MW00453399

Reference Data for Engineers:

Reference Data for Engineers:

Radio, Electronics, Computer, and Communications

Reference Data for Engineers:

Radio, Electronics, Computer, and Communications

Seventh Edition

A Revision of
Reference Data for Radio Engineers

Howard W. Sams & Co., Inc.
A Subsidiary of Macmillan, Inc.
4300 West 62nd Street, Indianapolis, Indiana 46268 U.S.A.

SEVENTH EDITION
FIRST PRINTING—1985

International Standard Book Number: 0-672-21563-2
Library of Congress Catalog Card Number: 43-14665

Printed in the United States of America

EDWARD C. JORDAN
EDITOR IN CHIEF

Professor Emeritus (and formerly Head) of Electrical Engineering, University of Illinois at Urbana-Champaign. Member, National Academy of Engineering; Member, NAE Telecommunications Committee; Fellow, IEEE; Past-Chairman and Honorary Life Member, IEEE Antennas and Propagation Society; Chairman (1967–1970), U.S. National Committee of the International Union of Radio Science (URSI); Past Chairman, USNC/URSI Commission VI. IEEE Education Medal, Ohio State University Centennial Award.

Editorial Board

Preface

Reference Data for Engineers: Radio, Electronics, Computer, and Communications is the seventh edition of a favorite handbook that originated more than forty years ago in an ITT affiliate company as a sixty-page brochure called *Reference Data for Radio Engineers*. The nature and extent of the remarkable changes that have occurred in this field in recent years are reflected in this new edition, and have necessitated the change of title. A review of the contents of the present edition will serve to indicate its enlarged scope, as well as to remind the reader how the pioneering field of ''radio'' has developed and branched into the modern fields of electronics, computers, and communications.

Approximately fifty percent of the seventh edition is new, and the remainder has been thoroughly revised and updated. The basic structure and organization remain the same, and the volume continues to be useful to the design engineer as a *handbook* with informative tables, graphs, and sample worked problems. The ''classical'' material that has survived the test of time has been retained, but much new material has been added. The availability of greatly improved electronic components, particularly semiconductor devices, and vastly more powerful computing methods and ''machines'' have resulted in a rapid evolution of engineering art and practice as applied to communications and control. Digital methods, computer organization and programming, digital communications, and computer communications networks are now essential background for the engineering practitioner and have been included. The pervasive influence of the computer has effected marked changes even in such traditional fields as antennas, filters, and measurements. Finally, it is clear that all of radio science and electronics is now different, in both substance and extent, because of the semiconductor and all that has followed from the transistor. Almost every electronic ''box'' is now driven or controlled by some form of semiconductor device or IC ''chip.'' Nothing in electronics has escaped the effect of the transistor and the computer chip that came from it. The result has been a major metamorphosis and expansion of the entire field far beyond informed predictions of only twenty years ago.

The topics treated in this edition were selected by a panel of eminent engineers representing the broad areas of radio, electronics, electronic devices, communications, and computer engineering. The specialized chapters have been written or revised by more than seventy contributors, each a recognized expert in his subject, many of whom have been involved directly or indirectly in the major developments that have occurred in the fields covered in this book. The result is a handbook that is both comprehensive and completely authoritative. The editor-in-chief expresses his pride and satisfaction in the quality of the authors' contributions and thanks them for their conscientious efforts.

E. C. JORDAN

Contents

Contributors

M. R. Aaron (Ch. 38)
Head, Digital Techniques Dept., AT&T—Bell Laboratories
Member, National Academy of Engineering; Fellow IEEE; Alexander Graham Bell Medal, 1978

Richard W. Avery (Ch. 13)
Member of Technical Staff, AT&T—Bell Laboratories

Fred J. Banzi, Jr. (Ch. 13)
Member of Technical Staff, AT&T—Bell Laboratories
Member IEEE

Pier L. Bargellini (Ch. 27)
Senior Scientist, COMSAT Laboratories
Fellow IEEE; Assoc. Fellow AIAA

David Bender (Ch. 48)
Of Counsel, White & Case (New York)
American Bar Association; American Patent Law Association; Member ACM

Richard E. Blahut (Ch. 25)
IBM Fellow, IBM Corporation
Fellow IEEE; Past-President IEEE/GIT

Robert A. Blaszczyk (Ch. 5)
Applications Engineer, VRN Division of Vernitron Corp.

John J. Bohrer (Ch. 5)
Consultant, Bohrcon
Senior Member IEEE; Past-Chairman ECC; Past-President AI Chem

John F. X. Browne (Ch. 35)
President, John F. X. Browne and Associates
Fellow SMPTE, Governor SMPTE; Senior Member IEEE; Past-President, Association of Federal Communications Consulting Engineers

Robert H. Brunner (Ch. 12)
General Manager, Instrument Marketing Group (retired), Hewlett-Packard Co.

Charles A. Cain (Ch. 48)
Professor of Electrical Engineering and Bioengineering, UIUC
Senior Member IEEE; Member, Board of Directors, Bioelectromagnetic Society

Bernard B. Carniglia (Ch. 13)
Member of Technical Staff, AT&T—Bell Laboratories

Dorothy H. Cerni (Ch. 2)
Technical Information Specialist, Institute for Telecommunications Sciences, National Telecommunications and Information Administration

Pallab K. Chatterjee (Ch. 20)
TI Fellow and Associate Director VLSI Design Laboratory, Texas Instruments, Inc.
Senior Member IEEE; Member APS

Gilbert Y. Chin (Ch. 4)
Director, Materials Research Laboratory, AT&T—Bell Laboratories
Member, National Academy of Engineering; Fellow, TMS-AIME, Mathewson Gold Medal 1974; Member IEEE Magnetics Society; Member A Cer S

Marvin Chodorow (Chs. 16 and 17)
Professor Emeritus of Applied Physics and Electrical Engineering, Ginzton Laboratory, Stanford University
Member, National Academy of Engineering; Fellow IEEE; W. R. G. Baker Award, 1962; Lamme Medal, 1982

Erich Christian (Ch. 9)
Manager of Advanced Engineering, ITT Telecom (retired); Professor NCSU
Senior Member IEEE

Robert Coackley (Ch. 12)
R&D Manager, Communications Division (Scotland); Information Networks Division, CA; Hewlett-Packard Co.
Fellow (British) IEE

Douglass D. Crombie (Chs. 1, 2, and 33)
Chief Scientist, National Telecommunications and Information Administration
Member National Academy of Engineering; Member IEEE; Member USNC/URSI; Member CCIR (U.S. Study groups 1, 6).

Frank Davidoff (Ch. 35)
Staff Consultant, Advanced Technology (retired) CBS Television Network
Fellow SMPTE; Progress and David Sarnoff Gold Medals; Senior Member IEEE; Past-Chairman Broadcasting Group Audio-Video Techniques Committee; Member EBU Technical Committee

Thijs de Haas (Ch. 2)
Networks and Systems Division, ITS, National Telecommunications and Information Administration

Georges A. Deschamps (Chs. 31 and 44)
Professor Emeritus of Electrical and Computer Engineering, UIUC
Member National Academy of Engineering; Fellow IEEE; Fellow AAAS; Member APS; Member Antennas and Propagation Society

Thomas A. DeTemple (Ch. 41)
Professor of Electrical and Computer Engineering, UIUC
Senior Member IEEE; Member APS

Michiel deWit (Ch. 20)
Senior Member Technical Staff, Texas Instruments Inc.
Member IEEE; Member APS

John J. Dupre (Ch. 12)
R&D Manager, Signal Analysis Division, Hewlett-Packard Co.
Member IEEE

John D. Dyson (Ch. 31)
Professor of Electrical and Computer Engineering, UIUC
Fellow IEEE; Member Antenna and Propagation Society; Member MTT, I and M, EMC; Member Antenna Standards Committee

Hermann Fickenscher (Ch. 13)
Supervisor, Magnetic Components Group, AT&T—Bell Laboratories
Senior Member IEEE, Member Electronics Transformers Technical Committee

Michael C. Fischer (Ch. 12)
Microwave Instruments R&D, Hewlett-Packard Co.
Member IEEE; Member AES

Arthur Fong (Ch. 12)
Manager, Corporate Engineering Design, Hewlett-Packard Co.
Fellow IEEE; Member IEC/TC-66 Committee on Microwave Measurements

Robert L. Gallawa (Ch. 22)
Professor Adjoint, Univ. of Colorado
Member IEEE; Member OSA

Chester S. Gardner (Ch. 22)
Professor of Electrical and Computer Engineering, UIUC
Fellow OSA, Vice Chairman for Remote Sensing OS Technical Council; Senior Member IEEE

Pierre J. Gerondeau (Ch. 5)
Senior Engineer, Western Electric Company
Member EIA, P.2.4 Committee

Robert C. Hansen (Ch. 32)
President, R. C. Hansen, Inc.
Fellow (British) IEE; Fellow IEEE; Past-President Antennas and Propagation Society (1964 and 1980); Chairman, Standards Subcommittee 2.5; Past-Chairman USNC/URSI Comm. VI

Walter S. Hayward, Jr. (Ch. 39)
Consultant, SPC Studies Center, AT&T—Bell Laboratories
Member IEEE; U.S. Member, Advisory Committee of International Teletraffic Congress

Charles F. Hempstead (Ch. 13)
Supervisor, Magnetics and Characterization, AT&T—Bell Laboratories
Member IEEE

Harold H. Hosack (Ch. 20)
Director, Charge Coupled Device Imagers and Cameras Project, Central Research Laboratories, Texas Instruments Inc.
Senior Member IEEE; Member APS

Bill J. Hunsinger (Ch. 28)
Professor of Electrical and Computer Engineering, UIUC
Senior Member IEEE

Tatsuo Itoh (Chs. 29 and 30)
Hayden Professor in Engineering University of Texas
Fellow IEEE; Editor, IEEE Trans. MTT; Chairman Microwave Field Theory Comm.

Horace G. Jackson (Ch.19)
Lecturer, Department of Electrical Engineering and Computer Science, Univ. of California at Berkeley

George Jacobs (Ch. 35)
President, George Jacobs Associates
Fellow IEEE; Marconi Memorial Gold Medal

W. Kenneth Jenkins (Ch. 28)
Professor of Electrical and Computer Engineering, UIUC
Fellow IEEE; Past Associate Editor, IEEE Transactions on Circuits and Systems

Amos E. Joel, Jr. (Chs. 38 and 39)
Switching Consultant, Bell Telephone Laboratories (retired)
Member National Academy of Engineering; Fellow IEEE; Alexander Graham Bell Medal, 1976; ITU Centennary Award, 1983

Durwood R. Kressler (Ch. 13)
Member of Technical Staff, AT&T—Bell Laboratories

Benjamin C. Kuo (Ch. 15)
Professor of Electrical and Computer Engineering, UIUC
Fellow IEEE

Kenneth R. Laker (Ch. 10)
Professor and Chair, Department of Electrical Engineering, University of Pennsylvania. (Formerly, Supervisor, Signal Processing Subsystems Group, AT&T—Bell Laboratories)
Fellow IEEE; Past-President Circuits and Systems Society.

Satwinder D. S. Mahli (Ch. 20)
Member Technical Staff, Texas Instruments Inc.
Member IEEE

Douglas L. Marriott (Ch. 45)
Associate Professor of Mechanical and Industrial Engineering, UIUC
Member ASME; Member ASQC; Member ASM

Edward J. McCluskey (Ch. 43)
Professor and Director, Center for Reliable Computing, Stanford University
Fellow IEEE; Member ACM; SIGARCH Director 1980-83; Past-President IEEE Computer Society

Richard A. McDonald (Ch. 38)
District Manager, Distribution Characterization and Engineering, Bell Communications Research
Member IEEE TRANSYSCOM Working Group on Loop Performance Standards

Eugene A. Mechtly (Chs. 3 and 4)
Associate Professor of Electrical and Computer Engineering, UIUC
Member ASTM Committee on Metric Practice E-43; Past-Chairman ASEE Metrication Coordinating Committee

Edward Mette (Ch. 5)
Resistive Products Division, TRW, Inc.

David C. Munson, Jr. (Ch. 7)
Associate Professor of Electrical and Computer Engineering, UIUC
Member IEEE

Arthur Olsen, Jr. (Ch. 13)
Supervisor, Magnetics and Data Group, AT&T—Bell Laboratories

Ronald E. Pratt (Ch. 12)
R&D Project Manager, Stanford Park Division, Hewlett-Packard Co.
Member IEEE

Rick Price (Ch. 5)
Capacitor Division, TRW, Inc.

Michael B. Pursley (Chs. 23 and 24)
Professor of Electrical and Computer Engineering, UIUC
Fellow IEEE; Past-President IEEE/GIT

Frederick M. Remley, Jr. (Ch. 35)
Director, Michigan Media, Univ. of Michigan Media Resources Center
Fellow SMPTE, Past-Chairman Video Recording Comm., Member IEEE, Member AES

E. A. Robinson (Ch. 37)
TI Fellow, Texas Instruments Inc.
Member, Radio Technical Commission for Maritime Services

Lawrence C. J. Roscoe (Ch. 38)
Department Head, AT&T—Bell Laboratories
Member IEEE

Douglas K. Rytting (Ch. 12)
Manager, R&D Laboratory Section, Networks Measurements Division, Hewlett-Packard Co.

Rolf Schaumann (Ch. 10)
Associate Professor of Electrical Engineering, Univ. of Minn.
Senior Member IEEE; Editor, *IEEE Transactions on Circuits and Systems*

Paul D. Schomer (Ch. 40)
Team Leader—Acoustics, U.S. Army Construction Engineering Research Laboratory
Fellow ASA; Member IEEE; Member, Institute of Noise Control Engineering; Chairman, NAE, U.S. National Counterpart Committee to CIB-W-51-Acoustics

W. Ford Shepherd (Ch. 9)
Director, Research and Development, Network Systems Division, ITT Telecom

Merril I. Skolnik (Ch. 36)
Superintendent Radar Division, Naval Research Laboratory
Fellow IEEE; Harry Diamond Award

Frank S. Stein (Chs. 3 and 4)
Manager, Semiconductor R&D, Delco Electronics Div., GMC
Senior Member IEEE; Past-Chairman JEDEC Solid State Products Engineering Council; ETA Board of Governors; Chairman Solid State Products Division

Kent W. Sternstrom (Ch. 13)
Member of Technical Staff, AT&T—Bell Laboratories

Gregory E. Stillman (Ch. 21)
Professor of Electrical and Computer Engineering, UIUC
Fellow IEEE; President, Electron Devices Society; Member APS; Member ECS; Member, National Academy of Engineering

Ben G. Streetman (Ch. 18)
Professor and J. S. Cockrell Centennial Chair, Department of Electrical Engineering, Univ. of Texas at Austin
Fellow IEEE; Member Electrochemical Society; ASEE Terman Award

Fouad A. Tobagi (Ch. 26)
Associate Professor of Electrical Engineering, Stanford Univ.
Fellow IEEE; Member ACM

Joseph A. Toro (Ch. 5)
Manager, Special Products, TRW Capacitor Division
Member IEEE

Fred G. Turnbull (Ch. 14)
Electronics Engineer, General Electric Co.
Senior Member IEEE

Joseph T. Verdeyen (Ch. 41)
Professor of Electrical and Computer Engineering, UIUC
Senior Member IEEE; Member APS; Chairman, Gaseous Electronics Conference

Wilbur R. Vincent (Ch. 34)
Professor of Electrical Engineering, Naval Postgraduate School, Monterey, CA; Staff Scientist, SRI International
Member IEEE

Theodore F. VonKampen (Ch. 5)
Engineering Manager, TRW Capacitor Division

John Wakerly (Ch. 42)
Director, System Architecture, David Systems, Santa Clara, CA
Member IEEE; Member ACM

Richard C. Walker (Ch. 13)
Member of Technical Staff, AT&T—Bell Laboratories

Alva L. Wallis (Ch. 48)
Meteorologist, National Climatic Data Center, Asheville, NC

Bruno O. Weinschel (Ch. 11)
President, Weinschel Engineering
Fellow and President IEEE; Chairman IEEE Standards Committee on Network Analyzers; Past-Chairman CPEM; Past-Chairman, U.S. Commission I of URSI

Acknowledgments

In addition to the 78 persons in the list of contributors, there were many others who contributed to the accuracy and completeness of the seventh edition by reviewing manuscripts or suggesting changes and additions. The help of the following persons is especially acknowledged:

George Anner, Paul Bourget, S. A. Bowhill, Milton Crothers, Willis Emery, Edward Ernst, Jack Farley, Jay Gooch, James Guilbeau, S. R. Hartshorn, Kim Jovanovich, George Keller, Sidney Metzger, Saburo Muroga, Carl Myers, Tom Purl, Andrew Reed, Leslie Smith, Timothy Trick, James Wait, Edward Wetherhold, Nick Yaru.

The seventh edition, although mostly new or revised, is built upon the work of contributors to earlier editions. The difficulty of identifying the authors of material carried over during the more than 40-year history of the handbook makes it impossible to list all their names despite the debt we owe them. Nonetheless, the name of W. L. McPherson who compiled the original 60-page brochure must be mentioned. Also, special acknowledgment is made of the valuable contributions of A. G. Kandoian as Chairman of the Editorial Boards for the third through fifth editions, and of H. P. Westman who edited the fourth and fifth editions. Contributors from outside the ITT system acknowledged in the sixth edition were:

A. F. Barghausen, R. C. Barker, Colin Cherry, M. S. Cord, W. Q. Crichlow, R. T. Disney, E. A. Guillemin, C. W. Haydon, J. W. Herbstreit, W. K. Kahn, J. D. Kraus, R. S. Lawrence, D. J. LeVine, L. J. Lidofsky, N. Marchand, N. Marcuvitz, J. A. Pierce, J. R. Ragazzini, P. L. Rice, H. R. Romig, C. Tamir

Known contributors to earlier editions from within the ITT system were:

E. Baguley, E. E. Benham, R. A. Bones, T. Brown, J. H. Brundage, F. X. Bucher, J. A. Budek, H. G. Busignies, A. Casabona, R. Clayton, D. K. Coles, C. R. Cook, A. E. Cookson, M. Dishal, S. H. Dodington, J. G. Dunn, E. Eberhardt, J. A. Fingerett, M. T. Fujita, D. S. Girling, F. F. Hall, I. W. Hammer, D. E. Herrington, J. L. Jatlow, P. King, J. Kylander, P. Lighty, W. Litchman, J. G. Litterick, C. W. Moody, J. M. Moore, H. G. Nordlin, J. E. Obst, J. A. O'Connell, C. P. Oliphant, J. Polyzou, M. C. Poylo, L. G. Rado, D. S. Ridler, L. Rosenberg, J. E. Schlaikjer, H. H. Smith, R. Smith, T. L. Squires, J. L. Storr-Best, J. G. Tatum, L. F. Turner, R. Vachss, J. M. Valentine, R. Weber, A. K. Wing, Jr., J. Youlios

The close and continuous cooperation of the publisher's technical-books management and editorial staff over a period of nearly four years was required to bring this edition to completion. Special acknowledgments are made to Charlie Dresser, who initiated work on the seventh edition and persuaded the editor in chief to undertake that task, to Louis Keglovits who guided the publication through its latter stages, and to technical editor James Moore whose background in electrical engineering enabled him to do a superb job of checking the technical as well as typographical correctness of the material.

The advice and guidance of members of the Editorial Board proved invaluable in producing a handbook covering such a wide range of topics; it is a pleasure to acknowledge their contributions. Finally, the always cheerful and competent help of Nancy Dimond as assistant to the editor-in-chief ensured the successful completion of a long and arduous task.

1 Frequency Data

Revised by
Douglass D. Crombie

WAVELENGTH-FREQUENCY CONVERSION

Fig. 1 permits conversion between frequency and wavelength. The f scale may be multiplied by a power of 10 if the λ scale is divided by the same power of 10.

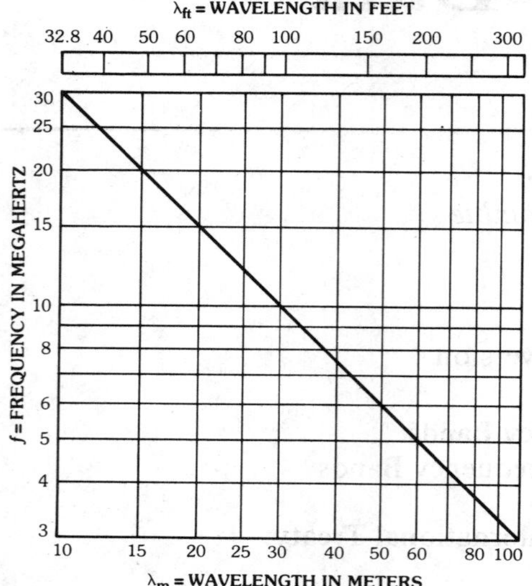

Fig. 1. Wavelength-frequency conversion.

Conversion Equations

Propagation velocity

$$c \approx 3\times10^8 \text{ meters/second}$$

Wavelength in meters

$$\lambda_m = \frac{300\,000}{f \text{ in kilohertz}} = \frac{300}{f \text{ in megahertz}}$$

Wavelength in centimeters

$$\lambda_{cm} = \frac{30}{f \text{ in gigahertz}}$$

Wavelength in feet

$$\lambda_{ft} = \frac{984\,000}{f \text{ in kilohertz}} = \frac{984}{f \text{ in megahertz}}$$

Wavelength in inches

$$\lambda_{in} = \frac{11.8}{f \text{ in gigahertz}}$$

1 angstrom unit, Å $= 3.937\times10^{-9}$ inch

$= 1\times10^{-10}$ meter

$= 1\times10^{-4}$ micrometer

1 micrometer, μm $= 3.937\times10^{-5}$ inch

$= 1\times10^{-6}$ meter

$= 1\times10^{4}$ angstrom units

(Note that the term "micrometer" has superceded the term "micron.")

Nomenclature of Frequency Bands

Table 1 is adapted from the Radio Regulations of the International Telecommunication Union, Article 2, 208, Geneva; 1982.

Letter Designations for Frequency Bands

Letter designations commonly used for microwave bands (particularly in references to radar equipment) are shown in Table 2. These designations have no official international standing, and various engineers have used limits for the bands and subbands other than those listed in the table.

Subband code letters should be used as subscripts in designating particular frequency ranges; for example, L_x indicates the band between 0.950 and 1.150 gigahertz.

FREQUENCY ALLOCATIONS BY INTERNATIONAL TREATY

The following information is adapted from the Radio Regulations of the ITU, Geneva, 1982. Some 400 footnotes describing special conditions pertaining to allocations within particular frequency bands and much other detailed information are not reproduced here. Copies of the Radio Regulations may be obtained from the Secretary General, International Telecommunication Union, Place des Nations, CH-1211, Geneva 20, Switzerland.*

For purposes of frequency allocations, the world has been divided into regions, as shown in Fig. 2.

See Article 8, Section 391–412 of the ITU Radio Regulations for definitions of the regions and of lines A, B, and C.

Frequency bands are allocated to services defined as follows:

Fixed: Radio communication between specified fixed points. Examples are point-to-point high-frequency circuits and microwave links.

Mobile: Radio communication between stations intended to be used while in motion or during halts at unspecified points or between such stations and fixed stations.

Aeronautical Mobile: Radio communication between a land station and an aircraft or between

*In the official documents of the ITU and FCC, the following terms are combined as single words: radiobeacon, radiocommunication, radiodetermination, radiolocation, radionavigation, radiorange, radiosonde, radiotelegraphy, and radiotelephony.

TABLE 1. NOMENCLATURE OF FREQUENCY BANDS

Band Number*	Frequency Range	Metric Subdivision	Adjectival Designation	
2	30 to 300 hertz	Megametric waves	ELF	Extremely low frequency
3	300 to 3000 hertz		VF	Voice frequency
4	3 to 30 kilohertz	Myriametric waves	VLF	Very low frequency
5	30 to 300 kilohertz	Kilometric waves	LF	Low frequency
6	300 to 3000 kilohertz	Hectometric waves	MF	Medium frequency
7	3 to 30 megahertz	Decametric waves	HF	High frequency
8	30 to 300 megahertz	Metric waves	VHF	Very high frequency
9	300 to 3000 megahertz	Decimetric waves	UHF	Ultra high frequency
10	3 to 30 gigahertz	Centimetric waves	SHF	Super high frequency
11	30 to 300 gigahertz	Millimetric waves	EHF	Extremely high frequency
12	300 to 3000 gigahertz or 3 terahertz	Decimillimetric waves		

* "Band Number N" extends from 0.3×10^N to 3×10^N hertz. The upper limit is included in each band; the lower limit is excluded.

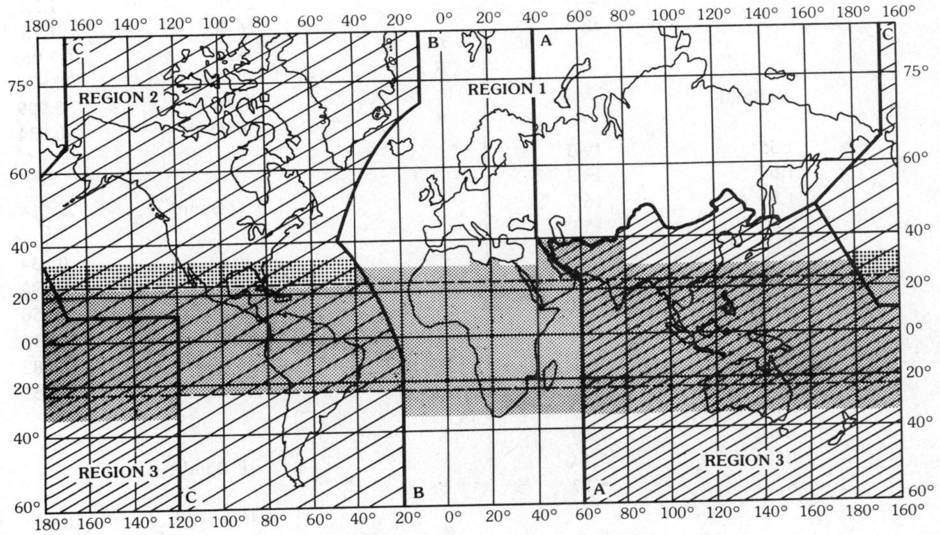

Fig. 2. Regions defined for frequency allocations. Shaded area represents tropical zone.

aircraft. (R indicates frequency bands for communication within regions. OR indicates bands for communication between regions.)

Maritime Mobile: Radio communication between a coast station and a ship or between ships.

Land Mobile: Radio communication between a base station and land mobile station or between land mobile stations. Examples are radio communication with taxicabs and police vehicles.

Radio Navigation: The determination of position for purposes of navigation by means of the propagation properties of radio waves. This includes obstruction warning. An example is loran.

Aeronautical Radio Navigation: A radio navigation service intended for the benefit of aircraft. Examples are VOR and Tacan systems, aeronautical radio beacons, instrument landing systems, radio altimeters, and airborne obstruction-indicating radar.

Maritime Radio Navigation: A radio navigation service intended for the benefit of ships. Examples

TABLE 2. LETTER DESIGNATIONS FOR MICROWAVE BANDS

Subband	Frequency in Gigahertz	Wavelength in Centimeters
	P Band	
	0.225	133.3
	0.390	76.9
	L Band	
	0.390	76.9
p		
	0.465	64.5
c		
	0.510	58.8
l		
	0.725	41.4
y		
	0.780	38.4
t		
	0.900	33.3
s		
	0.950	31.6
x		
	1.150	26.1
k		
	1.350	22.2
f		
	1.450	20.7
z		
	1.550	19.3
	S Band	
	1.55	19.3
e		
	1.65	18.3
f		
	1.85	16.2
t		
	2.00	15.0
c		
	2.40	12.5
q		
	2.60	11.5
y		
	2.70	11.1
g		
	2.90	10.3
s		
	3.10	9.67
a		
	3.40	8.32
w		
	3.70	8.10
h		
	3.90	7.69
*z**		
	4.20	7.14
d		
	5.20	5.77
	X Band	
	5.20	5.77
a		
	5.50	5.45
q		
	5.75	5.22
*y**		
	6.20	4.84
d		
	6.25	4.80
b		
	6.90	4.35
r		
	7.00	4.29
c		
	8.50	3.53
	X Band—*Continued*	
l		
	9.00	3.33
s		
	9.60	3.13
x		
	10.00	3.00
f		
	10.25	2.93
k		
	10.90	2.75
	K Band	
	10.90	2.75
p		
	12.25	2.45
s		
	13.25	2.26
e		
	14.25	2.10
c		
	15.35	1.95
u†		
	17.25	1.74
t		
	20.50	1.46
q†		
	24.50	1.22
r		
	26.50	1.13
m		
	28.50	1.05
n		
	30.70	0.977
l		
	33.00	0.909
a		
	36.00	0.834
	Q Band	
	36.0	0.834
a		
	38.0	0.790
b		
	40.0	0.750
c		
	42.0	0.715
d		
	44.0	0.682
e		
	46.0	0.652
	V Band	
	46.0	0.652
a		
	48.0	0.625
b		
	50.0	0.600
c		
	52.0	0.577
d		
	54.0	0.556
e		
	56.0	0.536
	W Band	
	56.0	0.536
	100.0	0.300

**C* Band includes S_z through X_y (3.90–6.20 gigahertz).
†K_l Band includes K_u through K_q (15.35–24.50 gigahertz).

are coastal radio beacons, direction-finding stations, and shipboard radar.

Radio Location: The determination of position for purposes other than those of navigation by means of the propagation properties of radio waves. Examples are land radars, coastal radars, and tracking systems.

Broadcasting: Radio communication intended for direct reception by the general public. Examples are amplitude-modulation broadcasting on medium and high frequencies, frequency-modulation broadcasting, and television.

Amateur: Radio communication carried on by persons interested in the radio technique solely with a personal aim and without pecuniary interest.

Space: Radio communication between space stations.

Earth-Space: Radio communication between earth stations and space stations. An example is between the earth and a satellite.

Radio Astronomy: Astronomy based on the reception of radio waves of cosmic origin.

Standard Frequency: Radio transmission of specified frequencies of stated high precision, intended for general reception for scientific, technical, and other purposes.

The allocations in Chart 1 apply to all three regions. They represent the primary services for each region. The order of listing in each band does not indicate relative priority.

DESIGNATION OF EMISSIONS

According to the ITU (Article 4; Radio Regulations, Geneva, 1982), emissions shall be designated according to their necessary bandwidth and their classification.

Necessary Bandwidth

Necessary bandwidth for a given class of emission is defined as the width of the frequency band that is just sufficient to ensure the transmission of information at the rate and with the quality required under specified conditions. Emissions needed for satisfactory functioning of the receiving equipment, such as the carrier in reduced-carrier systems or a vestigial sideband, are included in the necessary bandwidth.

The necessary bandwidth (the value of which is determined later) shall be expressed by three numerals and one letter. The letter occupies the position of the decimal point and represents the unit of bandwidth as shown below and in the subsequent examples.

Necessary bandwidths

- between 0.001 and 999 Hz shall be expressed in hertz (letter H);
- between 1.00 and 999 kHz shall be expressed in kilohertz (letter K);
- between 1.00 and 999 MHz shall be expressed in megahertz (letter M);
- between 1.00 and 999 GHz shall be expressed in gigahertz (letter G).

Examples:

0.002	Hz	= H002
0.1	Hz	= H100
25.3	Hz	= 25H3
400	Hz	= 400H
2.4	kHz	= 2K40
6	kHz	= 6K00
12.5	kHz	= 12K5
180.4	kHz	= 180K
180.5	kHz	= 181K
180.7	kHz	= 181K
1.25	MHz	= 1M25
2	MHz	= 2M00
10	MHz	= 10M0
202	MHz	= 202M
5.65	GHz	= 5G65

Classification

Classification of the signal is given by three additional symbols. The first symbol denotes the type of modulation of the main carrier; the second symbol denotes the nature of signal(s) modulating the main carrier; the third symbol denotes the type of information to be transmitted.

Modulation used only for short periods and for incidental purposes (such as, in many cases, for identification or calling) may be ignored provided that the necessary bandwidth as indicated is not thereby increased.

First Symbol—Type of Modulation of the Main Carrier:

(1.1) Emission of an unmodulated carrier N
(1.2) Emission in which the main carrier is amplitude-modulated (including cases where subcarriers are angle-modulated)
 (1.2.1) Double-sideband A
 (1.2.2) Single-sideband, full carrier H
 (1.2.3) Single-sideband, reduced or variable-level carrier R
 (1.2.4) Single-sideband, suppressed carrier J
 (1.2.5) Independent sidebands B
 (1.2.6) Vestigial sideband C
(1.3) Emission in which the main carrier is angle-modulated
 (1.3.1) Frequency modulation F
 (1.3.2) Phase modulation G

CHART 1. ALLOCATION TO SERVICES

Kilohertz

Region 1	Region 2	Region 3
Below 9 (Not allocated)		
9–14 Radionavigation		
14–19.95 Fixed/Maritime Mobile		
19.95–20.05 Standard Frequency and Time Signal (20 kHz)		
20.05–70 Fixed/Maritime Mobile		
70–72 Radionavigation	*70–90* Fixed/Maritime Mobile/Maritime Radio-Navigation	*70–72* Radionavigation
72–84 Fixed/Maritime Mobile/Radionavigation		*72–84* Fixed/Maritime Mobile/Radionavigation
84–86 Radionavigation		*84–86* Radionavigation
86–90 Fixed/Maritime Mobile/Radionavigation		*86–90* Fixed/Maritime Mobile/Radionavigation
90–110 Radionavigation		
110–112 Fixed/Maritime Mobile/Radionavigation	*110–130* Fixed/Maritime Mobile/Maritime Radio-navigation	*110–112* Fixed/Maritime Mobile/Radionavigation
112–117.6 Radionavigation		*112–117.6* Radionavigation
117.6–126 Fixed/Maritime Mobile/Radionavigation		*117.6–126* Fixed/Maritime Mobile/Radionavigation
126–129 Radionavigation		*126–129* Radionavigation
129–130 Fixed/Maritime Mobile/Radionavigation		*129–160* Fixed/Maritime Mobile/Radionavigation
130–148.5 Maritime Mobile/Fixed	*130–160* Fixed/Maritime Mobile	
148.5–255 Broadcasting	*160–190* Fixed	
	190–285 Aeronautical Radionavigation	
255–283.5 Broadcasting/Aeronautical Radio-navigation		
283.5–315 Maritime Radio-navigation/Aeronautical Radionavigation	*285–315* Maritime Radionavigation/Aeronautical Radionavigation	
315–325 Aeronautical Radionavigation	*315–325* Maritime Radionavigation	*315–325* Aeronautical Radionavigation/Maritime Radionavigation
325–405 Aeronautical Radionavigation		
405–415 Radionavigation		
415–435 Aeronautical Radionavigation/Maritime Mobile	*415–495* Maritime Mobile	
435–495 Maritime Mobile		
495–505 Mobile (Distress and calling)		
505–526.5 Maritime Mobile/Aeronautical Radionavigation	*505–510* Maritime Mobile	*505–526.5* Maritime Mobile/Aeronautical Radionavigation
	510–525 Mobile/Aeronautical Radionavigation	

CHART 1 (CONT). ALLOCATION TO SERVICES

Region 1	Region 2	Region 3
Kilohertz		
526.5–1 606.5 Broadcasting	*525–535* Broadcasting/Aeronautical Radionavigation	*526.5–1 606.5* Broadcasting
	535–1 605 Broadcasting	
1 606.5–1 625 Maritime Mobile/Fixed/Land Mobile	*1 605–1 625* Broadcasting	*1 606.5–1 800* Fixed/Mobile/Radiolocation/Radionavigation
1 625–1 635 Radiolocation	*1 625–1 705* Broadcasting/Fixed/Mobile	
1 635–1 800 Maritime Mobile/Fixed/Land Mobile	*1 705–1 800* Fixed/Mobile/Radiolocation/Aeronautical Radionavigation	
1 800–1 810 Radiolocation	*1 800–1 850* Amateur	*1 800–2 000* Amateur/Fixed/Mobile*/Radio-Navigation
1 810–1 850 Amateur		
1 850–2 045 Fixed/Mobile*	*1 850–2 000* Amateur/Fixed/Mobile*/Radiolocation/Radionavigation	
	2 000–2 065 Fixed/Mobile	
2 045–2 160 Maritime Mobile/Fixed/Land Mobile	*2 065–2 107* Maritime Mobile	
2 160–2 170 Radiolocation	*2 107–2 170* Fixed/Mobile	
2 170–2 173.5	Maritime Mobile	
2 173.5–2 190.5	Mobile (Distress and calling)	
2 190.5–2 194	Maritime Mobile	
2 194–2 300 Fixed/Mobile*	*2 194–2 300* Fixed/Mobile	
2 300–2 498 Fixed/Mobile*/Broadcasting	*2 300–2 495* Fixed/Mobile/Broadcasting	
2 498–2 501 Standard Frequency and Time Signal (2 500 kHz)	*2 495–2 501* Standard Frequency and Time Signal (2 500 kHz)	
2 501–2 502 Standard Frequency and Time Signal		
2 502–2 625 Fixed/Mobile*	*2 502–2 505* Standard Frequency and Time Signal	
2 625–2 650 Maritime Mobile/Maritime Radionavigation	*2 505–2 850* Fixed/Mobile	
2 650–2 850 Fixed/Mobile*		
2 850–3 025	Aeronautical Mobile (R)	
3 025–3 155	Aeronautical Mobile (OR)	
3 155–3 200	Fixed/Mobile*	
3 200–3 400	Fixed/Mobile*/Broadcasting	
3 400–3 500	Aeronautical Mobile (R)	

Continued on next page.

CHART 1 (CONT). ALLOCATION TO SERVICES

Kilohertz

Region 1	Region 2	Region 3
3 500–3 800 Amateur/ Fixed/Mobile*	*3 500–3 750* Amateur	*3 500–3 900* Amateur/ Fixed/Mobile
3 800–3 900 Fixed/ Aeronautical Mobile (OR)/Land Mobile	*3 750–4 000* Amateur/ Fixed/Mobile*	
3 900–3 950 Aeronautical Mobile (OR)		*3 900–3 950* Aeronautical Mobile/Broadcasting
3 950–4 000 Fixed/ Broadcasting		*3 950–4 000* Fixed/ Broadcasting
4 000–4 063	Fixed/Maritime Mobile	
4 063–4 438	Maritime Mobile	
4 438–4 650	Fixed/Mobile*	
4 650–4 700	Aeronautical Mobile (R)	
4 700–4 750	Aeronautical Mobile (OR)	
4 750–4 850 Fixed/ Aeronautical Mobile (OR)/Land Mobile/ Broadcasting	*4 750–4 850* Fixed/Mobile*/ Broadcasting	*4 750–4 850* Fixed/ Broadcasting
4 850–4 995	Fixed/Land Mobile/Broadcasting	
4 995–5 003	Standard Frequency and Time Signal (5 000 kHz)	
5 003–5 005	Standard Frequency and Time Signal	
5 005–5 060	Fixed/Broadcasting	
5 060–5 250	Fixed	
5 250–5 450	Fixed/Mobile*	
5 450–5 480 Fixed/ Aeronautical Mobile (OR)/ Land Mobile	*5 450–5 480* Aero- nautical Mobile (R)	*5 450–5 480* Fixed/ Aeronautical Mobile (OR)/ Land Mobile
5 480–5 680	Aeronautical Mobile (R)	
5 680–5 730	Aeronautical Mobile (OR)	
5 730–5 950 Fixed/ Land Mobile	*5 730–5 950* Fixed/ Mobile*	*5 730–5 950* Fixed
5 950–6 200	Broadcasting	
6 200–6 525	Maritime Mobile	
6 525–6 685	Aeronautical Mobile (R)	
6 685–6 765	Aeronautical Mobile (OR)	
6 765–7 000	Fixed	
7 000–7 100	Amateur/Amateur-Satellite	
7 100–7 300 Broadcasting	*7 100–7 300* Amateur	*7 100–7 300* Broadcasting
7 300–8 100	Fixed	
8 100–8 195	Fixed/Maritime Mobile	
8 195–8 815	Maritime Mobile	
8 815–8 965	Aeronautical Mobile (R)	
8 965–9 040	Aeronautical Mobile (OR)	
9 040–9 500	Fixed	

CHART 1 (CONT). ALLOCATION TO SERVICES

Region 1	Region 2	Region 3

Kilohertz

9 500–9 900	Broadcasting
9 900–9 995	Fixed
9 995–10 003	Standard Frequency and Time Signal (10 000 kHz)
10 003–10 005	Standard Frequency and Time Signal
10 005–10 100	Aeronautical Mobile (R)
10 100–11 175	Fixed
11 175–11 275	Aeronautical Mobile (OR)
11 275–11 400	Aeronautical Mobile (R)
11 400–11 650	Fixed
11 650–12 050	Broadcasting
12 050–12 230	Fixed
12 230–13 200	Maritime Mobile
13 200–13 260	Aeronautical Mobile (OR)
13 260–13 360	Aeronautical Mobile (R)
13 360–13 410	Fixed/Radio Astronomy
13 410–13 600	Fixed
13 600–13 800	Broadcasting
13 800–14 000	Fixed
14 000–14 250	Amateur/Amateur–Satellite
14 250–14 350	Amateur
14 350–14 990	Fixed
14 990–15 005	Standard Frequency and Time Signal (15 000 kHz)
15 005–15 010	Standard Frequency and Time Signal
15 010–15 100	Aeronautical Mobile (OR)
15 100–15 600	Broadcasting
15 600–16 360	Fixed
16 360–17 410	Maritime Mobile
17 410–17 550	Fixed
17 550–17 900	Broadcasting
17 900–17 970	Aeronautical Mobile (R)
17 970–18 030	Aeronautical Mobile (OR)
18 030–18 068	Fixed
18 068–18 168	Amateur/Amateur–Satellite
18 168–18 780	Fixed
18 780–18 900	Maritime Mobile
18 900–19 680	Fixed
19 680–19 800	Maritime Mobile
19 800–19 990	Fixed
19 990–19 995	Standard Frequency and Time Signal
19 995–20 010	Standard Frequency and Time Signal (20 000 kHz)
20 010–21 000	Fixed
21 000–21 450	Amateur/Amateur–Satellite

Continued on next page.

CHART 1 (CONT). ALLOCATION TO SERVICES

Region 1	Region 2	Region 3

Kilohertz

Frequency	Allocation (all regions)
21 450–21 850	Broadcasting
21 850–21 870	Fixed
21 870–21 924	Aeronautical Fixed
21 924–22 000	Aeronautical Mobile (R)
22 000–22 855	Maritime Mobile
22 855–23 200	Fixed
23 200–23 350	Aeronautical Fixed/Aeronautical Mobile (OR)
23 350–24 000	Fixed/Mobile*
24 000–24 890	Fixed/Land Mobile
24 890–24 990	Amateur/Amateur-Satellite
24 990–25 005	Standard Frequency and Time Signal (25 000 kHz)
25 005–25 010	Standard Frequency and Time Signal
25 010–25 070	Fixed/Mobile*
25 070–25 210	Maritime Mobile
25 210–25 550	Fixed/Mobile*
25 550–25 670	Radio Astronomy
25 670–26 100	Broadcasting
26 100–26 175	Maritime Mobile
26 175–27 500	Fixed/Mobile*
27 500–28 000	Meteorological Aids/Fixed/Mobile
28 000–29 700	Amateur/Amateur-Satellite
29 700–30 005	Fixed/Mobile

Megahertz

Region 1	Region 2	Region 3
30.005–30.01 Space Operation/Fixed/Mobile/Space Research (all regions)		
30.01–47 Fixed/Mobile (all regions)		
47–68 Broadcasting	*47–50* Fixed/Mobile	*47–50* Fixed/Mobile/ Broadcasting
	50–54 Amateur (Regions 2 and 3)	
	54–68 Broadcasting	*54–68* Fixed/Mobile/ Broadcasting
68–74.8 Fixed/Mobile*	*68–72* Broadcasting	*68–74.8* Fixed/Mobile
	72–73 Fixed/Mobile	
	73–74.6 Radio Astronomy	
	74.6–74.8 Fixed/Mobile	
74.8–75.2 Aeronautical Radionavigation (all regions)		
75.2–87.5 Fixed/Mobile*	*75.2–75.4* Fixed/Mobile (Regions 2 and 3)	
	75.4–76 Fixed/Mobile	*75.4–87* Fixed/Mobile
	76–88 Broadcasting	
87.5–100 Broadcasting	*88–100* Broadcasting	*87–100* Fixed/Mobile/ Broadcasting
100–108 Broadcasting (all regions)		
108–117.975 Aeronautical Radionavigation (all regions)		

CHART 1 (CONT). ALLOCATION TO SERVICES

Megahertz

<table>
<tr><th>Region 1</th><th>Region 2</th><th>Region 3</th></tr>
<tr><td colspan="3">117.975–137 Aeronautical Mobile (R)</td></tr>
<tr><td colspan="3">137–138 Space Operation[1]/Meteorological–Satellite[1]/Space Research[1]</td></tr>
<tr><td>138–143.6 Aeronautical Mobile (OR)</td><td>138–143.6 Fixed/Mobile/Radiolocation</td><td>138–143.6 Fixed/Mobile</td></tr>
<tr><td>143.6–143.65 Aeronautical Mobile (OR)/Space Research[1]</td><td>143.6–143.65 Fixed/Mobile/Space Research[1]/Radiolocation</td><td>143.6–143.65 Fixed/Mobile/Space Research[1]</td></tr>
<tr><td>143.65–144 Aeronautical Mobile (OR)</td><td>143.65–144 Fixed/Mobile/Radiolocation</td><td>143.65–144 Fixed/Mobile</td></tr>
<tr><td colspan="3">144–146 Amateur/Amateur–Satellite</td></tr>
<tr><td rowspan="2">146–149.9 Fixed/Mobile*</td><td>146–148 Amateur</td><td>146–148 Amateur/Fixed/Mobile</td></tr>
<tr><td colspan="2">148–149.9 Fixed/Mobile</td></tr>
<tr><td colspan="3">149.9–150.05 Radionavigation–Satellite</td></tr>
<tr><td>150.05–153 Fixed/Mobile*/Radio Astronomy</td><td colspan="2" rowspan="2">150.05–156.7625 Fixed/Mobile</td></tr>
<tr><td>153–156.7625 Fixed/Mobile*</td></tr>
<tr><td colspan="3">156.7625–156.8375 Maritime Mobile (Distress and calling)</td></tr>
<tr><td>156.8375–174 Fixed/Mobile*</td><td colspan="2">156.8375–174 Fixed/Mobile</td></tr>
<tr><td rowspan="4">174–230 Broadcasting</td><td>174–216 Broadcasting</td><td rowspan="2">174–223 Fixed/Mobile/Broadcasting</td></tr>
<tr><td>216–220 Fixed/Maritime Mobile</td></tr>
<tr><td>220–225 Amateur/Fixed/Mobile</td><td>223–230 Fixed/Mobile/Broadcasting/Aeronautical Radionavigation</td></tr>
<tr><td rowspan="2">225–235 Fixed/Mobile</td><td rowspan="2">230–235 Fixed/Mobile/Aeronautical Radionavigation</td></tr>
<tr><td>230–235 Fixed/Mobile</td></tr>
<tr><td colspan="3">235–272 Fixed/Mobile</td></tr>
<tr><td colspan="3">272–273 Space Operation[1]/Fixed/Mobile</td></tr>
<tr><td colspan="3">273–322 Fixed/Mobile</td></tr>
<tr><td colspan="3">322–328.6 Fixed/Mobile/Radio Astronomy</td></tr>
<tr><td colspan="3">328.6–335.4 Aeronautical Radionavigation</td></tr>
<tr><td colspan="3">335.4–399.9 Fixed/Mobile</td></tr>
<tr><td colspan="3">399.9–400.05 Radionavigation–Satellite</td></tr>
<tr><td colspan="3">400.05–400.15 Standard Frequency and Time Signal Satellite (400.1 MHz)</td></tr>
<tr><td colspan="3">400.15–401 Meteorological Aids/Meteorological–Satellite[1]/Space Research[1]</td></tr>
<tr><td colspan="3">401–402 Meteorological Aids/Space Operation[1]</td></tr>
<tr><td colspan="3">402–406 Meteorological Aids</td></tr>
<tr><td colspan="3">406–406.1 Mobile–Satellite[2]</td></tr>
<tr><td colspan="3">406.1–410 Fixed/Mobile*/Radio Astronomy</td></tr>
<tr><td colspan="3">410–430 Fixed/Mobile*</td></tr>
</table>

Continued on next page.

CHART 1 (CONT). ALLOCATION TO SERVICES

Megahertz

<table>
<tr><th>Region 1</th><th>Region 2</th><th>Region 3</th></tr>
<tr><td>430–440 Amateur/ Radiolocation</td><td colspan="2">430–440 Radiolocation</td></tr>
<tr><td colspan="3">440–450 Fixed/Mobile*</td></tr>
<tr><td colspan="3">450–470 Fixed/Mobile</td></tr>
<tr><td rowspan="5">470–790 Broadcasting</td><td rowspan="2">470–608 Broadcasting</td><td>470–585 Fixed/Mobile/ Broadcasting</td></tr>
<tr><td rowspan="2">585–610 Fixed/Mobile/ Broadcasting/Radio-navigation</td></tr>
<tr><td rowspan="2">608–614 Radio Astronomy</td></tr>
<tr><td rowspan="9">610–960 Fixed/Mobile/ Broadcasting</td></tr>
<tr><td rowspan="2">614–806 Broadcasting</td></tr>
<tr><td rowspan="2">790–862 Fixed/Broad-casting</td></tr>
<tr><td rowspan="2">806–890 Fixed/Mobile/ Broadcasting</td></tr>
<tr><td rowspan="5">862–960 Fixed/Mobile*/ Broadcasting</td></tr>
<tr><td>890–902 Fixed/Mobile*</td></tr>
<tr><td>902–928 Fixed</td></tr>
<tr><td>928–942 Fixed/Mobile*</td></tr>
<tr><td>942–960 Fixed</td></tr>
<tr><td colspan="3">960–1 215 Aeronautical Radionavigation</td></tr>
<tr><td colspan="3">1 215–1 260 Radiolocation/Radionavigation-Satellite[1]</td></tr>
<tr><td colspan="3">1 260–1 300 Radiolocation</td></tr>
<tr><td colspan="3">1 300–1 350 Aeronautical Radionavigation</td></tr>
<tr><td>1 350–1 400 Fixed/ Mobile/Radiolocation</td><td colspan="2">1 350–1 400 Radiolocation</td></tr>
<tr><td colspan="3">1 400–1 427 Earth Exploration Satellite[3]/Radio Astronomy/Space Research[3]</td></tr>
<tr><td colspan="3">1 427–1 429 Space Operation[2]/Fixed/Mobile*</td></tr>
<tr><td>1 429–1 525 Fixed/ Mobile*</td><td colspan="2">1 429–1 525 Fixed/Mobile</td></tr>
<tr><td>1 525–1 530 Space Operation[1]/Fixed</td><td>1 525–1 530 Space Operation[1]</td><td>1 525–1 530 Space Operation[1]/Fixed</td></tr>
<tr><td colspan="3">1 530–1 535 Space Operation[1]/Maritime Mobile Satellite[1]</td></tr>
<tr><td colspan="3">1 535–1 544 Maritime Mobile Satellite[1]</td></tr>
<tr><td colspan="3">1 544–1 545 Mobile Satellite[1]</td></tr>
<tr><td colspan="3">1 545–1 559 Aeronautical Mobile Satellite (R)[1]</td></tr>
<tr><td colspan="3">1 559–1 610 Aeronautical Radionavigation/Radionavigation Satellite[1]</td></tr>
<tr><td colspan="3">1 610–1 626.5 Aeronautical Radionavigation</td></tr>
<tr><td colspan="3">1 626.5–1 645.5 Maritime Mobile Satellite[2]</td></tr>
<tr><td colspan="3">1 645.5–1 646.5 Mobile Satellite[2]</td></tr>
<tr><td colspan="3">1 646.5–1 660 Aeronautical Mobile Satellite (R)[2]</td></tr>
<tr><td colspan="3">1 660–1 660.5 Aeronautical Mobile Satellite (R)[2]/Radio Astronomy</td></tr>
<tr><td colspan="3">1 660.5–1 668.4 Radio Astronomy/Space Research[3]</td></tr>
<tr><td colspan="3">1 668.4–1 670 Meteorological Aids/Fixed/Mobile*/Radio Astronomy</td></tr>
<tr><td colspan="3">1 670–1 690 Meteorological Aids/Fixed/Meteorological Satellite[1]/Mobile*</td></tr>
<tr><td colspan="3">1 690–1 700 Meteorological Aids/Meteorological Satellite[1]</td></tr>
</table>

CHART 1 (CONT). ALLOCATION TO SERVICES

Megahertz

Region 1	Region 2	Region 3
1 700–1 710 Fixed/ Meteorological Satellite[1]	*1 700–1 710* Fixed/ Meteorological Satellite[1]/ Mobile*	
1 710–2 290 Fixed	*1 710–2 290* Fixed/ Mobile	
2 290–2 300 Fixed/ Space Research[1]	*2 290–2 300* Fixed/ Mobile*/ Space Research[1]	
2 300–2 450 Fixed	*2 300–2 500* Fixed/ Mobile/ Radiolocation	
2 450–2 500 Fixed/ Mobile		
2 500–2 690 Fixed/ Mobile*/ Broadcasting Satellite	*2 500–2 655* Fixed/ Fixed Satellite[1]/ Mobile*/ Broadcasting Satellite	*2 500–2 535* Fixed/ Fixed Satellite[1]/ Mobile*/ Broadcasting Satellite
		2 535–2 655 Fixed/ Mobile*/ Broadcasting Satellite
	2 655–2 690 Fixed/ Fixed Satellite[1, 2]/ Mobile*/ Broadcasting Satellite	*2 655–2 690* Fixed/ Fixed Satellite[2]/ Mobile*/ Broadcasting Satellite
2 690–2 700	Earth Exploration Satellite[3]/ Radio Astronomy/ Space Research[3]	
2 700–2 900	Aeronautical Radionavigation	
2 900–3 100	Radionavigation	
3 100–3 400	Radiolocation	
3 400–4 200 Fixed/ Fixed Satellite[1]	*3 400–3 500* Fixed/ Fixed Satellite[1]	
	3 500–4 200 Fixed/ Fixed Satellite[1]/ Mobile*	
4 200–4 400	Aeronautical Radionavigation	
4 400–4 500	Fixed/ Mobile	
4 500–4 800	Fixed/ Fixed Satellite[1]/ Mobile	
4 800–4 990	Fixed/ Mobile	
4 990–5 000	Fixed/ Mobile*/ Radio Astronomy	
5 000–5 250	Aeronautical Radionavigation	
5 250–5 350	Radiolocation	
5 350–5 460	Aeronautical Radionavigation	
5 460–5 470	Radionavigation	
5 470–5 650	Maritime Radionavigation	
5 650–5 725	Radiolocation	
5 725–5 850 Fixed Satellite[2]/ Radiolocation	*5 725–5 850* Radiolocation	
5 850–7 075	Fixed/ Fixed Satellite[2]/ Mobile	
7 075–7 250	Fixed/ Mobile	
7 250–7 300	Fixed/ Fixed Satellite[1]/ Mobile	
7 300–7 450	Fixed/ Fixed Satellite[1]/ Mobile*	
7 450–7 550	Fixed/ Fixed Satellite[1]/ Meteorological Satellite[1]/ Mobile*	
7 550–7 750	Fixed/ Fixed Satellite[1]/ Mobile*	
7 750–7 900	Fixed/ Mobile*	

Continued on next page.

CHART 1 (CONT). ALLOCATION TO SERVICES

Region 1	Region 2	Region 3

Megahertz

<table>
<tr><td colspan="3">7 900–8 025 Fixed/Fixed Satellite[2]/Mobile</td></tr>
<tr><td>8 025–8 175 Fixed/Fixed Satellite[2]/Mobile</td><td>8 025–8 175 Earth Exploration Satellite[1]/Fixed/Fixed–Satellite[2]/Mobile</td><td>8 025–8 175 Fixed/Fixed Satellite[2]/Mobile</td></tr>
<tr><td>8 175–8 215 Fixed/Fixed Satellite[2]/Meteorological Satellite[2]/Mobile</td><td>8 175–8 215 Earth Exploration Satellite[1]/Fixed/Fixed Satellite[2]/Meteorological Satellite[2]/Mobile</td><td>8 175–8 215 Fixed/Fixed Satellite[2]/Meteorological Satellite[2]/Mobile</td></tr>
<tr><td>8 215–8 400 Fixed/Fixed Satellite[2]/Mobile</td><td>8 215–8 400 Earth Exploration Satellite[1]/Fixed/Fixed Satellite[2]/Mobile</td><td>8 215–8 400 Fixed/Fixed Satellite[2]/Mobile</td></tr>
<tr><td colspan="3">8 400–8 500 Fixed/Mobile*/Space Research[1]</td></tr>
<tr><td colspan="3">8 500–8 750 Radiolocation</td></tr>
<tr><td colspan="3">8 750–8 850 Radiolocation/Aeronautical Radionavigation</td></tr>
<tr><td colspan="3">8 850–9 000 Radiolocation/Maritime Radionavigation</td></tr>
<tr><td colspan="3">9 000–9 200 Aeronautical Radionavigation</td></tr>
<tr><td colspan="3">9 200–9 300 Radiolocation/Maritime Radionavigation</td></tr>
<tr><td colspan="3">9 300–9 500 Radionavigation</td></tr>
<tr><td colspan="3">9 500–9 800 Radiolocation/Radionavigation</td></tr>
<tr><td colspan="3">9 800–10 000 Radiolocation</td></tr>
</table>

Gigahertz

<table>
<tr><td>10–10.45 Fixed/Mobile/Radiolocation</td><td>10–10.45 Radiolocation</td><td>10–10.45 Fixed/Mobile/Radiolocation</td></tr>
<tr><td colspan="3">10.45–10.5 Radiolocation</td></tr>
<tr><td>10.5–10.55 Fixed/Mobile</td><td colspan="2">10.5–10.55 Fixed/Mobile/Radiolocation</td></tr>
<tr><td colspan="3">10.55–10.6 Fixed/Mobile*</td></tr>
<tr><td colspan="3">10.6–10.68 Earth Exploration Satellite[3]/Fixed/Mobile*/Radio Astronomy/Space Research[3]</td></tr>
<tr><td colspan="3">10.68–10.7 Earth Exploration Satellite[3]/Radio Astronomy/Space Research[3]</td></tr>
<tr><td>10.7–11.7 Fixed/Fixed Satellite[1,2]/Mobile*</td><td colspan="2">10.7–11.7 Fixed/Fixed Satellite[1]/Mobile*</td></tr>
<tr><td rowspan="4">11.7–12.5 Fixed/Broadcasting/Broadcasting Satellite</td><td>11.7–12.1 Fixed/Fixed Satellite[1]</td><td rowspan="2">11.7–12.2 Fixed/Mobile*/Broadcasting/Broadcasting Satellite</td></tr>
<tr><td rowspan="2">12.1–12.3 Fixed/Fixed Satellite[1]/Mobile*/Broadcasting/Broadcasting Satellite</td></tr>
<tr><td rowspan="2">12.2–12.5 Fixed/Mobile*/Broadcasting</td></tr>
<tr><td rowspan="2">12.3–12.7 Fixed/Mobile*/Broadcasting/Broadcasting Satellite</td></tr>
<tr><td rowspan="2">12.5–12.75 Fixed Satellite[1,2]</td><td rowspan="2">12.5–12.75 Fixed/Fixed Satellite[1]/Mobile*/Broadcasting Satellite</td></tr>
<tr><td>12.7–12.75 Fixed/Fixed Satellite[2]/Mobile*</td></tr>
</table>

CHART 1 (CONT). ALLOCATION TO SERVICES

Region 1	Region 2	Region 3

Gigahertz

<table>
<tr><td>12.75–13.25</td><td colspan="2">Fixed/Fixed Satellite[2]/Mobile</td></tr>
<tr><td>13.25–13.4</td><td colspan="2">Aeronautical Radionavigation</td></tr>
<tr><td>13.4–14</td><td colspan="2">Radiolocation</td></tr>
<tr><td>14–14.3</td><td colspan="2">Fixed Satellite[2]/Radionavigation</td></tr>
<tr><td>14.3–14.4 Fixed/Fixed Satellite[2]/Mobile*</td><td>14.3–14.4 Fixed Satellite[2]</td><td>14.3–14.4 Fixed/Fixed Satellite[2]/Mobile*</td></tr>
<tr><td>14.4–14.5</td><td colspan="2">Fixed/Fixed Satellite[2]/Mobile*</td></tr>
<tr><td>14.5–14.8</td><td colspan="2">Fixed/Fixed Satellite[2]/Mobile</td></tr>
<tr><td>14.8–15.35</td><td colspan="2">Fixed/Mobile</td></tr>
<tr><td>15.35–15.4</td><td colspan="2">Earth Exploration Satellite[3]/Radio Astronomy/Space Research[3]</td></tr>
<tr><td>15.4–15.7</td><td colspan="2">Aeronautical Radionavigation</td></tr>
<tr><td>15.7–17.3</td><td colspan="2">Radiolocation</td></tr>
<tr><td>17.3–17.7</td><td colspan="2">Fixed Satellite[2]</td></tr>
<tr><td>17.7–18.6</td><td colspan="2">Fixed/Fixed Satellite[1]/Mobile</td></tr>
<tr><td>18.6–18.8 Fixed/Fixed Satellite[1]/Mobile*</td><td>18.6–18.8 Earth Exploration Satellite[3]/Fixed/Fixed Satellite[1]/Mobile*/Space Research[3]</td><td>18.6–18.8 Fixed/Fixed Satellite[1]/Mobile*</td></tr>
<tr><td>18.8–19.7</td><td colspan="2">Fixed/Fixed Satellite[1]/Mobile</td></tr>
<tr><td>19.7–20.2</td><td colspan="2">Fixed Satellite[1]</td></tr>
<tr><td>20.2–21.2</td><td colspan="2">Fixed Satellite[1]/Mobile Satellite[1]</td></tr>
<tr><td>21.2–21.4</td><td colspan="2">Earth Exploration Satellite[3]/Fixed/Mobile/Space Research[3]</td></tr>
<tr><td>21.4–22</td><td colspan="2">Fixed/Mobile</td></tr>
<tr><td>22–22.21</td><td colspan="2">Fixed/Mobile*</td></tr>
<tr><td>22.21–22.5</td><td colspan="2">Earth Exploration Satellite[3]/Fixed/Mobile*/Radio Astronomy/Space Research[3]</td></tr>
<tr><td>22.5–22.55 Fixed/Mobile</td><td colspan="2">22.5–22.55 Fixed/Mobile/Broadcasting Satellite</td></tr>
<tr><td>22.55–23 Fixed/Intersatellite/Mobile</td><td colspan="2">22.55–23 Fixed/Intersatellite/Mobile/Broadcasting Satellite</td></tr>
<tr><td>23–23.55</td><td colspan="2">Fixed/Intersatellite/Mobile</td></tr>
<tr><td>23.55–23.6</td><td colspan="2">Fixed/Mobile</td></tr>
<tr><td>23.6–24</td><td colspan="2">Earth Exploration Satellite[3]/Radio Astronomy/Space Research[3]</td></tr>
<tr><td>24–24.05</td><td colspan="2">Amateur/Amateur Satellite</td></tr>
<tr><td>24.05–24.25</td><td colspan="2">Radiolocation</td></tr>
<tr><td>24.25–25.25</td><td colspan="2">Radionavigation</td></tr>
<tr><td>25.25–27</td><td colspan="2">Fixed/Mobile</td></tr>
<tr><td>27–27.5 Fixed/Mobile</td><td colspan="2">27–27.5 Fixed/Fixed Satellite[2]/Mobile</td></tr>
<tr><td>27.5–29.5</td><td colspan="2">Fixed/Fixed Satellite[2]/Mobile</td></tr>
<tr><td>29.5–30</td><td colspan="2">Fixed Satellite[2]</td></tr>
<tr><td>30–31</td><td colspan="2">Fixed Satellite[2]/Mobile Satellite[2]</td></tr>
<tr><td>31–31.3</td><td colspan="2">Fixed/Mobile</td></tr>
</table>

Continued on next page.

CHART 1 (CONT). ALLOCATION TO SERVICES

Region 1	Region 2	Region 3

Gigahertz

31.3–31.5	Earth Exploration Satellite[3]/Radio Astronomy/Space Research[3]
31.5–31.8	Earth Exploration Satellite[3]/Radio Astronomy/Space Research[3]
31.8–32	Radionavigation
32–33	Intersatellite/Radionavigation
33–33.4	Radionavigation
33.4–35.2	Radiolocation
35.2–36	Meteorological Aids/Radiolocation
36–37	Earth Exploration Satellite[3]/Fixed/Mobile/Space Research[3]
37–37.5	Fixed/Mobile
37.5–39.5	Fixed/Fixed Satellite[1]/Mobile
39.5–40.5	Fixed/Fixed Satellite[1]/Mobile/Mobile Satellite[1]
40.5–42.5	Broadcasting Satellite/Broadcasting
42.5–43.5	Fixed/Fixed Satellite[2]/Mobile*/Radio Astronomy
43.5–47	Mobile/Mobile Satellite/Radionavigation/Radionavigation Satellite
47–47.2	Amateur/Amateur Satellite
47.2–50.2	Fixed/Fixed Satellite[2]/Mobile
50.2–50.4	Earth Exploration Satellite[3]/Fixed/Mobile/Space Research[3]
50.4–51.4	Fixed/Fixed Satellite[2]/Mobile
51.4–54.25	Earth Exploration Satellite[3]/Space Research[3]
54.25–58.2	Earth Exploration Satellite[3]/Fixed/Inter–satellite/Mobile/Space Research[3]
58.2–59	Earth Exploration Satellite[3]/Space Research[3]
59–64	Fixed/Intersatellite/Mobile/Radiolocation
64–65	Earth Exploration Satellite[3]/Space Research[3]
65–66	Earth Exploration Satellite/Space Research
66–71	Mobile/Mobile Satellite/Radionavigation/Radionavigation Satellite
71–74	Fixed/Fixed Satellite[2]/Mobile/Mobile Satellite[2]
74–75.5	Fixed/Fixed Satellite[2]/Mobile
75.5–76	Amateur/Amateur Satellite
76–81	Radiolocation
81–84	Fixed/Fixed Satellite[1]/Mobile/Mobile Satellite[1]
84–86	Fixed/Mobile/Broadcasting/Broadcasting Satellite
86–92	Earth Exploration Satellite[3]/Radio Astronomy/Space Research[3]
92–95	Fixed/Fixed–Satellite[2]/Mobile/Radiolocation
95–100	Mobile/Mobile Satellite/Radionavigation/Radionavigation Satellite

CHART 1 (CONT). ALLOCATION TO SERVICES

Region 1	Region 2	Region 3

Gigahertz

Frequency	Allocation
100–102	Earth Exploration Satellite[3]/Fixed/Mobile/Space Research[3]
102–105	Fixed/Fixed Satellite[1]/Mobile
105–116	Earth Exploration Satellite[3]/Radio Astronomy/Space Research[3]
116–126	Earth Exploration Satellite[3]/Fixed/Inter-satellite/Mobile/Space Research[3]
126–134	Fixed/Intersatellite/Mobile/Radiolocation
134–142	Mobile/Mobile Satellite/Radionavigation/Radionavigation Satellite
142–144	Amateur/Amateur Satellite
144–149	Radiolocation
149–150	Fixed/Fixed Satellite[1]/Mobile
150–151	Earth Exploration Satellite[3]/Fixed/Fixed Satellite[1]/Mobile/Space Research[3]
151–164	Fixed/Fixed Satellite[1]/Mobile
164–168	Earth Exploration Satellite[3]/Radio Astronomy/Space Research[3]
168–170	Fixed/Mobile
170–174.5	Fixed/Intersatellite/Mobile
174.5–176.5	Earth Exploration Satellite[3]/Fixed/Inter-satellite/Mobile/Space Research[3]
176.5–182	Fixed/Intersatellite/Mobile
182–185	Earth Exploration Satellite[3]/Radio Astronomy/Space Research[3]
185–190	Fixed/Intersatellite/Mobile
190–200	Mobile/Mobile Satellite/Radionavigation/Radionavigation Satellite
200–202	Earth Exploration Satellite[3]/Fixed/Mobile/Space Research[3]
202–217	Fixed/Fixed Satellite[2]/Mobile
217–231	Earth Exploration Satellite[3]/Radio Astronomy/Space Research[3]
231–235	Fixed/Fixed Satellite[1]/Mobile
235–238	Earth Exploration Satellite[3]/Fixed/Fixed Satellite[1]/Mobile/Space Research[3]
238–241	Fixed/Fixed Satellite[1]/Mobile
241–248	Radiolocation
248–250	Amateur/Amateur Satellite
250–252	Earth Exploration Satellite[3]/Space Research[3]
252–265	Mobile/Mobile Satellite/Radionavigation/Radionavigation Satellite
265–275	Fixed/Fixed Satellite[2]/Mobile/Radio Astronomy
275–400	(Not allocated)

[1] Space to earth.
[2] Earth to space.
[3] Passive.
* Except aeronautical mobile.

(1.4) Emission in which the main carrier is amplitude- and angle-modulated either simultaneously or in a pre-established sequence D
(1.5) Emission of pulses[1]
 (1.5.1) Sequence of unmodulated pulses P
 (1.5.2) A sequence of pulses
  (1.5.2.1) modulated in amplitude K
  (1.5.2.2) modulated in width/duration L
  (1.5.2.3) modulated in position/phase M
  (1.5.2.4) in which the carrier is angle-modulated during the period of the pulse Q
  (1.5.2.5) which is a combination of the foregoing or is produced by other means V
(1.6) Cases not covered above, in which an emission consists of the main carrier modulated, either simultaneously or in a pre-established sequence, in a combination of two or more of the following modes: amplitude, angle, pulse W
(1.7) Cases not otherwise covered X

Second Symbol—Nature of Signal(s) Modulating the Main Carrier:
(2.1) No modulating signal 0
(2.2) A single channel containing quantized or digital information without the use of a modulating subcarrier[2] 1
(2.3) A single channel containing quantized or digital information with the use of a modulating subcarrier[2] 2
(2.4) A single channel containing analog information 3
(2.5) Two or more channels containing quantized or digital information 7
(2.6) Two or more channels containing analog information 8
(2.7) Composite system with one or more channels containing quantized or digital information, together with one or more channels containing analog information 9
(2.8) Cases not otherwise covered X

Third Symbol—Type of Information To Be Transmitted[3]:
(3.1) No information transmitted N
(3.2) Telegraphy—for aural reception A
(3.3) Telegraphy—for automatic reception B
(3.4) Facsimile C
(3.5) Data transmission, telemetry, telecommand D
(3.6) Telephony (including sound broadcasting) E
(3.7) Television (video) F
(3.8) Combination of the above W
(3.9) Cases not otherwise covered X

Two more optional symbols can be added to the basic characteristics described above for a more complete description of an emission (Appendix 6, Radio Regulations, ITU; Geneva, 1982). They are:
 Fourth symbol—Detail of signal(s)
 Fifth symbol—Nature of multiplexing
Where the fourth or the fifth symbol is not used, this should be indicated by a dash where the symbol would otherwise appear.

Fourth Symbol—Details of Signal(s):
(4.1) Two-condition code with elements of differing numbers and/or durations A
(4.2) Two-condition code with elements of the same number and duration without error correction B
(4.3) Two-condition code with elements of the same number and duration with error correction C
(4.4) Four-condition code in which each condition represents a signal element (of one or more bits) D
(4.5) Multicondition code in which each condition represents a signal element (of one or more bits) E
(4.6) Multicondition code in which each condition or combination of conditions represents a character F
(4.7) Sound of broadcasting quality (monophonic) G
(4.8) Sound of broadcasting quality (stereophonic or quadraphonic) H
(4.9) Sound of commercial quality (excluding categories given in subparagraphs 1.10 and 1.11) J
(4.10) Sound of commercial quality with the use of frequency inversion or band splitting K
(4.11) Sound of commercial quality with separate frequency-modulated signals to control the levels of demodulated signal L
(4.12) Monochrome M
(4.13) Color N
(4.14) Combination of the above W
(4.15) Cases not otherwise covered X

Fifth Symbol—Nature of Multiplexing:
(5.1) None N
(5.2) Code-division multiplex* C
(5.3) Frequency-division multiplex F

---

[1]Emissions where the main carrier is directly modulated by a signal which has been coded into quantized form (e.g., pulse code modulation) should be designated under (1.2) or (1.3).

[2]This excludes time-division multiplex.

[3]In this context the word "information" does not include information of a constant, unvarying nature such as is provided by standard-frequency emissions, continuous-wave and pulse radars, etc.

*This includes bandwidth expansion techniques.

network. These indicators consist of two letters signifying the country and its telegraph network (if more than one) followed by two letters signifying the town on that network. Examples: Vienna AUWI, Panama City (Tropical Radio) PAPA, Balboa (ITTCACR) PZBA, Stockholm SWSM, San Francisco (ITT Worldcom) UISF.

The CCITT has approved a worldwide numbering system for telex services. The telex destination code consists of two or three numerical digits signifying the country or network within the country. The destination code is followed by the telex subscriber's national number, also consisting of numerical digits.

The telex system provides also for *designation* codes, for identifying the country and network of the originator of a communication. The designation code consists of two letters, the same two letters that compose the first half of the message-retransmission-system *destination* indicator.

Examples of destination codes are:

North and Central America: 200 Cuba, 205 Puerto Rico (RCA), 206 Puerto Rico (ITTWC), 207 Puerto Rico (C & W), 21 Canada (except TWX), 22 Mexico, 25 USA (TWX), 271 Guatemala, 275 British Honduras, 290 Bermuda, 292 Virgin Islands.

South America: 304 Surinam, 305 Paraguay, 31 Venezuela, 36 Peru, 381 Brazil (Radio Brazil), 383 Brazil (PTT), 387 Argentina (ITTCM), 390 Netherlands Antilles, 391 Trinidad.

Europe: 400 Canary Islands, 403 Spain, 409 Algeria, 41 Germany, 46 Belgium, 492 Syria, 496 Kuwait, 501 Iceland, 51 United Kingdom, 57 Finland.

Eastern Europe: 601 Greece, 606 Israel, 61 Hungary, 64 USSR, 65 Romania.

Pacific: 702 Guam, 704 Hawaii (RCA), 705 Hawaii (ITTWC), 71 Australia, 72 Japan, 75 Philippines.

Asia: 801 Korea, 802 Hong Kong, 81 India, 85 China, 88 Iran.

Africa: 901 Libya, 907 Southern Rhodesia, 91 United Arab Republic, 94 Ghana, 95 South Africa, 972 Dahomey, 975 Niger, 981 Congo (Brazzaville), 982 Congo (Leopoldville), 991 Angola, 992 Mozambique.

CCITT AND TELEPHONY

International Country Codes

The addressing signals of worldwide automatic telephony consist of the national telephone number, as used for long-distance dialing within a country, prefixed by a country code. Country codes are grouped by continental regions; for example, the country codes of all South American countries begin with "5." Where the national numbering system includes more than one country, the country code may also include the countries included in the national system. Thus the country code for the United States—"1"—includes Canada and some other countries. The following are examples of some country codes, grouped by world numbering regions or zones, as assigned by the Third Plenary Assembly of the CCITT in Geneva in 1964.

Zone 1—Code 1: USA, Canada, Mexico and Central America, Bahamas, Bermuda, Jamaica, French Antilles, Netherlands Antilles.
Zone 2—Africa: 51 countries, 48 country codes (Algeria, Morocco, Tunisia, Libya in one group—the Maghreb— code 21). United Arab Republic 20, South Africa 27, 45 three-digit codes.
Zones 3 and 4—Europe, Iceland, Malta, Cyprus: 17 two-digit and 13 three-digit country codes. Examples: France 33, Spain 34, Italy 39, United Kingdom 44, Germany 49, Iceland 354, Finland 401, Hungary 402.
Zone 5—South America and Cuba: 6 two-digit and 8 three-digit country codes. Examples: Cuba 53, Argentina 54, Brazil 55, Chile 56, Columbia 57, Venezuela 58, Peru 596.
Zone 6—Southwestern Pacific: 6 two-digit and 14 three-digit country codes. Examples: Malaysia 60, Australia 61, Indonesia 62, Philippines 63, New Zealand 64, Thailand 66, Guam 682.
Zone 7—Country code 7: Soviet Union
Zone 8—Northwestern Pacific: 4 two-digit and 6 three-digit country codes. Examples: Japan 81, Korea 82, Vietnam 84, China (Formosa) 85, Hong Kong 852, Mongolia 854, Laos 856.
Zone 9—East: 5 two-digit and 15 three-digit country codes. Examples: India 91, Burma 95, Iran 98, Lebanon 961, Saudi Arabia 966, Israel 972.

TELEPHONE SIGNALING

CCITT signaling systems have been standardized for international use. General descriptions are given in Table 4, and some of the signaling characteristics are given in Table 5.

Signals in communications are used for passing information, for identifying the called subscriber or addressee (with resulting internal system signals concerned with the establishment of a connection), and for supervising and controlling the connection once it has been established.

Information signals may be analog (voice, telemetry, or facsimile) or digital (teleprinter or data).

Addressing signals may be dial pulse, multifrequency, or binary. They are not needed once a communication has been established.

(A) *Dial pulse* signals consist of a series of from one to ten pulses representing the corresponding numerical digits 1 to 9 and 0. The pulses are breaks

TABLE 4. CCITT SIGNALING SYSTEMS

No.	Systems
1	500/20-hertz system used in the international manual service (ringdown). Used until 1980.
2	600/750-hertz two-frequency systen. Never used in international service.
	International Automatic and Semiautomatic Systems
3	For unidirectional operation of circuits. Uses one in-band frequency (2280 hertz) for the transmission of both line and interregister signals; used for terminal traffic; in general not to be used for new installations. Used until late 1970s.
4	For undirectional operation of circuits (circuits seized from one end only). Uses two in-band frequencies (2040 and 2400 hertz) for the end-to-end transmission of both line and register signals; used for international intracontinental traffic; suitable for terminal and transit traffic; in the latter case two or three circuits equipped with System No. 4 may be switched in tandem. Suitable for submarine- or land-cable circuits and microwave radio circuits; not applicable to TASI-equipped systems. Capable of interworking with System No. 5
5	For both-way operation of circuits. Uses two in-band signaling frequencies (2400 and 2600 hertz) for the link-by-link transmission of line signals, and six in-band frequencies (700, 900, 1100, 1300, 1500, and 1700 hertz) in a two-out-of-six code (numerical information transmitted *en bloc*) for the link-by-link transmission of register signals; used for intercontinental traffic. Suitable for submarine- or land-cable circuits and microwave links, whether or not TASI is used; suitable for terminal or transit traffic—in the latter case, two or more circuits equipped with System No. 5 may be switched in tandem but are subject to possible undesirable delays if all are TASI-equipped. Capable of interworking with System No. 4
6	Based on principles of common channel signaling techniques in which the signaling is removed from the voice path; for both analog and digital transmission facilities; signaling link carriers information by serial mode of synchronous data transmission; link-by-link operation; can be associated, quasi-associated, or nonassociated mode; capable of interworking with No. 4, No. 5, and No. 5 bis. (Yellow Book, V1.3)
7	A common channel system specified in 1980; conceived for the digital environment; preferred system for integrated digital networks (IDN) for both telephony and circuit switched data; designed for interexchange signaling in integrated services digital network (ISDN); optimized for operation over 64 kb/s digital channels; employs packet-type information transfer. Flexible enough to evolve with future enhanced service requirements. (Yellow Book, VI.6)

in a continuous direct current on the line, usually lasting from 58 to 67 percent of the time interval between the starts of successive pulses. These breaks in direct current may have to be converted into pulses in a tone, or to frequency shifts between tones, in order to pass through some media or multiplexing systems. Dial pulse speeds are usually 10 pulses per second, although high-speed dials of nearly twice that speed are sometimes used.

(**B**) *Multifrequency* signals represent numerical digits by one pulse (or sometimes four pulses) of a specific frequency combination. The two-out-of-six and four-by-four multifrequency codes are shown in Tables 6 and 7. Inherent in both is the constant-ratio error-control principle (the simultaneous receipt of three or more, or one only, frequencies indicates an error). Table 8 shows the US Army two-out-of-five numerical code. The CCITT two-voice-frequency code, consisting of two frequencies sent one at a time in four pulses, is given in Table 9.

(**C**) *Binary* signals for addressing are usually in a numeric or alphanumeric code used also for information signals, such as in telegraph addresses or headings. Multifrequency signals are sometimes directly converted to binary signals by changing the two frequencies-out-of-six code to a corresponding two-time-slots-out-of-six synchronous six-element binary code. Another example of a binary code used only for addressing is the CCITT one-voice-frequency code, and four-element start-stop code given in Table 10. (The CCITT two-voice-frequency code is not truly binary since it uses a third condition—no tone—in addition to the two tones.)

Supervisory or Line Signals

Supervisory or line signals cannot be generated exclusively by registers because they are required during the entire use of the connection, after the registers that established the connection have been disconnected. Supervisory signals are an extension of the original basic signals of ringing and of closing a line loop to allow direct current to flow. Supervisory signals may be classified as spurt (discontinuous) and continuous.

Continuous signals are based on conditions of on-hook and off-hook, representing the condition of locked or flowing direct current on the subscriber's line, and their extension to trunk signaling is given in Table 11. Either condition is continuous and may be detected at any time. On the other hand, discontinuous signals must be recorded, and the condition represented is presumed to continue until a new signal is sent. Supervisory signaling in a backward direction (toward the calling end) is also needed in automatic working and is also described as on-hook or off-hook, although on two-wire metallic circuits the signaling condition is usually a

TABLE 5. LINE SIGNALS IN THREE CCITT SYSTEMS

Signal	Direction	CCITT No. 3 (*1VF*)	CCITT No. 4 (*2VF*)	CCITT No. 5 (*2VF*)
Seize, terminal	→	*X*	*PX*	*X*
Seize, transit	→		*PY*	
Start pulsing, terminal	←	*X*	*X*	*Y*
Start pulsing, transit	←		*Y*	
End of pulsing (ST)	→	250 ms*	*xSxSxSx**	1500+1700 hertz*
Busy	←	*XX*	*PX*	*Y*
Acknowledge	→			*X*
Answer	←	*XSX*	*PY*	*X*
Acknowledge	→			*X*
Clear back (on-hook)	←	*XX*	*PX*	*Y*
Acknowledge	→			*X*
Ring forward	→	*XSX*	*PYY*	*Y*(850±200 ms)
Clear forward (disconnect)	→	*XXSXX*	*PXX*	*X+Y*
Release guard (disconnect acknowledge)	←	*XXSXX*	*PYY*	*X+Y*
		X: 2280±6 hertz, 150±30 ms *XX*: 2280±6 hertz, 600±120 ms *S*: 100±20 ms silence	*X*: 2040±6 hertz, 100±20 ms *Y*: 2400±6 hertz, 100±20 ms *XX*: 350±70 ms *YY*: 350±70 ms *S*: 35±7 ms; *x*: 2040 hertz, 35 ms *P*: (2040 hertz, 2400 hertz), 150+30ms	*X*: 2400±6 hertz *Y*: 2600±6 hertz

* Combination No. 15 of address code.

TABLE 6. MULTIFREQUENCY NUMERICAL CODE USED BY CCITT (TWO-OUT-OF-SIX)

Digit	Frequencies	Weighting
1	700 + 900	0 + 1
2	700 + 1100	0 + 2
3	900 + 1100	1 + 2
4	700 + 1300	0 + 4
5	900 + 1300	1 + 4
6	1100 + 1300	2 + 4
7	700 + 1500	0 + 7
8	900 + 1500	1 + 7
9	1100 + 1500	2 + 7
0	1300 + 1500	4 + 7
Code 11	700 + 1700	0 + 11 } for inward
Code 12	900 + 1700	1 + 11 } operators
KP	1100 + 1700	2 + 11 start of pulsing
KP2	1300 + 1700	4 + 11 transit traffic
ST	:1500 + 1700	7 + 11 end of pulsing

TABLE 8. US ARMY TA-341/PT NUMERICAL CODE

Digit	Frequencies
1	2100 + 2300
2	2300 + 2500
3	1900 + 2700
4	1900 + 2100
5	2500 + 2700
6	2300 + 2700
7	2100 + 2500
8	1900 + 2300
9	2100 + 2700
0	1900 + 2500

reversal of flow of the direct current rather than an interruption.

Continuous supervisory signaling over longer distances is effected by use of signaling paths distinct from the voice path. These signaling paths may be telegraph legs of a composite telegraph system, simplexing of the voice pair, or special tones inside or outside of the voice channel. Whatever the paths, they are extensions of separate direct-current leads from the trunk circuits, known as *E* and *M* leads. The relation between the conditions of these leads and the on-hook or off-hook signaling conditions they represent is shown in Fig. 6. The usual method of extending the *E* and *M* leads over tone channels is shown in Fig. 7. The frequency of the tone used is preferably higher than 2000 hertz and is usually 2600 hertz on four-wire circuits. A 3825-hertz signaling system is known as "out-of-band" and must be built into the carrier system using it.

Spurt signaling avoids the necessity of using distinct signaling paths. For manual operation, ring-down signaling is usually satisfactory. For semiautomatic or automatic operation, however, more elaborate systems are required. Voice-frequency signals are used, and they are distinguished from

TABLE 7. NUMERICAL FOUR-BY-FOUR MULTIFREQUENCY CODE

Touch-Tone or Touch Calling

Low group (hertz)				
697	1	2	3	
770	4	5	6	
852	7	8	9	
941	spare	0	spare	
High group (hertz)	1209	1336	1477	(1633)

US Air Force 412L

Low group (hertz)				
1020	1	2	3	
1140	4	5	6	
1260	7	8	9	
1380		0		
High group (hertz)	1620	1740	1860	(1980)

Note: Each digit is composed of one frequency from the low group and one frequency from the high group. The frequencies have been chosen to minimize voice simulation.

TABLE 9. NUMERICAL CODE, TWO-VOICE-FREQUENCY SIGNALING SYSTEM, CCITT NO. 4

Digit	Successive Elements 1	2	3	4
1	*y*	*y*	*y*	*x*
2	*y*	*y*	*x*	*y*
3	*y*	*y*	*x*	*x*
4	*y*	*x*	*y*	*y*
5	*y*	*x*	*y*	*x*
6	*y*	*x*	*x*	*y*
7	*y*	*x*	*x*	*x*
8	*x*	*y*	*y*	*y*
9	*x*	*y*	*y*	*x*
0	*x*	*y*	*x*	*y*

Note: The two frequencies are sent one at a time, with a silent space between pulses. The duration of both frequency and silent periods is 35±7 milliseconds. Frequencies: x = 2040± 6 hertz; y = 2400±6 hertz. Power level: −9 decibels.

TABLE 10. NUMERICAL CODE, ONE-VOICE-FREQUENCY SIGNALING SYSTEM, CCITT NO. 3.

Digit	Time Elements Start	1	2	3	4	Stop
1	1				1	
2	1			1		
3	1			1	1	
4	1		1			
5	1		1		1	
6	1		1	1		
7	1		1	1	1	
8	1	1				
9	1	1			1	
0	1	1		1		

Note: "1" signifies frequency present. Length of each time element is 50 milliseconds±1 percent. Frequency: 2280±6 hertz. Power level: −6 decibels.

TABLE 11. ON-HOOK AND OFF-HOOK SIGNALS

Direct Current Telephone Line	Trunk
On-hook signifies loop is open to direct current supplied from other end. *Off-hook* signifies loop is closed, allowing relay at other end to operate. Signaling in reverse direction is ringdown.	If idle, signals on-hook to other end. Seizure at calling end signals off-hook to called end. While calling end awaits answer, called end signals on-hook to calling end. Answer results in signaling off-hook from called end. If called end is not ready to receive address signals when seized, it signals off-hook to calling end until ready.

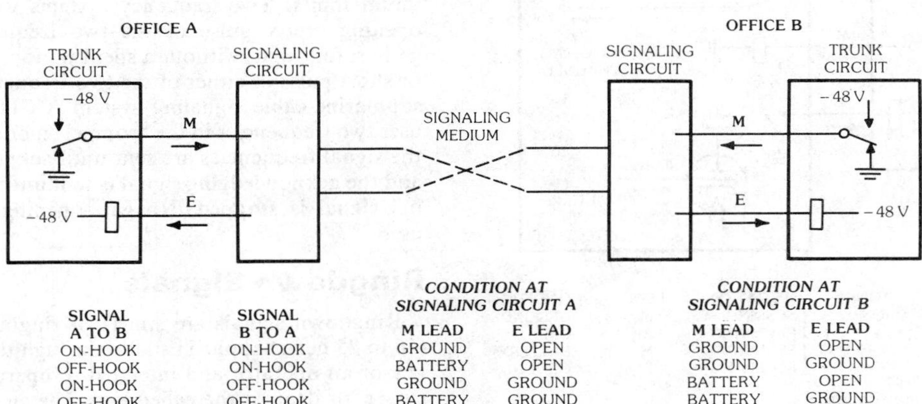

SIGNAL A TO B	SIGNAL B TO A	CONDITION AT SIGNALING CIRCUIT A: M LEAD	E LEAD	CONDITION AT SIGNALING CIRCUIT B: M LEAD	E LEAD
ON-HOOK	ON-HOOK	GROUND	OPEN	GROUND	OPEN
OFF-HOOK	ON-HOOK	BATTERY	OPEN	GROUND	GROUND
ON-HOOK	OFF-HOOK	GROUND	GROUND	BATTERY	OPEN
OFF-HOOK	OFF-HOOK	BATTERY	GROUND	BATTERY	GROUND

Fig. 6. *E* and *M* signaling. The *E* lead receives open or ground signals from the signaling circuit. The *M* lead sends ground or battery signals to the signaling circuit.

TABLE 12. TABLE OF PROPOSED STANDARD OF AUDIBLE TONES IN NORTH AMERICA (FROM CCITT DOCUMENT AP III-84)

Use	Frequencies§ (hertz) 350	440	480	620	Power per Frequency at Exchange Where Tone is Applied (dBm0)	Cadence
Dial tone	x	x			−13	Continuous
Busy tone			x	x	−24	0.5 second on 0.5 second off
Reorder tone*			x	x	−24	0.2 second on 0.3 second off or 0.3 second on 0.2 second off
Audible ringing tone		x	x		−16	2 seconds on 4 seconds off
High tone†			x		−16	Varies according to use
Preemption tone‡		x		x	−18	Single 200/500-ms pulse
Call-waiting tone		x			−13	Single 500-ms pulse

Notes:

* A possible alternative is the use of a call-failure tone, which would identify the office and type of condition that prevented the successful completion of the call.

† High tone is used in many ways. For example:

(A) Spurts of tone to indicate specific orders to operators in the manual service (order tones).

(B) To inform operators of lines that are temporarily out of service (permanent signal tone).

(C) To alert customers that their services are in a permanent off-hook condition.

‡ Preemption tones are used in certain private switched networks which may interconnect with the national networks.

§ Frequency limits are ±0.5 percent of nominal.

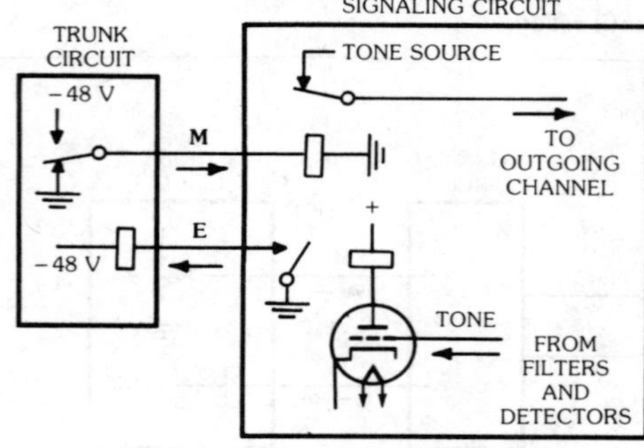

SIGNAL	TONE	OPERATION	LEAD	CONDITION
ON-HOOK	ON	SENDING	M	GROUND
		RECEIVING	E	OPEN
OFF-HOOK	OFF	SENDING	M	BATTERY
		RECEIVING	E	GROUND

Fig. 7. Tone signaling.

voice transmissions by filters and timing. Single-frequency systems use one or two long or short pulses of the specific signaling frequency, with the pulses carefully timed to within minimum and maximum limits. Two-frequency systems use a gate-opening prefix pulse of the two frequencies together, followed, without a silent period, by a long or short pulse of either of the two frequencies. The submarine-cable signaling system (CCITT No. 5) uses two frequencies in the "compel" mode, wherein the signal frequencies are sent until acknowledged, and the acknowledging signal is sent until the original signal is stopped. No gate-opening prefix is used.

Ringdown Signals

Ringdown signals are spurts of ringing current (16 to 25 hertz) applied usually through the ringing key of an operator and intended to operate a bell, ringer, or drop at the called end. The current may be generated by a manually operated magneto or by a ringing machine with or without automatically inserted silent periods. Ringing to telephone subscribers in automatic central offices is stopped or "tripped" automatically by relay action resulting from the subscriber's off-hook condition. Ringing signals may be converted to 500 or 1000 hertz, usually interrupted at a 20-hertz rate, to pass through voice channels of carrier equipment. A ringing signal to a manual switchboard usually lights a switchboard lamp, which can be darkened again only by local action and not by stopping or repeating the ringing signal. This characteristic makes ringdown operation unsuitable for fully automatic operation. Ringdown signaling over carrier circuits has the advantages of simplicity and of not requiring the distinct signaling channels of *E* and *M* systems.

Tones

A special case of signaling is that of information in the form of tones to the subscriber of a telephone system. The basic tones are dial tone, busy tone, and ring-back tone (representing the ringing of the called subscriber's line). The dial tone is generated at the subscriber's local switching center, but the busy tone and ring-back tone, plus special tones such as no-such-number and line-out-of-order, are generated at the called subscriber's switching center and should be standardized for universal intelligibility. A proposal by the American Telephone and Telegraph Company for standard tones is described in Table 12. On some international calls, the busy and ring-back tones are generated locally in accordance with spurt supervisory signals from the called end.

Alternative Routing

Switching systems in which the complete called number is recorded in the first center to which the subscriber is connected permit a translation of digits from those identifying the called subscriber to those most conveniently used by the switching mechanism to establish the desired connection. This translation of digits permits the controlling mechanism to use a variety of routes in such a way that if the first-choice route is occupied or disabled a second-choice route may be tried, etc. Safeguards are required to prevent doubling back of routes and dead-end choices. In the near future, many sophisticated plans may be expected that depend on rapid analysis of network possibilities by data-transmission means. The signals used for this purpose are interregister signals and are classed with addressing signals in that they are used in establishing a connection to the desired addressee, not being needed after a connection has been established.

3 Units, Constants, and Conversion Factors

Revised by
Eugene A. Mechtly

SYMBOLS FOR UNITS*

Unit symbols are letters, combinations of letters, or other characters that may be used in place of the names of the units. The following list covers symbols for *units* only. It does not cover letter symbols for physical *quantities.*

> Example: In the expression $I = 150$ mA, I is the symbol for a physical quantity (current), A is the symbol for the unit of current (ampere), and m is the symbol for the prefix milli. Together, m and A form the symbol for a submultiple unit of current, the milliampere.

The unit symbols listed in this section are documented in NBS Special Publication 330, 1981.

When an unfamiliar unit symbol is first used in text, it should be followed by its name in parentheses. Only the symbol need be used thereafter.

Symbols for units are written in lower-case (small) letters, except for the first letter if the name of the unit is derived from a proper name, and except for a very few that are not formed from letters. Every effort should be made to follow the distinction between upper- and lower-case letters, even if the symbols appear in applications where the other lettering is in upper-case style.

Symbols for units are printed in roman (upright) type. Their form is the same for both singular and plural, and they are not followed by a period. When there is risk of confusion in using the standard symbols, e.g., "l" for liter and "s" for second, the name of the unit should be spelled out.

When a compound unit is formed by multiplication of two or more other units, its symbol consists of the symbols for the separate units joined by a raised dot (for example, N · m for newton meter). The dot may be omitted in the case of familiar compounds if no confusion would result. For example, V · s and V s are both acceptable representations of the unit weber, Wb. When a unit symbol prefix is identical to a unit symbol, special care must be taken. For example, the symbol m · N indicates the product of the units *meter* and *newton,* while mN is the symbol for *millinewton.* Hyphens should not be used in symbols for compound units.

Positive and negative exponents may be used with unit symbols, but care must be taken in text to avoid confusion with superscripts that indicate footnotes or references.

When a compound unit is formed by division of one unit by another, its symbol consists of the symbols for the separate units either separated by a solidus (for example, m/s for meter per second) or multiplied using negative powers (for example, $m \cdot s^{-1}$ for meter per second). In simple cases, use of the solidus is recommended, but in no case should more than one solidus on the same line, or a solidus followed by a product, be included in such a combination unless parentheses are inserted to avoid ambiguity. In complicated cases, negative powers should be used.

Prefixes

The prefixes used to indicate multiples or submultiples of units are listed in Table 1.

TABLE 1. PREFIXES

Multiple	Prefix	Symbol
10^{18}	exa	E
10^{15}	peta	P
10^{12}	tera	T
10^{9}	giga	G
10^{6}	mega	M
10^{3}	kilo	k
10^{2}	hecto	h
10	deka	da
10^{-1}	deci	d
10^{-2}	centi	c
10^{-3}	milli	m
10^{-6}	micro	μ
10^{-9}	nano	n
10^{-12}	pico	p
10^{-15}	femto	f
10^{-18}	atto	a

Symbols for prefixes are printed in roman type, without space between the prefix and the symbol for the unit. The distinctions between upper- and lower-case letters must be observed.

Compound prefixes should not be used:

Use		Do not use	
tera	T	megamega	MM
giga	G	kilomega	kM
nano	n	millimicro	mμ
pico	p	micromicro	μμ

When a symbol representing a unit that has a prefix carries an exponent, this indicates that the multiple (or submultiple) unit is raised to the power expressed by the exponent. For example:

$$2\ \mathrm{cm}^3 = 2(\mathrm{cm})^3 = 2(10^{-2}\mathrm{m})^3 = 2 \cdot 10^{-6}\ \mathrm{m}^3$$

$$1\ \mathrm{ms}^{-1} = 1(\mathrm{ms})^{-1} = 1(10^{-3}\mathrm{s})^{-1} = 10^3\mathrm{s}^{-1}$$

Numerical Values

To facilitate the reading of numbers, the digits may be separated into groups of three, counting from the decimal sign toward the left and the right. The groups should be separated by a small space, but not by a comma or a point. In numbers of four digits, the space is usually not necessary. For example:

2.141 596   73 772   7372   0.133 47

* This section is a revision of a section in the sixth edition. Much of the material in that section was taken from ANSI Y10.19-1969.

TABLE 2. SI UNITS

Name	Symbol		Quantity
ampere (an SI base unit)	A		electric current
ampere per meter	A/m		magnetic field strength
ampere per square meter	A/m^2		current density
becquerel	Bq	s^{-1}	activity (of a radionuclide)
candela (an SI base unit)	cd		luminous intensity
candela per square meter	cd/m^2		luminance
coulomb	C	s·A	electric charge, quantity of electricity
coulomb per cubic meter	C/m^3		electric charge density
coulomb per kilogram	C/kg		exposure (x and gamma rays)
coulomb per square meter	C/m^2		electric flux density
cubic meter	m^3		volume
cubic meter per kilogram	m^3/kg		specific volume
degree Celsius	°C		Celsius temperature
farad	F	C/V	capacitance
farad per meter	F/m		permittivity
gray	Gy	J/kg	absorbed dose, specific energy imparted, kerma, absorbed dose index
gray per second	Gy/s		absorbed dose rate
henry	H	Wb/A	inductance
henry per meter	H/m		permeability
hertz	Hz	s^{-1}	frequency
joule	J	N·m	energy, work, quantity of heat
joule per cubic meter	J/m^3		energy density
joule per kelvin	J/K		heat capacity, entropy
joule per kilogram	J/kg		specific energy
joule per kilogram kelvin	J/(kg·K)		specific heat capacity, specific entropy
joule per mole	J/mol		molar energy
joule per mole kelvin	J/(mol·K)		molar entropy, molar heat capacity
kelvin (an SI base unit)	K		thermodynamic temperature
kilogram (an SI base unit)	kg		mass
kilogram per cubic meter	kg/m^3		density, mass density
lumen	lm		luminous flux
lux	lx	lm/m^2	illuminance
meter (an SI base unit)	m		length
meter per second	m/s		speed, velocity
meter per second squared	m/s^2		acceleration
mole (an SI base unit)	mol		amount of substance
mole per cubic meter	mol/m^3		concentration (of amount of substance)
newton	N		force
newton meter	N·m		moment of force
newton per meter	N/m		surface tension
number per meter	m^{-1}		wave number
ohm	Ω	V/A	electric resistance
pascal	Pa	N/m^2	pressure, stress
pascal second	Pa·s		dynamic viscosity
radian	rad		plane angle
radian per second	rad/s		angular velocity
radian per second squared	rad/s^2		angular acceleration
second (an SI base unit)	s		time
siemens	S	A/V	electric conductance
sievert	Sv	J/kg	dose equivalent, dose equivalent index
square meter	m^2		area
steradian	sr		solid angle
tesla	T	Wb/m^2	magnetic flux density
volt	V	W/A	electric potential, potential difference, electromotive force
volt per meter	V/m		electric field strength
watt	W	J/s	power, radiant flux
watt per meter kelvin	W/(m·K)		thermal conductivity
watt per square meter	W/m^2		heat flux density, irradiance, power density
watt per square meter steradian	$W \cdot m^{-2} \cdot sr^{-1}$		radiance
watt per steradian	W/sr		radiant intensity
weber	Wb	V·s	magnetic flux

American-English documents use a period for the decimal sign. British-English documents use a raised dot (·). For all other languages, the recommended decimal sign is a comma.

If the magnitude of a number is less than unity, the decimal sign should be preceded by a zero.

The sign of multiplication of numbers is a cross (×) or a raised dot (·).

SI Units

Table 2 contains only units of the International System of Units (Système International d'Unités), which is the name given in 1960 by the Conférence Générale des Poids et Mesures to the coherent system of units that has the following base units and quantities:

Unit	Quantity
meter	length
kilogram	mass
second	time
ampere	electric current
kelvin	thermodynamic temperature
mole	amount of substance
candela	luminous intensity

Units of this system are called SI units.*

Common multiples of the SI units are listed in Table 3. Miscellaneous non-SI units, most of which are deprecated by the IEEE, are listed in Table 4.

FUNDAMENTAL NUMERICAL CONSTANTS

$\pi = 3.141\,592\,653\,589$
e = 2.718 281 828 459

TEMPERATURE

The kelvin (symbol K) is the SI unit of thermodynamic temperature, T. The degree Celsius (symbol °C) is the commonly used unit of Celsius temperature, t. Celsius temperature is defined by the equation:

$$t = T - T_0$$

where $T_0 = 273.15$ kelvins.

Fahrenheit temperature, t_F (°F), and Rankine temperature, t_R (°R), are converted to kelvins or to degrees Celsius by the equations:

$$T = (5/9)(t_F + 459.67)$$
$$t = (5/9)(t_F - 32)$$
$$T = (5/9)t_R$$

* The International System of Units is becoming *the* system of units in most fields of science and engineering throughout the world, so where appropriate SI units have been introduced in the text. Nevertheless, non-SI units have been retained when required by working standards and/or current industrial practice.

TABLE 3. MULTIPLES OF SI UNITS IN FREQUENT USE

Unit	Symbol
centimeter	cm
cubic centimeter	cm^3
cubic meter per second	m^3/s
gigahertz	GHz
gram	g
kilohertz	kHz
kilohm	kΩ
kilojoule	kJ
kilometer	km
kilovolt	kV
kilovoltampere	kVA
kilowatt	kW
megahertz	MHz
megavolt	MV
megawatt	MW
megohm	MΩ
microampere	μA
microfarad	μF
microgram	μg
microhenry	μH
micrometer	μm
microsecond	μs
microwatt	μW
milliampere	mA
milligram	mg
millihenry	mH
millimeter	mm
millisecond	ms
millivolt	mV
milliwatt	mW
nanoampere	nA
nanofarad	nF
nanometer	nm
nanosecond	ns
nanowatt	nW
picoampere	pA
picofarad	pF
picosecond	ps
picowatt	pW

DECIBELS AND NEPERS

The decibel, symbol dB, is a dimensionless unit for representing the ratio of two values of power. The number, n, of decibels is 10 times the logarithm to the base 10 of the power ratio:

$$n(\text{dB}) = 10 \log_{10} (P_2/P_1)$$

In the electrical case of voltage, V, current, I, and impedance, Z (when Z is a pure resistance):

$$P_2/P_1 = (V_2/V_1)^2 (Z_1/Z_2) = (I_2/I_1)^2 (Z_2/Z_1)$$

Thus,

TABLE 4. MISCELLANEOUS NON-SI UNITS

(These units are deprecated unless marked with the symbol *.)

Unit	Symbol	Notes
ampere-hour	Ah	
ampere (turn)	At	Unit of magnetomotive force
angstrom	Å	1 Å = 10^{-10} m
apostilb	asb	1 asb = $(1/\pi)$ cd/m^2 A unit of luminance. The SI unit, candela per square meter, is preferred.
atmosphere:		
standard atmosphere	atm	1 atm = 101 325 N/m^2
technical atmosphere	at	1 at = 1kg$_f$/cm^2
*atomic mass unit (unified)	u	The (unified) atomic mass unit is defined as one-twelfth of the mass of an atom of the ^{12}C nuclide. Use of the old atomic mass unit (amu), defined by reference to oxygen, is deprecated.
bar	bar	1 bar = 100 000 N/m^2
barn	b	1 b = 10^{-28} m^2
baud	Bd	Unit of signaling speed equal to one element per second.
*bel	B	
*billion electronvolts		The name *billion electronvolts* is deprecated; see *giga-electronvolt* (GeV).
*bit	b	
British thermal unit	Btu	
candle		The unit of luminous intensity has been given the name *candela;* use of the word *candle* for this purpose is deprecated.
circular mil	cmil	1 cmil = $\pi/4) \cdot 10^{-6}$ in^2
cubic foot	ft^3	
cubic foot per minute	ft^3/min	
cubic foot per second	ft^3/s	
cubic inch	in^3	
cubic yard	yd^3	
curie	Ci	Unit of activity in the field or radiation dosimetry.
cycle	c	
cycle per second	c/s	Deprecated. Use hertz.
*decibel	dB	
*degree (plane angle)	°	
degree (temperature):		
degree Celsius	°C	
degree Fahrenheit	°F	Note that there is no space between the symbol ° and the letter. The use of the word *centigrade* for the Celsius temperature scale was abandoned by the Conférence Générale des Poids et Mesures in 1948.
degree Rankine	°R	
dyne	dyn	
*electronvolt	eV	
erg	erg	
erlang	E	Unit of telephone traffic.
foot	ft	
footcandle	fc	The name *lumen per square foot* (lm/ft^2) is recommended for this unit. Use of the SI unit of illuminance, the lux (lumen per square meter), is preferred.

Continued on next page.

TABLE 4 (CONT). MISCELLANEOUS NON-SI UNITS
(These units are deprecated unless marked with the symbol *.)

Unit	Symbol	Notes
footlambert	fL	If luminance is to be measured in English units, the candela per square inch (cd/in^2) is recommended. Use of the SI unit, the candela per square meter, is preferred.
foot per minute	ft/min	
foot per second	ft/s	
foot per second squared	ft/s^2	
foot pound-force	ft · lb_f	
gal	Gal	1 Gal = 1cm/s^2
gallon	gal	The gallon, quart, and pint differ in the US and the UK, and their use is deprecated.
gauss	G	The gauss is the electromagnetic cgs (centimeter gram second) unit of magnetic flux density. The SI unit, tesla, is preferred.
*gigaelectronvolt	GeV	
gilbert	Gb	The gilbert is the electromagnetic cgs (centimeter gram second) unit of magnetomotive force. Use of the SI unit, the ampere (or ampere-turn), is preferred.
grain	gr	
horsepower	hp	Use of the SI unit, the watt, is preferred.
*hour	h	Time may be designated as in the following example: $9^h46^m30^s$.
inch	in	
inch per second	in/s	
*kiloelectronvolt	keV	
kilogauss	kG	
kilogram-force	kg_f	In some countries the name *kilopond* (kp) has been adopted for this unit.
kilometer per hour	km/h	
kilopond	kp	See kilogram-force.
kilowatthour	kWh	
knot	kn	1 kn = 1 nmi/h
lambert	L	The lambert is the cgs (centimeter gram second) unit of luminance. The SI unit, candela per square meter, is preferred.
*liter	l	
*liter per second	l/s	
lumen per square foot	lm/ft^2	Use of the SI unit, the lumen per square meter, is preferred.
*lumen per square meter	lm/m^2	Unit of luminous exitance.
*lumen per watt	lm/W	Unit of luminous efficacy.
*lumen second	lm · s	Unit of quantity of light.
maxwell	Mx	The maxwell is the electromagnetic cgs (centimeter gram second) unit of magnetic flux. Use of the SI unit, the weber, is preferred.
*megaelectronvolt	MeV	
mho	mho	1 mho = 1 Ω^{-1} = 1 S
microbar	μbar	
micron		the name *micrometer* (μm) is preferred.
mil	mil	1 mil = 0.001 in.

Continued on next page.

TABLE 4 (CONT). MISCELLANEOUS NON-SI UNITS

(These units are deprecated unless marked with the symbol *.)

Unit	Symbol	Notes
mile		
nautical	nmi	
statute	mi	
mile per hour	mi/h	
millibar	mbar	mb may be used.
milligal	mGal	
*milliliter	ml	
millimeter of mercury	mm Hg	1 mm Hg = 133.322 N/m^2.
millimicron		The name *nanometer* (nm) is preferred.
minute (plane angle)	. . .′	
*minute (time)	min	Time may be designated as in the following example: $9^h46^m30^s$.
nautical mile	nmi	
*neper	Np	
nit	nt	1 nt = 1 cd/m^2
oersted	Oe	The oersted is the electromagnetic cgs (centimeter gram second) unit of magnetic field strength. Use of the SI unit, the ampere per meter, is preferred.
ounce (avoirdupois)	oz	
pint	pt	The gallon, quart, and pint differ in the US and the UK, and their use is deprecated.
pound	lb	
pound-force	lb_f	
pound-force foot	$lb_f \cdot ft$	
pound-force per square inch	lb_f/in^2	
pound per square inch		Although use of the abbreviation psi is common, it is not recommended. See pound-force per square inch.
poundal	pdl	
quart	qt	The gallon, quart, and pint differ in the US and the UK, and their use is deprecated.
rad	rd	Unit of absorbed dose in the field of radiation dosimetry.
rem	rem	Unit of dose equivalent in the field of radiation dosimetry.
revolution per minute	r/min	Although use of the abbreviation rpm is common, it is not recommended.
revolution per second	r/s	
roentgen	R	Unit of exposure in the field of radiation dosimetry.
second (plane angle)	. . .″	
square foot	ft^2	
square inch	in^2	
square yard	yd^2	
stilb	sb	1 sb = 1 cd/cm^2. Use of the SI unit,the candela per square meter, is preferred.
tonne	t	1 t = 1000 kg.
*(unified) atomic mass unit	u	See atomic mass unit (unified) .
var	var	Unit of reactive power.
*voltampere	V · A	SI unit of apparent power.
watthour	W · h	
yard	yd	

TABLE 5. EXAMPLES OF POWER, CURRENT, AND VOLTAGE RATIOS CONVERTED TO DECIBELS AND NEPERS

Power Ratio	Z-Matched Voltage or Current Amplitude Ratio	Decibels	Nepers	Power Ratio	Z-Matched Voltage or Current Amplitude Ratio	Decibels	Nepers
1.0233	1.0116	0.1	0.01	19.953	4.4668	13.0	1.50
1.0471	1.0233	0.2	0.02	25.119	5.0119	14.0	1.61
1.0715	1.0351	0.3	0.03	31.623	5.6234	15.0	1.73
1.0965	1.0471	0.4	0.05	39.811	6.3096	16.0	1.84
1.1220	1.0593	0.5	0.06	50.119	7.0795	17.0	1.96
1.1482	1.0715	0.6	0.07	63.096	7.9433	18.0	2.07
1.1749	1.0839	0.7	0.08	79.433	8.9125	19.0	2.19
1.2023	1.0965	0.8	0.09	100.00	10.0000	20.0	2.30
1.2303	1.1092	0.9	0.10	158.49	12.589	22.0	2.53
1.2589	1.1220	1.0	0.12	251.19	15.849	24.0	2.76
1.3183	1.1482	1.2	0.14	398.11	19.953	26.0	2.99
1.3804	1.1749	1.4	0.16	630.96	25.119	28.0	3.22
1.4454	1.2023	1.6	0.18	1000.0	31.623	30.0	3.45
1.5136	1.2303	1.8	0.21	1584.9	39.811	32.0	3.68
1.5849	1.2589	2.0	0.23	2511.9	50.119	34.0	3.91
1.6595	1.2882	2.2	0.25	3981.1	63.096	36.0	4.14
1.7378	1.3183	2.4	0.28	6309.6	79.433	38.0	4.37
1.8197	1.3490	2.6	0.30	10^4	100.000	40.0	4.60
1.9055	1.3804	2.8	0.32	1.5849×10^4	125.89	42.0	4.83
1.9953	1.4125	3.0	0.35	2.5119×10^4	158.49	44.0	5.06
2.2387	1.4962	3.5	0.40	3.9811×10^4	199.53	46.0	5.29
2.5119	1.5849	4.0	0.46	6.3096×10^4	251.19	48.0	5.52
2.8184	1.6788	4.5	0.52	10^5	316.23	50.0	5.76
3.1623	1.7783	5.0	0.58	1.5849×10^5	398.11	52.0	5.99
3.5481	1.8836	5.5	0.63	2.5119×10^5	501.19	54.0	6.22
3.9811	1.9953	6.0	0.69	3.9811×10^5	630.96	56.0	6.45
5.0119	2.2387	7.0	0.81	6.3096×10^5	794.33	58.0	6.68
6.3096	2.5119	8.0	0.92	10^6	1000.00	60.0	6.91
7.9433	2.8184	9.0	1.04	10^7	3162.3	70.0	8.06
10.0000	3.1623	10.0	1.15	10^8	10000.0	80.0	9.21
12.589	3.5481	11.0	1.27	10^9	31623	90.0	10.36
15.849	3.9811	12.0	1.38	10^{10}	100000	100.0	11.51

To convert:

Decibels to nepers, multiply by 0.1151
Decibels per statute mile to nepers per kilometer, multiply by 7.154×10^{-2}
Decibels per nautical mile to nepers per kilometer, multiply by $6.215 \times 10'$
Nepers to decibels, multiply by 8.686
Where the power ratio is less than unity, it is usual to invert the fraction and express the answer as a decibel loss.

$$n_V(dB) = 20 \log_{10}(V_2/V_1) + 10 \log_{10}(Z_1/Z_2)$$

and

$$n_I(dB) = 20 \log_{10}(I_2/I_1) + 10 \log_{10}(Z_2/Z_1)$$

Only in the special case of matched impedances ($Z_2 = Z_1$) does

$$n = n_{z\text{-matched}} = 20 \log_{10}(V_2/V_1) = 20 \log_{10}(I_2/I_1)$$

The neper, symbol Np, is a dimensionless unit for expressing the ratio of two values of amplitude. The number, n, of nepers is the natural logarithm of the amplitude ratio:

$$n(\text{Np}) = \ln (A_2/A_1)$$

Some examples of power ratios and the corresponding numbers of decibels and nepers are shown in Table 5.

CONVERSION FACTORS

Exact numerical values (based on definitions) are followed by *** ....

The symbol E marks a power-of-ten multiplier in the usual way for the entry of data into calculators and computers.

To Convert From	Multiply By		To Get Unit
abampere	1.*** *** *** *** ***	E 1	A
abcoulomb	1.*** *** *** *** ***	E 1	C
abfarad	1.*** *** *** *** ***	E 9	F
abhenry	1.*** *** *** *** ***	E– 9	H
abmho	1.*** *** *** *** ***	E 9	S
abohm	1.*** *** *** *** ***	E– 9	ohm
abvolt	1.*** *** *** *** ***	E– 8	V
acre	4.046 9–– ––– ––– –––	E 3	m^2
angstrom	1.*** *** *** *** ***	E–10	m
are	1.*** *** *** *** ***	E 2	m^2
astronomical unit (IAU)	1.495 978 70– ––– –––	E 11	m
astronomical unit (radio)	1.495 978 9–– ––– –––	E 11	m
atmosphere, standard	1.013 25* *** *** ***	E 5	Pa
atmosphere, technical, kg_f/cm^2	9.806 65* *** *** ***	E 4	Pa
bar	1.*** *** *** *** ***	E 5	Pa
barn	1.*** *** *** *** ***	E–28	m^2
barrel (petroleum, 42 gallons)	1.589 873 ––– ––– –––	E– 1	m^3
barye	1.*** *** *** *** ***	E– 1	Pa
board foot (12 cubic inches)	2.359 737 216 *** ***	E– 3	m^3
Btu (IST before 1956)	1.055 04– ––– ––– –––	E 3	J
Btu (IST after 1956)	1.055 056 ––– ––– –––	E 3	J
Btu (mean)	1.055 87– ––– ––– –––	E 3	J
Btu (thermochemical)	1.054 350 ––– ––– –––	E 3	J
Btu (39°F)	1.059 67– ––– ––– –––	E 3	J
Btu (60°F)	1.054 68– ––– ––– –––	E 3	J
bushel (US)	3.523 907 016 688 ***	E– 2	m^3
cable	2.194 56– ––– ––– –––	E 2	m
caliber (0.01 inch)	2.54* *** *** *** ***	E– 4	m
calorie (IST)	4.186 8** *** *** ***	E 0	J
calorie (mean)	4.190 02– ––– ––– –––	–––	J
calorie (thermochemical)	4.184 *** *** *** ***	E 0	J
calorie (15°C)	4.185 80– ––– ––– –––	–––	J
calorie (20°C)	4.181 90– ––– ––– –––	–––	J
calorie, kilogram (IST)	4.186 8** *** *** ***	E 0	J
calorie, kilogram (mean)	4.190 02– ––– ––– –––	E 3	J
calorie, kilogram (thermochemical)	4.184 *** *** *** ***	E 3	J
carat (metric)	2.*** *** *** *** ***	E– 4	kg
centimeter of mercury (0°C)	1.333 22– ––– ––– –––	E 3	Pa
centimeter of water (4°C)	9.806 38– ––– ––– –––	E 1	Pa
chain (engineer or Ramden)	3.048 *** *** *** ***	E 1	m
chain (surveyor or Gunter)	2.011 684 ––– ––– –––	E 1	m
circular mil	5.067 074 8–– ––– –––	E–10	m^2
clo	2.003 712 ––– ––– –––	E– 1	$K{\cdot}m^2/W$
cord	3.624 556 3–– ––– –––	–––	m^3
cubit	4.572 ––– ––– ––– –––	E– 1	m
cup	2.365 882 365 ––– –––	E– 4	m^3
curie	3.7** *** *** *** ***	E 10	Bq
day (mean solar)	8.64* *** *** *** ***	E 4	s (mean solar)
day (sidereal)	8.616 409 0–– ––– –––	E 4	s (mean solar)
degree (angle)	1.745 329 251 994 3––	E– 2	rad
denier (international)	1.*** *** *** *** ***	E– 7	kg/m
dram (avoirdupois)	1.771 845 195 312 5**	E– 3	kg
dram (troy or apothecary)	3.887 934 6** *** ***	E– 3	kg

Continued on next page.

To Convert From	Multiply By		To Get Unit
dram (US fluid)	3.696 691 195 312 5**	E- 6	m^3
dyne	1.*** *** *** *** ***	E- 5	N
electronvolt (eV)	1.602 189 2-- --- ---	E-19	J
emu of xxx (see ab---)			
erg	1.*** *** *** *** ***	E- 7	J
esu of xxx (see stat---)			
faraday (based on carbon 12)	9.648 70- --- --- ---	E 4	C
faraday (chemical)	9.649 57- --- --- ---	E 4	C
faraday (physical)	9.652 19- --- --- ---	E 4	C
fathom	1.828 804 --- --- ---	----	m
fermi (femtometer, fm)	1.*** *** *** *** ***	E-15	m
fluid ounce (US)	2.957 352 956 25* ***	E- 5	m^3
foot	3.048 *** *** *** ***	E- 1	m
foot (US survey)	1200/3937 *** *** ***	E 0	m
	3.048 006 096 --- ---	E- 1	m
foot of water (39.2°F)	2.988 98- --- --- ---	E 3	Pa
footcandle	1.076 391 0-- --- ---	E 1	lx
footlambert	3.426 259 --- --- ---	----	cd/m^2
free fall (standard, g)	9.806 65* *** *** ***	E 0	m/s^2
furlong	2.011 68- --- --- ---	E 2	m
gal	1.*** *** *** *** ***	E- 2	m/s^2
gallon (Canadian liquid)	4.546 090 --- --- ---	E- 3	m^3
gallon (UK dry)	4.404 883 770 86* ***	E- 3	m^3
gallon (UK liquid)	4.546 092 --- --- ---	E- 3	m^3
gallon (US dry)	4.404 884 --- --- ---	E- 3	m^3
gallon (US liquid)	3.785 411 784 *** ***	E- 3	m^3
gamma	1.*** *** *** *** ***	E- 9	T
gauss	1.*** *** *** *** ***	E- 4	T
gilbert	7.957 747 2-- --- ---	E- 1	A
gill (UK)	1.420 652 --- --- ---	E- 4	m^3
gill (US)	1.182 941 2-- --- ---	E- 4	m^3
grad (angle)	9.*** *** *** *** ***	E- 1	degree (angle)
	1.570 796 3-- --- ---	E- 2	rad
grain	6.479 891 *** *** ***	E- 5	kg
gram	1.*** *** *** *** ***	E- 3	kg
hand	1.016 *** *** *** ***	E- 1	m
hectare	1.*** *** *** *** ***	E 4	m^2
hogshead (US)	2.384 809 423 92* ***	E- 1	m^3
horsepower (boiler)	9.809 50- --- --- ---	E 3	W
horsepower (electric)	7.46* *** *** *** ***	E 2	W
horsepower (metric)	7.354 99- --- --- ---	E 2	W
horsepower (UK)	7.457 --- --- --- ---	E 2	W
horsepower (water)	7.460 43- --- --- ---	E 2	W
horsepower (550 $ft{\cdot}lb_f/s$)	7.456 998 7-- --- ---	E 2	W
hour (mean solar)	3.6** *** *** *** ***	E 3	s (mean solar)
hour (sidereal)	3.590 170 4-- --- ---	E 3	s (mean solar)
hundredweight (long)	5.080 234 544 *** ***	E 1	kg
hundredweight (short)	4.535 923 7** *** ***	E 1	kg
inch	2.54* *** *** *** ***	E- 2	m
inch of mercury (32°F)	3.386 389 --- --- ---	E 3	Pa
inch of mercury (60°F)	3.376 85- --- --- ---	E 3	Pa
inch of water (39.2°F)	2.490 82- --- --- ---	E 2	Pa
inch of water (60°F)	2.488 4-- --- --- ---	E 2	Pa
jansky	1.*** *** *** *** ***	E-26	$W/(m^2\ Hz)$
kayser	1.*** *** *** *** ***	E 2	m^{-1}
kilocalorie (IST)	4.186 8-- --- --- ---	E 3	J
kilocalorie (mean)	4.190 02- --- --- ---	E 3	J
kilocalorie (thermochemical)	4.184 *** *** *** ***	E 3	J

To Convert From	Multiply By		To Get Unit
kilogram force (kg_f)	9.806 65* *** *** ***	E 0	N
kilopond	9.806 65* *** *** ***	E 0	N
kip	4.448 221 615 260 5**	E 3	N
knot (international)	5.144 444 444 --- ---	E- 1	m/s
lambert	1/pi* *** *** *** ***	E 4	cd/m^2
	3.183 098 8-- --- ---	E 3	cd/m^2
langley	4.184 --- --- --- ---	E 4	J/m^2
lb_f (pound force, avoirdupois)	4.448 221 615 260 5**	E 0	N
lb_m (pound mass, avoirdupois)	4.535 923 7** *** ***	E- 1	kg
league	4.828 --- --- --- ---	E 3	m
league (international)	5.556 *** *** *** ***	E 3	m
league (US nautical)	5.559 552 *** *** ***	E 3	m
light year	9.460 55- --- --- ---	E 15	m
link (engineer or Ramden)	3.048 *** *** *** ***	E- 1	m
link (surveyor or Gunter)	2.011 68- --- --- ---	E- 1	m
liter	1.*** *** *** *** ***	E- 3	m^3
maxwell	1.*** *** *** *** ***	E- 8	Wb
meter (based on constant *c*)†			
meter (based on Kr 86)	1.650 763 73- --- ---	E 6	wavelengths
mho	1.*** *** *** *** ***	E 0	S
micron	1.*** *** *** *** ***	E- 6	m
mil	2.54* *** *** *** ***	E- 5	m
mile (international)	1.609 344 *** *** ***	E 3	m
mile (international nautical)	1.852 *** *** *** ***	E 3	m
mile (UK nautical)	1.835 184 *** *** ***	E 3	m
mile (US nautical)	1.852 *** *** *** ***	E 3	m
mile (US statute)	1.609 3-- --- --- ---	E 3	m
millibar	1.*** *** *** *** ***	E 2	Pa
millimeter of mercury (0°C)	1.333 224 --- --- ---	E 2	Pa
minute (angle)	2.908 882 086 66- ---	E- 4	rad
minute (mean solar)	6.*** *** *** *** ***	E 1	s (mean solar)
minute (sidereal)	5.983 617 4-- --- ---	E 1	s (mean solar)
month (mean calendar)	2.628 *** *** *** ***	E 6	s (mean solar)
nautical mile (international)	1.852 *** *** *** ***	E 3	m
nautical mile (UK)	1.853 184 *** *** ***	E 3	m
nautical mile (US)	1.852 *** *** *** ***	E 3	m
oersted	7.957 747 2-- --- ---	E 1	A/m
ounce (UK fluid)	2.841 307 --- --- ---	E- 5	m^3
ounce (US fluid)	2.957 352 956 25* ***	E- 5	m^3
ounce force (avoirdupois)	2.780 138 5-- --- ---	E- 1	N
ounce mass (avoirdupois)	2.834 952 312 5** ***	E- 2	kg
ounce mass (troy or apothecary)	3.110 347 68* *** ***	E- 2	kg
pace	7.62- --- --- --- ---	E- 1	m
parsec (IAU)	3.085 678 --- --- ---	E 16	m
peck (US)	8.809 767 541 72* ***	E- 3	m^3
pennyweight	1.555 173 84* *** ***	E- 3	kg
perch	5.029 2-- --- --- ---	----	m
phot	1.--- --- --- --- ---	E 4	lx
pica (printer's)	4.217 517 6** *** ***	E- 3	m
pint (US dry)	5.506 104 713 575 ***	E- 4	m^3
pint (US liquid)	4.713 764 73* *** ***	E- 4	m^3
point (printer's)	3.514 598 *** *** ***	E- 4	m
poise	1.*** *** *** *** ***	E- 1	N·s/m^2

†On October 20, 1983, the 17th CGPM adopted the following definition: "The meter is the length of the path traveled by light in vacuum during a time interval of 1/299 792 458 of a second." This new definition has the effect of giving a fixed value to the speed of light in vacuum of c = 299 792 458 m/s exactly. (Metrologia, *Vol. 19, p. 163, 1984.*)

Table continued on next page.

To Convert From	Multiply By		To Get Unit
pole	5.029 2-- --- --- ---	----	m
pound force (lb$_f$ avoirdupois)	4.448 221 615 260 5**	****	N
pound mass (lb$_m$ avoirdupois)	4.535 923 7** *** ***	E- 1	kg
pound mass (troy or apothecary)	3.732 417 216 *** ***	E- 1	kg
poundal	1.382 549 543 76* ***	E- 1	N
quad	1.055 --- --- --- ---	E 18	J
quart (US dry)	1.101 220 942 715 ***	E- 3	m^3
quart (US liquid)	9.463 529 46* *** ***	E- 4	m^3
rad (absorbed radiation dose)	1.*** *** *** *** ***	E- 2	Gy
rayleigh (photon emission rate)	1.*** *** *** *** ***	E 10	$m^{-2}\ s^{-1}$
rem (dose equivalent)	1.*** *** *** *** ***	E- 2	Sv
rhe	1.*** *** *** *** ***	E 1	$m^2/(N{\cdot}s)$
rod	5.029 2-- --- --- ---	----	m
roentgen	2.579 76* *** *** ***	E- 4	C/kg
rutherford	1.*** *** *** *** ***	E 6	Bq
scruple (apothecary)	1.295 978 2** *** ***	E- 3	kg
second (angle)	4.848 136 811 --- ---	E- 6	rad
second (ephemeris)	1.000 000 000 --- ---	----	s (SI atomic)
second (mean solar)	Consult American Ephemeris and Nautical Almanac.		s (ephemeris)
second (sidereal)	9.972 695 7-- --- ---	E- 1	s (mean solar)
section	2.589 998 --- --- ---	E 6	m^2
shake	1.--- --- --- --- ---	E- 8	s
skein	1.097 28* *** *** ***	E 2	m
slug	1.459 390 29- --- ---	E 1	kg
span	2.286 *** *** *** ***	E- 1	m
statampere	3.335 640 --- --- ---	E-10	A
statcoulomb	3.335 640 --- --- ---	E-10	C
statfarad	1.112 650 --- --- ---	E-12	F
stathenry	8.987 554 --- --- ---	E 11	H
statohm	8.987 554 --- --- ---	E 11	ohm
statute mile (US)	1.609 3-- --- --- ---	E 3	m
statvolt	2.997 925 --- --- ---	E 2	V
stere	1.*** *** *** *** ***	****	m^3
stilb	1.--- --- --- --- ---	E 4	cd/m^2
stokes	1.*** *** *** *** ***	E- 4	m^2/s
tablespoon	1.478 676 478 125 ***	E- 5	m^3
teaspoon	4.928 921 593 75* ***	E- 6	m^3
tex	1.*** *** *** *** ***	E- 6	kg/m
therm	1.055 --- --- --- ---	E 8	J
ton (assay)	2.916 666 6-- --- ---	E- 2	kg
ton (long)	1.016 046 908 8** ***	E 3	kg
ton (metric)	1.*** *** *** *** ***	E 3	kg
ton (nuclear equiv of TNT)	4.184 *** *** *** ***	E 9	J
ton (refrigeration)	3.516 800 --- --- ---	E 3	W
ton (register)	2.831 684 659 2** ***	****	m^3
ton (short, 2000 lb$_m$)	9.071 847 4** *** ***	E 3	kg
tonne	1.*** *** *** *** ***	E 3	kg
torr (0°C)	1.333 22- --- --- ---	E 2	Pa
township	9.323 994 --- --- ---	E 7	m^2
unit pole	1.256 637 --- --- ---	E- 7	Wb
yard	9.144 *** *** *** ***	E- 1	m
year (calendar, 365 days)	3.153 6** *** *** ***	E 7	s (mean solar)
year (sidereal)	3.155 815 0-- --- ---	E 7	s (mean solar)
year (tropical)	3.155 692 6-- --- ---	E 7	s (mean solar)
year 1900 (tropical, January, day 0, hour 12)	3.155 692 597 47* ***	E 7	s (ephemeris)
	3.155 692 597 47- ---	E 7	s (SI atomic)

ENERGY CONVERSION FACTORS*†

Quantity	Value*	Unit	Error (ppm)
1 kg	5.609538(24)	10^{29} MeV	4.4
1 amu	931.4812(52)	MeV	5.5
Electron mass	0.5110041(16)	MeV	3.1
Proton mass	938.2592(52)	MeV	5.5
Neutron mass	939.5527(52)	MeV	5.5
1 electronvolt	1.6021917(70)	10^{-19} J	4.4
		10^{-12} erg	
	2.4179659(81)	10^{14} Hz	3.3
	8.065465(27)	10^{5} m^{-1}	3.3
		10^{3} cm $^{+1}$	
	1.160485(49)	10^{4} K	42
Energy-wavelength conversion	1.2398541(41)	10^{-6} eV·m	3.3
		10^{-4} eV·cm	
Rydberg constant, R_∞	2.179914(17)	10^{-18} J	7.6
		10^{-11} erg	
	13.605826(45)	eV	3.3
	3.2898423(11)	10^{15} Hz	0.35
	1.578936(67)	10^{5} K	43
Bohr magneton, μ_B	5.788381(18)	10^{-5} eV T^{-1}	3.1
	1.3996108(43)	10^{10} Hz T^{-1}	3.1
	46.68598(14)	m^{-1}·T^{-1}	3.1
		10^{-2} cm^{-1}·T^{-1}	
		K T^{-1}	43
	0.671733(29)	10^{-8} eV T^{-1}	6.8
Nuclear magneton, μ_n	3.152526(21)	10^{6} Hz T^{-1}	5.5
	7.622700(42)	10^{-2} m^{-1}·T^{-1}	5.5
	2.542659(14)	10^{-4} cm^{-1}·T^{-1}	
	3.65846(16)	10^{-4} K T^{-1}	44
Gas constant, R_0	8.20562(35)	10^{-2} m^3·atm kmol^{-1}·K^{-1}	42
Standard volume of ideal gas, V_0	22.4136	m^3 kmol^{-1}	

*Compiled by B. N. Taylor, W. H. Parker, and D. N. Langenberg. Reprinted from *Reviews of Modern Physics, Vol. 41 (1969)*. The numbers in parenthesis are the standard-deviation uncertainties in the last digits of the quoted value, computed on the basis of internal consistency.

†For more recent values, see Cohen, E. R., and Taylor, B. N. "The 1973 Least-Squares Adjustment of the Fundamental Constants." *J. Phys. Chem. Ref. Data*, Vol. 2, No. 4 (1973), pp. 663-734. The International Committee on Data for Science and Technology, chaired by B. N. Taylor, is expected to publish a new adjustment in late 1984.

FUNDAMENTAL PHYSICAL CONSTANTS*†

Quantity	Symbol	Value	Error (ppm)	Units SI	Units cgs
Velocity of light	c	2.9979250(10)	0.33	10^{8} m s^{-1}	10^{10} cm s^{-1}
Fine structure constant, $[\mu_0 c^2/4\pi](e^2/\hbar c)$	α	7.297351(11)	1.5	10^{-3}	10^{-3}
	α^{-1}	137.03602(21)	1.5		
Electron charge	e	1.6021917(70)	4.4	10^{-19} C	10^{-20} emu
		4.803250(21)	4.4		10^{-10} esu
Planck's constant	h	6.626196(50)	7.6	10^{-34} J·s	10^{-27} erg·s
	$\hbar = h/2\pi$	1.0545919(80)	7.6	10^{-34} J·s	10^{-27} erg·s
Avogadro's number	N	6.022169(40)	6.6	10^{26} kmol^{-1}	10^{23} mol^{-1}
Atomic mass unit	amu	1.660531(11)	6.6	10^{-27} kg	10^{-24} g
Electron rest mass	m_e	9.109558(54)	6.0	10^{-31} kg	10^{-28} g
	m_e*	5.485930(34)	6.2	10^{-4} amu	10^{-4} amu
Proton rest mass	M_p	1.672614(11)	6.6	10^{-27} kg	10^{-24} g
	M_p*	1.00727661(8)	0.08	amu	amu
Neutron rest mass	M_n	1.674920(11)	6.6	10^{-27} kg	10^{-24} g
	M_n*	1.00866520(10)	0.10	amu	amu
Ratio of proton mass to electron mass	M_p/m_e	1836.109(11)	6.2		
Electron charge to mass ratio	e/m_e	1.7588028(54)	3.1	10^{11} C kg^{-1}	10^{7} emu g^{-1}
		5.272759(16)	3.1		10^{17} esu g^{-1}
Magnetic flux quantum, $[c]^{-1}(hc/2e)$	Φ_0	2.0678538(69)	3.3	10^{-15} T·m^2	10^{-7} G·m^2
	h/e	4.135708(14)	3.3	10^{-15} J·s C^{-1}	10^{-7} erg·s emu^{-1}
		1.3795234(46)	3.3		10^{-17} erg·s esu^{-1}
Quantum of circulation	$h/2m_e$	3.636947(11)	3.1	10^{-4} J·s kg^{-1}	erg·s g^{-1}
	h/m_e	7.273894(22)	3.1	10^{-4} J·s kg^{-1}	erg·s g^{-1}
Faraday constant, Ne	F	9.648670(54)	5.5	10^{7} C kmol^{-1}	10^{3} emu mol^{-1}
		2.892599(16)	5.5		10^{14} esu mol^{-1}

Rydberg constant, $[\mu_0 c^2/4\pi]^2\ (m_e e^4/4\pi\hbar^3 c)$	R_∞	1.09737312(11)	0.10	$10^7\ m^{-1}$	$10^5\ cm^{-1}$
Bohr radius, $[\mu_0 c^2/4\pi]^{-1}\ (\hbar^2/m_e e^2) = \alpha/4\pi R_\infty$	α_0	5.2917715(81)	1.5	10^{-11} m	10^{-9} cm
Classical electron radius, $[\mu_0 c^2/4\pi]\ (e^2/m_e c^2) = \alpha^3/4\pi R_\infty$	r_0	2.817939(13)	4.6	10^{-15} m	10^{-13} cm
Electron magnetic moment in Bohr magnetons	μ_e/μ_B	1.0011596389(31)	0.0031		
Bohr magneton, $[c](e\hbar/2m_e c)$	μ_B	9.274096(65)	7.0	$10^{-24}\ J\ T^{-1}$	$10^{-21}\ erg\ G^{-1}$
Electron magnetic moment	μ_e	9.284851(65)	7.0	$10^{-24}\ J\ T^{-1}$	$10^{-21}\ erg\ G^{-1}$
Gyromagnetic ratio of protons in H_2O	γ'_p $\gamma'_p/2\pi$	2.6751270(82) 4.257597(13)	3.1 3.1	$10^8\ rad\ s^{-1} \cdot T^{-1}$ $10^7\ Hz\ T^{-1}$	$10^4\ rad\ s^{-1} \cdot G^{-1}$ $10^3\ Hz\ G^{-1}$
γ'_p corrected for diamagnetism of H_2O	γ_p $\gamma_p/2\pi$	2.6751965(82) 4.257707(13)	3.1 3.1	$10^8\ rad\ s^{-1} \cdot T^{-1}$ $10^7\ Hz\ T^{-1}$	$10^4\ rad\ s^{-1} \cdot G^{-1}$ $10^3\ Hz\ G^{-1}$
Magnetic moment of protons in H_2O in Bohr magnetons	μ'_p/μ_B	1.52099312(10)	0.066	10^{-3}	10^{-3}
Proton magnetic moment in Bohr magnetons	μ_p/μ_B	1.52103264(46)	0.30	10^{-3}	10^{-3}
Proton magnetic moment	μ_p	1.4106203(99)	7.0	$10^{-26}\ J\ T^{-1}$	$10^{-23}\ erg\ G^{-1}$
Magnetic moment of protons in H_2O in nuclear magnetons	μ'_p/μ_n	2.792709(17)	6.2		
μ'_p/μ_n corrected for diamagnetism of H_2O	μ_p/μ_n	2.792782(17)	6.2		
Nuclear magneton, $[c](e\hbar/2M_p c)$	μ_n	5.050951(50)	10	$10^{-27}\ J\ T^{-1}$	$10^{-24}\ erg\ G^{-1}$
Compton wavelength of the electron, $h/m_e c$	λ_C $\lambda_C/2\pi$	2.4263096(74) 3.861592(12)	3.1 3.1	10^{-12} m 10^{-13} m	10^{-10} cm 10^{-11} cm
Compton wavelength of the proton, $h/M_p c$	$\lambda_{C,p}$ $\lambda_{C,p}/2\pi$	1.3214409(90) 2.103139(14)	6.8 6.8	10^{-15} m 10^{-16} m	10^{-13} cm 10^{-14} cm

Continued on next page.

Continued from preceding page.

Quantity	Symbol	Value	Error (ppm)	Units: SI	Units: cgs
Compton wavelength of the neutron, h/M_nc	$\lambda_{C,n}$	1.3196217(90)	6.8	10^{-15} m	10^{-13} cm
	$\lambda_{C,n}/2\pi$	2.100243(14)	6.8	10^{-16} m	10^{-14} cm
Gas constant	R_0	8.31434(35)	42	10^3 J $\text{kmol}^{-1}\cdot\text{K}^{-1}$	10^7 erg $\text{mol}^{-1}\cdot\text{K}^{-1}$
Boltzmann's constant R_0/N	k	1.380622(59)	43	10^{-23} J K^{-1}	10^{-16} erg K^{-1}
Stefan-Boltzmann constant, $\pi^2k^4/60\hbar^3c^2$	σ	5.66961(96)	170	10^{-8} W m^{-2} K^4	10^{-5} erg $\text{s}^{-1}\cdot\text{cm}^{-2}\cdot\text{K}^{-4}$
First radiation constant, $8\pi hc$	c_1	4.992579(38)	7.6	10^{-24} J·m	10^{-15} erg·cm
Second radiation constant, hc/k	c_2	1.438833(61)	43	10^{-2} m·K	cm·K
Gravitational constant	G	6.6732(31)	460	10^{-11} $\text{N}\cdot\text{m}^2\ \text{kg}^{-2}$	10^{-8} $\text{dyn}\cdot\text{cm}^2\ \text{g}^{-2}$
kx-unit-to-angstrom conversion factor, $\Lambda = \lambda(\text{Å})/\lambda(\text{kxu})$; $\lambda(\text{Cu}K\alpha_1) \equiv$ 1.537400 kxu	Λ	1.0020764(53)	5.3		
Å-to-angstrom conversion factor, $\Lambda = \lambda(\text{Å})/\lambda(\text{Å}^*)$; $\lambda(\text{W}K\alpha_1) \equiv$ 0.2090100 Å*	Λ^*	1.0000197(56)	5.6		

*Compiled by B. N. Taylor, W. H. Parker, and D. N. Langenberg. Reprinted from *Reviews of Modern Physics,* Vol. 41, 1969, p.375.

Note that the unified atomic mass scale $^{12}\text{C} \equiv 12$ has been used throughout, that amu = atomic mass unit, C = coulomb, G = gauss, Hz = hertz = cycles/s, J = joule, K = kelvin, T = tesla (10^4 G), V = volt, and W = watt. In cases where formulas for constants are given (e.g., $R\infty$), the relations are written as the product of two factors. The second factor, in parentheses, is the expression to be used when all quantities are expressed in cgs units, with the electron charge in electrostatic units. The first factor, in brackets, is to be included only if all quantities are expressed in SI units. With the exception of the auxiliary constants which have been taken to be exact, the uncertainties of these constants are correlated, and therefore the general law of error propagation must be used in calculating additional quantities requiring two or more of these constants. The numbers in parentheses are the standard-deviation uncertainties in the last digits of the quoted value, computed on the basis of internal consistency.

†For more recent values, see Cohen, E. R., and Taylor, B. N. "The 1973 Least-Squares Adjustment of the Fundamental Constants." *J. Phys. Chem. Ref. Data,* Vol. 2, No. 4 (1973), pp. 663–734. The International Committee on Data for Science and Technology, chaired by B. N. Taylor, is expected to publish a new adjustment in late 1984.

GREEK ALPHABET

Name	Capital	Small	Commonly Used to Designate
Alpha	A	α	Angles, coefficients, attenuation constant, absorption factor, area
Beta	B	β	Angles, coefficients, phase constant
Gamma	Γ	γ	Complex propagation constant (cap), specific gravity, angles, electrical conductivity, propagation constant
Delta	Δ	δ	Increment or decrement (cap or small), determinant (cap), permittivity (cap), density, angles
Epsilon	E	ϵ	Dielectric constant, permittivity, base of natural logarithms, electric intensity
Zeta	Z	ζ	Coordinates, coefficients
Eta	H	η	Intrinsic impedance, efficiency, surface charge density, hysteresis, coordinates
Theta	Θ	ϑ, θ	Angular phase displacement, time constant, reluctance, angles
Iota	I	ι	Unit vector
Kappa	K	κ	Susceptibility, coupling coefficient, thermal conductivity
Lambda	Λ	λ	Permeance (cap), wavelength, attenuation constant
Mu	M	μ	Permeability, amplification factor, prefix micro
Nu	N	ν	Reluctivity, frequency
Xi	Ξ	ξ	Coordinates
Omicron	O	o	
Pi	Π	π	3.1416
Rho	P	ρ	Resistivity, volume charge density, coordinates
Sigma	Σ	σ	Summation (cap), surface charge density, complex propagation constant, electrical conductivity, leakage coeffieicnt, deviation
Tau	T	τ	Time constant, volume resistivity, time-phase displacement, transmission factor, density
Upsilon	Υ	υ	
Phi	Φ	ϕ, φ	Scalar potential (cap), magnetic flux, angles
Chi	X	χ	Electric susceptibility, angles
Psi	Ψ	ψ	Dielectric flux, phase difference, coordinates, angles
Omega	Ω	ω	Resistance in ohms (cap), solid angle (cap), angular velocity

Note: Small letter is used where capital (cap) is indicated.

4 Properties of Materials

Revised by
Gilbert Y. Chin and Eugene A. Mechtly

GENERAL PROPERTIES OF THE ELEMENTS

Some properties of the elements are listed in Table 1.* Some of the listed quantities are defined as follows.

Atomic number Z represents the number of protons per atom.

Mass number $Z + N$ is equal to the number of protons Z plus the number of neutrons N present in the nuclei. Mass numbers of the most abundant isotopes are given in order of decreasing abundance. For example, Cadmium Cd-48 mass numbers 114–112 means that cadmium atoms of greater abundance (28.86%) have a mass number of 114; that is, the nucleus of Cd^{114} has $114 - 48 = 66$ neutrons while isotope Cd^{112} of lower abundance (24.07%) has $112 - 48 = 64$ neutrons.

Atomic radius values listed provide a comparison of sizes (deduced from interatomic spacing of bound atoms).

Gram atomic volume in cubic centimeters gives the volume occupied in the solid state by an atom at its melting point. The gram atomic volume contains the Avogadro number of atoms (6.0225×10^{23}).

Electronegativity represents the relative tendency of an atom to attract shared electron pairs. The highest electronegativity is assigned to fluorine with the value 3.90.

First ionization potential is the work in electron-volts required to pull 1 electron off an isolated neutral atom.

$$1 \text{ electronvolt} = 1.602 \times 10^{-19} \text{ joule}$$

Electron work function, expressed in electron-volts, represents the energy that must be supplied to an electron to cross over the surface barrier of a metal. That energy may be supplied by heat (thermionic work function), by light (photoelectric work function), or by contact with a dissimilar metal (contact potential).

Electrochemical equivalents are expressed in ampere-hours per gram liberated at the electrode.

*Tables 1 and 2, Chart 1, and Fig. 1 of this chapter are partly based on data from the following sources: *Handbook of Chemistry and Physics*, 55th ed., CRC Press, Inc.; 1974. *Fundamentals of Chemistry*, John Wiley & Sons. *The Encyclopedia of Electrochemistry*, Reinhold Publishing Corp.; 1964. *American Institute of Physics Handbook*, 3rd ed., McGraw-Hill Book Co.; 1972. *Lange's Handbook of Chemistry*, 11th ed., McGraw-Hill Book Co.; 1973.

PERIODIC CLASSIFICATION OF THE ELEMENTS

Fig. 1 is a periodic table of the elements.

Oxidation number is defined as the charge that an atom appears to have in a compound when electrons are counted according to certain rules:

(A) In the free elements each atom has an oxidation number of 0.

(B) Electrons shared between two unlike atoms are counted with the more electronegative atom.

(C) Electrons shared between the two like atoms are divided equally between sharing atoms.

PHYSICAL PROPERTIES OF THE ELEMENTS

Some of the physical properties of the elements are listed in Table 2.

GALVANIC SERIES IN SEA WATER

In sea water, two dissimilar metals connected by a conductor form a galvanic cell. If the two metals are in different groups of Chart 1 (separated by spaces), the metal coming first in the series—starting from corroded end to protected end—will be anodic (i.e., corroded by the metal contained in the group farther from the corroded end). If the two metals are in the same group, no appreciable corrosive action will take place.

TEMPERATURE-EMF CHARACTERISTICS OF THERMOCOUPLES*

Fig. 2 shows temperature-emf characteristics of thermocouples.

Electromotive Force and Other Properties

Electromotive force and other properties of thermocouples are listed in Tables 3 and 4.

CONDUCTING MATERIALS

Conducting materials (Tables 5–7) can be classified as follows:

Conductors: Resistive from 10^{-6} to 10^{-4} ohm-cm (1 to 100 microhm-cm). Conductivities from 10^4 to 10^6 S-cm^{-1}.

Semiconductors: Resistivities from 10^{-4} to 10^9 ohm-cm. Conductivities from 10^{-9} to 10^4 S-cm^{-1}.

*R. L. Weber, *Temperature Measurement and Control* (Philadelphia: Blakiston Co., 1941; pp. 68–71).

TABLE 1. PROPERTIES

	Symbol	Atomic Number Z	Mass Number $Z+N$	Relative Atomic Mass	Atomic Radius (Å)	Gram Atomic Volume (cm^3)
Actinium	Ac	89	227	227		
Aluminum	Al	13	27	26.98	1.25	10
Americium	Am	95	243	243		
Antimony	Sb	51	121-123	121.75	1.41	18
Argon	Ar or A	18	40	39.948	1.74	24
Arsenic	As	33	75	74.92	1.21	16
Astatine	At	85	210	210		
Barium	Ba	56	138	137.34	1.98	38
Berkelium	Bk	97	247	247		
Beryllium	Be	4	9	9.012	0.89	5
Bismuth	Bi	83	209	208.98	1.52	21
Boron	B	5	11	10.81	0.88	5
Bromine	Br	35	79-81	79.904	1.14	23
Cadmium	Cd	48	114-112	112.40	1.41	13
Calcium	Ca	20	40	40.08	1.74	26
Californium	Cf	98	251	251		
Carbon	C	6	12	12.011	0.77	5
Cerium	Ce	58	140	140.12	1.65	21
Cesium	Cs	55	133	132.905	2.35	71
Chlorine	Cl	17	35	35.453	0.99	19
Chromium	Cr	24	52	51.996	1.17	7
Cobalt	Co	27	59	58.933	1.16	7
Copper	Cu	29	63	63.546	1.17	7
Curium	Cm	96	247	247		
Dysprosium	Dy	66	164-162-163	162.50	1.59	19
Einsteinium	Es or E	99	254	254		
Erbium	Er	68	166-168-167	167.26	1.57	18
Europium	Eu	63	153-151	151.96	1.85	29
Fermium	Fm	100	257	257		
Fluorine	F	9	19	18.998	0.64	15
Francium	Fr	87	223	223		
Gadolinium	Gd	64	158-160-156	157.25	1.61	20
Gallium	Ga	31	69-71	69.72	1.25	12
Germanium	Ge	32	74-72-70	72.59	1.22	13
Gold	Au	79	197	196.967	1.34	10
Hafnium	Hf	72	180-178-177	178.49	1.44	13
Helium	He	2	4	4.003		32
Holmium	Ho	67	165	164.93	1.58	19
Hydrogen	H	1	1	1.008	0.37	13
Indium	In	49	115	114.82	1.50	16
Iodine	I	53	127	126.904	1.33	26
Iridium	Ir	77	193-191	192.22	1.26	9
Iron	Fe	26	56	55.847	1.17	7
Krypton	Kr	36	84-86	83.80	1.89	33
Lanthanum	La	57	139	138.905	1.69	22
Lawrencium	Lw	103	257	257		
Lead	Pb	82	208-206-207	207.2	1.54	18
Lithium	Li	3	7	6.940	1.23	13
Lutetium	Lu	71	175	174.97	1.56	18
Magnesium	Mg	12	24	24.305	1.36	14
Manganese	Mn	25	55	54.938	1.17	7

OF THE ELEMENTS

Electro-negativity, Relative Scale	First Ionization Potential (eV)	Electron Work Function			Electrochemical Equiv.	
		Thermionic	Photoelectric	Contact	Valence Involved	Amp-Hours per Gram
1.1	6.9				3	0.35
1.5	5.98		4.08	3.38	3	2.98
	6.05					
2.05	8.64		4.01	4.14	5	1.1
0	15.76				n*	0.67
2.0	9.81		5.11		5	1.79
2.2						
0.9	5.21	2.11	2.48	1.73	2	0.39
1.5	9.32		3.92	3.10	2	5.94
1.9	7.29		4.25	4.17	5	0.64
2.0	8.3		4.5		3	7.43
2.85	11.81				1	0.335
1.7	8.99		4.07	4.0	2	0.477
1.0	6.11	2.24	2.706	3.33	2	1.337
2.6	11.26	4.34	4.81		4	8.93
1.1	5.6	2.6	2.84		3	0.574
0.7	3.89	1.81	1.92	4.46	1	0.2
3.15	12.97				1	0.756
1.6	6.76	4.60	4.37	4.38	3	1.546
1.8	7.86	4.40	4.20	4.21	2	0.91
1.9	7.72	4.26	4.18	4.46	2	0.84
1.2	5.93				3	0.495
1.2	6.10				3	0.48
1.1	5.67				3	0.53
3.9	17.42				1	1.41
0.65						
1.1	6.16				3	0.513
1.6	5.99	4.12		3.80	3	1.15
1.9	7.89		4.5	4.5	4	1.48
2.4	9.22	4.32	4.82	4.46	3	0.41
1.3	7.0	3.53			4	0.600
0	24.59				n*	6.698
1.2	6.02				3	0.488
2.2	13.59				1	26.59
1.7	5.78				3	0.700
2.65	10.45		6.8		1	0.211
2.2	9.1	5.3		4.57	4	0.555
1.8	7.87	4.25	4.33	4.40	3	1.440
0	13.99				n*	0.32
1.1	5.61	3.3			3	0.579
1.8	7.42		4.05	3.94	4	0.517
1.0	5.39		2.35	2.49	1	3.862
1.2	6.15				3	0.46
1.2	7.64		3.68	3.63	2	2.204
1.5	7.43	3.83	3.76	4.14	4	1.952

Continued on next page.

TABLE 1 (CONT). PROPERTIES

	Symbol	Atomic Number Z	Mass Number $Z+N$	Relative Atomic Mass	Atomic Radius (Å)	Gram Atomic Volume (cm^3)
Mendelevium	Md or Mv	101	256	256		
Mercury	Hg	80	202-200-199	200.59	1.44	14
Molybdenum	Mo	42	98-96-92-95	95.94	1.29	9
Neodymium	Nd	60	142-144-146	144.24	1.64	21
Neon	Ne	10	20	20.179	1.31	17
Neptunium	Np	93	237	237.048		
Nickel	Ni	28	58	58.71	1.15	6
Niobium	Nb	41	93	92.906	1.34	11
Nitrogen	N	7	14	14.007	0.70	14
Nobelium	No	102	254	254		
Osmium	Os	76	192-190-189	190.2	1.26	9
Oxygen	O	8	16	15.999	0.66	11
Palladium	Pd	46	108-106-105	106.4	1.28	9
Phosphorus	P	15	31	30.974	1.10	17
Platinum	Pt	78	195-194-196	195.09	1.29	9
Plutonium	Pu	94	242	242		
Polonium	Po	84	209	210	1.53	
Potassium	K	19	39	39.098	2.03	46
Praseodymium	Pr	59	141	140.907	1.65	21
Promethium	Pm	61	145	145		
Protactinium	Pa	91	231	231.036		
Radium	Ra	88	226	226.025		45
Radon	Rn	86	222	222	2.14	50
Rhenium	Re	75	187-185	186.2	1.28	9
Rhodium	Rh	45	103	102.905	1.25	8
Rubidium	Rb	37	85-87	85.468	2.16	56
Ruthenium	Ru	44	102-104-101	101.07	1.24	8
Samarium	Sm	62	152-154-147	150.35	1.66	20
Scandium	Sc	21	45	44.956	1.44	15
Selenium	Se	34	80-78	78.96	1.17	16
Silicon	Si	14	28	28.086	1.17	12
Silver	Ag	47	107-109	107.868	1.34	10
Sodium	Na	11	23	22.99	1.57	24
Strontium	Sr	38	88	87.62	1.92	34
Sulfur	S	16	32	32.064	1.04	16
Tantalum	Ta	73	181	180.948	1.34	11
Technetium	Tc	43	99	98.906		
Tellurium	Te	52	130-128-126	127.60	1.37	21
Terbium	Tb	65	159	158.925	1.59	19
Thallium	Tl	81	205-203	204.37	1.55	17
Thorium	Th	90	232	232.038	1.65	20
Thulium	Tm	69	169	168.934	1.56	18
Tin	Sn	50	120-118	118.69	1.40	16
Titanium	Ti	22	48	47.90	1.32	11
Tungsten	W	74	184-186-182	183.85	1.30	10
Uranium	U	92	238	238.029	1.42	13
Vanadium	V	23	51	50.94	1.22	8
Xenon	Xe	54	132-129-131	131.30	2.09	43
Ytterbium	Yb	70	174-172-173	173.04	1.70	25
Yttrium	Y	39	89	88.906	1.62	21
Zinc	Zn	30	64-66-68	65.38	1.25	9
Zirconium	Zr	40	90-94-92	91.22	1.45	14

OF THE ELEMENTS

Electro-negativity, Relative Scale	First Ionization Potential (eV)	Electron Work Function			Electrochemical Equiv.	
		Thermionic	Photoelectric	Contact	Valence Involved	Amp-Hours per Gram
1.9	10.43		4.53	4.50	2	0.267
1.8	7.10	4.20	4.25	4.28	6	1.67
1.1	5.49	3.3			3	0.557
0	21.56				n*	1.33
1.3	5.8					
1.8	7.63	5.03	5.01	4.96	2	0.913
1.6	6.88	4.01	4.5			
3.05	14.53				5	9.57
2.2	8.7			4.55	4	0.56
3.5	13.62				2	3.35
2.2	8.33	4.99	4.97	4.49	4	1.005
2.15	10.48				5	4.33
2.2	9.0	5.32	5.22	5.36	4	0.549
	5.8					
2.0	8.43				6	0.766
0.8	4.34		2.24	1.60	1	0.685
1.1	5.42	2.7			3	0.571
1.1	5.55					
1.5					5	0.580
0.9	5.28				2	0.237
0	10.75				n*	0.121
1.9	7.87	5.1	5.0		7	1.007
2.2	7.46	4.80	4.57	4.52	4	1.042
0.8	4.18		2.09		1	0.314
2.2	7.37			4.52	4	1.054
1.1	5.63	3.2			3	0.535
1.3	6.54				3	1.783
2.45	9.75		4.8	4.42	6	2.037
1.9	8.15	3.59	4.52	4.2	4	3.821
1.9	7.57	3.56	4.73	4.44	1	0.248
0.9	5.14		2.28	1.9	1	1.166
1.0	5.69		2.74		2	0.612
2.6	10.36				6	5.01
1.3	7.88	4.19	4.14	4.1	5	0.741
1.9	7.28					
2.3	9.01		4.76	4.70	6	1.260
1.2	5.98				3	0.505
1.8	6.11		3.68	3.84	3	0.393
1.3	6.95	3.35	3.47	3.46	4	0.462
1.2	6.18				3	0.475
1.8	7.34		4.38	4.09	4	0.903
1.5	6.82	3.95	4.06	4.14	4	2.238
1.7	7.98	4.52	4.49	4.38	6	0.874
1.7	6.08	3.27	3.63	4.32	6	0.676
1.6	6.74	4.12	3.77	4.44	5	2.63
0	12.13				n*	0.204
1.2	6.25				3	0.465
1.3	6.38				3	0.904
1.6	9.39		3.73	3.78	2	0.820
1.6	6.84	4.21	3.82	3.60	4	1.175

*n = nonvalent

TABLE 2. PHYSICAL PROPERTIES

	Symbol	Atomic Number	Density at 20°C (g/cm^3)	Relative Hardness	Melting Point (°C)	Boiling Point (°C)
Actinium	Ac	89			1 050	3 200
Aluminum	Al	13	2.70	2.9	660	2 467
Americium	Am	95	13.67		994	2 600
Antimony	Sb	51	6.62	3	630.5	1 750
Argon	Ar or A	18	1.78*		−189.2	−185.7
Arsenic (gray)	As	33	5.73	3.5	820‡	615**
Astatine	At	85			302	337
Barium	Ba	56	3.5		725	1 640
Berkelium	Bk	97				
Beryllium	Be	4	1.82	3	1 278	2 970
Bismuth	Bi	83	9.80	2.5	271.3	1 560
Boron	B	5	2.46	9.5	2 300	2 550**
Bromine	Br	35	3.12		−7.2	58.8
Cadmuim	Cd	48	8.65	2.0	320.9	765
Calcium	Ca	20	1.54		842	1 487
Californium	Cf	98				
Carbon	C	6	2.22	10†	>3 500	4 827
Cerium	Ce	58	6.9	2.5	795	3 468
Cesium	Cs	55	1.87	0.2	28.5	690
Chlorine	Cl	17	3.21*		−100.98	−34.7
Chromium	Cr	24	7.14	9	1 890	2 672
Cobalt	Co	27	8.9	5	1 495	2 900
Copper	Cu	29	8.96	3	1 083	2 595
Curium	Cm	96	13.51		1 340	
Dysprosium	Dy	66	8.54		1 407	2 600
Einsteinium	Es or E	99				
Erbium	Er	68	9.05		1 497	2 900
Europium	Eu	63	5.26		826	1 439
Fermium	Fm	100				
Fluorine	F	9	1.69* ††		−220	−188
Francium	Fr	87			27	677
Gadolinium	Gd	64	7.89		1 312	3 000
Gallium	Ga	31	5.91	1.5	29.78	2 403
Germanium	Ge	32	5.36	6.2	937.4	2 830
Gold	Au	79	19.3	2.5	1 063	2 966
Hafnium	Hf	72	13.31		2 220	4 602
Helium	He	2	0.1664*		<−272§	−268.94
Holmium	Ho	67	8.803		1 461	2 600
Hydrogen	H	1	0.08375*		−259.14	−252.8
Indium	In	49	7.31	1.2	156	2 050
Iodine	I	53	4.93		113.5	184.3
Iridium	Ir	77	22.4	6.15	2 410	4 527
Iron	Fe	26	7.87	4	1 535	3 000
Krypton	Kr	36	3.448*		−156.6	−152.3
Lanthanum	La	57	6.15		920	3 469
Lawrencium	Lw	103				
Lead	Pb	82	11.34	1.5	327.4	1 744
Lithium	Li	3	0.53	0.6	179	1 336
Lutetium	Lu	71	9.84		1 652	3 327
Magnesium	Mg	12	1.74	2	651	1 100
Manganese	Mn	25	7.44	5.0	1 244	2 097
Mendelevium	Md or Mv	101				
Mercury	Hg	80	13.55	1.5	−38.87	356.9

OF THE ELEMENTS

Latent Heat of Fusion (cal/g)	Specific Heat at 20°C (cal/g°C)	Thermal Conductivity at 20°C (W/cm°C)	Linear Thermal Expansion per °C at 20°C (×10^{-6})	Elasticity Modulus (GN/m^2)	Tensile Strength (MN/m^2)
93	0.226	2.18	22.9	71.1	61.8
38.3	0.049	0.19	8.5—10.8	77.5	10.3
6.7	0.125	1.7×10^{-4}			
	0.082		4.7		
324	0.425	1.64	12	294	117
12.5	0.0294	0.084	13.3	31.4	
	0.307		2		
16.2	0.107				
13.2	0.055	0.91	29.8	53.9	70.6
	0.145		25	20.6	55.9
	0.165	0.24	0.6—4.3	4.9	
	0.042				88.8
3.8	0.052		97		
23	0.226	0.072×10^{-4}			
75.6	0.11	0.69	6.2		
58.4	0.1001	0.69	12.3	206	237
50.6	0.0921	3.94	16.5	108	221
10.1					
19.2	0.079		18		
	0.073				
16.1	0.031	2.96	14.2	71.6	113
	1.25	13.9×10^{-4}			
15	3.415	17×10^{-4}			
	0.057	0.24	33		2.9
15.8	0.052	43.5×10^{-4}	93		
33	0.032	1.4	6.5	515	
65	0.108	0.79	11.7	196	201
		0.89×10^{-4}			
	0.045				
6.3	0.030	0.35	28.7	17.7	13
159	0.79	0.71	56		
88	0.249	1.55	25.2	45.1	89.7
64.8	0.107		23	157	283
2.7	0.033	0.084			

continued on next page.

TABLE 2 (CONT). PHYSICAL

	Symbol	Atomic Number	Density at 20°C (g/cm^3)	Relative Hardness	Melting Point (°C)	Boiling Point (°C)
Molybdenum	Mo	42	10.2	6	2 610	4 800
Neodymium	Nd	60	7.05		1 024	3 027
Neon	Ne	10	0.8387*		−248.7	−245.9
Neptunium	Np	93	20.45		640	3 902
Nickel	Ni	28	8.9	5	1 453	2 732
Niobium	Nb	41	8.57		2 468	5 127
Nitrogen	N	7	1.1649*		−209.9	−195.8
Nobelium	No	102				
Osmium	Os	76	22.48	7.0	3 000	5 000
Oxygen	O	8	1.3318*		−218.4	−183
Palladium	Pd	46	12	4.8	1 552	2 927
Phosphorus	P	15	1.82		44.1	280
Platinum	Pt	78	21.45	4.3	1 769	3 827
Plutonium	Pu	94	19.82		639.5	3 235
Polonium	Po	84	9.2		254	962
Potassium	K	19	0.86	0.5	63.65	774
Praseodymium	Pr	59	6.63		935	3 127
Promethium	Pm	61			1 027	2 027
Protactinium	Pa	91	15.37		1 227	4 027
Radium	Ra	88	5		700	1 525
Radon	Rn	86	4.40*		−71	−61.8
Rhenium	Re	75	20		3 180	5 627
Rhodium	Rh	45	12.44	6	1 966	3 727
Rubidium	Rb	37	1.53	0.3	38.5	688
Ruthenium	Ru	44	12.2	6.5	2 250	3 900
Samarium	Sm	62	7.7		1 072	1 900
Scandium	Sc	21	2.5		1 539	2 727
Selenium	Se	34	4.81	2	217	685
Silicon	Si	14	2.4	7	1 410	2 355
Silver	Ag	47	10.49	2.7	960.8	2 212
Sodium	Na	11	0.97	0.4	97	892
Strontium	Sr	38	2.6	1.8	769	1 384
Sulfur	S	16	2.07	2.0	116	444.6
Tantalum	Ta	73	16.6	7	2 996	5 425
Technetium	Tc	43	11.49		2 200	4 700
Tellurium	Te	52	6.24	2.3	449.5	990
Terbium	Tb	65	8.27		1 356	2 800
Thallium	Tl	81	11.85	1.2	303.5	1 457
Thorium	Th	90	11.5		1 800	4 200
Thulium	Tm	69	9.33		1 545	1 727
Tin	Sn	50	7.3	1.8	231.89	2 270
Titanium	Ti	22	4.54	4	1 675	3 260
Tungsten	W	74	19.3	7	3 410	5 660
Uranium	U	92	18.7		1 133	3 818
Vanadium	V	23	5.68		1 890	3 400
Xenon	Xe	54	5.495*		−111	−107
Ytterbium	Yb	70	6.98		824	1 427
Yttrium	Y	39	5.51		1 495	2 927
Zinc	Zn	30	7.14	2.5	419.4	907
Zirconium	Zr	40	6.4	4.7	1 852	4 377

*g/liter †diamond ‡36 atm §26 atm **sublimes †† At 15°C

PROPERTIES OF THE ELEMENTS

Latent Heat of Fusion (cal/g)	Specific Heat at 20°C (cal/g°C)	Thermal Conductivity at 20°C (W/cm°C)	Linear Expansion per °C at 20°C (10^6)	Elasticity Modulus (GN/m^2)	Tensile Strength (MN/m^2)
69.0	0.065	1.46	4.9	343	1177
	0.045				
		4.57×10^{-4}			
73.8	0.112	0.9	13.3	206	317
68.0	0.064	0.52	7.1		
6.2	0.247				
34.0	0.031	0.61	5		
3.3	0.218				
34.2	0.059	0.70	11.8	117	137
5.0	0.177		125		
27.1	0.032	0.69	8.9	147	157
3.0	0.032	0.08	54		
14.5	0.177	0.99	83		
	0.458				
	0.035				
50.0	0.060	1.5	8.1	29.4	
6.1	0.080		90		
	0.061		9.1		
16.0	0.077	0.005	37		
430.0	0.176	0.84	2.8—7.3	108	
24.3	0.056	4.08	18.9	70.6	148
27.5	0.295	1.35	71		
25					
9.3	0.175	26.4×10^{-4}	6.4		
41.0	0.036	0.54	6.6	186	490
25.3	0.047	0.060	16.8	20.6	11.0
7.2	0.031	0.39	28		
17.0	0.028	0.41	11.1		549
14.4	0.054	0.64	23	403	13.7
100.0	0.142	0.2	8.5	83.4	
44.0	0.034	1.99	4.3	343	2650
12.0	0.028	0.25	13.4		
98.0	0.115	0.60	8		
		5.9×10^{-4}			
24.1	0.09	1.1	17.39	82.4	103
	0.066		5.6	73.5	294

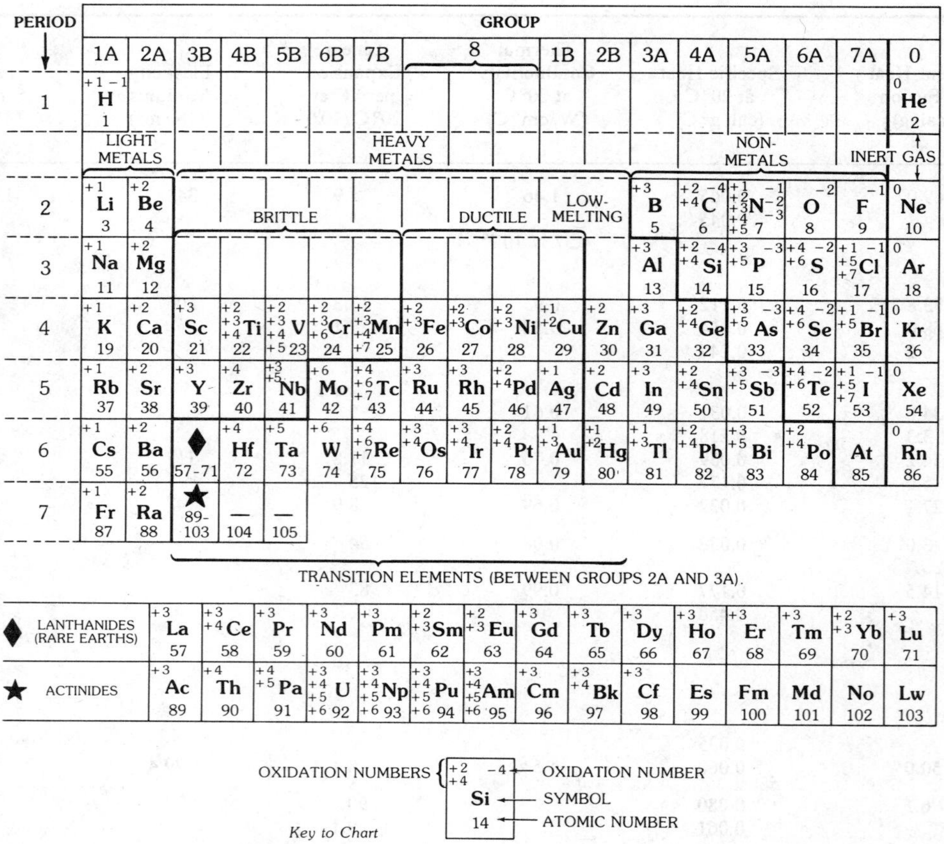

PERIOD	1A	2A	3B	4B	5B	6B	7B	8	8	8	1B	2B	3A	4A	5A	6A	7A	0
1	+1 −1 H 1																	0 He 2
2	+1 Li 3	+2 Be 4											+3 B 5	+2 +4 −4 C 6	+1 +2 +3 +4 +5 −1 −2 −3 N 7	−2 O 8	−1 F 9	0 Ne 10
3	+1 Na 11	+2 Mg 12											+3 Al 13	+2 +4 −4 Si 14	+3 +5 −3 P 15	+4 +6 −2 S 16	+1 +5 +7 −1 Cl 17	0 Ar 18
4	+1 K 19	+2 Ca 20	+3 Sc 21	+2 +3 +4 Ti 22	+2 +3 +4 +5 V 23	+2 +3 +6 Cr 24	+2 +3 +4 +7 Mn 25	+2 +3 Fe 26	+2 +3 Co 27	+2 +3 Ni 28	+1 +2 Cu 29	+2 Zn 30	+3 Ga 31	+2 +4 Ge 32	+3 +5 −3 As 33	+4 +6 −2 Se 34	+1 +5 −1 Br 35	0 Kr 36
5	+1 Rb 37	+2 Sr 38	+3 Y 39	+4 Zr 40	+3 +5 Nb 41	+6 Mo 42	+4 +6 +7 Tc 43	+3 Ru 44	+3 Rh 45	+2 +4 Pd 46	+1 Ag 47	+2 Cd 48	+3 In 49	+2 +4 Sn 50	+3 +5 −3 Sb 51	+4 +6 −2 Te 52	+1 +5 +7 −1 I 53	0 Xe 54
6	+1 Cs 55	+2 Ba 56	♦ 57-71	+4 Hf 72	+5 Ta 73	+6 W 74	+4 +6 +7 Re 75	+3 +4 Os 76	+3 +4 Ir 77	+2 +4 Pt 78	+1 +3 Au 79	+1 +2 Hg 80	+1 +3 Tl 81	+2 +4 Pb 82	+3 +5 Bi 83	+2 +4 Po 84	At 85	0 Rn 86
7	+1 Fr 87	+2 Ra 88	★ 89-103	— 104	— 105													

LIGHT METALS (groups 1A–2A); HEAVY METALS; BRITTLE; DUCTILE; LOW-MELTING; NON-METALS; INERT GAS

TRANSITION ELEMENTS (BETWEEN GROUPS 2A AND 3A).

♦ LANTHANIDES (RARE EARTHS)	+3 La 57	+3 +4 Ce 58	+3 Pr 59	+3 Nd 60	+3 Pm 61	+2 +3 Sm 62	+2 +3 Eu 63	+3 Gd 64	+3 Tb 65	+3 Dy 66	+3 Ho 67	+3 Er 68	+3 Tm 69	+2 +3 Yb 70	+3 Lu 71
★ ACTINIDES	+3 Ac 89	+4 Th 90	+4 +5 Pa 91	+3 +4 +5 +6 U 92	+3 +4 +5 +6 Np 93	+3 +4 +5 +6 Pu 94	+3 +4 +5 +6 Am 95	+3 Cm 96	+3 +4 Bk 97	+3 Cf 98	Es 99	Fm 100	Md 101	No 102	Lw 103

Key to Chart: OXIDATION NUMBERS {+2 +4 −4 ← OXIDATION NUMBER; Si ← SYMBOL; 14 ← ATOMIC NUMBER

Fig. 1. Periodic classification of the elements.

Insulators: Resistivities from 10^9 to 10^{25} ohm-cm. Conductivities from 10^{-25} to 10^{-9} S-cm^{-1}.

SEMICONDUCTING MATERIALS

Some properties of semiconductor materials are listed in Table 8.

INSULATING MATERIALS

The permittivity ϵ of an insulating material is defined by

$$\epsilon = \epsilon_0 + (P/E)$$

where,

ϵ_0 = permittivity of free space,
P = flux density from dipoles within the dielectric medium,
E = electric-field intensity,
P/E = electric susceptibility.

To deal with a dimensionless coefficient, it is customary to use the relative dielectric constant, ϵ_r, defined by

$$\epsilon_r = \epsilon/\epsilon_0$$

The relative dielectric constant, ϵ_r, is a function of temperature and frequency. From Table 9, which gives the values of ϵ_r as a function of frequencies at room temperature, it is easy to get

$$\epsilon = \epsilon_r\epsilon_0$$

In the mks rationalized system of units, the permittivity of vacuum is equal to

$$\epsilon_0 = 10^{-9}/36\pi = 8.854 \times 10^{-12} \text{ farad/meter}$$

and we have

Coulomb's Law

$$F = (1/4\pi\epsilon_0\epsilon_r)(q_1q_2/R^2)$$

Gauss's Law

$$\Phi = (\epsilon_0\epsilon_r)^{-1}\sum q_i$$

CHART 1. GALVANIC SERIES IN SEA WATER

Corroded end (anodic)

Magnesium
Magnesium alloys

Zinc
Galvanized steel
Galvanized wrought iron

Aluminum:
52SH, 4S, 3S, 2S, 53ST
Aluminum clad

Cadmium

Aluminum:
A17ST, 17ST, 24ST

Mild steel
Wrought iron
Cast iron

Ni-resist

13% chromium stainless steel (type 410—active)

50-50 lead-tin solder

18–8 stainless steel type 304 (active)
18–8–3 stainless steel type 316 (active)

Lead
Tin

Muntz metal
Manganese bronze
Naval brass

Nickel (active)
Inconel (active)

Yellow brass
Admiralty brass
Aluminum bronze
Red brass
Copper
Silicon bronze
Ambrac
70-30 copper-nickel
Comp. G, bronze
Comp. M, bronze

Nickel (passive)
Inconel (passive)

Monel

18–8 stainless steel type 304 (passive)
18–8–3 stainless steel type 316 (passive)

Protected end (cathodic or most noble)

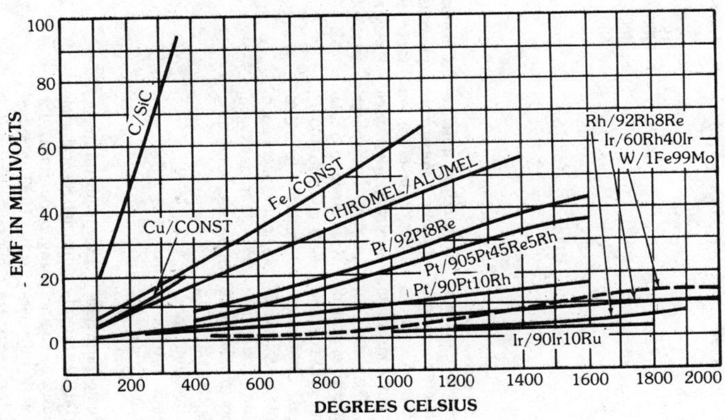

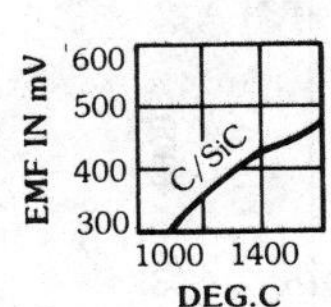

Fig. 2. Temperature-emf characteristics of thermocouples.

The dissipation factor of an insulating material (Table 9) is defined as the ratio of the energy dissipated to the energy stored in the dielectric per hertz, or as the tangent of the loss angle. For dissipation factors less than 0.1, the dissipation factor may be considered equal to the power factor of the dielectric, which is the cosine of the phase angle by which the current leads the voltage.

Many of the materials listed are characterized by a peak dissipation factor that occurs somewhere in

TABLE 3. THERMOCOUPLES AND THEIR CHARACTERISTICS

	Copper/Constantan	Iron/Constantan	Chromel/Constantan	Chromel/Alumel	Platinum/Platinum Rhodium (10)	Platinum/Platinum Rhodium (13)	Carbon/Silicon Carbide
Composition, percent	100 Cu/60 Cu 40 Ni	100 Fe/60 Cu 40 Ni	90 Ni 10 Cr/55 Cu 45 Ni	90 Ni 10 Cr/94 Ni 2 Al 3 Mn 1 Si	Pt/90 Pt 10 Rh	Pt/87 Pt 13 Rh	C/SiC
*Range of application, °C	− 200 to +300	− 200 to + 1100	0 to +1100	− 200 to +1200	0 to +1450	0 to +1450	0 to + 2000
Resistivity, microhm-cm	1.75 49	10 49	70 49	70 29.4	10 21		
Temperature coefficient of resistivity, per °C	0.0039 0.00001	0.005 0.00001	0.00035 0.0002	0.00035 0.000125	0.0030 0.0018		
Melting temperature, °C	1085 1190	1535 1190	1400 1190	1400 1430	1755 1700		3000 2700
emf in millivolts; reference junction at 0°C	100°C 4.24 mV 200 9.06 300 14.42	100°C 5.28 mV 200 10.78 400 21.82 600 33.16 800 45.48 1000 58.16	100°C 6.3 mV 200 13.3 400 28.5 600 44.3	100°C 4.1 mV 200 8.13 400 16.39 600 24.90 800 33.31 1000 41.31 1200 48.85 1400 55.81	100°C 0.643 mV 200 1.436 400 3.251 600 5.222 800 7.330 1000 9.569 1200 11.924 1400 14.312 1600 16.674	100°C 0.646 mV 200 1.464 400 3.398 600 5.561 800 7.927 1000 10.470 1200 13.181 1400 15.940 1600 18.680	1210°C 353.6 mV 1300 385.2 1360 403.2 1450 424.9
Influence of temperature and gas atmosphere	Subject to oxidation and alteration above 400°C due Cu, above 600° due constantan wire. Ni-plating of Cu tube gives protection in acid-containing gas. Contamination of Cu affects calibration greatly. Resistance to oxide. atm. good. Resistance to reducing atm. good. Requires protection from acid fumes.	Oxidizing and reducing atmosphere have little effect on accuracy. Best used in dry atmosphere. Resistance to oxidation good to 400°C. Resistance to reducing atmosphere good. Protect from oxygen, moisture, sulphur.	Chromel attacked by sulphurous atmosphere. Resistance to oxidation good. Resistance to reducing atmosphere poor.	Resistance to oxidizing atmosphere very good. Resistance to reducing atmosphere poor. Affected by sulphur, reducing or sulphurous gas, SO_2 and H_2S.	Resistance to oxidizing atmosphere very good. Resistance to reducing atmosphere poor. Susceptible to chemical alteration by As, Si, P vapor in reducing gas (CO_2, H_2, H_2S, SO_2). Pt corrodes easily above 1000°. Used in gas-tight protecting tube.		Used as tube element. Carbon sheath chemically inert.
Particular applications	Low temperature, industrial. Internal-combustion engine. Used as a tube element for measurements in steam line.	Low temperature, industrial. Steel annealing, boiler flues, tube stills. Used in reducing or neutral atmosphere.		Used in oxidizing atmosphere. Industrial. Ceramic kilns, tube stills, electric furnaces.	International Standard 630 to 1065°C.	Similar to Pt/PtRh (10) but has higher emf.	Steel furnace and ladle temperatures. Laboratory measurements.

*For prolonged use; can be used at higher temperature for short periods.

TABLE 4. THERMAL ELECTROMOTIVE FORCE OF PLATINUM-RHODIUM ALLOYS VERSUS PLATINUM*

	Electromotive Force (Millivolts)							
	Percent Rhodium							
Temp.(°C)	0.5	1.0	5.0	10.0	20.0	40.0	80.0	100.0
0	0.00	0.00	0.00	0.00	0.00	0.00	0.00	0.00
100	+ 0.10	+ 0.18	+ 0.54	+ 0.64	+ 0.63	+ 0.65	+ 0.62	+ 0.70
200	0.20	0.37	1.16	1.43	1.44	1.52	1.49	1.61
300	0.29	0.57	1.82	2.32	2.40	2.55	2.55	2.68
400	0.39	0.76	2.49	3.25	3.47	3.70	3.77	3.91
500	0.48	0.94	3.17	4.22	4.63	4.97	5.12	5.28
600	0.58	1.12	3.86	5.22	5.87	6.36	6.60	6.77
700	0.67	1.30	4.55	6.26	7.20	7.85	8.20	8.40
800	0.76	1.48	5.25	7.33	8.59	9.45	9.92	10.16
900	0.85	1.66	5.96	8.43	10.06	11.16	11.76	12.04
1000	0.94	1.84	6.68	9.57	11.58	12.98	13.73	14.05
1100	1.03	2.02	7.42	10.74	13.17	14.90	15.81	16.18
1200	1.13	2.20	8.16	11.93	14.84	16.91	17.99	18.42

*From *Smithsonian Physical Tables*, 9th revised edition, Vol. 120. (Washington, D.C.: Smithsonian Institution, 1969).

TABLE 5. RESISTIVITIES OF METALS AND ALLOYS

Material	Form	Resistivity ($\times 10^{-6}$ ohm-cm)	Temperature (°C)	Temperature Coefficient (°C^{-1})
Alumel	solid	33.3	0	0.0012
Aluminum	liquid	20.3	670	
	solid	2.62	20	0.0039
Antimony	liquid	123	800	
	solid	39.2	20	0.0036
Arsenic	solid	35	0	0.0042
Beryllium		4.57	20	
Bismuth	liquid	128.9	300	
	solid	115	20	0.004
Boron		1.8×10^{12}	0	
Brass (66 Cu 34 Zn)		3.9	20	0.002
Cadmium	liquid	34	400	
	solid	7.5	20	0.0038
Carbon	diamond	5×10^{20}	15	
	graphite	1400	20	− 0.0005
Cerium		78	20	
Cesium	liquid	36.6	30	
	solid	20	20	
		18.83	0	
Chromax (15 Cr, 35 Ni, balance Fe)		100	20	0.00031
Chromel	solid	70–110	0	0.00011–0.000054
Chromium		2.6	0	
Cobalt		9.7	20	0.0033
Constantan (55 Cu, 45 Ni)		44.2	20	+ 0.0002
Copper (commercial annealed)	liquid	21.3	1083	
	solid	1.7241	20	0.0039
Gallium	liquid	27	30	
	solid	53	0	
Gold	liquid	30.8	1063	
	solid	2.44	20	0.0034
		2.19	0	
Hafnium		32.1	20	
Indium	liquid	29	157	
	solid	9	20	0.00498
Iridium		5.3	20	0.0039

Continued on next page.

TABLE 5 (CONT). RESISTIVITIES OF METALS AND ALLOYS

Material	Form	Resistivity ($\times 10^{-6}$ ohm-cm)	Temperature (°C)	Temperature Coefficient (°C^{-1})
Iron		9.71	20	0.0052–0.0062
Kovar A (29 Ni, 17 Co, 0.3 Mn, balance Fe)		45–84	20	
Lead	liquid	98	400	
	solid	21.9	20	0.004
PbO_2		92		
Lithium	liquid	45	230	0.003
	solid	9.3	20	0.005
Magnesium		4.46	20	0.004
Manganese		5	20	
MnO_2		6 000 000	20	
Manganin (84 Cu, 12 Mn, 4 Ni)		44	20	± 0.0002
Mercury	liquid	95.8	20	0.00089
	solid	21.3	− 50	
Molybdenum		5.17	0	
		4.77	20	0.0033
Monel metal (67 Ni, 30 Cu, 1.4 Fe, 1 Mn)	solid	42	20	0.002
Neodymium	solid	79	18	
Nichrome (65 Ni, 12 Cr, 23 Fe)	solid	100	20	0.00017
Nickel	solid	6.9	20	0.0047
Nickel-silver (64 Cu, 18 Zn, 18 Ni)	solid	28	20	0.00026
Niobium		12.4	20	
Osmium		9	20	0.0042
Palladium		10.8	20	0.0033
Phosphor bronze (4 Sn, 0.5 P, balance Cu)		9.4	20	0.003
Platinum		10.5	20	0.003
Plutonium		150	20	
Potassium	liquid	13	62	
	solid	7	20	0.006
Praseodymium		68	25	
Rhenium		19.8	20	
Rhodium		5.1	20	0.0046
Rubidium		12.5	20	
Ruthenium		10	20	
Selenium	solid	1.2	20	
Silver		1.62	20	0.0038
Sodium	liquid	9.7	100	
	solid	4.6	20	
Steel (0.4–0.5 C, balance Fe)		13–22	20	0.003
Steel, manganese (13 Mn, 1 C, 86 Fe)		70	20	0.001
Steel, stainless (0.1 C, 18 Cr, 8 Ni, balance Fe)		90	20	
Strontium		23	20	
Sulfur		2×10^{23}	20	
Tantalum		13.1	20	0.003
Thallium		18.1	20	0.004
Thorium		18	20	0.0021
Tin		11.4	20	0.0042
Titanium		47.8	25	
Tophet A (80 Ni, 20 Cr)		108	20	0.00014
Tungsten		5.48	20	0.0045
W_2O_5		450	20	
WO_3		2×10^{11}	20	
Uranium		29	0	0.0021
Zinc	liquid	35.3	420	
	solid	6	20	0.0037
Zirconium		40	20	0.0044

TABLE 6. ELECTRICAL RESISTIVITY OF ROCKS AND SOILS*

	Resistivity (ohm-cm)
Igneous Rocks	
Granite	10^7–10^9
Lava flow (basic)	10^6–10^7
Lava, fresh	3×10^5–10^6
Quartz vein, massive	$> 10^6$
Metamorphic Rocks	
Marble	4×10^8
Marble, white	10^{10}
Marble, yellow	10^{10}
Schist, mica	10^7
Shale, bed	10^5
Shale, Nonesuch	10^4
Sedimentary Rocks	
Limestone	10^4
Limestone, Cambrian	10^4–10^5
Sandstone	10^5
Sandstone, eastern	3×10^3–10^4
Unconsolidated Materials	
Clay, blue	2×10^4
Clay, fire	2×10^5
Clayey earth	10^4–4×10^4
Gravel	10^5
Sand, dry	10^5–10^6
Sand, moist	10^5–10^6

* From *Smithsonian Physical Tables*, 9th revised edition, Vol. 120. Washington, D.C.: Smithsonian Institution, 1969.

TABLE 7. SUPERCONDUCTIVITY OF SOME METALS, ALLOYS, AND COMPOUNDS

Material	Critical Temperature (K)
NbC	10.1
Niobium	9.22
TaC	9.2
Pb-As-Bi	9.0
Pb-Bi-Sb	8.9
Pb-Sn-Bi	8.5
Pb-As	8.4
MoC	7.7
Lead	7.2
N_2Pb_5	7.2
Bi_6Tl_3	6.5
Sb_2Tl_7	5.5
Lanthanum	5.2
Tantalum	4.4
Vanadium	4.3
TaSi	4.2
Mercury	4.15
PbS	4.1
Hg_5Tl_7	3.8
Tin	3.71
Indium	3.38
ZrB	2.82
WC	2.8
Rhenium	2.57
Mo_2C	2.4
Thallium	2.4
W_2C	2.05
Au_2Bi	1.84
CuS	1.6
TiN	1.4
Thorium	1.32
VN	1.3
Aluminum	1.15
Gallium	1.12
TiC	1.1
Zinc	0.95
Uranium	0.75
Osmium	0.71
Zirconium	0.54
Cadmium	0.54
Titanium	0.53
Ruthenium	0.47
Hafnium	0.35

the frequency range, this peak being accompanied by a rapid change in the dielectric constant. These effects are the result of a resonance phenomenon occurring in polar materials. The position of the dissipation-factor peak in the frequency spectrum is very sensitive to temperature. An increase in the temperature increases the frequency at which the peak occurs, as illustrated qualitatively in Fig. 3. Nonpolar materials have very low losses without a noticeable peak; the dielectric constant remains essentially unchanged over the frequency range.

Another effect that contributes to dielectric losses is that of ionic or electronic conduction. This loss, if present, is important usually at the lower end of the frequency range only, and is distinguished by the fact that the dissipation factor varies inversely with frequency. Increase in temperature increases the loss due to ionic conduction because of increased ionic mobility.

The data given on dielectric strength are accompanied by the thickness of the specimen tested because the dielectric strength, expressed in volts/mil, varies inversely with the square root of thickness, approximately.

The direct-current volume resistivity of many materials is influenced by changes in temperature or humidity. The values given in the table may be reduced several decades by raising the temperature

TABLE 8. PROPERTIES OF SEMICONDUCTORS

Units	IV		III–V											II–VI					
	Si	Ge	AlP	AlAs	AlSb	GaN	GaP	$GaAs_6P_4$	GaAs	GaSb	InP	InAs	InSb	ZnS	ZnSe	ZnTe	CdS	CdSe	CdTe
E_g (eV)	1.11**I**	0.65**I**	2.45**I**	2.14**I**	1.62**I**	3.39**D**	2.26**I**	~1.93**D**	1.43**D**	0.70**D**	1.35**D**	0.356**D**	0.180**D**	3.66**D**	2.67**D**	2.25**D**	2.42**D**	1.74**D**	1.50**D**
E_g (μ)	1.117	1.907	0.506	0.579	0.765	0.366	0.548	0.642	0.867	1.77	0.918	3.48	6.89	0.339	0.464	0.551	0.512	0.712	0.861
$E_{g\,4.2K}$ (eV)	1.21	0.74	2.52	2.22	~1.70	3.47	2.34		1.52	0.81	1.42	0.409	0.235	3.80	2.80	2.38	2.58	1.84	1.60
$E_{CB2} - E_{VB}$ (eV)	~1.9**I**	0.81**D**	3.6**D**	3.14**D**	2.22**D**		2.81**D**	~2.03**I**	1.87**I**	0.78**I**	2.25**I**	1.82**I**	~0.8**I**			3.7**I**			3.4**I**
$\Delta E_{\text{Spin Orbit}}$ (eV)	0.038	0.29		0.29	0.75		0.08		0.34	0.80	0.11	0.41	0.82	0.07	0.45	0.93			0.81
a_0 (Å)	5.431	5.658	5.463	5.661	6.138	3.180	5.449	5.572	5.654	6.095	5.868	6.058	6.479	5.409	5.669	6.104	4.137	4.299	6.481
c_0 (Å)						5.166											6.716	7.015	
d (gm/cm^3)	2.328	5.323	2.40	3.73	4.26	6.10	4.129	4.88	5.316	5.613	4.787	5.667	5.775	4.09	5.26	5.64	4.82	5.81	5.86
Therm. Exp. (10^{-6}/°C)	2.56	5.92		5.20	4.88	†5.59/3.17	5.8	5.9	6.8	6.7	4.5	5.19	5.04	6.2	7	8	†4/2.1		5.5
Therm. Cond. (W/cm°C)	1.41	0.61	0.9	0.8	0.54		0.97	0.16	0.54	0.35	0.68	0.26	0.18	0.26	0.13	0.112	0.2	.063	.07
Hardness (Knoop)	1150	780	500	481	360		945		750	450	535	381	223	178	150	130	55	44	100
Melting Pt. (°C)	1415	958	>2000	1740	1080	1500	1467		1238	712	1070	943	525	1830	1520	1295	1475	1239	1098
Molecular Weight	28.09	72.60	57.95	101.90	148.74	83.73	100.70	127.06	144.64	191.48	145.79	189.74	236.58	97.45	144.34	192.99	144.48	191.37	240.02
m_e (m_0)	‡.19 .90	‡.082 1.64		0.35	0.39	0.19	0.35	0.089	0.065	0.049	0.078	0.023	0.014	0.34	0.17	0.09	0.20	0.13	0.11
m_{h1} (m_0)	0.15	0.042			0.11		0.14		0.087	0.056		0.025	.016			0.15	0.7	0.45	0.13
m_{h2} (m_0)	0.52	0.34			0.5	0.6	0.86		0.47	0.33	0.8	0.41	0.43		0.6	0.68	5	>1	2.1
*μ_{e300K} (cm^2/V·s)	1900	3600	80	280	200	150	190	2900	8800	6000	4700	22 600	8.2–10^4	140	600	340	350	650	1050
*μ_{e77K} (cm^2/V·s)	19 000	40 000	30	18	700		2700	17 000	2.1–10^5	10 000	60 000	1.2–10^5	1.2–10^6		7000		5000	5000	15 000
*μ_{h300K} (cm^2/V·s)	425	2300			400		120		400	800	150	260	1700	5	28	110	15		80
*μ_{h77K} (cm^2/V·s)	8000	40 000			4000		2000		7500	6000	1200	350	7000			900			500
Phonon LO/TO(me V)	51.0/57.4	28.1/33.3	62.0/54.6	49.8/45.1	42.1/39.5		49.9/45.5		36.2/33.3	29.8/26.8	42.8/37.7	30.2/27.1	24.2/22.6	36.9/28.4	31/26	25.5/22.3	36.8/32.1	26.2/20.6	21.2/17.4
**Ref. Index	3.5	4.1		3.1	3.4	2.1	3.3	3.4	3.4	3.8	3.1	3.5	3.9	2.4	2.8	3.1	2.5	2.6	2.8
E_∞	11.94	16.0		8.16	10.24	4	9.04	~10.2	10.9	14.44	9.52	11.8	15.7	5.07	6.1	7.3	†5.3/5.4	†6.2/6.3	6.7
E_0	11.94	16.0		10.06	14.4		11.1	~12.4	13.18	15.69	12.35	14.55	17.72	8.3	9.2	10.4	†8.9/8.5	†9.3/10	9.4

All values at 300K unless otherwise noted; compiled by D. E. Hill

*Highest experimental Hall mobilities

† ⊥/∥ to c axis

**Value near Eq on long λ side

‡m_L/m_T

D = "Direct" energy gap

I = "Indirect" energy gap

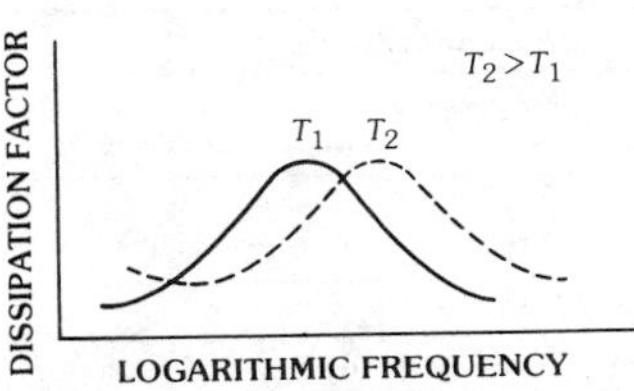

Fig. 3. Variation of dissipation factor with frequency.

toward the higher end of the working range of the material, or by raising the relative humidity of the air surrounding the material to above 90 percent.

MAGNETIC MATERIALS*

All materials that are magnetized by a magnetic field are called magnetic materials. Depending on the magnetic response, various kinds of magnetism are classified. Most magnetic materials of commercial importance today are ferromagnets or ferrimagnets. Iron, nickel, cobalt, and their alloys are examples of ferromagnets; spinel ferrites and garnets are examples of ferrimagnets. These exhibit spontaneous magnetism and develop a flux density B upon application of a magnetic field of strength H in accordance with

$$B = \mu_0 H + J$$

In this equation, $\mu_0 = 1.257 \times 10^{-6}$ henry per meter is the permeability of free space, and J is the magnetic polarization. The units of B and J are teslas (T) or webers per square meter (Wb/m^2), and H is in amperes per meter (A/m).

The permeability of a magnetic material is

$$\mu = B/H = \mu_0 + J/H$$

It is, however, customary to use the relative permeability, μ_r, defined by

$$\mu_r = \mu/\mu_0 = 1 + J/\mu_0 H$$

The permeability of a ferromagnet (or ferrimagnet) is a function of applied field, temperature, and frequency. For electronic transformer and inductor applications where a small field is impressed, the initial permeability, μ_i (often measured at $B = 4$ mT), is the most useful quantity. For power transformer applications, values at higher fields are more useful. Sometimes the value of the maximum permeability, μ_m, is quoted. This is a useful quantity for materials exhibiting a square hysteresis loop ideal for amplifier-type applications. Magnetic materials are classified as soft or hard (permanent) depending on the value of coercivity H_c, which is the field strength required to reduce the flux density to zero after the material has been magnetized. Soft magnets have values of H_c less than about 1 kA/m, and hard magnets have H_c greater than about 10 kA/m. Some applications such as reed contacts make use of semihard magnets with 1 kA/m $< H_c < 10$ kA/m.

* This section contributed by Gilbert Y. Chin.

Soft Magnetic Metals

Table 10 lists some typical commercial soft magnetic metals generally used in low-frequency transformers and inductors. They are often available as laminations, cut cores, and tape-wound cores. The nickel-iron alloys known as permalloys exhibit the highest initial permeability and lowest coercivity. The most commonly used materials in this category are the 48-percent nickel alloy with μ_i about 11 000 and the 80-percent nickel – 4-percent molybdenum alloy with μ_i about 70 000, the latter being more expensive. The nickel-iron alloys can also be processed to exhibit a square hysteresis loop, useful for amplifier-type applications, or a skewed loop, useful for unipolar pulse transformer designs. For common 50- or 60-Hz transformers used in the electrical utility industry, silicon steel is the most common material. Low-carbon steels are very inexpensive and widely used in small motors and generators. The cobalt-iron alloys, usually known as permendur, have the highest values of saturation polarization and Curie temperature and tend to be used in high-performance, lightweight applications such as airborne motors. Appearing on the market recently is a new class of soft magnetic alloys called amorphous magnets or metallic glasses. These materials are rapidly solidified from the melt into thin tapes such that the usual crystalline structure is absent. Some amorphous magnets have been prepared to have properties similar to the silicon steels but with substantially lower core losses, while others have been prepared with properties similar to the best grades of the permalloys.

Permanent-Magnet Materials

Table 11 lists some typical permanent-magnet materials. The alnicos are brittle and hence can be used only in cast or sintered form. The hexagonal ferrites are oxides having the general formula $MO{\cdot}6Fe_2O_3$, where M is barium or strontium. The magnets are prepared by ceramic techniques and are often called ceramic magnets. Large numbers of low-cost ferrite magnets are bonded in plastics and widely used in door catches, wall magnets, refrigerator door gaskets, and toys.

In recent years, a new class of high-performance permanent magnets made of cobalt rare earths, particularly those containing samarium, have been developed commercially. These magnets have the highest combination of coercivity and maximum energy product available on the market, with applications ranging from tiny wristwatch and earphone magnets to horsepower-size industrial dc and syn-

TABLE 9. CHARACTERISTICS

Material Composition	T°C	Dielectric Constant at (Frequency in hertz) 60	10^3	10^6	10^8	3×10^9	2.5×10^{10}
Ceramics:							
Aluminum oxide	25	—	8.83	8.80	8.80	8.79	—
Barium titanate†	26	1250	1200	1143	—	600	100
Calcium titanate	25	168	167.7	167.7	167.7	165	—
Magnesium oxide	25	—	9.65	9.65	9.65	—	—
Magnesium silicate	25	6.00	5.98	5.97	5.96	5.90	—
Magnesium titanate	25	—	13.9	13.9	13.9	13.8	13.7
Oxides of aluminum, silicon, magnesium, calcium, barium	24	—	6.04	6.04	—	5.90	—
Porcelain (dry process)	25	5.5	5.36	5.08	5.04	—	—
Steatite 410	25	5.77	5.77	5.77	5.77	5.7	—
Strontium titanate	25	—	233	232	232	—	—
Titanium dioxide (rutile)	26	—	100	100	100	—	—
Glasses:							
Iron-sealing glass	24	8.41	8.38	8.30	8.20	7.99	7.84
Soda-borosilicate	25	—	4.97	4.84	4.84	4.82	4.65
100% silicon dioxide (fused quartz)	25	3.78	3.78	3.78	3.78	3.78	3.78
Plastics:							
Alkyd resin	25	—	5.10	4.76	4.55	4.50	—
Cellulose acetate-butyrate, plasticized	26	3.60	3.48	3.30	3.08	2.91	—
Cresylic acid-formaldehyde, 50% α-cellulose	25	5.45	4.95	4.51	3.85	3.43	3.21
Cross-linked polystyrene	25	2.59	2.59	2.58	2.58	2.58	—
Epoxy resin (Araldite CN-501)	25	—	3.67	3.62	3.35	3.09	—
Epoxy resin (Epon resin RN-48)	25	—	3.63	3.52	3.32	3.04	—
Foamed polystyrene, 0.25% filler	25	1.03	1.03	1.03	—	1.03	1.03
Melamine—formaldehyde, α-cellulose	24	—	7.57	7.00	6.0	4.93	—
Melamine—formaldehyde, 55% filler	26	—	6.00	5.75	5.5	—	—
Phenol—formaldehyde (Bakelite BM 120)	25	4.90	4.74	4.36	3.95	3.70	3.55
Phenol—formaldehyde, 50% paper laminate	26	5.25	5.15	4.60	4.04	3.57	—
Phenol—formaldehyde, 65% mica, 4% lubricants	24	5.1	5.03	4.78	4.72	4.71	—
Polycarbonate	—	3.17	3.02	2.96	—	—	—
Polychlorotrifluoroethylene	25	2.72	2.63	2.42	2.32	2.29	2.28
Polyethylene	25	2.26	2.26	2.26	2.26	2.26	2.26
Polyethylene-terephthalate	—	3.16	3.12	2.98	—	—	—
Polyethylmethacrylate	22	—	2.75	2.55	2.52	2.51	2.5
Polyhexamethylene-adipamide (nylon)	25	3.7	3.50	3.14	3.0	2.84	2.73
Polyimide	—	—	3.5	3.4	—	—	—
Polyisobutylene	25	2.23	2.23	2.23	2.23	2.23	—
Polymer of 95% vinyl-chloride, 5% vinyl-acetate	20	—	3.15	2.90	2.8	2.74	—
Polymethyl methacrylate	27	3.45	3.12	2.76	—	2.60	—
Polyphenylene oxide	—	2.55	2.55	2.55	—	2.55	—
Polypropylene	—	2.25	2.25	2.55	—	—	—
Polystyrene	25	2.56	2.56	2.56	2.55	2.55	2.54
Polytetrafluoroethylene (teflon)	22	2.1	2.1	2.1	2.1	2.1	2.08
Polyvinylcyclohexane	24	—	2.25	2.25	2.25	2.25	—
Polyvinyl formal	26	3.20	3.12	2.92	2.80	2.76	2.7
Polyvinylidene fluoride	—	8.4	8.0	6.6	—	—	—
Urea-formaldehyde, cellulose	27	6.6	6.2	5.65	5.1	4.57	—
Urethane elastomer	—	6.7–7.5	6.7–7.5	6.5–7.1	—	—	—
Vinylidene-vinyl chloride copolymer	23	5.0	4.65	3.18	2.82	2.71	—
100% aniline-formaldehyde (Dilectene-100)	25	3.70	3.68	3.58	3.50	3.44	—
100% phenol-formaldehyde	24	8.6	7.15	5.4	4.4	3.64	—
100% polyvinyl-chloride	20	3.20	3.10	2.88	2.85	2.84	—
Organic Liquids:							
Aviation gasoline (100 octane)	25	—	—	1.94	1.94	1.92	—
Benzene (pure, dried)	25	2.28	2.28	2.28	2.28	2.28	2.28

OF INSULATING MATERIALS*

Dissipation Factor at (Frequency in hertz) 60	10^3	10^6	10^8	3×10^9	2.5×10^{10}	Dielectric Strength in Volts/Mil at 25°C	DC Volume Resistivity in Ohm-cm at 25°C	Thermal Expansion (Linear) in Parts/°C	Softening Point in °C	Moisture Absorption in Percent
—	0.00057	0.00033	0.00030	0.0010	—	—	—	—	—	—
0.056	0.0130	0.0105	—	0.30	0.60	75	10^{12}–10^{13}	—	1400–1430	0.1
0.006	0.00044	0.0002	—	0.0023	—	100	10^{12}–10^{14}	—	1510	<0.1
—	<0.0003	<0.0003	<0.0003	—	—	—	—	—	—	—
0.012	0.0034	0.0005	0.0004	0.0012	—	—	$>10^{14}$	9.2×10^{-6}	1350	0.1–1
—	0.0011	0.0004	0.0005	0.0017	0.0065	—	—	—	—	—
—	0.0019	0.0011	—	0.0024	—	—	—	7.7×10^{-6}	1325	—
0.03	0.0140	0.0075	0.0078	—	—	—	—	—	—	—
—	0.0030	0.0007	0.0006	0.00089	—	—	—	—	—	—
—	0.0011	0.0002	0.0001	—	—	100	10^{12}–10^{14}	—	1510	0.1
—	0.0015	0.0003	0.00025	—	—	—	—	—	—	—
—	0.0004	0.0005	0.0009	0.00199	0.0112	—	10^{10} at 250°	132×10^{-7}	484	Poor
—	0.0055	0.0036	0.0030	0.0054	0.0090	—	7×10^{7} at 250°	50×10^{-7}	693	—
0.0009	0.00075	0.0001	0.0002	0.00006	0.00025	410 (0.25″)	$>10^{19}$	5.7×10^{-7}	1667	—
—	0.0236	0.0149	0.0138	0.0108	—	—	—	—	—	—
0.0045	0.0097	0.018	0.017	0.028	—	250–400 (0.125″)	—	11–17×10^{-5}	60–121	2.3
0.098	0.033	0.036	0.055	0.051	0.038	1020 (0.033″)	3×10^{13}	3×10^{-5}	>125	1.2
0.0004	0.0005	0.0016	0.0020	0.0019	—	—	—	—	—	—
—	0.0024	0.019	0.034	0.027	—	405 (0.125″)	$>3.8\times10^{7}$	4.77×10^{-5}	109 (distortion)	0.14
—	0.0038	0.0142	0.0264	0.021	—	—	—	—	—	—
<0.0002	<0.0001	<0.0002	—	0.0001	—	—	—	—	85	low
—	0.0122	0.041	0.085	0.103	—	300–400	—	—	99 (stable)	0.4–0.6
—	0.0119	0.0115	0.020	—	—	—	—	1.7×10^{-5}	—	0.6
0.08	0.0220	0.0280	0.0380	0.0438	0.0390	300 (0.125″)	10^{11}	30–40×10^{-6}	<135 (distortion)	<0.6
0.025	0.0165	0.034	0.057	0.060	—	—	—	—	—	—
0.015	0.0104	0.0082	0.0115	0.0126	—	—	—	—	—	—
0.009	0.0021	0.010	—	—	—	364 (0.125″)	2×10^{16}	7×10^{-5}	135 (deflection)	—
0.015	0.0270	0.0082	—	0.0028	0.0053	—	10^{18}	—	—	—
<0.0002	<0.0002	<0.0002	0.0002	0.00031	0.0006	1200 (0.033″)	10^{17}	19×10^{-5} (varies)	95–105 (distortion)	0.03
0.0021	0.0047	0.016	—	—	—	4000 (0.002″)	—	—	—	—
—	0.0294	0.0090	—	0.0075	0.0083	—	—	—	60 (distortion)	low
0.018	0.0186	0.0218	0.0200	0.0117	0.0105	400 (0.125″)	8×10^{14}	10.3×10^{-5}	65 (distortion)	1.5
—	0.002	0.003	—	—	—	570	—	—	—	—
0.0004	0.0001	0.0001	0.0003	0.00047	—	600 (0.010″)	—	—	25 (distortion)	low
—	0.0165	0.0150	0.0080	0.0059	—	—	—	—	—	—
0.064	0.0465	0.0140	—	0.0057	—	990 (0.030″)	$>5\times10^{16}$	8–9×10^{-5}	70–75 (distortion)	0.3–0.6
0.0004	0.0003	0.0007	—	0.0011	—	500 (0.125″)	10^{17}	5.3×10^{-5}	195 (deflection)	—
<0.0005	<0.0005	<0.0005	—	—	—	650 (0.125″)	6×10^{16}	6–8.5×10^{-5}	99–116 (deflection)	—
<0.00005	<0.00005	0.00007	<0.0001	0.00033	0.0012	500–700 (0.125″)	10^{18}	6–8×10^{-5}	82 (distortion)	0.05
<0.0005	<0.0003	<0.0002	<0.0002	0.00015	0.0006	1000–2000 (0.005″–0.012″)	10^{17}	9.0×10^{-5}	66 (distortion) (stable to 300)	0.00
—	0.0002	<0.0002	<0.0002	0.00018	—	—	—	—	—	—
0.003	0.0100	0.019	0.013	0.0113	0.0115	860 (0.034″)	5×10^{16}	7.7×10^{-5}	190	1.3
0.049	0.018	0.17	—	—	—	260 (0.125″)	2×10^{14}	12×10^{-5}	148 (deflection)	—
0.032	0.024	0.027	0.050	0.0555	—	375 (0.085″)	—	2.6×10^{-5}	152 (distortion)	2
0.016	0.055	—	—		—	450–500 (0.125″)	2×10^{11}	10–20×10^{-5}	—	—
0.042	0.063	0.057	0.0180	0.0072	—	300 (0.125″)	10^{14}–10^{16}	15.8×10^{-5}	150	<0.1
0.0033	0.0032	0.0061	0.0033	0.0026	—	810 (0.068″)	10^{16}	5.4×10^{-5}	125	0.06–0.08
0.15	0.082	0.060	0.077	0.052	—	277 (0.125″)	—	8.3–13×10^{-5}	50 (distortion)	0.42
0.0115	0.0185	0.0160	0.0081	0.0055	—	400 (0.125″)	10^{14}	6.9×10^{-5}	54 (distortion)	0.05–0.15
—	—	—	0.0001	0.0014	—	—	—	—	—	—
<0.0001	<0.0001	<0.0001	<0.0001	<0.0001	<0.0001	—	—	—	—	—

Continued on next page.

TABLE 9 (CONT). CHARACTERISTICS

Material Composition	T°C	Dielectric Constant at (Frequency in hertz) 60	10^3	10^6	10^8	3×10^9	2.5×10^{10}
Organic Liquids (cont):							
Carbon tetrachloride	25	2.17	2.17	2.17	2.17	2.17	—
Ethyl alcohol (absolute)	25	—	—	24.5	23.7	6.5	—
Ethylene glycol	25	—	—	41	41	12	—
Jet fuel (JP-3)	25	—	—	2.08	2.08	2.04	—
Methyl alcohol (absolute analytical grade)	25	—	—	31	31.0	23.9	—
Methyl or ethyl siloxane polymer (1000 cs)	22	2.78	2.78	2.78	—	2.74	—
Monomeric styrene	22	2.40	2.40	2.40	2.40	2.40	—
Transil oil	26	2.22	2.22	2.22	2.20	2.18	—
Vaseline	25	2.16	2.16	2.16	2.16	2.16	—
Waxes:							
Beeswax, yellow	23	2.76	2.66	2.53	2.45	2.39	—
Dichloronaphthalenes	23	3.14	3.04	2.98	2.93	2.89	—
Polybutene	25	2.34	2.34	2.34	2.30	2.27	—
Vegetable and mineral waxes	25	2.3	2.3	2.3	2.3	2.25	—
Rubbers:							
Butyl rubber	25	2.39	2.38	2.35	2.35	2.35	—
GR-S rubber	25	2.96	2.96	2.90	2.82	2.75	—
Gutta-percha	25	2.61	2.60	2.53	2.47	2.40	—
Hevea rubber (pale crepe)	25	2.4	2.4	2.4	2.4	2.15	—
Hevea rubber, vulcanized (100 pts pale crepe, 6 pts sulfur)	27	2.94	2.94	2.74	2.42	2.36	—
Neoprene rubber	24	6.7	6.60	6.26	4.5	4.00	4.0
Organic polysulfide, fillers	23	—	2260	110	30	16	13.6
Silicone-rubber compound	25	—	3.35	3.20	3.16	3.13	—
Woods:‡							
Balsa wood	26	1.4	1.4	1.37	1.30	1.22	—
Douglas fir	25	2.05	2.00	1.93	1.88	1.82	1.78
Douglas fir, plywood	25	2.1	2.1	1.90	—	—	1.6
Mahogany	25	2.42	2.40	2.25	2.07	1.88	1.6
Yellow birch	25	2.9	2.88	2.70	2.47	2.13	1.87
Yellow poplar	25	1.85	1.79	1.75	—	1.50	1.4
Miscellaneous:							
Amber (fossil resin)	25	2.7	2.7	2.65	—	2.6	—
DeKhotinsky cement	23	3.95	3.75	3.23	—	2.96	—
Gilsonite (99.9% natural bitumen)	26	2.69	2.66	2.58	2.56	—	—
Shellac (natural XL)	28	3.87	3.81	3.47	3.10	2.86	—
Mica, glass-bonded	25	—	7.45	7.39	—	—	—
Mica, glass, titanium dioxide	24	—	9.3	9.0	—	—	—
Ruby mica	26	5.4	5.4	5.4	5.4	5.4	—
Paper, royalgrey	25	3.30	3.29	2.99	2.77	2.70	—
Selenium (amorphous)	25	—	6.00	6.00	6.00	6.00	6.00
Asbestos fiber-chrysotile paper	25	—	4.80	3.1	—	—	—
Sodium chloride (fresh crystals)	25	—	5.90	5.90	—	—	5.90
Soil, sandy dry	25	—	2.91	2.59	2.55	2.55	—
Soil, loamy dry	25	—	2.83	2.53	2.48	2.44	—
Ice (from pure distilled water)	−12	—	—	4.15	3.45	3.20	—
Freshly fallen snow	−20	—	3.33	1.20	1.20	1.20	—
Hard-packed snow followed by light rain	−6	—	—	1.55	—	1.5	—
Water (distilled)	25	—	—	78.2	78	76.7	34

* Mostly taken from *Tables of Dielectric Materials*, Vols. I-IV, prepared by the Laboratory for Insulation Research of the Massachusetts Institute of Technology, Cambridge, Massachusetts, January 1953; from *Dielectric Materials and Applications*, A. R. von Hippel, editor, John Wiley & Sons, New York, N. Y., 1954; and from *Modern Plastics Encyclopedia*, Joel Frados, editor, 1301 Avenue of the Americas, New York, N. Y., 1962. Materials listed are typical of a class. Further data should be sought for a particular material of interest.

OF INSULATING MATERIALS*

Dissipation Factor at (Frequency in hertz) 60	10^3	10^6	10^8	3×10^9	2.5×10^{10}	Dielectric Strength in Volts/Mil at 25°C	DC Volume Resistivity in Ohm-cm at 25°C	Thermal Expansion (Linear) in Parts/°C	Softening Point in °C	Moisture Absorption in Percent
0.007	0.0008	<0.00004	<0.0002	0.0004	—	—	—	—	—	—
—	—	0.090	0.062	0.250	—	—	—	—	—	—
—	—	-0.030	0.045	1.00	—	—	—	—	—	—
—	—	0.0001	—	0.0055	—	—	—	—	—	—
—	—	0.20	0.038	0.64	—	—	—	—	—	—
0.0001	0.00008	<0.0003	—	0.0096	—	—	—	—	—	—
0.01	0.005	<0.0003	—	0.0020	—	300 (0.100")	3×10^{12}	—	—	0.06
0.001	<0.00001	<0.0005	0.0048	0.0028	—	300 (0.100")	—	—	−40 (pour point)	—
0.0004	0.0002	<0.0001	<0.0004	0.00066	—	—	—	—	—	—
—	0.0140	0.0092	0.0090	0.0075	—	—	—	—	45–64 (melts)	—
0.10	0.0110	0.0003	0.0017	0.0037	—	—	—	—	35–63 (melts)	nil
0.0002	0.0003	0.00133	0.00133	0.0009	—	—	—	—	—	—
0.0009	0.0006	0.0004	0.0004	0.00046	—	—	—	—	57	—
0.0034	0.0035	0.0010	0.0010	0.0009	—	—	—	—	—	—
0.0008	0.0024	0.0120	0.0080	0.0057	—	870 (0.040")	2×10^{15}	—	—	—
0.0005	0.0004	0.0042	0.0120	0.0060	—	—	10^{15}	—	—	—
0.0030	0.0018	0.0018	0.0050	0.0030	—	—	—	—	—	—
0.005	0.0024	0.0446	0.0180	0.0047	—	—	—	—	—	—
0.018	0.011	0.038	0.090	0.034	0.025	300 (0.125")	8×10^{12}	—	—	nil
—	1.29	0.39	0.28	0.22	0.10	—	—	—	—	—
—	0.0067	0.0030	0.0032	0.0097	—	—	—	—	—	—
0.058	0.0040	0.0120	0.0135	0.100	—	—	—	—	—	—
0.004	0.0080	0.026	0.033	0.027	0.032	—	—	—	—	—
0.012	0.0105	0.0230	—	—	0.0220	—	—	—	—	—
0.008	0.0120	0.025	0.032	0.025	0.020	—	—	—	—	—
0.007	0.0090	0.029	0.040	0.033	0.026	—	—	—	—	—
0.004	0.0054	0.019	—	0.015	0.017	—	—	—	—	—
0.001	0.0018	0.0056	—	0.0090	—	2300 (0.125")	Very high	—	200	—
0.049	0.0335	0.024	—	0.021	—	—	—	9.8×10^{-5}	80–85	—
0.006	0.0035	0.0016	0.0011	—	—	—	—	—	155 (melts)	—
0.006	0.0074	0.031	0.030	0.0254	—	—	10^{16}	—	80	low after baking
—	0.0019	0.0013	—	—	—	—	—	—	—	—
—	0.0125	0.0026	—	0.0040	—	—	—	—	400	<0.5
0.005	0.0006	0.0003	0.0002	0.0003	—	3800–5600 (0.040")	5×10^{13}	—	—	—
0.010	0.0077	0.038	0.066	0.056	—	202 (0.125")	—	—	—	—
—	0.0004	<0.0003	<0.0002	0.00018	0.0013	—	—	—	—	—
—	0.15	0.025	—	—	—	—	—	—	—	—
—	<0.0001	<0.0002	—	—	<0.0005	—	—	—	—	—
—	0.08	0.017	—	0.0062	—	—	—	—	—	—
—	0.05	0.018	—	0.0011	—	—	—	—	—	—
—	—	0.12	0.035	0.0009	—	—	—	—	—	—
—	0.492	0.0215	—	0.00029	—	—	—	—	—	—
—	—	0.29	—	0.0009	—	—	—	—	—	—
—	—	0.040	0.005	0.157	0.2650	—	10^6	—	—	—

†Dielectric constant and dissipation factor on electrical field strength.
‡Field perpendicular to grain.

TABLE 10. PROPERTIES OF SOFT MAGNETIC METALS

Name	Composition, %	Permeability Initial	Permeability Maximum	Coercivity H_c (A/m)	Retentivity B_r (T)	B_{max} (T)	Resistivity ($\mu\Omega$-cm)
Ingot iron	99.8 Fe	150	5 000	80	0.77	2.14	10
Low carbon steel	99.5 Fe	200	4 000	100	—	2.14	12
Silicon iron, unoriented	3 Si, bal Fe	270	8 000	60	—	2.01	47
Silicon iron, grain oriented	3 Si, bal Fe	1 400	50 000	7	1.20	2.01	50
4750 alloy	48 Ni, bal Fe	11 000	80 000	2	—	1.55	48
4-79 Permalloy	4 Mo, 79 Ni, bal Fe	40 000	200 000	1	—	0.80	58
Supermalloy	5 Mo, 80 Ni, bal Fe	80 000	450 000	0.4	—	0.78	65
2V-Permendur	2V, 49 Co, bal Fe	800	8 000	160	—	2.30	40
Supermendur	2V, 49 Co, bal Fe	—	100 000	16	2.00	2.30	26
Metglas* 2605SC	$Fe_{81}B_{13.5}Si_{3.5}C_2$	—	210 000	14	1.46	1.60	125
Metglas* 2605S-3	$Fe_{79}B_{16}Si_5$	—	30 000	8	0.30	1.58	125

*Metglas is Allied Corporation's registered trademark for amorphous alloys.

TABLE 11. TYPICAL PROPERTIES OF PERMANENT-MAGNET MATERIALS

Name	Composition, %	Retentivity *Br* (T)	Coercivity *Hc*/*Hci* (kA/m)	Max. Energy Product (*BH*) max (kJ/m³)
Cast Alnico 2	10 Al, 19 Ni, 13 Co, 3 Cu, bal Fe	0.75	45	13.5
Cast Alnico 5	8 Al, 14 Ni, 24 Co, 3 Cu, bal Fe	1.28	51	44.0
Cast Alnico 5-7	8 Al, 14 Ni, 24 Co, 3 Cu, bal Fe	1.35	59	60.0
Cast Alnico 6	8 Al, 16 Ni, 24 Co, 3 Cu, 1 Ti, bal Fe	1.05	62	31.0
Cast Alnico 8	7 Al, 15 Ni, 35 Co, 4 Cu, 5 Ti, bal Fe	0.82	130	42.0
Cast Alnico 9	7 Al, 15 Ni, 35 Co, 4 Cu, 5 Ti, bal Fe	1.05	120	72.0
Sintered Alnico 2	10 Al, 19 Ni, 13 Co, 3 Cu, bal Fe	0.71	44	12.0
Sintered Alnico 5	8 Al, 14 Ni, 24 Co, 3 Cu, bal Fe	1.09	49	31.0
Sintered Alnico 8	7 Al, 15 Ni, 35 Co, 4 Cu, 5 Ti, bal Fe	0.74	120	32.0
Bonded ferrite	$BaO \cdot 6\ Fe_2O_2$ + organics	0.16	110/240	4.4
Bonded ferrite	$BaO \cdot 6\ Fe_2O_3$ + organics	0.24	170/215	11.0
Sintered ferrite	$BaO \cdot 6\ Fe_2O_3$	0.22	140/280	8.0
Sintered ferrite	$BaO \cdot 6\ Fe_2O_3$	0.32	240/290	20.0
Sintered ferrite	$BaO \cdot 6\ Fe_2O_3$	0.38	180/190	27.0
Sintered ferrite	$SrO \cdot 6\ Fe_2O_3$	0.36	250/290	24.0
Sintered ferrite	$SrO \cdot 6\ Fe_2O_3$	0.40	180/190	30.0
Lodex 31	16 Fe - 9 Co - 67.5 Pb - 7.5 Sb	0.63	90	27.0
Lodex 32	19.2 Fe - 10.8 Co - 63 Pb - 7 Sb	0.74	75	28.0
Lodex 41	16 Fe - 9 Co - 67.5 Pb - 7.5 Sb	0.44	79	11.0
Lodex 42	19.2 Fe - 10.8 Co - 63 Pb - 7 Sb	0.53	68	11.0
Cunife	20 Fe, 20 Ni, 60 Cu	0.54	44	12.0
Vicalloy I	39 Fe, 51 Co, 10V	0.84	19	7.0
Iron Chrome	28 Cr, 10.5 Co, bal Fe	0.98	30	16.0
Iron Chrome	23 Cr, 15 Co, 3V, 2 Ti, bal Fe	1.35	44	44.0
Cobalt Rare Earth	Co_5Sm	0.82	600/ > 2400	130.0
Cobalt Rare Earth	Co_5Sm	0.87	660/ > 1200	144.0
Cobalt Rare Earth	$(Co, Cu, Fe)_7\ Sm$	1.10	510/530	240.0

For US suppliers, contact Magnetic Materials Producers Association, 1717 Howard St., Evanston, IL 60202.

chronous motors and generators. Some are available in plastic bonded form.

An emerging class of permanent-magnet alloys is the iron chromes, containing iron, chromium, and cobalt. These have magnetic properties similar to the alnicos, but they are sufficiently ductile that they can be cold-rolled to thin strips, drawn into fine wires, or machined or punched into intricate shapes. They also contain less cobalt than the alnicos.

Ferrites

"Ferrite" is the common term applied to a wide range of different ceramic ferromagnetic materials. Specifically, the term applies to those materials with the spinel crystal structures having the general formula XFe_2O_4, where X is any divalent metallic ion having the proper ionic radius to fit in the spinel structure. Several ceramic ferromagnetic materials have been prepared that deviate stoichiometrically from the basic formula XFe_2O_4, but common usage has included them in the family of ferrite materials.

The behavior of the conductivity and dielectric constant of ferrites can be understood by considering them as grains (crystals) of fairly low-resistance material separated by thin layers of a relatively poor conductor. Therefore, the dielectric constant and conductivity show a relaxation as a function of frequency with the relaxation frequency varying from 1000 to several million hertz. Most ferrites have relatively high resistivities ($\sim 10^6$ ohm-centimeters) if they are prepared carefully so as to avoid the presence of any divalent iron ion in the material. However, if the ferrite is prepared with an appreciable amount of divalent iron, then both the conductivity and dielectric constant are very high. Relative dielectric constants as high as 100 000 and resistivities less than 1 ohm-centimeter have been measured in several ferrites having a small amount of divalent iron in their composition.

Since the electrical resistivity of ferrites is typically 10^6 times that of metals, ferrite components have much lower eddy-current losses and hence are used at frequencies above about 10 kHz. Table 12

TABLE 12. CHARACTERISTICS OF FERRITES

Material	MnZn Ferrites									NiZn Ferrites		
Code*	H5A	H5B	H5C2	H5E	H6F	H6H3	H6K	H7C1	H7C2	K5	K6A	K8
Practical freq. (MHz)	< 0.2	< 0.1	< 0.1	< 0.01	0.2–2.0	0.01–0.8	0.01–0.3	< 0.3	< 0.2	< 8	1–50	< 200
Initial permeability	3 300	5 000	10 000	18 000	800	1 300	2 200	2 500	3 900	290	25	16
Relative loss factor, tan $\delta/\mu_i \times 10^6$	< 2.5 10 kHz	< 6.5 10 kHz	< 7.0 10 kHz	—	< 17 1 MHz	< 1.2 100 kHz	< 3.5 100 kHz	—	—	< 28 1 MHz	< 150 10 MHz	< 250 100 MHz
Temp. coef of $\mu_i \times 10^6$, −30 to 20°C†	−0.5 to 2.0	−0.5 to 2.0	−0.5 to 1.5	−0.5 to 2.0	—	0.3 to 2.0	0.4 to 1.2	—	—	−4.0 to 2.0	—	—
Curie temp. (°C)	> 130	> 130	> 120	> 115	> 200	> 200	> 130	> 230	> 200	> 280	> 450	> 500
Saturation flux density (T)	0.41	0.42	0.40	0.44	0.40	0.47	0.39	0.51	0.48	0.33	0.30	0.27
Disaccommodation factor $\times 10^6$, 1–10 min.**	< 3	< 3	< 1	< 1	< 12	< 5	< 2	—	—	< 30	< 20	—
Resistivity (Ω·m)	1	1	0.15	0.05	4	25	8	10	2	20×10^5	2.5×10^5	1.0×10^5
Applications	Transformers				Inductors			Power Supplies		Inductors		

From Kirk-Othmer, *Encyclopedia of Chemical Technology*, Vol. 14, 3rd ed. New York: John Wiley & Sons, Inc.; p. 665, Table 7.

*TDK Data Book: Ferrite Cores-2, Aug., 1978.

† $(\mu_2 - \mu_1)/\mu_1^2(T_2 - T_1)$

** $(\mu_1 - \mu_2)\mu_1^2 \log(t_2/t_1)$

lists some of the pertinent information with respect to typical commercial ferrites used in transformers and inductors. As a general rule, manganese zinc (MnZn) ferrites are used at frequencies up to about 1 MHz, beyond which nickel zinc (NiZn) ferrites become more efficient as a result of greater electrical resistivity. Relative initial permeabilities can be as high as 20 000 in commercial MnZn ferrites for transformer applications, while the relative loss factor, $\tan \delta / \mu$, can be as low as 10^{-6} (at 100 kHz) for some low-loss-inductor MnZn ferrites. For filter applications, a low temperature coefficient of permeability and a low disaccommodation factor are often critical. Table 12 also contains data for some MnZn ferrites used in the design of switched-mode power supplies where a high saturation flux density is desirable.

For microwave applications, devices make use of the nonreciprocal propagation characteristics of ferrites close to or at a gyromagnetic resonance frequency between about 1 and 100 GHz. The most important of such devices are isolators and circulators. Table 13 lists the pertinent design values of several current commercial microwave ferrites, including saturation polarization J_s, resonance line width ΔH, dielectric constant ϵ', loss tangent $\tan \delta$, and Curie temperature T_c. Materials having a range of J_s are needed, depending on the operating frequency, ω, of the device, since for resonance, $J_s < \omega / \delta$, where δ is the gyromagnetic ratio. In the low-

TABLE 13. SELECTED MICROWAVE MATERIALS

Material	Saturation Polarization J_s (T)	Line Width ΔH (A/cm)	Dielectric Constant ϵ'	Loss Tangent $\tan\delta$	Curie Temperature T_c (°C)	Remarks
Garnets						
Y	0.180	36	15.0	2×10^{-4}	280	—
YAl	0.018	36	13.8	2×10^{-4}	105	Decreasing
⋮	⋮	⋮	⋮	⋮	⋮	aluminum
YAl	0.120	36	14.8	2×10^{-4}	220	content
YGd	0.073	160	15.4	2×10^{-4}	280	Decreasing Gd,
⋮	⋮	⋮	⋮	⋮	⋮	low $\Delta M_s / \Delta T$,
YGd	0.160	40	15.1	2×10^{-4}	280	const. T_c
YGdAl	0.040	52	14.2	2×10^{-4}	150	Similar to
⋮	⋮	⋮	⋮	⋮	⋮	YAl, but
YGdAl	0.140	40	15.1	2×10^{-4}	265	lower $\Delta M_s / \Delta T$
Spinels						
MgMnAl	0.075	96	11.3	2.5×10^{-4}	90	Decreasing
⋮	⋮	⋮	⋮	⋮	⋮	aluminum
MgMnAl	0.175	180	12.2	2.5×10^{-4}	225	content
MgMn	0.215	432	12.7	2.5×10^{-4}	320	—
MgMnZn	0.250	416	12.9	2.5×10^{-4}	275	Increasing
⋮	⋮	⋮	⋮	⋮	⋮	zinc
MgMnZn	0.280	432	13.1	2.5×10^{-4}	225	content
NiZn	0.400	272	12.3	2.5×10^{-3}	470	High $4\pi M_s$
NiZn	0.500	128	12.5	1.0×10^{-3}	375	High $4\pi M_s$
Li	0.375	520	15.0	2.5×10^{-3}	640	High T_c
LiZn	0.480	192	14.5	2.5×10^{-3}	400	High $4\pi M_s$
LiTi	0.100	240	18.0	2.5×10^{-3}	300	Decreasing
⋮	⋮	⋮	⋮	⋮	⋮	titanium
LiTi	0.290	440	15.2	2.5×10^{-3}	600	content

From Kirk-Othmer, *Encyclopedia of Chemical Technology*, Vol. 14, 3rd ed. New York: John Wiley & Sons, Inc.; p. 668, Table 8. (Source: Trans-Tech, Inc.)

frequency range (1–5 GHz), a class of ferrites called garnets (general formula $R_3Fe_5O_{12}$ with R being yttrium or other rare earth element) is used, whereas in the high-frequency range, the spinel ferrites are used.

MAGNETOSTRICTION

The static strain $\Delta l/l$ produced by a direct-current polarizing flux density B_0 is given by

$$\Delta l/l = cB_0^2$$

c being a material constant expressed in m^4/Wb^2.

If a small alternating-current driving field is superimposed on a large constant polarizing field B_0, we have

$$d(\Delta l/l) = 2cB_0B = \beta B$$

The magnetostriction stress constant Λ in newtons/weber is

$$\Lambda = \beta Y_0 = 2cB_0Y_0$$

Y_0 being the Young's modulus for a free bar.

Nickel contracts with increasing B, so Λ is negative. Permalloy and Alfer expand, and their Λ is positive (Fig. 4).

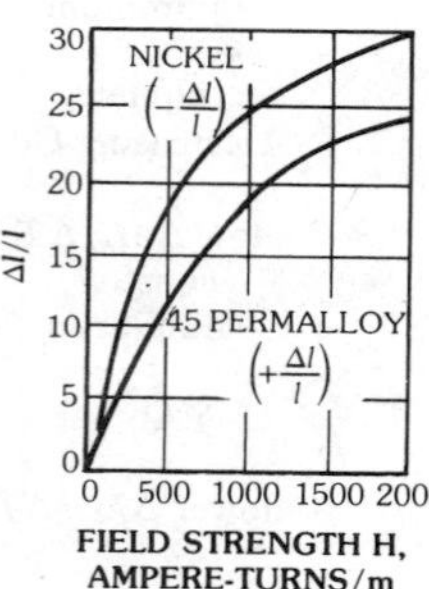

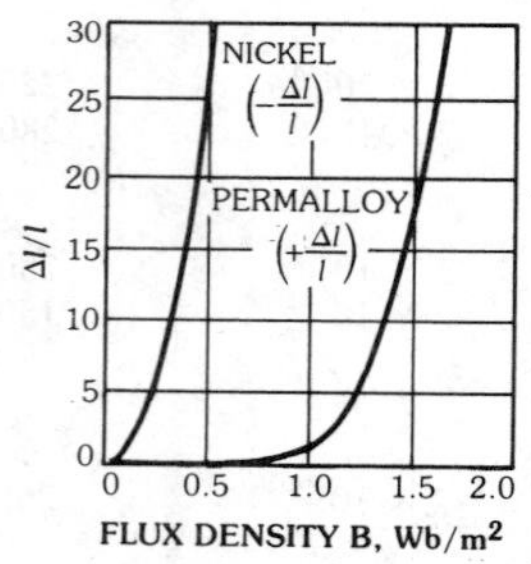

Fig. 4. Strain versus field strength (left) and versus flux density (right). (*From T. F. Hueter and R. H. Bolt, eds.,* Sonics. *New York: John Wiley & Sons, Inc., 1955; p. 173.*)

Table 14 gives values for the three important transducer materials in both SI and cgs units: annealed nickel, 45 Permalloy (45-percent nickel, 55-percent iron), and Alfer (an alloy of 13-percent aluminum and 87-percent iron). For nickel the values for two different polarizing conditions are given: 160 ampere-turns per meter and 1200 ampere-turns per meter, the latter appearing in parentheses. Table 15 compares properties of six magnetostriction materials.

PIEZOELECTRICITY

Table 16 lists piezoelectric strain coefficients d_{ij} which are ratios of piezoelectric polarization components to components of applied stress at constant electric field (direct piezoelectric effect) and also ratios of piezoelectric strain components to applied electric field components at constant mechanical stress (converse effect). The subscripts $i = 1$ to 3 indicate electric field components, and the subscripts $j = 1$ to 6 indicate mechanical stress or strain components. These components are referred to the crystallographic principal axes. For correlation of these to crystallographic axes, we follow Standards on Piezoelectric Crystals. For completeness, the full d matrix is listed in Fig. 5 for the various crystal classes.

In the monoclinic system, indices 2 and 5 refer to the symmetry (b) axis, in distinction from the older convention relating indices 3 and 6 to the symmetry axis. Crystal classes are designated by international (Hermann-Mauguin) symbols. A dash in place of a coefficient indicates that it is equal by symmetry from another listed coefficient; a blank space indicates that the coefficient is zero by symmetry. If the sign of a coefficient is not given, it is unknown, not necessarily positive.

Unit for $d_{ij} = 1/3 \times 10^{-12}$ coulomb/newton

$= 1/3 \times 10^{-12}$ meter/volt

Coupling factor k is defined practically by

$k^2 =$ (mechanical energy converted into electric energy)/(mechanical energy put into the crystal)

The converse effect is also true. The same type of relationship holds, and the coupling coefficient is numerically identical to what it was before, namely

$k^2 =$ (electrical energy converted into mechanical energy)/(electrical energy put into the crystal)

d is the measure of the deflection caused by an applied voltage or the amount of charge produced by a given force (units = meters per volt or coulombs per newton)

g denotes a field produced in a piezoelectric crystal by an applied stress unit:

$$\frac{\text{volts/meter}}{\text{newtons/square meter}}$$

Equations that relate g, d, and k are:

$$g = d/\epsilon_r\epsilon_0$$

and

$$k^2 = gdE$$

where,

ϵ_r = relative permittivity of the dielectric,

ϵ_0 = permittivity of free space = 8.85×10^{-12} farad/meter,

E = Young's modulus.

Constants of some piezoelectric materials are listed in Table 17.

TABLE 14. MAGNETOMECHANICAL COEFFICIENTS OF THREE IMPORTANT MAGNETOSTRICTIVE MATERIALS AT INTERNAL POLARIZING FIELD H_0*†

	Annealed Nickel**		45 Permalloy		Alfer (13% Al, 87% Fe)	
Coefficient	SI	CGS	SI	CGS	SI	CGS
H_0						
A/m	160 (1200)		600		800	
Oe		2 (15)		7.5		10
B_0						
V·s/m²	0.25 (0.51)		1.43		1.15	
G		2500 (5100)		14 300		11 500
μ	1250 (340)		1900		1150	
μ_i	137 (41)		230		190	
$\Delta l/l$ at H_0	-8×10^{-6} (-26×10^{-6})		14×10^{-6}		26×10^{-6}	
c						
m⁴/Wb²	-1×10^{-4}		6.9×10^{-6}		19.5×10^{-6}	
G^{-2}		-1×10^{-12}		6.9×10^{-14}		19.5×10^{-14}
Λ						
N/Wb	-4.8×10^{6} (-20×10^{6})		2.7×10^{6}		6.7×10^{6}	
dyn/G · cm³		-4.8×10^{3} (-20×10^{3})		2.7×10^{3}		6.7×10^{3}
Y_0						
N/m²	20×10^{10}		13.8×10^{10}		15×10^{10}	
dyn/m²		20×10^{11}		13.8×10^{11}		15×10^{11}
ρ						
kg/m³	8.7×10^{3}		8.25×10^{3}		6.7×10^{3}	
g/cm³		8.7		8.25		6.7
k_c% (electromechanical coupling factor)	14 (31)		12.4		27	
ρ_c						
Ω · m	7×10^{-8}		7×10^{-7}		9×10^{-7}	
Ω · cm		7×10^{-6}		7×10^{-5}		9×10^{-5}

* From T. F. Hueter and R. H. Bolt, eds., *Sonics* (New York: John Wiley & Sons, Inc., 1955; p. 175).

† The number of external ampere-turns per meter required to produce H_0 depends on the shape of the core. For closed magnetic loops H_{ext} is equal to H_0. For rod-shaped cores external fields larger than H_0 are necessary to compensate for the demagnetizing effect of the poles at the free ends.

** The values for two different polarizing conditions are given: 160 and 1200 ampere-turns/meter, the latter in parentheses.

TABLE 15. OTHER CONSTANTS FOR SOME MAGNETOSTRICTION MATERIALS

Composition	$\Delta l/l$ at Saturation of B	Λ (newtons/weber)	μ_i (henries/meter)	Young's modulus (newtons/m^2)	k_c (%)	Curie Temperature (°C)
99.9% Ni annealed	-33×10^{-6}	-20×10^{6}	4.3×10^{-6}	2.0×10^{11}	31	358
2V 49 Co 49 Fe (2V Permadur)	$+70\times10^{-6}$	—	—	1.7×10^{11}	20–37	980
45 Ni 55 Fe (45 Permalloy)	$+27\times10^{-6}$	2.7×10^{6}	2.9×10^{-4}	1.4×10^{11}	12	440
13 Al 87 Fe (13 Alfer)	$+40$	6.7×10^{6}	2.4×10^{-4}	1.5×10^{11}	27	500
Fe_3O_4	$+40$	-90×10^{6}	1.9×10^{-2}	1.8×10^{11}	3	190
Ferrite 7 Al	-28	-28 to -44×10^{6}	$4\text{–}5\times10^{-5}$	1.68 to 1.75×10^{11}	25–30	640

TABLE 16. PIEZOELECTRIC STRAIN COEFFICIENTS FOR VARIOUS MATERIALS*

(A) Cubic and Tetragonal Crystals	Composition	Class	d_{14}	d_{36}
Sphalerite	ZnS	$\bar{4}3m$	9.7	—
Sodium chlorate	$NaClO_3$	23	5.2	—
Sodium bromate	$NaBrO_3$	23	7.3	—
"ADP"	NH_4H_2PO	$\bar{4}2m$	−1.5	+48.0
"KDP"	KH_2PO_4	$\bar{4}2m$	+1.3	+21
"ADA"	$NH_4H_2AsO_4$	$\bar{4}2m$	+41	+31
"KDA"	KH_2AsO_4	$\bar{4}2m$	+23.5	+22

(B) Trigonal Crystals	Class	d_{11}	d_{14}	d_{15}	d_{22}	d_{31}	d_{33}
Quartz	32	+6.9	−2.0				
Tourmaline	3			+11.0	−0.94	+0.96	+5.4

(C) Orthorhombic Crystals	Class	d_{14}	d_{25}	d_{36}
Epsomite	222	−6.2	−8.2	−11.5
Iodic acid	222	57	46	70
Rochelle salt (30°C)	222	+1500†	−160	+35
$NaNH_4$ tartrate	222	+56	−150	+28
LiK tartrate	222	9.6	33.6	22.8
$LiNH_4$ tartrate	222	13.2	19.6	14.8
$(NH_4)_2$ oxalate	222	50	11	25

	d_{15}	d_{24}	d_{31}	d_{32}	d_{33}
K pentaborate	9.5	1.7	−5.4	0	+5.6

Body of table continued on next page.

Note: If the sign of a coefficient is not given, it is unknown (not necessarily positive).

*From *Smithsonian Physical Tables*, 9th revised edition, Vol. 120. (Washington, D.C.: Smithsonian Institution, 1969; p.432).

†The coefficient d_{14} of Rochelle salt is extremely dependent on temperature and amplitude. The ratio of d_{14} to dielectric constant ϵ_r is, however, nearly constant.

TABLE 16 (CONT). PIEZOELECTRIC STRAIN COEFFICIENTS FOR VARIOUS MATERIALS**

(D) Monoclinic Crystals (Class 2)	d_{14}	d_{16}	d_{21}	d_{22}	d_{23}	d_{25}	d_{34}	d_{36}
Lithium sulfate	+14.0	−12.5	+11.6	−45.0	−5.5	+16.5	−26.4	+10.0
Tartaric acid	+24.0	+15.8	−2.3	−6.5	−6.3	+1.1	−32.4	+35.0
K_2 tartrate (DKT)	−25	+6.5	−2.2	+8.5	−10.4	−22.5	+29.4	−66.0
$(NH_4)_2$ tartrate	+9.3	−8.5	+17.6	−26.2	+1.8	−5.9	−14.0	+5.6
EDT (ethylene diamine tartrate)	−31.1	−36.5	+30.6	+6.6	−33.8	−54.3	−51	−56.9
Cane sugar	−3.7	−7.2	+4.4	−10	+2.2	−2.6	−1.3	+1.3

(E) Polarized Polycrystalline Substance	d_{15}	d_{31}	d_{33}
Barium titanate ceramic $\epsilon_r = 1700$	750	−235	+570

**See notes for Table 16 on page 4-30.

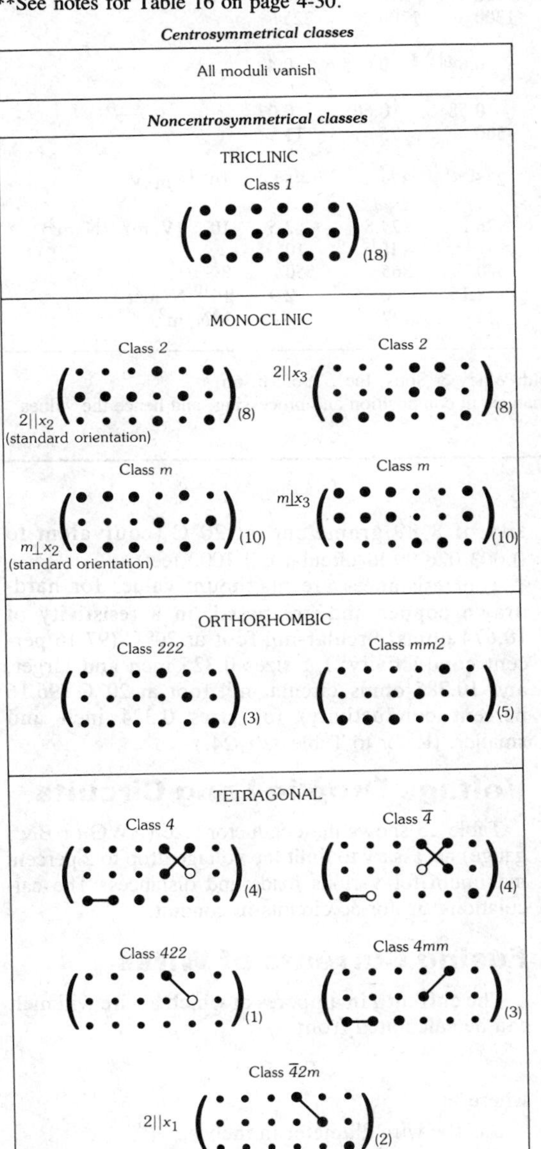

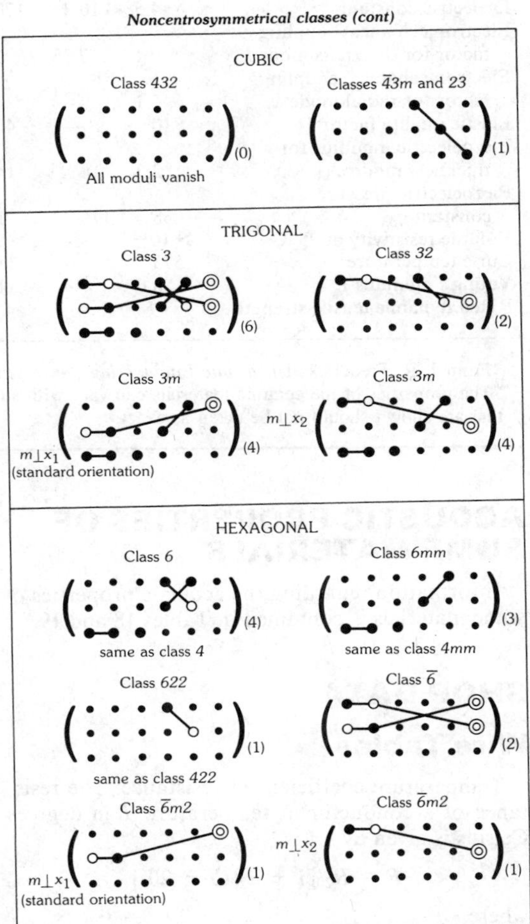

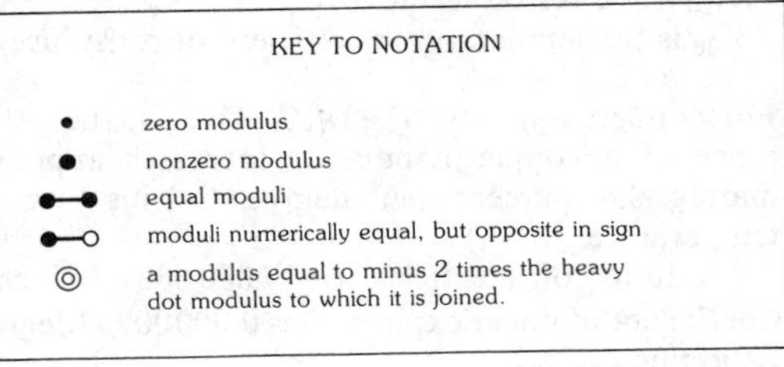

Fig. 5. Form of the (d_{ij}) matrix.

TABLE 17. CONSTANTS OF SOME PIEZOELECTRIC MATERIALS*†

Physical Property	Quartz O° X-cut	Lithium Sulfate O° Y-cut	Barium Titanate Type B	Lead Zirconate-/Titanate PZT-4	Lead Zirconate-/Titanate PZT-5	Lead Meta-/Niobate	Units
Density p	2.65	2.06	5.6	7.6	7.7	5.8	10^3 kg/m^3
Acoustic impedance *pc*	15.2	11.2	24	30.0	28.0	16	10^6 kg/m^{2}s
Frequency thickness constant *ft*	2870	2730	2740	2000	1800	1400	kHz·mm
Maximum operating temperature	550	75	70–90	250	290	500	°C
Dielectric constant	4.5	10.3	1700	1300	1700	225	—
Electromechanical coupling factor for thickness mode k_{33}	0.1	0.35	0.48	0.64	0.675	0.42	—
Electromechanical coupling factor for radial mode k_p	0.1	—	0.33	0.58	0.60	0.07	—
Elastic quality factor Q	10^6	—	400	500	75	11	—
Piezoelectric modulus for thickness mode d_{33}	2.3	16	149	285	374	85	10^{-12} m/V
Piezoelectric pressure constant g_{33}	58	175	14.0	26.1	24.8	42.5	10^{-3} (V/m)/(N/m^2)
Volume resistivity at 25°C	$> 10^{12}$	—	$> 10^{11}$	$> 10^{12}$	$> 10^{13}$	10^9	—
Curie temperature	575	—	115	320	365	550	°C
Young's modulus E	8.0	—	11.8	8.15	6.75	2.9	10^{10} N/m^2
Rated dynamic tensile strength	—	—	—	24	27	—	MN/m^2

*From J. R. Frederick, *Ultrasonic Engineering* (New York: John Wiley & Sons, Inc., 1965; p. 66).
†The properties of the ceramic materials can vary with slight changes in composition and processing, and hence the values that are shown should not be taken as exact.

ACOUSTIC PROPERTIES OF SOME MATERIALS

Information regarding the acoustic properties of some materials is contained in Tables 18 and 19.

SHOP DATA

Wire Tables

Temperature coefficient of resistance: The resistance of a conductor at temperature T in degrees Celsius is given by

$$R = R_{20}[1 + \alpha_{20}(T - 20)]$$

where,

R_{20} is the resistance at 20°C,

α_{20} is the temperature coefficient of resistance at 20°C.

For copper, $\alpha_{20} = 0.003\,93/°C$. That is, the resistance of a copper conductor increases approximately 0.4 percent per degree Celsius rise in temperature.

Modulus of elasticity is 17 000 000 lb/inch^2. Coefficient of linear expansion is 0.000 009 4/degree Fahrenheit.

In Tables 20 through 24 masses are based on a density of 8.89 grams/cm^3 at 20°C (equivalent to 0.003 026 99 lb/circular mil·1000 feet).

The resistances are maximum values for hard-drawn copper and are based on a resistivity of 10.674 ohms/circular-mil foot at 20°C (97.16 percent conductivity) for sizes 0.325 inch and larger, and 10.785 ohms/circular-mil foot at 20°C (96.16 percent conductivity) for sizes 0.324 inch and smaller. (Refer to Tables 20–24.)

Voltage Drop in Long Circuits

Table 25 shows the conductor size (AWG or B&S gauge) necessary to limit the voltage drop to 2 percent maximum for various loads and distances. The calculations are for ac circuits in conduit.

Fusing Currents of Wires

The current I in amperes at which a wire will melt can be calculated from

$$I = Kd^{3/2}$$

where,

d is the wire diameter in inches,

K is a constant that depends on the metal concerned.

Table 26 gives the fusing currents in amperes for

TABLE 18. SIMPLIFIED EQUATIONS FOR SOUND INTENSITY*

Material	Mode of Vibration	Crystal Cut	Effective† Piezo Modulus H (coulombs/m^2)	Sound Velocity c (m/s)	Density ρ (kg/m^3)	Sound Intensity Water for Airbacked Transducer $\mathcal{G}$ (watts/m^2)	Units
Quartz	Thickness	X	$H = e_{11}$ $= 0.17$ coulomb/m^2	5.72×10^3	2.65×10^3	$f_0^2V^2 \times 10^{-14}$	V (V) rms f (Hz)
	Longitudinal	X	$H = d_{11}/s'$ $= 0.18$ coulomb/m^2	5.44×10^3	2.65×10^3	$0.087 \times E^2 \times 10^{-6}$	E (V/m) rms
Ammonium dihydrogen phosphate (ADP)	Longitudinal	45°Z	$H = d_{36}/2s'$ $= 0.042$ coulomb/m^2	3.28×10^3	1.8×10^3	$0.6E^2 \times 10^{-6}$	E (V/m) rms
Rochelle salt (0°C)	Longitudinal	45°X	$H = d_{14}/2s'$ $= 5.67$ coulombs/m^2	3.4×10^3	1.77×10^3	$85E^2 \times 10^{-6}$	E (V/m) rms
Barium titanate (40°C)	Thickness	Polarized normal to thickness	$H = e_{33}$ $= 10$ to 17 coulombs/m^2	5×10^3	5.5×10^3	$0.005f_0^2V^2 \times 10^{-8}$ to $0.014f_0^2V^2 \times 10^{-8}$	V (V) rms f (Hz)

*From T. F. Hueter and R. H. Bolt, eds., *Sonics* (New York: John Wiley & Sons, Inc., 1955; p.125).
†The quantity s' in the relationship $H = d_{ih}/s'$ is the effective compliance in the direction of longitudinal vibration.

TABLE 19. VELOCITIES, DENSITIES, AND CHARACTERISTIC IMPEDANCES OF VARIOUS METALS*

	Velocities				
	Longitudinal				Characteristic Impedance ρc
Metals	Bulk (m/s)	Bar (m/s)	Shear (m/s)	Density ρ (kg/m^3)	Bulk (kg/m^{2}s)
	$\times 10^3$	$\times 10^3$	$\times 10^3$	$\times 10^3$	$\times 10^6$
Aluminum	6.40	5.15	3.13	2.7	17.3
Beryllium	12.89	...	8.88	1.8	23.2
Brass, 70-30	4.37	3.40	2.10	8.5	37.0
Cast iron	3.50–5.60	3.0–4.7	2.2–3.2	7.2	25.0–40.0
Copper	4.80	3.65	2.33	8.9	42.5
Gold	3.24	2.03	1.20	19.3	63.0
Iron	5.96	5.18	3.22	7.9	46.8
Lead	2.40	1.25	0.79	11.3	27.2
Magnesium	5.74	4.90	3.08	1.7	9.9
Mercury	1.45	...	...	13.6	19.6
Molybdenum	6.25	...	3.35	10.2	63.7
Nickel	5.48	4.70	2.99	8.9	48.5
Platinum	3.96	2.80	1.67	21.4	85.0
Steel, mild	6.10	5.05	3.24	7.9	46.7
Silver	3.70	2.67	1.70	10.5	36.9
Tin	3.38	2.74	1.61	7.3	24.7
Titanium	5.99	...	3.12	4.50	27.0
Tungsten	5.17	...	2.88	19.3	100.0
Uranium	3.37	...	2.02	18.7	63.0
Zinc	4.17	3.81	2.48	7.1	29.6
Zirconium	4.65	...	2.30	6.4	29.8
Other Solid Materials:					
Crown glass	5.66	5.30	3.42	2.5	14.0
Granite	...	3.95	...	2.75	...
Ice	3.98	...	1.99	0.9	3.6
Nylon	1.8–2.2	...	...	1.1–1.2	2.0–2.7
Paraffin, hard	2.2	...	...	0.83	1.8
Plexiglas or Lucite	2.68	1.8	1.32	1.20	3.2
Polystyrene	2.67	...	...	1.06	2.8
Quartz, fused	5.57	5.37	3.52	2.6	14.5
Teflon	1.35	...	...	2.2	3.0
Tungsten carbide	6.66	...	3.98	10.0–15.0	66.5–98.5
Wood, oak	...	4.1	...	0.8	...
Fluids:					
Benzene	1.32	...	...	0.88	1.16
Castor oil	1.54	...	...	0.95	1.45
Glycerine	1.92	...	...	1.26	2.5
Methyl iodide	0.98	...	...	3.23	3.2
Oil, SAE 20	1.74	...	...	0.87	1.5
Water, fresh	1.48	...	...	1.00	1.48

Note: The values that are shown should not be taken as exact values because of the effects of variations in composition and processing. They are adequate for most practical purposes, however.

*From J. R. Frederick, *Ultrasonic Engineering* (New York: John Wiley & Sons, Inc., 1965; p. 363).

TABLE 20. SOLID COPPER — COMPARISON OF GAUGES

American (B & S) Wire Gauge	Birmingham (Stubs') Iron Wire Gauge	British Standard (NBS) Wire Gauge	Diameter		Area			Mass	
			Mils	Millimeters	Circular Mils	Square Millimeters	Square Inches	per 1000 Feet in Pounds	per Kilometer in Kilograms
—	0	—	340.0	8.636	115 600	58.58	0.090 79	350	521
0	—	—	324.9	8.251	105 500	53.48	0.082 89	319	475
—	—	0	324.0	8.230	105 000	53.19	0.082 45	318	472
—	1	1	300.0	7.620	90 000	45.60	0.070 69	273	405
1	—	—	289.3	7.348	83 690	42.41	0.065 73	253	377
—	2	—	284.0	7.214	80 660	40.87	0.063 35	244	363
—	—	—	283.0	7.188	80 090	40.58	0.062 90	242	361
—	—	2	276.0	7.010	76 180	38.60	0.059 63	231	343
—	3	—	259.0	6.579	67 080	33.99	0.052 69	203	302
2	—	—	257.6	6.544	66 370	33.63	0.052 13	201	299
—	—	3	252.0	6.401	63 500	32.18	0.049 88	193	286
—	4	—	238.0	6.045	56 640	28.70	0.044 49	173	255
—	—	4	232.0	5.893	53 820	27.27	0.042 27	163	242
3	—	—	229.4	5.827	52 630	26.67	0.041 34	159	237
—	5	—	220.0	5.588	48 400	24.52	0.038 01	147	217
—	—	5	212.0	5.385	44 940	22.77	0.035 30	136	202
4	—	—	204.3	5.189	41 740	21.18	0.032 78	126	188
—	6	—	203.0	5.156	41 210	20.88	0.032 37	125	186
—	—	6	192.0	4.877	36 860	18.68	0.028 95	112	166
5	—	—	181.9	4.621	33 100	16.77	0.026 00	100	149
—	7	—	180.0	4.572	32 400	16.42	0.025 45	98.0	146
—	—	7	176.0	4.470	30 980	15.70	0.024 33	93.6	139
—	8	—	165.0	4.191	27 220	13.86	0.021 38	86.2	123
6	—	—	162.0	4.116	26 250	13.30	0.020 62	79.5	118
—	—	8	160.0	4.064	25 600	12.97	0.020 11	77.5	115
—	9	—	148.0	3.759	21 900	11.10	0.017 20	66.3	98.6
7	—	—	144.3	3.665	20 820	10.55	0.016 35	63.0	93.7
—	—	9	144.0	3.658	20 740	10.51	0.016 29	62.8	93.4
—	10	—	134.0	3.404	17 960	9.098	0.014 10	54.3	80.8
8	—	—	128.8	3.264	16 510	8.366	0.012 97	50.0	74.4
—	—	10	128.0	3.251	16 380	8.302	0.012 67	49.6	73.8
—	11	—	120.0	3.048	14 400	7.297	0.011 31	43.6	64.8
—	—	11	116.0	2.946	13 460	6.818	0.010 57	40.8	60.5
9	—	—	114.4	2.906	13 090	6.634	0.010 28	39.6	58.9
—	12	—	109.0	2.769	11 880	6.020	0.009 331	35.9	53.5
—	—	12	104.0	2.642	10 820	5.481	0.008 495	32.7	48.7
10	—	—	101.9	2.588	10 380	5.261	0.008 155	31.4	46.8
—	13	—	95.00	2.413	9 025	4.573	0.007 088	27.3	40.6
—	—	13	92.00	2.337	8 464	4.289	0.006 648	25.6	38.1
11	—	—	90.74	2.305	8 234	4.172	0.006 467	24.9	37.1
—	14	—	83.00	2.108	6 889	3.491	0.005 411	20.8	31.0
12	—	—	80.81	2.053	6 530	3.309	0.005 129	19.8	29.4
—	—	14	80.00	2.032	6 400	3.243	0.005 027	19.4	28.8
—	15	15	72.00	1.829	5 184	2.627	0.004 072	16.1	23.4

Continued on next page.

TABLE 20 (CONT). SOLID COPPER — COMPARISON OF GAUGES

American (B & S) Wire Gauge	Birmingham (Stubs') Iron Wire Gauge	British Standard (NBS) Wire Gauge	Diameter		Area			Mass	
			Mils	Millimeters	Circular Mils	Square Millimeters	Square Inches	per 1000 Feet in Pounds	per Kilometer in Kilograms
13	—	—	71.96	1.828	5 178	2.624	0.004 067	15.7	23.3
—	16	—	65.00	1.651	4 225	2.141	0.003 318	12.8	19.0
14	—	—	64.08	1.628	4 107	2.081	0.003 225	12.4	18.5
—	—	16	64.00	1.626	4 096	2.075	0.003 217	12.3	18.4
—	17	—	58.00	1.473	3 364	1.705	0.002 642	10.2	15.1
15	—	—	57.07	1.450	3 257	1.650	0.002 558	9.86	14.7
—	—	17	56.00	1.422	3 136	1.589	0.002 463	9.52	14.1
16	—	—	50.82	1.291	2 583	1.309	0.002 028	7.82	11.6
—	18	—	49.00	1.245	2 401	1.217	0.001 886	7.27	10.8
—	—	18	48.00	1.219	2 304	1.167	0.001 810	6.98	10.4
17	—	—	45.26	1.150	2 048	1.038	0.001 609	6.20	9.23
—	19	—	42.00	1.067	1 764	0.8938	0.001 385	5.34	7.94
18	—	—	40.30	1.024	1 624	0.8231	0.001 276	4.92	7.32
—	—	19	40.00	1.016	1 600	0.8107	0.001 257	4.84	7.21
—	—	20	36.00	0.9144	1 296	0.6567	0.001 018	3.93	5.84
19	—	—	35.89	0.9116	1 288	0.6527	0.001 012	3.90	5.80
—	20	—	35.00	0.8890	1 225	0.6207	0.000 962 1	3.71	5.52
—	21	21	32.00	0.8128	1 024	0.5189	0.000 804 2	3.11	4.62
20	—	—	31.96	0.8118	1 022	0.5176	0.000 802 3	3.09	4.60

five commonly used types of wire. Owing to the wide variety of factors that can influence the rate of heat loss, these figures must be considered as only approximations.

Physical Properties

Physical properties of several kinds of wire are listed in Table 27. Physical properties of stranded copper wire are listed in Table 28.

Adhesives

Some facts about adhesives are given in Table 29.

Machine Screws and Drill Sizes

Dimensions and other data concerning machine screws are given in Table 30. Head styles and the method of length measurement are illustrated in Fig. 6. Drill sizes are listed in Table 31.

Sheet-Metal Gauges

Materials are customarily made according to certain gauge systems. While materials can usually be had specially in any system, some usual practices are shown in Tables 32 and 33.

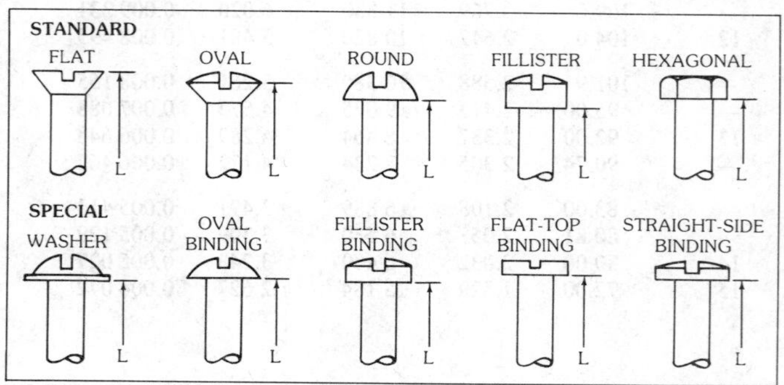

Fig. 6. Head styles and length measurement (L) for machine screws.

TABLE 21. ANNEALED COPPER

AWG B & S Gauge	Diameter in Mils	Cross Section: Circular Mils	Cross Section: Square Inches	Ohms per 1000 Ft at 20°C (68°F)	Lb per 1000 Ft	Ft per Lb	Ft per Ohm at 20°C (68°F)	Ohms per Lb at 20°C (68°F)
0000	460.0	211 600	0.1662	0.049 01	640.5	1.561	20 400	0.000 076 52
000	409.6	167 800	0.1318	0.061 80	507.9	1.968	16 180	0.000 121 7
00	364.8	133 100	0.1045	0.077 93	402.8	2.482	12 830	0.000 193 5
0	324.9	105 500	0.082 89	0.098 27	319.5	3.130	10 180	0.000 307 6
1	289.3	83 690	0.065 73	0.123 9	253.3	3.947	8 070	0.000 489 1
2	257.6	66 370	0.052 13	0.156 3	200.9	4.977	6 400	0.000 777 8
3	229.4	52 640	0.041 34	0.197 0	159.3	6.276	5 075	0.001 237
4	204.3	41 740	0.032 78	0.248 5	126.4	7.914	4 025	0.001 966
5	181.9	33 100	0.026 00	0.313 3	100.2	9.980	3 192	0.003 127
6	162.0	26 250	0.020 62	0.395 1	79.46	12.58	2 531	0.004 972
7	144.3	20 820	0.016 35	0.498 2	63.02	15.87	2 007	0.007 905
8	128.5	16 510	0.012 97	0.628 2	49.98	20.01	1 592	0.012 57
9	114.4	13 090	0.010 28	0.792 1	39.63	25.23	1 262	0.019 99
10	101.9	10 380	0.008 155	0.998 9	31.43	31.82	1 001	0.031 78
11	90.74	8 234	0.006 467	1.260	24.92	40.12	794	0.050 53
12	80.81	6 530	0.005 129	1.588	19.77	50.59	629.6	0.080 35
13	71.96	5 178	0.004 067	2.003	15.68	63.80	499.3	0.127 8
14	64.08	4 107	0.003 225	2.525	12.43	80.44	396.0	0.203 2
15	57.07	3 257	0.002 558	3.184	9.858	101.4	314.0	0.323 0
16	50.82	2 583	0.002 028	4.016	7.818	127.9	249.0	0.513 6
17	45.26	2 048	0.001 609	5.064	6.200	161.3	197.5	0.816 7
18	40.30	1 624	0.001 276	6.385	4.917	203.4	156.6	1.299
19	35.89	1 288	0.001 012	8.051	3.899	256.5	124.2	2.065
20	31.96	1 022	0.000 802 3	10.15	3.092	323.4	98.50	3.283
21	28.46	810.1	0.000 636 3	12.80	2.452	407.8	78.11	5.221
22	25.35	642.4	0.000 504 6	16.14	1.945	514.2	61.95	8.301
23	22.57	509.5	0.000 400 2	20.36	1.542	648.4	49.13	13.20
24	20.10	404.0	0.000 317 3	25.67	1.223	817.7	38.96	20.99
25	17.90	320.4	0.000 251 7	32.37	0.969 9	1 031.0	30.90	33.37
26	15.94	254.1	0.000 199 6	40.81	0.769 2	1 300	24.50	53.06
27	14.20	201.5	0.000 158 3	51.47	0.610 0	1 639	19.43	84.37
28	12.64	159.8	0.000 125 5	64.90	0.483 7	2 067	15.41	134.2
29	11.26	126.7	0.000 099 53	81.83	0.383 6	2 607	12.22	213.3
30	10.03	100.5	0.000 078 94	103.2	0.304 2	3 287	9.691	339.2
31	8.928	79.70	0.000 062 60	130.1	0.241 3	4 145	7.685	539.3
32	7.950	63.21	0.000 049 64	164.1	0.191 3	5 227	6.095	857.6
33	7.080	50.13	0.000 039 37	206.9	0.151 7	6 591	4.833	1 364
34	6.305	39.75	0.000 031 22	260.9	0.120 3	8 310	3.833	2 168
35	5.615	31.52	0.000 024 76	329.0	0.095 42	10 480	3.040	3 448
36	5.000	25.00	0.000 019 64	414.8	0.075 68	13 210	2.411	5 482
37	4.453	19.83	0.000 015 57	523.1	0.060 01	16 660	1.912	8 717
38	3.965	15.72	0.000 012 35	659.6	0.047 59	21 010	1.516	13 860
39	3.531	12.47	0.000 009 793	831.8	0.037 74	26 500	1.202	22 040
40	3.145	9.888	0.000 007 766	1049.0	0.029 93	33 410	0.9534	35 040

TABLE 22. HARD-DRAWN COPPER

AWG B & S Gauge	Wire Diameter in Inches	Breaking Load in lb_f	Tensile Strength in lb_f/in^2	Mass: Pounds per 1000 Feet	Mass: Pounds per Mile	Maximum Resistance (ohms per 1000 feet at 68°F)	Cross-Sectional Area: Circular Mils	Cross-Sectional Area: Square Inches
4/0	0.460 0	8 143	49 000	640.5	3 382	0.050 45	211 600	0.166 2
3/0	0.409 6	6 722	51 000	507.9	2 682	0.063 61	167 800	0.131 8
2/0	0.364 8	5 519	52 800	402.8	2 127	0.080 21	133 100	0.104 5
1/0	0.324 9	4 517	54 500	319.5	1 687	0.101 1	105 500	0.082 89
1	0.289 3	3 688	56 100	253.3	1 338	0.128 7	83 690	0.065 73
2	0.257 6	3 003	57 600	200.9	1 061	0.162 5	66 370	0.052 13
3	0.229 4	2 439	59 000	159.3	841.2	0.204 9	52 630	0.041 34
4	0.204 3	1 970	60 100	126.4	667.1	0.258 4	41 740	0.032 78
5	0.181 9	1 591	61 200	100.2	529.1	0.325 8	33 100	0.026 00
—	0.165 0	1 326	62 000	82.41	435.1	0.396 1	27 225	0.021 38
6	0.162 0	1 280	62 100	79.46	419.6	0.410 8	26 250	0.020 62
7	0.144 3	1 030	63 000	63.02	332.7	0.518 1	20 820	0.016 35
—	0.134 0	894.0	63 400	54.35	287.0	0.600 6	17 956	0.014 10
8	0.128 5	826.0	63 700	49.97	263.9	0.653 3	16 510	0.012 97
9	0.114 4	661.2	64 300	39.63	209.3	0.823 8	13 090	0.010 28
—	0.104 0	550.4	64 800	32.74	172.9	0.997 1	10 816	0.008 495
10	0.101 9	529.2	64 900	31.43	165.9	1.039	10 380	0.008 155
11	0.090 74	422.9	65 400	24.92	131.6	1.310	8 234	0.006 467
12	0.080 81	337.0	65 700	19.77	104.4	1.652	6 530	0.005 129
13	0.071 96	268.0	65 900	15.68	82.77	2.083	5 178	0.004 067
14	0.064 08	213.5	66 200	12.43	65.64	2.626	4 107	0.003 225
15	0.057 07	169.8	66 400	9.858	52.05	3.312	3 257	0.002 558
16	0.050 82	135.1	66 600	7.818	41.28	4.176	2 583	0.002 028
17	0.045 26	107.5	66 800	6.200	32.74	5.266	2 048	0.001 609
18	0.040 30	85.47	67 000	4.917	25.96	6.640	1 624	0.001 276

TABLE 23. TENSILE STRENGTH OF COPPER WIRE

AWG B & S Gauge	Wire Diameter in Inches	Hard Drawn: Minimum Tensile Strength lb_f/in^2	Hard Drawn: Breaking Load in Pounds	Medium-Hard Drawn: Minimum Tensile Strength lb_f/in^2	Medium-Hard Drawn: Breaking Load in lb_f	Soft or Annealed: Maximum Tensile Strength lb_f/in^2	Soft or Annealed: Breaking Load in lb_f
1	0.289 3	56 100	3 688	46 000	3 024	37 000	2 432
2	0.257 6	57 600	3 003	47 000	2 450	37 000	1 929
3	0.229 4	59 000	2 439	48 000	1 984	37 000	1 530
4	0.204 3	60 100	1 970	48 330	1 584	37 000	1 213
5	0.181 9	61 200	1 591	48 660	1 265	37 000	961.9
—	0.165 0	62 000	1 326	—	—	—	—
6	0.162 0	62 100	1 280	49 000	1 010	37 000	762.9
7	0.144 3	63 000	1 030	49 330	806.6	37 000	605.0
—	0.134 0	63 400	894.0	—	—	—	—
8	0.128 5	63 700	826.0	49 660	643.9	37 000	479.8
9	0.114 4	64 300	661.2	50 000	514.2	37 000	380.5
—	0.104 0	64 800	550.4	—	—	—	—
10	0.101 9	64 900	529.2	50 330	410.4	38 500	314.0
11	0.090 74	65 400	422.9	50 660	327.6	38 500	249.0
12	0.080 81	65 700	337.0	51 000	261.6	38 500	197.5

Antifreeze Solutions

Data regarding commercial antifreeze solutions are given in Table 34. For each type of antifreeze, the freezing point (top number) and specific gravity (bottom number) are given for several percentages of antifreeze in water.

TABLE 24. COPPER-CLAD STEEL

AWG		Cross-Sectional Area		Mass			Resistance Ohms/1000 ft at 68°F		Breaking Load Pounds		Attenuation in Decibels/Mile*				Characteristic Impedance*	
											40% Cond		30% Cond			
B & S Gauge	Diam Inch	Circular Mils	Square Inch	Pounds per 1000 Feet	Pounds per Mile	Feet per Pound	40% Conduct	30% Conduct	40% Conduct	30% Conduct	Dry	Wet	Dry	Wet	40% Cond	30% Cond
4	0.2043	41 740	0.032 78	115.8	611.6	8.63	0.633 7	0.844 7	3 541	3 934	—	—	—	—	—	—
5	0.1819	33 100	0.026 00	91.86	485.0	10.89	0.799 0	1.065	2 938	3 250	—	—	—	—	—	—
6	0.1620	26 250	0.020 62	72.85	384.6	13.73	1.008	1.343	2 433	2 680	0.078	0.086	0.103	0.109	650	686
7	0.1443	20 820	0.016 35	57.77	305.0	17.31	1.270	1.694	2 011	2 207	0.093	0.100	0.122	0.127	685	732
8	0.1285	16 510	0.012 97	45.81	241.9	21.83	1.602	2.136	1 660	1 815	0.111	0.118	0.144	0.149	727	787
9	0.1144	13 090	0.010 28	36.33	191.8	27.52	2.020	2.693	1 368	1 491	0.132	0.138	0.169	0.174	776	852
10	0.1019	10 380	0.008 155	28.81	152.1	34.70	2.547	3.396	1 130	1 231	0.156	0.161	0.196	0.200	834	920
11	0.0907	8 234	0.006 467	22.85	120.6	43.76	3.212	4.28	896	975	0.183	0.188	0.228	0.233	910	1 013
12	0.0808	6 530	0.005 129	18.12	95.68	55.19	4.05	5.40	711	770	0.216	0.220	0.262	0.266	1 000	1 120
13	0.0720	5 178	0.004 067	14.37	75.88	69.59	5.11	6.81	490	530						
14	0.0641	4 107	0.003 225	11.40	60.17	87.75	6.44	8.59	400	440						
15	0.0571	3 257	0.002 558	9.038	47.72	110.6	8.12	10.83	300	330						
16	0.0508	2 583	0.002 028	7.167	37.84	139.5	10.24	13.65	250	270						
17	0.0453	2 048	0.001 609	5.684	30.01	175.9	12.91	17.22	185	205						
18	0.0403	1 624	0.001 276	4.507	23.80	221.9	16.28	21.71	153	170						
19	0.0359	1 288	0.001 012	3.575	18.87	279.8	20.53	27.37	122	135						
20	0.0320	1 022	0.000 802 3	2.835	14.97	352.8	25.89	34.52	100	110						
21	0.0285	810.1	0.000 636 3	2.248	11.87	444.8	32.65	43.52	73.2	81.1						
22	0.0253	642.5	0.000 504 6	1.783	9.413	560.9	41.17	54.88	58.0	64.3						
23	0.0226	509.5	0.000 400 2	1.414	7.465	707.3	51.92	69.21	46.0	51.0						
24	0.0201	404.0	0.000 317 3	1.121	5.920	891.9	65.46	87.27	36.5	40.4						
25	0.0179	320.4	0.000 251 7	0.889	4.695	1 125	82.55	110.0	28.9	32.1						
26	0.0159	254.1	0.000 199 6	0.705	3.723	1 418	104.1	138.8	23.0	25.4						
27	0.0142	201.5	0.000 158 3	0.559	2.953	1 788	131.3	175.0	18.2	20.1						
28	0.0126	159.8	0.000 125 5	0.443	2.342	2 255	165.5	220.6	14.4	15.9						
29	0.0113	126.7	0.000 099 5	0.352	1.857	2 843	208.7	278.2	11.4	12.6						
30	0.0100	100.5	0.000 078 9	0.279	1.473	3 586	263.2	350.8	9.08	10.0						
31	0.0089	79.70	0.000 062 6	0.221	1.168	4 521	331.9	442.4	7.20	7.95						
32	0.0080	63.21	0.000 049 6	0.175	0.926	5 701	418.5	557.8	5.71	6.30						
33	0.0071	50.13	0.000 039 4	0.139	0.734	7 189	527.7	703.4	4.53	5.00						
34	0.0063	39.75	0.000 031 2	0.110	0.582	9 065	665.4	887.0	3.59	3.97						
35	0.0056	31.52	0.000 024 8	0.087	0.462	11 430	839.0	1 119	2.85	3.14						
36	0.0050	25.00	0.000 019 6	0.069	0.366	14 410	1 058	1 410	2.26	2.49						
37	0.0045	19.83	0.000 015 6	0.055	0.290	18 180	1 334	1 778	1.79	1.98						
38	0.0040	15.72	0.000 012 3	0.044	0.230	22 920	1 682	2 243	1.42	1.57						
39	0.0035	12.47	0.000 009 79	0.035	0.183	28 900	2 121	2 828	1.13	1.24						
40	0.0031	9.89	0.000 007 77	0.027	0.145	36 440	2 675	3 566	0.893	0.986						

*DP insulators, 12-inch wire spacing at 1000 hertz.

TABLE 25. CONDUCTOR SIZE FOR 2-PERCENT VOLTAGE DROP

Current in Amperes	Distance in Feet									Distance in Feet								
	25	50	75	100	150	200	300	400	500	25	50	75	100	150	200	300	400	500
	Single-phase—110 volts									Single-phase—220 volts								
1	—	—	—	—	—	—	14	12	10	—	—	—	—	—	—	—	—	14
1.5	—	—	—	—	14	14	12	10	10	—	—	—	—	—	—	14	14	12
2	—	—	—	—	14	12	10	10	8	—	—	—	—	—	—	14	12	12
3	—	—	14	14	12	10	8	8	6	—	—	—	—	14	14	12	10	10
4	—	—	14	12	10	10	8	6	6	—	—	—	—	14	12	10	10	8
5	—	14	12	12	10	8	6	6	4	—	—	—	14	12	12	10	8	8
6	—	14	12	10	8	8	6	4	4	—	—	14	14	12	10	8	8	6
7	—	14	12	10	8	8	6	4	2	—	—	14	14	12	10	8	8	6
8	—	12	10	10	8	6	4	2	2	—	—	14	12	10	10	8	6	6
9	—	12	10	8	8	6	4	2	2	—	14	14	12	10	8	8	6	4
10	14	12	10	8	6	6	4	2	2	—	14	12	12	10	8	6	6	4
12	14	10	8	8	6	4	2	1	1	—	14	12	10	8	8	6	4	4
14	14	10	8	8	6	4	2	0	0	—	14	12	10	8	8	6	4	2
16	12	10	8	6	4	4	2	0	00	—	12	10	10	8	6	4	4	2
18	12	8	8	6	4	2	1	00	00	14	12	10	8	8	6	4	2	2
20	12	8	6	6	4	2	1	00	000	14	12	10	8	6	6	4	2	2
25	10	8	6	4	2	2	0	000	0000	14	10	8	8	6	4	2	2	1
30	10	6	4	4	2	1	00	—	—	12	10	8	6	4	4	2	1	0
35	10	6	4	2	2	0	000	—	—	12	10	8	6	4	2	2	0	00
40	8	6	4	2	1	00	0000	—	—	12	8	6	6	4	2	1	00	0000
45	8	4	4	2	0	00	—	—	—	10	8	6	4	4	2	0	00	0000
50	8	4	2	2	0	000	—	—	—	10	8	6	4	2	2	0	000	0000
60	6	4	2	1	00	0000	—	—	—	10	6	4	4	2	1	00	0000	—
70	6	2	2	0	000	—	—	—	—	10	6	4	2	2	0	000	—	—
80	6	2	1	00	0000	—	—	—	—	8	6	4	2	1	00	0000	—	—
90	4	2	0	00	—	—	—	—	—	8	4	4	2	0	00	—	—	—
100	4	2	0	000	—	—	—	—	—	8	4	2	2	0	000	—	—	—
120	4	1	00	0000	—	—	—	—	—	6	4	2	1	00	0000	—	—	—
	Three-phase—220 volts									Three-phase—440 volts								
1	—	—	—	—	—	—	—	—	—	—	—	—	—	—	—	—	—	—
1.5	—	—	—	—	—	—	—	14	14	—	—	—	—	—	—	—	—	—
2	—	—	—	—	—	—	14	14	12	—	—	—	—	—	—	—	—	—
3	—	—	—	—	—	14	12	12	10	—	—	—	—	—	—	—	14	14
4	—	—	—	—	14	14	12	10	10	—	—	—	—	—	—	14	14	12
5	—	—	—	—	14	12	10	10	8	—	—	—	—	—	—	14	12	12
6	—	—	—	14	12	12	10	8	8	—	—	—	—	—	14	12	12	10
7	—	—	14	14	12	10	8	8	6	—	—	—	—	14	14	12	10	10
8	—	—	14	14	12	10	8	6	6	—	—	—	—	14	14	12	10	10
9	—	—	14	12	10	10	8	6	6	—	—	—	—	14	12	10	10	8
10	—	—	14	12	10	10	8	6	6	—	—	—	—	14	12	10	10	8
12	—	14	12	12	10	8	6	6	4	—	—	—	14	12	12	10	8	8
14	—	14	12	10	8	8	6	4	4	—	—	14	14	12	10	8	8	6
16	—	14	12	10	8	8	6	4	2	—	—	14	14	12	10	8	8	6
18	—	12	10	10	8	6	4	4	2	—	—	14	12	10	10	8	6	6
20	—	12	10	10	8	6	4	2	2	—	—	14	12	10	10	8	6	6
25	14	12	10	8	6	6	4	2	1	—	14	12	12	10	8	6	6	4
30	14	10	8	8	6	4	2	2	0	—	14	12	10	8	8	6	6	4
35	12	10	8	6	4	4	2	1	0	—	12	10	10	8	6	4	4	4
40	12	10	8	6	4	2	2	0	00	—	12	10	10	8	6	4	2	2
45	12	8	6	6	4	2	1	0	000	14	12	10	8	6	6	4	2	2
50	12	8	6	4	4	2	0	00	000	14	12	10	8	6	6	4	2	1
60	10	8	6	4	2	2	0	000	—	14	10	8	8	6	4	2	2	0
70	10	6	4	4	2	1	00	0000	—	12	10	8	6	4	4	2	1	0
80	10	6	4	2	2	0	000	—	—	12	10	8	6	4	2	2	0	00
90	8	6	4	2	1	0	0000	—	—	12	8	6	6	4	2	1	0	000
100	8	6	4	2	0	00	—	—	—	12	8	6	6	4	2	0	00	000
120	8	4	2	2	0	0000	—	—	—	10	8	6	4	2	2	0	000	0000

TABLE 26. FUSING CURRENTS IN AMPERES*

AWG B & S Gauge	Diam d in Inches	Diam d in mm	Copper ($K = 10\,244$)	Aluminum ($K = 7585$)	German Silver ($K = 5230$)	Iron ($K = 3148$)	Tin ($K = 1642$)
40	0.0031	0.079	1.77	1.31	0.90	0.54	0.28
38	0.0039	0.099	2.50	1.85	1.27	0.77	0.40
36	0.0050	0.127	3.62	2.68	1.85	1.11	0.58
34	0.0063	0.160	5.12	3.79	2.61	1.57	0.82
32	0.0079	0.201	7.19	5.32	3.67	2.21	1.15
30	0.0100	0.254	10.2	7.58	5.23	3.15	1.64
28	0.0126	0.320	14.4	10.7	7.39	4.45	2.32
26	0.0159	0.409	20.5	15.2	10.5	6.31	3.29
24	0.0201	0.511	29.2	21.6	14.9	8.97	4.68
22	0.0253	0.643	41.2	30.5	21.0	12.7	6.61
20	0.0319	0.810	58.4	43.2	29.8	17.9	9.36
19	0.0359	0.912	69.7	51.6	35.5	21.4	11.2
18	0.0403	1.02	82.9	61.4	42.3	25.5	13.3
17	0.0452	1.15	98.4	72.9	50.2	30.2	15.8
16	0.0508	1.29	117	86.8	59.9	36.0	18.8
15	0.0571	1.45	140	103	71.4	43.0	22.4
14	0.0641	1.63	166	123	84.9	51.1	26.6
13	0.0719	1.83	197	146	101	60.7	31.7
12	0.0808	2.05	235	174	120	72.3	37.7
11	0.0907	2.30	280	207	143	86.0	44.9
10	0.1019	2.59	333	247	170	102	53.4
9	0.1144	2.91	396	293	202	122	63.5
8	0.1285	3.26	472	349	241	145	75.6
7	0.1443	3.67	561	416	287	173	90.0
6	0.1620	4.12	668	495	341	205	107

*Courtesy of Automatic Electric Co., Chicago, Ill.

TABLE 27. PHYSICAL

Property		Copper Annealed	Copper Hard-Drawn	Aluminum 99 Percent Pure
Conductivity, Matthiessen's standard in percent		99 to 102	96 to 99	61 to 63
Ohms/mil-foot at 68°F = 20°C		10.36	10.57	16.7
Circular-mil-ohms/mile at 68°F = 20°C		54 600	55 700	88 200
Pounds/mile-ohm at 68°F = 20°C		875	896	424
Mean temp coefficient of resistivity/°F		0.002 33	0.002 33	0.002 2
Mean temp coefficient of resistivity/°C		0.004 2	0.004 2	0.004 0
Mean specific gravity		8.89	8.94	2.68
Pounds/1000 feet·circular mil		0.003 027	0.003 049	0.000 909
Density in pounds/inch^3		0.320	0.322	0.096 7
Mean specific heat		0.093	0.093	0.214
Mean melting point in °F		2 012	2 012	1 157
Mean melting point in °C		1 100	1 100	625
Mean coefficient of linear expansion/°F		0.000 009 50	0.000 009 50	0.000 012 85
Mean coefficient of linear expansion/°C		0.000 017 1	0.000 017 1	0.000 023 1
Solid wire	Ultimate tensile strength	30 000 to 42 000	45 000 to 68 000	20 000 to 35 000
	Average tensile strength	32 000	60 000	24 000
(Values in pounds/in^2)	Elastic limit	6 000 to 16 000	25 000 to 45 000	14 000
	Average elastic limit	15 000	30 000	14 000
	Modulus of elasticity	7 000 000 to 17 000 000	13 000 000 to 18 000 000	8 500 000 to 11 500 000
	Average modulus of elasticity	12 000 000	16 000 000	9 000 000
Concentric strand	Tensile strength	29 000 to 37 000	43 000 to 65 000	25 800
	Average tensile strength	35 000	54 000	—
	Elastic limit	5 800 to 14 800	23 000 to 42 000	13 800
(Values in pounds/in^2)	Average elastic limit	—	27 000	—
	Modulus of elasticity	5 000 000 to 12 000 000	12 000 000	Approx 10 000 000

TABLE 28. PHYSICAL PROPERTIES OF STRANDED COPPER (AWG)*

Circular Mils	AWG B & S Gauge	Number of Wires	Individual Wire Diam in Inches	Cable Diam in Inches	Area in Square Inches	Mass in lb per 1000 Ft	Mass in lb per Mile	*Maximum Resistance in Ohms/1000 ft at 20°C
211 600	4/0	19	0.1055	0.528	0.166 2	653.3	3 450	0.050 93
167 800	3/0	19	0.0940	0.470	0.131 8	518.1	2 736	0.064 22
133 100	2/0	19	0.0837	0.419	0.104 5	410.9	2 170	0.080 97
105 500	1/0	19	0.0745	0.373	0.082 86	325.7	1 720	0.102 2
83 690	1	19	0.0664	0.332	0.065 73	258.4	1 364	0.128 8
66 370	2	7	0.0974	0.292	0.052 13	204.9	1 082	0.164 4
52 640	3	7	0.0867	0.260	0.041 34	162.5	858.0	0.204 8
41 740	4	7	0.0772	0.232	0.032 78	128.9	680.5	0.258 2
33 100	5	7	0.0688	0.206	0.026 00	102.2	539.6	0.325 6
26 250	6	7	0.0612	0.184	0.020 62	81.05	427.9	0.410 5
20 820	7	7	0.0545	0.164	0.016 35	64.28	339.4	0.517 6
16 510	8	7	0.0486	0.146	0.012 97	50.98	269.1	0.652 8

Continued on next page.

PROPERTIES OF VARIOUS WIRES*

Iron (Ex BB)	Steel (Siemens-Martin)	Crucible Steel, High Strength	Plow Steel, Extra-High Strength	Copper-clad	
				30% Cond	40% Cond
16.8	8.7	—	—	29.4	39.0
62.9	119.7	122.5	125.0	35.5	26.6
332 000	632 000	647 000	660 000	187 000	140 000
4 700	8 900	9 100	9 300	2.775	2.075
0.002 8	0.002 78	0.002 78	0.002 78	0.002 4	—
0.005 0	0.005 01	0.005 01	0.005 01	0.004	0.004 1
7.77	7.85	7.85	7.85	8.17	8.25
0.002 652	0.002 671	—	—	0.002 81	0.002 81
0.282	0.283	0.283	0.283	0.298	0.298
0.113	0.117	—	—	—	—
2 975	2 480	—	—	—	—
1 635	1 360	—	—	—	—
0.000 006 73	0.000 006 62	—	—	0.000 007 2	0.000 007 2
0.000 012 0	0.000 011 8	—	—	0.000 012 9	0.000 012 9
50 000 to 55 000	70 000 to 80 000	—	—	—	—
55 000	75 000	125 000	187 000	60 000	100 000
25 000 to 30 000	35 000 to 50 000	—	—	—	—
30 000	38 000	69 000	130 000	30 000	50 000
22 000 000 to 27 000 000	22 000 000 to 29 000 000	—	—	—	—
26 000 000	29 000 000	30 000 000	30 000 000	19 000 000	21 000 000
—	74 000 to 98 000	85 000 to 165 000	140 000 to 245 000	70 000 to 97 000	—
—	80 000	125 000	180 000	80 000	—
—	37 000 to 49 000	—	—	—	—
—	40 000	70 000	110 000	—	—
—	12 000 000	15 000 000	15 000 000	—	—

*Reprinted by permission from *Transmission Towers* (Pittsburgh, Pa.: American Bridge Co., 1925; p. 169).

TABLE 28 (CONT). PHYSICAL PROPERTIES OF STRANDED COPPER (AWG)*

Circular Mils	AWG B & S Gauge	Number of Wires	Individual Wire Diam in Inches	Cable Diam in Inches	Area in Square Inches	Mass in lb per 1000 Ft	Mass in lb per Mile	*Maximum Resistance in Ohms/1000 ft at 20°C
13 090	9	7	0.0432	0.130	0.010 28	40.42	213.4	0.823 3
10 380	10	7	0.0385	0.116	0.008 152	32.05	169.2	1.038
6 530	12	7	0.0305	0.0915	0.005 129	20.16	106.5	1.650
4 107	14	7	0.0242	0.0726	0.003 226	12.68	66.95	2.624
2 583	16	7	0.0192	0.0576	0.002 029	7.975	42.11	4.172
1 624	18	7	0.0152	0.0456	0.001 275	5.014	26.47	6.636
1 022	20	7	0.0121	0.0363	0.000 802 7	3.155	16.66	10.54

*The resistance values in this table are trade maxima for soft or annealed copper wire and are higher than the average values for commercial cable. The following values for the conductivity and resistivity of copper at 20°C were used:

Conductivity in terms of International Annealed Copper Standard: 98.16 percent
Resistivity in pounds per mile-ohm: 891.58

The resistance of hard-drawn copper is slightly greater than the values given, being about 2 percent to 3 percent greater for sizes from 4/0 to 20 AWG.

TABLE 29. ADHESIVES CLASSIFIED BY CHEMICAL COMPOSITION*

	Natural	Thermoplastic	Thermosetting	Elastomeric	Alloys†	Miscellaneous
Types Within Group	Casein, blood albumin, hide, bone, fish, starch (plain and modified); rosin, shellac, asphalt; inorganic (sodium silicate, litharge-glycerin); soybean glue; dextrin	Polyvinyl acetate, polyvinyl alcohol, acrylic, cellulose nitrate, asphalt, oleo-resin, acrylate-vinyl acetate copolymer, acrylic-ethylene, acrylonitrile-butadiene-styrene, cellulose acetate, cellulose acetate-butyrate, cellulose caprate, chlorinated polyethylene, chlorinated polyvinyl chloride, cyanoacrylate, ethyl cellulose, hydroxyethyl cellulose, methyl cellulose, polyacrylate, polyacrylate (carboxylic), polyacrylic esters, polyamide, polyester, polyhydroxyether, polyimide, polymethylmethacrylate, polystyrene, polysulfone, polyvinyl acetal, polyvinyl alkyl ether, polyvinyl butyral, polyvinyl chloride, polyvinyl chloride (modified), polyvinyl ester, polyvinyl formal, vinyl acetate-ethylene copolymer, vinyl chloride-vinyl acetate copolymer, vinyl chloride vinylidene	Phenolic, resorcinol, phenol-resorcinol, epoxy, epoxy-phenolic, urea, melamine, alkyd, epoxy-alkyl ester, epoxy bisphenol A-based, epoxy bitumen, furan resin, melamine formaldehyde, phenol formaldehyde, phenol formaldehyde-resorcinol formaldehyde, polyester, polyethylene imine, polyisocyante, polyurethane, resorcinol formaldehyde, silicone resins, urea formaldehyde, urea formaldehyde-melamine formaldehyde	Natural rubber, reclaim rubber, butadiene-styrene (GR-S), neoprene, acrylonitrile-butadiene (Buna-N), silicone, butadiene-polyacrylate rubber, butyl rubber, chlorinated rubber, chlorobutyl rubber, cyclized rubber, depolymerized rubber, polybutadiene, polyisobutylene, polyisoprene, polysulfide, polyurethane rubber	Phenolic-polyvinyl butyral, phenolic-polyvinyl formal, phenolic-neoprene rubber, phenolic-nitrile rubber, modified epoxy, acrylic rubber-phenolic resin, neoprene-phenolic, nylon-epoxy resins, phenolic-silicone	Bitumen-asphaltic, ceramic, lacquer, bitumen-latex, varnishes
Most Used Form	Liquid, powder	Liquid, some dry film	Liquid, but all forms common	Liquid, some film	Liquid, paste, film	

Common Further Classifications	By vehicle (water emulsion is most common, but many types are solvent dispersions)	By vehicle (most are solvent dispersions or water emulsions)	By cure requirements (heat and/or pressure most common, but some are catalyst types)	By cure requirements (all are common); also by vehicle (most are solvent dispersions or water emulsions)	By cure requirements (usually heat and pressure except some epoxy types); by vehicle (most are solvent dispersions or 100% solids); and by type of adherends or end-service conditions
Bond Characteristics	Wide range, but generally low strength; good resistance to heat, chemicals; generally poor moisture resistance	Good to 200–500° F; poor creep strength; fair peel strength	Good to 200–500° F; good creep strength; fair peel strength	Good to 150–400° F; never melt completely; low strength; high flexibility	Balanced combination of properties of other chemcial groups depending on formulation; generally higher strength over wider temperature range
Major Type of Use**	Household, general purpose, quick set, long shelf life	Unstressed joints; designs with caps, overlaps, stiffeners	Stressed joints at slightly elevated temperature	Unstressed joints on lightweight materials; joints in flexure	Where highest and strictest end-service conditions must be met, sometimes regardless of cost, as military uses
Materials Most Commonly Bonded	Wood (furniture), paper, cork, liners, packaging (food), textiles, some metals and plastics. Industrial uses giving way to other groups	Formulation range covers all materials, but emphasis on nonmetallics—especially wood, leather, cork, paper, etc.	Epoxy-phenolics for structural uses of most materials; others mainly for wood; alkyds for laminations; most epoxies are modified (alloys)	Few used "straight" for rubber, fabric, foil, paper, leather, plastics, films; also as tapes. Most modified with synthetic resins	Metals, ceramics, glass, thermosetting plastics; nature of adherends often not as vital as design or end-service conditions (e.g., high strength, temperature)

*Expanded and updated from Clauser, et al, *Encyclopedia of Engineering Materials* (New York: Reinhold Publishing Corp., 1963). Additional practical information including trade names, suppliers, and types of adhesives may be found in the *Adhesives Red Book*, Atlanta, Ga.: Communication Channels, Inc.

†"Alloy," as used here, refers to formulations containing resins from two or more *different* chemical groups. There are also formulations which benefit from compounding two resin types from the same chemical group (e.g., epoxy-phenolic).

**Although some uses of the "nonalloyed" adhesives absorb a large percentage of the quantity of adhesives sold, the uses are narrow in scope; from the standpoint of diversified applications, by far the most important use of any group is the forming of adhesive alloys.

TABLE 30. MACHINE-SCREW DIMENSIONS AND OTHER DATA

Screw		Threads/in		Clearance Drill*		Tap Drill†			Head								Washer		
							Diameter		Round		Flat	Fillister		Hex Nut					
No.	Diam	Coarse	Fine	No.	Diam.	No.	Inches	mm	Max OD	Max Height	Max OD	Max OD	Max Height	Across Flat	Across Corner	Thick-ness	OD	ID	Thick-ness
0	0.060	—	80	52	0.064	56	0.047	1.2	0.113	0.053	0.119	0.096	0.059	0.156	0.171	0.046	—	—	—
1	0.073	64	72	47	0.079	53	0.060	1.5	0.138	0.061	0.146	0.118	0.070	0.156	0.171	0.046	—	—	—
2	0.086	56	64	42	0.094	50	0.070	1.8	0.162	0.070	0.172	0.140	0.083	0.187	0.205	0.062	1/4	0.093	0.032
3	0.099	48 / —	— / 56	37	0.104	47 / 45	0.079 / 0.082	2.0 / 2.1	0.187	0.078	0.199	0.161	0.095	0.187	0.205	0.062	1/4	0.105	0.020
4	0.112	40 / —	— / 48	31	0.120	43 / 42	0.089 / 0.094	2.3 / 2.4	0.211	0.086	0.225	0.183	0.107	0.250	0.275	0.093	5/16	0.125	0.032
5	0.125	40 / —	— / 44	29	0.136	38 / 37	0.102 / 0.104	2.6 / 2.6	0.236	0.095	0.252	0.205	0.120	0.312	0.344	0.109	3/8	0.140	0.032
6	0.138	32 / —	— / 40	27	0.144	36 / 33	0.107 / 0.113	2.7 / 2.9	0.260	0.103	0.279	0.226	0.132	0.312	0.344	0.109	5/16 / 3/8	0.156	0.026 / 0.046
8	0.164	32 / —	— / 36	18	0.170	29 / 29	0.136 / 0.136	3.5 / 3.5	0.309	0.119	0.332	0.270	0.156	0.344	0.373	0.125	3/8 / 7/16	0.186	0.032 / 0.046
10	0.190	24 / —	— / 32	9	0.196	25 / 21	0.150 / 0.159	3.8 / 4.0	0.359	0.136	0.385	0.313	0.180	0.375	0.413	0.125	7/16 / 1/2	0.218	0.036 / 0.063
12	0.216	24 / —	— / 28	2	0.221	16 / 14	0.177 / 0.182	4.5 / 4.6	0.408	0.152	0.438	0.357	0.205	0.437	0.488	0.156	1/2 / 9/16	0.250	0.063
1/4	0.250	20 / —	— / 28	—	17/64	7 / 3	0.201 / 0.213	5.1 / 5.5	0.472	0.174	0.507	0.414	0.237	0.437 / 0.500	0.488 / 0.577	0.203 / 0.250	9/16 / 5/8	0.281	0.040 / 0.063

All dimensions in inches except where noted.

*Clearance-drill sizes are practical values for use of the engineer or technician doing his own shop work.

†Tap-drill sizes are for use in hand tapping material such as brass or soft steel. For copper, aluminum, Norway iron, cast iron, Bakelite, or for very thin material, the drill should be a size or two larger in diameter than shown.

TABLE 31. DRILL SIZES*

Drill	Inches	Drill	Inches	Drill	Inches	Drill	Inches
0.10 mm	0.003 937	1.15 mm	0.045 275	2.75 mm	0.108 267	no 12	0.189 000
0.15 mm	0.005 905	no 56	0.046 500	7/64 in	0.109 375	no 11	0.191 000
0.20 mm	0.007 874	3/64 in	0.046 875	no 35	0.110 000	4.90 mm	0.192 913
0.25 mm	0.009 842	1.20 mm	0.047 244	2.80 mm	0.110 236	no 10	0.193 500
0.30 mm	0.011 811	1.25 mm	0.049 212	no 34	0.111 000	no 9	0.196 000
no 80	0.013 000	1.30 mm	0.051 181	no 33	0.113 000	5.00 mm	0.196 850
no 79½	0.013 500	no 55	0.052 000	2.90 mm	0.114 173	no 8	0.199 000
0.35 mm	0.013 779	1.35 mm	0.053 149	no 32	0.116 000	5.10 mm	0.200 787
no 79	0.014 000	no 54	0.055 000	3.00 mm	0.118 110	no 7	0.201 000
no 78½	0.014 500	1.40 mm	0.055 118	no 31	0.120 000	13/64 in	0.203 125
no 78	0.015 000	1.45 mm	0.057 086	3.10 mm	0.122 047	no 6	0.204 000
1/64 in	0.015 625	1.50 mm	0.059 055	1/8 in	0.125 000	5.20 mm	0.204 724
0.40 mm	0.015 748	no 53	0.059 500	3.20 mm	0.125 984	no 5	0.205 500
no 77	0.016 000	1.55 mm	0.061 023	3.25 mm	0.127 952	5.25 mm	0.206 692
0.45 mm	0.017 716	1/16 in	0.062 500	no 30	0.128 500	5.30 mm	0.208 661
no 76	0.018 000	1.60 mm	0.062 992	3.30 mm	0.129 921	no 4	0.209 000
0.50 mm	0.019 685	no 52	0.063 500	3.40 mm	0.133 858	5.40 mm	0.212 598
no 75	0.020 000	1.65 mm	0.064 960	no 29	0.136 000	no 3	0.213 000
no 74½	0.021 000	1.70 mm	0.066 929	3.50 mm	0.137 795	5.50 mm	0.216 535
0.55 mm	0.021 653	no 51	0.067 000	no 28	0.140 500	7/32 in	0.218 750
no 74	0.022 000	1.75 mm	0.068 897	9/64 in	0.140 625	5.60 mm	0.220 472
no 73½	0.022 500	no 50	0.070 000	3.60 mm	0.141 732	no 2	0.221 000
no 73	0.023 000	1.80 mm	0.070 866	no 27	0.144 000	5.70 mm	0.224 409
0.60 mm	0.023 622	1.85 mm	0.072 834	3.70 mm	0.145 669	5.75 mm	0.226 377
no 72	0.024 000	no 49	0.073 000	no 26	0.147 000	no 1	0.228 000
no 71½	0.025 000	1.90 mm	0.074 803	3.75 mm	0.147 637	5.80 mm	0.228 346
0.65 mm	0.025 590	no 48	0.076 000	no 25	0.149 500	5.90 mm	0.232 283
no 71	0.026 000	1.95 mm	0.076 771	3.80 mm	0.149 606	ltr A	0.234 000
no 70	0.027 000	5/64 in	0.078 125	no 24	0.152 000	15/64 in	0.234 375
0.70 mm	0.027 559	no 47	0.078 500	3.90 mm	0.153 543	6.00 mm	0.236 220
no 69½	0.028 000	2.00 mm	0.078 740	no 23	0.154 000	ltr B	0.238 000
no 69	0.029 000	2.05 mm	0.080 708	5/32 in	0.156 250	6.10 mm	0.240 157
no 68½	0.029 250	no 46	0.081 000	no 22	0.157 000	ltr C	0.242 000
0.75 mm	0.029 527	no 45	0.082 000	4.00 mm	0.157 480	6.20 mm	0.244 094
no 68	0.030 000	2.10 mm	0.082 677	no 21	0.159 000	ltr D	0.246 000
no 67	0.031 000	2.15 mm	0.084 645	no 20	0.161 000	6.25 mm	0.246 062
1/32 in	0.031 250	no 44	0.086 000	4.10 mm	0.161 417	6.30 mm	0.248 031
0.80 mm	0.031 496	2.20 mm	0.086 614	4.20 mm	0.165 354	ltr E } 1/4 in }	0.250 000
no 66	0.032 000	2.25 mm	0.088 582	no 19	0.166 000		
no 65	0.033 000	no 43	0.089 000	4.25 mm	0.167 322	6.40 mm	0.251 968
				4.30 mm	0.169 291		
0.85 mm	0.033 464	2.30 mm	0.090 551			6.50 mm	0.255 905
no 64	0.035 000	2.35 mm	0.092 519	no 18	0.169 500	ltr F	0.257 000
0.90 mm	0.035 433	no 42	0.093 500	11/64 in	0.171 875	6.60 mm	0.259 842
no 63	0.036 000	3/32 in	0.093 750	no 17	0.173 000	ltr G	0.261 000
no 62	0.037 000	2.40 mm	0.094 488	4.40 mm	0.173 228	6.70 mm	0.263 779
0.95 mm	0.037 401	no 41	0.096 000	no 16	0.177 000	17/64 in	0.265 625
no 61	0.038 000	2.45 mm	0.096 456	4.50 mm	0.177 165	6.75 mm	0.265 747
no 60½	0.039 000	no 40	0.098 000	no 15	0.180 000	ltr H	0.266 000
1.00 mm	0.039 370	2.50 mm	0.098 425	4.60 mm	0.181 102	6.80 mm	0.267 716
no 60	0.040 000	no 39	0.099 500	no 14	0.182 000	6.90 mm	0.271 653
no 59	0.041 000	no 38	0.101 500	no 13	0.185 000	ltr I	0.272 000
1.05 mm	0.041 338	2.60 mm	0.102 362	4.70 mm	0.185 039	7.00 mm	0.275 590
no 58	0.042 000	no 37	0.104 000	4.75 mm	0.187 007	ltr J	0.277 000
no 57	0.043 000	2.70 mm	0.106 299	3/16 in	0.187 500	7.10 mm	0.279 527
1.10 mm	0.043 307	no 36	0.106 500	4.80 mm	0.188 976	ltr K	0.281 000

Continued on next page.

TABLE 31 (CONT). DRILL SIZES*

Drill	Inches	Drill	Inches	Drill	Inches	Drill	Inches
9/32 in	0.281 250	8.80 mm	0.346 456	29/64 in	0.453 125	47/64 in	0.734 375
7.20 mm	0.283 464	ltr S	0.348 000	15/32 in	0.468 750	19.00 mm	0.748 030
7.25 mm	0.285 432	8.90 mm	0.350 393	12.00 mm	0.472 440	3/4 in	0.750 000
7.30 mm	0.287 401	9.00 mm	0.354 330	31/64 in	0.484 375	49/64 in	0.765 625
ltr L	0.290 000	ltr T	0.358 000	12.50 mm	0.492 125	19.50 mm	0.767 715
7.40 mm	0.291 338	9.10 mm	0.358 267	1/2 in	0.500 000	25/32 in	0.781 250
ltr M	0.295 000	23/64 in	0.359 375	13.00 mm	0.511 810	20.00 mm	0.787 400
7.50 mm	0.295 275	9.20 mm	0.362 204	33/64 in	0.515 625	51/64 in	0.796 875
19/64 in	0.296 875	9.25 mm	0.364 172	17/32 in	0.531 250	20.50 mm	0.807 085
7.60 mm	0.299 212	9.30 mm	0.366 141	13.50 mm	0.531 495	13/16 in	0.812 500
ltr N	0.302 000	ltr U	0.368 000	35/64 in	0.546 875	21.00 mm	0.826 770
7.70 mm	0.303 149	9.40 mm	0.370 078	14.00 mm	0.551 180	53/64 in	0.828 125
7.75 mm	0.305 117	9.50 mm	0.374 015	9/16 in	0.562 500	27/32 in	0.843 750
7.80 mm	0.307 086	3/8 in	0.375 000	14.50 mm	0.570 865	21.50 mm	0.846 455
7.90 mm	0.311 023	ltr V	0.377 000	37/64 in	0.578 125	55/64 in	0.859 375
5/16 in	0.312 500	9.60 mm	0.377 952	15.00 mm	0.590 550	22.00 mm	0.866 140
8.00 mm	0.314 960	9.70 mm	0.381 889	19/32 in	0.593 750	7/8 in	0.875 000
ltr O	0.316 000	9.75 mm	0.383 857	39/64 in	0.609 375	22.50 mm	0.885 825
8.10 mm	0.318 897	9.80 mm	0.385 826	15.50 mm	0.610 235	57/64 in	0.890 625
8.20 mm	0.322 834	ltr W	0.386 000	5/8 in	0.625 000	23.00 mm	0.905 510
ltr P	0.323 000						
8.25 mm	0.324 802	9.90 mm	0.389 763	16.00 mm	0.629 920	29/32 in	0.906 250
8.30 mm	0.326 771	25/64 in	0.390 625	41/64 in	0.640 625	59/64 in	0.921 875
21/64 in	0.328 125	10.00 mm	0.393 700	16.50 mm	0.649 605	23.50 mm	0.925 195
8.40 mm	0.330 708	ltr X	0.397 000	21/32 in	0.656 250	15/16 in	0.937 500
		ltr Y	0.404 000	17.00 mm	0.669 290	24.00 mm	0.944 880
ltr Q	0.332 000						
8.50 mm	0.334 645	13/32 in	0.406 250	43/64 in	0.671 875	61/64 in	0.953 125
		ltr Z	0.413 000	11/16 in	0.687 500	24.50 mm	0.964 565
8.60 mm	0.338 582	10.50 mm	0.413 385	17.50 mm	0.688 975	31/32 in	0.968 750
ltr R	0.339 000	27/64 in	0.421 875	45/64 in	0.703 125	25.00 mm	0.984 250
8.70 mm	0.342 519	11.00 mm	0.433 070	18.00 mm	0.708 660	63/64 in	0.984 375
11/32 in	0.343 750	7/16 in	0.437 500	23/32 in	0.718 750	1 in	1.000 000
8.75 mm	0.344 487	11.50 mm	0.452 755	18.50 mm	0.728 345		

*From *New Departure Handbook*.

TABLE 32. COMMON GAUGE PRACTICES

Material	Sheet	Wire
Aluminum	B&S	AWG(B&S)
Brass, bronze; sheet	B&S	—
Copper	B&S	AWG(B&S)
Iron, steel; band and hoop	BWG	—
Iron, steel; telephone and telegraph wire	—	BWG
Steel wire, except telephone and telegraph	—	W&M
Steel sheet	US	—
Tank steel	BWG	—
Zinc sheet	"Zinc gauge" proprietary	—

TABLE 33. COMPARISON OF GAUGES*

Gauge	AWG B&S	Birmingham or Stubs BWG	Wash. & Moen W&M	British Standard NBS SWG	London or Old English	United States Standard US	American Standard Preferred Thickness†
0000000	—	—	0.490	0.500	—	0.500 00	—
000000	0.580 0	—	0.460	0.464	—	0.468 75	—
00000	0.516 5	—	0.430	0.432	—	0.437 50	—
0000	0.460 0	0.454	0.393 8	0.400	0.454	0.406 25	—
000	0.409 6	0.425	0.362 5	0.372	0.425	0.375 00	—
00	0.364 8	0.380	0.331 0	0.348	0.380	0.343 75	—
0	0.324 9	0.340	0.306 5	0.324	0.340	0.312 50	—
1	0.289 3	0.300	0.283 0	0.300	0.300	0.281 25	—
2	0.257 6	0.284	0.262 5	0.276	0.284	0.265 625	—
3	0.229 4	0.259	0.243 7	0.252	0.259	0.250 000	0.224
4	0.204 3	0.238	0.225 3	0.232	0.238	0.234 375	0.200
5	0.181 9	0.220	0.207 0	0.212	0.220	0.218 750	0.180
6	0.162 0	0.203	0.192 0	0.192	0.203	0.203 125	0.160
7	0.144 3	0.180	0.177 0	0.176	0.180	0.187 500	0.140
8	0.128 5	0.165	0.162 0	0.160	0.165	0.171 875	0.125
9	0.114 4	0.148	0.148 3	0.144	0.148	0.156 250	0.112
10	0.101 9	0.134	0.135 0	0.128	0.134	0.140 625	0.100
11	0.090 74	0.120	0.120 5	0.116	0.120	0.125 000	0.090
12	0.080 81	0.109	0.105 5	0.104	0.109	0.109 375	0.080
13	0.071 96	0.095	0.091 5	0.092	0.095	0.093 750	0.071
14	0.064 08	0.083	0.080 0	0.080	0.083	0.078 125	0.063
15	0.057 07	0.072	0.072 0	0.072	0.072	0.070 312 5	0.056
16	0.050 82	0.065	0.062 5	0.064	0.065	0.062 500 0	0.050
17	0.045 26	0.058	0.054 0	0.056	0.058	0.056 250 0	0.045
18	0.040 30	0.049	0.047 5	0.048	0.049	0.050 000 0	0.040
19	0.035 89	0.042	0.041 0	0.040	0.040	0.043 750 0	0.036
20	0.031 96	0.035	0.034 8	0.036	0.035	0.037 500 0	0.032
21	0.028 46	0.032	0.031 75	0.032	0.031 5	0.034 375 0	0.028
22	0.025 35	0.028	0.028 60	0.028	0.029 5	0.031 250 0	0.025
23	0.022 57	0.025	0.025 80	0.024	0.027 0	0.028 125 0	0.022
24	0.020 10	0.022	0.023 00	0.022	0.025 0	0.025 000 0	0.020
25	0.017 90	0.020	0.020 40	0.020	0.023 0	0.021 875 0	0.018
26	0.015 94	0.018	0.018 10	0.018	0.020 5	0.018 750 0	0.016
27	0.014 20	0.016	0.017 30	0.0164	0.018 7	0.017 187 5	0.014
28	0.012 64	0.014	0.016 20	0.0148	0.016 5	0.015 625 0	0.012
29	0.011 26	0.013	0.015 00	0.0136	0.015 5	0.014 062 5	0.011
30	0.010 03	0.012	0.014 00	0.0124	0.013 72	0.012 500 0	0.010
31	0.008 928	0.010	0.013 20	0.0116	0.012 20	0.010 937 50	0.009
32	0.007 950	0.009	0.012 80	0.0108	0.011 20	0.010 156 25	0.008
33	0.007 080	0.008	0.011 80	0.0100	0.010 20	0.009 375 00	0.007
34	0.006 305	0.007	0.010 40	0.0092	0.009 50	0.008 593 75	0.006
35	0.005 615	0.005	0.009 50	0.0084	0.009 00	0.007 812 50	—
36	0.005 000	0.004	0.009 00	0.0076	0.007 50	0.007 031 250	—
37	0.004 453	—	0.008 50	0.0068	0.006 50	0.006 640 625	—
38	0.003 965	—	0.008 00	0.0060	0.005 70	0.006 250 000	—
39	0.003 531	—	0.007 50	0.0052	0.005 00	—	—
40	0.003 145	—	0.007 00	0.0048	0.004 50	—	—

*Courtesy Whitehead Metal Products Co., Inc.

†These thicknesses are intended to express the desired thickness in decimal fractions of an inch. They have no relation to gauge numbers; they are approximately related to the AWG sizes 3–34.

TABLE 34. FREEZING POINTS AND SPECIFIC GRAVITIES OF COMMERCIAL ANTIFREEZE SOLUTIONS

	Percent by Volume in Water				
	10	20	30	40	50
Typical Commercial Methanol Antifreeze	− 5.2°C	− 12.0°C	− 21.1°C	− 32.2°C	− 45.0°C
Sp. gr. at 15°C/15°C*	0.986	0.975	0.963	0.950	0.935
Typical Commercial Ethanol Antifreeze	− 3.3°C	− 7.7°C	− 14.2°C	− 22.0°C	− 30.6°C
Sp. gr. at 15°C/15°C*	0.988	0.977	0.967	0.955	0.938
Commercial Glycerine† Antifreeze	− 1.6°C	− 4.7°C	− 9.5°C	− 15.4°C	− 23.0°C
Sp. gr. at 15°C/15°C*	1.023	1.048	1.074	1.101	1.128
Typical Commercial Ethylene Glycol† Antifreeze	− 3.8°C	− 8.8°C	− 15.5°C	− 24.3°C	− 36.5°C
Sp. gr. at 15°C/15°C*	1.015	1.030	1.045	1.060	1.074

*Specific gravity is measured for mixture at 15°C referred to water at 15°C.

†Glycerine and ethylene glycol are practically nonvolatile. All types must be suitably inhibited to prevent cooling-system corrosion. Commercial antifreeze solutions based on ethylene glycol and on glycerine are in use at the present time.

5 Components or Parts

Revised by
John J. Bohrer, Robert Blaszczyk,
Pierre J. Gerondeau, Jack Isken, Edward Mette,
Rick Price, Joseph A. Toro, and Theodore E. VanKampen

GENERAL STANDARDS

Standardization of electronic components or parts is handled by several cooperating agencies.

In the US, the Electronic Industries Association (EIA)* and the American National Standards Institute (ANSI)† are active in the commercial field. Electron-tube and semiconductor-device standards are handled by the Joint Electron Device Engineering Council (JEDEC), a cooperative effort of EIA and the National Electrical Manufacturers Association (NEMA)‡. Military (MIL) standards are issued by the US Department of Defense or one of its agencies such as the Defense Electronics Supply Center (DESC).

International standardization in the electronics field is carried out by the various Technical Committees of the International Electrotechnical Commission (IEC)§. A list of the available IEC Recommendations is included in the ANSI Index (outside the US, consult the national standardization agency or the IEC). Documents from the IEC may be used directly, or their recommendations may be incorporated in whole or in part in national standards issued by the EIA or ANSI. A few broad areas may be covered by standards issued by the International Standards Organization (ISO).

These organizations establish standards for electronic components or parts (and in some cases, for equipments) to provide interchangeability among different products regarding size, performance, and identification; minimum number of sizes and designs; and uniform testing of products for acceptance. This chapter presents a brief outline of the requirements, characteristics, and designations for the major types of component parts used in electronic equipment. Such standardization offers economic advantages to both the parts user and the parts manufacturer, but is not intended to prevent the manufacture and use of other parts under special conditions.

Color Coding

The color code of Table 1 is used for marking electronic parts.

---

*EIA Engineering Dept., Washington, D.C. Index of standards is available. EIA was formerly Radio-Electronics-Television Manufacturers' Association (RETMA).

†ANSI, New York, New York. Index of standards is available. ANSI was formerly the USA Standards Institute (USASI).

‡NEMA, New York, New York. Index of standards is available.

§IEC, Central Office; Geneva, Switzerland. The US National Committee for the IEC operates within the ANSI.

Tolerance

The maximum deviation allowed from the specified nominal value is known as the tolerance. It is usually given as a percentage of the nominal value, though for very small capacitors the tolerance may be specified in picofarads (pF). For critical applications it is important to specify the permissible tolerance; where no tolerance is specified, components are likely to vary by ±20 percent from the nominal value.

Do not assume that a given lot of components will have values distributed throughout the acceptable range of values. A lot ordered with a ±20% tolerance may include *no* parts having values within 5% of the desired nominal value; these may have been sorted out before shipment. The manufacturing process for a given lot may produce parts in a narrow range of values only, not necessarily centered in the acceptable tolerance range.

Preferred Values

To maintain an orderly progression of sizes, preferred numbers are frequently used for the nominal values. A further advantage is that all parts manufactured are salable as one or another of the preferred values. Each preferred value differs from its predecessor by a constant multiplier, and the final result is conveniently rounded to two significant figures.

ANSI Standard Z17.1-1973 covers a series of preferred numbers based on $(10)^{1/5}$ and $(10)^{1/10}$ as listed in Table 2. This series has been widely used for fixed wirewound power-type resistors and for time-delay fuses.

Because of the established practice of using ±20-, ±10-, and ±5-percent tolerances, a series of values based on $(10)^{1/6}$, $(10)^{1/12}$, and $(10)^{1/24}$ has been adopted by the EIA, and is now an ANSI Standard (C83.2-1971) (EIA RS-385). It is widely used for such small electronic components as fixed composition resistors and fixed ceramic, mica, and molded paper capacitors. These values are listed in Table 2. (For series with smaller steps, consult the ANSI or EIA Standard.)

Voltage Rating

Distinction must be made between the breakdown-voltage rating (test volts) and the working-voltage rating. The maximum voltage that may be applied (usually continuously) over a long period of time without causing the part to fail determines the working-voltage rating. Application of the test voltage for more than a very few minutes, or even repeated applications of short duration, may result in permanent damage or failure of the part.

Characteristic

The term "characteristic" is frequently used to include various qualities of a part such as tempera-

TABLE 1. STANDARD COLOR CODE OF ELECTRONICS INDUSTRY

Color	Significant Figure	Decimal Multiplier	Tolerance in Percent*	Voltage Rating	Characteristic
Black	0	1	±20 (M)	—	A
Brown	1	10	±1 (F)	100	B
Red	2	100	±2 (G)	200	C
Orange	3	1 000	±3	300	D
Yellow	4	10 000	GMV‡	400	E
Green	5	100 000	±5(J)†, (0.50(D))§	500	F
Blue	6	1 000 000	±6(0.25(C))§	600	G
Violet	7	10 000 000	±12.5, (0.10(B))§	700	—
Gray	8	0.01†	±30, (0.05(N))§	800	I
White	9	0.1†	±10†	900	J
Gold	—	0.1	±5 (J), (0.50(E))‖	1 000	—
Silver	—	0.01	±10 (K)	2 000	—
No Color	—	—	±20	500	—

* Tolerance letter symbol as used in type designations has tolerance meaning as shown. ±3, ±6, ±12.5, and ±30 percent are tolerances for USA Std 40-, 20-, 10-, and 5-step series, respectively.

† Optional coding where metallic pigments are undesirable.

‡ GMV is −0 to +100-percent tolerance or Guaranteed Minimum Value.

§ For some film and other resistors only.

‖ For some capacitors only.

ture coefficient of capacitance or resistance, Q value, maximum permissible operating temperature, stability when subjected to repeated cycles of high and low temperature, and deterioration when it is subjected to moisture either as humidity or water immersion. One or two letters are assigned in EIA or MIL type designations, and the characteristic may be indicated by color coding on the part. An explanation of the characteristics applicable to a component or part will be found in the following sections covering that part.

ENVIRONMENTAL TEST METHODS

Since many component parts and equipments have the same environmental exposure, environmental test methods are becoming standardized. The principal standards follow.

EIA Standard RS-186-E (*ANSI C83.58-1978*): Standard Test Methods for Passive Electronic Component Parts.

IEC Publication 68: Basic Environmental Testing Procedures for Electronic Components and Electronic Equipment (published in multiple).

MIL-STD-202F: Military Standard Test Methods for Electronic and Electrical Component Parts.

MIL-STD-750B: Test Method for Semiconductor Device.

MIL-STD-810C: Military Standard Environmental Test Methods.

MIL-STD-883B: Test Methods and Procedures for Microelectronics.

MIL-STD-1344A: Test Methods for Electrical Connectors.

ASTM Standard Test Methods*—Primarily applicable to the materials used in electronic component parts.

Wherever the test methods in these standards are reasonably applicable, they should be specified in preference to other methods. This simplifies testing of a wide variety of parts, testing in widely separated locations, and comparison of data.

When selecting destructive environmental tests to determine the probable life of a part, distinguish between the environment prevailing during normal equipment operation and the environment used to accelerate deterioration. During exposure to the latter environment, the item may be out of tolerance with respect to its parameters in its normal operating-environment range. Accelerated tests are most meaningful if some relation between the degree of acceleration and component life is known. Such acceleration factors are known for many insulation systems.

STANDARD AMBIENT CONDITIONS FOR MEASUREMENT

Standard ambient conditions for measurement are listed in Table 3.

---

*ASTM = American Society for Testing and Materials; Philadelphia, Pa. Index of standards is available.

TABLE 2. PREFERRED VALUES*

	USA Standard Z17.1-1973†		USA Standard C83.2-1971 (R 1977)‡		
Name of Series	"5"	"10"	±20% (E6)	±10% (E12)	±5% (E24)
Percent step size	60	25	≈40	20	10
Step multiplier	$(10)^{1/5}=1.58$	$(10)^{1/10}=1.26$	$(10)^{1/6}=1.46$	$(10)^{1/12}=1.21$	$(10)^{1/24}=1.10$
Values in the series (Use decimal multipliers for smaller or larger values)	10	10	10	10	10
	—	12.5	—	—	11
	—	(12)	—	12	12
	—	—	—	—	13
	—	—	15	15	15
	16	16	—	—	16
	—	—	—	18	18
	—	20	—	—	20
	—	—	22	22	22
	—	—	—	—	24
	25	25	—	—	—
	—	—	—	27	27
	—	31.5	—	—	30
	—	(32)	—	—	—
	—	—	33	33	33
	—	—	—	—	36
	—	—	—	39	39
	40	40	—	—	—
	—	—	—	—	43
	—	—	47	47	47
	—	50	—	—	—
	—	—	—	—	51
	—	—	—	56	56
	—	—	—	—	62
	63	63	—	—	—
	—	—	68	68	68
	—	—	—	—	75
	—	80	—	—	—
	—	—	—	82	82
	—	—	—	—	91
	100	100	100	100	100

* ANSI Standard C83.2-1971 applies to most electronics components. It is the same as EIA Standard RS-385 (formerly GEN-102) and agrees with IEC Publication 63. ANSI Standard Z17.1-1973 covers preferred numbers and agrees with ISO 3 and ISO 497.

† "20" series with 12-percent steps ($(10)^{1/20}=1.122$ multiplier) and a "40" series with 6-percent steps ($(10)^{1/40}=1.059$ multiplier) are also standard.

‡ Associate the tolerance ±20%, ±10%, or ±5% only with the values listed in the corresponding column. Thus, 1200 ohms may be either ±10 or ±5, but not ±20 percent; 750 ohms may be ±5, but neither ±20 nor ±10 percent.

OTHER STANDARD ENVIRONMENTAL TEST CONDITIONS

Ambient Temperature

Dry heat, °C: + 30, +40, (+ 49), +55, (+ 68), +70, (+ 71), + 85, +100, +125, +155, +200 (values in parentheses not universally used).

Cold, °C: − 10, −25, −40, −55, −65.

Constant-Humidity Tests

40° C, 90 to 95% RH; 4, 10, 21, or 56 days.

66° C, ≈ 100% RH: 48, 96, or 240 hours (primarily for small items).

Cycling Humidity Tests

Fig. 1 shows a number of cycling humidity tests. (See applicable chart in standard for full details.) Preconditioning is customary before starting cycle series. RH = relative humidity.

High-Altitude Tests

Information regarding high-altitude tests is given in Table 4.

Vibration Tests

The purposes of vibration tests are:

(A) Search for resonance.

TABLE 3. STANDARD AMBIENT CONDITIONS FOR MEASUREMENT

	Standard	Temperature (°C)	Relative Humidity (%)	Barometric Pressure mm Hg	Barometric Pressure mbar
Normal range	RS-186-D	15–35	45–75	650–800	860–1060
	IEC-68	15–35	45–75	(645–795)	860–1060
	MIL-STD-202E	15–35	45–75	650–800	(866–1066)
	MIL-STD-810C	13–33	20–80	650–775	(866–1033)
Closely controlled range	IEC-68	20±1	65±2	(645–795)	860–1060
	IEC-68	23±1	50±2	(645–795)	860–1060
	MIL-STD-202E	23±1	50±2	650–800	(866–1066)
	MIL-STD-810C	23±1.4	50±5	650–775	(866–1033)
	IEC-68	27±1	65±2	(645–795)	860–1060
	RS-186-D	25±2	50±2	650–800	860–1060

Notes:

1. Use the closely controlled range only if the properties are sensitive to temperature or humidity variations, or for referee conditions in case of a dispute. The three temperatures 20°, 23°, and 27°C correspond to normal laboratory conditions in various parts of the world.

2. Rounded derived values are shown in parentheses ( ).

3. 25±2°C, 20 to 50% relative humidity (RH) has been widely used as a closely controlled ambient for testing electronics components.

(B) Determination of endurance (life) at resonance (or at specific frequencies).

(C) Determination of deterioration resulting from long exposure to swept frequency (or random vibration).

Recommended Frequency Ranges for Tests— Hertz: 1 to 10, 5 to 35, 10 to 55, 10 to 150, 10 to 500, 10 to 2000, 10 to 5000.

Recommended Combinations of Amplitude and Frequency—IEC Publication 68 recommends testing at constant amplitude below and constant acceleration above the crossover frequency (57 to 62 hertz). MIL-STD-202 and MIL-STD-810 also follow this principle but use different crossover points and low-frequency severities. The choice of frequency range and vibration amplitude or acceleration should bear some relation to the actual service environment. Successful completion of 10^7 vibration cycles indicates a high probability of no failures in a similar service environment. Resonances may make the equipment output unusable, although the mechanical life may be adequate.

COMPONENT VALUE CODING

Axial-lead and some other components are often color coded by circumferential bands to indicate

TABLE 4. HIGH-ALTITUDE TESTS

Pressure mbar	Pressure mm Hg	Pressure in. Hg	Approximate Corresponding Altitude feet	Approximate Corresponding Altitude meters	Standard
700	*525*	*20.67*	*7 218*	2 200	IEC
600	*450*	*17.72*	*11 483*	3 500	IEC
533	*400*	*15.74*	*14 108*	4 300	IEC
586	439	17.3	15 000	4 572	MIL-202
466	*349*	13.75	20 000	*6 096*	RS-186
300	*225*	*8.86*	*27 900*	8 500	IEC
300	226	8.88	30 000	9 144	MIL-202, RS-186
116	87.0	3.44	50 000	15 240	MIL-202, RS-186
85	*63.8*	*2.51*	*52 500*	16 000	IEC
44	*33.0*	*1.30*	*65 600*	20 000	IEC
44.4	33.0	1.31	70 000	21 336	MIL-202
20	*17.2*	*0.677*	*85 300*	26 000	IEC
10.6	8.00	0.315	100 000	30 480	MIL-202
1.28	1.09	0.043	150 000	45 720	MIL-202
3.18×10^{-6}	2.40×10^{-6}	9.44×10^{-8}	656 000	200 000	MIL-202

Notes:

1. The inconsistency in the pressure–altitude relation arises from the use of different model atmospheres. For testing purposes always specify the desired pressure rather than an elevation in feet or meters.

2. Values in italics are derived from the values specified in the associated standard.

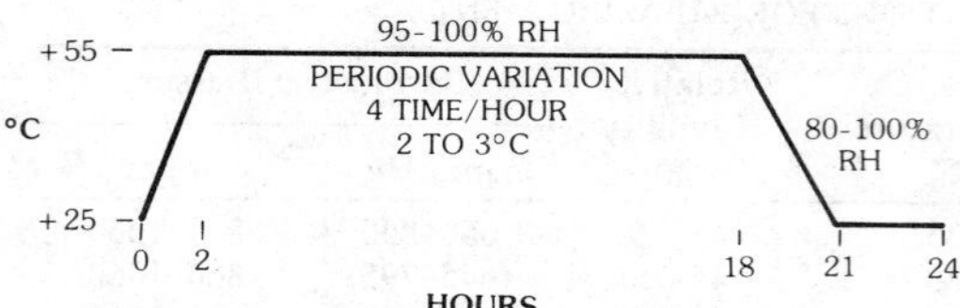

IEC-68—1, 2, or 6 24-Hour Cycles

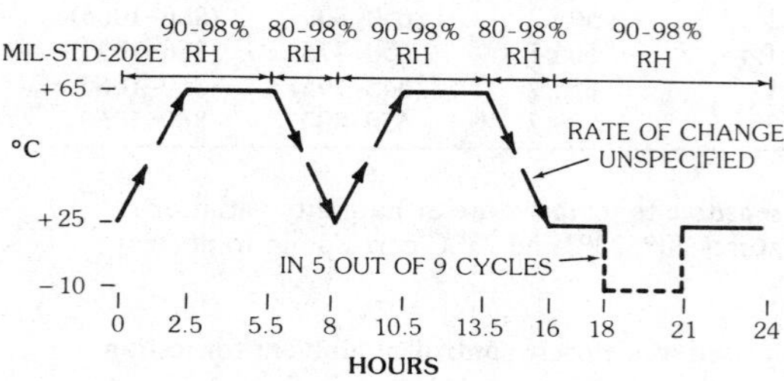

RS-186-D—4, 10, or 28 24-Hour Cycles
IEC-68, MIL-STD-202E—10 24-Hour Cycles

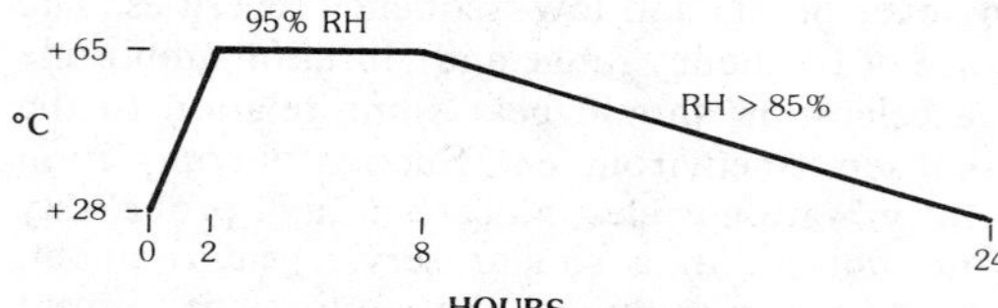

MIL-STD-810C—10 24-Hour Cycles

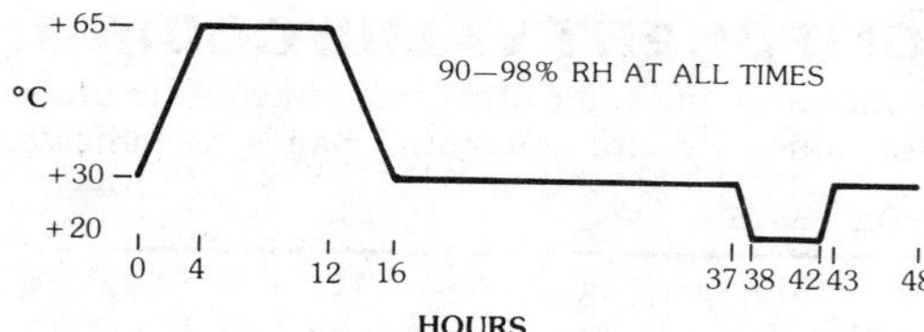

MIL-STD-810C—5 48-Hour Cycles or 5 48-Hour Cycles Plus 480 Hours at 30°C, 90-98% RH

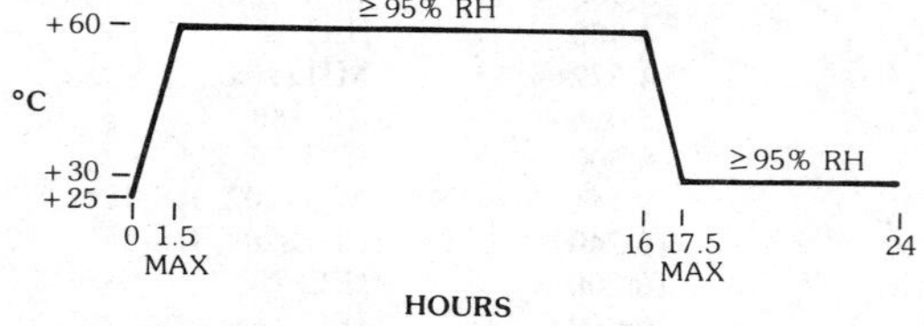

MIL-STD-810C—5 24-Hour Cycles

Fig. 1. Cycling humidity tests. Relative humidity for RS-186-D is 90–95% but may be uncontrolled during temperature changes.

the resistance, capacitance, or inductance value and its tolerance. Usually the value may be decoded as indicated in Fig 2 and Table 1.

Sometimes, instead of circumferential bands, colored dots are used as shown in Fig. 3 and Table 5.

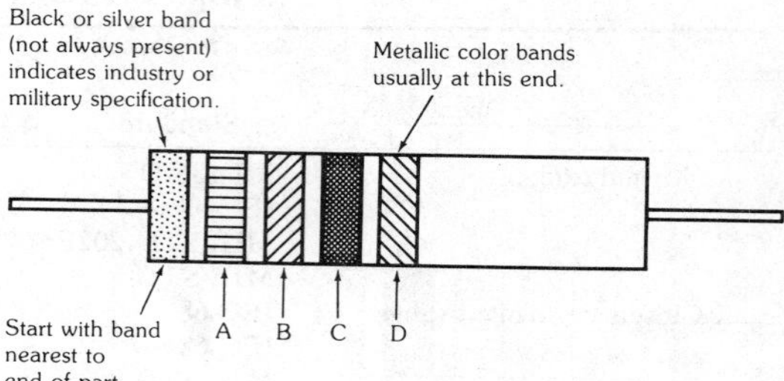

Fig. 2. Component value coding. The code of Table 1 determines values. Band A color = First significant figure of value in ohms, picofarads, or microhenries. Band B color = Second significant figure of value. Band C color = Decimal multiplier for significant figures. Band D color = Tolerance in % (if omitted, the broadest tolerance series of the part applies).

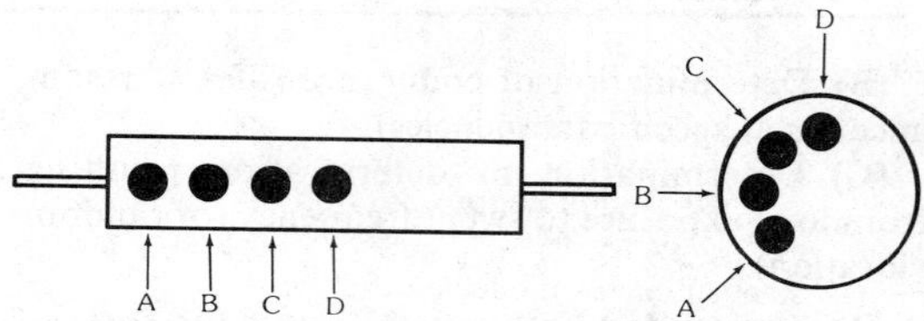

Fig. 3. Alternative methods of component value coding.

TABLE 5. COLOR-CODE EXAMPLES

Component Value	Band or Dot Color A	B	C	D
3300±20%	Orange	Orange	Red	Black or omitted
5.1±10%	Green	Brown	Gold or white	Silver
1.8 megohms ±5% (as applied to a resistor)	Brown	Gray	Green	Gold

Semiconductor-Diode Type-Number Coding

The sequential number portion (following the "1N" of the assigned industry type number) may be indicated by color bands* as shown in Fig. 4. Colors have the numerical significance given in Table 1.

Bands J, K, L, M represent the digits in the sequential number (for two-digit numbers, band J

*EIA Standard RS-236-B.

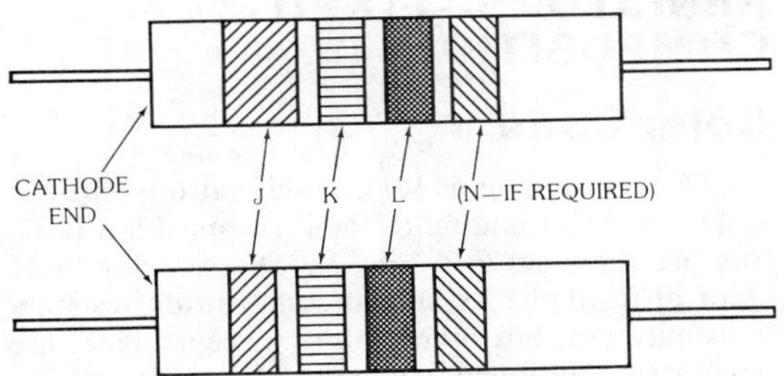

(A) 2- or 3-digit sequential number.

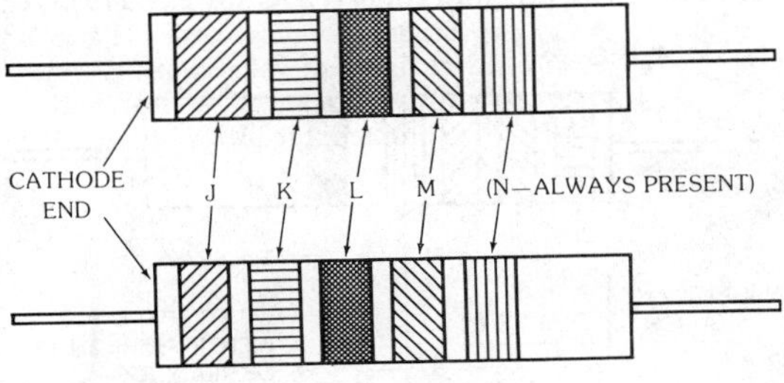

(B) 4-digit sequential number.

Fig. 4. Semiconductor-diode value coding.

is black). Band N is used to designate the suffix letter as shown in Table 6. Band N may be omitted

TABLE 6. DIODE COLOR CODE

Color	Suffix Letter	Number
Black	—	0
Brown	A	1
Red	B	2
Orange	C	3
Yellow	D	4
Green	E	5
Blue	F	6
Violet	G	7
Gray	H	8
White	J	9

in two- or three-digit number coding if not required, but it will always be present in 4-digit number coding (black if no suffix letter is required). See Table 7 for examples.

A single band indicates the cathode end of a diode or rectifier.

RESISTORS—DEFINITIONS

Wattage Rating: The maximum power that the resistor can dissipate, assuming (A) a specific life, (B) a standard ambient temperature, and (C) a stated long-term drift from its no-load value. Increasing the ambient temperature or reducing the allowable deviation from the initial value (more-stable resistance value) requires derating the allowable dissipation. With few exceptions, resistors are derated linearly from full wattage at rated temperature to zero wattage at the maximum temperature.

TABLE 7. EXAMPLES OF DIODE COLOR CODING

Band	Band Color		
J	Red	Red	Orange
K	Green	Green	Blue
L	Yellow	Yellow	Violet
M	—	—	Red
N	—	Red	Black
	1N254	1N254B	1N3672

Temperature Coefficient (*Resistance-Temperature Characteristic*): The magnitude of change in resistance due to temperature, usually expressed in percent per degree Celsius or parts per million per degree Celsius (ppm/°C). If the changes are linear over the operating temperature range, the parameter is known as "temperature coefficient"; if nonlinear, the parameter is known as "resistance-temperature characteristic." A large temperature coefficient and a high hot-spot temperature cause a large deviation from the nominal condition; e.g., 500 ppm/°C and 275°C result in a resistance change of over 12 percent.

Maximum Working Voltage: The maximum voltage that may be applied across the resistor (maximum working voltage) is a function of (A) the materials used, (B) the allowable resistance deviation from the low-voltage value, and (C) the physical configuration of the resistor. Carbon composition resistors are more voltage-sensitive than other types.

Noise: An unwanted voltage fluctuation generated within the resistor. Total noise of a resistor always includes Johnson noise, which depends only on resistance value and the temperature of the resistance element. Depending on type of element and construction, total noise may also include noise caused by current and noise caused by cracked bodies and loose end caps or leads. For adjustable resistors, noise may also be caused by jumping of the contact over turns of wire and by imperfect electrical path between contact and resistance element.

Hot-Spot Temperature: The maximum temperature measured on the resistor due to both internal heating and the ambient operating temperature. The allowable maximum hot-spot temperature is predicated on thermal limits of the materials and the design. Since the maximum hot-spot temperature may not be exceeded under normal operating conditions, the wattage rating of the resistor must be lowered if it is operated at an ambient temperature higher than that at which the wattage rating was established. At zero dissipation, the maximum ambient around the resistor may be at its maximum hot-spot temperature. The ambient temperature for a resistor is affected by surrounding heat-producing devices; resistors stacked together do not experience the ambient surrounding the stack except under forced cooling.

Critical Resistance Value: A resistor of specified power and voltage ratings has a critical resistance value above which the allowable voltage limits the permissible power dissipation. Below the critical resistance value, the maximum permitted voltage across the resistor is never reached at rated power.

Inductance and Other Frequency Effects: For other than wirewound resistors, the best high-frequency performance is secured if (A) the ratio of resistor length to cross section is a maximum, and (B) dielectric losses are kept low in the base material and a minimum of dielectric binder is used in composition types.

Carbon composition types exhibit little change in effective dc resistance up to frequencies of about 100 kHz. Resistance values above 0.3 megohm start to decrease in resistance at approximately 100 kHz. Above 1 MHz, all resistance values decrease.

Wirewound types have inductive and capacitive effects and are unsuited for use above 50 kHz, even when specially wound to reduce the inductance and capacitance. Wirewound resistors usually exhibit an increase in resistance at high frequencies because of "skin" effect.

Film types have the best high-frequency performance. The effective dc resistance for most resistance values remains fairly constant up to 100 MHz and decreases at higher frequencies. In general, the higher the resistance value the greater the effect of frequency.

Established-Reliability Resistors: Some resistor styles can be purchased with maximum-failure-rate guarantees. Standard-failure-rate levels are:

%/1000 hours—1.0; 0.1; 0.01; 0.001.

Resistance Value and Tolerance Choice: A calculated circuit-resistance nominal value should be checked to determine the allowable deviation in that value under the most unfavorable circuit, ambient, and life conditions. A resistor type, resistance value, and tolerance should be selected considering (A) standard resistance values (specials are uneconomical in most cases), (B) purchase tolerance, (C) resistance value changes caused by temperature, humidity, voltage, etc., and (D) long-term drift.

RESISTORS—FIXED COMPOSITION

Color Code

EIA-standard and MIL-specification requirements for color coding of fixed composition resistors are identical (see Fig. 2). The exterior body color of insulated axial-lead composition resistors is usually tan, but other colors (except black) are permitted. Noninsulated axial-lead composition resistors have a black body color.

If three significant figures are required, Fig. 5 shows the resistor markings (EIA Std RS-279).

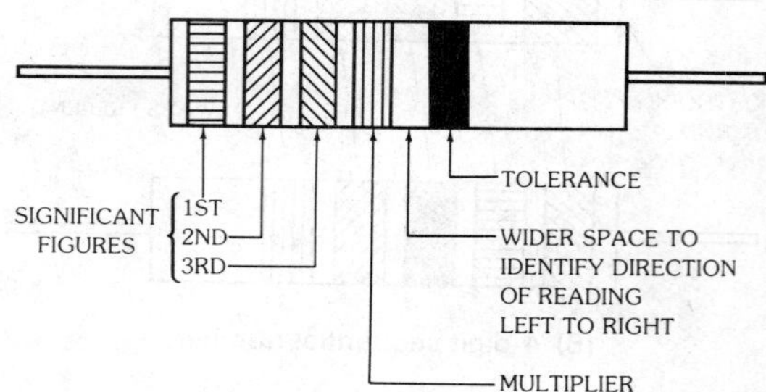

Fig. 5. Resistor value color code for three significant figures. Colors of Table 1 determine values.

Another form of resistor color coding (MIL-STD-1285A) is shown in Fig. 6. Colors have the significance shown in Table 8 for the fifth band.

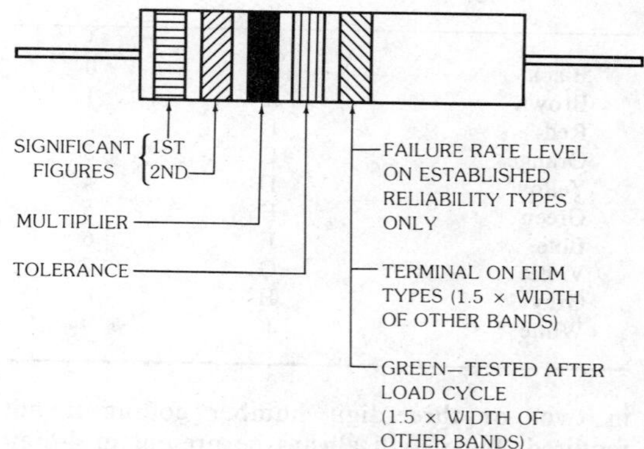

Fig. 6. Resistor color code per MIL-STD-1285A.

Tolerance

Standard resistors are furnished in ±20-, or ±10-, and ±5-percent tolerances, and in the preferred-value series of Table 2. "Even" values, such as 50 000 ohms, may be found in old equipment, but they are seldom used in new designs.

TABLE 8. COLOR CODE FOR FIFTH BAND (FIG. 6)

	Brown	Red	Orange	Yellow	Green	White
Failure Rate Level:						
Letter	M	P	R	S	—	—
Rate (%/1000 hrs)	1.0	0.1	0.01	0.001	—	—
Terminal	—	—	—	—	—	Weldable
Special	—	—	—	—	Fig. 6	—

Packaging Styles

In addition to the familiar axial-lead configuration, resistors are also available in SIP (single in-line package), DIP (dual in-line package), flat-pack, and leadless chip configurations.

The SIP and DIP types are designed for thru-hole or socket mounting on printed circuit boards. Flat-pack and chip configurations are surface mounted.

Flat-packs, SIPs, and DIPs contain multiple resistors, which may be internally connected to form networks or to have a common terminal pin.

The chip configuration allows increased component density and is autoinsertable. Since there are no leads, basic circuit reliability and high-frequency performance are improved.

Temperature and Voltage Coefficients

Resistors are rated for maximum wattage at an ambient temperature of 70° C; above these temperatures up to the maximum allowable hot-spot temperature of 130 or 150°C, it is necessary to operate at reduced wattage ratings. Resistance values are a function of voltage as well as temperature; present specifications allow a maximum voltage coefficient of resistance as given in Table 9 and permit a resistance-temperature characteristic as in Table 10.

A 1000-hour rated-load life test should not cause a change in resistance greater than 12% for 1/8-watt resistors and 10% for all other ratings. A severe cycling humidity test may cause resistance changes of 10% average and 15% maximum; 250 hours at 40° C and 95% relative humidity may cause up to 10% change. Five temperature-change cycles, − 55° C to +85° C, should not change the resistance value by more than 4% from the 25° C value. Soldering the resistor in place may cause a resistance change of 3%. Always allow 1/4-inch minimum lead length; use heat-dissipating clamps when soldering confined assemblies. The preceding summary indicates that close tolerances cannot be maintained over a wide range of load and ambient conditions.

Noise

Composition resistors above 1 megohm have high Johnson noise levels, precluding their use in critical applications.

TABLE 9. STANDARD RATINGS FOR COMPOSITION RESISTORS

Watts	Working Volts (Maximum)	Hot-Spot Temperature (°C) (Maximum)	Critical Resistance (Megohms)	Voltage Coefficient* (%/Volt) (Maximum)
1/8	150	150	0.22	0.05
1/4	250	130	0.25	0.035
1/2	350	130	0.25	0.035
1	500	130	0.25	0.02
2	500	130	0.12	0.02

* Applicable only to resistors of 1000 ohms and over.

TABLE 10. TEMPERATURE COEFFICIENT OF RESISTANCE FOR COMPOSITION RESISTORS

	Characteristic*	Percent Maximum Allowable Change from Resistance at 25 Degrees Celsius†					
At −55°C ambient	F	±6.5	±10	±13	±15	±20	±25
At +105°C ambient	F	±5	±6	±7.5	±10	±15	±15
Nominal resistance in ohms		0 to 1 000	>1 000 to 10 000	>10 000 to 0.1 meg	>0.1 meg to 1.0 meg	>1 meg to 10 meg	>10 meg

* Resistance-temperature.
† Up to 1 megohm, data also apply to MIL Established Reliability characteristic *G* (= former *GF*).

RF Effects

The end-to-end shunted capacitance effect may be noticeable because of the short resistor bodies and small internal distance between the ends. Operation at vhf or higher frequencies reduces the effective resistance because of dielectric losses (Boella effect).

Good Design Practice

Operate at one-half the allowable wattage dissipation for the expected ambient temperature. Provide an adequate heat sink. Mount no other heat-dissipating parts within one diameter. Use only in applications where a 15% change from the installed value is permissible or where the environment is controlled to reduce the resistance-value change.

RESISTORS—FIXED WIREWOUND

Fixed wirewound resistors are available as low-power insulated types, precision types, and power types.

EIA Low-Power Insulated Resistors†

These resistors are furnished with power ratings from 1 watt through 15 watts, in tolerances of ±5 and ±10 percent, and in resistance values from 0.1 ohm to 30 000 ohms in the preferred-value series of Table 2. They may be color coded as described in Fig. 2, but band A will be twice the width of the other bands. They may also be typographically marked in accordance with the EIA Standard.

The stability of these resistors is somewhat better than that of composition resistors, and they may be preferred except where a noninductive resistor is required.

EIA Precision Resistors‡

These resistors are furnished in ±1.0-, ±0.5-, ±0.25-, ±0.1-, and ±0.05-percent tolerances and in any value from 1.0 ohm to 1.0 megohm in the preferred-value series of Table 2. Power ratings range from 0.1 watt to 0.5 watt. The maximum ambient temperature for full wattage rating is 125°C. If the resistor is mounted in a confined area or may be required to operate in higher ambient temperatures (145°C maximum), the allowable dissipation must be reduced in accordance with the EIA Standard.

These resistors have an inherently low noise level, approaching the thermal agitation level, and their stability is excellent—the typical change in resistance for the lifetime of the resistor will not exceed 50 percent of the initial resistance tolerance when used within the specified design limits of the EIA Standard.

The temperature coefficient of resistance over the range − 55°C to +145°C, referred to 25°C, may have maximums as follows:

Value	EIA Standard
Above 10 ohms	±0.002%/°C
5 ohms to 10 ohms	±0.006%/°C
Below 5 ohms	±0.010%/°C

Where required, temperature coefficients of less than ±20 ppm/°C can be obtained by special selection of the resistance wire. Temperature coefficients of ±10ppm/°C may be obtained by limiting the range of temperatures for testing from −40°C to +105°C. The application of temperature coefficient to resistors should be limited, where possible, to the actual temperatures at which the equipment will operate.

†EIA Standard RS-344.

‡EIA Standard RS-229-A.

EIA Power Resistors*

These resistors are furnished in 3 styles (strip; tubular, open end; and axial lead) and 24 power ratings ranging from 1 watt to 210 watts in tolerances of ±1.0 percent and ±5 percent. Resistance values range from 1.0 ohm to 182 kilohms in the preferred-value series of Table 2.

Axial-lead types are available in two general inductance classifications—inductive winding and noninductive winding. The noninductive styles have a maximum resistance value of 1/2 the maximum resistance of inductive styles because of the special manner in which they are wound. The inductance of noninductive styles must not exceed 0.5 microhenry when measured at a test frequency of 1.0 megahertz ±5%. However, these resistors should not be used in very-high-frequency circuits where the inductance may affect circuit operation.

The maximum ambient temperature for full wattage rating for these resistors is 25°C. When the resistors are operated at ambient temperatures above 25°C, the wattage dissipation must be reduced in accordance with the EIA Standard.

RESISTORS—FIXED FILM

Film-type resistors use a thin layer of resistive material deposited on an insulating core. The low-power types are more stable than the usual composition resistors. Except for very high-precision requirements, film-type resistors are a good alternative to accurate wirewound resistors, being both

*EIA Standard RS-155-B.

smaller and less expensive and having excellent noise characteristics.

The power types are similar in size and performance to conventional wirewound power resistors. While their 200°C maximum operating temperature limits the power rating, the maximum resistance value available for a given physical size is much higher than that of the corresponding wirewound resistor.

Construction

For low resistance values, a continuous film is applied to the core, a range of values being obtained by varying the film thickness. Higher resistances are achieved by the use of a spiral pattern, a coarse spiral for intermediate values and a fine spiral for high resistance. Thus, the inductance is greater in high values, but it is likely to be far less than in wirewound resistors. Special high-frequency units having greatly reduced inductance are available.

Resistive Films

Resistive-material films presently used are microcrystalline carbon, boron-carbon, and various metallic oxides or precious metals.

Deposited-carbon resistors have a negative temperature coefficient of 0.01 to 0.05 percent/°C for low resistance values and somewhat larger for higher values. Cumulative permanent resistance changes of 1 to 5 percent may result from soldering, overload, low-temperature exposure, and aging. Additional changes up to 5 percent are possible from moisture penetration and temperature cycling.

The introduction of a small percentage of boron into the deposited-carbon film results in a more stable unit. A negative temperature coefficient of 0.005 to 0.02 percent/°C is typical. Similarly, a metallic dispersion in the carbon film provides a negative coefficient of 0.015 to 0.03 percent/°C. In other respects, these materials are similar to standard deposited carbon. Carbon and boron-carbon resistive elements have the highest random noise of the film-type resistors.

Metallic-oxide and precious-metal-alloy films permit higher operating temperatures. Their noise characteristics are excellent. Temperature coefficients are predominantly positive, varying from 0.03 to as little as 0.0025 percent/°C.

Applications

Power ratings of film resistors are based on continuous direct-current operation or on root-mean-square operation. Power derating is necessary for operation at ambient temperatures above the rated temperature. In pulse applications, the power dissipated during each pulse and the pulse duration are more significant than average power conditions. Short high-power pulses may cause instantaneous local heating sufficient to alter or destroy the film. Excessive peak voltages may result in flashover between turns of the film element. Derating under these conditions must be determined experimentally.

Film resistors are fairly stable up to about 10 megahertz. Because of the extremely thin resistive film, skin effect is small. At frequencies above 10 megahertz, it is advisable to use only unspiraled units if inductive effects are to be minimized (these are available in low resistance values only).

Under extreme exposure, deposited-carbon resistors deteriorate rapidly unless the element is protected. Encapsulated or hermetically sealed units are preferred for such applications. Open-circuiting in storage as the result of corrosion under the end caps has been reported in all types of film resistors. Silver-plated caps and core ends effectively overcome this problem.

Technical Characteristics

Some technical characteristics of film resistors are given in Tables 11, 12, and 13.

TABLE 11. STABLE EQUIVALENTS FOR COMPOSITION RESISTORS; AXIAL LEADS; DATA FOR MIL "RL" SERIES

Watts	1/4	1/2	1	2
Voltage rating	250	350	500	500
Critical resistance (megohms)	0.25	0.25	0.25	0.12

Maximum temp for full load—70°C; for 0 load—150°C

Resistance-temperature characteristic: ±200 ppm/°C maximum

Life-test resistance change: ±2% maximum

Moisture resistance test: ±1.5% maximum change

Resistance values: E24 series, same as composition resistors; tolerances 2% or 5%.

The MIL "RN" series of film resistors is more stable than the "RL" series and is available in a wider range of ratings. Commercial equivalents are also offered. Where stability and reliability are desired, the "RN" series is economically very competitive with the "RC" or "RL" series.

RESISTORS—ADJUSTABLE

Adjustable resistors may be divided into three separate and distinct categories, potentiometers, trimmers, and rheostats.

Potentiometers are control devices that are used

TABLE 12. HIGH-STABILITY FILM RESISTORS; AXIAL LEADS; DATA FOR MIL "RD" SERIES*

	Watts							
	$\frac{1}{20}$	$\frac{1}{10}$	$\frac{1}{8}$	$\frac{1}{4}$	$\frac{1}{2}$	$\frac{3}{4}$	1	2
Characteristic	Voltage Rating							
B	—	—	—	—	—	—	500	750
D	—	—	200	300	350	500	—	—
C,E	200	200	250	300	350	—	500	—

	Characteristic			
	B	C	D	E
Maximum temp (°C):				
Full load	70	125	70	125
0 load	150	175	165	175
Life-test resistance change (max)	±1%	±0.5%	±1%	±0.5%
Resistance-temperature characteristic (max ppm/°C)	±500	±50	+200 −500	±25
Moisture resistance test (max change)	±1.5%	±0.5%	±1.5%	±0.5%

Resistance Values: E96 series (E48 preferred): 1% tolerance;
E192 series (E96 preferred): 0.5%, 0.25%, 0.1% tolerances.

*Uninsulated commercial versions have lower temperature limits and greater resistance change. Color coding is the same as previously indicated for composition resistors.

TABLE 13. POWER-TYPE FILM RESISTORS, UNINSULATED

	Axial-Lead and MIL "RD" Series			Commercial Tab-Terminal Styles				
Watts	2	4	8	7	23	25	55	115
Voltage rating	350	500	750	525	1380	2275	3675	7875
Critical resistance (kilohms)	61	62	70	39	83	208	245	540
Maximum temperature (°C):								
Full load	25	25	25	25	25	25	25	25
0 load	235 275 (MIL)	235 275 (MIL)	235 275 (MIL)	235	235	235	235	235

Life-test resistance change: ±5% maximum
Resistance-temperature characteristic: ±500 ppm/°C maximum
Moisture resistance test: ±3% maximum change
Resistance values: E12 series (approx. for MIL)
Tolerances: Axial lead: $\frac{1}{2}$%, 1%, 2%, (MIL "RD" Series), 5%, 10%.
Tab lead: 1%, 2%, 5%, 10%, 20%.

where the frequency of adjustment is high. They may be operated manually by human effort or mechanically served by machine. Potentiometers are designed with a long mechanical life in view, generally from 10 000 to 100 000 cycles, with certain types having life capabilities in the millions of cycles. A cycle, or excursion, consists of wiper traverse from one limit of travel to the other limit and back.

Trimmers differ from potentiometers in that they are designed to be adjusted infrequently, sometimes only once, and normally exhibit greater setting stability once set. Their employment eliminates the use of expensive precision related components and provides an easily retunable vehicle to compensate for drift or aging in related parts. Normal life designs are rated at approximately 200 excursions. Some typical trimmer characteristics are listed in Table 14.

TABLE 14. TYPICAL CHARACTERISTICS OF ½-INCH TRIMMERS

	Resistance Ranges	Typical Resistance Tolerance	Wattage Rating	Rotational Life	Temperature Coefficient of Resistance
Wirewound	10 Ω - 50 kΩ	±5%	1 W at 70°C	200	±50 ppm/°C
Cermet	50 Ω - 2 MΩ	±10%	1 W at 70°C	200	±100 ppm/°C
Conductive Plastic	50 Ω - 2 MΩ	±10%	0.33 W at 50°C	1×10^6	±500 ppm/°C
Carbon Composition	100 Ω - 2.5 MΩ	±10% ±20%	0.33 W at 50°C	200	±10%/°C*

*Temperature characteristic.

Unlike potentiometers and trimmers, whose primary function is to control voltage, the rheostat is basically a current-controlling device. Rheostats are made in much the same manner as potentiometers and trimmers, but with more attention paid to wiper current-carrying ability and generally higher power ratings. Some rheostats are wirewound tubular ceramic power resistors with a track of exposed wire to allow for the adjustable feature. The wiper is in the form of a dimpled clamp band that is screw-tightened at the desired setting. Due to the heat generated by these units, care should be exercised in the circuit location.

Wattage ratings of adjustable resistors apply only when all the resistance is in the circuit. To avoid overloading any section, never exceed the maximum rated current based on total resistance.

Types of Adjustable Resistors

Wirewound Resistors—Wirewound resistor elements are made by winding a very fine resistance wire precisely around a mandrel. Most resistance wire is made from a nickel-chrome alloy with other elemental additives to enhance its electrical characteristics.

Wirewound adjustable resistors exhibit superior independent linearity characteristics, and it is for this reason that they are frequently specified for direct motion controls. Precision wirewounds are available with independent linearity ratings as low as 0.1%. They are very stable over a wide range of operating temperatures. Panel controls and precision potentiometers are recommended for normal operating temperature ranges from −65° to + 125°C, and most trimmers will perform satisfactorily from −55° to +150°C. Temperature coefficients of resistance as low as ±20 ppm/°C are available.

Wirewound elements, with rare exceptions, change value in steps as the wiper traverses each individual winding. Resolution may be improved by using multiturn adjustable resistors. This type of construction increases the winding length and decreases incremental resistance steps from one winding to the next.

Wirewound resistive elements are usually not suitable for frequency-sensitive rf circuits because of inductive and capacitive effects. They are impractical above a resistance value determined by the winding space available and the smallest resistance wire that can be space-wound. If infinite resolution is required, then a cermet or composition element must be substituted.

Cermet-Element Resistors—Cermets are a mixture of fine metal-oxide or precious-metal particles and glass in a viscous organic vehicle. This paste is screened onto a ceramic substrate and fired at vitrifying temperatures.

Cermet adjustable resistors are designed for low to moderate adjustment life. They feature infinite resolution and are generally available having temperature coefficients of resistance of ±100 ppm/°C. Sheet resistivities* are available from one ohm per square (1 Ω/□) to one megohm per square (1 MΩ/□). Ultimate resistance values are limited by substrate geometry.**

Carbon Composition Resistors—A mixture of carbon powders and a binder is molded under heat and pressure into a solid mass. In some constructions, the carbon composition is molded at the same time as the plastic substrate. This process is called comolding.

Carbon composition adjustable resistors are the least expensive and most common type of potentiometer for general electronics use. Their low noise makes them ideal for use in live audio controls.

In addition to a linear rotation-vs-resistance

*Sheet resistivity (Ω/□) is the resistance of a square sheet of material. It is independent of the units of length used because the resistance increases with length but decreases with width.

**For material of given volume resistivity ρ (ohm-meter), sheet resistivity given by ρ/d will increase as depth d is decreased.

characteristic, composition elements can have a wide range of nonlinear output curves. Standard logarithmic curves (tapers) are used as volume controls for radio, tv, etc.

Temporary resistance changes up to ±10% can be expected when these devices are operated near the extreme limits of a temperature range of −55°C to +120°C. Their use is not recommended for precision controls or in varying hostile environments.

Conductive Plastic—Conductive plastic is an ink formulated from carbon, other proprietary materials, a resin, and solvent. It is applied to a substrate by screening, dipping, or comolding. The low curing temperature (150 to 300°C) of the ink allows it to be applied to a wide variety of substrates.

Conductive-plastic potentiometers are most notable for their high rotational life, and they are used most frequently as machine-operated servo-controls. Another desirable feature is their low noise or output smoothness.

Resistance ranges in sheet resistivities of up to 50 000 Ω/□ are available with temperature coefficients of ± 500 ppm/°C. As with cermets, ultimate resistance values are limited by substrate geometry.

Terminal Identification

Industry standards have been developed for identification of potentiometer terminals. Most potentiometer terminals are either numbered or color coded as follows:

1 Yellow	Always the counterclockwise element limit.
2 Red	Wiper (or collector).
3 Green	Always the clockwise element limit.

For rotary potentiometers, clockwise is always defined by viewing the specified mounting end of the potentiometer.

Mounting Characteristics

Potentiometers usually must be accessible from the outside of a product, and they are often mounted on a panel by means of a threaded bushing. Most precision potentiometers are manufactured with both bushing and servo-mount options. Servo-mount units are secured by servo clamps to assure precise shaft alignment.

Trimmers are usually not accessible from outside the instrument. Most are circuit-board mounted by their terminals and are small in size to conserve space.

CAPACITORS—DEFINITIONS

Dielectric: A dielectric is a medium that can withstand high electric stress without appreciable conduction. When such stress is applied, energy in the form of an electric charge is held by the dielectric. Most of this stored energy is recovered when the stress is removed. The only perfect dielectric in which no conduction occurs and from which the whole of the stored energy may be recovered is a perfect vacuum.

Relative Capacitivity: The relative capacitivity or relative permittivity or dielectric constant is the ratio by which the capacitance is increased when another dielectric replaces a vacuum between two electrodes.

Dielectric Absorption: Dielectric absorption is the absorption of charge by a dielectric when subjected to an electric field by other than normal polarization. This charge is not recovered instantaneously when the capacitor is short-circuited, and a decay current will continue for many minutes. If the capacitor is short-circuited momentarily, a new voltage will build up across the terminals afterward. This is the source of some danger with high-voltage dc capacitors or with ac capacitors not fitted with a discharge resistor. The phenomenon may be used as a measure of dielectric absorption.

Tangent of Loss Angle: This is a measure of the energy loss in the capacitor. It is expressed as $\tan\delta$ and is the power loss of the capacitor divided by its reactive power at a sinusoidal voltage of specified frequency. (This term also includes power factor, loss factor, and dielectric loss. The true power factor is $\cos(90 - \delta)$.)

Insulation Resistance: This is a measure of the conduction in the dielectric. Because this conduction takes a very long time to reach a stable value, it is usually measured after 2 minutes of electrification for nonelectrolytic types and 3 minutes for electrolytics. It is measured preferably at the rated working voltage or at a standardized voltage.

The insulation resistance is usually multiplied by the capacitance to give the ohm-farad value, which is the apparent discharge time constant (seconds). This is a figure of merit for the dielectric, although for small capacitances a maximum value of insulation resistance is usually also specified.

In electrolytics, the conduction is expressed as leakage current at rated working voltage. It is calculated as $\mu A/\mu FV$, which is the reciprocal of the ohm-farad value. In this case, a maximum value of leakage current is specified for small capacitances.

Leakage Current: The current flowing between two or more electrodes by any path other than the interelectrode space is termed the leakage current, and the ratio of this to the test voltage is the insulation resistance.

Impedance: Impedance is the ratio of voltage to current at a specified frequency. At high frequencies, the inductance of leads becomes a limiting factor, in which case a transfer impedance method may be employed. This then measures the impedance of the shunt path only.

DC or AC Capacitor: A dc capacitor is designed to operate on direct current only. It is normally not suitable for use above 200 volts ac because of the occurrence of discharges in internal gas bubbles (corona). An ac capacitor is designed to have freedom from internal discharges and low tangent of loss angle to minimize internal heating.

Rated Voltage and Temperature: The rated voltage is the direct operating voltage that may be applied continuously to a capacitor at the rated temperature.

Category Voltage and Temperature: The category voltage is the voltage that may be applied to the capacitor at the maximum category temperature. It differs from the rated voltage by a derating factor.

Ripple Voltage: If alternating voltages are present in addition to direct voltage, the working voltage of the capacitor is taken as the sum of the direct voltage and the peak alternating voltages. This sum must not exceed the value of the rated voltage.

In electrolytics, the permissible ripple may be expressed as a rated ripple current.

Surge Voltage: This is a voltage above the rated voltage which the capacitor will withstand for a short time.

Voltage Proof Test (*Dielectric Strength*): This is the highest possible voltage that may be applied without breakdown to a capacitor during qualification approval testing to prove the dielectric. The repeated application of this voltage may cause failure.

Forming Voltage (*Electrolytics*): The voltage at which the anodic oxide has been formed. The thickness of the oxide layer is proportional to this voltage.

Burnout Voltage (*Metallized Types*): The voltage at which metallized types burn out during manufacture.

Self-Healing (*Metallized Types*): A momentary partial discharge of a capacitor resulting from a localized failure of the dielectric. Burning away the metallized electrodes isolates the fault and effectively restores the properties of the capacitor. The self-healing action is also called "clearing."

Equivalent Series Resistance (*ESR*): Equivalent series resistance (ESR) is a single resistive value that represents the sum of the ac losses (due to the leads, electrode plates, and junction terminations), the resistive losses due to leakage currents, and the resistive losses due to the inherent molecular polarization dielectric absorption factors of the base dielectric material.

Volt-Ampere Rating (*VA*): This is the reactive power in a capacitor when an ac voltage is applied. VA $\cos\theta$ gives the amount of heat generated in the capacitor. Since the amount of heat that can be dissipated is limited, the VA must also be limited and in some cases a VA rating is quoted. (Note that $\cos\theta = \cos(90 - \delta) \approx \tan\delta$, when δ is small.)

Scintillation: Minute and rapid fluctuations of capacitance formerly exhibited by silvered mica or silvered ceramic types but overcome by modern manufacturing techniques.

Corona Discharge: Partial discharge of a capacitor due to ionization of the gas in a bubble in the dielectric. On ac or pulse operation, this may occur in dielectric stressed above 200 volts and is a major cause of failure. On dc, such discharges are very infrequent and normally are not a cause of failure.

CLASSES OF CAPACITORS

Modern electronic circuits require the smallest possible capacitors, which are usually made with the thinnest possible dielectric material since they are for operation at low voltages. There are three broad classes of capacitors.

(A) Low-loss capacitors with good capacitance stability. These are usually of mica, glass, ceramic, or a low-loss plastic such as polypropylene or polystyrene.

(B) Capacitors of medium loss and medium stability, usually required to operate over a fairly wide range of ac and dc voltages. This need is met by paper, plastic film, or high-*K* ceramic types. The first two of these may have electrodes of metal foil or electrodes of evaporated metal which have a self-healing characteristic.

(C) Capacitors of the highest possible capacitance per unit volume. These are the electrolytics, which are normally made either of aluminum or tantalum. Both of these metals form extremely thin anodic oxide layers of high dielectric constant and good electrical characteristics. Contact with this oxide layer is normally by means of a liquid electrolyte that has a marked influence on the characteristics of the capacitor. In solid tantalum, the function of the electrolyte is performed by a manganese-dioxide semiconductor.

PLASTIC FILM CAPACITORS

Advances in organic chemistry have made it possible to produce materials of high molecular weight. These are formed by joining together a number of basic elements (monomers) to produce a polymer. Some of these have excellent dielectric characteristics.

Physically, they can be classified as thermoplastic or thermosetting. In the former case, the molecule consists of long chains with little or no branching, whereas in the latter the molecules are crosslinked. Thermosetting materials have no clearly defined melting point and are usually hard and brittle, making them unsuitable for the manufacture of plastic films. A cast film is usually amorphous, but by extrusion, stretching, and heat treatment, oriented crystalline films are produced with good flexibility and dielectric characteristics.

The electrical properties of the plastics depend on the structure of the molecule. If the molecule is not symmetrical, it will have a dipole moment giving increased dielectric constant. On the other hand, the dielectric constant and tanδ are then dependent on frequency. Generally speaking, nonpolar materials have electrical characteristics that are independent of frequency, while polar materials exhibit a decrease in capacitance with increasing frequency, and tanδ may pass through a maximum in the frequency range.

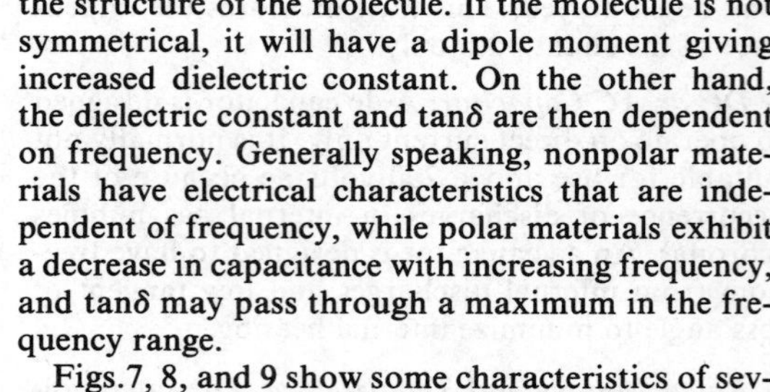

Figs. 7, 8, and 9 show some characteristics of several types of capacitors. At the present time, two classes of plastic film capacitors are recognized.

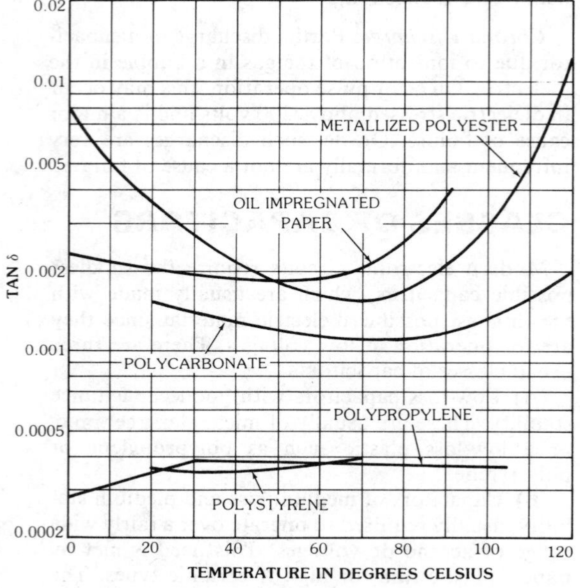

Fig. 7. Variation of tanδ with temperature for various plastic films compared with oil-impregnated paper.

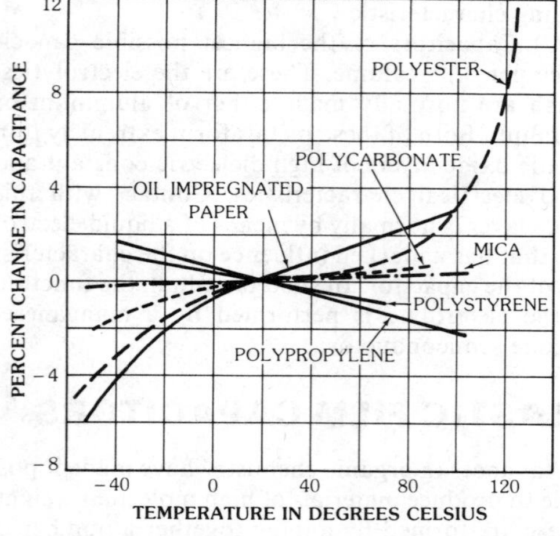

Fig. 8. Typical capacitance characteristics of various capacitors as a function of temperature. Measured on 0.1-microfarad capacitors at 1000 hertz.

(A) *Polystyrene and Polypropylene Capacitors.* Polystyrene and polypropylene are nonpolar plastics that have excellent electrical characteristics which are independent of frequency.

(B) *Polyester Films.* Strictly speaking, these are the polyethylene terephthalates (Mylar, Melinex, Hostaphan), but the polycarbonates are now included in this group because they have similar electrical characteristics.

Plastic films for capacitor manufacture are usually of the oriented crystalline type because of their good combinations of characteristics. One important feature of some of these films is that they tend to shrink back to their original shape after being heated. This fact is sometimes exploited in manufacturing the capacitor.

Moisture usually has little effect on the dielectric properties of plastic films, and capacitors made from them require less protection than paper or mica types. This, together with simple processing, has permitted them to be mass-produced at relatively low cost.

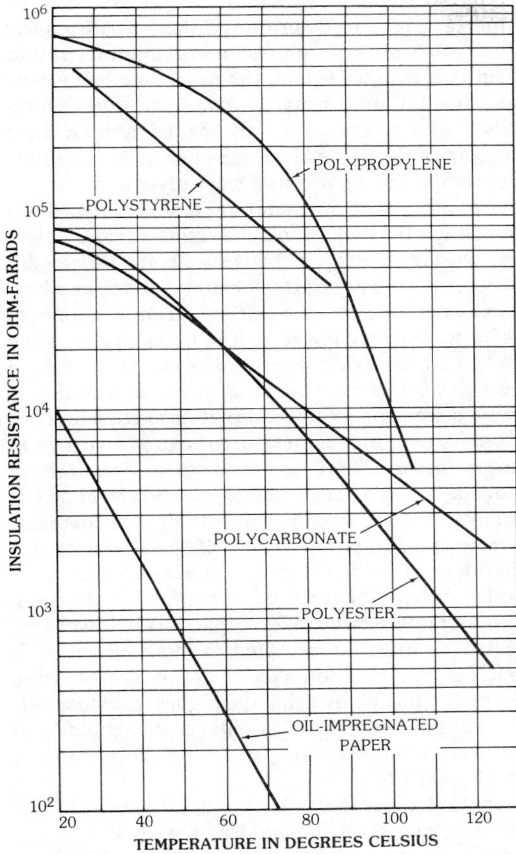

Fig. 9. Variation of insulation resistance with temperature for various plastic films compared with the resistance variation for oil-impregnated paper.

The electrical characteristics of capacitors made with these materials depend on the construction employed. The resistance of metallized electrodes and the shape and method of connection to the unit are particularly important.

Many plastic materials are being used in the manufacture of capacitors for which there are no internationally agreed specifications. In this case, it is necessary to obtain the relevant data from the manufacturer. There is no doubt that in the future the range of plastic film capacitors will be considerably extended.

Polystyrene Film Capacitors With Foil Electrodes

Polystyrene has excellent electrical characteristics. The film employed is of the oriented crystalline type, which makes it flexible and suitable for forming into thin films. On heat treatment, the film shrinks considerably, and this is used in the manufacturing process to obtain capacitance stability. The film is affected by greases and solvents, and care must be taken both in manufacture and in use to ensure that capacitors do not come into contact with these materials.

The power factor of polystyrene is low over the whole frequency range, but the resistance of the electrodes may result in an increase of power factor at high frequencies in the larger values, as shown in Table 15.

TABLE 15. POWER FACTOR OF POLYSTYRENE AT VARIOUS FREQUENCIES

Frequency (hertz)	Nominal Capacitance (pF) Up to 1 000	1 000 to 10 000	10 000 to 100 000	Above 100 000
800	0.0003	0.0003	0.0003	0.0003
10 000	0.0003	0.0003	0.0003	0.001
100 000	0.0003	0.0005	0.001–0.003	—
1 000 000	0.001	0.002	0.005–0.02	—

Polyester Film Capacitors With Foil Electrodes

The generic title "polyester" is usually used to apply to polyethylene terephthalate. It is a slightly polar plastic film suitable for operation up to temperatures of 125°C. Capacitors are available with foil electrodes.

Plastic Film Capacitors With Metallized Electrodes

Plastic film capacitors with metallized electrodes have superseded metallized paper capacitors in dc applications because of superior electrical characteristics, less tendency for self-healing to occur during service, higher and more stable insulation resistance, and approximately the same space factors. Three types of film are generally used, polyethylene terephthalate, polycarbonate, and polypropylene. For some purposes these are comparable, but polypropylene has a lower loss angle, and polycarbonate has a smaller change of capacitance with temperature. Polyester and polycarbonate are also available in thinner films, giving an advantage of space factor.

ELECTROLYTIC CAPACITORS

Electrolytic capacitors (Fig. 10) employ for at least one of their electrodes a "valve metal." This

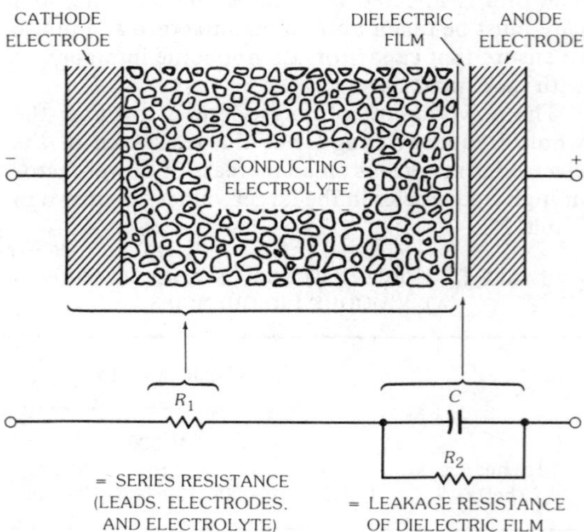

Fig. 10. Basic cell and simplified equivalent circuit for polar electrolytic capacitor.

metal, when operated in an electrolytic cell as the anode, forms a layer of dielectric oxide. The most commonly used metals are aluminum and tantalum. The valve-metal behavior of these metals was known about 1850. Tantalum electrolytic capacitors were introduced in the 1950s because of the need for highly reliable miniature capacitors in transistor circuits over a wide temperature range. These capacitors were made possible by improved refining and powder metallurgy techniques.

The term "electrolytic capacitor" is applied to any capacitor in which the dielectric layer is formed by an electrolytic method. The capacitor does not necessarily contain an electrolyte.

The oxide layer is formed by placing the metal in a bath containing a suitable forming electrolyte, and applying voltage between the metal as anode and another electrode as cathode. The oxide grows at a rate determined by the current, but this rate of growth decreases until the oxide has reached a limiting thickness determined by the voltage. For most practical purposes, it may be assumed that the thickness of the oxide is proportional to the forming voltage.

Properties of aluminum and tantalum and their oxides are shown in Table 16.

TABLE 16. ALUMINUM AND TANTALUM PROPERTIES

Metal	Density	Principal Oxide	Dielectric Constant	Thickness (Å/V)
Aluminum	2.7	Al_2O_3	8	13.5
Tantalum	16.6	Ta_2O_5	27.6	17

The structure of these oxide layers plays an important part in determining their performance. Ideally they are amorphous, but aluminum tends to form two distinct layers, the outer one being porous. Tantalum normally forms an amorphous oxide which, under conditions of a high field strength of the oxide layer, may become crystalline. Depending on the forming electrolyte and the surface condition of the metal, there is an upper limit of voltage beyond which the oxide breaks down. The working voltage is between 25 and 90 percent (according to type) of the forming voltage at which stable operation of the oxide layer can be obtained.

To produce a capacitor, it is necessary to make contact to the oxide layer on the anode, and there are two distinct methods of doing this. The first is to use a working electrolyte that has sufficient conductivity over the temperature range to give a good power factor. There are many considerations in choosing the working electrolyte, and the choice is usually a compromise between high- and low-temperature performance. The working electrolyte also provides a rehealing feature in that any faults in the oxide layer will be repaired by further anodization.

In aluminum electrolytic capacitors, the working electrolyte must be restricted to those materials in which aluminum and its oxide are inert. Corrosion can be minimized by using the highest possible purity of aluminum. This also reduces the tendency of the oxide layer to dissolve in the electrolyte, giving a better shelf life.

Tantalum, on the other hand, is very inert and therefore allows a wider choice of electrolyte. Since there is no gas evolution, better methods of sealing can be employed. The characteristics of aluminum and tantalum electrolytic capacitors are shown in Figs. 11 and 12.

A major problem with all electrolytic capacitors is to ensure that the electrolyte is retained within the case under all operating conditions. In the aluminum capacitor, allowance must be made for gas evolution on reforming. Even the tantalum capacitor usually must employ only organic materials for sealing, and these do not provide completely hermetic sealing. All organic materials have finite moisture transmission properties, and, therefore, at the maximum category temperature the high vapor pressure of the electrolyte results in some diffusion.

An elegant solution to this problem, using a semiconductor instead of an electrolyte, was found by Bell Telephone Laboratories. The semiconductor is manganese dioxide in a polycrystalline form and has a higher conductivity than conventional electrolyte systems. This material also provides a limited self-healing feature at a fault, resulting in oxidation of the tantalum and reduction of the manganese dioxide to a nonconducting form.

Electrolytic capacitors take many forms, and the

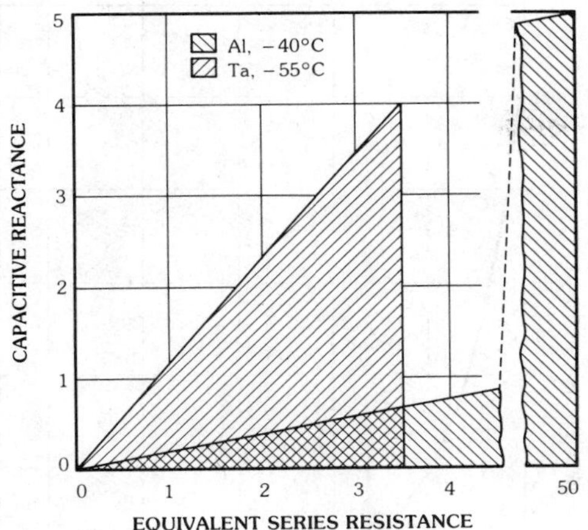

(A) Low temperature.

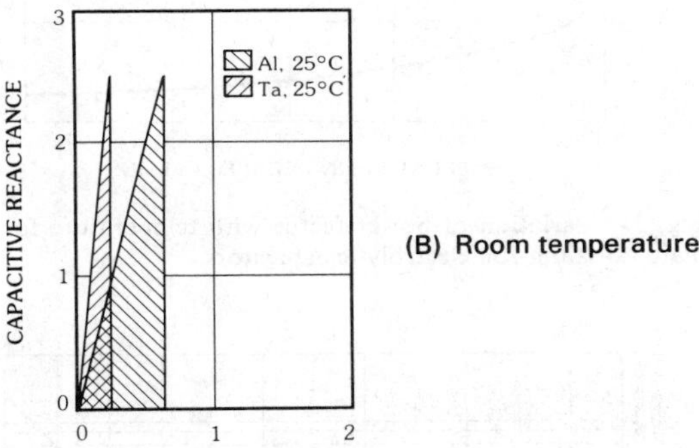

(B) Room temperature.

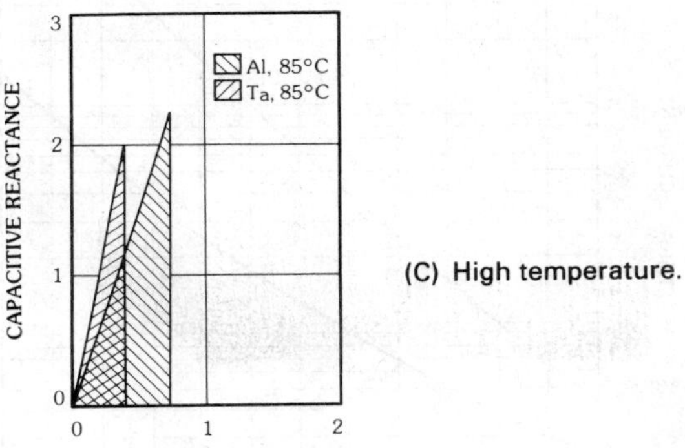

(C) High temperature.

Fig. 11. Typical 120-hertz impedance diagrams for aluminum (Al) and tantalum (Ta) plain-foil polar electrolytic capacitors of 150-volt rating.

anode may be of foil, wire, or a porous sintered body. The foil may be either plain or etched. The porous body may be made with fine or with coarse particles, and the body itself may be short and fat or long and thin. The aluminum-foil capacitor has a

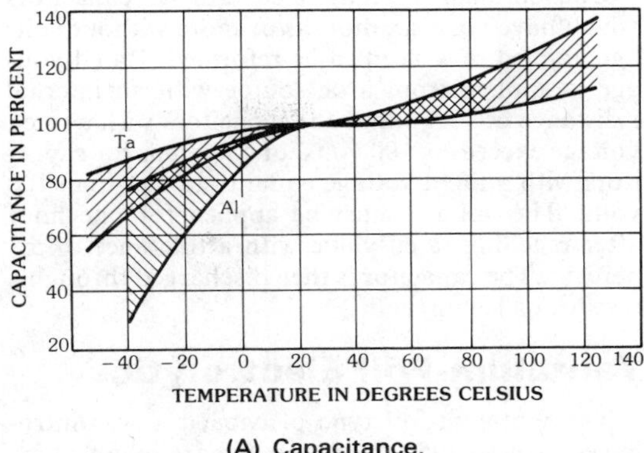

(A) Capacitance.

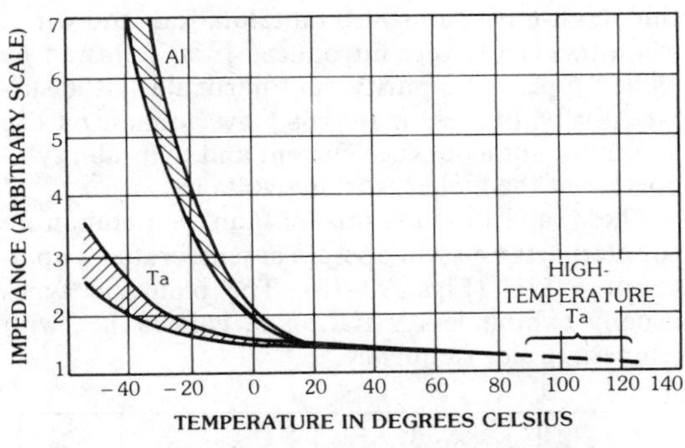

(B) Impedance.

Fig. 12. Capacitance and 120-hertz impedance as a function of temperature for aluminum (Al) and tantalum (Ta) electrolytic capacitors.

space factor about six times better than that of the equivalent paper capacitor, whereas tantalum capacitors are even smaller and enjoy a space factor up to 20 times better.

When electrolytic capacitors are operated in series-parallel, stabilizing resistors should be used to equalize the voltage distribution. It should also be noted that, even when the case is not connected to one terminal, a low resistance path exists between it and the electrodes. The case must be insulated from the chassis, particularly if the chassis and the negative terminal are not at the same potential.

Aluminum Electrolytics

The aluminum type is the most widely known electrolytic capacitor and is used extensively in radio and television equipment. It has a space factor about six times better than the equivalent paper capacitor. Types of improved reliability are now available using high-purity (better than 99.99%) aluminum.

Conventional aluminum electrolytic capacitors which have gone six months or more without voltage applied may need to be reformed. Rated voltage is applied from a dc source with an internal resistance of 1500 ohms for capacitors with a rated voltage exceeding 100 volts, or 150 ohms for capacitors with a rated voltage equal to or less than 100 volts. The voltage must be applied for one hour after reaching rated value with a tolerance of ±3 percent. The capacitor is then discharged through a resistor of 1 ohm/volt.

Tantalum-Foil Electrolytics

The tantalum-foil type of capacitor was introduced around 1950 to provide a more reliable type of electrolytic capacitor without shelf-life limitation. It was made possible by the availability of thin, high-purity annealed tantalum foils and wires. Plain-foil types were introduced first, followed by etched types. The purity, and particularly the surface purity, of these materials plays a major part in determining the leakage current and their ability to operate at the higher working voltages.

These capacitors are smaller than their aluminum counterparts and will operate at temperatures up to about 125°C (Figs. 13–15). The plain-foil types usually exhibit less variation of capacitance with temperature or frequency.

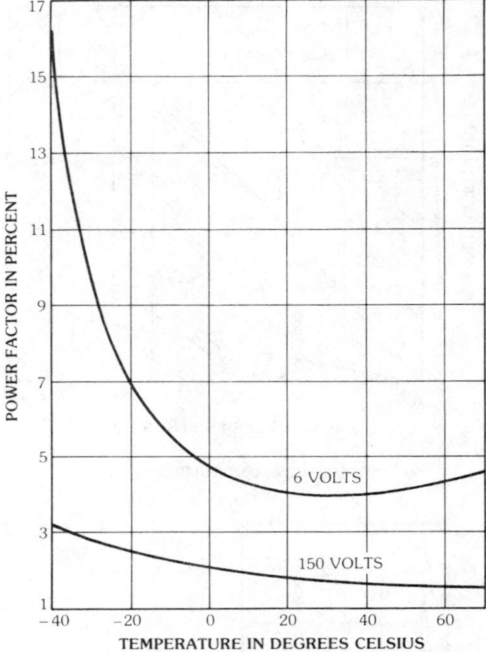

Fig. 14. Variation of power factor with temperature for plain tantalum-foil electrolytic capacitors.

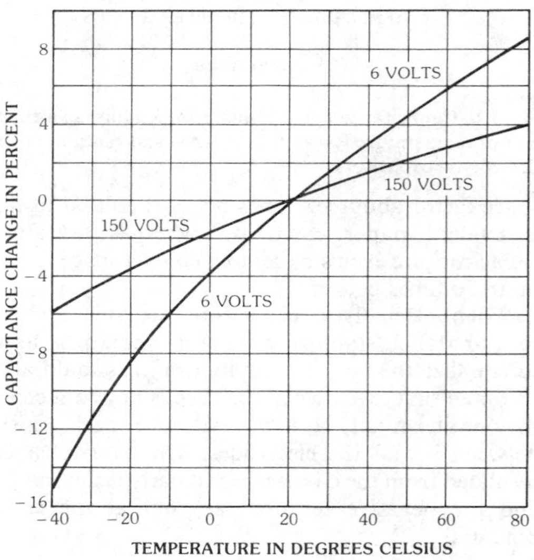

Fig. 13. Variation of capacitance with temperature for plain tantalum-foil electrolytic capacitors.

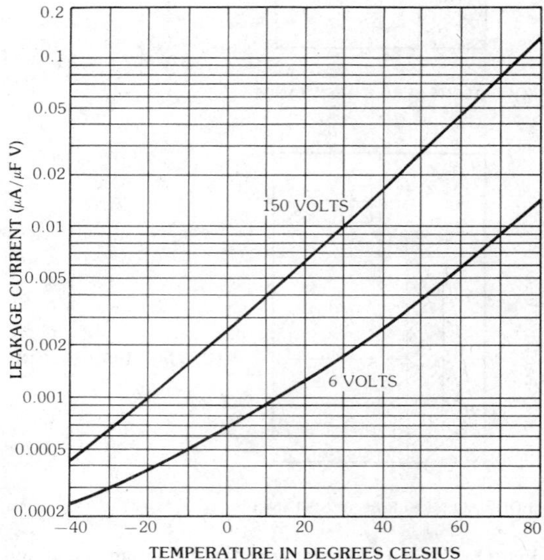

Fig. 15. Variation of leakage current with temperature for plain tantalum-foil electrolytic capacitors.

Tantalum Electrolytics With Porous Anode and Liquid Electrolyte

The tantalum electrolytic with porous anode and liquid electrolyte was the first type of tantalum electrolytic capacitor to be introduced and still has the best space factor. Types using sulphuric-acid electrolyte have excellent electrical characteristics up to about 70 working volts. Other types contain neutral electrolytes.

Basically, this type of capacitor consists of a sintered porous anode of tantalum powder housed in a silver or silver-plated container. The porous anode is made by pressing a high-purity tantalum powder into a cylindrical body and sintering in vacuum at about 2000°C.

Tantalum Electrolytics With Porous Anode and Solid Electrolyte

The so-called "solid" tantalum capacitor originally developed by Bell Telephone Laboratories developed from the porous-anode type with liquid electrolyte by replacing the liquid with a semiconductor. This overcame the problem of sealing common to all other types of electrolytic capacitors. Since there is no liquid electrolyte, it is possible to use a conventional hermetic seal.

CERAMIC CAPACITORS

Ceramic capacitors are defined in classes based on their distinct and inherent electrical properties.

Class I

Class I capacitors are stable temperature-compensating capacitors that have essentially linear characteristics with properties independent of frequency over the normal range. Materials are usually magnesium titanate for positive temperature coefficient of capacitance and calcium titanate for negative temperature coefficient of capacitance. Combinations of these and other materials produce a dielectric constant of 5 to 150 and temperature coefficient of capacitance of +150 to −4700 ppm /°C with tolerances of ±15 ppm/°C.

Low-*K* ceramics are suitable for resonant-circuit or filter applications, particularly where temperature compensation is a requirement. Disc and tubular types are the best forms for this purpose. Stability of capacitance is good, being next to that of mica and polystyrene capacitors.

Class II

Class II includes ceramic dielectrics suitable for fixed capacitors used for bypass, coupling, and decoupling. This class is usually divided into two subgroups for which the temperature characteristics define the characteristics.

(A) Embody stable *K* values of 250 to 2400 over a temperature range of −55°C to +125°C with maximum capacitance change 15% from a 25°C ambient.

(B) Embody combinations of materials, including titanates, with *K* values of 3000 to 12 000 made by keeping the Curie point near room ambient.

The high-*K* materials are the ferroelectrics. Because of their crystal structure, they sometimes have very high values of internal polarization, giving very high effective dielectric constants. In this way, these materials are comparable with ferromagnetic materials. Above the Curie temperature, a change of domain structure occurs that results in a change of electrical characteristics. This region is known as the paraelectric region. In common with the ferromagnetic materials, a hysteresis effect is apparent, and this makes the capacitance voltage-dependent.

The ferroelectrics are based on barium titanate, which has a peak dielectric constant of 6000 at the Curie point of 120°C. Additions of barium stanate, barium zirconate, or magnesium titanate reduce this dielectric constant but make it more uniform over the temperature range. Thus a family of materials can be obtained with a Curie point at about room temperature and with the dielectric constant falling off on either side. The magnitude of this change increases with increasing dielectric constant. These materials exhibit a decrease of capacitance with time and, as a result of the hysteresis effect, with increasing voltage.

Inductance in the leads and element causes parallel resonance in the megahertz region. Care is necessary in their application above about 50 megahertz for tubular styles and about 500 megahertz for disc types.

Class III

Class III includes reduced barium titanate in which a reoxidized layer or diffusion zone is the effective dielectric. Also included are internal grain boundary layer strontium titanate capacitors in which an internal insulating layer surrounds each grain. These capacitors are suitable for use in low-voltage circuits for coupling and bypass where low insulation resistance and voltage coefficient can be tolerated.

Color Code

The significance of the various colored dots for EIA Standard RS-198-B (ANSI C83.4-1972) fixed ceramic dielectric capacitors is explained by Figs. 16 and 17 and may be interpreted from Table 17.

Temperature Coefficient

Standard temperature coefficients of capacitance expressed in parts per million per degree Celsius are: +150, +100, + 33, 0, −33, −75, −150, −220, −330, −470, − 750, −1500, −2200, −3300, and −4700.

PAPER FOIL-TYPE CAPACITORS

In general, paper capacitors have been largely replaced by plastic film capacitors in both foil and

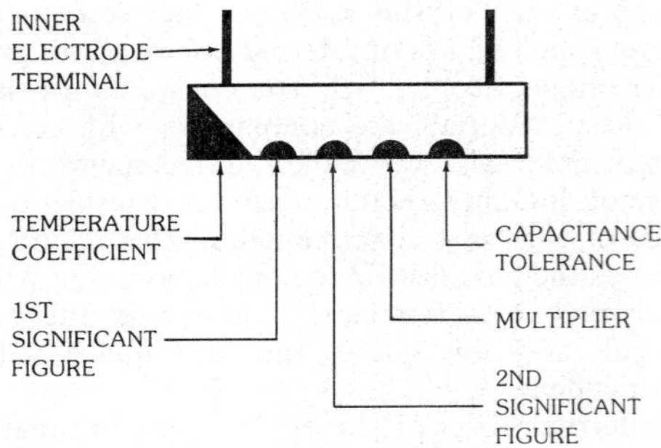

(A) Five-dot system.

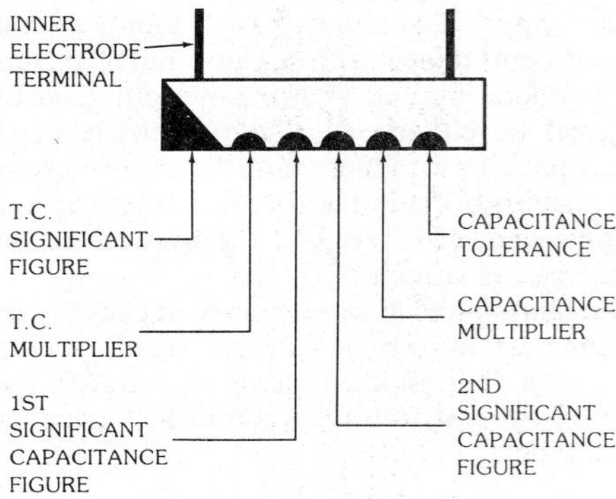

(B) Six-dot system.

Fig. 16. Color coding of EIA Class-I ceramic dielectric capacitors. See Table 17 for color code. Tubular style shown to illustrate identification of inner electrode. For disc or plate styles, color code will read from left to right as observed with lead wires downward.

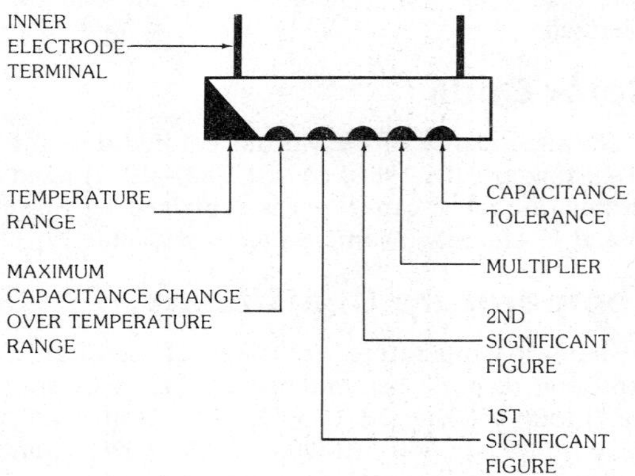

Fig. 17. Color coding of EIA Class-II ceramic dielectric capacitors. See Table 17 for color code. Tubular style shown to illustrate identification of inner electrode. For disc or plate styles, color code will read from left to right as observed with lead wires downward.

metallized constructions and in most dc electronic circuits. Impregnated paper capacitors for ac power applications are also being replaced with plastic film capacitors with both impregnated and dry constructions.

Construction usually consists of multiple layers of paper as the dielectric interleaved with aluminum or tin alloy foil electrodes. Termination is either tabs inserted during winding or leads welded to extended foils.

Paper capacitors are impregnated with stabilized waxes or oils. Most chlorinated materials, such as chlorinated diphenyls, have been banned by the EPA as impregnants. The impregnant usually has a dielectric constant of about 6 and is used for size reduction and improvement in corona start and to reduce internal discharge in ac power applications.

Included among dc applications are coupling, decoupling, bypass, smoothing filters, power-separating filters, energy-storage capacitors, etc. Included among ac applications are motor start, fluorescent lighting, interference suppression, power-factor correction, power-line coupling, distribution capacitors for high-voltage switching gear, capacitor voltage dividers for ac measurement, etc.

MICA CAPACITORS

Mica capacitors fall within the classification of low loss and good capacitance stability. Mica is one of the earliest dielectric materials used and has an unrivaled combination of physical and electrical characteristics. It is of mineral origin and, because of its monoclinic structure, can be readily split into thin plates. It has a dielectric constant of about 6 (largely independent of frequency) together with a very low loss.

Construction

Mica capacitor constructions of the eyelet, molded, bonded, and button styles have been largely replaced by precision polystyrene, polypropylene, and polyester plastic film capacitors and by ceramic capacitors. The main surviving style is the dipped epoxy-coated radial-leaded capacitor using a clamp-type silvered mica stack with tin electrodes.

Applications

Their low temperature coefficient of capacitance and good stability with temperature and frequency make mica capacitors a good choice for critical precision circuitry such as filter applications.

Type Designation

A comprehensive numbering system, the type designation, is used to identify mica capacitors. Type designations are of the form shown in Fig. 18.

TABLE 17. COLOR CODE FOR CERAMIC DIELECTRIC CAPACITORS, CLASSES I AND II*

			Class I						Class II		
			Capacitance Tolerance								
Color	Digit	Multiplier	10 pF or less (pF)	Over 10 pF (%)	Temperature Coefficient ppm/°C (5-Dot System)	Temperature Coefficient Significant Figure (6-Dot System)	Temperature Coefficient Multiplier (6-Dot System)	Capacitance Tolerance (%)	Temperature Range (°C)	Maximum Capacitance Change Over Temperature Range (%)
Black	0	1	±2.0	±20	0	0.0	−1	±20	—	±2.2
Brown	1	10	±0.1	±1	−33	—	−10	—	+10 to +85	±3.3
Red	2	100	—	±2	−75	1.0	−100	—	−55 to +125	±4.7
Orange	3	1000	—	±3	−150	1.5	−1000	—	+10 to +65	±7.5
Yellow	4	10 000	—	—	−220	2.2	−10 000	GMV	—	±10
Green	5	—	±0.5	±5	−330	3.3	+1	±5	—	±15
Blue	6	—	—	—	−470	4.7	+10	—	—	±22
Violet	7	—	—	—	−750	7.5	+100	—	—	+22, −33
Gray	8	0.01	±0.25	—	+150 to −1500	(−1000 to −5200 ppm/°C, With Black Multiplier)	+1000	+80, −20	—	+22, −56
White	9	0.1	±1.0	±10	+100 to −750	—	+10 000	±10	—	+22, −82
Silver	—	—	—	—	—	—	—	—	−30 to +85	±1.5
Gold	—	—	—	—	—	—	—	—	−55 to +85	±1

* EIA Standard RS-198-B (ANSI C83.4-1972). This standard classifies ceramic dielectric, fixed capacitors as follows:

Class I —Temperature compensating ceramics suited for resonant circuit or other applications where high Q and stability of capacitance characteristics are required.

Class II —Ceramics suited for bypass and coupling applications, or for frequency discriminating circuits where high Q and stability of capacitance characteristics are not of major importance.

Class III—Low-voltage ceramics specifically suited for transistorized or other electronic circuits for bypass, coupling, or frequency determination where dielectric losses, high insulation resistance, and capacitance stability are not of major importance.

Note: Where size permits, EIA Class-III ceramics are typographically marked as follows:
(1) Capacitance value in microfarads. (2) Rated voltage. (3) Manufacturer's mark or EIA source code.
(4) Capacitance value tolerance or appropriate code letter; either ±20% (Code M) or +80, −20% (Code Z).
(5) Temperature stability code (see the EIA or ANSI Standard).

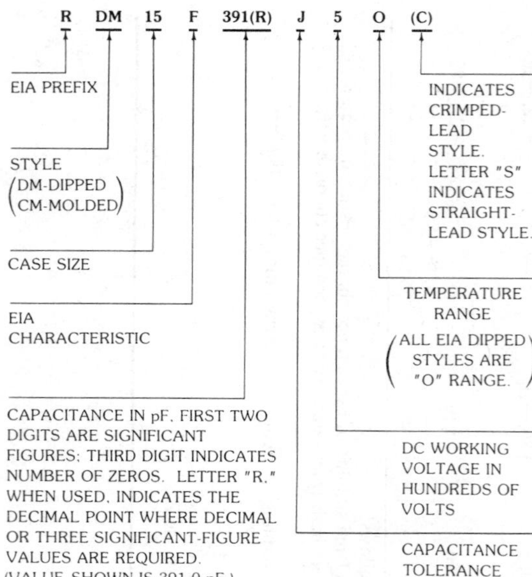

(A) EIA.

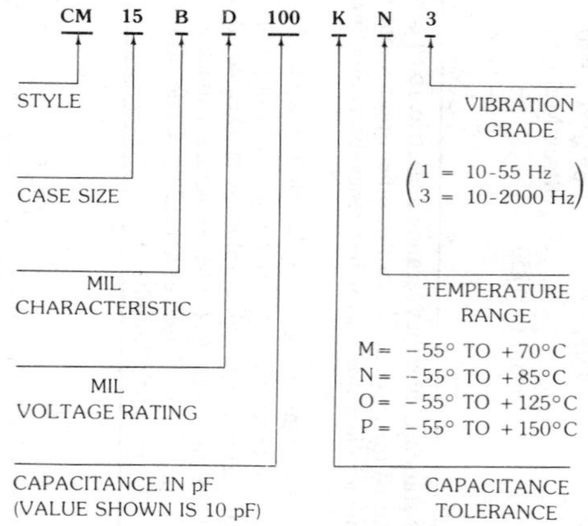

(B) MIL.

Fig. 18. Type designation for mica-dielectric capacitors.

MIL specifications now require type designation marking. Color coding is now used only for EIA standard capacitors.

Component Designation—Fixed mica-dielectric capacitors are identified by the symbol CM. For EIA, a prefix letter R is always included, and dipped types are identified by the symbol DM.

Case Designation—The case designation is a two-digit symbol that identifies a particular case size and shape.

Characteristic—The EIA or MIL characteristic is indicated by a single letter in accordance with Table 18.

TABLE 18. FIXED-MICA-CAPACITOR REQUIREMENTS BY EIA AND MIL CHARACTERISTIC

EIA or MIL Charac-teristic	Maximum Capacitance Drift	Maximum Range of Temperature Coefficient (ppm/°C)
B	Not Specified	Not Specified
C	±(0.5% + 0.1 pF)	±200
D	±(0.3% + 0.1 pF)	±100
E	±(0.1% + 0.1 pF)	−20 to +100
F	±(0.05% + 0.1 pF)	0 to +70

Capacitance Value—The nominal capacitance value in picofarads is indicated by a three-digit number. The first two digits are the first two digits of the capacitance value in picofarads. The final digit specifies the number of zeros that follow the first two digits. For EIA, if more than two significant figures are required, an additional digit is used, and the letter "R" is inserted to designate the decimal position.

Capacitance Tolerance—The symmetrical capacitance tolerance in percent is designated by a letter as shown in Table 1.

Voltage Rating—MIL voltage ratings are designated by a single letter as follows. A = 100, B = 250, C = 300, D = 500, E = 600, F = 1000, G = 1200, H = 1500, J = 2000, K = 2500, L = 3000, M = 4000, N = 5000, P = 6000, Q = 8000, R = 10 000, S = 12 000, T = 15 000, U = 20 000, V = 25 000, W = 30 000, and X = 35 000 volts. EIA dc working voltage is a number designating hundreds of volts.

Temperature Range—MIL specifications provide for four temperature ranges, all of which have a lower limit of −55°C; the upper limits are M = +70, N = +85, O = +125, and P = +150°C. The EIA uses only N and O, which are identical to the MIL standard.

Vibration Grade—The MIL vibration grade is a number, 1 corresponding to vibration from 10 to 55 hertz at 10g for 4.5 hours and 3 corresponding to 10 to 2000 hertz at 20g for 12 hours.

Capacitance

Capacitance is measured at 1 megahertz for capacitors of 1000 picofarads or smaller. Larger capacitors are measured at 1 kilohertz.

Temperature Coefficient

Measurements to determine the temperature coefficient of capacitance and the capacitance drift are based on one cycle over the following temperature values (all in degrees Celsius): +25, −55. −40, −10, +25, +45, +65, +70, +85, +125, +150, +25. Measurements at +85, +125, and +150 are not made if these values are not within the applicable temperature range of the capacitor.

Dissipation Factor

The EIA and MIL specifications require that for molded and dipped capacitors the dissipation factor not exceed the values shown in Fig. 19. For

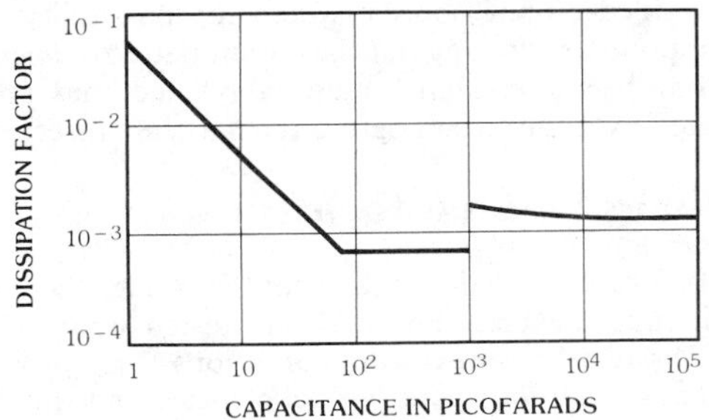

Fig. 19. EIA and MIL maximum dissipation factor at 1 megahertz for capacitance of 1000 picofarads or less and at 1 kilohertz for capacitance greater than 1000 picofarads.

potted and cast epoxy capacitors, the dissipation factor shall not exceed 0.35 percent from 1 to 1000 picofarads and 0.15 percent above 1000 picofarads.

High-Potential or Withstanding-Voltage Test

Molded or dipped mica capacitors are subjected to a test potential of twice their direct-current voltage rating.

Humidity and Thermal-Shock Tests

EIA Standard RS-153-B capacitors must withstand 5 cycles of −55, +25, +85, or +125 (as applicable), and +25 degrees Celsius thermal shock followed by a humidity test of 10 cycles (each of 24 hours) given for EIA Standard RS-186-D in Fig. 1. Units must pass a withstanding-voltage test. Capacitance may not change by more than 1.0 percent or 1.0 picofarad, whichever is greater. Insulation resistance must meet or exceed 30 percent of the initial requirements at 25°C (50 000 megohms for capacitances of 20 000 picofarads or less; 1000 ohm-farads for larger capacitances).

MIL Specification MIL-C-5D capacitors must withstand 5 cycles of −55; +25; +85, +125, or +150 (as applicable); and +25 degrees Celsius thermal shock followed by a humidity test of 10 cycles (each of 24 hours) given for MIL-STD-202E in Fig. 1. Units must pass a withstanding-voltage test. Capacitance may not change by more than ±(0.2 percent + 0.5 picofarad). Insulation resistance must meet or exceed 30 percent of the initial requirements at 25°C (100 000 megohms for 10 000 picofarads or less; 1000 megohm-microfarads for larger capacitances).

Life

Capacitors are given accelerated life tests at 85 degrees Celsius with 150 percent of rated voltage applied for 2000 hours for MIL specification or 250 hours for EIA standard. If capacitors are rated above +85°C, the test will be at their maximum rated temperature.

PRINTED CIRCUITS

A printed circuit is a conductive circuit pattern on one or both sides of an insulating substrate. Multilayer boards with tens of levels of circuitry are manufactured with conducting thru-holes to interconnect the circuitry levels. The conductive pattern can be formed by any of several techniques after which component lead holes are drilled or punched in the substrate and components are installed and soldered in place. Printed-circuit construction is ideal for assembly of circuits that employ miniature solid-state components. Its advantages over conventional chassis and point-to-point wiring include:

(A) Considerable space savings over conventional construction methods is usually a result.

(B) A complex circuit may be modularized by using several small printed circuits instead of a single larger one. Modularization simplifies trouble-shooting, circuit modification, and mechanical assembly in an enclosure.

(C) Soldering of component leads may be accomplished in an orderly sequence by hand or by dip or wave soldering.

(D) A more uniform product is produced because wiring errors are eliminated and because distributed capacitances are constant from one production unit to another.

(E) The printed-circuit method of construction lends itself to automatic assembly and testing.

(F) Using appropriate base metals, flexible cables or flexible circuits can be built.

(G) By using several layers of circuits (in proper registry) in a sandwich construction, with the conductors separated by insulating layers, relatively complex wiring can be provided.

Printed-Circuit Base Materials

Rigid printed-circuit base materials are available in thicknesses varying from 1/64 to 1/2 inch. The important properties of the usual materials are given in Table 19. For special applications, other rigid or flexible materials are available as follows:

(A) Glass-cloth Teflon (polytetrafluoroethylene, PTFE) laminate.
(B) Kel-F (polymonochlorotrifluoroethylene) laminate
(C) Silicone rubber (flexible)
(D) Glass-mat–polyester-resin laminate.
(E) Teflon film.
(F) Ceramic.

The most widely used base material is NEMA-XXXP paper-base phenolic.

Conductor Materials

Copper is used almost exclusively as the conductor material, although silver, brass, and aluminum also have been used. The common thicknesses of foil are 0.0014 inch (1 oz/ft^2) and 0.0028 inch (2 oz/ft^2). The current-carrying capacity of a copper conductor may be determined from Fig. 20.

Manufacturing Processes

The most widely used production methods are:

(A) Etching process, wherein the desired circuit is printed on the metal-clad laminate by photographic, silk-screen, photo-offset, or other means, using an ink or lacquer resistant to the etching bath. The board is then placed in an etching bath that removes all of the unprotected metal (ferric chloride is a commonly used mordant for copper-clad laminates). After the etching is completed, the ink or lacquer is removed to leave the conducting pattern exposed.

(B) Plating process, wherein the designed circuit pattern is printed on the unclad base material using an electrically conductive ink, and, by electroplating, the conductor is built up to the desired thickness. This method lends itself to plating through punched holes in the board for making connections from one side to the other.

(C) Other processes, including metal spraying and die stamping.

Circuit-Board Finishes

Conductor protective finishes are required on the circuit pattern to improve shelf-storage life of the circuit boards and to facilitate soldering. Some of the most widely used finishes are:

(A) Hot-solder coating (done by dip-soldering in a solder bath) is a low-cost method and gives good results where coating thickness is not critical.

(B) Silver plating used as a soldering aid but is subject to tarnishing and has a limited shelf life.

(C) Hot-rolled or plated solder coat gives good solderability and uniform coating thickness.

(D) Other finishes for special purposes are gold plate, for corrosion resistance and solderability, and electroplated rhodium over nickel, for wear resistance. Insulating coatings such as acrylic, polystyrene, epoxy, or silicone resin are sometimes applied to circuit boards to improve circuit performance under high humidity or to improve the anchorage of parts to the board. Conformal coatings are relatively thick and tend to smooth the irregular contour of the mounted items; they add less mass than encapsulation. A protective organic coating (unless excessively thick) will not improve the electrical properties of an insulating base material during long exposure to high humidity. On two-sided circuit boards, where the possibility of components shorting out the circuit patterns exists, a thin sheet of insulating material is sometimes laminated over the circuit before the parts are inserted.

Design Considerations

Before a printed-circuit layout is made, the circuit must be breadboarded and tested under the anticipated final operating conditions. This procedure will permit operating deficiencies and quirks to be detected and corrected before the time-consuming process of producing the circuit board is begun. It is important to note that certain circuits may operate differently on a printed-circuit board than on a breadboard, and appropriate corrective steps may be necessary. For example, inductive coupling between foil patterns may cause unwanted oscillation in high-frequency or amplifier circuits.

All features (terminal areas, contacts, board boundaries, holes, etc.) should be arranged to be centered at the intersections of a 0.100-, 0.050-, or 0.025-inch rectangular grid, with preference in the order stated. Many components are available with leads spaced to match the standard grids. Devices with circular lead configurations and a few other multilead devices are exceptions that require special attention and dimensioning. Following this grid-layout principle simplifies drafting and subsequent machine operations in board manufacture and assembly.

Drilled holes must be employed if the stated requirements for punched holes cannot be met, or if the material is not of a punching grade. Drilling is less detrimental to the laminate surrounding the hole; punching may cause crazing or separation of the laminate layers.

The diameter of punched holes in circuit boards should not be less than 2/3 the thickness of the base material.

The distance between punched holes or between holes and the edge of the material should not be less than the material thickness.

TABLE 19. PROPERTIES OF TYPICAL PRINTED-CIRCUIT DIELECTRIC BASE MATERIALS

Material	Comparable MIL Type	Punchability	Mechanical Strength	Moisture Resistance	Insulation	Arc Resistance	Abrasive Action on Tools	Max Temperature (°C)*
NEMA type XXXP paper-base phenolic	—	Good	Good	Good	Good	Poor	No	105
NEMA type XXXPC paper-base phenolic	—	Very good	Good	Very good	Good	Poor	No	105
NEMA type FR-2 paper-base phenolic, flame resistant	—	Very good	Good	Very good	Good	Poor	No	105
NEMA type FR-3 paper-base epoxy, flame resistant	PX	Very good	Very good	Very good	Very good	Good	No	105
NEMA type FR-4 glass-fabric-base epoxy, general purpose, flame resistant	GF	Fair	Excellent	Excellent	Excellent	Very good	Yes	130 (125)
NEMA type FR-5 glass-fabric-base epoxy, temperature and flame resistant	GH	Fair	Excellent	Excellent	Excellent	Very good	Yes	155 (150)
NEMA type G-10 glass-fabric-base epoxy, general purpose	GE	Fair	Excellent	Excellent	Excellent	Very good	Yes	130 (125)
NEMA type G-11 glass-fabric-base epoxy, temperature resistant	GB	Poor	Excellent	Excellent	Excellent	Very good	Yes	155 (150)
Glass-fabric-base polytetrafluoroethylene	GT	—	Good	Excellent	Excellent	Excellent	—	(150)
Glass-fabric-base fluorinated ethylene propylene	FEP	—	Good	Excellent	Excellent	Excellent	—	(150)

*MIL-STD-275C rating shown in parentheses if different from industry rating.

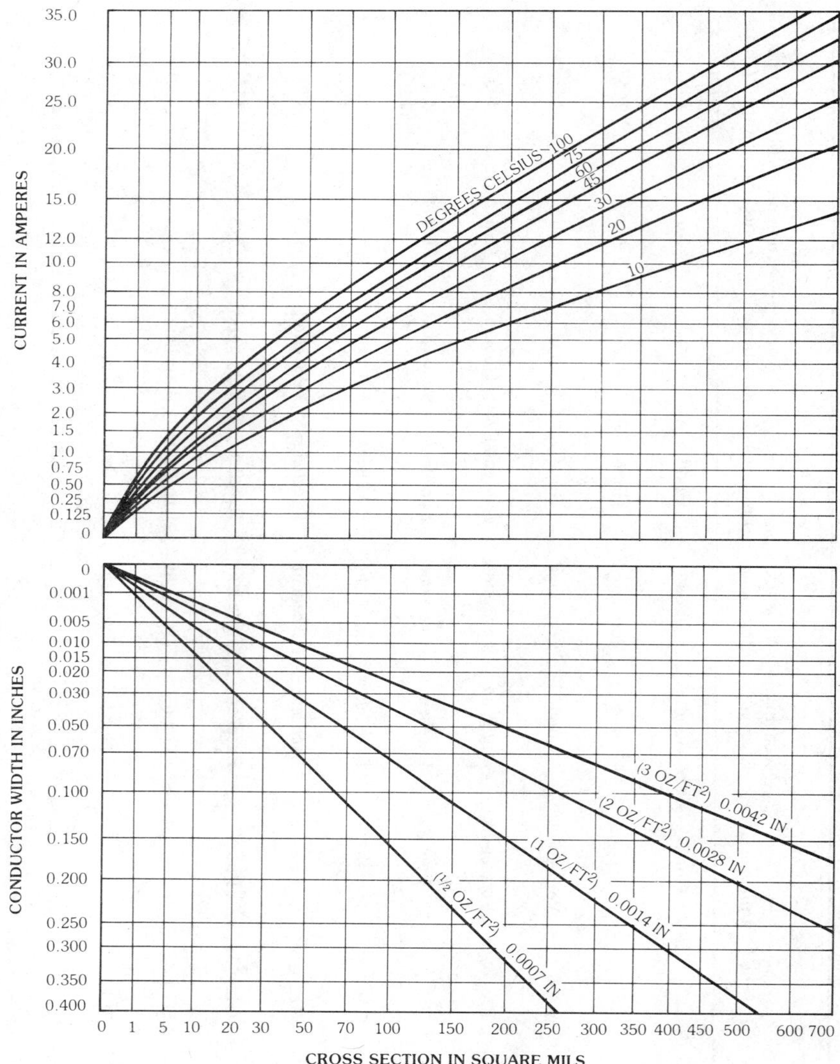

Fig. 20. Current-carrying capacity and sizes of etched copper conductors for various temperature rises above ambient. *From MIL-STD-275C, 9 January 1970.*

Punched-hole tolerance should not be less than ±0.005 inch on the diameters.

Hole sizes should not exceed by more than 0.020 inch the diameter of the wire to be inserted in the hole. With smaller holes, hand insertion of the wire is difficult. Machine insertion requires the larger allowance. Clinching of the lead is desirable if the clearance is larger.

Tolerances with respect to the true-grid location for terminal area centers and for locating edges of boards or other locating features (datums) should not exceed on the board: 0.014 inch diameter for conductor widths and spacings above 0.031 inch; 0.010 inch diameter for conductor widths and spacings 0.010 to 0.031 inch, inclusive. Tolerances on other dimensions (except conductor widths and spacings) may be larger. Closer tolerances may be needed if machine insertion is required.

Terminal area diameters should be at least (A) 0.020 inch larger than the diameter of the flange or projection of the flange on eyelets or standoff terminals, or the diameter of a plated-through hole, and (B) 0.040 inch larger than the diameter of an unsupported hole. Since the terminal area should be unbroken around the finished hole, the diameter should be further increased over the above minimum to allow for the permitted hole-position tolerance.

Conductor widths should be adequate for the current carried. See Fig. 20. For a given conductor-width and copper-thickness intersection, proceed vertically to the allowable temperature-rise line and then horizontally to the left to determine the permissible current. An additional 15% derating is recommended for board thicknesses of 1/32 inch or less, or for conductors thicker than 0.004 inch (3 oz). The normal ambient temperature surrounding the board plus the allowable temperature rise should not exceed the maximum safe operating temperature of the laminate. For ordinary work, copper conductor widths of 0.060 inch are convenient; with high-grade technique (extra cost), conductor widths as small as 0.010 inch can be readily produced.

Conductor spacing requirements are governed by the applied voltage, the maximum altitude, the conductor protective coating used, and the power-source size. The guide in Table 20 is suggested.

1. Hand application of opaque, permanent black ink
2. Pressure-sensitive tape
3. Hand-cut stencil made from self-adhesive opaque film
4. Preformed self-adhesive layout patterns

Artwork should be prepared to a scale that is two to five times oversize. Photographic reduction to final negative size should be possible, however, in one step.

Avoid the use of sharp corners when laying out the circuit. See Fig. 21.

The centers of holes to be manually drilled or punched in the circuit board should be indicated by a circle of 1/32-inch diameter (final size after reduction). See Fig. 22. This feature is not needed on each board if templates or numerically controlled

TABLE 20. CONDUCTOR SPACINGS*

Voltage Between Conductors	Minimum Spacing Between Conductors (Inches): Uncoated Boards, Sea Level to 10 000 Ft	Uncoated Boards, Over 10 000 Ft	Conformal Coated Boards, All Altitudes
0–30	0.025	0.025	0.010
0–50	0.025	0.025	—
0–150	0.025	—	—
31–50	0.025	0.025	0.015
51–100	0.025	0.060	0.020
101–170	—	0.125	0.030
101–300	—	—	0.030
151–300	0.050	—	0.030
171–250	0.050	0.250	0.030
251–500	—	0.500	—
301–500	0.100	0.500	0.060
Above 500	0.0002 per volt	0.001 per volt	0.00012 per volt

*From MIL-STD-275C

Preparation of Artwork

In preparing the master artwork for printed circuits, careful workmanship and accuracy are important. When circuits are reproduced by photographic means, much retouching time is saved if care is taken with the original artwork.

Artwork should be prepared on a dimensionally stable material. Tracing paper and bristol board are now outmoded, and specially treated (toothed) polyethylene terephthalate (Mylar, Cronar) base drafting films are used for most printed-circuit layouts. The layout pattern may be produced by one of the following methods:

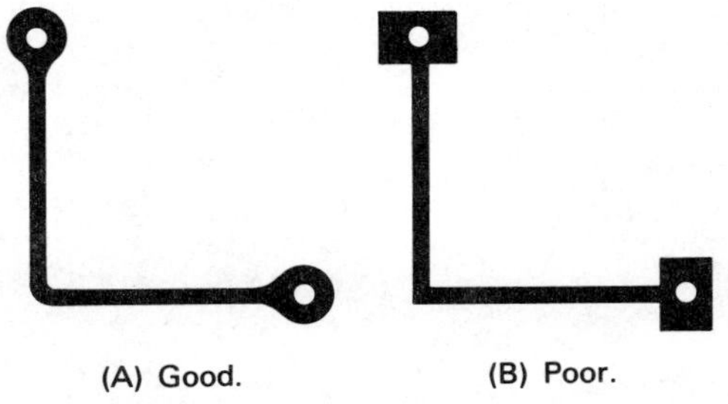

Fig. 21. Design of bends for printed-circuit conductors.

Fig. 22. Indication for hole.

machine tools are used for hole preparation; however, it is still a convenience for checking drawings, master artwork, and photographic negatives.

When drawing the second side of a printed-circuit board, corresponding centers should be taken directly from the back of the drawing of the first side.

In addition to the illustration of the circuit pattern, the trim line, registration marks, and two scale dimensions at right angles should be shown. Nomenclature, reference designations, operating instructions, and other information may also be added.

Assembly

All components should be inserted on one side of the board if practicable. In the case of boards with the circuit on one side only, the parts should be inserted on the side opposite the circuit. This allows all connections to be soldered simultaneously by dip-soldering.

Dip-soldering consists of applying a flux, usually a rosin-alcohol mixture, to the circuit pattern and then placing the board in contact with molten solder. Slight agitation of the board will ensure good fillets around the wire leads. In good present technique, the circuit board with its components assembled (on one side only) has its conductor pattern passed through the crest of a "wave" of molten solder; all junctions are soldered as the board progresses through the wave. The flux, board temperature, solder temperature, and immersion time are interrelated and must be adjusted for best results. Long exposure to hot solder is detrimental to the insulating material and to the adhesive that joins the copper foil to the insulation. For hand dipping, a 5-second dip in a 60/40 tin-lead solder bath maintained at a temperature of 450 degrees Fahrenheit will give satisfactory results.

After solder-dipping, the residual flux should be removed by a suitable solvent. Be sure the solvent is compatible with the materials used in the component parts mounted on the board; solvents frequently dissolve cements or plastics and marking inks, or cause severe stress cracking of plastics.

To secure the advantages of machine assembly:

(A) Components should be of similiar size and shape, or separate inserting heads will be required for each different shape of item.

(B) Components of the same size and shape must be mounted using the same terminal lead spacing at all points.

(C) Different values of a part, or even different parts of similiar shape and sizes (if axial-lead style) may be specially sequenced in a lead-taped package for insertion by one programmed insertion head.

(D) A few oddly sized or shaped components may be economically inserted by hand after the machine insertion work is completed.

Reference

An excellent reference on microelectronic printed-circuit techniques is Scarlett, J.A. *Printed Circuit Boards for Microelectronics.* New York: Van Nostrand Reinhold Co., 1970.

6 Fundamentals of Networks

INDUCTANCE OF SINGLE-LAYER SOLENOIDS

The approximate value of the low-frequency inductance of a single-layer solenoid is*

$$L = Fn^2d$$

where,

L = inductance in microhenries,
F = form factor, a function of the ratio d/l (value of F may be read from Fig. 1),
n = number of turns,
d = diameter of coil (inches) between centers of conductors,
l = length of coil (inches) = n times the distance between centers of adjacent turns.

The equation is based on the assumption of a uniform current sheet, but the correction due to the use of spaced round wires is usually negligible for practical purposes. For higher frequencies, skin effect alters the inductance slightly. This effect is not readily calculated, but is often negligibly small. However, it must be borne in mind that the equation gives approximately the true value of inductance. In contrast, the apparent value is affected by the shunting effect of the distributed capacitance of the coil.

Example: Required, a coil of 100 microhenries inductance, wound on a form 2 inches in diameter by 2 inches winding length. Then $d/l = 1.00$, and $F = 0.0173$ in Fig. 1.

$$n = (L/Fd)^{1/2}$$
$$= [100/(0.0173 \times 2)]^{1/2}$$
$$= 54 \text{ turns}$$

Reference to Table 1 will assist in choosing a desirable size of wire, allowing for a suitable spacing between turns according to the application of the coil. A slight correction may then be made for the increased diameter (diameter of form, plus two times radius of wire), if this small correction seems justified.

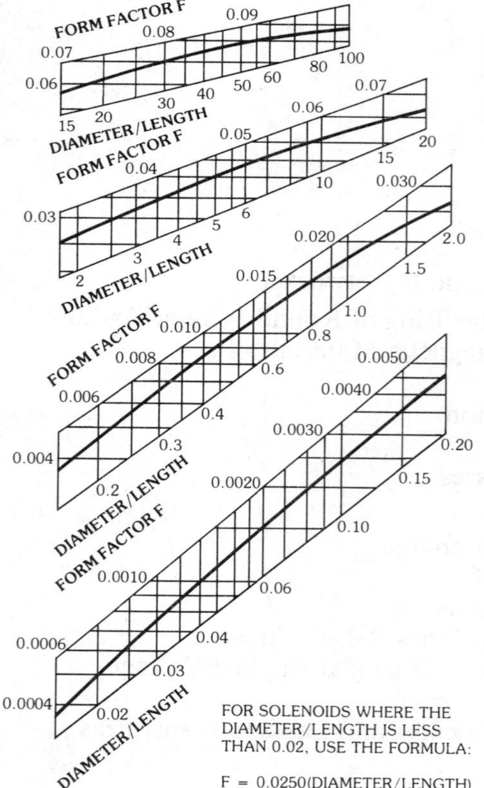

Fig. 1. Chart showing inductance of a single-layer solenoid, form factor = F.

Approximate Equation

For single-layer solenoids of the proportions normally used in radio work, the inductance in microhenries is given to an accuracy of about 1 percent by the formula

$$L = n^2[r^2/(9r + 10l)]$$

where $r = d/2$ and the other quantities are as defined for the previous inductance formula.

General Remarks

In the use of various charts, tables, and calculators for designing inductors, the following relationships are useful in extending the range of the devices. They apply to coils of any type or design.

(A) If all dimensions are held constant, inductance is proportional to n^2.

(B) If the proportions of the coil remain unchanged, then for a given number of turns the inductance is proportional to the dimensions of the coil. A coil with all dimensions m times those of a given coil (having the same number of turns) has m times the inductance of the given coil. That is, inductance has the dimensions of length.

Decrease of Solenoid Inductance by Shielding*

When a solenoid is enclosed in a cylindrical shield, the inductance is reduced by a factor given in Fig. 2. This effect has been evaluated by considering the shield to be a short-circuited single-turn secondary. The curves in Fig. 2 are reasonably

*Equations and Fig. 1 are derived from equations and tables in Bureau of Standards Circular No. C74.

*RCA Application Note No. 48; 12 June 1935.

TABLE 1. MAGNET-WIRE DATA

AWG B & S Gauge	Bare Nom Diam (in.)	Enam Nom Diam (in.)	SCC* Diam (in.)	DCC* Diam (in.)	SCE* Diam (in.)	SSC* Diam (in.)	DSC* Diam (in.)	SSE* Diam (in.)	Bare Min Diam (in.)	Bare Max Diam (in.)	Enameled Min Diam (in.)	Enameled Diam* (in.)
10	0.1019	0.1039	0.1079	0.1129	0.1104	—	—	—	0.1009	0.1029	0.1024	0.1044
11	0.0907	0.0927	0.0957	0.1002	0.0982	—	—	—	0.0898	0.0917	0.0913	0.0932
12	0.0808	0.0827	0.0858	0.0903	0.0882	—	—	—	0.0800	0.0816	0.0814	0.0832
13	0.0720	0.0738	0.0770	0.0815	0.0793	—	—	—	0.0712	0.0727	0.0726	0.0743
14	0.0641	0.0659	0.0691	0.0736	0.0714	—	—	—	0.0634	0.0647	0.0648	0.0664
15	0.0571	0.0588	0.0621	0.0666	0.0643	0.0591	0.0611	0.0613	0.0565	0.0576	0.0578	0.0593
16	0.0508	0.0524	0.0558	0.0603	0.0579	0.0528	0.0548	0.0549	0.0503	0.0513	0.0515	0.0529
17	0.0453	0.0469	0.0503	0.0548	0.0523	0.0473	0.0493	0.0493	0.0448	0.0457	0.0460	0.0473
18	0.0403	0.0418	0.0453	0.0498	0.0472	0.0423	0.0443	0.0442	0.0399	0.0407	0.0410	0.0422
19	0.0359	0.0374	0.0409	0.0454	0.0428	0.0379	0.0399	0.0398	0.0355	0.0363	0.0366	0.0378
20	0.0320	0.0334	0.0370	0.0415	0.0388	0.0340	0.0360	0.0358	0.0316	0.0323	0.0326	0.0338
21	0.0285	0.0299	0.0335	0.0380	0.0353	0.0305	0.0325	0.0323	0.0282	0.0287	0.0292	0.0303
22	0.0253	0.0266	0.0303	0.0343	0.0320	0.0273	0.0293	0.0290	0.0251	0.0256	0.0261	0.0270
23	0.0226	0.0238	0.0276	0.0316	0.0292	0.0246	0.0266	0.0262	0.0223	0.0228	0.0232	0.0242
24	0.0201	0.0213	0.0251	0.0291	0.0266	0.0221	0.0241	0.0236	0.0199	0.0203	0.0208	0.0216
25	0.0179	0.0190	0.0224	0.0264	0.0238	0.0199	0.0219	0.0213	0.0177	0.0181	0.0186	0.0193
26	0.0159	0.0169	0.0204	0.0244	0.0217	0.0179	0.0199	0.0192	0.0158	0.0161	0.0166	0.0172
27	0.0142	0.0152	0.0187	0.0227	0.0200	0.0162	0.0182	0.0175	0.0141	0.0144	0.0149	0.0155
28	0.0126	0.0135	0.0171	0.0211	0.0183	0.0146	0.0166	0.0158	0.0125	0.0128	0.0132	0.0138
29	0.0113	0.0122	0.0158	0.0198	0.0170	0.0133	0.0153	0.0145	0.0112	0.0114	0.0119	0.0125
30	0.0100	0.0108	0.0145	0.0185	0.0156	0.0120	0.0140	0.0131	0.0099	0.0101	0.0105	0.0111
31	0.0089	0.0097	0.0134	0.0174	0.0144	0.0109	0.0129	0.0119	0.0088	0.0090	0.0094	0.0099
32	0.0080	0.0088	0.0125	0.0165	0.0135	0.0100	0.0120	0.0110	0.0079	0.0081	0.0085	0.0090
33	0.0071	0.0078	0.0116	0.0156	0.0125	0.0091	0.0111	0.0100	0.0070	0.0072	0.0075	0.0080
34	0.0063	0.0069	0.0108	0.0148	0.0116	0.0083	0.0103	0.0091	0.0062	0.0064	0.0067	0.0071
35	0.0056	0.0061	0.0101	0.0141	0.0108	0.0076	0.0096	0.0083	0.0055	0.0057	0.0059	0.0063
36	0.0050	0.0055	0.0090	0.0130	0.0097	0.0070	0.0090	0.0077	0.0049	0.0051	0.0053	0.0057
37	0.0045	0.0049	0.0085	0.0125	0.0091	0.0065	0.0085	0.0071	0.0044	0.0046	0.0047	0.0051
38	0.0040	0.0044	0.0080	0.0120	0.0086	0.0060	0.0080	0.0066	0.0039	0.0041	0.0042	0.0046
39	0.0035	0.0038	0.0075	0.0115	0.0080	0.0055	0.0075	0.0060	0.0034	0.0036	0.0036	0.0040
40	0.0031	0.0034	0.0071	0.0111	0.0076	0.0051	0.0071	0.0056	0.0030	0.0032	0.0032	0.0036
41	0.0028	0.0031	—	—	—	—	—	—	0.0027	0.0029	0.0029	0.0032
42	0.0025	0.0028	—	—	—	—	—	—	0.0024	0.0026	0.0026	0.0029
43	0.0022	0.0025	—	—	—	—	—	—	0.0021	0.0023	0.0023	0.0026
44	0.0020	0.0023	—	—	—	—	—	—	0.0019	0.0021	0.0021	0.0024

* Nominal bare diameter plus maximum additions.
For additional data on copper wire, see Chapters 4 and 13.

accurate provided the clearance between each end of the coil winding and the corresponding end of the shield is at least equal to the radius of the coil. For square shield cans, take the equivalent shield diameter (for Fig. 2) as being 1.2 times the width of one side of the square.

Example: Let the coil winding length be 1.5 inches and its diameter 0.75 inch, while the shield diameter is 1.25 inches. What is the reduction of inductance due to the shield? The proportions are

$$(\text{winding length})/(\text{winding diameter}) = 2.0$$

$$(\text{winding diameter})/(\text{shield diameter}) = 0.6$$

Referring to Fig. 2, the actual inductance in the shield is 72 percent of the inductance of the coil in free space.

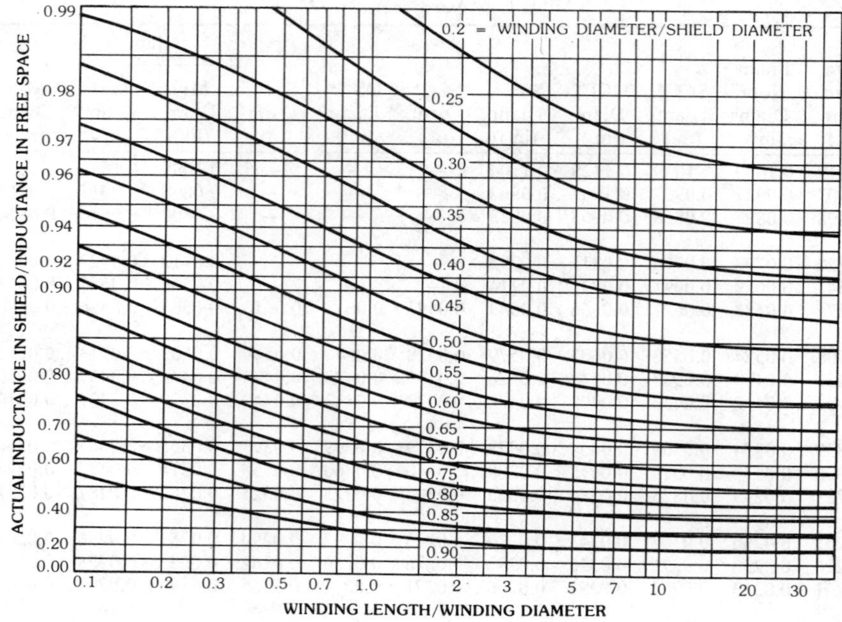

Fig. 2. Inductance decrease when a solenoid is shielded. *By permission of RCA, copyright proprietor.*

Q of Unshielded Solenoid

Fig. 3 can be used to obtain the unloaded Q of an unshielded solenoid.

REACTANCE CHARTS

Figs. 4, 5, and 6 give the relationships of capacitance, inductance, reactance, and frequency. Any one value may be determined in terms of two others by use of a straightedge laid across the correct chart for the frequency under consideration.

Example: Given a capacitance of 0.001 microfarad, find the reactance at 50 kilohertz and inductance required to resonate. Place a straightedge through these values and read the intersections on the other scales, giving approximately 3200 ohms and 11 millihenries. See Fig. 5.

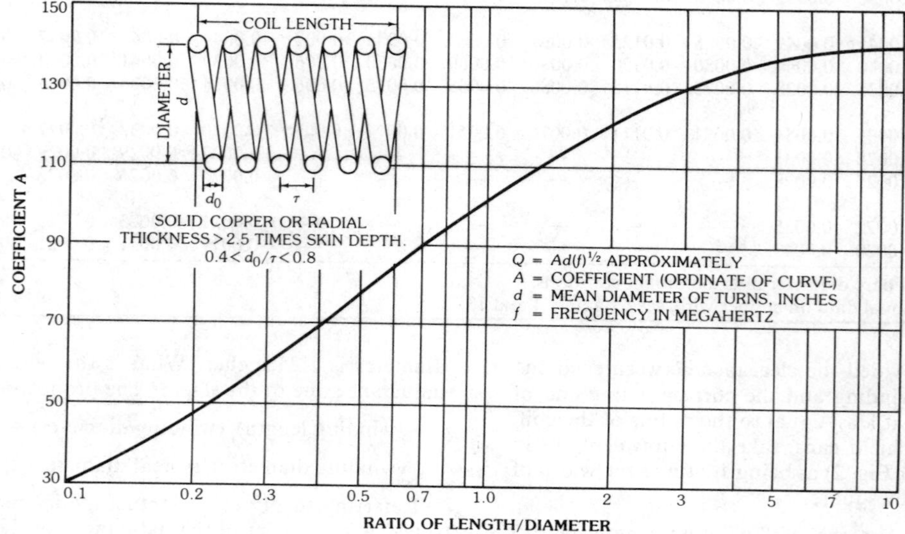

Fig. 3. Q of unshielded coil.

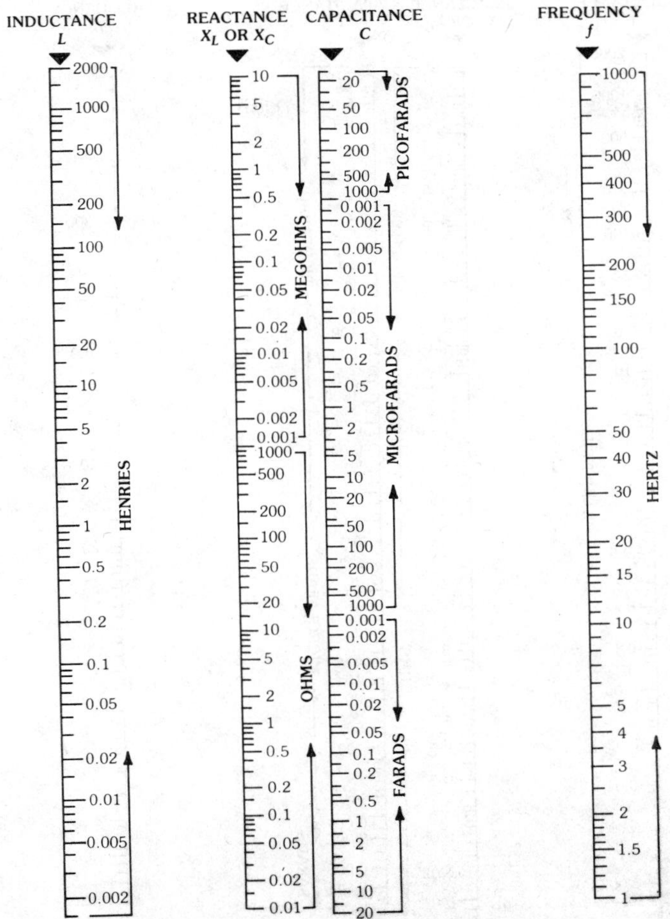

Fig. 4. Chart covering 1 hertz to 1000 hertz.

IMPEDANCE FORMULAS

Impedance and admittance formulas for various combinations of L, C, and R are given in Tables 2 and 3.

SKIN EFFECT

Symbols

A = correction coefficient
D = diameter of conductor in inches
f = frequency in hertz
R_{ac} = resistance at frequency f
R_{dc} = direct-current resistance
R_{sq} = resistance per square
T = thickness of tubular conductor in inches
T_1 = depth of penetration of current
δ = skin depth
λ = free-space wavelength in meters
μ_r = relative permeability of conductor material ($\mu_r = 1$ for copper and other nonmagnetic materials)
ρ = resistivity of conductor material at any temperature
ρ_c = resistivity of copper at 20°C = 1.724 microhms-centimeter

Skin Depth

The skin depth is that distance below the surface of a conductor where the current density has diminished to $1/e$ of its value at the surface. The thickness of the conductor is assumed to be several (perhaps at least three) times the skin depth. Imagine the conductor replaced by a cylindrical shell of the same surface shape but of thickness equal to the skin depth, with uniform current density equal to that which exists at the surface of the actual conductor. Then the total current in the shell and its resistance are equal to the corresponding values in the actual conductor.

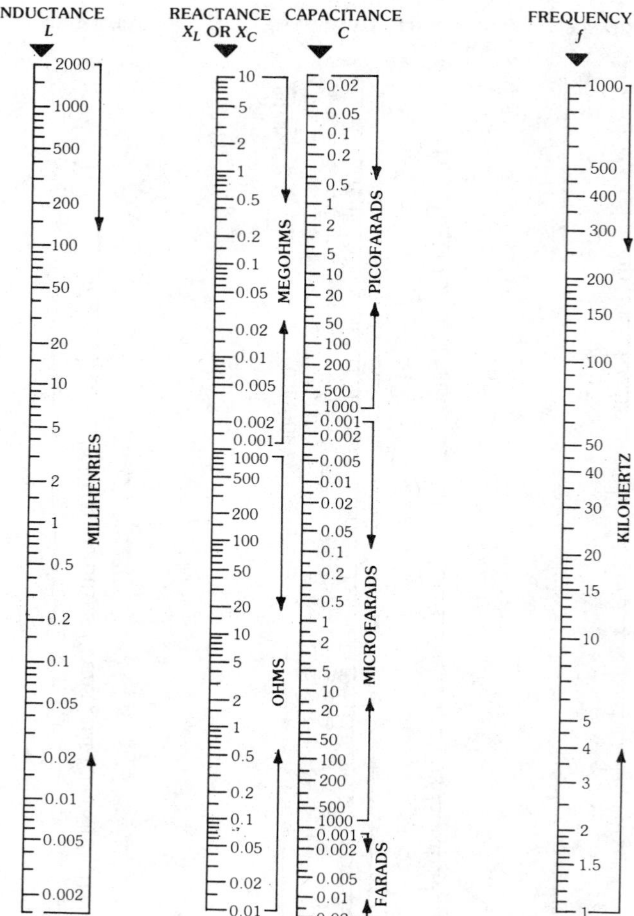

Fig. 5. Chart covering 1 kilohertz to 1000 kilohertz.

The skin depth and the resistance per square (of any size), in meter-kilogram-second (rationalized) units, are

$$\delta = (\lambda/\pi\sigma\mu c)^{1/2}$$

$$R_{sq} = 1/\delta\sigma$$

where,

δ = skin depth in meters,

R_{sq} = resistance per square in ohms,

c = velocity of light *in vacuo*

$= 2.998 \times 10^8$ meter/second,

$\mu = 4\pi \times 10^{-7}\ \mu_r$ henry/meter,

$1/\sigma = 1.724 \times 10^{-8}\ \rho/\rho_c$ ohm-meter.

For numerical computations

$$\delta = (3.82 \times 10^{-4}\lambda^{1/2})k_1$$

$$= (6.61/f^{1/2})k_1 \text{ centimeter}$$

$$\delta = (1.50 \times 10^{-4}\lambda^{1/2})k_1$$

$$= (2.60/f^{1/2})k_1 \text{ inch}$$

$$\delta_m = (2.60/f_{mc}{}^{1/2})k_1 \text{ mil}$$

$$R_{sq} = (4.52 \times 10^{-3}/\lambda^{1/2})k_2$$

$$= (2.61 \times 10^{-7}f^{1/2})k_2 \text{ ohm}$$

$$k_1 = [(1/\mu_r)\rho/\rho_c]^{1/2}$$

$$k_2 = (\mu_r\rho/\rho_c)^{1/2}$$

k_1, k_2 = unity for copper

Example: What is the resistance/foot of a cylindrical copper conductor of diameter D inches?

$$R = (12/\pi D)R_{sq}$$

$$= (12/\pi D) \times 2.61 \times 10^{-7}(f^{1/2})$$

$$= 0.996 \times 10^{-6}(f^{1/2})/D \text{ ohm/foot}$$

If $D = 1.00$ inch and $f = 100 \times 10^6$ hertz, then $R = 0.996 \times 10^{-6} \times 10^4 \approx 1 \times 10^{-2}$ ohm/foot.

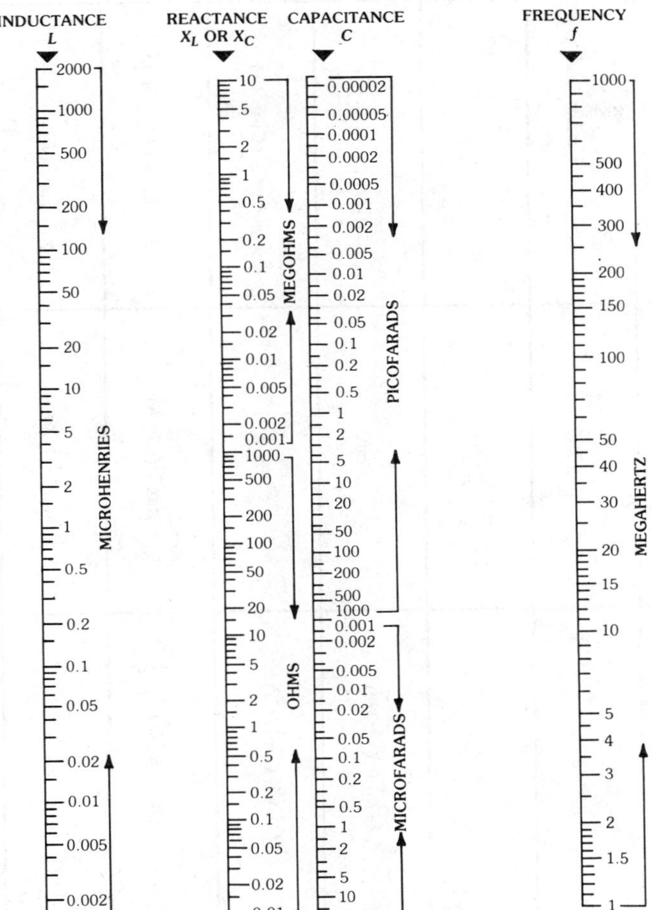

Fig. 6. Chart covering 1 megahertz to 1000 megahertz.

General Considerations

Fig. 7 shows the relationship of R_{ac}/R_{dc} versus $D(f^{1/2})$ for copper, or versus $D(f^{1/2})(\mu_r\rho_c/\rho)^{1/2}$ for any conductor material, for an isolated straight solid conductor of circular cross section. Negligible error in the equations for R_{ac} results when the conductor is spaced at least $10D$ from adjacent conductors. When the spacing between axes of parallel conductors carrying the same current is $4D$, the resistance R_{ac} is increased about 3 percent, when the depth of penetration is small. The equations are accurate for concentric lines due to their circular symmetry.

For values of $D(f^{1/2})(\mu_r\rho_c/\rho)^{1/2}$ greater than 40

$$R_{ac}/R_{dc} = 0.0960D(f^{1/2})(\mu_r\rho_c/\rho)^{1/2} + 0.26 \quad (1)$$

The high-frequency resistance in ohms/foot of an isolated straight conductor (either solid or tubular for $T < D/8$ or $T_1 < D/8$) is given in Eq. (2). If the current is along the inside surface of a tubular conductor, D is the inside diameter.

$$R_{ac} = A[(f^{1/2})/D][\mu_r(\rho/\rho_c)]^{1/2} \times 10^{-6} \quad (2)$$

The values of the correction coefficient A for solid conductors and for tubular conductors are given in Table 4.

The value of $T(f^{1/2})(\mu_r\rho_c/\rho)^{1/2}$ that just makes $A = 1$ indicates the penetration of the currents below the surface of the conductor. Thus, approximately

$$T_1 = [3.5/(f^{1/2})](\rho/\mu_r\rho_c)^{1/2} \quad (3)$$

where T_1 is in inches.

When $T_1 < D/8$, the value of R_{ac} as given by Eq. (2) (but not the value of R_{ac}/R_{dc} in Table 4, "Tubular Conductors") is correct for any value $T \geq T_1$.

Under the limitation that the radius of curvature of all parts of the cross section is appreciably greater than T_1, Eqs. (2) and (3) hold for isolated

TABLE 2. IMPEDANCE AND ADMITTANCE FORMULAS—SERIES AND PARALLEL COMBINATIONS

	Impedance $Z = R + jX$	Magnitude $\lvert Z \rvert = (R^2 + X^2)^{1/2}$	Phase Angle $\phi = \tan^{-1}(X/R)$	Admittance $Y = 1/Z$
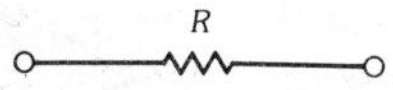 R	R	R	0	$1/R$
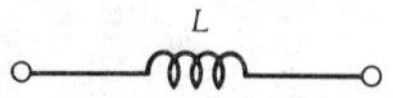 L	$j\omega L$	ωL	$+\pi/2$	$-j/\omega L$
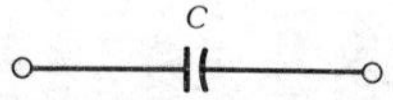 C	$-j/\omega C$	$1/\omega C$	$-\pi/2$	$j\omega C$
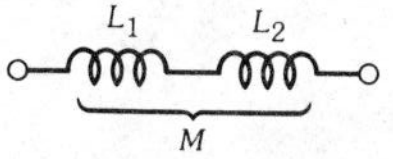 L_1 L_2 M	$j\omega\,(L_1 + L_2 \pm 2M)$	$\omega(L_1 + L_2 \pm 2M)$	$+\pi/2$	$-j/\omega\,(L_1 + L_2 \pm 2M)$
 C_1 C_2	$-(j/\omega)(1/C_1 + 1/C_2)$	$(1/\omega)(1/C_1 + 1/C_2)$	$-\pi/2$	$j\omega C_1 C_2/(C_1 + C_2)$
 R L	$R + j\omega L$	$(R^2 + \omega^2 L^2)^{1/2}$	$\tan^{-1}(\omega L/R)$	$(R - j\omega L)/(R^2 + \omega^2 L^2)$
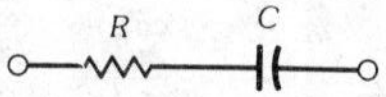 R C	$R - j/\omega C$	$(1/\omega C)(1 + \omega^2 C^2 R^2)^{1/2}$	$-\tan^{-1}(1/\omega CR)$	$(R + j/\omega C)/(R^2 + 1/\omega^2 C^2)$
 L C	$j(\omega L - 1/\omega C)$	$(\omega L - 1/\omega C)$	$\pm\pi/2$	$j\omega C/(1 - \omega^2 LC)$
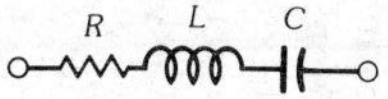 R L C	$R + j(\omega L - 1/\omega C)$	$[R^2 + (\omega L - 1/\omega C)^2]^{1/2}$	$\tan^{-1}[(\omega L - 1/\omega C)/R]$	$\dfrac{R - j(\omega L - 1/\omega C)}{R^2 + (\omega L - 1/\omega C)^2}$

TABLE 2 (CONT). IMPEDANCE AND ADMITTANCE FORMULAS—SERIES AND PARALLEL COMBINATIONS

Circuit	Impedance $Z = R + jX$	Magnitude $\lvert Z\rvert = (R^2 + X^2)^{1/2}$	Phase Angle $\phi = \tan^{-1}(X/R)$	Admittance $Y = 1/Z$
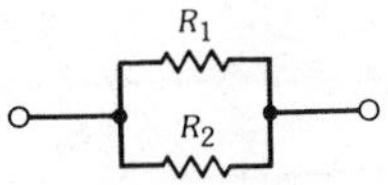 R_1 R_2	$R_1R_2/(R_1 + R_2)$	$R_1R_2/(R_1 + R_2)$	0	$(1/R_1 + 1/R_2)$
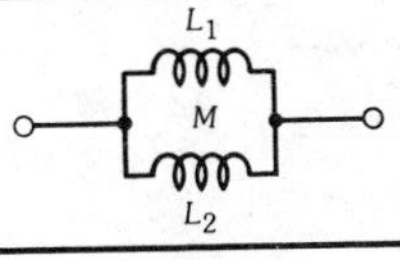 L_1 M L_2	$j\omega\left(\dfrac{L_1L_2 - M^2}{L_1 + L_2 \mp 2M}\right)$	$\omega\left(\dfrac{L_1L_2 - M^2}{L_1 + L_2 \mp 2M}\right)$	$+\pi/2$	$-(j/\omega)\left(\dfrac{L_1+L_2\mp 2M}{L_1L_2 - M^2}\right)$
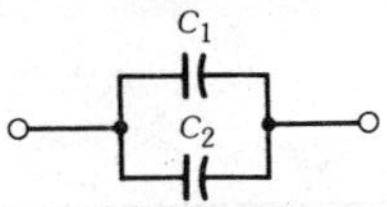 C_1 C_2	$-j/\omega(C_1 + C_2)$	$1/\omega(C_1 + C_2)$	$-\pi/2$	$j\omega(C_1 + C_2)$
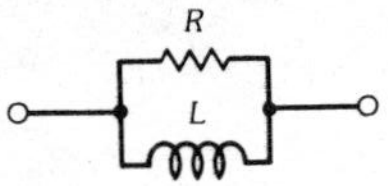 R L	$\omega LR\left(\dfrac{\omega L + jR}{R^2 + \omega^2L^2}\right)$	$\omega LR/(R^2 + \omega^2L^2)^{1/2}$	$\tan^{-1}(R/\omega L)$	$1/R - j/\omega L$
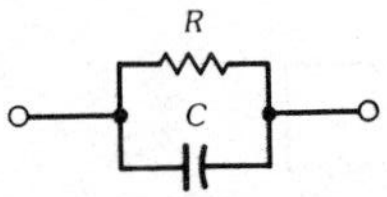 R C	$R(1 - j\omega CR)/(1 + \omega^2C^2R^2)$	$R/(1 + \omega^2C^2R^2)^{1/2}$	$-\tan^{-1}\omega CR$	$1/R + j\omega C$
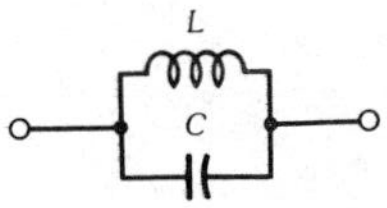 L C	$j\omega L/(1 - \omega^2 LC)$	$\omega L/(1 - \omega^2LC)$	$\pm\pi/2$	$j(\omega C - 1/\omega L)$
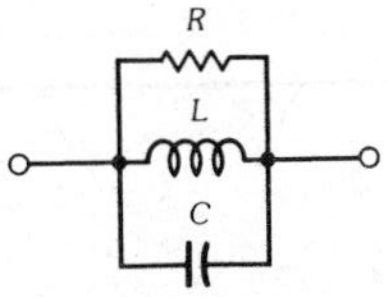 R L C	$\dfrac{1/R - j(\omega C - 1/\omega L)}{(1/R)^2 + (\omega C - 1/\omega L)^2}$	$[(1/R)^2 + (\omega C - 1/\omega L)^2]^{-1/2}$	$\tan^{-1} R(1/\omega L - \omega C)$	$1/R + j(\omega C - 1/\omega L)$

TABLE 3. IMPEDANCE AND ADMITTANCE FORMULAS—SERIES-PARALLEL COMBINATIONS

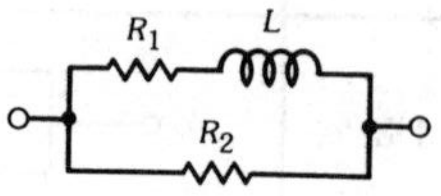	Impedance Z	$R_2\dfrac{R_1(R_1+R_2)+\omega^2L^2+j\omega LR_2}{(R_1+R_2)^2+\omega^2L^2}$
	Magnitude $\lvert Z\rvert$	$R_2\left[\dfrac{R_1{}^2+\omega^2L^2}{(R_1+R_2)^2+\omega^2L^2}\right]^{1/2}$
	Phase Angle ϕ	$\tan^{-1}\dfrac{\omega LR_2}{R_1(R_1+R_2)+\omega^2L^2}$
	Admittance Y	$\dfrac{R_1(R_1+R_2)+\omega^2L^2-j\omega LR_2}{R_2(R_1{}^2+\omega^2L^2)}$
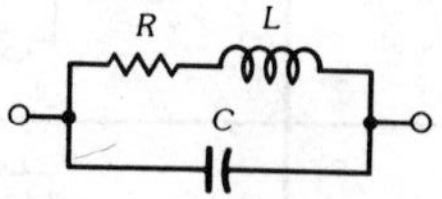	Impedance Z	$\dfrac{R+j\omega[L(1-\omega^2LC)-CR^2]}{(1-\omega^2LC)^2+\omega^2C^2R^2}$
	Magnitude $\lvert Z\rvert$	$\left[\dfrac{R^2+\omega^2L^2}{(1-\omega^2LC)^2+\omega^2C^2R^2}\right]^{1/2}$
	Phase Angle ϕ	$\tan^{-1}\left\{\omega[L(1-\omega^2LC)-CR^2]/R\right\}$
	Admittance Y	$\dfrac{R-j\omega[L(1-\omega^2LC)-CR^2]}{R^2+\omega^2L^2}$

TABLE 3 (CONT). IMPEDANCE AND ADMITTANCE FORMULAS—SERIES-PARALLEL COMBINATIONS

Circuit	Quantity	Formula
	Impedance Z	$X_1\dfrac{X_1R_2 + j[R_2^2 + X_2(X_1 + X_2)]}{R_2^2 + (X_1 + X_2)^2}$
	Magnitude $\|Z\|$	$X_1\left[\dfrac{R_2^2 + X_2^2}{R_2^2 + (X_1 + X_2)^2}\right]^{1/2}$
	Phase Angle ϕ	$\tan^{-1}\dfrac{R_2^2 + X_2(X_1 + X_2)}{X_1R_2}$
	Admittance Y	$\dfrac{R_2X_1 - j(R_2^2 + X_2^2 + X_1X_2)}{X_1(R_2^2 + X_2^2)}$
	Impedance Z	$\dfrac{R_1R_2(R_1 + R_2) + \omega^2L^2R_2 + R_1/\omega^2C^2}{(R_1 + R_2)^2 + (\omega L - 1/\omega C)^2} + j\dfrac{\omega LR_2^2 - R_1^2/\omega C - (L/C)(\omega L - 1/\omega C)}{(R_1 + R_2)^2 + (\omega L - 1/\omega C)^2}$
	Magnitude $\|Z\|$	$\left[\dfrac{(R_1^2 + \omega^2L^2)(R_2^2 + 1/\omega^2C^2)}{(R_1 + R_2)^2 + (\omega L - 1/\omega C)^2}\right]^{1/2}$
	Phase Angle ϕ	$\tan^{-1}\left[\dfrac{\omega LR_2^2 - R_1^2/\omega C - (L/C)(\omega L - 1/\omega C)}{R_1R_2(R_1 + R_2) + \omega^2L^2R_2 + R_1/\omega^2C^2}\right]$
Note: When $R_1 = R_2 = \sqrt{L/C}$, then $Z = R_1 = R_2$, a pure resistance at any frequency where the given conditions hold.	Admittance Y	$\dfrac{R_1 + \omega^2C^2R_1R_2(R_1 + R_2) + \omega^4L^2C^2R_2}{(R_1^2 + \omega^2L^2)(1 + \omega^2C^2R_2^2)} + j\omega\left[\dfrac{CR_1^2 - L + \omega^2LC(L - CR_2^2)}{(R_1^2 + \omega^2L^2)(1 + \omega^2C^2R_2^2)}\right]$

Continued on next page.

TABLE 3 (CONT). IMPEDANCE AND ADMITTANCE FORMULAS—SERIES-PARALLEL COMBINATIONS

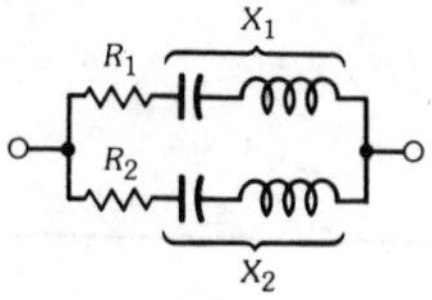	Impedance Z	$\frac{(R_1R_2 - X_1X_2) + j(R_1X_2 + R_2X_1)}{(R_1 + R_2) + j(X_1 + X_2)}$
	Magnitude $\vert Z\vert$	$\left[\frac{(R_1^2 + X_1^2)(R_2^2 + X_2^2)}{(R_1 + R_2)^2 + (X_1 + X_2)^2}\right]^{1/2}$
	Phase Angle ϕ	$\tan^{-1}(X_1/R_1) + \tan^{-1}(X_2/R_2) - \tan^{-1}[(X_1 + X_2)/(R_1 + R_2)]$
	Admittance Y	$1/(R_1 + jX_1) + 1/(R_2 + jX_2)$

$Z = R + jX$ $\qquad |Z| = (R^2 + X^2)^{1/2}$ $\qquad \phi = \tan^{-1}(X/R)$ $\qquad Y = 1/Z$

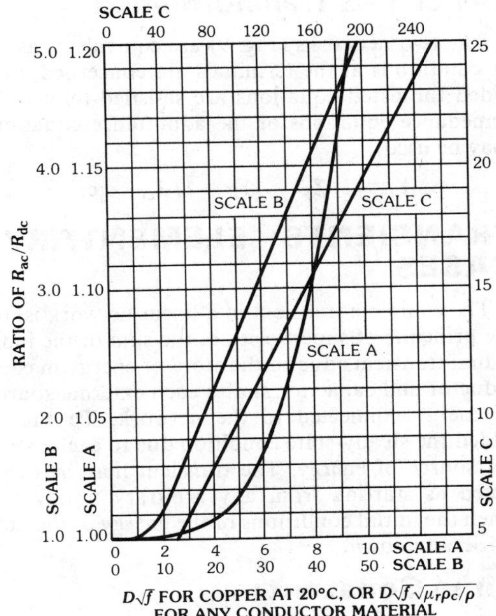

Fig. 7. Resistance ratio for isolated straight solid conductors of circular cross section.

straight conductors of any shape. In this case the term D = (perimeter of cross section)/π.

Examples:

(A) At 100 megahertz, a copper conductor has a depth of penetration $T_1 = 0.00035$ inch.

(B) A steel shield with 0.005-inch copper plate, which is practically equivalent in R_{ac} to an isolated copper conductor 0.005-inch thick, has a value of $A = 1.23$ at 200 kilohertz. This 23-percent increase in resistance over that of a thick copper sheet is satisfactorily low as regards its effect on the losses of the components within the shield. By comparison, a thick aluminum sheet has a resistance $(\rho/\rho_c)^{1/2} = 1.28$ times that of copper.

EQUATIONS FOR SIMPLE R, L, AND C NETWORKS*

Self-Inductance of Circular Ring of Round Wire at Radio Frequencies, for Nonmagnetic Materials

$$L = (a/100) \times [7.353 \log_{10}(16a/d) - 6.386]$$

*Many equations for computing capacitance, inductance, and mutual inductance will be found in Bureau of Standards Circular No. C74, obtainable from the Superintendent of Documents, Government Printing Office, Washington, D.C. 20402.

TABLE 4. SKIN-EFFECT CORRECTION COEFFICIENT A FOR SOLID AND TUBULAR CONDUCTORS

Solid Conductors

$D(f^{1/2})[\mu_r(\rho_c/\rho)]^{1/2}$	A
> 370	1.000
220	1.005
160	1.010
98	1.02
48	1.05
26	1.10
13	1.20
9.6	1.30
5.3	2.00
< 3.0	$R_{ac} \approx R_{dc}$

$$R_{dc} = (10.37/D^2)(\rho/\rho_c) \times 10^{-6} \text{ ohm/foot}$$

Tubular Conductors

$T(f^{1/2})[\mu_r(\rho_c/\rho)]^{1/2}$	A	R_{ac}/R_{dc}
$= B$ where $B > 3.5$	1.00	$0.384B$
3.5	1.00	1.35
3.15	1.01	1.23
2.85	1.05	1.15
2.60	1.10	1.10
2.29	1.20	1.06
2.08	1.30	1.04
1.77	1.50	1.02
1.31	2.00	1.00
$= B$ where $B < 1.3$	$2.60/B$	1.00

where,

L = self-inductance in microhenries,
a = mean radius of ring in inches,
d = diameter of wire in inches,
$a/d > 2.5$.

Capacitance

For Parallel-Plate Capacitor:

$$C = 0.0885\epsilon_r[(N-1)A]/t$$
$$= 0.225\epsilon_r[(N-1)A''/t'']$$

where,

C = capacitance in picofarads,
A = area of one side of one plate in square centimeters,
A'' = area in square inches,
N = number of plates,
t = thickness of dielectric in centimeters,
t'' = thickness in inches,
ϵ_r = dielectric constant relative to air.

This equation neglects "fringing" at the edges of the plates.

For Coaxial Cylindrical Capacitor (Fig. 8):

$$C = 2\pi\epsilon_r\epsilon_v/[\log_e(b/a)]$$
$$= \{(5 \times 10^6\epsilon_r)/[c^2\log_e(b/a)]\}$$

where,

C = capacitance per unit axial length in farads/meter,
c = velocity of light in vacuo, meters per second
$= 2.998 \times 10^8$,
ϵ_r = dielectric constant relative to air,
ϵ_v = permittivity of free space in farads/meter
$= 8.85 \times 10^{-12}$

Alternate forms of the equation are:

$C = 0.2416\epsilon_r/[\log_{10}(b/a)]$ picofarads/centimeter
$= 0.614\epsilon_r/[\log_{10}(b/a)]$ picofarads/inch
$= 7.36\epsilon_r/[\log_{10}(b/a)]$ picofarads/foot

When $1.0 < (b/a) < 1.4$, then with accuracy of 1 percent or better, the capacitance in picofarads/foot is:

$$C = 8.50\epsilon_r \frac{(b/a)+1}{(b/a)-1}$$

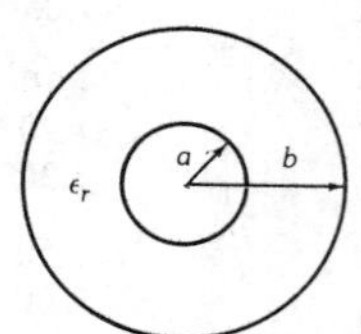

Fig. 8. Coaxial cylindrical capacitor.

T-π or Y-Δ Transformation

The two networks (Fig. 9) are equivalent, as far as conditions at the terminals are concerned, provided the listed equations are satisfied (either the impedance equations or the admittance equations may be used)

$$Y_1 = 1/Z_1 \qquad Y_c = 1/Z_c, \quad \text{etc.}$$

TRANSIENTS—ELEMENTARY CASES

The complete transient in a linear network is, by the principle of superposition, the sum of the individual transients due to the store of energy in each inductor and capacitor and to each external source of energy connected to the network. To this is added the steady-state condition due to each external source of energy. The transient may be computed as starting from any arbitrary time $t = 0$ when the initial conditions of the energy of the network are known.

Time Constant

The time constant (designated T) of the discharge of a capacitor through a resistor is the time $t_2 - t_1$ required for the voltage or current to decay to $1/\epsilon$ of its value at time t_1. For the charge of a capacitor, the same definition applies, the voltage "decaying" toward its steady-state value. The time constant of discharge or charge of the current in an inductor through a resistor follows an analogous definition.

Stored energy, in joules (watt-seconds), is:

Energy stored in a capacitor $= CE^2/2$

Energy stored in an inductor $= LI^2/2$

In the equations in this section, T and t are in

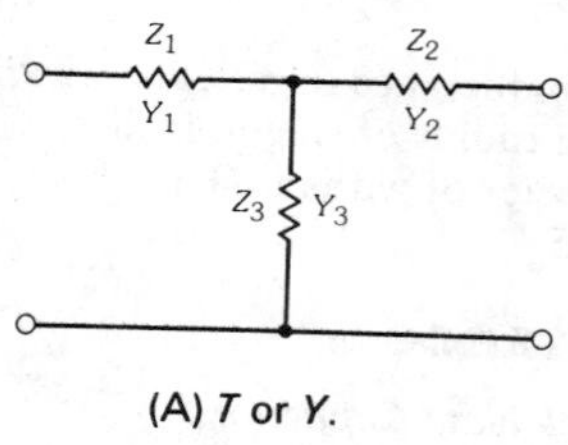

(A) T or Y.

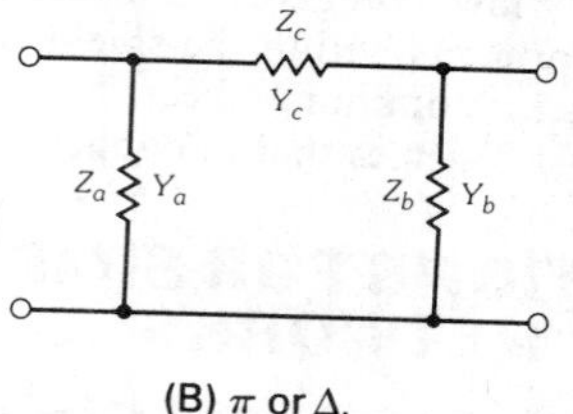

(B) π or Δ.

Impedance Equations:

$$Z_c = (Z_1Z_2+Z_1Z_3+Z_2Z_3)/Z_3$$
$$Z_a = (Z_1Z_2+Z_1Z_3+Z_2Z_3)/Z_2$$
$$Z_b = (Z_1Z_2+Z_1Z_3+Z_2Z_3)/Z_1$$

$$Z_1 = Z_aZ_c/(Z_a+Z_b+Z_c)$$
$$Z_2 = Z_bZ_c/(Z_a+Z_b+Z_c)$$
$$Z_3 = Z_aZ_b/(Z_a+Z_b+Z_c)$$

Admittance Equations:

$$Y_c = Y_1Y_2/(Y_1+Y_2+Y_3)$$
$$Y_a = Y_1Y_3/(Y_1+Y_2+Y_3)$$
$$Y_b = Y_2Y_3/(Y_1+Y_2+Y_3)$$

$$Y_1 = (Y_aY_b+Y_aY_c+Y_bY_c)/Y_b$$
$$Y_2 = (Y_aY_b+Y_aY_c+Y_bY_c)/Y_a$$
$$Y_3 = (Y_aY_b+Y_aY_c+Y_bY_c)/Y_c$$

(C) Equations.

Fig. 9. T or Y network and π or Δ network.

seconds, R is in ohms, L is in henries, C is in farads, E is in volts, and I is in amperes. Values of the numerical constants are:

$\epsilon = 2.718$ $1/\epsilon = 0.3679$ $\log_{10}\epsilon = 0.4343$

Capacitor Charge and Discharge

Closing of switch (Fig. 10) occurs at time $t = 0$.
Initial conditions (at $t=0$): Battery $= E_b$; $e_c = E_0$.
Steady state (at $t = \infty$): $i = 0$; $e_c = E_b$.
Transient:

$$i = [(E_b - E_0)/R]\exp(-t/RC) = I_0 \exp(-t/RC)$$

$$\log_{10}(i/I_0) = -(0.4343/RC)t$$

$$e_c = E_0 + C^{-1}\int_0^t i\,dt$$

$$= E_0 \exp(-t/RC) + E_b[1 - \exp(-t/RC)]$$

Time constant:

$$T = RC$$

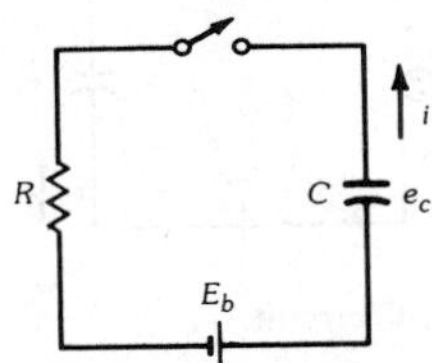

Fig. 10. Circuit for capacitor charge and discharge.

Fig. 11 shows current:
$$i/I_0 = \exp(-t/T)$$

Fig. 11 shows discharge (for $E_b = 0$):
$$e_c/E_0 = \exp(-t/T)$$

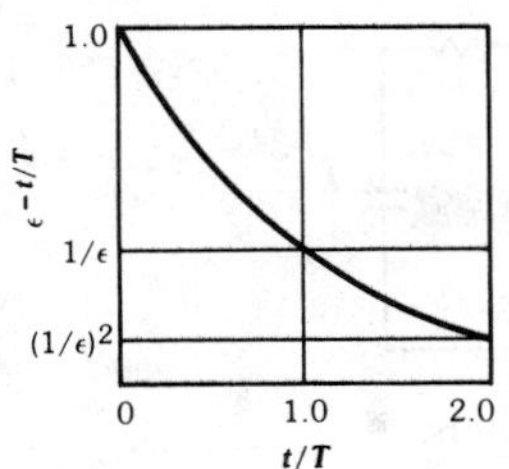

Fig. 11. Capacitor discharge.

Fig. 12 shows charge (for $E_0 = 0$):
$$e_c/E_b = 1 - \exp(-t/T)$$

These curves are plotted for a wider range in Fig. 13.

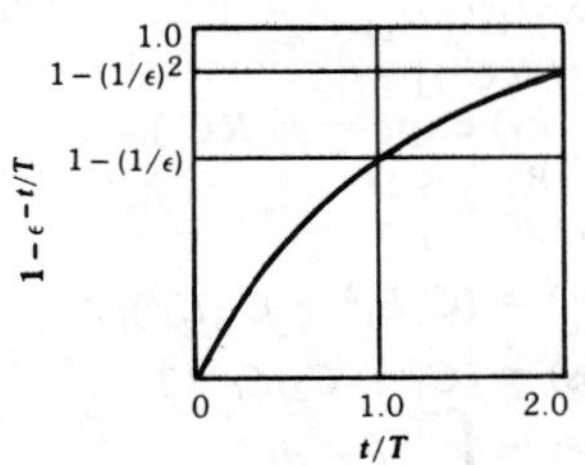

Fig. 12. Capacitor charge.

Two Capacitors

Closing of switch (Fig. 14) occurs at time $t = 0$.
Initial conditions (at $t = 0$):

$$e_1 = E_1; \quad e_2 = E_2$$

Steady state (at $t=\infty$):

Use exponential $\exp(-t/T)$ for charge or discharge of capacitor or discharge of inductor:

(current at time t)/(initial current)

Discharge of capacitor:

(voltage at time t)/(initial voltage)

Use exponential $1-\exp(-t/T)$ for charge of capacitor:

(voltage at time t)/(battery or final voltage)

Charge of inductor:

(current at time t)/(final current)

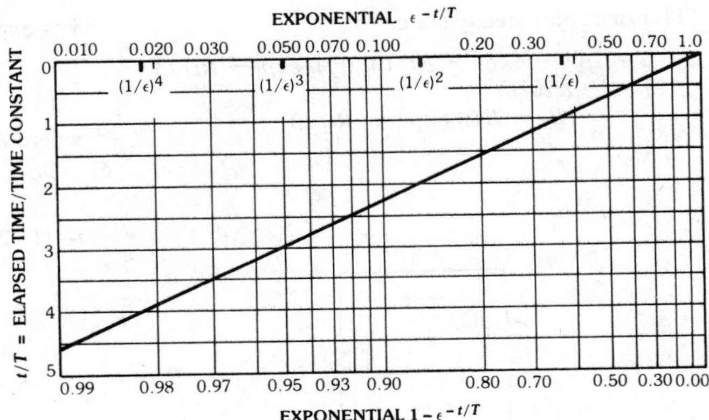

Fig. 13. Exponential functions $\exp(-t/T)$ and $1-\exp(-t/T)$ applied to transients in R-C and L-R circuits.

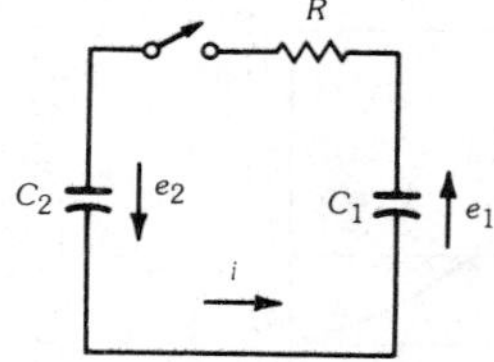

Fig. 14. Circuit for two capacitors.

$$e_1 = E_f;\quad e_2 = -E_f;\quad i = 0$$
$$E_f = (E_1C_1 - E_2C_2)/(C_1 + C_2)$$
$$C' = C_1C_2/(C_1 + C_2)$$

Transient:

$$i = [(E_1 + E_2)/R]\ \exp(-t/RC')$$
$$e_1 = E_f + (E_1 - E_f)\exp(-t/RC')$$
$$= E_1 - (E_1 + E_2)(C'/C_1) \times [1 - \exp(-t/RC')]$$
$$e_2 = -E_f + (E_2 + E_f)\exp(-t/RC')$$
$$= E_2 - (E_1 + E_2)(C'/C_2) \times [1 - \exp(-t/RC')]$$

Original energy (joules) $= (C_1E_1^2 + C_2E_2^2)/2$

Final energy (joules) $= (C_1 + C_2)E_f^2/2$

Loss of energy (joules) $= \int_0^\infty i^2R\,dt$

$= C'(E_1 + E_2)^2/2$

(Loss is independent of the value of R.)

Inductor Charge and Discharge

Initial conditions (at $t = 0$) in Fig. 15:

Battery $= E_b$; $i = I_0$

Steady state (at $t = \infty$):

$$i = I_f = E_b/R$$

Transient plus steady state:

$$i = I_f[1 - \exp(-Rt/L)] + I_0 \exp(-Rt/L)$$
$$e_L = -L\,di/dt$$
$$= -(E_b - RI_0)\exp(-Rt/L)$$

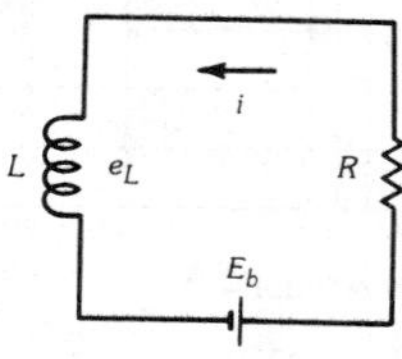

Fig. 15. Circuit for inductor charge and discharge.

Time constant:

$$T = L/R$$

Fig. 11 shows discharge (for $E_b = 0$):

$$i/I_0 = \exp(-t/T)$$

Fig. 12 shows charge (for $I_0 = 0$)

$$i/I_f = [1 - \exp(-t/T)]$$

These curves are plotted for a wider range in Fig. 13.

Charge and Discharge of Series R-L-C Circuit

Initial conditions (at $t = 0$) in Fig. 16:

Battery $= E_b$; $e_c = E_0$; $i = I_0$

Steady state (at $t = \infty$):

$$i = 0;\ e_c = E_b$$

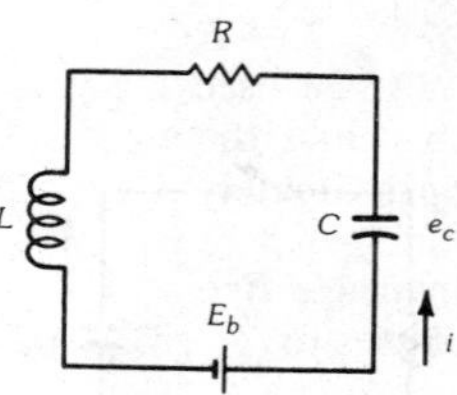

Fig. 16. Series R-L-C circuit.

Differential equation:

$$E_b - E_0 - C^{-1}\int_0^t i\,dt - Ri - L(di/dt) = 0$$

when

$$L(d^2i/dt^2) + R(di/dt) + (i/C) = 0$$

Solution of equation:

$$i = \exp(-Rt/2L)\left[\frac{2(E_b - E_0) - RI_0}{R(D^{1/2})}\sinh(Rt/2L) \times (D^{1/2}) + I_0\cosh(Rt/2L)(D^{1/2})\right]$$

where $D = 1 - (4L/R^2C)$.

Case 1: When L/R^2C is small

$$i = (1 - 2A - 2A^2)^{-1}\left\{\left[\frac{E_b - E_0}{R} - I_0(A + A^2)\right] \times \exp\left(-\frac{t}{RC}(1 + A + 2A^2)\right) + \left[I_0(1 - A - A^2) - \frac{E_b - E_0}{R}\right] \times \exp\left(-\frac{Rt}{L}(1 - A - A^2)\right)\right\}$$

where $A = L/R^2C$.

For practical purposes, the terms A^2 can be neglected when $A < 0.1$. The terms A may be neglected when $A < 0.01$.

Case 2: When $4L/R^2C < 1$ for which $D^{1/2}$ is real

$$i=\frac{\exp(-Rt/2L)}{D^{1/2}}\left\{\left[\frac{E_b-E_0}{R}-\frac{1}{2}I_0\,(1-D^{1/2})\right]\times\exp\left(\frac{Rt}{2L}\,D^{1/2}\right)+\left[\frac{1}{2}I_0(1+D^{1/2})-\frac{E_b-E_0}{R}\right]\exp\left(-\frac{Rt}{2L}\,D^{1/2}\right)\right\}$$

Case 3: When D is a small positive or negative quantity

$$i=\exp(-Rt/2L)\left\{\frac{2(E_b-E_0)}{R}\left[\frac{Rt}{2L}+\frac{1}{6}\left(\frac{Rt}{2L}\right)^3D\right]+I_0\left[1-\frac{Rt}{2L}+\frac{1}{2}\left(\frac{Rt}{2L}\right)^2D-\frac{1}{6}\left(\frac{Rt}{2L}\right)^3D\right]\right\}$$

This equation may be used for values of D up to ± 0.25, at which values the error in the computed current i is approximately 1 percent of I_0 or of $(E_b - E_0)/R$.

Case 3A: When $4L/R^2C = 1$ for which $D = 0$, the equation reduces to

$$i = \exp(-Rt/2L)\left[\frac{E_b-E_0}{R}\,\frac{Rt}{L}+I_0\left(1-\frac{Rt}{2L}\right)\right]$$

or $i = i_1 + i_2$, plotted in Fig. 17. For practical purposes, this equation may be used when $4L/R^2C = 1\pm 0.05$ with errors of 1 percent or less.

Case 4: When $4L/R^2C > 1$ for which $D^{1/2}$ is imaginary

$$i = \exp(-Rt/2L)\left[\left(\frac{E_b-E_0}{\omega_0 L}-\frac{RI_0}{2\omega_0 L}\right)\times\sin\omega_0 t + I_0\cos\omega_0 t\right]$$

$$= I_m \exp(-Rt/2L)\sin(\omega_0 t + \psi)$$

where,

$$\omega_0 = [(LC)^{-1} - (R^2/4L^2)]^{1/2}$$
$$I_m = (\omega_0 L)^{-1}\{[E_b - E_0 - (RI_0)/2]^2 + \omega_0^2L^2I_0^2\}^{1/2}$$
$$\psi = \tan^{-1}\{\omega_0 L I_0/[E_b - E_0 - (RI_0)/2]\}$$

The envelope of the voltage wave across the inductor is

$$\pm\exp(-Rt/2L)[\omega_0(LC)^{1/2}]^{-1}\times\{[E_b - E_0 - (RI_0)/2]^2 + \omega_0^2L^2I_0^2\}^{1/2}$$

Example: Relay with transient-suppressing capacitor (Fig. 18). The switch is closed until time $t = 0$, then opened.

Let $L = 0.10$ henry, $R_1 = 100$ ohms, and $E = 10$ volts.

Suppose we choose $C = 10^{-6}$ farad and $R_2 = 100$ ohms.

Then $R = 200$ ohms, $I_0 = 0.10$ ampere, $E_0 = 10$ volts, $\omega_0 = 3 \times 10^3$, and $f_0 = 480$ hertz.

Maximum peak voltage across L (envelope at $t = 0$) is approximately 30 volts. Time constant of decay of envelope is 0.001 second.

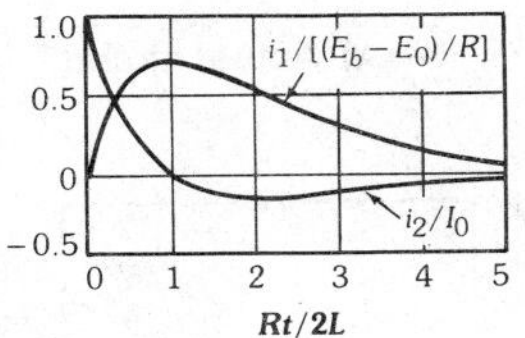

Fig. 17. Transients for $4L/R^2C = 1$.

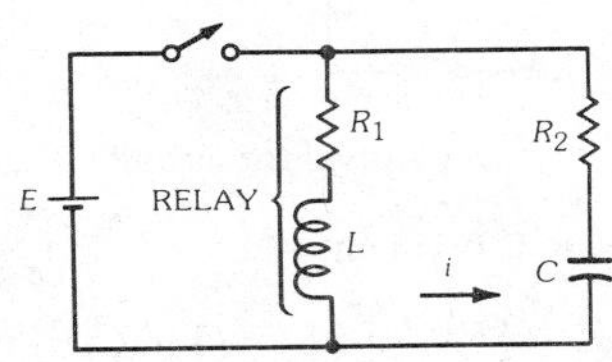

Fig. 18. Equivalent circuit of relay with transient-suppressing capacitor.

Nonoscillating Condition: It is preferable that the circuit be just nonoscillating (Case 3A) and that it present a pure resistance at the switch terminals for any frequency.

$$R_2 = R_1 = R/2 = 100 \text{ ohms}$$

$$4L/R^2C = 1$$

$$C = 10^{-5}\text{ farad} = 10\text{ microfarads}$$

At the instant of opening the switch, the voltage across the parallel circuit is $E_0 - R_2I_0 = 0$.

Series R-L-C Circuit With Sinusoidal Applied Voltage

By the principle of superposition, the transient and steady-state conditions are the same for the actual circuit and the equivalent circuit shown in Fig. 19, the closing of the switch occurring at time $t = 0$. In the equivalent circuit, the steady state is due to the source e acting continuously from time $t = -\infty$, while the transient is due to short-circuiting the source $-e$ at time $t = 0$.

Source:

$$e = E\sin(\omega t + \alpha)$$

Steady state:

$$i = (E/Z)\sin(\omega t + \alpha - \phi)$$

where,

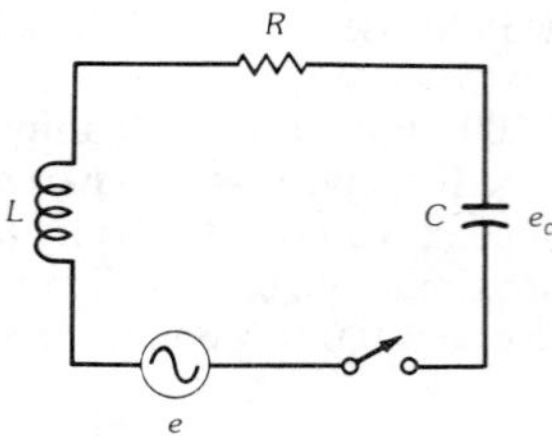

(A) Actual circuit.

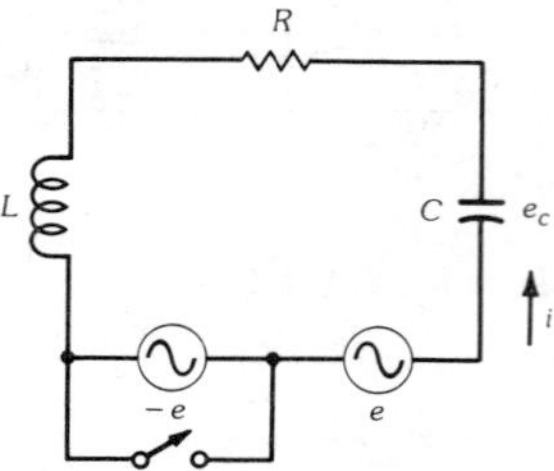

(B) Equivalent circuit.

Fig. 19. Series R-L-C circuit.

$$Z = \{R^2 + [\omega L - (1/\omega C)]^2\}^{1/2}$$
$$\tan\phi = (\omega^2 LC - 1)/\omega CR$$

The transient is found by determining current $i = I_0$ and capacitor voltage $e_c = E_0$ at time $t = 0$, due to the source $-e$. These values of I_0 and E_0 are then substituted in the equations of Case 1, 2, 3, or 4, above, according to the values of R, L, and C.

At time $t = 0$, due to the source $-e$:

$$i = I_0 = -(E/Z)\sin(\alpha - \phi)$$

$$e_c = E_0 = (E/\omega CZ)\cos(\alpha - \phi)$$

This form of analysis may be used for any periodic applied voltage e. The steady-state current and the capacitor voltage for an applied voltage $-e$ are determined, the periodic voltage being resolved into its harmonic components for this purpose, if necessary. Then the instantaneous values $i = I_0$ and $e_C = E_0$ at the time of closing the switch are easily found, from which the transient is determined. It is evident, from this method of analysis, that the waveform of the transient need bear no relationship to that of the applied voltage, depending only on the constants of the circuit and the hypothetical initial conditions I_0 and E_0.

TRANSIENTS—OPERATIONAL CALCULUS AND LAPLACE TRANSFORMS

Among the various methods of operational calculus used to solve transient problems, one of the most efficient makes use of the Laplace transform.

If we have a function $v = f(t)$, then by definition the Laplace transform is $\mathcal{L}[f(t)] = F(p)$, where

$$F(p) = \int_0^\infty \exp(-pt) f(t)\, dt \qquad (4)$$

The inverse transform of $F(p)$ is $f(t)$. Most of the mathematical functions encountered in practical work fall in the class for which Laplace transforms exist. Transforms of several functions are given in Chapter 46.

In the following, an abbreviated symbol such as $\mathcal{L}[i]$ is used instead of $\mathcal{L}[i(t)]$ to indicate the Laplace transform of the function $i(t)$.

The electrical (or other) system for which a solution of the differential equation is required is considered only in the time domain $t \geq 0$. Any currents or voltages existing at $t = 0$, before the driving force is applied, constitute initial conditions. Driving force is assumed to be 0 when $t < 0$.

Example

Take the circuit of Fig. 20, in which the switch is closed at time $t = 0$. Before the closing of the switch, suppose the capacitor is charged; then at $t = 0$, we have $v = V_0$. It is required to find the voltage v across capacitor C as a function of time.

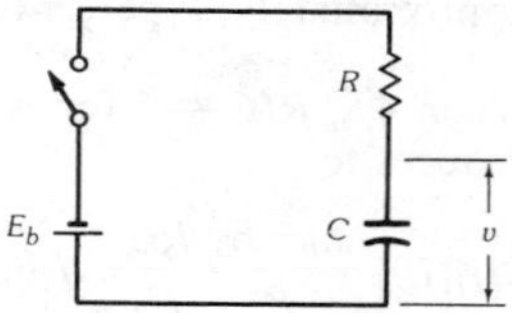

Fig. 20. Series R-C circuit.

Since $i = dq/dt = C(dv/dt)$, the differential equation of the circuit in terms of voltage is:

$$e(t) = v + Ri = v + RC(dv/dt) \qquad (5)$$

where $e(t) = E_b$.

Referring to the table of transforms, the applied voltage is E_b multiplied by unit step, or $E_b S_{-1}(t)$; the transform for this is E_b/p. The transform of v is $\mathcal{L}[v]$. That of $RC(dv/dt)$ is $RC[p\mathcal{L}[v] - v(0)]$, where $v(0) = V_0 =$ value of v at $t = 0$. Then the transform of Eq. (5) is:

$$E_b/p = \mathcal{L}[v] + RC[p\mathcal{L}[v] - V_0]$$

Rearranging and resolving into partial fractions:

$$\mathcal{L}[v] = \frac{E_b}{p(1 + RCp)} + \frac{RCV_0}{1 + RCp}$$
$$= E_b[p^{-1} - (p + 1/RC)^{-1}] + \frac{V_0}{p + 1/RC} \qquad (6)$$

Now we must determine the equation that would transform into Eq. (6). The inverse transform of $\mathcal{L}[v]$ is v, and those of the terms on the right-hand side are found in the table of transforms. Then, in the time domain $t \geq 0$

$$v = E_b[1 - \exp(-t/RC)] + V_0 \exp(-t/RC) \quad (7)$$

This solution is also well known by classical methods. However, the advantages of the Laplace transform method become more and more apparent in reducing the labor of solution as the equations become more involved.

Circuit Response Related to Unit Impulse

Unit impulse (see Laplace transforms) has the dimensions of time^{-1}. For example, suppose a capacitor of 1 microfarad is suddenly connected to a battery of 100 volts, with the circuit inductance and resistance negligibly small. Then the current is 10^{-4} coulomb multiplied by unit impulse.

The general transformed equation of a circuit or system may be written

$$\mathcal{L}[i] = \phi(p)\,\mathcal{L}[e] + \psi(p) \quad (8)$$

Here $\mathcal{L}[i]$ is the transform of the required current (or other quantity) and $\mathcal{L}[e]$ is the transform of the applied voltage or driving force $e(t)$. The transform of the initial conditions, at $t = 0$, is included in $\psi(p)$.

First considering the case when the system is initially at rest, $\psi(p) = 0$. Writing i_a for the current in this case

$$\mathcal{L}[i_a] = \phi(p)\,\mathcal{L}[e] \quad (9)$$

Now apply unit impulse $S_0(t)$ (multiplied by 1 volt-second), and designate the circuit current in this case by $B(t)$ and its transform by $\mathcal{L}[B]$. The Laplace transform of $S_0(t)$ is 1, so

$$\mathcal{L}[B] = \phi(p) \quad (10)$$

Equation (9) becomes, for any driving force

$$\mathcal{L}[i_a] = \mathcal{L}[B]\,\mathcal{L}[e] \quad (11)$$

Applying the convolution function (Laplace transform)

$$i_a = \int_0^t B(t-\lambda)\, e(\lambda)\, d\lambda$$

$$= \int_0^t B(\lambda)\, e(t-\lambda)\, d\lambda \quad (12)$$

To this there must be added the current i_0 due to any initial conditions that exist. From (8)

$$\mathcal{L}[i_0] = \psi(p) \quad (13)$$

Then i_0 is the inverse transform of $\psi(p)$.

Circuit Response Related to Unit Step

Unit step is defined and designated $S_{-1}(t) = 0$ for $t < 0$ and equals unity for $t > 0$. It has no dimensions. Its Laplace transform is $1/p$. Let the circuit current be designated $A(t)$ when the applied voltage is $e = S_{-1}(t) \times$ (1 volt). Then, the current i_a for the case when the system is initially at rest, and for any applied voltage $e(t)$, is given by any of

$$i_a = A(t)\, e(0) + \int_0^t A(t-\lambda)\, e'(\lambda)\, d\lambda$$

$$= A(t)\, e(0) + \int_0^t A(\lambda)\, e'(t-\lambda)\, d\lambda$$

$$= A(0)\, e(t) + \int_0^t A'(t-\lambda)\, e(\lambda)\, d\lambda$$

$$= A(0)\, e(t) + \int_0^t A'(\lambda)\, e(t-\lambda)\, d\lambda \quad (14)$$

where A' is the first derivative of A and similarly for e' of e.

As an example, consider the problem of Fig. 20 and Eqs. (5) to (7) above. Suppose $V_0 = 0$, and that the battery is replaced by a linear source

$$e(t) = Et/T_1$$

where T_1 is the duration of the voltage rise in seconds. By Eq. (7), setting $E_b = 1$

$$A(t) = 1 - \exp(-t/RC)$$

Then using the first equation in (14) and noting that $e(0) = 0$, and $e'(t) = E/T_1$ when $0 \leq t \leq T_1$, the solution is

$$v = (Et/T_1) - (ERC/T_1)[1 - \exp(-t/RC)]$$

This result can, of course, be found readily by direct application of the Laplace transform to Eq. (5) with $e(t) = Et/T_1$.

Heaviside Expansion Theorem

When the system is initially at rest, the transformed equation is given by Eq. (9) and may be written

$$\mathcal{L}[i_a] = [M(p)/G(p)]\,\mathcal{L}[e] \quad (15)$$

$M(p)$ and $G(p)$ are rational functions of p. In the following, $M(p)$ must be of lower degree than $G(p)$, as is usually the case. The roots of $G(p) = 0$ are p_r, where $r = 1, 2, \cdots, n$, and there must be no repeated roots. The response may be found by application of the Heaviside expansion theorem.

For a force $e = E_{max} \exp(j\omega t)$ applied at time $t = 0$

$$\frac{i_a(t)}{E_{max}} = \frac{M(j\omega)}{G(j\omega)} \exp(j\omega t) + \sum_{r=1}^{n} \frac{M(p_r)\exp(p_r t)}{(p_r - j\omega)G'(p_r)} \quad (16a)$$

$$= \frac{\exp(j\omega t)}{Z(j\omega)} + \sum_{r=1}^{n} \frac{\exp(p_r t)}{(p_r - j\omega)Z'(p_r)} \quad (16b)$$

The first term on the right-hand side of either form of (16) gives the steady-state response, and the second term gives the transient. When $e = E_{max}$

$\cos\omega t$, take the real part of (16), and similarly for $\sin\omega t$ take the imaginary part. $Z(p)$ is defined in Eq. (19). If the applied force is the unit step, set $\omega = 0$ in Eq. (16).

Application to Linear Networks

The equation for a single mesh is of the form

$$A_n(d^n i/dt^n) + \cdots + A_1(di/dt) + A_0 i + B\int i\,dt = e(t) \quad (17)$$

System Initially at Rest—Then, Eq. (17) transforms into

$$(A_n p^n + \cdots + A_1 p + A_0 + B_p{}^{-1})\ [i] = \ [e] \quad (18)$$

where the expression in parentheses is the operational impedance, equal to the alternating-current impedance when we set $p = j\omega$.

If there are m meshes in the system, we get m simultaneous equations like (17) with m unknowns $i_1, i_2, \cdots, i_m$. The m algebraic equations like (18) are solved for $\mathcal{L}\,[i_1]$, etc., by means of determinants, yielding an equation of the form of (15) for each unknown, with a term on the right-hand side for each mesh in which there is a driving force. Each such driving force may of course be treated separately and the responses added.

If any two meshes are designated by the letters h and k, the driving force $e(t)$ being in either mesh and the mesh current $i(t)$ in the other, then the fraction $M(p)/G(p)$ in (15) becomes

$$M_{hk}(p)/G(p) = 1/Z_{hk}(p) = Y_{hk}(p) \quad (19)$$

where $Y_{hk}(p)$ is the operational transfer admittance between the two meshes. The determinant of the system is $G(p)$, and $M_{hk}(p)$ is the cofactor of the row and column that represent $e(t)$ and $i(t)$.

System Not Initially at Rest—The transient due to the initial conditions is solved separately and added to the above solution. The driving force is set equal to zero in Eq. (17), $e(t) = 0$, and each term is transformed according to

$$\mathcal{L}[d^n i/dt^n] = p^n \mathcal{L}[i] - \sum_{r=1}^{n} p^{n-r}[d^{r-1}i/dt^{r-1}]_{t=0} \quad (20a)$$

$$\mathcal{L}\left[\int_0^t i\,dt\right] = p^{-1}\mathcal{L}[i] + p^{-1}\left[\int i\,dt\right]_{t=0} \quad (20b)$$

where the last term in each equation represents the initial conditions. For example, in Eq. (20b) the last term would represent, in an electrical circuit, the quantity of electricity existing on a capacitor at time $t = 0$, the instant when the driving force $e(t)$ begins to act.

Resolution into Partial Fractions—The solution of the operational form of the equations of a system involves rational fractions that must be simplified before the inverse transform is found. Let the fraction be $h(p)/g(p)$ where $h(p)$ is of lower degree than $g(p)$, for example $(3p+2)/(p^2 + 5p + 8)$. If $h(p)$ is of equal or higher degree than $g(p)$, it can be reduced by division.

The reduced fraction can be expanded into partial fractions. Let the factors of the denominator be $(p - p_r)$ for the n nonrepeated roots p_r of the equation $g(p) = 0$, and $(p - p_a)$ for a root p_a repeated m times.

$$\frac{h(p)}{g(p)} = \sum_{r=1}^{n} \frac{A_r}{p - p_r} + \sum_{r=1}^{m} \frac{B_r}{(p-p_a)^{m-r+1}} \quad (21a)$$

There is a summation term for each root that is repeated. The constant coefficients A_r and B_r can be evaluated by reforming the fraction with a common denominator. Then the coefficients of each power of p in $h(p)$ and the reformed numerator are equated and the resulting equations solved for the constants. More formally, they may be evaluated by

$$A_r = \frac{h(p_r)}{g'(p_r)} = \left[\frac{h(p)}{g(p)/(p-p_r)}\right]_{p=pr} \quad (21b)$$

$$B_r = [1/(r-1)!]\, f^{(r-1)}(p_a) \quad (21c)$$

where

$$f(p) = (p - p_a)^m [h(p)/g(p)]$$

and $f^{(r-1)}(p_a)$ indicates that the $(r-1)$th derivative of $f(p)$ is to be found, after which we set $p = p_a$.

Fractions of the form $(A_1 p + A_2)/(p^2 + \omega^2)$ or, more generally

$$\frac{A_1 p + A_2}{p^2 + 2ap + b} = \frac{A(p+a) + B\omega}{(p+a)^2 + \omega^2} \quad (22a)$$

where $b > a^2$ and $\omega^2 = b - a^2$ need not be reduced further. From the Laplace transforms the inverse transform of (22a) is

$$\exp(-at)(A\cos\omega t + B\sin\omega t) \quad (22b)$$

where

$$A = \frac{h(-a+j\omega)}{g'(-a+j\omega)} + \frac{h(-a-j\omega)}{g'(-a-j\omega)} \quad (22c)$$

$$B = j\left[\frac{h(-a+j\omega)}{g'(-a+j\omega)} - \frac{h(-a-j\omega)}{g'(-a-j\omega)}\right] \quad (22d)$$

Similarly, the inverse transform of the fraction

$$[A(p+a) + B\alpha]/[(p+a)^2 - \alpha^2]$$

is $\exp(-at)(A\cosh\alpha t + B\sinh\alpha t)$, where A and B are found by (22c) and 22d), except that $j\omega$ is replaced by α and the coefficient j is omitted in the expression for B.

7 Fourier Waveform Analysis

Revised by
David C. Munson, Jr.

FOURIER TRANSFORM OF A FUNCTION

The Fourier transform, F, of function f is defined by the integral (where x and y are real variables)

$$F(y) = \int_{-\infty}^{\infty} f(x) \exp(-j2\pi xy)\, dx$$

provided this integral exists.

A sufficient, but not necessary, existence condition is that f be absolutely integrable; that is

$$\int_{-\infty}^{\infty} |f(x)| dx < \infty$$

An important example of a function that has a Fourier transform even though it is not absolutely integrable is $(\sin x)/x$.

In general, F and f may be complex. Letting $f(x) = f_r(x) + jf_i(x)$, where f_r and f_i are real-valued, one has

$$F(y) = \int_{-\infty}^{\infty} [f_r(x) \cos 2\pi yx + f_i(x) \sin 2\pi yx] dx$$

$$-j\int_{-\infty}^{\infty} [f_r(x) \sin 2\pi yx - f_i(x) \cos 2\pi yx] dx$$

Conversely, the function f, whose Fourier transform is a given function F, is given by the integral (inverse Fourier transform)

$$f(x) = \int_{-\infty}^{\infty} F(y) \exp(j2\pi xy)\, dy$$

where it is assumed that, at points of discontinuity of the integral (if any), the function $f(x)$ is given the value

$$f(x) = [f(x^+) + f(x^-)]/2$$

The functions $f(x^+)$ and $f(x^-)$ are the limits of $f(x + t)$ as t approaches 0 through positive and negative values, respectively.

Letting $F(y) = F_r(y) + jF_i(y)$, one has

$$f(x) = \int_{-\infty}^{\infty} [F_r(y) \cos 2\pi yx - F_i(y) \sin 2\pi yx] dy$$

$$+j\int_{-\infty}^{\infty} [F_r(y) \sin 2\pi yx + F_i(y) \cos 2\pi yx] dy$$

In many engineering applications it is customary to denote the variable y as "frequency"; in most cases x represents time or space.

If the radian frequency $\omega = 2\pi y$ is introduced as a variable, the definitions of the Fourier transform and of its inverse are written as

$$F(\omega/2\pi) = F_1(\omega)$$

$$= \int_{-\infty}^{\infty} f(x) \exp(-j\omega x) dx$$

$$f(x) = (2\pi)^{-1} \int_{-\infty}^{\infty} F_1(\omega) \exp(j\omega x)\, d\omega$$

Properties of the Fourier transform are listed in Table 1. Table 2 contains normalized graphs of the Fourier transform (or its magnitude) for a number of common pulse shapes.

FOURIER SERIES

Real Form of Fourier Series

A periodic function with period T, defined by its values in the intervals $-T/2$ to $+T/2$ or 0 to T can be written in a Fourier series expansion as

$$f(x) = (1/2)A_0 + \sum_{n=1}^{\infty} [A_n \cos(n\omega_0 x) + B_n \sin(n\omega_0 x)$$

where $\omega_0 = 2\pi/T$, and the coefficients are given by

$$A_n = (2/T) \int_{-T/2}^{T/2} f(x) \cos(n\omega_0 x) dx$$

$$= (2/T) \int_0^T f(x) \cos(n\omega_0 x) dx$$

$$B_n = (2/T) \int_{-T/2}^{T/2} f(x) \sin(n\omega_0 x) dx$$

$$= (2/T) \int_0^T f(x) \sin(n\omega_0 x) dx$$

for $n = 0, 1, 2, \cdots$.

An alternate form of the above expansion is

$$f(x) = (1/2)C_0 + \sum_{n=1}^{\infty} C_n \cos(n\omega_0 x + \phi_n)$$

where

$$C_0 = A_0$$
$$C_n = (A_n^2 + B_n^2)^{1/2}$$
$$\cos\phi_n = A_n/C_n$$
$$\sin\phi_n = -B_n/C_n$$

Complex Form of Fourier Series

The Fourier series can be written more concisely as

$$f(x) = \sum_{n=1}^{\infty} D_n \exp(jn\omega_0 x)$$

where

$$D_n = T^{-1} \int_0^T f(x) \exp(-jn\, \omega_0 x) dx$$

and

$$D_0 = (1/2)A_0 = (1/2)C_0$$
$$D_n = (1/2)(A_n - jB_n)$$
$$D_{-n} = (1/2)(A_n + jB_n)$$

$n = 1, 2, \cdots$.

Average Power

The average power of the periodic waveform $f(x)$ is given by

TABLE 1. PROPERTIES OF FOURIER TRANSFORM*

	Function	Fourier Transform
1. Definition	$f(x)$	$F(y)=\int_{-\infty}^{+\infty} f(x)\exp(-2\pi jxy)dx$
2. Inverse transform	$f(x)=\int_{-\infty}^{+\infty} F(y)\exp(2\pi jxy)dy$	$F(y)$
3. Linearity	$af(x)$ $f_1(x)\pm f_2(x)$	$aF(y)$ $F_1(y)\pm F_2(y)$
4. Translation or shifting theorem	$g(x)=f(x-x_0)$, x_0=real const.	$G(y)=\exp(-2\pi jx_0y)F(x)$
	$g(x)=\exp(2\pi jy_0x)f(x)$, y_0=real const.	$G(y)=F(y-y_0)$
5. Change of scale	$g(x)=f(x/a)$, a=real const.	$G(y)=\|a\|F(ay)$
6. Frequency shifting and change of scale	$g(x)=\exp(2\pi jy_0x)f(x/a)$, y_0 and a=real const.	$G(y)=\|a\|F[a(y-y_0)]$
7. Interchange of function and transform	$g(x)=F(x)$	$G(y)=f(-y)$
8a. Convolution in x-space (product of Fourier transforms)	$h=f*g=g*f$ i.e., $h(x)=\int_{-\infty}^{+\infty} f(x-\tau)g(\tau)d\tau$ $=\int_{-\infty}^{\infty} f(\tau)g(x-\tau)d\tau$	$H=F\cdot G$
8b. Convolution in y-space (product of inverse Fourier transforms)	$h=f\cdot g$	$H=F*G$
9. Unit pulse (or Dirac function)	$\delta(x)$	$F(y)=1$ (for all y)
	$f(x)=1$ (for all x)	$\delta(y)$

Continued on next page.

TABLE 1 (CONT). PROPERTIES OF FOURIER TRANSFORM*

	Function	Fourier Transform
10. Periodic train of equal pulses	$A \sum_{n=-\infty}^{+\infty} \delta(x-nT)$	$(A/T) \sum_{n=-\infty}^{+\infty} \delta(y-n/T)$
11a. Derivative in x-space	$g(x)=d^n f/dx^n$	$G(y)=(2\pi jy)^n F(y)$, if $G(y)$ exists
11b. Derivative in y-space	$g(x)=(-2\pi jx)^n f(x)$	$G(y)=d^n F/dy^n$
12a. Integral in x-space	$g(x)=\int_{-\infty}^{x} f(x)\,dx$	$G(y)=[1/(2\pi jy)]F(y)+F(0)\delta(y)$
		where $F(0)=\int_{-\infty}^{\infty} f(x)\,dx$
12b. Integral in y-space	$g(x)=-[1/(2\pi jx)]f(x)$	$G(y)=\int_{-\infty}^{y} F(y)\,dy$
13. Symmetry	$g(x)=f(-x)$	$G(y)=F(-y)$
	f even: $f(x)=f(-x)$	F even: $F=2\int_{0}^{\infty} f(x)\cos(2\pi xy)\,dx$
	f odd: $f(x)=-f(-x)$	F odd: $F=-2j\int_{0}^{\infty} f(x)\sin(2\pi xy)\,dx$
14. Complex conjugate	$g(x)=f^*(x)$	$G(y)=F^*(-y)$
	Hence, if $f(x)$ is real	$F(-y)=F^*(y)$
15. Area under the curve	$\int_{-\infty}^{+\infty} f(x)dx=F(0)$	$\int_{-\infty}^{+\infty} F(y)dy=f(0)$
16a. Parseval's theorem	$\int_{-\infty}^{+\infty} f^*(x)g(x)dx$	$=\int_{-\infty}^{+\infty} F^*(y)G(y)dy$

Continued on next page.

TABLE 1 (CONT). PROPERTIES OF FOURIER TRANSFORM*

	Function	Fourier Transform
16b. Alternative forms	$\int_{-\infty}^{+\infty} f(x)g(x)\,dx$	$= \int_{-\infty}^{+\infty} F(-y)\,G(y)\,dy$
	$\int_{-\infty}^{+\infty} f(u)\,G(u)\,du$	$= \int_{-\infty}^{+\infty} F(u)g(u)\,du$
16c. "Energy" relation	$\int_{-\infty}^{+\infty} \lvert f(x)\rvert^2 dx$	$= \int_{-\infty}^{+\infty} \lvert F(y)\rvert^2 dy$
17. Initial value theorem	If $f(x) = 0$ for $x<0$, is real, and contains no pulses $f(0^+) = 2\int_{-\infty}^{\infty} F_r(y)\,dy$ where $F_r(y) = \mathrm{Re}[F(y)]$	
18. Relationships between $F_r(y)$ and $F_i(y)$	a) If $f(x) = 0$ for $x<0$, is real, and contains no pulses $\int_{-\infty}^{\infty} F_r^2(y)\,dy = \int_{-\infty}^{\infty} F_i^2(y)\,dy$ where $F_i(y) = \mathrm{Im}[F(y)]$ b) The following integral relationships apply (Hilbert transforms) $F_r(y) = 2\int_{-\infty}^{\infty} [F_i(\tau)/2\pi(y-\tau)]\,d\tau$ $F_i(y) = -2\int_{-\infty}^{\infty} [F_r(\tau)/2\pi(y-\tau)]\,d\tau$ in which the Cauchy principal values of the integrals are taken	

* In the table, functions of x are denoted by lower-case letters and their transforms by the corresponding capital letters.

TABLE 2. TIME AND FREQUENCY FUNCTIONS FOR COMMONLY ENCOUNTERED PULSE SHAPES*

Time Function	Frequency Function

A. Rectangular pulse

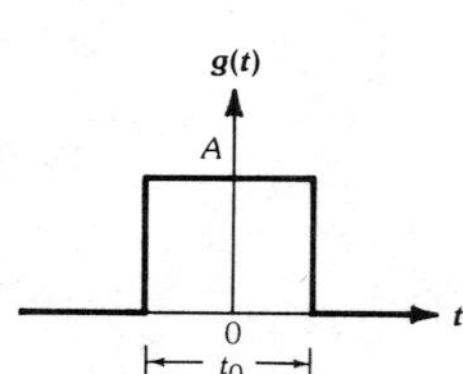

$g(t) = A$ for $-(1/2)t_0 < t < (1/2)t_0$

$= 0$ otherwise

Area $\mathcal{A} = At_0$

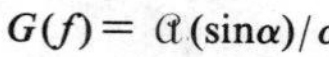

$G(f) = \mathcal{A}(\sin\alpha)/\alpha$

where $\alpha = \pi t_0 f$

[See curve $(\sin x)/x$ above.]

B. Isosceles-triangle pulse

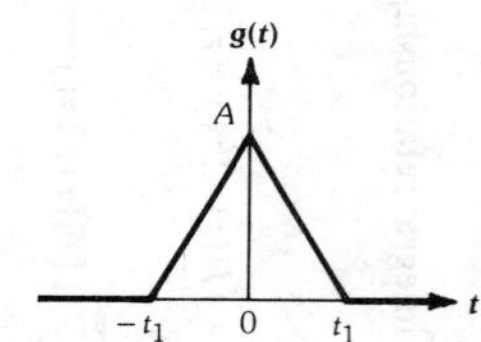

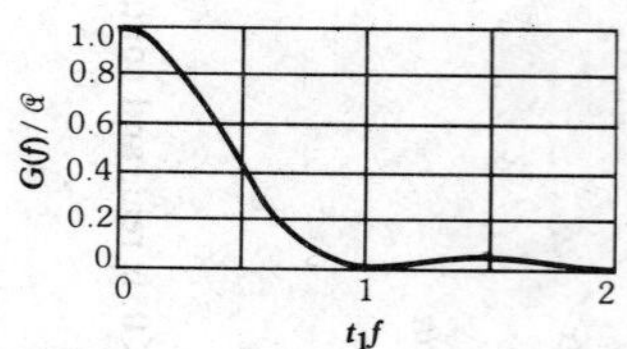

$g(t) = A[1-(t/t_1)],\ 0 \le t < t_1$

$= A[1+(t/t_1)],\ -t_1 < t \le 0$

$= 0$, otherwise

Area $\mathcal{A} = At_1$

$G(f) = \mathcal{A}[(\sin\alpha)/\alpha]^2$

where $\alpha = \pi t_1 f$

C. Sawtooth pulse

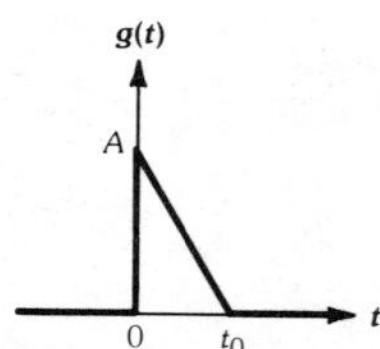

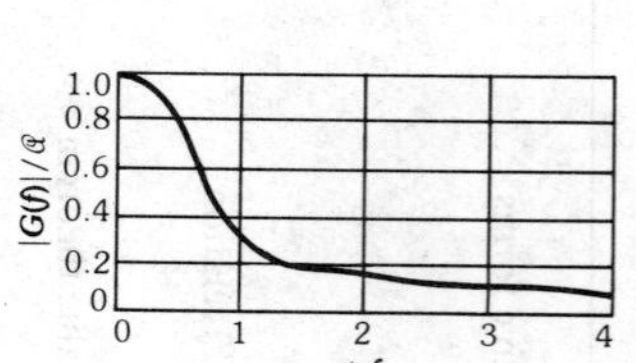

$g(t) = A[1-(t/t_0)],\ 0 < t < t_0$

$= 0$, otherwise

Area $\mathcal{A} = (1/2)At_0$

$$G(f) = \mathcal{A}\ (j/\alpha)\{[(\sin\alpha)/\alpha]\exp(-j\alpha) - 1\}$$

$$= \mathcal{A}\ \frac{1 - \exp(-2j\alpha) - 2j\alpha}{2\alpha^2}$$

where $\alpha = \pi t_0 f$

Continued on next page.

TABLE 2 (CONT). TIME AND FREQUENCY FUNCTIONS FOR COMMONLY ENCOUNTERED PULSE SHAPES*

Time Function	Frequency Function

D. Any pulse of polygonal form may be represented as a linear combination of waveforms such as **A**, **B**, and **C** above eventually after some shifts in time. The pulse spectrum is the same linear combination of the corresponding spectra (eventually modified according to property 4, Table 1).

E. Cosine pulse

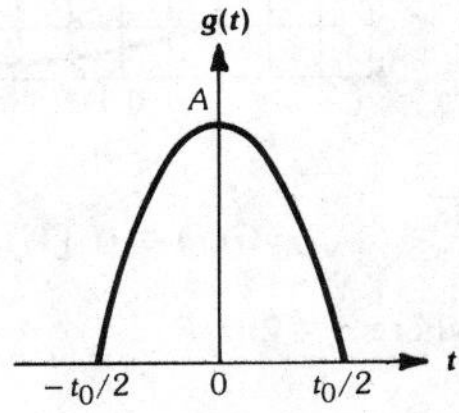

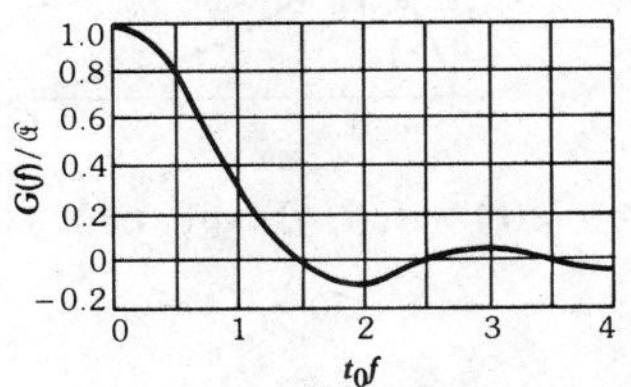

$g(t) = A\cos\pi(t/t_0),\ -(1/2)t_0 < t < (1/2)t_0$

$= 0$, otherwise

Area $\mathcal{A} = (2/\pi)At_0$

$G(f) = \mathcal{A}\{[\cos(\pi/2)\alpha]/(1-\alpha^2)\}$

where $\alpha = 2t_0f$

For $\alpha = 1$, $G(f) = \mathcal{A}\pi/4$

F. Cosine-squared pulse

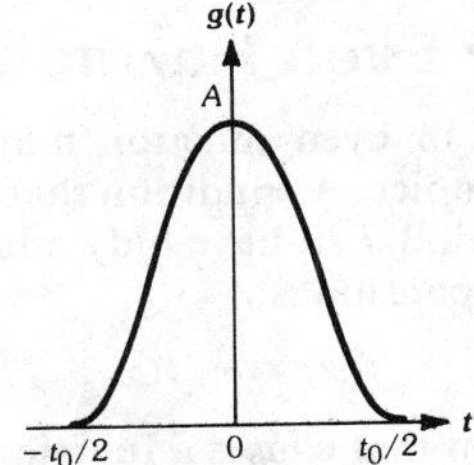

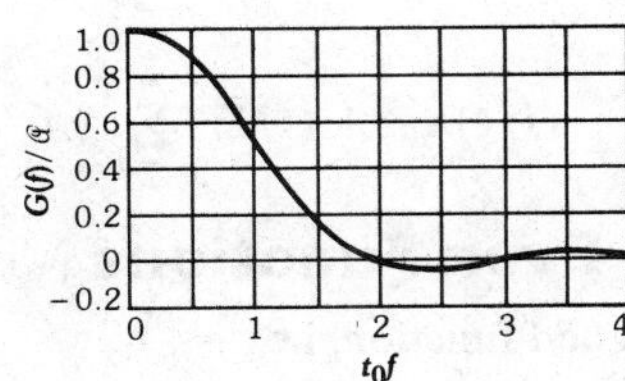

$$\left.\begin{aligned} g(t) &= A\cos^2\pi(t/t_0) \\ &= (1/2)A[1+\cos 2\pi(t/t_0)] \end{aligned}\right\} -(1/2)t_0 < t < (1/2)t_0$$

$= 0$, otherwise

Area $\mathcal{A} = (1/2)At_0$

$$G(f) = \mathcal{A}\,\frac{\sin\pi\alpha}{\pi\alpha(1-\alpha^2)}$$

where $\alpha = t_0f$

For $\alpha = 1$, $G(f) = (1/2)\,\mathcal{A}$

G. Gaussian pulse

Use curve of Fig. 1, Chapter 44 with standard deviation

$\sigma = t_1$

$= (2\ln 2)^{-1/2}t_{6dB};\ \Delta t_{6dB} \equiv 2t_{6dB}$

$g(t) = A\exp[-(1/2)(t/t_1)^2]$

$= A\exp[-(\ln 2)(t/t_{6dB})^2]$

Area $\mathcal{A} = (2\pi)^{1/2}At_1$

$= (1/2)(\pi/\ln 2)^{1/2}A\Delta t_{6dB}$

Use curve of Fig. 1, Chapter 44 with standard deviation

$\sigma = f_1 = 1/2\pi t_1$

$= (2\ln 2)^{1/2}/\pi\Delta t_{6dB};\ f_{3dB} = 2^{1/2}\ln 2/\pi\Delta t_{6dB}$

$G(f) = \mathcal{A}\exp[-(1/2)(f/f_1)^2]$

$= (\ln 2/2\pi)^{1/2}(A/f_{3dB})$

$\times\exp[-(1/2)(\ln 2)(f/f_{3dB})^2]$

$= (1/2)(\pi/\ln 2)^{1/2}A\Delta t_{6dB}$

$\times\exp[-(\pi^2/4\ln 2)(f\Delta t_{6dB})^2]$

Continued on next page.

TABLE 2 (CONT). TIME AND FREQUENCY FUNCTIONS FOR COMMONLY ENCOUNTERED PULSE SHAPES*

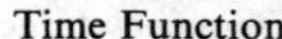

Time Function	Frequency Function
H. Critically damped exponential pulse	
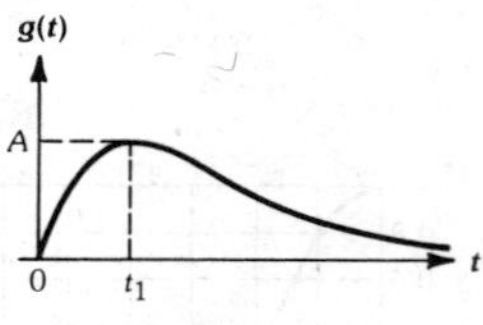	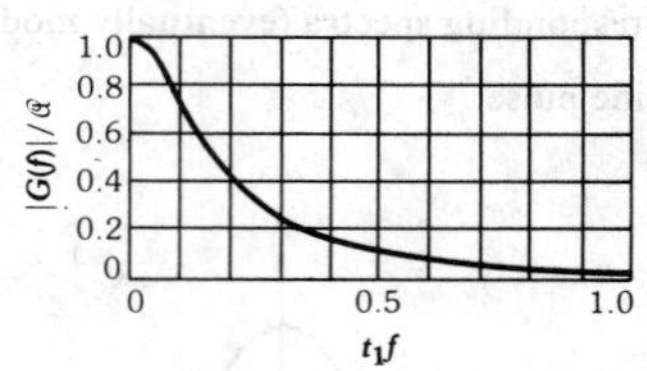
$g(t)=Ae(t/t_1)\exp(-t/t_1),\ t>0$	$G(f)=\mathcal{A}\,[1/(1+j\alpha)^2]$
$=0,\ t\leq 0$	where $\alpha=2\pi t_1 f$
$e=2.71828\cdots$	
Area $\mathcal{A}=Aet_1$	

*For an extensive tabulation of the Fourier transform and its inverse, see Campbell, G. A., and Foster, R. M. *Fourier Integrals for Practical Applications.* New York: D. Van Nostrand Co., Inc., 1948. See also Erdélyi, A., ed. *Tables of Integral Transforms*, Vol. 1, Bateman Manuscript Project. New York: McGraw-Hill Book Co., 1954.

$$T^{-1}\int_0^T |f(x)|^2 dx = \sum_{n=-\infty}^{+\infty} |D_n|^2$$

$$= (1/4)C_0^2 + (1/2)\sum_{n=1}^{\infty} C_n^2$$

$$= (1/4)A_0^2 + (1/2)\sum_{n=1}^{\infty} (A_n^2 + B_n^2)$$

Odd and Even Functions

If $f(x)$ is an odd function, i.e.,

$$f(x) = -f(-x)$$

then all the coefficients of the cosine terms (A_n) vanish, and the Fourier series consists of sine terms alone.

If $f(x)$ is an even function, i.e.,

$$f(x) = f(-x)$$

then all the coefficients of the sine terms (B_n) vanish, and the Fourier series consists of cosine terms alone, and a possible constant.

The Fourier expansions of functions in general include both cosine and sine terms. Every function capable of Fourier expansion consists of the sum of an even and an odd part.

$$f(x) = \underbrace{(1/2)A_0 + \sum_{n=1}^{\infty} A_n \cos n\omega_0 x}_{\text{even}} + \underbrace{\sum_{n=1}^{\infty} B_n \sin n\omega_0 x}_{\text{odd}}$$

To separate a general function $f(x)$ into its odd and even parts, use

$$f(x) \equiv \underbrace{(1/2)\,[f(x)+f(-x)]}_{\text{even}} + \underbrace{(1/2)\,[f(x)-f(-x)]}_{\text{odd}}$$

In some cases by suitable selection of the origin, the function may be made either odd or even, thus simplifying the expansion.

Odd or Even Harmonics

An odd or even function may contain odd or even harmonics. A condition that causes a function $f(x)$ of period T to have only odd harmonics in its Fourier expansion is

$$f(x) = -f(x+T/2)$$

A condition that causes a function $f(x)$ of period T to have only even harmonics in the Fourier expansion is

$$f(x) = f(x+T/2)$$

To separate a general function $f(x)$ into its odd and even harmonics, use

$$f(x) \equiv \underbrace{(1/2)[f(x)+f(x+T/2)]}_{\text{even harmonics}} + \underbrace{(1/2)[f(x)-f(x+T/2)]}_{\text{odd harmonics}}$$

A periodic function may sometimes be changed from odd to even (and vice versa) by a shift of the origin, but the presence of particular odd or even harmonics is unchanged by such a shift.

PULSE-TRAIN ANALYSIS

If the pulse defined by the function $g(t)$ is repeated every interval T, a periodic waveform

$$y(t) = \sum_{n=-\infty}^{+\infty} g(t-nT)$$

results with period T and repetition frequency $F = 1/T$ (see A and B in Table 3).

TABLE 3. THE SPECTRUM FOR PULSE TRAINS*

Waveform	Spectrum
A. Single pulse	
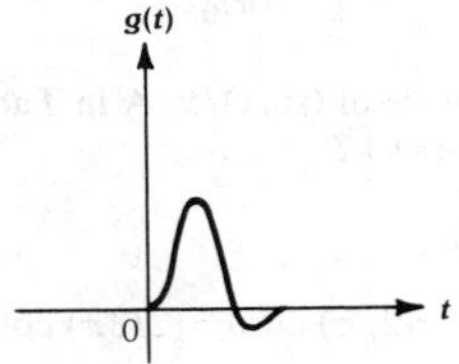	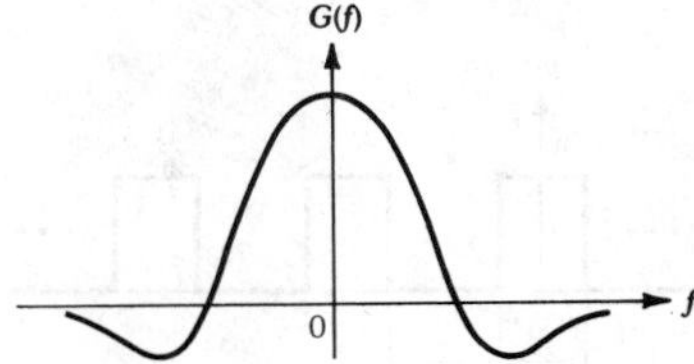
B. Infinite periodic pulse train	
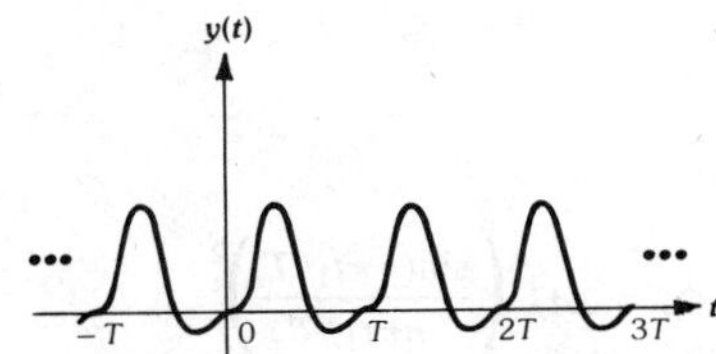	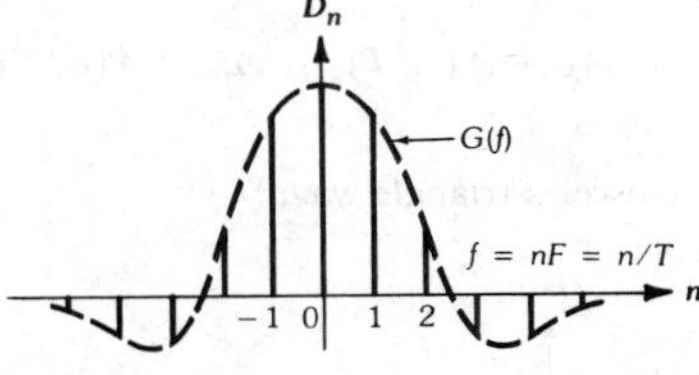
C. Limited pulse train	
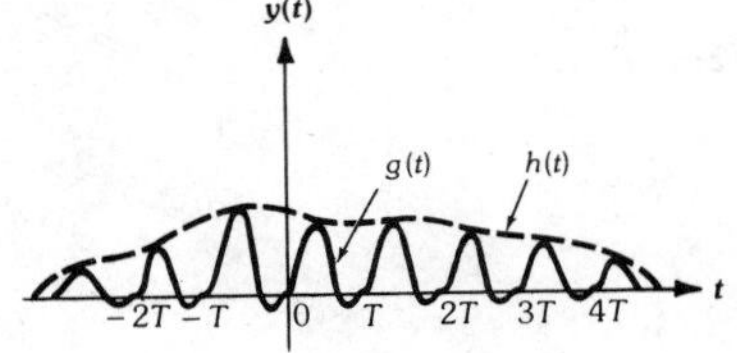	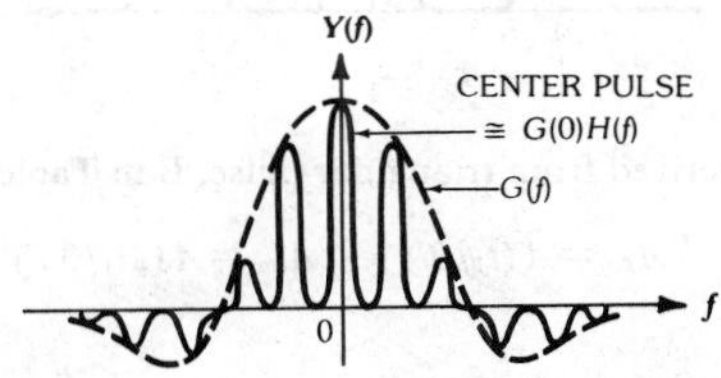

* Spectra are in general complex functions. They are represented here by real curves only to simplify the illustration.

This pulse train may be expressed as a convolution product

$$y(t) = [\sum_{n=-\infty}^{+\infty} \delta(t - nT)]*g(t)$$

and, applying properties 8 and 10 (Table 1), its Fourier transform is

$$Y(f) = (1/T)[\sum_{n=-\infty}^{+\infty} \delta(f - nF)]G(f)$$

The function $y(t)$ is represented by the Fourier series

$$y(t) = \sum_{n=-\infty}^{+\infty} D_n \exp(jn\omega_0 t)$$

where

$$D_n = (1/T)G(nF)$$

The coefficients D_n are obtained by sampling the pulse spectrum at frequencies that are multiples of the repetition frequency.

The amplitude C_n of the nth harmonic in the real representation (see section headed "Real Form of Fourier Series" earlier in this chapter) is

$$C_n = 2|D_n| = (2/T)|G(nF)|$$

The constant term of the series

$$D_0 = A_0/2 = C_0/2$$

is the average amplitude

$$A_{av} = \mathcal{a}T = G(0)/T$$

where

$$\mathcal{a} = \int_0^T g(t)dt$$

is the area under one pulse.

If the pulses do not overlap (i.e., if function $g(t)$ is zero outside of some period a to $a + T$), the energy in a pulse is

$$E = \int_a^{a+T} g^2(t)dt = \int_{-\infty}^{+\infty} |G(f)|^2 df$$

The root-mean-square amplitude is

$$A_{rms} = (E/T)^{1/2}$$

The average power of the pulse train is

Table 4. Periodic Waveforms and Fourier Series

Waveform	Coefficient of Fourier Series
A. Rectangular wave 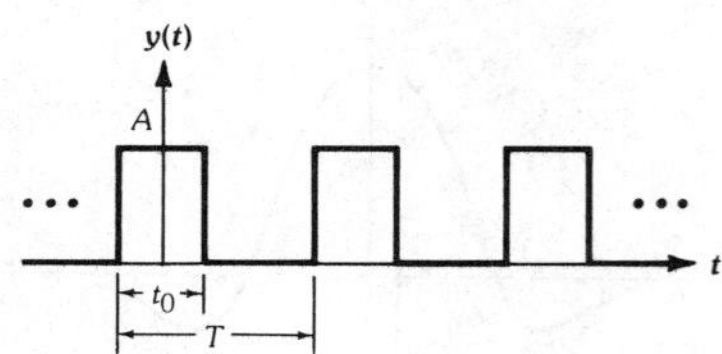 Derived from rectangular pulse, A in Table 2 $A_{Av}=A\,(t_0/T) \qquad A_{rms}=A\,(t_0/T)^{1/2}$	$C_n=2D_n=2A_{Av}\left\|\dfrac{\sin(n\pi t_0/T)}{n\pi t_0/T}\right\|$ Can be read off curve of $(\sin x)/x$, A in Table 2, by sampling at $n\pi t_0/T$ Example: If $T=2t_0$ $y(t)=2A_{Av}[{}^1\!/_2+(2/\pi)\cos\theta-(2/3\pi)\cos 3\theta+\cdots]$ where $\theta=2\pi t/T$
B. Isosceles-triangle wave 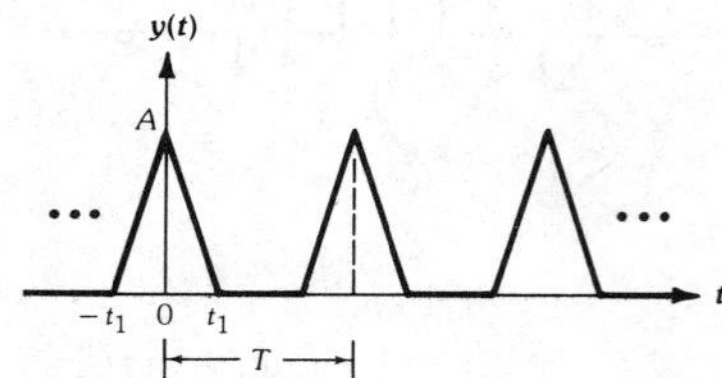 Derived from triangular pulse, B in Table 2 $A_{Av}=A\,(t_1/T) \qquad A_{rms}=A\,(2t_1/3T)^{1/2}$	$C_n=2A_{Av}\left(\dfrac{\sin(n\pi t_1/T)}{n\pi t_1/T}\right)^2$ Example: If $T=2t_1$ $y(t)=2A_{Av}[{}^1\!/_2+(2/\pi)^2\cos\theta+(2/3\pi)^2\cos 3\theta+\cdots]$ where $\theta=2\pi t/T$
C. Sawtooth wave 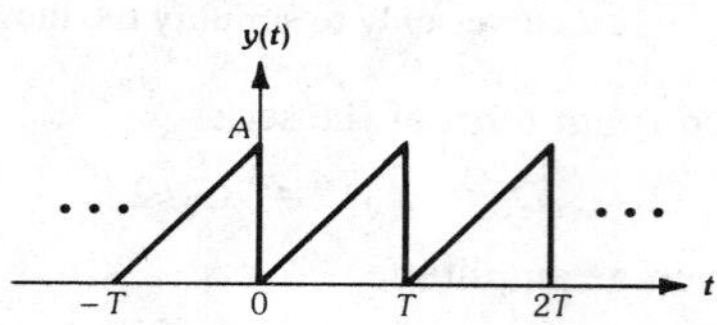 Derived from triangular pulse, C in Table 2 $A_{Av}=A/2 \qquad A_{rms}=A\,(3^{-1/2})$	$C_n=2A_{Av}(1/\pi n)$ $y(t)=2A_{Av}[{}^1\!/_2-(1/\pi)\sin\theta-(1/2\pi)\sin 2\theta-\cdots]$
D. Clipped sawtooth wave 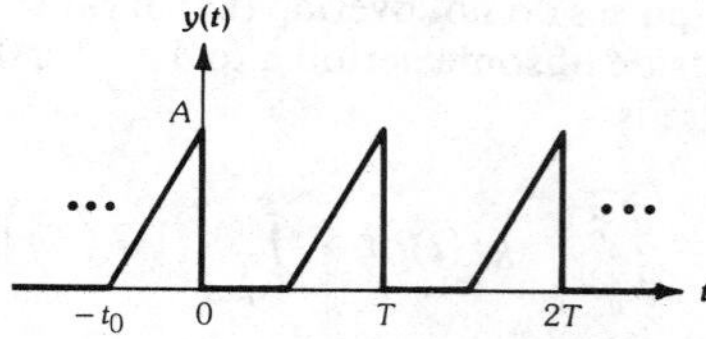 Derived from triangular pulse, C in Table 2 $A_{Av}=A\,(t_0/2T) \qquad A_{rms}=A\,(t_0/3T)^{1/2}$	$C_n=2A_{Av}(1/\alpha^2)[\sin^2\alpha+\alpha(\alpha-\sin 2\alpha)]^{1/2}$ where $\alpha=n\pi t_0/T$

Continued on next page.

TABLE 4 (CONT). PERIODIC WAVEFORMS AND FOURIER SERIES

Waveform	Coefficient of Fourier Series
E. Sawtooth wave 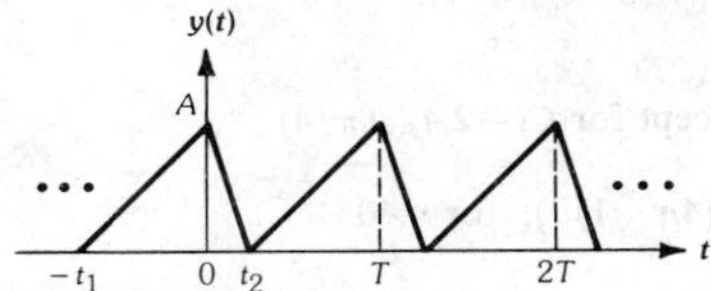 Derived from the sum of two triangular pulses, C in Table 2 $A_{Av}=A/2 \quad A_{rms}=A\,(3^{-1/2})$	$C_n=2A_{Av}(T^2/\pi^2n^2t_1t_2)\ \vert\sin(n\pi t_1/T)\vert$ where $t_1+t_2=T$
F. Symmetrical trapezoidal wave 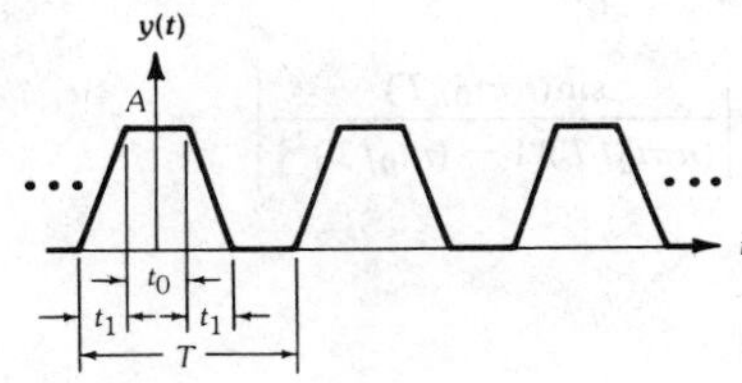 Derived as in D in Table 2 $A_{Av}=A[(t_0+t_1)/T]$ $A_{rms}=A[(3t_0+2t_1)/3T]^{1/2}$	$D_n=A_{Av}\dfrac{\sin(\pi nt_1/T)}{\pi nt_1/T}\dfrac{\sin[\pi n(t_1+t_0)/T]}{\pi n(t_1+t_0)/T}$ $C_n=2\ \vert D_n\vert$
G. Train of cosine pulses 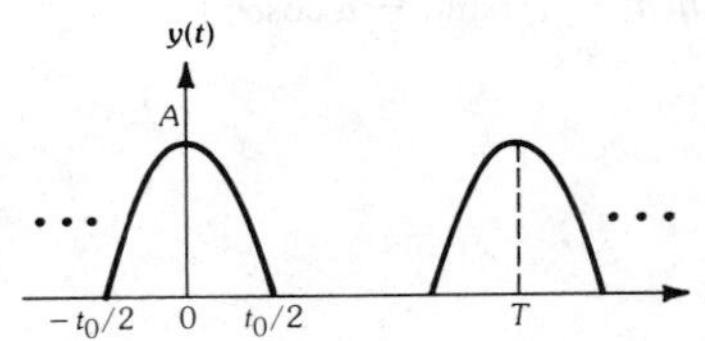 Derived from cosine pulse, E in Table 2 $A_{Av}=(2/\pi)A\,(t_0/T) \quad A_{rms}=A\,(t_0/2T)^{1/2}$	$C_n=2A_{Av}\left\vert\dfrac{\cos(n\pi t_0/T)}{1-(2nt_0/T)^2}\right\vert$ For $nt_0/T=1/2$, this becomes $\pi A_{Av}/2$
H. Full-wave-rectified sine wave 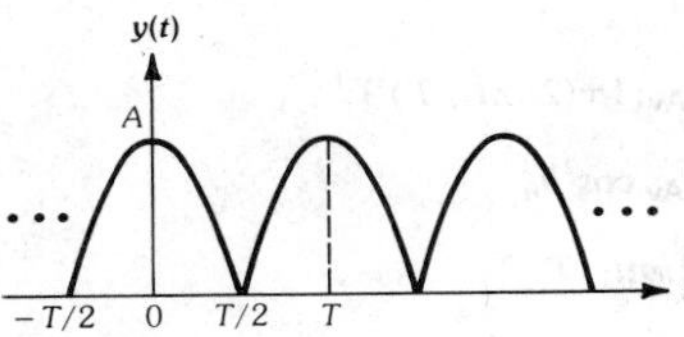 Derived from cosine pulse, E in Table 2 (same as G in Table 4 with $t_0=T$) $A_{Av}=(2/\pi)A \quad A_{rms}=A/(2^{1/2})$	$C_0=2A_{Av}$ $C_n=2A_{Av}(4n^2-1)^{-1}$, for $n\neq 0$ $y(t)=2A_{Av}[1/2+(1/3)\cos\theta-(1/15)\cos 2\theta+(1/35)\cos 3\theta\cdots-(-1)^n(4n^2-1)^{-1}\cos n\theta\cdots]$ where $\theta=2\pi t/T$

Continued on next page.

TABLE 4 (CONT). PERIODIC WAVEFORMS AND FOURIER SERIES

Waveform	Coefficient of Fourier Series

I. Half-wave-rectified sine wave

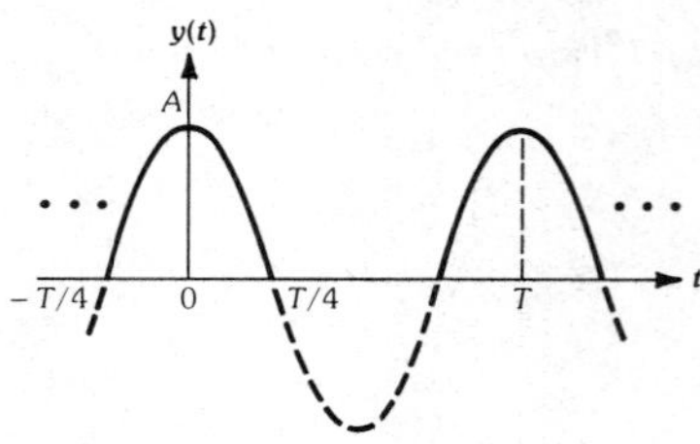

$C_0 = 2A_{Av}$

$C_{2n+1} = 0$, except for $C_1 = 2A_{Av}(\pi/4)$

$C_{2n} = 2A_{Av}(4n^2-1)^{-1})$, for $n \neq 0$

$y(t) = 2A_{Av}[1/2 + (\pi/4)\cos\theta + (1/3)\cos 2\theta - (1/15)\cos 4\theta + \cdots$

$-(-1)^n(4n^2-1)^{-1}\cos 2n\theta \cdots]$

Derived from cosine pluse, E in Table 2 (same as G in Table 4 with $t_0 = T/2$)

$A_{Av} = (1/\pi)A \qquad A_{rms} = A/2$

J. Train of cosine-squared pulses

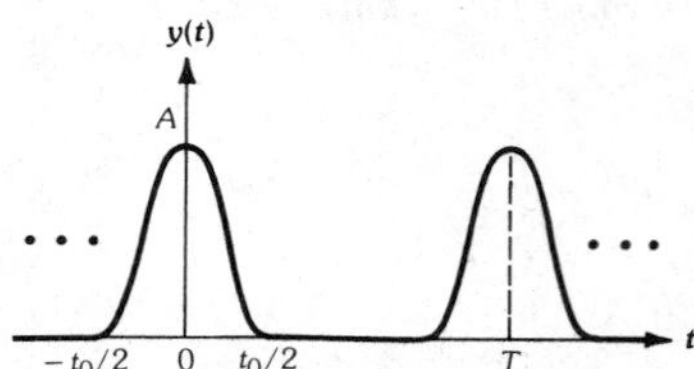

$$C_n = 2A_{Av}\left|\frac{\sin(n\pi t_0/T)}{(n\pi t_0/T)[1-(nt_0/T)^2]}\right|$$

Derived from cosine-squared pulse, F in Table 2

$A_{Av} = (1/2)A(t_0/T) \qquad A_{rms} = (1/2)4(3t_0/2T)^{1/2}$

K. Fractional sine wave

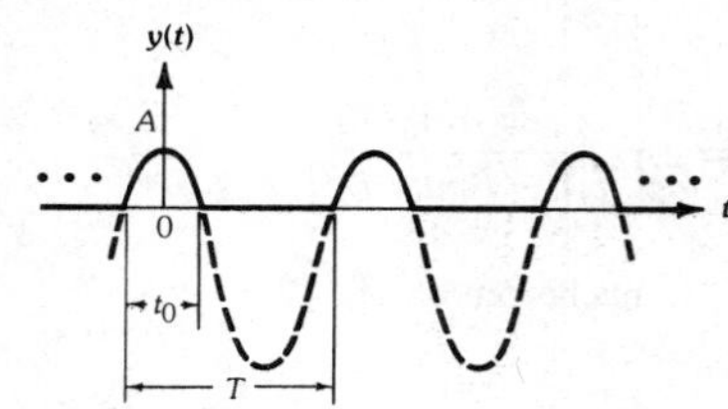

$$C_n = 2A_{Av}\left|\frac{\sin n\alpha \cos\alpha - n\sin\alpha\cos n\alpha}{n(n^2-1)(\sin\alpha - \alpha\cos\alpha)}\right|$$

$$A_{Av} = \frac{A}{\pi}\,\frac{\sin\alpha - \alpha\cos\alpha}{1-\cos\alpha}$$

when $n = 1$,

$$C_1 = A_{Av}\left|\frac{\alpha - \cos\alpha\sin\alpha}{\sin\alpha - \alpha\cos\alpha}\right|$$

$$A_{rms} = \frac{A}{(2\pi)^{1/2}}\,\frac{[2\alpha + \alpha\cos 2\alpha - (3/2)\sin 2\alpha]^{1/2}}{1-\cos\alpha}$$

where $\alpha = \pi t_0/T$

L. Critically damped exponential wave

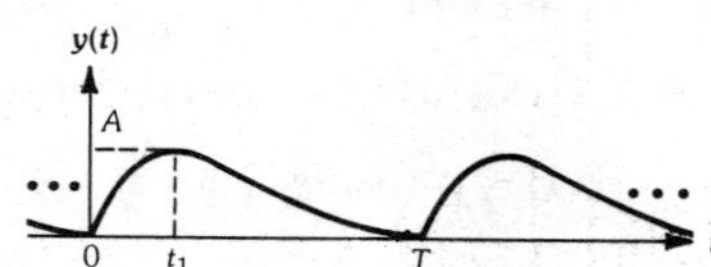

$C_n = 2A_{Av}[1 + (2\pi n t_1/T)^2]^{-1}$

$= 2A_{Av}\cos^2\theta_n$

where $\tan\theta_n = 2\pi n t_1/T$

Derived from exponential pulse, H in Table 2 (period $T >>$ period t_1 to make overlap negligible)

$A_{Av} = Ae(t_1/T) \qquad A_{rms} = (Ae/2)(t_1/T)^{1/2}$

$e = 2.71828\cdots$

$$E/T = A_{rms}^2 = \sum_{n=-\infty}^{+\infty} |D_n|^2$$

$$= (1/4)\,C_0^2 + (1/2) \sum_{n=1}^{+\infty} C_n^2$$

A pulse train of finite extent, where all pulses have the same shape and are spaced periodically, may be represented as a product

$$y(t) = h(t) \sum_{n=-\infty}^{+\infty} g(t - nT)$$

The function $h(t)$ defines the envelope of the pulse train.

The Fourier transform

$$Y(f) = (1/T) \sum_{n=-\infty}^{+\infty} G(nF)H(f - nF)$$

is given by a weighted sum of shifted Fourier transforms of $h(t)$. If $h(t) = 1$, then $H(f)$ is the δ function and the pulse train is a periodic waveform having a line spectrum as in B of Table 3. However, if $h(t)$ does not have infinite support, then this line spectrum is broadened. If $h(t)$ has support over a large number of periods T, so that $H(f)$ is narrow compared to F, then $Y(f)$ is approximately as shown in C of Table 3 where the envelope is $G(f)$ and each pulse is a scaled, shifted version of $H(f)$.

The Fourier series coefficients for a number of commonly encountered pulse trains are given in Table 4 (page 7-10).

SPECTRAL ANALYSIS

If $g(t)$ is band-limited and also nearly time-limited, then approximate samples of its Fourier transform $G(f)$ can be computed from samples of $g(t)$ by using the discrete Fourier transform (DFT). The DFT of the sampled sequence $\{g(nT_s)\}_{n=0}^{N-1}$ is defined by

$$G_m = \sum_{n=0}^{N-1} g(nT_s) \exp[-j(2\pi/N)nm]$$

$m = 0, \ldots, N-1$

If the sampling frequency $1/T_s$ is greater than twice the highest frequency contained in $g(t)$, and if $g(t)$ is nearly zero outside the interval $[0, (N-1)T_s]$, then it can be shown that

$$G_m \approx \begin{cases} (1/T_s)\, G\left(\dfrac{m}{NT_s}\right) & 0 \le m < N/2 \\[2ex] (1/T_s)\, G\left(\dfrac{m-N}{NT_s}\right) & N/2 \le m \le N-1 \end{cases}$$

where the argument of the Fourier transform G is in hertz. Ordinarily $g(t)$ is not time-limited, and only a finite-length section of $g(t)$ can be sampled and processed. This places a limit on achievable spectral resolution.

For additional discussion, including implementation of the DFT using the fast Fourier transform (FFT) algorithm, see Chapter 28.

8 Filters, Simple Bandpass Design

COEFFICIENT OF COUPLING

Several types of coupled circuits are shown in Table 1 (pages 8-4 and 8-5) and Figs. 1 and 2, together with equations for the coefficient of coupling. Also shown is the dependence of bandwidth on resonance frequency. This dependence is only a rough approximation to show the trend and may be altered radically if L_m, M, or C_m is adjusted in tuning to various frequencies.

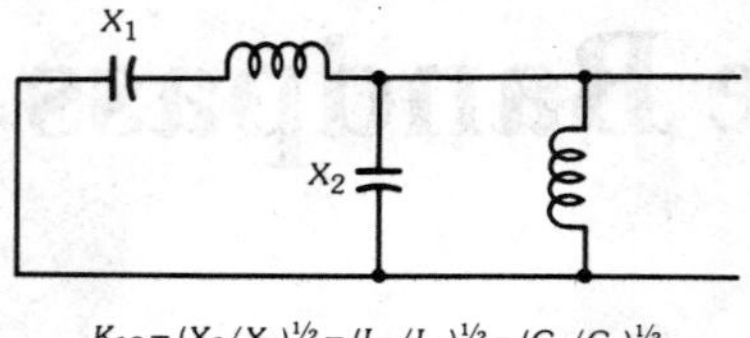

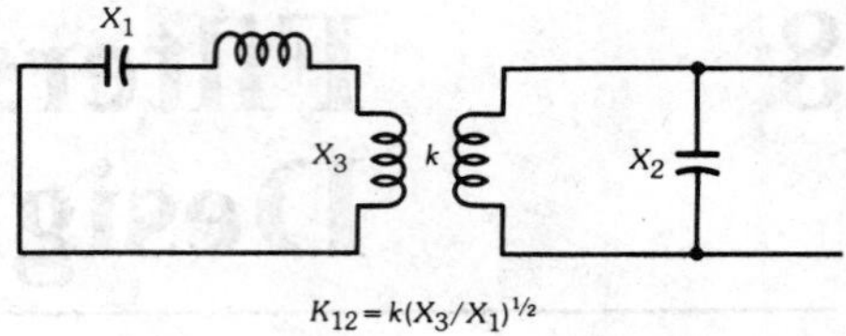

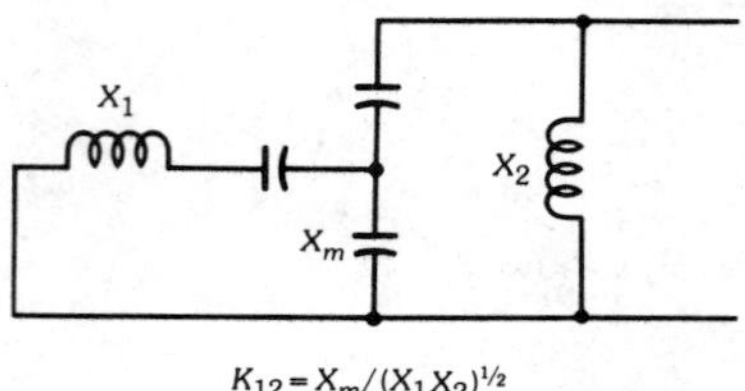

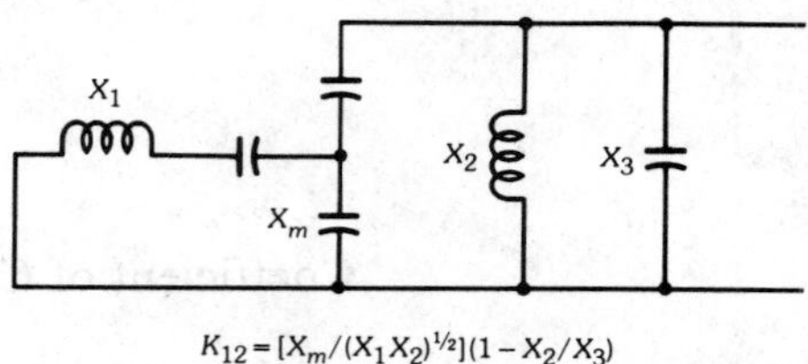

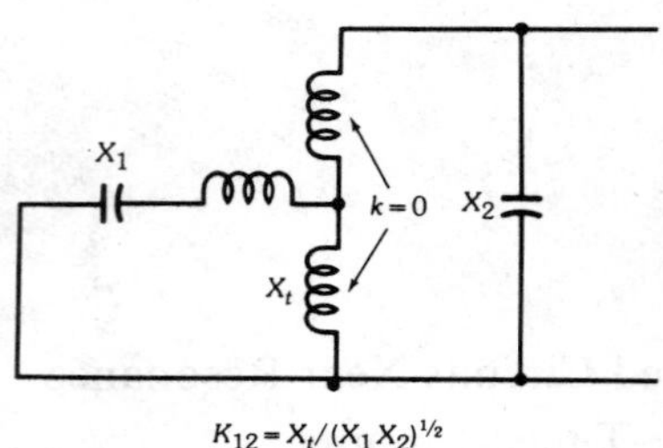

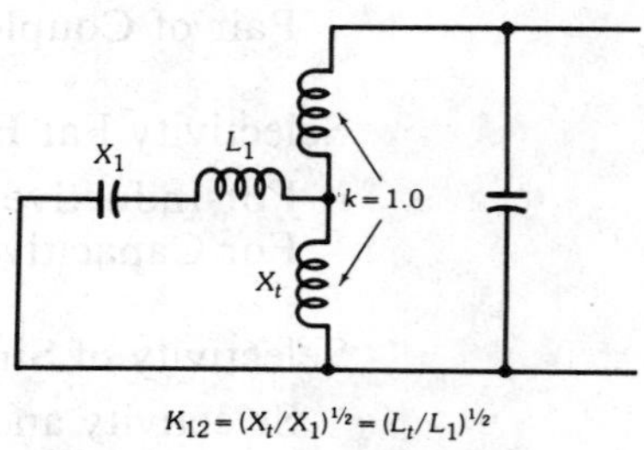

Fig. 1. Additional coefficient-of-coupling configurations (the node resonator is tuned to the desired midfrequency with the mesh resonator open-circuited, or the mesh resonator is tuned to this midfrequency with the node resonator short-circuited).

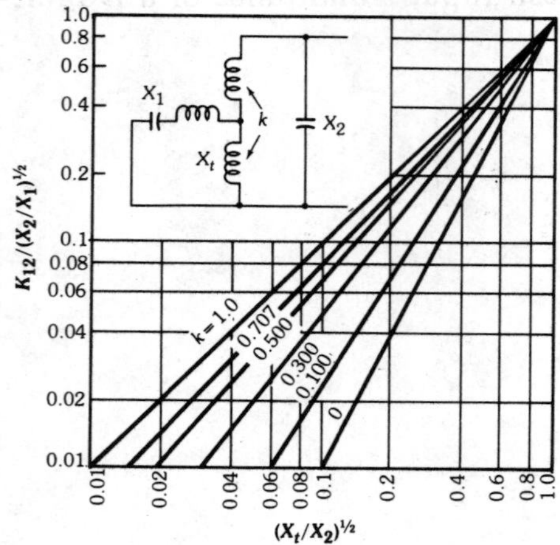

Fig. 2. Coefficient of coupling for configuration shown.

K_{12} = coefficient of coupling between resonant circuits

X_{10} = reactance of inductor (or capacitor) of first circuit at f_0

X_{20} = reactance of similar element of second circuit at f_0

$(bw)_C$ = bandwidth with capacitive tuning

$(bw)_L$ = bandwidth with inductive tuning

GAIN AT RESONANCE

Single Circuit

In Table 1A

$$E_0/E_g = -g_m|X_{10}|Q$$

where,

E_0 = output volts at resonance frequency f_0,

E_g = input volts to active device,
g_m = transconductance of active device.

Pair of Coupled Circuits

Pairs of coupled circuits are illustrated in Figs. 3 and 4.

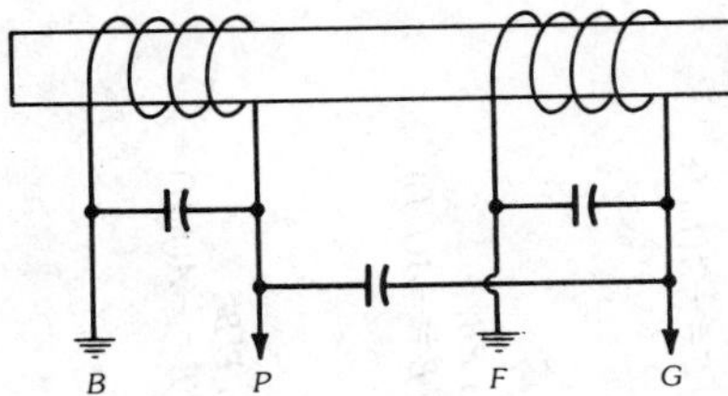

Fig. 3. Connection wherein k_m opposes k_c (k_c may be due to stray capacitance). Peak of attenuation is at $f = f_0(-k_m/k_c)^{1/2}$. Reversing connections or winding direction of one coil causes k_m to aid k_c.

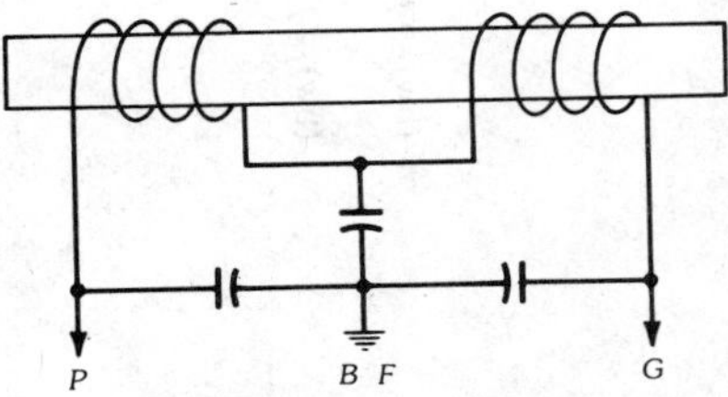

Fig. 4. Connection wherein k_m aids k_c. If mutual-inductance coupling is reversed, k_m will oppose k_c and there will be a transfer minimum at $f = f_0(-k_m/k_c)^{1/2}$.

In B through F in Table 1

$$E_0/E_g = jg_m(X_{10}X_{20})^{1/2}Q[K_{12}Q/(1 + K_{12}^2Q^2)]$$

This is maximum at critical coupling, where $K_{12}Q = 1$.

$Q = (Q_1Q_2)^{1/2}$ = geometric-mean Q for the two circuits, as loaded with active device input and output impedances

For circuits with critical coupling and overcoupling, the approximate gain is

$$|E_0/E_g| \approx 0.1g_m/(C_1C_2)^{1/2}(\text{bw})$$

where,

(bw) is the useful passband in megahertz,
g_m is in micromhos,
C is in picofarads.

SELECTIVITY FAR FROM RESONANCE

The selectivity curves of Fig. 5 are based on the presence of only a single type of coupling between the circuits. The curves are useful beyond the peak region treated in the section on selectivity near resonance.

In the equations for selectivity in Table 1

E = output volts at signal frequency f for same value of E_g as that producing E_0

For Inductive Coupling

$$A = \frac{Q^2}{1 + K_{12}^2Q^2}\left[\left(\frac{f}{f_0} - \frac{f_0}{f}\right)^2 - K_{12}^2\left(\frac{f}{f_0}\right)^2\right]$$

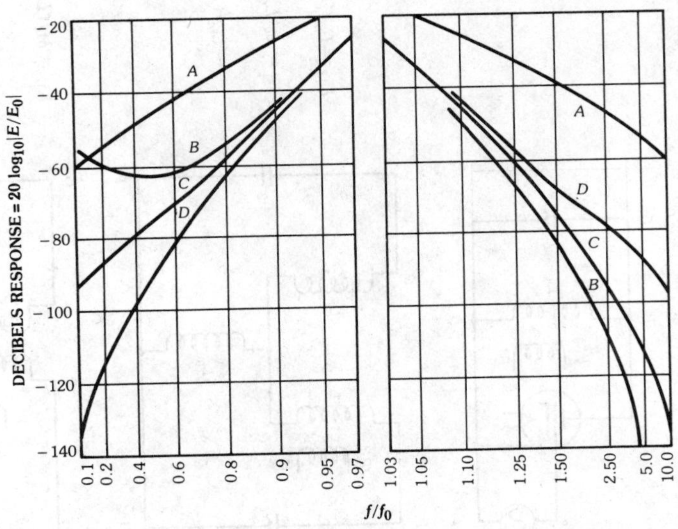

Fig. 5. Selectivity for frequencies far from resonance. Q = 100 and $|K_{12}|\ Q = 1.0$.

TABLE 1. SEVERAL TYPES OF COUPLED CIRCUITS, SHOWING COEFFICIENT OF COUPLING AND SELECTIVITY EQUATIONS

Diagram	Coefficient of Coupling	Approximate Bandwidth Variation With Frequency	Selectivity Far From Resonance: Equation*	Selectivity Far From Resonance: Curve in Fig. 5
A 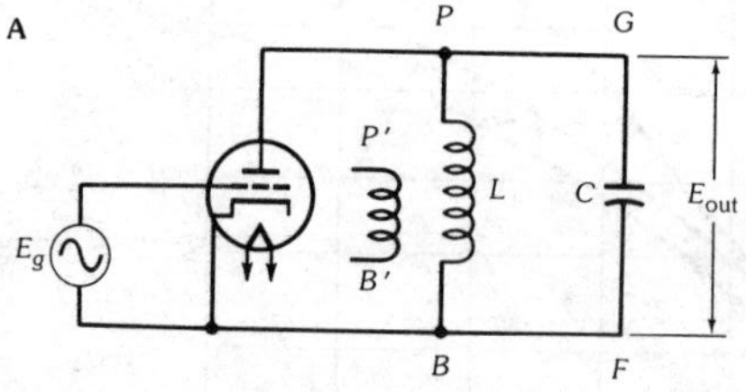			Input to PB or to $P'B'$: $E_0/E=jQ[(f/f_0)-(f_0/f)]$	A
B 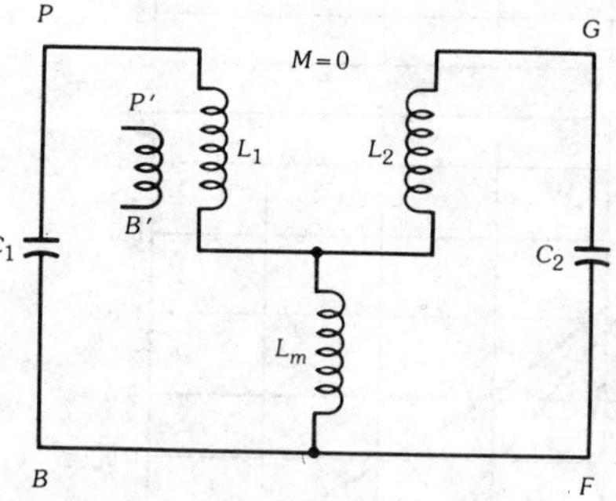	$K_{12}=L_m/[(L_1+L_m)(L_2+L_m)]^{1/2}$ $=\omega_0{}^2L_m(C_1C_2)^{1/2}$ $\approx L_m/(L_1L_2)^{1/2}$	$(\mathrm{bw})_C\propto f_0$ $(\mathrm{bw})_L\propto f_0{}^3$	Input to PB: $E_0/E=-A(f/f_0)$ Input to $P'B'$: $E_0/E=-A(f_0/f)$	C D
C 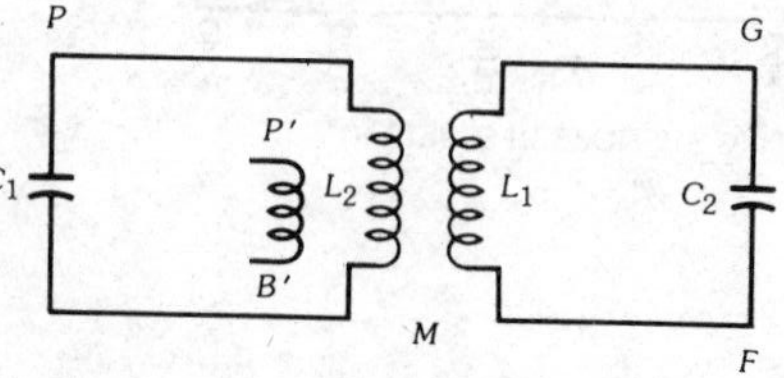	$K_{12}=M/(L_1L_2)^{1/2}$ $=\omega_0{}^2M(C_1C_2)^{1/2}$ M may be positive or negative	$(\mathrm{bw})_C\propto f_0$ $(\mathrm{bw})_L\propto f_0{}^3$	Input to PB: $E_0/E=-A(f/f_0)$ Input to $P'B'$: $E_0/E=-A(f_0/f)$	C D

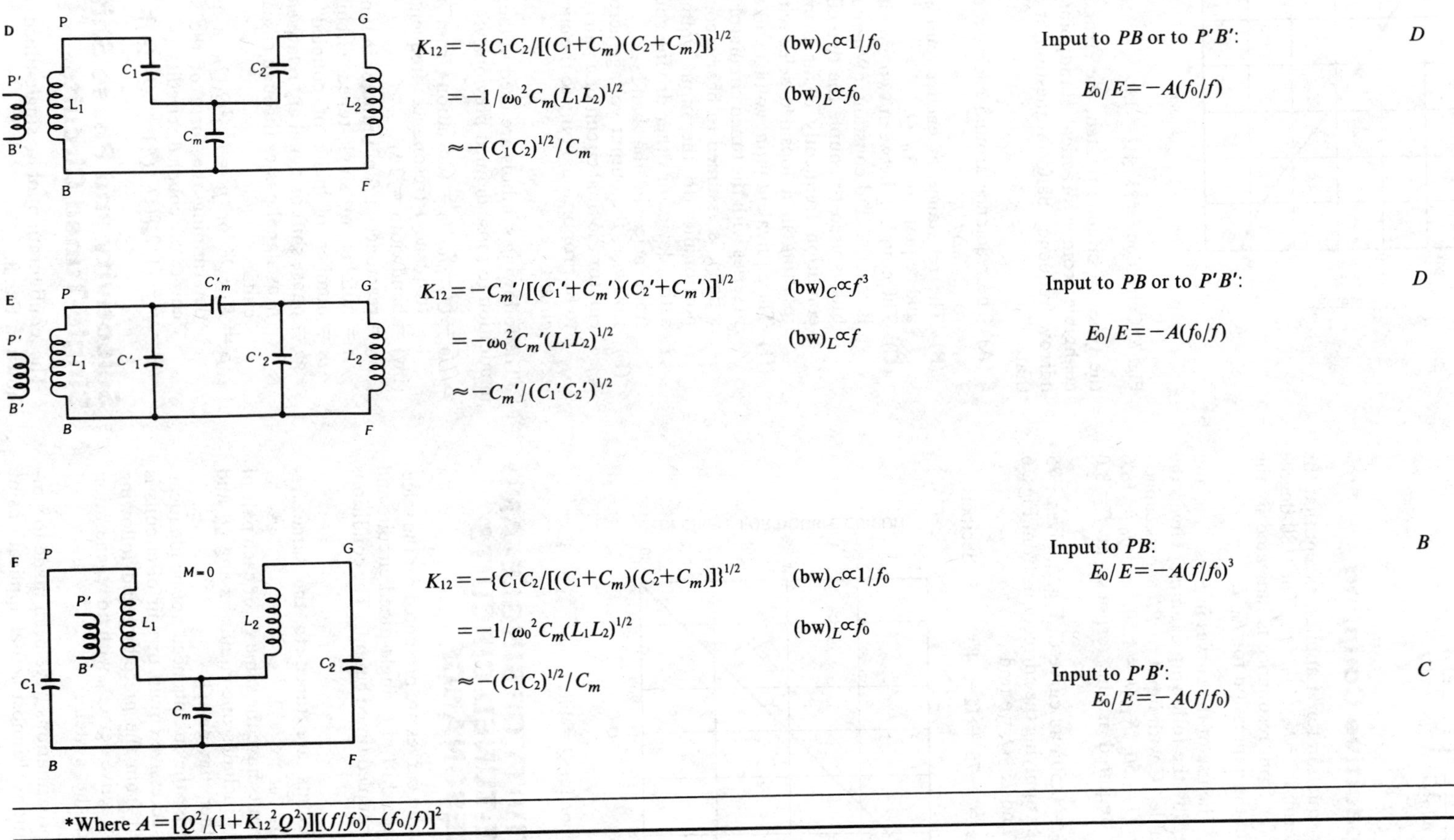

Circuit	Coupling coefficient	Bandwidth	Response*	Curve
D	$K_{12} = -\{C_1C_2/[(C_1+C_m)(C_2+C_m)]\}^{1/2}$ $= -1/\omega_0^2 C_m (L_1L_2)^{1/2}$ $\approx -(C_1C_2)^{1/2}/C_m$	$(\mathrm{bw})_C \propto 1/f_0$ $(\mathrm{bw})_L \propto f_0$	Input to PB or to $P'B'$: $E_0/E = -A(f_0/f)$	D
E	$K_{12} = -C_m'/[(C_1'+C_m')(C_2'+C_m')]^{1/2}$ $= -\omega_0^2 C_m'(L_1L_2)^{1/2}$ $\approx -C_m'/(C_1'C_2')^{1/2}$	$(\mathrm{bw})_C \propto f^3$ $(\mathrm{bw})_L \propto f$	Input to PB or to $P'B'$: $E_0/E = -A(f_0/f)$	D
F	$K_{12} = -\{C_1C_2/[(C_1+C_m)(C_2+C_m)]\}^{1/2}$ $= -1/\omega_0^2 C_m (L_1L_2)^{1/2}$ $\approx -(C_1C_2)^{1/2}/C_m$	$(\mathrm{bw})_C \propto 1/f_0$ $(\mathrm{bw})_L \propto f_0$	Input to PB: $E_0/E = -A(f/f_0)^3$ Input to $P'B'$: $E_0/E = -A(f/f_0)$	B C

*Where $A = [Q^2/(1+K_{12}^2Q^2)][(f/f_0)-(f_0/f)]^2$

$$\approx \frac{Q^2}{1+K_{12}{}^2Q^2}\left(\frac{f}{f_0}-\frac{f_0}{f}\right)^2$$

For Capacitive Coupling

A similar equation for A applies, except that the neglected term is $-K_{12}{}^2(f_0/f)^2$. The 180-degree phase shift far from resonance is indicated by the minus sign in the expression for E_0/E.

Example. The use of the curves in Figs. 5, 6, and 7 is indicated by the following example. Given the circuit of Table 1C with input to PB across capacitor C_1. Let $Q = 50$, $K_{12}Q = 1.50$, and $f_0 = 16.0$ megahertz. Required is the response at $f = 8.0$ megahertz.

Here, $f/f_0 = 0.50$ and curve C, Fig. 5, gives -75 decibels. Then applying the corrections from Figs. 6 and 7 for Q and $K_{12}Q$, we find

$$\text{Response} = -75 + 12 + 4 = -59 \text{ decibels}$$

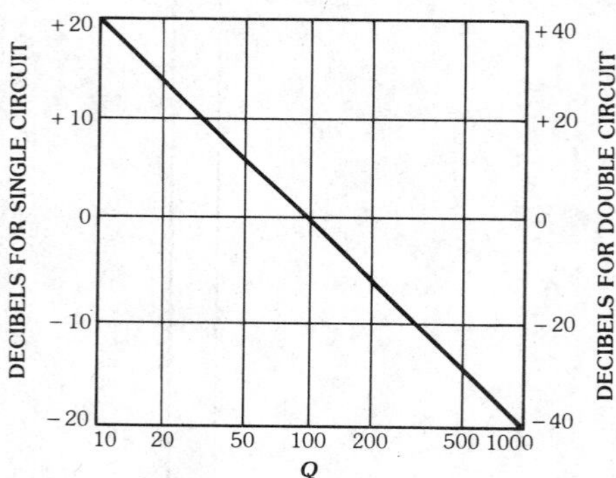

Fig. 6. Correction for $Q \neq 100$.

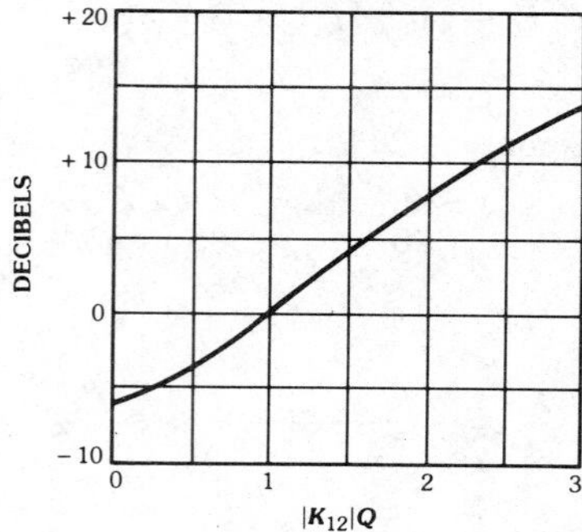

Fig. 7. Correction for $|K_{12}|Q \neq 1.0$.

SELECTIVITY OF SINGLE- AND DOUBLE-TUNED CIRCUITS NEAR RESONANCE

Equations and curves are presented for the selectivity and phase shift of n single-tuned circuits and of m pairs of coupled tuned circuits. The conditions assumed are

(A) All circuits are tuned to the same frequency, f_0.
(B) All circuits have the same Q, or each pair of circuits includes one circuit having Q_1 and the other having Q_2.
(C) Otherwise the circuits need not be identical.
(D) Each successive circuit or pair of circuits is isolated from the preceding and following ones by active devices, with no regeneration around the system.

Certain approximations have been made to simplify the equations. In most actual applications of the types of circuits treated, the error involved is negligible from a practical standpoint. Over the narrow frequency band in question, it is assumed that

(A) The reactance around each circuit is equal to $2X_0\Delta f/f_0$.
(B) The resistance of each circuit is constant and equal to X_0/Q.
(C) The coupling between two circuits of a pair is reactive and constant. (When an untuned link is used to couple the two circuits, this condition frequently is far from satisfied, resulting in a lopsided selectivity curve.)
(D) The equivalent input voltage, taken as being in series with the tuned circuit (or the first of a pair), is assumed to bear a constant proportionality to the input voltage of the active device or other driving source, at all frequencies in the band.
(E) Likewise, the output voltage across the circuit (or the final circuit of a pair) is assumed to be proportional only to the current in the circuit.

The following symbols are used in the equations in addition to those defined previously:

$\Delta f/f_0 = (f - f_0)/f_0 =$ (deviation from resonance frequency)/(resonance frequency)
(bw) = bandwidth = $2\Delta f$
$X_0 =$ reactance at f_0 of inductor in tuned circuit
$n =$ number of single-tuned circuits
$m =$ number of pairs of coupled circuits
$\phi =$ phase shift of signal at f relative to shift at f_0 as signal passes through cascade of circuits
$p = K_{12}{}^2Q^2$ or $p = K_{12}{}^2Q_1Q_2$, a parameter determining the form of the selectivity curve of coupled circuits
$B = p - (1/2)[(Q_1/Q_2) + (Q_2/Q_1)]$

Selectivity and Phase Shift of Single-Tuned Circuits

The configuration of a single-tuned circuit is shown in Fig. 8.

$$E/E_0 = \{[1 + (2Q\Delta f/f_0)^2]^{-1/2}\}^n$$

$$\Delta f/f_0 = \pm(2Q)^{-1}[(E_0/E)^{2/n} - 1]^{1/2}$$

Decibel response $= 20 \log_{10}(E/E_0)$

(dB response of n circuits) $= n \times$ (dB response of single circuit)

$$\phi = n \tan^{-1}(-2Q\,\Delta f/f_0)$$

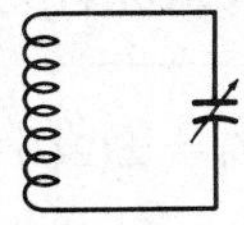

Fig. 8. Single-tuned circuit.

These equations are plotted in Figs. 9 and 10.

Example of the use of Figs. 9 and 10: Suppose there are three single-tuned circuits ($n = 3$). Each circuit has a Q of 200 and is tuned to 1000 kilohertz. The results are shown in Table 2.

Q Determination by 3-Decibel Points

For a single-tuned circuit, when

$E/E_0 = 0.707$ (3 decibels down)

$$Q = f_0/2\Delta f$$

$$= (\text{resonance frequency})/(\text{bandwidth})_{3\text{dB}}$$

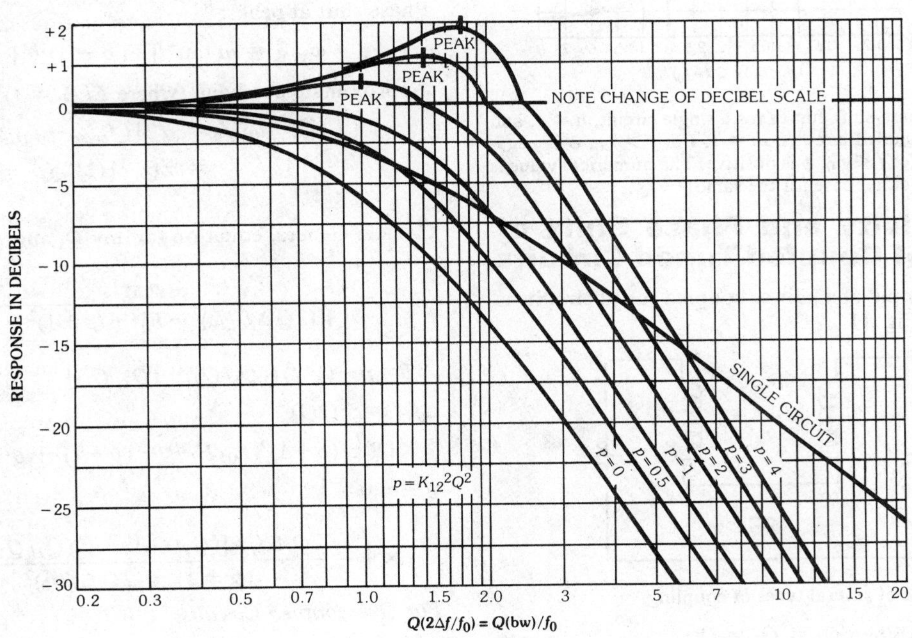

The selectivity curves are symmetrical about the axis $Q\Delta f/f_0 = 0$ for practical purposes.
Extrapolation beyond lower limits of chart:

Δ Response for Doubling Δf	Circuit	Useful Limit at (bw)/f_0	Useful Limit Error becomes
−6 dB	←single→	0.6	1 to 2 dB
−12 dB	←pair→	0.4	3 to 4 dB

Fig. 9. Selectivity curves showing response of a single circuit, $n = 1$, and a pair of coupled circuits, $m = 1$.

TABLE 2. EXAMPLE OF USE OF FIGS. 9 AND 10

Abscissa Q (bw)/f_0	Bandwidth (kilohertz)	Ordinate dB Response for n=1	Decibels Response for n=3	ϕ* for n=1	ϕ for n=3
1.0	5.0	−3.0	−9	∓45°	∓135°
3.0	15	−10.0	−30	∓71½°	∓215°
10.0	50	−20.2	−61	∓84°	∓252°

* ϕ is negative for $f > f_0$, and vice versa.

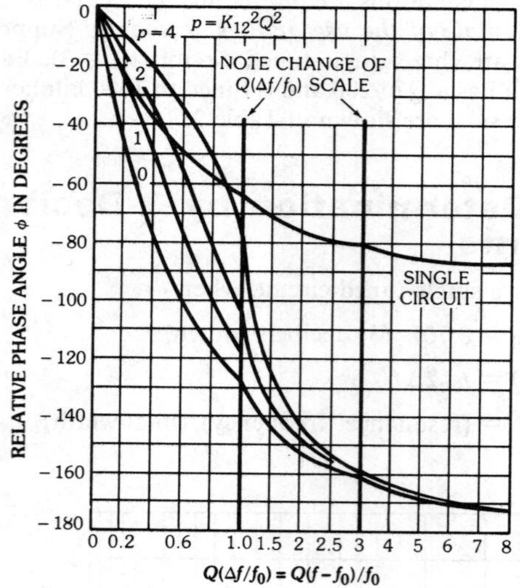

Fig. 10. Phase-shift curves for a single circuit, $n = 1$, and a pair of coupled circuits, $m = 1$. For $f > f_0$, ϕ is negative, while for $f < f_0$, ϕ is positive. The numerical value is identical in either case for the same $|f - f_0|$.

Selectivity and Phase Shift of Pairs of Coupled Tuned Circuits

An example of a pair of coupled tuned circuits is shown in Fig. 11.

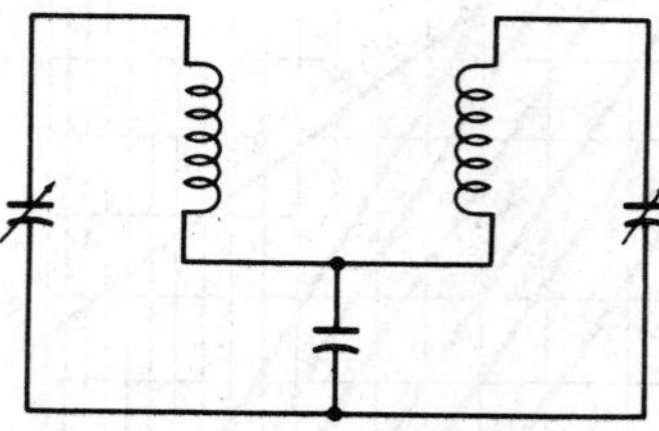

Fig. 11. One of several types of coupling.

CASE 1: When $Q_1 = Q_2 = Q$:

These equations can be used with reasonable accuracy when Q_1 and Q_2 differ by ratios up to 1.5 or even 2 to 1. In such cases, use the value $Q = (Q_1Q_2)^{1/2}$.

$$E/E_0 = \left[\frac{p+1}{\{[(2Q\,\Delta f/f_0)^2 - (p-1)]^2 + 4p\}^{1/2}}\right]^m$$

$$\Delta f/f_0 = \pm(2Q)^{-1}$$

$$\times\{(p-1)\pm[(p+1)^2(E_0/E)^{2/m} - 4p]^{1/2}\}^{1/2}$$

For very small values of E/E_0,

$$E/E_0 = [(p+1)/(2Q\Delta f/f_0)^2]^m$$

Decibel response $= 20\log_{10}(E/E_0)$

(dB response of m pairs of circuits) $= m \times$ (dB response of one pair)

$$\phi = m\tan^{-1}\left[\frac{-4Q\,\Delta f/f_0}{(p+1) - (2Q\,\Delta f/f_0)^2}\right]$$

As p approaches zero, the selectivity and phase shift approach the values for n single circuits, where $n = 2m$ (gain also approaches zero).

The above equations are plotted in Figs. 9 and 10.

For Overcoupled Circuits ($p > 1$):

Location of peaks:

$$(f_{\text{peak}} - f_0)/f_0 = \pm(2Q)^{-1}(p-1)^{1/2}$$

Amplitude of peaks:

$$E_{\text{peak}}/E_0 = [(p+1)/2(p^{1/2})]^m$$

Phase shift at peaks:

$$\phi_{\text{peak}} = m\tan^{-1}[\mp(p-1)^{1/2}]$$

Approximate passband (where $E/E_0 = 1$) is

$$(f_{\text{unity}} - f_0)/f_0 = \sqrt{2}\,[(f_{peak} - f_0)/f_0]$$
$$= \pm Q^{-1}[(1/2)(p-1)]^{1/2}$$

CASE 2: General equation for any Q_1 and Q_2:

$$E/E_0 = \left[\frac{p+1}{\{[(2Q\,\Delta f_0/f_0)^2 - B]^2 + (p+1)^2 - B^2\}^{1/2}}\right]^m$$

$$B = p - (1/2)[(Q_1/Q_2) + (Q_2/Q_1)]$$

$$\Delta f/f_0 = \pm(2Q)^{-1}$$

$$\times\{B\pm[(p+1)^2(E_0/E)^{2/m} - (p+1)^2 + B^2]^{1/2}\}^{1/2}$$

$$\phi = m\tan^{-1}$$

$$\times\left(-\frac{2Q\,\Delta f/f_0[(Q_1/Q_2)^{1/2} + (Q_2/Q_1)^{1/2}]}{(p+1) - (2Q\,\Delta f/f_0)^2}\right)$$

For Overcoupled Circuits:

Location of peaks:

$$(f_{\text{peak}} - f_0)/f_0 = \pm B^{1/2}/2Q$$
$$= \pm(1/2)[K_{12}^2 - (1/2)(1/Q_1^2 + 1/Q_2^2)]^{1/2}$$

Amplitude of peaks:

$$E_{\text{peak}}/E_0 = \{(p+1)/[(p+1)^2 - B^2]^{1/2}\}^m$$

CASE 3: Peaks just converged to a single peak:

Here $B = 0$ or $K_{12}^2 = (1/2)(1/Q_1^2 + 1/Q_2^2)$

$$E/E_0 = \{2/[(2Q'\Delta f/f_0)^4 + 4]^{1/2}\}^m$$

where,

$$Q' = 2Q_1Q_2/(Q_1 + Q_2)$$

$$\Delta f/f_0 = \pm(\sqrt{2}/4)(1/Q_1 + 1/Q_2)[(E_0/E)^{2/m} - 1]^{1/4}$$

$$\phi = m \tan^{-1}\left[-\frac{4Q'\,\Delta f/f_0}{2-(2Q'\Delta f/f_0)^2}\right]$$

The curves of Figs. 9 and 10 may be applied to this case, using the value $p = 1$ and substituting Q' for Q.

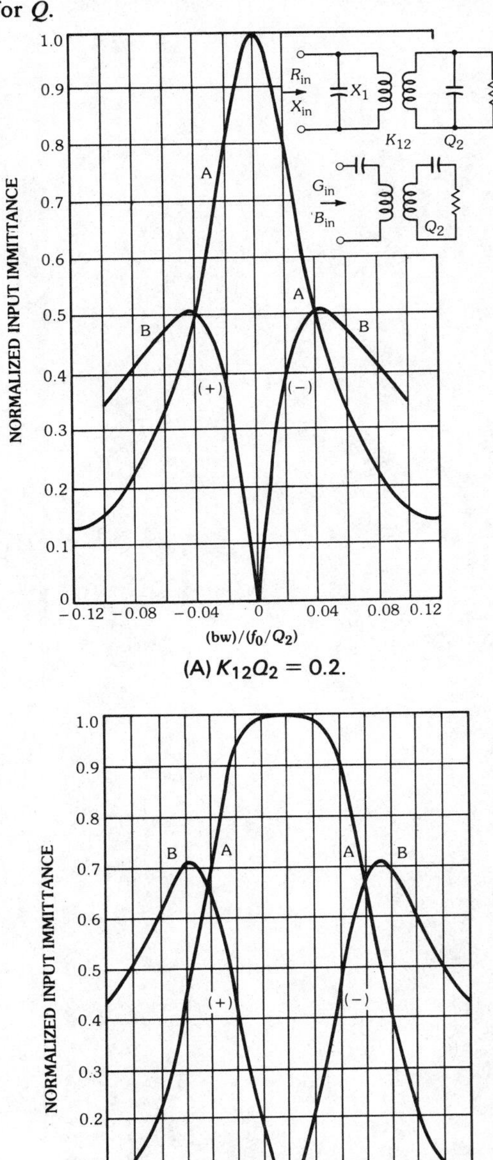

(A) $K_{12}Q_2 = 0.2$.

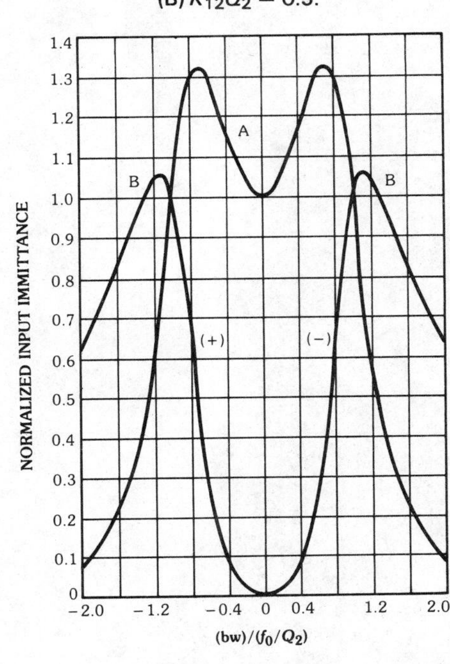

NODE INPUT IMPEDANCE OR MESH INPUT ADMITTANCE OF A DOUBLE-TUNED CIRCUIT

Fig. 12 gives the normalized input immittance versus the normalized frequency of double-tuned circuits.

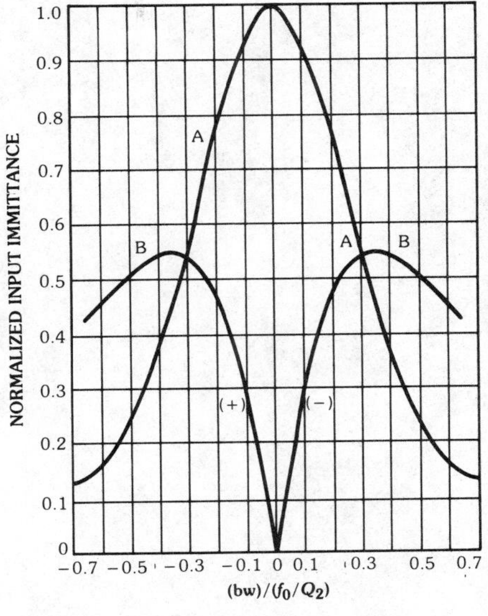

(B) $K_{12}Q_2 = 0.5$.

(C) $K_{12}Q_2 = 0.707$.

(D) $K_{12}Q_2 = 1.0$.

Fig. 12. Normalized input immittance versus normalized frequency of double-tuned circuits. $A = R_{in}/(X_1/K_{12}^2Q_2)$ or $G_{in}/(B_1/K_{12}^2Q_2)$. $B = X_{in}/(X_1/K_{12}^2Q_2)$ or $B_{in}/(B_1/K_{12}^2Q_2)$.

9 Filters, Modern-Network-Theory Design

Ford Shepherd

In contrast to filter design by image-parameter methods, the design of filters by the use of modern network theory is a domain for specialists with digital computers because of the complex calculations required. There are sufficient advantages of filter circuits computed by this method, however, to warrant making some application of them easy and straightforward. This chapter focuses on a very limited subset of the limitless possibilities for low-pass networks and simple transform methods to allow calculation of high-pass, bandpass, and band-stop circuits.

The design information is drawn from experience in the application of modern network theory to the design of electric wave filters. As stated above, only limited design results are supplied, and a concentrated study of the cited references is essential to gain a working knowledge of the synthesis process through which these results were computed. References 1, 4, and 6 provide details of the design theory. Reference 5 provides a concise summary of the theory with graphs and tables to enable an engineer to compute filter circuits with the help of a computer program.

Reference 2 provides a much larger tabulation of Cauer-parameter and Chebyshev filter networks up to degree 9, and Saal has also produced another volume that extends to degree 15. Reference 3 presents many practical ideas (drawn from a 25-year career in Europe and the USA) on designing, testing, and manufacturing filters and mentions two of the computer programs that are available. Many books and articles written on this subject since the work of Cauer and Darlington in the late 1930s are more than worthy of mention here. However, the scope of this chapter and the space available do not necessitate nor permit a detailed discussion. Reference 7 is an example of some of the work done toward practical implementation of these filter networks with standard-value capacitors.

No attempt has been made to present details of the theory and formulation involved, of the approximation of transfer polynomials to performance requirements, or the very useful but less frequently required topics such as zero and infinite terminations, phase and delay performance, and the effects of and compensation for the losses in real coils and capacitors.

INTRODUCTION

Filter networks continue to be of great importance in the design of electrical equipment, especially in communication engineering. Unlike previous methodologies to design spectrum-shaping networks, modern network theory enables the engineer to design filter networks that are based on the actual requirements for signal transmission. While image-parameter design is rather simplistic, only a very limited approximation of the specific requirements can be achieved. Today's methods are not so straightforward and are generally considered beyond the scope of nonspecialists, since the mathematical design process does not directly parallel physical conceptions and the calculations are complex and extensive. The development of digital computers has led to the capture of much of the knowledge of the specialist and the complex calculation algorithms into programs that allow many engineers to design some of their own networks. Still, the sophisticated requirements necessitate the special expert.

Many of the less complex requirements can be satisfied by designs that can be done with the procedures described here and without access to the computer programs. To accomplish this task, the requirements for all the filter types considered here (low-pass, high-pass, bandpass and band-stop) are transformed to a set of reference low-pass requirements. From the reference low-pass requirements, the network complexity is determined, and a normalized reference low-pass is selected. Then the suitable transformation is applied to this low-pass to arrive at the network to satisfy the initial requirements.

Fig. 1 shows a "reasonably typical" requirement for a filter network. To satisfy this requirement with the minimum network, it is necessary to use one of the available computer programs. However, it is possible to satisfy the modified requirements, as shown in Fig. 2, by the simple computations of this chapter. The requirements depicted by the tolerance plot of Fig. 2 can be transformed to the basic reference low-pass tolerance plot shown by Fig. 3 with $A_p = A_1$ and $A_s = A_6$. The algorithm for this

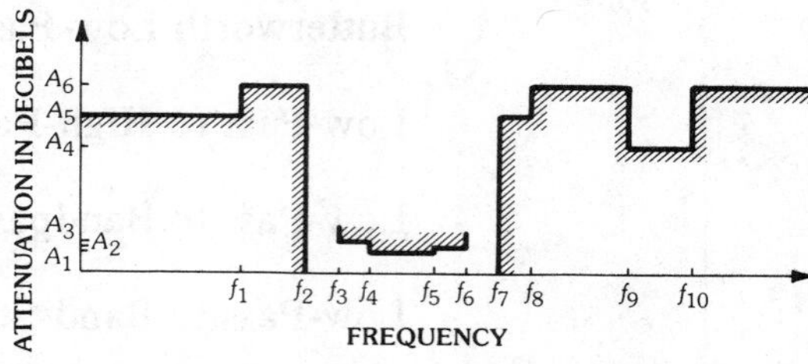

Fig. 1. Requirement for a filter network.

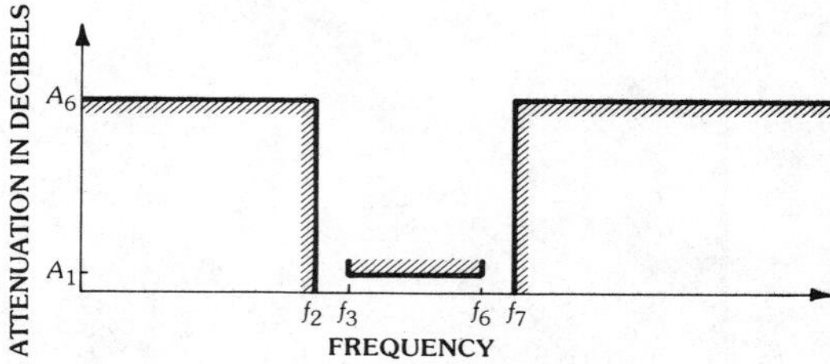

Fig. 2. Modified requirements for filter network.

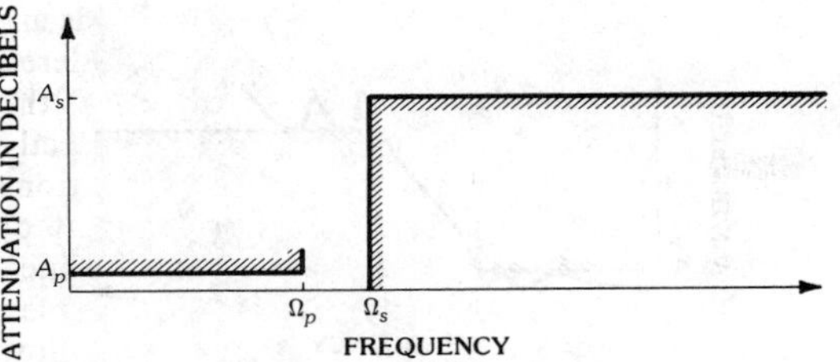

Fig. 3. Basic reference low-pass tolerance plot.

transform is covered in the section headed "The Reference Low-Pass."

NORMALIZATION

To simplify the calculations, frequencies are normalized with respect to some reference frequency, f_{ref}, and the impedances are normalized with respect to a reference resistance, R_{ref}. Hence, normalized frequencies are defined by $\Omega_i = f_i/f_{ref}$, and the normalized resistances are defined by $r_i = R_i/R_{ref}$.

In the case of low-pass filters, it is appropriate to use the upper edge of the passband (or cutoff frequency), f_p, as the reference frequency (hence, $\Omega_p = 1.0$) and use the input resistance, R_1, as the reference resistance ($r_1 = 1.0$).

Conversely, all normalized data of a circuit can be converted to any frequency range and any impedance level by selection of f_{ref} and R_{ref}. By defining

$$L_{ref} = R_{ref}/2\pi f_{ref}$$

and

$$C_{ref} = 1/2\pi f_{ref} R_{ref}$$

the entire circuit can be computed for the appropriate range of frequency and impedance.

$$R_i = r_i \cdot R_{ref} \text{ (ohms)}$$
$$f_i = \Omega_i \cdot F_{ref} \text{ (hertz)}$$
$$L_i = l_i \cdot L_{ref} \text{ (henries)}$$
$$C_i = c_i \cdot C_{ref} \text{ (farads)}$$

THE REFERENCE LOW-PASS

Fig. 4 shows a normal tolerance plot of the requirements for a low-pass filter. Also shown are the associated formulas to calculate the quantities for the reference low-pass of Fig. 3. With $f_{ref} = f_p$ ($\Omega_p = 1.0$), three parameters are required: (1) $\Omega_s = f_s/f_{ref}$; (2) A_p (decibels), the maximum acceptable ripple in the passband; and (3) A_s (decibels), the minimum attenuation in the stopband.

Fig. 5 shows a tolerance plot of requirements for a high-pass filter. Also shown are the equations to determine the reference low-pass parameters.

Fig. 6 shows a tolerance plot of requirements for a bandpass filter, together with the equations to resolve the parameters for the reference low-pass.

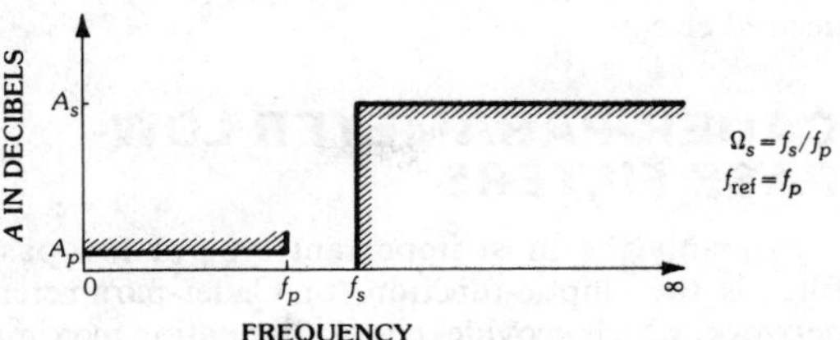

Fig. 4. Requirements for a low-pass filter.

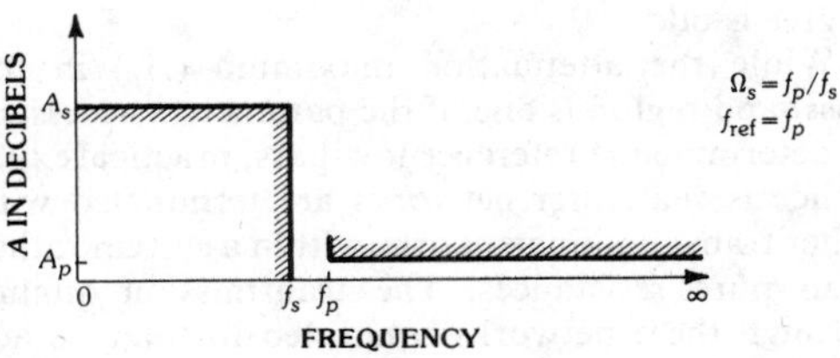

Fig. 5. Requirements for a high-pass filter.

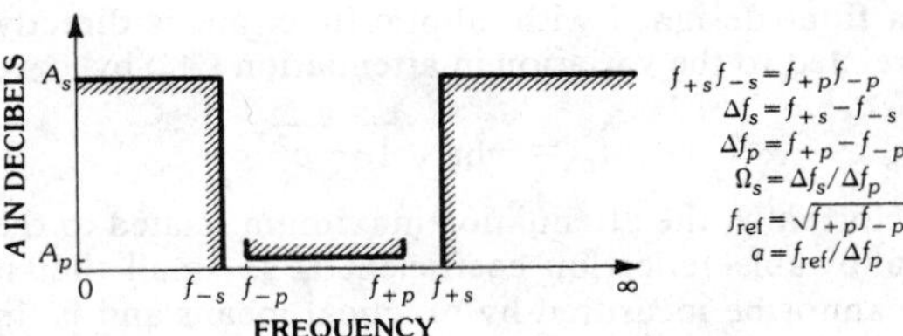

Fig. 6. Requirements for a bandpass filter.

Note that geometric symmetry is required for both stopband and passband limits. Generally, an "overdesign" will result for one or more of the frequencies.

Fig. 7 shows the tolerance plot of the requirements for a band-stop network, together with the associated equations to calculate the reference low-pass parameters. Again, symmetry is required as for the bandpass case.

In both the bandpass and band-stop equations, a transformation factor, *a*, is defined and is needed

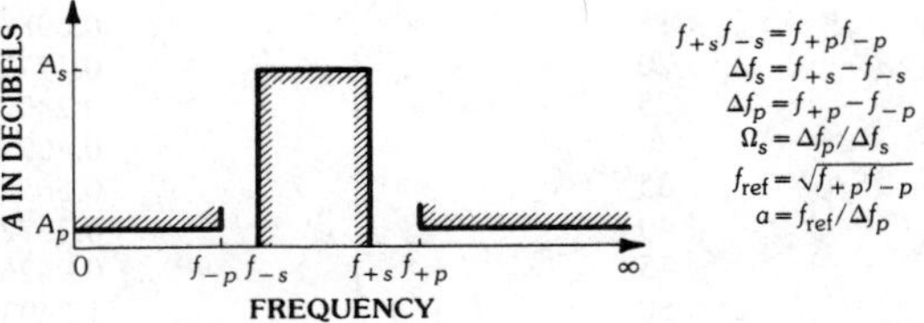

Fig. 7. Requirements for a band-stop network.

for transforming the reference low-pass into the desired circuit.

CAUER-PARAMETER LOW-PASS FILTERS

Probably the most important type of low-pass filter is the elliptic-function, or Cauer-parameter, network, which provides equal attenuation maxima in the passband region and equal attenuation minima in the stopband. Fig. 8 shows the attenuation versus frequency performance and the two possible circuit configurations for this type of filter when the degree is odd.

While the attenuation maximum (A_p) in the passband region is one of the parameters necessary to determine the reference low-pass, practical experience is that filter networks are terminated with other transmission networks within a system rather than pure resistances. The transmission quality through these networks connected to filters is adversely affected by excessive variations in the impedance of the filter network. The maximum variation of the input impedance, and consequently the related reflection coefficient (ρ), in the passband of a filter designed with elliptic functions is directly related to the variation in attenuation (A_p) by

$$A_p = -\ln \sqrt{1 - \rho^2}$$

Normally, the attenuation maximum related to the allowable reflection coefficient is so small that it cannot be measured by practical means and is, in fact, masked by the component losses. Table 1 shows the relationships among A_p, ρ, and return loss in networks with equal passband variations of attenuation.

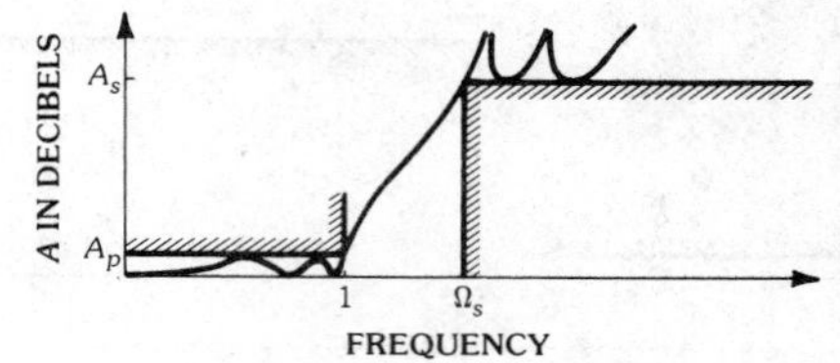

(A) Performance.

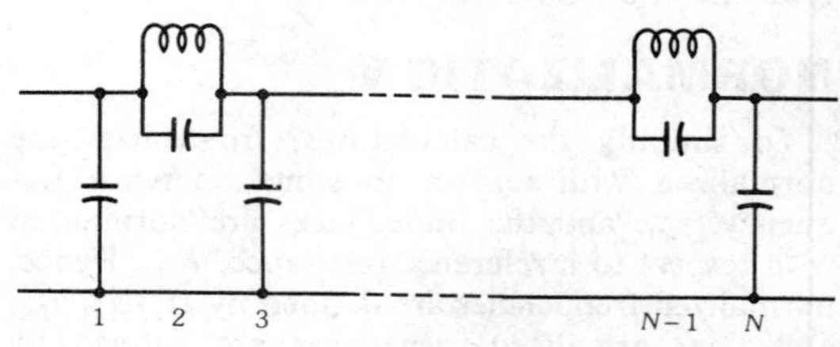

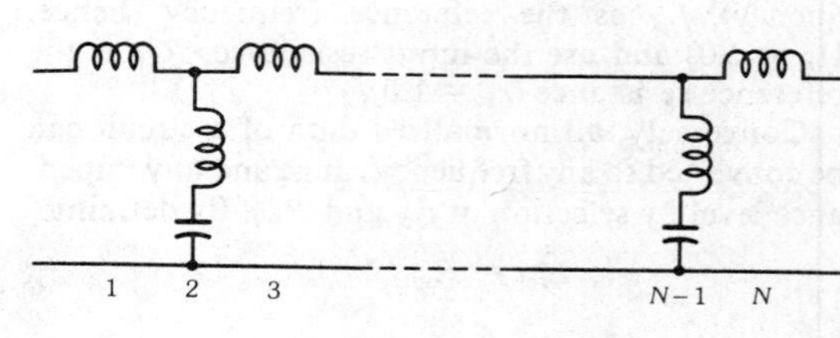

(B) Configurations.

Fig. 8. Cauer-parameter low-pass filters—odd degree.

TABLE 1. VALUES OF ρ, A_p, AND RETURN LOSS

ρ (%)	A_p (dB)	Return Loss (dB)
0.5	0.000109	46.021
1.0	0.000434	40.000
2.0	0.001738	33.979
3.0	0.003910	30.458
4.0	0.006954	27.959
5.0	0.01087	26.021
6.0	0.01566	24.437
8.0	0.02788	21.938
10	0.04365	20.000
12	0.06299	18.416
15	0.09883	16.478
20	0.17729	13.979
25	0.28029	12.041
30	0.40959	10.458
35	0.56753	9.119
40	0.75721	7.959
45	0.98269	6.936
50	1.24939	6.021
60	1.93820	4.437
70	2.92430	3.098

Tables 2 through 8 include normalized Cauer-parameter low-pass filters of degree 3 through 9 with reflection coefficients of 2, 5, 10, and 25 percent. For each reflection coefficient, ten different cutoff rates, and hence ten different stopband minima, are tabulated. In fact, any finer graduation of cutoff sharpness can be computed, and in several publications (Reference 2) the fine graduation is by integer degree of the modular angle, θ ($0° \leq \theta \leq 90°$). For odd degree, $\theta = \arcsin(1/\Omega_s)$.

TABLE 2. CAUER-PARAMETER LOW-PASS, DEGREE=3

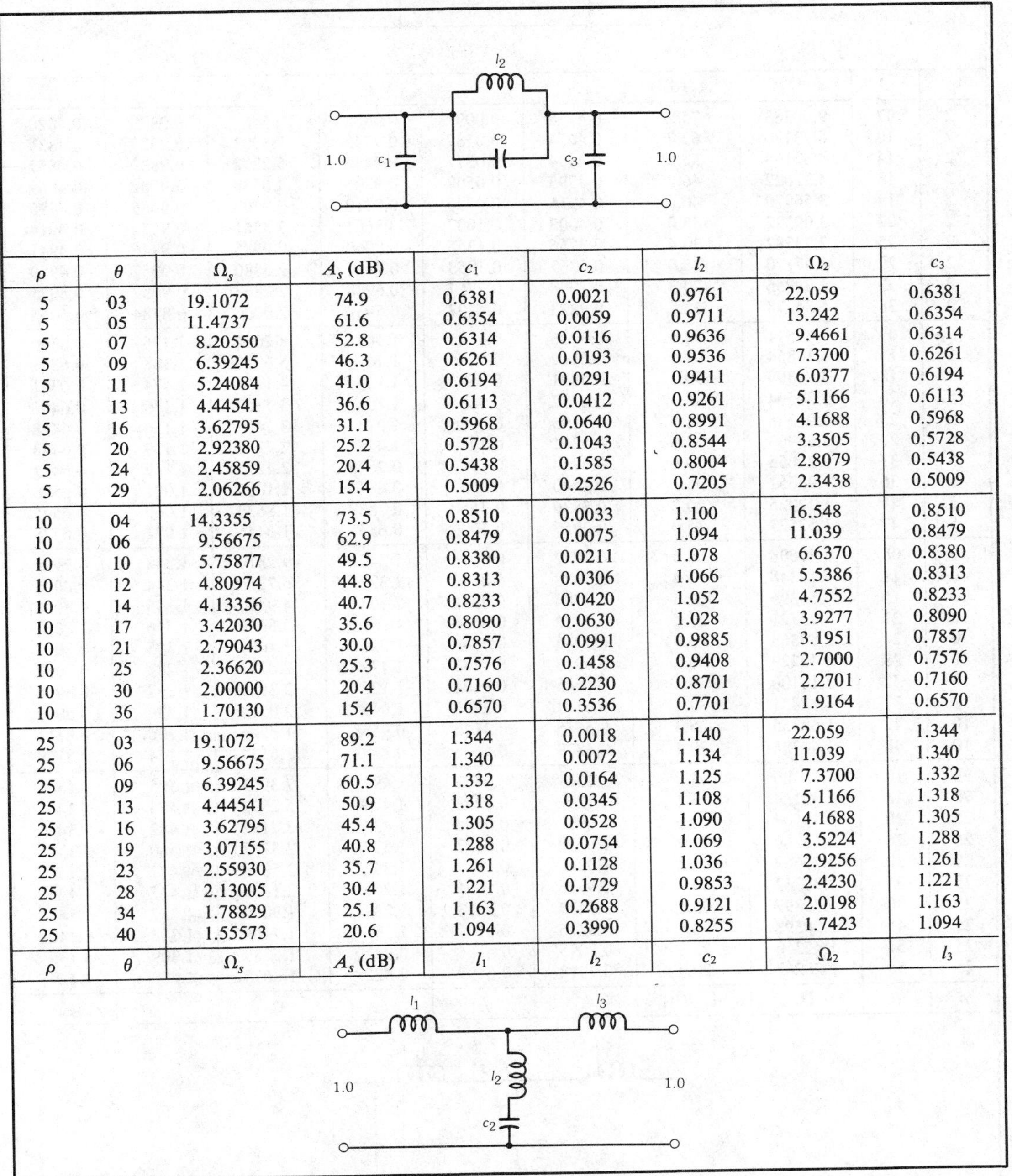

ρ	θ	Ω_s	A_s (dB)	c_1	c_2	l_2	Ω_2	c_3
5	03	19.1072	74.9	0.6381	0.0021	0.9761	22.059	0.6381
5	05	11.4737	61.6	0.6354	0.0059	0.9711	13.242	0.6354
5	07	8.20550	52.8	0.6314	0.0116	0.9636	9.4661	0.6314
5	09	6.39245	46.3	0.6261	0.0193	0.9536	7.3700	0.6261
5	11	5.24084	41.0	0.6194	0.0291	0.9411	6.0377	0.6194
5	13	4.44541	36.6	0.6113	0.0412	0.9261	5.1166	0.6113
5	16	3.62795	31.1	0.5968	0.0640	0.8991	4.1688	0.5968
5	20	2.92380	25.2	0.5728	0.1043	0.8544	3.3505	0.5728
5	24	2.45859	20.4	0.5438	0.1585	0.8004	2.8079	0.5438
5	29	2.06266	15.4	0.5009	0.2526	0.7205	2.3438	0.5009
10	04	14.3355	73.5	0.8510	0.0033	1.100	16.548	0.8510
10	06	9.56675	62.9	0.8479	0.0075	1.094	11.039	0.8479
10	10	5.75877	49.5	0.8380	0.0211	1.078	6.6370	0.8380
10	12	4.80974	44.8	0.8313	0.0306	1.066	5.5386	0.8313
10	14	4.13356	40.7	0.8233	0.0420	1.052	4.7552	0.8233
10	17	3.42030	35.6	0.8090	0.0630	1.028	3.9277	0.8090
10	21	2.79043	30.0	0.7857	0.0991	0.9885	3.1951	0.7857
10	25	2.36620	25.3	0.7576	0.1458	0.9408	2.7000	0.7576
10	30	2.00000	20.4	0.7160	0.2230	0.8701	2.2701	0.7160
10	36	1.70130	15.4	0.6570	0.3536	0.7701	1.9164	0.6570
25	03	19.1072	89.2	1.344	0.0018	1.140	22.059	1.344
25	06	9.56675	71.1	1.340	0.0072	1.134	11.039	1.340
25	09	6.39245	60.5	1.332	0.0164	1.125	7.3700	1.332
25	13	4.44541	50.9	1.318	0.0345	1.108	5.1166	1.318
25	16	3.62795	45.4	1.305	0.0528	1.090	4.1688	1.305
25	19	3.07155	40.8	1.288	0.0754	1.069	3.5224	1.288
25	23	2.55930	35.7	1.261	0.1128	1.036	2.9256	1.261
25	28	2.13005	30.4	1.221	0.1729	0.9853	2.4230	1.221
25	34	1.78829	25.1	1.163	0.2688	0.9121	2.0198	1.163
25	40	1.55573	20.6	1.094	0.3990	0.8255	1.7423	1.094
ρ	θ	Ω_s	A_s (dB)	l_1	l_2	c_2	Ω_2	l_3

Table 3. Cauer-Parameter Low-Pass, Degree=4

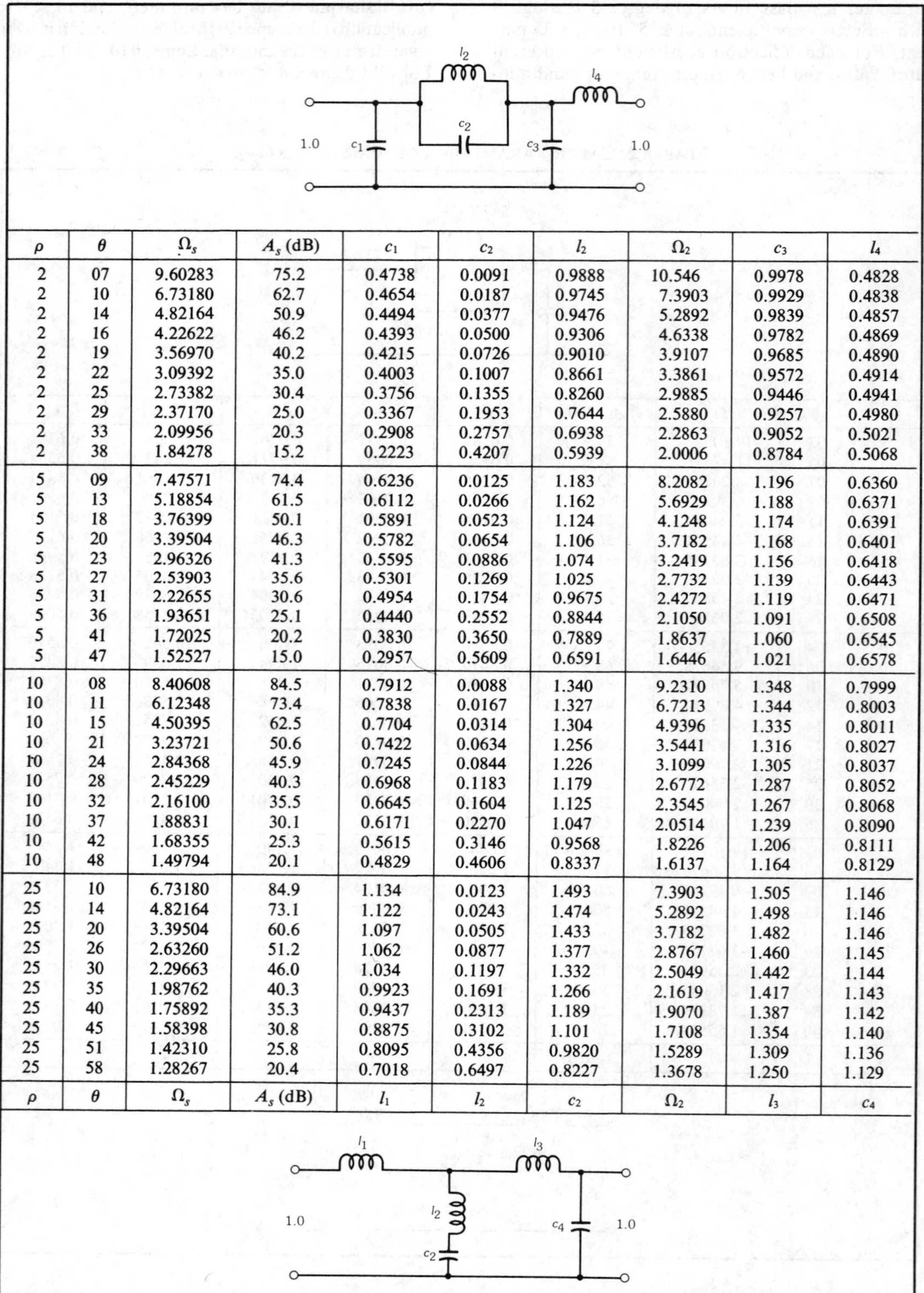

ρ	θ	Ω_s	A_s (dB)	c_1	c_2	l_2	Ω_2	c_3	l_4
2	07	9.60283	75.2	0.4738	0.0091	0.9888	10.546	0.9978	0.4828
2	10	6.73180	62.7	0.4654	0.0187	0.9745	7.3903	0.9929	0.4838
2	14	4.82164	50.9	0.4494	0.0377	0.9476	5.2892	0.9839	0.4857
2	16	4.22622	46.2	0.4393	0.0500	0.9306	4.6338	0.9782	0.4869
2	19	3.56970	40.2	0.4215	0.0726	0.9010	3.9107	0.9685	0.4890
2	22	3.09392	35.0	0.4003	0.1007	0.8661	3.3861	0.9572	0.4914
2	25	2.73382	30.4	0.3756	0.1355	0.8260	2.9885	0.9446	0.4941
2	29	2.37170	25.0	0.3367	0.1953	0.7644	2.5880	0.9257	0.4980
2	33	2.09956	20.3	0.2908	0.2757	0.6938	2.2863	0.9052	0.5021
2	38	1.84278	15.2	0.2223	0.4207	0.5939	2.0006	0.8784	0.5068
5	09	7.47571	74.4	0.6236	0.0125	1.183	8.2082	1.196	0.6360
5	13	5.18854	61.5	0.6112	0.0266	1.162	5.6929	1.188	0.6371
5	18	3.76399	50.1	0.5891	0.0523	1.124	4.1248	1.174	0.6391
5	20	3.39504	46.3	0.5782	0.0654	1.106	3.7182	1.168	0.6401
5	23	2.96326	41.3	0.5595	0.0886	1.074	3.2419	1.156	0.6418
5	27	2.53903	35.6	0.5301	0.1269	1.025	2.7732	1.139	0.6443
5	31	2.22655	30.6	0.4954	0.1754	0.9675	2.4272	1.119	0.6471
5	36	1.93651	25.1	0.4440	0.2552	0.8844	2.1050	1.091	0.6508
5	41	1.72025	20.2	0.3830	0.3650	0.7889	1.8637	1.060	0.6545
5	47	1.52527	15.0	0.2957	0.5609	0.6591	1.6446	1.021	0.6578
10	08	8.40608	84.5	0.7912	0.0088	1.340	9.2310	1.348	0.7999
10	11	6.12348	73.4	0.7838	0.0167	1.327	6.7213	1.344	0.8003
10	15	4.50395	62.5	0.7704	0.0314	1.304	4.9396	1.335	0.8011
10	21	3.23721	50.6	0.7422	0.0634	1.256	3.5441	1.316	0.8027
10	24	2.84368	45.9	0.7245	0.0844	1.226	3.1099	1.305	0.8037
10	28	2.45229	40.3	0.6968	0.1183	1.179	2.6772	1.287	0.8052
10	32	2.16100	35.5	0.6645	0.1604	1.125	2.3545	1.267	0.8068
10	37	1.88831	30.1	0.6171	0.2270	1.047	2.0514	1.239	0.8090
10	42	1.68355	25.3	0.5615	0.3146	0.9568	1.8226	1.206	0.8111
10	48	1.49794	20.1	0.4829	0.4606	0.8337	1.6137	1.164	0.8129
25	10	6.73180	84.9	1.134	0.0123	1.493	7.3903	1.505	1.146
25	14	4.82164	73.1	1.122	0.0243	1.474	5.2892	1.498	1.146
25	20	3.39504	60.6	1.097	0.0505	1.433	3.7182	1.482	1.146
25	26	2.63260	51.2	1.062	0.0877	1.377	2.8767	1.460	1.145
25	30	2.29663	46.0	1.034	0.1197	1.332	2.5049	1.442	1.144
25	35	1.98762	40.3	0.9923	0.1691	1.266	2.1619	1.417	1.143
25	40	1.75892	35.3	0.9437	0.2313	1.189	1.9070	1.387	1.142
25	45	1.58398	30.8	0.8875	0.3102	1.101	1.7108	1.354	1.140
25	51	1.42310	25.8	0.8095	0.4356	0.9820	1.5289	1.309	1.136
25	58	1.28267	20.4	0.7018	0.6497	0.8227	1.3678	1.250	1.129
ρ	θ	Ω_s	A_s (dB)	l_1	l_2	c_2	Ω_2	l_3	c_4

TABLE 4. CAUER-PARAMETER LOW-PASS, DEGREE=5

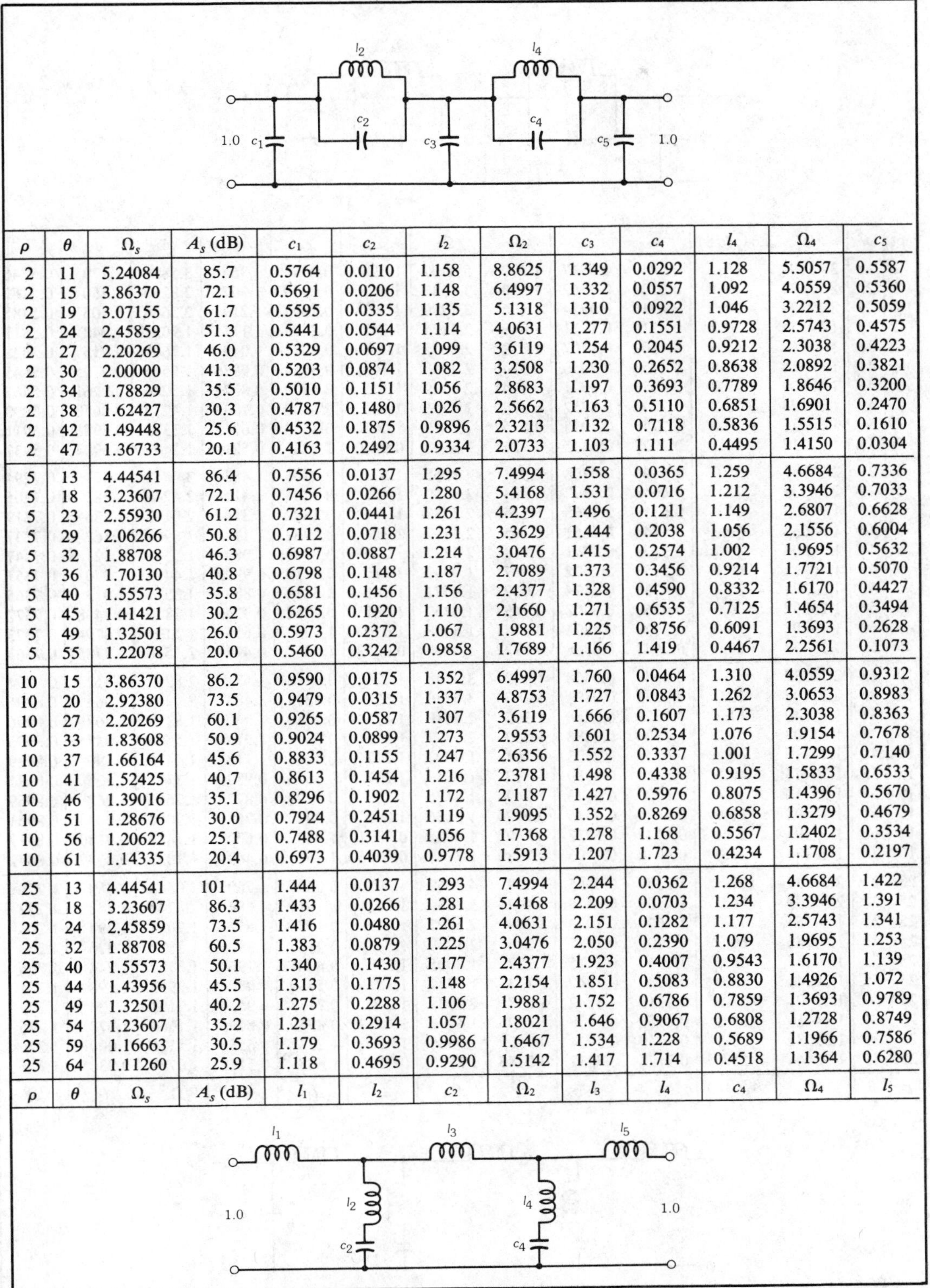

ρ	θ	Ω_s	A_s (dB)	c_1	c_2	l_2	Ω_2	c_3	c_4	l_4	Ω_4	c_5
2	11	5.24084	85.7	0.5764	0.0110	1.158	8.8625	1.349	0.0292	1.128	5.5057	0.5587
2	15	3.86370	72.1	0.5691	0.0206	1.148	6.4997	1.332	0.0557	1.092	4.0559	0.5360
2	19	3.07155	61.7	0.5595	0.0335	1.135	5.1318	1.310	0.0922	1.046	3.2212	0.5059
2	24	2.45859	51.3	0.5441	0.0544	1.114	4.0631	1.277	0.1551	0.9728	2.5743	0.4575
2	27	2.20269	46.0	0.5329	0.0697	1.099	3.6119	1.254	0.2045	0.9212	2.3038	0.4223
2	30	2.00000	41.3	0.5203	0.0874	1.082	3.2508	1.230	0.2652	0.8638	2.0892	0.3821
2	34	1.78829	35.5	0.5010	0.1151	1.056	2.8683	1.197	0.3693	0.7789	1.8646	0.3200
2	38	1.62427	30.3	0.4787	0.1480	1.026	2.5662	1.163	0.5110	0.6851	1.6901	0.2470
2	42	1.49448	25.6	0.4532	0.1875	0.9896	2.3213	1.132	0.7118	0.5836	1.5515	0.1610
2	47	1.36733	20.1	0.4163	0.2492	0.9334	2.0733	1.103	1.111	0.4495	1.4150	0.0304
5	13	4.44541	86.4	0.7556	0.0137	1.295	7.4994	1.558	0.0365	1.259	4.6684	0.7336
5	18	3.23607	72.1	0.7456	0.0266	1.280	5.4168	1.531	0.0716	1.212	3.3946	0.7033
5	23	2.55930	61.2	0.7321	0.0441	1.261	4.2397	1.496	0.1211	1.149	2.6807	0.6628
5	29	2.06266	50.8	0.7112	0.0718	1.231	3.3629	1.444	0.2038	1.056	2.1556	0.6004
5	32	1.88708	46.3	0.6987	0.0887	1.214	3.0476	1.415	0.2574	1.002	1.9695	0.5632
5	36	1.70130	40.8	0.6798	0.1148	1.187	2.7089	1.373	0.3456	0.9214	1.7721	0.5070
5	40	1.55573	35.8	0.6581	0.1456	1.156	2.4377	1.328	0.4590	0.8332	1.6170	0.4427
5	45	1.41421	30.2	0.6265	0.1920	1.110	2.1660	1.271	0.6535	0.7125	1.4654	0.3494
5	49	1.32501	26.0	0.5973	0.2372	1.067	1.9881	1.225	0.8756	0.6091	1.3693	0.2628
5	55	1.22078	20.0	0.5460	0.3242	0.9858	1.7689	1.166	1.419	0.4467	2.2561	0.1073
10	15	3.86370	86.2	0.9590	0.0175	1.352	6.4997	1.760	0.0464	1.310	4.0559	0.9312
10	20	2.92380	73.5	0.9479	0.0315	1.337	4.8753	1.727	0.0843	1.262	3.0653	0.8983
10	27	2.20269	60.1	0.9265	0.0587	1.307	3.6119	1.666	0.1607	1.173	2.3038	0.8363
10	33	1.83608	50.9	0.9024	0.0899	1.273	2.9553	1.601	0.2534	1.076	1.9154	0.7678
10	37	1.66164	45.6	0.8833	0.1155	1.247	2.6356	1.552	0.3337	1.001	1.7299	0.7140
10	41	1.52425	40.7	0.8613	0.1454	1.216	2.3781	1.498	0.4338	0.9195	1.5833	0.6533
10	46	1.39016	35.1	0.8296	0.1902	1.172	2.1187	1.427	0.5976	0.8075	1.4396	0.5670
10	51	1.28676	30.0	0.7924	0.2451	1.119	1.9095	1.352	0.8269	0.6858	1.3279	0.4679
10	56	1.20622	25.1	0.7488	0.3141	1.056	1.7368	1.278	1.168	0.5567	1.2402	0.3534
10	61	1.14335	20.4	0.6973	0.4039	0.9778	1.5913	1.207	1.723	0.4234	1.1708	0.2197
25	13	4.44541	101	1.444	0.0137	1.293	7.4994	2.244	0.0362	1.268	4.6684	1.422
25	18	3.23607	86.3	1.433	0.0266	1.281	5.4168	2.209	0.0703	1.234	3.3946	1.391
25	24	2.45859	73.5	1.416	0.0480	1.261	4.0631	2.151	0.1282	1.177	2.5743	1.341
25	32	1.88708	60.5	1.383	0.0879	1.225	3.0476	2.050	0.2390	1.079	1.9695	1.253
25	40	1.55573	50.1	1.340	0.1430	1.177	2.4377	1.923	0.4007	0.9543	1.6170	1.139
25	44	1.43956	45.5	1.313	0.1775	1.148	2.2154	1.851	0.5083	0.8830	1.4926	1.072
25	49	1.32501	40.2	1.275	0.2288	1.106	1.9881	1.752	0.6786	0.7859	1.3693	0.9789
25	54	1.23607	35.2	1.231	0.2914	1.057	1.8021	1.646	0.9067	0.6808	1.2728	0.8749
25	59	1.16663	30.5	1.179	0.3693	0.9986	1.6467	1.534	1.228	0.5689	1.1966	0.7586
25	64	1.11260	25.9	1.118	0.4695	0.9290	1.5142	1.417	1.714	0.4518	1.1364	0.6280
ρ	θ	Ω_s	A_s (dB)	l_1	l_2	c_2	Ω_2	l_3	l_4	c_4	Ω_4	l_5

TABLE 5. CAUER-PARAMETER LOW-PASS, DEGREE=6

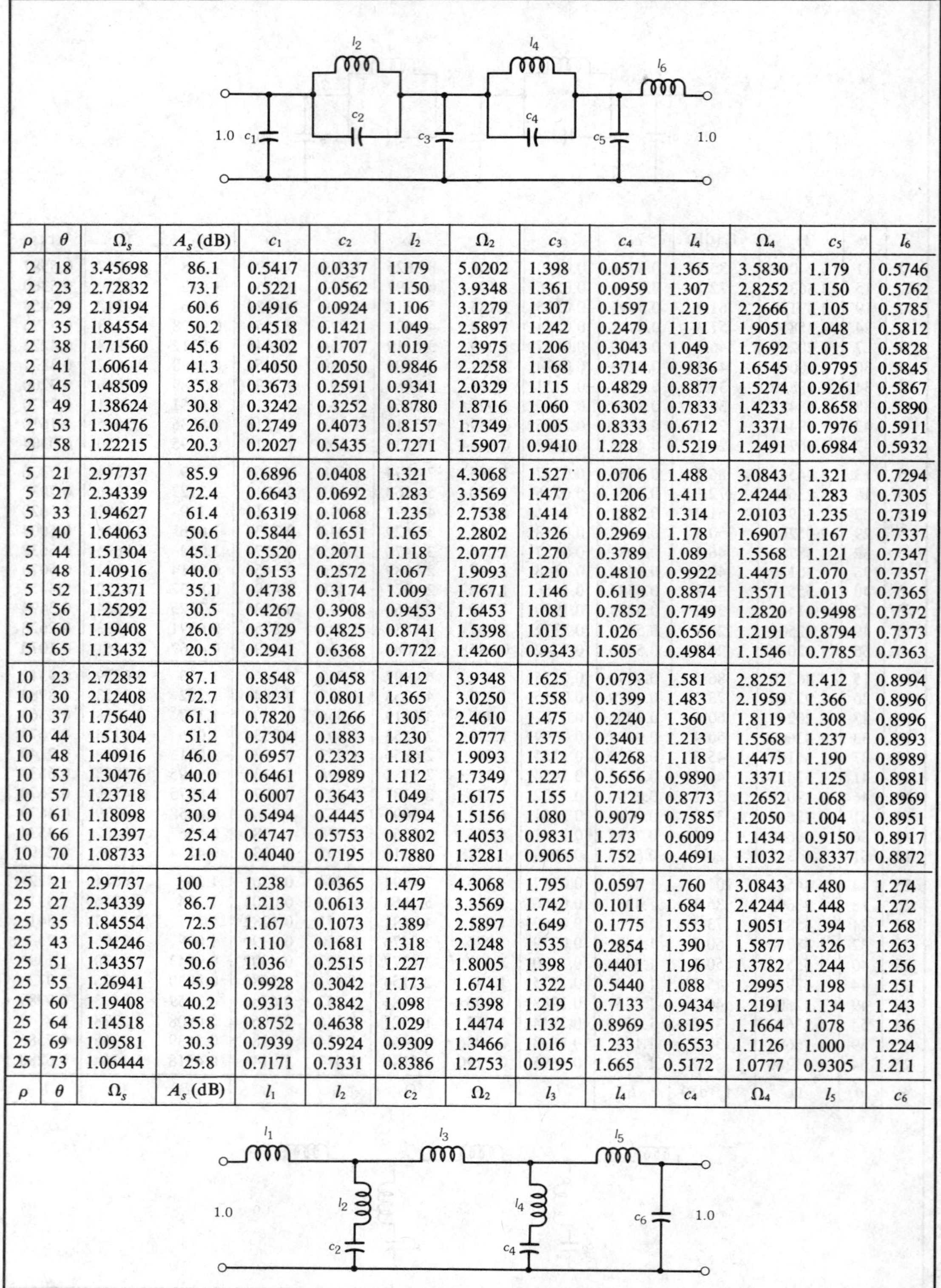

ρ	θ	Ω_s	A_s (dB)	c_1	c_2	l_2	Ω_2	c_3	c_4	l_4	Ω_4	c_5	l_6
2	18	3.45698	86.1	0.5417	0.0337	1.179	5.0202	1.398	0.0571	1.365	3.5830	1.179	0.5746
2	23	2.72832	73.1	0.5221	0.0562	1.150	3.9348	1.361	0.0959	1.307	2.8252	1.150	0.5762
2	29	2.19194	60.6	0.4916	0.0924	1.106	3.1279	1.307	0.1597	1.219	2.2666	1.105	0.5785
2	35	1.84554	50.2	0.4518	0.1421	1.049	2.5897	1.242	0.2479	1.111	1.9051	1.048	0.5812
2	38	1.71560	45.6	0.4302	0.1707	1.019	2.3975	1.206	0.3043	1.049	1.7692	1.015	0.5828
2	41	1.60614	41.3	0.4050	0.2050	0.9846	2.2258	1.168	0.3714	0.9836	1.6545	0.9795	0.5845
2	45	1.48509	35.8	0.3673	0.2591	0.9341	2.0329	1.115	0.4829	0.8877	1.5274	0.9261	0.5867
2	49	1.38624	30.8	0.3242	0.3252	0.8780	1.8716	1.060	0.6302	0.7833	1.4233	0.8658	0.5890
2	53	1.30476	26.0	0.2749	0.4073	0.8157	1.7349	1.005	0.8333	0.6712	1.3371	0.7976	0.5911
2	58	1.22215	20.3	0.2027	0.5435	0.7271	1.5907	0.9410	1.228	0.5219	1.2491	0.6984	0.5932
5	21	2.97737	85.9	0.6896	0.0408	1.321	4.3068	1.527	0.0706	1.488	3.0843	1.321	0.7294
5	27	2.34339	72.4	0.6643	0.0692	1.283	3.3569	1.477	0.1206	1.411	2.4244	1.283	0.7305
5	33	1.94627	61.4	0.6319	0.1068	1.235	2.7538	1.414	0.1882	1.314	2.0103	1.235	0.7319
5	40	1.64063	50.6	0.5844	0.1651	1.165	2.2802	1.326	0.2969	1.178	1.6907	1.167	0.7337
5	44	1.51304	45.1	0.5520	0.2071	1.118	2.0777	1.270	0.3789	1.089	1.5568	1.121	0.7347
5	48	1.40916	40.0	0.5153	0.2572	1.067	1.9093	1.210	0.4810	0.9922	1.4475	1.070	0.7357
5	52	1.32371	35.1	0.4738	0.3174	1.009	1.7671	1.146	0.6119	0.8874	1.3571	1.013	0.7365
5	56	1.25292	30.5	0.4267	0.3908	0.9453	1.6453	1.081	0.7852	0.7749	1.2820	0.9498	0.7372
5	60	1.19408	26.0	0.3729	0.4825	0.8741	1.5398	1.015	1.026	0.6556	1.2191	0.8794	0.7373
5	65	1.13432	20.5	0.2941	0.6368	0.7722	1.4260	0.9343	1.505	0.4984	1.1546	0.7785	0.7363
10	23	2.72832	87.1	0.8548	0.0458	1.412	3.9348	1.625	0.0793	1.581	2.8252	1.412	0.8994
10	30	2.12408	72.7	0.8231	0.0801	1.365	3.0250	1.558	0.1399	1.483	2.1959	1.366	0.8996
10	37	1.75640	61.1	0.7820	0.1266	1.305	2.4610	1.475	0.2240	1.360	1.8119	1.308	0.8996
10	44	1.51304	51.2	0.7304	0.1883	1.230	2.0777	1.375	0.3401	1.213	1.5568	1.237	0.8993
10	48	1.40916	46.0	0.6957	0.2323	1.181	1.9093	1.312	0.4268	1.118	1.4475	1.190	0.8989
10	53	1.30476	40.0	0.6461	0.2989	1.112	1.7349	1.227	0.5656	0.9890	1.3371	1.125	0.8981
10	57	1.23718	35.4	0.6007	0.3643	1.049	1.6175	1.155	0.7121	0.8773	1.2652	1.068	0.8969
10	61	1.18098	30.9	0.5494	0.4445	0.9794	1.5156	1.080	0.9079	0.7585	1.2050	1.004	0.8951
10	66	1.12397	25.4	0.4747	0.5753	0.8802	1.4053	0.9831	1.273	0.6009	1.1434	0.9150	0.8917
10	70	1.08733	21.0	0.4040	0.7195	0.7880	1.3281	0.9065	1.752	0.4691	1.1032	0.8337	0.8872
25	21	2.97737	100	1.238	0.0365	1.479	4.3068	1.795	0.0597	1.760	3.0843	1.480	1.274
25	27	2.34339	86.7	1.213	0.0613	1.447	3.3569	1.742	0.1011	1.684	2.4244	1.448	1.272
25	35	1.84554	72.5	1.167	0.1073	1.389	2.5897	1.649	0.1775	1.553	1.9051	1.394	1.268
25	43	1.54246	60.7	1.110	0.1681	1.318	2.1248	1.535	0.2854	1.390	1.5877	1.326	1.263
25	51	1.34357	50.6	1.036	0.2515	1.227	1.8005	1.398	0.4401	1.196	1.3782	1.244	1.256
25	55	1.26941	45.9	0.9928	0.3042	1.173	1.6741	1.322	0.5440	1.088	1.2995	1.198	1.251
25	60	1.19408	40.2	0.9313	0.3842	1.098	1.5398	1.219	0.7133	0.9434	1.2191	1.134	1.243
25	64	1.14518	35.8	0.8752	0.4638	1.029	1.4474	1.132	0.8969	0.8195	1.1664	1.078	1.236
25	69	1.09581	30.3	0.7939	0.5924	0.9309	1.3466	1.016	1.233	0.6553	1.1126	1.000	1.224
25	73	1.06444	25.8	0.7171	0.7331	0.8386	1.2753	0.9195	1.665	0.5172	1.0777	0.9305	1.211
ρ	θ	Ω_s	A_s (dB)	l_1	l_2	c_2	Ω_2	l_3	l_4	c_4	Ω_4	l_5	c_6

TABLE 6. CAUER-PARAMETER LOW-PASS, DEGREE=7

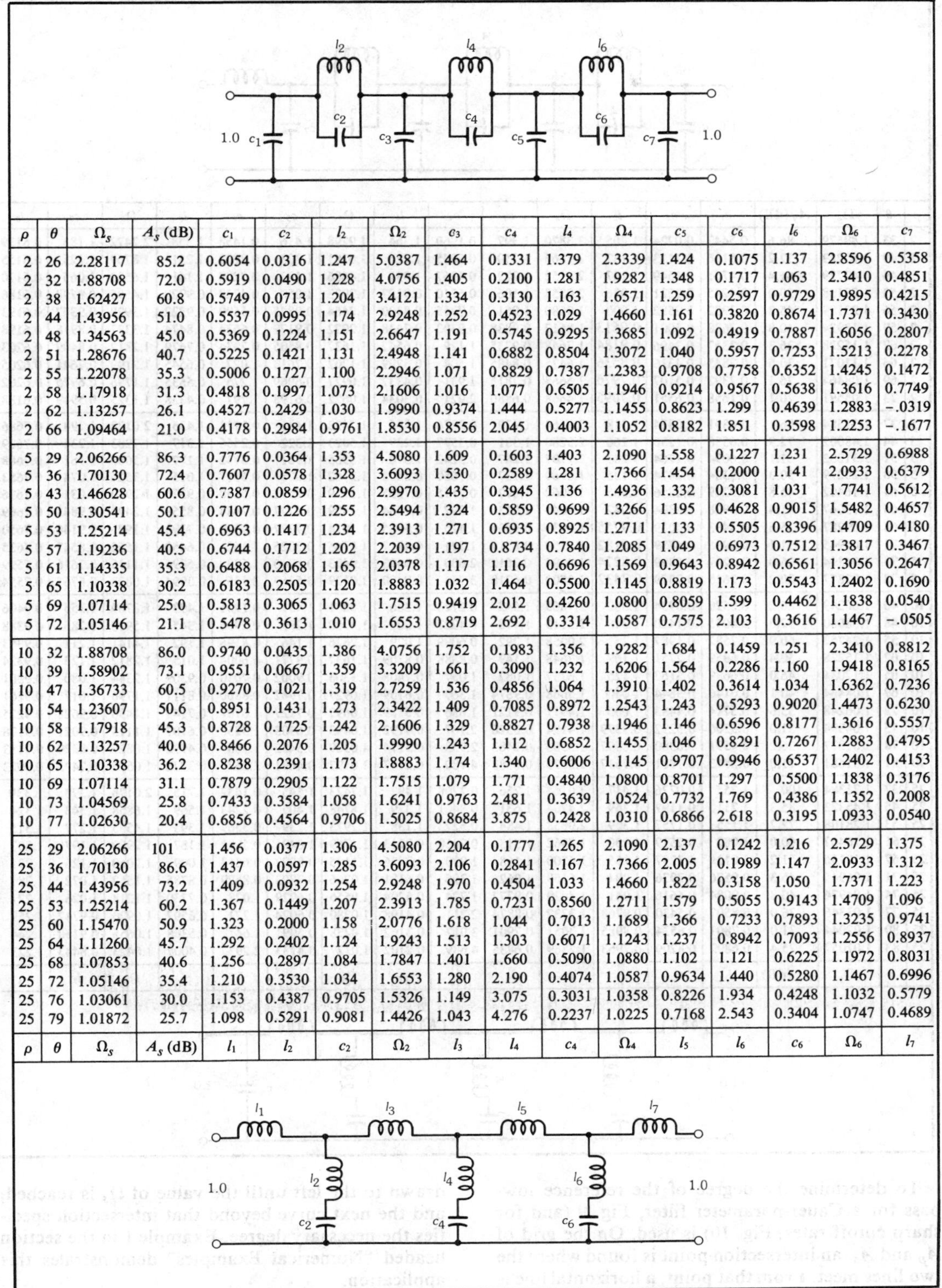

ρ	θ	Ω_s	A_s (dB)	c_1	c_2	l_2	Ω_2	c_3	c_4	l_4	Ω_4	c_5	c_6	l_6	Ω_6	c_7
2	26	2.28117	85.2	0.6054	0.0316	1.247	5.0387	1.464	0.1331	1.379	2.3339	1.424	0.1075	1.137	2.8596	0.5358
2	32	1.88708	72.0	0.5919	0.0490	1.228	4.0756	1.405	0.2100	1.281	1.9282	1.348	0.1717	1.063	2.3410	0.4851
2	38	1.62427	60.8	0.5749	0.0713	1.204	3.4121	1.334	0.3130	1.163	1.6571	1.259	0.2597	0.9729	1.9895	0.4215
2	44	1.43956	51.0	0.5537	0.0995	1.174	2.9248	1.252	0.4523	1.029	1.4660	1.161	0.3820	0.8674	1.7371	0.3430
2	48	1.34563	45.0	0.5369	0.1224	1.151	2.6647	1.191	0.5745	0.9295	1.3685	1.092	0.4919	0.7887	1.6056	0.2807
2	51	1.28676	40.7	0.5225	0.1421	1.131	2.4948	1.141	0.6882	0.8504	1.3072	1.040	0.5957	0.7253	1.5213	0.2278
2	55	1.22078	35.3	0.5006	0.1727	1.100	2.2946	1.071	0.8829	0.7387	1.2383	0.9708	0.7758	0.6352	1.4245	0.1472
2	58	1.17918	31.3	0.4818	0.1997	1.073	2.1606	1.016	1.077	0.6505	1.1946	0.9214	0.9567	0.5638	1.3616	0.7749
2	62	1.13257	26.1	0.4527	0.2429	1.030	1.9990	0.9374	1.444	0.5277	1.1455	0.8623	1.299	0.4639	1.2883	–.0319
2	66	1.09464	21.0	0.4178	0.2984	0.9761	1.8530	0.8556	2.045	0.4003	1.1052	0.8182	1.851	0.3598	1.2253	–.1677
5	29	2.06266	86.3	0.7776	0.0364	1.353	4.5080	1.609	0.1603	1.403	2.1090	1.558	0.1227	1.231	2.5729	0.6988
5	36	1.70130	72.4	0.7607	0.0578	1.328	3.6093	1.530	0.2589	1.281	1.7366	1.454	0.2000	1.141	2.0933	0.6379
5	43	1.46628	60.6	0.7387	0.0859	1.296	2.9970	1.435	0.3945	1.136	1.4936	1.332	0.3081	1.031	1.7741	0.5612
5	50	1.30541	50.1	0.7107	0.1226	1.255	2.5494	1.324	0.5859	0.9699	1.3266	1.195	0.4628	0.9015	1.5482	0.4657
5	53	1.25214	45.4	0.6963	0.1417	1.234	2.3913	1.271	0.6935	0.8925	1.2711	1.133	0.5505	0.8396	1.4709	0.4180
5	57	1.19236	40.5	0.6744	0.1712	1.202	2.2039	1.197	0.8734	0.7840	1.2085	1.049	0.6973	0.7512	1.3817	0.3467
5	61	1.14335	35.3	0.6488	0.2068	1.165	2.0378	1.117	1.116	0.6696	1.1569	0.9643	0.8942	0.6561	1.3056	0.2647
5	65	1.10338	30.2	0.6183	0.2505	1.120	1.8883	1.032	1.464	0.5500	1.1145	0.8819	1.173	0.5543	1.2402	0.1690
5	69	1.07114	25.0	0.5813	0.3065	1.063	1.7515	0.9419	2.012	0.4260	1.0800	0.8059	1.599	0.4462	1.1838	0.0540
5	72	1.05146	21.1	0.5478	0.3613	1.010	1.6553	0.8719	2.692	0.3314	1.0587	0.7575	2.103	0.3616	1.1467	–.0505
10	32	1.88708	86.0	0.9740	0.0435	1.386	4.0756	1.752	0.1983	1.356	1.9282	1.684	0.1459	1.251	2.3410	0.8812
10	39	1.58902	73.1	0.9551	0.0667	1.359	3.3209	1.661	0.3090	1.232	1.6206	1.564	0.2286	1.160	1.9418	0.8165
10	47	1.36733	60.5	0.9270	0.1021	1.319	2.7259	1.536	0.4856	1.064	1.3910	1.402	0.3614	1.034	1.6362	0.7236
10	54	1.23607	50.6	0.8951	0.1431	1.273	2.3422	1.409	0.7085	0.8972	1.2543	1.243	0.5293	0.9020	1.4473	0.6230
10	58	1.17918	45.3	0.8728	0.1725	1.242	2.1606	1.329	0.8827	0.7938	1.1946	1.146	0.6596	0.8177	1.3616	0.5557
10	62	1.13257	40.0	0.8466	0.2076	1.205	1.9990	1.243	1.112	0.6852	1.1455	1.046	0.8291	0.7267	1.2883	0.4795
10	65	1.10338	36.2	0.8238	0.2391	1.173	1.8883	1.174	1.340	0.6006	1.1145	0.9707	0.9946	0.6537	1.2402	0.4153
10	69	1.07114	31.1	0.7879	0.2905	1.122	1.7515	1.079	1.771	0.4840	1.0800	0.8701	1.297	0.5500	1.1838	0.3176
10	73	1.04569	25.8	0.7433	0.3584	1.058	1.6241	0.9763	2.481	0.3639	1.0524	0.7732	1.769	0.4386	1.1352	0.2008
10	77	1.02630	20.4	0.6856	0.4564	0.9706	1.5025	0.8684	3.875	0.2428	1.0310	0.6866	2.618	0.3195	1.0933	0.0540
25	29	2.06266	101	1.456	0.0377	1.306	4.5080	2.204	0.1777	1.265	2.1090	2.137	0.1242	1.216	2.5729	1.375
25	36	1.70130	86.6	1.437	0.0597	1.285	3.6093	2.106	0.2841	1.167	1.7366	2.005	0.1989	1.147	2.0933	1.312
25	44	1.43956	73.2	1.409	0.0932	1.254	2.9248	1.970	0.4504	1.033	1.4660	1.822	0.3158	1.050	1.7371	1.223
25	53	1.25214	60.2	1.367	0.1449	1.207	2.3913	1.785	0.7231	0.8560	1.2711	1.579	0.5055	0.9143	1.4709	1.096
25	60	1.15470	50.9	1.323	0.2000	1.158	2.0776	1.618	1.044	0.7013	1.1689	1.366	0.7233	0.7893	1.3235	0.9741
25	64	1.11260	45.7	1.292	0.2402	1.124	1.9243	1.513	1.303	0.6071	1.1243	1.237	0.8942	0.7093	1.2556	0.8937
25	68	1.07853	40.6	1.256	0.2897	1.084	1.7847	1.401	1.660	0.5090	1.0880	1.102	1.121	0.6225	1.1972	0.8031
25	72	1.05146	35.4	1.210	0.3530	1.034	1.6553	1.280	2.190	0.4074	1.0587	0.9634	1.440	0.5280	1.1467	0.6996
25	76	1.03061	30.0	1.153	0.4387	0.9705	1.5326	1.149	3.075	0.3031	1.0358	0.8226	1.934	0.4248	1.1032	0.5779
25	79	1.01872	25.7	1.098	0.5291	0.9081	1.4426	1.043	4.276	0.2237	1.0225	0.7168	2.543	0.3404	1.0747	0.4689
ρ	θ	Ω_s	A_s (dB)	l_1	l_2	c_2	Ω_2	l_3	l_4	c_4	Ω_4	l_5	l_6	c_6	Ω_6	l_7

TABLE 7. CAUER-PARAMETER LOW-PASS, DEGREE=8

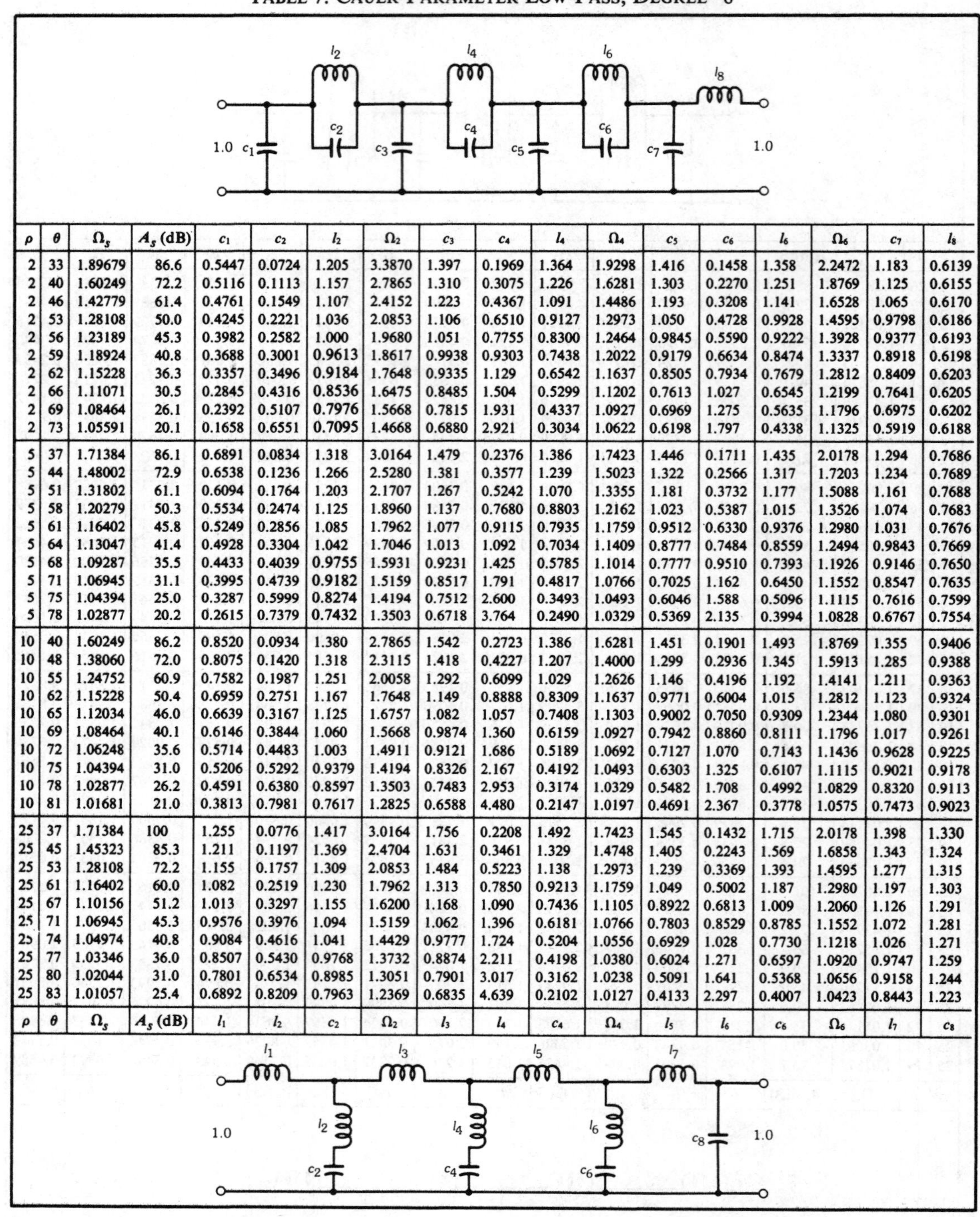

ρ	θ	Ω_s	A_s (dB)	c_1	c_2	l_2	Ω_2	c_3	c_4	l_4	Ω_4	c_5	c_6	l_6	Ω_6	c_7	l_8
2	33	1.89679	86.6	0.5447	0.0724	1.205	3.3870	1.397	0.1969	1.364	1.9298	1.416	0.1458	1.358	2.2472	1.183	0.6139
2	40	1.60249	72.2	0.5116	0.1113	1.157	2.7865	1.310	0.3075	1.226	1.6281	1.303	0.2270	1.251	1.8769	1.125	0.6155
2	46	1.42779	61.4	0.4761	0.1549	1.107	2.4152	1.223	0.4367	1.091	1.4486	1.193	0.3208	1.141	1.6528	1.065	0.6170
2	53	1.28108	50.0	0.4245	0.2221	1.036	2.0853	1.106	0.6510	0.9127	1.2973	1.050	0.4728	0.9928	1.4595	0.9798	0.6186
2	56	1.23189	45.3	0.3982	0.2582	1.000	1.9680	1.051	0.7755	0.8300	1.2464	0.9845	0.5590	0.9222	1.3928	0.9377	0.6193
2	59	1.18924	40.8	0.3688	0.3001	0.9613	1.8617	0.9938	0.9303	0.7438	1.2022	0.9179	0.6634	0.8474	1.3337	0.8918	0.6198
2	62	1.15228	36.3	0.3357	0.3496	0.9184	1.7648	0.9335	1.129	0.6542	1.1637	0.8505	0.7934	0.7679	1.2812	0.8409	0.6203
2	66	1.11071	30.5	0.2845	0.4316	0.8536	1.6475	0.8485	1.504	0.5299	1.1202	0.7613	1.027	0.6545	1.2199	0.7641	0.6205
2	69	1.08464	26.1	0.2392	0.5107	0.7976	1.5668	0.7815	1.931	0.4337	1.0927	0.6969	1.275	0.5635	1.1796	0.6975	0.6202
2	73	1.05591	20.1	0.1658	0.6551	0.7095	1.4668	0.6880	2.921	0.3034	1.0622	0.6198	1.797	0.4338	1.1325	0.5919	0.6188
5	37	1.71384	86.1	0.6891	0.0834	1.318	3.0164	1.479	0.2376	1.386	1.7423	1.446	0.1711	1.435	2.0178	1.294	0.7686
5	44	1.48002	72.9	0.6538	0.1236	1.266	2.5280	1.381	0.3577	1.239	1.5023	1.322	0.2566	1.317	1.7203	1.234	0.7689
5	51	1.31802	61.1	0.6094	0.1764	1.203	2.1707	1.267	0.5242	1.070	1.3355	1.181	0.3732	1.177	1.5088	1.161	0.7688
5	58	1.20279	50.3	0.5534	0.2474	1.125	1.8960	1.137	0.7680	0.8803	1.2162	1.023	0.5387	1.015	1.3526	1.074	0.7683
5	61	1.16402	45.8	0.5249	0.2856	1.085	1.7962	1.077	0.9115	0.7935	1.1759	0.9512	0.6330	0.9376	1.2980	1.031	0.7676
5	64	1.13047	41.4	0.4928	0.3304	1.042	1.7046	1.013	1.092	0.7034	1.1409	0.8777	0.7484	0.8559	1.2494	0.9843	0.7669
5	68	1.09287	35.5	0.4433	0.4039	0.9755	1.5931	0.9231	1.425	0.5785	1.1014	0.7777	0.9510	0.7393	1.1926	0.9146	0.7650
5	71	1.06945	31.1	0.3995	0.4739	0.9182	1.5159	0.8517	1.791	0.4817	1.0766	0.7025	1.162	0.6450	1.1552	0.8547	0.7635
5	75	1.04394	25.0	0.3287	0.5999	0.8274	1.4194	0.7512	2.600	0.3493	1.0493	0.6046	1.588	0.5096	1.1115	0.7616	0.7599
5	78	1.02877	20.2	0.2615	0.7379	0.7432	1.3503	0.6718	3.764	0.2490	1.0329	0.5369	2.135	0.3994	1.0828	0.6767	0.7554
10	40	1.60249	86.2	0.8520	0.0934	1.380	2.7865	1.542	0.2723	1.386	1.6281	1.451	0.1901	1.493	1.8769	1.355	0.9406
10	48	1.38060	72.0	0.8075	0.1420	1.318	2.3115	1.418	0.4227	1.207	1.4000	1.299	0.2936	1.345	1.5913	1.285	0.9388
10	55	1.24752	60.9	0.7582	0.1987	1.251	2.0058	1.292	0.6099	1.029	1.2626	1.146	0.4196	1.192	1.4141	1.211	0.9363
10	62	1.15228	50.4	0.6959	0.2751	1.167	1.7648	1.149	0.8888	0.8309	1.1637	0.9771	0.6004	1.015	1.2812	1.123	0.9324
10	65	1.12034	46.0	0.6639	0.3167	1.125	1.6757	1.082	1.057	0.7408	1.1303	0.9002	0.7050	0.9309	1.2344	1.080	0.9301
10	69	1.08464	40.1	0.6146	0.3844	1.060	1.5668	0.9874	1.360	0.6159	1.0927	0.7942	0.8860	0.8111	1.1796	1.017	0.9261
10	72	1.06248	35.6	0.5714	0.4483	1.003	1.4911	0.9121	1.686	0.5189	1.0692	0.7127	1.070	0.7143	1.1436	0.9628	0.9225
10	75	1.04394	31.0	0.5206	0.5292	0.9379	1.4194	0.8326	2.167	0.4192	1.0493	0.6303	1.325	0.6107	1.1115	0.9021	0.9178
10	78	1.02877	26.2	0.4591	0.6380	0.8597	1.3503	0.7483	2.953	0.3174	1.0329	0.5482	1.708	0.4992	1.0829	0.8320	0.9113
10	81	1.01681	21.0	0.3813	0.7981	0.7617	1.2825	0.6588	4.480	0.2147	1.0197	0.4691	2.367	0.3778	1.0575	0.7473	0.9023
25	37	1.71384	100	1.255	0.0776	1.417	3.0164	1.756	0.2208	1.492	1.7423	1.545	0.1432	1.715	2.0178	1.398	1.330
25	45	1.45323	85.3	1.211	0.1197	1.369	2.4704	1.631	0.3461	1.329	1.4748	1.405	0.2243	1.569	1.6858	1.343	1.324
25	53	1.28108	72.2	1.155	0.1757	1.309	2.0853	1.484	0.5223	1.138	1.2973	1.239	0.3369	1.393	1.4595	1.277	1.315
25	61	1.16402	60.0	1.082	0.2519	1.230	1.7962	1.313	0.7850	0.9213	1.1759	1.049	0.5002	1.187	1.2980	1.197	1.303
25	67	1.10156	51.2	1.013	0.3297	1.155	1.6200	1.168	1.090	0.7436	1.1105	0.8922	0.6813	1.009	1.2060	1.126	1.291
25	71	1.06945	45.3	0.9576	0.3976	1.094	1.5159	1.062	1.396	0.6181	1.0766	0.7803	0.8529	0.8785	1.1552	1.072	1.281
25	74	1.04974	40.8	0.9084	0.4616	1.041	1.4429	0.9777	1.724	0.5204	1.0556	0.6929	1.028	0.7730	1.1218	1.026	1.271
25	77	1.03346	36.0	0.8507	0.5430	0.9768	1.3732	0.8874	2.211	0.4198	1.0380	0.6024	1.271	0.6597	1.0920	0.9747	1.259
25	80	1.02044	31.0	0.7801	0.6534	0.8985	1.3051	0.7901	3.017	0.3162	1.0238	0.5091	1.641	0.5368	1.0656	0.9158	1.244
25	83	1.01057	25.4	0.6892	0.8209	0.7963	1.2369	0.6835	4.639	0.2102	1.0127	0.4133	2.297	0.4007	1.0423	0.8443	1.223
ρ	θ	Ω_s	A_s (dB)	l_1	l_2	c_2	Ω_2	l_3	l_4	c_4	Ω_4	l_5	l_6	c_6	Ω_6	l_7	c_8

To determine the degree of the reference low-pass for a Cauer-parameter filter, Fig. 9 (and for sharp cutoff rates, Fig. 10) is used. On the grid of A_p and A_s, an intersection point is found where the two lines meet. From that point, a horizontal line is drawn to the left until the value of Ω_s is reached, and the next curve beyond that intersection specifies the necessary degree. Example 1 in the section headed "Numerical Examples" demonstrates this application.

TABLE 8. CAUER-PARAMETER LOW-PASS, DEGREE=9

ρ	θ	Ω_s	A_s (dB)	c_1	c_2	l_2	Ω_2	c_3	c_4	l_4	Ω_4	c_5	c_6	l_6	Ω_6	c_7	c_8	l_8	Ω_8	c_9.
2	34	1.78829	101	0.6236	0.0332	1.289	4.8248	1.498	0.1723	1.423	2.0198	1.479	0.2254	1.353	1.8111	1.413	0.1236	1.162	2.6394	0.5415
2	40	1.55573	87.0	0.6123	0.0476	1.273	4.0634	1.438	0.2489	1.324	1.7423	1.368	0.3280	1.231	1.5740	1.325	0.1805	1.096	2.2480	0.4964
2	47	1.36733	72.9	0.5955	0.0689	1.249	3.4103	1.355	0.3674	1.188	1.5135	1.219	0.4907	1.068	1.3815	1.207	0.2707	1.002	1.9200	0.4306
2	54	1.23607	60.2	0.5736	0.0968	1.217	2.9140	1.257	0.5320	1.032	1.3498	1.053	0.7263	0.8855	1.2469	1.075	0.3999	0.8896	1.6766	0.3469
2	60	1.15470	50.0	0.5493	0.1282	1.182	2.5690	1.159	0.7331	0.8806	1.2446	0.8987	1.032	0.7158	1.1632	0.9530	0.5632	0.7753	1.5133	0.2558
2	63	1.12233	45.0	0.5345	0.1476	1.160	2.4161	1.105	0.8673	0.7990	1.2013	0.8192	1.250	0.6269	1.1297	0.8902	0.6752	0.7111	1.4432	0.2011
2	66	1.09464	40.0	0.5172	0.1704	1.135	2.2732	1.046	1.037	0.7131	1.1631	0.7389	1.540	0.5358	1.1009	0.8270	0.8193	0.6415	1.3793	0.1381
2	69	1.07114	35.1	0.4970	0.1978	1.106	2.1381	0.9828	1.259	0.6225	1.1296	0.6592	1.948	0.4430	1.0764	0.7643	1.013	0.5661	1.3208	0.0642
2	72	1.05146	30.1	0.4726	0.2315	1.070	2.0089	0.9141	1.569	0.5266	1.1002	0.5817	2.570	0.3490	1.0558	0.7039	1.287	0.4841	1.2669	−.0247
2	74	1.04030	26.6	0.4534	0.2592	1.041	1.9249	0.8650	1.857	0.4594	1.0827	0.5326	3.204	0.2864	1.0440	0.6662	1.546	0.4253	1.2332	−.0958
5	37	1.66164	102	0.7940	0.0370	1.385	4.4144	1.625	0.2023	1.415	1.8692	1.548	0.2644	1.337	1.6820	1.521	0.1360	1.247	2.4282	0.7041
5	44	1.43956	86.7	0.7798	0.0545	1.364	3.6664	1.546	0.3014	1.293	1.6018	1.402	0.3972	1.189	1.4554	1.404	0.2042	1.167	2.0480	0.6501
5	51	1.28676	73.5	0.7614	0.0773	1.336	3.1119	1.453	0.4347	1.151	1.4137	1.236	0.5807	1.021	1.2990	1.270	0.2973	1.070	1.7727	0.5826
5	58	1.17918	61.3	0.7374	0.1072	1.301	2.6775	1.343	0.6207	0.9884	1.2767	1.051	0.8480	0.8349	1.1884	1.120	0.4290	0.9531	1.5640	0.4977
5	64	1.11260	51.3	0.7103	0.1416	1.260	2.3675	1.234	0.8514	0.8322	1.1880	0.8812	1.201	0.6641	1.1196	0.9810	0.5941	0.8334	1.4212	0.4062
5	67	1.08636	46.3	0.6935	0.1631	1.235	2.2274	1.173	1.008	0.7480	1.1514	0.7931	1.457	0.5754	1.0923	0.9082	0.7071	0.7654	1.3592	0.3513
5	70	1.06418	41.4	0.6737	0.1890	1.206	2.0945	1.107	1.210	0.6594	1.1193	0.7037	1.804	0.4849	1.0691	0.8339	0.8531	0.6911	1.3023	0.2883
5	73	1.04567	36.3	0.6498	0.2208	1.171	1.9667	1.036	1.484	0.5658	1.0912	0.6141	2.309	0.3930	1.0497	0.7586	1.051	0.6092	1.2498	0.2142
5	76	1.03061	31.1	0.6202	0.2616	1.127	1.8420	0.9578	1.884	0.4665	1.0668	0.5256	3.114	0.3005	1.0337	0.6835	1.338	0.5181	1.2011	0.1245
5	79	1.01872	25.6	0.5821	0.3172	1.069	1.7175	0.8713	2.538	0.3601	1.0459	0.4412	4.597	0.2087	1.0209	0.6109	1.801	0.4159	1.1555	0.0102
10	40	1.55573	101	0.9894	0.0430	1.410	4.0634	1.754	0.2443	1.348	1.7423	1.618	0.3195	1.264	1.5740	1.626	0.1567	1.262	2.2489	0.8867
10	48	1.34563	85.0	0.9710	0.0651	1.383	3.3315	1.650	0.3753	1.205	1.4868	1.429	0.4957	1.092	1.3593	1.473	0.2419	1.168	1.8808	0.8199
10	55	1.22078	72.5	0.9498	0.0909	1.353	2.8522	1.543	0.5334	1.059	1.3303	1.241	0.7152	0.9223	1.2312	1.319	0.3452	1.068	1.6470	0.7465
10	62	1.13257	60.7	0.9218	0.1253	1.313	2.4659	1.417	0.7579	0.8936	1.2152	1.034	1.043	0.7376	1.1403	1.147	0.4917	0.9466	1.4658	0.6547
10	68	1.07853	50.7	0.8894	0.1657	1.267	2.1824	1.292	1.046	0.7351	1.1403	0.8439	1.494	0.5694	1.0841	0.9870	0.6783	0.8214	1.3397	0.5551
10	71	1.05762	45.8	0.8689	0.1920	1.238	2.0514	1.221	1.251	0.6496	1.1095	0.7454	1.837	0.4825	1.0622	0.9023	0.8091	0.7492	1.2844	0.4953
10	74	1.04030	40.6	0.8440	0.2244	1.203	1.9249	1.145	1.525	0.5592	1.0827	0.6454	2.328	0.3941	1.0440	0.8140	0.9838	0.6683	1.2332	0.4267
10	77	1.02630	35.3	0.8130	0.2661	1.159	1.8006	1.060	1.922	0.4634	1.0595	0.5447	3.093	0.3052	1.0291	0.7254	1.228	0.5793	1.1856	0.3424
10	79	1.01872	31.6	0.7873	0.3019	1.123	1.7175	0.9987	2.309	0.3958	1.0459	0.4781	3.903	0.2458	1.0209	0.6645	1.462	0.5124	1.1555	0.2761
10	82	1.00983	25.6	0.7369	0.3771	1.050	1.5893	0.8954	3.290	0.2874	1.0284	0.3807	6.205	0.1576	1.0113	0.5730	2.029	0.3983	1.1124	0.1503
25	44	1.43956	101	1.452	0.0572	1.301	3.6664	2.091	0.3414	1.141	1.6018	1.849	0.4468	1.057	1.4554	1.910	0.2068	1.153	2.0480	1.318
25	52	1.26902	85.9	1.428	0.0847	1.274	3.0437	1.958	0.5122	1.008	1.3914	1.610	0.6778	0.8994	1.2808	1.713	0.3101	1.066	1.7395	1.238
25	60	1.15470	72.2	1.397	0.1223	1.238	2.5690	1.800	0.7565	0.8534	1.2446	1.338	1.022	0.7235	1.1632	1.487	0.4562	0.9572	1.5133	1.137
25	67	1.08636	60.6	1.358	0.1686	1.196	2.2274	1.635	1.079	0.6991	1.1514	1.077	1.505	0.5570	1.0923	1.264	0.6444	0.8400	1.3592	1.024
25	73	1.04569	50.6	1.312	0.2258	1.145	1.9667	1.470	1.523	0.5514	1.0912	0.8373	2.230	0.4069	1.0497	1.052	0.8937	0.7163	1.2498	0.9005
25	76	1.03061	45.3	1.281	0.2653	1.111	1.8420	1.375	1.865	0.4712	1.0668	0.7127	2.836	0.3299	1.0337	0.9388	1.078	0.6431	1.2011	0.8245
25	78	1.02234	41.7	1.256	0.2982	1.084	1.7592	1.306	2.176	0.4149	1.0525	0.6282	3.423	0.2781	1.0248	0.8596	1.240	0.5887	1.1704	0.7662
25	81	1.01247	35.8	1.209	0.3637	1.031	1.6328	1.190	2.878	0.3251	1.0339	0.4995	4.859	0.2001	1.0142	0.7350	1.590	0.4956	1.1265	0.6615
25	83	1.00751	31.5	1.166	0.4258	0.9844	1.5447	1.102	3.665	0.2606	1.0233	0.4128	6.637	0.1481	1.0087	0.6467	1.959	0.4230	1.0985	0.5748
25	85	1.00382	26.7	1.108	0.5174	0.9194	1.4499	0.9994	5.088	0.1911	1.0143	0.3257	10.24	0.0968	1.0045	0.5537	2.578	0.3381	1.0711	0.4640
ρ	θ	Ω_s	A_s (dB)	l_1	l_2	c_2	Ω_2	l_3	l_4	c_4	Ω_4	l_5	l_6	c_6	Ω_6	l_7	l_8	c_8	Ω_8	l_9

CHEBYSHEV LOW-PASS FILTERS

If the passband requirement is for a variation of attenuation between zero and a maximum and the stop requirement is only to increase monotonically to infinity, a Chebyshev (sometimes spelled "Tchebyscheff") filter can be used. Fig. 11 shows this performance characteristic and the circuit configurations for both odd- and even-degree filters.

The discussion of impedance variation in the preceding section also applies for this network type.

Table 9 contains data for a number of normalized Chebyshev low-pass filters for degree 2 through 9. For each degree, one network is tabulated for reflection coefficients of 2, 5, 10, and 25 percent. Since stopband attenuation increases monotonically, there is only one circuit for a given A_p.

Figs. 12 and 13 are used in the same way as de-

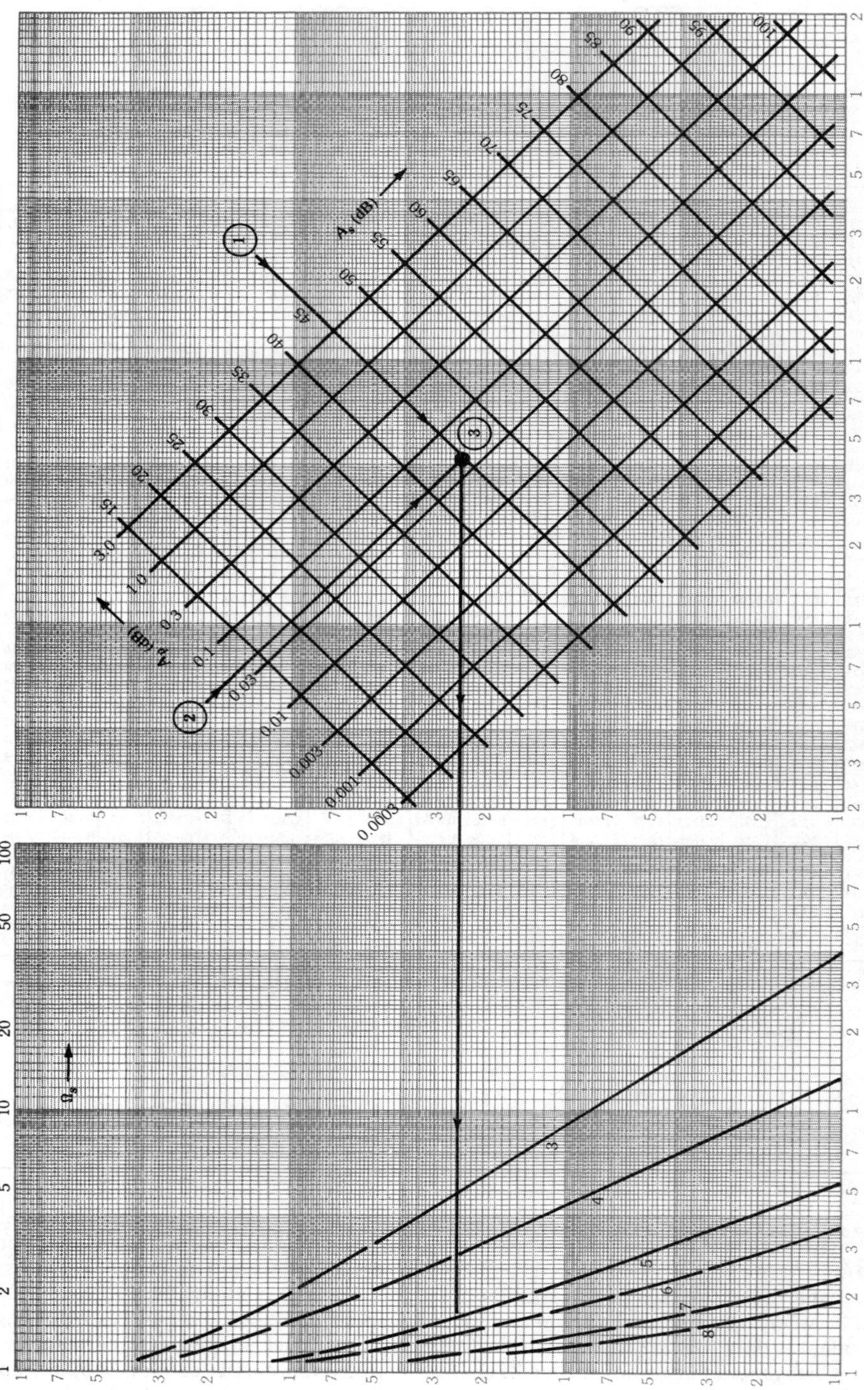

Fig. 9. Selecting the degree for a Cauer-parameter low-pass. (Example 1 is drawn in to illustrate use of the chart.)

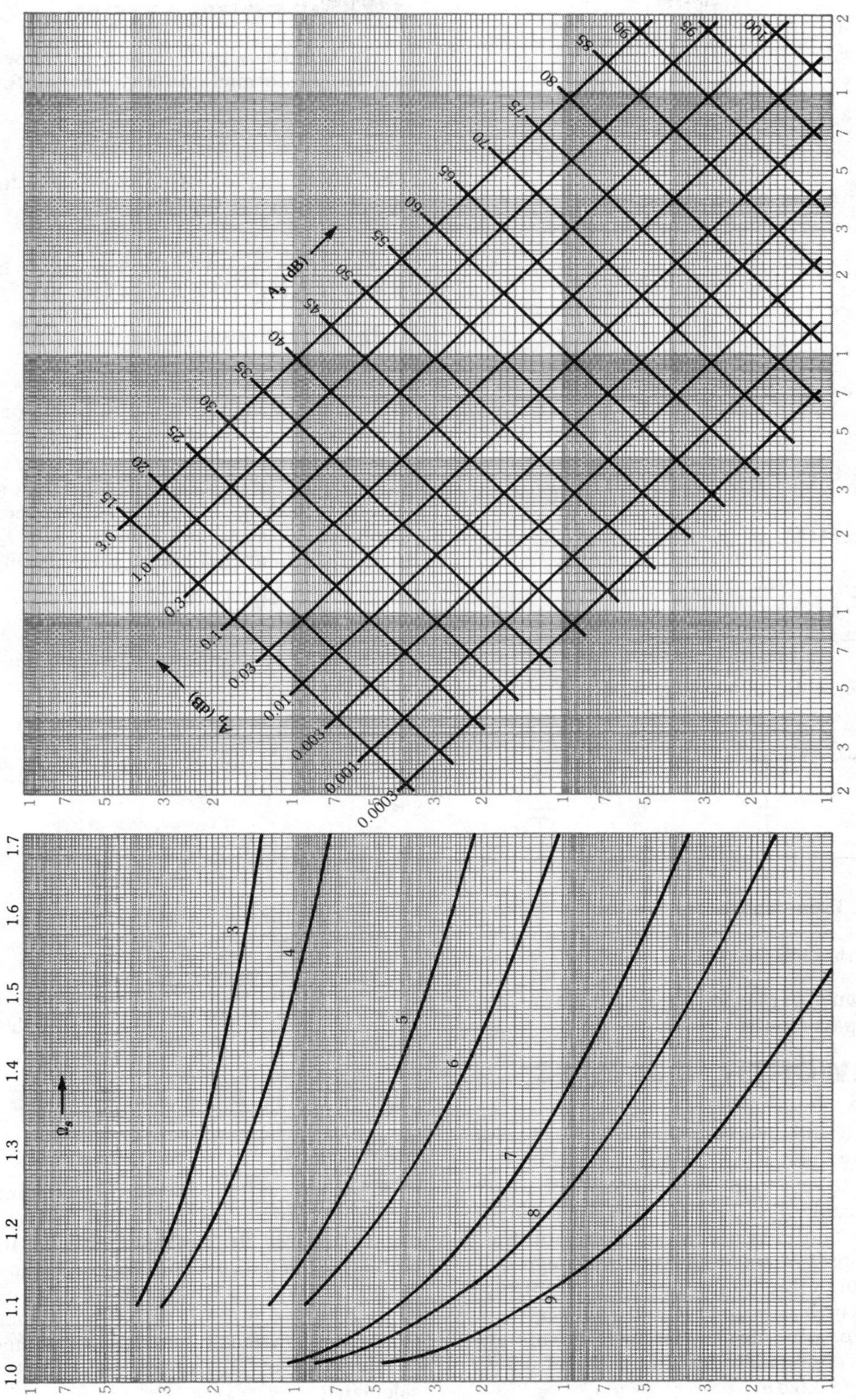

Fig. 10. Selecting the degree for a Cauer-parameter low-pass—expanded.

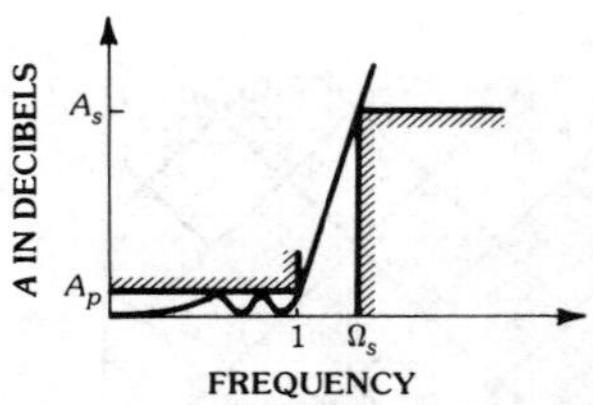

(A) Odd-degree performance.

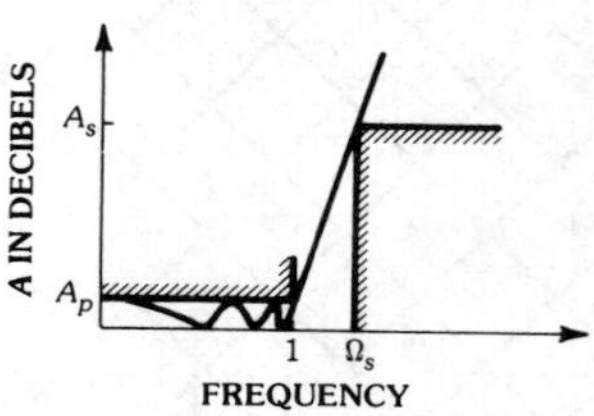

(B) Even-degree performance.

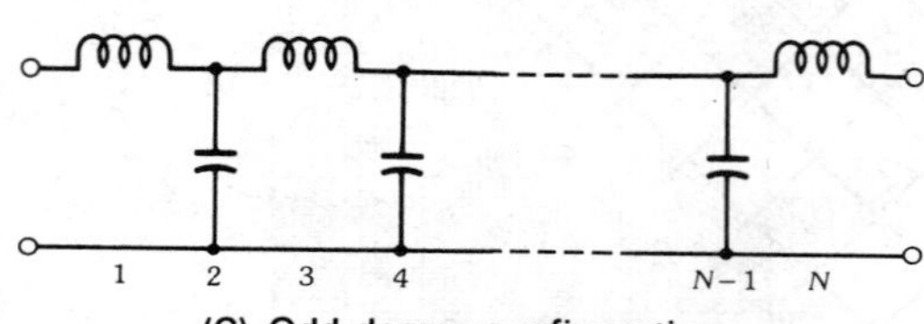

(C) Odd-degree configuration.

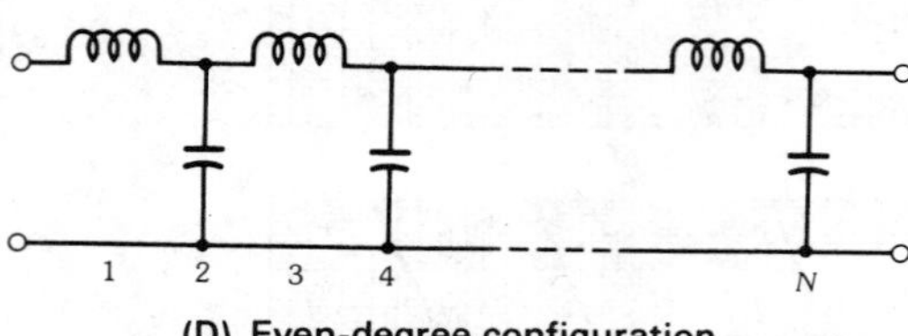

(D) Even-degree configuration.

Fig. 11. Chebyshev low-pass filters.

scribed for Cauer-parameter filters to ascertain the order of low-pass required.

BUTTERWORTH LOW-PASS FILTERS

Whenever a filter performance is desired with monotonically increasing attenuation from frequency equal to zero to infinity, Butterworth filters may be applicable. Fig. 14 shows the performance characteristic and circuit configurations for Butterworth low-pass filters of odd and even degree.

Table 10 contains data for a number of normalized Butterworth low-pass filters of degree 2 through 9. For each degree, circuit elements are tabulated for attenuation maxima at f_{ref} of 0.011, 0.044, 0.28, and 3.0 dB.

Figs. 15 and 16 are used in the same way as described for Cauer-parameter filters to determine the degree of the low-pass to satisfy the requirements.

LOW-PASS TO HIGH-PASS TRANSFORMATION

The frequency performance of a low-pass filter can be transformed into high-pass performance by the frequency transformation

$$\Omega_{HP} = 1/\Omega_{LP}$$

The transformation of normalized circuit-element values is shown in Fig. 17.

LOW-PASS TO BANDPASS TRANSFORMATION

By the transformation

$$\Omega = a(\eta - 1/\eta)$$

the frequency characteristic of a low-pass filter is converted into that of a bandpass filter. Two normalized bandpass frequencies, $\eta_p = f_{+p}/f_{ref}$ and $\eta_{-p} = 1/\eta_p = f_{-p}/f_{ref}$, correspond to the normalized low-pass Ω_p. Similarly, the geometrically symmetrical stopband limits, $\eta_s = f_s/f_{ref}$ and $\eta_{-s} = f_{-s}/f_{ref}$ are defined as shown in Fig. 6, and:

$$\Omega_p = a(\eta_{-p} + \eta_{+p}) = 1$$
$$\Omega_s = a(\eta_{-s} + \eta_s)$$

The relation between stopband limits and passband limits is

$$\Omega_s = (\eta_s - \eta_{-s})/(\eta_p - \eta_{-p})$$

The transformation of normalized circuit-element values is shown in Fig. 18.

LOW-PASS TO BAND-STOP TRANSFORMATION

The transformation of a low-pass filter into a band-stop filter is similar to the bandpass tranform.

$$\Omega = 1/a(\eta - 1/\eta)$$

The resulting transformation of circuit-element values is shown in Fig. 19.

NUMERICAL EXAMPLES

Example 1

A low-pass filter with input impedance of 600 ohms and output impedance less than or equal to 600 ohms is needed to pass frequencies up to 3.4 kHz with less than 0.05 dB attenuation and to attenuate frequencies at 8.0 kHz and above by at least 45 dB (Fig. 20). Design both a Cauer-parameter and a Chebyshev low-pass to meet the requirement.

The value of Ω_s for the reference low-pass is calculated

$$\Omega_s = 8000/3400 = 2.353$$

Cauer-Parameter Design—To determine the

TABLE 9. CHEBYSHEV LOW-PASS, DEGREE 2–9

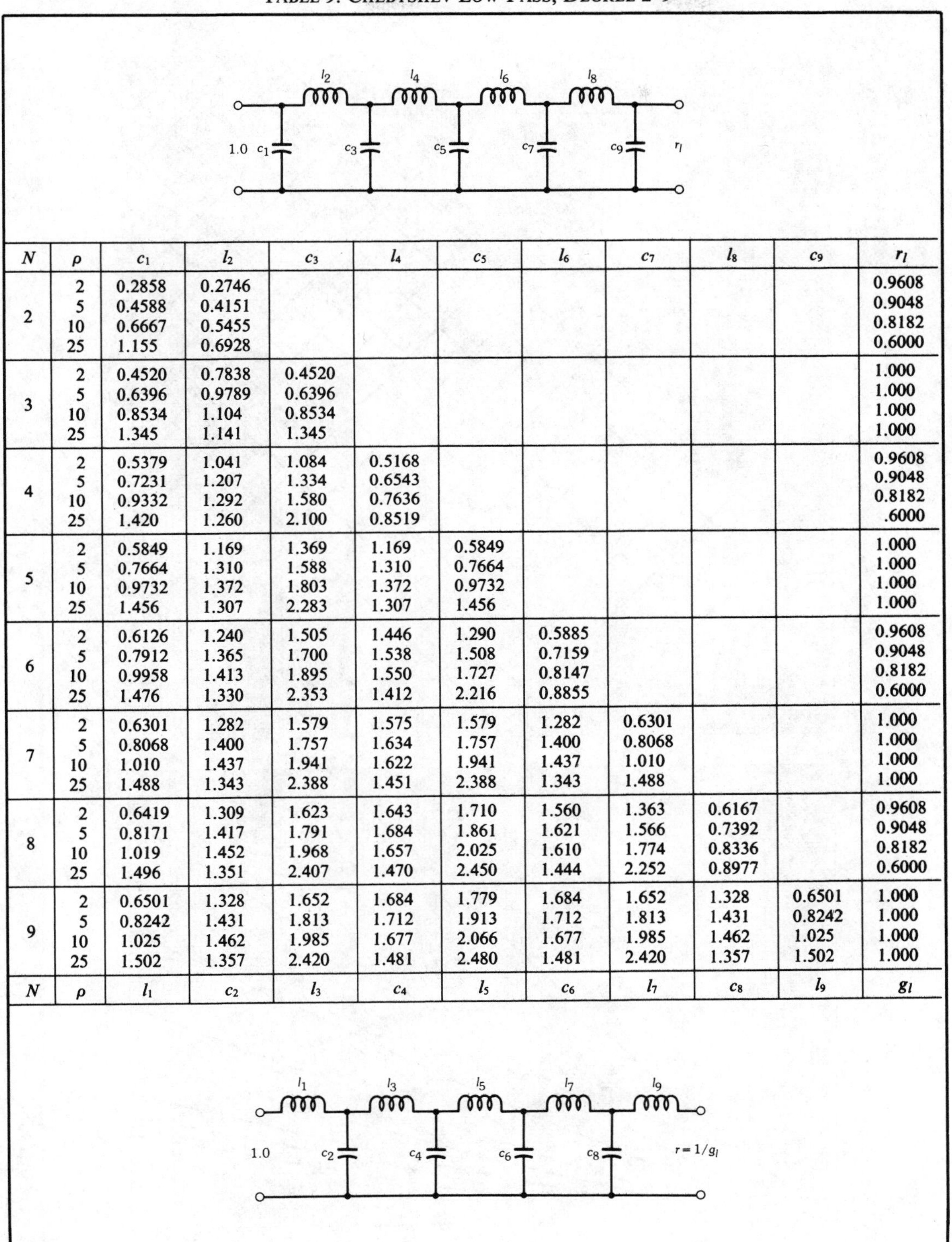

N	ρ	c_1	l_2	c_3	l_4	c_5	l_6	c_7	l_8	c_9	r_l
2	2	0.2858	0.2746								0.9608
	5	0.4588	0.4151								0.9048
	10	0.6667	0.5455								0.8182
	25	1.155	0.6928								0.6000
3	2	0.4520	0.7838	0.4520							1.000
	5	0.6396	0.9789	0.6396							1.000
	10	0.8534	1.104	0.8534							1.000
	25	1.345	1.141	1.345							1.000
4	2	0.5379	1.041	1.084	0.5168						0.9608
	5	0.7231	1.207	1.334	0.6543						0.9048
	10	0.9332	1.292	1.580	0.7636						0.8182
	25	1.420	1.260	2.100	0.8519						.6000
5	2	0.5849	1.169	1.369	1.169	0.5849					1.000
	5	0.7664	1.310	1.588	1.310	0.7664					1.000
	10	0.9732	1.372	1.803	1.372	0.9732					1.000
	25	1.456	1.307	2.283	1.307	1.456					1.000
6	2	0.6126	1.240	1.505	1.446	1.290	0.5885				0.9608
	5	0.7912	1.365	1.700	1.538	1.508	0.7159				0.9048
	10	0.9958	1.413	1.895	1.550	1.727	0.8147				0.8182
	25	1.476	1.330	2.353	1.412	2.216	0.8855				0.6000
7	2	0.6301	1.282	1.579	1.575	1.579	1.282	0.6301			1.000
	5	0.8068	1.400	1.757	1.634	1.757	1.400	0.8068			1.000
	10	1.010	1.437	1.941	1.622	1.941	1.437	1.010			1.000
	25	1.488	1.343	2.388	1.451	2.388	1.343	1.488			1.000
8	2	0.6419	1.309	1.623	1.643	1.710	1.560	1.363	0.6167		0.9608
	5	0.8171	1.417	1.791	1.684	1.861	1.621	1.566	0.7392		0.9048
	10	1.019	1.452	1.968	1.657	2.025	1.610	1.774	0.8336		0.8182
	25	1.496	1.351	2.407	1.470	2.450	1.444	2.252	0.8977		0.6000
9	2	0.6501	1.328	1.652	1.684	1.779	1.684	1.652	1.328	0.6501	1.000
	5	0.8242	1.431	1.813	1.712	1.913	1.712	1.813	1.431	0.8242	1.000
	10	1.025	1.462	1.985	1.677	2.066	1.677	1.985	1.462	1.025	1.000
	25	1.502	1.357	2.420	1.481	2.480	1.481	2.420	1.357	1.502	1.000
N	ρ	l_1	c_2	l_3	c_4	l_5	c_6	l_7	c_8	l_9	g_l

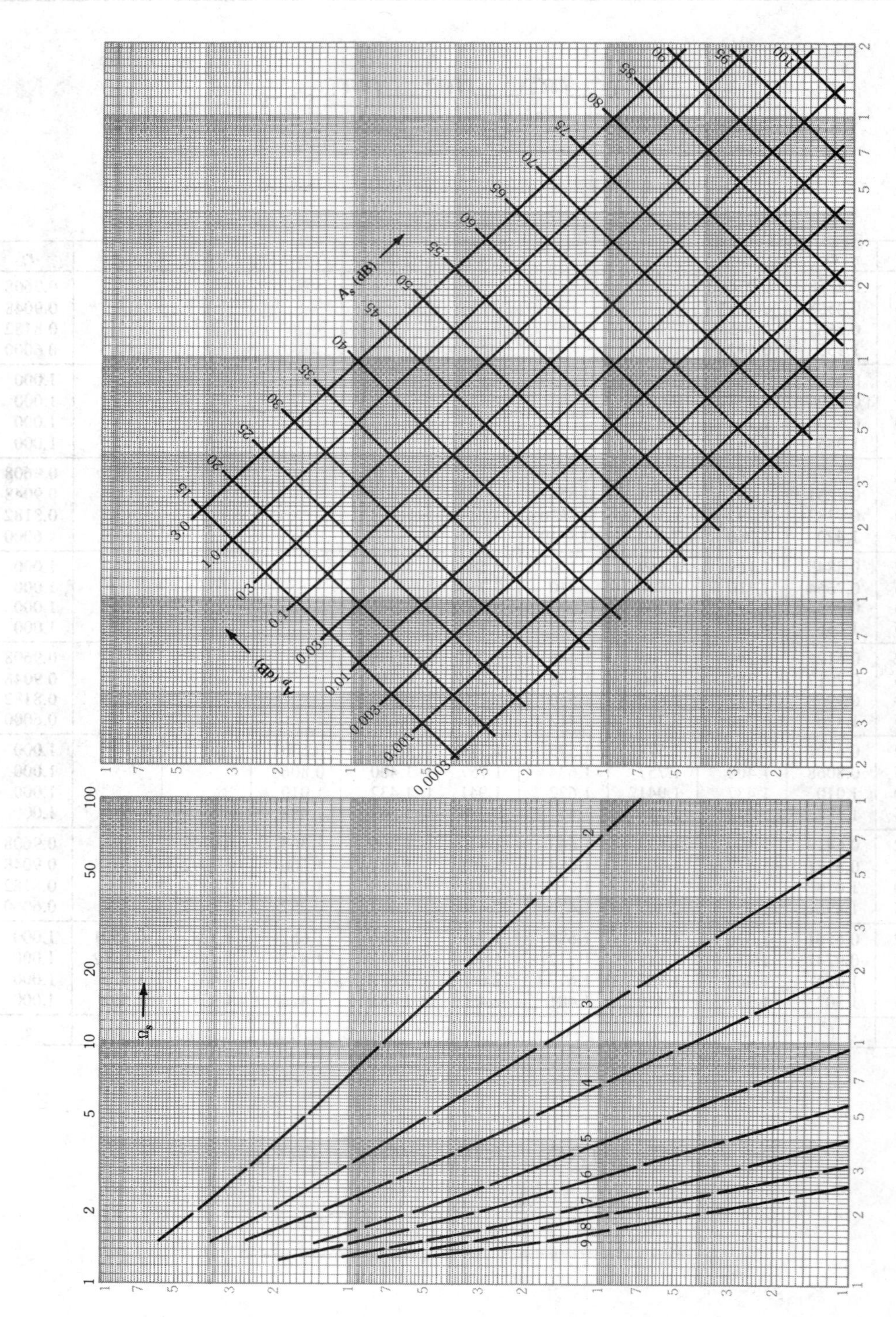

Fig. 12. Selecting the degree for a Chebyshev low-pass.

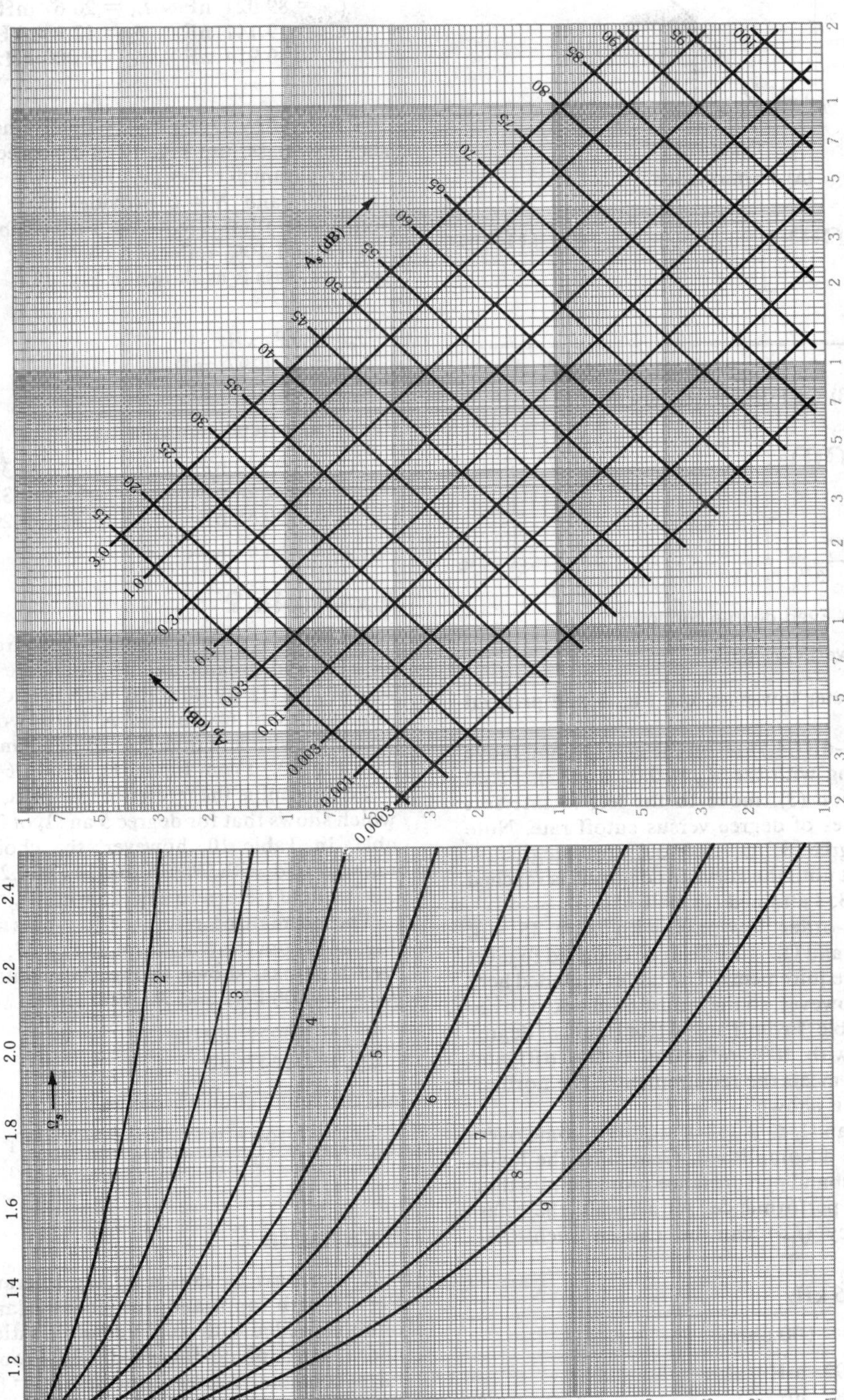

Fig. 13. Selecting the degree for a Chebyshev low-pass—expanded.

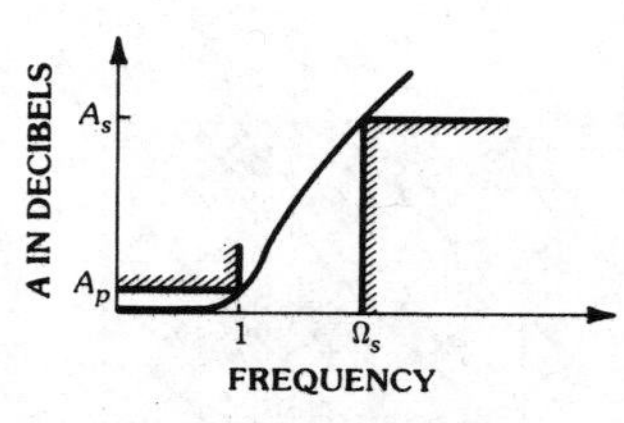

(A) Performance.

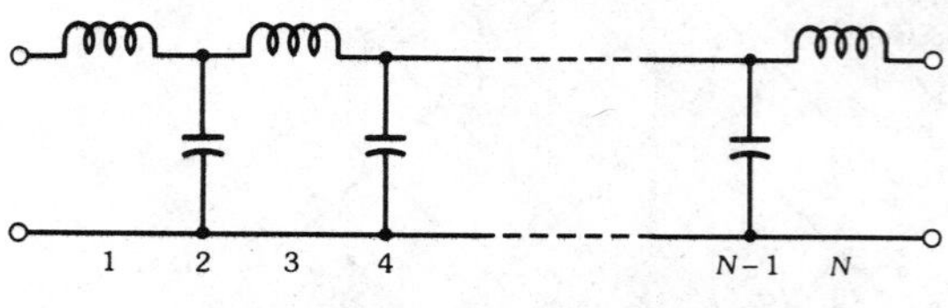

(B) Odd-degree configuration.

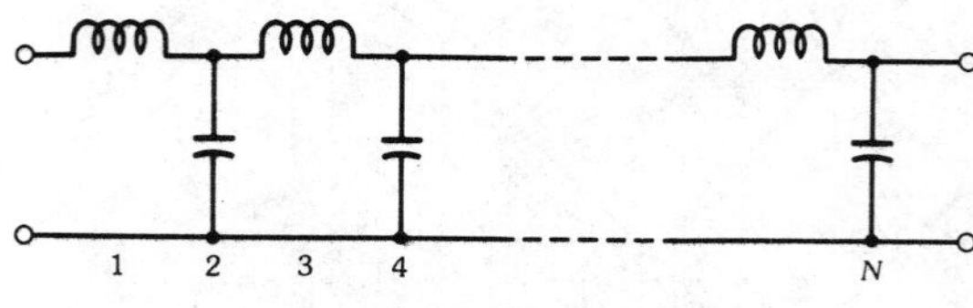

(C) Even-degree configuration.

Fig. 14. Butterworth low-pass.

Cauer-parameter low-pass, use Fig. 9 and proceed as follows.

Find the 45-dB (A_s) line on the grid (1), and move to the crossing with the $A_p = 0.05$ line (2). From this intersection (3), move to the left to intersect with the curves of degree versus cutoff rate. Note that for degree 3, Ω_s is $\cong 5.0$; for degree 4, $\Omega_s \cong 2.9$. Not until the curve for degree 5 is reached is an $\Omega_s < 2.353$ attained ($\Omega_s \cong 1.65$).

So a fifth-order low-pass with $A_s = 45$ dB, $A_p = 0.05$ dB and $\Omega_s = 1.65$ can be designed to meet the specified requirements. The tables included here are limited, however, so they must be checked to see what is available. From Table 4 (for filters of degree 5), $\theta = 37°$ with 10% reflection (0.044 dB) and $\Omega_s = 1.66164$ meets the requirements. The normalized low-pass is shown in Fig. 21.

Since only an $\Omega_s \leq 2.353$ is required, some overdesign can be accomplished. Generally, the losses erode the passband performance more rapidly than the stopband performance. As a result, the reference frequency (f_{ref}) can be chosen such that $f_s = 8000$ Hz.

$$f_{ref} = 4.6245 \text{ kHz}$$

and since $R_{ref} = 600$ ohms,

$$L_{ref} = 20.6492 \text{ mH}$$
$$C_{ref} = 57.3588 \text{ nF}$$

The actual low-pass circuit (Fig. 21) then is:

$C_1 = 50.665$ nF
$C_2 = 6.625$ nF $L_2 = 25.75$ mH
$C_3 = 89.021$ nF $L_4 = 20.67$ mH
$C_4 = 19.141$ nF $f_2 = 12.188$ kHz
$C_5 = 40.954$ nF $f_4 = 8.000$ kHz

Chebyshev Design—To determine the Chebyshev low-pass, use Fig. 12 and proceed as for the Cauer-parameter design. Here $N = 5$ results in $\Omega_s \cong 2.7$, and $N = 6$ provides $\Omega_s \cong 2.1$. Thus a sixth-order filter is required. From Table 9 for sixth degree and 10% reflection coefficient the normalized low-pass is as shown in Fig. 22.

With

$$f_{ref} = 3.8 \text{ kHz}$$
$$L_{ref} = 25.1297 \text{ mH}$$
$$C_{ref} = 69.8048 \text{ nF}$$

the actual circuit elements are:

$C_1 = 69.512$ nF $L_2 = 35.508$ mH
$C_3 = 132.28$ nF $L_4 = 38.951$ mH
$C_5 = 120.55$ nF $L_6 = 20.473$ mH

$R_l = 490.9$ ohms

Example 2

A high-pass filter is needed to pass frequencies of 64 kHz and greater and to attenuate frequencies below 16 kHz by 40.0 dB with input and output impedances of 600 ohms. A Butterworth filter is preferred with less than 0.4-dB passband variation. (See Fig. 23.) From Fig. 5, $\Omega_s = 64/16 = 4.0$.

To determine the reference low-pass, use Fig. 15, which shows that for degree 5 an Ω_s of 3.2 is attainable. In Table 10, however, the choice is more limited, and with $N = 5$ and $A_p = 0.28$, an Ω_s of 3.3 is found (return to Fig. 15 with $A_p = 0.28$ dB).

The normalized reference low-pass is shown in Fig. 24. From the transform of Fig. 17, the normalized high-pass shown in Fig. 25 is obtained. With $f_{ref} = 64.000$ kHz and R_{ref} 600 ohms,

$$L_{ref} = 1.4921 \text{ mH}$$
$$C_{ref} = 4.1447 \text{ nF}$$

and the actual circuit becomes

$L_2 = 1.209$ mH $C_1 = 8.791$ nF
$L_4 = 1.209$ mH $C_3 = 2.716$ nF
$C_5 = 8.791$ nF

Example 3

A bandpass as shown by the tolerance plot of Fig. 26 with input and output impedances of 300 ohms is needed with stopband attenuation of 58 dB and a reflection coefficient of 5%. From Fig. 6 we might compute a reference low-pass $\Omega_s = (104 - 72)/(96 - 80) = 2.0$. However, geometric symmetry is required for both passband and stopband, so more calculation is necessary. From Fig. 6, it can be found that:

TABLE 10. BUTTERWORTH LOW-PASS, DEGREE 2–9

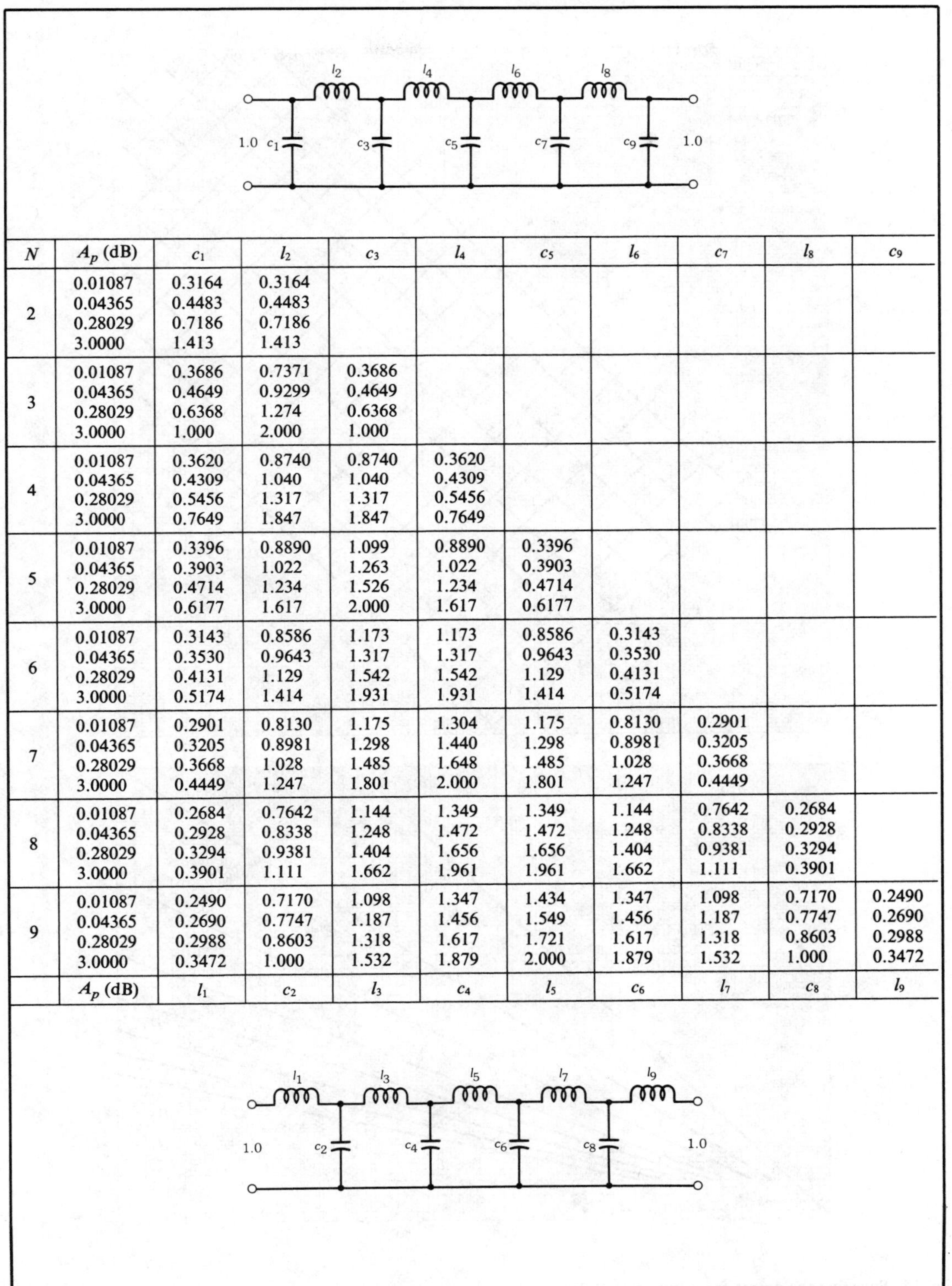

N	A_p (dB)	c_1	l_2	c_3	l_4	c_5	l_6	c_7	l_8	c_9
2	0.01087	0.3164	0.3164							
	0.04365	0.4483	0.4483							
	0.28029	0.7186	0.7186							
	3.0000	1.413	1.413							
3	0.01087	0.3686	0.7371	0.3686						
	0.04365	0.4649	0.9299	0.4649						
	0.28029	0.6368	1.274	0.6368						
	3.0000	1.000	2.000	1.000						
4	0.01087	0.3620	0.8740	0.8740	0.3620					
	0.04365	0.4309	1.040	1.040	0.4309					
	0.28029	0.5456	1.317	1.317	0.5456					
	3.0000	0.7649	1.847	1.847	0.7649					
5	0.01087	0.3396	0.8890	1.099	0.8890	0.3396				
	0.04365	0.3903	1.022	1.263	1.022	0.3903				
	0.28029	0.4714	1.234	1.526	1.234	0.4714				
	3.0000	0.6177	1.617	2.000	1.617	0.6177				
6	0.01087	0.3143	0.8586	1.173	1.173	0.8586	0.3143			
	0.04365	0.3530	0.9643	1.317	1.317	0.9643	0.3530			
	0.28029	0.4131	1.129	1.542	1.542	1.129	0.4131			
	3.0000	0.5174	1.414	1.931	1.931	1.414	0.5174			
7	0.01087	0.2901	0.8130	1.175	1.304	1.175	0.8130	0.2901		
	0.04365	0.3205	0.8981	1.298	1.440	1.298	0.8981	0.3205		
	0.28029	0.3668	1.028	1.485	1.648	1.485	1.028	0.3668		
	3.0000	0.4449	1.247	1.801	2.000	1.801	1.247	0.4449		
8	0.01087	0.2684	0.7642	1.144	1.349	1.349	1.144	0.7642	0.2684	
	0.04365	0.2928	0.8338	1.248	1.472	1.472	1.248	0.8338	0.2928	
	0.28029	0.3294	0.9381	1.404	1.656	1.656	1.404	0.9381	0.3294	
	3.0000	0.3901	1.111	1.662	1.961	1.961	1.662	1.111	0.3901	
9	0.01087	0.2490	0.7170	1.098	1.347	1.434	1.347	1.098	0.7170	0.2490
	0.04365	0.2690	0.7747	1.187	1.456	1.549	1.456	1.187	0.7747	0.2690
	0.28029	0.2988	0.8603	1.318	1.617	1.721	1.617	1.318	0.8603	0.2988
	3.0000	0.3472	1.000	1.532	1.879	2.000	1.879	1.532	1.000	0.3472
	A_p (dB)	l_1	c_2	l_3	c_4	l_5	c_6	l_7	c_8	l_9

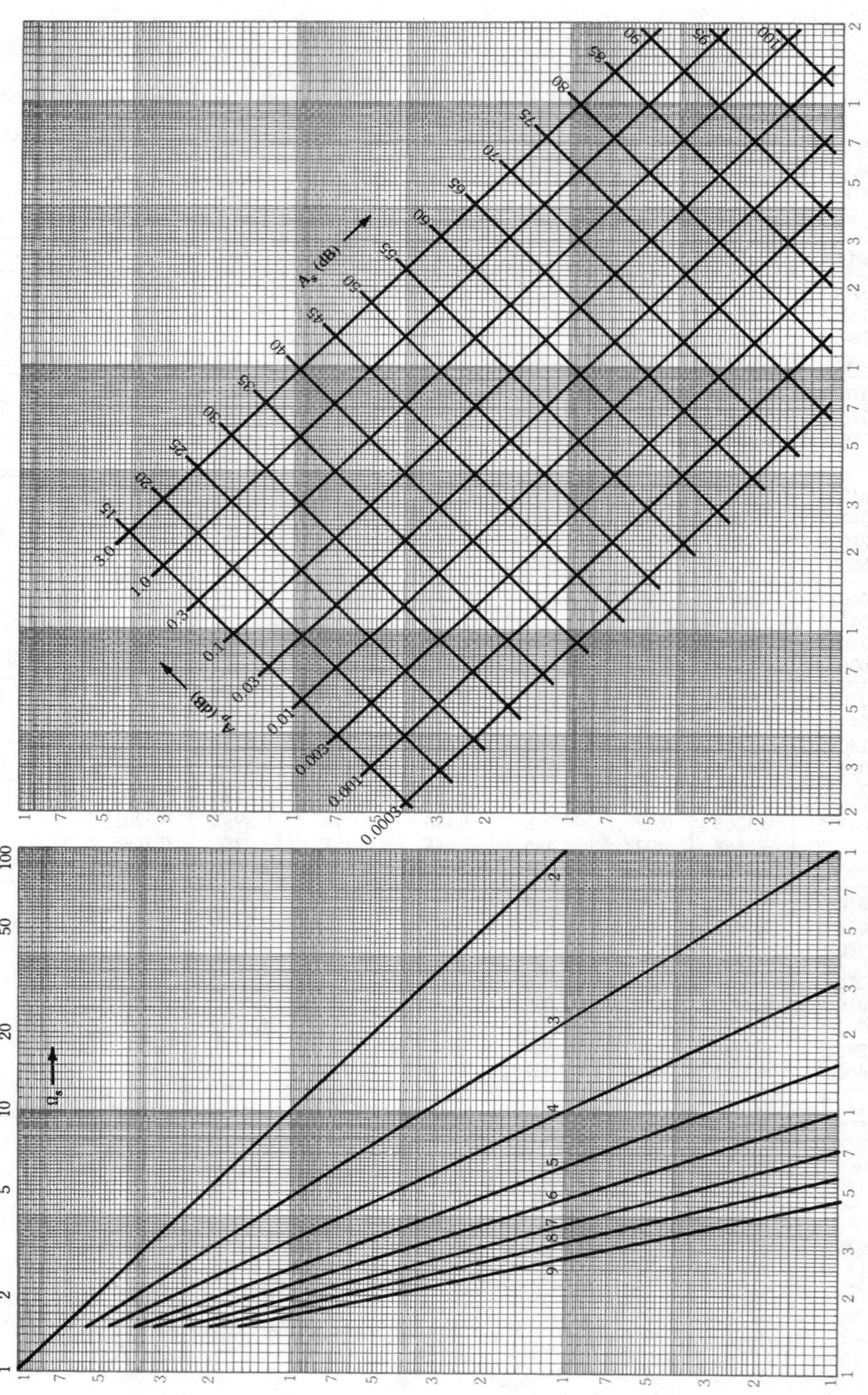

Fig. 15. Selecting the degree for a Butterworth low-pass.

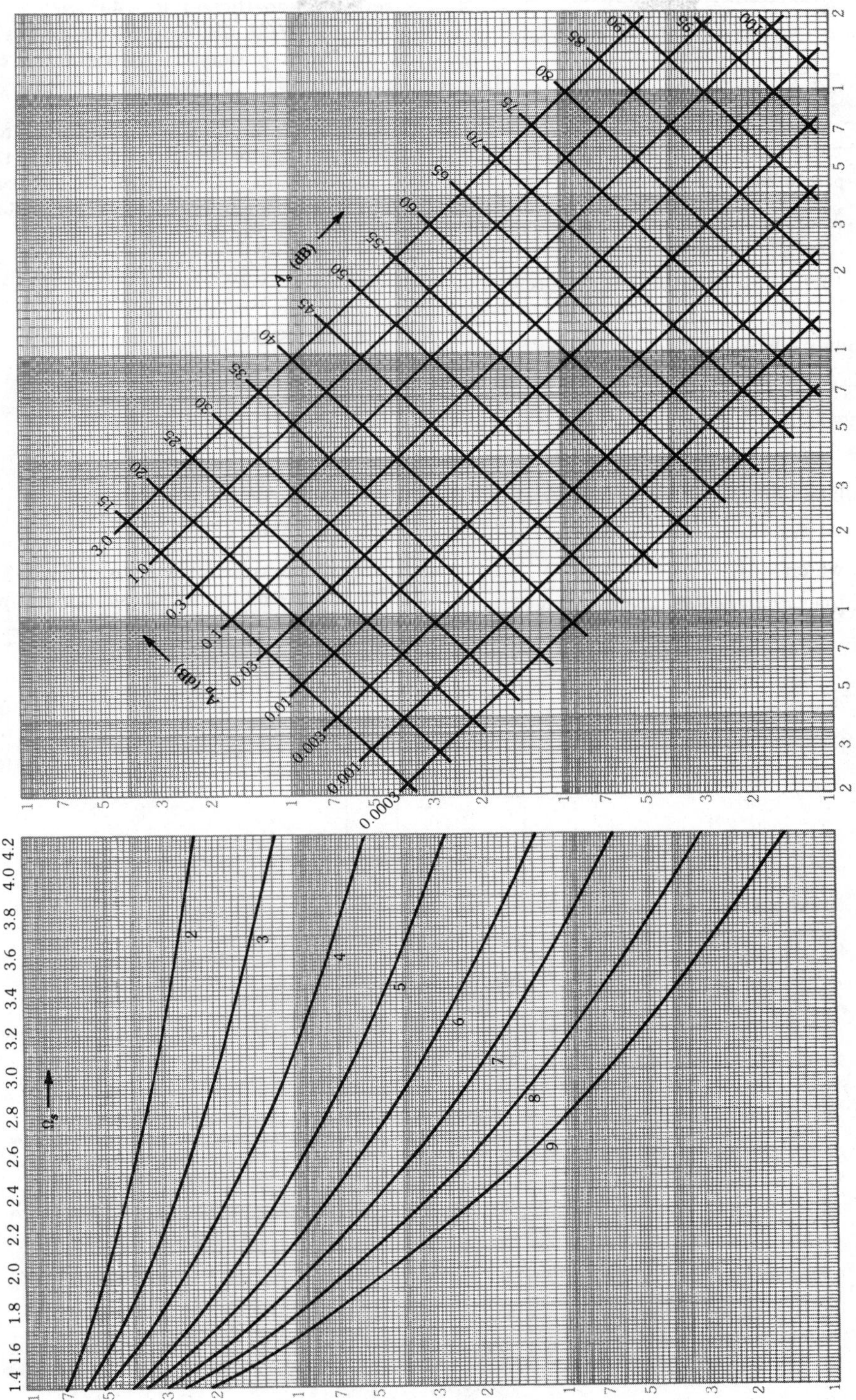

Fig. 16. Selecting the degree for a Butterworth low-pass—expanded.

LOW-PASS	HIGH-PASS	TRANSFORMATION
c_o (capacitor)	l (inductor)	$l=1/c_o$
l_o (inductor)	c (capacitor)	$c=1/l_o$

Fig. 17. Low-pass to high-pass transform.

$$f_{ref}^2 = f_{+s} f_{-s} = f_{+p} f_{-p}$$

There are, then, two choices for the reference low-pass:

(A) Calculate a new f_{-p}

$$f_{-p} = f_{-s} f_{+s}/f_{+p} = 78.0 \text{ kHz}$$

from which

$$\Omega_s = (104 - 72)/(96 - 78) = 1.778$$

(B) Calculate a new f_{-s}

$$f_{-s} = f_{+p} f_{-p}/f_{+s} = 73.846 \text{ kHz}$$

from which

$$\Omega_s = (104 - 73.846)/(96 - 80) = 1.885$$

Fig. 9 shows that a seventh-order filter is required to achieve an Ω_s as sharp as 1.885, and, in fact, $\Omega_s \cong 1.475$ can be achieved. From Table 6 with 5% reflection coefficient, $\theta = 43°$ yields $\Omega_s = 1.46628$ and $A_s = 60.6$ dB. (For filters of this complexity, only Cauer-parameter networks are generally considered.) The normalized reference low-pass is shown in Fig. 27.

Since this low-pass has a sharper cutoff rate than was required by either choice A or B, the bandpass transformation can be made with some overdesign throughout the tolerance plot. For instance, choose

$$f_{+p} = 97 \text{ kHz}$$

with $f_{ref} = \sqrt{104 \times 72} = 86.5332$ kHz

$$f_{-p} = f_{ref}^2/f_{+p} = 77.196 \text{ kHz}$$

$$\Delta f_p = 19.804 \text{ kHz}$$

$$\Delta f_s = \Omega_s \cdot \Delta f_p = 29.032 \text{ kHz}$$

Solving $f_{+s} f_{-s} = f_{ref}^2$ with $f_{+s} - f_{-s} = 29.032$

$$f_{+s} = 102.259 \text{ kHz}$$

$$f_{-s} = 73.226 \text{ kHz}$$

and

$$a = 4.369$$

With the transformations of Fig. 18, the normalized bandpass circuit becomes as shown in Fig. 28. (In Fig. 28, $\Omega_i = 1/\sqrt{c_i l_i}$.) Since $R_{ref} = 300$ ohms and $f_{ref} = 86.5332$ kHz,

LOW-PASS	BANDPASS	TRANSFORMATION
c_o (series capacitor)	l and c in parallel, η	$\eta = 1.0$ $c = 1/l = ac_o$
l_o (series inductor)	l and c in series, η	$\eta = 1.0$ $l = 1/c = al_o$
l_o parallel c_o, Ω_o	l_1, c_1 (η_1) in parallel; in series with l_2, c_2 (η_2) in parallel	$c_1 = 1/l_2 = ac_o(1+\eta_2^2)$ $c_1 = 1/l_1 = ac_o(1+\eta_1^2)$
		$\eta_{1,2} = \sqrt{(\Omega_o/2_a)^2+1} \pm \Omega_o/2a$
l_o series c_o, Ω_o	l_1, c_1 (η_1) in series; in parallel with l_2, c_2 (η_2) in series	$l_1 = 1/c_2 = al_o(1+\eta_2^2)$ $l_2 = 1/c_1 = al_o(1+\eta_1^2)$

Fig. 18. Low-pass to bandpass transform.

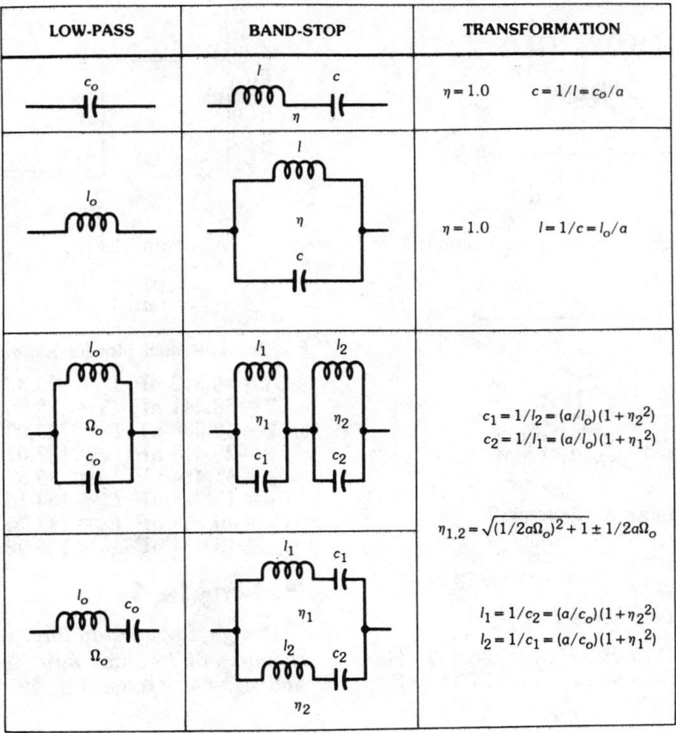

Fig. 19. Low-pass to band-stop transform.

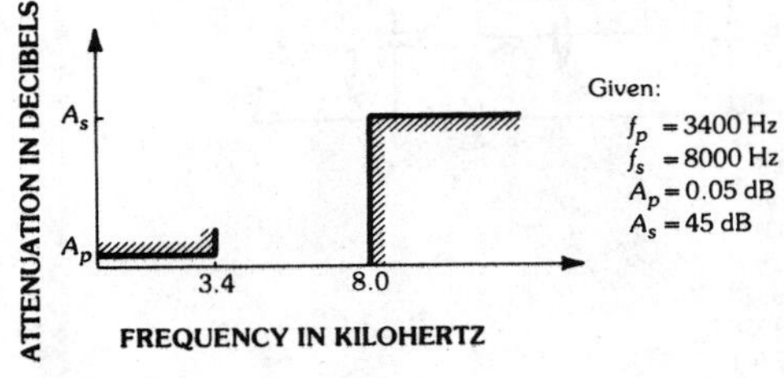

Fig. 20. Requirement for Example 1.

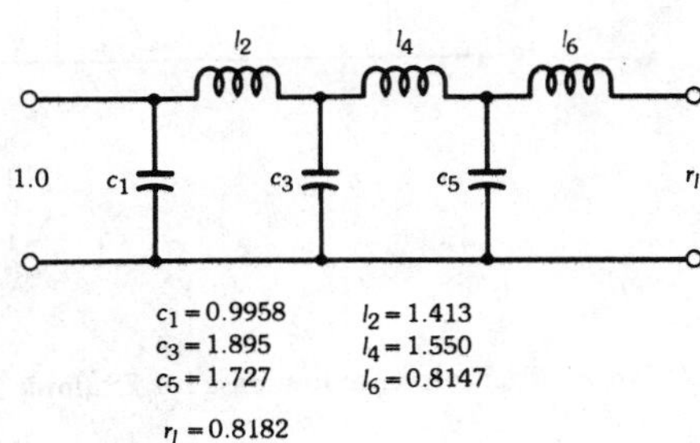

Fig. 22. Normalized Chebyshev low-pass for Example 1.

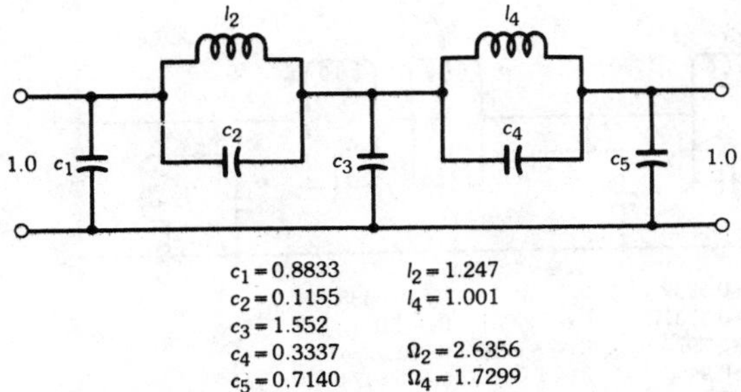

Fig. 21. Normalized Cauer-parameter low-pass for Example 1.

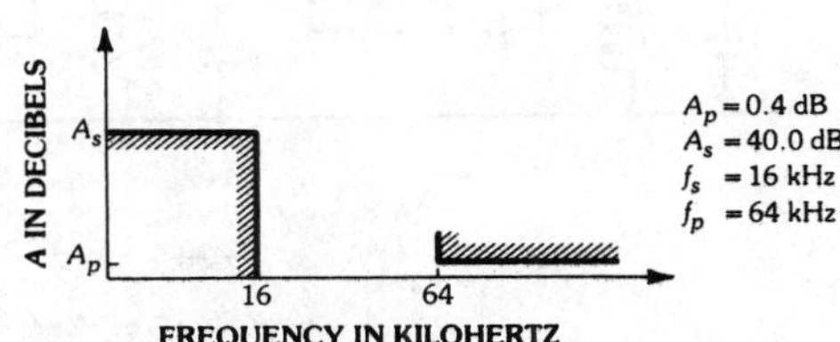

Fig. 23. Requirements for Example 2.

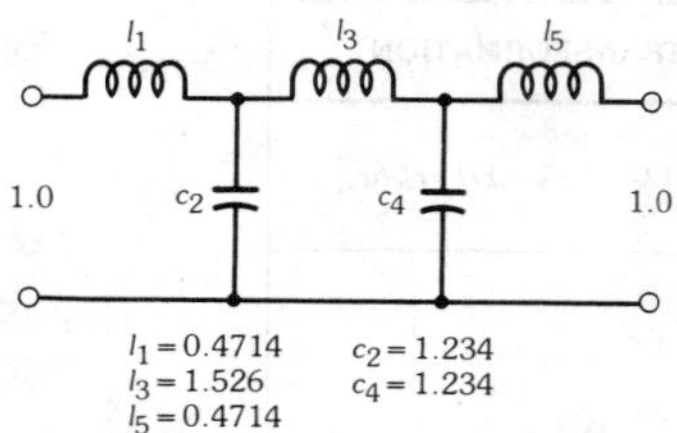

Fig. 24. Normalized reference low-pass for Example 2.

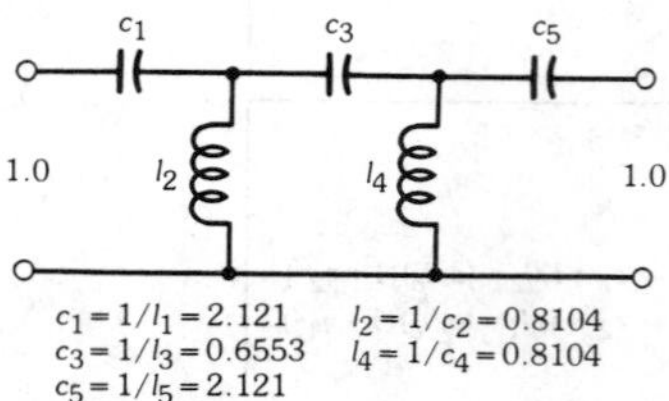

Fig. 25. Normalized high-pass for Example 2.

$L_{ref} = 551.77\ \mu H$
$C_{ref} = 6.131\ nF$

Then the actual circuit elements are:

$C_1 = 19.788$ nF $L_1 = 170.95\ \mu$H $f_1 = 86.533$ kHz
$C_2 = 3.475$ nF $L_2 = 496.59\ \mu$H $f_2 = 121.157$ kHz

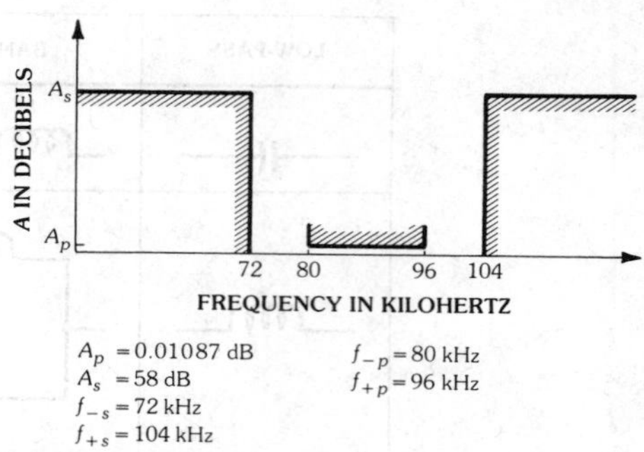

Fig. 26. Tolerance plot for Example 3.

$C_3 = 6.812$ nF $L_3 = 973.48\ \mu$H $f_3 = 61.804$ kHz
$C_4 = 38.441$ nF $L_4 = 88.00\ \mu$H $f_4 = 86.533$ kHz
$C_5 = 18.088$ nF $L_5 = 133.09\ \mu$H $f_5 = 102.578$ kHz
$C_6 = 25.418$ nF $L_6 = 187.01\ \mu$H $f_6 = 72.998$ kHz
$C_7 = 35.414$ nF $L_7 = 95.52\ \mu$H $f_7 = 86.533$ kHz
$C_8 = 13.768$ nF $L_8 = 164.16\ \mu$H $f_8 = 105.866$ kHz
$C_9 = 20.607$ nF $L_9 = 245.70\ \mu$H $f_9 = 70.731$ kHz
$C_{10} = 15.034$ nF $L_{10} = 225.02\ \mu$H $f_{10} = 86.533$ kHz

Example 4

Design a band-stop filter to operate between resistances of 75 ohms with 5% reflection coefficient and $A_s = 40$ dB, per Fig. 29.

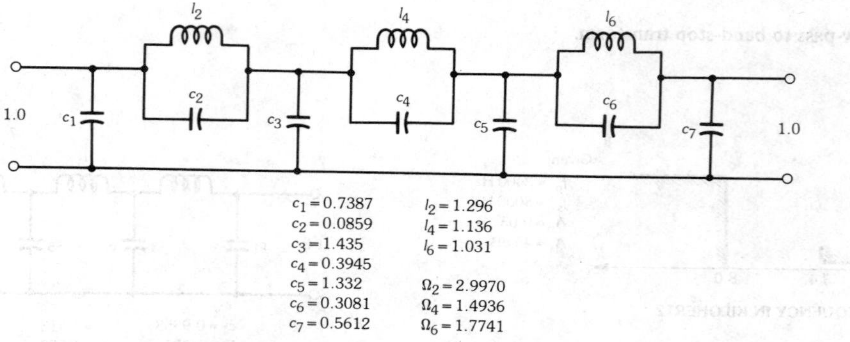

Fig. 27. Normalized reference low-pass for Example 3.

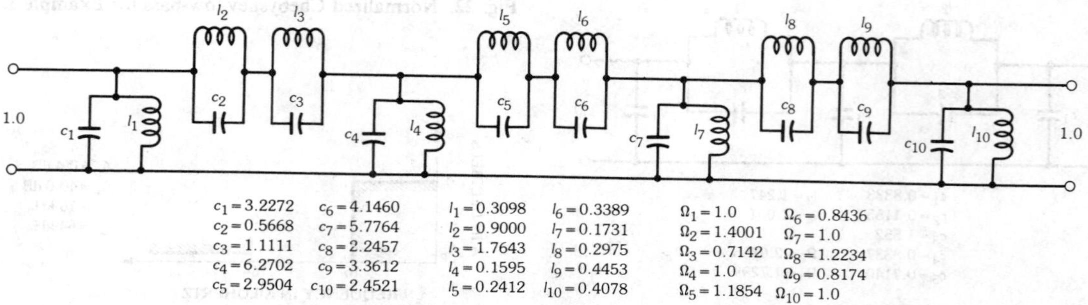

Fig. 28. Normalized bandpass circuit for Example 3.

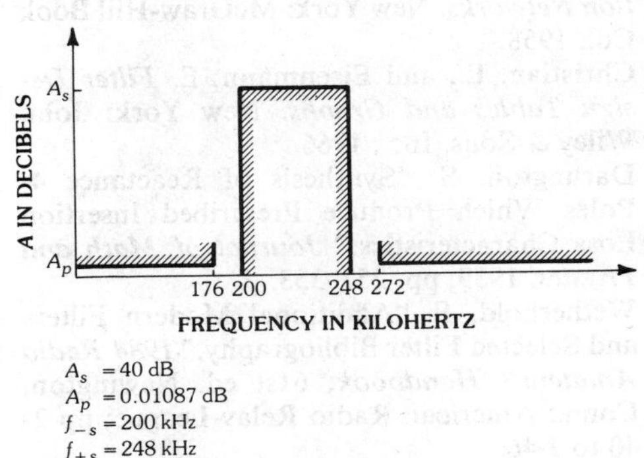

Fig. 29. Tolerance plot for Example 4.

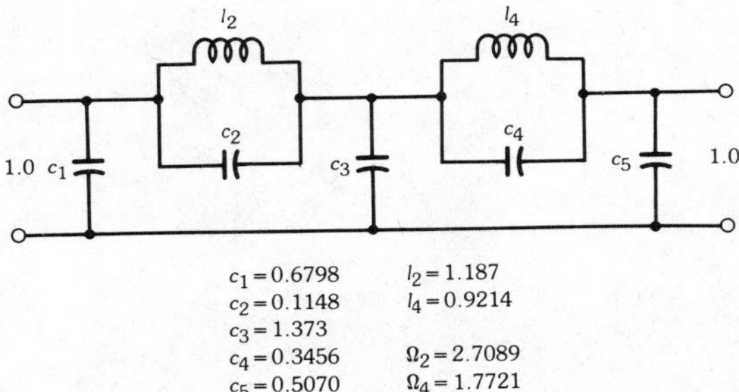

Fig. 30. Normalized reference low-pass for Example 4.

Two options exist for computing the reference low-pass (see Fig. 7) with geometric symmetry:

(A) Calculate a new f_{-p}:

$$f_{-p} = (200 \times 248)/272 = 182.35 \ \text{kHz}$$

from which

$$\Omega_s = (272 - 182.35)/(248 - 200) = 1.867$$

(B) Calculate a new f_{-s}:

$$f_{-s} = (272 - 176)/248 = 193.03 \ \text{kHz}$$

from which

$$\Omega_s = (272 - 176)/(248 - 193.03) = 1.7465$$

Fig. 9 indicates that a fifth-order filter is sufficient to meet the requirements. In fact, $\Omega_s \cong 1.70$ is achievable. From Table 4, for a reflection coefficient of 5%, $\theta = 36°$ produces $\Omega_s = 1.7013$ and $A_s = 40.8$ dB. The normalized reference low-pass is shown in Fig. 30. Selecting $f_{ref} = \sqrt{200 \times 248} = 222.711$ kHz and $f_s = 249.000$ kHz gives:

$$f_{-s} = f_{ref}^2/f_s = 199.197 \ \text{and} \ \Delta f_s = 49.803 \ \text{kHz}$$
$$\Delta f_p = \Delta f_s \cdot \Omega_s = 84.730 \ \text{kHz}$$

and solving $f_{-p} f_{+p} = f_{ref}^2$ with $\Delta f_p = f_{+p} - f_{-p}$ gives:

$$f_{-p} = 184.339 \ \text{kHz}$$
$$f_{+p} = 269.069 \ \text{kHz}$$
$$a = 2.6285$$

After application of the transformation of Fig. 19, the normalized bandstop circuit is as shown in Fig. 31.

With R_{ref} = 75 ohms and f_{ref} = 222.711 kHz

$$L_{ref} = 53.597 \ \mu\text{H}$$
$$C_{ref} = 9.5284 \ \text{nF}$$

and the actual circuit elements are:

$L_1 = 207.26\ \mu$H	$C_1 = 2.464$ nF	$f_1 = 222.711$ kHz
$L_2 = 11.254\ \mu$H	$C_2 = 39.437$ nF	$f_2 = 238.898$ kHz
$L_3 = 12.950\ \mu$H	$C_3 = 45.378$ nF	$f_3 = 207.620$ kHz
$L_4 = 102.58\ \mu$H	$C_4 = 4.978$ nF	$f_4 = 222.711$ kHz
$L_5 = 8.391\ \mu$H	$C_5 = 49.121$ nF	$f_5 = 247.897$ kHz
$L_6 = 10.397\ \mu$H	$C_6 = 60.859$ nF	$f_6 = 200.083$ kHz
$L_7 = 277.85\ \mu$H	$C_7 = 1.838$ nF	$f_7 = 222.711$ kHz

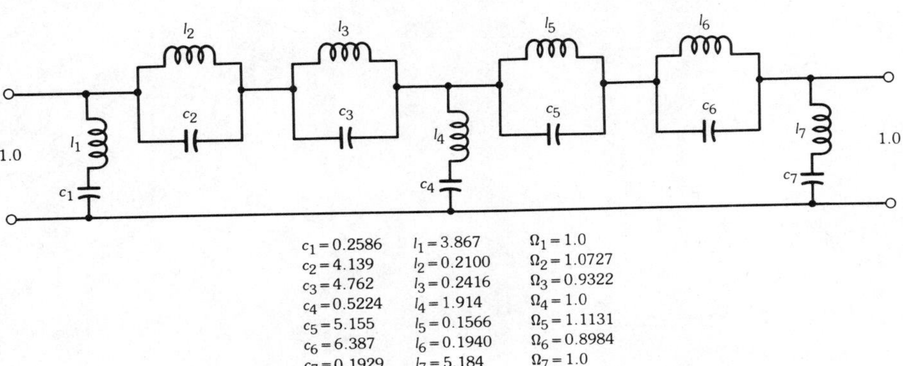

Fig. 31. Normalized band-stop circuit for Example 4.

REFERENCES

1. Saal, R., and Ulbrich, E. "On the Design of Filters by Synthesis." *IRE Transactions on Circuit Theory,* December, 1958, pp. 284–327.
2. Saal, R. *The Design of Filters Using the Catalog of Normalized Lowpass Filters.* Telefunken G. M. B. H., Fachbereich Anlagen Weitverkehr und Kabeltechnik, 1963.
3. Christian, E. *LC-Filters Design, Testing and Manufacturing.* New York: John Wiley & Sons, Inc., 1983.
4. Cauer, W. *Synthesis of Linear Communication Networks.* New York: McGraw-Hill Book Co., 1958.
5. Christian, E., and Eisenmann, E. *Filter Design Tables and Graphs.* New York: John Wiley & Sons, Inc., 1966.
6. Darlington, S. "Synthesis of Reactance 4-Poles Which Produce Prescribed Insertion Loss Characteristics." *Journal of Math and Physics,* 1939, pp. 257–353.
7. Wetherhold, E. "Additional Modern Filters and Selected Filter Bibliography." *1984 Radio Amateur's Handbook*, 61st ed. Newington, Conn.: American Radio Relay League; pp 2-40 to 2-46.

10 Active Filter Design

Kenneth R. Laker and Rolf Schaumann

The information in this chapter enables the engineer to design a wide variety of practical active filters for operation in the audio-frequency range. The equations presented permit the user to complete the design and arrive at a fairly comprehensive evaluation of the performance to be expected from the filter, without requiring complicated mathematics. Out of the countless different filters proposed in the technical literature, only those few circuits that have been proven to be practical, state-of-the-art designs are discussed in this chapter. Given the limited space available in a reference volume such as this, enough information can be provided only for the design of filters of relatively simple specifications; if system requirements are very stringent, the reader should consult the many excellent books (references 1–6) or papers referred to in the text.

INTRODUCTION

The technology of hybrid and monolithic integrated circuits has profoundly influenced the design and implementation of audio-frequency filters. Integration has allowed the realization of filters that are small in size, inexpensive, and mass-producible. During the past ten years, active R-C networks, typically comprising resistors, capacitors, and operational amplifiers, have been the primary means of integrated audio-filter implementation. Active R-C filters have eliminated the need for the bulky, expensive inductors required in passive implementations, and tuning is simplified and involves the adjustment of only resistors. Also, tuning can be automated in manufacture, using commercial laser trimming systems. In addition, active R-C filters have provided opportunities for standardization and modularity that significantly simplify design and fabrication.

More recently, switched-capacitor (SC) networks have allowed audio-frequency active filters to be realized with the metal-oxide-semiconductor (MOS), large-scale-integration (LSI) technologies associated with digital networks. Switched-capacitor filters typically contain capacitors, FET switches, and operational amplifiers. The switches are operated by clock signals that are digitally derived from a stable frequency source such as a crystal-controlled oscillator. The characteristics of the filter are then determined by capacitor ratios and the clock frequency, both inherently precise and stable parameters. Hence, SC filters rarely require trimming. The most important attribute of SC filters is that their implementation in silicon is compatible with digital-circuit integration. Hence, digital and analog circuitry can coexist on the same LSI chip.

To design active filters, whether active R-C or SC, which are used extensively in instrumentation and communication systems, one must first understand what an active filter is and how its performance requirements are specified.

An electric filter is a network that transforms an input signal in some specified way into a desired output signal. Although many applications exist where filter requirements are set in terms of time-domain specifications, the majority of filters are designed to satisfy certain frequency-domain criteria. Thus, as shown in Fig. 1, a filter is a two-port

Fig. 1. A general filter representation.

network with input voltage V_1 and output voltage V_2; the circuit response is described by a transfer function $H(s)$ defined by

$$H(s) = N(s)/D(s) = V_2(s)/V_1(s) \qquad (1)$$

where, in steady-state,

$s = j\omega$ is the frequency parameter,
$\omega = 2\pi f$ is the radian frequency (rad/s),
f is the frequency in hertz (Hz).

As indicated, $H(s)$ is a ratio of two polynomials $N(s)$ and $D(s)$. The roots of $N(s)$ are the transmission zeros of the filter, i.e., points of infinite attenuation; the roots of $D(s)$ are its poles. The transfer function is a complex quantity that may be expressed as

$$H(j\omega) = |H(j\omega)| \exp[j\phi(\omega)] \qquad (2)$$

where,

$|H(j\omega)|$ is the magnitude,
$\phi(\omega)$ is the phase.

Thus, to specify a transfer function completely, both magnitude and phase must be given at a sufficient number of frequency points. In many cases, the magnitude response is the dominant specification with the phase response either loosely specified or unspecified. In this case, a minimum-phase filter is designed to meet the magnitude specification, and whatever phase the design provides is accepted. When both magnitude and phase are specified, one widely accepted design procedure is to design first a minimum-phase filter to meet the magnitude response, as previously mentioned, and then to design a tandem nonminimum-phase all-pass filter, which, when cascaded with the minimum-phase filter, meets the desired phase specification. This nonminimum-phase all-pass network is often referred to as a phase or delay equalizer.

Filtering implies that certain frequency components of the input signal, those in the passband or passbands, are transmitted or passed to the output, whereas those in the stopband(s) are not transmitted. The most frequently used method of identifying the location of passbands and stopbands on the frequency axis is by specifying, versus frequency, the magnitude characteristic via the loss curve in decibels (dB), defined as

$$\alpha(\omega) = -20 \log |H(j\omega)| = -20 \log |V_2/V_1| \quad (3)$$

In the stopbands, where $|V_2| << |V_1|$, $|H(j\omega)|$ is small and the loss α is large, for example $|H(j\omega)| < 0.01$ or $\alpha > 40$ dB. In the passbands, $|V_2| \simeq |V_1|$ or even $|V_2| > |V_1|$, so that $|H(j\omega)| \simeq 1 (\alpha = 0$ dB) or $|H(j\omega)| > 1 (\alpha < 0$; i.e., the circuit provides gain, something an active filter can do, whereas a passive filter always provides a loss).

If the phase response is of prime importance, then $\phi(\omega)$ is specified directly in degrees or radians; alternatively, and perhaps more frequently, one prescribes the delay $T(\omega)$ in seconds, defined as

$$T(\omega) = -(d/d\omega)\,[\phi(\omega)] \quad (4)$$

For best, distortion-free transmission, delay should be constant, $T(\omega) = T_0$; i.e., the phase should be linear, $\phi(\omega) = -\omega T_0$, over the frequency range of interest.

Some additional criteria of practical interest in active-filter design are sensitivity, dynamic range, noise, power dissipation, number and range of components, method of fabrication, and cost. All of these specifications place limitations and constraints on the acceptable design. In more cases than one would like, the specifications conflict so that engineering tradeoffs have to be made to resolve the conflict.

In the following sections, the components used for active filters, some important design criteria, and several state-of-the-art practical active filters will be discussed in detail.

CIRCUIT ELEMENTS

Active filters are constructed from resistors, capacitors, and, usually, operational amplifiers (op amps). For more detailed information about these elements, refer to Chapters 5 and 20. However, a few comments will be made about these components, especially about the op amp because of its serious effect on filter performance.

Resistors*

Resistors used in active-filter design are carbon composition, metal or carbon film, thin or thick film, wirewound, and diffused. The selection depends on cost, on the technology used to implement the filter, and on filter requirements. Carbon composition resistors are the least expensive, but they have large tolerances and temperature coefficients. Further, tracking is not very good so that composition resistors should be used only for uncritical applications. Metal-film and wirewound resistors are better than composition types in all respects, although more expensive, and are the most frequently used resistors in active-filter design today. Wirewound resistors have somewhat larger parasitics (L and C) than metal-film resistors.

* Reference 3.

Capacitors*

Of the numerous different types of capacitors available, those commonly used in active filters are ceramic disc, Mylar, polystyrene, Teflon, and thin-film capacitors. Once again, the selection depends on factors such as cost, available range, tolerances, temperature coefficients, and dissipation factor (loss). Ceramic and Mylar capacitors are the least expensive types and have the highest loss; they are used only for uncritical applications. Teflon, thin-film, and especially polystyrene (or for small values, mica) capacitors are more expensive but have much lower dissipation factors and are therefore better suited for critical filter designs.

In setting filter parameters, apart from tolerances and temperature coefficient, the dissipation factor (DF) or quality factor (Q_c) is of some importance. If loss is modeled by means of a resistor R_c in parallel with capacitor C, Q_c and DF are defined as

$$Q_c = 1/DF = \omega C R_c \quad (5)$$

where ω is some critical frequency of interest. Also, it should be remembered that Q_c is a strong function of temperature. Typical values of Q_c range from less than 100 (ceramic) up to several thousand (polystyrene).

Operational Amplifiers

The active element used in the vast majority of all active filters is the operational amplifier. The "op amp" is an integrated circuit with five or more terminals, three of which are used for handling the signal (Fig. 2): the inverting (V^-) and noninverting (V^+) input terminals, at which the input voltages are applied, and the output terminal (V_o). The remaining terminals are for power supply and, in some models, for offset compensation and frequen-

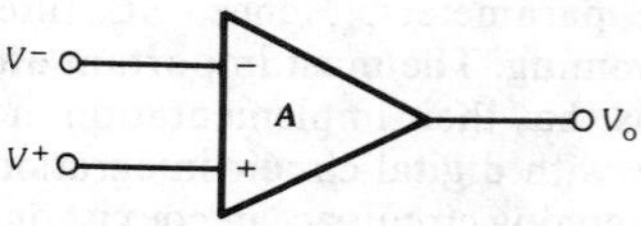

Fig. 2. Op-amp symbol.

* Reference 3.

cy compensation. The function of the op amp is described by

$$V_o = A(s)(V^+ - V^-) \tag{6}$$

where $A(s)$ is the open-loop gain.

In the ideal op amp, A is assumed to be infinite (i.e., $A \to \infty$), and, furthermore, all input impedances are infinite and the output impedance is zero. Thus, the ideal op amp is an ideal voltage-controlled voltage source with infinite gain; the input currents into the inverting and noninverting input terminals are zero. Further, since $A \to \infty$ and V_o must remain finite in practice, it follows that the input voltage $V_i = V^+ - V^-$ is zero; the operation is defined such that $V_o = 0$ when $V_i = 0$.

To get a feel for their operation, op-amp circuits often are first analyzed or designed under the assumption of ideal amplifiers ($A \to \infty$); it should, however, be strongly emphasized at this point that, except for uncritical applications at *very* low frequencies ($f < 1$ kHz), the operation of an active filter will rarely be satisfactory in practice if its design is based on ideal op amps. The main reason is that the op-amp gain $A(s)$ is a strong function of frequency. Specifically, for most op amps used in active-filter design, the gain decreases throughout the useful frequency range by 20 dB/decade. This frequency response is required for stability reasons and is achieved by use of internal or external compensation capacitors.

Thus, the most widely used and quite realistic op-amp model is

$$A(s) = A_0\sigma/(s + \sigma) = \omega_t/(s + \sigma) \tag{7}$$

where,

- $s = j\omega$ (as defined previously),
- σ is the open-loop -3 dB frequency (usually $\lesssim 2\pi \cdot 10$ Hz),
- A_0 is the dc gain (usually > 100 dB),
- $\omega_t = A_0\sigma$ is the gain-bandwidth product ($\simeq 2\pi \cdot 1$ MHz).

In most practical applications, $|s| >> \sigma$, so that, instead of Eq. (7), a commonly used model is

$$A(s) \simeq \omega_t/s \tag{8}$$

Analyzing an active filter with op amps represented by Eq. (7) or (8) rather than by a simple constant-gain model increases the degree of the describing network function [Eq. (1)] by one for each amplifier used. Thus, the filter acquires parasitic poles and zeros in addition to suffering shifts of the nominal pole-zero locations. Stated differently, the two polynomials $N(s)$ and $D(s)$ in Eq. (1) change from their ideal form. It is for this reason that filters designed with the assumption of ideal amplifiers ($\omega_t \to \infty$) do not normally have a satisfactory frequency response but show a potentially rather large deviation from their nominal performance due to finite ω_t.

Occasionally, a filter will behave unpredictably, possibly oscillate, in spite of a design based on the op-amp model of Eq. (7) or (8). The reason may often be found in the fact that, whereas the representation of Eq. (7) describes the op-amp gain very well, the phase shift is larger than indicated by Eq. (7). An adequate remedy is to multiply Eq. (7) by an "excess phase factor," $\exp(-j\,\omega/\omega_2)$, where ω_2 is a normalizing frequency of value $\omega_2 \simeq 3\omega_t$, and to use this augmented model in the analysis. Since for all practical purposes ω_2 is much larger than operating frequency ω, it has been found convenient for keeping the algebra manageable to set $\exp(-s/\omega_2) \simeq 1 - s/\omega_2$; i.e., a very accurate op-amp model for use in highly selective filters at fairly high frequencies is

$$A(s) = (\omega_t/s)(1 - s/\omega_2) \tag{9}$$

In most applications, however, the use of Eq. (8) leads to entirely satisfactory results.

Further op-amp characteristics of some concern to filter designers are slew rate and dc offset voltage.

Slew rate SR, given in volts per microsecond, refers to the maximum rate of change of a signal voltage that the amplifier can maintain at its output. Violating slew-rate limitations results in gross signal and/or transfer-function distortion and should be avoided. Thus, if $v_o(t) = V_o \sin \omega t$ is the amplifier output voltage, one needs to observe $V_o < \mathrm{SR}/\omega$. For example, if SR $= 0.7$ V/μs (a typical value) and the signal frequency is 15 kHz, the signal amplitude must satisfy $V_o \lesssim 7.2$ V.

There are two main reasons* for offset voltage. One is the need for dc input bias currents into the op-amp input stage; the second is imbalances in the input stage. To be able to provide the input bias, dc paths must exist from the inverting and noninverting input terminals to ground. To minimize offset, the resistances seen from these two terminals back into the network ought to be equal; then the voltage drops caused by the two bias currents at the op-amp inputs are equal and the direct differential input voltage is zero, resulting in zero contribution to offset. In practice, things are not quite so simple because due to imbalances in the op-amp input stage the bias currents are not exactly equal, causing a finite differential direct input voltage V. This voltage is multiplied by the dc closed-loop gain (see below), resulting in an output offset voltage that, however, frequently can be reduced to zero by means of a potentiometer connected to the offset-adjust terminals of the op amp.

All amplifier parameters, especially the important terms A_0 and ω_t, are strong functions of bias voltages and temperature, and in general are not

* Reference 3.

well determined or predictable from unit to unit. Therefore, one strives to minimize filter dependence on these parameters and always uses op amps in closed-loop feedback configurations (Fig. 3) where

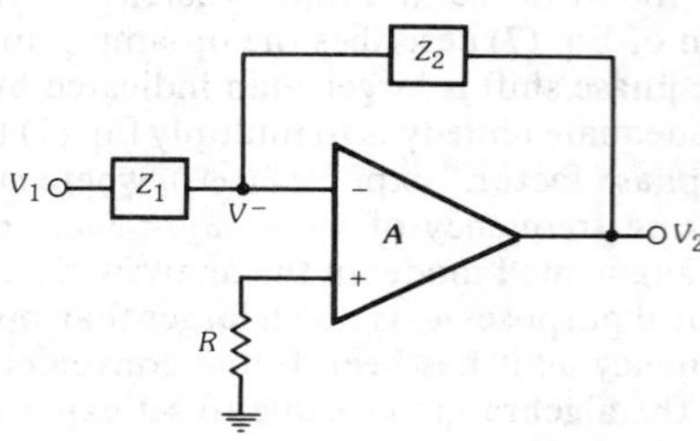

(A) Inverting gain.

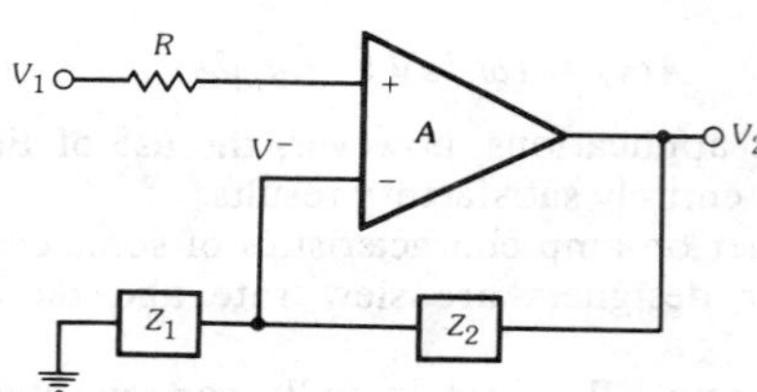

(B) Noninverting gain.

Fig. 3. Closed-loop feedback configurations.

the gain dependence on device parameters is reduced. Straightforward analysis, remembering that op-amp input currents are zero and that $V^- = -V_2/A$, yields

$$V_2/V_1 = -(Z_2/Z_1)/[1 + (1 + Z_2/Z_1)/A] \quad (10)$$

for Fig. 3A and

$$V_2/V_1 = +(1 + Z_2/Z_1)/[1 + (1 + Z_2/Z_1)/A] \quad (11)$$

for Fig. 3B. In both circuits, R is chosen to equal $1/[Y_1(0) + Y_2(0)]$ in line with the above discussion about offset minimization so that both op-amp inputs see the same dc impedance back into the circuit. Note, though, that resistor R does not affect the signal gain because no signal current flows through it. Thus, for simplicity, in the remaining discussion in this chapter, R will be neglected; i.e., $R = 0$ is assumed.

For ideal amplifiers, or, in practice, in the frequency range where $|A(j\omega)| >> |1 + Z_2(j\omega)/Z_1(j\omega)|$, Eqs. (10) and (11) reduce to

$$V_2/V_1 = -Z_2(s)/Z_1(s) \quad (12)$$

and

$$V_2/V_1 = 1 + Z_2(s)/Z_1(s) \quad (13)$$

respectively. The closed-loop gain functions are then independent of amplifer parameters, as desired, and are determined only by presumably accurately adjustable and stable external impedances. The quantities $Z_1(s)$ and $Z_2(s)$ are finally chosen to yield the desired frequency dependence of the gain. For example, setting $Z_1 = R_1$ and $Z_2 = KR_1$ results, of course, in the well-known amplifiers of inverting gain $-K$ (Fig. 3A) and noninverting gain $1 + K$ (Fig. 3B). More will be said later about these important building blocks.

FUNDAMENTALS AND TECHNIQUES OF ACTIVE-FILTER DESIGN

The Transfer Function

In Eq. (1) the transfer characteristic of a filter was introduced as a ratio of polynomials, i.e.,

$$H(s) = N(s)/D(s) = \frac{n_m s^m + n_{m-1} s^{m-1} + \ldots n_1 s + n_0}{s^r + d_{r-1} s^{r-1} + \ldots d_1 s + d_0} \quad (14)$$

where,

$m \leq r$,
r is the order of $H(s)$,
n_i and d_i are real coefficients, with $d_i > 0$.

It is customary to scale the frequency parameter by some convenient normalizing frequency ω_n; i.e., s in Eq. (14) is a normalized frequency

$$s = j\omega/\omega_n \quad (15)$$

Similarly, since a voltage transfer function is independent of impedance level, one scales all components (R and C) in the filter by a suitable normalizing resistance R_n so that the circuit is composed of dimensionless normalized resistors R/R_n and capacitors $\omega_n CR_n$. The advantages of this step are that all parameters in Eq. (14) are dimensionless and, more importantly, that the element values of the filter are dimensionless quantities, scaled for easier numerical computation.

In the following, it shall be assumed that $H(s)$ is given and our concern will be how to design an active filter to realize this prescribed transfer function. Space limitations do not permit us to discuss how to obtain $H(s)$ so that it approximates some desired magnitude or phase frequency response. This topic is treated in great detail in many excellent textbooks (references 1–5).

General Realization Methods

If a high-order function such as Eq. (14) with $r \geq 4$ is prescribed, the engineer has to decide how to find a network structure and element values such that the measured voltage ratio V_2/V_1 is indeed as prescribed in $H(s)$. Numerous different techniques have been developed for this purpose, but they fall essentially into two different groups. One method attempts to break Eq. (14) into simpler functions of lower order, usually first-order

$$T_1(s) = (as + b)/(s + c) \qquad c > 0 \quad (16)$$

and second-order sections

$$T_2(s) = \frac{as^2 + bs + c}{s^2 + s\omega_0/Q + \omega_0^2} \qquad Q > 0,\ \omega_0 > 0 \quad (17)$$

which are then interconnected in a configuration suitable to implement Eq. (14). In Eq. (17), the denominator coefficients have been expressed in terms of the pole frequency ω_0 and the pole quality factor Q in order to keep with standard nomenclature.

The second method draws on the vast experience in and the known excellent performance of passive L-C ladder filters. In this case, either the structure or the equations describing the L-C ladder are simulated via active R-C filters; the inconvenient inductors are thereby avoided, but the positive properties of L-C ladders are retained.

The details of high-order-function design are discussed in a later section of this chapter. Various popular and practically proven techniques are presented so that well-performing filters of reasonable complexity can be designed. In preparation for this step, realization methods for a number of elementary building blocks, such as first- and second-order sections, summers, integrators, and simulated inductors, are first discussed in the next two sections because they form the components of high-order filters besides being useful in their own right.

For trying to decide which of the high-order design methods or which of the numerous available second-order filter sections might be best for a given application, it is necessary to establish practically useful performance criteria that permit the designer to make an informed selection. Apart from obvious points, such as ease of design effort, number of active and passive components needed, spread and size of element values, and required power consumption, all of which reflect themselves in the final price of the filter, the generally accepted most important criterion for a good filter is that of low sensitivity.

Sensitivity

The response or performance parameters of networks depend in general on some or all of the components in the circuit. For example, the gain of the closed-loop amplifier in Fig. 3A, Eq. (10), depends on impedances Z_1 and Z_2 and on op-amp gain $A(s)$. In practice, the components cannot be expected to have or even maintain their ideal, calculated, nominal values. Rather, due to factors such as fabrication tolerances, aging, and temperature drifts, the element values will vary, and, consequently, so will the circuit response.

Clearly, the expected magnitude of the response deviation due to element variations is of great interest to the filter designer before the circuit is ever fabricated so that judgments can be made about the likelihood of the initial and continued operation of the circuit within specifications. The theory of sensitivity addresses this question.

Assume that a circuit response depends on the N elements or parameters k_i, $i = 1, \ldots N$, i.e., $H(j\omega) = H(j\omega, k_1, k_2, \ldots k_N)$. Then the total variation of H, when the elements change from k_i to $k_i + \Delta k_i$ with Δk_i a small change, is obtained from

$$\Delta H \simeq (\partial H/\partial k_1)\,\Delta k_1 + (\partial H/\partial k_2)\,\Delta k_2 + \ldots (\partial H/\partial k_N)\,\Delta k_N \quad (18)$$

where higher-order terms are neglected (Δk_i is small!) and $\partial H/\partial k_i$ is the partial derivative of H with respect to k_i, evaluated at the nominal point. Upon normalizing, Eq. (18) can be rewritten as

$$\Delta H/H \simeq \sum_{i=1}^{N} S_{k_i}{}^{H}(\Delta k_i/k_i) \quad (19)$$

where,

$$S_{k_i}{}^{H} = (k_i/H)(\partial H/\partial k_i) = (\partial H/H)/(\partial k_i/k_i) = [d(\ln H)]/[d(\ln k_i)] \quad (20)$$

is the classical definition of the single-parameter small-change sensitivity; i.e., $S_{k_i}{}^{H}$ is a measure that indicates the percentage deviation dH/H caused by the percentage variation dk_i/k_i of the element k_i. Thus, with $S_{k_i}{}^{H}$ and the component tolerance known, the expected response deviation is, from Eqs. (19) and (20), for a change in a single parameter

$$\Delta H/H = S_{k_i}{}^{H}(\Delta k_i/k_i) \quad (21)$$

Equation (19) of course gives an indication of the total change in H due to variations in all N parameters k_i. As an example, consider the op-amp stage with positive gain G in Fig. 3B, assuming for now $Z_i(s) = R_i$. From Eq. (11),

$$G \equiv V_2/V_1 = (1 + R_2/R_1)/[1 + (1 + R_2/R_1)/A]$$

Using (20) it is easy to show that

$$S_{R_2}{}^{G} = -S_{R_1}{}^{G} = \frac{R_2/R_1}{(1 + R_2/R_1)[1 + (1 + R_2/R_1)/A]}$$

and

$$S_A{}^{G} = \frac{1 + R_2/R_1}{1 + (1 + R_2/R_1)/A} \cdot \frac{1}{A}$$

Thus, if the desired value of G is, say, 50 and if in the frequency range of interest $|A(j\omega)| >> 50$ can be assumed, then $R_2/R_1 = 49$ and

$$S_{R_2}{}^{G} = -S_{R_1}{}^{G} \simeq 49/50 \simeq 1 \quad (22a)$$

$$S_A{}^{G} \simeq 50/A << 1 \quad (22b)$$

Using Eqs. (19), (21), and (22),

$$\Delta G/G \simeq -\Delta R_1/R_1 + \Delta R_2/R_2 + (50/A)(\Delta A/A) \quad (23)$$

These numbers indicate that, as long as $|A(j\omega)| >> G$, changes in A have a negligible effect on ΔG, but that the percentage deviation of gain equals that of R_2 for constant R_1 and equals the negative of that of R_1 for constant R_2. Note also that if R_1 and R_2 track, i.e., if they increase or decrease by the same amount, say, due to temperature changes in a thin-film hybrid circuit, then $\Delta G/G \simeq 0$, because the ratio R_2/R_1 stays constant.

Several further observations can be made at this point. First note that since $A(j\omega)$ is a function of frequency, as for example in Eq. (8), G and all sensitivities also are functions of frequency and the sensitivity to $A(s)$ will increase with increasing ω because $|A(j\omega)| \simeq \omega_t/\omega$ decreases. Second, note the frequency limitation that is imposed on the validity of Eq. (23); the requirement $|A| >> G = 50$ implies $\omega << \omega_t/50$ (e.g., $f << 20$ kHz for a 741-type op amp). This re-emphasizes the point made earlier that active-network designs based on assumptions of ideal op amps can be expected to give reasonable results only at very low frequencies. Finally, remember that op-amp gain is a very unreliable parameter: A_0 and ω_t often vary by more than 50% from unit to unit. Thus, this simple example indicates already that the circuit response, here the gain G, is very sensitive [Eq. (23)] to A variations unless strong feedback is applied, i.e., $G << |A|$.

A number of interesting relationships that simplify sensitivity calculations can be derived.* Two of the more useful ones will be discussed. First, if a network response function depends on a parameter k_1 that in turn depends on a second variable k_2,

$$H = H(k_1)$$

where,

$$k_1 = k_1(k_2)$$

then it is easy to show from Eq. (20) that

$$S_{k_2}{}^{H} = S_{k_1}{}^{H} \cdot S_{k_2}{}^{k_1} \quad (24)$$

For example, if k_1 is a resistor and k_2 the temperature T, i.e., $R = R(T)$, Eq. (24) permits an easy evaluation of $S_T{}^H$.

The second relationship makes use of Eq. (2); if Eq. (2) is inserted in Eq. (20), it should be clear that $S_K{}^H$ is a complex number equal to

$$S_k{}^{H(j\omega)} = S_k{}^{|H(j\omega)|} + j\frac{1}{k}\frac{\partial\phi(\omega)}{\partial k} \quad (25)$$

Thus, the real part of the sensitivity of the transfer function $H(j\omega)$ to a parameter k equals the sensitivity of the magnitude (gain) of $H(j\omega)$, and its imaginary part specifies the absolute variation of phase due to the relative error in k. Thus, gain and phase deviations can easily be estimated.

Equation (21) describes the change to be expected in $H(s)$ when only one circuit parameter varies. To minimize this change, $\Delta k_i/k_i$ has to be minimized (implying components of tight tolerances and therefore high cost) and/or a design has to be found that minimizes $S_{k_i}{}^H$. Thus, the importance of low-sensitivity filter circuits should be clear.

Although useful, the insight provided by single-parameter sensitivities is somewhat limited in many cases because the effect of the remaining element variations is not taken into account. Thus, for a more realistic picture, the deterministic multiparameter sensitivity measure, Eq. (19), ought to be consulted and a performance criterion, such as

$$\left[\sum_{i=1}^{N} |S_{k_i}{}^H|^2\right]^{1/2} \rightarrow \text{Min} \quad (26)$$

should be evaluated for a given selection of circuits in order to arrive at a best choice. Since $\Delta k_i/k_i$ and also $S_{k_i}{}^H$ in general can be of either sign, some cancellations can be expected when $\Delta H/H$ is calculated from Eq. (19). Thus, Eq. (26) gives a somewhat pessimistic worst-case picture of circuit performance. Further, this measure does not take into account that in many modern filter technologies, such as in hybrid or monolithic integrated circuits, the element variations are statistically related and frequently highly correlated. For example, all resistors may track and increase and all capacitors track and decrease with changes in temperature or during fabrication. Treatment of this case is beyond the scope of this chapter; the interested reader is referred to Reference 1.

Although the above sensitivity discussion has been concentrated solely on deviations of the transfer function $H(s)$ due to component changes, it should be understood that sensitivity calculations apply to any network parameter whose value depends on variable circuit components. Thus, for example, the center frequency and the selectivity of a bandpass function, or the cutoff frequency of a low-pass filter, depend in general on several resistors and capacitors and possibly on op-amp gain. Any shift in these parameters due to component variations can be estimated by calculating the corresponding sensitivities. Similarly, the precise location of transmission zeros or filter poles depends, of course, on the correct element values; any variations or tolerances in the latter cause shifts in these critical frequencies and therefore transfer-function errors that can readily be evaluated by sensitivity calculations. All second- and higher-order filters discussed in the remaining two sections of this chapter have been evaluated extensively as to their sensitivity performance and are recognized as the best designs available. If the reader for some reason wishes to use different active R-C filters out of the numerous topologies presented in the literature, he

* References 1 and 2.

is well advised to investigate the sensitivities of the contemplated circuit to make sure that the design will work to his satisfaction in practice. It is noted again that simple post-design tuning to eliminate fabrication tolerances will in general not suffice because circuit components cannot be expected to retain their values under environmental stresses, such as aging or temperature fluctuations. Thus, low sensitivity is a necessary requirement for any circuit. When evaluating a filter, the reader ought to keep in mind that the sensitivity results presented are valid for *small* changes in element values because of the linearizations involved in the analysis: Note from Eq. (18) that the function $H(j\omega,k)$ has been replaced by its slope in the nominal point $k = k_0$ in order to estimate changes caused by varying k from k_0 to $k_0 + \Delta k$. In particular, this means that a circuit with the desirable property $S_k{}^H = 0$ may not be entirely independent of the parameter k; it only says that at the nominal value $k = k_0$ the slope dH/dk is zero, usually implying a quadratic dependence on k. Large variations Δk can still cause unacceptable performance errors!

Also, it is emphasized that sensitivity expressions are usually functions of frequency so that the results ought to be evaluated in the frequency range of interest (normally in the passband or at the passband edges) when different possible designs are compared.

Finally, a last point worth noting: Sensitivity is only an intermediate result and by itself can still create a misleading picture of circuit performance. In the ultimate analysis, it is the deviation or variability $\Delta H/H$, i.e., by Eq. (18) or (21), the sensitivity multiplied by the expected component tolerances, that is important. Thus, relatively large sensitivities are acceptable when the relevant "component" (e.g., a resistor ratio in a thin-film hybrid realization) can be expected to be very accurate and stable. On the other hand, when a component varies strongly (e.g., op-amp gain-bandwidth product), very low sensitivities must be insisted upon.

FREQUENTLY USED BUILDING BLOCKS

Active filters are generally constructed by interconnecting a number of well understood elementary building blocks. Understanding the performance of those blocks enables the designer to assemble a better working and more reliable filter. Thus, in this section, the most important functional blocks are being introduced: the summer, the integrator, the general impedance converter (GIC), the frequency-dependent negative resistor (FDNR), and a circuit for realizing a first-order voltage transfer function. A separate section is devoted to the realization of second-order biquadratic transfer functions because of their special importance.

Summers

As the name implies, a summer is used to add different signals in a filter to form a desired sum signal. The circuit is essentially an extension of Fig. 3A in which various input signals V_i, $i = 1, \ldots n$ are fed through weighting resistors to the virtual ground node V^- of the op amp (Fig. 4). Simple analysis yields

$$V_o = -\left(\sum_{i=1}^{n} a_i V_i\right)/\left[1 + \left(1 + \sum_{i=1}^{n} a_i\right)/A\right]\Big|_{A\to\infty}$$

$$= -\sum_{i=1}^{n} a_i V_i \tag{27}$$

Thus, for ideal amplifiers, $A \to \infty$, the output voltage is simply the negative sum of the input voltages, weighted by resistor ratios a_i. If the op-amp gain is finite and frequency-dependent, e.g., via Eq. (8), the summing coefficients a_i exhibit errors (specifically low-pass behavior) as is evident from Eq. (27). Whether or not this effect causes unacceptable deviations in a given filter depends on several factors and must be investigated from case to case; no general rule can be given.

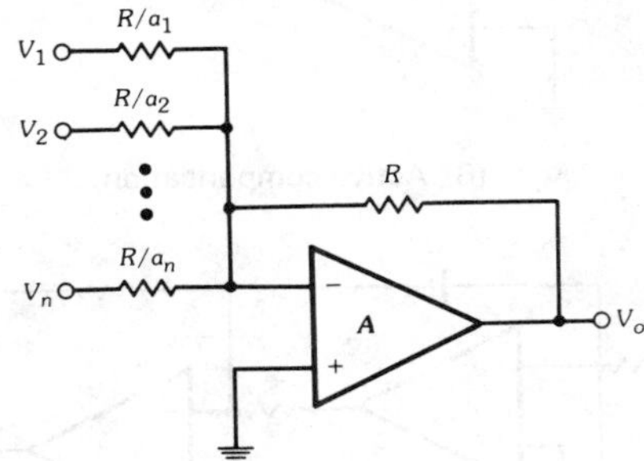

Fig. 4. Resistive op-amp summer.

Integrators

One of the most basic functions in analog signal processing is that of integration; i.e., an output signal is obtained by integrating an input signal over time. In the frequency domain, this results in the transfer function

$$V_2/V_1 = \pm 1/(s\tau) \tag{28}$$

where τ is the integrator time constant. Equation (28) can be implemented by use of Fig. 3A with $Z_1 = R$ and $Z_2 = 1/(sC)$; i.e., assuming ideal op amps, the simple inverting integrator realizes

$$V_2/V_1 = -1/(sRC) \tag{29}$$

so that $\tau = RC$. Constructing filters by use of these very elementary, so-called Miller integrators results invariably in large errors, even at frequencies as low as 1 kHz, because of the overly ideal assumptions made about the op-amp gain. The problem can be appreciated by substituting the more realistic model of Eq. (8) into Eq. (10) with the result, for $\tau >> 1/\omega_t$

$$V_2/V_1 = -1/(s\tau + s^2\tau/\omega_t) \qquad (30)$$

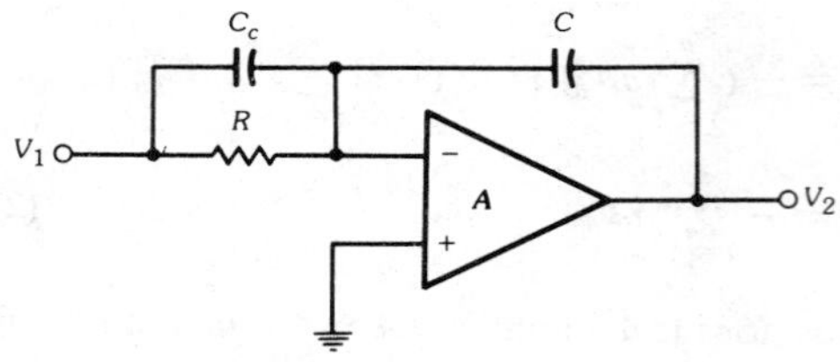

(A) Passive compensation.

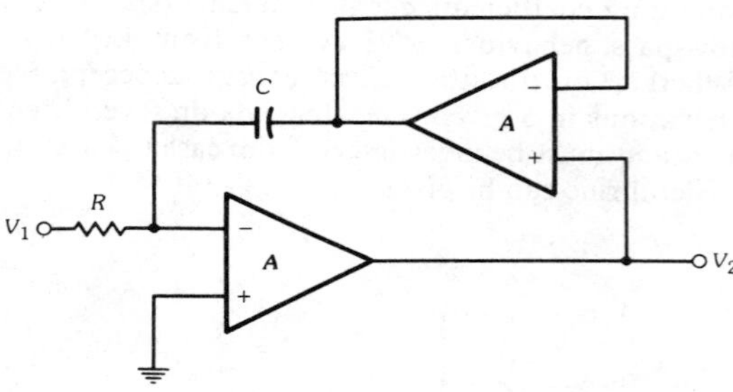

(B) Active compensation.

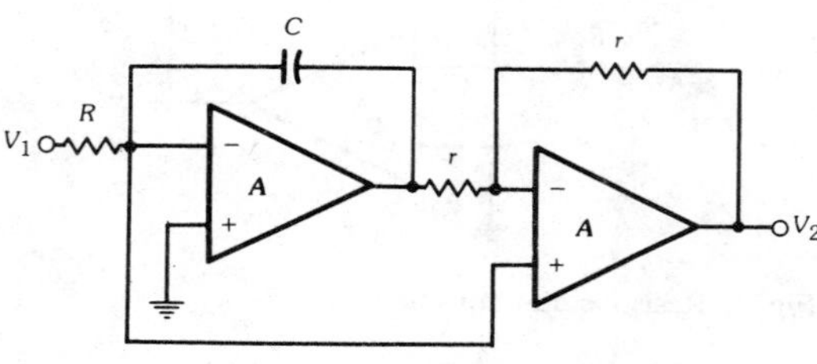

(C) Noninverting, Q negative.

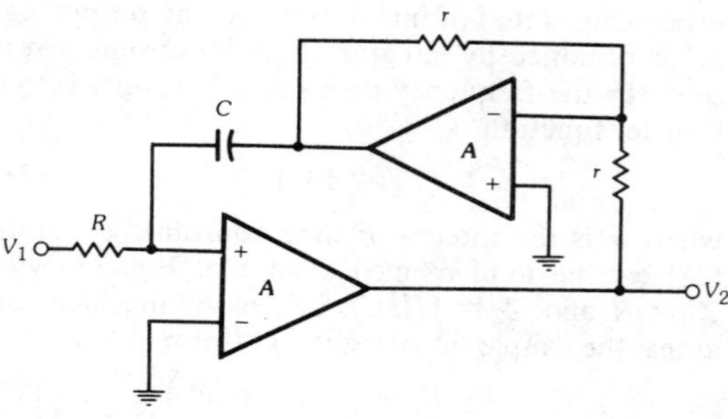

(D) Noninverting, Q positive.

Fig. 5. Different phase-compensated integrator circuits.

Thus, compared with the ideal case of Eq. (29), the integrator transfer function is no longer purely imaginary (it has acquired a parasitic pole), and its phase is not $\pi/2$. In analogy to the quality factor of capacitors and inductors, the ratio of imaginary to real part of the denominator of Eq. (30) is defined as the integrator Q-factor

$$Q_I = -\omega_t/\omega = -|A(j\omega)| \qquad (31)$$

That is, the Q-factor of the Miller integrator is the negative of the op-amp gain at the frequency of interest.

It is apparent from Eq. (30) that the op amp introduces an excess phase lag into the integrator transfer function when compared to the ideal. Since filter behavior is usually particularly sensitive to such phase errors, procedures for their elimination have been devised; these procedures can be classified as passive and active compensation methods, respectively. Fig. 5A shows a Miller integrator with passive phase compensation. Adding a phase-lead compensation capacitor of value $C_c \simeq 1/(R\omega_t) << C$ approximately just cancels the phase lag introduced by the op amp. The difficulty is that ω_t is not well determined and is variable, so compensation is difficult to maintain. A better, though more expensive, approach is active compensation (Fig. 5B), which attempts to cancel the errors by use of a second op amp. For well matched amplifiers, routine analysis shows that the integrator of Fig. 5B has a Q-factor of value $Q \simeq -(\omega_t/\omega)^3 = -|A(j\omega)|^3$, a significant improvement. Two further integrators are shown in Figs. 5C and D; they are useful for those applications where a *non*inverting [*plus* sign in Eq. (28)] integrator is required. The circuit in Fig. 5C again has a Q-factor of value $Q \simeq -|A(j\omega)|$, whereas Fig. 5D can be shown to have $Q \simeq +|A(j\omega)|$. It is noted here especially that the Q-factor of Fig. 5D is positive; i.e., the integrator has a *leading* phase error. This fact is quite significant because active-filter structures often consist of two-integrator loops where the total phase error can be virtually eliminated by using the circuit in Fig. 5D together with the regular Miller integrator in Fig. 5A (with $C_c = 0$). The result is much improved filter performance in spite of the relatively large phase error in each individual integrator.*

The General Impedance Converter (GIC)

One of the motivations for active filters is the elimination of inductors. Thus, finding methods for inductance simulation is an important topic in active-filter design. The best op-amp–based circuit for inductance simulation is the general impedance

* Reference 2.

converter (GIC) shown in Fig. 6. It realizes between terminal n and ground the input impedance

$$Z_{in}(s) = sCR_1R_3R_4/R_2 \quad (32)$$

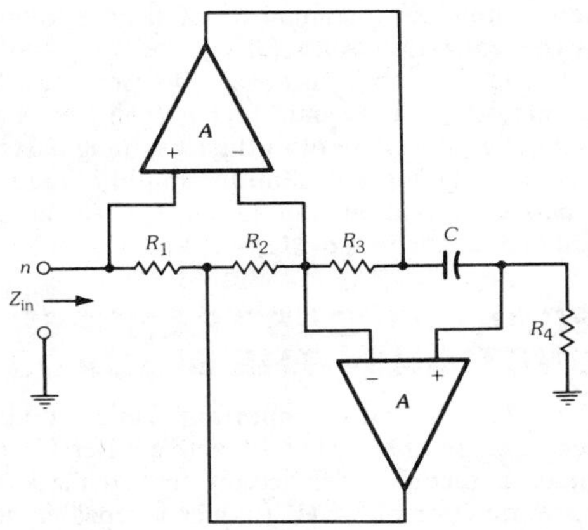

Fig. 6. Inductor-simulation circuit.

That is, the circuit simulates a grounded inductor of value $L_0 = CR_1R_3R_4/R_2$. In order to minimize the dependence of L_0 on op-amp performance (finite ω_t), the circuit is carefully optimized by setting $R_2 = R_3$ and $R_4 = 1/(\omega_c C)$ where ω_c is some critical frequency value chosen in the filter passband or at the passband edge. For these conditions, $L_0 = R_1/\omega_c$; R_1 is chosen to set the inductor value, and the inductor error and the inductor quality factor can be shown to be

$$\Delta L_0/L_0|_{\omega=\omega_c} \simeq 4/|A(j\omega_c)| \quad (33)$$

$$Q \simeq |A(j\omega_c)|/2\epsilon_r \quad (34)$$

where ϵ_r is the fractional mismatch between R_2 and R_3; i.e., $\epsilon_r = |R_2 - R_3|/R_2$. Note that $Q = \infty$ can be obtained by tuning the circuit such that $\epsilon_r = 0$.

Fig. 6 simulates a grounded inductor. Although several procedures exist for simulating floating inductors, none has ever been found to be satisfactory in practice. Thus, alternative procedures had to be invented for those (e.g., low-pass) filters whose passive equivalent requires floating inductors. One such method makes use of impedance transformations and the so-called frequency-dependent negative resistor (FDNR).

The FDNR Element

The use of and motivation for FDNR elements in designing active filters based on passive L-C prototypes will be discussed in a later section. Here, only the building block is to be introduced. The input impedance between terminal n and ground in the circuit of Fig. 7 is

$$Z_{in}(s) = (1/s^2)(R_2/C^2R_1R_3) \quad (35)$$

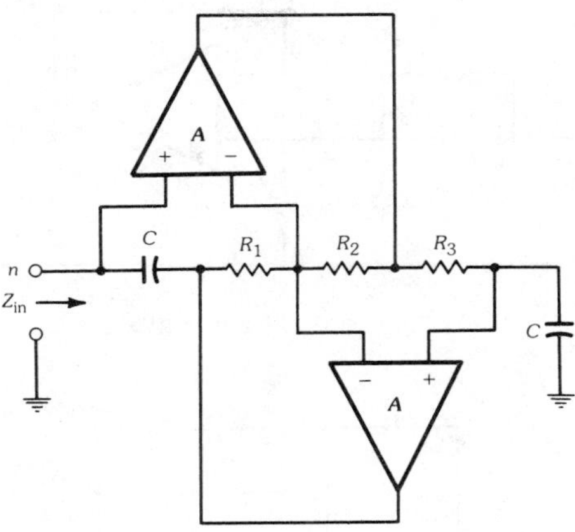

Fig. 7. FDNR element.

For $s = j\omega$, Z_{in} is a frequency-dependent negative resistance of value $-1/(\omega^2D_0)$ with $D_0 = C^2R_1R_3/R_2$. Again, to minimize the dependence of D_0 on op-amp parameters (finite ω_t), the circuit is carefully optimized by choosing $R_1 = R_2$ and $R_3 = 1/(\omega_c C)$ where ω_c is a critical frequency value chosen in the passband of the filter or at the band edge. The fractional deviation of D_0 is then

$$\Delta D_0/D_0 \simeq 4/|A(j\omega_c)| \quad (36)$$

and its quality factor, defined as the ratio of real to imaginary part of $Z_{in}(j\omega)$, is

$$Q \simeq |A(j\omega_c)|/2\epsilon_r \quad (37)$$

where ϵ_r is the fractional mismatch between R_1 and R_2.

First-Order Transfer Functions

It was stated earlier that one technique for realizing a high-order transfer function involves breaking the function into first- and/or second-order subfunctions that are then realized by suitable networks that are interconnected in an appropriate fashion. The realization of first-order voltage transfer functions is relatively simple and will be treated in the following. Second-order functions are discussed in the next section.

Fig. 8 shows different realizations of a function of the form of Eq. (16), repeated here as Eq. (38):

$$T(s) = (as + b)/(s + c) \qquad c > 0 \quad (38)$$

Fig. 8A is appropriate if $0 \leq a \leq 1$ and $0 \leq b \leq c$. Note that in this case $T(s)$ is realized by a passive R-C network. The unity-gain buffer amplifier with very high input and zero output impedance, a special case of Fig. 3B for $Z_2 = 0$ and $Z_1 = \infty$, has been included for the case in which the load for the R-C network is significant; otherwise it can be omitted. Fig. 8B realizes $-T(s)$; the minus sign, a 180° phase shift, is immaterial in many applica-

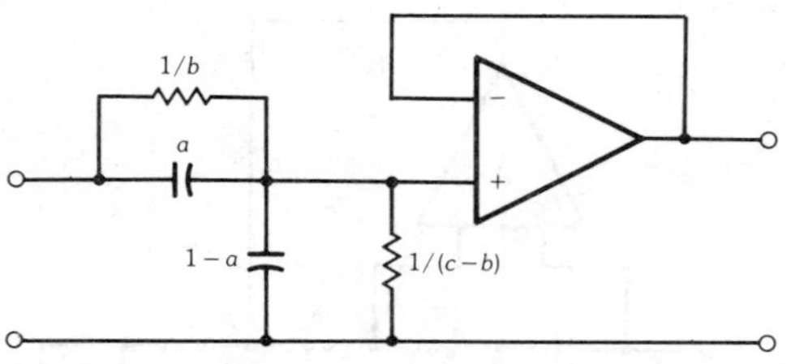

(A) For $0 \leq a \leq 1$ and $0 \leq b \leq c$.

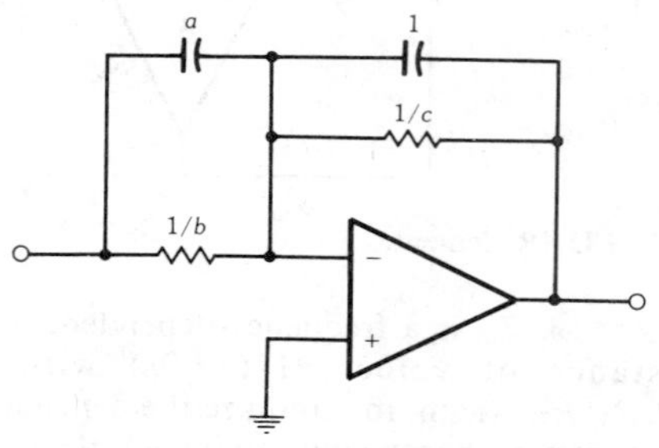

(B) Realizes $-T(s)$.

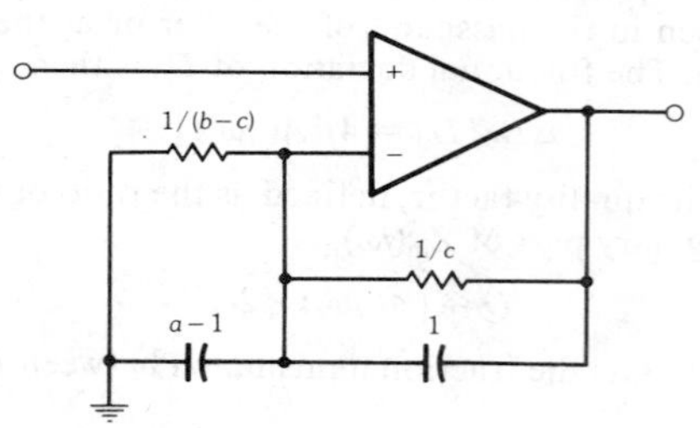

(C) For $a \geq 1$ and $0 \leq c \leq b$.

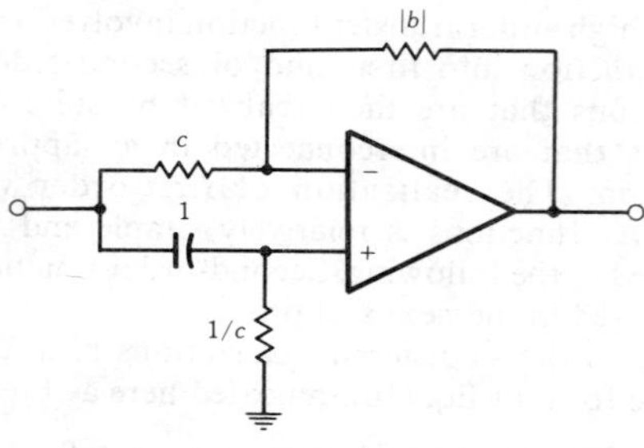

(D) For $a = 1$ and $b < 0$.

Fig. 8. Different realizations of $T(s)$ in Eq. (38). The element values are normalized resistors and capacitors.

tions. Fig. 8C realizes $T(s)$ for the case in which $a \geq 1$ and $0 \leq c \leq b$. Finally, Fig. 8D takes care of the case when $a = 1$ and $b < 0$, important for all-pass functions (see below).

It is noted here that the circuits in Figs. 8B, C, and D realize Eq. (38) under the assumption of ideal amplifiers; if the op amp is described by Eq. (8), discrepancies from the ideal in the frequency response must be expected as discussed earlier. Fortunately, however, the majority of all cases require only the realization of Eq. (38) with either $a = 0$ or $b = 0$. Fig. 8A can then always be used, possibly after factoring out a multiplying (gain) constant that can be taken care of in the remaining parts of the circuit. The unity-gain buffer amplifier requires for perfect operation only $|A(j\omega)| >> 1$ as can readily be seen from Eq. (11).

SECOND-ORDER TRANSFER FUNCTIONS—THE BIQUADS

Probably the most important basic building block used in the design of active filters is the biquad, a second-order circuit that realizes the transfer function of Eq. (17), which is repeated here as Eq. (39):

$$T(s) = V_2/V_1 = N(s)/D_2(s) = (as^2 + bs + c)/(s^2 + s\omega_0/Q + \omega_0^2) \quad Q > 0,\ \omega_0 > 0 \qquad (39)$$

It is standard practice to identify the coefficients of the second-order denominator polynomial $D_2(s)$ via the pole frequency ω_0 and the pole quality factor Q. The coefficients a, b, and c of the numerator polynomial $N(s)$ specify the type of transfer function realized (Fig. 9): If $N(s) = c$, i.e., $a = b = 0$, $T(s)$ is a low-pass (LP) function; if $N(s) = as^2$, i.e.,

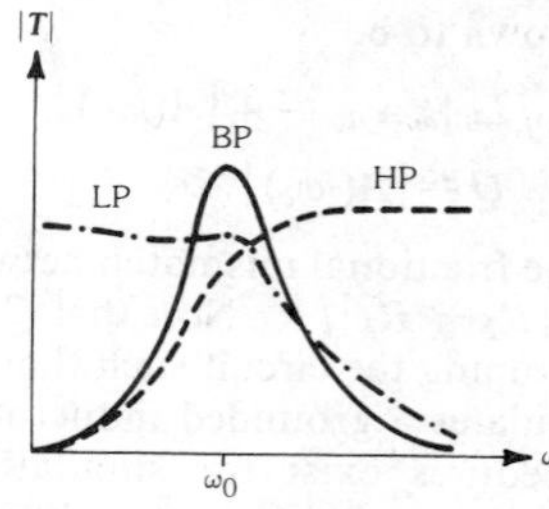

(A) Pass characteristics.

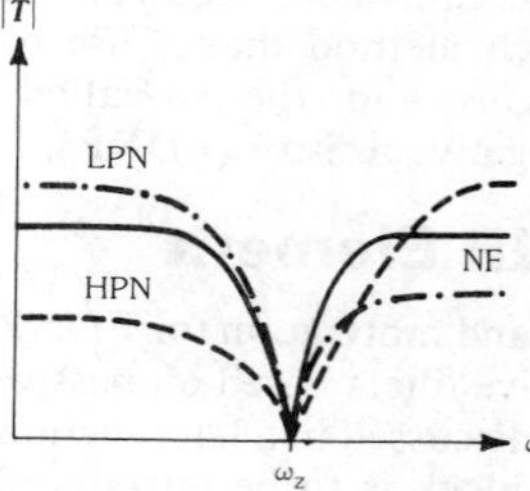

(B) Notch characteristics.

Fig. 9. Typical plots of the basic second-order transfer functions.

$b = c = 0$, $T(s)$ is a high-pass (HP) function; if $N(s) = bs$, i.e., $a = c = 0$, $T(s)$ reduces to a bandpass (BP) function. A finite transmission zero at $\omega = \omega_z$ is obtained by setting $b = 0$; i.e., $N(s) = as^2 + c = a(s^2 + \omega_z^2)$ where $\omega_z^2 = c/a$. Such a circuit is called a "notch" filter (NF) and is distinguished further between a high-pass notch (HPN) if $\omega_z \leq \omega_0$ and a low-pass notch (LPN) if $\omega_z \geq \omega_0$. Finally, if $N(s)$ can be brought into the form $N_2(s) = a(s^2 + sb/a + c/a) = a(s^2 - s\omega_0/Q + \omega_0^2)$, i.e., $a \neq 0$, $c > 0$, $b < 0$, then $|N(j\omega)| = |D_2(j\omega)|$, so that $|T(j\omega)| = 1$. Such an "all-pass" (AP) filter passes signals of all frequencies without attenuation but imposes a frequency-dependent phase shift of value

$$\phi = -2 \arctan [(\omega\omega_0/Q)/(\omega_0^2 - \omega^2)] \quad (40)$$

To realize the function of Eq. (39) or any of the mentioned special cases, the technical literature contains literally hundreds of different circuits. Fortunately, the choice is simplified for the engineer because closer examination reveals that the vast majority of all circuits have some drawback or another, usually excessive sensitivities, so that only two or three structures are used for practical active filters. These will be discussed in the following.

Single-Amplifier Filters*

Fig. 10 contains three state-of-the-art single-amplifier active filters that can be used to realize the function in Eq. (39) with any of its special cases. The circuits are related to each other by complementary and by *RC:CR* dual transformations. The transfer-function pole sensitivities, i.e., the ω_0 and Q sensitivities, can be shown to be the same for all three circuits. Specifically, the sensitivities of ω_0 to the passive elements are at their theoretical minima, $S_{k_i}^{\omega_0} = -0.5$, where k_i stands for any of the resistors and capacitors, and the passive sensitivities of Q satisfy

$$|S_{k_i}^{Q}| \leq Q/q - 0.5 \quad (41)$$

where q is a free design parameter, defined as a resistor ratio in Fig. 10. Further, it can be shown that ω_0 and Q deviations caused by the finite op-amp gain-bandwidth product ω_t are

$$\Delta\omega_0/\omega_0 \simeq -q(\omega_0/\omega_t) \quad (42)$$

$$\Delta Q/Q \simeq q(\omega_0/\omega_t)[1 - 2QK(\omega_0/\omega_t)] \quad (43)$$

where K is defined in Fig. 10. The value of free parameter q is arrived at through a compromise dictated by the technology chosen to implement the filter: It is apparent from Eqs. (41), (42), and (43) that increasing q will reduce the passive Q sensitivities but will increase the ω_0 and Q deviations caused by the active element via finite ω_t. Taking into consideration that in the design, Eq. (48) below, the

* Reference 7.

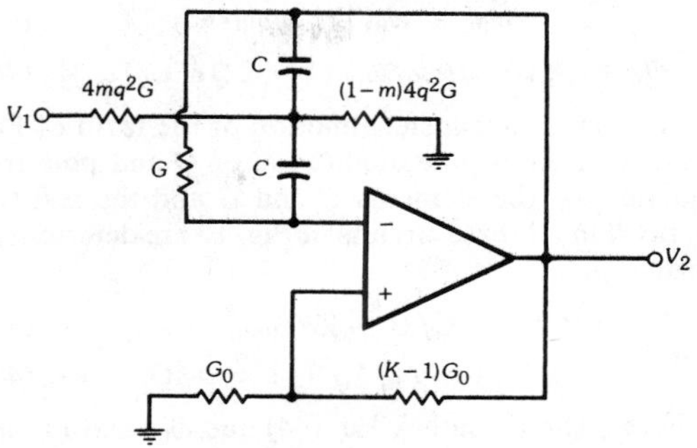

(A) Bandpass biquad.

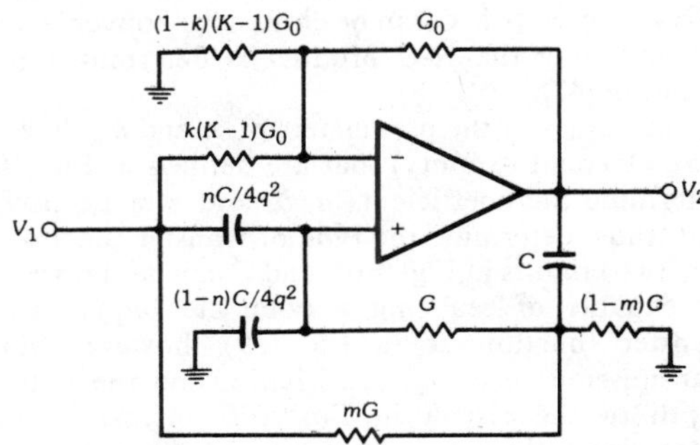

(B) Low-pass biquad.

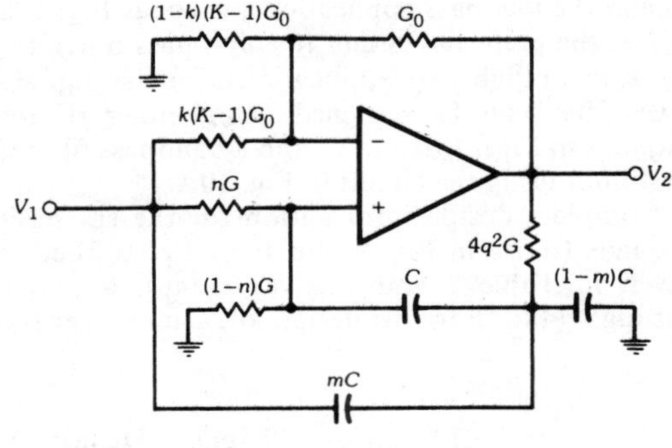

(C) High-pass biquad.

Fig. 10. Single-amplifier biquads.

value of K depends on the choice of q, it can be shown that for sensitivity and realizability reasons [see Eq. (48)] the value of q must be restricted by

$$\sqrt{6}/2 \leq q \leq Q \quad (44)$$

In practice, q in the range of 3 to 5 is a reasonable tradeoff between active and passive sensitivities.

The initial error caused by finite ω_t can be eliminated by predistortion, i.e., by applying the negative of the deviations, Eqs. (42) and (43), to the nominal values of ω_0 and Q and by basing the design on the predistorted values

$$\omega_{0p} = \omega_0(1 + q\omega_0/\omega_t) \quad (45)$$

$$Q_p = Q[1 - q(\omega_0/\omega_t)(1 - 2QK\,\omega_0/\omega_t)] \quad (46)$$

To realize a transfer function of the form of Eq. (39) with given pole quality factor Q and pole frequency ω_0, the elements C and G and the resistor ratio K in all three circuits in Fig. 10 are determined from

$$C/G = 2q/\omega_{0p} \quad (47)$$

$$K = 1 + (1/2q^2)(1 - q/Q_p) \quad (48)$$

where q should satisfy Eq. (44) and ω_{0p} and Q_p are replaced by ω_0 and Q if predistortion is not used. Resistor $R_0 = 1/G_0$ is arbitrary, because only the ratio $K - 1$ enters the transfer function. Further, either C or $R = 1/G$ can be chosen for convenience because only the R-C product is determined by equation (47).

The values of the parameters m, n and k (all less than or equal to unity) that are defined in Fig. 10 determine the coefficients a, b, and c in Eq. (39) and, thus, determine the type of transfer function. The two circuits in Figs. 10B and C can be shown to be capable of realizing a complete biquadratic transfer function as in Eq. (39); however, the transmission zero (ω_z) sensitivities and the notch depth of the circuit in Fig. 10C increase with increasing ratio ω_z/ω_0. The opposite is true for the circuit in Fig. 10B. Thus, Fig. 10B gives the preferred circuit for low-pass-notch, $\omega_z \geq \omega_0$, (see Fig. 9) and the low-pass applications, whereas Fig. 10C shows the preferred circuit for high-pass-notch ($\omega_z \leq \omega_0$) and high-pass applications. Further, an all-pass filter can be designed using either of the biquads in Figs. 10B and C, and a bandpass filter is best built using the circuit in Fig. 10A.

Complete design information for the six basic biquads (those in Fig. 9 plus the all-pass filter) is given in Table 1 and Eqs. (39), (40), and (44) through (48). For the design of a more general biquadratic transfer function, the reader is referred to Reference 7, on which the information in this subsection is based.

A couple of examples will illustrate the use of the information presented. First, assume the bandpass function

$$T(s) = -2s_n/(s_n^2 + 0.1s_n + 1) \quad (49)$$

has to be realized, where the frequency parameter s_n is normalized with respect to $\omega_0 = 2\pi\cdot 3$ kHz, i.e., $s_n = s/\omega_0$. Comparing Eq. (49) with Eq. (39) results in $\omega_0 = 2\pi\cdot 3$ kHz, $Q = 10$, $a = c = 0$, and $b = -2\omega_0$. Note that at $\omega = \omega_0$, i.e., $|s_n| = 1$, $T(s_n) = -20$; the midband gain of the filter is inverting with a magnitude of 20, or 26 dB. From Eq. (44), q should satisfy $1.23 \leq q \leq 10$; a good choice, as stated earlier, is $q = 4$. Assuming the circuit will be built using a 741-type op amp with $\omega_t \simeq 2\pi\cdot 900$ kHz, the expected sensitivities and deviations are, from Eqs. (41) through (43), $|S_{k_i}{}^Q| \leq 2$, $\Delta\omega_0/\omega_0 \simeq -0.013$, and $\Delta Q/Q \simeq 0.013(1 - 0.07K)$ where, from Eq. (48) with $Q_p = Q$, $K = 1.01875$. Thus $\Delta Q \simeq 0.12$, i.e., a 1.2% error. To reduce this error, use predistortion; i.e., from Eqs. (46) and (48) the circuit is designed for $Q = Q_p = 9.88$, resulting in $K - 1 = 0.0186$. Further, from Eq. (45), $\omega_{0p} = 2\pi\cdot 3.04$ kHz so that, from Eq. (47), $RC = 418.83$ μs. Choosing $C = 10$ nF gives R = 41.9 kΩ. Finally, the bandpass design is completed by calculating, from Table 1, $m = |b|/(2Kq\,\omega_0) = 2/(2Kq) = 0.2454$ (note that in this example $b < 0$), and choosing the resistance level $R_0 = 1$ kΩ. The resulting bandpass filter is shown in Fig. 11.

As the next example, design a low-pass notch filter realizing, with the maximum possible high-frequency gain H,

$$T(s) = H(s_n^2 + 2.3)/(s_n^2 + 0.12s_n + 1) \quad (50)$$

where s_n is normalized with respect to $\omega_0 = 2\pi\cdot 2.4$ kHz. Comparing Eq. (50) with Eq. (39) results in ω_0

TABLE 1. DESIGN DATA FOR CIRCUITS IN FIG. 10

Filter Type	$N_2(s)$	k	m	n	Best Circuit
Low-Pass	c	0	$c/(K\omega_0^2)$	0	Fig. 10B
High-Pass	as^2	0	a/K	0	Fig. 10C
Bandpass	$-bs$	0	$b/(2Kq\omega_0)$	0	Fig. 10A
Low-Pass Notch	$a(s^2+\omega_z^2)$ $\omega_z \geq \omega_0$	$a/(1-q/Q)$	$k[(K-1)/K] \times[1+2q^2(\omega_z/\omega_0)^2]$	$k(1-q/KQ)$	Fig. 10B
High-Pass Notch	$a(s^2+\omega_z^2)$ $\omega_z \leq \omega_0$	$a(\omega_z/\omega_0)^2/(1-q/Q)$	$k[(K-1)/K] \times[1+2q^2(\omega_z/\omega_0)^2]$	$k(1-q/KQ)$	Fig. 10C
All-Pass [a=(Q−q)/(Q+q) is fixed]	$a[s^2-s(\omega_0/Q)+\omega_0^2]$	1	$1-2/K(1+Q/q)$	$1-2/K(1+Q/q)$	Either Fig. 10B or Fig. 10C

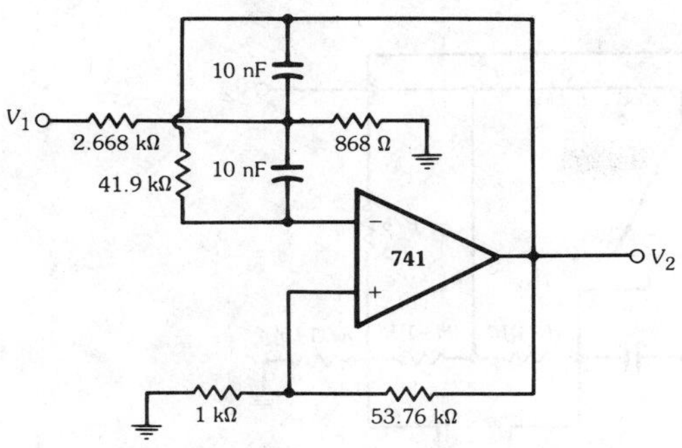

Fig. 11. Bandpass circuit of Fig. 10A realizing the example of Eq. (49).

$= 2\pi \cdot 2.4$ kHz, $Q = 1/0.12 = 8.333$, $a = H$, $b = 0$, and $c = 2.3\ \omega_0^2 H$, i.e., $\omega_z = 1.5166\ \omega_0$. Choosing $q = 3.5$ and assuming $\omega_t = 2\pi \cdot 900$ kHz gives $|S_{k_i}{}^Q| \leq 1.88$, $\Delta\omega_0/\omega_0 \simeq -0.009$, $\Delta Q/Q \simeq 0.009$, and $K = 1.0237$, where Eqs. (41) through (43) and (48) were used. The deviations are quite small, so predistortion appears unnecessary; thus, from Eq. (47), $R = 2q/(\omega_0 C) = 46.24$ kΩ, where $C = 10$ nF was assumed. The function will be realized via the circuit in Fig. 10B; thus from Table 1: $k = 1.724H$, $m = 2.289H$, and $n = 1.0167H$. Since k, m, and n must be less than or equal to 1, $H_{max} = 0.4369$; then $k = 0.7532$, $m = 1$, and $n = 0.4442$. Finally, R_0 is chosen equal to 1 kΩ, and the final low-pass notch circuit is as shown in Fig. 12.

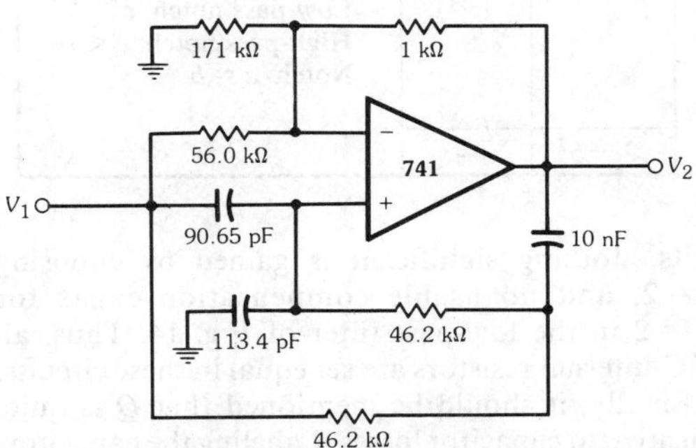

Fig. 12. Realization of low-pass notch function of Eq. (50) using circuit of Fig. 10B.

Two-Amplifier Filters*

The literature contains several multiamplifier second-order filters whose main advantage over single-amplifier circuits is that sensitivities both to the passive components and to the finite gain-bandwidth product ω_t of the op amps are reduced. Based on a sensitivity comparison, only very few of these circuits have been shown to result in practical designs; among these, the biquad based on inductance simulation using an alternative version of the general impedance converter (GIC) of Fig. 6 has emerged as the best biquad due to the reduced dependence of filter paramaters on ω_t.

The circuit capable of realizing a general biquadratic transfer function (except low-pass) is shown in Fig. 13. Simple analysis yields, for $\omega_t \to \infty$,

$$\begin{aligned} T(s) &= V_2/V_1 = N(s)/D_2(s) \\ &= (a_2 s^2 + a_1 s + a_0)/(s^2 + s\omega_0/Q + \omega_0^2) \end{aligned} \tag{51}$$

with

$$\omega_0 = 1/RC \tag{52a}$$

and

$$a_0 = b\omega_0^2 \tag{52b}$$

$$a_1 = [H(a/Q - bR/R_c) + (1 - H)b/Q]\ \omega_0 \tag{52c}$$

$$a_2 = cH + b(1 - H) \tag{52d}$$

Different types of transfer functions are realized by an appropriate choice of the parameters a, b, and c that are defined in Fig. 13. Table 2 gives the details. Of course, a, b, and c must be equal to or less than 1. The circuit of Fig. 13 cannot realize a low-pass function because a_1 and a_2 cannot be set to zero without forcing a_0 to zero also. Thus, a low-pass filter is built with a slightly modified circuit as shown in Fig. 14; it realizes, with $\omega_0 = 1/CR$,

$$T(s) = V_2/V_1 = 2\omega_0^2/(s^2 + s\omega_0/Q + \omega_0^2) \tag{53}$$

A few observations are in order about the performance of GIC-based filters: As was the case with the GIC in Fig. 6, the circuit is optimally insensitive to ω_t if all GIC-internal resistors are equal. In Fig. 13, this clearly requires $H = 2$, which according to Table 2 implies that the midband gain of the bandpass filter and the high-frequency gain of the high-pass filter are fixed at $H = 2$. However, in many applications, especially in high-order filters based on second-order building blocks (see next section), the gain must be adjustable in order to maximize dynamic range.

If the gain required is less than 2, one simply splits the lead-in element (QR for the bandpass filter, C for the high-pass filter) into a voltage divider of an appropriate ratio without having to upset the desirable value $H = 2$. However, $H > 2$ can be achieved only by setting the two resistors $(H - 1)R$ in Fig. 13 to the appropriate value. The effect of $H > 2$ can be shown to result in a very significant enhancement of Q in addition to a decrease of ω_0 with increasing ω_0/ω_t. The ω_0 error

$$\begin{aligned} \Delta\omega_0/\omega_0 &= (\omega_a - \omega_0)/\omega_0 \\ &\simeq -0.5[H^2/(H - 1)]\,\omega_0/\omega_t \end{aligned} \tag{54}$$

(where ω_a is the actually realized pole frequency) has its minimum at $-2\omega_0/\omega_t$ for $H = 2$, and can be corrected only by predistortion. The Q deviation, expressed in

$$\begin{aligned} Q_a \simeq Q/\{1 + [1 - 2Q(2 - H)/H]\ \Delta\omega_0/\omega_0 \\ + 2(3 - 4/H)Q(\Delta\omega_0/\omega_0)^2\} \end{aligned} \tag{55}$$

* References 2 and 8.

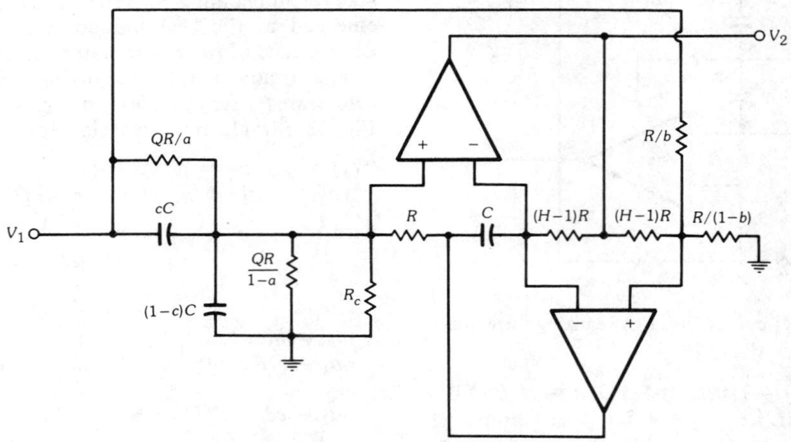

Fig. 13. A GIC-based biquad.

TABLE 2. DESIGN DATA FOR CIRCUIT IN FIG. 13

Filter Type	$N_2(s)$	a	b	c	Gain at $\omega =$		Comments
Low-Pass	a_0	Not realizable with this circuit.					
Bandpass	$a_1 s$	1	0	0	H	ω_0	Set $H = a_1 Q/\omega_0$ R_c From Eq. (56)
High-Pass	$a_2 s^2$	0	0	1	H	∞	Set $H = a_2$ R_c From Eq. (56)
All-Pass	$s^2 - s(\omega_0/Q) + \omega_0^2$	0	1	1	1	All	Set $H = 2$, $R_c = \infty$
Notch:	$a_2(s^2 + \omega_z^2)$ $\omega_z^2 = b\omega_0^2/(2c-b)$	$(1/2)\, b$	b	c	b $2c-b$	0 ∞	Set $H = 2$, $R_c = \infty$ Low-pass notch: $c > b$ High-pass notch: $c < b$ Notch: $c = b$

(where Q_a is the actually realized pole quality factor) also is minimized for $H = 2$ and can be shown to be reduced to zero, i.e., $Q_a = Q$, by connecting a compensation resistor

$$R_c \simeq R(1 + m)^2/m(H - 2 - m) \qquad (56)$$

where,

$$m = [H/(H - 1)]\omega_0/\omega_t \qquad (56a)$$

in shunt with the GIC input, as indicated in Fig. 13. Equation (56) is valid for pole frequencies satisfying

$$\omega_0 \lesssim 0.3\, \omega_t/H \qquad (57)$$

For larger values of ω_0, Eq. (56) becomes increasingly inaccurate, but Q compensation can still be achieved by functional tuning of R_c.

The function of R_c is only to eliminate ω_t-caused Q enhancement for $H > 2$; otherwise, it has no effect on the realized transfer function apart from the slight dependence of a_1 on R_c shown by Eq. (52c). As is seen from Eq. (56), R_c is eliminated, i.e., $R_c \rightarrow \infty$, for $H = 2$. In the all-pass and notch circuits, nothing significant is gained by choosing $H \neq 2$, and no usable compensation exists for $H \neq 2$ in the low-pass filter of Fig. 14. Thus, all GIC-internal resistors are set equal in these circuits.

Finally, it should be mentioned that Q is quite sensitive to capacitor losses. Labeling the capacitor-loss resistor R_L and using Eqs. (5) and (52a), we obtain $Q_c = \omega_0 C R_L = R_L/R$. The actual Q factor, Q_a, can be shown to be

$$Q_a = Q/(1 + 2Q/Q_c) \qquad (58)$$

Thus, high-quality capacitors should be used to build these filters.

Further details about GIC-filter performance can be found in References 2 and 8, on which most of this discussion is based.

As a design example, assume that the bandpass function

$$T(s) = 0.8 s_n/(s_n^2 + 0.1 s_n + 1) \qquad (59)$$

with $s_n = s/\omega_0$ and $\omega_0 = 2\pi\cdot 6$ kHz has to be realized. Assume further that op amps with ω_t

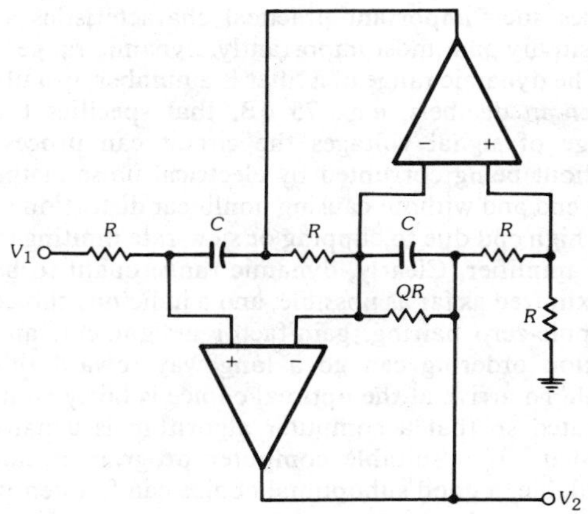

Fig. 14. GIC low-pass filter.

$= 2\pi \cdot 900$ kHz are available. From Eq. (59), pole frequency, quality factor, and midband gain are, respectively, $f_0 = 6$ kHz, $Q = 10$, and $H = 8$. From Table 2 for a bandpass filter: $a = 1, b = c = 0$. From Eq. (57) it follows that $f_0 = 6 \text{ kHz} < 33.7$ kHz so that Eq. (56) applies. Thus, $R_c = 22.24R$ will compensate for ω_t-caused Q enhancement that by Eq. (55) would be 79%; i.e., without R_c, $Q_a = 1.79Q = 17.9$. Further, from Eq. (54), $\Delta f_0 \simeq -183$ Hz, a −3.0% error. To eliminate this nominal deviation, the filter is designed with predistortion to realize $\omega_0 = 2\pi \cdot 6.189$ kHz so that a −3% error gives $\omega_a = 2\pi \cdot 6$ kHz. Thus, from Eq. (52a) with $C = 10$ nF, $R = 2572\ \Omega$; therefore, $QR = 25.7$ kΩ and $(H - 1)R = 18.0$ kΩ. The circuit is shown in Fig. 15.

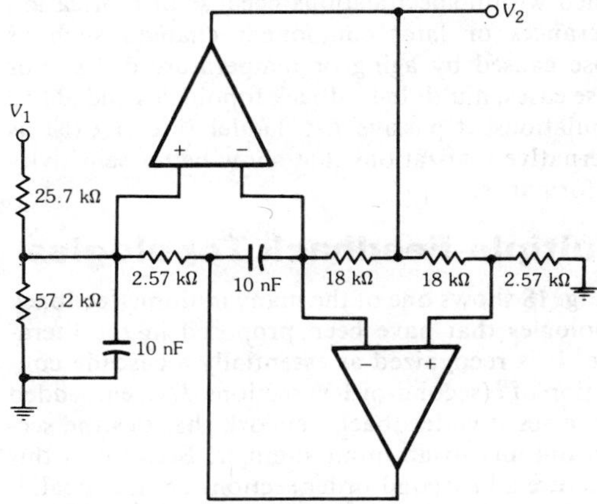

Fig. 15. GIC filter realizing Eq. (59).

HIGH-ORDER TRANSFER FUNCTIONS

As indicated earlier, high-order transfer functions are realized by one of three methods: (1) by cascading first- and/or second-order sections, (2) by connecting second-order sections in a suitable feedback topology, and (3) by simulating the operation of the elements of a lossless LC ladder. The primary advantage of methods 2 and 3 is their lower sensitivity to component variations than in the cascade approach; the price paid for this benefit is a more complicated design and more difficult tuning.

In the following, cascade and "follow-the-leader feedback" (FLF) designs, in addition to a ladder-simulation approach, are discussed in sufficient detail to permit construction of reliable high-order filters. Derivations, proofs, and further details about these and several other approaches are contained in many excellent textbooks and papers (see references 1 through 6).

Cascade Realization

The easiest, and for many applications entirely adequate, realization of a high-order function $H(s)$ is via a cascade connection of second-order filter sections as shown in Fig. 16. Assuming no interaction between the sections T_i, $i = 1, \ldots l$, i.e., each filter block $T_i(s)$ has a low output and/or high input impedance, the overall voltage gain of the circuit in Fig. 16 is

$$H(s) = V_0/V_i = (V_1/V_i)(V_2/V_1)\ldots(V_{l-1}/V_{l-2})(V_0/V_{l-1}) = T_1 \cdot T_2 \ldots T_l \quad (60)$$

To this end, Eq. (14) is factored into the product of l second-order transfer functions of the form of Eq. (39),

$$T_i(s) = N_i(s)/D_i(s) = (a_i s^2 + b_i s + c_i)/(s^2 + s\omega_{0i}/Q_i + \omega_{0i}^2) \qquad i = 1, \ldots l \quad (61)$$

such that $N(s) = \prod_{i=1}^{l} N_i(s)$ and $D(s) = \prod_{i=1}^{l} D_i(s)$.

This step is accomplished simply by finding the in general complex roots of $N(s)$ and $D(s)$ and keeping conjugate complex terms together so that all coefficients in $T_i(s)$ are real.

The notation in Eqs. (14), (60), and (61) has tacitly assumed that $H(s)$ is of even order, i.e., that r is even. Of course, if r is odd, one of the transfer functions $T_i(s)$ in Eq. (60) must be of first order; that is, it must be of the form of Eq. (38), whose realizations are given in Fig. 8. Such a first-order function can always be cascaded with the remainder

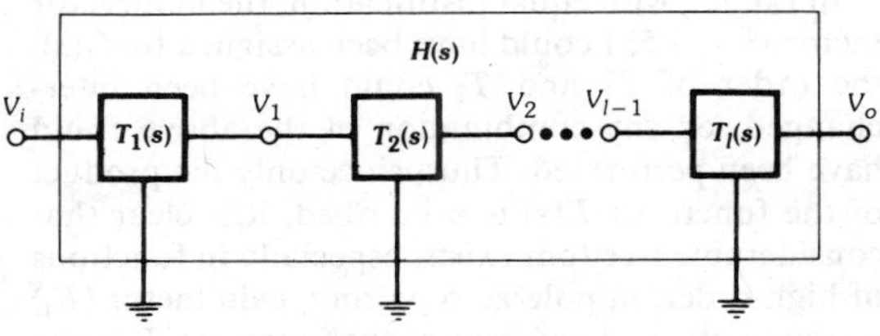

Fig. 16. Cascade connection of blocks $T_i(s)$.

of the network. Thus, it should be clear that the discussion can concentrate on the synthesis of even-order functions only; i.e., the question is how to realize Eq. (14) as expressed in Eq. (60) with Eq. (61). This in turn implies that practical methods have to be found to realize biquadratic transfer functions of the form of Eq. (61), a topic discussed in the previous section of this chapter.

As a simple example for the steps discussed, assume the function

$$H(s) = \frac{K(s^2+3.3551)}{s^4+2.7555s^3+3.7964s^2+3.0995s+1.4246} \tag{62}$$

a low-pass function with the frequency response sketched in Fig. 17, has to be realized. In Eq. (62), s is normalized with respect to the low-pass cutoff frequency $f_{3dB} = 4.55$ kHz; i.e., the filter passband is in $0 \leq f \leq f_{3dB}$, and $s = j\omega/(2\pi \cdot f_{3dB})$. Realizing Eq. (62) as a cascade of two second-order sections requires factoring the denominator and writing $H(s)$ as a product of two functions, T_1 and T_2. Simple algebra (root finding) results in

$$H(s) = \frac{K_1}{s^2+2.0905s+1.3544} \cdot \frac{K_2(s^2+3.551)}{s^2+0.6650s+1.0518}$$
$$= T_1(s) \cdot T_2(s) \tag{63}$$

where $K_1K_2 = K$. The low-pass and low-pass-notch functions $T_1(s)$ and $T_2(s)$ can then be realized as described in the previous section, e.g., by the circuit in Fig. 10B.

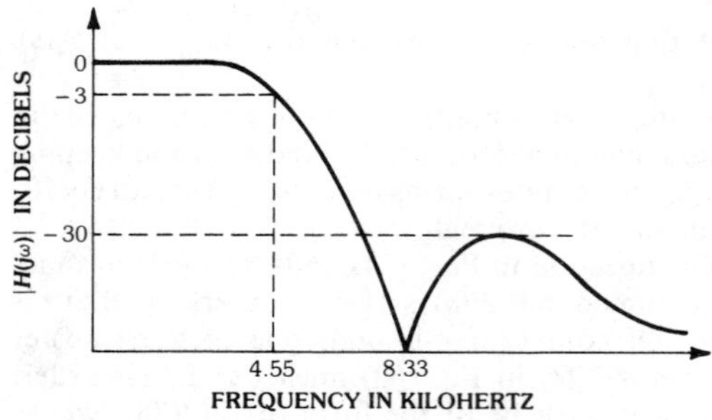

Fig. 17. Frequency-response plot, Eq. (62).

In Eq. 63, with equal justification, the numerator factor $s^2 + 3.551$ could have been assigned to $T_1(s)$, the order of T_1 and T_2 could have been interchanged, or any combination of the above could have been performed. Thus, since only the product of the functions $T_i(s)$ is prescribed, it is clear that considerable freedom exists, especially in functions of high order, in pole-zero pairing, gain factor (K_i) assignments, and ordering of the functions T_i in the cascade. The final choice of these factors determines such important practical characteristics as sensitivity and, most importantly, dynamic range.

The dynamic range of a filter is a number, usually given in decibels, e.g., 75 dB, that specifies the range of signal voltages the circuit can process without being corrupted by electrical noise at the low end and without causing nonlinear distortion at the high end due to clipping or slew-rate limiting in the amplifier. Clearly, dynamic range ought to be maximized as far as possible, and a judicious choice of pole-zero pairing, gain-factor assignment, and section ordering can go a long way toward this goal. To arrive at the optimal choice is fairly complicated so that a computer algorithm is usually needed.* If a suitable computer program is not available, a good suboptimal choice can frequently be obtained by

(1) assigning the poles with the highest Q factor to the closest transmission zero,
(2) choosing a low-pass or bandpass section as the first block and a high-pass or bandpass section as the last block in the cascade,
(3) ordering the remaining sections such that the voltages at the intermediate section outputs in the passband are as "flat" (independent of frequency) as possible, and, finally,
(4) assigning the gain constants K_i such that the voltage maxima at all section outputs are equal.

For critical filter specifications, requiring a transfer function of order 6 to 8 or higher, it has been found in some cases that cascade realizations are too sensitive to element variations; that is, the filter response cannot be tuned correctly or maintained within specifications because of fabrication tolerances or later component changes such as those caused by aging or temperature drifts. For these cases, multiple-feedback topologies and active simulations of passive *L-C* ladder filters† exist as alternative realizations that show better sensitivity performance.

Multiple-Feedback Topologies

Fig. 18 shows one of the many multiple-feedback topologies that have been proposed in the literature. It is recognized as essentially a cascade connection of l (second-order) sections $T(s)$, embedded into a resistive feedback network that ties the section outputs to an input summer. Because in this structure all l second-order sections are identical, it has been labeled the primary-resonator-block (PRB) topology.

The transfer function of the circuit in Fig. 18 can be shown, for an ideal summing amplifier, to be equal to

* References 2, 6, and 9.
† References 1, 2, and 6.

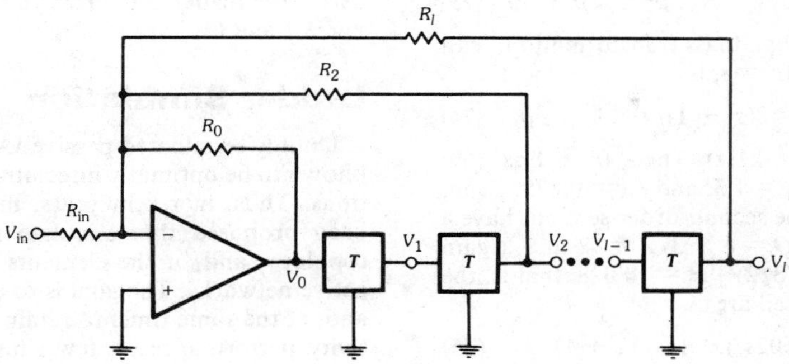

Fig. 18. The FLF topology for the special case of all identical sections $T(s)$.

$$H(s) = V_l/V_{in} = -F_0T^l(s)/[1 + \sum_{i=2}^{l} F_iT^i(s)] \tag{64}$$

where $F_i = R_i/R_{in}$, i = 0, 2, . . . l.

Such multiple-feedback structures are most convenient for the realization of symmetrical all-pole bandpass transfer functions (Butterworth and Chebishev filters, see Chapter 9) of even order that are most easily designed by transforming a prescribed bandpass transfer function

$$H_{BP}(s) = -Ks^{r/2}/(s^r + D_{r-1}s^{r-1} + \ldots D_1s + 1) \quad (r = 2l = \text{even}) \tag{65}$$

via the frequency transformation $p = \hat{Q}(s^2 + 1)/s$ into "a prototype low-pass function"

$$H_{LP}(p) = -KQ^l/(p^l + d_{l-1}p^{l-1} + \ldots d_1p + d_0) \tag{66}$$

In Eqs. 65 and 66, $r = 2l$ = degree of desired bandpass; l = degree of prototype low-pass filter; $K\hat{Q}^l = K_0$ is the desired midband gain of the bandpass filter; p is the frequency parameter of the low-pass filter, normalized to the corner frequency of the passband; and $s = j\omega/\omega_0$, the normalized frequency parameter of the bandpass function. Further, ω_0 is the center frequency of the bandpass filter, and $\hat{Q} = \omega_0/\Delta\omega_0$, where $\Delta\omega_0$ is the desired bandwidth. For this situation, the blocks T in the low-pass equivalent of Fig. 18 are

$$T(p) = Mk/(p + k) \tag{67}$$

To design the circuit, one substitutes Eq. (67) into Eq. (64), brings the resulting equation into the form of Eq. (66), and compares coefficients with the prescribed function. If the definition

$$f_i = F_iM^i \quad (i = 2, \ldots l) \tag{68}$$

is used, this process yields the following solution:

$$k = d_{l-1}/l \tag{69a}$$

$$f_2 = d_{l-2}/k^2 - l(l-1)/2! \tag{69b}$$

$$f_i = d_{l-i}/k^i - l!/(l-i)!i! - [1/(l-i)!]\sum_{j=2}^{i-1} f_j(l-j)!/(i-j)! \quad (i = 3, \ldots l) \tag{69c}$$

Applying now the frequency transformation $p = \hat{Q}(s^2 + 1)/s$ to Eq. (67) results in $T(s)$, the blocks in Fig. 18, being second-order bandpass functions.

$$T(s) = (M/Q)s/(s^2 + s/Q + 1) \tag{70}$$

where $Q = \hat{Q}l/d_{l-1}$ is the quality factor of $T(s)$, i.e., ω_0 divided by the 3-dB bandwidth of T. Note that all blocks $T(s)$ are tuned to the same ω_0 because $s = j\omega/\omega_0$. The gain constant M of $T(s)$ should be selected for best dynamic range, such that all filter-internal signal maxima are equal. In general, this requires a computer routine and results in different M factors for each section*; a good suboptimal choice, however, which at the same time retains all identical sections, is simply

$$M = [1 + (Q/\hat{Q})^2]^{1/2} \tag{71}$$

With M and f_i, $i = 2, \ldots l$, known from Eqs. (69) and (71), the actual feedback factors $F_i = R_i/R_{in}$ can then be determined from Eq. 68. Finally, the PRB bandpass design is completed by setting

$$F_0 = R_0/R_{in} = K_0M^{-l}(Q/\hat{Q})^l \tag{72}$$

so that the prescribed gain constant K_0 is realized correctly.

An example will illustrate the process: To be designed is a sixth-order bandpass filter with a Butterworth magnitude characteristic, center frequency $f_0 = 4.8$ kHz, center-frequency gain $K_0 = 5$ (14 dB), and 3-dB bandwidth $\Delta f = 600$ Hz. For this case, the prototype third-order Butterworth low-pass transfer function, corresponding to Eq. (66), is

* Reference 10.

$$H_{LP}(p) = -5/(p^3 + 2p^2 + 2p + 1) \quad (73)$$

and the low-pass-to-bandpass transformation, with $\hat{Q} = 4.8\text{ kHz}/0.6\text{ kHz} = 8$, is

$$p = 8(s^2 + 1)/s \quad (74)$$

where $s = j\omega/(2\pi \cdot 4.8\text{ kHz})$. Then, from Eqs. (69) and (73), $k = 2/3$, $f_2 = 1.5$, and $f_3 = 0.875$. From Eqs. (70) and (71), the second-order sections have a pole quality factor $Q = 8 \times 3/2 = 12$ and a gain constant $M = [1 + (3/2)^2]^{1/2} = 1.8028$; that is, the functions to be realized are

$$T(s) = 0.1502s/(s^2 + s/12 + 1) \quad (75)$$

Finally, the feedback resistor ratios F_i are, from Eq. (68), $F_2 = f_2/M^2 = 0.4615$, $F_3 = f_3/M^3 = 0.1493$, and, from Eq. (72), $F_0 = 2.880$.

The single-amplifier bandpass filter of Fig. 10A will be used to realize the function of Eq. (75). The design is completed as follows: From Eqs. (48), (45), and (46) with a choice of $q = 5$, one obtains $K = 1.01149$ and the predistorted values, if $f_t = 1$ MHz, $f_{0p} = 4.915$ kHz and $Q_p = 11.75$. If C = 12 nF is chosen, Eq. (47) yields $R = 27.0$ kΩ. Finally, from Table 1, with $|b| = 0.1502\omega_{0p}$, $m = 0.0149 = 1/67.3$. The complete circuit realizing the prescribed bandpass filter is shown in Fig. 19. Note that an inverter is needed in the outer feedback loop because the bandpass sections have an inverting gain and the total loop gain must be negative to assure stability.

For the use of multiple-feedback structures for realizing bandpass filters with finite transmission zeros, the reader is referred to the literature (references 1 and 6).

Ladder Simulation

Doubly terminated passive L-C ladder filters are known to be optimally insensitive to element variations. Thus, in recent years, many methods have been proposed that simulate the operation, the topology, and/or the elements of L-C ladders by active networks. The goal is to eliminate inductors and, at the same time, to retain the superior sensitivity performance. A few simple procedures are described in the following; for more details and for the design of very demanding filter circuits, the reader is referred to the literature (references 1, 2, and 6).

Leapfrog Topology—The leapfrog method is an operational simulation of an L-C ladder and owes its name to the resulting circuit structure. The process is best explained by an example from which the general design procedure should become clear.

Consider the simple L-C low-pass filter in Fig. 20. Analysis yields the equations

$$V_o = I_3/(sC_4 + G_L) \quad (76a)$$
$$I_3 = (V_c - V_o)/sL_3 \quad (76b)$$
$$V_c = (I_1 - I_3)/sC_2 \quad (76c)$$
$$I_1 = (V_i - V_c)/(sL_1 + R_s) \quad (76d)$$

where the currents and voltages are interpreted as signals to be represented in a signal-flow graph. Since all signals ought to be voltages, Eqs. (76) are

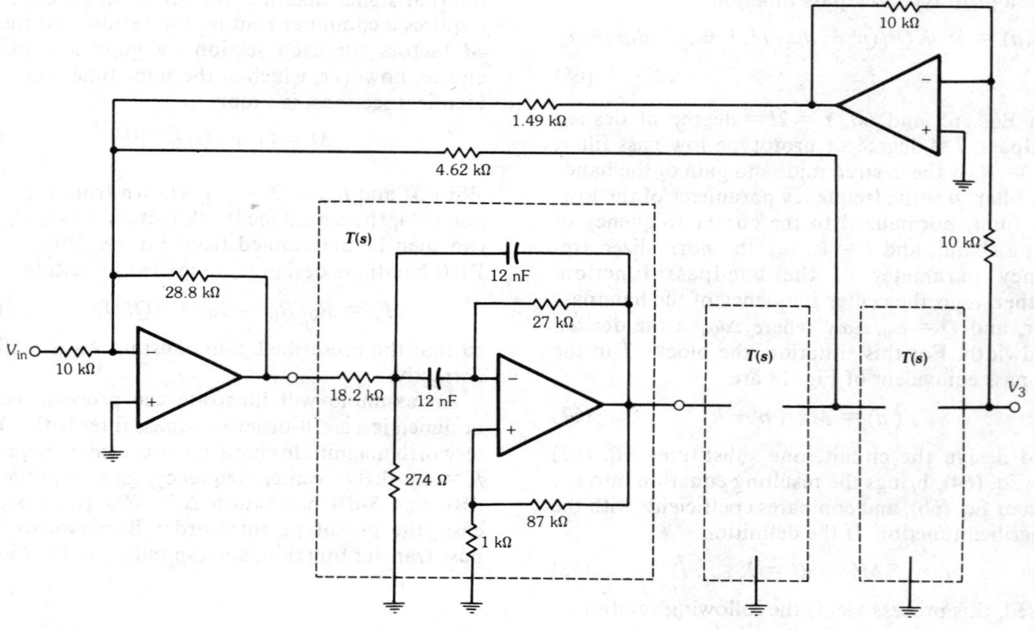

Fig. 19. Realization of the example.

then normalized by an arbitrary resistor R, and the notation $R_s/R = r_s$; $R_L/R = r_L$; $I_mR = V_m$, $m = 1, 3$; $L_j/R = l_j$, $j = 1, 3$; $C_kR = c_k$, $k = 2, 4$, is introduced. For example:

$$V_o = I_3R/(sC_4R + G_LR)$$

becomes

$$V_o = V_3/(sc_4 + g_L)$$

Using this notation, Eqs. (76) are rewritten as follows:

$$V_o = V_3/(sc_4 + g_L) \quad (77a)$$
$$V_3 = (V_c - V_o)/sl_3 \quad (77b)$$
$$V_c = (V_1 - V_3)/sc_2 \quad (77c)$$
$$V_1 = (V_i - V_c)/(sl_1 + r_s) \quad (77d)$$

These equations are represented in the flow graph of Fig. 21, which demonstrates that the operation of the ladder in Fig. 20 is simulated by use of summers and inverting and noninverting integrators. These two functions are easy to realize with active circuits, as has been shown earlier in this chapter. Note that all internal integrators are ideal, i.e., lossless, reflecting that the L-C prototype ladder is lossless, but that the two end sections are realized via lossy integrators, thereby taking care of the resistive source and load terminations. Note further that each loop contains one inverting and one noninverting integrator (compare Fig. 5), thereby assuring inverting loop gains for stability reasons. Stability will always be assured if all inductors are simulated by noninverting and all capacitors by inverting integrators. Lossy integrators are implemented by connecting a resistor in parallel with the integrator capacitor, and the summing operation indicated in Fig. 21 is realized by combining the summer of Fig. 4 and the integrator of Fig. 5 into one circuit. These operations are implemented in Fig. 22 and described by Eq. (78).

$$V_o = -V_1/(sCR_1 + R_1/R_3) - V_2/(sCR_2 + r_2/R_3) \quad (78)$$

Clearly, if $R_3 = \infty$ the summing integrator is lossless. Of course, to obtain a noninverting summing integrator operation, the circuit of Fig. 5C or D is used instead of a simple Miller integrator.

The actual circuit implementing the L-C ladder of Fig. 20 is shown in Fig. 23; the similarity of this diagram to the flow diagram in Fig. 21 should be apparent. To minimize phase errors caused by the finite gain-bandwidth product ω_t of the op amps, each loop consists of a Miller integrator with $Q_I < 0$ [see Eq. (31)] and a phase-lead integrator with $Q_I > 0$ (Fig. 5D). For convenience, all capacitors are selected to be equal. To determine resistors R_2 and R_9 in Fig. 23, the time constants of the corresponding integrators are set equal to the time constants of the input and output branches of the L-C prototype in Fig. 20. Thus:

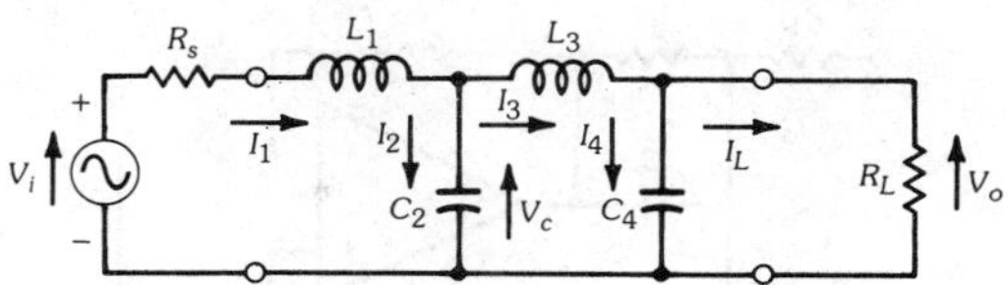

Fig. 20. L-C low-pass ladder.

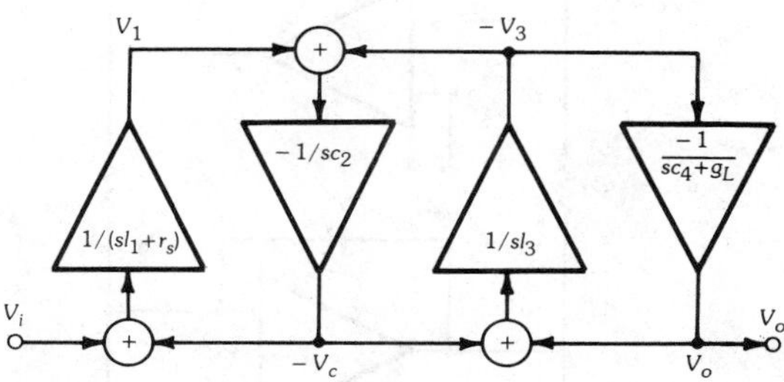

Fig. 21. Signal-flow diagram representing the L-C filter shown in Fig. 20.

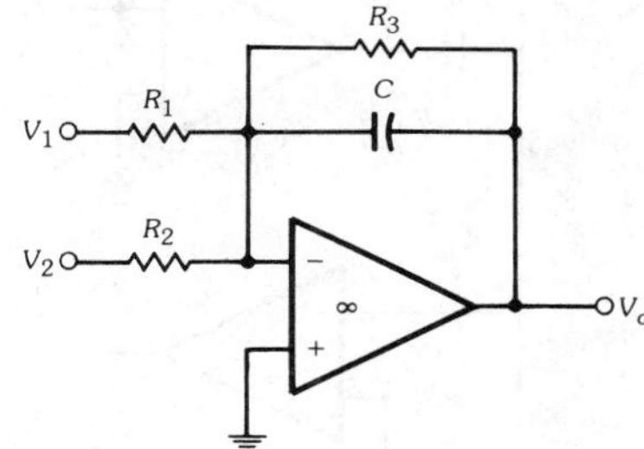

Fig. 22. A summing lossy integrator.

$$CR_2 = L_1/R_s \quad (79a)$$
$$C_4R_L = CR_9 \quad (79b)$$

or, for some suitable choice of C,

$$R_2 = L_1/(CR_s) \quad (80a)$$
$$R_9 = (C_4/C)R_L \quad (80b)$$

The "internal" resistors R_i, $i = 3, \ldots 8$, for Fig. 23 are obtained by equating the loop gains in the active circuit, i.e., the product of any two adjacent integrator time constants, to the product of the two corresponding reactances in the L-C prototype. Thus:

$$L_1C_2 = CR_3 \cdot CR_4 \quad (81a)$$
$$C_2L_3 = CR_5 \cdot CR_6 \quad (81b)$$
$$L_3C_4 = CR_7 \cdot CR_8 \quad (81c)$$

So, with the reactances of the L-C ladder known and some suitable choice of C, as for Eq. (80), one has, for example,

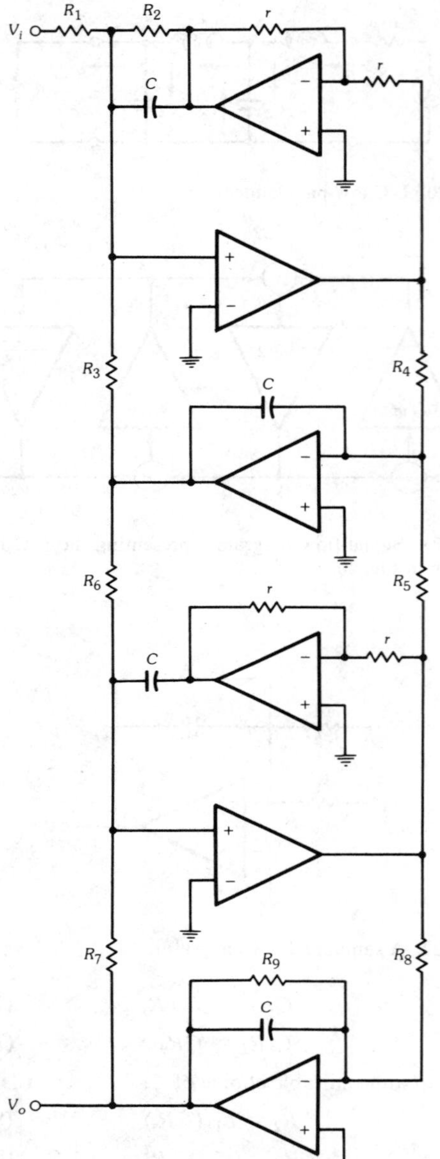

Fig. 23. Active simulation of the L-C ladder of Fig. 20.

$$R_3 = R_4 = \sqrt{L_1 C_2 C^{-2}} \quad \text{(82a)}$$

$$R_5 = R_6 = \sqrt{C_2 L_3 C^{-2}} \quad \text{(82b)}$$

$$R_7 = R_8 = \sqrt{L_3 C_4 C^{-2}} \quad \text{(82c)}$$

Finally, resistor R_1 determines the desired dc gain of the filter. In the active network, this can be chosen freely, whereas in the passive prototype it is fixed at $R_L/(R_s + R_L)$. Let the dc gain be called K; then it can be shown that

$$R_1 = (R_s/R)(R_D/K) \quad \text{(83a)}$$

if the L-C low-pass ladder starts with a series inductor (as in Fig. 20) and

$$R_1 = (R/R_s)(R_D/K) \quad \text{(83b)}$$

if the L-C ladder starts with a shunt capacitor. In Eqs. (83), R_D is the damping resistor of the first integrator (R_2 in Fig. 23), and R is the value of the normalizing resistor that was chosen in connection with Eqs. (77).

In a similar fashion, any type of L-C filter can be simulated as an active R-C filter by going through a signal-flow graph. For a detailed discussion, see reference 2. If the desired filter function is an all-pole bandpass function obtained via the low-pass–to-bandpass transformation from a prototype low-pass function, as illustrated in connection with the transformation from Eq. (66) to Eq. (65), the design is particularly simple: One only needs to simulate the low-pass filter, as shown in Fig. 21 for a fourth-order case, and then use the low-pass–to-bandpass transformation to transform each integrator of the form $T_i\,(p) = a/(bp + c)$ into a second-order bandpass section

$$T_i\,(s) = (sa/b\hat{Q})/(s^2 + sc/b\hat{Q} + 1) \quad \text{(84)}$$

similar to the transformation from Eq. (67) to Eq. (70) above. For ideal integrators, coefficient $c = 0$, implying a bandpass of infinite Q* for all internal sections of the simulated ladder (compare Fig. 21). After a suitable realization for the second-order sections is obtained as discussed earlier, these are then interconnected in a "leapfrog" topology as shown in Fig. 24, where the resistors of the feedback network are adjusted for the correct loop gains. For details of the design of these very well

* In practice, a very large but finite Q is sufficient.

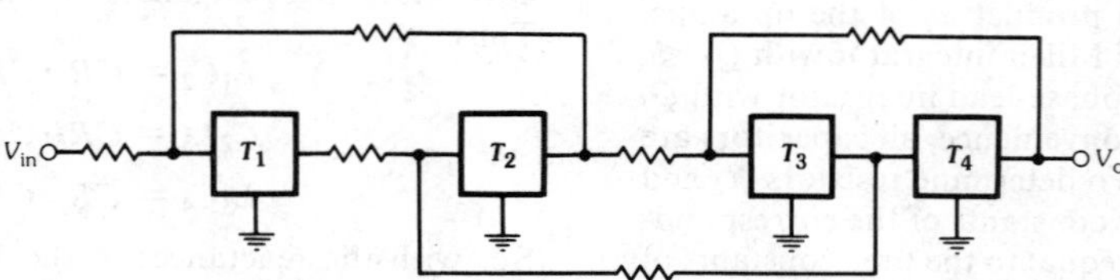

Fig. 24. Leapfrog bandpass filter.

behaved active ladder simulations, reference 1 or 6 should be consulted.

Ladders Using Inductor Simulation—It was pointed out earlier that high-quality grounded inductors can be simulated by the use of general impedance converters implemented with two op amps. Fig. 6 shows the circuit realizing, for $R_2 = R_3$ and $R_4 = 1/(\omega_c C)$, an inductor of value $L = R_1/\omega_c$, where ω_c is a frequency in or at the edge of the filter passband. It stands to reason, therefore, that an L-C ladder can be simulated by simply replacing all inductors by the circuit in Fig. 6, *provided* the inductors are grounded. An illustration of this rather straightforward procedure is given in Fig. 25. Fig. 25A shows an L-C ladder, which may have been obtained from tables and/or filter design handbooks. The filter has only grounded inductors and a finite transmission zero where C_2 and L_2 resonate (at $\omega_z = (C_2L_2)^{-1/2}$. If the signal spectrum does not extend beyond upper audio frequencies, a range through which GIC circuits are known to behave well, replacing inductors by optimized GICs [see earlier section, "The General Impedance Converter (GIC)."] results in the circuit of Fig. 25B, which will perform as reliably and predictably as the passive prototype.

A difficulty arises if the L-C filter contains—as in all low-pass and in many bandpass circuits—floating inductors, because no economical method exists to date that can be used to simulate high-quality floating inductors. Often, one of many ingenious network transformations can be used to eliminate all ungrounded inductors* so that no floating impedance converters are required. A procedure that works particularly well for low-pass filters makes use of frequency-dependent negative resistors.

Ladders Using FDNR Circuits—If the normalized impedances of a circuit are scaled (multiplied) by an impedance scaling factor $1/s$, the result is that a resistor of value R_i ohms becomes a capacitor of value $(1/R_i)$ farads, an inductor of L_i henries becomes a resistor of L_i ohms, and a capacitor of C_i farads is converted into a frequency-dependent negative resistor (FDNR) of value $D_0 = C_i$ as introduced in the section "The FDNR Element." Thus, $Z = R + sL + 1/sC$ is transformed into

$$\begin{aligned} Z_t &= Z/s \\ &= R/s + L + 1/s^2C \triangleq 1/sC_t + R_t + 1/s^2D_0 \end{aligned} \tag{85}$$

* References 2 and 5.

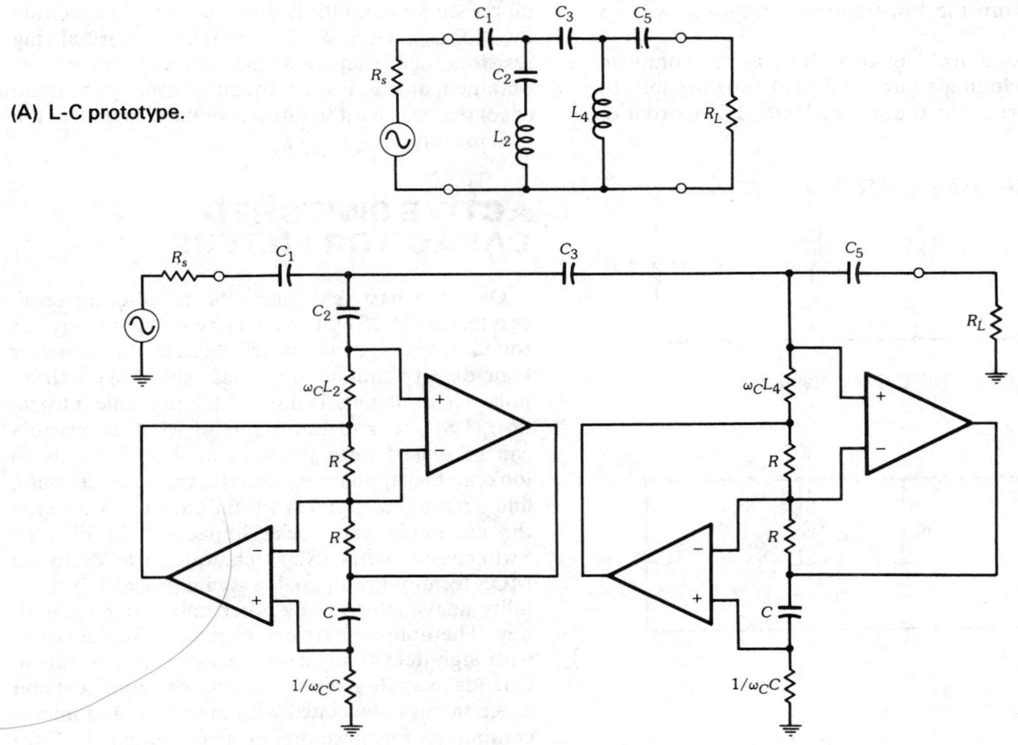

(B) Active simulation using GICs.

Fig. 25. Simulation of an L-C ladder.

The voltage transfer function, being a dimensionless quantity, is of course unaffected by impedance scaling so that the transformed circuit has the same response as the original filter. Hence, according to Eq. (85), inductors are eliminated, and FDNR elements are introduced that can readily be realized as discussed earlier in connection with Fig. 7. As with the inductor-simulation approach, the one limitation is that good floating FDNRs are very difficult to realize. Thus, this approach is most useful for passive L-C prototype filters that have no floating capacitors. In this sense, the FDNR and inductor simulation methods are complementary.

An example will illustrate the procedure: Fig. 26A shows an L-C low-pass filter with finite transmission zeros at $\omega_1 = (L_2C_2)^{-1/2}$ and $\omega_2 = (L_4C_4)^{-1/2}$. Impedance scaling by a factor $1/s$ transforms the circuit into Fig. 26B, which shows also the commonly used symbol for FDNR elements. Two practical problems become apparent in Fig. 26B. First, with reference to the FDNR circuit in Fig. 7, it is seen that there is no dc path to the noninverting input of one op amp of each of the two elements D_2 and D_4 in Fig. 26B; hence there can be no dc bias current. This difficulty is solved by shunting C_s and C_L by *large* resistors. (Note that in the L-C prototype these resistors correspond to large parasitic inductors shunting R_s and R_L, and that therefore the low-frequency response will be affected!)

The second problem arises because the complete circuit, including source and load resistors, has to be transformed by the scaling factor $1/s$ in order to leave the transfer function undisturbed. The terminations, however, may be imposed by the system so that they cannot be transformed into capacitors. This difficulty is circumvented by isolating the transformed filter between two unity-gain buffer amplifiers so that arbitrary terminations can be connected without affecting the response of the circuit. Making use of these ideas, the complete final active simulation of the L-C filter of Fig. 26A is shown in Fig. 27. The two bias resistors R_{B1} and R_{B2} should be chosen such that

$$R_{B1}, R_{B2} >> (\text{all } L_i)$$

and

$$R_{B2} = (R_{B1} + R_{B2} + \Sigma L_{\text{series}})\, H(0)$$

where $H(0) < 1$ is the prescribed dc gain of the prototype L-C filter. Further, as indicated, the *numerical* values of the resistors L_i are equal to those of the inductors in the L-C filter; the "source and load capacitors" G_s and G_L are equal to the source and load conductances, respectively; C_2 and C_4 in the FDNR blocks are equal to C_2 and C_4 in the L-C prototype; and the FDNR resistors are $R_2 = 1/C_2$ and $R_4 = 1/C_4$. Remember that all elements are normalized dimensionless quantities; the final design step consists of denormalizing the frequency by ω_n and the impedance level by R_n (i.e., all resistors are multiplied by R_n and all capacitors by $1/R_n\omega_n$). Here R_n is a suitable denormalizing resistor such that practical element values are obtained, and ω_n is the frequency (usually the band edge) that was used to normalize the elements of the L-C prototype.

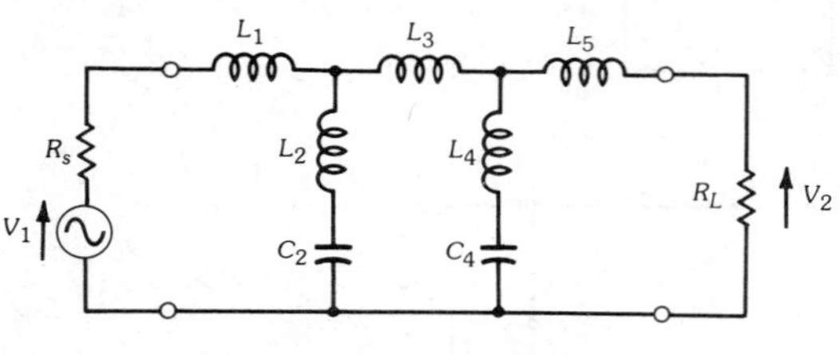

(A) L-C prototype.

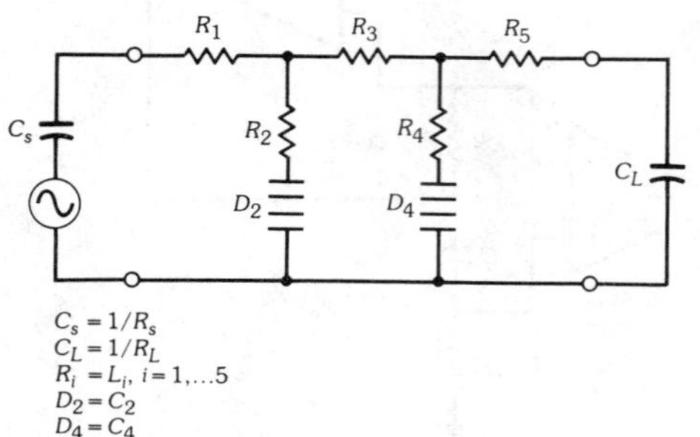

(B) Transformed circuit.

Fig. 26. FDNR ladder implementation.

ACTIVE SWITCHED-CAPACITOR FILTERS

Over the past few years, the metal-oxide-semiconductor (MOS) integrated circuit technology has found wide usage in industry because of its superior logic density compared to that achievable with bipolar technologies. Today, with large-scale integration (LSI), tens of thousands of MOS transistors can be placed on a single chip. We need only to look at the appliances, entertainment electronics, and personal computers within our homes to sense the economic and social impact of MOS LSI. Switched-capacitor (SC) techniques provide to the MOS technology an analog signal processing capability unavailable in any other monolithic technology. The unique marriage of analog SC functions with high-density logic on the same piece of silicon extends to analog/digital systems the same cost and space savings associated with memories and microcomputers. Furthermore, properly designed SC filters realize characteristics that usually require no trimming and that are inherently stable over process and environment variations.

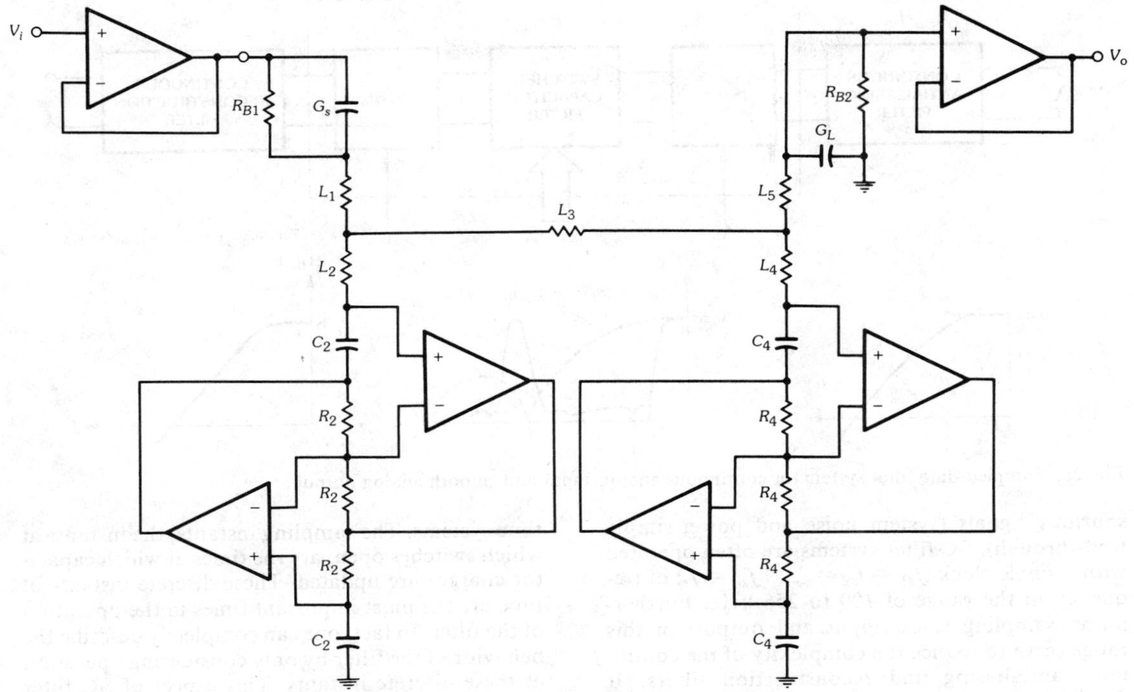

Fig. 27. FDNR realization of Fig. 26A.

Switched-capacitor filters,* made up of MOS capacitors, switches, and op amps, realize infinite impulse response (IIR) analog sampled-data filters, similar topologically to the active R-C filters described in previous sections. Therefore, narrow and flat passbands can be realized efficiently. Unlike their continuous-time active R-C counterparts, SC filters are sampled-data systems, which complicates their use and design. The sampled-data character of SC filters will be discussed in a later section. These filters do indeed take full advantage of the inherent precision achieved by MOS processing. As we will show later, the transfer functions are completely determined by precise crystal controlled clocks and ratioed capacitors. It has been demonstrated† that capacitor ratios can be held to about 0.3 percent and, with appropriate circuit techniques,‡ capacitances as small as 0.5 pF can be used. Furthermore, MOS capacitors are nearly ideal, with very low dissipation factors and good temperature stability. These properties can be achieved with either NMOS or CMOS processing. However, CMOS, with its added flexibility for realizing high-gain, low-noise op amps and low power dissipation, is recognized as the technology of choice.

* References 1, 6, 11, and 12.
† Reference 11.
‡ References 1 and 12.

Sampled-Data Filter Systems

Let us consider a sampled-data filter system that is suitable for use in a continuous analog (i.e., analog input/analog output) environment. This represents, from a hardware point of view, the most severe environment for an SC filter, or for that matter, any sampled-data filter. The system given in Fig. 28 shows the analog signal being passed through a continuous antialiasing filter (low-pass); an input sample-and-hold circuit, $(S/H)_i$, which samples the band-limited analog input at intervals of $1/f_{s1}$; the switched-capacitor filter; an output sample-and-hold circuit, $(S/H)_o$, which resamples the output of the SC filter at intervals $1/f_{sn}$; and a final continuous reconstruction filter (low-pass), which serves to smooth the sharp transitions in the sampled-data output waveform. The SC filter is shown, in general, to be controlled by clocks of multiple frequencies (f_{s2} through f_{sn-1}). Since the capacitor ratios (hence silicon area) required to realize a given transfer function scale in proportion to the ratio of the clock frequency to the pole (zero) frequencies, silicon area can be minimized by employing multiple clocks.* However, for simplicity and to minimize the aliasing of out-of-band

* Reference 1.

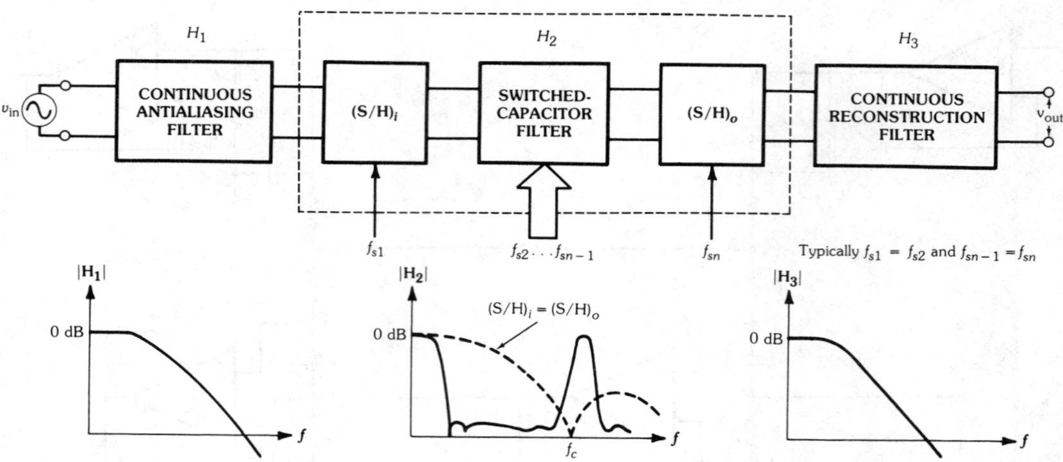

Fig. 28. Sampled-data filter system for continuous analog input and smooth analog output.

spurious signals (system noise and power-supply feed-through), SC filter systems are often operated with a single clock ($f_{s1} = f_{s2} = \ldots f_{sn} = f_s$) of frequency in the range of 100 to 256 kHz. Furthermore, sampling rates (input and output) in this range serve to reduce the complexity of the continuous antialiasing and reconstruction filters. It should be noted that in many SC filters, particularly low-pass filters, the sample-and-hold operations shown in Fig. 28 are inherently performed by the SC filter. In these cases, the SC filter includes the three blocks enclosed by the dashed box in Fig. 28. Typical frequency responses for the various blocks (shown for a low-pass SC filter) in Fig. 28 are shown below each system block for the typical case where $f_{s1} = f_{s2} = \ldots = f_{sn} = f_s$.

It is noted that when an input or output is interfaced with digital or sampled-data circuitry, such as d/a and a/d converters, some of this hardware is no longer needed. For example, when the output is to be interfaced to another sample-data function, the reconstruction filter is not required, and a sample-and-hold circuit is typically included at the input of this function. Although the need for continuous filtering is reduced, interfacing to digital and sampled-data circuits requires careful synchronization between the clocks that control the SC filter and those that control the external sampling operations. This is accomplished by passing synchronization pulses between the SC network and the external samplers. One reason for synchronization is to ensure that the SC network output is sampled after all transients have settled and the output is steady.

Another consequence of the sampled-data character of SC networks is the mathematical convenience of the z transform in analyzing or specifying SC filters. The z transform, where $z = e^{s\tau}$ and τ is the clock period, is covered in Chapter 28 as a mathematical tool for digital filters and discrete-time systems. The sampling instants, the instants at which switches open, are the times at which capacitor charges are updated. These discrete instants of time are the most important times in the operation of the filter. In fact, one can completely describe the behavior of the filter by only considering operation at these discrete instants. This aspect of SC filter behavior is analogous to the operation of digital filters; hence, the mathematical analysis follows in a similar manner.

It has been mentioned that the input and output of an analog sampled-data system are analog or continuous signals. We note that the antialiasing filter serves to band-limit the input spectrum to $f_c < f_s/2$ so that the signal can be reconstructed without error. Here, f_c refers to the highest allowed component in the input spectrum, and f_s is the sampling frequency. The antialiasing filter will also serve to band-limit high-frequency input noise, which would otherwise be aliased back into the baseband. Inherent in all analog sampled-data systems is the means to provide some sort of analog reconstruction. The simplest form of reconstruction is the zero-order hold or sample-and-hold (S/H). The impulse response of the zero-order hold is (with $\tau = 1/f_s$)

$$h_0(t) = \begin{cases} 1/\tau & \text{for } 0 \le t \le \tau \\ 0 & \text{elsewhere} \end{cases} \tag{86}$$

The sample-and-hold impulse response is sketched in Fig. 29, and an S/H reconstructed signal $x_r(t)$ is shown in Fig. 30. The transfer function for this sample-and-hold is

$$H_0(s) = (1 - e^{-s\tau})/(s\tau) \tag{87}$$

and its frequency response is

$$\begin{aligned} H_0(j\omega) &= (1 - e^{-j\omega\tau})/j\omega\tau \\ &= e^{-j\omega\tau/2}(\sin \omega\tau/2)/(\omega\tau/2) \end{aligned} \tag{88}$$

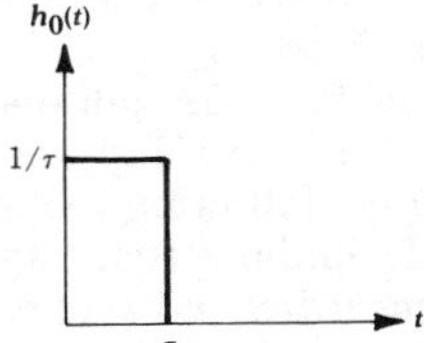

Fig. 29. Impulse response for sample-and-hold.

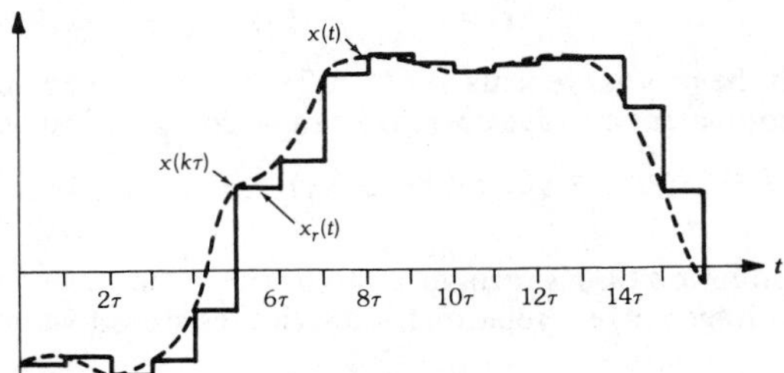

Fig. 30. Reconstruction with sample-and-hold.

It can be seen that $H_0(j\omega)$ has the $\sin(x)/x$ response shown in Fig. 31A and the phase characteristic shown in Fig. 31B.

Let us examine what happens to the spectrum of a sampled signal $X^{\#}(j\omega)$ when the sampled signal is reconstructed with a sample-and-hold. Note that $X^{\#}(j\omega)$ is precisely the spectrum predicted by z-transform analysis. The spectrum of reconstructed signal $X_r(j\omega)$ is simply the product of the sampled signal spectrum $X^{\#}(j\omega)$ and the sample-and-hold spectrum $H_0(j\omega)$, as shown in Fig. 32. Note that the

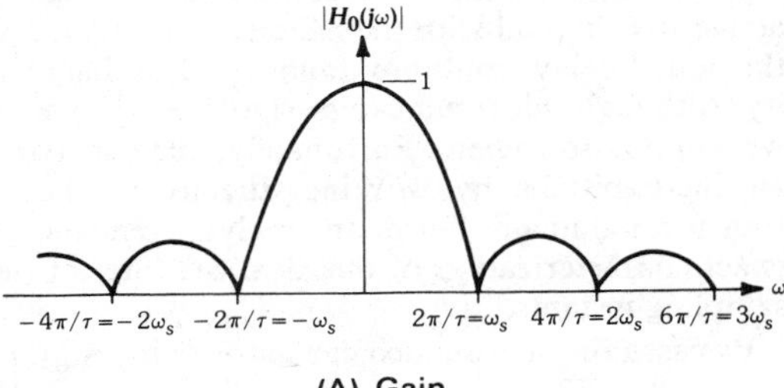

(A) Gain.

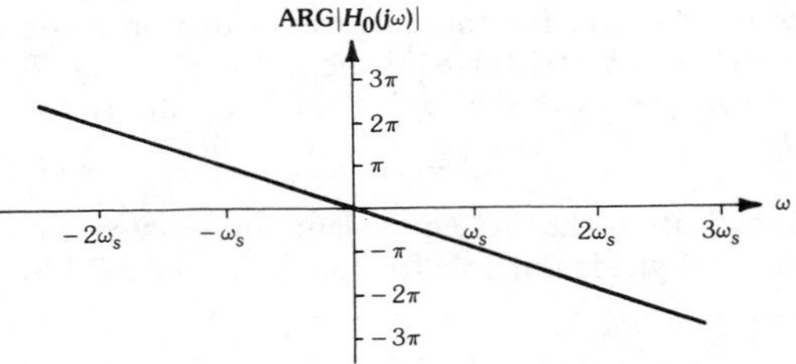

(B) Phase.

Fig. 31. Sample-and-hold responses.

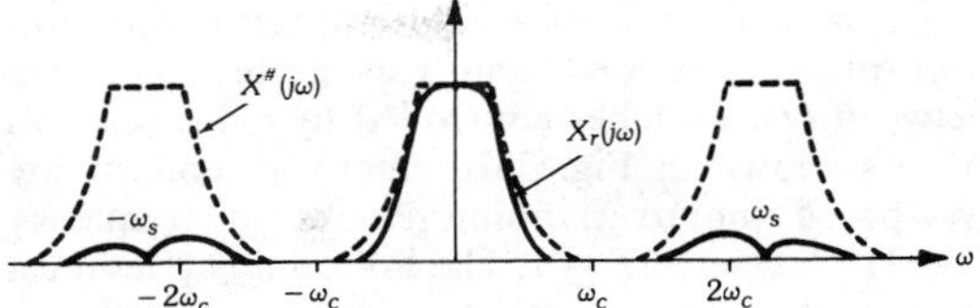

Fig. 32. Spectrum of reconstructed sampled signal.

high-frequency content of the sampled signal has been substantially reduced and the baseband spectrum is slightly altered near the band edge. When $f_s/f_c >> 1$, as is usually the case in SC filters, this baseband distortion is negligible and

$$X_r(j\omega) = X^{\#}(j\omega) \text{ for } \omega < \omega_s/2 = \pi f_s \quad (89)$$

For lower sampling rates, the sampled-data filter that operates on signal X is designed with a peak to compensate for the S/H-related band-edge droop. To further smooth the $X_r(j\omega)$ (i.e., further attenuate the high-frequency components of X_r), the analog sampled-data system can be followed by a continuous low-pass filter as shown in Fig. 28.

Now that we have discussed in general terms the concept of analog sampled-data systems and the necessary hardware to support the sampled system in a non–band-limited analog environment, let us focus our attention on the SC-filter portion of this system.

The Operation of Ideal SC Filters

Consider now the operation of an ideal SC filter, made up of ideal capacitors, ideal switches, and ideal op amps (infinite gain and infinite bandwidth). It is noted that MOS op amps that settle to within 0.1 percent of final value in less than 2 μs and achieve dc gains of greater than 60 dB have been designed.* One subtle difference between SC filters and their active R-C counterparts is the effect of op-amp dynamics on filter behavior. Although related, op-amp settling time, rather than op-amp frequency response, is the important op-amp characteristic for SC filters. In fact, it is op-amp settling time that establishes the upper limit for the sampling rate. It has been demonstrated that settling errors can be minimized by carefully orchestrating† the timing among the various samples within the filter. Hence, for sampling rates f_s of no more than 256 kHz and careful timing, a practical MOS op amp usually can be modeled as a voltage-controlled voltage source with gain equal to the dc gain of the op amp.

---

* Reference 11.

† Reference 13.

Consider for the moment the simple SC circuit in Fig. 33A, consisting of a single capacitor with bottom plate grounded and top plate connected between two switches controlled by clocks ϕ^a and ϕ^b. As shown in Fig. 33B, clocks ϕ^a and ϕ^b are two-phase, nonoverlapping clocks of frequency $f_s = 1/2T$, where $2T = \tau$. The clock duty cycles must be less than 50 percent to achieve nonoverlap. However, to allow maximum time for op-amp settling, duty cycles of greater than 35 percent are typically used. Most SC filters are insensitive to duty cycle, but nonoverlap is required for the circuit to operate correctly. The fact that op-amp settling typically establishes the upper limit for f_s has already been mentioned. It was also mentioned in the previous section that high f_s results in large capacitor ratios and low f_s results in complex antialiasing/reconstruction filters. Other factors that impact the selection of a sampling rate are noise and capacitor leakage. It should be evident that the choice of sampling rate involves a compromise among several conflicting factors and no general rule can be stated. One can find in the literature* SC filters with sampling frequencies that range from 256 kHz to as low as 8 kHz.

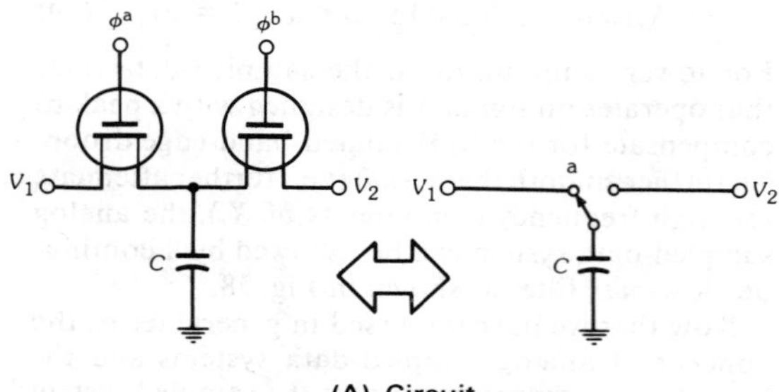

(A) Circuit.

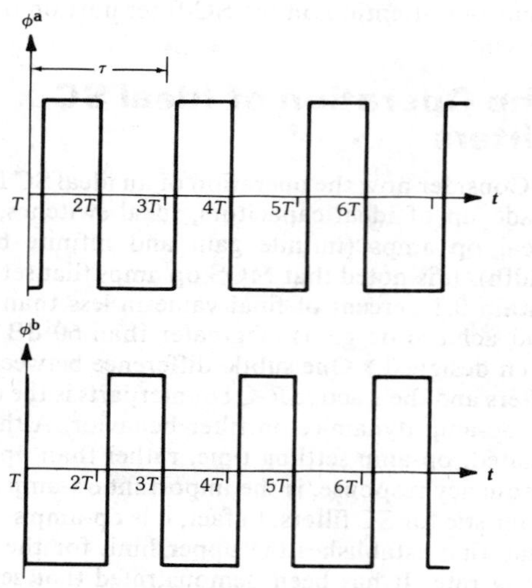

(B) Biphase nonoverlapping clocks.

Fig. 33. Simple SC circuit.

* References 1 and 11.

The circuit in Fig. 33A, although simple, provides a rather interesting and important function. Let us consider the following simple analysis of this circuit. Assume initially that capacitor C is uncharged. When switch ϕ^a closes and switch ϕ^b opens, at $t = 0$, capacitor C charges to voltage V_1. When $t = T$ (or $\tau/2$), switch ϕ^a opens and switch ϕ^b closes, and capacitor C discharges to voltage V_2. The amount of charge that flows into (or out of) V_2 is given by

$$\Delta Q = C(V_2 - V_1) \tag{90}$$

If the process repeats at $t = \tau, 2\tau, \ldots$ etc., then an equivalent average current I can be computed to be

$$I = \Delta Q/\tau = (C/\tau)(V_2 - V_1) = Cf_s(V_2 - V_1) \tag{91}$$

Equation (91) is recognized to be the instantaneous voltage-current relation for a series resistor of value

$$R = 1/Cf_s \tag{92}$$

Equations (90)–(92) establish the well known switched capacitor-resistor equivalence. An oversimplified application of Eq. (92) is to derive SC filters from active R-C prototypes by replacing each resistor with an equivalent switched capacitor (with $C = 1/Rf_s$). However, a careful analysis of Fig. 33A reveals that Eq. (92) is an approximation in that the charge transfer described in Eq. (90) involves a delay of T seconds. The approximation improves as T becomes small (f_s becomes large). Hence, the SC filter synthesis method of replacing resistors in an active R-C prototype with SCs is not a recommended procedure. In fact, if one were to replace the resistor in the feedback loop of a high-Q active R-C biquad with the SC circuit in Fig. 33A, the added delay would no doubt result in instability. Although this result can be rigorously proven,* we will not do so here. Fortunately, one can avoid all these pitfalls by working directly in the z-transform domain. Such an analysis renders an exact characterization of the ideal SC filter at the sampling instants.

By cascading a grounded capacitor to the SC circuit in Fig. 33A, we construct a simple one-pole low-pass filter. This circuit is shown in Fig. 34. Let us once again, for the purpose of demonstration, resort to the resistor equivalence in Eq. (92). The pole frequency of the equivalent R-C filter is

$$\omega_p = 1/R_1C_2 \tag{93}$$

Substituting the SC equivalent for resistor R_1 in Eq. (93) yields the pole frequency for the SC filter, i.e.,

$$\omega_p \approx f_s(C_1/C_2) \text{ for } f_s >> 2\pi f_p \tag{94}$$

From Eq. (94), one immediately observes the inher-

* Reference 1.

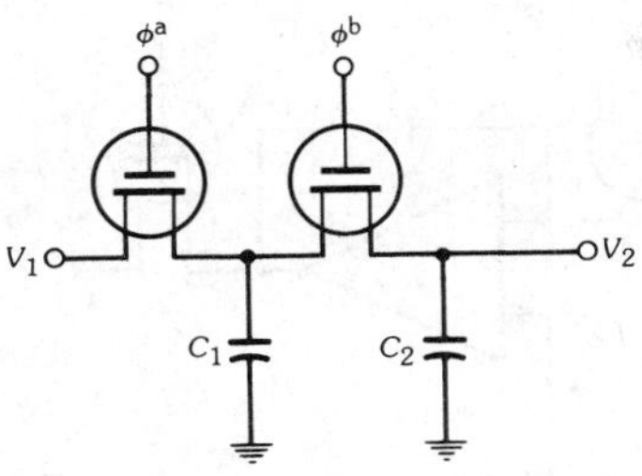

Fig. 34. One-pole low-pass SC filter.

ent accuracy of the SC implementation. That is, ω_p no longer depends on the product of an R and a C, but on the ratio of two Cs and f_s. It was mentioned previously that MOS capacitor ratios can be held to tight tolerances and f_s is derived from a very precise crystal-controlled master clock. In addition, note that the capacitor ratio $C_2/C_1 = f_s/\omega_p$. That is, for constant ω_p, C_2/C_1 scales in direct proportion to f_s. As a consequence, the silicon area required to realize the SC filter in Fig. 34 (for a fixed ω_p and minimum capacitance C_1) scales with f_s. For IC realizations, minimum capacitances on the order of 1 pF are typically used.

SC Integrators

Most practical SC filters employ SC integrators as basic building blocks. Moreover, SC filters are realized with one op amp per pole (and zero). As we will soon demonstrate, SC filters implemented in this manner can be made insensitive to the unavoidable parasitic capacitances that occur in an IC realization.

Let us first consider the inverting SC integrator in Fig. 35. Also shown are the parasitic capacitances associated with each node of the circuit. Capacitors (C_1 and C_2 in Fig. 35) are typically implemented as a thin (about 1000 Å) SiO_2 dielectric sandwiched between two polysilicon plates. The interconnecting lines are typically polysilicon and metal that lie on top of a thick layer of SiO_2 (referred to as field oxide). The switches are usually either polysilicon gate CMOS or enhancement NMOS devices. In the process of constructing these features in an integrated circuit, parasitic capacitors are simultaneously created with the reverse-biased substrate. The parasitic capacitances are not controllable, do not track the intended capacitors, and sometimes are nonlinear. Hence, it is important to reduce significantly or eliminate the effect of parasitic capacitances on the transfer characteristic of the SC filter.

Returning to the integrator in Fig. 35, let us consider the roles played by each of the parasitic capacitances, C_{p1} through C_{p4}. First, parasitics C_{p1} and C_{p4} shunt low-impedence voltage sources. Also, parasitic C_{p3} shunts the "virtual" ground of a very high-gain op amp. These parasitics have no effect on circuit performance. Parasitic C_{p2}, on the other hand, parallels C_1 and adds directly to C_1. In the absence of parasitics, the z-domain transfer function* for the inverting SC integrator in Fig. 35 is

$$H(z) = -(C_1/C_2)z^{-1/2}/(1 - z^{-1}) \quad (95)$$

(Due to space limitations, z-domain transfer functions are given without derivation. The interested reader is referred to the literature* for detailed derivations.) If C_{p2} is included, the transfer function becomes*:

$$H(z) = -(C_1/C_2)(1 + C_{p2}/C_1)z^{-1/2}/(1 - z^{-1}) \quad (96)$$

From Eq. (96), the parasitic capacitance is seen to result in a gain error of $1 + C_{p2}/C_1$.

Let us now consider the alternative SC inverting integrator shown in Fig. 36. Also shown in Fig. 36 are the two important parasitic capacitances. All other parasitic capacitances shunt either a voltage source or the op-amp virtual ground and have been shown to have no effect. To determine the effect of C_{pl} and C_{p2}, let us examine the circuit as switches ϕ^a and ϕ^b turn on and off. With ϕ^a on and ϕ^b off, parasitics C_{pl} and C_{p2} shunt a voltage source and the op-amp virtual ground, respectively. When ϕ^a turns off and ϕ^b turns on, both C_{pl} and C_{p2} are shorted harmlessly to ground. Hence, this SC inverting integrator operates without error. This realization in Fig. 36 is said to be parasitic insensitive. The transfer function for this integrator* is

* Reference 1.

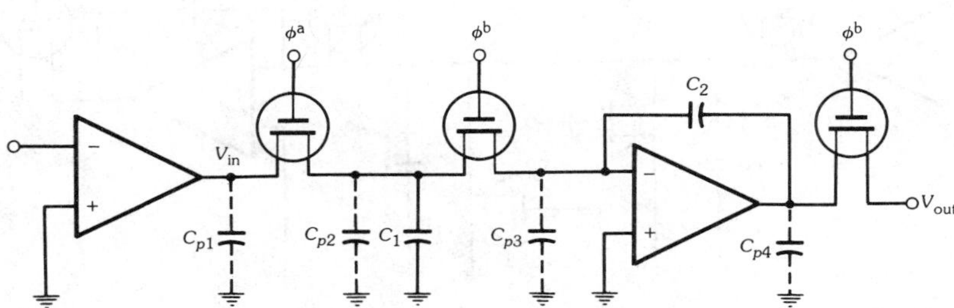

Fig. 35. Parasitic-sensitive SC inverting integrator with associated parasitic capacitances.

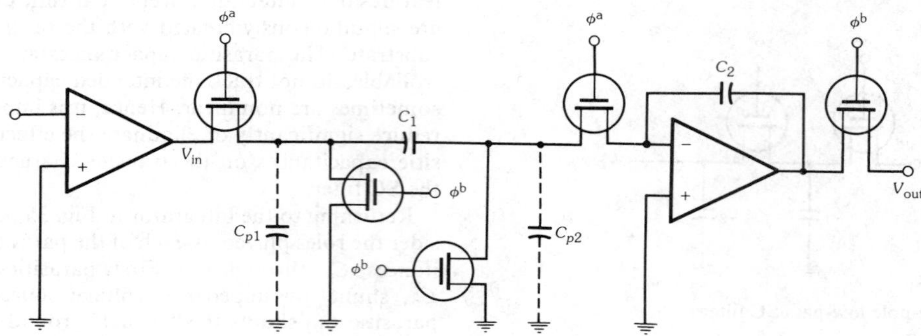

Fig. 36. Parasitic-insensitive SC inverting integrator with associated parasitic capacitances.

$$V_{out}/V_{in} = -(C_1/C_2)z^{-1/2}/(1 - z^{-1}) \quad (97)$$

Note the presence of the $z^{-1/2}$ (one-half clock period delay) term in Eqs. (95)–(97). In the digital and SC filter literature, integrators of the form $Kz^{-1/2}/(1 - z^{-1})$ are known * as lossless discrete integrators (LDI).

Switched-capacitor techniques allow the efficient realization of noninverting integrators. As shown in Fig. 37, an SC noninverting integrator can be realized with a single op amp. In the active R-C world, noninverting integrators require two op amps. With the switching arrangement shown for switched capacitor C_1, the necessary second inversion is elegantly performed by C_1. It can be verified that this circuit, like that in Fig. 36, is parasitic insensitive. The transfer function for the noninverting integrator in Fig. 37 is

$$H(z) = (C_1/C_2)z^{-1/2}/(1 - z^{-1}) \quad (98)$$

With the inverting and noninverting integrators in Figs. 36 and 37, parasitic insensitive SC filters of all transfer-function types can be realized. Parasitic insensitivity is not achieved at the expense of design generality. In fact, there is no reason to implement SC filters in any other way.

* Reference 1.

SC *z*-Domain Biquadratic Transfer Functions

To design an SC filter, one must first derive a *z*-domain transfer function from the filter requirements. This can be done by first obtaining the *s*-domain transfer function according to classic methods. The desired *z*-domain transfer function can then be obtained from the *s*-domain function by using the well known bilinear transform,

$$s = (2/\tau)(1 - z^{-1})/(1 + z^{-1}) \quad (99)$$

where τ represents one clock period. The bilinear transform is treated in Chapter 28 of this handbook; hence, it will not be discussed in further detail here. Alternatively, *z*-domain transfer functions can be obtained directly from specified filter requirements by using any one of several computational techniques (see Chapter 28).

It was noted in previous sections that active R-C biquads have played a dominant role in the realization of high-order active R-C filters. Although SC filters provide the opportunity to implement much higher-order filters on a single chip, the advantage of the biquadratic building block approach remains substantial. It lends regularity to the design process and can considerably reduce the chip layout effort. As active R-C biquads realize biquadratic trans-

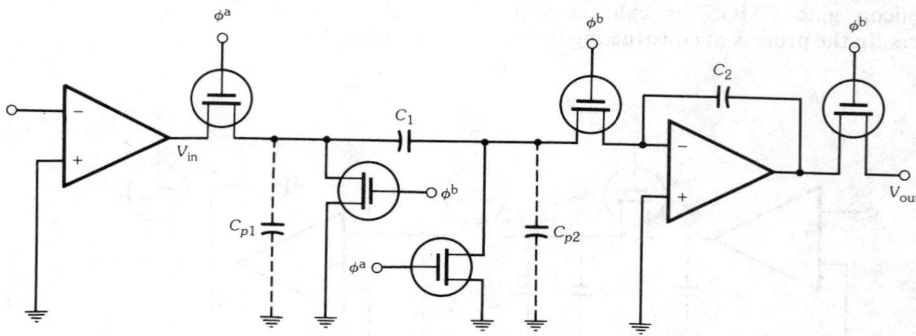

Fig. 37. Parasitic-insensitive SC noninverting integrator with associated parasitic capacitances.

fer functions in the s-domain, SC biquads realize biquadratic z-domain transfer functions. Thus, the transfer functions to be realized are of the form:

$$H(z) = N(z)/D(z) = (\gamma + \epsilon z^{-1} + \delta z^{-2}) / (1 + \alpha z^{-1} + \beta z^{-2}) \quad (100)$$

The well known generic forms of Eq. (100), namely low-pass (LP), high-pass (HP), bandpass (BP), low-pass notch (LPN), high-pass notch (HPN), and all-pass (AP), can be derived by applying the bilinear transform to the well known s-domain generic biquadratic functions.

One important property of the bilinear transfer function that we need to consider at this point is the property that the s-plane zeros at infinity map into z-plane zeros at the one-half sampling frequency (i.e., $z = -1$). Such zeros appear in LP and BP functions. Although bilinear BP and LP provide desirable steep cutoff in the vicinity of the half-sampling frequency, they may not offer the most economical SC realization. For the high sampling rates typically used in SC filters, the attenuation provided at the half-sampling frequency is of little importance and diminishes in importance as $\omega_p T$, $\omega_z T$ become small (ω_p and ω_z refer to the biquad pole and zero frequencies). When this relation is taken into account, several alternative LP and BP biquadratic transfer functions can be derived by replacing the zeros at $z = -1$ (i.e., the factors $1 + z^{-1}$) with either 2 or $2z^{-1}$.

In Table 3, the numerators $N(z)$ are listed for the various z-domain generic biquadratic forms. The LP and BP forms are referred to in this table as LP IJ and BP IJ, where I and J may have the value 0, 1, or 2. The suffixes I and J denote, respectively, the number of $1 + z^{-1}$ factors and the number of z^{-1} factors.

The recommended design procedure is first to obtain $D(z)$ and $N(z)$ via the bilinear transform. This will ensure the proper placement of the poles and zeros (other than those at $z = -1$). One or more of the zeros at $z = -1$ can then be replaced by either 2 or $2z^{-1}$. To place the gain level accurately, the gain constant K may require slight alteration.

TABLE 3. GENERIC BIQUADRATIC TRANSFER FUNCTIONS

Generic Form	Numerator $N(z)$
LP 20 (bilinear transform)	$K(1+z^{-1})^2$
LP 11	$Kz^{-1}(1+z^{-1})$
LP 10	$K(1+z^{-1})$
LP 02	Kz^{-2}
LP 01	Kz^{-1}
LP 00	K
BP 10 (bilinear transform)	$K(1-z^{-1})(1+z^{-1})$
BP 01	$Kz^{-1}(1-z^{-1})$
BP 00	$K(1-z^{-1})$
HP	$K(1-z^{-1})^2$
LPN	$K(1+\epsilon z^{-1}+z^{-2})$, $\epsilon > \alpha/\sqrt{\beta}$, $\beta > 0$
HPN	$K(1+\epsilon z^{-1}+z^{-2})$, $\epsilon < \alpha/\sqrt{\beta}$, $\beta > 0$
AP	$K(\beta+\alpha z^{-1}+z^{-2})$
General	$\gamma+\epsilon z^{-1}+\delta z^{-2}$

Parasitic-Insensitive SC Biquads

A general SC biquad that realizes Eq. (100) and its special cases in Table 3 was introduced by Fleischer and Laker.* This parasitic-insensitive biquad, shown in Fig. 38, is widely used throughout the industry. Due to space limitations, in this section we can only highlight the salient features of this biquad. The interested reader is referred to references 1 and 12 for detailed derivations and demonstrations of individual features.

It simplifies the general analysis of this biquad to assume the input (V_{in}) is sampled and held for the full clock period. This condition is not necessary in every special-case implementation of Fig. 38, but the condition is readily arranged. Furthermore, if the input to a cascade of SC biquads of the form of that in Fig. 38 is presented with a full-clock-period sampled-and-held signal, the switch timing in the biquads will propagate this condition through the entire filter. This condition is provided inherently by LP 02 and BP 01 biquads for any properly band-limited input.

Since, depending on the application, the desired output may be either V or V', both corresponding transfer functions are of interest. These are given as follows:

* References 1 and 12.

$$H=\frac{V}{V_{\text{in}}}=-\frac{D(I+K)+(AG+AL-DI-DJ-2DK)z^{-1}+(DJ+DK-AH-AL)z^{-2}}{D(F+B)+(AC+AE-DF-2BD)z^{-1}+(DB-AE)z^{-2}} \quad (101a)$$

$$H'=\frac{V'}{V_{\text{in}}}=\frac{(IC+IE+KC+KE-GF-GB-LF-LB)+(FH+BH+BG+FL+2BL-JC-JE-IE-KC-2KE)z^{-1}+(EJ+EK-BH-BL)z^{-2}}{D(F+B)+(AC+AE-DF-2BD)z^{-1}+(DB-AE)z^{-2}} \quad (101b)$$

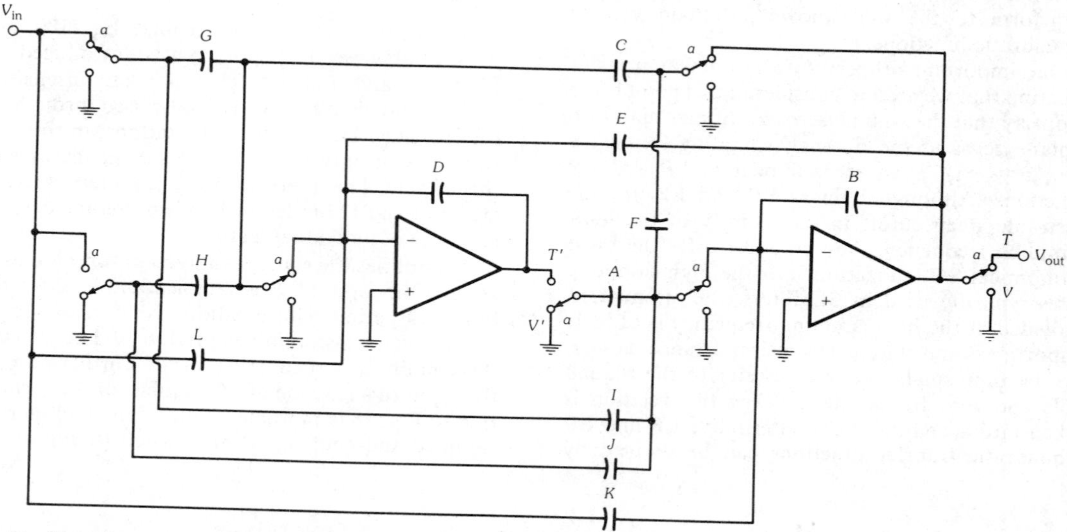

Fig. 38. General active SC biquad topology.

From Fig. 38 and Eqs. (101), it will be recognized that the transfer-function poles are determined by the feedback loop, consisting of capacitors A, B, C, D, E, and F. Further, the transfer-function zeros are determined by the two feed-forward paths, consisting, respectively, of capacitors G, H, L and I, J, K. As expected, both H and H' share the same denominator; however, the numerators are quite different.

The properties of this general biquad circuit are much more easily discerned if certain simplifications are introduced. Thus, it may be readily observed that the general circuit of Fig. 38 is not minimal, with redundancy occurring in both the feedback and feed-forward paths. In the feedback paths, capacitor E and switched capacitor F provide two means for damping the transfer-function poles. Thus, in practice either E or F is used, but not both. Accordingly, it is useful to define an "E-circuit" in which $E\neq 0$ and $F=0$ and an "F-circuit" in which $E=0$ and $F\neq 0$.

Also, as noted before, the two zero-forming feed-forward paths consist of six elements. At most, four of these six elements, that is two elements for each path, are required to realize arbitrary zero locations. Consequently, during the initial design of a biquad, it is convenient to assign $K=L=0$. This degree of freedom is readily restored to the design by using the element equivalences* shown in Fig. 39. That is, after an initial design is completed, these equivalences are employed to modify the circuit until an acceptable design is obtained. The z-domain validity of the equivalences relies on terminals 1 and 2 being connected to a voltage source (independent voltage source or op-amp output) and virtual ground, respectively.*

Since the transfer function of switched-capacitor filters depends only on capacitor ratios, one capacitor in each stage may be arbitrarily chosen. It is convenient to exercise these degrees of freedom by setting $B=D=1$. In addition, it may be shown that one can initially set $A=1$. The effect of this choice is to relinquish temporarily control of the gain constant associated with the transfer function to the "secondary" output. Once the initial design is completed, the stages may be properly rescaled to restore full generality to the biquad.

In view of the above, it is convenient to set $K=L=0$ and $A=B=D=1$ in Eqs. (101) and further to consider the E-circuit and the F-circuit separately. This results in the following useful equations:

---

* References 1 and 12.

$$H_E = -\frac{I+(G-I-J)z^{-1}+(J-H)z^{-2}}{1+(C+E-2)z^{-1}+(1-E)z^{-2}} \tag{102a}$$

$$H'_E = \frac{(IC+IE-G)+(H+G-JC-JE-IE)z^{-1}+(EJ-H)z^{-2}}{1+(C+E-2)z^{-1}+(1-E)z^{-2}} \tag{102b}$$

$$H_F = -\frac{I+(G-I-J)z^{-1}+(J-H)z^{-2}}{(F+1)+(C-F-2)z^{-1}+z^{-2}} \tag{103a}$$

$$H'_F = -\frac{(GF+G-IC)+(JC-FH-H-G)z^{-1}+Hz^{-2}}{(F+1)+(C-F-2)z^{-1}+z^{-2}} \tag{103b}$$

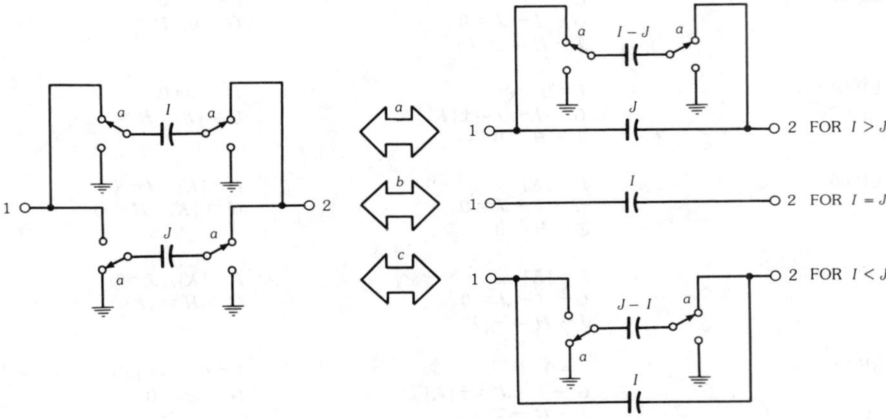

Fig. 39. SC element equivalences.

The "hats" are placed on the F-circuit elements in order to distinguish them from the E-circuit elements.

The synthesis equations for the biquad can be readily derived from Eqs. (100), (102), and (103). To avoid repetition, design equations will be given only for the more often used H_E and H_F functions. A complete set of design equations can be found in reference 12.

For pole placement, the synthesis equations for E, C and F, C in terms of the z-domain transfer-function coefficients α and β can be stated as follows:

For the "E-circuit"

$$E = 1 - \beta \tag{104a}$$

$$C = 1 + \alpha + \beta \tag{104b}$$

For the "F-circuit"

$$\hat{F} = (1 - \beta)/\beta \tag{105a}$$

$$\hat{C} = (1 + \alpha + \beta)/\beta \tag{105b}$$

The synthesis equations for zero placement can be derived in terms of the z-domain transfer-function coefficients γ, ϵ, and δ by comparing Eqs. (102) and (103) with the generic numerator forms in Table 3. Since in Eqs. (102) and (103) those numerator coefficients are determined by four capacitors, the solution is nonunique. Hence, in Table 4 a complete set of design equations is given for each generic case. For each of the cases a "simple" solution is also offered.

The synthesis equations given in the previous paragraphs result in unscaled capacitor values. To complete the synthesis, some scaling is required. The first order of business is to adjust the voltage level at the "secondary" output. If this voltage is too high, overloads will result; if it is too low, unnecessary noise penalties may be taken.

Although the voltage levels may be obtained by using analysis techniques,* the simplest procedure is to simulate the unscaled circuit on an analysis program. This also serves as a confirmation of the correctness of the design.

To adjust the voltage level V', i.e., the flat gain of H', without affecting H, only capacitors A and D

* Reference 1.

TABLE 4. ZERO PLACEMENT FORMULAS FOR T_E AND T_F

Filter Type	Design Equations	Simple Solution
LP 20	$I=\|K\|$ $G-I-J=2\|K\|$ $J-H=\|K\|$	$I=J=\|K\|$ $G=4\|K\|,\ H=0$
LP 11	$I=0$ $G-I-J=\pm\|K\|$ $J-H=\pm\|K\|$	$I=0,\ J=\|K\|$ $G=2\|K\|,\ H=0$
LP 10	$I=\|K\|$ $G-I-J=\|K\|$ $J-H=0$	$I=\|K\|,\ J=0$ $G=2\|K\|,\ H=0$
LP 02	$I=0$ $G-I-J=0$ $J-H=\pm\|K\|$	$I=J=0$ $G=0,\ H=\|K\|$
LP 01	$I=0$ $G-I-J=\pm\|K\|$ $J-H=0$	$I=J=0$ $G=\|K\|,\ H=0$
LP 00	$I=\|K\|$ $G-I-J=0$ $J-H=0$	$I=\|K\|,\ J=0$ $G=\|K\|,\ H=0$
BP 10	$I=\|K\|$ $G-I-J=0$ $J-H=-\|K\|$	$I=\|K\|,\ J=0$ $G=H=\|K\|$
BP 01	$I=0$ $G-I-J=\pm\|K\|$ $J-H=\mp\|K\|$	$I=0,\ J=\|K\|$ $G=H=0$
BP 00	$I=\|K\|$ $G-I-J=-\|K\|$ $J-H=0$	$I=\|K\|,\ J=0$ $G=H=0$
HP	$I=\|K\|$ $G-I-J=-2\|K\|$ $J-H=\|K\|$	$I=J=\|K\|$ $G=H=0$
HPN and LPN	$I=\|K\|$ $G-I-J=\|K\|\epsilon$ $J-H=\|K\|$	$I=J=\|K\|$ $G=\|K\|\{2+\epsilon\},\ H=0$
AP ($\beta>0$)	$I=\|K\|\beta$ $G-I-J=\|K\|\alpha$ $J-H=\|K\|$	$I=\|K\|\beta,\ J=\|K\|$ $G=\|K\|(1+\beta+\alpha)=\|K\|C$ $H=0$
General ($\gamma>0$)	$I=\gamma$ $G-I-J=\epsilon$ $J-H=\delta$	$I=\gamma$ $J=\delta+x$ $G=\gamma+\delta+\epsilon+x$ $H=x\geq 0$

Note: $\hat{G}=G(1+\hat{F})$, $\hat{H}=H(1+\hat{F})$, $\hat{I}=I(1+\hat{F})$, and $\hat{J}=J(1+\hat{F})$.

need be scaled. More precisely, if it is desired to modify the gain constant associated with V' according to

$$H' \rightarrow \mu H' \tag{106}$$

then it is only necessary to scale A and D as

$$(A, D) \rightarrow (1/\mu)A, (1/\mu)D \tag{107}$$

The gain constant associated with H remains invariant under this scaling. The correctness of this procedure follows directly from signal-flow graph concepts.

In a similar fashion, it can be shown that if the flat gain associated with V is to be modified, i.e.,

$$H \rightarrow vH \tag{108}$$

the following capacitors must be scaled:

$$(B,\ C,\ E,\ F) \rightarrow (1/v)B,\ (1/v)C,\ (1/v)\ E,\ (1/v)F \tag{109}$$

Once satisfactory gain levels have been obtained at both outputs, it is convenient to scale the admittances associated with each stage so that the minimum capacitance value in the circuit becomes unity. This makes it easier to observe the maximum capacitance ratios required to realize a given circuit and also serves to "standardize" different designs so that the total capacitance required can be readily observed. The two groups of capacitors that are to be scaled together are listed below:

Group 1: $(C,\ D,\ E,\ G,\ H,\ L)$

Group 2: $(A,\ B,\ F,\ I,\ J,\ K)$

Note that capacitors in each group are distinguished by the fact that they are all incident on the same input node of one of the operational amplifiers.

This completes the design process for synthesizing practical SC-biquad networks. In the next section, a detailed example is given to demonstrate each step of the design.

Low-Pass Notch Example

The transfer function to be realized will be based on the s-domain transfer function

$$H(s)=\frac{0.891975s^2+(1.140926\times 10^8)}{s^2+356.047s+(1.140926\times 10^8)} \tag{110}$$

This transfer function provides a notch frequency of $f_z = 1800$ Hz, a peak corresponding to a quality factor $Q_p = 30$ at $f_p = 1700$ Hz, and 0 dB dc gain. The assigned sampling frequency is 128 kHz; i.e., $\tau = 7.8125\ \mu$s.

The z-domain transfer function is conveniently obtained via the bilinear transformation shown in Eq. (99). Because the band-edge frequency of 1700 Hz is much less than the sampling rate, it is not necessary to prewarp* the $H(s)$ given in Eq. (110). Applying the bilinear transformation to Eq. (110) yields, after some algebraic manipulations:

$$H(z)=0.89093\,\frac{1-1.99220z^{-1}+z^{-2}}{1-1.99029z^{-1}+0.99723z^{-2}} \tag{111}$$

Note that in obtaining this transfer function a high degree of numerical precision is required. However, this does not result in high sensitivities, since the capacitor ratios define only the departures from -2 and $+1$ in the above terms.

Only the H_E and H_F realizations of the above circuit will be given here, as these circuits are more economical in the number of capacitors required for realization. The synthesis itself is straightforward. Capacitors C, E or $\hat{C}$, $\hat{F}$ are determined from Eqs. (104) or (105), respectively, and capacitors G, H, I, J or $\hat{G}$, $\hat{H}$, $\hat{I}$, $\hat{J}$ are obtained from the "simple solution" entry in Table 4. Finally, of course, A, B, D or $\hat{A}$, $\hat{B}$, $\hat{D}$ are set equal to unity. The resulting unscaled capacitor values are given in the appropriate columns of Table 5. Note that in this table the hats are omitted from the F-circuit capacitors for notational convenience. Also note that since $I = J$ these two switched capacitors are replaced by the unswitched capacitor K ($K = I = J$) in accordance with Fig. 39.

At this point, the unscaled E- and F-circuits were computer simulated. The results confirmed that the H_E and H_F were both correct. In particular, the maximum gain in both these realizations was approximately 10.56 dB. However, the maximum gains for H'_E and H'_F were very low. It was decided to increase these gains also to a maximum of 10.56 dB. In this way, the first stage is no more susceptible to overloads than the second stage. Specifically, it was found that

* See Chapter 28.

TABLE 5. LOW-PASS NOTCH REALIZATION

Capacitor (pF)	*E*-Circuit Initial	*E*-Circuit Dynamic Range Adjusted	*E*-Circuit Final	*F*-Circuit Initial	*F*-Circuit Dynamic Range Adjusted	*F*-Circuit Final
A	1.0000	0.08308	1.0000	1.0000	0.08395	30.1895
B	1.0000	1.0000	12.0365	1.0000	1.0000	359.629
C	0.00694	0.00694	2.5035	0.00696	0.00696	1.0000
D	1.0000	0.08308	29.9613	1.0000	0.08395	12.0591
E	0.00277	0.00277	1.0000	—	—	—
F	—	—	—	0.00278	0.00278	1.0000
G	0.00694	0.00694	2.5035	0.00696	0.00696	1.0000
H	—	—	—	—	—	—
I	—	—	—	—	—	—
J	—	—	—	—	—	—
$K(I=J)$	0.89093	0.89093	10.7238	0.89340	0.89340	321.293
ΣC (pF)	—	—	59.7	—	—	726.1

$H'_{E\,MAX} \approx -11.05$ dB, $H'_{F\,MAX} \approx -10.96$ dB

Therefore, in accordance with Eq. (106)

$$\mu = 12.0365, \hat{\mu} = 11.9124$$

Using these factors to rescale A, D, and $\hat{A}$, $\hat{D}$, respectively, as given in Eq. (107), yields the "dynamic range adjusted" capacitor values shown in Table 5. Finally, the capacitors associated with each operational-amplifier stage are separately rescaled so that the minimum capacitance value becomes 1 pF. These "final" values are also shown in Table 5.

In comparing the "final" realizations, we note that the F-circuit requires roughly 12 times the total capacitance of the E-circuit, in spite of the fact that the initial values were almost identical. Thus, alternative designs must be carried to completion before they can be meaningfully compared. It should be noted that other practical examples exist in which the F-circuit designs are dramatically more efficient than the corresponding E-circuit designs. The sensitivities for both designs are found to be equivalent.

REFERENCES

1. Ghausi, M. S., and Laker, K. R. *Modern Filter Design: Active RC and Switched Capacitor.* Englewood Cliffs, N.J.: Prentice Hall, Inc., 1981.
2. Sedra, A. S., and Brackett, P. O. *Filter Theory and Design: Active and Passive.* Portland, Ore.: Matrix Publishers, Inc., 1978.
3. Lindquist, C. S. *Active Network Design with Signal Filtering Applications.* Long Beach, Calif.: Steward & Sons, 1977.
4. Van Valkenburg, M. E. *Analog Filter Design.* New York: Holt, Rinehart and Winston, 1982.
5. Temes, G. C., and LaPatra, J. W. *Introduction to Circuit Synthesis and Design.* New York: McGraw-Hill Book Co., 1977.
6. Schaumann, R., Soderstrand, M. S., and Laker, K. R., eds. *Modern Active Filter Design*, IEEE Press Selected Reprint Series, 1981.
7. Sedra, A. S., Zharab, M. A., and Martin, K. "Optimum Configuration of Single-Amplifier Biquadratic Filters." *IEEE Transactions on Circuits and Systems*, Vol. CAS-27, 1980, pp. 1155–1163, 1980.
8. Chiou, C.-F., and Schaumann, R. "Performance of GIC-Derived Active RC Biquads with Variable Gain." *Proc. IEE*, Vol. 128, Part G, Electronic Circuits and Systems, No. 1, pp. 46–52, 1981.
9. Snelgrove, W. M., and Sedra, A. S. "Optimization of Dynamic Range in Cascade Active Filters." *Proc. IEEE Int. Symp. on Circuits and Systems*, pp. 151–155, 1978.
10. Chiou, C.-F., and Schaumann, R. "Comparison of Dynamic Range Properties of High-Order Active Bandpass Filters." *Proc. IEE*, Vol. 127, Part G, Electronic Circuits and Systems, No. 3, pp. 101–108, 1980.
11. Gray, P. R., Hodges, D. A., and Broderson, R. W., eds. *Analog MOS Integrated Circuits*, IEEE Press Selected Reprint Series, 1980.
12. Fleischer, P. E., and Laker, K. R. "A Family of Active Switched-Capacitor Biquad Building Blocks." *The Bell System Technical Journal*, Vol, 58, No. 10, December 1979. (Reprinted in Reference 6.)
13. Martin, K., and Sedra, A. S. "Effects of Op Amp Finite Gain and Bandwidth on the Performance of Switched Capacitor Filters." *IEEE Transactions on Circuits and Systems*, Vol. CAS28, August 1981, pp. 822–829.

11 Attenuators

Revised by
Bruno O. Weinschel

DEFINITIONS

An attenuator is a network that reduces the input power by a predetermined ratio. The ratio of input power to output power is expressed in logarithmic terms such as decibels (dB).

$$\text{Attenuation in dB} = 10 \log_{10} P_{in}/P_{out}$$
$$\equiv 20 \log_{10} E_{in}/E_{out}$$

NOTE: $Z_{source} = Z_{load} = Z_{attenuator}$
All resistive, matched

Examples:

1) $P_{in}/P_{out} = 13.18 = 10 \times 1.318$
$10 \log_{10} P_{in}/P_{out}$ dB $= 10 (\log_{10} 10 + \log_{10} 1.318)$ dB
$= 10(1 + 0.1199)$ dB $\simeq 11.2$ dB
2) $E_{in}/E_{out} = 3.630$
$20 \log_{10} E_{in}/E_{out}$ dB $= 20 \log_{10} 3.630$ dB
$= 20 \times 0.560$ dB $= 11.2$ dB

To convert attenuation in decibels into power or voltage ratio:

$$P_{in}/P_{out} = \log_{10}^{-1} \text{dB}/10 = 10^{\text{dB}/10}$$
$$E_{in}/E_{out} = \log_{10}^{-1} \text{dB}/20 = 10^{\text{dB}/20}$$

Examples:

3) 11.2 dB $P_{in}/P_{out} = 10^{11.2/10} = 10^{1.12} = 13.18$
4) 11.2 dB $E_{in}/E_{out} = 10^{11.2/20} = 10^{0.56} = 3.630$

Table 1 lists a few decibel values together with the corresponding power and voltage ratios.

TABLE 1. DECIBELS VERSUS POWER AND VOLTAGE RATIO

dB	P_{in}/P_{out}	E_{in}/E_{out}
10	10	—
20	10^2	10
30	10^3	—
40	10^4	10^2
50	10^5	—
60	10^6	10^3
70	10^7	—
80	10^8	10^4
90	10^9	—
100	10^{10}	10^5
110	10^{11}	—
120	10^{12}	10^6

The power ratio used in the *characteristic insertion loss* relation (page 82, IEEE Standard 100—1972) is shown in Fig. 1. The ratio specified is equal to the characteristic insertion loss since source and load are reflectionless. Usually, one is interested in resistive attenuators of nominally equal impedance as the source and load impedance. Resistive devices are most common. Attenuators include absorptive and reflective devices having at least two ports.

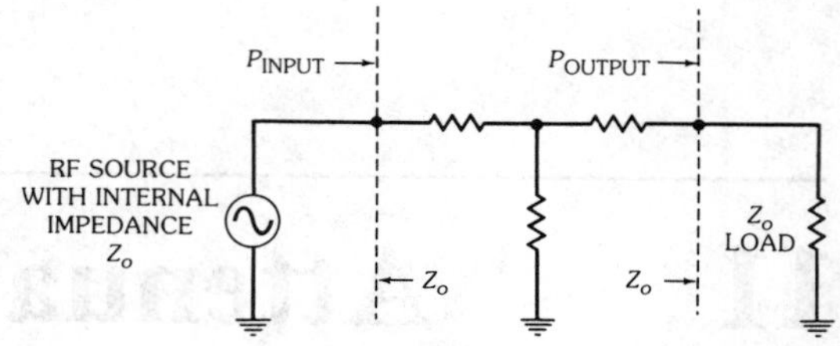

Fig. 1. Definition of characteristic insertion loss.

There are fixed attenuators, continuously variable attenuators, and step attenuators. This chapter treats principally Z_o *matched, two-port* attenuators, for use in a Z_o *system*. They are reversible except for a high-power attenuator designed for unilateral power flow. Minimum-loss L-pads for matching between two different impedances are also treated.

TYPICAL DESIGNS OF RESISTIVE ATTENUATORS

Simple wirewound resistors are used to upper audio frequencies. Nonreactive wirewound resistors (mica card, Ayrton-Perry winding, woven resistors) can be used to higher frequencies for low characteristic impedance values of, for example, 50 ohms. Preferred US impedance values are 600 ohms for audio and 50 or 75 ohms for coaxial video, rf, and microwave transmissions. For coaxial applications to over 26.5 GHz, thin-film resistors on low-loss insulators are used as attenuating elements. Up to a few gigahertz, lumped rod and disc resistors can be used as series and shunt elements in coaxial lines. The rods should be shorter than $\lambda/8$ at the maximum frequency (Figs. 2 and 3). For higher frequen-

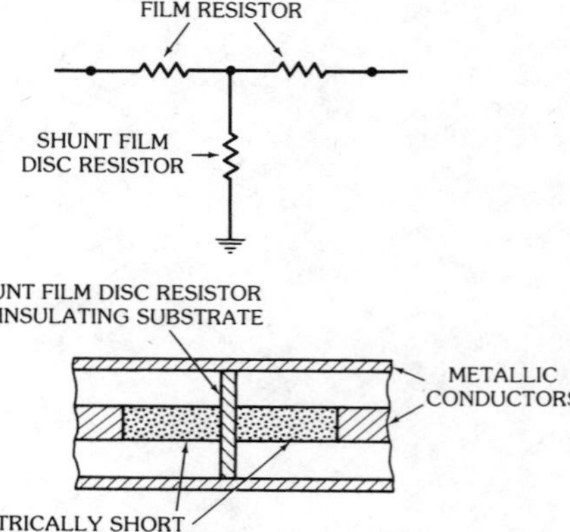

Fig. 2. Coaxial tee section.

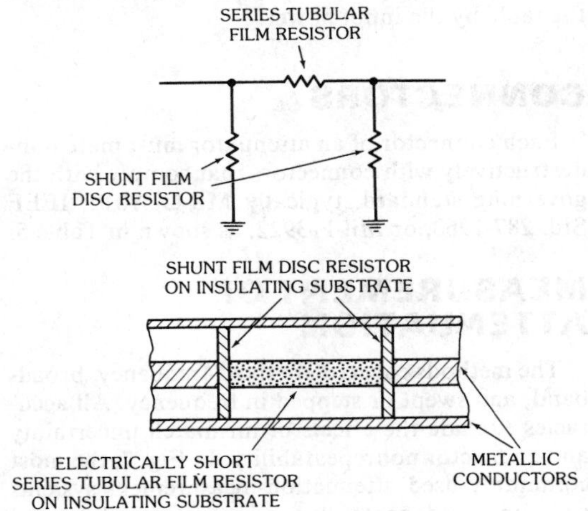

Fig. 3. Coaxial pi section.

cies, distributed resistive films on suspended substrates are used (Fig. 4).

Resistance Networks for Attenuators

Four types of symmetrical pads are shown in Fig. 5. The formulas for the resistance values in ohms for these pads when $R_o = 1$ ohm are given below. If R_o has a value other than 1 ohm, multiply each of the resistance values (a, b, c, $1/a$, $1/b$, and $1/c$) by R_o.

$$a = (10^{dB/20} - 1)/(10^{dB/20} + 1)$$
$$b = 2.10^{dB/20}/(10^{dB/10} - 1)$$
$$c = 10^{dB/20} - 1$$

An unsymmetrical matching pad is shown in Fig. 6. The formulas for this pad are:

$$j = R_1 - kR_2/(k + R_2)$$
$$k = [R_1 R_2^2/(R_1 - R_2)]^{1/2}, \text{ where } R_1 > R_2$$

$$\text{Minimum loss in dB} = 20 \log_{10}\{[(R_1 - R_2) / R_2]^{1/2} + (R_1/R_2)^{1/2}\}$$

Typical values for the pads in Fig. 5 are listed in Table 2. Typical values for the pad in Fig. 6 are listed in Table 3.

Balanced pads are used in balanced transmission lines. For voice, Z_o is typically 600 ohms; for vhf tv, Z_o is typically 300 ohms. Bridged tee pads are common in low-frequency step attenuators. Lumped-element pi pads are short and easily matched up to about 1.5 GHz, with 75- or 50-ohm impedance. Coaxial tee pads of 50-ohm impedance can work above 10 GHz if the series elements are short in comparison with the wavelength. For low frequency sensitivity, thin resistive films (Nichrome, cracked carbon, tantalum nitride) on an insulating

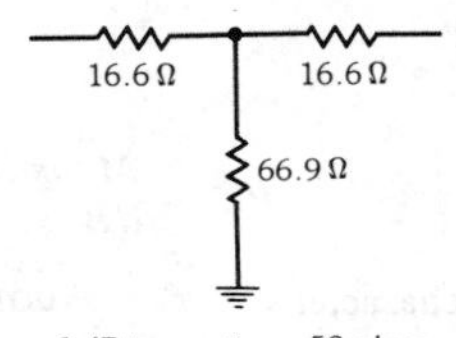

6 dB tee section—50 ohms.

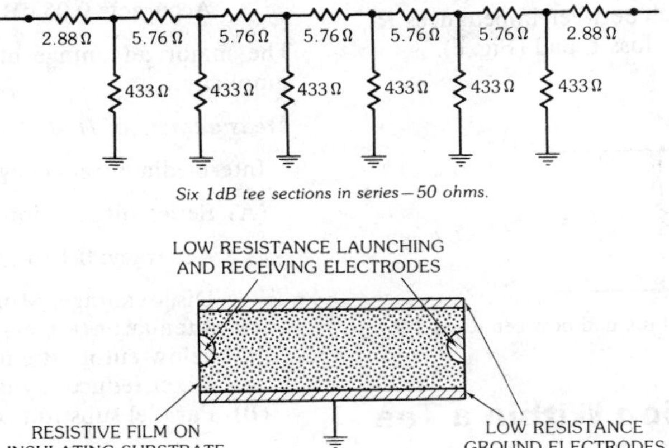

Distributed series and shunt resistors.

Fig. 4. Distributed series and shunt resistive attenuator element on suspended substrate in coaxial line.

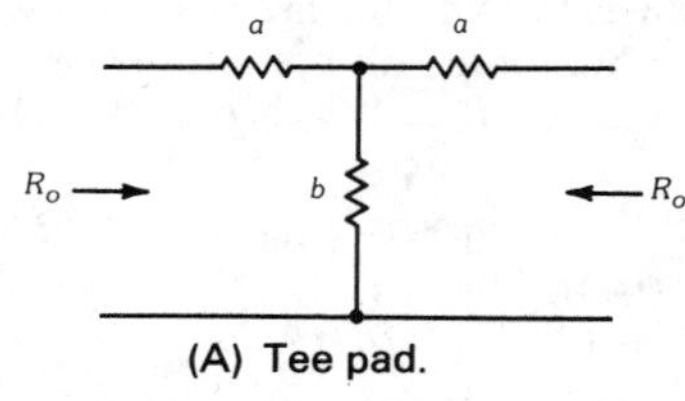

(A) Tee pad.

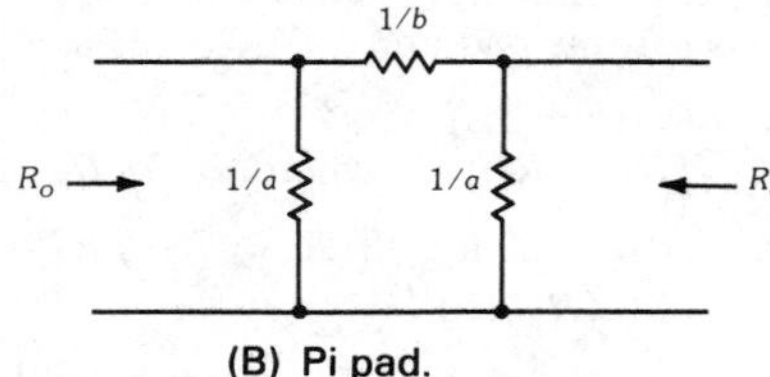

(B) Pi pad.

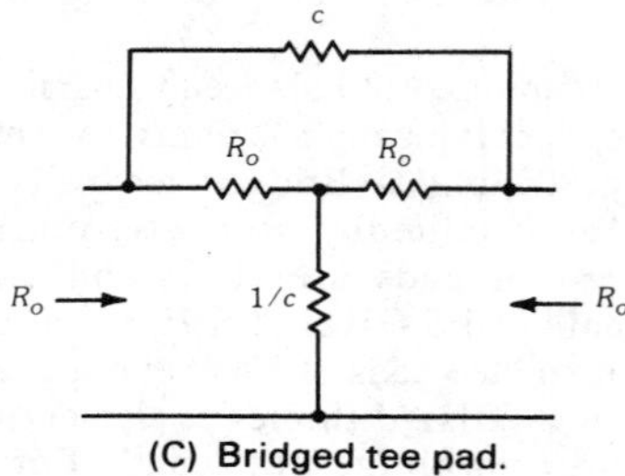

(C) Bridged tee pad.

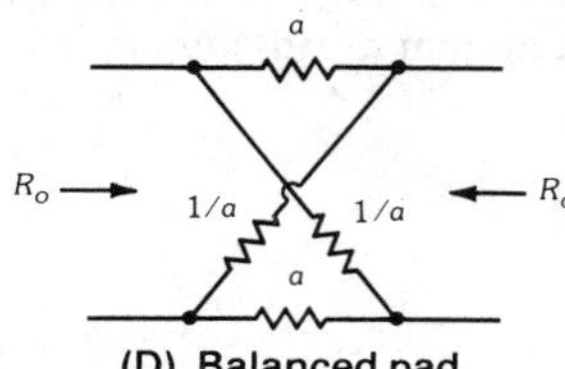

(D) Balanced pad.

Fig. 5. Symmetrical pads for use between equal characteristic impedances.

substrate (mica, ceramic, quartz) are used. Figs. 2, 3, and 4 illustrate coaxial cross sections.

For a broad-band match between impedances R_1 and R_2, use the minimum-loss L pad (Fig. 6).

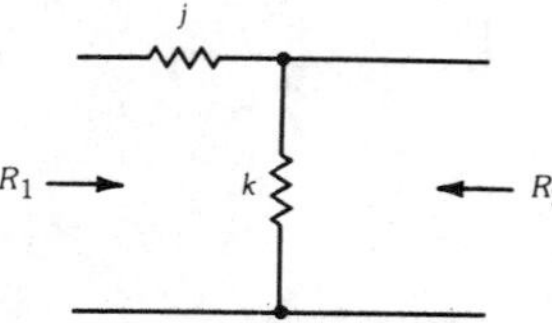

Fig. 6. Unsymmetrical matching pad between R_1 and R_2.

Power Dissipation Within a Tee Pad

Table 4 lists values of power dissipation within a tee pad. The values in the table are for an input of 1 watt; for other input powers, multiply the values in the table by the input power.

CONNECTORS

Each connector of an attenuator must mate nondestructively with connectors that comply with the governing standard, typically Mil-C-39012, IEEE Std. 287-1968, or Mil-F-3922, as shown in Table 5.

MEASUREMENT OF ATTENUATION

The methods covered are fixed-frequency, broadband, and swept or stepped in frequency. All accuracies exclude the effects of mismatch uncertainty and connector nonrepeatability. In Fig. 7, the most commonly used attenuation-measurement systems are compared as to dynamic range and typical accuracy.

Fixed-Frequency

Measurement of Small Attenuation ($<$ 5 dB)

Dual-channel audio substitution (Fig. 8)

Accuracy: 0.01 dB

Advantage: Single radio-frequency source

Attenuation standards: Audio attenuator (resistive)
Ratio transformer (inductive)

Measurement of Moderate Attenuation (1 to 30 dB)

Audio substitution

(A) Single channel (Fig. 9)
Accuracy: 0.1 db

(B) Dual channel (Fig. 8)
Accuracy: 0.05 dB

The major advantage of systems (A) and (B) is simplicity.

Measurement of High Attenuation (30 to $>$ 100 dB)

Intermediate-frequency substitution

(A) Series substitution (Fig. 10)

Accuracy: 0.1 to 1 or 2 dB

Disadvantage: Minimum loss of standard attenuator (typically 30 dB for a waveguide-below-cutoff attenuator) in series with the mixer reduces dynamic range.

(B) Parallel substitution (Fig. 11)

Accuracy: 0.1 to 1 or 2 dB

Advantage: Minimum loss of standard intermediate-frequency attenuator does not

TABLE 2. RESISTANCE VALUES FOR ATTENUATOR PADS WHEN R_o=1 OHM[1]

	Tee Pad		Pi Pad		Bridged Tee Pad		Balanced Pad	
dB	a	b	$1/b$	$1/a$	c	$1/c$	a	$1/a$
0.1	0.0057567	86.853	0.011514	173.71	0.011580	86.356	0.0057567	173.71
0.2	0.011513	43.424	0.023029	86.859	0.023294	42.930	0.011513	86.859
0.3	0.017268	28.947	0.034546	57.910	0.035143	28.455	0.017268	57.910
0.4	0.023022	21.707	0.046068	43.438	0.047128	21.219	0.023022	43.438
0.5	0.028775	17.362	0.057597	34.753	0.059254	16.877	0.028775	34.753
0.6	0.034525	14.465	0.069132	28.965	0.071519	13.982	0.034525	28.965
0.7	0.040274	12.395	0.080678	24.830	0.083927	11.915	0.040274	24.830
0.8	0.046019	10.842	0.092234	21.730	0.096478	10.365	0.046019	21.730
0.9	0.051762	9.6337	0.10380	19.319	0.10918	9.1596	0.051762	19.319
1.0	0.057501	8.6668	0.11538	17.391	0.12202	8.1954	0.057501	17.391
2.0	0.11462	4.3048	0.23230	8.7242	0.25893	3.8621	0.11462	8.7242
3.0	0.17100	2.8385	0.35230	5.8481	0.41254	2.4240	0.17100	5.8481
4.0	0.22627	2.0966	0.47697	4.4194	0.58489	1.7097	0.22627	4.4194
5.0	0.28013	1.6448	0.60797	3.5698	0.77828	1.2849	0.28013	3.5698
6.0	0.33228	1.3386	0.74704	3.0095	0.99526	1.0048	0.33228	3.0095
7.0	0.38248	1.1160	0.89604	2.6145	1.2387	0.80727	0.38248	2.6145
8.0	0.43051	0.94617	1.0569	2.3229	1.5119	0.66143	0.43051	2.3229
9.0	0.47622	0.81183	1.2318	2.0999	1.8184	0.54994	0.47622	2.0999
10.0	0.51949	70 273*	1.4230	1.9250	2.1623	46 248*	0.51949	1.9250
20.0	0.81818	20 202*	4.9500	1.2222	9.0000	11 111*	0.81818	1.2222
30.0	0.93869	6330.9*	15.796	1.0653	30.623	3265.5*	0.93869	1.0653
40.0	0.980198	2000.2*	49.995	1.0202	99.000	1010.1*	0.980198	1.0202
50.0	0.99370	632.46*	158.11	1.0063	315.23	317.23*	0.99370	1.0063
60.0	0.99800	200.00*	500.00	1.0020	999.00	100.10*	0.99800	1.0020
70.0	0.99937	63.246*	1581.1	1.0006	3161.3	31.633*	0.99937	1.0006
80.0	0.99980	20.000*	5000.0	1.0002	9999.0	10.001*	0.99980	1.0002
90.0	0.99994	6.3246*	15 811	1.0001	31 622	3.1633*	0.99994	1.0001
100.0	1.0000	2.0000*	50 000	1.0000	99 999	1.0000*	1.0000	1.0000

* These values have been multiplied by 10^5.

NOTES:

1. If $R_o \neq 1$ ohm, multiply all values by R_o.
2. For other decibel values, use formulas in text.

affect system sensitivity; gain stability of intermediate-frequency amplifier does not influence accuracy.

Disadvantage: More complex

All intermediate-frequency systems have an inherent limitation due to the range of power conversion linearity of the mixer. Beyond 90 dB, corrections may have to be made for linearity deviation. Partial radio-frequency substitution is often used to extend range, but mismatch and leakage errors are prevalent.

TABLE 3. RESISTANCE VALUES AND ATTENUATION FOR MINIMUM-LOSS L PAD*

R_1/R_2	j	k	dB
20.0	19.49	1.026	18.92
16.0	15.49	1.033	17.92
12.0	11.49	1.044	16.63
10.0	9.486	1.054	15.79
8.0	7.484	1.069	14.77
6.0	5.478	1.095	13.42
5.0	4.472	1.118	12.54
4.0	3.469	1.155	11.44
3.0	2.449	1.225	9.96
2.4	1.833	1.310	8.73
2.0	1.414	1.414	7.66
1.6	0.9798	1.633	6.19
1.2	0.4898	2.449	3.77
1.0	0	∞	0

*For R_2=1 ohm and $R_1>R_2$. If $R_2 \neq 1$ ohm, multiply values by R_2. For ratios not in the table, use the formulas in the text.

Examples of use of table:

If R_1=50 ohms and R_2=25 ohms, then R_1/R_2=2.0, and $j=k$ =1.414×25 ohms=35.35 ohms.

If R_1/R_2=1.0, minimum loss=0 dB.

For R_1/R_2=2.0, the insertion loss with the use of j and k for matching is 7.66 dB above that for R_1/R_2=1.0.

Broad-Band

Measurement of Small Attenuation (Less Than 5 dB)

Dual-channel audio substitution (Fig. 8)

Accuracy: 0.01 dB

Measurement of Attenuation Above 1 dB

Single- or dual-channel audio (Figs. 8 and 9)

Accuracy: 0.01 to 0.05 dB

Advantage: Single radio-frequency source

Disadvantage: Limited dynamic range (about 30 dB) without partial radio-frequency substitution

The major difficulty in broad-band techniques is maintaining a Z_o-matched source and terminating impedance. The increased mismatch uncertainty due to impedance variations of source and load with frequency may degrade the above accuracy figures by an order of magnitude. A technique producing an equivalent Z_o-matched source impedance is discussed in Reference 9.

Swept or Stepped in Frequency

Measurement of Attenuation Less than 30 dB

(1) Audio-substitution dual channel (Fig. 8 with addition of level-stabilized swept radio-frequency source)

Accuracy: 0.05 to 0.1 dB

Advantage: Potentially most accurate system

Disadvantage: Impedance variation

(2) Audio ratio technique (Fig. 12)

Accuracy: 0.1 to 0.3 dB

Advantage: Rapid measurement, simplicity

Disadvantage: Curvilinear reference lines which can be straightened through the use of memory circuits

(3) Sampling (Fig. 13)

Accuracy: 0.1 to 0.5 dB dependent on uncertainty of system precalibration

Advantages: (A) Digital technique permits use of logic circuitry for automatic correction of measurement errors

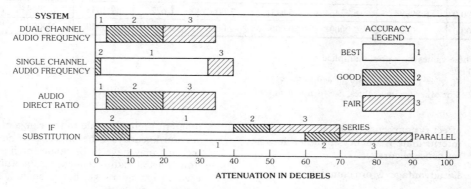

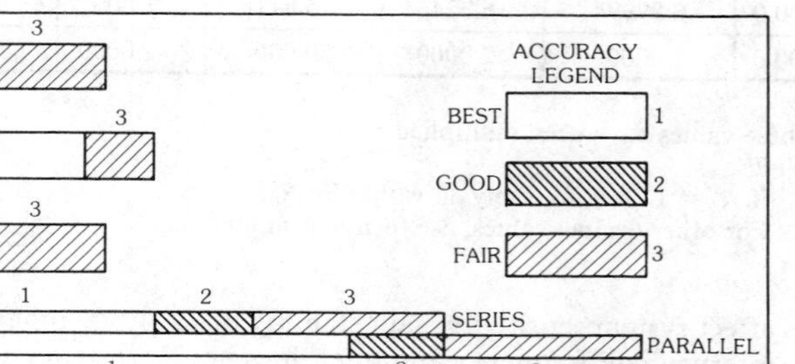

Fig. 7. Comparison of most commonly used attenuation measurement systems.

TABLE 4. POWER DISSIPATION WITHIN TEE PAD*

dB	Watts Input Series Resistor	Watts Shunt Resistor	Watts Output Series Resistor
0.1	0.00576	0.0112	0.005625
0.3	0.0173	0.0334	0.016113
0.5	0.0288	0.0543	0.025643
0.7	0.0403	0.0743	0.034279
0.9	0.0518	0.0933	0.0421
1.0	0.0575	0.1023	0.0456
1.2	0.0690	0.120	0.0523
1.4	0.0804	0.1368	0.0582
1.6	0.0918	0.1525	0.0635
1.8	0.103	0.1672	0.0679
2.0	0.114	0.1808	0.0718
2.2	0.126	0.1953	0.0758
2.4	0.137	0.2075	0.0787
2.6	0.149	0.2205	0.0818
2.8	0.160	0.232	0.0839
3.0	0.170998	0.242114	0.085698
3.2	0.182	0.2515	0.0870
3.4	0.193	0.2605	0.0882
3.6	0.204	0.2695	0.0890
3.8	0.215	0.2775	0.0897
4.0	0.226	0.285	0.0898
5	0.280	0.3145	0.0884
6	0.332	0.332	0.0833
7	0.382	0.341	0.0761
8	0.430	0.343	0.0681
9	0.476218	0.33794	0.0599527
10	0.519	0.328	0.0519
12	0.598	0.3005	0.0377
14	0.667	0.266	0.0266
16	0.726386	0.23036	0.0182460
18	0.776	0.1955	0.0123
20	0.818	0.1635	0.0100
30	0.938	0.0593	0.0010
40	0.980	0.0196	0.0001

*For 1-watt input and matched termination. If input ≠ 1 watt, multiply values by P_{in}.

(B) Extremely rapid measurement technique

Disadvantage: Very complex

Measurement of Attenuation Greater than 30 dB

Sampling (Fig. 13)

Accuracy: 0.5–2 dB to 50 or 60 dB dependent on uncertainty of system precalibration

Advantages: (A) Digital technique permits use of logic circuitry for automatic correction of measurement errors

(B) Extremely rapid measurement technique

Disadvantage: Very complex

The aforementioned measurement techniques are not intended to be all-inclusive. These are representative of current practices only. No mention has been made of specialized power ratio, modulated

TABLE 5. LIST OF POPULAR COAXIAL CONNECTORS

Connector Type	Contact Type	Specifications
N	Female	Mil-C-39012/2
N	Male	Mil-C-39012/1
C	Male	Mil-C-39012/6
C	Female	Mil-C-39012/7
SC	Male	Mil-C-39012/39
SC	Female	Mil-C-39012/36
TNC	Male	Mil-C-39012/26
TNC	Female	Mil-C-39012/27
BNC	Male	Mil-C-39012/16
BNC	Female	Mil-C-39012/17
SMA	Male	Mil-C-39012/55
SMA	Female	Mil-C-39012/57
7-mm GPC	Sexless	IEEE Std. 287-1968
14-mm GPC	Sexless	IEEE Std. 287-1968
Waveguide	Choke or cover flange	Mil-F-3922

References—Mil-C-39012 on Coaxial Connectors; Mil-F-3922 on Waveguide Flanges, General Purpose; IEEE Std. 287-1968 on Precision Coaxial Connectors

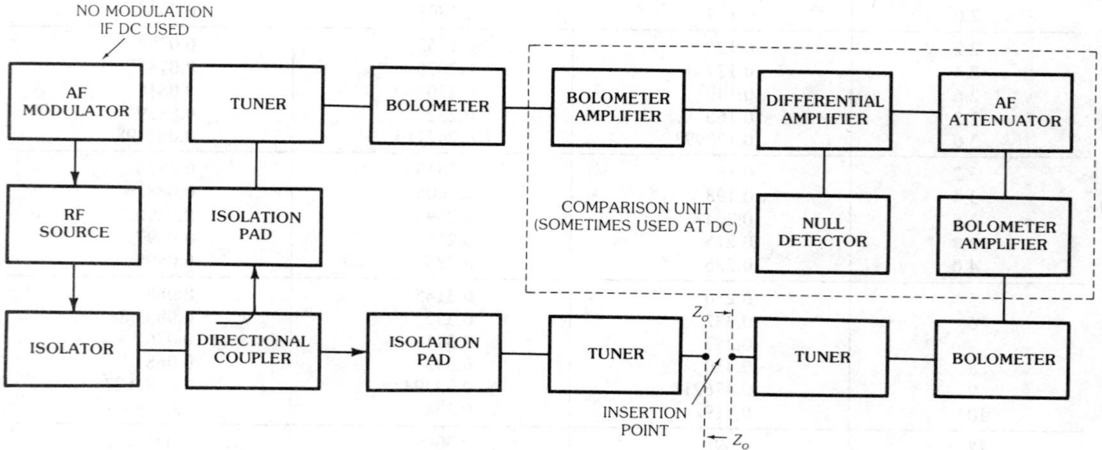

Fig. 8. Dual-channel audio substitution system (Reference 1).

subcarrier, radio-frequency substitution, homodyne, impedance, or self-calibrating methods, or range extension by using a supervised change of incident power level.

REFERENCES ON ATTENUATION MEASUREMENT

1. Sorger, G.U., and Weinschel, B. O., "Comparison of Deviations from Square Law for RF Crystal Diodes and Barretters," *IRE Transactions on Instrumentation*, Vol. I-8, Dec. 1959, pp. 103-111.
2. Weinschel, B. O., "Measurement of Microwave Parameters by the Ratio Method," *Microwave Journal,* Aug. 1969, pp. 69–73.
3. Terman, F. E., "Linear Detection of Heterodyne Signals," *Electronics,* Nov. 1930, pp. 386–387.
4. Harnett, D. E., and Case, N. P., "The Design and Testing of Multi-Range Receivers," *Proceedings of the IRE,* Vol. 23, June 1935, pp. 578–593.
5. Gainsborough, G. F., "A Method of Calibrating Standard Signal Generators and Radio Frequency Attenuators," *Journal of the IEE,* Vol. 94, Pt. III, May 1947, pp. 203–210.
6. Weinschel, B. O., Sorger, G. U., and Hedrich, A. L., "Relative Voltmeter for VHF/UHF Signal Generator Attenuator Calibration," *IRE Transactions on Instrumentation,* Vol. I-8, Mar. 1959, pp. 22–31.
7. Schafer, G. E., and Rumfelt, A. Y., "Mismatch Errors in Cascade-Connected Variable

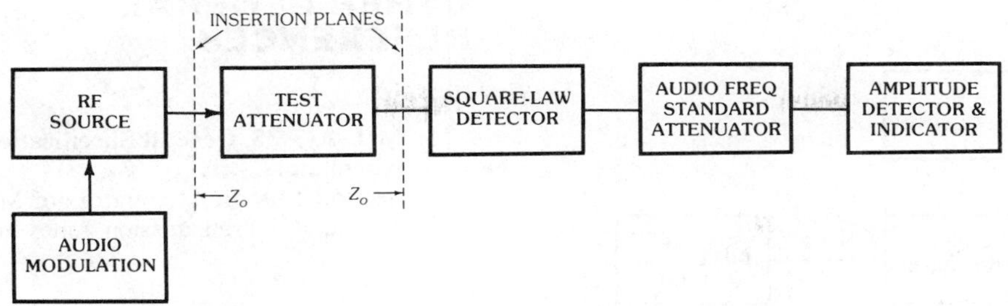

Fig. 9. Single-channel audio substitution system (Reference 1).

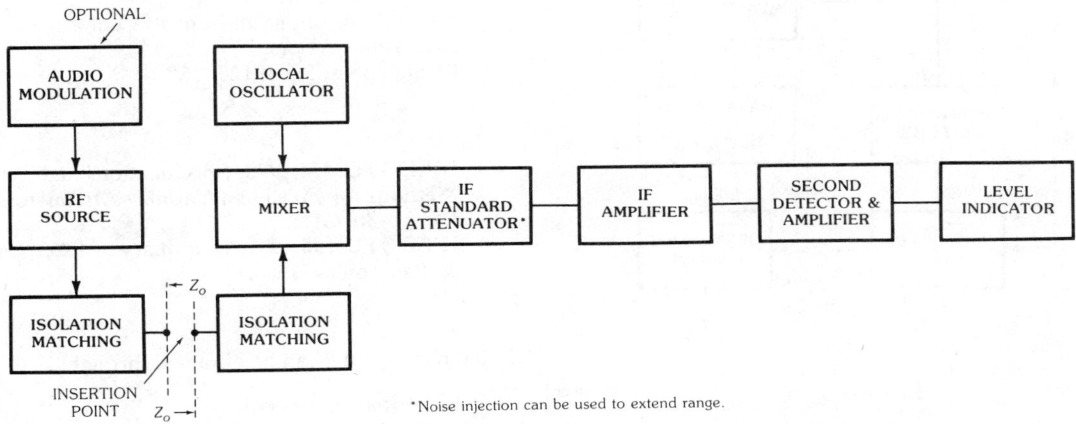

Fig. 10. Series type intermediate-frequency substitution system (References 3 and 4).

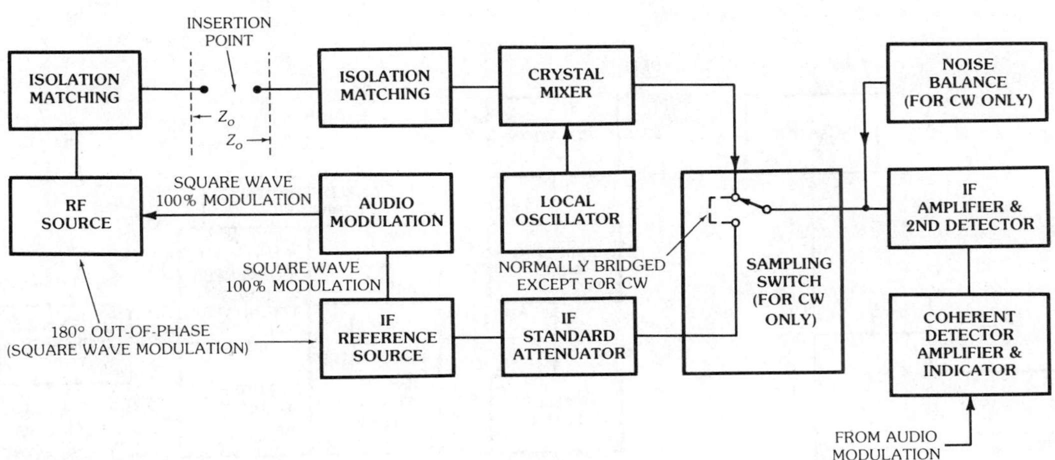

Fig. 11. Parallel if substitution system with noise injection and coherent detection (References 5 and 6).

Attenuators," *IRE Transactions on Microwave Theory and Techniques,* Vol. MTT-7, Oct. 1959, pp. 447–453.

8. Adam, S. F., "A New Precision Automatic Microwave Measurement System," *IEEE Transactions on Instrumentation and Measurement,* Vol. IM-17, Dec. 1968, pp. 308–313.
9. Engen, G. F., "Amplitude Stabilization of a Microwave Signal Source," *IRE Transactions on Microwave Theory and Techniques,* Vol. MTT-6, Apr. 1958, pp. 202–206.

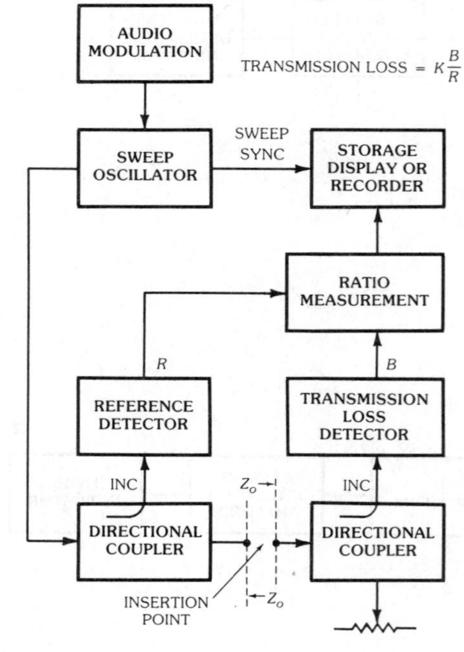

AUDIO MODULATION
TRANSMISSION LOSS = $K\frac{B}{R}$
SWEEP OSCILLATOR
SWEEP SYNC
STORAGE DISPLAY OR RECORDER
RATIO MEASUREMENT
R
B
REFERENCE DETECTOR
TRANSMISSION LOSS DETECTOR
INC
Z_o
INC
DIRECTIONAL COUPLER
DIRECTIONAL COUPLER
INSERTION POINT
Z_o

Fig. 12. Diagram of broad-band direct-ratio system (References 2 and 9).

OTHER GENERAL REFERENCES

1. MIL-A-3933, General Specification for Fixed Attenuators.
2. Section 14A on Attenuators of MIL-HDBK-216 on RF Transmission Lines and Fittings.

References 1 and 2 can be obtained through:

Commanding Officer
Naval Publications and Forms Center
5801 Tabor Avenue
Philadelphia, PA 19120

3. IEEE STD-474-1973, Specifications and Test Methods for Fixed and Variable Attenuators, DC to 40 GHz.
4. IEEE STD-100-1977, Dictionary of Electrical & Electronics Terms

References 3 and 4 can be obtained through:

IEEE Service Center
445 Hoes Lane
Piscataway, NJ 08854

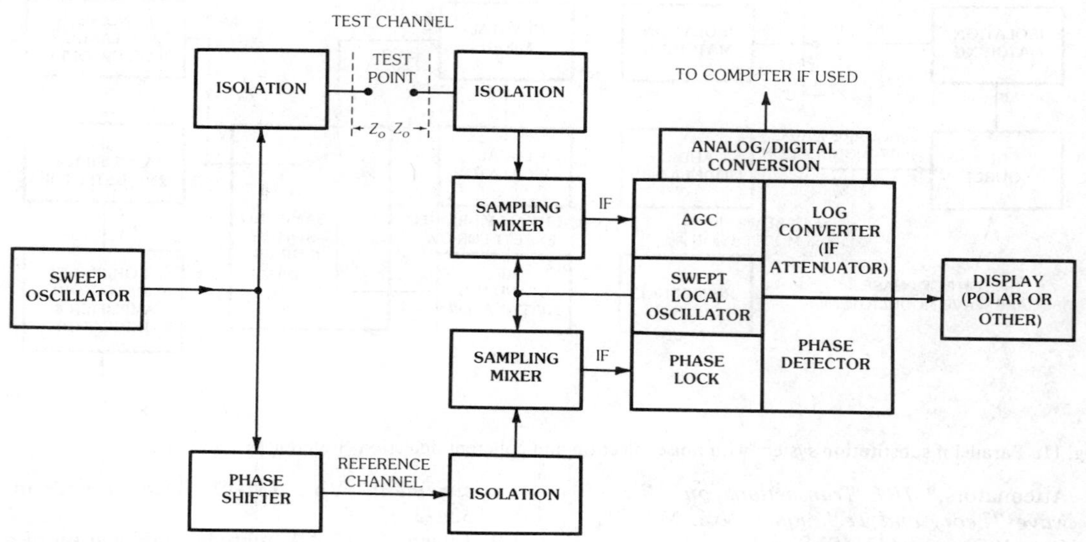

Fig. 13. Sampling conversion method of attenuation measurement (Reference 8).

12 Measurements and Analysis

Arthur Fong, Robert Coackley, John J. Dupre,
Michael C. Fischer, Ronald E. Pratt, and Douglas K. Rytting

The aim of this chapter is to provide a condensed description of methods of electronic measurements. Science and technology would indeed be vague without the ability to measure. Lord Kelvin cautioned that "Knowledge not expressible in numbers is of a meager and unsatisfactory kind"; he was identifying an essential aspect of scientific knowledge.

In the past, measurements of voltage, frequency, impedance, or power were made by using bridges or substitution methods with prototype standards derived from primary standards. Now, using digital techniques, one can create signals with precise voltage, frequency, and phase and can measure signals with the same precision. Many measurements are now made by stimulating the network or system with a precise signal, measuring the response, and using a microprocessor to compute the impedance, gain, phase, or other required results.

Another method for determination of measured quantities uses the technique of ratio measurements. For example, the complex ratio of E/I can be determined very precisely for a given frequency; then a microprocessor instantaneously calculates the impedance or response. In addition, the sequential steps of measurements, computation, and final display can also be controlled by the same microprocessor. Touch and read instruments were created by these technologies.

These measurements topics will be covered:

Impedance
Networks
Signals
Time and Frequency
Power
Microwave Links
IEEE 488 Data Bus
EMI/EMC

It is assumed that the reader is already familiar with such fundamental topics as voltage, current, gain, phase, distortion, etc. Specialized measurements for linear systems, digital signals, time domains, fields, magnetics, etc., are left for specialized publications. (Some chapters in this book covering specialized topics do contain information on the associated measurements.)

IMPEDANCE BRIDGES

In the diagrams of bridges in this section, the source (generator) and the detector (headphones) may be interchanged as dictated by the location of grounds. For all but the lowest frequencies, a shielded transformer is required at either the input or output (but not usually both) terminals of the bridge. The detector is chosen according to the frequency of the source. When insensitivity of the ear makes direct use of headphones impractical, a simple radio receiver or its equivalent is essential. Some selectivity is desirable to discriminate against harmonics, for the bridge is often frequency sensitive. The source may be modulated to obtain an audible signal, but greater sensitivity and discrimination against interference are obtained by the use of a continuous-wave source and a heterodyne detector. An oscilloscope is sometimes preferred for observing nulls. In this case, it is convenient to have an audible output signal available for the preliminary setup and for locating trouble, since much can be deduced from the quality of the audible signal that would not be apparent from observation of amplitude only.

Fundamental Alternating-Current or Wheatstone Bridge

Refer to Fig. 1. The balance condition is $Z_x = Z_s Z_a / Z_b$. Maximum sensitivity exists when Z_d is the conjugate of the bridge output impedance and Z_g is the conjugate of its input impedance. Greatest sensitivity exists when the bridge arms are equal; for example, for resistive arms

$$Z_d = Z_a = Z_b = Z_x = Z_s = Z_g$$

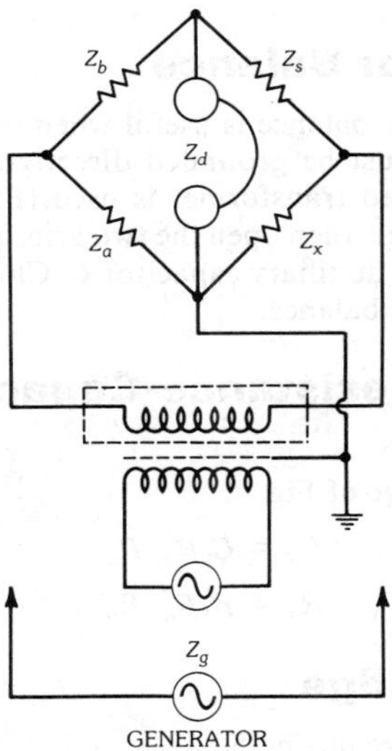

Fig. 1. Fundamental ac bridge.

Wagner Ground Connection

None of the bridge elements (Fig. 2) is grounded directly. First balance the bridge with the switch to B. Throw the switch to G, and rebalance by means of R and C. Recheck the bridge balance and repeat as required. The capacitor balance C is necessary only when the frequency is above the audio range. The transformer may have only a single shield as shown, with the capacitance of the secondary to the shield kept to a minimum.

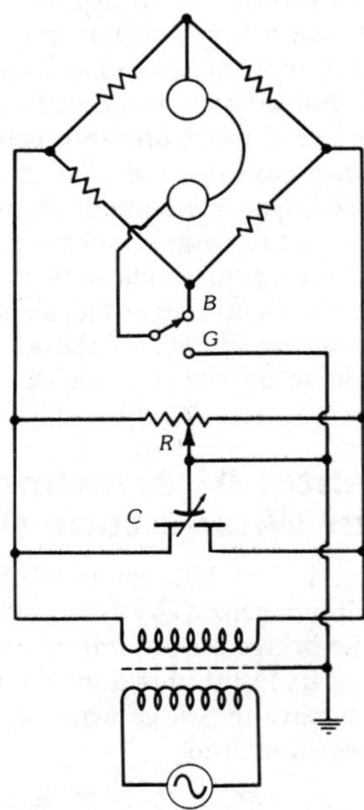

Fig. 2. Wagner ground connection.

Capacitor Balance

A capacitor balance is useful when one point of the bridge must be grounded directly and only a simple shielded transformer is used (Fig. 3). Balance the bridge, then open the two arms at P and Q. Rebalance by auxiliary capacitor C. Close P and Q and check the balance.

Series-Resistance-Capacitance Bridge

In the bridge of Fig. 4:

$$C_x = C_s R_b / R_a$$
$$R_x = R_s R_a / R_b$$

Wien Bridge

In the bridge of Fig. 5:

$$C_x C_s = (R_b / R_a) - (R_s / R_x)$$
$$C_s / C_x = 1/\omega^2 R_s R_x$$

For measurement of frequency, or in a frequency-selective application, if we make $C_x = C_s$, $R_x = R_s$, and $R_b = 2R_a$, then

$$f = (2\pi C_s R_s)^{-1}$$

Owen Bridge

In the bridge of Fig. 6:

$$L_x = C_b R_a R_d$$
$$R_x = (C_b R_a / C_d) - R_c$$

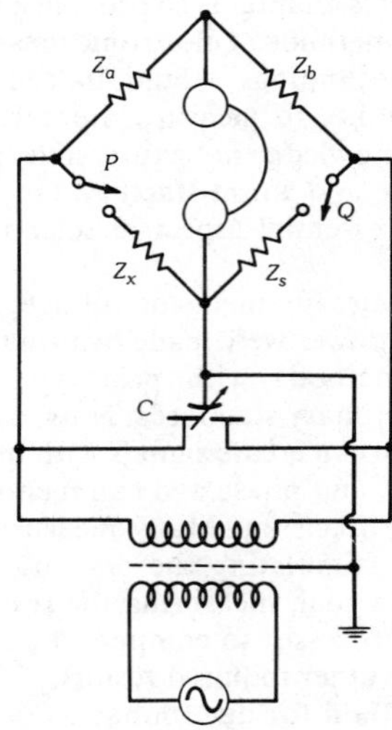

Fig. 3. Capacitor balance.

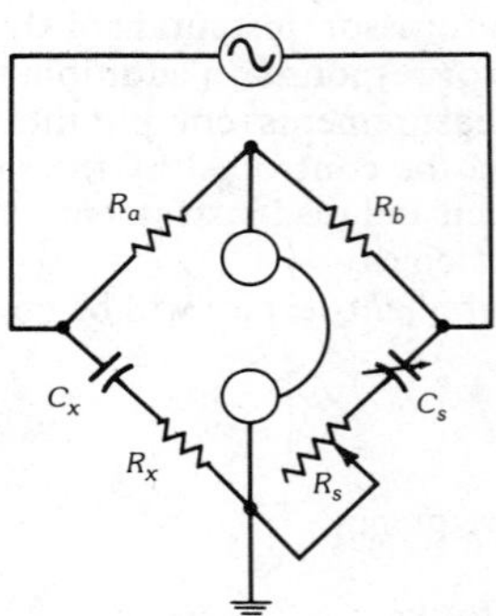

Fig. 4. Series-resistance-capacitance bridge.

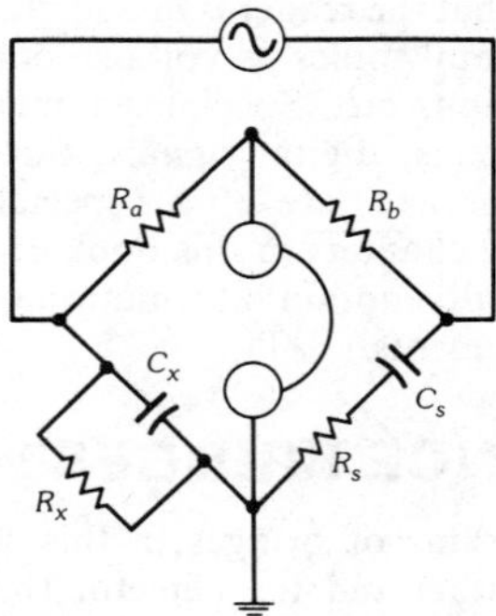

Fig. 5. Wien bridge.

Resonance Bridge

In the bridge of Fig. 7:

$$\omega^2 LC = 1$$
$$R_x = R_s R_a / R_b$$

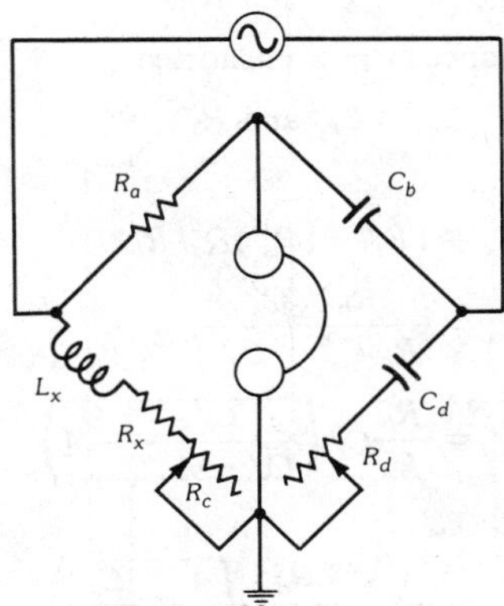

Fig. 6. Owen bridge.

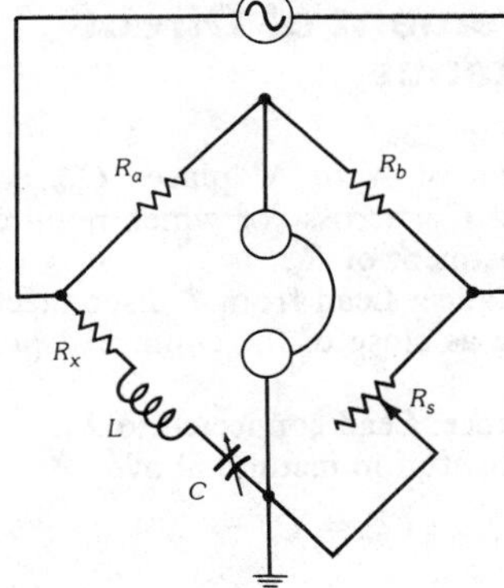

Fig. 7. Resonance bridge.

Maxwell Bridge

In the bridge of Fig. 8:

$$L_x = R_a R_b C_s$$
$$R_x = R_a R_b / R_s$$
$$Q_x = \omega(L_x / R_x) = \omega C_s R_s$$

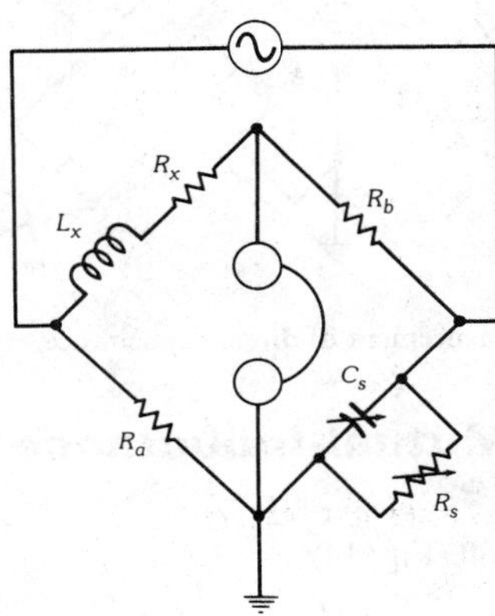

Fig. 8. Maxwell bridge.

Hay Bridge

The bridge of Fig. 9 is for the measurement of large inductance.

$$L_x = R_a R_b C_s / (1 + \omega^2 C_s^2 R_s^2)$$
$$Q_x = \omega L_x / R_x = (\omega C_s R_s)^{-1}$$

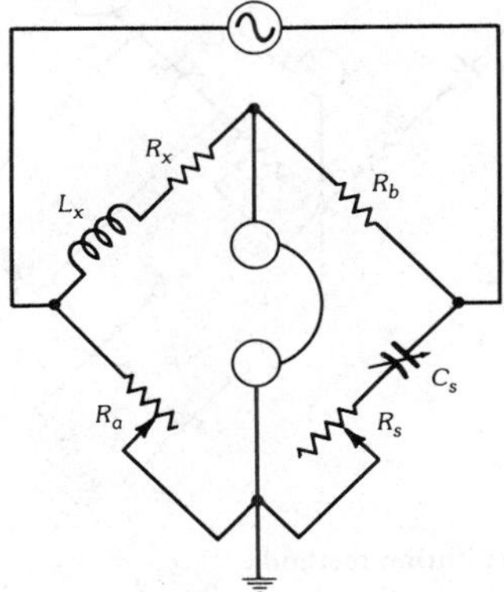

Fig. 9. Hay bridge.

Schering Bridge

In the bridge of Fig. 10:

$$C_x = C_s R_b / R_a$$
$$1/Q_x = \omega C_x R_x = \omega C_b R_b$$

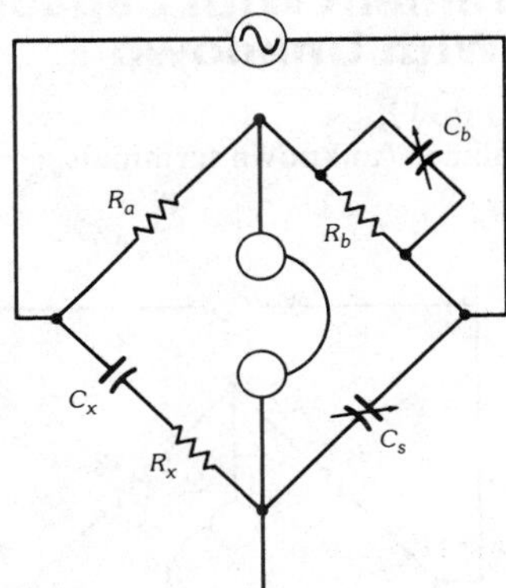

Fig. 10. Schering bridge.

Substitution Method for High Impedances

Refer to Fig. 11.

Initial balance (unknown terminals $x - x$ open):

$$C_s' \text{ and } R_s'$$

Final balance (unknown connected to $x - x$):

$$C_s'' \text{ and } R_s''$$

Then when $R_x > 10/\omega C_s'$, there results, with error < 1 percent

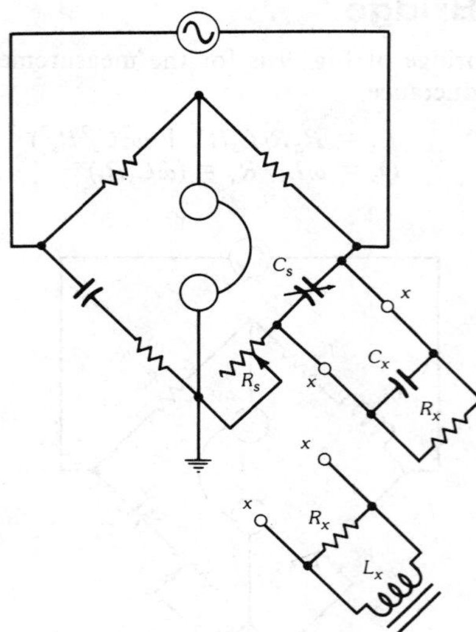

Fig. 11. Substitution method.

$$C_x = C_s' - C_s''$$

The parallel resistance is

$$R_x = [\omega^2 C_s'^2 (R_s' - R_s'')]^{-1}$$

If unknown is an inductor

$$L_x = -(\omega^2 C_x)^{-1} = [\omega^2 (C_s'' - C_s')]^{-1}$$

Measurement With Capacitor in Series With Unknown

Refer to Fig. 12.

Initial balance (unknown terminals $x - x$ short-circuited):

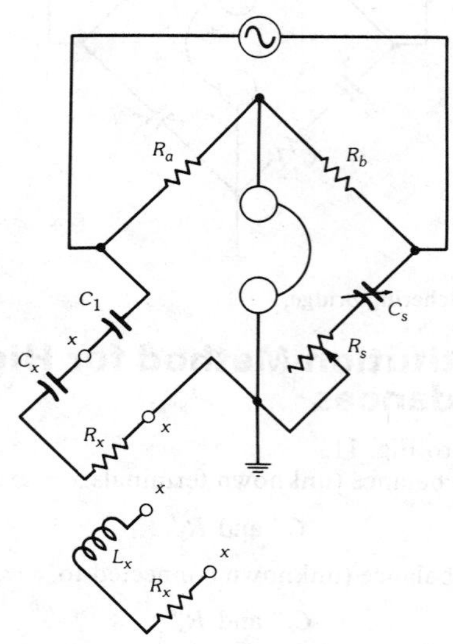

Fig. 12. Measurement with capacitor in series with unknown.

$$C_s' \text{ and } R_s'$$

Final balance ($x - x$ unshorted):

$$C_s'' \text{ and } R_s''$$

Then

$$R_x = (R_s'' - R_s')R_a/R_b$$

$$C_x = \frac{R_b C_s' C_s''}{R_a(C_s' - C_s'')}$$

$$= \frac{R_b}{R_a} C_s' \left(\frac{C_s'}{C_s' - C_s''} - 1 \right)$$

When $C_s'' > C_s'$

$$L_x = \frac{1}{\omega^2} \; \frac{R_a}{R_b C_s'} \left(1 - \frac{C_s'}{C_s''} \right)$$

Measurement of Direct Capacitance

Refer to Fig. 13.

Connection of N to N' places C_{nq} across the detector and C_{np} across R_b, which requires only a small readjustment of R_s.

Initial balance: Lead from P disconnected from X_1 but lying as close to the connected position as practical.

Final balance: Lead connected to X_1.

By the substitution method above

$$C_{pq} = C_s' - C_s''$$

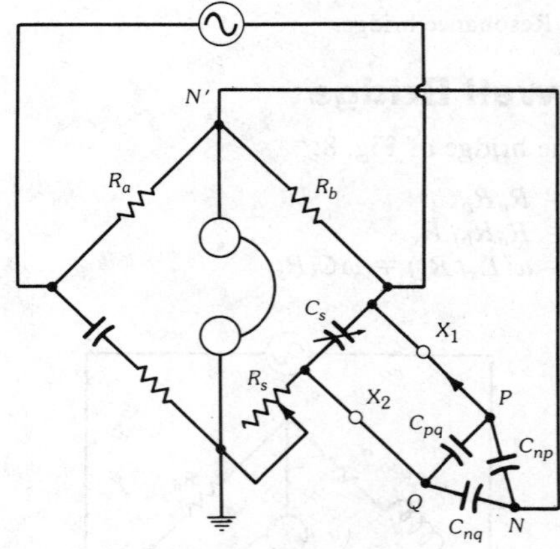

Fig. 13. Measurement of direct capacitance.

Felici Mutual-Inductance Balance

At the null (Fig. 14):

$$M_x = -M_s$$

This is useful at lower frequencies where capaci-

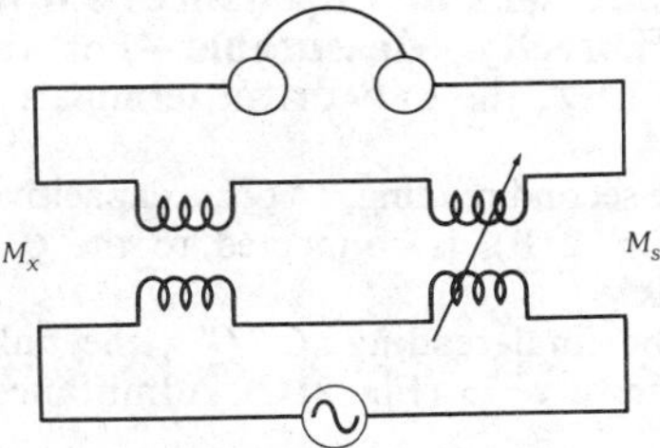

Fig. 14. Felici mutual-inductance balance.

tive reactances associated with windings are negligibly small.

Mutual-Inductance Capacitance Balance

With a low-loss capacitor (Fig. 15), at the null:

$$M_x = 1/\omega^2 C_s$$

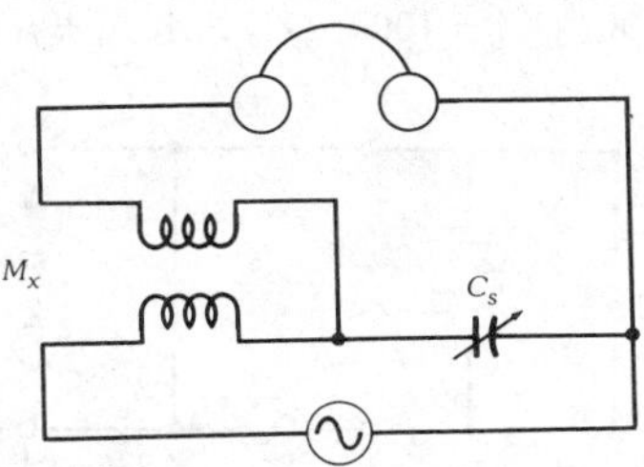

Fig. 15. Mutual-inductance capacitance balance.

Hybrid-Coil Method

At the null (Fig. 16):

$$Z_1 = Z_2$$

The transformer secondaries must be accurately matched and balanced to ground. This is useful at audio and carrier frequencies.

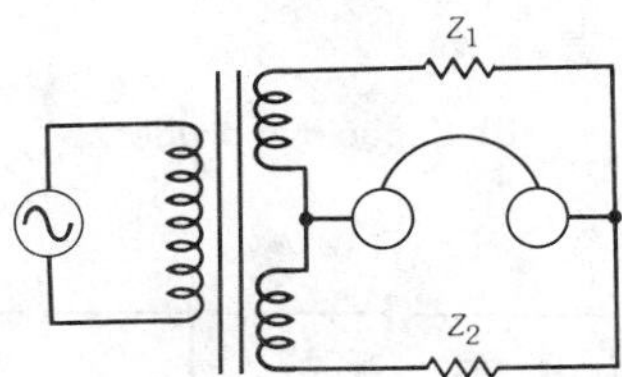

Fig. 16. Hybrid-coil method.

Q of Resonant Circuit by Bandwidth

The method of Fig. 17 may be used to evaluate Q by finding the 3-dB, or half-power, points. The source should be loosely coupled to the circuit. Adjust the frequency to each side of resonance, noting the bandwidth between the points where $V = 0.71 \times (V \text{ at resonance})$. Then

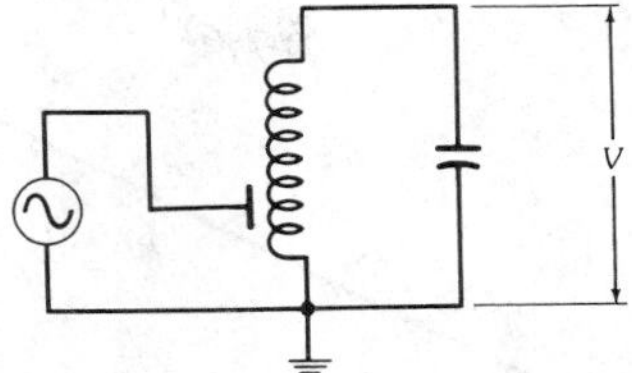

Fig. 17. Measurement of Q by bandwidth.

$$Q = (\text{resonance frequency})/(\text{bandwidth})$$

Q Meter (Hewlett-Packard 4342A)

Refer to Fig. 18. In this circuit, T_1 is a wideband transformer with n turns in the primary and one turn in the secondary. The secondary impedance is 0.001 ohm. The combination $L_xR_xC_0$ represents an unknown coil plugged into the COIL terminals for measurement; V_2 is a very high impedance voltmeter. With this arrangement,

$$Q = nV_2/V_1$$

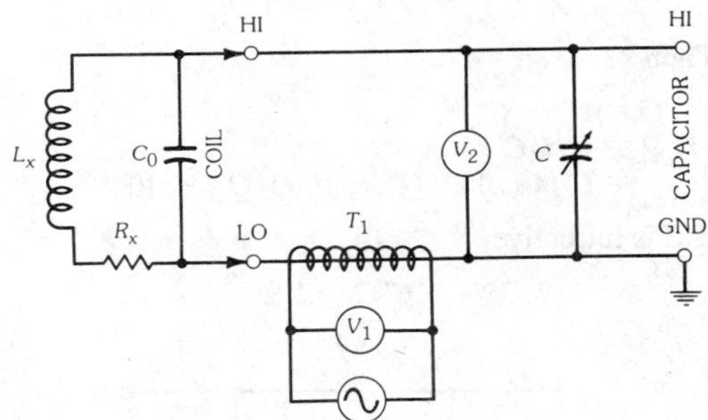

Fig. 18. Q meter.

Correction of Q Reading—The value of Q corrected for distributed capacitance C_0 of the coil is given by

$$Q_{\text{true}} = Q[(C + C_0)/C]$$

where,

Q = reading of Q-meter (corrected for internal resistors R_1 and R_2 if necessary),
C = capacitance reading of Q-meter.

Measurement of C_0 and True L_x—The plot of $1/f^2$ against C is a straight line (Fig. 19).

$$L_x = \text{true inductance} = \frac{1/f_2^2 - 1/f_1^2}{4\pi^2(C_2 - C_1)}$$

C_0 = negative intercept
f_0 = natural frequency of coil

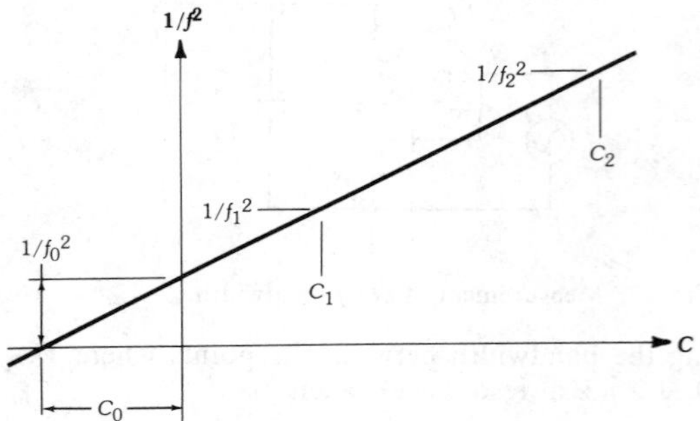

Fig. 19. Plot of C and $1/f^2$.

When only two readings are taken and $f_1/f_2 = 2.00$

$$C_0 = (C_2 - 4C_1)/3$$

With values in microhenries, megahertz, and picofarads

$$L_x = 19\,000/f_2^2(C_2 - C_1)$$

Measurement of Admittance—An initial reading, $C'Q'$, is taken as in Fig. 20A (LR_p is any suitable coil). For the final reading, $C''Q''$ (Fig. 20B):

$$1/Z = Y = G + jB = 1/R_p + j\omega C$$

Then

$$C = C' - C''$$
$$1/Q = G/\omega C$$
$$= C'/C(1000/Q'' - 1000/Q') \times 10^{-3}$$

If Z is inductive

$$C'' > C'$$

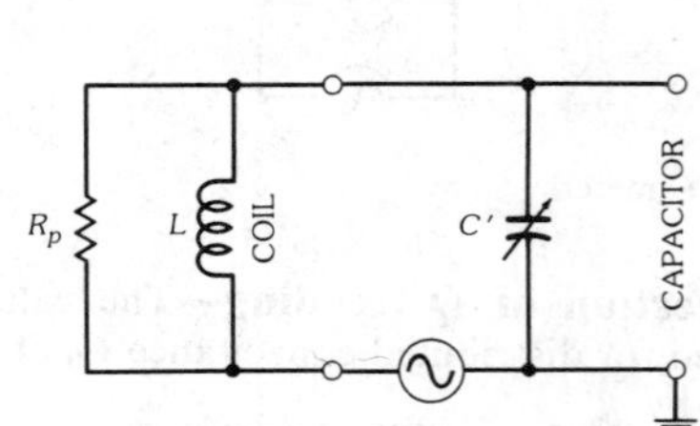

(A) Initial reading.

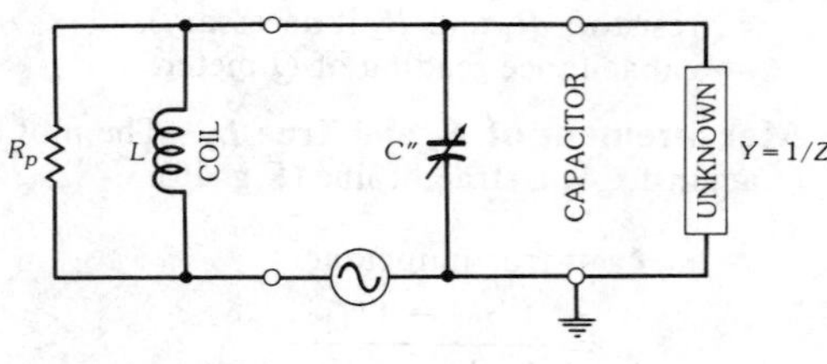

(B) Final reading.

Fig. 20. Measurement of admittance.

Measurement of Impedances Lower Than Those Directly Measurable—For the initial reading, $C'Q'$, the CAPACITOR terminals are open (Fig. 21A).

On the second reading, $C''Q''$, a capacitive divider, C_aC_b (Fig. 21B), is connected to the CAPACITOR terminals.

For the final reading, $C'''Q'''$, the unknown is connected to $x - x$ (Fig. 21C). Admittances Y_a and Y_b are

$$Y_a = G_a + j\omega C_a \qquad Y_b = G_b + j\omega C_b$$

with G_a and G_b not shown in the diagrams.

Then the unknown impedance is

$$Z = [Y_a/(Y_a + Y_b)]^2(Y''' - Y'')^{-1} - (Y_a + Y_b)^{-1} \text{ ohms}$$

where, with capacitance in picofarads and $\omega = 2\pi \times$ frequency in megahertz

$$(Y''' - Y'')^{-1} = \frac{10^6/\omega}{C'(1000/Q''' - 1000/Q'') \times 10^{-3} + j(C'' - C''')}$$

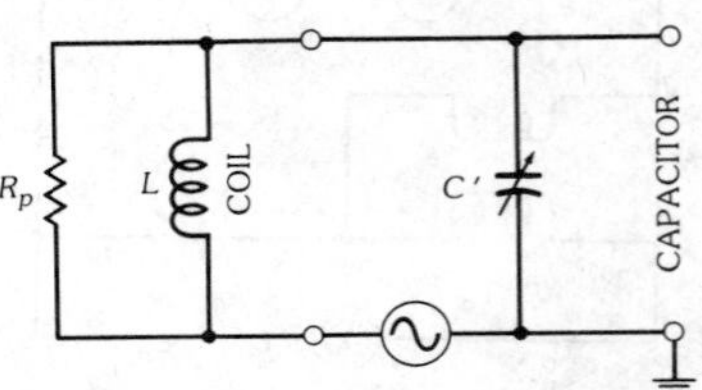

(A) Initial reading.

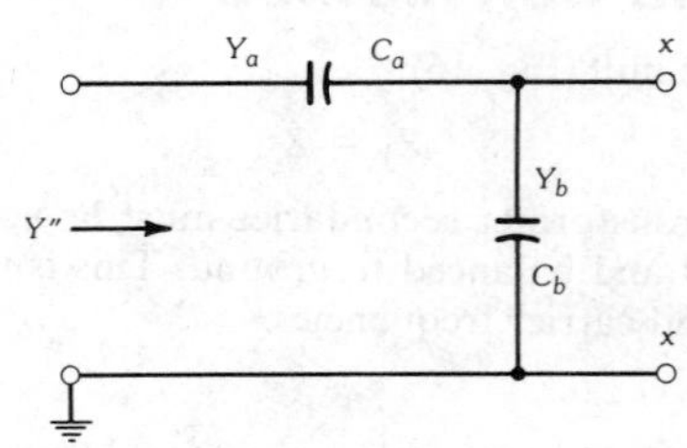

(B) Capacitive divider.

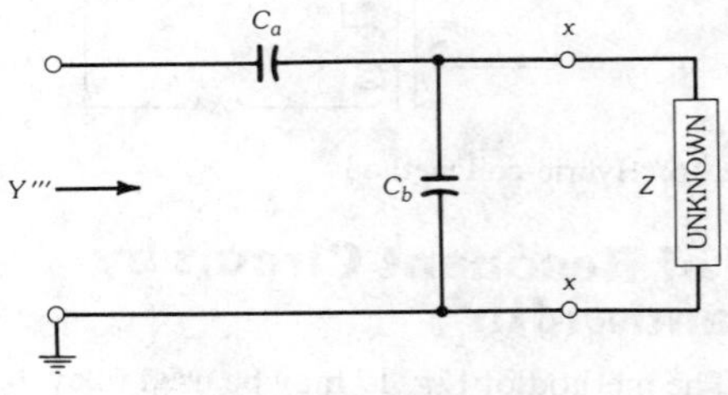

(C) Connection of unknown.

Fig. 21. Measurement of low impedances.

Usually G_a and G_b may be neglected; then there results

$$Z=\left(\frac{1}{1+C_b/C_a}\right)^2 (Y'''-Y'')^{-1} + j\frac{10^6}{\omega(C_a+C_b)} \text{ ohms}$$

For many measurements, C_a may be 100 picofarads. Capacitance $C_b = 0$ for very low values of Z and for highly reactive values of Z. For unknowns that are principally resistive and of low or medium value, C_b may take sizes up to 300 to 500 picofarads. When $C_b = 0$

$$Z=(Y'''-Y'')^{-1}+j(10^6/\omega C_a) \text{ ohms}$$

and the "second" reading above becomes the "initial," with $C' = C''$ in the equations.

Measurement of Coupling Coefficient of Loosely Coupled Coils—The coefficient of coupling

$$k = M/(L_1L_2)^{1/2}$$

between two high-Q coils can be obtained by measuring the inductance L with S_1 closed and again with S_1 open (Fig. 22). From these two measurements

$$k=(1-L_{\text{closed}}/L_{\text{open}})^{1/2}$$

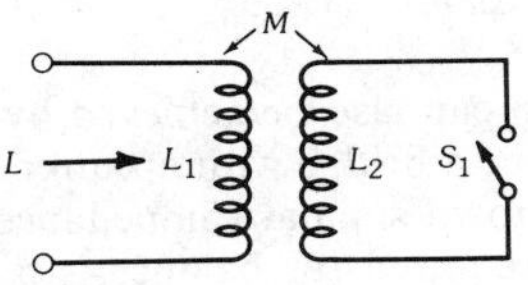

Fig. 22. Method of determining coefficient of coupling between two coils.

When the coil self-inductances are known, a measurement of L_a and L_b (Fig. 23) yields

$$k=(L_a-L_b)/4(L_1L_2)^{1/2}$$

If $L_1 = L_2$

$$k=(L_a-L_b)/(L_a+L_b)$$

Neither of the above methods provides adequate precision when two high-Q coils are only about critically coupled. In that case, the Q of each coil is measured with the other coil open-circuited. Then the coupled coils (L_1R_1 and L_2R_2) and a low-loss adjustable capacitor (C_2) are connected to the Q-meter as shown in Fig. 24, C_2 is disconnected, and

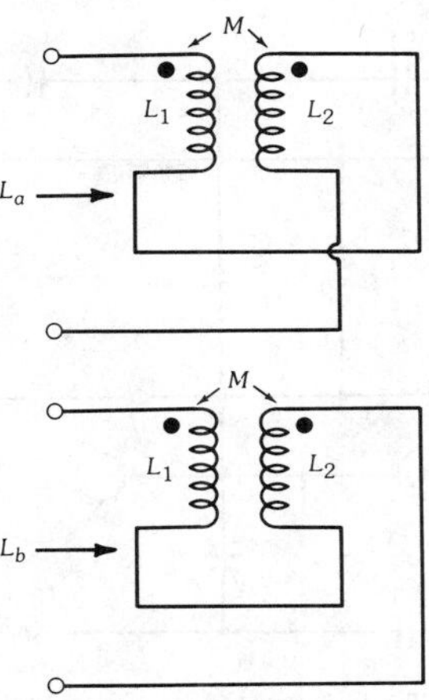

Fig. 23. Method of determining coefficient of coupling when self-inductances are known.

C is adjusted to maximize the Q-meter reading; C_2 is then connected and adjusted to minimize the reading. If the final reading is Q_0

$$K=[1/Q_2(1/Q_0-1/Q_1)]^{1/2}$$

If the final reading is too small to be read accurately, Q_2 may be reduced by inserting a small resistance in series with L_2R_2.

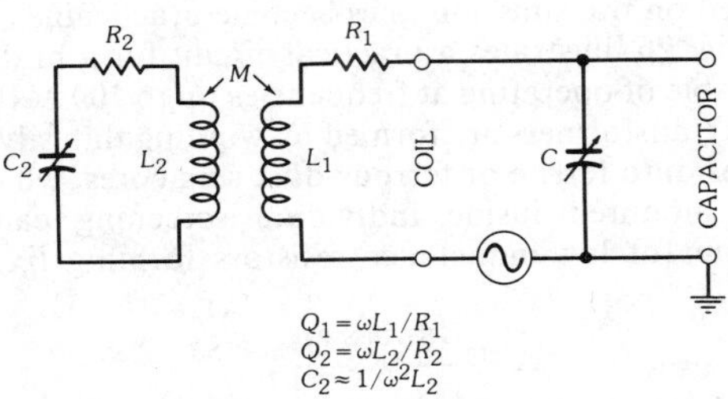

Fig. 24. Method of determining coefficient of coupling when coils are only about critically coupled.

Twin-T Admittance-Measuring Circuit (General Radio Type 821-A)

The circuit in Fig. 25 may be used for measuring admittances in a range somewhat exceeding 400 kilohertz to 40 megahertz. It is applicable to the special measuring techniques described above for the Q-meter.

Conditions for a null in the output are

$$G+G_l = R\omega^2C_1C_2(1+C_g/C_3)$$
$$C+C_b = 1/\omega^2L - C_1C_2(1/C_1+1/C_2+1/C_3)$$

With the unknown disconnected, call the initial balance C_b' and C_g'. With the unknown connected,

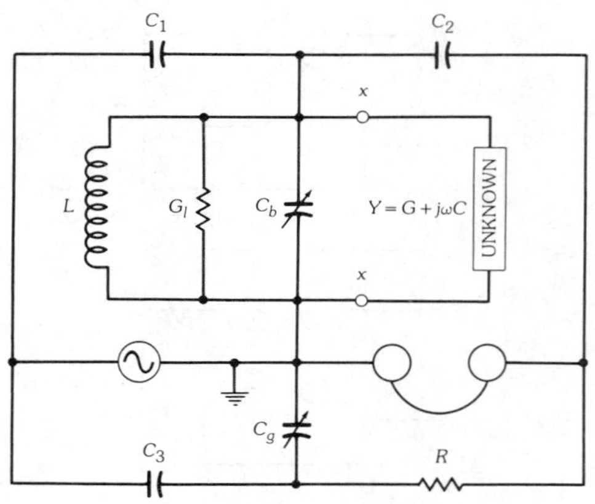

Fig. 25. Twin-T admittance-measuring circuit.

the final balance is C_b'' and C_g''. Then the components of the unknown, $Y = G + j\omega C$, are

$$C = C_b' - C_b''$$
$$G = (R\omega^2 C_1 C_2 / C_3)(C_g'' - C_g')$$

Ratio-Arm Bridges (Wayne Kerr)*

Transformer ratio-arm bridges can be designed to operate at radio frequencies up to about 250 MHz. Beyond that point, other forms of bridges based on transmission lines become practicable.

Fig. 26 illustrates a practical circuit for a bridge capable of operating at frequencies up to 100 MHz. The transformers are formed by winding thin silver tapes onto ferrite or ferrous-dust ring cores, which are mounted inside individual screening cans. Drums of low-inductance resistors forming fixed conductance standards are arranged to engage with spring contacts. A variable conductance for interpolation is formed by means of resistor R, which is fed with a voltage derived from a resistive potential divider, P.

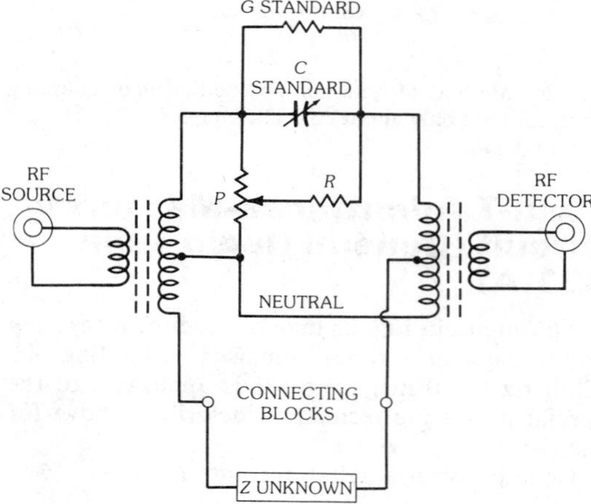

Fig. 26. Ratio-arm bridge.

Radio-frequency bridge measurements require that considerable care should be taken in setting up the apparatus. Any leakage of power from the source to the detector that bypasses the bridge network will give errors.

Automatic Impedance Meters

The availability of digital voltmeter technology has created a generation of touch-and-read impedance meters. One basic technique is to impress a known current or voltage upon an impedance and measure the magnitude and phase of the voltage or current resulting from it (Fig. 27). In its simplest form, the ratio of the complex voltage to the complex current is the value of the impedance:

$$Z\angle\theta = E\angle\theta_1 / I\angle\theta_2$$

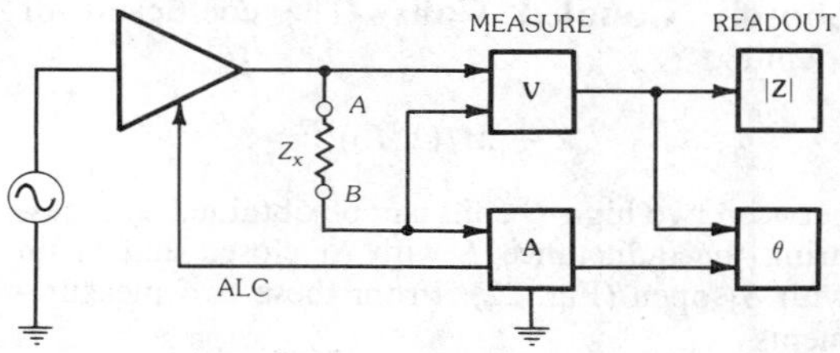

Fig. 27. Vector impedance meter. (*From Alonzo, G. J., et al. "Direct-Reading Vector Impedance Meters."* Hewlett-Packard Journal, *January 1967. © 1967 Hewlett-Packard, used with permission.*)

Automation can also be achieved by using feedback to balance a bridge with an unknown impedance and a known standard impedance. The complex voltage required for balance is a measure of the unknown impedance.

At frequencies above 10 MHz, it is convenient to compare an impedance to a precise 50-ohm coaxial transmission line and measure the reflection coefficient. The impedance is calculated using microprocessor techniques.

An example of each technique is shown below.

Four-Terminal-Pair Bridge, 5 Hz–13 MHz—Refer to Fig. 28. A four-terminal-pair configuration is used to avoid errors caused by mutual coupling between leads. The bridge provides a complex voltage, V_x, across the device under test (DUT) and another complex voltage V_r proportional to the current through the DUT at bridge balance. The bridge is balanced by a heterodyne method through the frequency range of 5 Hz to 13 MHz.†

* From R. Calvert, *The Transformer Ratio-Arm Bridge.* Bognor Regis, Sussex, England: Wayne Kerr Co. Ltd.

† Y. Narimatsu, et al, "A Versatile LF Impedance Analyzer," *H-P Journal*, Sept. 1981.

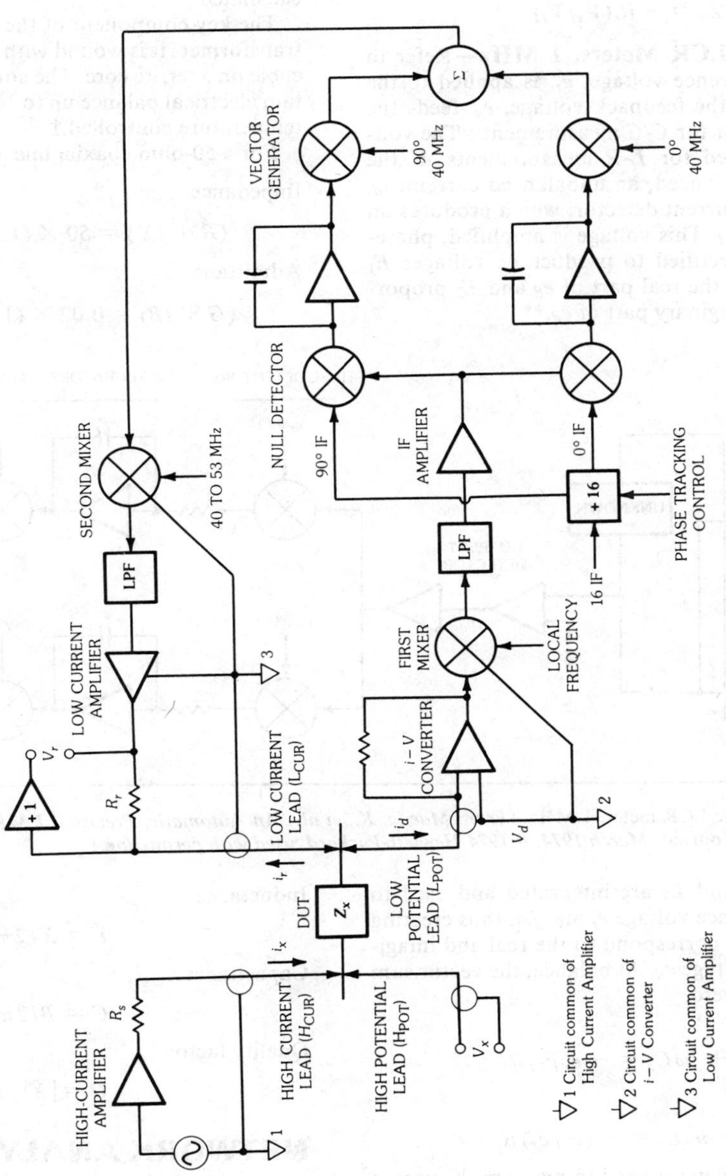

Fig. 28. 5-Hz to 13-MHz bridge. (*From Narimatsu, Y., et al. "A Versatile LF Impedance Analyzer."* Hewlett-Packard Journal, September 1981. *© 1981 Hewlett-Packard, used with permission.*)

A vector ratio detector is used to measure V_x/V_r. Since

$$V_x/Z_x = i_x = i_r = -V_r/R_r$$

therefore,

$$Z_x = -R_r(V_x/V_r)$$

Automatic LCR Meters, 1 MHz—Refer to Fig. 29. A reference voltage, e_r, is applied to the unknown, and the feedback voltage, e_v, feeds the standard resistor for *C-G* measurements. The voltages are reversed for *L-R* measurements. If the bridge is not balanced, an unbalanced current, i_d, flows into the current detector, which produces an error voltage, e_d. This voltage is amplified, phase-detected, and rectified to produce dc voltages E_1 proportional to the real part of e_d and E_2 proportional to the imaginary part of e_d.**

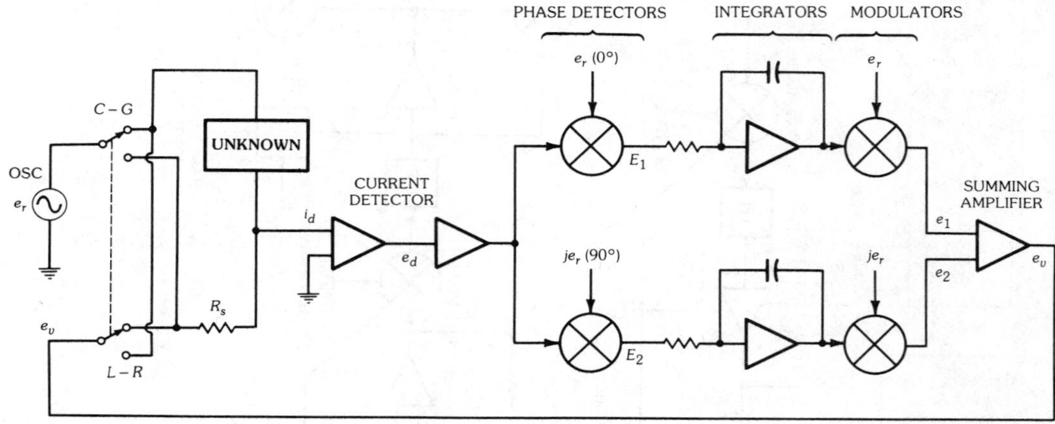

Fig. 29. Automatic LCR meter, 1 MHz. (*From Maeda, K., et al. "An Automatic, Precision 1 MHz Digital LCR Meter."* Hewlett-Packard Journal, *March 1974. © 1974 Hewlett-Packard, used with permission.*)

Voltages E_1 and E_2 are integrated and used to modulate reference voltage e_r and je_r, thus creating e_1 and e_2, which correspond to the real and imaginary parts of voltage e_d. At balance, the vector sum e_v reduces i_d to zero.

At balance

$$G_x + j\omega C_x = -e_v/e_rR_s$$

or

$$R_x + j\omega L_x = -(e_v/e_r)R_s$$

A digital voltmeter is used to measure the vector ratio of e_v and e_r.

Vector Impedance Analyzer, to 1000 MHz—Refer to Fig. 30. Outputs from a directional rf bridge channel and a compensated reference channel go to a receiver for synchronous detection and processing by a dual-slope integrator, thus providing a vector ratio that is a measure of the reflection coefficient, Γ. With a microprocessor, the impedance, admittance, inductance, capacitance, or Q of the device under test can be calculated.

The key component of the rf bridge is the balun transformer. It is wound with fine semirigid coaxial cable on a ferrite core. The construction must maintain electrical balance up to 1000 MHz and must be temperature controlled.‡

For a 50-ohm coaxial line, calculations are:

Impedance

$$(R + jX) = 50 \times (1 + \Gamma)/(1 - \Gamma)$$

Admittance

$$(G + jB) = 0.02 \times (1 - \Gamma)/(1 + \Gamma)$$

Inductance

$$L = X/2\pi f$$

Capacitance

$$C = B/2\pi f$$

Quality factor

$$Q = |X|/R$$

NETWORK ANALYSIS

A network analyzer consists of the five basic blocks shown in Fig. 31. At low frequencies, voltage and current can be measured by probes to determine the Z, Y, or h parameters (Fig. 32). Typically, a network analyzer is used to characterize small signal parameters. For higher frequencies (up

** K. Maeda, et al, "An Automatic, Precision 1-MHz Digital LCR Meter," *H-P Journal,* March 1974.

‡ T. Ichino, et al, "Vector Impedance Analysis at 1000 MHz," *H-P Journal,* Jan. 1980.

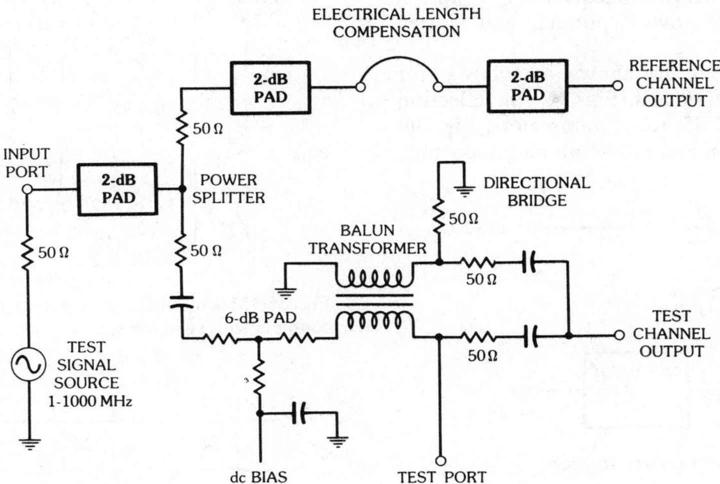

Fig. 30. Rf impedance analyzer. (*From Ichino, T., et al. "Vector Impedance Analysis at 1000 MHz."* Hewlett-Packard Journal, *January 1980. © 1980 Hewlett-Packard, used with permission.*)

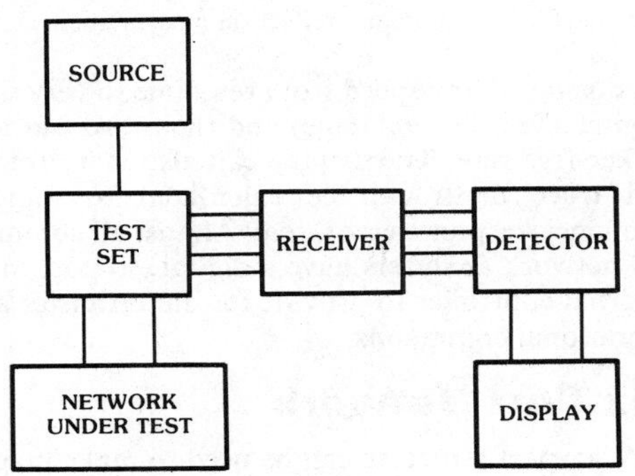

Fig. 31. Block diagram of a network analyzer.

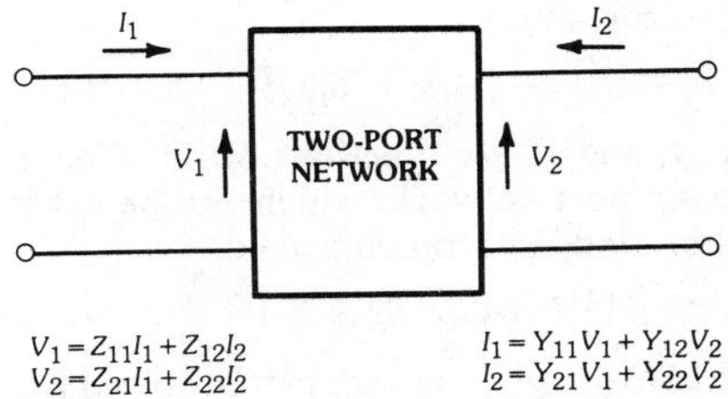

Fig. 32. *Z* and *Y* parameters typically used at low frequency.

through microwave frequencies), network analyzers primarily characterize the magnitude and phase of reflection coefficients, transmission coefficients, and *S* parameters of networks (Fig. 33). (See Chapter 31 for a definition of *S* parameters.)

At microwave frequencies, it is difficult to build perfect structures that have no frequency-response, directivity, or port-match errors. It is widely ac-

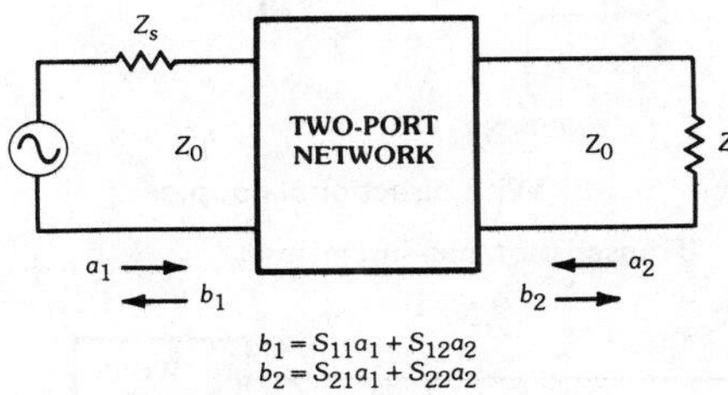

Fig. 33. Parameters used at higher frequencies.

cepted practice to reduce these errors with a mathematical error-correction procedure implemented with a computer.

Sources

The most common source used for network analysis is a sweep oscillator. This provides a real-time frequency response of the network characteristics. When greater frequency accuracy is required, synthesizers are used. However, the switching times can be long (10 ms to 50 ms). These synthesizers provide the best phase accuracy and repeatability, but at the expense of fast real-time measurements.

Receivers

There are three common receiver techniques, the diode detector, homodyne or self-mixing, and the heterodyne approach (both fundamental and harmonic), usually with multiple inputs.

Test Sets

The test-set portion provides connection of the source and receiver to the network under test. At low frequencies, voltage and current probes are sufficient, but at microwave frequencies, couplers, directional bridges, power splitters, and slotted lines are required.

Test configurations are shown as follows: for transmission measurements, Fig. 34; for reflection measurements, Fig. 35; for S parameters, Fig. 36; and for transmission and reflection measurements, Fig. 37.

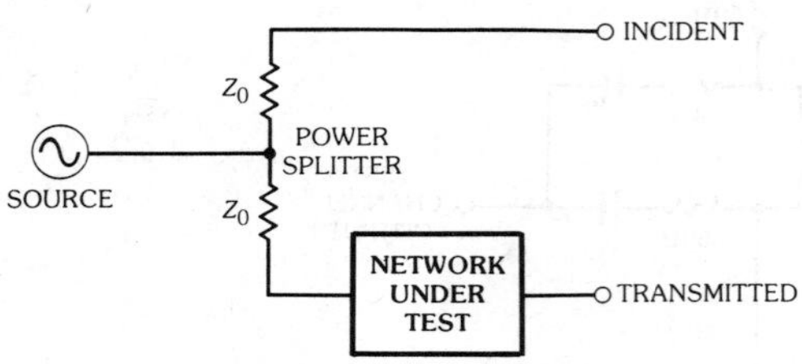

(A) With power splitter.

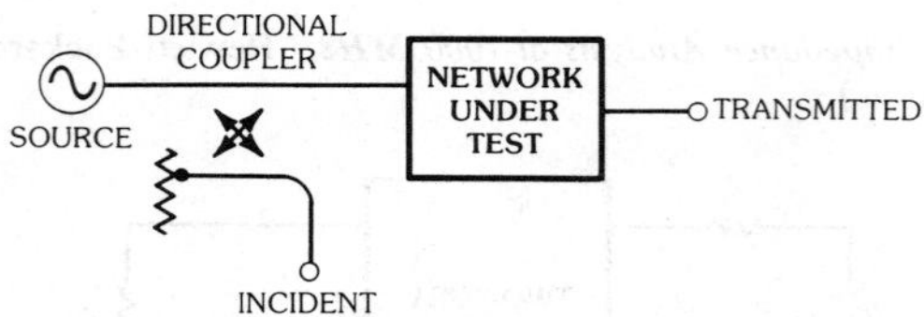

(B) With directional coupler.

Fig. 34. Transmission measurements.

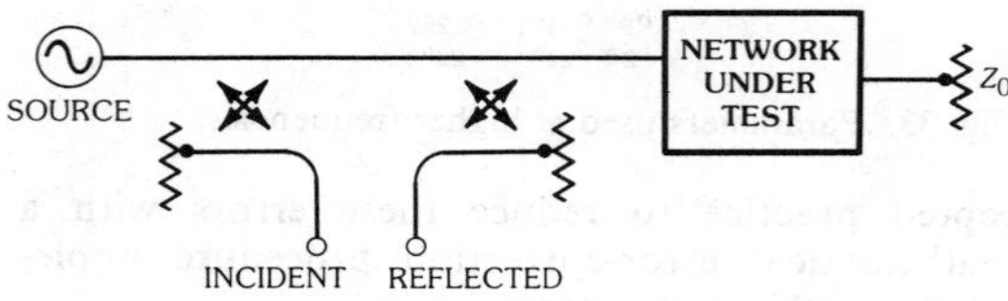

(A) With dual directional coupler.

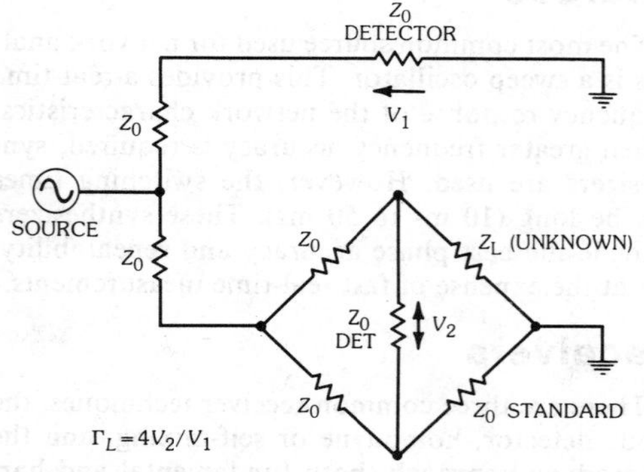

(B) With directional bridge.

Fig. 35. Reflection measurements.

Displays and Output

The most useful display is the digital storage crt system (Fig. 38). Data can be written into the dig-

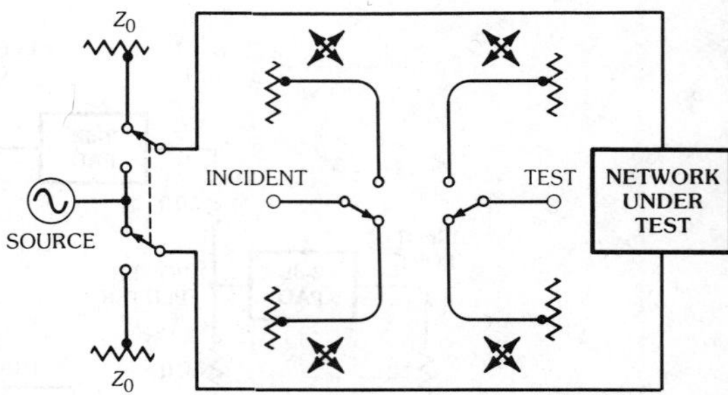

Fig. 36. Measurement of S parameters, using directional couplers with switching.

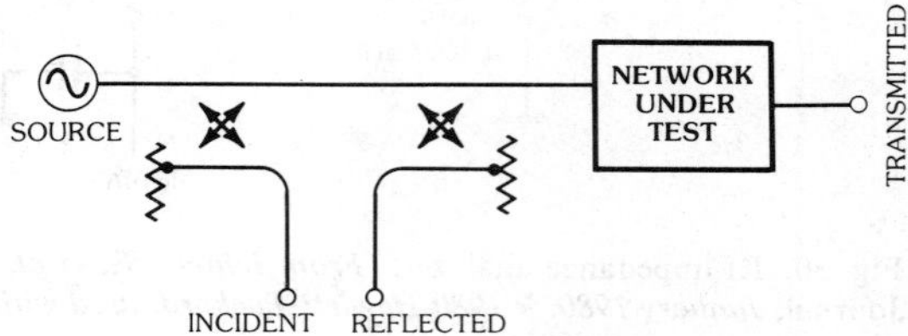

Fig. 37. Transmission and reflection measurements.

ital display at any speed from real time to very slow (digital sweep for example) and then read out at a flicker-free rate. The display can also store traces, and trace math can be calculated to remove frequency-response errors, etc. Almost all commercial network analyzers have a digital i/o port to an external controller to provide for data transfer and operational commands.

Six-Port Network

A six-port structure can be used to make magnitude and phase measurements using magnitude-only detectors such as the diode, thermistor, or bolometer. The general solution is shown below (refer to Fig. 39).

$$\rho_i = |b_i|^2 = |A_i a + B_i b|^2 \qquad i = 3 \text{ to } 6$$

where A_i and B_i are functions of the S parameters of the six-port network, which can be determined by calibration. Rearranging yields

$$\rho_i = |B_i|^2 |b|^2 |(A_i/B_i)\Gamma + 1|^2 \qquad i = 3 \text{ to } 6$$

The above four equations can be solved for Γ, the reflection coefficient of the device under test:

$$\Gamma = \frac{\Sigma C_i \rho_i + j\Sigma S_i \rho_i}{\Sigma \beta_i \rho_i} \qquad i = 3 \text{ to } 6$$

where C_i, S_i, and β_i are functions of A_i and B_i. Note that Γ consists of a sum of noncomplex terms.

A dual six-port structure can be used to measure the S parameters of a two-port network provided that the restriction $S_{12} = S_{21}$ is applied (Fig. 40). The mathematical approach of calibrating and measuring with the six-port structure is divided into

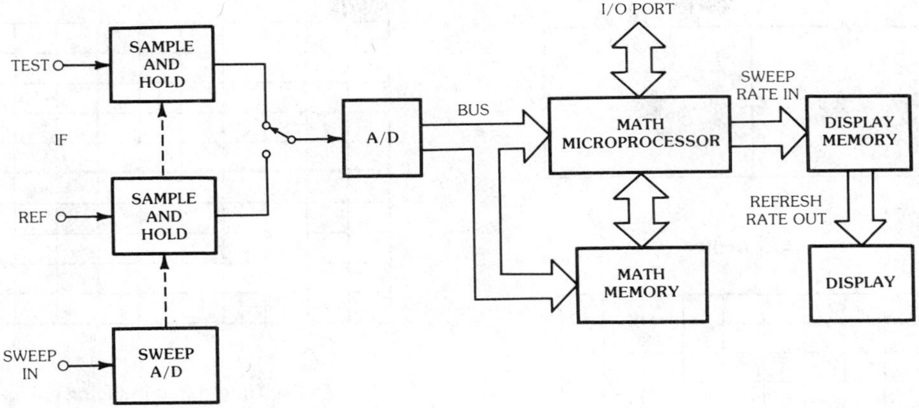

Fig. 38. Digital storage crt system.

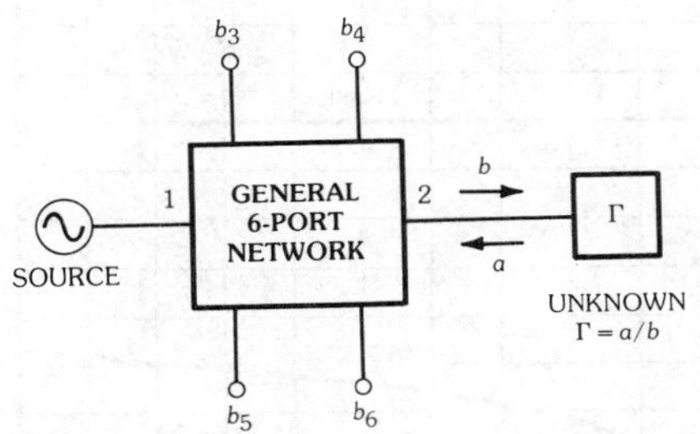

Fig. 39. Measurements with six-port system.

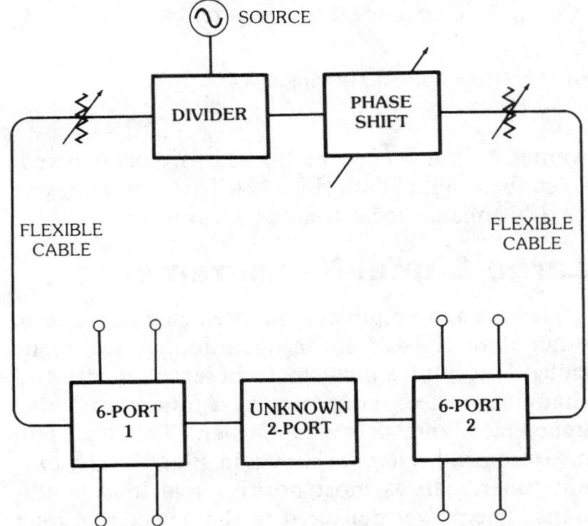

Fig. 40. A dual six-port structure used to measure a two-port network.

two steps. Step one is the calibration of the six-port structure to determine phase from the magnitude-only measurements. Step two is the same as the error-correction procedure for the four-port network analyzers. There has been some excellent work done on six-port networks by Glen Engen and Cletus Hoer of the National Bureau of Standards.

The approach can be realized in all frequency bands from audio to greater than 100 GHz. Great stability of the measurement system is required to achieve accurate results. There are also limitations in measuring active devices where power setting flexibility is needed and when S_{21} does not equal S_{12}. Typically, data gathering time is slow but accuracy is very good.

Error Correction

There are error-correction procedures by which the linear time invariant errors of the network analyzer can be characterized and then, by an inverse mathematical procedure, be removed from the measured data. See references 1, 2, and 3 (listed at the end of this chapter) for descriptions of these error-correction procedures.

Frequency and Time-Domain Relationships

With the advent of the modern computer, measuring data in the frequency domain and then converting to the time domain by means of the inverse-Fourier transform has become practical. This ability to observe measurements in the time domain adds additional insight to microwave measurements. Many times in microwave measurements, circuit discontinuities are separated by lengths of transmission lines. Taking the Fourier transform of such a circuit allows us to isolate the various impulse responses to the circuit discontinuities (see Fig. 41C).

Also, by mathematically generating a step stimulus, a traditional time-domain reflectometer (TDR) display with high resolution and stability is achieved (see Fig. 41D). With the time data isolated, it is possible to filter out unwanted time-domain responses and then transform back to the frequency domain. By means of this technique, connector discontinuities, launches, and other un-

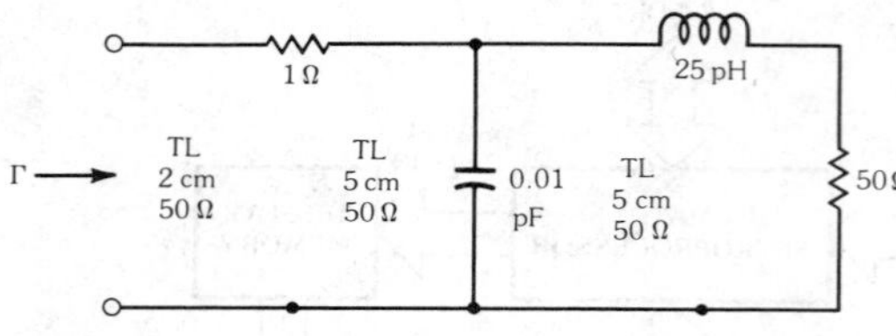

(A) Circuit.

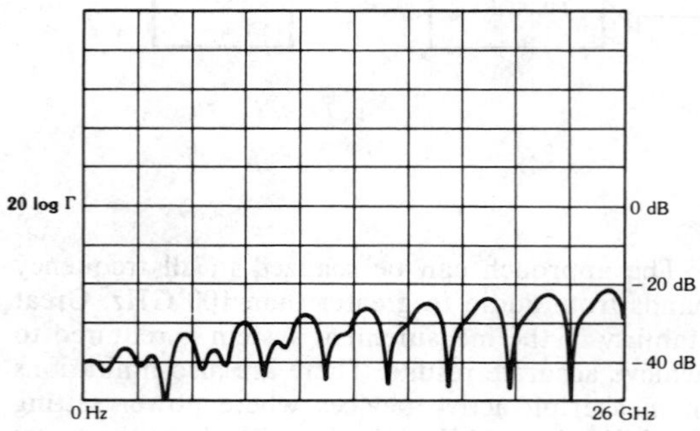

(B) Frequency response.

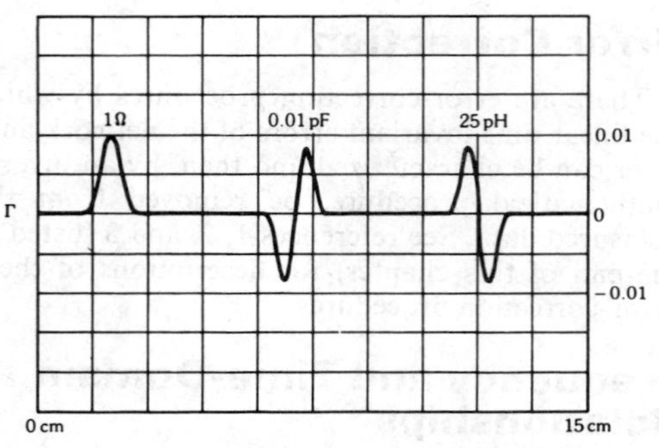

(C) Impulse time domain.

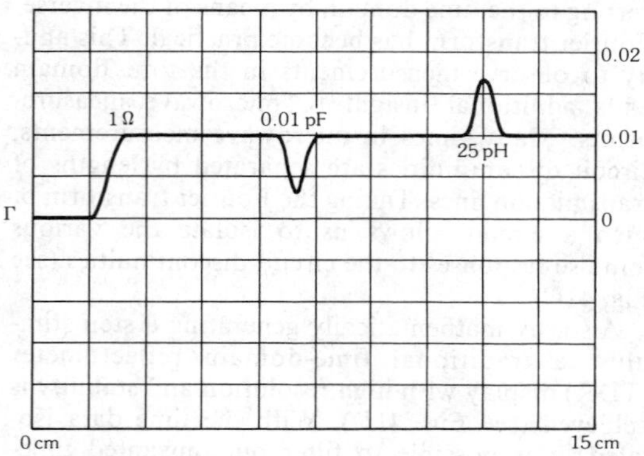

(D) Step time domain or time-domain reflectometer.

Fig. 41. Frequency and time relationships.

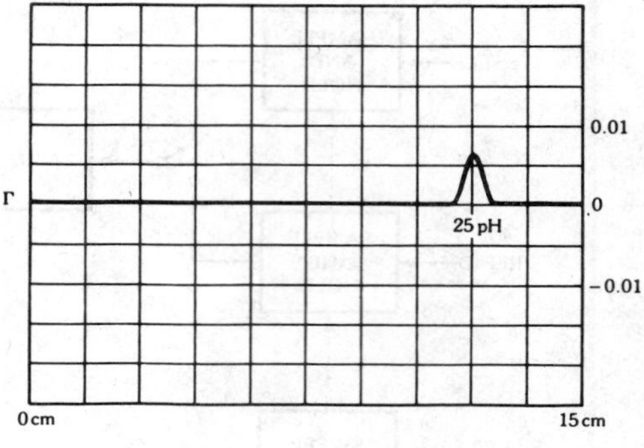

(A) Time-gated inductor.

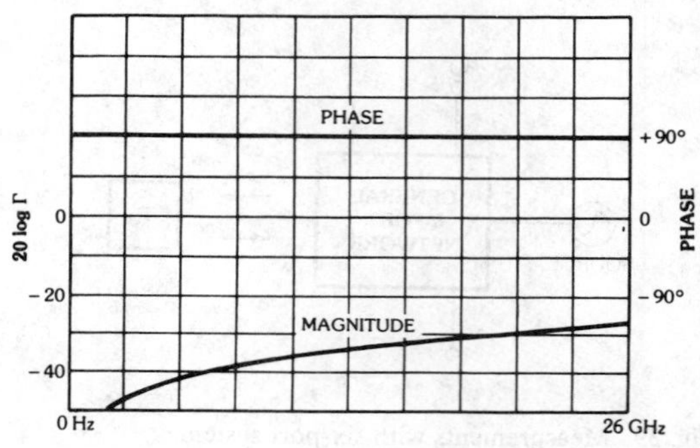

(B) Frequency response after gating.

Fig. 42. Time-domain filtering.

wanted responses can be filtered (or deconvolved) from the original data (Fig. 42). This technique can greatly enhance measurement accuracy.

Large-Signal Measurements

Devices and amplifiers are often characterized in other than class-A linear operation. A technique called load-pull is one way to observe the effect of changing device parameters as a function of load impedance and delivered power. The load-pull measurement system is shown in Fig. 43. The output tuner realizes most positive real load conditions. The power delivered to the device and load can be measured by power meters connected to the coupler side arms. The impedance of the load can be measured with a network analyzer. Fig. 44 shows contours of constant delivered power as a function of load impedance. These plots can be generated for the fundamental or the harmonics to give good insight into the nonlinear behavior. With these data, matching networks can be designed that will optimize the delivered power under nonlinear operating conditions.

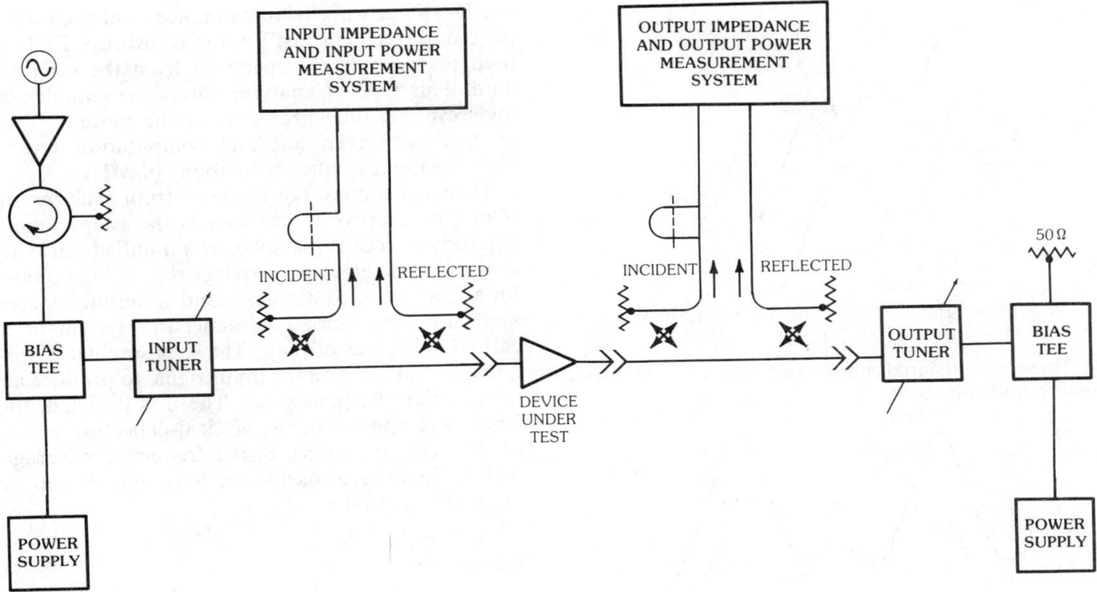

Fig. 43. Large-signal measurement system utilizing load-pull technique.

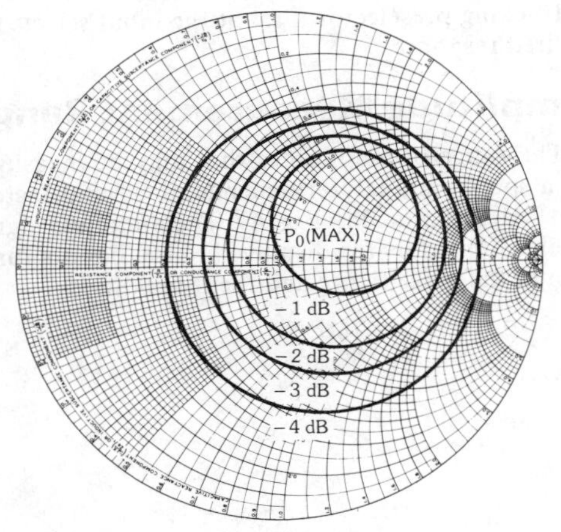

Fig. 44. Contours of constant delivered power vs load impedance under large-signal conditions.

SIGNAL ANALYSIS

Analysis of signals in the frequency domain is an important technique to determine the characteristics of audio, rf, and microwave signals. While the oscilloscope is the basic tool for time-domain analysis, the spectrum analyzer provides a display of signal amplitude as a function of frequency and is the basic tool for frequency-domain analysis. The relationship between the time and frequency domains is illustrated in Fig. 45 for the case of two sine waves. Signal amplitude vs time and frequency is shown in the three-dimensional representation of Fig. 45A. In the time-domain representation of Fig. 45B, the two signal components add at each instant of time, producing the composite as would be viewed on an oscilloscope. Fig. 45C is the frequency-domain view. Each frequency is represented by a vertical line, with the height representing the amplitude and the horizontal position representing the frequency. Although the information about the signal is the same in either domain, the frequency domain provides a tool to observe small but important signal characteristics such as a small amount of harmonic distortion.

Conceptually, a spectrum analyzer could consist of a parallel bank of rectangular bandpass filters each followed by an envelope detector as in Fig. 46. The filters cover adjacent frequency bands and do not overlap. The detector outputs are sequentially scanned to provide an amplitude-vs-frequency display on a cathode-ray tube. The resolution of the spectrum analyzer, that is, its ability to resolve closely spaced spectral lines, is determined by the width of the filter bandpass. Obviously, a very large number of filters would be required to obtain wide frequency spans with narrow resolution. Practical filters would have long time constants, thereby requiring a long time for their outputs to reach a steady-state value.

Another type of spectrum analyzer uses digital signal-processing techniques. Called a Fourier analyzer or dynamic signal analyzer, it is shown in Fig.

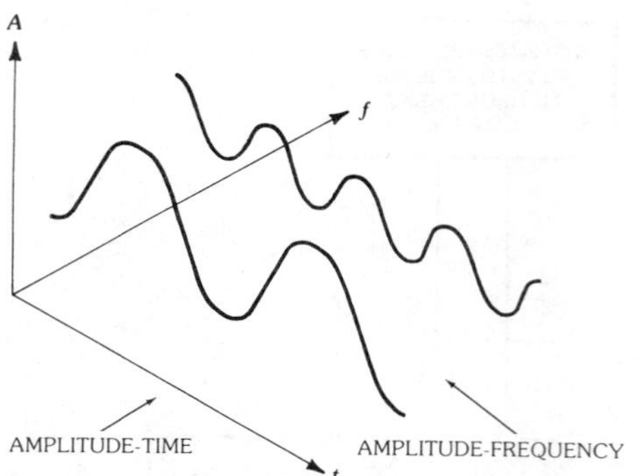

(A) Three-dimensional coordinates showing time, frequency, and amplitude.

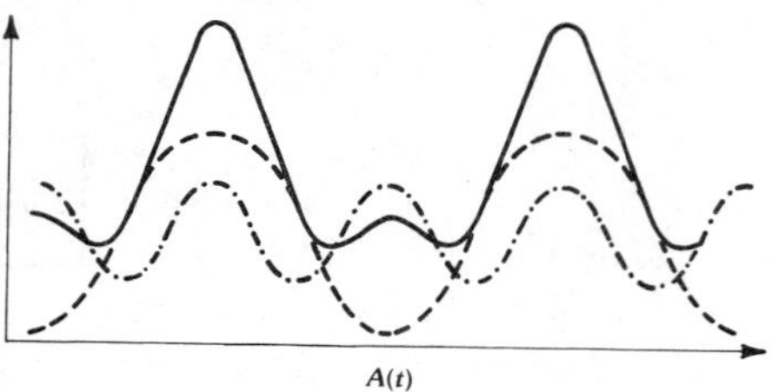

(B) View seen in the time domain.

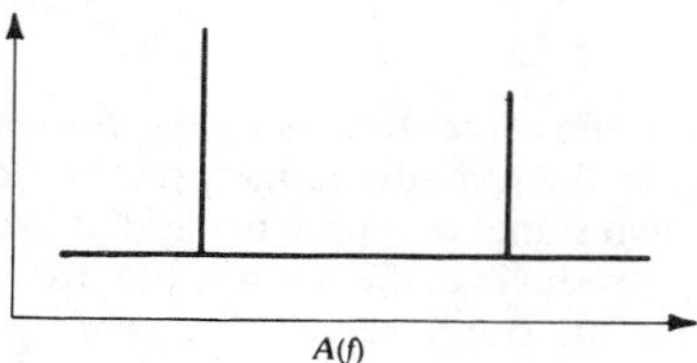

(C) View seen in the frequency domain.

Fig. 45. Relationship between time and frequency domains. (*From Hewlett-Packard Application Note No. 150 on Spectrum Analysis. © Hewlett-Packard, used with permission.*)

47. The input waveform is sampled, and each sample is digitized. The fast Fourier transform (FFT) is used to compute the spectrum from the sampled data. This type of analyzer solves the complexity and response-time problems of the parallel filter analyzer; however, adc and computation speeds limit the input frequency to about 10 MHz.

The most common form of spectrum analyzer for rf and microwave frequencies is the swept superheterodyne receiver shown in simplified form in Fig. 48. A sweep ramp provides the tuning voltage for a voltage-tuned oscillator and is simultaneously applied to the horizontal-deflection system of a cathode-ray-tube display. The local-oscillator frequency is mixed with the input signal to produce an intermediate frequency (if). The if is detected, filtered, and applied to the vertical-deflection plates of the crt. To achieve broad frequency coverage with a single local oscillator, harmonic mixing is often used so that

$$f_s = nf_{LO} \pm f_{if}$$

where,

f_s = signal frequency,
f_{LO} = local-oscillator frequency,
f_{if} = intermediate frequency,
$n = 1, 2, 3 \ldots.$

A tracking preselector filter at the input selects the desired response.

Amplitude Measurement Range

The range of amplitudes that may be measured on a spectrum analyzer depends on several factors and is diagrammed in Fig. 49. The minimum signal level is limited by the displayed noise that is a function of the noise figure and the if bandwidth:

$$N = (F - 1)\ kT_0B$$

where,

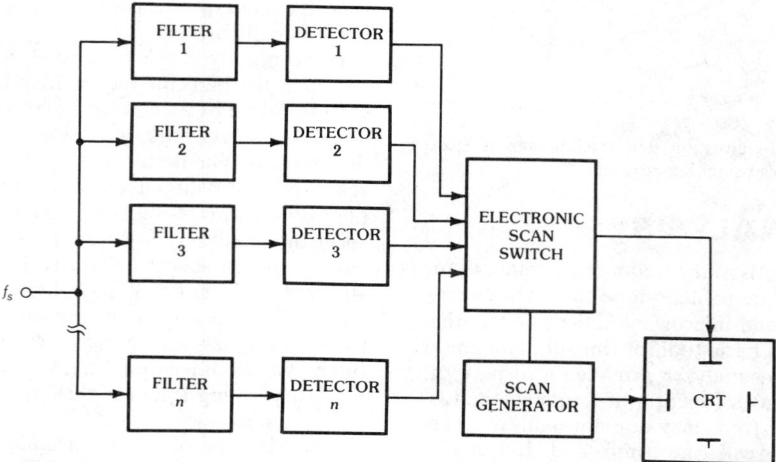

Fig. 46. A multiple-filter spectrum analyzer. (*From Hewlett-Packard Application Note No. 150 on Spectrum Analysis. © Hewlett-Packard, used with permission.*)

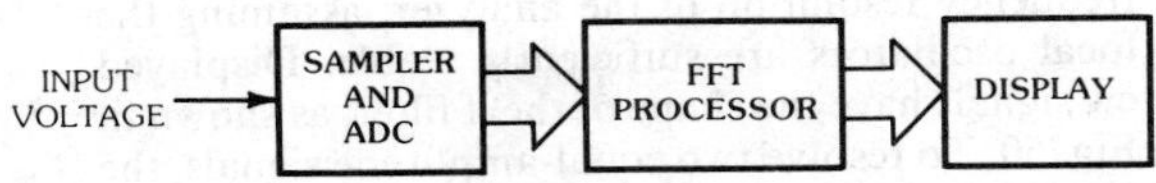

Fig. 47. Block diagram of a Fourier analyzer.

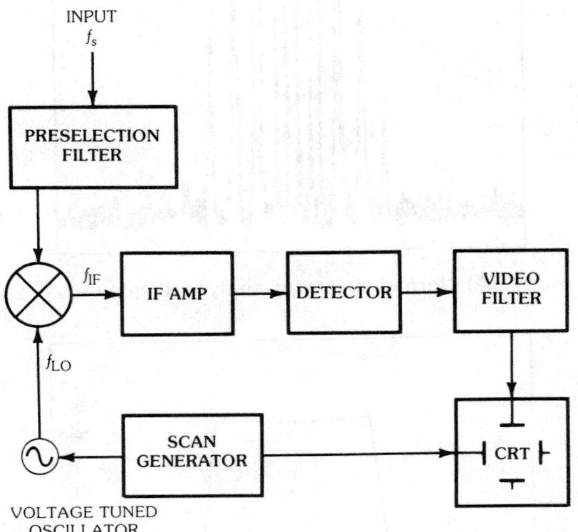

Fig. 48. Swept superheterodyne spectrum analyzer.

N = displayed noise level, in watts,
T_0 = absolute temperature, in kelvins,
k = Boltzmann's constant,
F = spectrum-analyzer noise figure,
B = if bandwidth.

For example, a noise factor of 100 (20 dB) and an if bandwidth of 10 kHz give a displayed noise level of 41×10^{-15} watt ($-$ 114 dBm).

The maximum signal level is the damage level of the input circuitry, which is typically +13 dBm for an input mixer and +30 dBm for an attenuator. Below that level, the input mixer compresses the input signal, causing an amplitude inaccuracy. The maximum signal at the input mixer for less than 1 dB of gain compression is typically 0 dBm (1 milliwatt).

Distortion products produced within the spectrum analyzer pose yet another constraint on maximum signal amplitude. For a specified spurious level, the signal level to the mixer must not exceed a given level. The dynamic range is the ratio of the largest to the smallest signal amplitudes that can be displayed simultaneously with no internally generated spurious products being present. Dynamic range depends on the display range, the distortion characteristics, and the noise level.

Signal-Analysis Characteristics as Determined by the IF Filter, Detector, and Video Amplifier

The input mixer and local oscillator translate the input frequency to a convenient intermediate frequency. Typically, several conversions are used to reduce spurious responses. The final if filter, detector, video amplifier, and display-processing circuitry largely determine the signal-analysis characteristics of a spectrum analyzer.

The if-filter bandwidth and shape determine the

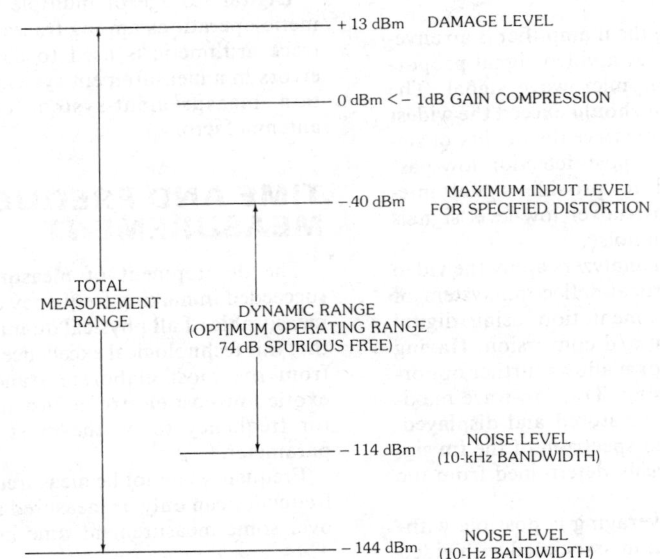

Fig. 49. Typical spectrum-analyzer input signal range.

frequency resolution of the analyzer, assuming the local oscillators are sufficiently stable. Displayed cw signals have the shape of the if filter, as shown in Fig. 50. To resolve two equal-amplitude signals, the if 3-dB bandwidth should be less than the signal separation. Resolving unequal-amplitude signals requires narrower bandwidths or rectangularly shaped filters. Filters approximating a rectangular shape, however, have poor pulse response and limit the scan time. Filters approximating a Gaussian shape allow the shortest scan times without introducing amplitude errors due to overshoot or ringing. The fastest scan rate for Gaussian-shaped filters is approximately given by:

$$R = BW^2$$

where,

R = maximum scan rate in Hz/s,
BW = 3-dB if filter bandwidth.

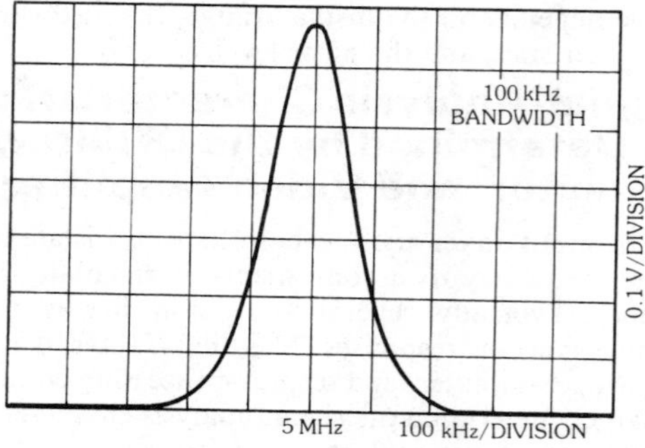

Fig. 50. Displayed cw signal has shape corresponding to if filter bandpass.

The detector following the if amplifier is an envelope detector that provides a video signal proportional to the input rf or microwave signal. The bandwidth of the detector should exceed the widest resolution bandwidth to preserve the fidelity of signal modulation. However, post-detection low-pass filtering is often useful to provide signal integration—for example, to recover low-level signals in the presence of random noise.

While analog spectrum analyzers apply the video signal directly to the vertical-deflection system of the display, newer instrumentation using digital displays first performs an a/d conversion. Having the video data in digital form allows further opportunities for signal processing. Trace-to-trace maximums or minimums can be stored and displayed. For instance, the occupied spectrum of the fm signal shown in Fig. 51 is easily determined from the "max hold" trace.

Trace-to-trace signal averaging is possible without the sweep-time limitation associated with video filtering. Fig. 52 shows the result of digital averaging on a noise-modulated signal. The signal power

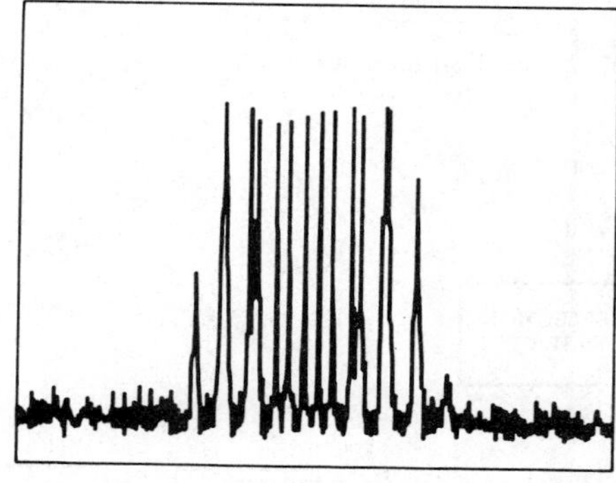

(A) Normal spectrum-analyzer display.

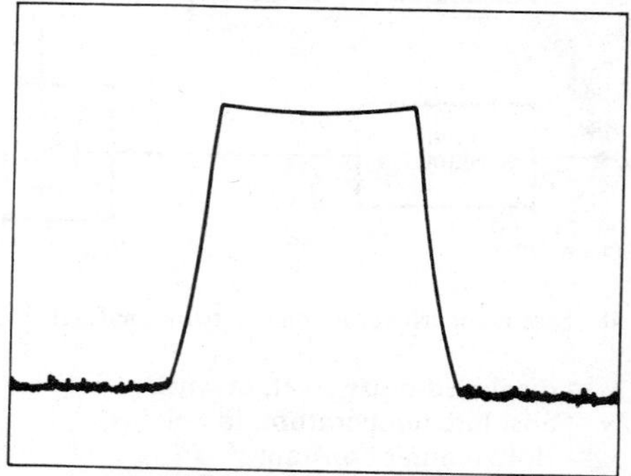

(B) Digitally processed "max hold" trace.

Fig. 51. Occupied spectrum of an fm signal.

distribution and average frequency are much more apparent.

Digital storage of multiple traces allows arithmetic operations among traces. Fig. 53 shows how trace arithmetic is used to correct for amplitude errors in a measurement system. Trace B may contain measurement-system correction factors, antenna factors, etc.

TIME AND FREQUENCY MEASUREMENT

The development of measurement science has succeeded in making frequency the most accurately measurable of all physical quantities. It is fortunate that this technological excellence scales conveniently from the most elaborate standards to the least exotic low-cost electronic instruments. It is typical for frequency to be the most accurately known parameter.*

Frequency cannot be measured at a point in time; frequency can only be measured as an average value over some measurement time interval. This meas-

* Reference 12.

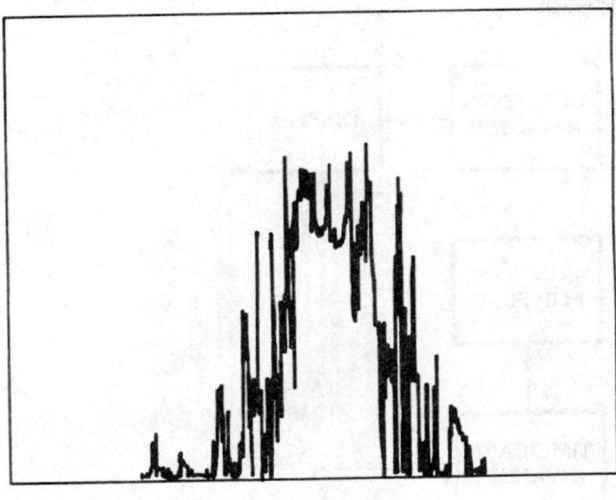

(A) Normal trace.

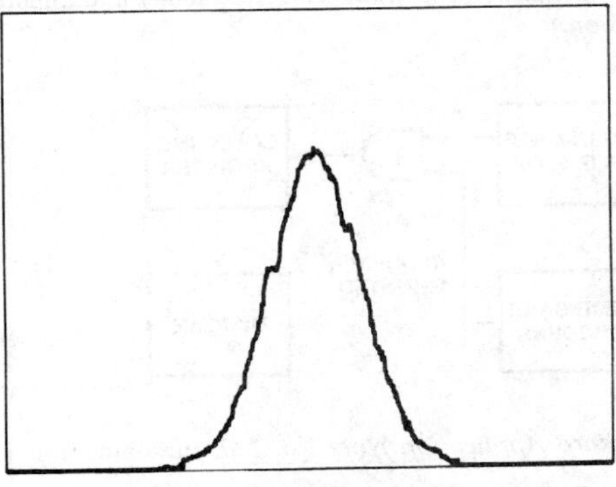

(B) 100 digitally averaged traces.

Fig. 52. Noise-modulated fm signal.

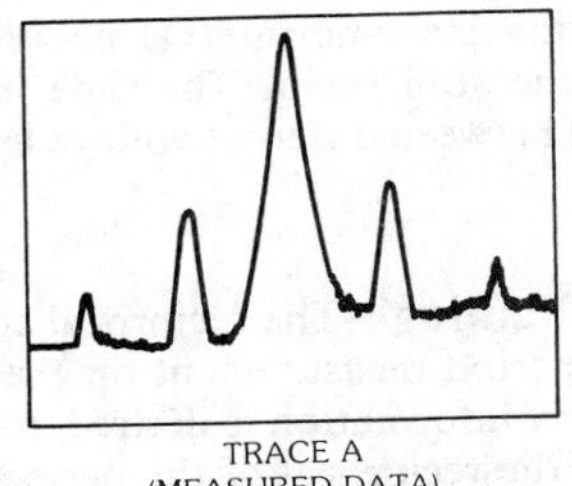

TRACE A (MEASURED DATA)

−

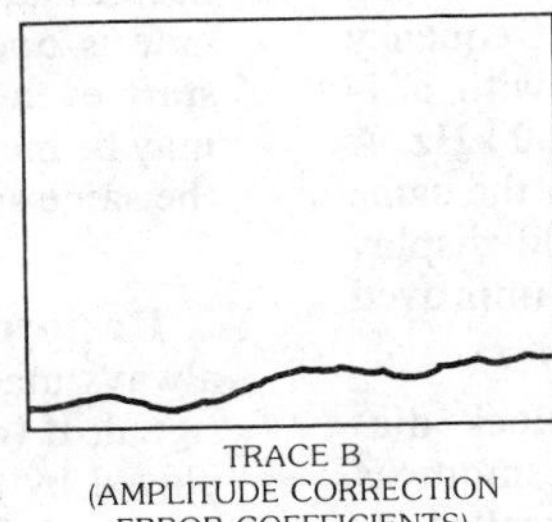

TRACE B (AMPLITUDE CORRECTION ERROR COEFFICIENTS)

=

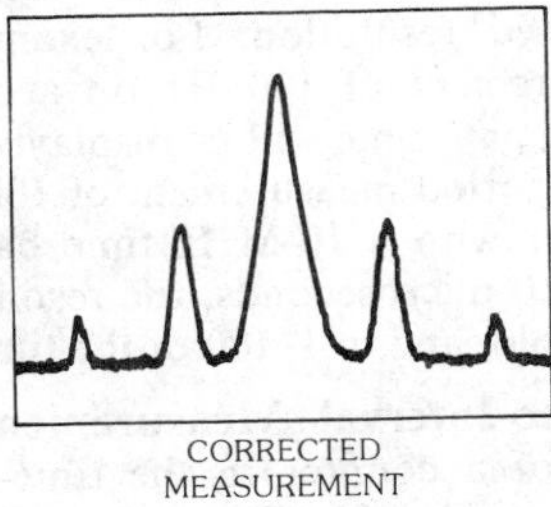

CORRECTED MEASUREMENT

Fig. 53. Error correction by arithmetic operations on multiple traces.

urement interval is generally known as averaging time, τ.

Frequency-multiplication techniques can be used to allow a frequency measurement to be made with an averaging time that is a small fraction of the period of the original signal. Frequency multiplication is usually performed by either a distort-and-filter process or a phase-locked loop. In either case, the frequency of the multiplied signal that is being measured will represent the average of the input signal over the most recent few cycles,

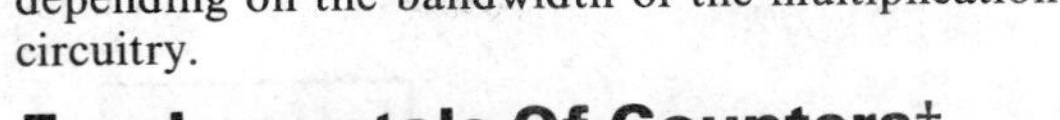

depending on the bandwidth of the multiplication circuitry.

Fundamentals Of Counters†

The counter is a digital electronic device that measures the frequency of an input signal. It may also perform related measurements such as the period of the input signal, ratio of the frequencies of two input signals, time interval between two events, and totalizing of events.

Frequency Measurement—The frequency, *f*, of a repetitive signal is measured by counting the number of cycles, *n*, and dividing this number by the time interval, *t*. The basic block diagram of a counter in its frequency mode of measurement is shown in Fig. 54.

The input signal is conditioned to form a digital pulse train in which each pulse corresponds to one cycle or event of the input signal. While the main gate is open, pulses are allowed to pass and be totalized by the counting register. The time, *t*, from the opening to the closing of the main gate, or gate time, is controlled by the time base, which usually employs a quartz-crystal oscillator at 5 or 10 MHz, where long-term stability is best.

Period Measurement—The period, *t*, of an input signal is the inverse of its frequency. The period of a signal is therefore the time taken for the signal to complete one cycle of oscillation. If the time is measured over several input cycles, then the average period of the repetitive signal is determined. This is called multiple-period averaging.

The block diagram of a counter in the period-measurement mode is shown in Fig. 55. The time during which the main gate is open is controlled by the period of the input signal. The counting register counts the output pulses from the time base dividers for one or more periods of the input signal.

The conditioned input signal may also be divided so that the gate is open for decade multiples of the input signal rather than for a single period. This is

† Abstracted from reference 8.

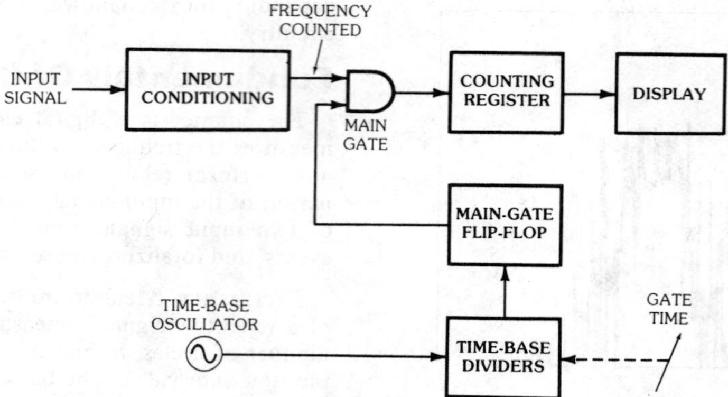

Fig. 54. Counter in frequency-measurement mode. (*From Hewlett-Packard Application Note No. 200*, Fundamentals of the Electronic Counters. © *1978 Hewlett-Packard, used with permission.*)

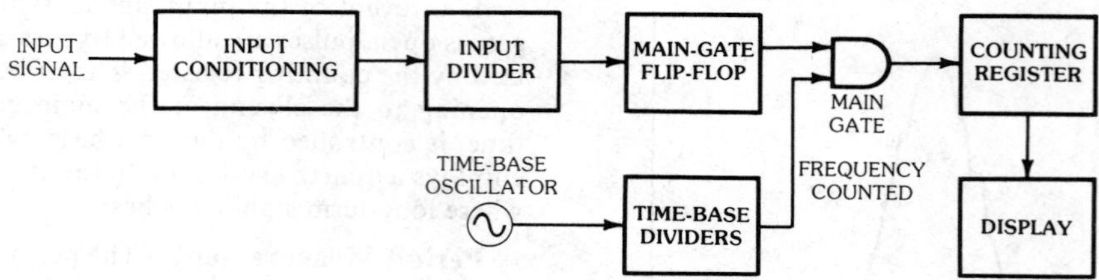

Fig. 55. Counter in period-measurement mode. (*From Hewlett-Packard Application Note No. 200*, Fundamentals of the Electronic Counters. © *1978 Hewlett-Packard, used with permission.*)

the basis of the multiple-period averaging technique.

Period measurement yields more accurate measurement of low-frequency signals by allowing increased resolution. For example, a frequency measurement of 100 Hz on a counter with a 1-second gate time will be displayed as 0.100 kHz. A single-period measurement of 100 Hz on the same counter with a 10-MHz time base would display 10 000.0 microseconds, the resolution is improved 1000-fold, and in 1/100 of the time.

Time-Interval Measurement—A block diagram of a counter in the time-interval mode is shown in Fig. 56. The main gate is controlled by two independent inputs, the start input, which opens the gate, and the stop input, which closes it.

Clock pulses from the time-base dividers are accumulated over the time duration for which the gate is open, giving the time interval between the start event and the stop event. The time interval may be measured between different voltage levels of the same signal.

Reciprocal Counting—The reciprocal counter always makes a period measurement on the input signal. If frequency information is desired, it is displayed by taking the reciprocal of the period measurement. This technique offers two features:

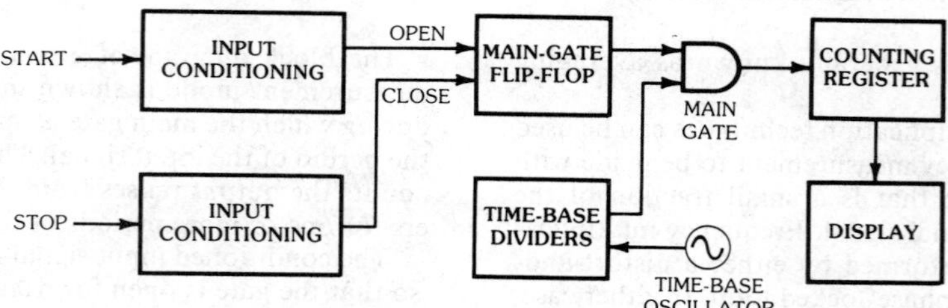

Fig. 56. Time-interval measurement mode. (*From Hewlett-Packard Application Note No. 200*, Fundamentals of the Electronic Counters. © *1978 Hewlett-Packard, used with permission.*)

1. The ±1-count quantization error is independent of the input signal frequency.
2. The period counting characteristic of the reciprocal technique allows external enabling of the main gate.

The ±1-count quantization error in a period measurement is always smaller than in a corresponding frequency measurement for all input frequencies less than that of the counted clock for a given measurement time. Assuming negligible trigger and time-base errors, the period measurement always has a higher resolution than a corresponding frequency measurement for all input frequencies less than that of the counted clock. The corollary to this is that the reciprocal technique can achieve the same resolving capability as the conventional frequency-measurement approach with a significantly smaller measurement time.

The block diagram of a reciprocal counter (Fig. 57) is similar to that of the conventional counter except for the fact that the counting is done in separate registers for time and event counts. The contents of these registers are read and a quotient computed to obtain either the period or frequency information, which is displayed directly.

The inherent high resolving power of period counting, plus the ability to initiate a measurement at any point in real time via external arming, allows frequency profiling. This is useful on frequency-agile, pulse-compression, and Doppler radar systems.

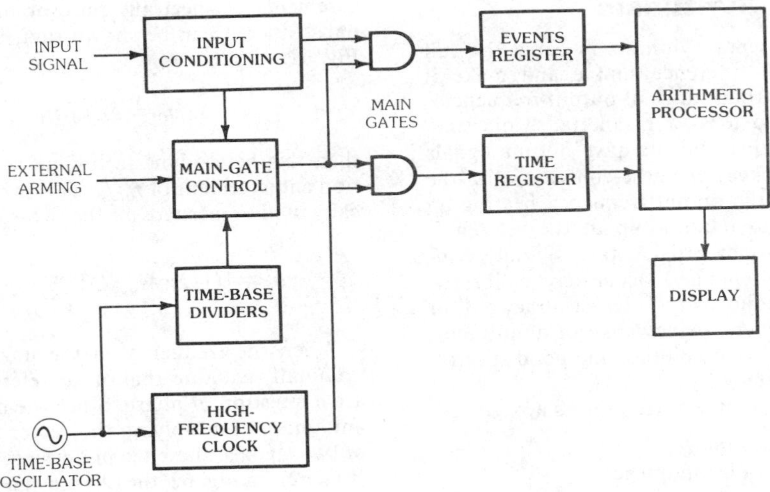

Fig. 57. Reciprocal or multiple-register counter.

Microwave Frequency Counters—Because a frequency counter, being a digital instrument, is limited in its frequency range by the speed of its logic circuitry, it must use some form of down-conversion. Four techniques are in use: prescaling, heterodyne down-conversion, transfer-oscillator down-conversion, and harmonic heterodyne down-conversion.

Prescaling, with a range to the lower microwave frequencies, results in a lower-frequency signal that can be counted in digital circuitry. A prescaler is a digital frequency-divider stage that runs continuously, ungated; its only output is a carry pulse every n cycles of the input.

Heterodyne down-conversion uses a mixer to beat the incoming microwave frequency with a high-stability local-oscillator signal, resulting in a difference frequency that is within the bandwidth of the conventional counter. In the block diagram of Fig. 58, the down-converter section is enclosed by the dashed line.

The transfer oscillator uses the technique of phase locking a harmonic of a low-frequency oscillator to the microwave input signal (Fig. 59). The frequency of the oscillator can then be measured in a conventional counter. To determine the harmonic relationship between that frequency and the input, a parallel channel with an offset oscillator is used.

The harmonic heterodyne converter is a hybrid of the previous two techniques. A counter such as the one represented by Fig. 60 acquires the input microwave frequency in the manner of the transfer oscillator, but it then makes a frequency measurement in the manner of a heterodyne converter.

The acquisition routine for this down-converter consists of tuning the synthesizer f_s until the signal detector finds a video signal, f_{if}, of the appropriate frequency range (defined by the bandpass filter). Next, the harmonic number N must be determined by the method of a second sampler loop, as with the transfer oscillator (Fig. 59), or by stepping the synthesizer back and forth between two closely spaced frequencies and observing the differences in counter readings. The value of N is then calculated.

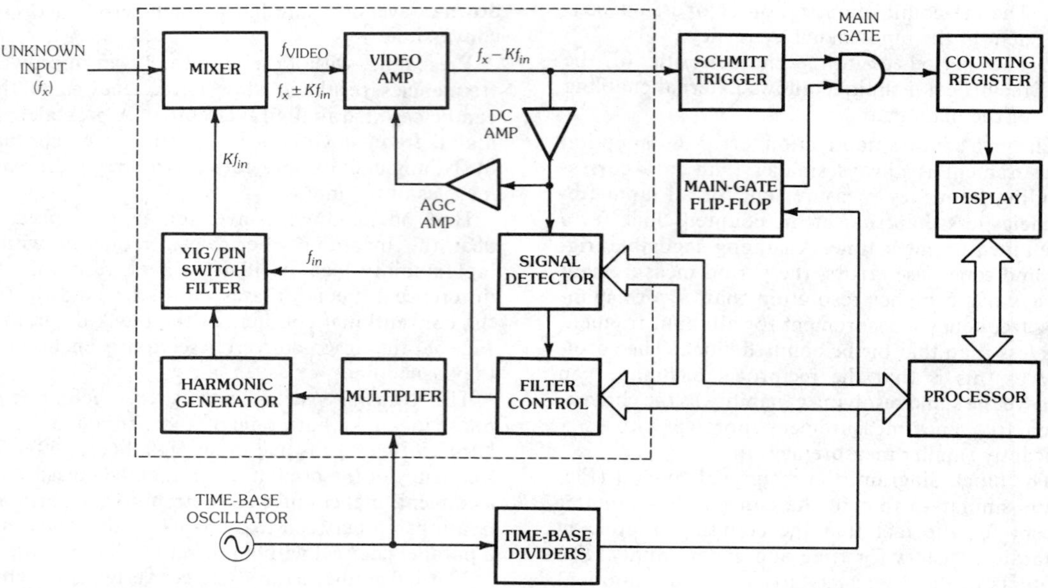

Fig. 58. Heterodyne down-conversion. (*From Hewlett-Packard Application Note No. 200*, Fundamentals of the Electronic Counters. © *1978 Hewlett-Packard, used with permission.*)

Frequency Synthesis

Coherent frequency synthesizers carry the full accuracy of their reference source and make it available over a broad range of output frequencies from millihertz to tens of gigahertz. "Coherent" means that these instruments have output signals that are phase-locked, or phase-coherent, with the reference source, the output frequency bearing an exact integer ratio relationship to the reference. Synthesizers also can provide fast digitally controlled frequency changes, accurately calibrated phase-increment adjustments, high-accuracy output level controlled over a wide range of amplitudes, various modulation capabilities, and good spectral purity of their outputs.*

Basic frequency-synthesis techniques are:

A. Division by an integer
B. Multiplication by an integer
C. Mixing with another frequency and filtering to obtain the sum or difference
D. Phase-locked loop with both input and feedback frequency division, yielding a ratio of integers times the input

Phase Noise and Stability Measurement by High-Resolution Systems

Mixers (usually double-balanced) are widely used in arrangements to extend the resolution of measurements of spectrally pure or highly stable signals. Since it can be shown that, in the system of Fig. 61,

$$\Delta f/f_0 = \Delta\tau f_B{}^2/f_0$$

then this heterodyne technique achieves a resolution enhancement of f_b/f_0. This is the method typically used to measure σ_y, the two-sample deviation:

$$\sigma_y^2 = [(1/2)(M-1)] \sum_{k=1}^{M=1} (y_{k+1} - y_k)^2$$

When the frequency of the unknown signal is nominally equal to that of the reference, the errors are a question of phase. Since the mixing function amounts to multiplying the vector representations of two signals, these techniques are phase sensitive (Fig. 62). In Fig. 62, the (X) may go by many different names: mixer, phase detector, phase comparator, phase meter, or time-interval counter. The phase-detector output is a voltage proportional to phase which may be analyzed statistically to obtain $\Delta\phi$, phase modulation; or σ_x, rms time error. For example, a phase meter or a modest counter can, by multiple time-interval average, achieve 100-picosecond resolution. A pair of such phase measurements, separated by a 100-second elapsed time, offers 10^{-12} fractional frequency resolution.

This arrangement is also widely used to measure phase modulation in the frequency domain by connecting a spectrum analyzer (typically low frequency) to the voltage output described above. This

* Reference 11.

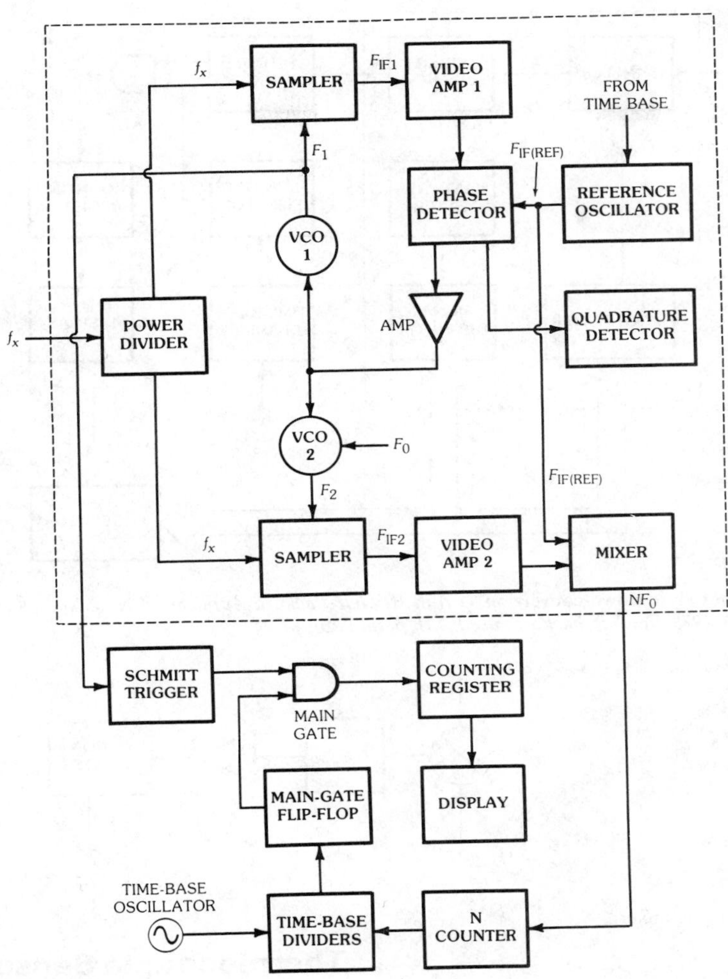

Fig. 59. Transfer-oscillator down-conversion. (*From Hewlett-Packard Application Note No. 200*, Fundamentals of the Electronic Counters. © *1978 Hewlett-Packard, used with permission.*)

yields S_ϕ or $\mathscr{L}$, the spectral density of phase modulation.*

RF AND MICROWAVE POWER MEASUREMENTS

A common technique for measuring power at high frequencies is to employ a sensing element that converts the rf power to a measurable dc or low-frequency signal. The sensing element is often designed to form a termination that is matched to the characteristic impedance (Z_0) of the input transmission line. Various types of sensing elements are used.

* References 4, 6, 7, 10, 13, and 16.

Thermistor Sensors

Thermistor sensors provide a change of resistance. The typical power range is 1 μW to 10 mW; the maximum frequency is greater than 100 GHz. Fig. 63A shows a typical power sensor employing thermistors. The thermistors form the termination for the rf input, and dc or audio power from the self-balancing bridge in Fig. 63B raises the temperature of the thermistors until they each have a resistance of $2Z_0$. The rf impedance then becomes equal to Z_0.

Since the bridge keeps the thermistor resistance constant, any heat added by the rf power causes a corresponding reduction in bias power. The rf power level is determined by measuring this change in bias power.

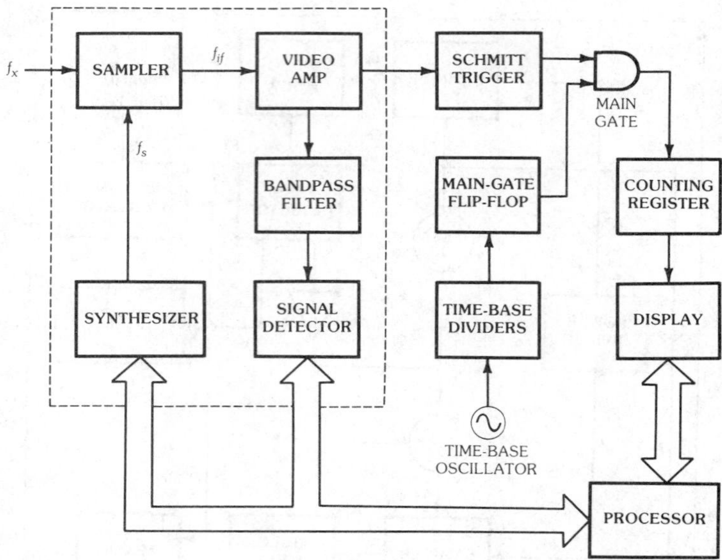

Fig. 60. Harmonic-heterodyne down-conversion. (*From Hewlett-Packard Application Note No. 200*, Fundamentals of the Electronic Counters. *© 1978 Hewlett-Packard, used with permission.*)

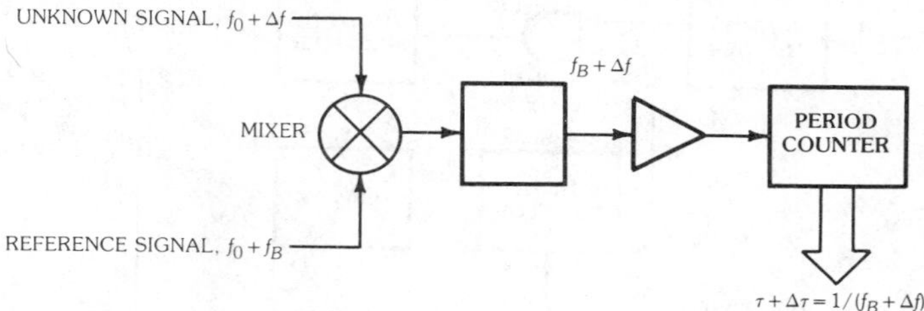

Fig. 61. Arrangement using mixer.

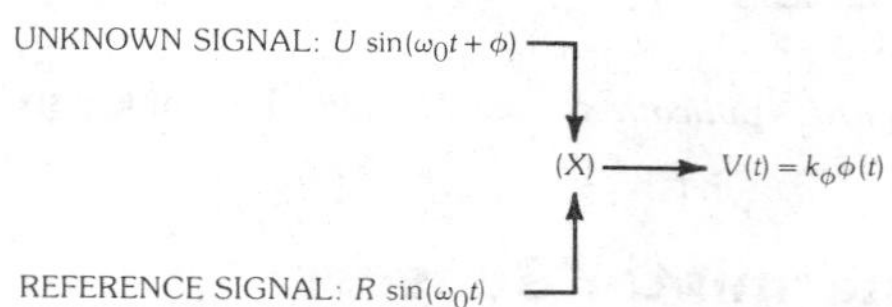

Fig. 62. Mixing function.

Thermal Converters

Thermal converters provide a dc voltage (less than 10 mV). The typical power range is 0.1 to 100 mW; the maximum frequency is less than 1 GHz. Thermal converters employ a number of thermocouples (thermopile) mounted with good thermal contact to the rf termination (Fig. 64). The rf power heats the termination, and the thermopile output voltage is proportional to the amount of power dissipated. The converter can be calibrated by applying a precisely known dc or rf power level at the input. This calibration yields the value of C.

Thermocouple Sensors

Thermocouple sensors provide a dc voltage (less than 10 mV). The typical power range is 0.1 μW to 100 mW; the maximum frequency is greater than 100 GHz. Thermoelectric sensors differ from thermal converters in that the thermocouples are used as the terminating resistors (Fig. 65). This type of sensor must be calibrated with a precise rf power level to determine the value of C.

Diode Sensors

Diode sensors provide a dc voltage (approximately 1 V at 10 mW). The typical power range is 0.1 nW to 10 mW; the maximum frequency is greater than 18 GHz. Diode power sensors (Fig. 66) use point-contact or Schottky barrier diodes to detect the rf signal. If the rf voltage is less than 20 mV, the diode output follows the square of the applied voltage, so the dc voltage is a function of rf power. At higher levels, the rectified output gradually changes to the more familiar peak-detection mode, and harmonics in the signal can cause errors

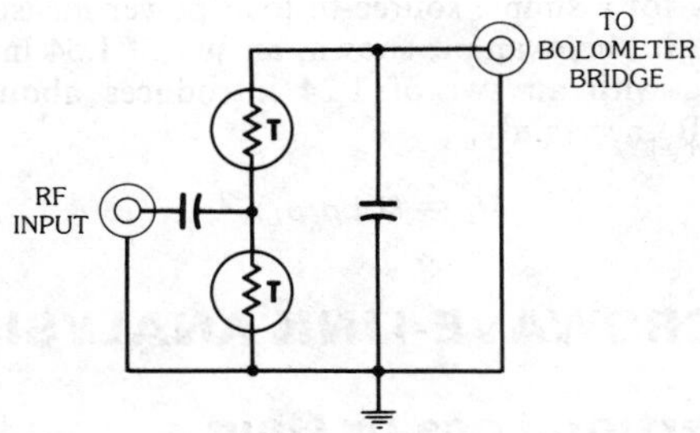

(A) Thermistor sensor.

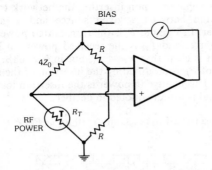

(B) Self-balancing bridge.

Fig. 63. Power measurement with thermistor sensor.

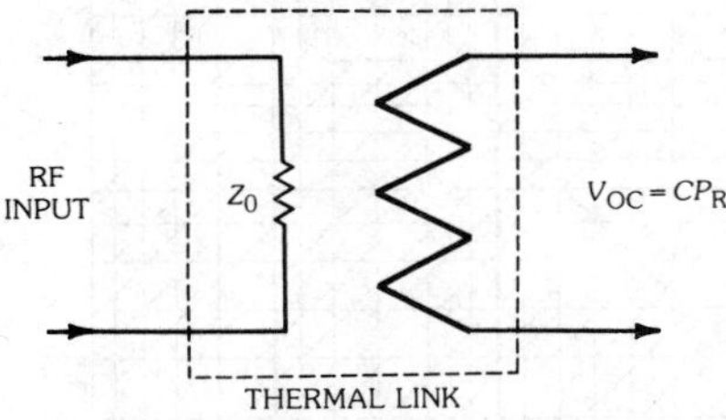

Fig. 64. Thermal converter.

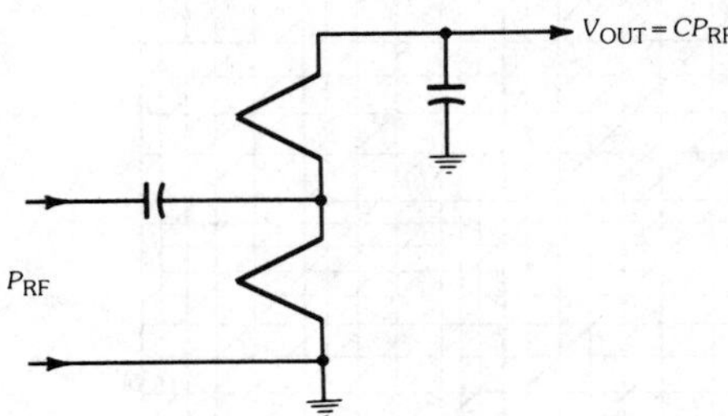

Fig. 65. Thermocouple sensor.

in the power reading. Diode sensors must be calibrated with a precise rf signal level to determine the value of C.

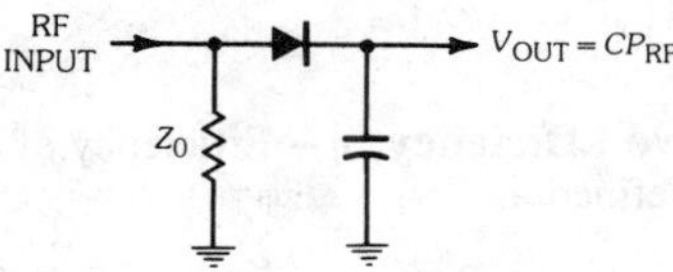

Fig. 66. Diode power sensor.

Power-Measurement Definitions

Maximum Available Power, P_{avs}—The power obtainable from a source when it is terminated in a load whose impedance is the complex conjugate of the source impedance. This condition is usually obtained by installing a tuning device between the source and the power sensor and adjusting the tuner for a peak in the power reading.

Z_0 Available Power, P_{z0}—The power obtainable from a source when it is terminated with a load matched to the Z_0 of the transmission line. The Z_0 available power is related to the maximum available power by:

$$P_{z0} = P_{avs}(1 - \rho_g^2)$$

where $\rho_g = |\Gamma_g|$

Power Incident Upon the Load, P_i—The amount of power the source transmits toward the load is usually measured by use of a directional coupler connected such that its coupled output is governed by the wave emerging from the source. The incident power is related to the Z_0 available power by:

$$P_i = P_{z0}/|1 - \Gamma_g\Gamma_L|^2$$

Power Reflected by the Load, P_r—This power is usually measured by using a directional coupler connected such that its coupled output is governed by the amount of power reflected by the load. The reflected power is related to the incident power by:

$$P_r = \rho^2 P_i$$

Power Dissipated in a Load, P_L—The power dissipated in the load is related to the other power levels by:

$$P_L = P_i - P_r = P_i(1 - \rho^2)$$

$$= P_{avs}\frac{(1 - \rho_g^2)(1 - \rho_L^2)}{|1 - \Gamma_g\Gamma_L|^2}$$

Source and Load Reflection Coefficients, Γ_g and Γ_L—Some of the foregoing equations refer to the match of the source and load impedance to the characteristic impedance of the transmission line. Values of reflection coefficients Γ_g and Γ_L are given by:

$$\Gamma_g = (Z_g - Z_0)/(Z_g + Z_0)$$
$$\Gamma_L = (Z_L - Z_0)/(Z_L + Z_0)$$

$\rho_g = |\Gamma_g|$
$\rho_L = |\Gamma_L|$

Effective Efficiency, η—Efficiency of a power sensor is defined as

$$\eta = P_{sub}/P_0$$

where,

P_{sub} is the dc substituted or rf calibration power,
P_0 is the amount of power dissipated in the sensor.

Calibration Factor, K_B—Calibration factor is related to efficiency by:

$$K_B = P_{sub}/P_i = \eta(1 - \rho^2)$$

where ρ is the magnitude of the reflection coefficient of the sensor.

Multiple Mismatch Error, M_u—The precise amount of power dissipated in the sensor is a function of many variables, the largest one being the interaction of the source and load reflections. Fig. 67 can be used to estimate the worst-case uncertainty for a simple source-to-load power measurement. In the example shown, an swr of 1.54 interacting with an swr of 1.24 introduces about a 0.2-dB uncertainty.

$$M_u = (1 \pm \rho_g \rho_L)^{-2}$$

MICROWAVE-LINK ANALYSIS

Insertion Loss or Gain

Insertion loss or gain is defined as the loss or gain that is apparent upon inserting the network to be measured between a given source and a given receiver. In Fig. 68, P_s is the transmitted power of the source, and P_R is the received power. In Fig. 68A, it is clear that $P_R = P_s$, but in Fig. 68B, the value of P_R is modified by the insertion of the network. This change in power is the insertion loss or gain and is usually quoted in decibels:

$$\text{Insertion Loss or Gain} = 10 \log_{10}(P_R/P_s)$$

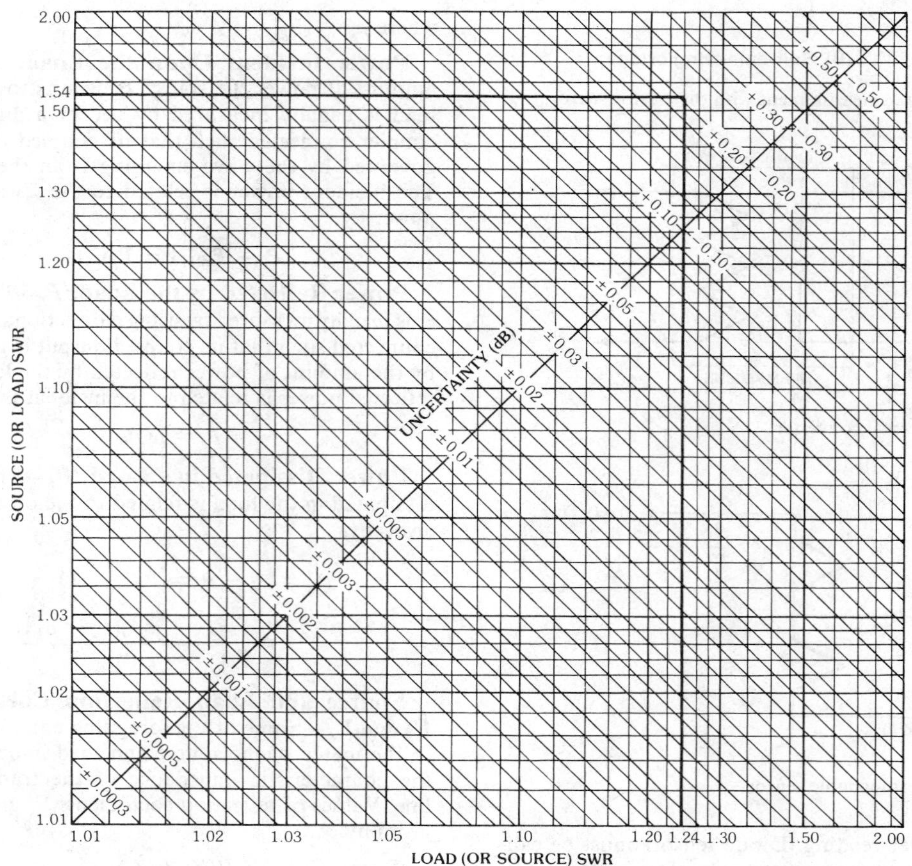

Fig. 67. Mismatch uncertainty limits.

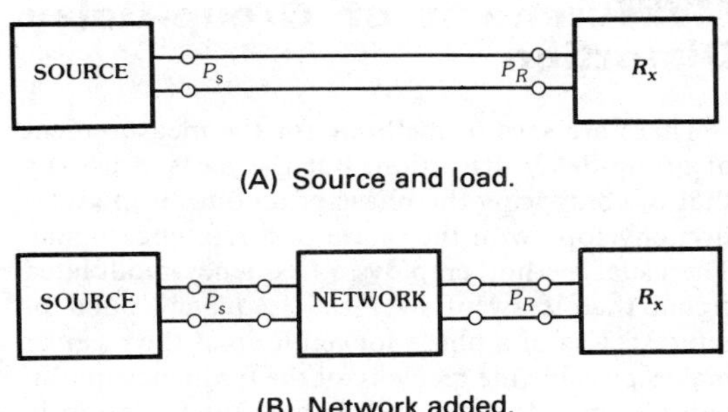

Fig. 68. Insertion loss or gain.

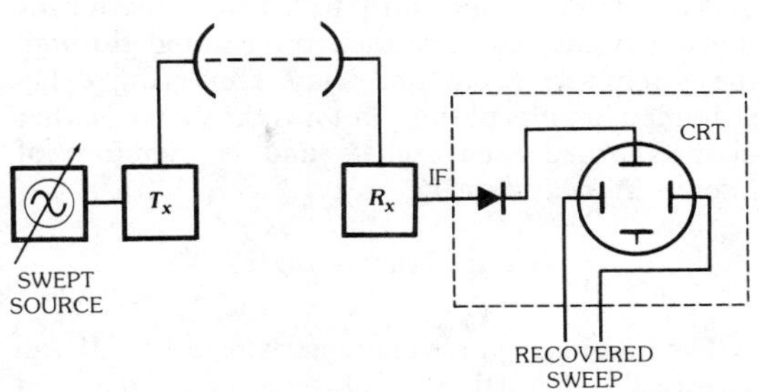

Fig. 69. Amplitude-response measurement.

Amplitude Response

Amplitude response is sometimes referred to as "flatness" or frequency response and is the variation in gain or loss with change in frequency over a defined frequency band. While it may be thought that this parameter is not important for frequency-modulated radio links, it generally must be well controlled, especially for high-capacity systems. Amplitude response is also very important for the rf and if sections of digital radio links.

In order to measure amplitude response accurately, it is important that harmonics of the transmitted signal not be included in the measurement of received power. Thus the ideal measurement technique would be one using a tunable receiver. In practice, however, the measurement of radio links using intermediate frequencies of 70 MHz or 140 MHz requires a swept frequency range that allows a simple low-pass filter to remove harmonics. Additionally, errors due to the use of a broad-band detector are generally low. This, then, allows a simple measurement principle to be used that lends itself to remote or "straightaway" measurements (Fig. 69).

Measurement receivers usually incorporate a frequency-tracking loop that reduces the effective sweep range in the analyzer where the benefits arising from a reduced sweep may be used for group-delay and/or linearity measurements. The frequency-tracking loop, however, also gives a recovered sweep signal that is used to drive the *X* deflection of the display and makes the "straightaway" measurement possible (Fig. 70).

The swept range of the input signal is reduced by the loop gain of the arrangement shown in Fig. 70, where the derived low-frequency signal used to drive the vco is a facsimile of the frequency variation against time characteristic of the source signal.

These techniques are quite satisfactory when used to measure amplitude response of if or rf sections of radio links. They are, however, generally inadequate for the measurement of baseband amplitude response, where a selective receiver should be used. This implies that the receiver should be capable of following the frequency changes of the input signal; i.e., it must "track" the input frequency. Various techniques have been developed for this, even for those measurements where frequency "stepping" synthesizers are used. These methods usually require that a receiver be programmed to tune to its next frequency only after having successfully completed a measurement at the current frequency. The method therefore usually relies on the generator stepping at a slower rate than the receiver.

Envelope-Delay Distortion (EDD)

Envelope-delay distortion is often termed group-delay distortion and is effectively the variation in the derivative of the phase-vs-frequency response. For zero group-delay distortion, the phase/fre-

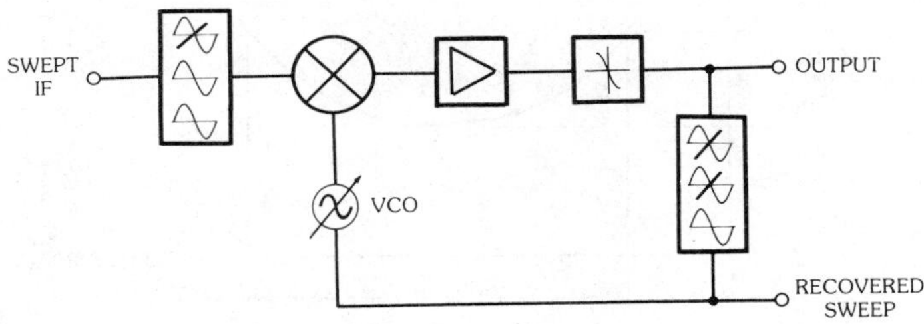

Fig. 70. Frequency-tracking loop.

quency characteristic must be linear. This means that all frequencies are then transmitted through the system with equal time delay. Thus group delay is defined in units of time, but it is the variation that distorts broad-band signals and is therefore of interest in measurement.

$$\text{Group Delay} = d\phi/df$$

The nonlinear phase characteristic of Fig. 71 will give rise to distortion of a broad-band signal that will manifest itself in the form of noise and intermodulation for analog systems. The group-delay response derived from Fig. 71 will be of the form shown in Fig. 72. As can be seen, there is a mean level of group delay that is not very important; it is the variation of distortion that must be equalized.

Equalization is achieved by means of networks that give inverse group-delay variations. The most common types required are those that compensate for linear group-delay slope or parabolic group-delay distortion. It is important that equalizers be connected as close as possible to the source of group-delay distortion or at least that no nonlinear networks be connected before the equalizer. This is because nonlinearities that introduce amplitude-to-phase-modulation conversion will produce effects that cannot be removed by group-delay equalization.

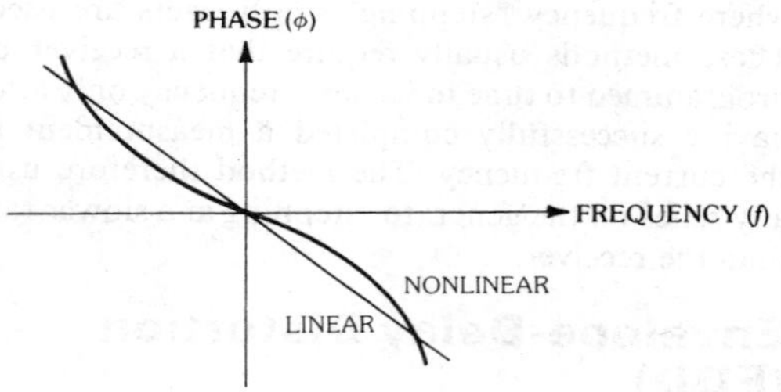

Fig. 71. Phase versus frequency slope.

Measurement of Group-Delay Distortion

There are several methods for the measurement of group-delay distortion, but the basis is usually that of comparing the phase of an output modulation envelope with the phase of a reference signal. The usual method employs a frequency-modulated signal that is swept over the frequency band of interest. Use of a phase-locked loop at the receiver makes possible the recovery of the frequency modulation, which then is used to give the variation in phase as the input signal is swept (Fig. 73).

It was shown earlier that a frequency-tracking loop could be used to recover the sweeping signal for the *X* deflection of the crt. Additionally, the tracking loop reduces the swept range of the received if signal so that distortions arising from the nonideal nature of networks in the receivers are minimized. For example, a signal that is used to test a radio link may sweep over a band of more than 30 MHz but will be sweep reduced in the analyzer to yield a sweep range of less than 100 kHz. Consequently, the errors introduced by the analyzer are minimized. The phase-locked loop in the group-delay measurement then uses the demodulated baseband frequency to control the phase of its reference oscillator, which will be held at the mean phase of the modulating frequency. Variations in phase are then available at point B in Fig. 73.

Another way of looking at this is to say that if point A represents the reference phase of the modulating signal, then going backward around the double integrating loop means that the signal at point B is the derivative of this phase, or $d\phi/dt$. In this system, however, time is the same as frequency, because as time proceeds, the if is swept; hence the instantaneous value of the dc voltage at B represents the group delay $(d\phi/df)$ at the instantaneous value of the if.

The reason for using a phase-tracking detector is that it is always working very close to zero phase difference between the inputs. This means that the

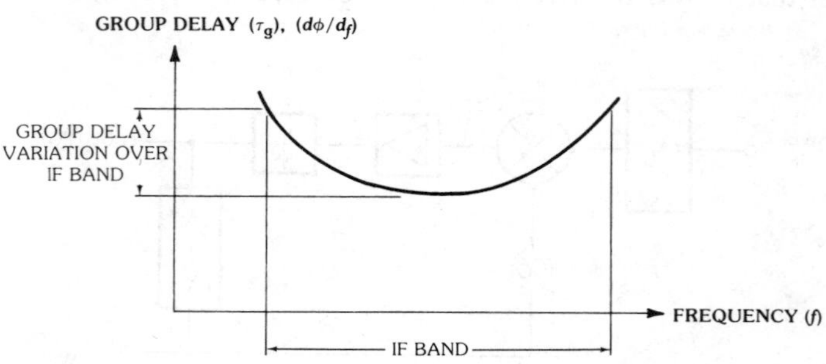

Fig. 72. Group-delay distortion.

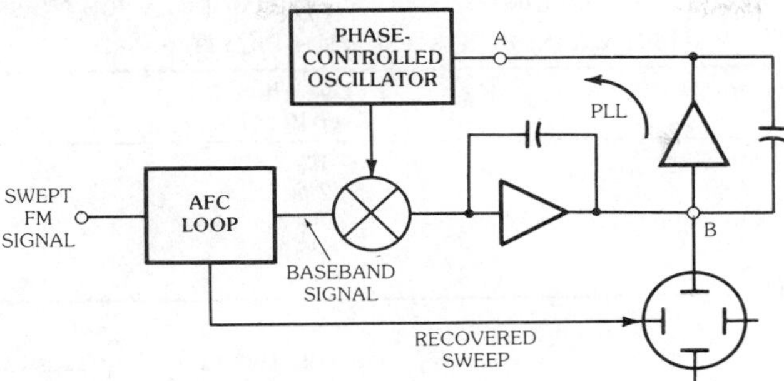

Fig. 73. Measurement of group-delay distortion.

phase detector is always working on the central part of its dc/phase conversion slope, and nonlinearity of this response does not affect the measurement.

The measurement of group-delay distortion involves the careful selection of several test parameters:

Sweep range
Sweep rate
Modulation frequency (test tone)
Modulation index
Post-detection bandwidth

It is important to select a sweep range appropriate for the device or system under test. Some consideration should be given to the spectrum of the modulating signal, since the device under test will be subjected to the total spectrum. This is usually important only where a high-frequency test tone (> 1 MHz) is used, but it can be important for testing components such as narrow-band filters with lower test-tone frequencies.

The sweep rate employed is often in the range of 50 to 100 Hz, but this range may not be suitable for systems such as satellite communication links where use of a lower sweep rate will allow a narrow bandwidth to be selected for the post-detection bandwidth. This will then enhance the measurement resolution by reducing noise power.

The modulation frequency used is a compromise between two conflicting effects:

1. Use of too high a frequency will tend to conceal rapid fluctuations in group delay such as the ripple produced by imperfect impedance matching.
2. Use of too low a frequency will produce a low voltage at the output of the group-delay detector, and the signal-to-noise ratio of the display will be poor.

It is therefore usual to select frequencies between 50 kHz and 500 kHz to give an appropriate compromise. There are cases, however, where much higher modulation frequencies can be used. This is usually true when either television or broad-band telephone systems are carried and the test signal has to reveal low values of nonlinearity that produce intermodulation. In these cases, the measurement sensitivity is enhanced by using a modulation frequency between 1 MHz and 12 MHz. In using these high frequencies, it is not appropriate to refer to the measurement as a group-delay measurement, but rather as a measurement of "differential phase."

In practice, this measurement is not as important as the measurement of "differential gain," which can be very useful in tracing problems of amplitude-to-phase modulation conversion.

Television systems are often measured with a test-tone frequency equal to the television color subcarrier frequency. The measurement of differential phase is then used to define system performance.

Differential gain is an extension of the measurement of linearity that uses techniques similar to those for group delay, but, whereas this reveals how the phase for the modulation envelope varies, linearity shows how the amplitude of the envelope varies. The same measurement principle is applied to modulators and demodulators to measure nonlinearity of the voltage/frequency characteristic. It is interesting to note that for a nonlinearity occurring in a discriminator or modulator, the measured nonlinearity will be relatively independent of the modulation frequency employed. This contrasts with the case for if nonlinearities where the distortion value changes with the square of the change in test-tone frequency. This property may then be used in analyzing system deficiencies.

It can be valuable to be able to separate key characteristics that contribute significantly to intermodulation distortion. Table 1 shows the relative effects of distortion styles to give an approximate indication of which shapes are most serious.

Table 1 does not show the effects of the "coupled" responses where amplitude-to-phase modulation conversion interacts with distortions in other parameters, but these interactions can have a serious effect in high-capacity radio systems. Additionally, nonflatness can be important, especially for digital radio systems.

TABLE 1. DISTORTION FOR 100-pW INTERMODULATION NOISE (1800-CHANNEL SYSTEM WITH PREEMPHASIS)

Parameter	Distortion Over 10 MHz	Test Tone
Differential-Gain Slope	1%	2.4 MHz
Group-Delay Slope	2 ns	—
Differential-Phase Slope	3% rad	2.4 MHz
Group Delay, Cubic	3.5 ns	—
Group Delay, Parabolic	14 ns	—

Return Loss

In the alignment of microwave radio links, it is important that the impedance match of the various sections be well maintained. This is especially important where cabling is used between a source and a load, since any mismatch will produce time-delayed reflections that may impair link performance. The normal way of describing the mismatch of a source and load is by using the term "return loss." Return loss is the measure of the ratio between the transmitted and reflected signals:

$$\text{Return loss} = 20 \log_{10} |E_i/E_r|$$

where,

E_i is the incident signal,

E_r is the reflected signal.

Return loss is a measure of the magnitudes of incident and reflected signals and does not take account of phase relationships.

Measurement of Return Loss

Long-Cable Method—In this method, a "long cable" is connected to the termination under test as shown in Fig. 74. When the swept if signal is applied to the long cable and its termination, a series of ripples will appear on the crt display. If the test termination is removed, the open circuit produces a large amplitude of ripple that is then adjusted by the attenuator to equal the level when the termination is connected. The return loss is equal to 2 times the attenuation inserted.

Note that the cable must be long enough to produce at least one ripple over the swept range, but more than ten are required to observe variations across the band. Thus, for normal if measurements, a length in excess of 20 meters is required.

Standard-Mismatch Method—This method (Fig. 75) relies on the measurement of power passed from a hybrid when it is terminated by a known mismatch and then by the test item. A typical mismatch of 17 dB is used to calibrate the power meter/detector, after which the actual return loss may be measured directly.

Baseband Measurements

Most of the above measurements are made at the intermediate frequency (if) of the radio system or at the microwave carrier frequency. Obviously, some measurements such as modulator linearity also involve the direct use of baseband inputs, but there is an additional class of measurements that define system performance and are carried out at baseband level. The most obvious of these is the measurement of baseband amplitude/frequency response, but checks for spurious frequencies should also be made. These measurements can be made with special instruments that provide spectrum analysis.

For multichannel telephony systems, the prime concern is to minimize the level of noise in each channel. There can be many sources of noise, but they are mainly either thermal or caused by cross talk or intermodulation. Thermal noise is not affected by the traffic level (loading) of the system, whereas intermodulation distortion is sensitive to loading. The system designer therefore attempts to define a loading level at which intermodulation noise is low and yet the ratio of signal level to thermal noise is also satisfactory.

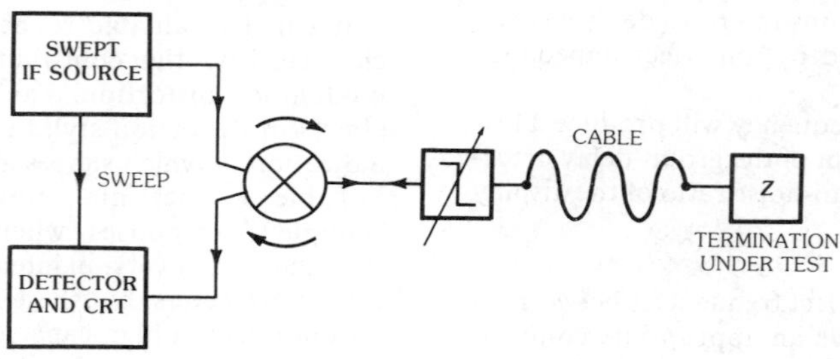

Fig. 74. Return-loss measurement by long-cable method.

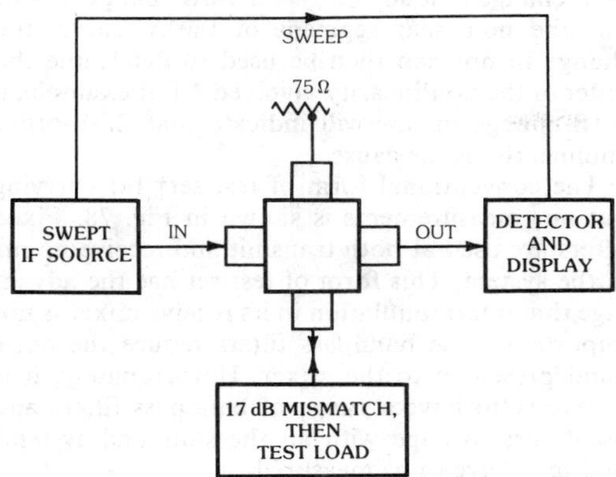

Fig. 75. Measurement of return loss by standard-mismatch method.

In order to verify that the correct balance of thermal and intermodulation noise is obtained, it is necessary to be able to simulate the traffic load and then measure the effects in channels throughout the baseband. This is achieved by using a band of thermal noise to represent the traffic.

Noise Loading—Traffic simulation using a band of thermal noise is possible because the amplitude distribution of such noise is Gaussian and the power-versus-frequency spectrum is uniform. This latter characteristic is analogous to white light, so the band of noise used to simulate traffic is often referred to as "white noise."

In order to simulate a given traffic load, the formula normally used to calculate the noise power required is

$$\text{Mean noise power} = -15 + 10\log_{10}N$$

where,

Mean noise power is in dBm0 (power in dBm referred to a point of zero relative transmission level),

N is the number of telephone channels.

For this formula to be valid, N must be greater than 240, because as the noise bandwidth is reduced to simulate high-capacity systems, the signal departs significantly from a Gaussian distribution. The formula then used is

$$\text{Mean noise power} = -1 + 4\log_{10}N$$

where,

Mean noise power is in dBm0,

N is less than 240.

Modern systems may, of course, be loaded with data. In this case, a modified formula is used:

$$\text{Mean noise power} = -10 + 10\log_{10}N$$

where,

Mean noise power is in dBm0,

Data loading N is greater than 12 channels.

It is necessary to limit the noise bandwidth to represent the bandwidth of the system. Filters for controlling noise bandwidth are prescribed by the various telecommunications authorities.

Noise Power Ratio—When the system has been loaded with noise to simulate traffic, it is then necessary to measure intermodulation noise. This is done by introducing a quiet "slot" into the noise band. This slot is then inspected to determine the level of noise introduced by intermodulation. This is known as the measurement of noise power ratio (npr), which may be defined as the ratio of the noise power in a measurement channel with the baseband fully loaded to the noise power in that channel when the baseband is noise-loaded except for the measurement channel. In Fig. 76, P_1 (Fig. 76A) represents the power in the measurement channel due to the loading signal, P_2 (Fig. 76B) would equal zero for the ideal case where there is no thermal noise or intermodulation noise, and P_3 (Fig. 76C) shows the effect that these noise components will have. Thus,

$$\text{npr} = P_1/P_3$$

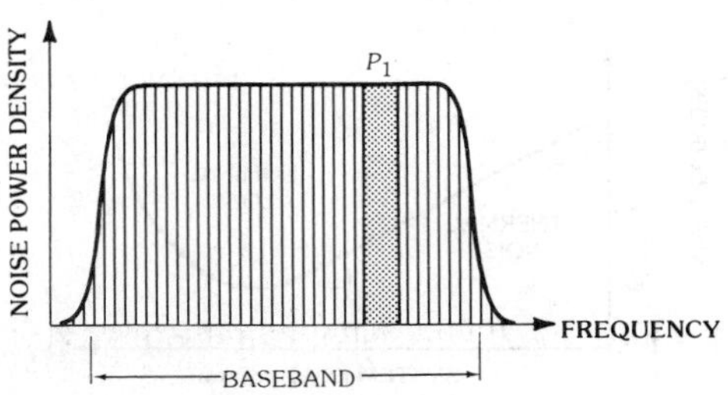

(A) Measurement channel.

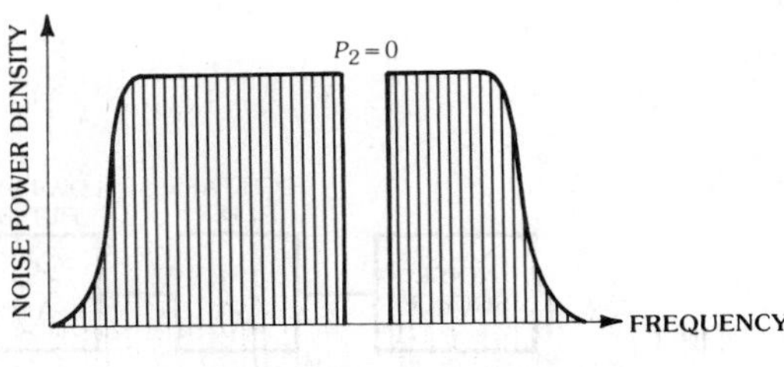

(B) Quiet "slot" introduced.

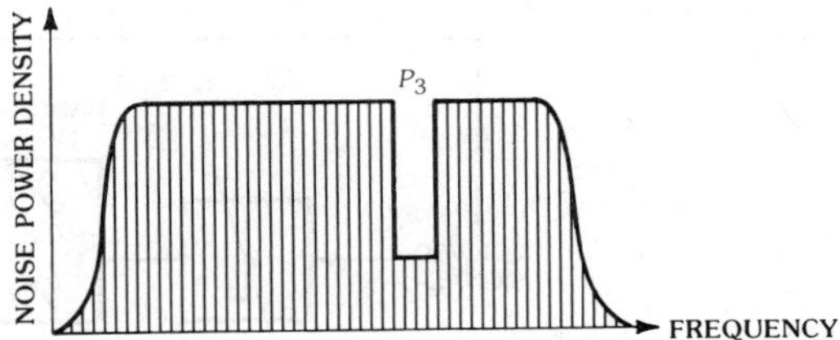

(C) Effect of thermal and intermodulation noise.

Fig. 76. Principle of npr measurement.

If the noise load is removed, the thermal noise component in the measurement channel can be measured. (Thermal noise is sometimes referred to as basic noise or intrinsic noise.) Then if the noise load is added, the additional noise introduced by intermodulation and cross talk can be measured.

By varying the level of the baseband noise loading signal, it is possible to explore the combined effects of thermal noise and intermodulation noise to determine the optimum operating point for the system.

The npr curve is often referred to as a "*V*" curve because of the characteristic shape (Fig. 77). At low loading levels, thermal noise will predominate, and npr will simply be proportional to the loading power level. Thus this part of the characteristic is linear. As intermodulation noise rises, however, the noise power ratio will fall rapidly and give the shape of Fig. 77. It is possible to gain troubleshooting information by inspecting the *V* curve for slots placed at different frequencies throughout the baseband. The method employed is to select a slot and observe the change in npr for a 1-dB change in noise loading. In the linear region of the curve, a

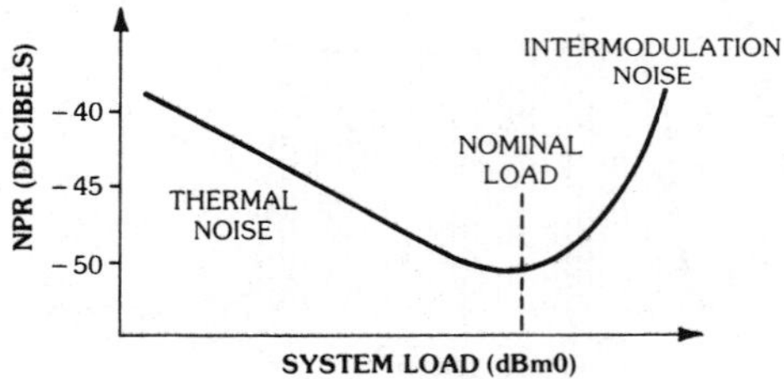

Fig. 77. Npr curve for a multichannel radio system.

1-dB change in load will give a 1-dB change in npr. For the nonlinear segment of the V curve, the change in npr can then be used to determine the order of the nonlinearity involved.* For example, a 2-dB change in npr will indicate that third-order nonlinearity is the cause.

The conventional form of test set† for carrying out npr measurements is shown in Fig. 78. Fixed filters are used at both transmit and receive points in the system. This form of test set has the advantage that intermodulation in its receive mixer is not important since bandpass filters reduce the noise band presented to the mixer. Unfortunately, it is necessary to have a range of bandpass filters and oscillators to cope with all the slots and systems that may have to be measured.

The modern design of highly linear mixers, however, allows a selective receiver to make the npr measurement. Such receivers use a synthesizer as the local oscillator, so the requirement for separate filters and oscillators is avoided.

Conversions—It is sometimes necessary to convert from npr to other measures, such as signal-to-noise ratio (snr). The following relations are used.

1. Weighted signal-to-noise ratio in decibels:

$$\text{Weighted snr} = \text{npr} + 10 \log_{10}(f/3.1) - P + W$$

where,

f = bandwidth of baseband (kHz),
P = nominal load (dBm0),
W = weighting factor (dB).

When applied to systems having greater than 240 channels and using *C*-message weighting, this simplifies to

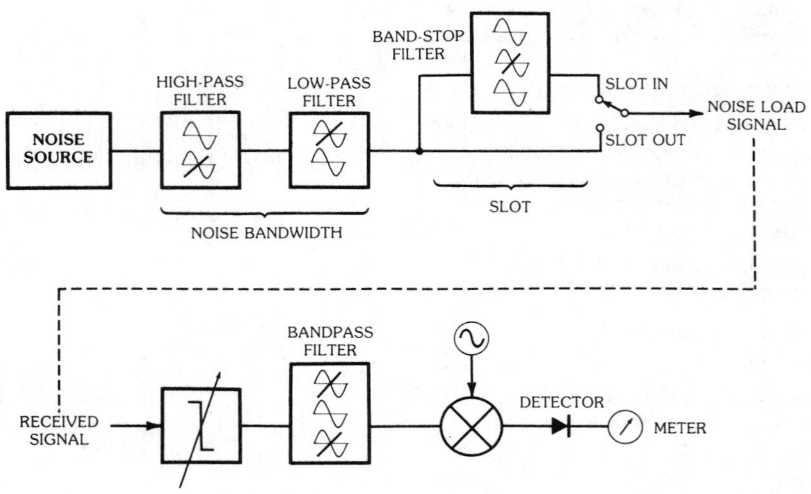

Fig. 78. Npr measurement.

* References 18 and 19. † Reference 18.

$$\text{snr} = \text{npr} + 17.8$$

2. The notation dBrnC0 means decibels of noise with respect to reference noise of −90 dBm and with *C* message weighting, corrected to a point of zero transmission level. Conversion to dBrnC0 for systems that have more than 240 channels is given by

$$\text{dBrnC0} = 71.7 - \text{npr}$$

Note that the European psophometric weighting and *C* message differ by 1 dB, but in practice the above formula is often used for psophometric weighting.

Carrier/Noise Measurement

For digital radio systems, the normal overall performance measurement is based on "bit error rate," where a pseudo-random-sequence bit pattern is applied to the system and errors are indicative of system imperfections. A more detailed analysis is made possible by inspection of the so-called "eye" diagram at the sampling stage of digital signal recovery, but measurement of error rate as the carrier-to-noise ratio is varied can be a more revealing method of analyzing system performance.

Diagnostic Measurements and System Performance

The relationships between diagnostic tests and system performance are often very complex. Thus it is difficult to relate measurements of envelope-delay distortion to the intermodulation that will be produced. It is possible to make such predictions,* but this generally requires the use of a computer. It is generally more useful to derive a clear understanding of how these measurements can best be used to optimize system performance, or at least to be able to separate those measurement responses that have a significant effect on system performance from those that have a lesser effect.

IEEE 488 OR IEC 625-1 GENERAL PURPOSE INTERFACE BUS (GPIB)**

The GPIB provides a means by which system components communicate with each other. It is analogous to the telephone system that allows one person to exchange verbal information with another person—or, in the case of a conference call, with several other people.

Every participating device in a GPIB system performs at least one of three roles: talker, listener, or controller. A talker transmits data to other devices via the bus. A listener receives data from other devices. A controller manages communications on the bus, primarily by designating which devices are to send or receive data during each measurement sequence. The controller may also interrupt and command specific actions within devices.

Many devices are both talkers and listeners. For example, a programmable multimeter or electronic counter listens when receiving its program instructions, and it talks when sending its measurements to another system component such as a printer or computer. There can be several listeners, but to avoid confusion, only one talker can be active on the bus at any one time.

In its simplest form, a GPIB system can consist of only one talker and one listener—for example, an electronic counter linked to a printer. In this application, there is no need for a controller; the counter "talks" and the printer "listens."

The full versatility of the GPIB becomes apparent when there is more stimulative interaction among the interconnected devices. To achieve this requires a controller that can schedule measurement tasks, set up instruments to perform specified tests and measurements, monitor processes on-line, and process data and analyze and interpret the results. A desktop computer, or "computing controller," is often used as a system manager to control the bidirectional data flow of the GPIB.

Interface circuitry is part of each interfaced instrument or computing controller—the GPIB cable itself is passive. Inside the cable, 16 active signal lines are grouped into three sets, according to function (Fig. 79).

The eight data lines carry coded messages—such as addresses, program data, measurements, and status bytes—among as many as 15 devices interconnected with a single bus. The same data lines are used for both input and output, and the messages are in bit-parallel, byte-serial form. Data are exchanged asynchronously, for maximum compatibility among a wide variety of devices.

For unambiguous, intelligible communication between instrument and computer devices, some rules or protocol must apply to the communication process itself. Thus, the exchange of data is controlled by the second set of signal lines, the three data byte transfer control lines. All listeners addressed on the GPIB send a signal when they are "ready for data." The talker then initiates data-byte transfer by signaling that "valid data" are on the data lines. When each listening device successfully receives the message, it acknowledges "data accepted."

When two or more devices are listening, all must acknowledge "data accepted" before new data can be transmitted on the bus. Thus, data are transferred at the rate of the slowest listener participating in the particular conversation.

High-speed and slow devices can be connected on

---

* Reference 17.

** From Hewlett-Packard publication 5952-0050.© Hewlett-Packard, used with permission.

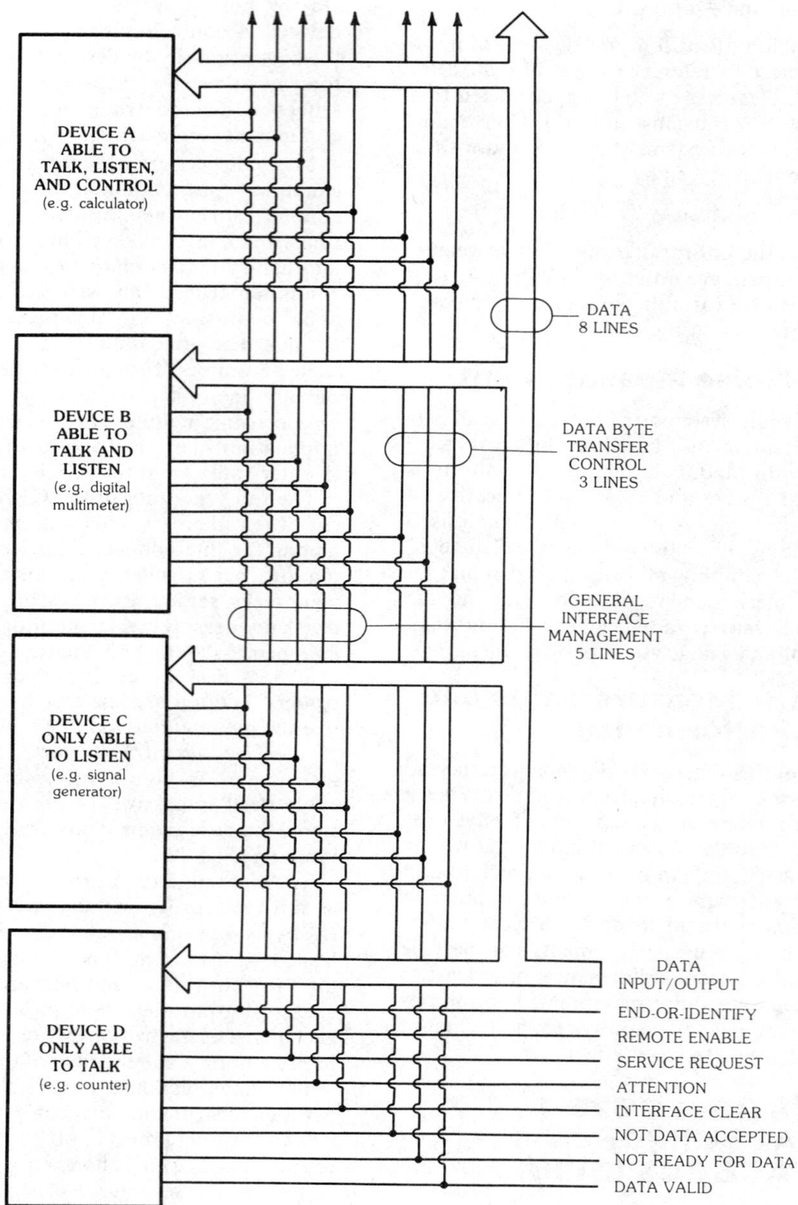

Fig. 79. General purpose interface bus. (*From Hewlett-Packard publication 5952-0050. © Hewlett-Packard, used with permission.*)

the same bus. Data transfer between high-speed devices is not adversely affected, as long as the slow devices are not addressed or participating in the conversation.

The remaining five general interface management lines are used for such things as activating all the connected devices at once, clearing the interface, and so forth.

Up to 15 GPIB devices can be linked together on one bus. The GPIB cable connects them in parallel functionally, in either star or linear fashion physically. Total cable transmission path length on any one bus should not exceed 20 meters or 2 meters per device, whichever is less, except where the distance is extended by devices such as a common carrier interface.

A single computing controller can manage more than one bus instrument cluster at once. This is particularly useful for achieving concurrent actions at different test stations in a production test measurement application.

Interface circuitry for controllers and instruments is commercially available in integrated-circuit packages.

ELECTROMAGNETIC COMPATIBILITY, INTERFERENCE, AND SUSCEPTIBILITY

EMC/EMI

Electromagnetic compatibility (EMC) is the ability of items of electronic equipment to function properly together in the electronic environment. Most electronic equipment is designed to perform an assigned task. However, when improperly designed or poorly maintained, a device may emit unintentional radiation to interfere with the proper functioning of other equipment. Sometimes equipment is designed to emit radiation at designated frequencies and with prescribed bandwidths but may inadvertently emit at other frequencies or harmonics. These situations create electromagnetic interference (EMI) to other equipment.

Engineers would prefer to design equipment to survive under the most severe electromagnetic interference,* but they are often constrained by size, weight, and cost requirements. Designing equipment to work at prescribed interference levels is designing for acceptable susceptibility.

Regulations**

In the United States, the Federal Communications Commission is the body responsible for the control of EMI. Part 15 (for radio-frequency devices) and Part 18 (for medical, scientific, and industrial equipment) of the FCC Rules and Regulations contain sections on the control of interference. Devices intended for commercial and scientific applications with limited usage are governed by class A limits. Devices intended for home or consumer use are covered by class B limits, which are up to 22 dB lower in level of conducted interference and 10 dB lower in level of radiated interference. Certification from the FCC is required before equipment may be sold in the USA.

Another source of EMI regulations, particularly for builders of military equipment is MIL-STD-461. Test methods for the frequency range 30 Hz to 10 GHz are covered in MIL-STD-462.

European countries are following the lead of the West German (Federated Republic) VDE 0871 (Verband Deutscher Electrotechniker) for interference levels and measurements. These emission limits, as well as the FCC limits, were based on the IEC (International Electrotechnical Commission) recommendations. Certification to these regulations must

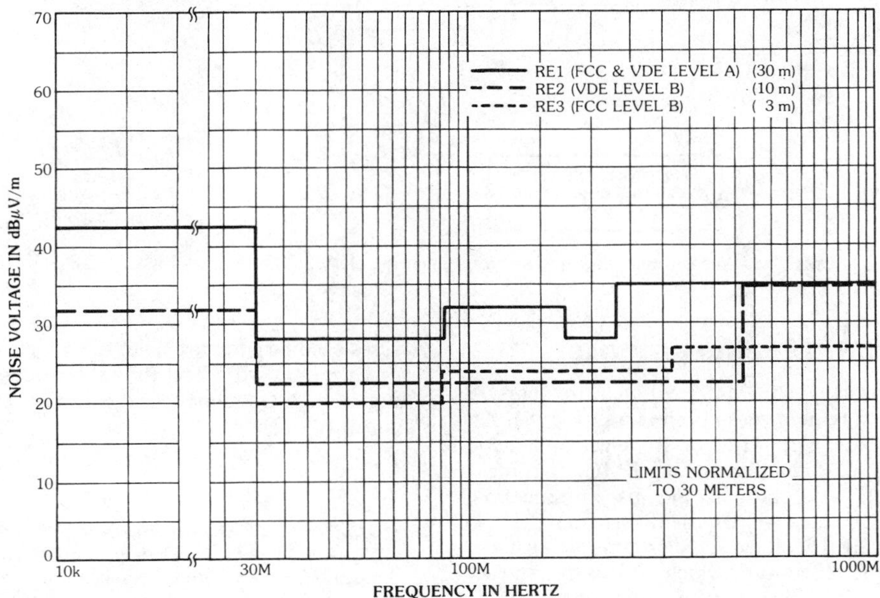

Fig. 80. Typical emission limits.

* Ott, Henry W., *Noise Reduction Techniques in Electronic Systems* (New York: Wiley-Interscience).

** White, D. R. J., *EMI Specifications, Standards, and Regulations,* Electromagnetic Compatibility Series, Vol. 6 (Gainesville, Va.: Don White Consultants, Inc.)

be performed before the equipment is offered for sale.

Other international bodies provide detailed limits of EMI for specialized equipment. In general, compliance with their standards is voluntary, but compliance will often mean better chances of success in interfacing with existing equipment. These bodies are the International Special Committee on Radio Interference (CISPR) and the International Telecommunication Union (ITU).

Typical emission limits are shown in Fig. 80.

Inexpensive techniques for the testing of EMI or EMS (electromagnetic susceptibility) are offered by tem cells (transmission lines) if the equipment is small relative to the dimensions of the cell. For EMI measurements, a spectrum analyzer* is connected as the load to the cell, and for EMS measurements, a signal generator is connected as the source. The tem cell** is completely shielded and does not have the problem of multiple reflections. However, for larger equipment the tem cell becomes prohibitive in size and cost. Also, as the cell dimensions become comparable to the wavelength, higher-order modes will affect the calibrations, thus creating a high-frequency usage limit known as the cutoff frequency.

CHART 1. TYPICAL EMC/EMI MEASUREMENTS

Unwanted Emissions	Unwanted Susceptibility
Conducted	Conducted
Power lines	Power lines (including dropouts & transients)
Control leads	Receiving antenna
Signal leads	
Antenna leads	
Radiated	Radiated
Magnetic fields	Magnetic fields
Electric fields	Electric fields

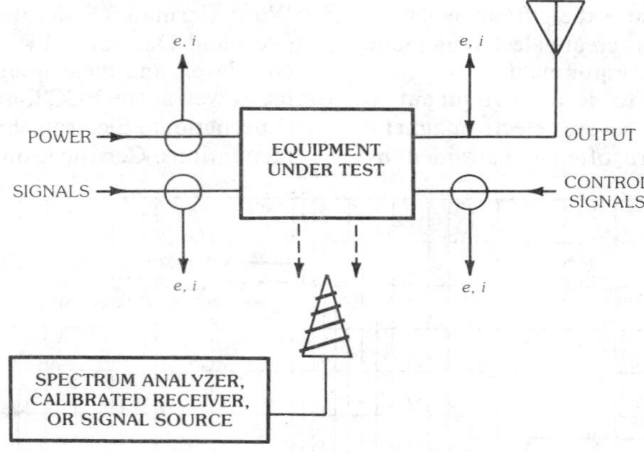

Fig. 81. Typical EMC/EMI measurement setup. (*From* Engineering Design Handbook—EMC, *DARCOM-P 706-410, US Army; Chapter 7.*)

EMC/EMI Measurements

Chart 1 lists typical EMC/EMI measurements. A typical measurement setup is shown in Fig. 81. An open field site with little or no electromagnetic ambient noise is ideal for making EMC/EMI measurements.† Next best would be an anechoic chamber, a shielded room with absorbent materials lining the walls.‡ Unlined screen rooms are afflicted with multiple reflections, which add to or subtract from the true magnitude of the radiation.

† Madzy, T.M., and Nordby, K.S., *IBM Endicott EMI Range,* IEEE International Symposium on EMC 1981; p. 17.

‡ *EMC Design Guide,* NAVAIR AD 1115 (Dept. of Navy; Chapter 9).

* *Modern EMI Measurements,* Application Note 63E (Palo Alto, Calif.: Hewlett-Packard, Oct. 1969). *EMI Measurement Procedure,* Application Note 142 (Palo Alto, Calif.: Hewlett-Packard, July 1972).

** Crawford, M. L., *Measurement of EM Radiation From Electronic Equipment Using TEM Cells,* US National Bureau of Standards Report NBSIR 73-306, 1973. Other TEM reports: IEEE Trans. EMC-16, Nov. 1974, pp. 189–195; IEEE 1976 International Symposium, IEEE Cat. No. 76-CH01104-9 EMC, pp. 369–374.

Strip-line cells*** are handy for testing small electronic modules. They are inexpensive, easy to build, and economical of bench space. Unfortunately, they are not shielded, and their use is therefore limited to preliminary testing and troubleshooting.

REFERENCES

Error Correction

1. Kruppa, W., and Sodomsky, K. F. "An Explicit Solution for the Scattering Parameters of a Linear Two-Port Measurement of an Imperfect Test Set." *IEEE Trans. MTT,* Jan. 1971, pp. 122–123.
2. Fitzpatrick, J. "Error Models for Systems Measurement." *Microwave Journal,* May 1978, pp. 63–66.
3. Gupta, K. C., Ramesh, G., and Rakesh, C. *Computer-Aided Design of Microwave Circuits.* Dedham, Mass.: Artech House Inc., 1981, pp. 307–317.

Time and Frequency Measurement

4. Barnes, J. A., et al. "Characterization of Frequency Stability." *IEEE Trans. Instrum. Meas.,* Vol. IM-20, No. 2 (May 1971), pp. 105–120.
5. "Standard Frequencies and Time Signals." Vol. VII, CCIR XIII Plenary Assembly. Geneva: International Telecommunication Union, 1975, p. 18.
6. Fischer, Michael C. "Frequency Stability Measurement Procedures." Proceedings, Eighth Annual PTTI Meeting, Goddard Space Flight Center, Code 250, Greenbelt, Md., 1976, pp. 575–618.
7. Fischer, Michael C. "Analyze Noise Spectra With Tailored Test Gear." *Microwaves,* Vol. 18, No. 7 (July 1979), pp. 66-75.
8. *Fundamentals of the Electronic Counters,* Application Note 200. Palo Alto, Calif.: Hewlett-Packard, 1978.
9. *Fundamentals of Time and Frequency Standards,* Application Note 52-1. Palo Alto, Calif.: Hewlett-Packard, 1974.
10. Kartaschoff, Peter. *Frequency and Time.* New York: Academic Press, Inc., 1978.
11. Manassewitsch, Vadim. *Frequency Synthesizers Theory and Design.* 2nd ed. New York: John Wiley & Sons, Inc., 1980.
12. Oliver, B. M., and Cage, J. M., eds. *Electronic Measurements and Instrumentation.* New York: McGraw-Hill Book Co., 1971, Ch. 6.
13. Rutman, Jacques. "Characterization of Phase and Frequency Instabilities in Precision Frequency Sources: Fifteen Years of Progress." *Proceedings of the IEEE,* Vol. 66, No. 9 (Sept. 1978), pp. 1048–1075.
14. *Timekeeping and Frequency Calibration,* Application Note 52-2. Palo Alto, Calif., Hewlett-Packard, 1979.
15. Vessot, Robert F. C., et al. "Research With a Cold Atomic Hydrogen Maser." *Proceedings 33rd Frequency Control Symposium.* Washington, D.C.: Electronic Industries Association, 1979, pp. 511–514.
16. Winkler, Gernot M. R. "A Brief Review of Frequency Stability Measures." Proceedings, Eighth Annual PTTI Meeting, Goddard Space Flight Center, Code 250, Greenbelt, Md., 1976, pp. 489–527.

Microwave-Link Analysis

17. *Differential Phase & Gain at Work,* Application Note 175-1. Palo Alto, Calif.: Hewlett-Packard, Nov. 1975.
18. Tant, M. J. *The White Noise Book.* White Crescent Press Ltd., 1974.
19. "An Integrated Test Set for Microwave Radio Link Baseband Analysis." *Hewlett-Packard Journal,* April 1982.
20. Smith, Emerson C. *Glossary of Communications.* Telephony Publishing Corp., 1971.
21. Preston, C. L. *Broadband Transmission System,* a volume from the series Noise Analysis Theory and Technique. New York: American Telephone and Telegraph Corp., 1972.

*** *Engineering Design Handbook—EMC,* DARCOM-P 706-410 (US Army; pp. 7–49).

13 Magnetic-Core Transformers and Reactors

Revised by
Charles F. Helpstead, Durwood R. Kressler, Richard W. Avery,
Fred J. Banzi, Bernard B. Carniglia,
Kent W. Sternstrom, and Richard C. Walker

INTRODUCTION

Definition of Transformer and Inductor

Magnetic-core transformers are static devices containing magnetically coupled windings. They are used in power systems to change values of voltage and current at a single frequency. In communications circuits, often over a wide band of frequencies, they are used to provide direct-current isolation, signal splitting and combining functions, specific current or voltage ratios, impedance matching, and phase inversion.

The Institute of Electrical and Electronics Engineers, Inc. (IEEE) has defined a transformer as follows: "A static device consisting of a winding, or two or more coupled windings, with or without a magnetic core, for introducing mutual coupling between circuits. Note: Transformers are extensively used in electric power systems to transfer power by electromagnetic induction between circuits at the same frequency, usually with changed values of voltage and current."*

Magnetic-core inductors and reactors are static devices containing one or more windings to introduce inductance into an electric circuit. Reactors are used in power circuits primarily to filter alternating current from direct current. Inductors are used in communications systems primarily in frequency-selective circuits.

In this chapter, only those devices having magnetic cores will be considered. The type of core material is known as soft magnetic material, which is defined as ferromagnetic material which, once having been magnetized, is very easily demagnetized (i.e., requires only a slight coercive force to remove the resultant magnetism). A ferromagnetic material usually has relatively high values of specific permeability, and it exhibits hysteresis. The principal ferromagnetic materials are iron, nickel, cobalt, and certain of their alloys.†

Transformer Types and Frequency Ranges

The major types of transformers for both power and communications applications are listed below, along with the general operating frequencies for each type.

Power

Power transformers	50, 60, and 400 Hz
Ferroresonant transformers	50, 60, and 400 Hz
Converter transformers	100 Hz to 150 kHz

Communications

Audio-frequency transformers	20 Hz to 20 kHz
Carrier-frequency transformers	20 kHz to 20 MHz
High-frequency transformers	20 MHz to 1000 MHz
Pulse transformers	repetition rates to 4 MHz

*Reference 1.

†Reference 2.

Inductor Types and Frequency Ranges

The major types of reactors or inductors are rectifier-filter reactors, alternating-current reactors, audio-frequency inductors, and precision adjustable inductors for filters. Inductors are used from 20 Hz to 1000 MHz or higher.

Generalized Equivalent Circuit for a Transformer

Fig. 1 shows the equivalent circuit for a generalized transformer having two windings. Commonly accepted nomenclature * is as follows:

a = turns ratio = N_p/N_s
C_p = primary equivalent shunt capacitance
C_s = secondary equivalent shunt capacitance
E_g = root-mean-square generator voltage
E_{out} = root-mean-square output voltage
k = coefficient of coupling
L_p = primary inductance
l_p = primary leakage inductance
l_s = secondary leakage inductance
R_c = core-loss equivalent shunt resistance
R_g = generator impedance
R_l = load impedance
R_p = primary-winding resistance
R_s = secondary-winding resistance

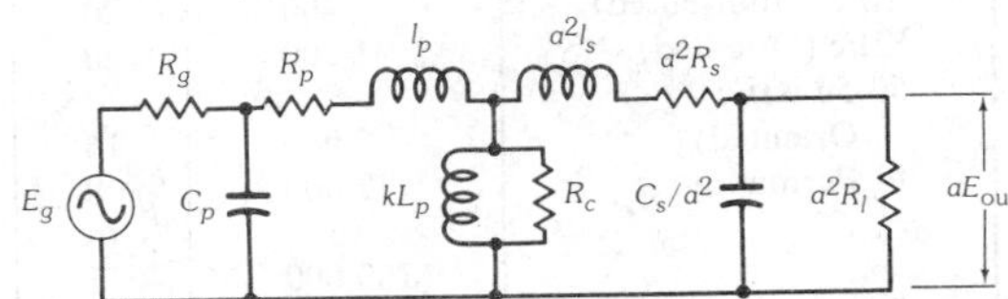

Fig. 1. Equivalent network of a transformer.

POWER TRANSFORMERS

Power transformers operate from a low source impedance at a low frequency. Depending on the source of power, the frequency may vary as much as ±5.5% at 50 or 60 Hz, and as much as ±20% at 400 Hz.

Types of Magnetic Cores

The magnetic cores used for power transformers are usually E and I laminations stamped from

*Reference 3.

silicon-iron sheet when low cost is of primary importance. When minimum size or low loss is of greatest concern, wound-cut "*C*" cores of oriented silicon steel or the more expensive Supermendur could be considered.

Table 1 lists basic properties of soft magnetic materials. Table 2 gives the maximum-flux-density operating conditions for various core materials at 60 and 400 hertz. Two types of laminations are listed as typical, although there are more types that are of both a higher and lower grade. The 0.014 M-6 material is of a higher grade than the 0.0185 M-19 material, but the cost of the former may be 50% more per pound than the cost of the latter, depending on the lamination size chosen. On the other hand, less of the better-grade material is required for the same performance. In a recent study,* it was shown that transformers and reactors constructed of the higher-grade materials were always smaller, as expected, but that in some cases the total material cost of the structures having the better-grade laminations was the lowest.

Wound-cut "*C*" cores are wound in tape form on a rectangular mandrel, impregnated, and cut into halves. These halves are then banded around the wound transformer coil.

The sheet-form magnetic materials may also be made in thin strip form and supplied as toroidal tape-wound bobbin cores for high frequencies. The 79 permalloy in 0.000125-inch tape thickness on a bobbin core may be used at frequencies as high as 1 MHz. These cores are not cut into halves.

*Reference 4.

Tables 1 and 2 make reference to "Metallic Glass." METGLAS† is Allied Corporation's registered trademark for amorphous alloys of metals. This material is new to the list of available core materials. Its advantage is low core loss and exciting volt-amperes. It is currently manufactured in ribbon form with a nominal thickness of 0.0011 inch and widths of 1.0 to 4.0 inches. The material can be stamped into laminations, but the form lends itself best to tape-wound cores. METGLAS material was originally developed for low-frequency power transformers at a commercially competitive price compared with the standard Si-Fe material. The general-purpose alloy with the formulation $Fe_{81}\,B_{13.5}\,Si_{3.5}\,C_2$, known as Alloy 2605SC, is specified in Tables 1 and 2. Other alloys are available.

The approximate number of exciting volt-amperes is largely dependent on the air gap or equivalent air gap in series with the magnetic path. The values specified in Table 2 are for the material exclusive of the air gap. Information on core loss and exciting volt-amperes for specific lamination sizes may be obtained from Thomas and Skinner.**

†Reference 5.
**Reference 6.

TABLE 1. PROPERTIES OF SOFT MAGNETIC MATERIALS

Material	Initial Permeability	B_S (kG)	Specific Gravity (g/cm^2)	Curie Temp. (°C)	Resistivity (Ω-cm)	Operating Frequency (Hz)
Sheet Form						
SiFe (Unoriented)	400	20	7.65	740	47×10^{-6}	60 to 1 000
SiFe (Oriented)	1 500	20	7.65	740	50×10^{-6}	60 to 1 000
50-50 NiFe (Oriented)	2 000	16	8.25	360	40×10^{-6}	60 to 1 000
79 Permalloy	12 000 to 100 000	8 to 11	8.74	450	55×10^{-6}	1k to 75k
Supermendur ($Co_{49}\,Fe_{49}\,V_2$)	800	23	8.15	980	26×10^{-6}	60 to 5 000
Metallic Glass ($Fe_{81}\,B_{13.5}\,Si_{3.5}\,C_2$)	2 000	16	7.32	370	125×10^{-6}	60 to 100k
Bonded Powder Form						
Permalloy Powder	14 to 550	3		450	1.0	10k to 200k
Iron Powder	5 to 80	10		770	10^4	100k to 100M
Ferrite—MnZn	750 to 15 000	3 to 5	4.5 to 5.2	100 to 300	10 to 100	10k to 2M
Ferrite—NiZn	10 to 1 500	3 to 5	3.7 to 5.3	150 to 450	10^6	200k to 100M

Sources: Allegheny Ludlum, Allied Corp., Armco, Arnold Engineering, Indiana General, Magnetics

TABLE 2. MAXIMUM-FLUX-DENSITY OPERATING CONDITIONS FOR CORE MATERIALS AT 60 AND 400 Hz

Freq. (Hz)	Material Thickness (Inches)	Core Material	Core Flux Density B_m (kG)	Approx. Core Loss (W/lb)	Approx. Exciting VA* (VA/lb)
60	0.0185	M-19 Nonoriented Laminations	12	1.0	3.3
60	0.014	M-6 Oriented Laminations	15	0.65	0.83
60	0.012	Silectron Wound-Cut Core	16	1.1	4.0
60	0.004	Supermendur Wound-Cut Core	21	1.1	4.0
60	0.0011	Metallic-Glass Wound-Cut Core	14	0.14	0.16
400	0.004	Silectron Wound-Cut Core	15	10.0	15.0
400	0.004	Supermendur Wound-Cut Core	20	7.0	30.0
400	0.0011	Metallic-Glass Wound-Cut Core	14	1.6	1.82

Sources: Allied Corp., Arnold Engineering, U.S. Steel
*Exciting VA values are for the material exclusive of the air gap.

Similar information for cut-cores may be obtained from catalogs of the Arnold Engineering Co.

Design of Power Transformers for Rectifiers

The design of transformers for rectifiers was chosen as an example because it represents one of the more difficult transformer design problems. The reason it is difficult is that all parts of all windings are not operating continuously with a sinusoidal current. The result is that the capability of the transformer structure is not as fully utilized as it would be if the transformer supplied a nonrectified resistive load. An early paper by R. W. Armstrong* gave the effective, or rms, values of voltages and currents for the primary and secondary windings of a transformer that provides the power for various types of single-phase and polyphase rectification followed by a reactor-input filter. Table 3 lists single-phase rectifier transformer ratings for both inductive (L) and resistive (R) filter inputs. A more comprehensive listing of all the circuit parameters of these single-phase as well as polyphase rectifier circuits is found in Chapter 14 of this book. Should the input to the filter be capacitive, higher rms values of current appear in the secondary winding, resulting in a higher volt-ampere rating. A detailed analysis of the effect of the capacitor-input filter on the transformer-winding voltage and current requirements was made by O. H. Schade in 1943.† More recent analyses have been made by R. Lee** and N. R. Grossner.‡

*Reference 7. **Reference 10.
†Reference 9. ‡Reference 8.

The first step in designing a transformer is to determine its volt-ampere (VA) rating. In the case of the transformer for a rectifier,

$$VA = VA_T E_{dc}' I_{dc}$$

where,

VA_T may be obtained from Table 3,
E_{dc}' is the sum of the output dc voltage (E_{dc}), the diode drop (E_{dd}), and the reactor drop (E_{rd}).

In the case of the bridge rectifier, the current passes through two diodes on each half cycle, and it is necessary to use the value of two diode drops in the calculation of E_{dc}'.

At this point, it is handy to have a table of VA ratings versus lamination sizes, such as Table 4. Note the constraints of the 60-Hz frequency and the 75°C winding temperature rise in an ambient of 75°C. If a lower frequency or temperature rise is required, then the VA average will be lower for a given structure, with other parameters also being affected. It might also be noted that if a bobbin type of construction is used on the smaller structures instead of a layer-wound construction with interlayer insulation, then up to 30% more copper may be used in a given structure, resulting in up to 30% more VA average at the same frequency and temperature rise. The above calculated VA rating may be used to select a suitable structure from Table 4.

In order to determine the number of turns and wire size of each winding, it is first necessary to determine the voltage and current for each winding. Table 3 gives the relationships between the rms voltages and currents and the dc voltages and currents.

TABLE 3.* SINGLE-PHASE RECTIFIER TRANSFORMER RATINGS**

	Half-Wave	Full-Wave, Center-Tapped		Bridge	
	R Input	R Input	L Input	R Input	L Input
Ripple frequency	f	$2f$	$2f$	$2f$	$2f$
Ripple V_{rms}	1.11	0.471	0.471	0.471	0.471
Ripple V_{rms} total‡	1.21	0.482	0.482	0.482	0.482
Primary E_p	2.22	1.11	1.11	1.11	1.11
Primary I_p	1.21	1.11	1.00	1.11	1.00
Primary VA	2.69	1.23	1.11	1.23	1.11
Secondary E_s	2.22	1.11†	1.11†	1.11	1.11
Secondary I_s	1.57	0.785	0.707	1.11	1.00
Secondary VA	3.49	1.74	1.57	1.23	1.11
Average VA_T	3.09	1.49	1.34	1.23	1.11
PIV per Diode	3.14	3.14	3.14	1.57	1.57
I_{pk} per Diode	3.14	1.57	1.00	1.57	1.00
I_{av} per Diode	1.00	0.50	0.50	0.50	0.50
I_{rms} per Diode	1.57	0.785	0.707	0.785	0.707

* From N. R. Grossner, *Transformers and Electronic Circuits* (New York: McGraw-Hill Book Co., 1967).

** Voltages, currents, and VA are based on unity dc output voltage and current. Voltage drops in transformer, diode, and inductor are neglected.

† Secondary voltage at each side of center tap.

‡ Includes second and third harmonics of ripple frequency.

The value of E_{dc}' calculated above is used for the dc voltage. The primary voltage, E_p, is determined by the specified input voltage. The primary rms current, I_p, is determined as follows:

$$I_p = [(\text{VA})\ (\text{Primary VA})]\ /\ [(\text{Efficiency})(E_p)]$$

The quantities "primary VA" and "efficiency" are obtained from Tables 3 and 4, respectively. The secondary rms voltage, E_s, and the secondary rms current, I_s, are obtained from the following relationships:

$$E_s = (\text{Secondary } E_s)(E_{dc}')$$
$$I_s = (\text{Secondary } I_s)(I_{dc})$$

The quantities "Secondary E_s" and "Secondary I_s" are obtained from Table 3.

Now the number of turns in each winding can be calculated. The number of primary-winding turns, N_p, is obtained as follows:

$$N_p = (E_p \times 10^5)\ /\ (4.44 \times f \times A_c \times B_m)$$

where f is the frequency of the applied voltage, which is assumed to be sinusoidal. If the applied voltage has other than a sinusoidal waveform, then the 4.44 constant must be changed. An ASTM standard explains this.* The quantity A_c is the area of the core, and B_m is the core flux density. If the metric unit the kilogauss (kG) is used, as listed in Table 4, then A_c must be in square centimeters. If B_m is in kilolines per square inch, then the units of A_c must be square inches. The A_c of the structure selected from Table 4 may be computed by squaring the specified tongue width (in the appropriate units) and multiplying by the core manufacturer's recommended stacking factor. Although Table 4 structures are all square stacks (i.e., the stack height equals the tongue width), the stack heights are often increased or decreased for a larger or smaller VA rating.

The number of secondary-winding turns, N_s, is obtained as follows:

$$N_s = N_p(E_s/\ E_p)[1 + (\text{Copper Regulation}\ /\ 100)]$$

where "Copper Regulation" (in percent) is obtained from Table 4.

The wire sizes for each winding are determined by dividing the current calculated for the winding by the current density specified in Table 4. This determines the cross-sectional area needed for the conductor. A wire table, such as Table 5, will give the bare area for each wire size. This table also gives the thickness of insulation necessary to support each layer of a given wire size in the layer-wound construction, as well as the maximum diameter of the wire over its film insulation.

The next step is to design the coil that will result in the configuration shown in Fig. 2. This wound coil will then be laminated to form the finished product as shown in Fig. 3.

In designing the coil, one first selects the core

*Reference 11.

TABLE 4.* VA RATINGS VS LAMINATION SIZES—60 HZ, 75°C TEMPERATURE RISE**

Lamination E-I Type	Tongue Width, in	Area Product	VA Average	Flux Density, Kilogauss	Current Density, A/in^2	Efficiency, %	Core Loss, Watts	Copper Loss, Watts	Copper Regulation, %	Weight, lb	
										Iron	Copper
625	5/8	0.114	9.1	14.0	4 060	63.5	0.5	4.7	52	0.37	0.098
75	3/4	0.237	20.3	14.2	3 480	73.2	0.88	6.6	32.6	0.63	0.182
87	7/8	0.441	40.0	14.4	3 040	80.9	1.47	8.7	22	1.00	0.32
100	1	0.750	72.5	14.6	2 580	84.5	2.3	11.0	15.1	1.49	0.46
125	1 1/4	1.825	163	14.8	2 220	88.6	4.7	16.2	9.9	2.91	1.0
138	1 3/8	2.66	229	14.8	2 130	90.2	6.2	19.0	8.3	3.88	1.44
150	1 1/2	3.80	298	14.8	2 000	91.0	8.1	21.8	7.3	5.05	1.75
175	1 3/4	7.0	524	14.8	1 845	92.8	12.8	27.8	5.3	8.0	2.86
212	2 1/8	15.3	1 050	14.8	1 550	94.7	22.6	37.4	3.6	14.2	5.18
250	2 1/2	29.3	1 823	14.8	1 335	95.7	35.5	47.5	2.6	22.2	8.5
251	2 1/2	68.8	3 551	14.8	935	96.5	49.8	79.2	2.23	31.1	26.6

* *Electro-Technol.*, Vol. 67, No. 1, p. 61, January, 1961. Copyright C-M Technical Publications Corp., 1961.

** Table based on 29-gauge, grain-oriented (M6) silicon steel, square stack. Exciting VA/input VA is 23.5 percent (EI625) to 12.7 percent (EI251). Operating temperature = 75°C (amb) + 75°C (rise) = 150°C. Copper weight will ordinarily be less than values in the table.

TABLE 5. WIRE TABLE FOR TRANSFORMER DESIGN*

AWG Size†	Nom. Bare Area ($in^2 \times 10^{-3}$)	Maximum Overall Diameter Single Build (inches)	Maximum Overall Diameter Heavy Build (inches)	Layer Factor	Nominal Resist.** (Ω/1000 ft)	Nominal Weight (lb/1000 ft)	Minimum Margin m (inches)	Kraft Layer Insulation (inches)
14	3.227	0.0666	0.0682	0.90	2.52	12.44	0.1875	0.0100
15	2.561	0.0594	0.0609	0.90	3.18	9.87	0.1562	0.0100
16	2.027	0.0531	0.0545	0.90	4.02	7.812	0.1562	0.0100
17	1.612	0.0475	0.0488	0.90	5.05	6.213	0.1562	0.0070
18	1.276	0.0424	0.0437	0.90	6.39	4.914	0.125	0.0070
19	1.012	0.0379	0.0391	0.90	8.05	3.900	0.125	0.0070
20	0.804	0.0339	0.0351	0.90	10.13	3.099	0.125	0.0050
21	0.638	0.0303	0.0314	0.90	12.77	2.459	0.125	0.0050
22	0.503	0.0270	0.0281	0.90	16.20	1.937	0.125	0.0050
23	0.401	0.0243	0.0253	0.90	20.30	1.546	0.125	0.0050
24	0.317	0.0217	0.0227	0.90	25.67	1.223	0.125	0.0020
25	0.252	0.0194	0.0203	0.90	32.37	0.970	0.125	0.0020
26	0.1986	0.0173	0.0182	0.89	41.02	0.7650	0.125	0.0020
27	0.1584	0.0156	0.0164	0.89	51.44	0.6101	0.125	0.0020
28	0.1247	0.0140	0.0147	0.89	65.31	0.4806	0.125	0.0015
29	0.1003	0.0126	0.0133	0.89	81.21	0.3866	0.125	0.0015
30	0.0785	0.0112	0.0119	0.89	103.71	0.3025	0.0938	0.0015
31	0.0622	0.0100	0.0108	0.88	130.9	0.2398	0.0938	0.0015
32	0.0503	0.0091	0.0098	0.88	162.0	0.1937	0.0938	0.0013
33	0.0396	0.0081	0.0088	0.88	205.7	0.1526	0.0938	0.0013
34	0.0312	0.0072	0.0078	0.88	261.3	0.1201	0.0938	0.0010
35	0.0246	0.0064	0.0070	0.88	330.7	0.0949	0.0938	0.0010
36	0.0196	0.0058	0.0063	0.87	414.8	0.07569	0.0938	0.0010
37	0.0159	0.0052	0.0057	0.87	512.1	0.06128	0.0938	0.0010
38	0.01257	0.0047	0.0051	0.87	648.2	0.04844	0.0625	0.0010
39	0.00962	0.0041	0.0045	0.86	846.6	0.03708	0.0625	0.0007
40	0.00755	0.0037	0.0040	0.86	1079.2	0.02910	0.0625	0.0007
41	0.00616	0.0033	0.0036	0.85	1323.	0.02374	0.0625	0.0007
42	0.00491	0.0030	0.0032	0.85	1659.	0.01892	0.0625	0.0005
43	0.00380	0.0026	0.0029	0.85	2143.	0.01465	0.0625	0.0005
44	0.00314	0.0024	0.0027	0.85	2593.	0.01210	0.0625	0.0005

* Data for this table are courtesy of NEMA (ref. 12) and Phelps Dodge (ref. 13).

† Square or rectangular wire is recommended for wire sizes heavier than AWG 14. Single build is not recommended for these sizes.

**Resistance values are at 20°C.

tube thickness (J in Fig. 2). This thickness may vary from 0.025 to 0.050 inch for lamination stack heights (p in Fig. 3) of 0.5 to 2 inches. Next, calculate the coil-build (a in Fig. 3) for each winding, using the number of turns and wire sizes previously calculated.

$$a = 1.1[n_l(D + t) - t + t_c]$$

where,

n_l is the number of layers in the winding,

D is the diameter of the insulated wire,

t_c is the thickness of the insulation under and over the winding.

The numeric 1.1 allows for a 10% bulge factor. This factor may be reduced, depending on the winding equipment and methods of winding used. The total coil-build should not exceed 85–90% of the window width.

Compute the mean length per turn (MLT) of each winding from the geometry of the core and windings (Fig. 2).

$$(\text{MLT})_1 = 2(r + J) + 2(s + J) + \pi a_1$$
$$(\text{MLT})_2 = 2(r + J) + 2(s + J) + \pi(2a_1 + a_2)$$

where,

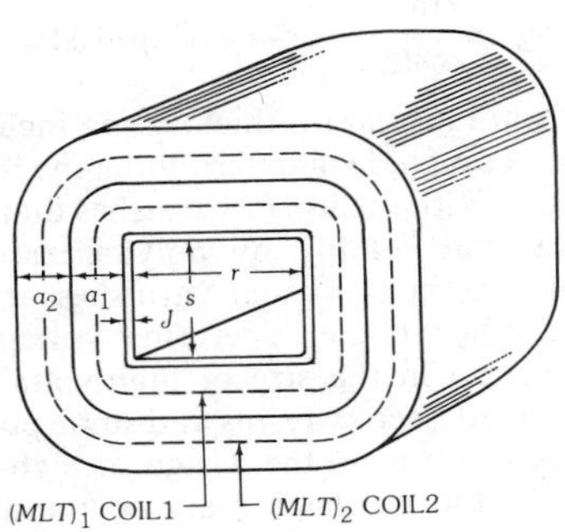

Fig. 2. Dimensions relating to coil mean length of turn (MLT).

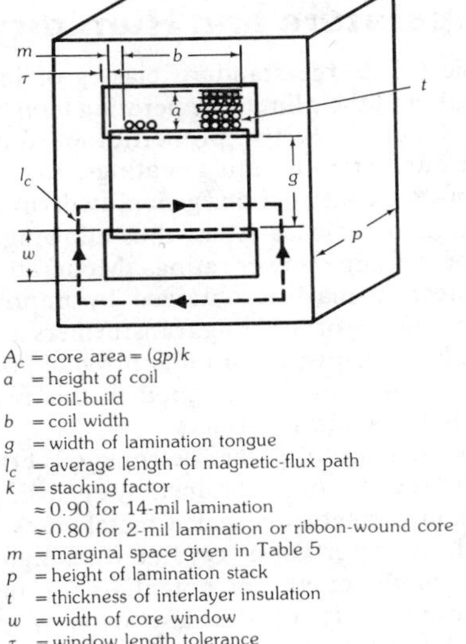

A_c = core area = $(gp)k$
a = height of coil
= coil-build
b = coil width
g = width of lamination tongue
l_c = average length of magnetic-flux path
k = stacking factor
≈ 0.90 for 14-mil lamination
≈ 0.80 for 2-mil lamination or ribbon-wound core
m = marginal space given in Table 5
p = height of lamination stack
t = thickness of interlayer insulation
w = width of core window
τ = window length tolerance
= 1/16 inch, total

Fig. 3. Dimensions relating to the design of a transformer coil-build and core.

a_1 is the build of the first winding,
a_2 is the build of the second winding,
J is the thickness of the winding form,
r and s are the winding-form dimensions.

Calculate the total length and resistance of each winding, and determine the IR drop and I^2R loss for each winding.

Make corrections, if required, in the number of turns of the windings to allow for the IR drops, so as to have the required E_s.

$$E_s = (E_p - I_pR_p)(N_s/N_p) - I_sR_s$$

Compute the core loses from the weight of the core and Table 2 or reference 6.

Determine the percent efficiency η and voltage regulation (vr) from

$$\eta = \frac{W_{out} \times 100}{W_{out} + (\text{core loss}) + (\text{copper loss})}$$

$$(\text{vr}) = \frac{I_s[R_s + (N_s/N_p)^2R_p]}{E_s}$$

For a more accurate evaluation of voltage regulation, determine leakage-reactance drop $= I_{dc}\omega l_{sc}/2\pi$, and add to the above (vr) the value of $I_{dc}\omega l_{sc}/2\pi E_{dc}$. Here, l_{sc} = leakage inductance viewed from the secondary; see "Methods of Winding Transformers" (in this chapter) to evaluate l_{sc}.

Bring out all terminal leads. Use the wire of the coil, insulated with suitable sleevings, for all sizes of wire heavier than 21. Use 7-30 stranded and insulated wire for smaller sizes.

Effect of Duty Cycle on Design

If a transformer is operated at different loads according to a regular duty cycle, the equivalent volt-ampere (VA) rating is

$$(\text{VA})_{eq} = \left[\frac{(\text{VA})_1{}^2t_1 + (\text{VA})_2{}^2t_2 + (\text{VA})_3{}^2t_3 + \ldots + (\text{VA})_n{}^2t_n}{t_1 + t_2 + t_3 + \ldots t_n}\right]^{1/2}$$

where $(\text{VA})_1$ = output during time t_1, etc.

Example: 5 kilovolt-ampere output, 1 minute on, 1 minute off.

$$(\text{VA})_{eq} = \left[\frac{(5000)^2(1) + (0)^2(1)}{1+1}\right]^{1/2} = \left[\frac{(5000)^2}{2}\right]^{1/2}$$

$$= 5000/(2)^{1/2} = 3535 \text{ volt-amperes}$$

Methods of Winding Transformers

The most common methods of winding transformers are shown in Fig. 4. Leakage inductance is

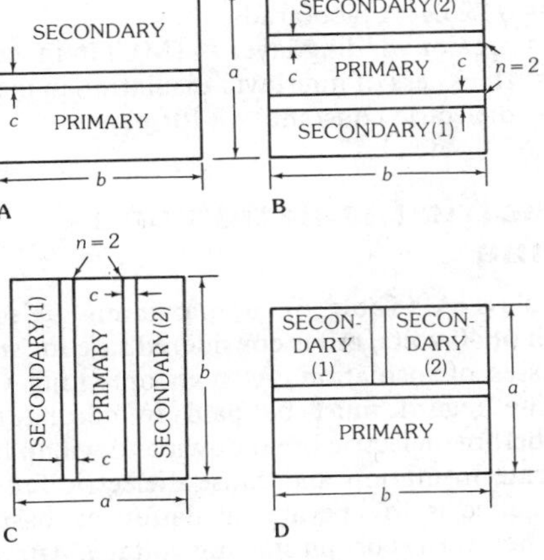

Fig. 4. Methods of winding transformers.

reduced by interleaving, i.e., by dividing the primary or secondary coil into two sections and placing the other winding between the two sections. Interleaving may be accomplished by concentric and by coaxial windings, as shown in Figs. 4B and C. Reduction of leakage inductance is computed from

$$l_{sc} = \frac{10.6N^2(\text{MLT})(2nc + a)}{n^2 b \times 10^9}$$

where l_{sc} is the leakage inductance (referred to the winding having N turns) in henries and the dimensions are in inches and to be the same for Figs. 4B and C.

Means of reducing leakage inductance are:

(**A**) Minimize turns by using high-permeability core.
(**B**) Reduce build of coil.
(**C**) Increase winding width.
(**D**) Minimize spacing between windings.
(**E**) Use bifilar windings.

Means of minimizing capacitance are:

(**A**) Increase dielectric thickness t.
(**B**) Reduce winding width b and thus area A.
(**C**) Increase number of layers.
(**D**) Avoid large potential differences between winding sections, as the effect of capacitance is proportional to applied potential squared.

Note: Leakage inductance and capacitance requirements must be compromised in practice since corrective measures are opposites.

Effective interlayer capacitance of a winding may be reduced by sectionalizing it as shown in Fig. 4D. This can be seen from

$$C_e = (4C_l/3n_l)(1 - 1/n_l)$$

where,

C_e = effective capacitance in picofarads,
n_l = number of layers,
C_l = capacitance of one layer to another $= 0.225A\epsilon/t$ picofarads,
A = area of winding layer = (MLT)b inches^2,
t = thickness of interlayer insulation in inches,
ϵ = dielectric constant ≈ 3 for paper.

Dielectric Insulation and Corona

For class A (Table 6), a maximum dielectric strength of 40 volts/mil is considered safe for small thicknesses of insulation. At high operating voltages, due regard must be paid to corona that occurs before dielectric breakdown and will in time deteriorate insulation and cause dielectric failure. Best practice is to operate insulation at least 25 percent below the corona starting voltage. Approximate 60-hertz rms corona voltage V is

$$\log \frac{V\text{ (in volts)}}{800} = (2/3)\log(100t)$$

where t = total insulation thickness in inches. This may be used as a guide in determining the thickness of insulation. With the use of varnishes that require no solvents, but solidify by polymerization, the bubbles present in the usual varnishes are eliminated, and much higher operating voltages and, hence, reduction in the size of high-voltage units may be obtained. Epoxy resins and some polyesters belong in this group. In the design of high-voltage transformers, the creepage distance required between wire and core may necessitate the use of insulating channels covering the high-voltage coil, or taping of the latter. For units operating at 10 kilovolts or higher, oil insulation will greatly reduce creepage and, hence, the size of the transformer.

Temperature and Humidity

Table 6 lists the standard classes of insulating materials and their limiting operating temperatures. Table 7 compares the properties of four high-temperature wire-insulating coatings.

Open-type constructions generally permit greater cooling than enclosed types, thus allowing smaller sizes for the same power ratings. Moderate humidity protection may be obtained by impregnating and dip-coating or molding transformers in polyester or epoxy resins; these units provide good heat dissipation but are not as good in this respect as completely open transformers.

Protection against the detrimental effects of humidity is commonly obtained by enclosing transformers in hermetically sealed metal cases. This is particularly important if very fine wire, high output voltage, or direct-current potentials are involved. Heat conductivity to the case exterior may be improved by the use of asphalt or thermosetting resins as filling materials. Best conductivity is obtained with high-melting-point silica-filled asphalts or resins of the polyester or epoxy types. Coils impregnated with these resins dissipate heat best, since voids in the heat path may be eliminated.

Immersion in oil is an excellent means of removing heat from transformers. An air space or bellows must be provided to accommodate expansion of oil when heated.

FERRORESONANT TRANSFORMERS

Ferroresonant transformers, also known as constant-voltage transformers or ferroresonant voltage regulators, make use of an alternating-current phenomenon involving at least one magnetic component in combination with a suitable capacitor to provide stabilized voltage to a load when the source voltage fluctuates. A magnetic component must have a magnetization characteristic that is sharply

TABLE 6. CLASSIFICATION OF ELECTRICAL INSULATING MATERIALS†

Class	Insulating Material	Limiting Insulation Temperature (Hottest Spot) in °C
O	Materials or combinations of materials such as cotton, silk, and paper without impregnation*	90
A	Materials or combinations of materials such as cotton, silk, and paper when suitably impregnated or coated or when immersed in a dielectric liquid*	105
B	Materials or combinations of materials such as mica, glass fiber, asbestos, etc., with suitable bonding substances*	130
F	Same as for Class B	155
H	Materials or combinations of materials such as silicone elastomer, mica, glass fiber, etc., with suitable bonding substances such as appropriate silicone resins*	180
C	*	220
Over C	Materials consisting entirely of mica, porcelain, glass, quartz, and similar inorganic materials*	Over 220

* (Other) materials or combinations of materials may be included in this class if by experience or accepted tests they can be shown to have comparable thermal life at the temperature given in the right-hand column.

These temperatures are, and have been in most cases over a long period of time, benchmarks descriptive of the various classes of insulating materials, and various accepted test procedures have been or are being developed for use in their identification. They should not be confused with the actual temperatures at which these same classes of insulating materials may be used in the various specific types of equipment nor with the temperatures on which specified temperature rise in equipment standards are based.

In the above definitions the words "accepted tests" are intended to refer to recognized test procedures established for the thermal evaluation of materials by themselves or in simple combinations. Experience or test data, used in classifying insulating materials, are distinct from the experience or test data derived for the use of materials in complete insulation systems. The thermal endurance of complete systems may be determined by suitable test procedures.

A material that is classified as suitable for a given temperature may be found suitable for a different temperature, either higher or lower, by an insulation system test procedure. For example, it has been found that some materials suitable for operation at one temperature in air may be suitable for a higher temperature when used in a system operated in an inert gas atmosphere. Likewise some insulating materials when operated in dielectric liquids will have lower or higher thermal endurance than in air.

It is important to recognize that other characteristics (in addition to thermal endurance) such as mechanical strength, moisture resistance, and corona endurance are required in varying degrees in different applications for the successful use of insulating materials.

† From "Insulation class ratings," *IEEE Standard Dictionary of Electrical and Electronics Terms,* IEEE Std. 100-1972

nonlinear, and it must operate above the "knee" of the characteristic, where the change in voltage across it is small compared to the change in current through it. A set of design equations for commonly used ferroresonant regulators and regulated rectifiers are derived in a paper by Hart and Kakalec.*

*Reference 14.

These equations are for the basic open-loop type of ferroresonant voltage regulator.

A more recent development is the closed-loop, or feedback-controlled, ferroresonant voltage regulator.† Whereas the regulating function of the open-loop type depends on the stability of the frequency

†Reference 15.

of the source and the magnetic characteristics of the core, the regulating function of the closed-loop type is dependent on the voltage stability of the reference in the control circuit, and also on the gain of the feedback loop. Regulation better than 0.5% for line, load, frequency, and temperature changes can be attained. The circuit of reference 15 requires the use of a magnetic component, an inductor, in the control circuit. A later development** eliminates the need for this inductor as a separate component by the use of a double-shunt feedback ferroresonant transformer.

CONVERTER/INVERTER TRANSFORMERS

The IEEE‡ defines a converter as a machine that changes alternating-current power to direct-current power or vice versa, or from one frequency to another. It also defines an inverter as a machine, device, or system that changes direct-current power to alternating-current power.

Transformers are used in static dc-to-dc converters and dc-to-ac inverters. The core materials in these types of transformers may or may not be driven to a saturation flux density, depending on the control circuits involved. Also, the core may operate in a double-ended (bipolar) or single-ended (unipolar) mode, again dependent on the external circuitry. Bipolar is also known and defined by ASTM§ as a symmetrically cyclically magnetized condition (SCM); unipolar is defined as a cyclically magnetized condition (CM). The latter has flux-current loops that are not symmetrical with respect to the origin of the axes.

Details of how to design the circuitry and the magnetics (transformer and inductor) may be found in reference 17. Details on the selection of magnetic materials for static converter and inverter transformers may be found in reference 18.

TABLE 7. COMPARISON OF FOUR WIRE INSULATING COATINGS*

NEMA Std. Designation	MW 15C	MW 28C	MW 35C	MW 16C
Thermal Class	105°C	130°C	200°C	220°C
Insulating Material	Polyvinyl formal, modified	Polyurethane, followed by Nylon	Modified Polyester, followed by Polyamide Imide	Aromatic Polyimide
Min. Dielectric Strength at Rated Temperature of Single-Film Coated 36 AWG	1900 V	1725 V	1900 V	1900 V
Scrape Resistance (Min. Grams-to-Fail of Single-Film Coated 30 AWG)	250	250	250	160
Min. Thermoplastic Flow Temperature of Single-Coated 36 AWG	180°C	170°C	300°C	400°C
Trade Names				
Anaconda	Formvar	Nylac	AP 2000	ML
Phelps Dodge	Formvareze	Nyleze	Armored Polythermaleze	ML
Rea	Formvar	Nysol	Thermamid	Pyre-ML
Westinghouse	Formvar	Nythane	Omegaklad	ML

* From NEMA Standards Publication No. MW 1000-1981, Revised July 1982.

**Reference 16.

‡Reference 1.

§Reference 11.

AUDIO-FREQUENCY TRANSFORMERS

Audio-frequency transformers are used mainly for matching impedances and transmitting audio frequencies. They also provide isolation from direct currents and present balanced impedances to lines or circuits.

Types of Magnetic Cores

The magnetic core for this type of transformer is usually an EI or EE type using either audio-grade silicon steel or nickel-alloy steel (refer to Tables 1 and 10). High-permeability nickel-alloy tape cores in toroidal form are used for extreme bandwidths. High-permeability ferrite cores with highly polished mating surfaces are also used to obtain wide bandwidths.

If there is no direct current in the windings, it is possible to design a small, high-quality audio transformer with a high-permeability ferrite core. Material with initial permeability above 10 000 is available. To minimize the air gap in the assembled core, the mating surfaces are ground flat and polished to a mirror finish. Because of the small core volume, when these transformers are used at low frequencies care must be taken to avoid core saturation. The maximum flux density in a core is given by:

$$B_{max} = (3.49 \times 10^6)\ E/fNA_c$$

where,

B_{max} = maximum flux density (gauss),
E = rms volts,
f = frequency (hertz),
N = number of winding turns,
A_c = cross section of core (square inches).

Design of Audio-Frequency Transformers

Important Parameters—Important parameters are generator and load impedances, R_g and R_l, respectively; generator voltage E_g; frequency band to be transmitted; harmonic distortion; and operating voltages (for adequate insulation). See Fig. 1. Refer to the section on power-transformer design for details about physical design, cores, winding, and so forth.

Midband Frequencies—The relative low- and high-frequency responses are taken with reference to midband frequencies where

$$aE_{out}/E_g = [(1 + R_s/R_l) + R_l/a^2R_l]^{-1}$$

Low Frequencies—At low frequencies, the equivalent unity-ratio network of a transformer becomes approximately as shown in Fig. 5.

$$\text{Amplitude} = [1 + (R'_{par}/X_m)^2]^{-1/2}$$

$$\text{Phase angle} = \tan^{-1}(R'_{par}/X_m)$$

where,

$R'_{par} = (R_1R_2a^2)/(R_1 + R_2a^2)$
$R_1 = R_g + R_p$
$R_2 = R_l + R_s$
$X_m = 2\pi fL_p$

In a good output transformer, R_p, R_s, and R_c may be neglected. In input or interstage transformers, R_c may be omitted.

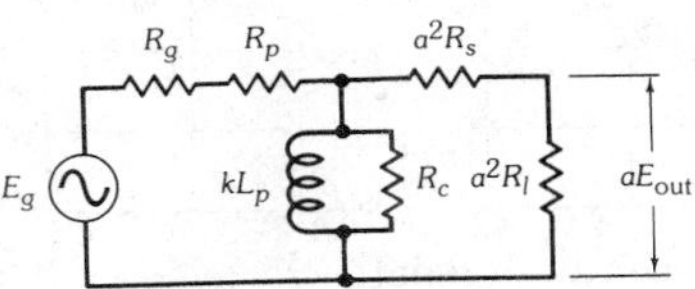

Fig. 5. Equivalent network of an audio-frequency transformer at low frequencies.

High Frequencies—At high frequencies, neglecting the effect of winding and other capacitances, the equivalent unity-ratio network becomes approximately as in Fig. 6.

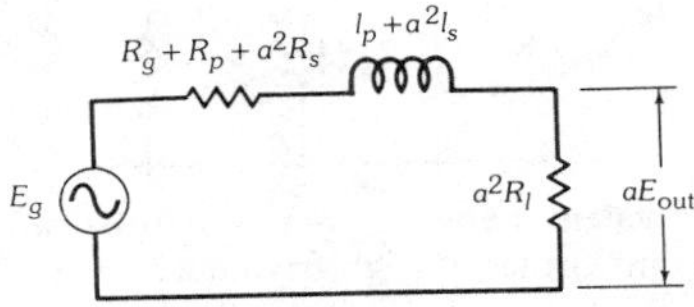

Fig. 6. Equivalent network of an audio-frequency transformer at high frequencies, neglecting the effect of the winding shunt capacitances.

$$\text{Amplitude} = [1 + (X_l/R'_{se})^2]^{-1/2}$$

$$\text{Phase angle} = \tan^{-1}(X_l/R'_{se})$$

where,

$R'_{se} = R_1 + R_2a^2$,
$X_l = 2\pi fl_{scp}$,
l_{scp} = inductance measured across primary with secondary short-circuited $= l_p + a^2l_s$.

The low- and high-frequency responses are shown by the curves of Fig. 7.

If at high frequencies the effect of winding and other capacitances is appreciable, the equivalent network on a 1:1-turns-ratio basis becomes as shown in Fig. 8. In a step-up transformer, C_2 = equivalent shunt capacitances of both windings. In a step-down transformer, C_2 shunts both leakage inductances and R_2. The relative high-frequency response of this network is given by

$$\frac{(R_1 + R_2)/R_2}{[(R_1/X_c + X_l/R_l)^2 + (X_l/X_c - R_g/R_l - 1)^2]^{1/2}}$$

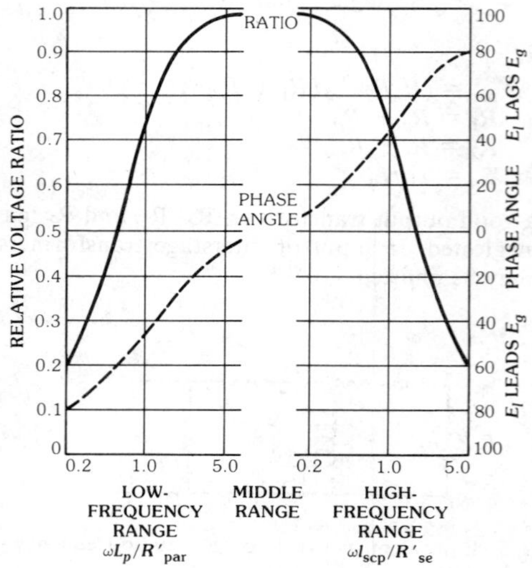

Fig. 7. Universal frequency and phase response characteristics of output transformers. (*Courtesy McGraw-Hill Book Co.*)

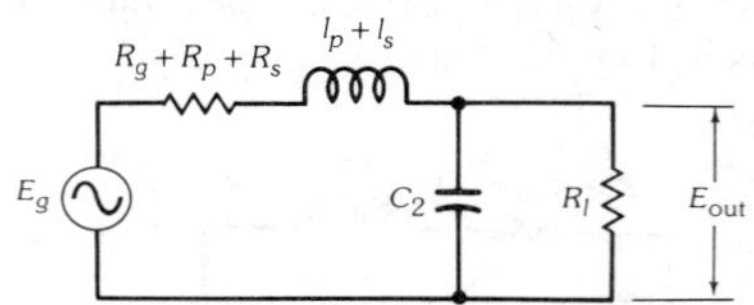

Fig. 8. Equivalent network of a 1:1-turns-ratio audio-frequency transformer at high frequencies when effect of winding shunt capacitances is appreciable.

This high-frequency response is plotted in Fig. 9 for $R_2 = R_1$ (matched impedances), based on simplified equivalent networks as indicated. At frequency f_r, $X_l = X_c$ and $B = X_c/R_l$.

Harmonic Distortion—Harmonic distortion requirements may constitute a deciding factor in the design of transformers. Such distortion is caused by either variations in load impedance or nonlinearity of magnetizing current. The percent harmonic voltage appearing in the output of a loaded transformer is given by

$$(\text{Percent harmonics}) = 100E_h/E_f = 100I_h/I_f(R'_{par}/X_m)[1-(R'_{par}/4X_m)]$$

where $100I_h/I_f$ = percent of harmonic current measured with a zero-impedance source (values in Table 8 are for a 4-percent silicon-steel core).*

Insertion Loss—Insertion loss is the loss introduced into the circuit by addition of the transformer. At midband, the loss is caused by winding resistance and core loss. Frequency discrimination adds to this at low and high frequencies. Insertion loss is input divided by output expressed in decibels, or (in terms of measured voltages and impedance)

$$(\text{dB insertion loss}) = 10 \log[(E_g^2 R_l)/(4E_0^2 R_g)]$$

Impedance Match—For maximum power transfer, the reflected load impedance should equal the generator impedance. Winding resistance should be included in this calculation: For matching

$$R_g = a^2(R_l + R_s) + R_p$$

Also, in a properly matched transformer

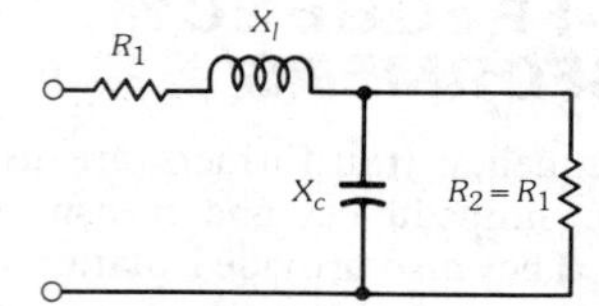

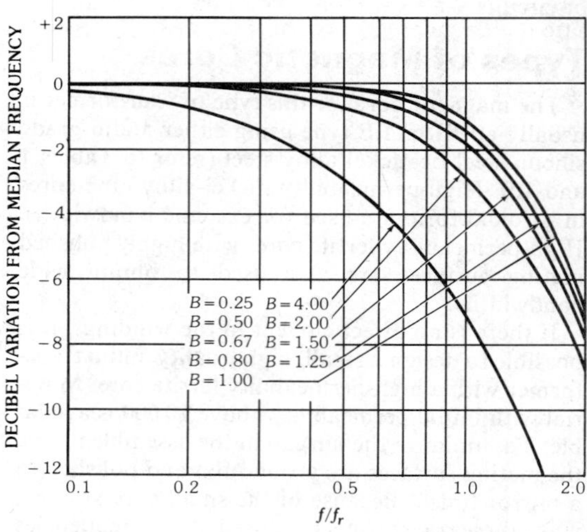

Fig. 9. Transformer characteristics at high frequencies for matched impedances. At frequency f_r, $X_l = X_c$ and $B = X_c/R_1$. (*Reprinted from* Electronic Transformers and Circuits, *by R. Lee, 2nd ed., p. 151, 1955; by permission, John Wiley & Sons, New York.*)

Table 8. Harmonics Produced by Various Flux Densities B_m in a 4-Percent Silicon-Steel-Core Audio Transformer

B_m	Percent 3rd Harmonic	Percent 5th Harmonic
100	4	1.0
500	7	1.5
1 000	9	2.0
3 000	15	2.5
5 000	20	3.0
10 000	30	5.0

*N. Partridge, "Harmonic Distortion in Audio-Frequency Transformers," *Wireless Engineer,* Vol. 19; September, October, and November 1942.

$$R_g = a^2 R_l = (Z_{oc} \times Z_{sc})^{1/2}$$

where,

Z_{oc} = transformer primary open-circuit impedance,
Z_{sc} = transformer primary impedance with secondary winding short-circuited.

If more than one secondary is used, the turns ratio to match impedances properly depends on the power delivered from each winding.

$$N_s / N_p = [(R_n / R_g)(W_n / W_p)]^{1/2}$$

Example: Using Fig. 10

$$N_2 / N_p = [(10/600)(10/16)]^{1/2} = 0.102$$
$$N_3 / N_p = [(50/600)(5/16)]^{1/2} = 0.161$$
$$N_4 / N_p = [(100/600)(1/16)]^{1/2} = 0.102$$

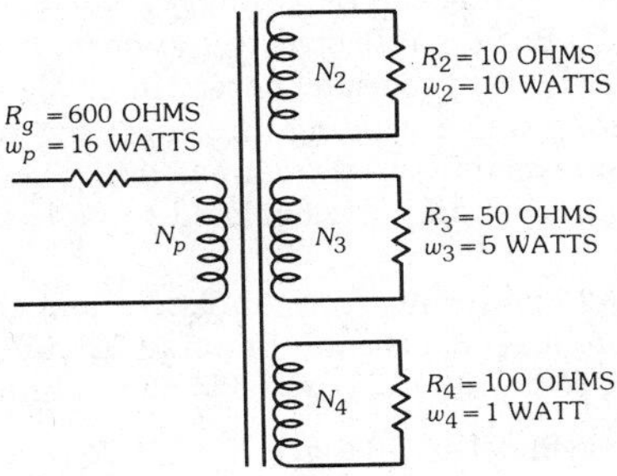

Fig. 10. Multisecondary audio transformer.

WIDEBAND TRANSFORMERS

Wideband transformers operate over a range of one to three or more frequency decades without significant attenuation. The frequency band of interest can be considered to lie between 20 kHz and 1 GHz. This frequency range can be broken into two subranges: carrier frequency (20 kHz to 20 MHz) and high frequency (20 MHz to 1 GHz). Wideband transformers are used to provide impedance matching, voltage or current ratios, dc isolation, connection of balanced to unbalanced circuits, power splitting, and phase inversion.

Maximum bandwidth is achieved when the ratio of shunt inductance to leakage inductance is maximized, neglecting the effect of shunt capacitance. To increase this ratio, we must use the highest μ_e possible and make the ratio of winding width to core length approach unity. The latter occurs for a toroidal core, but is smaller for spool-wound structures. Modern wideband transformers utilize ferrite cores. Some commonly used core shapes are EE, EI, EP, RM, pot, Q, toroid, and X.

Core-Material Considerations

Carrier-frequency transformers often use cores of MnZn ferrite having material permeability (μ) in the 1000 to 20 000 range. High-frequency transformers sometimes use NiZn ferrite cores having material permeability as low as 15, especially when low core loss is important. Most high-frequency transformers operate at relatively low impedance levels. To obtain effective permeability above about 10 000, either toroidal cores or split cores having extremely small air gaps are needed. Highly polished, clean, and coplanar core mating surfaces can achieve air gaps as small as 0.2 μm. Even at such small air gaps, μ_e can be considerably reduced from μ and is given by:

$$\mu_e = l_e \mu / [l_e + (\mu - 1) l_g]$$

where,

μ_e = effective permeability,
μ = material permeability,
l_e = effective core magnetic path length,
l_g = length of air gap.

Electrical loss in ferrite material is usually expressed as a relative loss coefficient, $\tan \delta / \mu = \omega L_p / R_p$. It has three components: hysteresis loss, eddy-current loss, and residual loss. The effect in transformer cores can be more conveniently expressed as a parallel resistance per turn squared in ohms (R_c / N^2) and plotted as a function of frequency. Ferrite core loss is a frequency-dependent nonlinear resistance; hence it cannot be equalized easily by a simple reactive network, should its effects be significant.

The Curie temperature of the core material can be a consideration when transformers are operated at high ambient temperatures. High-permeability materials tend to have lower Curie temperatures. Core temperature and disaccommodation factors are usually not as critical in wideband transformers as in stable filter-class inductors.

Design Example for Carrier Frequencies

Assume the following problem: Design a 75-to-150-ohm unbalanced transformer to operate over a frequency range of 50 kHz to 10 MHz. In-band loss shall be less than 0.1 dB, and loss shaping at band ends shall be less than 0.1 dB. The transformer will not carry any direct current and should be small in size (less than 0.75 in^2 in area and less than 0.60 in high).

The initial step is to determine the minimum inductance for low-frequency loss shaping.

$$\text{Loss in dB} = 10 \log_{10} [1 + (R/\omega L_p)^2]$$

where,

$R = (R_g R')/(R_g + R')$,
R_g = 75 ohms,
R_l = 150 ohms,

$R' = a^2 R_l$,

a = turns ratio = N_p/N_s

For a loss of 0.1 dB maximum

$$1 + (37.5/\omega L_p)^2 = 10^{0.1/10}$$

From this, it is found that $\omega L_p = 245.7$ ohms minimum at 50 kHz, and L_p therefore must be greater than 0.782 millihenries.

A minimum value for L_p could also be determined from a return-loss requirement:

$$\text{Return loss in dB} = 10 \log_{10} [1 + (2\omega L_p/R')^2]$$

The minimum inductance previously selected for loss shaping would provide a return loss relative to R' of only 16.4 dB. If a greater return loss were specified, L_p minimum would have to be larger. Return-loss requirements for wideband transformers frequently cause more severe constraints than transformer loss shaping. Transformer loss is related to return loss by the expression:

$$\text{Transformer loss} = -10 \log_{10} (1 - 10^{-RL/10})$$

The expression is applicable when the equivalent circuit is considered to contain pure reactances.

The EP13 core is a small, efficient core that meets the size constraints of this example. It is available with an A_L value of 3500 nH/N^2. With no direct current, no air gap is required between the core halves, and the full 3500 nH/N^2 may be utilized.

$$N_p = (782/3.5)^{1/2} = 14.9 \text{ turns minimum}$$
$$Ns/Np = (150/75)^{1/2} = 1.41$$

Primary turns of 17 and secondary turns of 24 will closely approximate the 1.41 ratio. This number of primary turns results in a minimum shunt inductance of 1.012 millihenries ($17^2 \times 3.5\ \mu\text{H}/N^2$), providing an inductance margin to allow for variations in assembly.

High-frequency loss shaping is controlled by both leakage inductance and effective shunt capacitance. Interleaved windings reduce leakage inductance but cause increased shunt capacitance. For a low-impedance transformer, the leakage inductance dominates, while in a high-impedance transformer the shunt capacitance is controlling. Windings usually can be oriented to balance and minimize the total contributions to shaping loss from leakage inductance and shunt capacitance. As a starting point, allocate half the requirement to each effect; thus, assume 0.05 dB caused by the leakage inductance.

The leakage inductance of the primary winding with the secondary winding shorted (l_{sc}) is that of both primary and secondary windings referred to the primary.

$$\text{Loss in dB} = 10 \log_{10} [1 + (\omega l_{sc})^2/(R_g + R')^2]$$
$$1 + (\omega l_{sc}/150)^2 = 10^{0.05/10}$$
$$\omega l_{sc} = 16.14 \text{ ohms maximum}$$
$$l_{sc} = 0.257\ \mu\text{H max. at 10 MHz}$$

By singly interleaving the secondary around the primary, leakage can be reduced to a sufficiently low value. The primary is a single layer of 34 AWG wire wound as a parallel pair to achieve a smooth full single layer. Each secondary section is wound as a single layer of 31 AWG wire. Two layers of 2.5-mil Mylar tape are applied between windings. From the winding geometry, the leakage inductance referred to the primary is calculated to be 0.18 microhenry.

The loss contribution from effective shunt capacitance is also allowed to be 0.05 dB.

$$\text{Loss in dB} = 10 \log_{10}[1 + (\omega RC)^2]$$
$$1 + (\omega RC)^2 = 10^{0.05/10}$$
$$\omega RC = 0.1076$$
$$C = 45.7 \text{ pF max. at 10 MHz}$$

The direct capacitance between each secondary interleaved winding and the primary is calculated to be 32.8 pF. Each is reflected to the primary as (C direct)/3 with appropriate corrections made for turns ratio, relative winding direction, and reversing or nonreversing circuit connection. The effective shunt capacitance is calculated to be less than 40 pF.*

Midband loss results from both core and copper losses. Core loss (R_c) is a minimum of 22 ohms/N^2. Therefore, R_c is ($17^2 \times 22$), or 6358 ohms minimum.

$$\text{Loss in dB} = 20 \log_{10} (1 + R/R_c) = 0.051 \text{ dB max.}$$

The series ac resistances of the primary and secondary windings are R_p and a^2R_s when referred to the primary winding. The value of R_p is calculated to be 0.177 ohm maximum, and the value of a^2R_s is calculated to be 0.062 ohm maximum. These values lead to a copper loss of

$$\text{Loss in dB} = 20 \log_{10} [1 + (R_p + a^2R_s)/(R_g + R')] = 0.014 \text{ dB max.}$$

The total midband loss should be less than 0.065 decibel.

This transformer design was realized. The loss-frequency characteristic is shown in Fig. 11B. Loss shaping at the band ends, 50 kHz and 10 MHz, is 0.04 dB and 0.07 dB, respectively. Midband loss is 0.035 dB.

High-Frequency Wideband Transformers

Circuit impedances are usually low at high frequencies (20 MHz to 1 GHz). Transformer designs typically require only a few turns because shunt inductances are small and leakage inductances must be kept low. Transformers are small in size, which is compatible with low leakage inductance and low effective shunt capacitance. Winding dc resistances

*Reference 19.

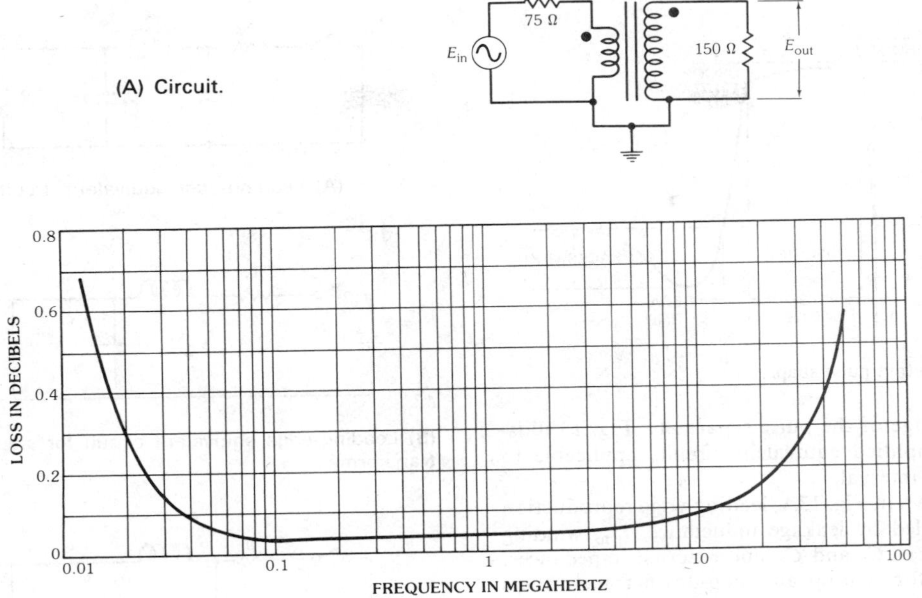

(B) Loss-frequency characteristic, 20 $\log_{10} E_{out}/E_{in}$.

Fig. 11. Response of transformer in example.

are usually negligible, but core loss can be important for low turns and some core materials. Small toroidal cores are often used because of their ability to achieve low leakage inductance.

Transmission-line techniques can be applied to transformer windings at high frequencies by tightly coupling the primary and secondary windings to form a transmission line having a particular characteristic impedance. The overall characteristic impedance (Z) of a transformer is approximately equal to $(Z_pZ_s)^{1/2}$. This impedance is matched as nearly as possible to a transmission-line winding whose characteristic impedance (Z_0) is equal to $(L/C)^{1/2}$. In this expression, L is the inductance of the transmission line when the far ends are shorted, and C is the effective capacitance of the transmission line with the far ends open. Windings are typically wound tightly coupled as parallel or twisted groups of two, three, or four wires to obtain a particular ratio. Only a few integer turns ratios are easily obtainable; ratios of 1:1, 1:2, and 1:3 are the most common. Tightly coupled windings produce low leakage inductance. By controlling wire size, wire insulation, and degree of twisting, various uniform characteristic impedances can be obtained. Values of Z_0 between 50 and 100 ohms are most easily made.

Autotransformers provide very low leakage inductances and lend themselves well to a distributed transmission-line analysis. They are preferred in applications where direct-current isolation of windings is not a factor.

Use of NiZn cores, which have lower permeability than MnZn cores, will restrict low-end response considerably. However, NiZn cores usually have lower core loss at high frequencies, which is important when only a few turns are required.

PULSE TRANSFORMERS

Pulse transformers are designed to transmit rectangular waves or trains of pulses while maintaining as closely as possible the original shape. Functions performed by pulse transformers are similar to those of broad-band analog transformers. Examples include impedance conversion, dc isolation, coupling between balanced and unbalanced circuits, voltage transformation, and phase inversion. Analytical design is usually done in the time domain instead of the frequency domain, and transient considerations become important.

Pulse waveshapes contain a wide range of frequencies. Lower frequencies relate to pulse duration and repetition rate, while higher frequencies determine the shape of pulse edges. Core and winding parameters limit both extremes of the frequency response of a transformer. Fig. 12 portrays a typical transformer output pulse compared with the corresponding input pulse. (In the strictest sense, pulse rise and decay times are measured between the 10- and 90-percent values; width is measured between the 50-percent values.) Pulse transformers can be analyzed by considering the leading edge, top, and

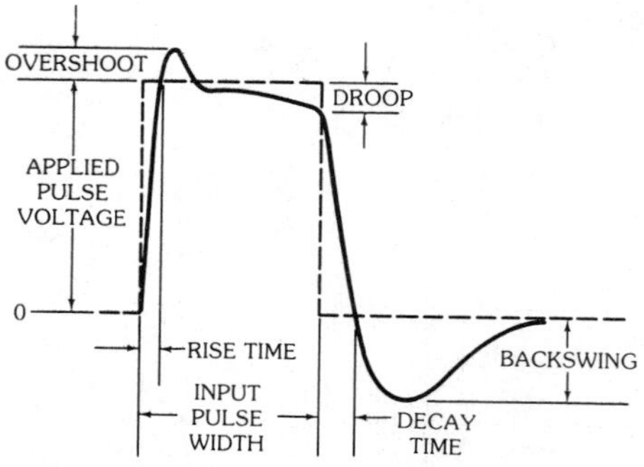

Fig. 12. Output pulse shape.

trailing edge of the pulse separately. Fig. 13 illustrates simplified equivalent circuits applicable to each time interval.

As shown in Fig. 13A, leading-edge reproduction is controlled by leakage inductance, l_{scp}, winding capacitances C_p and C_s, and external impedances. Analysis for step-up and step-down transformers varies slightly, as shown in Figs. 13B and C. Leakage inductance and winding capacitance must be minimized to achieve a sharp rise; however, output voltage may overshoot input voltage (Fig. 12), and oscillation may be encountered where very abrupt rise times are invoked. Rise time T_r may be related to the high-frequency 3-dB loss point f_h of a wideband analog transformer as follows:*

$$f_h = 0.382/T_r$$

A graphical solution of rise time may be found in additional references.†

Pulse-top response (droop, Fig. 12) depends on the magnitude of open-circuit inductance L_p and external impedances as shown in Fig. 13D. The circuit is similar to the low-frequency equivalent circuit for wideband analog transformers. However, the value of open-circuit inductance is proportional to the average slope of the part of the *B-H* loop traversed and may differ when measured under continuous-wave and pulse conditions. A cw input signal generally is sinusoidal, and the core *B-H* loop traversed is symmetrical about the origin. As shown in Fig. 14A, a unipolar pulse traverses a smaller, asymmetrical portion of the *B-H* loop, resulting in a lower average slope. Fig. 14B shows the *B-H* loop generated with bipolar pulse excitation (see reference 20). A graphical solution of pulse-top response is also shown in reference 10.

Trailing-edge response is controlled by the dissipation of energy stored in the transformer after the supporting pulse voltage has been removed. It depends on the open-circuit inductance, secondary

*Reference 20.
†Reference 10.

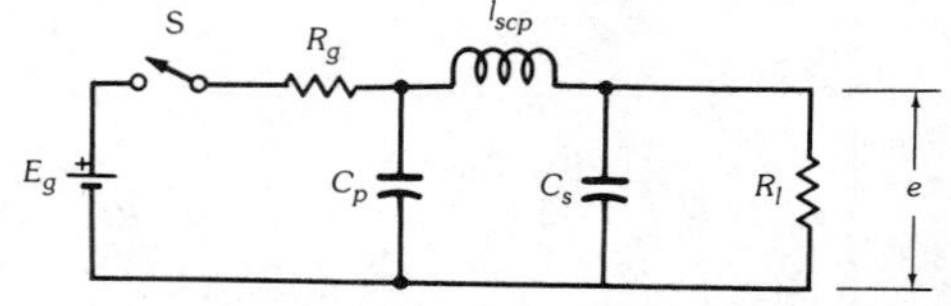

(A) Leading-edge equivalent circuit.

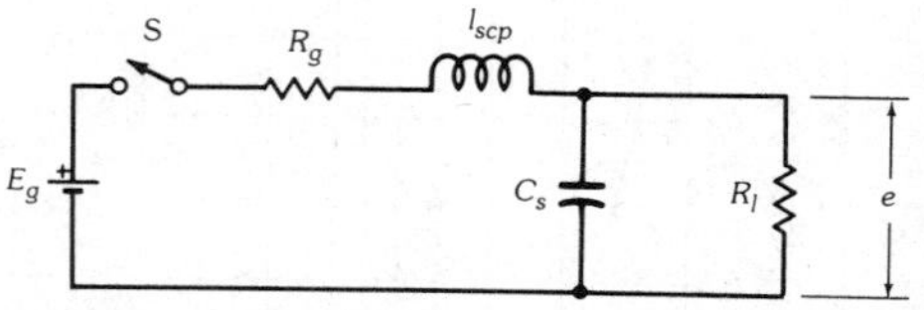

(B) Leading-edge equivalent circuit for step-up-ratio transformer.

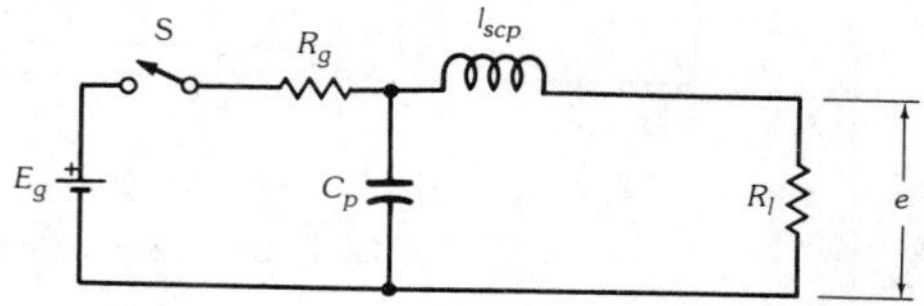

(C) Leading-edge equivalent circuit for step-down-ratio transformer.

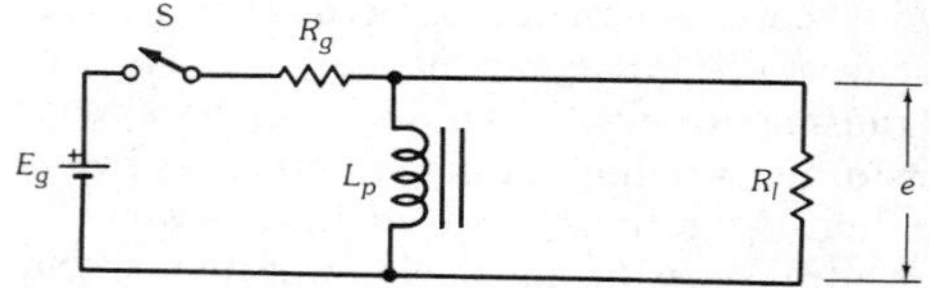

(D) Top-of-pulse equivalent circuit.

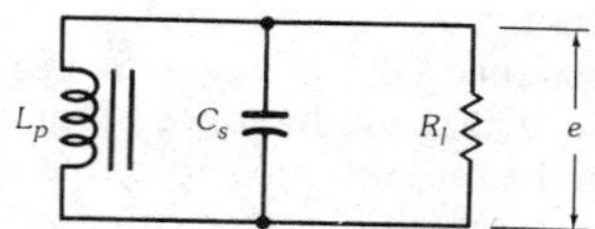

(E) Trailing-edge equivalent circuit.

Fig. 13. Pulse-transformer equivalent circuits.

winding capacitance, and external impedances as shown in Fig. 13E. A lower capacitance results in a faster rate of voltage decay. Negative backswing is proportional to the magnitude of the transformer magnetizing current. A graphical analysis of trailing-edge response can be found in reference 10.

Choice of a pulse-transformer core is usually determined by the *Et* product, which refers to the pulse voltage *E* and the duration *t* in microseconds. The core should not saturate, so *Et* is proportional to the flux swing traversed on the *B-H* loop just

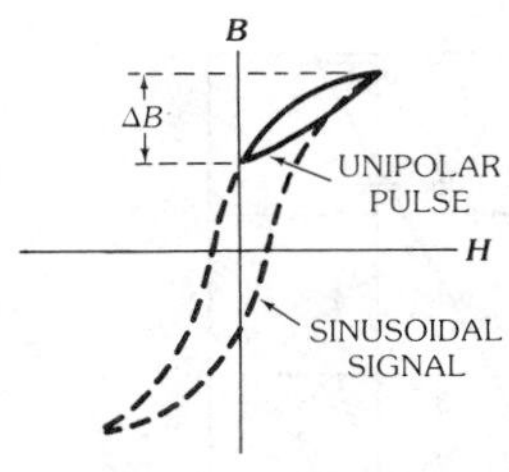

(A) Unipolar pulse.

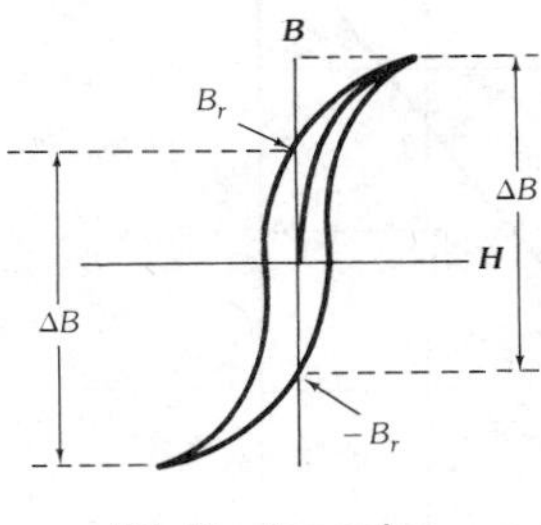

(B) Bipolar pulse.

Fig. 14. Traverses of *B-H* loop.

below saturation and to core cross-sectional area.*

Ferrite materials, specified in terms of pulse excitation, are typically used for pulse-transformer cores. Gaps may be used to provide greater flux swing for unipolar pulse applications. Core configurations may be toroidal where self-shielding and low leakage inductance are paramount, "E" shape for typically lower-cost bobbin winding techniques, and cup-core or RM type for greater magnetic efficiency with bobbin windings.

MAGNETIC-CORE REACTORS

The purpose of a reactor is to introduce reactance into a circuit. Inasmuch as this reactance is inductive, this device is also called an inductor. Reactors consist of one or more windings and may or may not have a magnetic core. This section is concerned with those reactors that have magnetic cores. A filter reactor is defined† as ". . . a reactor used to reduce harmonic voltage in alternating-current or direct-current circuits."

Rectifier-Filter Reactors

The rectifier-filter reactor is used mainly in direct-current power supplies to smooth the output ripple voltage. It carries all the direct current of the rectifier filter and must be designed not to saturate with direct current in the reactor winding.

Optimum design data may be obtained from Hanna curves, Fig. 15. These curves relate direct-current energy stored in the core per unit volume, LI_{dc}^2/V, to magnetizing field NI_{dc}/l_c (where l_c = average length of flux path in core), for an appropriate air gap. Heating is seldom a factor, but direct-current-resistance requirements affect the design; however, the transformer equivalent volt-ampere ratings of chokes given in Table 9 should be useful in determining their sizes. This is based on the empirical relationship $(VA)_{eq} = 188LI_{dc}^2$.

As an example, take the design of a choke that is to have an inductance of 10 henries with a superimposed direct current of 0.225 ampere and a direct-current resistance ≤ 125 ohms. This reactor is to be used for suppressing harmonics of 60 hertz, where the alternating-current ripple voltage (second harmonic) is about 35 volts.

(A) $LI^2 = 0.51$. Based on the data of Table 9, try a 4% silicon-steel core, type EI-125 lamination, with a core buildup of 1.5 inches.

(B) From Table 9: $V = (11.4/1.25) \times 1.5 = 13.7$ in^3, $l_c = 7.50$, $LI^2/V = 0.51/13.7 = 0.037$. From Fig. 15: $NI/l_c = 88$ ampere-turns per inch, $N = 88l_c/I = (88 \times 7.5)/0.225 = 2930$ turns, $l_g/l_c = 0.0032$, the length of air gap $l_g = 0.0032 \times 7.5 = 0.024$ inch.

(C) From Table 9, coil (MLT) = (7.21 + 0.5)/12 = 0.643 foot, and length of coil = $N \times$ (MLT) = 2930 × 0.643 = 1884 feet. Since the maximum resistance is 125 ohms, the maximum ohms/ft = 125/1884 = 0.0663, or 66.3 ohms/1000 ft. From Table 5, the nearest size of wire is No. 28.

(D) Now see if 2930 turns of No. 28 single-insulated wire will fit in the window space of the core. (Determine turns per layer, number of layers, and coil-build, as explained in the design of power transformers.)

(E) This is an actual coil design; in case the lamination window space is too small (or too large), change stack of laminations, or size of laminations, so that the coil meets the electrical requirements and the total coil-build ≈ 0.85 to 0.90 × (window width).

Note: To allow for manufacturing variations in permeability of cores and resistance of wires, use at least 10-percent tolerance.

Swinging reactors are used where the direct current in the rectifier circuit varies. These reactors are designed to saturate under full load current while providing adequate inductance for filtering. At light load current, higher inductance is available to perform proper filtering and prevent "capacitor effect." The equivalent reactor size is determined from:

$$LI^2 = (L_{max} \times L_{min})^{1/2} I_{dc(max)}^2$$

The design is similar to that of a normal reactor and is based on meeting both L and I_{dc} extremes. The typical swing in inductance is 4:1 for a current swing of 10:1.

*Reference 10.

†Reference 1.

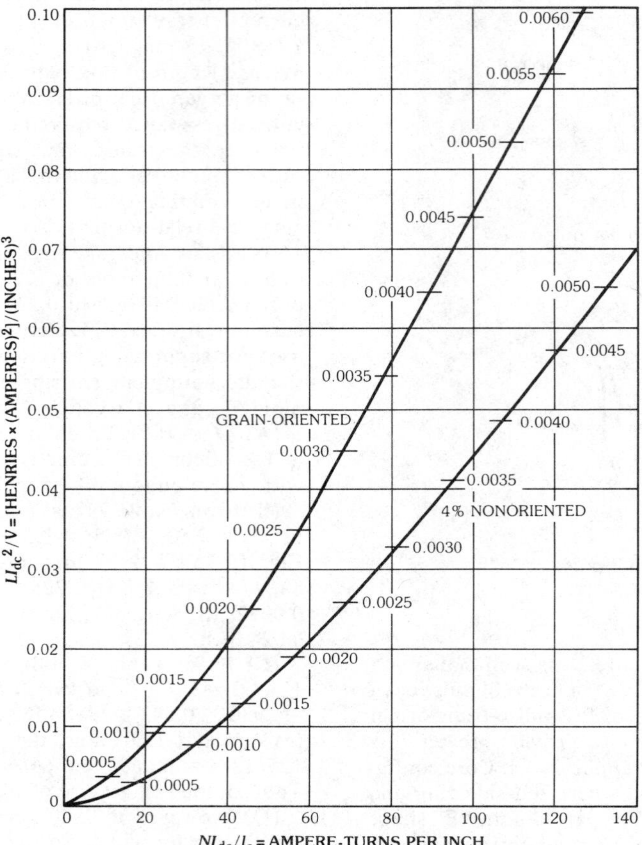

Fig. 15. Hanna curves for silicon steel. The numbers on the curves represent length of air gap l_g/ length of flux path l_c.

AC-Filter Reactors

In the rectifier-filter reactor, the amount of ac flux in the magnetic core is small in comparison to the dc flux. With the ac-filter reactor, there is no dc flux, and the ac flux is large. In the rectifier-filter reactor, the direct current through the coil causes the core to operate in a unipolar, or CM, mode, which necessitates the use of an air gap as calculated from the Hanna curves. In the ac-filter reactor, the core operates in a bipolar, or SCM, mode, but the peak values of the alternating current are large, which also necessitates the use of an air gap in the core to prevent it from saturating.

A method for designing ac-filter reactors may be found in Chapter VIII of reference 18.

MAGNETIC-CORE INDUCTORS

Magnetic-core inductors used in frequency-selective filters must provide a specific inductance value, a high Q value (ratio of reactance to effective resistance), and stable parameter values in the operating environment. The inductance value depends on the core inductance coefficient, $A_L = L/N^2$, and the number of turns in the winding. The A_L values for specific cores are a function of the effective permeability, μ_e, and the core geometry; core manufacturers tabulate these data in their catalogs. Achieving high Q is more complex. The inductor is modeled with a set of independent loss tangent ($1/Q$) values that together describe all the loss mechanisms for the inductor. An attempt is made to minimize the major contributors at a given frequency. At the same time, the designer must also attempt to minimize distributed capacitance, μ_e, and mechanical stresses on the cores in order to promote stable performance. The completed structure must be mechanically sound to minimize the effects of shock and vibration on the parameter values.

Audio-Frequency Inductors

Audio-frequency inductors operate in the frequency band of about 200 Hz to 20 kHz. At the

TABLE 9. EQUIVALENT LI^2 RATING OF FILTER REACTOR FOR RECTIFIERS*

LI^2†	Current Density† (A/in²)	EI-Type Lamination‡	Stack Height p (in)	Core Volume V (in³)	Magnetic Path Length l_c (in)	Average Copper Mean Length per Turn (MLT) (in)
0.0195	3200	EI-21	0.5	0.80	3.25	3.12
0.0288	2700	EI-625	0.625	1.45	3.75	3.62
0.067	2560	EI-75	0.75	2.51	4.50	4.33
0.088	2560	EI-75	1.00	3.35	4.50	4.83
0.111	2330	EI-11	0.875	3.88	5.25	5.04
0.200	2130	EI-12	1.00	5.74	6.00	5.71
0.300	2030	EI-12	1.50	8.61	6.00	6.71
0.480	1800	EI-125	1.25	11.4	7.50	7.21
0.675	1770	EI-125	1.75	16.0	7.50	8.21
0.850	1600	EI-13	1.50	19.8	9.00	8.63
1.37	1500	EI-13	2.00	26.4	9.00	9.63
3.70	1200	EI-19	1.75	39.4	13.0	12.8

* L=inductance in henries, I=direct current in amperes. The rating is based on power-supply frequencies up to 400 hertz and 50°C temperature rise above ambient. The LI^2 values should be reduced for lower temperature rises and high-voltage operation.

† From *Radio Components Handbook* (Cheltenham, PA.: Technical Advertising Associates, 1948; page 92).

‡ Lamination designation and constants per Allegheny Ludlum Corp., Pittsburgh, Pa.

lowest frequencies, laminated cores are generally used, although recent designs have used ferrite. Where low-cost, nonadjustable inductors are needed in the range 300 Hz to 10 kHz, permalloy powder toroids are often used. For frequencies above 10 kHz, or for applications requiring adjustable inductors, ferrite pot cores are preferred (see Table 10). Pot-core inductors are discussed in the section on precision adjustable inductors.

To design stable, high-Q inductors, it is necessary to understand the equivalent circuit of an inductor (Fig. 16) and its application to the parallel-tuned mesh (Fig. 17) and series-tuned mesh (Fig. 18). The distributed capacitance, C_d, is very important in series-tuned meshes, because it affects both the inductance and the Q. If possible, it is desirable to keep the distributed capacitance to one tenth of the tuning capacitance, C_s. Means of minimizing C_d are discussed in succeeding paragraphs.

The copper loss in the coil winding, R_w, is made up of the dc resistance of the wire and the eddy-current losses generated in the winding by stray flux from the core that cuts the winding. This eddy-current loss can be minimized at high frequencies by dividing the wire into many small strands. This type of wire is called litz wire.

The loss due to the magnetic core reflected in series with the coil winding is represented by R_c.

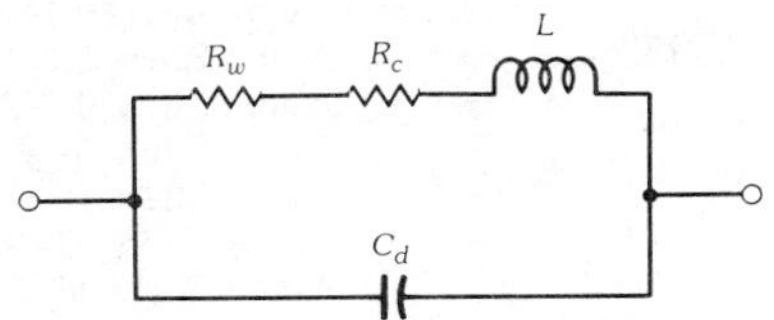

L = calculated inductance based on turns and magnetic-core constant
C_d = distributed capacitance of winding
R_w = copper losses in winding
R_c = magnetic-core losses reflected in series with winding
Q = quality factor = $\omega L/(R_w + R_c)$, $\omega = 2\pi f$

Fig. 16. Equivalent network of an audio filter coil.

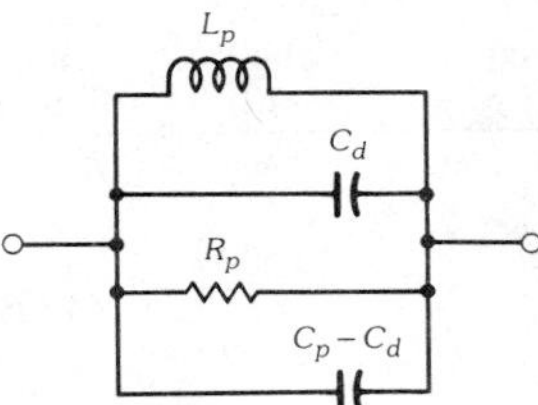

L_p = equivalent parallel inductance
R_p = equivalent parallel resistance
C_d = distributed capacitance of winding
C_p = parallel tuning capacitance
Q_p = effective parallel Q
$L_p = L$, for $Q > 10$
$R_p = (R_w + R_c)(Q^2 + 1)$, for $Q_p = Q$

Fig. 17. Parallel-tuned mesh.

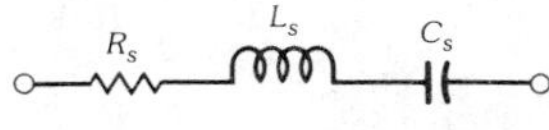

L_s = equivalent series inductance
R_s = equivalent series resistance
C_s = series tuning capacitance
Q_s = effective series Q

For $Q > 10$
$L_s = L/(1 - \omega^2 LC_d)$
$R_s = (R_w + R_c)/(1 - \omega^2 LC_d)^2$
$Q_s = Q(1 - \omega^2 LC_d)$
$L = L_s/(1 + \omega^2 L_s C_d)$

Fig. 18. Series-tuned mesh.

The calculation of this factor is discussed in succeeding paragraphs.

Laminated-Core Audio-Filter Inductors— Externally, a laminated-core audio-filter inductor resembles a transformer. The laminations are usually "F" in shape to create an air gap in the center leg of the core. For high Q, nickel-steel laminations either 14 or 6 mils thick are used (Table 10). The air gap in the core reduces the effective permeability, μ_e, which reduces the effective core losses and stabilizes the inductance.

The winding is usually layer wound as described in the section on methods of winding transformers. The techniques for reducing capacitance described in that section can be used to reduce the distributed capacitance of the winding. A detailed method for designing this type of inductor can be found in *High Q Reactors for Low Frequencies,* Bulletin A10, Magnetic Metals Co., Camden, N.J.

Toroidal Core Types—Toroidal coils are doughnut shaped, with the winding covering the entire core. The core is usually made of pressed molybdenum-permalloy powder, although some cores are made of carbonyl powdered iron for very-high-frequency applications (Table 10).

Permalloy-powder toroidal cores are made in various sizes from 0.140 to 2.25 inches od and with ten material compositions to change the effective permeability, μ_e, from 14 to 550 for use at different frequency ranges.

Fig. 20 gives the Q-vs-frequency characteristics of four sizes of toroids with five different effective permeabilities. In addition, the inductance in millihenries per 1000 turns, A_L, is given for each size of core.

Fig. 21 gives the number of turns of various sizes of enameled wire that can be placed on the toroid as well as the mean length of turn (MLT) for the respective toroidal cores of Fig. 20. These curves

TABLE 10. CHARACTERISTICS OF SOME CORE MATERIALS FOR AUDIO-FILTER COILS

Material or Alloy	Initial Perme-ability (μ_0)	Resistivity (ohm-cm)	Hysteresis Coefficient ($a\times10^6$)	Residual Coefficient ($c\times10^6$)	Eddy-Current Coefficient ($e\times10^9$)	Gauge (mils)	Application and Frequency Range (kilohertz)
4% silicon steel	400	60×10^{-6}	120	75	870	14	Rectifier filters
Low nickel	3 500 to 10 000	44×10^{-6}	0.4	14	1 550	14	Audio filters up to 0.2
					284	6	Audio filters up to 10
High nickel	10 000 to 20 000	57×10^{-6}	0.05	0.05	950	14	Audio filters up to 0.2
					175	6	Audio filters up to 10
Molybdenum permalloy powder	550†	1.0	1.5	88	27	—	Audio filters 0.1–6
	200†	1.0	0.7	21	25	—	Audio filters 0.1–7
	160†	1.0	0.9	25	17	—	Audio filters 0.1–10
	125†	1.0	0.9	32	15	—	Audio filters 0.2–20
	60†	1.0	1.5	50	7.5	—	Audio filters 5–50
	25†	1.0	4.0	96	7.0	—	Audio filters 15–60
	26†	1.0	4.0	96	7.0	—	Audio filters 15–60
	14†	1.0	7.0	143	6.5	—	Audio filters 40–150
Carbonyl types:							
C	55	—	9	80	7	—	High-frequency filters
P	26	—	3.4	220	27	—	High-frequency filters
Th	16	—	2.5	80	8	—	High-frequency filters
Ferrites‡:							
3B7	2 300	100	§	§	§	—	Audio filters 0.2–300
3B9	1 800	100	§	§	§	—	Audio filters 0.2–300
3D3	750	10^5	§	§	§	—	HF filters 200–2 500
4C4	125	10^5	§	§	§	—	HF filters 1 000–20 000

$R_c/(\mu_0 Lt) = aB_m + c + ef$, where R_c = series resistance in ohms due to core loss.*

* Data and coefficients *a*, *c*, and *e* are from V. E. Legg and F. J. Given, "Compressed Powdered Molybdenum Permalloy for High Quality Inductance Coils," *Bell System Technical Journal*, Vol. 19, No. 3, July 1940; pp. 385–406.

† Data from Catalog PC303T, Magnetics, Inc., Butler, Pa.

‡ Data from Bulletin 220-C, Ferroxcube Corporation of America, Saugerties, N.Y.

§ See Fig. 19.

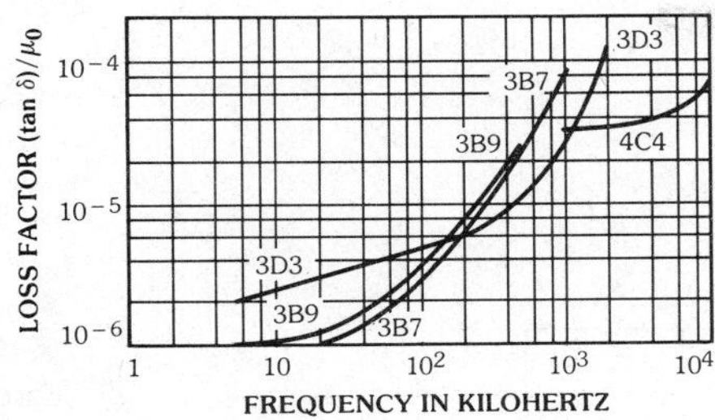

Fig. 19. Loss factor as a function of frequency. (*From Ferrite Pot Cores, Bulletin 220-C, Ferroxcube Corp. of America, Saugerties, N.Y.*)

are based on a single winding wound with 180° traverse using commercially available toroidal winding machines.

Fig. 22 gives the core-loss factor in ohms per millihenry for various effective permeabilities and frequencies. This factor is multiplied by the inductance in millihenries to determine R_c. The curves are based on a magnetic induction of 100 gauss. For other levels, R_c can be calculated from the Legg coefficients given in Table 10.

Table 11 gives an approximate value of distributed capacitance C_d that can be expected for the different sizes of cores and winding methods.

When maximum temperature stability of inductance is required, most manufacturers of permalloy toroidal cores can provide various types of stabilized cores. The stabilizations most used are listed in Table 12.

Since the distributed capacitance changes rapidly with temperature, it should be kept to a minimum to avoid changing the inductance.

Fig. 23 illustrates the most common methods of winding toroids. In Fig. 23A the toroid is rotated over a 360° arc for every layer of the winding (called 360° traverse winding). In Fig. 23B, the toroid is rotated over a 180° arc for each layer until half the coil is wound. The other half of the winding is similarly wound. This is called 180° traverse winding. In Fig. 23C the toroid is rotated over only 90° until one quarter is wound. The other three quarters are wound in the same manner. This winding method is called 90° traverse or quadrature winding.

Since most toroid cores are made with a tolerance of ±8% on the A_L (millihenries/1000 turns), it is usually necessary to adjust the inductance after winding. This is done by winding about 5% more turns than calculated on the core and removing turns until the inductance reaches the desired value. Coils for series-tuned meshes should be adjusted to the resonance frequency of the mesh with the tuning capacitor in series to eliminate the effect of distributed capacitance. Coils for parallel tuned meshes should be adjusted at low frequency in such a way that the tuning capacitance is 1000 times the distributed capacitance for 0.1% accuracy.

Toroidal Core Design Example—It is desired to design an inductor of 100 millihenries for a series-tuned mesh that resonates at 10 kilohertz. The Q must be 150 minimum and the size as small as possible.

(A) Consulting the Q curves of Fig. 20B shows that a 1.06-inch od toroid core with $\mu_e = 125$ is the smallest core that will meet the Q requirements. This has an A_L value of 157 millihenries per 1000 turns.

(B) From Table 11, $C_d \approx 200$ picofarads. Calculate L from Fig. 18:

$$L = \frac{L_s}{1 + \omega^2 L_s C_d}$$

$$= \frac{100}{1 + (2\pi 10\ 000)^2 \times 0.1 \times 200 \times 10^{-12}}$$

$$= 92.68 \text{ millihenries}$$

(C) Compute the number of turns required from

$$N = 1000(L_{\text{millihenries}}/A_L)^{1/2}$$
$$= 1000(92.68/157)^{1/2}$$
$$= 768 \text{ turns}$$

(D) Fig. 21 gives the maximum size of wire and mean length of turn. Use No. 30 heavy enameled wire for 768 turns. On a 1.06-inch od core the (MLT) is 1.57 inches.

(E) Calculate R_w from (MLT), N, and Table 5.

$$R_w = \frac{(\text{MLT}) \times N \times \text{ohms}/1000 \text{ ft}}{12\ 000}$$

$$= \frac{1.57 \times 768 \times 103.2}{12\ 000}$$

$$= 10.36 \text{ ohms}$$

(F) Calculate R_c from R/L values of Fig. 22 for $\mu_e = 125$ and $f = 10$ kilohertz. $R/L = 0.23$ ohm/millihenry. $R_c = (R/L)L = 0.23 \times 92.68 = 21.32$ ohms.

(G) Calculate Q per Fig. 16.

$$Q = \frac{\omega L}{R_w + R_c} = \frac{2\pi 10 \times 92.68}{10.36 + 21.32} = 183.8$$

(H) Calculate Q_s per Fig. 18.

$$Q_s = Q(1 - \omega^2 L C_d)$$
$$= 183.8[1 - (2\pi 10000)^2 \times 0.09268 \times 200 \times 10^{-12}]$$
$$= 170.4$$

This should approximate the measured Q.

(I) The coil is wound using 360° traverse winding, and, since no special temperature stability is required, a standard unstabilized core is used.

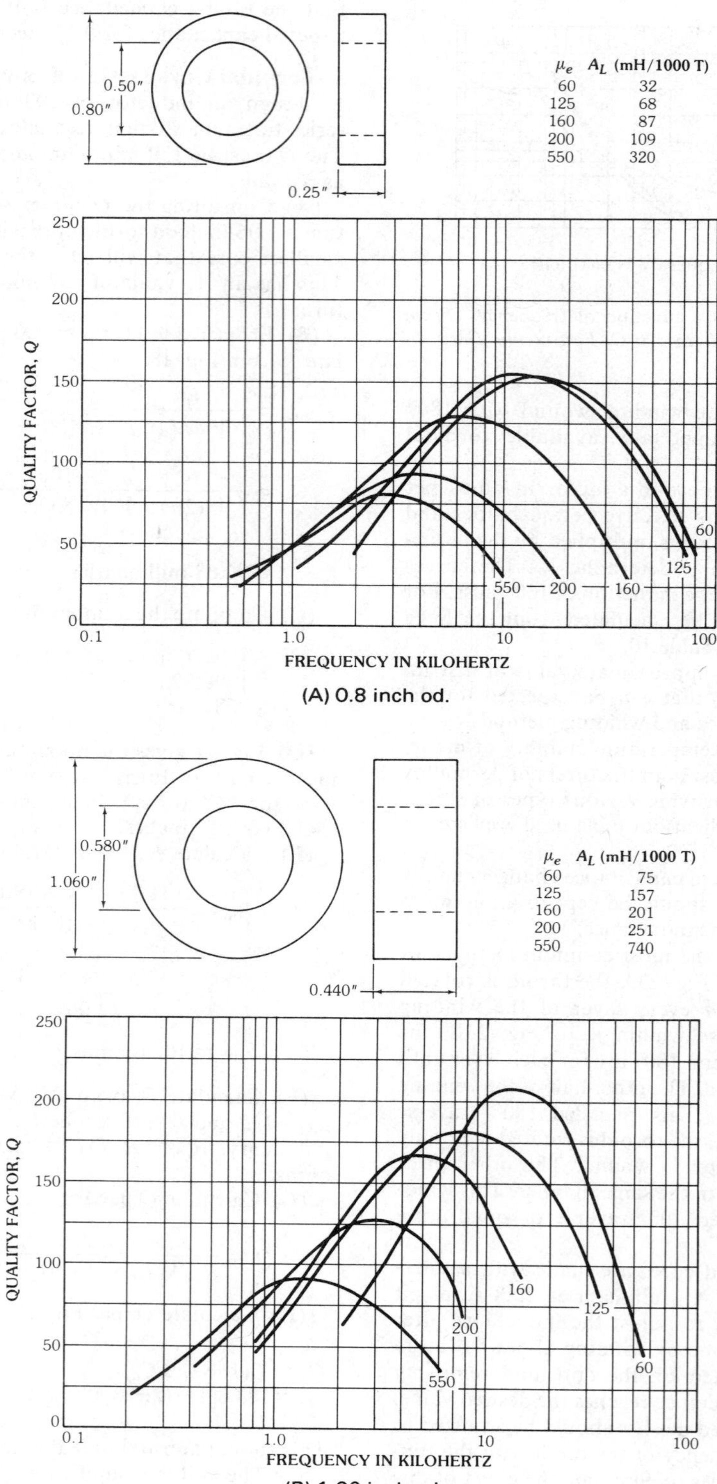

Fig. 20. Quality factor Q as a function of frequency for several sizes of permalloy-dust toroids. (*Data replotted from*

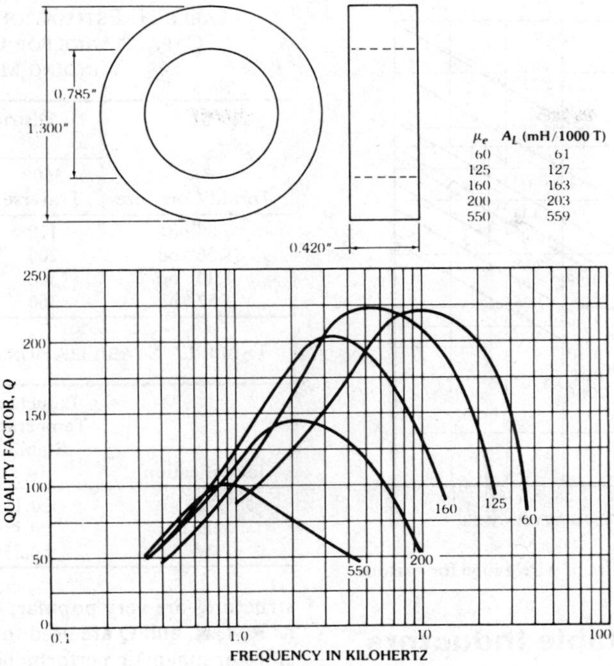

μ_e	A_L (mH/1000 T)
60	61
125	127
160	163
200	203
550	559

(C) 1.3 inch od.

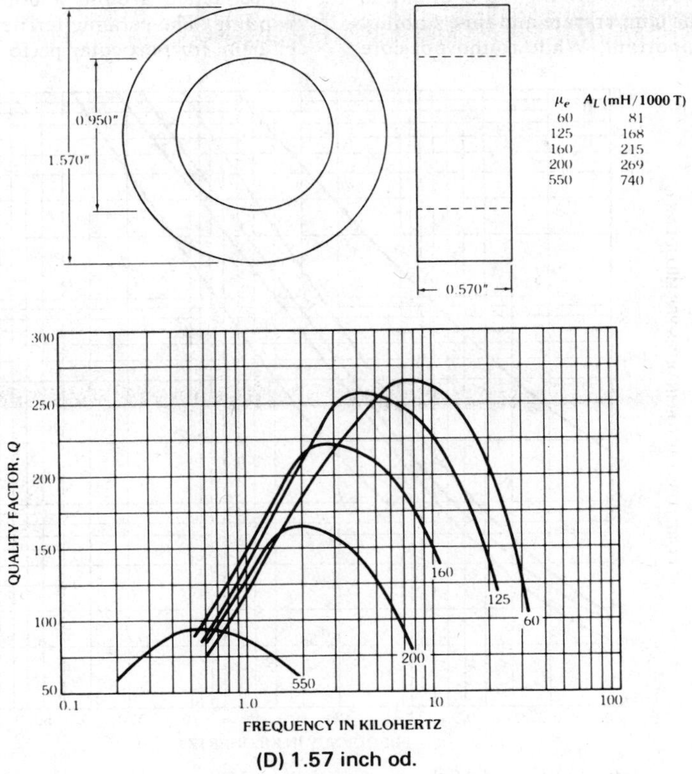

μ_e	A_L (mH/1000 T)
60	81
125	168
160	215
200	269
550	740

(D) 1.57 inch od.

Permalloy Powder Cores, *Catalog PC-303T, Magnetics, Inc., Butler, Pa.)*

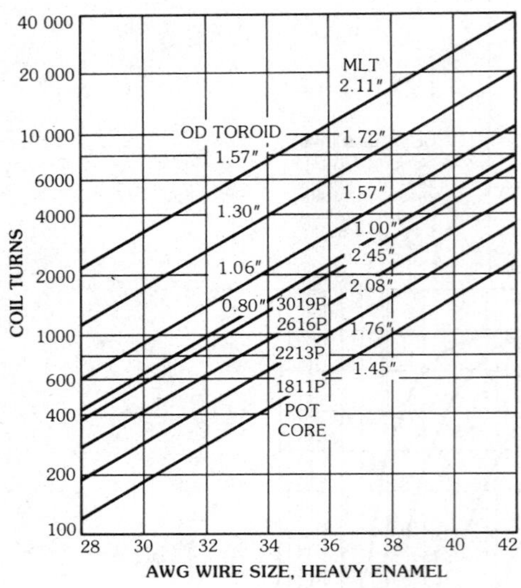

Fig. 21. Coil turns as a function of wire gauge for various core sizes and types.

Precision Adjustable Inductors

Pot Cores—Filter-class inductors are precision, adjustable devices designed to provide high Q values at a particular frequency. Although Q is often the greatest concern, the temperature and time stability are usually also important. While round pot-core structures are very popular, other core shapes such as RM, X, and Q are used to improve packing densities or magnetic performance.

A cross-section of a typical pot-core inductor is shown in Fig. 24. The two core halves are clamped or cemented around a bobbin that contains the winding. The ceramic ferrite cores are chosen specifically for particular performance characteristics.

TABLE 11. ESTIMATION OF DISTRIBUTED CAPACITANCE FOR CORE SIZE AND WINDING METHOD

	Distributed Capacitance C_d (picofarads)		
Toroid Core Size	360° Traverse	180° Traverse	90° Traverse
0.8″ od	120	60	30
1.06″ od	200	100	50
1.30″ od	300	150	75
1.57″ od	360	180	90

TABLE 12. STABILIZATION OF TOROIDAL CORES

Identification	Inductance Temperature Stability (%)	Temperature Range (°C)
B	±0.1	+13 to +35
D	±0.1	0 to +55
W	±0.25	−55 to +85

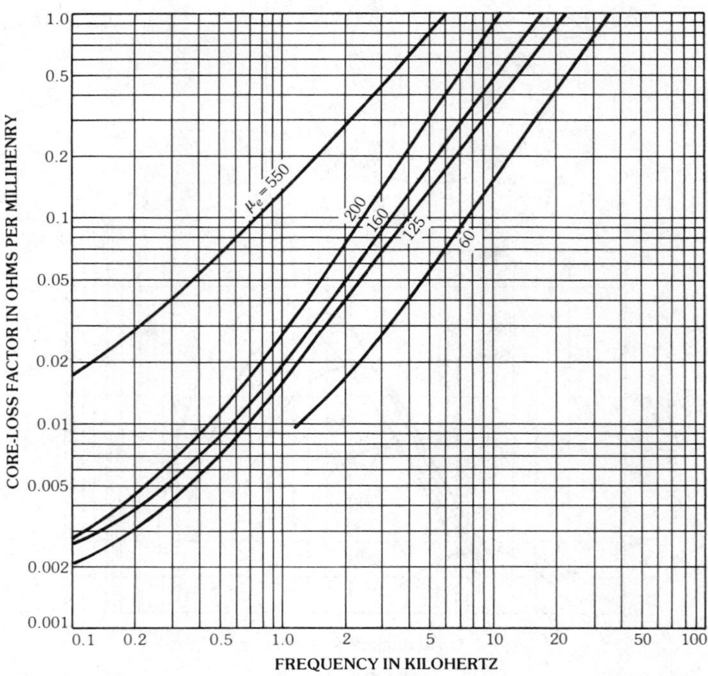

Fig. 22. Molybdenum permalloy-dust-core loss characteristics. B_m = 100 gauss. (*Data replotted from* Permalloy Powder Cores, *Catalog MPP-303T, Magnetics, Inc., Butler, Pa.*)

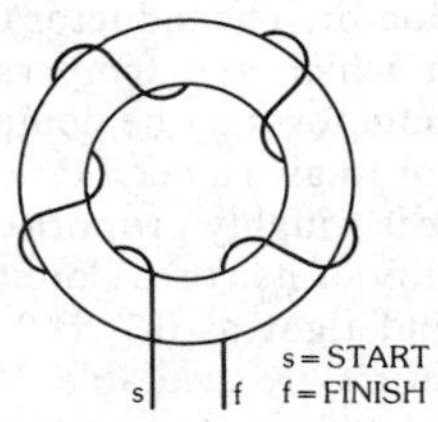

(A) 360° traverse.

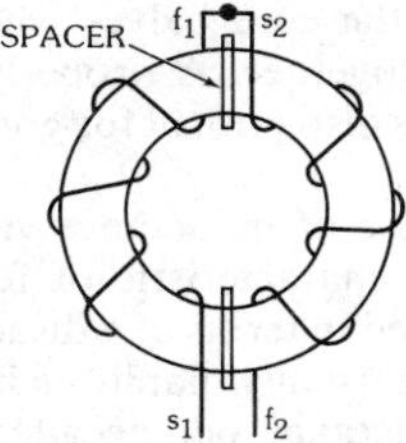

(B) 180° traverse.

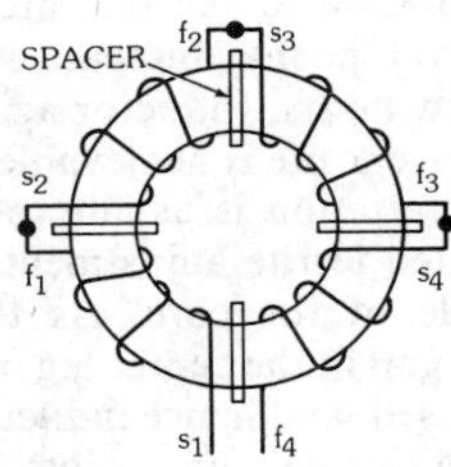

(C) 90° traverse.

Fig. 23. Methods of winding toroid cores.

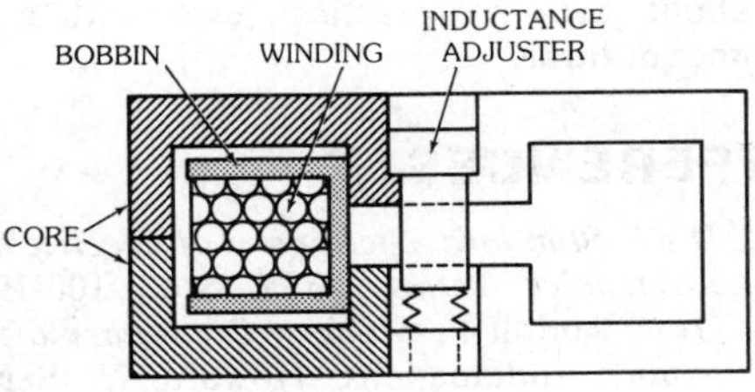

Fig. 24. Typical pot-core coil.

A wide range of permeabilities, loss characteristics, and stability performance exists. The information is available in ferrite-supplier catalogs.

The cores have a recessed center post so that when they are mated, an air gap exists. Although the gap reduces the effective permeability, μ_e, more importantly it reduces the core loss and temperature coefficient of inductance. The gap is used to control the A_L value (in nanohenries per turn squared), a most important inductor design parameter. Inductors operating at low frequencies have relatively high values of μ_e and A_L. As the operating frequency is increased, lower values of these parameters usually are adequate.

Loss Mechanisms and *Q*—Inductor Q, the ratio of reactance to effective resistance at a given frequency, is controlled by four loss factors: core loss, winding resistance loss, capacitance loss, and proximity-effect loss. These are expressed in terms of loss tangents. Inductor Q is the reciprocal of the sum of these four loss tangents. Many inductor-loss equations are given by Snelling,* but only four equations are needed in practice. Optimization equations are given by Banzi.†

Most manufacturers of inductor cores list the core-material loss, $\tan \delta_m$, which has been normalized with respect to the core-material permeability μ_i. To calculate the effective core-loss tangent, it is necessary only to multiply the advertised value by the effective permeability, μ_e, listed separately in the core literature for specific geometries and inductance factors (A_L values). The effective magnetic core-loss equation is

$$\tan \delta_{me} = (\tan \delta_m)(\mu_e/\mu_i) = \mu_e/\mu_i Q_m$$

where Q_m is the core-material Q. While it is apparent that a low ratio of μ_e/μ_i (implying a large air gap) would reduce the effective core loss, a large gap results in higher winding losses because of flux fringing at the gap.

Although coils generally have associated capacitances that are distributed in nature, a single lumped capacitance shunting the winding may be considered a sufficiently accurate model for most Q calculations. The capacitive loss is the product of a ratio and a constant, as shown in the equation below. The ratio is the parallel winding capacitance, C_p, divided by the capacitance, C_r, necessary to resonate the inductor at the frequency of interest. The value of C_p can be measured, or it can be calculated based on the winding geometry.‡ Constant K_D depends on the dielectric constant of the wire insulating material. The capacitive loss-tangent formula is

$$\tan \delta_c = (C_p/C_r)K_D$$

Experimental work has shown the value of K_D to be 0.02 for nylon-served litz wire (groups of individually insulated wires twisted into a bundle that is then wrapped in nylon yarn) and 0.01 for polyurethane-coated solid wire. It is not necessarily advantageous to use solid wire to minimize capacitive losses. For multiple-layer windings, litz wire usually results in lower capacitance than does solid wire because litz-wire insulation provides more separation between layers. The high loss associated with the nylon serving, then, tends to be offset by the lower value of C_p. For single-layer windings,

*Reference 19.
†Reference 21.
‡Reference 8.

though, there may be very little difference in capacitance between the two wire types, depending on the number of turns used, so that here solid wire presents an advantage in minimizing capacitive loss.

The winding-resistance loss tangent, $\tan \delta_r$, is the ratio of the ac resistance of the winding, R_{ac}, to the coil reactance, ωL, at the frequency of interest:

$$\tan \delta_r = R_{ac}/\omega L$$

Because the dc resistance can be determined, it would be convenient to calculate the ac resistance in terms of the dc resistance. The skin-effect factor, SEF, is the ratio of ac to dc resistance. The SEF versus d/Δ (wire diameter/skin depth) characteristic is used in conjunction with the skin depth versus frequency characteristic. These characteristics are plotted by Snelling.* The ac resistance is thus the dc resistance times SEF. This can be rewritten in more useful terms:

$$\tan \delta_r = [K_R (N/n)/\omega L] (\text{SEF})$$

where,

$K_R =$ dc resistance per strand turn of wire,
$N =$ number of turns of wire,
$n =$ number of wire strands,
SEF = skin-effect factor.

The proximity-effect loss-tangent formula is

$$\tan \delta_{pe} = K_E f N n d^4 (\text{PEF})/A_L$$

where,

$K_E =$ a constant that is geometry dependent,
$f =$ frequency of operation in hertz,
$N =$ number of winding turns,
$n =$ number of strands of wire,
$d =$ bare diameter of an individual strand of wire in millimeters,
PEF = proximity-effect factor, which is frequency dependent,
$A_L =$ inductance factor in nH/N^2

The value of K_E depends on the winding configuration and core type and must be determined experimentally; it varies slowly with frequency. Snelling gives typical values of K_E as well as means for obtaining PEF.*

Combining these effects:

$$Q = 1/(\tan \delta_{me} + \tan \delta_c + \tan \delta_r + \tan \delta_{pe})$$

For details see reference 21.

Inductor Q values as high as 1000 are obtainable with proper design and at frequencies in the range around 100 kHz. Modern designs use cores with a high A_L value and low mean length of turn to achieve high Q; unfortunately, this choice leads to higher effective permeability, which adversely affects stability of inductance.

Stability Considerations—Temperature stability of inductance may be critical for a precision adjustable inductor. The inductor is part of a tuned *LC* circuit in which the temperature coefficient (TC) of the inductor may be compensated by that of the capacitor to give a net TC near zero. The TC of an inductor is roughly proportional to the effective permeability of its core. Core temperature factors as low and tight as 0.7 ± 0.3 ppm/°C near room temperature are available. With an effective permeability of 100 for its core, an inductor would have a TC of 70 ± 30 ppm/°C. Ferrite cores exhibit magnetostrictive effects, so care must be taken to avoid stress on the core halves when bonding or clamping them together. A properly designed adjuster assembly is also critical for good temperature stability

A slow decrease of inductance value with time, calling aging, is characteristic of ferrite-cored inductors. Described in terms of a disaccommodation factor, this aging usually manifests itself as a linear decrease in inductance per decade of time, on a semilog plot. Manufacturers usually measure disaccommodation factor between 10 and 100 minutes, but an extrapolation to years of life is risky, since the longer-term slope may increase above that measured over a few hours. Inductor aging of less than 0.5% over a 20-year life is achievable.

Inductance variation is usually obtained by the use of a threaded ferrite adjustment mechanism in the center hole of the core. As the ferrite slug bridges the air gap in the center leg of the core (Fig. 24), μ_e is increased, and hence inductance increases. In many applications, an adjustment range of greater than ± 8%, with a sensitivity of 0.1%, is possible. Most ferrite inductor cores have preadjusted air gaps that result in an A_L tolerance of ±3%; thus guaranteed inductor adjustment ranges of about ±2% are practical, even with a small number of turns.

*Reference 19

REFERENCES

1. *IEEE Standard Dictionary of Electrical and Electronics Terms,* IEEE Std., 100-1972.
2. Graf, Rudolf F. *Modern Dictionary of Electronics.* Indianapolis: Howard W. Sams & Co., Inc., 1968.
3. *IEEE Standard for High-Power Wide-Band Transformers (100 Watts and Above),* IEEE Std. 264-1977.
4. Workman, T.J., of Thomas & Skinner, Inc. "Transformer Lamination Study." *Proceedings of Coil Winding/Electrical Manufacturing Expo '81.*
5. *Catalogue 15M-10/81.* Parsippany, N.J.: Allied Corp., 1981.
6. *Mini-Log IV.* Indianapolis: Thomas & Skinner, Inc.
7. Armstrong, R. W. "Polyphase Rectification Special Connections." *Proceedings of the In-*

stitute of Radio Engineers, Vol. 19, No. 1, January 1931.

8. Grossner, N. R. *Transformers for Electronic Circuits.* New York: McGraw-Hill Book Co., 1967. (Second edition published in 1983)
9. Schade, O. H. "Analysis of Rectifier Operation." *Proceedings of the IRE,* July 1943.
10. Lee, R. *Electronic Transformers and Circuits.* 2nd ed. New York: John Wiley & Sons, Inc., 1961.
11. *Standard Definitions of Terms, Symbols and Conversion Factors Relating to Magnetic Testing,* ANSI/ASTM A 340-77.
12. NEMA Standards Publication No. MW-1000-1981. Washington, D.C.: National Electrical Manufacturers Association, October 1981.
13. *Engineering Data for Film Insulations.* Fort Wayne, Ind.: Phelps Dodge Copper Products, January 1, 1961.
14. Hart, H. P., and Kakalec, R. J. "The Derivation and Application of Design Equations for Ferroresonant Voltage Regulators and Regulated Rectifiers." *IEEE Transactions on Magnetics,* Vol. Mag-7, No. 1, March 1971.
15. Kakalec, R. J. "A Feedback-Controlled Ferroresonant Voltage Regulator." *IEEE Transactions on Magnetics,* Vol. Mag-6, No. 1, March 1970.
16. Hart, H. P., and Kakalec, R. J. "A New Feedback Controlled Ferroresonant Regulator Employing a Unique Magnetic Component." *IEEE Transactions on Magnetics,* Vol. Mag-7, No. 3, September 1971.
17. Pressman, A. I. *Switching and Linear Power Supply, Power Converter Design.* Rochelle Park, New Jersey: Hayden Book Co., Inc., 1977.
18. McLyman, W. T. *Transformer and Inductor Design Handbook.* New York: Dekker Inc., 1978.
19. Snelling, E. C. *Soft Ferrites.* The Chemical Rubber Co., 1969.
20. *IEEE Standard for Low-Power Pulse Transformers,* IEEE Std., 390-1975.
21. Banzi, Fred J., Jr., "Higher *Q* From Pot Core Inductors." *IEEE Trans. Parts, Hybrids, and Packaging,* Vol. PHP-13, No. 4, December 1977.

14 Rectifiers, Filters, and Power Supplies

Revised by Fred G. Turnbull

CHARACTERISTICS OF POWER SEMICONDUCTOR DEVICES

This section briefly describes the terminal properties of various types of power semiconductor devices to aid in understanding their operation in power conversion equipment. Power semiconductor devices discussed are rectifiers, including fast recovery and Schottky; bipolar and field-effect transistors; thyristors and triacs; and voltage transient clippers.

Rectifiers

Voltage and Current Ratings—Silicon-rectifier ratings* are generally expressed in terms of reverse-voltage ratings and of mean-forward-current ratings in a half-wave circuit operating from a 60-hertz sinusoidal supply and into a purely resistive load.

There are three reverse-voltage ratings of importance:

Peak transient reverse voltage	V_{RM}
Maximum repetitive reverse voltage	$V_{RM(\mathrm{rep})}$
Working peak reverse voltage	$V_{RM(\mathrm{wkg})}$

Peak transient reverse voltage (V_{RM}) is the rated maximum value of any nonrecurrent surge voltage, and this value must not be exceeded under any circumstances, even for a microsecond. Maximum repetitive reverse voltage ($V_{RM(\mathrm{rep})}$) is the maximum value of reverse voltage that may be applied recurrently, e.g., in every cycle, and will include any circuit oscillatory voltage that may appear on the sinusoidal supply voltage. Working peak reverse voltage ($V_{RM(\mathrm{wkg})}$) is the crest value of the sinusoidal voltage of the supply at its maximum limit. The manufacturer generally recommends a $V_{RM(\mathrm{wkg})}$ that has an appreciable safety margin in relation to the V_{RM} to allow for the commonly experienced transient overvoltages on power mains.

Three forward-current ratings are similarly of importance:

Nonrecurrent surge current	$I_{FM(\mathrm{surge})}$
Repetitive peak forward current	$I_{FM(\mathrm{rep})}$
Average forward current	$I_{F(\mathrm{av})}$

Silicon diodes have comparatively small thermal mass, and care must be taken to ensure that short-term overload currents are limited. The nonrecurrent surge current is sometimes given as a single value that must not be exceeded at any time, but it is more generally given in the form of a graph of permissible surge current versus time. It is important to observe whether the surge-current scale is marked in peak, rms, or average value in order that the data be correctly interpreted. The repetitive peak forward current is the peak value of the forward current reached in every cycle and excludes random peaks due to transients. Its relation to the average forward current depends on the circuit used and on the load that is applied. For example, the repetitive peak is about three times the average for a half-wave or bridge circuit working into a resistive load; it may be many times greater when the same circuits work into capacitive loads.

Forward Characteristics—The manufacturer generally supplies curves of instantaneous forward voltage versus instantaneous forward current at one or more operating temperatures; a typical characteristic is shown by the solid curve in Fig. 1. Such curves are not exact for all rectifiers of a given type but are subject to normal production spreads. They are of particular importance in determining the power dissipated by the rectifier under given working conditions.

Calculation of power dissipation from the voltage-current curves need not be done in every instance, since the manufacturer gives curves of power dissipation versus forward current for a limited number of commonly used circuits. However, cases do arise for which the particular form of circuit or load is not covered, and it is then necessary to calculate the dissipation for these particular

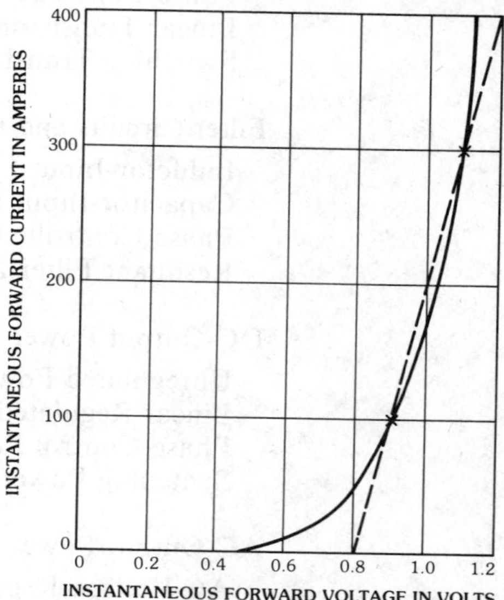

Fig. 1. Instantaneous forward voltage-current characteristic for a typical 100-ampere diode operated at 100°C junction temperature. The ideal threshold forward voltage V_T is the value where the broken line intercepts zero forward current.

*For a complete list of silicon-rectifier ratings, refer to the EIA-JEDEC *Recommendations for Letter Symbols, Abbreviations, Terms and Definitions for Semiconductor Device Data Sheets and Specifications,* published by Electronic Industries Association, 2001 I Street, N.W., Washington, D.C. 20006.

conditions. The calculation can be greatly simplified, with little loss of accuracy in most cases, by approximating the actual *V-I* characteristic curve to a straight line, as shown by the broken line in Fig. 1. The approximate characteristic corresponds to that of a fixed voltage (the threshold voltage) plus a fixed resistance (the slope resistance). For any shape of current waveform, the power dissipated at constant voltage is the product of the average current and this fixed voltage, while the power dissipated at constant resistance is the product of the square of the rms current and this fixed resistance. Thus, the following simple equation can be used:

$$P = I_{F(\mathrm{av})} \times V_T + I_{F(\mathrm{rms})}{}^2 \times R_S \qquad (1)$$

where,

- P is the forward power dissipation,
- $I_{F(\mathrm{av})}$ is the average forward current through the rectifier, averaged over one complete cycle,
- $I_{F(\mathrm{rms})}$ is the rms value of the forward current through the rectifier,
- V_T is the threshold voltage,
- R_S is the slope resistance.

For the best accuracy, the straight-line approximation should be drawn through points on the current curve corresponding to 50% and 150% of the peak current at which the rectifier is to be used. Thus, in Fig. 1 the broken line would correspond to a peak working current of 200 amperes.

Carrier Storage—On switching from forward conduction to reverse blocking, a silicon diode cannot immediately revert to its blocking state because of the presence of the stored carriers in the junction. These have the effect of allowing current to flow in reverse, as through a forward-biased junction, when reverse voltage is applied. The current is limited only by the external voltage and circuit parameters. However, the carriers are rapidly removed from the junction both by internal recombination and by the sweep-out effect of the reverse current, and when this has happened, the diode reverts to its blocking condition in which only a low leakage current flows. This sudden cessation of a large reverse current can cause objectionable voltage transients if there is appreciable circuit inductance and surge-suppression components have not been included. The reverse current due to carrier storage is not excessive in normal operation of power rectifier circuits and does not in itself constitute a hazard; however, its effect can sometimes lead to complications in switching arrangements. For example, in an inductively loaded circuit, the current will "free-wheel" through the diodes after the supply has been removed until the inductive energy has been discharged. Should the supply be reapplied while this process is going on, some of the diodes will be required to conduct in a forward direction, but others will be required to block; while the latter are recovering from the carrier storage injected by the free-wheeling current, the short-circuit across the supply can cause a damaging surge current to flow.

A technique for reducing this problem is to use "fast-recovery" rectifiers. These rectifiers are tailored to operate in high-frequency circuits with reduced and specified amounts of recovered charge. The devices change from conducting reverse current to leakage current in a "nonabrupt" manner. The finite range of change in current reduces the transient voltages that appear on the power semiconductors and other circuit elements.

Schottky Rectifiers

The Schottky rectifier has the same terminal characteristics as do conventional and fast-recovery rectifiers; that is, it conducts current in one direction and blocks voltage of one polarity. Rather than rely on a pn junction for rectification, the Schottky diode uses a metal-to-semiconductor contact. Depending on the metal, the forward voltage drop is typically one-half that of a conventional silicon pn rectifier. The reverse voltage rating is lower than that of a pn rectifier, being on the order of 50 to 100 volts. The rectifying action of the Schottky rectifier depends upon the majority carriers, so there is very little reverse current caused by minority-carrier recombination. Therefore, these devices are characterized by very low recovered charge and as such are suitable for high-frequency rectification, up to typically 200 kHz. The devices are used extensively as rectifiers on the low-voltage windings of high frequency switching mode power supplies. The output voltage of these systems is typically 5 volts dc, and they operate typically at 20 kHz. The Schottky rectifiers operate in these systems with considerably lower conduction and switching losses than silicon pn rectifiers.

Zener Diodes

"Zener" is the name given to a class of silicon diodes having a sharp turnover characteristic at a particular reverse voltage, as shown in Fig. 2. If such a diode is operated on this part of its characteristic, no breakdown (in the sense of dielectric breakdown) occurs, and the process is reversible without damage. The steepness of the reverse part of the current-voltage characteristic in the turnover region makes these diodes excellent elements for voltage reference and voltage regulation.

The temperature coefficient for a typical range of zener diodes is shown in Fig. 3. It will be seen that the coefficient changes from negative to positive in the region of 5 volts. Use is sometimes made of this phenomenon to match diodes of opposite coefficient to produce a series pair having a low effective temperature coefficient in combination.

Zener diodes are used to provide stable reference

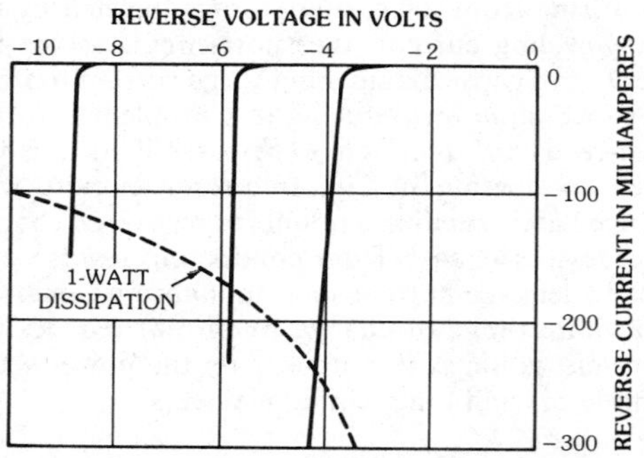

Fig. 2. Typical reverse characteristics for three low-voltage zener diodes of 1-watt rating. (*From J. M. Waddell and D. R. Coleman, "Zener Diodes—Their Properties and Applications,"* Wireless World, *Vol. 66, No. 1, p. 18, Fig. 2; January 1960.©1959, Iliffe Electrical Publications, Ltd., London, England.*)

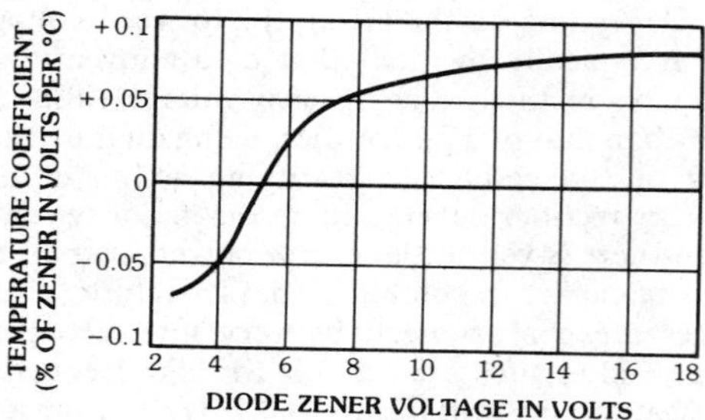

Fig. 3. Temperature coefficient for a typical range of low-voltage zener diodes. (*From J. M. Waddell and D. R. Coleman, "Zener Diodes—Their Properties and Applications,"* Wireless World, *Vol. 66, No. 1, p. 18, Fig. 4; January 1960.©1959, Iliffe Electrical Publications, Ltd., London, England.*)

voltages for electronic control circuits and as a voltage reference for closed-loop regulating systems in which the actual voltage is compared to the reference voltage in order to develop an error signal. They are available in voltage ratings from 3 to 200 volts and power ratings from less than a watt to 50 watts. The voltage tolerance can be specified as ±20%, ±10%, ±5%, and, with further selection, to closer tolerances. Special units designed for transient voltage clipping are discussed in the section on transient-voltage suppressors.

Power Transistors

Power transistors are three-terminal semiconductor devices that are widely used in power supplies and other power conversion equipment. The addition of a third terminal allows the device to have the capability of electronic control of its impedance to the flow of current. The device impedance can range from a very low forward voltage drop, called *saturation voltage*, with the current limited by the external load to a very low leakage current during its off condition. The transistor can operate in its linear region with a simultaneous high voltage across the device and a high current flowing through the device. This linear mode is used in the linear regulated power supplies described in a later section. The power transistor can also be operated in the switching mode, where it is either on or off with the shortest possible times to accomplish the turn-on and turn-off. In order to increase the gain defined as the ratio of collector current to base current, two transistors are arranged in a Darlington connection in which the collector current of one transistor provides the base current to the output transistor. This connection increases the collector-to-emitter saturation voltage. Silicon high-voltage transistors useful for high-frequency switching circuits are characterized by a set of voltage, current-gain, and switching characteristics that are briefly discussed below.

Breakdown Sustaining Voltages—Power transistors during conditions of turn-off are required to withstand immediately a forward collector-to-emitter voltage. Because of the rapid change from forward conduction to forward blocking, the sustaining voltages are less than the steady-state forward blocking voltage ratings.

The collector-to-emitter sustaining-voltage rating is specified as a function of the base-to-emitter voltage conditions:

$V_{CEO(SUS)}$ = collector-emitter sustaining voltage, base open

$V_{CER(SUS)}$ = collector-emitter sustaining voltage, base-emitter resistor

$V_{CES(SUS)}$ = collector-emitter sustaining voltage, base-emitter reverse-bias voltage

The $V_{CEO(SUS)}$ rating is the lowest of the three ratings and is specified at a low level of collector current. It is important in switching inductive loads, and this voltage should not be exceeded during the switching interval.

DC Current Gain and Saturation Voltage—The collector-emitter saturation voltage is the voltage drop that occurs when the transistor is carrying current. It is specified at a given collector current and a given base current or as a gain that is the ratio of collector current to base current. The saturation voltage is also a function of junction temperature. The gain has a peak value at nominal current levels and decreases at both lower and higher values of collector current. The typical value of saturation voltage for single transistors is one volt. This voltage drop times the collector current represents a major part of the conduction losses in the transistor. The collector-emitter saturation-voltage versus collector-current characteristics of a Darlington power transistor are plotted in Fig. 4 for a 500-volt, 50-ampere npn power transistor.

Forward- and Reverse-Biased Second Breakdown—When transistors are being turned on and off, the possibility for simultaneous high currents and high voltages exists. This results in localized high power dissipation in the transistor, leading to increasing temperature rise and potential device failure. This failure is described as "second breakdown." Power-transistor manufacturers present a family of curves outlining a safe operating area for the device. The actual device switching locus should remain within the designated area. Two sets of curves are provided, one with forward bias on the base and the other with reverse bias on the base (turn-off). The forward-biased safe operating area for a 500-volt, 100-ampere Darlington transistor is shown in Fig. 5. The curves are bounded by peak current limits, peak voltage limits, thermal limits, and second breakdown limits. The safe operating area increases as the switching times decrease. The reverse-biased safe operating area for a 500-volt, 100-ampere Darlington transistor is shown in Fig. 6. The curve shown is for turn-off with a voltage clamp on the maximum collector-to-emitter voltage.

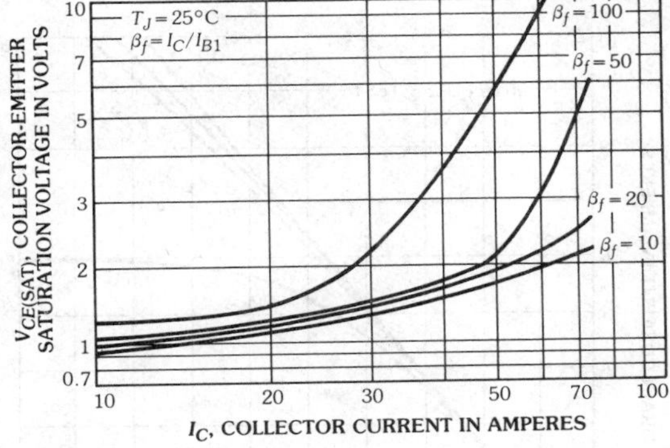

Fig. 4. Plots of $V_{CE(SAT)}$ versus I_C, $T_J = 25°C$. (*From* Transistors-Diodes. *Auburn, N.Y.: General Electric Co., Semiconductor Products Dept., 1982.*)

Switching Times—The switching times, both on and off, are important during high-frequency switching because both voltage and current are high, resulting in high peak power dissipation. Reduction in these switching times results in less average power dissipation due to switching losses. The turn-on time is composed of two parts, delay time and rise time. With inductive loads, these times are quite short, typically less than 1 microsecond. The turn-off time is also composed of two parts, a storage time and a fall time. Fall time is the most important parameter because the peak device power dissipation is high and the heat generated is nonuniform due to current crowding. The switching times are functions of turn-on base current, turn-off base current, base-to-emitter reverse voltage, junction temperature, and device design. Typical values are from 0.1 to 15 microseconds.

Reverse Voltage Operation—If the emitter is biased positive with respect to the collector and current is supplied to the base, then the device will

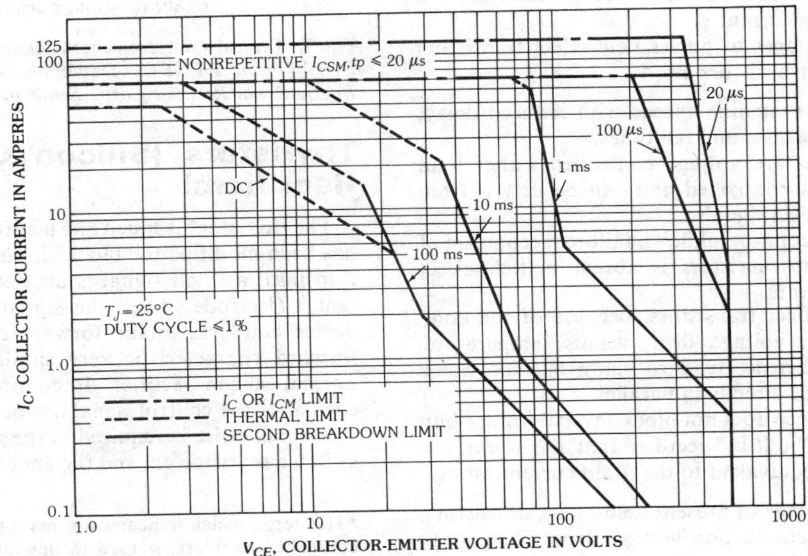

Fig. 5. Forward-bias safe operating area for a 100-ampere power transistor. (*From* Transistors-Diodes. *Auburn, N.Y.: General Electric Co., Semiconductor Products Dept., 1982.*)

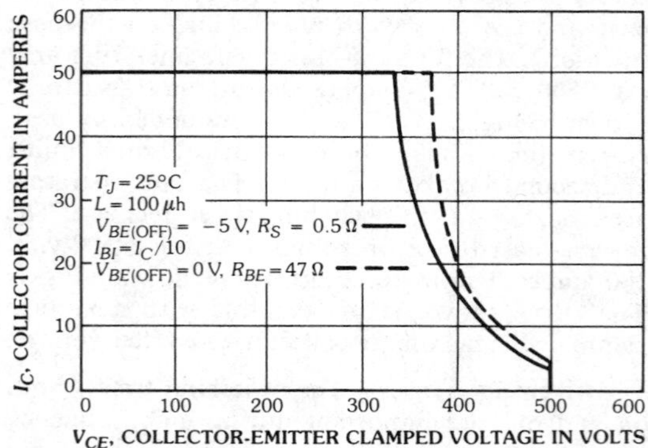

Fig. 6. Reverse-bias safe operating area (clamped) for a 100-ampere power transistor. (*From* Transistors-Diodes. *Auburn, N.Y.: General Electric Co., Semiconductor Products Dept., 1982.*)

conduct in the "inverted mode." The gain is low, and there exists a possibility that the current capability of the device is low. In normal practice, a rectifier is placed in inverse parallel with the collector and emitter of the transistor so that the rectifier instead of the transistor carries the reverse current.

Field-Effect Transistors

The field-effect transistor is also used as a controlled switch in high-voltage and high-frequency power circuits. The three terminals, drain, gate, and source, in an n-channel device bear the same relationship as the collector, base, and emitter in an npn bipolar transistor. That is, a positive signal from gate to source causes the device to conduct a positive drain current.

The advantages of power field-effect transistors over bipolar transistors are:

(1) Faster switching speeds with reduced delay, rise, storage, and fall times.
(2) Devices are voltage controlled rather than current controlled and can be driven from logic-level signals.
(3) The second-breakdown failure mechanism of bipolar transistors is absent in field-effect transistors.
(4) Field-effect transistors, because of the conduction voltage drop versus temperature characteristic, tend to share current when operated directly in parallel.
(5) The device does not block reverse voltage but has a "built-in" rectifier that has a current rating equivalent to the drain current rating.

A disadvantage of present field-effect transistors is the higher conduction voltage drop when compared with a bipolar transistor of the same current rating. The value of "on" resistance is a function of the drain-source voltage rating of the device. Higher-voltage devices have higher on resistances and therefore lower drain currents for the same temperature rise. The voltage drop can be comparable with the voltage drop of a Darlington transistor.

At present, the voltage and current ratings of field-effect transistors are not as high as those available in bipolar transistors. Field-effect transistors have replaced some bipolar transistors in switching power supplies at generally higher operating frequencies, typically over 50 kHz. Fig. 7 shows the drain-to-source voltage, during saturation, versus drain current characteristic for a 500 volt, 10 ampere n-channel field-effect transistor. The safe operating region for the same field-effect transistor is plotted on Fig. 8.

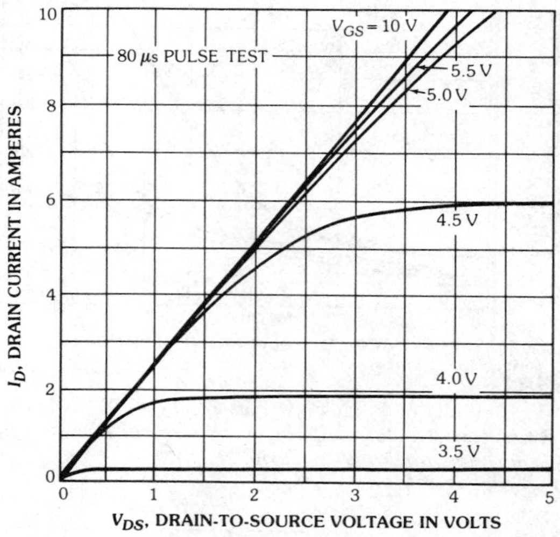

Fig. 7. Typical saturation characteristics for a 10-ampere FET. (*From* HEXFET Databook. *El Segundo, Cal.: International Rectifier Corp., Semiconductor Div., 1981.*)

Thyristors (Silicon Controlled Rectifiers)

The thyristor* is much like a normal rectifier that has been modified to "block" in the forward direction until a small signal is applied to the control (gate) electrode. After the signal is applied, the device conducts in the forward direction with a forward characteristic very similar to that of a normal silicon rectifier. It continues to conduct even after the control signal has been removed.

A small pulse is required at the gate electrode to switch a thyristor on, and the anode supply must be

*This term, which indicates a general class of solid-state controlled rectifiers, is used throughout this section instead of the term "silicon controlled rectifier" (a four-layer pnpn device that is the most common member of the class).

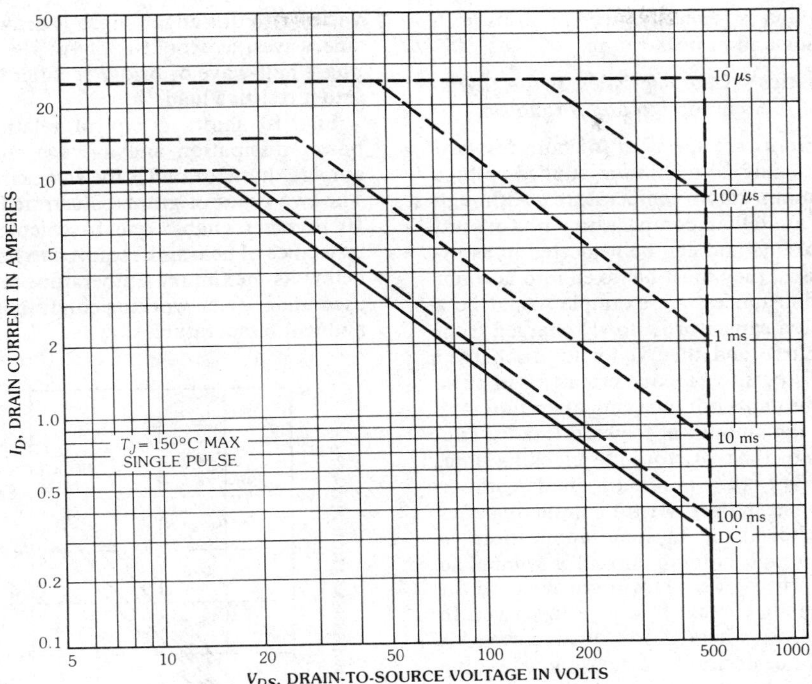

Fig. 8. Maximum safe operating area for a 10-ampere FET. (*From* HEXFET Databook. *El Segundo, Cal.: International Rectifier Corp., Semiconductor Div., 1981.*)

removed, reduced, or reversed to switch the thyristor off. If proportional control is required, means must be provided for adjusting the phase of the trigger pulse with respect to the supply to control the proportion of the cycle during which the thyristor is permitted to conduct. In ac circuit applications, the turn-off is obtained by the natural reversal of the supply voltage every half-cycle; in dc circuits it is usual to charge a "commutating" capacitor during the "on" period and to apply this charge in negative polarity between the anode and cathode when it is desired to turn the thyristor off.

Ratings—Voltage and current ratings are generally expressed in terms similar to those for silicon rectifiers, as discussed in a previous section. It is necessary, however, to add the following ratings.

Peak forward blocking voltage is the maximum safe value that may be applied, under recurrent or nonrecurrent transient conditions, while the thyristor is in the blocking state. The thyristor may break over into a conducting state regardless of gate drive if either (A) too high a positive voltage is applied between anode and cathode or (B) a positive anode-cathode voltage is applied too quickly (*dv/dt* firing). Even small voltage pulses, if their leading edges are sufficiently steep at the anode, can turn the thyristor on. Firing by condition (A) is generally avoided by making the peak forward blocking voltage lower than the breakover voltage of any thyristor of a particular type and by seeing that this voltage is not exceeded in practice. Trouble from *dv/dt* effects may be minimized by locating the gate wires to avoid stray coupling between anode and gate, by use of negative bias on the gate during blocking, and by use of R-C damping circuits between anode and cathode to slow down the rate of change of applied voltage.

Continuous forward blocking voltage covers operation under dc conditions, with forward voltage from anode to cathode.

Peak forward gate voltage is quoted for the anode positive with respect to the cathode and for the anode negative with respect to the cathode. The voltage rating is quite low in the latter case (typically 0.25 V), since the reverse voltage rating is reduced by forward gate current.

Peak reverse gate voltage is generally the same whether the anode is positive or negative with respect to the cathode.

Peak forward gate current: Forward gate impedance is a finite value subject to quite large variations between samples and over a temperature range. It is usually necessary to plot a load line on the gate current-voltage characteristics to determine the gate current that may flow due to a given external gate voltage and source resistance. Care must be taken that the rating is not exceeded with all known spreads of gate-cathode characteristic and temperature.

Gate dissipation is generally given in terms both of average rating and of peak rating.

Characteristics—Characteristics of the thyristor important for circuit design are as follows:

Leakage currents are specified for both forward and reverse blocking, at maximum applied voltage and at maximum rated temperature. Although these currents are low in comparison with forward conducting currents and can be neglected in assessing power losses, they must be taken into account in certain circumstances. An example would be a circuit in which a capacitor is slowly charged from an external source and then suddenly discharged through a thyristor into a second circuit (as in pulse modulators); the capacitor charging operation may be affected by the amount of forward leakage current conducted by the thyristor in its blocking state.

Holding current is the minimum anode-cathode current that will keep the thyristor conducting after it has been switched on. In some applications, a thyristor with a high holding current is wanted so that it can be turned off easily without the need for reducing the anode current to a very low level. In other applications, where a low load current is normal, it might be desirable to have a low holding current to ensure that the thyristor latches on reliably with light loading.

Forward voltage drop is important in assessing power loss. The same methods of assessing power loss in terms of the forward current-voltage characteristic apply as in the case of silicon rectifiers. It is impracticable to measure the junction temperature under working conditions, and therefore the manufacturers list maximum values of stud or case temperature related to the forward current. This relationship is expressed in the form of a graph, as shown in Fig. 9.

Note that since the ratio of rms to average forward current varies with the angle of conduction, the power dissipation for any average current also varies with this angle. Fig. 9 is drawn for fractional sine waves, as would apply to the cases of single-phase half-wave or bridge rectifier circuits working into a resistive load.

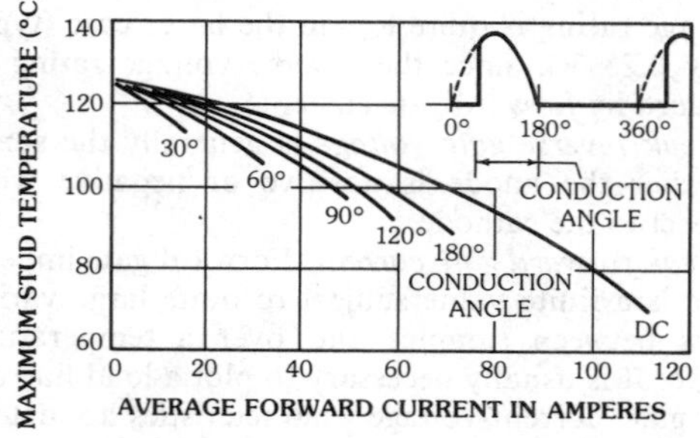

Fig. 9. Maximum permitted stud temperature vs average thyristor forward current in single-phase half-wave or bridge circuit feeding a resistive load. (70-ampere average rated thyristor.)

Fig. 10 shows a typical relationship between power dissipation and average current for a 70-ampere thyristor under the same circuit conditions. The two types of graphs illustrated by Figs. 9 and 10 together enable one to calculate the thermal resistance of heat sink required to keep the thyristor below its maximum temperature ratings when it is used under given working conditions of current and ambient temperature.

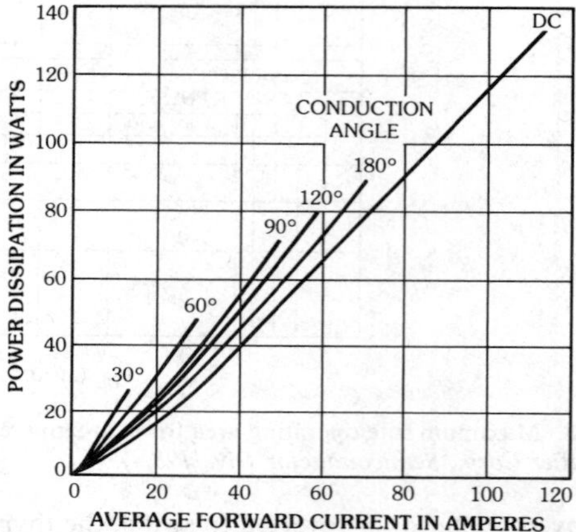

Fig. 10. Power dissipation vs average thyristor forward current in single-phase half-wave or bridge circuit feeding a resistive load. (70-ampere average rated thyristor.)

Gate trigger sensitivity is specified in terms of a minimum voltage and/or current that must be applied to ensure that all samples of a particular type of thyristor will be triggered into conduction. The minimum voltage is not temperature sensitive, but the minimum trigger current varies considerably with temperature, more current being required to turn on at low temperature than at high. The basic requirements of a gate drive circuit are therefore that the driving voltage and source resistance must be such that either the minimum voltage or the minimum current (or both) is exceeded but that the rated gate dissipation is not exceeded.

Switching times of importance are the turn-on and turn-off times, the latter generally being at least one order of magnitude greater than the former. When a gate signal is applied to the thyristor, there is a finite delay time during which the anode current remains at its normal blocking level; this is followed by a "rise time" during which the anode current increases from its blocking level to a value determined by the external load circuit. Turn-on time is the sum of these two times. For a given thyristor, the turn-on time is influenced by the magnitude of gate drive, the load current to be achieved, and, to a lesser extent, the applied anode supply voltage. The time is reduced by high gate drive, low load current, and high anode supply voltage. Turn-off time is similarly composed of two individual periods; the first is a storage time, analogous to that obtained with a saturated transistor, and the second is a recovery time. Forward voltage may not be reapplied before the completion of both phases of the turn-off process, or the thyristor may conduct load current again. After this period, however, forward voltage may be applied, and the thyristor will remain in its blocking state provided that the rate of rise of anode voltage is not allowed to exceed the specified maximum dv/dt, as already discussed.

Triacs

A triac is a three-terminal ac semiconductor switch that is triggered into conduction by a gate signal much as a thyristor is. The triac was developed to provide a single device that could control current flow in both directions with the application of a gate signal. This allowed for the replacement of two inverse-parallel thyristors and the complex gate-drive circuit they require with a single device that has a single gate.

The terms "anode" and "cathode" are not used; instead, the power terminals are numbered MT_1 and MT_2. The gate terminal is associated with MT_1. The triggering characteristics are such that with MT_2 positive with respect to MT_1, either positive or negative gate current will trigger the device into conduction. With MT_1 positive with respect to MT_2, again either positive or negative gate current will trigger the device into conduction. However, positive gate current should be used only if needed, because the gain is lower.

The other major difference from two inverse thyristors is in the commutating dv/dt rating. When two thyristors are connected in inverse parallel and operating with a lagging-power-factor load, one device ceases to conduct, and the voltage rises on the previously conducting device in the reverse or non-turn-on direction. The other device sees a positive dv/dt in the turn-on direction, but it has had an entire half-cycle for turn-off or commutation. In a triac, the only period for turn-off occurs when the current goes through zero and a pulse of recovery current is conducted through the device. The voltage rises to the circuit voltage level in the turn-on direction very rapidly, and the device does not have an entire half-cycle to regain its forward blocking capability. A suitable R-C snubber is generally provided.

Triacs are available for standard ac voltages (120, 240, and some higher voltages) and up to 40–100 amperes operating at 50, 60, or 400 hertz. The remainder of their ratings are consistent with thyristor ratings.

Transient-Voltage Suppressors

Transient-voltage suppressors have a terminal characteristic such that, above a certain voltage level, the voltage drop across the device increases slightly for large changes in current magnitude. Some of these devices exhibit voltage clamping characteristics for both voltage polarities, whereas others exhibit clamping characteristics for only one voltage polarity. Two types are currently available. One is similar to a silicon zener diode, selected for a high-pulse-power, low-average-power duty cycle. The second type is composed of a polycrystalline nonlinear resistive material. Both of these categories of devices come in a range of voltage and power ratings. They are used as transient-voltage clippers on incoming ac or dc utility lines and as transient-voltage limiters for semiconductors and other voltage-sensitive equipment. Fig. 11 shows a comparison of several types and sizes of devices as a function of standardized pulse current waveforms and supply voltages.

AC-DC CONVERTER CIRCUITS

This section describes ac-to-dc converter circuits, both unregulated and electronically regulated, using rectifiers, phase-controlled thyristors, linear transistors, and switching-mode transistors. The basic design equations together with some of the technical advantages, disadvantages, and common applications are given and discussed.

Rectifier Circuits

Table 1 shows seven of the most commonly used power-rectifier circuits and general design information for each type. Their advantages, disadvantages, and common applications follow.

Single-Phase Half-Wave Rectifier—Since only half of the input wave is used, the efficiency is low and the regulation is relatively poor. Capacitors are commonly used in half-wave circuits to increase the output voltage and decrease the voltage ripple. The output voltage and degree of filtering are determined by the value of capacitance used in relation to the load current. Transformer design is complicated, and the unidirectional secondary current causes core saturation and poor regulation. Most half-wave circuits operate either directly from ac lines or at a high voltage with a relatively low current.

Single-Phase Full-Wave Center-Tap Rectifier—The efficiency is good, but the transformer ac voltage is approximately 2.2 times the dc output voltage. The circuit requires a larger transformer than an equivalent bridge rectifier, with the added complication of a center tap. Each arm of the center-tap circuit must block the full terminal vol-

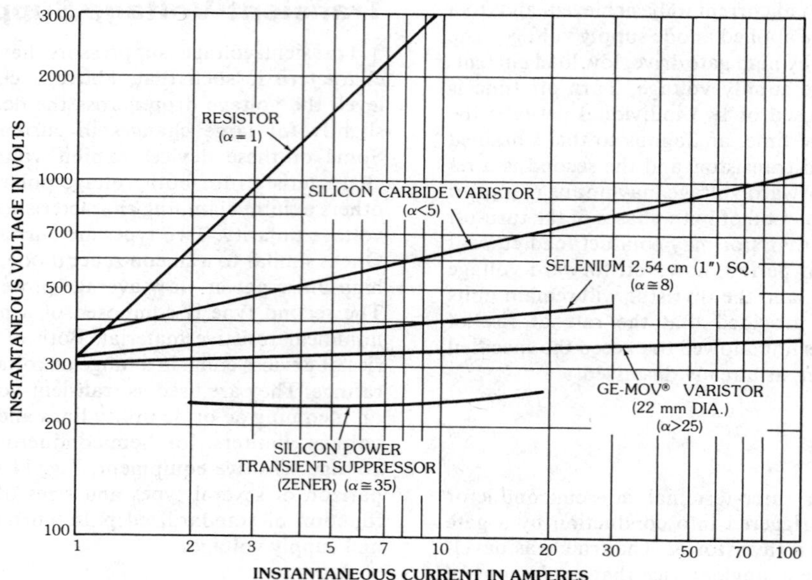

Fig. 11. Voltage-current characteristics of four transient-suppressor devices. (*From* Transient Voltage Suppression, *3rd ed. Auburn, N.Y.: General Electric Co., Semiconductor Products Dept., 1982.*)

tage of the transformer. Because of this, center-tap connections are economical only in voltage ranges where not more than one rectifier per arm is required. If series units must be used to obtain the required output voltage, a bridge circuit is preferable. This circuit is used with low-voltage logic-level power supplies where the efficiency is improved because there is only one diode in series with the output.

Single-Phase Full-Wave Bridge Rectifier— If single-phase full-wave output is required, a bridge circuit is commonly used. Efficiency is good and transformer design is easy. Filtering is simplified because the ripple frequency is twice the input frequency.

Three-Phase Wye (or Star) Half-Wave Rectifier—This circuit is commonly used if dc output-voltage requirements are relatively low and current requirements are moderately large. The dc output voltage is approximately equal to the phase voltage. However, each of the three arms must block the line-to-line voltage, which is approximately 2.5 times the phase voltage. For this reason, it is desirable to use a three-phase half-wave connection only where one series unit per arm will provide the required dc output. The transformer design and utilization are somewhat complicated because there is a tendency to saturate the core with unidirectional current in each winding.

Three-Phase Full-Wave Bridge Rectifier— This circuit is commonly used if high dc power is required and if efficiency must be considered. The ripple component in the load is 4.2% at a frequency six times the input frequency, so additional filtering is not required in most applications. The dc output voltage is approximately 25% higher than the phase voltage, and each arm must block only the phase voltage. Transformer utilization is good. This is the most common three-phase rectifier connection.

Three-Phase Diametric Half-Wave Rectifier—The characteristics of this circuit approximate those of the three-phase double-wye circuit without an interphase transformer. Popular applications include requirements for very high dc load currents in low-to-medium voltage ranges (approximately 6 to 125 volts dc).

Three-Phase Double-Wye Half-Wave Rectifier—A three-phase double-wye connection is recommended if a very high direct current is required at a relatively low dc voltage. Each arm is required to block the full phase voltage of the secondary windings. The dc output current rating is double that of a three-phase bridge or half-wave connection. However, the output voltage is only 75% of the phase voltage. The transformer design is complicated by additional connections and extra insulation, and an interphase transformer (or balance coil) is required.

Single-Phase Voltage Multipliers

These circuits use the principle of charging capacitors in parallel from the ac input and adding the voltages across them in series to obtain dc voltages higher than the source voltage. Filtering must be of the capacitor-input type.

Conventional and Cascade Voltage Doublers—In the conventional circuit (Fig. 12), capacitors C_1 and C_2 are each charged, during alternate half-cycles, to the peak value of the alternating input voltage. The capacitors are discharged in series into load R_L, thus producing an output across the load of approximately twice the ac peak voltage.

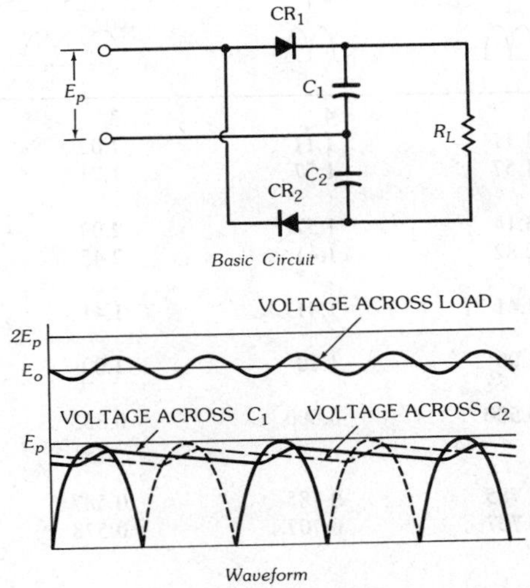

Fig. 12. Conventional voltage doubler.

In the cascade circuit (Fig. 13), C_1 is charged to the peak value of the ac input voltage through rectifier CR_2 during one half-cycle, and during the other half-cycle it discharges in series with the ac source through CR_1 to charge C_2 to twice the ac peak voltage.

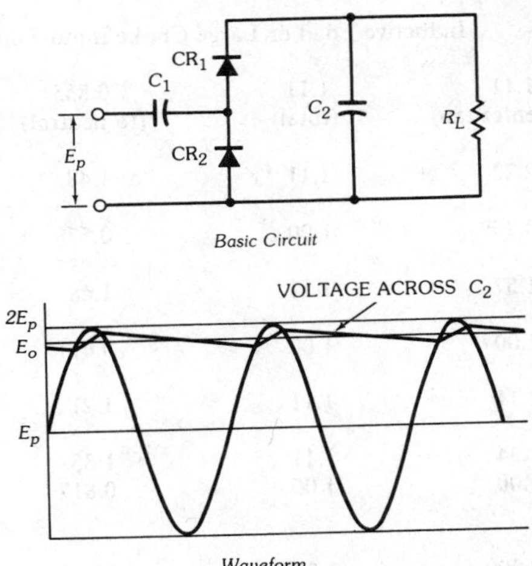

Fig. 13. Cascade voltage doubler.

The "conventional" circuit has slightly better regulation, and, since the ripple frequency is twice the supply frequency, the output is easier to filter, the percentage ripple being approximately the same in both circuits. In addition, both capacitors are rated at the peak ac voltage, whereas C_2 in the cascade circuit must be rated at twice this value. With both circuits, the peak inverse voltage across each rectifier is twice the ac peak. The cascade circuit, however, has the advantage of a common input and output terminal and, therefore, permits the combination of units to give higher-order voltage multiplications. The regulation of both circuits is poor, so that only small load currents can be drawn.

Bridge Rectifier or Voltage Doubler—If SW in Fig. 14 is open, the circuit is a bridge rectifier; if the peak ac applied voltage is $2E_p$, the dc output voltage is $2E_p$. If SW is closed, the circuit is a voltage doubler; an ac input of E_p gives a dc output of $2E_p$. This circuit is used if the same dc voltage is desired and the input voltage can be either 115 or 230 volts ac.

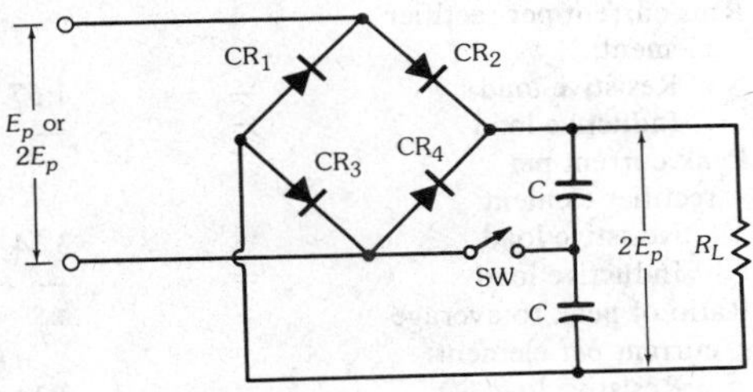

Fig. 14. Bridge rectifier or voltage doubler.

Further Voltage Multiplication—The cascade voltage doubler shown in Fig. 13 can be combined several times to obtain higher dc voltages, as shown in Fig. 15. The voltage ratings of all the capacitors and rectifiers are twice the ac peak voltage, but the capacitors must have the values shown. The value of C will be the same as that for the cascade voltage doubler (Fig. 13), which is the basic unit for the circuit in Fig. 15. The load current must be small. The increasing size of capacitors and the deterioration in regulation limit the voltages that can be obtained from this type of circuit.

Phase-Controlled Thyristor Circuits

As described earlier, the thyristor can block forward voltage until it is triggered into conduction. The delay angle of gating the thyristor into conduction is designated as alpha (α). With $\alpha = 0$, the thyristor behaves like a rectifier. As the delay angle is increased, the average dc output voltage de-

TABLE 1. RECTIFIER CIRCUIT CHART. THE DATA ASSUME ZERO FORWARD DROP AND ZERO

Type of Circuit→		Single-Phase Half-Wave	Single-Phase Center Tap	Single-Phase Bridge	Three-Phase Star (Wye)
Primary→ / Secondary→ / One Cycle Wave of Rectifier Output Voltage (No Overlap)		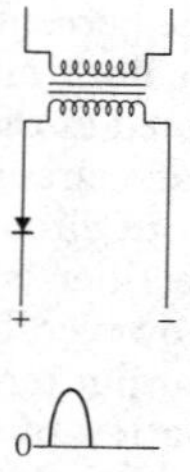	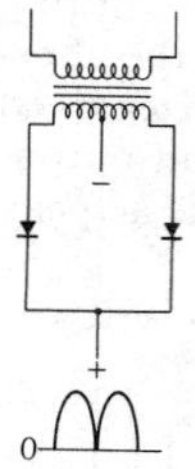	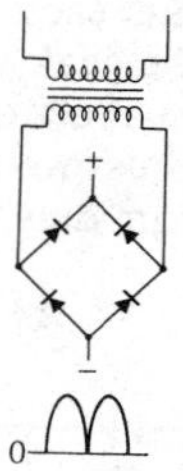	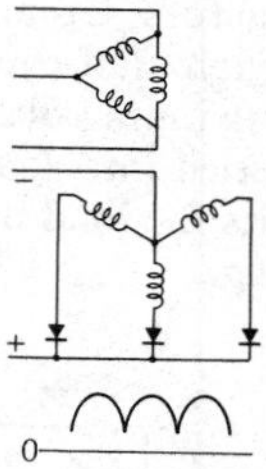
Number of rectifier elements	=	1	2	4	3
Rms dc volts output	=	1.57	1.11	1.11	1.02
Peak dc volts output	=	3.14	1.57	1.57	1.21
Peak reverse volts per rectifier element	=	3.14	3.14	1.57	2.09
	=	1.41	2.82	1.41	2.45
	=	1.41	1.41	1.41	1.41
Average dc output current	=	1.00	1.00	1.00	1.00
Average dc output current per rectifier element	=	1.00	0.500	0.500	0.333
Rms current per rectifier element:					
Resistive load	=	1.57	0.785	0.785	0.587
Inductive load	=	—	0.707	0.707	0.578
Peak current per rectifier element:					
Resistive load	=	3.14	1.57	1.57	1.21
Inductive load	=	—	1.00	1.00	1.00
Ratio of peak to average current per element:					
Resistive load	=	3.14	3.14	3.14	3.63
Inductive load	=	—	2.00	2.00	3.00
% Ripple (rms of ripple/average output voltage)	=	121%	48%	48%	18.3%
Ripple Frequency	=	1	2	2	3
		Resistive Load	Inductive Load or Large Choke Input Filter		
Transformer secondary rms volts per leg	=	2.22	1.11 (to center tap)	1.11 (total)	0.855 (to neutral)
Transformer secondary rms volts line-to-line	=	2.22	2.22	1.11	1.48
Secondary line current	=	1.57	0.707	1.00	0.578
Transformer secondary volt-amperes	=	3.49	1.57	1.11	1.48
Transformer primary rms amperes per leg	=	1.21	1.00	1.00	0.471
Transformer primary volt-amperes	=	2.69	1.11	1.11	1.21
Average of primary and secondary volt-amperes	=	3.09	1.34	1.11	1.35
Primary line current	=	1.21	1.00	1.00	0.817
Line power factor	=	—	0.900	0.900	0.826

REVERSE CURRENT IN RECTIFIERS AND NO ALTERNATING-CURRENT LINE OR SOURCE REACTANCE

Note: Assumes perfect rectifiers and zero reactance of ac line and source

Three-Phase Bridge	Six-Phase Star (Three-Phase Diametric)	Three-Phase Double Wye With Interphase Transformer		To Determine Actual Value of Parameter in Any Column, Multiply Factor Shown by Value of:
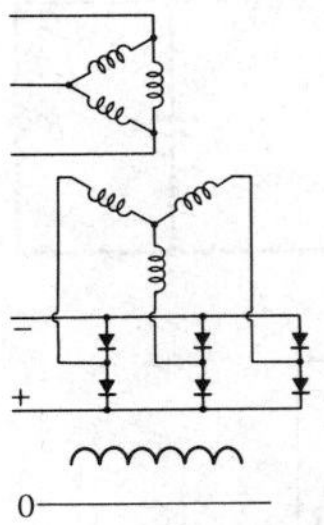	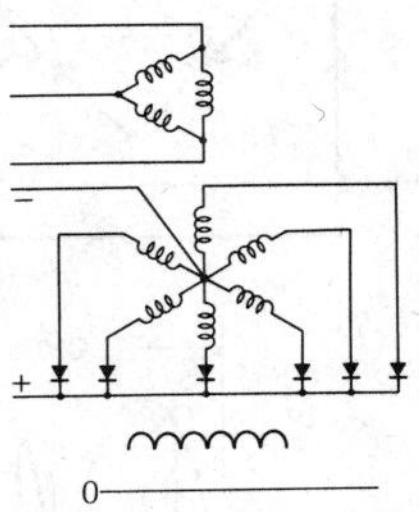	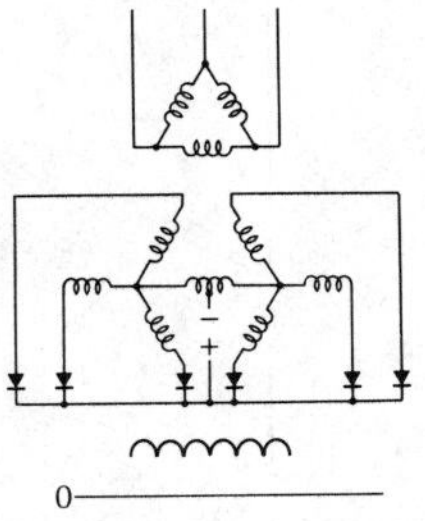		
6	6	6		
1.00	1.00	1.00	×	Average dc voltage output
1.05	1.05	1.05	×	Average dc voltage output
1.05	2.09	2.42	×	Average dc voltage output
2.45	2.83	2.83	×	Rms secondary volts per transformer leg
1.41	1.41	1.41 (diametric)	×	Rms secondary volts line-to-line
1.00	1.00	1.00	×	Average dc output current
0.333	0.167	0.167	×	Average dc output current
0.579	0.409	0.293	×	Average dc output current
0.578	0.408	0.289	×	Average dc output current
1.05	1.05	0.525	×	Average dc output current
1.00	1.00	0.500	×	Average dc output current
3.15	6.30	3.15		
3.00	6.00	3.00		
4.2%	4.2%	4.2%		
6	6	6	×	Line frequency, f
Inductive Load or Large Choke Input Filter				
0.428 (to neutral)	0.740 (to neutral)	0.855 (to neutral)	×	Average dc voltage output
0.740	1.48 (max)	1.71 (max-no load)	×	Average dc voltage output
0.816	0.408	0.289	×	Average dc output current
1.05	1.81	1.48	×	Dc watts output
0.816	0.577	0.408	×	Average dc output current
1.05	1.28	1.05	×	Dc watts output
1.05	1.55	1.26	×	Dc watts output
1.41	0.817	0.707	×	(Avg. load current × sec. leg voltage)/primary line voltage
0.955	0.955	0.955		

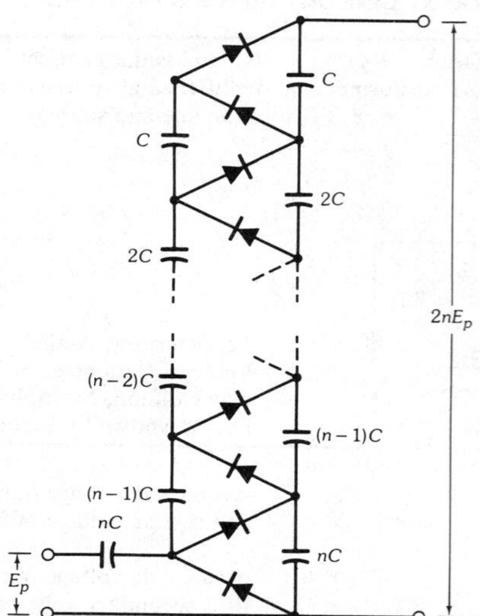

Fig. 15. Circuit for high-order voltage multiplication.

creases, and therefore these circuits provide electronic regulation of the dc output voltage.

Figs. 16 through 21 are basic circuits of thyristors used as controlled rectifiers. In many applications, it is not necessary for all rectifier elements to be controllable, and it is common to find bridges composed of thyristors and ordinary diodes in equal numbers; such circuits are called *half-controlled,* whereas those containing only thyristors are termed *full-controlled.*

Figs. 16 and 17 show a full-controlled single-phase thyristor bridge, the thyristor circuit being the same whether a resistive or inductive load is being driven. The voltage waveforms are different, however, in the two cases. The principal difference of practical importance is that the range of firing-pulse phase control is required to be different in the two cases. For the resistive load, phase control over the range 0° to 180° is necessary to obtain full control from maximum output voltage down to zero;

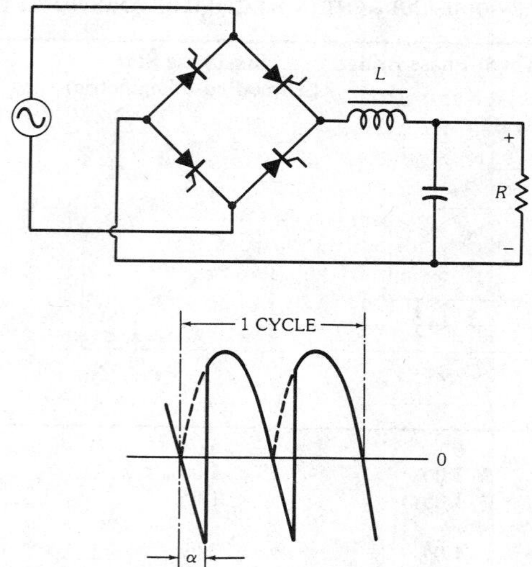

Fig. 17. Full-controlled single-phase thyristor bridge working into an inductive load where $\omega L >> R$. Delay range required is 0° to 90°.

for an almost pure inductive load (*i.e.*, a very high $\omega L/R$ ratio), full-phase control is obtained with a range of only 0° to 90°.

Fig. 18 shows a half-controlled single-phase rectifier driving an inductive load. For such loads a bypass diode,* CR_3, must be added at the output; at the end of a voltage half-cycle, current still flows in the choke, but in this circuit the current is transferred at the end of the voltage half-cycle to the bypass diode. The two important effects of this diode are (A) that the output voltage is clamped to zero while this inductively maintained current flows, so that the output waveform is the same as that of Fig. 16 (having the characteristic flat portion) and (B) that the transfer of load current from the thyristor to the diode turns the thyristor off. If CR_3 were

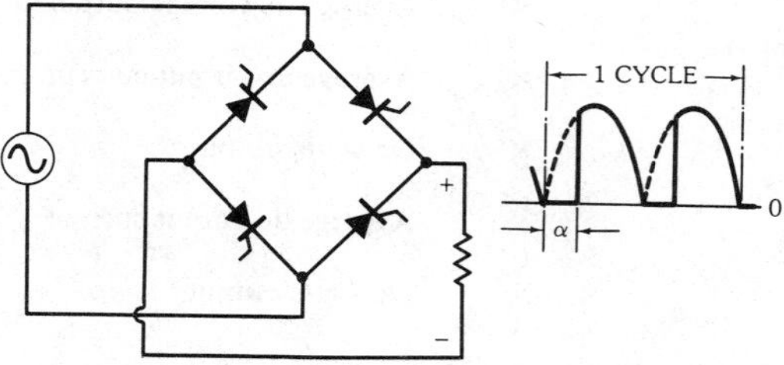

Fig. 16. Full-controlled single-phase thyristor bridge. The broken lines indicate the path of the normal waveform at full conduction, *i.e.*, where $\alpha = 0°$ and the circuit behaves like a diode rectifier. The quantity α is the firing angle delay. Delay range required for this circuit is 0° to 180°.

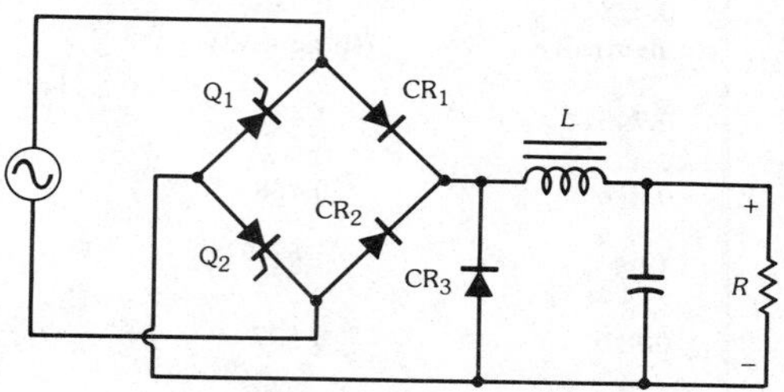

Fig. 18. Half-controlled single-phase thyristor bridge working into inductive load. Common-anode thyristor connection. Waveform and delay range as for Fig. 16.

*Sometimes called a "commutating," "freewheel," or "flyback" diode.

not present, the waveform would still be the same, since a zero voltage clamp would exist through a series combination CR_1-Q_1 or CR_2-Q_2, depending on which thyristor was conducting during the previous half-cycle. However, this flow of uncontrolled current through the thyristor, bypassing the supply, is undesirable; for example, should it be required that the output voltage be turned off by removal of gate pulses, this action may prove to be impossible. The thyristor can be held on continuously through the negative half-cycle by inductive circulating current and is then ready to conduct on the next positive half-cycle. Thus the gate has lost control, and a continuous half-wave output is produced. The bypass diode overcomes these difficulties by ensuring thyristor turn-off at the end of each voltage half-cycle.

Fig. 19 is a push-pull controlled rectifier circuit. Fig. 20 is a three-phase full-controlled rectifier circuit. Fig. 21 shows the circuit of a three-phase half-controlled rectifier, the bypass diode being necessary only for inductive loads.

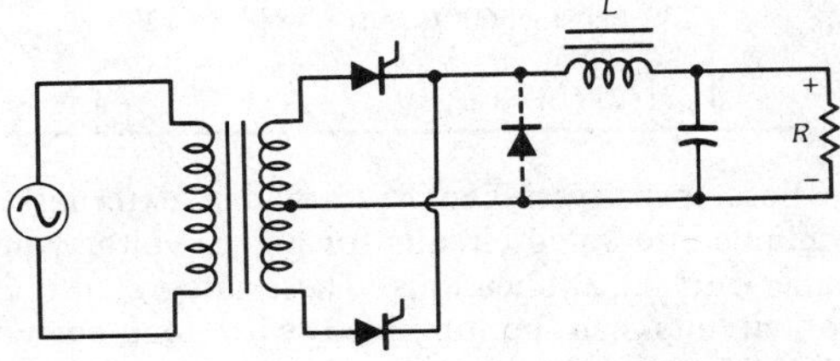

Fig. 19. Full-controlled single-phase push-pull thyristor rectifier working into inductive load. With diode connected, waveform and delay range are as for Fig. 16. Without diode, waveform and delay range are as for Fig. 17 (if $\omega L >> R$). Addition of diode reduces critical inductance of L for continuous current.

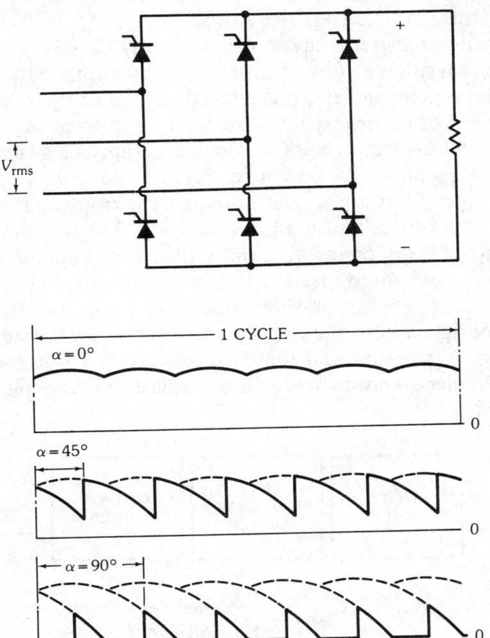

Fig. 20. Full-controlled three-phase thyristor bridge working into resistive load; waveforms of voltage output are shown for three values of α. Waveform develops characteristic flat portion when $\alpha \geq 60°$. Delay range required is 0° to 120°.

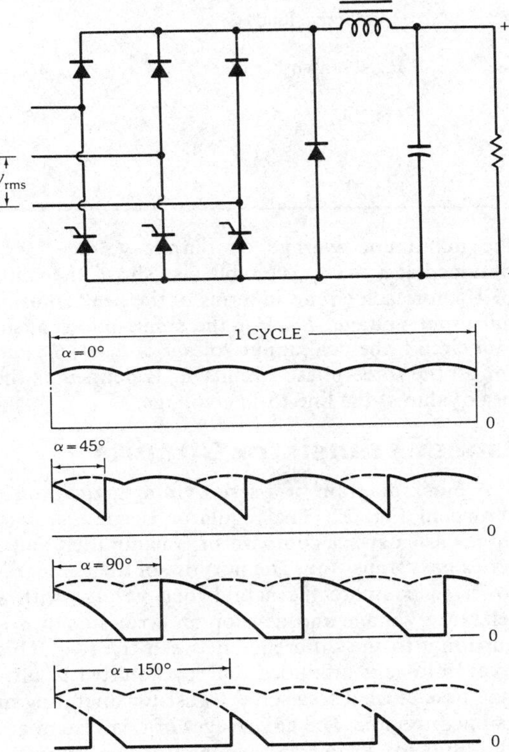

Fig. 21. Half-controlled three-phase thyristor bridge. The bypass diode is necessary only when feeding an inductive load. The waveforms are the same for resistive or inductive load. Delay range required is 0° to 180°. Note that the three thyristors could have been put in the positive bridge arms; it is more usual to put them in the negative arms because a common anode connection permits the use of a common heat sink when thyristors with anode studs are used to implement the circuit.

In all the cases illustrated in which a bypass diode is used, this diode must be rated continuously for a maximum average current equal to the load current if the full load current is to be drawn when the average output voltage is reduced almost to zero. In practice, a larger diode is often used in this position than in the bridge arms, or several diodes may be used in parallel.

For each of the circuits illustrated in Figs. 16 through 21, Table 2 gives equations for the average dc output voltage, $V_{d\alpha}$, at any angle α in terms of

TABLE 2. MEAN DC OUTPUT VOLTAGE FOR THYRISTOR-CONTROLLED RECTIFIERS

Circuit	V_{do}	$V_{d\alpha}$
Fig. 16	$2E_p/\pi$	$V_{do}(1/2)(1+\cos\alpha)$
Fig. 17	$2E_p/\pi$	$V_{do}\cos\alpha$
Fig. 18	$2E_p/\pi$	$V_{do}(1/2)(1+\cos\alpha)$
Fig. 19, no diode and inductive load	$2E_p/\pi$	$V_{do}\cos\alpha$
Fig. 19, all other cases	$2E_p/\pi$	$V_{do}(1/2)(1+\cos\alpha)$
Fig. 20	$3E_p/\pi$	$V_{do}\cos\alpha$, for $\alpha=0°$ to $60°$ $V_{do}[1+\cos(\alpha+60°)]$, for $\alpha=60°$ to $120°$
Fig. 21	$3E_p/\pi$	$V_{do}(1/2)(1+\cos\alpha)$

the maximum average dc output voltage, V_{do}, obtained at $\alpha = 0°$. The table also shows the value of V_{do} for each circuit in terms of the peak sinusoidal input voltage, E_p. For the single-phase push-pull circuit, the peak input voltage is E_p–0–E_p, and for all the three-phase circuits E_p is defined as the peak value of the line-to-line voltage.

Linear Transistor Circuits

A block diagram of a series voltage regulator is shown in Fig. 22. The regulator is composed of three basic parts, a comparator, an amplifier, and a series pass transistor. The purpose of the comparator is to compare the actual load voltage with a reference voltage and develop an error signal proportional to the difference between the two. This error voltage is amplified and level shifted to provide base current to a series transistor operating in its linear region. The advantages of this system are its simplicity, its current availability as a complete integrated circuit from several manufacturers, its fast response, and its low switching noise. Its disadvantage is its power dissipation, since the series transistor operates as a linear resistive element. Series pass transistors can be provided external to available integrated circuits for higher-voltage and higher-current applications. These integrated control circuits can be provided with load-current limit, remote start-up and shut-down, and over-temperature limits.

Fig. 22. Block diagram of linear series transistor voltage regulator.

Switching Transistor Circuits

To achieve higher output power capability, transistor switching-mode power circuits have been developed. The transistor is operated either in its saturated state or its cutoff state. At present, switching frequencies are typically 5–100 kHz. Load-voltage control is provided by adjustment of the repetition rate, pulse width, or on-to-off ratio. Field-effect transistors can be used at even higher switching frequencies. A block diagram of a switching regulator is shown in Fig. 23. An input and output filter are shown to reduce the ripple current in the source and load, respectively. The transistor either is on, connecting the source and the load, or is cut off, disconnecting the source and the load.

There are four basic dc-to-dc switching converter configurations: the step-down, the step-up, the step-down/step-up, and the Cuk converter. These four circuits do not provide ohmic isolation between the

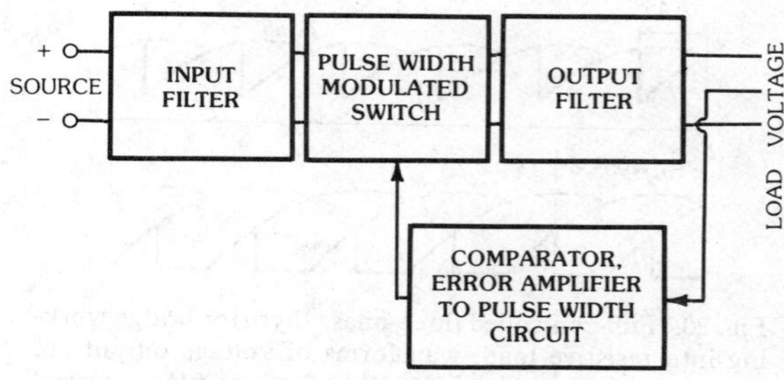

Fig. 23. Block diagram of switching-mode power supply.

input and the output. Five circuits have been developed that do provide transformer isolation between input and output. They are the flyback converter, the forward converter, the half-bridge, the full-bridge, the push-pull, and the Cuk converter with isolation transformer. Table 3 shows these circuits, their schematic diagrams, transfer functions, device rating equations, circuit waveforms, and advantages and disadvantages.

Integrated circuits are available to control these various circuits properly. These signal-level integrated circuits contain reference voltages, voltage-error amplifiers, current-error amplifiers, linear timing ramps, error-voltage–to–pulse-width circuits, remote on-off, synchronization, and power-transistor base-drive signals. Only a few external components are required to set the basic operating frequency and voltage scaling in order to provide a complete control circuit. Isolation between the output voltage and the input voltage can be provided with optical couplers. Isolation of the power-transistor base-drive signals can be accomplished by pulse transformers with energy-storage capacitors to provide a reverse pulse of base current for fast turn-off of the power transistor or field-effect transistor.

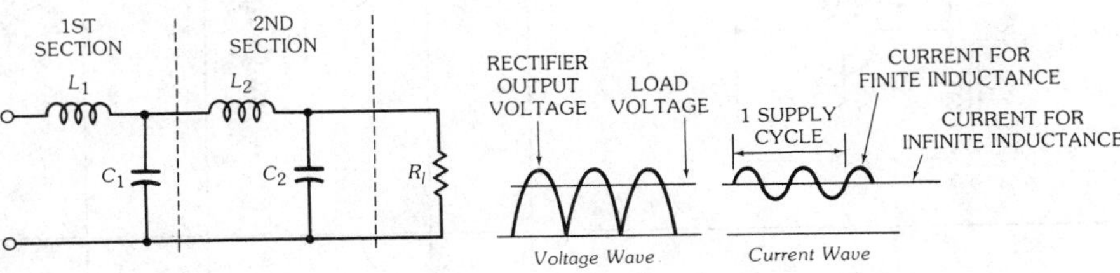

Fig. 24. Inductor-input filter.

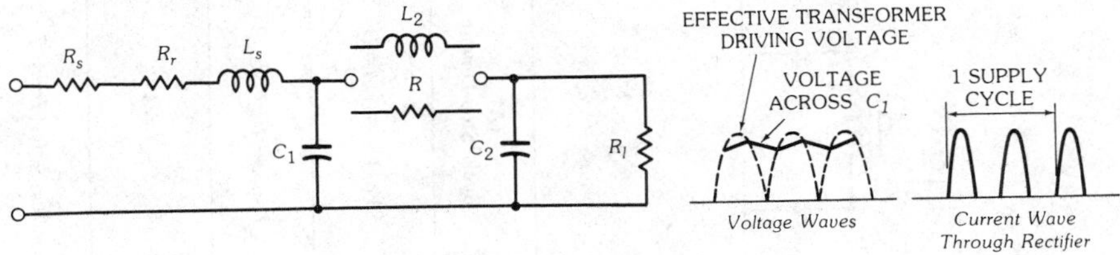

Fig. 25. Capacitor-input filter. C_1 is the input capacitor. $R_s = 1/2 \times$ (secondary-winding resistance). L_s = leakage inductance viewed from 1/2 secondary winding. R_r = equivalent resistance of IR drop in rectifier element.

FILTER CIRCUITS AND DESIGN

Rectifier filters may be classified into three types:

(1) *Inductor Input* (Fig. 24): Have good voltage regulation, high transformer utilization factor, and low rectifier peak currents, but also give relatively low output voltage.

(2) *Capacitor Input* (Fig. 25): Have high output voltage, but poor regulation, poor transformer utilization factor, and high peak currents. Used mostly in television and radio receivers.

(3) *Resistor Input* (Fig. 26): Used for low-current and low-power applications.

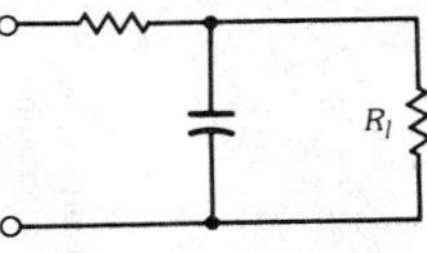

Fig. 26. Resistor-input filter.

Inductor-Input Filter Design

The constants of Fig. 24 are determined from the following considerations:

(A) There must be sufficient inductance to ensure continuous operation of rectifiers and good voltage regulation. When this critical value of inductance is increased by a 25-percent safety factor, the minimum value becomes

$$L_{min} = (K/f_s)R_l \text{ henry} \tag{2}$$

where,

f_s = frequency of source in hertz,
R_l = maximum value of total load resistance in ohms,
K = 0.060 for full-wave single-phase circuits
= 0.0057 for full-wave two-phase circuits
= 0.0017 for full-wave three-phase circuits.

TABLE 3. POWER TRANSISTOR AND DIODE REQUIREMENTS FOR SWITCHING POWER SUPPLIES[1]

Circuit Configuration	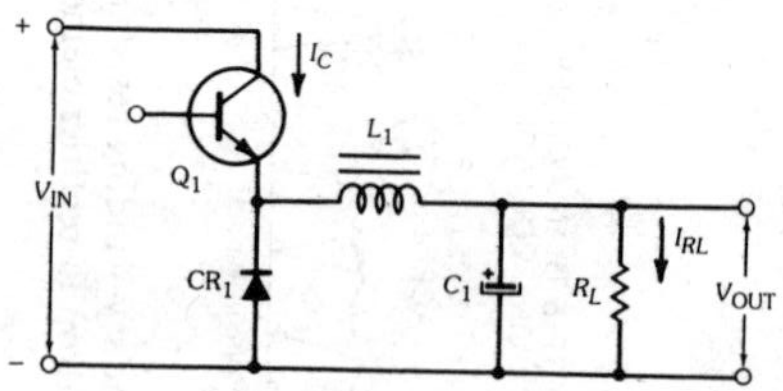	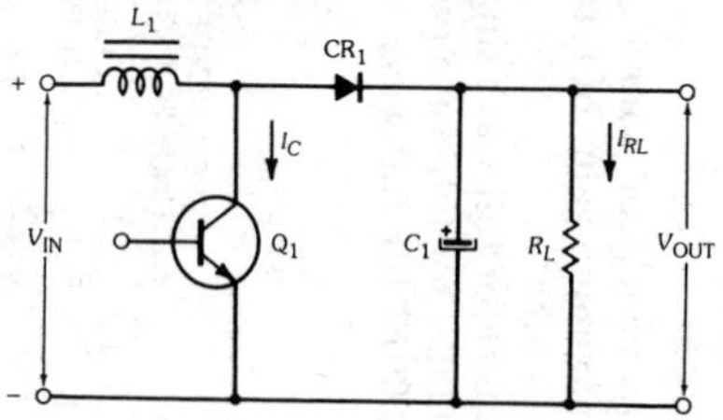	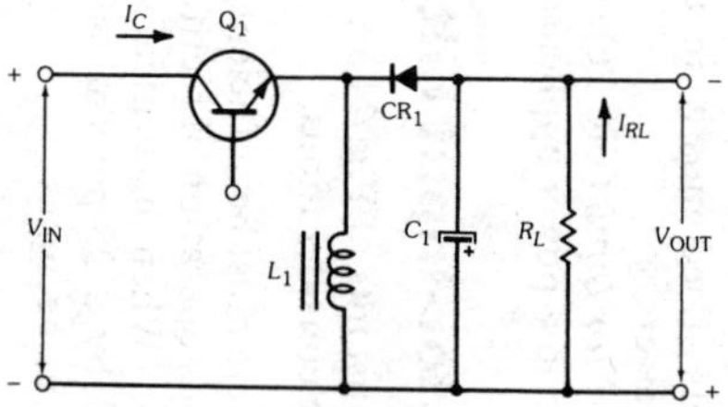
Type of Converter	(A) Buck (Step-Down)	(B) Boost (Step-Up)	(C) Buck-Boost
Ideal Transfer Function	$V_O/V_{IN} = \tau/T = D$	$V_O/V_{IN} = T/(T-\tau)$	$V_O/V_{IN} = -\tau/(T-\tau)$
*Collector Current(i_c)	$I_{CMAX} = I_{RL} + \Delta I_{L1}/2$	$I_{CMAX} = I_{RL}[T/(T-\tau)] + \Delta I_{L1}/2$	$I_{CMAX} = I_{RL}[T/(T-\tau)] + \Delta I_{L1}/2$
*Collector Voltage Rating	$V_{CEO} = V_{IN}$	$V_{CEO} > V_O + 1$	$V_{CEO} > V_{IN} + V_O$
*Diode Currents	$I_{CR1} = I_{RL}(T-\tau)/T$	$I_{CR1} = I_{RL}$	$I_{CR1} = I_{RL}$
*Diode Voltages (V_{RM})	$V_{CR1} = V_{IN}$	$V_{CR1} = V_O$	$V_{CR1} = V_O + V_{IN}$

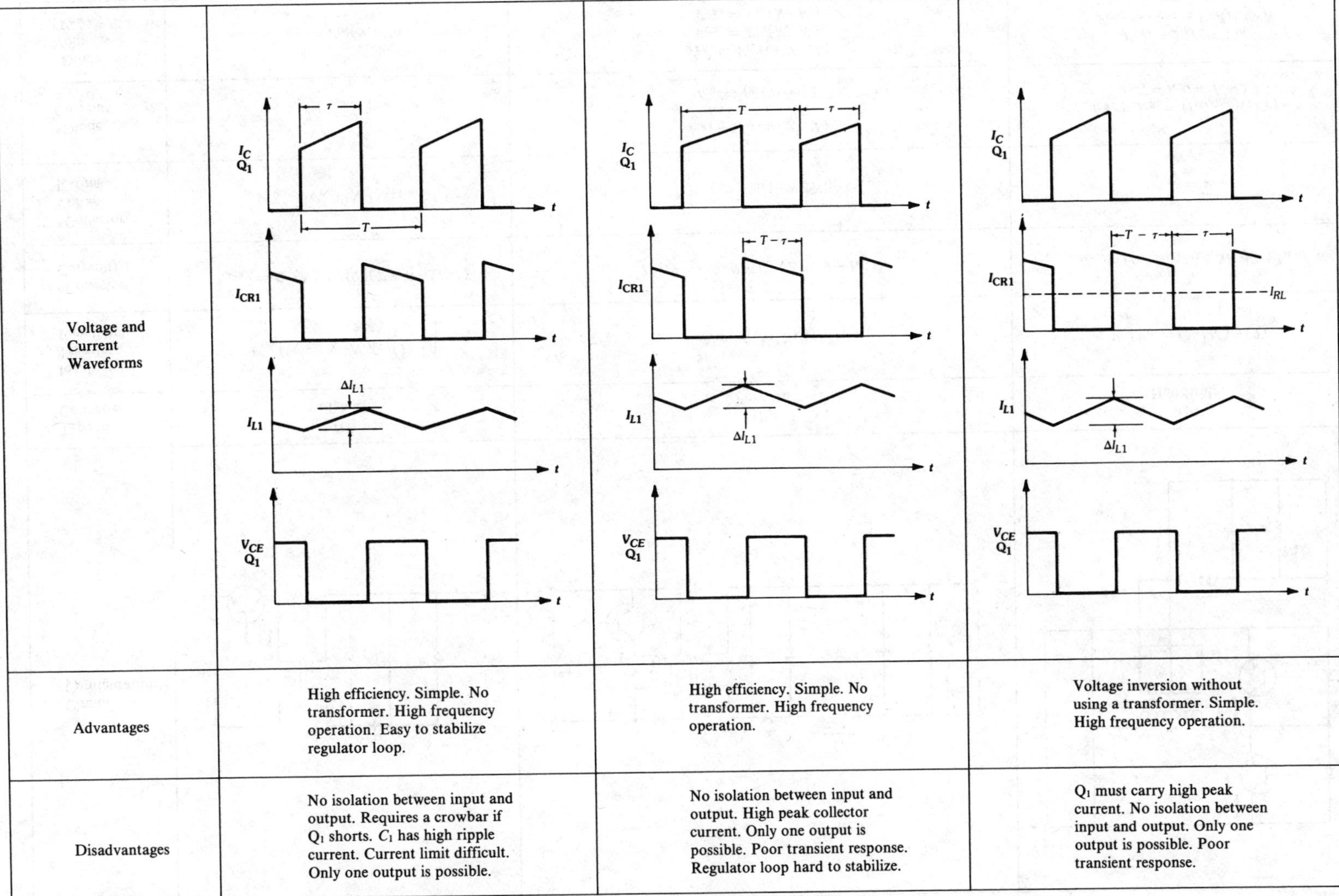

Voltage and Current Waveforms			
Advantages	High efficiency. Simple. No transformer. High frequency operation. Easy to stabilize regulator loop.	High efficiency. Simple. No transformer. High frequency operation.	Voltage inversion without using a transformer. Simple. High frequency operation.
Disadvantages	No isolation between input and output. Requires a crowbar if Q_1 shorts. C_1 has high ripple current. Current limit difficult. Only one output is possible.	No isolation between input and output. High peak collector current. Only one output is possible. Poor transient response. Regulator loop hard to stabilize.	Q_1 must carry high peak current. No isolation between input and output. Only one output is possible. Poor transient response.

1. From Application Note 200.87. Auburn, N.Y.: General Electric Co., Semiconductor Products Dept., 1979.

* For reliable operation, it is suggested and recommended that all voltage and current ratings be increased to 125% of the required maximum.

Continued on next page.

TABLE 3 (CONT). POWER TRANSISTOR AND DIODE REQUIREMENTS FOR SWITCHING POWER SUPPLIES[1]

Circuit Configuration	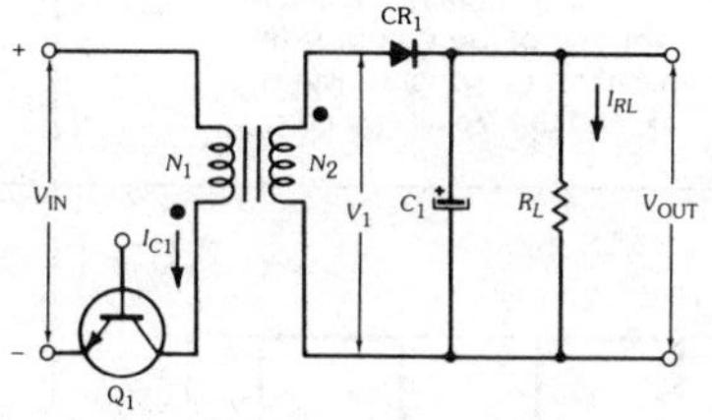	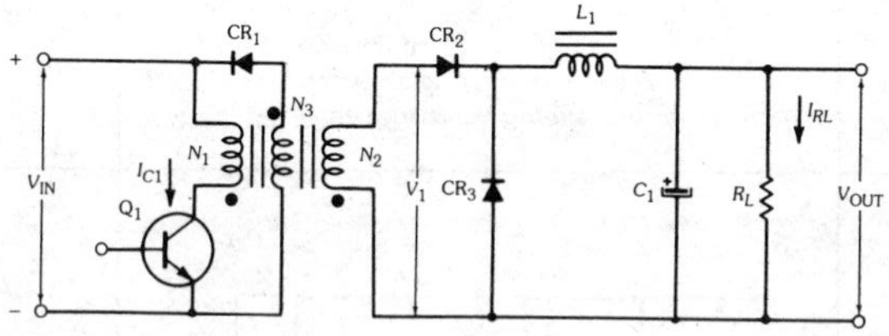	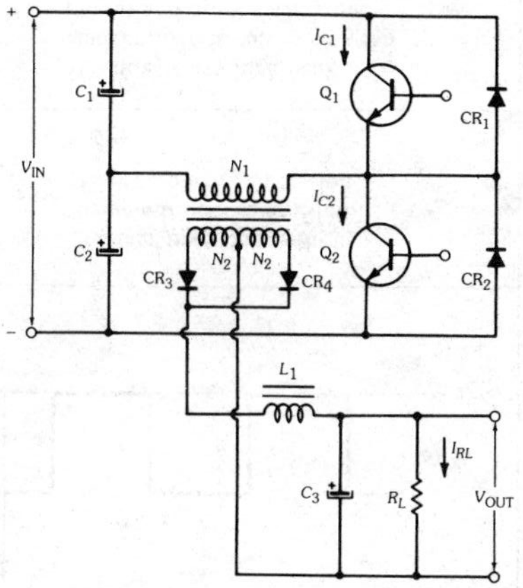
Type of Converter	(D) Flyback	(E) Forward	(F) Half Bridge
Ideal Transfer Function	$V_O/V_{IN} = (N_2/N_1)[\tau/(T-\tau)]$	$V_O/V_{IN} = (N_2/N_1)(\tau/T)$	$V_O/V_{IN} = (N_2/N_1)(\tau/T)$
*Collector Current(i_c)	$I_{CMAX} = I_{RL}(N_2/N_1)[T/(T-\tau)]+\Delta I_{L1}/2$	$I_{CMAX} = (N_2/N_1)(I_{RL}+\Delta I_{L1}/2)+\hat{I}_{MAG}$	$I_{CMAX} = (N_2/N_1)(I_{RL}+\Delta I_{L1}/2)+\hat{I}_{MAG}$
*Collector Voltage Rating	$V_{CEO} > V_{IN}+(N_1/N_2)V_{OUT}$	$V_{CEO} > V_{IN}(1+N_1/N_3)$	$V_{CEO} = V_{IN}$
*Diode Currents	$I_{CR1} = I_{RL}$	$I_{CR1} = (\hat{I}_{MAG}/2)(\tau/T)$ $I_{CR2} = I_{RL}(\tau/T)$ $I_{CR3} = I_{RL}(T-\tau)/T$	$I_{CR1} = I_{CR2} = (\hat{I}_{MAG}/2)(\tau/T)$ $I_{CR3} = I_{CR4} = I_{RL}/2$
*Diode Voltages (V_{RM})	$V_{CR1} = V_{IN}(N_2/N_1)$	$V_{CR1} = V_{IN}(1+N_3/N_1)$ $V_{CR2} = V_{IN}(N_2/N_3)$ $V_{CR3} = V_{IN}(N_2/N_1)$	$V_{CR1} = V_{CR2} = V_{IN}$ $V_{CR3} = V_{CR4} = V_{IN}(N_2/N_1)$

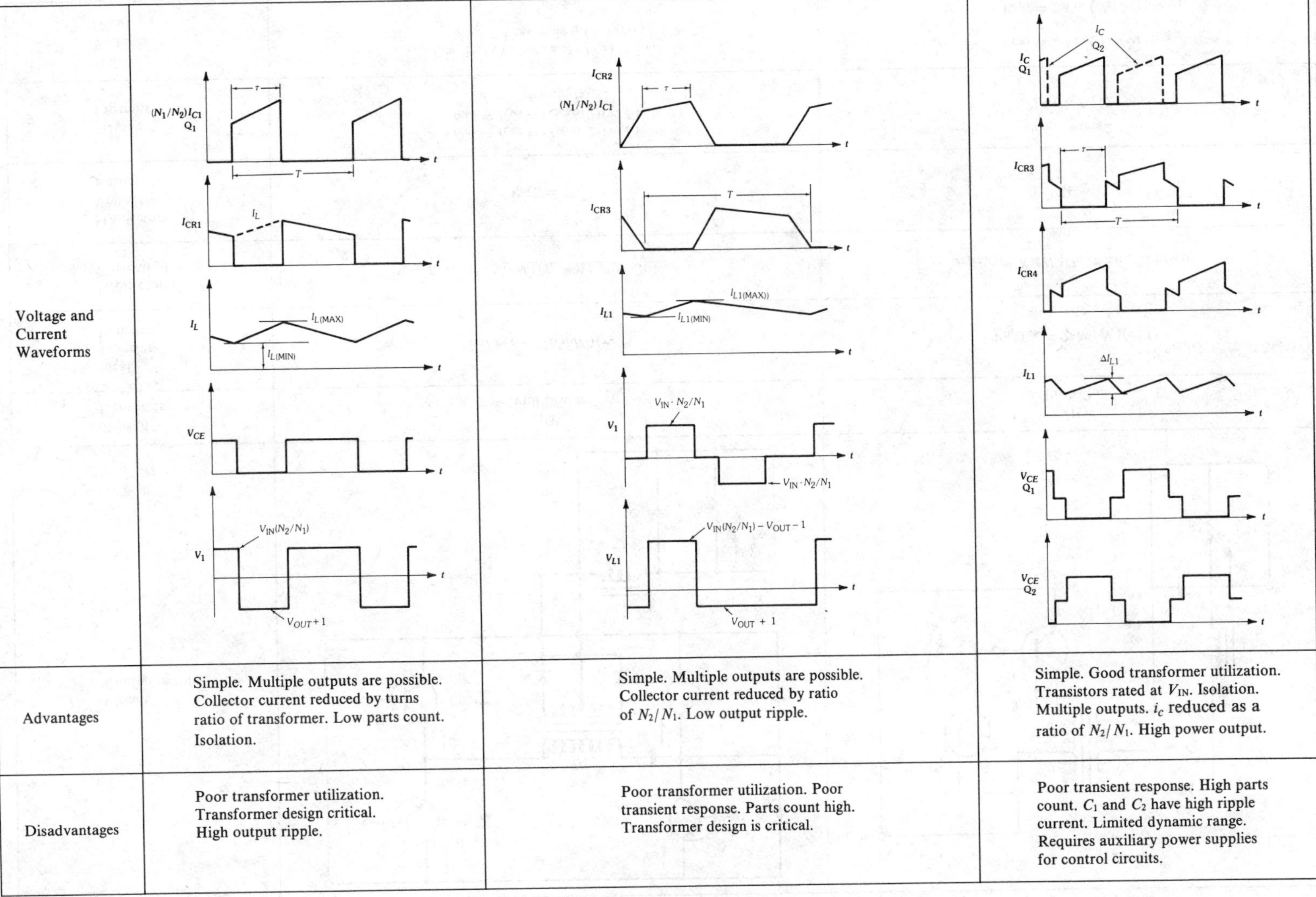

Advantages	Simple. Multiple outputs are possible. Collector current reduced by turns ratio of transformer. Low parts count. Isolation.	Simple. Multiple outputs are possible. Collector current reduced by ratio of N_2/N_1. Low output ripple.	Simple. Good transformer utilization. Transistors rated at V_{IN}. Isolation. Multiple outputs. i_c reduced as a ratio of N_2/N_1. High power output.
Disadvantages	Poor transformer utilization. Transformer design critical. High output ripple.	Poor transformer utilization. Poor transient response. Parts count high. Transformer design is critical.	Poor transient response. High parts count. C_1 and C_2 have high ripple current. Limited dynamic range. Requires auxiliary power supplies for control circuits.

1. From Application Note 200.87. Auburn, N.Y.: General Electric Co., Semiconductor Products Dept., 1979.

* For reliable operation, it is suggested and recommended that all voltage and current ratings be increased to 125% of the required maximum.

Continued on next page.

TABLE 3 (CONT). POWER TRANSISTOR AND DIODE REQUIREMENTS FOR SWITCHING POWER SUPPLIES[1]

Circuit Configuration	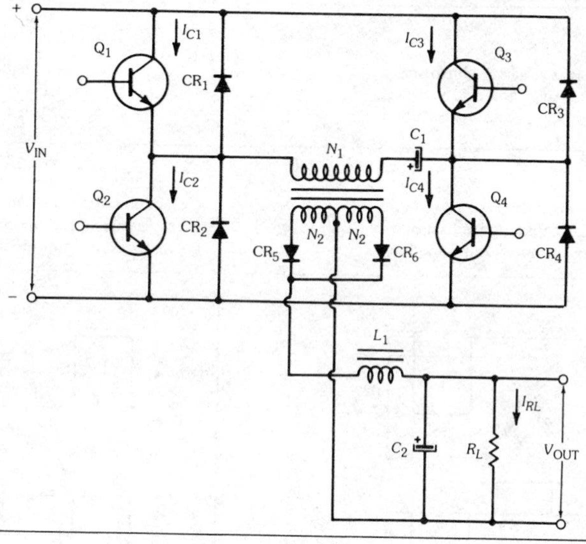	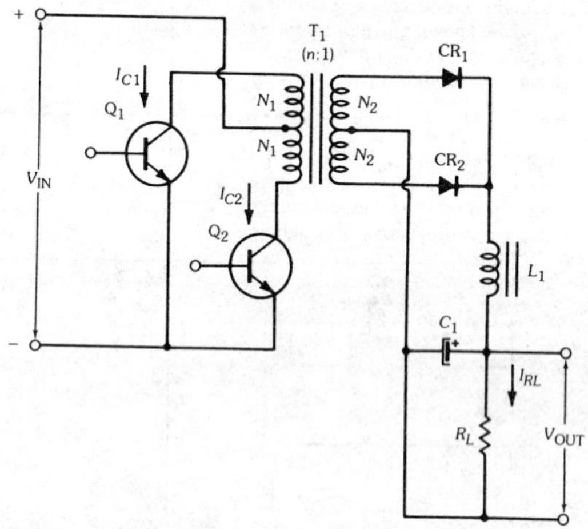
Type of Converter	(G) Full Bridge	(H) Push-Pull
Ideal Transfer Function	$V_O/V_{IN} = 2(N_2/N_1)(\tau/T)$	$V_O/V_{IN} = 2(N_2/N_1)(\tau/T)$
*Collector Current(i_c)	$I_{CMAX} = (N_2/N_1)(I_{RL} + \Delta I_{L1}/2) + \hat{I}_{MAG}$	$I_{CMAX} = (N_2/N_1)(I_{RL} + \Delta I_{L1}/2) + \hat{I}_{MAG}$
*Collector Voltage Rating	$V_{CEO} = V_{IN}$	$V_{CEO} = 2\ V_{IN}$
*Diode Currents	$I_{CR1} = I_{CR2} = (\hat{I}_{MAG}/2)(\tau/T)$ $I_{CR3} = I_{CR4} = (\hat{I}_{MAG}/2)(\tau/T)$ $I_{CR5} = I_{CR6} = I_{RL}/2$	$I_{CR1} = I_{RL}/2$ $I_{CR2} = I_{RL}/2$
*Diode Voltages (V_{RM})	$V_{CR1} = V_{CR2} = V_{CR3} = V_{CR4} = V_{IN}$ $V_{CR5} = V_{CR6} = 2V_{IN}(N_2/N_1)$	$V_{CR1} = 2\ V_{IN}\ (N_2/N_1)$ $V_{CR1} = 2\ V_{IN}\ (N_2/N_1)$

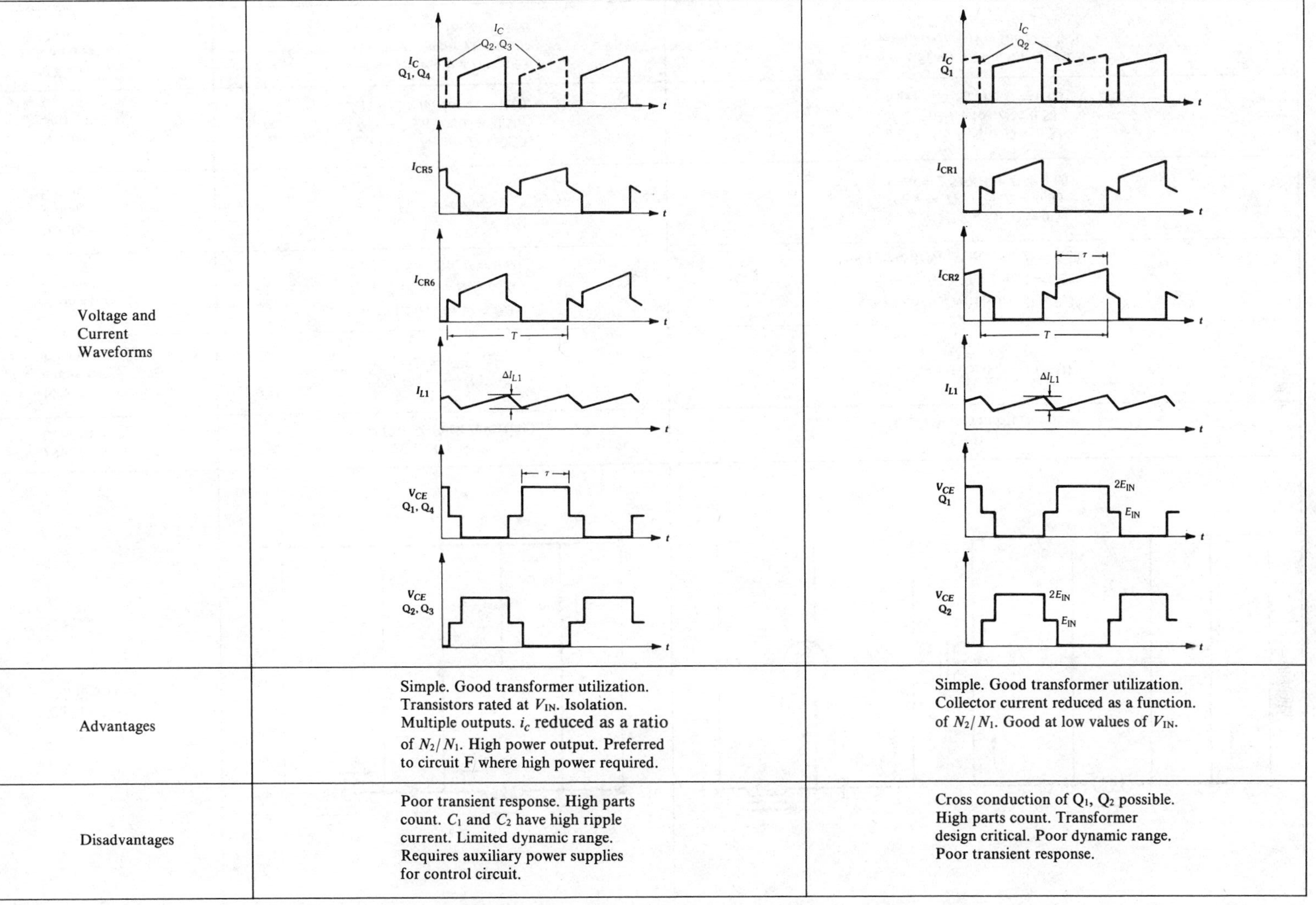

Voltage and Current Waveforms		
Advantages	Simple. Good transformer utilization. Transistors rated at V_{IN}. Isolation. Multiple outputs. i_c reduced as a ratio of N_2/N_1. High power output. Preferred to circuit F where high power required.	Simple. Good transformer utilization. Collector current reduced as a function. of N_2/N_1. Good at low values of V_{IN}.
Disadvantages	Poor transient response. High parts count. C_1 and C_2 have high ripple current. Limited dynamic range. Requires auxiliary power supplies for control circuit.	Cross conduction of Q_1, Q_2 possible. High parts count. Transformer design critical. Poor dynamic range. Poor transient response.

1. From Application Note 200.87. Auburn, N.Y.: General Electric Co., Semiconductor Products Dept., 1979.

* For reliable operation, it is suggested and recommended that all voltage and current ratings be increased to 125% of the required maximum.

Continued on next page.

TABLE 3 (CONT). POWER TRANSISTOR AND DIODE REQUIREMENTS FOR SWITCHING POWER SUPPLIES[1]

Circuit Configuration	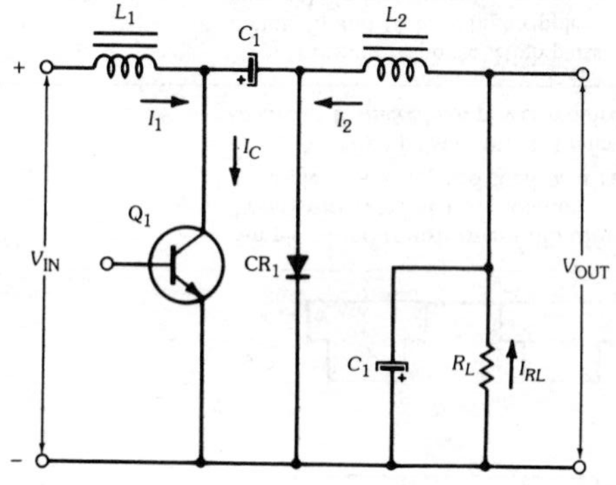	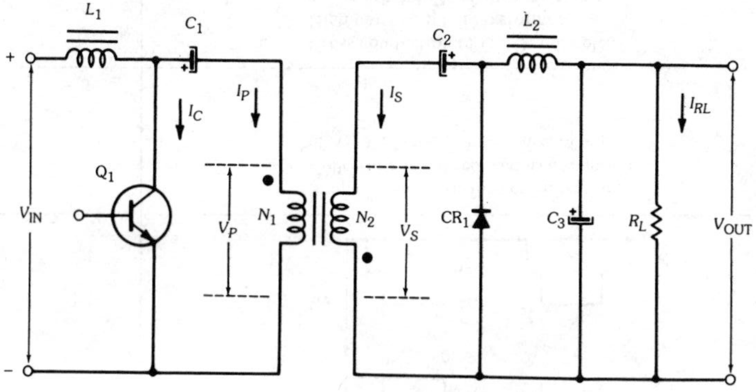
Type of Converter	(I) Cuk (Boost-Buck Inverting)	(J) Cuk (With Transformer)
Ideal Transfer Function	$V_O/V_{IN} = -\tau/(T-\tau)$ $D = \tau/T$	$V_O/V_{IN} = (N_2/N_1)[\tau/(T-\tau)]$ $D = \tau/T$
*Collector Current(i_c)	$I_C = 1.5I_{RL}$ for $D = 0.33$ $I_C = 2I_{RL}$ for $D = 0.50$ $I_C = 2.5I_{RL}$ for $D = 0.60$	$I_C = 1.5(N_2/N_1)I_{RL}$ for $D = 0.33$ $I_C = 2(N_2/N_1)I_{RL}$ for $D = 0.50$ $I_C = 2.5(N_2/N_1)I_{RL}$ for $D = 0.60$
*Collector Voltage Rating	$V_{CEO} \geq 2\ V_{IN}$	$V_{CEO} = 1.5V_{IN}$ for $D = 0.33$ $V_{CEO} = 2V_{IN}$ for $D = 0.50$ $V_{CEO} = 2.5V_{IN}$ for $D = 0.60$
*Diode Currents	$I_{CR1} = 1.5\ I_{RL}$ for $D = 0.33$ $I_{CR1} = 2\ I_{RL}$ for $D = 0.50$ $I_{CR1} = 2.5\ I_{RL}$ for $D = 0.60$	$I_{CR1} = 1.5\ I_{RL}$ for $D = 0.33$ $I_{CR1} = 2\ I_{RL}$ for $D = 0.50$ $I_{CR1} = 2.5\ I_{RL}$ for $D = 0.60$
*Diode Voltages (V_{RM})	$V_{CR1} = 1.5\ V_{IN}$ for $D = 0.33$ $V_{CR1} = 2\ V_{IN}$ for $D = 0.50$ $V_{CR1} = 2.5\ V_{IN}$ for $D = 0.60$	$V_{CR1} = 1.5(N_2/N_1)V_{IN}$ for $D = 0.33$ $V_{CR1} = 2(N_2/N_1)V_{IN}$ for $D = 0.50$ $V_{CR1} = 2.5(N_2/N_1)V_{IN}$ for $D = 0.60$

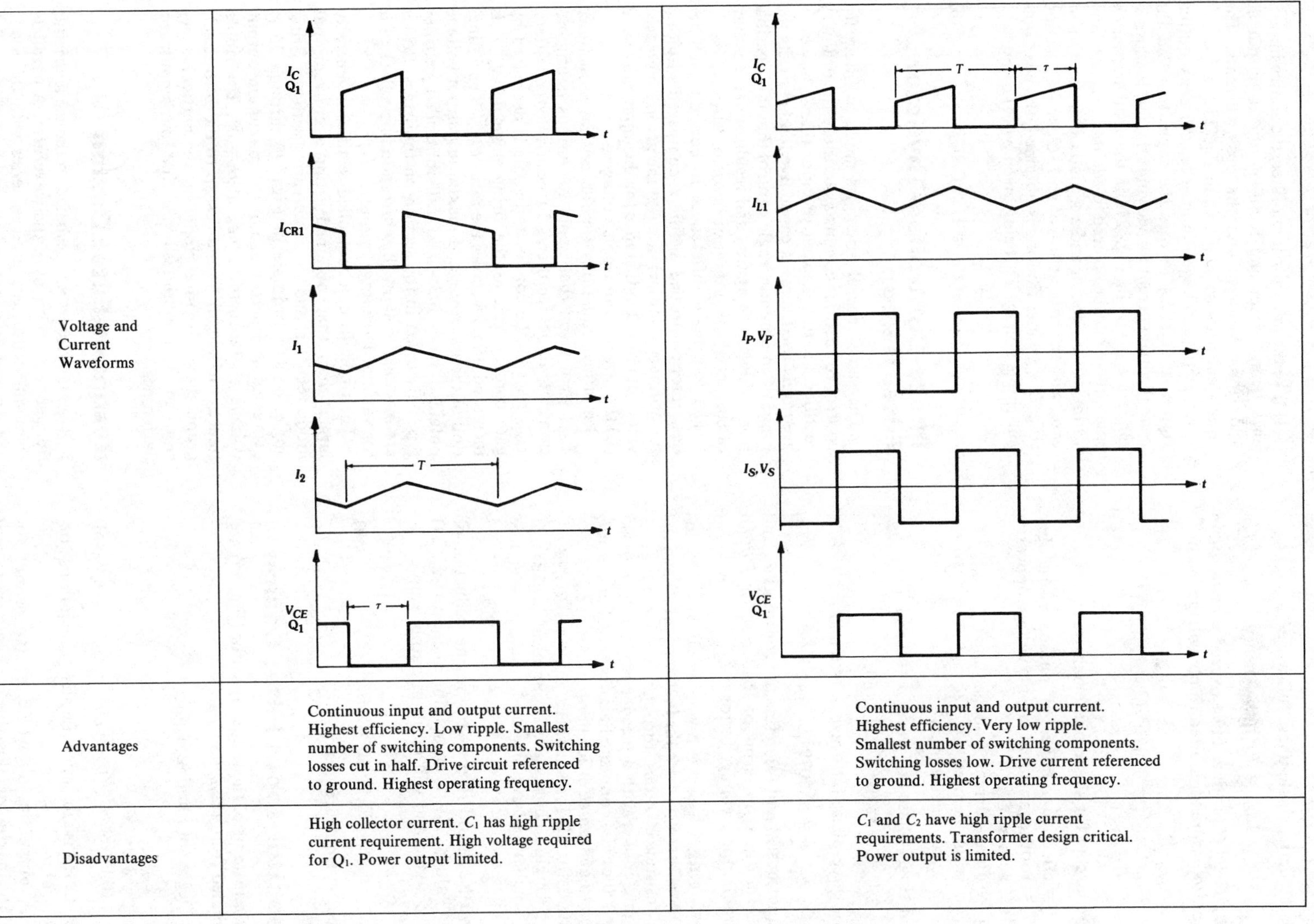

Voltage and Current Waveforms	(waveforms)	(waveforms)
Advantages	Continuous input and output current. Highest efficiency. Low ripple. Smallest number of switching components. Switching losses cut in half. Drive circuit referenced to ground. Highest operating frequency.	Continuous input and output current. Highest efficiency. Very low ripple. Smallest number of switching components. Switching losses low. Drive current referenced to ground. Highest operating frequency.
Disadvantages	High collector current. C_1 has high ripple current requirement. High voltage required for Q_1. Power output limited.	C_1 and C_2 have high ripple current requirements. Transformer design critical. Power output is limited.

1. From Application Note 200.87. Auburn, N.Y.: General Electric Co., Semiconductor Products Dept., 1979.

* For reliable operation, it is suggested and recommended that all voltage and current ratings be increased to 125% of the required maximum.

At 60 hertz, single-phase full-wave

$$L_{min} = R_l/1000 \quad \text{henry} \tag{2a}$$

(B) The *LC* product must exceed a certain minimum, to ensure a required ripple voltage factor

$$\begin{aligned} r &= E_r/E_{dc} \\ &= [\sqrt{2}/(p^2-1)][10^6/(2\pi f_s p)^2 L_1 C_1] \\ &= K'/L_1C_1 \end{aligned} \tag{3}$$

where, except for single-phase half-wave,

p = effective number of phases of rectifier,
E_r = root-mean-square ripple voltage appearing across C_1,
E_{dc} = direct-current voltage on C_1,
L_1 is in henries,
C_1 is in microfarads.

For single-phase full-wave, $p = 2$ and

$$r = (0.83/L_1C_1)(60/f_s)^2 \tag{3a}$$

For three-phase full-wave, $p = 6$ and

$$r = (0.0079/L_1C_1)(60/f_s)^2 \tag{3b}$$

Equations(2) and (3) define the constants L_1 and C_1 of the filter, in terms of load resistor R_l and allowable ripple voltage factor *r*.

Swinging Chokes—Swinging chokes have inductances that vary with the load current. When the load resistance varies through a wide range, a swinging choke, with a bleeder resistor R_b (10 000 to 20 000 ohms) connected across the filter output, is used to guarantee efficient operation; i.e., $L_{min} = R_l'/1000$ for all loads, where $R_l' = (R_lR_b)/(R_l + R_b)$. Swinging chokes are economical because of their smaller relative size and result in adequate filtering in many cases.

Two-Section Filters—For further reduction of ripple voltage E_{r1}, a smoothing section (Fig. 24) may be added and will result in output ripple voltage E_{r2}.

$$E_{r2}/E_{r1} \approx 10^6/(2\pi f_r)^2 L_2C_2 \tag{4}$$

where f_r = ripple frequency.

Capacitor-Input Filter Design

The constants of the input capacitor (Fig. 25) are determined from the following.

(A) Degree of filtering required.

$$\begin{aligned} r &= E_r/E_{dc} \\ &= \sqrt{2}/2\pi f_r C_1 R_l \\ &= (0.00188/C_1R_l)(120/f_r) \end{aligned} \tag{5}$$

where C_1R_l is in microfarads × megohms or farads × ohms.

(B) A maximum allowable C_1 (so as not to exceed the maximum allowable peak-current rating of the rectifier).

Unlike the inductor-input filter, the source impedance (transformer and rectifier) affects output dc and ripple voltages and the peak currents. The equivalent network is shown in Fig. 25.

Neglecting leakage inductance, the peak output ripple voltage E_{r1} (across the capacitor) and the peak rectifier current for varying effective load resistance are given in Fig. 27. If the load current is small, there may be no need to add the L-section consisting of an inductor and a second capacitor. Otherwise, with the completion of an L_2C_2 or RC_2 section (Fig. 25), greater filtering is obtained, the peak output-ripple voltage E_{r2} being given by (4) or by

$$E_{r2}/E_{r1} = 1/2\pi f_r RC_2 \tag{6}$$

Phase-Controlled-Thyristor Filter Design

The same general principles apply to filters for controlled rectifier circuits as to those for ordinary rectifier circuits. Capacitive-input filters, however, are rarely used, since a capacitive load restricts the range of conduction-angle control that it is possible to obtain with thyristors and results in large values of current at the instant of thyristor turn-on.

The two main differences to consider in filter design for controlled rectifiers, compared with that for ordinary rectifiers, are (A) the greatly increased values of critical inductance for the input choke and (B) the larger input ripple voltages.

Empirical equations for the critical inductance L_{min} were given that apply to various configurations of diode rectifiers. In controlled rectifiers, L_{min} rises as the conduction angle is decreased by gate control, *i.e.*, as the firing-angle delay, α, is increased. Fig. 28 shows the ratio L_{min}/R as a function of the percentage of maximum output voltage obtained from half- and full-controlled rectifier types discussed in the section on thyristors. Scales are included to show the corresponding values of α for various percentage outputs.

The ripple from a controlled rectifier is generally larger than that obtained from a conventional diode rectifier, and its value varies with the firing angle, α. The ordinate in Figs. 29 through 35 is expressed as a percentage of the maximum output voltage of the rectifier, *i.e.*, at $\alpha = 0°$. The ripple figures include all harmonics up to the 24th. Each graph shows the total ripple for the particular circuit and the amplitudes of the first four significant harmonics.

Resonant-Filter Design

Resonant filters are designed to reduce a specific frequency component of the waveform. A parallel-resonant section is placed in series with the load, and a series-resonant section is placed in parallel with the load. Fig. 36 shows a resonant filter. Both

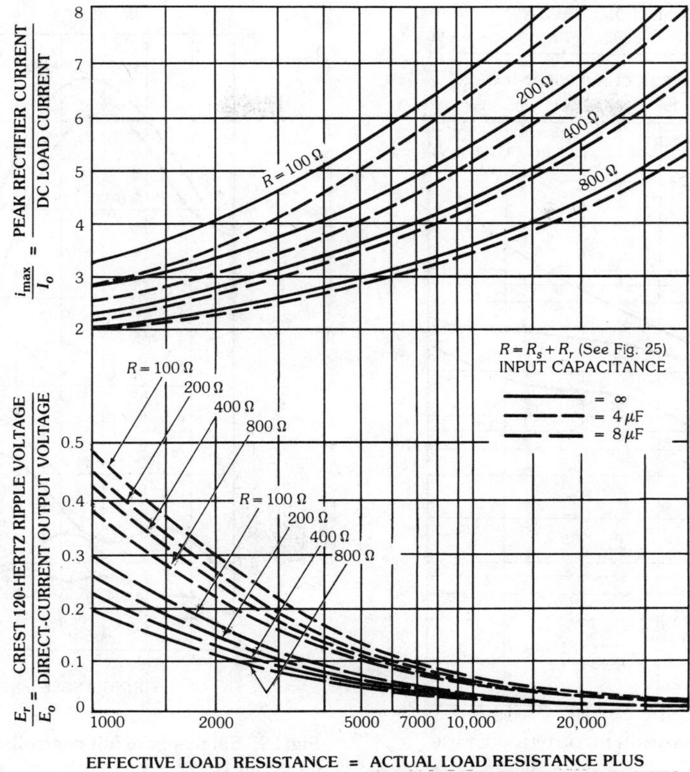

Fig. 27. Performance of capacitor-input filter for 60-hertz full-wave rectifier, assuming negligible leakage-inductance effect. (*Adapted from* Radio Engineers Handbook, *by F. E. Terman, 1st ed., p. 603; 1943. By permission McGraw-Hill Book Co., New York, N.Y.*)

L_1 and C_1 and L_2 and C_2 are selected to resonate at the unwanted frequency. These systems are applied to power-system filters where additional losses cannot be tolerated.

DC-OUTPUT POWER SUPPLIES

The previous sections described the various types of semiconductor devices, ac-to-dc converter circuits, and filter circuits. This section assembles these subsystems into complete dc power supplies. Block diagrams illustrate the various approaches to dc power supplies.

Unregulated Power Supplies

The most common type of unregulated power supply is the ac-to-dc rectifier with uncontrolled rectifiers followed by a suitable filter. Polyphase power produces less dc ripple voltage, less input harmonic current, and reduced dc voltage regulation. A power transformer allows for voltage matching between the incoming ac and the dc load voltage. If the load requires high-voltage dc, then a voltage-multiplier circuit can be chosen, since the design of the step-up transformer is more difficult with a large step-up ratio and severe insulation requirements. Load-voltage adjustment can be provided by transformer taps or an adjustable-voltage transformer.

Linear Regulated Power Supplies

A block diagram of a low-voltage linear regulated power supply is shown in Fig. 37. A utility-frequency transformer provides voltage matching between the incoming ac voltage and the desired load voltage. A diode rectifier converts the secondary ac voltage to dc voltage. A filter provides coarse filtering and fills in the gaps in the dc voltage caused by the rectification of single-phase power. The series transistor, operating in its linear region, absorbs the difference between its input voltage and the load voltage. The load voltage is sensed and fed back to the closed-loop voltage regulator. Load current limiting can also be provided to protect the series pass transistor from excessive power dissipation. Integrated and hybrid control circuits provide these and other functions in a single package. Instead of sensing output voltage, the output current can be sensed, and the system can provide a

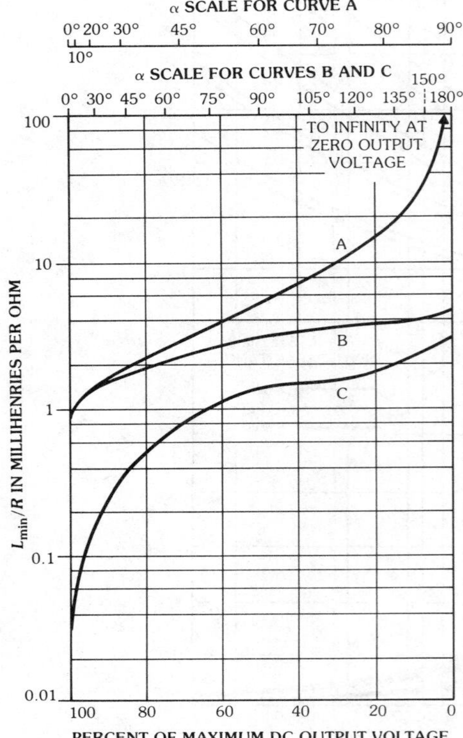

Fig. 28. L_{min}/R as a function of the percentage of maximum dc output voltage for thyristor rectifiers. The values of L_{min}/R given by the curves apply to a supply frequency of 60 hertz. Curve A is for full-controlled single-phase rectifiers such as those in Fig. 17 and in Fig. 19 with the bypass diode omitted. Curve B is for half-controlled single-phase rectifiers such as those in Fig. 18 and in Fig. 19 with the bypass diode included. Curve C is for a half-controlled three-phase rectifier such as that in Fig. 21.

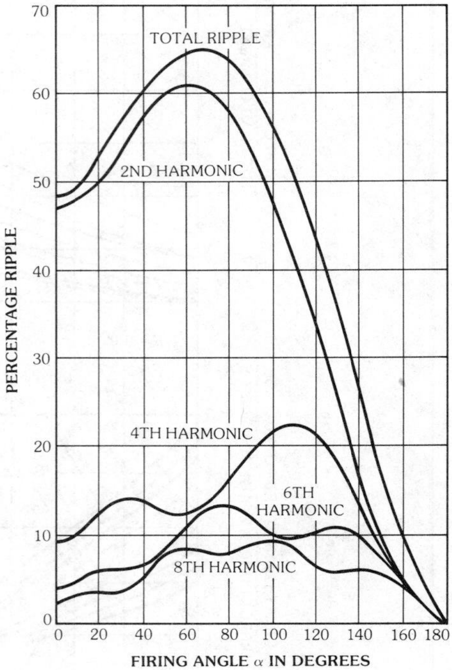

Fig. 29. Single-phase full-controlled rectifiers with bypass diode on resistive or inductive loads, or without bypass diode on resistive load; also applies to half-controlled rectifiers with or without bypass diode on resistive or inductive loads. The waveform is as illustrated by the example of Fig. 16, and typical circuits are Figs. 16, 18, and 19 with the diode inserted. (*From R. Smith, "Harmonic Voltages in the Outputs of Controlled Rectifier Circuits,"* Electronic Engineering, *Vol. 36, No. 442, p. 833, Fig. 1; © December 1964, Morgan Brothers (Publishers) Ltd., London, England.*)

constant current to a variable load. A diode is sometimes provided to prevent load-voltage reversals due to inductive energy in the load. Linear power supplies are used at low power (less than 50–100 watts); where multiple output voltages must be regulated; and where low noise, low ripple voltage, and fast response are essential.

Phase-Control Power Supplies

Two basic dc power-supply systems are configured with phase-controlled thyristors either on the primary or the secondary side of the isolation transformer. Figs. 38 and 39 show these two versions. The choice between the two systems is dependent on the cost of thyristors as a function of their voltage and current ratings. The usual configuration is for the ac input to be high voltage and the dc load to be low voltage and high current. Placing the thyristors in the primary allows high-voltage, low-current thyristors to provide the control of the output voltage. If the thyristors are located on the low-voltage side of the transformer, they must be rated at high current and low voltage. The efficiencies of the two systems are comparable, as the double power conversion of one system balances the losses associated with a high-current, low-voltage phase-controlled rectifier. Control of the transformer primary voltage requires that the gating signals be symmetrical to avoid transformer saturation.

Systems without isolation transformers are used to supply adjustable dc voltage to dc motor armatures and field supplies. Isolation transformers are provided in general-purpose laboratory power supplies in the range of 1–50 kW.

Switching Power Supplies

Switching type power supplies are currently used for general-purpose dc power supplies in the 100–1500-watt range. It is expected that this range will increase in the future due to the technical fea-

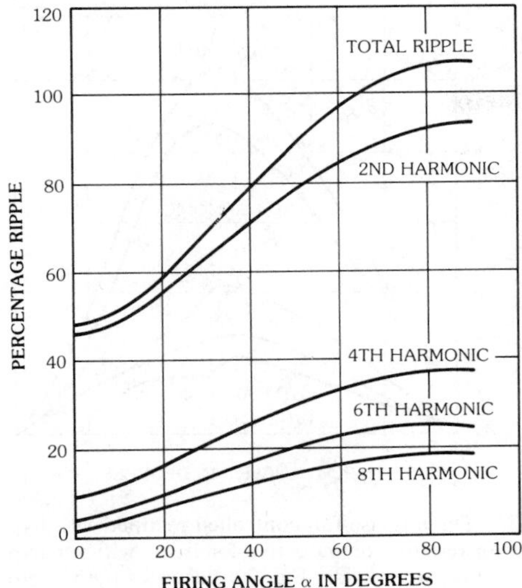

Fig. 30. Single-phase full-controlled rectifier without bypass diode on infinitely inductive load. The waveform is as illustrated by the example of Fig. 17, and typical circuits are Fig. 17 and Fig. 19 with the diode omitted. (*From R. Smith, "Harmonic Voltages in the Outputs of Controlled Rectifier Circuits,"* Electronic Engineering, *Vol. 36, No. 442, p. 833, Fig. 2; © December 1964, Morgan Brothers (Publishers) Ltd., London, England.*)

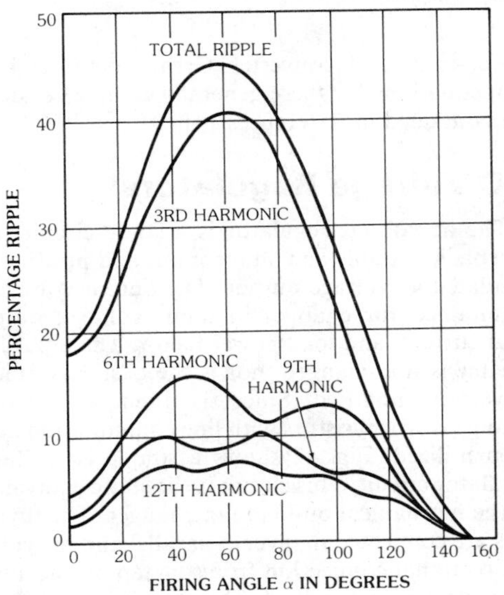

Fig. 31. Three-phase half-wave full-controlled rectifier with bypass diode on resistive or inductive loads, or without bypass diode on resistive load. (*From R. Smith, "Harmonic Voltages in the Outputs of Controlled Rectifier Circuits,"* Electronic Engineering, *Vol. 36, No. 442, p. 833, Fig. 3; © December 1964, Morgan Brothers (Publishers) Ltd., London, England.*)

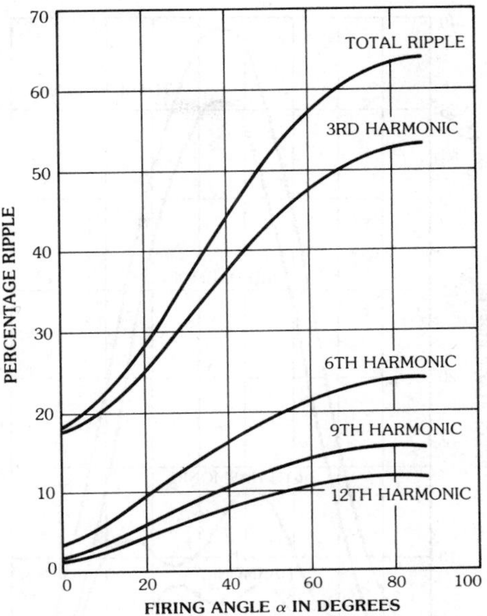

Fig. 32. Three-phase half-wave full-controlled rectifier without bypass diode on infinitely inductive load. (*From R. Smith, "Harmonic Voltages in the Outputs of Controlled Rectifier Circuits."* Electronic Engineering, *Vol. 36, No. 442, p. 833, Fig. 4; © December 1964, Morgan Brothers (Publishers) Ltd., London, England.*)

tures and cost reductions associated with this approach. Table 3 provides the design equations and device requirements for the most common types of dc-to-dc converters. A 60-hertz isolation transformer is not provided, in order to achieve small size and light weight. An uncontrolled diode rectifier and small capacitor filter provide an approximately 120-volt dc supply from 115-volt single-phase ac power. At light load, this voltage can increase to approximately 170–200 volts. A block diagram of this approach is shown in Fig. 40. A high-frequency pulse-width-modulated chopper or inverter is provided to regulate the voltage applied to the high-frequency transformer. The high-frequency transformer (20–50 kHz) provides voltage isolation and voltage matching at low weight, small size, and high efficiency. An uncontrolled ac-to-dc rectifier, at low voltages using Schottky diodes, converts from ac to dc. A low-pass filter removes the high-frequency (40–100 kHz) ripple from the output voltage. A closed-loop voltage regulator is provided by sensing the output voltage, comparing it to a reference, and controlling the on-off ratio of the primary transistors. Isolation of the feedback signal can be provided by optical couplers or voltage-to-frequency converters, pulse transformers, and frequency-to-voltage converter circuits.

A second approach is to separate the regulation and transformer functions into two separate circuits. A block diagram is shown in Fig. 41. Voltage

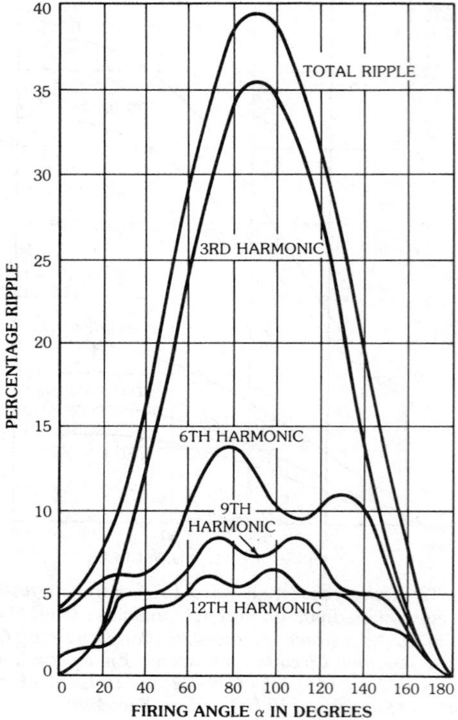

Fig. 33. Three-phase half-controlled rectifier with or without bypass diode on resistive or inductive loads. The waveform and the circuit are as shown by the example of Fig. 21. (*From R. Smith, "Harmonic Voltages in the Outputs of Controlled Rectifier Circuits,"* Electronic Engineering, *Vol. 36, No. 442, p. 834, Fig. 5; © December 1964, Morgan Brothers (Publishers) Ltd., London, England.*)

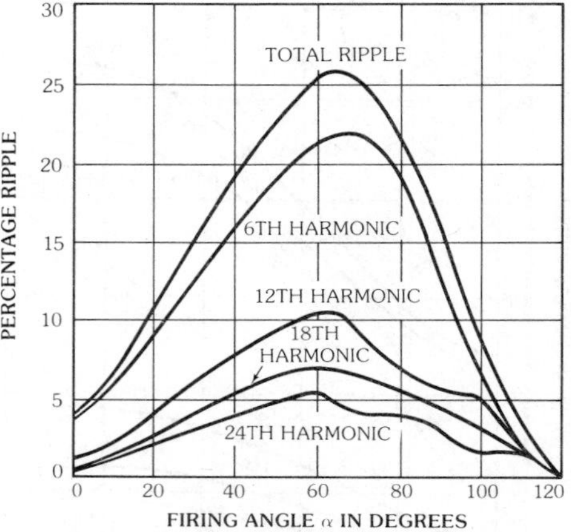

Fig. 34. Three-phase full-controlled rectifier with bypass diode on resistive or inductive loads, or without bypass diode on resistive loads. Fig. 20 illustrates the waveform and also the circuit without bypass diode for resistive load. The curve also applies to a double-star circuit with interphase transformer connections with bypass diode on resistive or inductive loads and without bypass diode on resistive loads. For the double-star circuit, α is measured from the normal output-voltage commutation point, not from the commutation point of the individual three-phase half-wave rectifiers. (*From R. Smith, "Harmonic Voltages in the Outputs of Controlled Rectifier Circuits,"* Electronic Engineering, *Vol. 36, No. 442, p. 834, Fig. 6; © December 1964, Morgan Brothers (Publishers) Ltd., London, England.*)

control is provided by a series dc/dc chopper operating at high frequency followed by a low-pass filter. The dc-to-ac inverter operates with a fixed 180° waveform that allows the use of a lightweight non-air-gap transformer. After rectification, the output filter needs to provide only a small amount of high-frequency filtering. This system is useful when the output voltage is low and the current is high, so that a large filter inductor is indicated.

AC-OUTPUT POWER SUPPLIES

The previous sections have discussed the aspects of converter and filter circuits and dc-output power supplies. This section describes some of the various types of power supplies that provide an alternating voltage and alternating current output. Some of the specific problem areas with an alternating output are associated with the power factor of ac loads, ac voltage magnitude control, polyphase output, adjustable-frequency output, and reverse power flow. Three basic ac-output power supplies are described, the ac voltage regulator, the dc-to-ac inverter, and the ac-to-ac cycloconverter. Greater detail and design equations for these generally complex systems are contained in References 1–17.

AC Voltage Regulators

The ac voltage regulator is a static circuit that accepts a variable ac voltage input and produces a regulated ac voltage output. The output voltage is maintained for changes in input voltage, output load current, and load power factor. The input frequency is not changed, nor is the basic sinusoidal waveform modified. Since the input and output voltage is ac, thyristors with line commutation have been utilized. Fig. 42 shows a simple ac voltage-regulator circuit. This circuit is the static equivalent of the mechanical on-load tap changer. The thyristors are connected in inverse-parallel and provide a bidirectional connection from the tap on the autotransformer to the load. One tap is at a voltage lower than the input voltage; the other tap is at a higher voltage. In this manner, the input voltage can be either increased or decreased to produce a fixed voltage output. Output voltage control is provided by a closed-loop regulator that adjusts the firing angle of the two thyristors connected to the higher-voltage transformer tap. The gating on of

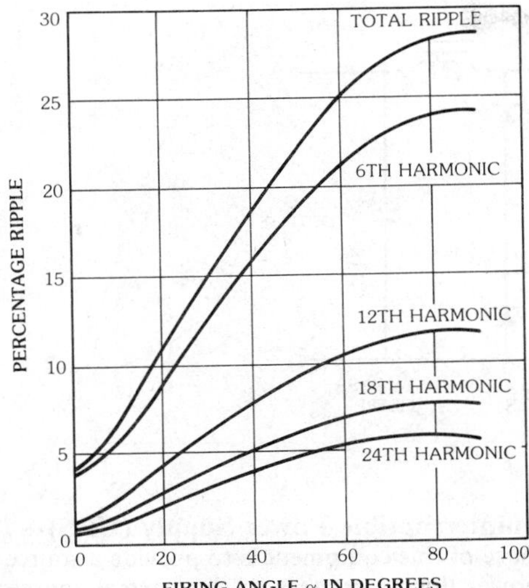

Fig. 35. Three-phase full-controlled rectifier without bypass diode on infinitely inductive load. The curve applies to the double-star circuit with interphase transformer when there is no bypass diode and the load is infinitely inductive. The same remarks as in Fig. 34 apply to the measurement of α. (*From R. Smith, "Harmonic Voltages in the Outputs of Controlled Rectifier Circuits,"* Electronic Engineering, *Vol. 36, No. 442, p. 834, Fig. 7; © December 1964, Morgan Brothers (Publishers) Ltd., London, England.*)

these thyristors reverse biases the lower two thyristors, causing them to cease conducting the load current and transferring the load current to the higher-tap thyristors. More than two taps can be provided to develop a regulated output voltage with reduced harmonic content. The gating of the lower two thyristors is accomplished at the zero crossing of the load current. In this manner, the load current is

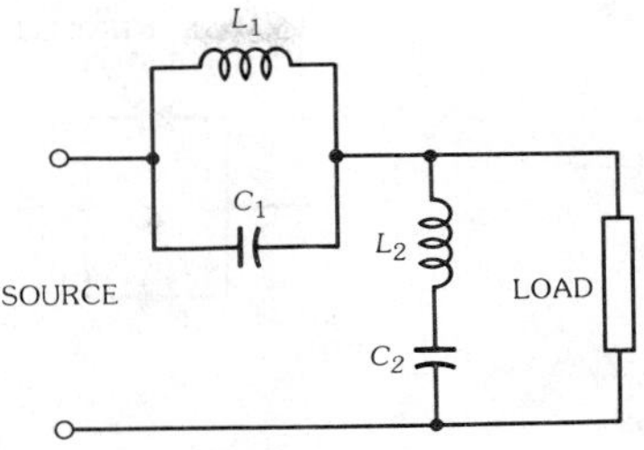

Fig. 36. Resonant filter circuit.

initially in the lower tap irrespective of the load power factor. The transfer of load current at current zero allows the upper-tap thyristors to turn off and allow the lower-tap thyristors to conduct the load current. This transfer from upper to lower tap is at load-current zero crossing and is dependent on the load power factor. Therefore, the output voltage waveform and magnitude change with respect to the load power factor, putting an additional burden on the ac voltage regulator. Both three-phase and single-phase circuits with full isolation transformers have been fabricated. Where low harmonic distortion is required, an output filter can be provided.

DC-AC Inverters

This broad class of circuits is characterized by a dc input voltage and an ac output voltage. Since the input is dc, commutation of thyristors must be provided by other techniques than the reversal of the input voltage. Transistors are widely used in lower-power inverters because of their self-commutation capability. Four major classes of inverters are described in this section: first, a system called an "uninterruptible power supply (UPS)," which provides a fixed-voltage, fixed-frequency output; second, an

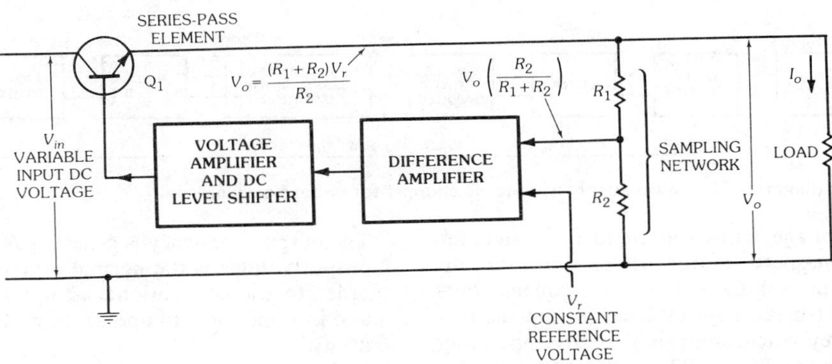

Fig. 37. Basic series-pass voltage regulator. Transistor Q_1 is an electronically controlled variable resistance in series with the load. (*From Pressman, Abraham I.* Switching and Linear Power Supply, Power Converter Design. *Rochelle Park, N.J.: Hayden Book Co., Inc., 1977.*)

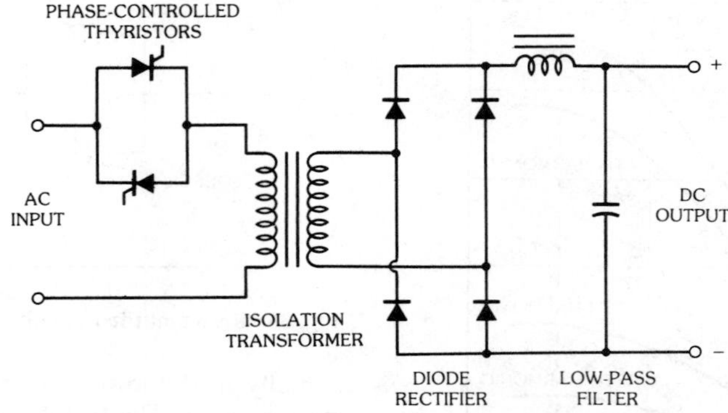

Fig. 38. Adjustable dc power supply with primary voltage control.

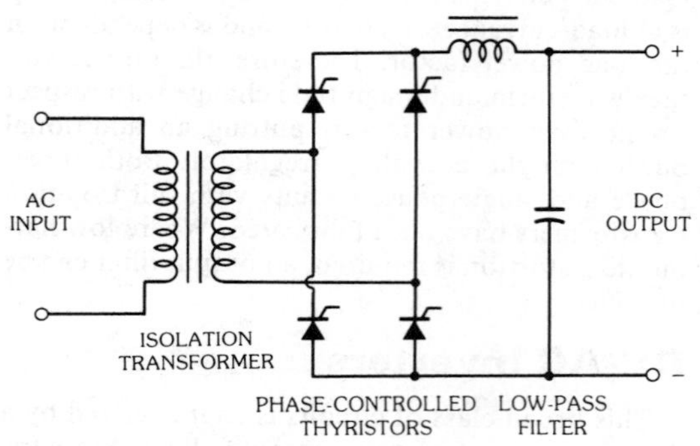

Fig. 39. Adjustable dc power supply with secondary voltage control.

Uninterruptible Power Supply (UPS)—The purpose of this equipment is to provide a source of ac power during outages of the normal source of utility supply. Uninterruptible power supplies are used in computer installations where power outages can mean loss of stored data (for example, in on-line reservations systems). Lower-power systems are provided to maintain continuous power to critical instrumentation (for example, a boiler-flame detector in a power plant). The source of power for these UPS installations is a battery that is kept charged from the utility. When the utility voltage is lost, the battery supplies power to the inverter and the connected load. These installations range in size from 1 kVA single-phase to 1000 kVA three-phase.

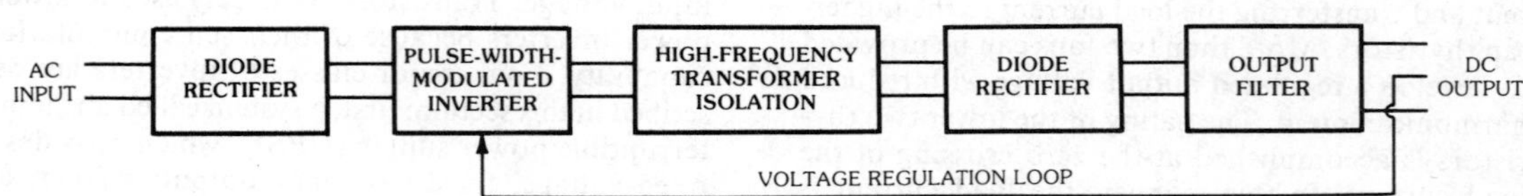

Fig. 40. Block diagram of dc power supply with pulse-width-modulated inverter for voltage control.

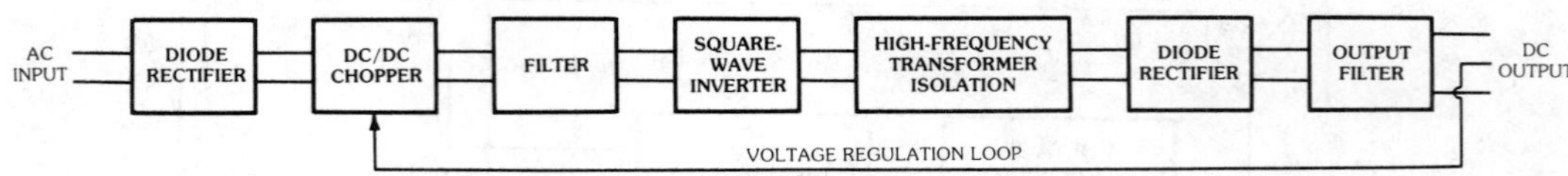

Fig. 41. Block diagram of dc power supply with dc/dc chopper for voltage control.

adjustable-voltage, adjustable-frequency system for an adjustable-speed motor drive; third, an induction-heating supply with a high-frequency output; and fourth, an adjustable-ac-current, adjustable-frequency synchronous-motor drive operating a leading power factor. These constitute typical applications for dc-to-ac inverters. Design equations and discussions of operating conditions in greater depth can be obtained in References 5–17.

The output frequency is generally 60 hertz, and the output voltage is the normal utilization voltage in order to use conventional ac-operated equipment and in some cases to operate in parallel with the ac utility.

Fig. 43 shows a block diagram of a large UPS installation. The separate battery charger, battery, inverter, utility bypass connection, and critical loads are shown. Large systems use thyristors with forced

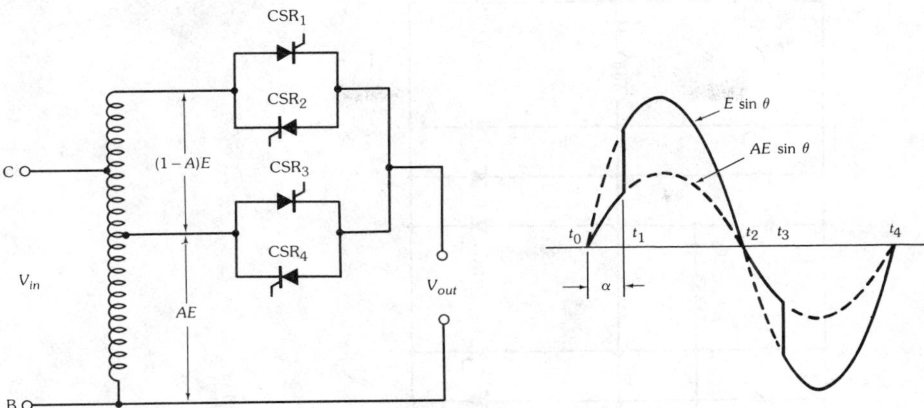

Fig. 42. Circuit diagram and waveforms for single-phase ac voltage regulator. (*From Mazda, F. F.* Thyristor Control. *New York: John Wiley & Sons, Inc., 1973.*)

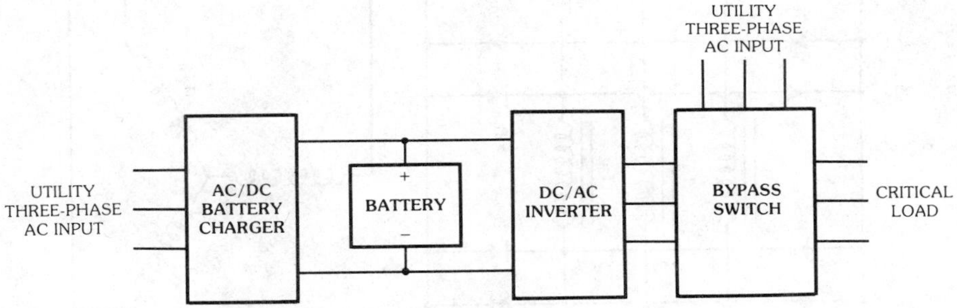

Fig. 43. Block diagram of UPS with utility bypass switch.

commutation to convert the dc voltage to ac voltage. Transformers are provided for voltage matching and isolation between the dc battery voltage and the ac utilization voltage. An example of a three-phase high-power thyristor bridge is shown on Fig. 44. Main thyristors T_1–T_{12} are turned on in sequence to generate an ac voltage on the primary of the three-phase transformer. Feedback diodes D_1–D_{12} conduct the reactive component of the load current and allow the load voltage to be independent of load power factor. Commutation or turn-off of the main thyristors is accomplished by auxiliary thyristors TA_1–TA_{12}, capacitors C_1–C_6, and reactors L_1–L_6. A filter is provided at the output to reduce the output harmonic voltages to within specified levels. Control of the voltage magnitude is provided by adjusting the phase shift between the two three-phase bridge inverters. In order to keep the output phase angle fixed, one bridge is phased forward, and the second bridge is phased back the same amount. This inverter circuit is called auxiliary thyristor commutation because of the use of auxiliary thyristors. The design equations for this commutation circuit are given in Reference 6.

Drive for Adjustable-Speed AC Motor—A second application for an ac-output system is an adjustable-voltage, adjustable-frequency inverter operating from a dc source and providing power to an adjustable-speed ac motor. To operate at high efficiency over a wide range of speed, the ac motor must be provided with adjustable-voltage and adjustable-frequency power. The voltage and the frequency are not totally independent and are generally varied in such a way as to maintain a constant ratio of voltage to frequency. That is, higher ac voltage and higher frequency result in higher motor speed. Since the source is dc and induction motors operate at lagging power factor, forced commutation must be provided for the switching devices used in the inverter. If thyristors are used, they must be provided with auxiliary components to reduce the thyristor current to zero and provide a reverse voltage on the previously conducting device. Transistors with their self-commutation capability are becoming popular in inverters below 50 horsepower. The inverter operates from a relatively fixed source of dc voltage; therefore control of the output voltage must be provided in the inverter. The pre-

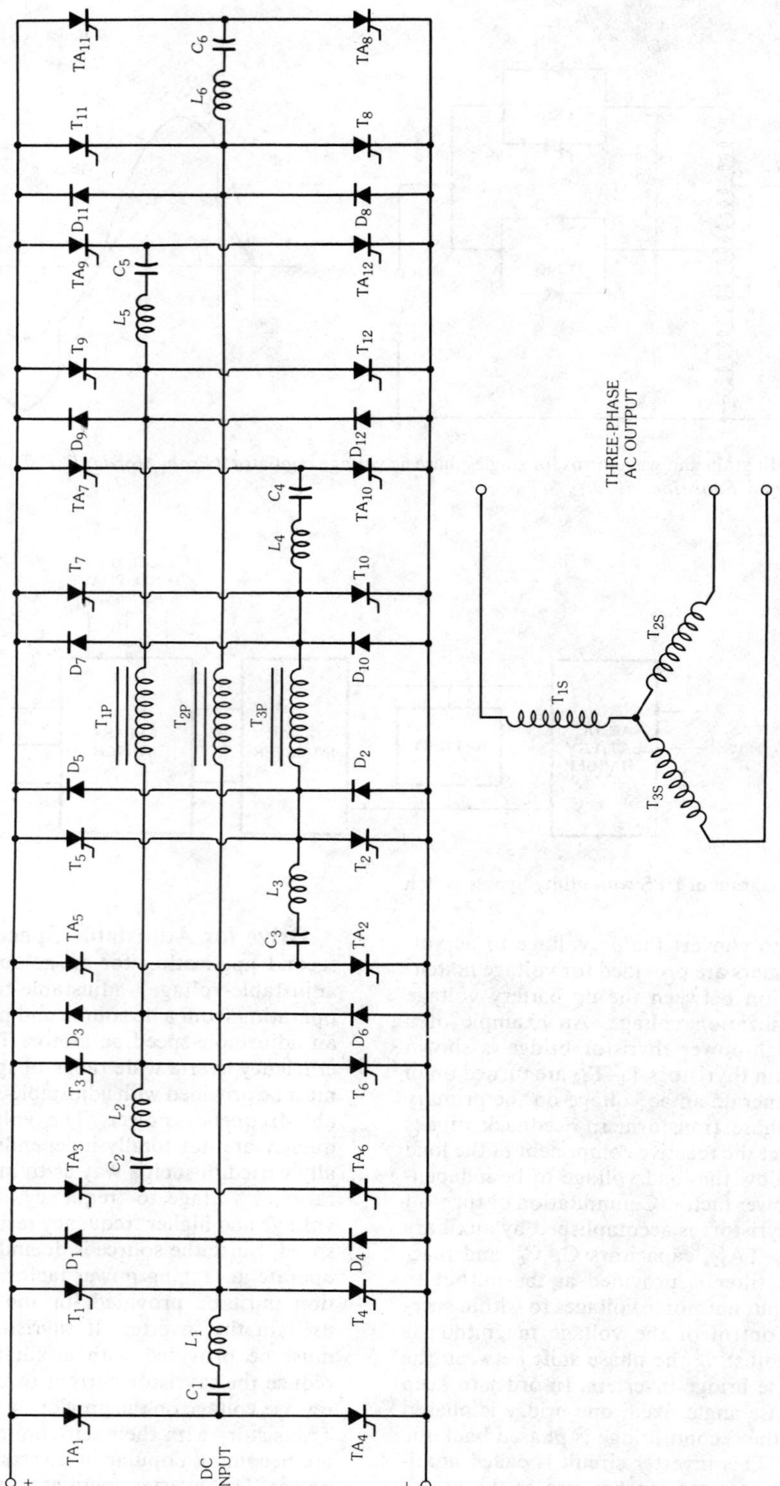

Fig. 44. Twelve-thyristor, auxiliary-impulse-commutated inverter.

ferred technique is to pulse-width modulate the output voltage by operating the transistors in an on-off mode many times during a given half-cycle of the motor fundamental frequency. Power transistors, with turn-on and turn-off times measured in microseconds, can be switched on and off many times during one half-cycle of a 60-hertz voltage, corresponding to a 3600-rpm synchronous speed with a two-pole induction motor. Various forms of pulse-width-modulation (pwm) have been proposed, with a form of sine-wave shaping being a preferred approach. Rapid advancements in microprocessor-based control systems and stored-program memory devices have reduced their cost and complexity to the point where they are used to generate the pulse-width-modulation switching times.

A three-phase full-bridge inverter is shown in Fig. 45. Transistors T_1–T_6, together with feedback diodes D_1–D_6, operate in a pulse-width-modulated manner to generate a load voltage as shown in Fig. 46. The motor leakage reactance acts as a filter for the motor current. For industrial applications, the source of dc voltage shown in Fig. 45 is an uncontrolled diode rectifier for either a single-phase or polyphase ac supply.

Induction-Heating Supply—In this application, a source of high-frequency ac power is converted to localized heating for melting and heat treating metals and ferrous parts. The frequency is relatively fixed and is dependent on the application. Low frequency, 60–180 hertz, is used for large ferrous heating and melting applications, whereas 10–50 kHz is used for surface hardening of metallic parts and in ultrasonic cleaning tanks. This application requires that the ac power be coupled into the load with an induction coil or transducer. Because of the lagging power factor of the work coil, a capacitor, either fixed or adjustable in steps, is provided to increase the power factor. High-power applications require the use of thyristors, and forced commutation is required because the operating frequency is higher than 60 hertz. The inductance of the work coil, an external capacitance, and the load

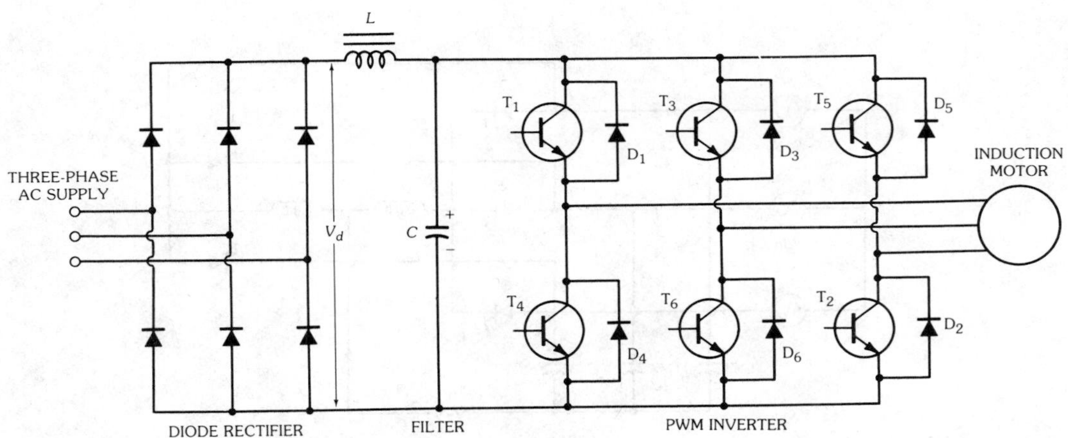

Fig. 45. Six-transistor pwm inverter/induction-motor drive. (*From Bose, Bimal K., ed.* Adjustable Speed AC Drive Systems. *New York: IEEE Press, 1981.*)

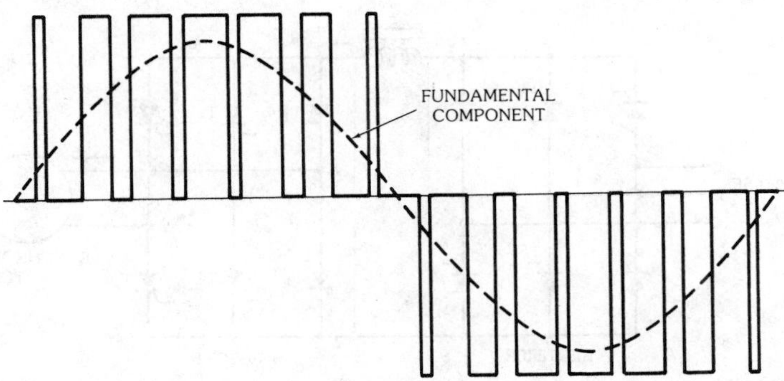

Fig. 46. Output-voltage waveform of a pulse-width-modulated inverter with sinusoidal modulation of the pulse width. (*From Murphy, J. M. D.* Thyristor Control of A.C. Motors. *New York: Pergamon Press, 1973.*)

resistor are selected to form an underdamped resonant circuit. Changes in the load resistance only change the damping ratio and barely affect the resonant frequency. The sinusoidal current waveform in the thyristors allows them to operate at higher frequencies with reduced losses. Fig. 47 shows an induction-heating inverter operating from a three-phase ac supply and supplying a single-phase load. High-power systems use this technique to balance the load on the three-phase supply.

Synchronous-Motor Drive—This system illustrates a fourth commutation technique for thyristors. In this case, the synchronous machine is operated at a leading fundamental-frequency power factor provided by overexciting the field of the machine. An adjustable-speed drive system of this type is shown in Fig. 48. The source of supply is an adjustable dc current supply. Thyristors T_1–T_6 steer the dc current into the motor windings in sequence to provide the ac-motor line current. A suitable control system turns on the thyristors at the proper time with respect to the back electromotive force in the motor to provide a reverse voltage for commutation of the thyristors. Shaft-position sensors or electrical signal processing is used to determine the correct time to gate the thyristors. The source of adjustable dc current for the inverter is an ac-to-dc phase-controlled rectifier. Special commutating circuits or turn-off intervals in the dc supply current are necessary to provide thyristor commutation during initial start-up and low-speed operation of the motor when the back emf is insufficient to commutate the thyristors. These drives have been fabricated to supply large synchronous machines up to 10 000 horsepower.

Cycloconverters

The preceding sections have described power-conversion systems that operate from dc power and produce an ac output. In industrial applications, the source is usually ac, thereby requiring a two-step conversion process from ac to dc to ac. The cycloconverter allows for a one-step conversion process from ac of a given voltage and frequency to

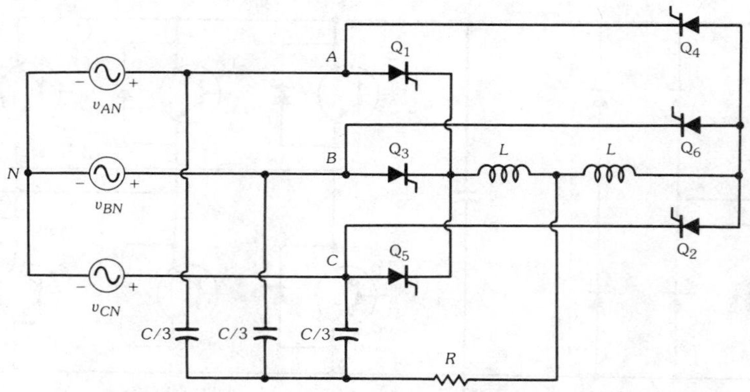

Fig. 47. Modified frequency multiplier with load commutation. (*From Dewan, S. B., and Straughen, A.* Power Semiconductor Circuits. *New York: John Wiley & Sons, Inc., 1975.*)

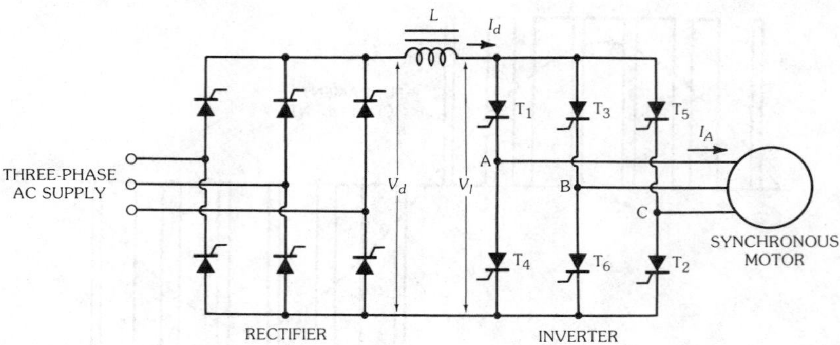

Fig. 48. Six-thyristor load-commutated inverter/synchronous-motor drive. (*From Bose, Bimal K., ed.* Adjustable Speed AC Drive Systems. *New York: IEEE Press, 1981.*)

adjustable-voltage ac (within limits) and adjustable-frequency ac (also within limits). This system uses the reversal of the line voltage to commutate the thyristors. The output voltage can be controlled by retarding the thyristor firing angle as in the phase-controlled rectifier. A second set of thyristors provides a path for the negative half-cycle of the load current. A three-phase to single-phase cycloconverter is shown in Fig. 49. The firing angle of the thyristors is modulated in a sinusoidal manner to generate a cycle of the desired output frequency. The inverse-parallel connection of the thyristors provides a bidirectional path for current flow at any power factor. Since the commutation of the thyristors is provided by the input frequency, the maximum output frequency must be less than the input frequency. The maximum output voltage is also less than the input voltage because of the need to generate a sinusoidal output voltage waveform.

There are two major applications for cycloconverters. The first is very-high-horsepower, very-low-speed, multiple-pole induction and synchronous motors. The maximum frequency for these motors can be less than 5 hertz in the case of an 8000-hp motor enclosing a rotary cement ball mill. The second application is for variable-speed constant-frequency (vscf) systems in aircraft. The speed of the aircraft generator, and hence its generated frequency, is a function of the engine speed. A cycloconverter is provided at the output of the generator to convert the varying frequency into a fixed 400-hertz frequency for the aircraft power system. The adjustment of the thyristor gating times generates a sinusoidal output voltage waveform with minimum filtering. Output voltage regulation is provided by both the adjustment in thyristor gating times and adjustment of the field excitation on the synchronous generator.

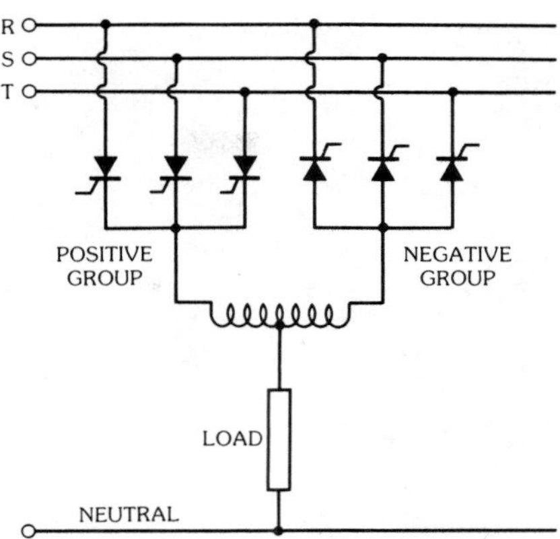

Fig. 49. Three-phase to single-phase cycloconverter power circuit. (*From Murphy, J. M. D.* Thyristor Control of A.-C. Motors. *New York: Pergamon Press, 1973.*)

REFERENCES

1. Grafham, D. R., and Golden, F. B., eds. *SCR Manual.* 6th ed. Auburn, N.Y.: General Electric Co., 1979.
2. Gentry, F., Gutzwiller, F., Holonyak, N., and Von Zastrow, E. E. *Semiconductor Controlled Rectifiers.* Englewood Cliffs, N.J.: Prentice-Hall, Inc., 1964.
3. Hnatek, E. R. *Design of Solid-State Power Supplies.* 2nd ed. New York: Van Nostrand Reinhold Co., 1981.
4. Pressman, A. I. *Switching and Linear Power Supply, Power Converter Design.* Rochelle Park, N.J.: Hayden Book Co., 1977.
5. Dewan, S. B., and Straughen, A. *Power Semiconductor Circuits.* New York: John Wiley & Sons, Inc., 1975.
6. Bedford, B. D., and Hoft, R. G. *Principles of Inverter Circuits.* New York: John Wiley & Sons, Inc., 1964.
7. Ramshaw, R. S. *Power Electronics—Thyristor Controlled Power for Electric Motors.* London: Chapman and Hall, 1973.
8. Mazda, F. F. *Thyristor Control.* New York: John Wiley & Sons, Inc., 1973.
9. Murphy, J. M. D. *Thyristor Control of A.C. Motors.* Oxford: Pergamon Press, 1973.
10. Sen, P. C. *Thyristor DC Drives.* New York: John Wiley & Sons, Inc., 1981.
11. Kusko, A. *Solid State DC Motor Drives.* Cambridge, Mass.: M.I.T. Press, 1969.
12. Möltgen, G. *Line Commutated Thyristor Converters.* London: Pitman Publishing, 1972.
13. Pelly, B. R. *Thyristor Phase-Controlled Converters and Cycloconverters.* New York: John Wiley & Sons, Inc., 1971.
14. McMurray, W. *The Theory and Design of Cycloconverters.* Cambridge, Mass.: M.I.T. Press, 1972.
15. Gyugyi, L., and Pelly, B. R. *Static Power Frequency Changers.* New York: John Wiley & Sons, Inc., 1976.
16. Harnden, J. D. Jr., and Golden, F. B., eds. *Power Semiconductor Applications.* Vol. 1 and Vol. 2. New York: IEEE Press, 1972.
17. Bose, B. K., ed. *Adjustable Speed AC Drive Systems.* New York: IEEE Press, 1981.

15 Feedback Control Systems

Revised by
Benjamin C. Kuo

Control systems are found in all sectors of the industry. They are used for such purposes as the quality control of manufactured products, machine-tool control, transportation systems, space technology and weapon systems, computer control, power systems control, robotics, and many others. Even such problems as inventory control and social and economic systems control may be approached from the control system standpoint. Therefore, in general, a control system usually consists of a process that is to be controlled in a certain desirable manner. Fig. 1 gives the simple block-diagram representation of a controlled process. For example, in a motor-speed control problem, the controlled process is the motor, the actuating signal is the voltage or current input to the motor, and the output in this case is the motor speed.

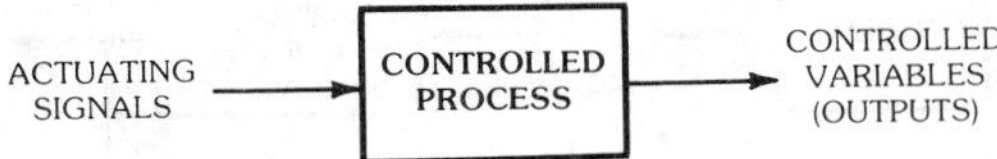

Fig. 1. Block diagram of a controlled process.

Quite often, the desired performance of the system demands that additional controllers be used to generate the actuating signal. Furthermore, other requirements such as accuracy and sensitivity of the system dictate that a feedback configuration be used. Thus, a practical control system often contains the elements shown in the block diagram of Fig. 2.

TYPES OF CONTROL SYSTEMS

Control systems may be classified according to the following classifications: (A) methods of analysis and design, and (B) types of signals.

Methods of Analysis and Design

Control systems may be classified into linear systems and nonlinear systems, time varying and time invariant. Strictly, all physical systems are nonlinear and have time-varying parameters to some degree. Where the effect of the nonlinearity is very small or the time-varying parameters are slow with respect to time, linear time-invariant analysis of the system is adequate for engineering purposes.

The most important property of a linear system is that the principle of superposition applies. Thus, in a linear system the shape of the response is independent of the magnitude of the input. A linear time-invariant dynamic system is generally described by a linear constant-coefficient differential equation.

In a control system where the system components possess such properties as saturation, limiting, backlash, or hysteresis, which are nonlinear characteristics, the principle of superposition does not hold. The system is generally described by nonlinear differential equations whose analytical solutions are rare, and the analysis of the nonlinear system must be carried out by numerical or graphical methods.

Types of Signals

According to the type of signals found in a control system, the latter may be classified into continuous-data (analog) systems and discrete-data (digital) systems. A continuous-data control system is one in which the signals at various parts of the system are all functions of the continuous time variable t. Discrete-data control systems contain signals that are in the form of either a pulse train or a digital code. For example, the system shown in Fig. 2 may be considered as a continuous-data control system if all the signals are functions of t, such as $r(t)$, $e(t)$, and $c(t)$. On the other hand, if the controller in Fig. 2 is replaced by a digital controller, the system becomes a discrete-data, or digital, control system. The input and output signals of the controller are digital and are represented by the sampling operation, as shown in Fig. 3.

CONTROL-SYSTEM COMPONENTS

Error-Measuring Systems: Potentiometers, Synchros, Incremental Encoders

Commonly used error-measuring systems, or comparators, are shown in Fig. 4. Rotary potentiometers are available commercially in single-revolution or multirevolution form. Some of the potentiometers have limited motion, such as one or more revolutions, and some have unlimited rotational motion. The potentiometers are commonly made with wirewound or conductive plastic resistance elements. Linear-motion potentiometers are also available for measuring linear motion properties.

The input-output relation of a potentiometer error detector is written

$$e(t) = K_s[r(t) - c(t)]$$

where,

$e(t)$ is the output voltage,
$r(t)$ is the reference input,
$c(t)$ is the controlled variable,
K_s is the gain of the error detector.

Synchros are used in control systems as detectors and encoders due to their ruggedness of construction and high reliability. A synchro is basically a rotary device that operates on the same principle as a transformer and produces a correlation between an angular position and a voltage or set of voltages.

For synchros whose primary excitation is 115 volts, the error sensitivity is approximately 1 volt/

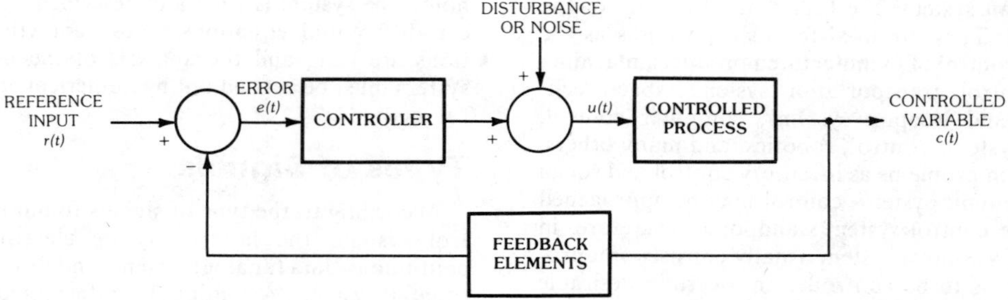

Fig. 2. Block diagram of feedback control system.

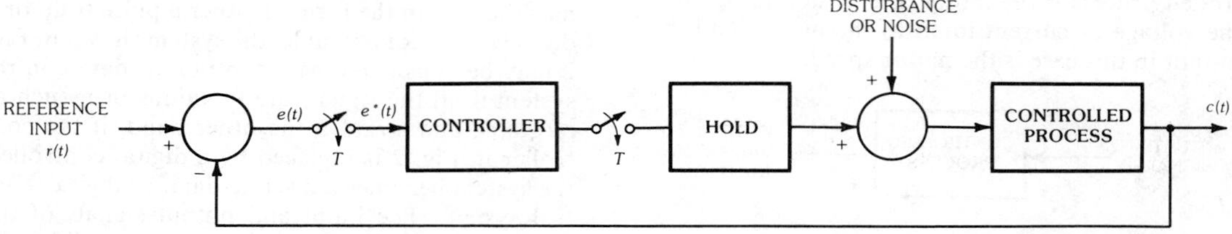

Fig. 3. Block diagram of a discrete-data, or digital, control system.

degree for a load resistance of 10 000 ohms across the control-transformer rotor.

The static error of a synchro transmitter and control transformer combination is of the order of 18 minutes maximum and is a function of the rotor position. In some precision units, this error may be reduced to a few minutes of arc. In synchro control transformers, a very undesirable characteristic is the presence of residual voltages at the null position. In well-designed units, this voltage will be less than 30 millivolts.

Synchro errors can be materially reduced by the use of double-speed systems. Such systems consist of a dual set of synchro units whose shafts are geared in such a manner as to provide a "fine" and a "coarse" control. The synchro error can be effectively reduced by the factor of the gear ratio employed. Synchronizing networks are employed to provide for proper switching between the two sets of synchros.

Incremental encoders are available in rotary or linear forms. These devices convert linear or rotary displacement into digitally coded or pulse signals. The outputs of the encoder are generally rectangular or sinusoidal in shape.

DC Motors

Types of dc motors include straight-series motors, split-series motors, shunt motors, compound motors (series-shunt fields), and permanent-magnet (pm) motors. Due to the advancements made in permanent-magnet materials, pm dc motors are one of the most widely used prime movers in industry today. The speed-torque characteristics of a pm dc motor are quite linear. However, the speed-torque characteristics of other types of dc motors are usually nonlinear.

The following variables and parameters are defined:

θ_m = motor angular position in radians
θ_l = load angular position in radians
ω = angular velocity in radians/s = $d\theta/dt$
T_m = motor-developed torque in newton-meters
J_m = motor moment of inertia in kilogram-meters^2
J_l = load moment of inertia in kilogram-meters^2
E_m = applied voltage in volts
K_m = motor torque constant in newton-meters/ampere
K_b = motor back emf constant in volts/rad/s
B_m = motor viscous-friction coefficient in Nm/rad/s
B_l = load viscous-friction coefficient in Nm/rad/s
N = load-to-motor gear ratio = θ_l/θ_m
B_{me} = viscous-friction coefficient reflected to motor shaft = $B_m + N^2 B_l$
J_{me} = inertia reflected to motor shaft = $J_m + N^2 J_l$
R_a = armature resistance of motor in ohms
L_a = armature inductance of motor in henries

The transfer function between the motor displacement and input voltage is

$$\theta_m(s)/E_m(s) = K_m/\{s[L_a J_{me} s^2 + (R_a J_{me} + B_{me} L_a)s + (K_b K_m + R_a B_{me})]\}$$

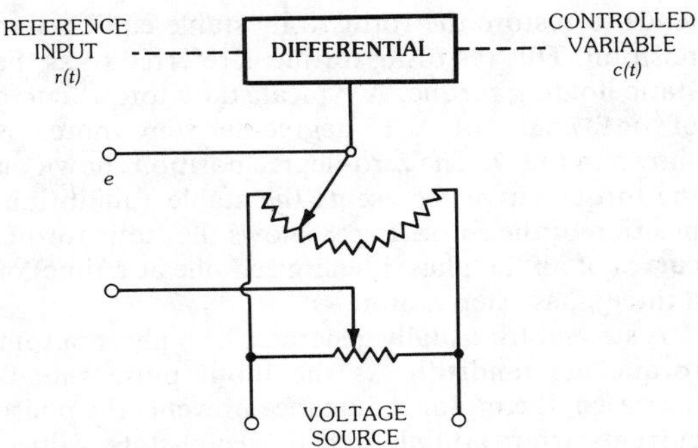

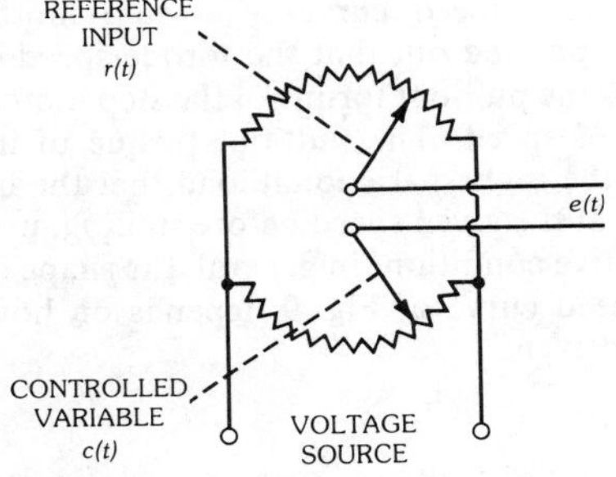

(A) Potentiometer systems.

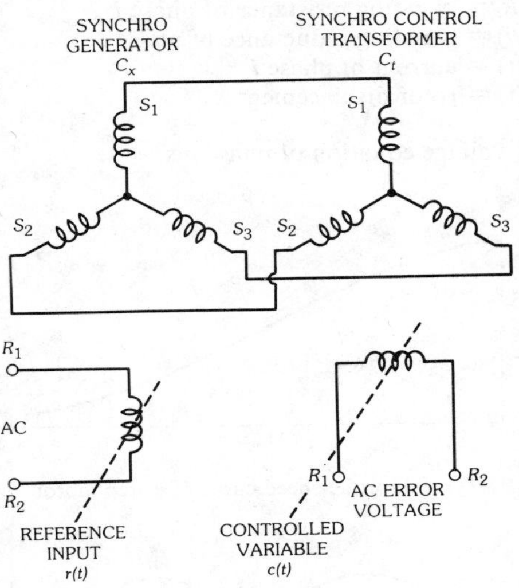

(B) Synchro system.

Fig. 4. Error-measuring systems.

Two-Phase Servomotors

For low-power applications in control systems, ac motors are sometimes used because of their rugged construction. Most ac motors used in control systems are of the two-phase induction type. One of the two phases is excited from a constant-voltage source (the reference winding). The speed-torque curves shown in Fig. 5 are linearized for analytical purposes. Let

$$k = \frac{\text{Blocked-rotor torque at rated } E_m}{\text{Rated control voltage } E_1}$$

$$= T_0/E_1$$

$$m = -\frac{\text{Blocked-rotor torque}}{\text{No-load speed}} = -T_0/\Omega_0$$

The transfer function between the control voltage and the motor displacement is

$$\theta_m(s)/E_m(s) = K_m/[s(1 + \tau_m s)]$$

where,

$$K_m = k/(B_{me} - m)$$

$$\tau_m = J_{me}/(B_{me} - m)$$

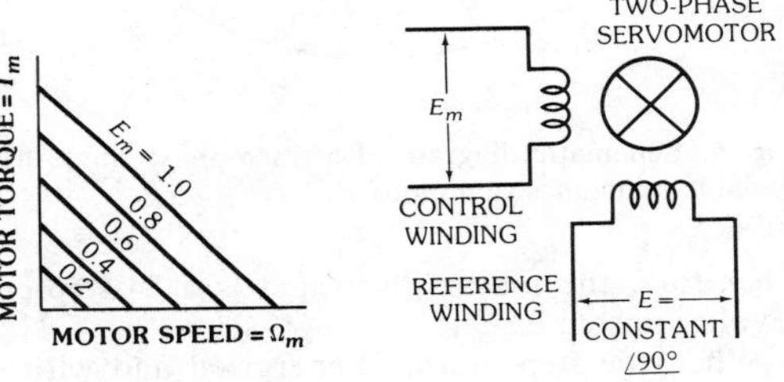

Fig. 5. Diagram of two-phase servomotor and idealized torque-speed curves.

Step Motors

Step motors are electromechanical incremental actuators that convert digital pulse inputs to analog output shaft motion. The advantage of a step motor is that the motor can be driven directly by digital inputs through the power driver, and no interface of d/a conversion is necessary. In a rotary step motor, the output shaft of the motor rotates in equal increments in response to a train of input pulses.

The three most popular types of step motors are the variable-reluctance motor, the permanent-magnet motor, and the hybrid permanent-magnet motor. The latter two types have a permanent magnet in the rotor assembly.

Fig. 6 shows the schematic diagram of a single-stack three-phase variable-reluctance step motor. The stator in this case has 12 teeth, and the rotor has 8 teeth. There are four teeth per phase, and only the windings of phase A are shown. The rotor is shown to be at the detent position when phase A is energized with a dc current. If the dc excitation is shifted to the windings of phase B, the rotor will rotate 15 degrees in the clockwise direction. If, instead, phase C is energized, the rotor will make a 15-degree step in the counterclockwise direction.

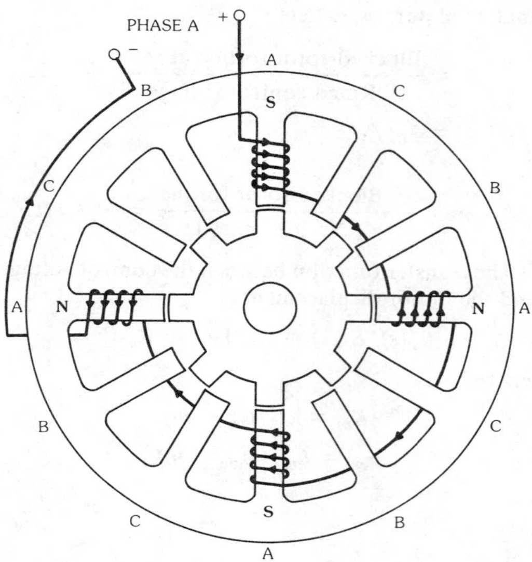

Fig. 6. Schematic diagram of a three-phase single-stack variable-reluctance step motor.

Therefore, the motor illustrated is a 24-step-per-revolution step motor.

When the step motor is energized and with its rotor at the equilibrium position of the energized phase, no torque is developed on the rotor shaft. When the rotor is displaced from the equilibrium position, a restoring torque is developed which tends to restore the rotor to its stable equilibrium position. This restoring torque is referred to as the static holding torque. A typical static torque curve of one phase of a 15-degree-per-step motor is shown in Fig. 7. The zero-degree position shown on the torque curve represents the stable equilibrium position of the rotor. Fig. 8 shows the static torque curves of all the phases (energized one at a time) of a three-phase step motor.

A step motor usually generates its highest output torque at standstill. As the input pulse rate is increased, the motor inductance prevents the phase currents from attaining their steady-state values, and the motor torque decreases. Fig. 9 illustrates a typical torque-speed curve of a step motor. It should be pointed out that the torque-speed curve represents the pull-out torque of the step motor as a function of speed. The pull-out torque of a step motor is the highest frictional load that the motor can drive at the given speed before stalling, under a specific drive condition. In general, the shape of the torque-speed curve of Fig. 9 depends on how the motor is driven.

Let

$e_i(t) =$ applied voltage of phase i, $i =$ A, B, C, . . . , N, for an N-phase motor
$R_i =$ winding resistance of phase i
$L_i(\theta) =$ winding inductance of phase i
$i_i(t) =$ current of phase i
$\theta(t) =$ rotor displacement

The voltage equation of phase i is

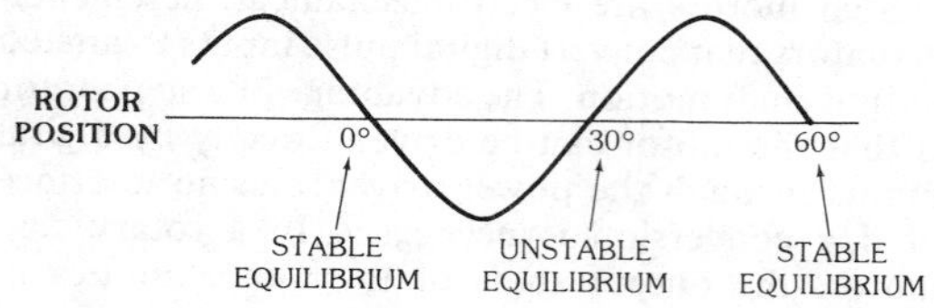

Fig. 7. Typical static holding torque of one phase of a step motor.

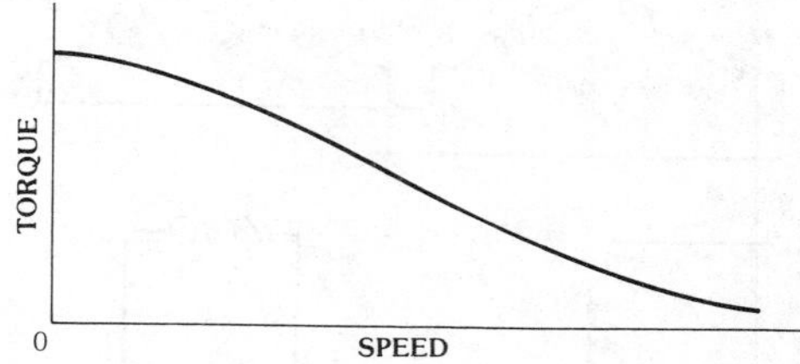

Fig. 9. Typical torque-speed curve of a step motor.

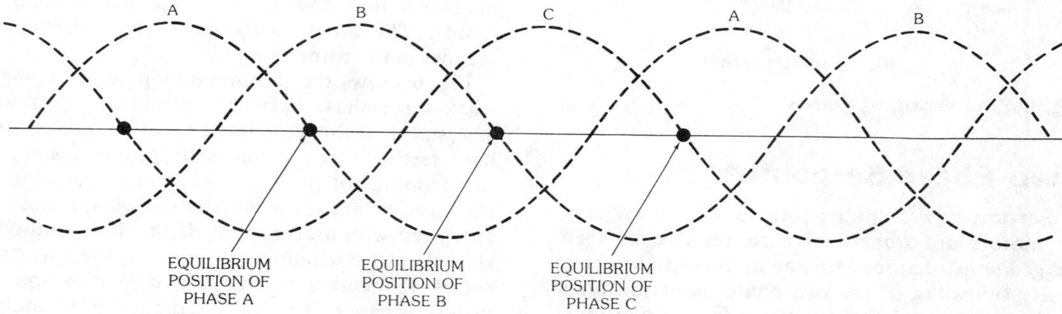

Fig. 8. Individual static holding torque curves of a three-phase step motor.

$$e_i(t) = R_i i_i(t) + L_i(\theta)\frac{di_i}{dt} + i_i \frac{d}{d\theta} L_i(\theta) \frac{d\theta}{dt}$$

where,

$R_i i_i(t) = IR$ drop in the phase windings,

$L_i(\theta)\dfrac{di_i}{dt}$ = transformer electromotive force,

$i_i \dfrac{d}{d\theta} L_i(\theta) \dfrac{d\theta}{dt}$ = back emf due to rotation of rotor.

The developed torque of phase i is written as

$$T_i = \frac{1}{2}\, i_i^2(t) \frac{d}{d\theta} L_i(\theta)$$

For a three-phase motor,

$$L_A(\theta) = L_1 + L_2 \cos N_r\theta$$
$$L_B(\theta) = L_1 + L_2 \cos (N_r\theta - 120°)$$
$$L_C(\theta) = L_1 + L_2 \cos (N_r\theta - 240°)$$

where,

N_r is the number of rotor teeth,

L_1 and L_2 are inductance components that are generally dependent on the phase currents.

Thus, for phase A,

$$T_A = -\frac{1}{2} L_2 N_r i_A^2(t) \sin N_r\theta$$
$$= -K i_A^2(t) \sin N_r\theta$$

which agrees with the torque curve in Fig. 7.

In general, as demonstrated by the above equations, the dynamics of a step motor are highly nonlinear. Therefore, transfer functions are not used to model step motors, except in certain very simple cases.

Fig. 10 illustrates the schematic diagrams of a hybrid permanent-magnet step motor. The permanent magnet located in the rotor creates a magnetic field in addition to the magnetic field caused by the excitations of the stator windings. As shown in Fig. 10, the teeth on the two rotor end sections are displaced by one-half of the rotor tooth pitch with respect to each other.

Rate Generators

A rate generator (or tachometer generator) is a precision electromechanical component resembling a small motor and having an output voltage proportional to its shaft rotational speed. Rate generators have extensive applications both as computing instruments and as stabilizing components of feedback control systems. An example of the latter is illustrated in Fig. 11. The use of the rate generator produces an effective viscous damping and also tends to linearize the servomechanism by inserting damping of a linear nature and of such magnitude that it swamps out the rather large nonlinear damping of the motor. To eliminate the backlash between rate generators and servomotors, they are often constructed as integral units having a common shaft. These units are available for dc or ac (either 400- or 60-hertz) operation.

TYPES OF LINEAR SYSTEMS

The various types of feedback control systems can be described most effectively in terms of the simple closed-loop direct feedback system. Fig. 12 shows such a system; $R(s)$, $C(s)$, and $E(s)$ are the Laplace transforms of the reference input, controlled variable, and error signal, respectively.

For a typical linear system, $G(s)$ might appear as

$$C(s)/E(s) = G(s) = [K(T_1 s + 1)(T_3 s + 1)]/[s^n(T_2 s + 1)(T_4 s + 1)]$$

The value of exponent n, an integer, designates the type of the system. This in turn reveals the nature of the steady-state performance of the system as outlined below.

Type-0 System

A constant value of the controlled variable requires a constant error signal under steady-state conditions. A feedback control system of this type is generally referred to as a regulator system.

Type-1 System

A constant rate of change of the controlled variable requires a constant error signal under steady-state conditions. A type-1 feedback control system is generally referred to as a servomechanism system. For reference inputs that change with time at a constant rate, a constant error is required to produce the same steady-state rate of the controlled variable. When applied to positions control, type-1 systems may also be referred to as "zero-displacement-error" systems. Under steady-state conditions, it is possible for the reference signal to have any desired constant position or displacement and the feedback signal or controlled variable to have the same displacement.

Type-2 System

A constant acceleration of the controlled variable requires a constant error under steady-state conditions for a type-2 system. Since these systems can maintain a constant value of controlled variable and a constant controlled variable speed with no actuating error, they are sometimes referred to as "zero-velocity-error" systems.

STATE-VARIABLE ANALYSIS OF LINEAR SYSTEMS

State-variable methods represent a modern approach to the analysis and design of control sys-

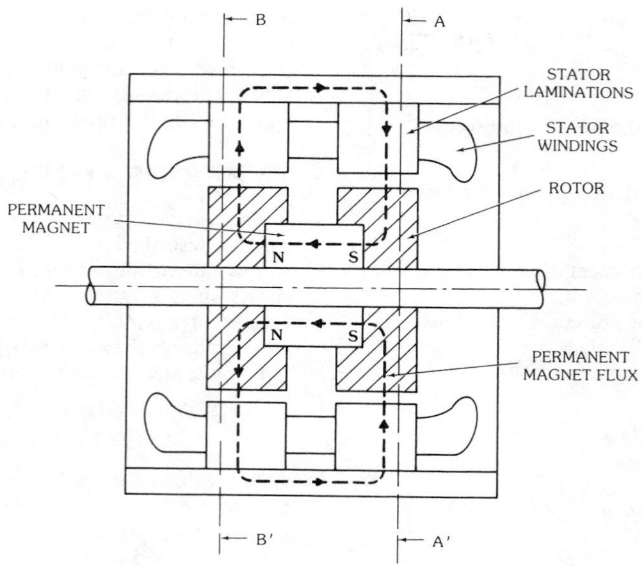

(A) Axial view.

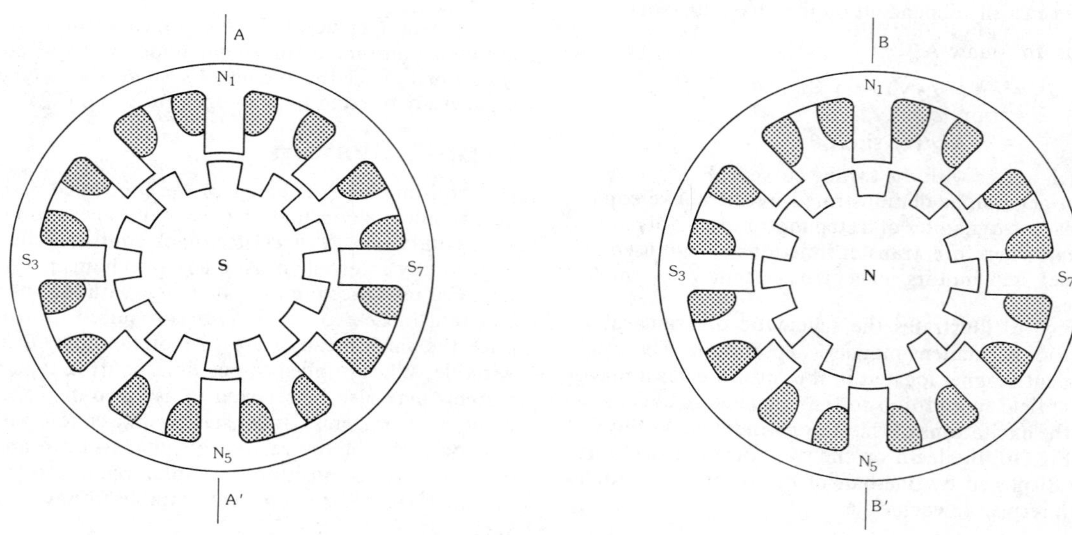

(B) Cross-sectional views.

Fig. 10. Diagrams of a hybrid pm step motor.

tems. For an nth order dynamic system, the n state equations may be written as

$$\frac{dx_i(t)}{dt} = f_i[x_1(t), x_2(t), \ldots, x_n(t), r_1(t), r_2(t), \ldots, r_p(t)]$$

where,

$i = 1, 2, \ldots, n,$

$x_1(t), x_2(t), \ldots, x_n(t)$ denotes the n state variables,

$r_1(t), r_2(t), \ldots, r_p(t)$ denotes the p inputs of the system.

Let $c_1(t), c_2(t), \ldots, c_q(t)$ be the q output variables of the system. Then, the output equations are

$$c_j(t) = g_j[x_1(t), x_2(t), \ldots, x_n(t), r_1(t), r_2(t), \ldots, r_p(t)]$$

where $j = 1, 2, \ldots, q$.

The state of n state equations and the q output equations together form the so-called dynamic equations.

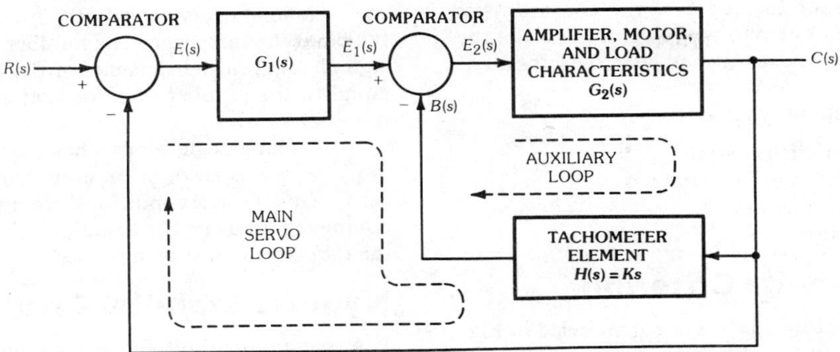

Fig. 11. Positioning-type servo.

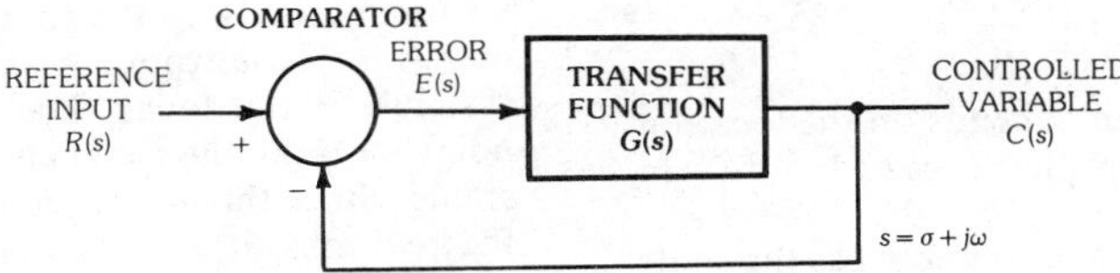

Fig. 12. Single-loop system.

For linear time-invariant systems, the dynamic equations may be written in the following vector-matrix form:

State equations

$$d\mathbf{x}(t)/dt = \mathbf{A}\mathbf{x}(t) + \mathbf{B}\mathbf{r}(t)$$

Output equations

$$\mathbf{c}(t) = \mathbf{D}\mathbf{x}(t) = \mathbf{E}\mathbf{r}(t)$$

where,

$$\mathbf{x}(t) = \begin{bmatrix} x_1(t) \\ x_2(t) \\ \vdots \\ x_n(t) \end{bmatrix} = \text{state vector}$$

$$\mathbf{r}(t) = \begin{bmatrix} r_1(t) \\ r_2(t) \\ \vdots \\ r_p(t) \end{bmatrix} = \text{input vector}$$

$$\mathbf{c}(t) = \begin{bmatrix} c_1(t) \\ c_2(t) \\ \vdots \\ c_q(t) \end{bmatrix} = \text{output vector}$$

$$\mathbf{A} = \begin{bmatrix} a_{11} & a_{12} \ldots a_{1n} \\ a_{21} & a_{22} \ldots a_{2n} \\ \vdots & \vdots \\ a_{n1} & a_{n2} \ldots a_{nn} \end{bmatrix} \; n \times n \text{ coefficient matrix}$$

$$\mathbf{B} = \begin{bmatrix} b_{11} & b_{12} \ldots b_{1p} \\ b_{21} & b_{22} \ldots b_{2p} \\ \vdots & \vdots \\ b_{n1} & b_{n2} \ldots b_{np} \end{bmatrix} \; n \times p \text{ coefficient matrix}$$

The solutions of the state equations are written as

$$\mathbf{x}(t) = \boldsymbol{\phi}(t - t_0)\mathbf{x}(t_0) + \int_{t_0}^{t} \boldsymbol{\phi}(t - \tau)\mathbf{B}\mathbf{r}(\tau)d\tau$$

where the input $\mathbf{r}(t)$ is applied at $t = t_0$, and $\mathbf{x}(t_0)$ denotes the initial state vector. The $n \times n$ matrix $\boldsymbol{\phi}(t)$ is defined as the state transition matrix, and is written as

$$\boldsymbol{\phi}(t) = \mathbf{I} + \mathbf{A}t + \frac{\mathbf{A}^2 t^2}{2!} + \ldots \frac{\mathbf{A}^k t^k}{k!} + \ldots$$

where $\mathbf{I}$ is the $n \times n$ identity matrix.

The state transition matrix has the following properties:

$$\boldsymbol{\phi}(0) = \mathbf{I}$$
$$\boldsymbol{\phi}^{-1}(t) = \text{matrix inverse of } \boldsymbol{\phi}(t) = \boldsymbol{\phi}(-t)$$
$$\boldsymbol{\phi}(t_2 - t_1)\boldsymbol{\phi}(t_1 - t_0) = \boldsymbol{\phi}(t_2 - t_0)$$

for any t_0, t_1, t_2

$$\boldsymbol{\phi}(t) = \mathcal{L}^{-1}[s\mathbf{I} - \mathbf{A})^{-1}]$$

where $\mathcal{L}^{-1}$ denotes the "inverse Laplace transform of." The characteristic equation of $\mathbf{A}$ is defined as

$$\Delta = \text{determinant of } s\mathbf{I} - \mathbf{A} = |s\mathbf{I} - \mathbf{A}|$$

STABILITY OF LINEAR TIME-INVARIANT SYSTEMS

A linear system is stable if its output is bounded for any bounded input. The stability criteria can be stated in terms of the roots of the characteristic equation. For a system to be stable, the roots of the characteristic equation must all lie in the left half of the s-plane.

Methods of stability analysis:

1. Routh-Hurwitz criterion
2. Nyquist criterion
3. Root locus diagram
4. Bode diagram

Routh-Hurwitz Criterion

The stability of the linear system modeled in Fig. 12 can be investigated by referring to the closed-loop transfer function

$$\frac{C(s)}{R(s)} = \frac{G(s)}{1 + G(s)}$$

The stability of the system depends on the location of the poles of $C(s)/R(s)$ or the zeros of $1 + G(s)$ in the complex s-plane.

The zeros of $1 + G(s)$ are also known as the roots of the characteristic equation, which can be written

$$D = \sum_{i=0}^{n} a_i s^i$$

where all the coefficients are real.

The necessary conditions for the last equation to have no roots on the imaginary axis or in the right half of the s-plane are that all the coefficients of the equation must be of the same sign and that none of the coefficients is zero.

To check the necessary and sufficient conditions, we form the following tabulation:

$$\begin{array}{cccccccc}
a_n & a_{n-2} & a_{n-4} & a_{n-6} & \cdot & \cdot & \cdot \\
a_{n-1} & a_{n-3} & a_{n-5} & a_{n-7} & \cdot & \cdot & \cdot \\
b_1 & b_2 & b_3 & b_4 & \cdot & \cdot & \cdot \\
c_1 & c_2 & c_3 & c_4 & \cdot & \cdot & \cdot \\
d_1 & d_2 & d_3 & \cdot & \cdot & \cdot & \cdot \\
e_1 & e_2 & \cdot & \cdot & \cdot & \cdot & \cdot \\
f_1 & \cdot & \cdot & \cdot & \cdot & \cdot & \cdot \\
\cdot & \cdot & \cdot & \cdot & \cdot & \cdot & \cdot \\
\cdot & \cdot & \cdot & \cdot & \cdot & \cdot & \cdot \\
\cdot & \cdot & \cdot & \cdot & \cdot & \cdot & \cdot
\end{array}$$

where,

$$b_1 = (a_{n-1}a_{n-2} - a_{n-3}a_n)/a_{n-1}$$
$$b_2 = (a_{n-1}a_{n-4} - a_{n-5}a_n)/a_{n-1}$$
$$b_3 = (a_{n-1}a_{n-6} - a_{n-7}a_n)/a_{n-1}$$
$$c_1 = (b_1a_{n-3} - b_2a_{n-1})/b_1$$
$$c_2 = (b_1a_{n-5} - b_3a_{n-1})/b_1$$
$$c_3 = (b_1a_{n-7} - b_4a_{n-1})/b_1$$
$$d_1 = (c_1b_2 - b_1c_2)/c_1$$
$$d_2 = (c_1b_3 - b_1c_3)/c_1$$
$$d_3 = (c_1b_4 - b_1c_4)/c_1$$
$$\vdots$$

The table will consist of n rows.

The system is stable, *i.e.*, the polynominal has no zeros on the imaginary axis or in the right half of the s-plane, if every entry in the first column of the table has the same sign. The number of consecutive sign changes in the elements of the first column is equal to the number of zeros that are in the right half-plane.

A singular case develops when the first element in any one row is zero, or an entire row contains all zeros. Under such conditions, the table cannot be completed by using the equations given above, and the tabulation must be modified.

Nyquist Stability Criterion

A second method for determining stability is known as the Nyquist stability criterion. This method consists in obtaining the locus of the transfer function $G(s)$ in the complex G plane for values of $s = j\omega$ for ω from $-\infty$ to $+\infty$. For single-loop systems, if the locus thus described encloses the point $-1 + j0$, the system is unstable; otherwise it is stable. Since the locus is always symmetrical about the real axis, it is sufficient to draw the locus for positive values of ω only. Fig. 13 shows loci for several simple systems. Curves A and C represent stable systems and are typical of the type-1 system; curve B is an unstable system. Curve D is conditionally stable; that is, for a particular range of values of gain K it is unstable. The system is stable both for larger and smaller values of gain. *Note:* It is unstable as shown.

Phase margin θ_p and gain margin g are also illustrated in Fig. 13A. The former is the angle between the negative real axis and $G(j\omega)$ at the point where the locus intersects the unit-gain circle. It is positive when measured as shown.

Gain margin g is the negative dB value of $G(j\omega)$ corresponding to the frequency at which the phase angle is 180° (i.e., where $G(j\omega)$ intersects the negative real axis). The gain margin is often expressed in decibels, so that $g = -20 \times \log_{10} G(j\omega)$. Typical satisfactory values are -10 dB for g and an angle of 30° for θ_p. These values are selected on the basis of a good compromise between speed of response and reasonable overshoot. Note that for conditionally stable systems, the terms gain margin and phase margin are without their usual significance.

Logarithmic Plots (Bode Diagrams)

The transfer function of a feedback control system can be described by separate plots of attenuation and phase versus frequency. This provides a very simple method for constructing a Nyquist diagram from a given transfer function. Use of a logarithmic frequency scale permits simple straight-line (asymptotic) approximations for each curve. Fig. 14 illustrates the method for a transfer function with a single time constant. A comparison between approximate and actual values is included.

Transfer functions of the form $G = (1 + j\omega T)$

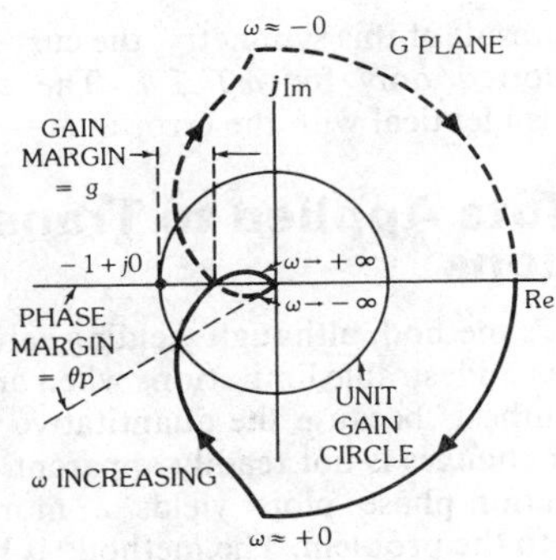

(A) Stable system.

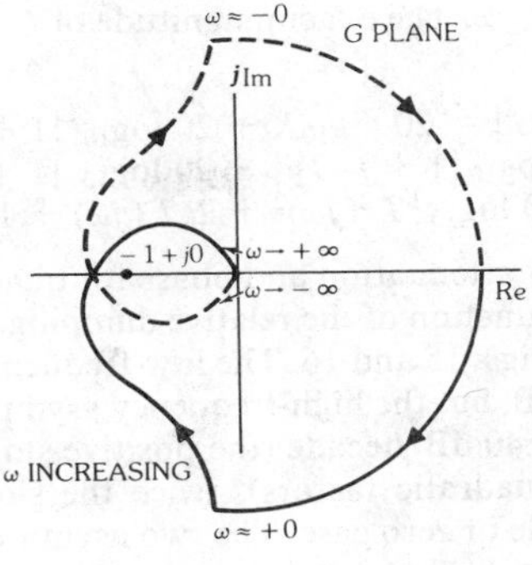

(B) Unstable system.

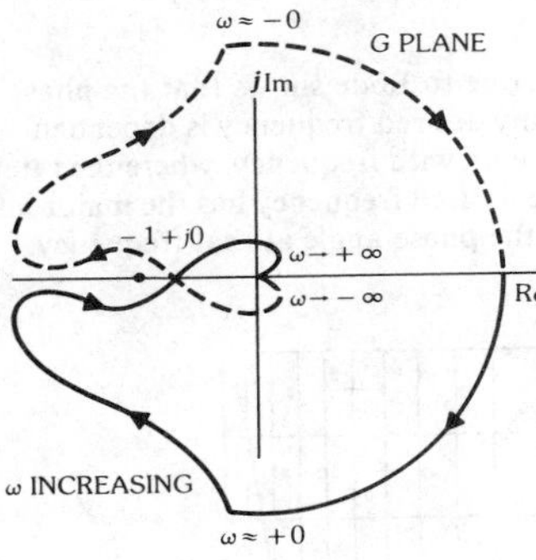

(C) Stable system.

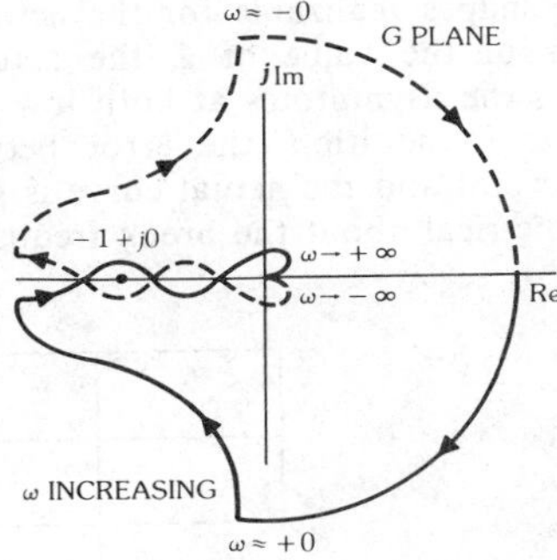

(D) Conditionally stable system.

Fig. 13. Typical Nyquist loci.

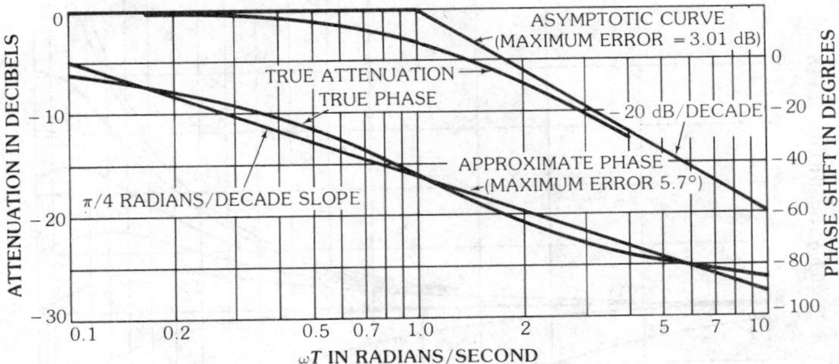

Fig. 14. Transfer-function plot. $G(j\omega) = 1/(1 + j\omega T)$.

have similar approximations except that the attenuation curve slope is inverted upward (+ 20 dB/decade) and the values of phase shift are positive.

The transfer function of feedback control systems can often be expressed as a fraction with the numerator and denominator each composed of linear factors of the form $(Ts + 1)$. Certain types of control systems, such as hydraulic motors where compressibility of the oil in the pipes is appreciable or some steering problems where the viscous damping is small, give rise to transfer functions in which quadratic factors occur in addition to the linear factors. The process of taking logarithms (as in making a dB plot) facilitates computation because only the addition of product terms is involved. The associated phase angles are directly additive.

For example,

$$G(j\omega) = \frac{K(1 + j\omega T_2)}{[T^2(j\omega)^2 + 2\zeta T(j\omega) + 1]\,(1 + j\omega T_1)\,(1 + j\omega T_3)}$$

where $s = j\omega$. The exact magnitude of G in decibels is

$$20 \log_{10} |G| = 20 \log_{10} K + 20 \log_{10} |1 + j\omega T_2| - 20 \log_{10} |1 + j\omega T_1| - 20 \log_{10} |1 + j\omega T_3| - 20 \log_{10} |T^2(j\omega)^2 + 2\zeta T(j\omega) + 1|$$

Plots of attenuation and phase for quadratic factors as a function of the relative damping ratio ζ are given in Figs. 15 and 16. The low-frequency asymptote is 0 dB, but the high-frequency asymptote has a slope of ±40 dB/decade (the positive slope applies to zero quadratic factors), twice the slope of the simple pole or zero case. The two asymptotes intersect at $\omega = 1/T$.

The difference between the asymptotic plot and the actual curves depends on the value of ζ with a variety of shapes realizable for the actual curve. Regardless of the value of ζ, the actual curve approaches the asymptotes at both low and high frequencies. In addition, the error between the asymptotic plot and the actual curve is geometrically symmetrical about the break frequency $\omega = 1/T$. As a result of this symmetry, the curves of Fig. 15 are plotted only for $\omega T \leq 1$. The error for $\omega = \alpha/T$ is identical with the error at $\omega = 1/\alpha T$.

Log Plots Applied to Transfer Functions

Nyquist's method, although yielding satisfactory results, has undesirable limitations when applied to system synthesis because the quantitative effect of parameter changes is not readily apparent. The use of attenuation-phase plots yields a more direct approach to the problem. The method* is based on the relation between phase and the rate of change of gain with frequency of networks. As a first approximation, which is valid for simple systems, a gain

* A theorem due to Bode shows that the phase angle of a network at any desired frequency is dependent on the rate of change of gain with frequency, where the rate of change of gain at the desired frequency has the major influence on the value of the phase angle at that frequency.

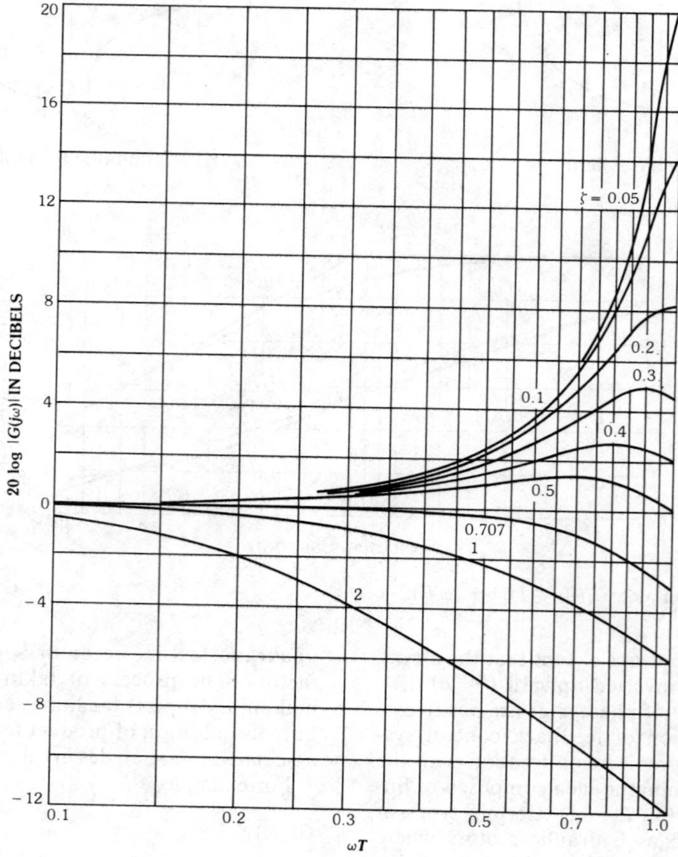

Fig. 15. Attenuation curve for quadratic factor. $G(j\omega) = 1/[T^2(j\omega^2) + 2\zeta T(j\omega) + 1]$. (*By permission from* Automatic Feedback Control System Synthesis, *by J. G. Truxal. Copyright 1955, McGraw-Hill Book Co.*)

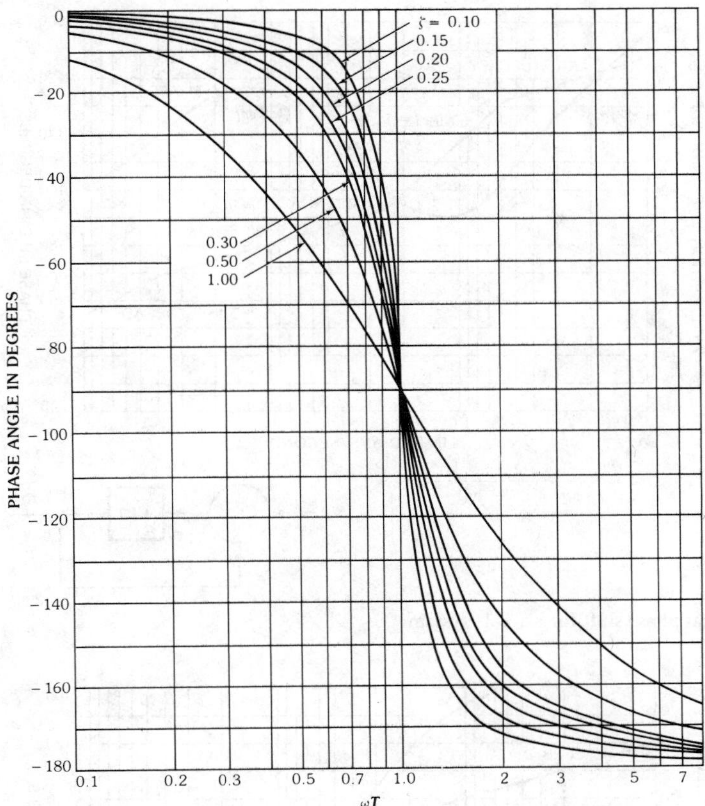

Fig. 16. Phase characteristic. (*By permission from* Theory of Servomechanisms, *by H. M. James, N. B. Nichols, and R. S. Phillips. Copyright 1947, McGraw-Hill Book Co.*)

rate of change of 20 dB/decade corresponds to a phase shift of 90°. Since the stability of a system can be determined from its phase margin at unity gain (0 dB), simple criteria for the slope of the attenuation curve can be established. Thus it is obvious that to avoid instability, the slope of the attenuation curve at unity gain must be appreciably less than −40 dB/decade (commonly about −33 dB/decade).

The design procedure is to construct asymptotic attenuation-phase curves as a first approximation. From this it can be determined whether the stability requirements are met. Refinements can be made by using the actual instead of asymptotic values for the curve as outlined in Fig. 14.

Figs. 17 and 18 are examples of transfer functions plotted in this manner. In Fig. 17, a positive phase margin exists, and the system is stable. Associated with the first-order pole at the origin is a uniform (low-frequency) slope of − 20 dB/decade and −90° phase shift. This may be considered characteristic of the integrating action of a type-1 control system. Fig. 18 represents an unstable system. It has a negative phase margin (as a result of the steep slope of the attenuation curve). The former is stable, the latter is unstable.

Root-Locus Method

Root locus is a method of design due to Evans, based on the relation between the poles and zeros of the closed-loop system function and those of the open-loop transfer function. The rapidity and ease with which the loci can be constructed form the basis for the success of root-locus design methods, in much the same way that the simplicity of the gain and phase plots (Bode diagrams) makes design in the frequency domain so attractive. The root-locus plots can be used to adjust system gain, guide the design of compensation networks, or study the effects of changes in system parameters.

In the usual feedback control system, $G(s)$ is a rational algebraic function, the ratio of two polynomials in s; thus

$$G(s) = m(s)/n(s)$$

From Fig. 12

$$(C/R)\ (s) = G(s)/[1 + G(s)]$$

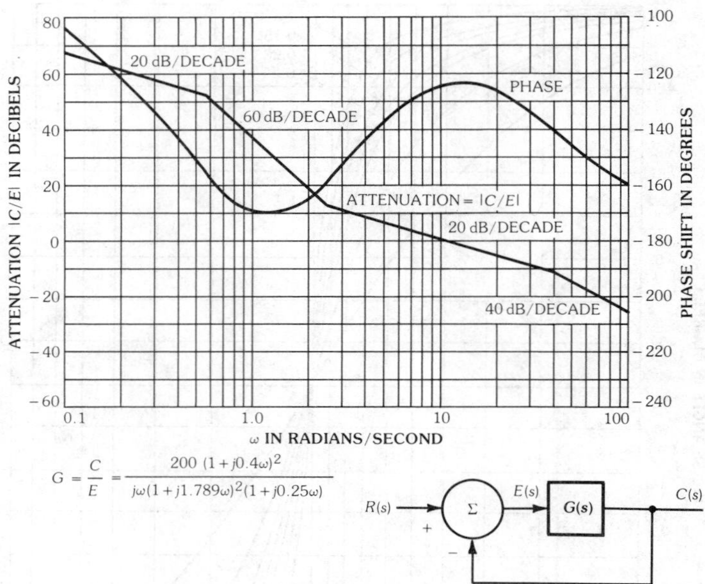

Fig. 17. Attenuation and phase shift for a stable system.

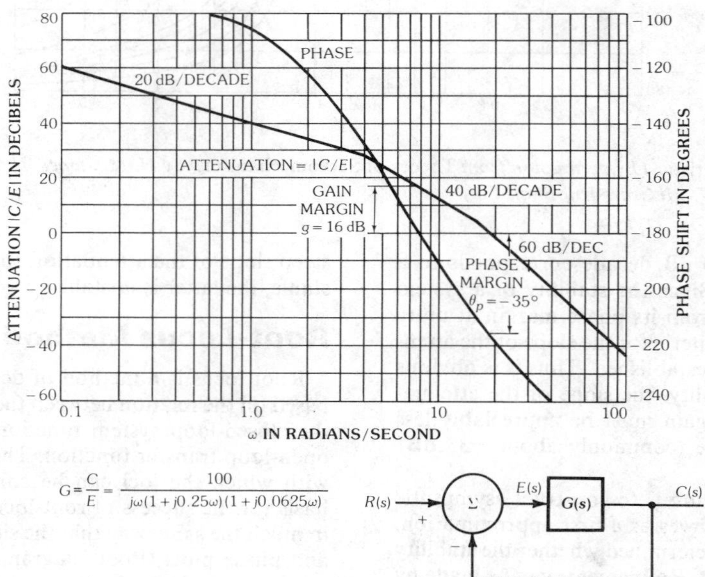

Fig. 18. Attenuation and phase shift for an unstable system.

$$= \frac{m(s)/n(s)}{1 + [m(s)/n(s)]}$$
$$= m(s)/[m(s) + n(s)]$$

The zeros of the closed-loop system are identical with those of the open-loop system function.

The closed-loop poles are the values of s at which $m(s)/n(s) = -1$. The root-locus method is a graphic technique for determination of the zeros of $m(s) + n(s)$ from the zeros of $m(s)$ and $n(s)$. Root loci are plots in the complex s plane of the variations of the poles of the closed-loop system function with changes in the open-loop gain. For the single-loop system of Fig. 12, the root loci constitute all s-plane points at which

$$\angle \mathbf{G(s)} = 180° + n\,360°$$

where n is any integer including zero. For a type-1 feedback control system

$$G(s) = \frac{K(s + z_1)(s + z_2)}{s(s + p_1)(s + p_2)(s + p_3)}$$

A graphic interpretation is given in Fig. 19. Examples are given in Figs. 20 and 21.

Gain K_1, Fig. 21, produces the case of critical damping. An increase in gain somewhat beyond this value causes a damped oscillation to appear. The latter increases in frequency (and decreases in damping) with further increase in gain. At gain K_3, a sustained oscillation will result. Instability exists for gain greater than K_3, as at K_4. This corresponds to poles in the right half of the s plane for the closed-loop transfer function.

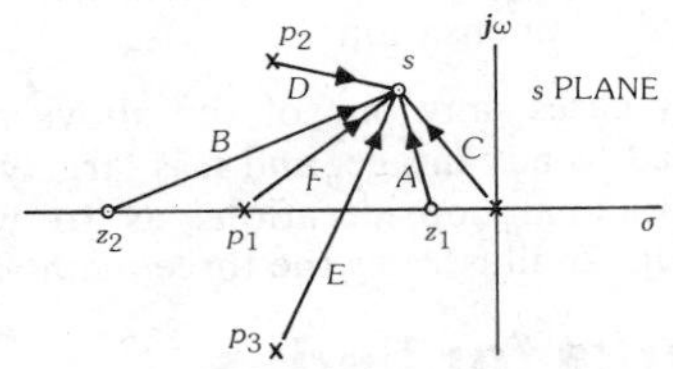

Fig. 19. Graphic interpretation of $G(s)$.

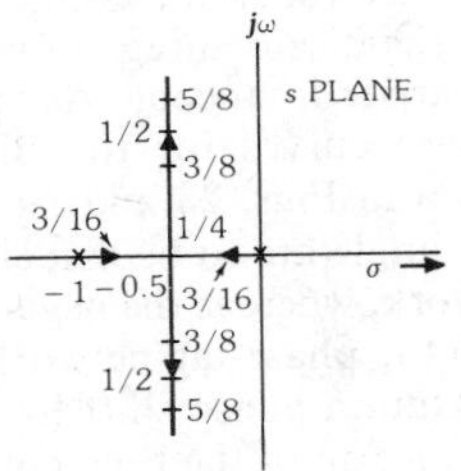

Fig. 20. Root loci for $G(s) = K/[s(s + 1)]$. Values of K as indicated by fractions.

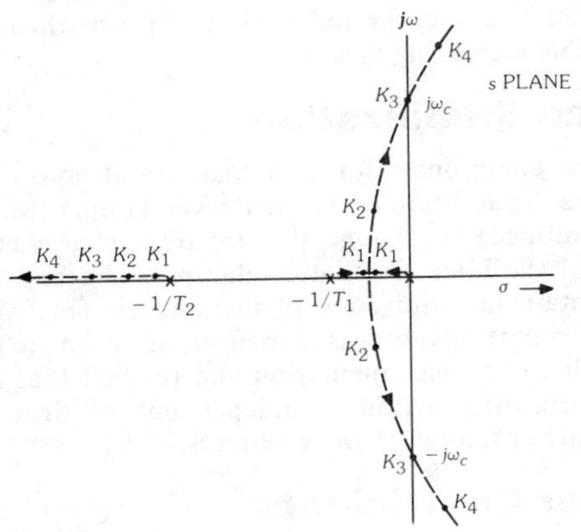

Fig. 21. Root loci for $G(s) = K/[s(T_1s + 1)(T_2s + 1)]$.

Aids in Sketching Root-Locus Plots

Intervals Along the Real Axis—The simplest portions of the plot to establish are the intervals along the negative real ($-\sigma$) axis, because then all angles are either 0° or 180°.

Complex pairs of zeros or poles contribute no net angle for points along the real axis.

Along the real axis, the locus will exist for intervals that have an *odd* number of zeros and poles to the right of the interval (Fig. 22).

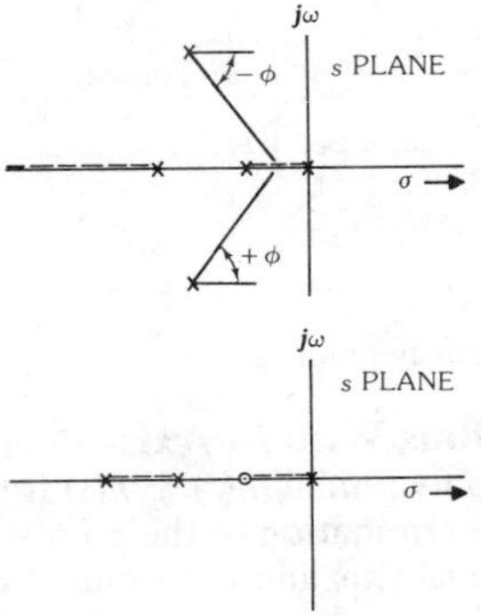

Fig. 22. Root-locus intervals along the real axis.

Asymptotes—For very large values of s, all angles are essentially equal. The locus will thus finally approach asymptotes at the angles (Fig. 23) given by the expression

$$\frac{180 + n\,360^\circ}{(\text{poles}) - (\text{zeros})}$$

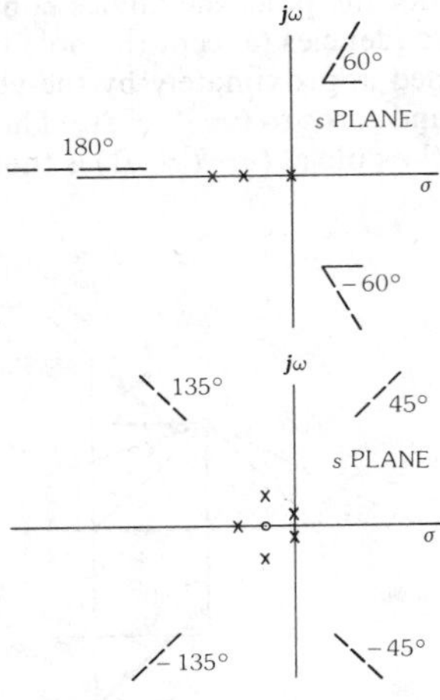

Fig. 23. Final asymptotes for root loci. Top, 60° asymptotes for system having three poles. Bottom, 45° asymptotes for system having an excess of four poles over zeros.

These asymptotes meet at a point s_1 (on the negative real axis) given by

$$s_1 = \frac{\sum(\text{poles}) - \sum(\text{zeros})}{(\text{finite poles}) - (\text{finite zeros})}$$

Breakaway Points—Breakaway points from the real axis occur where the net change in angle caused by a small vertical displacement is zero. In Fig. 24, point p satisfies this condition at $1/x_0 = (1/x_1) + (1/x_2)$.

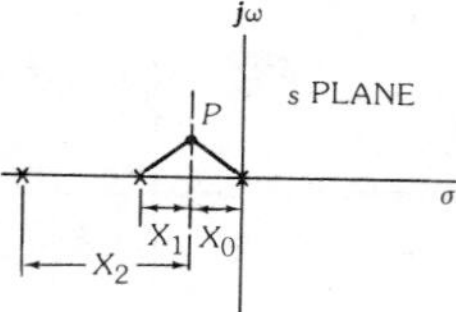

Fig. 24. Breakaway point.

Intersections With $j\omega$ Axis—Routh's test applied to the polynomial $m(s) + n(s)$ frequently permits rapid determination of the points at which the loci cross the $j\omega$ axis and the value of gain at these intersections.

Angles of Departure and Arrival—The angles at which the loci leave the poles and arrive at the zeros are readily evaluated from

$$\underline{\sum\angle\text{vectors from zeros to } s} - \underline{\sum\angle\text{vectors from poles to } s} = 180° + n\ 360°$$

For example, consider Fig. 25. The angle of departure of the locus from the pole at $(-1 + j1)$ is desired. If a test point is assumed only slightly displaced from the pole, the angles contributed by all critical frequencies (except the pole in question) are determined approximately by the vectors from these poles and zeros to $(-1 + j1)$. The angle contributed by the pole at $(-1 + j1)$ is then just sufficient to make the total angle 180°. In the example shown in the figure, the departure angle is found from the relation

$$\underbrace{+45°}_{s+2} - (\underbrace{135°}_{s} + \underbrace{90°}_{s+1+j1} + \underbrace{26.6°}_{s+3} + \underbrace{\theta}_{s+1-j1}) = 180° + n\,360°$$

Hence, $\theta = -26.6°$, the angle at which the locus leaves $(-1 + j1)$.

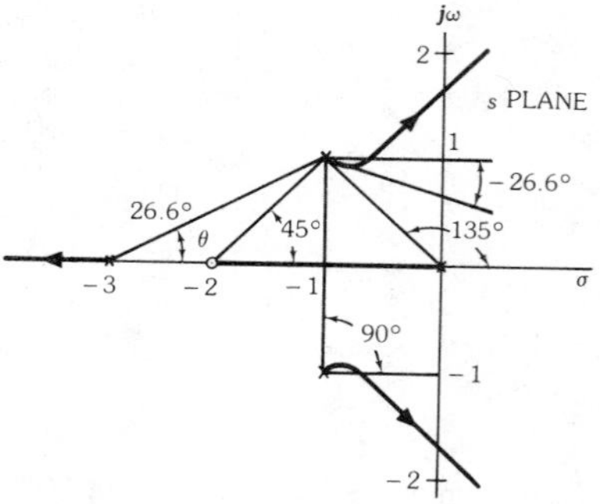

Fig. 25. Loci for $G(s) = K(s + 2)/[s(s + 3)(s^2 + 2s + 2)]$.

METHODS OF STABILIZATION

Methods of stabilization for improving feedback-control-system response fall into the following basic categories:

(A) Series (cascade) compensation
(B) Feedback (parallel) compensation
(C) Load compensation

In many cases, any one of the above methods may be used to advantage, and it is largely a question of practical considerations as to which is selected. Fig. 26 illustrates the three methods.

Networks for Series Stabilization

Common networks for stabilization are shown in Fig. 27 with the transfer functions. The bridged-*T* network can be used for stabilization of ac systems, although it has the disadvantage of requiring close control of the carrier frequency. Asymptotic attenuation and phase curves for the first three networks are shown in Figs. 28 and 29. The positive values of phase angle are to be associated with the phase-lead network, whereas the negative values are to be applied to the phase-lag network. Fig. 30 is a plot of the maximum phase shift for lag and lead networks as a function of the time-constant ratio.

Instead of direct feedback, the feedback connection may contain frequency-sensitive elements. Typical of such frequency-sensitive elements are tachometers or other rate- or acceleration-sensitive devices that may be fed back directly or through suitable stabilizing means.

Load Stabilization

The commonest form of load stabilization involves the addition of an oscillation damper (tuned or untuned) to change the apparent characteristics of the load. Oscillation dampers can be used to obtain the equivalent of tachometric feedback. The primary advantages of load stabilization are the simplicity of instrumentation and the fact that the compensating action is independent of drift of the carrier frequency in ac systems.

Error Coefficients

Of major importance in feedback control systems, along with stability, is system accuracy. *Static*

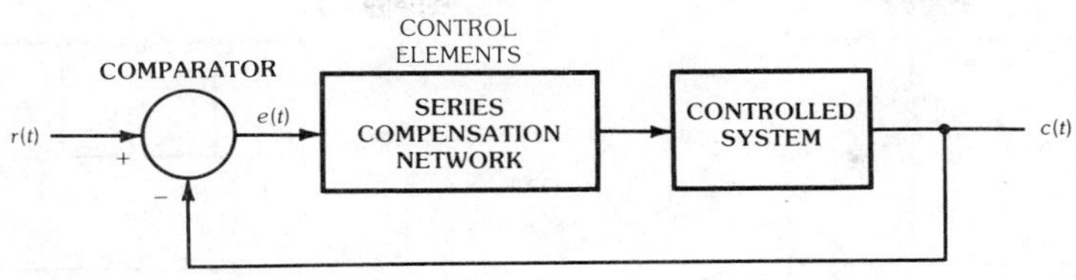

(A) Series compensation.

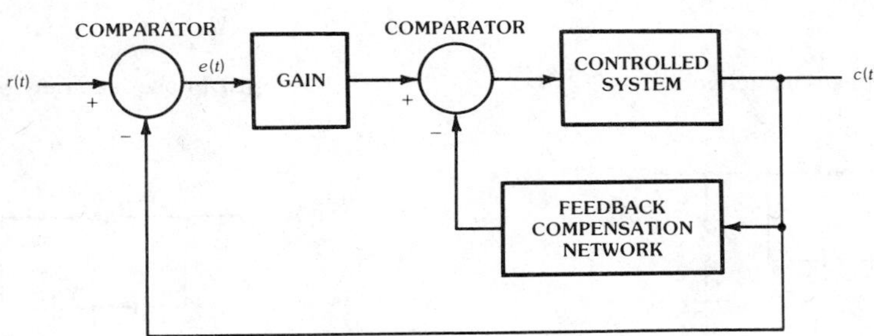

(B) Feedback compensation.

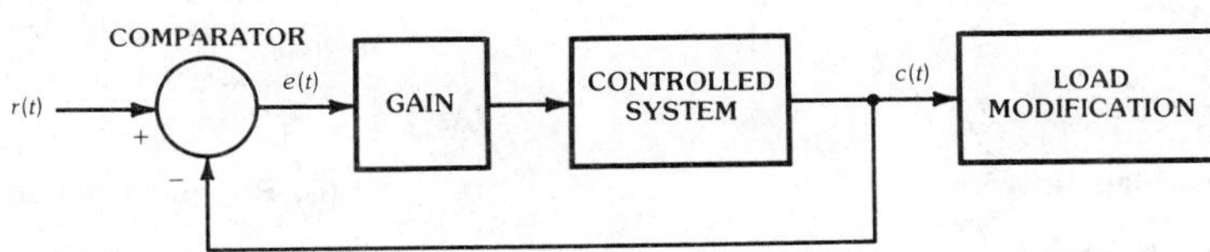

(C) Load compensation.

Fig. 26. Simple schemes for compensation.

accuracy refers to the accuracy of a system after the steady state is reached and is ordinarily measured with the system input constant or slowly varying. *Dynamic accuracy* refers to the ability of the system to follow rapid changes of the input. The following refers to a system such as Fig. 12.

Static-Error Coefficients

Position Error Constant—

$$K_p = \lim_{s \to 0} [C(s)/E(s)] = \lim_{s \to 0} G(s)$$

$$= \text{(controlled variable)}/\text{(actuating error)}$$

for a constant value of controlled variable.

Velocity Error Constant—

$$K_v = \lim_{s \to 0} [sC(s)/E(s)] = \lim_{s \to 0} sG(s)$$

$$= \frac{\text{(velocity of controlled variable)}}{\text{(actuating error)}}$$

for a constant velocity of controlled variable.

Acceleration Error Constant—

$$K_a = \lim_{s \to 0} [s^2C(s)/E(s)] = \lim_{s \to 0} s^2G(s)$$

$$= \frac{\text{(acceleration of controlled variable)}}{\text{(actuating error)}}$$

for constant acceleration of the controlled variable.

MULTIPLE INPUTS AND LOAD DISTURBANCES

Frequently, systems are subjected to unwanted signals entering the system at points other than the input. Examples are load-torque disturbances, noise generated at a point within the system, etc. These may be represented as additional inputs to the system. Fig. 31 is a block diagram of such a condition.

For linear operation

(A) $C/R = G_1G_2/(1 + HG_1G_2)$

(B) $C/U = G_2/(1 + HG_1G_2)$

Combining **(A)** and **(B)**

$$C/U = (1/G_1)(C/R)$$

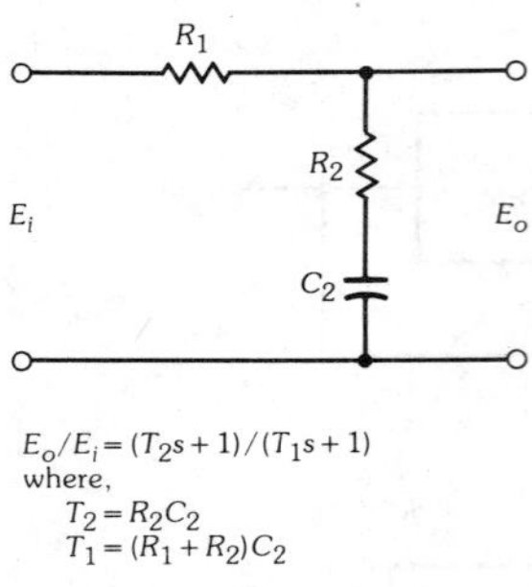

(A) Phase-lag network.

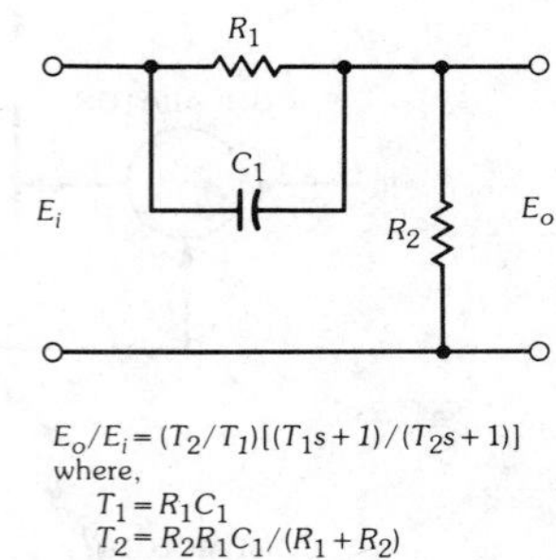

R_1
C_1
E_i
R_2
E_o
$E_o/E_i = (T_2/T_1)[(T_1 s + 1)/(T_2 s + 1)]$
where,
$T_1 = R_1 C_1$
$T_2 = R_2 R_1 C_1/(R_1 + R_2)$

(B) Phase-lead network.

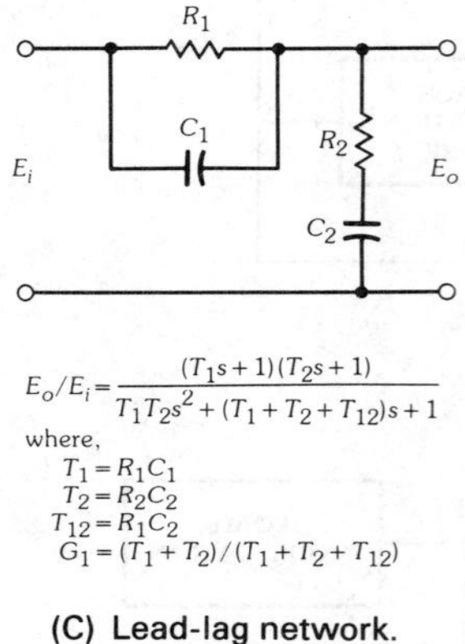

(C) Lead-lag network.

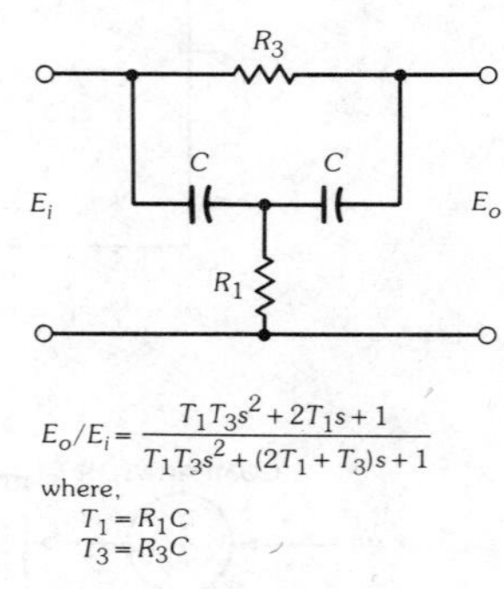

(D) Bridged-T network.

Fig. 27. Networks for stabilization.

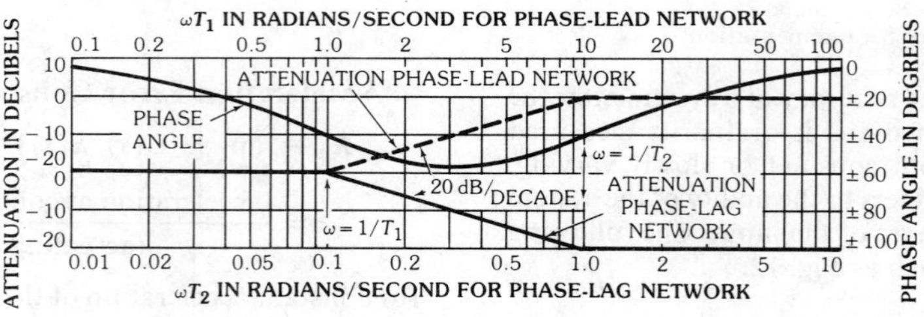

Fig. 28. Phase and attenuation for phase-lead and phase-lag networks. $T_1 = 10T_2$.

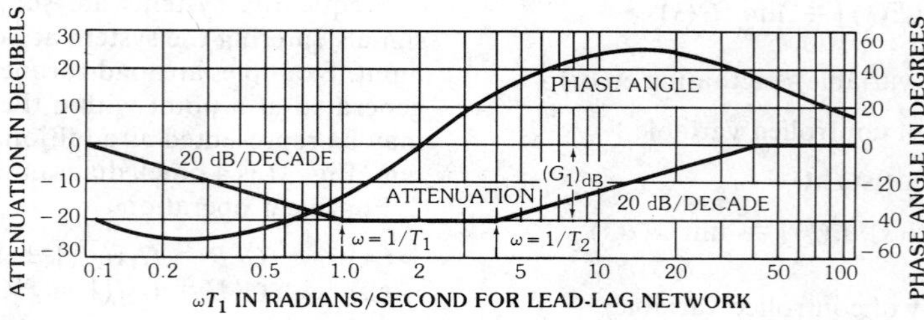

Fig. 29. Phase and attenuation for lead-lag network. $G_1 = (T_1 + T_2)/(T_1 + T_2 + T_{12})$. $T_2 = T_1/4$ and $T_{12} = 11.25T_1$.

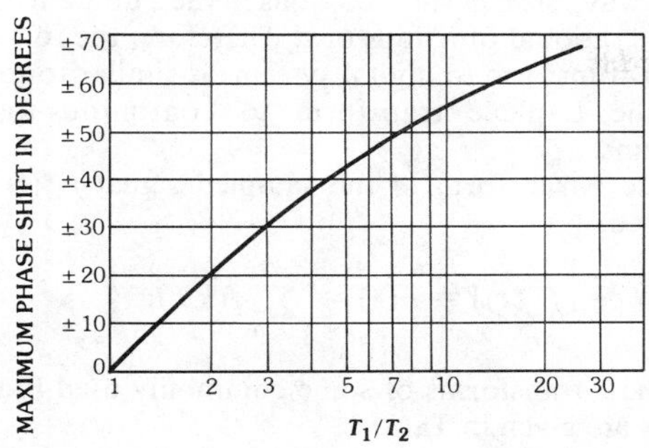

Fig. 30. Maximum phase shift for phase-lead (use positive angles) and phase-lag (negative angles) networks.

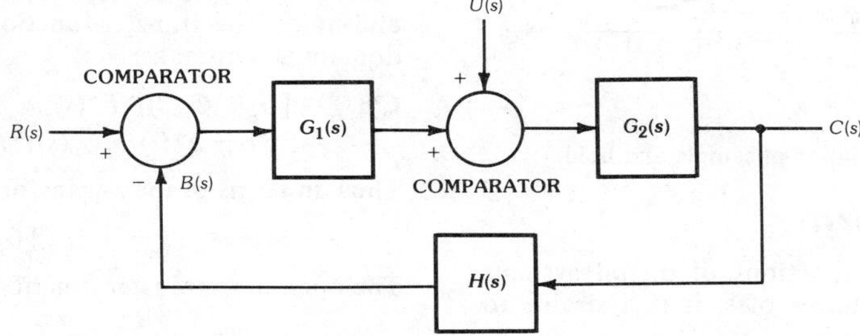

Fig. 31. Multiple-input control system.

If it is desired that the sum of R and U be reproduced in the output (controlled variable), then G_1 should be equal to unity. If U is a disturbance to be minimized, then G_1 should be as large as possible. An example of such a disturbance is the torque produced on a radar antenna by wind forces.

DIGITAL CONTROL SYSTEMS

Digital control systems differ from the conventional continuous-data, or analog, systems in that the signals found in one or more parts of the system are in the form of either pulses or digitally coded signals. Due to the advances made in microcomputers, digital controllers have become very popular in control systems.

Fig. 32 shows the block diagram of a typical digital control system. Typically, the controlled process may be an analog device, but the controller is digital. In this case, the reference input is shown to be an analog signal. The analog-to-digital converter (a/d) converts the analog signal into a digitally coded signal. The output of the digital controller is a digitally coded (such as binary-coded) signal. The digital-to-analog converter (d/a) converts the digital signal into an analog one for the controlled process.

From the analytical standpoint, the a/d operation can be represented by a *sampler* that opens and closes every T second, where T is the sampling period. The d/a operation can be represented by a sample-and-hold device. The block-diagram representation of the sample-and-hold is shown in Fig. 33. The hold device simply holds the output of the sampler for one sampling period, T. Fig. 34 illustrates a typical set of waveforms of the inputs and the outputs of the sampler and the hold device.

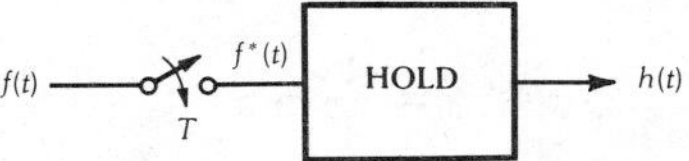

Fig. 33. Sample-and-hold.

The sampler simply defines the values of the function $f(t)$ at the sampling instants. In the time domain, the input-output relation of the sampler is written:

$$f^*(t) = \sum_{k=0}^{\infty} f(kT)\,\delta(t - kT)$$

where $\delta(t)$ is the unit impulse function.

The Laplace transform of the last equation is

$$F^*(s) = \sum_{k=0}^{\infty} f(kT)e^{-kTs}$$

The transfer function of the hold device is

$$G_h(s) = (1 - e^{-Ts})/s$$

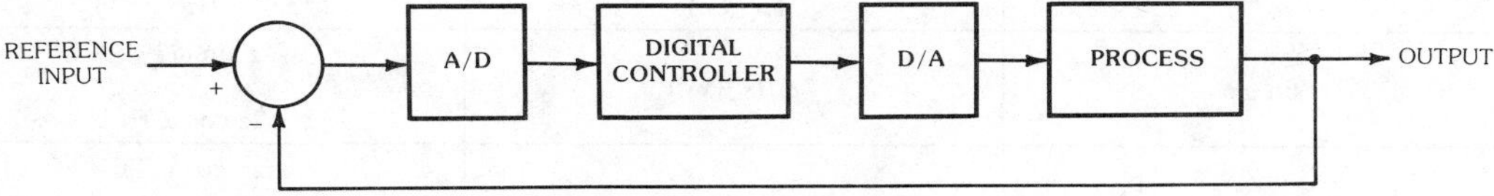

Fig. 32. Typical digital control system.

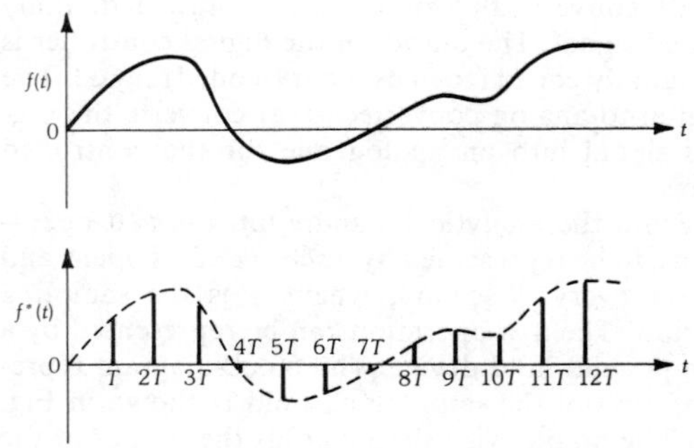

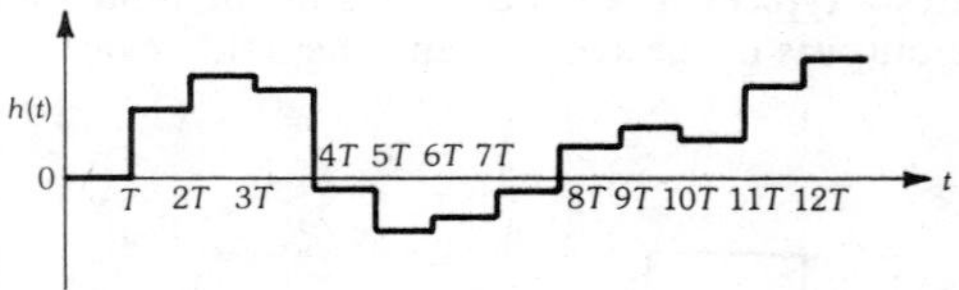

Fig. 34. Input-output relation of sample-and-hold.

The *z*-Transform

Since the transfer functions of digital systems contain exponential terms of s, it is desirable to introduce the z-transform:

$$z = e^{Ts}$$

or

$$s = \frac{1}{\mathrm{T}} \ln z$$

This way, the transfer functions in the z domain will all be rational functions of z. Therefore, the role of the z-transform to digital systems is similar to that of the Laplace transform to continuous-data systems.

The z-transform of the sampled signal $f^*(t)$ is written:

$$\mathfrak{z}\,[f^*(t)] = F(z) = \sum_{k=0}^{\infty} f(kT) z^{-k}$$

The z-transforms of some commonly used functions are given in Table 1.

The *z*-Transfer Function

Fig. 35 shows the block diagram of a typical open-loop digital control system with a sample-and-hold. The transfer-function relation in the s-domain is written as:

$$C^*(s) = [G_h(s)G_p(s)]^* F^*(s)$$
$$= \{[(1 - e^{-Ts})/s]G_p(s)\}^* F^*(s)$$

Thus, in terms of the z-transform,

$$C(z) = (1 - z^{-1})\,\mathfrak{z}\,[G_p(s)/s]F(z)$$

The open-loop transfer function is defined as

$$\frac{C(z)}{F(z)} = (1 - z^{-1})\,\mathfrak{z}\,[G_p(s)/s]$$

For the closed-loop digital control system shown in Fig. 36, the closed-loop transfer function is

$$C(z)/\mathrm{R}(z) = G(z)/[1 + G(z)]$$

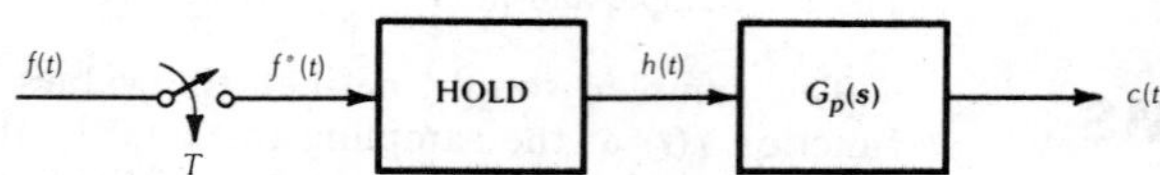

Fig. 35. Open-loop digital control system.

TABLE 1. *z*-TRANSFORMS

Time Function	Laplace Transform	*z*-Transform
$u(t)$ Unit-step function	$1/s$	$z/(z-1)$
t	$1/s^2$	$Tz/(z-1)^2$
$t^2/2$	$1/s^3$	$\frac{T^2z(z+1)}{2(z-1)^3}$
e^{-at}	$1/(s+a)$	$z/(z-e^{-aT})$
te^{-at}	$1/(s+a)^2$	$(Tze^{-aT})/(z-e^{-aT})^2$
$\sin \omega t$	$\omega/(s^2+\omega^2)$	$\frac{z \sin \omega T}{z^2-2z \cos \omega T+1}$
$\cos \omega t$	$s/(s^2+\omega^2)$	$\frac{z(z-\cos \omega T)}{z^2-2z \cos \omega T+1}$

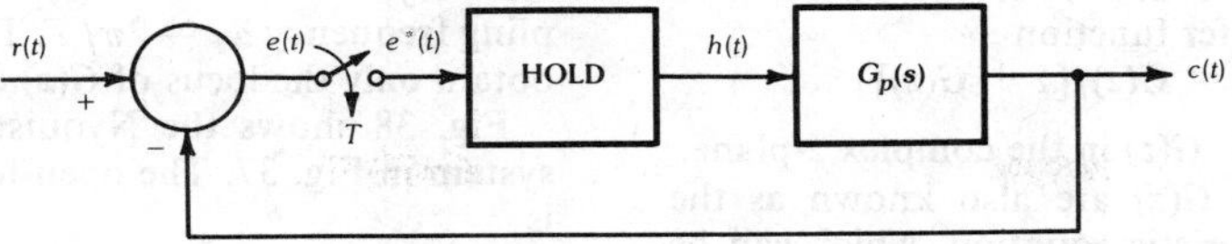

Fig. 36. Closed-loop digital control system.

where,

$$G(z) = (1 - z^{-1})\, \mathfrak{z}\, [G_p(s)/s]$$

The Inverse z-Transform

There are three methods of evaluating the inverse z-transform. These are the partial-fraction expansion method, the power-series method, and the inversion formula method.

Partial-Fraction Expansion Method—The function $F(z)/z$ is expanded into the form

$$F(z)/z = K_1/(z+a) + K_2/(z+b) + \cdots$$

by partial-fraction expansion. Then

$$F(z) = K_1 z/(z+a) + K_2 z/(z+b) + \cdots$$

The inverse z-transform of $F(z)$ is then taken term by term.

Power-Series Method—The function $F(z)$ is expanded into a power series in z^{-1}; *i.e.,*

$$F(z) = \sum_{k=0}^{\infty} f(kT) z^{-k}$$

Then, by the definition of the z-transform, the coefficient of z^{-k}, $k = 0, 1, 2, 3, \ldots$ is $f(kT)$.

The Inversion Formula Method—The inverse z-transform of $F(z)$ can be expressed as the inversion formula

$$f(kT) = \frac{1}{2\pi j} \oint F(z) z^{k-1} dz$$

where the integral is taken over a circle that encloses all the singularities of $F(z)z^{k-1}$. Using the residue theorem of complex variable theory:

$f(kT) = \Sigma$ Residues of $F(z)z^{k-1}$ at the poles of $F(z)z^{k-1}$

State Variable Analysis of Digital Control Systems

The state equations for continuous-data systems can be applied directly to digital systems with the input defined as

$$\mathbf{r}(t) = \mathbf{r}(kT) \qquad kT \le t < (k+1)T$$

due to the action of the sample-and-hold. Thus, the state equations of an nth-order digital system are written as

$$\frac{d\mathbf{x}(t)}{dt} = \mathbf{A}\mathbf{x}(t) + \mathbf{B}\mathbf{r}(kT) \qquad kT \le t < (k+1)T$$

The solution of the state equation, the state transition equation, is

$$\mathbf{x}(t) = \boldsymbol{\phi}(t-kT)\mathbf{x}(kT) + \int_{kT}^{t} \boldsymbol{\phi}(t-\tau)\mathbf{B}\, d\tau \cdot \mathbf{r}(kT)$$

Let

$$\boldsymbol{\theta}(t-kT) = \int_{kT}^{t} \boldsymbol{\phi}(t-\tau)\mathbf{B}\, d\tau$$

Then,

$$\mathbf{x}(t) = \boldsymbol{\phi}(t-kT)\mathbf{x}(kT) + \boldsymbol{\theta}(t-kT)\mathbf{r}(kT)$$

To describe the state variables only at the sampling instants, let $t = (k+1)T$. The last equation becomes

$$\mathbf{x}[(k+1)T] = \boldsymbol{\phi}(T)\mathbf{x}(kT) + \boldsymbol{\theta}(T)\mathbf{r}(kT)$$

which is in the form of a vector-matrix difference equation. The solution of these different equations is found to be

$$\mathbf{x}(NT) = [\boldsymbol{\phi}(T)]^N \mathbf{x}(0) + \sum_{k=0}^{N-1} \boldsymbol{\phi}[(N-k-1)T]\boldsymbol{\theta}(T)\mathbf{r}(kT)$$

$$= \boldsymbol{\phi}(NT)\mathbf{x}(0) + \sum_{k=0}^{N-1} \boldsymbol{\phi}[N-k-1)T]\boldsymbol{\theta}(T)\mathbf{r}(kT)$$

where,

$$\boldsymbol{\phi}(NT) = \boldsymbol{\phi}(T)\cdot\boldsymbol{\phi}(T)\cdots\boldsymbol{\phi}(T) = [\boldsymbol{\phi}(T)]^N$$

and

$$\boldsymbol{\phi}(T) = e^{\mathbf{A}T} = \mathbf{I} + \mathbf{A}T + \frac{\mathbf{A}^2T^2}{2!} + \frac{\mathbf{A}^3T^3}{3!} + \ldots$$

Stability of Linear Time-Invariant Digital Systems

The stability-analysis methods devised for linear continuous-data systems can all be extended to the stability study of digital systems. Since the z-transformation $z = e^{Ts}$ maps the imaginary axis in the s-plane onto the unit circle, $|z| = 1$, in the z-plane, the stability criterion of linear time-invariant digital systems is that all the roots of the characteristic equation must be found inside the unit circle in the z-plane.

The stability of the digital control system shown

in Fig. 36 depends on the location of the poles of the closed-loop transfer function

$$C(z)/R(z) = G(z)/[1 + G(z)]$$

or of the zeros of $1 + G(z)$ in the complex z-plane.

The zeros of $1 + G(z)$ are also known as the roots of the characteristic equation, which can be written as

$$F(z) = a_n z^n + a_{n-1} z^{n-1} + \cdots + a_2 z^2 + a_0 = 0$$

where all the coefficients are real.

The tabulation in Chart 1 can be formed to check the necessary and sufficient conditions for the roots of the characteristic equation to be inside the unit circle.

Actually, the locus of $G(z)$ repeats for every sampling frequency $\omega_s = 2\pi/T$. Thus, it is necessary to obtain only the locus of $G(z)$ for $0 \le \omega \le \omega_s$.

Fig. 38 shows the Nyquist loci of $G(z)$ of the system in Fig. 37. The open-loop transfer function is

$$G(z) = \frac{T^2 K_p (z+1)}{2J_v z^2 + (2K_r T - 4J_v)z + 2J_v - 2K_r T}$$

with $J_v = 41822$, $K_r = 317000$, and $T = 0.1$ sec. Or,

$$G(z) = \frac{1.2 \times 10^{-7} K_p (z+1)}{(z-1)(z-0.242)}$$

CHART 1. CHART FOR CHECKING CONDITIONS FOR ROOTS OF CHARACTERISTIC EQUATION TO BE INSIDE UNIT CIRCLE

a_n	a_{n-1}	$a_{n-2} \ldots a_2$	a_1	a_0	$k_a = a_0/a_n$
$a_0 k_a$	$a_1 k_a$	$a_2 k_a \ldots a_{n-2} k_a$	$a_{n-1} k_a$		
b_0	b_1	$b_2 \ldots b_{n-2}$	b_{n-1}		$k_b = b_{n-1}/b_0$
$b_{n-1} k_b$	$b_{n-2} k_b$	$b_{n-3} k_b \ldots b_1 k_b$			
c_0	c_1	$c_2 \ldots c_{n-2}$			$k_c = c_{n-2}/c_0$
$c_{n-2} k_c$	$c_{n-3} k_c$	$c_{n-4} k_c \ldots$			
....					
p_0	p_1	p_2			$k_p = p_2/p_0$
$p_2 k_p$	p_1/k_p				
q_0	q_1				$k_q = q_1 q_0$
$q_1 k_q$					
r_0					

The conclusions are:

Number of *positive* calculated elements in the first column, ($b_0, c_0, \ldots p_0, q_0, r_0$) = number of roots inside the unit circle.

Number of *negative* calculated elements in the first column, ($b_0, c_0, \ldots p_0, q_0, r_0$) = number of roots outside the unit circle.

The Nyquist stability criterion can be applied directly to determine the stability of digital control systems. The method involves the construction of the Nyquist locus of $G(z)$, with $z = e^{j\omega T}$, in the complex $G(z)$ plane for values of ω from $-\infty$ to $+\infty$.

The Bode diagram of a digital control system can be obtained by use of the bilinear transformation

$$z = (1 + W)/(1 - W)$$

where W is a complex variable. For $z = e^{j\omega T}$,

$$W = j \tan(\omega T/2) = \sigma_W + j\omega_W$$

Thus,

$$\omega_W = \tan(\omega T/2)$$

Then the Bode plot of $G(z)$ can be made in the

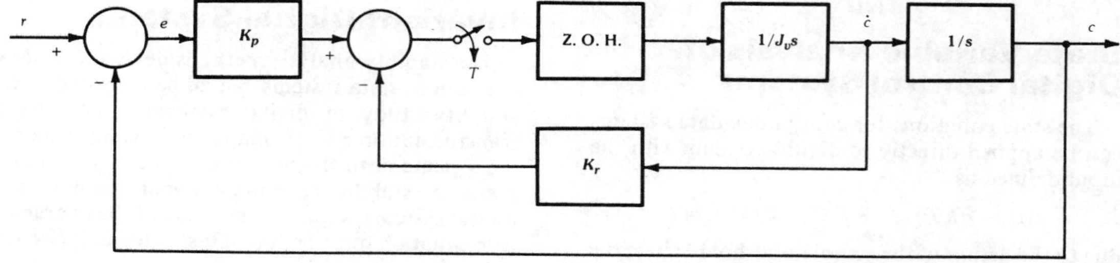

Fig. 37. Closed-loop digital control system.

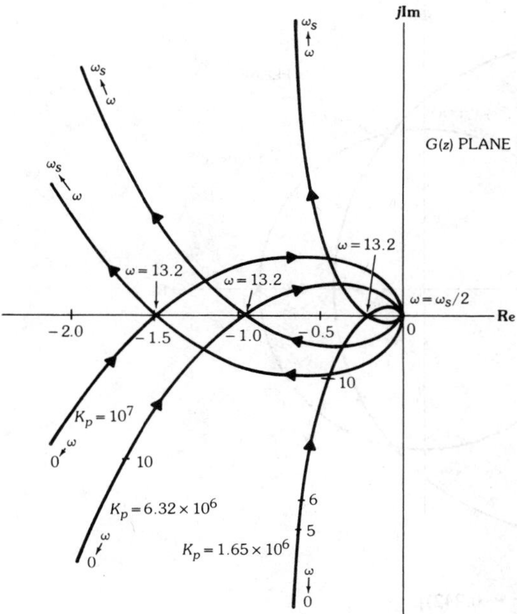

Fig. 38. Typical Nyquist loci of digital control systems.

logarithmic coordinates using the definition of ω_W. For the system shown in Fig. 37,

$$G(j\omega_W) = \frac{1.583 \times 10^{-7} K_p(1 - j\omega_W)}{j\omega_W(1 + j\,1.636\omega_W)}$$

The Bode diagram of $G(j\omega_W)$ is shown in Fig. 39.

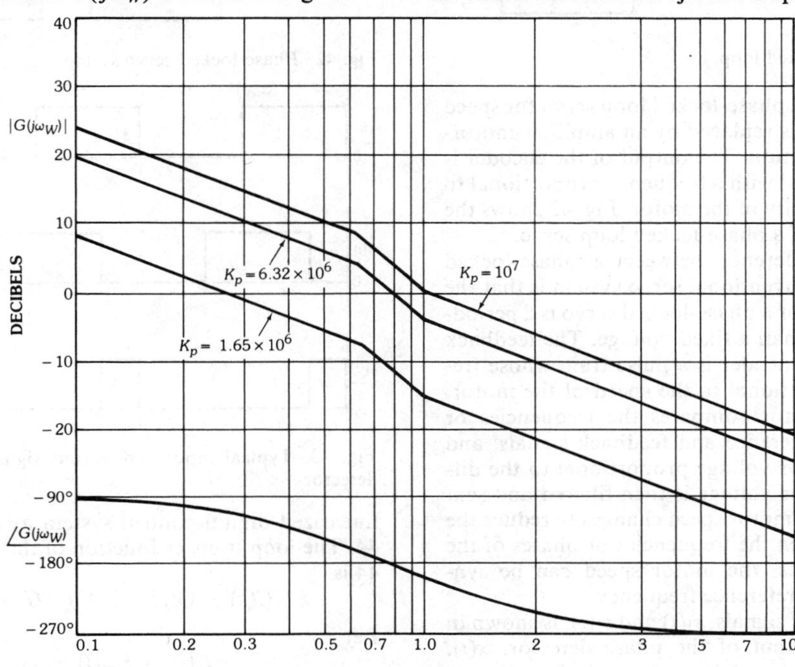

Fig. 39. Bode diagram of the digital control system in Fig. 37.

The rules on the construction of root loci in the s-plane for continuous-data systems can be applied directly to the root loci in the z-plane. The only difference is that in the z-plane the stability of the digital control system must be investigated with respect to the unit circle $|z| = 1$.

For the digital control system in Fig. 37, the open-loop transfer function $G(z)$ has a zero at $z = -1$, and two poles at $z = 1$ and $z = 0.242$. The root loci of the characteristic equation for $0 \leq K_p < \infty$ are constructed as shown in Fig. 40.

PHASE-LOCKED LOOP SERVO SYSTEMS

A phase-locked loop servo is a closed-loop control system that is used widely in communication systems for frequency demodulation and bit synchronization. Phase-locked loop servos are also used for velocity control, especially when a high degree of speed regulation and accuracy is desired.

A phase-locked loop in its basic form is represented by the block diagram of Fig. 41. The input signal and the feedback signal are sinusoidal, and, upon locking, both signals will have the same frequency and a constant phase difference. Any deviation from the desired phase difference is detected by the phase detector and is transmitted to the voltage-controlled oscillator (vco) to correct the error. Since the phase difference is constant, the frequencies of the input and the feedback signals are the same. This principle can be applied to servo systems for which the control objective is speed regulation.

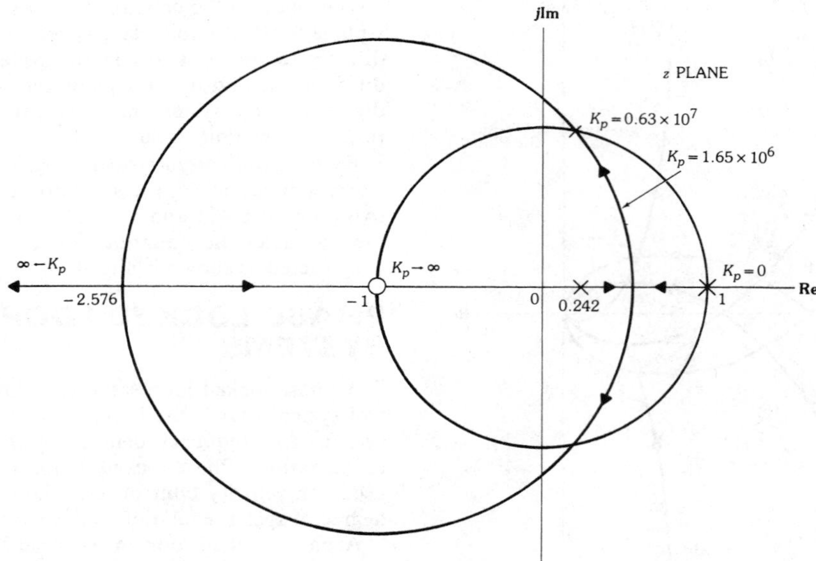

Fig. 40. Root loci for $G(z) = [(1.2 \times 10^{-7})(z + 1)K_p]/[(z - 1)(z - 0.242)]$.

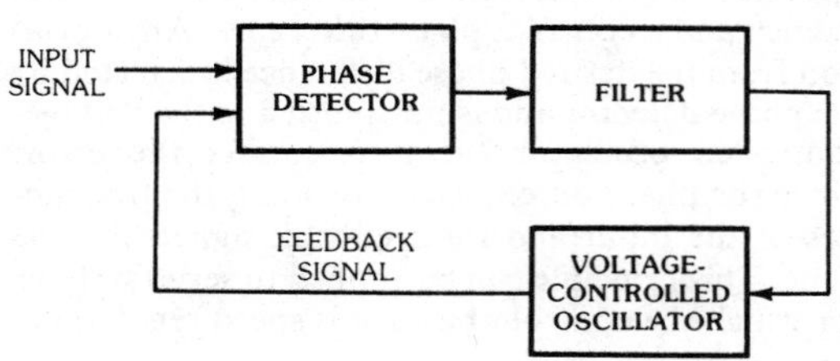

Fig. 41. Phase-locked loop.

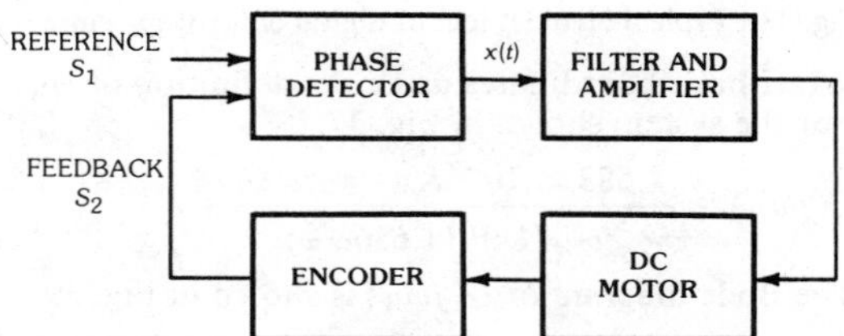

Fig. 42. Phase-locked servo system.

In the case of a phase-locked loop servo for speed control, the vco is replaced by an amplifier-motor-encoder combination. The output of the encoder is a sinusoidal signal with a frequency proportional to the angular velocity of the motor. Fig. 42 shows the block diagram of a phase-locked loop servo.

The main difference between a phase-locked servo and the conventional servo system is that the reference input for a phase-locked servo is a periodic signal rather than a fixed voltage. The feedback signal from the encoder is a pulse train whose frequency is proportional to the speed of the motor. The phase detector compares the frequencies or phases of the reference and feedback signals, and generates an error voltage proportional to the difference. The error voltage is then filtered and sent to the motor; the motor speed changes to reduce the difference between the frequencies or phases of the two signals. Thus, the motor speed can be synchronized to the reference frequency.

A typical set of signals, $s_1(t)$ and $s_2(t)$, is shown in Fig. 43. The output of the phase detector, $x(t)$, switches between 0 and V_s as shown in Fig. 43.

The phase-locked loop servo can be modeled as a

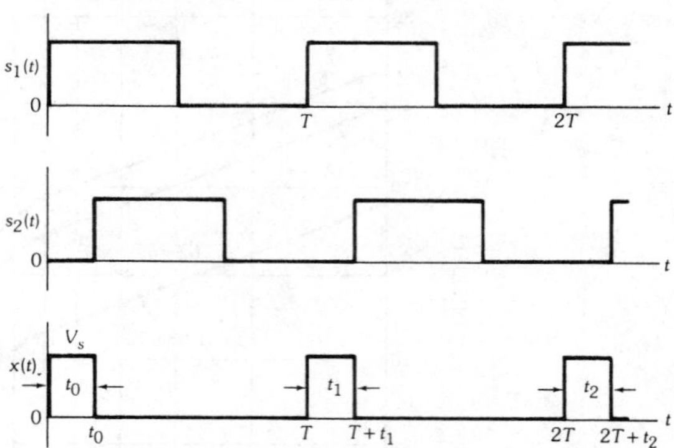

Fig. 43. Typical input and output signals of the phase detector.

linearized digital control system as shown in Fig. 44. The loop transfer function of the system in Fig. 44 is

$$G(z) = (V_s Tn/2\pi)\ \mathcal{Z}\ [G_p(s)/s]$$

Let

$$G_p(s) = K_m/(1 + \tau s)$$

Then,

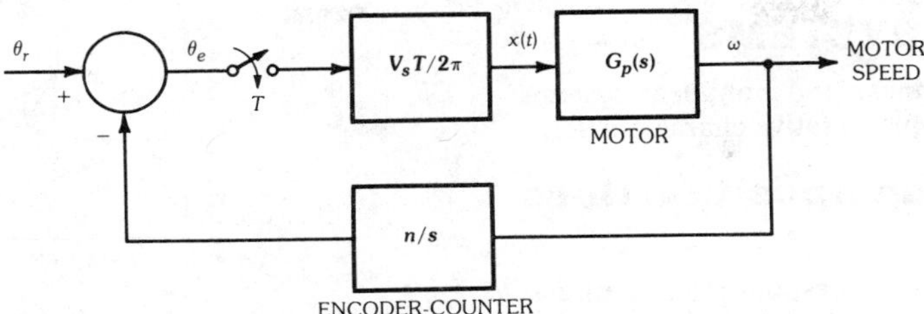

Fig. 44. Linearized model for a phase-locked loop servo.

$$\mathfrak{z}\,[G_p(s)/s] = \mathfrak{z}\,\{K_m/[s(1+\tau s)]\} = K_m z/(z-1) - K_m z/(z - e^{-T/\tau})$$

Let

$$K = (V_s T n/2\pi)K_m$$

Then,

$$G(z) = \frac{Kz(1 - e^{-T/\tau})}{(z-1)\,(z - e^{-T/\tau})}$$

The characteristic equation of the digital phase-locked loop servo is

$$z^2 + (-1 - e^{-T/\tau} + K - Ke^{-T/\tau})z + e^{-T/\tau} = 0$$

Applying the stability test to the last equation yields the condition of stability:

$$K < 2(1 + e^{-T/\tau})/(1 - e^{-T/\tau})$$

NONLINEAR SYSTEMS

All physical systems have nonlinearities and time-varying parameters to some degree. This is justified by the fact that any element may physically break down or exhibit deterioration as a result of time.

Linear systems, as described in the preceding sections, have a linear relationship between the variables described by a linear differential equation; the theory of superposition also applies. Analysis and synthesis techniques applicable to any linear system have been thoroughly investigated and developed.

Nonlinear systems have no general methods of analysis and synthesis. Therefore, for reasons of simplicity, they are often treated with linear approximations and, in many cases of small nonlinear effects, satisfactory results have been obtained. In general, however, linear methods become restrictive in their application and quite often unrealistic.

There are many different ways of solving nonlinear systems that may be applicable to a certain type of system but not to all. Among these methods, two techniques have proved useful in the study of nonlinear systems.

The describing-function technique was first applied to the analysis of nonlinear feedback control systems by Kochenburger.* It is the object of the describing-function method of analysis to reduce the representation of the nonlinearity to an equivalent linear gain and phase angle. The representation of the nonlinearity is described in terms of the fundamental component of the distorted output waveform in response to a sinusoidal input. The result of the describing-function analysis is a representation of the system in the frequency domain; however, the correlation between the time and frequency domains in comparison with linear system analysis is much less precise in nonlinear system analysis. The synthesis of nonlinear systems can be carried out with the describing-function technique in much the same way as is done with linear systems. The describing-function technique is therefore most useful in complex systems of relatively high order and where the effect of the nonlinearity is small but significant.

The phase-plane method, on the other hand, is a representation of the behavior of the first- and second-order systems portrayed on the phase plane. The phase-plane diagrams can be interpreted directly in terms of time domain; they thus are a useful tool in the study of transient response to any initial condition and in some cases to step and ramp inputs. The phase-plane method does not readily indicate the steps required to correct system performance, and synthesis is carried out by trial-and-error procedures.

This method can be extended to higher-order systems in the so-called phase space. However, an *n*th-order system in the *n*th-dimensional phase space compared with the phase plane is difficult to envision and interpret and may be less effective in its usefulness. In summary, this method is useful for large nonlinearities of the second-order systems.

The two methods are in a sense complementary and constitute the principal tools for the study of nonlinear control problems.

* Kochenburger, R.J. "A Frequency Response Method For Analyzing and Synthesizing Contractor Servomechanisms." *AIEE Transactions* Vol. 69 (1950), pp. 270-284.

CHARACTERISTICS OF NONLINEAR SYSTEMS

A comparison of linear and nonlinear systems yields the following representative characteristics.

Principle of Superposition Does Not Apply

In nonlinear systems, the response to a combination of individual signals at the input will not be the same as the response to the sum of those same signals. The impulse response (weighting function) describing the frequency response is therefore not applicable. The familiar Laplace transform operation ($s = \sigma + j\omega$) and the transfer-function concept used extensively in linear systems cannot be directly applied without some modification. In fact, transfer functions are noncommutative in nonlinear systems. That is, in a linear system transfer, functions $G_1(s)$ and $G_2(s)$ can be cascaded as $G_1(s) \cdot G_2(s)$ or $G_2(s) \cdot G_l(s)$, whereas in nonlinear systems $N_1 \cdot N_2$ is not the same as $N_2 \cdot N_1$.

Nonlinear Response Is Dependent On Input Signal

In linear systems, the system response is strictly a function of the system parameters. A system proved to possess a good stable response to one type of input will behave similarly to any other type of input. Nonlinear systems, on the other hand, are dependent on the input-signal size and initial conditions as well as the system parameter. A stable response for one input signal may be unstable for another.

In nonlinear systems, phenomena that are ordinarily unexplainable by linear analytical methods (generation of new frequencies, jump resonance, limit cycles, etc.) exist. The following are examples of such unique nonlinear phenomena.

Jump Resonance

The phenomenon called jump resonance is observed in certain closed-loop systems with saturation, where the input-output amplitude ratio and phase angle as a function of frequency exhibit sudden discontinuities. The typical closed-loop gain characteristics of a saturating system with jump resonance are shown in Fig. 45. Amplitude ratio $|\theta_o/\theta_i|$ is plotted as a function of frequency for a fixed amplitude θ_i. As the frequency is increased from zero, the frequency response follows the curve along points A, B, and C. At point C a sudden discontinuous jump to D is observed with an incremental increase in frequency. Further increase in frequency leads to point E along the curve. If the frequency is reversed, the response retraces the path E, D and continues to point F, at which a sudden jump to point G occurs; it then continues on through B and A of the gain curve. The phase-angle response behaves similarly. The overall response curve exhibits a hysteresis-type property or jump resonance.

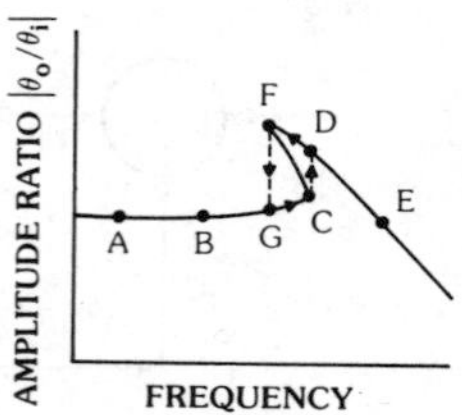

(A) Amplitude ratio.

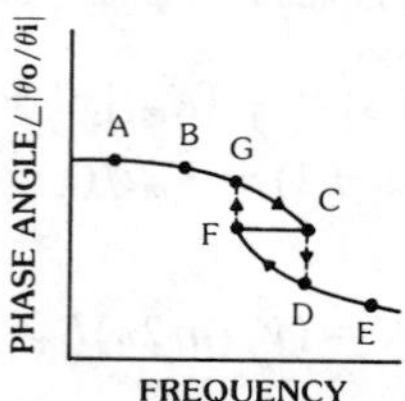

(B) Phase angle.

Fig. 45. Jump resonance.

Limit Cycle

Limit cycle is a phenomenon of oscillation peculiar to nonlinear systems. The oscillatory behavior, unexplainable in terms of linear theory, is characterized by a constant amplitude and frequency determined by the nonlinear properties of the system. Limit cycles are distinguishable from linear oscillation in that their amplitude of oscillation is independent of initial conditions. For instance, if a system has a stable limit cycle, the system will tend to fall into the limit cycle, with the output approaching the amplitude of that limit cycle regardless of the initial condition and forcing function. A limit cycle is easily recognized in the phase plane as an isolated closed path as shown in Fig. 46.

Often the system falls into a limit cycle in the

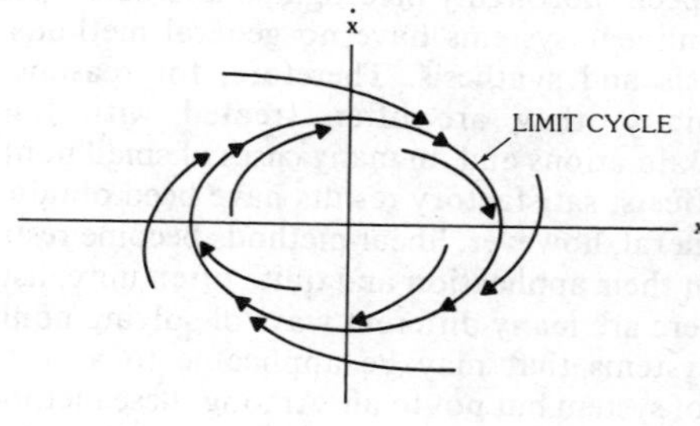

Fig. 46. Limit cycle.

presence of very small excitation or disturbances. This behavior is termed *soft self-excitation.* Conversely, for systems requiring forced excitation above a certain minimum amplitude or appropriate initial condition before entering a limit cycle, the term *hard self-excitation* is used. In any event, the presence of limit cycles in control systems is undesirable and makes isolation of these limit cycles important to the analysis of nonlinear systems. The phase-plane method described later in this chapter provides techniques for investigating limit cycles.

Generation of New Frequencies

In a nonlinear system the output of the nonlinear device contains harmonic and subharmonic frequencies of the input signal. For instance, application of two sine waves of different frequencies f_1 and f_2 to the input will produce components corresponding to the input frequencies f_1, f_2, their sum and difference $f_1 \pm f_2$, their higher harmonics mf_1, nf_2, and various combinations of sums and differences $mf_1 \pm nf_2$. In a linear system only the components of the input frequencies f_1, f_2 will be reflected in the output. This exemplifies why the frequency-response concept pertaining to linear systems must be modified for nonlinear applications.

Hysteresis

Multivalued functions exist when two or more function values correspond to the value of the variable. Multivalued functions are intrinsically nonlinear. Some examples of this are the hysteresis curves of magnetic materials and the backlash of a gear train.

TYPES OF NONLINEAR ELEMENTS

Because of the infinite variety of nonlinearities, specific classification becomes impossible. Two generally accepted classifications do exist; they are the *incidental* and *intentional* nonlinearities.

Incidental nonlinearities are extraneous to system design and are usually undesirable. Saturation, dead zone, and backlash are examples of nonlinearities that may lead to inaccurate and poor response or even instability. There are cases, however, where incidental nonlinearities aid and improve the system response.

Intentional nonlinearities are functions purposely introduced into the system to compensate and improve performance. The relay servo system is typical of this type. Other classifications are the *single-valued* function and *multivalued* function. These functions are transfer characteristics describing the input-to-output relationship of the nonlinear element. Single-valued functions have only one output value corresponding to the input value, while the multivalued functions have two or more output values. Several typical nonlinear characteristics commonly encountered in control systems are presented in Fig. 47.

In addition to the classifications described above, other functions to be found may include large and small values of continuous and discontinuous types of nonlinearities. More than one classification may apply to any nonlinear function.

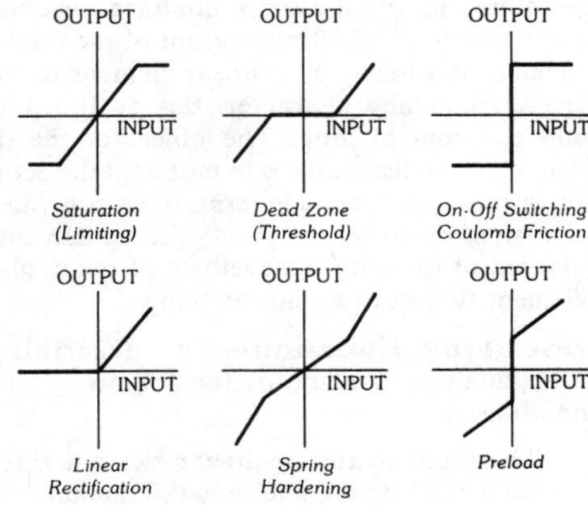

(A) Single-valued functions.

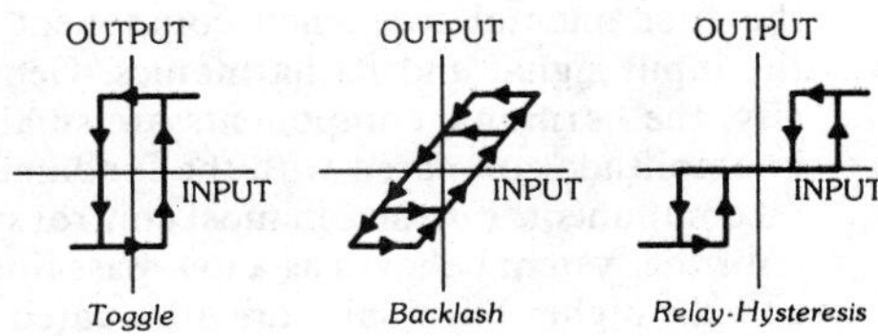

(B) Multivalued functions.

Fig. 47. Types of nonlinear elements.

ANALYTICAL METHODS FOR SOLVING NONLINEAR SYSTEMS

The analytical methods for solving nonlinear systems described in this section will be concentrated mainly on graphical methods for describing-function techniques and phase-plane methods. Other methods of analysis are:

A. Direct Solution to the Nonlinear Differential Equation. There are certain nonlinear differential equations of lower order that are analytically solvable or integrable; however, they are very rare.

B. Numerical Method. The numerical method is a step-by-step process obtaining the solution to the differential equation as a table of corresponding values of independent and dependent variables. In theory any equation can be solved numerically, although the process may be quite complex.

Neither (A) nor (B) is discussed here, but further information may be found in references 1, 8, and 12 of the bibliography.

Describing-Function Technique

This technique is valuable in the analysis and design of an important class of nonlinear feedback control systems, in which the output of the nonlinear element is filtered by a linear element having low-pass frequency characteristics as it travels around the control loop. The object of the describing-function technique is to represent the actual nonlinearity of the system in terms of an equivalent linear system by considering only the fundamental component of the output waveform of the nonlinear element subject to a sinusoidal input.

Describing Function—The describing-function analysis is made of the following basic assumptions.

A. The input to the nonlinear element n is a sinusoidal signal, and only the fundamental component of the output of n contributes to the input. The output response of a nonlinear element to a periodic signal consists of the fundamental frequency component of the input signal and its harmonics. Generally, the harmonic components are smaller in amplitude compared with the fundamental component. Further, in most control systems the system behaves as a low-pass filter, and the higher harmonics are attenuated. If the higher harmonics are sufficiently small, they can be neglected, and the equivalent linear approximation may be justified.

B. There is only one nonlinear element in the system. All nonlinearities in the system are lumped into one single nonlinear element n. Fig. 48 shows a block diagram of a closed-loop system containing a nonlinear element n.

C. The output of the lumped nonlinear element is a function only of the present value and past history of the input; *i.e.*, n is not a function of time.

The describing function of a nonlinear element is defined as the ratio of the fundamental-frequency component of the output as a complex quantity

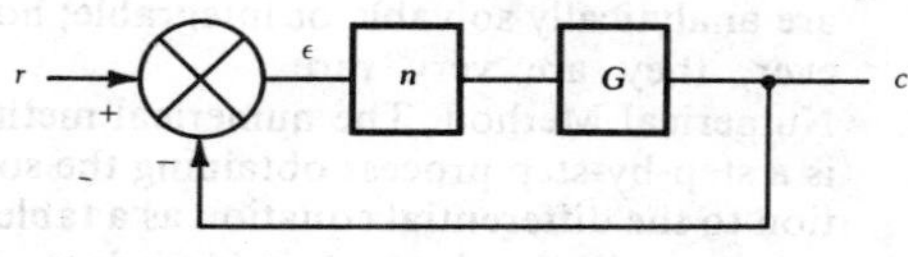

Fig. 48. Block diagram of nonlinear closed-loop system.

(amplitude and phase angle) to the amplitude of the sinusoidal input signal. If the input signal as applied to the nonlinear element n is described by

$$e_{in}(t) = X \sin\omega t$$

the output response $e_o(t)$ may generally take the form of

$$e_o(t) = (a_0X/2) + a_1X \sin\omega t + b_1X \cos\omega t$$
$$+ \sum_{n=2}^{\infty} a_nX \sin n\omega t + \sum_{n=2}^{\infty} b_nX \cos n\omega t + \text{subharmonics}$$

The $(a_0/2)$ term is the dc component; a_n and b_n are the harmonic components.

The fundamental-frequency component of the output may be expressed in terms of amplitude and phase angle as

$$e_{o1} = A(\omega, X)\, X \sin[\omega t + \phi(\omega,X)]$$

In this expression, $A(\omega,X)X$ is the amplitude, and $\phi(\omega, X)$ is the phase angle of the fundamental component. Both amplitude and phase angle are a function of the frequency and amplitude of the input signal. The describing function $N(\omega, X)$ by definition is

$$\begin{aligned} N(\omega, X) &= \{A(\omega, X)X \exp[j\phi(\omega, X)]/X\} \\ &= A(\omega, X) \exp[j\phi(\omega, X)] \\ &= A(\omega, X) \cos\phi(\omega, X) \\ &\quad + jA(\omega, X) \sin\phi(\omega, X) \end{aligned} \qquad (1)$$

The describing function $N(\omega, X)$ may be purely real or contain a phase angle depending on the type of nonlinearity. For single-valued nonlinear functions N is real, whereas for multivalued functions phase shift exists, generally lagging.

Calculation of a Describing Function—Calculation of the describing function involves performing a conventional Fourier analysis on the output waveform to obtain the fundamental component. The Fourier series expansion of the output waveform to an input sinusoidal $X \sin\omega t$ may be expressed as

$$e_o(t) = (a_0X/2) + a_1X \sin\omega t + b_1X \cos\omega t + a_2X \sin 2\omega t + b_2X \cos 2\omega t + \cdots$$

For the describing function, only the coefficients of the fundamental-frequency component are required. The coefficients may be obtained from the integrals

$$a_1 = (\pi X)^{-1} \int_0^{2\pi} f_0(t) \sin\omega t \cdot d(\omega t) \qquad (2)$$

$$b_1 = (\pi X)^{-1} \int_0^{2\pi} f_0(t) \cos\omega t \cdot d(\omega t) \qquad (3)$$

where $f_0(t)$ is the exact output of the nonlinear element expressed as a function of time. The describing function is then

$$|N(\omega, X)| = |a_1 + jb_1| = (a_1^2 + b_1^2)^{1/2}$$
$$\angle N(\omega, X) = \tan^{-1}(b_1/a_1)$$

Where the exact output function $f_0(t)$ is known, the above method is applicable. If the function is not known, a graphical Fourier expansion can be performed on the output waveform. Two examples describing the procedure for calculation of the describing function using the graphical method are given.

Example 1: Saturation-Type Nonlinearity— A nonlinear element with saturation is shown in Fig. 49. Output y is held constant for input values greater than S. This region is called saturation or limiting. For input values less than S, the output behaves linearly with the input. The input-output relationship can be expressed by

(A) $y = kx$ for $-S < x < S$
(B) $y = kS$ for $x > S$
(C) $y = -kS$ for $x < -S$

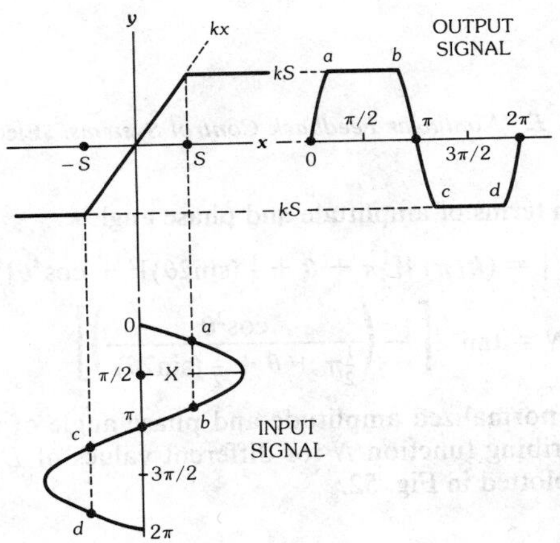

Fig. 49. Graphic representation of saturation.

The output is an odd function, and thus only the sine term of the fundamental equation need be calculated. Furthermore, because of symmetry only the first quarter of the integration need be evaluated as follows:

$$a_1 = (4/\pi X)\int_0^{\pi/2} f_0(t)\sin\omega t \cdot d(\omega t)$$

If the input is $x = X\sin\omega t$, then the output $f_0(t) = y$ is expressed by

$$f_0(t) = kX\sin\omega t \qquad X < S$$

and for X greater than S

$$f_0(t) = kX\sin\omega t \qquad 0 < \omega t < \sin^{-1}(S/X)$$
$$= kS \qquad \sin^{-1}(S/X) < \omega t < \pi/2$$

Therefore the coefficients a_1 are

$$a_1 = k \qquad \text{for } X < S$$

and for X greater than S

$$a_1 = (4/\pi X)\int_0^{\sin^{-1}(S/X)} kX\sin^2(\omega t)\cdot d(\omega t) + (4/\pi X)\int_{\sin^{-1}(S/X)}^{\pi/2} kS\sin\omega t\cdot d(\omega t)$$
$$= (2k/\pi)[\phi + (\sin 2\phi)/2]$$

where $\phi = \sin^{-1}(S/X)$.

The describing function N is given by

$N = k \qquad X < S$
$N = k(2/\pi)[\phi + (\sin 2\phi)/2] \qquad X > S$

The variation of amplitude of N with respect to X/S is plotted in Fig. 50. The phase angle is zero over the entire range.

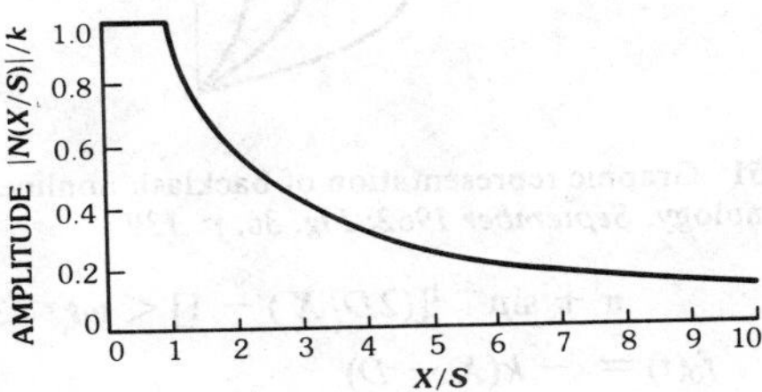

Fig. 50. Describing function for saturation (normalized amplitude).

Example 2: Backlash-Type Nonlinearity— For the second example, a simple backlash-type nonlinearity (Fig. 51) is evaluated. The backlash is a multivalued nonlinearity where the input-output relationship follows a different path dependent on the input-signal amplitude (curves 1, 2, and 3). After the steady state is established, the output $f_0(t)$ corresponding to different values of X of an input signal $x = X\sin\omega t$ are:

For $X < D$ (curve 1):

$$f_0(t) = 0$$

For $D < X < 2D$ (curve 2):

$$f_0(t) = -k(X - D) \qquad 0 < \omega t < \sin^{-1}[(2D/X) - 1]$$

$$f_0(t) = k(X\sin\omega t - D) \qquad \sin^{-1}[(2D/X) - 1] < \omega t < \pi/2$$

$$f_0(t) = k(X - D) \qquad \pi/2 < \omega t < \pi + \sin^{-1}[(2D/X) - 1]$$

$$f_0(t) = k(X\sin\omega t + D)$$

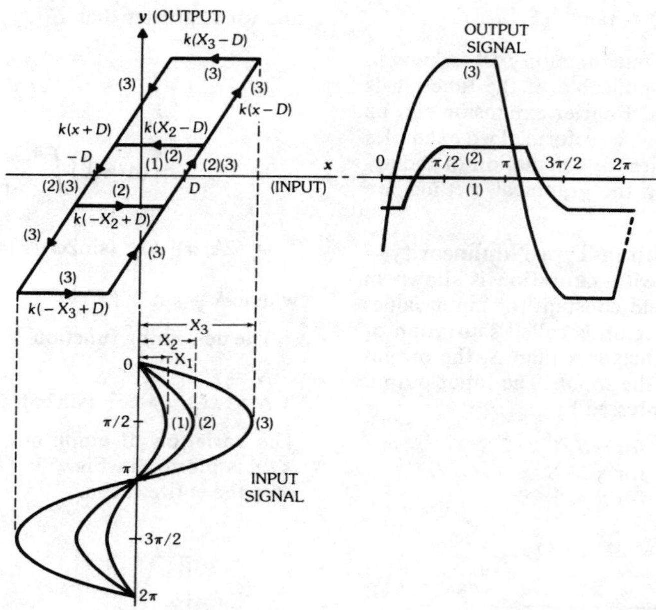

Fig. 51. Graphic representation of backlash nonlinearity. (*Levinson, E. "Nonlinear Feedback Control Systems."* Electro-Technology, *September 1962; Fig. 36, p. 139.*)

$$\pi + \sin^{-1}[(2D/X) - 1] < \omega t < 3\pi/2$$

$$f_0(t) = -k(X - D)$$

$$3\pi/2 < \omega t < 2\pi$$

For $X > 2D$ (curve 3):

$$f_0(t) = k(X \sin\omega t - D)$$

$$0 < \omega t < \pi/2$$

$$f_0(t) = k(X - D)$$

$$\pi/2 < \omega t < \pi - \sin^{-1}[1 - (2D/X)]$$

$$f_0(t) = k(X \sin\omega t + D)$$

$$\pi - \sin^{-1}[1 - (2D/X)] < \omega t < 3\pi/2$$

$$f_0(t) = -k(X - D)$$

$$3\pi/2 < \omega t < 2\pi - \sin^{-1}[1 - (2D/X)]$$

$$f_0(t) = k(X \sin\omega t - D)$$

$$2\pi - \sin^{-1}[1 - (2D/X)] < \omega t < 2\pi$$

Solving for a_1 and b_1 of equations (2) and (3), for the three conditions above, yields the describing-function terms

$a_1 = 0$ $\quad X < D$

$a_1 = (k/\pi)[\pi/2 + \theta + (\sin 2\theta)/2]$ $\quad X > D$

$b_1 = 0$ $\quad X < D$

$b_1 = (-k/\pi)\cos^2\theta$ $\quad X > D$

where,

$$\theta = \sin^{-1}[1 - (2D/X)]$$

or in terms of amplitude and phase angle

$$|N| = (k/\pi)\{[\tfrac{1}{2}\pi + \theta + \tfrac{1}{2}(\sin 2\theta)]^2 + \cos^4\theta\}^{1/2}$$

$$\angle N = \tan^{-1}\left[-\left(\frac{\cos^2\theta}{\frac{1}{2}\pi + \theta + \frac{1}{2}(\sin 2\theta)}\right)\right]$$

The normalized amplitude and phase angle of the describing function N for different values of D/X are plotted in Fig. 52.

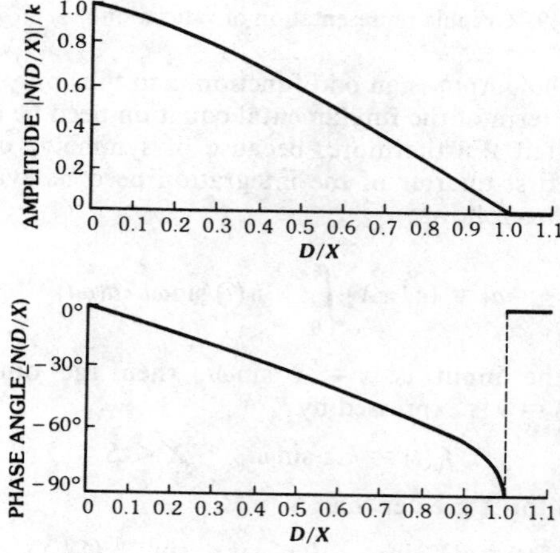

Fig. 52. Describing function for backlash (normalized amplitude and phase angle).

The describing function is calculated simply by determining the fundamental output component of the nonlinear element. The third-harmonic component may be obtained to estimate the accuracy of the describing-function analysis. Describing functions for some of the common nonlinear elements are given in Table 2.

Stability Analysis—The describing function N of the nonlinearity can be used to determine the stability of the system, providing the harmonics are sufficiently attenuated. In general, the describing function is a function of both frequency and amplitude of the input signal.

The closed-loop transfer function of the nonlinear feedback system in Fig. 53 is given by

$$(c/r)\,(\omega, X) = \frac{N(\omega, X)\,G(j\omega)}{1 + N(\omega, X)\,G(j\omega)}$$

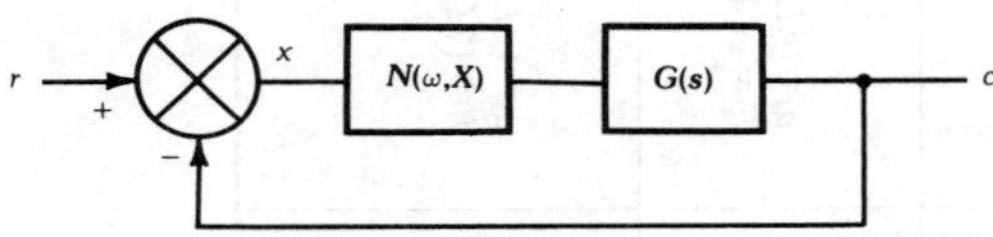

Fig. 53. Single-loop nonlinear system.

The characteristic equation of the system is

$$1 + N(\omega, X)G(j\omega) = 0 \tag{4}$$

or

$$G(j\omega) = -\,[1/N(\omega, X)]$$

The condition of (4) must be satisfied for sustained oscillation of the output with zero input. Since $N(\omega, X)$ is a function of both frequency and amplitude, various combinations of ω and X can be found for oscillation. If there are no possible combinations satisfying the oscillatory condition, the system is stable. In the case of sustained oscillation, the oscillatory mode may be either stable or unstable. If a slight disturbance in amplitude or frequency occurs and the oscillation returns to its original value, the oscillation is stable (stable limit cycle). If the oscillation amplitude increases or decreases from the original value, the oscillation is unstable (unstable limit cycle). The stability of the closed-loop system may be evaluated analytically by directly solving the characteristic equation by any one of the modified linear graphical methods.

Polar Plot (Nyquist Diagram)—The conventional Nyquist diagram must be modified to apply the Nyquist stability criteria to the frequency-response plot. In a linear system, the critical point on the Nyquist diagram is -1. For nonlinear systems the $-[1/N(\omega, X)]$ locus corresponds to the critical point -1. To evaluate the stability of the system, both $-[1/N(\omega, X)]$ and the $G(j\omega)$ function are plotted on the polar plane. The describing function $N(\omega, X)$ generally is a function of both ω and X. If N is only a function of X, there will be one locus $-[1/N(x)]$ plotted as a function of X. If N is also a function of ω, a family of constant-frequency loci are plotted for different values of ω (see Fig. 54).

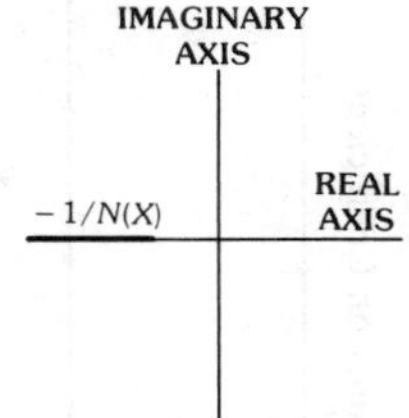

(A) Plot of $-1/N(X)$, zero phase angle.

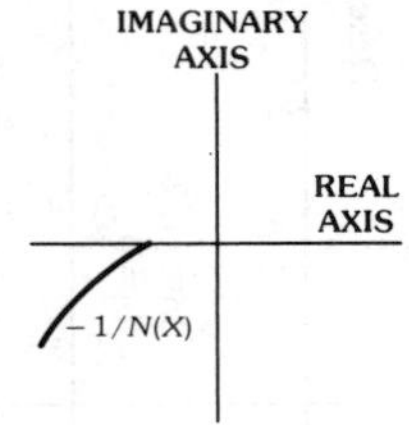

(B) Plot of $-1/N(X)$, with phase angle.

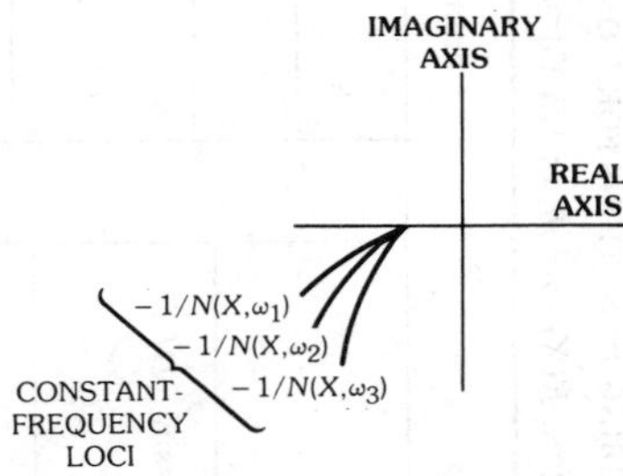

(C) Plot of $-1/N(X)$, as a function of constant ω.

Fig. 54. Typical polar plots of various $N(\omega,X)$.

The stability of the system is determined by the following relationship between the $-[1/N(\omega, X)]$ locus and the $G(j\omega)$ plot (Fig. 55). If the $-[1/N(\omega, X)]$ lies to the left of the $G(j\omega)$ plot or is not enclosed, the system is *stable*. Conversely, if the $-[1/N(\omega, X)]$ lies to the right of the $G(j\omega)$ plot or is enclosed, the system is *unstable*. If the $-[1/N(\omega, X)]$ locus intersects with the $G(j\omega)$ plot, the system may have a *sustained oscillation*. In the case where N is a function of ω, the condition for sustained oscillation is satisfied if the ω of the $G(j\omega)$ plot at the intersecting point is the same ω of the $-[1/N(\omega, X)]$ locus (see Fig. 56).

The oscillation may be either stable or unstable. If the $G(j\omega)$ intersects with the $-[1/N(\omega, X)]$ locus at one point only, the oscillation is stable (stable limit cycle). If more points of intersection exist, the limit cycle may be either stable or unstable. The

Table 2. Describing Functions for Common Nonlinear Elements

Describing Function: $N(X)=a_1+jb_1$ or $|N(X)|=(a_1^2+b1^2)^{1/2}$. $\angle N(X)=\tan^{-1}(b_1/a_1)$

Characteristic	Describing-Function Coefficients	
A 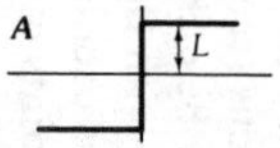	$a_1=4S/\pi X$ $b_1=0$	
B 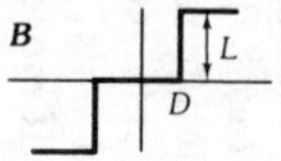	$a_1=(4S/\pi X)\cos\theta$ $b_1=0$	$\theta=\sin^{-1}(D/X)$
C 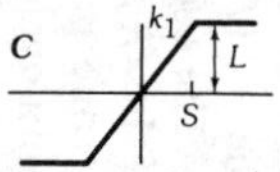	$a_1=(2k_1/\pi)[\theta+(1/2)(\sin 2\theta)]$ $b_1=0$	$\theta=\sin^{-1}(S/X)$
D 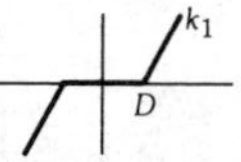	$a_1=(2k_1/\pi)[\pi/2-\theta-(1/2)(\sin 2\theta)]$ $b_1=0$	$\theta=\sin^{-1}(D/X)$
E 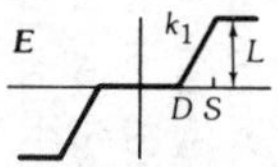	$a_1=(2k_1/\pi)[\psi-\theta+(1/2)(\sin 2\psi)-(1/2)(\sin 2\theta)]$ $b_1=0$	$\psi=\sin^{-1}(S/X)$ $\theta=\sin^{-1}(D/X)$
F 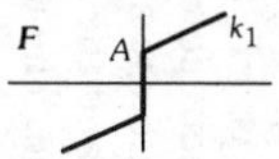	$a_1=k_1+(4A/\pi X)$ $b_1=0$	
G 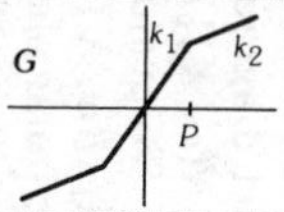	$a_1=k_2-[(k_2-k_1)/\pi](2\theta+\sin 2\theta)$ $b_1=0$	$\theta=\sin^{-1}(P/X)$
H 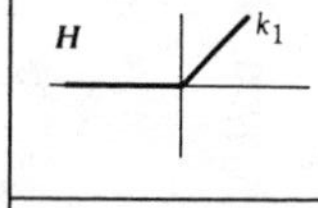	$a_1=k_1/2$ $b_1=0$	
I 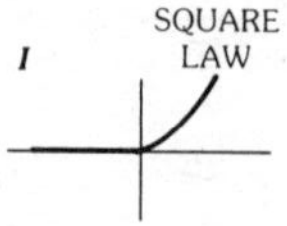 (SQUARE LAW)	$a_1=4X/3\pi$ $b_1=0$	
J 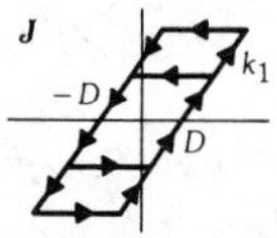	$a_1=(k_1/\pi)[\pi/2+\theta+(1/2)(\sin 2\theta)]$ $b_1=-(k_1/\pi)\cos^2\theta$	$\theta=\sin^{-1}[1-(2D/X)]$
K 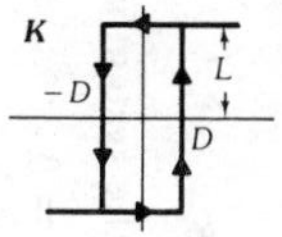	$a_1=(4L/\pi X)\cos\theta$ $b_1=-(4L/\pi X)\sin\theta$	$\theta=\sin^{-1}(D/X)$
L 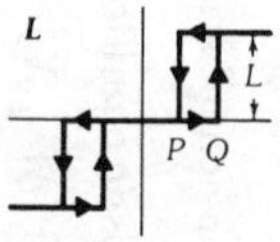	$a_1=(2L/\pi X)(\cos\theta+\cos\psi)$ $b_1=(2L/\pi X)(\sin\psi-\sin\theta)$	$\psi=\sin^{-1}(P/X)$ $\theta=\sin^{-1}(Q/X)$

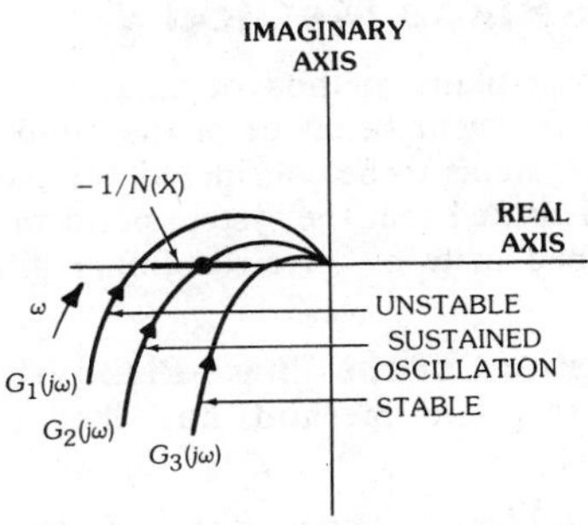

Fig. 55. Polar-plot stability criteria.

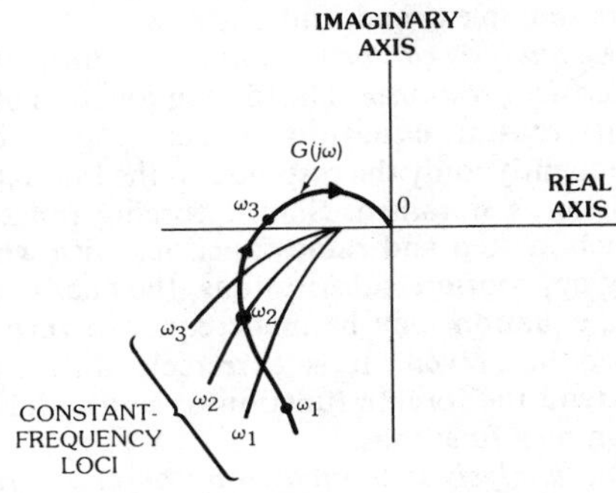

Fig. 56. Typical polar plot of $G(j\omega)$ and $-[1/N(\omega,X)]$ as a function of ω.

stability of the limit cycle is determined by the direction of the two loci at the crossover point.

By establishing the $G(j\omega)$ locus pointing in the direction of increasing frequency as a reference, if the $-[1/N(X)]$ locus pointing in the direction of increasing amplitude X crosses the $G(j\omega)$ locus from right to left, the limit cycle is stable. If the crossover occurs from left to right, the limit cycle is unstable. A polar plot with both stable and unstable limit cycles is shown in Fig. 57.

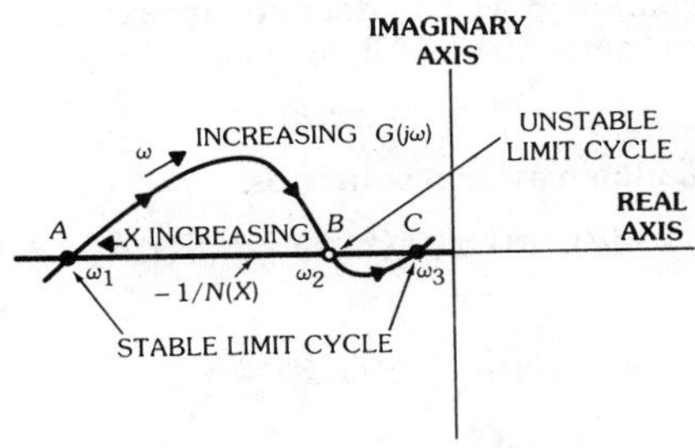

Fig. 57. Polar plot of stable and unstable limit cycles.

Gain-Phase Plot—The gain-phase plot is the direct transfer of the polar plot from the polar coordinate to the rectangular coordinate. The ordinate is the gain in decibels, and the abscissa is the phase angle in degrees.

The gain and phase angle of the two functions $G(j\omega)$ and $-[1/N(\omega, X)]$ are for $G(j\omega)$:

Gain	$20 \log \vert G(j\omega) \vert$
Phase angle	$\angle G(j\omega)$

and for $N(X, \omega)$:

Gain	$-20 \log_{10} \vert N(\omega, X) \vert$
Phase angle	$-180° - \angle N(\omega, X)$

Typical gain-phase plots for various types of $N(\omega, X)$ are given in Fig. 58.

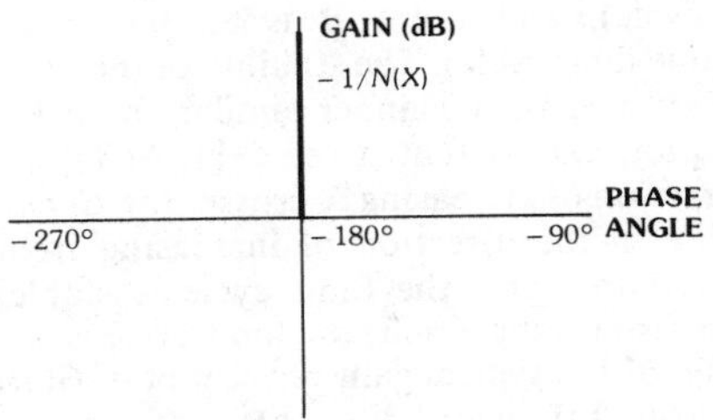

(A) Plot of $-1/N(X)$, zero phase angle.

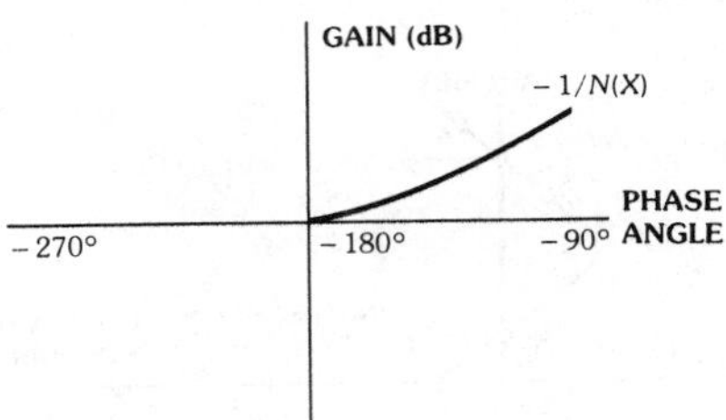

(B) Plot of $-1/N(X)$, with phase angle.

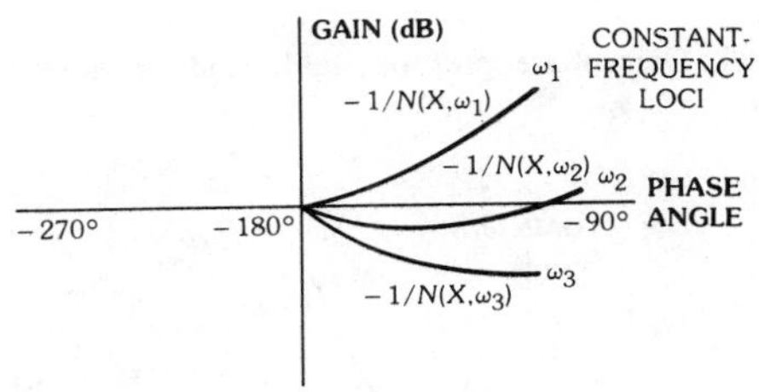

(C) Plot of $-1/N(X)$ as a function of constant ω.

Fig. 58. Typical gain-phase plots of various $N(\omega,X)$.

The system is stable if the $-(1/N)$ locus does not intersect with the $G(j\omega)$ plot. If the $-(1/N)$ locus intersects with the $G(j\omega)$ plot, the system has a sustained oscillation (Fig. 59).

In the case of sustained oscillation, there may be more than one point of intersection, as shown in

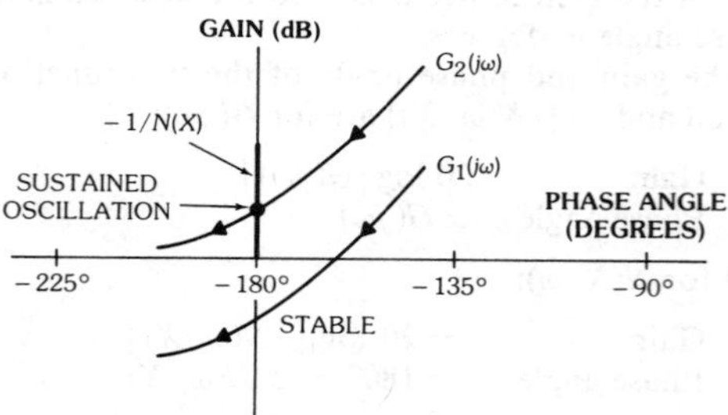

Fig. 59. Stability criteria for gain-phase plot.

Fig. 60. Points A and C are stable points (stable limit cycle), and point B is an unstable point (unstable limit cycle). The stability of the limit cycle is determined in a manner similar to that of the polar plot, except that if the $-[1/N(X)]$ locus in the direction of increasing X crosses the $G(j\omega)$ locus pointing in the direction of increasing frequency from left to right, the limit cycle is stable; if it crosses from right to left, the limit cycle is is unstable. Fig. 61 is a typical gain-phase plot of $G(j\omega)$ and $-[1/N(\omega, X)]$, where $N(\omega, X)$ is a function of ω. The family of $-(1/N)$ plots are the constant-frequency loci. Point A is the location for sustained oscillation.

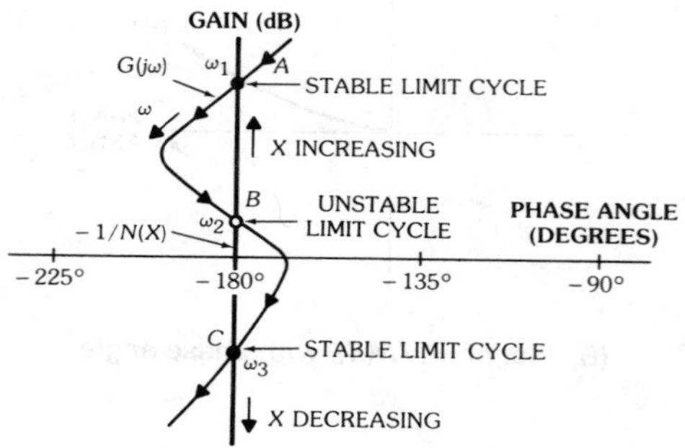

Fig. 60. Gain-phase plot of stable and unstable limit cycles.

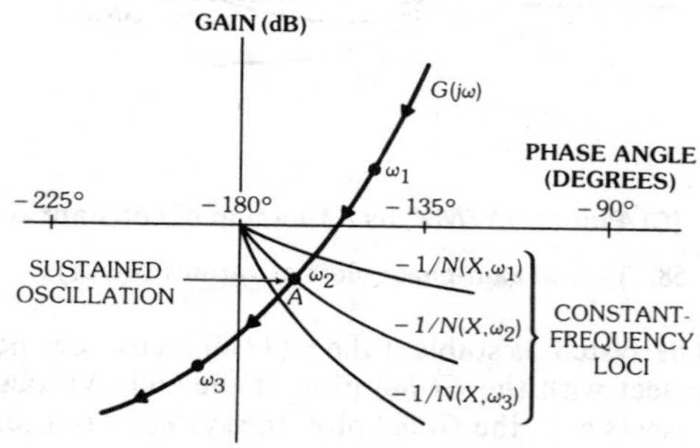

Fig. 61. Typical gain-phase plot of $G(j\omega)$ and $-[1/N(\omega,X)]$ as a function of ω.

Phase-Plane Method

The phase-plane method of analysis is used to study the transient behavior of the nonlinear system. The systems to be considered are assumed to be so constituted that the system performance can be described in terms of an ordinary differential equation.

Restrictions of the Phase-Plane Method— The phase-plane method has the following restrictions.

A. *The analysis is limited to systems described by the first and second order.* Differential equations for systems of higher order may be solved in the phase space; however, the results are complex and unwieldy.

B. *The analysis can be used only for study of the transient response.* The forcing function of the differential equation is zero, and, consequently, only the response to the initial condition is obtained. Simple forcing functions such as step and ramp functions with which, by appropriate substitutions, the characteristic equation may be made equal to zero can also be solved. It is extremely difficult to extend the forcing function to sinusoidal and complex functions.

C. *The analysis is limited to autonomous functions.* That is, the coefficients of the derivatives must be functions of x and $\dot{x}$ and not of time explicitly.

Phase Plane—The differential equation describing a second-order system may be expressed by

$$f_1(x,dx/dt,t)\,(d^2x/dt^2) + f_2\,(x,\,dx/dt,\,t)\,(dx/dt) + f_3(x,\,dx/dt,\,t)x = g(t)$$

The type of equations that can be evaluated in the phase plane is of the form

$$f_1(x,\,dx/dt)\,(d^2x/dt^2) + f_2\,(x,\,dx/dt)\,(dx/dt) + f_3(x,\,dx/dt)x = 0 \quad (7)$$

The equation in which t does not appear explicitly is called "autonomous." By defining

$$\dot{x} = dx/dt$$

the equation may be rewritten as

$$f_1(x,\,\dot{x})\,(d\dot{x}/dt) + f_2\,(x,\,\dot{x})\,\dot{x} + f_3(x,\,\dot{x})x = 0$$

or

$$dx/dt = P\,(x,\dot{x}) = \dot{x} \quad (8)$$

$$d\dot{x}/dt = Q(x,\,\dot{x}) = -\,[f_2(x,\dot{x})\dot{x} + f_3(x,\,\dot{x})x]/f_1(x,\dot{x}) \quad (9)$$

and further into the form of

$$\frac{(d\dot{x}/dt)}{(dx/dt)} = d\dot{x}/dt$$

$$= Q(x,\dot{x})/P(x,\dot{x})$$
$$= -[f_2(x,\dot{x})\dot{x} + f_3(x,\dot{x})x]/\dot{x}f_1(x,\dot{x}) \quad (10)$$

The second-order differential equation with respect to time is reduced to a first-order equation of x and $\dot{x}$.

The phase-plane diagram has the $\dot{x}$ as its ordinate and x as its abscissa. The plot of $\dot{x}$ as a function of x on the phase-plane diagram is termed *phase trajectory*. A family of phase trajectories is called the *phase portrait*.

The phase trajectory originates at a point corresponding to the initial condition (x_0, $\dot{x}_0$) and moves to a new location at each increment of time. Generally, the increments of time are not portrayed on the trajectory and must be obtained by other means described in a later section. If the value of time at each point on the trajectory is obtained, the time response of $\dot{x}(t)$ and $x(t)$ can be plotted. The phase trajectory has a definite direction associated with time. When $\dot{x}$ is positive the trajectory moves from left to right, and for negative values of $\dot{x}$ all paths move from right to left. If the trajectory approaches the origin or some finite point on the phase plane as time goes to ∞, the system is stable. If the trajectory goes to ∞ with time, the system is unstable. If the trajectory approaches an enclosed path in the phase plane, the system has sustained oscillation. The enclosed path is called the limit cycle.

Construction of the Phase Portrait, Method of Isoclines—The slope of $d\dot{x}/dx$ of equation (10) is simply the slope of the trajectory in the phase plane. The locus of constant $d\dot{x}/dx$ is termed an *isocline* corresponding to the slope α; that is

$$\alpha = d\dot{x}/dx$$
$$= -[f_2(x,\dot{x})/f_1(x,\dot{x})] - [f_3(x,\dot{x})/f_1(x,\dot{x})](x/\dot{x})$$
$$= -g(x,\dot{x}) - h(x,\dot{x})(x/\dot{x})$$

The phase portrait is constructed by plotting a large number of isoclines corresponding to the various slopes of the trajectory on the phase plane. All points located on the same isocline have the same slope α. Beginning at the location of the initial condition (x_0, $\dot{x}_0$), the trajectory traverses in the clockwise direction, crossing each isocline at an angle corresponding to that slope α. Fig. 62 shows the isocline for a damped, linear, second-order system. Isoclines for first- and second-order linear differential equations are straight lines.

Determination of Time on the Phase Plane—*TIME FROM RECIPROCAL PLOT*: This method is based on the relationship of time and the reciprocal plot of $\dot{x}$. Since

$$\dot{x} = dx/dt$$

dt may be expressed as

$$dt = dx/\dot{x}$$

Integrating both sides yields

$$t = \int_{x0}^{x1} (1/\dot{x})dx$$

Since $\dot{x}$ as a function of x is known from the phase plane, the reciprocal $1/\dot{x}$ may be plotted as a function of x, and the integral under the curve between any two points is the time required for the trajectory to change from one point to the other.

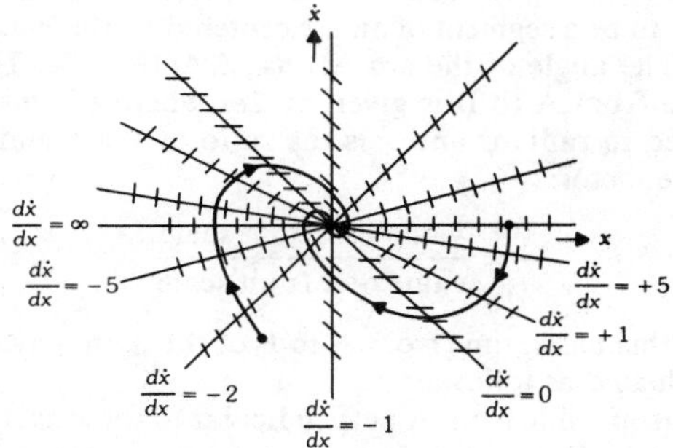

Fig. 62. Isocline method.

A typical reciprocal plot of $\dot{x}$ to determine time is shown in Fig. 63. The integration is in the direction of increasing time from right to left in the lower half-plane and vice versa in the upper half-plane.

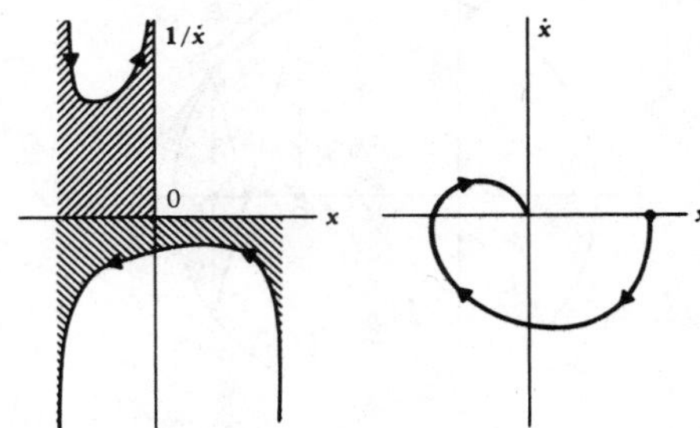

Fig. 63. Reciprocal plot of $\dot{x}$ for determining time.

In the vicinity of the x axis, the function $1/\dot{x}$ approaches infinity. Even with the integral unbounded the integral is finite; however, it is not readily evaluated graphically.

As an alternative method, the value of time may be evaluated from integration involving $\dot{x}$ instead of x. From the original differential equation (9)

$$d\dot{x}/dt = Q(x,\dot{x})$$

dt is written as

$$dt = [1/Q(x,\dot{x})]d\dot{x}$$

and therefore

$$t = \int_{\dot{x}_1}^{\dot{x}_2} [1/Q(x, \dot{x})] d\dot{x}$$

The expression $1/Q(x, \dot{x})$ may be plotted as a function of $\dot{x}$, and graphic integration may be performed.

The two methods may be used alternately if the integral goes to infinity.

GRAPHIC CONSTRUCTION: This method is based on the approximation of the phase trajectory by a series of circular arcs centered on the x axis (Fig. 64A). Consider the section of the path $\overline{AB}$ to be a segment of an arc centered on the x-axis O. The angle of the arc $\overline{AB}$ is $\angle AOB = 2\epsilon$. The time from A to B is given by $2\epsilon\tau$ where ϵ is measured in radians and τ is the ratio of the x and $\dot{x}$ scale factor:

$$\tau = \frac{\text{value of } x/\text{unit scale}}{\text{value of } \dot{x}/\text{unit scale}}$$

On this basis, time from A to B of the path may be evaluated as follows.

Drop a line from A perpendicular to the x axis at point D. Drop a line from B perpendicular to the x axis, and scale a distance equal to $\overline{BE}$ on the opposite side of the x-axis ($\overline{CE}$). Draw a line $\overline{AC}$ connecting points A and C. Measure the angle $\angle$ CAD in radians. The time from point A to B is then

$$t = 2\angle \text{CAD} \cdot \tau$$

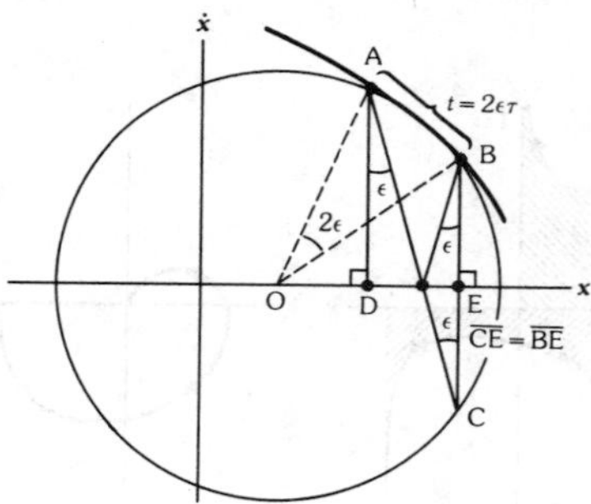

(A) Procedure.

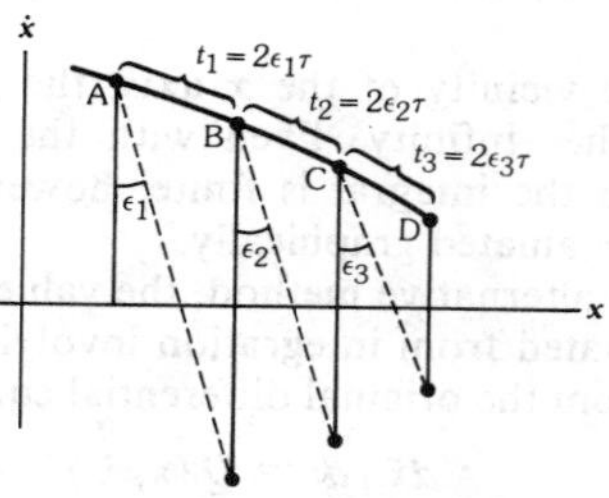

(B) Example.

Fig. 64. Graphic construction for determining time.

Repeat the procedure as the points move along the trajectory. A typical example is shown in fig. 64B.

SINGULAR POINTS: In a second-order system, the differential equation of the system may be described by two variables x and $\dot{x}$ in the following form:

$$dx/dt = P(x, \dot{x})$$

$$d\dot{x}/dt = Q(x, \dot{x})$$

The points where dx/dt and $d\dot{x}/dt$ vanish are called *singular points.* At a singular point, the system is in a state of equilibrium.

The importance of a singular point in the phase plane is how the trajectories of the phase portrait behave in the vicinity of the singular point. When the trajectory converges to the singular point the system is stable, whereas if it diverges the system is unstable. Typical singular points are described below.

TYPES OF SINGULAR POINTS: Besides stable and unstable equilibrium, the singular points may be classified into node, focus, center, and saddle points.

Consider, for example, a singular point at $x = a$ and $\dot{x} = b$ of equations (8) and (9). At a singular point the derivatives dx/dt and $d\dot{x}/dt$ are both zero, and the location may be solved in the phase plane by setting (8) and (9) equal to zero. A singular point exists at $x = a$, $\dot{x} = b$, and the functions P and Q can be expressed in terms of the Taylor series about those points; then

$$dx/dt = c_1(x-a) + c_2(\dot{x}-b) + c_3(x-a)^2 + c_4(x-a)(\dot{x}-b) + c_5(\dot{x}-b)^2 + \cdots$$

$$d\dot{x}/dt = d_1(x-a) + d_2(\dot{x}-b) + d_3(x-a)^2 + d_4(x-a)(\dot{x}-b) + d_5(\dot{x}-b)^2 + \cdots$$

If a sufficiently small region around the singular point is taken, the derivatives are dominated by the linear terms and hence quantities c_1, c_2, d_1, and d_2. By changing the variables, the singular point may be moved to the origin. Then the system equation may be rewritten as

$$dx/dt = p_1x + p_2\dot{x} = 0$$

$$d\dot{x}/dt = q_1x + q_2\dot{x} = 0$$

and the characteristic equation is

$$\lambda^2 - (p_1 + q_2)\lambda + (p_1q_2 - p_2q_1) = 0$$

The roots of the characteristic equation determine the nature of the critical points. The roots are

$$\lambda = \tfrac{1}{2}\{(p_1 + q_2) \pm [(p_1 + q_2)^2 - 4(p_1q_2 - p_2q_1)]^{1/2}\}$$

There are six possible cases for the six types of singular points.

A. The roots are real and are both negative. If the initial condition is $x_0 \neq 0, \dot{x}_0 = 0$, the trajectory approaches the singular point without an overshoot or oscillation and is called a *stable node* (Fig. 65A).

B. The roots are complex conjugate with negative real parts. The trajectory displays a spiraling response as it converges to the singular point as shown in Fig. 65B. This is called the *stable-focus* type of singularity.

C. The roots are conjugate and pure imaginary. The response exhibits a sustained oscillatory motion with the amplitude dependent on the initial condition. The trajectory displays a family of ellipses about the singular point, and is termed *center* (Fig. 65C).

D. The roots are both positive real. The response in the time domain increases exponentially and is unstable. The portrait is the same as the stable node except the trajectory diverges from the singular point. This is termed *unstable node* (Fig. 65D).

E. The roots are complex conjugate with positive real parts. The phase portrait is the same as the stable focus except the trajectory diverges from the singular point. This is termed the *unstable focus* (Fig. 65E).

F. The roots are real with one negative and the other positive. The phase portrait consists of a family of curves of the hyperbolic type having $k_1 = (\lambda_1 - p_1)/p_2$ and $k_2 = (\lambda_2 - p_1)/p_2$ for its asymptotes. The direction of the paths is toward the singular point on the negative asymptote and away from the singular point on the positive asymptote. Singular points of this type are called *saddle points* and are unstable (Fig. 65F).

The six types of singularities correspond to the six regions of Fig. 66.

BIBLIOGRAPHY

1. Cunningham, W. J. *Introduction to Nonlinear Analysis.* New York: McGraw-Hill Book Co., 1958.
2. Evans, W. R. "Graphical Analysis of Control Systems." *Trans. AIEE,* Vol. 67 (1948), pp. 547–551.

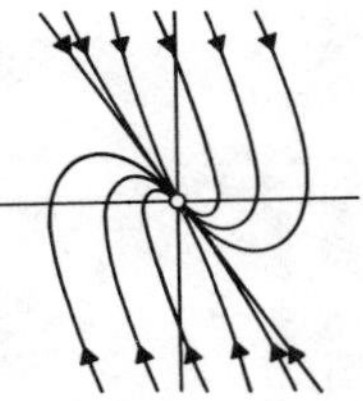

(A) Stable node.

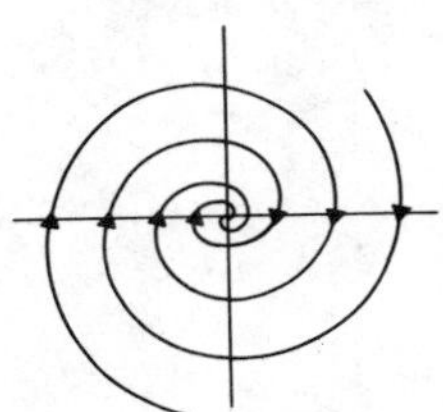

(B) Stable focus.

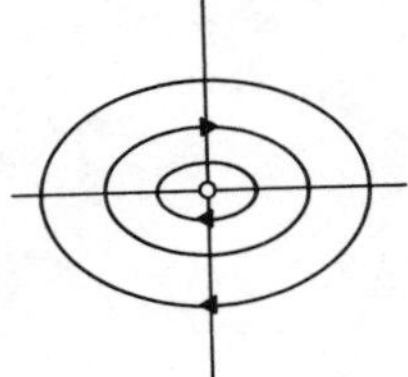

(C) Center.

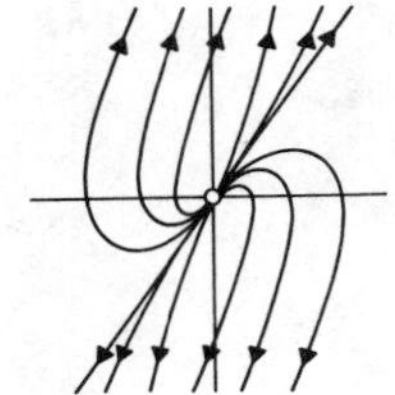

(D) Unstable node.

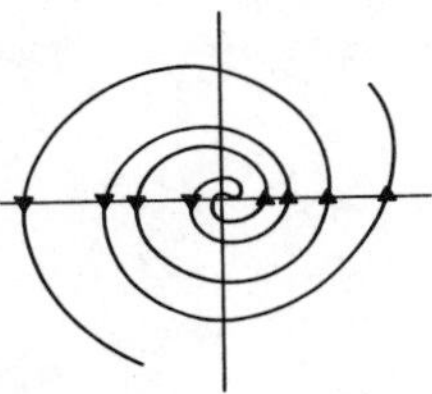

(E) Unstable focus.

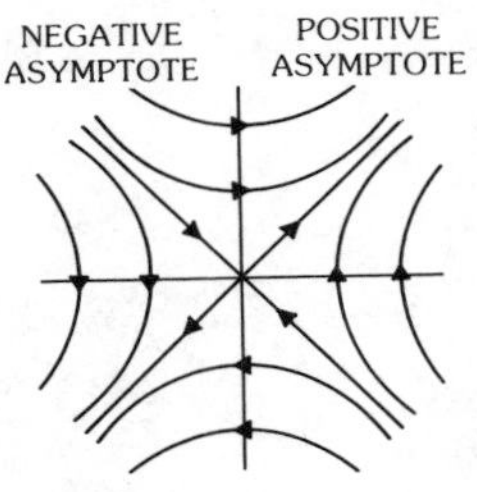

(F) Saddle point.

Fig. 65. Types of singular points.

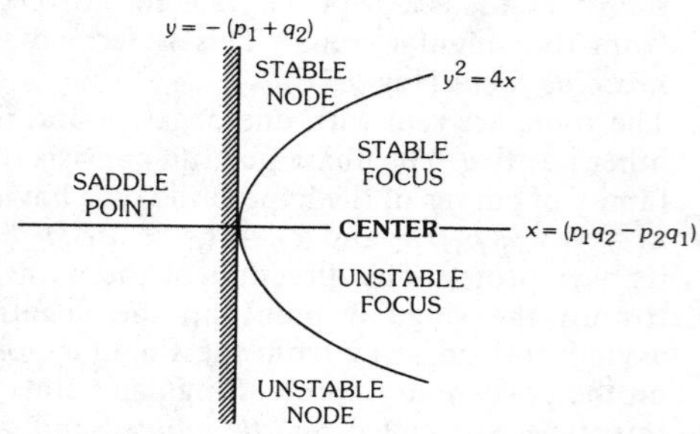

Fig. 66. Regions of various singular points.

3. Evans, W. R. "Control System Synthesis by Root Locus Method." *Trans. AIEE,* Vol. 69 (1950), pp. 66–69.
4. Evans, W. R. *Control System Dynamics.* New York: McGraw-Hill Book Co., 1954.
5. Gibson, J. E. *Nonlinear Automatic Control.* New York: McGraw-Hill Book Co., 1963.
6. Graham, D. and McRuer, D. *Analysis of Nonlinear Control Systems.* New York: John Wiley & Sons, Inc., 1961.
7. Hsu, J. C. and Meyer, A. U. *Modern Control Principles and Applications.* New York: McGraw-Hill Book Co., 1968.
8. Ince, E. L. *Ordinary Differential Equations.* New York: Dover Publications, 1953.
9. Kuo, B. C., ed. *Incremental Motion Control, Vol. 2, Step Motors and Control Systems.* Champaign, Ill.: SRL Publishing Co., 1979.
10. Kuo, B. C. *Digital Control Systems.* New York: Holt, Rinehart and Winston, 1980.
11. Kuo, B. C. *Automatic Control Systems.* 4th ed. Englewood Cliffs, N.J.: Prentice-Hall, Inc., 1982.
12. McLachlan, N. W. *Ordinary Nonlinear Differential Equations in Engineering and Physical Sciences.* Oxford: University Press, 1955.
13. Moore, A. W. "Phase-Locked Loops for Motor Speed Control." *IEEE Spectrum,* April 1973, pp. 61–67.
14. Tal, J. and Rushforth, K. "Phase-Locked Servo Systems," Chap. 14, *Incremental Motion Control, Vol. 1, DC Motors and Control Systems.* B. C. Kuo, ed. Champaign, Ill.: SRL Publishing Co., 1978.
15. Truxal, J. G. *Automatic Feedback Control System Synthesis.* New York: McGraw-Hill Book Co., 1955.

16 Electron Tubes

Revised by
Marvin Chodorow

ELECTRON EMISSION

All electron tubes* depend for their operation on the flow of electrons within the tube, through either high vacuum or an ionized gas. The electrons are emitted from a cathode surface as a result of one of four processes that are distinguished on the basis of the mechanism by which the electrons are enabled to leave the surface. These processes are elevated temperature (thermionic or primary emission); bombardment by other particles, generally electrons (secondary emission); the action of a high electric field (field emission); or the incidence of photons (photoemission).

Thermionic Emission

Thermionic emission occurs when the electrons in the cathode material have enough thermal energy to overcome the forces at the surface and escape.

The thermal emission of electrons from metals obeys the Richardson-Dushman equation

$$J_0 = AT^2 \exp(-11600\phi_0/T)$$

where,

J_0 is emission density in amperes/cm^2,
A is a constant [amperes/cm^2(K)^2],
ϕ_0 is the work function (electronvolts),
T is temperature in kelvins.

Both A and ϕ_0 are characteristic of the specific material.

The current density given by this equation is usually referred to as the saturation emission current density. Typical constants are given in Table 1 for several commonly used cathode materials.

The maximum current of which a cathode is capable at the operating temperature is known as the saturation current and is normally taken as the value at which the current first fails to increase as the three-halves power of the voltage causing the current. Thoriated-tungsten filaments for continuous-wave operation are usually assigned an available emission of approximately half the saturation value. Oxide-coated emitters do not have a well-defined saturation point and are designed empirically. The available emission from the cathode must be at least equal to the sum of the peak currents drawn by all the electrodes.

Fig. 1 gives a plot of saturation current as a function of temperature for several types of emitters in common use. The shaded blocks at the bottom of the figure show the normal operating range for three of the cathodes. The curves labeled (A) through (F) are explained as follows. (A) *The oxide-coated cathode.* Curve A_1 gives the saturation emission current density under pulsed conditions. Curve A_2 gives the direct-current saturation emission density. The position of this curve may vary substantially with environmental conditions. Direct-current densities much in excess of 0.5 A/cm^2 lead to relatively short cathode life. (B) *The pressed nickel cathode.* Curve B shows the direct-current saturation emission current density obtained from a pressed nickel cathode. (C) *The impregnated nickel cathode.* Curve C shows the saturation emission current obtained from the impregnated nickel cathode. The measurements were taken with 40-microsecond pulses and a repetition rate of 60 pulses per second. (D) *Pressed and impregnated tungsten cathodes.* Curve D shows the saturation emission density obtained from pressed and impregnated tungsten cathodes based on $A = 2.5$ A/cm^2 (K)^2 and $\phi_0 = 1.67$ electronvolts. (E) *The thoriated-tungsten cathode.* Curve E shows

TABLE 1. COMMONLY USED CATHODE MATERIALS

Type	A	ϕ_0	Efficiency (milliamperes/watt)	Specific Emission (amperes/cm^2)	Emissivity (watts/cm^2)	Operating Temperature (K)	Resistance Ratio (hot/cold)
Bright tungsten (W)	70	4.50	5–10	0.25–0.7	70–84	2500–2600	14/1
Thoriated tungsten (Th-W)	4	2.65	40–100	0.5–3.0	26–28	1950–2000	10/1
Tantalum (Ta)	37	4.12	10–20	0.5–1.2	48–60	2380–2480	6/1
Oxide coated (Ba-Ca-Sr)	*	1.0–1.3	50–150	0.5–2.5	3–5	1000–1150	2.5 to 5.5/1
Impregnated	2.4	1.65		1.8–5.4	2.6–3.8	1300–1400	

* The Richardson-Dushman equation does not apply to a composite surface of this type.

*J. W. Gewartowski and H. A. Watson, *Principles of Electron Tubes* (Princeton, N.J.: Van Nostrand, 1965). K. R. Spangenberg, *Vacuum Tubes,* 1st ed. (New York: McGraw-Hill Book Co., 1948). A. H. W. Beck, *Thermionic Valves, Their Theory and Design* (London: Cambridge University Press, 1953). *Standards on Electron Tubes: Definitions of Terms* (New York: Institute of Radio Engineers, 1950).

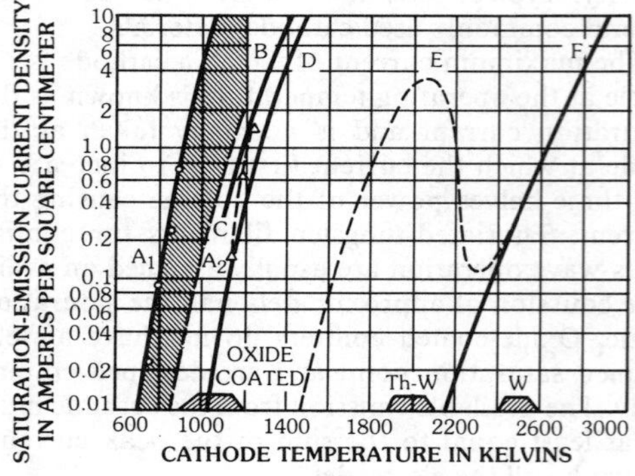

Fig. 1. Emission current density vs cathode temperature for several types of thermionic emitters. (*From J. W. Gewartowski and H. A. Watson,* Principles of Electron Tubes, *1965; p. 42. Courtesy D. Van Nostrand Co., Inc.*)

the measured saturation emission current density of an uncarburized thoriated-tungsten filament. (F) *Tungsten filaments.* Curve F shows the saturation emission current density of a tungsten filament based on $A = 70\ \mathrm{A/cm^2\ (K)^2}$ and $\phi_0 = 4.5$ electronvolts.*

Thoriated-tungsten and oxide-coated emitters should be operated close to specified temperature. A customary allowable heating-voltage deviation is ±5 percent. Bright-tungsten emitters may be

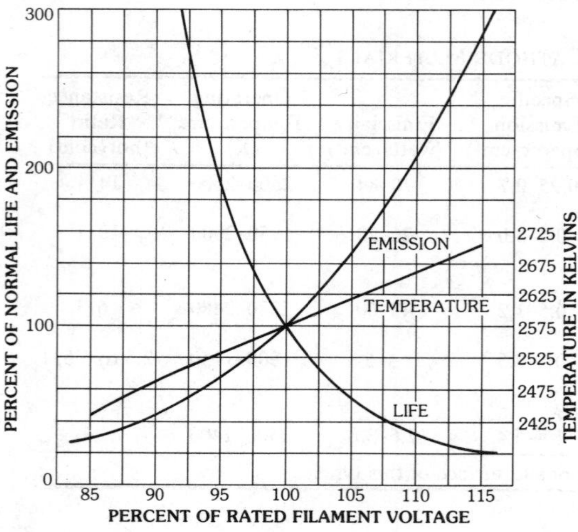

Fig. 2. Effect of change in filament voltage on the temperature, life, and emission of a bright-tungsten filament (based on 2575-K normal temperature of filament).

*J. W. Gewartowski and H. A. Watson, *Principles of Electron Tubes* (Princeton, N.J.: D. Van Nostrand Co., Inc., 1965).

operated at the minimum temperature that will supply required emission as determined by power-output and distortion measurements. Life of a bright-tungsten emitter is lengthened by lowering the operating temperature. Fig. 2 shows a typical relationship between filament voltage and temperature, life, and emission.

Mechanical stresses in filaments due to the magnetic field of the heating current are proportional to I_f^2. Current through a cold filament should be limited to 150 percent of the normal operating value for large tubes, and 250 percent for medium types. Excessive starting current may easily warp or break a filament.

Secondary Emission

When the surface of a solid is bombarded by charged particles having appreciable velocity, electrons are emitted from the solid. This is the process of secondary emission,* the most important case being when the bombarding particles are also electrons. One then differentiates between incident and emitted electrons by the terms primary and secondary, respectively. The latter term commonly describes all electrons collected from a secondary emitter; these electrons comprise three groups: (A) true secondaries, (B) inelastically reflected primaries, and (C) elastically reflected primaries. True secondaries are considered to be those of the solid which have been excited above the energy level required for escape across the surface barrier. The three groups are separable to a degree on the basis of energy as indicated in the energy distribution curve of Fig. 3. True secondaries constitute the bulk of

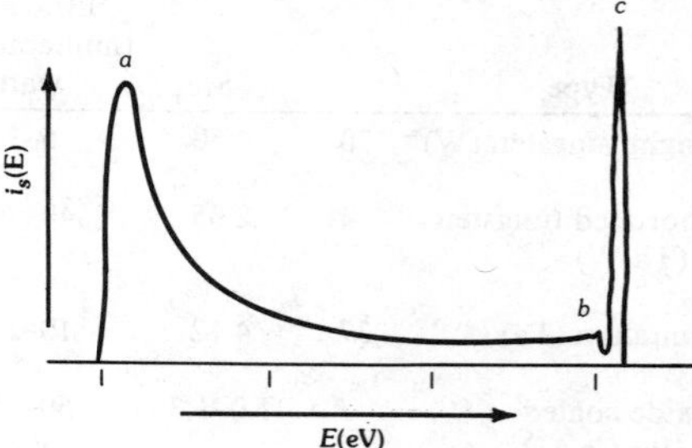

Fig. 3. Total energy distribution of secondary electrons.

*H. Bruining, *Physics and Applications of Secondary Electron Emission* (New York: McGraw-Hill Book Co., 1954). O. Hackenberg and W. Brauer, *Secondary Electron Emission from Solids,* Advances in Electronics and Electron Physics, Vol. XI (New York: Academic Press, 1959). A. J. Dekker, *Secondary Electron Emission,* Solid State Physics, Vol. 6 (New York: Academic Press, 1958). R. Kollath, *Sekundarelektronen-Emission fester Korper bei Bestrahlung mit Elektronen,* Handbuch der Physik, Band XXI (Berlin: Springer-Verlag, 1956).

emitted electrons at moderate primary energies and have a mode energy of at most a few electronvolts. Their distribution is almost independent of primary energy. Electrons in the relatively flat interval within *a, b* constitute a mixture of true secondaries and inelastically reflected primaries. It has become customary to arbitrarily designate those emitted electrons having energies less than 50 electronvolts as true secondaries.

Total secondary yield δ, defined as the ratio of secondary to primary electron current, is independent of primary current but strongly dependent on primary energy as indicated in Fig. 4. The shape of the yield curve follows from generating and escape mechanisms, the former leading to an initial rise in yield with primary energy, and the latter causing an eventual reduction owing to increased penetration of primaries and a greater mean depth of escape of secondaries. Significant points of the yield curve are first and second crossover at which yield becomes unity, the maximum yield δ_m, and the primary energy eV_m at which the maximum occurs. For most insulators, first crossover occurs between 15 and 25 eV primary energy. Insulators generally exhibit higher yields than conductors, a property attributed to the absence of conduction electrons which tend to reduce the mean energy and the escape probability of secondaries through collision losses within the solid. The yield of insulators decreases noticeably as temperature is increased, owing to increasing electron-phonon interaction.

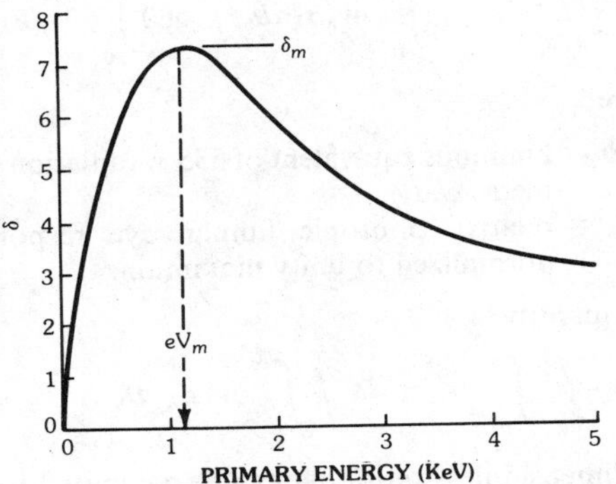

Fig. 4. Secondary-emission yield curve.

Secondary yield increases with angle of primary incidence, the effect being most pronounced at high primary energies. Yield is also a function of surface structure and may be minimized by employing physical trapping such as provided by a porous surface. Lowest yields are obtained for porous carbon deposits and highest yields for single crystal insulators having low electron affinity. Secondary yield may also be influenced by internal electric fields which tend to assist or retard the escape of secondaries. If such fields are strongly dependent on charge transport within the bombarded material, yield may become dependent on primary current which can in turn give rise to anomalous time-dependent effects. Barring such effects, it appears that the interaction time for the secondary-emission process is of the order of 10^{-12} second.

When the rate of bombardment by primary electrons becomes very low, as in single-electron counting, the statistical nature of the secondary-emission process becomes evident. The probability of obtaining 0, 1, 2, ···, *n* secondaries per incident primary is given by the Poisson distribution.

Commonly used secondary-emission materials are silver-magnesium or copper-beryllium alloy processed to provide a high-yield partly conductive surface film. Typically such surfaces exhibit yields of 2.5 to 4 at 100 eV primary energy.

Secondary emission is employed advantageously in the operation of many electron devices, such as camera tubes, storage tubes, and image intensifiers. A most important application lies in secondary-electron multiplication, which provides a means for amplifying very weak electron currents as in photomultiplier tubes. A conventional electron multiplier consists of a number of secondary-emitting dynodes operated at progressively higher potentials and terminated by an electron-collecting electrode. Electrons incident on the first dynode are multiplied, the resultant secondaries are accelerated to the second dynode where the process is repeated, and so on throughout the multiplier structure.

If the dynodes exhibit uniform yield characteristics, the overall amplification G obtained from a multiplier having n dynodes is

$$G = g^n$$

where g is the gain per dynode. The actual gain g may be slightly less than the secondary yield δ because of multiplier geometry. In the absence of appreciable space-charge effects, maximum gain is realized when the available potential is uniformly distributed across the dynode chain. An empirical relation for g may be obtained by approximating the initial portion of the yield curve; that is

$$g \cong \delta \cong A \Delta V^m$$

In this equation, ΔV is the interdynode potential (primary energy), and A and m are empirical constants. Using this approximation, one obtains

$$g_{opt} = \epsilon^m$$
$$n_{opt} = A^{1/m}(V/\epsilon)$$

(ϵ = base of natural logarithms), where g_{opt} and n_{opt} are the optimum values of gain/dynode and number of dynodes, respectively, for maximizing overall amplification, given a total potential V available for distribution across the dynodes. The number of dynodes n is taken as the closest integer to the calculated value. In some cases, deviation from a uniform potential distribution and optimum gain conditions is advantageous, for example to reduce space-charge effects and improve time-delay and time-dispersion properties. Increasing the energy of electrons incident on the first dynode improves signal-to-noise ratio and single-electron-counting capability.

In addition to the conventional discrete dynode multiplier, a number of novel arrangements have been devised including crossed-field strip multipliers and tubular multipliers which commonly employ a continuous semiconductive dynode surface for potential distribution and multiplication. Tubular multipliers, when formed into a parallel array of small-diameter elements, may be employed for electron image intensification. Another form of multiplier commonly used for this purpose is the transmission secondary-emission multiplier, wherein secondaries exit from the side opposite primary incidence. The structure normally takes the form of a thin-film or porous supported layer, having the side of primary incidence made electrically conductive.

Field Emission

If an electric field of sufficient magnitude is offered to the surface of a metal, the potential barrier at the surface will be lowered, allowing the escape of electrons, and field emission* will result. The current has been found to vary with the applied field in accordance with

$$J = CE^2 \exp(-D/E)$$

where,

J is the current density in A/cm^2,
E is the electric field at the surface,
C and D are approximately constant coefficients.

Coefficient D is determined mainly by the work function. Field emission must be taken into account in the design of very-high-voltage tubes and apparatus, and it is a factor in the operation of cold-cathode gas tubes. Although development is being carried on, there has been little use made yet of field emission in high-vacuum tubes.

Photocathode Response

Response to Spectrally Distributed Sources— The total photocurrent I in amperes emitted by a photocathode subjected to a spectrally distributed input flux is given by

$$I = w_{\lambda\max} s_{\lambda\max} \int_0^\infty w_\lambda \sigma_\lambda \, d\lambda$$

$$= W_{\lambda_1\lambda_2} s_{\lambda\max} \left(\int_0^\infty w_\lambda \sigma_\lambda \, d\lambda \Big/ \int_{\lambda_1}^{\lambda_2} w_\lambda d_\lambda \right)$$

where,

$w_{\lambda\max}$ = peak input flux spectral density (watts/m),
$s_{\lambda\max}$ = peak monochromatic radiant sensitivity of the photocathode (A/watt),
w_λ = relative spectral distribution of the input flux,
σ_λ = relative spectral distribution of the radiant sensitivity of the photocathode,
λ = wavelength,
$W_{\lambda_1\lambda_2}$ = flux in watts within the wavelength interval $\lambda_1 \leq \lambda \leq \lambda_2$.

Typical values of the dimensionless "spectral matching factor" ratio

$$\int_0^\infty w_\lambda \sigma_\lambda \, d\lambda \Big/ \int_{\lambda_1}^{\lambda_2} w_\lambda \, d\lambda$$

appearing in the above relationship and describing the comparative spectral match between various photocathode and input-flux spectral distributions are shown in Table 2 evaluated for $\lambda_1 = 0$ and $\lambda_2 = 1.2\ \mu m = 1.2 \times 10^{-6}$ meter.

If the input flux is measured in lumens L instead of watts, the resultant ratio I/L of emitted current I to flux input is designated as the photocathode luminous sensitivity S, and is given by

$$S = I/L = s_{\lambda\max} \int_0^\infty w_\lambda \sigma_\lambda \, d\lambda \Big/ 680 \int_0^\infty w_\lambda E_\lambda \, d\lambda$$

where,

680 = luminous equivalent of 555 Å radiation (lumens/watt),
E_λ = relative photopic human eye response, normalized to unity maximum.

The quantity

$$\int_0^\infty w_\lambda \sigma_\lambda \, d\lambda \Big/ \int_0^\infty w_\lambda E_\lambda \, d\lambda$$

is a dimensionless ratio that can be computed from the spectral-matching-factor data in Table 2 for various photocathode and input-flux spectral distributions by dividing the spectral-matching-factor data for

$$\int_0^\infty w_\lambda \sigma_\lambda \, d\lambda \Big/ \int_{\lambda_1}^{\lambda_2} w_\lambda \, d\lambda$$

by the spectral-matching-factor data for

$$\int_0^\infty w_\lambda E_\lambda \, d\lambda \Big/ \int_{\lambda_1}^{\lambda_2} w_\lambda \, d\lambda$$

In the special case where the input spectral distri-

*R. H. Fowler and L. Nordheim, *Royal Society Proceedings*, Vol. 119, p. 173; 1928.

TABLE 2. SPECTRAL MATCHING FACTORS

			$\int_0^\infty w_\lambda \sigma_\lambda d\lambda \Big/ \int_0^{1.2\mu} w_\lambda\, d\lambda$			and $\int_0^\infty w_\lambda E_\lambda\, d\lambda \Big/ \int_{\lambda 1}^{\lambda 2} w_\lambda\, d\lambda$
			Photocathode Type			
Source	λ_1	λ_2	S1	S11	S20	Photopic Eye
2854 K lamp	0	1200	0.52	0.060	0.112	0.071
5000 K blackbody	0	1200	0.53	0.26	0.34	0.140
Mean solar flux	0	1200	0.54	0.32	0.36	0.197
P1 phosphor	0	∞	0.28	0.28	0.69	0.768
P4 phosphor	0	∞	0.31	0.67	0.73	0.402
P11 phosphor	0	∞	0.22	0.91	0.88	0.201
P20 phosphor	0	∞	0.39	0.42	0.58	0.707
NaI(Th)	0	∞	0.53	0.88	0.90	0.046

Note: λ_1 and λ_2 are in nanometers.

bution w_λ, designated as w_λ (2854), corresponds to a standard 2854 K color-temperature tungsten lamp, the luminous sensitivity S, designated as S(2854), and often used as a specification on photocathode sensitivity, is given by

$$S(2854) = s_{\lambda \max}$$

$$\times \int_0^\infty w_\lambda(2854)\, \sigma_\lambda\, d\lambda \Big/ 681 \int_0^\infty w_\lambda(2854)\, E_\lambda\, d\lambda$$

This equation relates the peak cathode monochromatic radiant sensitivity $s_{\lambda \max}$ (also commonly used to specify cathode sensitivity) to the standard luminous sensitivity S(2854). Since the wavelength at which the cathode quantum efficiency has its peak value does not correspond, in general, to the wavelength at which the radiant sensitivity has its peak value, no general relationship exists between the peak radiant sensitivity $s_{\lambda \max}$ and the peak quantum efficiency $q_{\lambda \max}$.

Photocathode Response With Optical Filter— The ratio of (*A*) the emitted photocurrent, I(filter), with a filter inserted between a given flux source and the photocathode to (*B*) the current without the filter, I(no filter), is called the filter factor $T(t_\lambda, w_\lambda, \sigma_\lambda)$ and is given by

$$T(t_\lambda, w_\lambda, \sigma_\lambda) = I(\text{filter}) / I(\text{no filter})$$

$$= \int_0^\infty t_\lambda w_\lambda \sigma_\lambda\, d\lambda \Big/ \int_0^\infty w_\lambda \sigma_\lambda\, d\lambda$$

where t_λ is the transmission of the filter at a given wavelength λ, and the notation $T(t_\lambda, w_\lambda, \sigma_\lambda)$ indicates that the filter factor is a function not only of the filter transmission t_λ but also of the detector response σ_λ and the source distribution w_λ. Typical filter factors are given in Table 3.

The ratio of emitted photocurrent with the filter, I(filter), to the flux in lumens, L(2854), incident on the filter (not on the cathode) from a 2854 K source is designated as S(photocathode + filter) and is given by

$$S(\text{photocathode} + \text{filter}) = I(\text{filter}) / L(2854)$$

$$= S(2854)\, T(t_\lambda, 2854, \sigma_\lambda)$$

The magnitude of the luminous sensitivity, S(photocathode + filter), in amperes per lumen is used to specify cathode sensitivity, or more pre-

TABLE 3. TYPICAL FILTER FACTORS

Filter					Filter Factor $\int_0^\infty t_\lambda w_\lambda \sigma_\lambda\, d\lambda \Big/ \int_0^\infty w_\lambda \sigma_\lambda\, d\lambda$, $w_\lambda = w_\lambda(2854)$			
					Photocathode Type			
Manufacturer	Glass Number	Thickness	Color Series	Description	S1	S4	S11	S20
Corning	2540	stock	CS 7-56	Infrared	0.108	0.000	0.000	0.000
Corning	5113	½ stock	CS 5-58	Blue	0.004	0.126	0.103	0.055
Corning	2403	stock	CS 2-58	Deep red	0.788	0.114	0.112	0.257

cisely, cathode-plus-filter sensitivity, over a selected spectral region, where the filter is chosen to restrict the flux incident on the photocathode to the desired region. The sensitivity, *S*(photocathode + filter), is then designated as the "infrared" sensitivity, or "red" sensitivity, or "blue" sensitivity, et cetera, depending on the predominant spectral region passed by the filter.

ELECTRODE DISSIPATION

After the electron stream has given up the useful component of its energy, the remainder is dissipated as heat in some suitable part of the tube. Five processes are commonly used to remove this heat. The amount that can be removed depends on the area available, the temperature differential, and, in the cases of forced cooling, the coolant flow.

In computing cooling-medium flow, a minimum velocity sufficient to assure turbulent flow at the dissipating surface must be maintained. The figures for specific dissipation (Table 4) apply to clean cooling surfaces and may be reduced to a small fraction of the values shown by heat-insulating coatings such as scale or dust.

TABLE 4. TYPICAL OPERATING DATA FOR COMMON TYPES OF COOLING

Type	Average Cooling Surface Temperature (°C)	Specific Dissipation of Cooling Surface (watts/cm²)	Cooling-Medium Supply
Radiation	400–1000	4–10	
Water	30–150	30–110	0.25–0.5 gallon/minute/kilowatt
Forced air	150–200	0.5–1	50–150 feet³/minute/kilowatt
Evaporative	100–120	80–125	Water-, air-, or convection-cooled condenser. A water-cooled condenser would require 0.07–0.1 gallon/minute/kilowatt
Conduction	100–250	5–30	Heat sink operating at 50–100°C

TABLE 5. TOTAL THERMAL EMISSIVITY ϵ_t OF ELECTRON-TUBE MATERIALS

Material	Temperature (K)	Thermal Emissivity
Aluminum	450	0.1
Anode graphite	1000	0.9
Copper	300	0.07
Molybdenum	1300	0.13
Molybdenum, quartz-blasted	1300	0.5
Nickel	600	0.09
Tantalum	1400	0.18
Tungsten	2600	0.30

Radiation Cooling

In a radiation-cooled system, that portion of the tube on which the heat is dissipated is allowed to reach a temperature such that the heat is radiated to the surroundings. The amount of heat that can be removed in this manner is given by the equation

$$P = \epsilon_t \sigma (T^4 - T_0^4)$$

where,

P = radiated power in watts/centimeter²,
ϵ_t = total thermal emissivity of the surface,
σ = Stefan-Boltzmann constant = 5.67×10^{-12} W cm^{-2} K^{-4},
T = temperature of radiating surface in kelvins,
T_0 = temperature of surroundings in kelvins.

Total thermal emissivity varies with the degree of roughness of the surface of the material and the temperature. Values for typical surfaces are in Table 5.

Water Cooling

For water cooling, the water is circulated through a suitably designed structure. The amount of heat that can be removed by this process is given by

$$P = 264 Q_W (T_2 - T_1)$$

where,

P = power in watts,
Q_W = flow in gallons per minute,
T_2, T_1 = outlet and inlet water temperatures, respectively, in kelvins.

This same relationship is given in the nomogram of Fig. 5 with the temperature rise in degrees Fahrenheit or Celsius and the power in kilowatts.

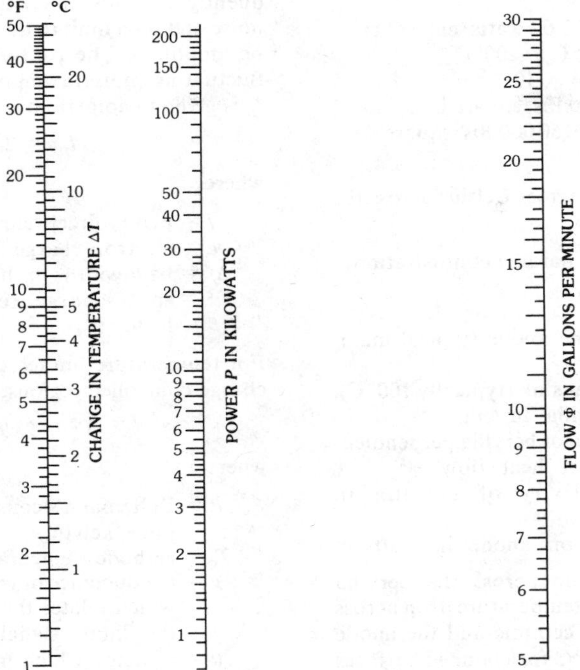

Fig. 5. Heat transfer in cooling water. $P = 0.1466\Phi\Delta T$, for T in Fahrenheit. (*Courtesy Clyde G. Haehnle*)

Forced-Air Cooling

With forced-air cooling, a stream of air is forced past a suitable radiator. The heat that can be removed by this process is given by

$$P = 169Q_A[(T_2/T_1) - 1]$$

where Q_A = air flow in feet^3/minute, other quantities as above.

Evaporative Cooling

A typical evaporative-cooled system consists of a tube with a specially designed anode immersed in a boiler containing distilled water. When power is dissipated on the anode, the water boils and the steam is conducted upward through an insulating pipe to a condenser. The condensate is then gravity fed back to the boiler, thus eliminating the pump required in a circulating water system.

For some transmitter applications, the steam is directed downward to leave the space above the tube available for other components. Such a system requires a pump to return the condensate to the boiler, but even then the pump has to handle only about 0.05 of the amount of water required for a water-cooled system because of the exploitation of the latent heat of steam.

The size of the heat-exchanger equipment for an evaporative-cooled system is less than one-third of that required for a water-cooled system because of the greater mean temperature differential between the cooled liquid and the secondary coolant. Typical temperature differentials for the two systems are 75° C and 30° C, respectively.

The anode dissipation should not exceed 135 watts per square centimeter of external anode surface because at this point, often referred to as the "Leidenfrost" or "calefaction" point, the surface becomes completely covered with a sheath of vapor, and the thermal conductivity between the anode and the cooling liquid drops to 30 watts per square centimeter, with resultant overheating of the anode. Special designs of the external anode surface (such as the "pineapple") allow up to 500 watts to be dissipated per square centimeter of internal anode surface.

Conduction Cooling

When an external heat sink is available, heat may be removed from the tube by conduction. Since the electrode where the heat appears is usually at an elevated potential, it is often necessary to conduct the heat through an electrical insulator.

Because of its relatively high thermal conductivity, beryllia ceramic can be used as a common insulator and thermal conductor between the anode of a tube and a heat sink.

Properties of beryllia:

Breakdown strength = 10 kV/mm

Dielectric constant = 6–8
Thermal conductivity = 2.62 watts/cm/°C at 20°C, 1.75 watts/cm/°C at 200°C
Dielectric loss factor = 4×10^{-5}
Tensile strength = 18 000 lbs/square inch
Compressive strength = 150 000 lbs/square inch

The temperature drop in degrees Celsius across the beryllia ceramic is given by

$$t_1 - t_2 = dW_a/KA \text{ (for a parallel configuration)}$$

where,

t_1 = temperature of tube anode (typical maximum 250°C),
t_2 = temperature of heat sink (typically 100°C),
d = thickness of beryllia in cm,
A = cross-sectional area of beryllia perpendicular to direction of heat flow,
K = thermal conductivity of beryllia in watts/cm/°C,
W_a = power dissipated on anode in watts.

To the temperature drop across the beryllia ceramic must be added the temperature drop across the interfaces between the ceramic and the anode and heat sink, typically 20°C for clamped surfaces at a loading of 25 watts/cm^2.

Because of its toxic nature, care must be taken in handling and disposal of beryllia ceramic.

Grid Temperature

Operation of grids at excessive temperatures will result in one or more harmful effects: liberation of gas, high primary (thermal) emission, contamination of the other electrodes by deposition of grid material, and melting of the grid. Grid-current ratings should not be exceeded, even for short periods.

NOISE IN TUBES

There are several sources of noise in electron tubes,* some of which are associated with the nature of electron emission and some of which are caused by other effects in the tube.

Shot Effect

The electric current emitted from a cathode consists of a large number of electrons and consequently exhibits fluctuations that produce tube noise and set a limit to the minimum signal that can be amplified. The root-mean-square value of the fluctuating (noise) component of the plate current I_n is given in amperes by

$$I_n^2 = 2eI\Gamma^2\Delta f$$

where,

I = plate direct current in amperes,
e = electron charge = 1.6×10^{-19} coulomb,
Δf = bandwidth in hertz,
Γ^2 = space-charge reduction or smoothing factor.

For temperature-limited cases, $\Gamma^2 = 1$. For space-charge-controlled regions

$$\Gamma^2 = 2kT_cg\theta/\sigma eI$$

where,

k = Boltzmann's constant = 1.380×10^{-23} joule/kelvin,
T_c = cathode temperature in kelvins,
g = conductance or transconductance in mhos, which relates the output signal current to the input signal voltage,*
θ = a factor which in most practical cases is nearly equal to its asymptotic value of $3[1 - (\pi/4)] = 0.644$,
σ = a tube parameter, related to the amplification factor and electrode spacings, which has a value of unity for diodes and varies between 0.5 and 1.0 for negative-grid tubes.

Partition Noise

Excess noise appears in multicollector tubes because of fluctuations in the division of the current between the different electrodes. In a grid-controlled tube, these fluctuations in current division reduce the effectiveness of the space-charge smoothing of the shot noise in the plate current. For a screen-grid tube, the root-mean-square noise currents in the cathode lead, the screen-grid lead, and the plate lead (I_{nk}, I_{nc2}, and I_n, respectively) are given by

$$I_{nk}^2 = 2eI_k\Gamma^2\Delta f$$
$$I_{nc2}^2 = 2eI_{c2}[(\Gamma^2 I_{c2} + I)/I_k]\Delta f$$
$$I_n^2 = 2eI[(\Gamma^2 I + I_{c2})/I_k]\Delta f$$

where I_k and I_{c2} are the cathode and screen-grid currents, respectively.

*S. Goldman, *Frequency Analysis, Modulation and Noise* (New York: Dover Publications, 1967). A. van der Ziel, *Noise* (Englewood Cliffs, N.J: Prentice-Hall, 1954). L. D. Smullin and H. A. Haus, eds., *Noise in Electron Devices* (The Technology Press of Massachusetts Institute of Technology and John Wiley & Sons, New York, 1959). D.H. Bell, *Electron Noise* (London: D. Van Nostrand Co., 1960). W. R. Bennett, *Electrical Noise* (New York: McGraw-Hill Book Co., 1960). D. K. C. MacDonald, *Noise and Fluctuations: An Introduction* (New York: John Wiley & Sons, 1962).

*For diodes, g is the conductance; for triode and pentode amplifiers, g is the transconductance g_m; and for triode or pentode mixers and converters, g is the conversion conductance g_c.

Flicker Effect

The mechanism involved in the flicker effect is not completely understood but appears to depend on the field distribution in the surface layer of the cathode due to its porous structure. Because this same field distribution also will influence the cathode activity and temperature, flicker noise will depend on cathode activity and temperature in a complicated manner.

The flicker noise spectrum is usually of the form $f^{-\alpha}$ with α close to unity and thus is important only at low frequencies. The sensitivity of audio, subaudio, and direct-current amplifiers is limited by the flicker noise generated in the first tube.

Collision Ionization

Free gas ions can be generated by collisions with the electron stream. The electrons thus liberated and collected by the anode will appear as noise in the anode circuit. The ions that travel to the cathode will travel slowly through the potential minimum and reduce the space charge, which in turn will reduce the space-charge smoothing effect. This also will increase the noise in the anode circuit.

Induced Noise

At high frequencies it is not necessary for electrons to reach an electrode for induced current to flow in the electrode leads. This noise is an important consideration in miniature tubes above 15 megahertz and becomes the principal limiting factor in low-noise amplifier design above about 100 megahertz. For microwave tubes, this is the dominant method by which beam noise is coupled to the output circuit.

Miscellaneous Noise

Other noise may be present due to microphonics, hum, leakage, charges on insulators, poor contacts, and secondary emission.

Microwave Tubes

The noise appearing in the output circuit of a microwave tube is due in part to induced noise from the beam. Also, some of the electrons may be intercepted by the radio-frequency structure (microwave cavity, slow-wave circuit, et cetera) giving rise to partition noise. In well-designed low-noise tubes, however, the latter effect is kept negligibly small.

For lossless linear beam tubes (traveling-wave amplifiers, klystron amplifiers, backward-wave amplifiers), the minimum obtainable noise figure F_{min} for one-dimensional single-velocity small-signal theory and high gain has been found to be given by

$$F_{min} = 1 + (2\pi/kT_0)(S - \pi)$$

where $S - \pi$ is the basic noise parameter and is established in the region of the potential minimum of the beam. If certain assumptions concerning the potential minimum are made, such as full shot noise and uncorrelated current and velocity fluctuations,* then values for S and π can be obtained. They are given as

$$\pi = 0$$
$$S = [1 - (\pi/4)]^{1/2}(kT_c/\pi)$$

therefore

$$F_{min} = 1 + (4 - \pi)^{1/2}(T_c/T_0)$$

For $T_c/T_0 = 4$, $F_{min} \approx 4$. The assumptions made are not entirely valid, as shown by the fact that noise figures of less than 4 have been obtained experimentally. At the present time values of S and π/S are obtained by measurement.

LOW-, MEDIUM- AND HIGH-FREQUENCY TUBES

This section applies particularly to triodes and multigrid tubes operated at frequencies where electron-inertia effects are negligible. Traditionally, the vacuum envelope of such tubes has been of glass with metal, usually copper, for the anode in larger sizes. In recent years, the trend has been toward ceramic in place of glass for the external insulating portions of such tubes. Fig. 6 shows a typical construction of a medium-power transmitting tube.

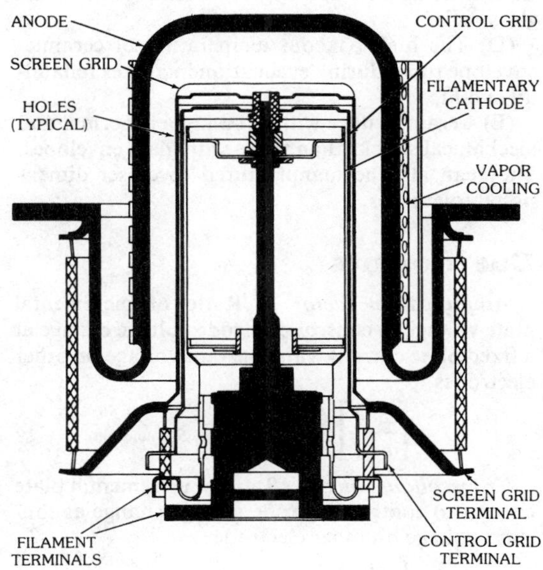

Fig. 6. Typical medium-power transmitting tube.

---

*J. R. Pierce, "A Theorem Concerning Noise in Electron Streams," *Journal of Applied Physics,* Vol. 25, p. 931; 1954.

Ceramic-envelope tubes have the following advantage over glass tubes.

(A) The radio-frequency loss P_{rf} in the seals of a tube is given by

$$P_{rf} = Kf^{5/2}R^{1/2}\mu^{1/2}$$

where,

K = constant,
f = frequency,
R = resistivity of the conducting material,
μ = permeability of the conducting material.

In glass-to-metal seals, the metal is normally of a magnetic material such as Kovar. As Kovar has high resistivity and permeability, the radio-frequency losses at the seals are therefore high, and at high frequencies cracking and/or glass suck-in near the seals can result. With ceramic-to-metal seals, this problem is minimized because the radio-frequency circulating currents at the seals flow through the metallizing and plating on the ceramic. The resistivity is low, and the permeability is unity.

(B) Ceramics have a lower dielectric loss than glass. Furthermore, the loss factor of glass rapidly rises with temperature. This leads to a "runaway" condition, glass suck-in, and hence severe limitation of maximum frequency of operation of glass tubes.

(C) The safe operating temperature of a ceramic-to-metal seal may be between 220 and 250 degrees Celsius as against 180 degrees Celsius for Kovar-glass seals.

(D) The high bakeout temperature of ceramic-envelope tubes during evacuation increases reliability and life.

(E) Ceramic tubes withstand higher thermal and mechanical shocks than those with glass envelopes. They can also be manufactured to closer dimensional tolerances.

Coefficients

Amplification factor μ: Ratio of incremental plate voltage to control-electrode voltage change at a fixed plate current with constant voltage on other electrodes

$$\mu = \left[\frac{\delta e_b}{\delta e_{c1}}\right]_{\substack{I_b,\, E_{c2}, \ldots, E_{cn} \text{ constant} \\ rl = 0}}$$

Transconductance s_m: Ratio of incremental plate current to control-electrode voltage change at constant voltage on other electrodes

$$s_m = \left[\frac{\delta i_b}{\delta e_{c1}}\right]_{\substack{E_b,\, E_{c2}, \ldots, E_{cn} \text{ constant} \\ rl = 0}}$$

When electrodes are plate and control grid, the ratio is the mutual conductance g_m

$$g_m = \mu / r_p$$

Variational (*ac*) *plate resistance* r_p: Ratio of incremental plate voltage to current change at constant voltage on other electrodes

$$r_p = \left[\frac{\delta e_b}{\delta i_b}\right]_{\substack{E_{c1}, \ldots, E_{cn} \text{ constant} \\ rl = 0}}$$

Total (dc) *plate resistance* R_p: Ratio of total plate voltage to current for constant voltage on other electrodes

$$R_p = \left[\frac{E_b}{I_b}\right]_{\substack{E_{c1}, \ldots, E_{cn} \text{ constant} \\ rl = 0}}$$

A useful approximation of these coefficients may be obtained from a family of anode characteristics, Fig. 7. Relationships between the actual geometry of a tube and its coefficients are given roughly in Table 6.

When the operating frequency is increased, the operation of triodes and multigrid tubes is affected by electron-inertia effects. The design features that distinguish the high-frequency tube shown in Fig. 9 from the lower-frequency tube (Fig. 6) are: reduced cathode-to-grid and grid-to-anode spacings, high emission density, high power density, small active and inactive capacitances, heavy terminals, short support leads, and adaptability to a cavity circuit.

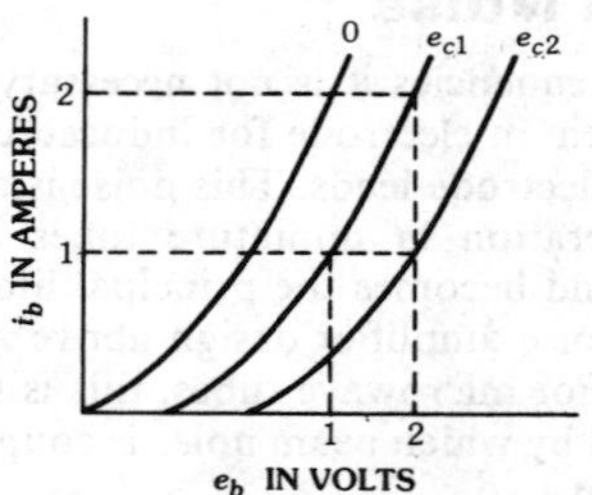

Amplification factor $\mu = (e_{b2} - e_{b1})/(e_{c2} - e_{c1})$
Mutual conductance $g_m = (i_{b2} - i_{b1})/(e_{c2} - e_{c1})$
Total plate resistance $R_p = e_{b2}/i_{b2}$
Variational plate resistance $r_p = (e_{b2} - e_{b1})/(i_{b2} - i_{b1})$

Fig. 7. Graphic method of determining coefficients.

MATERIALS AND STRUCTURES

Cathodes

Tubes for power levels up to about 10 kW often contain oxide-coated cathodes (barium oxide + calcium oxide + strontium oxide on nickel is typical). Operating temperatures are 750–800° C. Structures are plain cylinders or discs.

At higher power levels, thoriated tungsten, carburized, is most commonly found. Operating temperatures are 1600–1650° C. Structures are usually meshes or parallel wires.

Grids

Grids are conductors, semitransparent to electrons, usually in the form of meshes or parallel

TABLE 6. TUBE CHARACTERISTICS FOR UNIPOTENTIAL CATHODE AND NEGLIGIBLE SATURATION OF CATHODE

Function	Parallel-Plane Cathode and Anode	Cylindrical Cathode and Anode
Diode anode current (amperes)	$G_1 e_b^{3/2}$	$G_1 e_b^{3/2}$
Triode anode current (amperes)	$G_2[(e_b+\mu e_c)/(1+\mu)]^{3/2}$	$G_2[(e_b+\mu e_c)/(1+\mu)]^{3/2}$
Diode perveance G_1	$2.3\times10^{-6}(A_b/d_b^2)$	$2.3\times10^{-6}(A_b/\beta^2 r_b^2)$
Triode perveance G_2	$2.3\times10^{-6}(A_b/d_b d_c)$	$2.3\times10^{-6}(A_b/\beta^2 r_b r_c)$
Amplification factor μ	$2.7d_c[(d_b/d_c)-1]/[\rho \log(\rho/2\pi r_g)]$	$(2\pi d_c/\rho)[\log(d_b/d_c)/\log(\rho/2\pi r_g)]$
Mutual conductance g_m	$1.5G_2[\mu/(\mu+1)](E'_g)^{1/2}$	$1.5G_2[\mu/(\mu+1)](E'_g)^{1/2}$
	$E'_g=(E_b+\mu E_c)/(1+\mu)$	$E'_g=(E_b+\mu E_c)/(1+\mu)$

A_b=effective anode area in square centimeters; d_b=anode-cathode distance in centimeters; d_c=grid-cathode distance in centimeters; β=geometric constant (a function of ratio of anode-to-cathode radius), $\beta^2\approx1$ for $r_b/r_k>10$ (Fig. 8); ρ=pitch of grid wires in centimeters; r_g= grid-wire radius in centimeters; r_b=anode radius in centimeters; r_k=cathode radius in centimeters; r_c=grid radius in centimeters.

Note:

These equations are based on theoretical considerations and do not provide accurate results for practical structures; however, they give a fair idea of the relationship between the tube geometry and the constants of the tube.

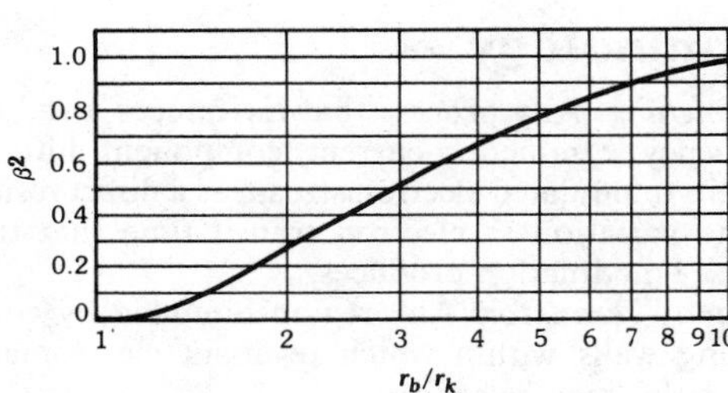

Fig. 8. Values of β^2 for values of $r_b/r_k < 10$.

bars. Their function is to control or accelerate electrons by establishing appropriate electric fields. An ideal grid would intercept no electrons; in practice, interception is minimized by careful design but still occurs, causing heating that usually sets the upper limit to power output of the tube.

In tubes with oxide cathodes, grids may be tungsten or molybdenum wire structures, coated with gold to reduce primary emission caused by deposition of material evaporated from the cathode during operation. In tubes with thoriated tungsten cathodes, grids may be made of tungsten or molybdenum coated with proprietary compounds to reduce primary emission. Relatively recently, pyrolytic graphite has been found to be an excellent grid material because of its combination of good ther-

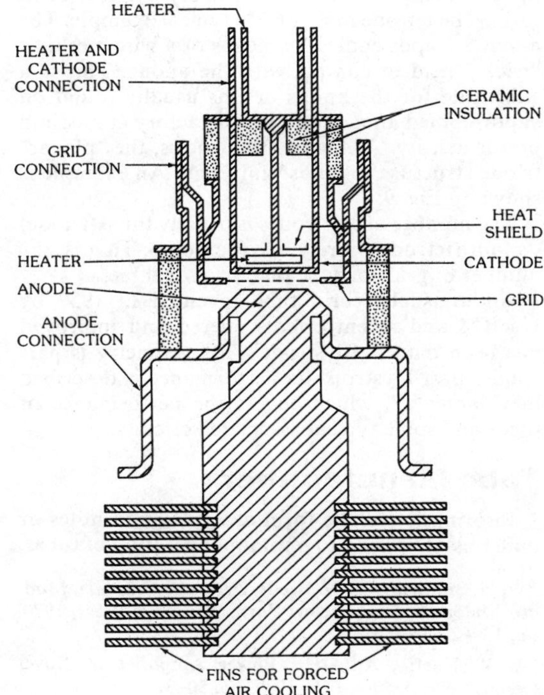

Fig. 9. Electrode arrangement of a small high-frequency external-anode triode.

mal and mechanical properties and its low primary and secondary emission.* Grids made of this material behave elastically up to very high temperatures and as a result are more dimensionally stable than refractory metal wire grids. This material is anisotropic, is formed by chemical vapor deposition at high temperature (2000°C or higher), and should not be confused with ordinary graphite, which is isotropic.

Anodes

Anodes are usually made of copper, cooled by air or water. Some glass-envelope tubes have molybdenum internal anodes. Carbon has been used. Insulating structures inside the vacuum envelope are usually high-purity aluminum oxide or beryllium oxide. Conductors are molybdenum, tungsten, nickel, Kovar, cupro-nickel, Hastalloy, or copper.

Getters

Getters are used to maintain a high vacuum during the life of the tube. They are so called because they "get" or trap and hold gases that may evolve. Typical materials are zirconium, cerium, barium, and titanium.

Tube Geometry

Gridded tubes at all power levels for frequencies up to about 1 GHz are invariably cylindrical in form. The tetrode in Fig. 6 is a typical example. The anode is vapor cooled by means of a wire mesh, or "wick," held in contact with the anode. This is a substitute for the knobs or fins usually found on vapor-cooled anodes and is satisfactory at medium power density. At higher frequencies, the "planar" triode structure is almost universal. An example is shown in Fig. 9.

The number of electrodes is usually three (triode) or four (tetrode), rarely five (pentode). That is, the number of grids may be one, two, or three.

An unusual hybrid tube invented in 1939 by Haeff†§ and recently rediscovered and improved has been named Klystrode™. Its geometry is part triode, part klystron. Its performance is described in Chapter 17, which covers the performance of tubes and some typical associated circuits.

Tube Manufacturers

Information in the form of application notes or bulletins pertaining to optimum operation of tubes, circuit and cooling arrangements, etc., is available from some of the following manufacturers:

Varian/EIMAC, San Carlos, CA and Salt Lake City, UT
R.C.A., Lancaster, PA
ITT, Easton, PA
Amperex, Hicksville, NY
Westinghouse, Elmira, NY
General Electric Co., Owensboro, KY
Thompson-CSF, Paris, France
Siemens, Munich, Germany
English Electric Valve Co., Chelmsford, England

MICROWAVE TUBES*

The reduced performance of space-charge control tubes in the microwave region has fostered the development of other types of tubes for use as oscillators and amplifiers at microwave frequencies. Such tubes generally function on the basis of the modulation of the velocity of an electron stream rather than of its density. They may be roughly divided simply into linear-beam devices and crossed-field devices. In the former, the electron stream flows essentially linearly, often with a collimating magnetic field to counteract space-charge spreading; in the latter, the electron stream follows a curved path under the action of orthogonal electric and magnetic fields. The linear-beam devices are often referred to as *O*-type, while the crossed-field devices are referred to as *M*-type.

Terminology

Bunching: Any process that introduces a radio-frequency conduction-current component into a velocity-modulated electron stream as a direct result of the variation in electron transit time that the velocity modulation produces.

Cavity resonator: Any region bounded by conducting walls within which resonant electromagnetic fields may be excited.

Circuit efficiency: The ratio of (A) the power of the desired frequency delivered to the output terminals of the circuit of an oscillator or amplifier to (B) the power of the desired frequency delivered by the electron stream to the circuit.

Coherent-pulse operation: Method of pulse operation in which the phase of the radio-frequency wave is maintained through successive pulses.

Drift space: In an electron tube, a region substan-

---

*W. H. Smith and D. H. Leeds, *Pyrolytic Graphite,* Modern Materials, Vol. 7 (New York: Academic Press, 1970; pp. 139–221).

†A. V. Haeff, "An UHF Power Amplifier of Novel Design," *Electronics,* Feb. 1939, pp. 30–32.

§D. Priest and M. B. Shrader, "The Klystrode—An Unusual Transmitting Tube With Potential for UHF-TV," *Proc. IEEE,* Nov. 1982.

*A. H. W. Beck, *Space-Charge Waves and Slow Electromagnetic Waves* (New York: Pergamon Press, 1958). R. G. E. Hutter, *Beam and Wave Electronics in Microwave Tubes,* (Princeton, N.J.: D. Van Nostrand Co., 1960). W. J. Kleen, *Electronics of Microwave Tubes* (New York: Academic Press, 1958). J. C. Slater, *Microwave Electronics* (Princeton, N.J.: D. Van Nostrand Co., 1950). J. F. Hull, "Microwave Tubes of the Mid-Sixties," 1965 *IEEE International Convention Record,* IEEE, New York.

tially free of externally applied alternating fields in which a relative repositioning of the electrons is determined by their velocity distributions and the space-charge forces.

Duty cycle: The product of the pulse duration and the pulse repetition rate. It is also the ratio of the average power output to the peak power output.

External Q: The reciprocal of the difference between the reciprocals of the loaded Q and the unloaded Q.

Frequency pulling of an oscillator is the change in the generated frequency caused by a change of the load impedance.

Frequency pushing of an oscillator is the change in frequency due to change in anode current (or in anode voltage).

Loaded Q of a specific mode of resonance of a system is the Q when there is external coupling to that mode. Note: When the system is connected to the load by means of a transmission line, the loaded Q is customarily determined when the line is terminated in its characteristic impedance.

Mode: One of the components of a general configuration of a vibrating system. A mode is characterized by a particular geometric pattern of the electromagnetic field and a resonant frequency (or propagation constant).

Noise Figure: The ratio in decibels of the total available output noise from an amplifier to the available noise which would be present at the output if the amplifier itself were noiseless, assuming a source temperature of 290 K.

Perveance: The ratio of electron beam current to the $^3/_2$ power of the beam voltage ($I/V^{3/2}$), which is an invariant for a particular electron-gun design.

Pulling figure of an oscillator is the difference in megahertz between the maximum and minimum frequencies of oscillation obtained when the phase angle of the load-impedance reflection coefficient varies through 360 degrees, while the absolute value of this coefficient is constant and is normally equal to 0.20.

Pulse: Momentary flow of energy of such short time duration that it may be considered as an isolated phenomenon.

Pushing figure of an an oscillator is the rate of frequency pushing in megahertz per ampere or megahertz per volt.

Q: The Q of a specific mode of resonance of a system is 2π times the ratio of the stored electromagnetic energy to the energy dissipated per cycle when the system is excited in this mode.

Reflector: Electrode whose primary function is to reverse the direction of an electron stream. It is also called a *repeller*.

Reflex bunching: Type of bunching that occurs when the velocity-modulated electron stream is made to reverse its direction by means of an opposing direct-current field.

Slow-wave structure: A microwave circuit, as used in beam-type microwave tubes, capable of propagating radio-frequency waves with phase velocities appreciably less than the velocity of light.

Linear-Beam Tubes

The principal types of linear-beam tubes are the klystron, the traveling-wave amplifier, and the backward-wave oscillator.

Klystrons—A klystron* is an electron tube in which the following processes may be distinguished:

(A) Periodic variations of the longitudinal velocities of the electrons forming the beam in a region confining a radio-frequency field.

(B) Conversion of the velocity variation into conduction-current modulation by motion in a region free from radio-frequency fields.

(C) Extraction of the radio-frequency energy from the beam in another confined radio-frequency field.

The transit angles in the confined fields are made short ($\delta \doteq \pi/2$) so that there is no appreciable conduction-current variation while traversing them.

Several variations of the basic klystron exist. These include the two-cavity amplifier or oscillator, the reflex klystron, and the multicavity high-power amplifier.

Two-Cavity Klystron Amplifiers: An electron beam is formed in an electron gun and passed through the gaps associated with the two cavities (Fig. 10). After emerging from the second gap, the electrons pass to a collector designed to dissipate the remaining beam power without the production of secondary electrons. In the first gap, the electron

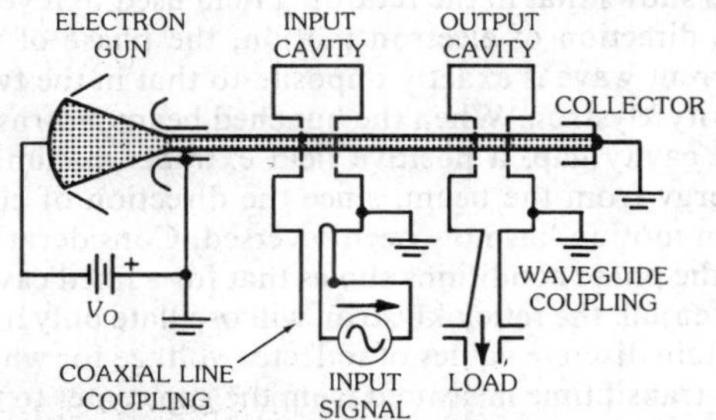

Fig. 10. Two-cavity klystron amplifier. (*From J. W. Gewartowski and H. A. Watson,* Principles of Electron Tubes, *1965; p. 296. Courtesy D. Van Nostrand Co., Inc.*)

*D. R. Hamilton, J. K. Knipp, and J. B. H. Kuper, *Klystrons and Microwave Triodes* (New York: McGraw-Hill Book Co., 1948). A. H. W. Beck, *Velocity-Modulated Thermionic Valves* (London: Cambridge University Press, 1948). A. H. W. Beck, *Thermionic Valves, Their Theory and Design* (London: Cambridge University Press, 1953). A. H. W. Beck, *Space-Charge Waves and Slow Electromagnetic Waves* (New York: Pergamon Press, 1958).

beam is alternately accelerated and decelerated in succeeding half-periods of the radio-frequency cycle, the magnitude of the change in speed depending on the magnitude of the alternating voltage impressed on the cavity. The electrons then move in a drift space where there are no radio-frequency fields. Here, the electrons that were accelerated in the input gap during one half-cycle catch up with those that were decelerated in the preceding half-cycle, and a local increase of current density occurs in the beam. Analysis shows that the maximum of the current-density wave occurs at the position, in time and space, of those electrons that passed the center of the input gap as the field changed from negative to positive. There is therefore a phase difference of $\pi/2$ between the current wave and the voltage wave that produced it. Thus at the end of the drift space, the initially uniform electron beam has been altered into a beam showing periodic density variations. This beam now traverses the output gap, and the variations in density induce an amplified voltage wave in the output circuit, phased so that the negative maximum corresponds with the phase of the bunch center. The increased radio-frequency energy has been gained by conversion from the direct-current beam energy.

The two-cavity amplifier can be made to oscillate by providing a feedback loop from the output to the input cavity. A much simpler but less efficient structure results if the electron-beam direction is reversed by a negative electrode, termed the reflector.

Reflex Klystrons*: A schematic diagram of a reflex klystron is shown in Fig. 11. The velocity-modulation process takes place as before, but analysis shows that in the retarding field used to reverse the direction of electron motion, the phase of the current wave is exactly opposite to that in the two-cavity klystron. When the bunched beam returns to the cavity gap, a positive field extracts maximum energy from the beam, since the direction of electron motion has now been reversed. Consideration of the phase conditions shows that for a fixed cavity potential, the reflex klystron will oscillate only near certain discrete values of reflector voltage for which the transit time measured from the gap center to the reflection point and back is given by

$$w\tau = 2\pi(N + 3/4)$$

where N is an integer called the mode number.

By varying the reflector voltage around the value corresponding with the mode center, it is possible to vary the oscillation frequency by a small percentage. This fact is made use of in providing automatic frequency control or in frequency-modulation transmission.

*J. R. Pierce and W. G. Shepherd, "Reflex Oscillators," *Bell System Technical Journal,* Vol. 26, pp. 460-681; July 1947.

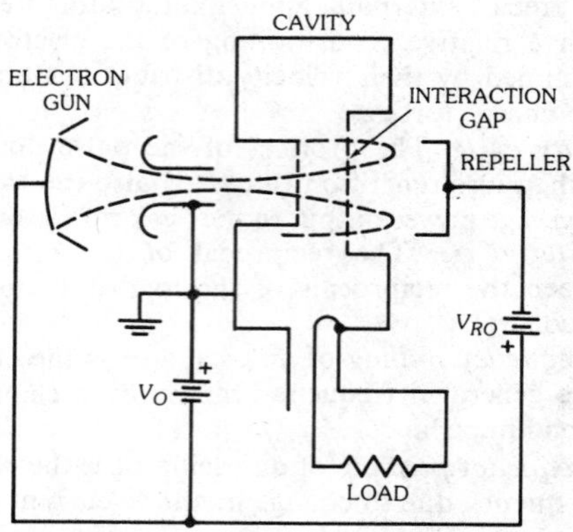

Fig. 11. Schematic diagram of reflex klystron with power supply. (*From J. W. Gewartowski and H. A. Watson,* Principles of Electron Tubes, *1965; p. 311. Courtesy D. Van Nostrand Co., Inc.*)

Reflex-Klystron Performance Data: The performance data for a reflex klystron are usually given in the form of a reflector-characteristic chart. This chart displays power output and frequency deviations as a function of reflector voltage. Several modes are often displayed on the same chart. A typical chart is shown in Fig. 12.

Table 7 shows typical reflex-klystron performance.

Multicavity Klystrons: Multicavity klystrons* have been perfected for use in two rather different fields of application: applications requiring extremely high pulse powers, and continuous-wave

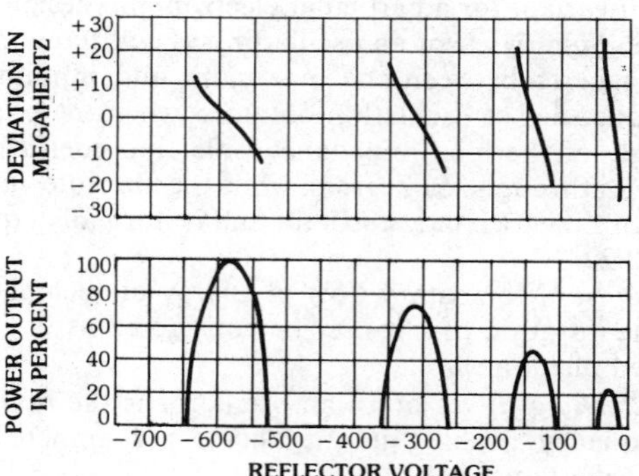

Fig. 12. Klystron reflector-characteristic chart. (*Courtesy Sperry Gyroscope Co.*)

*M. Chodorow, E. L. Ginzton, I. R. Neilson, and S. Sonkin, "Design and Performance of a High-Power Pulsed Klystron," *Proceedings of the IRE,* Vol. 41, pp. 1584–1602; November 1953. D. H. Priest, C. E. Murdock, and J. J. Woerner, "High-Power Klystrons at UHF," *Proceedings of the IRE,* Vol. 41, pp. 20–25; January 1953. A. Staprans, E. McCune, and J. Ruetz, "High-Power Linear Beam Tubes," *Proceedings of the IEEE,* Vol. 61, pp. 299–329; March 1973.

TABLE 7. CLASSES OF REFLEX KLYSTRONS

Frequency (MHz)	Power Output (mW)	Useful Mode Width Δf_{3dB} (MHz)	Operating Voltage
	Local Oscillators		
3 000	150	40	300
9 000	40	40	350
24 000	35	120	750
35 000	>15	50	2 000
50 000	10–20	60–140	600
	Maser pumps		
35 000	500–1 500	70	2 000
45 000	500–1 000	80	2 000
	Frequency-modulation transmitters		
4 000	10 000	40	1 100
7 000	10 000	37	750
9 000	6 000	60	500

systems in which a few kilowatts to hundreds of kilowatts are required. Examples of the first application are power sources for nuclear-particle acceleration and radar; ultra-high-frequency television, troposcatter, satellite-ground-station, and space-communication transmitters are examples of the latter.

A multicavity klystron amplifier is shown schematically in Fig. 13. The example shown has three cavities, all coupled to the same beam, although as many as seven cavities are sometimes used. The radio-frequency input modulates the beam as before. The bunched beam induces an amplified voltage across the second cavity, which is tuned to the operating frequency. This amplified voltage remodulates the beam with a certain phase shift, and the now more strongly bunched beam excites a highly amplified wave in the output circuit. It is found that the optimum power output is obtained when the second cavity is slightly detuned. When four or more cavities are used, optimum efficiency is obtained with both the penultimate and the antipenultimate cavities tuned above the operating frequency. Moreover, when increased bandwidth is required, the earlier cavities may be loaded. Modern multicavity klystrons use magnetically focused, high-perveance beams, and under these conditions

INTERMEDIATE CAVITY
ELECTRON GUN
INPUT CAVITY
OUTPUT CAVITY
COLLECTOR
V_O
INPUT SIGNAL
OUTPUT SIGNAL

Fig. 13. Three-cavity klystron amplifier. (*From J. W. Gewartowski and H. A. Watson,* Principles of Electron Tubes, *1965; p. 340. Courtesy D. Van Nostrand Co., Inc.*)

high gains, large power output, and reasonable values of efficiency are readily obtained.

Fig. 14 shows the power output and tuning range versus frequency for a number of typical cw klystrons. Efficiencies from 30% to 75% are typical. The lower efficiencies apply for high-perveance ($3 \times 10^{-6}\ A/V^{3/2}$) stagger-tuned, broad-band klystrons, and the higher efficiencies apply for narrow-band, lower-perveance ($0.5 \times 10^{-6} A/V^{3/2}$) klystrons. Average output power scales as about λ^3 (λ = wavelength) at constant perveance if cavity losses limit output power. At the lowest frequencies, the limit on power is imposed by single-mode-waveguide, output window capability, and power

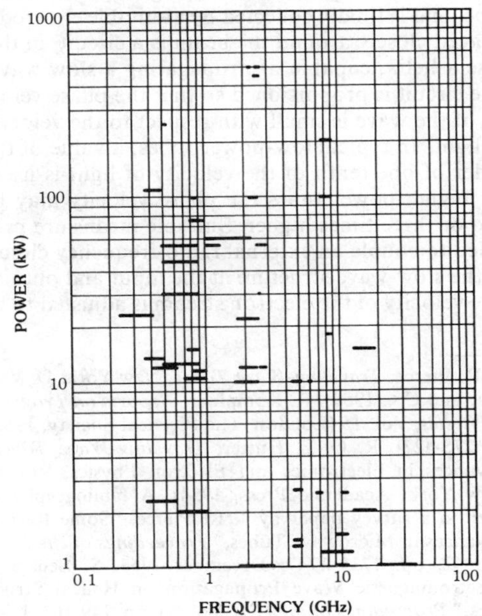

Fig. 14. Output power vs frequency for a representative group of cw klystrons.

scales as λ. At the highest frequencies, cathode current and cavity losses impose the limit, and the power scales as $\lambda^{9/2}$.

Fig. 15 shows the power output and tuning range versus frequency for pulsed klystrons. Because most pulse klystrons use higher-perveance beams (1.5 to $3 \times 10^{-6}\ A/V^{3/2}$) in order to keep the beam voltage low, efficiencies range from 30% to 50%. As much as 11.8% bandwidth has been achieved from a 10-MW klystron with a 1000-ohm "beam impedance." Bandwidth scales inversely with the ratio of dc beam voltage to dc beam current, or "beam impedance," in both cw and pulse klystrons. Peak output power from pulsed klystrons is usually limited by the beam voltage that can be applied to the electron gun without arcing problems and by the current that can be put through the beam tunnel while keeping the cathode current density reasonable. For this reason, maximum peak power from klystrons is relatively constant up to about 3000 MHz and then falls as λ^2.

Traveling-Wave Tubes—The traveling-wave tube* differs from the klystron in that the radio-frequency field is not confined to a limited region but is distributed along a wave-propagating structure. A longitudinal electron beam interacts continuously with the field of a wave traveling along this wave-propagating structure. In its most common form, it is an amplifier, although there are related types of tubes that are basically oscillators.

The principle of operation may be understood by reference to Fig. 16. An electron stream is produced by an electron gun, travels along the axis of the tube, and is finally collected by a suitable electrode. Spaced closely around the beam is a circuit, in this case a helix, capable of propagating a slow wave. The circuit is proportioned so that the phase velocity of the wave is small with respect to the velocity of light. In typical low-power tubes, a value of the order of one-tenth of the velocity of light is used; for higher-power tubes the phase velocity may be two or three times higher. Suitable means are provided to couple an external radio-frequency circuit to the slow-wave structure at the input and output. The velocity of the electron stream is adjusted to be approximately the same as the axial phase velocity of the wave on the circuit.

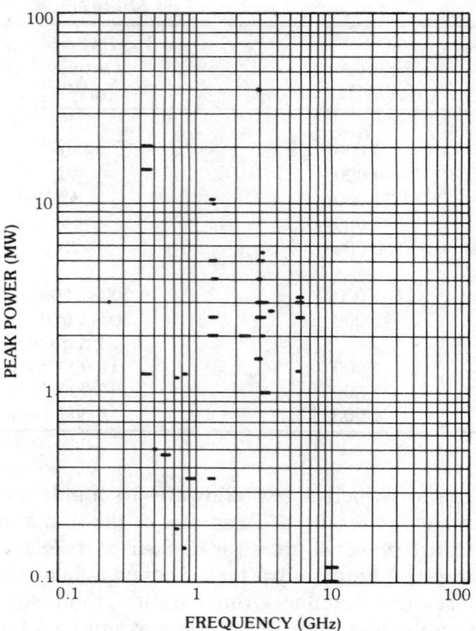

Fig. 15. Output power vs frequency for a representative group of pulsed klystrons.

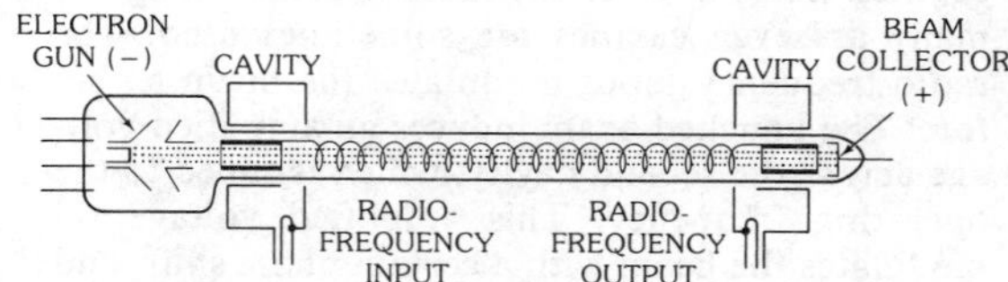

Fig. 16. Basic helix traveling-wave tube. The magnetic beam-focusing system between input and output cavities is not shown.

When a wave is launched on the circuit, the longitudinal component of its field interacts with the electrons traveling in approximate synchronism with it. Some electrons will be accelerated and some decelerated, resulting in a progressive rearrangement in phase of the electrons with respect to the wave. The electron stream, thus modulated, in turn induces additional waves on the helix. This process of mutual interaction continues along the length of the tube with the net result that direct-current energy is given up by the electron stream to the circuit as radio-frequency energy, and the wave is thus amplified.

By virtue of the continuous interaction between a wave traveling on a broad-band circuit and an electron stream, traveling-wave tubes do not suffer the gain-bandwidth limitation of ordinary types of electron tubes. By proper circuit design, such tubes are made to have bandwidths of an octave in frequency, and even more in special cases.

*J. R. Pierce, *Traveling-Wave Tubes* (New York: D. Van Nostrand Co., 1950). R. Kompfner, *Reports on Progress in Physics*, Vol. 15 (London: The Physical Society, 1952; pp. 275-327). R. G. E. Hutter, *Traveling-Wave Tubes,* Advances in Electronics and Electron Physics, Vol. 6, (New York: Academic Press, 1954). A bibliography is given in a survey paper by J. R. Pierce, "Some Recent Advances in Microwave Tubes," *Proceedings of the IRE,* Vol. 42, pp. 1735–1747; December 1954. S. Sensiper, "Electromagnetic Wave Propagation on Helical Structures," *Proceedings of the IRE,* Vol. 43, pp. 149-161; February 1955. A. H. W. Beck, *Space-Charge Waves and Slow Electromagnetic Waves* (New York: Pergamon Press, 1958). D. A. Watkins, *Topics in Electromagnetic Theory* (New York: John Wiley & Sons, 1958).

The helix is an extremely useful form of slow-wave circuit because the impedance that it presents to the wave is relatively high and because, when properly proportioned, its phase velocity is almost independent of frequency over a wide range.

An essential feature of this type of tube is the approximate synchronism between the electron stream and the wave. For this reason, the traveling-wave tube will operate correctly over only a limited range in voltage. Practical considerations require that the operating voltages be kept as low as is consistent with obtaining the necessary beam input power; the voltage, in turn, dictates the phase velocity of the circuit. The electron velocity v in centimeters/second is determined by the accelerating voltage V in accordance with

$$v = 5.93 \times 10^7 V^{1/2}$$

Fig. 17 shows a typical relationship between gain and beam voltage. The gain G, in decibels, of a traveling-wave tube is given approximately by

$$G = A + BCN$$

where,

A is the initial loss due to the establishment of the modes on the helix and lies in the range from −6 to −9 decibels,
B is a gain coefficient that accounts for the effect of circuit attenuation and space charge,
C is a gain parameter that depends on the impedances of the circuit and the electron stream,
N is the number of active wavelengths in the tube.

$$C = \{[E^2/(\omega/v)^2P] \times (I_0/8V_0)\}^{1/3}$$

and

$$N = (l/\lambda_0)(c/v)$$

where,

I_0 = beam current,
V_0 = beam voltage,
l = axial length of the helix,
λ_0 = free-space wavelength,
v = phase velocity of wave along tube,
c = velocity of light.

The term $E^2/(\omega/v)^2P$ is a normalized wave impedance that may be defined in a number of ways.

In practice, the attenuation of the circuit will vary along the tube, and consequently the gain per unit length will not be constant. The total gain will be a summation of the gains of various sections of the tube.

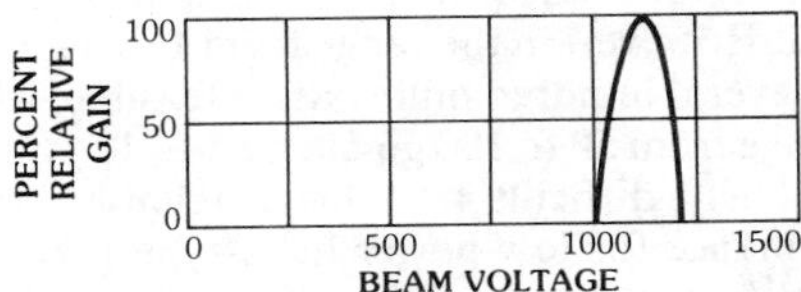

Fig. 17. Traveling-wave-tube gain vs accelerating voltage.

Commonly, C is of the order of 0.02 to 0.2 in helix traveling-wave tubes. The gain of low- and medium-power tubes varies from 20 to 70 decibels with 30 decibels being a common value. The gain in a tube designed to produce appreciable power will vary somewhat with signal level when the beam voltage is adjusted for optimum operation. Fig. 18 shows a typical characteristic.

To restrain the physical size of the electron stream as it travels along the tube, it is necessary to provide a focusing field, either magnetic or electrostatic, of a strength appropriate to overcome the space-charge forces that would otherwise cause the beam to spread. Until fairly recently, a longitudinal magnetic field supplied by a solenoid electromagnet was used for this purpose. Continuing demands for improved efficiency and reliability, and for weight and size reduction, however, have forced the development of permanent-magnet focusing structures. At the present time, focusing by periodically reversing magnetic fields produced by permanent-magnet structures is rapidly becoming predominant.

Several techniques for electrostatic containment of the electron stream have also been developed. Typical is the use of a bifilar helix slow-wave structure where an appropriate voltage difference between the helices provides, in effect, a distributed Einzel lens. Because of voltage-breakdown problems as well as increased power-supply requirements, electrostatic focusing has not yet proved practical for use in linear-beam tubes.

Other types of slow-wave circuits in addition to the helix are possible, including a number of periodic structures. In general, such designs are capable of operation at higher power levels but at the expense of bandwidth.

Traveling-Wave-Tube Performance Data—Traveling-wave tubes are designed to emphasize particular inherent characteristics for specific applications. Three general classes are distinguished.

Low-Noise Amplifiers: Tubes of this class are intended for the first stage of a receiver and are proportioned to have the best possible noise figure. This requires that the random variations in the elec-

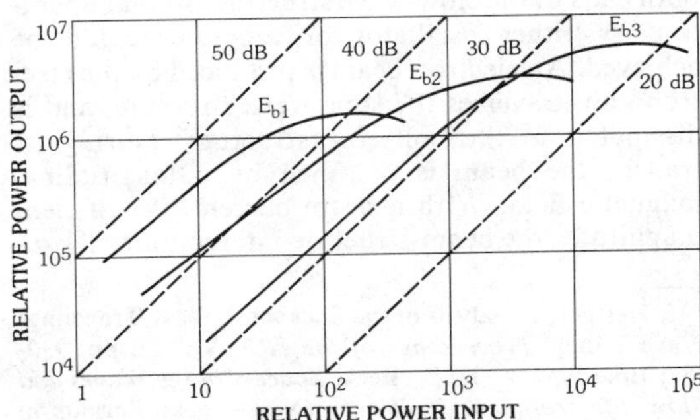

Fig. 18. Gain of traveling-wave tube as a function of input level and beam voltage. $E_{b1} < E_{b2} < E_{b3}$.

tron stream be minimized and that steps be taken also to minimize partition noise. Tubes have been made for commercial use with noise figures as low as 3 decibels in S-band and 6 decibels or less over the entire range from 1000 to 12 000 megahertz. Gains of from 20 to 35 decibels are typical. The maximum power output is generally not more than a few milliwatts. Performance of this order can be achieved with either permanent-magnet or electromagnet focusing structures. Recently a new class of tubes has been developed which offers medium-power performance (1 to 5 watts) with reasonable low-noise performance (noise figures of 10 to 14 decibels).

Intermediate Power Amplifiers: These tubes are intended to provide power gain under conditions where neither noise nor large values of power output are important. Gains of 30 or more decibels are customary, and the maximum output power is usually in the range from 100 milliwatts to 1 watt.

Power Amplifiers: For this class of tubes, the application is usually the output stage of a transmitter; the power output, either continuous-wave or pulsed, is of primary importance. Much active development continues in this area, and the values of power obtainable are steadily increasing. At present, continuous-wave powers range from tens of kilowatts in the ultra-high-frequency region to more than 100 watts at 10 000 megahertz. Tubes especially designed for pulsed operation provide considerably higher power. Several megawatts of peak power have been achieved at 3000 megahertz. Efficiencies in excess of 30 percent have been obtained, and this may be further enhanced by recently developed collector-depression techniques. Power gains of 30 to 50 decibels are normal.

Backward-Wave Oscillators—A member of the traveling-wave-tube family, the *O*-type backward-wave oscillator* makes use of the interaction of the electron stream with a radio-frequency-circuit wave whose phase and group velocities are 180° apart. The group velocity, and thus the direction of energy flow, is directly opposed to the direction of electron motion. Fig. 19 shows schematically a backward-wave tube with connection to both ends of the slow-wave structure, so that operation as either oscillator or amplifier could be achieved. An electron beam is produced by the electron gun, traverses the slow-wave structure, and is dissipated in the collector structure. During its transit, the beam is confined by a longitudinal magnetic field. With a beam current of sufficient magnitude, the beam-structure interaction will produce oscillations, and microwave power will be delivered from the end of the structure adjacent to the electron gun. At beam-current levels below the "start-oscillation" value, a radio-frequency signal may be introduced at the collector end of the device, and the tube will operate as an amplifier.

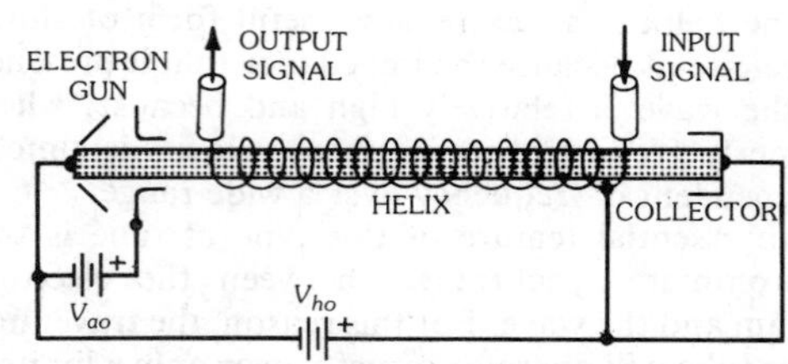

Fig. 19. A traveling-wave tube in operation as a backward-wave amplifier. A separate power supply connected to the anode permits beam-current control independent of the helix voltage. (*From J. W. Gewartowski and H. A. Watson,* Principles of Electron Tubes, *1965; p. 398. Courtesy D. Van Nostrand Co., Inc.*)

To improve interaction efficiency, electron beams with hollow cross sections are usually used. This places all the electrons as close as possible to the slow-wave structure in the region of maximum radio-frequency field. The reason is that the strength of the −1 space harmonic field goes to zero on the axis. To produce this hollow-cross-section beam, it is necessary to use magnetically confined electron flow from the cathode, and thus the electron gun is entirely immersed in the magnetic field.

An *O*-type backward-wave device is voltage tunable, with the frequency being proportional to the $^1/_2$ power of the cathode-helix voltage as well as dependent on the dimensions of the structure. Typically, tuning over a full octave range is possible, and in special cases a range of two or more octaves can be achieved. However, where confined limits are desired on power variation or other special characteristics, more restricted frequency ranges may be necessary. Where full octave tuning is used, power output variation of 6 to 10 decibels across the range is usual. In most cases, a separate control element in the gun permits adjustment of beam-current amplitude and thus provides control of power output. Oscillators of this type have very low pulling figures, but the pushing figure is often substantial. Frequency stability is generally excellent, with the achievable value normally depending on power-supply capabilities rather than inherent tube limitations.

Generally, *O*-type backward-wave oscillators are low-power devices, with 10 to 50 milliwatts being typical. However, in the range from 1 to 4 gigahertz up to several hundred milliwatts is feasible, while in the range from 50 to 100 gigahertz 5 to 10 milliwatts is relatively difficult to achieve reliably. Typical performance for low-power helix-type permanent-magnet-focused backward-wave oscillators is listed in Table 8.

* H. Heffner, "Analysis of the Backward-Wave Traveling-Wave Tube," *Proceedings of the IRE,* Vol 42, pp. 930-937; June 1954. A. H. W. Beck, *Space-Charge Waves and Slow Electromagnetic Waves* (New York: Pergamon Press, 1958; pp. 241-255). R. Kompfner and N. T. Williams, "Backward-Wave Tubes," *Proceedings of the IRE,* Vol. 41, pp. 1602-1611; November 1953.

TABLE 8. PERFORMANCE OF TYPICAL LOW-POWER BACKWARD-WAVE OSCILLATORS

Frequency Range (GHz)	Tuning Voltage (V)	Cathode Current (mA)	Minimum Power Output (mW)
1.0–2.0	250–1150	15	100
2.0–4.0	300–1800	10	100
4.0–8.0	250–2400	12	25
5.3–11.0	245–2400	10	25
8.0–12.4	550–2400	10	25

Crossed-Field Tubes

Microwave crossed-field tubes are often called *M*-type devices to distinguish them from linear-beam tubes that are often called *O*-type devices. The earliest type of crossed-field tube* was the magnetron oscillator. Crossed-field amplifiers (CFA) and voltage-tunable crossed-field oscillators have been developed more recently. Crossed-field tubes generally operate with higher electronic conversion efficiency for dc power to microwave power than linear-beam tubes, so they are especially attractive for high-power applications.

Magnetrons—A magnetron is a high-vacuum tube containing a cathode and an anode, the latter usually divided into two or more segments. A constant magnetic field modifies the space-charge distribution and the current-voltage relations. In modern usage, the term "magnetron" refers to the magnetron oscillator in which the interaction of the electronic space charge with the resonant system converts direct-current power into alternating-current power, usually at microwave frequencies.

Many forms of magnetrons have been made in the past, and several kinds of operation have been employed. The type of tube that is now almost universally employed is the multicavity magnetron generating traveling-wave oscillations. It possesses the advantages of good efficiency at high frequencies, capability of high outputs either in pulse or continuous-wave operation, moderate magnetic-field requirements, and good stability of operation. A section through the basic anode structure of a typical fixed-frequency, conventional magnetron is shown in Fig. 20.

In magnetrons, the operating frequency is determined by the resonant frequency of the separate cavities arranged around the central cylindrical cathode and parallel to it. A high direct-current potential is placed between the cathode and the cavities, and radio-frequency output in the conventional type of magnetron is brought out through a suitable transmission line or waveguide, usually coupled to one of the resonator cavities. Under the action of the radio-frequency voltages across these resonators and the axial magnetic field, the elec-

*E. Okress, ed., *Crossed-Field Microwave Devices* (New York: Academic Press, 1961).

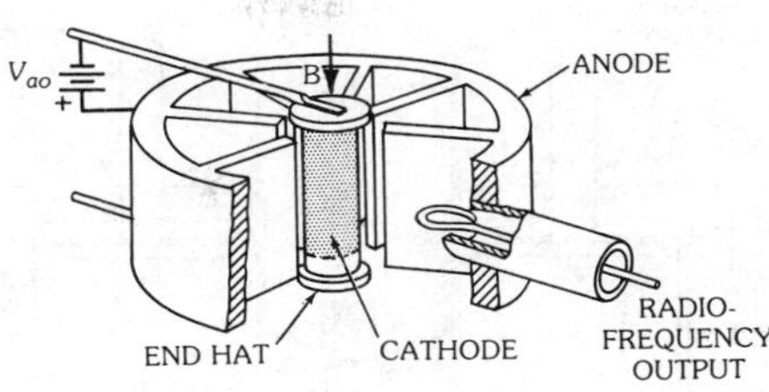

Fig. 20. Magnetron oscillator. (*From J. W. Gewartowski and H. A. Watson,* Principles of Electron Tubes, *1965; p. 428. Courtesy D. Van Nostrand Co., Inc.*)

trons from the cathode form a bunched space-charge cloud that rotates around the tube axis, exciting the cavities and maintaining their radio-frequency voltages.

Pulsed magnetrons have been developed to operate with peak power output from a few kilowatts to several megawatts and at frequencies from a few hundred megahertz to frequencies having millimeter wavelengths. The primary application for these tubes is the rf power generator for radar transmitters. Continuous-wave magnetrons have also been developed with power output of several hundred watts at frequencies from a few hundred megahertz to about 10–12 GHz. The principal use of many of these cw tubes is in countermeasure equipment used for electronic jamming. In addition, many fixed-frequency cw magnetrons at 2450 MHz have been used in microwave ovens.

The output frequency of conventional magnetrons can be altered in an undesired fashion by variations in the microwave load impedance (frequency pulling) or by changes in cathode current (frequency pushing) caused by fluctuations in voltage applied to the magnetron. Significant improvement in operational stability for the magnetron is obtained by coupling to a high-Q stabilizing cavity. A convenient method for doing this is shown in Fig. 21. The stabilizing cavity is made an integral part of the magnetron by surrounding the magnetron anode with a resonant coaxial cavity—hence, the name "coaxial magnetron." Alternate resonators of the anode circuit are slotted to provide coupling to excite the TE_{011}, circular electric mode in the coaxial cavity. Power output from a coaxial magnetron is obtained through a slot in the outer wall of the cavity.

The circulating space charge in both conventional and coaxial magnetrons contains wideband noise-frequency components that can couple to the output. In conventional magnetrons, this spurious noise can couple directly to the output waveguide. Spurious noise power measured in a 1-MHz bandwidth is typically greater than 30 to 40 dB below the carrier. The coaxial cavity in the coaxial magnetron provides some isolation between the spurious noise coupled to the vanes and the output waveguide. The spurious noise power from coaxial magnetrons is typically 10 to 20 dB lower than the noise power

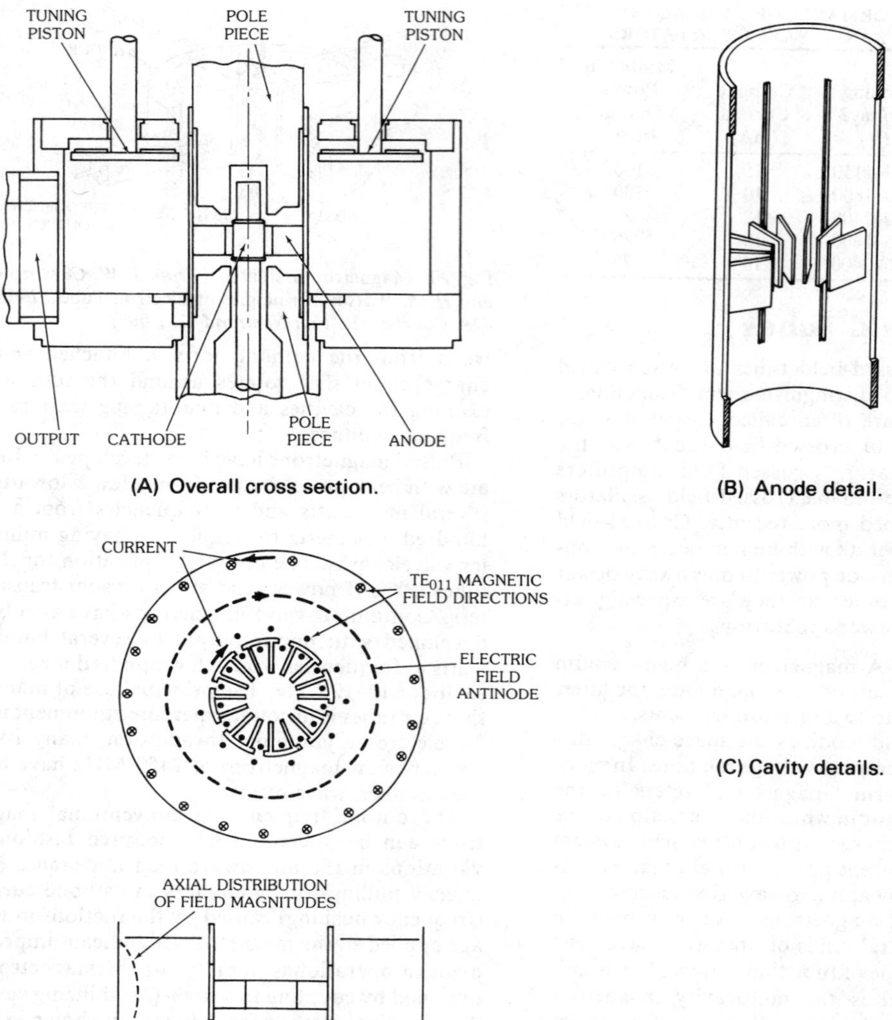

Fig. 21. Geometry of the composite anode. (*From E. Okress,* Crossed-Field Microwave Devices, *Vol. 2. New York: Academic Press, Inc., 1961; p. 125.*)

from conventional magnetrons of comparable peak power level.

Magnetron frequencies can be tuned by various means. One method used for conventional magnetrons employs metallic probes that are inserted uniformly into the multiple resonant cavities of the anode. This changes the resonant frequency of the cavities that control the magnetron frequency. Coaxial magnetrons are usually tuned by changing the resonant frequency of the coaxial stabilizing cavity by moving an end plate in the cavity. Other methods are also used in each magnetron type.

Many conventional and coaxial magnetrons are designed for very rapid tuning. Fast frequency operation is employed in radar transmitters to improve radar signal detection and/or as an electronic countermeasure to avoid jamming. Fast tuning is performed by several methods. Some tubes have tuners that are driven mechanically by an electromagnet in a manner similar to the voice coil of a radio speaker. Tuning rates, tuning range, and oscillator frequency-versus-time profiles are determined by the waveform of the voltage applied to the electromagnet. Other tubes have a mechanical linkage between the tuner and a rotating motor and produce approximately sinusoidal variation in frequency-time profile. Tuning rates and range are limited by acceleration and deceleration forces in the reciprocating tuner. For example, X-band, 200-kW tubes with narrow-band, rapid tuning of 30–50

MHz are tuned through this range at 200 Hz. Wider-band tuner excursions of 250–500 MHz are cycled at 25–40 Hz. Other tuning mechanisms in conventional and coaxial magnetrons utilize continuously rotating tuners. More rapid frequency excursions through a fixed tuning range can be obtained than from reciprocating tuners, but the rotational inertia of the tuners prohibits rapid changes in tuning profiles.

Voltage-Tuned Crossed-Field Oscillators— There are two types of voltage-tunable, crossed-field oscillators. These are carcinotron and voltage-tuned magnetrons (vtm's).

The carcinotron is an *M*-type backward-wave oscillator in which the electron stream traverses the tube and interacts with the fields on the slow-wave structure under conditions where the electric and magnetic fields are perpendicular to each other. Fig. 22 shows schematically a linear version of the carcinotron. In the electron gun, current is drawn from the cathode when the accelerator voltage is applied. Because of the presence of the magnetic field, directed as shown, the electron paths are curved approximately 90° so that they enter the interaction region between the slow-wave structure and the sole. If the voltages and the magnetic field strength are proper, the electrons will travel along a path approximately parallel to the structure until they reach the collector.

Although Fig. 22 shows a linear arrangement, carcinotrons are conventionally designed in a circular arrangement to conserve magnet size and weight. In this arrangement, the sole approximates the appearance of the cathode of a magnetron, and the slow-wave structure is in the position of the magnetron anode, but neither the sole nor the structure is reentrant.

The carcinotron gives performance similar to that of the *O*-type backward-wave oscillator, but it offers several of the advantages of crossed-field devices. High-efficiency operation is possible, with values of 20 to 30% being readily obtained. This efficiency capability makes the carcinotron useful as a high-power device with continuous-wave capabilities of hundreds of watts through X-band. Its construction is such as to permit direct scaling to very high frequencies, with several milliwatts of power having been achieved at frequencies beyond 300 gigahertz.

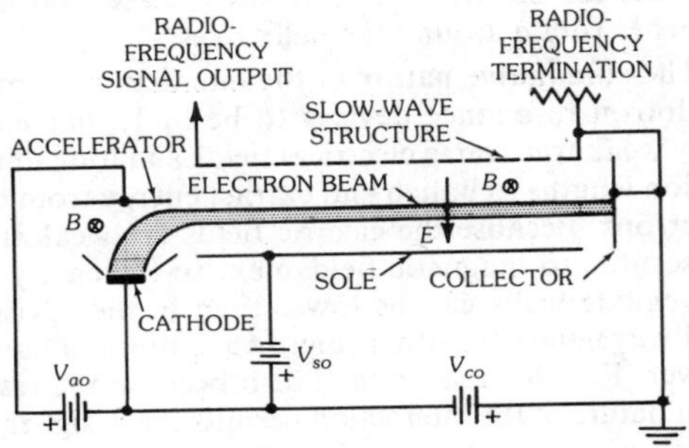

Fig. 22. Linear version of an *M*-carcinotron oscillator. (*From J. W. Gewartowski and H. A. Watson,* Principles of Electron Tubes, *1965; p. 459. Courtesy D. Van Nostrand Co., Inc.*)

The carcinotron, like the *O*-type backward-wave oscillator, is voltage tunable with the oscillation frequency being approximately directly proportional to the voltage between the cathode and the slow-wave structure. This linear relationship simplifies the associated electronic tuning circuit considerably. Frequency pushing is considerably lower than in *O*-type backward-wave oscillators. The *M*-type carcinotron has the disadvantage, however, that it is relatively noisy, with spurious power output often not more than 10 to 15 decibels below the main signal output.

Voltage-tuned magnetrons (vtm's) are distinguished from conventional magnetrons in two critical ways:

1. The output is coupled very heavily to the circuit, so the loaded Q is very low, typically between 1 and 10.
2. The normal cathode is replaced by a nonemitting post, and a cylindrical beam is injected into the interaction space from an emitting filament at one end; the injected current is controlled by the voltage on a control electrode surrounding the filament.

Under these conditions, the oscillation frequency is no longer controlled by the resonator, but by the ratio of electric to magnetic fields in the interaction space. The magnetic field is held constant by a permanent magnet so that the frequency becomes directly proportional to the applied voltage, which is normally between 1 and 4 kV dc. The control-electrode voltage is normally about 20% of the cathode-anode voltage.

Power output of vtm's ranges from a few watts to a few hundred watts at frequencies from about 1 GHz to 5 GHz. Tuning range is more than one octave at low powers (under 10 watts), but reduces to about 30% at 100 watts.

Efficiency is remarkably high, exceeding 60% in the higher-power vtm's. The tubes are more compact and more efficient that carcinotrons, but they do not reach as high an absolute power level or as high a frequency.

Because the heavy coupling would make the vtm very susceptible to load changes, a ferrite isolator is normally made an integral part of the vtm package.

The application of high-power, voltage-tuned, crossed-field oscillators is exclusively in the field of electronic countermeasures. Low-power vtm's are also used as test oscillators, capable of extremely high modulation rates.

Crossed-Field Amplifiers—Crossed-field amplifiers (cfa's) can be divided into two general classes, injected-beam tubes and distributed-emission (or sometimes emitting-sole) tubes. The former are similar in appearance to the carcinotron in that they employ separate electron guns, interaction regions, and beam-collector elements. For amplifiers, the slow-wave circuit is provided with both input and output terminals. The rf input is close to the electron gun, and the rf wave is amplified as it travels along the slow-wave circuit to the output while maintaining synchronism with the electron stream. Most injected-beam cfa's are fabricated in a linear format and are often compared to beam-type traveling-wave tubes. The principal application of injected-beam cfa's is in electronic-countermeasures equipment.

Distributed-emission cfa's are similar to magnetrons in that electron current for interaction is obtained from the cathode (or sole electrode) throughout the interaction space. Most distributed-emission tubes are fabricated in circular format like magnetrons. As may be seen from Fig. 23, the major difference is that the slow-wave structure is not reentrant, whereas in the magnetron both the beam and the circuit are reentrant. In Fig. 23, voltage and magnetic field are applied as for the magnetron. A radio-frequency signal is applied to the structure and progresses in a clockwise direction toward the output terminal. Current spokes, produced in the cathode-circuit region by the radio-frequency electric fields, also progress in a clockwise direction synchronously with the circuit wave. The interaction between the beam and circuit wave results in a growing of the circuit wave and thus gain. If desired, interaction with a backward mode may also be accomplished with this device.

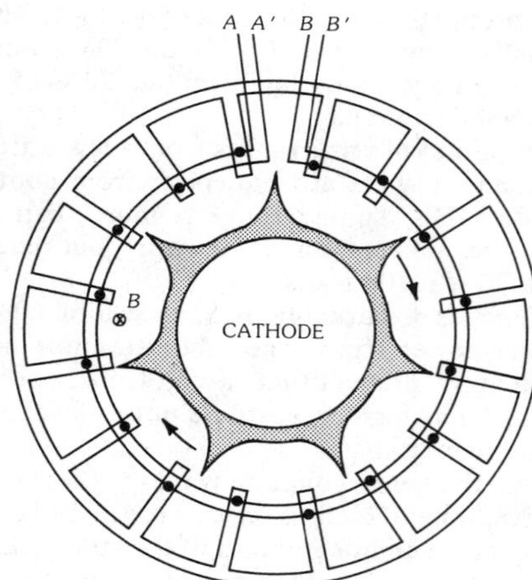

Fig. 23. Schematic drawing of a crossed-field amplifier. (*From J. W. Gewartowski and H. A. Watson,* Principles of Electron Tubes, *1965; p. 449. Courtesy D. Van Nostrand Co., Inc.*)

Since the beam is reentrant, the crossed-field amplifier will oscillate if the circuit gain becomes high. Gain is usually limited to 10 to 15 decibels. If only a portion of the circumference is used for the slow-wave structure and a drift area is left between the two ends of the structure, the feedback mechanism is disrupted, and gains of 15 to 20 decibels may be realized.

The power output of the crossed-field amplifier is essentially independent of the radio-frequency drive signal, and the device thus operates as a saturated amplifier. This characteristic makes it unsuitable for amplifying amplitude-modulated signals.

Distributed-emission crossed-field amplifiers offer the advantage of relatively high efficiency, 40 to 60% or even higher, and they may be designed to provide very high peak output powers. Their disadvantages are their low gain, limited bandwidth, high noise, and saturated-amplifier characteristic.

CYCLOTRON RESONANCE MICROWAVE TUBES*

In cyclotron resonance devices (gyrotrons, gyro klystrons, gyro traveling-wave tubes, and gyro backward-wave oscillators), the electrons in the electron beam have substantial motion perpendicular to the axis of the beam and the focusing magnetic field. They interact with radio-frequency electric fields perpendicular to the magnetic focusing field. As the electrons rotate and the fields alternate in synchronism, there is a cumulative interaction. Some electrons gain energy, and other electrons lose energy. The electrons that gain energy undergo a relativistic-mass increase, and the ones that lose energy undergo a relativistic-mass decrease. Because the cyclotron frequency is equal to eB/m, in which m is the relativistic mass of the electrons, electrons which gain energy and mass slow down in angular rotation, and electrons that lose energy and mass speed up in angular rotation. Hence, the electrons gather into rod-like bunches parallel to the axis of the helical electron trajectories. These rod-like bunches rotate around the helix axes.

The cumulative nature of the interaction permits cyclotron resonance devices to be built that have very weak transverse electrical fields and long interaction lengths to bunch and extract energy from the electrons. Because the electric fields are weak, the losses due to magnetic fields next to the cavity or waveguide walls can be lower than in the cavities and waveguides of other microwave tubes of equal power. For this reason, and also because the resonant nature of the interaction permits the designer to select an interaction with a higher-order cavity

*R. S. Symons and H. R. Jory, "Cyclotron Resonance Devices," *Advances in Electronics,* Vol. 55, pp. 1–75 (New York: Academic Press, Inc., 1980).

mode without necessarily having interference with lower-order modes, cyclotron resonance devices will produce higher powers at higher frequencies than will other microwave tubes.

Cyclotron resonance devices that have been built so far fall into four major categories: (A) gyrotron oscillators, single-cavity devices that oscillate because of the negative interaction impedance of the electron beam with the cavity; (B) gyro klystrons, in which azimuthal bunching of the helically traveling electrons is achieved by passing them through one cavity and energy extraction is achieved in a second cavity; (C) gyro-traveling-wave tubes (twt's), in which the electrons travel through a uniform waveguide supporting a transverse electric mode; and (D) gyro backward-wave oscillators, which are similar in structure to gyro-twt's but in which the group velocity of the wave in the uniform waveguide is opposed to electron flow and power comes out of the circuit at the end where the electron beam enters.

Fig. 24 shows the output power of a number of cw and pulse gyrotron oscillators compared with that of other, more conventional microwave tubes. Gyro devices have been built primarily for plasma heating in fusion reactors, although they could equally well be used for millimeter-wave radar.

Gyro traveling-wave tubes are under development and show possibility of producing pulse powers on the order of 100 kilowatts peak and 10 kilowatts average at a frequency of 100 GHz. They behave as coherent linear amplifiers of the input signal.

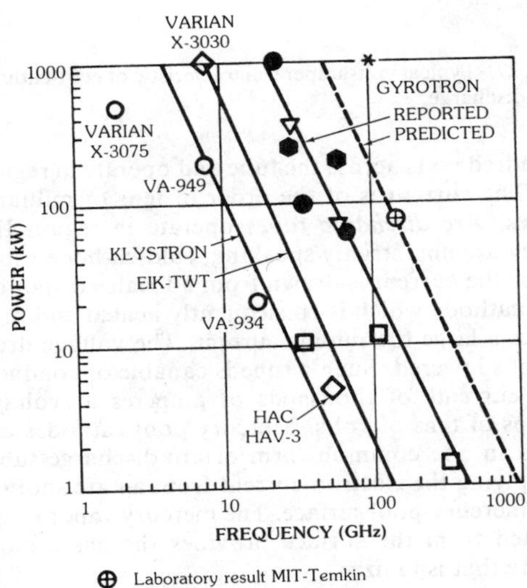

Fig. 24. Power output of microwave tubes.

GAS TUBES

A gas tube* is an electron tube in which the pressure of the contained gas is such as to affect substantially the electrical characteristics of the tube. Such effects are caused by collisions between moving electrons and gas atoms. These collisions, if of sufficient energy, may dislodge an electron from the atom, thereby leaving the atom as a positive ion. The electron space charge is effectively neutralized by these positive ions, and comparatively high free-electron densities are easily created.

Table 9 gives the energy in electronvolts necessary to produce ionization. The column headed P_c contains the kinetic-theory collision probability per centimeter of path length for an electron in a gas at 15° Celsius at a pressure of 1 millimeter of mercury. The collision frequency is given by

$$\nu_c = P_c p_o v$$

where,

ν_c = collisions per second,
P_c = collision probability in collisions per centimeter per torr of pressure,
p_o = reduced gas pressure in torr,
v = electron velocity.

Characteristics of Gas Tubes

Gas tubes may be generally divided into two classes, depending on whether the cathode is hot or cold and thus on the mechanism by which electrons are supplied.

Hot-Cathode Gas Tubes—The electrons in the hot-cathode gas tube are produced thermionically. The voltage drop across such tubes is that required to produce ionization of the gas and is generally a few tens of volts. The current conducted by the tube depends primarily on the emission capability of the cathode. Fig. 25 shows the effect of the ionized gas on the voltage distribution in a hot-cathode tube.

Cold-Cathode Gas Tubes—The electrons in a cold-cathode tube are produced by bombardment of the cathode by ions and/or by the action of a localized high electric field. The voltage drop across such a tube is higher than in the hot-cathode tube because of this mechanism of electron generation, and the current is limited. Fig. 26 shows the effect

* B. E. Cherrington, *Gaseous Electronics and Gas Lasers* (Elmsford, N. Y.: Pergamon Press, 1979). B. N. Chapman, *Glow Discharge Processes* (New York: John Wiley & Sons, Inc., 1980).

TABLE 9. IONIZATION PROPERTIES OF GASES

Gas	Ionization Energy (electronvolts)	Collision Probability P_c
Helium	24.5	12.7
Neon	21.5	17.5
Nitrogen	16.7	37.0
Hydrogen (H_2)	15.9	20.0
Argon	15.7	34.5
Carbon monoxide	14.2	23.8
Oxygen	13.5	34.5
Krypton	13.3	45.4
Water vapor	13.2	55.2
Xenon	11.5	62.5
Mercury	10.4	67.0

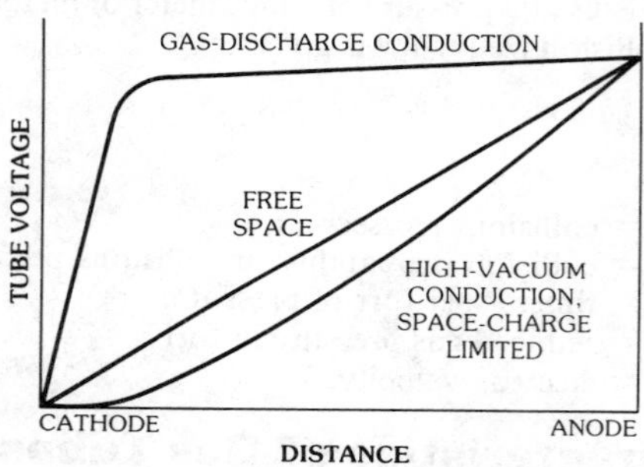

Fig. 25. Voltage distribution between plane parallel electrodes showing effect of space-charge neutralization in a hot-cathode gas tube.

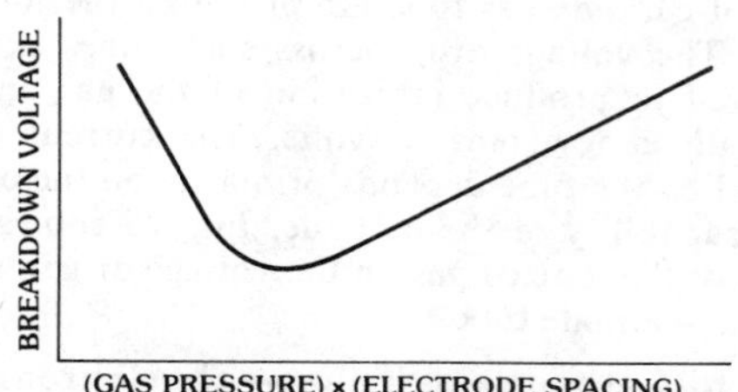

Fig. 26. Effect of gas pressure and tube geometry on gap voltage required for breakdown to occur in a cold-cathode gas tube.

of tube geometry and gas pressure on the voltage required to initiate the discharge.

Fig. 27 shows a typical volt-ampere characteristic of a cold-cathode discharge. Cold-cathode gas tubes may be divided into two categories, depending on the region of this characteristic in which they operate. *Glow discharge tubes* require a drop of several

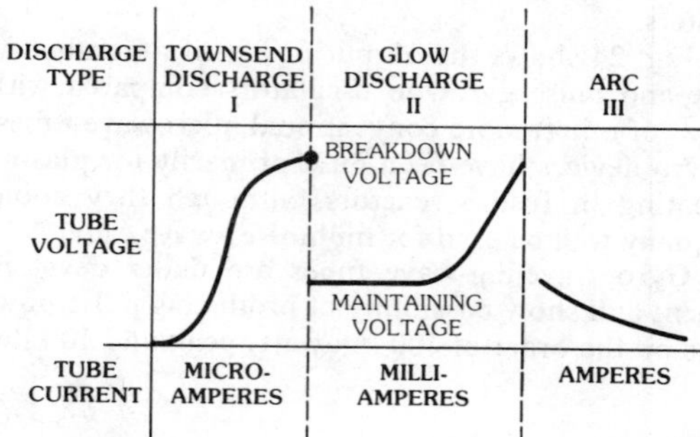

Fig. 27. Typical volt-ampere characteristic of cold-cathode gas discharge.

hundred volts across the tube and operate in region II. The current is of the order of tens of milliamperes. *Arc discharge tubes* operate in region III. They are not, strictly speaking, cold-cathode tubes since the current is drawn from a localized spot on the cathode which is consequently heated and provides a large thermionic current. The voltage drop is thus lowered. Such a tube is capable of conducting currents of thousands of amperes at voltage drops of tens of volts. Mercury-pool cathodes are used in one common form of arc discharge tube, supplying the electron current from an arc spot on the mercury-pool surface. The mercury vapor evaporated from the surface provides the gas atmosphere that is ionized.

Power Applications of Gas Tubes

Power rectifier and control tubes include mercury-vapor rectifiers, thyratrons, and ignitrons. These tubes employ the very high current-carrying capacity of gas discharge tubes with low power losses for rectification and control in high-power equipment. The operation of mercury-vapor tubes depends on temperature insofar as tube voltage drop and peak inverse voltages are concerned. (See Fig. 28.)

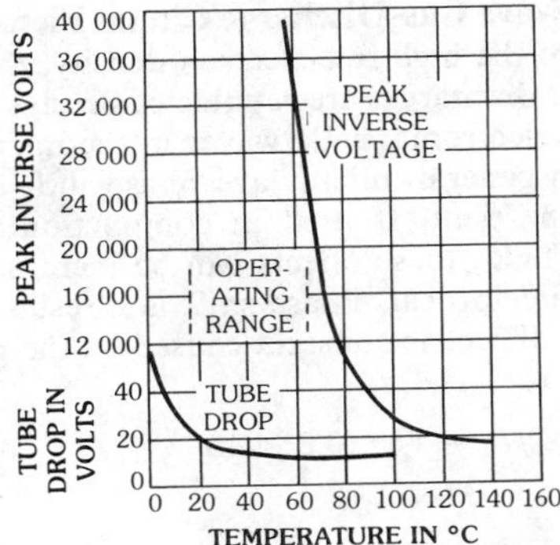

Fig. 28. Tube drop and arc-back voltages as a function of the condensed mercury temperature in a hot-cathode mercury-vapor tube. (*Courtesy McGraw-Hill Book Co.*)

Hydrogen thyratrons are hot-cathode hydrogen-filled triodes designed for use as electronic switching devices where short anode delay time is important. In pulsing service, they are capable of switching tens of megawatts at voltages of tens of kilovolts. Anode delay time and time jitter are in the nanosecond range, and the tubes do not depend on ambient temperature for proper operation. Hydrogen thyratrons are also used in crowbar applications to protect other circuit components against fault voltages or currents and are capable of handling peak currents of several thousand amperes.

Triggered spark gaps are cold-cathode gas tubes operating in the arc discharge region (region III). The gaps contain two high-power electrodes and a trigger electrode, which is generally fired through a step-up pulse transformer by a simple low-energy pulse. The gaps are used as electronic switching devices for peak currents of tens of thousands of amperes and voltages of tens of kilovolts. They can discharge stored energies of several thousand joules and are used for energy transfer in exploding-bridge-wire circuits, gas plasma discharges, spark chambers, and Kerr cells. They are also used in crowbar applications for fast-acting protection of other circuit components against fault voltages and currents. Before conduction, the gap presents a low capacitance and a very high impedance to the circuit. After triggering, when the gap is conducting, the impedance drops to a few ohms or less.

Voltage regulators of the glow-discharge type take advantage of the volt-ampere characteristic in region II, where the voltage is nearly independent of the current. They operate at milliamperes and up to a few hundred volts.

Voltage regulators of the corona-discharge type operate at currents of less than a milliampere and at voltages up to several thousand volts.

Microwave Applications of Gas Tubes

Noise Sources—Gas-discharge devices possess a highly stable and repeatable effective noise temperature when in the fired condition. This feature provides a convenient and accurate means for determining noise figure. The microwave energy radiated from a gas-discharge plasma is coupled into a radio-frequency transmission line with which it is used. The amount of radio-frequency power available from a gas-discharge tube depends mainly on the nature of the gas fill, the geometric characteristic of the discharge tube, and the electron temperature of the positive column or plasma. The design parameter that most strongly determines the noise temperature is the type of gas employed. Any of the noble gases may be used in a noise source. In practice, however, only two or three are normally used:

Gas	$F = \text{ENR(dB)}$
Helium	21.0
Neon	18.5
Argon	15.3

In referring to a noise source or generator, the ratio of its noise power output to thermal noise power is called the Excess Noise Ratio (ENR).

$$F = \text{ENR} = \frac{[(T_2/T_0) - 1] - Y[(T_1/T_0) - 1]}{Y - 1}$$

where,

Y = ratio of the noise output power of the receiver with the noise generator on to that with the noise generator off,
T_0 = 290 K,
T_1 = temperature (in kelvins) of the termination,
T_2 = effective noise temperature (in kelvins) of the noise generator in the fired condition.

The expression $[(T_2/T_0) - 1]$ is termed the excess noise power of the noise source. When $T_1 = T_0 = 290$ K

$$\text{ENR} = [(T_2/T_0) - 1]/(Y - 1)$$

$$\text{ENR(dB)} = 10 \log_{10} \text{ENR}$$

The effective temperature of the noise source is equal to the temperature of the discharge only if the coupling of the transmission line to the discharge is complete. Otherwise, there is a reduction in the noise power output that can best be determined by measuring the fired and unfired insertion loss of the unit at the frequency of interest. The relation between these factors is given by

$$[(T_e/T_0) - 1]/[(T_2/T_0) - 1] = 1 - (L_u/L_f)$$

where,

$[(T_e/T_0) - 1]$ is the effective excess noise power of the generator,

$[(T_2/T_0) - 1]$ is the excess noise power,

L_u and L_f are the insertion losses in the unfired and fired conditions, respectively.

This correction should be subtracted from the apparent measured noise figure.

Noise figure is always measured with reference to a standard temperature of 290 K (T_0). If the ambient temperature (T_1) of the noise-generator termination differs from the standard temperature, the noise figure calculated must be corrected. To find the correction factor, substitute the ambient temperature of the noise-generator termination for T_1 in the following equation, and add the temperature factor (F_T) to the noise figure calculated.

$$F_T = [Y/(Y-1)][(T_1/T_0) - 1]$$

TR Tubes—Transmit-receive tubes are gas-discharge devices designed to isolate the receiver section of radar equipment from the transmitter during the period of high power output. A typical TR tube and its circuit are illustrated in Fig. 29. The cones in the waveguide form a transmission cavity tuned to the transmitter frequency, and the tube conducts received low-power-level signals from the antenna to the receiver. When the transmitter is operated, however, the high-power signal causes gas ionization between the cone tips, which detunes the structure and reflects all the transmitter power to the antenna. The receiver is protected from the destructively high level of power, and all of the available transmitter power is useful output.

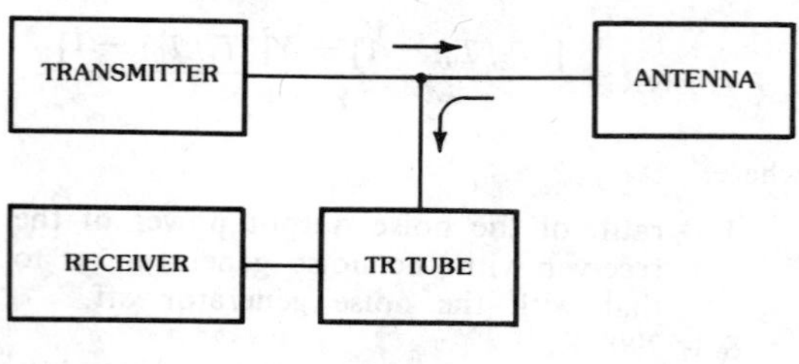

(A) Use of tr tube.

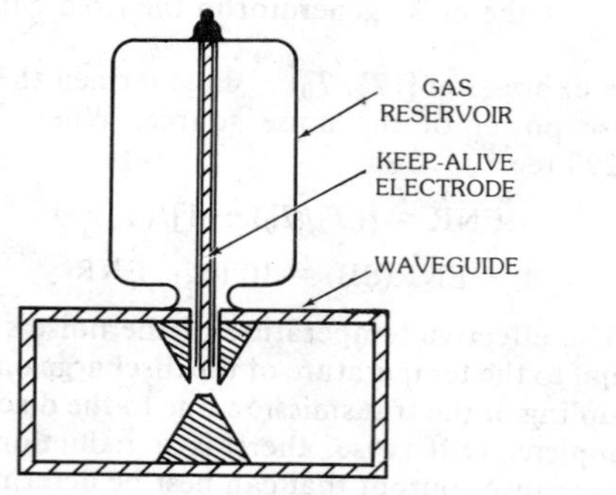

(B) Diagram of tr tube.

Fig. 29. Transmit-receive tube.

Microwave Gas-Discharge Circuit Elements—Because of the high free-electron density, the plasmas of gas discharges are capable of strong interaction with electromagnetic waves in the microwave region. In general, microwave phase shift and/or absorption result. If used in conjunction with a magnetic field, these effects can be increased and made nonreciprocal. Phase shift is a result of the change in dielectric constant caused by the plasma according to

$$\epsilon_p/\epsilon_0 = 1 - (0.8 \times 10^{-4} N_0/f_s^2)$$

where,

ϵ_p = dielectric constant in plasma,

ϵ_0 = dielectric constant in free space,

N_0 = electron density in electrons/centimeter^3,

f_s = signal frequency in megahertz.

Absorption of microwave energy results when electrons, having gained energy from the electric field of the signal, lose this energy in collisions with the tube envelope or neutral gas molecules. This absorption is a maximum when the frequency of collisions is equal to the signal frequency and the absolute magnitude is proportional to the free-electron density.

LIGHT-SENSING AND -EMITTING TUBES

Radiometry and Photometry*

Radiometric and photometric† systems are generally based on the concept of radiated flux, where flux is defined as the total amount of radiation passing through a unit area per unit time.

If a flux is measured in terms of its thermal heating ability, the most common unit is the watt, and the resultant measurement system is referred to as radiometry.

If a flux is measured in terms of its ability to stimulate the standard photopic human eye, the resultant unit is the lumen, and the resultant measurement system is called photometry.

A third choice for the measurement of flux is the number of photons per unit time.

These three choices, in conjunction with the mks system of units, lead to the three mutually compatible systems of units shown in Table 10. Table 11 gives equivalents between units in different photometric measurement systems.

*See also Chapter 21.

†J. W. T. Walsh, *Photometry* (London: Constable and Co., Ltd., 1958).

TABLE 10. COMPATIBLE SYSTEMS OF RADIATION UNITS

Parameter	Radiometric System	Photometric System	Photon System
Flux	watt	lumen	photon s^{-1}
Source intensity	watt sr^{-1}	lumen sr^{-1}	photon s^{-1} sr^{-1}
Incidence	watt m^{-2} (irradiance)	lumen m^{-2} (illuminance)	photon s^{-1} m^{-2}
Excitance	watt m^{-2} (emittance)	lumen m^{-2} (emittance)	photon s^{-1} m^{-2}
Sterance	watt sr^{-1} m^{-2} (radiance)	lumen sr^{-1} m^{-2} (luminance)	photon s^{-1} sr^{-1} m^{-2}
Energy	watt second	lumen second	photon

Note: The terms in parentheses are often used to characterize a measurement as either radiometric or photometric.

TABLE 11. PHOTOMETRIC EQUIVALENTS

Photometric Unit	Equivalent Unit Based on the Lumen (lm) as the Unit of Flux	Equivalent Lumen-MKS Unit
	Source Intensity, C	
1 candela	1 lm sr^{-1}	1 lm sr^{-1}
1 Hefner candle	0.92 lm sr^{-1}	0.92 lm sr^{-1}
	Surface Luminance, B	
1 candle cm^{-2}	1 lm sr^{-1} cm^{-2}	10^4 lm sr^{-1} m^{-2}
1 candle m^{-2}	1 lm sr^{-1} m^{-2}	1 lm sr^{-1} m^{-2}
1 candle in^{-2}	1 lm sr^{-1} in^{-2}	1.55×10^3 lm sr^{-1} m^{-2}
1 candle ft^{-2}	1 lm sr^{-1} ft^{-2}	10.8 lm sr^{-1} m^{-2}
1 nit	10^{-4} lm sr^{-1} cm^{-2}	1 lm sr^{-1} m^{-2}
1 stilb	1 lm sr^{-1} cm^{-2}	10^4 lm sr^{-1} m^{-2}
1 apostilb	π^{-1} lm sr^{-1} m^{-2}	π^{-1} lm sr^{-1} m^{-2}
1 lambert	π^{-1} lm sr^{-1} cm^{-2}	10^4 π^{-1} lm sr^{-1} m^{-2}
1 millilambert	10^{-3} π^{-1} lm sr^{-1} cm^{-2}	$10\pi^{-1}$ lm sr^{-1} m^{-2}
1 footlambert	π^{-1} lm sr^{-1} ft^{-2}	$10.8\pi^{-1}$ lm sr^{-1} m^{-2}
	Illuminance of a Surface, I_L	
1 lux	1 lm m^{-2}	1 lm m^{-2}
1 phot	1 lm cm^{-2}	10^4lm m^{-2}
1 milliphot	10^{-3} lm cm^{-2}	10 lm m^{-2}
1 footcandle	1 lm ft^{-2}	10.8 lm m^{-2}
	Energy, U	
1 talbot	1 lm s	1 lm s

Flux Units

The number of lumens dL_λ and the number of photons per second dN_λ associated with a monochromatic flux dW_λ in watts are given by

$$dL_\lambda = 680E_\lambda dW_\lambda \quad \text{and} \quad dN_\lambda = (\lambda / hc)dW_\lambda$$

where,

- $680 =$ number of lumens per watt of radiation at the peak photopic eye response,
- $E_\lambda =$ normalized (to unity maximum) photopic human eye response (Fig. 30),
- $\lambda =$ wavelength of the monochromatic radiation (m),
- $h =$ Planck's constant $\simeq 6.6 \times 10^{-34}$ (J·s),
- $c =$ velocity of light $\simeq 3.0 \times 10^8$ (m/s).

The number of lumens L and the number of photons per second N between the wavelength units of λ_3 to λ_4 associated with a distributed spectral radiation source having a wattage W between the wavelength limits λ_1 and λ_2 are given by

$$L/W = 680 \int_0^\infty E_\lambda w_\lambda d\lambda \Big/ \int_{\lambda_1}^{\lambda_2} w_\lambda \, d\lambda$$

and

$$Nhc/W = \int_{\lambda_3}^{\lambda_4} \lambda w_\lambda \, d\lambda \Big/ \int_{\lambda_1}^{\lambda_2} w_\lambda \, d\lambda$$

where,

$$W = w_{\lambda\,\max} \int_{\lambda_1}^{\lambda_2} w_\lambda \, d\lambda$$

and where,

$w_{\lambda\,\max}$ = maximum spectral density in watts per unit wavelength in the spectral band between λ_1 and λ_2,

w_λ = relative spectral distribution of the radiation source on thermal-energy basis, normalized to a maximum value of unity.

Typical w_λ spectral distributions are in Fig. 30.

Optical Imaging

In an optical lens system of flux-gathering diameter D_f in meters, focal length f in meters, and optical transmittance T, the ratio $f/D_f = n_f$ is called the f-number of the lens. If the surface of an object of radiance or luminance B in flux units per steradian per meter^2 is imaged by this system with a linear magnification m, and assuming Lambertian emittance characteristics over the solid angle subtended by the optical system, the image will be subjected to an irradiance or illuminance I_L in flux units per meter^2 given by

$$I_L = \pi BT/[4n_f^2(m+1)^2 + m^2]$$

For objects at infinity, $m = 0$, and

$$I_L \text{ (object at infinity)} = \pi BT/4n_f^2$$

If the irradiance (or illuminance) I_L in flux units per meter^2 is allowed to fall on a nonabsorbing Lambertian diffusing surface, the resultant image radiance (or luminance) B_i in flux units per steradian per meter^2 is given by $\Pi B_i = I_L$.

Any desired method of measuring flux units, such as watts, lumens, or photons/second (Table 10), can be selected for expressing the object radiance (or luminance) B in flux units steradian^{-1} meter^{-2} and the irradiance (or illuminance) I_L in flux units meter^{-2} in these relationships. Thus, a radiance B in watts steradian^{-1} meter^{-2} would be paired with an irradiance I_L in watts meter^{-2}, a luminance B in lumens steradian^{-1} meter^{-2} with an illuminance I_L in lumens meter^{-2}, and a radiance B in photons second^{-1} steradian^{-1} meter^{-2} with an irradiance I_L in photons second^{-1} meter $^{-2}$.

Any spectral distribution modifications, if present, would be included in the numerical magnitude of the lens transmission T, defined as the ratio of the total output flux from the optical system to the corresponding input flux.

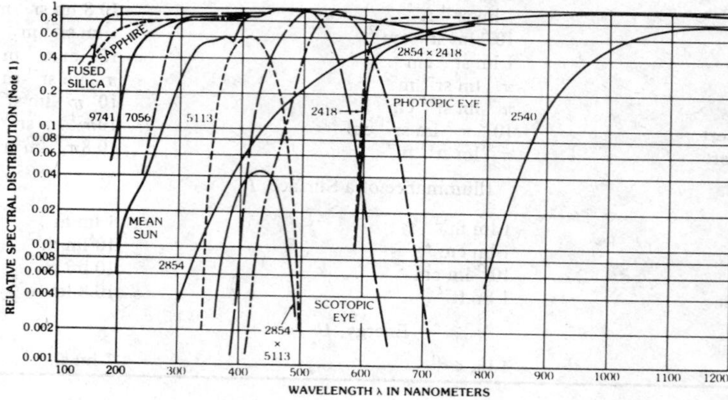

Fused silica: transmission through polished "Suprasil" (1 millimeter thick)
Sapphire: transmission through polished Sapphire (1 millimeter thick)
9741: transmission through polished Corning 9741 glass (1 millimeter thick)
7056: transmission through polished Corning 7056 glass (1 millimeter thick)
Mean sun: mean solar distribution at Earth's surface
5113: transmission through polished Corning 5113 filter (CS-5-58) (half-stock thickness)
2854 × 5113: product of 5113 curve and 2854 curve
2854: spectral density distribution of 2854 K color-temperature tungsten lamp
Scotopic eye: relative response of dark-adapted eye
Photopic eye: standard eye response
2418: transmission through polished Corning 2418 filter (CS-2-62) (stock thickness)
2854 × 2418: product of 2854 and 2418 curves
2540: transmission through polished Corning 2540 filter (CS-7-56) (stock thickness)

Note 1: "Relative spectral distribution" designates the relative radiant-energy density distribution w_λ for sources, the relative visual stimulation for equi-energy inputs for the eye response, and the spectral transmission t_λ for windows and filters. The transmission characteristics of individual filter and window samples can be expected to depart appreciably from these typical values.

Fig. 30. Useful spectral distributions.

Selection of appropriate alternative pairs of luminance and illuminance units when the flux units are not explicitly stated (first column of Table 11) must be made with care. Thus candle centimeter^{-2} (or stilb) would be paired with phot, candle meter^{-2} (or nit) with lux, and candle foot^{-2} with footcandle. Even greater difficulty arises when the factor Π in the preceding relationships is absorbed or included in the units of luminance. Thus the product ΠB in apostilbs would be paired with I_L in lux, the product ΠB in lamberts with I_L in phots, the product ΠB in millilamberts with I_L in milliphots, and the product ΠB in footlamberts with I_L in footcandles. These difficulties are avoided by the use of the compatible systems of radiation units shown in Table 10.

Typical Approximate Illumination Values at the Surface of the Earth

Sun at zenith $\simeq 10^4$ footcandles
$\simeq 10^5$ lumens meter^{-2}
Full moon $\simeq 3 \times 10^{-2}$ footcandles
$\simeq 3 \times 10^{-1}$ lumens meter^{-2}

Typical Approximate Brightness Values

	footlamberts	lm sr^{-1} m^{-2}
Highlights, 35-millimeter movie	$\simeq 4$	$\simeq 100$
Page brightness for reading fine print	$\simeq 10$	$\simeq 3 \times 10^2$
November football field	$\simeq 50$	$\simeq 1.5 \times 10^3$
Surface of moon seen from Earth	$\simeq 1.5 \times 10^3$	$\simeq 5 \times 10^4$
Summer baseball field	$\simeq 3 \times 10^3$	$\simeq 10^5$
Surface of 40-watt frosted lamp bulb	$\simeq 8 \times 10^3$	$\simeq 2.5 \times 10^5$
Crater of carbon arc	$\simeq 4.5 \times 10^7$	$\simeq 10^9$
Sun seen from Earth	$\simeq 5.2 \times 10^8$	$\simeq 1.5 \times 10^{10}$

LIGHT-SENSING TUBES

Image Tubes and Image Intensifiers

An image tube* is an optical-image-in to optical-image-out electron-tube device, combining an input photocathode and an output phosphor screen such that photoelectrons emitted from each point on the photocathode subsequently excite a corresponding individual image "point" on the phosphor screen.

Various focusing means, including magnetic and electrostatic electron lenses, may be used to assure maximum point-to-point correlation between the input and output images. The principal operating requirements are a lens to form the input image and a high-voltage supply, typically 5–25 kilovolts, to provide sufficient electron-beam energy to excite the output phosphor screen.

If means are provided within the image tube to amplify the photoelectrons before they strike the output phosphor screen, or if the tube without such means produces a much brighter output image than the input image would produce on a diffusing screen, the tube is commonly called an image-intensifier tube.

Image-intensifier tubes are used to amplify the brightness of a faint input image for better visual or photographic viewing, whereas image tubes without amplification are used to convert radiation from one spectral region to another (image conversion) or to perform such control operations as optical shuttering by programming the applied high voltage.

The total output flux dW_o in watts exiting (through 2Π steradians) from the phosphor-screen faceplate of an image-intensifier tube for an input monochromatic flux dW_λ in watts at a wavelength λ is given by

$$dW_o = s_\lambda G \mathcal{E}_w (V - V_k)\, dW_\lambda$$
$$= G_\lambda dW_\lambda$$

where,

$G_\lambda =$ monochromatic wattage gain of the image-intensifier tube at a wavelength $\lambda =$ ratio of the total output flux dW_o in watts to the input monochromatic flux dW_λ in watts,

$s_\lambda =$ radiant sensitivity of the input photocathode in amperes per watt (see Fig. 13, Chapter 21),

$G =$ internal current gain ratio of the image-intensifier tube $=$ ratio of the current bombarding the output phosphor screen to the corresponding photocurrent leaving the input photocathode,

$\mathcal{E}_w =$ absolute phosphor efficiency $=$ ratio of the total radiated flux in watts to the exciting electron-beam power in watts dissipated in the particles of the output phosphor screen,

$V =$ energy of the electron beam in volts bombarding the output phosphor screen,

$V_k =$ extrapolated knee voltage of the output phosphor screen.

If the phosphor screen radiates flux according to Lambert's Law (usually only approximately valid), the corresponding output image radiance R_o in watts steradian^{-1} meter^{-2} is given by

$$R_o = G_{\lambda 1} I_{\lambda 1} / \Pi m^2$$

*"Photo-Electric Image Devices," *Advances in Electronics and Electron Physics,* Vols. 12, 16, 22A, and 22B (New York and London: Academic Press, 1960, 1962, and 1966). H. V. Soule, *Electro-Optical Photography at Low Illumination Levels* (New York: John Wiley & Sons, 1968).

where,

$I_{\lambda 1}$ = input image irradiance on the photocathode expressed in watts meter^{-2} at the wavelength λ_1,

m = differential magnification ratio of the image tube = output incremental image size divided by the corresponding input incremental image size.

For a spectrally distributed input flux having a known relative spectral distribution w_λ and a known total radiated power $W_{\lambda 1\lambda 2}$ in watts between the wavelength limits λ_1 and λ_2, the resulting total output flux W_o in watts exiting from the image tube is given by

$$W_o = s_{\lambda\max}\left(\int_0^\infty \sigma_\lambda w_\lambda\, d\lambda \Big/ \int_{\lambda 1}^{\lambda 2} w_\lambda\, d\lambda\right) \times G\mathcal{E}_w(V - V_k)W_{\lambda 1\lambda 2}$$

$$= G_{\lambda 1\lambda 2}W_{\lambda 1\lambda 2}$$

where,

$s_{\lambda\max}$ = peak radiant sensitivity of the input photocathode in amperes per watt,

σ_λ = relative radiant sensitivity of the input photocathode as a function of wavelength λ normalized to unity maximum,

w_λ = relative spectral distribution of the power density spectrum of the input flux normalized to unity maximum,

$G_{\lambda 1\lambda 2}$ = wattage gain of the image tube for the relative spectral distribution w_λ and the wavelength limits λ_1 and λ_2.

Typical values for the magnitude of the dimensionless spectral-matching-factor ratio

$$\int_0^\infty \sigma_\lambda w_\lambda\, d\lambda \Big/ \int_{\lambda 1}^{\lambda 2} w_\lambda\, d\lambda$$

are found in Table 2.

The total output flux L_o in lumens exiting from an image tube, corresponding to the total output flux W_o in watts, can be computed from the flux conversion relationships given in the section on radiometry and photometry, or from the following relationship

$$L_o = s_{\lambda\max}\left(\int_0^\infty \sigma_\lambda w_\lambda\, d\lambda \Big/ \int_0^\infty E_\lambda w_\lambda\, d\lambda\right) \times G\mathcal{E}_w\left(\int_0^\infty E_\lambda w_{o\lambda}\, d\lambda \Big/ \int_0^\infty w_{o\lambda}\, d\lambda\right) (V - V_k)L_i$$

$$= G_L L_i$$

where,

G_L = luminous gain of the image intensifier tube = ratio of the output flux in lumens to the corresponding input flux in lumens for the spectral input distribution w_λ,

E_λ = standard tabulated average relative photopic eye response (Table 2),

$w_{o\lambda}$ = relative spectral density distribution of the output flux,

L_i = input flux in lumens.

The typical values of the dimensionless spectral matching factors given in Table 2 can be used to determine the magnitude of the dimensionless integral ratios appearing in these relationships.

For the special case where the input flux $L_i(2854)$ in lumens is generated by a 2854 K color-temperature tungsten-filament lamp, the output flux L_o (2854) in lumens is given by

$$L_o(2854) = S(2854)G\mathcal{E}_w\left(\int_0^\infty w_{o\lambda}E_\lambda\, d\lambda \Big/ \int_0^\infty w_{o\lambda}\, d\lambda\right) \times (V - V_k)L_i(2854)$$

$$= G_L(2854)L_i(2854)$$

where,

$G_L(2854)$ = luminous gain of the image intensifier for 2854 K tungsten-lamp radiation,

$S(2854)$ = luminous sensitivity of the input photocathode for 2854 K tungsten-lamp radiation.

The magnitude of the luminous gain $G_L(2854)$ is commonly used to characterize the image intensification properties of an image-intensifier tube.

If the output phosphor screen radiates flux according to Lambert's Law (usually only approximately valid), the output image luminance (or brightness) B_o in lumens steradian^{-1} meter^{-2} is given by

$$B_o = G_L I_i / \pi m^2$$

where,

G_L = luminous gain of the image intensifier for input spectral distribution w_λ,

I_i = input illuminance (or illumination) on the photocathode in lumens meter^{-2} for the spectral distribution w_λ,

m = differential magnification ratio of the image tube = output incremental image size divided by the corresponding input incremental image size.

Internal current gain G of an image-intensifier tube can be obtained by the use of an internal sandwich electrode, in which an auxiliary or sandwich photocathode is mounted in close proximity to and following an auxiliary or sandwich phosphor. Photoelectrons from the input photocathode of the tube are then imaged onto this sandwich phosphor screen, and the flux from this screen is coupled to the sandwich photocathode, generating an enhanced photocurrent. The current gain ratio G of this sandwich phosphor-photocathode combination, defined as the ratio of output photocurrent to input photocurrent, is given by

$$G = s_{\lambda \text{ max, sand}} \left(\int_0^\infty w_{\lambda \text{ (sand)}} \sigma_{\lambda \text{ (sand)}} \, d\lambda \Big/ \int_0^\infty w_{\lambda \text{ (sand)}} \, d\lambda \right) \mathcal{E}_{w, \text{ sand}} (V_{\text{sand}} - V_{k, \text{ sand}}) \gamma$$

where,

$s_{\lambda \text{ max, sand}}$ = peak monochromatic responsivity of the sandwich photocathode in amperes watt^{-1},

$w_{\lambda \text{ (sand)}}$ = relative spectral distribution of the flux emitted by the sandwich phosphor screen,

$\sigma_{\lambda \text{ (sand)}}$ = the relative spectral distribution of the sandwich photocathode,

$\mathcal{E}_{w, \text{ sand}}$ = absolute efficiency of the sandwich phosphor screen in watts watt^{-1},

V_{sand} = electron beam energy in volts bombarding the sandwich phosphor screen,

$V_{k, \text{ sand}}$ = extrapolated knee voltage in volts for the sandwich phosphor screen,

γ = optical coupling efficiency of the sandwich electrode = ratio of the flux falling onto the sandwich photocathode to the corresponding flux emitted by the sandwich phosphor screen.

Typical values of the dimensionless ratio of the two integrals appearing in this relationship are given in Table 2.

The combination of a phosphor screen and a photocathode to produce current gain G can also be achieved by optically coupling the output flux from one image tube to the input of a second tube.

Resolution in image tubes and image-intensifier tubes is a subjective parameter describing the number of pairs of equally spaced illuminated and unilluminated bars per unit distance at the photocathode imaged onto the input photocathode surface which can just be distinguished visually by a trained observer under stated test conditions.

Distortion is a parameter describing any change in the geometric shape of the output image compared with the input image. Radially increasing magnification leads to "pincushion" distortion; radially decreasing magnification leads to "barrel" distortion; and radially changing image rotation leads to "S" distortion.

Gas Photodiodes

In diode phototubes not containing a high vacuum, ionization by collision of electrons with neutral molecules may occur so that more than one electron reaches the anode for each originally emitted photoelectron. This "gas amplification factor" has a value of between 3 and 5; a higher factor causes instabilities. Gas-tube operation is restricted to frequencies below about 10 000 hertz.

Image Orthicons

The image orthicon* is a camera tube that is widely used for commercial television. This fact derives from its high sensitivity, its close spectral-sensitivity match to the human eye, and its relatively fast response. Good-quality commercial television pictures can be generated by an image orthicon viewing a 5-to-20-footlambert ($\simeq$ 15–50 lumen steradian^{-1} meter^{-2}) scene through an f/5.6 lens. The image orthicon is generally available with either S-10 or S-20 spectral response (see Fig. 13, Chapter 21) and is capable of 500 picture elements per raster height (9.9 line pairs/millimeter) at 30-percent video-amplitude response.

Principle of Operation—Fig. 31 is a diagram of an image orthicon. In the image section, a light image incident on the translucent photocathode liberates photoelectrons into the adjacent vacuum region in proportion to the light intensity (gamma is unity) on each element of the cathode. These photoelectrons are accelerated toward and magnetically focused onto the surface of a thin semiconducting target. Electrons strike this target with sufficient energy to liberate a larger number of secondary electrons (typically 5) for each incident primary. The secondary electrons are collected by a mesh closely spaced from the target membrane. Hence, by depletion of electrons from the thin membrane, incremental areas become positive in proportion to the number of photoelectrons striking each element. In cases of high-light-level operation, parts of the target may become charged to target (collector) mesh potential, and saturation charge results. This phenomenon accounts for the so-called "knee" in the signal-vs-illumination transfer curve (Fig. 32).

Because the target membrane is very thin, of the order of microns, a charge distribution pattern formed on the image-section surface appears nearly simultaneously and identically on the scanning-section surface.

In the scanning section, an electron gun generates a highly apertured electron beam from a fraction to

*A. Rose, P. K. Weimer, and H. B. Law, "The Image Orthicon—A Sensitive Television Pickup Tube," *Proceedings of the IRE,* Vol. 34, No. 7, pp. 424–432; July 1946.

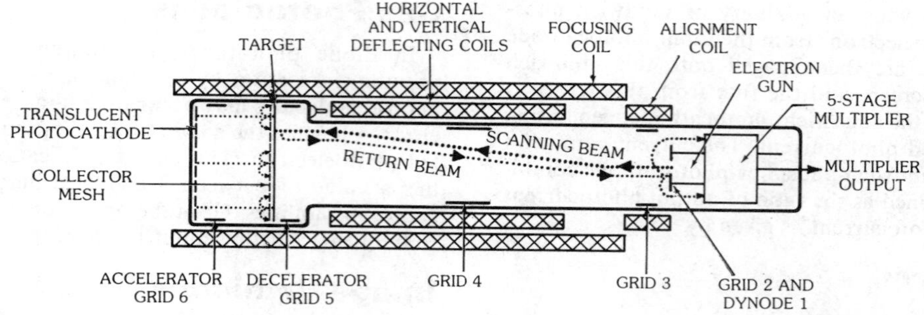

Fig. 31. Image orthicon. (*By permission of RCA, copyright proprietor.*)

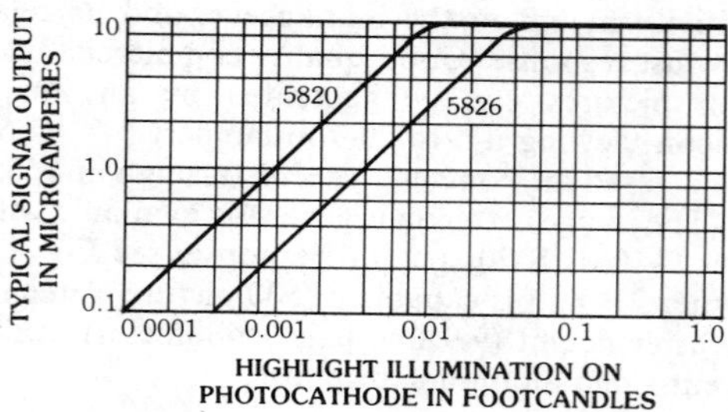

Fig. 32. Basic light-transfer characteristic for type 5820 and 5826 image orthicons. The curves are for small-area highlights illuminated by tungsten light, white fluorescent light, or daylight. (*By permission of RCA, copyright proprietor.*)

tens of microamperes in intensity. A solenoidal magnetic-focus coil and saddle-type deflection coils surrounding the scan section focus this beam on the insulator target and move it across the target. Scan-beam electrons impinge on the target at very low velocity, giving rise to relatively few secondary electrons. The target acts somewhat as a retarding-field electrode and reflects a large number of the beam electrons that have less than average axial velocity. These two phenomena, small but finite secondary emission and reflection of slow beam electrons, limit scan-beam modulation to a maximum of about 30 percent at high light levels, and to 2 orders less at threshold. As will be shown later, the large unmodulated return beam current is the primary source of noise in the image orthicon.

Another problem created by the retarding-field aspect of low-velocity target scanning appears when the deflected beam does not strike the target normally. Since the entire beam-velocity component normal to the surface is now reduced by the cosine of the angle of incidence, the effective beam impedance is greatly increased. To overcome this problem, the decelerating field between grids 4 and 5 is shaped such that the electron beam always approaches normal to the plane of the target at a low velocity. If the elemental area on the target is positive, then electrons from the scanning beam deposit until the charge is neutralized. If the elemental area is at cathode potential (corresponding to a dark picture area), no electrons are deposited. In both cases, the excess beam electrons are turned back and focused into a 5-stage electron multiplier. The charges existing on either side of the semiconductive target membrane will, by conductivity, neutralize each other in less than one frame time. Electrons turned back at the target form a return beam that has been amplitude-modulated in accordance with the charge pattern of the target.

The return beam is redirected by the deflection and focus fields toward the electron gun where it originated. Atop the electron gun, and forming the final aperture for that gun, is a flat secondary-emitting surface comprising the first dynode of the electron multiplier. The return beam strikes this surface, generating secondary electrons in a ratio of approximately 4:1.

Grid 3 facilitates a more complete collection by dynode 2 of the secondary electrons emitted from dynode 1. The gain of the multiplier is high enough that in operation the limiting noise is the shot noise of the returned electron beam rather than the input noise of the video amplifier.

Signal and Noise—Typical signal output currents for tube types 5820 and 5826 are shown in Fig. 32. The tubes should be operated so that the highlights on the photocathode bring the signal output slightly over the knee of the signal-output curve.

The spectral response of type 5820 and 5826 image orthicons is shown in Fig. 33. Note that when a Wratten 6 filter is used with the tube, a spectral curve closely approximating that of the human eye is obtained.

From the standpoint of noise, the total television system can be represented as shown in Fig. 34, where I_s = signal current, I_n = total image-orthicon noise current, E_{nt} = thermal noise in R_1, E_{ns} = shot noise in the input amplifier tube, R_1 = input load, C_1 = total input shunt capacitance, and R_t = shot-noise equivalent resistance of the input amplifier = $2.5/g_m$ for triode or cascode input = $[I_b/(I_b + I_c)][(2.5/g_m) + (20I_{c2}/g_m^2)]$ for pentode

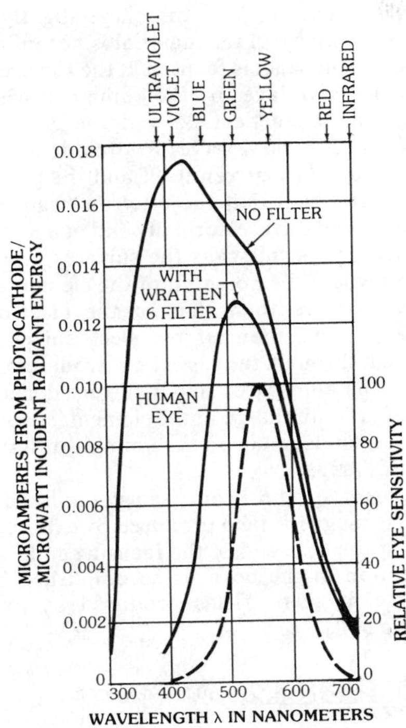

Fig. 33. Spectral sensitivity of image orthicon. (*By permission of RCA, copyright proprietor.*)

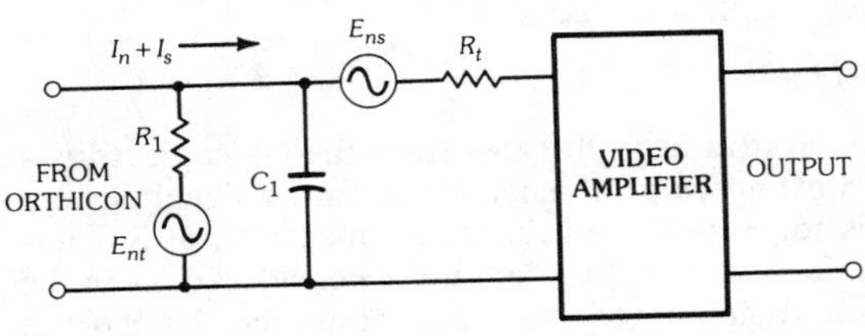

Fig. 34. Equivalent circuit for noise in orthicon and first amplifier stage.

input, with g_m = transconductance of input tube or cascode combination, I_b = amplifier direct plate current, and I_c = amplifier direct screen-grid current.

The noise added per stage is

$$\Delta n = [\sigma/(\sigma - 1)]^{1/2}$$

where σ = stage gain in the multiplier. For a total multiplier noise figure to be directly usable, it must be referred to the first-dynode current; therefore, for five multiplier stages

$$\overline{\Delta N} = \Delta n^2 + \frac{\Delta n^2}{\sigma^2} + \frac{\Delta n^2}{\sigma^4} + \frac{\Delta n^2}{\sigma^6} + \frac{\Delta n^2}{\sigma^8}$$

where ΔN = electron-multiplier noise factor referred to multiplier input.

After combining all noise sources

$$\frac{S}{N} = \frac{I_s}{\left\{F\left[2eIk_m{}^2 + 4kT\left(\frac{1}{R_1} + \frac{R_t}{R_1{}^2} + \frac{\omega^2 C_1{}^2 R_t}{3}\right)\right]\right\}^{1/2}}$$

where,

S/N = signal-to-noise ratio,
F = bandwidth in hertz,
e = electron charge = 1.6×10^{-19} coulomb,
I = image-orthicon beam current,
k_m = electron-multiplier noise factor, referred to multiplier output = $m \, \Delta \, N$,
k = Boltzmann's constant = 1.38×10^{-23} joule/kelvin,
T = absolute temperature in kelvins,
$\omega = 2\pi f$ in hertz.

The signal current is an alternating-current signal superimposed on a larger direct beam current. This can be thought of as a modulation of the beam current. Properly adjusted tubes obtain as much as 30-percent modulation.

$$I_s = mMI$$

where,

m = multiplier gain,
M = percentage modulation.

If S/N is now rewritten,

$$\frac{S}{N} = \frac{I_s}{\left[4kTF\left(\frac{2eI_s m\overline{\Delta N}^2}{4kTM} + \frac{1}{R_1} + \frac{R_t}{R_1{}^2} + \frac{\omega^2 C_1{}^2 R_t}{3}\right)\right]^{1/2}}$$

In typical television operation, the thermal noise of the load resistor and the shot noise of the first amplifier can be neglected.

Focusing and Scanning Fields—The electron optics of the scanning section of the tube are quite complicated, and space does not permit the inclusion of the complete equations. A simple relationship between the strength of the magnetic focusing field and the magnetic deflection field is given below.

The image orthicon is usually operated with multiple-node focus in the scanning section. Working at a multiple-node focus not only demands more focus current but also more deflection current. Note the deflection path in Fig. 35. Let H = horizontal dimension of scanned area or target, L = effective length of horizontal deflection field, H_d = horizontal deflection field (peak-to-peak value), and H_f = focusing field, Then

$$H_d = H_f H / L$$

For the image orthicon, $H \approx 1.25$ inches, and $L \approx 4$ inches. Thus $H_f \approx 75$ gauss, and $H_d \approx 23$ gauss.

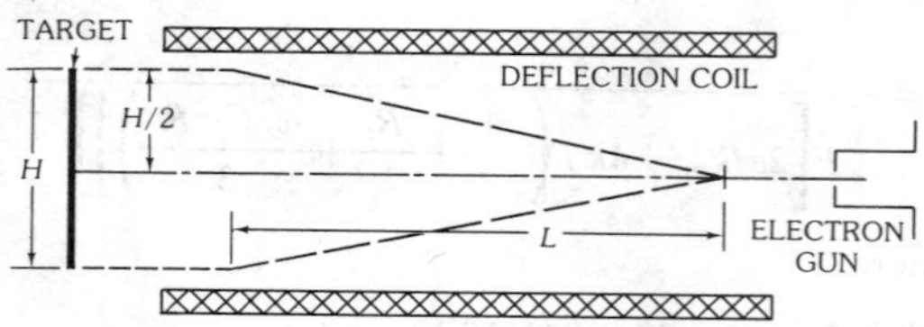

Fig. 35. Deflection in image orthicon.

Vidicons

The vidicon* is a small television camera tube that is used primarily for industrial television, space application, and studio film pickup because of its small size and simplicity.

As shown in Fig. 36, the tube consists of a signal electrode composed of a transparent conducting film on the inner surface of the faceplate, a thin layer (a few micrometers) of photoconductive material deposited on the signal electrode, a fine mesh screen (grid 4) located adjacent to the photoconductive layer, a focusing electrode (grid 3) connected to grid 4, and an electron gun.

imaged on the opposite surface of the layer. Even those areas that are dark discharge slightly, since the dark resistivity of the material is not infinite.

The electron beam is focused at the surface of the photoconductive layer by the combined action of the uniform magnetic field and the electrostatic field of grid 3. Grid 4 serves to provide a uniform decelerating field between itself and the photoconductive layer such that the electron beam always approaches the surface normally and at a low velocity. When the beam scans the surface, it deposits electrons where the potential of the elemental area is more positive than that of the electron-gun cathode. At this moment the electrical circuit is completed through the signal-electrode circuit to ground. The amount of signal current depends on the amount of discharge in the elemental capacitor, which in turn depends on the amount of light falling on this area.

Alignment of the beam is accomplished by a transverse magnetic field produced by external coils located at the base end of the focusing coil.

Deflection of the beam is accomplished by the transverse magnetic fields produced by external deflecting coils.

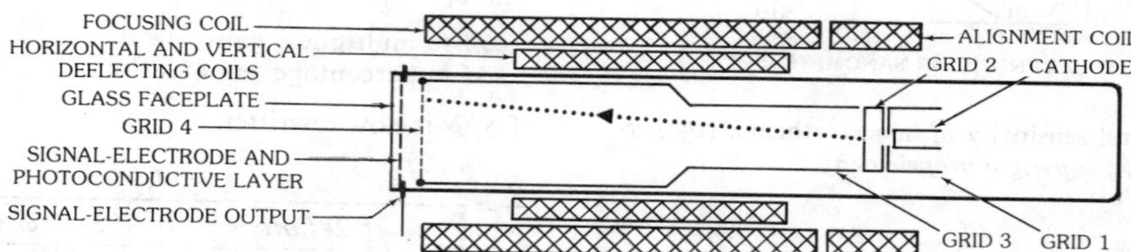

Fig. 36. Vidicon construction. (*By permission of RCA, copyright proprietor.*)

Principle of Operation—Each elemental area of the photoconductor can be likened to a leaky capacitor with one plate electrically connected to the signal electrode that is at some positive voltage (usually about 20 volts) with respect to the thermionic cathode of the electron gun and the other plate floating except when commutated by the electron beam. Initially, the gun side of the photoconductive surface is charged to cathode potential by the electron gun, thus leaving a charge on each elemental capacitor. During the frame time, these capacitors discharge in accordance with the value of their leakage resistance, which is determined by the amount of light falling on each elemental area. Hence, there appears on the gun side of the photoconductive surface a positive-potential pattern corresponding to the pattern of light from the scene

*B. H. Vine, R. B. Janes, and F. S. Veith, "Performance of the Vidicon—A Small Developmental Camera Tube," *RCA Review,* Vol. 13, No. 1, pp. 3-10; March 1952. P. Weimer, S. Forgue, and R. Goodrich, "The Vidicon Photoconductive Camera Tube," *Electronics,* Vol. 23, No. 5, pp. 70-73; May 1950.

Signal and Noise—Since the vidicon acts as a constant-current generator as far as signal current is concerned, the value of the load resistor is determined by bandpass and noise considerations in the input circuit of the video amplifier. Unlike the image orthicon, the vidicon has the signal current removed at the target, and only that portion of the scan beam actually involved in the target discharge contributes shot noise. Moreover, electron-beam contributions to noise are minimal for low-light portions of the scene.

The primary noise associated with vidicon operation is seldom scan-beam shot noise. Where the signal current is less than 1 microampere and the bandpass is relatively wide, the principal noise in the system is contributed by the input circuit and the first stage of the video amplifier. To minimize the thermal noise of the load resistor, its resistance is made much higher than flat-bandpass considerations would indicate, since signal voltage increases directly and noise voltage increases as the square root. To correct for attenuation of the signal with increasing frequency, the amplitude response of the video amplifier frequently employs high-frequency boost of the following form, where C_1 and R_1 refer to Fig. 37:

$$G = G_0(1 + 4\pi^2 F^2 C_1^2 R_1^2)^{1/2} / R_1$$

A representative plot of amplitude response as a function of the number of television lines (per raster height) is shown in Fig. 38.

The vidicon has somewhat more lag or image persistence than the image orthicon. This is the result of two factors. To obtain high-sensitivity surfaces, the photoconductive decay time is made as long as tolerable, since quantum efficiency is limited by the ratio of effective carrier lifetime to carrier transit time across the photoconductor. A second source of lag is simply the RC time constant of the target recharging circuit; that is, the target capacitance and the beam impedance.

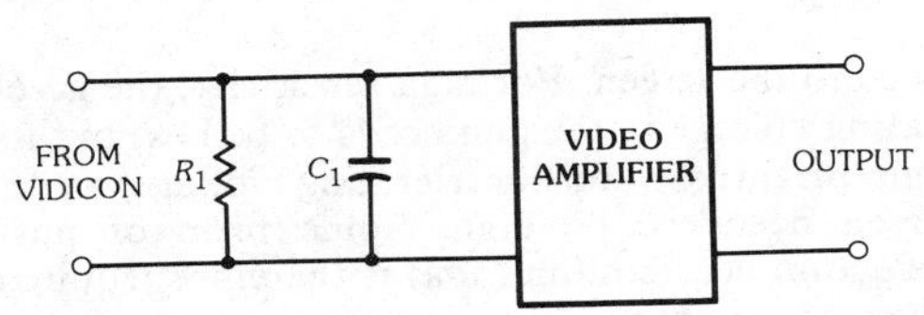

Fig. 37 Equivalent input circuit for first-stage amplifier in vidicon circuit.

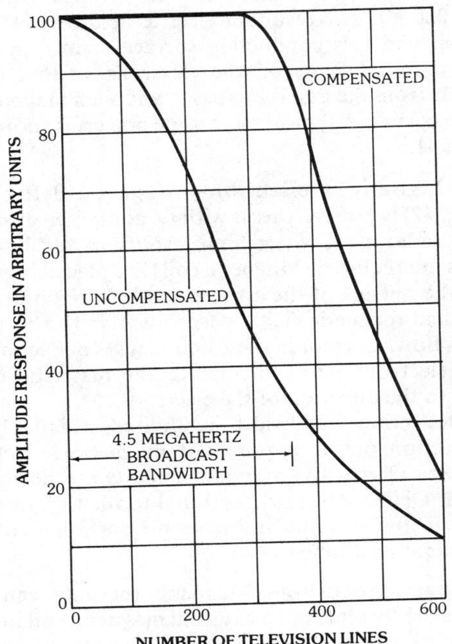

Fig. 38. Vidicon resolution, showing uncompensated and compensated horizontal responses. Highlight signal-electrode microamperes = 0.35; test pattern = transparent square-wave resolution wedge; 80 television lines = 1-megahertz bandwidth. (*By permission of RCA, copyright proprietor.*)

The spectral response of most commercial vidicons, designated S-18, is more actinic than that of the human eye. Fig. 39 compares these responses with the spectrum of a 2854 K tungsten source.

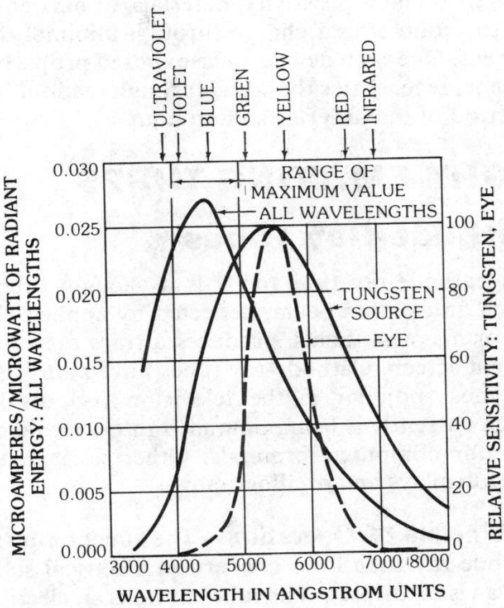

Fig. 39. Spectral response of vidicon. (*By permission of RCA, copyright proprietor.*)

Variations of the Vidicon

Interest in optical guidance and surveillance from aircraft and spacecraft has given rise to a wide variety of vidicon camera tubes. To treat these variations in detail becomes encyclopedic, but the following gives some indication of the choices now available to the user.

Effective Sensitivity—True photoconductive tubes now offer sensitivities of 150–200 nanoamperes for 1/2-footcandle illumination with 20 nanoamperes dark current. Improved methods of deposition of photoconductors have made possible higher-voltage operation without objectionable dark shading. Special devices using junction effects promise even better sensitivity.

Spectral Response—Available photoconductors, taken as a whole, provide sensitivity over the entire visible range with usual (7056) glass windows. Quartz-window tubes offer useful sensitivity to below 2000 angstrom units. Numerous applications of direct excitation of photoconductors by X-radiation have been reported. High-velocity electron excitation (bombardment-induced conductivity) is also in use.

Size and Deflection—Vidicons are available in sizes ranging from 1/2 inch to 2 inches in diameter.

Various combinations of deflection and focus are available.

Storage—A number of manufacturers have produced vidicons with long-storage characteristics. Many are merely long-lag tubes; however, a few rely on high-resistivity materials, or barrier layers to retain stored charge through minimal dark current. One such device, once exposed properly to a scene, regenerates the scene through readout over a period of the order of half an hour.

LIGHT-EMITTING TUBES

Cathode-Ray Tubes

A cathode-ray tube (crt)* is a vacuum tube in which an electron beam, deflected by applied electric or magnetic fields, produces a trace on a fluorescent screen. Cathode-ray tubes have many uses; the most popular is the television (tv) display, which currently is being upgraded into a visual display for computer terminals. Other uses include radar displays and oscillography.

Principle of Operation—The function of the cathode-ray tube is to convert an electrical signal into a visual display. The tube contains an electron-gun structure (to provide a narrow beam of electrons) and a phosphor screen. The electron beam is directed to the phosphor screen and strikes it, causing light to be emitted in a small area or spot in proportion to the intensity of the electron beam. The beam intensity varies as a function of the beam-control element in the electron gun. In a tv or computer display crt, the electrical signal that controls the beam intensity corresponds to the desired picture information and is referred to as the video signal. Although one spot or picture element is not enough to reproduce a picture, by moving the spot over the entire screen in a systematic manner, the complete picture can be reproduced.

For oscilloscope use, the deflection of the beam at the screen is proportional to the voltages applied to the deflection electrodes within the tube, and a visual picture of time-dependent waveforms can be produced on the screen in the same manner as a graph is drawn on paper.

Electrostatic Deflection—Electrostatic deflection is generally used in oscillography where high-frequency signals need to be analyzed. Electrostatic deflection consumes low power, and deflectors can be designed to perform at extremely high frequencies. Electrodes placed within the tube (Fig. 40) form an electric field perpendicular to the electron path. Deflection is determined by the voltage across the electrodes, the accelerating voltage, and the distance to the screen. For high sensitivity, the accelerating voltage in the gun needs to be low, but for high brightness, the accelerating voltage at the screen needs to be high. Some form of post-deflection acceleration (pda) is therefore required. The most common form of pda uses a dome-shaped mesh at the end of the electron gun to form a scan-expansion lens. The mesh is maintained at gun voltage (2 000 V), and the conductive coating in the glass envelope is maintained at a higher voltage (20 000 V). The resultant electric field between the mesh and this conducting surface causes an outward radial force on the electrons so that their angle from the gun is increased and scan magnification occurs. A typical mesh-type pda crt is shown in Fig. 41.

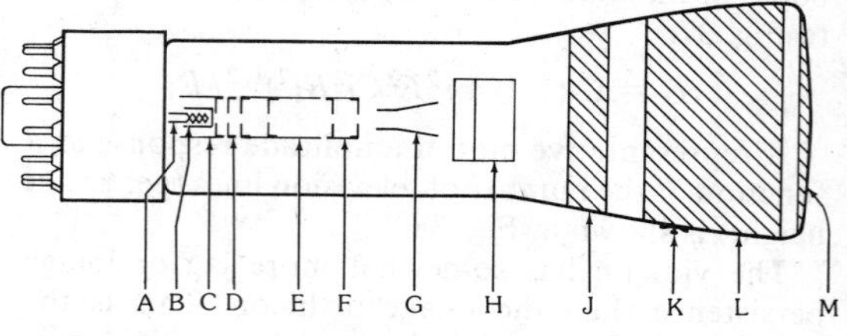

A—Heater
B—Cathode
C—Control Electrode
D—Screen grid or preaccelerator
E—Focusing electrode
F—Accelerating electrode
G—Deflection-plate pair
H—Deflection-plate pair
J—Conductive coating connected to accelerating electrode
K—Intensifier-electrode terminal
L—Intensifier electrode (conductive coating on glass)
M—Fluorescent screen

Fig. 40. Electrode arrangement of typical electrostatic-focus and -deflection cathode-ray tube.

Magnetic Deflection—Magnetic deflection (Fig. 42) is generally used with tv, computer-display, and radar crt's, where high resolution and brightness are required. Magnetic coils are placed in pairs on the outside of the crt to provide horizontal and vertical magnetic fields perpendicular to the electron flow. Current in these coils causes deflection of the electrons perpendicular to the magnetic field and to the direction of the electrons.

Deflection bandwidth is limited by the high deflection power required of magnetic-deflection systems. Repetitive resonant circuits are normally used. Deflection is proportional to the flux or current in the coil and inversely proportional to the accelerating voltage.

Beam Focusing—Magnetic focusing can be achieved by placing an external magnetic coil in the form of a short solenoid on the outside of the crt over the electron gun (Fig. 43). Because of the disadvantages of weight and the difficulty in aligning the coil to the beam, most modern crt's use electrostatic focusing. Cylindrical electrodes at differing voltages form electron lenses that focus the beam to a fine spot at the phosphor screen.

*K. R. Spangenberg, *Vacuum Tubes,* 1st ed. (New York: McGraw-Hill Book Co., 1948).

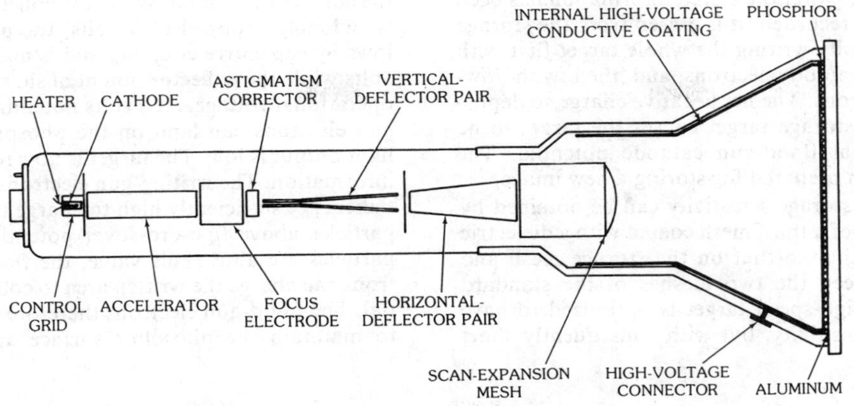

Fig. 41. Crt with mesh-type post-deflection acceleration.

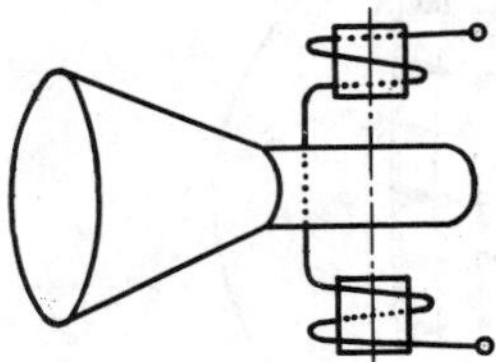

Fig. 42. Magnetic deflection.

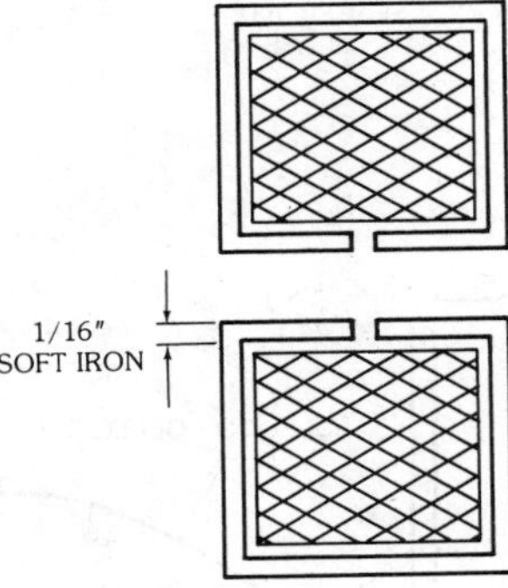

Fig. 43. Magnetic focusing.

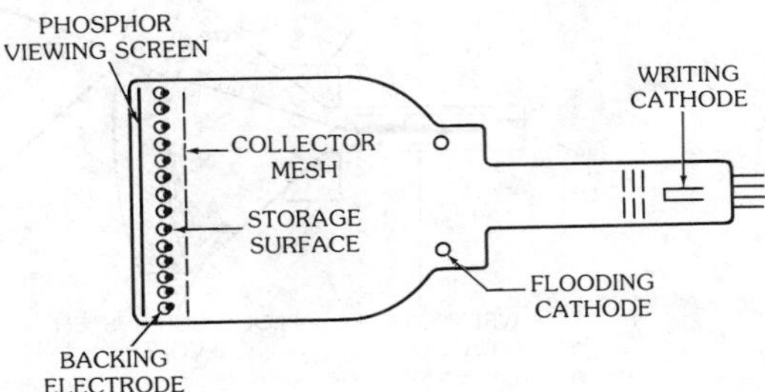

Fig. 44. Construction of storage cathode-ray tube.

Storage Cathode-Ray Tubes

The storage cathode-ray tube* produces a visual display of controllable duration. The tube has two electron guns, a phosphor viewing screen, and two or more fine-mesh screens. One of the electron guns is referred to as the writing gun and the other as the flood gun. The screen nearest to the guns is the collector mesh. The other mesh is the storage mesh and is coated with a thin dielectric material to form a surface on which electrons store information. A typical storage tube is shown in Fig. 44.

*M. Knoll and B. Kazan, *Storage Tubes and Their Basic Principles* (New York: John Wiley & Sons, Inc., 1952).

The writing gun emits a pencil-like electron beam, which can be modulated by the information to be stored. This information is in the form of an electrical input signal that can be applied to the control grid for intensity modulation or to the deflection electrodes for spatial modulation. The storage surface is scanned by this high-resolution beam, which actually strikes the surface. A positive-charge image corresponding to the input signal pattern is created on the storage surface by secondary-emission effects. The image remains on the storage surface until it decays from the neutralizing action of gas ions or is erased intentionally. The storage screen acts as an array of elemental electron guns, with each mesh hole acting as a control element of one of the guns. The desired information is stored on the storage mesh by the action of the writing gun, and the entire surface is flooded by electrons from the flood gun. The value of the positive charge deposited at each mesh aperture controls the amount of flood-gun current that can pass through the mesh aperture to the phosphor screen. The current that passes through the mesh strikes the phosphor, and light is emitted in proportion to the current density arriving and to the landing energy of the electrons. In other words, a gray scale is reproduced in the stored image. After the stored information has been observed or recorded, it is erased from the storage surface by fully writing the whole target first with high-energy flood electrons and then with low-energy electrons. The net negative charge so deposited on the storage target causes the target to be maintained at flood-gun cathode potential. The target is then prepared for storing a new image.

Increased storage sensitivity can be obtained by the addition of a third mesh coated with a dielectric material similar to that on the storage mesh and placed between the two meshes of the standard tube. This high-speed target is optimized to have very high sensitivity, but with consequently short retention (view time). The charge pattern written on this surface is quickly transferred to the front storage mesh, on which the coating has been optimized for long retention time and low sensitivity. Such crt's are generally referred to as transfer storage tubes and are capable of storing signals with bandwidth in excess of 400 MHz.

Bistable Storage Tube

The bistable storage tube allows storage of charge in a crt without the complication and expense of added meshes. Two guns are used as in the mesh storage tubes, but the phosphor screen is structured to have collector islands surrounded by a phosphor layer deposited on a transparent film (Fig. 45). The phosphor itself acts as both the dielectric storage surface and the light emitter. The potential of the phosphor can be maintained at two stable potentials by the action of secondary emission from the phosphor layer.

To prepare the phosphor surface for storage, the collector electrode is pulsed with the waveform shown in Fig. 46. The flood electrons strike the whole target, and since the target potential is above first crossover, the target charges positively toward the collector potential. When the collector potential is suddenly dropped to 0 volts, the phosphor follows by capacitive coupling and is maintained at 0 volts while the collector potential slowly rises to its operational voltage. At 0 volts potential, few flood-gun electrons can land on the phosphor, and the light output is low. The target is now ready to store information. The writing-gun electrons have potential energy sufficiently high to charge the phosphor particles above first-crossover potential. Once the particles are above this value, the flood-gun electrons can charge the written area to collector potential. The flood-gun electrons therefore can be made to maintain the phosphor surface at two stable

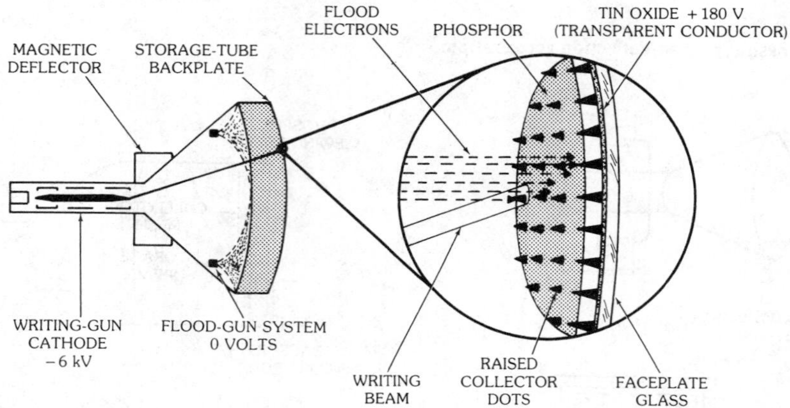

Fig. 45. Typical bistable storage tube.

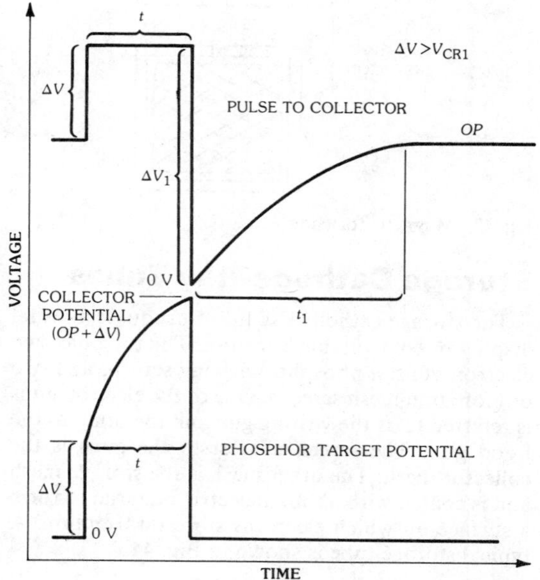

Fig. 46. Erasure waveforms.

potentials, flood-gun cathode and collector potential, hence the term "bistable storage." At collector potential, the flood-gun electrons arrive with sufficient energy to excite the phosphor, and the stored image is clearly visible and remains as long as desired or until erased.

These crt's have enjoyed wide use in computer graphics terminals. Very high resolution can be obtained on large display tubes of 19 and 25 inches (diagonal). The 19-inch tube is capable of displaying 8500 characters, and the 25-inch crt is capable of over 15 000 characters. Since these characters are actually stored on the face of the crt until erased, no solid-state memory is required to refresh them.

Another advantage of this type of crt is that when a line is scanned with the writing gun at low beam current, the collector current will be a function of whether the area scanned is written or not. This signal current can be amplified and processed to produce an electronic image of the written areas. This image can be fed into a copy device for producing a hard copy of the display.

17 Power Grid-Tube Circuits

Marvin Chodorow

It is common practice to differentiate between types of vacuum-tube circuits, particularly amplifiers, on the basis of the operating regime of the tube.

Class-A: Grid bias and alternating grid voltages such that plate current flows continuously throughout electrical cycle ($\theta_p = 360°$).

Class-AB: Grid bias and alternating grid voltages such that plate current flows appreciably more than half but less than entire electrical cycle ($360° > \theta_p > 180°$).

Class-B: Grid bias close to cutoff such that plate current flows only during approximately half of electrical cycle ($\theta_p \approx 180°$).

Class-C: Grid bias appreciably greater than cutoff so that plate current flows for appreciably less than half of electrical cycle ($\theta_p < 180°$).

A further classification between circuits in which positive grid current is conducted during some portion of the cycle and those in which it is not is denoted by subscripts 2 and 1, respectively. Thus a class-AB_2 amplifier operates with a positive swing of the alternating grid voltage such that positive electronic current is conducted and accordingly in-phase power is required to drive the tube.

GENERAL DESIGN

For quickly estimating the performance of a tube from catalog data, or for predicting the characteristics needed for a given application, the ratios given below may be used.

Table 1 gives correlating data for typical operation of tubes in the various amplifier classifications. If the maximum ratings of a tube are known, the maximum power output, currents, voltages, and corresponding load impedance may be estimated from the table. Take for example a type F-124-A water-cooled transmitting tube operated as a class-C radio-frequency power amplifier and oscillator (the constant-current characteristics are shown in Fig. 1). Published maximum ratings are as follows.

Dc plate voltage:

$$E_b = 20\,000 \text{ volts}$$

Dc grid voltage:

$$E_c = 3000 \text{ volts}$$

Dc plate current:

$$I_b = 7 \text{ amperes}$$

Rf grid current:

$$I_g = 50 \text{ amperes}$$

Plate input:

$$P_i = 135\,000 \text{ watts}$$

Plate dissipation:

$$P_p = 40\,000 \text{ watts}$$

Maximum conditions may be estimated as follows. For $\eta = 75$ percent

$$P_i = 135\,000 \text{ watts}$$
$$E_b = 20\,000 \text{ volts}$$

Power output $P_o = \eta P_i = 100\,000$ watts.

Average dc plate current $I_b = P_i/E_b = 6.7$ amperes

From a tabulated typical ratio ${}^M i_b/I_b = 4$, the instantaneous peak plate current ${}^M i_b = 4I_b = 27$ amperes.*

The rms plate alternating-current component, taking the ratio $I_p/I_b = 1.2$, is

$$I_p = 1.2 I_b = 8 \text{ amperes}$$

The rms value of the plate alternating-voltage component from the ratio $E_p/E_b = 0.6$ is $E_p = 0.6E_b = 12\,000$ volts.

The approximate operating load resistance, R_l, is now found from

$$R_l = E_p/I_p = 1500 \text{ ohms}$$

An estimate of the grid drive power required may be obtained by reference to the constant-current characteristics of the tube and determination of the peak instantaneous positive grid current ${}^M i_c$ and the corresponding instantaneous total grid voltage ${}^M e_c$. If the value of grid bias for the given operating condition is E_c, the peak alternating grid drive voltage is

$${}^M E_g = ({}^M e_c - E_c)$$

*In this discussion, the superscript M indicates the use of the maximum or peak value of the varying component; i.e., ${}^M i_b$ = maximum or peak value of the alternating component of the plate current.

TABLE 1. TYPICAL AMPLIFIER OPERATING DATA (MAXIMUM-SIGNAL CONDITIONS, PER TUBE)

Function	Class A	Class B af (p-p)	Class B rf	Class C rf
Plate efficiency η (percent)	20–30	35–65	60–70	65–85
Peak instantaneous to dc plate-current ratio ${}^M i_b/I_b$	1.5–2	3.1	3.1	3.1–4.5
Rms alternating to dc plate-current ratio I_p/I_b	0.5–0.7	1.1	1.1	1.1–1.2
Rms alternating to dc plate-voltage ratio E_p/E_b	0.3–0.5	0.5–0.6	0.5–0.6	0.5–0.6
Dc to peak instantaneous grid-current $I_c/{}^M i_c$		0.1–0.25	0.1–0.25	0.1–0.25

from which the peak instantaneous grid drive power can be determined:

$$^{M}P_c = {}^{M}E_g{}^{M}i_c$$

An approximation to the average grid drive power, P_g, necessarily rough due to neglect of negative grid current, is obtained from the typical ratio of dc to peak value of grid current, $I_c/{}^{M}i_c = 0.2$. The result is

$$P_g = I_cE_g = 0.2{}^{M}i_cE_g \text{ watt}$$

Plate dissipation P_p may be checked with published values since

$$P_p = P_i - P_o$$

It should be borne in mind that combinations of published maximum ratings as well as each individual maximum rating must be observed. Thus, for example in this case, the maximum dc plate operating voltage of 20 000 volts does not permit operation at the maximum dc plate current of 7 amperes since this exceeds the maximum plate input rating of 135 000 watts.

Plate load resistance R_l may be connected directly in the tube plate circuit as in the resistance-coupled amplifier, through impedance-matching elements as in audio-frequency transformer coupling, or effectively represented by a loaded parallel-resonant circuit as in most radio-frequency amplifiers. In any case, calculated values apply only to effectively resistive loads, such as are normally closely approximated in radio-frequency amplifiers. With appreciably reactive loads, operating currents and voltages will in general be quite different, and their precise calculation is quite difficult.

The physical load resistance present in any given setup may be measured by audio-frequency or radio-frequency bridge methods. In many cases, the proper value of R_l is ascertained experimentally as in radio-frequency amplifiers that are tuned to the proper minimum dc plate current. Conversely, if the circuit is to be matched to the tube, R_l is determined directly as in a resistance-coupled amplifier or as

$$R_l = N^2R_s$$

in the case of a transformer-coupled stage, where N is the primary-to-secondary voltage transformation ratio. In a parallel-resonant circuit in which the output resistance R_s is connected directly in one of the reactance legs

$$R_l = X^2/R_s = L/Cr_s = QX$$

where,

X is the leg reactance at resonance (ohms),
L and C are leg inductance in henries and capacitance in farads, respectively,
$Q = X/R_s$.

GRAPHIC DESIGN METHODS

When accurate operating data are required, more precise methods must be used. Because of the nonlinear nature of tube characteristics, graphic methods usually are most convenient and rapid. Examples of such methods are given below.

A comparison of the operating regimes of class A, AB, B, and C amplifiers is given in the constant-current–characteristics graph of Fig. 1. The lines corresponding to the different classes of operation are the loci of instantaneous grid voltage e_c and plate voltage e_b, corresponding to their respective load impedances.

For radio-frequency amplifiers and oscillators having tuned circuits that give an effectively resistive load, plate and grid tube and load alternating voltages are sinusoidal and in phase (disregarding transit time), and the loci become straight lines.

For amplifiers having nonresonant resistive loads, the loci are in general nonlinear except in the distortionless case of linear tube characteristics (constant r_p), for which they are again straight lines.

Thus, for determination of radio-frequency performance, the constant-current chart is convenient. For solution of audio-frequency problems, however, it is more convenient to use the i_b-e_c transfer characteristics of Fig. 2, on which a dynamic load line may be constructed.

Methods for calculation of the most important cases are given below.

Class-C RF Amplifier or Oscillator

Draw a straight line from A to B (Fig. 1) corresponding to the chosen dc operating plate and grid voltages, and to the desired peak alternating plate and grid voltage excursions. The projection of AB on the horizontal axis thus corresponds to ${}^{M}E_p$. Using Chaffee's 11-point method of harmonic analysis, lay out on AB points

$$e_p' = {}^{M}E_p$$
$$e_p'' = 0.866{}^{M}E_p$$
$$e_p''' = 0.5{}^{M}E_p$$

to each of which correspond instantaneous plate currents, i_b', i_b'', and i_b''' and instantaneous grid currents i_c', i_c'', and i_c'''. The operating currents are obtained from

$$I_b = [i_b' + 2i_b'' + 2i_b''']/12$$
$$I_c = [i_c' + 2i_c'' + 2i_c''']/12$$
$${}^{M}I_p = [i_b' + 1.73i_b'' + i_b''']/6$$
$${}^{M}I_g = [i_c' + 1.73i_c'' + i_c''']/6$$

Substitution of the above in the following gives the desired operating data.

Power output $P_o = ({}^{M}E_p{}^{M}I_p)/2$
Power input $P_i = E_bI_b$

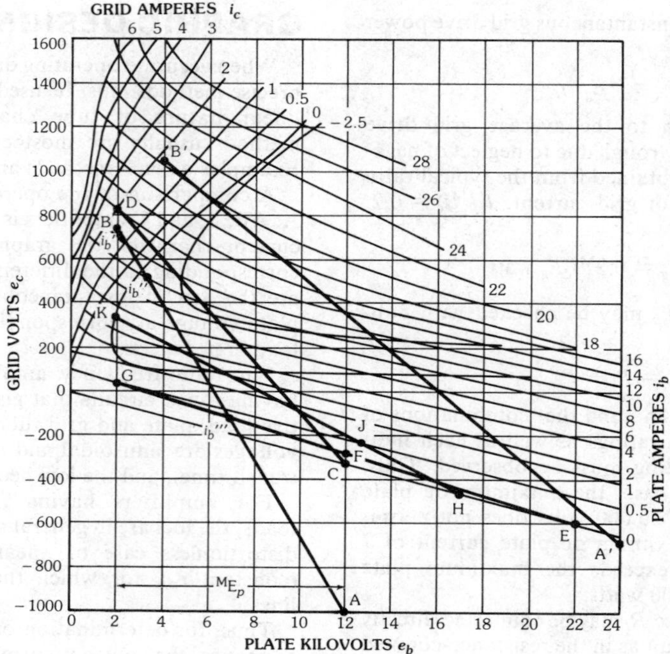

Fig. 1. Constant-current characteristics of type F-124-A tube with typical load lines: AB—class C, CD—class B, EFG—class A, HJK—class AB.

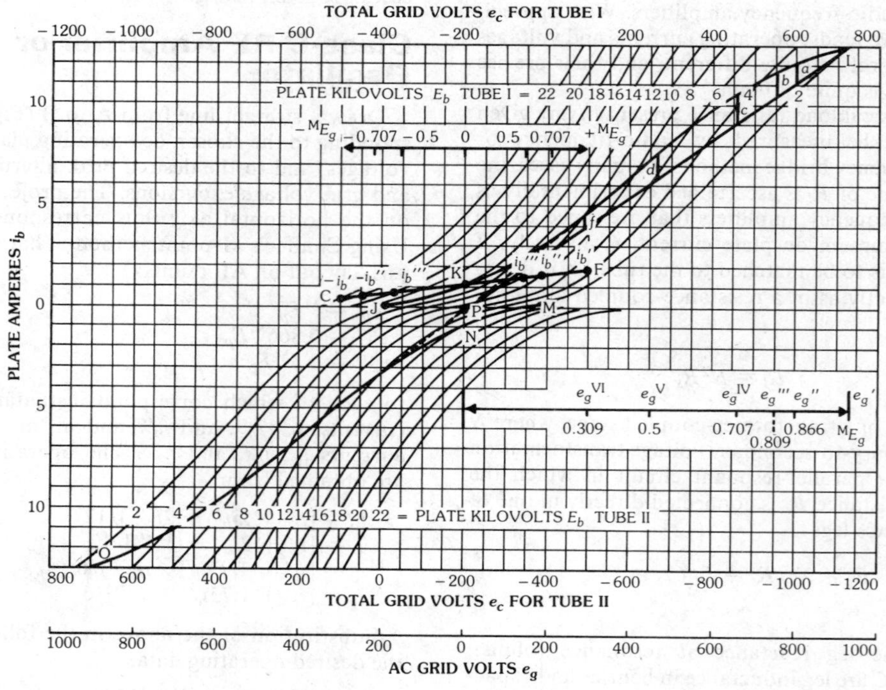

Fig. 2. Transfer characteristics i_b versus e_c with load lines: CKF—class A_2 and OPL—class B.

Average grid excitation power $P_g = ({}^ME_g{}^MI_g)/2$
Peak grid excitation power ${}^MP_c = {}^ME_g i_c'$
Plate load resistance $R_l = {}^ME_p/{}^MI_p$
Grid bias resistance $R_c = E_c/I_c$
Plate efficiency $\eta = P_o/P_i$
Plate dissipation $P_p = P_i - P_o$

The above procedure may also be applied to plate-modulated class-C amplifiers. Take the above data as applying to carrier conditions, and repeat the analysis for ${}^{crest}E_b = 2E_b$ and ${}^{crest}P_o = 4P_o$, keeping R_l constant. After a cut-and-try method has given a peak solution, it will often be found that combination fixed and self grid biasing as well as grid modulation are indicated to obtain linear operation.

To illustrate the preceding exposition, a typical amplifier calculation is given below.

Operating requirements (carrier condition):

$$E_b = 12\,000 \text{ volts}$$
$$P_o = 25\,000 \text{ watts}$$
$$\eta = 75 \text{ percent}$$

Preliminary calculation (refer to Tables 1 and 2):

$$E_p/E_b = 0.6$$
$$E_p = 0.6 \times 12\,000 = 7200 \text{ volts}$$
$${}^ME_p = 1.41 \times 7200 = 10\,000 \text{ volts}$$
$$I_p = P_o/E_p \quad 25\,000/7200 = 3.48 \text{ amperes}$$
$${}^MI_p = 4.9 \text{ amperes}$$
$$I_p/I_b = 1.2$$
$$I_b = 3.48/1.2 = 2.9 \text{ amperes}$$
$$P_i = 12\,000 \times 2.9 = 35\,000 \text{ watts}$$
$${}^Mi_b/I_b = 4.5$$
$${}^Mi_b = 4.5 \times 2.9 = 13.0 \text{ amperes}$$
$$R_l = E_p/I_p = 7200/3.48 = 2060 \text{ ohms}$$

Complete Calculation: Lay out carrier operating line AB on the constant-current graph, Fig. 1, using values of E_b, ${}^ME_p$, and ${}^Mi_b$ from the preliminary calculated data. Operating carrier bias voltage E_c is chosen somewhat greater than twice the cutoff value (1000 volts) to locate point A.

The following data are taken along AB.

$$i_b' = 13 \text{ amperes}$$
$$i_b'' = 10 \text{ amperes}$$
$$i_b''' = 0.3 \text{ ampere}$$
$$i_c' = 1.7 \text{ amperes}$$
$$i_c'' = -0.1 \text{ ampere}$$
$$i_c''' = 0 \text{ ampere}$$
$$E_c = -1000 \text{ volts}$$
$$e_c' = 740 \text{ volts}$$
$${}^ME_p = 10\,000 \text{ volts}$$

From the equations, complete carrier data as follows are calculated.

$${}^MI_p = [13 + 1.73 \times 10 + 0.3]/6 = 5.1 \text{ amperes}$$
$$P_o = (10\,000 \times 5.1)/2 = 25\,500 \text{ watts}$$
$$I_b = [13 + 2 \times 10 + 2 \times 0.3]/12 = 2.8 \text{ amperes}$$
$$P_i = 12\,000 \times 2.8 = 33\,600 \text{ watts}$$
$$\eta = (25\,500/33\,600) \times 100 = 76 \text{ percent}$$
$$R_l = (10\,000/5.1) = 1960 \text{ ohms}$$
$$I_c = [1.7 + 2(-0.1)]/12 = 0.125 \text{ ampere}$$
$${}^MI_g = [1.7 + 1.7(-0.1)]/6 = 0.255 \text{ ampere}$$
$$P_g = (1740 \times 0.255)/2 = 220 \text{ watts}$$

Operating data at 100-percent positive modulation crests are now calculated based on the fact that here

$$E_b = 24\,000 \text{ volts}$$
$$R_l = 1960 \text{ ohms}$$

and for undistorted operation

$$P_o = 4 \times 25\,500 = 102\,000 \text{ watts}$$
$${}^ME_p = 20\,000 \text{ volts}$$

The crest operating line A'B' is now located by trial so as to satisfy the above conditions, by the use

TABLE 2. CLASS-C RF AMPLIFIER DATA FOR 100-PERCENT PLATE MODULATION

Symbol	Preliminary Carrier	Detailed Carrier	Detailed Crest
E_b (volts)	12 000	12 000	24 000
${}^ME_p$ (volts)	10 000	10 000	20 000
E_c (volts)	—	−1 000	−700
${}^ME_g$ (volts)	—	1 740	1 740
I_b (amperes)	2.9	2.8	6.4
${}^MI_p$ (amperes)	4.9	5.1	10.2
I_c (amperes)	—	0.125	0.083
${}^MI_g$ (amperes)	—	0.255	0.183
P_i (watts)	35 000	33 600	154 000
P_o (watts)	25 000	25 500	102 000
P_g (watts)	—	220	160
η (percent)	75	76	66
R_l (ohms)	2 060	1 960	1 960
R_c (ohms)	—	7 100	7 100
E_{cc} (volts)		−110	−110

of the same equations and method as for the carrier condition.

It is seen that to obtain full-crest power output, in addition to doubling the alternating plate voltage, the peak plate current must be increased. This is accomplished by reducing the crest bias voltage with a resultant increase of current conduction period but lower plate efficiency.

The effect of grid secondary emission to lower the crest grid current is taken advantage of to obtain the reduced grid-resistance voltage drop required. By use of combination fixed and grid-resistance bias, proper variation of the total bias is obtained. The value of grid resistance required is given by

$$R_c = -(E_c - {}^{\text{crest}}E_c)/(I_c - {}^{\text{crest}}I_c)$$

and the value of fixed bias by

$$E_{cc} = E_c - (I_c R_c)$$

Calculations at carrier and positive crest together with the condition of zero output at negative crest give sufficiently complete data for most purposes. If accurate calculation of audio-frequency harmonic distortion is necessary, the above method may be applied to the additional points required.

Class-B RF Amplifiers

A rapid approximate method is to determine by inspection from the tube i_b-e_b characteristics the instantaneous current, i_b', and voltage, e_b', corresponding to the peak alternating voltage swing from operating voltage E_b.

Ac plate current:

$${}^{M}I_p = i_b'/2$$

Dc plate current:

$$I_b = i_b'/\pi$$

Ac plate voltage:

$${}^{M}E_p = E_b - e_b'$$

Power output:

$$P_o = [(E_b - e_b')i_b']/4$$

Power input:

$$P_i = E_b i_b'/\pi$$

Plate efficiency:

$$\eta = (\pi/4)[1 - (e_b'/E_b)]$$

Thus $\eta \approx 0.6$ for the usual crest value of ${}^{M}E_p \approx 0/8E_b$.

The same method of analysis used for the class-C amplifier may also be used in this case. The carrier and crest-condition calculations, however, are now made from the same E_b, the carrier condition corresponding to an alternating-voltage amplitude of ${}^{M}E_p/2$ such as to give the desired carrier power output.

For greater accuracy than the simple check of carrier and crest conditions, the radio-frequency plate currents ${}^{M}I_p'$, ${}^{M}I_p''$, ${}^{M}I_p'''$, ${}^{M}I_p{}^0$, $-{}^{M}I_p'''$, $-{}^{M}I_p''$, and $-{}^{M}I_p'$ may be calculated for seven corresponding selected points of the audio-frequency modulation envelope $+{}^{M}E_g$, $+0.707{}^{M}E_g$, $+0.5{}^{M}E_g$, 0, $-0.5{}^{M}E_g$, $-0.707{}^{M}E_g$, and $-{}^{M}E_g$, where the negative signs denote values in the negative half of the modulation cycle. If the designations

$$S' = {}^{M}I_p' - (-{}^{M}I_p')$$
$$D' = {}^{M}I_p' + (-{}^{M}I_p') - 2{}^{M}I_p{}^0$$

are used, the fundamental and harmonic components of the output audio-frequency current are obtained as

$${}^{M}I_{p1} = (S'/4) + [S''/2(2)^{1/2}] \text{ (fundamental)}$$
$${}^{M}I_{p2} = (5D'/24) + (D''/4) - (D'''/3)$$
$${}^{M}I_{p3} = (S'/6) - (S'''/3)$$
$${}^{M}I_{p4} = (D'/8) - (D''/4)$$
$${}^{M}I_{p5} = (S'/12) - [S''/2(2)^{1/2}] + (S'''/3)$$
$${}^{M}I_{p6} = (D'/24) - (D''/4) + (D'''/3)$$

This detailed method of calculation of audio-frequency harmonic distortion may, of course, also be applied to calculation of the class-C modulated amplifier, as well as to the class-A modulated amplifier.

Class-A and -AB AF Amplifiers

Approximate equations assuming linear tube characteristics:

Maximum undistorted power output

$${}^{M}P_o = ({}^{M}E_p{}^{M}I_p)/2$$

when plate load resistance

$$R_l = r_p\left[\frac{E_c}{({}^{M}E_p/\mu) - E_c} - 1\right]$$

and negative grid bias

$$E_c = ({}^{M}E_p/\mu)[(R_l + r_p)/(R_l + 2r_p)]$$

giving maximum plate efficiency

$$\eta = {}^{M}E_p{}^{M}I_p/8E_bI_b$$

Maximum maximum undistorted power output

$${}^{MM}P_o = {}^{M}E^2{}_p/16r_p$$

when

$$R_l = 2r_p$$
$$E_c = \tfrac{3}{4}({}^{M}E_p/\mu)$$

An exact analysis may be obtained by use of a dynamic load line laid out on the transfer characteristics of the tube. Such a line is CKF of Fig. 2, which is constructed about operating point K for a given load resistance r_l from

$$i_b{}^S = [(e_b{}^R - e_b{}^S)/R_l] + i_b{}^R$$

where R, S, etc., are successive conveniently spaced construction points.

Using the seven-point method of harmonic analysis, plot instantaneous plate currents i_b', i_b'', i_b''', i_b, $-i_b'''$, $-i_b''$, and $-i_b'$, corresponding to $+{}^ME_g$, $+0.707{}^ME_g$, $+0.5{}^ME_g$, 0, $-0.5{}^ME_g$, $-0.707{}^ME_g$, and $-{}^ME_g$, where 0 corresponds to operating point K. In addition to the equations given under class-B radio-frequency amplifiers

$$I_b \text{ average} = I_b + (D'/8) + (D''/4)$$

from which complete data may be calculated.

Class-AB and -B AF Amplifiers

Approximate equations assuming linear tube characteristics give (referring to Fig. 1, line CD) for a class-B audio-frequency amplifier

$$\begin{aligned} {}^MI_p &= i_b' \\ P_o &= {}^ME_p{}^MI_p/2 \\ P_i &= (2/\pi)E_b{}^MI_p \\ \eta &= (\pi/4)\ ({}^ME_p/E_b) \\ R_{pp} &= 4({}^ME_p/i_b') = 4R_l \end{aligned}$$

An exact solution may be derived by use of dynamic load line JKL on the i_b-e_c characteristic of Fig. 2. This line is calculated about the operating point K for the given R_l (in the same way as for the class-A case). However, since two tubes operate in phase opposition in this case, an identical dynamic load line MNO represents the other half cycle, laid out about the operating bias abscissa point but in the opposite direction (see Fig. 2).

Algebraic addition of instantaneous current values of the two tubes at each value of e_c gives the composite dynamic characteristic OPL for the two tubes. Inasmuch as this curve is symmetrical about point P, it may be analyzed for harmonics along a single half-curve PL by use of the Mouromtseff 5-point method. A straight line is drawn from P to L, and ordinate plate-current differences a, b, c, d, f between this line and the curve, corresponding to e_g'', e_g''', $e_g{}^{IV}$, $e_g{}^V$, and $e_g{}^{VI}$, are measured. Ordinate distances measured upward from curve PL are taken positive.

Fundamental and harmonic current amplitudes and power are found from

$$\begin{aligned} {}^MI_{p1} &= i_b' - {}^MI_{p3} + {}^MI_{p5} - {}^MI_{p7} + {}^MI_{p9} - {}^MI_{p11} \\ {}^MI_{p3} &= 0.4475(b+f) + (d/3) - 0.578d - \tfrac{1}{2}\,{}^MI_{p5} \\ {}^MI_{p5} &= 0.4(a-f) \\ {}^MI_{p7} &= 0.4475(b+f) - {}^MI_{p3} + 0.5{}^MI_{p5} \\ {}^MI_{p9} &= {}^MI_{p3} - \tfrac{2}{3}\,d \\ {}^MI_{p11} &= 0.707c - {}^MI_{p3} + {}^MI_{p5} \end{aligned}$$

Even harmonics are not present due to the symmetry of the dynamic characteristic. The direct-current and power-input values are found by the 7-point analysis from curve PL and doubled for two tubes.

CIRCUIT CLASSIFICATION

The classification of amplifiers in classes A, B, and C is based on the operating conditions of the tube. Another classification can be used, based on the type of circuits associated with the tube.

A tube can be considered as a four-terminal network with two input terminals and two output terminals. One of the input terminals and one of the output terminals are usually common; this common junction or point is usually called "ground."

When the common point is connected to the filament or cathode of the tube, we can speak of a grounded-cathode circuit (the most conventional type of vacuum-tube circuit). When the common point is the grid, we can speak of a grounded-grid circuit; and when the common point is the plate or anode, we can speak of a grounded-anode circuit. This last type of circuit is most commonly known by the name "cathode-follower."

A fourth and most general class of circuit is obtained when the common point or ground is not directly connected to any of the three electrodes of the tube. This is the condition encountered at uhf where the series impedances of the internal tube leads make it impossible to ground any of them. It is also encountered in such special types of circuits as the phase-splitter, in which the impedance from plate to ground and the impedance from cathode to ground are made equal to obtain an output between plate and cathode balanced with respect to ground.

Design information for the first three classifications is given in Table 3, where

Z_2 = load impedance to which output terminals of amplifier are connected,
$\mathbf{E}_1$ = phasor input voltage to amplifier,
$\mathbf{E}_2$ = phasor output voltage across load impedance Z_2,
$\mathbf{A}$ = voltage gain of amplifier = $\mathbf{E}_2/\mathbf{E}_1$,
Y_1 = input admittance to input terminals of amplifier,
$\omega = 2\pi \times$ (frequency of excitation voltage $\mathbf{E}_1$),
$j = (-1)^{1/2}$.

RF AMPLIFIER CIRCUITS

The power grid tube requires external circuits. Examples are shown schematically below for each kind of tube.

Triodes

The triode has three electrodes: the thermionic cathode, which emits electrons; the control grid; and the anode, which collects most of the electrons. If the grid is "biased" to a sufficiently high negative potential (cutoff bias), no current flows. As the grid potential becomes less negative, more current flows to the anode. When the grid becomes positive with respect to the cathode, both grid and anode draw current. At some value of positive grid potential,

TABLE 3. DESIGN INFORMATION FOR THREE CLASSES OF AMPLIFIERS

Grounded-Cathode	Grounded-Grid	Grounded-Plate or Cathode-Follower
	Circuit Schematic	
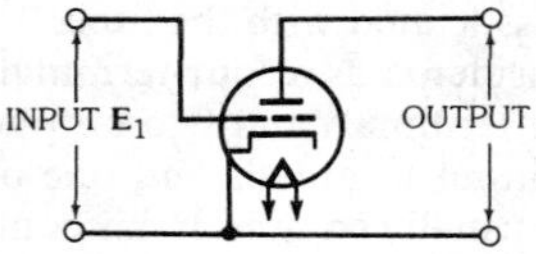 INPUT E_1 — OUTPUT E_2	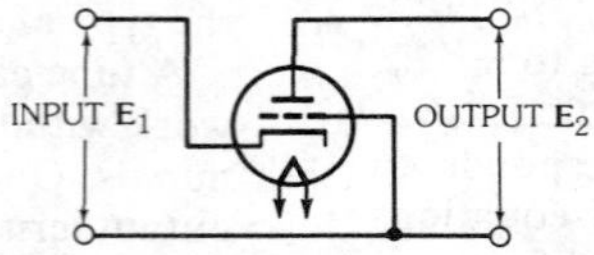 INPUT E_1 — OUTPUT E_2	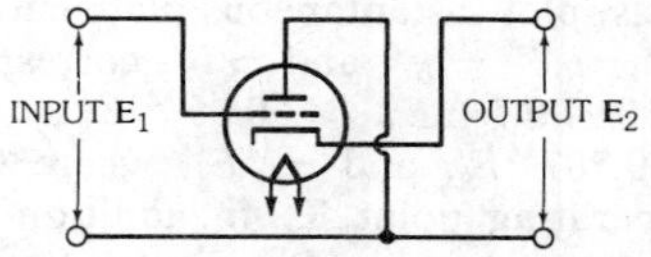 INPUT E_1 — OUTPUT E_2
	Equivalent Circuit, Alternating-Current Component, Class-A Operation	
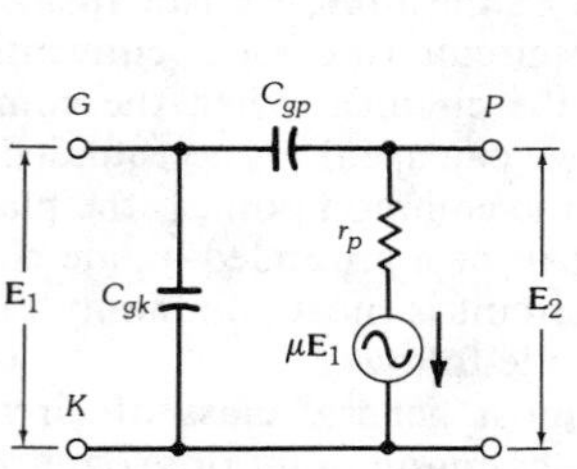 G, C_{gp}, P, r_p, E_1, C_{gk}, μE_1, E_2, K	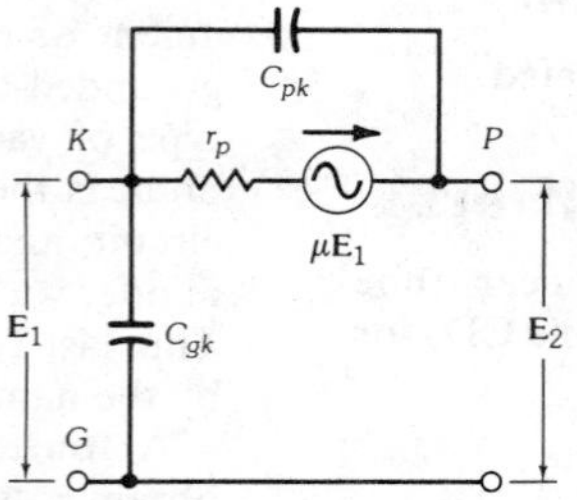 C_{pk}, K, r_p, P, μE_1, E_1, C_{gk}, E_2, G	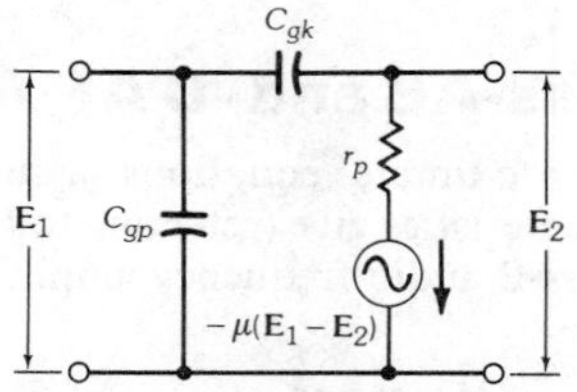 C_{gk}, r_p, E_1, C_{gp}, E_2, $-\mu(E_1-E_2)$
	Voltage Gain A for Output Load Impedance Z_2; $A=E_2/E_1$	
$A=-\mu Z_2/(r_p+Z_2)$ $=-g_m[r_p Z_2/(r_p+Z_2)]$	$A=(1+\mu)[Z_2/(r_p+Z_2)]$	$A=\mu Z_2/[r_p+(1+\mu)Z_2]$
Neglecting C_{gp}	Neglecting C_{pk}	Neglecting C_{gk}
(Z_2 includes C_{pk})	(Z_2 includes C_{gp})	(Z_2 includes C_{pk})
	Input Admittance; $Y_1=I_1/E_1$	
$Y_1=j\omega[C_{gk}+(1-A)C_{gp}]$	$Y_1=j\omega[C_{gk}+(1-A)C_{pk}]$ $+[(1+\mu)/(r_p+Z_2)]$	$Y_1=j\omega[C_{gp}+(1-A)C_{gk}]$
	Equivalent Generator Seen by Load at Output Terminals	

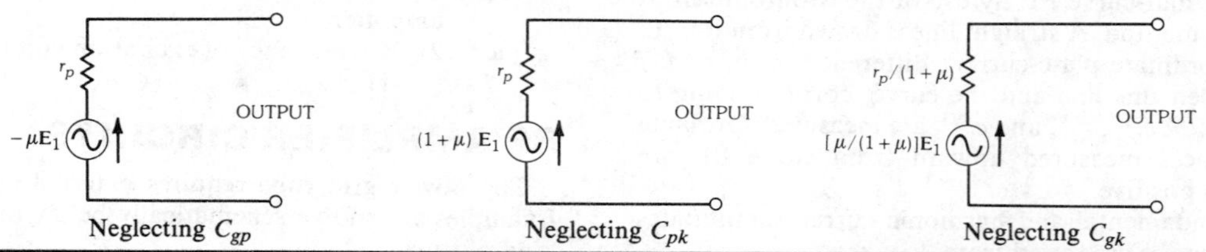

Neglecting C_{gp}	Neglecting C_{pk}	Neglecting C_{gk}

the total space current starts to exceed the emitting capability of the cathode (cathode saturation) or the product of the grid current and grid-cathode voltage (grid dissipation) exceeds the limit above which the grid will emit electrons (primary emission). Excessive grid dissipation interferes with the desired operation of the tube or results in mechanical distortion due to excessive temperature. There is also a limit to the power dissipation of the anode, depending on the cooling method used.

In operation as an rf power amplifier, the triode must be either "neutralized" as in Fig. 3A or operated "grounded-grid" ("cathode-driven") as in Fig. 3B; otherwise, the internal capacitance between grid and anode produces positive feedback that may cause self-oscillation at a frequency close to the operating frequency. The triode may be operated as an efficient oscillator by optimizing the feedback through the addition of extra capacitance or by other means. Oscillators are used for rf heating of materials in industrial operations where precise control of frequency is not required.

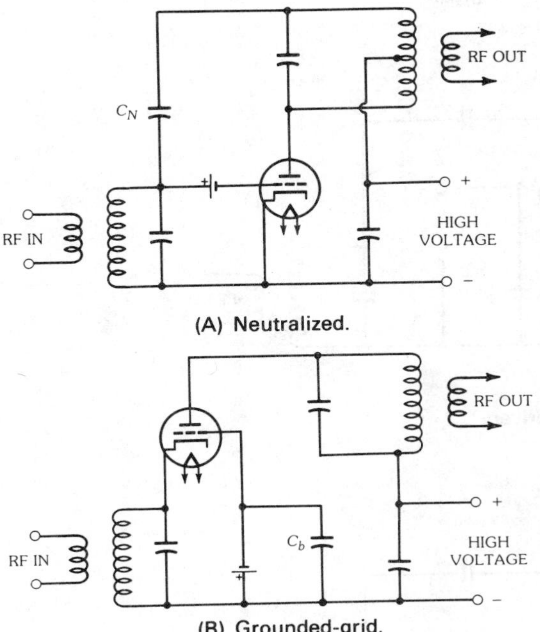

(A) Neutralized.

(B) Grounded-grid.

Fig. 3. Triode amplifier circuits.

Small planar triodes are used at uhf and microwave frequencies up to about 4 GHz, especially in pulse service where peak-to-average power ratios are 100–1000. The anode supply voltage is typically 1 kV or more.

Small cylindrical triodes are used mainly at vhf and uhf where cw power of a few hundred watts or pulse powers of tens of kilowatts are required. Modern triodes are designed with beam-forming cathode and control-grid geometry to allow the simplicity of design and circuit advantages of a triode with the gain of a tetrode.

Tetrodes and Pentodes

The tetrode and pentode have four or five electrodes, respectively. A tetrode has a cathode, a control grid, a screen grid, and an anode. The screen grid greatly reduces the capacitance between the anode and control grid and makes neutralization unnecessary or easy to accomplish. The pentode has an additional "suppressor grid" to control secondary electrons. Modern tetrodes accomplish this control in other ways and have almost completely displaced pentodes.

Two tetrode circuits for rf amplification are shown in Fig. 4. The arrangement in Fig. 4A is commonly used at frequencies below about 30 MHz for very large tubes and below about 400 MHz for small tubes. A power gain of 1000 (30 dB) is typical at lower frequencies.

At vhf and uhf, the cathode-lead inductance, L_k, of Fig. 4A produces excessive negative feedback, which can be overcome by incorporating the inductance into the input resonant circuit as shown in Fig. 4B. In this arrangement, both the control grid and screen grid are maintained at rf ground potential by the bypassing capacitors (C_b). In this circuit, the power gain is reduced compared to that of Fig. 4A, but 10–15 dB is typical at frequencies below about 1000 MHz.

Tetrode tubes are available in various sizes for different power levels and frequencies. Applications are radio broadcasting (am and fm), television (vhf and uhf), communications, radar, navigational aids, and high-energy physics, including particle accelerators and thermonuclear-fusion machines. Functions are rf power generation, modulation (am), and switching for pulse service.

UHF Operation

When the transit time of the electrons from cathode to anode is an appreciable fraction of one radio-frequency cycle:

(*A*) Input conductance due to reaction of electrons with the varying field from the grid becomes appreciable. This conductance, which increases as the square of the frequency, results in lowered gain, an increase in driving-power requirement, and loading of the input circuit.

(*B*) Grid-anode transit time introduces a phase lag between grid voltage and anode current. In oscillators, the problem of compensating for the phase lag by design and adjustment of a feedback circuit becomes difficult. Efficiency is reduced in both oscillators and amplifiers.

(*C*) Distortion of the current pulse in the grid-anode space increases the anode-current conduction angle and lowers the efficiency.

In amplifiers, the effect of cathode-lead inductance is to introduce a conductance component in the grid circuit. This effect is serious because the loading of the input circuit by the conductance current limits the gain of the stage. Cathode-grid and grid-anode capacitive reactances are of small magnitude at ultrahigh frequencies, and heavy currents flow as a result of these reactances. Tubes must be designed to carry these currents without serious loss. Coaxial cavities are used to resonate with the tube reactances and to minimize resistive and radiation losses.

Klystrode™ Amplifiers

At frequencies between about 100 and 1000 MHz, a hybrid tube, part triode, part klystron, has promise. Called a Klystrode™, it was invented in 1939 and demonstrated at low power levels, but lay dormant for over 40 years. It now appears capable of high power, somewhat greater conversion efficiency, and considerably more power gain than the conventional tetrode. It is physically shorter than a

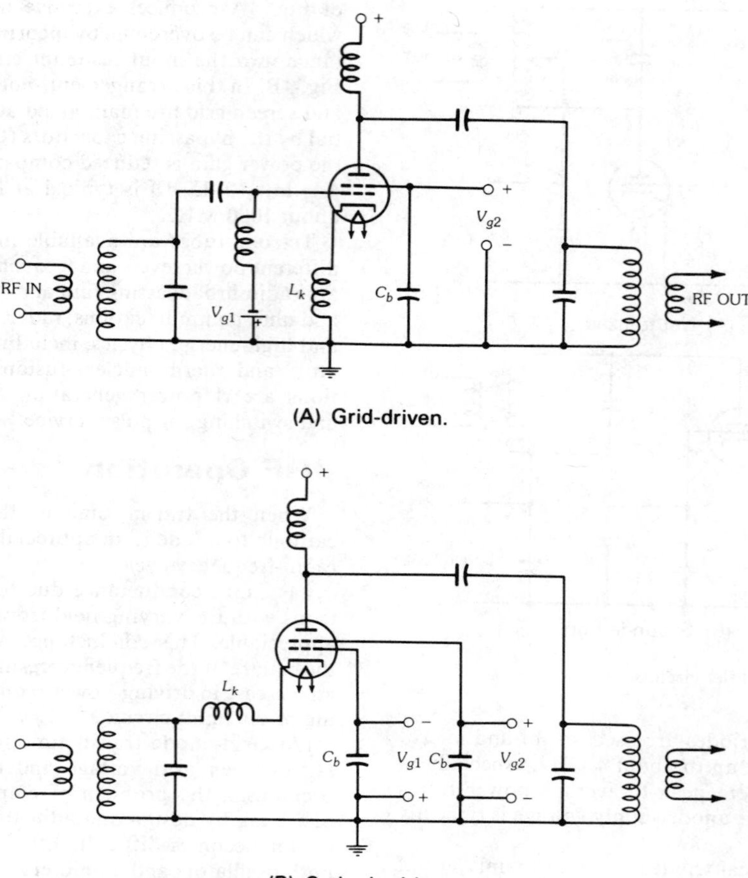

Fig. 4. Tetrode amplifier circuits.

klystron because of the absence of a multicavity buncher. The Klystrode requires a resonant output cavity; in this respect it resembles the klystron. At frequencies much above 1000 MHz, the power output falls off rapidly with increasing frequency, because the device is basically a "density-modulated" or grid-controlled tube limited by electron transit time and grid limitations. A diagram of a Klystrode is shown in Fig. 5, and expected power versus frequency is shown in Fig. 9.

Circuits of Special Interest

High-Efficiency Circuits for RF Amplification—Increasing energy costs have led to attempts to improve the efficiency of conversion from dc to rf, especially in very high-power equipment. Most of these attempts are based on wave shaping. It can be shown that the power lost as heat at the anode of a tetrode, for example, can be reduced as the waveforms of voltage and current approach square waves rather than sine waves or portions of sine waves. One method due to Tyler* still in use in am broadcast transmitters is to add harmonics in the correct phase and amplitude to the original sine functions. The harmonics are generated in tuned circuits that are resonant at the harmonic frequencies and placed in series with the circuits tuned to the fundamental. The current and voltage waveforms approach square waves.

Another approach not involving extra circuits is simply to restrict the time duration of the current (class-C operation). This, however, reduces the power output or increases the cathode current density.

For a power amplifier at 1.5 MHz, typical efficiencies are: Class B, 75%; Class C, 80%–85%; "Tyler," 93%.

Circuits That Provide High Linearity—In some applications, for example tv visual channels, it is necessary to achieve rf power amplification

*Reference 1.

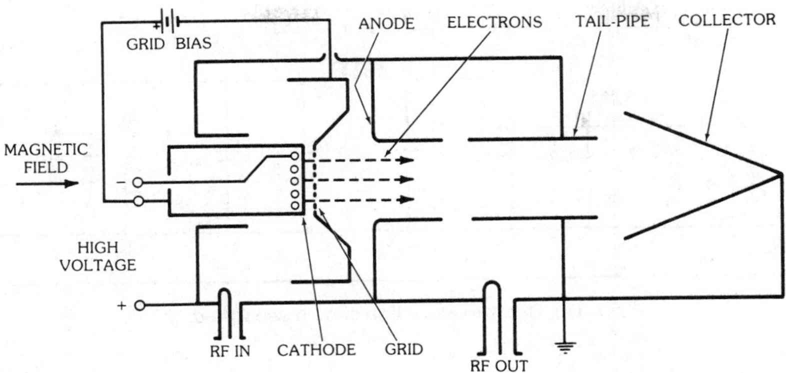

Fig. 5. Klystrode™ amplifier schematic. (*Courtesy Varian/Eimac*)

with very low distortion of the modulation content, measured either as total harmonic distortion or, more often, as intermodulation distortion. Present practice is to use triodes as class-A amplifiers or tetrodes as class-B or class-AB linear amplifiers. In class-B service, grid bias is adjusted to the value that just reduces the anode current to zero. In class-AB, a small "idling" current is allowed to flow. By careful design of the tube and circuit combination, the modulated rf output power can be a faithful copy of the input over the bandwidth required for transmission of the information contained in the modulation. In this system, the average conversion efficiency over a period of time long compared to the modulation period or cycle can be reasonably high, because power input to the tube falls when the modulation level is low (the carrier is suppressed). In contrast, an amplifier operating in class A, such as a tetrode biased so that a sinusoidal anode current waveform is produced, would draw the same dc current whether the modulation were 100% or zero. The same is true of a klystron or twt amplifier in which the beam current does not vary with the rf drive level.

In the linear amplifier with suppressed carrier, the efficiency varies as the square root of the modulation depth. For example, if with 100% modulation the efficiency is 50%, with 30% modulation the efficiency would be $50\% \times \sqrt{0.3} = 27\%$. The corresponding klystron efficiency would be $50\% \times 0.3 = 15\%$.

Another tube-circuit combination of interest is the high-efficiency Doherty linear amplifier* shown schematically in Fig. 6. In this configuration, modulation is accomplished by causing two linear amplifiers, both driven by modulated rf grid-cathode voltages, to feed power into a common load impedance. The tubes are essentially in parallel. They are biased differently so that with no modulation the carrier power is supplied only by one (the "carrier" tube). The other ("peaking" tube) is cut off. At full modulation, both tubes deliver power. A quarter-wavelength transmission line is connected between the tubes. The impedance inverting properties of this line allow each tube to "see" a load impedance that varies with modulation depth. This arrangement is found in some am broadcast transmitters in which the carrier and both sidebands are transmitted. This circuit has the advantage that efficiency tends to be high at all modulation levels, no high-level modulation transformer is required, and the peak voltage on the tubes is much lower than with conventional anode modulation.

Modulators—Modulation of the signal from a tetrode rf amplifier may be accomplished by:

A. Variation of rf drive, described above
B. Variation of screen-grid voltage, keeping drive and anode voltage constant
C. Variation of screen-grid and anode voltage together
D. Variation of control-grid voltage only

Each of these systems has advantages and disadvantages. Most transmitters today use method A or C. Pulse transmission, where essentially square pulses are required, can be accomplished by method D, with variation of the grid-bias voltage between cutoff and the value required for efficient operation when current flows.

Power grid tubes are frequently used to pulse-modulate velocity-modulated microwave tubes such as klystrons, twt's, magnetron oscillators, and cross-field amplifiers. Simple circuits can be used because a single gridded tube can perform the pulsing function. An interesting example at high power level is the "linear beam switch tube" series modulator. This tube is a form of tetrode in which the pseudo-screen grid is an apertured electrode and the final anode, or collector, is a reentrant chamber designed to minimize the production of secondary electrons, which would be accelerated back to the screen grid

*References 2 and 3.

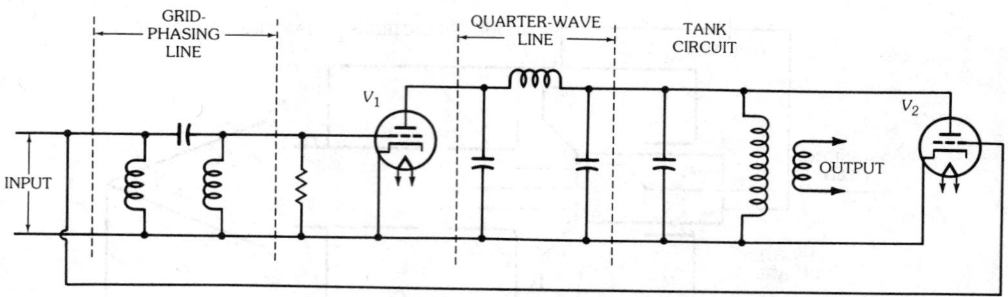

(A) Quarter-wave line shown separated.

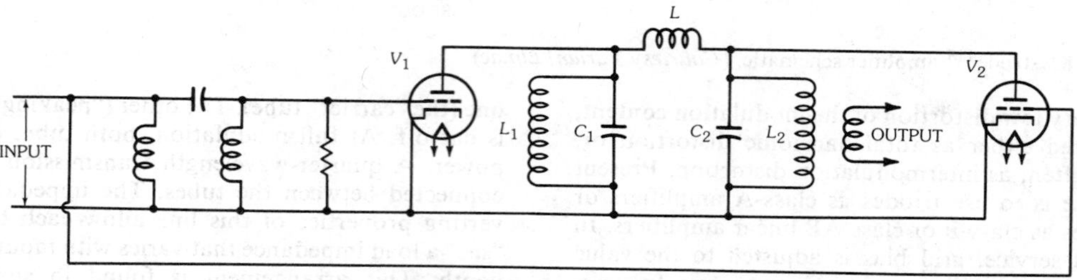

(B) Quarter-wave line incorporated into tank circuits.

Fig. 6. Schematic diagram of high-efficiency linear amplifier. (*After F.E. Terman*, Radio Engineers Handbook. *New York: McGraw-Hill Book Co., 1943; p. 456.*)

because of its higher potential. This arrangement is especially suitable for modulation of a high-power klystron that requires a high beam voltage negative in polarity with respect to ground. Fig. 7 shows this arrangement schematically. The switch tube itself uses a convergent electron beam from a concave spherical cathode and control grid in an electron gun that is similar to that of the klystron.

A relatively new modulation technique that is finding increasing use in am broadcast transmitters is pulse duration modulation (pdm)*. An rf power amplifier is anode-modulated by a series modulator tube. The control grid of the modulator tube is pulsed by a train of rectangular pulses at a repetition frequency well above the highest modulation frequency required. The pulses are arranged to drive the modulator tube to its maximum anode current and to zero current alternately. The width of the pulses is varied at the modulation frequency. A filter tuned to the pulse repetition frequency is placed between the modulator tube and the rf amplifier tube so that only the required modulating signal is applied to the rf tube. The result is a relatively efficient system at all modulation levels because the modulator tube is either fully on or fully off, and because the efficiency of the rf tube is constant over a wide range of applied anode voltage.

Broad-Band Distributed Amplifiers—For some applications, an amplifier that covers several octaves of bandwidth is required. At microwave frequencies, a single twt can be made to do this. At lower frequencies, from dc to several hundred megahertz, an arrangement of tetrode tubes and lumped-constant transmission lines known as a "distributed amplifier" can be used, as shown in Fig. 8. The transmission lines are terminated by load resistances of magnitude equal to their characteristic impedances. A growing wave of current is present on the output transmission line, each tube

*Reference 4.

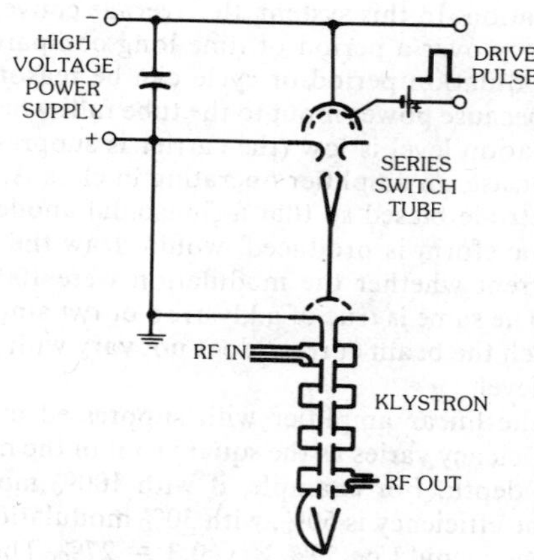

Fig. 7. Klystron modulated by linear beam switch tube. (*Courtesy Varian*)

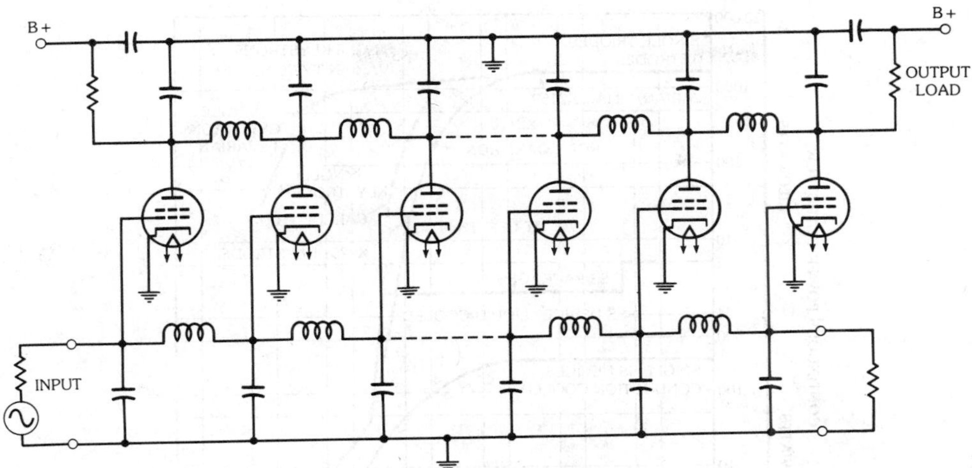

Fig. 8. Basic distributed amplifier circuit. (*Courtesy IEEE*)

providing its contribution of current in the correct phase. Such amplifiers are quite inefficient. Typically eight to sixteen tubes are used. Tube requirements are high input impedance (there must be no grid current) and high anode dissipation capability, dictating the use of tetrodes with grounded cathodes. The upper cutoff frequency is limited mainly by cathode and grid lead inductances and grid-cathode capacitance of the tubes.*

PERFORMANCE OF ELECTRON POWER TUBES; COMPARISON WITH SOLID-STATE DEVICES

Power grid tubes are traditionally used for generation and amplification of power at radio frequencies; for modulation, switching, and rectification; and, formerly, for low-power-level applications including receivers and early computers. Solid-state devices and packages have eliminated tubes from all low-power-level applications where information processing is the objective, and they are becoming widely used where up to 2 kW of cw power is required for radio transmission below about 2 GHz. At much higher power levels, power tubes remain the economical choice and are likely to remain so for the forseeable future.

Fig. 9 shows the rf power obtainable from various devices as a function frequency. Data are taken from manufacturers' catalogs and other published information. It is clear that in terms of maximum cw power obtainable, a single power-tube device is many orders of magnitude more powerful than a single solid-state device over the whole frequency range. This situation is not likely to change. It exists because of the fundamental physical distinction between the properties of electrons moving in a vacuum and electrons moving in solid material, and the properties of the media themselves. This can be seen from the following discussion.

Radio-frequency generators are really converters of dc to rf power. The key elements are:

A. A dc power supply
B. An rf resonant circuit with Q greater than approximately 5
C. A source of electrons
D. A means of "bunching" electrons and accelerating the bunches
E. A means for interaction between the bunches of electrons and the resonant circuit so that energy is extracted from the electrons and transferred via the circuit (which, like a flywheel, stores energy) to a useful load such as an antenna or a substance to be heated. (See Fig. 10.)

More specifically, typical configurations are shown in Fig. 11. Simplified circuit diagrams are essentially the same for a triode tube and a transistor, both of which illustrate the principles involved.

Fig. 12, which illustrates current and voltage as functions of time, shows that the voltage appearing across the output region of the device is small when current is flowing, but peaks at a value roughly equal to twice the dc supply voltage half a period later. The device must withstand this voltage without internal arcing. The current waveform shown in Fig. 12 implies that the time of transit of electrons between the source (cathode or emitter) and the final electrode (anode or collector) is very small compared to the period of the radio frequency. If the transit time is large, the current waveform will depart from that shown as the solid line in Fig. 12 and will tend to be shown by the dash line. The rf

*Reference 5.

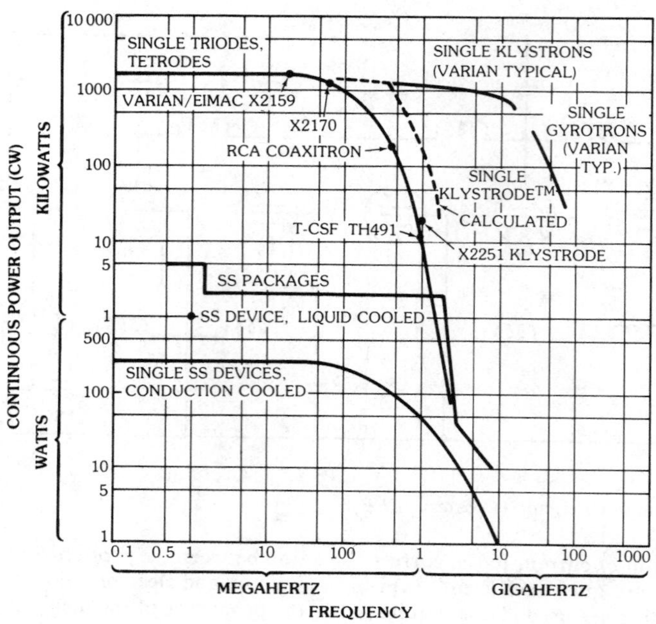

Fig. 9. Maximum cw power versus frequency from rf power sources, 1982. Available except where noted.

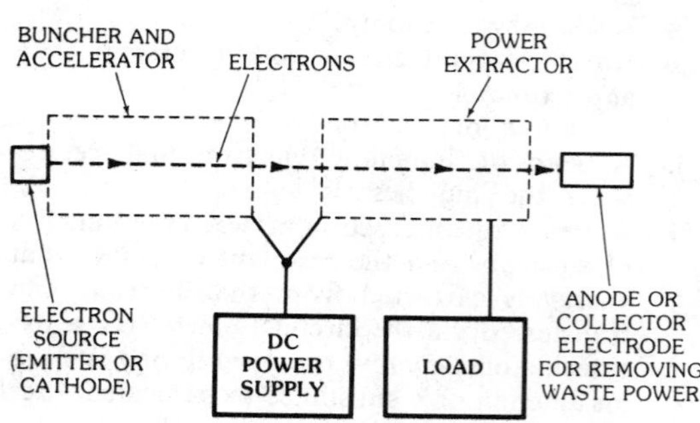

Fig. 10. Basic schematic diagram of rf amplifier using electron device.

component of this current will decrease as the transit time increases, and the conversion efficiency of the device will fall, eventually to zero. In quantitative terms, it can be shown that, other factors being equal, the parameter v/fd (where v is electron velocity across the interelectrode gap, f is frequency, and d is gap length in the direction of electron flow) must be held constant for a given performance.

The above is true for electrons both in a vacuum and in the solid state, but if we now consider the actual values of v and d, we note a fundamental difference.

$$v \text{ in vacuum} \propto \phi^{1/2}$$

where ϕ is the potential difference through which the electron has passed. For example, if $\phi = 2000$ volts, as commonly encountered in a high-power tube, $v \simeq 2 \times 10^9$ cm/s. On the other hand, in silicon, for example, v_{max} is a constant 6×10^6 cm/s.

Assume that for efficient energy extraction an electron must pass through the output gap in a time $t = (1/2\pi) \times$ (period of radio frequency). (The output gap is grid to anode in a tube or base to collector in a transistor.) At a frequency of 100 MHz, for example, $t = (1/2\pi)10^{-8}$ s. Then the output gap length must be less than the product vt, or 10^{-2} cm for the transistor compared to 3 cm for the tube. At 1000 MHz, these dimensions would be 10^{-3} cm for the transistor and 0.3 cm for the tube.

The maximum applied voltage is determined by the above dimensions and the dielectric strength of the medium. For silicon, the breakdown voltage is about 2×10^5 V/cm. In a vacuum, the situation is more complicated, but under typical conditions in a high-power tube the same number can be used. This leads to the conclusion that at 1000 MHz the maximum voltage across the tube could be $2 \times 10^5 \times 3 \times 10^{-1} = 60\,000$ V and the maximum voltage across the transistor could be 200 V. In both devices, when used in the circuits shown in Fig. 11, the peak applied voltage occurs during the half cycle when no current flows and will reach a value of about twice the dc power-supply voltage. Therefore, the dc voltage would ideally be limited, at 1000 MHz, to 100 volts for the transistor and 30 000 volts for the tube. In practice, the transistor dc voltage tends to be much lower, typically 28 volts, because the base-to-collector distance is in fact less

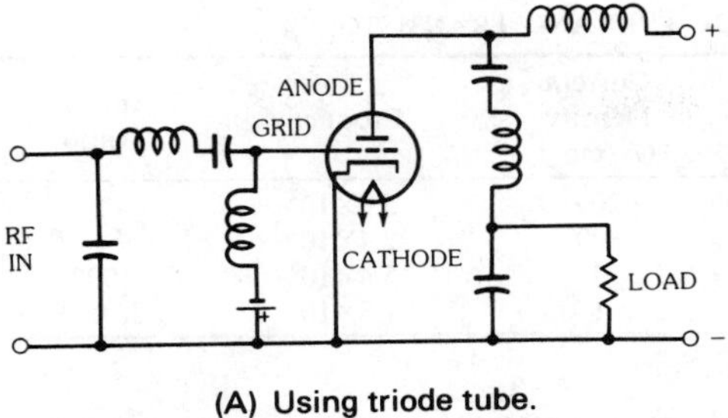

(A) Using triode tube.

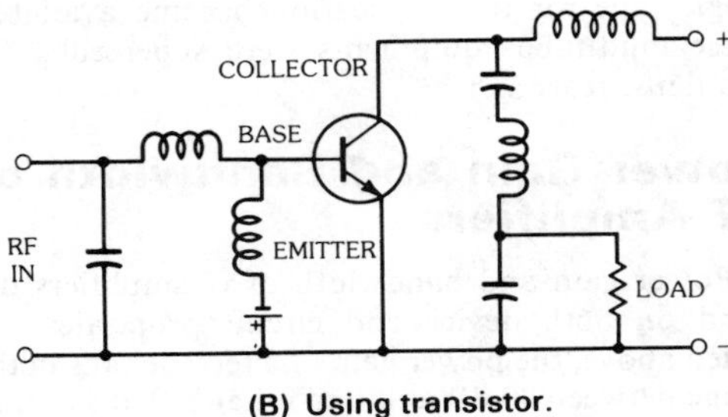

(B) Using transistor.

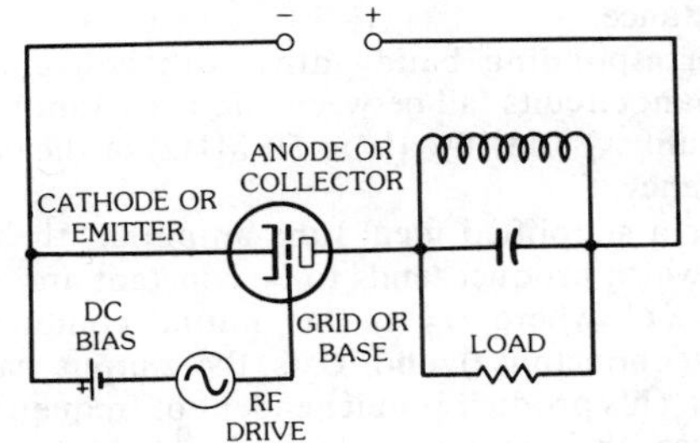

(C) Equivalent circuit for both types of device.

Fig. 11. Simplified rf amplifier circuit diagrams.

than the distance limit determined by electron transit time (for reasons given below), and the maximum usable voltage remains constant as the frequency is reduced. In the tube, the voltage ideally could rise inversely with frequency, but it is limited in practice because of the cost of high-voltage power supplies and the problem of insulation outside the vacuum.

Summarizing, then, it appears that for reasons of breakdown voltage alone, the relative power levels of transistors and gridded power tubes would be in the ratio of 1 to 300, assuming the same current in both and no other limitations. In practice, the ratio is higher because the problem of heat removal in the transistor sets an upper limit to the voltage that tends to be constant at about 60 volts rather than 200 volts, so that power-supply voltages of 28 volts are typical. The heat-removal problem is severe because, unlike the case in a vacuum, heat is generated by the passage of current through the semiconductor due to collisions between electrons and atoms (the reason for the velocity limit referred to above) and also by the thermal vibration of the atoms in the rf field (dielectric loss). In a vacuum, this is zero. The heat generated must be conducted through the semiconductor to the heat sink at the collector terminal, so the thermal conductivity of the material is critical. To add to the difficulties, semiconductor performance deteriorates as temperature increases. The heat-removal problem in fact sets an upper limit to power output well below the more basic limitation due to electron transit time.

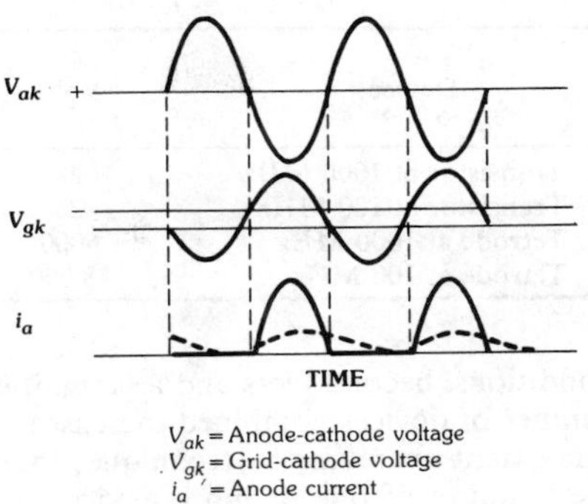

Fig. 12. Waveforms of voltage and current in class-B radio-frequency amplifier.

The remaining factors that determine the power output of a device are its cross-sectional area and current density. Here again is a great distinction. The size of the solid-state device is limited by fabrication technology and cost to the order of 0.1 cm^2. The gridded power tube is limited by the more fundamental fact that its efficiency falls off if one dimension perpendicular to electron flow exceeds approximately one-tenth the operating wavelength. The other dimension may be larger. A cylindrical tube at 1000 MHz (30 cm wavelength) may be $3 \times 10 = 30$ cm^2 in area. At lower frequencies, it can be correspondingly larger, as will be seen later. The maximum current density on peaks of rf achievable in the transistor is about 200 A/cm^2 at uhf and 800 A/cm^2 at 100 MHz. In the tube, it is about 3 A/cm^2. The current per device is the product of the current density and the area.

In summary, the power output per device is proportional to applied voltage times current density times cross-sectional area times conversion efficiency. If efficiency is neglected, Table 4 is a rough guide to relative performance. From the table, it is evident that large numbers of transistors must be combined to give the power output obtainable from a single tube. Combination of transistors has been done to the extent indicated by Fig. 9, in which available power from combinations is shown.

It should be noted that the major advantage of the solid-state device, inherent long life under nor-

TABLE 4. RELATIVE PERFORMANCE OF TUBE AND TRANSISTOR

Device	Volts	Area (cm^2)	Current Density (A/cm^2)	Product	Approx Ratio
Transistor at 1000 MHz	28	0.1	200	5.6×10^2	1
Transistor at 100 MHz	28	0.1	800	22.4×10^2	4
Tetrode at 1000 MHz	6000	30	3	5.4×10^5	1000
Tetrode at 100 MHz	18 000	1000	1	1.8×10^7	30 000

mal conditions, becomes less and less significant as the number of devices combined increases. This is to some extent overcome by techniques that allow complete failure of one or more devices to occur without failure of the system but with only a degradation in performance that can often be tolerated ("graceful degradation"). It is also not surprising that some performance criteria—bandwidth, for example—tend to fall off as the number of devices combined increases. This happens because the input and output impedances of the combination, which with simple paralleling are inversely proportional to the number of devices, become very low compared to the impedances of the circuits with which the combination is associated. Also, it should be noted that paralleling requires low-voltage regulated power supplies with high current capability. This is by no means easy to achieve and is very costly. One form of degradation, hardly graceful, is total failure of all devices due to a transient overcurrent or overvoltage phenomenon such as the electromagnetic pulse (emp) that can result from nearby lightning or a nuclear explosion. A less dramatic but equally fatal circumstance producing the same result is an unplanned increase in load impedance that causes the voltage to rise to the breakdown limit. It is interesting that the solid-state device and the tube device react differently to this event because of their different physical nature. In the solid-state device, the effect is a permanent breakdown, or short-circuit. In the vacuum device, the effect is a temporary breakdown followed by a resumption of full performance a short time after the transient has subsided.

In summary, then, it appears that both tubes and solid-state devices have fundamental limitations. Both have reached a stage of development where subsequent improvements will probably be small relative to those already made. Essentially, the transistor is a low-power device. The art of combining transistors may progress so that higher-power packages may become available, but at extra cost. In nonmilitary systems, it is ultimately the cost in terms of amortized capital cost plus operating and maintenance costs per unit time that governs the choice of device type.

(It is of historical interest that before the advent of high-power velocity-modulated tubes, serious and successful attempts were made to combine triodes and tetrodes to increase rf power output at vhf and uhf, and many transmitters were built this way. As soon as tubes giving sufficient power from a single type for the application became available, these multitube equipments were superceded for economic reasons.)

Power Gain and Bandwidth of RF Amplifiers

Power gain and bandwidth of rf amplifiers depend on both device and circuit properties. As noted above, the power gain of a tetrode falls in the range between 30 dB at lf and mf and 10 dB at uhf. Power gain at uhf is limited by electron-transit-time effects and by the inevitable effect of cathode-lead inductance.

Corresponding bandwidths with conventional resonant circuits fall between about 1% (am broadcast, uhf tv) and 10% (tv at 50 MHz) of the carrier frequency.

For a simplified ideal tube amplifier, the gain-bandwidth product tends to be constant and equal to G_m/C, where G_m is the mutual conductance (transconductance) and C is the output capacitance. This product is independent of frequency. In practice, the product is always less because of the presence of the external circuits that store energy and because the bandwidth of the input circuit may be smaller than that of the output circuit. Furthermore, the performance of the amplifier may be largely determined by the amount of feedback present. It is nevertheless true that the gain-bandwidth product will always increase with G_m/C.

This factor is related to interelectrode spacings and is approximately inversely proportional to the grid-cathode spacing. The achievement of small spacing, which implies fine grid wires and structures of high dimensional stability with respect to change of temperature, has always been a major challenge in tube design.

Power Output Under Pulse Conditions

Fig. 9 shows cw power available versus frequency. Under pulse conditions, gridded tubes can be made to give many times this power, depending on the pulse width and duty factor (ratio of pulse width to the interval between pulses). Tube manufacturers will supply data on the exact amount of power. A

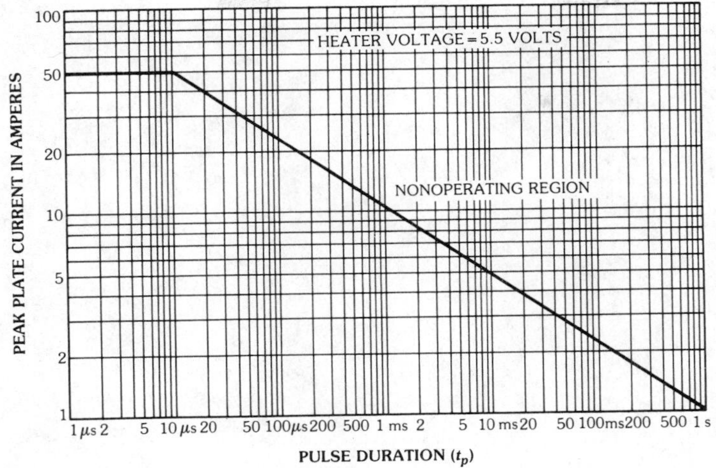

Fig. 13. Pulse-derating data for triode type 3CPX1500A7, pulse modulator or regulator service. (*Courtesy Varian/ Eimac*)

typical pulse derating chart is shown in Fig. 13. The power is limited by the thermal capacity of the grids, by the breakdown voltage between electrodes, and by the emission capability of the cathode.

Life Expectancy of Tubes

The principal factor governing tube life is the temperature of the thermionic cathode. This is related to emission and therefore to output power. In general, longer life results from lower cathode temperature. In a well designed tube, the available emission at the specified cathode heating power is many times that required for satisfactory operation. It is therefore often possible to reduce the heating power and increase the life of the tube.

Another factor affecting life is the relationship between maximum ratings and operating conditions. Exceeding ratings causes overheating, which tends to shorten life. Careful attention must be paid to manufacturers' ratings and recommendations.

Life expectancy for a highly stressed oxide-cathode tube at uhf may be 2500 hours. For a thoriated-tungsten-filament tube in broadcast service, 50 000 hours is not uncommon.

Development Trends in Gridded Power Tubes

Improved performance of power grid tubes, including longer life, can be anticipated from the incorporation of improved materials and fabrication techniques. For example, the use of pyrolytic graphite for grids is becoming standard practice. Impregnated tungsten dispenser cathodes commonly found in microwave velocity-modulated tubes may find their way into gridded power tubes.

In addition, improved structural forms such as modular construction may allow the manufacture of very-high-power tubes (capable of 10 MW of cw power output) having higher gain-bandwidth products because of smaller grid-cathode spacings.

At vhf and uhf, developments of the Klystrode™ amplifier may appear in applications where the combination of small size compared to a klystron; relatively high efficiency, especially with amplitude modulation; high power capability; and interesting power gain makes it more attractive than a klystron or a conventional tetrode.

REFERENCES

1. Tyler, V. J. "A New High-Efficiency Power Amplifier." *Marconi Review*, Vol. 21, No. 130 (1958), pp. 6–109.
2. Doherty, W. H. "A New High-Efficiency Power Amplifier for Modulated Waves." *Proc. IRE*, Vol. 24 (Sept. 1936), p. 1163.
3. Terman, F. E. *Radio Engineers Handbook*. New York: McGraw-Hill Book Co., 1943; pp. 455–458.
4. Swanson, H. "The Pulse Duration Modulator, a New Method of High Level Modulation in Broadcast Transmitters." *IEEE Trans. on Broadcasting*, Vol. BC17, No. 4 (Dec. 1971), pp. 89–92.
5. Ginzton, Hewlett, Jasberg, and Noe. "Distributed Amplifications." *Proc. IRE*, 1948, pp. 956–969.

18 Semiconductors and Transistors

Ben G. Streetman

The basic building blocks of any electronic circuit or system are electronic devices—the transistors, diodes, and other elements that collectively allow the system to perform its function. These devices may be individual transistors, diodes, and passive elements soldered in appropriate interconnection. Or, the system may be a collection of complex integrated circuits, each composed of thousands of active and passive elements on a silicon chip (see Chapter 20). The system may even involve information transfer by a modulated light beam and detector (see Chapter 21). In any case, a great variety of electronic elements is available for performing the desired function, and those who design or use electronic equipment have a wide range of devices from which to choose.

Most modern electronic devices are made with semiconductor materials. The unusual properties of these solids are responsible for a revolution in electronics since the invention of the transistor. This chapter* deals with semiconductors and their applications in pn junctions, transistors, and other devices. Circuit models of transistors are discussed in Chapter 19, and fabrication methods are discussed in Chapter 20.

SEMICONDUCTORS

Semiconductors are a group of materials that have electrical conductivities intermediate between those of metals and insulators. It is significant that the conductivity of these materials can be varied considerably by changes in temperature, optical excitation, and impurity content. This variability of electrical properties makes possible the wide range of modern electronic devices.

Semiconductor Materials

Semiconductors are found in column IV and neighboring columns of the periodic table (Table 1). The column-IV semiconductors silicon (Si) and germanium (Ge) are called *elemental* semiconductors because they are composed of single species of atoms. In addition to the elemental materials, compounds of column-III and column-V atoms, as well as certain combinations from columns II and VI, make up the *intermetallic*, or *compound*, semiconductors.

As Table 1 indicates, there are numerous semiconductor materials. Among these, silicon is used for the majority of semiconductor devices; rectifiers, transistors, and integrated circuits are now made mostly in silicon, although GaAs is also used in high-speed integrated circuits. The compounds are most widely used in devices requiring the emission or absorption of light. For example, semiconductor light-emitting diodes commonly are made of such compounds as GaAs, GaP, and alloys such as GaAsP. An important microwave device, the gunn diode, is usually made in GaAs. Thus, the wide range of semiconductor materials offers considerable variety in properties and provides experts in electronic circuits and systems with much flexibility in the design of electronic functions.

The electronic and optical properties of semiconductor materials are strongly affected by impurities, which may be added in precisely controlled amounts. Such impurities are used to vary the con-

*Much of this discussion is from Ben G. Streetman, *Solid State Electronic Devices,* 2nd ed., © 1980. Reprinted by permission of Prentice-Hall, Inc., Englewood Cliffs, N.J.

TABLE 1. COMMON SEMICONDUCTOR MATERIALS

The Portion of the Periodic Table Where Semiconductors Occur

II	III	IV	V	VI
	B	C		
	Al	Si	P	S
Zn	Ga	Ge	As	Se
Cd	In		Sb	Te

Elemental and Compound Semiconductors

Elemental	IV Compounds	III-V Compounds	II-VI Compounds
Si	SiC	AlP	ZnS
Ge		AlAs	ZnSe
		AlSb	ZnTe
		GaP	CdS
		GaAs	CdSe
		GaSb	CdTe
		InP	
		InAs	
		InSb	

ductivities of semiconductors, and even to alter the nature of the conduction processes from conduction by negative charge carriers to conduction by positive charge carriers. For example, an impurity concentration of one part per million can change a sample of silicon from a poor conductor to a good conductor of electric current. This process of controlled addition of impurities, called *doping*, will be discussed below.

Virtually all semiconductor devices require single crystals of extremely pure materials. Since small concentrations of impurities can radically alter their electrical properties, device-grade semiconductor crystals are grown with greater perfection than any other materials.

Energy Bands and Charge Carriers

The atomic arrangement in most semiconductor crystals is similar to the diamond lattice, in which each atom is surrounded by four nearest neighbors. The atomic bonding in the crystal is largely covalent (as in the H_2 molecule). That is, the sharing of electrons between adjacent atoms in covalent bonds holds the crystal together. In the column-IV semiconductor silicon, for example, each atom has four valence electrons shared with four nearest neighbors.

Electrons in isolated atoms are restricted to certain discrete energy levels, predictable generally from the Bohr model and more precisely from the results of quantum mechanics. In a similar fashion, electrons in solids are restricted to certain energies and are not allowed at other energies. The basic difference between the case of an electron in a solid and that of an electron in an isolated atom is that in the solid the electron has a *range*, or *band*, of available energies. General features of the energy bands for insulators, metals, and semiconductors are shown in Fig. 1. In an insulator or semiconductor, a lower (valence) band is filled with electrons at low temperatures, and an upper (conduction) band is empty of electrons. The separation between the valence and conduction bands is called the *band gap* E_g. Insulators have a wider band gap than do semiconductors, and metals have overlapping bands.

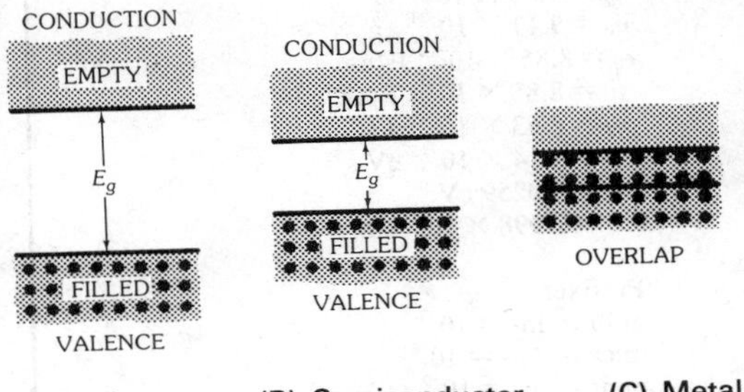

Fig. 1. Energy bands for typical materials.

The variety of electrical conductivities of metals, insulators, and semiconductors arises primarily from their band structures (Fig. 1). If electrons are to experience acceleration in an electric field, they must be able to move into new energy states. In a completely filled band, there are no available empty states, and therefore no net charge transport (no current) can take place. The same is true for an empty band, which contains no electrons to move. As a result, a perfect insulator does not conduct current. Metals, on the other hand, have electrons mixed with unfilled allowable energy states. Electrons are therefore available to participate in current flow, and there are plenty of energy states for the electrons to occupy in response to a field. This accounts for the high conductivity of metals.

Although semiconductors and insulators have similar band structures, there is an important difference: The band gap E_g of an insulator is several electronvolts wide,* whereas E_g for a typical semiconductor is sufficiently small (about one eV) to allow excitations of electrons from the valence band to the conduction band. These excitations can result from the application of thermal or optical energy. For example, at room temperature in a perfect silicon crystal ($E_g = 1.1$ eV), the equilibrium concentration of thermally generated electrons in the conduction band is about 1.5×10^{10} per cubic centimeter (cm^{-3}). Of course, there is an equal number of empty energy states in the valence band. These unoccupied valence-band states are referred to as *holes*. Electron-hole pairs (EHP) can also be created by optical excitation; for example, a quantum of light (*photon*) with energy greater than the band gap E_g can be absorbed by the semiconductor crystal, exciting a valence-band electron to the conduction band. The result is a conduction-band electron and a valence-band hole. As discussed below, this optical generation of electron-hole pairs is basic to the operation of semiconductor photoconductors and photodiodes.

At thermal equilibrium, the excitation of electrons to the conduction band is balanced by electrons falling back to the valence band (recombination). This balance of electron-hole generation and recombination results in a predictable concentration of EHP at a given temperature. Similarly, a steady optical excitation causes a new steady-state balance between generation and recombination, resulting in a higher density of electrons and holes.

It is clear that electrons which have been excited

*The electronvolt (eV) is a particularly convenient unit of measure for the energy of an electron. One eV is defined as the energy acquired by an electron moving through a potential of one volt. Thus, 1 eV = 1.6×10^{-19} joule.

to the conduction band of a semiconductor can participate in current conduction, since ample unoccupied energy states are available in that band. A less obvious but important feature of semiconductors is charge transport involving holes in the valence band. The presence of an empty state (hole) in the otherwise filled valence band allows electrons in the band to move, with a resulting net motion of charge, or current. The approach for calculating this charge transport is to sum the contributions of all electrons in a filled valence band (which results in zero net current) and then subtract the contribution of the missing electron. The result is equivalent to considering the hole as a positive charge carrier (i.e., a particle with charge $+q$, where $-q$ is the charge on an electron).* For example, a semiconductor with 10^{10} electron-hole pairs per cubic centimeter contains a concentration of 10^{10} cm^{-3} negatively charged electrons in the conduction band, which drift opposite to an applied electric field, and 10^{10} cm^{-3} positively charged holes in the valence band, which drift in the direction of the field. The two current components add in the field direction to give the total current. Both electrons and holes are important charge carriers in semiconductors.

In addition to the direct excitation of electrons and holes in pairs, each type of charge carrier can be introduced by appropriate doping of the semiconductor with impurities. Impurities can be added to the crystal in the growth process or introduced later during device fabrication. By doping, a crystal can be altered so that it has a predominance of either electrons or holes. Thus there are two types of doped semiconductors, n-type (mostly electrons) and p-type (mostly holes). The designation n and p is chosen to reflect the dominance of negative or positive charge carriers.

When impurities are introduced into an otherwise perfect crystal, additional levels are created in the energy-band structure, usually within the band gap. For example, an impurity from column V of the periodic table (phosphorus, P; arsenic, As; antimony, Sb) introduces an energy level very near the conduction band in germanium or silicon. This level is filled with electrons at 0 K, and very little thermal energy is required to excite these electrons to the conduction band (Fig. 2A). Thus at room temperature virtually all of the electrons in the impurity level are "donated" to the conduction band. Such an impurity level is called a *donor* level, and the column-V impurities in germanium or silicon are called donor impurities. Material doped with donor impurities can have a considerable number of electrons in the conduction band, even when the temperature is too low for the thermally generated EHPs to be appreciable. Thus for semiconductors doped with a significant number of donor atoms, the electron concentration n will be much greater than the hole concentration p. This is n-type material.

Atoms from column III (boron, B; aluminum, Al; gallium, Ga; indium, In) introduce impurity levels in germanium or silicon near the valence band (Fig. 2B). These levels are empty of electrons at 0 K. At higher temperatures, enough thermal energy is available to excite electrons from the valence band into the impurity level, leaving behind holes in the valence band. Since this type of impurity level "accepts" electrons from the valence band, it is called an *acceptor* level, and the column-III

*Table 2 lists the magnitude of the electronic charge and other information.

TABLE 2.* PHYSICAL CONSTANTS AND CONVERSION FACTORS

Avogadro's number	$N_A = 6.02 \times 10^{23}$ molecules/mole
Boltzmann's constant	$k = 1.38 \times 10^{-23}$ J/K
	$= 8.62 \times 10^{-5}$ eV/K
Electronic charge (magnitude)	$q = 1.60 \times 10^{-19}$ C
Electronic rest mass	$m_0 = 9.11 \times 10^{-31}$ kg
Permittivity of free space	$\epsilon_0 = 8.85 \times 10^{-14}$ F/cm
	$= 8.85 \times 10^{-12}$ F/m
Planck's constant	$h = 6.63 \times 10^{-34}$ J s
	$= 4.14 \times 10^{-15}$ eV s
Room temperature value of kT	$kT = 0.0259$ eV
Speed of light	$c = 2.998 \times 10^{10}$ cm/s
	Prefixes:
1 Å (angstrom) $= 10^{-8}$ cm	milli-, m- $= 10^{-3}$
1 μm (micron) $= 10^{-4}$ cm	micro-, μ- $= 10^{-6}$
1 mil $= 10^{-3}$ in	nano-, n- $= 10^{-9}$
2.54 cm = 1 in	pico-, p- $= 10^{-12}$
1 eV $= 1.6 \times 10^{-19}$ J	kilo-, k- $= 10^{3}$
	mega-, M- $= 10^{6}$
	giga-, G- $= 10^{9}$

A wavelength λ of 1 μm corresponds to a photon energy of 1.24 eV.

* From Ben G. Streetman, *Solid State Electronic Devices*, 2nd ed., © 1980. Reprinted by permission of Prentice-Hall, Inc., Englewood Cliffs, N.J.

impurities are acceptor impurities in germanium and silicon. Doping with acceptor impurities can create a semiconductor with a hole density much greater than the conduction-band electron density (p-type material).

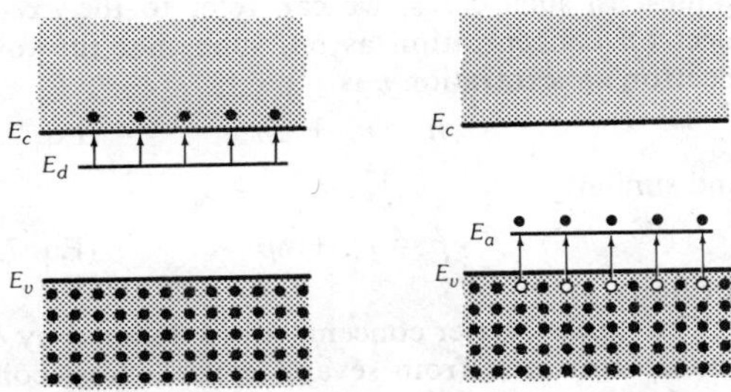

(A) Electrons are thermally excited from a donor level to the conduction band.

(B) Valence-band electrons are excited to E_a, leaving holes in the valence band.

Fig. 2. Donor and acceptor levels (E_d and E_a) in a semiconductor.

A semiconductor without doping impurities is called *intrinsic* material, and the *intrinsic carrier concentration* n_i is the concentration of thermally generated electron-hole pairs. In silicon, for example, the intrinsic concentration n_i is about 10^{10} cm^{-3} at room temperature. If silicon is doped with 10^{15} antimony atoms/cm^3, the conduction electron concentration n increases from 10^{10} to 10^{15} cm^{-3}, a change of five orders of magnitude. As a result, the resistivity of silicon changes from about 2×10^5 ohm-cm to 5 ohm-cm with this doping.

When a semiconductor is doped n-type or p-type, one type of carrier dominates. In the example given above, the conduction-band electrons outnumber the holes in the valence band by many orders of magnitude. We refer to the small number of holes in n-type material as *minority carriers* and the relatively large number of conduction electrons as *majority carriers*. Similarly, electrons are the minority carriers in p-type material and holes are the majority carriers.

Carrier Concentrations

In calculating semiconductor electrical properties and analyzing device behavior, it is often necessary to know the concentration of charge carriers in the material. The majority-carrier concentration is usually obvious in heavily doped material, since one majority carrier is obtained for each impurity atom. The concentration of minority carriers is not obvious, however, unless we consider details of the electron distribution in the solid. An important result of semiconductor statistics is that the *electron-hole product np* at equilibrium is a constant for a given material at a given temperature, whether the material is doped or not:

$$np = n_i^2 \qquad \text{(Eq. 1)}$$

where n and p are the equilibrium electron and hole concentrations, respectively. Thus, if we know the majority-carrier concentration (e.g., from the doping density) we can find the minority-carrier concentration from Eq. 1. For example, if a silicon sample ($n_i = 1.5 \times 10^{10}$ cm^{-3}) is doped with 10^{15} donors/cm^3, the electron concentration is essentially 10^{15} and the hole concentration is only 2.25×10^5 cm^{-3} at room temperature. It is interesting to note from Eq. 1 that as the majority-carrier concentration increases, the minority-carrier concentration must decrease.

If a semiconductor contains both donors and acceptors, the exact relationship among the electron, hole, donor, and acceptor concentrations can be obtained by considering the requirements for *space-charge neutrality*. If the material is to remain electrostatically neutral, the sum of the positive charges (holes and ionized donor atoms N_d^+) must balance the sum of the negative charges (electrons and ionized acceptor atoms N_a^-):

$$p + N_d^+ = n + N_a^- \qquad \text{(Eq. 2)}$$

If the material is strongly n-type ($n >> p$) and all of the impurities are ionized, we can approximate Eq. 2 by $n \cong N_d - N_a$.

Drift of Carriers

The electrons and holes in a semiconductor are in constant motion due to their thermal energy. At thermal equilibrium, the movement is random, and carriers are scattered from lattice atoms, impurities, and defects. Since the scattering is completely random, there is no net motion of the group of carriers over a period of time. On the other hand, if an electric field is applied to the sample, a net drift of carriers is superimposed on the random thermal motion. Thus, a field $\mathcal{E}_x$ applied in the x direction results in a net force $q\mathcal{E}_x$ on each hole and $-q\mathcal{E}_x$ on each electron. The current resulting from this net drift is just the number of carriers crossing a given area A per unit time (number of carriers times velocity) multiplied by the charge of the carrier:

$$I = qApv_p - qAn(-v_n) \qquad \text{(Eq. 3)}$$

In this equation, v_p is the average drift velocity of holes in the x direction, and $-v_n$ is the average electron velocity, which is in the opposite direction. The resulting electrical conductivity is usually written in terms of the *mobility* of each charge carrier. This quantity measures the ease with which carriers drift in a given material. The hole mobility μ_p is the average drift velocity per unit applied field $v_p/\mathcal{E}$, and similarly for electron mobility μ_n. Thus the current is

$$I = qA\,(\mu_p p + \mu_n n)\,\mathcal{E} \equiv \sigma \mathcal{E} A \qquad \text{(Eq. 4)}$$

where the conductivity of the sample σ has the units $(\text{ohm-cm})^{-1}$. Conductivity is determined by electron and hole concentration (n, p), and therefore by the doping of the sample, and also by the carrier mobilities.

If a semiconductor bar contains both types of carriers, holes drift as a group in the direction of the electric field, and electrons drift in the opposite direction. Both components of current are in the direction of the $\mathcal{E}$ field, however, since conventional current is positive in the direction of hole flow and opposite to the direction of electron flow.

If a magnetic field is applied perpendicular to the direction in which holes drift in a p-type bar, the path of the holes tends to be deflected (Fig. 3). In

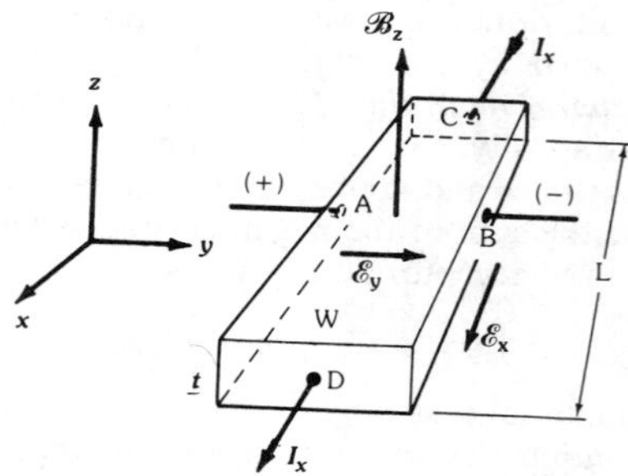

Fig. 3. A p-type semiconductor bar in a magnetic field, resulting in the production of a Hall-effect voltage. (*From Ben G. Streetman,* Solid State Electronic Devices, *2nd ed., © 1980. Reprinted by permission of Prentice-Hall, Inc., Englewood Cliffs, N.J.*)

vector notation, the total force on a single hole due to the combined electric and magnetic fields is

$$\mathbf{F} = q(\mathbf{E} + \mathbf{v} \times \mathbf{B}) \qquad \text{(Eq. 5)}$$

In the y direction the force is

$$F_y = q\,(\mathcal{E}_y - v_x \mathcal{B}_z) \qquad \text{(Eq. 6)}$$

An electric field $\mathcal{E}_y$ builds up in the y direction as holes are displaced laterally, and in steady state this field is sufficient to just balance the force due to the magnetic field ($\mathcal{E}_y = v_x\mathcal{B}_z$). The resulting voltage between A and B can be measured with a high-impedance voltmeter. This is called the Hall effect, and V_{AB} is called the Hall voltage. This effect can be used to measure magnetic fields and in a variety of other applications. An important use for the Hall effect in semiconductor research is that the Hall voltage can be easily related to the concentration of majority carriers in the sample. Although the discussion here relates to p-type material, similar results obtain for n-type samples, in which the majority carriers are electrons. By combining Hall measurements with conductivity measurements, the majority-carrier concentration and mobility can be obtained for a semiconductor sample.

Excess Carriers; Diffusion

To this point, we have discussed electron and hole carrier concentrations at thermal equilibrium or with steady fields applied. In each case, the carrier concentrations have been the equilibrium values (let us call them n_o and p_o). However, most semiconductor devices operate by the creation of carrier concentrations greater than the equilibrium values. In such cases, we can refer to the *excess* electron concentration as δn, such that the total electron concentration n is

$$n = n_o + \delta n \qquad \text{(Eq. 7A)}$$

and similarly

$$p = p_o + \delta p \qquad \text{(Eq. 7B)}$$

for holes.

The excess carrier concentrations denoted by δn and δp can arise from several causes, most commonly from *optical excitation* or from *carrier injection,* which is characteristic of pn junctions. Optical excitation is the basis of semiconductor light detectors.

When excess carriers are created at a rate g_{op} by a steady light, a steady-state balance is established between generation and recombination. That is, in steady state the EHPs recombine as fast as they are generated; otherwise, there would be a buildup of EHPs with time. Thus, the recombination rate equals the generation rate in steady state. However, the electron and hole concentrations in the illuminated sample are greater than the equilibrium values. In particular,

$$\delta n = g_{op}\tau_n, \qquad \delta p = g_{op}\tau_p \qquad \text{(Eq. 8)}$$

where τ_n and τ_p are called the electron and hole *lifetimes,* respectively. The lifetime of a carrier is the average time it spends in its respective band before recombination. Thus $\delta n/\tau_n$ is the electron recombination rate, and $\delta p/\tau_p$ is the rate at which holes are generated and recombine in steady state.

The complementary mechanism to optical absorption is *radiative recombination.* In some semiconductors, the energy lost by an electron in recombining with a hole is given off by a photon of light. If the recombination takes place directly, without involving an intermediate state in which a carrier is temporarily captured, the energy of the emitted photon is equal to the band gap (see Table 3). For example, recombination of excess carriers in GaAs results in light emission with photon energies equal to the band-gap energy of about 1.4 eV. Such emission is called *luminescence.* Recombination in many semiconductors (such as silicon and germanium) takes place indirectly; for example, an electron is trapped at an impurity or lattice defect, and then a hole is captured to complete the recombination process. In such cases, the energy of the electron is given up as heat to the lattice, and no light is emitted.

If a current is passed through an illuminated semiconductor sample, we find that the conductivity has increased due to the presence of the excess carriers. From Eqs. 4 and 7, the conductivity is

$$\begin{aligned}\sigma &= q(\mu_p p + \mu_n n)\\ &= q[\mu_p(p_o + \delta p) + \mu_n (n_o + \delta n)] \qquad \text{(Eq. 9)}\end{aligned}$$

The change in conductivity between its dark value and its value with the generation rate g_{op} is

$$\begin{aligned}\Delta\sigma &= q(\mu_p \delta p + \mu_n \delta n)\\ &= q\, g_{op}(\tau_p \mu_p + \tau_n \mu_n) \qquad \text{(Eq. 10)}\end{aligned}$$

This increase in conductivity for the illuminated sample is called *photoconductivity*.. We notice from Eq. 10 that a sensitive photoconductive detector should have reasonably long carrier lifetimes and high carrier mobilities. Carrier lifetime is determined for most semiconductors by crystal quality, doping, and other material properties.

If the excess carrier concentrations δn and δp vary with position in the sample, *diffusion* occurs. Diffusion is the familiar process by which particles migrate from regions where their concentration is high to regions of lower concentration. A simple example of this process is the opening of a perfume bottle in a closed room with very still air. The scented air molecules soon disperse throughout the room, even if no air currents exist. The diffusion is due simply to the random thermal motion of the air molecules. The diffusion continues until the particles are evenly distributed in space. This process is described mathematically by *Fick's first law of diffusion*

$$\phi(x) = -D[dN(x)/dx] \qquad \text{(Eq. 11)}$$

for a one-dimensional problem. This relation states that the rate of particle flow in the x direction [the particle flux density $\phi(x)$] is proportional to the negative gradient of the particle concentration $N(x)$ at each point x. The proportionality constant D is called the *diffusion coefficient* (cm^2/s). Particles diffuse in the direction of decreasing particle concentration; the rate of flow depends on how steeply the concentration profile is graded; and no external driving force besides random thermal motion is involved.

The same diffusion mechanism applies to charge carriers if their concentration varies spatially, and the result is a diffusion current

$$\begin{aligned}J_n(\text{diff.}) &= -(-q)\, D_n\, [dn(x)/dx]\\ &= +q\, D_n\, [dn(x)/dx] \qquad \text{(Eq. 12A)}\\ J_p(\text{diff.}) &= -(+q)\, D_p\, [dp(x)/dx]\\ &= -q\, D_p\, [dp(x)/dx] \qquad \text{(Eq. 12B)}\end{aligned}$$

In these equations J(diff.) is the current density (A/cm^2) due to diffusion, and the n and p subscripts refer to electrons and holes, respectively. The current density J is related to the particle flux density ϕ by the charge on the carrier. As a result of the opposite charge on the two types of carriers, Eqs. 12 indicate that electrons and holes diffusing

TABLE 3.* PROPERTIES OF SEMICONDUCTOR MATERIALS (300 K)

	E_g (eV)	μ_n ($cm^2/V s$)	μ_p ($cm^2/V s$)	Lattice	a (Å)	ϵ_r	Density (g/cm^3)	Melting point (°C)
Si	1.11	1350	480	*D*	5.43	11.8	2.33	1415
Ge	0.67	3900	1900	*D*	5.66	16	5.32	936
SiC(α)	2.86	500		*W*	3.08	10.2	3.21	2830
AlP	2.45	80		*Z*	5.46		2.40	2000
AlAs	2.16	180		*Z*	5.66	10.9	3.60	1740
AlSb	1.6	200	300	*Z*	6.14	11	4.26	1080
GaP	2.26	300	150	*Z*	5.45	11.1	4.13	1467
GaAs	1.43	8500	400	*Z*	5.65	13.2	5.31	1238
GaSb	0.7	5000	1000	*Z*	6.09	15.7	5.61	712
InP	1.28	4000	100	*Z*	5.87	12.4	4.79	1070
InAs	0.36	22600	200	*Z*	6.06	14.6	5.67	943
InSb	0.18	10^5	1700	*Z*	6.48	17.7	5.78	525
ZnS	3.6	110		*Z, W*	5.409	8.9	4.09	
ZnSe	2.7	600		*Z*	5.671	9.2	5.65	
ZnTe	2.25		100	*Z*	6.101	10.4	5.51	
CdS	2.42	250	15	*W, Z*	4.137	8.9	4.82	1475
CdSe	1.73	650		*W*	4.30	10.2	5.81	1258
CdTe	1.58	1050	100	*Z*	6.482	10.2	6.20	1098
PbS	0.37	575	200	*H*	5.936	161	7.6	1119
PbSe	0.27	1000	1000	*H*	6.147	280	8.73	1081
PbTe	0.29	1600	700	*H*	6.452	360	8.16	925

Definitions of symbols: *a* is lattice constant; *D* is diamond; *Z* is zinc blende; *W* is wurtzite; *H* is halite (NaCl). Values of mobility are for material of available purity. These values are considered approximate. Most of the values in this table were taken from publications of the Electronic Properties Information Center (EPIC), Hughes Aircraft Co., Culver City, California; also, M. Neuberger, *III-V Semiconducting Compounds—Data Tables*, Plenum Publishing Corp., 1970.

together in a particle gradient give rise to currents in opposite directions.

In most cases, electric fields are present along with gradients in the carrier concentrations. We can include drift in the field along with diffusion:

$$J_n(x) = q\mu_n n(x)\ \mathcal{E}(x) + qD_n[dn(x)/dx] \quad \text{(Eq. 13A)}$$

drift diffusion

$$J_p(x) = q\mu_p p(x)\ \mathcal{E}(x) - qD_p[dp(x)/dx] \quad \text{(Eq. 13B)}$$

The total current density is the sum of the electron and hole components:

$$J(x) = J_n(x) + J_p(x) \quad \text{(Eq. 14)}$$

At equilibrium, there is no net current in a semiconductor. Thus, any fluctuation that would begin a diffusion current sets up an electric field that redistributes carriers by drift. An examination of the requirements for equilibrium indicates that the diffusion coefficient and mobility must be related by

$$D/\mu = kT/q \quad \text{(Eq. 15)}$$

for either carrier type. This important equation is called the *Einstein relation.* It allows us to measure either D or μ and calculate the other.

In many carrier-diffusion problems, a steady-state distribution is maintained by a constant generation of excess carriers at some point in the semiconductor. For example, in Fig. 4, a steady excess hole concentration Δp is maintained by the injection of holes at $x = 0$ in a long semiconductor bar. Clearly, these excess holes will diffuse into the semiconductor; but they will also recombine with a characteristic lifetime τ_p. Thus the excess hole concentration $\delta p(x)$ decreases with distance into the semiconductor. This process is governed by the *steady-state diffusion equations:*

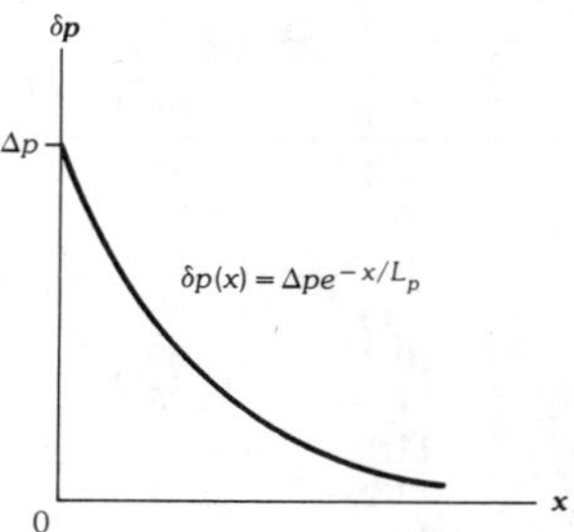

Fig. 4. Excess holes injected into a long n-type semiconductor diffuse and recombine, giving an exponential distribution.

$$d^2\delta n/dx^2 = \delta n/D_n\tau_n \equiv \delta n/L_n^2 \quad \text{(Eq. 16A)}$$

$$d^2\delta p/dx^2 = \delta p/D_p\tau_p \equiv \delta p/L_p^2 \quad \text{(Eq. 16B)}$$

where,

$L_n \equiv \sqrt{D_n\tau_n}$ is called the electron *diffusion length,*

L_p is the diffusion length for holes.

For the example of Fig. 4, the solution to Eq. 16B is

$$\delta p(x) = C_1 e^{x/L_p} + C_2 e^{-x/L_p} \quad \text{(Eq. 17)}$$

We can evaluate C_1 and C_2 from the boundary conditions. Since recombination must reduce $\delta p(x)$ to zero for large values of x, $\delta p = 0$ at $x = \infty$, and therefore $C_1 = 0$. Similarly, the condition $\delta p = \Delta p$ at $x = 0$ gives $C_2 = \Delta p$, and the solution is

$$\delta p(x) = \Delta p e^{-x/L_p} \quad \text{(Eq. 18)}$$

for the steady-state hole distribution. The injected excess hole concentration dies out exponentially in x due to recombination, and the diffusion length L_p represents the average distance a hole diffuses before recombining.

The steady-state distribution of excess holes causes diffusion, and therefore a hole current, in the direction of decreasing concentration. From Eqs. 12B and 18 we have

$$J_p(x) = -qD_p(dp/dx) = -qD_p(d\delta p/dx)$$
$$= q(D_p/L_p)\Delta p e^{-x/L_p} = q(D_p/L_p)\delta p(x) \quad \text{(Eq. 19)}$$

Since $p(x) = p_o + \delta p(x)$, the space derivative involves only the excess holes. Notice that since $\delta p(x)$ is proportional to its derivative for an exponential distribution, the diffusion current at any x is just proportional to the excess density δp at that position.

Although this example seems rather restricted, it is typical of carrier injection in pn junctions. The injection of minority carriers across a junction often leads to exponential distributions as in Eq. 18, with the resulting diffusion current of Eq. 19.

pn JUNCTIONS

Most semiconductor devices contain at least one junction between p-type and n-type material. Such junctions are responsible for the injection and collection of charge carriers necessary for the operation of diodes, transistors, and other devices. Junctions are typically formed by crystal growth processes, by alloying, or by ion implantation or diffusion of doping impurities into crystals of the opposite conductivity type. We will consider here an abrupt junction between a uniformly doped p region and a uniformly doped n region of a single crystal semiconductor.

A Junction at Equilibrium

Let us consider separate regions of p- and n-type semiconductor material, brought together to form a junction (Fig. 5). Since the n material has a large

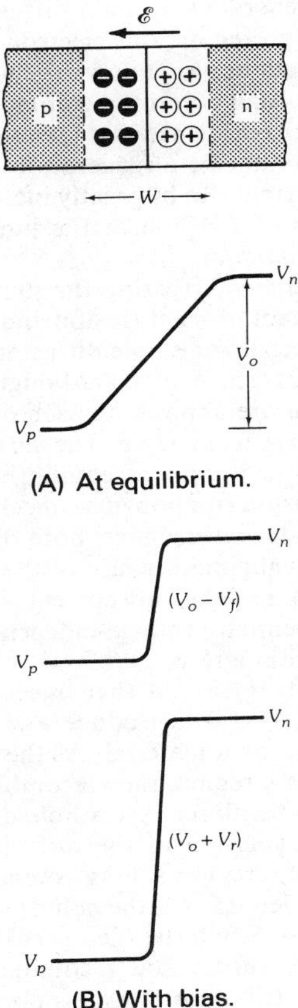

Fig. 5. A pn junction.

concentration of electrons and few holes, and conversely for the p material, we expect considerable diffusion of carriers across the junction. Thus holes diffuse from the p side into the n side, and electrons diffuse from n to p.

Uncharged particles would diffuse throughout the material, resulting in a homogeneous distribution. This cannot occur in the case of the charged particles in a pn junction because of the development of space charge and the electric field ℰ in the neighborhood of the junction. Electrons diffusing from n to p leave behind uncompensated* donor ions (N_d^+) in the n material, and holes leaving the p region leave behind uncompensated acceptors (N_a^-); thus a region of positive space charge develops near the n side of the junction, and negative charge develops near the p side. The resulting electric field is directed from the positive charge toward the negative charge. The influence of ℰ for each type of carrier is in the direction opposite to that of diffusion current. Therefore the field creates a drift component of current from n to p, opposing the diffusion current.

Since no *net* flow of electrons or holes across the junction can take place at equilibrium, the current due to the drift of each type of carrier in the ℰ field must exactly cancel the diffusion current in Eq. 13. Therefore, the electric field ℰ builds up to the point that the net current is zero at equilibrium. The electric field appears in some region W about the junction, and there is an equilibrium potential difference V_o across W. We assume the electric field is zero in the neutral regions outside W. Thus there is a constant potential V_n in the neutral n material, a constant V_p in the neutral p material, and a potential difference $V_o = V_n - V_p$ between the two. The region W is called the *transition region* (also called the *depletion region*), and the potential V_o is called the *contact potential.*

If we consider the junction to be made up of material with N_a acceptors per cubic centimeter on the p side and N_d donors on the n side, we can show that the contact potential is

$$V_o = (kT/q) \ln N_a N_d / n_i^2 \qquad \text{(Eq. 20)}$$

Forward and Reverse Bias

One of the useful features of a pn junction is the fact that current flows quite freely in the p-to-n direction when the p region has a positive external voltage bias relative to n (forward bias and forward current), whereas virtually no current flows when p is made negative relative to n (reverse bias and reverse current). This current asymmetry makes the pn junction diode very useful as a rectifier of ac signals, and forms the basis for many other applications of junctions.

Assume an applied voltage bias V appears across the transition region of the junction rather than in the neutral n and p regions. In typical devices, the neutral regions have low resistance, and the space charge regions about the junction take up most of the applied voltage. Since an applied voltage changes the electrostatic potential barrier and thus the electric field within the transition region, we would expect changes in the various components of current at the junction.

The *electrostatic potential barrier* at the junction is lowered by a forward bias, V_f, from the equilibrium contact potential, V_o, to the smaller value $V_o - V_f$. This lowering of the potential barrier occurs because a forward bias (p positive with respect to n) raises the electrostatic potential on the p side relative to the n side (Fig. 5B). For a reverse bias

*Neutrality is maintained in the bulk materials of Fig. 5 by the presence of one electron for each ionized donor ($n = N_d^+$) in the n material, and one hole for each ionized acceptor ($p = N_a^-$) in the p material (neglecting minority carriers). Thus, if electrons leave n, some of the positive donor ions near the junction are left uncompensated. The donors and acceptors are fixed in the lattice, in contrast to the mobile electrons and holes.

($V = -V_r$) the opposite occurs; the electrostatic potential of the p side is depressed relative to the n side, and the potential barrier at the junction becomes larger ($V_o + V_r$).

The *electric field* within the transition region decreases with forward bias, since the applied electric field opposes the built-in field. With reverse bias, the field at the junction is increased by the applied field.

The change in electric field at the junction calls for a change in the *transition region width, W*, since it is necessary that a proper number of positive and negative charges (in the form of uncompensated donor and acceptor ions) be exposed for a given value of the $\mathcal{E}$ field. Thus, we expect the width, W, to decrease under forward bias (smaller $\mathcal{E}$, fewer uncompensated charges) and to increase under reverse bias.

The *diffusion current* is composed of majority-carrier electrons on the n side surmounting the potential-energy barrier to diffuse the p side, and holes surmounting their barrier from p to n. With forward bias, the barrier is lowered from V_o to $V_o - V_f$, and many more electrons in the n-side conduction band have sufficient energy to diffuse from n to p over the smaller barrier. Therefore, the electron diffusion current can be quite large with forward bias. Similarly, more holes can diffuse from p to n under forward bias because of the lowered barrier. For reverse bias, the barrier becomes so large ($V_o + V_r$) that virtually no electrons in the n-side conduction band or holes in the p-side valence band have enough energy to surmount it. Therefore, the diffusion current is usually negligible for reverse bias.

The *drift current* is relatively insensitive to the height of the potential barrier. The reason for this is that the drift current is limited *not* by *how fast* carriers are swept down the barrier, *but* rather *how often*. For example, minority-carrier electrons on the p side which wander into the transition region will be swept down the barrier of the $\mathcal{E}$ field, giving rise to the electron component of drift current. However, this current is small not because of the size of the barrier, but because there are very few minority electrons in the p side to participate. Every p-side electron that diffuses to the transition region will be swept down the potential energy hill, whether the hill is large or small. Similar comments apply regarding the drift of minority holes from the n side to the p side of the junction. To a good approximation, therefore, the electron and hole drift currents at the junction are independent of the applied voltage.

The supply of minority carriers on each side of the junction required to participate in the drift component of current is generated by thermal excitation of electron-hole-pairs (EHPs). For example, an EHP created near the junction on the p side provides a minority electron in the p material. If the EHP is generated within a diffusion length L_n of the transition region, this electron can diffuse to the junction and be swept down the barrier to the n side. The resulting current is commonly called the *generation current* since its magnitude depends entirely upon the rate of generation of EHP. This generation current can be greatly increased by optical excitation of EHPs near the junction (the pn junction *photodiode*).

The *total current* crossing the junction is composed of the sum of the diffusion and drift components. The electron and hole diffusion currents are both directed from p to n (although the particle flow directions are opposite to each other), and the drift currents are from n to p. The *net* current crossing the junction is zero at equilibrium, since the drift and diffusion components cancel for each type of carrier. Under reverse bias, both diffusion components are negligible because of the large barrier at the junction, and the only current is the relatively small (and essentially voltage-independent) generation current from n to p.

We expect from Fig. 4 that injection of excess holes from p to n will produce a *distribution* of excess holes in the n material. As the holes diffuse deeper into the n region, they recombine with electrons, and the resulting excess hole distribution is obtained as a solution of the diffusion equation, Eq. 16B. If the n region is long compared with the hole diffusion length, L_p, the solution is exponential, as in Fig. 4. Similarly, the injected electrons in the p material diffuse and recombine, giving an exponential distribution of excess electrons. These excess carrier distributions must be accounted for in switching a pn junction from forward bias to reverse bias.

By evaluating the diffusion currents of electrons and holes across the junction, it can be shown that the current-voltage characteristic of the diode is given by

$$I = qA[(D_p/L_p)p_n + (D_n/L_n)\,n_p](e^{qV/kT} - 1) \quad \text{(Eq. 21)}$$

Eq. 21 is called the *diode equation*, which describes the total current through the diode for either forward or reverse bias. We can calculate the current with reverse bias (commonly called the reverse saturation current) by letting $V = -V_r$

$$I = qA[(D_p/L_p)p_n + (D_n/L_n)\,n_p](e^{-qV_r/kT} - 1) \quad \text{(Eq. 22)}$$

If V_r is larger than a few kT/q, the total current is just the reverse saturation current

$$I = -qA[(D_p/L_p)p_n + (D_n/L_n)\,n_p] \quad \text{(Eq. 23)}$$

The *I-V* characteristic for the pn junction is shown in Fig. 6. It exhibits conduction for forward bias (V positive), negligible current for moderate reverse bias, and avalanche breakdown at a large reverse bias (V_{br}).

Reverse Breakdown

The small reverse saturation current typical of reverse bias in a junction is valid until a critical reverse bias is reached, for which *reverse breakdown* occurs (Fig. 6). At this critical voltage (V_{br}),

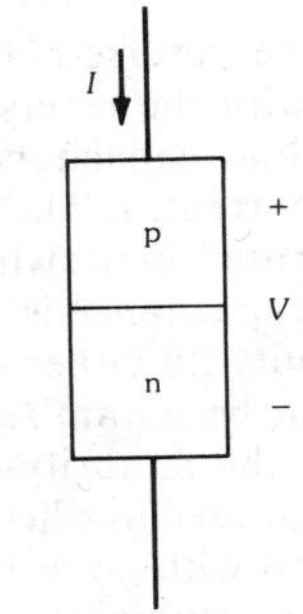

(A) Definitions of directions.

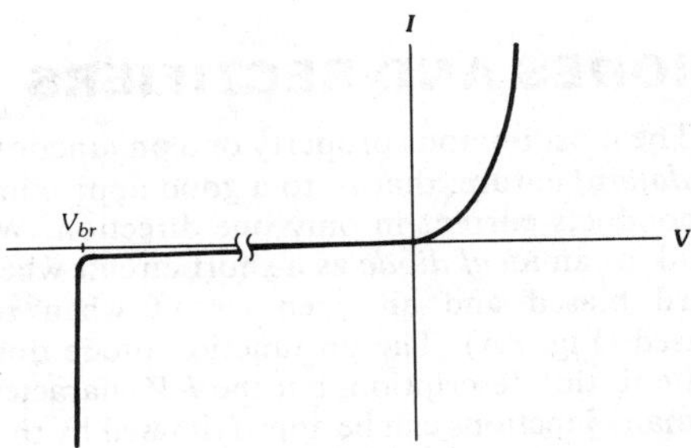

(B) *I-V* characteristic.

Fig. 6. Current and voltage in a pn junction.

the reverse current through the diode increases sharply, and relatively large currents can flow with little further increase in voltage. The existence of a critical breakdown voltage introduces almost a right-angle appearance to the reverse characteristic of most diodes.

There is nothing inherently destructive about reverse breakdown. If the current is limited to a reasonable value by the external circuit, the pn junction can be operated safely in the reverse-breakdown condition. Useful devices called *breakdown diodes* are designed to operate in the reverse-breakdown regime of their characteristics.

Reverse breakdown can occur by two mechanisms, each of which requires a critical electric field in the junction transition region. The first mechanism, called the *zener effect*, is operative at low voltages (up to a few volts reverse bias). If the breakdown occurs at higher voltages (from a few volts to hundreds of volts), the mechanism is *avalanche breakdown.*

Zener breakdown occurs in heavily doped junctions in which the transition from the p side to the n side is very abrupt. Under these conditions, the potential barrier separating carriers on the two sides of the junction is very narrow, and a quantum-mechanical process called *tunneling* can take place. In effect, electrons from the p-side valence band are able to tunnel across the barrier into the empty states in the n-side conduction band when a small reverse bias is applied. The result is a strong current from n to p in the diode, causing zener breakdown. We shall discuss tunneling again in relation to tunnel diodes in a later section.

For diodes with lighter doping or with a graded junction, the avalanche mechanism is the operative breakdown process. In this effect, the electric field in the transition region becomes strong enough under reverse bias to accelerate carriers to quite high kinetic energy. For example, if a minority hole is generated thermally on the n side and wanders into the transition region, it is swept by the junction field to the p side. If the field is high enough, the hole can collide with the lattice with sufficient energy to create an electron-hole pair. As a result, the newly created hole is swept to the p side and the electron is swept to the n side. With one carrier entering the transition region, we in effect have three carriers leaving. This *carrier multiplication* adds to the reverse current. Furthermore, if on the average each carrier (including carriers created by ionizing collisions) can create one EHP during its transit across the transition region, a runaway (avalanche) process develops. Avalanche breakdown usually occurs at a sharply defined breakdown voltage which corresponds to the establishment of a critical field in the transition region. This critical field is that necessary to cause a series of ionizing collisions by carriers in transit across the transition region.

Capacitance and Transient Behavior

Since many pn-junction devices are used in ac circuits or for switching applications, it is important to consider their time-dependent properties. The transient behavior of a junction is influenced strongly by the presence of stored charge. For example, the charge (Q) due to excess minority carriers in a forward-biased junction must be changed as the current varies. Since this charge cannot be altered instantaneously, current and voltage may vary in a complex fashion for the transient case.

The problem of stored charge is particularly important in applications that call for switching a diode from forward conduction to the reverse-biased state. In this case, the distribution of excess minority carriers is swept across the junction, giving rise to a brief pulse of reverse current after the applied voltage has been switched negative (Fig. 7). The time this reverse current ($-I_r$) lasts is called the *storage delay time, t_{sd}*. This time is proportional

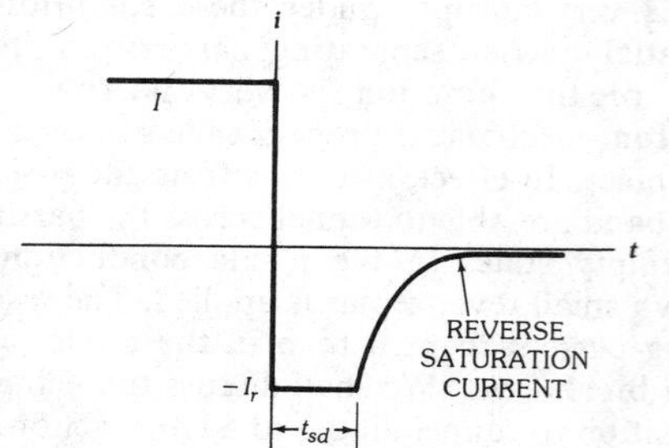

Fig. 7. Reverse recovery transient for a diode switched from forward conduction (I) to reverse bias.

to the minority-carrier lifetime, τ, in the region where the charge is stored. After the charge distribution has properly rearranged to that expected for a reverse-biased junction, the small reverse saturation current is reached. However, the relatively large reverse current during switching must be accounted for in circuit designs. Special switching devices can be constructed which store very little charge in forward bias, and therefore are able to switch from the forward to the reverse state with a minimum of time delay or reverse current.

When ac signals are applied to a junction, the necessity for buildup and depletion of stored charge causes the voltage to lag behind the current. The resulting capacitance is an important junction property. For a reverse-biased junction, the capacitance is due to the changing charge in the depletion region (Fig. 5). As the reverse bias increases, W must increase to accommodate the increased charge on each side of the junction. We calculate the junction capacitances, C_j, from the rate of change of charge within W as the voltage is varied. The result is equivalent to the usual parallel-plate capacitor formula

$$C_j = \epsilon A / W \qquad \text{(Eq. 24)}$$

where,

W is the depletion width,
ϵ is the permittivity of the semiconductor.

Since W depends on the reverse bias, C_j is a *voltage-variable capacitance.* This property can be used in tuning circuits, as discussed below.

For forward bias, ac variations impressed on a steady forward current I (dc) call for variations in stored charge of the minority carrier distribution (Fig. 4). A forward-biased pn diode in which hole injection dominates responds to ac signals with an equivalent RC time constant τ_p required for redistribution of the charge due to holes stored in the n region. The current in a forward-biased junction ($V >> kT/q$) is

$$I = I_s\, e^{qV/kT} \qquad \text{(Eq. 25)}$$

where I_s is the magnitude of the saturation current, described in Eq. 23.

If we take the derivative with respect to V, we find the ac conductance to be

$$G_{ac} = (q/kT)\, I_s\, e^{qV/kT} = (q/kT)\, I \qquad \text{(Eq. 26)}$$

Thus charge-storage capacitance $C_s = \tau_p G$ is

$$C_s = (q\, \tau_p / kT)\, I \qquad \text{(Eq. 27)}$$

In summary, the reverse-bias junction capacitance, C_j, varies with the reverse bias voltage, V_r, and the forward bias capacitance, C_s, varies with the forward bias current, I. The storage delay time is an important limitation in switching circuits, and the forward-bias capacitance is important in high-frequency ac circuits. In either case, the effects of charge storage can be minimized by reducing the carrier lifetime in the minority-carrier storage region. One approach used in silicon devices involves doping the junction with gold, which introduces a high density of recombination centers, thereby reducing carrier lifetime.

DIODES AND RECTIFIERS

The most obvious property of a pn junction is its *unilateral* nature; that is, to a good approximation it conducts current in only one direction. We can think of an *ideal diode* as a short circuit when forward biased and an open circuit when reverse biased (Fig. 8A). The pn junction diode does not quite fit this description, but the *I-V* characteristics of many junctions can be approximated by the ideal diode in series with other circuit elements to form an equivalent circuit. For example, most forward-biased diodes exhibit an *offset voltage,* E_o, which can be approximated in a circuit model by a battery in series with the ideal diode (Fig. 8B). The series battery in the model keeps the ideal diode turned off for applied voltages less than E_o. In some cases, the approximation to the actual diode characteristic is improved by adding a series resistor, R. The values of E_o and R depend on the particular diode, and can be obtained from the actual *I-V* characteristic of the device. The circuit approximations illustrated in Fig. 8B are called *piecewise-linear equivalents,* since the approximate characteristics are linear over specific ranges of voltage and current.

An ideal diode can be placed in series with an ac voltage source to provide *rectification* of the signal. Since current can flow only in the forward direction through the diode, only the positive half-cycles of an input sine wave are passed to the load. During the negative half-cycles, the diode is nonconducting, and the resulting voltage across a load resistor is zero. The output voltage is a *half-rectified sine wave.* The rectified signal has an average value, and therefore contains a dc component. By appropriate filtering, this dc level can be extracted from the rectified signal.

The unilateral nature of diodes is useful for many

other circuit applications that require *waveshaping*. This involves alteration of ac signals by passing only certain portions of the signal while blocking other portions.

Other properties of pn junctions can be exploited for electronic-device applications. For example, the *breakdown diode* (sometimes called *zener diode*) makes use of the abrupt reverse breakdown shown in Fig. 6. The breakdown voltage is strongly dependent on doping, and junctions can be made with accurately chosen values of V_{br}. Breakdown diodes can be used as *voltage regulators* in circuits with varying inputs. Such a device can also be used as a *reference diode*; since the breakdown voltage of a particular diode is known, the voltage across it during breakdown can be used as a reference in circuits that require a known value of voltage.

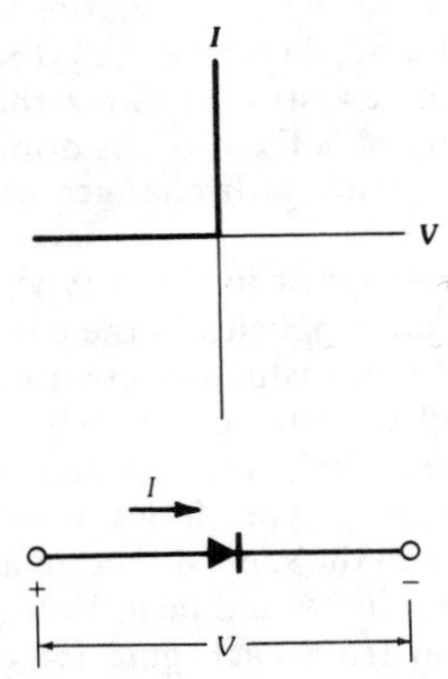

(A) Behavior of an ideal diode.

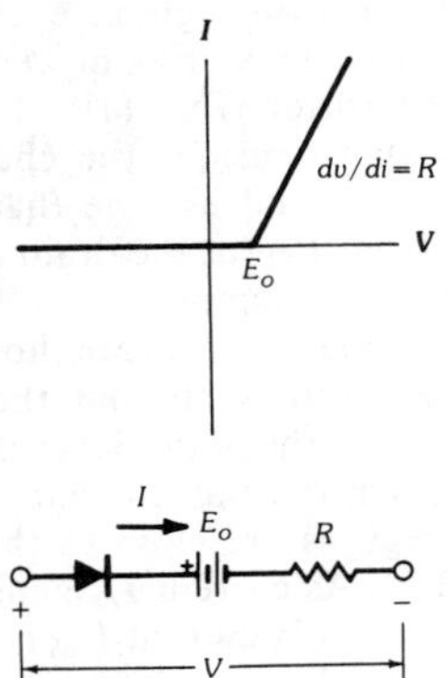

(B) Piecewise linear model of pn junction.

Fig. 8. Semiconductor diode.

The junction capacitance described by Eq. 24 can be put to use in a device called a *varactor*. Since C_j is inversely proportional to the depletion width, W, and W varies with the reverse bias, V_r, the junction is a *voltage-variable capacitor*. For an abrupt junction, C_j is proportional to $V_r^{-1/2}$. Thus it is possible to vary C_j by choosing values of the applied bias. A set of varactors can be used, for example, to replace variable-plate capacitors in the tuning section of a radio receiver. By proper adjustment of doping profiles in the junction, the voltage dependence of C_j can be tailored for the specific application.

In abrupt pn junctions that are very heavily doped on both sides, current can be carried by the quantum-mechanical process called tunneling. In reverse bias, a tunnel diode passes current at very small voltage, and the same occurs for a fraction of a volt of forward bias (Fig. 9). However, at a critical forward voltage (V_p), the tunneling decreases, and eventually the usual forward diode current dominates. As Fig. 9 indicates, a range of the *I-V* characteristic exists in which the tunneling current decreases with increasing voltage, giving rise to a *negative resistance* region. Such a negative resistance can be used in various switching, oscillation, amplification, and other circuit functions.

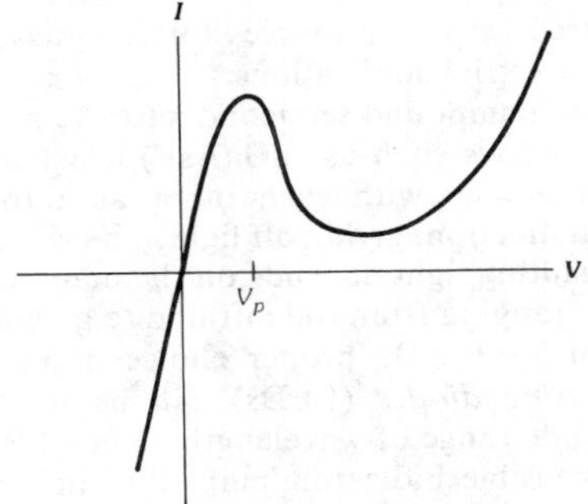

Fig. 9. *I-V* characteristic for a tunnel diode.

When light shines on a pn junction, electron-hole pairs are generated by optical excitation. The collection of these optically generated charge carriers across the junction results in a net excess current from n to p (i.e., *I-V* curve A in Fig. 10 is moved down to curve B). The shift of the *I-V* curve is proportional to the optical generation rate of excess electrons and holes, and therefore to the intensity of the light. In the third quadrant (III), the junction can be used as a *photodiode*, in which the reverse current at a fixed reverse bias voltage is proportional to the intensity of the light. When the junction operates in the fourth quadrant of the *I-V* curve (IV), power is delivered from the junction to the external load. (Notice that at IV, the diode current is negative while the junction voltage is positive. Thus, the current in this case flows from minus to plus relative to the voltage, as in a battery.) In this case, the pn junction can be used as a *solar cell* (or *photovoltaic junction*). Such solar cells are

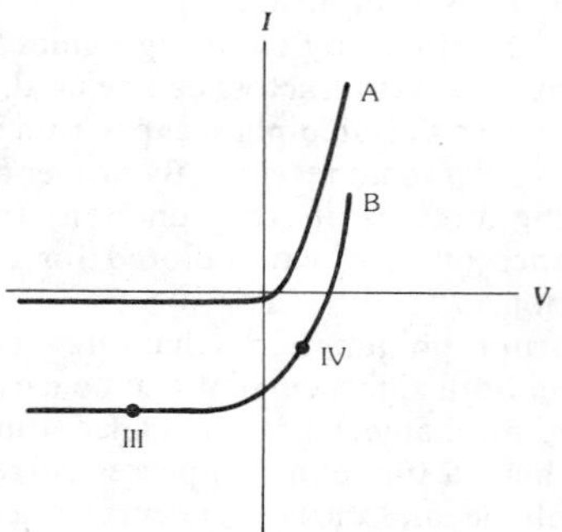

Fig. 10. *I-V* characteristics for a photodiode in the dark (A) and with light applied (B).

widely used as power sources for satellites and in certain terrestrial applications.

In many compound semiconductors (e.g., GaAs, InP, and alloys such as InGaAsP), electrons and holes recombine with each other at a forward-biased pn junction, giving off light. The wavelength of the resulting light depends on the material used, generally ranging from the infrared (e.g., GaAs) to the green (GaP). By proper choice of materials, *light-emitting diodes* (LEDs) can be made that cover a wide range of wavelengths. Such LEDs are useful as visible indicator lamps and in alphanumeric displays. In addition, infrared LEDs can be used in optical communication systems in which light signals are sent over optical fibers and are detected at the other end with a photodiode. Such optoelectronic systems can be made very efficient if the compound semiconductor diode is made in the form of a laser and the photodetector operates in the reverse-bias avalanche condition. For further discussion, see Chapters 21 and 22.

TRANSISTORS

The basic building block of modern electronics is the transistor. This device, which has replaced the vacuum tube in all but a few special applications, is used in most circuits requiring amplification or switching. Complex integrated circuits are also based on the transistor. This section deals with the operation of three transistor types—the junction field-effect transistor, the MOS field-effect transistor, and the bipolar junction transistor. We concentrate here on the internal operation of these devices, reserving treatments of their circuit operation for later chapters. Transistor fabrication methods are discussed in Chapter 20.

As an amplifying device, the transistor converts weak time-varying signals into strong signals. As a switching element, it can be changed from a conducting state to a nonconducting state quickly with the application of very little control power. In the field-effect transistor, current through two terminals is varied by voltage applied to a third terminal. In the bipolar transistor, the current through two terminals is controlled by a small current applied to a third terminal.

Junction Field-Effect Transistors

In Fig. 5, the width of the depletion region (W) under reverse bias is controlled by the voltage applied to the junction. This property is the basis for the operation of the *junction field-effect transistor (JFET)*. In a JFET, the voltage-variable depletion-region width of a junction is used to control the effective cross-sectional area of a conducting *channel*. In the device of Fig. 11, current I_D flows through an n-type channel between two p regions. A reverse bias between these p regions and the channel causes the depletion regions to intrude into the n material, and therefore the effective width of the channel can be restricted. Since the resistivity of the channel region is fixed by its doping, the channel resistance varies with changes in the effective cross-sectional area.

In Fig. 11, electrons in the n-type channel drift from right to left, opposite to the current. The end of the channel from which electrons flow is called the *source*, and the end toward which they flow is called the *drain*. The p regions are called *gates*. If the channel were p-type, holes would flow from source to drain, in the same direction as the current, and the gate regions would be n. Voltage V_{GS} refers to the potential from each gate region, G, to the source, S. Since the conductivity of the heavily doped gate regions is high, we can assume that the potential is uniform throughout each gate. In the lightly doped channel material, however, the potential varies with position (Fig. 11B). For low values of current, we can consider the channel to be a distributed resistor, and assume that voltage V_{xS} varies linearly from V_{DS} at the drain end to zero at the source end of the channel.

In Fig. 12, assume the gates are short-circuited to the source ($V_{GS} = 0$), such that the potential at $x = L$ is the same as the potential everywhere in the gate regions. For very small currents, the widths of the depletion regions are close to the equilibrium values (Fig. 11A). As current I_D is increased, however, it becomes important that V_{xS} is large near the drain end and small near the source end of the channel. Since the reverse bias across each point in the gate-to-channel junction ($-V_{Gx}$) is simply V_{xS} when V_{GS} is zero, we can estimate the shape of the depletion regions as in Fig. 12A. The reverse bias is relatively large near the drain ($-V_{GD} = V_{DS}$) and decreases toward zero near the source. As a result, the depletion region intrudes into the channel near the drain, and the effective channel area is constricted. Since the resistance of the constricted channel is higher, the *I-V* plot for the channel be-

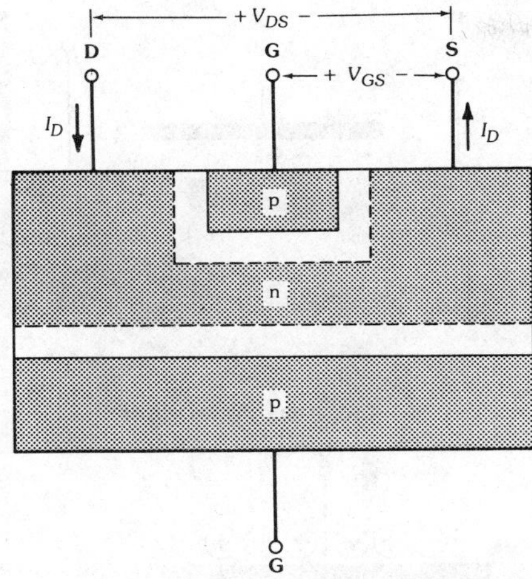

(A) Transistor geometry.

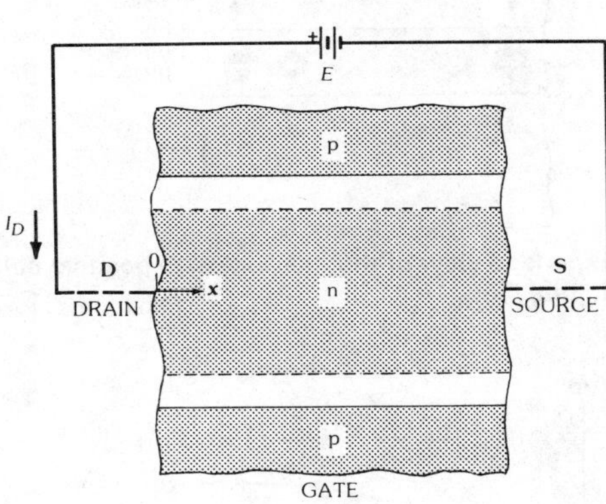

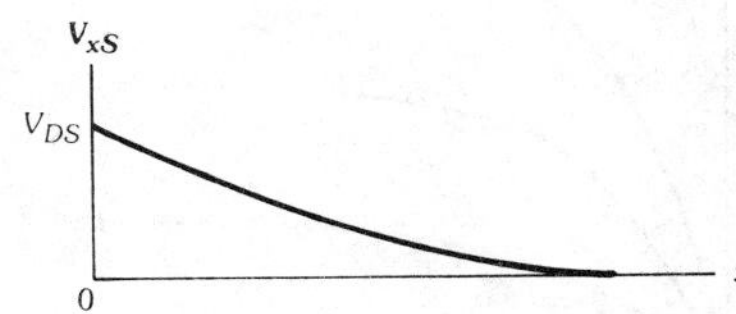

(B) Detail of channel and voltage variation along channel with $V_{GS} = 0$ and small I_D.

Fig. 11. Simplified cross-sectional view of a junction FET. (*From Ben G. Streetman,* Solid State Electronic Devices, *2nd ed.,* © *1980. Reprinted by permission of Prentice-Hall, Inc., Englewood Cliffs, N.J.*)

gins to depart from the straight line that was valid at low current levels. As V_{DS} is increased, there must be some bias voltage at which the depletion regions meet near the drain and essentially *pinch off* the channel. When this happens, current I_D cannot increase significantly with further increase in V_{DS}. For higher voltages, the current is *saturated* approximately at its value at pinch-off.

The effect of a negative gate bias, $-V_{GS}$, is to increase the resistance of the channel and induce pinch-off at a lower value of current (Fig. 12B). Since the depletion regions are larger with V_{GS} negative, the slopes of the I_D-versus-V_{DS} curves below pinch-off become smaller. As current I_D increases, the pinch-off condition is reached at a lower drain-to-source voltage, and the saturation current is lower than for the case of zero gate bias. As V_{GS} is varied, a family of curves is obtained for the *I-V* characteristic of the channel, as in Fig. 12B. By varying the gate bias, we can obtain amplification of an ac signal or switch the device from its off ($I_D = 0$) to its on condition. Since the input control voltage, V_{GS}, appears across the reverse-biased gate junctions, the input impedance of the device is relatively high.

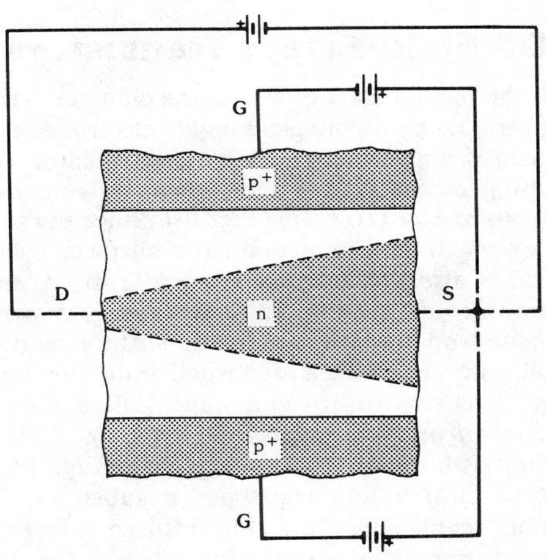

(A) Variation of depletion-region widths.

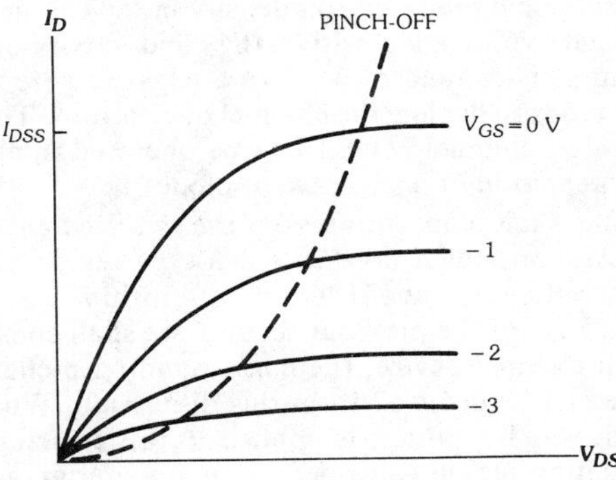

(B) Family of current-voltage curves for the channel as V_{GS} is varied.

Fig. 12. Effects of negative gate bias in the JFET.

The JFET is generally used in the saturation region of its *I-V* characteristic (i.e., beyond pinch-off). The value of I_D at saturation (called I_{DS}) depends on V_{GS}, as Fig. 12B indicates. To a good approximation, I_{DS} beyond pinch-off can be written as follows:

$$I_{DS} \simeq I_{DSS}(1 + V_{GS}/V_p)^2 \qquad (V_{GS}\text{ negative}) \qquad \text{(Eq. 28)}$$

In this expression, I_{DSS} is the saturation value with the gate shorted to the source (i.e., with $V_{GS} = 0$). The term V_p is the pinch-off voltage defined as the gate bias required to pinch off the channel with $I_D = 0$. The quantity V_p is dictated by the device geometry and doping. For an n-channel device, V_{GS} is a negative number. Thus, Eq. 28 indicates that the saturation drain current, I_{DS}, is greatest (I_{DSS}) when V_{GS} is zero, and decreases to zero when V_{GS} becomes as negative as pinch-off voltage V_p is positive.

MOS Field-Effect Transistors

In the second class of FETs, the channel current is controlled by a voltage at a gate electrode that is isolated from the channel by an insulator. The resulting device is called an *insulated-gate field-effect transistor* (*IGFET*) (Fig. 13). Since the channel exists at the semiconductor surface, such a device is also called a *surface FET*. In the most common configuration, an oxide layer is grown or deposited on the semiconductor surface, and the metal gate electrode is deposited onto this oxide layer. This structure is commonly called a *metal-oxide-semiconductor transistor* (*MOST*).

The n source and drain regions of Fig. 13 are diffused into a high-resistivity p substrate. The channel region may be a thin diffused n layer, or more commonly an *induced inversion region*. If an n-type diffused channel is included between source and drain, the effect of the field is to raise or lower the conductance of the channel by either depleting or enhancing the electron density in the channel. If the gate voltage is positive, the conductivity of the channel is enhanced, whereas a negative gate voltage tends to deplete the channel of electrons. Thus a diffused-channel MOST can be operated in either the *depletion* or *enhancement* mode.

We shall concentrate on the *induced-channel MOST*, in which no diffused n-type region exists between source and drain at equilibrium. As with the JFET of the previous section, we shall consider an n-channel device; the corresponding p-channel case can be deduced from this discussion. When a positive gate voltage is applied to this structure, a depletion region is formed in the p material, and a thin layer of mobile electrons is drawn from the source and drain into the channel (Fig. 14A). Where the mobile electrons dominate, the material is effectively n-type. This is called an *inversion layer*, since that material was originally p-type. Once an inversion layer is formed near the semiconductor surface, a conducting channel exists from source to drain. The operation of the device is

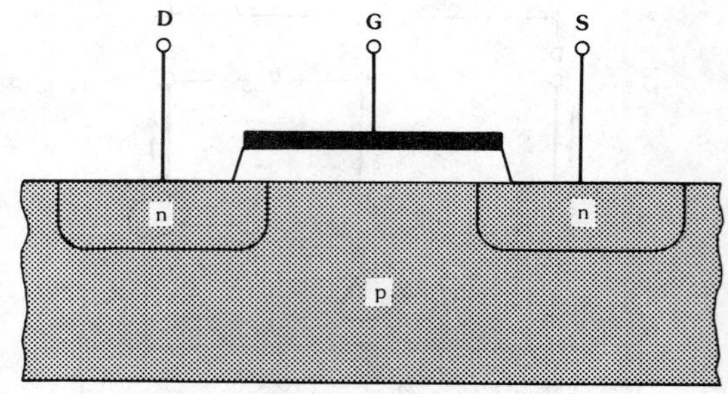

(A) Cross-sectional view.

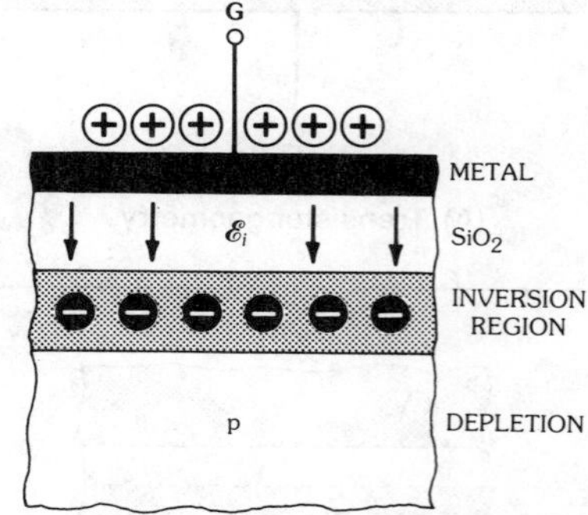

(B) Expanded view of the channel with positive polarity of gate bias.

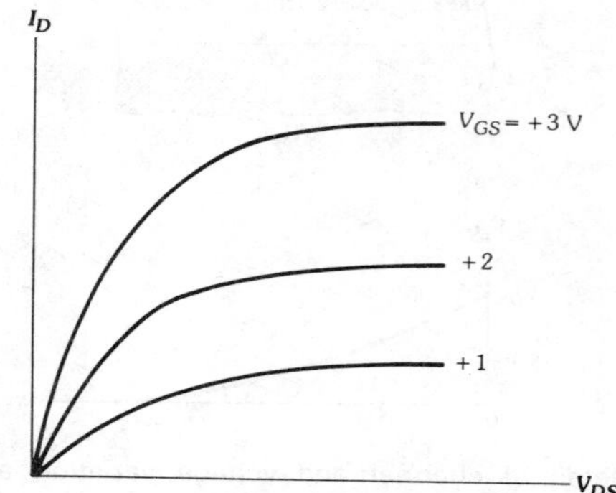

(C) *I*-*V* characteristics as gate bias is varied.

Fig. 13. The induced-channel MOST.

then quite similar to that of the JFET. The channel conductance is controlled by field $\mathcal{E}_i$ in the insulator, but the magnitude of this field varies along the channel as V_{Gx} varies from $V_{GS} - V_{DS}$ at the drain to V_{GS} at the source (Fig. 14C); this corresponds qualitatively to the variation in junction reverse bias along the channel for the JFET. Since a positive voltage is required between the gate and each point x in the channel to maintain inversion, a large enough value of V_{DS} can cause $\mathcal{E}_i$ to go to zero at the drain. This corresponds to the pinch-off condition. Once pinch-off is reached, the saturation cur-

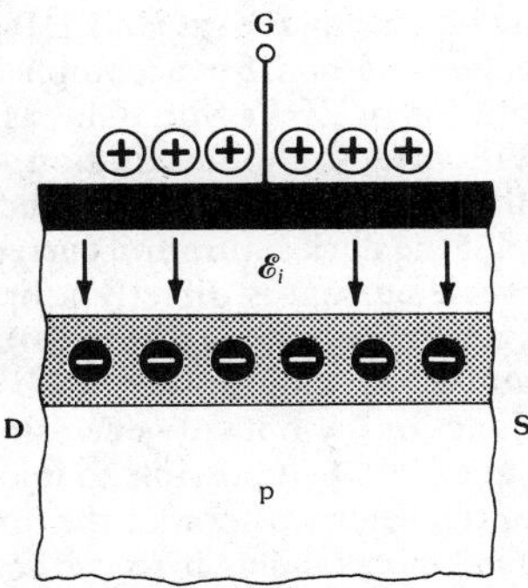

(A) Channel with positive gate bias, low current.

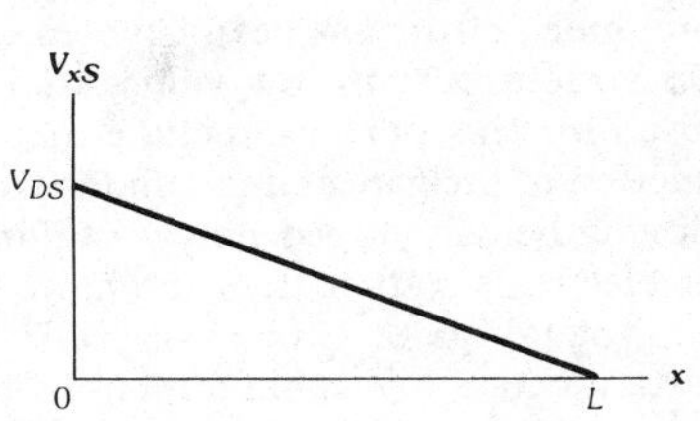

(B) Voltage variation along channel for low I_D.

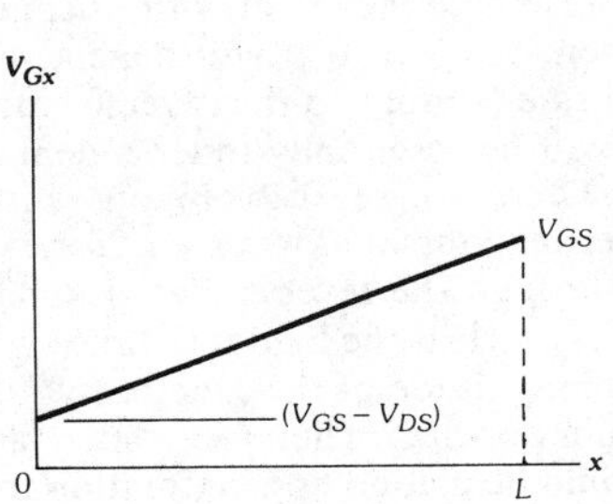

(C) Voltage from gate to channel as a function of *x* near pinch-off.

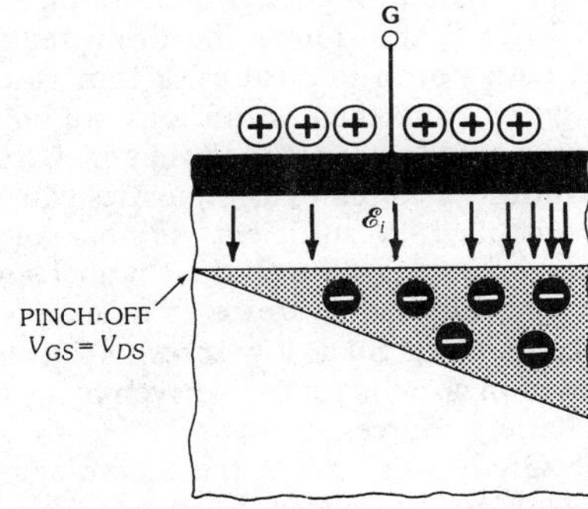

(D) View of channel at pinch-off.

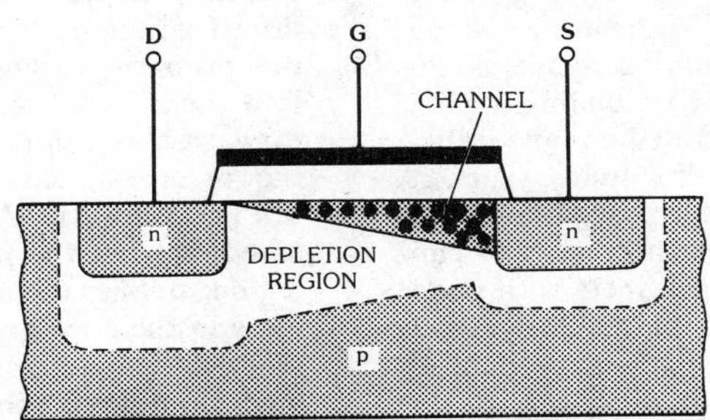

(E) Shape of channel and depletion region at pinch-off.

Fig. 14. Properties of an induced-channel MOST.

rent remains essentially constant, as in the JFET.

The saturation drain current for the MOST is described by an equation of the same form as Eq. 28 for the JFET. For the MOS transistor with an induced channel,

$$I_{DS} = (\mu_n C_o / 2L^2)(V_{GS} - V_T)^2 \qquad \text{(Eq. 29)}$$

where,

μ_n is the electron mobility in the channel,
C_o is the capacitance between gate and channel,
L is the length of the channel from source to drain,
V_T is the *threshold voltage*, the gate bias that must be applied to induce a channel in the underlying semiconductor.

The value of the threshold voltage is dependent on several properties of the MOS structure. For example, differences in work function between the metal gate and the semiconductor must be balanced by the applied gate voltage. Furthermore, trapped charges at the semiconductor-oxide interface and in the oxide layer may induce image charges in the channel region that must also be overcome by the applied gate voltage. In many cases, several volts must be applied to the gate to create an inversion region (and therefore a conducting channel) in the semiconductor. The existence of such a threshold voltage can be a serious limitation in circuits that are designed to operate with low-voltage batteries. Therefore, much effort has been devoted to designing MOS structures with low values of V_T. Techniques used for this purpose include matching the work function of the gate and the underlying silicon by employing heavily doped polycrystalline silicon as a gate electrode, rather than the more common aluminum metal gate. Silicon crystals oriented along certain crystallographic directions that minimize surface-charge effects are also used to reduce V_T. One of the most successful methods for controlling the threshold voltage is the use of ion implantation in the channel region. For example, in a p-channel MOST, the source and drain regions are diffused with boron impurities, a thin gate oxide layer is grown, and then boron ions are implanted through the oxide region into the channel. Since the concentration of implanted impurities can be precisely controlled, a sufficient number of boron atoms can be implanted into the channel such that the gate bias required to create an inversion region is greatly reduced. Such low-threshold transistors can be used in circuits that are driven by quite low-voltage battery sources.

The MOS transistor has the advantage of extremely high input impedance between the gate and source electrodes, since these terminals are separated by an oxide layer. The input impedance of a MOST can be on the order of 10^{14} ohms, making this device very useful in amplifying signals delivered by circuits that are sensitive to loading.

The MOST is useful in integrated circuits utilizing silicon planar technology. For many years, it was difficult to control the surface charge and other effects in MOS structures, but these devices can now be manufactured in large numbers with good reproducibility, as discussed in Chapter 20.

Bipolar Junction Transistors

The BJT is a current-controlled device, in which the current through two terminals is controlled by a relatively small current in a third terminal. We shall begin the discussion by considering the reverse-biased pn junction. As discussed earlier, the reverse saturation current through a junction depends on the rate at which minority carriers are generated in the neighborhood of the junction. For example, the reverse current due to holes being swept from n to p is essentially independent of the size of the junction $\mathcal{E}$ field and hence independent of the reverse bias. The reason for this is that the hole current depends on how often minority holes are generated by EHP creation within a diffusion length of the junction, and not upon how fast a particular hole is swept across the depletion layer by the field. As a result, it is possible to increase the reverse current through the diode by increasing the rate of EHP generation. One convenient method for accomplishing this is optical excitation of EHPs with light, as in the photodiode. With steady photoexcitation, the reverse current will still be essentially independent of bias voltage, and if the dark saturation current is negligible, the reverse current is directly proportional to the optical generation rate, g_{op} (Fig. 10).

The example of external control of current through a junction by optical generation raises an interesting question: Is it possible to inject minority carriers into the neighborhood of the junction *electrically* instead of optically? If so, we could control the junction reverse current by simply varying the rate of minority-carrier injection. For example, let us consider a hypothetical hole-injection device that can inject holes at a predetermined rate into the n side of the junction. The effect on the junction current will resemble the effects of optical generation. The current from n to p will depend on the hole injection rate (similar to the reverse current in Fig. 10) and will be essentially independent of the bias voltage. There are several obvious advantages to such external control of a current; for example, the current through the reverse-biased junction would change very little if the load resistance were altered, since the magnitude of the junction voltage is relatively unimportant. Therefore, such an arrangement should be a good approximation to a controllable constant-current source.

A convenient hole-injection device is a forward-biased p^+n junction (p^+ refers to a very heavily doped p region). The current in such a junction is due primarily to holes injected from the p^+ region into the n material. If we make the n side of the forward-biased junction the same as the n side of the reverse-biased junction, the p^+np structure of Fig. 15 results. With this configuration, injection of holes from the p^+n junction into the center n region supplies the minority-carrier holes to participate in the current through the reverse-biased np junction. Of course, it is important that the injected holes do not recombine in the n region before they can diffuse to the depletion layer of the reverse-biased junction. Thus we must make the n region narrow compared with a hole diffusion length.

The structure we have described is a pnp bipolar junction transistor. The forward-biased junction that injects holes into the center n region is called the *emitter junction*, and the reverse-biased junction that collects the injected holes is called the *collector junction*. The p^+ region that serves as the source of injected holes is called the *emitter*, and the p region into which the holes are swept by the reverse-biased junction is called the *collector*. The center n region is called the *base*. The biasing arrangement of Fig. 15 is called the *common-base* configuration, since the base electrode (B) is common to the emitter and collector circuits.

To have a good pnp transistor, almost all of the holes injected by the emitter into the base should be

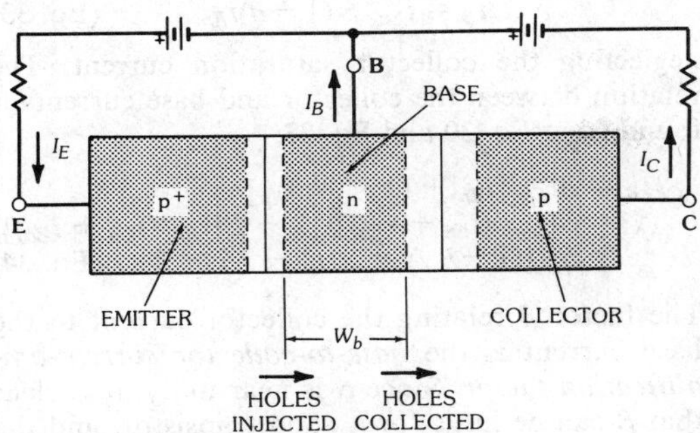

(A) Schematic representation of pnp device with forward-biased emitter junction and reverse-biased collector junction.

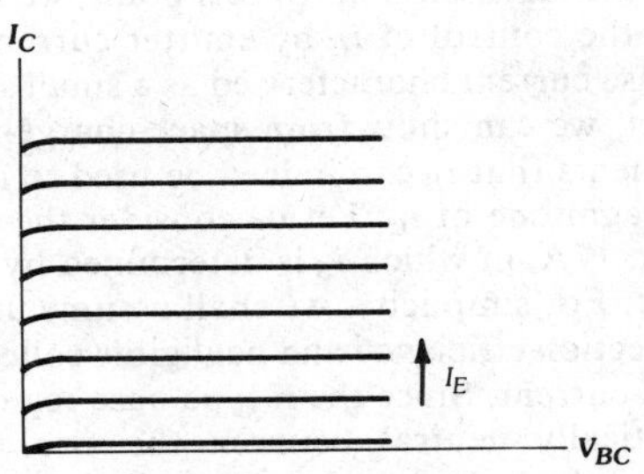

(B) *I-V* characteristics of the reverse-biased np junction as a function of emitter current.

Fig. 15. A pnp transistor. (*From Ben G. Streetman,* Solid State Electronic Devices, *2nd ed., © 1980. Reprinted by permission of Prentice-Hall, Inc., Englewood Cliffs, N.J.*)

collected. Thus the n-type base region should be narrow, and the hole lifetime, τ_p, should be long. This requirement is summed up by specifying $W_b << L_p$, where W_b is the length of the *neutral* n material of the base (measured between the depletion regions of the emitter and collector junctions) and L_p is the diffusion length for holes in the base, $(D_p\tau_p)^{1/2}$. With this requirement satisfied, an average hole injected at the emitter junction will diffuse to the depletion region of the collector junction without recombination in the base. A second requirement is that current I_E crossing the emitter junction should be composed almost entirely of holes injected into the base, rather than electrons crossing from base to emitter. This requirement is satisfied by doping the base region lightly compared with the emitter, so that the p^+n emitter junction of Fig. 15 results.

It is clear that current I_E flows into the emitter of a properly biased pnp transistor and that I_C flows out at the collector, since the direction of hole flow is from emitter to collector. However, base current I_B requires a bit more thought; I_B flows out since this current supplies electrons to the base region. In a good transistor, the base current will be very small since I_E is essentially hole current and the collected hole current, I_C, is almost equal to I_E. There must be some base current, however, due to requirements of electron flow into the n-type base region (Fig. 16). We can account for I_B physically by three dominant mechanisms:

(A) There must be some recombination of injected holes with electrons in the base, even with $W_b << L_p$. The electrons lost to recombination must be resupplied through the base contact.

(B) Some electrons will be injected from n to p in the forward-biased emitter junction, even if the emitter is heavily doped compared to the base. These electrons must also be supplied by I_B.

(C) Some electrons are swept into the base at the reverse-biased collector junction due to thermal generation in the collector. This small current reduces I_B by supplying electrons to the base.

The dominant mechanism in the base current is usually recombination, and we can often approximate the base current by calculating the recombination rate in the base. In a well-designed transistor, I_B will be a very small fraction (perhaps one-hundredth) of I_E.

In an npn transistor, the three current directions are reversed, since electrons flow from emitter to collector and holes must be supplied to the base.* The physical mechanism for operation of the npn device can be understood simply by reversing the roles of electrons and holes in the pnp discussion.

The BJT is useful in amplifiers because the currents at the emitter and collector are controllable by the relatively small base current. The essential mechanisms are easy to understand if various secondary effects are neglected. We shall use total current (dc plus ac) in this discussion, with the understanding that the simple analysis applies only to dc and to small-signal ac at low frequencies. We can relate the terminal currents of the transistor (i_E, i_B, and i_C) by several important factors. In this introduction, we shall neglect the saturation current at the collector (3 in Fig. 16) and such effects as recombination in the transition regions. Under these assumptions, the collector current is made up entirely of those holes injected at the emitter that are not lost to recombination in the base. Thus i_C is proportional to the hole component of the emitter current i_{Ep}:

$$i_C = Bi_{Ep} \qquad \text{(Eq. 30)}$$

Proportionality factor B is simply the fraction of

*In the present discussion of device operation, we show the directions of the currents as they actually flow in normal operation. In circuit analysis, however, it is common to define all currents as flowing *into* the transistor, and incorporate minus signs where they are needed.

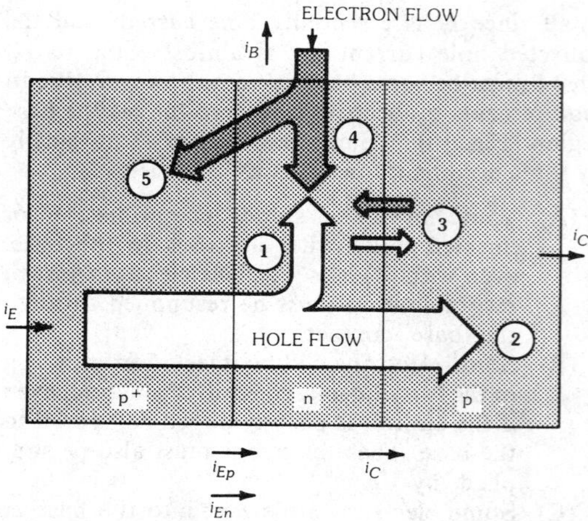

1 Injected holes lost to recombination in the base
2 Holes reaching the reverse-biased collector junction
3 Thermally generated electrons and holes making up the reverse saturation current of the collector junction
4 Electrons supplied by the base contact for recombination with holes
5 Electrons injected across the forward-biased emitter junction

Fig. 16. Hole and electron flow in a pnp transistor with normal biasing. (*From Ben. G. Streetman,* Solid State Electronic Devices, *2nd ed.,* © *1980. Reprinted by permission of Prentice-Hall, Inc., Englewood Cliffs, N.J.*)

injected holes that make it across the base to the collector; B is called the *base transport factor*. The total emitter current, i_E, is made up of the hole component, i_{Ep}, and an electron component, i_{En}, due to electrons injected from base to emitter (5 in Fig. 16). The *emitter injection efficiency*, γ, is

$$\gamma = i_{Ep}/(i_{En} + i_{Ep}) \qquad \text{(Eq. 31)}$$

For an efficient transistor, B and γ should each be very near unity; that is, the emitter current should be due mostly to holes ($\gamma \simeq 1$), and most of the injected holes should eventually participate in the collector current ($B \simeq 1$). The relation between the collector and emitter currents is

$$i_C/i_E = Bi_{Ep}/(i_{En} + i_{Ep}) = B\gamma \qquad \text{(Eq. 32)}$$

The product $B\gamma$ is defined as the factor α, which represents the emitter-to-collector current amplification. There is no real amplification between these currents, since α is smaller than unity. On the other hand, the relation between i_C and i_B is more promising for amplification.

In accounting for the base current, we must include the rates at which electrons are lost from the base by injection across the emitter junction (i_{En}) and the rate of electron recombination with holes in the base. In each case, the lost electrons must be resupplied through the base current, i_B. If the fraction of injected holes making it across the base *without* recombination is B, then it follows that $(1-B)$ is the fraction *recombining* in the base. Thus the base current is

$$i_B = i_{En} + (1-B)i_{Ep} \qquad \text{(Eq. 33)}$$

neglecting the collector saturation current. The relation between the collector and base currents is found from Eq. 30 and Eq. 33:

$$\begin{aligned} i_C/i_B &= Bi_{Ep}/[i_{En} + (1-B)\, i_{Ep}] \\ &= B[i_{Ep}/(i_{En} + i_{Ep})]/\{1 - B[i_{Ep}/(i_{En} + i_{Ep})]\} \\ &= B\gamma/(1-B\gamma) = \alpha/(1-\alpha) \equiv \beta \qquad \text{(Eq. 34)} \end{aligned}$$

The factor β relating the collector current to the base current is the *base-to-collector current-amplification factor*. Since α is near unity, it is clear that β can be large for a good transistor, and the collector current is large compared with the base current.

It remains to be shown that collector current i_C can be controlled by variations in the small current i_B. In the discussion up to this point, we have indicated the control of i_C by emitter current i_E, with the base current characterized as a small side effect. In fact, we can show from space-charge-neutrality arguments that i_B can indeed be used to determine the magnitude of i_C. Let us consider the transistor of Fig. 17A, in which i_B is determined by a biasing circuit. For simplicity, we shall assume unity emitter injection efficiency and negligible collector saturation current. Since the n-type base region is electrostatically neutral between the two transition regions, the presence of excess holes in transit from emitter to collector calls for compensating excess electrons from the base contact. However, there is an important difference in the times which electrons and holes spend in the base. The average hole spends a time τ_t, defined as the *transit time* from emitter to collector. Since the base width W_b is made small compared with L_p, this transit time is much less than the average hole lifetime, τ_p, in the base. On the other hand, an average excess electron supplied from the base contact spends τ_p seconds in the base (for simple recombination and equal excess carrier densities, τ_n and τ_p are equal). While the average electron waits τ_p seconds for recombination, many holes can enter and leave the base region, each with an average transit time τ_t. In particular, for each electron entering from the base contact, τ_p/τ_t holes can pass from emitter to collector while maintaining space-charge neutrality. Thus the ratio of collector current to base current is simply

$$i_C/i_B = \beta = \tau_p/\tau_t \qquad \text{(Eq. 35)}$$

for $\gamma = 1$ and negligible collector saturation current.

If the electron supply to the base (i_B) is restricted, the traffic of holes from emitter to base is correspondingly reduced. This can be argued in a simple way by supposing that the hole injection does continue despite a reduction of electrons from the base contact. The result would be a net buildup of positive charge in the base and a loss of forward bias (and therefore a loss of hole injection) at the emitter junction. Clearly, the supply of electrons through i_B

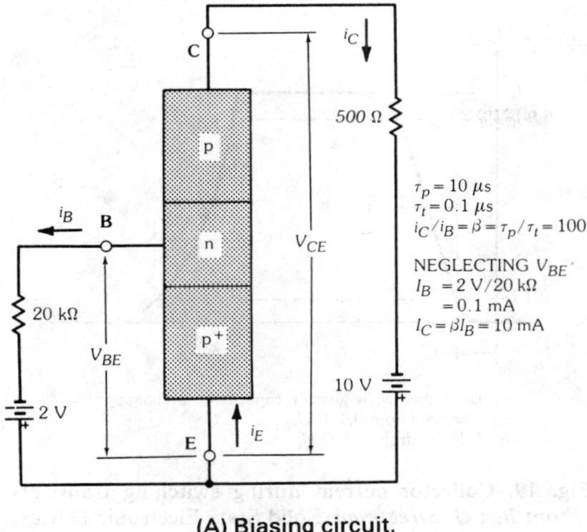

(A) Biasing circuit.

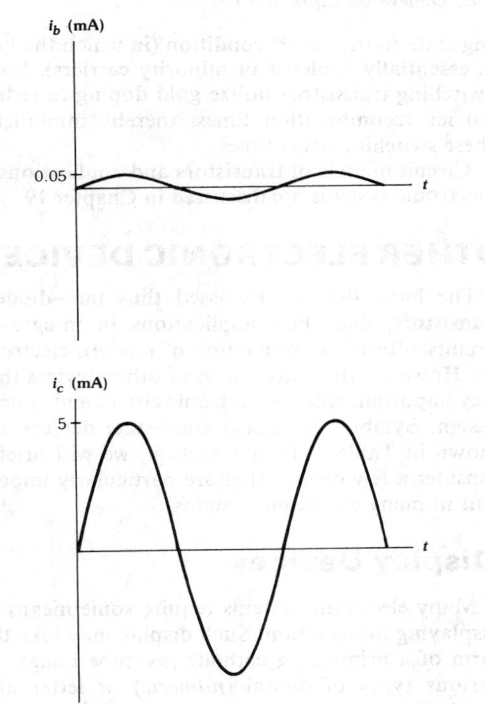

(B) Addition of ac variation of base current.

Fig. 17. Example of amplification in a common-emitter transistor circuit.

can be used to raise or lower the hole flow from emitter to collector.

The base current is controlled independently in Fig. 17A. This is called a *common-emitter* circuit, since the emitter electrode is common to the base and collector circuits. The emitter junction is clearly forward biased by the battery in the base circuit. The voltage drop in the forward-biased emitter junction is small, however, so that almost all of the voltage from collector to emitter appears across the reverse-biased collector junction. Since v_{BE} is small for the forward-biased junction, we can neglect it and approximate the base current as 2 V/20 kΩ = 0.1 mA. If we assume $\tau_t = 0.1\ \mu s$ and $\tau_p = 10\ \mu s$, β for the transistor is 100, and the collector current i_C is 10 mA. It is important to notice that i_C is determined by β and the base current, rather than by the battery and resistor in the collector circuit (as long as these are of reasonable values to maintain a reverse-biased collector junction). In this example 5 V of the collector-circuit battery voltage appears across the 500 Ω resistor, and 5 V serves to reverse-bias the collector junction.

If a small ac current, i_b, is superimposed on the steady-state base current of Fig. 17, a corresponding ac current, i_c, appears in the collector circuit (Fig. 17B). The time-varying portion of the collector current will be larger than i_b by the factor β, and current gain results.

In a switching operation, a transistor is usually controlled in two conduction states that can be loosely referred to as the *on* state and the *off* state. Ideally, a switch should appear as a short circuit when turned on and an open circuit when turned off. Furthermore, it is desirable to switch the device from one state to the other with no lost time in between. Transistors do not fit this ideal description of a switch, but they can serve as a useful approximation in practical electronic circuits. The two states of a transistor in switching can be seen in the simple common-emitter example of Fig. 18. In this figure, collector current i_C is controlled by base current i_B over most of the family of characteristic curves. The *load line* specifies the locus of allowable (i_C, $-v_{CE}$) points for the circuit. The load line is obtained by writing a loop equation around the collector circuit, which results in a linear equation with an intercept on the $-V_{CE}$ axis ($I_C = 0$) and an intercept on the i_C axis ($V_{CE} = 0$).

If I_B is such that the operating point lies somewhere between the two end points of the load line (Fig. 18B), the transistor operates in the *normal active mode*. That is, the emitter junction is forward biased and the collector is reverse biased, with a reasonable value of I_B flowing out of the base. On the other hand, if the base current is zero or negative, the cutoff point (C in Fig. 18B) is reached at the bottom end of the load line, and the collector current is negligible. This is the *off* state of the transistor, and the device is said to be operating in the *cutoff* regime. If the base current is positive and sufficiently large, the device is driven to the *saturation* regime, marked S. This is the *on* state of the transistor, in which a large value of I_C flows with only a very small voltage drop V_{CE}. The saturation regime corresponds to the loss of reverse bias across the collector junction. In a typical switching operation, the base current swings from positive to negative, thereby driving the device from saturation to cutoff and vice versa.

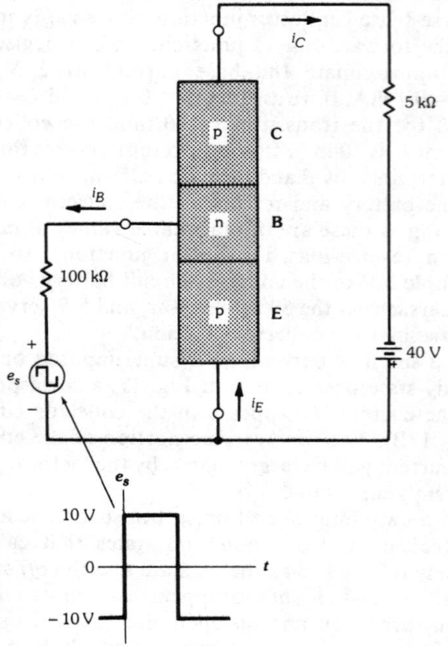

(A) Biasing circuit.

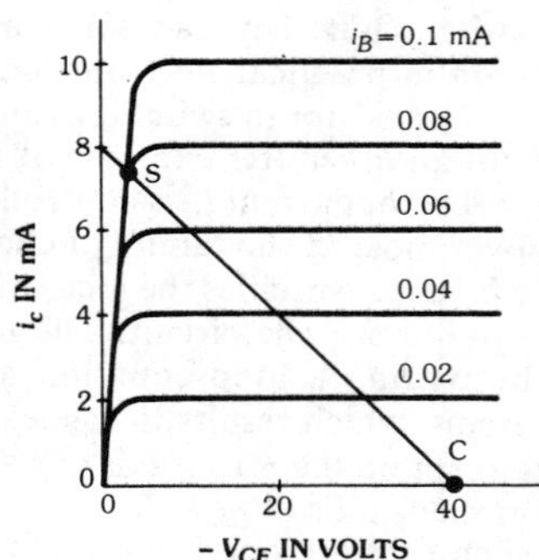

(B) Collector characteristics and load line for circuit.

Fig. 18. Simple common-emitter switching circuit for a transistor. (*From Ben G. Streetman,* Solid State Electronic Devices, *2nd ed., © 1980. Reprinted by permission of Prentice-Hall, Inc., Englewood Cliffs, N.J.*)

The response of a BJT to switching signals is not instantaneous. Junction capacitance and charge-storage effects, discussed earlier, cause delays in switching response that must be taken into account in circuit design. A typical time response of collector current to the input square wave of Fig. 18 is shown in Fig. 19. An initial delay, t_d, is required for charging of the junction space-charge regions. Then a rise time, t_r, is required before the minority-carrier distribution is established in the base region. Similarly, a fall time, t_f, characterizes the decay of the minority-carrier distribution from the conducting state to the cutoff condition (in which the base is essentially depleted of minority carriers). Some switching transistors utilize gold doping to reduce carrier recombination times, thereby minimizing these switching delay times.

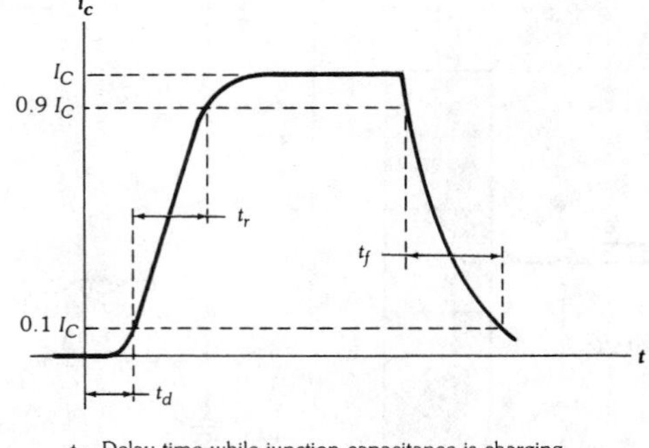

Fig. 19. Collector current during switching transients. (*From Ben G. Streetman,* Solid State Electronic Devices, *2nd ed., © 1980. Reprinted by permission of Prentice-Hall, Inc., Englewood Cliffs, N.J.*)

Circuit models of transistors and applications in electronic systems are discussed in Chapter 19.

OTHER ELECTRONIC DEVICES

The basic devices discussed thus far—diodes, transistors, and their applications in integrated circuits—form the foundation of modern electronics. However, there are a host of other devices that play important roles in electronic circuit and system design. Symbols for many solid-state devices are shown in Table 4. In this section, we will briefly consider a few devices that are particularly important in many electronic systems.

Display Devices

Many electronic systems require some means of displaying information. Such display may take the form of a printout, a cathode-ray-tube image, or various types of digital (*numeric*) or letter and number (*alphanumeric*) displays. Even analog displays, such as the indicator needle on a meter scale, are designed to present information from a circuit or system to the viewer. Most of these display methods are familiar and need no further elaboration. We will discuss here only a few of the newer displays that are currently being used.

Liquid crystal (LC) materials are made up of molecules that have considerable freedom of movement and can be aligned in various configurations by application of an electric field. Depending on the molecular alignment, light can be passed through the LC or can be scattered within it. A

TABLE 4.* COMMON CIRCUIT SYMBOLS FOR SOLID STATE DEVICES †

Diodes

pn diode, pin, IMPATT, Schottky barrier diode — A K

Breakdown ("zener") diode — B

Bidirectional zener diode — B

Varactor

Tunnel diode

Photodiode, solar cell

Light-emitting diode (LED)

Transistors

Bipolar pnp — E C B

Bipolar npn

Unijunction (n-base) — E B_1 B_2

Unijunction (p-base)

JFET (n-channel) — G D S

JFET (p-channel)

IGFET (n-channel) — Depletion: G D B (Bulk) S; Enhancement

Continued on next page.

TABLE 4*(CONT). COMMON CIRCUIT SYMBOLS FOR SOLID STATE DEVICES †

Transistors (Continued)

IGFET (p-channel)

Four-Layer and Related Devices

pnpn (Shockley) diode

SCR

A K G

Bilateral switch

TRIAC

SCS

G_1 A K G_2

A K G_1 G_2

* From Ben G. Streetman, *Solid State Electronic Devices*, 2nd ed., © 1980. Reprinted by permission of Prentice-Hall, Inc., Englewood Cliffs, N.J.

† This table gives many of the commonly used symbols for semiconductor devices. Standardization is incomplete, however, and other symbols are often used in the electronics literature.

typical display cell is made up of the liquid crystal held between two glass plates with conductive coatings. The top plate is transparent, and the bottom plate is either reflecting or absorbing. By the application of appropriate voltages across the plates, incident light may be either scattered within the LC or transmitted through the cell and absorbed or reflected from the back plate. As a result, an array of LC cells can be made to reflect light in certain segments but not others. Typically, the array is in the form of a series of bars that can be selectively addressed to display numbers or letters.

Liquid-crystal displays require little power for their operation and can be made in rather large arrays. Certain LC systems can be made to reflect colors selectively, so that colored displays are possible. Disadvantages of the LC devices include the requirement of incident light, sensitivity to temperature variations, and relatively slow response time (milliseconds) to the control signal.

One of the most versatile display systems is the *light-emitting diode* (LED). As discussed earlier, carriers injected across a forward-biased pn junction recombine within approximately one diffusion length of the junction. If the semiconductor is a luminescent material, the energy given up when an electron-hole pair recombines results in a photon of light. For example, when an electron in the conduction band of GaAs recombines with a hole in the valence band, a photon is emitted with energy equal to the GaAs band gap (1.4 eV). A photon with this energy is in the infrared portion of the spectrum. Thus in a properly constructed GaAs junction, forward bias results in considerable carrier injection, recombination, and the resulting emission of infrared light. The intensity of the light can be varied (with nanosecond response time) by changing the diode current; thus electrical signals can be converted into optical signals by the LED. Used in conjunction with a photodiode, the LED can send information from one point to another optically. Another application of the LED-photodiode is in an *isolator*. Since the signal transmission is optical, complete electrical isolation can be achieved. Such an isolator pair can be mounted on an insulating substrate and packaged to form a device that allows signal transfer between input and output while maintaining electrical isolation.

If the information receiver is the human eye, the display device must emit light in the visible part of the spectrum (approximately 4000 to 7000 Å in wavelength, or 3.1 to 1.8 eV in photon energy). Visible LEDs are built that cover the range from red to green. For example, the compound GaAsP has a band gap that can be varied from that of GaAs (1.4 eV) to that of GaP (2.3 eV) by choosing the appropriate mixture. When the phosphorus content is about 40% of the column V constituent, the band gap is 2 eV, in the red. This is the most common material for LED display fabrication. By using segments made up of GaAsP strips, a numeric display can be made that emits light in the appropriate segments to display numbers or letters. Such an LED display operates at very low voltage, has a long operating lifetime, and is highly reliable.

Microwave Devices

The use of transistors at high frequencies is generally limited to the range of a few gigahertz by capacitance and transit-time effects. Therefore, generation and amplification of microwave signals usually depend on special devices that can deliver high-frequency ac power to a resonant cavity or waveguide. Power is delivered to an ac signal if there is an increase in the motion of charge through a region where the field is changing such as to retard such motion. For charges in a solid or gas, this is called *negative differential conductivity.* As an example, suppose holes are drifting down a typical semiconductor bar in the direction of an electric field. If the field varies with time, the holes speed up or slow down as the field changes, and power is extracted from the field. However, if the arrival of holes at a certain point were to increase during a time interval in which the field at that point decreases, power would be delivered to the field, and the apparent conductivity $dJ/d\mathcal{E}$ would be negative. The object of microwave oscillators is to modulate the rate of arrival of electrons or holes at a point where the electric field is varying, such that the arrival coincides with the retarding half-cycle of the ac field.

A number of solid-state microwave devices have been introduced in recent years. Most of these devices are variations of the *IMPATT diode* and the *gunn diode.* In each of these devices, a negative conductance allows current oscillations that can be used to generate microwave signals.

The negative conductance in an IMPATT diode results from avalanche multiplication and transit-time effects. A simple version of the IMPATT structure is the n^+pip^+ device shown in Fig. 20A. In this diode, the n^+ and p^+ regions are heavily doped, and the i region is essentially intrinsic. Assume the diode is biased such that avalanche multiplication at the n^+p junction begins at $t = 0$ in the voltage cycle (Fig. 20B). Then a positive ac voltage superimposed on the dc bias causes more multiplication. The avalanche-generated holes drift to the right and into the i region in the diagram of Fig. 20C. Since the hole pulse grows as long as the critical field, $\mathcal{E}_a$, is exceeded, the pulse reaches its maximum at $\omega t = \pi$ (i.e., the hole pulse grows during the entire positive half-cycle of the voltage). Then as v enters its negative half-cycle, avalanche ceases at the n^+p junction (Fig. 20D). However, the hole pulse drifts through the i region from left to right while the ac terminal voltage is negative. Since the holes are collected during the negative half-cycle of voltage, the ac conductance is negative during this period. If length L is chosen such that the pulse drifts through the i region for the full negative half-cycle of voltage, negative conductance will be obtained for the entire half-cycle. The IMPATT device can be placed in a resonant cavity tuned to the appropriate frequency, and microwave generation results.

Current pulses are created in a GaAs gunn diode by a mechanism that transfers electrons from one region of the conduction band to another. The conduction electrons in GaAs normally reside in a band 1.43 eV above the valence band. In addition, a subsidiary conduction band lies about 0.3 eV above the first. Of course, this higher-lying band is generally of little interest, since conduction-band electrons are usually found only in the lower-energy band. It is possible, however, to excite electrons into the upper conduction band by applying a sufficiently large electric field. The reason this transfer is of interest is that the mobility of electrons in the upper band is much smaller than the usual mobility of electrons in the lower conduction band. Thus, when electrons are transferred to the low-mobility subsidiary band, they actually *slow down.* At a critical threshold field, $\mathcal{E}_{th}$, electrons begin to transfer to the upper band, where their velocity is smaller. This transfer results in a negative differential conductance that can be used in microwave generation.

Switching Devices

Many electronic applications call for a device that can be switched from a nonconducting "off" state to a conducting "on" state. Several devices can be used in switching applications, and selection of the appropriate device depends on requirements of power level, switching time, and other factors. A common electronic switch is the bipolar transistor, which can be driven from cutoff to saturation by controlling the small base current, or the FET, which can be switched by controlling the gate voltage.

The most widely used controllable switch for large currents is the *semiconductor controlled rectifier (SCR).* This is a four-layer pnpn device (Fig. 21) with terminals attached to the anode (A) and the cathode (K) and a third terminal attached to one of the central regions. This third terminal is called the gate (G). When the gate is left open, the SCR has the characteristic of Figs. 21A-C. With a negative applied voltage, junctions j_1 and j_3 are reverse biased (Fig. 21A), and current through the device is effectively blocked, even at high reverse voltage. With a positive voltage (A positive with respect to K), junctions j_1 and j_3 are forward biased,

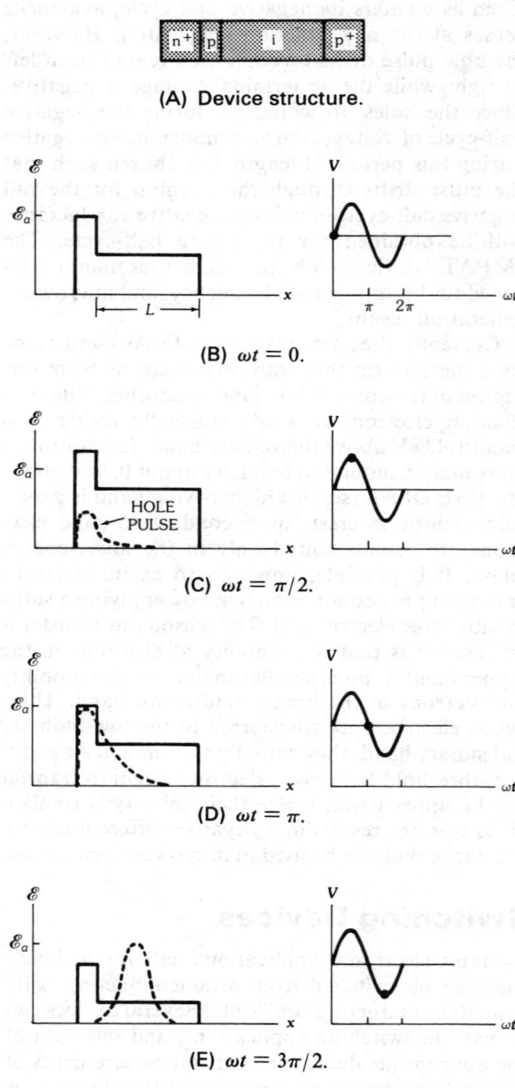

Fig. 20. Time dependence of the growth and drift of holes during a cycle of applied voltage for the n⁺pip⁺ IMPATT diode. (*From Ben G. Streetman,* Solid State Electronic Devices, *2nd ed.,* © *1980. Reprinted by permission of Prentice-Hall, Inc., Englewood Cliffs, N.J.*)

while j_2 is reverse biased (Fig. 21B). Initially, the device current is restricted to the small saturation current of j_2. This is called the *forward blocking state* and corresponds to the "off" condition of the switch. The SCR can be thought of as two coupled transistors ($p_1n_1p_2$ and $n_1p_2n_2$) with a common collector junction, j_2. If transistor action is initiated, holes injected into n_1 from the forward-biased emitter junction, j_1, can be transported across the base of the pnp structure into p_2. Such transistor action in effect feeds holes into the base of the npn structure, thereby increasing electron injection from n_2 into p_2. Such injected electrons can then be collected across j_2 into n_1 by transistor action, and the process continues. The result is a combination of two saturated transistors, typified by high current at low voltage (the *forward conducting state*), as shown in Fig. 21C.

The initiation of transistor action (and therefore switching) can occur as a result of raising the bias to a critical value (V_p). At this voltage, avalanche multiplication at j_2 and base-width narrowing* in n_1 and p_2 combine to initiate transport of minority carriers across the two base regions. Alternatively, a small gate current i_g can supply sufficient base current to the npn structure to initiate transistor action (Fig. 21D). The latter switching method is the most common type for an SCR. In this mode of operation, the device is nonconducting at forward voltages until a small pulse of current is applied to the gate. Such a pulse initiates transistor action in the device and switches it into the conducting state. The SCR remains in the conducting state until current I is dropped below a value called the *holding current* required to maintain transistor action. In addition, some SCR devices can be turned off by applying a negative current to the gate, thereby extracting carriers and terminating the transistor action within the device.

REFERENCES

General

Milnes, A. G. *Semiconductor Devices and Integrated Electronics.* New York: Van Nostrand Reinhold Co., 1980.

Muller, R. S., and Kamins, T. I. *Device Electronics for Integrated Circuits.* New York: John Wiley & Sons, Inc., 1977.

Pierret, R. F., and Neudeck, G. W. *Modular Series on Solid State Devices.* Reading, Mass.: Addison-Wesley Publishing Co., Inc., 1983.

Streetman, B. G. *Solid State Electronic Devices.* 2nd ed. Englewood Cliffs, N.J.: Prentice-Hall, Inc., 1980.

Sze, S. M. *Physics of Semiconductor Devices.* 2nd ed. New York: John Wiley & Sons, Inc., 1981.

Yang, E. S. *Fundamentals of Semiconductor Devices.* New York: McGraw-Hill Book Co., 1978.

Semiconductor Physics

Kittel, C. *Introduction to Solid State Physics.* 5th ed. New York: John Wiley & Sons, Inc., 1976.

*Base-width narrowing occurs as the depletion region about j_2 grows due to the increased reverse bias. As the effective widths of n_1 and p_2 become smaller, base transport factor B increases.

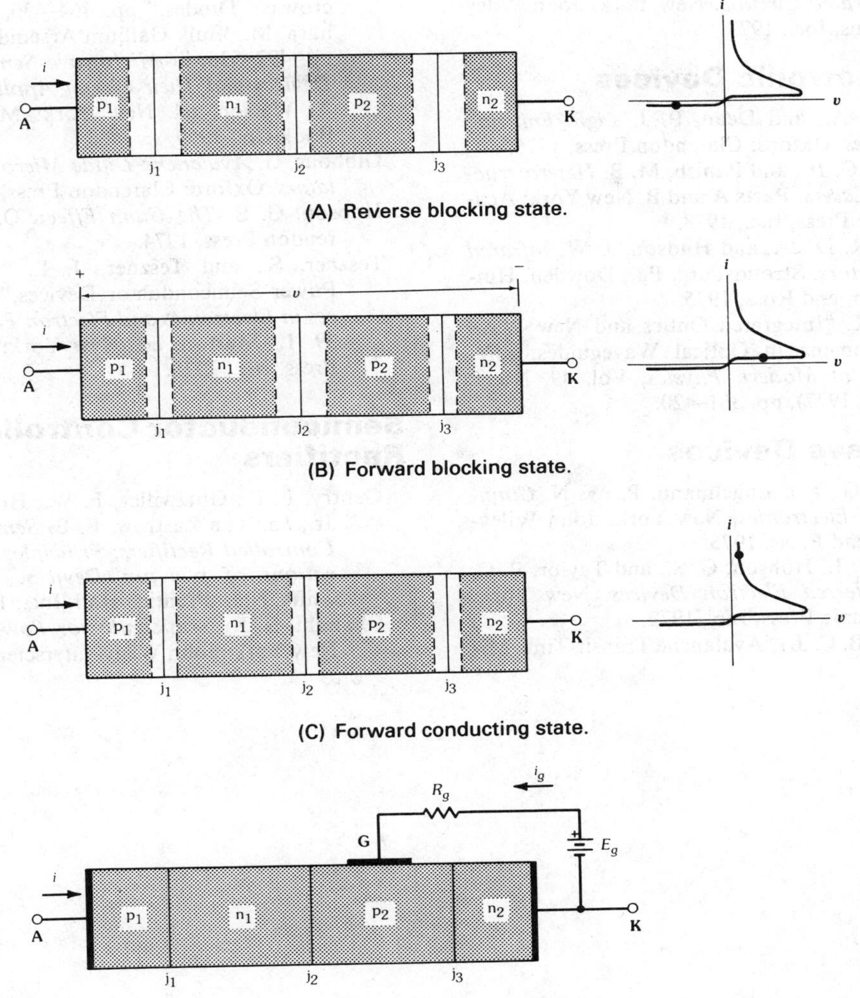

Fig. 21. Biasing of a pnpn device. (*From Ben G. Streetman*, Solid State Electronic Devices, *2nd ed.*, © *1980. Reprinted by permission of Prentice-Hall, Inc., Englewood Cliffs, N.J.*)

McKelvey, J. P. *Solid State and Semiconductor Physics*. New York; Harper & Row Publishers, Inc., 1966.

Pankove, J. I. *Optical Processes in Semiconductors*. Englewood Cliffs, N.J.: Prentice-Hall, Inc., 1971.

van der Ziel, A. *Solid State Physical Electronics*. 3rd ed. Englewood Cliffs, N.J.: Prentice-Hall, Inc., 1976.

Wolf, H. F. *Semiconductors*. New York: John Wiley-Interscience, 1977.

Junctions and Transistors

American Micro-Systems Engineering Staff. *MOS Integrated Circuits*. W. M. Penney and L. Lau, eds. New York: Van Nostrand Reinhold Co., 1972.

Glaser, A. B., and Subak-Sharpe, G. E. *Integrated Circuit Engineering*. Reading, Mass.: Addison-Wesley Publishing Co., Inc., 1977.

Grove, A. S. *Physics and Technology of Semiconductor Devices*. New York: John Wiley & Sons, Inc., 1967.

Milnes, A. G., and Feucht, D. L. *Heterojunction and Metal-Semiconductor Junctions*. New York: Academic Press, Inc., 1972.

Nicollian, E. H., and Brews, J. R., *MOS (Metal Oxide Semiconductor) Physics and Technology*. New York: John Wiley & Sons, Inc., 1982.

Richman, P. *MOS Field-Effect Transistors and*

Integrated Circuits. New York: John Wiley & Sons, Inc., 1973.

Optoelectronic Devices

Bergh, A. A., and Dean, P. J. *Light-Emitting Diodes*. Oxford: Clarendon Press, 1976.

Casey, H. C. Jr., and Panish, M. B. *Heterostructure Lasers*, Parts A and B. New York: Academic Press, Inc., 1978.

Hudson, R. D. Jr., and Hudson, J. W. *Infrared Detectors*. Stroudsburg, Pa.: Dowden, Hutchison, and Ross, 1975.

Tien, P. K. "Integrated Optics and New Wave Phenomena in Optical Waveguides." *Reviews of Modern Physics*, Vol. 49, No. 2 (April 1977), pp. 361–420.

Microwave Devices

Bosch, B. G., and Engelmann, R. W. N. *Gunn-Effect Electronics*. New York: John Wiley-Halstead Press, 1975.

Bulman, P. J., Hobson, G. S., and Taylor, B. C. *Transferred Electron Devices*. New York: Academic Press, Inc., 1972.

DeLoach, B. C. Jr. "Avalanche Transit-Time Microwave Diodes," pp. 464–496, and Uenohara, M. "Bulk Gallium Arsenide Devices," pp. 497–536, in *Microwave Semiconductor Devices and Their Circuit Applications*. H. A. Watson, ed. New York: McGraw-Hill Book Co., 1969.

Gibbons, G. *Avalanche-Diode Microwave Oscillators*. Oxford: Clarendon Press, 1973.

Hobson, G. S. *The Gunn Effect*. Oxford: Clarendon Press, 1974.

Teszner, S., and Teszner, J. L. "Microwave Power Semiconductor Devices," in *Advances in Electronics and Electron Physics*, Vol. 39. L. Marton, ed. New York: Academic Press, Inc., 1975.

Semiconductor Controlled Rectifiers

Gentry, F. E., Gutzwiller, F. W., Holonyak, N. Jr., and von Zastrow, E. E. *Semiconductor Controlled Rectifiers: Principles and Applications of p–n–p–n Devices*. Englewood Cliffs, N.J.: Prentice-Hall, Inc., 1964.

Ghandhi, S. K. *Semiconductor Power Devices*. New York: John Wiley-Interscience, 1977.

19 Transistor Circuits

Horace G. Jackson

This chapter gives condensed descriptions of many types of circuits in which transistors are used. Also presented is design information that makes possible the determination of the various circuit parameters. In accordance with the accepted practice, upper-case variables with upper-case subscripts (V_{CE}) are used to indicate the static, or large-signal, quantities, and lower-case variables with lower-case subscripts (v_{ce}) are used to indicate the dynamic, or small-signal, values.

The overwhelming majority of transistor circuits are made with silicon devices; therefore, this is the assumed technology in this chapter. However, germanium and gallium-arsenide devices are also available to a limited degree.

DEVICE MODELS AND EQUATIONS*

This section presents the basic large-signal circuit models and equations for three common semiconductor active devices, namely:

(1) Bipolar junction transistor (BJT, or simply transistor)
(2) Junction field-effect transistor (JFET or FET)
(3) Metal-oxide-semiconductor field-effect transistor (MOSFET or MOS)

The first two are readily found in both discrete and integrated transistor circuits. Except for some high-power applications, the third is restricted to integrated circuits only.

The symbol and large-signal model for an npn transistor are illustrated in Fig. 1. For a pnp transistor, the polarities of the terminal voltages, V_{BC}, V_{BE} and V_{CE}, must be reversed; the direction of the junction diodes must be reversed; and the direction of all the currents must also be reversed. The model equations are listed in Chart 1. Typical values for the device parameters are given in Chart 2.

The symbol and large-signal model for an n-channel JFET are given in Fig. 2. For a p-channel device, the polarities of the terminal voltages, V_{GD}, G_{GS}, and V_{DS}; the direction of the junction diodes; and the direction of all the currents must be reversed. The model equations, given in Chart 3, assume that the JFET approximates a square-law device. Typical parameter values are listed in Chart 4.

The symbol and large-signal model for an n-channel MOSFET are given in Fig. 3. For a p-channel device, the polarities of the five terminal voltages, V_{GS}, V_{GD}, V_{DS}, V_{BS}, and V_{BD}; the two junction diodes; and the currents must all be reversed. The model equations and parameter values are given in Chart 5 and Chart 6, respectively.

* References 1, 13, and 17.

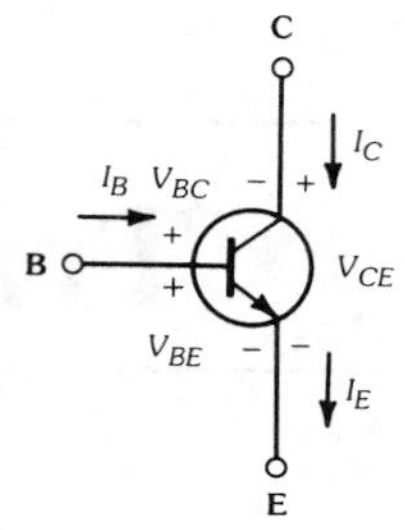

(A) Symbol and nomenclature.

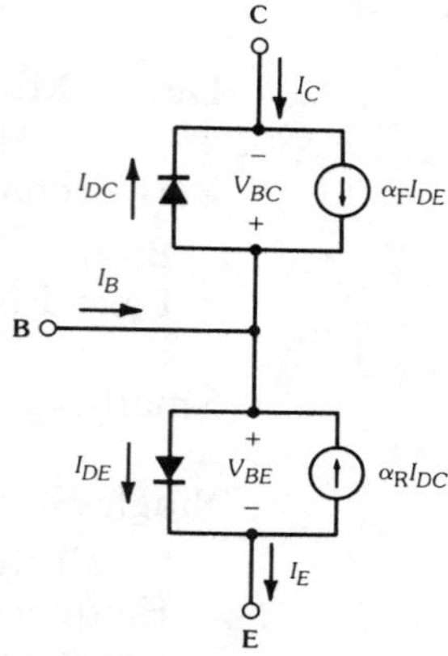

(B) Large-signal model.

Fig. 1. Symbol, nomenclature, and large-signal model for npn transistor (BJT).

Note that a MOSFET may be one of two types:

A. Enhancement type: with $V_{GS} = 0$ V there is no conducting channel, and the drain current is zero.
B. Depletion type: with $V_{GS} = 0$ V there is a conducting channel, and the drain current is finite.

The FET terms I_{DSS} and V_T are graphically defined in the transfer characteristic of Fig. 4.

A summary of the operating modes for JFETs and MOSFETs is presented in Table 1.

BIAS TECHNIQUES†

Bipolar Transistor

As an amplifier, the BJT is normally operated in the forward-active region. That is, the base-emitter junction is forward-biased, and the base-collector junction is reverse-biased. Thus, with $V_{BE} >> 4V_T$ and $V_{BC} << -4V_T$ the equations in Chart 1 reduce to

$$I_E = (\beta_F + 1)\,(I_B + I_{CO})$$

$$I_C = \beta_F I_B + (\beta_F + 1) I_{CO}$$

† References 6 and 14.

CHART 1. MODEL EQUATIONS FOR AN NPN TRANSISTOR

$$I_E = I_{ES}(e^{V_{BE}/V_T} - 1)(1 + V_{BC}/V_A) - \alpha_R I_{CS}(e^{V_{BC}/V_T} - 1)(1 + V_{BE}/V_B)$$

$$I_C = \alpha_F I_{ES}(e^{V_{BE}/V_T} - 1)(1 + V_{BC}/V_A) - I_{CS}(e^{V_{BC}/V_T} - 1)(1 + V_{BE}/V_B)$$

$$I_B = I_E - I_C$$

Also,

$$\alpha_F I_{ES} = \alpha_R I_{CS}$$

where,

α_F = Forward current gain,

α_R = Reverse current gain,

I_{ES} = Emitter junction saturation current,

I_{CS} = Collector junction saturation current,

V_A = Forward-mode basewidth modulation factor,

V_B = Reverse-mode basewidth modulation factor,

V_T = Thermal voltage* = kT/q,

k = Boltzmann's constant = 1.38×10^{-23} J/K,

T = Absolute temperature = [273 + temp (°C)] K,

q = Electronic charge = 1.60×10^{-19} C.

* The thermal voltage, used with bipolar devices, is not to be confused with the threshold voltage (V_T) used with field-effect devices.

CHART 2. TYPICAL VALUES OF THE DEVICE PARAMETERS FOR A LOW-POWER SILICON BJT

$\alpha_F = 0.98$	$\beta_F = \alpha_F/(1-\alpha_F) = 50$
$\alpha_R = 0.49$	$\beta_R = \alpha_R/(1-\alpha_R) = 1$
$I_{ES} = 1 \times 10^{-14}$ A	V_A = 20 to 200 V
$I_{CS} = 2 \times 10^{-14}$ A	V_B = 10 to 100 V
	V_T = 26 mV at 27°C

where

$$I_{CO} = (1 - \alpha_F \alpha_R) I_{CS}$$

Good design of the bias circuit for a BJT requires that two deficiencies of the transistor be overcome. They are:

1. The transistor is a temperature sensitive device, in particular with respect to V_{BE} and β_F.

$$\Delta V_{BE}(\mathrm{T}) \approx -2.5\ \mathrm{mV}/°\mathrm{C}$$
$$\Delta \beta_F(\mathrm{T}) \approx +0.7\%/°\mathrm{C}$$

Also note that I_{CO} doubles for every 8° C rise in temperature. But with silicon devices I_{CO} is usually so small that it may be neglected.

2. The parameters of a transistor are subject to process variation, in particular with respect to β_F. Typically β_F = 30 to 300 (or greater).

Generally, a constant collector current I_C and constant collector-emitter voltage V_{CE} are required. That is, the operating, or quiescent, point—the Q-point—of the transistor is designed to be insensitive to temperature and

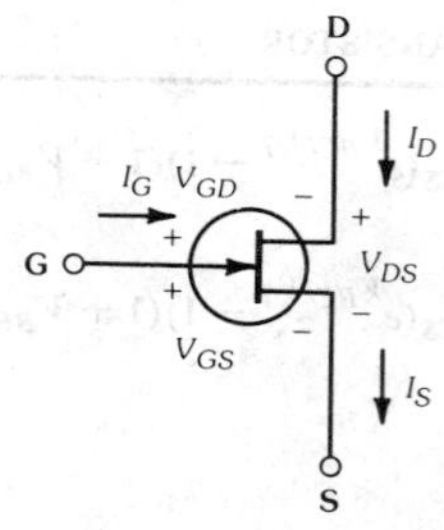

(A) Symbol and nomenclature.

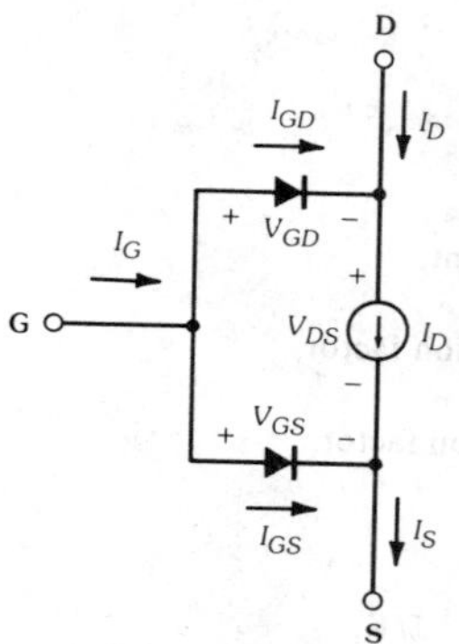

(B) Large-signal model.

Fig. 2. Symbol and large-signal model for n-channel junction field-effect transistor (JFET).

process variations. A self-bias, or emitter bias, circuit is illustrated in Fig. 5A. The circuit in Fig. 5B is equivalent.

Given I_C, V_{CE}, and V_{BE}, the design procedure is as follows:

A. Choose R_E, where $R_E \geq 5\Delta V_{BE}/I_E$. Here, ΔV_{BE} is the change in V_{BE} over the temperature range of interest. This tends to desensitize the operating point to temperature variations of V_{BE}.

B. Determine R_B, where $R_B \leq \beta_F R_E/5$. Here, β_F is the nominal value. This tends to make the operating point independent of variations in the value of β_F.

C. Find R_C, from $R_C = [(V_{CC} - V_{CE})/I_C] - R_E$.

D. Using nominal values for V_{BE} and β_F, solve for

$$V_{BB} \approx V_{BE} + I_C[(R_B + \beta_F R_E)/\beta_F]$$

In an amplifier design, R_E is usually paralleled by a capacitor or by a series resistor-capacitor combination.

Field-Effect Transistor

The JFET is also a temperature sensitive device.

$$\Delta I_{DSS}\,(\mathrm{T}) \approx -0.5\%/^\circ\mathrm{C}$$
$$\Delta V_T\,(\mathrm{T}) \approx -2\ \mathrm{mV}/^\circ\mathrm{C}$$

Due to process variations, I_{DSS} may range 3:1 and V_T may vary 2:1.

For a MOSFET, the parameters k and V_T have temperature and process dependence similar to those of I_{DSS} and V_T for a JFET.

In designing for a constant I_D and constant V_{DS} in the saturation region, the self-bias circuit shown in Fig. 6A can be used with JFETs or depletion-type MOSFETs. Since the gate current is negligible

$$R_S = -V_{GS}/I_D$$
$$R_D = [(V_{DD} - V_{DS})/I_D] - R_S$$

The circuit of Fig. 6A is unsuitable for biasing an

CHART 3. MODEL EQUATIONS FOR AN N-CHANNEL JFET

Nonsaturation region,* $V_{DS} < (V_{GS} - V_T)$:

$$I_D = \frac{I_{DSS}}{V_T^2}[2(V_{GS} - V_T)V_{DS} - V_{DS}^2](1 + \lambda V_{DS})$$

Saturation region,** $V_{DS} \geq (V_{GS} - V_T)$:

$$I_D = \frac{I_{DSS}}{V_T^2}[V_{GS} - V_T]^2(1 + \lambda V_{DS})$$

where,

I_{DSS} = Zero-bias saturation drain current, *i.e.*, $V_{GS} = 0$ V, $V_{DS} > (V_{GS} - V_T)$,

V_T = Threshold voltage,†

λ = Channel-length modulation factor.

* Also termed the linear, ohmic, or triode region.
** Also the constant current, pinchoff, or pentode region.
† Also the pinchoff voltage (V_P) or $V_{GS(\mathrm{off})}$. Not to be confused with the thermal voltage (V_T) used with bipolar devices.

CHART 4. TYPICAL DEVICE PARAMETERS FOR A LOW-POWER SILICON JFET

	N-Channel	P-Channel
$\lvert I_{DSS}\rvert$	1 to 10 mA	1 to 10 mA
V_T	−2 to −6 V	+2 to +6 V
λ	0.1 to 0.01 V^{-1}	0.1 to 0.01 V^{-1}

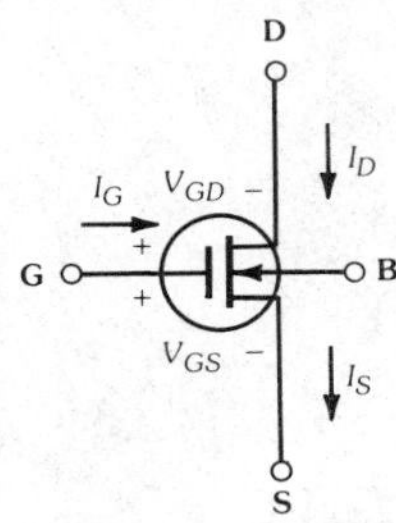

(A) Symbol and nomenclature.

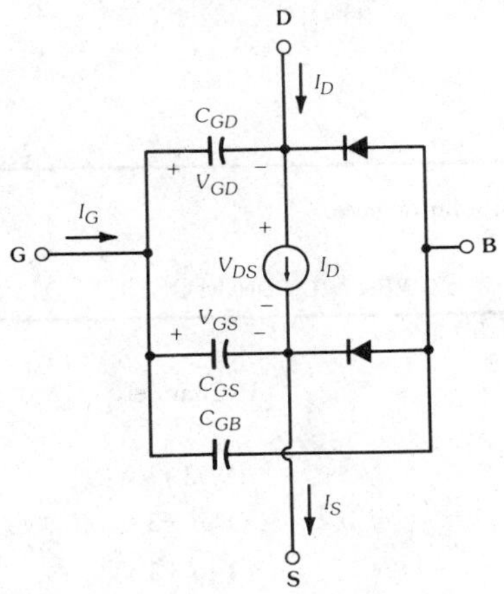

(B) Large-signal model.

Fig. 3. Symbol and large-signal model for n-channel MOSFET.

enhancement-type MOSFET, since for an n-channel device V_T is positive and it is required that $V_{GS} > V_T$. Two alternative circuits for biasing an enhancement-type MOSFET are included in Fig. 6. In Fig. 6B, with negligible gate current $V_{GS} = V_{DS}$ so that the transistor is operating in the saturation region.

$$R_D = (V_{DD} - V_{DS})/I_D$$
$$R_G \geq 10R_D$$

When it is required that $V_{GS} \neq V_{DS}$, the circuit in Fig. 6C may be used.

As in the BJT case, in most amplifier applications R_S is paralleled by a capacitor.

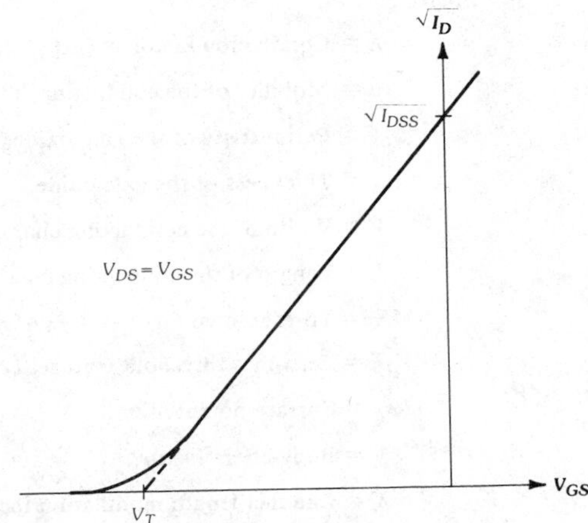

Fig. 4. Typical JFET transfer characteristic.

SMALL-SIGNAL MODELS AND EQUATIONS *

For a BJT biased in the forward-active region, the basic small-signal model for the device is illustrated in Fig. 7. Included in the model are the bulk ohmic effects of the neutral base, collector, and emitter regions, designated r_b, r_c, and r_e. Typically, these are about 100 ohms, 50 ohms, and 1 ohm, respectively. The forward-biased base-emitter junction is modeled by r_π and C_π. The reverse-biased base-collector junction is modeled by C_μ, and r_μ is due to the basewidth modulation effect. That the collector current is a function of the base-emitter voltage is modeled by current source $g_m v_\pi$ with output resistance r_{ce}. How these small-signal parameters relate to the operating point of the transistor is indicated in Chart 7.

The small-signal model for a JFET biased in the saturation region is given in Fig. 8. The ohmic resistances of the neutral gate, drain, and source regions are modeled by the linear resistors r_g, r_d, and r_s. Typically, these are 1 ohm, 10 ohms, and 10 ohms, respectively. In most applications they may be neglected.

The reverse-biased gate-source and gate-drain junctions of the JFET are modeled by r_{gs}-C_{gs} and

* References 2, 4, and 5.

CHART 5. MODEL EQUATIONS FOR AN N-CHANNEL MOSFET

Nonsaturation region, $V_{DS} < (V_{GS} - V_T)$:

$$I_D = \frac{k}{2}[2(V_{GS} - V_T)\, V_{DS} - V_{DS}^2](1 + \lambda V_{DS})$$

Saturation region, $V_{DS} \geq (V_{GS} - V_T)$:

$$I_D = \frac{k}{2}[V_{GS} - V_T]^2 (1 + \lambda V_{DS})$$

where

k = Conduction factor = $(\mu \epsilon_{ox}/t_{ox})\,(W/L)$,

μ = Mobility of the conducting carriers,

ϵ_{ox} = Permittivity of the gate oxide,

t_{ox} = Thickness of the gate oxide,

W = Width of the conducting channel,

L = Length of the conducting channel,

V_T = Threshold voltage* = $V_{T0} + \gamma[\sqrt{|2\varphi_F| + |V_{BS}|} - \sqrt{|2\varphi_F|}\,]$,

V_{T0} = Zero-bias threshold voltage, *i.e.*, $V_{BS} = 0$ V,

$2\varphi_F$ = Surface potential $\approx$ 0.6 V,

γ = Body-effect factor,

λ = Channel-length modulation factor.

* Not to be confused with the thermal voltage (V_T) used with bipolar devices.

CHART 6. TYPICAL DEVICE PARAMETERS FOR A LOW-POWER SILICON MOSFET

	N-Channel	P-Channel
k	20(W/L) μA/V^2	7(W/L) μA/V^2
V_{TEO}	+1 to +3 V	−1 to −3 V
V_{TDO}	−1 to −3 V	+1 to +3 V
$\lvert 2\varphi_F \rvert$	−0.6 to −0.8 V	+0.6 to +0.8 V
γ	+0.3 to +0.6 V	−0.3 to −0.6 V
λ	0.1 to 0.01 V^{-1}	0.1 to 0.01 V^{-1}

r_{gd}-C_{gd}, respectively. The voltage-dependent drain-current generator is modeled by $g_m v_{gs}$ with an output resistance of r_{ds}. The equations for the latter two parameters are found in Chart 8.

For low-frequency applications, the charge-storage elements, C_{gs} and C_{gd}, may be neglected. Because r_{gs} and r_{gd} are generally very large, they also may be neglected in many applications.

The small-signal model for a MOSFET biased in the saturation region is given in Fig. 9. The ohmic effects of the gate, drain, source, and body regions are modeled by the linear resistors r_g, r_d, r_s, and r_b. Typical values are 1 ohm, 1 ohm, 1 ohm, and 10 ohms, respectively. In most applications they may be neglected.

Charge stored between the insulated gate and the source, drain, and body regions is modeled by the capacitors C_{gs}, C_{gd}, and C_{gb}, respectively. The reverse-biased source-body and drain-body junctions are included in the model by r_{sb}-C_{sb} and r_{db}-C_{db}. However, generally the resistance values are so large that their effect is usually neglected.

As with the JFET, the voltage-dependent current source is modeled by $g_m v_{gs}$ with an output resis-

TABLE 1. SUMMARY OF OPERATING MODES FOR FET DEVICES

	JFET		MOSFET		
	N-Channel	P-Channel	N-Channel	P-Channel	N-Channel Depletion
Symbol					
V_T Polarity	Negative	Positive	Positive	Negative	Negative
V_{GS} Polarity	Negative	Positive	Positive	Negative	Neg or Pos
V_{DS} Polarity	Positive	Negative	Positive	Negative	Positive
Cutoff Region	$V_{GS}<V_T$	$V_{GS}>V_T$	$V_{GS}<V_T$	$V_{GS}>V_T$	$V_{GS}<V_T$
Nonsaturation Region	$V_{DS}<(V_{GS}-V_T)$	$V_{DS}>(V_{GS}-V_T)$	$V_{DS}<(V_{GS}-V_T)$	$V_{DS}>(V_{GS}-V_T)$	$V_{DS}<(V_{GS}-V_T)$
Saturation Region	$V_{DS}\geq(V_{GS}-V_T)$	$V_{DS}\leq(V_{GS}-V_T)$	$V_{DS}\geq(V_{GS}-V_T)$	$V_{DS}\leq(V_{GS}-V_T)$	$V_{DS}\geq(V_{GS}-V_T)$

tance of r_{ds}. The equations for these parameters are found in Chart 9.

The general ranges of the small-signal parameters for each of the devices considered in this section are compared in Table 2.

SINGLE-STAGE AMPLIFIERS*

Small-Signal Characteristics

The three possible circuit configurations for a single-transistor amplifier are illustrated in Fig. 10. For clarity, the bias networks have been omitted. It is assumed that the bipolar transistor is operating in the forward-active region and the field-effect transistor is operating in the saturation region. The small-signal design equations listed in Table 3 are for the input resistance (r_i), output resistance (r_o), voltage gain ($a_v = v_o/v_i$), and current gain ($a_i = i_o/i_i$). The small-signal transducer power gain is obtained from $a_p = a_v a_i$. The second-order terms due to the bulk ohmic effects have been omitted. The equations for the FET apply for both the JFET and the MOSFET. Numerical values for each of the amplifier configurations are compared in Table 4.

* References 2, 8, and 14.

Frequency Response

In Figs. 11 through 13 are shown simple circuit schematics for each of the single-stage amplifier configurations. Included in each figure are the small-signal equivalent circuits appropriate for the three ranges of frequencies, low, middle, and high. The corresponding small-signal transfer functions derived from the equivalent circuits are listed in Tables 5 through 7.

For the common-emitter and common-source amplifiers, the low cutoff frequency (− 3 dB) is due to the greater of ω_{l1} or ω_{l2}, provided the ratio of the two is greater than 10. In the high-frequency region, normally $\omega_{zh} >> \omega_{h2} >> \omega_{h1}$, so that ω_{h1} is the high cutoff frequency (− 3 dB).

For the emitter-follower and the source-follower, there is only one low break frequency. In the high-frequency region, typically ω_{zh} is only slightly greater than ω_{h1}, though ω_{h2} is much greater than either. Hence, the high cutoff frequency is difficult to determine accurately. However, it is greater than ω_{h1}, which is approximately equal to ω_T of the transistor.

For the common-base amplifier, the determination of the low cutoff frequency is similar to the case of the common-emitter amplifier. However, note that for the common-gate amplifier there is only one low break frequency. In the high-fre-

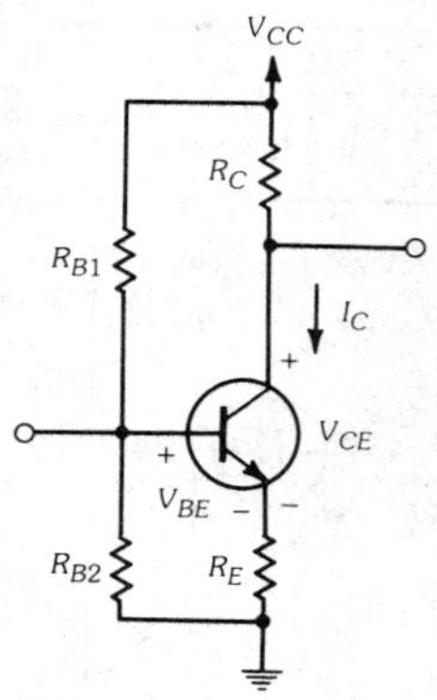

(A) Single voltage supply.

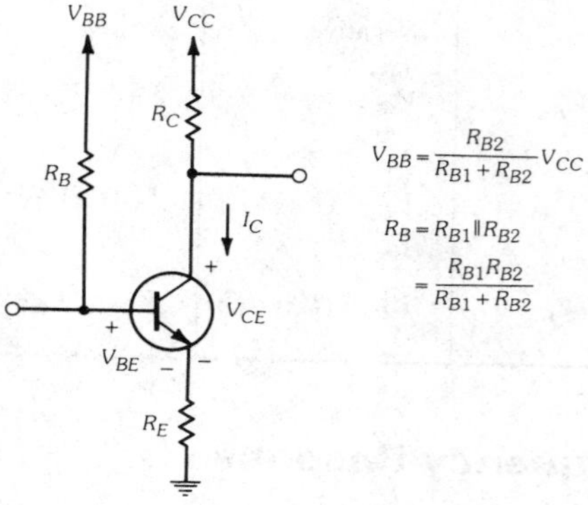

(B) Two-voltage-supply equivalent.

Fig. 5. Bias circuits for bipolar transistors.

quency region, typically $\omega_{h2} >> \omega_{h1}$ so that ω_{h1} is the high cutoff frequency.

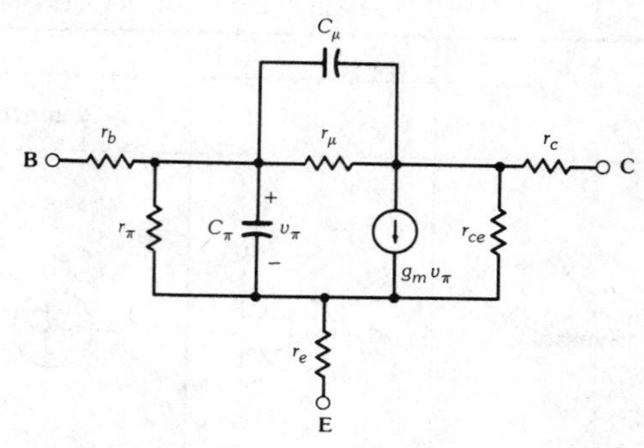

Fig. 7. Small-signal model for BJT.

CHART 7. SMALL-SIGNAL PARAMETERS FOR A BJT

$$g_m = i_c/v_\pi = \left.\frac{dI_C}{dV_{BE}}\right|_{OP} = I_C/V_T$$

$$r_\pi = v_\pi/i_b = \left.\frac{dV_{BE}}{dI_B}\right|_{OP} = \beta_F V_T/I_C = \beta_F/g_m$$

$$r_{ce} = v_{ce}/i_c = \left.\frac{dV_{CE}}{dI_C}\right|_{OP} = (V_A + V_{CE})/I_C$$

$$r_\mu = v_{cb}/i_b = \left.\frac{dV_{CB}}{dI_B}\right|_{OP} = (V_A + V_{CB})/I_B$$

$$C_\pi = (g_m/2\pi f_T) - C_\mu$$

Large-Signal Characteristics

For large-signal operation, both bipolar and field-effect transistor amplifiers are classified as one of the following.

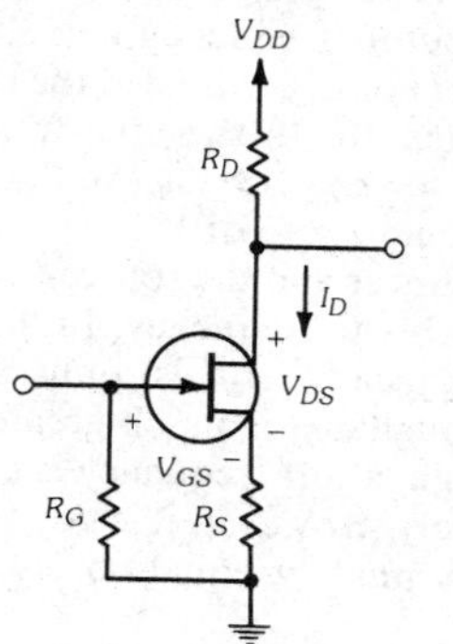

(A) For JFET and depletion-type MOSFET.

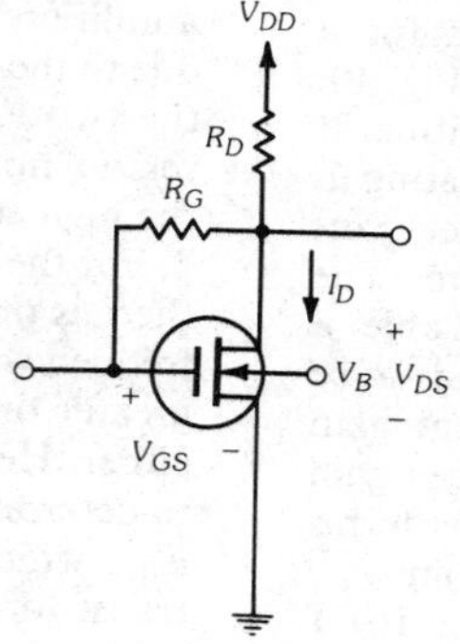

(B) For enhancement-type MOSFET with $V_{DS} = V_{GS}$.

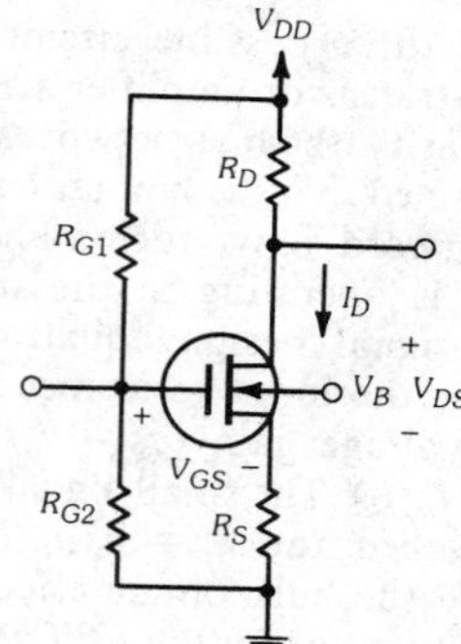

(C) Alternative circuit for enhancement MOSFET.

Fig. 6. Bias circuits for FETs.

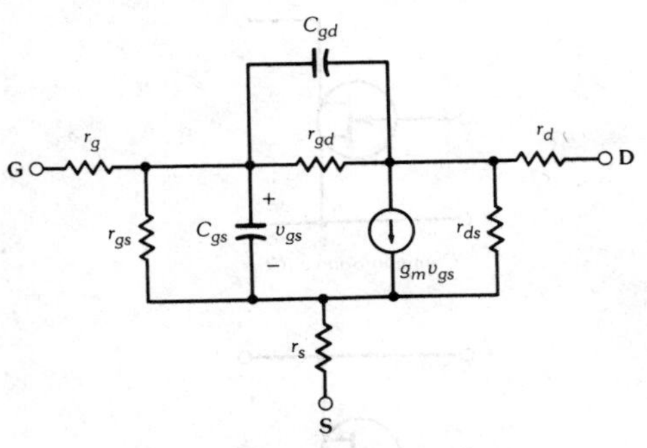

Fig. 8. Small-signal model for JFET.

CHART 8. SMALL-SIGNAL PARAMETERS FOR JFET

$$g_m = i_d / v_{gs} = \left.\frac{dI_D}{dV_{GS}}\right|_{OP} = (2I_{DSS}/V_T)\,(1 - V_{GS}/V_T)$$

$$= (2/V_T)\,(I_D I_{DSS})^{1/2}$$

$$r_{ds} = v_{ds}/i_d = \left.\frac{dV_{DS}}{dI_D}\right|_{OP} = (V_A + V_{DS})/I_D$$

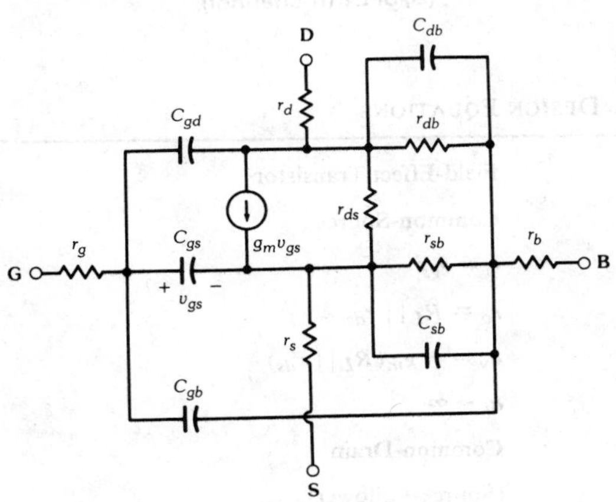

Fig. 9. Small-signal model for MOSFET.

CHART 9. SMALL-SIGNAL PARAMETERS FOR MOSFET

$$g_m = i_c / v_{gs} = \left.\frac{dI_C}{dV_{GS}}\right|_{OP} = k(V_{GS} - V_T)$$

$$= (2kI_D)^{1/2}$$

$$r_{ds} = v_{ds}/i_d = \left.\frac{dV_{DS}}{dI_D}\right|_{OP} = (V_A + V_{DS})/I_D$$

Class A—The transistor is conducting at all times. Essentially this describes a transistor operating in the linear region of the characteristic curves.

Class B—The transistor is conducting only one-half of the operating cycle, *i.e.*, for 180° of a sine-wave input. This operation is used in output stages to obtain a higher power-conversion efficiency than class A provides.

Class AB—Operation is similar to class B, but in the absence of an input signal the transistor is conducting a small quiescent current.

Class C—The transistor is conducting less than 180° of a sine-wave input. Class-C operation is used mostly in high-power amplifiers and oscillators.

Output Stages

The simple emitter-follower of Fig. 14A illustrates a class-A output stage. From the maximum undistorted output voltage swing, where $|V_{max}| = |V_{min}|$, and the maximum output current swing, the maximum output power is $V_{rms} I_{rms} = V_{max} I_{max}/2$. For the ideal case ($V_{CE\,(sat)} = 0$ V for the output transistor and the current source is effective with 0 V across it) $V_{max} = V_{CC+}$ and $V_{min} = V_{CC-}$. With $|V_{CC+}| = |V_{CC-}| = V_{CC}$, the maximum output power is as listed in Table 8. Also listed in Table 8 is the average supply power for the output stage. The ratio of the output power to the supply power

TABLE 2. COMPARISON OF SMALL-SIGNAL PARAMETERS

BJT	JFET	MOSFET
$g_m = 1$ to 100 mA/V	$g_m = 0.1$ to 10 mA/V	$g_m = 0.1$ to 10 mA/V
$r_\pi = 1$ to 100 kΩ	$r_{gs} \geq 10^9$ Ω	$r_{gs} \geq 10^{10}$ Ω
$r_{ce} = 0.01$ to 1 MΩ	$r_{ds} = 0.01$ to 1 MΩ	$r_{ds} = 0.01$ to 1 MΩ
$C_\pi = 10$ to 100 pF	$C_{gs} = 1$ to 10 pF	$C_{gs} = 1$ to 10 pF
$C_\mu = 1$ to 10 pF	$C_{gd} = 1$ to 10 pF	$C_{gd} = 1$ to 10 pF

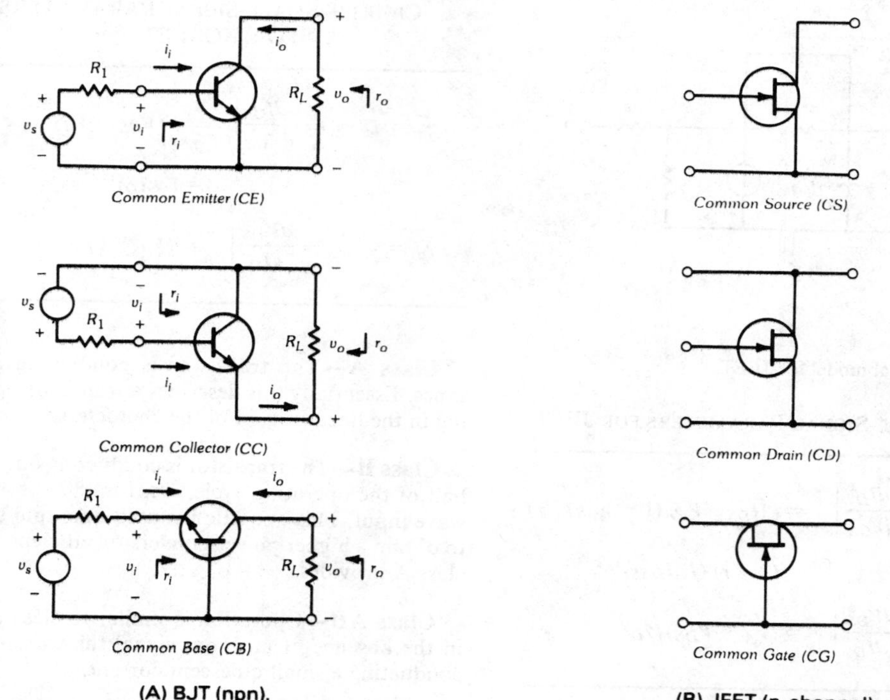

Fig. 10. Single-transistor amplifier configurations.

TABLE 3. SMALL-SIGNAL DESIGN EQUATIONS

Bipolar Transistor

Common-Emitter

$r_i = r_\pi$

$r_o = R_L || r_{ce}$

$a_v = - g_m(R_L || r_{ce})$

$a_i = \beta_0$

Common-Collector

(Emitter-Follower)

$r_i = r_\pi + (\beta_0 + 1) R_L$

$r_o = R_L || [(R_1 + r_\pi)/(\beta_0 + 1)]$

$$a_v = \frac{(\beta_0 + 1) R_L}{r_\pi + (\beta_0 + 1) R_L}$$

$a_i = \beta_0 + 1$

Common-Base

$r_i = r_\pi/(\beta_0 + 1) \approx 1/g_m$

$r_o = R_L || r_{ce}(1 + g_m R_1)$

$$a_v = \alpha_0 \frac{[R_L || r_{ce}(1 + g_m R_1)]}{r_\pi/(\beta_0 + 1)}$$

$a_i = \alpha_0$

Field-Effect Transistor

Common-Source

$r_i = r_{gs}$

$r_o = R_L || r_{ds}$

$a_v = - g_m(R_L || r_{ds})$

$a_i \to \infty$

Common-Drain

(Source-Follower)

$r_i = r_{gs}$

$r_o = R_L || 1/g_m$

$a_v = R_L/(1/g_m + R_L)$

$a_i \to \infty$

Common-Gate

$r_i = 1/g_m$

$r_o = R_L || r_{ds}(1 + g_m R_1)$

$a_v = g_m[R_L || r_{ds}(1 + g_m R_1)]$

$a_i \to \infty$

TABLE 4. COMPARISON OF SMALL-SIGNAL AMPLIFIER PARAMETERS

	BJT			FET		
	CE	CC	CB	CS	CD	CG
r_i	1.3 kΩ	256 kΩ	26 Ω	10^9 Ω	10^9 Ω	224 Ω
r_o	5 kΩ	120 Ω	5 kΩ	5 kΩ	214 Ω	5 kΩ
a_v	−192	0.99	188	−22	0.96	22
a_i	50	51	0.98	—	—	1.0
Circuit Data	$R_1 = 5$ kΩ, $R_C = 5$ kΩ $I_C = 1$ mA $\beta_0 = 50$, $V_A = 50$ V			$R_1 = 5$ kΩ, $R_D = 5$ kΩ $I_D = 1$ mA $I_{DSS} = 5$ mA, $V_T = -1$ V, $V_A = 50$ V		

gives the power-conversion efficiency of the circuit. For a class-A amplifier this is 25% maximum, but a typical figure is about 15%.

A class-B output stage is shown in Fig. 14B. For this circuit, in the quiescent state neither transistor is conducting. On the positive half-cycle of an input sine wave, the npn transistor conducts and acts as a source of current to the load. On the negative half-cycle, the pnp transistor conducts and sinks current from the load. Hence, this configuration is known as a push-pull output stage. Since each transistor conducts only during one-half cycle of the input sine wave, the average supply power is as given in Table 8. For the class-B amplifier, the maximum power-conversion efficiency is 79%, but a typical figure is about 65%.

A problem with the class-B push-pull output stage is that each base-emitter junction must be forward biased before current can flow to the load. This gives rise to crossover distortion—a deadband of about $2V_{BE}$ around the zero axis of the output waveform. This problem can be avoided with the class-AB stage illustrated in Fig. 14C. The addition of a current source and two diodes permits a controlled quiescent current to flow in the two output transistors. The crossover distortion is improved at the cost of a small decrease in power-conversion efficiency.

Harmonic Distortion

For a common-emitter amplifier, the small-signal output voltage is given as

$$v_o = -g_m R_L v_i$$

But for large-signal class-A operation (Fig. 15),

$$V_o = -R_L(I_S e^{V_i/V_T} - I_Q)$$

where,

$$I_Q = I_S e^{V_{BE}/V_T}$$

and

$$V_i = V_s + V_{BE}$$

Therefore

$$V_o = -R_L I_Q (e^{V_s/V_T} - 1)$$

Expanding this in a power series

$$V_o = -R_L I_Q [(V_s/V_T) + (V_s/V_T)^2/2 + (V_s/V_T)^3/6 + \dots]$$
$$= a_1 V_s + a_2 V_s^2 + a_3 V_s^3 \dots$$

where,

$$a_1 = -R_L I_Q / V_T$$
$$a_2 = -R_L I_Q / 2V_T^2$$
$$a_3 = -R_L I_Q / 6V_T^3$$

Note that in this equation, with $V_s/V_T << 1$ the first term dominates and the circuit is essentially linear.

For a sine-wave input

$$V_s = \hat{V}_s \sin \omega t$$

Then

$$V_o = a_1 \hat{V}_s \sin \omega t + a_2 \hat{V}_s^2 \sin^2 \omega t + a_3 \hat{V}_s^3 \sin^3 \omega t + \dots$$
$$= a_1 \hat{V}_s \sin \omega t + (a_2 \hat{V}_s^2/2)(1 - \cos 2\omega t) + (a_3 \hat{V}_s^3/4)(3 \sin \omega t - \sin 3\omega t) + \dots$$

Hence, with large-signal operation, harmonics are introduced into the output voltage waveform due to the exponential relationship of the transfer characteristic for the bipolar transistor.

Equations for the second (HD_2) and third (HD_3) harmonic distortion for each of the bipolar transis-

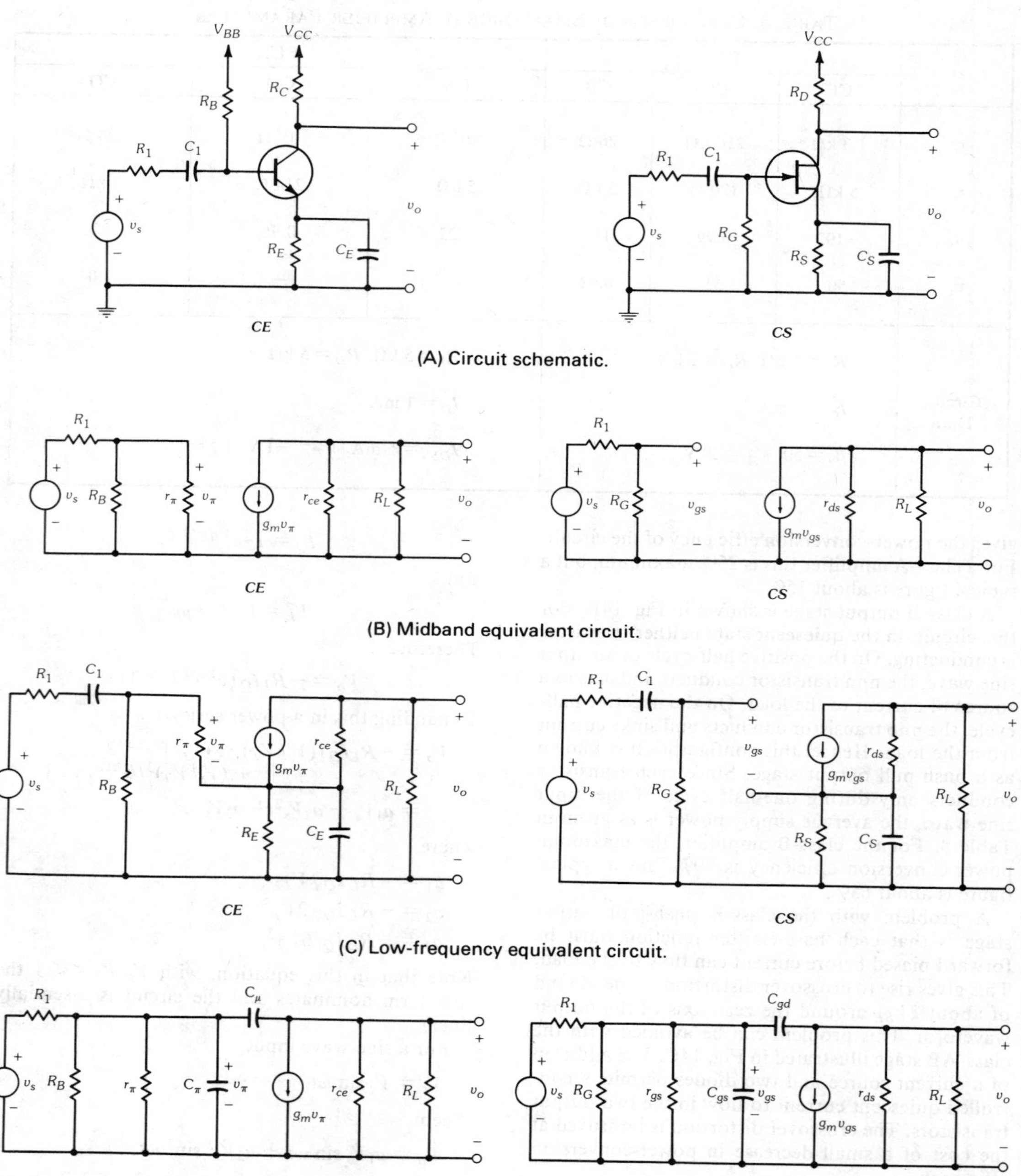

Fig. 11. Circuit schematic and small-signal equivalent circuits for common-emitter and common-source amplifier.

tor configurations are given in Table 9. Note, for example, that for the CE stage with $\hat{V}_s = 10$ mV, $HD_2 = 10\%$ and $HD_3 = 0.62\%$.

In an ideal push-pull amplifier stage, the positive half-cycle of the output waveform exactly matches the negative half-cycle. The result is that all even harmonic terms are balanced out, leaving the third harmonic as the prime source of harmonic distortion.

The transfer characteristic of an FET has a

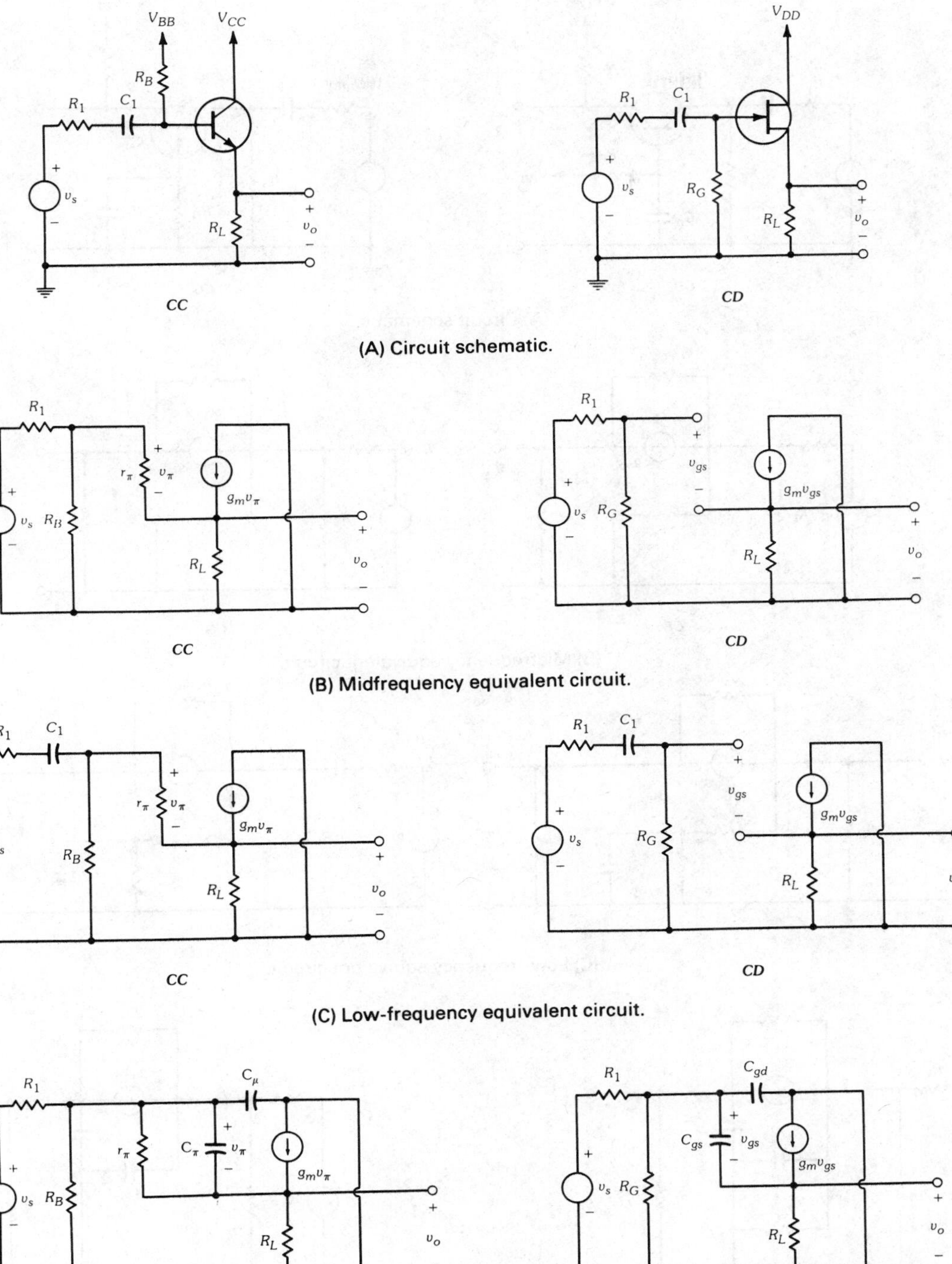

Fig. 12. Circuit schematic and small-signal equivalent circuits for common-collector (emitter-follower) and common-drain (source-follower) amplifier.

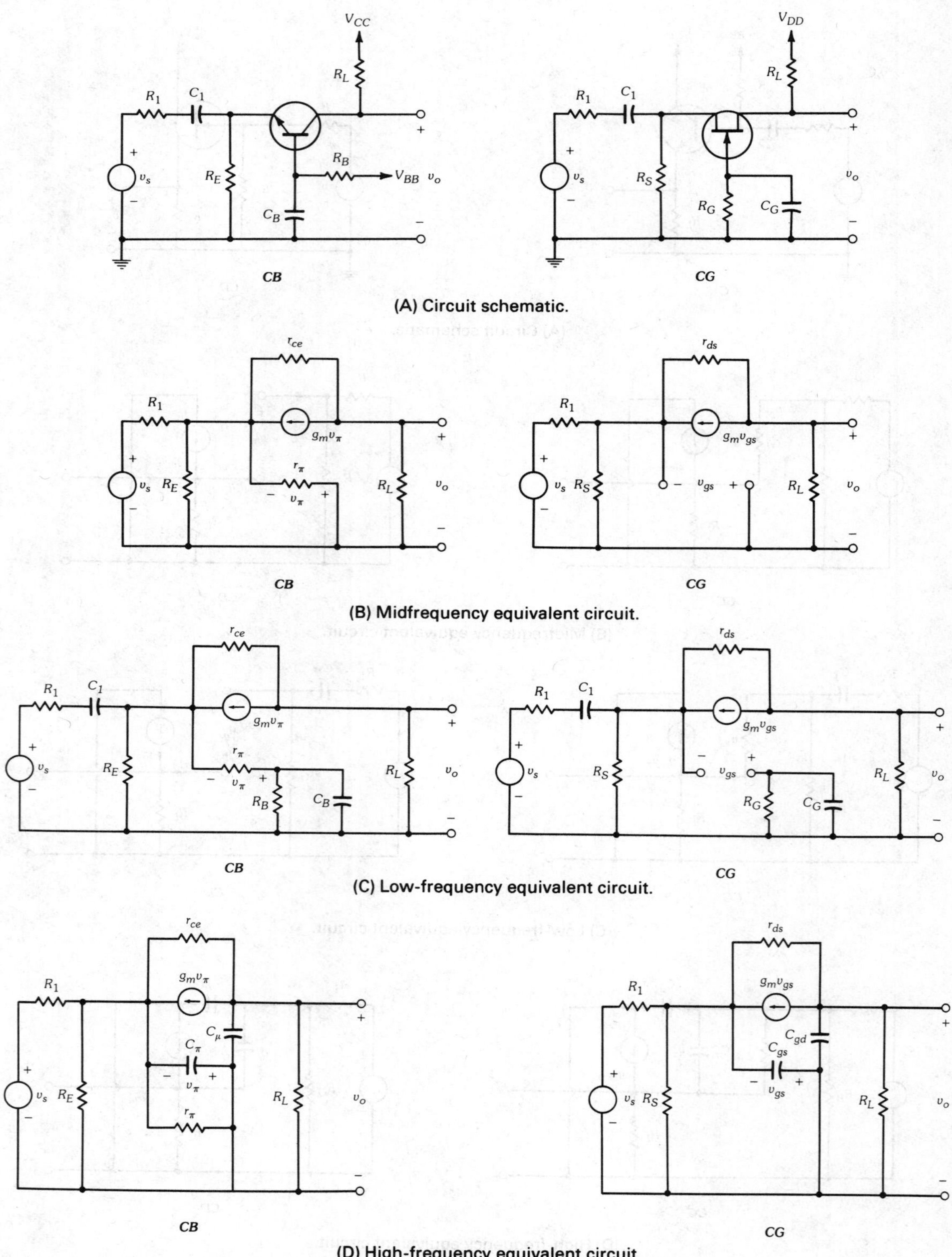

Fig. 13. Circuit schematic and small-signal equivalent circuits for common-base and common-gate amplifier.

TABLE 5. SMALL-SIGNAL TRANSFER FUNCTIONS: COMMON-EMITTER AND COMMON-SOURCE

Bipolar Transistor	Field-Effect Transistor
Midfrequency	Midfrequency
$a_0 = v_o/v_s = -g_m(R_L\|\|r_{ce})[r_\pi/(R_1 + r_\pi)]$	$a_0 = v_o/v_s = -g_m(R_L\|\|r_{ds})[R_G/(R_1 + R_G)]$
Low Frequency 1. Due to C_1 (assume that C_E is a short-circuit)	Low Frequency 1. Due to C_1 (assume that C_S is a short-circuit)
$a_v(j\omega) = a_0/(1 - j\omega_{l1}/\omega)$	$a_v(j\omega) = a_0/(1 - j\omega_{l1}/\omega)$
where, $\omega_{l1} \approx 1/(R_1 + r_\pi)C_1$	where, $\omega_{l1} \approx 1/(R_1 + R_G)C_1$
2. Due to C_E (assume that C_1 is a short-circuit)	2. Due to C_S (assume that C_1 is a short-circuit)
$a_v(j\omega) = a_0(1 + j\omega/\omega_{zl})/(1 + j\omega/\omega_{l2})$	$a_v(j\omega) = a_0(1 + j\omega/\omega_{zl})/(1 + j\omega/\omega_{l2})$
where, $\omega_{zl} \approx 1/R_EC_E$	where, $\omega_{zl} \approx 1/R_SC_S$
$\omega_{l2} \approx 1/\{R_E\|\|[(R_1 + r_\pi)/(\beta_0 + 1)]\}C_E$	$\omega_{l2} \approx 1/[R_S\|\|(1/g_m)]C_S$
High Frequency	High Frequency
$a_v(j\omega) = a_0 \dfrac{(1 + j\omega/\omega_{zh})}{(1 + j\omega/\omega_{h1})(1 + j\omega/\omega_{h2})}$	$a_v(j\omega) = a_0 \dfrac{(1 + j\omega/\omega_{zh})}{(1 + j\omega/\omega_{h1})(1 + j\omega/\omega_{h2})}$
where, $\omega_{zh} \approx 1/(C_\mu/g_m)$	where, $\omega_{zh} \approx 1/(C_{gd}/g_m)$
$\omega_{h1} \approx 1/C_T(R_1\|\|r_\pi)$	$\omega_{h1} \approx 1/C_T(R_1\|\|R_G)$
$C_T = C_\pi + C_\mu[1 + g_m(R_L\|\|r_{ce})]$	$C_T = C_{gs} + C_{gd}[1 + g_m(R_L\|\|r_{ds})]$
$\omega_{h2} \approx g_m/C_\pi + 1/C_\pi(R_1\|\|R_L) + 1/C_\mu R_L$	$\omega_{h2} \approx g_m/C_{gs} + 1/C_{gd}(R_1\|\|R_L) + 1/C_{gd}R_L$

Note: In this analysis it is assumed
1. For the BJT circuit that $R_1 << R_B >> r_\pi$.
2. For the FET circuit that $R_1 << R_G << r_{gs}$.

square-law relationship. Hence, ideally, only second-harmonic distortion appears at the output of a field-effect transistor amplifier. In practice, all FETs deviate somewhat from the ideal transfer characteristic and exhibit a third-order term. However, this term is usually very small. For a common-source amplifier

$$HD_2 = (1/4)(\hat{V}_s/V_T)(I_{DSS}/I_Q)^{1/2}$$

The second harmonic in other FET single amplifier configurations follows from Table 9.

DIFFERENTIAL AMPLIFIERS*

A simple differential amplifier is exemplified by the BJT emitter-coupled pair illustrated in Fig. 16. A source-coupled FET pair could readily replace the bipolar transistors. Fig. 17 shows the small-signal difference and common-mode input voltages of the circuit. The relationship between the difference and common-mode operation is given in Table 10. The design equations are listed in Table 11.

* References 8 and 10.

CURRENT SOURCES†

A useful component in many circuits, but especially with differential amplifiers, is a transistor current source.

Current Mirrors

The concept of current mirrors is widely used in linear integrated circuits, where the matching and tracking of transistor characteristics are well controlled. The idea may also be used with discrete

† References 9 and 11.

TABLE 6. SMALL-SIGNAL TRANSFER FUNCTIONS: COMMON-COLLECTOR AND COMMON-DRAIN

Bipolar Transistor	Field-Effect Transistor
Midfrequency	Midfrequency
$a_0 = v_o/v_s = [(\beta_0 + 1)R_L]/[R_1 + r_\pi + (\beta_0 + 1)R_L]$	$a_0 = v_o/v_s = R_L/(1/g_m + R_L)$
Low Frequency	Low Frequency
$a_v(j\omega) = a_0/(1 - j\omega_{l1}/\omega)$	$a_v(j\omega) = a_0/(1 - j\omega_{l1}/\omega)$
where, $\omega_{l1} \approx 1/[R_1 + r_\pi + (\beta_0 + 1)R_L]C_1$	where, $\omega_{l1} \approx 1/(R_1 + R_G)C_1$
High Frequency	High Frequency
$a_v(j\omega) = a_0 \dfrac{(1 + j\omega/\omega_{zh})}{(1 + j\omega/\omega_{h1})(1 + j\omega/\omega_{h2})}$	$a_v(j\omega) = a_0 \dfrac{(1 + j\omega/\omega_{zh})}{(1 + j\omega/\omega_{h1})(1 + j\omega/\omega_{h2})}$
where, $\omega_{zh} \approx 1/(C_\pi/g_m)$	where, $\omega_{zh} \approx 1/(C_{gs}/g_m)$
$\omega_{h1} \approx 1/\{[(R_1 + R_L)/(1 + g_m R_L)]C_\pi + R_1 C_\mu\}$	$\omega_{h1} \approx 1/\{[(R_1 + R_L)/(1 + g_m R_L)]C_{gs} + R_1 C_{gd}\}$
$\omega_{h2} \approx g_m/C_\pi + 1/(R_1 \| \| R_L)C_\mu$	$\omega_{h2} \approx g_m/C_{gs} + 1/(R_1 \| \| R_L)C_{gd}$

Note: In this analysis it is assumed
1. For the BJT circuit that $R_1 << R_B >> r_\pi$.
2. For the FET circuit that $R_1 << R_G << r_{gs}$.

dual transistors that have the same matching and tracking characteristics as in ICs.

Four simple, but effective, current sources are shown in Fig. 18. Their characteristics are listed in Table 12.

In Fig. 18A, with $V_{BE1} = V_{BE2}$ and the characteristics of Q_1 closely matching those of Q_2, to a first order $I_R = I_{C1} = I_{C2}$. The effect of finite base currents is included in the equations given in Table 12, where, provided $\beta_F >> 2$, $I_{C2} = I_R$. The effects of base current are even further reduced in the improved circuit of Fig. 18B.

The output resistance of a transistor current source can be improved, as in the Widlar current source of Fig. 18C. For this circuit, note that typically $r_{\pi 2} > (1/g_{m1} + R_2)$; then the output resistance is simply given as $r_o = r_{ce}(1 + g_{m2}R_2)$. The solution of the transcendental equation for I_{C2} can be done readily by trial and error with one or two iterations, or with a programmable calculator. An even higher output resistance can generally be obtained with the Wilson current source of Fig. 18D, where, due to negative feedback, $r_o \approx \beta_0 r_{ce2}/2$.

The Widlar circuit is useful for obtaining small output currents, and the Wilson circuit is useful for obtaining high output resistance and low sensitivity to transistor base currents.

Each of these circuits is also viable with matching dual JFETs or MOSFETs.

Active Loads

Another useful application for current sources is as active loads in amplifier circuits. The output resistance of the current source provides a high collector resistance with a relatively small voltage drop. A basic circuit is illustrated in Fig. 19A.

Neglecting base currents, in this circuit

$$I_{C1} = I_{C2} = I_{C3} = I_{C4} = I_{C5} = I_{EE}/2$$

The effective load resistance for differential input signals is

$$R_{L1} = R_{L2} = r_{ce1} \| r_{ce3}$$

The output resistance of the current source may be further increased by adding resistances, $R_{E3} = R_{E4}$, in the emitter leads of Q_3 and Q_4, respectively. The effective load resistance then becomes

$$R_{L1} = R_{L2} = r_{ce1} \| [r_{ce3}(1 + g_{m3}R_{E3})]$$

Fig. 19B shows an amplifier with a differential input signal but a single-ended output signal. In this circuit, not only does Q_4 effectively provide a large collector resistance for Q_2, but also since the collector current of Q_4 mirrors that of Q_3, the small-signal change in the collector current of Q_1 is added to that of Q_2 to provide the full differential voltage gain of the stage at the single-ended output, v_o.

Each of these circuits may be designed with JFETs or MOSFETs. Also, many times it is advan-

TABLE 7. SMALL-SIGNAL TRANSFER FUNCTIONS: COMMON-BASE AND COMMON-GATE

Bipolar Transistor	Field-Effect Transistor
Midfrequency	Midfrequency
$a_0 = v_o/v_s = \alpha_0 \dfrac{R_L\|\|r_{ce}[1 + gm(R_1)]}{R_1 + r_\pi/(\beta_0 + 1)}$	$a_0 = v_o/v_s = \dfrac{R_L\|\|r_{ds}[1 + gm(R_1)]}{R_1 + 1/g_m}$
Low Frequency 1. Due to C_1 (assume that C_B is a short-circuit)	Low Frequency 1. Due to C_1 (assume that C_G is a short-circuit)
$a_v(j\omega) = a_0/(1 - j\omega_{l1}/\omega)$	$a_v(j\omega) = a_0/(1 - j\omega_{l1}/\omega)$
where, $\omega_{l1} \approx 1/(R_1 + 1/g_m)C_1$	where, $\omega_{l1} \approx 1/(R_1 + 1/g_m)C_1$
2. Due to C_B (assume that C_1 is a short-circuit)	2. Due to C_G (assume that C_1 is a short-circuit)
$a_v(j\omega) = a_0(1 + j\omega/\omega_{zl})/(1 + j\omega/\omega_{l2})$	$a_v(j\omega) = a_0$
where, $\omega_{zl} \approx 1/R_BC_B$	
$\omega_{z2} \approx 1/\{R_B\|\|[r_\pi + (\beta_0 + 1)R_1]\}C_B$	
High Frequency	High Frequency
$a_v(j\omega) = a_0/(1 + j\omega/\omega_{h1})(1 + j\omega/\omega_{h2})$	$a_v(j\omega) = a_0/(1 + j\omega/\omega_{h1})(1 + j\omega/\omega_{h2})$
where, $\omega_{h1} \approx 1/(C_\mu/g_m + R_LC_\mu)$	where, $\omega_{h1} \approx 1/(C_{gs}/g_m + R_LC_{gd})$
$\omega_{h2} \approx g_m/C_\pi + 1/R_LC_\mu$	$\omega_{h2} \approx g_m/C_{gs} + 1/R_LC_{gd}$

Note: In this analysis it is assumed
1. For the BJT circuit that $R_1 << R_E >> r_\pi$.
2. For the FET circuit that $R_1 << R_S << r_{gs}$.

tageous to use JFETs as the amplifier devices and BJTs for active loads.

FEEDBACK AMPLIFIERS*

Negative feedback is widely used in amplifier design because it produces several benefits. Among these are:

- Desensitivity against parameter changes
- Improved input/output characteristics
- Reduction in harmonic distortion
- Increased frequency response

Unfortunately there are some disadvantages:

- Reduced gain
- Instability problems

Basic Properties

The basic properties of negative feedback are presented with the aid of the block diagram in Fig. 20. In this diagram, the gain of the basic amplifier, *a*, is controlled by undependable transistor parameters that are subject to considerable variation. It is assumed that gain *a* is much larger than needed. It is also assumed in the block diagram that there is no loading of one block by another block and the signal path is only in the direction of the arrows; *i.e.*, the gain of each block is unilateral.

With *positive feedback:*

$$s_i = s_s + s_f$$

That is, the signals *add* and $s_i > s_s$.

With *negative feedback:*

$$s_i = s_s - s_f$$

That is, the signals *subtract* and $s_i < s_s$.

Also from the block diagram:

$$s_o = as_i$$
$$s_f = fs_o$$

Chart 10 presents the basic equation for negative feedback, along with other important definitions. Note from the basic equation that with negative

* References 7 and 15.

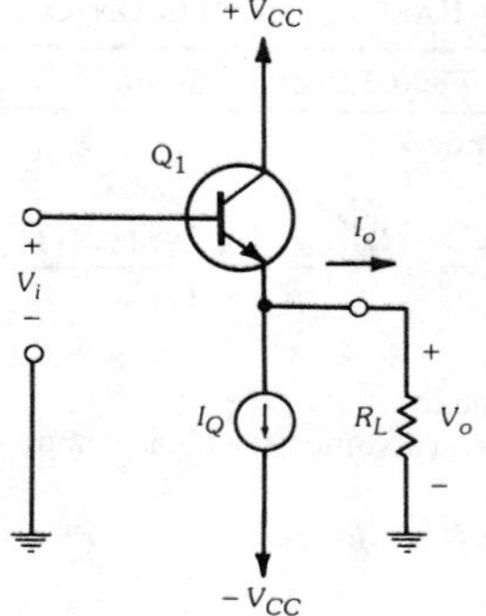

(A) Emitter-follower (class A).

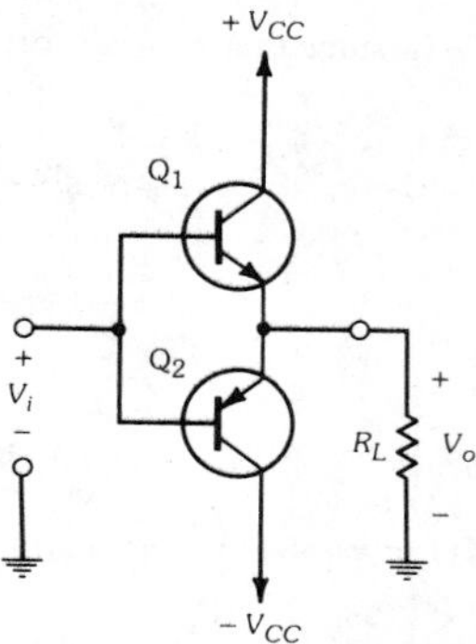

(B) Push-pull (class B).

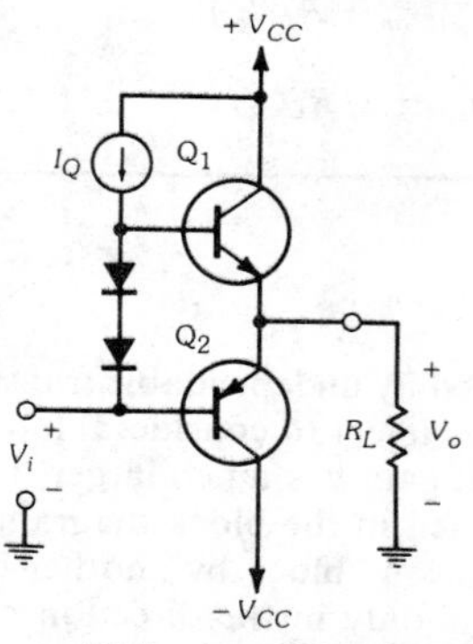

(C) Push-pull (class AB).

Fig. 14. Output stages.

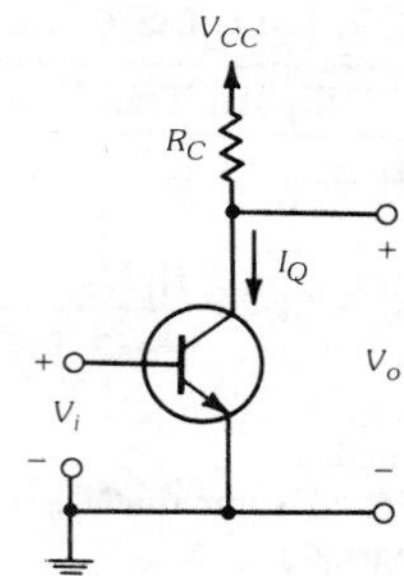

Fig. 15. Large-signal operation of common-emitter amplifier circuit.

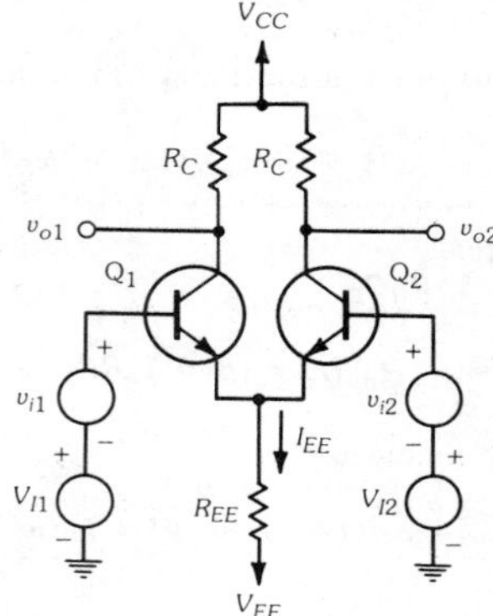

Fig. 16. Differential amplifier pair.

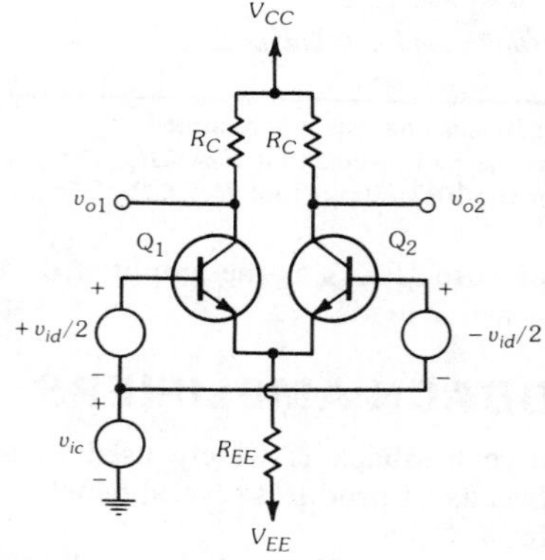

Fig. 17. Differential amplifier pair showing small-signal difference and common-mode input voltages.

feedback the open-loop gain is reduced by the factor $1 + T$. Further, if $T >> 1$ the closed-loop gain becomes $1/f$. It is common for the feedback network to be made up of stable, high-precision, passive components; the value of f is then well defined, and so is the overall amplifier gain.

Basic Feedback Circuit Topologies

Preliminary to the presentation of the feedback topologies, consider the circuit diagrams of four basic classes of amplifier (Fig. 21). The classification is based on the magnitude of the input and output impedance relative to the source and load impedance. The ideal characteristics for each class of amplifier are given in Table 13.

A feedback amplifier is described in terms of the way in which the feedback network is connected to the basic amplifier. There are four basic feedback circuit topologies. These are illustrated in the block diagrams of Fig. 22. Clearly, in Fig. 22A the feed-

TABLE 8. OUTPUT CHARACTERISTICS FOR LARGE-SIGNAL AMPLIFIERS

	Class A	Class B
Output voltage swing	$V_{max}-V_{min}$	$V_{max}-V_{min}$
Output current swing	$I_{max}-I_{min}$	$I_{max}-I_{min}$
Maximum output power	$V_{max}I_{max}/2 = V_{CC}I_Q/2$	$V_{max}I_{max}/2 = V_{CC}I_Q/2$
Average supply power	$2V_{CC}I_Q$	$2V_{CC}I_Q/\pi$
Maximum conversion efficiency	25%	79%

Note: This analysis assumes that $\|V_{CC+}\| = \|V_{CC-}\| = V_{CC}$

TABLE 9. HARMONIC DISTORTION IN BIPOLAR TRANSISTOR AMPLIFIERS

Parameter	CE	CC	CB
HD_2	$\frac{1}{4}(\hat{V}_s/V_T)$	$\frac{1}{4}(\hat{V}_s/V_T)/(1+g_mR_L)$	$\frac{1}{4}(\hat{V}_s/V_T)/(1+g_mR_E)$
HD_3	$\frac{1}{24}(\hat{V}_s/V_T)^2$	$\frac{1}{24}(\hat{V}_s/V_T)^2\,\frac{1-3g_mR_L/(1+g_mR_L)}{1+g_mR_L}$	$\frac{1}{24}(\hat{V}_s/V_T)^2\,\frac{1-3g_mR_E/(1+g_mR_E)}{1+g_mR_E}$

TABLE 10. RELATIONSHIP FOR DIFFERENCE AND COMMON-MODE OPERATION

	Difference Mode	Common Mode
Input	$v_{id} = v_{i1} - v_{i2}$	$v_{ic} = (v_{i1} + v_{i2})/2$
Output	$v_{od} = v_{o1} - v_{o2}$	$v_{oc} = (v_{o1} + v_{o2})/2$
Gain	$a_{dm} = v_{od}/v_{id}$	$a_{cm} = v_{oc}/v_{ic}$
Input	$v_{i1} = v_{ic} + v_{id}/2$ $v_{i2} = v_{ic} - v_{id}/2$	
Output	$v_{o1} = a_{cm}v_{ic} + a_{dm}v_{id}/2$ $v_{o2} = a_{cm}v_{ic} - a_{dm}v_{id}/2$	

TABLE 11. DIFFERENTIAL AMPLIFIERS: SMALL-SIGNAL DESIGN EQUATIONS

Difference Mode	Common Mode
$a_{dm} = v_{od}/v_{id} = -g_mR_C$	$a_{cm} = v_{oc}/v_{ic} = -g_mR_Cr_\pi/[r_\pi + (\beta_0+1)2R_{EE}]$
$r_{id} = v_{id}/i_{id} = 2r_\pi$	$r_{ic} = v_{ic}/i_{ic} = r_\pi + (\beta_0+1)2R_{EE}$
$r_{od} = 2R_C$	$r_{oc} = 2R_C$
CMRR (dB) $= 20 \log \|a_{dm}/a_{cm}\|$	

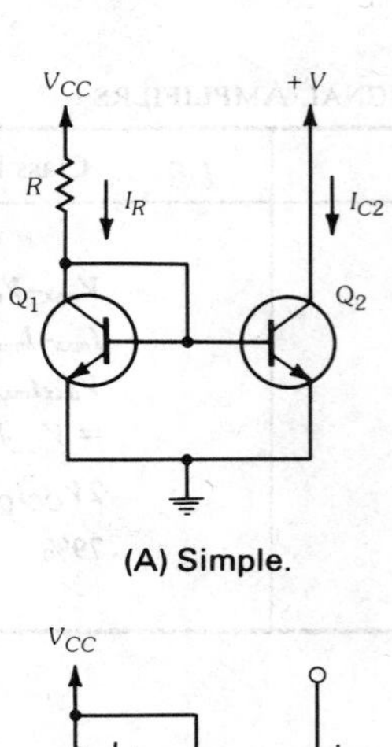

(A) Simple.

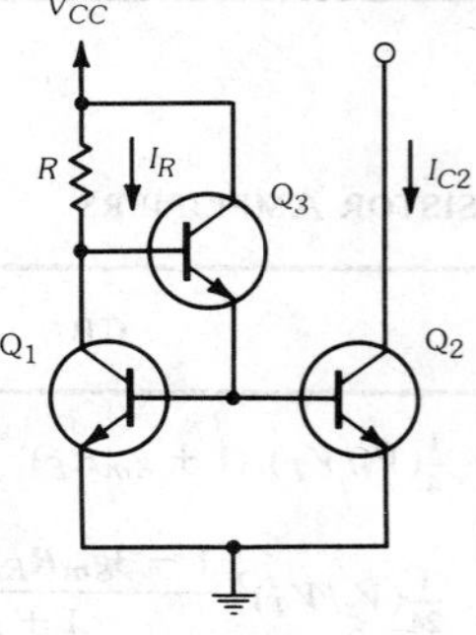

(B) Improved.

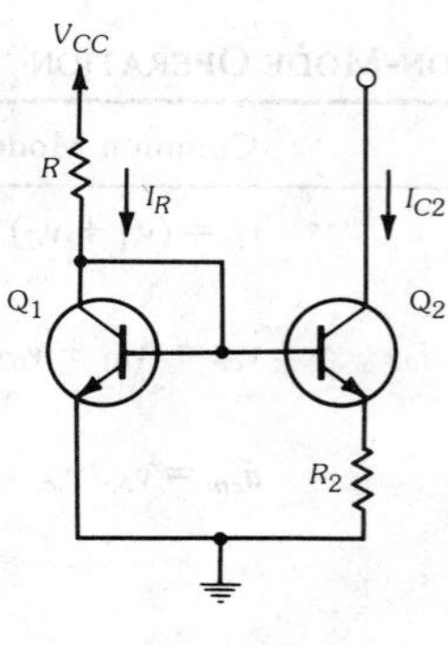

(C) Widlar.

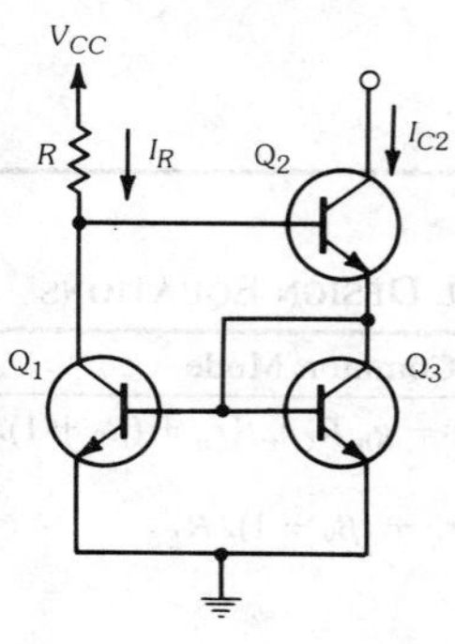

(D) Wilson.

Fig. 18. Current-mirror circuits.

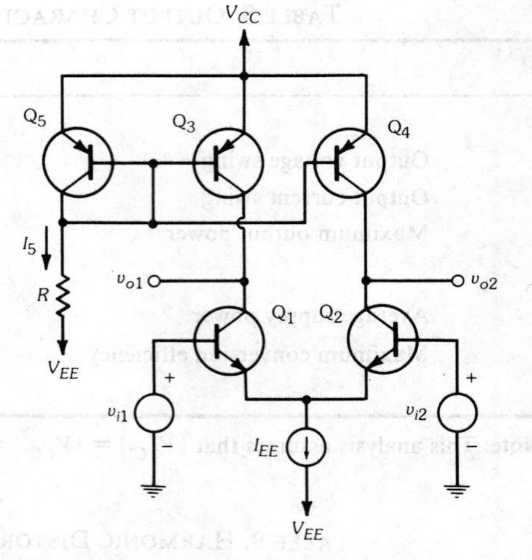

(A) Differential amplifier pair with active load.

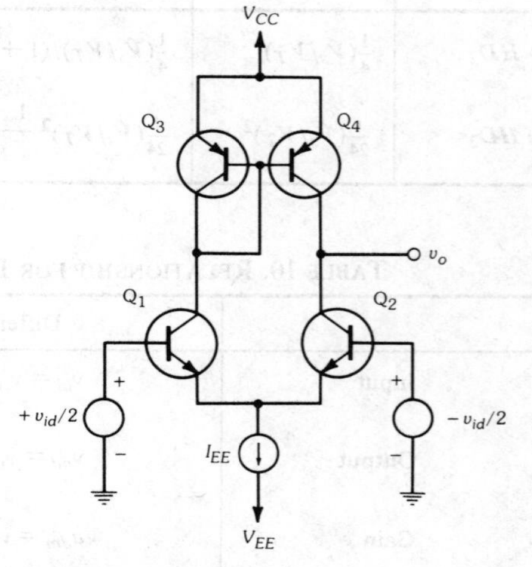

(B) Modified active load for differential amplifier pair.

Fig. 19. Active loads.

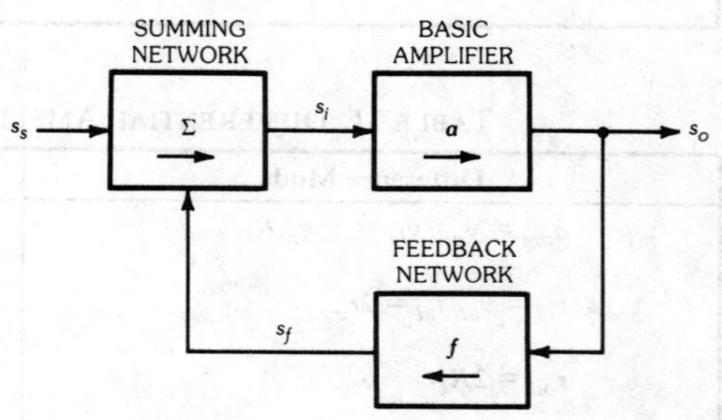

Fig. 20. Ideal feedback block diagram.

CHART 10. BASIC EQUATIONS FOR NEGATIVE FEEDBACK

A	$= a/(1 + af) = a/(1 + T)$	$= a/D_s$
A	= closed-loop gain	$= s_o/s_s$
a	= open-loop gain	$= s_o/s_i$
f	= feedback factor	$= s_f/s_o$
T	= loop gain = af	$= s_f/s_i$
D_s	= desensitivity factor	$= 1 + T$

Note: The basic equation for negative feedback is often written as $A_{CL} = A_{OL}/(1 + A_{OL}\beta)$

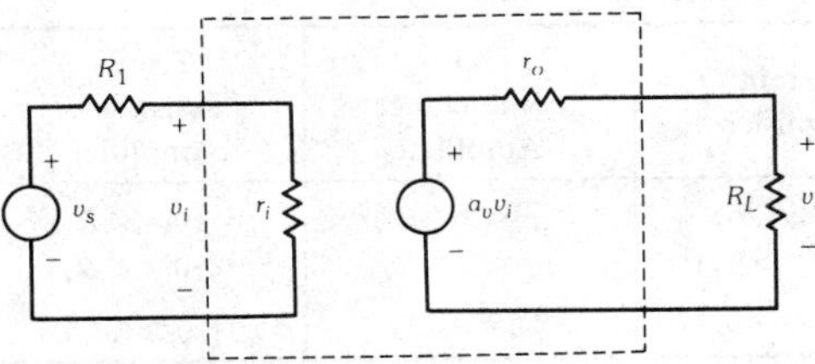

(A) Voltage amplifier ($v_{in} \rightarrow v_{out}$).

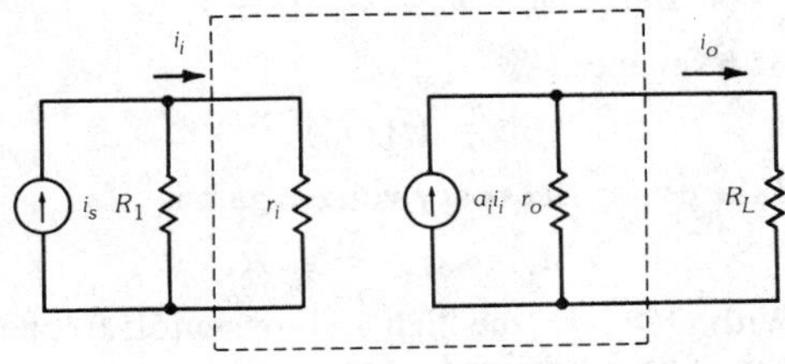

(B) Current amplifier ($i_{in} \rightarrow i_{out}$).

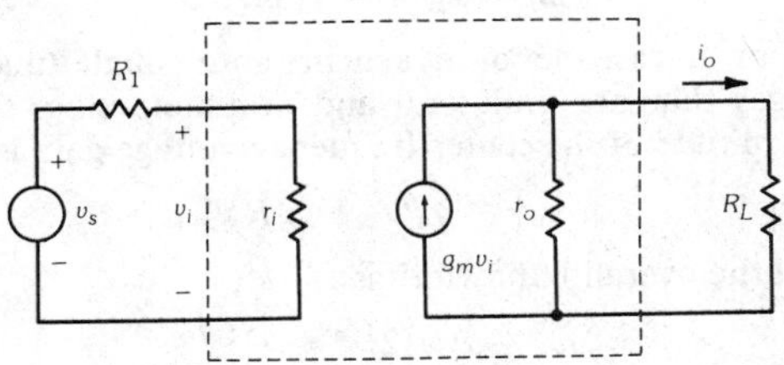

(C) Transconductance amplifier ($v_{in} \rightarrow i_{out}$).

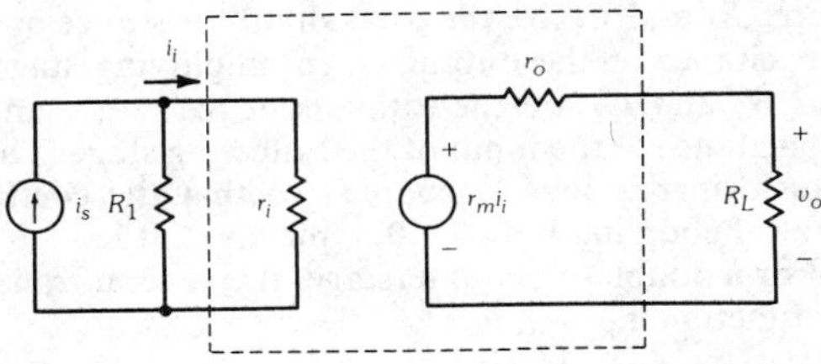

(D) Transresistance amplifier ($i_{in} \rightarrow v_{out}$).

Fig. 21. Amplifier classifications.

back network is connected in *series* with the input terminals of the basic amplifier and in *shunt* with the output terminals. Notice:

With series feedback at the input, voltages v_s and v_f are algebraically summed.
With shunt feedback at the input, currents i_s and i_f are algebraically summed.
With series feedback at the output, a current i_o is sampled.
With shunt feedback at the output, a voltage v_o is sampled.

Included in Fig. 22 are simple examples of each feedback connection, implemented with bipolar transistors. Especially note the correspondence between each circuit schematic and the related block diagram. To avoid complexity, all biasing resistors have been omitted from the circuit diagrams, but it is assumed that all transistors are biased in the forward-active region to yield a high-gain amplifier.

Presented below is a method of analysis of feedback amplifiers. The design of a feedback amplifier would follow a similar procedure.

1. Identify the feedback topology:
   - A. Is feedback signal s_f applied in series (v_f) or in shunt (i_f) with the signal source s_s?
   - B. Is sampled signal s_o obtained at the output node (v_o) or from the output loop (i_o)?
2. Draw the basic amplifier circuit with the feedback set to zero; that is
   - A. For the correct input circuit:
     - (a) With shunt sampling, short-circuit the output nodes to set $v_o = 0$.
     - (b) With series sampling, open-circuit the output loop to set $i_o = 0$.
   - B. For the correct output circuit:
     - (a) With series summing, open-circuit the input loop to set $v_f = 0$.
     - (b) With shunt summing, short-circuit the input nodes to set $i_f = 0$.
3. Indicate s_f and s_o, and solve for the feedback factor ($f = s_f/s_o$).

TABLE 12. CURRENT SOURCES

Type	Reference Current (I_R)	Output Current (I_{C2})	Output Resistance (r_o)
Simple	$(V_{CC} - V_{BE1})/R$	$I_R/(1 + 2/\beta_F)$	r_{ce2}
Improved	$(V_{CC} - V_{BE3} - V_{BE1})/R$	$I_R/[1 + 2/(\beta_F^2 + \beta_F)]$	r_{ce2}
Widlar	$(V_{CC} - V_{BE1})/R$	$(V_T/R_2)\ln(I_{C1}/I_{C2})$	$r_{ce2}[1 + \beta_0 R_2/(1/g_{m1} + r_{\pi 2} + R_2)]$
Wilson	$(V_{CC} - V_{BE2} - V_{BE3})/R$	$I_R[1 - 2/(\beta_F^2 + 2\beta_F + 2)]$	$\beta_0 r_{ce2}/2$

TABLE 13. IDEAL AMPLIFIER CHARACTERISTICS

Parameter	Voltage Amplifier	Current Amplifier	Transconductance Amplifier	Transresistance Amplifier
Input resistance (r_i)	$\to\infty(>>R_1)$	$\to 0(<<R_1)$	$\to\infty(>>R_1)$	$\to 0(<<R_1)$
Output resistance (r_o)	$\to 0(<<R_L)$	$\to\infty(>>R_L)$	$\to\infty(>>R_L)$	$\to 0(<<R_L)$
Transfer characteristic (a)	$a_v = v_o/v_s$	$a_i = i_o/i_s$	$g_m = i_o/v_s$	$r_m = v_o/i_s$

Note: Transconductance (g_m) used here refers to an amplifier parameter, not necessarily just a device parameter.

4. Evaluate the open-loop gain function (a).
5. From a and f, find T, D_s, A, R_i, and R_o.

Information to aid in the analysis and design of feedback amplifiers is summarized in Table 14. Notice that the effect of negative feedback is to modify the open-loop parameters of an amplifier so that the closed-loop performance approaches the ideal characteristics as listed in Table 13.

BANDPASS AMPLIFIERS *

A bandpass amplifier selectively amplifies a narrow band of frequencies around a center frequency. The selectivity is indicated by

$$Q = \omega_0/\Delta\omega = \omega_0/(\omega_h - \omega_l)$$

where,
ω_0 is the center frequency,
$\Delta\omega$ is the bandwidth,
ω_h and ω_l are the high and low cutoff frequencies (−3 dB).

For a bandpass amplifier, typically $Q > 10$.

Single-Tuned Interstage

A single-tuned interstage is modeled in Fig. 23. Resistor R is the total shunt resistance at the output of an amplifying stage including the input resistance of the following stage. Similarly, capacitor C is the total shunt capacitance.

For a single-tuned interstage the general transfer function is

$$a_v(j\omega) = v_o/v_i = -g_m R/[1 + jQ(\omega/\omega_o - \omega_o/\omega)]$$

and

$$\Delta\omega = \omega_h - \omega_l = \omega_o/Q = 1/RC$$

At resonance:

$$\omega_o = 1/(LC)^{1/2}$$

and the center-frequency voltage gain is

$$a_v(j\omega_o) = -g_m R$$

With $Q >> 1$, the high and low cutoff frequencies may be determined from

$$\omega_h = \omega_o(1 + 1/2Q)$$
$$\omega_l = \omega_o(1 - 1/2Q)$$

For a cascade of n synchronous single-tuned stages that are unilateral and noninteracting, the magnitude of the center-frequency voltage gain is

$$|A_v| = a_v^n = (g_m R)^n$$

and the overall bandwidth is

$$\Delta\omega_n = \Delta\omega(2^{1/n} - 1)^{1/2}$$

Double-Tuned Interstage

A double-tuned interstage is modeled in Fig. 24. Here, R_1 and C_1 are the total shunt resistance and capacitance at the output of an amplifying stage, and R_2 and C_2 are the total shunt resistance and capacitance at the input of the following stage. The transformer is loosely coupled so that the coefficient of coupling $k << 1.0$, typically < 0.1.

For a double-tuned interstage, the general transfer function is

$$a_v(j\omega) = v_o/v_i$$

$$= -g_m(R_1R_2)^{1/2}\frac{s}{[(1+s^2)^2 - 2(s^2 - b/2)a^2 + a^4]^{1/2}}$$

* References 18 and 20.

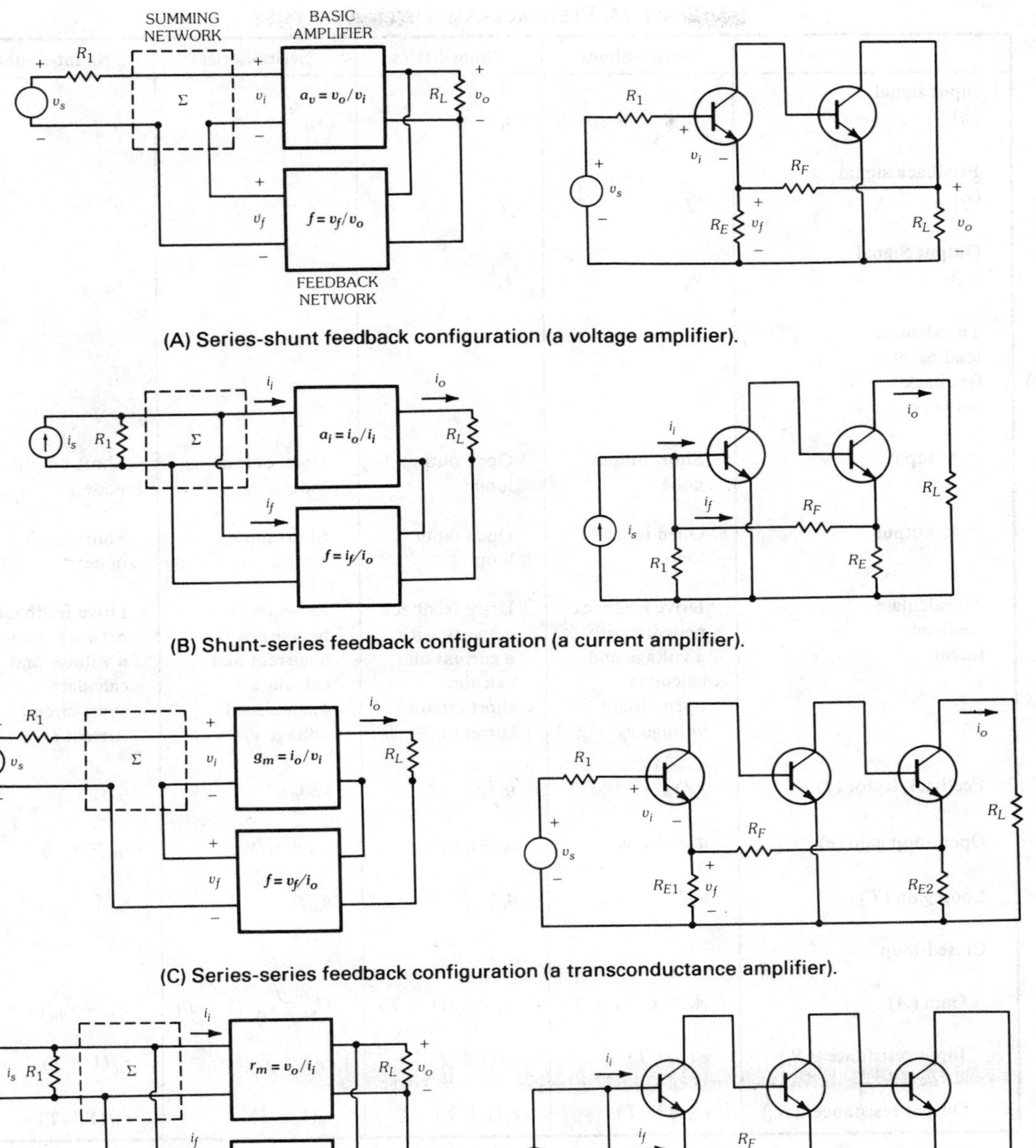

(A) Series-shunt feedback configuration (a voltage amplifier).

(B) Shunt-series feedback configuration (a current amplifier).

(C) Series-series feedback configuration (a transconductance amplifier).

(D) Shunt-shunt feedback configuration (a transresistance amplifier).

Fig. 22. Feedback-amplifier topologies.

where,

$s = k(Q_1 Q_2)^{1/2}$
$p = Q_1/Q_2$
$b = (Q_1/Q_2 + Q_2/Q_1)$
$a = (Q_1 Q_2)^{1/2}(\omega/\omega_0 - \omega_0/\omega)$

At resonance:

$$\omega_0 = 1/(L_1C_1)^{1/2} = 1/(L_2C_2)^{1/2}$$

The gain of the interstage is at a maximum with $s = 1$; the transformer is then critically coupled, and

$$a_v(j\omega_0) = - g_m(R_1R_2)^{1/2}/2$$

TABLE 14. FEEDBACK-AMPLIFIER ANALYSIS

	Series-Shunt	Shunt-Series	Series-Series	Shunt-Shunt
Input signal (s_i)	v_i	i_i	v_i	i_i
Feedback signal (s_f)	v_f	i_f	v_f	i_f
Output Signal (s_o)	v_o	i_o	i_o	v_o
To calculate loading of feedback network				
At input	Short output node	Open output loop	Open output loop	Short output node
At output	Open input loop	Open input loop	Short input node	Short input node
To calculate feedback factor	Drive feedback network with a voltage and calculate open-circuit voltage v_f	Drive feedback network with a current and calculate short-circuit current i_f	Drive feedback network with a current and calculate open-circuit voltage v_f	Drive feedback network with a voltage and calculate short-circuit current i_f
Feefback factor (f)	v_f/v_o	i_f/i_o	v_f/i_o	i_f/v_o
Open-loop gain (a)	$a_v = v_o/v_i$	$a_i = i_o/i_i$	$g_m = i_o/v_i$	$r_m = v_o/i_i$
Loop gain (T)	$a_v f$	$a_i f$	$g_m f$	$r_m f$
Closed-loop				
Gain (A)	$A_v = a_v/(1+T)$	$A_i = a_i/(1+T)$	$G_m = g_m/(1+T)$	$R_m = r_m/(1+T)$
Input resistance (R_i)	$r_i(1+T)$	$r_i/(1+T)$	$r_i(1+T)$	$r_i/(1+T)$
Output resistance (R_o)	$r_o/(1+T)$	$r_o(1+T)$	$r_o(1+T)$	$r_o/(1+T)$

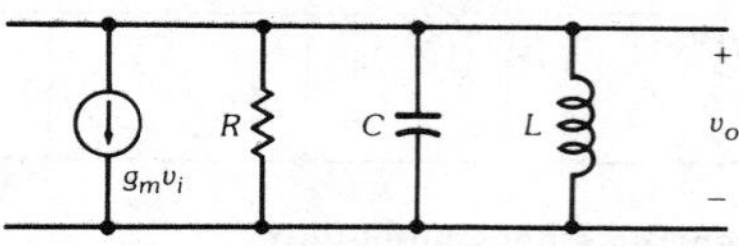

Fig. 23. Single-tuned interstage.

with

$$k_c = 1/(Q_1 Q_2)^{1/2}$$

In practice, it is customary to use not the value of k which gives the maximum gain, but that value which gives the flattest selectivity curve. The circuit is then said to be transitionally coupled.

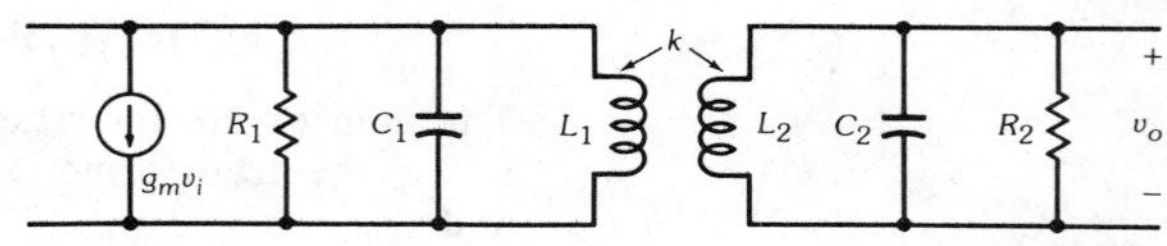

Fig. 24. Double-tuned interstage.

$$k_t = [(1/Q_1^2 + 1/Q_2^2)/2]^{1/2}$$

For a transitionally coupled circuit

$$a_v(j\omega_0) = -g_m(R_1R_2)^{1/2}[2p(1+p^2)]^{1/2}/(1+p)^2$$

and the bandwidth is

$$\Delta\omega = [(1+p)/\sqrt{2}\,p]\,\omega_0/Q_2$$

With equal Q's, $Q_1 = Q_2 = Q_m$, it follows that $k_t = k_c = k_m$ and

$$a_v(j\omega_0) = -g_m(R_1R_2)^{1/2}/2$$

$$\Delta\omega = \sqrt{2}\,\omega_0/Q_m$$

and

$$k_m = \Delta\omega/\sqrt{2}\omega_0 = 1/Q_m$$

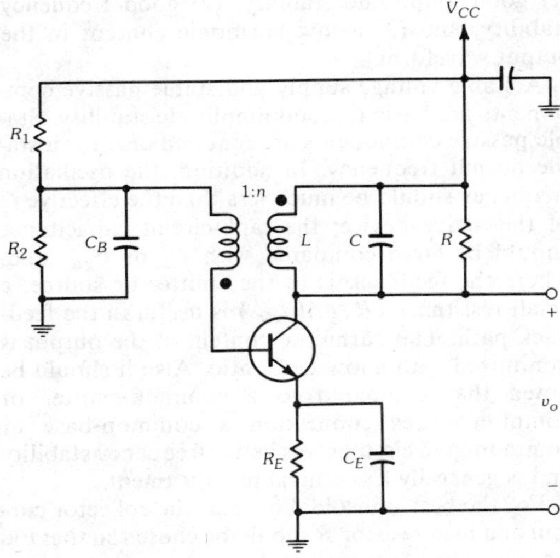

(A) Bipolar junction transistor.

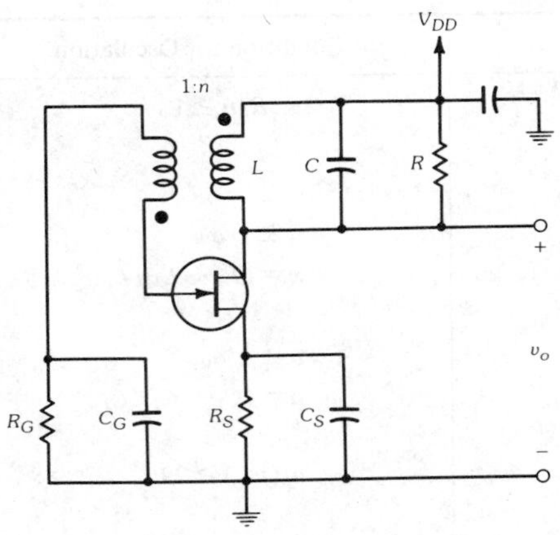

(B) Junction field-effect transistor.

Fig. 25. Transformer-coupled feedback oscillators.

Comparing a double-tuned interstage (with $n = 1$) and a single-tuned interstage yields the same center-frequency gain, but the double-tuned circuit provides a squarer response over the passband.

For a cascade of n synchronous double-tuned stages that are unilateral and noninteracting, the magnitude of the center-frequency voltage gain is

$$|A_v| = a_v{}^n = [g_m(R_1R_2)^{1/2}/2]^n$$

and the overall bandwidth is

$$\Delta\omega_n = \Delta\omega(2^{1/n} - 1)^{1/4}$$

SINUSOIDAL OSCILLATORS*

In the section on feedback amplifiers, *negative* feedback was used to advantage to produce near-ideal amplifiers. In this section, *positive* feedback is used to produce near-ideal sinusoidal oscillators. Fig. 25 illustrates two transformer-coupled feedback oscillator circuits. In Fig. 26, the popular Colpitts and Hartley circuits are shown. These are LC tuned-circuit feedback oscillators. Two RC oscilla-

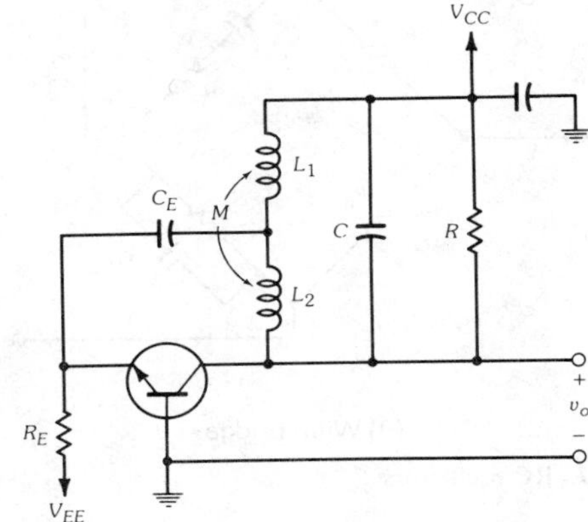

(A) Hartley.

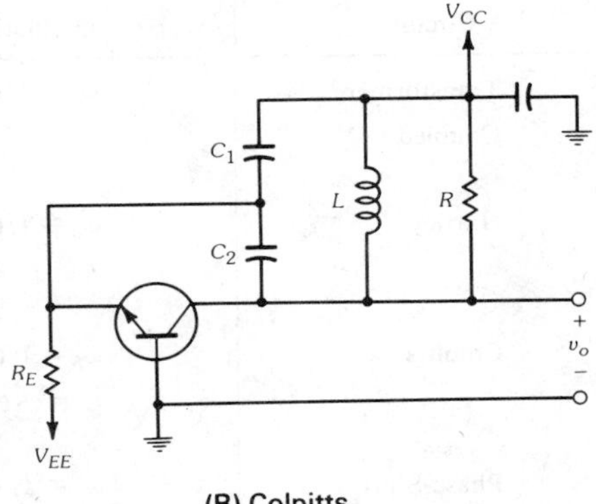

(B) Colpitts.

Fig. 26. Tuned-circuit feedback oscillators.

* References 3 and 19.

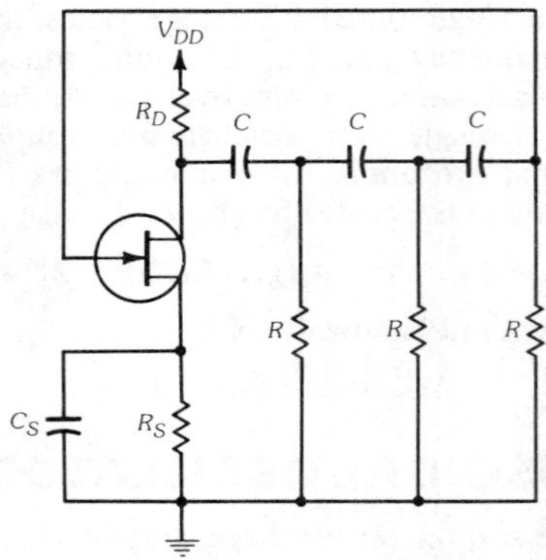

(A) Phase-shift.

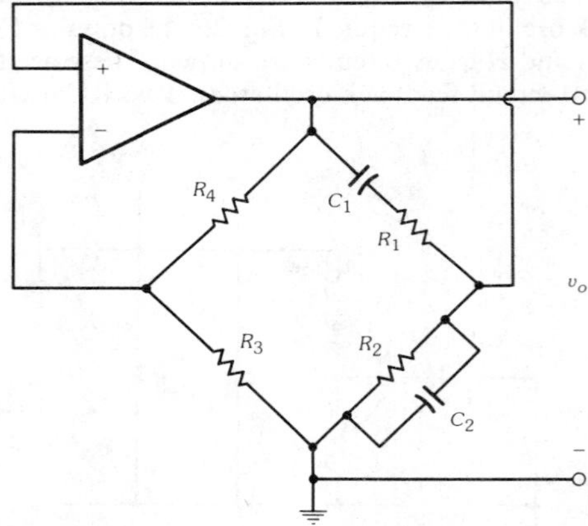

(B) Wien bridge.

Fig. 27. RC oscillators.

tor circuits are presented in Fig. 27. Equations for the oscillation frequency and conditions for oscillation are listed in Table 15.

Tight coupling is used in the transformer-coupled oscillators so that coefficient of coupling k is close to 1.0. To sustain oscillations, but to prevent gross distortion in the output waveform, the ratio $g_m R/n$ is typically chosen to be about 3. The turns ratio for the transformer is not critical; a value of about 10 is appropriate. For the Hartley and Colpitts oscillators, "turns ratio" n is typically about 3.

For the RC oscillators, loop gain T is equal to 1 at ω_0, provided that the voltage gain of the basic amplifier is as listed in Table 15. The RC oscillators are used at low frequencies, where the size of the inductor in an LC oscillator would be too large.

Prime requirements of sinusoidal oscillators are (1) good amplitude stability, (2) good frequency stability, and (3) a low harmonic content in the output waveform.

A stable voltage supply and stable passive components are basic to good amplitude stability. Stable passive components are required also for a stable output frequency. In addition, the oscillation frequency should be much less than the effective f_T of the active device; the tank-circuit capacitance should be large compared with C_μ or C_{gd}; and, where the feedback is to the emitter or source, a small resistance ($R \approx 10/g_m$) is useful in the feedback path. The harmonic content at the output is minimized with a low L/C ratio. Also it should be noted that compared to a common-emitter or common-source connection a common-base or common-gate circuit gives better frequency stability and is generally less critical in adjustment.

For the best sinusoidal output, the collector current and load resistor R should be chosen so that the transistor just cuts off but does not saturate on the

TABLE 15. OSCILLATOR EQUATIONS

Circuit	Oscillation Frequency	Condition for Oscillation
Transformer Coupled	$\omega_o = 1/(LC)^{1/2}$	$g_m R/n \geq 1$
Hartley	$\omega_o = 1/(LC)^{1/2}$ $L = L_1 + L_2$	$n \geq 1/\alpha_0$ $n = (L_1 + L_2)/L_1$
Colpitts	$\omega_o = 1/(LC)^{1/2}$ $C = C_1C_2/(C_1 + C_2)$	$n \geq 1/\alpha_0$ $n = (C_1 + C_2)/C_2$
Phase-Shift	$\omega_o = 1/RC$	$a_v(j\omega_o) \geq 29$
Wien Bridge	$\omega_o = 1/R_1C_1 = 1/R_2C_2$	$a_v(j\omega_o) = 3$ i.e., $R_4 = 2R_3$

peaks of the waveform. However, this class-A operation does lead to a low power-conversion efficiency. An oscillator operated near class A has good frequency stability and a low harmonic content, but for a high power-conversion efficiency class-C operation is preferred.

The RC oscillators are normally used at frequencies below 1 MHz. The Hartley is a good general-purpose oscillator to about 10 MHz. The Colpitts oscillator has good stability and a low harmonic output. It is used almost exclusively at frequencies above 10 MHz.

PULSE CIRCUITS†

Pulse Shaping

The shaping of pulses by means of passive RC and RL circuits is common. Fig. 28 shows the response of each of these simple networks to a step-voltage input and a sine-wave input. For a step input, $v_i = V$. With a step-voltage at the input of a simple high-pass network, the fall time between the 0.9 and 0.1 points on the output waveform is $t_f = 2.2\tau$. Similarly, for a simple low-pass network the rise time between the 0.1 and 0.9 points of the output waveform is $t_r = 2.2\tau$. The high-pass network also serves as a voltage differentiator, since $v_o(t) = \tau(dv_i/dt)$. The low-pass network serves as a voltage integrator, since $v_o(t) = (1/\tau)\int v_i\,dt$. The response of these shaping networks to an exponential voltage input, $v_i = V(1 - e^{-t/\tau})$, is illustrated in Fig. 29.

Fig. 30 shows the response of an RLC circuit (Fig. 31) to a step-voltage input. Parameter k in these curves is related to capacitance C or resistance R by the following equation:

$$k = (1/2R)\sqrt{L/C}$$

For the critically damped case, $k = 1$ and

$$v_o(t)/V = (4Rt/L)e^{-2Rt/L}$$

† References 12 and 16.

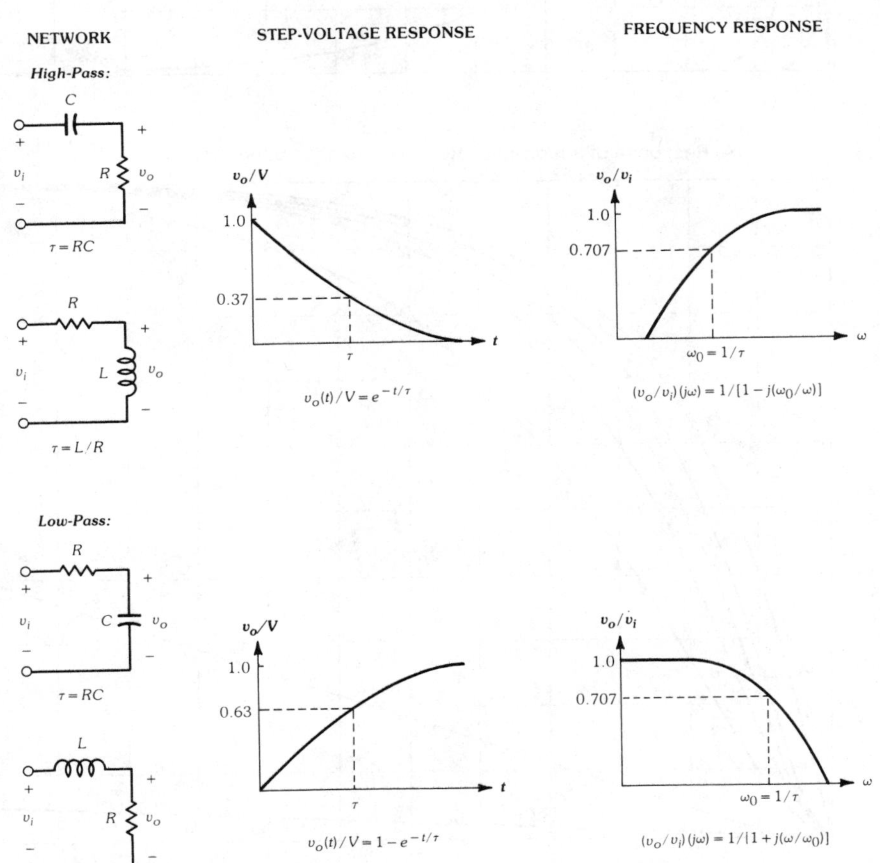

Fig. 28. Response of single RC and RL networks.

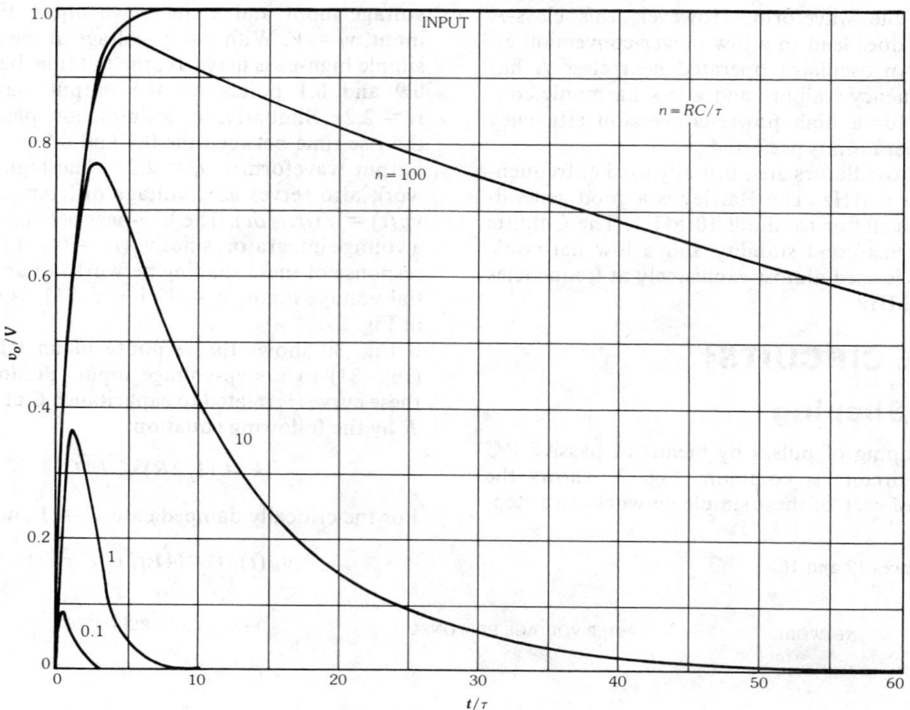

(A) Response of a high-pass RC circuit to an exponential input.

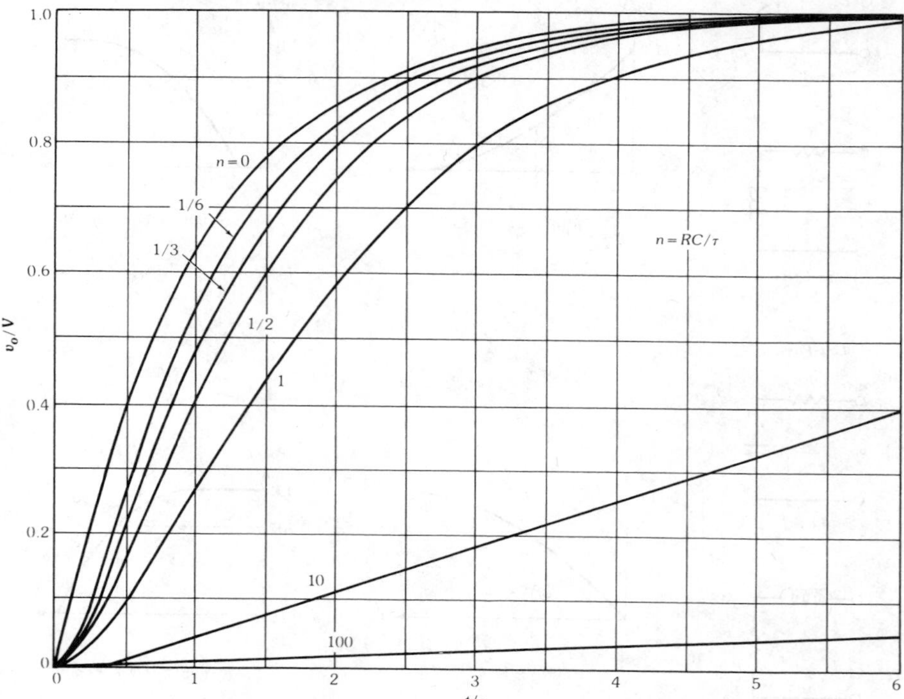

(B) Response of a low-pass RC circuit to an exponential input.

Fig. 29. Response curves for RC networks.

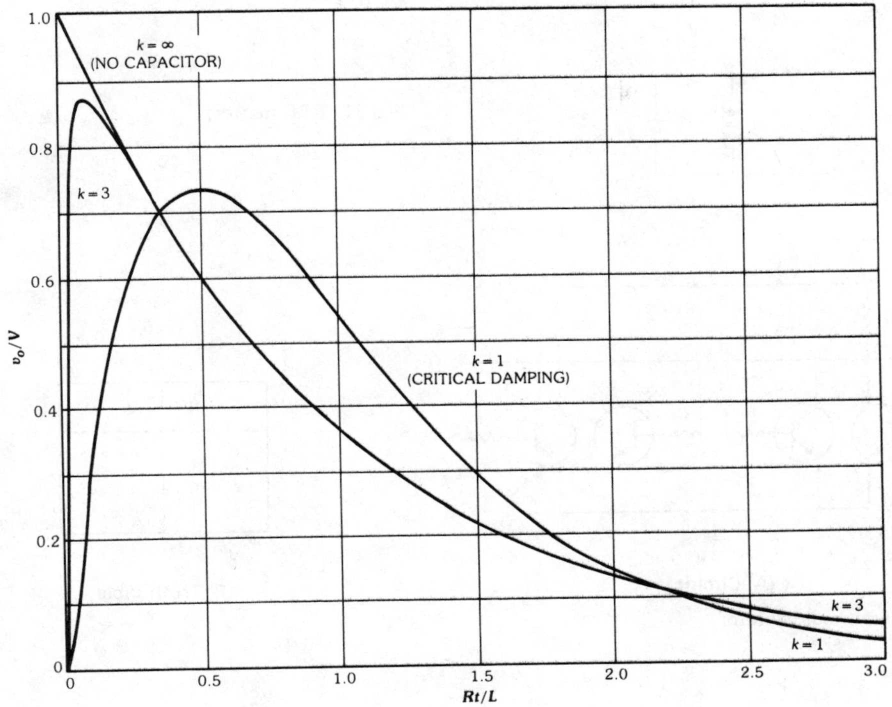

(A) Critically damped and overdamped cases for fixed values of *R* and *L*.

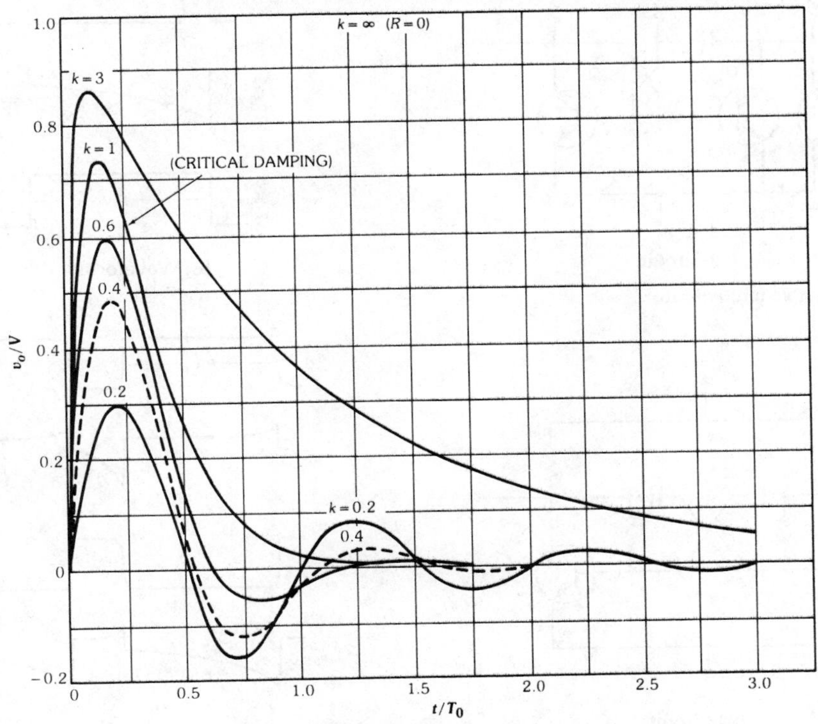

(B) Fixed values of *L* and *C*.

Fig. 30. Response curves for the RLC circuit of Fig. 31.

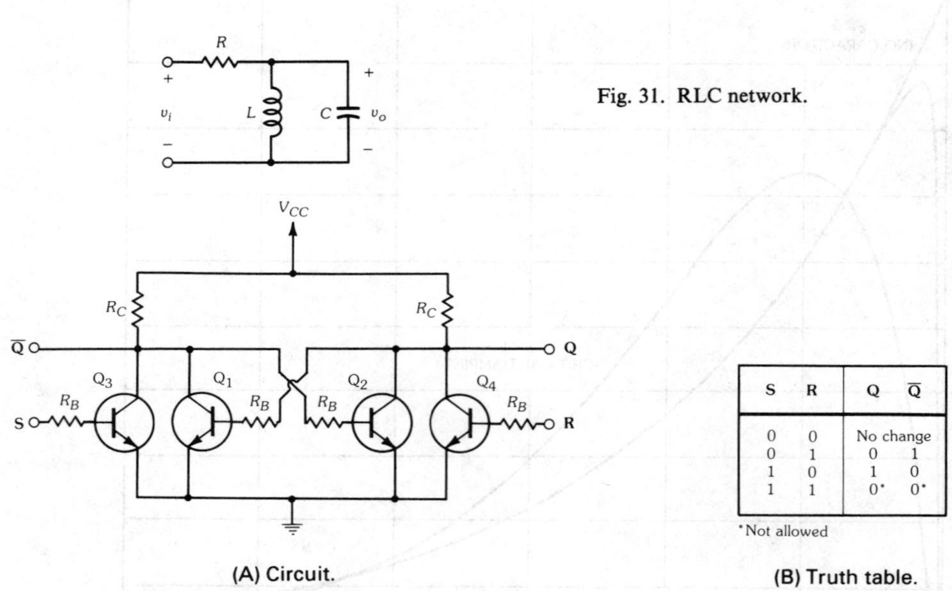

Fig. 31. RLC network.

S	R	Q	Q̄
0	0	No change	
0	1	0	1
1	0	1	0
1	1	0*	0*

*Not allowed

(A) Circuit.

(B) Truth table.

Fig. 32. Bistable multivibrator.

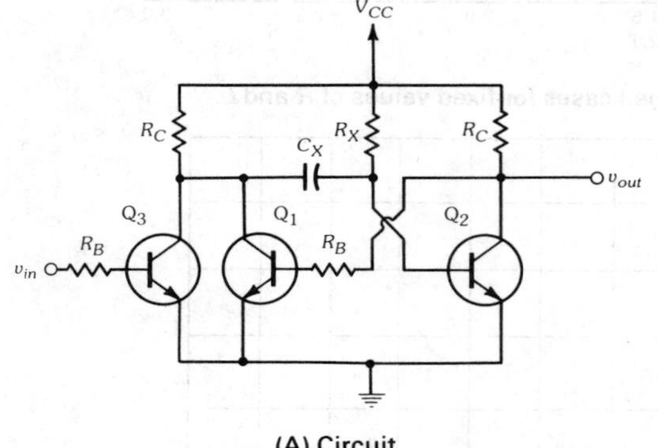

(A) Circuit.

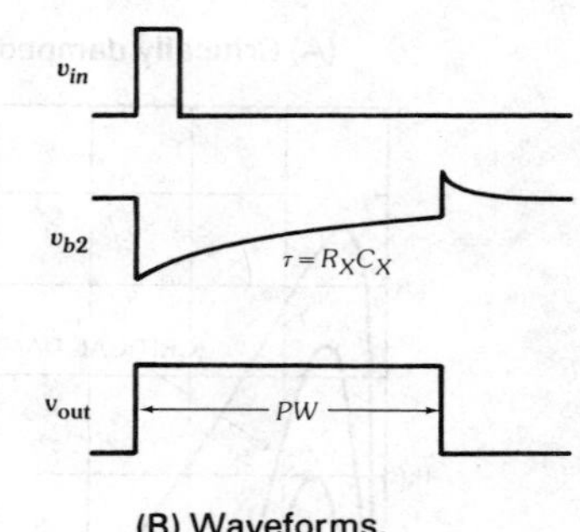

(B) Waveforms.

Fig. 33. Monostable multivibrator.

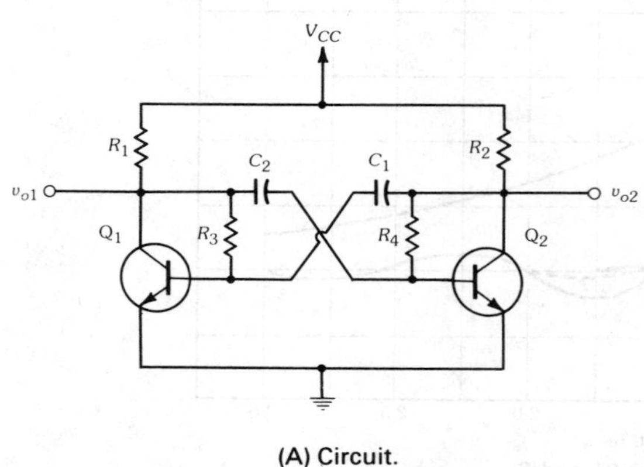

(A) Circuit.

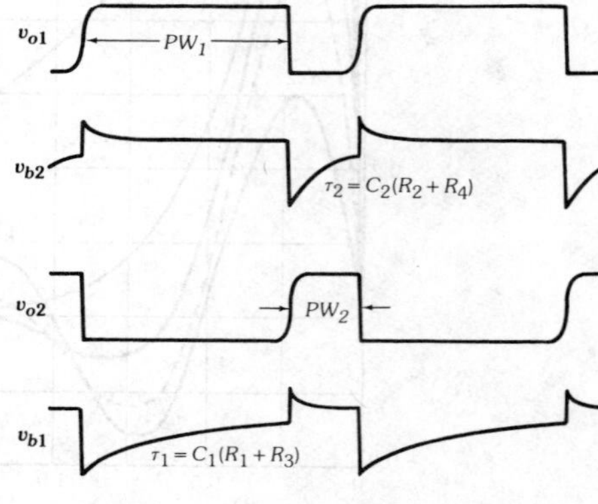

(B) Waveforms.

Fig. 34. Astable multivibrator.

In the overdamped case $k > 1$, and with $T_0 \equiv 2\pi\sqrt{LC}$

$$v_o(t)/V = e^{-\pi t/kT_0} - e^{-4\pi kt/T_0} \approx e^{-Rt/L}$$

In the underdamped case, $k < 1$ and

$$v_o(t)/V = [2k/(1-k^2)^{1/2}]\,e^{-2\pi t/T_0} \sin 2\pi(1-k^2)^{1/2}t/T_0$$

Multivibrators

The three forms of a multivibrator circuit are bistable, monostable, and astable. Each of these operations is best performed with digital integrated circuits. However, discrete transistor versions of these circuits may be used, and some simple examples using saturating bipolar transistors are shown in Figs. 32–34.

For the bistable circuit of Fig. 32, typically $R_B \approx 10R_C$. Included in the figure is the truth table for this bistable latch circuit. Note that since $\bar{Q}$ is defined as the inverse of Q, the $R = S = 1$, $Q = \bar{Q} = 0$ state is not allowed.

In the monostable circuit of Fig. 33, a rectangular pulse is generated at v_{out} due to the trigger pulse at v_{in}. The output pulse width is given by

$$PW \approx 0.7R_xC_x$$

The width of the input pulse should be much less than the output pulse width.

The feedback biasing scheme in the astable circuit of Fig. 34 ensures that the astable will oscillate when power is initially applied to the circuit. For this circuit

$$PW_1 \approx 0.7C_1(R_1 + R_3)$$

$$PW_2 \approx 0.7C_2(R_2 + R_4)$$

With $PW_1 = PW_2$ the output at each collector of the astable is a square wave with a frequency of $1/2PW_1$.

The timing of these circuits may be improved if the transistors are prevented from saturating by connecting a Schottky-barrier diode, or other fast-recovery low forward-voltage diode, from the base to the collector. The anode of the diode is connected to the base terminal.

Finally, each of these multivibrator circuits may be implemented with field-effect transistors, especially complementary MOS(CMOS).

REFERENCES

1. Carr, W. N., and Mize, J. P. *MOS/LSI Design and Application*, Chapter 1. New York: McGraw-Hill Book Co., 1972.
2. *Ibid.*, Chapter 9.
3. Chirlian, P. M. *Analysis and Design of Integrated Circuits*, Chapter 18. New York: Harper and Row Publishers, Inc., 1981.
4. Getreu, I. *Modeling the Bipolar Transistor*. Beaverton, Ore.: Tektronix Inc., 1976.
5. Gray, P. E., and Searle, C. L. *Electronic Principles—Physics, Models, and Circuits*, Chapter 11. New York: John Wiley and Sons, Inc., 1969.
6. *Ibid.*, Chapter 13.
7. *Ibid.*, Chapter 18.
8. Gray, P. R., and Meyer, R. G. *Analysis and Design of Analog Integrated Circuits*, Chapter 3. New York: John Wiley and Sons, Inc., 1977.
9. *Ibid.*, Chapter 4.
10. Grinich, V. H., and Jackson, H. G. *Introduction to Integrated Circuits*, Chapter 8. New York: McGraw-Hill Book Co., 1975.
11. Hamilton, D. J., and Howard, W. G. *Basic Integrated Circuit Engineering*, Chapter 9. New York: McGraw-Hill Book Co., 1975.
12. Hodges, D. A., and Jackson, H. G. *Analysis and Design of Digital Integrated Circuits*, Chapter 8. New York: McGraw-Hill Book Co., 1983.
13. Holt, C. A. *Electronic Circuits—Digital and Analog*, Chapters 2 and 3. New York: John Wiley and Sons, Inc., 1978.
14. Millman, J. *Microelectronics*, Chapter 11. New York: McGraw-Hill Book Co., 1979.
15. *Ibid.*, Chapter 12.
16. Millman, J., and Taub, H. *Pulse Digital and Switching Waveforms*, Chapter 2. New York: McGraw-Hill Book Co., 1965.
17. Muller, R. S., and Kamins, T. I. *Device Electronics for Integrated Circuits*, Chapters 3 and 5. New York: John Wiley and Sons, Inc., 1977.
18. Pederson, D. O. *Electronic Circuits*, Chapter 12. New York: McGraw-Hill Book Co., 1965.
19. *Ibid.*, Chapter 14.
20. Valley, G. E., and Wallman, H. *Vacuum Tube Amplifiers*, Chapters 4 and 5. New York: McGraw-Hill Book Co., 1948.

20 Integrated Circuits

Pallab K. Chatterjee, Satwinder D. S. Mahli,
Michiel deWit, and Harold Hosack

Progress in the complexity and usefulness of integrated circuits over the past two decades has been extremely rapid. The degree of miniaturization provided by advances in silicon technology has spurred on this progress at an exponential rate. The applications of integrated circuits have been quite pervasive. They have allowed major advances in computers and entertainment electronics and have impacted the way we live. In this regard, the explosive nature of this growth is comparable in impact to the industrial revolution. A similar rate of progress is expected through the decade of the 80s. This chapter describes some of the basic aspects of the integrated-circuit technology, designs, and a few applications.

DEFINITIONS AND TERMINOLOGY

access time: 1. Address-to-read output access time—The time delay in random-access memories from the application of appropriate address signal levels to the presence of valid data signal levels at the output. The signal to enable the memory chip enable is present. 2. Chip-enable-to-read output access time—With an appropriate address signal level present, the time delay in random-access memories from the application of appropriate signal levels to the enable inputs to the presence of valid data signal levels at the output.

adder: Switching circuit that combines binary bits to generate the sum and carry of these bits. Takes the bits from two binary numbers to be added (addend and augend), plus the carry from the preceding less-significant bit, and generates the sum and carry.

APCVD: Atmospheric pressure chemical vapor deposition.

ARC: Antireflection coating used in resist processing.

ash: Isotropic dry etching of resist with O_2.

avalanche breakdown: Reverse-voltage breakdown due to electrons gaining sufficient speed to dislodge valence electrons and thus create more current carriers.

backgrind: Process of removal of material from back side of wafers by grinding.

ball bond: Connecting wire from package to IC chip.

barrier layer: A thin layer deposited on contacts before metallization that prevents metal spikes from penetrating through thin junctions and causing unwanted shorts.

beta ratio: For all standard p-channel and n-channel MOS structures, the coefficient $(\mu\epsilon_{ox})/(2t_{ox})$ for the driver device is equal to that of the load device. In this equation, μ = majority carrier mobility, ϵ_{ox} = permittivity of the oxide, and t_{ox} = thickness of the oxide over the channel. Hence, the beta ratio simply becomes the ratio of $(W/L)_{\text{driver}}$ to $(W/L)_{\text{load}}$, where W = width of the channel and L = length of the channel in the direction of the current. Some engineers use the symbol K_R to denote this same ratio.

bias, forward: An external voltage applied in the conducting direction of a pn junction. This is accomplished by connecting the positive terminal of the source to the p-type region and the negative terminal to the n-type region.

bias, reverse: An external voltage applied in the nonconducting direction of a pn junction. The connections are opposite to those for forward bias.

binary coded decimal (bcd): A binary numbering system for coding decimal numbers in groups of four bits. The binary value of these four-bit groups ranges from 0000 to 1001 and codes the decimal digits 0 through 9. To count to 9 takes four bits; to count to 99 takes two groups of four bits; to count to 999 takes three groups of four bits; etc.

binary logic: Digital logic elements that operate with two distinct states. The two states are variously called true and false, high and low, on and off, or 1 and 0. In computers they are represented by two different voltage levels. The level that is more positive (or less negative) than the other is called the high level; the other is called the low level. If the true (1) level is the most positive voltage, such logic is referred to as "positive true" or "positive logic."

bipolar technology: Semiconductor fabrication involving an active semiconductor device with two or more pn junctions. Conduction is by a flow of electrons (in n-type material) or positive holes (in p-type material) across the junction of the two materials.

bird beak: Edge feature produced during local oxidation.

bistable element: Another name for flip-flop. A

circuit in which the output has two stable states (output levels 0 and 1) and can be caused to go to either of these states by input signals, but remains in that state permanently after the input signals are removed. This differentiates the bistable element from a gate, which also has two output states but which requires the retention of the input signals to stay in a given state. The characteristic of two stable states also differentiates the bistable element from a monostable element, which keeps returning to a specific state, and an astable element, which keeps changing from one state to the other.

bit: Binary digit.

boat: Wafer holder generally made of quartz or polysilicon used during furnace processes.

BTS: Bias temperature stress.

buffer memory: A memory system, usually of small capacity compared to a main-frame memory, that provides a buffer between two digital activities.

bus: Long signal-carrying line.

byte: Eight bits.

CAD/CAE/CAM: Computer-aided design/computer-aided engineering/computer-aided manufacture.

CDI: Collector diffusion isolation. A bipolar semiconductor fabrication process that uses an epitaxial layer for the base structure (eliminating a base-diffusion step) and combines the deep collector diffusion contact and isolation steps.

Cell: A collection of devices that is repeated in a circuit. Mostly used to describe unit memory structures.

character generation: A design technique for integrated circuits that utilizes a fixed program in a storage element as the means for generating a dot matrix to represent alphanumeric characters on a crt display.

charge-coupled device (CCD): A type of semiconductor device in which the presence or absence of charge represents the information. Charge-coupled devices store minority-carrier charges in potential wells created at the silicon surface and transport these charge packets along the surface by moving the potential wells.

chip enable: The control signal that activates a complete storage element. No reading or writing of data occurs unless this signal activates the storage element.

class N: Measure of particulate contamination in clean rooms.

clock: A timing control signal required by storage elements and memory systems.

CMOS: *See* complementary MOS.

C_o: Gate capacitance corresponding to the oxide layer over the channel area; $C_o = A(\epsilon_{ox}/t_{ox})$. A = channel area, ϵ_{ox} = permittivity of oxide, t_{ox} = thickness of oxide over channel.

column decoder: Circuits within the storage elements designed to route data from a column of storage cells.

complementary MOS: An MOS fabrication process that combines both p-channel and n-channel transistors on the same substrate. The n-channel transistor is usually the driver device, and the p-channel transistor is the load. Only one transistor of the pair is normally on, except during the switching mode.

complexity: The number of equivalent gates in a circuit.

current, forward: The net current that flows across the forward-biased pn junction.

CVD: Chemical vapor deposition of thin films.

cycle time: 1. Read cycle time—The total time required between the application of address information for reading data number 1 and the application of address information for reading data number 2 with the memory in the read mode. 2. Write cycle time—Same as read cycle time, but for writing of data and memory in the write mode. 3. System cycle time—The time between memory cycle initiations. It usually is determined by the longest of the read or write access or read or write cycle times.

Czochralski: Crystal-pulling process from a melt using a single crystal seed.

decoder: A conversion circuit that accepts digital input information (in the memory case, binary address information) that appears as a small number of lines and selects and activates one of a large number of output lines.

deglaze: Removal of thin layer of SiO_2 by the use of weak HF.

depletion region: The region in a semiconductor in which the mobile-carrier charge density is insufficient to neutralize the net charge density of the fixed donor and acceptor ions.

descum: Anisotropic plasma removal of resist residues after development.

die-by-die alignment: Alignment mechanism in which each die is aligned separately.

diffusion current: The current produced when charges move by diffusion.

diffusion length: The average distance excess minority carriers diffuse between injection and recombination.

diode isolation: A method of producing a high electrical resistance between an integrated-circuit element and the substrate by surrounding the element with a reverse-biased pn junction. The method is also called junction isolation (JI).

diode, semiconductor: A two-electrode semiconductor device that conducts current more easily in one direction than in the other.

DIP: Dual in-line package.

direct mapping: A design technique, used in virtual memory design, that determines the way data are organized and transferred from primary memory to the buffer memory. Data are mapped directly as blocks.

dope: Introduction of different atoms (dopant) in a semiconductor structure, Q.

dose: Number of implanted or diffused ions per unit area.

dot matrix: A matrix of dots that is used to identify alphanumeric characters.

dRAM/sRAM: Dynamic (refreshed storage cells), static (V_{cc} only) RAM.

drift current: A current that is produced when the carriers move under the influence of an applied voltage, i.e., due to a voltage gradient.

drive-in: Deeper thermal diffusion of dopants introduced during predeposition.

DSW: Direct step on wafer for step and repeat projection optical lithography.

dynamic storage elements: Storage elements that contain storage cells that must be refreshed at appropriate time intervals to prevent the loss of information content.

EAROM: Electrically alterable ROM. A fixed-program semiconductor storage element whose program can be altered by the application of external electrical or optical means.

E-beam: Electron beam (exposes resist to pattern masks or wafers).

E_C: Conduction-band energy level.

ECL circuits: Bipolar emitter-coupled logic, also called current-mode logic circuits.

ECL storage cell: A type of bipolar storage cell that is like the multiple-emitter cell but is used for ECL interface.

E_F: Fermi energy level.

E_g: Band-gap energy $= E_C - E_v$.

E_i: Intrinsic energy level; assumed to be at the center of the band gap.

electromigration: Metal line failure due to mass transport of metal under large dc electric field.

electron-beam lithography: Use of focused electron beam to delineate patterns in resist. Commonly used for mask making.

epitaxial growth: The deposition of a monocrystalline layer of material onto a substrate material such that the layer thus formed has the same crystal orientation as the substrate.

EPROM, EEPROM: Electrically programmable, electrically erasable and programmable ROM.

etch filament: Thin residue left along steps during anisotropic dry etching.

etch selectivity: Ability of an etchant to remove layers of certain material with minimal attack on others.

E_v: Valence-band energy level.

excess minority carriers: The number of minority carriers that exceeds the normal equilibrium number in a semiconductor.

factory-programmed ROM: A fixed-program

semiconductor storage element that has been programmed at the factory with a unique bit pattern.

failure rate: The number of components that fail in a given amount of time, usually expressed as a percentage per 1000 hours.

fall time: A measure of the time required for the output voltage of a circuit to change from a high-voltage level to a low-voltage level once a level change has started.

fan-out: The number of loads connected to the output of a logic stage. (A load normally consists of the input impedance of a logic circuit.)

fiducial: Markers on reticles for aid in alignment in projection printers.

filming: Lifting of resist due to lack of adhesion.

fixed-program storage (read-only memory or ready-only store): A special application of random-access storage in which storage is fixed after programming.

flip-flop: A storage element; a circuit having two stable states and the capability of changing from one state to another with the application of a control signal and remaining in that state after removal of signals. (*See* bistable element.)

flip-flop, D: The letter D stands for delay. A flip-flop whose output is a function of the input that appeared one pulse earlier; for example, if a 1 appeared at the input, the output after the next clock pulse would be a 1.

flip-flop, JK: A flip-flop having two inputs, designated J and K. At the application of a clock pulse, a 1 on the J input and a 0 on the K input will set the flip-flop to the 1 state; a 1 on the K input and a 0 on the J input will reset it to the 0 state; and 1 simultaneously on both inputs will cause it to change state regardless of the previous state. J = 0 and K = 0 will prevent change.

flip-flop, RS: A flip-flop consisting of two cross-coupled NAND gates and having two inputs designated R and S. A 1 on the S input and 0 on the R input will reset (clear) the flip-flop to the 0 state; 1on the R input and 0 on the S input will set it to the 1 state. It is assumed that 0 will never appear simultaneously at both inputs. If both inputs have 1, the flip-flop will stay as it was. A 1 is considered nonactivating. A similar circuit can be formed with NOR gates.

flip-flop, RST: A flip-flop having three inputs designated R, S, and T. This unit works like the RS flip-flop, except that the T input is used to cause the flip-flop to change states.

flip-flop, T: A flip-flop having only one input. A pulse appearing on the input will cause the flip-flop to change states. Used in ripple counters.

float zone: Crystal growth process by repetitive melting and slow solidification.

fusible metallization: Utilization of a fusible metal, such as Nichrome, in the fabrication of semiconductor memory circuits. The metal interconnection is severed by a high current to program the storage element.

gate, AND: A logic circuit in which all inputs must have 1-level signals at the input to produce a 1-level output (assuming positive logic).

gate, NAND: A logic circuit in which all inputs must have 1-level signals at the input to produce a 0-level output (assuming positive logic).

gate, NOR: A logic circuit in which any one input or more than one input having a 1-level signal will produce a 0-level output (assuming positive logic).

gate, OR: A logic circuit in which any one input or more than one input having a 1-level signal will produce a 1-level output (assuming positive logic).

gate, XOR: A logic circuit in which any odd number of inputs having a 1-level signal will produce a 1-level output (assuming positive logic).

glass spray: Quartz particles generated by friction of boat with furnace tube during loading and unloading.

global alignment: Alignment mechanism in which only two geometries per wafer are aligned automatically, assuring requisite alignment of others.

g_m: Transconductance = $\delta I_D/\delta V_G\,|_{V_D}$.

GOI: Gate oxide integrity against applied voltage.

hillocks: Protrusions in deposited thin films.

Hilton haze: Hazy look on thin films due to departure from stoichiometry, with appearance akin to Hilton Hotel symbol.

hold time: The time that address information and data information must be maintained after write-enable to guarantee successful writing of data in the memory.

hydrophilic: Surface readily wetted by water.

hydrophobic: Surface not wetted by water.

IGFET: Insulated-gate field effect transistor.

input loading factor (ILF): The load that an input line presents to a driver. It may be expressed as a current, voltage, or impedance.

insulator: A material in which the outer (valence) electrons are tightly bound to the atom and are not free to move. No current can flow when a voltage less than breakdown is applied across the material.

integrated circuit: "The physical realization of a number of electrical elements inseparably associated on or within a continuous body of semiconductor material to perform the functions of a circuit." (EIA definition)

intrinsic concentration: The number of free electrons (or holes) per cubic centimeter in an undoped semiconductor at thermal equilibrium.

ion-beam lithography: Use of focused ion beams for pattern delineation.

ion implantation: An MOS semiconductor fabrication process often used to adjust threshold voltage values by implantation of dopant ions in the gate region, after source and drain formation. The implanted doping level is controlled by the ion accelerator beam current and implant time.

ion milling: Micromachining of thin flims using mechanical motion of energetic ions.

isoplanar: A bipolar semiconductor fabrication process in which the p-diffused isolation regions are replaced by selectively grown oxide isolation.

junction avalanche: Utilization of an avalanche junction in the fabrication of fixed-program (read-only) semiconductor circuits. The avalanched junction forms a connection by high voltage and current to program the storage element.

junction barrier: The opposition to the diffusion of majority carriers across a pn junction due to the charge of fixed donor and acceptor ions.

junction capacitor: A capacitor utilizing the capacitance of a reverse-biased pn junction.

junction, pn: The region of transition between p-type and n-type materials.

junction transistor: An active semiconductor device with a base electrode and two or more junction electrodes.

KOOI effect: Inhibition of subsequent oxidation of silicon underlying Si_3N_4 due to NH_3 formed during wet local oxidation.

L: Effective channel length (in direction of current) in a MOSFET.

large-scale integration (LSI): The simultaneous realization of large-area chips and optimum component packing density, resulting in cost reduction by maximizing the number of system connections done at the chip level. Circuit complexity above 100 gates.

laser anneal: Use of high-energy laser beam for local melting and recrystallization of semiconductors.

lifetime: The average time interval between the introduction and recombination of minority carriers.

loading factors: Specifically used here for memory systems. A numerical measure of the load that must be supplied to drive lines in a memory system.

load line: A line drawn on the family of collector characteristic curves of a transistor showing how the transistor collector voltage changes as the current through the transistor and load resistance changes.

LOCOS: Local oxidation of silicon. This refers to the use of silicon nitride to protect against oxidation.

logic swing: The voltage difference between the two logic levels, 1 and 0.

LPCVD: Low pressure chemical vapor deposition.

mainframe memory: The main memory of the digital system.

mechanically programmable semiconductor ROM: A fixed-program (read-only) semiconductor storage element that can be programmed by breaking interconnection by mechanical means (wiping metal away).

medium-scale integration (MSI): The realization of circuit complexities between 12 and 100 equivalent gates.

MESFET: Metal gate Schottky field-effect transistor.

metallization: A thin-film pattern of conductive material (usually aluminum) deposited on a substrate to interconnect electronic components or to provide conductive contacts to which interconnecting wires may be bonded.

MINIMOS: MOSFET simulation program (available from University of Vienna).

mobile charge: Alkali ions that move through oxide under the influence of applied electric field.

mobility: The average velocity attained by a charge carrier under the influence of a unit electric field.

monolithic integrated circuit: An electronic circuit that has been fabricated as an inseparable assembly of circuit elements in a single structure that cannot be divided without permanently destroying its intended electronic function.

MOS capacitor: A capacitor formed by depositing a silicon-oxide dielectric layer and then a metal top electrode on the surface of a semiconductor region that forms the bottom electrode. The use of silicon as one of the capacitor plates makes the capacitance a function of applied voltage.

MOSFET (metal-oxide-semiconductor field-effect transistor): An active semiconductor device in which a conducting channel is induced in the region between two electrodes by a voltage applied to an insulated electrode on the surface of the region.

MOS, MIS: Metal-oxide-semiconductor, metal-insulator-semiconductor.

MTBF: Mean time between failures of an electronic system.

multilevel oxide: Insulating layer between two conductive layers.

multiple-emitter cell: A type of bipolar storage cell that uses a multiple-emitter transistor control for coupling to the bit lines.

N_A, N_D, N: Doping levels. (A = acceptor, D = donor.)

negative logic: Logic in which the more negative voltage represents the 1 state and the less negative voltage represents the 0 state. (*See* binary logic.)

nibble: Four binary digits.

NMOS, PMOS: Designations for n-channel, p-channel MOSFET.

noise immunity: A measure of the insensitivity of a logic circuit to triggering or reacting to spurious or undesirable electrical signals or noise, largely determined by the signal swing of the logic. Noise can occur in either of two directions, positive or negative.

nondestructive readout: Semiconductor memory designed so that readout does not affect the content stored. It is not necessary to perform a write after every read operation.

NOT: A Boolean logic operation indicating negation. Actually an inverter. If input is 1, output is *NOT* 1 = 0; if input is 0, output is *NOT* 0 = 1. Graphically represented by a bar over a Boolean symbol: $\bar{A}$. $\bar{A}$ means "when A is not 1."

n-type semiconductor: A semiconductor in which electric conduction is due to the presence of more free electrons than holes.

NVRAM: Nonvolatile RAM.

OC curve: Operating characteristic curve. For a particular sampling plan, the graph of the probability of acceptance for all values of percent defective. Completely describes the risks involved in using the plan.

ODE: Orientation dependent etch.

ohmic contact: A resistive contact area that permits aluminum to be used as interconnecting metal from one high-resistivity n-type region to another.

OR: Oxide removal (for contact openings).

oxidation: A process that converts the surface of a silicon wafer to silicon dioxide. This is accomplished by subjecting the wafer to an oxygen or steam atmosphere at very high temperatures.

P: Heavy p-type diffusion.

parallel: The technique for handling a binary data word that has more than one bit. All bits are acted upon simultaneously. It is like the line of a football team; upon a signal, all linemen act.

parallel operation: The organization of data manipulation within computer circuitry wherein all the digits of a word are transmitted simultaneously on separate lines in order to speed up operation, as opposed to serial operation.

parasitics: Stray components associated with the desired components diffused into an integrated circuit. Such parasitics may consist of capacitances, resistances, diodes, or transistors effectively in series or in shunt with the diffused components. They tend to limit the performance of the desired components in a circuit unless compensated for in device and circuit design.

passivation: Protection against penetration by impurity atoms. A silicon surface is passivated by covering it with a thin layer of variously doped films. This layer cannot easily be penetrated by impurities at normal processing and operating temperatures and, therefore, provides the necessary protection.

passive elements: Electronic components, such as resistors and capacitors, that simply introduce resistance or reactance into an electrical circuit but cannot change the waveform of an applied sine wave.

photomasking: A semiconductor-fabrication process in which a photographic negative is used to delineate selective chemical change to portions of the semiconductor surface.

pinholes: Weak spots in thin films due to defects or contamination.

Pirahna: A cleaning agent consisting of H_2O_2 and H_2SO_4.

PLA: Programmable logic array. An integrated circuit that employs ROM matrices to combine sum and product terms of logic networks.

planar transistor: A diffused-junction transistor in which the emitter, base, and collector regions all come to the same plane surface, with the junctions between the regions protected at the surface by a layer of material such as silicon oxide.

plasma: Ionized gas in which concentrations of positive and negative charge carriers are almost equal.

plasma etch: Use of a plasma of reactive ions to etch thin layers.

plug bar: Assortment of test structures added on IC mask for model extraction and diagnostic purposes.

pn junction: The region of transition between p-type and n-type semiconductor materials.

poly: Polycrystalline silicon.

potential barrier: The difference in potential across a pn junction.

precharge time (reset time): The timing pulse width within a memory cycle that is used for charging node capacitances to particular starting-point voltage levels.

PREDEP: Predeposition of dopants.

process flow: The detailed step-by-step sequence of a fabrication schedule.

propagation delay: The time required for a change in logic level to be transmitted through an element or a chain of elements.

PSG: Phosphosilicate glass.

p-type semiconductor: A semiconductor in which electric conduction is due to the presence of more holes than free electrons.

purge: Extended exposure to only a desired gas or a mixture of gases.

ϕ_F: Fermi potential; the amount the Fermi level is displaced from the intrinsic level or the center of the gap (as measured in the bulk). Units are volts.

ϕ_s: Surface potential; the amount the intrinsic Fermi level, at the surface, has been shifted with respect to the bulk Fermi level.

q: Electronic charge; 1.6×10^{-19} coulomb.

random-access memory (RAM): A memory from which information can be obtained at the output with approximately the same time delay by choosing an address randomly and without first searching through a vast amount of irrelevant data.

range: Depth of peak of implanted ion distribution.

ratio inverter: An inverter whose logic swing is determined by the beta ratio (β_R) of the load and driver devices.

ratioless inverter: An inverter whose logic swing closely approximates the power-supply voltage. ("Ratioless" because the load and driver device do not conduct simultaneously.)

ratioless-type shift register: Current does not flow through the inverter when the clock and data inputs are simultaneously at the logic 1 level.

ratio-type shift register: Current flows through the inverter when the clock and data inputs are simultaneously at logic 1.

reactive ion etching: Etching of films using plasma and mechanical motion of reactive ions.

reflow: High-temperature treatment of phosphosilicate glass that causes its mass flow.

refresh: Method that restores capacitance charge that deteriorates because of leakage.

register: Temporary storage for digital data.

reox: Oxide removal from heavily implanted region and its thermal oxidation for sake of passivation.

repair frequency: The rate at which an electronic system must be repaired, i.e., once/day, once/week, once/four weeks, etc.

reset: Also called "clear." Similar to set except it is the input through which the Q output can be made to go to 0.

resist: Photosensitive organic or inorganic resins.

resistivity: The (volume) resistivity ρ is the electric field E required to produce a unit current density J. That is, $\rho = E/J$.

rinse cascade: A series of containers with running deionized water used for wafer rinsing after chemical treatment.

ROM: Read-only memory.

row decoder: Circuits within the storage elements designed to route data from a row of storage cells.

SAG (self-aligned gate): An MOS fabrication process using a self-aligning gate formed from deposited silicon.

SAMPLE: Lithography and etch simulation program available from the University of Berkeley.

SATO (self-aligned thick oxide) process: An MOS fabrication process using nitride as the self-aligning gate material. Thus, the gate metallization does not have to withstand diffusion temperature.

Schmitt trigger: An input circuit with hysteresis. There is a higher threshold for positive-going inputs than for negative-going inputs.

Schottky barrier diode: Metal-semiconductor barrier diode.

segregation: Preferential accumulation of dopants in either silicon or oxide during thermal oxidation.

self-refresh: A circuit-design technique that incorporates the refresh method in the storage-element circuitry so that external refresh circuitry is not required.

SEM: Scanning electron microscope.

semiconductor: A material with conductivity roughly midway between that of conductors and insulators, and in which the conductivity increases with temperature over a certain temperature range.

sense amplifier: A sensitive amplifier accepting linear voltage or current signals and producing logic-level outputs.

sequentially accessed memory: A memory from which information is received at the output in varying time delays from a reference point depending on the position of the data in a time sequence.

serial accumulator: A register that receives data bits in sequence and temporarily holds the data for future use.

serial operation: The organization of data manipulation within computer circuitry wherein the digits of a word are transmitted one at a time along a single line. The serial mode of operation is slower than parallel operation, but utilizes less complex circuitry.

sheet resistance: The resistance per square of a sheet of material.

SIM: Secondary ion microscope.

single crystal: A piece of material in which all the basic groups of atoms have the same crystallographic orientation.

sinter: Annealing treatment after metal etch to promote ohmic contacts.

slice: A single wafer cut from a silicon ingot, forming a thin substrate on which all active and passive elements for multiple integrated circuits have been fabricated by semiconductor epitaxial growth, diffusion, passivation, masking, photoresist, and metallization technologies. A completed slice generally contains hundreds of individual circuits, called chips or bars.

SOI: Silicon on insulator.

solid-state diffusion: The introduction of atoms of an impurity element into the surface regions of a solid semiconductor wafer.

SOS: Silicon on sapphire.

SPICE 2: Version of circuit simulation computer program (University of Berkeley).

state: The condition of an input or output of a circuit as to whether it is logic 1 or logic 0. The state of a circuit (gate or flip-flop) refers to its output. A flip-flop is said to be in the 1 state when its Q output is 1. A gate is in the 1 state when its output is 1.

static storage elements: Storage elements that contain storage cells that retain their information as long as power is applied unless the information is altered by external excitation.

step coverage: Ability of thin films to maintain thickness when going over feature steps.

stored-charge programmable semiconductor ROM: Utilizing a charge stored on a floating gate or a dielectric as a means of programming an MOS fixed-program (read-only) semiconductor storage element.

straggle: Standard deviation around the range of implanted ions.

substrate: The physical material on which an integrated circuit is fabricated. Its primary function is mechanical support, but it may serve some electrical function also.

SUPREM: Process simulation program available from Stanford University.

TEM: Transmission electron microscope.

thermal compression bond: A commonly used method for attaching a very fine wire to a point (usually a bonding pad) on an integrated-circuit chip.

thermal generation: The creation of a hole and a free electron by freeing a bound electron through the addition of heat energy.

three-state output: An output condition that has a low on impedance for driving to the high state, a low on impedance for driving to the low state, and an intermediate high-impedance off state.

throughput: Rate of material processed per hour.

TIRAM: Taper isolated random-access memory, an advanced dynamic RAM cell.

toggle: To switch between two states, as in a flip-flop.

t_{ox}: Oxide thickness.

trigger: A timing pulse used to initiate the transmission of logic signals through the appropriate circuit signal paths.

truth table: A chart that tabulates and summarizes all the combinations of possible states of the inputs and outputs of a circuit. It tabulates what will happen at the output for a given input combination.

TTL: Transistor-transistor logic multiple emitter bipolar semiconductor circuit. Bipolar semiconductor transistor-transistor coupled logic circuits.

two-level main memory: A memory system featuring two separate memories. One memory is a buffer store, or cache memory. The other memory is the primary storage. Coupled together, these form a virtual memory with the capacity of the primary memory and the speed of the buffer memory.

two-level metallization: A semiconductor fabrication process in which there are two levels of interconnecting metal on the surface of the integrated circuit.

ULSI: Ultralarge-scale integration; circuit complexity above 1 million transistors.

vacancy: Unoccupied position in bond structure receptive to substitutional occupancy.

V_{BG}: Back-gate bias.

V_{cc}, V_{dd}, V_{ss}: High, intermediate, low supply-voltage levels.

vector generation: A design technique for generating the coordinates for positioning the beam in a crt display that uses only the changes in coordinates of the beam rather than the absolute coordinates.

VFB: Flatband voltage.

V_{GG}: Gate supply voltage.

V_{GS}: Gate-to-source voltage.

VHPIC: Very-high-performance integrated circuits (British program).

VHSIC: Very-high-speed integrated circuits.

VLSI: Very large-scale integration; circuit complexity above 100 000 transistors.

V_{TD}: Threshold voltage of the driver device.

V_{TL}: Threshold voltage of the load device.

V_T (V_{BG}): Threshold voltage as a function of back-gate bias.

W: Effective channel width (perpendicular to current) in a MOSFET.

wafer flat: A flat portion of an otherwise circular wafer for orientation identification.

wafer stepper: Resist exposure system in which each die is focused and exposed separately.

word: Sixteen bits.

write enable: Also called read/write or R/W. The control signal to a storage element or a memory that activates the write mode or operation. When the device is not in the write mode, the read mode is active.

write time: 1. Address-to-write time—The time delay in random-access memories from the application of appropriate address signal levels until the write-mode control signal of an appropriate level is applied. 2. Chip-enable-to-write time—The time delay, with appropriate address signal levels present, from the application of appropriate signal levels to the enable inputs until the write-mode control signal of an appropriate level is applied. 3. The time that the appropriate level must be maintained on the write-enable line and that data must be present to guarantee successful writing of data in the memory.

yield: The percentage of acceptable circuits (chips) produced by a particular process, process step, inspection, or test.

zener: Breakdown in diodes due to tunneling.

IC FABRICATION

The discipline of integrated silicon processing is one that has been continuously and rapidly evolving, especially in the last decade. This section discusses processes that currently are or are about to be in widespread production use. New processes continue to be invented and developed, and to keep abreast of current understanding, the reader is well advised to refer to pertinent technical journals.* Since silicon-based circuits are the mainstream of IC technology, repeated references are made to silicon in the following discussion; however, the concepts presented apply equally to processing of other materials, such as gallium arsenide.

Substrate-Cleaning Procedures

The performance characteristics of semiconductor devices are sensitive to cleanliness in processing. The removal of unwanted impurities from the wafer surface is important because such impurities may diffuse into the semiconductor during high-temperature processing, altering its bulk and surface properties. Many of the spurious impurities may be either donor or acceptor dopants, directly affecting the device characteristics. Other impurities may cause surface or bulk defects such as traps, stacking faults, or dislocations. Surface contaminants such as oil, grease, or other organic matter may lead to poor film adhesion. This requires a careful chemical cleaning of wafers at the initiation of the process and appropriate cleaning at various steps during processing.

The initial cleanup generally starts with wafer scrubbing to remove loose particulate contaminants. This is followed by treatment with organic chemicals to get rid of any possible organic impurities such as hydrocarbons and greases that may be

* For example, references 1, 2, and 3.

TABLE 1. CLEANING REAGENTS USED IN IC PROCESSING

Cleaning Agent	Boiling Point (°C)	Purpose
Ethanol	78.3	Remove organic contaminants
Trichloroethylene	87.2	
Acetone	56.2	
p-Xylene	138.4	
5 H_2O: 1 H_2O_2: 1 NH_4OH	—	Remove inorganic contaminants (heavy metals)
6 H_2O: 1 H_2O_2: 1 HCl	—	
1 H_2SO_4: 1 H_2O_2	—	
1 HCl: 1 HNO_3	—	

remnants from the wafer-grinding process. Organic solvents such as methanol and ethanol are suitable for this purpose. The final cleanup consists of a variety of inorganic chemicals to remove heavy metals, etc. Most of these chemical mixtures are strong oxidants,† forming a thin oxide at the wafer surface. This oxide is then stripped so that impurities absorbed therein are removed. Table 1 lists some cleaning reagents commonly used in IC processing.

Thin films

All IC processes involve the use of a number of thin films—insulators, semiconductors, and conductors. These films may form an essential part of a particular structure or just be of use in the implementation of a process step. The following means of achieving thin films are of widespread use.

Thermal Growth—This technique involves heating the substrate in a furnace at precisely controlled temperature and gas ambient. Generally, a high temperature, in the range from 800°C to 1200°C for silicon processing, is used to promote chemical reaction between the ambient gases and the substrate. The prominent example is the growth of SiO_2 on silicon in O_2 ambient. It is also possible to grow Si_3N_4 in an N_2 or, preferably, NH_3 atmosphere. It is possible to exercise extremely good control on absolute thickness as well as thickness uniformity across the wafer. Figs. 1 and 2 show the relationship between oxidation time and resultant oxide thickness in dry O_2 and steam, respectively.**

Chemical Vapor Deposition—The gas-phase reduction of highly reactive chemicals under low pressure forms a convenient way of obtaining very uniform thin films. This method is widely used for deposition of oxides, nitrides, and polycrystalline semiconductors. A conformal deposition around sharp edges is an important attribute of this technique. Table 2 lists constituent gases and suitable temperatures for the deposition of some common films.‡

† Reference 4.

** Reference 5.

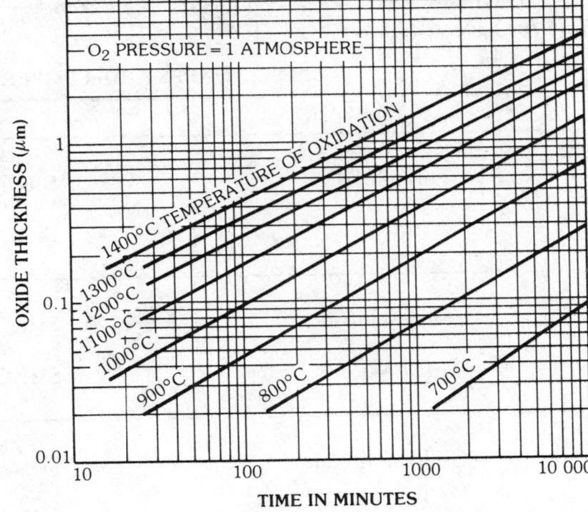

Fig. 1. Silicon dioxide growth rate in dry oxygen. (*From A. M. Smith, "Experimental Measurements,"* in Burger and Donovan, eds.,* Fundamenals of Silicon Integrated Device Technology, *Vol. I. Englewood Cliffs, N.J.: Prentice-Hall, Inc.; Fig. 6-42.)*

Plasma Deposition—The production of thin films by electric discharge or plasma depends on the capability of a nonequilibrium but sustained plasma to generate chemically reactive species at

* References: B. I. Boltaks and H. Shih-yin, "Diffusion, Solubility and the Effect of Silver Impurities on Electrical Properties of Silicon," *Soviet Phys. Solid State* 2, May 1961, p. 2303. F. A. Trumbore, "Solid Solubilities of Impurity Elements in Germanium and Silicon," *Bell System Tech. J.* 39, January 1960, pp. 205—233.

‡ Reference 6.

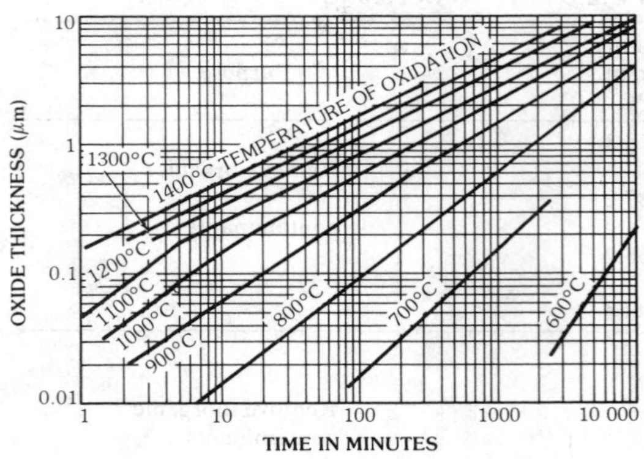

Fig. 2. Silicon dioxide growth rate in steam. (*From A. M. Smith, "Experimental Measurements,"* in Burger and Donovan, eds.,* Fundamentals of Silicon Integrated Device Technology, *Vol. I. Englewood Cliffs, N.J.: Prentice-Hall, Inc.; Fig. 6-42.*)

TABLE 2. GASES AND TEMPERATURES FOR DEPOSITION OF FILMS

Film	Gases	Temperature (°C)
Polysilicon	SiH_4, N_2	650 – 700
(Si)	SiH_4, H_2	850 – 950
Silicon nitride	SiH_4, NH_3, N_2	750 – 800
(Si_3N_4)	SiH_2Cl_2, NH_3, N_2	750 – 900
Silicon dioxide	SiH_4, N_2O, N_2	750 – 850
(SiO_2)	SiH_4, CO_2, H_2	950 – 1000

TABLE 3. GASES USED FOR PLASMA DEPOSITIONS

Film	Gases
Amorphous silicon (Si)	SiH_4, Ar
Silicon dioxide (SiO_2)	SiH_4, N_2O
Silicon nitride (Si_3N_4)	SiH_4, NH_3

low temperature. The plasma is typically sustained at 0.1 to several torr and exhibits free electron temperatures of tens of thousands of kelvins, while the temperature of the translational or rotational modes of atoms, radicals, or molecules is only hundreds of kelvins. Thus, deposition can be made at very low temperatures. The deposition scheme is in use for oxides, nitrides, carbides, and amorphous semiconductors. Table 3 list gases used.§

Evaporation—Evaporation involves film deposition by vaporizing the material on heating it past its melting point under vacuum to produce enough vapor pressure. Either resistive heating or E-beam heating is used to bring about melting. This technique is used mainly to deposit metals such as aluminum. A list of temperatures and support materials used to evaporate various elements is in Table 4.†

Sputter Deposition—In this scheme, the material to be deposited is bombarded with positive inert ions with kinetic energy far exceeding the heat of sublimation of the target material. This results in dislodging of target atoms and their ejection into the gas phase succeeded by deposition on the substrate, which may or may not be biased negatively. In widespread use are sputter deposition of metals and metal silicides.

Spin-On Deposition—The material to be deposited is mixed with a suitable solvent and spun coated on the substrate. The resulting thickness is a function of spin speed and viscosity of the solution used. Subsequently, an oven-bake drives out the

* Boltaks and Shih-yin, *loc. cit.*

§ Reference 7.

† Reference 8.

TABLE 4. TEMPERATURES AND SUPPORT MATERIALS USED TO EVAPORATE VARIOUS ELEMENTS†

Element and Predominant Vapor Species	Temp (°C)		Support Materials		Remarks
	mp	$p^* = 10^{-2}$ Torr	Wire, Foil	Crucible	
Aluminum (Al)	659	1220	W	C, BN TiB_2–BN	Wets all materials readily and tends to creep out of containers. Alloys with W and reacts with carbon. Nitride crucibles preferred.
Antimony (Sb_4, Sb_2)	630	530	Mo, Ta, Ni	Oxides, BN, metals, C	Polyatomic vapor, $\alpha v = 0.2$. Requires temperatures above mp. Toxic.
Arsenic (As_4, As_2)	820	300	......	Oxides, C	Polyatomic vapor, $\alpha v = 5\times10^{-5}$–$5\times10^{-2}$. Sublimates but requires temperatures above 300°C. Toxic.
Barium (Ba)	710	610	W, Mo, Ta, Ni, Fe	Metals	Wets refractory metals without alloying. Reacts with most oxides at elevated temperatures.
Beryllium (Be)	1283	1230	W, Mo, Ta	C, refractory oxides	Wets refractory metals. Toxic, particularly BeO dust.
Bismuth (Bi, Bi_2)	271	670	W, Mo, Ta, Ni	Oxides, C, metals	Vapors are toxic.
Boron (B)	2100 ±100	2000	......	C	Deposits from carbon supports are probably not pure boron.
Cadmium (Cd)	321	265	W, Mo, Ta, Fe, Ni	Oxides, metals	Film condensation requires high supersaturation. Sublimates. Wall deposits of Cd spoil vacuum system.
Calcium (Ca)	850	600	W	Al_2O_2	
Carbon (C_3, C_1, C_2)	3700	2600	......	......	Carbon-arc or electron-bombardment evaporation. $\alpha v < 1$

(Continued on next page.)

solvent, leaving behind a stable layer. This procedure is used for deposition of layers of resists and doped or undoped oxides.

Lithography

Lithography is the aspect of the IC fabrication process that deals with transferring onto a substrate the detailed features associated with individual components that collectively comprise an integrated circuit. The following lithographic steps must be implemented for each level of an IC fabrication schedule.

Mask Generation—A mask, or reticle, is a flat plate or a membrane that features a geometrical pattern with areas that are selectively transparent or opaque to a wavelength or a band of wavelengths used in a particular lithographic system. The pattern dimensions may be the same as the final size required in the circuit, or they may be larger by a factor N in an N:1 exposure system, where generally $N = 1$, 5, or 10. If $N > 1$, then the resolution requirements during the process of mask generation are relaxed, and less exacting equipment and techniques may be utilized to reduce costs.

The first step during mask making is the circuit layout. The goals of a good layout are to transform all designed components associated with a circuit

TABLE 4 (CONT). TEMPERATURES AND SUPPORT MATERIALS USED TO EVAPORATE VARIOUS ELEMENTS†

Element and Predominant Vapor Species	Temp (°C) mp	Temp (°C) p* = 10^{-2} Torr	Support Materials Wire, Foil	Support Materials Crucible	Remarks
Chromium (Cr)	1900	1400	W, Ta	......	High evaporation rates without melting. Sublimation from radiation-heated Cr rods preferred. Cr electrodeposits are likely to release hydrogen.
Cobalt (Co)	1495	1520	W	Al_2O_3, BeO	Alloys with W, charge should not weigh more than 30% of filament to limit destruction. Small sublimation rates possible.
Copper (Cu)	1084	1260	W, Mo, Ta	Mo, C, Al_2O_3	Practically no interaction with refractory materials. Mo preferred for crucibles because it can be machined and conducts heat well.
Gallium (Ga)	30	1130	......	BeO, Al_2O_3	Alloys with refractory metals. The oxides are attacked above 1000°C.
Germanium (Ge)	940	1400	W, Mo, Ta	W, C, Al_2O_3	Wets refractory metals but low solubility in W. Purest films by electron-gun evaporation.
Gold (Au)	1063	1400	W, Mo	Mo, C	Reacts with Ta, wets W and Mo. Mo crucibles last for several evaporations.
Indium (In)	156	950	W, Mo	Mo, C	Mo boats preferred.
Iron (Fe)	1536	1480	W	BeO, Al_2O_3, ZrO_2	Alloys with all refractory metals. Charges should not weigh more than 30% of W filament to limit destruction. Small sublimation rates possible.
Lead (Pb)	328	715	W, Mo, Ni, Fe	Metals	Does not wet refractory metals. Toxic.
Magnesium (Mg)	650	440	W, Mo, Ta, Ni	Fe, C	Sublimates.
Manganese (Mn)	1244	940	W, Mo, Ta	Al_2O_2	Wets refractory metals.

(Continued on next page.)

into a geometrical layout that achieves the required packing density while keeping the parasitics small. The parasitics are those electrical circuit components that are not designed in but that inevitably originate due to some features of a particular layout. Also, the performance of certain designs critically depends on well matched components; therefore, such circuit elements must be laid out identically and close together.

First, the layout rules are established with the capabilities of a particular technology kept in mind. The layout may be accomplished by means of a variety of computer-aided design methodologies. All schemes utilize advanced pattern generating capabilities. Various levels are denoted by differing colors. The geometries may be laid out explicitly or implicitly with a symbolic representation that is later converted into required geometries.* After the layout is completed, this information is fed into the mask-generation system. The mask generation in itself must utilize various lithographic steps to be discussed in detail below.

* Reference 9.

TABLE 4 (CONT). TEMPERATURES AND SUPPORT MATERIALS USED TO EVAPORATE VARIOUS ELEMENTS†

Element and Predominant Vapor Species	Temp (°C)		Support Materials		Remarks
	mp	$p^* = 10^{-2}$ Torr	Wire, Foil	Crucible	
Molybdenum (Mo)	2620	2530	......	......	Small rates by sublimation from Mo foils. Electron-gun evaporation preferred.
Nickel (Ni)	1450	1530	W, W foil lined with Al_2O_3	Refractory oxides	Alloys with refractory metals; hence charge must be limited. Small rates by sublimation from Ni foil or wire. Electron-gun evaporation preferred.
Palladium (Pd)	1550	1460	W, W foil lined with Al_2O_3	Al_2O_3	Alloys with refractory metals. Small sublimation rates possible.
Platinum (Pt)	1770	2100	W	ThO_2, ZrO_2	Alloys with refractory metals. Multistrand W wire offers short evaporation times. Electron-gun evaporation preferred.
Rhodium (Rh)	1966	2040	W	ThO_2, ZrO_2	Small rates by sublimation from Rh foils. Electron-gun evaporation preferred.
Selenium (Se_2, Se_n: $n = 1–8$)	217	240	Mo, Ta, stainless steel 304	Mo, Ta, C, Al_2O_3	Wets all support materials. Wall deposits spoil vacuum system. Toxic. $\alpha v = 1$
Silicon (Si)	1410	1350	......	BeO, ZrO_2, ThO_2, C	Refractory oxide crucibles are attacked by molten Si, and films are contaminated by SiO. Small rates by sublimation from Si filaments. Electron-gun evaporation gives purest films.
Silver (Ag)	961	1030	Mo, Ta	Mo, C	Does not wet W. Mo crucibles are very durable sources.
Strontium (Sr)	770	540	W, Mo, Ta	Mo, Ta, C	Wets all refractory metals without alloying.

(Continued on next page.)

Resist Casting—The pattern on a mask is replicated on a substrate through the use of resists. Resists are organic or inorganic resins that are sensitive to a wavelength or a band of wavelengths used in a lithographic exposure system and, when exposed, undergo a chemical transformation so that selective removal can be accomplished. A resist is referred to as a positive resist if when exposed it is removed during development; it is referred to as a negative resist if when not exposed it is removable on development.

The resist application involves a thorough cleaning of the substrate to rid it of undesirable contaminants. After careful drying, the substrate is covered by a layer of resist, typically less than 3 μm thick, by spin-coating, spraying, or immersion.

The resist thickness uniformity is critical to obtaining high resolution in a lithographic system. With the advent of very large scale integration (VLSI), a tight line-width control has become of supreme importance. Often, the requirements of obtaining good line-width control, high resolution, and good step coverage are difficult to meet simultaneously. Good step coverage demands a thicker resist, whereas a thinner resist is necessary for good resolution. This is true for both positive and negative, organic and inorganic resists. The problem is shown schematically in Fig 3. The resist thickness diminishes when the resist goes over steps. The effect becomes more pronounced in VLSI due to a larger ratio of line height to line width.

Regardless of resist type, a thin coating and a flat

TABLE 4 (CONT). TEMPERATURES AND SUPPORT MATERIALS USED TO EVAPORATE VARIOUS ELEMENTS†

Element and Predominant Vapor Species	Temp (°C) mp	Temp (°C) $p^* = 10^{-2}$ Torr	Support Materials Wire, Foil	Support Materials Crucible	Remarks
Tantalum (Ta)	3000	3060	......	......	Evaporation by resistance heating of touching Ta wires, or by drawing an arc between Ta rods. Electron-gun evaporation preferred.
Tellurium (Te_2)	450	375	W, Mo, Ta	Mo, Ta, C, Al_2O_3	Wets all refractory metals without alloying. Contaminates vacuum system. Toxic. $\alpha v = 0.4$
Tin (Sn)	232	1250	W, Ta	C, Al_2O_3	Wets and attacks Mo.
Titanium (Ti)	1700	1750	W, Ta	C, ThO_2	Reacts with refractory metals. Small sublimation rates from resistance-heated rods or wires. Electron-gun evaporation preferred.
Tungsten (W)	3380	3230	......	......	Evaporation by resistance heating of touching W wires, or by drawing an arc between W rods. Electron-gun evaporation preferred.
Vanadium (V)	1920	1850	Mo, W	Mo	Wets Mo without alloying. Alloys slightly with W. Small sublimation rates possible.
Zinc (Zn)	420	345	W, Ta, Ni	Fe, Al_2O_3, C, Mo	High sublimation rates. Wets refractory metals without alloying. Wall deposits spoil vacuum system.
Zirconium (Zr)	1850	2400	W	......	Wets and slightly alloys with W. Electron-gun evaporation preferred.

† From Maissel and Glang, *Handbook of thin Film Technology*. New York: McGraw-Hill Book Co., 1970; pp. 1-37 and 1-38.

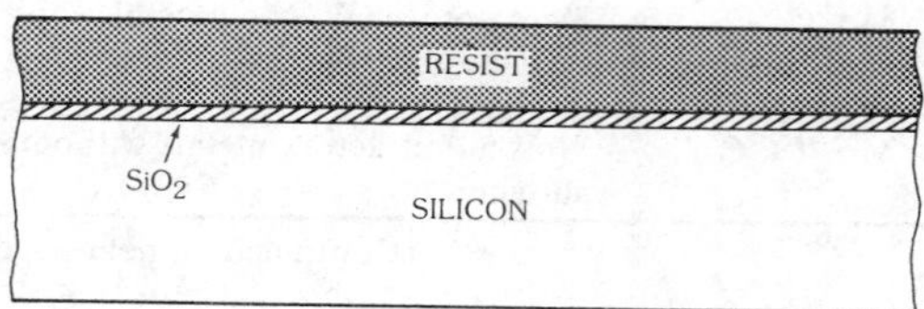

(A) The resist is uniform when the wafer does not have any topology.

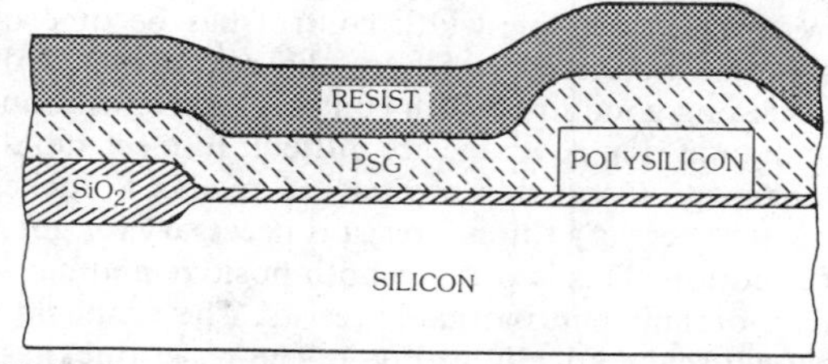

(B) Resist thins as it goes over steps in the presence of surface features.

Fig. 3. Schematic cross-sections showing details of resist-coated wafers.

resist surface are needed for high resolution and good line-width control. Multilevel resist systems are being investigated to fulfill this need.† In a typical double-level resist system, a thick organic resist is first spun. This resist conforms to the wafer surface and is planar at the top. Then an intermediate layer such as SiO_2 is deposited. Now another thin coating of resist is made. This resist is capable of achieving high resolution and good line-width control. This is shown in Fig 4.

After the resist coating is complete, an oven bake, called prebake, is carried out to drive out the solvents, increase sensitivity, and improve resistance to mechanical handling. This finishes the procedure of resist casting.

Resist Exposure—After prebake, the resist is ready for exposure. A wide variety of exposure systems is currently in production use or under development (Fig 5).

† Reference 10.

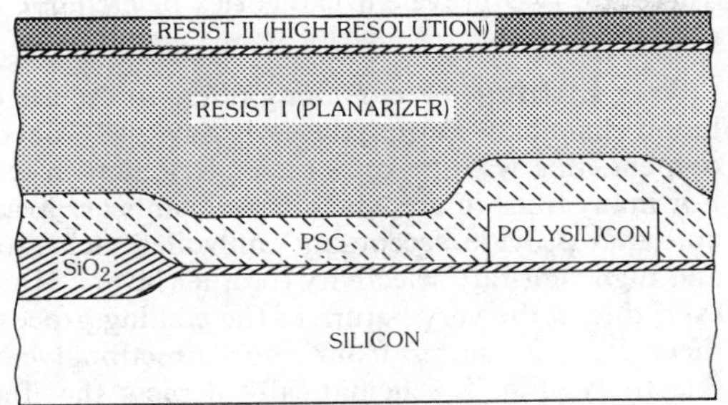

Fig. 4. Schematic cross-section of a wafer coated with double-level resist system.

mask are made to overlap the corresponding structure on the wafer within reasonable bounds, a good alignment is said to have been made. The detailed features of alignment mechanisms differ, and the superiority of a particular system is partly determined by its capability to achieve alignment within tighter bounds.

Resist Development—After exposure, the resist is ready for development. Development refers to selective removal of resist as determined by the mask. It is accomplished either by immersing the wafer in a bath of developer or by spraying fine jets of developer on the wafer surface. Control of both the development time and temperature is important

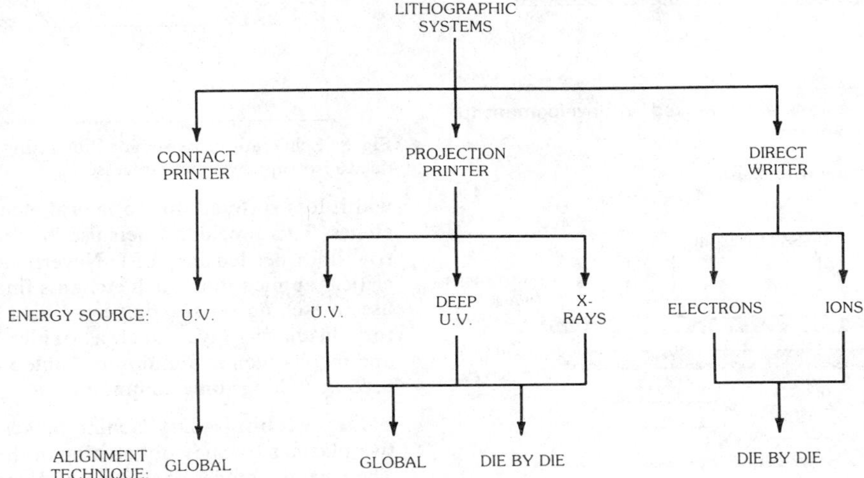

Fig. 5. Current lithographic systems.

The contact printer refers to the scheme in which the mask is in intimate contact with the resist-covered wafer when exposure is made. Contact printers have been the predominant lithographic exposure tool in production in the seventies. They can be used reliably down to 3–4 μm geometries. In a projection printer, the mask or reticle is placed between the exposure source and the wafer to be exposed and does not make contact with the wafer. The mask image is reduced by a factor N and focused on the wafer. The scheme is suitable for high resolution. In a direct-write system, the need for a mask is completely eliminated. The resist is exposed by a finely focused particle beam that can be electrostatically scanned and blinked to form the required exposure pattern. This is inherently a slower procedure, and throughput therefore is adversely affected. Such systems, however, are uniquely suitable for mask generation.

A key feature of an exposure system is the capability to align a given level to an existing level on the wafer. This is done with the help of specially designed geometrical patterns (fiducials), which are placed on each level. When these patterns on the

for reproducible line-width control. Fig. 6 shows schematically a positive-resist film after exposure and development. After development, the wafer is rinsed to remove all unwanted resist residues. Then the wafer is oven baked to make the resist more resistant to subsequent chemical etching (postbake). In multilevel-resist systems, only the top high-resolution resist is developed. With this resist used as a mask, the intermediate layer and thick planarizing resist must be dry etched to reveal the underlying films, as shown in Fig. 7.

Etching

All components of an integrated circuit are made up of a selective arrangement of variously doped semiconductors, insulators, and conductors. This requires the capability to etch certain layers selectively without damaging others. The important features that determine the preferential choice of an etchant are:

(A) Etch rate
(B) Etch rate selectivity
(C) Directional selectivity

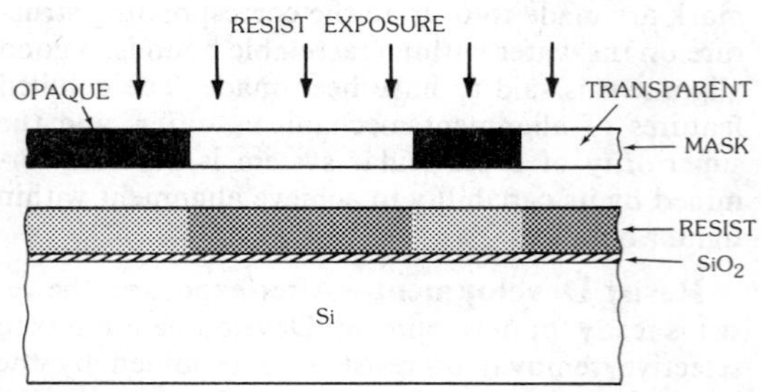

(A) Wafer coated with positive resist is exposed through mask.

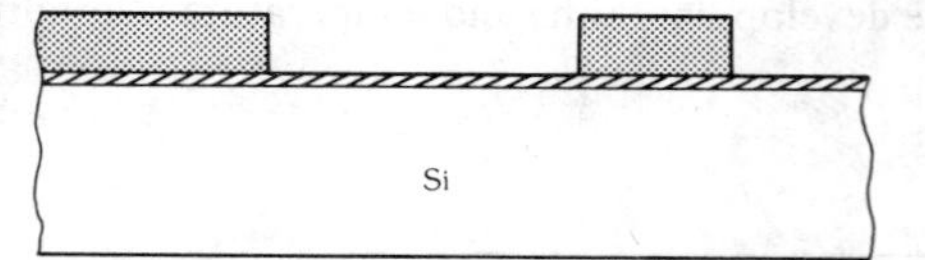

(B) Exposed regions are removed on development to leave desired pattern.

Fig. 6. Patterning with resist.

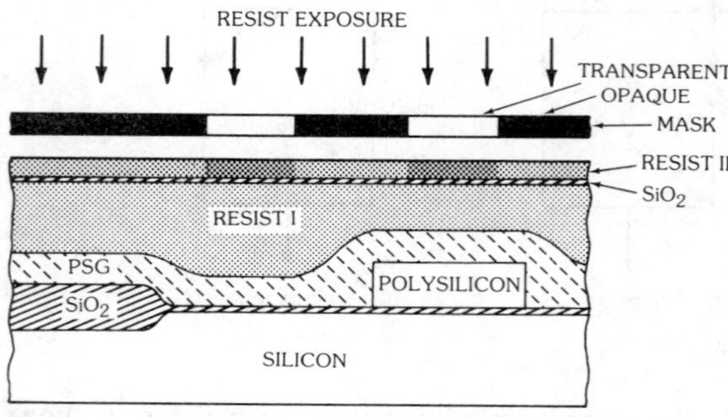

(A) After resist-II exposure.

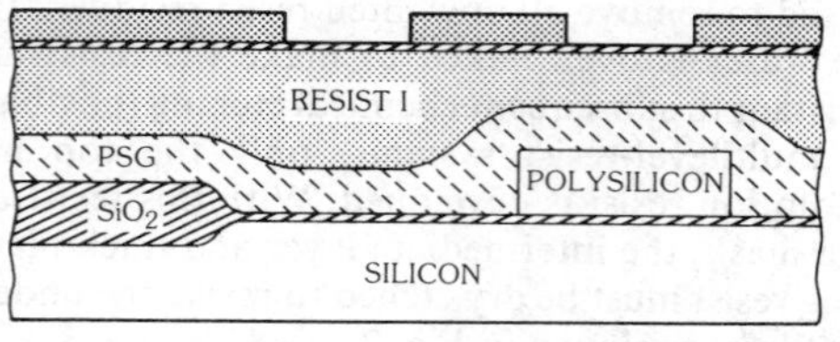

(B) After resist-II development.

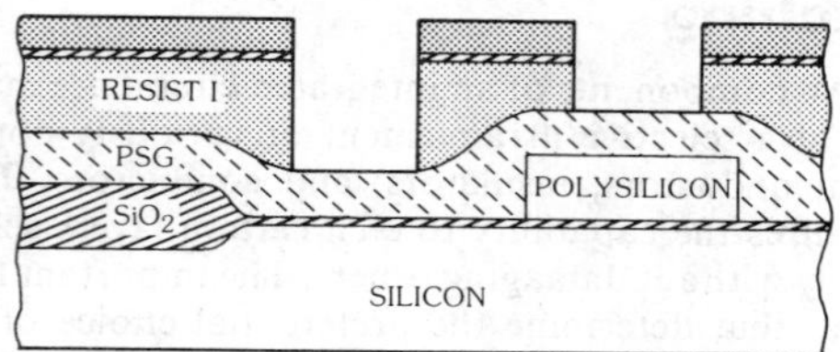

(C) After intermediate-layer and resist-I dry etching.

Fig. 7. Double-level resist systems.

There are two prevalent categories of etchants in production use.

Wet Etching—Wet etching refers to the use of liquid chemicals. During the early days of IC development, this was the only kind of etchant in use. For many films in use, chemical etchants are available that provide acceptably controlled etch rates and high etch-rate selectivity to other layers. However, due to the very nature of the etching process, these etchants suffer from poor directional etch selectivity. Fig. 8 schematically depicts the line-width loss suffered due to lateral etching with wet etches. This precludes their use in etching the narrow lines needed in VLSI. Nevertheless, for noncritical applications, such etchants find widespread use in etching resists, polycrystalline semiconductors, insulating layers such as oxides and nitrides, and metals such as aluminum. Table 5 lists etchants used in etching some common films.

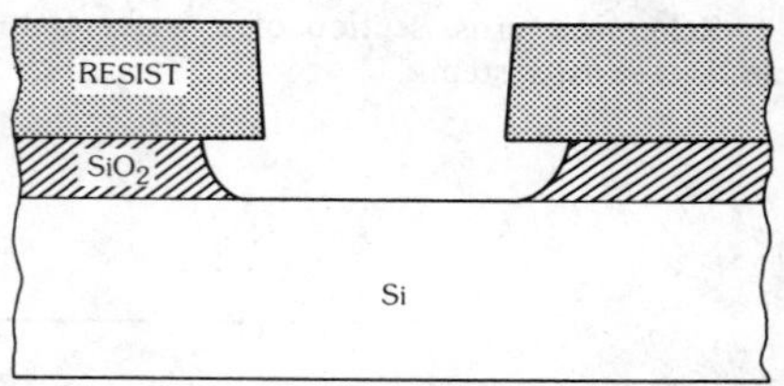

Fig. 8. Schematic cross-section illustrating line-width loss due to isotropic wet etch processes.

Dry Etching—Dry etching makes use of reactive plasmas to carry out etching in the gas phase.* The key advantages of this methodology are highly directional etch anisotropy and the facility to penetrate small resist openings; these make it suitable for etching the small geometries encountered in VLSI. By suitably adjusting the constituents and proportions of etchant species, a good etch-rate selectivity to other layers is achieved. Plasma etching of resists, SiO_2, Si_3N_4, Al, Poly-Si, and metal silicides is in extensive use in the industry today. Fig. 9 shows schematically small geometries etched through a resist mask. A list of gases commonly used for plasma etching is given in Table 6.†

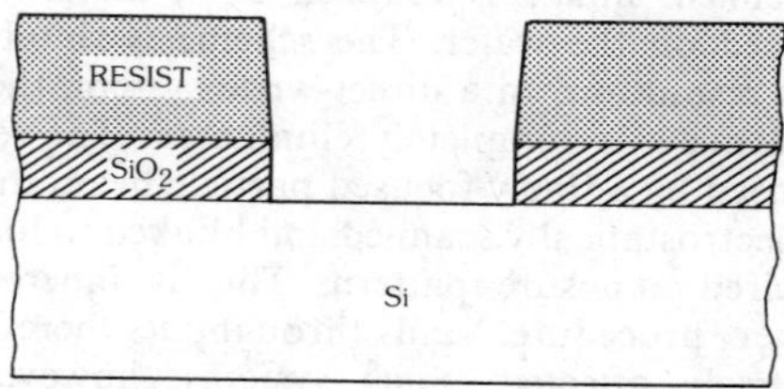

Fig. 9. Due to high anisotropy in etch rate, almost vertical walls are obtainable with plasma etching without any line-width loss.

* Reference 11.
† Reference 7.

TABLE 5. ETCHANTS USED WITH COMMON FILMS

Film	Etch	Composition	Temp. (°C)
Silicon	Planar etch	2 ml HF, 15 ml HNO_3 5 ml CH_3COOH	25
Polysilicon	Iodine etch	50 ml HF, 100 ml HNO_3 110 ml CH_3COOH, 0.3 g I_2	25
Silicon dioxide	Bell 2 etch	54 ml H_2O, 36 ml NH_4F 10 ml HF	25
Silicon nitride	Hot phosphoric etch	H_3PO_4	165
Aluminum	Phosphoric etch	55 ml H_3PO_4, 11 ml CH_3OOH 4.5 ml HNO_3, 2 ml H_2O	25
Titanium	—	90 ml H_2O, 10 ml HF	25
Tantalum	—	20 ml HNO_3, 10 ml HF 10 ml H_2O	25
Molybdenum	Dalton etch	92 g $K_3[Fe(CN)_6]$ 20 g KOH 300 ml H_2O	25
Resists	J100	*	100

* Unknown, manufactured by Indust-Ri-Chem Lab., Richardson, Tex.

TABLE 6. GASES USED FOR PLASMA ETCHING

Film	Gases
Silicon dioxide	SiF_4 CF_4 C_3F_8
Silicon	CF_4, O_2 CCl_4, HCl
Silicon nitride	CF_4
Tantalum, titanium, tungsten, molybdenum, vanadium	CF_4
Chrome, chrome oxide	CCl_4
Aluminum	CCl_4 BCl_3
Resists	Ar, O_2

Substrate Doping Procedures

All semiconductor devices rely on selective doping of various areas either n-type or p-type to a required concentration and depth. There are basically three predominant techniques of introducing dopants in a controlled manner.

Predeposition and Drive-In—The geometries to be doped are lithographically defined on the substrate in a mask layer such as oxide or nitride that is impervious to the given dopant. Then the wafers are exposed to the dopant source, which may be a gas, solid, or liquid, in a well controlled furnace at high temperature. An inert gas such as nitrogen or argon is used as a carrier for the dopant species.

The doping density in the substrate is a function of predeposition temperature and time and is given by the equation

$$N(x,t_p) = N_s \operatorname{erfc}[x/2(D_p t_p)^{1/2}] \quad (1)$$

where,

N is the doping density at a distance x below the surface,
t_p is the predeposition time,
D_p is the diffusion constant,
N_s is the surface concentration.

The surface concentration is usually equal to the solid solubility. The solid solubility and diffusion constant of common impurities in silicon are shown in Figs. 10 and 11.*

The total number of dopants introduced in the substrate per unit area is obtained, by integrating Equation 1 over the depth, to be the dose

$$Q_p = 2N_s (D_p t_p/\pi)^{1/2} \quad (2)$$

After the predeposition cycle, wafers are generally loaded in a different furnace at a higher temperature to drive in the impurities to obtain a required junction depth. A passivating layer such as an oxide is initially either grown or deposited on the wafers to prevent the escape of impurities. A redistribution of impurities takes place, and the doping density as a function of drive-in time is given by

$$N(x,t_d) = [Q_p/(D_d t_d)^{1/2}] \exp(-x^2/4D_d t_d) \quad (3)$$

where,

t_d is the drive-in time,
D_d is the diffusion constant at the drive-in temperature.

Equation (3) can be used to calculate the resulting junction depth if the background concentration is known.

Diffusion From Doped Oxides—Because of the solubility of the dopant at the surface, the standard two-step diffusion process is not suitable for obtaining shallow junctions of low doping concentration, as required in a number of applications. One alternative is to use doped glass as a diffusion source.§ Doped glass containing the required doping can be either chemically deposited at low temperature and low pressure or spun-coated on the wafer surface. The impurities are then driven in at a higher temperature. By limiting the dopant concentration in the doped oxide, shallow junctions with low doping density can be realized.

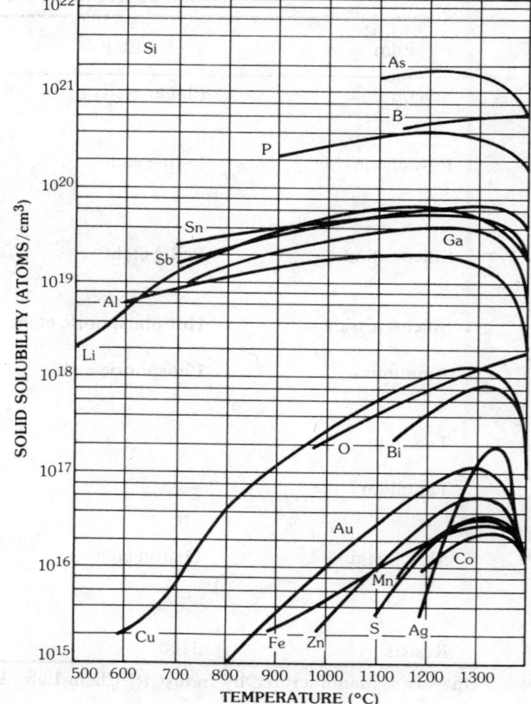

Fig. 10. Solid solubility of elements in silicon. (*From F. A. Trumbore, "Solid Solubilities of Impurity Elements in Germanium and Silicon,"* The Bell System Technical Journal, *Vol. 39, Jan. 1960, pp. 205–233. © 1960, AT&T; reprinted with permission.*)

Doping by Ion Implantation—There are certain limitations associated with the diffusion processes that can be overcome by ion implantation. Ion implantation is a technique of extracting dopant species from a source, separating the required ions from other spurious particles, accelerating them to the required energy, and embedding them in the substrate. The implantation energy determines the impurity concentration. To a good approximation, the implanted ions settle with a gaussian distribution given by reference 15:

* References 12 and 13.

§ Reference 14.

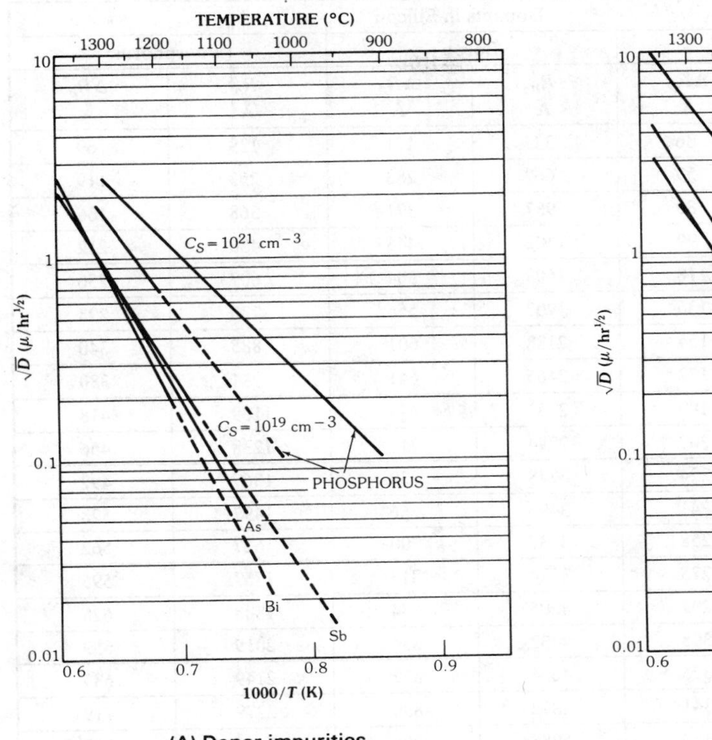

(A) Donor impurities.

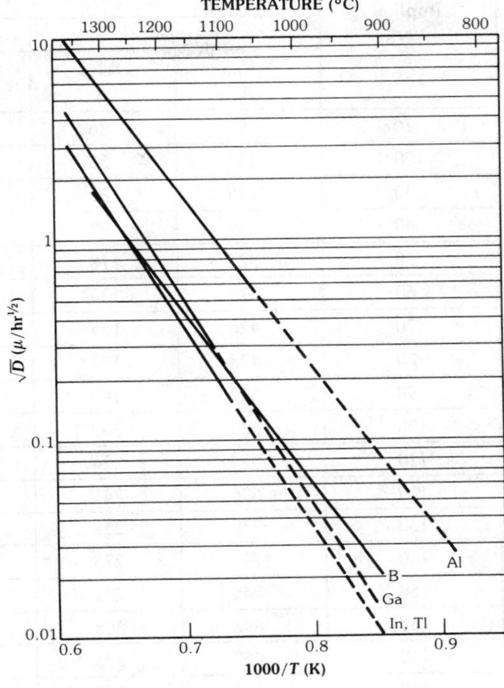

(B) Acceptor impurities.

Fig. 11. Diffusion constant in silicon. (*From A. S. Grove,* Physics and Technology of Semiconductor Devices, *New York: John Wiley & Sons, Inc., 1967.*)

$$N(x) = (Q/\sqrt{2\pi}\Delta R_p)\exp -\tfrac{1}{2}[(R_p - x)/\Delta R_p]^2 \quad (4)$$

where,

$N(x)$ is the dopant concentration at a distance x below the surface,
Q is the ion dose,
R_p is the ion range,
ΔR_p is the range straggle.

The values for R_p and ΔR_p as a function of implant energy are available in published tables. Table 7 gives values for usual dopants in silicon.**

With this technique, it is possible to exercise a tight control on both the depth and the concentration of dopants, which is of prime importance in VLSI. By choosing multiple implant energies and doses, a variety of intended profiles can be approximated. The ion implantation is followed by a thermal anneal, which activates the dopants and anneals out the crystal damage produced by the process.

** Reference 16.

An IC Process Example—NMOS Process

The information presented above can best be consolidated by illustrating an IC fabrication process. For this purpose, we have chosen to discuss NMOS technology, whose typical basic process flow can be illustrated with the help of Fig. 12.

The starting material is p-type silicon in wafers that are carefully cleaned and thermally oxidized in steam to obtain an SiO_2 thickness of 300 Å. An LPCVD Si_3N_4 is now deposited to a thickness of 2000 Å. Next, the isolation regions are defined by using a moat mask as shown in Fig. 12A. Sequentially, dry plasma etching of nitride and oxide is carried out, with resist protecting the areas not to be etched. Resist is removed by J100 wet etch, and a low-energy boron implant is made that raises the p-type doping density in the opened windows but does not penetrate through the nitride and oxide stack, as illustrated in Fig. 12B. Slices are then cleaned and thermally oxidized in steam to oxidize the exposed regions to obtain an SiO_2 thickness of 7000 Å; nitride prohibits oxidation of other areas, as shown in Fig. 12C. Nitride is now wet etched in

TABLE 7. VALUES OF R_p AND ΔR_p

Implant Energy keV	Dopants in Silicon					
	Arsenic		Boron		Phosphorus	
	R_p Å	ΔR_p Å	R_p Å	ΔR_p Å	R_p Å	ΔR_p Å
10	97	36	333	171	139	69
20	159	59	662	283	253	119
30	215	80	987	371	368	166
40	269	99	1302	443	486	212
50	322	118	1608	504	607	256
60	374	136	1903	556	730	298
70	426	154	2188	601	885	340
80	478	172	2465	641	981	380
90	530	189	2733	677	1109	418
100	582	207	2994	710	1238	456
110	634	224	3248	739	1367	492
120	686	241	3496	766	1497	528
130	739	258	3737	790	1627	562
140	791	275	3774	813	1757	595
150	845	292	4205	834	1888	628
160	898	308	4432	854	2019	659
170	952	325	4654	872	2149	689
180	1005	341	4872	890	2279	719
190	1060	358	5086	906	2409	747
200	1114	374	5297	921	2539	775

hot phosphoric acid, and an unmasked boron implant is made to adjust the threshold voltage of enhancement-mode devices, as shown in Fig. 12D.

The depletion-mode-device region is defined by a second mask in order to adjust the threshold voltage of depletion-mode devices. An arsenic implant dopes the exposed silicon areas n-type near the surface. This is shown in Fig. 12E. The slices are now carefully cleaned, a high-quality 500 Å thermal oxide is grown in dry oxygen to serve as the gate insulator, and a 5000 Å LPCVD polycrystalline silicon layer is deposited. This polysilicon layer is doped heavily n-type (n+) by phosphorus diffusion; the wafer cross-section at this step is shown in Fig. 12F. The polysilicon gate regions are patterned with a third mask and dry etched, as shown in Fig. 12G. The resist is wet etched, and a heavy arsenic implant is made that converts regions not covered by polysilicon or thick oxide to n+ type, as depicted in Fig. 12H.

A 7000 Å phosphosilicate glass (PSG) is deposited on cleaned wafers and patterned with a fourth mask, and the doped glass is dry etched to open contact areas, as shown in Fig. 12I. Resist is removed by wet etching, and the wafers are cleaned and subjected to a temperature greater than 900° C, which causes the doped glass to flow so that all edges are smooth. Now a 10 000 Å aluminum film is evaporated or sputter deposited (Fig. 12J). A fifth mask patterns the aluminum interconnect, aluminum is dry etched, and resist is removed by wet etch (Fig. 12K). After careful clean-up, the aluminum is sintered at 450° C in H_2 to promote good ohmic contacts. A 3000 Å layer of plasma nitride is now deposited (Fig. 12L). A sixth mask defines the bonding-pad regions, where the coated nitride is plasma etched to allow access to the bonding pads. The wafers are cleaned, and this finishes the fabrication process.

The circuits are tested for functionality and then diced. Good dies are bonded in appropriate packages, retested, and shipped to customers.

Yield Statistics

Only a fraction of a large number of chips on a silicon wafer are completely functional. Defects in the masks, dust particles on wafer surfaces, nonideality of the basic silicon material, and short or open circuits in the wiring all cause some of the circuits to be nonfunctional. With present design techniques, any single defect of sufficient size will kill an entire chip.

The simplest model for the yield, or the fraction of chips that do not have defects, assumes a random defect distribution across the wafer. If there are D fatal defects per unit area and the area of an individual chip is A, then the probability that a chip has n flaws is, in the simplest case, given by the Poisson distribution $P_n(DA)$. The probability of a good chip is:

$$P_0(DA) = e^{-DA} \tag{5}$$

While this equation is not rigorously applicable to fabrication processes, it is a good approximate model for estimating the yield of various design alternatives.

IC DESIGN*

The design of integrated circuits requires the synthesis and analysis of a large number of active elements. The classical form of analysis can be extended to integrated circuits, whereas the focus of synthesis goes beyond the classical notions inasmuch as the geometric layout of the circuit topography for integrated circuits is a major fraction of the circuit synthesis. The electrical aspects of circuit design derive from the considerations that are laid out in Chapter 18, with certain constraints based on the scale of integration. The design of the circuit topography is unique to integrated circuits and has evolved as a major discipline.

Modeling and Simulation

The design of integrated circuits requires the electrical analysis of circuits that contain a large number of elements. For LSI or VLSI circuits, this involves the simulation of over 100 000 circuit elements if the total circuit response is to be examined. The evaluation of this class of circuits is extremely computer-intensive. In most cases, it is not practical to simulate the circuit with classical time- or frequency-domain analysis at the transistor level. The current practice for simulation and analysis of such classes of circuits is to adopt a hierarchical procedure with different levels of abstraction at each level. The design of an LSI logic circuit would involve a typical modeling hierarchy such as that in Chart 1.

CHART 1. DESIGN HIERARCHY FOR LSI CIRCUITS

Architecture simulation
Behavioral simulation
Functional simulation
Logic simulation
Transistor simulation

The top of the hierarchy is an abstract definition of the architecture of the circuit, which is used to provide a guideline for the various ways of accomplishing the objective of the circuit. It trades off, for example, the use of pipeline processing versus serial processing. The behavioral level of simulation actually involves the definition of the major blocks of the circuit and their interaction, with the details of the overall data or control flow to accomplish the objective of the chip being examined. The functional level actually describes the overall logical response of the major blocks, relating the logical inputs and outputs with no details of internal realization of logic in the block. The logic-simulation level details the realization of each block at the gate level, provides logic minimization, and in some cases introduces the notion of relative timing. The transistor-level simulation considers the transient response of the circuit, including the detailed simulation of all elements of the circuit. The key to the usefulness of this hierarchical simulation is the ability to mix the different levels of abstraction in order to examine the performance of the entire circuit with focus on one block at a time. Such mixed-mode simulators are being evolved, and common hardware-description languages that operate on a unified data base that is accessed by any level of the hierarchy are now available.

Topological Realization of Circuits

The design of integrated circuits differs from the design of board-level circuits in the importance of the actual physical realization of the active elements and interconnections. Typical LSI circuits have many thousands of active elements and interconnections that must be topologically related to each other. This requires the interaction of the device physics, process technology constraints, and topological constraints. In order to obtain a practical solution to this complex interactive design environment, it is generally accepted practice to describe the process constraints in the form of a design rule package. This design rule package is a simple description of the lateral spatial relationship of the various active-element forming geometries as well as the wiring. A simple example of design rules is shown in Fig. 13. The original concept of the design rules was governed by the constraints of the technology only. However, the design data base for the geometrical description of the circuit for VLSI circuits is very large, and some constraints that limit the size of this data base have been introduced into the design rules.* This is simply a quantization of the minimum spatial distance describing any

* Reference 9.

* Reference 9.

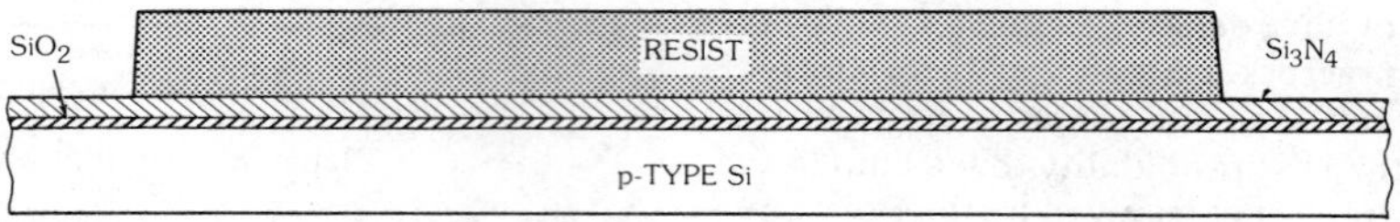

(A) After field pattern.

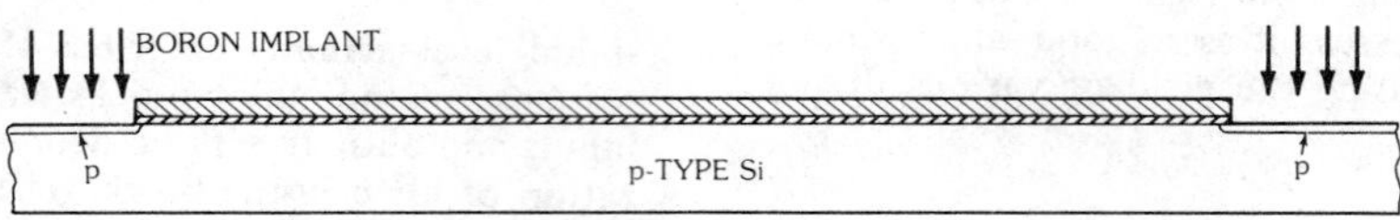

(B) After field implant.

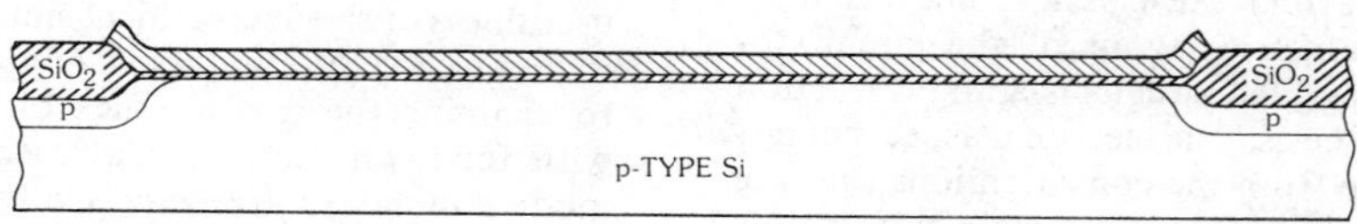

(C) After field oxidation.

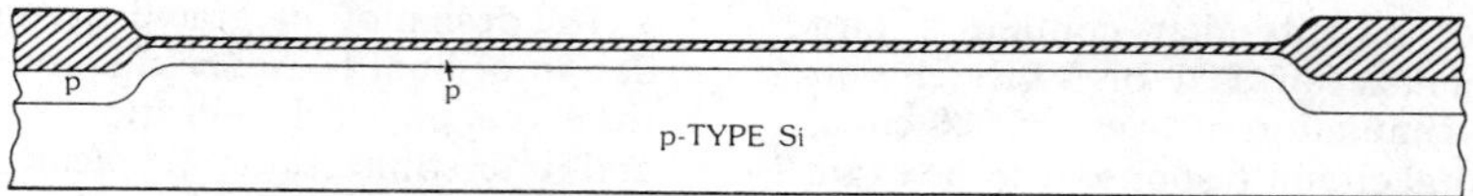

(D) After enhancement threshold implant.

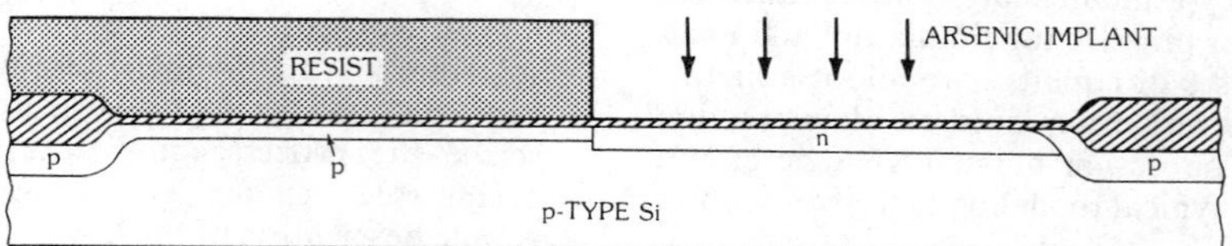

(E) After pattern and depletion implant.

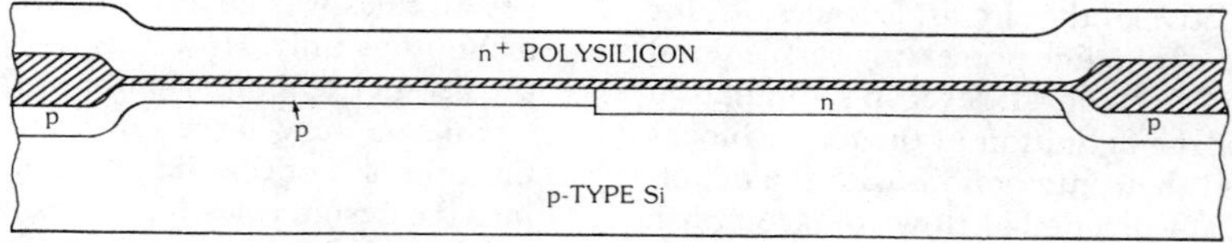

(F) After gate oxidation, poly deposition, and doping.

Fig. 12. Schematic cross-sections illustrating the purpose and

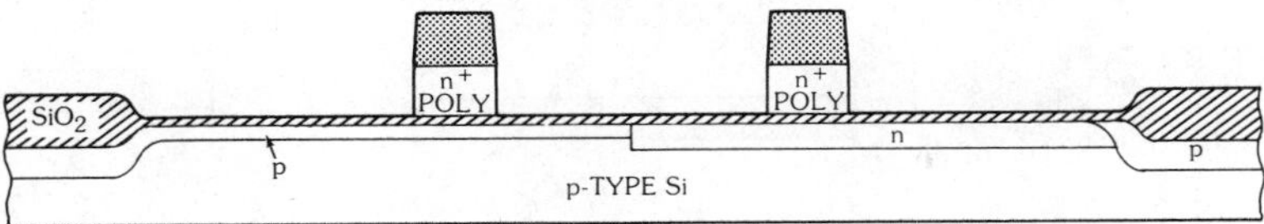

(G) After gate pattern and etch.

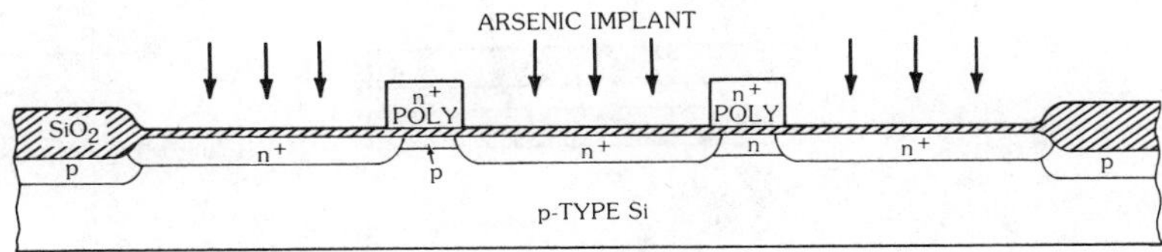

(H) After source-drain implant.

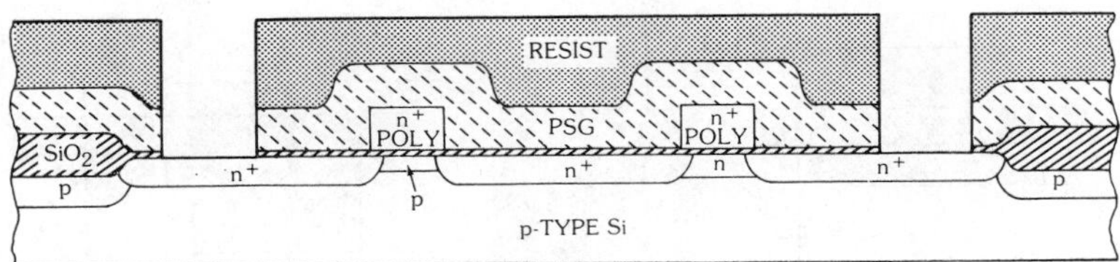

(I) After PSG deposition, contact pattern, and etch.

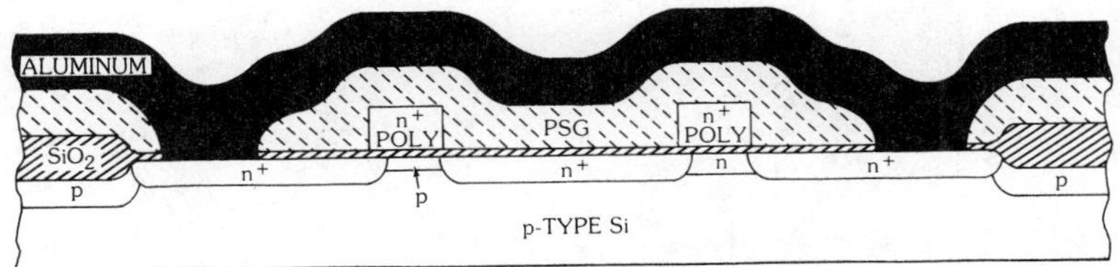

(J) After aluminum deposition.

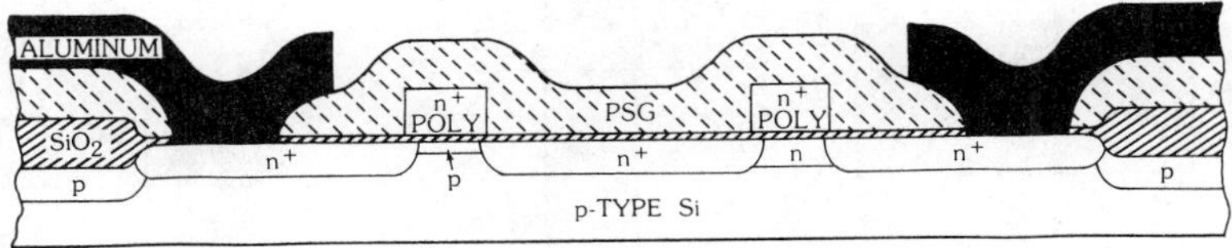

(K) After aluminum etch.

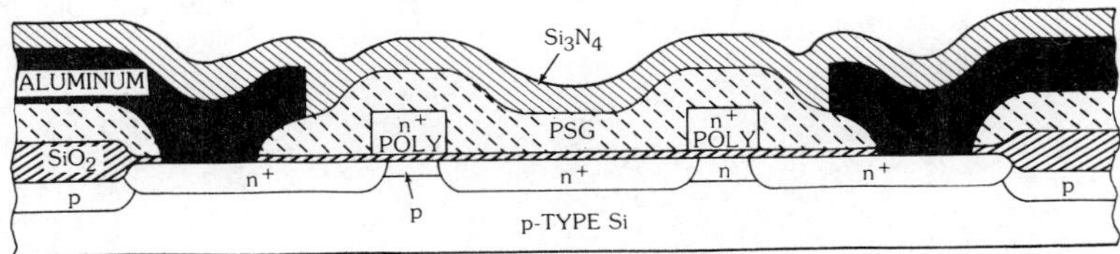

(L) After nitride overcoat.

consequence of various steps involved in an NMOS process.

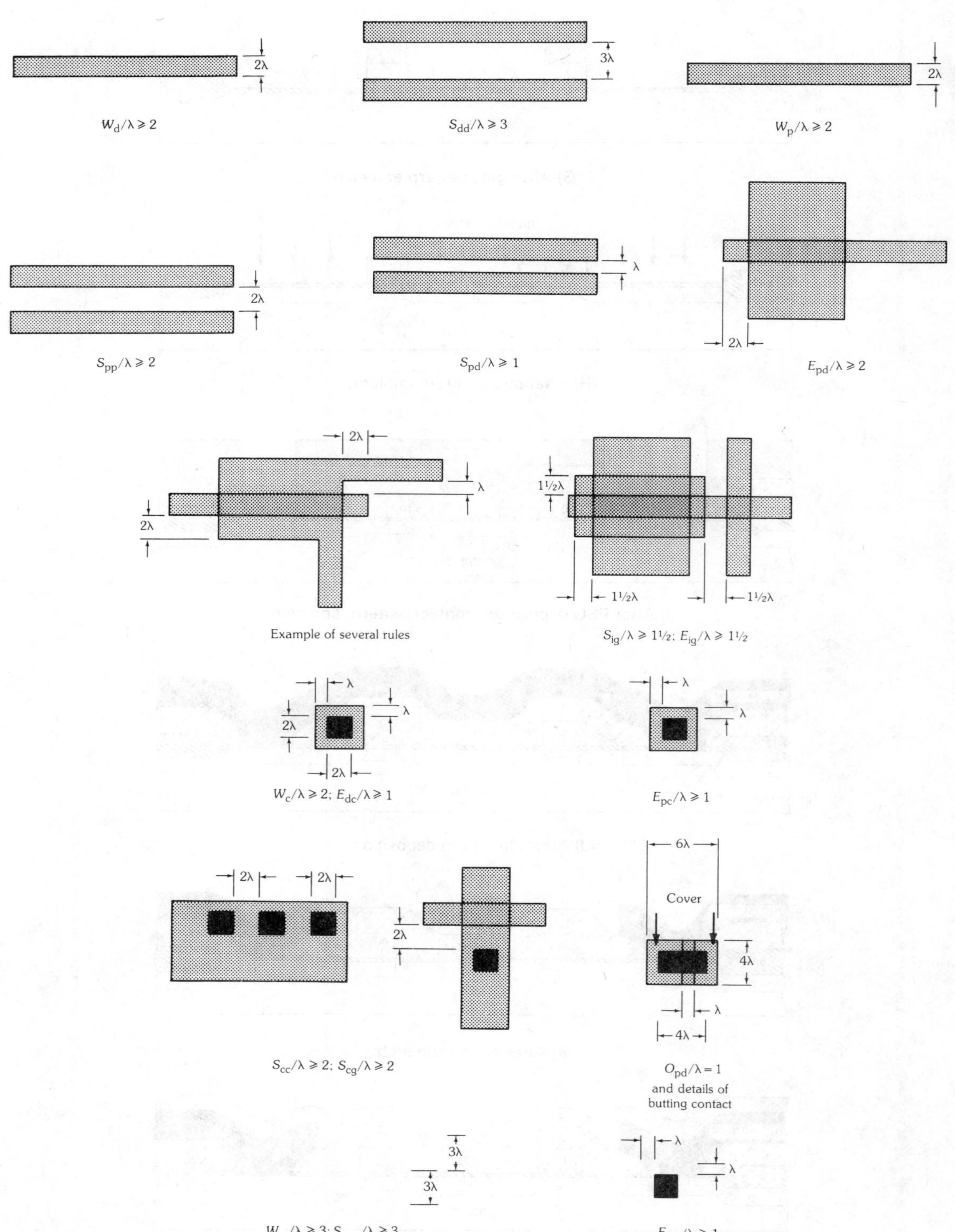

Fig. 13. An example of a design rule set. (*From C. Mead and L. Conway,* Introduction to VLSI Systems. *Reading, Mass.: Addison-Wesley Publishing Co., Inc., 1980.*)

technology constraint. This kind of quantization is equivalent to the definition of the finest grid on which a geometry must fall.

There are two major approaches to the realization of the geometrical data base for an IC. The first is the classical approach in which the complete drawing of each individual section is introduced into the data base in detail. This is a tedious and inefficient way of generating data. The preferred approach to data entry is through a symbolic description of the layout in which the definitions of the often-used elements are built into the system and the designer calls up such elements and strings them together on a graphic entry system. These symbols are a modified description of the circuit elements much like the drawing of a classical circuit schematic (Fig. 14). The design rules are then used to generate automatically the full two-dimensional geometrical description from the symbolic description. The symbolic layout system is very efficient and suffers only from a 10–20% reduction of the packing density of circuit elements compared to a detailed customized input of the complete geometries. In an attempt to reduce this gap, software compaction of the design data base has been developed. The connection of geometries on a layout data base is another aspect of integrated-circuit design that has been traditionally inefficient. Automatic wire-routing software developed for multilayered printed circuit boards has been adapted for this purpose and is being used. The use of artificial-intelligence concepts in the realization of more efficient routing algorithms is under investigation.

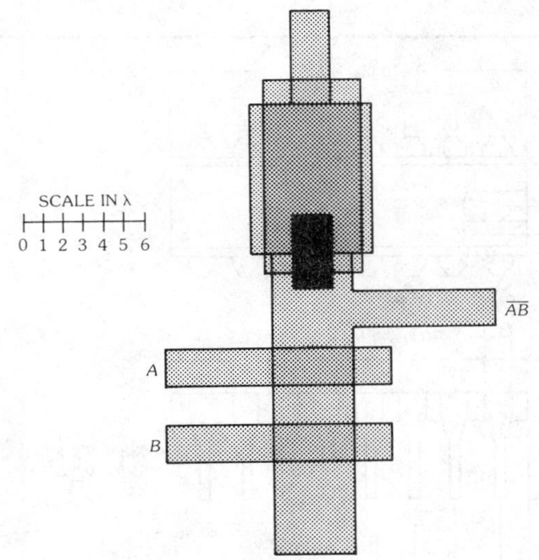

(A) NAND-gate layout geometry.

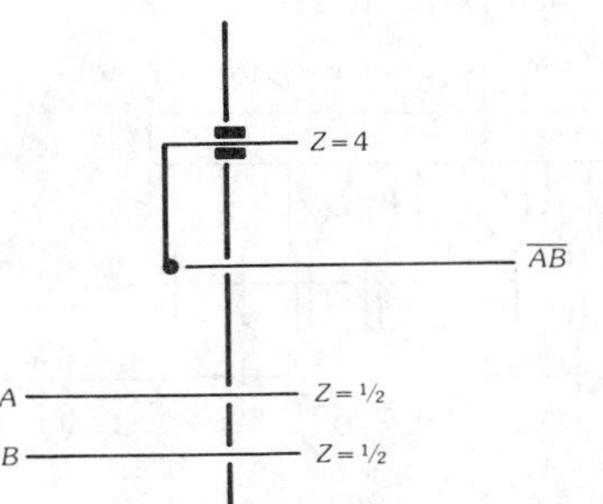

(B) NAND-gate topology using symbolic layout.

Fig. 14. Symbolic layout and its topological equivalent. (*From C. Mead and L. Conway,* Introduction to VLSI Systems. *Reading, Mass.: Addison-Wesley Publishing Co., Inc., 1980.*)

Once the design data base has been created, the design task is to verify the validity of the data base geometrically and electrically. The geometric verification is a checking of all geometries for design-rule violations. This is achieved routinely by all layout systems through software analysis of the data base. The electrical validity of the geometrical data base involves the extraction of the electrical circuit schematic and the comparison of the electrical schematic to the various levels of hierarchy of the initial design. This upward feedback through the hierarchy is of crucial importance to the design cycle and requires the design and geometric data bases to be compatible. It is currently common practice to extract electrical schematics and parasitic resistances and capacitances from the layout data base and compare them to the design data base. The comparison occurs at the transistor level.

Various integrated design systems with the above functions are currently available.

PACKAGING INTEGRATED CIRCUITS

The standards for packaging integrated circuits and for allocation of functions to pins for the purposes of interchangeability are set by international committees EIA/JEDEC (Electronic Industries Association) and IEC (International Electrotechnical Commission). There are two major kinds of specifications for each type of package: mechanical and thermal. A variety of package types exist, the most popular of which will be described.*

Plastic Packages

Plastic dual-in-line packages consist of a circuit mounted on a 16-, 18-, 20-, 24-, or 28-pin lead frame and encapsulated within an electrically nonconductive plastic compound. The compound will withstand soldering temperature with no deformation, and circuit performance characteristics remain sta-

* Details of packages can be found in integrated-circuit data books.

ble when the device is operated in high-humidity conditions. These are the lowest-cost packages for integrated circuits. An example of a 16-pin plastic package is shown in Fig 15.

Ceramic Packages

Another type of dual-in-line package is hermetically sealed and consists of a ceramic base, a ceramic cap, and a 16-, 18-, 20-, 24-, 28-, or 48-lead frame. Hermetic sealing is accomplished with glass. These devices are divided into two categories: side-braze and frit seal. An example of the mechanical dimensions of a 16-pin dual-in-line ceramic package is shown in Fig. 16.

Ceramic Flat Package

Hermetically sealed ceramic flat packages consist of an electrically nonconductive ceramic base and cap and a 16- or 24-pin lead frame. An example of the mechanical dimensions of such a package is shown in Fig. 17.

Leadless Ceramic Chip Carrier

Leadless ceramic chip-carrier packages are emerging as the most popular for LSI and VLSI chips due to their board-level density advantage. An example of a 36-pad leadless ceramic chip carrier is shown in Fig. 18.

Thermal Resistance

The most important system specification of an IC package is its ability to conduct heat, since the IC dissipates power in the package. The thermal properties of a package are specified as a thermal resistance. Junction-to-ambient thermal-resistance values of dual-in-line packaging systems are shown in

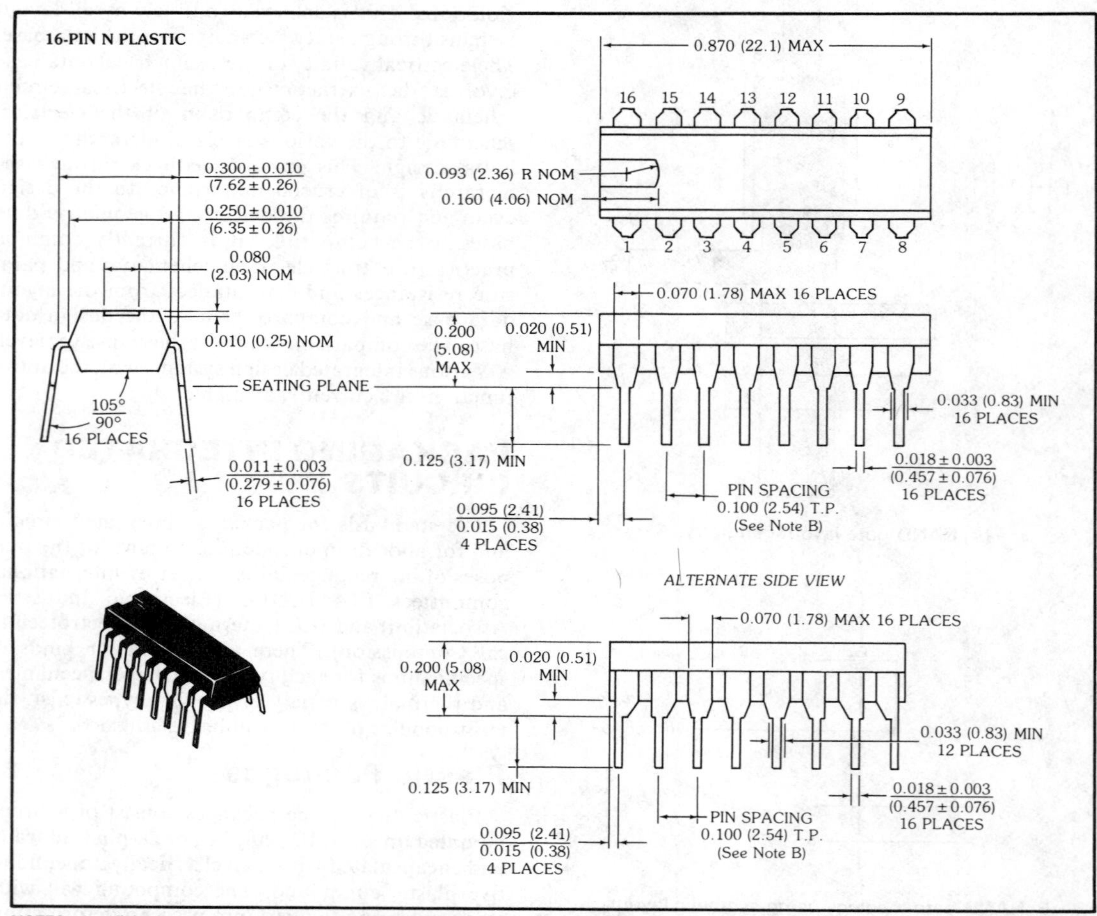

NOTES: A. All dimensions are shown in inches (and parenthetically in millimeters for reference only). Inch dimensions govern.
B. Each pin centerline is located within 0.010 (0.26) of its true longitudinal position.

Fig. 15. Dual-in-line plastic packaging for integrated circuits. (*From* TTL Data Book, *2nd ed. Dallas, Tex.: Texas Instruments, Inc., 1976; p. 4-6.*)

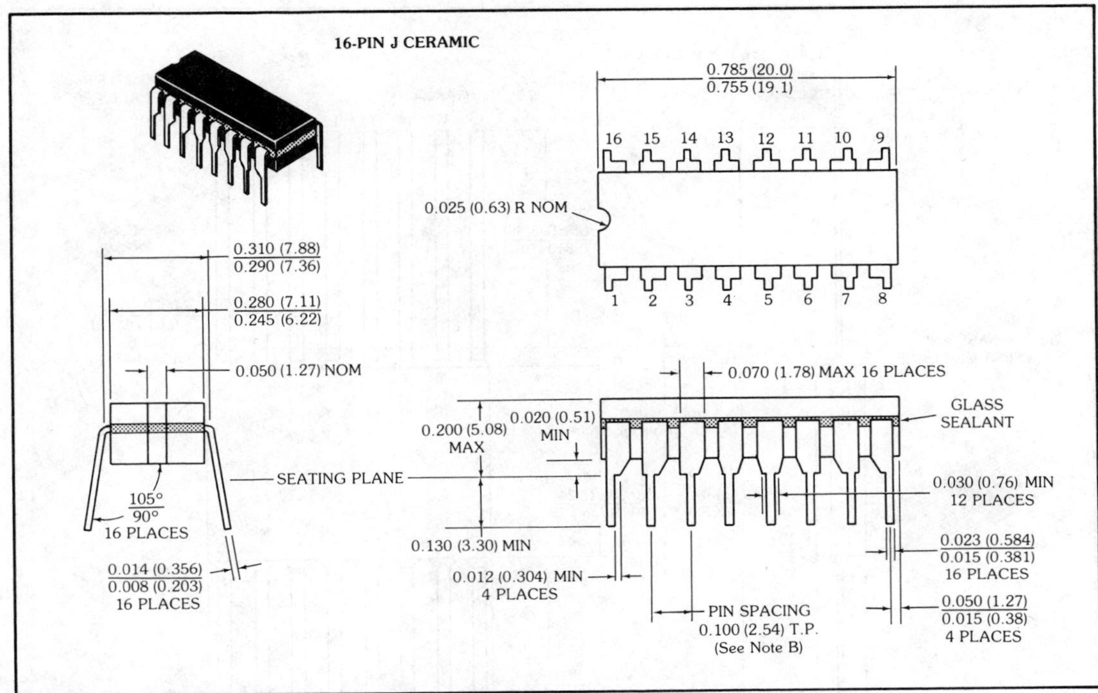

NOTES: A. All dimensions are shown in inches (and parenthetically in millimeters for reference only). Inch dimensions govern.
B. Each pin centerline is located within 0.010 (0.26) of its true longitudinal position.

Fig. 16. Dual-in-line ceramic packaging for integrated circuits. (*From* TTL Data Book, *2nd ed. Dallas, Tex.: Texas Instruments, Inc., 1976; p. 4-4.*)

Fig. 19. This figure of merit is used to relate the internal chip temperature to the ambient system temperature.

DIGITAL INTEGRATED CIRCUITS

Digital logic functions can be realized in many different configurations. The traditional logic realization used resistors or diodes in conjunction with a transistor. This approach has been replaced in the current integrated circuits by the use of active transistors mostly, since these can be realized most efficiently in the technology. The logic realization is dependent on the property of the transistor used and will be categorized differently for bipolar and field-effect transistors.

The implementation of logic integrated circuits requires the definition of the voltage, current, and timing standards. These have been defined in Table 8. A very large variety of digital logic circuits spanning SSI, MSI, LSI, and VLSI is commercially available. The various manufacturers' data books and applications books should be consulted for detailed information.

Bipolar Logic Families

Bipolar logic functions are commercially available as SSI and MSI circuits, and they are being developed for LSI and VLSI circuits. The different families described here are graded from the most popular SSI and MSI standard logic to the more advanced families used for LSI and VLSI.

Transistor-Transistor Logic (TTL)*—The most widely used logic family for SSI and MSI logic is transistor-transistor logic. A broad spectrum of TTL circuits are available that allow logic designers to optimize all portions of a system cost effectively.

The basic schematic of generic TTL logic is shown in Fig. 20. This arrangement utilizes the base-emitter diodes and collector-base diode of the multiemitter transistor for logic and the output transistor for drive. This is very easily realized in planar bipolar technology. Although the basic TTL circuit consists of two transistors, additional components are necessary to increase circuit speed and fan-out capability (Fig. 21). Two major classes of TTL logic exist. These are called 54XX and 74XX

* Reference 17.

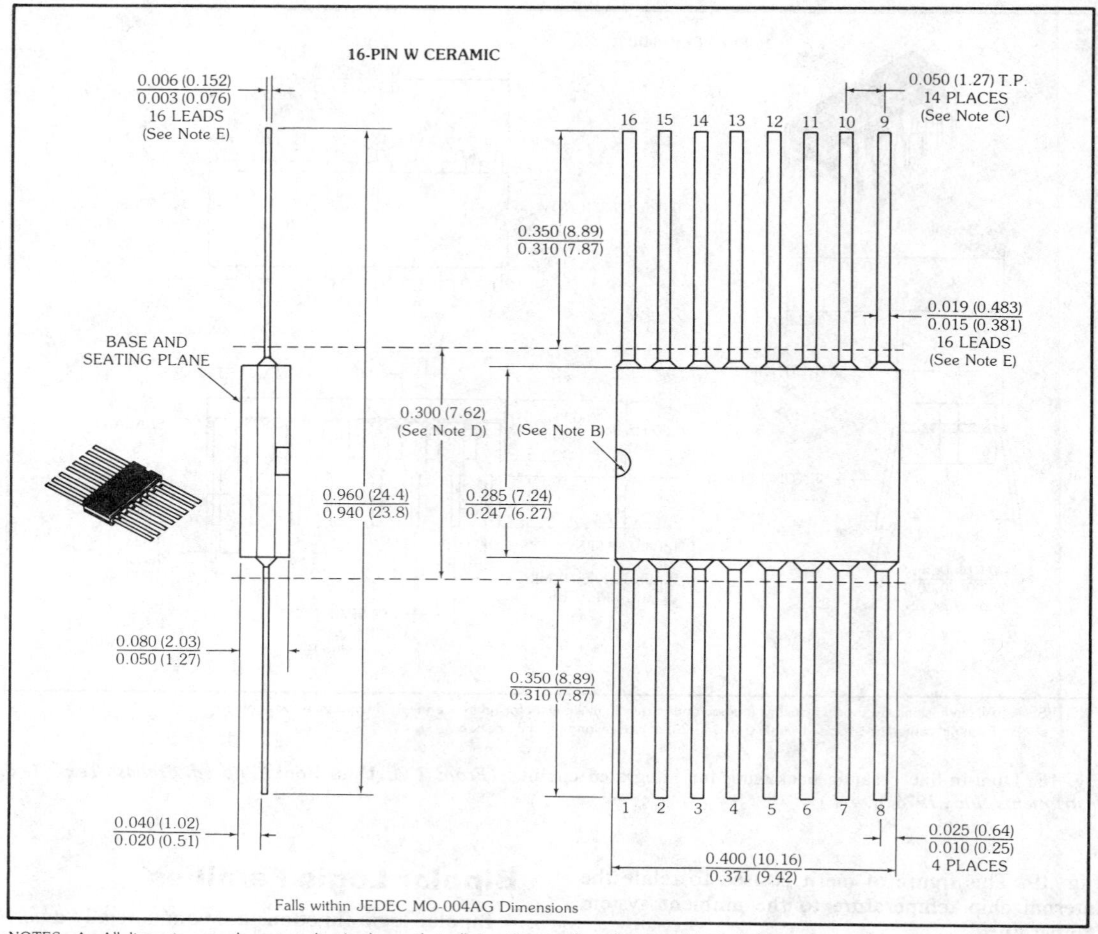

NOTES: A. All dimensions are shown in inches (and parenthetically in millimeters for reference only). Inch dimensions govern.
B. Index point is provided on cap for terminal identification only.
C. Leads are within 0.005 (0.13) radius of true position (T.P.) at maximum material condition.
D. This dimension determines a zone within which all body and lead irregularities lie.
E. Not applicable for solder-dipped leads.
F. When solder-dipped leads are specified, dipped area extends from lead tip to within 0.050 (1.27) of package body.

Fig. 17. Ceramic flat packaging for integrated circuits. (*From* TTL Data Book, *2nd ed., Dallas, Tex.: Texas Instruments, Inc., 1976; p. 4-10.*)

series TTL. The 74XX series is a limited-temperature (0°C to +70°C) IC, and the 54XX series is a full-military-temperature (−55°C to +125°C) IC. A classification by low power and high speed in addition to the standard series exists.

An improvement in the TTL performance is achieved by the addition of a Schottky barrier diode that permits the transistor to operate in the nonsaturated mode; this reduces storage time and improves speed. Initially, the Schottky TTL was developed for high speed. Further evolution of the Schottky technology led to the low-power Schottky TTL family in which Schottky diodes replace the multiemitter transistor as the input element. The Schottky TTL circuits are shown in Fig. 22.

Table 9 defines the voltage and current standards of various TTL logic families. Table 10 shows the performance characteristics of different TTL families.

Emitter-Coupled Logic (ECL)—The highest-speed, lowest-noise bipolar logic family is emitter-coupled logic. It is also known as current-mode logic (CML), since the circuit works on a current-steering principle. It is designed as a nonsaturating form of logic, which eliminates transistor storage time as a speed-limiting characteristic. The typical ECL gate shown in Fig. 23 comprises a differential-amplifier input, an internal bias reference, and an emitter-follower output to restore dc levels. High-

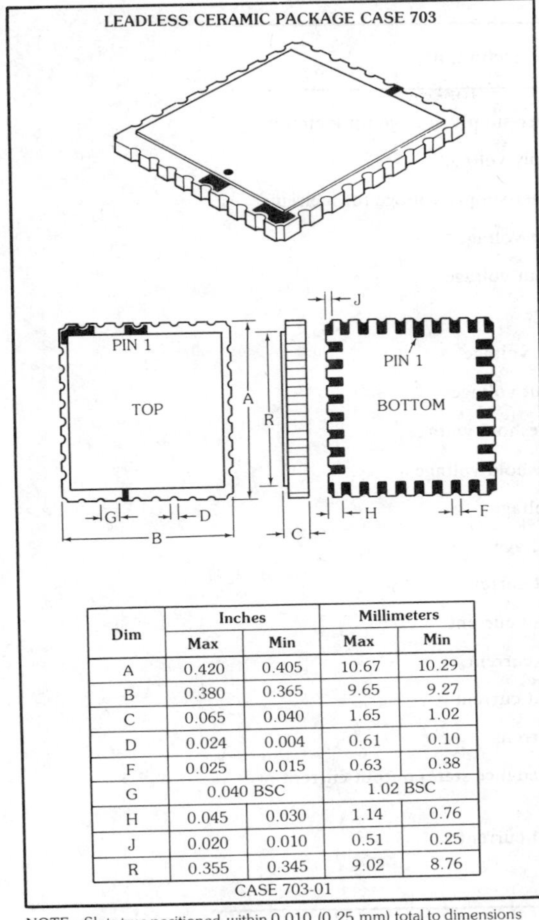

Dim	Inches		Millimeters	
	Max	Min	Max	Min
A	0.420	0.405	10.67	10.29
B	0.380	0.365	9.65	9.27
C	0.065	0.040	1.65	1.02
D	0.024	0.004	0.61	0.10
F	0.025	0.015	0.63	0.38
G	0.040 BSC		1.02 BSC	
H	0.045	0.030	1.14	0.76
J	0.020	0.010	0.51	0.25
R	0.355	0.345	9.02	8.76

CASE 703-01

NOTE: Slots true positioned within 0.010 (0.25 mm) total to dimensions A and B at maximum material condition.

Fig. 18. Leadless ceramic chip-carrier package. (*From* Motorola CMOS Data, *1978; p. g-14.*)

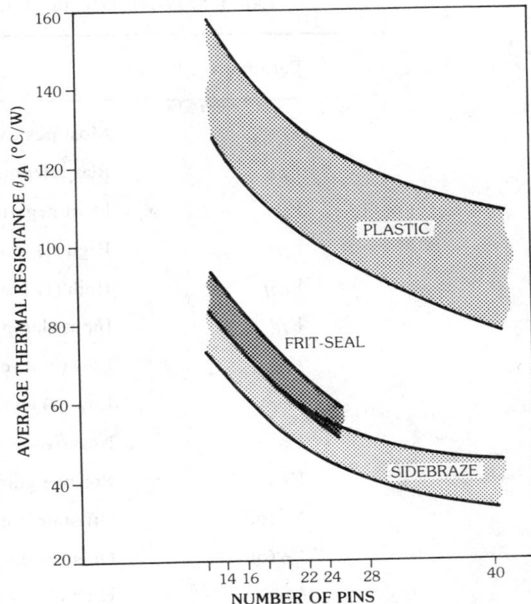

Fig. 19. Typical thermal resistance for various packages. (*From* Motorola CMOS Data, *1978; p. g-3.*)

fan-out operation is possible because of the high input impedance of the differential amplifier and the low output impedance of the emitter follower. Power-supply noise is virtually eliminated by the nearly constant current drain of the differential amplifier even during transition time. The logic-voltage standards for ECL are shown in Table 11. Note that CML circuits tend to adopt TTL voltage standards. Typical gate speed for ECL logic ICs is 2 ns at 100 mW power dissipation for commercially available circuits.

Integrated Injection Logic (I^2L)*—Integrated injection logic (I^2L), also known as merged transistor logic (MTL), is a bipolar logic comparable in density and power dissipation with dynamic MOS logic. It is an LSI- or VLSI-caliber bipolar logic family. The basic I^2L gate is shown in Fig. 24. The basic cell is made up of a multiemitter npn transistor operated in the inverted mode and a lateral pnp transistor whose base and collector are common to the collector and base of the first transistor. For logic operation, the pnp transistor is used both as an injector of base current for the npn transistor in the same cell and as a current-source load for an npn transistor in an adjacent cell. The operation of the basic cell as an inverter can be understood as follows. When V_{in} is high, all the current from the pnp transistor is used up by npn transistor T_1 operating in the saturated mode, and no current is left for the base of npn transistor T_2. Hence V_{O1} is low and V_{O2} is high. When V_{in} is low, the opposite situation applies. Typical gate delays of 5 ns per gate at 0.2 mW have been achieved at 2-micron geometries. Many advances based on self-aligned structures are being made that are enhancing the performance of this logic family.

Schottky Transistor Logic (STL)†—Schottky transistor logic, also called Schottky coupled transistor logic (SCTL) and complementary constant current logic (C^3L), is a generic diode transistor logic (DTL) implemented with Schottky logic diodes and a single npn switch transistor. Fig. 25 shows a three-output STL gate. For relatively high-gain transistors ($h_{fe} > 10$) and fan-out limited to one per logic diode, the signal-voltage swing is shown to be the difference in the Schottky-diode

* Reference 18.

† Reference 19.

TABLE 8. DEFINITIONS OF LOGIC STANDARD TERMINOLOGY

Parameter	Definition
V_{CC}	Most positive power-supply voltage for a circuit
V_{BB}	Bias reference supply voltage
V_{EE}	Most negative power-supply voltage for a circuit
V_{IH}	High (1) level input voltage
V_{OH}	High (1) level output voltage
V_{IK}	Input clamp voltage
V_{IL}	Low (0) level input voltage
V_{OL}	Low (0) level output voltage
V_{T-}	Negative-going threshold voltage
V_{T+}	Positive-going threshold voltage
V_O (off)	Off-state output voltage
V_O (on)	On-state output voltage
I_{IH}	High (1) level input current
I_{OH}	High (1) level output current
I_{IL}	Low (0) level input current
I_{OL}	Low (0) level output current
I_O (off)	Off-state output current
I_{OZ}	Off-state (high-impedance state) output current of three-state output
I_{OS}	Short-circuit output current
I_{CC}	Supply current

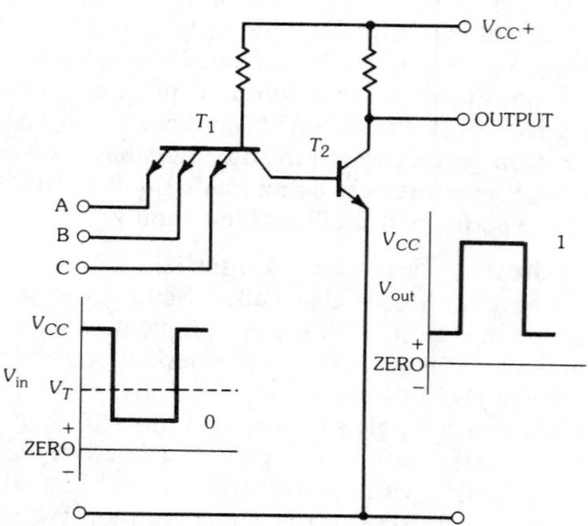

Fig. 20. The basic TTL circuit. (*From* Designing With TTL Integrated Circuits. *Dallas, Tex.: Texas Instruments, Inc., 1971; Fig. 1-13, p. 12.*)

forward voltages. The STL characteristics are compatible with the supply voltages, power, density, and speed requirements of VLSI circuits. Typical on-chip gate speeds of less than 2 ns at 10 μW of power are achievable with 2-micron technology.

MOS Logic Families §

There are two classes of MOS logic families, based on the operation of the logic: (A) static, or ratioed, logic and (B) dynamic, or ratioless, logic. The realization of these two classes of logic is further described by technology type: p-channel (PMOS), n-channel (NMOS), and complementary (CMOS). The unique feature of a MOS technology is the symmetric nature of the MOSFET. This has been utilized in logic circuits by the use of a pass transistor as shown in Fig. 26. The signal flow can be interrupted by the use of such a series switch

§ Reference 9.

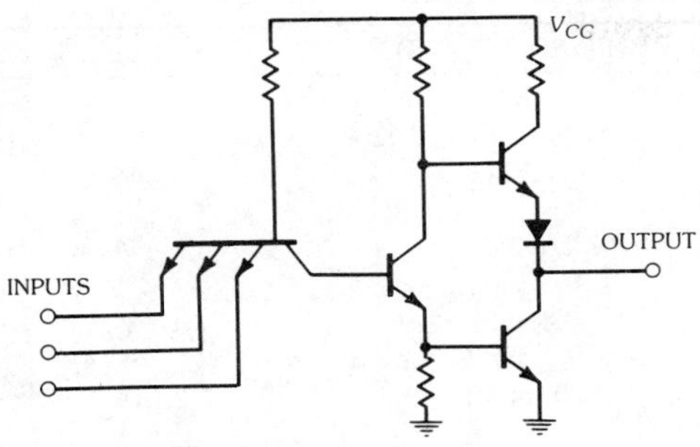

(A) Standard, or low-power, circuit.

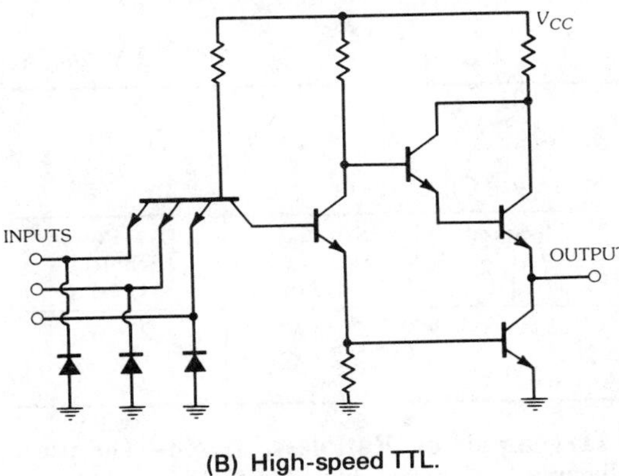

(B) High-speed TTL.

Fig. 21. Diagrams of typical implementations of TTL circuits. (*From* Designing With TTL Integrated Circuits. *Dallas, Tex.: Texas Instruments, Inc., 1971; Fig. 2-1, p. 18, Fig. 2-3, p. 19.*)

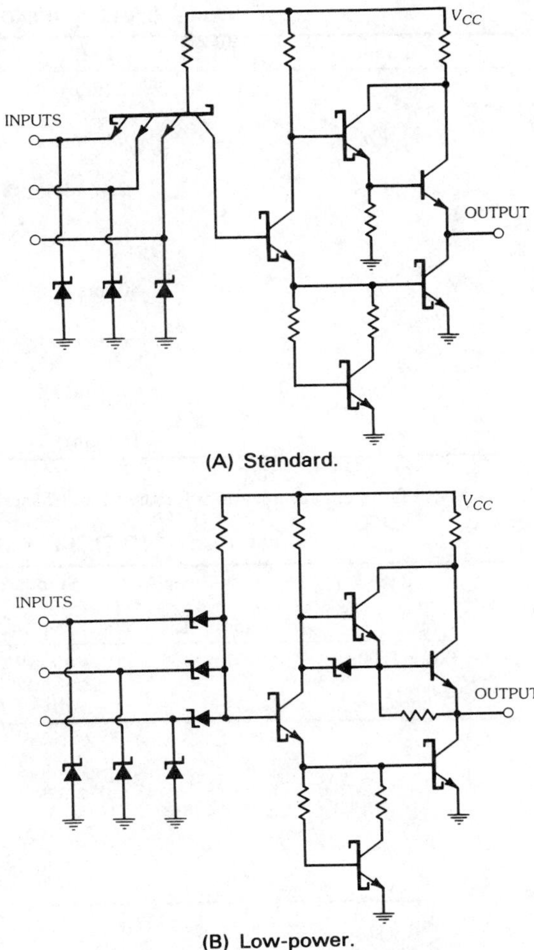

(A) Standard.

(B) Low-power.

Fig. 22. Schottky TTL circuits. (*From G. D. Kraft and W. N. Toy,* Mini/Micro Computer Hardware Design. *Englewood Cliffs, N.J.: Prentice-Hall, Inc., 1979; p. 71.*)

when the switch is turned off. In the on state, the switch allows signal flow in either direction.

Static, or Ratioed, Logic—The generalized schematic of a static, or ratioed, logic gate is shown in Fig. 27A. It consists of a load device and a driver device or several driver devices that can be switched to perform a logic function. The ideal characteristic for a load device is a constant-current source. The typical device realization of the load in a PMOS or NMOS technology is an active depletion-type transistor with the gate and source connected as shown in Fig. 27B. The typical load line achieved by this technique is shown in Fig. 28. Note that the nonideal behavior is due to the change in the depletion-mode-device threshold voltage with source-to-substrate bias, commonly known as the body effect. This static circuit has a logic high level determined by the power-supply voltage, since the driver device is switched off in this state and only draws subthreshold current. The logic low level is determined by the ratio of the load current and the on current of the driver device. This ratio is a function of the threshold and aspect ratio of the load and the driver device defined as the beta ratio of the inverter. This is the most important design parameter for ratioed static logic.

$$\text{Beta ratio} = \frac{(\text{Width/Length})_{\text{driver}}}{(\text{Width/Length})_{\text{load}}}$$

The load current is constantly turned on in this circuit for the depletion load realization as in PMOS or NMOS. This causes a static power dissipation in one logic state. The static logic has the advantage of being totally asynchronous and requires no clock pulses. The power dissipation is drastically reduced by the use of CMOS technology (Fig. 27C). The load is a p-channel transistor, and the driver is an n-channel transistor. The CMOS inverter or logic gate draws power only during state transition. Note also that the CMOS static inverter has an inherently large ratio since one device is in the subthreshold regime in each state.

TABLE 9. TTL VOLTAGE AND CURRENT STANDARDS*

Parameter	TTL	Schottky TTL(LS)	Units
V_{IH}	2 (min)	2 (min)	V
V_{IL}	0.8 (max)	0.8 (max)	V
V_{OH}	2.4 (min)	2.5–2.7†	V
V_{OL}	0.4 (max)	0.4–0.5†	V
I_{IH}	40 (max)	20 (max)	μA
I_{IL}	−1.6 (max)	−0.36 (max)	mA
I_{OH}	−400 (min)	−400 (min)	μA
I_{OL}	16 (min)	4—8†	mA

*($V_{CC} = 5\text{ V} \pm 10\%$)
† Different specifications for the 54 and 74 series

TABLE 10. 54/74 TTL TYPICAL PERFORMANCE CHARACTERISTICS

	Low-Power 54/74L	Standard 54/74	High-Speed 54/74H	Schottky 54/74S	Low-Power Schottky 54/74LS
Power/Gate	1 mW	10 mW	22 mW	19 mW	2 mW
Delay/Gate	33 ns	10 ns	6 ns	3 ns	7 ns

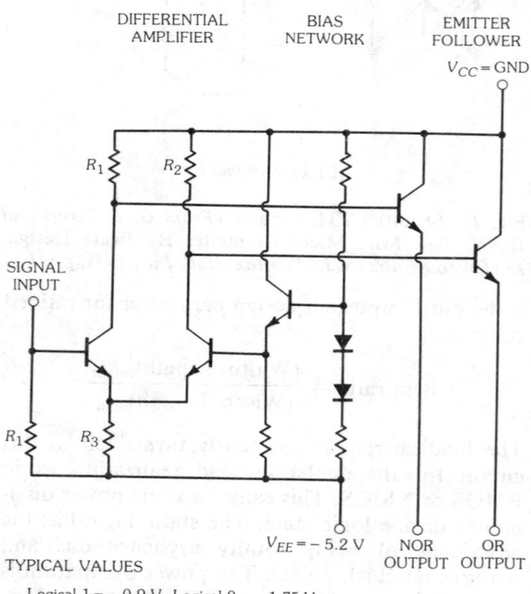

Fig. 23. Diagram of a typical ECL gate. (*From* Signetics ECL Data Book.)

Dynamic, or Ratioless, Logic—The use of dynamic, or clocked, logic for MOS circuits is very attractive for low-power synchronous applications. Examples of dynamic logic gates realized in NMOS and CMOS technology are shown in Fig. 29. The use of two-phase clocks in synchronous systems is standard practice. The timing of these clocks is shown in Fig. 29A. The ϕ_A clock turns on the active load and precharges the output node to the supply voltage independent of the state of the input. The ϕ_B clock discharges the output only if the input logic combination is appropriate. This time-multiplexed logic is possible if the leakage rate through the clocked transistors does not discharge the output stage within a clock period. This scheme requires the generation of a two-phase clock for NMOS and PMOS circuits. However, for the CMOS circuit (Fig. 29B), single-clock operation is possible. The advantages of dynamic logic are the low power required and synchronous operation. It is assumed that the clock propagation speed on-chip is faster than the clock period. This requirement tends to limit the overall gate speed for a given technology.

Microprocessors and Microcomputers*

The implementation of logic in integrated circuits in the decade of the sixties and part of the seventies

* References 20 and 21.

TABLE 11. LOGIC STANDARDS FOR 10000 SERIES ECL*

Parameter	10 100 10 200	10 500 10 600
V_{IL} (min)	−1.85 V	−1.85 V
V_{IH} (max)	−0.81 V	−0.72 V
V_{OL} (min)	−1.85 V	−1.85 V
V_{OL} (max)	−1.65 V	−1.62 V
V_{OH} (min)	−0.96 V	−0.93 V
V_{OH} (max)	−0.98 V	−0.95 V
V_{BB} (typical)	−1.29 V	−1.29 V

* 25°C, V_{EE} = 5.2 V, V_{CC} = Gnd

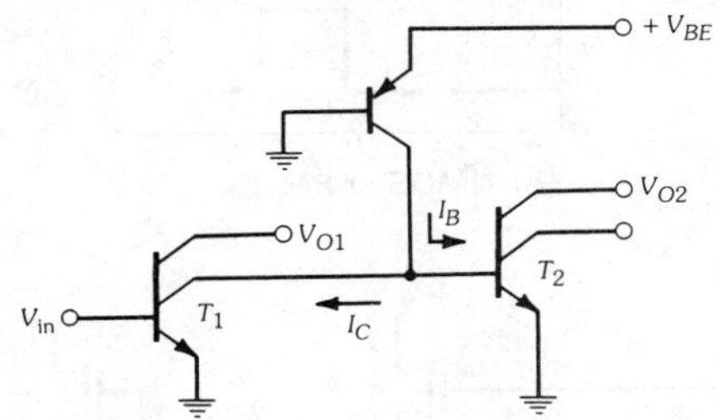

Fig. 24. The integrated-injection-logic, I²L, gate. (*From K. Hart and A. Slob, "Integrated Injection Logic,"* IEEE Trans. Solid State Circuits, *SC-7 (1972), p. 346.)*

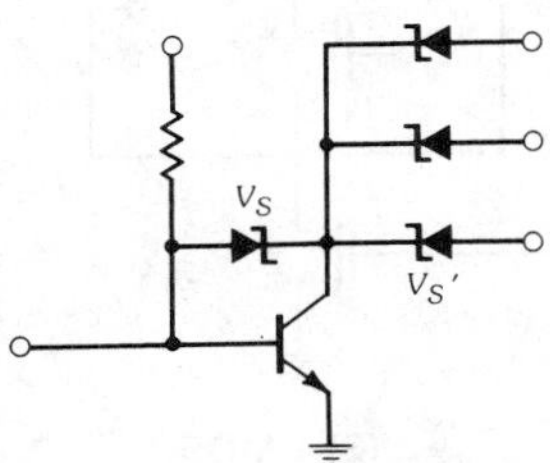

Fig. 25. Three-output STL gate. (*From K. Hart and A. Slob, "Integrated Injection Logic,"* IEEE Trans. Solid State Circuits, *SC-7 (1972), p. 346.)*

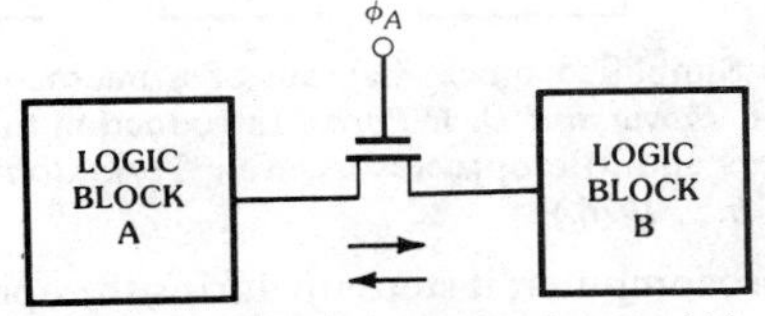

Fig. 26. Pass transistor in MOS logic.

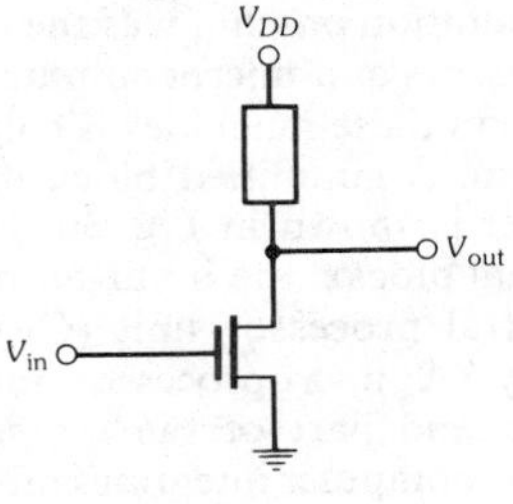

(A) Generic logic.

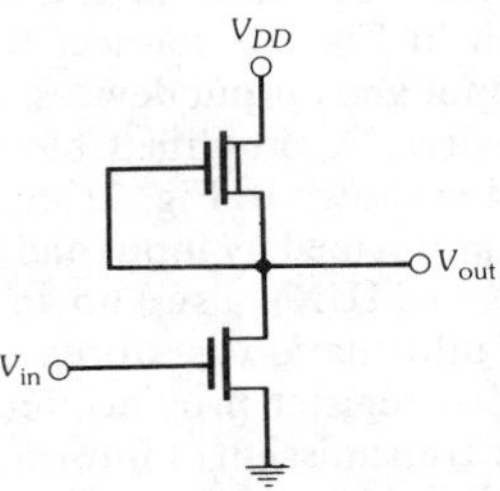

(B) PMOS or NMOS realization with depletion load.

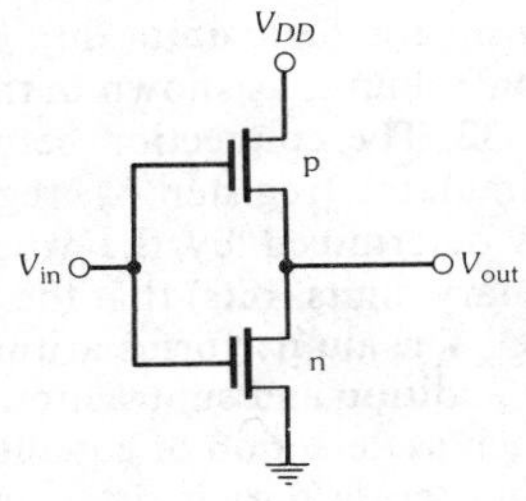

(C) CMOS inverter.

Fig. 27. Static, or ratioed, logic.

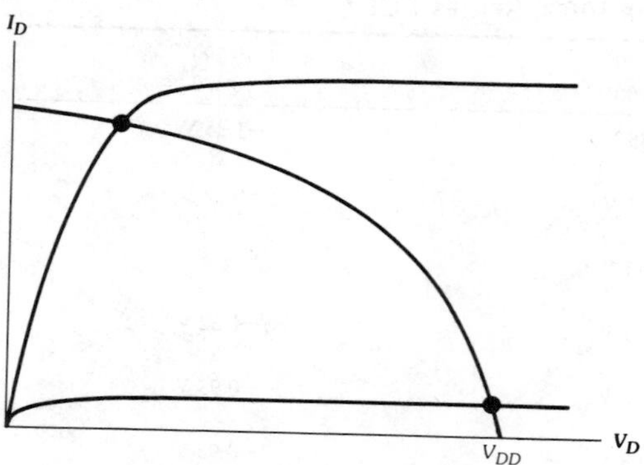

Fig. 28. Typical load line achieved with depletion load.

relied on providing building blocks that integrated unique logic functions. The major breakthrough in logic implementation on-chip was the development of microprocessors and microcomputers. The logical operations in these machines is controlled by a stored program. A simplified block diagram of a microcomputer is shown in Fig. 30. It consists of three functional blocks: the input-output (i/o) section, the central processor unit (CPU), and the main memory. A microprocessor integrates the CPU function and part of the i/o section only, whereas a microcomputer integrates all three blocks shown in Fig. 30.

I/O Section—The lines at the left of the i/o section shown in Fig. 30 connect the microcomputer to the input and output devices, also known as peripheral devices. A simplified block diagram of an i/o section is shown in Fig. 31. Selection of the i/o devices is performed by input and output multiplexers (MPX or MUX), also known as data selectors. Output information is stored in the output buffers. The i/o register provides temporary storage during the transmission of information between the CPU and i/o section.

The Central Processor Unit—The structure of the central processor unit varies widely among the various microcomputers. A simple CPU consists of an arithmetic-logic unit (alu), several registers, and a control unit, as shown in the block diagram of Fig. 32. The connection between the alu and the accumulator (register A), register B, and register M is determined by the word length or number of binary digits (bits) that the alu can process in parallel. The alu performs arithmetic operations, such as addition and subtraction, and logical operations, such as detection of equality. The CPU uses several registers as data registers, working registers, and scratch-pad memory; these help in storing intermediate results (registers A and B) and communicating with the i/o section and main memory (register M). Register P is the program counter that determines the operation sequence of the microcomputer; it steps up during the operation of the program unless otherwise commanded. The control unit in the CPU provides suitable direction of computer operation.

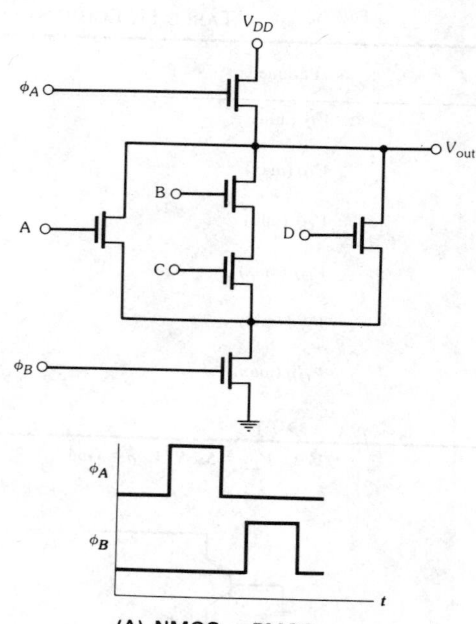

(A) NMOS or PMOS.

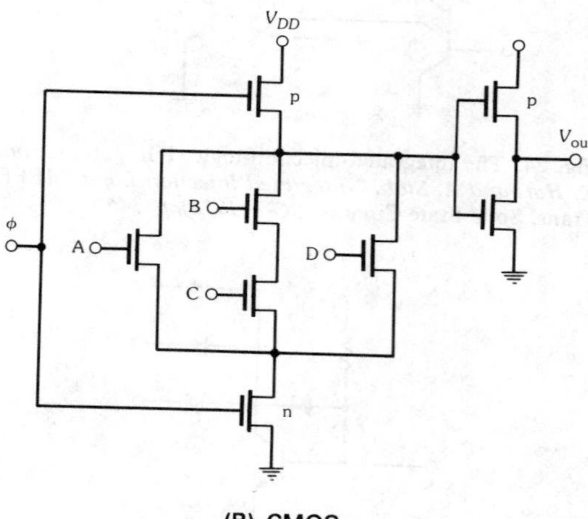

(B) CMOS.

Fig. 29. Dynamic logic.

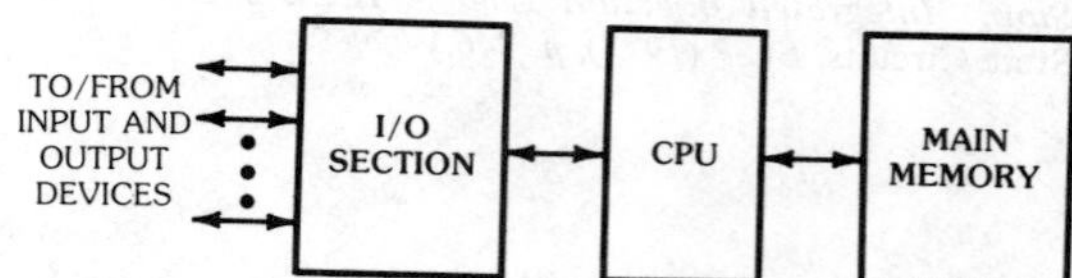

Fig. 30. Simplified block diagram of a microcomputer. (*From A. Barna and D. I. Porat,* Introduction to Microcomputers and Microprocessors. *New York: John Wiley & Sons, Inc., 1976.*)

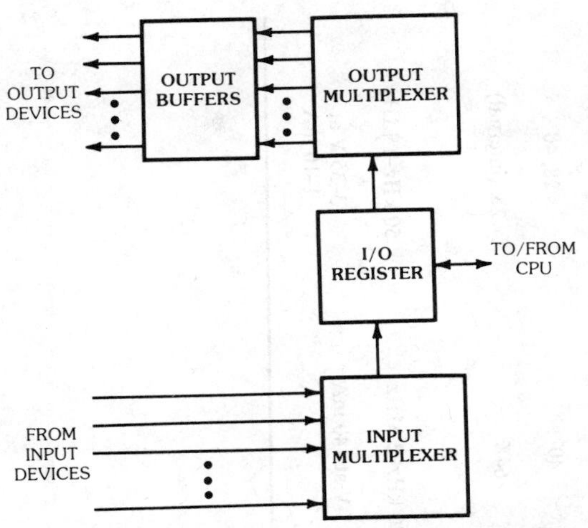

Fig. 31. The i/o section. (*From A. Barna and D. I. Porat,* Introduction to Microcomputers and Microprocessors. *New York: John Wiley & Sons, Inc., 1976.*)

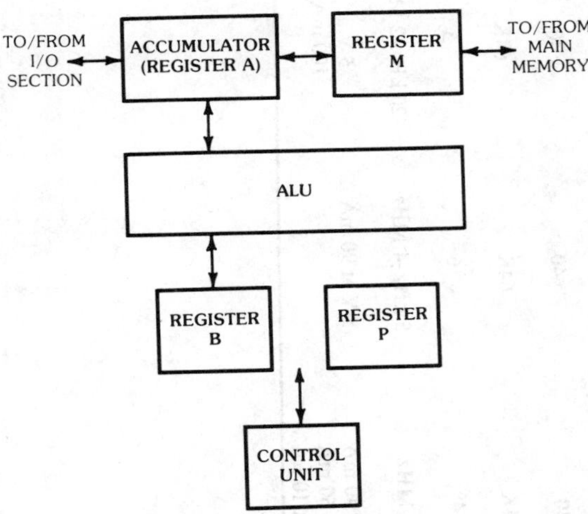

Fig. 32. The CPU. (*From A. Barna and D. I. Porat,* Introduction to Microcomputers and Microprocessors. *New York: John Wiley & Sons, Inc., 1976.*)

The Main Memory—The i/o section and CPU contain several temporary storage registers for digital information. The majority of data storage in a microcomputer is in the main memory. A part of the memory can also be used to store often-used sequences of operations or instructions, to help more complex programs to be executed efficiently. The details of memory functions will be described in the section on memory integrated circuits.

Available Microprocessors—The early microprocessors were direct mappings of early-generation CPU architectures, and early microcomputers were used for hand-held calculator applications. However, the design of microprocessor integrated circuits has evolved into the use of structured design and architecture that take advantage of the best topological layout and conserve silicon area. Tables 12 and 13 list some of the popular microprocessors available currently and their capabilities.

Gate Arrays

Another direction in the integration of random logic functions on-chip in a standard configuration is a gate array, also known as a master slice or an uncommitted logic array. The concept of this class of logic circuits is to provide a structured array of simple logic gates, typically three-input NOR gates, and implement custom logic by automatic connection of these gates through software programs that translate logical gate connections to physical wire routing over wiring channels provided on-chip between gates. This is a direct miniaturization of the printed-circuit-board concept with the scale of wiring channels decreased by many orders of magnitude. The design of the basic gate used in the array is determined by the loading due to a maximum wiring channel length. Gate arrays have been implemented in both CMOS and bipolar technologies.

Standard Cell Library

Logic designers are traditionally used to building special-purpose logic systems out of a choice of simple logic building blocks. These building blocks have been available in single-package units as medium-scale integrated logic. In the application of LSI and VLSI, these blocks are offered as software data bases with complete performance specifications that can be integrated together in a topological data base and fabricated as an LSI or VLSI chip.

MEMORY INTEGRATED CIRCUITS*

The most explosive growth of integrated circuits has perhaps been in the area of semiconductor memories. The major advantages of semiconductor memories are that they can utilize the most advanced technology and they are required in very large volumes in all systems so that they provide the economies of scale in the learning process for the maturing of the technology in time. The large volume of memory integrated circuits thus acts as a catalyst in the timely development of the technology it uses and thereby allows increases in yield and decreases in unit costs, which in turn propels an increase in volume. The progress in the density of memories has been phenomenal because of this synergism between the system pull for more

* Reference 22.

TABLE 12. 8-BIT MICROPROCESSORS

	8080	Z80	6800	6502	TMS 1000 (4-Bit Microcomputer)
Manufacturer	Intel	Zilog	Motorola	MOS Technology	TI
Second source	AMD, NEC, TI	Mostek	AMI	Rockwell, Synertek	
Technology	NMOS	NMOS	NMOS	NMOS	NMOS, CMOS, PMOS
Number of basic instructions	78	158	72	56	54
Number of registers	10	14 (duplicated)	6	6	7
Pin count	40	40	40	40	28, 40
Direct addressing range (bytes)	64K	64K	64K	64K	2K (internal)
Number of addressing modes	4	6	5	8	
Basic clock frequency	0.5–4 MHz	5 kHz–4 MHz	20 kHz–2 MHz	20 kHz–2 MHz	50 kHz–1 MHz
Power supply	12 V at 40 mA 5 V at 60 mA −5 V at 10 μA	5 V at 90 mA	5 V at 100 mA	5 V at 140 mA	3–35 V at 1–10 mA

TABLE 13. SPECIFICATIONS OF 16-BIT MICROPROCESSORS

	8086	Z8000	68000	16008/16016	16032
Year of commercial introduction	1978	1979	1981	1981	1981
Number of basic instructions	95	110	61	100	100
Number of general-purpose registers	14	16	16	8	8
Pin count	40	48/40	64	40	48
Direct address range (Bytes)	1M	48M	16M/64M	64K/16M	16M
Number of addressing modes	24	6	14	9	9
Basic clock frequency	5 MHz (4–8 MHz)	2.5–3.9 MHz	5–8 MHz	10 MHz	10 MHz

memory at low cost and the technology push via geometry scaling that allows improved bit density, improved performance, and decreased cost. Fig. 33 shows the unit volumes of one class of random-access memory as a function of time. The increase in volumes through the decade of the seventies has been at the rate of four times every two years, with a somewhat slower pace noted in the present generation. The slowing of pace is due to a complex interaction of manufacturing economics and the precision demanded by the micron-sized geometries in the technology. Many different memories are available that fall into the categories described here. The various manufacturers' data books and applications books should be consulted for detailed information.

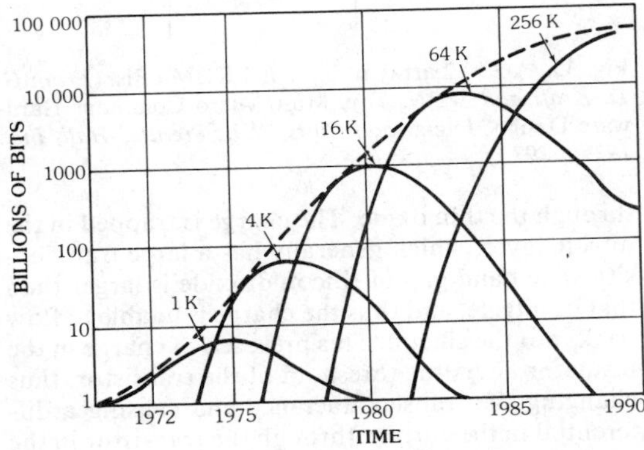

Fig. 33. Unit volumes of memory as a function of time.

Semiconductor memories are classified according to their functions as read-only memories (ROM) or random-access memories (RAM). The use of serial-access memory integrated circuits has been attempted, but it has not become common due to various cost-related factors; therefore, this type of memory will not be considered further here. Externally, memory chips typically require the application of a binary address input to locate the information. A set of control or mode-setting signals is needed to tell the memory to read or write, and the memory circuit provides the desired information after a time interval that is called the access time of the memory. The typical memories of today are organized to provide either one bit of data at a time or one byte of data at a time. The choice of the width of the data word is a system partitioning function. In addition, other system options such as power-down modes are common. Density and performance of state-of-the-art semiconductor memories are compared in Table 14.

Read-Only Memories (ROM)

Read-only memories have information programmed into them during manufacture. They act as tables of data that can be accessed by the system at any time. This class of memories is extremely useful in the storage of programs, operating systems, fixed utilities for a system, etc.

The read-only memory function is realized by the use of single transistors as memory cells connected in an *X-Y* matrix as shown in Fig. 34. The gate of each transistor is turned on by an *X* address signal, and the current through the transistor is monitored through the *Y* address line. The programming of the memory results in the presence or absence of the transistor in a specific location. This may be done at various points in the fabrication process. The presence of current through the transistor is detected as a 1, and the absence of current is detected as a 0. Note that the memory cell in a ROM is a single transistor. Further, only a read function is required. For this reason, ROMs are the densest semiconductor memories.

Nonvolatile Read-Only Memories

Nonvolatile read-only memories may be classified as "read-mostly memories." They are programmable after manufacturing. They are referred to as nonvolatile because they retain data even

TABLE 14. COMPARISON OF MEMORY CIRCUITS*

Type	Bits/Chip	Cell Size	Access Time	Power Dissipation
NMOS ROM	256K	40 μm^2	250 ns	300 mW
Bipolar PROM	64K	300 μm^2	50 ns	800 mW
EPROM	128K	100 μm^2	150 ns	300 mW
EEPROM	32K	260 μm^2	90 ns	300 mW
NMOS sRAM	64K	300 μm^2	35 ns	450 mW
CMOS sRAM	64K	450 μm^2	65 ns	200 mW
Bipolar sRAM	16K	900 μm^2	25 ns	800 mW
dRAM	256K	75 μm^2	150 ns	150 mW

* Data represent typical devices of each class.

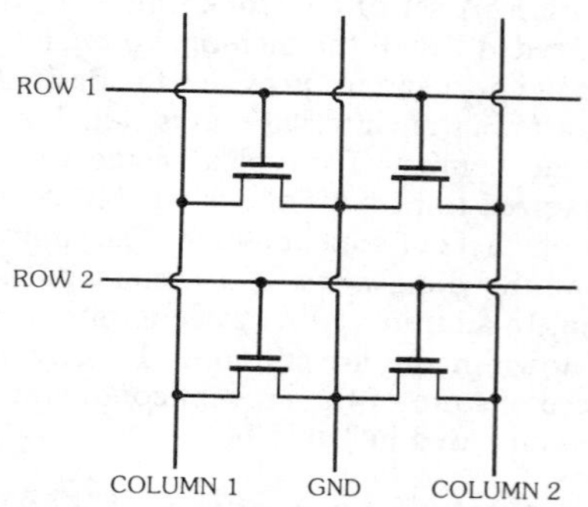

Fig. 34. A 2 × 2 ROM memory-cell array.

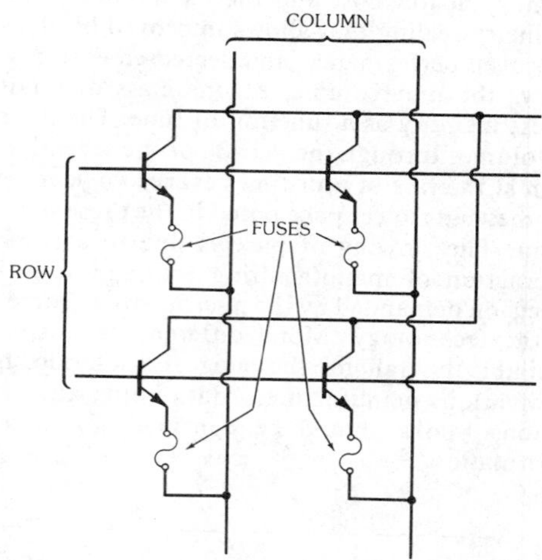

Fig. 35. A 2 × 2 array of bipolar PROM cells. (*From G. D. Kraft and W. N. Toy,* Mini/Micro Computer Hardware Design. *Englewood Cliffs, N.J.: Prentice-Hall, Inc., 1979; p. 97.*)

when no power is applied. Three subclasses of nonvolatile ROMs exist. The first is a programmable read-only memory (PROM), which can be electrically programmed. The second is an erasable and programmable read-only memory (EPROM), which can be programmed by electrical signals but erased only by exposure to ultraviolet light. The third is an electrically erasable and programmable read-only memory (EEPROM).

Generally, PROMs are constructed with bipolar technology. A simple 2 × 2 array of bipolar PROM cells is shown in Fig. 35. Programming is accomplished by applying a sequence of 20–30 mA current pulses to blow a fusible link.

Both EPROMs and EEPROMs are generally constructed by the use of MOS technology. The physical mechanism by which these memory cells store data is a storage of charge in the gate insulator of a MOSFET. This has been accomplished by two categories of transistor structures, shown in Fig. 36. The first structure incorporates multiple dielectric layers (Fig. 36A). The most common version of the multiple-insulator structure is the MNOS (metal-nitride-oxide-silicon) transistor. The oxide layer in contact with the silicon surface is very thin (2–3 nm). The charge is injected from the substrate into the nitride layer, by way of hot electron tunneling through the thin oxide. The charge is trapped in the nitride layer, which generally has a large trap density. The band gap of silicon dioxide is larger than that of nitride, and thus the charge is unable to flow back into the silicon. This presence of charge in the insulator alters the threshold of the transistor, thus changing the transconductance and causing a differential in the current through the transistor in the presence or absence of charge. This is read much the same way as the ROM cell.

The second structure that achieves the same purpose is the floating-gate transistor (Fig. 36B). The charge is injected into the floating gate from the silicon by the application of the proper electric field. The charge on the floating gate provides the change in threshold voltage. The floating-gate structure has been used in both EPROMs and EEPROMs, since the floating gate acts as an equipotential surface and can be biased with a reversed

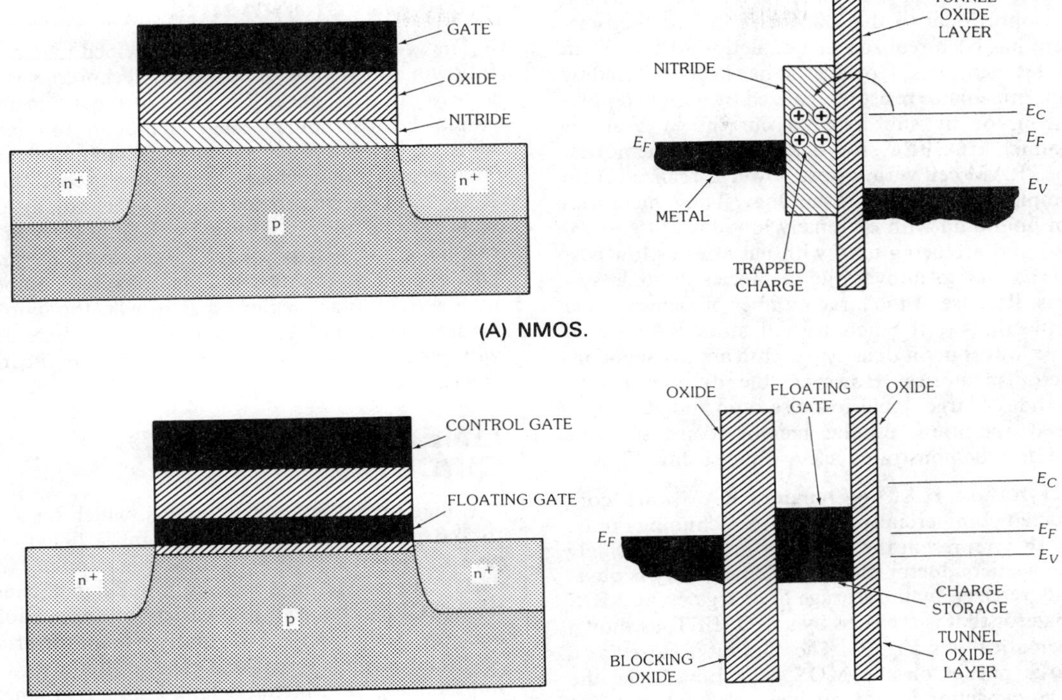

Fig. 36. Transistors for EPROM and EEPROM memories.

electric field to result in an erase function. The control of the floating-gate process is easier than the control of the MNOS process, and most of the nonvolatile memories are fabricated this way. There are various subclasses of the floating-gate structure that are based on the biasing scheme or the injection mechanism used for the program and the erase modes.

Random-Access Memories (RAM)

A random access memory is one in which data may be written in or read out at very high speed. The bit or set of bits into which data may be written is chosen randomly in an *X-Y* matrix. There are two types of random-access memories, static and dynamic. Static memories retain data indefinitely while power is applied to the chip; dynamic memories retain data for a short period of time, after which they must be refreshed. The system overhead associated with the refreshing of the memories is offset by the fact that dynamic memories are four times denser than static memories at the same design rules.

Static RAM—Static RAMs are realized from a memory cell with a bistable circuit. A simple flip-flop is commonly used for the memory cell, as shown in Fig. 37. Note that this can be constructed with either bipolar or MOS technology. The bipolar versions have traditionally been faster than the MOS versions. However the bit density of MOS sRAMs has been larger than for bipolar versions. The important elements of the cell are the load (M_1, M_2) and the cross-coupled driver devices (M_3, M_4) that form the bistable latch. The access transistors (M_5, M_6) are used to address the cell. The major considerations in the design of an sRAM cell are

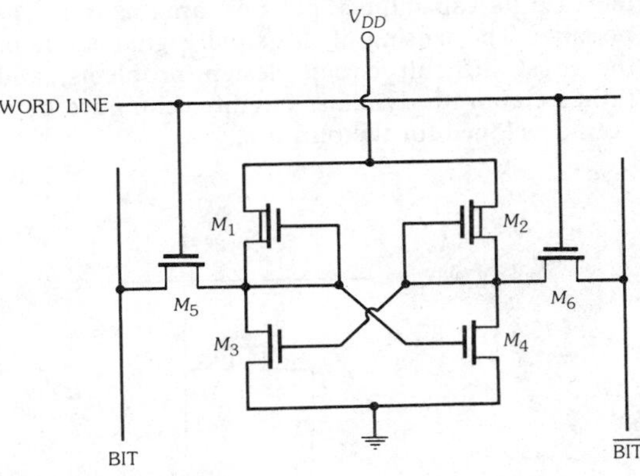

Fig. 37. A static RAM cell.

power, speed, and data hold. The power in the cell is proportional to the load current. The load element has been realized as a depletion MOSFET in 4K-bit memories. However, it has been replaced by high-impedance resistors realized by undoped polysilicon, or by subthreshold current in a short-channel MOSFET, for higher-density memories. The sRAM cell with lowest power is realized from complementary MOS technology. These memories can hold data with extremely low power (50–1000 nW) and are being used with batteries as low-cost alternatives to nonvolatile memories in some systems. Because of the large number of elements and connections to the memory cell, static RAMs tend to be lowest in bit density per chip among semiconductor memories. However, the differential cell provides a large signal-to-noise ratio and very high-speed operation. Bipolar memories are all static and have demonstrated access times below 10 ns.

Dynamic RAM—Dynamic RAMs are constructed commercially with MOS technology only, and they represent the largest-volume, most widely used semiconductor memory. The memory is physically realized by the storage of charge on an MOS capacitor that is accessed by a MOSFET, as shown schematically in Fig. 38. The data bit is stored as a charge packet on the MOS capacitor. Since the MOS capacitor is not an ideal element and has nonzero leakage, it is discharged through these leakage mechanisms as a function of time. The design of dynamic RAMs is thus critically dependent on the reduction of leakage in the MOS capacitor. This requires the realization of very long carrier lifetimes in the semiconductor and very high crystalline quality. The signal-to-noise ratio depends on the ratio of the storage capacitor to the capacitance of the digit or bit line on which the charge is dumped when the cell is accessed. As the bit density per chip grows, this ratio is degraded, since the cell area is decreased and more cells are added on the bit line. It is thus important to store as high a voltage as possible on the cell capacitor and increase its capacitance per unit area as much as possible. The sensing of this small signal is one of the most difficult circuit design problems, and rather elaborate dynamic circuit techniques have been developed for the purpose.

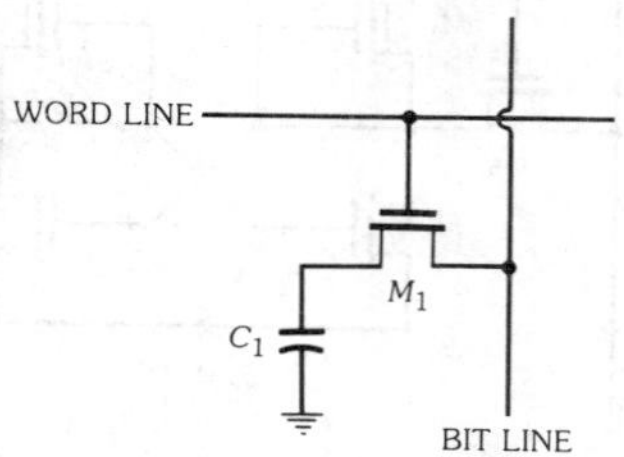

Fig. 38. The one-transistor/capacitor dRAM cell.

New Developments

The growth of all the above-described memories following the scaling of MOSFET technology is on an evolutionary path. The state-of-the-art memories in the development stage include up to 4 megabits of ROM and 4 megabits of dynamic RAM. The memory market for all sectors except static RAM has been dominated by MOS technology and will continue to be so. However, the bipolar memories have had a strong edge in speed in the past. One-micron MOS static RAMs are speed-competitive with bipolar memories. However, the bipolar memories are significantly more radiation hard and will tend to have major applications in hostile environments.

LINEAR INTEGRATED CIRCUITS*

Unlike digital integrated circuits, which respond to and produce two-state logic signals, linear ICs give an output signal that can be made to vary linearly with respect to a varying input signal. Since linear ICs can be used in a variety of applications and new designs appear regularly, standardization of circuit elements as in the case of digital ICs is impractical. Several important families of linear ICs have evolved. They include the device categories described below. The various manufacturers' data books and applications books should be consulted for detailed information.

Differential Amplifier

The basic differential amplifier consists of two identical input transistors connected to respond to the difference between two input signals while simultaneously blocking the identical part of these signals. This common-mode rejection is useful in noisy environments. The amplifier may consist of a single stage or two stages and usually has a low-impedance output stage. The output is also differential. These amplifiers are typified by a wide bandwidth (dc to vhf), moderate gain (less than 1000 times) and moderate common-mode rejection ratio (cmrr) (50 dB). Internal feedback is provided for linearity control, and external feedback is difficult to apply without severely degrading the bandwidth. Applications include linear amplification, mixers, product detectors, amplitude modulators, frequency multipliers, voltage-controlled oscillators, and disk or tape memory read amplifiers.

Operational Voltage Amplifier

The basic operational amplifier, or op-amp, is a dual-input differential amplifier followed by one or more direct-coupled gain stages and a low-imped-

* References 23, 24, 25, 26, and 27.

ance output stage. The typical op-amp exhibits very high voltage gain at dc (80 to 120 dB) and a unity-gain bandwidth of 1 to 20 MHz. The op-amp may be internally compensated, i.e., 100% negative feedback can be applied externally without oscillation problems, or pins may be provided to tailor the compensation for the application. Pins may also be provided to adjust the input offset voltage. Op-amps are commonly made with a bipolar process but may have JFET or MOSFET input stages. Complete CMOS designs are also available.

Any op-amp can be connected as shown in Fig. 39 to provide a linear voltage gain equal to R_f/R_{in}, provided the op-amp is ideal and properly compensated. The "ideal" op-amp would have the following characteristics:

1. A negligibly small differential input voltage can produce any desired output.
2. The input currents are negligibly small.
3. The output impedance is zero.

Application of these rules to negative-feedback circuits such as in Fig. 39 immediately leads to the ideal transfer function. No existing op-amp has these ideal characteristics, but many come close to meeting one or more of them. A variety of op-amps are available with emphasis on perfecting different properties such as low noise, low input-voltage offset, low input current, high slew rate, single-power-supply operation, and large power-supply-range operation.

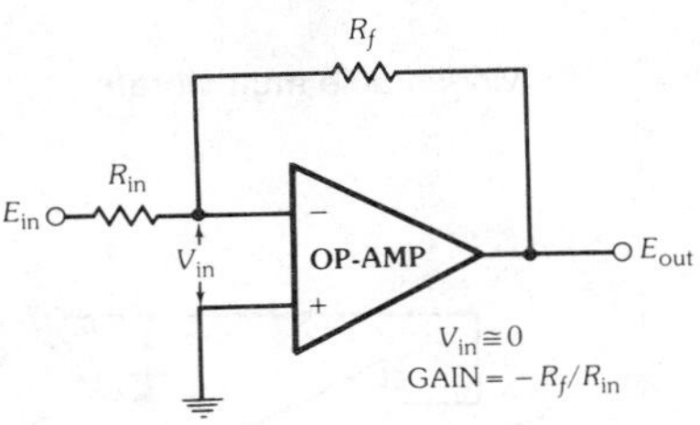

Fig. 39. Inverting dc amplifier.

The op-amp was originally designed to perform such mathematical operations as integration, differentiation, comparison, summation, multiplication, and others. Though op-amps are still used in these applications, complete ICs performing these functions are available. The use of various types of external feedback networks permits a vast array of additional applications. Accordingly, the op-amp is the most versatile of linear ICs, and its circuit applications include analog-to-digital conversion, peak detection, averaging, function generation, oscillator and pulse circuits, biological function amplification, voltage-controlled oscillator circuits, automatic gain control, voltage comparators, active filters, precision diodes, sample-and-hold circuits, and numerous others. Many of these complete circuits in turn are available in IC form.

Fig. 40 summarizes some of the elementary op-amp circuit configurations and gives ideal formulas that describe the circuit performance. The op-amp is used in a number of modes besides the basic operational one, such as a high-impedance buffer to minimize loading in RC filters or to compress the nonlinearity of other devices by the op-amp open-loop gain for voltage comparison.

Other Amplifiers

The operational transconductance amplifier (OTA) provides an output current in response to a voltage input. The OTA bias current controls the transconductance, and considerable control over the performance of an OTA is possible. The operating characteristics make the OTA usable in a wide variety of circuits such as multiplexers, sample-and-hold circuits, gain controls, modulators, multipliers, comparators, multistable circuits, and (with the addition of an output stage) a high-gain op-amp.

The current-differencing amplifier, or "Norton" amplifier, provides a voltage output proportional to the difference of two input currents. It is especially suited for single-power-supply operation without loss of common-mode range. Again, a large variety of circuit functions can be obtained.

Voltage Reference

Although classified under linear ICs, the voltage reference is a highly nonlinear device and provides a constant output voltage almost independent of input voltage, load current, temperature, and time. Output voltages range typically from 1.22 volts to 10 volts with a precision of 0.05% to 5%. Temperature coefficients range from 100 parts per million down to 0.5 ppm. Output current capabilities are less than 10 mA. Both fixed-output-voltage and programmable-output-voltage devices are available.

Reference diodes also come in IC form and simulate the characteristic of a zener diode, but with a much sharper breakdown characteristic.

Some applications are amplifier biasing for temperature independence, constant-current-source circuits, level detectors, and low-voltage regulators.

Voltage Regulators

Bipolar voltage-regulator ICs incorporate a voltage reference and a sense amplifier and maintain the output voltage at a value almost independent of load, input voltage and ripple, and temperature. Both fixed- and programmable-output-voltage units are made. Voltages up to 50 volts can be regulated. Current capability is usually on the order of tens of milliamperes, and the IC is intended as the driver for large pass transistors that can accommodate the higher currents and operating temperatures that occur in power supplies. External components for current limiting, noise reduction, and

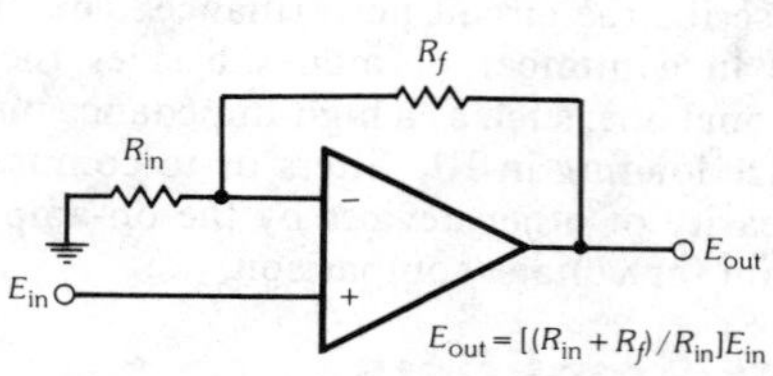

(A) Dc amplifier (noninverting).

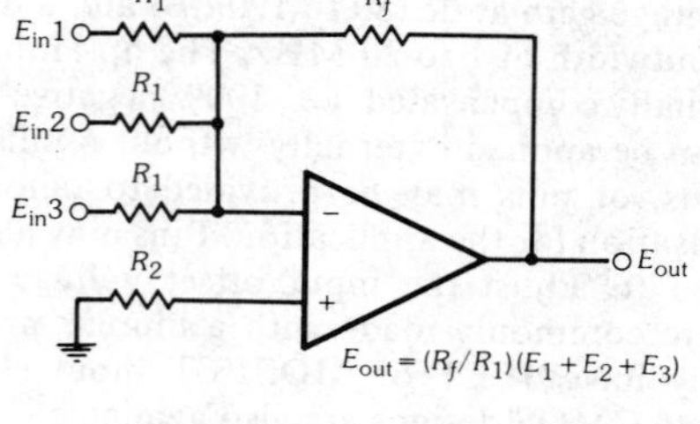

(B) Summing amplifier.

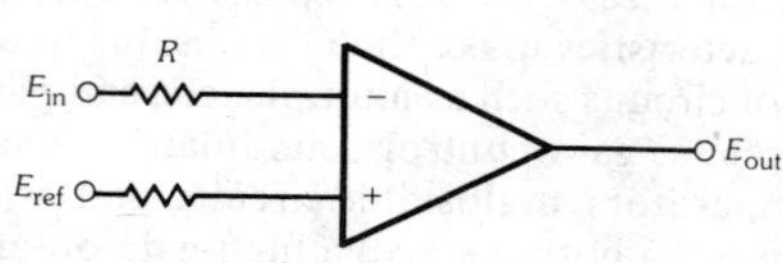

(C) Analog-to-digital converter.

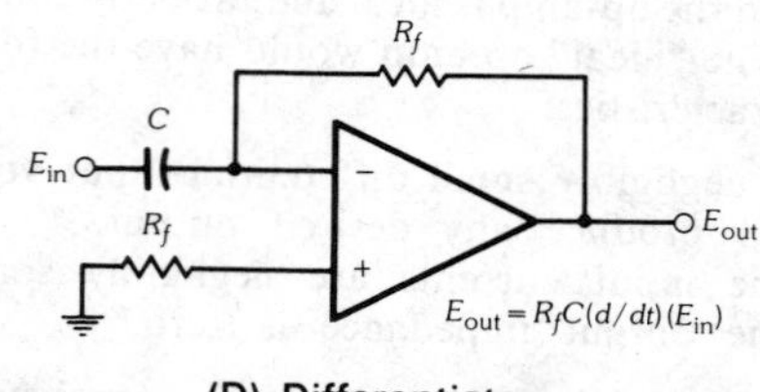

(D) Differentiator.

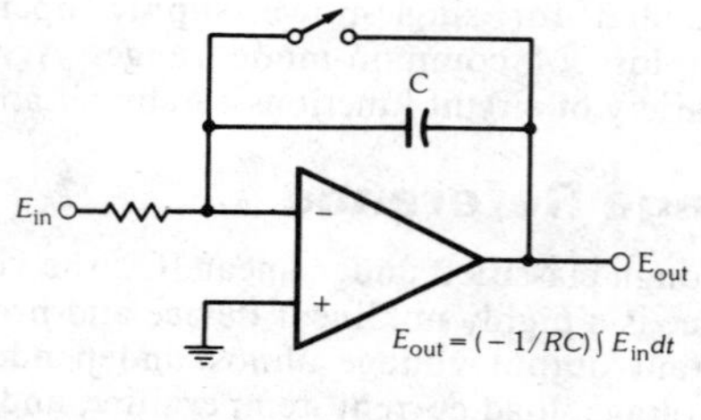

(E) Integrator.

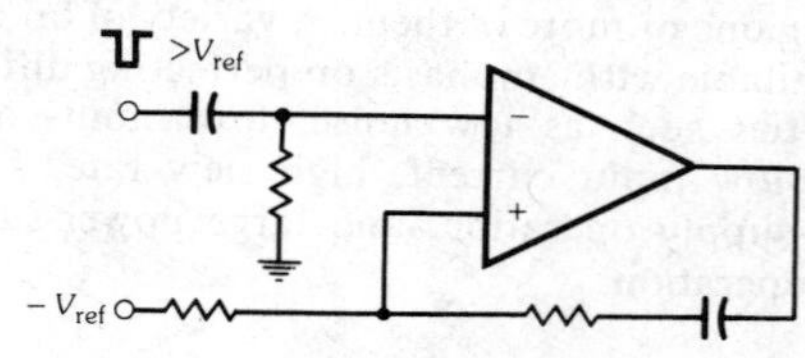

(F) Monostable multivibrator.

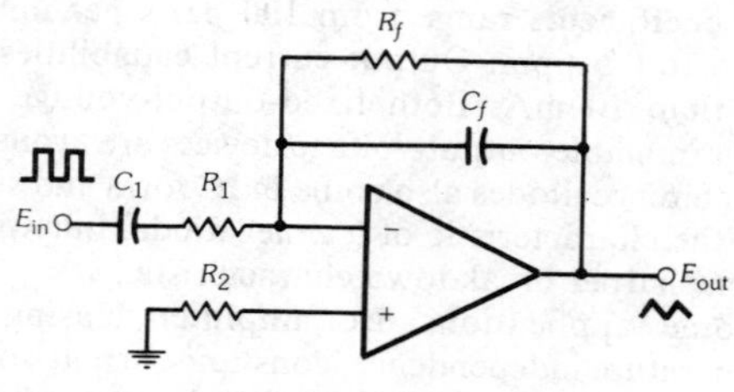

(G) Sweep generator.

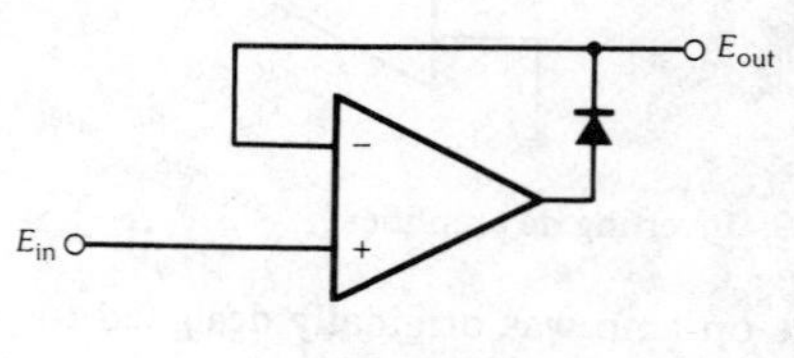

(H) Precision diode.

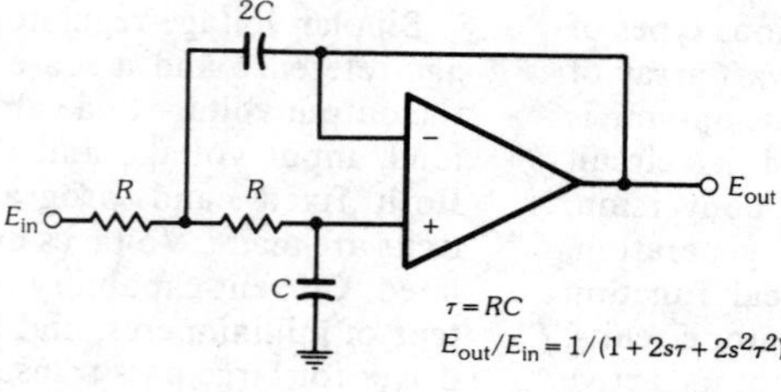

(I) Low-pass filter.

Fig. 40. Simple ideal op-amp circuits.

compensation are desirable additions in the design of a complete power supply.

Comparators

The voltage comparator is a differential amplifier designed with a small delay time between the application of a differential input signal and the output transition. The output swing is made compatible with TTL inputs. Emphasis is placed on low input bias current and its offset and low input offset voltage, and the usual amplifier specifications are of lesser importance. Common applications are high-speed analog-to-digital converters, fast zero-crossing detectors, tape- and disk-file read channels, and differential line receivers.

Special-Purpose Linear Integrated Circuits

Since any circuit that can be assembled from discrete semiconductor components can almost always be duplicated or simulated with monolithic IC technology, a great many special-purpose ICs exist. These include custom designs for use in various kinds of commercial products as well as a wide range of off-the-shelf numbered units. The commercial units include such devices as music synthesizers, phase-locked loops, tone decoders and encoders, function generators, programmable filters, fm and video demodulators, if detectors, subcarrier regenerators, agc, fm stereo demultiplexers, am-receiver functional blocks, timers, audio amplifiers and power amplifiers, voltage-to-frequency converters, analog-to-digital and digital-to-analog converters, instrumentation amplifiers, and sample-and-hold circuits.

Though many of the ICs are dedicated to a specific application or function, the addition of a few external components can often result in a wide range of additional applications. The list of special-purpose linear ICs grows constantly, and the design engineer is well advised to consult current manufacturers' data and application literature.

Miscellaneous Linear ICs

Other useful building blocks are timers, transistor-diode arrays, digitally controlled analog pass-gates, optical isolators, bar or dot LED-display drivers, and analog shift registers.

TRENDS IN INTEGRATED CIRCUITS

Scaling and Miniaturization

The explosive growth of integrated circuits in the sixties and seventies has been fueled by the ability to scale the minimum lithographic dimensions of an integrated circuit. This results in a threefold advantage:

1. Increase in density of circuit elements per chip.
2. Increase in circuit performance due to increased device gain and reduced load capacitance.
3. Decrease in cost per function, which provides economic incentive.

The most significant improvement in integration complexity has been achieved in MOS circuits. An exponential growth of the number of circuit elements in time has been noted. This has been made possible by the simple scaling laws that relate the scaling of vertical and lateral dimensions to the scaling of doping and voltages. Table 15 shows three sets of scaling laws that have been used to various degrees. The first assumes that electric field in the device must be held constant. The second is based on complying with currently set voltage standards, and the third proposes a change in the voltage standards only when the electric field is high enough to cause problems due to secondary effects. Another scaling theory has been proposed on the basis of the off-state switching behavior of MOS devices. Scaling of bipolar devices does not proceed along scaling laws since the base width that is the critical dimension is much smaller than all other dimensions. Scaling of bipolar technology is aimed at reduction of parasitic device capacitance. The prospects of scaling to submicron dimensions have been shown to be technically feasible, with the reduction of parasitic effects and statistical fluctuation, both of which are expensive problems to solve, being the dominant deterrents. It is recognized that the improvement of circuit performance in the submicron regime is incremental compared to the current 2-micron circuits. The integration complexity can also be increased by using larger silicon area and not reducing feature size. The trend to bigger chips is currently noted, with fault-tolerant architectures being implemented. The decision between reducing feature size and increasing chip area is expected to be totally cost-based.

Image-Sensing ICs

The application of solid-state devices to high-resolution imaging began in earnest with the invention of the charge coupled device (CCD) at Bell Laboratories in 1969. Many approaches to the solid-state imaging problem have been undertaken. These approaches may generally be divided into three distinct areas: (1) memory arrays, (2) charge injection devices (CIDs), and (3) charge coupled devices (CCDs). Memory arrays, as the name implies, are simply solid-state memory devices that have been packaged in a format that allows an image to be placed on the area that is the usual memory array. The primary advantage of the memory-array approach to solid-state imaging is the fact that these devices are *X-Y* addressed as well

TABLE 15. DEFINITION OF SCALING LAWS

Scaling Law	Constant Field	Constant Voltage	Quasi-Constant Voltage
Dimensions (λ)	λ	λ	λ
Gate Oxide (λ_O)	λ	$\sqrt{\lambda}$	λ
Doping (λ_N)	λ	λ	λ
Voltage (λ_V)	λ	1	$\sqrt{\lambda}$

Full-scale (λ = 1) values of $L = 3\ \mu m$, $t_{ox} = 500$ Å, $N_A = 2.5 \times 10^{15}\ cm^{-3}$ are used.

as being process compatible with other memory components. The primary disadvantage of this approach to solid-state imaging is the fact that memory-cell architectures as normally designed for digital applications have relatively high-capacitance sense nodes, and therefore they have relatively low sensitivity and high noise, which precludes these devices from being used for most high-performance applications.

Charge injection devices employ construction similar to memory arrays in that they are *X-Y* addressed, but the charge detection scheme is different. The device in its modern form consists of an array of charge-coupled pairs of capacitors as illustrated in Fig. 41. The readout is achieved by applying a capacitively coupled pulse to the row while a certain column is selected by lowering its potential.

The most widely explored architecture for solid-state imaging is the CCD. These devices store charge in a potential well and transfer the charge from one electrode area to an adjacent electrode area by application of voltages with appropriate time and phase relationships, as illustrated in Fig. 42. The initial CCD devices employed three separately clocked phases to store and transfer charge in an MOS structure. The CCD concept has been extended, however, to include four-phase, three-phase, two-phase, and virtual-phase (one-phase) architectures, with significant advantages for devices with fewer gates. The construction of these different structures is illustrated in Fig. 43.

Solid-state imaging devices provide the combined advantages of high performance, low power, high reliability, and ease of use. These features are available in a wide variety of imagers fabricated as memory arrays, CIDs, and CCDs, and in a wide spectrum of formats and performance ranges. These features will combine to make solid-state imaging the mainstay of advanced imaging systems for future applications.

Speech-Synthesizer IC*

Speech is generated with these chips by the excitation of a time-varying digital filter. The excitation and filter parameters are stored in ROM, EPROM, RAM, or disk memory or are generated by a program. Overall control is provided by a microprocessor. The process is entirely digital up to the analog-to-digital converter, which provides the analog signal to drive a speaker.

An alternative to the variable-filter approach is to store digitized speech and sample the data at an 8-kHz rate, which leads to a digital data rate of about 100 kHz. Logarithmically compressed amplitude data could be used, analogous to digital tele-

* Reference 28.

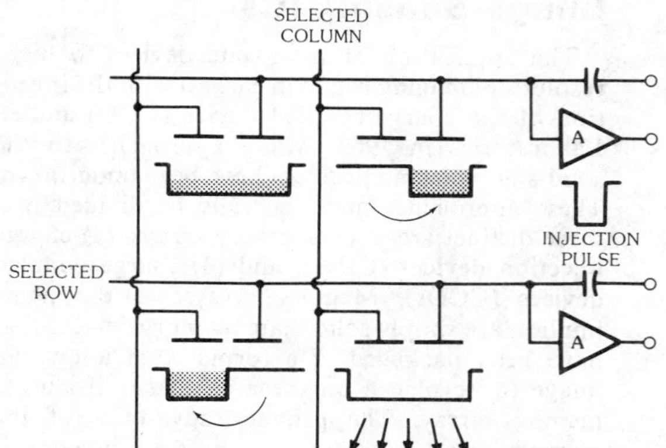

Fig. 41. Basic operation of a charge injection device.

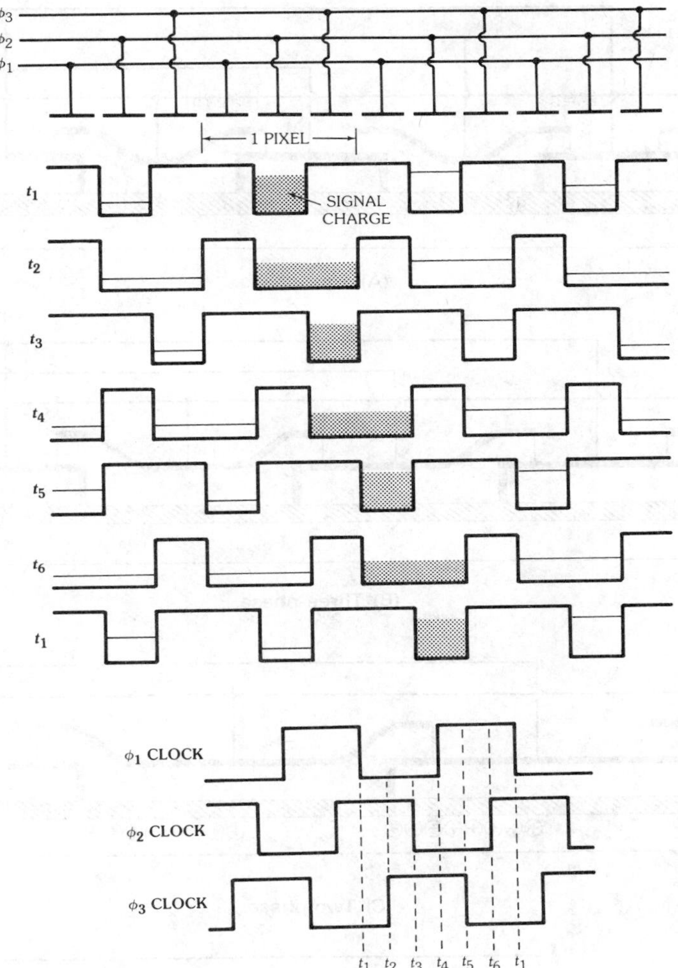

Fig. 42. Basic operation of a charge coupled device.

phone systems, which results in a rate of 64 kilobits/second with very good quality. The time-varying filter techniques provide acceptable speech quality but at a much lower digital input data rate, down to an average rate of 1200 bits/second for a ten-pole filter derived from a linear prediction model of speech.

The low data rates are made possible because of the redundancy in speech and by using a simplified simulator of the human speech-generating system. The vocal tract is simulated by a dozen or so connected pipes of differing diameter, and the excitation is represented by a pulse stream at the vocal-chord rate for voiced sound or a random noise source for the unvoiced parts of speech. The reflection coefficients at the junctions of the pipes can be obtained from a linear prediction analysis of the speech waveform,† and the names associated with the ICs, such as LPC or PARCOR, refer to the particular analysis used to obtain the coefficients for the equivalent electrical digital filter.

REFERENCES

1. *Journal of Electrochemical Society,* The Electrochemical Society, Inc., Manchester, N.H.
2. *Journal of Vacuum Science and Technology,* published for the American Vacuum Society by the American Institute of Physics, New York.
3. *Semiconductor International,* Cahners Publishing Co., Chicago.

† Reference 28.

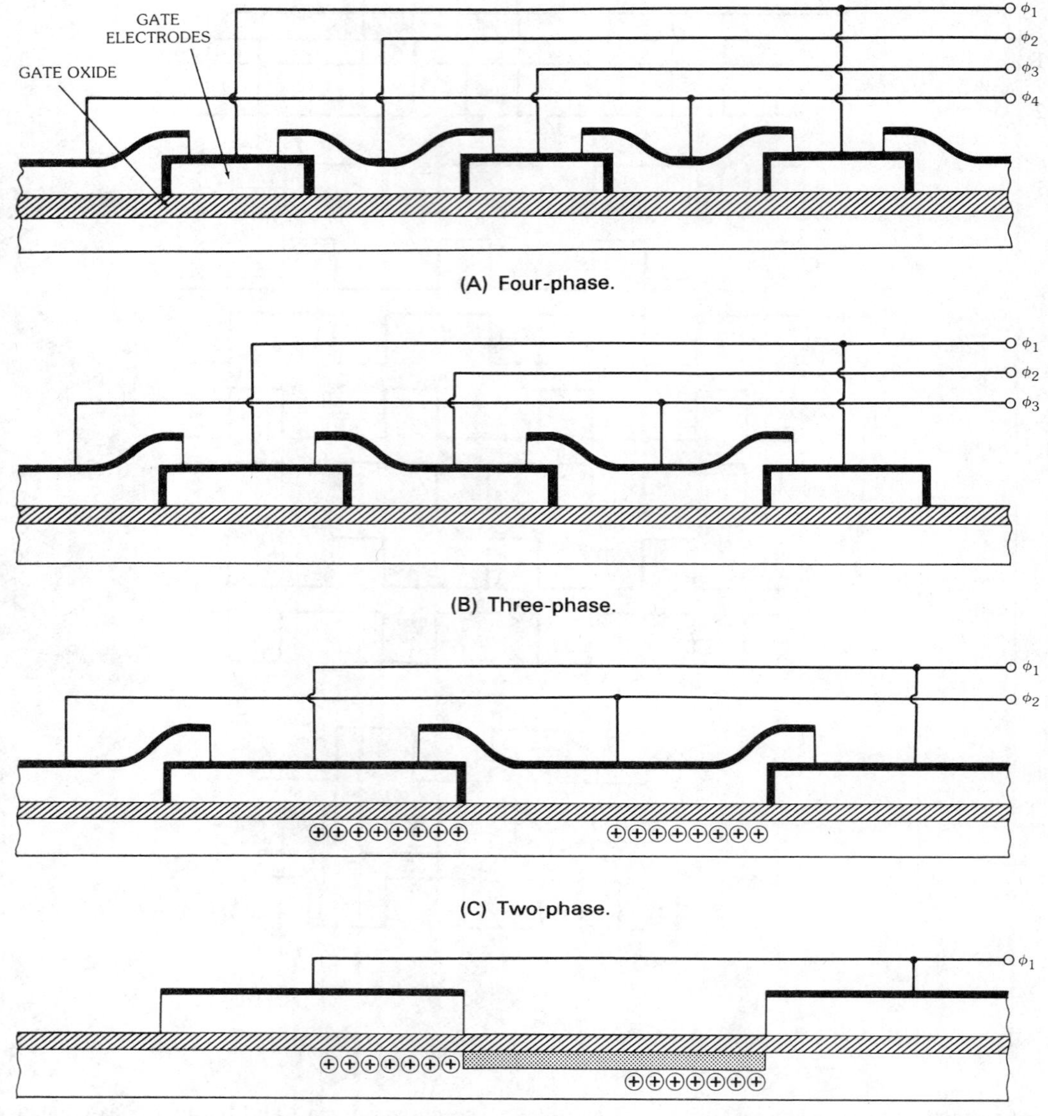

Fig. 43. Alternate structures for CCDs.

4. Kern, W., and Puotinen, D. "Cleaning Solutions Based on Hydrogen Peroxide for Use in Silicon Semiconductor Technology." *RCA Review*, June 1970, pp 187–206.
5. Burger, R. M., and Donovan, R. P. *Fundamentals of Silicon Integrated Device Technology,* Vol. 1. Englewood Cliffs, NJ: Prentice-Hall, Inc., 1967, p. 254.
6. Rosler, R. S. "Low Pressure Production Processes for Poly, Nitride and Oxide." *Solid State Technology*, April 1977, p. 63.
   Kern, W. and Ban, V. S. "Chemical Vapor Deposition of Inorganic Thin Films." *Thin Film Processes,* T. L. Vossen and W. Kern, eds. New York: Academic Press, Inc., 1978.
7. Hollahan, J. R., and Rosler, R. S. "Plasma Deposition of Inorganic Thin Films." *Thin Film Processes*, T. L. Vossen and W. Kern, eds. New York: Academic Press, Inc., 1978.
8. Maissel, L. I., and Glang, R., Eds. *Handbook of Thin Film Technology*. New York: McGraw-Hill Book Co., 1970, pp 1–37.
9. Mead, C., and Conway, L. *Introduction to VLSI Systems.* Reading, Mass.: Addison-Wesley Publishing Co., Inc., 1980.
10. Moran, J. M., and Maydan, D. "High Resolu-

tion, Steep Profile Resist Patterns." *J. Vac. Sci. Technol.*, Vol. 16 (1979), p. 1620.

11. Poulsen, B. "Plasma Etching—A Review." *J. Vac Sci. Technol.*, Vol. 14 (1977), p. 266.
12. Trumbore, F. A. "Solid Solubilities of Impurity Elements in Germanium and Silicon." *Bell Sys. Tech. J*, Vol. 39 (1960), p. 205.
13. Research Triangle Institute Technical Report, "Integrated Silicon Device Technology." *RTI Review-Reports*, Vol. 4 ASD-TDR-63-316, 1964.
14. Brown, D. M., and Kennicott, P. R. "Glass Source B Diffusion in Si and SiO_2." *J. Electrochem. Soc.*, 1971, p. 293.
15. Meyer, J. W., Ericksson, L., and Davies, J. A. *Ion Implantation in Semiconductors.* New York: Academic Press, Inc., 1970.
16. Gibbons, T. F., Johnson, W. S., and Mylroie, S. W. *Projected Range Statistics—Semiconductors and Related Materials.* 2nd ed. New York: Halsted Press, 1975.
17. Morris, Robert L., and Miller, John R. *Designing with TTL Integrated Circuits.* Texas Instruments Electronics Series. New York: McGraw-Hill Book Co., 1971.
18. Hart, K., and Slob, A. "Integrated Injection Logic." *IEEE Trans Solid State Circuits,* SC-7 (1972), p. 346.
19. Sloan, B. J. "STL Technology." *IEEE IEDM Tech Digest,* 1979, p. 324.
20. Barna, Arpad, and Porat, Dan I. *Introduction to Microcomputers and Microprocessors.* New York, New York: John Wiley and Sons, Inc., 1976.
21. Kraft, George D., and Toy, Wing N. *Mini /Microcomputer Hardware Design.* Englewood Cliffs, New Jersey: Prentice-Hall, Inc., 1979.
22. Elmasry, Mohammed I., ed. *Digital MOS Integrated Circuits.* IEEE Press, John Wiley & Sons, Inc., distributor, 1981.
23. Roberge, James K. *Operational Amplifiers: Theory and Practise,* New York: John Wiley & Sons, Inc., 1975.
24. Tobey, G. E., Graeme, J. G., and Huelsman, L. P. *Operational Amplifiers: Design and Applications.* New York: McGraw-Hill Book Co., 1971.
25. Graeme, J. G. *Applications of Operational Amplifiers: Third-Generation Techniques.* New York: McGraw-Hill Book Co., 1973.
26. Wong, Y. J., and Ott, W. E. *Function Circuits: Design and Applications.* New York: McGraw-Hill Book Co., 1976.
27. Graeme, J. G. *Designing with Operational Amplifiers: Application Alternatives.* New York: McGraw-Hill Book Co., 1977.
28. Markel, J. D., and Gray, A. H. Jr., *Linear Prediction of Speech.* New York: Springer-Verlag, 1976.

21 Optoelectronics

Revised and Expanded by
Gregory E. Stillman

Optoelectronics is the technological marriage of the fields of optics and electronics. It includes the generation and evaluation of electromagnetic radiation in the optical wavelength range and its conversion into electrical current or signals, the interaction of light with matter, radiometry, and the characteristics of sources and detectors.

THE OPTICAL SPECTRUM

The optical spectrum is generally defined to encompass electromagnetic radiation with wavelengths in the range from 10 nm to 10^3 μm, or frequencies in the range from 300 GHz to 3000 THz (Fig. 1).

Other units are often used to describe the optical spectrum:

Wavelength λ:

$$1\ \mu\text{m} = 10^{-3}\ \text{mm} = 10^3\ \text{nm} = 10^4\ \text{Å}$$

Frequency (ν or f):

$$1\ \text{cm}^{-1}\ \text{(wave number)} = 30\ \text{GHz} = 3 \times 10^{10}\ \text{Hz}$$

$$\nu\ (\text{or } f) = \frac{c}{\lambda} = \frac{2.998 \times 10^{14}}{\lambda\ (\mu\text{m})}\ \text{Hz}$$

$$\bar{\nu}\ \text{(wave number)} = \frac{\nu}{c} = \frac{1}{\lambda}\ (\text{cm}^{-1})$$

Photon energy (E):

$$E = h\nu = \frac{hc}{\lambda} = \frac{1.2399}{\lambda\ (\mu\text{m})}\ \text{eV},$$

where the velocity of light in free space is

$$c = (2.99792458 \pm 0.000000012) \times 10^{10}\ \text{cm/s}$$

and Planck's constant h is

$$h = (6.626176 \pm 0.000036) \times 10^{-34}\ \text{J}$$

The optical spectrum is divided into three major categories, as follows.

Ultraviolet: Wavelengths shorter than those in the visible spectrum and longer than for x-rays are collectively designated ultraviolet (uv). Ultraviolet is classified according to wavelength as extreme (100–2000 Å), far (2000–3000 Å), or near (3000–3700 Å). Ultraviolet is also sometimes designated as short-wave or long-wave.

Visible: Those wavelengths in the approximate range 3700–7500 Å can be perceived by the human eye and are therefore collectively designated as visible light. Visible light is classified according to the various colors its wavelengths elicit in the mind of a standard observer. The major color categories are violet (3700–4550 Å), blue (4560–4920 Å), green (4930–5770 Å), yellow (5780–5970 Å), orange (5980–6220 Å), and red (6230–7500 Å).

Infrared: Those wavelengths longer than those in the visible spectrum and shorter than microwaves are collectively designated infrared (ir). Infrared is classified according to its wavelength as near (0.75–3 μm), middle (3–6 μm), far (6–15 μm), and extreme or submillimeter (15 μm–1 mm), although different authors often use slightly different wavelength ranges for these classifications.

RADIOMETRY

Terms and Definitions

Radiometry is the science of measurement of optical radiation at any wavelength, based simply on physical measurements. Radiant energy cannot be measured quantitatively directly, but must always be converted into some other form such as thermal, electrical, or chemical. Radiometry applies over the entire electromagnetic spectrum, not just the optical region. The terms used to describe radiant power in radiometry are summarized in

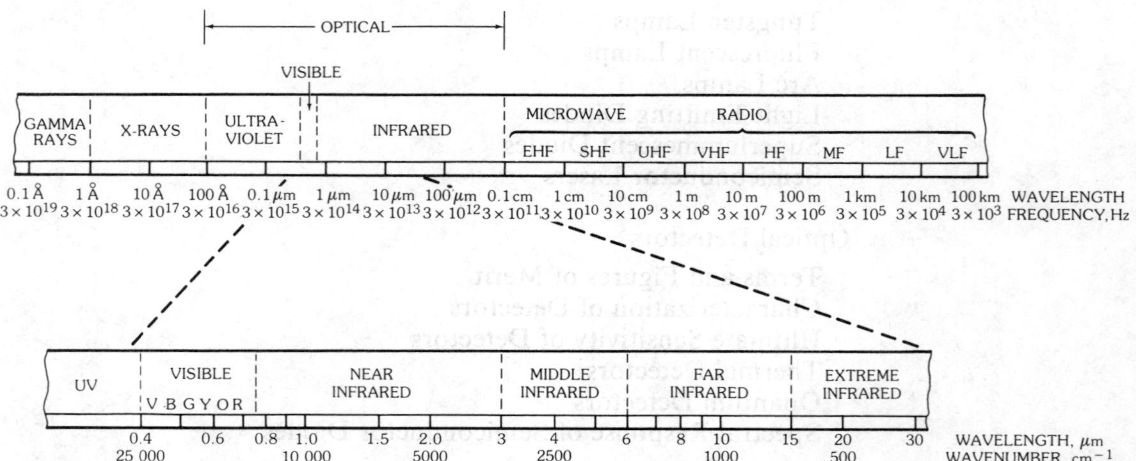

Fig. 1. The electromagnetic spectrum.

Table 1. All the terms used in radiometry are defined in terms of energy. When certain quantities are considered as a function of wavelength or frequency, the adjective "spectral" is used to modify the term, and the symbol for the quantity is followed by the appropriate spectral symbol, λ, ν, or $\bar{\nu}$, in parentheses. When the spectral concentration of a quantity is considered, it is also described by the adjective spectral, but in this case the symbol is subscripted with the appropriate spectral quantity; for example the spectral radiant excitance can be given as W_λ in units of W m^{-2} µm^{-1}.

Blackbody Radiation

Any surface of a body with a temperature greater than absolute zero ($T = 0$ K) is a source of radiation. The term "blackbody" applies to a thermal radiator that absorbs completely all incident electromagnetic radiation regardless of the wavelength, the direction of incidence, or the polarization. Such a radiator also has the maximum emission possible for any wavelength and in any direction for a thermal radiator in thermal equilibrium at a given temperature. Kirchhoff's law states that for any body (all materials) in an isothermal enclosure at temperature T, the ratio of the radiant excitance (emittance), W, to the absorptance, α, is equal to the radiant excitance of a blackbody, W_{bb}, at the same temperature:

$$W(T)/\alpha = W_{bb}(T)$$

Planck's law gives the blackbody spectral power for unpolarized radiation emitted at temperature T between the wavelengths λ and $\lambda + d\lambda$ as

$$W_\lambda(\lambda,T)d\lambda = 2\pi c^2 h\lambda^{-5}(e^{hc/\lambda kT} - 1)^{-1}d\lambda$$
$$= C_1 d\lambda/\lambda^5(e^{C_2/\lambda T} - 1)$$

where,

$k = (1.380662 \pm 0.000044) \times 10^{-23}$ J K^{-1} is Boltzmann's constant,

$C_1 = 2\pi c^2 h = (3.741832 \pm 0.000020) \times 10^4$ W cm^{-2} µm^4,

$C_2 = ch/k = (1.438786 \pm 0.000045)\ 10^4$ µm K.

Fig. 2A shows the variation of $W_\lambda(\lambda,T)$ with λ for various values of T. Fig. 2B shows the value of $N_\lambda(\lambda,T)$ which is equal to $W_\lambda(\lambda,T)/\pi$, since a blackbody is a Lambertian radiator. The integral of $W_\lambda(\lambda,T)d\lambda$ over all wavelengths gives the Stefan-Boltzmann law for total blackbody radiant excitance $W_{bb}(T)$ in W/cm^2,

$$W_{bb}(T) = \int_0^\infty W_\lambda(\lambda,T)\,d\lambda = (2\pi^5k^4/15c^2h^3)T^4$$
$$= (\pi^4C_1/15C_2^4)T^4 = \sigma T^4$$

where $\sigma = (5.67032 \pm 0.00071) \times 10^{-12}$ W cm^{-2} K^{-4} is the Stefan-Boltzmann constant. The blackbody spectrum has a pronounced maximum at a particular wavelength λ_m for a given temperature T, and the relationship

$$\lambda_m T = C_2/4.9651 = 2897.8 \pm 0.4 \text{ µm K}$$

is known as Wien's displacement law.

Planck's law can also be specified in terms of frequency or wave number instead of wavelength, and these relations result in slightly different forms of Wien's displacement law. For calculations it is often convenient to use the dimensionless quantity

$$x = hc/\lambda kT = h\nu/kT = C_2/\lambda T$$

for which Planck's law becomes

$$W_x(x,T)dx = (15\sigma T^4/\pi^4)[x^3dx/(e^x - 1)]$$

The energy flux per unit area for wavelengths lying between x_1 and x_2 can easily be calculated using the series expansions below:

$$W_{0-\lambda_0}(T) = \int_0^{\lambda_0} W_\lambda(\lambda,T)d\lambda$$
$$= (15\sigma T^4/\pi^4)\int_{x_0}^\infty x^3dx/(e^x - 1)$$
$$= (15\sigma T^4/\pi^4)\sum_{m=1}^\infty (e^{-mx_0}/m^4)\,[(mx_0)^3 + 3(mx_0)^2 + 6(mx_0) + 6]$$

where

$$x_0 = hc/\lambda_0kT = C_2/\lambda_0T$$

and the energy flux per unit area for wavelengths between λ_2 and λ_1 is given by

$$W_{\Delta\lambda} = (15\sigma T^4/\pi^4)$$
$$\sum_{n=1}^\infty (e^{-nx}/n^4)\,[(nx)^3 + 3(nx)^2 + 6nx + 6]\Big|_{x_2}^{x_1}$$

for $x_i \geq 2$.

A similar expression can be derived for values of $x_i < 2$ using

$$W_{0-\lambda_0}(T) = \sigma T^4\{1 - (15/\pi^4)x_0^3\,[1/3 - x_0/8 + x_0^2/60 - x_0^4/5040 + x_0^6/272\,160 - x_0^8/13\,305\,600 + \ldots]\}$$

which converges rapidly for $x_0 < 2$.

Similar relations for the number of quanta or photons emitted by a body at temperature T can be obtained by dividing $W_\lambda(\lambda,T)$ by the energy of the quanta at wavelength λ, hc/λ, as

$$Q_\lambda(\lambda,T)d\lambda = (C_3/\lambda^4)[d\lambda/(e^{c_2/\lambda T} - 1)]$$

where $C_3 = C_1/hc = 2\pi c$. The variation of $Q_\lambda(\lambda,T)$ with λ for several temperatures is shown in Fig. 2C. The relation corresponding to Wien's displacement law relating the wavelength of maximum quanta emission λ'_m to the temperature is

$$\lambda'_m T = 3669.73 \text{ µm K}$$

TABLE 1. RECOMMENDED RADIOMETRIC TERMINOLOGY

Symbol	Term	Description	Units	Equation
U	Radiant energy	Energy transferred by electromagnetic waves	J	
u	Radiant energy density	Radiant energy per unit volume	J cm^{-3}	$u=dU/dV$
P	Radiant flux	Rate of transfer of radiant energy	W	$P=dU/dt$
W	Radiant emittance	Radiant flux emitted per unit area of a source	W cm^{-2}	$W=dP/dA$
Q	Radiant photon emittance	Number of photons emitted per second per unit area	Photons s^{-1} cm^{-2}	
J	Radiant intensity	Radiant flux per unit solid angle	W sr^{-1}	$J=dP/dQ$
N	Radiance	Radiant flux per unit solid angle per unit area	W cm^{-2} sr^{-1}	$N=W/\pi$
H	Irradiance	Radiant flux incident per unit area	W cm^{-2}	$H=dP/dA$
P_λ	Spectral radiant flux	Radiant flux per unit wavelength interval at a particular wavelength	W μm^{-1}	$P_\lambda=dP/d\lambda$
W_λ	Spectral radiant emittance	Radiant emittance per unit wavelength interval at a particular wavelength	W cm^{-2} μm^{-1}	$W_\lambda=dW/d\lambda$
Q_λ	Spectral radiant photon emittance	Radiant photon emittance per unit wavelength interval at a particular wavelength	Photons s^{-1} cm^{-2} μm^{-1}	$Q_\lambda=dQ/d\lambda$
J_λ	Spectral radiant intensity	Radiant intensity per unit wavelength interval at a particular wavelength	W sr^{-1} μm^{-1}	$J_\lambda=dJ/d\lambda$
N_λ	Spectral radiance	Radiance per unit wavelength interval at a particular wavelength	W cm^{-2} sr^{-1} μm^{-1}	$N_\lambda=dN/d\lambda$
H_λ	Spectral irradiance	Irradiance per unit wavelength interval at a particular wavelength	W cm^{-2} μm^{-1}	$H_\lambda=dH/d\lambda$
ϵ	(Radiant) emissivity	Ratio of radiant emittance of a source to that of a blackbody at the same temperature	(Numeric)	
α	(Radiant) absorptance	Ratio of absorbed radiant flux to incident radiant flux	(Numeric)	
θ	(Radiant) reflectance	Ratio of reflected radiant flux to incident radiant flux	(Numeric)	
τ	(Radiant) transmittance	Ratio of transmitted radiant flux to incident radiant flux	(Numeric)	

for λ'_m in micrometers, and the Stefan-Boltzmann law for the total number of photons emitted per unit area in blackbody radiation is

$$Q_{bb}(T) = \sigma T^4/2.75kT = \sigma' T^3$$

where $\sigma' = 1.49341 \times 10^{11}\ s^{-1}\ cm^{-2}\ K^{-3}$. These results are summarized in Table 2.

The results given above apply only to a blackbody. The emissivity of actual bodies can be accounted for by multiplying $W_\lambda d\lambda$, for example, by $\epsilon(\lambda,T)$, the variation of the emissivity of a particular material with a particular surface preparation or condition with wavelength and temperature. If $\epsilon(\lambda,T) = \epsilon(T) = \text{constant} < 1$, the body is referred to as a "graybody." The relationship between the emissivity and spectral radiant emittance is shown in Fig. 3.

INTERACTION OF OPTICAL WAVES WITH MATTER

An optical wave may interact with matter by being reflected, refracted, absorbed, or transmitted. The interaction normally involves two or more of these effects.

Reflectance

Some of the optical radiation impinging upon a surface is usually reflected from the surface. Reflectance varies according to the properties of the surface and the wavelength, and in real circumstances may range from more than 98% (smoked MgO at visible wavelengths) to less than 1% (lampblack at visible wavelengths). Reflection from a surface may be either diffuse, specular, or both. A diffuse reflector has a surface that is rough when compared to the wavelength of the impinging radiation. Lambert's law specifies a perfectly diffuse surface as one having a constant radiance independent of the viewing angle according to

$$N = W/\pi$$

where,

N is radiance ($W/m^2/sr$),
W is the radiant excitance (W/m^2).

In practice, the reflectance of real reflectors varies with the cosine of the viewing angle. A specular reflector has a surface that is smooth when compared to the wavelength of the impinging radiation. A perfect specular reflector will reflect an oncoming beam without altering the divergence of the beam. A narrow beam of optical radiation impinging upon a specular reflector obeys two rules:

1. The angle of reflection is equal to the angle of incidence.
2. The incident ray and the reflected ray lie in the same plane as a normal line extending perpendicularly from the surface.

Fig. 4 illustrates both diffuse and specular reflection.

Absorptance

Some of the optical radiation impinging upon any substance is absorbed by the substance. Absorptance varies according to the properties of the substance and the wavelength, and in real circumstances may range from a low of less than 2 dB/km for certain ultrapure fused silica glasses to more than 98% for lampblack.

Transmittance

Some of the optical radiation impinging upon a substance is transmitted into the substance. The penetration depth may be slight, in which case the transmittance is zero. Certain ultrapure silica glasses may have a transmittance of greater than 75% at certain wavelengths over a distance of 1 km. The reflectance, absorptance, and transmittance of a substance must satisfy

$$\rho + \alpha + \tau = 1$$

where,

ρ is reflectance,
α is absorptance,
τ is transmittance.

Refraction

A ray of optical radiation passing from one medium to another is bent at the interface of the two mediums if the angle of incidence is unequal to 90°. The index of refraction for a substance is the sine of the angle of incidence divided by the sine of the angle of refraction. Refractive index varies with wavelength and ranges from 1.002914 (air at 656 nm) to 2.7 (crystalline titanium oxide).

OPTICAL SOURCES

Important optoelectronic sources are tungsten lamps, fluorescent lamps, glow-discharge lamps, electroluminescent diodes, and lasers.

Tungsten Lamps*

Fig. 5 shows the spectral output of a tungsten lamp at a temperature of 2800 K. A typical tungsten lamp is an efficient optical source, but only about 5% of its radiant flux falls within the visible wavelengths. The addition of a halogen gas to a tungsten lamp can increase the life and efficiency of the lamp. At high operating temperature, the halogen (iodine or bromine) vaporizes and combines chemically with the evaporated tungsten on the quartz

---

* W. L. Wolfe and G. J. Zessis, eds, *The Infrared Handbook* (Ann Arbor, Mich.: Environmental Research Institute of Michigan, 1978).

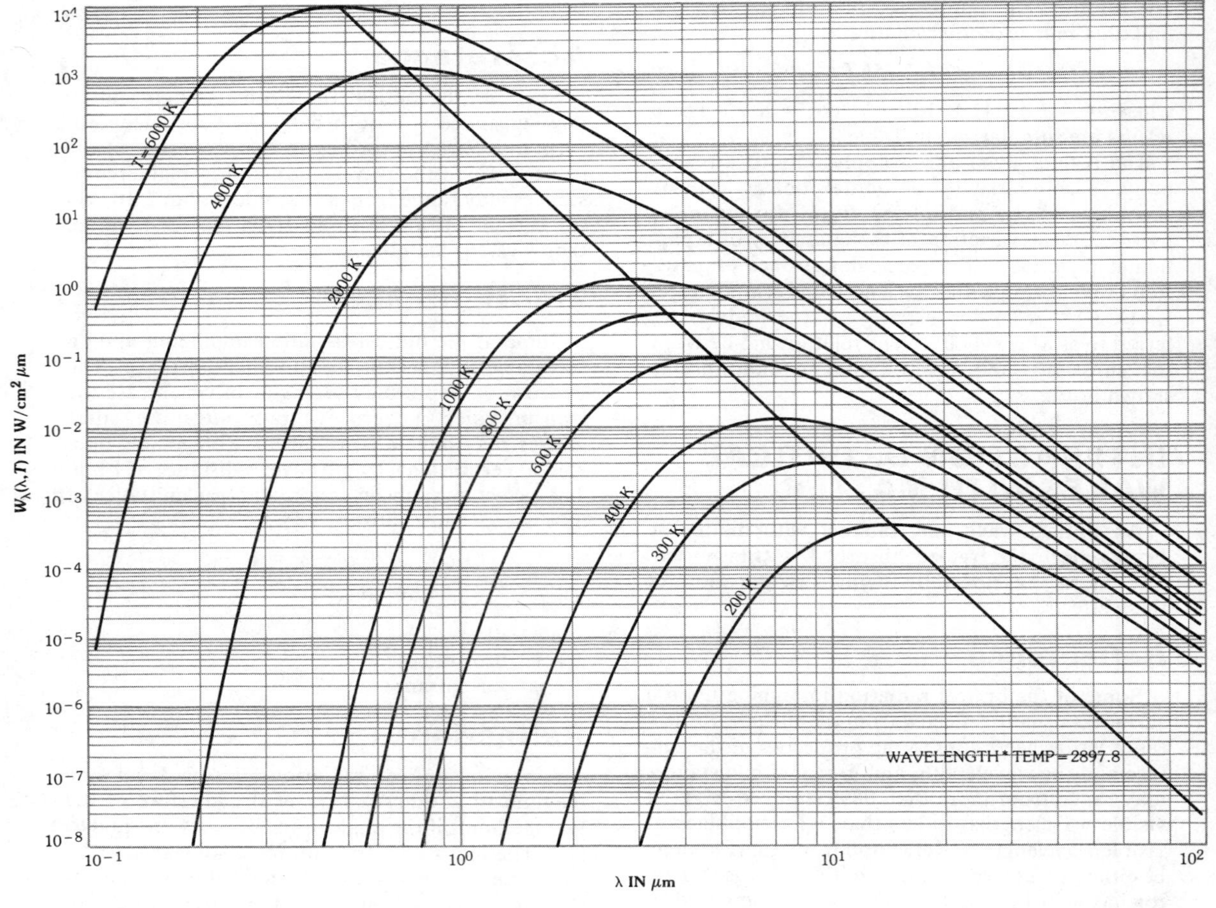

(A) $W_\lambda(\lambda,T)$ versus λ.

Fig. 2. Blackbody

envelope and in the gas. The resulting tungsten-halogen gas migrates back to the filament, where the very high temperatures decompose it so that tungsten is redeposited on the filament and the halogen repeats the cycle. This permits uniform and constant output throughout the lamp life.

Fluorescent Lamps†

A typical fluorescent lamp is a sealed glass tube filled with argon gas and containing a small amount of mercury. When an electrical discharge is established in the tube, ultraviolet radiation is produced, which causes a phosphor coating on the inside wall of the tube to fluoresce with a bright white glow. Fluorescent lamps provide high efficiency.

† F. Grum and Richard J. Becherer, *Optical Radiation Measurements,* Vol. 1, *Radiometry* (New York: Academic Press, Inc., 1979; p. 142).

Arc Lamps*

Arc lamps operate at considerably higher temperatures than other sources and, with the exception of certain lasers, are the most brilliant artificial sources. A representative arc lamp consists of a heavy-walled quartz envelope filled with gas (mercury, xenon, krypton, etc.) at a typical pressure of 20–40 atmospheres at operating temperature. The emission from a typical high-pressure arc lamp originates from the gaseous discharge, from the incandescence of the electrodes, and from the envelope.

Light-Emitting Diodes*

The operation of light-emitting diodes (LEDs) is dependent on the radiative recombination of elec-

* Wolfe and Zessis, *loc. cit.*

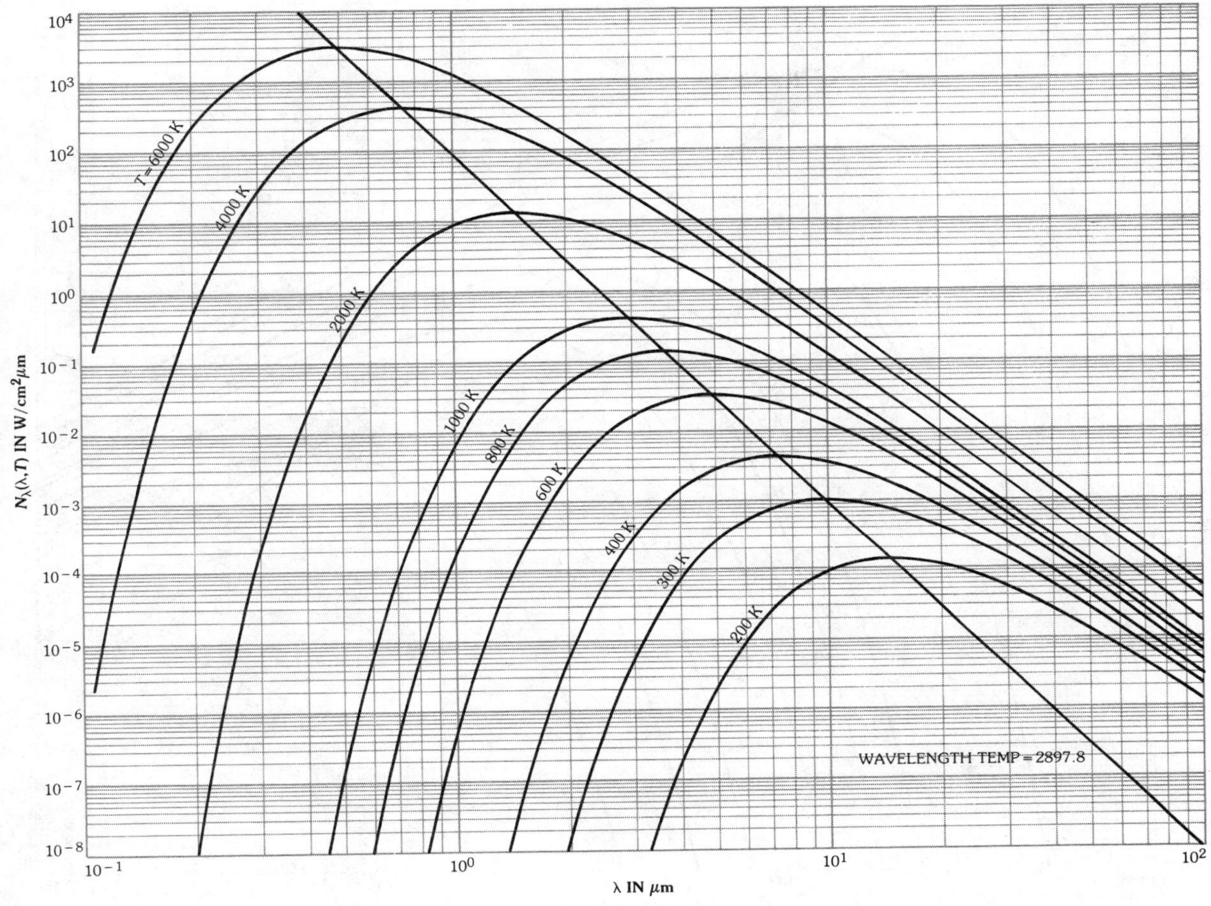

(B) $N_\lambda(\lambda, T)$ versus λ.

Continued on next page.

radiation.

tron-hole pairs at a forward-biased pn junction. The recombination of holes and electrons that occurs when current is injected into a pn junction results in the release of energy. This energy, which corresponds to the band gap, may be in the form of a photon (radiative recombination), a series of phonons (lattice vibrations, nonradiative), or both. In some cases, the energy may be transferred to another electron. Light-emitting diodes are electroluminescent diodes in which radiative recombination is significant. Recombination radiation emitted by LEDs is peaked at or near the band-gap energy, and in practical devices ranges from 0.55 μm to 34 μm. The spectral emission width of a representative device may be 25–30 nm, and this provides sufficient monochromaticity for the production of discrete wavelength bands (including the visible colors green, yellow, orange, and red).

Direct band-gap semiconductors make more efficient LEDs than indirect band-gap semiconductors (those in which electron-hole recombination is accompanied by emission of a photon and phonons). Silicon and germanium, both indirect band-gap materials, produce far too little recombination radiation for practical use, but other direct and indirect band-gap materials or direct-indirect band-gap alloys, particularly GaAs, GaP, and solid solution alloys of GaAs and GaP, GaAs and AlAs, and InAs and GaP, can be tailored to produce relatively efficient radiative recombination.

Gallium arsenide (GaAs) LEDs and similar direct band-gap LEDs in other materials exhibit internal quantum efficiencies (ratio of emitted photons per injected electrons) very nearly unity. Due to such factors as internal absorption, contact shadows, and refractive-index-induced surface reflectance, the external quantum efficiency of practical LEDs is usually much lower (less than about 0.1). Both

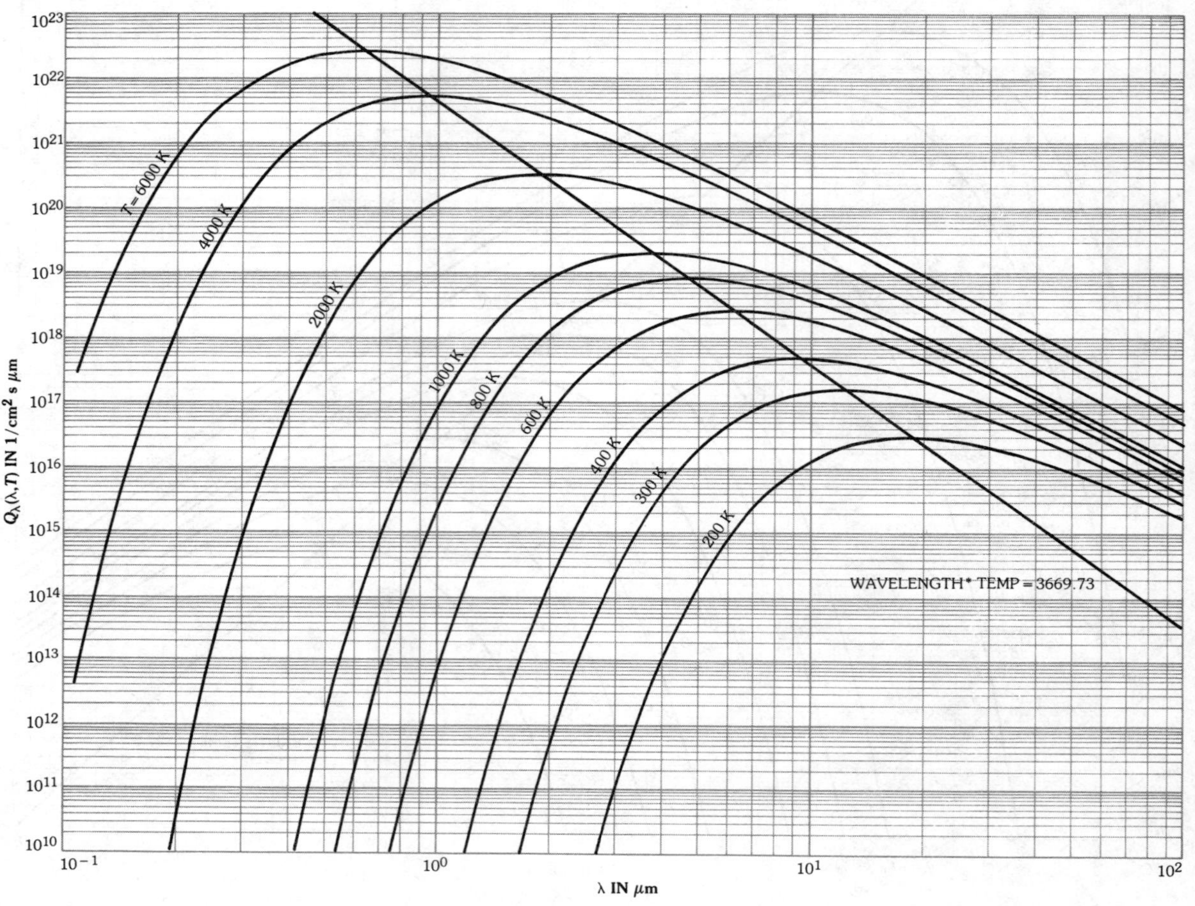

(C) $Q_\lambda(\lambda, T)$ versus λ.

Fig. 2. (cont). Blackbody radiation.

quantum efficiency and recombination radiation wavelength are affected by temperature, and the external quantum efficiency of specially fabricated LEDs may exceed 0.4 at 20 K. Wavelength is directly related to junction temperature and typically varies about 0.25 nm per degree Celsius. Therefore, changes in either the ambient temperature or junction heating will alter the emitted wavelength.

A wide range of economical long-lived plastic- and metal-packaged LEDs is commercially available. It is common practice to install the diode chip inside a miniature directional reflector and encapsulate the entire assembly in an index-matching epoxy to enhance the radiation extraction efficiency. This fabrication technique improves the external efficiency by a factor of 2 or more. Visible emitters suitable for use as visual indicators and other display roles include: GaP and GaP:N (green—550 nm); $GaAs_{0.25}P_{0.75}$:N (yellow—610 nm); $GaAs_{0.6}P_{0.4}$ (red—660 nm); $Al_{0.3}Ga_{0.7}As$ (red—675 nm); and GaP:Zn,O (red—690 nm). Several of these materials, particularly $GaAs_{0.6}P_{0.4}$, are in widespread use as display devices for digital clocks, watches, calculators, and electronic instruments. The most common infrared-emitting materials are GaAs (905 nm) and GaAs:Si (940 nm). These materials offer much higher efficiencies than visible emitters and are useful in such applications as optical communications and ranging, position sensing, object detection, and electro-optical isolation. Ordinary GaAs LEDs are characterized by a reasonable power output (about 1.5 mW @ 100 mA I_F) and fast turn-on time (about 1 ns), whereas special GaAs diodes doped with Si are more efficient (about 10 mW @ 100 mA I_F) but slower (t_{on} of

TABLE 2. BASIC LAWS AND CONSTANTS OF BLACKBODY RADIATION

Stefan-Boltzmann Law	Wien's Displacement Law	Planck's Equation
$W_{bb}(T)=\sigma T^4$	$\lambda_m T=b$	$W_\lambda(\lambda, T)=C_1\lambda^{-5}(e^{c^2/\lambda T}-1)^{-1}$
$Q_{bb}(T)=\sigma' T^3$	$\lambda_m' T=b'$	$Q_\lambda(\lambda, T)=C_3\lambda^{-4}(e^{c^2/\lambda T}-1)^{-1}$

$W_{bb}(T)$ = Total blackbody radiant emittance, W cm^{-2}, for temperature T

$Q_{bb}(T)$ = Total blackbody photon emittance, photons cm^{-2} s^{-1}

σ = Stefan-Boltzmann constant

$=(5.67032\pm0.00071)\times10^{-12}$ W cm^{-2} K^{-4}

σ' = Photon Stefan-Boltzmann constant
$=1.49341\times10^{11}$ s^{-1} cm^{-2} K^{-3}

λ_m = Wavelength of maximum radiant emittance

λ_m' = Wavelength of maximum radiant photon emittance

λ = Radiant wavelength

$b=(2897.8\pm0.4)$ μm K

$b'=(3669.7\pm0.4)$ μm K

$h=(6.626176\pm0.000036)\times10^{-34}$ W s
Planck's constant

$W_\lambda(\lambda, T)$ = Blackbody spectral radiant emittance, W cm^{-2} μm^{-1}

$Q_\lambda(\lambda, T)$ = Blackbody spectral photon emittance, photons s^{-1} cm^{-2} μm^{-1}

$c=(2.99792458\pm.000000012)\times10^{10}$ cm s^{-1}

$C_1=2\pi c^2 h=(3.741832\pm0.000020)\times10^4$ W cm^{-2} μm^4

$C_2=ch/k=(1.438786\pm0.000045)\times10^4$ μm K

$C_3=C_1/hc=2\pi c$

$k=(1.380662\pm0.000044)\times10^{-23}$ K^{-1}
Boltzmann's constant

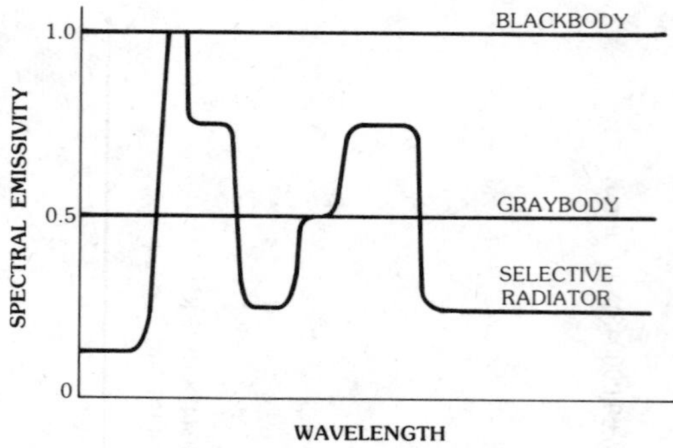

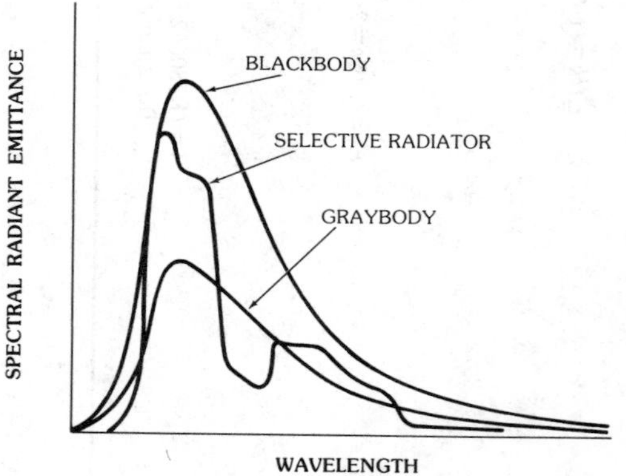

Fig. 3. Spectral emissivity and spectral radiant emittance of three types of radiators. (*From Hudson, R. D. Jr.,* Infrared System Engineering. *New York: John Wiley & Sons, Inc., 1969.*)

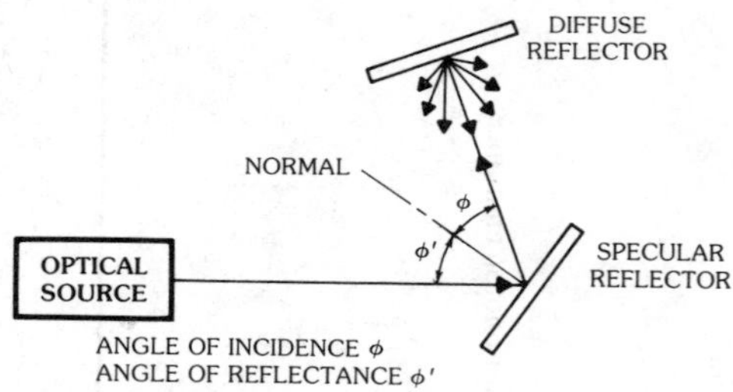

Fig. 4. Specular and diffuse (lambertian) reflectance. (*From F. M. Mims III,* Optoelectronics. *Indianapolis: Howard W. Sams & Co., Inc., 1975; p. 19.*)

about 300 ns). Since the radiant emission from electroluminescent diodes is generally linear with respect to applied current below the saturation region, they can be easily and directly pulse or analog modulated by simply controlling the forward bias. Electroluminescent diodes of GaAs, GaAs:Si, GaAs-$Al_xGa_{1-x}As$, and related materials emitting in the near infrared (approximately 750–950 nm) are finding wide use in fiber-optical communication applications, but because of lower fiber losses in the 1.3- to 1.6-μm wavelength range, InGaAsP and

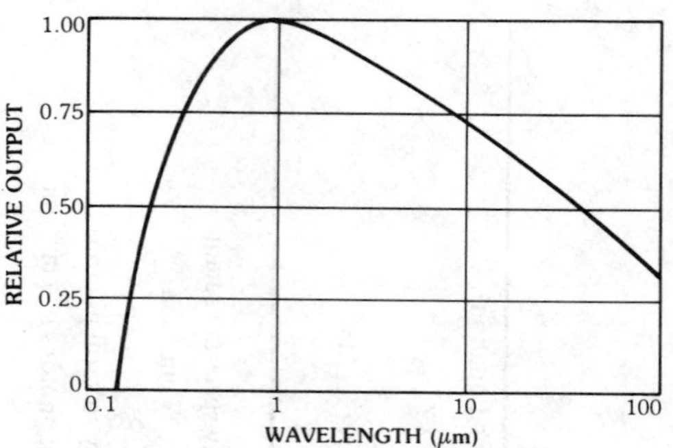

Fig 5. Spectral output of a tungsten lamp at 2800 K.

InGaAs LEDs are being rapidly developed for use at these longer wavelengths.

Electroluminescent diodes normally emit radiation in a relatively broad pattern and therefore cannot be coupled to optical waveguide fibers as efficiently as the laser sources described below. Nevertheless, reasonable coupling efficiency can be obtained by utilizing special structures such as the Burrus diode structure shown in Fig. 6. This structure has a typical emission area of 2×10^{-5} cm^2 and a radiance of 100 W/sr/cm^2 at a forward bias of 150 mA. The output wavelength can be peaked anywhere between 750 and 905 nm by adjusting the aluminum concentration in the $Al_xGa_{1-x}As$ alloy making up the diode. About 2 mW can be coupled into a suitable optical waveguide fiber when this structure is used. This and most other electroluminescent diodes exhibit operating lifetimes of more

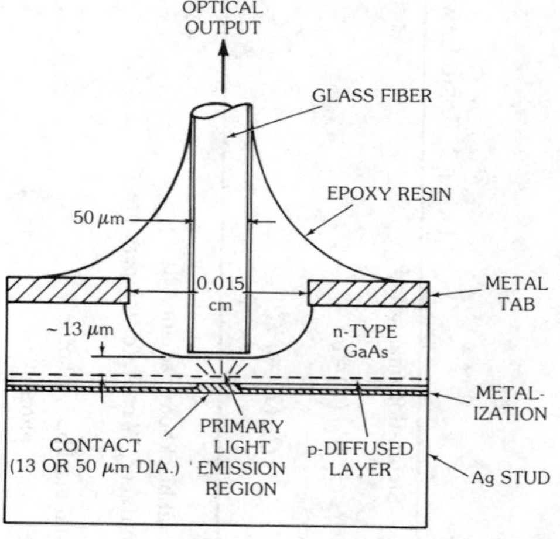

Fig. 6. GaAs electroluminescent diode specifically designed for direct coupling to an optical-fiber waveguide. (*After C. A. Burrus and R. W. Dawson, "Small-Area High-Current-Density GaAs Electroluminescent Diodes and a Method of Operation for Improved Degradation Characteristics,"* Applied Physics Letters, *1 August 1970; pp. 97–99.*)

than 10^4 hours at the relatively high current densities required to obtain significant optical output.

Numerous electroluminescent-diode configurations have been utilized in atmospheric optical communication links. A typical device structure is shown schematically in Fig. 7. Typically, approximately 20% of the radiation emitted by such a structure can be collected and collimated by a simple $f/1$ lens. A typical cw power output of about 7 mW at 100 mA forward bias and a pulsed power output of more than 90 mW at several amperes forward bias can be obtained from economical commercial devices. These relatively high power levels are obtained from silicon compensated GaAs diodes having a peak spectral wavelength of about 940 nm. The high peak powers are obtained at the expense of modulation bandwidth, since the rise and fall times of GaAs:Si diodes are approximately 300 and 200 ns, respectively, as compared to about 10 ns for typical GaAs devices. Somewhat lower optical powers (e.g., a few milliwatts at 100 mA forward bias) can be obtained from ordinary GaAs electroluminescent diodes.

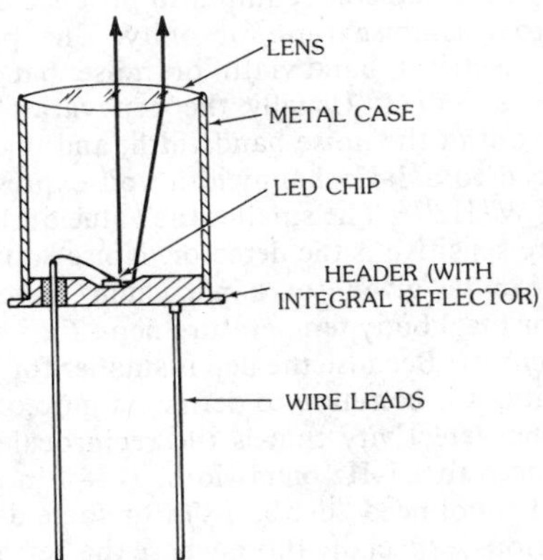

Fig. 7. Construction of a typical GaAs electroluminescent diode of the type employed in atmospheric communication links.

Superluminescent Diodes

Specially fabricated GaAs-$Al_xGa_{1-x}As$ double heterostructure and other electroluminescent diodes can be made to emit both spontaneous (quantum noise) and stimulated (amplified quantum noise) radiation. As in the case of a laser, stimulated emission implies a narrower spectral emission width and higher radiance than spontaneous emission. Fig. 8 shows the construction of a typical superluminescent diode (SLD). The structure of the device in the figure is virtually identical to that of a stripe-geometry injection laser with the major exception being an incomplete upper electrode. The incomplete electrode eliminates current injection near one of the mirrors, and this effectively isolates the mirror from the optical wave propagating along the plane of the junction and suppresses laser action. A similar effect can be had by angling one mirror of an injection laser a few degrees away from the normal or by coating one of the end mirrors of an injection laser with an antireflective film. The SLD is less efficient than most other electroluminescent sources, but its narrow spectral width and high radiance make it well suited for optical-waveguide links. Pulsed radiation of 50 mW has been coupled into an optical fiber from an SLD source.

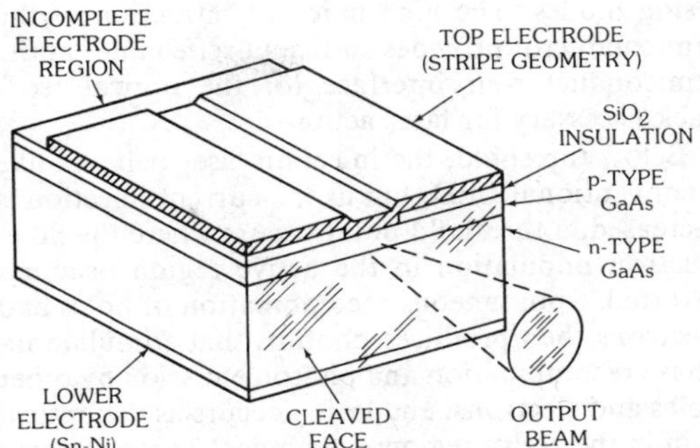

Fig. 8. Stripe-geometry semiconductor injection laser modified with an incomplete upper contact electrode to achieve superluminescent emission. (*After T. Lee, C. A. Burrus, Jr., and B. I. Miller, "A Stripe-Geometry Double-Heterostructure Amplified-Spontaneous-Emission* [*Superluminescent*] *Diode,"* IEEE Journal of Quantum Electronics, *August 1973; pp. 820–828.*)

Semiconductor Lasers*

Several fundamental modifications of the basic pn-junction electroluminescent diode exist, and chief among these is the semiconductor injection laser. In its simplest form, the injection laser is a direct band-gap LED having an exceptionally flat and uniform junction (the active region) bounded on facing sides by two parallel mirrors perpendicular to the plane of the junction, which provide a Fabry-Perot resonant cavity. The mirrors are usually produced by cleaving the semiconductor chip along parallel planes to produce perfectly parallel and flat surfaces. The remaining two sides of the chip perpendicular to the junction plane are intentionally roughened during the sawing process, which separates bars of material into individual chips. This surface roughening suppresses off-axis

* H. Kressel and J. K. Bulter, *Semiconductor Lasers and Heterojunction LEDs* (New York: Academic Press, Inc., 1977); H. C. Casey, Jr. and M. B. Panixh, *Heterostructure Lasers* (New York: Academic Press, Inc., 1978); G. H. B. Thompson, *Physics of Semiconductor Laser Devices* (Chichester: John Wiley & Sons, Inc., 1980).

lasing modes. The high index of refraction of the semiconductor provides sufficient reflectance at the semiconductor-air interface for the optical feedback necessary for laser action.

Below threshold, the injection laser behaves like a conventional LED, but as the current injection is increased, a threshold point occurs where the hole-electron population in the active region becomes inverted. Spontaneous recombination of holes and electrons then produces photons that stimulate in-phase recombination and photon emission by other holes and electrons, and lasing occurs as the optical gain in the active region overcomes absorption and other losses. The mirrors on either end of the active region provide the optical feedback necessary to sustain laser action, and a small fraction of the wave propagating between the mirrors emerges from each on each pass. One end facet on many commercial lasers is overcoated with a reflective gold film to cause all the radiation to emerge from only one end of the device and thus enhance collection efficiency.

The most common and best developed injection lasers utilize GaAs and AlGaAs (820–905 nm), though many other semiconductors have been used to produce wavelengths ranging from as short as 630 nm ($Al_xGa_{1-x}As$) to as long as 34 μm (PbSnSe). The high current density required to achieve lasing in broad-area homojunction injection lasers (8000 A/cm^2 or more) precludes continuous operation above temperatures greater than about 77 K. Room-temperature operation of a homojunction laser requires short current pulses (10–100 amperes for typical diodes) no more than 200 ns wide and with a duty cycle of 0.1%. Peak pulse power outputs at 300 K of up to 100 watts from single diodes and several kilowatts from diode arrays have been obtained from commercial devices. Injection lasers whose GaAs active region is sandwiched between two $Al_xGa_{1-x}As$ layers to produce a double heterojunction, which confines both the recombination and the optical wave to a very thin layer, have been fabricated with a threshold current density of less than 1000 A/cm^2. These devices are now commercially available and can be operated continuously at 300 K with appropriate heatsinking. Individual diodes may emit up to a few tens of milliwatts at 500 mA forward bias.

Injection-laser recombination radiation is characterized by a narrow spectral bandwidth (approximately 0.3 nm), beam directionality, and moderate beam divergence. The beam directionality permits collection of up to 90% of the radiation by simple optics in most cases. The beam divergence, which may be 10–15° in the plane of the junction (half-power measurement) and 25–45° perpendicular to the plane of the junction, is a result of the very narrow emission region of the laser. Essentially a diffraction limited slit, the emission region and internal lasing modes both contribute to the various diffraction patterns and structures seen in the far-field pattern of the beam from most injection lasers.

OPTICAL DETECTORS

Detectors are designed to convert optical radiation into a current or voltage. Associated with both the "signal" radiation and the background radiation incident on the detector are fluctuations in the photon arrival rate, and these fluctuations induce fluctuations, or noise, in the current or voltage produced by the detector. There are other sources of noise, such as thermal noise (Johnson or Nyquist noise), due to random thermal motion in any resistance in the circuit, and generation-recombination noise in photoconductive detectors, due to random generation and recombination of free carriers, that can limit the sensitivity of an optical detector. There are several figures of merit used to evaluate the performance of detectors.

Terms and Figures of Merit

Noise equivalent power: (Abbreviated nep.) The rms value of sinusoidally modulated radiant energy falling on the detector required to produce an rms-signal–to–rms-noise ratio of unity. The post-detection electrical bandwidth or noise bandwidth must be specified. Usually the nep varies as the square root of the noise bandwidth, and the nep is referenced to a 1-Hz bandwidth and expressed in units of W/Hz$^{1/2}$. The smaller the value of the nep, the more sensitive is the detector. Noise equivalent power can be given for a particular wavelength (nep_λ) or blackbody temperature nep (T).

Detectivity: Because the nep is smaller for higher sensitivity, it is common to define a figure of merit called the detectivity that is the reciprocal of the nep reduced to a 1-Hz bandwidth, $D = 1/\text{nep}$.

D^*: (Pronounced "dee star.") For some detector applications, especially those where the nep and/or D is limited by background radiation, the nep increases with the square root of the detector area, A, and to permit comparison of the performance of similar detectors with different areas, a normalized detectivity

$$D^* = A^{1/2}B^{1/2}/\text{nep} \quad \text{cm Hz}^{1/2}\ \text{W}^{-1}$$

is defined, where B is the electrical noise bandwidth.

Because of the way infrared detectors are characterized, it is common to specify the D^* as D^*_{bb} or D^*_λ to indicate measurement of the D^* with a blackbody or with a monochromatic source. Usually, D* is given as $D^*_{bb}\ (T, f, \Delta f)$ for a blackbody temperature T, chopping frequency f, and bandwidth Δf, and $D^*_\lambda\ (\lambda, f, \Delta f)$ for measurements at wavelength λ, chopping frequency f, and bandwidth Δf.

D^{**}: (Pronounced "dee-double star.") The detectivity of a background-limited detector depends not only on the temperature of the background but also on the field of view. As the field of view is decreased, the background radiation is reduced, and the detectivity is increased. To permit the comparison of different detectors of the same kind under different background conditions, a figure of merit is defined as

$$D^{**} = (\Omega/2\pi)^{1/2} D^*$$

where for circular geometry the solid angle Ω subtended by the background for a detector with an angle of view of θ is $\Omega = 2\pi \sin^2(\theta/2)$. Thus,

$$D^{**} = D^* \sin(\theta/2) \quad \text{cm Hz}^{1/2}\,\text{sr}^{1/2}\,\text{W}^{-1}$$

For $\theta/2 = \pi/2$ or $\theta = \pi$, and $\Omega = 2\pi$ sr,

$$D^{**} = D^*$$

Responsivity: The responsivity is defined as the rms signal voltage or rms signal current per unit rms radiant power incident on the detector in units of volts/watt or amperes/watt. The voltage responsivity must be specified for a given load resistance or referenced to an open-circuit condition. The responsivity is also specified as R_{bb} *or* R_λ, denoting a blackbody responsivity or spectral responsivity as for D^*.

Spectral response: The spectral responsivity variation for ideal photon detectors and thermal detectors is shown in Fig. 9. For photon detectors, the responsivity $R(\lambda)$, defined in terms of incident power, increases linearly with increasing wavelength up to the long-wavelength cutoff or the minimum energy for carrier excitation and then decreases abruptly to zero, whereas the responsivity for an ideal thermal detector is independent of wavelength. For photon detectors, a related figure of merit is the quantum efficiency η, the ratio of the number of electrons collected to the incident number of photons.

Response time: Another important measure of the performance of optical detectors is the response time. This is often specified in terms of a simple time constant, a rise time, or a cutoff modulation frequency.

Characterization of Detectors

The characterization of detectors is important because the results of such measurements make it possible to select the best possible detector for a particular application. The determination of nep, D, D^*, or D^{**} described above each requires measurements of the responsivity, or the rms signal for a given incident power, and the noise of the detector being characterized.

Responsivity—The power responsivity of a photon detector is usually determined by the following procedure:

1. The relative spectral power responsivity is measured at all wavelengths to which the detector responds by using a monochromator source and thermal detector that is assumed to have a flat or constant spectral power responsivity.
2. The absolute spectral power responsivity is then determined by either: (A) measuring the absolute power responsivity at one or a few wavelengths by using a laser or calibrated source of known power output, or (B) measuring the blackbody responsivity of the detector and, from this result and the relative spectral responsivity, determining the responsivity at the peak of the relative spectral response curve by means of the technique described below.

Fig. 10 shows a schematic diagram of an experimental arrangement for the measurement of the relative spectral power responsivity. The monochromator provides a nearly monochromatic source of radiation that is modulated by the chopper at some suitable frequency (in the range from about 10 Hz to about 1000 Hz) depending on the type of detector being characterized. This chopped radiation is divided by a 50% beam splitter that is flat over the wavelength range of interest so that equal intensities of radiation fall on both the detector being characterized and a thermocouple or other reference detector that is assumed to be spectrally flat. Alternatively, the radiation can be switched back and forth from the detector being characterized and the reference detector at a low frequency (about 10 Hz) compatible with the thermal detector by means of mirrored chopper blades. The signal from the thermal detector is then detected synchronously with the mirrored chopper. The output sig-

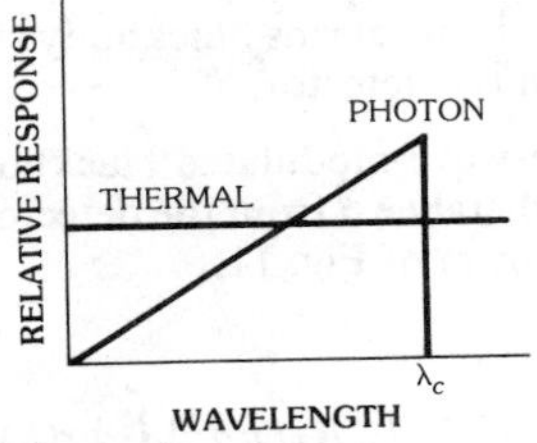

(A) Constant flux per unit wavelength interval.

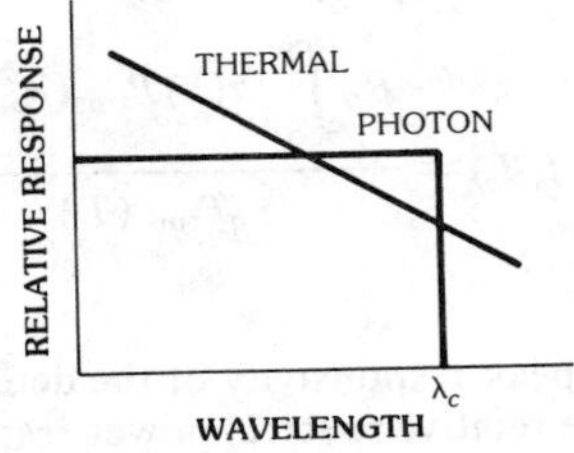

(B) Constant photons per unit wavelength interval.

Fig. 9. Idealized spectral response curves for photon and thermal detectors. (*From Grum, F., and Becherer, R. J.,* Optical Radiation Measurements, *Vol. 1,* Radiometry. *New York: Academic Press, Inc., 1979.*)

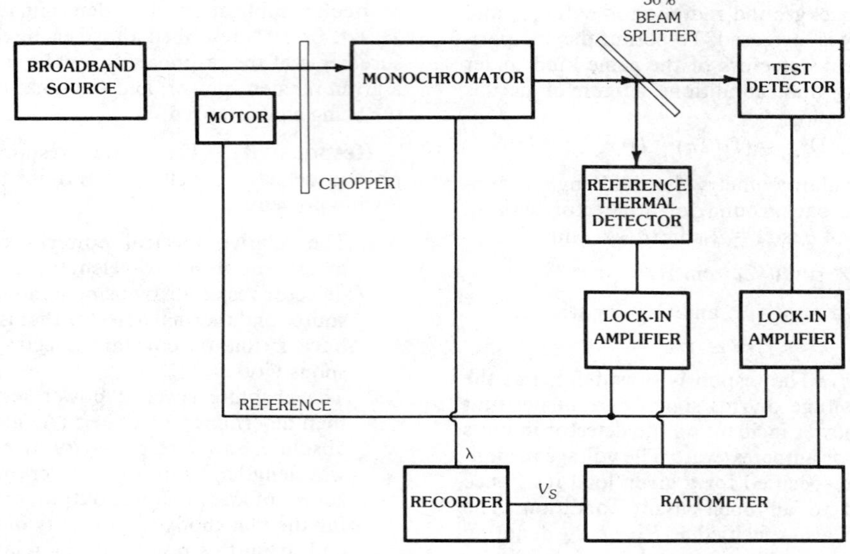

Fig. 10. Measurement of relative spectral power responsivity.

nals from the detector being measured and the thermocouple detector are then ratioed by the ratiometer to give a reading proportional to the power responsivity of the test detector. By recording this signal as the wavelength of the output radiation of the monochromator is scanned over the desired range, the relative spectral response of the test detector, $r(\lambda)$, can be obtained.

The absolute spectral responsivity of near-infrared detectors is usually determined by measuring the absolute responsivity at one or two wavelengths by using lasers with emission in the spectral range of interest and a calibrated output power. From these values of responsivity at specified wavelength, the relative spectral power responsivity curves can be made absolute.

For detectors in the wavelength range where calibrated lasers or reference detectors are not available, the spectral power responsivity is usually determined by using blackbody radiation. Fig. 11 shows an experimental arrangement for this process. The blackbody responsivity is

$$R_{bb}(f, T) = \frac{R_p \int_0^\infty r(\lambda) P_{rms}(\lambda, T) d\lambda}{P_{rms}(T)}$$

where,

R_p is the peak responsivity of the detector,

$r(\lambda)$ is the relative spectral power responsivity of the detector ($0 \leqslant r(\lambda) \leqslant 1$),

$P_{rms}(\lambda, T)$ is the rms value of the incident radiation at modulation frequency f and wavelength λ,

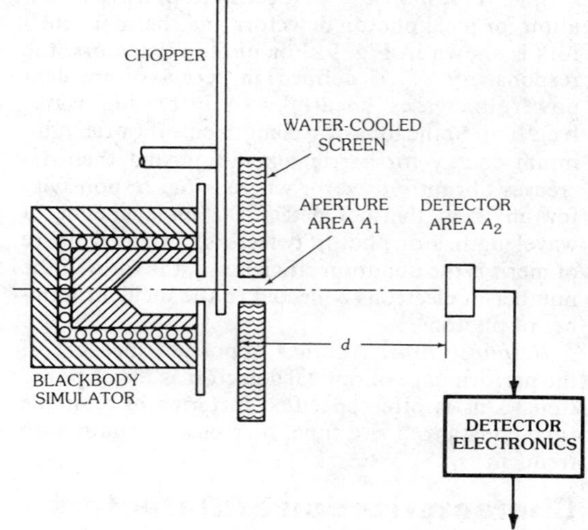

Fig. 11. Determination of spectral power responsivity with blackbody radiation.

$P_{rms}(T)$ is the total rms blackbody radiation falling on the detector.

For a square-wave-modulated blackbody source of area A_1 at a distance d from the detector of area A_2, as in the diagram of Fig. 11,

$$R_{bb}(f, T) = \frac{[(0.45 A_1 A_2) R_P / \pi d^2] \int_0^\infty r(\lambda) W(\lambda, T) d\lambda}{[(0.45 A_1 A_2)/\pi d^2] \int_0^\infty W(\lambda, T) d\lambda}$$

$$= \frac{R_p \int_0^\infty r(\lambda)\, W(\lambda, T) d\lambda}{\sigma T^4}$$

But, the experimentally determined value of $R_{bb}(f,T)$ is

$$R_{bb}(f,T) = V_{rms}(f,T)/(0.45 A_1 A_2/\pi d^2)\, \sigma T^4$$

where $V_{rms}\,(f,T)$ is the experimentally measured value of the rms voltage, for example, generated by the detector. The peak value of responsivity R_p is given in terms of the experimental blackbody responsivity $R_{bb}(f,T)$ by

$$R_p = \sigma T^4 R_{bb}(f,T) / \int_0^\infty r(\lambda)\, W(\lambda,T) d\lambda$$

$$= R_{bb}(f,T)/\gamma(T)$$

where $\gamma(T)$ is called the effectiveness factor. For an ideal thermal detector, $r(\lambda) = 1$, $\gamma = 1$, and $R_p = R_{bb}(f,T)$. The units of R_p are volts/watt or amperes/watt.

Noise Measurements—In order to determine the nep and related figures of merit for the detectors described above, it is necessary to measure the rms noise voltage or current characteristics of the detector being evaluated. The rms noise can be measured with rms-responding meters, lock-in amplifiers, spectrum analyzers with special noise-measuring capability, or tuned voltmeters with average-responding meters. When the latter are used, it should be noted that the meter reading for the rms value of a sinusoid must be multiplied by 1.128 in order to obtain the rms value of Gaussian noise. The nep can then be calculated from the responsivity and the rms noise as $\text{nep}_\lambda = V_n/R_\lambda$, where V_n is the rms noise voltage, for example. The *D*, *D**, etc., can be calculated similarly.

Ultimate Sensitivity of Detectors

There are many noise sources that can limit the detection capability of any detection system. These noise sources can generally be categorized as noise in the detector, noise in the amplifier, and fluctuations in the radiating background and the signal itself. The noise in the detector comes from fluctuations in the concentration or motion of the current carriers. The amplifier noise consists of thermal noise as well as excess noise due to the same type of fluctuations as in the detector. The background noise or photon noise is due to the quasi-random absorption of photons from the surroundings of the detector; for thermal detectors it also comes from the quasi-random emission of photons from the detector. It is this background noise that limits the ultimate performance of infrared detectors.

Johnson, Nyquist, or Thermal Noise—This is the random fluctuation in the current or voltage at the electrical contacts of a resistor due to the random thermal motion of electrons in the resistor. The rms values of these fluctuations are given by

$$\overline{v_J^2}^{\,1/2} = (4kTBR_L)^{1/2}$$
$$= 1.287 \times 10^{-10} R_L^{1/2}\,(T/300)^{1/2} B \text{ volts}$$

and

$$\overline{i_J^2}^{\,1/2} = (4kTB/R_L)^{1/2}$$
$$= 1.287 \times 10^{-10} R_L^{-1/2}\,(T/300)^{1/2} B \text{ amperes}$$

where,

R_L is the resistance in ohms,
T is the absolute temperature,
B is the electrical bandwidth.

This noise source is independent of current for a given temperature.

Shot Noise—Shot noise results whenever the fluctuation arises because of a series of independent events occurring at random, such as the emission of electrons by a temperature-limited thermionic cathode, or a photocathode, or the crossing of a junction by electrical carriers (electrons or holes) in a pn junction or transistor. This is also the source of noise when transitions occur between two energy levels as for generation and recombination of carriers in a semiconductor or when photons are emitted by a laser. The root-mean-square current fluctuations for the process can be expressed in terms of the average dc current as

$$\overline{i_S^2}^{\,1/2} = (2qI_{dc}B)^{1/2}$$
$$= 5.657 \times 10^{-10} (I_{dc})^{1/2} B^{1/2} \text{ amperes}$$

where,

I_{dc} is the average dc current in amperes,
B is the electrical bandwidth in hertz.

Generation-Recombination Noise—Although the noise that results from fluctuations in the instantaneous values of the free carrier concentration in semiconductors can be described as shot noise for some cases, for photoconductive detectors it is more meaningful to describe this noise source in terms of the generation, recombination, and trapping processes that cause the fluctuations in the free carrier concentrations. The G-R noise for a homogenous n-type extrinsic semiconductor is

$$\overline{i_{GR}^2}^{\,1/2} = \{(4I_{dc}^2/N_0)\,[\tau_0/(1 + \omega^2\tau_0^2)]B\}^{1/2} \text{ amperes}$$

where,

I_{dc} is the average dc current,
N_0 is the total number of free carriers in the semiconductor,
τ_0 is the carrier lifetime.

1/f Noise—There are several types of noise that increase as the frequency decreases. Because of the different, and not necessarily well understood, sources of this noise, there is no simple expression that describes $1/f$ noise accurately. However, empirical expressions like

$$\overline{i_{1/f}^2}^{1/2} = (KI_{dc}^{\alpha} B/f^{\beta})^{1/2}$$

where K is a proportionality factor, $\alpha \approx 2$, and $\beta \approx 1$ are frequently used. In this expression, I_{dc} is the average current, and B is the bandwidth.

Photon Noise—Also called radiation noise, photon noise results from fluctuations in the rate of absorption (and for thermal detectors, emission) of photons by the detector. For photons from a laser or nonblackbody source, this noise can be calculated accurately by the shot noise of the average current induced by the photon flux. For blackbody radiation, the noise is greater than that estimated by the shot-noise formula because blackbody photons, described by Bose-Einstein statistics, do not arrive randomly in time but, instead, are clustered together. Thus, the mean square fluctuation in the arrival rate of photons, which for independent events would simply be the average photon arrival rate, $\bar{n}$, is increased by the factor $e^{h\nu/kT}/(e^{h\nu/kT}-1)$. For practical applications, this distinction is generally not important.

Temperature Noise—Temperature noise is the fluctuation in temperature of a thermal detector that is in contact with its surroundings by both conduction and radiation. If the detector is only in contact with its surroundings through radiation, the temperature noise reduces to blackbody photon noise due to both absorption and emission of photons.

Amplifier Noise—There is always the thermal Johnson noise of some equivalent load resistance associated with a detector circuit, but in addition, with real amplifiers there is some additional added noise called amplifier noise. This noise, as well as the Johnson noise of the load resistor, is usually referred to the input of the amplifier. The rms noise voltage referred to the amplifier input is described by the equation

$$\overline{v_a^2}^{1/2} = [4k(T_L + T_A)R_L B]^{1/2}$$

where,

T_L is the noise temperature of the equivalent load resistance, R_L,

T_A is the amplifier noise temperature,

B is the electrical bandwidth.

The noise figure, F, is another useful measure of amplifier performance and is the ratio of the signal-to-noise ratio at the amplifier input to the signal-to-noise ratio at the output for a room-temperature load resistor ($T_L = 290$ K). In terms of T_A and T_L,

$$F = 1 + T_A/T_L$$

The noise figure is usually expressed as a logarithmic function,

$$\text{nf} = 10 \log_{10} F \quad \text{decibels}$$

In this notation,

$$\overline{v_a^2} = 4kFT_L R_L B$$

It should be noted that F for a given amplifier usually depends on both R_L and the frequency, f.

The ultimate sensitivity of a detector is that limited by the shot noise of the signal itself. For photon detectors this is given by

$$\text{nep}_\lambda = 2hcB/\lambda\eta(\lambda)$$

where $\eta(\lambda)$ is the detector quantum efficiency at wavelength λ. In practice, this limit is seldom if ever achieved, because of the various noise sources described above. However, because of the large amount of background radiation present in much of the infrared spectral region, it is common for the sensitivity of good infrared detectors to be limited by the photon noise due to background radiation. The mean square noise power spectral density of the photon flux from blackbody radiation at temperature T incident on a thermal detector of area A is given by

$$S_p(f) = \int_0^\infty \frac{2A2\pi h^2\nu^4}{c^2} \frac{e^{h\nu/kT} d\nu}{(e^{h\nu/kT}-1)^2} = 8AkT\sigma T^4$$

and for a thermal detector, in equilibrium with and at the same temperature as the background (and with unity emissivity), this fluctuation must be doubled because of the fluctuations due to the photons emitted by the detector; so

$$S_p(T) = 16AkT\sigma T^4$$

Thus, the mean square noise power in a bandwidth B is

$$\overline{P_b^2} = 16AkT\sigma T^4 B$$

and the D^* for such a thermal detector is

$$D^* = A^{1/2}B^{1/2}/\overline{P_b^2}^{1/2} = 1/4(kT\sigma T^4)^{1/2}$$
$$= 5.5171 \times 10^{-11}(T/300)^{5/2} \text{ W Hz}^{-1/2}$$

The D^* for an ideal photon detector with a long-wavelength cutoff of $\lambda_0 = c/\nu_0$ for a signal at ν_s within the spectral range of the detector is given by

$$D_\lambda^* = \frac{c\eta(\nu_s)}{2h\nu_s\pi^{1/2}\left[\int_{\nu_0}^\infty \frac{\eta(\nu)\nu^2 \exp(h\nu/kT)\, d\nu}{[\exp(h\nu/kT)-1]^2}\right]^{1/2}}$$

and this can be readily evaluated for the peak detectivity of an ideal detector for ν_s and ν_0, assuming $\eta(n)$ is independent of wavelength, by using a programmable calculator and the series solution

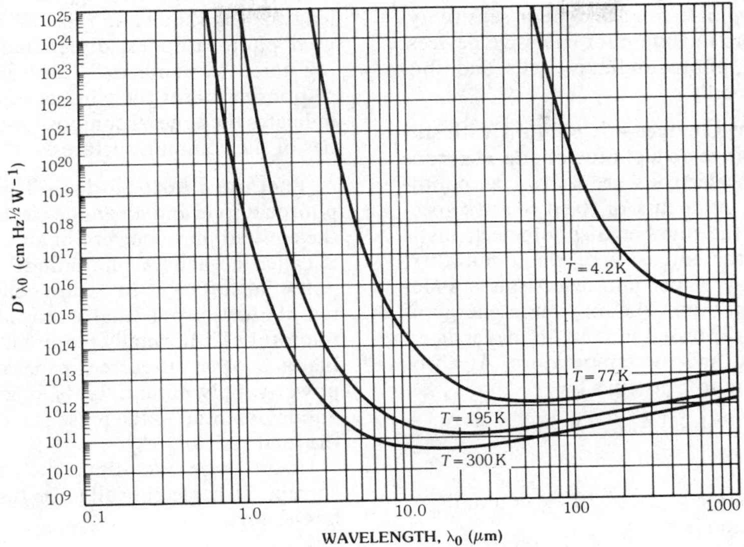

Fig. 12. Background-limited $D^*_{\lambda 0}$ at peak wavelength, λ_0.

$$D_{\lambda 0}^* = \frac{c\,[\eta(\nu_0)]^{1/2}}{2\pi^{1/2} h^{1/2} \nu_0^2 (kT)^{1/2}}$$
$$\left\{ \sum_{m=1}^{\infty} \exp(-mh\nu_0/kT)\,[1 + 2kT/mh\nu_0 + 2(kT/mh\nu_0)^2] \right\}^{-1/2}$$

The background-limited $D^*_{\lambda 0}$ at the peak wavelength, λ_0, is shown in Fig. 12 for several values of the background temperature. This result applies to photoemission and photovoltaic detectors where only the background fluctuations contribute to the noise. For $T = 300$ K, the minimum $D^*_{\lambda 0}$ occurs at about 14 μm. The rapid increase in $D^*_{\lambda 0}$ for detectors with short cutoff wavelengths occurs because these detectors respond to less of the photon noise of this background. The small increase in $D^*_{\lambda 0}$ for longer wavelength occurs because the increase in the photon noise received from the background is more than compensated by the increase in the number of photons per watt of signal. For photoconductive photon detectors, there is the additional fluctuation due to fluctuations in the recombination of carriers in addition to the excitation of free carriers, so for photoconductive detectors the photon noise is larger by $\sqrt{2}$ and the background-limited D^* is smaller by $\sqrt{2}$.

The various types of thermal and photon or quantum detectors are described briefly below.

Thermal Detectors

Thermal detectors respond to optical-radiation-induced temperature variations and are therefore well-suited for broad-band detection throughout the optical spectrum. Thermal detectors include the bolometer, thermocouple, thermopile, thermopneumatic cell, and pyroelectric detector.

Bolometer—The bolometer changes its resistance in response to thermal energy resultant from impinging radiant energy. The most common bolometric detector is the thermistor.

Thermocouple—A thermocouple is a junction of two dissimilar metals that, upon absorbing thermal energy, produces an emf.

Thermopile—The thermopile is an array of thermocouples. Miniature thermopiles made with thick film deposition techniques are commonly used in infrared detection applications.

Thermopneumatic Cell—The thermopneumatic cell senses the presence of thermal energy by means of a sealed cell that expands or contracts in response to variations in applied radiant energy. The magnitude of the expansion can be detected by means of interferometric techniques.

Pyroelectric Detector—The pyroelectric detector is a temperature-sensitive current source. In a typical detector, a thin wafer of a ferroelectric crystal such as triglycine sulfate or lithium tantalate forms a capacitor whose capacitance is altered by thermal energy. Because of the ferroelectric effect, this detector will not detect constant unmodulated incident radiation.

Quantum Detectors

Quantum detectors respond to variations in the number of incident photons. Quantum detectors have a more limited spectral sensitivity range than thermal detectors but are generally characterized by relatively fast response time and high sensitivity. Quantum detectors include photoemissive devices, photovoltaic cells, photoconductive cells, and photoelectromagnetic cells.

Photoemissive Devices—If photons with sufficient energy impinge on a photocathode, electrons are emitted. Such electrons are known as photoelectrons. For an input flux of fixed relative spectral distribution, the number of photoelectrons is proportional to the intensity of the input flux, whereas the energy of the photoelectrons is independent of this intensity. The maximum energy of emitted electrons expressed in volts, V, depends on the wavelength, λ, and the temperature. At absolute zero, according to Einstein's law

$$e(V + \phi) = hc/\lambda$$

where,

e = electron charge = 1.6×10^{-19} coulomb,
ϕ = work function in volts,
h = Planck's constant = 6.6×10^{-34} joule-second,
c = velocity of light in meters/second,
λ = wavelength in meters.

If a threshold wavelength, λ_0, is defined by

$$e\phi = hc/\lambda_0$$

then V is seen to be zero (except for thermal velocities) at the wavelength λ_0; for $\lambda > \lambda_0$, there is no photoelectric emission at absolute zero. At temperatures above absolute zero, there is always a finite probability of some photoemission at all wavelengths due to the thermalization of the electron distribution.

Photocathode Response to Monochromatic Radiation—The output current dI_λ in amperes, generated by a photocathode subjected to a monochromatic input flux dW_λ in watts, is given by

$$dI_\lambda = s_\lambda dW_\lambda$$

where s_λ is the monochromatic radiant responsivity of the photocathode in amperes/watt defined by this equation. Similarly, the number of electrons/second, dn_λ, generated by an input flux of dN_λ photons/second is given by

$$dn_\lambda = \eta_\lambda dN_\lambda$$

where η_λ is the monochromatic quantum efficiency of the photocathode in electrons/photon defined by this equation.

The monochromatic radiant responsivity, s_λ, is related to the monochromatic quantum efficiency, η_λ, by

$$s_\lambda = e\lambda\eta_\lambda/hc = 8.06 \times 10^5 \lambda\eta_\lambda$$

Typical values of the monochromatic radiant responsivity, s_λ, and corresponding monochromatic quantum efficiency, η_λ, as a function of wavelength, λ, are shown in Fig. 13 for some commonly used photocathodes, designated by their JEDEC registered "S numbers." Table 3 gives typical peak responsivities for the various surfaces, and Table 4 indicates the general composition and other properties of the common surfaces.

Vacuum Photodiodes—The combination of a photocathode and an anode electrode for collecting the emitted photocurrent in an evacuated envelope is called a vacuum photodiode. A positive anode potential sufficient to assure collection of all emitted photoelectrons (that is, to "saturate" the diode phototube) is normally required, the tube then acting as a constant-current generator (Fig. 14). The power-supply potential, V_B, must assure sufficient anode potential in the presence of a voltage drop in the load resistor, R_L.

Under these conditions the total anode output current, I_a, neglecting all noise fluctuations, is given by the equation

$$I_a = I_s + I_b + I_d + I_L$$

where,

I_s = emitted photocathode signal current,
I_b = emitted photocathode photocurrent due to stray background flux,
I_d = photocathode thermionic dark current,
I_L = residual dark current (leakage, etc.).

The instantaneous value of signal current I_s will follow the instantaneous signal flux input magnitude from direct current up to an upper frequency limit (commonly 0.2–2 GHz) set by the transit-time spread of the electrons crossing the gap between cathode and anode, and including induced displacement currents during transit.

For steady-state or slowly varying input flux rates, the total noise current output, i_n, from the diode and load resistor is given by

$$\overline{i_n^2} = 2qB(I_s + I_b + I_d) + (4kTB/R_L) + i_L^2$$

where,

q = the electron charge = 1.6×10^{-19} coulomb,
B = noise-current measurement bandwidth,
k = Boltzmann's constant = 1.38×10^{-23} joule per K,
T = absolute temperature of the load resistor (K),
R_L = load resistance (ohms),
i_L = residual dark noise current (from leakage, stray pickup, etc.).

To increase the absolute level of the noise voltage generated by noise current i_n so that tube noise predominates over the noise voltage of the subsequent amplifiers, and to suppress load-resistor noise relative to tube noise, large values of the load resistor (of the order of 10^7–10^9 ohms) are commonly used when response time is not a limitation.

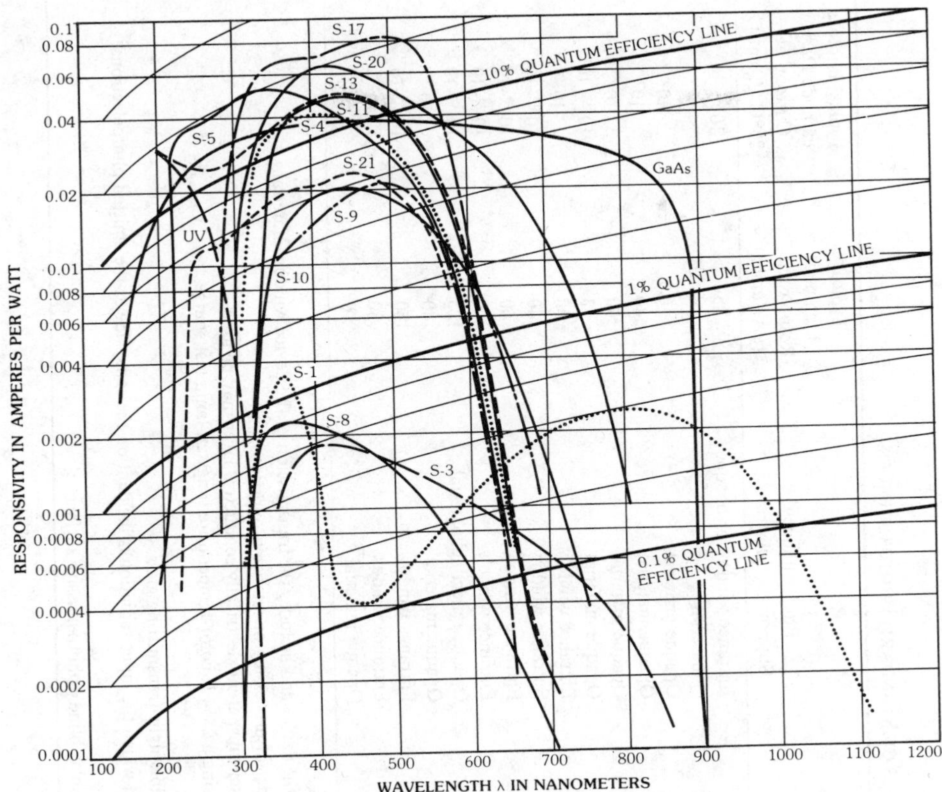

Fig. 13. Typical absolute spectral response characteristics of photoemissive devices.

TABLE 3. TYPICAL PEAK PHOTOCATHODE RESPONSIVITIES

	Radiant Responsivity		Quantum Efficiency	
S Number	$s_{\lambda\,max}$ (A W^{-1})	λ_{max} (nano-meters)	$\eta_{\lambda\,max}$ (electron photon^{-1})	λ_{max} (nano-meters)
S1	0.0025	800*	0.004	770*
S3	0.0019	420	0.0058	400
S4	0.042	420	0.13	380
S5	0.052	330	0.21	320
S8	0.0024	360	0.0082	350
S9	0.023	490	0.056	480
S10	0.021	440	0.061	420
S11	0.048	450	0.14	400
S13	0.048	450	0.20	<200
S17	0.083	500	0.26	350
S20	0.066	420	0.20	400
S21	0.024	460	0.07	370

* Neglecting short wavelength peak

For a plane-parallel vacuum photodiode, the space-charge-limited output current, $I_{a(\max)}$, in amperes for a given applied cathode-to-anode potential difference, V, in volts is given by

$$I_{a(\max)} = 2.33 \times 10^{-6}\, A V^{3/2}/d^2$$

where,

A = uniformly emitting emission area (meters^2),
d = anode-cathode spacing (meters).

In practice, linear output currents up to approximately half of this maximum limit can be obtained.

For a plane-parallel vacuum phototube, the output anode current, I_A, as a function of time, t, for an ultrashort exciting light pulse is given by

$$I_A = (2QRC/T^2)[(t/RC) + \exp(-t/RC) - 1] \quad \text{for } 0 < t < T$$

$$I_A = (2QRC/T^2)[(T/RC) + \exp(-T/RC) - 1]\,\exp[-(t-T)/RC] \quad \text{for } t \geq T$$

and

$$T = d(2m/qV)^{1/2} = (3.37 \times 10^{-6})d/(V)^{1/2} \text{ seconds}$$

where,

T = transit time of the charge from cathode to anode (seconds),
V = cathode-to-anode potential (volts),
C = total capacitance including external circuit capacitance,

TABLE 4. CHARACTERISTICS OF STANDARD PHOTOSURFACES

S Number[1]	Principal Photocathode Components[2]	Entrance Window Material	Photocathode Supporting Substrate[3]	Typical Luminous Responsivity[4] (μA/lumen)	Typical Photocathode Dark Current[5] at 25°C (A/cm²)
S1	Ag-O-Cs	Visible-light-transmitting glass[6]	Entrance window or opaque material[7]	25	10^{-11}–10^{-13}
S3	Ag-O-Rb	Visible-light-transmitting glass[6]	Opaque material[7]	6.5	10^{-12}
S4	Cs-Sb	Visible-light-transmitting glass[6]	Opaque material[7]	40	10^{-14}
S5	Cs-Sb	Ultraviolet-transmitting glass	Opaque material[7]	40	10^{-14}
S8	Cs-Bi	Visible-light-transmitting glass[6]	Opaque material[7]	3	10^{-14}–10^{-15}
S9	Cs-Sb	Visible-light-transmitting glass[6]	Entrance window	30	10^{-14}
S10	Ag-Bi-O-Cs	Visible-light-transmitting glass[6]	Entrance window	40	10^{-13}–10^{-14}
S11	Cs-Sb	Visible-light-transmitting glass[6]	Entrance window	60	10^{-14}–10^{-15}
S13	Cs-Sb	Fused silica	Entrance window	60	10^{-14}–10^{-15}
S17	Cs-Sb	Visible-light-transmitting glass[6]	Opaque reflecting material[7]	125	10^{-14}–10^{-15}
S19	Cs-Sb	Fused silica	Opaque material[7]	40	10^{-14}
S20	Sb-K-Na-Cs	Visible-light-transmitting glass[6]	Entrance window	150	10^{-15}–10^{-16}
S21	Cs-Sb	Ultraviolet-transmitting glass	Entrance window	30	10^{-14}
UV[8]	Cs-Te	Sapphire	Opaque material[7]	0	—

Notes:

1. The S number is the designation of the spectral response characteristic of the device and includes the transmission of the device window material.
2. Principal components of the photocathode are listed without regard to order of processing or relative proportions.
3. When the supporting substrate is the entrance window, an intermediate semitransparent electrically conductive layer may be used.
4. Corresponding to the specific absolute response curves shown in Fig. 13 using a 2854 K color-temperature tungsten-lamp test source.
5. Specific dark current excludes direct-current leakage.
6. Lime glass and Kovar sealing borosilicate glass are commonly used for visible-light-transmitting glass.
7. The opaque material used as the supporting substrate for photocathodes in which the input radiation is incident on the same side as the emitted photoelectrons is usually metallic in nature.
8. An S number designation has not yet been assigned to this experimental "solar blind" photoemissive surface.

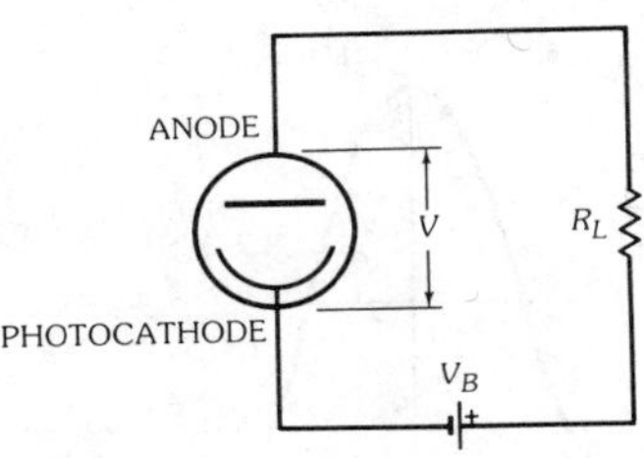

Fig. 14. Photodiode circuit.

R = load resistance,
Q = total charge,
m = electron mass.

Photomultipliers—The combination of a photocathode and a secondary-emission electron multiplier is called a photomultiplier. Emitted photoelectrons from the photocathode are directed under the influence of a suitable electrode, often called the "focus electrode," to the surface of a secondary-emitting electrode, the "first dynode." Subsequently emitted secondary electrons, increased in number by the effective secondary-emission ratio, σ_1, are then directed to the secondary emitting surface of a subsequent dynode by an appropriate electric field for further multiplication. Continuing this process for n successive dynodes and collecting the multiplier charge at an output electrode called the "anode" or "collector" leads to a charge or current amplification, G, given by

$$G = \sigma^n, \quad \text{for } \sigma_1 = \sigma_2 = \sigma_3 \ldots = \sigma$$

Gains as high as 10^5–10^9 are commonly achieved in 10-stage to 16-stage multipliers.

Disregarding all noise fluctuations of the output current, and assuming operation within the usual linear-response region, a photomultiplier acts as a constant-current source generating an output current, I_o, given by

$$I_o = I_s + I_b + I_d = G\epsilon I_{ks} + G\epsilon I_{kb} + G\epsilon I_{kd} + I_{ad}$$

where,

I_s = anode signal current due to an incident signal flux to be detected,
I_b = anode current due to any background flux simultaneously present on the photocathode,
I_d = anode dark current,
G = current gain of the electron multiplier,
ϵ = collection efficiency (ratio of current entering the electron multiplier to emitted photocathode current),
I_{ks} = photocathode signal current due to an incident signal flux to be detected,
I_{kb} = photocathode current generated by any background flux simultaneously present on the photocathode,
I_{kd} = photocathode dark current,
I_{ad} = component of anode dark current I_d not originating from the photocathode.

The output signal current, I_s, follows the instantaneous value of the input signal flux from direct current up to an upper frequency limit (typically 20–200 megahertz) established by the response time or "transit-time spread" (typically 1–10 nanoseconds).

Noise fluctuations of the output current in photomultipliers can be divided into two classes: dark noise, occurring in the absence of input flux, and noise-in-signal, including "quantum" noise resulting from the inherent quantum nature of the input flux as well as uncontrolled fluctuations of that flux. The presence of an appreciable, in fact often predominant, noise-in-signal current component in photomultipliers depending on the instantaneous signal current magnitude requires caution in applying noise concepts to photomultipliers and may lead to erroneous conclusions regarding photomultiplier behavior, particularly for a modulated flux input.

For a steady-state unmodulated flux input, the total noise current, i_n, in the load resistance, R, is given by

$$i_n^2 = 2eGK\Delta f(I_s + I_b + G\epsilon I_{kd}) + i_r^2 + (4kT\Delta f/R)$$

where,

e = electron charge,
K = photomultiplier noise factor,
Δf = noise bandwidth of the noise-current measuring circuits (Hz),
i_r = residual photomultiplier anode dark noise current, excluding dark-current emission from the photocathode (amperes),
$(4kT\Delta f/R)^{1/2}$ = Johnson-Nyquist noise current in load resistance R (amperes),
k = Boltzmann's constant, 1.38×10^{-23} joule/K,
T = absolute temperature of the load resistance (kelvins).

For photomultipliers with a constant gain per stage, σ, in the first few stages of the electron multiplier, the noise factor, K, may be estimated from

$$K = \sigma/(\sigma - 1)$$

Photomultipliers as Scintillation and Single-Electron Counters—In combination with suitable scintillating material, typically thallium-activated NaI crystals, photomultipliers are extensively used to detect the single flashes of light generated by the scintillating material on bombardment by a single triggering particle, typically X-rays or a gamma ray from a nuclear disintegration process. If the scintillating material generates an average of N photons per disintegration incident on the effective photocathode of peak quantum efficiency, $Y_{\max}$, the resultant average charge pulse, Q_A, appearing in the anode circuit (disregarding all photons or electrons producing no output charge) will be given by the equation

$$Q_A = NY_{max}\alpha Ge$$

where α is a spectral matching factor describing the relative match between the scintillator spectral output and cathode sensitivity.

Because of the random statistical fluctuations of cathode quantum efficiency Y_{max} and electron-multiplier gain G, as well as in the number of effective photons, N, generated by the scintillator, the anode charge, Q_A, will vary in magnitude from pulse to pulse, introducing ambiguity in the determination of the average magnitude of N, which in turn is used to determine the energy of the triggering input particle, for example the gamma ray. The ratio of the spread of the amplitude of individually observed values of charge Q_A at half maximum to the most probable value $Q_{A\,(max)}$ is called the "energy resolution" of the photomultiplier-plus-scintillator combination and is commonly 7–10% minimum (see Fig. 15).

If the input flux has no time-coherent groups of photons, as it does in scintillation detection, photoelectrons are emitted singly at random emission times from the photocathode and also generate an average output charge, Q_A, given by

$$Q_A = Gq$$

where all photons or electrons generating no output pulse are disregarded in measuring G and computing Q_A. Assuming sufficiently large gain G and sufficiently low generation of dark pulses of similar charge amplitude, the individual anode pulses of charge amplitude Gq can be detected and counted individually, the photomultiplier then acting as a single-electron counter.

Photoconductivity—Photoconductivity is the increase in electrical conductivity of a material that takes place when the material is illuminated with infrared, visible, or ultraviolet light.

The absorption of light is a quantum process in which electrons are excited to higher energy levels. Ordinarily, the excited electrons are more mobile than unexcited electrons. Photoconductivity is commonly analyzed in terms of the number and mobility of the excited electrons in an electron conduction band and of holes in a lower-energy valence band. To maintain a steady current, both types of current carriers must be generated in the volume of the material, or else charge carriers must enter the photoconductor at one of the electrodes. Many high-resistivity photoconductors make "ohmic" contacts with their electrodes. These serve as practically unlimited reservoirs of mobile electrons, free to enter the photoconductor volume. Even in these photoconductors, the steady dark current is usually limited to a low value by a build-up of a space-charge-potential barrier in the photoconductor.

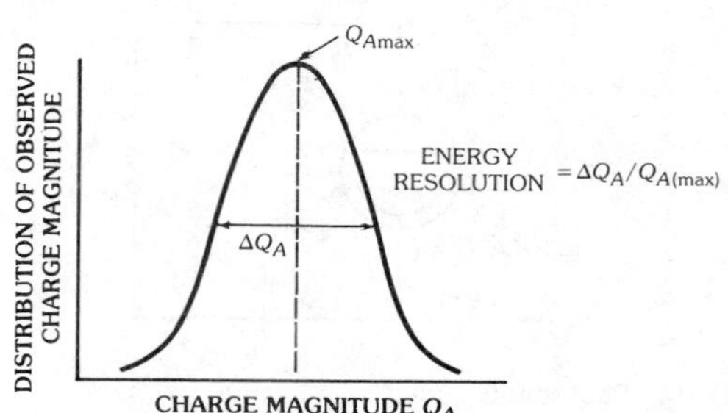

Fig. 15. Energy resolution of photomultiplier.

At the same time that mobile photoelectrons are excited (thermally or optically) in the interelectrode volume, positive charges must also be generated; these compensate the charge of the photoelectrons in such a way that a "photocurrent" can be superimposed on the small space-charge-limited "dark" current originating at the electrode. If the positive charges are immobile, then long after the photoelectron has passed through the photoconductor into the anode, the immobile positive charges may remain to support a photocurrent of electrons, drawing on the reservoir of electrons at the cathode. This will continue until the immobile "holes" or impurity centers are neutralized by recombination with some of the mobile electrons. Since the recombination lifetime may be much longer than the electron transit time between electrodes, the number of "photoelectrons" transported across the photoconductor may be much larger than the rate of generation of photoelectrons in the photoconductor volume. This ratio of photocurrent to generation rate is called the photoconductive gain.

The photoconductive gain of a pure material can often be greatly increased by addition of localized traps lying near the conducting band. Since these are in thermal equilibrium with the conducting band, they serve as an additional reservoir of the charge carriers. This can increase both the response time and the sensitivity by a large factor.

Practically all materials are photoconductors in the sense that light of the correct wavelengths will generate current carriers. However, in many materials the photoconductivity is not detectable by ordinary measurements, either because of very short carrier lifetimes or because of a large dark current. The useful photoconductors, characterized by comparatively long lifetimes and low dark currents, have most of their charge carriers immobile (in the dark). Light of the proper energy can excite these carriers through the forbidden energy regions into the conduction bands. The long-wavelength limit of photoconductivity at low temperatures is given approximately by

$$\lambda_{max} = hc/E_g$$

where,

E_g is the forbidden band gap,
h is Planck's constant,
c is the velocity of light.

For wavelengths longer than 5 micrometers, this equation gives a band gap smaller than 1/4 electron-volt. Photoconductors with such small energy gaps are usually cooled to reduce the dark conductivity due to thermal excitation of carriers across the gap.

Values of D^*, a commonly used figure of merit for photoconductors, for several typical photoconductors at room temperature are shown in Fig. 16. The photoconductors with long-wavelength cutoffs will be considerably more sensitive if they are cooled below room temperature.

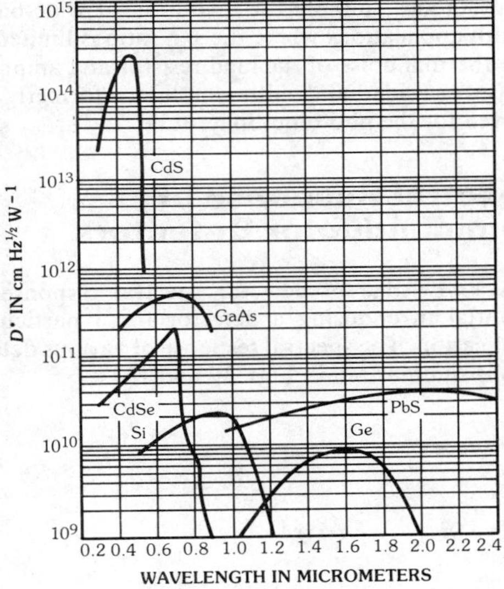

Fig. 16. Detectivity for some typical photoconductors at room temperature.

Photodiodes† — When photons irradiate a semiconductor having a band gap less than the energy of the photons, hole-electron pairs are produced. This phenomenon is best exploited by forming a pn junction in a semiconductor. The resulting photodiode may have a quantum efficiency (ratio of photoelectrons to impinging photons) of from 0.2 to 0.7.

Photodiodes have two primary operating modes, photovoltaic and photoconductive. In the photovoltaic mode, the unbiased junction is illuminated to stimulate the production of hole-electron pairs. Charge separation then occurs in the field of the junction, and a current and/or voltage results.

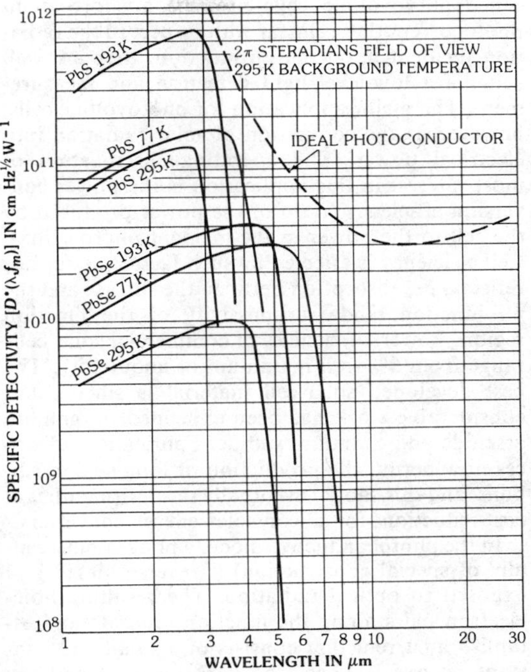

Fig. 17. Spectral detectivities of several thin-film detectors at frequency f_m. (*From Santa Barbara Research Center.*)

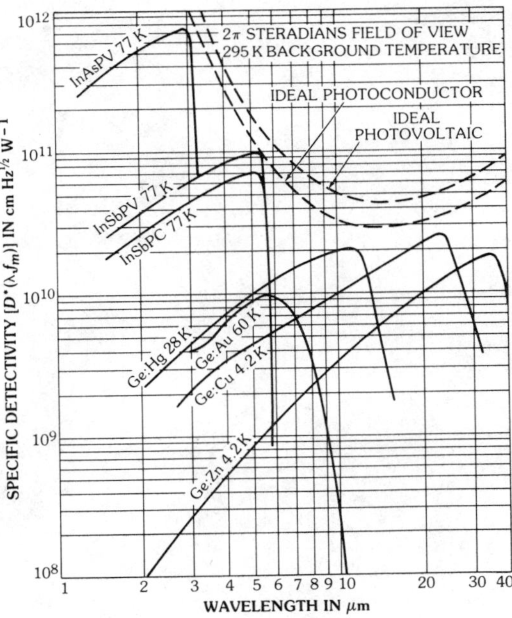

Fig. 18. Spectral detectivities of several crystal detectors at frequency f_m. (*From Santa Barbara Research Center.*)

† G. E. Stillman and C. M. Wolfe, "Avalanche Photodiodes," in *Semiconductor and Semimetals,* Vol. 12 (New York: Academic Press, Inc., 1977; pp. 291–393).

Since photovoltaic cells produce a short-circuit photocurrent that is linear with respect to the radiation incident upon the pn junction, they are well suited for low-level light detection and measurement. The major application for photovoltaic cells, however, is the conversion of solar radiation into electrical power. The operating specification of most interest in this application is the power conversion efficiency (ratio of the power produced by the cell to the power in the incident photon flux). Cell efficiency is affected by such factors as surface reflectance, absorption between the surface and the pn junction, and transmissivity of the junction region. Power efficiencies of commercial solar cells range from 5% to a maximum of about 15%. The best developed solar-cell material is silicon, but considerable work has been expended on gallium arsenide and cadmium sulfide. Fabrication advances now permit the production of long silicon ribbons, and this should eventually result in significant cost reductions for silicon solar-energy converters.

In the photoconductive mode, a photodiode (usually of special construction) is reverse biased and exposed to optical radiation. The resulting hole-electron pairs along the junction separate and establish a current that consists of a small and constant reverse leakage (the dark current) and the photocurrent (the signal). The photocurrent is linear with respect to the incident radiation. The dark current establishes a minimum noise level below which operation in the presence of a dc signal is impractical. An important advantage of photoconductive operation over the photovoltaic mode is response time. A photodiode resembles a parallel-plate capacitor. Unbiased operation is slowed by high junction capacitance, whereas reverse-biased operation reduces junction capacitance and enhances response time.

A photodiode of major importance is the avalanche photodiode. This device is a modified photodiode biased just below the avalanche breakdown region. Photons impinging upon the junction region create hole-electron pairs that initiate an avalanche multiplication of carriers across the junction. The resulting internal gain (which may be a few hundred) and response time make avalanche photodiodes suitable for operation in wide-bandwidth applications where the s/n ratio is limited by the thermal noise of the load resistor and amplifier noise of wide-bandwidth amplifiers formerly reserved for the photomultiplier tube.

Spectral Response of Semiconductor Detectors

A knowledge of detector spectral response is essential in choosing a detector for a particular application. The spectral response of various detectors is summarized in Figs. 17 and 18.

22 Optical Communications

Chester S. Gardner and Robert L. Gallawa

During the past decade, advances in device technology have significantly enhanced the capabilities of optical communication systems. High-data-rate systems have been constructed for space applications, and fiber-optic systems are finding wide use in telephone and data communications. In this chapter, the design of both fiber and free-space optical communication systems is discussed. Systems aspects are emphasized in this chapter. More detailed information on device technology can be found in Chapter 21 on Optoelectronics.

OPTICAL DETECTORS AND NOISE

Optical detectors can be classified into two general categories. Photomultiplier tubes (PMT) and vacuum photodiodes are photoemissive devices that emit electrons into a vacuum or gas. Avalanche photodiodes (APD), pin photodetectors, and phototransistors are solid-state devices in which the excited charge is transported in the solid by holes and electrons. Photomultiplier tubes, APDs, and pin photodetectors are the most widely used in optical communication systems.

Photomultiplier tubes are typically used when high sensitivity and bandwidth are required. Because their internal gain can be as high as 10^7, the output signal level is large compared to the thermal noise of the succeeding amplifiers. Receivers employing PMTs are, therefore, shot-noise limited. The disadvantages of PMTs are that they are relatively bulky and fragile compared to solid-state detectors and they require high-voltage power supplies. Solid-state detectors are small, reliable, and inexpensive, and they exhibit high quantum efficiency through the visible to near-infrared portions of the spectrum. Avalanche photodiodes are characterized by high gain and bandwidth.

Optical detectors are square-law devices because they respond to intensity rather than amplitude. The detection process involves the interaction of incident photons with the detector material. In the case of vacuum photodiodes and PMTs, the incident photons generate photoelectrons that are ejected from the photocathode and flow to the anode. Electron-hole pairs are created in the lattice structure of solid-state detectors by the incident photons. The quantum efficiency (η) of a detector is the fraction of incident photons that produce a photoelectron or electron-hole pair. The quantum efficiencies of photocathodes in PMTs range from a few tenths of a percent to about 30 percent. Silicon photodetectors can have quantum efficiencies exceeding 80 percent. In either case, each detected photon causes an impulse of current to flow in the detector load resistor. The photocurrent is equal to the number of electrons emitted per second times the electron charge ($e = 1.6 \times 10^{-19}$ C). Photodetector efficiency is sometimes expressed in terms of responsivity, which is defined as the photocurrent divided by the incident optical power. Responsivity (R) can be written in terms of the detector quantum efficiency, electron charge, and photon energy (hf)

$$R = \eta e / hf \tag{1}$$

where,

h is Planck's constant (6.63×10^{-34} J·S),
f is the optical frequency.

The detector output current can be modeled by using the following formula:

$$i(t) = \sum_{j=1}^{N(t)} G_j g(t - \tau_j) + i_{Th}(t) \tag{2}$$

where,

$N(t)$ = number of photons that have been detected during the time interval $(-\infty, t)$,
τ_j = arrival time of the jth photon,
G_j = detector gain for the jth photon,
$g(t)$ = detector impulse response,
$i_{Th}(t)$ = detector thermal noise current.

In the above expression, N is a random process whose statistics are related to the optical intensity. If the intensity is deterministic, N is Poisson distributed

$$\text{Prob}\,(N = k) = (\mu^k / k!) e^{-\mu} \tag{3}$$

where,

$$\mu = (\eta / hf) \int_{-\infty}^{t} P(\tau)\, d\tau, \tag{4}$$

$P(t)$ is the optical power.

The expected value and variance of N equal μ.

$$E(N) = \text{Var}(N) = (\eta / hf) \int_{-\infty}^{t} P(\tau)\, d\tau \tag{5}$$

The internal amplification processes of PMTs and APDs are noisy so that G is also random. This effect introduces additional noise into the detection process. Amplification noise is most significant in APDs and should be considered in analyzing receiver performance. However, if the gain is assumed to be constant, the mean and variance of the detector output current are given by Campbell's theorem

$$E[i(t)] = G(\eta / hf) g(t) * P(t) \tag{6}$$

$$\text{Var}[i(t)] = G^2 (\eta / hf) g^2(t) * P(t) + \sigma^2_{Th} \tag{7}$$

The first term on the right-hand side of Eq. 7 is called shot noise. It is caused by the statistical nature of the photon detection process. The term σ^2_{Th} is the thermal noise variance, which is given by

$$\sigma^2_{Th} = 4kT_L B / R_L \tag{8}$$

where,

R_L = detector load resistance,
k = Boltzmann's constant,
T_L = equivalent temperature of the detector load resistor (kelvins),
B = detector bandwidth.

Typically, the detector bandwidth will be larger than or equal to the bandwidth of $P(t)$. In this case, the convolutions in Eqs. 6 and 7 can be approximated by

$$g(t) * P(t) \cong P(t) \int_{-\infty}^{\infty} g(t)\, d\tau = eP(t) \quad (9)$$

$$g^2(t) * P(t) \cong P(t) \int_{-\infty}^{\infty} g^2(\tau)\, d\tau = 2e^2BP(t) \quad (10)$$

In deriving Eqs. 9 and 10, it was assumed that $g(t)$ is an ideal low-pass filter, with bandwidth B. Consequently, the mean and variance of the output current are given by

$$E[i(t)] = (\eta eG/hf)(P_s + P_b) \quad (11)$$

$$\mathrm{Var}[i(t)] = (2\eta e^2G^2B/hf)(P_s + P_b) + 4kT_LB/R_L \quad (12)$$

where,

P_s is the signal power,
P_b is the optical background noise power.

For atmospheric systems operating during the daytime, P_b can be quite large. For fiber systems, P_b is negligible.

Eq. 12 is valid for PMTs and detectors with no internal gain ($G = 1$). For APDs, the random variation of the internal gain must be considered when calculating the variance of the output current. The details are complicated and need not be repeated here. The variance of the output current for an APD is given by*

$$\mathrm{Var}[i(t)] = [2\eta e^2G^2F(G)B/hf](P_s + P_b) + 4kT_LB/R_L \quad (13)$$

where,

$$F(G) = \rho G + (2 - 1/G)(1 - \rho), \quad (14)$$

G is the mean gain,
ρ is the ratio of hole collision ionization probability to electron collision ionization probability.

For silicon APDs, ρ is between 0.02 and 0.03, and G is typically less than 200. For noise-free avalanche gain, $F(G)$ would be equal to 1.

The current variance is proportional to the noise power at the detector output. The signal power is proportional to the square of the mean signal current. Therefore, the signal-to-noise power ratio (snr) at the detector output is given by

$$\mathrm{snr} = \frac{[(\eta eG/hf)P_s]^2}{(2\eta e^2G^2B/hf)(P_s + P_b) + 4kT_LB/R_L} \quad (15)$$

Thermal-noise-limited detection occurs when the internal gain of the detector and the optical signal power are low. In this case, thermal noise dominates the detection process, and the snr becomes

$$\mathrm{snr} = [(\eta eG/hf)P_s]^2/(4kT_LB/R_L) \quad \text{Thermal noise limited} \quad (16)$$

If the internal gain of the detector is high so that detection is shot-noise limited, the snr can be written as

$$\mathrm{snr} = (\eta/2hfB)[P^2{}_s/(P_s + P_b)] \quad \text{Shot noise limited} \quad (17)$$

MODULATION TECHNIQUES

An optical carrier can be modulated with any of the conventional amplitude, phase, and frequency techniques employed at radio frequencies. However, intensity modulation and polarization modulation are the most widely used methods for optical systems. In digital communications, the polarization state of the optical carrier can be used to represent the value of a data bit. Bits can be represented as either right or left circular polarization or as any two orthogonal linear polarization states. Lithium niobate ($LiNbO_3$) and lithium tantalate ($LiTaO_3$) are two of the more commonly used electro-optic crystals for polarization modulators. Intensity modulation can be used for analog and digital signals. Analog signals are sometimes used to modulate rf subcarriers by conventional phase or frequency modulation. The optical intensity is then modulated by the rf subcarrier. Intensity modulation can be accomplished directly in laser diodes and LEDs by varying the diode forward current. External acousto-optic and electro-optic modulators can also be used to vary the optical intensity continuously.

Pulse-Position Modulation

Pulse-position modulation (ppm) is a very effective digital intensity-modulation technique. In M-ary ppm, each time slot is divided into M equal intervals. A single pulse of constant energy is transmitted during one of the M intervals. During a single time slot, one of M different messages can be transmitted. If $M = 2^k$, this corresponds to a binary word of k bits. This modulation technique is a form of block coding. In general, the optimum detection strategy can be quite complicated. However, when the optical detector is shot-noise limited, the maximum-likelihood receiver reduces to a particularly

* Reference 1.

simple form. For each time slot, the receiver counts the number of detected photons in each of the M intervals. The transmitted message is assumed to be the message corresponding to the interval with the largest count. The probability of word error (pwe) for this case is*

$$\text{pwe} = 1 - \{\exp[-(N_s + MN_b)]\}/M - \sum_{k=1}^{\infty} [(N_s + N_b)^k/k!] \exp[-(N_s + N_b)] \times \left[\sum_{j=1}^{k-1} (N_b^j/j!)\, e^{-N_b}\right]^{M-1} (1/aM)[(1+a)^M - 1] \quad (18)$$

where,

$$a = (N_b{}^k/k!)/\sum_{i=0}^{\infty} N^i{}_b/i! \quad (19)$$

$$N_s = (\eta/hf)J_s \quad (20)$$

$$N_b = (\eta/hf)J_b \quad (21)$$

In these expressions, N_s is the expected signal photocount; N_b is the expected background noise count per pulse interval; and J_s and J_b are, respectively, the signal and background noise energies per pulse interval. In Fig. 1, pwe is plotted versus N_s for several values of M. The plots indicate that pwe increases as M increases. It is misleading to use these plots to compare M-ary systems at different values of M. As M increases, more bits are transmitted per pulse. Also, for a fixed time-slot width, the data rate increases with M. Comparisons of M-ary systems should be made only on a bit-error-rate basis for a fixed data rate. In Fig. 2, the equivalent probability of bit error (pbe) is plotted versus N_s for several values of M. Notice that for N_s sufficiently large, the pbe continually decreases with increasing M. When N_s is above the crossover point, there is a significant advantage to increasing M. However, it should be noted that the peak power of the optical transmitter and the bandwidths of the modulator and receiver must also increase.

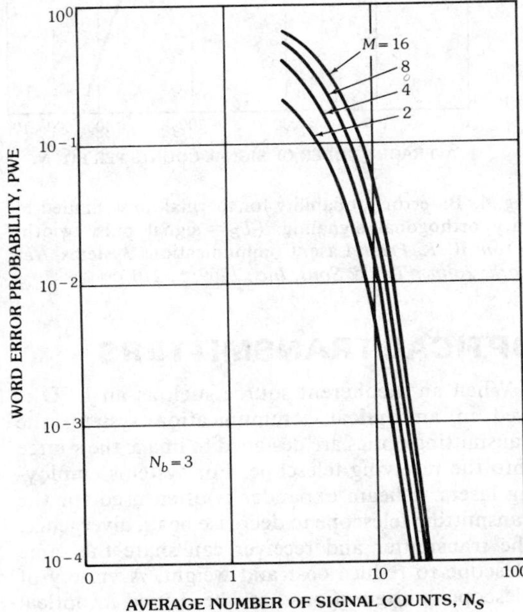

Fig. 1. Word error probability for block-coded pulse-position modulation. (*From R. Gagliardi and S. Karp.* Optical Communications. *New York: John Wiley & Sons, Inc., 1976; p. 264.*)

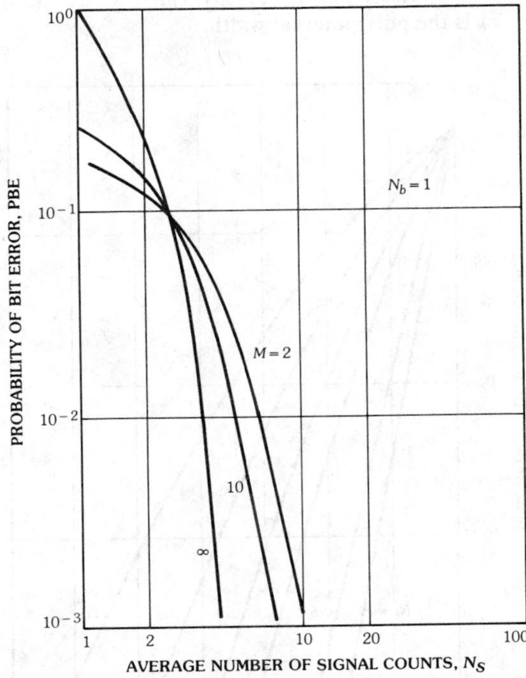

Fig. 2. Equivalent bit-error probability for block-coded pulse-position modulation. (*From R. Gagliardi and S. Karp,* Optical Communications. *New York: John Wiley & Sons, Inc., 1976; p. 271.*)

Binary Orthogonal Signaling

When $M = 2$, we have binary modulation. In this case, the pbe is given by*

$$\text{pbe} = \tfrac{1}{2}\{1 + Q[\sqrt{2N_b}, \sqrt{2(N_s + N_b)}] - Q[\sqrt{2(N_s + N_b)}, \sqrt{2N_b}]\} \quad (22)$$

* Reference 3.

* Reference 4.

where,

$$Q(a,b) = \int_b^{\infty} \exp[-(a^2 + x^2)/2]\, I_o(ax)\, x\, dx \qquad (23)$$

is Marcum's Q function. This expression for the pbe is also valid for other types of binary orthogonal signaling such as polarization modulation.† In Fig. 3, the pbe is plotted versus N_s for several values for N_b. For thermal-noise-limited detection, the pbe can be expressed in terms of the error function and the snr:

$$\text{pbe} = \tfrac{1}{2}[1 - \text{erf}(\sqrt{\text{snr}}/2)] \qquad (24)$$

where,

$$\text{snr} = [G(eN_s/\tau_B)]^2/(4kT_L/R_L\tau_B) \qquad (25)$$

τ_B is the pulse interval width.

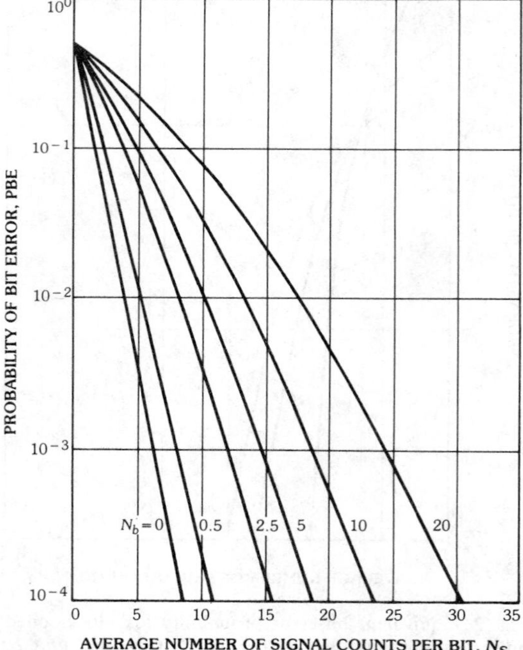

Fig. 3. Bit-error probability for shot-noise-limited binary orthogonal signaling. (*From W. K. Pratt,* Laser Communications Systems. *New York: John Wiley & Sons, Inc., 1969; p. 209.*)

Eq. 25 was obtained from Eq. 16 by replacing the detector bandwidth by $1/\tau_B$ and P_s by N_s/τ_B. The pbe for the thermal-noise-limited detection is plotted in Fig. 4. In general, the signal levels required to realize a given pbe are much higher for thermal-noise-limited detection than for shot-noise-limited detection. This is expected because, by definition, the thermal noise power is much larger than the shot noise power in a thermal-noise-limited receiver.

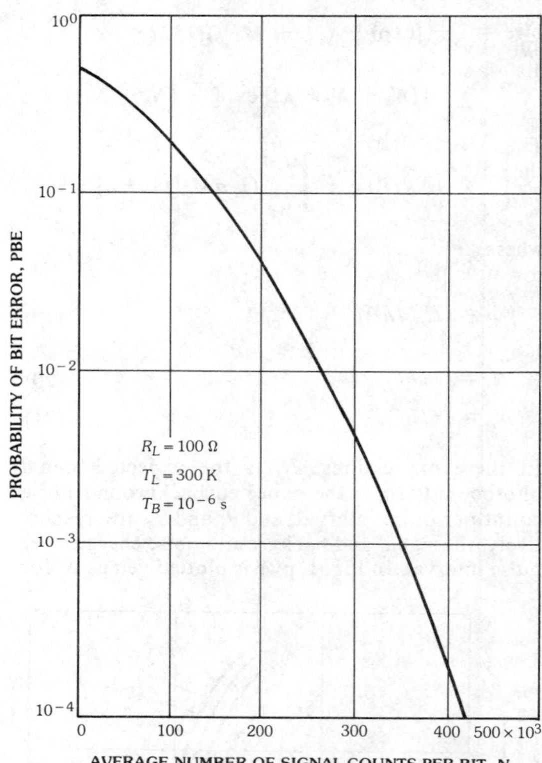

Fig. 4. Bit-error probability for thermal-noise-limited binary orthogonal signaling. (T_B = signal pulse width). (*From W. K. Pratt,* Laser Communications Systems. *New York: John Wiley & Sons, Inc., 1969; p. 210.*)

OPTICAL TRANSMITTERS

When an incoherent source such as an LED is used in an optical communication system, the transmitting optics are designed to image the source onto the receiving telescope. For systems employing lasers, a beam expander is often used for the transmitting telescope to decrease beam divergence. The transmitter and receiver can share the same telescope to reduce cost and weight. A variety of telescope configurations have been used in optical communication systems. The most common are the Cassegrain reflector and the simple refracting telescope. Because of the high cost of low-loss lens materials, reflecting optics are utilized almost exclusively in systems operating at the middle and far ir wavelengths. Both reflecting and refracting telescopes are used for near ir and visible systems.

† Reference 5.

Incoherent Sources

A simple imaging telescope for an LED source is diagrammed in Fig. 5. The objective lens is designed to project an image of the LED onto the receiver. The image intensity depends on the image magnification and the power radiated by the LED. The image magnification (m) is the ratio of the image distance (z) to the object distance (Z_o)

$$m = z/Z_o \tag{26}$$

The image and object distances are related through the lens-maker's equation

$$1/f_L = 1/z + 1/Z_o \tag{27}$$

where f_L is the lens focal length. By solving Eq. 27 for the object distance, we obtain

$$Z_o = f_L/(1 - f_L/z) \tag{28}$$

Usually, the image distance is large compared to the lens focal length so that $Z_o = f_L$ and $m = z/f_L$. The image diameter (d_i) is equal to the LED diameter (d_{LED}) multiplied by the magnification

$$d_i = (z/f_L)d_{\text{LED}} \tag{29}$$

The total optical power in the image is equal to the power collected by the transmitting telescope objective multiplied by the atmospheric transmittance. To compute the power collected by the objective, we need to know the radiation pattern of the LED. If the LED has no lens, the radiation is Lambertian and is proportional to the cosine of the angle between the observation point and the optical axis. Many LEDs are packaged with small lenses to collimate the radiation and reduce the beamwidth. In most cases, the radiation pattern can be fairly accurately modeled by the formula

$$P_L(\theta,\phi) = [(\xi + 1)/2\pi]P_L \cos^{\xi} \theta \tag{30}$$

where,

$$\xi = -\ln 2/\ln \cos(\theta_B/2) \tag{31}$$

θ and ϕ are the spherical coordinate angles,
P_L is the total power emitted by the LED into the hemisphere,
θ_B is the full-width at half-maximum (FWHM) beamwidth.

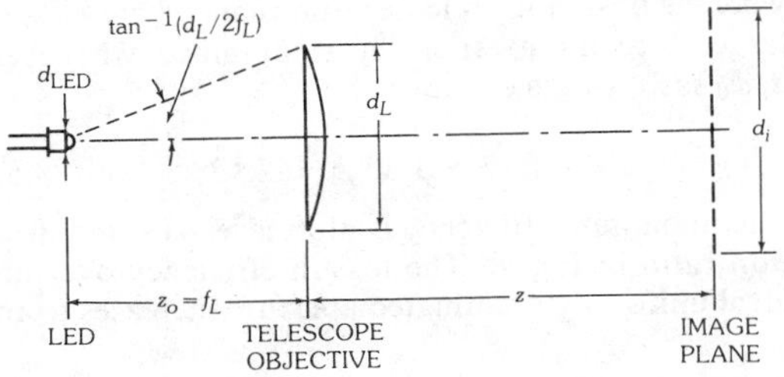

Fig. 5. Transmitting telescope for an LED source.

For a Lambertian source, $\theta_B = 120°$ and $\xi = 1$. For LEDs, θ_B can be as small as 10° and ξ can approach 200.

The power collected by the objective lens is calculated by integrating the LED radiation pattern over the solid angle of the lens.

$$P_T = \int_0^{2\pi}\int_0^{\tan^{-1}(d_L/2f_L)} \{(\xi + 1)/2\pi\} P_L \cos^{\xi} \theta \cdot \sin\theta \, d\theta \, d\phi$$
$$= P_L[1 - (1 + d_L^2/4f_L^2)^{-(\xi+1)/2}] \tag{32}$$

The quantity P_T is the total power transmitted by the telescope, and d_L is the diameter of the telescope objective lens. The maximum value of P_T is P_L, the total power emitted by the LED. The bracketed term is a factor describing the efficiency of the LED-telescope combination. The efficiency increases with decreasing beamwidth (increasing ξ) and decreasing f/number (f_L/d_L) of the objective lens. This is illustrated in Fig. 6, where telescope efficiency is plotted versus the LED beamwidth for several values of the objective-lens f/number. For a Lambertian source, Eq. 32 reduces to

$$P_T = P_L d_L^2/(d_L^2 + 4f_L^2) \tag{33}$$

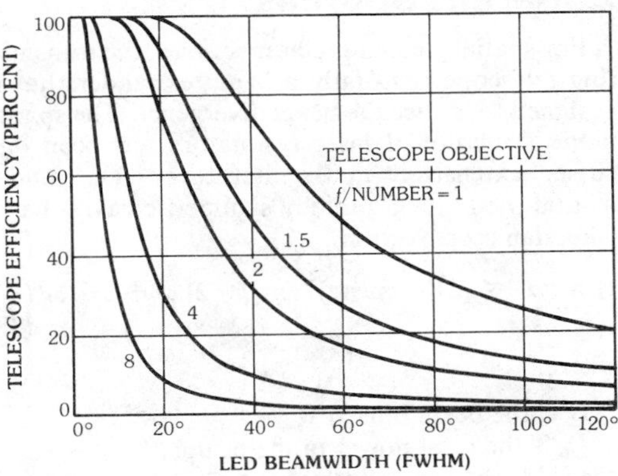

Fig. 6. Variation of the transmitting-telescope efficiency with LED beam width.

If we assume that the image is uniformly bright, then the optical power density or signal intensity (I_s) in the receiver plane is equal to the total power in the image divided by the image area. For a circular LED source, the signal intensity is given by

$$I_s = (4f_L^2/\pi d^2_{\text{LED}})(T_a P_T/z^2) \tag{34}$$

where T_a is the atmospheric transmittance. As expected, the signal intensity decreases inversely with the square of the distance between the transmitter and receiver. Although the telescope efficiency decreases with increasing f_L, the f_L^2 dependence in Eq. 34 dominates so that I_s increases with increasing f_L. However, this does not always mean that it is better to have a large focal length. The magnification and image size decrease with increasing f_L. Smaller images require more precise pointing and tracking. The maximum permitted pointing error (θ_E) is approximately equal to the arctangent of the image radius divided by the image distance

$$\theta_E = \tan^{-1}(d_i/2z) = \tan^{-1}(d_{\text{LED}}/f_L) \quad (35)$$

This is the condition for the edge of the image to lie on the optical axis of the receiving telescope. The transmitting telescope must be pointed with an accuracy less than the angle given by Eq. 35. Small changes in the pointing angle can be obtained by moving the LED perpendicular to the optical axis of the transmitting telescope.

In direct analogy with conventions established in microwave-communication systems analysis, it is sometimes convenient to express the signal power density in terms of the transmitter or source power and an effective antenna gain. In this case, Eq. 34 is written as

$$I_S = (G_T/4\pi)\,(T_a P_L/z^2) \quad (36)$$

where,

$$G_T = (16 f_L^2/d_{\text{LED}}^2)\,[1 - (1 + d_L^2/4f_L^2)^{-(\xi+1)/2}] \quad (37)$$

Coherent Sources

For spatially coherent laser sources, the transmitting telescope is usually a beam expander that is designed to reduce the beam divergence. The spatial mode structure of laser resonators has been discussed extensively in the literature.* The fundamental mode for a radially symmetric cavity has a Gaussian cross section

$$I_L(x,y,z) = [2P_L/\pi\omega^2(z)]\exp[-2(x^2+y^2)/\omega^2(z)] \quad (38)$$

where,

I_L is the beam intensity,
P_L is the total power in the beam,
$\omega(z)$ is the beamwidth at the e^{-2} intensity point.

In Eq. 38, the optical axis of the laser is assumed to be the z axis. The laser is located at the origin, and the beam is propagating in the positive z direction. The factor $\omega(z)$ is a function of the beamwidth and phase-front curvature at the laser. If the receiver is in the far field, then $\omega(z)$ can be written as

$$\omega(z) = z\tan\theta_L \quad (39)$$

where θ_L is the beam divergence angle. Usually, θ_L is a few milliradians or less. The smallest beam is obtained whenever the laser radiation is focused onto the receiver plane. If the receiver is in the far field, this is equivalent to collimating the beam. For a collimated beam, $\omega(z)$ is given by

$$\omega(z) = \omega_0(1 + \lambda^2 z^2/\pi^2\omega_0^4)^{1/2}$$

or

$$\omega(z) = \omega_0[1 + (\lambda z/\pi\omega_0^2)^2]^{1/2} \quad (40)$$

where,

λ is the optical wavelength,
ω_0 is the beamwidth at the output of the beam expander.

The divergence angle for a collimated beam is

$$\theta_L = \lambda/\pi\omega_0 \quad (41)$$

In the far field, the on-axis signal intensity is calculated by substituting Eq. 40 into Eq. 38 and evaluating the result at $x = y = 0$:

$$I_S = (2\pi\omega_0^2/\lambda^2)\,(T_a P_L/z^2) \quad (42)$$

where T_a is the atmospheric transmittance. This expression is similar to Eq. 34 for incoherent sources. The beam divergence decreases and the on-axis intensity increases with increasing beamwidth, ω_0.

It is sometimes convenient to express the on-axis intensity in terms of an equivalent antenna gain:

$$I_S = (G_T/4\pi)\,(T_a P_L/z^2) \quad (43)$$

where

$$G_T = 8\pi^2\omega_0^2/\lambda^2 \quad (44)$$

Because a Gaussian beam is infinite in extent, Eq. 44 is the maximum gain that would be obtained with an infinitely large beam expander. However, in practice, the beam will be truncated by the finite extent of the transmitting aperture. In addition, if a Cassegrain reflecting telescope is used to expand the beam, the central obscuration caused by the secondary mirror will also reduce the effective gain. It is necessary to match the beamwidth to the telescope to minimize the gain loss.† This is illustrated in Fig. 7, where the relative far-field axial gain is plotted versus the ratio of the aperture radius to beamwidth for several values of obscuration ratio ($\gamma = b/a$). The radii of the primary and secondary mirrors are denoted, respectively, by a and b. The gain is plotted relative to the gain for an unobscured uniformly illuminated aperture, $4\pi^2a^2/\lambda^2$. A uniformly illuminated aperture has the maximum possible on-axis gain. Thus, the data in Fig. 7 represent the efficiency of the transmitting antenna. Truncation losses dominate when a/ω_0 is small, whereas obscuration losses dominate when a/ω_0 is large. Maximum efficiency is obtained whenever a/ω_0 satisfies the equality

$$a/\omega_0 = 1.12 - 1.30\gamma^2 + 2.12\gamma^4 \quad (45)$$

The maximum efficiency is plotted versus obscuration ratio in Fig. 8. The loss in efficiency over an ideal uniformly illuminated aperture increases from

* Reference 6.

† Reference 7.

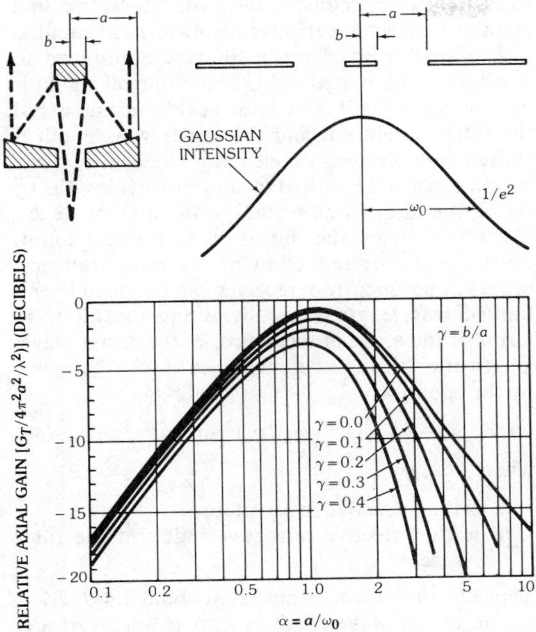

Fig. 7. Far-field axial gain of a centrally obscured transmitting telescope as a function of the ratio of the telescope radius to the laser beam width. (*From B. J. Klein and J. J. Degnan, "Optical Antenna Gain. 1. Transmitting Antennas,"* Appl. Opt., *13, September 1974; p. 2137.*)

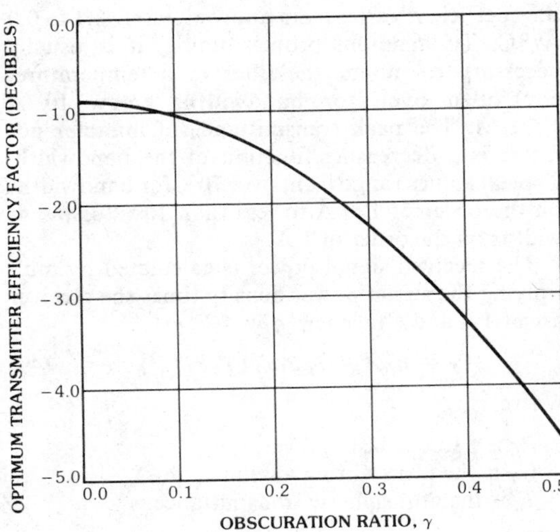

Fig. 8. Optimum telescope efficiency relative to an unobscured uniformly illuminated aperture $(4\pi^2a^2/\lambda^2)$ as a function of obscuration ratio. (*From B. J. Klein and J. J. Degnan, "Optical Antenna Gain. 1. Transmitting Antennas,"* Appl. Opt., *13, September 1974; p. 2139.*)

approximately 1 dB to almost 5 dB as the obscuration ratio increases from 0 to 0.5. As a consequence, these effects must be taken into account, particularly for systems designed with a low gain margin.

In some cases, the spider support structure for the secondary mirror can also have a measurable effect on the axial gain.* However, the effects are usually significant only when the obscuration ratio is small ($\gamma < 0.2$).

The pointing and tracking requirements become more severe as the beamwidth increases and beam divergence decreases. Small pointing errors can result in substantial loss of signal. This is illustrated in Fig. 9, where the loss due to transmitter pointing error is plotted versus pointing error for several values of the obscuration ratio. These curves correspond to the optimum antenna configuration where Eq. 45 is satisfied. The loss is approximately 10 dB when the pointing error is $\lambda/2a$ radians.

Because of the geometry of the emitting stripe in a semiconductor laser diode, the output beam has an elliptical cross section that is highly asymmetric. Therefore, the transmitting telescope must be designed to expand the minor axis of the beam more than the major axis to produce a symmetric cross section. This can be accomplished by using an anamorphic prism pair or cylindrical telescope.

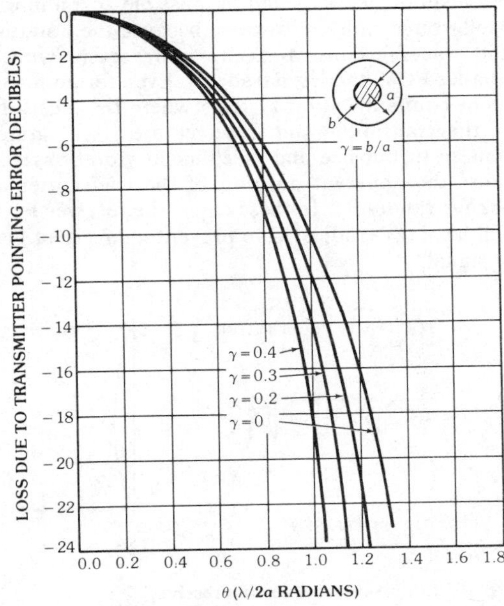

Fig. 9. The decibel loss due to transmitter pointing error as a function of the angular pointing error for optimum antenna configuration (maximum far-field gain) for several values of the obscuration ratio. (*From B. J. Klein and J. J. Degnan, "Optical Antenna Gain. 1. Transmitting Antennas,"* Appl. Opt., *13, September 1974; p. 2139.*)

OPTICAL RECEIVERS

Receiving telescopes are designed to focus the optical signal onto the photodetector and to reject

---

* Reference 8.

as much of the background radiation as is practical. Interference filters are employed to eliminate background radiation that is not the same wavelength as the optical signal. Field stops are used to reject radiation that is not emitted from the region surrounding the transmitting telescope. Direct-detection receivers respond to the signal intensity and are the most widely used in communication systems. In heterodyne-detection receivers, the optical signal is combined with a local-oscillator beam, and then both signals are focused onto the same detector. Heterodyne-detection receivers respond to signal amplitude and are used primarily in the far infrared to overcome limitations imposed by poor detector sensitivity.

Direct-Detection Receivers

A simple direct-detection receiver is diagrammed in Fig. 10. Because of cost, the objective is usually a reflector such as a Cassegrain system whenever the aperture is greater than about 10 cm. The field-stop iris limits the field of view (FOV) of the telescope to a small region surrounding the transmitter. The FOV should be as small as possible to minimize background noise. However, because the pointing and tracking constraints are more severe with a smaller FOV, there is a tradeoff. Even for point-to-point communication systems where the locations of the transmitter and receiver are fixed, atmospheric turbulence and mechanical vibrations can cause the apparent position of the transmitter to change randomly. In these cases, the receiver FOV will have to be adjusted to prevent inadvertent loss of signal.

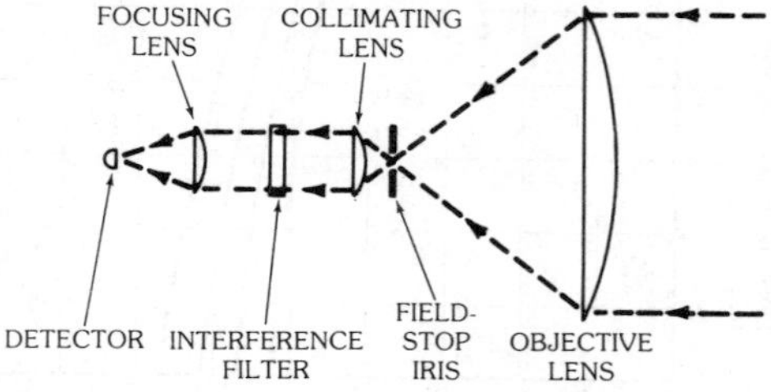

Fig. 10. Simple direct-detection receiver.

The planar (θ_R) and solid-angle (Ω_R) fields of view can be calculated from simple geometric considerations. If f_R and d_I denote, respectively, the focal length of the receiver objective and the field-stop iris diameter, then we have

$$\theta_R = \tan^{-1}(d_I/2f_R) \tag{46}$$

$$\Omega_R = 4\pi \sin^2(\theta_R/2) \tag{47}$$

These expressions for the FOV are useful for calculating the amount of background noise received by the telescope.

Interference filters are multilayered thin-film devices. They are constructed with dielectric and metallic layers and can have bandwidths as small as 1 Å. Usually, interference filters are designed to work with collimated light. If the optical signal is not adequately collimated, the peak transmission of the filter is reduced, and the center wavelength is shifted to a shorter wavelength. The center wavelength can also be shifted to shorter wavelengths by tilting the filter from normal to oblique incidence. For small angles, the shift is without distortion of the passband or reduction of the peak transmittance. Consequently, a receiver can be tuned over a limited wavelength range by tilting the filter. In terms of the angle of incidence, ϕ, the center wavelength at small angles of incidence ($\phi < 5°$) is given by the equation

$$\lambda = \lambda_{max}\,[1 - (n_o/n_e)^2 \sin^2 \phi]^{1/2} \tag{48}$$

where,

n_o is the refractive index of air,
n_e is the effective refractive index of the filter spacer.

Typical values for n_e range from about 1.4 to 2.1.

The center wavelength is also temperature dependent because the layer thicknesses and refractive indices change with temperature. The center wavelength shifts to longer wavelengths as the temperature increases. The thermal coefficient of wavelength shift is a function of wavelength and filter construction. For visible and near ir wavelengths, the coefficient varies from approximately 0.1 to 0.3 Å/°C. To maintain proper tuning, it is usually necessary to mount the filter in a temperature-controlled oven for bandwidths below 10 Å FWHM. The peak transmittance of intereference filters is a decreasing function of the bandwidth. Typical values range from over 70% for bandwidths on the order of 100 Å to less than 10% for bandwidths on the order of 1 Å.

The received signal power is calculated by multiplying the signal power density times the receiver area (A_R) and efficiency (η_R):

$$P_S = \eta_R A_R (G_T/4\pi)\,(T_a P_L/z^2) \tag{49}$$

where,

P_L is LED or laser power,
G_T is the transmitting antenna gain,
T_a is the atmospheric transmittance.

Efficiency η_R includes the effects of reflective and absorptive losses in the receiver optics including the interference filter. Low-loss optical materials and antireflection coatings can reduce losses to a few percent or less for each lens or mirror. Because losses are wavelength dependent, the optics should be coated for the specific wavelength of interest to obtain optimum performance.

Equation 49 can be expressed in terms of an equivalent receiving antenna gain (G_R):

$$P_S = (\lambda/4\pi z)^2 T_a \eta_T \eta_R G_T G_R P_L \quad (50)$$

where,

$$G_R = 4\pi A_R/\lambda^2 \quad (51)$$

A transmitting-antenna efficiency factor (η_T) has been included in Eq. 50 to account for losses in the transmitter optics. The transmitting antenna gain is given by Eq. 37 for LED sources and by Eq. 44 for laser sources with a Gaussian cross section. The factor $(\lambda/4\pi z)^2$ is called the free-space loss and represents the effect of path length on the signal strength. The snr for a direct-detection receiver is calculated by substituting Eq. 50 into Eq. 15. The background noise power can be computed from the data presented in the section on background radiation.

Heterodyne-Detection Receivers

A simple heterodyne-detection receiver telescope is diagrammed in Fig. 11. The local oscillator (lo) beam is combined with the optical signal, and both are focused onto the same detector. Because the objective and collimating lenses demagnify the signal beam, the receiving telescope illustrated in Fig. 11 is optically equivalent to combining the signal and lo in front of the objective lens. Optical mixing of the signal and lo beams produces an intermediate-frequency (if) signal in the detector output that is proportional to the signal amplitude. The if is the difference between the signal and lo frequencies.

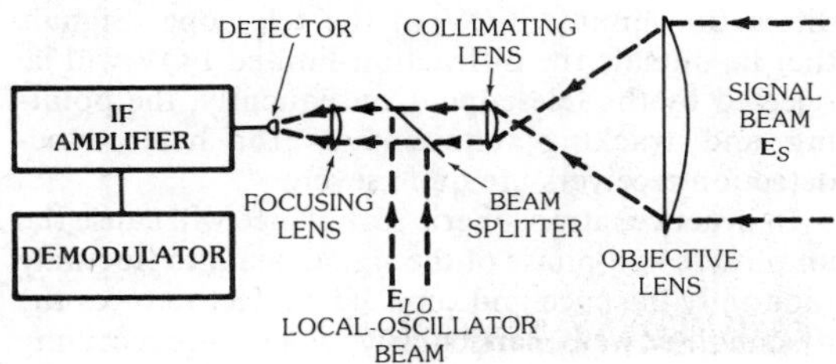

Fig. 11. Simple heterodyne-detection receiver.

The electric field vectors for the signal ($\mathbf{E}_S$) and lo ($\mathbf{E}_{LO}$) can be written in the form

$$\mathbf{E}_S = \mathbf{A}_S \cos(2\pi f_s t + \phi_s) \quad (52)$$

$$\mathbf{E}_{LO} = \mathbf{A}_{LO} \cos(2\pi f_{LO} t + \phi_{LO}) \quad (53)$$

where,

$\mathbf{A}_S$ and $\mathbf{A}_{LO}$ are the vector amplitudes,
f_s and f_{LO} are the optical frequencies,
ϕ_S and ϕ_{LO} are the phases of the signal and lo, respectively.

The amplitudes and phases of both beams can depend on time and position. The detector responds to the total incident power, which is calculated by integrating over the receiver aperture the magnitude squared of the total field:

$$P_D = \int_{\substack{\text{Receiver}\\\text{Aperture}}} \overline{|\mathbf{E}_S + \mathbf{E}_{LO}|^2}\, dxdy = P_S + P_{LO} + S_{IF} \quad (54)$$

The bar in Eq. 54 represents a time average over an interval that is short compared to the inverse of the detector bandwidth but is long compared to the optical periods of the signal and lo. Terms P_S and P_{LO} are, respectively, the total signal and lo powers focused onto the detector by the telescope. Term S_{IF} is the if signal, which is given by

$$S_{IF} = \int_{\substack{\text{Receiver}\\\text{Aperture}}} \mathbf{A}_S \cdot \mathbf{A}_{LO} \cos(2\pi f_{IF} t + \phi_S - \phi_{LO})\, dxdy \quad (55)$$

where,

$$f_{IF} = f_s - f_{LO} \quad (56)$$

The receiver sensitivity and S_{IF} are maximum when the polarization, amplitude distribution, and phase of the lo beam are matched to the signal beam.* If the receiver is in the far field and the effects of atmospheric turbulence are neglected, the signal beam can be approximated as a uniform plane wave. Then for uniform lo and signal beams, $\mathbf{A}_S$ and $\mathbf{A}_{LO}$ are constant throughout the receiver aperture and can be moved outside the integral sign in Eq. 55. To obtain maximum sensitivity, the signal and lo beams must be properly aligned. Small pointing errors can significantly reduce the if signal strength. To illustrate, the phase fronts of the signal beam are assumed to be tilted with respect to the lo. The lo phase is constant, and the signal phase is

$$\phi_S = \phi_0 + (2\pi/\lambda) y \sin\theta_R \quad (57)$$

For simplicity, the coordinate system was chosen so that the tilt (θ_R) is with respect to the y-axis. Angle θ_R may be regarded as the angular pointing error of the receiver.

After substituting Eq. 57 into Eq. 55 and carrying out the integration for an unobscured circular aperture, we obtain

$$S_{IF} = \frac{2J_1[(2\pi a/\lambda)\sin\theta_R]}{(2\pi a/\lambda)\sin\theta_R} A_R \mathbf{A}_S \cdot \mathbf{A}_{LO} \cos(2\pi f_{IF} t + \phi_0 - \phi_{LO}) \quad (58)$$

where,

A_R is the receiver aperture area (πa^2),
a is the aperture radius,
J_1 is a Bessel function of the first kind.

* Reference 3.

The factor involving the Bessel function is a gain-reduction factor related to pointing error. When θ_R is zero (i.e., perfect lo and signal beam alignment), this factor is one. The if signal is proportional to the lo amplitude. By increasing the lo power, the if signal can be made much larger than the receiver thermal noise and the signal and background shot noises. Typically, the limiting noise for heterodyne detection is shot noise contributed by the lo. There is an if signal contribution due to background noise. However, for typical if filter bandwidths (B_{IF} less than approximately 1 GHz), this background-noise contribution is small compared to the if laser signal and can be neglected.

The detector output is processed by an if amplifier and filter and final demodulator. The expected current at the if-amplifier output due to the optical signal is calculated by substituting Eq. 58 into Eq. 11 and is

$$E[i_{IF}(t)] = (\eta eG/hf)S_{IF} \tag{59}$$

If the polarization of the lo is matched to the signal, the if signal power is given by

$$P_{IF} = 2(\eta eG/hf)^2 \left(\frac{2J_1(2\pi a/\lambda)\sin\theta_R}{(2\pi a/\lambda)\sin\theta_R} \right)^2 P_{LO}P_S \tag{60}$$

where,

$$P_S = A_R A_S^2/2 \tag{61}$$
$$P_{LO} = A_R A_{LO}^2/2 \tag{62}$$

The noise power is calculated from Eq. 12 with B replaced by the if bandwidth (B_{IF}) and P_{LO} added to the signal and background shot-noise power. Therefore, when the pointing error is zero, the if snr is given by

$$\text{snr}_{IF} = \frac{2(\eta eG/hf_s)^2 P_{LO}P_S}{(2\eta e^2G^2B_{IF}/hf_s)(P_s + P_{LO} + P_b) + 4kT_LB_{IF}/R_L} \tag{63}$$

If the lo power is large, the thermal noise power and the signal plus background shot-noise power will be negligible compared to the lo shot-noise power. In this case, the snr reduces to

$$\text{snr}_{IF} = \eta P_S/hf_sB_{IF} \tag{64}$$

The signal power can be calculated from Eq. 50, which was derived for a direct-detection receiver. The background noise power can be computed from the data presented in the section on background radiation.

The pointing error can have a significant effect on the if snr. This is illustrated in Fig. 12, where the if snr loss in decibels is plotted versus normalized pointing error for an unobscured circular aperture and uniform lo beam. To prevent significant loss,

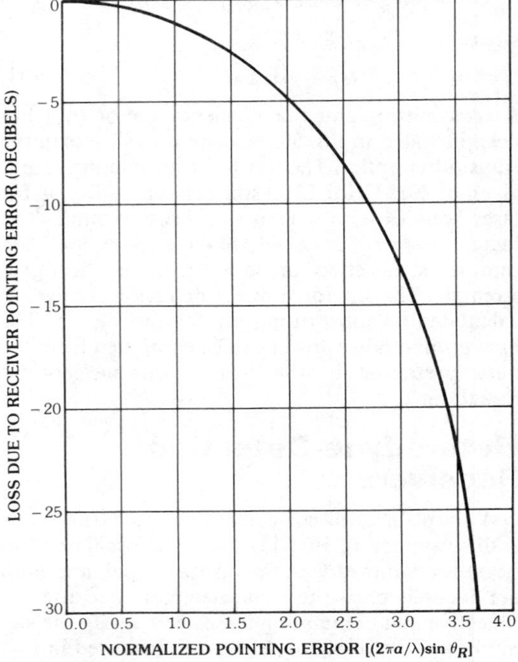

Fig. 12. Intermediate-frequency signal power loss due to receiver pointing error.

the pointing error must be very small. The condition for negligible loss is

$$\theta_R << \lambda/2\pi a \tag{65}$$

Eq. 65 is the condition for pointing within the diffraction-limited FOV of the telescope. Signals that lie outside the diffraction-limited FOV will be rejected by the telescope. Consequently, the pointing and tracking requirements for heterodyne-detection receivers are quite severe.

In practice, atmospheric turbulence will cause the amplitude and phase of the signal beam to fluctuate randomly in space and time. The effect reduces the if signal power, particularly when the receiving aperture is large. The if signal strength is maximum when the lo and signal beams add coherently across the aperture. Turbulence reduces the spatial coherence of the signal beam and causes the signal phase to vary randomly across the aperture. As the diameter of the aperture increases, the additional signal may add incoherently or out of phase with the lo beam, resulting in either no increase or a reduction in the snr. This phenomenon is discussed in more detail in the section on atmospheric effects.

BACKGROUND RADIATION

During the day, the dominant background-noise source is the sun, whereas at night, radiation from man-made sources, the moon, the stars, and the planets becomes important. Background radiation may enter the receiving telescope directly, when the source falls within the receiver FOV, or indirectly, when the radiation is reflected or scattered. The spectral distribution of the background radiation depends on many factors, including the spectral distribution of the source, absorption characteristics of the atmosphere, and reflection characteristics of the earth and objects near the transmitter. Uniform background sources are conveniently described by their spectral radiance, $N(\lambda)$, which is defined as the power radiated at wavelength λ per unit bandwidth into a unit solid angle per unit source area.* The total background noise power collected by the receiving telescope is

$$P_b = \begin{cases} N(\lambda)\ \Delta\lambda A_R\ \Omega_{FOV} & \text{if } \Omega_{FOV}\ \epsilon\Omega_S \\ N(\lambda)\ \Delta\lambda A_R\ \Omega_S & \text{if } \Omega_S\ \epsilon\Omega_{FOV} \end{cases} \tag{66}$$

where,

$\Delta\lambda$ is the optical bandwidth of the receiver,
A_R is the receiver area,
Ω_{FOV} is the receiver solid-angle FOV,
Ω_S is the solid angle subtended by the source when viewed from the receiver.

When $\Omega_{FOV}\ \epsilon\Omega_S$, the source completely fills the receiver FOV, and only a fraction of the total power within the receiver optical bandwidth collected by the aperture is focused onto the detector. This is the case for sky background. When $\Omega_S\ \epsilon\ \Omega_{FOV}$, the complete image of the source is focused onto the detector. In this case, Ω_S must be known to evaluate Eq. 66. This is not particularly convenient for small sources such as stars and planets. Consequently, the radiation from point sources is usually described in terms of the spectral irradiance, $H(\lambda)$, which is the power per unit bandwidth per unit receiver area.* The spectral irradiance is related to spectral radiance by

$$H(\lambda) = \Omega_S N(\lambda) \tag{67}$$

Expressed in terms of the spectral irradiance, the received background noise power is

$$P_b = H(\lambda)\ \Delta\lambda A_R \tag{68}$$

Often the spectral radiance of a background source can be approximated by a blackbody curve

$$N(\lambda) = (2hc^2/\lambda^5)\ [\exp\ (hc/\lambda k T_b) - 1]^{-1} \tag{69}$$

where,

h is Planck's constant (6.63×10^{-34} J·s),
k is Boltzmann's constant (1.38×10^{-23} J/K),
T_b is the source temperature,
c is the velocity of light.

Most earth objects have temperatures between 200 K and 300 K so that their spectral radiance peaks within the wavelength region 9 to 15 μm. The moon has an equivalent blackbody temperature of 373 K. Radiation from the sun and stars peaks at the near ir and visible wavelengths. The equivalent blackbody temperture of the sun is 5900 K.

Absorption in the atmosphere of the earth can significantly alter the spectral radiance of a background source. This is illustrated in Fig. 13, where the spectral irradiance of the sun outside the atmosphere of the earth and at sea level is plotted. Also noted in the figure are the atmospheric constituents that are responsible for some of the major absorption bands. Fig. 14 illustrates the measured spectral radiance of the clear daytime sky. (The zenith angle of the sun is 45°, and the visibility is excellent.) The radiance for sunlit clouds is approximately one order of magnitude larger. On a clear day, the color temperature of the sky is approximately 20 000 K to 25 000 K.

The spectral irradiances of planets and stars also approximate blackbody radiation curves. Fig. 15 shows the calculated spectral irradiances outside the atmosphere of the earth from the planets. It is likely that many stars of widely differing temperatures will be within the receiver FOV. The probable spectral irradiance from a one-square-degree star field near the galactic plane is plotted in Fig. 16.

If the background radiation is due to the reflection of the sun and sky, knowledge of the reflectance of the terrain surrounding the transmitter is required. In Fig. 17, the typical reflectances of water, snow, soil, and vegetation are plotted versus wavelength.

ATMOSPHERIC EFFECTS

Absorption, scattering, and turbulence affect any optical signal that propagates through the atmosphere of the earth. Absorption by minor constituents such as water vapor, carbon dioxide, and ozone can significantly attenuate the signal beam. Rayleigh scattering by air molecules and Mie scattering by larger particles such as aerosols, dust, and clouds also attenuate the signal. Turbulence is caused by small temperature fluctuations in the atmosphere that give rise to random variations in the refractive index. The optical signal is scattered by these refractive inhomogeneities. Portions of the signal that are scattered by different inhomogeneities interfere at the receiver aperture. The amplitude and phase of the total signal vary randomly in space and time. This loss of signal coherence results in reduced gain and fading in both direct-detection and heterodyne-detection receivers.

The transmittance of the atmosphere (T_a) over a path length z for radiation of wavelength λ is given by Bouguer's law*

* References 2 and 4.

* References 9 and 10.

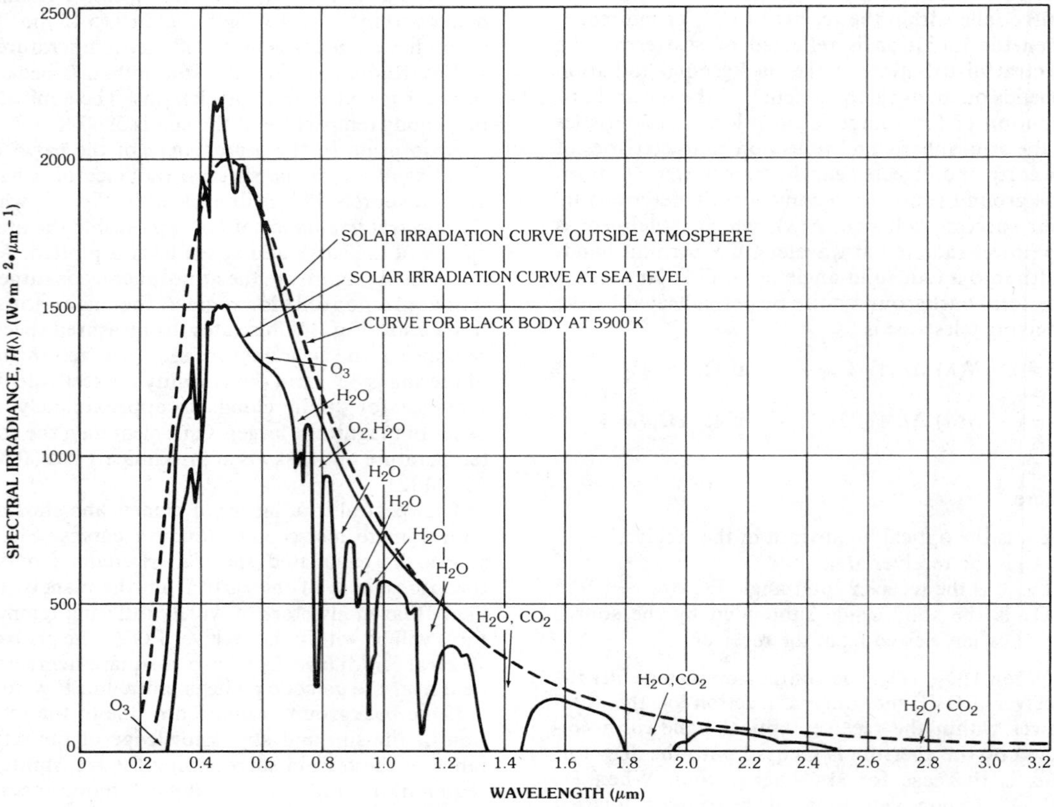

Fig. 13. Solar spectral irradiance with sun at zenith. (*From P. R. Gait, "Solar Spectral Irradiance," in* Handbook of Geophysical and Space Environment, *Sec. 16.1, S. L. Valley, ed. Cambridge, Mass., 1965*)

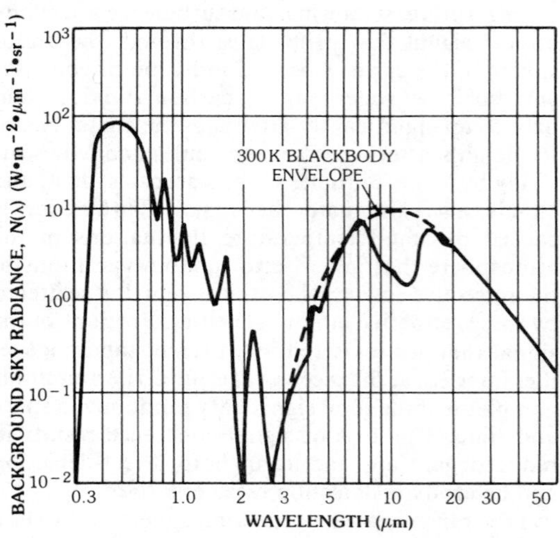

Fig. 14. Diffuse component of the typical background radiance from the sky viewed at sea level. (*From W. K. Pratt,* Laser Communications Systems. *New York: John Wiley & Sons, Inc., 1968; p. 121.*)

$$T_a = \exp\left[-\int_0^z \alpha(\lambda,r)dr\right] \quad (70)$$

where $\alpha(\lambda,r)$ is the attenuation or extinction coefficient and the integral is taken over the propagation path. The attenuation coefficient has units of inverse meters. The integral of $\alpha(\lambda,r)$ in Eq. 70 is called the optical thickness of the propagation path. Optical thickness is dimensionless. Because the atmosphere attenuates through absorption and scattering by both gases and particles, the attenuation coefficient can be written as the sum of four terms

$$\alpha = \alpha_{g,a} + \alpha_{g,s} + \alpha_{p,a} + \alpha_{p,s} \quad (71)$$

where the subscripts g, p, a, and s denote, respectively, gases, particles, absorption, and scattering. Absorption by gases and scattering by particles are the most significant at optical wavelengths.

Absorption

The main absorbing gases in the atmosphere for ultraviolet, visible, and infrared wavelengths in

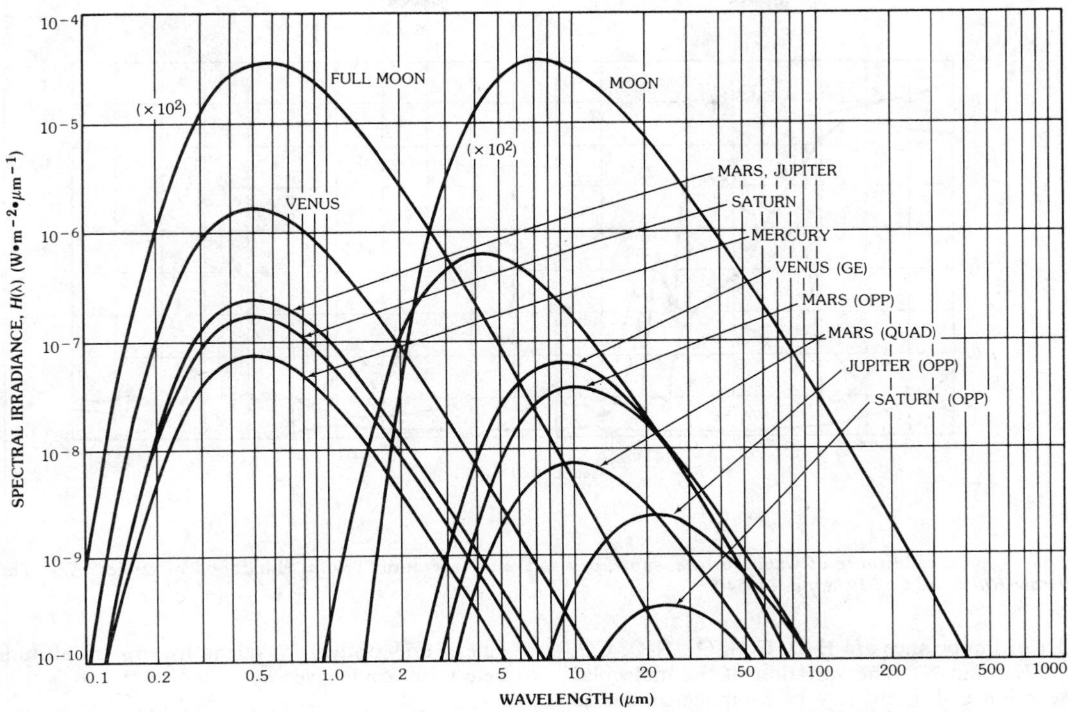

Fig. 15. Calculated planetary and lunar spectral irradiance outside the terrestrial atmosphere. (*From W. K. Pratt,* Laser Communications Systems. *New York: John Wiley & Sons, Inc., 1968; p. 123.*)

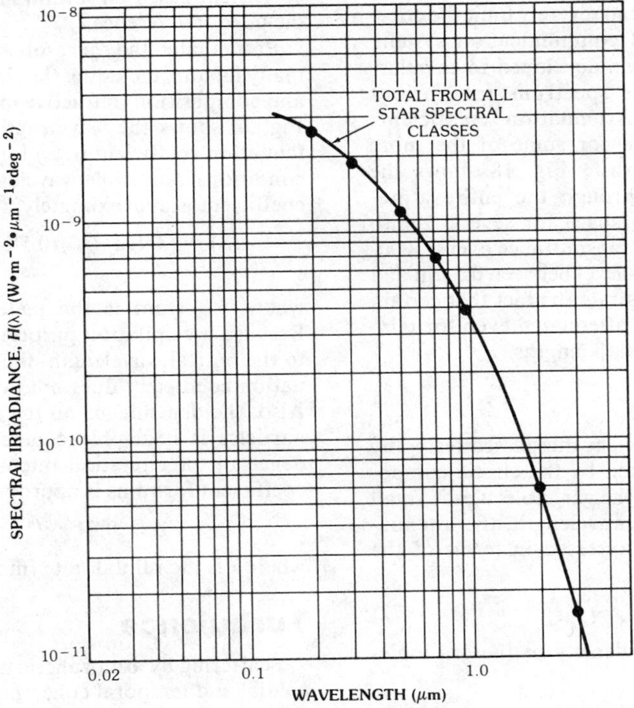

Fig. 16. Probable spectral irradiance from a one-square-degree star field near the galactic plane. (*From* RCA Electro-Optics Handbook. *RCA Commercial Engineering, 1974; p. 69.*)

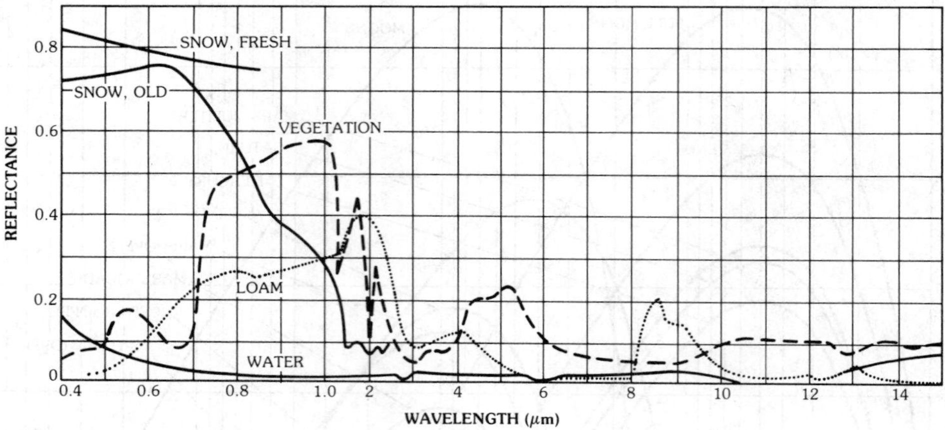

Fig. 17. Typical reflectance of water surface, snow, dry soil, and vegetation. (*From* Handbook of Optics. *New York: McGraw-Hill Book Co., 1978; pp. 14–54.*)

order of importance are H_2O, CO_2, O_3, N_2O, CO, O_2, CH_4, and N_2. The variation of the molecular absorption coefficient is very complicated due to the presence of various absorption band complexes. Strong absorption lines occur most frequently at ultraviolet ($\lambda < 0.3$ μm) and infrared ($\lambda > 0.9$ μm) wavelengths, where they can severely limit the effective range of an optical communication system. Computer codes have been developed to calculate the detailed absorption spectrum for various atmospheric models.† Attenuation coefficients have also been tabulated for some of the more common laser wavelengths.§ Fig. 18 shows the spectral transmittance through the entire atmosphere from sea level to space for several zenith angles. Fig. 19 shows the transmittance over a 1000-foot horizontal path at sea level between 0.5 μm and 25 μm. These figures illustrate the fact that certain wavelengths are strongly attenuated even for relatively short propagation path lengths.

Scattering

The attenuation coefficient due to scattering by gas molecules is dominated by the elastic component. Because the size of the gas molecules is small compared to the optical wavelength, the attenuation coefficient can be expressed in terms of the Rayleigh cross section σ_R

$$\alpha_{g,s} = N_g \sigma_R \tag{72}$$

where N_g is the number density of the gas molecules. For optical wavelengths, σ_R at altitudes below 100 km is given by**

$$\sigma_R = 4.59[\lambda\ (\mu\text{m})/0.55]^{-4} \times 10^{-27}\ \text{cm}^2 \tag{73}$$

Extensive tabulations of the Rayleigh attenuation coefficient based on several model atmospheres are included in reference 12.

Particulate and aerosol scattering depend on many factors, including the size distribution, shape, and composition (refractive index) of the scatterers. Fig. 20 shows the wavelength variation of the attenuation coefficient ($\alpha_{p,s}$) for various atmospheric conditions. At visible wavelengths, the attenuation coefficient is approximately**

$$\alpha(\lambda) \simeq (3.912/V_M)(0.55\ \mu\text{m}/\lambda)^{0.585\, V_M{}^{1/3}} \tag{74}$$

where V_M (km) is the meteorological visibility. Because precipitation particles are large compared to the optical wavelength, the corresponding attenuation coefficient does not depend on wavelength. Also, the dependence on the microstructure of the particles is negligible compared with the dependence on precipitation intensity. The attenuation coefficient for rains is approximately**

$$\alpha \cong 0.21\ r^{0.74}\ \text{km}^{-1} \tag{75}$$

where r is the rainfall rate (mm/hr).

Turbulence

Scattering by atmospheric turbulence reduces the spatial and temporal coherence of the signal beam.

† Reference 11.
§ Reference 12.

** Reference 10.

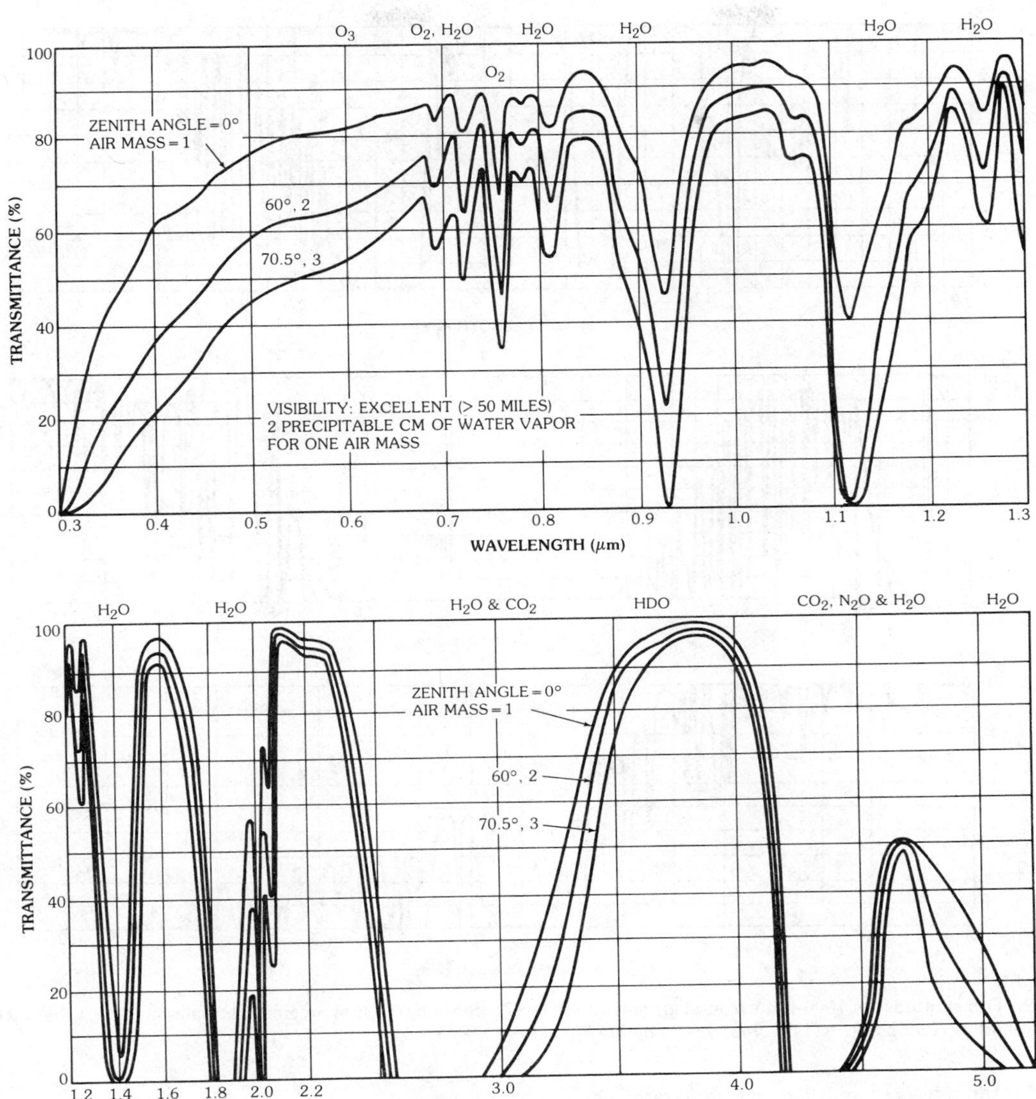

Fig. 18. Spectral transmittance from the atmosphere of the earth for varying optical air masses. (*From* RCA Electro-Optics Handbook. *RCA Commercial Engineering, 1974; p. 83.*)

The effect causes fading or scintillation in the signal power and a reduction in the snr in heterodyne-detection receivers. In weak turbulence, the probability density of the intensity fluctuations is very nearly log-normal*

$$p(I) = [1/(2\pi)^{1/2} I\sigma] \exp\{-[\ln(I/<I>) + (1/2)\,\sigma^2]^2/2\sigma^2\} \quad (76)$$

where,

$$\sigma^2 = \ln(1 + \sigma_I{}^2/<I>^2) \quad (77)$$

* Reference 13.

$\sigma_I{}^2$ is the intensity variance,
$<I>$ is the mean intensity.

In very strong turbulence, the intensity probability density approaches the exponential distribution

$$p(I) = (I/<I>)\exp(-I/<I>) \quad (78)$$

The intensity variance is a function of wavelength, path length, and turbulence strength. For weak turbulence, the intensity variance is given by

$$\sigma_I^2/<I>^2 = \sigma_1{}^2 \quad (79)$$

where,

$$\sigma_1{}^2 = 1.23(2\pi/\lambda)^{7/6}\, C_n{}^2 z^{11/16} \quad (80)$$

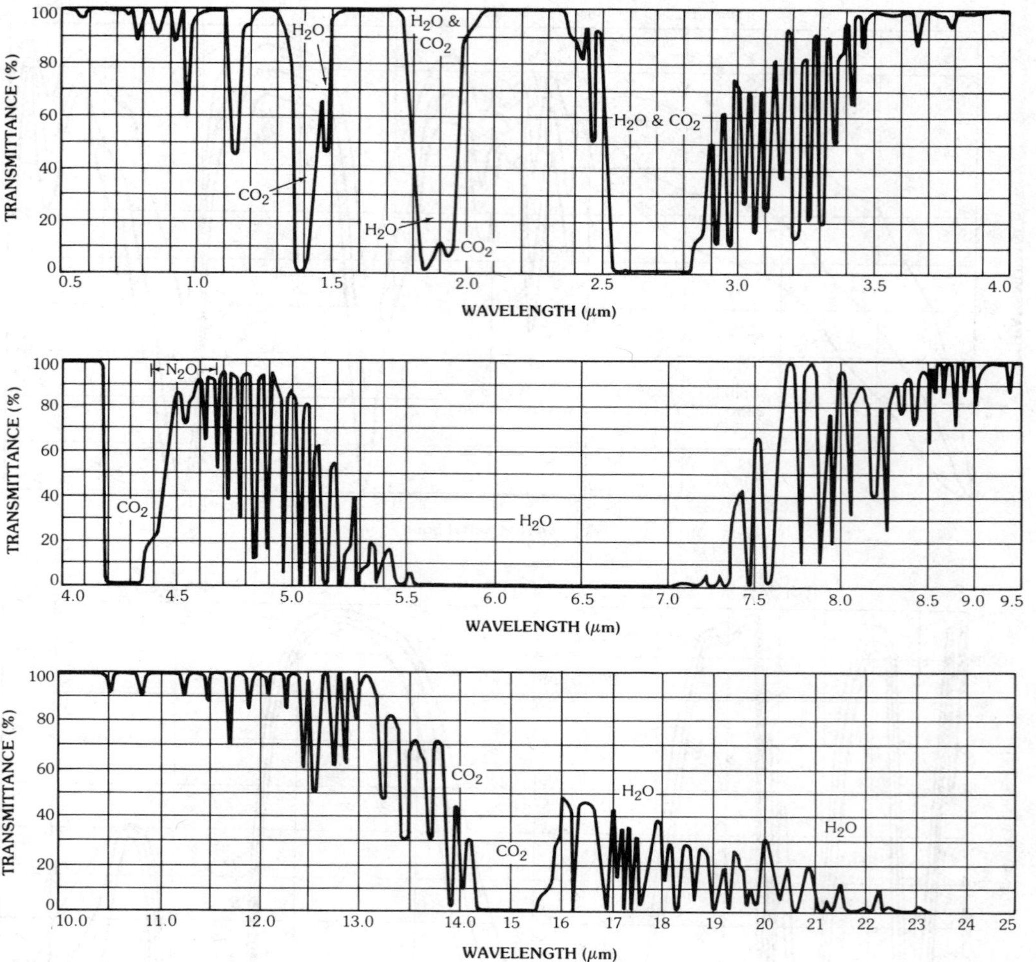

Fig. 19. Transmittance of 1000-ft horizontal air path at sea level containing 5.7 mm precipitable water at 79° F. (*From* RCA Electro-Optics Handbook. *RCA Commercial Engineering, 1974; p. 84.*)

C_n^2 is the refractive index structure parameter,
z is the path length.

The value of C_n^2 varies with altitude, meteorological conditions, and time of day. Typically, C_n^2 is maximum near noon, reaching values on the order of 10^{-14} to 10^{-13} $m^{-2/3}$. At night, the value ranges between 10^{-16} and 10^{-15} $m^{-2/3}$. Eq. 79 is valid for values of σ_1^2 up to approximately 0.3. In stronger turbulence, the normalized intensity variance is less than σ_1^2. For very strong turbulence, $\sigma_I^2/<I>^2$ approaches a value near one.

Eqs. 76 through 79 describe the statistics of the signal-intensity fluctuations at a point. They can also be used to characterize the signal-power fluctuations for a direct-detection receiver if the aperture diameter is smaller than the intensity correlation length, ρ_c. The intensity correlation length is given approximately by

$$\rho_c \simeq \begin{cases} (\lambda z)^{1/2} & \sigma_1^2 \lesssim 0.3 \\ 0.36(\lambda z)^{1/2}/(\sigma_1^2)^{3/5} & \sigma_1^2 >> 1 \end{cases} \quad (81)$$

Because of aperture averaging, power fluctuations are reduced in direct-detection receivers if the aperture area is large compared to the intensity coherence area ($\pi\rho_c^2$). This is illustrated in Fig. 21, where the ratio of the signal-power variance for a receiver of diameter $2a$ to the variance for a point aperture is plotted versus aperture diameter. In both strong and weak turbulence, there is a significant reduction in the power fluctuations whenever the aperture diameter exceeds ρ_c.

Temporal fluctuations of signal intensity are caused by the movement and breakup of refractive inhomogeneities within the propagation path. The

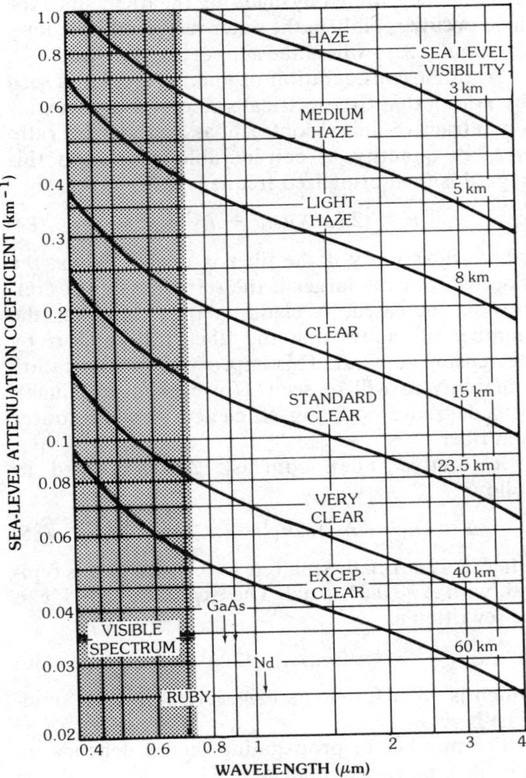

Fig. 20. Approximate variation of attenuation coefficient with wavelength at sea level for various atmospheric conditions. Absorption by water vapor and carbon dioxide is neglected. (*From* RCA Electro-Optics Handbook. *RCA Commercial Engineering, 1974; p. 89.*)

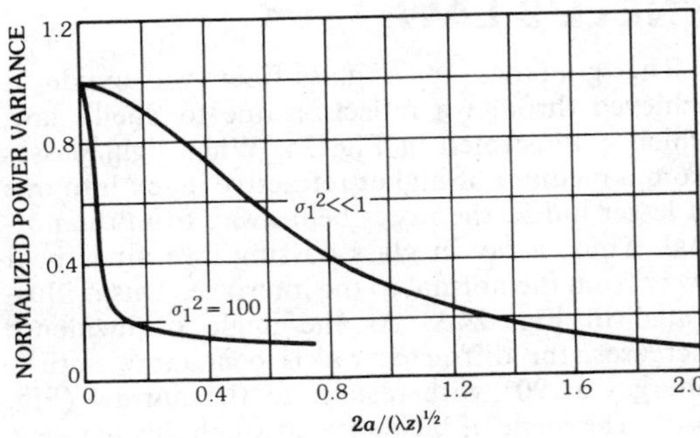

Fig. 21. Aperture averaging in direct-detection receivers for strong and weak turbulence. The received power variance is normalized to the variance for a point aperture; a = aperture radius. (*From R. L. Fante, "Electromagnetic Beam Propagation in Turbulent Media,"* Proc. IEEE, *63, December 1975; p. 1680.*)

intensity temporal power spectrum has a cutoff frequency (f_c) that is related to the wind velocity normal to the propagation path ($V_\perp$) and the intensity correlation length:

$$f_c \simeq V_\perp / \rho_c \qquad (82)$$

For weak turbulence, f_c is typically 10 to 100 Hz, whereas for strong turbulence, the cutoff frequency is approximately an order of magnitude higher. The inverse of f_c is the intensity coherence time and can be used to estimate the fade time.

Heterodyne-detection receivers are most affected by the signal phase fluctuations. Areas of the signal beam that are out of phase with the lo will add destructively at the detector and reduce the if signal amplitude. This results in a reduction of the if snr. For plane-wave sources that are propagating horizontally, the phase correlation length (r_0) is given by the equation†

$$r_0 = [2.36\,(2\pi/\lambda)^2 C_n^2 z]^{-3/5} \qquad (83)$$

The loss of phase coherence reduces the effective area of the receiver aperture. This is illustrated in Fig. 22, where the effective aperture area is plotted versus aperture diameter ($2a$). The effective area never exceeds the phase coherence area (πr_0^2). Because r_0 can be as small as 10 to 20 cm, this effect can severely limit receiver performance. Spatial diversity can be used to overcome some of these problems.§

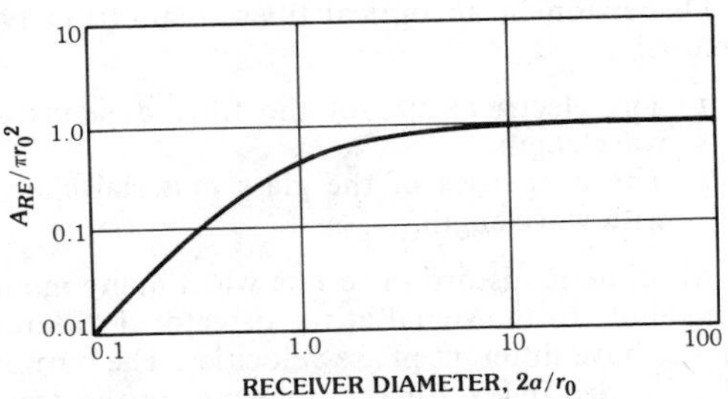

Fig. 22. Effective heterodyne receiver area A_{RE}; a is the aperture radius, and r_0 is the phase correlation length of received signal. (*From R. Gagliardi and S. Karp,* Optical Communications. *New York: John Wiley & Sons, Inc., 1976; p. 199.*)

FIBER WAVEGUIDE TRANSMISSION

The performance of any waveguide depends on four principal factors:

1. Modal characteristics
2. The geometry and size of the waveguide
3. Manufacturing tolerances
4. Materials used

These factors influence signal degradation through:

1. Delay distortion, due to waveguide dispersion
2. Signal distortion due to multimode operation

† Reference 14.
§ Reference 15.

3. Signal attenuation due to absorption, mode conversion, or scattering
4. Signal distortion due to material dispersion

Delay distortion is the result of the nonlinear phase characteristics of the waveguide. For any cylindrical waveguide, the axial propagation characteristics are describable as

$$\exp(-\gamma z) = \exp(-ihz - \alpha z)$$

Then, unless α and h are independent of frequency, the various frequency components of the signal travel at different velocities and with different attenuations. Normally, the distortion due to the variation in the attenuation, $\alpha(\omega)$, is negligible compared to the distortion due to the nonlinearity of the phase term, $h(\omega)$.

Dispersion is a term used to describe the chromatic or wavelength dependence of a parameter. The term is used, for example, to describe the process by which an electromagnetic signal is distorted because the various frequency (i.e., wavelength) components of that signal have different propagation characteristics. Thus, dispersion is one source of distortion. Multimode operation also causes signal distortion, but this is not a dispersive mechanism.

Dispersion in an optical fiber stems from two factors:

1. The electrical size of the fiber depends on wavelength.
2. The properties of the glass material change with wavelength.

Multimode distortion results when many modes contribute to the signal at the detector if different modes have different phase velocities. The purpose of a graded-index fiber is to equalize the travel times of the various modes.

A step-index glass-fiber waveguide is shown in Fig. 23. The radius of the core (or cladding) is designated a (or b), having refractive index n_1 (or n_2). (If the refractive index changes gradually from n_1 to n_2, the fiber is called a graded-index fiber.

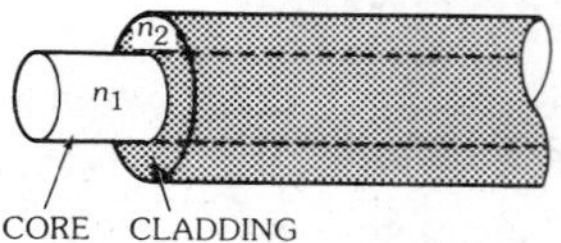

Fig. 23. Step-index optical waveguide (fiber).

Such fibers are discussed later.) The structure will act as a waveguide for $n_1 > n_2$. The cladding is desirable, but not necessary. There is good reason to make $n_2 > 1$. Since the waveguide is "open," the electromagnetic fields are not constrained to remain within the dielectric region. The fiber must be supported, and that support will cause reflections and radiation if it intersects the electromagnetic field. This can be avoided by cladding the fiber, since the field decays rapidly in the outer region and becomes essentially zero where the support is attached.

Another consideration in making $n_2 > 1$ is seen by noting that the electrical size of a fiber depends on refractive-index contrast, as well as the ratio a/λ. A quantity of considerable interest in this regard is the normalized frequency, V:

$$V = (2\pi a/\lambda)(n_1^2 - n_2^2)^{1/2} \quad (84)$$

where n_2 is unity if the fiber is unclad. Thus, the fiber is optically larger if the refractive-index contrast is increased. A cladding thus decreases the number of modes that the fiber will support by decreasing contrast. This improves communication capability as will be seen. However, decreasing V also degrades coupling efficiency between source and fiber.

Refractive index contrast, Δ, is defined as follows:

$$\Delta = (n_1^2 - n_2^2)/2n_1^2 \cong (n_1 - n_2)/n_1 \quad (85)$$

where Δ is normally small: $\Delta = 0.004$ to 0.05 is typical, with $\Delta < 0.01$ usual. The expression for V can be rewritten as

$$V \cong (2\pi a/\lambda)n_1(2\Delta)^{1/2} \quad (86)$$

which is valid for fibers of practical interest (contrast is small).

The number of propagating modes depends on the fiber cross-sectional area (V^2). The cladding allows the physical size to be increased while maintaining reasonably small V by adjusting n_2 to make Δ small. Small V (small number of propagating modes) enhances the communication capability of the fiber.

SNELL'S LAW*

The guidance in a glass-fiber waveguide is achieved through a reflection due to Snell's law, which is illustrated in Fig. 24. When light passes from a medium of higher refractive index into one of lesser index, the ray is bent away from the normal. Thus, a ray in glass passing into air is bent away from the normal to the interface. This is illustrated in Fig. 24A. As the angle of incidence increases, the diffracted ray is bent more until it emerges at 90° with respect to the normal (Fig. 24B). The angle of incidence at which this happens is called the critical angle. When the angle of incidence exceeds the critical angle, as shown in Fig. 24C, light is totally reflected. This is because there is no real angle in the upper medium into which the ray can emerge. The critical angle occurs when $\zeta_2 = 90°$. Then $\sin \zeta_c = n_2/n_1$.

The guidance in a glass-fiber waveguide is accomplished through the same reflection process.

* Reference 16.

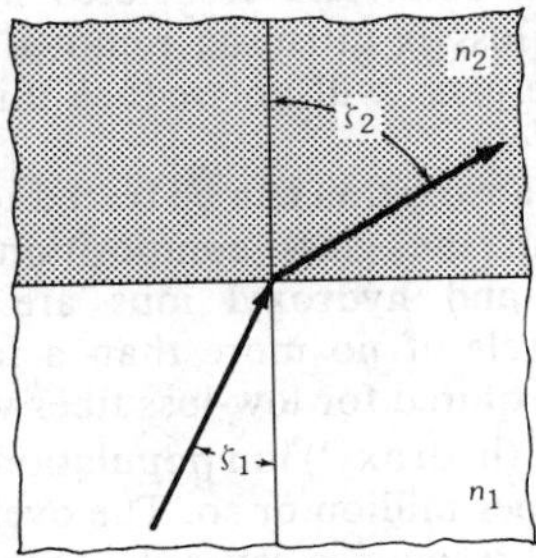

(A) Angle of incidence is less than critical angle.

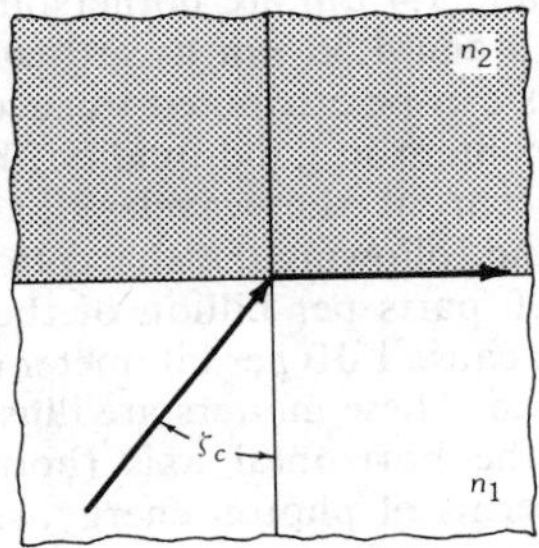

(B) Angle of incidence equals critical angle.

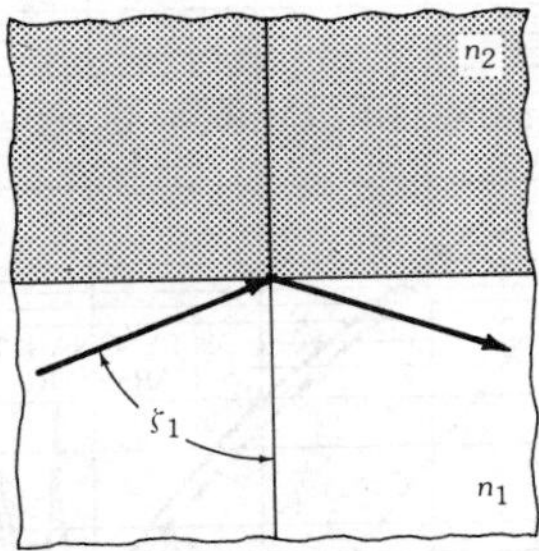

(C) Total internal reflection; incident angle exceeds critical angle.

Fig. 24. Snell's law of reflection.

Reflection at a surface having azimuthal curvature (the core-clad interface) complicates the picture, but the principles are still appropriate. We distinguish between rays that are confined to a single plane (meridional rays) and rays that are not so confined (skew rays). The concepts discussed above, with respect to reflection at a plane surface, carry over easily to meridional rays. Since the skew rays are not confined to a single plane, they are more difficult to track.

Fig. 25 shows rays entering a clad fiber and being reflected or refracted at the core-clad interface. The numerical aperture (NA) of the fiber† is:

† Reference 17.

$$NA = (n_1^2 - n_2^2)^{1/2} \qquad (87)$$

The numerical aperture is a measure of the light-gathering properties of the fiber, but it does not include the effect of skew rays. The numerical aperture can be adjusted via contrast without relying on the fiber dimensions: if a dielectric rod of index $n_1 > \sqrt{2}$ is immersed in free space ($n_2 = 1$), all rays incident at the entrance will be trapped.

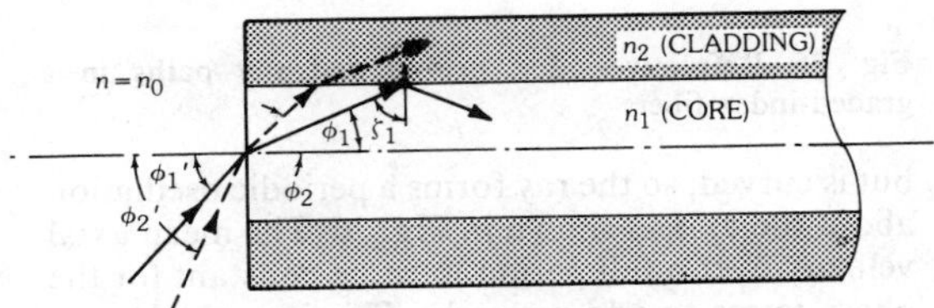

Fig. 25. Rays being reflected or refracted at the core-clad boundary.

GRADED-INDEX FIBERS*

It is common knowledge that continuous variations in refractive index can cause focusing that has a constructive effect on communication. This leads to speculation on the advantages of a continuously varying refractive index in the glass fiber. Intermodal delay is the dominant cause of pulse distortion in a step-index multimode fiber, and that distortion results from differences in axial travel time (group velocity) between the modes. The lower-order modes (small ray angles with respect to the axis) arrive at the detector before high-order modes (large ray angle), a condition that results in interference and pulse spreading. This interference completely masks the other causes of pulse spread. This multimode group delay can be eliminated or greatly reduced by grading the refractive index of the fiber. This is illustrated in Fig. 26, which shows how rays that suffer greater excursions from the axis have increased propagation velocity because of the decreased refractive index. The various optical path lengths are constrained to be constant:

$$\int n(r)ds = \text{constant} \qquad (88)$$

where,

ds is an element of length along the ray path,
r is the distance from the axis,
n(r) is the refractive index.

This equation, or variations of it, has been used to determine the appropriate profile $n(r)$ to minimize distortion.†

The path of a ray in such a fiber is not straight,

* Reference 18.
† Reference 19.

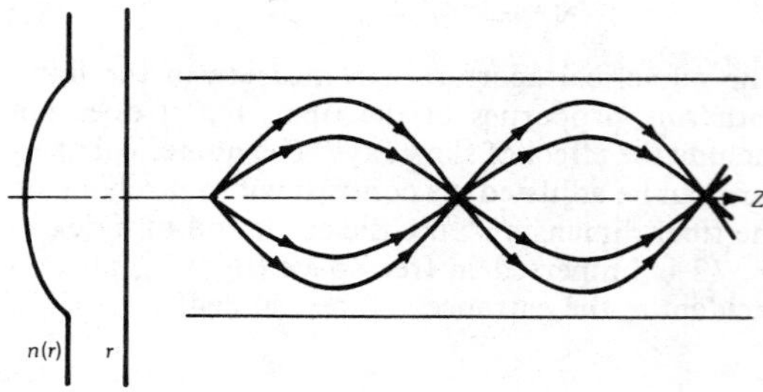

Fig. 26. Refractive-index profile and ray paths in a graded-index fiber.

but is curved, so the ray forms a periodic oscillation about the axis as shown in Fig. 26. The mean axial velocity over a full period must be constant for the phase terms to add properly. This is equivalent to requiring a constant optical path length, as expressed in the above equation.

Refractive-index profiles of interest belong to a class described by a power-law equation,§ as follows, where g is the profile parameter:

$$n(r) = \begin{cases} n_1[1 - 2\Delta(r/a)^g]^{1/2} & r \leq a \\ n_1[1 - 2\Delta]^{1/2} & r \geq a \end{cases} \tag{89}$$

Good quality fibers have $g \cong 2$ (near parabolic). A step-index fiber corresponds to large g. Curves given later will show how fiber bandwidth changes with profile parameter.

WAVEGUIDE LOSSES*

Fiber losses have several causes:

1. Absorption
2. Material scattering
3. Waveguide scattering
4. Losses at bends

Additional loss is due to so-called "microbends." If the core-cladding interface is not smooth, energy is coupled into radiation modes and lost. This microbending loss depends on the rms deviation of the core diameter and the correlation length of the inhomogeneities.

The spectacular decrease in fiber attenuation in recent years is due primarily to the reduction of absorption losses. There are three operating wavelength ranges of interest, corresponding to relative minima in total waveguide loss: 0.8 to 0.9 μm, 1.1 to 1.3 μm, and 1.5 to 1.6 μm. In each case, the low-loss window is attributable to crossover from one dominant loss mode to another.

Several absorption mechanisms are important. The glass constituents, usually oxides of silicon, germanium, or iron cause absorption by virtue of the chemical bond and associated natural vibrations. This loss is intrinsic; removal of contaminants in the glass will not reduce this basic ionic absorption loss.

Extrinsic absorption can be traced to defects and impurities in the glass composition. Transition metal ions and hydroxyl ions are responsible. Impurity levels of no more than a few parts per billion are required for low-loss fiber waveguides.

The OH^- (hydroxyl) ion population must be less than a part per million or so. The overtones of the fundamental resonance cause the absorption problem in the wavelength range of interest. The fundamental vibration is at 2.8 μm. The overtones at 0.75, 0.97, and 1.4 μm are bothersome. The emergence of 1.3 and 1.55 μm as attractive operating wavelengths can be attributed to success in reducing OH^- content in good-quality fibers. The two windows occur at the wavelengths immediately preceding and following the absorption peak at 1.4 μm. Only 30 parts per billion of the OH^- ion is sufficient to cause 1 dB per kilometer of absorption loss at 1.4 μm. These matters are illustrated in Fig. 27, where the horizontal axis (bottom scale) is labeled in terms of photon energy, since it is this

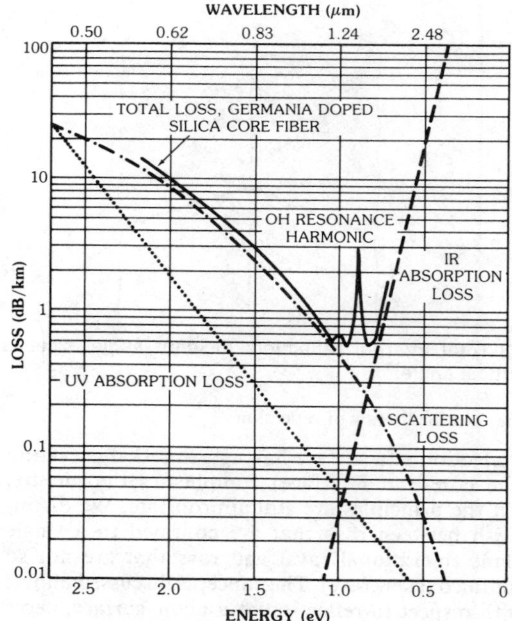

Fig. 27. Total loss and contributors in a germania doped silica core fiber (typical).

energy level that determines loss. Conversion to wavelength (top scale) is made with the basic relationship

$$\lambda\ (\mu m) = 1.24/eV$$

§ Reference 19.

* References 20, 21, and 22.

DISPERSION†

Pulse spread in fibers is attributable to four distinct factors. First, as frequency changes, the dimensions of the waveguide, in wavelengths, change. This causes the phase term to vary with frequency and leads to waveguide dispersion. Material dispersion, the second contributor, is due to the frequency dependence of the electrical properties of the waveguide material. At radio frequencies, the material properties are sensibly constant; at optical frequencies, the refractive index varies considerably, since ionic interactions with the electromagnetic wave become important.

Fig. 28 is useful in illustrating three forms of dispersion in a waveguide. The figure shows phase velocity variation with V (i.e., with normalized frequency) by plotting effective refractive index, n_e; phase velocity is related to n_e as shown in the figure. The plot is for a typical guided mode.

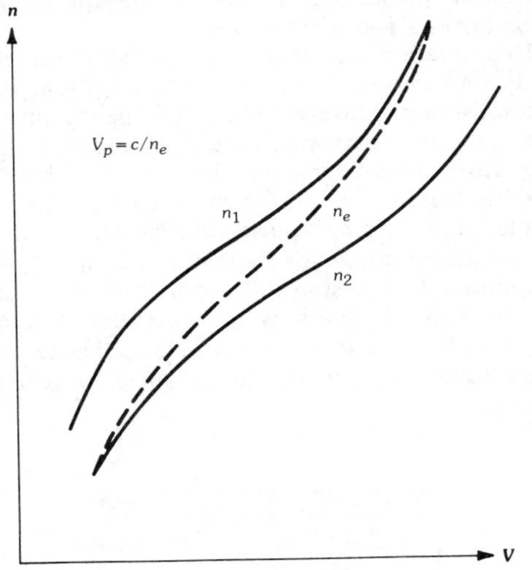

Fig. 28. Symbolic representation of material, waveguide, and profile dispersion in an optical fiber.

The figure shows that modal phase velocity is always such that

$$c/n_1 \leq V_p \leq c/n_2 \tag{90}$$

The value of n_e changes continuously from n_2 to n_1 as V increases. The resulting wavelength dependence of phase velocity is known as waveguide dispersion. The fact that n_1 and n_2 change with wavelength (i.e., with V) is known as material dispersion.

Fig. 28 shows another subtle cause of pulse distortion. Since the difference between n_1 and n_2 is a function of V, the contrast, Δ, is a function of V. Thus, as depicted in the figure,

$$d\Delta/d\lambda \neq 0 \tag{91}$$

leading to a contributor called profile dispersion.

Finally, signal distortion occurs if more than one mode propagates. This factor is important when V is large; it arises because the different modes travel at different velocities. The successive arrival of modes has the effect of spreading the pulse in the time domain. This third contributor to signal distortion is called intermodal or multimode distortion. Graded-index fibers minimize this contribution to pulse spread.

Material Dispersion

We expect material dispersion to be substantial for very short wavelengths and to decrease as wavelength increases. At the short wavelengths, the interactions of the electromagnetic field and the glass render the refractive index more dependent on frequency. On the other hand, waveguide dispersion is small at very short wavelengths. At short wavelengths, the signal is influenced most by the core, since the phase velocity approaches c/n_1; as it does, the phase velocity becomes independent of frequency unless n_1 is a function of frequency. As wavelength increases, waveguide dispersion increases but material dispersion decreases.

The group delay of a plane wave in a dispersive medium of refractive index n is

$$\tau = (1/c)[n - \lambda(dn/d\lambda)] = N/c \tag{92}$$

The quantity N is called the group index. If the spectral width of the optical source is σ_s, then the group delay produces pulse spreading; the pulse width after unit distance is

$$(\Delta\tau) = (d\tau/d\lambda)\sigma_s \tag{93}$$

A series expansion is normally used to deduce N for the wavelength range between 0.7 and 1.5 μm for glasses of interest for optical fibers. Derivatives are taken in accordance with Eq. 93 to determine pulse spread. The term multiplying σ_s on the right-hand side of Eq. 93 is usually called material dispersion, denoted M. Fig. 29 shows M for three candidate materials for glass fibers. Eq. 93 and the units given on the vertical axis of Fig. 29 clearly show the importance of rms source spectral width σ_s, defined as follows:*

$$\sigma_s = \left[\int_0^\infty (\lambda - \lambda_p)^2 S(\lambda) d\lambda\right]^{1/2} \tag{94}$$

where,

$S(\lambda)$ is the spectral distribution of the source,
λ_p is the mean value of source wavelength:

$$\lambda_p = \int_0^\infty S(\lambda)\lambda d\lambda \tag{95}$$

† References 23, 24, 25.

* Reference 23.

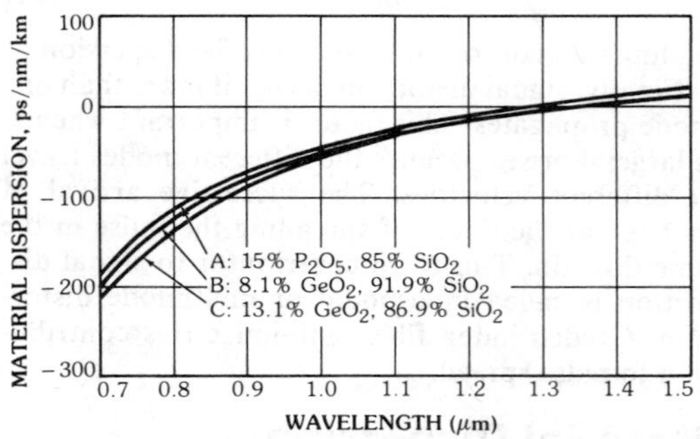

Fig. 29. Material dispersion for typical fiber materials.

$$\int_0^\infty S(\lambda)d\lambda = 1 \tag{96}$$

Since light-emitting diodes (LEDs) have a source spectral width about ten times that of a laser diode (LD), the LED is less desirable than the LD source when material dispersion is a dominant source of pulse spread. This will not be true if operation is at about 1.3 μm, where a point of inflection on the $N(\lambda)$ curve causes M to go through zero. Thus, LEDs become more attractive as sources for fiber systems at 1.3 μm. This happens also to correspond to a relative minimum in fiber attenuation, as seen from Fig. 27 (1.3-μm wavelength corresponds to 0.95 eV).

The wavelength at which material dispersion vanishes cannot easily be inferred. The amount of germania (or other) dopant plays a key role. For silica-based glasses, the wavelength ranges from about 1.27 μm for pure silica to about 1.35 μm with 16% germania, but its exact value depends on doping.

Intermodal Distortion†

In most cases, intermodal distortion is the dominant cause of signal degradation, and the bit-rate capability of the system is limited accordingly. There are two special operational cases for which intermodal distortion does not dominate:

1. When the fiber is monomode, there is only one propagating mode, and there can be no intermodal distortion.
2. The refractive-index profile can be optimized to reduce the intermodal term to a tolerable level.

Other terms then become important in limiting communication capability. Graded-index fibers are discussed in the following section.

The worst-case magnitude of intermodal distortion is seen through a ray treatment of an overmoded step-index fiber. In this approximation, the distortion is caused by the difference in propagation time for rays of different angles. Fig. 30 shows that different rays (at angles ϕ_1 and ϕ_2) require different times to travel axial distance L. The time delay varies as

$$\tau = (n_1/c)L/\cos\phi \tag{97}$$

The limiting delay is the difference between the delays for the axial ray and a ray incident at the critical angle. Thus,

$$\tau_m L = (\Delta t) = (L/c)(n_1/n_2)(n_1 - n_2) \tag{98}$$

where τ_m (s/km) is the delay difference per kilometer. The maximum pulse rate is taken as the reciprocal of this delay:

$$R_m = (cn_2/n_1)/(n_1 - n_2)L = 1/\tau_m L \tag{99}$$

where c is the velocity of light. Note that maximum pulse rate varies inversely with L and it is independent of fiber size. This approximation predicts a maximum pulse rate of tens of megapulses per second over a 1-kilometer fiber.

This equation is useful only to the extent that it is easily understood, being based on intuition. It is not satisfying, however, since it depends on the fiber only in a gross way. Indeed, n_1 and n_2 are the only fiber parameters included. A more precise analysis would account for pulse overlap, for example, and original pulse width. The analysis should also account for the amount of waveguide and material dispersion suffered by each mode and for the fact that various modes suffer different degrees of attenuation. Nevertheless, this expression is useful in predicting the worst-case maximum pulse rate.

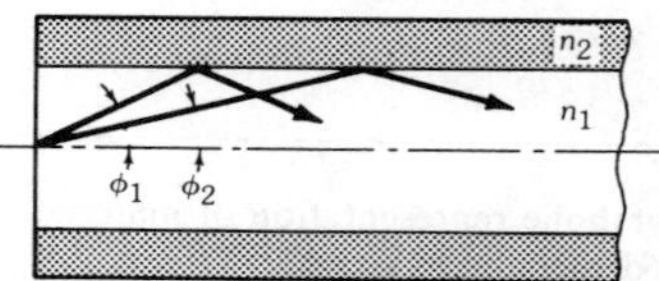

Fig. 30. Ray paths in a step-index fiber.

Pulse Spread in a Graded-Index Fiber§

The success of graded-index fibers is traced to equalization in travel time of the various rays. Those rays that take a short physical path close to the axis, where refractive index is high, traverse an optical path comparable to the longer physical paths through a lower refractive index. The discussion here will assume the power-law profile discussed earlier, with g the profile parameter of Eq. 89. We introduce a term called profile dispersion, y,

† Reference 17.

§ Reference 23.

which describes the change in contrast (Δ) with wavelength:

$$y = -2(n_1/N_1)(\lambda/\Delta)(d\Delta/d\lambda) \quad (100)$$

where n_1 and N_1 are, as before, refractive index and group index of the core:

$$N_1 = n_1 - \lambda dn_1/d\lambda \quad (101)$$

The prediction of pulse spread requires knowledge of the total number of propagating modes, M, and the propagation constant of each. A WKB approach yields

$$M = [g/(g+2)](kan_1)^2\Delta = [g/(g+2)](V^2/2) \quad (102)$$

and the phase term for mode m is

$$h_m = n_1k[1 - 2\Delta(m/M)^{g/(g+2)}]^{1/2} \quad (103)$$

where g is the profile parameter. Actually, m is not a mode label but merely a counting index that tracks the number of propagating modes between the allowed extremes.

$$n_1k < h < h_m$$

Using these equations yields a prediction of rms intermodal pulse broadening:

$$\begin{aligned}\sigma_{er} &= \sigma_{\text{intermodal}} \\ &= (LN_1\Delta/2c)[g/(g+1)][(g+2)/(3g+2)]^{1/2} \\ &\quad \cdot\{C_1^2 + 4C_1C_2\Delta(g+1)/(2g+1) \\ &\quad + [2C_2\Delta(2g+2)]^2/(5g+2)(3g+2)\}^{1/2} \quad (104)\end{aligned}$$

where terms of the order Δ^2 have been neglected,

$$C_1 = (g - 2 - y)/(g+2) \quad (105)$$

$$C_2 = (3g - 2 - 2y)/2(2+g) \quad (106)$$

and L is distance in kilometers.

The optimum value of g is

$$g_o = 2 + y - \Delta(4+y)(3+y)/(5+2y) \quad (107)$$

If profile dispersion is neglected ($y = 0$), the optimum profile is close to parabolic:

$$g_o = 2 - 2.4\Delta \quad (108)$$

The intramodal term is handled similarly. Contributions to rms pulse broadening include material dispersion and waveguide dispersion, but the latter is small if the profile parameter is close to optimum.*

$$\begin{aligned}\sigma_{ra} = \sigma_{\text{intramodal}} &= (L\sigma_s/c\lambda)\{(-\lambda^2n_1'')^2 \\ &\quad - 2\lambda^2n_1''N_1\Delta[(g-2-y)/(g+2)] \\ &\quad \cdot[2g/(2g+2)] + (N_1\Delta)^2[(g-2-y)/ \\ &\quad (g+2)]^2[2g/(3g+2)]\}^{1/2} \quad (109)\end{aligned}$$

Total rms pulse width is

$$\sigma_t = (\sigma_{er}^2 + \sigma_{ra}^2)^{1/2} \quad (110)$$

The expressions given here are quite general; they indicate that the optimum profile parameter is a function of operating wavelength. To utilize fully the capability of a graded-index fiber, then, the operating wavelength must be chosen carefully. This is illustrated in Fig. 31, which shows the optimum value of profile parameter as a function of wavelength for a 3.1% GeO_2-doped silica fiber core and fused silica cladding.

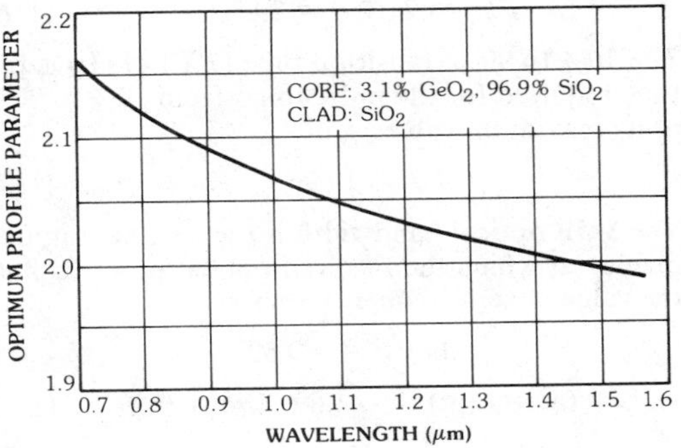

Fig. 31. Optical profile parameter variation with operating wavelength.

BANDWIDTH AND TRANSITION TIME FOR THE GAUSSIAN PULSE†

The Gaussian pulse is important in fiber systems because temporal outputs are approximated closely by such functions. The two most important engineering parameters associated with such pulse shapes are bandwidth and transition or rise time. Caution is in order, however, since optical and electrical bandwidths are not equivalent. Furthermore, electronic circuits (familiar RC circuits) act as low-pass filters that have exponential impulse response.

The frequency-domain function corresponding to the Gaussian temporal function,

$$f(t) = (1/\sigma\sqrt{2\pi})\exp(-t^2/2\sigma^2) \quad (111)$$

is

$$F(\omega) = \exp(-\omega^2\sigma^2/2) \quad (112)$$

where σ is the rms pulse width. Once the bandwidth criterion is established, Eq. 112 can be used to specify the electrical or optical bandwidth. Similarly, rise time or pulse duration can be specified in terms of σ, once a definition of rise time or pulse width is

* Reference 23.

† Reference 24, Appendix 5.

established. From these equations, the full duration of the pulse at half maximum (FDHM), a criterion frequently used, is

$$\text{FDHM} = 2T; \quad f(T) = 1/2 f(0)$$
$$\text{FDHM} = 2.35\,\sigma \tag{113}$$

Also,

$$T_e = \sqrt{2}\,\sigma \tag{114}$$

is the half width of the pulse at the $1/e$ point; i.e.,

$$f(T_e) = f(0)/e \tag{115}$$

and the corresponding full duration is

$$FD_e = 2\sqrt{2}\,\sigma = 2.83\sigma \tag{116}$$

The 10% to 90% transition time (RT) of $f(t)$ is the time required for the pulse to go from 10% to 90% of its maximum value:

$$RT = 1.69\sigma \tag{117}$$

The 3-dB optical bandwidth is the modulation frequency at which the received optical power is half the value at zero frequency. Since

$$\ln 0.5 = -0.693,$$
$$\text{BW (3 dB, opt)} = \sqrt{1.39}/2\pi\sigma = 0.19/\sigma \text{ Hz} \tag{118}$$

The relationship between optical bandwidth and rms pulse duration is shown in Fig. 32.

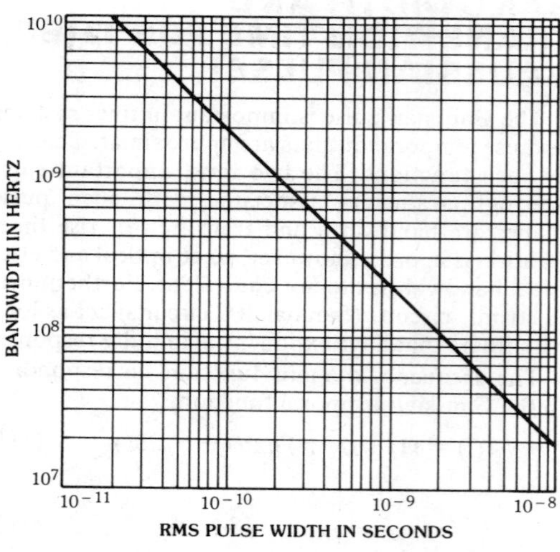

Fig. 32. Relationship of rms pulse width and bandwidth.

The 3-dB electrical bandwidth is $1/\sqrt{2}$ times this value by virtue of the square-law nature of optical detection; i.e.,

$$\text{BW (3 dB, elec)} = 0.19/\sigma\sqrt{2} \text{ Hz} \tag{119}$$

and is the frequency for which the optical power is $1/\sqrt{2}$ of the value at zero frequency. These two definitions arise because the photocurrent of the optical detector, not the electrical power, is proportional to incident optical power. Thus, the electrical power generated by the photocurrent is proportional to the square of the optical power. When the optical power has fallen to $1/\sqrt{2}$ of its peak value, the electrical power has dropped to 1/2 of its peak value. If Eq. 111 is integrated over the time interval 4σ (from $t = -2\sigma$ to $t = +2\sigma$), we find that about 95% of the pulse energy is contained within that interval. This is a convenient and often used criterion for minimum pulse period, the reciprocal of which is taken as the maximum pulse rate, R_m:

$$R_m = 1/4\sigma \tag{120}$$

Other Pulse Shapes*

The Gaussian pulse is of singular importance and has the advantage of being easily analyzed. Other pulse shapes are of practical interest, however. The geometrical optics approach to wave propagation sometimes assumes that all rays suffer the same attenuation and contain the same energy. In that case, the fiber response is a square wave, and the pulse shape of interest in an intensity-modulated system is

$$f(t) = \begin{cases} P_0 & \text{for} \quad 0 \le t \le T \\ 0 & \text{for} \quad t \ge T \end{cases} \tag{121}$$

where T is the envelope pulse width. The associated rms pulse width is

$$\sigma = T/\sqrt{12} = 0.29\,T \tag{122}$$

Maximum pulse rate is $1/T$, or

$$R_m = 0.29/\sigma \tag{123}$$

This equation lends credence to the rule of thumb mentioned earlier: maximum pulse rate is approximately $1/4\sigma$, where σ is rms pulse width.

FIBER BANDWIDTH†

The preceding sections form the basis for predictions of fiber bandwidth as a function of fiber parameters. We arbitrarily assume a fiber whose core is 3.1% GeO_2 and 96.9% SiO_2, and whose cladding is silica. The three-term Sellmeier equation is used to approximate $n(\lambda)$. By adjusting the constants to allow a best fit to measured data, the result can be used to calculate derivatives, allowing prediction of pulse spread and associated bandwidth. Fig. 33 shows n_1, n_2, and N, the group index.

Figs. 34 and 35 show fiber range-bandwidth product under various operating conditions. That the

* Reference 26.

† References 27 and 28.

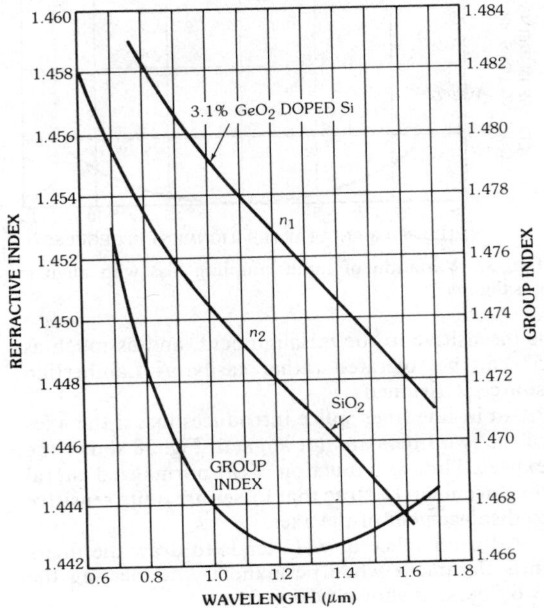

Fig. 33. Variation of refractive index and group index as a function of wavelength.

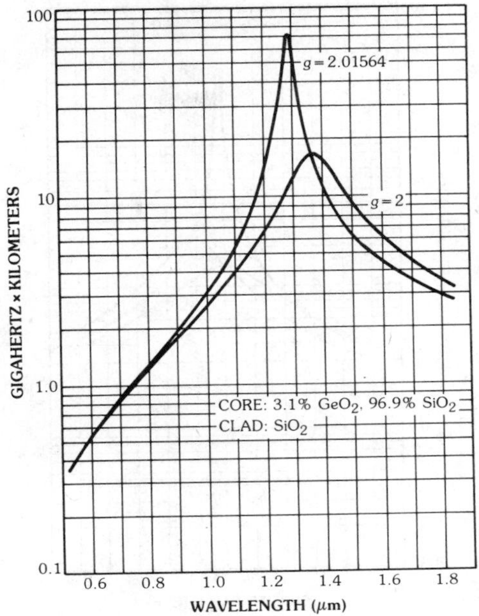

Fig. 34. Range-bandwidth product variation as a function of wavelength.

fiber capacity should be measured as the product of bandwidth and distance is apparent from Eqs. 104 and 109, which show that pulse spread increases linearly with distance. The importance of operating wavelength and profile parameter is apparent from Figs. 34 and 35, which are typical of good quality fibers. At about 1.3 μm, the optimum value of profile parameter is 2.01564. Fig. 34 shows the sensitivity of optimum profile parameter to wavelength and the sensitivity of bandwidth-distance product to profile parameter. Fig. 35 puts this into perspective by showing bandwidth-distance product for this fiber for two popular wavelengths and for two values of source spectral width. For operation at 0.85 μm and source spectral width of 50 nm, the material dispersion term obviously dominates pulse spread. A graded-index fiber offers theoretical improvement over the step-index counterpart on the order of $10/\Delta^2$, or a factor of about 1000.

A monomode fiber waveguide is capable of extremely high bandwidths, if the operating wavelength is chosen properly. A discussion of the theory is beyond the scope of this chapter, so we simply state the result that pulse spread can be reduced to a term that is proportional to σ_s^3:

$$\tau \cong (L/c)(\sigma_s^3/3!)(d^3N/d\lambda^3) \qquad (124)$$

Predictions require the evaluation of $d^3N/d\lambda^3$; range-bandwidth products of several tens of gigabit-kilometers are theoretically possible. The monomode fiber is the obvious candidate for long-haul, high-bit-rate systems.

DETECTOR NOISE LIMITED OPERATION*

When pulse amplitude limits the maximum pulse rate, the important engineering parameter is the power required at the detector to provide the desired bit-error rate (BER). If the energy required per bit is fixed for a given BER, the average power per bit must increase as the bit rate increases. The power required depends not just on the detector characteristics, but on the amplifier as well. At rates less than about 25 Mb/s, silicon field-effect transistors (FETs) are used immediately following the detector. At higher rates, silicon bipolar transistors are used.

Fig. 36 displays a typical design curve showing received power for a pin detector with FET amplifier and, at the higher rates, the bipolar transistor amplifier. Also shown is power required for a silicon APD. The figure assumes operation at 0.85 μm, with a detector having quantum efficiency of 75%; BER $= 10^{-9}$ is assumed. The received optical pulses are taken to be half-duty-cycle rectangular pulses, and the equalizing circuit yields raised cosine pulses.

At 25 Mb/s, the power required at the detector is 3.25×10^{-8} watt when the detector has no avalanche current gain. Under optimum gain conditions, the required power is 1.61×10^{-9} watt, and the value of optimum gain is 57. Optimum gain is

* Reference 29.

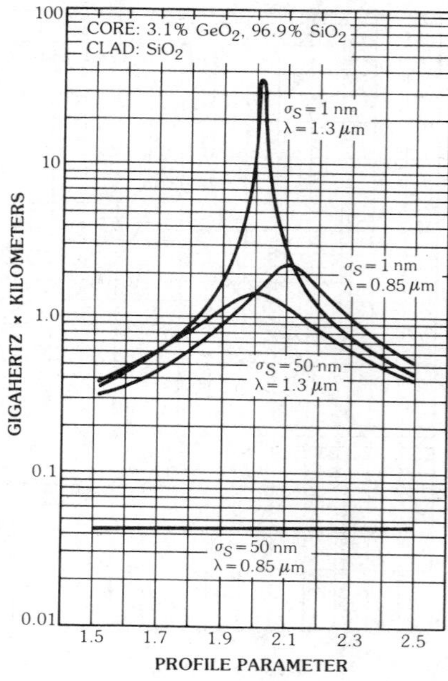

Fig. 35. Range-bandwidth product variation as a function of profile parameter.

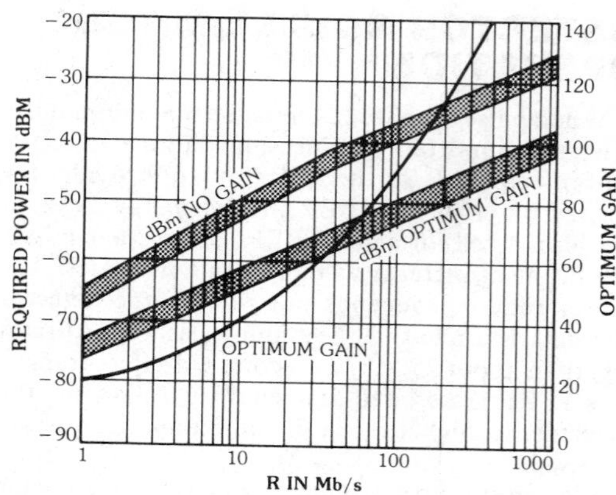

Fig. 36. Relationship among power required, optimum gain, and bit rate.

defined as the gain that minimizes the optical pulse energy required to yield the specified BER.

COUPLING AND SPLICE LOSS†

The optical axis of an optical source may be misaligned relative to the expected axis (usually the mechanical axis). The result is a power loss between the source and the fiber. Fig. 37 shows the expected loss owing to such angular misalignment. The loss

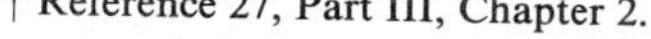

† Reference 27, Part III, Chapter 2.

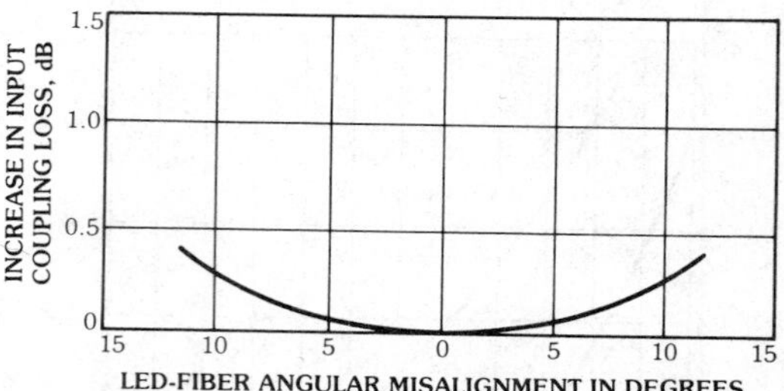

Fig. 37. Variation of input coupling loss with angular misalignment.

is insensitive to the misalignment, and as much as 5° can be tolerated rather easily. A Lambertian source is assumed.

An in-line fiber splice introduces loss if the axes of the two fibers are not aligned. Fig. 38 shows the expected loss as a function of the normalized lateral displacement, *d*. Note that losses are quite sensitive to displacement of the axes.

A fusion splice or weld tends to draw the fibers into alignment while permanently connecting the two fibers, as shown in Fig. 39.

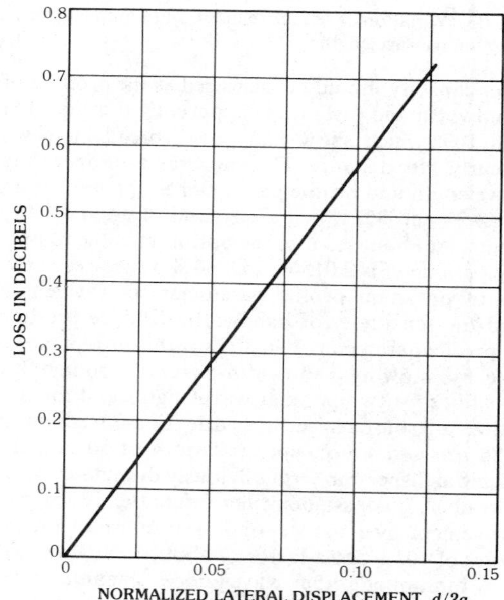

Fig. 38. Coupling-loss variation with lateral displacement of two fibers.

The fusion splice introduces only a small loss and is useful even on single-mode fibers. A loss of 0.2 dB per splice is typical. Insofar as end preparation can be performed in the field and reasonable care is used to maintain cleanliness, the technique is suitable for field use.

A fusion technique is used for power-splitting functions also. Fig. 40 shows two fibers being

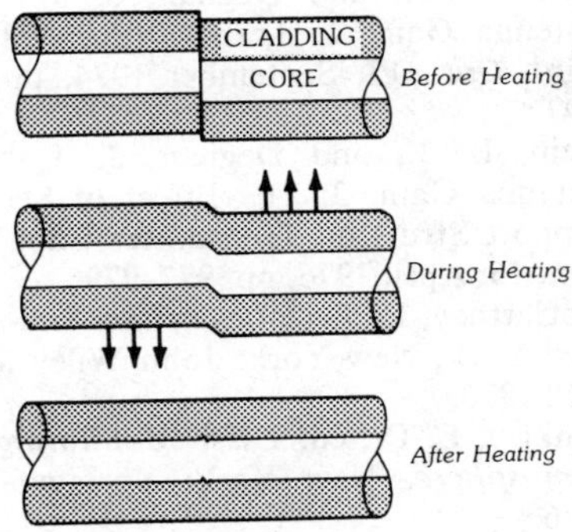

Fig. 39. The fusion splice concept.

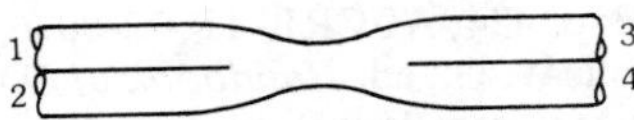

Fig. 40. A four-channel power splitter.

joined to accommodate the splitting of signals among four channels. By heating and tensing the fibers, the coupler is formed to have a variety of coupling levels.

The core radius of a fiber is subject to variation, even in the most refined drawing process. If fibers of different core radii or different NA are joined, additional loss is introduced. These are additive, and an in-line joint may introduce loss due to each of the effects. Fig. 41 shows the expected additional loss due to the mismatch between two fibers (subscripts 1 and 2).

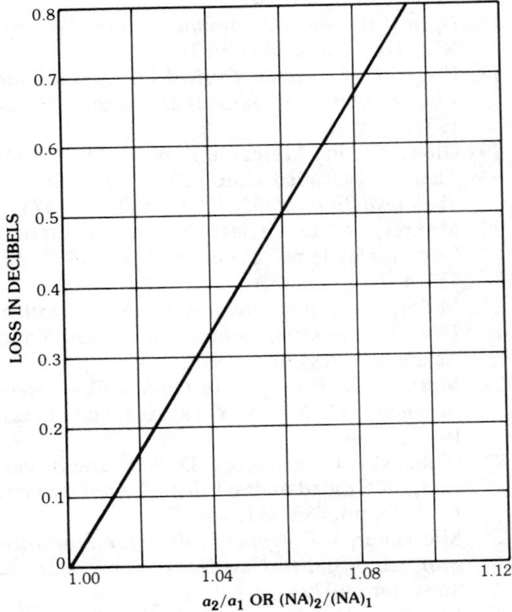

Fig. 41. Coupling loss between two fibers having different core parameters.

An in-line splice of two graded-index fibers introduces loss if the two fibers have different profiles. Fig. 42 shows the expected loss, in decibels, for an in-line splice of two fibers having different values of g; subscripts 1 and 2 refer to the two fibers, and power flow is from fiber 2 to fiber 1. As a point of interest, a splice between a parabolic profile and a step-index fiber introduces a 3-dB loss.

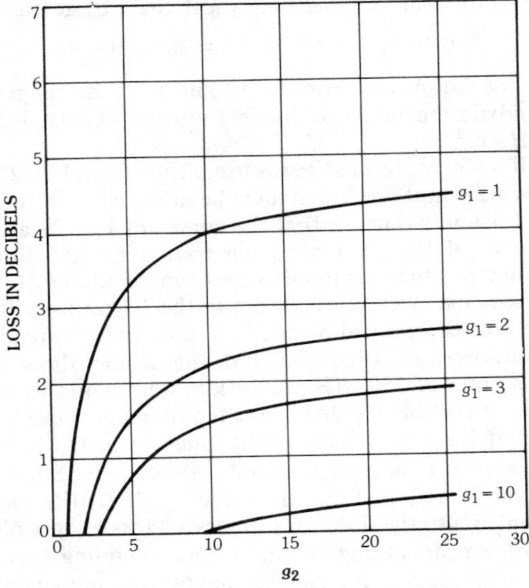

Fig. 42. Coupling loss between two fibers having different profile parameters.

SOURCES AND DETECTORS FOR FIBER SYSTEMS§

Semiconductor laser diodes (LDs) and light-emitting diodes (LEDs) are obvious choices for fiber systems. For long-haul, high-data-rate applications, LDs are normally chosen because of speed and relative high power coupled into the fiber. Light-emitting diodes are more robust, but they suffer from a broad radiation pattern (decreased coupling efficiency) and a broad spectral width; LEDs have rms spectral width about ten times that of an LD. Accordingly, material dispersion is intolerable when LEDs are used, except when the link is short, or the pulse rate is low, or the operation is at about 1.3 μm, where material dispersion is dramatically reduced.

The advantage (in decibels) of using an avalanche photodetector (APD) can be inferred from Fig. 36.

§ Reference 27, Part II.

The disadvantage is in circuit complexity and bias-voltage requirements. At high data rates (greater than 100 Mb/s), the APD has significant advantage in required detector power, and the complexities of using an APD may be justified.

SUMMARY AND DISCUSSION

The following example illustrates the use of concepts, equations, and curves given above. Consider a 10-km link that must pass 50 Mb/s without a repeater. If the on-off keying format is used, this is equivalent to 50 Mp/s (p = pulses). Eqs. 118 and 120 show that the fiber bandwidth must be at least

$$0.19 \times 4 \times 50 \times 10^6 = 38 \times 10^6$$

in one kilometer. For the 10-km link, the range-bandwidth product of the fiber must be at least 380 MHz-km.

If a 400 MHz-km fiber is used, then from Fig. 29 it is clear that operation must be in the range of 1.2 to 1.5 μm, assuming that the curves of Fig. 29 are typical of the fiber used. Operation at 0.85 μm would produce material dispersion of about 100 ps/km/nm, yielding 1 ns/nm in the 10-km link. If the source spectral width is 1 nm, pulse spread would be 1 ns. From Eq. 120, this would allow a pulse rate of 250 MHz in 10 km, and would thus not satisfy the 380 MHz-km requirement. Operation at 1.3 μm gives acceptable pulse spread.

From Fig. 36 used as a guide, about −40 dBm in power is required at the detector. If 0 dBm is coupled into the fiber, this suggests a loss budget of 4 dB/km, including splices, output coupling loss, and degradation. If a detector having optimum gain (about 80) is used, only −55 dBm of optical power is required, and the loss budget for the 10-km length increases to 5.5 dB/km.

REFERENCES

1. Personick, S. D. "Photodetectors for Fiber Systems." *Fundamentals of Optical Fiber Communications*. M. K. Barnoski, ed. New York: Academic Press, Inc., 1981.
2. *RCA Electro-Optics Handbook*. Tech. Series EOH-11, RCA Commercial Engineering, 1974.
3. Gagliardi, R. M., and Karp, S. *Optical Communications*. New York: John Wiley & Sons, Inc., 1976.
4. Pratt, W. K. *Laser Communications Systems*. New York: John Wiley & Sons, Inc., 1969.
5. Peterson, G. D., and Gardner, C. S. "Cross-Correlation Interference Effects in Multiaccess Optical Communications." *IEEE Trans. Aerospace Elect. Sys.*, AES-17, March 1981, pp. 199–207.
6. Kogelnik, H., and Li, T. "Laser Beams and Resonators." *Proc. IEEE*, 54, October 1966, pp. 1312–1329.
7. Klein, B. J., and Degnan, J. J. "Optical Antenna Gain. 1: Transmitting Antennas." *Appl. Opt.*, 13, September 1974, pp. 2134–2141.
8. Klein, B. J., and Degnan, J. J. "Optical Antenna Gain. 3: The Effect of Secondary Support Struts on Transmitter Gain." *Appl. Opt.*, 15, April 1976, pp. 977–979.
9. McCartney, E. J. *Optics of the Atmosphere*. New York, New York: John Wiley & Sons, Inc., 1976.
10. Hinkley, E. D., ed. *Laser Monitoring of the Atmosphere*. New York: Springer-Verlag, 1976.
11. McClatchey, R. A., et al. "AFCRL Atmospheric Absorption Line Parameter Compilation." Air Force Camb. Res. Lab. Environ. Res. Pap. 434, AFCRL-TR-73-0096 (1973).
12. Driscoll, W. G., ed. *Handbook of Optics*. New York: McGraw-Hill Book Co., 1978.
13. Fante, R. L. "Electromagnetic Beam Propagation in Turbulent Media." *Proc. IEEE*, 63, December 1975, pp. 1669–1692.
14. Fried, D. L., and Mevers, G. E. "Evaluation of r_o For Propagation Down Through the Atmosphere." *Appl. Opt.*, 13, November 1974, pp. 2620–2622.
15. Churnside, J. H., and McIntyre, C. M. "Heterodyne Receivers for Atmospheric Optical Communications." *Appl. Opt.*, 19, 15 February 1980, pp. 582–590.
16. Young, M. *Optics and Lasers, An Engineering Physics Approach*, Ch 8. New York: Springer Verlag, 1977.
17. *Optical Waveguide Communication Glossary*, NBS Handbook 140 (1982).
18. Unger, H. G. *Planar Optical Waveguides and Fibers*. Oxford: Oxford University Press, 1977.
19. Gloge, D., and Marcatili, E. A. J. "Multimode Theory of Graded Core Fibers." *Bell System Technical Journal*, 52, 1973, pp. 1563–1578.
20. Maurer, R. D. "Glass Fibers for Optical Communications." *Proc. IEEE*, 61, 1973, pp. 452–462.
21. Miller, S. E., and Chynoweth, A. G. *Optical Fiber Telecommunications*, Ch 11. New York: Academic Press, Inc., 1979.
22. Marcuse, D. *Principles of Optical Fiber Measurements*, Ch 5. New York: Academic Press, 1981.
23. Olshansky, R., and Keck, D. B. "Pulse Broadening in Graded-Index Fibers." *Appl. Optics*, 15, 1976, pp. 483–491.
24. Midwinter, J. E. *Optical Fibers for Transmission*, Ch 5 and 6. New York: John Wiley & Sons, Inc., 1979.
25. Marcuse, D. *Theory of Dielectric Optical Waveguides*, Ch 2. New York: John Wiley & Sons, Inc., 1974.

26. Gaskill, J. D. *Linear Systems, Fourier Transforms and Optics*, Ch 3. New York: John Wiley & Sons, Inc., 1978.
27. CSELT Technical Staff (Torino), *Optical Fiber Communication*, Part I, Ch 2. New York: McGraw-Hill Book Co., 1981.
28. Danielson, B. L., Day, G. W., Franzen, D. L., Kim, E. M., and Young, M. *Optical Fiber Characterization: Backscatter, Time Domain Bandwidth, Refracted Near Field*, and *Interlaboratory Comparisons*. NBS Special Publications 637, Vol. I (1982).
29. Personick, S. D. *Optical Fiber Transmission Systems*. New York: Plenum Press, 1982.

23 Analog Communications

Revised by
Michael B. Pursley

Modulation is a process whereby certain characteristics of a wave (often called a carrier) are varied or selected in accordance with a message signal. Modulation can be divided into continuous modulation, in which the modulated wave is always present, and pulsed modulation, in which no signal is present between pulses. Digital data modulation is discussed in Chapter 24.

PART 1—CONTINUOUS MODULATION

In continuous modulation* the modulated carrier can be given by the expression $s(t) = A(t) \cos\theta(t)$, where $A(t)$ is the *instantaneous amplitude* and $\theta(t)$ is the *instantaneous phase*. For a sinusoidal carrier of angular frequency ω_c, this expression reduces to $s(t) = A(t) \cos[\omega_c t + \phi(t)]$, where $\phi(t)$ is the carrier phase. When the instantaneous amplitude $A(t)$ is varied linearly by the message function and the carrier phase is constant, the process is called *amplitude modulation*; when the carrier phase angle $\phi(t)$ is modulated by the message function, the process is called *angular or phase modulation.*

The concept of a rotating vector can be used to represent a sinusoidal vector modulated in both amplitude and phase as shown in Fig. 1, where $s(t)$ is represented as the projection of a rotating vector on a fixed reference axis.

$$s(t) = A(t)\cos[\omega_c t + \phi(t)] = \mathrm{Re}(A(t)\exp\{j[\omega_c t + \phi(t)]\})$$

In these expressions, $A(t)$ represents the envelope of the modulated carrier, and $\phi(t)$ is the modulated phase. The vector rotates with an instantaneous angular frequency $\omega_i(t)$ given by

$$\omega_i(t) = \omega_c + [d\phi(t)/dt]$$

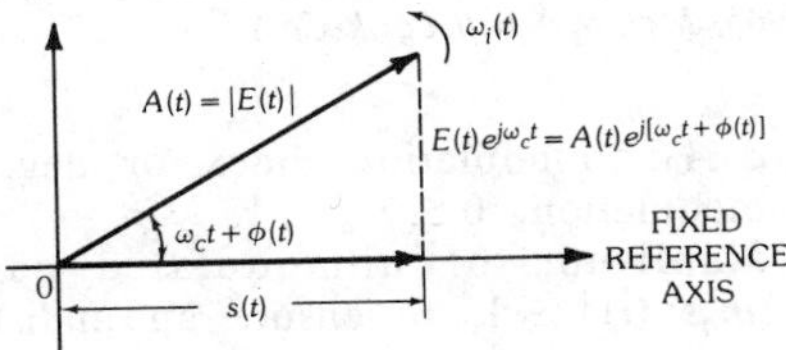

Fig. 1. Fixed-reference vector diagram. (*From P. F. Panter,* Modulation, Noise, and Spectral Analysis, *Fig. 2-7, © 1965, McGraw-Hill Book Co.*)

In amplitude modulation, only the amplitude changes, and the general expression reduces to

$$s(t) = \mathrm{Re}[A(t)\exp(j\phi_0)\cdot\exp(j\omega_c t)]$$

* P. F. Panter, *Modulation, Noise, and Spectral Analysis* (New York: McGraw-Hill Book Co., 1965; Ch. 5 and 6).

In phase modulation, only the phase changes so that

$$s(t) = \mathrm{Re}\{A_c\exp[j\phi(t)]\cdot\exp(j\omega_c t)\}$$

where A_c is constant.

ANALYTIC SIGNAL REPRESENTATION OF MODULATED WAVEFORMS

A real signal

$$s(t) = A(t)\cos[\omega_c t + \phi(t)]$$

may be expressed either as

$$s(t) = \mathrm{Re}(A(t)\exp\{j[\omega_c t + \phi(t)]\})$$

or as

$$s(t) = \mathrm{Re}[\psi(t)]$$

where $\psi(t)$ is the analytic signal defined by

$$\psi(t) = s(t) + j\hat{s}(t)$$

The function $s(t)$ is the Hilbert transform of $s(t)$, namely

$$\hat{s}(t) = \pi^{-1}\int \frac{s(t)}{t-\tau}\,d\tau$$

Basically, the analytic signal $\psi(t)$ is a complex function of a real variable whose real and imaginary parts form a Hilbert pair. The analytic signal is simply a formalized version of the "rotating vector" discussed above. If $S(j\omega)$ is the Fourier transform of $s(t)$, then $\Psi(j\omega)$, the Fourier transform of $\psi(t)$, can be written in terms of $S(j\omega)$ as

$$\begin{aligned}\Psi(j\omega) &= 2S(j\omega) && \omega > 0\\ &= S(j\omega) && \omega = 0\\ &= 0 && \omega < 0\end{aligned}$$

Also, $\hat{S}(j\omega)$, the Fourier transform of $s(t)$, is given by the equation

$$\hat{S}(j\omega) = -j(\mathrm{sgn}\omega)S(j\omega)$$

where

$$\begin{aligned}\mathrm{sgn}x &= 1 && x > 0\\ &= 0 && x = 0\\ &= -1 && x < 0\end{aligned}$$

and sgn$x$ is the signum function.

AMPLITUDE MODULATION

In amplitude modulation, the frequency components of the modulating signal are translated to occupy a different position in the spectrum. It is essentially a multiplication process in which the time functions that describe the modulating signal and carrier are multiplied together. The following amplitude-modulation systems are discussed.

(A) Double-sideband suppressed carrier (DSB-SC), also called DSB
(B) Conventional amplitude modulation (AM)
(C) Vestigial sideband
(D) Single sideband (SSB)

Double Sideband (DSB)

In DSB modulation, the message signal $g(t)$, whose Fourier transform is $G(j\omega)$, is considered to have zero dc component. The product

$$e(t) = A_c g(t) \cos\omega_c t$$

represents a double-sideband suppressed-carrier signal, and A_c = amplitude of unmodulated carrier. The radio-frequency envelope follows the waveform of the modulating signal $g(t)$ as shown in Fig. 2. The spectral components of the DSB signal $e(t)$ are given by its Fourier transform

$$E(j\omega) = \tfrac{1}{2}G[j(\omega - \omega_c)] + \tfrac{1}{2}G[j(\omega + \omega_c)]$$

as shown in Fig. 3. Note that the upper and lower sidebands are translated symmetrically $\pm\omega_c$ about the origin.

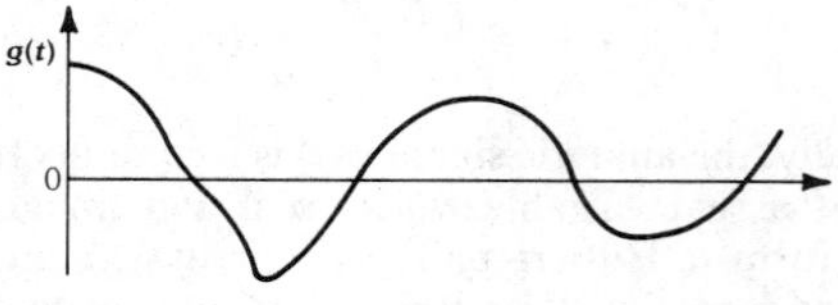

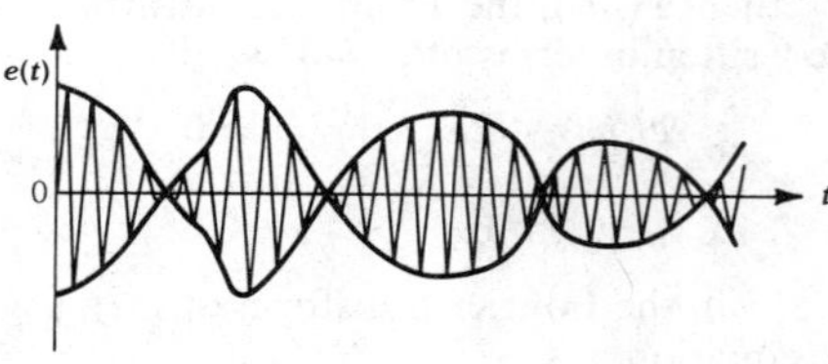

Fig. 2. Double-sideband waveforms. (*From P. F. Panter,* Modulation, Noise, and Spectral Analysis, *Fig. 5-3, © 1965, McGraw-Hill Book Co.*)

Conventional Amplitude Modulation (AM)

In amplitude modulation, a dc term is added to the modulating signal $g(t)$. The resulting waveform, shown in Fig. 4, is given by

$$e(t) = [A_0 + as(t)]\cos\omega_c t$$
$$= A_0\,[1 + m_a s(t)]\cos\omega_c t$$

where,

a = maximum amplitude of modulating function, $g(t) = as(t)$, $|s(t)| \leq 1$,

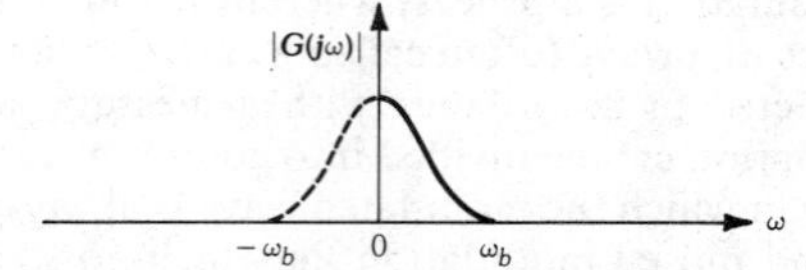

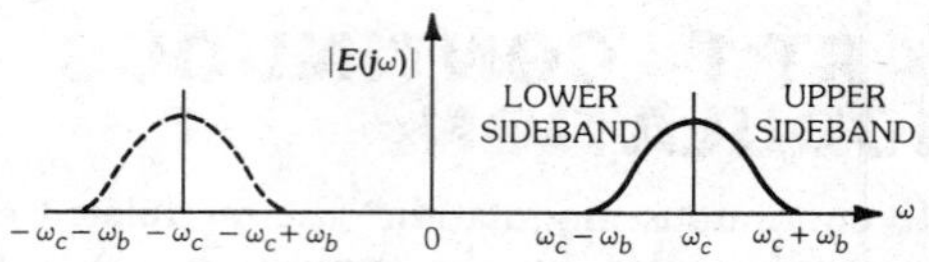

Fig. 3. Baseband signal and double-sideband spectra. (*From P. F. Panter,* Modulation, Noise, and Spectral Analysis, *Fig. 5-2, © 1965, McGraw-Hill Book Co.*)

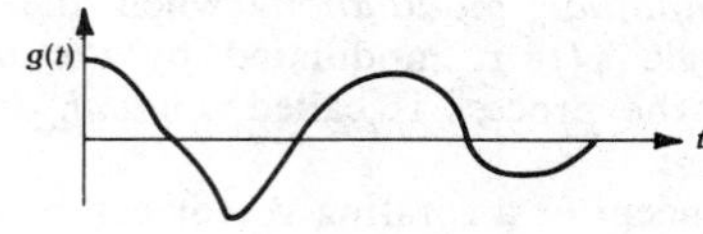

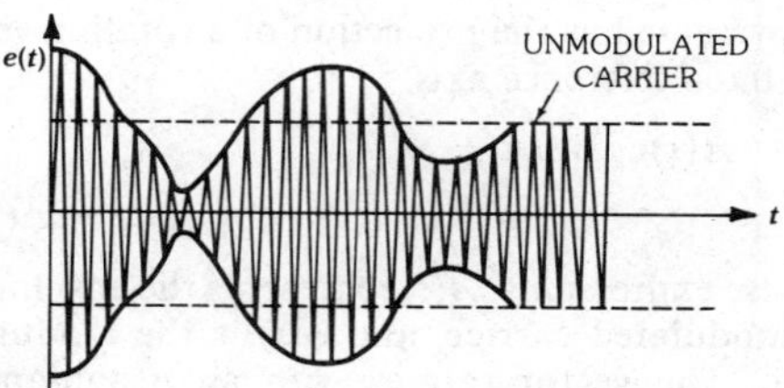

Fig. 4. Amplitude modulation. The modulating signal is at top and the modulated carrier at bottom. (*From P. F. Panter,* Modulation, Noise, and Spectral Analysis, *Fig. 5-4, © 1965, McGraw-Hill Book Co.*)

$m_a = a/A_0$ = modulation index or degree of modulation, $0 \leq m_a \leq 1$,

A_0 = amplitude of unmodulated carrier, $|m_a s(t)| \leq 1$, to ensure an undistorted envelope.

Vestigial Sideband

Vestigial-sideband modulation is derived from a DSB signal by passing the output of the product modulator through a filter whose transfer function is $H_v(j\omega)$, as shown in Fig. 5. The transfer function $H_v(j\omega)$ of the filter treats the two sidebands of the DSB signal in such a manner as to attenuate one sideband differently from the other. The process of vestigial-sideband modulation by the use of the filter network $H_v(j\omega)$ may be replaced by the

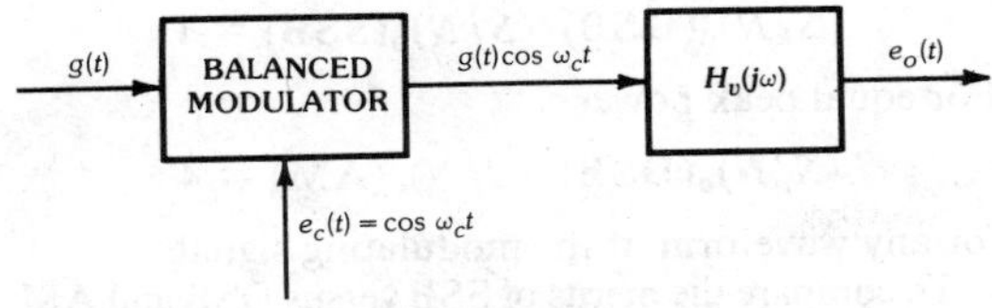

Fig. 5. Vestigial-sideband transmission system. (*From P. F. Panter,* Modulation, Noise, and Spectral Analysis, *Fig. 5-7, © 1965, McGraw-Hill Book Co.*)

equivalent vestigial system shown in Fig. 6, where the transfer functions $H_i(j\omega)$ and $H_q(j\omega)$ are given by

$$H_i(j\omega) = \tfrac{1}{2}\{H_v[j(\omega - \omega_c)] + H_v[j(\omega + \omega_c)]\}$$
$$H_q(j\omega) = (1/2j)\{H_v[j(\omega - \omega_c)] - H_v[j(\omega + \omega_c)]\}$$

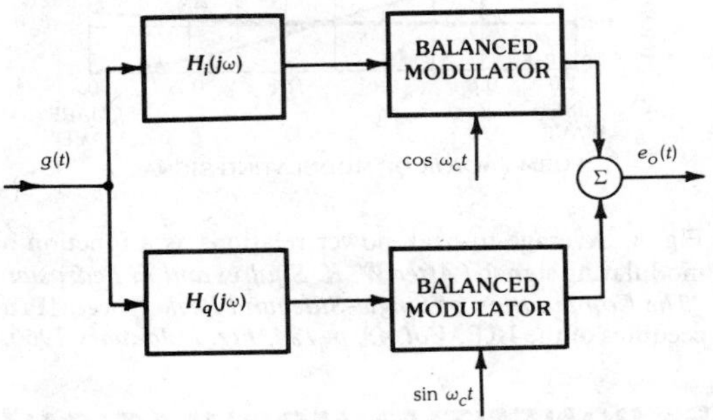

Fig. 6. Equivalent vestigial-sideband transmission system. (*From P. F. Panter,* Modulation, Noise, and Spectral Analysis, *Fig. 5-8, © 1965, McGraw-Hill Book Co.*)

Single Sideband (SSB)

Single-sideband transmission may be produced in the same manner as vestigial sideband by using a high-pass filter $H_s(j\omega)$ which completely eliminates all signals on one side of the carrier frequency. The transfer function $H_s(j\omega)$ of the ideal high-pass filter is defined by

$$H_s(j\omega) = [\tfrac{1}{2} + \tfrac{1}{2}\,\mathrm{sgn}(\omega - \omega_c)] + [\tfrac{1}{2} - \tfrac{1}{2}\,\mathrm{sgn}(\omega - \omega_c)]$$

where sgnω is the signum function. The output spectrum $E_s(j\omega)$ is given by

$$\begin{aligned} E_s(j\omega) &= H_s(j\omega)E(j\omega) \\ &= \tfrac{1}{2}G[j(\omega - \omega_c)]\,[\tfrac{1}{2} + \tfrac{1}{2}\,\mathrm{sgn}(\omega - \omega_c)] \\ &\quad + \tfrac{1}{2}G[j(\omega + \omega_c)]\,[\tfrac{1}{2} - \tfrac{1}{2}\,\mathrm{sgn}(\omega + \omega_c)] \end{aligned}$$

The SSB signal can also be regarded as the resultant of quadrature modulation of a carrier by a pair of signals in phase quadrature. The modulated wave

$$e_s(t) = s(t)\cos\omega_c t - \sigma(t)\sin\omega_c t$$

represents an upper-sideband signal with no spectral components below the carrier angular frequency ω_c, where $s(t)$ is an arbitrary message function and $\sigma(t)$ its harmonic conjugate.

This equation can be written in the form

$$\begin{aligned} e_s(t) &= [s^2(t) + \sigma^2(t)]^{1/2} \\ &\quad \cos\{\omega_c t + \tan^{-1}[\sigma(t)/s(t)]\} \\ &= \alpha(t)\cos[\omega_c t + \phi(t)] \end{aligned}$$

regarding the single-sideband signal as a hybrid amplitude-modulated and phase-modulated wave. The envelope $\alpha(t)$ and phase $\phi(t)$ are related by the analytic signal

$$\psi(t) = s(t) + j\sigma(t) = \alpha(t)\exp[j\phi(t)]$$

where $\sigma(t) = \hat{s}(t)$, the Hilbert transform of $s(t)$. The amplitude and phase of the complex signal $\psi(t)$ are identical to the envelope and phase of the single-sideband wave. The Fourier transform of the analytic signal $\psi(t)$ is

$$\begin{aligned} \Psi(j\omega) &= S(j\omega) + j\hat{S}(j\omega) \\ &= S(j\omega) + S(j\omega) = 2S(j\omega), \qquad \omega > 0 \\ &= S(j\omega) - S(j\omega) = 0, \qquad \omega < 0 \end{aligned}$$

Thus, a study of single sideband can be made through the analytic signal without reference to the arbitrary carrier frequency ω_c.

DEMODULATION, OR DETECTION, OF AMPLITUDE MODULATION

The process of separating the modulating signal from a modulated carrier is called demodulation or detection. In DSB or SSB detection, the detector must be supplied with a carrier wave that is synchronized with the wave used at the transmitter. This method of detection is called coherent or synchronous detection. In conventional amplitude-modulation systems, coherent detection is not necessary, and the modulating signal may be recovered by the use of envelope detection; e.g., the modulated carrier is applied to a half-wave rectifier whose output is then filtered to provide the desired modulating signal.

DSB Detection

In DSB reception, the incoming signal $e_r(t)$ is multiplied by a locally generated signal that is phase-synchronized with the carrier component of the received signal $e_r(t)$, as shown in Fig. 7. The detected output after filtering is given by

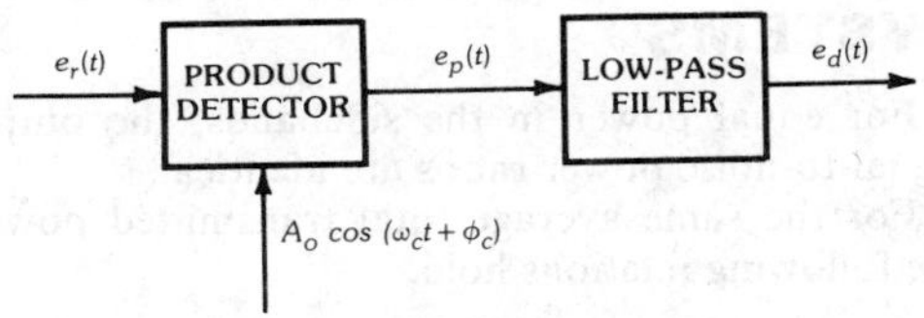

Fig. 7. Block diagram of double-sideband (dsb) receiver. (*From P. F. Panter,* Modulation, Noise, and Spectral Analysis, *Fig. 6-1, © 1965, McGraw-Hill Book Co.*)

$$e_d(t) = kg(t)\cos(\phi_c - \phi_0), \qquad k = \text{constant}$$

where $(\phi_c - \phi_0)$ represents the phase difference between the transmitted carrier and the locally generated oscillator. When the local carrier is in phase with the incoming carrier, the detected signal is maximum. The output signal-to-noise ratio $(S/N)_o$ is related to the input signal-to-noise ratio $(S/N)_i$ by the expression

$$\frac{(S/N)_o}{(S/N)_i} = 2\cos^2(\phi_c - \phi_0)$$

where the noise in each case is measured in a band occupied by the signal. This represents a maximum improvement of 3 decibels when the local oscillator is in phase with the incoming carrier.

AM Detection

Synchronous Detection:

$$\frac{(S/N)_o}{(S/N)_i} = \frac{2m_a^2\langle g\rangle^2(t)\cos^2(\phi_c - \phi_0)}{1 + m_a^2\langle g\rangle^2(t)}, \ |g(t)| \leq 1$$

where $\langle g\rangle^2(t)$ equals the mean-square value of the message function, which is maximum for $m_a = 1$ and $\phi_c = \phi_0$.

Envelope Detection: In case of a carrier much stronger than the noise (high input carrier-to-noise ratio), we have

$$\frac{(S/N)_o}{(S/N)_i} = \frac{2m_a^2\langle g\rangle^2(t)}{1 + m_a^2\langle g\rangle^2(t)}$$

which is identical to the case of synchronous detection with $\phi_c = \phi_0$.

In case of poor input carrier-to-noise ratio, the message function $g(t)$ may be lost in the noise, which results in a threshold effect. This effect exists only in envelope detection and does not exist if synchronous or coherent detection is used.

SSB Detection

$$(S/N)_o/(S/N)_i = \cos^2(\phi_c - \phi_0)$$

where the signal component of the output is measured by the correlation of the detected output with the transmitted signal.

COMPARISON OF AMPLITUDE-MODULATION SYSTEMS

For equal power in the sidebands, the output signal-to-noise power ratios are identical.

For the same average total transmitted power, the following relations hold.

$$(S/N)_o(\text{DSB})/(S/N)_o(\text{AM}) = 1 + r^{-1}$$

where r equals the ratio of the mean-square power of the message function to its peak power, and

$$(S/N)_o(\text{DSB})/(S/N)_o(\text{SSB}) = 1$$

For equal peak power

$$(S/N)_o(\text{DSB})/(S/N)_o(\text{AM}) = 4$$

for any waveform of the modulating signal.

To compare the merits of SSB versus DSB and AM on the basis of the signal-to-noise ratio, the waveform of the modulating signal must be specified. This is illustrated in Fig. 8 for a modulating signal $\sin^\nu x$, $0 \leq \nu \leq 1$.

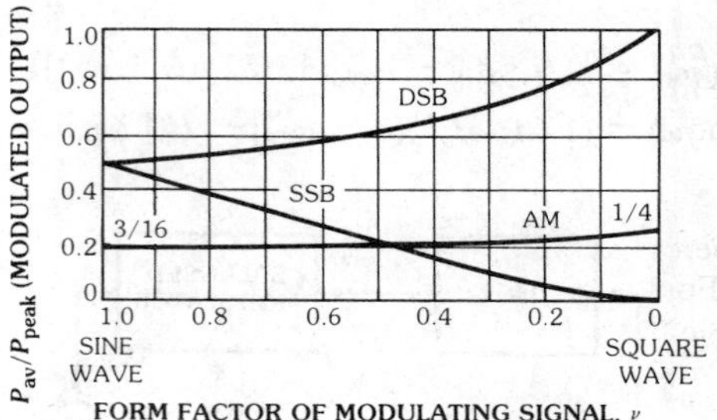

Fig. 8. Average-to-peak power relations as a function of modulating signal. (*After W. K. Squires and E. Bedrosian, "The Computation of Single-Sideband Peak Power,"* Proceedings of the IRE, *Vol. 48, p. 124, Fig. 2; January 1960.*)

EXPONENTIAL MODULATION

In exponential or angular modulation, the carrier analytic signal $A_c\exp[j(\omega_c t + \phi_c)]$ is multiplied by the transformed message function $\exp[j\psi(t)]$ to produce an angle-modulated carrier analytic signal.

$$\begin{aligned} e(t) &= \text{Re}\{A_c\exp[j(\omega_c t + \phi_c)]\cdot\exp[j\psi(t)]\} \\ &= \text{Re}\{A_c\exp[j\phi(t)]\} \end{aligned} \tag{1}$$

where,

A_c = amplitude of unmodulated carrier,
ω_c = angular frequency of unmodulated carrier,
ϕ_c = carrier phase angle,
$\phi(t) = [\omega_c t + \phi_c + \psi(t)]$,
= instantaneous phase angle modulated by the message function, $g(t)$.

Expanding Eq. 1 in powers of $\psi(t)$, we have

$$\begin{aligned} e(t) = \text{Re}\{A_c\exp[j(\omega_c t + \phi_c)] \times [1 + j\psi(t) \\ - (1/2!)\psi^2(t) - j(1/3!)\psi^3(t) + \cdots]\} \end{aligned}$$

When $|\psi(t)|_{\max} >> 1$, we have nonlinear modulation, since the carrier is multiplied by higher powers of $\psi(t)$. In case $|\psi(t)|_{\max} << 1$, the exponential modulation is approximately linear and is given by

$$e(t) \cong \text{Re}\{A_c[1 + j\psi(t)]\exp[j(\omega_c t + \phi_c)]\}$$

Note that for amplitude modulation we have

$$e_{\text{AM}}(t) = \text{Re}\{A_c[1 + m_a g(t)]\exp[j(\omega_c t + \phi_c)]\}$$

Expressing Eq. 1 in the real form, we obtain

$$e(t) = A_c[\cos\omega_c t + \phi_c + \psi(t)]$$

where for phase modulation

$$\psi(t) = m_p g(t),\ m_p = \text{constant}$$

and for frequency modulation

$$\psi(t) = m_f \int_0^t g(\tau)d\tau,\ m_f = \text{constant}$$

The instantaneous frequency $\omega_i(t)$ is defined by

$$\omega_i(t) = \left[\frac{d\phi(t)}{dt}\right] = \left[\omega_c + \frac{d\psi(t)}{dt}\right]$$

In *phase modulation,* the instantaneous phase of the modulated signal varies proportionally with the modulating signal $g(t)$

$$e_{PM}(t) = A_c \cos[\omega_c t + m_p g(t)]$$

where ϕ_c has arbitrarily been set to zero.

For single-tone sinusoidal modulation, $g(t) = \cos\omega_m t$, we have

$$e_{PM}(t) = A_c \cos(\omega_c t + m_p \cos\omega_m t)$$

where $m_p = \Delta\theta$, and the peak phase deviation is independent of ω_m.

The instantaneous frequency

$$\omega_i(t) = d\phi(t)/dt$$
$$= \omega_c - m_p\omega_m \sin\omega_m t$$

and the peak frequency deviation $\Delta\omega = m_p\omega_m$ is proportional to the modulating frequency ω_m.

In *frequency modulation,* the instantaneous frequency of the modulated signal is proportional to $g(t)$

$$\omega_i(t) = \omega_c + m_f g(t)$$

or

$$e_{FM}(t) = A_c \cos\left[\omega_c t + m_f \int_0^t g(\tau)d\tau\right]$$

For single-tone sinusoidal modulation

$$\omega_i(t) = \omega_c + \Delta\omega \cos\omega_m t$$
$$e_{FM}(t) = A_c \cos[\omega_c t + (m_f/\omega_m)\sin\omega_m t]$$

The peak frequency deviation $\Delta\omega \equiv m_f$ is independent of ω_m, whereas the peak phase deviation $\Delta\theta = \Delta\omega/\omega_m$ is inversely proportional to ω_m; $\Delta\theta$ (in radians) is the modulation index, often denoted by β. For broad-band application, $\Delta\omega << \omega_c$ and $\beta >> 1$.

Frequency Spectrum of Single-Tone Angular Modulation

Small Phase Deviation (Narrow-Band PM):

$$e(t) = A_c \cos(\omega_c t + \beta \sin\omega_m t), \qquad \beta << 1$$

$$e(t) \cong A_c(\cos\omega_c t - \beta \sin\omega_m t)$$

$$= \underbrace{A_c \cos\omega_c t}_{\text{carrier}} \underbrace{- \tfrac{1}{2}(A_c\beta)\cos(\omega_c - \omega_m)t}_{\text{lower sideband}} \underbrace{+ \tfrac{1}{2}(A_c\beta)\cos(\omega_c + \omega_m)t}_{\text{upper sideband}}$$

The corresponding equation for am is

$$e_{AM}(t) = A_c \cos\omega_c t + \tfrac{1}{2}(A_c m_a)\cos(\omega_c - \omega_m)t + \tfrac{1}{2}(A_c m_a)\cos(\omega_c + \omega_m)t$$

Large Phase Deviation (Wideband PM):

$$e(t) = A_c \cos(\omega_c t + \beta \sin\omega_m t), \quad \beta >> 1$$
$$= A_c[\cos\omega_c t \cos(\beta \sin\omega_m t) - \sin\omega_c t \sin(\beta \sin\omega_m t)]$$

$$= A_c[\cos\omega_c t \sum_{n=-\infty}^{\infty} J_n(\beta)\cos n\omega_m t - \sin\omega_c t \sum_{n=-\infty}^{\infty} J_n(\beta)\sin n\omega_m t]$$

The waveform for wideband modulation is given by the equation

$$e(t) = A_c\{J_0(\beta)\cos\omega_c t - J_1(\beta)[\cos(\omega_c - \omega_m)t - \cos(\omega_c + \omega_m)t] + J_2(\beta)[\cos(\omega_c - 2\omega_m)t + \cos(\omega_c + 2\omega_m)t] - J_3(\beta)[\cos(\omega_c - 3\omega_m)t - \cos(\omega_c + 3\omega_m)t] + \cdots\}$$

$$= A_c \sum_{n=-\infty}^{\infty} J_n(\beta)\cos(\omega_c + n\omega_m)t$$

as shown in Fig. 9.

In practical application, the required bandwidth is finite, for—beyond a certain frequency range from the carrier, depending on the magnitude of β—the sideband amplitudes, which are proportional to $J_n(\beta)$, are negligibly small (see Fig. 10). Note that at $\beta = 2.404$, $J_0(\beta) = 0$ and the carrier amplitude is zero.

The average power in an angle-modulated wave is constant

$$P = \tfrac{1}{2}(A_c^2) \sum_{-\infty}^{\infty} J_n^2(\beta)$$
$$= \tfrac{1}{2}(A_c^2)$$

Multitone Angle Modulation

For *two-tone angle modulation* with ω_1 *and* ω_2, the instantaneous frequency is given by

$$\omega_i(t) = \omega_c + \Delta\omega_{c1} t \cos\omega_1 t + \Delta\omega_{c2}\cos\omega_2 t$$

where $\Delta\omega_{c1}$ and $\Delta\omega_{c2}$ denote the corresponding frequency deviations of ω_c, and the FM signal is

$$e(t) = A_c \cos(\omega_c t + \beta_1 \sin\omega_1 t + \beta_2 \sin\omega_2 t)$$

where $\beta_1 = \Delta\omega_{c1}/\omega_1$ and $\beta_2 = \Delta\omega_{c2}/\omega_2$.

The spectral components are as follows.

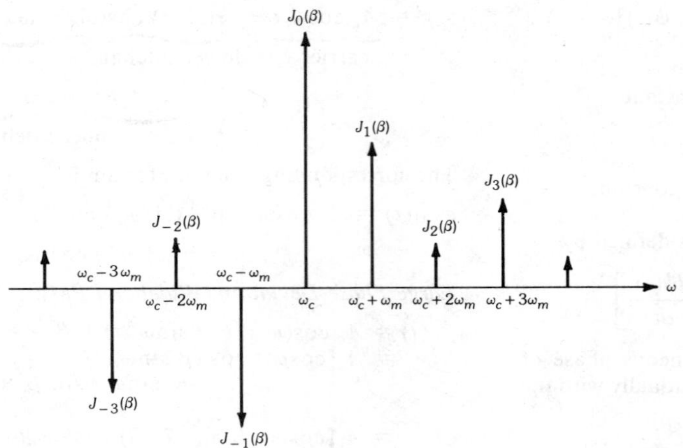

Fig. 9. Composition of fm wave into sidebands. (*From P. F. Panter,* Modulation, Noise, and Spectral Analysis, *Fig. 7-6, © 1965, McGraw-Hill Book Co.*)

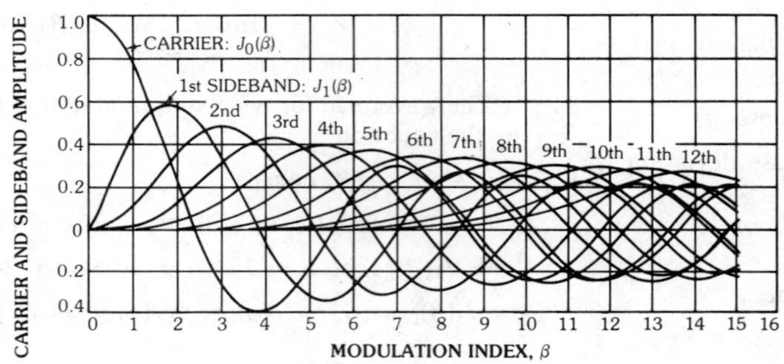

Fig. 10. Plot of Bessel functions of first kind as a function of argument β. (*From P. F. Panter,* Modulation, Noise, and Spectral Analysis, *Fig. 7-8, © 1965, McGraw-Hill Book Co.*)

(A) Carrier:

$$J_0(\beta_1)\, J_0(\beta_2) A_c \cos\omega_c t$$

(B) Sidebands due to ω_1:

$$J_n(\beta_1) J_0(\beta_2) A_c \cos(\omega_c \pm n\omega_1)t, \qquad n = 1, 2, 3$$

(C) Sidebands due to ω_2:

$$J_m(\beta_2) J_0(\beta_1) A_c \cos(\omega_c \pm m\omega_2)t, \qquad m = 1, 2, 3$$

(D) Beat frequencies at $\omega_c \pm n\omega_1 \pm m\omega_2$:

$$J_n(\beta_1) J_m(\beta_2) A_c \cos(\omega_c \pm n\omega_1 \pm m\omega_2)t$$

Spectral Distribution of an FM/FM Signal

Let ω_c = carrier angular frequency, ω_s = subcarrier, and ω_m = modulating angular frequency. The instantaneous frequency of the carrier wave is

$$\omega_i(t) = \omega_c + \Delta\omega \cos[\omega_s t + \phi_s + \beta_s \sin(\omega_m t + \phi_m)]$$

where,

$\Delta\omega$ = peak frequency deviation of carrier,
$\beta_s = \Delta\omega_s/\omega_m$ = peak phase deviation of subcarrier.

The spectral distribution is given by

$$e(t) = A_c \sum_{p=-\infty}^{\infty} \sum_{q=-\infty}^{\infty} J_p(\beta) J_q(p\beta_s) \times \cos[(\omega_c + p\omega_s + q\omega_m)t + \phi_c + p\phi_s + q\phi_m]$$

where $\beta = \Delta\omega/\omega_s$ = peak phase deviation of carrier.

Bandwidth Considerations in Multitone FM

An estimate of the IF bandwidth required for transmission of FM carrier by a complex modulating signal is given by

$$\beta_{IF} = 2(\Delta F + 2f_m) = 2\Delta F(1 + 2/\beta)$$

where,

ΔF = peak frequency deviation for the system,
f_m = highest baseband frequency (see Fig. 11).

Signal-to-Noise Improvement in FM Systems

The performance of a conventional FM receiver in the presence of random fluctuation noise is commonly judged on the basis of the variation of the output signal-to-noise $(S/N)_o$ power ratio as a function of the carrier-to-noise power ratio $(C/N)_i$ measured at the input to the limiter. This relationship is

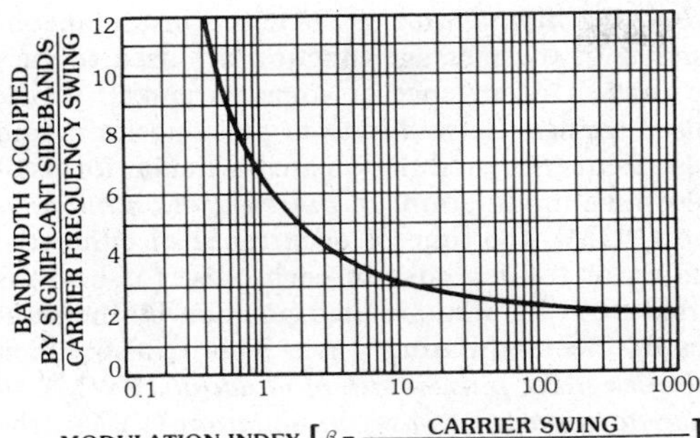

Fig. 11. Significant bandwidth (normalized) vs modulation index β. (*From C. E. Tibbs and G. G. Johnstone,* Frequency Modulation Engineering, *John Wiley & Sons, Inc., New York. Courtesy of Chapman & Hall, Ltd., London.*)

shown graphically in Fig. 12. The threshold of full improvement occurs when $(C/N)_i$ is about 12 decibels. For all values of the carrier greater than the threshold, the output $(S/N)_o$ is proportional to the input $(C/N)_i$. The signal-to-noise improvement ratio for a single-channel FM system is given by

$(S/N)_o/(C/N)_i = (\Delta\Phi)^2$, using a phase detector

where $\Delta\Phi$ = peak phase deviation, and

$$(S/N)_o/(C/N)_i = 3(\Delta F/f_m)^2$$
$$= 3\beta^2, \text{ using a frequency discriminator}$$

where,

ΔF = peak frequency deviation,
f_m = highest modulating frequency.

The signal-to-noise improvement ratio for a particular channel of a multiplex system is given by

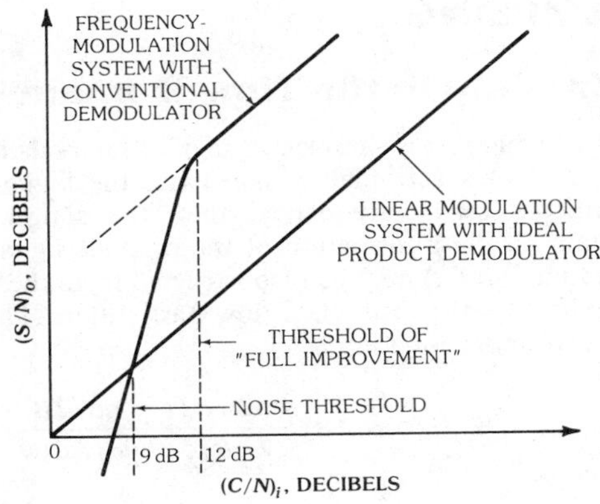

Fig. 12. Noise performance of conventional fm receiver. (*From P. F. Panter,* Modulation, Noise, and Spectral Analysis, *Fig. 14-2, © 1965, McGraw-Hill Book Co.*)

$$(S/N)_o/(C/N)_i = (B/B_c)(\Delta F_m/f_n)^2$$

where,

$2B$ = IF bandwidth,
B_c = channel bandwidth,
ΔF_m = peak channel frequency deviation,
f_n = midband channel frequency in the nth channel.

In the nonlinear region, when the noise is larger than the carrier there exists a signal-suppression effect, the average amplitude of the discriminator output is reduced, and for $(C/N)_i \ll 1$ we have $(S/N)_o \propto (C/N)_i^2$.

Signal-to-Noise Improvement Through De-Emphasis

The $(S/N)_o$ ratio of the high frequency end of the baseband can be increased by passing the modulating signal (at the transmitting end) through a pre-emphasis network (Figs. 13A and B), which em-

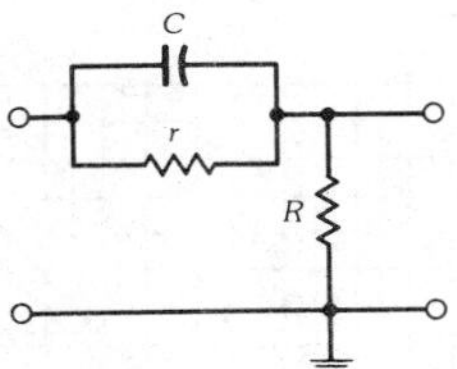

(A) Pre-emphasis network ($r \gg R$, $rC = 75\ \mu s$).

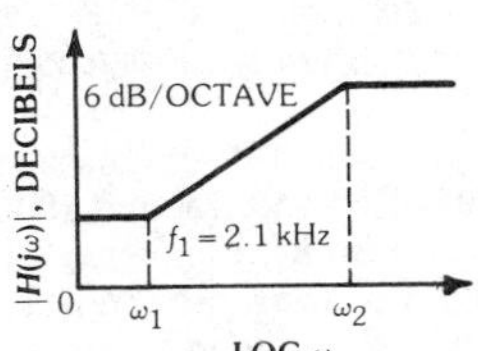

(B) Asymptotic response ($\omega_1 = 1/rC$, $\omega_2 \doteq 1/RC$).

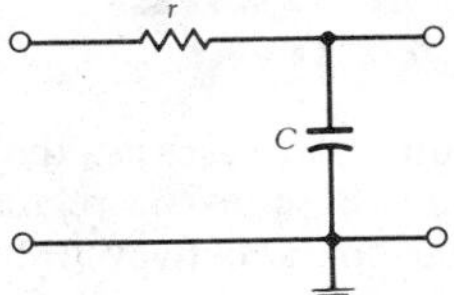

(C) De-emphasis network ($rC = 75\ \mu s$).

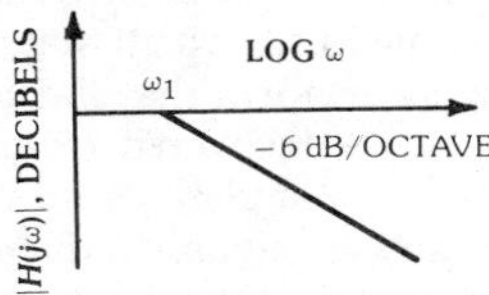

(D) Asymptotic response ($f_1 = 2.1$ kHz).

Fig. 13. Pre-emphasis and de-emphasis networks. (*From P. F. Panter,* Modulation, Noise, and Spectral Analysis, *Figs. 14-6 and 14-7, © 1965, McGraw-Hill Book Co.*)

phasizes the higher signal frequencies, and then passing the output of the discriminator through a de-emphasis network (Figs. 13C and D) to restore the original signal-power distribution. Typical pre-emphasis and de-emphasis circuit responses for general time constants τ are shown in Fig. 14.

The improvement factor ρ_{FM} is given by

$$\rho_{FM} = \frac{(2\pi f_m \tau)^3}{3(2\pi f_m \tau - \tan^{-1} 2\pi f_m \tau)}$$

where f_m denotes the highest baseband frequency.

For narrow-band FM

$$\rho_{FM} \to 1$$

For wideband FM, f_m is large, and

$$\rho_{FM} \to (2\pi f_m \tau)^2/3$$

The mean $(S/N)_o$ ratio for FM with pre-emphasis is given by

$$(S/N)_o = \rho_{FM} \cdot 3\beta^2 (C/N)_i$$

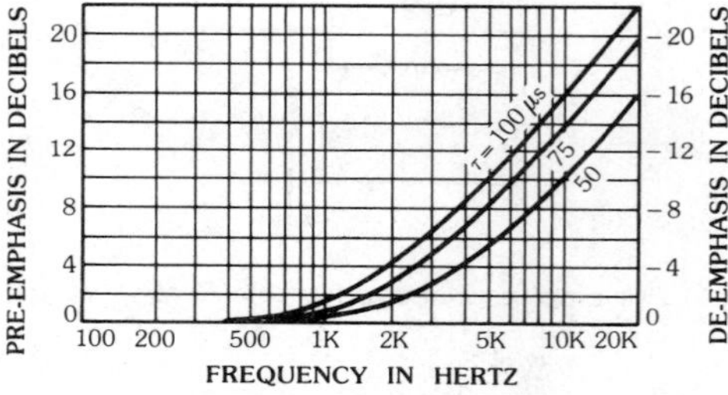

Fig. 14. Pre-emphasis and de-emphasis circuit response, for time constants of $\tau = 50$, 75, and 100 μs. (*From C. E. Tibbs and G. G. Johnstone,* Frequency Modulation Engineering, *John Wiley & Sons, Inc., New York. Courtesy of Chapman & Hall, Ltd., London.*)

PART 2—PULSE MODULATION

In pulse-modulation systems, the unmodulated carrier is usually a series of regularly recurrent pulses; information is conveyed by modulating some parameter of the transmitted pulses such as the amplitude, duration, time of occurrence, or shape of pulse. This type of modulation is based on the "sampling principle," which states that a continuous message waveform that has a spectrum of finite width could be recovered from a set of discrete instantaneous samples whose rate is higher than twice the highest signal frequency. This discrete set of periodic samples of the message function is used to modulate some parameter of the carrier pulses. In *pulse-amplitude modulation* (PAM), the series of periodically recurring pulses is modulated in amplitude by the corresponding instantaneous samples of the message function. In *pulse-time modulation* (PTM), the instantaneous samples of the message function are used to vary the time of occurrence of some parameter of the pulsed carrier. Pulse-duration, pulse-position, and pulse-frequency modulation are particular forms of pulse-time modulation. In *pulse-duration modulation* (PDM), the time of occurrence of either the leading or trailing edge of each pulse (or both) is varied from its unmodulated position by the samples of the modulating wave. This is also called *pulse-length* or *pulse-width modulation* (PWM), In *pulse-position (or phase) modulation* (PPM), the samples of the modulating wave are used to vary the position in time of a pulse, relative to its unmodulated time of occurrence. Pulse-position modulation is essentially the same as PDM, except that the variable edge is now replaced by a short pulse. In *pulse-frequency modulation* (PFM), the samples of the message function are used to modulate the frequency of the series of carrier pulses.

The pulse-modulation systems enumerated so far are examples of uncoded pulse systems. In *pulse-code modulation* (PCM), the modulating signal waveform is sampled at regular intervals as in conventional pulse modulation. However, in PCM, the samples are first quantized into discrete steps; i.e., within a specified range of expected sample values, only certain discrete levels are allowed, and these are transmitted over the system by means of a code pattern of a series of pulses.

Another example of a code-modulation system is *delta modulation.* As in PCM, the range of signal amplitudes is quantized, and binary pulses are produced at the sending end at regular intervals. However, in delta-modulation systems, instead of the absolute quantized signal amplitude being transmitted at each sampling, the transmitted pulses carry the information corresponding to the derivative of the amplitude of the modulating signal.

SAMPLING

Sampling in the Time Domain

If a signal $f(t)$ is sampled at regular intervals of time and at a rate higher than twice the highest significant signal frequency, then the samples contain all the information of the original signal. The function $f(t)$ may be reconstructed from these samples by the use of a low-pass filter. The reconstruction equation is

$$f(t) = \alpha \sum_{n=-\infty}^{\infty} f(n\alpha/2B) \frac{\sin 2\pi B(t - n\alpha/2B)}{2\pi B(t - n\alpha/2B)},$$

$$0 < \alpha \leq 1$$

where $f(t)$ is band-limited to B hertz, and the samples are taken at sampling intervals $\alpha/2B$ seconds apart.

Sampling in the Frequency Domain

A time-limited signal $f(t)$ that is zero outside the range $t_1 < t < t_2$ is completely determined by the values of the spectrum function $F(j\omega)$ at the angular-frequency sampling points given by

$$\omega_n = n[2/(t_2 - t_1)]$$

The function $f(t)$ expressed in terms of its sampling values in the frequency domain is given by the reconstruction equation

$$f(t) = \sum_{n=-\infty}^{\infty} (t_2 - t_1)^{-1} F\left(j\,\frac{2\pi n}{t_2 - t_1}\right) \times \exp[j2\pi nt/(t_2 - t_1)]$$

Sampling of a Bandpass Function (B_0, $B_0 + B$)

The reconstruction equation for $f(t)$ in terms of its sampled values is

$$f(t) = 2BT \sum_{n=-\infty}^{\infty} f(nT)\,\frac{\sin \pi B(t - nT)}{\pi B(t - nT)} \times \cos 2\pi B_c(t - nT)$$

where $B_c = B_0 + (B/2)$, the center frequency of the band-pass signal, and the permissible values of T are given by

$$m/2B_0 \leq T \leq [(m+1)/2(B_0 + B)],\ m = 0, 1, 2, \cdots$$

provided $B_0 \neq 0$.

The minimum sampling frequency for a band-limited signal of width B is illustrated in Fig. 15.

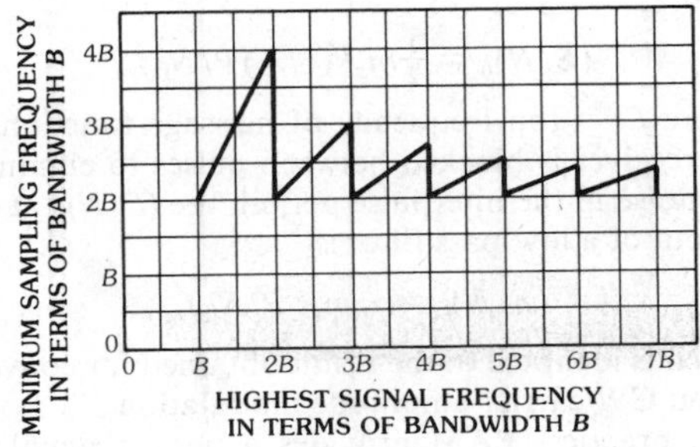

Fig. 15. Minimum sampling frequency for band of width *B*. (*From P. F. Panter,* Modulation, Noise, and Spectral Analysis, *Fig. 17-13, © 1965, McGraw-Hill Book Co.*)

PULSE-AMPLITUDE MODULATION (PAM)

In PAM, the samples of the message function are used to amplitude modulate the successive carrier pulses. When the modulated pulses follow the amplitude variation of the sampled time function during the sampling interval, the process is called *natural sampling* or *top sampling*. In contrast with natural sampling is instantaneous or square-topped sampling, where the amplitude of the pulses is determined by the instantaneous value of the sampled time function corresponding to a single instant (i.e., center or edge) of the sampling time interval. Pulse-amplitude modulation can be instrumented by two distinct methods. The first produces a variation of the amplitude of a pulse sequence about a fixed nonzero value or pedestal and constitutes double-sideband amplitude modulation (Figs. 16A and B). In the second method, the pedestal is zero, and the output signal consists of double-polarity modulated pulses and constitutes double-sideband suppressed-carrier modulation (Figs. 16C and D).

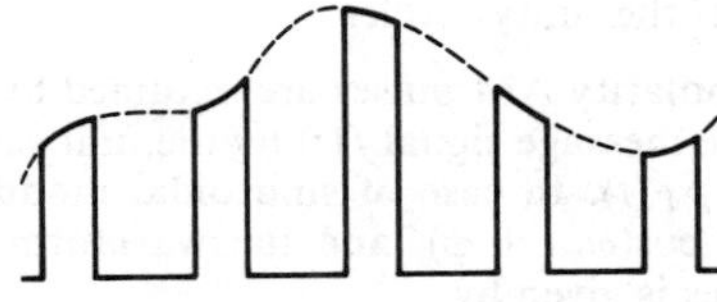

(A) Single-polarity pulses.

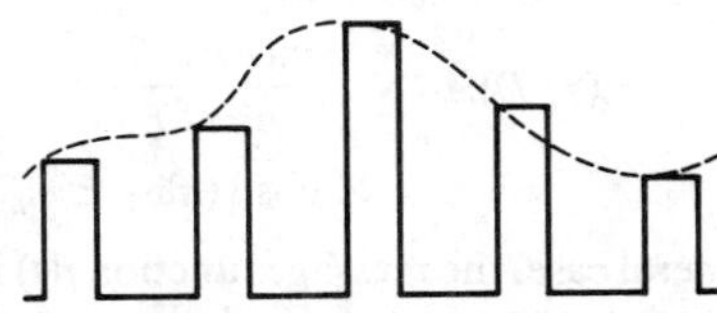

(B) Single-polarity flat-top pulses.

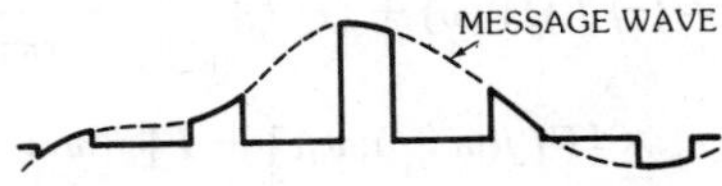

(C) Double-polarity pulses.

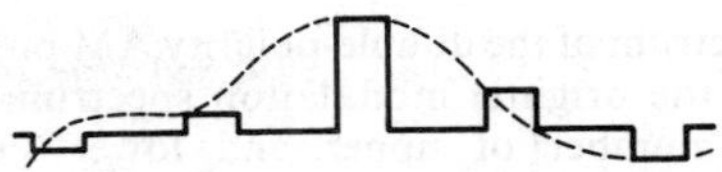

(D) Double-polarity flat-top pulses.

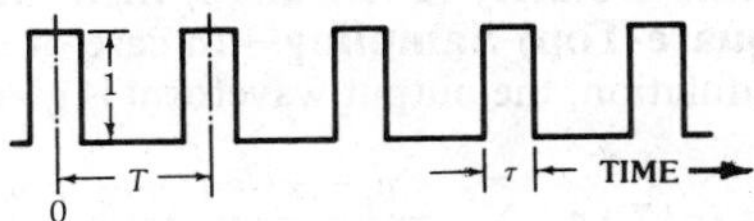

(E) Unit sampling function.

Fig. 16. Various shapes of amplitude-modulated pulses. (*From H. S. Black,* Modulation Theory. *Courtesy D. Van Nostrand Co., Inc., Princeton, N. J.*)

Spectra of Amplitude-Modulated Pulses

Double-Polarity AM Pulses, Natural (or Top) Sampling—In the process of natural sampling (or exact scanning), the modulated pulses follow the sampled time function during the sampling interval. The unit sampling function (Fig. 16E) consists of a train of unmodulated periodic pulses of unit amplitude given by

$$p_T(t) = (\tau/T) \sum_{n=-\infty}^{\infty} \frac{\sin(n\pi\tau/T)}{n\pi\tau/T} \exp(jn\omega_0 t)$$

where,

$\omega_0 = 2\pi f_0 = 2\pi/T$ is the fundamental angular frequency of the pulse train,
τ is the duration of the pulse,
τ/T is the duty cycle.

Double-polarity AM pulses are obtained by multiplying the message signal $f(t)$ by the unit sampling function $p_T(t)$. In case of sinusoidal modulation, $f(t) = A\cos(\omega_m t + \phi)$, and the waveform of the AM pulses is given by

$$f_{s1}(t) = f(t)p_T(t)$$
$$= (\tau/T)A\cos(\omega_m t + \phi)$$
$$+ (\tau/T)A \sum_{n=1}^{\infty} \frac{\sin(n\pi\tau/T)}{n\pi\tau/T}$$
$$\times \cos[(n\omega_0 \pm \omega_m)t \pm \phi]$$

In the general case, the message function $f(t)$ is band-limited, and its spectrum is $F(j\omega)$. The output spectrum is

$$F_{s1}(j\omega) = (\tau/T)F(j\omega) + (\tau/T) \sum_{n=1}^{\infty} \frac{\sin(n\pi\tau/T)}{n\pi\tau/T}$$
$$\times \{F[j(\omega - n\omega_0)] + F[j(\omega + n\omega_0)]\}$$
$$= (\tau/T) \sum_{n=-\infty}^{\infty} \frac{\sin(n\omega_0\tau/2)}{n\omega_0\tau/2} F[j(\omega - n\omega_0)]$$

The spectrum of the double-polarity AM pulses consists of the original modulation spectrum and an infinite number of upper and lower sidebands around ω_0 and its harmonics.

Double-Polarity AM Pulses, Instantaneous (or Square-Top) Sampling—In case of sinusoidal modulation, the output waveform is given by

$$f_{s2}(t) = (\tau/T)A \sum_{n=-\infty}^{\infty} \frac{\sin[\pi(\tau/T)(n\omega_0 + \omega_m)/\omega_0]}{[\pi(\tau/T)(n\omega_0 + \omega_m)/\omega_0]}$$
$$\times \cos[(n\omega_0 + \omega_m)(t - \tfrac{1}{2}\tau) + \phi]$$

In the general case, the output spectrum is

$$F_{s2}(j\omega) = (\tau/T) \frac{\sin(\omega\tau/2)}{\omega\tau/2} \sum_{n=-\infty}^{\infty} F[j(\omega - n\omega_0)]$$

Single-Polarity AM Pulses, Natural Sampling—For sinusoidal modulation

$$f_{s3}(t) = [1 + m_a\cos(\omega_m t + \phi)]$$
$$\times (\tau/T) \sum_{n=-\infty}^{\infty} \frac{\sin(n\pi\tau/T)}{n\pi\tau/T} \exp(jn\omega_0 t)$$
$$= p_T(t) + f_{s1}(t)$$

where m_a is the modulation index. In the general case

$$F_{s3}(j\omega) = P(j\omega) + F_{s1}(j\omega)$$

where $P(j\omega)$ is the Fourier transform of $p_T(t)$.

Single-Polarity AM Pulses, Instantaneous Scanning—For sinusoidal modulation

$$f_{s4}(t) = p_T(t) + f_{s2}(t)$$

In the general case

$$F_{s4}(j\omega) = P(j\omega) + F_{s2}(j\omega)$$

Signal-to-Noise Ratio in PAM

The signal to noise ratio for PAM is

$$(S/N)_i = \tfrac{1}{2} m_a^2 P/N_0 B$$

where,

P = average power of unmodulated radio-frequency pulse train,
N_0 = noise-power density in watts/hertz,
B = channel (RF) bandwidth.

Also

$$(S/N)_o = \tfrac{1}{2} m_a^2 (\tau/T) P/N_0 f_m$$

where f_m = top frequency of message function. If the receiver is blocked between pulses to eliminate the noise in the interpulse period, the $(S/N)_o$ at the output of a low-pass filter is

$$(S/N)_o = \tfrac{1}{2} m_a^2 P/N_0 f_m$$

which is identical to the result obtained for conventional CW carrier amplitude modulation.

In practice, PAM provides a poorer signal-to-noise ratio than conventional AM, because the receiver is unblocked for rather longer than the pulse-duration time owing to the sloping sides of the pulse.

PULSE-TIME MODULATION (PTM)

The improvement in signal-to-noise ratio obtained by the use of time-modulated pulses of constant amplitude instead of amplitude-modulated pulses led to the development of systems using pulse-duration and pulse-position modulation. The sampling associated with pulse modulation may be either natural or uniform (periodic). Natural sampling may be defined as a process of sampling in which the time of sampling coincides with the time of appearance of the time-modulated pulse as shown in Fig. 17A. In the process of natural sampling, the pulse duration τ_n corresponds to the value of the modulating signal $M(t_n)$ at that instant, and consequently the sampling intervals t_n are not equal but depend on the modulation level. Uniform sampling may be defined as a process of sampling in which the variation in the parameter of the pulse is proportional to the modulating signal at uniformly spaced sampling times. This is illustrated in Fig. 17B, where the width of the pulses is proportional to the modulating values $M(t_n)$ which are sampled at equal intervals $t_n = nT_r$ and are independent of the modulation process.

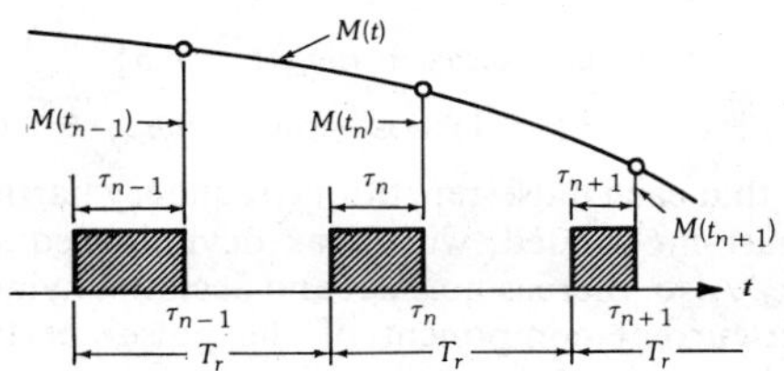

(A) Natural sampling.

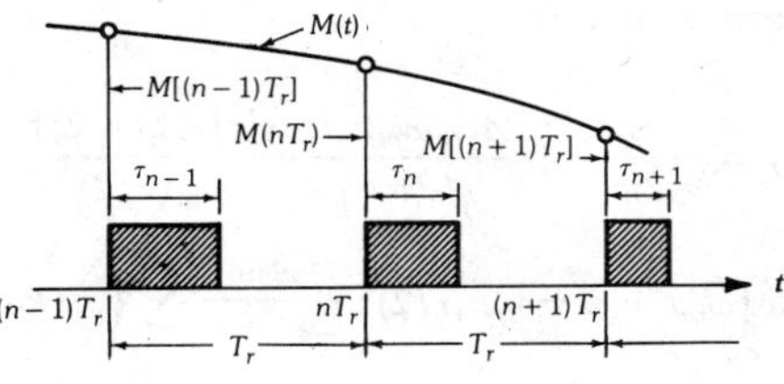

(B) Uniform sampling.

Fig. 17. Pulse-duration modulation. (*From P. F. Panter, Modulation, Noise, and Spectral Analysis, Fig. 18-14, © 1965, McGraw-Hill Book Co.*)

Spectra of Time-Modulated Pulses

The spectra of PTM pulses can be derived with reference to Fig. 18, where the two cosine waves, A and B, of angular frequency ω_r are displaced relative to each other by an amount τ, the width of the unmodulated pulse. The positive and negative steps that give rise to the pulse train are assumed to occur at the peaks of waveforms A and B, respectively. In the absence of modulation, the time of occurrence of the positive and negative steps is given by

$$\omega_r(t + \tau/2) = 2n\pi$$

and

$$\omega_r(t - \tau/2) = 2n\pi$$

With natural modulation, the time of occurrence of the positive and negative steps is given by

$$\omega_r(t + \tau/2) + \beta \sin(\omega_m t + \phi) = 2n\pi$$

and

$$\omega_r(t - \tau/2) + \beta \sin(\omega_m t + \phi) = 2n\pi$$

Similarly, with uniform modulation, the time of occurrence or the position of the leading and trailing edges of the pulses is determined by

$$\omega_r(t + \tau/2) + \beta \sin(\omega_m t + \phi) = 2n\pi$$

$$\omega_r(t - \tau/2) + \beta \sin(\omega_m \overline{t - \tau} + \phi) = 2n\pi$$

where,

ω_m is the modulating frequency,

β is the modulation index.

Pulses whose moments of occurrence satisfy these equations are said to be time modulated. In pulse-frequency modulation, $\beta = \Delta\omega/\omega_m$, and in pulse-phase (or pulse-position) modulation, β is constant independent of the modulating frequency.

Pulse-Frequency Modulation, Natural Sampling—A useful expression for an infinite train of unmodulated pulses is in the form

$$p_T(t) = (A/2\pi j) \sum_{k=-\infty}^{\infty} k^{-1} \{\exp[jk\omega_r(\tau/2)] - \exp[-jk\omega_r(\tau/2)]\} \exp(jk\omega_r t)$$

where,

A is the amplitude of the pulses,

ω_r is the pulse repetition frequency.

Frequency modulation can be taken into account by substituting for $\omega_r\tau/2$ in the expressions for the leading and trailing edges in the last equation, the expressions

$$\tfrac{1}{2}(\omega_r\tau) + \beta \sin(\omega_m t + \phi)$$

and

$$\tfrac{1}{2}(\omega_r\tau) - \beta \sin(\omega_m t + \phi)$$

The frequency-modulated pulse train is then

$$\begin{aligned} p_m(t) &= (A/2\pi j) \\ &\times \sum_{k=-\infty}^{\infty} k^{-1}(\exp\{j[k\omega_r(\tau/2) + k\beta \sin(\omega_m t + \phi)]\} \\ &\quad - \exp\{-j[k\omega_r(\tau/2) - k\beta \sin(\omega_m t + \phi)]\}) \\ &\quad \times \exp(jk\omega_r t) \\ &= \frac{A\omega_r\tau}{2\pi} + \frac{A\omega_r\tau}{\pi} \sum_{k=1}^{\infty} \frac{\sin[k\omega_r(\tau/2)]}{k\omega_r(\tau/2)} \\ &\quad \times (J_0(k\beta) \cos k\omega_r t + \sum_{n=1}^{\infty} J_n(k\beta) \\ &\quad \times \{\cos[(k\omega_r + n\omega_m)t + n\phi] \\ &\quad + (-1)^n \cos[(k\omega_r - n\omega_m)t - n\phi]\}) \end{aligned}$$

This expression may be compared with that for the spectrum of a frequency-modulated continuous wave given by

$$e_{FM}(t) = AJ_0(\beta)\cos\omega_r t + A\sum_{n=1}^{\infty} J_n(\beta)\{\cos[(\omega_r + n\omega_m)t + n\phi] + (-1)^n\cos[(\omega_r - n\omega_m)t - n\phi]\}$$

The conclusions reached are as follows.

(A) With pulse-frequency modulation using natural sampling, the direct-current component of the pulse spectrum has no sideband of the modulating frequency.

(B) The kth harmonic of the pulse-repetition frequency is frequency modulated, the modulation index being $k\beta$.

Pulse-Frequency Modulation, Uniform Sampling—In this type of modulation, the displacement of waveform B of Fig. 18 from its unmodulated position at any instant of time t will depend on the value of the modulating voltage at $(t - \tau)$. The expression for the modulated pulse train becomes

$$p_m(t) = (A/2\pi j) \times \sum_{k=1}^{\infty} k^{-1}(\exp\{j[k\omega_r(\tau/2) + k\beta\sin(\omega_m t + \phi)]\} - \exp\{-j[k\omega_r(\tau/2) - k\beta\sin(\omega_m \overline{t - \tau} + \phi)]\}) \times \exp(jk\omega_r t)$$

$$= \frac{A\omega_r\tau}{2\pi} + A\left(\frac{\Delta\omega}{2\pi}\right)\tau\frac{\sin[\omega_m(\tau/2)]}{\omega_m(\tau/2)} \times \cos[\omega_m t + \phi - (\omega_m\tau/2)] + \frac{A\omega_r\tau}{\pi}\sum_{k=1}^{\infty}\left(J_0(k\beta)\frac{\sin[k\omega_r(\tau/2)]}{k\omega_r(\tau/2)} \times \cos k\omega_r t + \sum_{k=1}^{\infty} J_n(k\beta)\left\{\frac{\sin(k\omega_r - n\omega_m)(\tau/2)}{k\omega_r(\tau/2)} \times \cos[(k\omega_r + n\omega_m)t + n\phi - n\omega_m(\tau/2)] + (-1)^n\frac{\sin(k\omega_r - n\omega_m)(\tau/2)}{k\omega_r(\tau/2)} \times \cos[(k\omega_r - n\omega_m)t - n\phi + n\omega_m(\tau/2)]\right\}\right)$$

The conclusions reached are as follows.

(A) The direct-current component of the pulse spectrum has a sideband of the modulating frequency of amplitude

$$(A\Delta\omega\tau/2\pi)\{\sin[\omega_m(\tau/2)]/\omega_m(\tau/2)\}$$

Modulation can therefore be recovered by means of a low-pass filter.

(B) The upper and lower sidebands of the kth harmonic of the pulse-repetition frequency are not equal in amplitude, whereas in the case of natural sampling they are equal.

Pulse-Position (or Pulse-Phase) Modulation—The waveform of pulse-phase modulation can be directly derived from that for pulse-frequency modulation by substituting $\omega_r\tau_d$ for β, where $\omega_r\tau_d$ represents the peak phase deviation of waveforms A and B, which is constant independent of the modulation frequency ω_m. The resulting waveform is the following:

Natural sampling:

$$p_m(t) = \frac{A\omega_r\tau}{2\pi} + \frac{A\omega_r\tau}{\pi}\sum_{k=1}^{\infty}\frac{\sin[k\omega_r(\tau/2)]}{k\omega_r(\tau/2)} \times \left(J_0(k\omega_r\tau_d)\cos k\omega_r t + \sum_{n=1}^{\infty} J_n(k\omega_r\tau_d) \times \{\cos[(k\omega_r + n\omega_m)t + n\phi] + (-1)^n\cos[(k\omega_r - n\omega_m)t - n\phi]\}\right)$$

Note that each pulse-repetition-frequency harmonic is phase-modulated, with peak deviation equal to $k\omega_r\tau_d$. Also, there is no sideband accompanying the direct-current component of the pulse spectrum, and hence modulation cannot be recovered by means of a low-pass filter.

Uniform sampling:

$$p_m(t) = \frac{A\omega_r\tau}{2\pi} + \frac{A\omega_r\omega_m\tau_d\tau}{2\pi}\frac{\sin[\omega_m(\tau/2)]}{\omega_m(\tau/2)} \times \cos[\omega_m t + \phi - (\omega_m\tau/2)] + \frac{A\omega_r\tau}{\pi}\sum_{k=1}^{\infty}\left(J_0(k\omega_r\tau_d) \times \frac{\sin[(k\omega_r(\tau/2)]}{k\omega_r(\tau/2)}\cos k\omega_r t + \sum_{n=1}^{\infty} J_n(k\omega_r\tau_d) \times \left\{\frac{\sin(k\omega_r + n\omega_m)(\tau/2)}{k\omega_r(\tau/2)}\cos[(k\omega_r + n\omega_m)t + n\phi - n\omega_m(\tau/2)] + (-1)^n\frac{\sin(k\omega_r - n\omega_m)(\tau/2)}{k\omega_r(\tau/2)} \times \cos[(k\omega_r - n\omega_m)t - n\phi + n\omega_m(\tau/2)]\right\}\right)$$

This is an equation very similar to that for pulse-frequency modulation.

Signal-to-Noise Improvement Ratio in PTM

In PDM, the noise manifests itself as jitter in the leading and trailing edges of the recovered pulses,

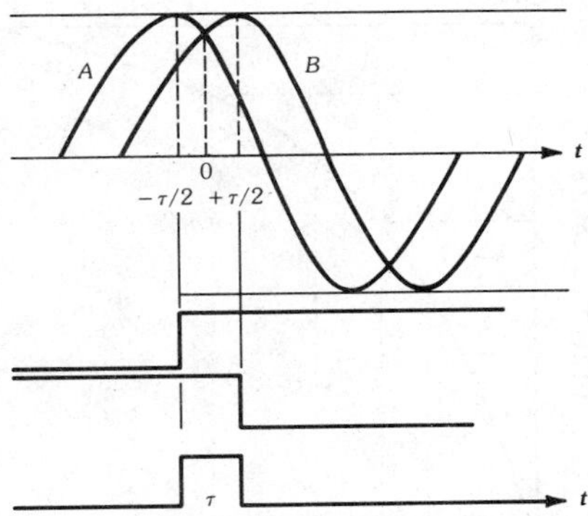

Fig. 18. Modulation process (modified). (*From P. F. Panter,* Modulation, Noise, and Spectral Analysis, *Fig. 17-14, © 1965, McGraw-Hill Book Co.*)

and the slopes of the pulse edges influence noise reduction. A PPM system is affected by noise in the same manner as a PDM system. For trapezoidal pulses (Figs. 19 and 20), the S/N power ratio at the demodulator output is

$$(S/N)_o = \frac{1}{2}(t_0/\tau_r)^2(A_c/\sigma)^2$$

The peak pulse power to mean noise power ratio is

$$(C/N)_i = (A_c/\sigma)^2$$

Hence

$$(S/N)_o = \frac{1}{2}(t_0/\tau_r)^2(C/N)_i$$

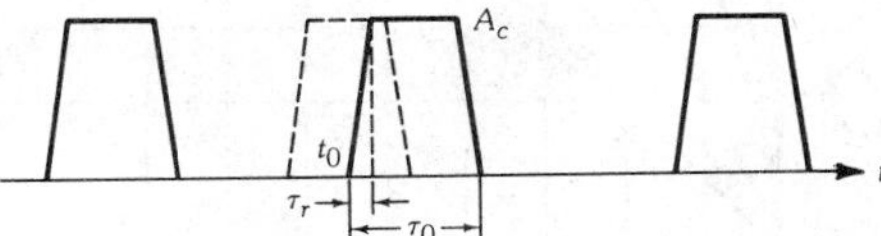

Fig. 19. Pulse-position modulation of trapezoidal pulses. (*From P. F. Panter,* Modulation, Noise, and Spectral Analysis, *Fig. 18-26, © 1965, McGraw-Hill Book Co.*)

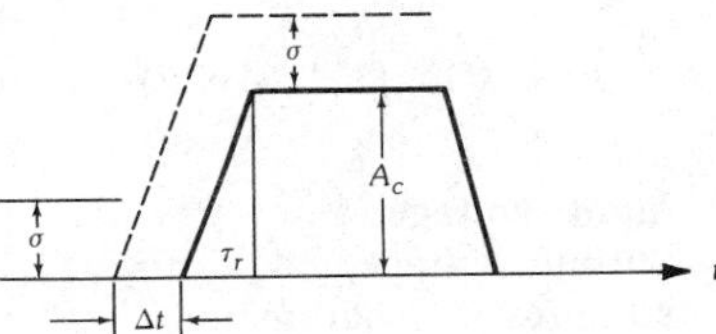

Fig. 20. Variation in pulse position due to the presence of noise or interference. (*From P. F. Panter,* Modulation, Noise, and Spectral Analysis, *Fig. 18-27, © 1965, McGraw-Hill Book Co.*)

The $(S/N)_o$ can be improved by decreasing the pulse rise τ_r or correspondingly by widening the transmission bandwidth. For $B \cong 1/\tau_r$

$$(S/N)_o = \frac{1}{2}t_0{}^2B^2(C/N)_i$$

For $B \cong 1/2\tau_r$

$$(S/N)_o = 2t_0{}^2B^2(C/N)_i$$

As in the case of FM, the $(S/N)_o$ ratio cannot be improved indefinitely by widening the bandwidth, because the noise power introduced at the receiver increases with bandwidth and eventually becomes comparable to the signal and "takes over" the system. A threshold level thus also exists just as in the FM case. This threshold level is usually taken as $A_c/\sigma = 2$, or $(C/N)_i = 4$ (6 dB).

PULSE-CODE MODULATION (PCM)

In PCM, several pulses are used as a code group to describe the quantized amplitude of a single sample. For example, a code group of n on-off pulses (binary code) can represent 2^n discrete amplitudes or levels, including zero level. In general, in an s-ary PCM system, the number of quantized amplitude levels the code group can express (including zero level) is given by

$$M = s^n$$

If a stands for 0 or 1, the binary notation with n digits, $a_1, a_2, \cdots, a_n$, represents the number

$$a_12^0 + a_22^1 + a_32^2 + \cdots + a_n2^{n-1}$$

In the ternary number system, a stands for the pulse amplitude 0, 1, 2, and the code group of n digits represents the number

$$a_13^0 + a_23^1 + \cdots + a_n3^{n-1}$$

Table 1 shows how the 64 numbers from 0 through 63 are represented in binary, quaternary, and octonary notation.

Quantization Noise in a PCM System

Representing the message signal by certain discrete allowed levels or steps is called quantizing. It inherently introduces an initial error in the amplitude of the samples, giving rise to quantization noise.

Uniform Spacing of Levels—In this case, the quantizing interval, or step, Δv is constant, and the quantizing noise power is given by

$$N_q = (\Delta v)^2/12$$

assuming that the quantization noise is uniformly distributed between $\pm \Delta v/2$. Assuming that the amplitudes of the samples are uniformly distributed, the signal power recovered from the quantized samples is

$$S_q = [(M^2 - 1)/12](\Delta v)^2$$

where M is the number of discrete levels assigned to

TABLE 1. ENCODING INTO BINARY, QUATERNARY, AND OCTONARY NUMBERS*

Decimal No.	Binary No.	Quaternary No.	Octonary No.
0	000000	000	00
1	000001	001	01
2	000010	002	02
3	000011	003	03
4	000100	010	04
5	000101	011	05
6	000110	012	06
7	000111	013	07
8	001000	020	10
9	001001	021	11
10	001010	022	12
11	001011	023	13
12	001100	030	14
⋮	⋮	⋮	⋮
62	111110	332	76
63	111111	333	77

*From P. F. Panter, *Modulation, Noise, and Spectral Analysis,* Table 20-1, © 1965, McGraw-Hill Book Co., New York.

the message signal. The ratio of the signal power to the quantizing noise power is

$$S_q/N_q = M^2 - 1 \cong M^2, \qquad M >> 1$$

Nonuniform Spacing of Levels—Quantization noise can be reduced by the use of nonuniform spacing of levels, to provide smaller steps for weaker signals and coarser quantization near the peak of large signals. Quantization noise can be minimized by an optimum level distribution that is a function of the probability density of the signal. The optimum level spacing Δv_k is given by

$$[p(v_k)]^{1/3}\,\Delta v_k = k/M, \qquad k = \text{constant}$$

With optimum level spacing, the total minimum error power is

$$(N_q)_{\min} = (2/3M^2)\left\{\int_0^V [p(v)]^{1/3}dv\right\}^3$$

where $p(v)$ is the probability density of the message signal, and the nonuniform levels are symmetrically disposed about zero level in the amplitude range $(-V, V)$.

In practice, nonuniform quantization is realized by compression, followed by uniform quantization as in Fig. 21. The logarithmic compression curve shown in Fig. 22 renders the distortion largely independent of the signal and is relatively easy to obtain in practice.

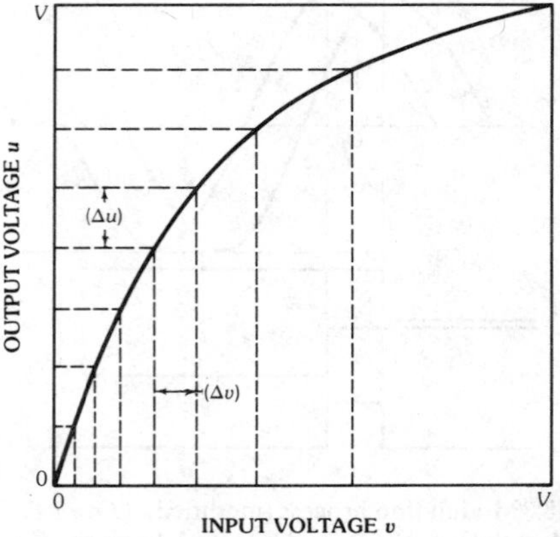

Fig. 21. Compression characteristic of "compressor." (*From P. F. Panter,* Modulation, Noise, and Spectral Analysis, *Fig. 20-10, © 1965, McGraw-Hill Book Co.)*

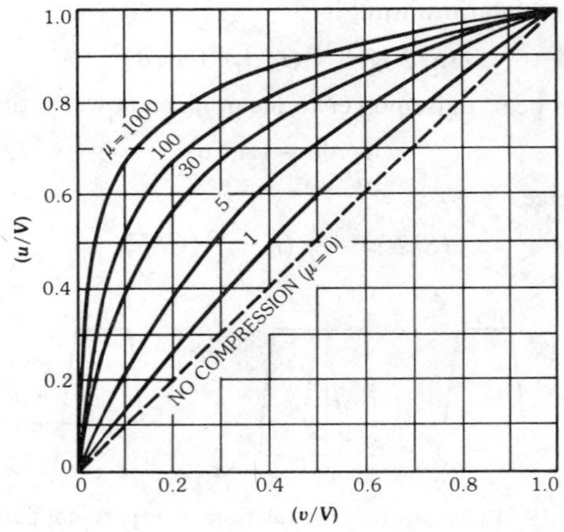

Fig. 22. Logarithmic compression characteristics. (*From P. F. Panter,* Modulation, Noise, and Spectral Analysis, *Fig. 20-11, © 1965, McGraw-Hill Book Co.)*

$$u = k\log[1 + (\mu v/V)]$$

where,

$v =$ input voltage,
$u =$ output voltage,
$\mu =$ compression parameter,
$k =$ undetermined constant.

If the maximum values of the input and the compressed signals are adjusted to be equal, this gives

$$u = \frac{V\log(1 + \mu v/V)}{\log(1 + \mu)}, \qquad 0 \le v \le V$$

and

$$u = \frac{-V\log(1 - \mu v/V)}{\log(1 + \mu)}, \qquad -V \le v \le 0$$

The quantizing noise power with logarithmic compression is

$$N_q = (\alpha^2/12)\,(V^2 + \mu^2 S)$$

where

$$\alpha = \frac{2\log(1+\mu)}{\mu M}$$

and

$$S = \int_{-V}^{V} v^2 p(v)\,dv = \text{average signal power}$$

False-Pulse Noise in a PCM System

In addition to quantization noise, a PCM system is characterized by *false-pulse noise,* which originates primarily at the receiving end of the system and is caused by noise spikes breaking through the threshold. This type of noise decreases rapidly as the signal power is increased above threshold. The effect of the false pulses introduced in the code group is to introduce an error in the decoded samples. The mean-square error introduced in the decoded signal is defined as the false-pulse noise. The output signal-to-noise ratio at the decoder is

$$(S/N)_o = (1/4pq) - 1$$

where,

- p is the probability of sending out state one and receiving state zero and vice versa,
- q is the probability that no transmission fault occurs $(p + q = 1)$.

The output signal-to-noise ratio drops from infinity with a noiseless channel $(p = 0,\ q = 1)$ to zero in the case of an infinitely large channel noise $(p = q = 1/2)$.

The output signal-to-noise ratio for K links in tandem is given by

$$(S/N)_{o,(K)} = \frac{(q-p)^{2K}}{1-(q-p)^{2K}}$$

These expressions for the $(S/N)_o$ in a PCM system are given in terms of the probability of false pulses in the code group due to channel noise. The following expressions relate the output signal-to-noise ratio to the input carrier-to-noise ratio for one link.

$$(S/N)_o = (\pi/8)^{1/2}\,(V_0/2\sigma)\,\exp[\tfrac{1}{2}(V_0/2\sigma)^2]$$

For unipolar or on-off binary system

$$(S/N)_o = (\pi/16)^{1/2}\,[(C/N)_i]^{1/2}\ \exp[\tfrac{1}{4}(C/N)_i]$$

For bipolar binary system

$$(S/N)_o = (\pi/8)^{1/2}\,[(C/N)_i]^{1/2}\ \exp[\tfrac{1}{2}(C/N)_i]$$

For K links in tandem

$$(S/N)_{o(K)} = (\mathrm{erf}\,x)^{2K}/[1-(\mathrm{erf}\,x)^{2K}]$$

where $x = V_0/2\sigma$. For high $(c/n)_i$, $x \gg 1$, and

$$\mathrm{erf}\,x \cong 1 - (2/\pi)^{1/2}\,\frac{\exp(-x^2/2)}{x}$$

The rapid improvement in $(S/N)_o$ for small increases in $(C/N)_i$ is illustrated in Fig. 23 for various links in tandem.

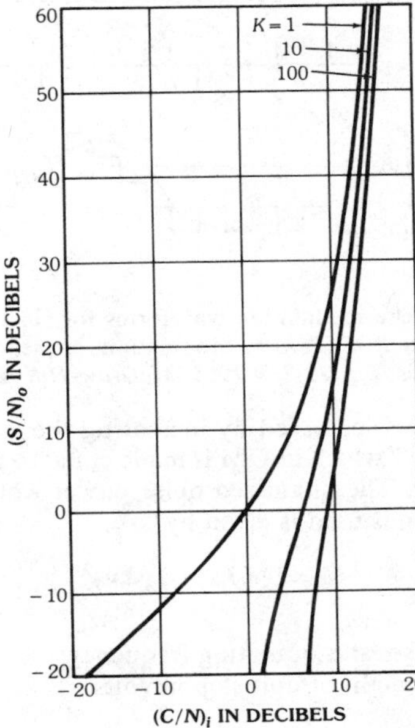

Fig. 23. Output signal-to-noise ratio for PCM. (*From H. F. Mayer, "Principles of Pulse Code Modulation,"* Advan. Electron., *Vol. 3, © 1951, Academic Press, Inc., New York.*)

DELTA MODULATION (DM)

In a DM system, instead of the absolute signal amplitude being transmitted at each sampling, only the changes in signal amplitude from sampling instant to sampling instant are transmitted. As shown in Fig. 24, the transmitted pulse train $e_2(t)$ of positive and negative pulses at the output of the encoder can be assumed to be generated at a constant clock rate. The transmitted pulses from the pulse generator are positive if the change in signal amplitude is positive; otherwise the transmitted pulses are negative. In the decoder, the delta-modulated pulse train $e_2(t)$ is integrated into the voltage $e_1(t)$, which consists of the original message function plus noise components due to sampling. These are eliminated by a low-pass filter so that the reconstructed signal of the final output is a close replica of the original modulating signal $e_0(t)$.

Signal-to-Noise Ratio in DM

The difference between the original and reconstructed signals gives rise to a "quantizing noise"

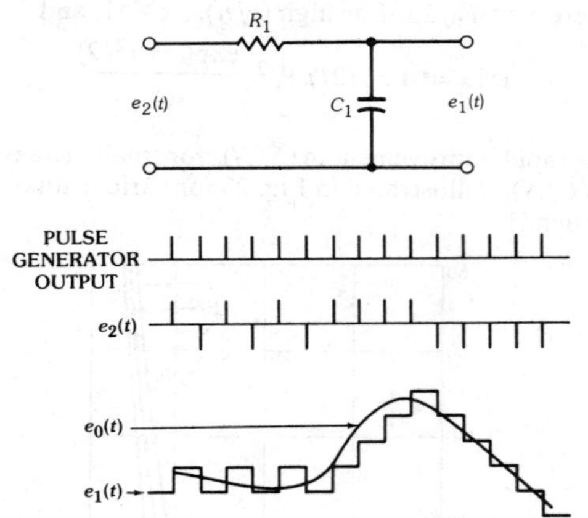

Fig. 24. Delta-modulation waveforms for single integration. (*From P. F. Panter,* Modulation, Noise, and Spectral Analysis, *Fig. 22-2, © 1965, McGraw-Hill Book Co.*)

that can be decreased by increasing the "sampling frequency," which in DM is made equal to the pulse frequency. The quantized noise power when single integration is used is given by

$$N_0 = \tfrac{2}{3}(f_m/f_s)(\Delta v)^2$$

where,

f_m = highest modulating frequency,
Δv = height of unit step in volts.

A DM system has no fixed maximum signal amplitude limitation but overloads when the slope of the signal is too large. The largest slope the system can reproduce is one that changes by one level or step every pulse interval, so that the maximum signal power depends on the type of signal. The signal power in the calculation of signal-to-noise ratio is taken as the power of the sinusoidal tone that is just below the overload point. The maximum amplitude of such a sinusoidal signal of frequency f that can be transmitted with single integration without overloading is

$$A = f_s(\Delta v)/2\pi f$$

The average signal power is

$$S_0 = f_s^2(\Delta v)^2/4\pi^2 f^2$$

so that the signal-to-noise ratio for single integration is

$$(S/N)_o = \tfrac{3}{2} r^3 (f_m/\pi f)^2$$

where $r = f_s/2f_m$ = bandwidth expansion factor. The signal-to-noise ratio for double integration is

$$(S/N)_o = \tfrac{3}{2} r^5 (f_m/\pi f)^4$$

Thus, the improvement in signal-to-noise ratio varies with f_s^3 for the system with single integration, whereas it varies with f_s^5 for double integration.

24 Digital Communications

Michael B. Pursley

A general model for a digital communication system is illustrated in the block diagram of Fig. 1. The source, encoder, and modulator are part of the transmitter, and the demodulator, decoder, and destination are in the receiver. In this chapter, we are primarily concerned with the *digital data channel*, which consists of the modulator, channel, and demodulator; the encoder and decoder are discussed in Chapter 25. To focus attention on the digital data channel, consider the model of Fig. 2. In this model, the source and the encoder are combined into a single element called the message source, and the decoder is absorbed into the destination.

are selected according to various criteria such as average probability of error.

The set of allowable decisions can be different from the message set, such as when we permit the receiver to erase symbols. For example, the binary erasure channel has a message set {0,1}, but the set of allowable decisions is {0,1,*e*}, where "*e*" denotes the erasure. This is in contrast to the binary "hard-decision" channel, which has the set {0,1} as both the message set and the set of allowable decisions.

The number of waveforms available to represent the messages is typically reflected in the names of the elements associated with the communications system (including the system itself). For $M = 2$, the

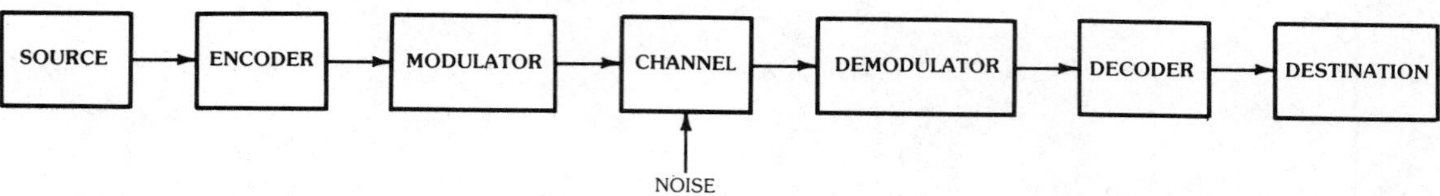

Fig. 1. General model for a communication system.

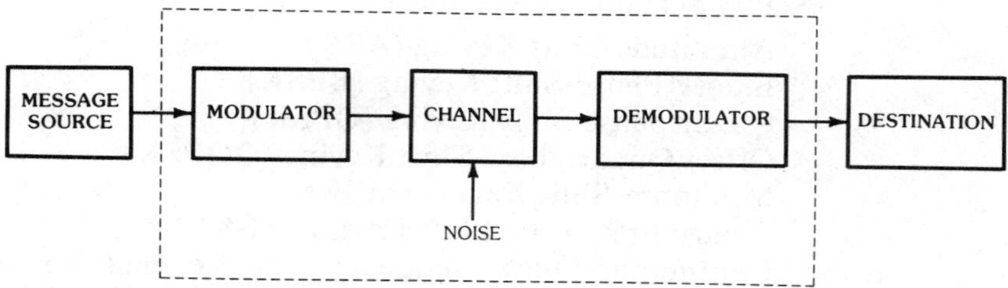

Fig. 2. Model for a digital data channel.

The information to be conveyed to the destination consists of a sequence of elements called *messages*. The *message set* is a set of M elements that are indexed by the integers $0, 1, \ldots, M-1$. The modulator produces signals that are used to represent these messages on the channel. During each transmission interval, the message source produces one of the M messages, and the corresponding waveform is transmitted over the channel to the receiver.

Rather than attempting to reproduce the transmitted waveform (as in an analog communication system), the goal of the demodulator in a digital communication system is to determine which of the M messages was sent. It accomplishes this by processing the received signal, which is a distorted, noisy version of the transmitted signal. The demodulator first produces a real number or vector called the *decision statistic*. This statistic is the input to a decision device (e.g., a threshold device), which bases its decision on a predetermined strategy referred to as a *decision rule*. Decision rules

signal sets, modulation techniques, and communications systems are referred to as *binary*; for $M = 3$, they are called *ternary*; and for $M = 4$, they are called *quaternary*. In this chapter, the term *M-ary* refers to nonbinary signaling (i.e., it is implicit that $M > 2$).

All of the digital modulation schemes described here use time-limited waveforms. A waveform $v(t)$ is *time limited* if $v(t) = 0$ for values of t outside some finite interval, and we say it is of *duration T* if this interval is $[0,T]$. The energy is perhaps the most important parameter of the transmitted signal. In general, the *energy* E_v in the waveform $v(t)$ is

$$E_v = \int_{-\infty}^{\infty} v^2(t)\, dt$$

If $v(t)$ has duration T,

$$E_v = \int_0^T v^2(t)\, dt$$

BASEBAND SIGNAL SETS

A baseband signal set is a collection of M baseband signals $\{s_k:k = 0, 1, \ldots, M-1\}$, each of which is of duration T. The kth signal $s_k(t)$ represents the message k, and a sequence of messages is sent to the receiver by transmitting the corresponding sequence of signals.

The energy in the kth signal is denoted by E_k. If all of the signals in the set have the same energy, the common value is denoted by E. Given two signals $s_k(t)$ and $s_j(t)$, their *inner product* (s_k,s_j) is defined by

$$(s_k, s_j) = \int_{-\infty}^{\infty} s_k(t)s_j(t)dt$$

Two types of signal sets are of particular interest: orthogonal signal sets and binary antipodal signal sets. In general, a signal set is said to be an *orthogonal* set if $(s_k,s_j) = 0$ for all $k \neq j$. A binary signal set is *antipodal* if $s_0(t) = -s_1(t)$ for all t in the interval $[0,T]$. Antipodal signals have equal energy E, and their inner product is $(s_0,s_1) = -E$.

Many of the baseband signal sets of interest can be defined in terms of a single waveform $v(t)$. To describe such sets, it is convenient to define a set of data variables $\{d_k:k = 0, 1, \ldots, M-1\}$ which are real numbers that represent the messages. Given a set of data variables and a pulse waveform $v(t)$ of duration T, a baseband signal set can then be defined as follows.

The kth signal in the set is given by

$$s_k(t) = Ad_kv(t)$$

where the constant A represents the signal amplitude due to amplification and attenuation (e.g., antenna gains and propagation losses). In many cases, A is the peak signal voltage at the receiver. Since $v(t)$ has duration T, each signal in the set $\{s_k:k = 0, 1, \ldots, M-1\}$ also has duration T. The energy in the kth signal is

$$E_k = A^2d_k^2E_v$$

Each antipodal signal set can be defined in this way with $M = 2$, $d_1 = -d_0$, and $E_1 = E_0$.

Four examples of binary baseband signal sets are illustrated in Fig. 3. Each of these signal sets can be obtained from a single waveform $v(t)$. For the signal sets shown in Figs. 3A and B, the waveform is a rectangular pulse of duration T. This waveform arises so frequently that it is convenient to give it a special notation. We denote the *rectangular pulse* of duration T by $p_T(t)$; it is defined by

$$p_T(t) = \begin{cases} 1 \text{ for } 0 \leq t < T \\ 0 \text{ otherwise} \end{cases}$$

Thus for the signal sets shown in Figs. 3A and B, $v(t) = p_T(t)$. The basic waveform for the baseband signals shown in Fig. 3C is the split-phase or Manchester-coded pulse. For the signal set shown in Fig. 3D, the basic waveform is the *sine pulse,* which is one-half of a full period of a sine wave. That is, $v(t) = \sin(\pi t/T)$ for $0 \leq t < T$.

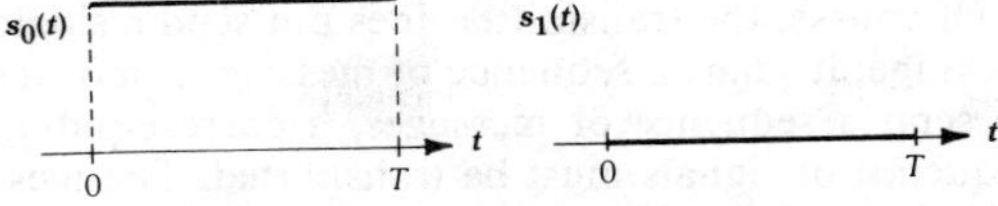

(A) Basic waveform: rectangular pulse.

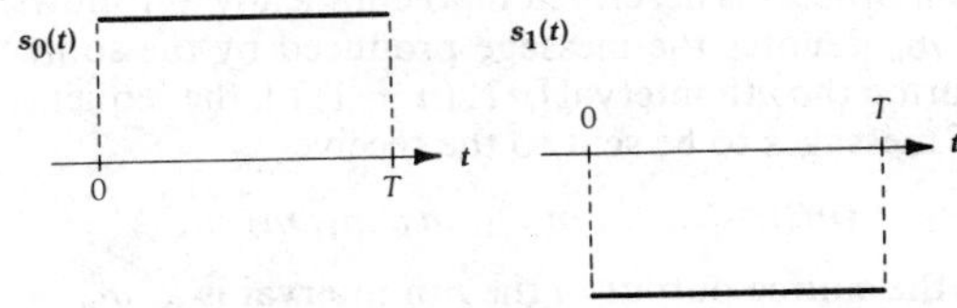

(B) Basic waveform: rectangular pulse.

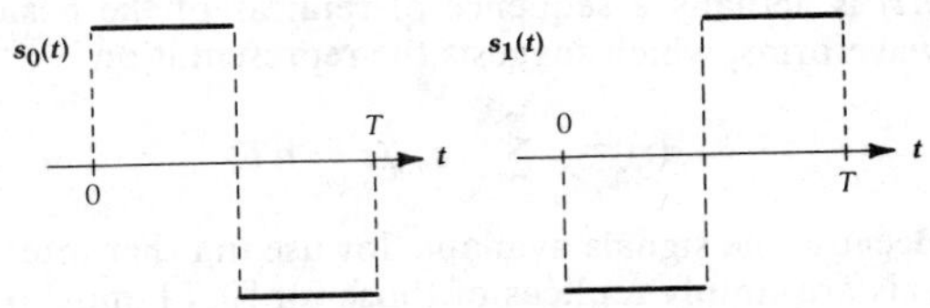

(C) Basic waveform: split-phase pulse.

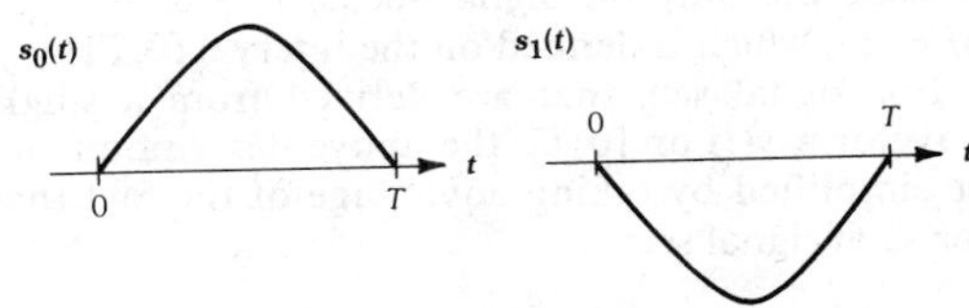

(D) Basic waveform: sine pulse.

Fig. 3. Four examples of baseband signal sets.

The data symbols for the signal set of Fig. 3A are $d_0 = 1$ and $d_1 = 0$, so this is an orthogonal signal set. For the other three signal sets, $d_0 = +1$ and $d_1 = -1$; thus each of these is an antipodal signal set.

There are binary and M-ary baseband signal sets that cannot be described in terms of a single waveform of duration T. If M is a power of 2, for example, a set of M orthogonal signals can be obtained by letting the signals be sequences of pulses (each pulse is of duration T/M) with amplitudes determined by the rows of an M by M Hadamard matrix.* Other M-ary orthogonal signal sets of this type can be designed from maximal-length linear feedback shift-register sequences. A biorthogonal signal set* consists of the signals from an orthogonal set together with all of their negatives. Two signals from a biorthogonal set are either orthogonal or antipodal.

---

* Reference 1.

Of course, the transmitter does not send a single message; it sends a sequence of messages. In order to send a sequence of messages, a corresponding sequence of signals must be transmitted. The message k is sent in the nth time interval by transmitting s_k during this interval. The baseband modulation process is described mathematically as follows. If m_n denotes the message produced by the source during the nth interval $[nT, (n + 1)T]$, the sequence of messages to be sent to the receiver is

$$(m_n) = \ldots, m_{-1}, m_0, m_1, m_2, \ldots$$

If the source output in the nth interval is k, $m_n = k$ and the transmitted signal $s(t)$ is equal to $s_k(t - nT)$ for $nT \leq t < (n + 1)T$. The signal $s(t)$ is actually a sequence of replicas of the basic waveforms, which suggests the representation

$$s(t) = \sum_{n=-\infty}^{\infty} s_{m_n}(t - nT)$$

Because the signals available for use in other intervals are simply replicas of those for $[0,T]$, most of the key features of a particular baseband modulation technique can be described and analyzed by considering only the signal set $\{s_k : k = 0, 1, \ldots, M - 1\}$, which is defined on the interval $[0,T]$.

For signal sets that are derived from a single waveform $v(t)$ on $[0,T]$, the above description can be simplified by taking advantage of the fact that for such signal sets

$$s_k(t) = Ad_k v(t)$$

The transmitted signal is defined as follows. First, the message sequence (m_n) is represented by the corresponding sequence (b_n) of data variables; the latter sequence is defined by $b_n = d_k$ if $m_n = k$. Next, the transmitted signal can be written as

$$s(t) = \sum_{n=-\infty}^{\infty} Ab_n v(t - nT)$$

It follows that the signal set can be described and analyzed by considering only the waveform $v(t)$ and the set of data variables $\{d_k : k = 0, 1, \ldots, M - 1\}$.

As an example, consider the waveform $v(t) = \sin(\pi t/T)$, which is shown in Fig. 3D. For binary communications, the message set is $\{0,1\}$. In order to use antipodal signals, we can let $d_0 = +1$ and $d_1 = -1$, so that a binary 0 is transmitted as a positive sine pulse and a binary 1 is transmitted as a negative sine pulse. Message sequences for which $m_0 = 0, m_1 = 1, m_2 = 1$, and $m_3 = 0$ correspond to data sequences with $b_0 = +1$, $b_1 = -1$, $b_2 = -1$, and $b_3 = +1$. The resulting signal $s(t)$ is shown in Fig. 4.

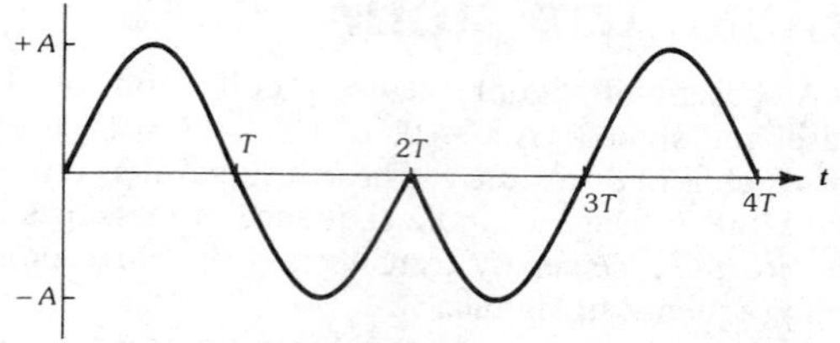

Fig. 4. Transmitted signal $s(t)$ for the message sequence 0110.

SIGNAL SETS FOR RF CHANNELS

The most general RF signals that are considered in this section are of the form

$$s(t) = A\, a(t) \cos[2\pi f_c t + \theta(t) + \phi]$$

where $a(t)$ and $\theta(t)$ are baseband signals of the type described in the preceding section. The signal $a(t)$ represents *amplitude modulation*, and $\theta(t)$ represents *phase modulation*. The variable A is a real number that represents the signal amplitude at the receiver. Physically, the parameter A may be the peak signal voltage, such as when $|a(t)| = 1$. The variable ϕ is the phase of the rf carrier at the time $t = 0$ in the absence of any phase modulation (i.e., $\theta(t) = 0$). This phase angle may or may not be known to the receiver.

A variety of RF signal sets can be obtained by employing the baseband waveforms of Fig. 3, or other elementary waveforms, as amplitude and phase modulation in the formulation given above. The most common RF signal sets that can be generated in this manner are described below.

The signal energy values given in the descriptions that follow are based on the assumption that either $2f_cT$ is an integer and both $|a(t)|$ and $\theta(t)$ are constant on $[0,T]$ (in which case the value is exact) or $f_cT >> 1$ (in which case the value is a good approximation). These conditions imply that the double-frequency component of the square of the signal is negligible. It is also assumed that the signals $a(t)$ and $\theta(t)$ use baseband waveforms of duration T. For such signals the *energy per data pulse* is defined as follows. If the kth message is being sent in the interval $[0,T]$, the energy is

$$E_k = \int_0^T s^2(t)\, dt$$

For several of the modulation schemes, $\theta(t)$ is equal to a constant θ_k for $0 \leq t < T$ if the kth message is sent during the interval $[0,T]$. In this case, the energy is

$$E_k = (A^2/2) \int_0^T a^2(t)\, dt$$

Because of the assumption concerning f_cT, the energy does not depend on the value of θ_k. It does depend on the energy in the waveform $a(t)$, so E_k may depend on the message k that is being sent.

Amplitude-Shift Keying (ASK)

The signal structure for amplitude-shift keying (ASK) is

$$s(t) = A\, a(t) \cos(2\pi f_c t + \phi)$$

where $a(t)$ is a sequence of baseband pulses whose amplitudes are modulated to represent the messages. This type of signaling is also commonly referred to as digital AM or multiamplitude signaling. The baseband signal $a(t)$ can be written as

$$a(t) = \sum_{n=-\infty}^{\infty} b_n\, v(t - nT)$$

where $b_n = d_k$ if the kth message is being sent during the nth interval. The set $\{d_k{:}k = 0, 1, \ldots, M - 1\}$ of data variables is the set of amplitudes for the sequence of pulses. The simplest case is binary ASK modulation with $d_0 = 1$ and $d_1 = 0$; this is known as *on-off keying* (*OOK*).

Another important special case results if $M = 2$, $v(t)$ is the rectangular pulse of duration T, $d_0 = +1$, and $d_1 = -1$. This is just the amplitude-modulation representation for BPSK (see next subsection).

In general, the energy per data pulse for ASK depends on the message being sent. For the kth message, the energy is

$$E_k = (A\, d_k)^2 E_v / 2$$

where E_v is the energy in the waveform $v(t)$, as defined at the beginning of this chapter.

A modification of ASK is *quadrature ASK* (*QASK*), which is also known as quadrature AM (QAM). The QASK signal is of the form

$$s(t) = A\,\{a_1(t) \cos[2\pi f_c t + \phi] + a_2(t) \sin[2\pi f_c t + \phi]\}$$

The baseband signals $a_1(t)$ and $a_2(t)$ are sequences of pulses of duration T_s with amplitudes from the set $\{d_k{:}k = 0, 1, \ldots, M - 1\}$. The QASK signal $s(t)$ can be written as

$$s(t) = s_I(t) + s_Q(t)$$

where

$$s_I(t) = A\, a_1(t) \cos[2\pi f_c t + \phi]$$

is the *in-phase* component of $s(t)$ and

$$s_Q(t) = A\, a_2(t) \sin[2\pi f_c t + \phi]$$

is the *quadrature* component. Each of these two components of the QASK signal is an ASK signal with pulse duration T_s.

In some applications, $a_1(t)$ and $a_2(t)$ are sequences of binary pulses derived from a single binary source (e.g., odd-numbered bits go to the in-phase channel and even-numbered bits go to the quadrature channel). If the source produces binary digits at the rate of one bit every T seconds, then $T_s = 2T$. If $a_1(t)$ and $a_2(t)$ are binary ($M = 2$), the in-phase and quadrature signals are binary ASK signals of rate $1/T_s$ bits per second. This gives a total data rate of $2/T_s = 1/T$ bits per second for the QASK signal.

Binary Phase-Shift Keying (BPSK)

One of the most commonly used binary signal sets is obtained by shifting the phase of the RF carrier by $+\pi/2$ radians or $-\pi/2$ radians, depending on whether the data bit is a 0 or a 1. Binary phase-shift keying can be viewed as binary phase modulation or binary amplitude modulation; the only requirement is that, during each signaling interval, $s(t)$ is one of two sinusoidal signals that differ in phase by π radians.

Suppose that the binary digit k is to be sent during the nth time interval. Viewed as phase modulation, the corresponding BPSK signal is

$$A \cos[2\pi f_c t + d_k(\pi/2) + \phi]$$

for $nT \leq t < (n + 1)T$ where $|d_0 - d_1| = 2$. The two most common choices for d_k are $d_0 = 0$ and $d_1 = 2$ or $d_0 = +1$ and $d_1 = -1$. Based on the latter choice for d_k, the BPSK signal can be defined by

$$s(t) = A\, a(t) \cos[2\pi f_c t + \theta(t) + \phi]$$

where $a(t) = 1$ for all t and

$$\theta(t) = (\pi/2) \sum_{n=-\infty}^{\infty} b_n p_T(t - nT)$$

The data symbol b_k is d_k if k is to be sent in the nth time interval. This formulation corresponds to phase modulation by a sequence of rectangular pulses of duration T with amplitudes $(\pi/2)b_n$. The resulting signal $s(t)$ is illustrated in Fig. 5A for the data sequence $b_0 = -1$, $b_1 = +1$, and $b_2 = +1$. This illustration is for $f_c T = 2$ and $\phi = 0$.

Viewed as binary amplitude modulation, the BPSK signal can be expressed in a form that reveals its connection with binary ASK. Let $\theta(t) = 0$ for all t and use the amplitude to convey the information by defining

$$a(t) = \sum_{n=-\infty}^{\infty} b_n p_T(t - nT)$$

Thus, 0 is sent by transmitting

$$s(t) = A \cos[2\pi f_c t + \phi]$$

and 1 is sent by transmitting

$$s(t) = -A \cos[2\pi f_c t + \phi]$$

This formulation of binary PSK corresponds to amplitude modulation by a sequence of rectangular pulses of duration T with amplitudes $+1$ and -1. However, it is equivalent to phase modulation of the form

$$s(t) = A \cos[2\pi f_c t + d_k(\pi/2) + \phi]$$

with $d_0 = 0$ and $d_1 = 2$.

Although the modulator implementations suggested by these two formulations are quite differ-

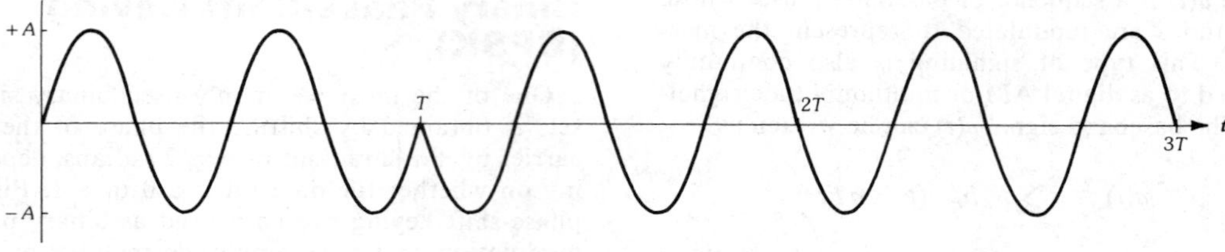

(A) Binary PSK, data rate $1/T$ bits per second.

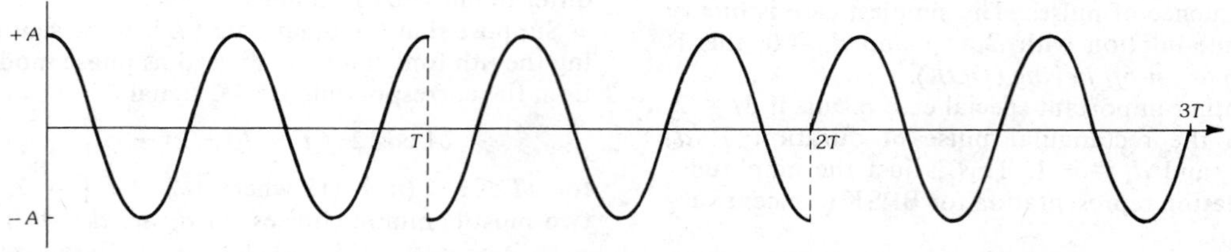

(B) Example of QPSK, data rate $2/T$ bits per second.

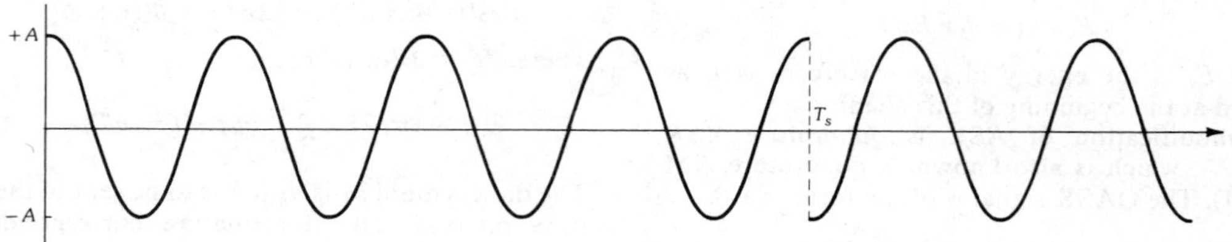

(C) Example of QPSK, data rate $1/T$ bits per second.

Fig. 5. Various forms of phase-shift keying.

ent, the resulting signal sets are mathematically equivalent. In particular, it is clear from either of these formulations that the BPSK signals representing the binary digits 0 and 1 form an antipodal signal set. Notice also that for each of these representations the energy per data pulse is $E = A^2T/2$.

Quadriphase Shift Keying (QPSK)

Binary PSK can be generalized by allowing the phase waveform to take on more than two values. The QPSK signal is of the same basic form as BPSK, except that quaternary modulation is employed instead of binary modulation. Thus the transmitted signal for the interval $[nT, (n+1)T]$ can be written as

$$s(t) = A\cos[2\pi f_c t + \theta_k + \phi]$$

where k denotes the message to be sent in the nth interval ($k = 0, 1, 2,$ or 3). The four different values for θ_k must be separated by $\pi/2$ radians. For example, we can let $\{\theta_k : k = 0, 1, 2, 3\}$ be the set $\{0, \pi/2, \pi, 3\pi/2\}$ or the set $\{\pi/4, 3\pi/4, 5\pi/4, 7\pi/4\}$. As with BPSK, the values themselves are unimportant, but the differences are constrained.

The QPSK signal can be expressed as

$$s(t) = A\cos[2\pi f_c t + \theta(t) + \phi]$$

The phase modulation $\theta(t)$ is a sequence of rectangular pulses of duration T and amplitudes $(\pi/2)\, b_n$ with $b_n = 0, 1, 2,$ or 3. This signal is illustrated in Fig. 5B for the data sequence $b_0 = 0$, $b_1 = 2$, and $b_2 = 3$. As in Fig. 5A, $f_c T = 2$ and $\phi = 0$.

The QPSK signal can also be viewed as quadrature amplitude modulation by writing

$$s(t) = A\cos[2\pi f_c t + \theta(t) + \phi]$$

as

$$s(t) = A\{\cos[\theta(t)]\cos[2\pi f_c t + \phi] - \sin[\theta(t)]\sin[2\pi f_c t + \phi]\}$$

If $\theta(t)$ takes on values in the set $\{\pi/4, 3\pi/4, 5\pi/4, 7\pi/4\}$, then $\cos[\theta(t)]$ is either $+1/\sqrt{2}$ or $-1/\sqrt{2}$, and $\sin[\theta(t)]$ takes on the same two values. Thus $s(t)$ can be written as

$$s(t) = (A/\sqrt{2})\,\{a_1(t)\cos[2\pi f_c t + \phi] + a_2(t)\sin[2\pi f_c t + \phi]\}$$

where $a_1(t)$ and $a_2(t)$ are sequences of positive and negative unit-amplitude rectangular pulses of duration T. This formulation of QPSK clearly exhibits its relationship to QASK. If we define the *in-phase* and *quadrature* signals by

$$s_I(t) = (A/\sqrt{2})\,a_1(t)\cos[2\pi f_c t + \phi]$$

and

$$s_Q(t) = (A/\sqrt{2})\,a_2(t)\sin[2\pi f_c t + \phi]$$

respectively, the QPSK signal can be written as

$$s(t) = s_I(t) + s_Q(t)$$

Each of the signals $a_1(t)$ and $a_2(t)$ has the same number of pulses per unit of time as the signal $a(t)$ that appears in the amplitude-modulation formulation of BPSK. Consequently, QPSK signaling has twice the data rate of BPSK. This can also be observed from the original quaternary-phase-modulation formulation by noticing that each pulse of duration T takes on one of four values, and this gives $\log_2(4) = 2$ bits of data per pulse.

The quadrature-amplitude-modulation formulation of QPSK displays the signal as the sum of two quadrature binary ASK signals. A QPSK signal can also be viewed as two quadrature BPSK signals. For example, we can write the QPSK signal as

$$s(t) = (A/\sqrt{2})\,\{\cos[2\pi f_c t + \theta_1(t) + \phi] + \sin[2\pi f_c t + \theta_2(t) + \phi]\}$$

where $\theta_1(t)$ and $\theta_2(t)$ take on the values 0 and π.

The power in the QPSK signal is $A^2/2$, just as for BPSK, and so the energy per pulse is $E = A^2T/2$. Since each pulse represents two bits of data, the energy per bit for QPSK is $E_b = A^2T/4$, which is one-half of the energy per bit for BPSK for the same pulse duration T. Because the pulse shape and duration are the same for BPSK and QPSK, the two signals have the same bandwidth.

For most purposes, it is better to compare BPSK and QPSK signals that have *the same energy per data bit and the same data rate* (rather than the same pulse duration as above). To do this, simply replace T by T_s in the QPSK expressions, and let $T_s = 2T$. That is, the signal $s(t)$ is given by

$$s(t) = A\cos[2\pi f_c t + \theta(t) + \phi]$$

as before, but now the phase modulation $\theta(t)$ is a sequence of rectangular pulses of duration $T_s = 2T$, rather than T. The amplitudes of these pulses are still given by $(\pi/2)b_n$ with $b_n = 0$, 1, 2, or 3; however, the sequence (b_n) is now transmitted at the rate of one quaternary symbol every T_s seconds. This signal is illustrated in Fig. 5C for $b_0 = 0$ and $b_1 = 2$, with $f_cT = 2$ and $\phi = 0$.

Because a QPSK pulse of duration T_s provides two bits of data, the data rate of a QPSK signal that uses pulses of duration T_s is

$$R_{\text{QPSK}} = 2/T_s = 1/T$$

bits per second. If the digital data consists of a single sequence of binary digits, half of the data bits can be modulated onto the in-phase component and half onto the quadrature component.

Since $1/T$ is also the data rate for a BPSK signal with pulse duration T, these two forms of PSK modulation can now be compared on the basis of the same data rate. In this case, the energy per pulse for QPSK is given by

$$E = A^2T_s/2$$

which is equivalent to an energy per bit of

$$E_b = A^2T/2$$

the same as BPSK. The pulse shapes for QPSK and BPSK are the same, but the pulse duration for QPSK is $T_s = 2T$, twice that for BPSK. This means the QPSK signal has one-half the bandwidth of the BPSK signal.

In summary, if QPSK and BPSK are compared for the *same data rate,* BPSK requires twice the bandwidth of QPSK. If the two signals have the *same data rate* and *equal power,* they also have the same energy per data bit.

Offset Quadriphase Shift Keying (OQPSK)

The OQPSK signal can be defined by replacing $\theta_1(t)$ with $\theta_1(t + T)$ in the quadrature-phase-modulation formulation of QPSK. That is, OQPSK can be represented in terms of two quadrature BPSK signals as

$$s(t) = (A/\sqrt{2})\,\{\cos[2\pi f_c t + \theta_1(t + T) + \phi] + \sin[2\pi f_c t + \theta_2(t) + \phi]\}$$

where $\theta_1(t)$ and $\theta_2(t)$ are sequences of pulses of duration $T_s = 2T$ as in the preceding subsection. Since the amount of the offset is one-half the pulse duration, the phase transitions for the in-phase and quadrature components of the OQPSK signal are separated in time by T seconds.

The OQPSK signal can be represented as quadrature amplitude modulation with the baseband signal of the in-phase component offset by T seconds relative to the baseband signal of the quadrature component. Thus, the amplitude-modulation representation of OQPSK is

$$s(t) = (A/\sqrt{2})\,\{a_1(t + T)\cos[2\pi f_c t + \phi] + a_2(t)\sin[2\pi f_c t + \phi]\}$$

where $a_1(t)$ and $a_2(t)$ are as in the preceding subsection. Notice that the baseband signal for the in-phase component can change polarity at times nT for odd integers n only, but the transitions of the baseband signal for the quadrature component are at times nT for even values of n only. This is an important feature of OQPSK for band-limited channels with nonlinearities. A band-limited version of QPSK or BPSK has an envelope that may go to zero when the in-phase and quadrature signals switch polarity at the same time; the offset in OQPSK leads to a more nearly constant envelope for the band-limited signal.

If the OQPSK signal is expressed in the form

$$s(t) = A \cos[2\pi f_c t + \theta(t) + \phi]$$

the phase $\theta(t)$ can change every T seconds, whereas the phase shifts in QPSK occur only at $2T$-second intervals. However, for OQPSK the magnitudes of the phase shifts are limited to 0 or $\pi/2$ only. Phase shifts for QPSK can have magnitudes 0, $\pi/2$, or π.

Minimum-Shift Keying (MSK)

The MSK signal can be considered as a special case of offset quadrature amplitude modulation in which the baseband waveform is the sine pulse $v(t) = \sin(\pi t/2T)p_{2T}(t)$. Thus the MSK signal can be written as

$$s(t) = A\{a_1(t+T)\cos[2\pi f_c t + \phi] + a_2(t)\sin[2\pi f_c t + \phi]\}$$

where $a_1(t)$ and $a_2(t)$ are sequences of sine pulses of duration $2T$. That is, for each i ($i = 1$ or $i = 2$)

$$a_i(t) = \sum_{n=-\infty}^{\infty} b_{i,n} v(t - 2nT)$$

where $(b_i) = \ldots, b_{i,0}, b_{i,1}, b_{i,2}, \ldots$ is a binary data sequence. Minimum-shift keying can also be viewed as continuous-phase frequency-shift keying (discussed in a later subsection).

Binary Frequency-Shift Keying (FSK)

Binary FSK signals are of the form

$$s(t) = A \cos[2\pi f_c t + \theta(t)]$$

where the modulation $\theta(t)$ is defined as follows. If 0 is the bit to be transmitted in the nth interval,

$$\theta(t) = 2\pi f_d t + \phi_0$$

and if 1 is sent,

$$\theta(t) = -2\pi f_d t + \phi_1$$

Thus, the signal

$$s(t) = A \cos[2\pi(f_c + f_d)t + \phi_0]$$

represents a 0 for the nth interval, and the signal

$$s(t) = A \cos[2\pi(f_c - f_d)t + \phi_1]$$

represents a 1. The quantity f_d is called the *frequency deviation,* and the parameter $h = 2f_dT$ is called the *deviation ratio* or *modulation index* for the FSK signal set. The FSK signals may be generated by switching between two oscillators or by applying a binary baseband signal at the input of a voltage-controlled oscillator (VCO).

In general, the phase angles ϕ_k are arbitrary, and ϕ_0 need not be related to ϕ_1 in any way. If the FSK signals are obtained by switching between two oscillators, one at frequency $f_c + f_d$ and one at frequency $f_c - f_d$, the phase angles ϕ_0 and ϕ_1 represent the phases of these oscillators at time $t = 0$. Alternatively, the FSK signals may be generated by applying a baseband signal to a VCO (or another frequency-modulation circuit), in which case the phase angles ϕ_0 and ϕ_1 may be related.

The signals in FSK systems are often referred to as "tones," and the two signals are distinguished by calling one of them "*mark*" and the other "*space*." The convention followed here is to refer to the signal at frequency $f_c + f_d$ as the space and the signal at frequency $f_c - f_d$ as the mark (the binary digit 0 is transmitted as a space, and the binary digit 1 is transmitted as a mark).

If $2f_cT$ and $2f_dT$ are integers, the mark and space signals form an *orthogonal* signal set for *all* values of ϕ_0 and ϕ_1. It is often the case that $f_cT >> 1$. Under this condition, the signals are approximately orthogonal if $2fdT$ is an integer or if $f_dT >> 1$. If $2f_cT$ and $2f_dT$ are integers or if $(f_c - f_d)T >> 1$, the energy per data bit is $A^2T/2$.

For certain applications, it is necessary to generalize the above formulation of FSK by introducing a phase angle $\phi(n)$ that depends on the interval in which the signal is transmitted. To send a 0 in the nth interval, the transmitted signal for $nT \le t < (n+1)T$ is

$$s(t) = A \cos[2\pi(f_c + f_d)t + \phi(n) + \phi_0]$$

and to send a 1 in this same interval, the signal is

$$s(t) = A \cos[2\pi(f_c - f_d)t + \phi(n) + \phi_1]$$

This generalization is required in order to characterize FSK signals in frequency-hopped spread-spectrum systems, for example, where the phase $\phi(n)$ represents a phase shift introduced by noncoherent frequency hopping.* This generalization is also required for the description of certain continuous-phase FSK modulation schemes.

Continuous-Phase Frequency-Shift Keying (CPFSK)

Continuous-phase FSK signals are of the form

$$s(t) = A \cos[2\pi f_c t + \theta(t)]$$

just as for other types of FSK signals. However, for CPFSK the phase modulation $\theta(t)$ is a continuous function of t. In order to describe CPFSK modulation, it is sufficient to consider only the case in

* Reference 21.

which $\phi_0 = \phi_1 = 0$. This is because the phase angles ϕ_0 and ϕ_1 can always be absorbed in the phase modulation $\theta(t)$ for CPFSK signals.

The signal $\theta(t)$ is defined as follows. Suppose b_n is the data symbol to be sent in the nth interval, and b_n is either +1 or −1. The phase modulation is

$$\theta(t) = 2\pi b_n f_d t + \phi(n)$$

for $nt \leq t < (n+1)T$. The phase angles $\phi(n)$ are such that the phase is continuous from one interval to the next. In order to make the phase continuous at time nT, the phase angles $\phi(n)$ and $\phi(n-1)$ must satisfy

$$2\pi b_n f_d nT + \phi(n) = 2\pi b_{n-1} f_d nT + \phi(n-1)$$

which is equivalent to the condition

$$\phi(n) = 2\pi(b_{n-1} - b_n) f_d nT + \phi(n-1)$$

In other words, $\phi(n) = \phi(n-1)$ if the two successive data symbols b_{n-1} and b_n are the same, but $\phi(n)$ and $\phi(n-1)$ differ by $4\pi f_d nT$ radians if these two data symbols are different.

The simplest CPFSK signal is obtained for the case in which the modulation index $h = 2f_d T$ is an integer. For this case, $4\pi f_d nT$ is an integer multiple of 2π, and $\phi(n) = \phi(n-1)$ modulo 2π, regardless of whether the successive data symbols are the same or different. But phase angles that differ by integer multiples of 2π are the same for our purposes, so we can let $\phi(n) = \phi$, where ϕ is a constant phase (independent of n). No phase changes are required in order to make $\theta(t)$ continuous provided the modulation index is an integer.

Another modulation index of considerable interest for CPFSK is 1/2. For $h = 1/2$, the frequency deviation is $f_d = 1/4T$, and the condition for continuous phase is

$$\phi(n) = \pi n[(b_{n-1} - b_n)/2] + \phi(n-1)$$

If the successive data bits b_{n-1} and b_n are the same, the phase angles $\phi(n-1)$ and $\phi(n)$ are also the same. If the successive data bits are different, $b_{n-1} - b_n$ is either +2 or −2, and so the phase angles must differ by an integer multiple of π radians. It follows that if $\phi(0) = 0$ and $2f_c T$ is an integer, or if $f_c T >> 1$, the mark and space signals are orthogonal. It is important to realize that orthogonality is obtained for $h = 1/2$ only if the phases in successive intervals are controlled.

From the relationship

$$\theta(t) = 2\pi b_n f_d t + \phi(n)$$

which holds for $nT \leq t < (n+1)T$, it follows that

$$\theta(nT) = 2\pi b_n f_d nT + \phi(n)$$

Making the substitution for $\phi(n)$ from the last equation into the expression for $\theta(t)$, we find that

$$\theta(t) = 2\pi b_n f_d (t - nT) + \theta(nT)$$

for $nT \leq t < (n+1)T$. This fact and the continuity condition imply

$$\theta((n+1)T) = 2\, b_n f_d (n+1)T + \theta(nT)$$

If we denote $\theta(kT)$ by θ_k and let $h = 1/2$ in this expression, it can be written as

$$\theta_{n+1} = (\pi/2) b_n + \theta_n$$

Similarly, the expression for $\theta(t)$ becomes

$$\theta(t) = (\pi/2T) b_n (t - nT) + \theta_n$$

which is valid for $nT \leq t < (n+1)T$.

The CPFSK signal is

$$s(t) = A\,\{\cos[\theta(t)]\cos[2\pi f_c t] - \sin[\theta(t)]\sin[2\pi f_c t]\}$$

The above relationships for $\theta(t)$, θ_n, and θ_{n+1} imply that if n is an even integer and $h = 1/2$,

$$-\sin\theta(t) = -\sin[\theta_{n+1}]\sin[\pi(t - nT)/2T]$$

for $nT \leq t < (n+2)T$. Similarly, if n is an odd integer and $h = 1/2$,

$$\cos\theta(t) = -\cos[\theta_{n+1}]\sin[\pi(t - nT)/2T]$$

for $nT < t < (n+2)T$. Define the data sequence (B_n) by

$$B_n = -\sin\theta_{n+1}$$

if n is an even integer and by

$$B_n = -\cos\theta_{n+1}$$

if n is an odd integer. The data sequence (B_n) is related to the original data sequence (b_n) by

$$b_n = -B_n \cos\theta_n$$

if n is an even integer and by

$$b_n = B_n \sin\theta_n$$

if n is an odd integer.

From the above development, we conclude that for $h = 1/2$ the CPFSK signal $s(t)$ is of the form

$$s(t) = A\,\{a_1(t + T)\cos[2\pi f_c t] + a_2(t)\sin[2\pi f_c t]\}$$

The baseband signals $a_1(t)$ and $a_2(t)$ are given by

$$a_1(t) = B_{2k-1}\sin[\pi(t - 2kT)/2T]$$

and

$$a_2(t) = B_{2k}\sin[\pi(t - 2kT)/2T]$$

for $2kT \leq t < 2(k+1)T$. A comparison of these last three equations with analogous equations in the subsection on minimum-shift keying shows that CPFSK with modulation index $h = 1/2$ is the same as MSK.

M-ary Signaling and Multiple Frequency Shift Keying (MFSK)

The general M-ary RF signal set has signals of the form

$$s_k(t) = A\,a_k(t)\cos[2\pi f_c t + \theta_k(t) + \phi_k]$$

for $k = 0, 1, \ldots, M-1$. The signals $a_k(t)$ and $\theta_k(t)$ are baseband signals with bandwidths much smaller than the carrier frequency f_c. The messages are indexed by the integers $0, 1, \ldots, M-1$, and in order to send the kth message during the nth interval, the signal $s(t) = p_T(t - nT)\, s_k(t)$ is transmitted. (The rectangular pulse of duration T is defined by $p_T(u) = 1$ for $0 \leq u < T$ and $p_T(u) = 0$ otherwise.)

The RF signal set is characterized by the inner products

$$(s_k,s_j) = \int_0^T s_k(t)\, s_j(t)\, dt$$

Since the amplitude modulation $a_k(t)$ and phase modulation $\theta_k(t)$ are narrow-band signals, the inner products are (at least approximately) given by

$$(s_k,\, s_j) = A^2 \int_0^T a_k(t)\, a_j(t) \cos[\theta_{k,j}(t) + \phi_{k,j}]dt$$

where

$$\theta_{k,j}(t) = \theta_k(t) - \theta_j(t)$$

and

$$\phi_{k,j} = \phi_k - \phi_j$$

If $(s_k,s_j) = 0$ for all $j \neq k$, the signals form an *orthogonal* signal set. One type of orthogonal signal set is obtained by letting $\theta_k(t) = 0$ for all t and using a set of orthogonal baseband signals for the amplitude modulation $a_k(t)$.

Multiple frequency shift keying (*MFSK*), which is also known as M-ary frequency shift keying, is a direct extension of binary FSK to modulation with more than two frequency tones. It is also a special case of M-ary signaling in which $a_k(t) = 1$ and

$$\theta_k(t) = (f_k - f_c)t$$

The resulting signals are of the form

$$s_k(t) = A\cos[2\pi f_k t + \phi_k]$$

In order to send the kth message in the nth interval, the transmitted signal is

$$s(t) = A p_T(t - nT)\cos[2\pi f_k t + \phi_k]$$

If the frequencies f_k are such that $f_k T = n_k$ for some integers $n_0, n_1, \ldots, n_{M-1}$, the M different signals are orthogonal on each interval $[nT, (n+1)T]$. The signals are at least approximately orthogonal if $|f_k - f_j|T >> 1$ for each choice of $j \neq k$. The phases $\phi_0, \phi_1, \ldots, \phi_{M-1}$ are arbitrary and, in general, not related to each other in any way.

Spectra of PSK and MSK Signals

By introducing appropriate random time delays and phase angles, we can model the signal $s(t)$ as a wide-sense stationary random process. The Fourier transform of the autocorrelation function for this random process is called the *power spectral density,* or simply the *spectrum.* The spectrum of a signal gives a measure of the distribution of its power as a function of frequency. For example, the bandwidth of a signal is usually defined in terms of its spectrum (e.g., see reference 11). The four digital modulation techniques considered are three forms of phase-shift keying (PSK) and MSK. The three PSK schemes are BPSK, QPSK, and OQPSK. The data rate in each case is assumed to be $R = 1/T$ bits per second.

The spectrum of the BPSK signal is

$$S_1(f) = (A^2T/4)\{G_1[(f - f_c)T] + G_1[(f + f_c)T]\}$$

where the function G_1 is defined by

$$G_1(x) = [\text{sinc}(x)]^2 = [\sin(\pi x)/\pi x]^2$$

The spectrum of a QPSK signal with data rate $R = 1/T$ bits per second is given by

$$S_2(f) = (A^2T/4)\{G_2[(f - f_c)T] + G_2[(f + f_c)T]\}$$

where

$$G_2(x) = 2\, G_1(2x)$$

The spectrum of the OQPSK signal is the same as the spectrum of the QPSK signal. The spectrum of the MSK signal is

$$S_3(f) = (A^2T/4)\{G_3[(f - f_c)\, T] + G_3[(f + f_c)T]\}$$

where the function G_3 is defined by

$$G_3(x) = \{[(4/\pi)\cos(2\pi x)]/(1 - 16x^2)\}^2$$

The power in each of the PSK and MSK signals is $P = A^2/2$. Since the power is equal to the integral of the power spectral density,

$$\int_{-\infty}^{\infty} S_i(f)\, df = (A^2/2)\int_{-\infty}^{\infty} G_i(x)\, dx = P$$

for each i. The comparisons below of the spectra of PSK and MSK signals are made on the basis of equal power in the signals. It is also common to compare the signals on the basis of equal values of $S_i(f_c)$, the spectral density at the carrier frequency (e.g., reference 11), but in this case the signals have different power levels.

Notice that the spectra $S_1(f)$, $S_2(f)$, and $S_3(f)$ are given in the form

$$S_i(f) = (A^2T/4)\{G_i[(f - f_c)T] + G_i[(f + f_c)T]\}$$

so that comparisons of these spectra can be made by comparing the corresponding functions $G_1(x)$, $G_2(x)$, and $G_3(x)$. In fact, if $f_cT >> 1$ the two components of the spectrum do not overlap significantly; that is, for each integer i,

$$G_i[(f - f_c)T] >> G_i[(f + f_c)\, T]$$

for $f > 0$, and the reverse inequality holds for $f < 0$. It follows that $S_i(f)$ is approximately

$$S_i(f) = G_i[(f - f_c)T]$$

for $f > 0$ and

$$S_i(f) = G_i[(f + f_c)T\}$$

for $f < 0$. Consequently, for $f_cT >> 1$, the shape of spectrum $S_i(f)$ for $f > f_c$ is just the shape of $G_i(x)$ for $x > 0$. Also notice that $G_i(x) = G_i(-x)$.

The spectra of BPSK, QPSK, OQPSK, and MSK are compared in Figs. 6 and 7. In Fig. 6, $G_i(x)$ is shown as a function of x for all three values of i. The parameter x can be thought of as *normalized frequency*, since it is frequency (in hertz) divided by the data rate R (in bits per second). Note that x is therefore a dimensionless parameter.

Specifications on signal spectra (e.g., bandwidths) are often given in terms of the power spectral densities in decibels (dB). Fig. 7 shows $G_i(x)$ in decibels as a function of x. Since $G_i(x)$ is a power density, the conversion is $10 \log_{10}[G_i(x)]$. The *3-dB bandwidth* of $S_i(f)$ is the width of a frequency band outside of which the power spectral density is at least 3 dB below $S_i(f_c)$, its value at the center of the band. Typically, bandwidths are normalized by dividing by R, which is equivalent to specifying the bandwidth in terms of the normalized frequency. Since $10 \log_{10}(0.5)$ is approximately -3 dB, the 3-dB bandwidth is also termed the *half-power bandwidth*. In terms of the function G_i, the 3-dB, or half-power,

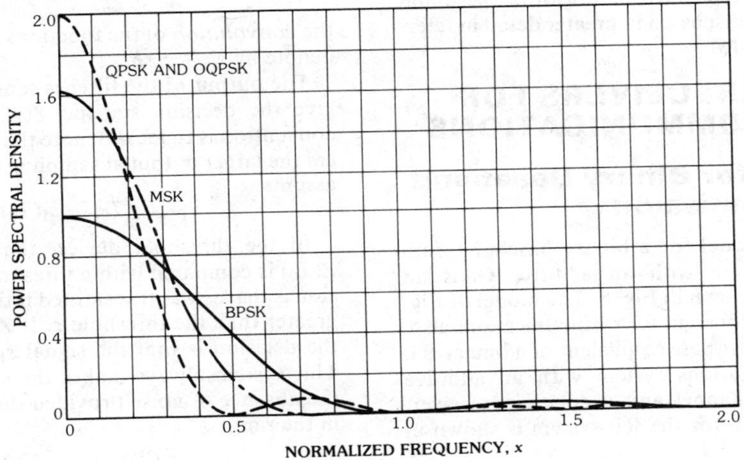

Fig. 6. Power spectral densities for PSK and MSK.

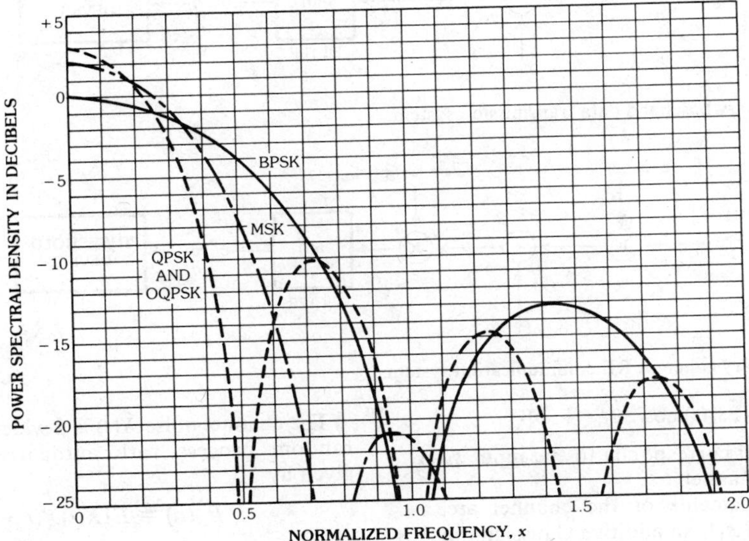

Fig. 7. Power spectral densities in decibels.

bandwidth is the largest value of y for which $G_i(y/2) = 0.5\ G_i(0)$.

Analogous definitions are given for other power-density levels (e.g., the 10-, 35-, or 50-dB bandwidths) in reference 11. The 3-dB bandwidths for the PSK and MSK signals are 0.88 for BPSK, 0.44 for QPSK and OQPSK, and 0.59 for MSK; and the 35-dB bandwidths are 35.12, 17.56, and 3.24, respectively.* The *null-to-null bandwidth* is the smallest value of y for which $G_i(y/2) = 0$. As can be seen in Figs. 6 and 7, the null-to-null bandwidths are 2.0 for BPSK, 1.0 for QPSK and OQPSK, and 1.5 for MSK. Notice that although MSK has lower side lobes and smaller 35-dB bandwidth than QPSK, QPSK has a narrower main lobe (i.e., smaller null-to-null bandwidth) and a smaller 3-dB bandwidth. These and related issues are discussed in greater detail in references 11 through 15.

OPTIMUM RECEIVERS FOR DIGITAL COMMUNICATIONS

Receivers for Binary Baseband Data Transmission

The general model for a binary baseband data transmission system with an additive Gaussian noise channel is shown in Fig. 8. The model of Fig. 8 is employed for coherent RF communications also, because it is the low-pass equivalent of a binary RF digital communications system with an additive Gaussian noise channel and a coherent receiver. The block diagram for the RF system is shown in Fig. 9. The input,

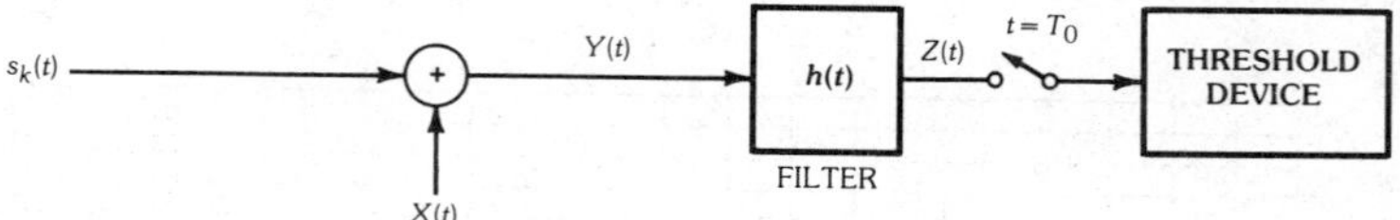

Fig. 8. Model of binary baseband data transmission system.

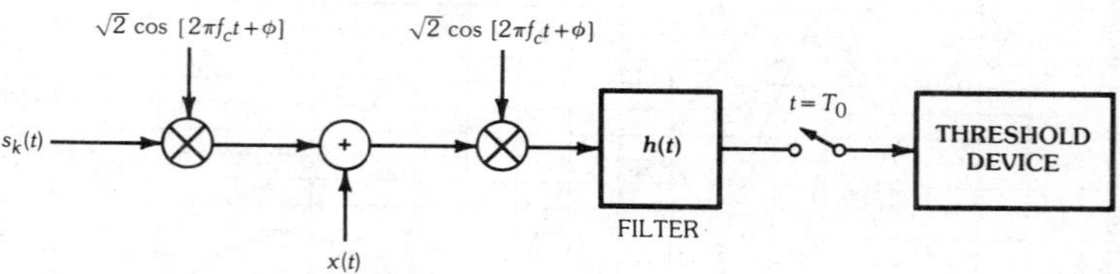

Fig. 9. Model of binary coherent RF communication system.

$$s(t) = \sqrt{2}s_k(t)\cos[2\pi f_c t + \phi]$$

for the RF channel corresponds to the input $s_k(t)$ for the baseband channel.

The principal elements of the channel are a binary signal set $\{s_0, s_1\}$, an additive Gaussian noise process $X(t)$, a linear filter with impulse response $h(t)$, a sampler, and a threshold device. The binary digit k is sent during the nth time interval by transmitting the signal $s_k(t - nT)$ for $nT < t < (n+1)T$. For convenience, we consider the first time interval (i.e., $n = 0$) in all that follows. For this time interval, the channel output $Y(t)$ is the sum of the signal $s_k(t)$ plus the noise $X(t)$.

The channel output is the input to the linear time-invariant filter. The filter output $Z(t)$ is the sum of a filtered version of the signal, which is denoted by $\hat{s}_k(t)$, and a filtered version of the noise. The signal component of the filter output is given by

$$\hat{s}_k(t) = \int_{-\infty}^{\infty} s_k(u)\, h(t-u)\, du$$

the *convolution* of the functions s_k and h, which we denote by $\hat{s}_k = s_k{*}h$.

The output of the filter is sampled at time T_0 to give the decision statistic $Z(T_0)$. Our notational convention is to let s_0 denote the signal that produces the larger output at sampling time T_0; that is, we assume

$$\hat{s}_0(T_0) > \hat{s}_1(T_0)$$

In the threshold device, the decision statistic $Z(T_0)$ is compared with a threshold z, and the decision is that s_0 was transmitted if the statistic $Z(T_0)$ is greater than the threshold z. If $Z(T_0)$ is less than z, the decision is that the signal s_1 was transmitted. This system always makes the correct decision in the absence of noise provided that the threshold is in the range

$$\hat{s}_0(T_0) > z > \hat{s}_1(T_0)$$

The channel noise $X(t)$ is a wide-sense stationary random process with autocorrelation function given by

$$R_X(u) = E\{X(t)X(t+u)\}$$

where $E\{Y\}$ denotes the expected value of a random variable Y. Without loss of generality, the channel

* Reference 11.

noise process can be assumed to have zero mean (the mean value can always be absorbed in the signal set). The autocorrelation function for the noise at the output of the filter is given by

$$R(u) = \int_{-\infty}^{\infty} R_X(v) f(u - v)\, dv$$

where the function f is defined by

$$f(t) = \int_{-\infty}^{\infty} h(v) h(v + t)\, dv$$

These two expressions are equivalent to $R = R_X * f$ and $f = h * g$, respectively, where the function g is defined by $g(t) = h(-t)$. That is, g is the time reverse of the impulse response of the filter. Since $f(t) = f(-t)$, $h(v + t)$ can be replaced by $h(v - t)$ in the definition of the function f, but the form given above is usually easier to evaluate.

Let $P_{e,k}$ denote the probability that an error is made by the receiver when the signal $s_k(t)$ is the input to the channel. The error probability $P_{e,k}$ can be thought of as a *conditional probability of error* given that $Y(t) = s_k(t) + X(t)$. That is, if H_k denotes the event that the signal $s_k(t)$ is transmitted, the probability $P_{e,k}$ is the conditional probability of error given H_k.

When the signal $s_0(t)$ is transmitted, the receiver makes an error if $Z(T_0)$ is less than the threshold z. Thus,

$$P_{e,0} = P[Z(T_0) < z \mid H_0]$$

Given that $s_0(t)$ is the transmitted signal, $Z(T_0)$ is a Gaussian random variable that has mean $\hat{s}_0(T_0)$ and variance $\sigma^2 = R(0)$. Consequently,

$$P_{e,0} = Q[\{\hat{s}_0(T_0) - z\}/\sigma]$$

where the function Q is defined by

$$Q(x) = \int_{x}^{\infty} (2\pi)^{-1/2} \exp(-y^2/2)\, dy$$

The function $1 - Q(x)$ is the distribution function of a zero mean, unit variance, Gaussian random variable. A discussion of the function Q and its properties can be found in reference 7.

Similarly, if $s_1(t)$ is transmitted, $Z(T_0)$ is a Gaussian random variable with mean $\hat{s}_1(T_0)$ and variance $\sigma^2 = R(0)$, and an error occurs if $Z(T_0)$ is greater than the threshold z. Therefore, the error probability $P_{e,1}$ is

$$P_{e,1} = P[Z(T_0) > z \mid H_1] = Q[\{z - \hat{s}_1(T_0)\}/\sigma]$$

Optimum Threshold for General Gaussian Noise—The choice of the threshold depends on the performance measure under consideration. One measure of performance that is often used is the maximum of the error probabilities,

$$P_{e,m} = \max\{P_{e,0}, P_{e,1}\}$$

The *minimax criterion* for selecting the threshold is to choose the threshold z to give the smallest possible value of $P_{e,m}$. The value of z that minimizes $P_{e,m}$ is given by

$$z_m = [\hat{s}_0(T_0) + \hat{s}_1(T_0)]/2$$

and this is referred to as the *minimax threshold.* Notice that the minimax threshold depends on the signal set $\{s_0, s_1\}$, the impulse response of the filter, and the sampling time T_0.

It is convenient to introduce the signal

$$s_d(t) = \{s_0(t) - s_1(t)\}/2$$

and let $\hat{s}_d(t)$ denote the output of the filter when $s_d(t)$ is the input (i.e., $\hat{s}_d = s_d * h$). If the minimax threshold is employed, the error probabilities $P_{e,0}$ and $P_{e,1}$ are given by

$$P_{e,0} = P_{e,1} = Q[\hat{s}_d(T_0)/\sigma]$$

Whenever these two probabilities are equal, we denote the common value by P_e.

As an example, consider an antipodal signal set (discussed earlier in this chapter). Since $s_0(t) = -s_1(t)$, the signal $s_d(t)$ is given by $s_d(t) = s_0(t)$. The error probability for the minimax threshold is

$$P_e = Q[\hat{s}_0(T_0)/\sigma]$$

There are other criteria for selecting the threshold z, but they typically require additional information about the data sequence that is to be transmitted. For instance, if the data sequence is modeled as a random process with a known probability distribution, the average probability of error can be employed as a criterion. If p_k denotes the *a priori probability* that the binary digit k is transmitted, the *average probability of error* is

$$P_{e,A} = P_{e,0}\, p_0 + P_{e,1}\, p_1$$

The threshold that minimizes $P_{e,A}$ is given by

$$z_A = z_m + 2\,\sigma^2 [\hat{s}_d(T_0)]^{-1} \ln(p_1/p_0)$$

If the data bits 0 and 1 are equally probable (i.e., $p_0 = p_1$), the thresholds z_A and z_m are the same. Otherwise, the second term on the right-hand side of the expression for z_A is a nonzero term that biases the decision in favor of the data bit with the largest a priori probability.

For applications in which unequal costs are associated with the two types of errors and the a priori probabilities p_i are known, it may be desirable to select the threshold to minimize the average cost. This is the well-known *Bayes criterion,* which is discussed in references 2, 5, and 6.

The Matched Filter—A white Gaussian noise process $X(t)$ has spectral density $S_X(f) = N_0/2$ for all f. This is just the Fourier transform of a delta function that has area $N_0/2$. If the channel noise process has this flat spectral density, the autocorrelation function of the noise at the output of the filter is given by

$$R(u) = (N_0/2) f(u)$$

where $f = h*g$, the convolution of the impulse response and its time-reverse. Letting $u = 0$, we find

$$\sigma^2 = R(0) = N_0 \, ||h||^2/2$$

The norm of the function h is defined by

$$||h|| = \{ \int_{-\infty}^{\infty} h^2(t) \, dt \}^{1/2}$$

Consider first the minimax criterion. The optimum threshold depends on the filter impulse response $h(t)$, and for each choice of the filter impulse response, the corresponding minimax threshold gives the smallest possible value of $P_{e,m}$. Let P_e denote this minimum value of $P_{e,m}$ (recall that $P_e = P_{e,0} = P_{e,1}$ if the minimax threshold is used).

The value of P_e depends on the impulse response of the filter. For a system with an *additive white Gaussian noise channel,* the minimum possible value of P_e is achieved by the impulse response

$$h_M(t) = s_d(T_0 - t)$$

Multiplication of the filter impulse response by a positive constant c does not change the error probability if the corresponding minimax threshold is used, so $h_M(t)$ can be any positive constant multiple of $s_d(T_0 - t)$. A filter with this impulse response is called a *matched filter.*

Given a binary signal set $\{s_0,s_1\}$, let E_i denote the energy in the signal s_i, and define E_A by

$$E_A = [E_0 + E_1]/2$$

The signal correlation r' is the inner product

$$r' = (s_0,s_1) = \int_{-\infty}^{\infty} s_0(t) \, s_1(t) \, dt$$

and $r = r'/E_A$ is the correlation coefficient for the given signal set. The value of $|r|$ is never greater than one; moreover, $r = -1$ only for antipodal signals, and $r = +1$ only if $s_0(t) = s_1(t)$.

The norm of signal s_d is related to the energy by

$$||s_d||^2 = E_A(1 - r)/2$$

The matched filter impulse response $h_M(t) = cs_d(T_0 - t)$ has norm

$$||h_M||^2 = c^2 \, ||s_d||^2$$

If this filter is employed in the receiver, the minimax threshold is $z_m = c(E_0 - E_l)/4$, and the signal $\hat{s}_d(T_0)$ is given by

$$\hat{s}_d(T_0) = ||s_d|| \; ||h_m|| = c \; ||s_d||^2$$

It follows that the error probability for a receiver with a matched filter and minimax threshold is

$$P_e = Q\{[E_A(1 - r)/N_0]^{1/2}\}$$

This is the smallest error probability that can be achieved (if a single bit is transmitted) with *binary* baseband data transmission or *binary* coherent rf communications over an additive white Gaussian noise channel.

Two important points should be made concerning the above expression for the minimum error probability. First, the result does not depend on the sampling time T_0. This is because the matched filter automatically compensates for the sampling time by incorporating a delay in its impulse response. As a consequence of this, the sampling time can be selected for convenience of implementation. In particular, if the signal set $\{s_0,s_1\}$ is *time limited* to the interval $[0,T]$, the matched filter is causal for any choice of T_0 not less than T. For such a signal set, the customary choice for the sampling time is $T_0 = T$.

The second point is that the minimum probability of error depends on the three parameters E_A, N_0, and r. The energy E_A can be increased only by increasing the power in the received signal, and the noise density can be decreased only by lowering the noise level in the receiver. Both of these require increased cost and complexity in the communication system. However, the parameter r can be decreased by proper signal design. As previously mentioned, r can be no smaller than -1, and $r = -1$ if and only if the signals are antipodal. Antipodal signals have equal energy: $E_A = E_0 = E_1$. As a consequence, the subscripts on E can be omitted, and the error probability for antipodal signaling, an additive white Gaussian noise channel, and a matched-filter receiver is given by

$$P_e = Q\{[2E/N_0]^{1/2}\}$$

where E is the energy per data bit.

The Correlation Receiver—The matched-filter receiver is shown in Fig. 10. The decision statistic $Z(T_0)$ for this receiver can be expressed in terms of the channel output $Y(t)$ by

$$Z(T_0) = \int_{-\infty}^{\infty} Y(T_0 - u) \, h_M(u) \, du$$

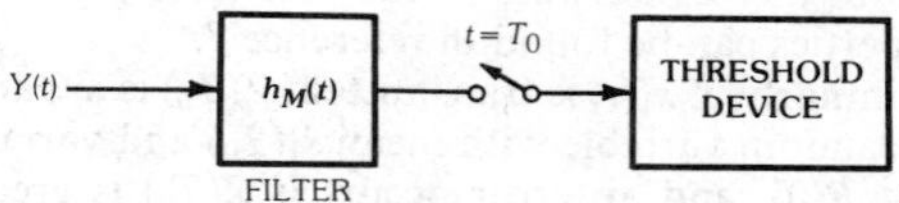

Fig. 10. Matched filter receiver.

The impulse response of the matched filter is defined in terms of the binary signal set $\{s_0,s_1\}$ by

$$h_M(t) = \{s_0(T_0 - t) - s_1(T_0 - t)\}/2$$

In terms of the signal

$$s_d(t) = \{s_0(t) - s_1(t)\}/2$$

the matched filter impulse response is

$$h_M(t) = s_d(T_0 - t)$$

The statistic $Z(T_0)$ can therefore be expressed as

$$Z(T_0) = \int_{-\infty}^{\infty} Y(T_0 - u)\, s_d(T_0 - u)\, du$$

If the signals s_0 and s_1 are time limited to $[0, T]$ and the sampling time is $T_0 = T$, this expression reduces to

$$Z(T) = \int_0^T Y(t)\, s_d(t)\, dt$$

Thus, $Z(T)$ can be obtained by multiplying the channel output $Y(t)$ by the signal $s_d(t)$ and integrating from 0 to T, as shown in Fig. 11. The receiver that processes the channel output in this manner is known as the *correlation receiver*. The correlation receiver gives the same error probability as the matched-filter receiver.

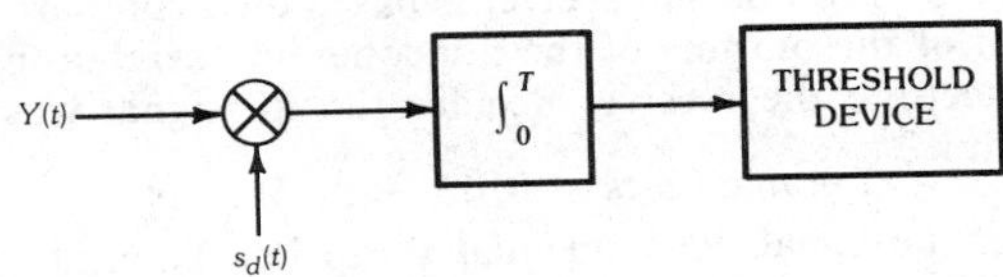

Fig. 11. Correlation receiver.

Coherent Receivers for Binary and Quaternary RF Signals

For binary RF communications with signals of the form

$$s(t) = \sqrt{2}\, s_k(t) \cos[2\pi f_c t + \phi]$$

the matched-filter receiver for an additive white Gaussian noise channel is the receiver portion of the system shown in Fig. 9 with $h(t) = h_M(t)$. As in baseband systems, the correlation receiver can be substituted for the matched filter in coherent RF communication systems. The correlation receiver for coherent RF communications is shown in the diagram of Fig. 12.

Notice from Fig. 12 that the receiver must know the carrier phase ϕ. In practice, it suffices to have a good estimate of this phase, and there are many different kinds of tracking loops that can be employed in the receiver to provide such an estimate. A presentation of phase-tracking loops and related synchronization devices may be found in reference 3.

Binary phase shift keying (BPSK) employs binary RF signals with $s_d(t) = A/\sqrt{2}$ for $0 \leq t < T$ (see subsection on BPSK). The optimum correlation receiver for coherent reception of BPSK is shown in Fig. 13. This is a special case of the receiver of Fig. 12. The square of the BPSK signal contains an unmodulated carrier at frequency $2f_c$ and phase 2ϕ, and this double-frequency carrier can be used to provide an estimate of the phase ϕ (this is the squaring loop*). This estimate is employed as the phase reference for coherent demodulation. An alternative method for coherent demodulation is to have the detection filter in the loop such as with a Costas loop or an *I-Q* loop.*

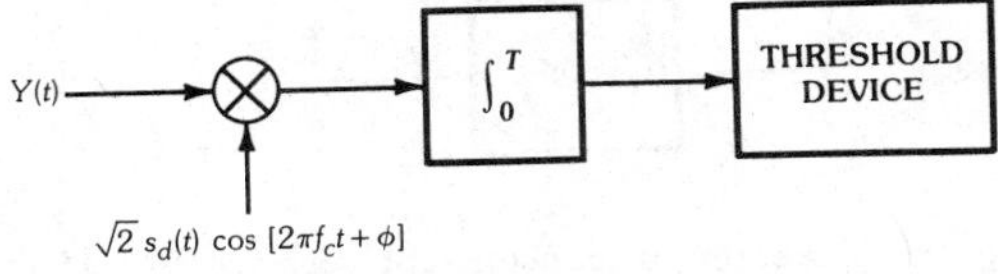

Fig. 12. Coherent correlation receiver for RF signals.

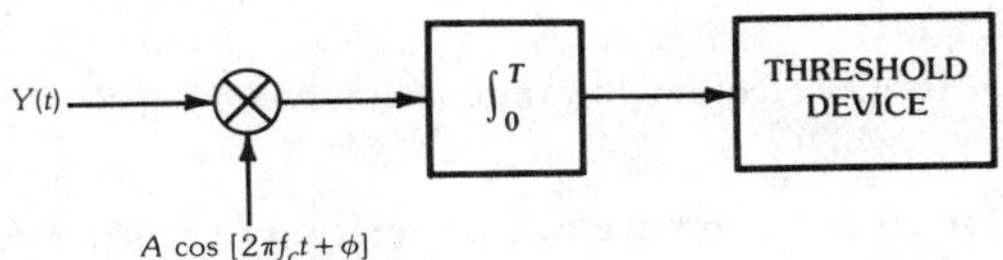

Fig. 13. Coherent correlation receiver for BPSK.

The coherent receivers for QPSK, OQPSK, and MSK are of the form shown in Fig. 14. These receivers consist of two branches, and each branch is a correlation receiver for an appropriate binary signal. The coherent references for the two branches differ in phase by $\pi/2$ radians, and in general there is a timing offset T' between the in-phase and quadrature branches. If the data rate is $R = 1/T$ bits per second in each case, then $T' = T$ for OQPSK and MSK, but $T' = 0$ for QPSK. The waveform $v(t)$ is the rectangular pulse $p_{2T}(t)$ for QPSK and OQPSK, and it is the sine pulse $\sin\,(\pi t/2T)\, p_{2T}(t)$ for MSK.

Noncoherent Receivers

The signals employed in noncoherent communications are binary and M-ary signals of the type described in the subsections on FSK and MFSK. The signal set consists of M equal-energy signals of the form

$$s_k(t) = A a_k(t) \cos[2 f_c t + \theta_k(t) + \phi_k]$$

where $a_k(t)$ and $\theta_k(t)$ are baseband signals with bandwidths much smaller than the carrier frequency f_c.

For noncoherent demodulation, it is assumed that the receiver has no knowledge of the phase ϕ_k and that the phases for different signals are unrelated (e.g., ϕ_k and ϕ_j are statistically independent for $k \neq j$). Moreoever, for some applications, a time-varying phase shift is introduced by the channel, so the phase of the received signal may change from one interval to the next.

The key component of the optimum noncoherent receiver for the signal set given above and an additive white Gaussian noise channel is the *noncoherent correlation detector* (*NCD*), which is also

---

* Reference 3.

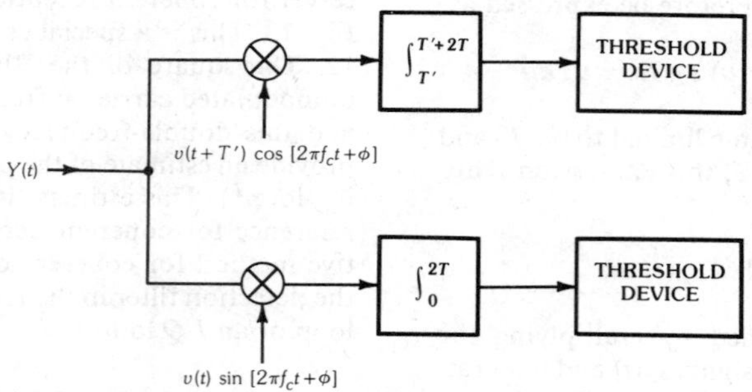

Fig. 14. Coherent receiver for quaternary RF signals.

referred to as an envelope correlation detector.* A block diagram of the noncoherent correlation detector for the signal $s_k(t)$ is shown in Fig. 15. This detector consists of in-phase and quadrature correlation receivers followed by squaring devices and a summing device. Thus the output of the noncoherent correlation detector is the square-law combination of the outputs of two correlation receivers whose phase references differ by $\pi/2$ radians. For convenience, let NCD/k denote the noncoherent correlation detector for the kth signal in the set.

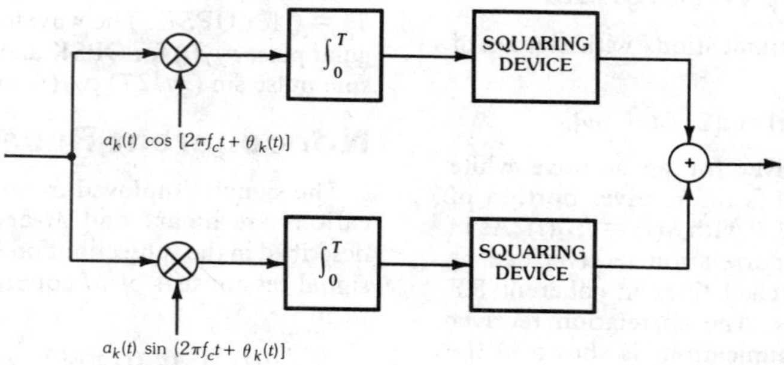

Fig. 15. Noncoherent correlation detector (NCD/k).

Let $P_{e,k}$ denote the probability that an error is made by the receiver when the signal $s_k(t)$ is sent. Two performance measures of interest are the maximum probability of error

$$P_{e,m} = \max\{P_{e,k}:k = 0, 1, \ldots, M-1\}$$

and the average probability of error

$$P_{e,A} = P_{e,0}p_0 + P_{e,1}p_1 + \cdots + P_{e,M-1}p_{M-1}$$

In the expression for the average probability of error, p_k denotes the a priori probability that the signal $s_k(t)$ is sent. The receiver that gives the minimum possible value of $P_{e,m}$ also gives the minimum possible value of $P_{e,A}$ for equally probable signals (i.e., $p_k = 1/M$ for each k). The block diagram for this optimum receiver is shown in Fig. 16. The decision of the receiver is based on a comparison of the outputs of the noncoherent correlation detectors: the receiver decides the ith signal was sent if

$$Z_i = \max\{Z_k:k = 0, 1, \ldots, M-1\}$$

If the signals have unequal a priori probabilities, some additional processing of the outputs of the noncoherent correlation detectors is necessary in order to give the minimum average probability of

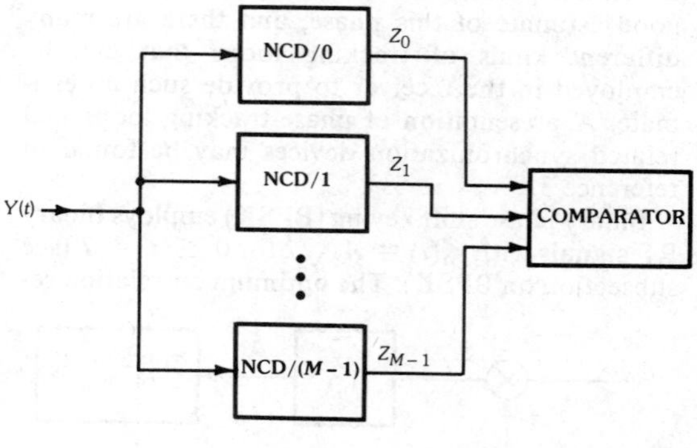

Fig. 16. Diagram of noncoherent receiver for M-ary communications.

* Reference 6.

error.* This is also necessary for signals with unequal energy.

Since MFSK is a special case of M-ary signaling, an optimum receiver for MFSK can be obtained from Figs. 15 and 16. The signals for MFSK are

$$s_k(t) = A \cos[2\pi f_k t + \phi_k]$$

so the noncoherent correlation detector for MFSK uses reference signals $\cos(2\pi f_k t)$ and $\sin(2\pi f_k t)$ as shown in Fig. 17. An optimum noncoherent receiver for MFSK is obtained by employing this noncoherent correlation detector in the system of Fig. 16. An optimum noncoherent receiver for binary FSK is this system with $M = 2$.

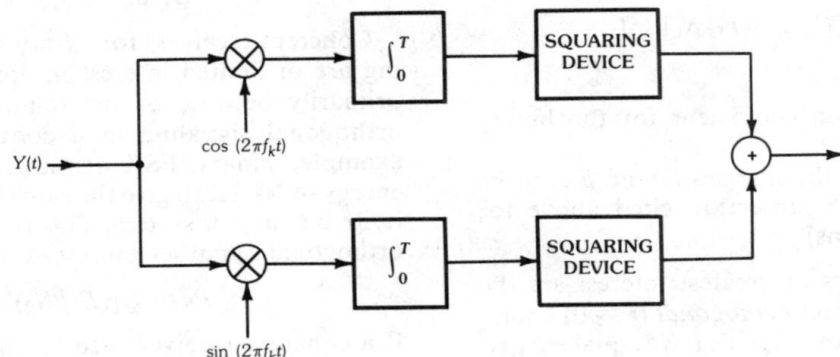

Fig. 17. Noncoherent correlation detector (NCD/ k) for MFSK signal set.

If the channel noise is a band-limited white Gaussian noise process and the noise bandwidth is small compared to the tone frequencies f_k, there is an alternative implementation of the optimum noncoherent receiver that utilizes an *envelope detector*. In order to describe the response of the envelope detector to narrow-band inputs, define the signal

$$x_k(t) = v(t) \cos(2\pi f_k t) + w(t) \sin(2\pi f_k t)$$

in terms of baseband signals $v(t)$ and $w(t)$ that are narrow-band with respect to the frequency f_k. If $x_k(t)$ is the input to an envelope detector, the output is

$$e(t) = [v^2(t) + w^2(t)]^{1/2}$$

The signal $e(t)$ is called the *envelope* of the narrowband signal $x_k(t)$.

Let $h_k(t)$ be the impulse response of a filter matched to the kth tone of the MFSK signal set; that is, let

$$h_k(t) = \cos(2\pi f_k t)$$

In contrast to the matched filters of the preceding subsection, this filter is *noncoherently matched* to the signal

$$s_k(t) = A \cos[2\pi f_k t + \phi_k]$$

since no phase information is utilized in $h_k(t)$. The output of the system of Fig. 18 at time T is equivalent to the output of the noncoherent correlation detector. Thus the matched filter/envelope detector can be substituted for the noncoherent correlation detector in the system of Fig. 16 to obtain an alternative form of the optimum receiver.

ERROR PROBABILITIES FOR DIGITAL COMMUNICATION SYSTEMS

In the preceding section, the optimal receivers are given for the most commonly used digital modulation techniques. Although there are a large number of different modulation techniques, many of them give the same performance. Throughout this section, performance will be measured by the *bit error probability* P_b as a function of the *bit energy to noise density ratio* E_b/N_0. Issues such as bandwidth and complexity of implementation will not be considered (see the subsection titled "Spectra of PSK and MSK Signals" for some comments on bandwidth of PSK and MSK signals).

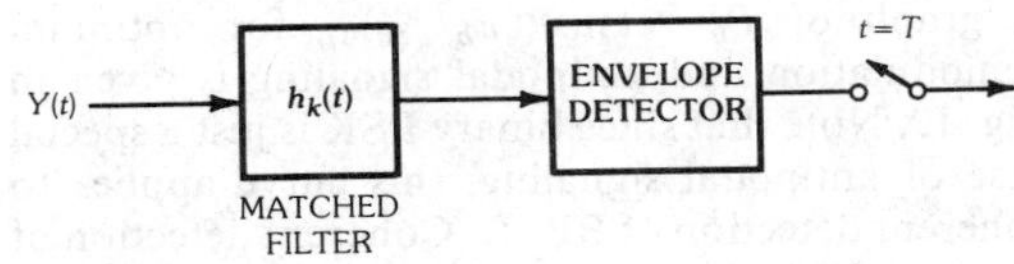

Fig. 18. Noncoherent matched filter/envelope detector.

Most binary communication systems of practical interest fall into two categories: coherent detection of antipodal signals and noncoherent detection of orthogonal signals. Moreover, certain quaternary signal sets such as QPSK, OQPSK, and MSK have the same bit error probability as binary antipodal signaling. For M-ary modulation, the primary case of interest is orthogonal signaling with noncoherent demodulation.

* Reference 6.

Coherent Systems With Additive White Gaussian Noise Channels

As discussed in the subsection titled "Receivers for Binary Baseband Data Transmission," binary coherent RF communication systems can be analyzed in terms of their low-pass equivalent systems. Thus, the results of that subsection give the error probabilities for optimum coherent demodulation of binary RF signaling over additive white Gaussian noise channels. If the channel noise has spectral density $N_0/2$, the error probability for binary signaling (baseband or RF) with an optimum receiver is

$$P_b = Q\{[E_A(1 - r)/N_0]^{1/2}\}$$

where,

- r is the correlation coefficient for the binary signal set,
- E_A is the average of the energies E_0 and E_1 for the two signals (see subsection cited above for formal definitions).

The binary signal sets of greatest interest are the *antipodal* ($r = -1$) and *orthogonal* ($r = 0$) equal-energy signals. If the two signals have equal energy, $E_A = E_0 = E_1$, and thus the *energy per bit* is E_A regardless of which bit is transmitted. For equal energy signals (binary or M-ary), it is customary to specify the performance in terms of the energy per bit, which is denoted by E_b.

It follows from the above comments that the bit error probability for optimum coherent detection of binary antipodal signals is

$$P_b = Q[(2E_b/N_0)^{1/2}]$$

For most applications, it is convenient to give the bit energy to noise density ratio in decibels. For this purpose, we define

$$(E_b/N_0)_{dB} = 10 \log_{10}(E_b/N_0)$$

A graph of P_b versus $(E_b/N_0)_{dB}$ for optimum demodulation and antipodal signaling is given in Fig. 19. Note that since binary PSK is just a special case of antipodal signaling, this curve applies to coherent detection of BPSK. Coherent detection of binary ASK with antipodal signals is another special case.

Many of the quaternary communication systems have the same bit error probability as binary antipodal signaling with an optimum receiver. The bit error probability for optimum coherent demodulation of QPSK, OQPSK, or MSK is given by

$$P_b = Q[(2E_b/N_0)^{1/2}]$$

exactly the same as for optimum coherent demodulation of BPSK. Consequently, the performance curve for QPSK, OQPSK, and MSK is the curve in Fig. 19 that corresponds to binary antipodal signaling.

Because the error probability $Q[(2E_b/N_0)^{1/2}]$ arises so frequently in the analysis of the performance of coherent communication systems, a short table of its values is given in Table 1. The table lists $(E_b/N_0)_{dB}$, E_b/N_0, and the error probability $Q[(2E_b/N_0)^{1/2}]$.

If the phase reference is not perfect, such as when it is estimated by some kind of tracking loop, the error probability for BPSK and related modulation schemes will be larger than given above. The *phase error* ϕ_e is the difference between the reference phase and the phase of the received signal. If ϕ_e is constant for the duration of the data bit, the bit error probability for BPSK is

$$P_b = Q[(2E_b/N_0)^{1/2} \cos(\phi_e)]$$

Coherent receivers for M-ary orthogonal signaling are of limited interest in applications. This is primarily because of the relative inefficiency of orthogonal signaling in a coherent system. For example, binary FSK requires twice the signal energy of BPSK to give the same bit error probability in a coherent system. This is true of any binary orthogonal signal set, since $r = 0$ implies

$$P_b = Q[(E_b/N_0)^{1/2}]$$

If a coherent receiver is to be employed, it makes sense to use a modulation scheme that is more efficient than orthogonal signaling.

The error probabilities for coherent demodulation of M-ary orthogonal signals are given in reference 3. The error probabilities for noncoherent demodulation of M-ary orthogonal signals are given in the next subsection. For large values of M, the performance of the coherent system is not much better than the performance of the noncoherent system (especially for low error probabilities). This is another reason there is little interest in implementing the more complex coherent receiver if M-ary orthogonal modulation is employed. For further discussion of this and related topics see reference 3.

Noncoherent Systems With Additive White Gaussian Noise Channels

The performance curve for binary FSK can be obtained by considering binary, orthogonal, equal-energy signals. For such signals, the bit error probability for optimum noncoherent detection is

$$P_b = 0.5 \exp(-E_b/2N_0)$$

which is also shown in Fig. 19.

Closely related to binary orthogonal signaling, although not always viewed as such, is *differential phase-shift keying* (*DPSK*). In DPSK, binary digits are transmitted at a rate of $1/T$ bits per second by phase transitions at the ends of successive T-second intervals. A phase change of 180 degrees can represent a binary 0, and the absence of a phase

TABLE 1. VALUES OF ERROR PROBABILITY $Q[(2E_b/N_0)^{1/2}]$

$(E_b/N_0)_{dB}$	E_b/N_0	$Q[(2E_b/N_0)^{1/2}]$
5.0	3.16	5.95×10^{-3}
5.5	3.55	3.86×10^{-3}
6.0	3.98	2.39×10^{-3}
6.5	4.47	1.40×10^{-3}
7.0	5.01	7.73×10^{-4}
7.5	5.62	3.99×10^{-4}
8.0	6.31	1.91×10^{-4}
8.5	7.08	8.40×10^{-5}
9.0	7.94	3.36×10^{-5}
9.5	8.91	1.21×10^{-5}
10.0	10.00	3.88×10^{-6}
10.5	11.22	1.09×10^{-6}
11.0	12.59	2.62×10^{-7}
11.5	14.13	5.34×10^{-8}
12.0	15.85	9.03×10^{-9}

change can represent a binary 1. This is simply differentially encoded PSK, and it can be demodulated either coherently or noncoherently.

The optimum noncoherent demodulator for DPSK is based on a noncoherent correlation detector (Fig. 15) in which the integration time is $2T$ seconds rather than T seconds. The optimality follows from the fact that DPSK is a form of orthogonal signaling on the interval $[0,2T]$: the signal corresponding to a 180-degree phase change is orthogonal to the signal corresponding to no phase change. Since the integration time is $2T$, the effective signal energy is $2E_b$. For optimum noncoherent demodulation of orthogonal signals of energy $2E_b$, the error probability is

$$P_b = 0.5 \exp(-E_b/N_0)$$

which is shown in Fig. 19. A more detailed discussion of DPSK is given in references 4 and 6.

Next we consider M-ary signaling with equal-energy, orthogonal signals. That is, the signal set $\{s_k: k = 0, 1, \ldots, M-1\}$ is such that $\|s_k\|^2 = E_s$ for all k, and $(s_k, s_j) = 0$ whenever $k \neq j$. The energy in the signal s_k is called the *energy per symbol* (denoted by E_s). A symbol error occurs if the receiver decides that s_j was sent when in fact s_k was sent and $k \neq j$. For an additive white Gaussian noise channel and an optimum noncoherent receiver, the symbol error probability P_s is given by

$$P_s = M^{-1} \sum_{n=2}^{M} \binom{M}{n} (-1)^n \exp\{-[(n-1)/n]E_s/N_0\}$$

This can be evaluated on a digital computer for values of M up to 32, although double precision computation may be required for values of M larger than about 16. The symbol error probability can be computed for larger values of M by numerical integration of an alternative expression for P_s; for example, see equation (10-15) on p. 489 of reference 3.

If M-ary modulation is employed to transmit binary data, M should be a power of 2. If $M = 2^m$, each M-ary symbol can represent a unique sequence of m binary digits. The energy per bit is $E_b = E_s/m$. A symbol error can cause as few as one or as many as m bit errors. On the average, the number of bit errors per symbol error is $mM/2(M-1)$ for M-ary orthogonal signaling with optimum noncoherent demodulation. Thus the bit error probability P_b is related to the symbol error probability by $P_b = MP_s/2(M-1)$. Making the appropriate substitutions in the above expression for P_s, we find that the bit error probability is given in terms of the energy per bit by

$$P_b = (2M-2)^{-1} \sum_{n=2}^{M} \binom{M}{n} (-1)^n \exp\{-[(n-1)/n]mE_b/N_0\}$$

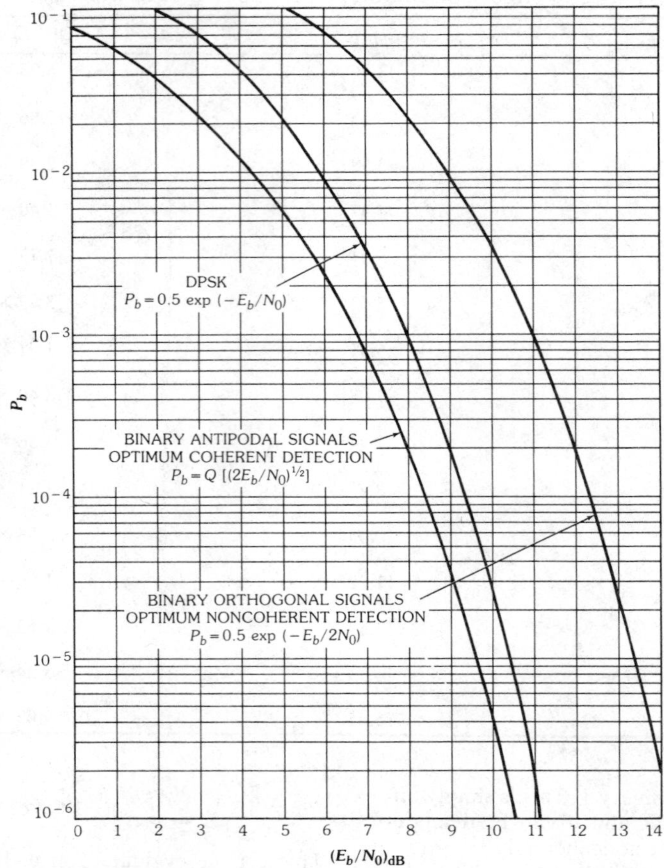

Fig. 19. Bit error probabilities.

A table of values of the symbol error probability is given in Table 2 for $M = 2$, 8, and 32. The bit error probabilities can be obtained from this table by multiplying the symbol error probabilities by $M/2(M-1)$. Curves of the bit error probability versus the bit energy to noise density ratio are shown in Fig. 20 for $M = 2, 4, 8, 16$, and 32. It can be shown that $P_b \rightarrow 0$ as $M \rightarrow \infty$ provided that

TABLE 2. SYMBOL ERROR PROBABILITIES FOR OPTIMUM NONCOHERENT DEMODULATION OF M-ARY ORTHOGONAL SIGNALING

$(E_b/N_0)_{dB}$	Symbol Error Probability, P_s $M = 2$	$M = 8$	$M = 32$
3.0	1.84×10^{-1}	1.05×10^{-1}	5.03×10^{-2}
4.0	1.42×10^{-1}	5.38×10^{-2}	1.71×10^{-2}
5.0	1.03×10^{-1}	2.25×10^{-2}	4.03×10^{-3}
6.0	6.83×10^{-2}	7.25×10^{-3}	6.03×10^{-4}
7.0	4.08×10^{-2}	1.67×10^{-3}	5.08×10^{-5}
8.0	2.13×10^{-2}	2.53×10^{-4}	2.10×10^{-6}
9.0	9.42×10^{-3}	2.27×10^{-5}	3.64×10^{-8}
10.0	3.37×10^{-3}	1.06×10^{-6}	2.15×10^{-10}

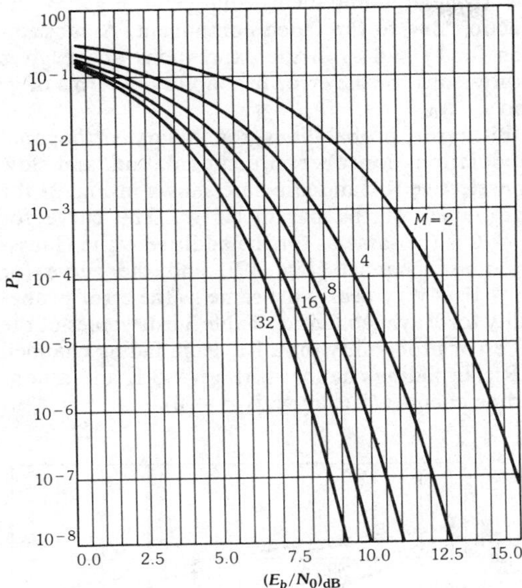

Fig. 20. Bit error probabilities for M-ary orthogonal signals and optimum noncoherent demodulation.

E_b/N_0 is greater than ln 2 (where ln denotes the natural logarithm).

Noncoherent Systems With Nonselective Fading Channels

In some communication channels, there are several propagation paths between the transmitter and the receiver. The signal components arriving at the receiver from different paths add destructively or constructively depending on the difference in the propagation times and on any phase shifts introduced by the medium (e.g., reflections from objects). Thus the signal strength depends on the relative phasing of the components of the received signal, and this gives rise to the condition known as *fading*.

If the differential propagation times for the paths are small compared with the data symbol duration, and the path strengths, propagation times, and phase shifts are nearly constant for the duration of the symbol, then there is no significant dispersion of the signal. In this case, the fading is referred to as *slow nonselective fading*. If the differential propagation times are large compared with the data symbol duration, signal dispersion will occur. Such channels are known as *time-dispersive* or *frequency-selective* fading channels. These channels may produce significant levels of intersymbol interference. A more complete discussion of fading and its effects on communication systems is given in reference 4, and the derivations of the results given in this section can be found in references 8, 9, and 10.

For *slow nonselective Rician fading*, the received signal is the sum of a nonfaded version of the transmitted signal and a slow nonselective Rayleigh faded version. The difference in propagation times for the two components is small enough that the channel is nonselective. The nonfaded component of the received signal is called the *specular component*, and the Rayleigh faded component is sometimes called the *random* or *scatter component*. The specular component may result from a direct path between the transmitter and receiver, and the faded component may arise from a large number of reflections.

Consider the transmission of a unit-amplitude RF signal over this Rician fading channel. The received signal is the sum of a deterministic component of amplitude A and a random component of amplitude V. The quantities A and V represent the gains associated with the specular path and the scatter path, respectively. The random variable V has the *Rayleigh distribution,* and its probability density function is

$$f_V(v) = (2v/m_f)\exp\{-v^2/m_f\}$$

for $v > 0$, where m_f denotes the *second moment* or *mean-square value* of the amplitude of the faded component (i.e., $m_f = E\{V^2\}$). The phase between the two components is random and uniformly distributed on the interval $[0, 2\pi]$. The amplitude of the sum of the specular and faded components is a random variable Y which has a *Rician distribution.* The probability density function for the Rician distribution is

$$f_Y(y) = (2y/m_f)\exp\{(y^2 + A^2)/m_f\}\, I_0(2yA/m_f)$$

for $y > 0$, where I_0 is the zero-order modified Bessel function. The *second moment* for this distribution is $E\{Y^2\} = A^2 + m_f$. Notice that for $A = 0$ (no specular component), the Rician density reduces to the Rayleigh density, because $I_0(0) = 1$.

Let E' denote the transmitted energy per symbol in an M-ary noncoherent communication system. The energy per symbol in the received signal is Y^2E'. The average energy per symbol in the received signal is

$$E_s = E\{Y^2\}\,E' = (A^2 + m_f)\,E'$$

Let $e_1 = A^2E'/N_0$ and $e_2 = m_fE'/N_0$ be the symbol energy to noise density ratios for the specular and faded components, respectively. The conditional probability of symbol error given $Y = y$ is (see preceding subsection)

$$P_s(y) = M^{-1}$$

$$\sum_{n=2}^{M} \binom{M}{n}(-1)^n \exp\{-[(n-1)/n]\,y^2E'/N_0\}$$

The *average probability of symbol error* is

$$P_s = \int_0^\infty P_s(y)\,f_Y(y)\,dy$$

$$= M^{-1} \sum_{n=2}^{M} \binom{M}{n} (-1)^n \, G[(n-1)/n, e_1, e_2]$$

where the function G is defined by

$$G(c,e_1,e_2) = (1 + ce_2)^{-1} \exp\{-ce_1/(1 + ce_2)\}$$

The *bit error probability* is obtained by multiplying the symbol error probability P_s *by* $M/2(M-1)$.

Two special cases are of interest. First, for the Rayleigh fading channel, $A = 0$ and $e_1 = 0$. It follows that $e_2 = E_s/N_0$. Moreover,

$$G(c,0,e_2) = (1 + ce_2)^{-1}$$

so that the error probability for noncoherent reception of M-ary orthogonal signals with nonselective Rayleigh fading is given by

$$P_s = M^{-1} \sum_{n=2}^{M} \binom{M}{n} (-1)^n \{1 + [(n-1)/n]E_s/N_0\}^{-1}$$

The second important special case is the *nonfading* channel. If the received signal has no faded component, then $m_f = 0$, $e_2 = 0$, and $e_1 = E_s/N_0$. Since

$$G(c,e_1,0) = \exp\{-ce_1\}$$

the error probability is

$$P_s = M^{-1} \sum_{n=2}^{M} \binom{M}{n} (-1)^n \exp\{-[(n-1)/n]E_s/N_0\}$$

as given in the preceding subsection.

For binary orthogonal signaling, each symbol represents one bit, so $P_s = P_b$. The resulting expression for the bit error probability in a Rician fading channel is

$$P_b = 0.5\ G(0.5,e_1,e_2)$$
$$= (2 + e_2)^{-1} \exp\{-e_1/(2 + e_2)\}$$

For Rayleigh fading, $e_1 = 0$ and $e_2 = E_s/N_0$. Since $E_b = E_s$ for binary signals, the bit error probability is

$$P_b = [2 + (E_b/N_0)]^{-1}$$

It is customary to present numerical data for the error probability in a Rician fading channel in terms of the ratio of the power in the two components of the received signal. Let $g^2 = m_f/A^2$, so that $e_2 = g^2 e_1$. The signal-to-noise ratios e_1 and e_2 are related to E_s/N_0 by

$$e_1 = [1/(1 + g^2)]\ E_s/N_0$$

and

$$e_2 = g^2 e_1 = [g^2/(1 + g^2)]\ E_s/N_0$$

The Rician channel can be specified by giving either e_1 and e_2 or g^2 and E_s/N_0. In the latter specification, E_s/N_0 is a measure of the average signal-to-noise ratio at the receiver, and g^2 specifies how the signal power is distributed between its two components. Specifically, $1/(1 + g^2)$ is the fraction due to the specular component, and $g^2/(1 + g^2)$ is the fraction due to the faded component. A substitution for e_1 and e_2 in the expressions above gives expressions for the error probability in terms of g^2 and E_s/N_0.

Bit error probabilities for binary orthogonal modulation, noncoherent demodulation, and slow nonselective Rician fading are shown in Fig. 21 for five values of the parameter g^2. The curve for $g^2 = 0$ is the same as for the additive white Gaussian noise channel (Fig. 19), and the curve for $g^2 = 10$ is very nearly the same as the error probability for Rayleigh fading. Table 3 lists values of the bit error probability for a Rayleigh fading channel, a Rician fading channel with $g^2 = 0.1$, and a nonfading channel (Rician with $g^2 = 0$).

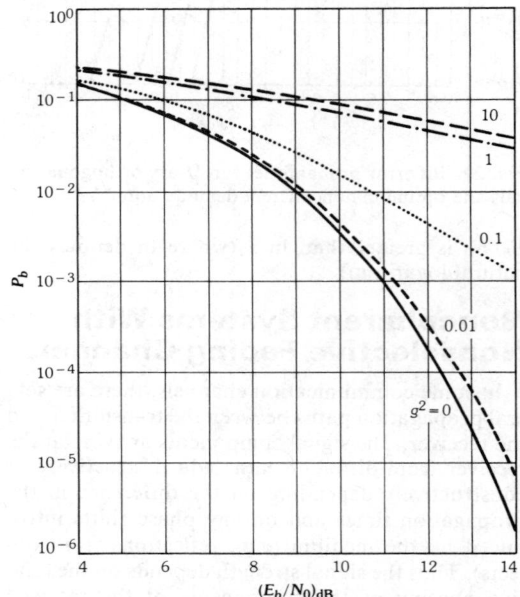

Fig. 21. Bit error probability for Rician fading.

SPREAD-SPECTRUM COMMUNICATIONS

A common trade-off in a communication system is performance versus bandwidth. That is, various aspects of communications performance can be improved at the expense of increased RF bandwidth. One example of this is the reduction of error probability in MFSK systems as the number of tones increases. More generally, the use of error-correcting coding gives lower error probabilities but requires a larger bandwidth than for uncoded signaling with the same information rate (see Chapter 25, Information Theory and Coding).

Spread-spectrum modulation is a very general method for improving other measures of system

TABLE 3. BIT ERROR PROBABILITIES FOR BINARY ORTHOGONAL SIGNALING, SLOW NONSELECTIVE FADING, AND NONCOHERENT DEMODULATION

(E_b/N_0) dB	Rayleigh	Bit Error Probability $g^2 = 0.1$	No Fading
4.0	2.22×10^{-1}	1.61×10^{-1}	1.42×10^{-1}
6.0	1.67×10^{-1}	9.15×10^{-2}	6.83×10^{-2}
8.0	1.20×10^{-1}	4.18×10^{-2}	2.13×10^{-2}
10.0	8.33×10^{-2}	1.51×10^{-2}	3.37×10^{-3}
12.0	5.60×10^{-2}	4.41×10^{-3}	1.81×10^{-4}
14.0	3.69×10^{-2}	1.13×10^{-3}	1.76×10^{-6}
20.0	9.80×10^{-3}	2.48×10^{-5}	9.64×10^{-23}

performance through the use of wideband signals. A properly designed spread-spectrum communication system can operate reliably in the presence of various types of radio-frequency interference (RFI) including multipath interference, multiple-access interference, and hostile jamming. In addition, when the signal power is spread over a large bandwidth, the signal has very small average power in any narrow-band slot. This means that the spread-spectrum system can share a frequency band with several narrow-band systems. It also means that it is difficult to detect the presence of the spread-spectrum signal by use of narrow-band equipment. The basic modulation formats and receiver structures are described below. Further details on the design and performance of spread-spectrum systems are given in references 16 through 24.

Direct-Sequence Spread-Spectrum Communications

The type of spread spectrum known as *binary direct-sequence spread spectrum* employs signals of the form

$$s(t) = A\, a_s(t)\, a_d(t) \cos[2\pi f_c t + \phi]$$

where,

$a_d(t)$ is a binary baseband data signal,

$a_s(t)$ is a baseband *spectral-spreading signal* with a bandwidth that is large compared to the data rate.

These binary baseband signals are of the form described in the section on baseband signal sets. Although only binary direct-sequence modulation is described here, there is a natural generalization to *quaternary direct sequence,* which is described in reference 16.

The data signal $a_d(t)$ consists of a sequence of positive and negative rectangular pulses; that is, it can be expressed as

$$a_d(t) = \sum_{n=-\infty}^{\infty} b_n b_T(t - nT)$$

where,

$p_T(t)$ is the rectangular pulse duration T,

b_n is either $+1$ or -1, depending on the data bit to be sent in the nth interval.

Similarly, the spectral-spreading signal can be written as

$$a_s(t) = \sum_{j=-\infty}^{\infty} a_j v(t - jT_c)$$

where $v(t)$ is the *chip waveform,* a time-limited pulse of duration T_c. The parameter T_c is called the *chip duration,* and the sequence

$$(a_j) = \ldots, a_{-1}, a_0, a_1, a_2, \cdots$$

is called the *signature sequence.* For reasonable choices of signature sequence and chip waveform, $1/T_c$ is a rough estimate of the bandwidth of the spread-spectrum signal. For most direct-sequence systems, $T_c << T$, so the bandwidth of the spread-spectrum signal is much larger than that of the data signal.

Usually, the data pulse duration T is an integer multiple of the chip duration. If $T = NT_c$ for some integer N, there are N chips per data pulse, and the bandwidth of the spread-spectrum signal is roughly N times the data rate. Usually the signature sequence is periodic, and the period p is equal to N or else $p >> N$, depending on the nature of the application.

The optimum receiver for binary direct-sequence spread-spectrum modulation and an additive white Gaussian noise channel (i.e., thermal noise is the only interference) is the correlation receiver shown in Fig. 22 (cf. Fig. 13). The received signal $Y(t)$ is multiplied by

$$A\, a_s(t) \cos[2\pi f_c t + \phi]$$

and the product is integrated over an appropriate interval of length T (this is illustrated for the interval $[0, T]$ in Fig. 22). Since the signals corresponding to $b_n = +1$ and $b_n = -1$ are antipodal, the bit error probability for binary direct-sequence spread-

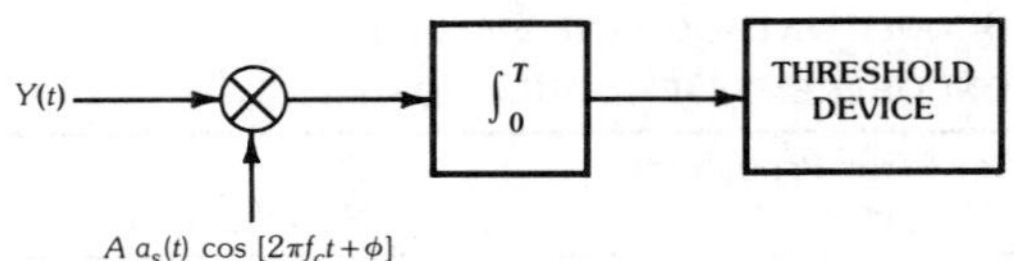

Fig. 22. Coherent correlation receiver for binary direct-sequence spread spectrum.

spectrum modulation and optimum coherent demodulation is

$$P_b = Q[(2E_b/N_0)^{1/2}]$$

the same as for binary BPSK with optimum coherent demodulation.

The motivation for considering direct-sequence spread spectrum is not performance improvement for channels in which the primary interference is thermal noise. Rather, it is the fact that the spectral-spreading signal can be selected to give the transmitted signal certain desirable properties and enable the receiver to discriminate against forms of interference other than thermal noise. For example, if the spread-spectrum signal is corrupted by additive RFI, and if we ignore the thermal noise for the moment, the received signal can be written as $Y(t) = s(t) + x(t)$, where $x(t)$ is an additive interference signal. The effects of this interference can be reduced by choosing the spectral-spreading signal to minimize

$$\int_0^T x(t)\, a_s(t) \cos[2\pi f_c t + \phi]\, dt$$

One application in which additive RFI is encountered is in a direct-sequence spread-spectrum *multiple-access* communication system.* In such a system, signals from the various transmitters have the same carrier frequency f_c. The spread-spectrum signals from other transmitters appear as additive RFI to a receiver that attempts to demodulate the signal $s(t)$. Let $s'(t)$ denote one of these other signals, and denote its spectral-spreading signal by $a'_s(t)$ and its signature sequence by (a'_j). Since the signal $s'(t)$ is from a different transmitter than $s(t)$, it arrives at the receiver with a different propagation delay and phase angle. If u denotes the propagation delay and ϕ' denotes the phase for $s'(t)$, the quantity to be minimized is

$$\int_0^T a'_s(t-u)\, a_s(t)\, dt \cos(\phi - \phi')$$

Since this should be small for all values of the phase angles, it is necessary to minimize the magnitude of

$$\int_0^T a'_s(t-u)\, a_s(t)\, \mathrm{dt}$$

for all values of u. The goal is to select the signature sequences to accomplish this minimization (see references 16 and 20).

* Reference 16.

Similar considerations arise for channels with specular multipath, narrow-band interference, hostile jamming, and other forms of RFI. For complete discussions of the problems of signature sequence selection see references 16, 19, and 20. The problems of obtaining phase and timing references for direct-sequence spread-spectrum communication systems are discussed in reference 18. The performance of direct-sequence spread spectrum with various forms of RFI is considered in references 16, 17, and 22 through 24.

Frequency-Hop Spread-Spectrum Communications

Signals with very large RF bandwidths can be generated by a method known as *frequency hopping* in which the carrier frequency of a digital communication signal is changed, or "hopped," over a wide range of frequencies. If the digital communication signal is

$$c(t) = A\, a(t) \cos[2\pi f_c t + \theta(t) + \phi]$$

the resulting *frequency-hopped* signal is

$$s(t) = A\, a(t) \cos[2\pi f(t)t + \theta(t) + \phi]$$

The function $f(t)$, which describes the carrier frequency as a function of time, is called the (frequency) *hopping pattern.* The hopping pattern is generated by applying a random or pseudorandom sequence of inputs to a frequency synthesizer. Typically, the available RF bandwidth is partitioned into q nonoverlapping frequency intervals called *slots*, and the q different frequencies generated by the frequency hopper are the center frequencies for these slots.

A frequency-hop (FH) spread-spectrum signal with hopping rate R_h hops per second is a signal that has the form of $s(t)$ above for which the frequency $f(t)$ can change every $1/R_h$ seconds. The frequency is constant on intervals of length $T_h = 1/R_h$. The parameter T_h is called the *frequency dwell time* or *hop interval.* In contrast to direct-sequence spread-spectrum signals, which occupy the full RF bandwidth at all times, FH spread-spectrum signals occupy only a small fraction of the RF bandwidth during a given hop interval.

Fast FH spread-spectrum systems have hopping rates that are larger than the data rate, and so the duration of a data pulse is larger than the hop interval. Since the transmission of a data pulse utilizes more than one of the q frequency slots, frequency diversity is obtained with fast FH spread-spectrum signaling. *Slow FH spread-spectrum systems* have hopping rates that are smaller than the data rate, and thus the hop interval is greater than the data symbol duration.

The total RF bandwidth of a slow FH spread-spectrum signal is approximately q times the bandwidth of the digital communication signal $c(t)$; it is virtually independent of the hopping rate. The total bandwidth of a fast FH spread-spectrum signal depends on the number of frequency slots and the hopping rate, but it does not depend very much on the data rate.

The multiple-access capability of frequency-hop spread spectrum is due to the fact that each signal occupies only the fraction $1/q$ of the bandwidth during each hop interval. Even for totally asynchronous operation of a larger number of transmitters,* the hopping patterns can be designed such that the probability of interference between the signals from any two given transmitters during a given hop interval is no more than $2/q$. Since errors occur with high probability whenever the signals interfere, some form of error-control coding is necessary for typical multiple-access systems. In fact, error-control coding is virtually a requirement for frequency-hop spread-spectrum communication in the presence of any form of partial-band or pulsed interference. Convolutional codes and Reed-Solomon block codes (see Chapter 25) appear to be the most suitable error-correcting codes for use in frequency-hop spread-spectrum communication systems.

REFERENCES

General References for Digital Communications

1. Golomb, S. W., ed. *Digital Communications With Space Applications.* Englewood Cliffs, N.J.: Prentice-Hall, Inc., 1964.
2. Helstrom, C. W. *Statistical Theory of Signal Detection.* 2d ed. New York: Pergamon Press, Inc., 1968.
3. Lindsey, W. C., and Simon, M. K. *Telecommunications Systems Engineering.* Englewood Cliffs, N. J.: Prentice-Hall, Inc., 1973.
4. Stein, S. Part III of *Communication Systems and Techniques.* New York: McGraw-Hill Book Co., 1966.
5. Van Trees, H. L. *Detection, Estimation, and Modulation Theory,* Part I. New York: John Wiley & Sons, Inc., 1968.
6. Weber, C. L. *Elements of Detection and Signal Design.* New York: McGraw-Hill Book Co., 1968.
7. Wozencraft, J. M., and Jacobs, I. M. *Principles of Communication Engineering.* New York: John Wiley & Sons, Inc., 1965.

Noncoherent Communication Over Nonselective Fading Channels

8. Turin, G. L. "Error Probabilities for Binary Symmetric Ideal Reception Through Nonselective Slow Fading and Noise." *Proceedings of the IRE,* Vol. 46, September 1958, pp. 1603–1619.
9. Lindsey, W. C. "Error Probabilities for Rician Fading Multichannel Reception of Binary and N-ary Signals." *IEEE Transactions on Information Theory,* October 1964, pp. 339–350 (reprinted in reference 10).
10. Brayer, K., ed. *Data Communications Via Fading Channels.* New York: IEEE Press, 1975.

Bandwidth-Efficient Modulation (QPSK, OQPSK, MSK, CPFSK)

11. Amoroso, F. "The Bandwidth of Digital Data Signals." *IEEE Communications Magazine,* November 1980, pp. 13–24.
12. Amoroso, F., and Kivett, J. A. "Simplified MSK Signaling Technique." *IEEE Transactions on Communications,* Vol. COM-25, April 1977, pp. 433–441.
13. de Buda, R. "Coherent Demodulation of Frequency-Shift Keying With Low Deviation Ratio." *IEEE Transactions on Communications,* Vol. COM-20, June 1972, pp. 429–435.
14. Gronemeyer, S. A., and McBride, A. L. "MSK and Offset QPSK Modulation." *IEEE Transactions on Communications,* Vol. COM-24, August 1976, pp. 809–820.
15 Pasupathy, S. "Minimum Shift Keying: A Spectrally Efficient Modulation." *IEEE Communications Magazine,* July 1979, pp. 14–22.

Spread-Spectrum Communications

16. Pursley, M. B. "Spread-Spectrum Multiple-Access Communications." In *Multi-User Communication Systems.* G. Longo, ed. Vienna and New York: Springer-Verlag, 1981; pp. 139–199.
17. Pickholtz, R. L., Schilling, D. L., and Milstein, L. B. "Theory of Spread-Spectrum Communications—A Tutorial." *IEEE Transactions on Communications,* Vol. COM-30, May 1982, pp. 855–884.
18. Holmes, J. K., *Coherent Spread Spectrum Systems.* New York: John Wiley & Sons, Inc., 1982.
19. MacWilliams, F. J., and Sloane, N. J. A. "Pseudo-Random Sequences and Arrays." *Proceedings of the IEEE,* Vol. 64, December 1976, pp. 1715–1729.
20. Sarwate, D. V, and Pursley, M. B., "Crosscorrelation Properties of Pseudorandom and Related Sequences." *Proceedings of the IEEE,* Vol. 68, May 1980, pp. 593–619.
21. Geraniotis, E. A., and Pursley, M. B. "Error

* Reference 21.

Probabilities for Slow-Frequency-Hopped Spread-Spectrum Multiple-Access Communications Over Fading Channels." *IEEE Transactions on Communications,* Vol. COM-30, May 1982, pp. 996–1009.

22. Special Issue on Spread-Spectrum Communications, *IEEE Transactions on Communications* ,Vol. COM-25, August 1977.

23. Special Issue on Mobile Spread-Spectrum Communications, *IEEE Transactions on Vehicular Technology,* Vol. VT-30, February 1981.

24. Special Issue on Spread-Spectrum Communications, *IEEE Transactions on Communications,* Vol. COM-30, May 1982.

25 Information Theory and Coding

Richard E. Blahut

Information theory is a discipline centered around a common mathematical approach to the study of the collection and manipulation of information. It studies the theoretical basis of such activities as observation, measurement, data compression, data storage, communication, estimation, decision making, and pattern recognition. Many complex and expensive systems are built for automating or expanding these operations. Information theory attempts to guide the development of such systems based on a study of the possibilities and limitations inherent in mathematics and probability theory.

The communication problem, represented in Fig. 1, is the archetypal problem of information theory; much of the underlying structure and semantics of information theory is suggested by the communication channel. A source of information is to be connected to a user of information by a channel. A communication system is provided to prepare the source output for the channel and to prepare the channel output for the user. It consists of a device between the source and the channel called the encoder/modulator, and another device between the channel and the user called the demodulator/decoder.

It is conventional to partition the major functions of a modern communication system in the manner shown in the block diagram of Fig. 2. Data from the data source are first processed by a source encoder, whose purpose is to represent the source data more compactly. A block of source data is represented by a sequence of symbols, usually binary, called the "source codeword." The data then are processed by the channel encoder, which transforms the symbols from a sequence of source codewords into another sequence of symbols called the "channel codeword." The channel codeword is a new, longer sequence that has more redundancy than the source codeword. Each symbol in the channel codeword might be represented by a bit or a group of bits. Next, the modulator represents each symbol of the channel codeword by its corresponding analog signal from a finite set of possible analog symbols. The sequence of analog symbols, called a "waveform," is transmitted through the channel. Because the channel is subject to various types of noise, distortion, and interference, the channel output differs from the channel input. The demodulator converts the noisy received sequence of analog symbols, possibly mutually interfering, into a sequence of discrete symbols of the channel codeword alphabet (based on a best estimate of the transmitted signal). The demodulated sequence of

Fig. 1. The communication problem.

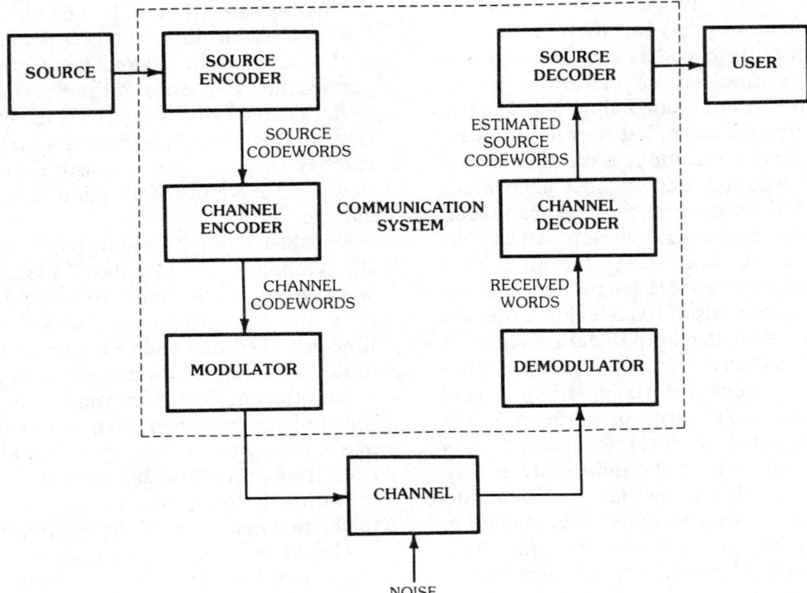

Fig. 2. Block diagram of a digital communication system.

symbols is called the "received word." Because of channel noise, the demodulator sometimes makes errors, so the symbols of the received word do not always match those of the channel codeword.

The function of the channel decoder is to use the redundancy in a channel codeword to correct the errors in the received word, and then to produce an estimate of the source codeword from it. If all errors are corrected, the estimated source codeword matches the original source codeword. The source decoder performs the inverse operation of the source encoder and delivers its output to the user. The source encoder and decoder are studied under the terms "data compaction" and "data compression." The channel encoder and decoder are commonly split into two functions: implementing error control to negate the effects of channel noise, and preparing the sequence of transmitted symbols to be compatible with channel constraints. These are studied under the terms "error control codes" and "constrained channel codes." The modulator and the demodulator are studied under the term "modulation theory."

Although we use the terminology of communication theory, the model is general and applies to a great variety of situations. One can interpret many other information-handling systems, such as mass storage systems, in terms of this model. It only is necessary to identify the boundaries between the boxes. This is arbitrary and depends on the goals of a particular analysis. Usually, the source, channel, and user are identified with those parts of the system that are fixed, and the encoder/decoder and modulator/demodulator are identified with those parts of the system that are subject to design. Therefore, in different circumstances, the identification of these functions may be different.

The operation of the encoders and decoders is to map strings of symbols from one alphabet into strings of symbols from a second alphabet. The two alphabets are often the same, but they need not be. A *block code* breaks the input data stream into blocks of fixed length k and encodes each block into a codeword of fixed length n; these are concatenated to form the output data stream. A *variable-to-fixed-length block code* breaks the input data stream into blocks of variable length and encodes the blocks into codewords of fixed length n that are concatenated to form the output data stream. A *fixed-to-variable-length block code* breaks the input data stream into blocks of fixed length k and encodes these into codewords of variable length that are concatenated to form the output data stream. A *tree code* breaks the input data stream into frames of length k_0 that are encoded into codeword frames of length n_0 with the encoding map depending on the previous m input data frames. The codeword frames are concatenated to form the output data stream. A tree code is called a variable-to-fixed-length tree code or a fixed-to-variable-length tree code when the input or output frames, respectively, are of variable length. A tree code with a finite encoding memory of m frames is called a *sliding block code* if the encoding operation is time invariant, and it is called a *convolutional code* if the encoding operation is both linear and time invariant.

Introductory information-theory textbooks intended for engineers are listed as references 1 through 4 at the end of this chapter. Other books and papers devoted to special topics will be cited in the appropriate section only if the topic is not treated within the general textbooks of the field.

CODING FOR NOISELESS CHANNELS

A discrete channel is a system by means of which an arbitrarily long sequence of symbols, each chosen from a finite set of I symbols $\{a_0, \ldots, a_{I-1}\}$, can be transmitted from one point to another. The transmission of symbol a_i requires a certain time duration, t_i seconds, which is not necessarily the same for all i. A noiseless channel is one in which the output is completely determined by the input—errors do not occur. It is not always true that all possible sequences of symbols from the set $\{a_i\}$ can be transmitted through the channel. Some channels, called constrained channels, forbid certain sequences of symbols from being transmitted.

Teletypewriters and telegraphy are two simple examples of discrete channels that are historically important. In the teletypewriter case, there are 32 symbols, each of the same duration, and any sequence of the 32 symbols is allowed. Each symbol can be used to represent five bits of information. If the system transmits r symbols per second, it is natural to say that the channel has a capacity of $5r$ bits per second. This does not mean that the teletypewriter channel will always be transmitting information at this rate. Whether or not the actual rate reaches this maximum possible rate depends on how the source of information is connected to the channel.

For the telegraphy channel, convention has fixed the symbols as a dot, a dash, a letter space, and a word space. We formalize these symbols as follows: (1) a dot, consisting of line closed for one unit of time and then line open for one unit of time; (2) a dash, consisting of three time units of closure and one unit open; (3) a letter space, consisting of three time units of line open; (4) a word space, consisting of six time units of line open. We also impose the restrictions on allowable sequences that no space may directly follow another space. This we take as the formal definition of the telegraphy channel.

The Morse code is one system of encoding information for this channel. However, one may properly question the efficiency of the Morse code. Is there a limit on the information that can be conveyed through the telegraphy channel, and does the Morse code achieve this limit? These questions are answered by information theory.

Capacity of Discrete Noiseless Channels

The capacity, C (in units of bits/second), of a discrete noiseless channel is defined by

$$C = \lim_{T \to \infty} (1/T) \log_2 N(T)$$

where $N(T)$ is the number of allowed sequences of symbols of duration T.

The limit in the definition will exist and be finite in most cases of interest. From the definition, it is clear that about 2^{CT} different messages can be transmitted through the channel in T seconds for large enough T. We say that the channel can transmit C bits per second. It is easily seen that in the teletypewriter example this definition of capacity reduces to the previous result of $5r$ bits per second.

The evaluation of capacity is more difficult if the symbols are of different length, as in the telegraphy channel, or if certain sequences are forbidden, also as in the telegraphy channel. Suppose first that all sequences of the symbols $a_0, \ldots, a_{I-1}$ are allowed and these symbols have durations $t_0, \ldots, t_{I-1}$ that are integer multiples of one time unit. Let $N(t)$ represent the number of sequences of duration t. We can set up a recursive equation to find $N(t)$. A sequence of length $N(t)$ can be produced from a sequence of length $N(t - t_0)$ by appending symbol a_0, or it can be produced from a sequence of length $N(t - t_1)$ by appending symbol a_1, and so on. That is

$$N(t) = N(t - t_0) + N(t - t_1) + \ldots + N(t - t_{I-1})$$

According to a well-known result in the study of finite-difference equations, $N(t)$ is asymptotic for large t to λ^t, where λ is the largest real root of the characteristic equation:

$$x^{-t_0} + x^{-t_1} + \ldots + x^{-t_{I-1}} = 1$$

Then the capacity, C, is equal to $\log_2 \lambda$.

Even when there are restrictions on allowed sequences, we may still be able to write down a different equation by inspection. In the case of the telegraphy channel:

$$N(t) = N(t-2) + N(t-4) + N(t-5) + N(t-7) + N(t-8) + N(t-10)$$

as we see by counting sequences of symbols according to the last or next-to-last symbol occurring. Hence, C equals $\log_2 \lambda$ where λ is the largest real zero of the polynomial $x^{-10} + x^{-8} + x^{-7} + x^{-5} + x^{-4} + x^{-2} = 1$. Solving this, we find $C = 0.539$ bit per unit of time. This is the maximum rate at which information can be conveyed by the telegraphy channel. One may devise many codes whose rates are close to the capacity of the channel, but never greater.

The Morse code shown in Table 1 is a widely used code for transmitting written text over the telegraphy channel. The Morse code combines both source encoding, described in the next section in terms of the *entropy* of the source, and channel encoding, described above. The Morse code does not exploit the Markov structure of natural language. We will evaluate the code for use with a memoryless source whose 27 output letters (including the space) occur with the same probabilities as in English text.

The entropy of this memoryless model of English is 4.03 bits per letter, and the telegraphy channel capacity is 0.539 bit per unit time. Hence, an optimum code uses an average of 7.48 units of signaling time per source output letter. The Morse code in Table 1 uses an average of 9.296 units of signaling time per source output letter, which is 124 percent of the time needed by the optimum code. This establishes the amount by which the Morse code could be improved for use with the memoryless model of English text. Of course, an optimum code may be too complex for an operator to learn. The Morse code is an excellent compromise between performance and simplicity.

State Diagrams and Trellises

A constrained channel is one that does not accept input sequences containing any of a certain collection of forbidden subsequences. Such a channel can be described by a state diagram. Each state, $s_0, s_1, \ldots, s_{m-1}$, corresponds to a recent past history of channel inputs. Loosely, we say that the channel is in one of these states, but we mean that its past history is described by that state. For each state, only certain symbols from the set $a_0, \ldots, a_{I-1}$ can be transmitted next. When one of these has been transmitted, the state changes to a new state depending both on the old state and on the particular symbol transmitted.

Sometimes the state diagram is augmented by a time axis so that one can see how the channel state changes with time. It is then called a *trellis*.

The telegraphy channel gives a simple illustration of a state diagram. There are two states. The state specifies whether or not a space was the last symbol transmitted. In state s_0, only a dot or a dash can be sent next, and the state always changes. In state s_1, any symbol can be transmitted, and the channel changes state if a space is sent; otherwise it remains in the same state. The conditions are indicated in the state diagram shown in Fig. 3.

The trellis diagram for the telegraphy channel is shown in Fig. 4. The Morse code is a collection of some of the paths through this trellis. An optimum code would use all of the paths, but the mapping

TABLE 1. THE MORSE CODE

Character	Per Letter Probability p_i	International Morse	Letter Duration l_i
Space	0.1859		6
A	0.0642	•–	9
B	0.0127	–•••	13
C	0.0218	–•–•	15
D	0.0317	–••	11
E	0.1031	•	5
F	0.0208	••–•	13
G	0.0152	––•	13
H	0.0467	••••	11
I	0.0575	••	7
J	0.0008	•–––	17
K	0.0049	–•–	13
L	0.0321	•–••	13
M	0.0198	––	11
N	0.0574	–•	9
O	0.0632	–––	15
P	0.0152	•––•	15
Q	0.0008	––•–	17
R	0.0484	•–•	11
S	0.0514	•••	9
T	0.0796	–	7
U	0.0228	••–	11
V	0.0083	•••–	13
W	0.0175	•––	13
X	0.0013	–••–	15
Y	0.0164	–•––	17
Z	0.0005	––••	15
	$\Sigma_i p_i \log p_i = 4.03$ Bits		$\Sigma_i p_i l_i = 9.296$ Time Units

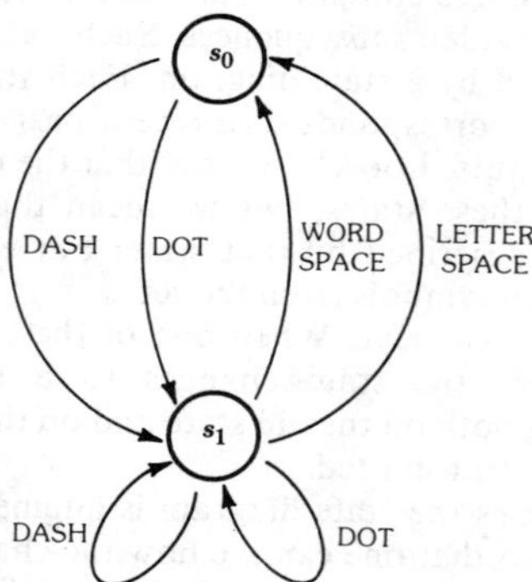

Fig. 3. State diagram for the telegraphy channel.

from the set of source output sequences to the set of trellis paths might be a complicated one.

The digital magnetic-recording channel is another important example of a constrained channel. The channel input alphabet is binary, and a typical set of channel constraints, called run-length constraints, are that at least r zeros follow every one and that no more than s zeros occur in sequence. These constraints are imposed to minimize intersymbol interference and to facilitate clock recovery at the receiver.

In another kind of application, certain bit sequences are used for higher-level protocols or punctuation in the communication system and so must be forbidden in the data. A code must be used to translate from the set of unconstrained binary sequences into the set of sequences that satisfy the constraints.

Fig. 5 is a state diagram for a channel with $r = 2$ and $s = 7$ run-length constraints. Fig. 6 is the trellis diagram for the same channel; the trellis is drawn with state s_2 as the initial state. The characteristic equation for this state diagram is found to be $x^8 - x^5 - x^4 - x^3 - x^2 - x - 1 = 0$. The capacity, C, of the channel, equal to the base-two logarithm of the largest zero of the characteristic polynomial, is 0.518 bit/bit. Codes exist that record 0.518 bit of information for each channel bit recorded, but no codes can do better. Practical codes in use fall a little short of the channel capacity. The Franaszek code* of Fig. 7 is a variable-to-variable-length code whose rate is 0.5. Hence, it achieves 96 percent of the channel capacity and yet is quite simple. A code that comes closer to channel capacity may be quite complex.

* Reference 5.

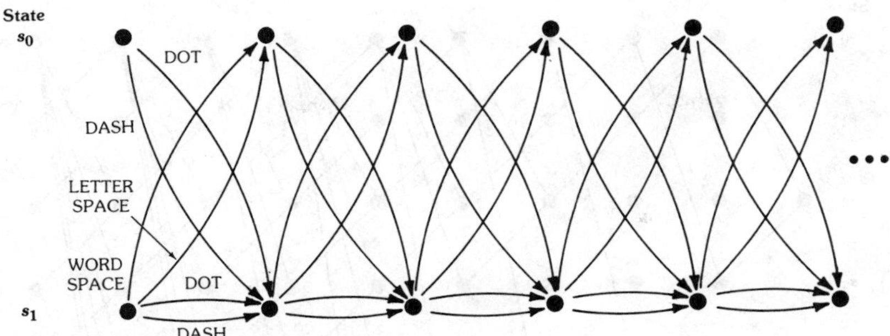

Fig. 4. Trellis diagram for the telegraphy channel.

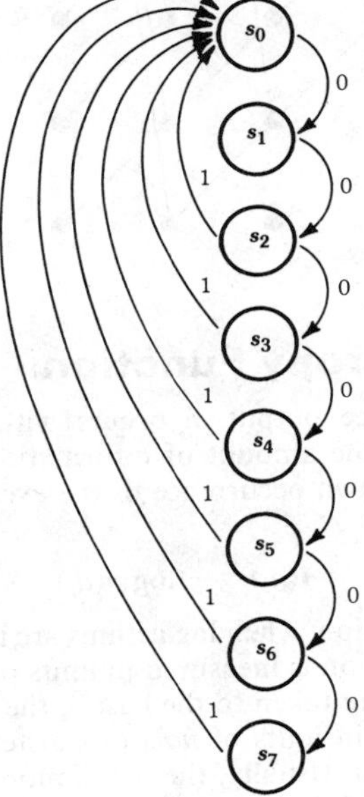

Fig. 5. State diagram of $r = 2$, $s = 7$ run-length-limited channel.

By inspection, one can see that any codeword of the Franaszek code followed by any other codeword satisfies the channel run-length constraints, and the codewords are uniquely decodable without punctuation. This last condition, known as the prefix condition, is due to the fact that no codeword looks like the beginning of any other codeword. This is an important requirement of a variable-length block code since the decoder must be able to break into codewords any concatenated string that is not punctuated.

Codes more elaborate than the Franaszek code are known, but they cannot be understood without developing more theory. The Adler-Hassner finite-state tree codes* encode a continuous bit sequence into a continuous bit sequence without word boundaries. Fig. 8 shows the encoder and decoder for an Adler-Hassner code for the $r = 1$, $s = 3$ run-length-limited channel. This channel has a channel capacity of 0.5515 bit/input. The rate one-half Adler-Hassner code implemented in Fig. 8 achieves 91 percent of the channel capacity. Binary symbols are clocked from the encoder into the channel and from the channel into the decoder at twice the source rate. Another Adler-Hassner code, say a rate 11/20 code, could achieve a larger fraction of a channel capacity, but the implementation would be much more complex.

* Reference 6.

SOURCE COMPACTION CODES

Source compaction codes are used to represent the output of a data source more efficiently. Data sources such as facsimile, voice, digital recording, data tables, or word text can produce many millions or billions of bits. However, in their natural form, the data from these sources can be highly redundant. Practical data compaction codes are now available that can reduce considerably the number of bits needed to encode many such sources.

Source Models

An information source produces messages by generating a sequence of letters from a fixed alphabet of permitted symbols called the source alphabet. The alphabet may be finite, in which case the source is called a discrete source, or the source alphabet may be continuous, such as the set of real numbers, in which case the source is called a continuous source. A source might also put out continuous functions on the time axis. Sampling techniques are available to make this source into a time-discrete source.

The output of a discrete information source is a random sequence of symbols from a finite alphabet containing J symbols given by $\{a_0, a_1, \ldots, a_{J-1}\}$. The sequence is produced according to some probability rule. The sources that are the easiest to study

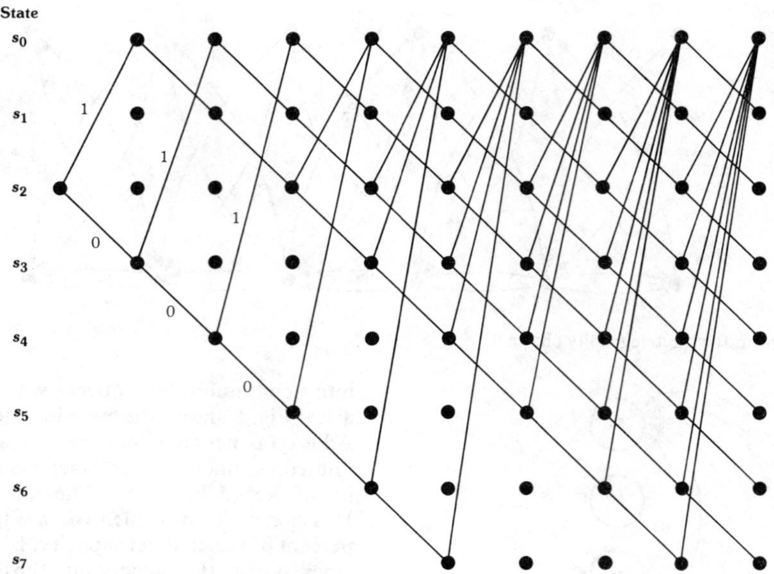

Fig. 6. Trellis diagram of $r = 2$, $s = 7$ run-length-limited channel.

mathematically are those known as ergodic sources. An *ergodic source* is one in which every sequence of symbols produced by the source is the same in statistical properties. If it is observed long enough, such a source will produce, with probability approaching one, a sequence of symbols that is "typical." In simple terms, this means that a sufficiently long sequence from the source will almost always produce the same frequency of occurrence of symbols and symbol combinations.

An information source is said to be *memoryless* if successive symbols generated by the source are statistically independent. That is, a source is memoryless if each symbol is selected without influence from all previous symbols. If previously selected symbols influence the selection of a symbol, then the source is said to possess memory. If the selection of a symbol is influenced only by the immediately preceding symbol, the source is known as a *Markov source.* If the selection is influenced by the m previously selected symbols, the source is called an *mth-order Markov* source.

INFORMATION WORDS	CODEWORDS
11	0100
10	1000
000	000100
010	001000
011	100100
0010	00001000
0011	00100100

Example:

1101010. . . ⟶ 01000010001000. . .

Fig. 7. The $r = 2$, $s = 7$ Franaszek rate one-half code.

The Entropy Function

If a source output a_j occurs with probability $p(a_j)$, then the amount of information associated with the known occurrence of the event is defined to be

$$I(a_j) = -\log p(a_j)$$

In the definition, when logarithms are to the base 2, the information is measured in units of *bits.* When logarithms are taken to the base *e,* the information is measured in units of *nats* (a shortened form of natural units). Usually, the nat is more convenient for theoretical developments, and the bit is more convenient for the final result. The conversion factor is: one nat equals 1.443 bits.

If the probabilities of selecting the source symbols are $p(a_0)$, $p(a_1)$, . . . , $p(a_{J-1})$, respectively, then the information generated each time a symbol is selected is $-\log_2 p(a_j)$ bits. Since the symbol a_j will, on average, be selected $np(a_j)$ times in a total of n selections, the average amount of information obtained from n selections is

$$np(a_0)\log_2 p(a_0)^{-1} + \ldots + np(a_{J-1})\log_2 p(a_{J-1})^{-1}$$

bits. Divide by n to obtain the average amount of information per symbol selection. This is known as the *average information,* the *uncertainty,* or the *entropy,* $H(p)$.

$$H(p) = -\sum_{j=0}^{J-1} p(a_j)\log_2 p(a_j) \text{ bits/symbol}$$

or

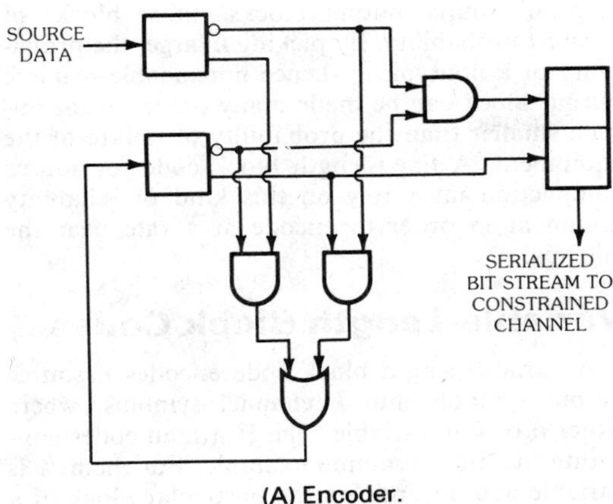

(A) Encoder.

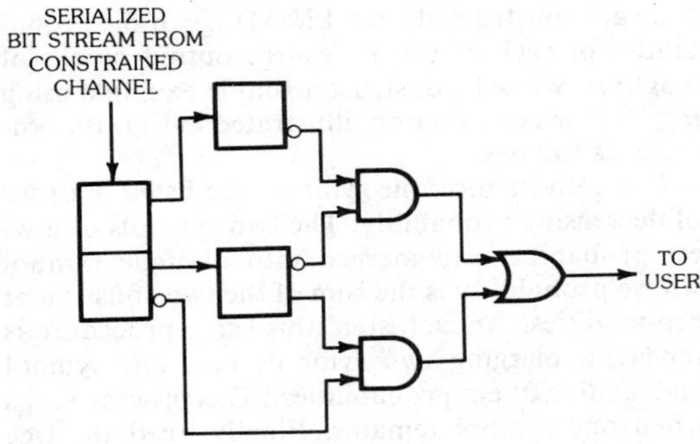

NOTES:
1. Every 1 followed by a 0.
2. At most three consecutive 0's.
3. Code rate equal to one half.
4. Channel symbol rate twice source symbol rate.

(B) Decoder.

Fig. 8. Encoder and decoder for an Adler-Hassner code.

$$H(p) = -\sum_{j=0}^{J-1} p_j \log_2 p_j$$

where p_j is an abbreviation of $p(a_j)$.

The entropy associated with n selections from the statistically independent set is equal to n times the entropy per single selection. The term entropy is used since the function is the same as that derived in statistical mechanics for the thermodynamic quantity entropy.

For a first-order Markov source, the memory can be expressed by a conditional probability distribution. This conditional probability may be written as $p(a_j\,|\,a_k')$ where a_j is any of the possible source symbols and a_k' is the source symbol emitted at the previous time instant.

The entropy is defined as an average

$$H(p) = -\sum_{k=0}^{J-1} p_k \sum_{j=0}^{J-1} p_{j\,|\,k} \log_2 p_{j\,|\,k}$$

where $p_{j\,|\,k}$ is an abbreviation of $p(a_j\,|\,a_k)$. This can also be rewritten

$$H(p) = -\sum_{k=0}^{J-1} \sum_{j=0}^{J-1} p_{jk} \log_2 p_{j\,|\,k}$$

where $p_{jk} = p_k p_{j\,|\,k}$.

The entropy function can be justified by its role in source compaction coding theorems, by its role in combinatorics, and by its intuitive properties as a measure of uncertainty. These properties are as follows:

1. The entropy function is continuous in p. This is a reasonable property, since it says that small changes in the probability distribution only make small changes in the uncertainty.
2. $H \geq 0$ and $H = 0$ if and only if all the p_j are zero except one which is unity. This a reasonable property of a measure of uncertainty, since if the outcome of a selection is sure, there is no uncertainty.
3. For a given n, $H(p) \leq \log n$ and is equal to $\log n$ if and only if the p_j are equal to $1/n$. This is a reasonable property which says that the more symbols available for selection without preference, the larger the uncertainty.

A binary source is one that contains only two symbols, a_0 and a_1; their probabilities are given by p and $1 - p$, respectively. The binary entropy function

$$H_2(p) = -p \log_2 p - (1-p) \log_2 (1-p)$$

is shown in Fig. 9.

The output of a binary source is a binary digit. The distinction between the bit used as a measure of information and the bit used as a binary output symbol should be carefully noted. Fig. 9 shows that on average the amount of information provided by a binary source is always equal to or less than 1 bit/bit (one bit of information per data bit). The binary source provides one bit of information for each selected symbol only when the two symbols are equiprobable.

Source Encoding

The entropy function can be interpreted as the average amount of information necessary to specify which symbol has been produced by the source. If a source selects n symbols, where n is a very large number, then with high probability, it will select a sequence from the set of 2^{nH} different typical sequences each having a probability of occurrence of about $(1/2)^{nH}$. This is a direct physical interpretation of H. It means that, theoretically, a very long typical sequence of n q-ary symbols selected by the

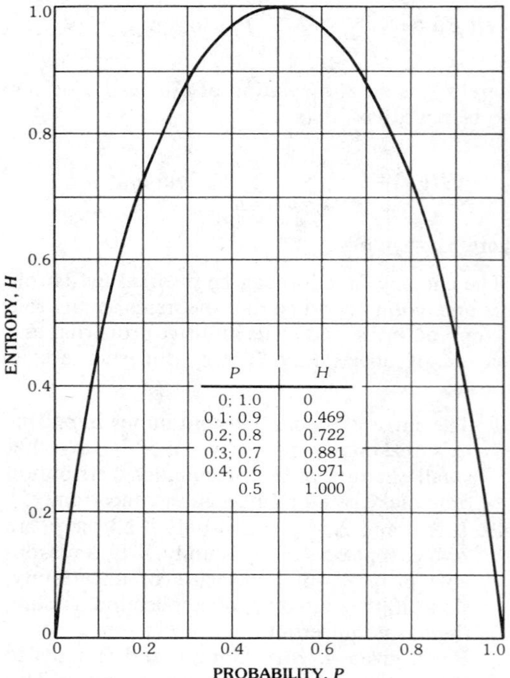

Fig. 9. The binary entropy function.

source can be encoded and retransmitted using only nH binary digits, each carrying one bit of information.

Usually, the binary codeword will be embedded in some longer binary text. Sometimes punctuation symbols are used and are not counted as part of the codeword. However, the more enlightened view is to incorporate the punctuation symbols into the code alphabet and devise an even more compact code. Then the code must be a self-punctuating code. Symbols to punctuate the code become implicit to the codeword.

Fixed-Length Block Codes

A fixed-length block code needs no punctuation. Compaction of a source with J symbols encodes n source output symbols into a codeword of length k bits, where n and k are fixed. There are J^n possible source output blocks the length n, so the block-length of a binary code must be $\log_2 J^n$ bits if every possible output block is to be encoded. This requires $\log_2 J$ codeword bits per source output symbol. In general, this is much greater than the source entropy rate.

A fixed-length block code can encode at the entropy rate but then must allow the possibility of an error. A source compaction block code with rate a little above the entropy provides only $n(H(p) + \epsilon)$ bits, so $2^{n(H(p)+\epsilon)}$ codewords are assigned to the "typical" source output blocks—those blocks of greatest probability. By picking n large, the probability of a nontypical—hence noncodable—source output block can be made many orders of magnitude smaller than the probability of failure of the equipment. A fixed-length block code for source compaction must rely on this kind of reliability argument in order to encode at a rate near the entropy.

Variable-Length Block Codes

A variable-length block code encodes n source output symbols into k channel symbols, where either n or k is variable. The Huffman codes constitute the most common example. For them, k is variable and depends on the particular block of n symbols observed, while n is fixed.

The variable-length codewords of a Huffman code are constructed from knowledge of the probabilities of each of the J^n source output blocks of length n. We will construct a simple example using $n = 1$. The construction, illustrated in Fig. 10, proceeds as follows.

For convenience, the symbols are listed in order of decreasing probability. The two symbols of lowest probability are merged into a single symbol whose probability is the sum of the two constituent probabilities. At each step, this same procedure is repeated, merging two symbols into one symbol and adding their probabilities. The process stops when one symbol remains. Finally, read the tree from right to left, labeling the branches that leave each node either with a zero or with a one. The codeword is the string of labels from the rightmost node back to the original symbol.

A Huffman code can be formed just as easily for blocks of length n. Simply replace the source symbols in the construction with blocks of source symbols, and replace the symbol probabilities with probabilities of blocks. By choosing a large enough n—usually an n quite small will do—the average

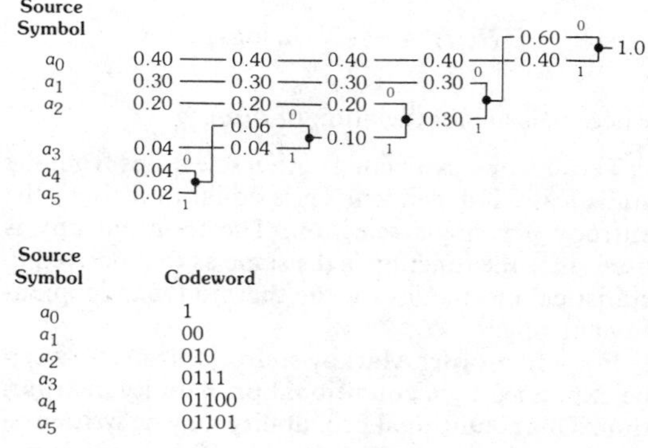

Average codeword length = 2.06 binary digits/symbol.
The entropy of the source = 1.999 bits/symbol.

Fig. 10. Example of Huffman encoding.

codeword length can be made as close to the source entropy as desired, and so a nearly optimal Huffman code can be constructed.

A Huffman code is a prefix code, sometimes said to be self-punctuating. It needs no explicit punctuation. Successive n blocks of source symbols can be encoded one after the other and concatenated. By decoding the codewords in a first-in, first-out fashion, the source symbols are uniquely recovered.

Variable-Length Tree Codes

A variable-length tree code for source compaction encodes indefinitely long strings of source output symbols into indefinitely long strings of code symbols. The encoding operation has a sliding structure whereby, as a few source symbols enter the encoder, a few codeword bits leave the encoder. The relationship between the number of symbols entering the encoder and the number of bits leaving it is variable, depending on the particular source symbols to be encoded.

One of the earliest variable-length tree codes for source encoding is the Elias code. Because of its simplicity, it is an excellent tutorial example. We describe an Elias code for a binary source with binary source alphabet $\{a_0, a_1\}$ and probability distribution $p = \{0.7, 0.3\}$. The entropy of this source is 0.88 bit. Imagine that the semi-infinite sequence of source output bits is a binary representation of a real number, r, in the interval [0, 1]. Refer to Fig. 11. The first symbol tells whether r is in the interval [0, 0.7) or in the interval [0.7, 1]. This interval is itself subdivided in the same proportion, and the next source bit selects one of these two intervals. At

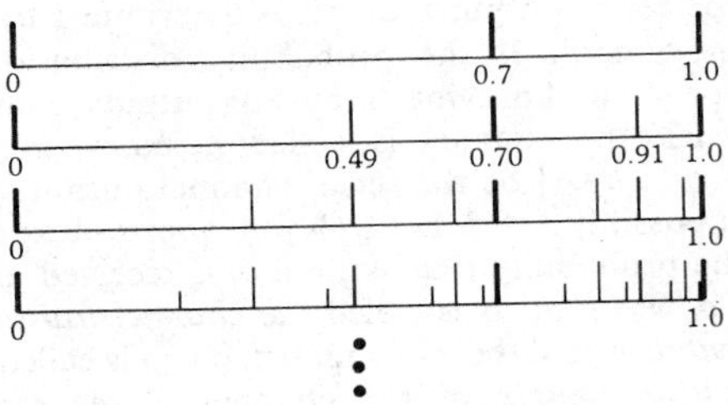

(A) Description of r by sourceword.

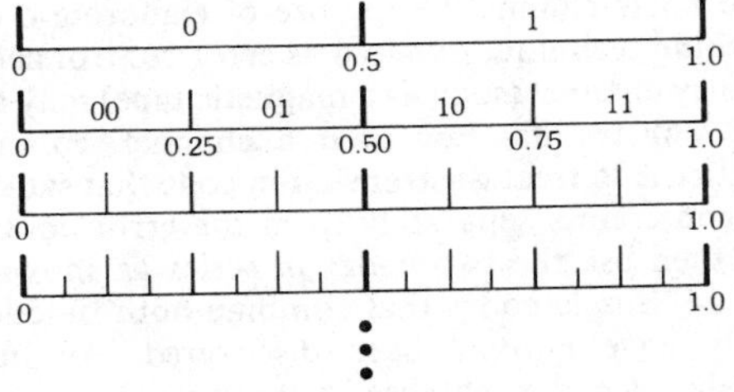

(B) Description of r by codeword.

Fig. 11. Construction of an Elias code.

each iteration, the process repeats, and, in this way, the process continues indefinitely.

The codeword, on the other hand, is a conventional binary representation of point r. As soon as enough source symbols are received to determine whether r is in the interval [0, 0.5) or the interval [0.5, 1], a codeword bit can be transmitted. The encoding is variable length since the number of source symbols needed to produce one codeword bit is random.

The decoder can begin its task after receiving only a few codeword bits. For example, if the binary sequence starts with 011 . . . , the point represented must lie between 0.375 and 0.50; hence the first symbol from the source must be a_0. If the binary sequence starts with 0110, the point represented must lie between 0.375 and 0.4375; hence the first three symbols from the source must be $a_0a_0a_1$.

The Elias code is not practical because of precision problems. Errors in early calculations, no matter how small, will eventually cause encoding and decoding errors. A practical variable-length tree code, due to Pasco and Rissanen,* is available, but the description is more difficult.

Universal Codes

A source compaction code usually requires a probabilistic model of the source in order to attain its best performance. The encoder will encode to near the entropy of the source only if it is given a satisfactory model of the source. Some source compaction codes, called *universal source codes,* implicitly construct their own model of the source as they go along (assuming that past source sequences are representative of future sequences) and so encode near the entropy of whatever source they are given.

The Lempel-Ziv code is a universal variable-to-fixed-length source-compaction block code that does not require an externally constructed source model. The technique, illustrated in Fig. 12, is to break up the string of source output symbols into substrings of variable length but not longer than a largest allowed length. The encoder has a buffered copy of the raw data that has already been encoded. At each iteration, it searches for a prefix of the data yet waiting to be encoded within the data already encoded. The longest prefix of which a copy can be found among the symbols recently encoded, together with one more innovation symbol, becomes the next substring.

A substring is encoded in three parts: a binary-encoded pointer telling where a copy of the substring begins within recently encoded data; a binary-encoded number giving the length of the substring; and the value of the innovation symbol. By including an innovation symbol, the encoder cannot degenerate into an unencodable situation.

* Reference 5.

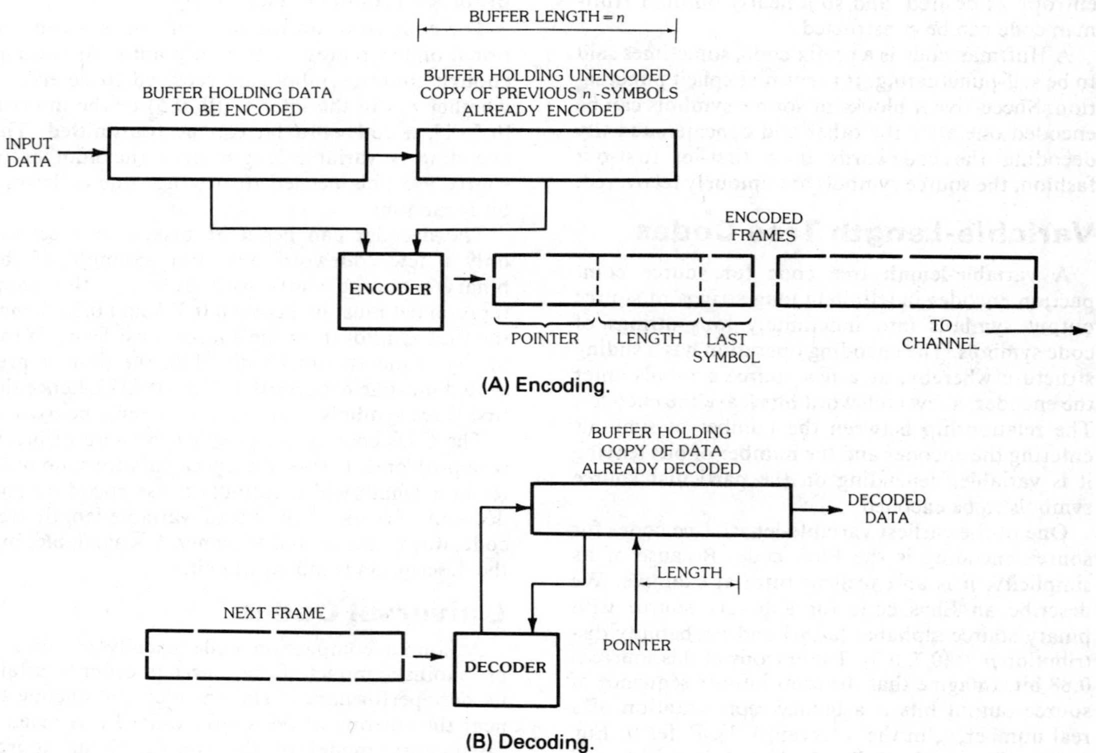

Fig. 12. Lempel-Ziv encoder and decoder.

The Lempel-Ziv decoder must keep a copy of recently decoded data in which it looks to recreate subsequent substrings that it must decode. The decoder contains a finite-length buffer equal in length to the encoder finite-length buffer. In decoding a substring, the decoder buffer contains the same symbols that the encoder buffer contained when that substring was encoded. Hence, the decoder can use the pointer and the substring length to reconstruct the next substring.

CODING FOR DISCRETE NOISY CHANNELS

The deep-space Gaussian noise channel and the noisy binary channel are two important examples of noisy channels. The first example is a continuous-time, continuous-amplitude channel that is usually made into a discrete channel by the modulator/demodulator; the second is a discrete channel.

A *discrete channel* is a system in which a sequence of letters chosen from a finite set of symbols $\{a_0 \ldots a_{I-1}\}$ can be transmitted from one point to another. A *noisy channel* is one for which the output symbol is not completely determined by the input symbol; only some probability distribution on the set of output symbols is determined by the input symbol. If the probability distribution is independent of previous inputs or outputs from the channel, the channel is called *memoryless.* Let $\{b_0, \ldots, b_{J-1}\}$ be the set of channel output symbols; possibly $J \neq I$. Let $Q(b_j | a_i)$, abbreviated $Q_{j|i}$, be the probability that symbol b_j is received given that a_i was sent. It is called the *channel transition probability,* and the J by I matrix $\{Q_{j|i}\}$ is called the *transition matrix* of the channel. Some simple channels are shown in Fig. 13.

Information can be sent reliably through a discrete noisy channel by the use of elaborate cross-checking techniques known as error control codes. A noisy channel (such as a magnetic tape) may also have constraints. For such a channel, common practice is to treat the translation code that satisfies the constraints separately from the error control, and then use the two codes in series as shown in Fig. 14. Single codes that combine both functions neatly have not yet been discovered. An inner encoder/decoder matches its input symbols to the channel. When the channel makes an error, the inner decoder makes an error. The outer encoder/decoder has no constraints on the input sequence.

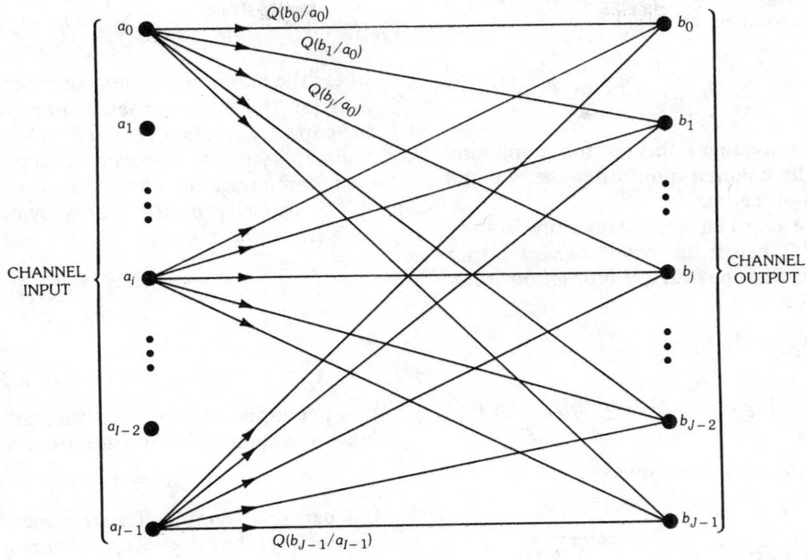

(A) The general case.

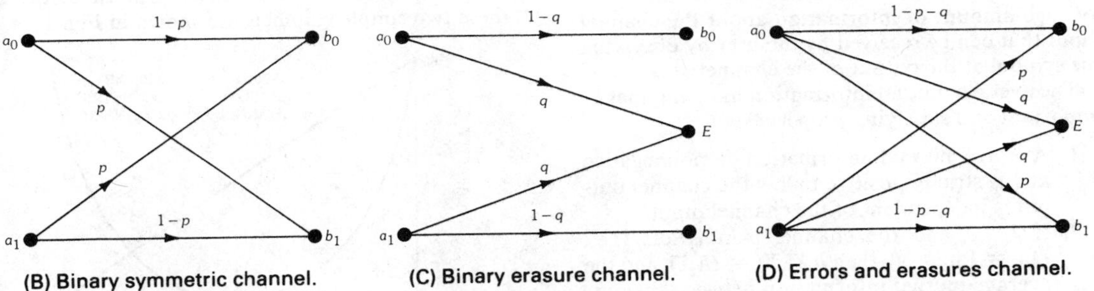

(B) Binary symmetric channel. (C) Binary erasure channel. (D) Errors and erasures channel.

Fig. 13. Some discrete memoryless channels.

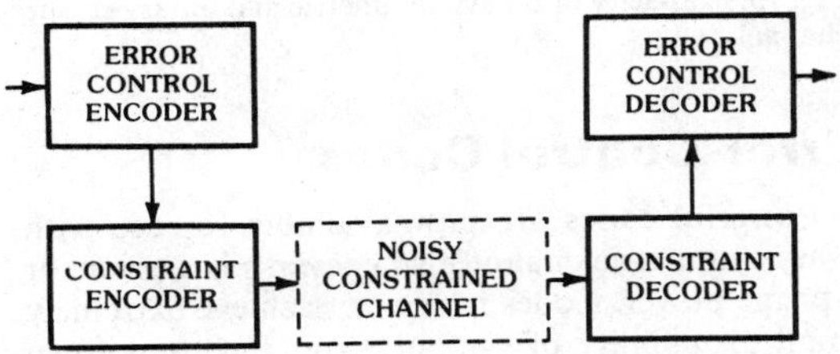

Fig. 14. Channel coding for noisy constrained channels.

Mutual Information

Let $p(a_i)$, abbreviated p_i, be the probability that input symbol a_i is sent through the channel. The probability, $q(b_j)$, abbreviated q_j, of symbol b_j being the channel output is

$$q_j = \sum_{i=0}^{I-1} p_i Q_{j|i}$$

Given output b_j, the probability, $P(a_i|b_j)$, abbreviated $P_{i|j}$, that symbol a_i was transmitted is given by the Bayes' rule

$$P_{i|j} = \frac{p_i Q_{j|i}}{\sum_i p_i Q_{j|i}}$$

The I by J matrix $\{P_{i|j}\}$ is called the *backward transition matrix* of the channel.

The average information per symbol carried by the input random variable, X, which takes value a_i with a probability $p(a_i)$, is the entropy.

$$H(X) = -\sum_{i=0}^{I-1} p_i \log_2 p_i$$

Conditional on b_j being received, the average information per symbol carried by input random variable X is

$$H(X|b_j) = -\sum_{i=0}^{I-1} P_{i|j} \log_2 P_{i|j}$$

The conditional input entropy, or *equivocation*, is then defined as an expectation

$$H(X \mid Y) = -\sum_{j=0}^{J-1} q_j \sum_{i=0}^{I-1} P_{i|j} \log P_{i|j}$$

bits/symbol. On average, this is the remaining uncertainty in the channel input after the channel output has been observed.

The difference between the average uncertainty in channel input X before and after channel output Y is received is called the *average mutual information* of the channel.

$$I(X;\, Y) = H(X) - H(X \mid Y)$$

$$= -\sum_{i=0}^{I-1} p_i \log p_i + \sum_{j=0}^{J-1} \sum_{i=0}^{I-1} q_j P_{i|j} \log P_{i|j}$$

By the Bayes' formula, this becomes

$$I(X;\, Y) = \sum_{i=0}^{I-1} \sum_{j=0}^{J-1} p_i Q_{j|i} \log \frac{Q_{j|i}}{\sum_i p_i Q_{j|i}}$$

The average mutual information is a measure of the average amount of information about the channel input that can be received by the user by observing the symbol at the output of the channel.

The average mutual information has a number of important and satisfying properties:

1. Average mutual information is nonnegative and is strictly positive unless the channel output is independent of the channel input.
2. If $I = J$ and the channel is noiseless (i.e., $Q_{j|i} = 1$ if $j = i$), then $I(X;\, Y) = H(X)$, and the average mutual information between the output and the input of the channel is equal to the average information into the channel.
3. Average mutual information is symmetric in X and Y. That is $I(X;\, Y) = I(Y;\, X)$ where

$$I(Y;\, X) = \sum_{j=0}^{J-1} \sum_{i=0}^{I-1} q_j P_{i|j} \log \frac{P_{i|j}}{\sum_j q_j P_{i|j}}$$

Channel Capacity

The *channel capacity*, C, is defined to be the maximum rate at which information can be transmitted through a channel. The *fundamental theorem of information theory* says that at any rate below channel capacity, an error control code can be designed whose probability of error is arbitrarily small. Intuitively, in a well-designed message, an isolated channel input symbol a_i should occur with a probability p_i such that the average mutual information is maximized.

For the noisy channel, with transition matrix Q, the channel capacity, C, is expressed mathematically by

$$C = \max_p \sum_{i=0}^{I-1} \sum_{j=0}^{J-1} p_i Q_{j|i} \log \frac{Q_{j|i}}{\sum_i p_i Q_{j|i}}$$

where the maximum is over all probability distributions on the input alphabet. For simple channels, the capacity can be evaluated by finding the maximum analytically. For more difficult channels, efficient computational algorithms exist.

The capacity of the binary symmetric channel with transition matrix

$$\begin{bmatrix} 1-p & p \\ p & 1-p \end{bmatrix}$$

is

$$C = 1 + p \log_2 p + (1-p) \log_2 (1-p)$$

bits per input symbol. The capacity of the binary erasure channel with erasure probability q is

$$C = 1 - q$$

bits per input symbol. This is somewhat surprising, since it says that the erasures cause a loss in capacity exactly equal to the fraction of symbols erased despite the fact that the encoder does not know which nq symbols will be erased. The capacities of these two simple channels are shown in Fig. 15.

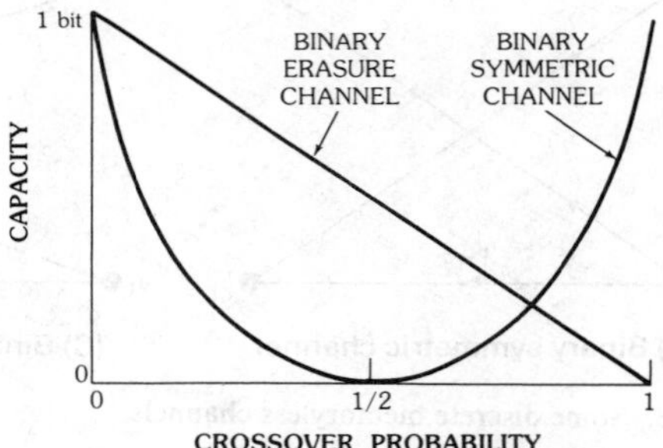

Fig. 15. Capacity of binary symmetric and binary erasure channels.

Error-Control Codes

Powerful codes are now available for use with q-ary symmetric channels whenever q is a prime or a prime power. Codes in use can achieve extremely small probability of symbol error, but the known codes are not good enough to achieve code rate near the channel capacity. Error control codes are routinely used for many purposes: to reduce required link signal-to-noise ratio; to protect against various forms of pulsed and other interference; to protect against hostile interference in military systems; to protect against noise and media defects in magnetic recording systems; and so forth.

Error-control codes in common use work with a code alphabet of size 2^m where m is a positive integer. For m equal to eight, the code alphabet consists of the set of all eight-bit bytes. It is common to use byte-organized codes because errors in practical systems tend to occur in bursts and because the hardware is simpler.

Error-control codes can be used for error detection or error correction. When they are used for error detection, a request for retransmission (called *automatic repeat request* (*ARQ*)) is fed back to the transmitter. When they are used for error correction (called *forward error correction* (*FEC*)), the link operates without interruption, and the error control is transparent to other levels of the system. Forward error correction can operate with noisy links and low transmitted power, whereas ARQ cannot; ARQ will play a reduced role in future systems.

English-language textbooks on error-control codes include Berlekamp (reference 8), Blahut (reference 9), Clark and Cain (reference 10), Lin and Costello (reference 11), and Peterson and Weldon (reference 12).

Block Codes

An (n, k) block code for error control encodes a block of k information symbols into a block of n codeword symbols. Each symbol is an m-bit byte. The rate, R, of the code is equal to k/n.

The rules for constructing good block codes for error control make use of the arithmetic of Galois fields. A Galois field with 2^m elements, denoted $GF(2^m)$, is an arithmetic system containing the operations of addition, subtraction, multiplication, and division defined in such a way that most arithmetic and algebraic procedures are valid. The arithmetic operations themselves are unconventional but have the enormous advantage that there is no overflow nor round-off error. Since the error control code is used to process bit packages but not to do real computations, it does not matter that the arithmetic rules are unconventional. Chart 1 shows addition and multiplication tables for several simple Galois fields. Notice that $GF(2)$ and $GF(3)$ are modulo 2 and modulo 3 arithmetic, respectively, but $GF(4)$ is *not* modulo 4 arithmetic. Modulo 4 arithmetic cannot form a field because $2 \cdot 1 = 2 \cdot 3 \pmod 4$, so division by 2 does not behave properly. Large fields such as $GF(256)$ are very important in practice, but the multiplication tables are too large to show here.

Chart 2 shows the set of codewords of the Hamming (7, 4) code, a binary code—symbols in $GF(2)$—that can correct a single bit error. This simple code has sixteen codewords. Chart 2 also shows some of the codewords of the Reed-Solomon (7, 5) code, an octal code—symbols are in $GF(8)$—that can correct a single octal symbol in error. This code has 8^5, or 32 768, codewords. This second example is actually a very small code although the number of codewords is already too great to enumerate. This is why it is important to use the computational structure of a Galois field to construct the encoders and decoders. Reed-Solomon codes as

CHART 1. EXAMPLES OF FINITE FIELDS

$GF(2)$

+	0	1
0	0	1
1	1	0

·	0	1
0	0	0
1	0	1

$GF(3)$

+	0	1	2
0	0	1	2
1	1	2	0
2	2	0	1

·	0	1	2
0	0	0	0
1	0	1	2
2	0	2	1

$GF(4)$

+	0	1	2	3
0	0	1	2	3
1	1	0	3	2
2	2	3	0	1
3	3	2	1	0

·	0	1	2	3
0	0	0	0	0
1	0	1	2	3
2	0	2	3	1
3	0	3	1	2

large as a (256 224) code over $GF(2^8)$ are now quite practical. This code consists of 224 information symbols followed by 32 parity symbols and can correct 16 symbol errors; a symbol error is a symbol that is wrong in any possible way. A symbol in $GF(2^8)$ may be used to represent eight bits by the user, yet on the channel may be represented by a 256-ary symbol.

The most important block codes in practice are those known as the *BCH codes* and the *Reed-Solomon codes.* They are important because efficient decoding algorithms exist for them. These decoders are based either on the *Berlekamp-Massey algorithm* or on the *Euclidean algorithm,* and they have a hardware cost proportional to nt where n is the blocklength and t is the number of errors to be corrected.

Convolutional Codes

A convolutional code encodes a stream of information symbols into a stream of codeword symbols. The duration of the stream is so long that it is effectively infinite and does not enter into the design of the encoder and decoder. An information sequence is shifted into the encoder beginning at time zero and continuing indefinitely into the future. The stream of incoming information symbols is broken into segments of k_0 symbols called information frames, which often may be as short as one symbol in practice. The encoder can store m frames. During each frame time, a new information

CHART 2. EXAMPLES OF BLOCK CODES

Hamming (7, 4) Code

Information				Parity		
0	0	0	0	0	0	0
0	0	0	1	0	1	1
0	0	1	0	1	1	0
0	0	1	1	1	0	1
0	1	0	0	1	1	1
0	1	0	1	1	0	0
0	1	1	0	0	0	1
0	1	1	1	0	1	0
1	0	0	0	1	0	1
1	0	0	1	1	1	0
1	0	1	0	0	1	1
1	0	1	1	0	0	0
1	1	0	0	0	1	0
1	1	0	1	0	0	1
1	1	1	0	1	0	0
1	1	1	1	1	1	1

Reed-Solomon (7, 5) Code

Information					Parity	
0	0	0	0	0	0	0
0	0	0	0	1	6	3
0	0	0	0	2	7	6
0	0	0	0	3	1	5
			⋮			
0	0	0	1	0	1	1
0	0	0	1	1	7	2
0	0	0	1	2	6	7
0	0	0	1	3	0	4
			⋮			
0	0	0	7	0	7	7
0	0	0	7	1	1	4
0	0	0	7	2	0	1
0	0	0	7	3	6	2
			⋮			
0	0	1	0	0	7	3
0	0	1	0	1	1	0
0	0	1	0	2	0	5
0	0	1	0	3	6	6
			⋮			

frame is shifted into the encoder, and the oldest information frame is shifted out and discarded. At the end of any frame time, the encoder has stored the most recent m frames, a total of mk_0 information symbols. At the beginning of a frame, from the new incoming information frame and the m previously stored information frames, the encoder computes a single codeword frame of length n_0 symbols. This codeword frame is shifted out of the encoder as the next information frame is shifted in. Hence, the channel must transmit n_0 codeword symbols for each k_0 information symbols. The rate, R, of the convolutional code is defined as $R = k_0/n_0$.

The *constraint length*, ν, of a convolutional code is defined as the number of memory stages in a minimum encoder. The complexity of a decoder is often q^ν. Binary convolutional codes used in practice may have a constraint length in the range of about seven to forty. Fig. 16 shows an encoder for a much simpler convolutional code, one of constraint length 2 with $k_0 = 1$ and $n_0 = 2$. A trellis description of the code is shown in Fig. 17. The convolutional code is the set of all semi-infinite binary words that may be read off any path through the trellis. A "zero" information bit entering the encoder at any node is encoded into the upward path out of that node, and a "one" information bit is encoded into the lower path. This convolutional code is able to correct any number of error events each containing two bit errors provided the error events are spaced far enough apart for the decoder to clear one error event before it sees the next.

Convolutional codes in use have a larger con-

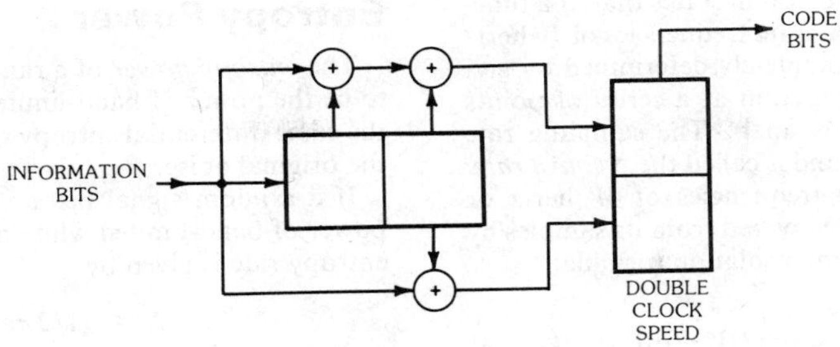

Fig. 16. Encoder for a convolutional code with constraint length 2.

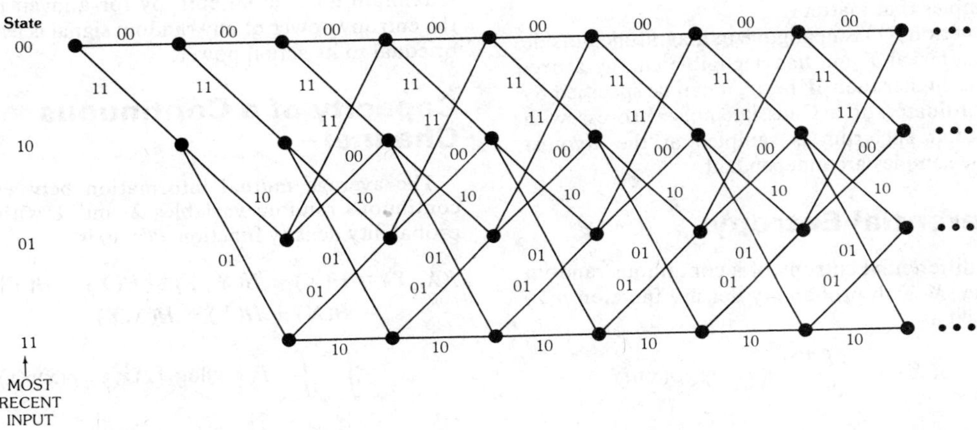

Fig. 17. Trellis diagram for a convolutional code with constraint length 2.

straint length. The most important decoders for convolutional codes either use the *Viterbi algorithm* or use *sequential decoding*. The Viterbi algorithm has a hardware cost proportional to 2^ν and so is practical only for small ν. Sequential decoding, using either the *Fano algorithm* or the *stack algorithm,* has a hardware cost that grows only slowly with ν, but has a random decoding delay and will occasionally overflow any finite-sized decoding buffer.

CONTINUOUS CHANNELS AND SOURCES

A discrete information source generates information at a finite rate; the entropy rate, which measures the information generated, is finite. A continuous information source can assume any one of an infinite number of amplitude values and so requires an infinite number of binary digits for its exact specification. The entropy rate is infinite. An immediate consequence of this is that in order to transmit the output of a continuous information source and recover it exactly, a channel of infinite capacity is required. Since, in practice, every continuous channel is perturbed by noise and therefore has a finite capacity, it is not possible to transmit the output of a continuous source over a channel and recover it exactly; there is always some distortion.

We distinguish among a discrete source or channel, a continuous source or channel defined on discrete time instants, and a waveform source or channel defined on a continuum of time points.

A waveform channel can be made into a discrete channel by choosing a finite set of modulation waveforms for the channel. The relationship between waveform design and information theory is discussed in Wozencraft and Jacobs.* A waveform source can be made into a continuous source by sampling, provided the bandwidth is finite, and into a discrete source by sampling and quantization provided a small amount of distortion can be introduced.

The Sampling Theorem

The sampling theorem is an important aid in the design and analysis of communication systems involving the use of continuous time functions of

* Reference 13.

finite bandwidth. The theorem states that, if a function of time, $f(t)$, contains no frequencies of W hertz or higher, then it is completely determined by giving the value of the function at a series of points spaced $(2W)^{-1}$ seconds apart. The sampling rate of $2W$ samples per second is called the *Nyquist rate.*

If $f(t)$ contains no frequencies of W hertz or higher, then it can be recovered from its samples by the Nyquist-Shannon interpolation formula:

$$f(t) = \sum_{n=-\infty}^{+\infty} f(n/2W)\{[\sin \pi(2Wt-n)]/\pi(2Wt-n)\}$$

The sampling theorem makes no mention of the time origin of the samples; it is only the spacing of the samples that matters.

If function $f(t)$ is negligible in magnitude outside a time interval T and has negligible energy at frequencies higher than W hertz, it can be specified by $2TW$ ordinates. If a Gaussian noise process with rectangular spectrum is sampled at the Nyquist rate, the samples are independent.

Differential Entropy

The differential entropy of a continuous random variable, X, with probability density function $p(x)$ is defined as

$$H(X) = -\int_{-\infty}^{+\infty} p(x)\log_2 p(x)dx$$

The differential entropy is *not* the limiting case of the entropy; the entropy of a continuous distribution is infinite. The differential entropy is not invariant under coordinate transformations. By itself, it has no fundamental physical meaning, but it occurs often enough to have a name.

The Gaussian probability distribution

$$p(x) = [1/(\sqrt{2\pi}\,\sigma)]\exp(-x^2/2\sigma^2)$$

plays a major role in information problems. Of all probability distributions with variance σ^2, the Gaussian distribution has the largest differential entropy, given by

$$H(X) = \tfrac{1}{2}\log_2(2\pi e\sigma^2)$$

The differential entropy of a vector of length n whose components are independent identically distributed Gaussian random variables is

$$H(X) = \tfrac{1}{2}n\log_2(2\pi e\sigma^2)$$

Consequently, if a Gaussian noise process of ideal rectangular spectrum is sampled at the Nyquist rate of $2W$ samples per second, we say the differential entropy rate of the source is $W\log_2(2\pi e\sigma^2)$. A Gaussian noise process whose spectrum is constant for frequencies below W hertz and is zero for larger frequencies is called *band-limited white noise.*

Entropy Power

The *entropy power* of a random signal is defined to be the power of band-limited white noise having the same differential entropy rate and bandwidth as the original noise.

If a random signal has an entropy H (bits), the power of band-limited white noise having the same entropy rate is given by

$$N_e = (1/2\pi e)2^{2H}$$

The power N_e is the entropy power of the random signal.

It should be noted that since white noise has the maximum differential entropy for a given power, the entropy power of any random signal is less than or equal to its actual power.

Capacity of a Continuous Channel

The average mutual information between two continuous random variables X and Y with joint probability density function $P(x, y)$ is

$$\begin{aligned} I(X;Y) &= H(X) - H(X\,|\,Y) = H(Y) - H(Y\,|\,X) \\ &= H(X) + H(Y) - H(X,Y) \\ &= \int_{-\infty}^{\infty}\int_{-\infty}^{\infty} P(x,y)\log_2[P(x,y)/p(x)p(y)]dydx \\ &= \int_{-\infty}^{\infty} p(x)\int_{-\infty}^{\infty} Q(y\,|\,x) \\ &\quad \log_2 \frac{Q(y\,|\,x)}{\int_{-\infty}^{\infty} p(x)\,Q(y\,|\,x)\,dx}\,dy\,dx \end{aligned}$$

bits. Even though the differential entropy terms appearing here are not invariant under a coordinate transformation, their difference is. The mutual information has a fundamental significance.

A continuous channel usually has some constraint on the input probability distribution; an average power constraint or a peak power constraint is the most common. Therefore, $p(x)$ must satisfy one or more constraint equations of the form

$$\int_{-\infty}^{\infty} e(x)p(x)dx \leq S$$

where $e(x)$ is some nonnegative function. The average power constraint is

$$\int_{-\infty}^{\infty} x^2p(x)dx \leq S$$

The capacity of the continuous channel is the maximum value of the average mutual information

$$C = \max_{p(x)} I(X; Y)$$

where the maximum is taken over the set of probability distributions that satisfy the constraints.

The Additive Gaussian Noise Channel

Calculation of the capacity can be a difficult task requiring a computer, but the important case of an additive Gaussian noise channel subject to an average power constraint can be solved analytically.

The channel output, Y, is given by $Y = X + Z$ where X is the channel input and Z is the noise. Since X and Z are independent, $H(Y|X) = H(Z)$, and $I(X; Y) = H(Y) - H(Z)$. For Gaussian noise, $H(Z) = \frac{1}{2}\log_2 2\pi eN$ where N is the noise power. Then

$$C = \max_{p(x)} 2W[H(Y) - \tfrac{1}{2}\log_2 2\pi eN]$$

Factor $2W$ converts the units of capacity to bits/second. Output Y has variance $S + N$ where S is the average power constraint. Entropy $H(Y)$ is largest if Y is Gaussian, which will be true if X is Gaussian of variance S. Then

$$\begin{aligned} C &= W\log_2 2\pi e(S + N) - W\log_2 2\pi eN \\ &= W\log_2 (1 + S/N) \end{aligned}$$

This is Shannon's formula for the capacity of the additive Gaussian noise channel with an average power constraint. The channel input signal that achieves channel capacity has an amplitude distribution described by the Gaussian function.

If a channel has additive but non-Gaussian noise and a power constraint, then the capacity may be difficult to calculate exactly. However, upper and lower bounds on the capacity are easy to obtain. At average transmitted power S, the capacity, C, in bits/second, is bounded by the inequalities

$$W\log_2 [(S + N)/N_e] \geq C \geq W\log_2 [(S + N)/N]$$

where,

W is the bandwidth,
N is the average noise power,
N_e is the entropy power of the noise.

Waveform Channels

An additive Gaussian noise waveform channel with a general transfer function is shown in Fig. 18. The capacity of the waveform channel under Gaussian noise is equal to the capacity one would obtain if the transfer function were approximated by many thin ideal rectangular tranfer functions of different amplitudes. The capacity is given by the so-called "water-pouring" formulas, written parametrically in terms of θ.

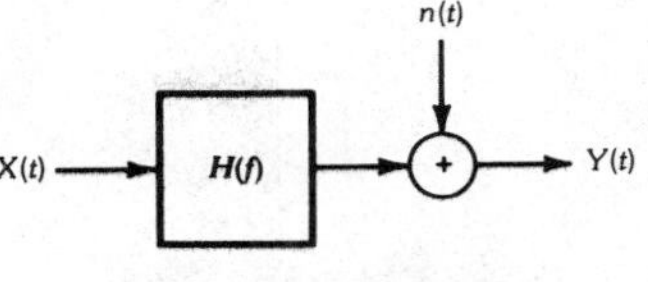

(A) Conventional model.

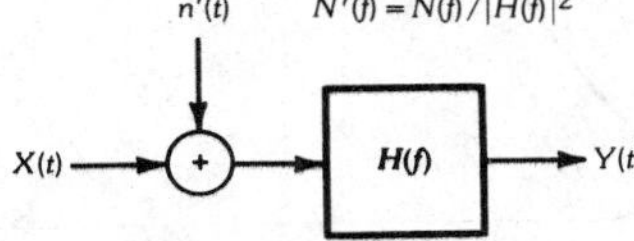

(B) Equivalent channel.

Fig. 18. Gaussian-noise waveform channel.

$$C(S, \theta) = \tfrac{1}{2}\int_{-\infty}^{\infty} \max[0, \log \frac{\theta}{N(f)/|H(f)|^2}]df$$

$$S(\theta) = \int_{-\infty}^{\infty} \max[0, \theta - N(f)/|H(f)|^2]df$$

The reason these are called the water-pouring formulas can be understood from Fig. 19. The input information is imagined as "poured" into a vessel whose shape is defined by $H(f)$ and $N(f)$. This produces the optimum spectral shape of input waveform $S(f) = \max[0, \theta - N(f)/|H(f)|^2]$ that achieves the channel capacity. The important lesson given by the water-pouring principle is that optimum waveforms put most of their energy in the spectral region where the channel is good and little or no energy in the spectral region where the channel is poor. The optimum strategy is exactly the opposite of an often-used equalization strategy that "boosts" the skirts of the channel by adding extra gain there.

Bit Energy and Bit Error Rate

The performance of a digital communication system is measured by the probability of bit error, also called the *bit error rate* (BER). On an additive Gaussian noise channel, the bit error rate can always be reduced by increasing transmitted power, but it is by the performance at low transmitted power that one judges the quality of a digital communication system. The better of two systems, otherwise the same, is the one that can achieve a desired bit error rate with the lower transmitted power.

Given a message, $s(t)$, of duration T containing K information bits, the bit energy, E_b, is given by

$$E_b = E_m/K$$

where

$$E_m = \int_0^T s(t)^2\, dt$$

is the message energy.

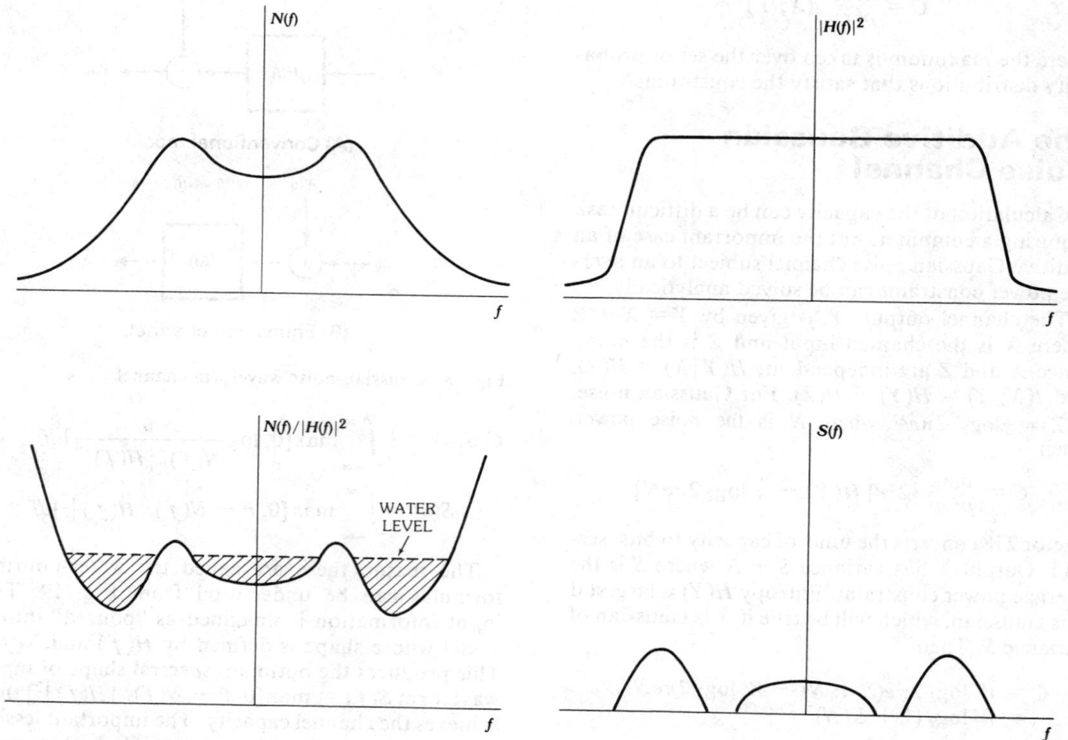

Fig. 19. Water pouring.

Bit energy E_b is calculated from the message energy and the number of information bits at the input to the encoder/modulator. At the input to the channel, one may find a message structure in which he perceives a larger number of bits. The extra symbols may be parity symbols for error control, or symbols for frame synchronization or channel protocol. These symbols do not represent transmitted information, and their energy must be amortized over information bits. Only information bits are used in calculating E_b.

For an infinite-length message of rate R information bits/second, E_b is defined by

$$E_b = S/R$$

where S is the message average power.

In addition to the message energy, the receiver also sees a white noise signal of one-sided spectral density N_o watts/hertz. Only the ratio E_m/N_o or E_b/N_o affects the bit error rate because the reception of the signal cannot be affected if both the signal and the noise are doubled. Signaling schemes are compared by comparing their respective graphs of BER versus required E_b/N_o.

It is possible to make precise statements about values of E_b/N_o for which good waveforms exist; these are a consequence of the channel capacity formula for the ideal rectangular bandpass channel in additive Gaussian noise. Let the signal power be $S = E_bR$ and the noise power be $N = N_oW$. Then

$$C/W = \log_2 (1 + RE_b/N_oW)$$

Define the spectral bit rate, r (measured in bits per second per hertz), by

$$r = R/W$$

The spectral bit rate, r, and E_b/N_o are the two most important figures of merit of a digital communication system.

Since the rate, R, is less than but can be made arbitrarily close to the capacity, C, the capacity formula becomes

$$E_b/N_o > (2^r - 1)/r$$

but E_b/N_o can be arbitrarily close to the bound by designing a sufficiently sophisticated digital communication system. This inequality, shown in Fig. 20, tells us that increasing the bit rate per unit bandwidth increases the required energy per bit. This is the basis of the energy/bandwidth trade of digital communication theory where increasing bandwidth at a fixed information rate can reduce power requirements.

Every communication system can be described by a point lying below the curve of Fig. 20. Any communication system that attempts to operate

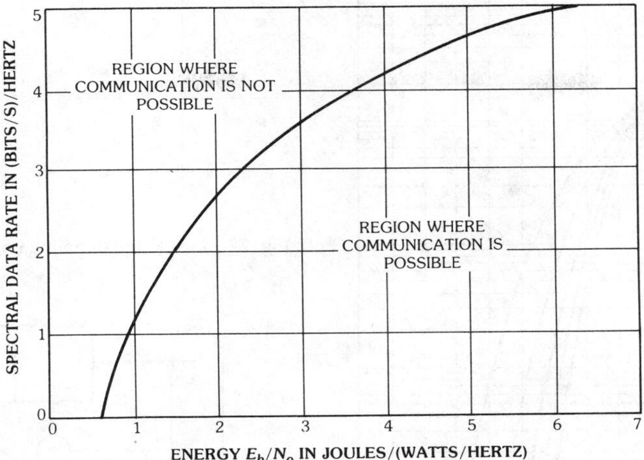

Fig. 20. Capacity of baseband Gaussian noise channel.

above the curve will lose enough data through errors so that its actual data rate will lie below the curve. By the fundamental theorem of information theory, for any point below the curve one can design a communication system that has as small a bit error rate as one desires. The history of digital communications can be described in part as a series of attempts to move ever closer to this limiting curve with systems that have very low bit error rate. Such systems employ both modem techniques and error-control techniques.

If bandwidth W is a plentiful resource but energy is scarce, then one should let W go to infinity, or r to zero. Then we have

$$E_b/N_o \geq \log_e 2 = 0.69$$

This is a fundamental limit. Ratio E_b/N_o is never less than -1.6 dB, and by a sufficiently expensive system one can communicate with any E_b/N_o larger than -1.6 dB.

Signaling Without Bandwidth Constraints

If the bandwidth is much larger than the data rate, then in principle one can signal arbitrarily closely to channel capacity by using an M-ary signaling alphabet. An M-ary signaling alphabet is a collection of M sufficiently distinct waveforms. Usually M is chosen equal to 2^k for some k. The modulator maps each k-bit word from the channel encoder into one of the 2^k waveforms in the signaling alphabet. The demodulator compares the received signal in each signaling time interval with each of the 2^k possible transmitted waveforms and chooses the most likely.

There are a great many sets of M-ary signal alphabets that can be used. Those with the best performance known are a type known as *simplex waveforms*, but orthogonal waveforms are almost as good and are usually used in practice. The advantage of M-ary signaling can be understood as the resultant of two opposing forces. Since the modulator puts k bits into one waveform, it has k times as much energy as it would for a binary waveform. Since the demodulator must make a 2^k-way decision rather than a binary decision, it needs more energy in the waveform to preserve a small error probability. However, the additional energy needed grows with k more slowly than the additional energy available. The net effect is a system that can operate at a very low E_b/N_o; however, the waveforms occupy a bandwidth much greater than the bit rate. Fig. 21 shows the performance of a 2^k-ary family of orthogonal waveforms. The figure shows that E_b/N_o of -1.6 dB is attainable for large k. Practical systems are usually limited to a k of about six. In the figure, E_b is the energy per data bit delivered; because some of these data bits will be incorrect, the energy per information bit is larger.

Signaling With a Bandwidth Constraint

When bandwidth is expensive, one no longer attempts to transmit with very small E_b/N_o. Now it becomes important to transmit with large spectral bit rate, r (bits/second)/hertz. Good performance can be achieved by careful combination of modulation and error control.

The Ungerboeck convolutional codes are a class of convolutional codes that provide good combinations of spectral bit rate and E_b/N_o. The rate 2/3 codes are designed to work with a set of eight fixed complex numbers, called a signaling constellation, which represent phase/amplitude modulation patterns. Two such signaling constellations are shown in Fig. 22. An encoder for the constraint-length-

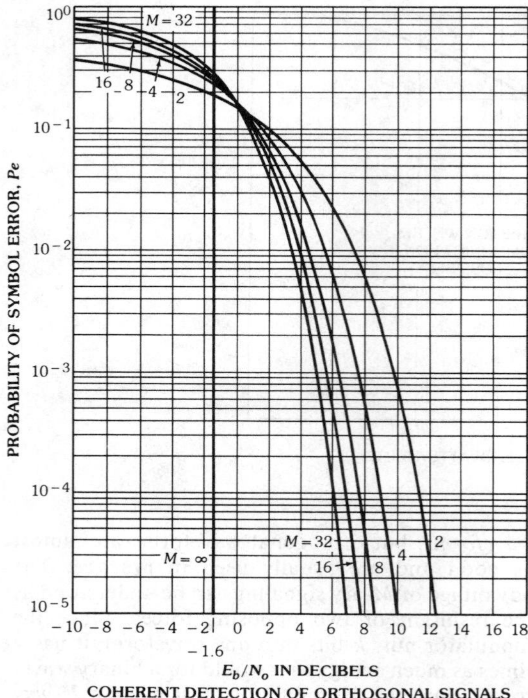

Fig. 21. Performance of *M*-ary signaling.

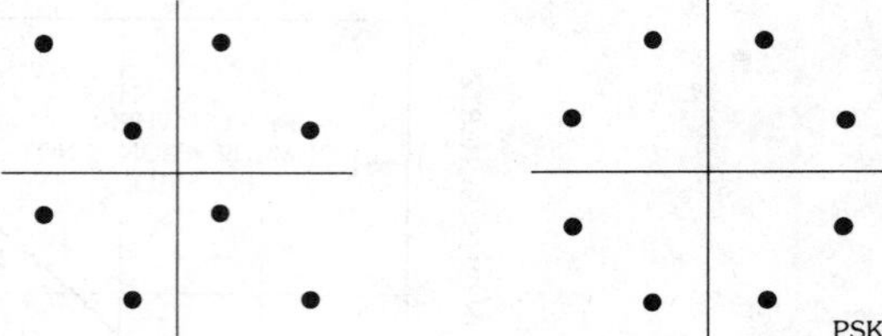

Fig. 22. Two 8-ary signaling constellations.

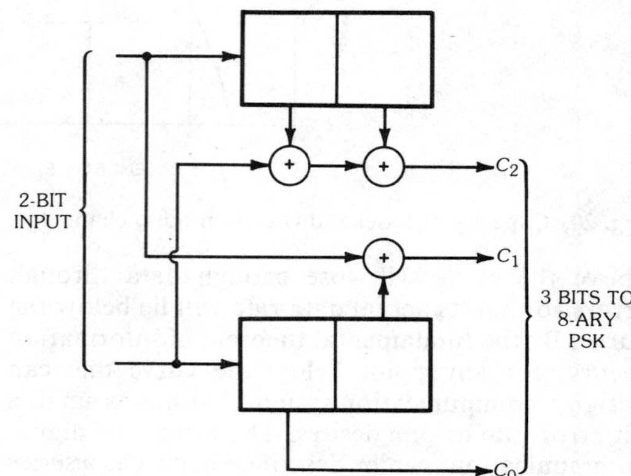

Fig. 23. Diagram of encoder for a constraint-length-four Ungerboeck code.

four Ungerboeck code for the PSK signaling constellation is shown in Fig. 23. This code can be used as a plug-in replacement for the popular uncoded four-phase PSK modulator. The information rate is still two bits per symbol. There is no change in the channel symbol rate, so the coded system has the same bandwidth as the uncoded system and transmits the same number of information bits per symbol. Hence, the user of the system is unaware of the presence of the code. However, the system now can run at a lower E_b/N_o or signal-to-noise ratio; the constraint-length-four code has a coding gain of 5.7 decibels.

DECISION THEORY AND ESTIMATION THEORY

Decision theory is concerned with the problem of deciding between a set of hypotheses when given a collection of imperfect measurements. Estimation theory is concerned with the problem of selecting the best value of a parameter from a continuum of possible values when given a collection of imperfect measurements. Estimation theory is also concerned with the selection of a best waveform from a collection of waveforms when given a collection of imperfect measurements.

General discussion of decision theory and estimation theory can be found in Davenport and Root* and Van Trees.†

Hypothesis Testing

The simplest problem of decision theory, called hypothesis testing or detection, is to decide between two mutually exclusive hypotheses. In the case of a radar, the hypotheses are *target present* or *target absent*. In the case of signal acquisition, the hypotheses are *signal present* or *signal absent*. In the general case, the hypotheses are called the *null hypothesis*, H_0, and the *alternate hypothesis*, H_1. The problem is to decide which hypothesis is correct by collecting data and processing them. The data are randomly distributed with a probability distribution that depends on the true hypothesis. The set of data may actually be quite extensive, but it is enough to think of it as a simple measurement whose outcome can only be an element of a finite set of K elements called the measurement space, and indexed by k. The theory applies to arbitrary data sets simply by replacing the scalar measurement by an appropriate vector measurement.

---

* Reference 14.

† Reference 15.

Associated with each hypothesis is a probability distribution on the measurement space. If H_0 is true, then q_{0k} gives the probability that k will be the measurement outcome; if H_1 is true, then q_{1k} gives this probability. A simple measurement consists of an observation of a realization of the random variable. A measurement is observed, and the problem is to decide whether hypothesis H_0 or hypothesis H_1 is true. A hypothesis testing rule is a partition of the measurement space into two disjoint sets $\mathcal{U}_0$ and $\mathcal{U}_1 = \mathcal{U}_0{}^c$. If the measurement is an element of $\mathcal{U}_0$, we decide that H_0 is true; if it is an element of $\mathcal{U}_1$, we decide that H_1 is true. Each hypothesis testing rule can be described as such a partition.

Accepting hypothesis H_0 when H_1 actually is true is called a type-one error, and the probability of this event is denoted by α. Accepting hypothesis H_1 when H_0 actually is true is called a type-two error, and the probability of this event is denoted by β. Obviously,

$$\alpha = \sum_{k \epsilon \mathcal{U}_0} q_{1k} \qquad \beta = \sum_{k \epsilon \mathcal{U}_1} q_{0k}$$

A method for finding the optimum decision regions is given by the Neyman-Pearson theorem. This theorem expresses the decision regions in terms of a parameter, T, called the threshold, and a function called the log-likelihood ratio given by

$$l(k) = \log\,(q_{0k}/q_{1k})$$

The Neyman-Pearson theorem says that the sets parametrized by T,

$$\mathcal{U}_0(T) = \{k \mid l(k) \geq T\}$$
$$\mathcal{U}_1(T) = \{k \mid l(k) < T\}$$

are an optimum family of decision rules in the sense that no decision rule can have both type-one error and type-two error better than any of these rules. The type-one and type-two error probabilities can be traded by varying the threshold, T.

The log-likelihood ratio for a block $v = (k_1, \ldots, k_n)$ of independent identically distributed measurements of length n can be written as a sum of per-letter log-likelihood ratios

$$l(v) = \log\,[q_0(v)/q_1(v)] = \sum_{l=1}^{n} \log\,[q_{0k_l}/q_{1k_l}]$$

because the probability distributions on v are products of the single letter probability distributions

$$q_0(v) = \prod_{l=1}^{n} q_{0k_l} \qquad q_1(v) = \prod_{l=1}^{n} q_{1k_l}$$

The Neyman-Pearson decision regions for the block measurement are expressed in terms of $l(v)$ by the inequalities $l(v) < T$ and $l(v) \geq T$.

The same description applies even when the measurements are not independent or identically distributed. However, the log-likelihood ratio, $l(v)$, then cannot be written as the sum of per-letter log-likelihood ratios.

The expected value of the log-likelihood ratio with respect to q_0 is the function known as the *discrimination*.

$$L(q_0,q_1) = \sum_{k=0}^{K-1} q_{0k} \log\,(q_{0k}/q_{1k})$$

The discrimination is closely related to the entropy and the average mutual information. It is useful for forming bounds on the probability of error.

Estimation Theory

Estimation theory is concerned with the problem of finding a best value for an unknown parameter when only imperfect data are available. The estimation of the unknown parameter depends on measurements that are random variables. The quality of the estimate is limited by the quality of the measurements. These limitations can be expressed by means of information-theoretic bounds; one such bound is known as the Cramer-Rao inequality.

The simplest estimation problem involves an unknown real parameter, θ, to be determined, and a random variable, X. Understanding this problem leads to understanding of the more general problems: the estimation of several parameters based on the observation of a finite number of random variables; the estimation of several parameters based on observing a sample waveform of a stochastic process; and the estimation of a function on an interval based on observing a sample waveform of a stochastic process.

The random variable has a continuous probability distribution, $q(x \mid \theta)$, conditional on the parameter θ. The unknown θ must be estimated based upon an observation of X. The estimate of θ given the measurement X is a function $\hat{\theta}(x)$. The estimate, $\hat{\theta}$, is a random variable, since it is a function of the random variable X. Estimation theory studies various criteria for making good estimates, that is, for selecting the function $\hat{\theta}(x)$.

The quality of an estimator is judged by its bias

$$\bar{\theta} = E[\hat{\theta}(x)]$$

and by its variance

$$\sigma_\theta^2 = E\{[\hat{\theta}(x) - \bar{\theta}]^2\}$$

Intuitively, one hopes to choose the estimator, $\hat{\theta}\,(x)$, so that its bias is zero and the variance is as small as possible. Such an estimator is called a *minimum-variance unbiased estimator*. Sometimes, however, the minimum-variance unbiased estimator does not exist, or another estimator has some advantage, and so another estimator is used.

The Cramer-Rao bound is a lower bound on the variance of any estimator. For an unbiased estimator of a single parameter θ, the most common form of the bound is

$$\sigma_\theta^2 \geq \{E[(\partial/\partial\theta)\log q(x|\theta)]^2\}^{-1}$$

The same bound holds even if X is a vector of measurements. The Cramer-Rao bound can also be expressed in terms of the discrimination

$$\sigma_\theta{}^2 \geq \left\{ \lim_{\theta' \to \theta} \frac{L(q_\theta;q_{\theta'})}{(\theta-\theta')^2} \right\}^{-1}$$

where,

$$L(q_\theta;q_{\theta'}) = \int q(x|\theta)\log[q(x|\theta)/q(x|\theta')]\,dx$$

The Matched-Filter Estimator

Suppose that a known waveform, $s(t)$, has spectrum $S(f)$ and that additive Gaussian noise has spectral density $N(f)$. The received noisy signal with an unknown time of arrival τ is

$$v(t) = s(t-\tau) + n(t)$$

The Cramer-Rao bound says that any estimator of the time of arrival satisfies

$$\sigma_\tau^2 \geq \left[\int_{-\infty}^{\infty} \frac{|2\pi f S(f)|^2}{N(f)}\,df\right]^{-1}$$

The optimal estimator, if the noise is white and the signal-to-noise ratio is sufficiently high, is the matched-filter estimator. It achieves the Cramer-Rao bound. The matched-filter estimator passes $v(t)$ through a filter with impulse response $s^*(-t)$ (with some fixed delay to make it realizable) and estimates τ from the peak of the filter output. If the signal-to-noise ratio is not sufficiently high, then the optimal estimator is not known.

Similar remarks apply to the problem of estimating the frequency offset, ϕ, of an otherwise known bandpass waveform

$$\begin{aligned} v(t) = s(t)\cos[2\pi(f_0+\phi)t+\theta] \\ + n_R(t)\cos 2\pi f_0 t - n_I(t)\sin 2\pi f_0 t \end{aligned}$$

where $s(t)$ has spectrum $S(f)$ and is a known and finite energy waveform whose bandwidth is much less than the known carrier frequency, f_0. A small offset in frequency ϕ is to be estimated, and $n_R(t)$, $n_I(t)$ are independent covariance stationary Gaussian processes with identical power density spectrum $N(f)$, which is known. The Cramer-Rao bound says that any estimator of frequency ϕ has an error variance satisfying

$$\sigma_\phi^2 \geq \left[\int_{-\infty}^{\infty} \frac{|S'(f)|^2}{N(f)}\,df\right]^{-1}$$

where $S'(f)$ is the derivative of $S(f)$. The matched-filter estimator of frequency consists of a bank of matched filters. Each filter has impulse response of the form $s^*(-t)\cos 2\pi(f_0+\Delta)t$ for Δ in a range of interest. The Δ of the filter with the largest output provides the estimate of ϕ. If the noise is white, and the signal-to-noise ratio is sufficiently high, the matched-filter estimator achieves the Cramer-Rao bound, and so is optimal. This matched-filter estimator of frequency offset often can be approximated by a phase-locked loop. If the signal-to-noise ratio is not sufficiently high, the optimal estimator of frequency is not known.

Maximum Entropy and Minimum Discrimination

The *Jaynes maximum entropy principle* of data reduction says that when reducing a set of data into the form of an underlying model, one should be maximally noncommittal with respect to missing data. If one must estimate a probability distribution subject to constraints on it, then one should choose the probability distribution of maximum entropy consistent with the constraints. For example, if $q(x)$ is a probability density function of a single variable, x, and the mean and variance of $q(x)$ are known and otherwise $q(x)$ is unknown, then one should estimate that $q(x)$ is a Gaussian probability density function with the given mean and variance.

The *Kullback principle of minimum discrimination* is a more general principle that applies when one is given a distribution, $p(x)$, that is a prior estimate of $q(x)$ and a set of constraints on $q(x)$. The principle states that, of the distributions that satisfy the constraints, one should choose as the new estimate that $q(x)$ that minimizes the discrimination

$$L(q,p) = \int q(x)\log[q(x)/p(x)]\,dx$$

If the probability distribution is discrete and $p(x)$ is a uniform distribution, then this reduces to the maximum entropy principle.

Spectral Estimation

Spectral estimation is the problem of estimating the power spectrum of a stochastic process given partial data, usually only a finite number of samples of the autocorrelation function of limited accuracy. Most spectral estimation methods in use ignore the noise on the autocorrelation samples and treat only the problem of dealing with the missing autocorrelation samples.

The simplest and most popular spectral estimation procedure is to choose as the estimated spectrum the Fourier transform of the known values of the autocorrelation function, possibly tapering the known values of the autocorrelation function or padding them with some zero components. The Fourier transform techniques implicitly augment the measured components of the autocorrelation function with artificial components: either a periodic continuation of the known values or some zero components.

The Jaynes maximum entropy principle applied to spectral estimation yields an alternative procedure known as *maximum-entropy spectral estima-*

tion. If the correlation samples are given at a set of equispaced lags, then the maximum entropy estimate of the stochastic process turns out to be

$$v_i = -\sum_{j=1}^{L} h_j v_{i-j} + n_i$$

which is the output of an *autoregressive filter* excited by discrete-time white noise. The filter coefficients are chosen so that the correlation coefficients of this process agree with the known correlation coefficients. Hence the filter coefficients are the solution of the matrix equation

$$\sum_{j=1}^{L} h_j R_{i-j} = -R_i$$

where the correlation coefficients, R_i, for $i = -L, \ldots, L$, are known and R_{-i} equals R_i. An efficient algorithm for solving this set of equations is the *Levinson algorithm.*

SOURCE COMPRESSION CODES

The average amount of information required to describe a source output symbol is equal to the entropy of the source. Sometimes it is not convenient or practical to retain all this information. It then is no longer possible to maintain an exact reproduction of the source. An analog source has infinite entropy, so distortion must always be present when the source output is passed through a channel of finite capacity.

Data compression is the practice of intentionally reducing the information content of a data record. This should be done in such a way that the least-distorted reproduction is obtained. Information theory finds the performance of the optimum compression of a random source of data; a naive encoder will have greater distortion.

The output of a discrete information source is a random sequence of symbols from a finite alphabet containing J symbols given by $\{a_0, a_1, \ldots, a_{J-1}\}$. A memoryless source produces the jth letter with probability p_j where p_j is strictly positive. The source output is to be reproduced in terms of a second alphabet called the reproducing alphabet, often identical to the source alphabet, but not always so. For example, the reproducing alphabet might consist of the union of the source alphabet and a single new element denoting "data erased."

A distortion matrix is a J by K matrix with nonnegative elements ρ_{jk} for $j = 0, \ldots, J-1$ and $k = 0, \ldots, K-1$ that specifies the distortion associated with reproducing the jth source letter by the kth reproducing letter. Without loss of generality, it can be assumed that for each source letter a_j there is at least one reproducing letter b_k such that the resulting distortion ρ_{jk} equals zero. Usually the distortion in a block is defined as the arithmetic average of the distortions of each letter of the block. This is called a per-letter fidelity criterion.

An important distortion matrix is the probability-of-error distortion matrix. For this case the alphabets are identical. For example, take $J = K = 4$ and

$$\rho = \begin{bmatrix} 0 & 1 & 1 & 1 \\ 1 & 0 & 1 & 1 \\ 1 & 1 & 0 & 1 \\ 1 & 1 & 1 & 0 \end{bmatrix}$$

This distortion matrix says that each error is counted as one unit of distortion. A different data-compression problem with the same source alphabet and reproducing alphabet is obtained if one takes the distortion matrix

$$\rho = \begin{bmatrix} 0 & 1 & 2 & 1 \\ 1 & 0 & 1 & 2 \\ 2 & 1 & 0 & 1 \\ 1 & 2 & 1 & 0 \end{bmatrix}$$

This distortion matrix says that, modulo four, an error of two units is counted as two units of distortion.

A source compression block code of blocklength n and size M is a set consisting of M sequences of reproducing letters, each sequence of length n. The source compression code is used as follows. Each source output block of length n is mapped to that one of the M codewords that results in the least distortion.

The entropy of the output of the data compressor is less than that of the original source and therefore can be encoded into a smaller number of bits.

The Distortion-Rate Function

Data compression is a deterministic process. The same block of source symbols always produces the same block of reproducing letters. Nevertheless, if attention is restricted to a single source output symbol without knowledge of the previous or subsequent symbols, then the reproducing letter is not predetermined. The letter, b_k, into which source letter a_j is encoded becomes a random variable even though the block encoding is deterministic. This random variable can be described by a transition matrix, $Q_{k|j}$. Heuristically, we think of $Q_{k|j}$ as describing an artificial channel that approximates the data compression. Each time the source produces letter a_j, it is reproduced by letter b_k with probability $Q_{k|j}$.

To obtain the greatest possible compression, it seems that this conditional probability should, on average, result in the smallest possible mutual information between the source and the reproduction provided that the average distortion is less than the allowable average distortion. This heuristic discussion motivates the following definition.

The distortion-rate function, $D(R)$, is given by

$$D(R) = \min \Sigma_j \Sigma_k p_j Q_{k|j} \rho_{jk}$$

where the minimum is over all probability transition matrices Q connecting the source alphabet and the reproducing alphabet that satisfy $I(p;Q) \leq R$. The definition is justified by the source compression theorems of information theory. Intuitively, if rate R bits per source letter is specified, then any compression must provide a distortion of at least $D(R)$. Conversely, compression to a level arbitrarily close to $D(R)$ is possible by appropriate selection of the compression scheme.

The distortion-rate function can be evaluated analytically for simple sources and distortion matrices. Fig. 24 shows $D(R)$ for the simplest example, a binary memoryless source with equiprobable zeros and ones, and a probability-of-error distortion matrix. This figure shows that if only one-half bit per bit is used to represent the output of a binary equiprobable source, then at least 11 percent of the reproduced bits must be in error. Fig. 25 shows $D(R)$ for Gaussian sources with mean-square distortion measure. This figure shows that the conventional equispaced quantizer is not optimal, but the performance loss is quite small. Source compression codes are discussed further in Berger.*

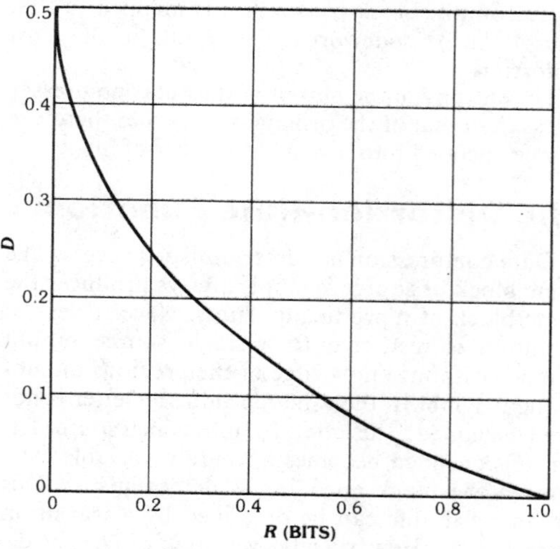

Fig. 24. Distortion-rate function for a binary symmetric source.

MULTITERMINAL INFORMATION NETWORKS

Information networks involving several sources or several channels can have conflicting requirements imposed by the several users. An efficient system tries to satisfy several goals at once, perhaps to interlock several independent messages into one efficient waveform, perhaps to share a multiple-access channel with several users, or perhaps to break a message into two distorted replicas that together contain enough information to reconstruct the original message without distortion. Problems of these kinds are not efficiently implemented in today's communication technology. Multiterminal information theory studies how to solve such problems efficiently. Most such problems are only partially solved. Further discussion of multiterminal information theory can be found in the survey articles by van der Meulen,* Berger,† and El Gamal and Cover.**

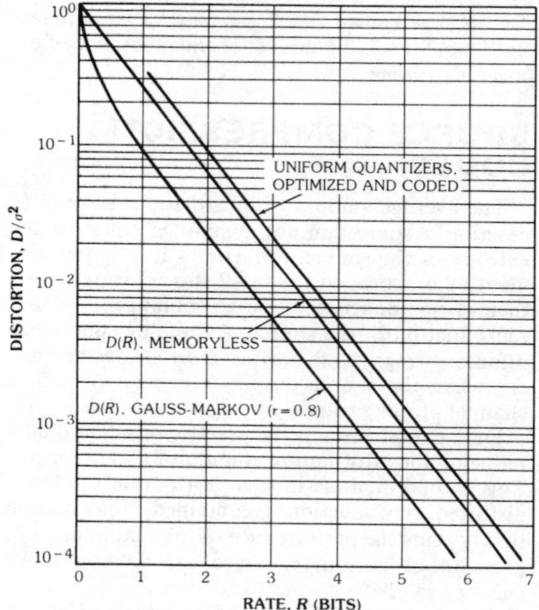

Fig. 25. $D(R)$ for Gaussian sources.

Two-Way Channels

A two-way channel has two terminals, as shown in Fig. 26. Each terminal attempts to get a message across to the other terminal through the two-way channel, but the transmission in one direction interferes with the transmission in the other direction. The problem is to design the terminals to achieve high data rates in both directions simultaneously.

Many practical channels are intrinsically two-way channels, but the designer elects to break them into two one-way channels using a technique such

* Reference 16.

* Reference 17.
† Reference 18.
** Reference 19.

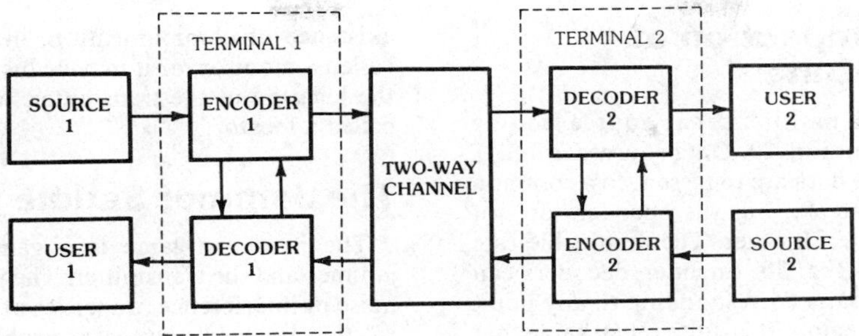

Fig. 26. The two-way channel.

as time division or frequency division. This is a simple and workable solution, but one wishes to know if there is a performance penalty. What is the two-way capacity region, and what sort of communication scheme is optimal? This problem, one of the most difficult ones in multiterminal information theory, has been open for twenty years. Some highly simplified models have been solved or partially solved. These solutions show that time division and frequency division are not optimal.

Broadcast Channels

A broadcast channel consists of a single transmitter and multiple receivers as shown in Fig. 27. If the same message is to be sent to all receivers, then the task of designing the communication waveform is no different than when there is a single receiver. However, if a different message is to be sent to each receiver, the problem is more complex. Conventionally, the channel is divided according to the needs of the receivers, using time-division or frequency-division signaling techniques. Optimum techniques are more tightly interlocked and do better. Time-division or frequency-division multiaccess signaling cannot achieve the capacity region of a broadcast channel.

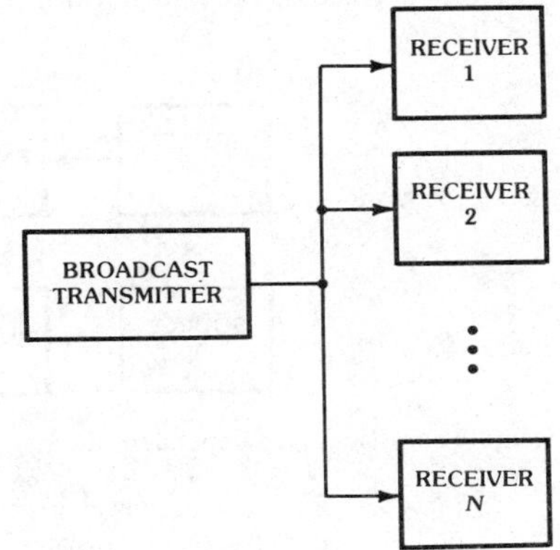

Fig. 27. Broadcast channel.

Degraded Diversity Systems

A *diversity system* sends the same information to a user twice through two channels so that if one channel is broken, the message will still arrive. A *degraded diversity system* is more subtle. It sends half the message through each of two channels but in such a way that either half suffices to reconstruct a degraded copy of the message. A high-fidelity reproduction is obtained if both channels are intact; a low-fidelity reproduction is obtained if only one channel is intact. The simplest model of a degraded diversity system is shown in Fig. 28. The theory shows that the data can be encoded so that either side channel can reconstruct the data from the half-rate code with a probability of error of 20.7 percent, and no better encoding exists.

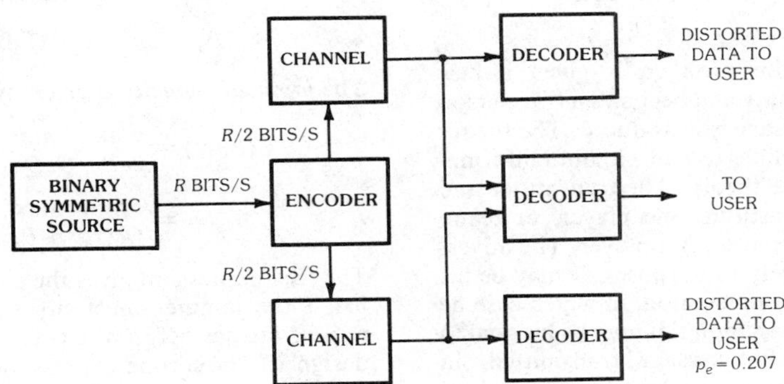

Fig. 28. A degraded diversity system.

Remote Compaction of Dependent Data

A fundamental multiterminal source coding problem is shown in Fig. 29. Data sources X and Y are dependent. The data are to be sent to a common site at rates R_x and R_y, but the encoders do not communicate with each other. The admissible rate region is shown in Fig. 30. Encoder/decoders can be built with rate pairs corresponding to any point in the upper-right region. It is quite remarkable that this region would not be any larger even if the two encoders could each see the output of both sources.

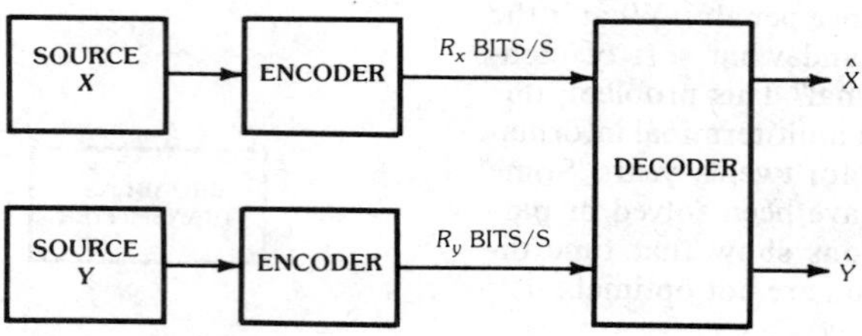

Fig. 29. Remote compaction of dependent data.

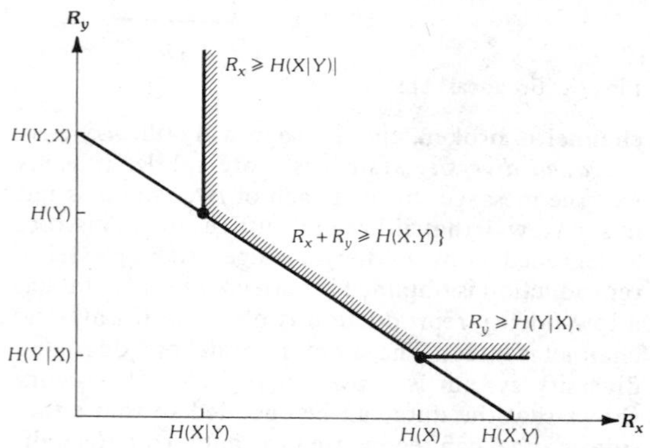

Fig. 30. The Slepian-Wolf admissible rate region.

THE COMMUNICATION GAMES

The communication problem is given a new dimension of complexity and becomes a two-person game when an adversary is introduced. The theory that is then developed has its roots in both information theory and game theory. The transmitter and receiver together constitute one player, or team, and the adversary is the second player. The adversary may have a variety of purposes. It may be his goal to interrupt communication, in which case he is called a jammer or a spoofer; it may be his goal to determine the specific message transmitted, in which case he is called a wiretapper or eavesdropper; it may be his goal simply to detect the existence of communication, in which case he is called a monitor; or it may be his goal to determine the location of the transmitter, in which case he is called a locator.

The Jammer Saddle Point

The jamming game is a game played by the jammer and the transmitter. The jammer selects the noise or interference properties of the channel. That is, the jammer selects a probability transition matrix, Q, connecting the input and output alphabets. The transmitter designs a communication waveform without knowledge of which channel noise was selected by the jammer. Variations are possible wherein the jammer has partial knowledge about the transmitted signal, but these variations will not be considered here. The jammer and the transmitter each must operate within certain constraints. The most important case is one in which the jammer and transmitter are each limited in average power, and this is the case we discuss.

The set of channels that the jammer can use is restricted in that he has a limited average transmitter power. Let Q_N denote the set of channel noise distributions that satisfy power constraints on the jammer,

$$Q_N = \{ Q\text{: } \int z^2 Q(z)\, dz \leq N\}$$

and let P_S denote the set of probability distributions on the channel input that satisfy average power constraints on the transmitter.

$$P_S = \{p\text{: } \int x^2 p(x)\, dx \leq S\}$$

The *jammed capacity* is given by:

$$C(S, N) = \max_{p\epsilon P_S} \min_{Q\epsilon Q_N} I(p;Q)$$

$$= \min_{Q\epsilon Q_N} \max_{p\epsilon P_S} I(p;Q)$$

The first expression gives the capacity of the channel if the jammer must first choose the channel noise statistics before the transmitted waveform is designed. The second expression gives the capacity if the transmitted waveform is designed before the jammer chooses the channel noise. The well-known minimax theorem of game theory says that the two expressions are equal.

For the discrete-time, continuous-amplitude channel with transmitter and jammer both limited in average power, the minimax jammer strategy is to transmit Gaussian noise, and the minimax transmitter strategy is to encode assuming Gaussian noise. From the point of view of the jammer, the jammed capacity is the data rate to which the jammer can hold the transmitter. From the point of view of the transmitter, the jammed capacity is the data rate that the transmitter can achieve in the presence of the jammer.

The most important application is the band-limited waveform channel in which the transmitter and the jammer are each limited in average power. The jammer saddle point is

$$C(S, N) = W \log (1 + S/N)$$

This expression is linear in the bandwidth, W. If the signal power, S, and the jammer power, N, are fixed, the capacity can be increased by increasing W. This is the basis for the practice of spectrum spreading discussed below. If the spectral power density, N_o (watts/hertz), is fixed rather than N, then

$$C(S, N_o) = W \log (1 + S/WN_o)$$

The capacity is bounded as W increases. Spectrum spreading is useful only when the jammer's total power is limited. It is worthless if his spectral power density is limited instead. Spectrum spreading does not improve performance against broad-band white noise.

Spectrum Spreading

A *spread-spectrum* system is one that takes a signal of small bandwidth, say γW, where γ is much smaller than one, and converts it into a signal of high bandwidth, W. This is done as shown in Fig. 31 by modulating the information waveform with a wideband waveform that is not known to the jammer. The wideband waveform may be noise-like (hence called pseudo-noise, or PN), or a time-varying carrier frequency (called frequency hopping). The wideband signal is transmitted through the channel, received, and then compressed back into the original low-bandwidth signal by demodulating the information waveform from the wideband waveform. This scheme requires that the transmitter and receiver share knowledge of the bandspreading waveform through a secure channel. If the jammer knows nothing about the wideband waveform, his "best" strategy is to use wideband noise of bandwidth W. After the signal spectrum is compressed, the jammer signal will still have bandwidth W, but the signal will have bandwidth γW. A bandpass filter will reduce the jammer noise power by the factor γ. The reciprocal of the factor γ is called the processing gain of the spread-spectrum system.

A spread-spectrum system does not make optimum use of the wide bandwidth. This can be seen by looking at the channel capacity. The channel capacity of the available wideband channel of bandwidth W and jammer power N is

$$C = W \log (1 + S/N)$$

However, the channel capacity of the spread-spectrum channel of compressed bandwidth γW and processed jammer power γN is

$$C_{SS}(\gamma) = \gamma W \log (1 + S/\gamma N)$$

whereas if only the narrow bandwidth, γW, is used, there is no reduction of jammer power, and the channel capacity is

$$C_{NB}(\gamma) = \gamma W \log (1 + S/N)$$

If the channel must transmit R bits per second, the wideband channel requires a signal-to-noise ratio of at least

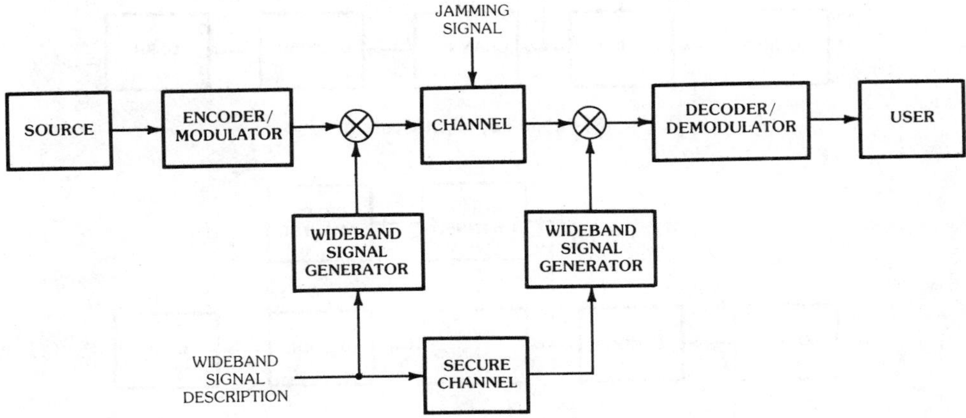

Fig. 31. Spread-spectrum signaling.

$$S/N = 2^{(R/W)} - 1$$

while the spread-spectrum channel requires a signal-to-noise ratio of

$$S/N = \gamma(2^{(R/\gamma W)} - 1)$$

which can be considerably larger than what is needed in the wide bandwidth but is smaller than

$$S/N = 2^{(R/\gamma W)} - 1$$

which is required if there is no spectrum spreading at all. Spread-spectrum by itself improves performance, but much better improvement is theoretically possible using the wide bandwidth.

The use of M-ary orthogonal signaling, spread-spectrum, and error-control codes in combination will make better use of the wide bandwidth. In addition, error-control codes are necessary in a spread-spectrum system to protect against partial-time or partial-band jamming tactics. A well-designed spread-spectrum system will use an error-control code and a modulation waveform that operate at a small E_b/N_o and then spread the spectrum of this waveform to fill out the available frequency band.

Cryptography

Three major kinds of cryptosystem are shown in Fig. 32. The first two kinds require some type of key; the third kind relies on noise in the channel to hide the message. While at first it may seem that the

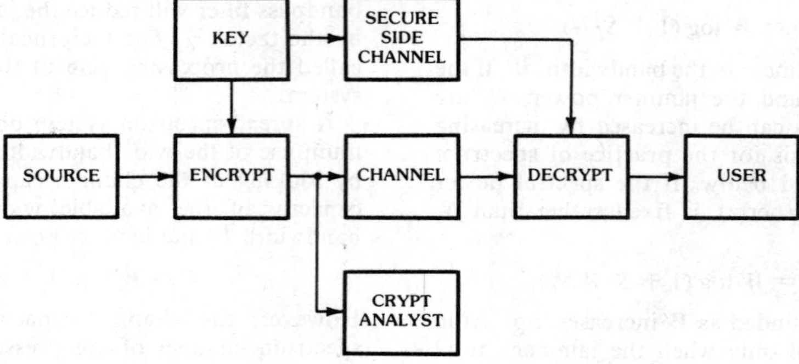

(A) Conventional cryptosystem.

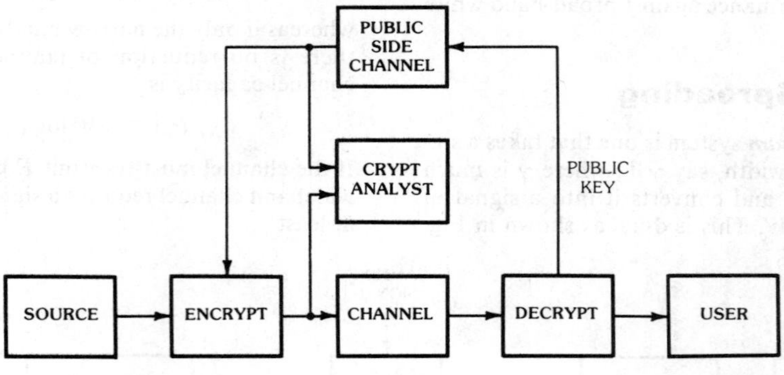

(B) Public-key cryptosystem.

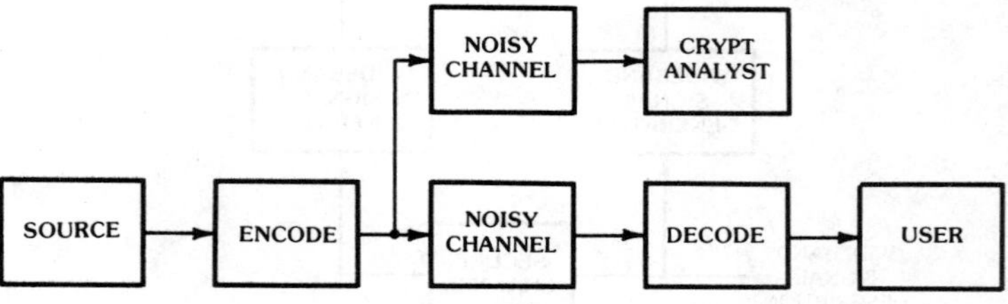

(C) Wire-tap channel.

Fig. 32. Types of cryptosystem.

wire-tap channel (Fig. 32C) is not a true cryptosystem, in fact it would be misleading to ignore it. All cryptosystems have internal compartments where the message or the key exists in the clear. The designer relies on the fact that any portion that leaks out will be buried in the ambient noise. Hence the wire-tap channel is present in every cryptosystem. A secure side channel for transmitting a key is of this type.

A conventional cryptosystem (Fig. 32A) is secure from a direct attack by the cryptanalyst if the key is nearly as long as the message. A so-called "one-time pad" has one random bit in the key for each bit of the message, and it is not reused. The one-time pad can be penetrated only by an attack on the secure side channel. Since it is difficult to transmit and store long keys, most cryptosystems have short keys. Then there is redundancy in the encrypted message that the cryptanalyst can exploit.

A public-key cryptosystem (Fig. 32B) differs from a conventional cryptosystem in that it relies on one-to-one functions, $f(x)$, whose inverses, $f^{-1}(x)$, are very difficult—even computationally intractable—to deduce. These are sometimes called one-way functions. Hence, the user randomly makes up an $f^{-1}(x)$, finds $f(x)$, and publicly announces that he can decrypt messages that are encrypted with $f(x)$. The cryptanalyst always knows the encoding key, but is unable to convert this into a decoding key.

REFERENCES

1. Gallager, R. G. *Information Theory and Reliable Communication*. New York: John Wiley & Sons, Inc., 1968.
2. McEliece, R. J. *The Theory of Information and Coding*. Reading, Mass.: Addison-Wesley Publishing Co., Inc., 1977.
3. Ash, R. B. *Information Theory*. Interscience, 1965.
4. Viterbi, A. J., and Omura, J. K. *Principles of Digital Communication and Coding*. New York: McGraw-Hill Book Co., 1979.
5. Franaszek, P. A. "Sequence-State Coding for Digital Transmission." *Bell System Tech. J.*, 1968, pp. 113–157.
6. Adler, R. L., Coppersmith, D., and Hassner, M. "Algorithms for Sliding Block Codes." *IEEE Transactions on Information Theory*, Vol. IT-29, 1983; pp. 5–22.
7. Rissanen, J., and Langdon, G. G. Jr. "Universal Modeling and Coding." *IEEE Transactions on Information Theory*, Vol. IT-27, 1981, pp. 12–23.
8. Berlekamp, E. R. *Algebraic Coding Theory*. New York: McGraw-Hill Book Co., 1968.
9. Blahut, R. E. *Theory and Practice of Error-Control Codes*. Reading, Mass.: Addison-Wesley Publishing Co., Inc., 1983.
10. Clark, G. C., and Cain, J. Bibb. *Error-Correction Coding for Digital Communications*. New York: Plenum Press, 1981.
11. Lin, S., and Costello, D. J. Jr. *Introduction to Error Correction Codes*. Englewood Cliffs, N.J.: Prentice-Hall, Inc., 1983.
12. Peterson, W. W., and Weldon, E. J. Jr. *Error Correcting Codes*, 2nd ed. Cambridge, Mass.: The M. I. T. Press, 1971.
13. Wozencraft, J. M., and Jacobs, I. M. *Principles of Communication Engineering*. New York: John Wiley & Sons, Inc., 1965.
14. Davenport, W. B. Jr., and Root, W. L. *Random Signals and Noise*. New York: McGraw-Hill Book Co., 1958.
15. Van Trees, H. L. *Detection, Estimation, and Modulation Theory*, Part 1. New York: John Wiley & Sons, Inc., 1968.
16. Berger, T. *Rate Distortion Theory: A Mathematical Basis for Data Compression*. Englewood Cliffs, N. J.: Prentice-Hall, Inc., 1971.
17. Van der Meulen, E. C. "A Survey of Multi-Way Channels in Information Theory: 1961–1976." *IEEE Transactions on Information Theory*, Vol. IT-23, 1977, pp. 1–37.
18. Berger, T. "Multiterminal Source Coding," in *The Information Theory Approach to Communications*, G. Longo, ed. New York: Springer-Verlag, 1977.
19. El Gamal, A., and Cover, T. M. "Multiple User Information Theory." *Proceedings IEEE*, Vol. 68, 1980, pp. 1066–1083.
20. Special Issue on Spread Spectrum, *IEEE Transactions on Communication Theory*, Vol. COM-30, May, 1982.

26 Computer Communications Networks

Fouad A. Tobagi

About two decades ago, a great challenge to many computer system designers was to enhance the processing power of computers and make them available to a large number of users on a time-sharing basis. The design of fast arithmetic and control units as well as the design of complex operating systems were identified among the key tasks needed to accomplish this objective. As time-sharing progressed, it was soon realized that such large resources would not be effectively utilized unless the problem of connecting remote user terminals to the central computing facility was adequately solved, thus allowing a large population of users to share the facility. With this problem, the field of computer communications came into existence. The early communications systems were merely *terminal access networks.* The next stage in the evolution consisted of the creation of the so-called distributed *resource-sharing networks.* The goal here is to interconnect computers and their users at various geographically distributed sites in order to allow the sharing of hardware and software resources developed at all sites by all users connected to the network. Today the term *computer communications* has a much broader meaning. Not only does it refer to communications among computers and their users, but it also refers to all kinds of communications applications (among humans and among machines) that make use of the computer as a tool. Examples of such applications are voice communications, electronic mail, facsimile, image transfer, process control, etc. The goals are also somewhat more diverse. Instead of communications per se, the driving force may very well be that of cost or reliability of computing power. High reliability may be achieved by having alternative sources of computing made available via a communications network. Lower cost may be achieved via distributed computing architectures that are simpler to design, cheaper to build, and easier to maintain, and of which communication is an intrinsic part. Thus, while in the seventies one might have characterized the computer communications field as having been in its research and development phase, the eighties will be marked by the wide use of this relatively new technology for a large repertoire of applications.

In the following, the structure of computer communications networks is examined and their building blocks identified. First, the various types of networks in existence and the switching techniques in use are described. Then the general function of a network needed in establishing a communications path among remote users is described, and the organization of these functions into a standard layered architecture called the *Open Systems Interconnection reference model* is discussed. Using this reference model as a guide, we then examine in more detail the functions in each of the layers and highlight standards whenever applicable.

THE STRUCTURE OF COMPUTER NETWORKS

A computer-communication network typically comprises a collection of computing resources called *hosts,* a collection of users, some of which are associated with the hosts, and a so-called *communication subnet* that connects them (Fig. 1.) The communication subnet consists of two basic components: the communication channels and the switching elements (or switching nodes). Depend-

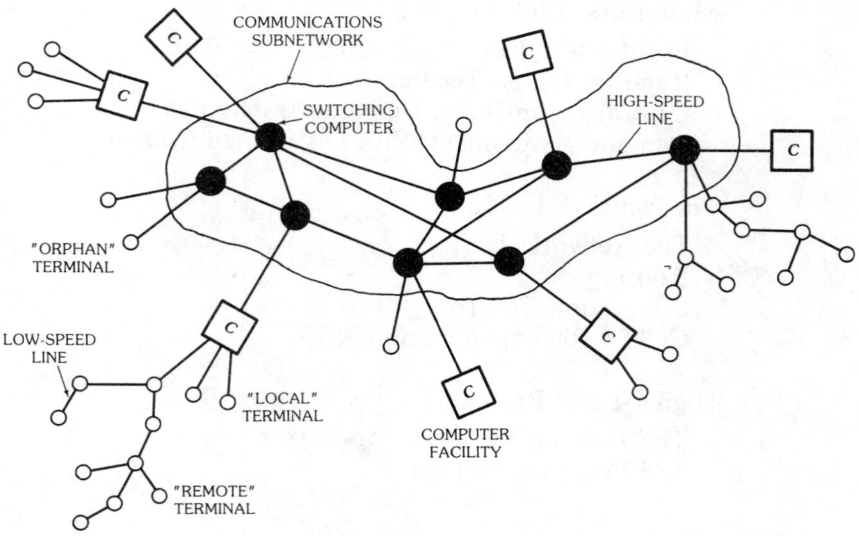

Fig. 1. The structure of a computer-communication network. (*From Leonard Kleinrock,* Computer Applications, *Vol. II of* Queueing Systems, *© 1976 by John Wiley & Sons, Inc., New York.*)

ing on (1) the physical medium used for the communication channels, (2) the subnet topology according to which the communication channels and the switching elements are interconnected to form a network, and (3) the switching technique used in providing a physical path among two or more communicating parties, several computer network types may be identified.

Switching Techniques

There are two basic types of switching techniques: *circuit switching* and *message switching.* In circuit switching, a total path of connected lines is set up from the origin to the destination at the time the call is made, and the path remains allocated to the source-destination pair (whether used or not) until it is released by the communicating parties. The switches, called circuit switches (or office exchange in telephone jargon), have no capability of storing or manipulating users' data on their way to the destination. The circuit is set up by a special signaling message that finds its way through the network, seizing channels in the path as it proceeds. Once the path is established, a return signal informs the source to begin transmission. Direct transmission of data from source to destination can then take place without any intervention on the part of the subnet.

In message switching, the transmission unit is a well defined block of data called a *message.* In addition to the text to be transmitted, a message comprises a *header* and a *checksum.* The header contains information regarding the source and destination addresses as well as other control information; the *checksum* is used for error control purposes. The switching element is a computer referred to as a *message processor,* with processing and storage capabilities. Messages travel independently and asynchronously, finding their own way from source to destination. First the message is transmitted from the host to the message processor to which it is attached. Once the message is entirely received, the message processor examines its header, and accordingly decides on the next outgoing channel on which to transmit it. If this selected channel is busy, the message waits in a queue until the channel becomes free, at which time transmission begins. At the next message processor, the message is again received, stored, examined, and transmitted on some outgoing channel, and the same process continues until the message is delivered to its destination. This transmission technique is also referred to as the *store-and-forward transmission* technique.

A variation of message switching is *packet switching.* Here the message is broken up into several pieces of a given maximum length, called *packets.* As with message switching, each packet contains a header and a checksum. Packets are transmitted independently in a store-and-forward manner.

With circuit switching, there is always an initial connection cost incurred in setting up the circuit. It is cost-effective only in those situations where once the circuit is set up there is a guaranteed steady flow of information transfer to amortize the initial cost. This is certainly the case with voice communication in the traditional way, and indeed circuit switching is the technique used in the telephone system. Communication among computers, however, is characterized as *bursty.* Burstiness is a result of the high degree of randomness encountered in the message-generation process and the message size, and of the low delay constraint required by the user. The users and devices require the communication resources relatively infrequently; but when they do, they require a relatively rapid response. If a fixed dedicated end-to-end circuit were to be set up connecting the end users, then one must assign enough transmission bandwidth to the circuit in order to meet the delay constraint with the consequence that the resulting channel utilization is low. If the circuit of high bandwidth were set up and released at each message transmission request, then the set-up time would be large compared to the transmission time of the message, resulting again in low channel utilization. Therefore, for bursty users (which can also be characterized by high peak-to-average data rate requirements), store-and-forward transmission techniques offer a more cost-effective solution, since a message occupies a particular communications link only for the duration of its transmission on that link; the rest of the time it is stored at some intermediate message switch and the link is available for other transmissions. Thus the main advantage of store-and-forward transmission over circuit switching is that the communication bandwidth is dynamically allocated, and the allocation is done on the fine basis of a particular link in the network and a particular message (for a particular source-destination pair).

Packet switching achieves the benefits discussed so far and offers added features. It provides the full advantage of the dynamic allocation of the bandwidth, even when messages are long. Indeed, with packet switching, many packets of the same message may be in transmission simultaneously over consecutive links of a path from source to destination, thus achieving a "pipelining" effect and reducing considerably the overall transmission delay of the message as compared to message switching. It tends to require smaller storage allocation at the intermediate switches. It also has better error characteristics and leads to more efficient error recovery procedures, as it deals with smaller entities. Needless to say, packet switching presents design problems of its own, such as the need to reorder packets of a given message that may arrive at the destination node out of sequence.

Fig. 2 illustrates the three switching modes for an example of a communication subnet involving four nodes and three transmission links. The figure shows the advantage of packet switching over message switching and of message switching over circuit switching. Clearly, this comparison involves a number of tradeoffs, and it is not hard to imagine situations where the conclusion may be reversed. This depends on a number of factors, among others the number of hops from source to destination, the length of the message, the amount of overhead incurred in the header and checksum of each message and packet, the circuit set-up delay, etc. For example, fast circuit switching accomplished with solid-state implementation of the switches can achieve a set-up time on the order of a fraction of a second (a few milliseconds per switch), which renders this technique a viable solution. In most cases, however, the picture shown in Fig. 2 reflects the real situation.

The advantages of store-and-forward transmission are obtained at the expense of higher processing and storage capabilities at all switches. Overall, however, the present economics are in favor of

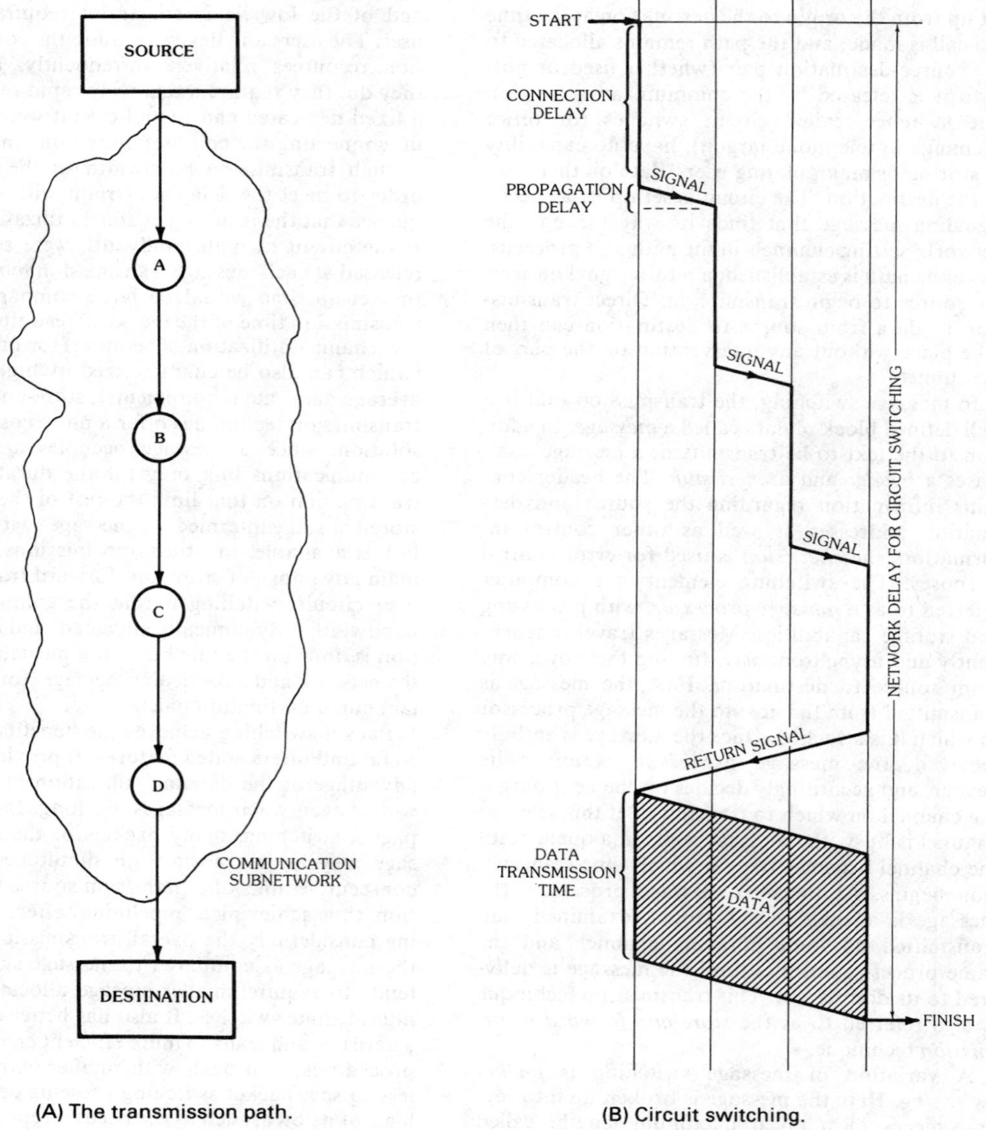

(A) The transmission path.

(B) Circuit switching.

Fig. 2. Comparison of network delay for circuit, message, and packet switching. (*From Leonard Kleinrock,*

store-and-forward. Roberts* has shown that since the early 60s, the incremental cost of computing needed to send one megabit of data through a nationwide network has been decreasing at a much faster rate than the incremental cost of the communication land lines in a national net, with a crossover point having occurred about 1969. Fig. 3 summarizes Roberts' findings concerning these cost trends (the reader is referred to the portion of the figure corresponding to the range up to 1974).

* Reference 1.

In the development that follows, the three terms message switching, packet switching, and store-and-forward are used interchangeably unless the distinction is made explicitly.

Types of Computer Communications Networks

A distinction is first made between *circuit-switched networks* and *packet communication networks.* Clearly, circuit-switched networks are those that use circuit switching. When used for local communication, these networks have a star topol-

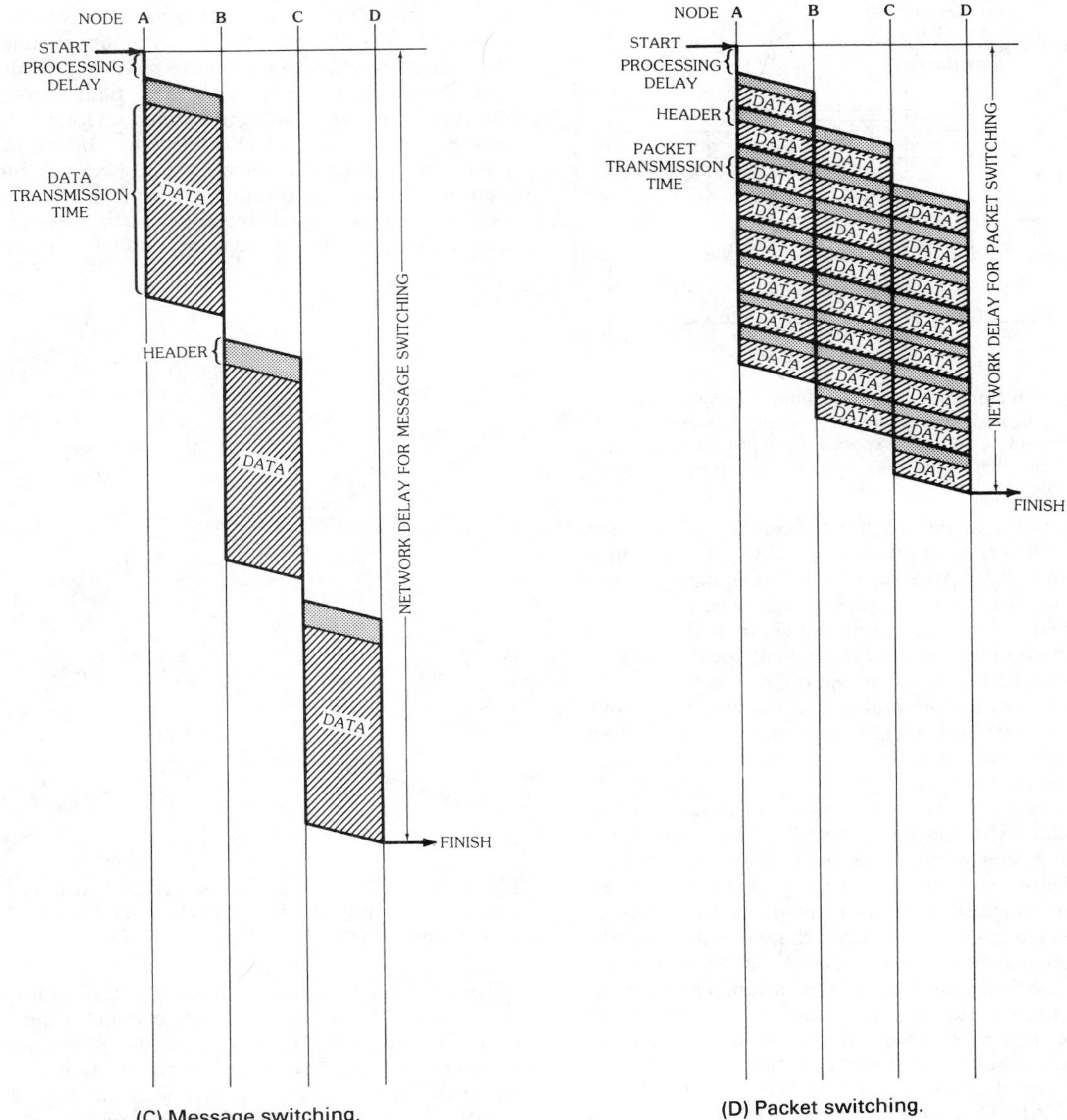

(C) Message switching.

(D) Packet switching.

Computer Applications, *Vol. II of* Queueing Systems, © *1976 by John Wiley & Sons, Inc., New York.)*

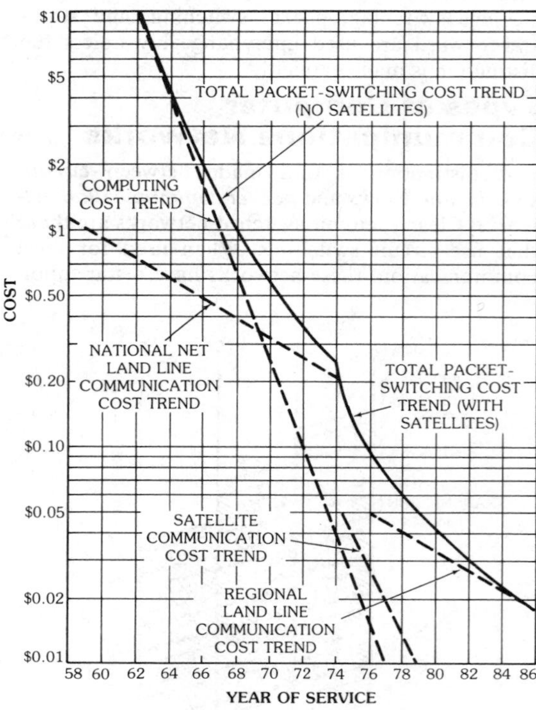

Fig. 3. Incremental cost for sending 1 kilopacket (1 megabit) through a nationwide network. (*From Leonard Kleinrock*, Computer Applications, *Vol. II of* Queueing Systems, © *1976 by John Wiley & Sons, Inc., New York.*)

ogy with a single switch in the center, and they may often be privately owned (e.g., CBX). Long-haul or nationwide networks, on the other hand, have a hierarchical mesh toplogy: subscribers are connected to local central offices via two-wire connections to form a star local network; local offices are connected to toll offices via toll connecting trunks; toll offices are interconnected via very high bandwidth intertoll trunks and other intermediate switching offices to form a highly redundant mesh topology.

Packet communication networks are all based on packet switching and may be of three types. The first is referred to as the *point-to-point store-and-forward* type. In this type, packet switches are interconnected by point-to-point data channels to form a mesh topology. Each channel is used only by the two switches adjacent to it, one in each direction; thus there exists no contention. The channels are usually full duplex, allowing transmission to take place in both directions simultaneously. Examples of such networks are the ARPANET, the Cigale Network, TELENET, TYMNET, DATAPAC, TRANSPAC, EURONET, etc.* This type is usually for large geographical scope, although it is also used for terminal access and local area communication; in the latter case, the network typically has a star topology with a packet switch at its center.

The second type of packet communication network is the *multiaccess-broadcast* type. This type of network consists of a *single* transmission medium that is shared by all subscribers; they access the medium according to some multiaccess scheme, each access being for the duration of a single packet. Any user's transmission is heard by all other users, hence the broadcast attribute. The single-hop broadcast nature of these networks achieves full connectivity at a very small cost: each subscriber is connected to the common channel through an interface that listens to all transmissions and accepts packets addressed to it. Examples of this type of network can be found in both long-haul and local-area communication systems. One example is that of a *satellite channel* (Fig. 4): a satellite transponder in a geostationary orbit above the earth can receive signals from any earth station in its coverage pattern and can transpond these signals

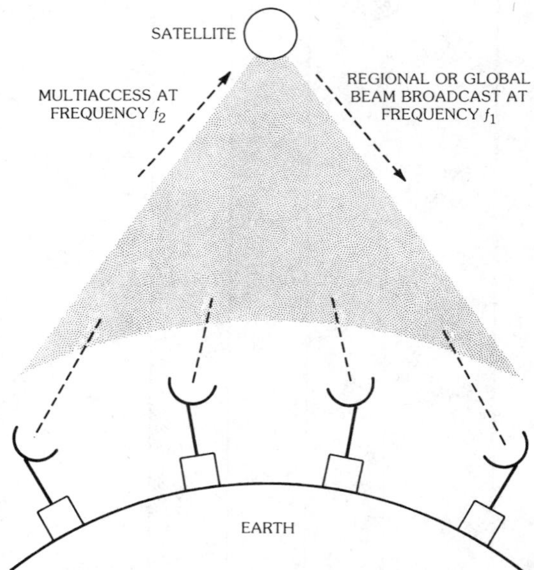

Fig. 4. A satellite channel. (*From Leonard Kleinrock*, Computer Applications, *Vol. II of* Queueing Systems, © *1976 by John Wiley & Sons, Inc., New York.*)

to all such earth stations (unless the satellite uses spot beams). Broad-band satellite channels offer a cost-effective alternative to store-and-forward networks for long-haul communication, as can be easily seen from Roberts' projections in Fig. 3. Another example is that of *ground packet radio* systems where the radio medium is used for terminal access and local-area communication among a

* References 2 and 3.

large population of terminals, possibly mobile (Fig. 5). The multiaccess and broadcast attributes are achieved when all users share a common radio frequency, employ omnidirectional antennas (thus facilitating communication among the mobile

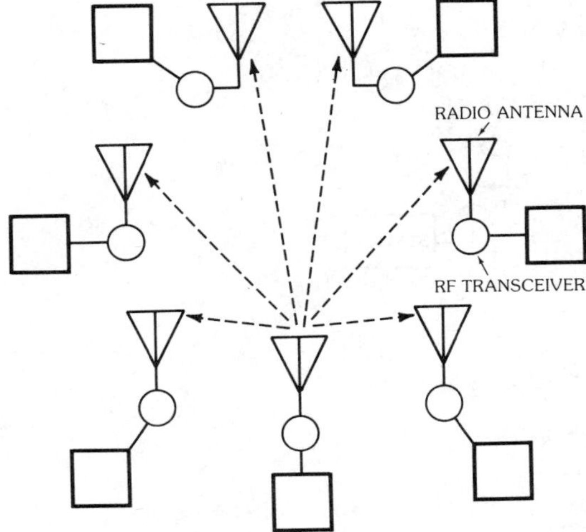

Fig. 5. A single-hop fully connected ground packet radio network.

users), and are in line-of-sight and within range of each other. Yet another example is the so-called *broadcast bus* system in which all users are connected to a single cable via simple passive taps (Fig. 6). These systems prove to be ideal for short-distance local-area communication within a building (up to a few kilometers) involving a large and often variable number of inexpensive devices requiring interconnection. The simple topology and simple and inexpensive connection interfaces of broadcast bus systems provide great flexibility in accommodating the high degree of variability in such environments while achieving a desired level of reliability. A final example that falls into this category is the *ring* network (Fig. 7). Each node is connected only to two neighbors via point-to-point unidirectional links. Messages (or packets) are circulated around the ring (hence achieving the broadcast feature), incurring a delay of only one or more bit times within each of the intermediate nodes. The multiaccess schemes available to operate multiaccess-broadcast networks are discussed in more detail below in the section on multiaccess link control.

A third type of packet network can be identified. It is the (multihop) *store-and-forward multiaccess/broadcast* type, which combines the features exhibited in the two types mentioned above. The best (and perhaps only) example is DARPA's packet radio network (PRNET). (See Fig. 8.) Its goal is to provide direct communication by a ground radio

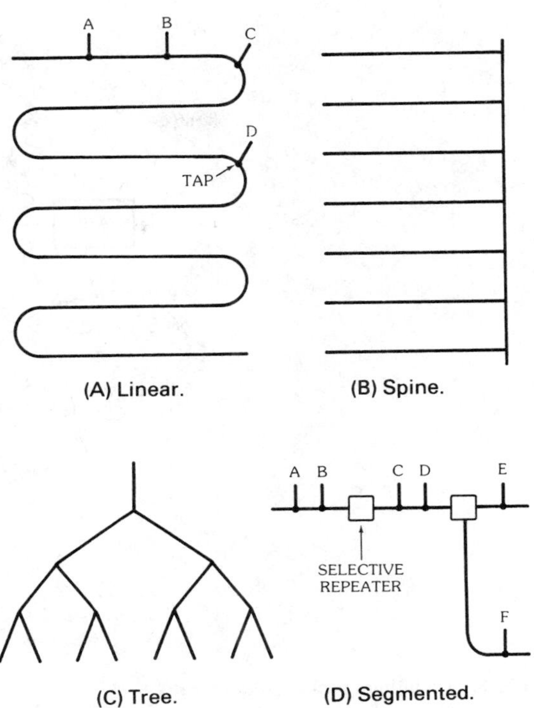

Fig. 6. Cable topologies. (*From Andrew S. Tanenbaum,* Computer Networks, *© 1981. Used by permission of Prentice-Hall, Inc., Englewood Cliffs, N.J.*)

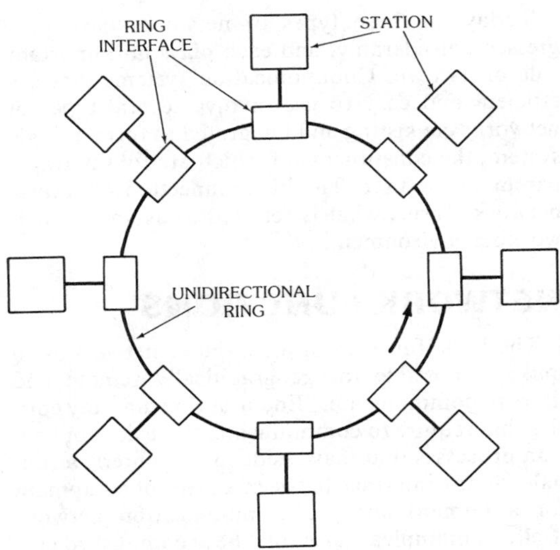

Fig. 7. A ring network. (*From Andrew S. Tanenbaum,* Computer Networks, *© 1981. Used by permission of Prentice-Hall, Inc., Englewood Cliffs, N.J.*)

network among mobile users over wide geographical areas. This requires store-and-forward switches called repeaters to become integral components of the system. Here too, for easy communication

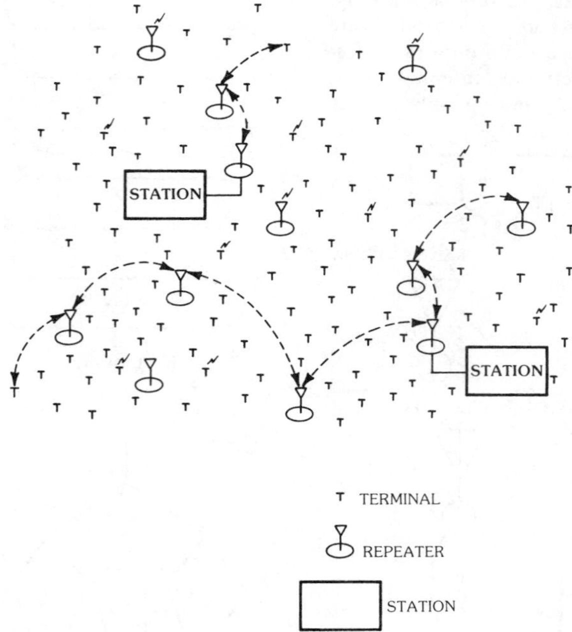

Fig. 8. A multihop packet radio network.

among mobile users and for rapid deployment in military applications, all users employ omnidirectional antennas and share a high-speed radio channel.

Today all these types of networks have progressed considerably, and each plays an important role of its own. Communication systems that are efficient and easy to use involve several types of networks coexisting and integrated to form a single system, the constituents of which are totally transparent to the user. The interconnection of several networks forms what is referred to as an internetworking environment.

NETWORK FUNCTIONS

The basic function of a computer network is to make it possible for geographically remote end users to communicate. "End users" means any entities that require to communicate; these are application processes that may reside in computers, terminals, or any interface between a piece of equipment (or a human) and the communication network itself. Communication should be accomplished efficiently, by making efficient use of the communications resources, and in a way that is responsive to the specific needs of the end users. It should also be made possible in spite of various types of errors that may occur in the transmission process, and in spite of the differences that exist between the end users with respect to the data formats used, the data rates supported, the patterns of intermittency, etc.

As this global function is a very complex one, it is best accomplished by dividing it into a sequence of more elementary functions that are organized in some fashion. The precise definition and structure of this aggregate set of functions is called a network *architecture*. Most typically, network architectures follow a linearly hierarchical model; i.e., the functions are organized into a linear succession of so-called *layers*. Such architectures are referred to as *layered* architectures. At the lowest level of the hierarchy, for example, are those functions that concern themselves with providing a physical connection between the end users and with the transmission of raw bits (i.e., regardless of their meaning) across the physical transmission resources. These functions constitute the physical layer. At the highest level, we find those portions of the application processes residing at the end users' machines that are the originators and destinations of the communication requests, and which directly serve the end user as far as the global communication function is concerned. The other intermediate layers comprise all other functions ranging from detecting and correcting errors to achieving efficient utilization of the resources; performing routing functions; preventing congestion; regulating the flow of data; preventing unfairness; supporting the patterns of intermittency and specific service requirements of the end users; and finally resolving the differences among the users pertaining to formats, character codes, device control procedures, and the like. In such a hierarchical structure, each layer is considered to be wrapping the lower layers and isolating them from the higher layers. Each layer offers a well specified service that it provides to the adjacent layer above it. In supplying that service, the layer obtains service from the adjacent layer just below it and performs its own functions. The rules for performing the functions at a given layer, once specified, are called *protocols.* The interactions between adjacent layers (at their boundaries) are called *interfaces.* The rules for achieving interactions and communication between adjacent layers, once specified, are called *interface protocols.* Also essential to a layered architecture is the concept of *peer protocols* or *peer interaction.* The processes implementing protocols within a layer at one machine (host or intermediate mode) communicate only with the processes implementing protocols within the same layer at another machine. Compatibility between protocols of the same level at (virtually) adjacent machines must be guaranteed. This layered approach to network functions for their description and design is now universal, and most (if not all) existing networks follow a layered architecture.

Layered architectures offer a number of advantages. The first is the prevention of ad hoc methods of network design, which, due to the intricacies involved, would certainly lead to a proliferation of designs difficult to understand, implement, or interact with. This is particularly beneficial when systems designed by different organizations or manufacturers are to be interconnected. The second advantage is that of modularity: If one specifies the interfaces between adjacent layers in a general way that is independent of the particular protocols designated at the interacting layers, then it becomes possible to modify the protocols at some layer without affecting the rest of the system. This is particularly beneficial in order to take advantage of newly emerging technologies, and in order to experiment with new protocols without the need for a major redesign of the entire system. For these reasons, the International Organization for Standardization* (ISO) and other standards organizations were urged to come up with standards for "Open Systems Interconnections" (OSI). The objective is that by conforming to those international standards, a system will be capable of interacting with all other systems obeying the same standards throughout the world. The standard architecture adopted by ISO, also the best known one, is a layered architecture with seven layers. It is only an architecture, in the sense that it only defines the services to be performed by a layer for the next higher layer, independent of how these services are performed. Accordingly, it is often referred to as ISO's OSI reference model. It is the first stage toward a complete standardization of network functions, although very few protocols and interfaces have been standardized so far.

The number of layers decided upon is the result of long debates and of a number of principles followed in the process. Some of these principles are:

1. To collect "similar" functions into the same layer and separate "manifestly different" functions into separate layers.
2. To create a layer of those functions that are easily localized and that may be totally redesigned in the future when taking advantage of new technological advances, without the need to change the services or the interfaces with the adjacent layers.
3. To create boundaries at those service points where the interactions across the boundaries are minimized, those which past experience has demonstrated to be successful, and those where it may be useful at some point in time to standardize the corresponding interface.

Another principle calls for a limitation on the number of layers so as to keep the engineering task of describing them and integrating them a simple one, but it allows further subgrouping of functions within a layer so as to form sublayers that may be bypassed if the corresponding services are not needed.

The seven layers of the ISO OSI reference model (Fig. 9) are: (1) the physical layer, (2) the data link layer, (3) the network layer, (4) the transport layer, (5) the session layer, (6) the presentation layer, and (7) the application layer. Specific protocols and their description are considered in subsequent sections.

While the ISO reference model is used here as a guide to describe network functions and protocols, it is to be noted that there are a few existing network architectures, either designed for experimental networks, such as the ARPANET, or supplied by computer manufacturers, such as the Digital Equipment Corporation DECNET and the IBM Systems Network Architecture (SNA). These architectures are layered but do not correspond exactly to the ISO reference model. See Table 1. For more information on these architectures, the reader is referred to references 2 and 3.

THE PHYSICAL LAYER

The physical layer is concerned first of all with the transparent transmission of a bit stream

* The International Organization for Standardization (ISO) is a voluntary nontreaty group, the membership of which includes the principal standardization body of each represented nation. The U. S. member body is the American National Standards Institute (ANSI). ANSI is a nonprofit, nongovernmental organization. It serves as the national clearing house and coordinating activity for voluntary standards in the U.S.

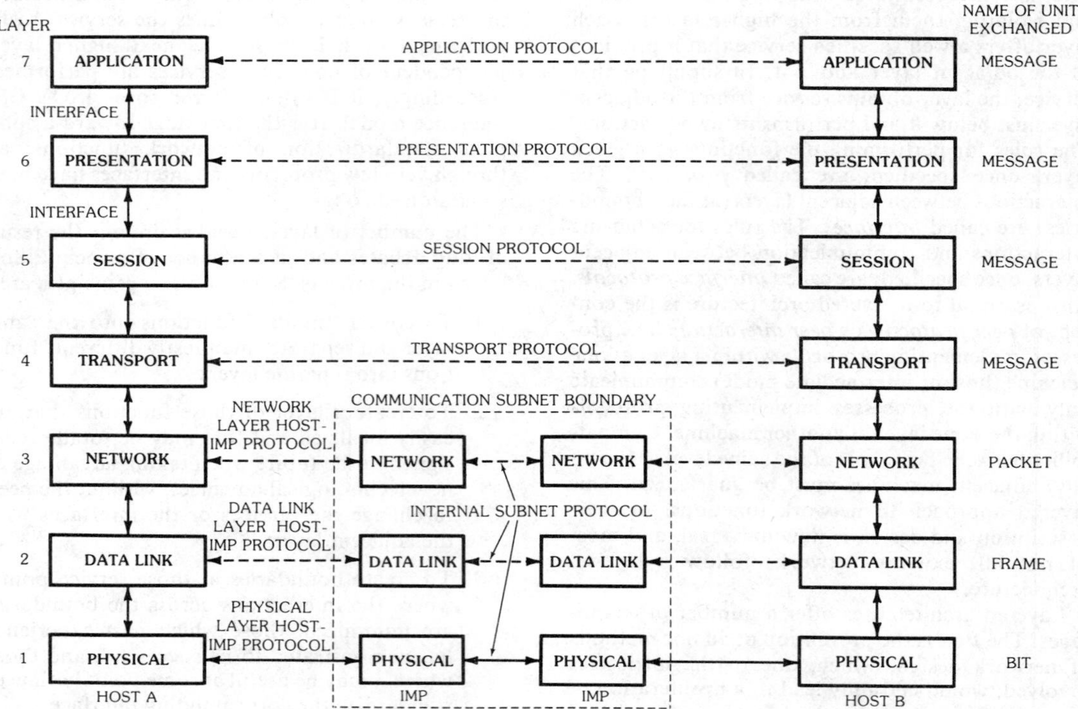

Fig. 9. The ISO OSI reference model. (*From Andrew S. Tanenbaum,* Computer Networks, *© 1981. Used by permission of Prentice-Hall, Inc., Englewood Cliffs, N.J.*)

(regardless of its meaning) across physical communication resources. In local networks, the physical medium may be a twisted pair, a coaxial cable, an optical fiber, or radio, and it may be privately owned. For long-haul links, the medium may be either copper wires or optical fibers as in terrestrial links, or radio as in satellite links; it is supplied by a common carrier. The methods used for the transmission of bits across the physical medium depend on the type of medium used. The description of these techniques is out of the scope of this chapter, but suffice it to mention here that bit transmission is done either in analog form by means of *modems* or in digital form by means of *line drivers,* and it

TABLE 1. APPROXIMATE CORRESPONDENCES AMONG THE VARIOUS NETWORK ARCHITECTURES*

Layer	ISO	ARPANET	SNA	DECNET
7	Application	User	End user	Application
6	Presentation	Telnet, FTP	NAU services	
5	Session	(None)	Data flow control	(None)
4	Transport	Host-host	Transmission control	Network services
3	Network	Source to destination IMP	Path control	Transport
2	Data link	IMP-IMP	Data link control	Data link control
1	Physical	Physical	Physical	Physical

* From Andrew S. Tanenbaum, *Computer Networks,* © 1981. Used by permission of Prentice-Hall, Inc., Englewood Cliffs, N.J.

involves determining such attributes as data encoding, timing, voltage levels, data rates, type of operation (half-duplex or full-duplex, synchronous or asynchronous), etc.

The second concern of the physical layer is to provide a physical interface between the end-user machine—which may be a terminal, a computer, or any other data-processing box, and which is referred to by the telecommunications administrations as the *Data Terminal Equipment* (DTE)—and the termination point of the communications circuit—that is the modem, line driver, etc., referred to as the *Data Circuit-Terminating Equipment* (DCE). (See Fig. 10.) The definition of such an interface requires the determination of four important characteristics: the mechanical, electrical, functional, and procedural characteristics.

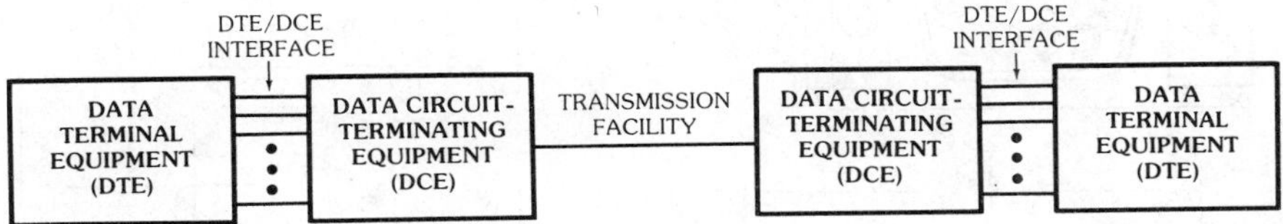

Fig. 10. DTE/DCE interface. (*From Paul E. Green, Jr.*, Computer Network Architectures and Protocols, © *1982. Used by permission of Plenum Publishing Corp., New York.*)

The mechanical aspects pertain to the point of demarcation, which most typically consists of a pluggable connector. They include specifications of the connector, its latching and mounting arrangements, its location with respect to the DCE, etc. Fig. 11 illustrates the various connectors known, along with their ISO identification numbers. The number of pins per connector ranges from 9 to 37.

The electrical aspects pertain to the electrical characteristics of the generators and receivers. They specify such parameters as the range of the signal voltage level, rise-time characteristics of the generator, data signaling rates as a function of the interconnecting cable distances, generator and receiver impedances, etc. They also include specifications regarding the reference with respect to which signal levels are measured; two cases exist: the "unbalanced" case where a single common return lead is used (or perhaps one common return for each direction), and the "balanced" case where each interchange circuit uses a pair of wires creating a differential signal. A balanced configuration is more desired because it permits longer distances and higher data rates.

The functional characteristics specify the number of interchange circuits used and their assigned functions. The functions are classified into four categories: data, control, timing, and grounds. Timing circuits are used for bit (and sometimes byte) synchronization, whereas control circuits are used for the exchange of status information, commands, and responses. Some interfaces employ only one function per interchange circuit (e.g., CCITT V.24*), whereas others multiplex DCE control functions and data over a single "data" interchange circuit in each direction, resulting in a considerably more compact interface (e.g., CCITT X.24).

The procedural characteristics consist of the set of procedures for using the interchange circuits, in order to achieve the transmission of bit streams as well as to provide maintenance test loops.

Three examples of widely used physical interface protocols are considered here to illustrate the concepts and physical characteristics discussed above. These are EIA RS-232-C,† EIA RS-449, and CCITT X.21. They are all serial data DTE/DCE interfaces. The dominant interface in use today between terminals and modems is RS-232-C. The connector is the 25-pin ISO 2110 (Fig. 11). Its electrical characteristics are compatible with CCITT recommendation V.28. This specifies a single-ended generator that produces a 5 to 15 volt signal with respect to signal ground, negative for binary 1 and positive for binary 0; a single common return lead is used for all interchange circuits (thus unbalanced); the generator rise time is fast (1 ms to cross ±3 V); the data signaling rates are limited to below 20 Kbits/s; the cable distances are limited to within 15 m. The functional characteristics of RS-232-C are compatible with CCITT recommendation V.24. Recommendation V.24 defines 43 interchange circuits, one function per circuit; RS-232-C uses 21 of the 43 interchange circuits; not all 21 circuits are needed in every application (for example, timing circuits are omitted for asynchronous applications). The procedural characteristics of RS-232-C are

* The International Telegraph and Telephone Consultative Committee (CCITT) is a committee of the International Telecommunication Union (ITU), a specialized agency of the United Nations Organization. The CCITT work on data communications is focused in two study groups. CCITT Study Group XVII is responsible for data communications over telephone facilities. Its work is contained in V-series recommendations. CCITT Study Group VII is responsible for data communications over data networks. Its work is contained in X-series recommendations.

† The Electronic Industries Association (EIA) is a trade association that represents manufacturers in the U.S. electronics industry. EIA standards on data communications are published in the RS-series.

many and complex. They contain procedures equivalent to those in recommendation V.24 describing the interrelationships between interchange circuits, as well as those in recommendation X.20bis and X.21bis, which specify asynchronous and synchronous operation on a public data network for DTEs designed to interface with V-series asynchronous and synchronous modems, respectively. (As indicated earlier in a footnote, the V-series recommendations correspond to data communications over telephone facilities.) For a detailed description of the procedural characteristics of RS-232-C, see reference 3.

In 1973, EIA began work on a new interface with improved performance over RS-232-C (longer cable distance and higher maximum data rates) and with additional interface functions such as maintenance loopback testing. The result was RS-449, published in November 1977. The new interface was carefully designed so as to allow interoperability with existing RS-232-C equipments. The RS-449 interface uses connectors of the same family as that used for

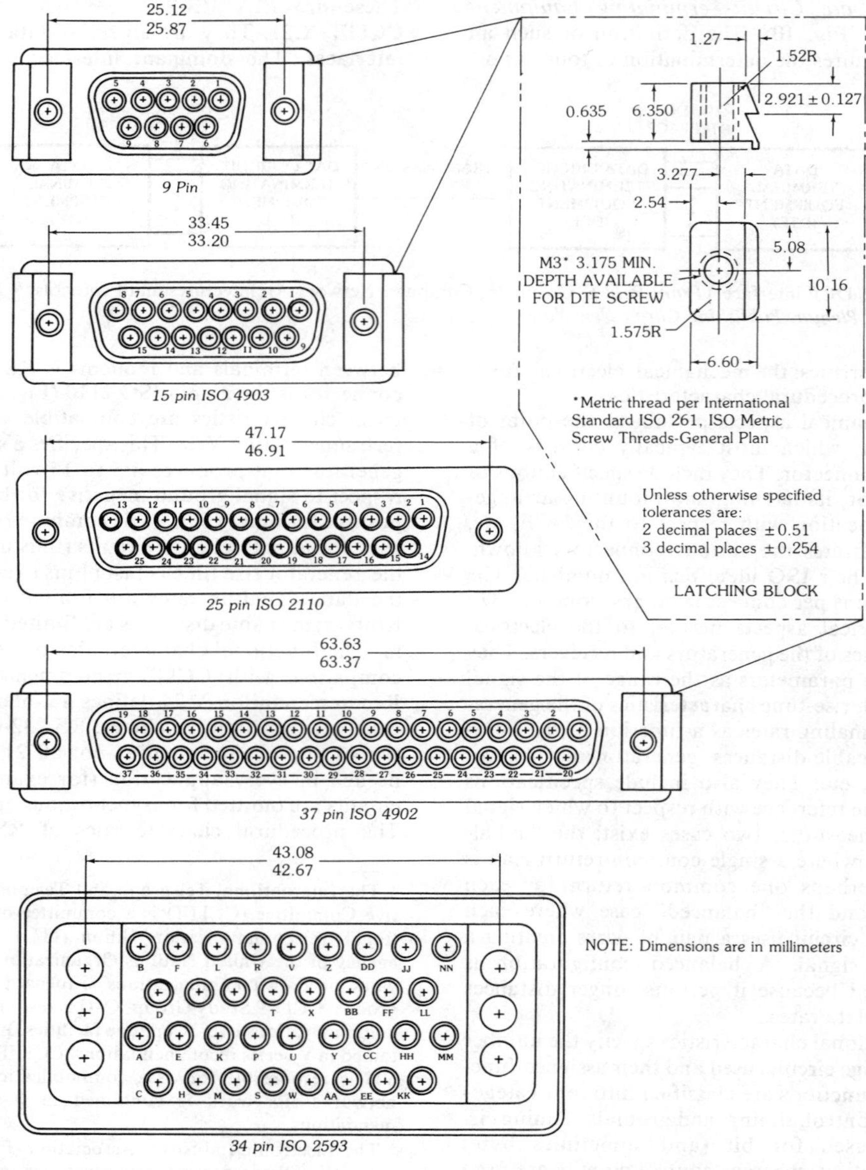

Fig. 11. DCE connectors. (*From Paul E. Green, Jr.*, Computer Network Architectures and Protocols, © *1982. Used by permission of Plenum Publishing Corp., New York.*)

RS-232-C: it uses a 37-pin connector (ISO 4902) for the basic interface, and a separate 9-pin connector if a ''secondary channel'' operation is in use (Fig. 11). Furthermore, a careful pin assignment plan is chosen to minimize cross talk in multipair cables and to facilitate the design of an adapter to RS-232-C. As in RS-232-C, the functional characteristics of RS-449 are compatible with V.24, one function per interchange circuit. There are 10 new circuits with respect to RS-232-C, among which we note the following: send common (SC) and receive common (RC), which are the common return leads for all unbalanced interchange circuits employing one wire in the direction toward the DCE and the DTE, respectively; terminal in service (IS), which indicates to the DCE whether the DTE is operational or not; local loopback (LL), remote loopback (RL), and test mode (TM), which are used in checking and testing; select standby (SS) and standby indicator (SB), which are used when standby facilities are in use in order to facilitate the rapid restoration of service when a failure has occurred. The electrical characteristics are defined in RS-423-A (compatible with V.10/X.26) and RS-422-A (compatible with V.11/X.27). Standard RS-423-A, referred to as the "new unbalanced" electrical characteristics, specifies a single-ended generator that produces 4 to 6 volts with respect to the common return; a single common return is used for each direction; data signaling rates are up to 3 Kbits/s over cable distances of 1000 meters and can be as high as 300 Kbits/s for distances up to 10 meters. Standard RS-422-A, referred to as the "new balanced" characteristics, specifies a balanced generator producing a 2 to 6 volt differential signal; each interchange circuit employs two wires; data rates are up to 100 Kbits/s over cable distances of 1000 meters and can be as high as 10 Mbits/s over 10 meters. Standard RS-449 uses the new unbalanced characteristics, RS-423-A, for all interchange circuits when the data rate is below 20 Kbits/s. This permits direct interoperability with RS-232-C since RS-423-A (V.10/X.26) is interoperable with both V.28 (RS-232-C) and RS-422-A (V.11/X.27). At data rates above 20 Kbits/s and up to 2 Mbits/s, the new balanced EIA RS-422-A must be used on 10 specific interchange circuits (send data, receive data, send timing, . . .) employing two wires per circuit, while RS-423-A is used for all other circuits.

As for the procedural characteristics of RS-449, all those procedures defined in RS-232-C carry over to RS-449. The newly added test and standby functions are conceived on the basis of action-reaction and thus are simple. For example, after the DTE turns on the local loopback (LL), it waits until the DCE responds with test mode on. The DTE can then proceed with test data transmission on the send data circuit, expecting to receive them back on the receive data circuit. Deactivation follows a similar procedure.

The physical characteristics for a general-purpose DTE/DCE interface for synchronous digital transmission on public data networks are defined in CCITT recommendation X.21. The DCEs are linked by means of real digital circuits, possibly through circuit-switching equipment. When circuit-switched services are provided, X.21 also specifies the call control procedures, i.e., the protocols by which to establish (and later to disconnect) a physical connection between two DTEs. Since the establishment of a circuit clearly requires knowledge of the address of the remote DTE (or its corresponding DCE) and the understanding of the various call progress signals, and since these requirements are supposed to be met only at the network layer (as discussed below), there is a general consensus that the call control procedure span the first three layers with the resulting service being offered by the network layer to the transport layer. Thus, these issues are dealt with in the network layer. The benefit of such a viewpoint is that of consistency in the network architecture, whereby the classification of functions is the same regardless of whether the data network is of the circuit-switched type or the packet-switched type. Indeed, in packet networks, the network layer takes care of opening a "connection," called the *"virtual circuit,"* with the remote DTE before the flow of packets takes place.

Simplicity and enhanced performance were prime objectives in the design of X.21. It permits interface operation over distances considerably greater than that available with V.28, and for the synchronous data rates specified by X.1—that is, 600, 2400, 4800, 9600, and 48 000 bits/s. Accordingly, the new balanced electrical characteristics (Recommendation X.27, introduced above as EIA RS-422-A) are specified for the DCE side of the interface. To allow flexibility in DTE design at the data rates 600, 2400, 4800, and 9600 bits/s, the DTE is permitted to use either the new balanced or the new unbalanced (X.26 or RS-423-A) electrical characteristics. The mechanical interface for X.21 is the 15-pin ISO 4903 connector. There is a careful assignment of interchange circuits to connector pin numbers, a pair of pins for each interchange circuit. The number of interchange circuits is reduced to 5 (Fig. 12). A transmit (T) circuit and a receive (R) circuit are used to convey both user data and network control information, depending on the state of the control (C) circuit and the indication (I) circuit; bit timing is continuously provided by a signal element timing (S) circuit; a sixth interchange circuit that provides byte timing information is optional; a signal ground circuit is also provided.

Among all the procedures specified in X.21, those associated with the quiescent phase are agreed upon to be within the physical level. Two quiescent signals are defined for the DCE: *DCE not ready*, indicating that no service is available, and *DCE ready*, indicating that the DCE (network) is ready

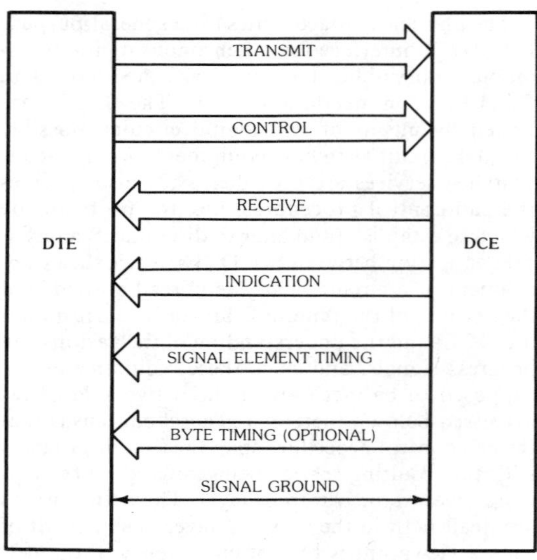

Fig. 12. CCITT Recommendation X.21 DTE/DCE interface. (*From Paul E. Green, Jr.*, Computer Network Architectures and Protocols, *© 1982. Used by permission of Plenum Publishing Corp., New York.*)

to enter the operational phase. Three quiescent signals are defined for the DTE: *DTE uncontrolled not ready,* indicating that the DTE is unable to enter operational phases because of abnormal conditions, *DTE controlled not ready,* indicating that the DTE is operational but is temporarily unable to enter operational phases, and *DTE ready,* indicating that the DTE is ready to enter operational phases. Fig. 13 shows the various combinations of quiescent states of the X.21 interface and the possible transitions between these states.

THE DATA LINK CONTROL LAYER

The data link control layer offers the capability for error-free conveyance of messages between machines that are physically connected by a communication channel. It consists of two sublayers. The upper sublayer deals with such issues as synchronization, error control, and link management and is present in all cases. The lower sublayer deals with the multiaccess link control and is present only when there is a shared channel that provides an any-to-any topological connectivity between stations. The focus in this section is on the upper sublayer. Multiaccess link control is described in the following section.

Asynchronous DLC Protocols

The DLC protocols used in early teleprinters are of the *asynchronous* type: the amount of data transmitted at a time is one character (5 to 8 bits—Fig. 14); when no transmission is taking place, the channel is in the idle state, which is represented by a continuous 1. When a character is to be transmitted, the terminal transmits a 0 (the start bit) followed by the character (of fixed length), and then followed by a 1 (the idle state) for a minimum period of time called the stop interval. The receiver determines the initial sampling from the 1 to 0 transition, using a clock sixteen times the transmission rate. The sampling period is known from the bit transmission rate. Bit synchronism is maintained only during the transmission time of a character. Anytime following the stop interval, another character can be transmitted.

The start-stop protocols are simple but present several disadvantages, among others the overhead of a start bit and a stop interval associated with each character, and the lack of any inherent link control capability.

Synchronous DLC Protocols

To improve the efficiency of a data link and provide link control capability, *synchronous* protocols were introduced. A block of several characters referred to as a *frame* is transmitted as a whole, with bit synchronism maintained during the entire transmission time of the frame. These protocols provide the following functions: (1) frame synchronization (a mechanism needed to delimit the beginning and end of transmission blocks, and to acquire bit synchronization), (2) error control (mechanisms needed to detect errors, acknowledge correctly received blocks, request the retransmission of incorrectly received blocks, and control the sequence of blocks to identify lost and duplicate blocks), (3) link management (mechanisms needed to establish a data link over a communication facility that has been idle and, in the case of multipoint facilities, to identify sender and receiver), (4) flow control (mechanisms needed to regulate the flow of information), and (5) functions to permit the recovery from abnormal conditions such as illegal sequences, loss of response, etc. Two types of synchronous DLC protocols exist: character-oriented and bit-oriented.

Character-Oriented DLC Protocols—These protocols are suitable for two-way alternate operation on full-duplex or half-duplex multipoint, switched, or dedicated link configurations. These protocols started with IBM's binary synchronous communication (BSC) protocol in the late 1960s; following it, there has been a proliferation of different protocols by various manufacturers, which are optimized for specific implementations. Although of the same type (all are character-oriented), these are incompatible with each other. A standardization effort was undertaken by ANSI and ISO. Today, the most widely used protocols are IBM's

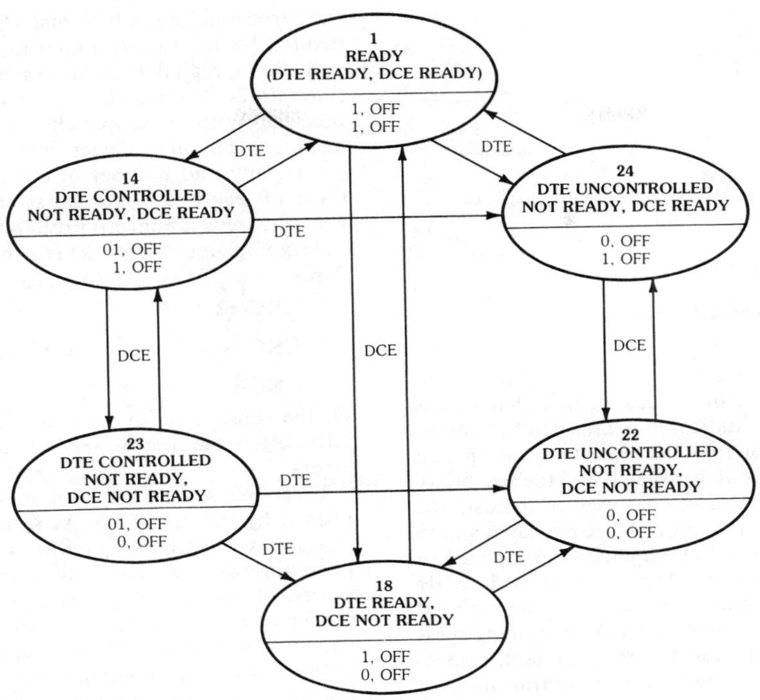

LEGEND: Each state is represented by an ellipse wherein the state name and number are indicated, together with the signals on the four interchange circuits which represent that state. Each state transition is represented by an arrow, and the equipment responsible for the transition (DTE or DCE) is indicated beside that arrow.

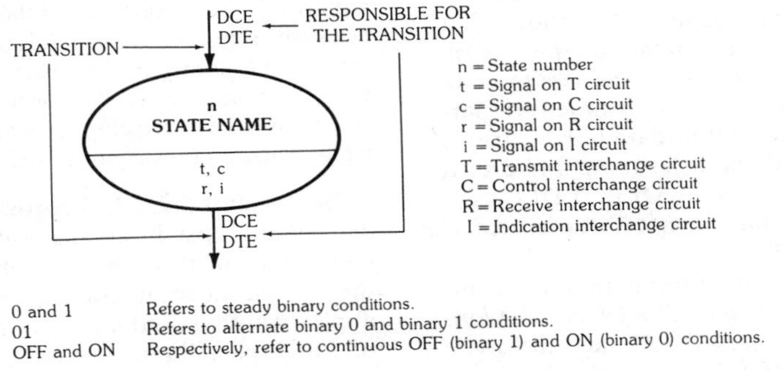

0 and 1	Refers to steady binary conditions.
01	Refers to alternate binary 0 and binary 1 conditions.
OFF and ON	Respectively, refer to continuous OFF (binary 1) and ON (binary 0) conditions.

Fig. 13. CCITT Recommendation X.21 quiescent states. (*From Paul E. Green, Jr.*, Computer Network Architectures and Protocols, *© 1982. Used by permission of Plenum Publishing Corp., New York.*)

BSC, ANSI X3.28 (1971, updated 1976), ISO's IS/745 (1975), and DEC's DDCMP. To illustrate this type of DLC protocol, ANSI 3.28 is used here.

In this protocol, the definitions of a message and a block are as follows: While a message is an ordered sequence of characters arranged to convey information from the originator user to the destination user (including a header portion), a block is a group of characters arranged for technical or logical reasons that is transmitted as a unit. A block may contain an entire message or a portion of a message.

Ten ASCII characters are designated as communication control characters. The first is SYN (synchronization character). It is used at the beginning of a data stream (following an idle time) to allow the receiver to establish byte synchronization. A minimum of two SYN characters are usually used to guarantee proper operation. This character is also used following the data stream (during the idle time) to allow the receiver to maintain synchronization, if this is desired. Two characters, ENQ (enquiry) and EOT (end of transmission), are used to manage the link. Using ENQ, a station

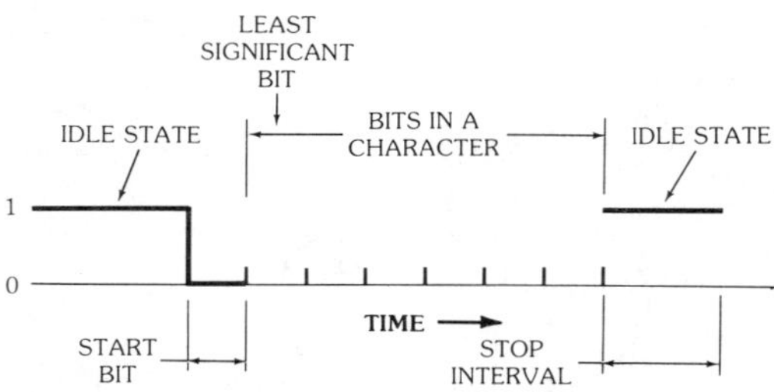

Fig. 14. Asynchronous DLC protocol.

solicits a response (such as status or identification) from another user. In particular, this character is used by a master station in a nonswitched multipoint configuration to poll a secondary station, and thus to establish a link between it and the secondary station. The EOT character is used to indicate the end of transmission of one or more messages and to relinquish the data link. Four characters are used to delimit blocks that are being transmitted: SOH (start of heading) and STX (start of text) are used to designate beginning of a block containing message header information and message text, respectively; ETB is used to designate end of transmission block; and ETX (end of text) is used at the end of the last block of a message. Two characters, ACK and NAK, are used as affirmative response and negative response (e.g., a block received in error) to a sender, respectively. Since all these characters have a specific control meaning, they should not normally appear in the user's data. To remove such a restriction and allow control characters to be in the user's data stream (that is, to achieve transparency), a tenth character, DLE (data link escape) is added; for any control character to be interpreted as such, it must be preceded by DLE. If DLE itself is to be transmitted as data, then it is transmitted as the sequence DLEDLE.

For error detection, three techniques are available: the *vertical redundancy check* (VRC), the *longitudinal redundancy check* (LRC), and the *cyclic redundancy check* (CRC). The vertical redundancy check consists of appending one parity bit to each character; thus it can only detect an odd number of bit errors in a character. The longitudinal redundancy check consists of computing parity on the entire sequence of successive characters in a block, and transmitting the result as an additional check character; thus it is vulnerable to double bit errors in the row of characters. With CRC, the bit stream constituting a block or a packet is treated as a polynomial, $M(x)$, with the least significant bit considered to be the coefficient of x^0. With k bits in the block, the polynomial is of degree $k - 1$. A generator polynomial, $G(x)$, of degree r is agreed upon by both sender and receiver. After appending r zero bits to the low-order end of the message to form the corresponding polynomial $x^r M(x)$, the latter is divided by $G(x)$ (polynomial division modulo 2), and the remainder is subtracted from $x^r M(x)$ (modulo 2). The result, $T(x)$, is the checksummed message. With the proper choice of $G(x)$, the CRC technique can safely detect single bit errors, double errors, any odd number of errors, and any burst error of length $\leq r$. For burst errors longer than $r + 1$, there is a nonzero probability of a bad message going undetected. Three international standards for a CRC generator exist. These are

$$\text{CRC-12} = x^{12} + x^{11} + x^3 + x^2 + x^1 + 1$$

$$\text{CRC-16} = x^{16} + x^{15} + x^2 + 1$$

$$\text{CRC-CCITT} = x^{16} + x^{12} + x^5 + 1$$

Of the three techniques (VRC, LRC, and CRC), CRC offers the best protection against undetected errors.

Error-free transmission of messages is accomplished by the use of the ACK and NAK control characters in conjunction with an error-detection procedure and the retransmission of blocks with errors. Moreover, timers are provided at both the sender and the receiver in order to recover from abnormal conditions. The *response timer* (typically 2–3 seconds) at the sending station protects against invalid or missing response, and the receive timer (about 500 ms) protects against failure to receive or recognize ETB or ETX. With these definitions and features, ANSI 3.28 specifies a number of specific protocols, each suitable to a particular scenario (or situation). For example, "subcategory 2.4" is for two-way alternate transmission on a nonswitched multipoint configuration with centralized operation; "subcategory D1" is for message-independent blocking, with cyclic checking, alternating acknowledgements, and transparent heading and text.

Bit-Oriented DLC Protocols—A number of deficiencies exist in character-oriented DLC protocols. First of all, such protocols are best suited for the two-way alternate transmission mode (i.e., half-duplex transmission) and thus make inefficient use of full-duplex lines. Secondly, error checking is done only on text, leaving control sequences unprotected, and thus leading to complex recovery procedures. Thirdly, a single data link function is performed with each transmission (e.g., either data transmission, or acknowledgement, or polling command, etc.), which leads to a larger number of turnarounds and to an unsatisfactory ratio of data transfer exchange to control exchange. Finally, character-oriented DLC protocols are too rigid and do not allow for easy expansion.

The new bit-oriented DLC protocols have been designed to overcome all the above limitations. Examples are ANSI's Advanced Data Communication Control Protocols (ADCCP), ISO's High-Level Data Link Control (HDLC), and IBM's Synchronous Data Link Control (SDLC). The

ADCCP and HDLC protocols are compatible. In this discussion, ADCCP is used to illustrate bit-oriented DLC protocols.

Bit-Oriented DLC protocols have been designed to support all possible environments: point-to-point and multipoint configurations using two-way alternate or two-way simultaneous operation over switched and nonswitched transmission lines, with both terrestrial and satellite connections. Three data transfer modes have been defined: the normal response mode (NRM) and the asynchronous response mode (ARM) for use in point-to-point and multipoint configurations, and the asynchronous balanced mode (ABM) for use in point-to-point configurations. In NRM and ARM, a so-called *primary station* controls the operation of the data link, while one or more *secondary stations* act as subservient to the primary. The primary issues commands and receives responses; the secondary station receives commands and issues responses in accordance with the command received and the mode of operation. In NRM, a secondary station initiates transmission only as a result of receiving explicit permission to do so. This mode is suitable for polled multipoint operation between a central location and a number of outlying stations. In ARM, a secondary station can initiate transmission without waiting for an explicit permission. It is suitable for a single primary and a *single* activated secondary wishing to transmit freely to one another without the overhead of polling. The ABM mode provides a *balanced* type of data transfer between two stations referred to as *logically equal* or *combined* stations. Each station operates as a primary for its data transfer, and is thus capable of initializing the link, activating the other combined station, and logically disconnecting the link. Typically, each combined station may be a host computer, an intelligent network node (e.g., a packet-switching node), or a highly intelligent terminal that has the capability to control the data link.

In bit-oriented DLC protocol, the basic transmission unit is called a *frame*. In order to achieve enhanced capabilities over character-oriented DLC protocols, a well defined frame format as shown in Fig. 15 is used, including an address field, a control field, and a frame check sequence (FCS).

The flag sequence (F) is the unique eight-bit pattern 01111110 used to delimit the start and end of each frame, and to fill idle time between frames during the transmission of multiple frames. In order to prohibit the occurrence of the flag in the address, control, information, and FCS fields, and thus achieve transparency, a technique called bit-stuffing is used. The transmitter inserts a 0 bit following any five contiguous 1 bits encountered in the above mentioned fields; the receiver deletes the 0 bit following five contiguous 1 bits following a 0 bit anywhere before receiving the closing flag sequence.

The address field in a command frame identifies the station (secondary or combined) that is to receive the command; in a response frame, it identifies the station (secondary or combined) that is sending the frame. Thus in NRM and ARM, the address always identifies the secondary station; in ABM, it identifies the response-generating portion of a combined station. There are two mutually exclusive address-field options: either a single octet accommodating up to 256 stations, or multiple octets recursively extendable by having the first bit of each octet be a 1 bit if the octet is the last one in the address or a 0 bit otherwise. Broadcast addressing is also possible by using the all-zero address.

The control field identifies the function and purpose of the frame. Three different formats exist, defining the three types of frames: information transfer type (I frame), supervisory type (S frame), and unnumbered type (U frame). See Fig. 16. The P/F bit in the control field provides for a checkpointing mechanism allowing a response frame to be logically associated with the appropriate initiating command frame. In NRM, the primary station sets the P/F bit to 1 to poll a secondary station. The secondary station sets P/F to 1 in the last frame sent in response to a poll. In ARM and ABM, receipt of a frame with the P/F bit set to 1 causes the responding station to set the P/F bit to 1 in the next appropriate frame.

The I frames are used to transfer user information across a data link. A so-called *window mechanism* is used to guarantee sequential error-free delivery of the information without degrading the performance of the link. Successive frames are numbered sequentially (modulo some number 2^n), where the numbers range from 0 to the maximum, $2^n - 1$. The frame sequence number is transmitted in the control field $N(S)$. The sender maintains a list of consecutive sequence numbers called the *sending window* corresponding to frames it is permitted to send. Likewise, the receiver maintains a list of consecutive sequence numbers called the *receiving window* for frames it is permitted to receive. Most acknowledgements in these DLC protocols are handled in a *piggyback* fashion. A receiver that has

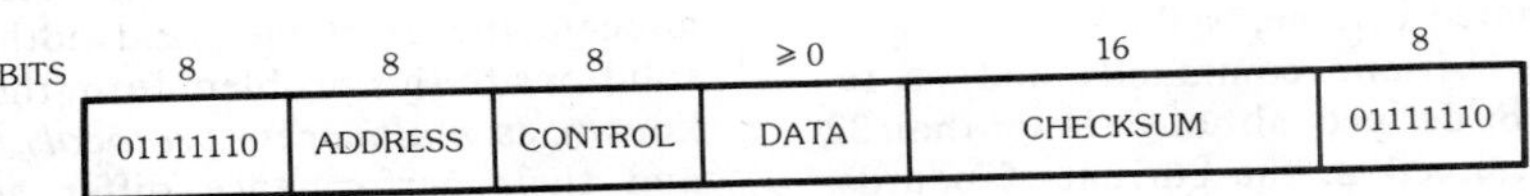

Fig. 15. Frame format for bit-oriented protocols. (*From Andrew S. Tanenbaum,* Computer Networks, © *1981. Used by permission of Prentice-Hall, Inc., Englewood Cliffs, N.J.*)

CONTROL FIELD FOR	CONTROL FIELD BITS							
	1	2	3	4	5	6	7	8
INFORMATION TRANSFER COMMAND/ RESPONSE (I FRAME)	0	N(S)			P/F	N(R)		
SUPERVISORY COMMANDS/ RESPONSES (S FRAME)	1	0	S	S	P/F	N(R)		
UNNUMBERED COMMANDS/ RESPONSES (U FRAME)	1	1	M	M	P/F	M	M	M

Fig. 16. Control field formats in ADCCP. (*From Paul E. Green, Jr.*, Computer Network Architectures and Protocols, *© 1982. Used by permission of Plenum Publishing Corp., New York.*)

correctly received all frames numbered up to some number, k, will send in the $N(R)$ field of the next I frame the number $N(R) = k + 1$; i.e., the sequence number of the next frame the receiver is expecting. In case an I frame is not ready for transmission, the receiver will send an S frame with $N(R) = k + 1$. Upon receiving an acknowledgement, the sender advances the lower edge of its sending window to $N(R)$ and the upper edge to $N(R) + W_S$. The sequence numbers of frames already transmitted that are within the sender window represent frames sent but as yet not acknowledged. The window size, W_S, represents the maximum number of such frames. It is determined as a function of the frame transmission time and the end-to-end propagation delay so as to "fill the pipe" and achieve high throughput. This feature is particularly important when dealing with wideband satellite channels where the ratio of round-trip delay to frame transmission time is high. The receive window, W_R, is determined as a function of the buffer space available at the receiver. The sending window, W_S, and receive window, W_R, need not be the same. The window mechanism protects against the loss of a frame and the loss of acknowledgement, detects and filters out duplicates, and guarantees that the network level receives the frames in sequential order. It is important, however, that the condition $W_S + W_R \leq 2^n$ be satisfied. In the ADCCP and HDLC standards, two sequence ranges are available: the unextended case, which is modulo 8, and the extended case, which is modulo 128.

There are four supervisory frames. Receive ready (RR) is used to indicate that the receiver is ready to receive I frames; receive not ready (RNR) is used to indicate a temporary busy condition. Both RR and RNR S frames acknowledge I frames already received. Reject (REJ) is used to request retransmission of all I frames starting with N(R); selective reject (SREJ) is used to request the retransmission of a single designated I frame, N(R).

There are 32 U-frame commands and/or responses that can be defined, although fewer than 32 are in fact determined in the current standards. They do not contain a sequence number and thus are not accounted for in the normal flow of frames. They are mainly used in the control of the data link. Examples are: the set mode commands (SNRM, SARM, SABM, etc.), disconnect command (DISC), an unnumbered poll command, an unnumbered information (UI) command/response, an unnumbered acknowledgement (UA), and a frame reject (FRMR). (The frame reject is used to indicate that a received frame was in error and cannot be corrected by retransmission, such as, for example, an information field greater than the maximum established length, receipt of a control field that is either invalid or not implemented, etc.)

The FCS consists of the 16-bit checksum obtained from the CRC error detection technique using the CRC-CCITT generator. An optional 32-bit FCS is also available if a higher degree of error detection is needed. The corresponding generator polynomial is $x^{32} + x^{26} + x^{23} + x^{22} + x^{16} + x^{12} + x^{11} + x^{10} + x^8 + x^7 + x^5 + x^4 + x^2 + x + 1$. Just as with character-oriented DLC protocols, classes of procedures are defined in order to provide organization and direction for the application of the bit-oriented data link control procedures. Certain commands belong to all the classes, while others are viewed as being optional. Examples are: UNC, 3, 4—unbalanced NRM with SREJ and UI capabilities; BAC, 2, 8—balanced ABM with REJ and the restriction on the use of I frames as commands only.

MULTIACCESS LINK CONTROL

Multiaccess protocols arise whenever a communication channel is shared by many independent contending users. Two major factors contribute to a multiaccess situation: the need to share an expensive communication channel in order to achieve its efficient utilization, and the need to provide a high degree of connectivity for communication among independent subscribers. The main issue of concern is how to control access to the common channel to allocate the available bandwidth efficiently. The solutions to this problem form the set of protocols known as *multiaccess protocols.* These protocols and their performance differ according to the environment and the specific requirements to be satisfied. In satellite channels, there is an inherent long propagation delay of approximately 0.25 s, a delay that is usually long compared to the transmission time of a packet given the bandwidth usually available. In ground radio environments, the opposite is true: the transmission radius is relatively small due to a limited power supply at the radio units and a heavy signal attenuation, and the data rate is relatively low, so that the propagation delay is relatively short compared to the transmission time of a packet. Finally, local area environments are characterized by a large and often variable number of inexpensive devices; they call for networks with simple topologies and inexpensive connection interfaces; the propagation delay is short but the bandwidth can be high; the propagation delay may or may not be larger than the packet transmission time.

Multiaccess schemes are evaluated according to various criteria: bandwidth utilization, message delay, the ability to support traffic of different types simultaneously (with different priorities, variable message lengths, and differing delay constraints), and robustness (defined here as the insensitivity to errors resulting in misinformation).

Multiaccess protocols differ by the static or dynamic nature of the bandwidth allocation algorithm, the centralized or distributed nature of the decision-making process, and the degree of adaptivity of the algorithm to changing needs. Accordingly, these protocols can be grouped into several classes described hereafter.

Fixed Assignment Techniques

Fixed assignment techniques consist of allocating the channel to the users, independently of their activity, by partitioning the time-bandwidth space into slots that are assigned in a static predetermined fashion. These techniques take two common forms: *orthogonal,* such as frequency division multiple access (FDMA) or synchronous time division multiple access (TDMA), and "*quasi-orthogonal,*" such as code division multiple access (CDMA).

The FDMA technique consists of assigning to each user a fraction of the bandwidth and confining its access to the allocated subband. Orthogonality is achieved in the frequency domain. The TDMA technique consists of assigning fixed predetermined channel time slots to each user; the user has access to the entire channel bandwidth, but only during its allocated slots. Here, signaling waveforms are orthogonal in time. The TDMA approach is more complex to implement than FDMA, but an important advantage is the connectivity that results from the fact that all receivers listen to the same channel while senders transmit on the same common channel at different times. Accordingly, with TDMA, many network realizations, both in ground and satellite environments, are easier to accomplish.

The CDMA technique allows overlap in transmission both in the frequency and time coordinates. It achieves orthogonality by the use of different signaling codes in conjunction with matched filters (or equivalently, correlation detection) at the intended receivers. Each user is assigned a particular code sequence, which is modulated on the carrier with the digital data modulated on top of that. Two common forms exist: the frequency-hopped SSMA and the phase-coded SSMA. In the former, the frequency is periodically changing according to some known pattern; in the latter, the carrier is phase modulated by the digital data sequence and the code sequence. Multiple orthogonal codes are obtained at the expense of increased bandwidth requirements (in order to spread the waveforms). With CDMA, there is also a lack of flexibility in interconnecting all users (unless, of course, matched filters corresponding to all codes are provided at all receivers). However, CDMA has the advantage of allowing the coexistence of several systems in the same band, as long as different codes are used for different systems.

Random Access Techniques

Since data traffic in computer communication is characterized as bursty, a more advantageous approach than fixed assignment is to provide a single sharable high-speed channel to the large number of users. The strong law of large numbers then guarantees that, with a very high probability, the demand at any instant will be approximately equal to the sum of the average demands of that population. When dealing with shared channels in a packet-switched mode, one must be prepared to resolve conflicts that arise when more than one demand is placed on the channel. For example, in packet-switched channels, whenever a portion of the transmission of one user overlaps with the transmission of another user, then the two collide and "destroy" each other (unless a code division multiple-access scheme is used). The existence of some positive acknowledgement scheme permits the transmitter to determine if a transmission is successful or not.

ALOHA—Pure ALOHA permits a user to transmit any time it desires. If a user transmits a packet, and within some appropriate time-out period following its transmission it receives an acknowledgement from the destination, then it knows that no conflict occurred. Otherwise, it assumes that a collision occurred and it must retransmit. To avoid continuously repeated conflicts, the retransmission delay is randomized across the transmitting devices, thus spreading the retry packets over time. A slotted version, referred to as *slotted ALOHA,* is obtained by dividing time into slots of duration equal to the transmission time of a single packet (assuming constant-length packets). Each user is required to synchronize the start of transmission of its packets to coincide with the slot boundary. When two packets conflict, they will overlap completely rather than partially, providing an increase in channel efficiency over pure ALOHA. Due to conflicts and idle channel time, the maximum channel efficiency available with ALOHA is less than 100 percent, 18 percent for pure ALOHA and 36 percent for slotted ALOHA. Note that, although the maximum achievable channel utilization is low, the ALOHA schemes are superior to fixed assignment schemes when there is a large population of bursty users and low packet delay is of the essence. (See reference 3.)

Carrier Sense Multiple Access (CSMA)—In the CSMA technique, an attempt is made to avoid collisions by listening to the carrier due to transmission from another user before transmitting, and inhibiting transmission if the channel is sensed busy. This is advantageous when the propagation delay between any source-destination pair is small compared to the packet transmission time. Many CSMA protocols exist; they differ according to the action that a terminal takes to transmit a packet after sensing the channel. In all cases, however, when a terminal learns that its transmission has incurred a collision, it reschedules the transmission of the packet according to a randomly distributed delay. At this new point in time, the transmitter senses the channel again and repeats the algorithm dictated by the protocol. In *nonpersistent* CSMA, a ready terminal senses the channel and operates as follows. If the channel is sensed idle, the terminal transmits the packet. If the channel is sensed busy, then the terminal schedules the retransmission of the packet to some later time according to the retransmission delay distribution. At this new point in time, it senses the channel and repeats the algorithm described.

In the *1-persistent* CSMA protocol, a ready terminal senses the channel and operates as follows. If the channel is sensed idle, it transmits the packet with probability one. If the channel is sensed busy, it waits until the channel goes idle and then immediately transmits the packet with probability one. In 1-persistent CSMA, whenever two or more terminals become ready during a packet transmission period, they wait for the channel to become idle (at the end of that transmission), and then they all transmit with probability one. A conflict will also occur with probability one. Randomizing the starting time of transmission of packets accumulating at the end of a transmission period reduces interference and improves performance. The *p-persistent* scheme involves including an additional parameter p, the probability that a ready packet persists ($1 - p$ being the probability of delaying transmission by τ seconds, where τ is the maximum propagation delay among all pairs). Parameter p is chosen to reduce the level of interference while keeping the idle periods between any two consecutive nonoverlapped transmissions as small as possible.

The CSMA technique has been applied to ground radio (e.g., PRNET), and to local area communications (e.g., ETHERNET). In ETHERNET, CSMA is used on a tapped coaxial cable to which all the communicating devices are connected. On the coaxial cable, in addition to sensing carrier, it is possible for the transceivers to detect collisions. This is achieved by having each transmitting device compare the bit stream it is transmitting to the bit stream it sees on the channel. When transmitting users detect interference among several transmissions (including their own), they abort the transmission of colliding packets. This variation of CSMA is referred to as *carrier sense multiple access with collision detection* (CSMA-CD).

The performance of CSMA is heavily dependent on the ratio, a, of propagation delay to packet transmission time. The maximum throughput of a CSMA protocol degrades significantly as a gets larger. For a ratio $a = 0.01$, nonpersistent CSMA achieves a channel utilization equal to 0.815, a significant improvement over the ALOHA schemes.

While until recently most of the concepts described in this section had been realized in experimental systems (namely, the ALOHA System, PRNET, and Xerox's experimental ETHERNET), it is important to note that today many contention systems of the ETHERNET type are available on the market. Examples are the Hyperchannel and the Hyperbus of Network Systems Corporation, Z-Net of Zilog, Omninet of Corvus, and ETHERNET itself. The latter has been announced as a product made available jointly by Xerox Corporation, Digital Equipment Corporation, and INTEL. Complete specifications of the data link and physical link protocols have been issued and constituted the basis of a standard for the IEEE Computer Society Project 802 on the standardization of local networks. A key feature that distinguishes this product from other already available systems is the LSI implementation of many of the data link and physical link protocols. The LSI implementation of network protocols clearly marks a trend in the evolution of computer networking, a trend that is indicative of the existence of a wide market and the need to provide reasonably priced components.

Busy-Tone Multiple Access (BTMA)—In ground radio environments, it is possible for two terminals to be within range of the intended receiver, but out of range of each other or separated by some physical obstacle opaque to radio signals. The existence of hidden terminals in such an environment signficantly degrades the performance of CSMA. The hidden-terminal problem can be eliminated by frequency dividing the available bandwidth into two separate channels, a busy-tone channel and a message channel, thus giving rise to *busy-tone multiple access* (BTMA). As long as a node senses carrier on the message channel, it transmits a (sine wave) busy-tone signal on the busy-tone channel. It is by sensing carrier on the busy-tone channel that nodes determine the state of the message channel. The action that a node takes pertaining to the transmission of the packet is again prescribed by the particular protocol being used, similar to those described for CSMA.

Capture—Capture in narrow-band channels can be defined as the ability of a receiver to receive a packet successfully (with nonzero probability) although the packet is partially or totally overlapped by another packet transmission. Capture is mainly due to a discrepancy in receive power between two signals that allows the receiver to receive the stronger correctly; both distance and transmit power contribute to this discrepancy. Clearly, capture improves the overall network performance, and, by means of adaptive transmit power control, it allows one to achieve either fairness to all users or intentional discrimination.

Spread-Spectrum Multiple Access (SSMA)—The SSMA technique is considered here to be a CDMA scheme. In one form of SSMA for packet radio, all transmitters employ the same code. Security, coexistence with other systems, and the ability to counteract the effects of multipath and capture are key benefits of SSMA. Contrary to the case of nonspread systems, the effect of interference in SSMA is minimized by the "capture effect," defined as the ability of the receiver to "lock on" one packet while all other overlapping packets appear as noise. The receiver locks on a packet by correctly receiving the preamble appended to the front of the transmitted packet. As long as the preambles of different packets do not overlap in time, and the signal strength of the late packets is not too high, capture of the earliest packet occurs with a high probability. In essence, SSMA allows a packet to be captured at the receiver, while CSMA allows a user to capture the channel. It is possible to use CSMA in conjunction with SSMA, but the channel sensing is more difficult. This mode will have the benefit of keeping away all users within hearing distance of the transmitter and thus help keep the capture effect and antijamming capability of the system at the desired level.

Centrally Controlled Demand Assignment

Demand assignment techniques require that explicit information regarding the need for the communication resource be exchanged. These techniques may be either centralized, whereby a central scheduler performs the assignment, or distributed, whereby all stations take active part in the assignment. Centrally controlled techniques are addressed in the present subsection.

Circuit Oriented Systems—In circuit oriented systems, the bandwidth is divided into FDMA or TDMA subchannels that are assigned on demand. The satellite SPADE system, for example, has a pool of FDMA subchannels that are allocated on request. It uses one subchannel operated in a TDMA fashion with one slot per frame permanently assigned to each user to handle the requests and releases of FDMA circuits. Intelsat's MAT-1 system uses the TDMA approach. The TDMA subchannels are periodically reallocated to meet the varying needs of earth stations.

The Advanced Mobile Phone Service (AMPS), introduced by Bell Laboratories, is another example of a centrally controlled FDMA system. The uniqueness of this system, however, lies in an efficient management of the spectrum based on space division multiple access (SDMA). That is, each subchannel in the pool of FDMA channels is allocated to different users in separate geographical areas, thus considerably increasing the spectrum utilization. To accomplish space division, the AMPS system has a cellular structure and uses a centralized handoff procedure (executed by a central office) that reroutes the telephone connections to other available subchannels as the mobile users move from one cell to another.

Polling Systems—In packet oriented systems, polling consists of having a central controller send polling messages to the terminals, one by one, asking the polled terminal to transmit. If the polled terminal has something to transmit, it goes ahead; if not, a negative reply (or absence of reply) is received by the controller, which then polls the next terminal in sequence. Polling requires this constant exchange of control messages between the controller and the terminals and is efficient only if (1) the round-trip propagation delay is small, (2) the overhead due to polling messages is low, and (3) the user population is not a large bursty one.

Adaptive Polling or Probing—The primary limitation of polling in lightly loaded systems is the high overhead incurred in determining which of the terminals have messages. A modified polling technique called *probing,* based on a tree searching algorithm, helps decrease this overhead. This technique assumes that the central controller can broadcast signals to all terminals. First the controller interrogates all terminals, asking if any of them has a message to transmit, and repeats this question until some terminals respond by putting a signal on the line. When a response is received, the central station divides the population into subsets (according to some tree structure) and repeats the question to each of the subsets. The process is continued until the terminals having messages are identified. When a single terminal is interrogated, it transmits its message. This probing technique can be made adaptive by having the controller start a cycle by probing groups of smaller size as the probability of terminals having messages to transmit increases.

Split-Channel Reservation Multiple Access (SRMA)—An attractive alternative to polling is the use of explicit reservation techniques. In dynamic reservation systems, it is the terminal that makes a request for service on some channel whenever it has a message to transmit. The central scheduler manages a queue of requests and informs the terminal of its allocated time. In SRMA, the available bandwidth is divided into two channels, one used to transmit control information and the second used for the data messages themselves. The request channel is operated in a random access mode (ALOHA or CSMA). Upon correct reception of the request packet, the scheduling station computes the time at which the backlog on the message channel will empty and transmits back to the terminal an answer packet containing the address of the terminal and the time at which it can start transmission.

Demand Assignment With Distributed Control

There are two reasons why distributed control is desirable. The first is reliability; with distributed control the system is not dependent on the proper operation of a central scheduler. The second is improved *performance,* especially when dealing with systems with long propagation delays, such as those using satellite channels. The basic element underlying all distributed algorithms is the need to exchange control information among the users, either explicitly or implicitly. Using this information, all users then execute independently the same algorithm, with the result that there is some coordination in their actions.

Reservation-ALOHA—Reservation-ALOHA for a satellite channel is based on a slotted time axis where the slots are organized into frames of equal size. The duration of a frame must be greater than the satellite propagation delay. A user that has successfully accessed a slot in a frame is guaranteed access to the same slot in the succeeding frame, and this continues until the user stops using it. "Unused" slots, however, are free to be accessed by all users in a slotted ALOHA contention mode. A slot in a frame is an unused slot if in the *preceding* frame it either was idle or contained a collision. Users need simply to maintain a history of the usage of each slot for just one frame duration. Since no request is explicitly issued by the user, this scheme has been referred to as an *implicit reservation* scheme. Clearly, Reservation-ALOHA is effective only if the users generate stream type traffic or long multipacket messages. Its performance will degrade significantly with single packet messages, since every time a packet is successful the corresponding slot in the following frame is likely to remain empty.

A First-in First-out (FIFO) Reservation Scheme—In this scheme, reservations are made explicitly. Time division is used to provide a reservation subchannel. The channel time is slotted as before, but every so often a slot is divided into V small slots that are used for the transmission of reservation packets (as well as possibly acknowledgments and small data packets); these packets contend on the V small slots in a slotted ALOHA mode. All other slots are data slots and are used on a reservation basis, free of conflict. To execute the reservation mechanism properly, each station must maintain information on the number of outstanding reservations (the "queue in the sky") and the slots at which its own reservations begin. These are determined by the FIFO discipline based on the successful reservations received. To maintain synchronization of control information at the proper time and to acquire the correct count of packets in the queue if out-of-sync conditions do occur, each station sends information regarding the status of its queue in its data packet. This information is also used by new stations that need to join the queue. The robustness of this system is achieved by a proper encoding of the reservation packets to increase the probability of their correct reception at *all* stations.

A Round-Robin (RR) Reservation Scheme—The basis of this scheme is fixed TDMA assignment, but with the major difference that "unused" slots are assigned to the active stations on a round-robin basis. This is accomplished by organizing packet slots into equal-size frames of duration greater than the propagation delay and such that the number of slots in a frame is larger than the number of stations. One slot in each frame is permanently assigned to each station. To allow other stations to know the current state (used or unused) of its own slot, each station is required to transmit information regarding its own queue of packets piggybacked in the data packet header (transmitted in the previous frame). A zero count indicates that the slot in question is free. All stations maintain a table of the queue lengths of all stations, allowing them to allocate among themselves unassigned slots in the current frame. Round-robin or other scheduling disciplines can be used. A station recovers its slot by deliberately causing a conflict in that slot, which other users detect.

Distributed Tree Retransmission Algorithms—Tree algorithms are based on the observation that a contention among several active sources is completely resolved if and only if all the sources are somehow subdivided into groups such that each group contains at most one active source. (See probing in the section on centrally controlled demand assignment.) Each source corresponds to a leaf on a *binary* tree. The channel time axis is slotted, and the slots are grouped into pairs. Each slot in a pair corresponds to one of the two subtrees of the node being visited. Starting with the root node of the tree, we let all terminals in each of the two subtrees of the root transmit in their corresponding slots. If any of the two slots contains a collision, then the algorithm proceeds to the root of the subtree corresponding to the collision and repeats itself. This continues until all the leaves are separated into sets such that each of them contains at most one packet. This is known to all users, as the outcome of the channel is either a successful transmission or an idle slot. Collisions caused by the left subtree (first slot of a pair) are resolved prior to resolving collisions in the right subtree. This scheme provides a maximum throughput of 0.347 packets/slot. Clearly, a binary tree is not always optimum. If, each time a return to the root node is made, the tree is reconfigured according to the current traffic conditions, it can be shown that the optimum tree is binary everywhere except for the root node, whose optimum degree depends on traffic conditions.

The preceding four schemes have been proposed for satellite channels. All assumed fixed-size slots, and thus can be implemented in systems that have been built for synchronous TDMA. If used in systems with small propagation delay, such as ground radio, then they will perform significantly better. Due to the inherent small propagation delay in ground radio and local environments, other access modes with distributed control are also possible if all devices are in line-of-sight and within range of each other. A description of these follows.

Minislotted Alternating Priorities (MSAP)—The MSAP technique is a "carrier-sense" version of polling with distributed control. The time axis is slotted with the slot size again equal to the maximum propagation delay (and referred to hereafter as a minislot). All users are synchronized and may start transmission only at the beginning of a minislot. Users are considered to be ordered from 1 to M. When a packet transmission ends, the alternating priorities (AP) rule assigns the channel to the same user that transmitted the last packet (say user i) if it is still busy; otherwise the channel is assigned to the next user in sequence (i.e., user $[i \,(\mathrm{mod}\, M) + 1]$). The latter (and all other users) detects the end of transmission of user i by sensing the absence of carrier over one minislot. At this new point in time, either user $[i(\mathrm{mod}\, M) + 1]$ starts transmission of a packet (which will be detected by all other users) or it is idle, in which case a minislot is lost and control of the channel is handed to the next user in sequence. The overhead at each poll in this scheme is one minislot. Scheduling rules other than AP are also possible, such as round-robin and random order.

The Assigned-Slot Listen-Before-Transmission Protocol—Time is minislotted and divided into frames, each containing an equal number of minislots (say L). To each minislot of a frame is assigned a given subset of M/L users. A user with a packet ready for transmission in a frame can sense the channel only in its assigned minislot. If the channel is sensed idle, transmission takes place; if not, the packet is rescheduled for transmission in a future frame. Parameter M/L is adjusted according to the load placed on the channel. For high throughput, $M/L = 1$ is found to be optimum, and the scheme becomes a conflict-free one that approaches MSAP. For very low throughput, $M/L = M$ (i.e., $L = 1$) is found to be optimum; this corresponds to pure CSMA. In between the two extreme cases, intermediate values of M/L are optimum.

The URN Scheme—The time axis is divided into packet slots, and all users are synchronized. Assuming that all users know the exact number, n, of busy users, the scheme consists of giving full access right (i.e., the right to transmit with probability 1) to some number, k, of users. A successful transmission will result if there is exactly one busy user among these k. The probability of such an event is maximized when $k = \lfloor M/n \rfloor$, where $\lfloor M/n \rfloor$ denotes the integer part of M/n. Assume the system is lightly loaded (for instance $n = 1$). A large number of users are given access right (in the example $n = 1$, the number is $k = M$), but only a few and hopefully only one will make use of it (in the example $n = 1$, a successful transmission takes place). As the load increases, k decreases and the access right is gradually restricted. For the extreme case of $n = M$, $k = 1$ and the scheme converges to TDMA. One possible scheme for estimating n with good accuracy is to include a single reservation minislot at the beginning of each data slot. An idle user that turns busy sends a standard reservation message of few bits. All users are able to detect the following three events: no new busy users, one new busy user, and more than one new busy user (termed an erasure). As it is impossible with this minimal overhead to estimate the exact number of new busy users when the latter is greater than one, errors in estimation result; however analysis and simulation have shown that this error is negligible and, furthermore, that the scheme is insensitive to small perturbations in n. This last statement is even more important with respect to the robustness of the scheme, since it means that all users need not have exactly the same estimate for n. As for coordinating the selection of the k users, an effective mechanism is the use of synchronized pseudorandom generators at all users, which allow them to draw the same k pseudorandom numbers. Another mechanism, referred to as a round-robin slot sharing window mechanism, consists of having a window of size k move over the population space.

When a collision occurs, the window stops and decreases in size. When there is no collision, the tail of the window is advanced to the head of the previous window, and the size is again set to k as determined by n.

Distributed Control Algorithms in Local Area Networks—In addition to the random-access schemes described previously, all above algorithms are also applicable to local-area (broadcast) *bus* networks, as these exhibit the required characteristics of small propagation delay and full connectivity. But in local-area communication, a slightly different topology, namely the *ring* (or loop), has also been widely considered. As described previously in connection with Fig. 7, in the ring topology messages are not broadcast but rather passed from node to node along unidirectional links, until they return to the originating node. A simple scheme suitable for a ring consists of passing the access right sequentially from node to node around the ring. (Note that in a ring, the physical locations of the nodes define a natural ordering among them.) One implementation of this scheme is exemplified by the Distributed Computing System's network where an 8-bit *control token* is passed sequentially around the ring. Any node with a ready message may, upon receiving the control token, remove the token from the ring, send the message, and then pass on the control token. Another implementation consists of providing a number of *message slots* that are continuously transmitted around the ring. A message slot may be empty or full; a node with a ready message waits to see an empty slot pass by, marks it as full, and uses it to send its message. A still different strategy is known as the *register insertion* technique. Here, a message to be transmitted is first loaded into a shift register. If the ring is idle, the shift register is just transmitted. If not, the register is inserted into the network loop at the next point separating two adjacent messages; the message to be sent is shifted out onto the ring while an incoming message is shifted into the register. The shift register can be removed from the network loop when the transmitted message has returned to it.

Priority-Oriented Demand Assignment (PODA)—In the context of a satellite channel, PODA has been proposed as the ultimate scheme that attempts to incorporate all the properties and advantages seen in many of the previous schemes. It has provision for both implicit and explicit reservations, thus accommodating both stream and packet-type traffic. It may also integrate the use of both centralized and distributed control techniques, thus achieving a high level of robustness. Channel time is divided into two basic subframes, an information subframe and a control subframe. The information subframe contains scheduled packets and packet streams that also contain, piggybacked, control information such as reservations and acknowledgements. The control subframe is used exclusively to send reservations that cannot be sent in the information subframe in a timely manner. In order to achieve integration of centralized and distributed assignments, the information subframe is further divided into two sections, one for each type. Access to the control subframe (which is divided into slots accommodating fixed-size control packets) can take any form that is suitable to the environment. It can be by *fixed assignment* (TDMA) if the number of stations is small (giving rise to the so-called FPODA), or by *contention* as in ALOHA if the stations have a low duty cycle (giving rise to CPODA), or a combination of both. The boundary between the control subframe and the information subframe is not fixed, but varies with the demand placed on the channel. As in the FIFO and RR reservation schemes, distributed control is achieved by having all stations involved in this type of control keep track of their queue length information. Priority scheduling can thus be achieved. For stream traffic, a reservation is made only once and is retained by each station in a stream queue. Centralized assignment may be used when delay is not the crucial element. This scheme has been proposed in the context of a satellite channel but may be applied to other environments as well.

THE NETWORK LAYER

The network layer consists of those functions that control the transportation of data from source-host to destination-host. It serves directly transport entities residing at the network hosts, relieving them from any concern about network issues such as switching, routing, and congestion control. The network-layer functions are implemented at all switching nodes of a network. This layer is the highest one that resides at the switching nodes. It makes use of the data-link layer to accomplish the error-free transmission of data over individual links.

The Network Services

The nature of the services provided by the network layer varies considerably depending on the switching technique used in the communication subnet and the transport-layer requirements pertaining to the delivery of data from one end to the other.

A network-layer protocol for a circuit-switched network is found in CCITT Recommendation X.21 (1972). As stated earlier, X.21 is a general-purpose interface between DTE and DCE for synchronous operation on public data networks. When circuit-switched services are provided, X.21 includes a data-link-layer function and a network-layer function needed for call establishment. The data-link layer is character oriented and includes only the minimum elements necessary for basic operation, namely character synchronization (using two or more SYN characters) and error detection (using odd parity). The network layer clearly defines the procedures used in processing the various phases of call requests, incoming calls, facility requests, call progress, and call clearing. The reason these procedures belong to the network layer as opposed to the physical layer (although the end result is a direct physical connection between the two end hosts) is that the network layer receives the remote DTE (or DCE) address from the transport layer and makes use of the data-link layer in processing the calls.

With packet switched networks, two types of network services exist: the *datagram service* and the *virtual circuit* (VC) *service.* In the VC model, the network layer provides the transport layer with a "perfect" connection: no errors, no duplicates, and all packets are delivered in order. In the datagram model, the network layer accepts messages from the transport layer and simply makes the best effort to deliver them independently (and not necessarily in order). The implementation of the datagram mode is simple; it merely consists of a routing algorithm that attempts delivery of the messages to their destination. For VC, in addition to routing the messages, error control and sequencing must be implemented at the end nodes. In theory, packets may travel on different routes and arrive in any order. Resequencing at the destination node would then be required. In practice, the implementation of a VC in the subnet is by establishing a route between the source and destination end nodes at connection time, and by continuously using the same route for all packets belonging to the same virtual circuit. The implementation requires the packet to carry a virtual circuit number, and each node to contain a table with an entry for each VC traversing it, relating incoming packets from an adjacent neighbor with a VC number to an outgoing link and a VC number on that link. Forwarding packets to the destination node is then straightforward. All nodes use the first-come-first-served service discipline, thus preserving the order of packets for each VC. End-to-end reliability and sequencing are achieved by means of a window mechanism, which also provides flow control. It is to be noted, however, that the network layer does not achieve complete end-host to end-host reliability, since it is subject to node and link failures; depending on the environment and on the application, higher level functions (at the transport layer) must exist to guarantee that reliability.

Routing

In multihop store-and-forward networks, the network layer includes a routing algorithm that is responsible for deciding on which output link a packet should be transmitted. Although one primary objective is that each packet reaches its destination, there are several other objectives that are also very important, such as to minimize packet transit times, to avoid congestion and deadlocks, to maximize the network throughput, etc. It is also desirable that the algorithm be simple, robust, stable, and fair to all users. There is a broad spectrum of routing algorithms. They vary according to various attributes pertaining to the nature of decision-making (centralized or distributed), the degree of adaptivity, the frequency of updates, etc. Several routing techniques are described below.

Directory Routing—In directory routing, each node maintains a table with one row for each destination. The row gives one or several outgoing links together with relative weights assigned to them. Upon receipt of a packet with a given destination address, the node simply performs a table look-up and chooses one of the alternatives, using the relative weights as probabilities. The selection of routes and their weights may be based on the number of hops. If the source-destination traffic requirement of the network is stationary, then it is possible to use routes that minimize the average message delay in the network. For a given destination, the routes from an intermediate node to that destination are entirely determined by the tables and independent of the source. This is not restrictive, since if node j is on the optimal path from source i to destination k, then the optimal path from j to k should follow the same route. This is known as the optimality principle. As a result, the set of routes from all sources to a given destination form a tree with the destination as a root; such a tree is called the *sink tree* for that destination.

Hierarchical Routing—If the size of the network is large, then hierarchical routing is used. The network is partitioned into regions; node addresses are hierarchical and contain a region number and a node number within the region. In the table at each node, there is an entry for each destination in the region in which the node is located, and an entry for each of the other regions. Hierarchical routing decreases the overhead incurred in terms of storage and processing requirement. If the network is very large, a hierarchy with more than two levels may be needed.

Static Versus Dynamic Routing—Static routing refers to the case in which the table content is fixed. Static routing is adequate if the topology and traffic conditions do not change much. Dynamic routing refers to the case in which table contents change as the network condition changes. This is also referred to as *adaptive routing.* For example, if routing tables are based on minimizing message delay, then the routes are modified as the traffic pattern changes. There is a wide range of adaptivity depending on the frequency of changes, the type and amount of information used, and the means for implementing the changes. For example, static routing may be used, but changed only when there is a failure. To construct the best routing tables in the nodes at all times, information is needed about the instantaneous state of the network and its traffic. Unfortunately, it is not possible for the nodes to have complete and up-to-date information about the entire network. To provide it would also constitute too great an overhead. Several practical alternatives are presented below.

Centralized Routing—In centralized routing, a node is designated as the *routing control center* (RCC). Each node periodically sends status information to the RCC. The RCC thus acquires global information, based upon which it computes optimal routes. New routing tables are periodically distributed to the nodes in the network. While centralized routing may achieve global optimal and relieve the node from the task of routing computation, it has some drawbacks: the information collected at the RCC may be old due to the delay in the network; the communication overhead incurred in collecting status information and distributing routing information may be substantial; the reliability of the entire network rests on the proper operation of the control center.

Isolated Routing—Isolated routing is the most extreme case of decentralized routing. In isolated routing, each node makes its own routing decision based on information it has at hand. The *hot potato* technique consists of passing the packet on as quickly as possible, by sending (or queueing) it on the outgoing link with the shortest queue. Variations of this shortest-queue routing are obtained by applying various biases. For example, the link selected may be determined by a combination of the weights assigned to the static alternate routes and the queue size at each. Another isolated technique is the *backward learning* technique. It consists of having a node attempt to estimate the number of hops (or delay) of a route going from it to some destination (starting with an outgoing link) by measuring the number of hops (or delay) incurred by packets arriving from that destination on that route (i.e., on that outgoing link). To implement this technique, using the delay measure for example, each packet is time-stamped when it sets off on its journey, and from this time-stamp each node compiles a table of information about delays. One main problem with that implementation is that the delays measured are incurred by packets traveling in the direction opposite to that of concern.

Delta Routing (or Hybrid Routing)—The delta routing algorithm consists of using both central and local decisions. Using information periodically sent to it by the nodes, the RCC computes the k best paths for each pair of nodes, where only paths that differ in the initial line are considered. The RCC then sends to each node all equivalent paths (i.e., those with cost or delay differing by less than some number δ) for each of its possible destinations. In routing a packet, the node may choose any of the equivalent paths either at random, or by choosing the line with the smallest current cost (or delay). By adjustment of k and δ, the scheme can be made more or less centralized. Transpac, the French public packet switching network, uses delta routing.

Distributed Algorithms—In this class of algorithms, the nodes exchange information about delays by sending control messages to one another. To keep the overhead low, this information is exchanged only among adjacent nodes. Each node communicates to its neighbors its estimate of the minimum delay to every other node of the network. When receiving such estimates from its neighbors, a node adds to them its own delay to reach each of the neighbors, and selects the best outgoing link for each destination. Information exchange may take place either periodically at regular intervals or asynchronously, such as when the estimates change by more than some amount. The old ARPANET routing algorithm was of this type.

Session Routing and Logical Circuit Routing—In the routing algorithms discussed above, the routing decision is made for different packets independently. In session routing, the route is chosen when a session is established. All packets in the session go on the same path. In logical circuit routing, the route is chosen by means of a route set-up packet when the virtual circuit is established. The route setup packet finds its way to the destination using any of the schemes described above. All packets belonging to a virtual circuit are then transmitted on the same route. Packets must only carry the virtual circuit number. Routing in the subnet is implemented as described in the subsection headed "The Network Services." Different logical circuits between the same pair of hosts may take different routes.

Broadcast Routing—Broadcast routing refers to those techniques by which to deliver a packet originating at some source to all possible destinations in the network. Of course, this can be accomplished by sending multiple copies of the same packet, one for each destination. However, more efficient techniques exist. In the *multidestination routing* scheme, a packet is issued with a list of destinations. At an intermediate node, a copy of the packet is sent out on an outgoing line if the latter is the best route for some destination (i.e., on the sink tree for that destination); the copy will then contain the list of destinations that are to use that line. After some number of hops, each copy will contain a single address and is treated as a normal packet. A more efficient technique makes use of the sink tree associated with the source of broadcast. Assuming knowledge of the spanning sink tree in both directions, each node will broadcast the packet on all links belonging to the spanning tree except the one on which it arrived. Normally, the spanning sink tree is known only in the direction toward the sink. In that case, an approximation of the above algorithm is as follows. Copies arriving at some node on a link belonging to the spanning sink tree are repeated on all links except the one on which they came. Copies arriving on all other links are discarded. This algorithm is called *reverse path forwarding*.

Congestion Control

Congestion in an uncontrolled network is inevitable due to the fact that all resources (line capacities, buffer space, and processing capability) are limited. It is also often due to the protocols in use, such as, for example, the need to retransmit packets that are in error, the need for sequencing, etc. Some types of congestion may be relieved by the routing element that, if made dynamic, would attempt to route traffic on underutilized paths. But unfortunately, routing is not sufficient to prevent congestion altogether; it merely helps reduce it or delay it. Other flow control procedures are needed to prevent congestion. Many of these functions are present in the network layer and thus are presented here. There are, however, other flow control procedures that reside at other layers of the hierarchy of protocols. Fig. 17 illustrates the various levels at which flow control is exercised. As will be clear from the following discussion, flow control helps prevent loss of efficiency, deadlocks, and unfair allocation of the resources.

Hop-Level Flow Control—Channel-Queue-Limit Flow Control: The *direct store-and-forward deadlock* occurs in nonlossy networks when all buffers in some node A are destined to an adjacent node B and vice versa. Channel-queue-limit flow-control techniques are used to prevent such deadlocks. They consist of partitioning the buffers available at a node among the various outgoing channels. Complete partitioning is such a scheme that does not allow any sharing of buffer among the channels. Sharing with maximum queues, with minimum queues, or with maximum and minimum queues are variants. In addition to preventing deadlocks, these techniques achieve more efficient and fairer operation of the network by avoiding hogging.

Structured Buffer Pool Flow Control: Given that a channel-queue-limit flow control is implemented, thus preventing the direct store-and-forward deadlock, another type of deadlock, called the *indirect store-and-forward deadlock,* might still occur. Consider a closed chain of consecutively adjacent nodes $N_0, N_1, \ldots N_{n-1}, N_0$. Assume all buffers on the outgoing queue from N_i to N_{i+1} are occupied by packets destined to N_{i+2}, for $i = 0, 1, \ldots, n-1$ (addition on indices is modulo n). This situation clearly leads to a deadlock. A solution to this problem is given by the structured buffer pool technique. Packets arriving at an outgoing link queue are classified according to the number of hops they have already traveled. With N nodes in the network, the maximum number of hops is $N - 1$. The buffers available for that queue are allocated to the various classes (either in a fixed fashion or adjustable as needed). A packet of class k can use all buffers available to classes 1 through k. It has been shown that such a technique avoids completely cycles such as those described above that lead to the indirect S/F deadlock.

Hop-Level Flow Control With Virtual Circuits: The above techniques are used with both datagram and virtual circuit modes. In virtual circuit modes, it is also possible to exercise selective flow control on individual virtual circuit streams. An example consists of setting a maximum limit, M, on the number of packets for each VC in transit at each intermediate node. Limit M may be fixed or dynamically adjusted. This scheme is effective in slowing down VCs that directly feed into congested areas; by backpressure the control is exercised at the source, which in response reduces its input. In the absence of this control, congestion is bound to spread to other areas of the network and affects other sources originally not responsible for the congestion.

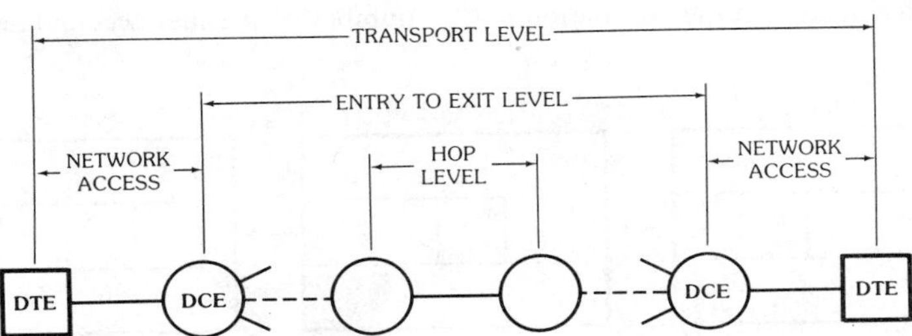

Fig. 17. Flow control levels. (*From Paul E. Green, Jr.*, Computer Network Architectures and Protocols, © *1982. Used by permission of Plenum Publishing Corp., New York.*)

Entry-to-Exit Flow Control—This level of flow control is primarily aimed at preventing buffer congestion at the exit node that may be due to conditions of overload of local lines connecting the exit node to the host. If the exit node must reassemble the messages before handing them to the host, then the problem is further complicated by reassembly and resequence deadlocks. An example illustrating a reassembly buffer deadlock is shown in Fig. 18. In this figure, A_i, $i = 1, 2, 3, 4$ are packets belonging to the same message. Similarly, B_j, $j = 1, 2, 3, 4$ and C_k, $k = 1, 2, 3$ represent packets of two other messages. Node 3 is the exit node. It is currently assembling message $A_1A_2A_3A_4$. Packet A_2 is missing. The deadlock occurs because packet A_2 cannot make it through node 2, where the buffers are full. An example illustrating resequence deadlocks is given in Fig. 19. Here A, B, . . . , K are consecutive packets. The exit node (node 3) must deliver packets to host 1 in sequence. The deadlock takes place because node 3 cannot deliver any of its packets, and packet A cannot reach node 3 via node 2.

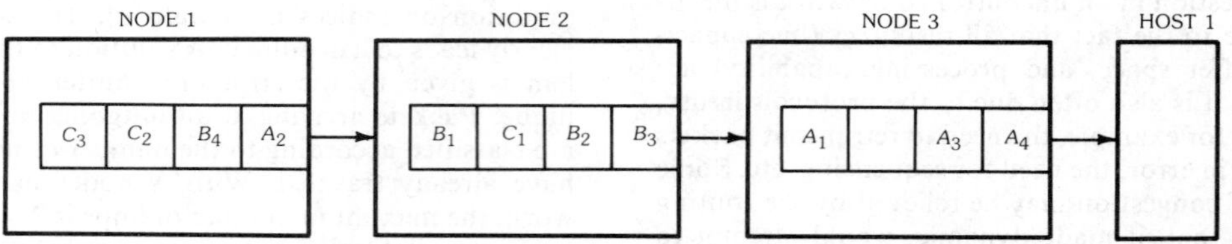

Fig. 18. Reassembly buffer deadlock. (*From Paul E. Green, Jr.*, Computer Network Architectures and Protocols, © *1982. Used by permission of Plenum Publishing Corp., New York.*)

The most common solution to reassembly deadlocks is to reserve a reassembly buffer for each message entering the network. The solution to the sequence deadlock is simply to discard out-of-sequence messages at the destination. The ARPANET applies this technique. A logical pipe is assumed to exist for all messages from some host A to some host B. Each pipe is individually flow controlled by a window mechanism of size $W = 8$. Messages arriving out-of-range are discarded. A request for next message (RFNM) is issued by the destination node to the source node to permit the source node to issue the next message. Buffer allocation for message reassembly at the destination is guaranteed prior to message transmission by the source node. In GMDNET, a network based on virtual circuits, entry-to-exit flow control is exercised individually on each VC. A window mechanism is used for each VC with the window size large enough to attain efficient utilization. However, the window size is variable: it is reduced if the destination is slow in accepting packets; it is increased otherwise.

Network Access Flow Control—Network access flow control is aimed at throttling the external input to the network with a view toward preventing overall internal network congestion. This control is based on measurements of internal network congestion, where the measures collected may be local (e.g., buffer occupancy at the entry node), global (e.g., total number of empty buffers available in the network), or selective (e.g., some measure of congestion on the path leading to the destination). Three techniques are described hereafter.

The Isarithmic Scheme: This scheme consists of setting an upper limit on the number of packets that can circulate in the network at any one time. It is based on the concept of *permits*. A number of permits are present in the network. Each packet offered to the network must secure a permit at the entry node before being admitted. Once a permit is so secured, it remains in use until the corresponding packet is delivered to its destination and hence exits from the network. The permit becoming free is added to the pool of permits at the node where it is freed. This scheme as described functions adequately if the traffic pattern is uniform and balanced. Otherwise, one must avoid having a large number of permits accumulate in certain parts of

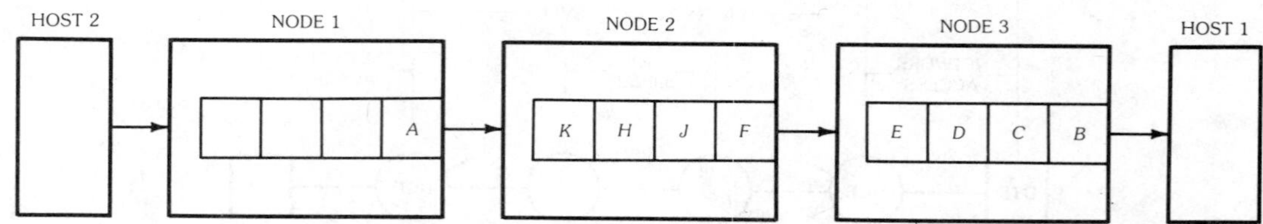

Fig. 19. Resequence deadlock. (*From Paul E. Green, Jr.*, Computer Network Architectures and Protocols, © *1982. Used by permission of Plenum Publishing Corp., New York.*)

the network at the expense of other parts. This is accomplished by setting limits on the number of permits at each node. The question of how to distribute the permits in the network then arises. The performance of the network under this scheme is sensitive to the permit distribution algorithm in use. A few such algorithms have been investigated analytically and by simulation.

The Input Buffer Limit Scheme: At each node, a distinction is made between (external) input traffic and transit traffic. The input is throttled based on buffer occupancy data at the entry node. This scheme rests on the observation that congestion at the entry node is a good indication of global congestion due to the back-pressure effect propagating internal network congestion back to the source nodes. Different versions exist. The structured buffer pool technique discussed previously falls into this category. Input traffic (which is in the 0-hop class) is discarded when all class-zero buffers are occupied. The number of class-zero buffers has been shown to have a great effect on the network performance under heavy load. Another version defines a limit, N_I, smaller than the total number of buffers, N, and new packets are limited to N_I while transit packets have no limitation imposed upon them. The performance here is sensitive to the ratio N_I/N. Another version yet consists of discarding an input packet if the total number of packets in the entry node exceeds a given threshold.

Choke Packet Scheme: In this scheme, a link is said to be congested if its utilization over some history window is beyond some threshold. A path is said to be congested if any of its links is congested. Along with routing information, congestion information is propagated in the network. When a packet is received by a node for some destination whose path is congested, then (1) if the packet is new external input, it is dropped; (2) otherwise, it is forwarded and a choke packet is sent back to the source, slowing the source. The path to the destination is unblocked gradually as no choke packets are received during a specified period of time.

CCITT Recommendation X.25

Recommendation X.25 is a widely accepted standard entitled "Interface Between a DTE and a DCE for Terminals Operating in the Packet Mode on Public Data Networks." It is a set of three peer protocols: a physical level, which is the same as in Recommendation X.21 (discussed in the section "The Physical Layer"); a link control level, which is essentially HDLC; and a network layer, which is discussed in the present subsection.

Recommendation X.25 is based on the concept of virtual circuits. These virtual circuits can be either temporary or permanent. A temporary virtual circuit referred to as a virtual call requires three phases: setting up the call, the data-transfer phase, and disconnecting the call. In setting up a virtual call, a logical channel number is assigned to the call. With a permanent virtual call, the numbers are assigned when the customer leases the facility from the common carrier. In such a case, the two phases of setting up and disconnecting the circuit do not exist.

It is important to note that Recommendation X.25 describes only the DTE/DCE interface. It is not concerned about how the network operates and how virtual circuits within the network are implemented. Recommendation X.25 defines the services rendered to the transport layer of the ISO model that resides in the user's equipment. Thus X.25 is known as a network access protocol standard. In many cases, the DTE is a terminal that handles characters rather than packets and that uses the start-stop asynchronous DLC protocol. In such cases, common carriers operating X.25 networks provide interface machines to which these terminals are connected. These interface machines, called PAD (Packet Assembly/Disassembly), are extensions to X.25 but not part of it.

The network layer of X.25 defines explicitly the formats of packets and the operational phases of a virtual circuit connection. Two types of packets exist: data packets and control packets. A data packet has a variable-length information field to carry user data. A control packet can be of different types, many of which are discussed below.

When a DTE sets up a virtual call, it selects a free logical channel number from the set allocated to it. Such a number consists of two parts: a logical channel group number and a logical channel number. A DTE has available to it up to 15 logical channel groups (addressed with 4 bits) and up to 255 logical channels within each group (addressed with 8 bits). Using a CALL REQUEST packet, the DTE sends the selected logical number along with its own address, the address of the destination, and other facility codes to the DCE to which it is attached. (For efficiency purposes, the CALL REQUEST packet may also carry user data up to 16 bytes.) The request to establish a connection arrives at the remote DTE from its attached DCE as an INCOMING CALL packet. The called DTE decides whether to accept or refuse the call. If it accepts it, then a CALL ACCEPTED packet is sent back to the DCE. The indication of acceptance of the call is achieved via a CALL CONNECTED packet, which is sent by the local DCE to its calling DTE. The disconnection of a call follows similar procedures. It uses the control packets CLEAR REQUEST, CLEAR CONFIRMATION, and CLEAR INDICATION. Once a connection is established, the flow of data packets can take place. Data packets are sequenced, and the flow is regulated according to a window mechanism as described in the section "The Data Link Control Layer." As with ADCCP and HDLC, the two sequence ranges of 8 and 128 are available. The control packets RECEIVE READY and RECEIVE NOT READY, similar to those described for ADCCP and HDLC, are used in a supervisory mode to acknowledge data packets. Other packets such as INTERRUPT, RESET, RESTART, and their confirmations are used for further control and to recover from certain minor and major problems. The INTERRUPT packet contains only one byte of user data and has no sequence number. It is transmitted as quickly as possible, preempting the queues of normal data packets. It is delivered (out of sequence) even when data packets are not being accepted. It is normally used to convey the fact that a terminal user has hit the break key to stop a flow of data from a distant computer. An INTERRUPT packet is acknowledged by the local DCE with an INTERRUPT CONFIRMATION packet. The RESET commands are used to reinitialize the window parameters corresponding to a particular virtual circuit to zero. Both DTEs and DCEs can initiate a reset. Reset packets contain up to two bytes of data explaining the reasons for the reset. These may be: the remote DTE is out of order, the network is congested or one of its nodes has failed, etc. Following a reset, the DTE has no knowledge about the status of outstanding (unacknowledged) packets. Recovery must be accomplished by higher levels. The *restart* condition provides a mechanism to recover from major failures. It has the effect of clearing all the virtual calls that a DTE has connected and resetting the permanent virtual calls, bringing the user/network interface to the state it was in when the service was initiated.

Since X.25 is an interface between a DTE and the local DCE to which it is attached, the receipt of an acknowledgement by a DTE from the DCE means that the latter has received the packet, but does not imply that the remote DTE has received it. To provide some level of end-to-end acknowledgement, a bit called the delivery confirmation bit, or D-bit, has been provided in the packet format. The DTE sets D to 1 to request confirmation of delivery to the remote DTE. An acknowledgement for this packet by the DCE would then have to be a guarantee of its delivery to the remote DTE.

In addition to the standard features of X.25, there are also some optional features that may be requested by a user when a call is set up. Examples are: reverse charging, closed user group (where members of a closed group can communicate only with one another), packet retransmission (where a DTE can ask its DCE to retransmit one or several data packets, and this is accomplished by the use of the REJECT command as in ADCCP), etc.

HIGHER-LEVEL PROTOCOLS

This section is concerned with layers four through seven of the ISO model. The brevity of this section is due to three factors: (1) many of the concepts and techniques needed at these higher levels, such as sequencing, acknowledging, and flow control, are similar to those used in lower layers and have already been described and discussed in previous sections; (2) many of the functions residing at these layers have not yet been well defined, and there are no internationally accepted standards yet; and (3) some of the functions (especially concerning the presentation and application layers) are not strictly speaking communication functions, and thus are considered out of the scope of this chapter.

The Transport and Session Layers

Internally to a host computer, application programs are represented by processes. Communication between such processes across a network is accomplished by means of "reliable connections" called *sessions* between specific pairs of processes, and it is the role of the session layer to establish and maintain such connections. It is possible to imagine that each such connection is achieved as a virtual circuit that the network layer provides. In that case, the session layer is trivial, the transport layer is nonexistent, and most of the work is done by the network layer. However, there are many problems with that design. The first problem arises when the network layer provides only a datagram service; the session layer will then have to be constructed appropriately so as to overcome the deficiencies of the network. In such a design, the session layer will also have to interface to the networks differently, depending on their type. It will have to be concerned with managing the buffers at the host where the processes reside, guaranteeing end-to-end reliability between the end hosts (which the network does not guarantee even when it is providing the virtual circuit service, since crashes, resets, and restarts can occur), and controlling the flow. As the protocols for many of these functions again differ depending on the network, it is best to render the session layer independent from network issues by the introduction of the transport layer. Thus, while the transport layer provides truly reliable host-to-host communication, the session layer connects two processes together in a session without worrying about any of the implementation details of the actual network. It merely requests from the transport layer its services. The functions of a session layer are thus considerably reduced. At session set-up time, it establishes the conventions necessary for the connection, conventions relating to the data-transfer mode (half-duplex versus full-duplex), the character codes, the flow control window sizes, how to recover from failures at the transport layer, etc. It also allows processes to refer to destinations using symbolic names by doing itself the mapping onto transport addresses.

The transport layer within a host defines a set of *transport addresses* or *sockets* through which connections are established. A transport address consists of a *network* number, a *host* number, and a *port* number assigned by the host. A CCITT numbering plan exists which uses 14-digit numbers. The first three identify the country; the fourth identifies the network within the country; the last ten digits are assigned by each network operator, some to indicate hosts, the remaining ones to indicate ports. There is a distinct difference between a transport address and a network layer address (which defines a virtual circuit), and the mapping need not be one-to-one. The transport layer uses network addresses to create transport connections. If the mapping is not one-to-one, then *multiplexing* is said to be in effect. In *upward multiplexing*, several transport connections are multiplexed onto the same network virtual circuit connection. In *downward multiplexing,* a single transport connection uses several virtual circuits. The choice of one type of multiplexing or the other is a function of the charging procedure used in the public network, the traffic volume exercised by a transport connection, and the window size of virtual circuits within the network.

The transport layer functions to establish and close connections as well as to control the flow across each connection. The techniques are similar to those described for lower levels. However, due to the ultimate end-to-end reliability that is to be guaranteed, several problems need to be solved. First of all, it is important to guarantee that old packets that for some reason have been delayed in the subnet but whose sequence numbers fall within the current receive window of the destination are not delivered to the destination. This is done by limiting the amount of time that a packet can exist in the network and by using a sequence space so large (e.g., 32 bits) that no packet can live for a complete cycle. For this method to be successful, however, it is important to have each connection initialize its sequence numbers to a value higher than the previous end sequence number by some margin so as to guarantee it is higher than any existing packet number. The second problem to deal with is to guarantee synchronization between the source host and destination host with regard to their respective initial sequence numbers. The solution to this problem is the *three-way handshake* mechanism. It consists of the following. The sender sends a CALL REQUEST packet with initial sequence number *S*. The destination responds with a CALL ACCEPTED with a sequence number *A* along with the source sequence number, *S*. Upon receipt of this packet, the source sends a packet with *S* and *A*. Thus, only when source and destination have confirmation of the combined sequence numbers would they consider the connection to have been established. The control of flow over a connection between source and destination is achieved by using a window mechanism similar to that of lower levels. The window sizes must be chosen appropriately. To be effective, it is important in some cases to distinguish acknowledgements from the destination confirming correct reception of packets from so-called *credits,* which are sent by the destination to inform the sender that receive buffers are available. Acknowledgement of a packet allows the sender to free its own buffer, while receipt of a credit allows the sender to advance its send window.

The Presentation Layer

The presentation layer performs transformation on the data generated by or destined to application programs before they are sent to or after they arrive from the session layer, respectively. The objectives and nature of such transformations are diverse. In order to save on communication bandwidth, the presentation layer compresses text and reduces the amount to be transmitted across the network. In order to guarantee the security of data, the presentation layer encrypts it at the source and decrypts at the destination. Many compression and encryption techniques are available. They are out of the scope of this chapter, and thus are not described here.

Another objective of the presentation layer is to resolve the differences that exist among the various equipments connected to the network. Advances of this nature have been made in two areas: (1) resolving the differences that exist among the various types of terminals in the networks leading to the so-called *virtual-terminal protocols,* and (2) resolving the differences that exist among the various hosts so that files stored in one host can be transferred to another host. This has led to the so-called *file transfer protocols.*

Terminals are basically of three types: *scroll-mode* terminals, which have no intelligence and generally use the start-stop protocol; *page-mode* terminals, which handle about 25 lines of 80 characters each at a time and have some local editing capabilities; and *form-mode* terminals, which are micro-based, have more intelligence, and handle specific forms that the operator fills out using local editing facilities. Besides the differences that exist among the different types of terminals, many differences exist within each class with regard to such attributes as the character set, line length, carriage-return and tab rules, cursor addressing, blinking, local editing capabilities, etc. Virtual-terminal protocols attempt to hide the differences by mapping real terminals onto a hypothetical network virtual terminal. For *scroll-mode* terminals, such a conversion is straightforward. Each time a new version of such a terminal type is to be supported, the corresponding conversion rules are added to the presentation layer at the host computer. For the more sophisticated types, a *data structure model* has proven useful and effective. Each end of the session has a data structure representing the state of the virtual terminal. Every time the application program changes the data structure on its machine, the presentation layer sends a message to the remote machine telling it how to change its data structure, and vice versa.

While the terminal conversion problem has been adequately solved, file transfer has proven to be much more difficult, and no general solutions exist. In particular, the concept of a network standard file format is not attractive due to the great difficulties involved in converting a file from one machine format to another. There are, however, several ad hoc file-transfer protocols particular to specific networks, e.g., ARPANET's FTP (ref. 2).

REFERENCES

The material presented in this chapter is primarily based on the content of the books listed below. By consulting these, the reader will be able to obtain more detailed information about this field and trace the original contributions and contributors.

1. Kleinrock, Leonard. *Computer Applications.* Vol. II of *Queueing Systems.* New York: John Wiley & Sons, Inc., 1976.
2. Tanenbaum, Andrew S. *Computer Networks.* Englewood Cliffs, N. J.: Prentice-Hall, Inc., 1981.
3. Green, Paul E. Jr., ed. *Computer Network Architectures and Protocols.* New York: Plenum Publishing Corp., 1982.
4. Martin, James. *Computer Networks and Distributed Processing: Software, Techniques, and Architecture.* Englewood Cliffs, N. J.: Prentice-Hall, Inc., 1981.

27 Satellite and Space Communications

Pier L. Bargellini

Advances in rocketry and microwave engineering inspired early proposals for communications satellites. Before the advent of artificial satellites, moon-reflection techniques were used for radar and communications purposes. In space systems, earth stations operate in conjunction with an orbiting spacecraft that probes the space environment, the earth as observable from space, the moon, a planet, or any other celestial body. In satellite communications systems, two or more stations located on or near the earth communicate via satellites that serve as relay stations in space. In both instances, control and monitoring of the spacecraft require that telemetry and command links be added to the main function of the mission.

Space systems include terrestrial missions, e.g., earth and/or sea surface observations of different kinds, weather satellites, and navigation satellites. Beyond the earth, space systems can be classified in terms of the mission range (i.e., cislunar, lunar, translunar, or planetary) as well as in terms of the specific nature of the observations to be carried out.

Communications satellite systems are classified in terms of their territorial coverage—e.g., global, regional, or national (domestic)—or in terms of the type of services offered—e.g., fixed, mobile, maritime, aeronautical, etc., or point-to-point, broadcasting, commercial, military, amateur, experimental, etc.

The environment of space affects the design of communications systems in several ways that make it different from the design of terrestrial systems. Major differences are:

A. Space and satellite communications systems cover distances far exceeding those encountered on earth.
B. As spacecraft power availability is limited and expensive, tradeoffs of space and earth segment design characteristics affect the overall system cost.
C. As the conditions along the signal paths are much more time invariant in space than on earth, it is possible to design space communications systems with great precision.

Three categories characterized by different environmental constraints can be identified, as follows.

Spacecraft-to-Spacecraft: In principle, the designer has maximum freedom in the choice of the operating frequency. The major difficulty resides in maintaining track between spacecraft.

Earth-to-Spacecraft (Up-Link): The choice of the operating frequency is primarily determined by the availability of spectral windows in the signal path. The window boundaries are dictated by absorption and dispersion phenomena in the ionosphere and atmosphere and also by the spectral and spatial distributions of natural noise sources. On earth, the easy availability of electrical power and the relatively benign environment make it possible to use large amounts of transmitter power enhanced by high-gain, large antennas that must be precisely aimed at the spacecraft. Launch-vehicle limitations restrict the size of the spacecraft receive antenna, and the spacecraft receiver is affected by the background noise of the earth.

Spacecraft-to-Earth (Down-Link): Spectral windows dominate the choice of the operating frequencies. The limited spacecraft transmitter power and antenna size (gain) are compensated by large earth receive antennas and low-noise receivers. The trend toward narrow-beam spacecraft antennas has increased the requirements for precise spacecraft positioning, stationkeeping, and antenna pointing with a consequent increase in the amount of onboard fuel required for the above-mentioned functions. The very stringent reliability requirements necessary to guarantee spacecraft design lifetime are achieved by painstaking selection of the equipment for all subsystems and provision for redundancy of critical components.

SATELLITE ORBITS

Satellites move around the earth as planets do around the sun. Kepler's laws apply:

First Law: Planetary orbits are elliptical with the sun at a focus.

Second Law: The radius vector from the sun to a planet sweeps equal areas in equal times.

Third Law: The ratio of the square of the period of revolution and the cube of the ellipse semimajor axis is the same for all planets.

These laws can be mathematically derived in terms of Newtonian mechanics and universal gravitation.

When the satellite mass is negligible with respect to the mass of the earth, the balance between gravitational and centrifugal force leads to the vector differential equation:

$$d^2\vec{r}/dt^2 + \mu\vec{r}/r^3 = 0 \qquad (1)$$

in which $\vec{r}$ is the radius vector from the center of the earth to the satellite and

$$\mu = GM = gR^2 = 3.99 \times 10^{14} \ m^3/s^2$$

where,

$G = 6.67 \times 10^{-11}$ N m^2/kg^2
$M = 5.98 \times 10^{24}$ kg
$g = 9.81$ m/s^2
$R = 6.38 \times 10^6$ m

As the centrifugal and gravitational forces are aligned and opposed, the satellite moves on a plane (orbit plane). Lengthy manipulations of Eq. 1 lead to Kepler's laws. The first law is verified by showing that the satellite moves along a conic on the orbit plane, the equation for which is:

$$r = p/(1 + e \cos\nu) \qquad (2)$$

where,

p is a parameter,

e is the eccentricity,
v is the central angle (true anomaly).

The value of the eccentricity determines the type of the conic.

$$e \begin{cases} = 0 & \text{circle} \\ < 1 & \text{ellipse} \\ = 1 & \text{parabola} \\ > 1 & \text{hyperbola} \end{cases}$$

With reference to Fig. 1, the following relationships, which hold for elliptical orbits, are frequently used:

$$x = r \cos v = a(\cos E - e)$$

$$y = r \sin v = a \sin E\, (1 - e^2)^{1/2}$$

Eccentricity:

$$e = c/a = (r_a - r_p)/(r_a + r_p)$$

Semimajor axis:

$$a = (r_a + r_p)/2$$

Apogee distance:

$$r_a = a + c = a(1 + e)$$

Perigee distance:

$$r_p = a - c = a(1 - e)$$

Locus parameter:

$$p = a\,(1 - e^2) = 2r_a r_p/(r_a + r_p)$$

Semiminor axis:

$$b = a\,(1 - e^2)^{1/2} = (r_a r_p)^{1/2}$$

Kepler's second and third laws can be derived from Eq. 1 through manipulations based on conservation of energy and angular momentum considerations whereby:

$$v^2/2 - \mu/2 = \text{constant} \quad (3)$$

and

$$\vec{r} \times \vec{v} = \vec{h} \quad (4)$$

The orbit period, P, is then

$$P = 2\pi a^{3/2}/\mu^{1/2} \quad (5)$$

Angle E (eccentric anomaly) is the central angle measured from the x axis to the vertical projection of the satellite point over the circle of radius a. The true anomaly, v, and the eccentric anomaly, E, are related by any of the expressions:

$$\cos E = (e + \cos v)/(1 + e \cos v)$$

$$\cos v = (e - \cos E)/(e \cos E - 1)$$

$$\tan (v/2) = [(1 + e)/(1 - e)]^{1/2} \tan (E/2)$$

For the hypothetical case of a uniform motion on the circle circumscribing the ellipse, the fictitious angle M (mean anomaly) can be defined:

$$M = 2\pi\, t/P \quad (6)$$

Angles M and E are related by the equation

$$M = E - e \sin E \quad (7)$$

It can also be shown that

$$v^2 = 2\mu/r - \mu/a \quad (8)$$

and that the orbit velocity vector has two constant components respectively normal to the radius vector and the major axis of the ellipse.

In order to define the satellite position in inertial space, three additional angles are required. These are: the inclination, i, of the orbit plane with respect to the equatorial plane of the earth; the right ascension (longitude), Ω, of the ascending node [i.e., the angle be-

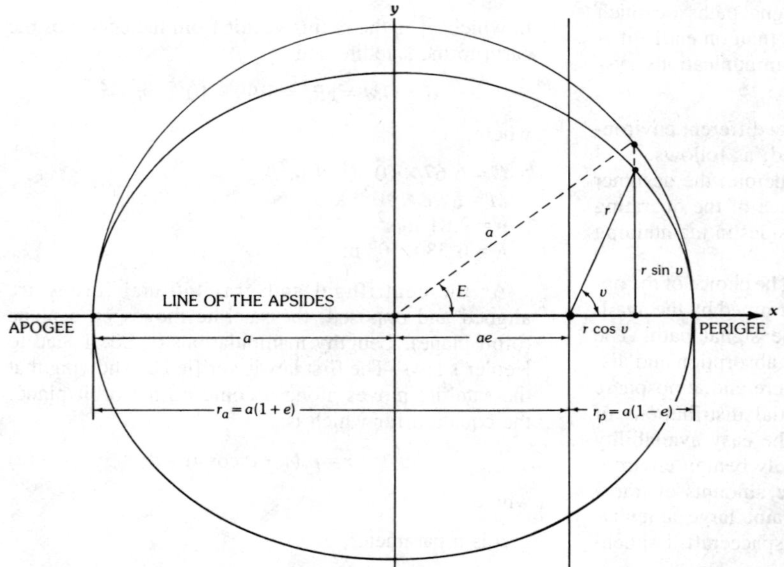

Fig. 1. Elliptical-orbit geometry.

tween the direction, γ, of the vernal equinox (point of Aries) and the intersection of the orbit and equatorial planes]; and finally the argument of perigee, ω, i.e., the angle measured in the orbit plane between the direction of the ascending node and that of the perigee.

The nonsphericity of the earth, the nonuniform distribution of masses of its surface, plus the effects of lunar and solar gravitation, make the actual motion of satellites depart from that described so far, which is based on the simple two-body problem.

The gravitational potential of the earth, which is no longer simply proportional to the inverse of the distance, can be expressed as a series of spherical functions. Two motions known as *secular variations* result from the above-mentioned causes; these are nodal regression and rotation of the line of the apsides.

Nodal regression is a rotation of the orbit plane in the direction opposite to the satellite motion around the axis of rotation of the earth. The rate of this motion (in degrees per day) is

$$\dot{\Omega} = -[10/(1-e^2)^2](R/a)^{7/2} \cos i \qquad (9)$$

Rotation of the line of the apsides is rotation of the ellipse major axis around the center of the earth on a fixed orbit plane. Apogee and perigee move then with respect to the earth, and the rate of motion of the argument of perigee is also a function of the orbit inclination angle, i, or:

$$\dot{\omega} = [5/(1-e^2)^2](R/a)^{7/2}(5 \cos^2 i - 1)\ °/\text{day} \qquad (10)$$

The critical value $i = \cos^{-1}(1/\sqrt{5}) = 63°24'$ stops this motion. Smaller values of i make the ellipse rotate in the same sense as the satellite motion along it; higher values of i make the ellipse rotate in the opposite sense. The effect becomes smaller for higher orbits.

Gravitational forces due to the moon and the sun also influence satellite motion. For low-altitude orbits, the gravitational field of the earth is preponderant, and solar and lunar effects can be neglected. For high-altitude orbits, gravitational disturbances of the moon and the sun cannot be neglected. In the case of a geostationary orbit, the sun and moon perturbation forces, respectively, amount to about 1/37 and 1/6000 of the gravitational force of the earth. The combined effect would change the orbital inclination of a satellite originally placed in a perfectly equatorial orbit by about 1 degree per year.

INFORMATION TRANSMISSION IN SPACE

In a link between two points in space, A and B, separated by the distance r, a transmitter located at A of power P_t, radiated by an antenna having a gain G_t produces at point B a power flux density

$$\phi = P_t G_t / 4\pi r^2 \qquad (11)$$

A receiving antenna of effective aperture A_r at B intercepts a signal power

$$P_r = P_t G_t A_r / 4\pi r^2 \qquad (12)$$

which equals the product of the rate of information transmission, R, and the energy, E, required to transmit one bit of information, whence:

$$R = BP_t G_t A_r / 4\pi \beta N r^2 \qquad (13)$$

where β is the ratio of the energy required to transmit one bit of information and the noise power density (i.e., $\beta = E/N_o$). This parameter, which is a function of the ratio of the channel bandwidth, B, and the information rate, R, characterizes the efficiency of modulation/demodulation schemes.

Equation 13 was derived by assuming an antenna of given gain at one end of the link and an antenna of given aperture at the other. In this case, frequency, or wavelength, does not appear explicitly in the equation, although the noise power density, N_o, is in general frequency dependent.

By assuming antennas of given gain at both ends, and using the well-known relationship

$$A = (\lambda^2/4\pi)G$$

the following result is obtained:

$$R = (\lambda^2/16\pi^2)(BP_t G_t G_r / \beta N r^2) \qquad (14)$$

On the other hand, assuming antennas of given aperture at both ends of the link yields

$$R = BP_t A_t A_r / \lambda^2 \beta N r^2 \qquad (15)$$

The above equations allow the information transmission rate, R, of a single space communications link to be computed. Since the noise encountered in space may be regarded as Gaussian, white, and additive, the Shannon model yields an upper bound of transmission rate. Equation 13 can be rewritten as:

$$R = BP_t G_t A_r / 4\pi r^2\ \beta N = (B/\beta) \cdot (P_r/N) \qquad (16)$$

The two factors on the right-hand side relate to the two cases of bandwidth- and power-limited transmission.

The lower bound of parameter $\beta = E/N_o$ has the value $\log_e 2 = 0.693$ in the limit case of infinite bandwidth and ideal (Shannon) modulation/demodulation processes. Then the signal-to-noise ratio goes to zero, and the information rate takes the value

$$\lim_{B \to \infty} R = 0.693\ P_r/N_o \qquad (17)$$

In physical systems, limited bandwidth and departure from ideal modulation/demodulation processes require higher values of E/N_o and P_r/N. The ratio of the actual information transmission rate and channel capacity (in the Shannon sense) becomes smaller than unity, and compromises among signal-to-noise ratio, bandwidth-to-information transmission rate ratio, and message error probabilities (or signal quality in analog systems) must be made.

ACTIVE VERSUS PASSIVE SATELLITES

Active satellites in geosynchronous orbit had been predicted in 1945, but as the early rockets could place only modest payloads in earth orbits a few hundred kilometers high and because reliable space-qualified receivers and transmitters were unavailable, the first space communications experiments were carried out with passive artificial satellites in low earth orbits. A passive satellite does not carry devices for signal amplification, but acts simply as a scatterer of electromagnetic waves.

Two approaches were pursued. In the first (ECHO I and ECHO II missions), the scatterer was a large lightweight metallized Mylar sphere. In the second (WESTFORD project), it was attempted to distribute a multitude of tiny dipoles in an orbital belt around the earth. Part of the energy radiated by an earth station impinges upon the sphere or the dipoles, and a small fraction of it is scattered back to earth. Thus the received power is

$$P_r = P_t G_t \Omega A_r / 4\pi r_1^2 r_2^2 \qquad (18)$$

where,

P_t is the transmitter power,
G_t is the gain of the transmit antenna,
Ω is the scattering area,
A_r is the effective area of the receive antenna,
r_1 and r_2 are the distances of the transmit and receive earth stations to and from the scatterer.

As $r_1 \simeq r_2$, the received power decreases with the fourth power of the distance. Thus even with very large amounts of transmitter power, large spheres in orbit, very large receive antennas, and very low-noise receivers, the communications capacity of passive satellites is inherently limited. In active satellites, the onboard amplifiers restore the signal power level, and as the received power at an earth station decreases only with the square of the distance, very substantial amounts of communications capacity can be obtained. Since space-qualified reliable, long-life electronic equipment became available, only active satellites have been used.

THE CHOICE OF THE ORBIT

The orbit height influences the communications range, the transmission delay, and the duration of the connection, which is determined by the time interval between satellite rise and set for points on the surface of the earth separated by a given angle measured from the center of the earth. The actual coverage is also a function of the minimum elevation angle of earth-station antennas. With reference to Fig. 2, the following relationship can be used to compute the coverage angle:

$$R/(R+h) = \cos(\beta+\theta)/\cos\theta \qquad (19)$$

where,

R = earth radius,
h = orbit height,
β = coverage angle,
θ = elevation angle.

The transmission delay can reach a maximum

$$t_d = [2(R+h)/c](\sin\beta/\cos\theta) \qquad (20)$$

Curves for the period, P, the earth coverage, and the maximum transmission delay, t_d, are shown in Fig. 3 for circular orbits.

Satellite systems can be classified in terms of three classes of orbits, i.e., in terms of the inclination angle, i, between the orbit plane and the equatorial plane:

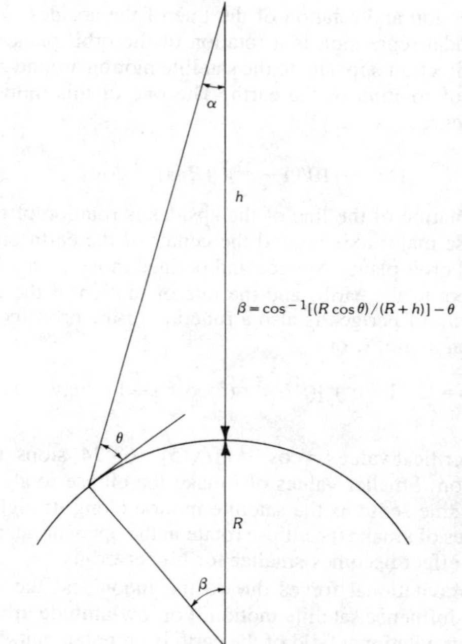

Fig. 2. Coverage and elevation angles.

$i = 90°$ polar orbits
$0 < i < 90°$ inclined orbits
$i = 0$ equatorial orbits

If the orbit is circular and the orbital height is 35.863 $\times 10^6$ m, the period equals a sidereal day (23 hr 56 min 4 s), and if the orbital plane coincides with that of the equator of the earth, the satellite becomes geostationary. In the absence of perturbations, such satellites hover over fixed points of the equator and subtend the earth with an angle of 17°27′. Communications can thus be established between any two points "visible" from the satellite over an area that amounts to about four-tenths of the entire surface of the earth. In this case, the handover problem, i.e., the passing of traffic from one satellite to another, which would be necessary with nonstationary satellites, is avoided. Other significant advantages are simplifi-

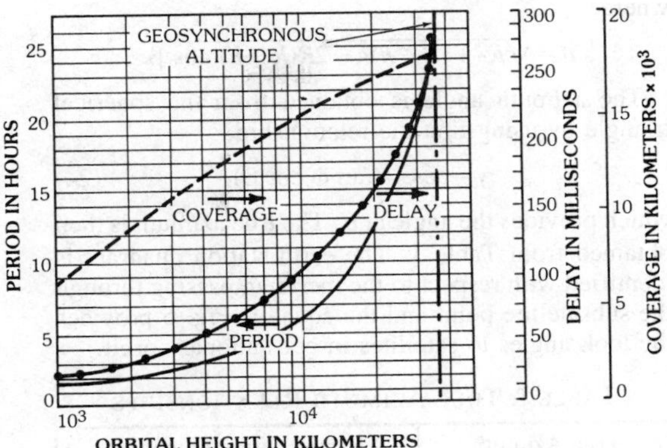

Fig. 3. Orbit period, earth coverage, and transmission delay versus orbit height.

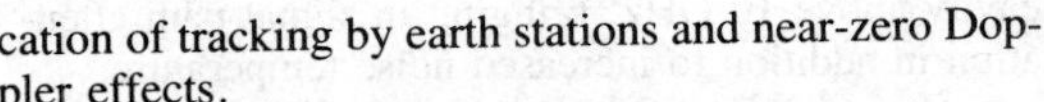

cation of tracking by earth stations and near-zero Doppler effects.

The transmission delay, which ranges from 0.238 to 0.275 second for single-hop circuits, has been found acceptable for telephone communications provided that the echo at times generated in terrestrial telephone plants at the transition points between 4- and 2-wire circuits is kept under control. This can be done to some extent by using improved echo suppressors or preferably echo cancellers.

Since the necessary rockets and know-how related to the complex in-orbit injection maneuvers were developed, the majority of communications satellites have been of the geostationary type. However, because the elevation angle at an earth station drops at higher latitudes, geostationary satellites cannot serve the near-polar regions. From Eq. 20, the maximum latitude that can be served is

$$\beta = \cos^{-1}\,[(R\cos\theta)/(R+h)] - \theta \qquad (21)$$

Hence, for an elevation angle $\theta = 5°$, $\beta = 76°21'$.

Thus, inclined orbits are required to serve near-polar regions. On the basis of energy considerations, it can be shown that for a given rocket, the lower the perigee the higher the apogee. The slower satellite motion around apogee results in the dual benefit of making communications possible over greater distances on earth and for longer periods of mutual visibility; the tracking problems are also eased.

As previously mentioned, an orbit inclination of 63.5° is advantageous because of the zero rotation of the line of the apsides for this critical angle. Polar orbits are also used for noncommunications-type missions such as earth observations, weather, surveillance, etc.

ELEVATION AND AZIMUTH ANGLES

For the case of a geostationary satellite and an earth station having latitude ϕ and longitude λ *relative to the subsatellite point on the equator*, the elevation angle (i.e., the angle above the horizon) and the azimuth angle (i.e., the angle measured clockwise from the direction of true North) can be computed by using standard spherical and plane trigonometry relationships. With reference to Fig. 4 and the spherical triangle $S'EB$ rectangle at B, it is

$$\cos\beta = \cos\phi\,\cos\lambda \qquad (22)$$

Then considering plane triangle OES, the elevation angle, θ, is

$$\theta = \cos^{-1}\{[(R+h)\sin\beta]/d\} \qquad (23)$$

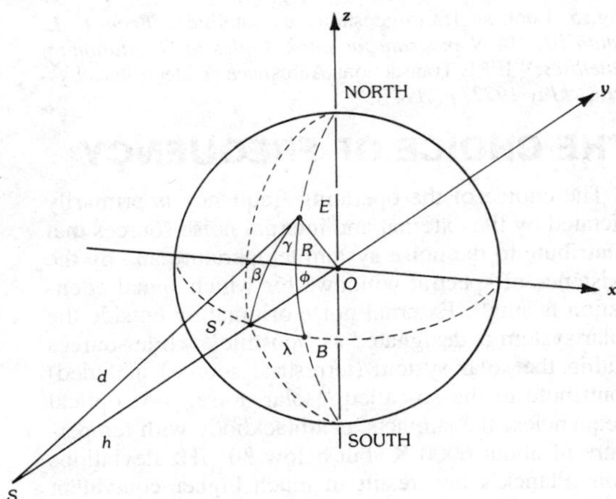

Fig. 4. Geometry of elevation and azimuth angles.

where,

$$d = \sqrt{R^2 + (R+h)^2 - 2R(R+h)\cos\beta}$$

The azimuth angle is obtained from the spherical triangle by using first the relationship

$$\gamma = \cos^{-1}(\tan\phi \cot\beta) \quad (24)$$

which provides the angle at E. The true azimuth is then obtained from Table 1. The earth-station quadrant is identified with respect to the meridian passing through the subsatellite point and the equator. Fig. 5 provides the look angles to satellites in geostationary orbit.

TABLE 1. TRUE AZIMUTH RELATIONSHIPS

True Azimuth	Earth-Station Quadrant
$180 - \gamma$	NW
$180 + \gamma$	NE
γ	SW
$360 - \gamma$	SE

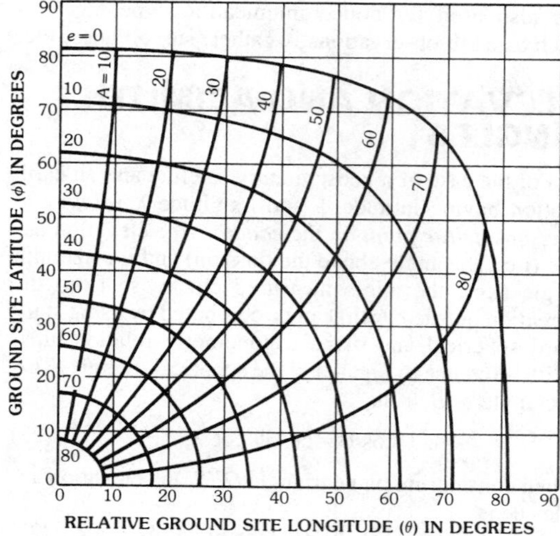

Fig. 5. Look angles to geostationary satellites. (*From F. L. Smith III. "A Nomogram for Look Angles to Geostationary Satellites,"* IEEE Transactions Aerospace & Electronics Systems, *May 1972, p. 394.*)

THE CHOICE OF FREQUENCY

The choice of the operating frequency is primarily dictated by the external and internal noise sources that contribute to the noise system temperature and by the existence of spectral windows for which signal attenuation is small. External noise originating outside the solar system is designated as "cosmic" while sources within the solar system (terrestrial sources included) contribute to the so-called "solar noise." At optical frequencies, the sun acts as a blackbody with temperature of about 6000 K, but below 30 GHz deviations from Planck's law result in much higher equivalent blackbody temperatures. These deviations depend on the solar activity during the 11-year sunspot cycles, with increases up to six orders of magnitude. A strong frequency dependency is observed as the photosphere, chromosphere, and corona contribute to the apparent temperature. Jupiter contributes up to 50 000 K at 440 MHz, a much higher value than that calculated from its surface temperature, the difference being attributable to synchrotron radiation. Venus contributes about 600 K to system noise temperature, and the moon about 300 K.

The ionosphere, the troposphere, and in particular the atmosphere, contribute as "terrestrial" sources. Lightning discharges produce very large values of noise temperature at lf and vlf. Ionospheric effects are secondary above 5 GHz. Hydrometeors contribute heavily above 10 GHz, bringing in signal-path attenuation in addition to increased noise temperature.

As shown in Fig. 6, the "clear sky" noise temperature observable at the surface of the earth has a broad minimum from 0.8 to 8 GHz and increases rapidly at the lower frequencies on account of terrestrial contributions, and at frequencies higher than 10 GHz on account of atmospheric phenomena. Fig. 6 applies to clear sky conditions; rain and other hydrometeors introduce additional noise as well as signal attenuation.

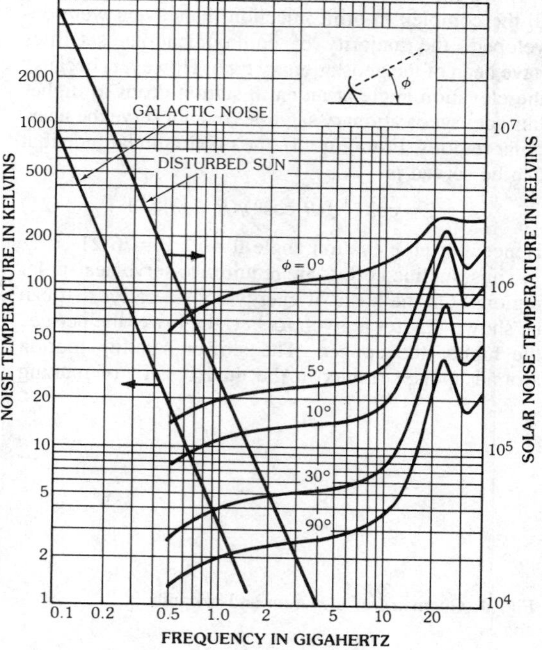

Fig. 6. Noise temperature versus frequency.

The effects of varying the antenna beam inclination angle are also shown. For low values of the elevation angle, larger amounts of noise power are collected by the antenna main beam in the atmosphere, and, in addition, the antenna side lobes pick up noise radiated by the surface of the earth and that part of the atmospheric noise reflected by it.

In conclusion, an optimum region for communications systems involving earth-to-space paths exists between 0.8 and 8 GHz. As it is relatively easy to build low-noise, high-gain electronic amplifiers for these frequencies, space communications systems have been preferentially designed to operate in this portion of the spectrum. The expansion of satellite communications systems has resulted in a crowding of these frequencies and has forced the opening of higher frequency bands. Allocations to fixed satellite service (FSS), intersatellite service (ISS), broadcasting satellite service (BSS), mobile satellite service (MSS), and the maritime mobile and aeronautical mobile satellite services can be found in Chapter 1.

LINK BUDGETS

Fig. 7 shows a model of an earth-space-earth communications system that comprises an up-link from an earth station to a spacecraft and a down-link from the spacecraft to earth. In the case of space systems, the information rates in the up- and down-links are different because the up-link carries only command and control information, while the down-link carries the information gathered by the spacecraft instruments. However, in satellite communications systems, because the spacecraft functions as a relay station in space, the information flow is the same in the up-link

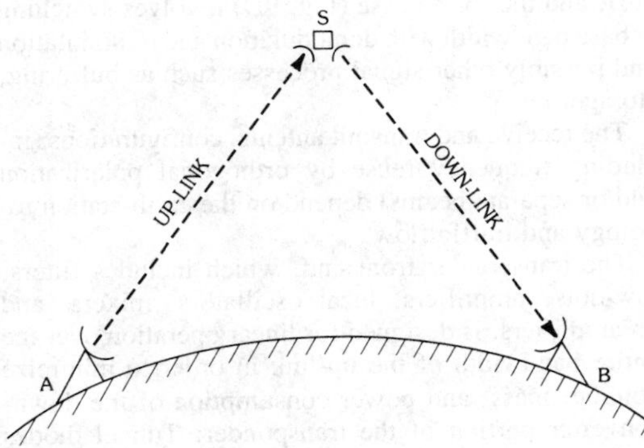

Fig. 7. Earth-space-earth links.

and the down-link. Messages (voice, video, data, etc.) generated by individual users and converted into electrical signals in the terrestrial network are routed to satellite earth stations, where, after suitable signal processing, modulated rf carriers are beamed to the satellite by a high-gain antenna. Conversely, the signals received from the satellite are processed, fed into the terrestrial network, and eventually distributed to the users.

Although a direct computation of the information transmission rate can be done in terms of Eq. 16 for the up- or down-link, or for their combination, it is convenient to derive relationships for the signal-to-noise power ratio in the two links taken separately. In this manner, specific contributions to the link budgets can be identified. In a successive step, the overall signal-to-noise power ratio is obtained by taking into account any additional noise contributions, such as interference or intermodulation. Finally, the number of channels (telephone, tv, or other) is computed for specific combinations of multiplexing, modulation, and multiple access techniques.

For the up-link, the power flux density at the satellite is given by

$$\phi = P_t G_t / 4\pi r_u^2 \tag{25}$$

Then the received signal power is:

$$\begin{aligned} P_u &= \phi A_{su} = \phi G_{su} \lambda^2{}_u / 4\pi \\ &= P_t G_t G_{su} (\lambda_u / 4\pi r_u)^2 \end{aligned} \tag{26}$$

where,

A_{su} = effective area of the satellite receive antenna,
G_{su} = gain of the satellite receive antenna,
λ_u = up-link wavelength.

The signal is received in the presence of thermal noise contributed by the receiver (internal noise) and the background (external noise). Both contributions can be assumed Gaussian and white; hence the total noise power density is $N_o = kT_s$, where $k = 1.38 \times 10^{-23}$ J/K is Boltzmann's constant and T_s is the system temperature. Thus the noise power in the rf transmission bandwidth, B, is $N = N_o B$. Substitution in Eq. 26 yields the up-link signal-to-noise power ratio:

$$P_u / N_u = (P_t G_t G_{su} / k T_s B)(\lambda_u / 4\pi r_u)^2 \tag{27}$$

and, in decibels,

$$\begin{aligned} \underset{(\mathrm{dB})}{P_u/N_u} &= 10 \log \underset{(\mathrm{dBW})}{P_t G_t} - 20 \log \underset{(\mathrm{dB})}{(\lambda_u / 4\pi r_u)} \\ &- 10 \log \underset{(\mathrm{dB \cdot Hz})}{B} + 10 \log \underset{(\mathrm{dB/K})}{(G/T)|_s} - 10 \log \underset{(\mathrm{dB/Hz \cdot K})}{k} \end{aligned} \tag{28}$$

The first term represents the earth station eirp; the second, the so-called "free space loss"; the third, the rf channel bandwidth; the fourth, the satellite G/T ratio; and the fifth, Boltzmann's constant.

A similar procedure yields the down-link budget:

$$P_d / N_d = (P_s G_{sd} / k T_e B)(\lambda_d / 4\pi r_d)^2 \tag{29}$$

and, in decibels,

$$\begin{aligned} \underset{(\mathrm{dB})}{P_d/N_d} &= 10 \log \underset{(\mathrm{dBW})}{(P_s G_{sd})} - 20 \log \underset{(\mathrm{dB})}{(\lambda_d / 4\pi r_d)} \\ &- 10 \log \underset{(\mathrm{dB \cdot Hz})}{B} + 10 \log \underset{(\mathrm{dB/K})}{(G/T)|_e} - 10 \log \underset{(\mathrm{dBW/Hz \cdot K})}{k} \end{aligned} \tag{30}$$

The numerical value of the Boltzmann's-constant term in Eqs. 28 and 30 is

$$10 \log (1.38 \times 10^{-23}) = -228.6 \text{ dBW/Hz·K}$$

and the "free-space loss" expressed in decibels is:

$$\underset{(\mathrm{dB})}{L_{fs}} = 92.44 + 20 \log \underset{(\mathrm{km})}{r} + 20 \log \underset{(\mathrm{GHz})}{f} \quad (31)$$

A plot of free-space loss versus frequency for the geostationary orbit appears in Fig. 8. If signal-power–to–system-noise-temperature ratios (dBW/K) are used

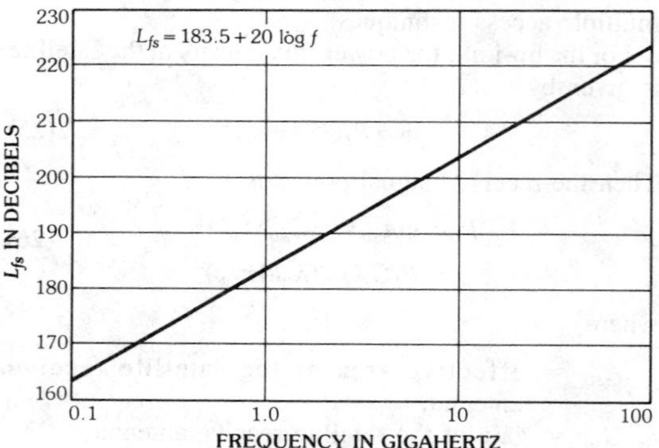

Fig. 8. Free-space loss versus frequency.

in lieu of signal-to-noise power ratios (dB), the Boltzmann's-constant term disappears in Eqs. 27 through 30. The rf channel bandwidth also disappears when signal-power–to–noise-power-density ratios (dB·Hz) are used. The choice of the notation is immaterial, but consistency is required in the use of units. Whenever angle modulation is employed, the symbol C (for carrier power) can be used in lieu of P (for signal power), as these two quantities have equal values.

SPACECRAFT ARCHITECTURE

The communications function is supported by the following subsystems:

A. Main structure
B. Primary propulsion subsystem, i.e., apogee/perigee motor(s)
C. Auxiliary propulsion subsystem for stationkeeping and orientation
D. Thermal control subsystem
E. Power subsystem
F. Telemetry, command, and control subsystem

The communications subsystem consists of antennas, receivers, and transmitters. Its mass plus that of the power subsystem defines the useful payload mass, M_u. The ratio of this mass and the total mass, M, of the spacecraft in orbit determines the spacecraft utilization factor:

$$u = M_u/M \quad (32)$$

Typical values of u range from 0.35 to 0.60. The difference,

$$M_b = M - M_u = M(1-u) \quad (33)$$

represents the mass of the bus (items A, C, D, and F of the above-mentioned support subsystems). Finally, the spacecraft mass at launch, M_l, is

$$M_l = M + M_b \quad (34)$$

where M_b represents the mass of the primary propulsion subsystem.

The three satellite parameters—transmit power, antenna gain, and bandwidth—that contribute to communications capacity are proportional to the total spacecraft mass. Designers' efforts aim at obtaining values as high as possible of the utilization factor (defined by Eq. 32), and for a given communications capacity, the total spacecraft mass should be kept as small as possible in order to keep launch costs down. Ever increasing communications capacity can be achieved not just by constructing bigger and more powerful satellites but also by the introduction of advanced technologies in all spacecraft subsystems as well as in the transmission system design.

TRANSPONDERS

Transponders are microwave repeaters carried by communications satellites. Four possible configurations are shown in Fig. 9. The first two cases (Figs. 9A and B), in which no signal processing takes place other than heterodyning from the up- to the down-link frequencies and amplification, represent a class of transponders designated as "transparent." These can handle any signal whose format can fit in the transponder bandwidth. The third case (Fig. 9C) involves switching at rf, and the fourth case (Fig. 9D) involves switching at base bandwidth with demodulation and remodulation and possibly other signal processes such as buffering, storage, etc.

The receive and transmit antenna configurations (including frequency reuse by orthogonal polarization and/or separate beams) depend on the earth-station topology and traffic flow.

The transponder front end, which includes filters, low-noise amplifiers, local oscillators, mixers, and preamplifiers, is designed for linear operation over the entire bandwidth of the up-link in order to minimize volume, mass, and power consumption of the down-converter portion of the transponder. Tunnel diodes and field-effect transistors are commonly used as active circuit elements. Distortionless transmission is ensured by flat amplitude and group-delay (linear phase) responses over the passband. Equalizers (mostly for compensation of group-delay distortion) are also used. However, limiters are used when antijam protection is needed.

Various contributions enter into the system noise temperature. If the receive antenna input port is taken as the reference point for the configuration shown in Fig. 10, the system noise temperature is

$$T_s = T_{ANT} + [(L-1) + L(F_{pr}-1) + L(F_r - 1)/G_{pr}]T_{ref} \quad (35)$$

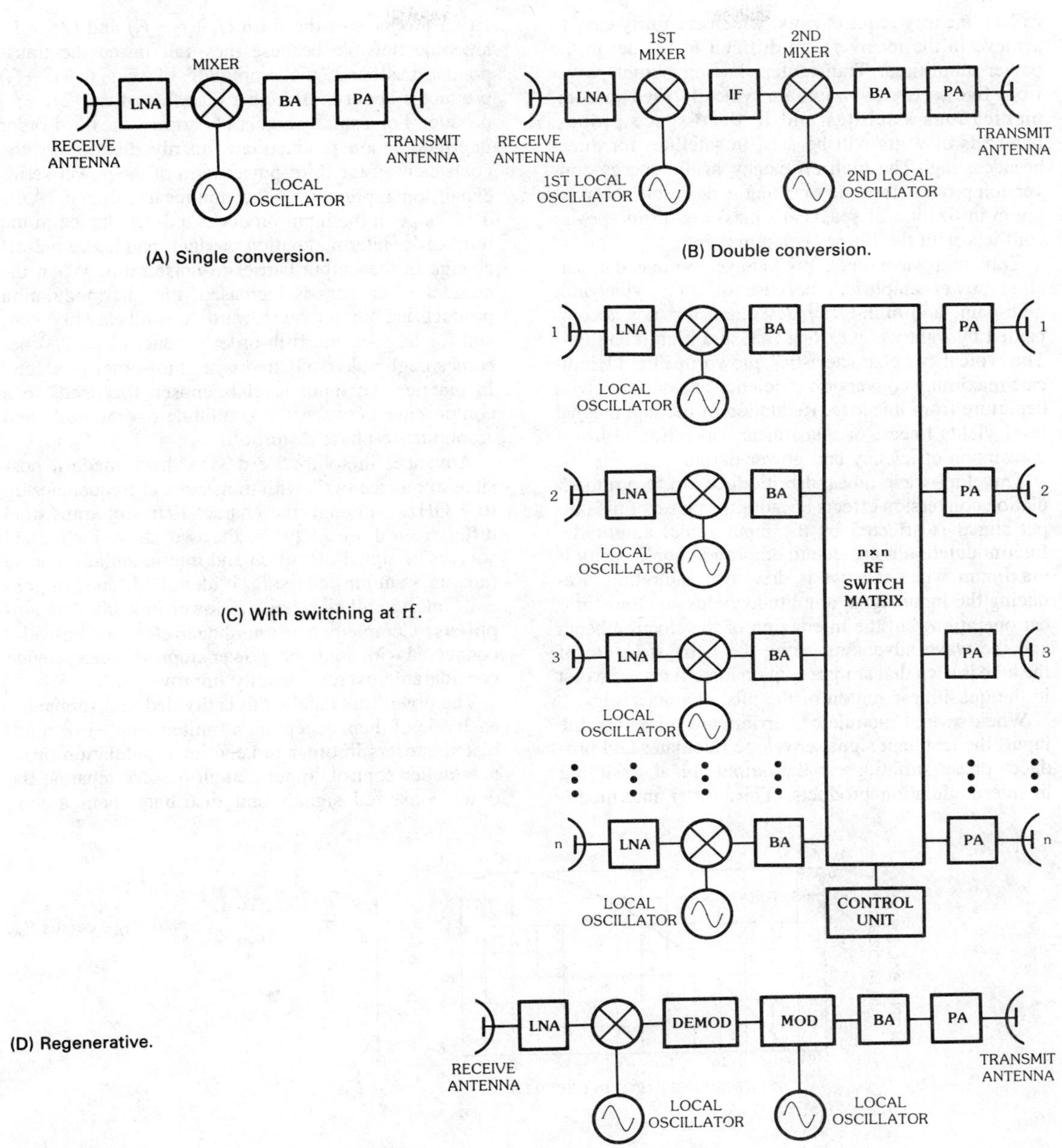

(A) Single conversion.

(B) Double conversion.

(C) With switching at rf.

(D) Regenerative.

Fig. 9. Transponders.

where,

T_{ANT} = antenna noise temperature,
L = transmission line (waveguide) loss,
F_{pr} = preamplifier noise factor,
G_{pr} = preamplifier power gain,
F_r = receiver (down-converter) noise figure,
T_{ref} = 290 K.

For a satellite antenna looking at the earth, the contribution of the term T_{ANT} is near 290 K. Low-loss waveguides and high-gain, low-noise preamplifiers are clearly desirable. After conversion to the down-link frequency and preamplification, the signal level needs to be raised to the required rf power output. Redundant receivers are used to ensure survivability over the planned spacecraft lifetime.

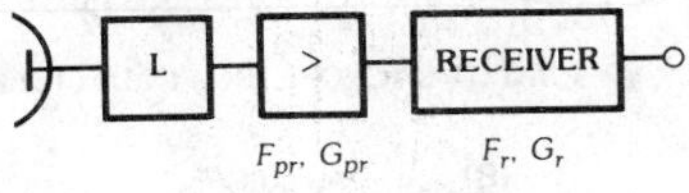

Fig. 10. Receiver front end.

The linearity requirements, which are fairly easy to achieve in the receiver, are difficult to be met in the power amplifiers. Transmitters having output power from five to twenty watts are typically used in communications satellites and in deep-space probes; hundreds of watts will be used in satellites for direct broadcasting. The high efficiency in the energy-conversion process from dc to rf that is desirable for maximum utilization of spacecraft mass and prime power conflicts with the linearity requirements.

Traveling-wave tubes (twt's) have dominated as satellite power amplifiers because of their wideband, high-gain, and high-efficiency characteristics accompanied by light weight, long life, and high reliability. The typical twt characteristics shown in Fig. 11 indicate maximum conversion efficiency at saturation with departure from linearity. Reduction of the input signal level yields linear, or quasilinear, operation at lower conversion efficiency and power output.

Traveling-wave tubes also produce am-to-pm modulation conversion effects because the phase of the output signal is affected by the input signal amplitude. Intermodulation noise due to amplitude nonlinearity is maximum when the twt is driven at saturation. Reducing the input signal amplitude yields an almost linear operation, but the interaction of the electron beam and the wave advancing along the helix structure of the tube is such that am/pm conversion effects are worst in the quasilinear region of the tube characteristics.

When several modulated carriers are present at the input, the resultant signal envelope fluctuates and produces phase variations in the output signal, resulting in intermodulation products. Third-order intermodulation products of the form $(f_1+f_2-f_3)$ and $(2f_1-f_2)$ are objectionable because they fall inside the transponder passband; the amplitude of the $(f_1+f_2-f_3)$ product is about 6 dB higher than that of the $(2f_1-f_2)$ product. For a small number of carriers, the third-order intermodulation products are mainly dictated by the coefficient of the third-power term of the power series expansion representing the nonlinear element. A 1-dB change in the input produces a 3-dB change in the third-order intermodulation product, and hence a 2-dB change in the output carrier-to-noise ratio. When the number of carriers is increased, the intermodulation products are compressed toward the limit case of $n=\infty$, and for large n, the fifth-order product $(3f_1-2f_2)$ becomes negligible compared with third-order products. In practice, an input level is chosen that leads to a compromise between the amplitude-to-amplitude and amplitude-to-phase distortions.

Advances in solid-state devices have made it possible to replace twt's with transistors at frequencies up to 4 GHz. Although the characteristics of transistors differ from those of twt's, the two above-mentioned sources of signal distortion and intermodulation noise (am/am - am/pm) can still be identified, and, in general, intermodulation noise is lower in solid-state amplifiers. Complementary nonlinear elements cascade-connected with nonlinear power amplifiers can provide considerable overall linearity improvement.

The down-link bandwidth is divided into subbands, each one of them accepting a limited number of modulated carriers in order to keep intermodulation products under control. Input demultiplexers separate the down-converted signals and distribute them among

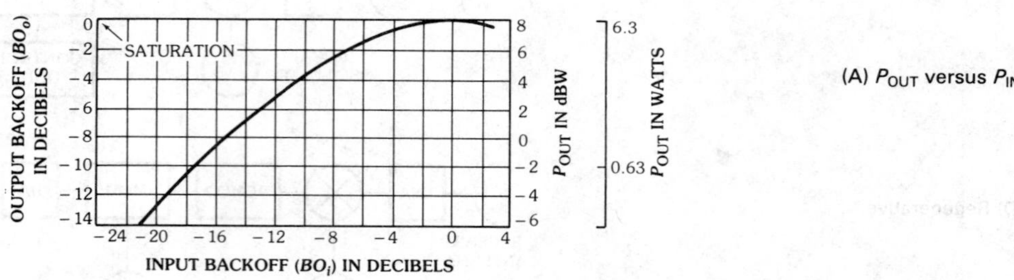

(A) P_{OUT} versus P_{IN}.

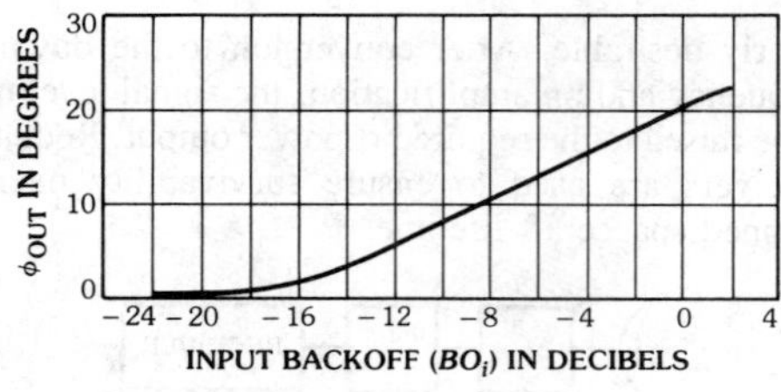

(B) ϕ_{OUT} versus P_{IN}.

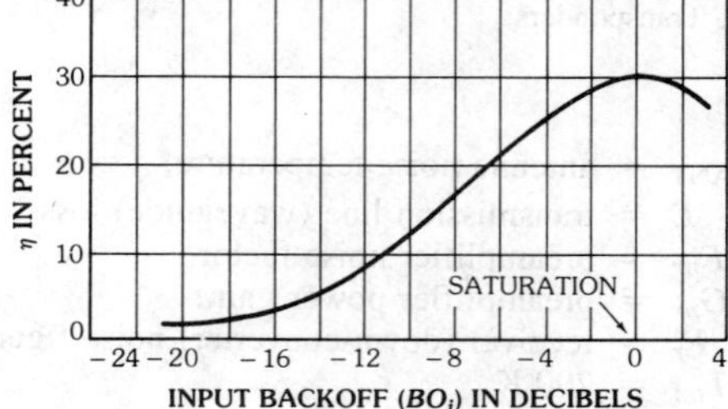

(C) Efficiency versus P_{IN}.

Fig. 11. Twta characteristics.

several transmitters, each one using a fraction of the entire down-link bandwidth. Output multiplexers are used at the interface between the transmit power amplifiers and the transmit antenna-feed waveguides.

Transponders are usually counted by the number of separate transmit channels in the down-link and the corresponding portions of the up-link wideband receivers.

For given amounts of power and mass available for the communications subsystem, the number of transponders and their bandwidth depends on tradeoffs of power, bandwidth, and intermodulation noise. With a fraction of the bandwidth lost to guard bands, the total usable bandwidth decreases when the number of transponders increases. A similar situation arises in regard to the available power per transponder, on account of the direct arithmetical proportion and also because of the increased complexity of the distribution network.

Transponders of equal bandwidth are advantageous for fixed allocations of traffic but do not provide flexibility when a network is reconfigured within a given coverage area or when a satellite is moved to serve an area characterized by different network topology and traffic requirements.

Until the mid 70s, the prevalent use of fdm/fm transmission led to a standardization of transponder bandwidth around 36 MHz and to a number of transponders per spacecraft around 12 for single-polarization satellites and 20–24 for double-polarization satellites. The introduction of the 11/14-GHz frequency bands and of high-speed digital transmission has accelerated the use of wider-band transponders (80 and 200 MHz) in addition to those having 36–40 MHz nominal bandwidth.

The increasing number of transponders needed to provide ever greater amounts of communications capacity has required advances in the design of flight-qualified microwave filters. Light weight has been achieved by using thin-wall Invar or graphite-fiber-reinforced plastic cavities. The requirement for flat amplitude and group time delay necessary to maintain low signal distortion and cross talk contrasts with the high skirt selectivity needed for efficient spectrum utilization. The design has evolved from conventional single-mode cascade rectangular-cavities Chebychev filters to longitudinal cross-coupled dual-mode circular-cavities filters as shown in Figs. 12 and 13, improved skirt selectivity being obtained at the expense of a finite but

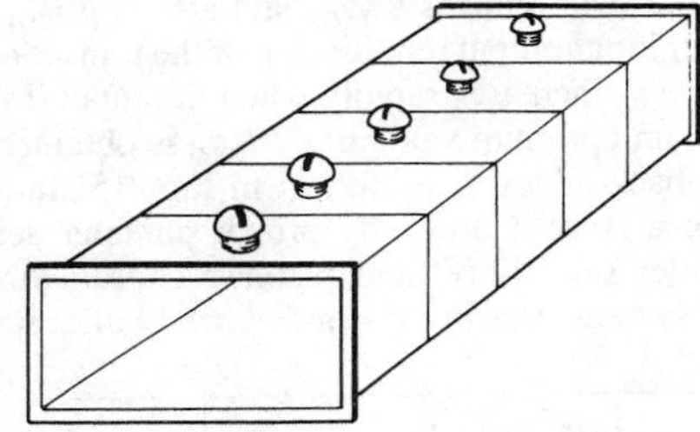

Fig. 12. Single-mode cascaded-cavities filter.

sufficiently large out-of-band attenuation. True elliptical response can be achieved in the canonical dual-mode filter shown in Fig. 14, in which the same physical cavity provides the input and output ports. This arrangement is also mechanically advantageous because of volume reduction.

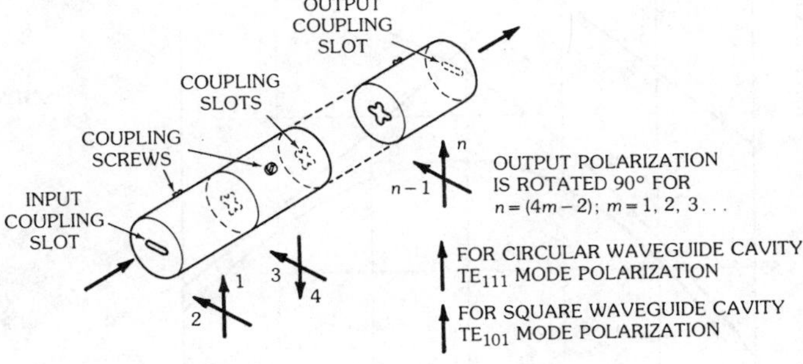

Fig. 13. Coupled dual-mode circular-cavities filter.

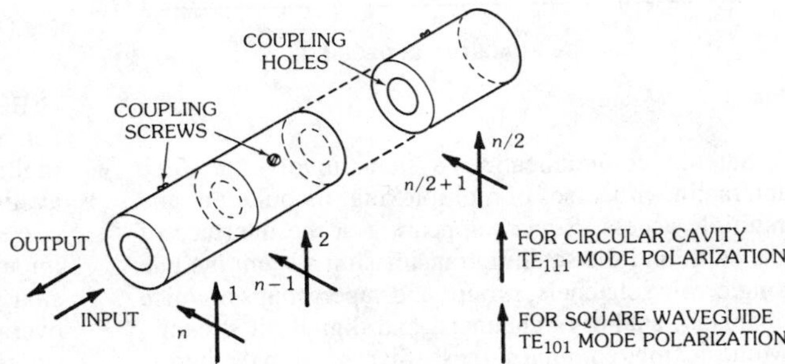

Fig. 14. Canonical dual-mode filter.

OVERALL TRANSMISSION SYSTEM CONSIDERATIONS

With transparent transponders and analog transmission, the total signal (carrier)-to-noise power ratio can be computed from the expression

$$(C/N_{\text{total}})^{-1} = (C/N_{\text{up}})^{-1} + (C/N_{\text{down}})^{-1} + (C/N_{\text{im}})^{-1} + (C/N_{\text{in}})^{-1} \quad (36)$$

where, in addition to the up- and down-link noise, the third term represents intermodulation noise, and the fourth term represents interfering signals. In the presence of intermodulation noise, C/N_{total} can be maximized by reducing the transponder input drive. Backing off the twta reduces C/N_{up} and also C/N_{down} (via the power input/output relationship of the transponder), but as C/N_{im} increases rapidly when the input drive is reduced, an optimum value of C/N_{total} is obtained at a specific backoff level, as shown in Fig. 15. Interference noise is kept down by proper antenna design, transponder selectivity, and planned coordination of satellite systems sharing given frequency allocations.

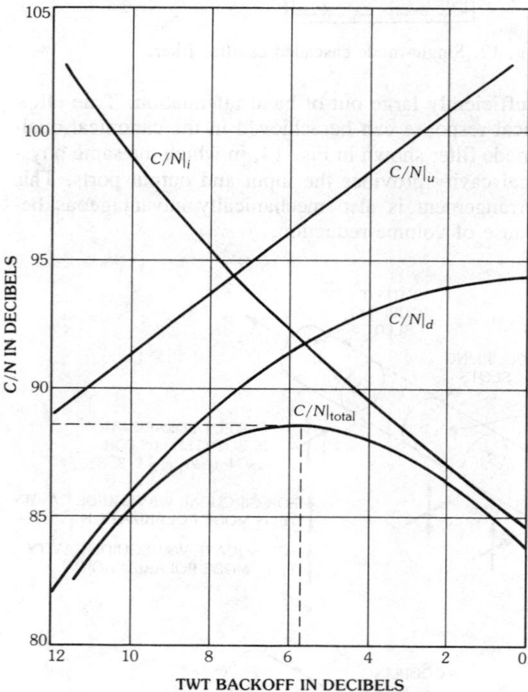

Fig. 15. *C/N* ratios.

Satellite communications systems involve the three interacting processes of multiplexing, modulation, and multiple access. Signals appearing at the interface of the earth and the satellite transmission system include single voice channels, groups and supergroups of voice channels, analog tv channels, and digital bit streams. Multiplexing combines these diverse signals into a composite baseband signal with individual inputs distinguishable in the frequency or time domain. Modulation can be either analog (i.e., continuous) or discrete (digital), and in the latter case, sampling and quantization occur prior to the modulation process itself. Amplitude modulation has been little used; analog phase or frequency modulation or digital bpsk and qpsk modulation have been widely used.

Multiple access is a unique feature of satellite communications due to the fact that when *m* earth stations are "visible" from a satellite, transponders can be shared by several earth-station pairs. The number of possible connections is then

$$n = m(m-1)/2$$

This "*n*-port" network feature is very attractive in comparison with the inflexibility of the "2-port" configuration encountered in most earth-based communications systems.

Individual accesses are kept separate in a physical domain such as frequency, time, or space (i.e., by separate antenna beams) or by encoding. The accesses can be preassigned in a static mode or dynamically assigned on demand. Fig. 16 illustrates three principal modes of multiple access, and Table 2 outlines their major features.

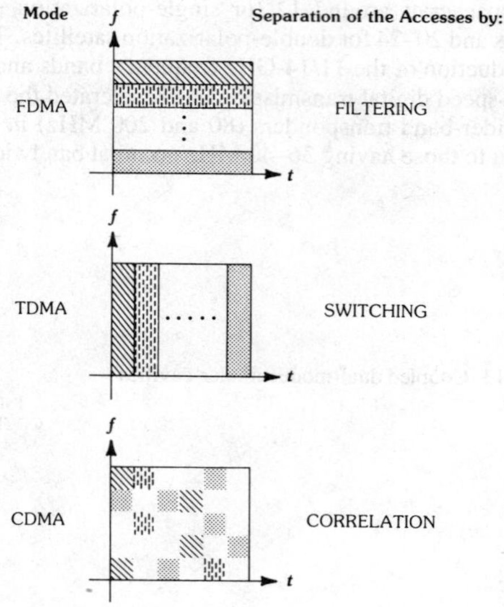

Fig. 16. Multiple-access schemes.

Because modulation and multiple access are interactive processes, the choice of a specific method of multiple access depends on the best utilization of the available power and bandwidth. The number of accesses should be made as large as possible while signal impairments due to the imperfections of the transmission channel are kept within specified margins. The overall system design is also influenced by the characteristics of the terrestrial interface. Analog voice and

TABLE 2. FEATURES OF MULTIPLE-ACCESS SCHEMES

Type	Characteristics	Advantages	Disadvantages
FDMA	Constant envelope signals Angle modulation (analog or digital) Separation by filtering	Easy interfacing with terrestrial analog facilities No need for synchronization Simple earth-station equipment	Intermodulation noise in nonlinear amplifiers reduces communications capacity Backoff required Up-link power coordination is required Transmission plan is difficult to reconfigure
TDMA	Bursts from and to different stations do not overlap Only one burst present at a given time Separation by time gating	Fairly easy interfacing with terrestrial digital facilities Power amplifiers operate near saturation at higher efficiency Highest communications capacity No need for power control and coordination Transmission plan is flexible	Need for coordinated synchronization More complex earth-station equipment A/d and d/a conversion required
CDMA	Constant envelope signals Each access takes the whole transponder bandwidth Separation by correlation	Spread spectrum provides protection against jamming	Synchronization is required Communications capacity is limited Signal quality worsens when number of accesses increases Power coordination is needed

video transmission systems predominant in the middle 60s in terrestrial systems led to the adoption of fdm/fm/fdma techniques in most commercial satellite systems, notwithstanding the loss of communications capacity resulting from the nonlinear channel characteristics when several carriers occupy a transponder. The expansion of digital systems, accelerated by the advances in solid-state devices that have made possible rugged, reliable, and cost-effective signal-processing equipment, is leading to a wider use of tdma. Future satellites providing on-board signal-processing functions, in addition to amplification, will further stimulate the use of digital techniques.

Digital transmission offers the following major advantages:

1. Guarantee of error control
2. Reduced sensitivity to channel nonlinearity
3. Efficient tradeoff of power and bandwidth
4. Flexibility with regard to multiplexing diverse signals
5. Easy combination of the functions of transmission switching and routing
6. Capability of signal regeneration
7. Implementation with rugged hardware

In satellite systems, points 1, 2, 3, 6, and 7 are especially meaningful. The up- and down-link can be separately designed to provide specified amounts of bit error rate (BER). For a given type of modulation, BER is primarily a function of the carrier-to-noise power ratio, C/N, (or the energy-per-bit–to–noise-power-density ratio, E/N_o) in the channel. Assuming independent noise performance of the links, the BER of their combination is the sum of the individual BERs. This condition is satisfied with repeaters capable of providing signal regeneration in the case of "transparent repeaters." Otherwise, the up-link BER must be much smaller than the down-link BER.

Digital methods are especially attractive in demand assigned multiple access (dama) and time division multiple access (tdma) systems.

ANALOG TRANSMISSION SYSTEMS

In satellite systems, multiplexing, modulation, and multiple-access methods and the channel characteristics interact in the determination of communications capacity. With analog transmission, the quality of the signal delivered to the user is a function of the post-detection signal-to-noise ratio, S/N, through a general functional relationship of the form

$$S/N = f(C/N_o, \text{CH, SIG, MUX, MOD, MA}) \quad (37)$$

where,

C/N_o = carrier-to-noise power density ratio in the rf channel,
CH = Rf channel characteristics (amplitude and phase, nonlinearity),
SIG = original signal descriptors,
MUX = type of multiplexing,
MOD = type of modulation,
MA = type of multiple access.

System performance can be assessed with reference to the Shannon bound (rate distortion theory and communications capacity) and, in practice, by the number of telex, telephone, or video channels that can be handled.

FDM/FM (Single-Carrier Case)

When a transponder handles a single carrier modulated by a baseband signal resulting from frequency-division multiplexing of numerous voice channels and with no interference, Eq. 36 simplifies to

$$(C/N|_{tot})^{-1} = (C/N|_{up})^{-1} + (C/N|_{down})^{-1} \quad (38)$$

The signal-to-noise power ratio in the worst (top) voice channel at the output of the fm demodulator can be computed by using the appropriate fm equations.

First, the weighted signal-to-noise ratio for a 1-mW test tone is computed by taking into account the CCIR recommendation on satellite systems channel noise performance, which requires the average noise power not to exceed 10 000 pW psophometrically weighted. Allowing 1000 pWp for the earth-station equipment noise, 1000 pWp for interference noise, and 500 pWp for the earth-station out-of-band noise, the balance of 7500 pWp is the noise allowed for the combination of the up- and down-links. Hence,

$$S/N = 10^{-3}/(7500 \times 10^{-12}) = 1.33 \times 10^5 \quad (39)$$

or

$$S/N_{dB} = 51.25 \text{ dB} \quad (40)$$

The relationship between the signal-to-noise ratios at the output and the input of the fm detector operating beyond threshold is*

$$S/N = (C/N)B/b(f_r/f_m)^2 pW \quad (41)$$

where,

S/N = weighted signal-to-noise power ratio at 1-mW test-tone level,
C/N = carrier-to-noise power ratio over the rf channel bandwidth,
B = rf channel bandwidth,
b = audio channel bandwidth (3.1 kHz),
f_r = rms test-tone frequency deviation,
f_m = maximum baseband frequency $\simeq 4.2 \times n$ kHz, with n the number of voice channels,
p = psophometric factor (1.78 or 2.5 dB),
W = preemphasis factor (2.5 or 4.0 dB).

Carson's rule defines the rf channel bandwidth as

$$B = 2(f_p + f_m) \quad (42)$$

* For a limiter-discriminator fm detector, the threshold is often taken as C/N = 10 to 13 dB. Fmfb and pll detectors can provide up to 3 to 5 dB of threshold extension.

where,

f_p is the peak frequency deviation,

f_m is the maximum frequency component of the baseband modulating signal.

Because this signal is Gaussianly distributed, the peak frequency deviation is undefined; however, a peak factor ρ, which is the ratio of the peak to the rms frequency deviations, can be assumed with values from 3.16 (10 dB) to 8.5 (18.6 dB), the lower value corresponding to large n.

Considering the load factor

$$L = \begin{cases} -15 + 10 \log n, & n \geq 240 \\ -1 + 4 \log n, & 12 < n < 240 \end{cases} \qquad (43)$$

which yields the average power of the baseband signal in dBm0, the quantity

$$g = \text{antilog } (L/20) \qquad (44)$$

is the ratio of the rms multichannel frequency deviation to the test-tone deviation. By introducing both factors in Eq. 42, the value of the rms test-tone deviation is determined:

$$f_r = (B/2 - f_m)/\rho g \qquad (45)$$

The value of B being fixed for a given transponder design, the number of voice channels per transponder is computed by an iterative procedure. Starting with an estimate for n, computed values of f_m and f_r are introduced into Eq. 41 together with the rf channel C/N power ratio. If the resulting S/N turns out to be greater (or smaller) than the prescribed 51.25 dB, the estimate for n was too low (or too high) and needs to be changed until the correct value of S/N is found.

In practice, each voice channel requires from 40 to 50 kHz of rf bandwidth; hence the channel density falls in the range from 20 to 25 voice channels per megahertz. The use of companded fdm/fm permits greater capacity; typically a doubling of the capacity can be achieved for a given transponder bandwidth.

FDM/FM/FDMA (Multicarrier Case)

As transponder bandwidth and power are shared among the modulated carriers, the 7500 pWp must include the intermodulation noise arising from the transponder nonlinearity. Although the intermodulation noise is reduced by backing off the traveling-wave-tube power amplifier, the ultimate result is a reduction of the number of voice channels that can be handled by a transponder in proportion to the number of the accesses. The smaller value of the loading factor contributes also to reducing communications capacity. Fig. 17 illustrates the performance of a typical commercial satellite (INTELSAT IV global beam) using FDM/FM/FDMA.

Within the constraints of the available bandwidth and power, and by taking into consideration interfer-

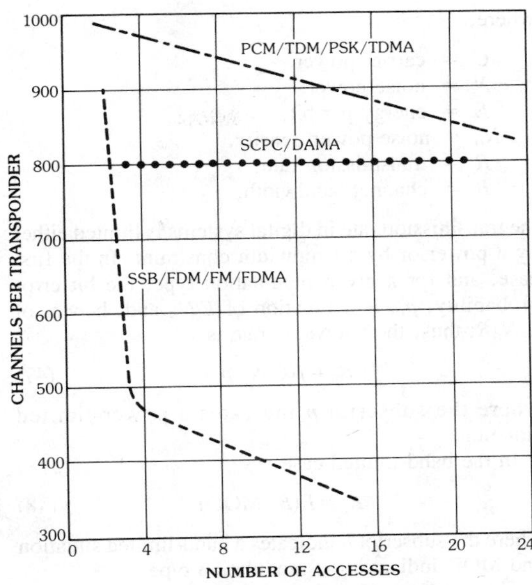

Fig. 17. Multiple-access systems performance.

ence and frequency-coordination problems, the actual value of the maximum transponder capacity depends on tradeoffs among the following causes of signal impairment:

A. Spacecraft twta impairments, which include in-band intermodulation products arising from both amplitude and phase nonlinearity, and intelligible cross talk due to am-pm conversion.

B. Frequency-modulation transmission impairments not attributable to the twta characteristics. These impairments include adjacent-channel interference due to spectral overlap, which produces convolution and impulse noise at baseband; dual path between transponders on a given spacecraft; interference produced by intermodulation in adjacent transponders; cochannel interference in frequency reuse systems; and earth-station rf out-of-band emission.

At the earth stations, the baseband signals frequency modulate preassigned multidestination carriers, which are transmitted via satellite to various receiving stations. After demodulation of the rf signals and demultiplexing of the baseband, the individual voice channels are recovered. This transmission method is rather inflexible with respect to changing traffic requirements, but it has been widely used for high-density trunks.

DIGITAL TRANSMISSION SYSTEMS

Since

$$C/N = (E/N_o)(R/B) \qquad (46)$$

where,

C = carrier power,
N = noise power,
E = energy per bit,
N_o = noise power density,
R = transmission rate,
B = channel bandwidth,

the transmission rate in digital systems is limited either by a power or by a bandwidth constraint. In the first case, and for a given modulation type, the bit error probability, p_e, is a function of E/N_o and, hence, of C/N_oR; thus, the allowable rate is

$$R_p = f(C/N_o, p_e) \quad (47)$$

where the subscript p indicates a power-limited situation.

In the band-limited case

$$R_b = F(B, \text{MOD}) \quad (48)$$

where the subscript b indicates a band-limited situation and MOD indicates the modulation type.

Thus, in general

$$R_{\max} = \min(R_p, R_b) = f(\text{MOD}, C/N_o, B, p_e) \quad (49)$$

Expressing bit error probability as a function of E/N_o determines modulation performance. In general,

$$p_e = f(E/N_o) = (1/2)\ \text{erfc}\ [(E/2N_o)(1-\rho)]^{1/2} \quad (50)$$

where

$$\text{erfc}(x) = (2/\sqrt{\pi}) \int_x^{\infty} e^{-t^2} dt$$

and

$$-1 \leq \rho = (1/E) \int_{-\infty}^{+\infty} s_1(t) s_2(t) dt \leq +1$$

the normalized correlation coefficient of the waveforms $s_1(t)$ and $s_2(t)$ used to represent zeros and ones.

The most frequently used digital modulation systems are binary phase shift keying (bpsk) and quaternary phase shift keying (qpsk), for both of which:

$$p_e = (1/2)\ \text{erfc}\ [E/N_o]^{1/2} \quad (51)$$

since in qpsk the symbol energy is twice the bit energy. Various forms of modulation characterized by constant envelope have been proposed to achieve a more efficient spectrum utilization and to avoid the amplitude changes occurring in bpsk and qpsk, which are objectionable in band-limited nonlinear systems. Plots of the bit error probability, p_e, as a function of E/N_o and various types of modulation appear in Fig. 18. The R/B ratio, which is a function of E/N_o, provides a measure of bandwidth utilization efficiency and, since R cannot exceed Shannon's channel capacity

$$C = B \log_2 (1 + C/N_oB)$$

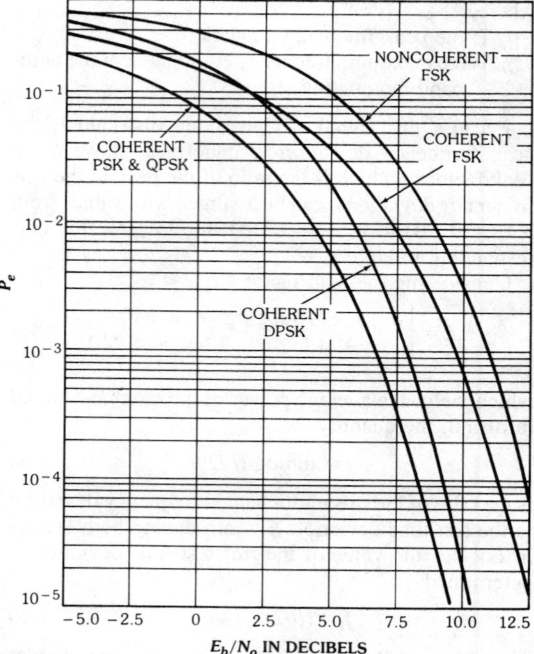

Fig. 18. Curves of BER versus E/N_o (*From Harry L. Van Trees, ed.*, Satellite Communications. *New York: IEEE Press, 1979; Fig. 12, p. 76.*)

a lower bound for E/N_o is

$$\frac{E}{N_o} \geq [\exp(0.69R/B) - 1]\ /\ (R/B)$$

In practice, since $R/C << 1$, E/N_o must be well above 0 dB. In power-limited situations, when the desired transmission rate cannot be achieved with a prescribed value of bit error probability, channel coding (forward error control) can be helpful. Error control in the channel requires the addition of redundant bits to the information bits and, hence, an increase of the overall rate and of the required bandwidth. Under certain conditions, the receiver can exploit the redundancy and reduce the error probability of the recovered message.

Various kinds of codes have been proposed and used in satellite systems; the two major distinct approaches are block and convolutional codes. Although the former may be attractive in certain cases, the latter have been found preferable because of ease of implementation and availability of efficient schemes such as Viterbi and sequential decoding.

In the power-limited case, the link budget is calculated from Eq. 46:

$$R_p = (C/N_o) \cdot (N_o/E) \quad (52)$$

which is often written in logarithmic form as

$$R_p|_{\text{dB}} = C/N_o|_{\text{dB}\cdot\text{Hz}} - E/N_o|_{\text{dB}} = \text{e.i.r.p.} + G/T - \text{free space loss} + 228.6 - E/N_o|_{\text{dB}} \quad (53)$$

For the bandwidth-limited case:

$$R_b = B \cdot \log_2 m\, R_s/B \qquad (54)$$

where,

m = number of bits/symbol,
R_s = channel symbol rate (symbols/s or bauds),

and using logarithmic units

$$R_b|_{dB} = B|_{dB} + 10 \log_{10} (\log_2 m) - B/R_s|_{dB} \qquad (55)$$

The lower bound of Eqs. 54 and 55 is the actual communications capacity.

The two operations of modulation and coding can be combined in a process known as "coded trellis signaling." By mating FEC coding to the modulation signal space, the minimum Euclidean distance between "words" can be maximized with the result of enhanced bit error rate performance versus E/N_o. In this unified approach, redundancy for FEC coding is achieved without sacrificing bandwidth by first doubling the dimension of the modulation signal space to an M-ary alphabet ($M = 2^n$). Then a convolutional code of rate $(n-1)/n$ is used to obtain the coding gain that provides better utilization of the available power.

In practice, the two functions of modulation and coding are performed separately for the sake of convenience.

DEMAND-ASSIGNED MULTIPLE ACCESS (DAMA)

In dama systems, satellite circuits are assigned to earth-station pairs upon request. Each voice channel is transmitted on a separate carrier, which is activated by the speech input. Thus, transponder power and bandwidth are more efficiently used than in preassigned systems. Modulation can be analog or digital, the latter being preferred, with either pcm or Δm voice encoding combined to bpsk or qpsk. Carriers are taken from a pool of frequencies that can be shared by the entire network of earth stations (fully variable dama) or limitedly assigned to certain destinations or origins (semivariable dama). The choice of the approach impacts on the complexity of the earth-station equipment.

An example of Dama techniques is the SPADE system of INTELSAT. The terminal shown in Fig. 19 comprises a terrestrial interface unit, several transmit/receive units, a signaling and switching unit, and an if subsystem. Through the terrestrial interface unit, calls originated in the local transmit center activate the signaling and switching unit, which selects a pair of frequencies and alerts the destination. Frequency synthesizers provide the outgoing carrier and the local-oscillator frequency for the received channel. Modulation/demodulation and a/d and d/a functions are performed by psk modems and pcm codecs. The if subsystem interfaces the earth-station up- and down-converters; it handles the outgoing and incoming modulated carriers and also the carrier for the common signaling system; single channel per carrier transmission (scpc) is used.* The common signaling system operates in a tdma broadcast mode at 128 kb/s with

* Scpc/fm/fdma transmission is also used in the INTELSAT preassigned network for small earth terminals.

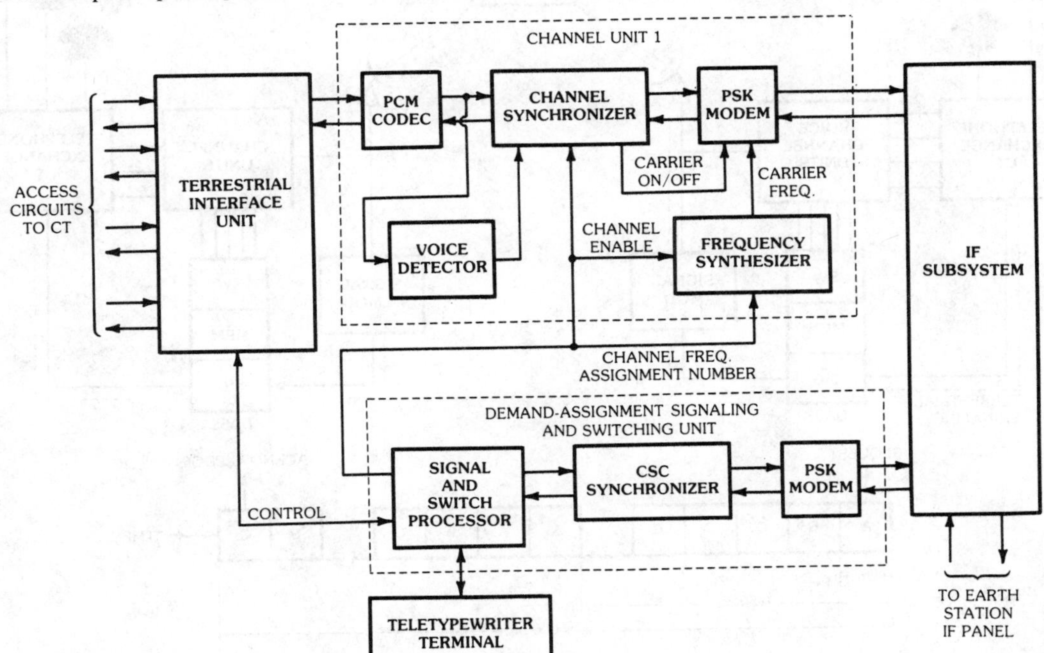

Fig. 19. SPADE terminal block diagram. (*From* COMSAT Technical Review, *Vol. 2, No. 1, p. 226 Spring 1972.*)

bpsk modulation at an error rate of 10^{-7}, which is three orders of magnitude better than that of the 64-kb/s qpsk communications channel. A variety of signaling and switching systems as used by different countries can be accommodated. The multichannel frequency allocation is shown in Fig. 20; Fig. 21 shows the signaling flow, and Fig. 22 shows the communications flow.

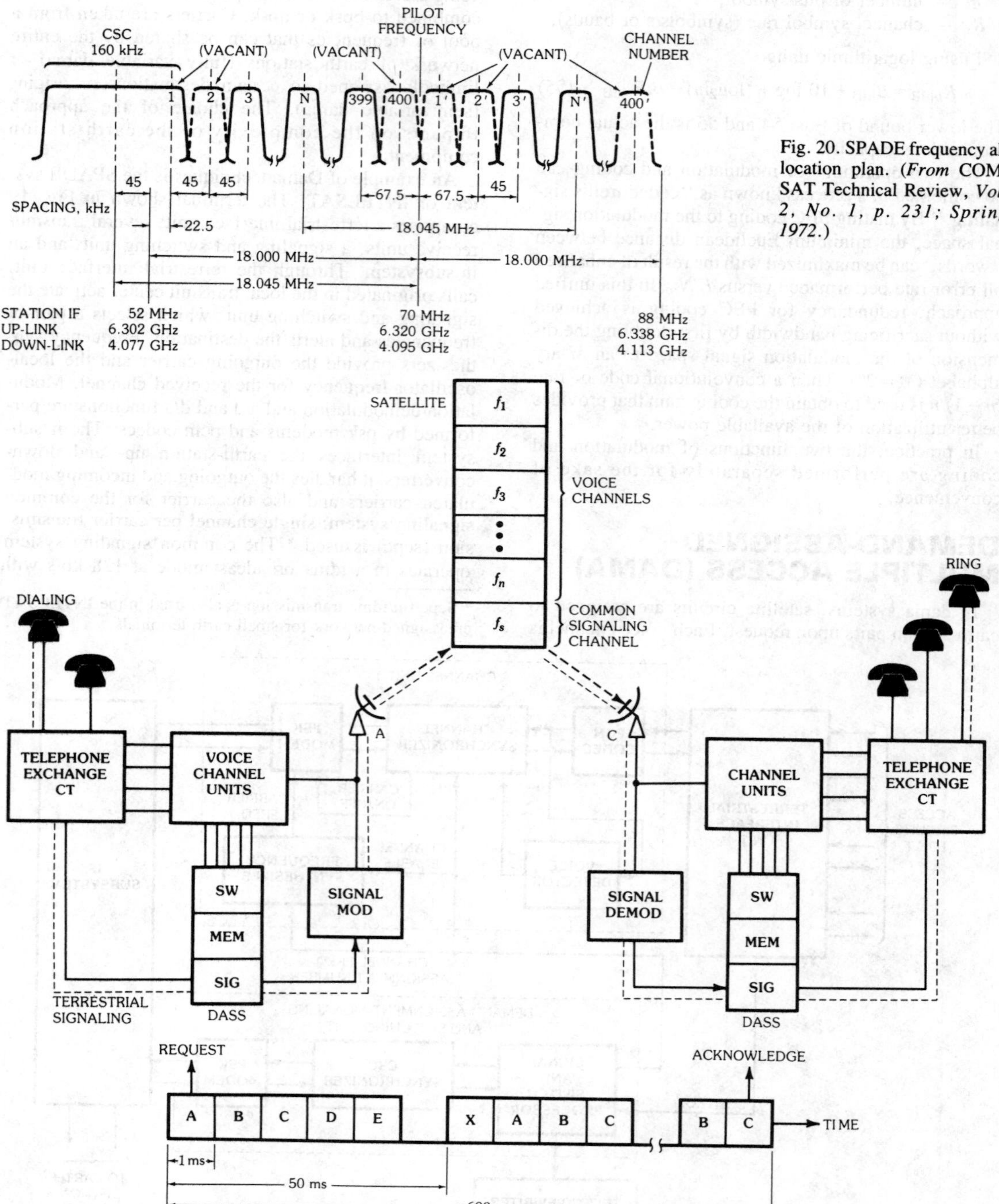

Fig. 20. SPADE frequency allocation plan. (*From* COMSAT Technical Review, *Vol. 2, No. 1, p. 231; Spring 1972.*)

Fig. 21. SPADE signaling flow. (*From* COMSAT Technical Review, *Vol. 2, No. 1, p. 229; Spring 1972.*)

The SPADE system provides 800 voice channels with an INTELSAT IV type transponder connected to a global beam antenna independently of the number of accesses. A voice activation factor of 0.4 implies that when all 800 channels are in use to provide up to 400 two-way conversations, only 320 channels are simultaneously active 90 percent of the time; thus, a power savings of 800/320 = 2.5, or 4 dB, is achieved.

TIME-DIVISION MULTIPLE ACCESS (TDMA)

In time-division multiple access (tdma), transponder power and bandwidth are shared by several earth stations. Each station transmits rf bursts at the same carrier frequency at different times; the carriers are modulated by signals coming from different sources. Because all stations are synchronized, only one burst occupies the transponder at any given time, and bursts from different stations never overlap. Power amplifiers can be operated at saturation, i.e., at maximum efficiency. Consequently, as intermodulation noise is absent, communications capacity is higher than that achievable with analog fdm/fm/fdma. The bursts are amplified by the satellite transponder and beamed down to earth to various earth stations. The down-link bursts are received and detected, and the demultiplexed signals are delivered to their destinations. Other advantages of tdma are compatibility with terrestrial digital systems and flexibility in accommodating transmission systems growth. The requirements of up-link power control, which are very stringent in fdm/fm/fdma systems, are greatly reduced in tdma systems. However, when bandwidth is only slightly greater than the signal symbol rate, the combined effect of transmit-side filtering and nonlinearities produces intersymbol interference and signal distortion that cannot be eliminated by linear filtering at the receive side. In such a case, backoff of the power amplifier is still required to keep signal degradation within limits.

Earth stations have parallel steady inputs, some already in digital form and others in analog form that need to be converted to digital format. Since each station transmits periodically in bursts at a rate much higher than the bit rate of any of the input signals, memory circuits are needed for data buffering. The burst periodicity determines the tdma frame duration; thus, for voice inputs (f_{max} = 4 kHz), the minimum value of the tdma frame duration corresponds to a Nyquist interval of 125 μs. Multiples of this minimum duration can be used, leading to increased communications capacity through tradeoffs of tdma frame duration and amounts of memory capacity. An upper bound to frame duration is determined by the transmission delay from earth station to earth station (about 270 ms for geostationary satellites at 10° elevation).

Table 3 illustrates trends of operational and planned tdma systems. Fig. 23 shows the tdma frame structure adopted in field trials conducted by INTELSAT during 1978–79. The transmission rate of 60 Mb/s (935 voice channels) achieved with pcm/qpsk via a 40-MHz transponder yields a channel density of 23 channels/MHz.

A reference burst consisting of 30 symbols for carrier and clock recovery, 10 symbols for unique word, and 3 symbols for station identification transmitted by a primary reference station (two standbys were provided) permits all other stations to transmit their bursts in a proper sequence, i.e., without mutual overlap. Transmission rate and channel density can be doubled with the use of digital speech interpolation.

The number of voice channels in a tdma system is

$$n = (1/r)\,(R - NP/T) \tag{56}$$

where,

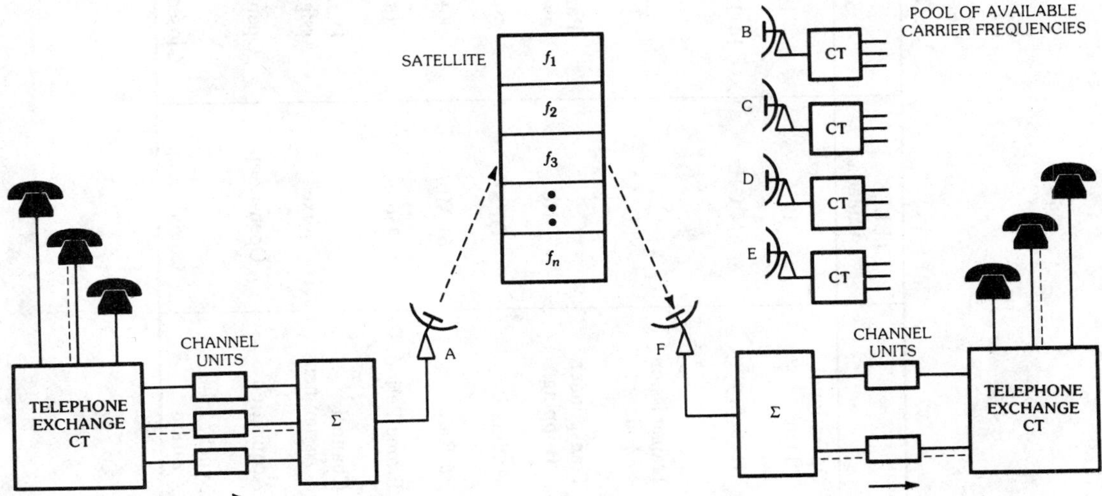

Fig. 22. SPADE communications flow. (*From* COMSAT Technical Review, *Vol. 2, No. 1, p. 229; Spring 1972.*)

TABLE 3. TDMA SYSTEMS CHARACTERISTICS

	TELESAT Canada	SBS	INTELSAT V	Advanced WESTAR	TELECOM 1	INTELSAT VI
Operational Date	1975	1981	1983	1983	1984	1986
Frequency	6/4 GHz	14/12 GHz	6/4 GHz and 14/11 GHz	14/12 GHz	14/12 GHz	6/4 GHz
Transmission Mode	TDMA	TDMA	TDMA	SS-TDMA	TDMA	SS-TDMA
Transponder Hopping	No	Parallel Down-link	Yes	No	Down-link Switching	Yes
Modulation	QPSK	QPSK	QPSK	QPSK	BPSK	QPSK
Bit Rate	61 Mb/s	48 Mb/s	121 Mb/s	250 Mb/s	25 Mb/s	121 Mb/s
Frame Period	250 μs	15 ms	2 ms	125–750 μs	20 ms	2 ms
Channel Assignment	Fixed	Demand Assignment	Fixed	Fixed	Demand Assignment	Fixed
Acquisition	Open-Loop	Open-Loop	Open-Loop	?	Open-Loop	Open-Loop
Synchronization	Closed-Loop	Closed-Loop	Feedback	?	Feedback	Feedback

r = voice channel bit rate,
R = satellite channel bit rate (power or band limited),
N = number of bursts in a frame,
P = number of digits in the preamble,
T = frame period.

Synchronization and acquisition in tdma systems can be implemented via closed- or open-loop techniques. In the closed-loop case, each earth station can monitor its own signals returned by the satellite via global or wide coverage area beam antennas. However, when narrow beam antennas (which provide greater eirp and communications capacity) are used, open-loop techniques must be employed.

Notwithstanding the loss of communications capacity encountered with fdm/fm/fdma when several accesses share a transponder, its simplicity and its "natural" interfacing with analog terrestrial networks made it almost universally the primary transmission method of the first decade of commercial satellite systems. With the growth of spacecraft available power and in view of bandwidth limitations, more complex transmissions techniques such as tdma have been used.

Fig. 17 illustrates the multiple-access performance of the three systems—fdm/fm/fdma, scpc/dama, and tdm/pcm/psk/tdma—when used with an INTELSAT IV transponder with global coverage. Clearly, bandwidth is more efficiently used by the latter two methods, which require, however, more complex equipment at the earth stations. Time division multiple access (tdma), whose potential had been identified many years ago, is now competitive with fdma and is already used operationally in certain systems.

The trend toward digital communications in terrestrial networks is a factor favoring the use of tdma in satellite systems.

SATELLITE-SWITCHED TIME-DIVISION MULTIPLE ACCESS (SS-TDMA)

Narrowing satellite antenna beamwidth increases communications capacity but decreases connectivity among coverage areas. By introducing on-board switching, connectivity can be restored. Fig. 24 illustrates the basic concept of satellite-switched time-division multiple access. Spacecraft so equipped operate not only as repeaters but also as switchboards. With time division multiplexing and digital modulation, the separation of the accesses is most effectively implemented in the time domain. It is also possible to use ss-fdma techniques, but their more complex implementation would require heavier spacecraft.

Switching can be done at rf or at baseband; in the first case, up- and down-links are interconnected via a microwave switching matrix whose output signals are amplified and eventually fed to the antennas. The switching matrix is steered by a distribution control unit (dcu), and an acquisition and synchronization unit provides the time references necessary for tdma operation. Cross-bar configurations have been designed to provide bandwidth up to 500 MHz at 4 GHz.

Fig. 25 shows a block diagram of an ss-tdma satellite carrying a 6 × 6 rf switch matrix. Data controlling the programmable cyclic switching states are transmitted via the telemetry, tracking, and command channel for

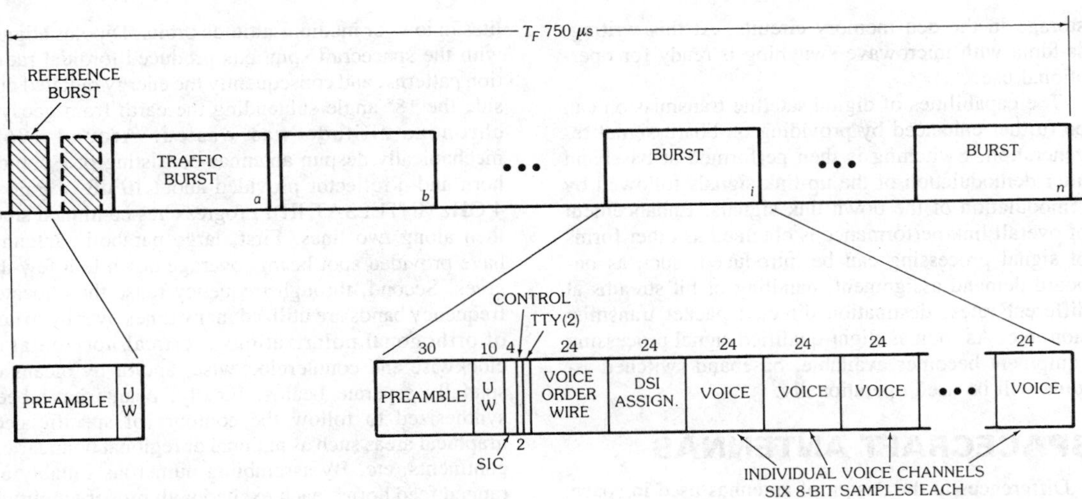

Fig. 23. Tdma frame structure.

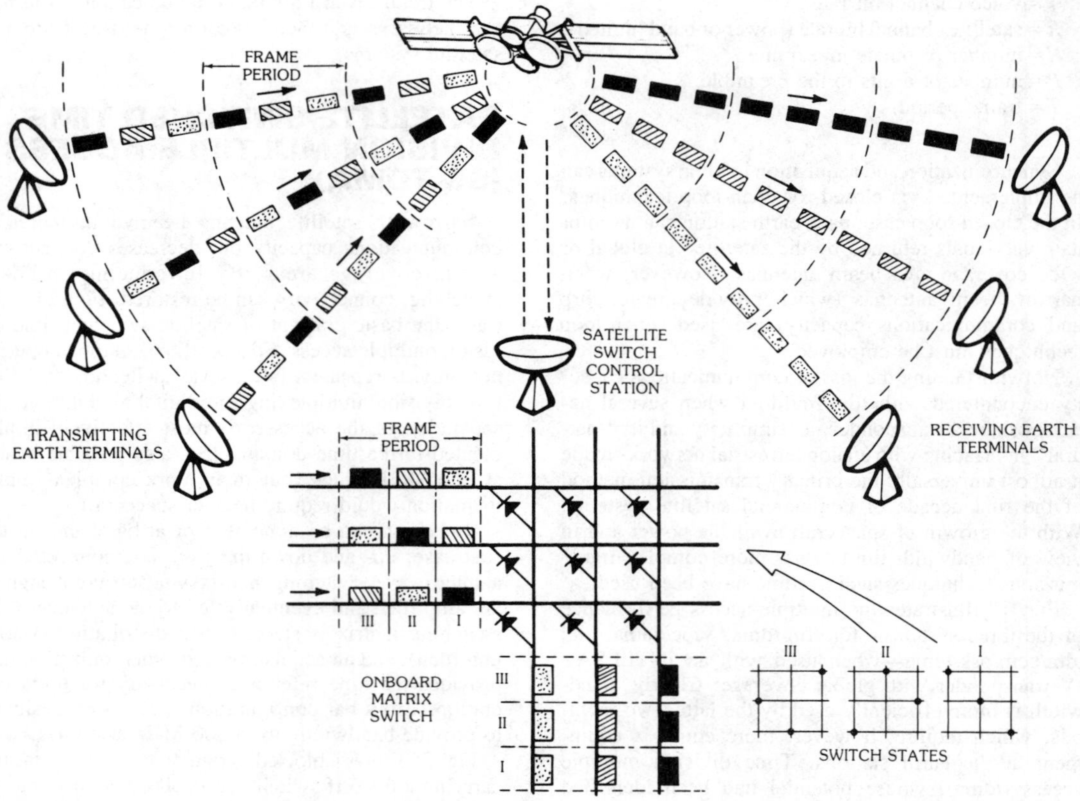

Fig. 24. Basic concept of ss-tdma.

storage in the dcu memory circuits. At this writing, ss-tdma with microwave switching is ready for operational use.

The capabilities of digital satellite transmission can be further enhanced by providing on-board signal regeneration. Switching is then performed at baseband after demodulation of the up-link signals followed by remodulation of the down-link signals. Enhancement of overall link performance is obtained as other forms of signal processing can be introduced, such as on-board demand assignment, handling of bit streams at different rates, destination directed packet transmission, etc. As soon as flight-qualified signal processing equipment becomes available, baseband switched ss-tdma will be used operationally.

SPACECRAFT ANTENNAS

Differences in the design of antennas used in space and on earth were quite marked in early communications satellite systems. Large apertures and very low noise system temperatures were then necessary at the earth stations because satellites were very much power-limited. Satellite antennas were small and had little or no directivity; this was especially the case with satellites in low- or medium-altitude orbits. Dipoles aligned with the spacecraft spin axis produced toroidal radiation patterns, and consequently the energy radiated outside the 18° angle subtending the earth from geosynchronous altitude was wasted. Around 1968, mechanically despun antennas consisting of a conical horn and a reflector provided about 19 dB of gain at 4 GHz (INTELSAT III). Progress has continued since then along two lines. First, large parabolic antennas have provided spot beam coverage down to a few degrees. Second, through frequency reuse the allocated frequency bands are utilized many times over by means of orthogonal polarizations (vertical/horizontal or clockwise and counterclockwise) and/or by means of spatially separate beams. Finally, beams have been synthesized to follow the contours of specific geographical areas such as national or regional boundaries, continents, etc. By assembling numerous suitably arranged feed horns, each excited with proper amplitude and phase, the radiated energy impinges upon a reflector and illuminates the desired areas on earth.

In the design of contour-shaped beams and frequency-reuse systems, it is difficult to maintain isolation between dually polarized and/or spatially separate beams, especially when the number of beams is

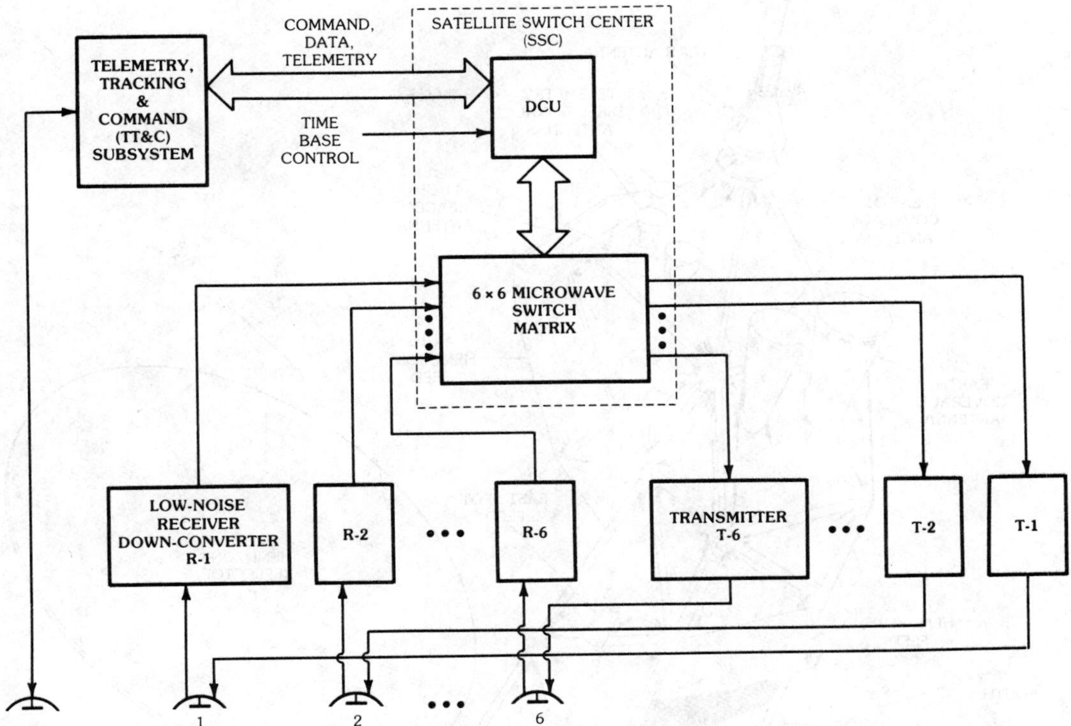

Fig. 25. Simplified block diagram of an ss-tdma satellite.

large. Until recently, parabolic reflectors with offset feed assemblies in the focal region have been adequate. The major role played by the antenna subsystem is exemplified by the design shown in Fig. 26 (INTELSAT V) with the coverage areas in the Atlantic Ocean Region shown in Fig. 27. In addition to beacon and tt&c antennas, circularly polarized conical horns provide global coverage at 6 GHz (receive) and 4 GHz (transmit). The 18° transmit horn is steerable up to ±2°; the receive horn, which has a wider beamwidth (22°), is fixed. Two parabolic reflectors of 2.44- and 1.54-m diameter provide hemispherical and zone area coverage for reception and transmission at 6 and 4 GHz. The offset hemi/zone antenna feeds are clusters of square horns with excitation in amplitude and phase producing the required shaped beams. Hemispherical and zone coverage are simultaneously obtained with opposite-sense circular polarizations. Feed characteristics, reflector size, and focal length are selected to yield high gain, sharp beam edges, low side lobes, and the high polarization purity required to achieve isolation between beams up to 27 dB.

Nominal 1-m diameter mechanically steerable parabolic receive and transmit antennas with linear orthogonal polarizations are used at 11/14 GHz to provide circular 1.6° beams in the west spot area and elliptical 1.8° by 3.2° beams in the east spot area. Spatial discrimination at 6/4 and 14/11 GHz and polarization discrimination at 6/4 GHz provide a fourfold frequency reuse of portions of the allocated spectrum and a total maximum usable bandwidth of 2137 MHz. Thus, it is clear that communications capacity can be effectively increased by adding sophistication to the antenna subsystem.

As more beams (i.e., greater amounts of frequency reuse) and better isolation among beams will be required, two approaches appear possible. Since feed blockage limits the use of single offset reflectors, dual offset reflectors with Gregorian or Cassegrain feeds could be used to allow better control of the illumination and permit the use of larger reflectors without incurring excessive focal lengths. Lens-type antennas or combinations of lenses and reflectors or phased arrays and reflectors can also be used. Ultimately, size and mass limitations are encountered. The launch vehicle payload volume geometry limits the maximum diameter of fixed antennas. This limitation can be circumvented by using deployable apertures, although structural problems arise in regard to surface tolerances.

PROPAGATION

Electromagnetic waves traveling along an earth-space path encounter four distinct regions: the troposphere, inner free space, the ionosphere, and outer free space. Free space is lossless and has unity refractive

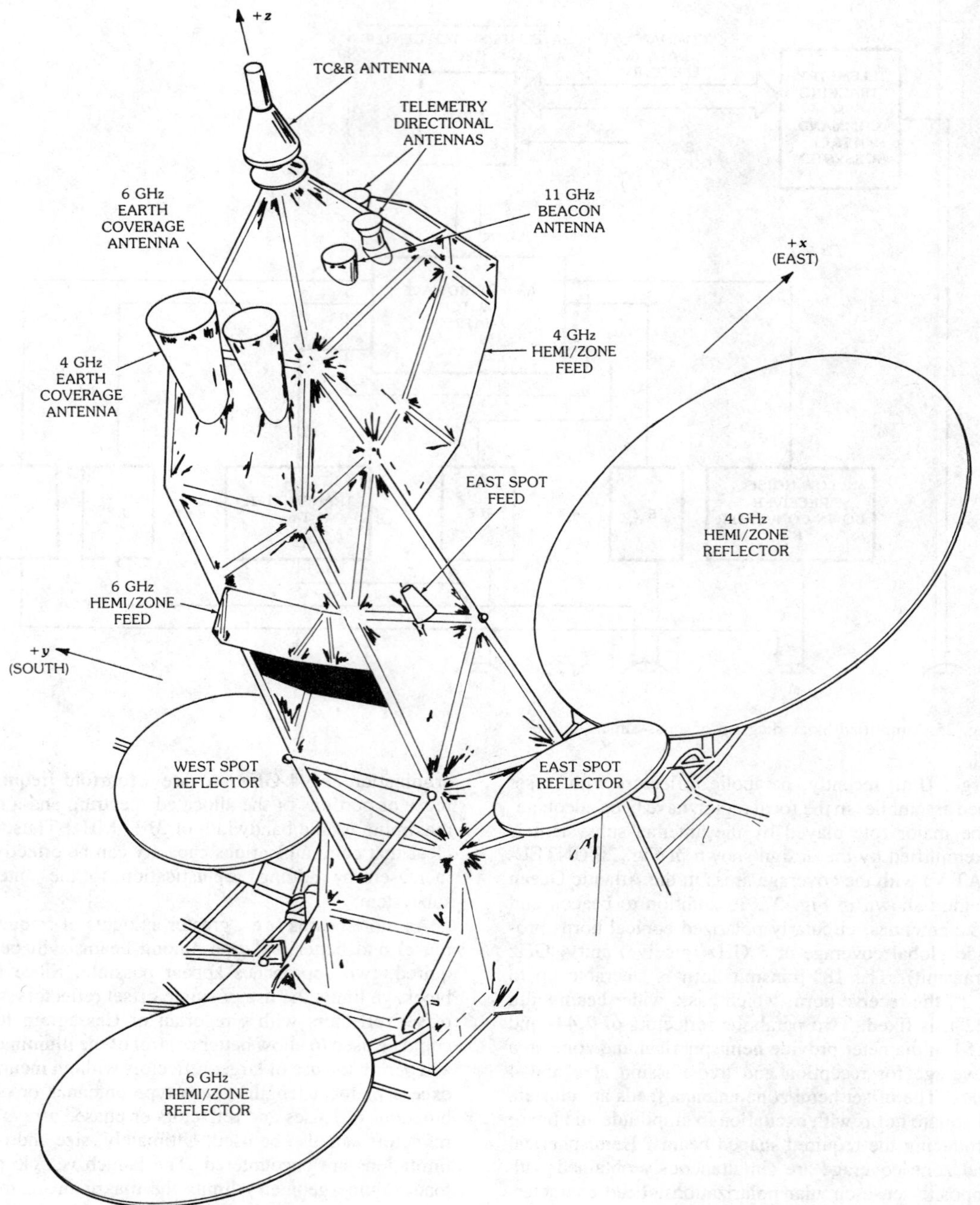

Fig. 26. INTELSAT V antenna subsystem. (*Courtesy Ford Aerospace and Communications Corp.*)

index, and the troposphere and the ionosphere have refractive indices respectively greater and smaller than unity; hence, refraction and absorption phenomena arise. In addition, the ionosphere, a magnetoionic medium, induces Faraday rotation. All these phenomena affect space and satellite communications systems. Propagation delays resulting from ray bending influence the performance of satellite navigation systems.

Faraday Rotation

A linearly polarized wave can be regarded as the sum of two counterrotating circularly polarized waves.

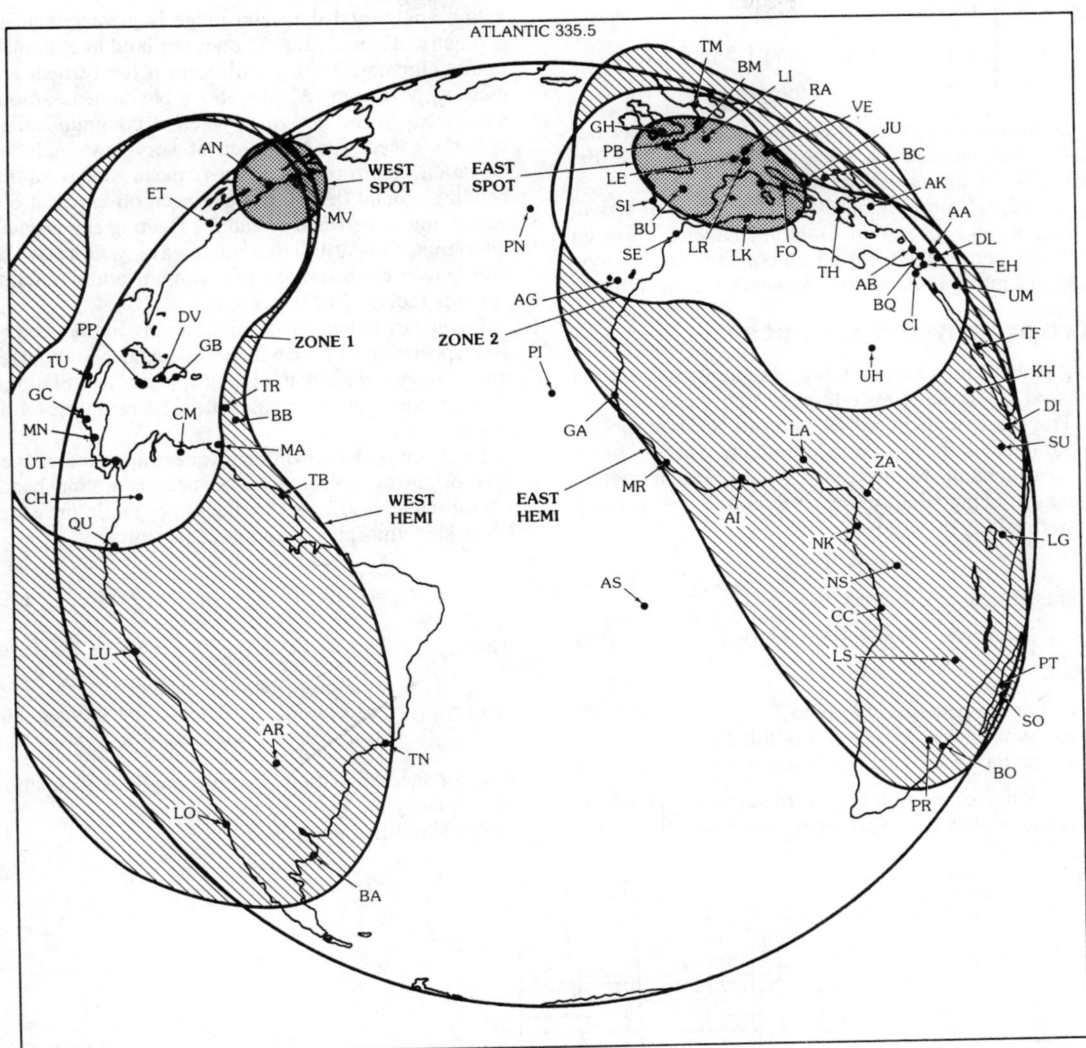

Fig. 27. INTELSAT V coverage of the Atlantic Ocean Region. (*From* COMSAT Technical Review, *Vol. 7, No. 1, p. 314; Spring 1977*)

Because the phase velocities of the two waves differ in a magnetoionic medium, the polarization plane rotates. Faraday rotation effects are negligible above 10 GHz; below 1 GHz they can be circumvented by using circular polarization. Faraday rotation can, however, degrade cross-polarization discrimination in frequency-reuse systems.

Ionospheric Scintillations

Strong short-term (1 to 15 s) variations of the amplitude, phase, polarization angle, and angle of arrival of electromagnetic waves reaching the earth from space observed up to microwave frequencies are known under the general name of ionospheric scintillations. These are attributable to fluctuations of the electron density in the sporadic E layer and to the spreading of the F layer. Earth-station location, season of the year, local time, and amount of solar activity influence the magnitude and occurrence of ionospheric scintillations, geomagnetic latitude and local time being the most influential factors.

Within $\pm 20°$ from the geomagnetic equator, heavy scintillations occur for a few hours before midnight with magnitude proportional to the number of sunspots and maxima around the equinoxes.

The scintillation index, S_4, defined as

$$S_4 = (1/R^2)\ \{(R^2 - \overline{R^2})^2\}^{1/2} \qquad (57)$$

has an $f^{-\alpha}$ frequency dependence with

$$\alpha = \begin{cases} 1/2 \text{ to } -1 & \text{at vhf} \\ -1 & \text{from 1.5 to 4.0 GHz} \\ -2 & \text{above 4 GHz} \end{cases}$$

Scintillations become less intense at intermediate latitudes but increase at higher latitudes beyond the boundary of the auroral region.

Additional ionospheric effects are angle-of-arrival variations and absorption. Both phenomena follow an f^{-2} law; hence, their impact on communications systems operating at gigahertz frequencies is negligible.

Tropospheric Effects

Ray bending, scintillation, attenuation, and increased sky noise temperature are major effects.

Tropospheric bending is opposite to ionospheric bending; it is frequency independent, whereas ionospheric bending follows an f^{-2} law. The refractive index, n, being slightly greater than one, the quantity N (refractivity) is used

$$N=(n-1)\times 10^6$$

It is empirically given as

$$N=(77.6/T)\,[p+4810\epsilon/T] \tag{58}$$

where,

T = air temperature in kelvins,
p = atmospheric pressure in millibars,
ϵ = partial water vapor pressure in millibars.

Since N decreases with height, the apparent elevation of a space object is greater than the geometrical elevation angle, and the radio range is greater than the geometric distance. Eq. 58 does not hold in extremely humid climates; ducting will occur if the vertical gradient of N is high. All the above phenomena are not significant in the design of satellite communications systems except for operation at very low angles of elevation. In tropical climates, mean values of ray bending around 0.5°–0.6° have been observed at elevation angles between 1° and 2°. Ducting can produce interference to earth-based microwave radio relays by high-power earth stations of communications satellite systems radiating at low angles.

Frequency independent scintillations are induced by atmospheric turbulence with time dependence about ten times greater than that of ionospheric scintillations. The phenomenon is strongly dependent on elevation angle.

Under clear sky conditions, water vapor and oxygen give origin to molecular resonance absorption bands whose width is affected by the atmospheric pressure. Clear sky atmospheric loss is of the form

$$L_a=\int_0^l \{\gamma_{O_2}(r,f)+\gamma_{H_2O}(r,f)\}dr \tag{59}$$

where,

l = total propagation distance,
γ_{O_2}, γ_{H_2O} = absorption coefficients for oxygen and water vapor in dB/km.

Plots of the attenuation per unit length as a function of frequency are given in Fig. 28.

The simplified expression

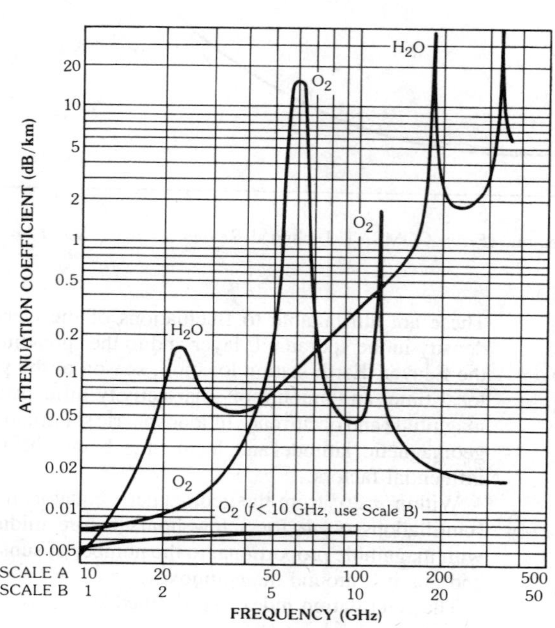

Fig. 28. Clear sky attenuation per unit length versus frequency. (*From K. Miya, ed.,* Satellite Communications Technology. *Tokyo: KDD Engineering and Consulting, Inc., 1982.*)

$$L_a = \gamma'_{O_2} L_{O_2} + \gamma'_{H_2O} L_{H_2O} \quad \text{in dB} \qquad (60)$$

may be used for horizontal paths, assuming that the absorption coefficients, γ', at the surface and the effective path length, L, for oxygen and water vapor are known. It is generally assumed that

$$L_{O_2} \cong 4 \text{ km}$$

$$L_{H_2O} \cong 2 \text{ km}$$

For a vertical path, total clear sky loss versus frequency is shown in Fig. 29. For slant paths, a factor about equal to the cosecant of the elevation angle must be introduced.

Hydrometeors

On account of the basic interaction of electromagnetic waves with water in liquid form, raindrops cause absorption, scattering, and depolarization phenomena. The first two result in signal attenuation (not to be confused with either water-vapor attenuation or attenuation by rain clouds) and increase in sky noise temperature. These effects are quite noticeable above 10 GHz. Depolarization is of little or no harm to communications systems using only one polarization, but it worsens the performance of dual-polarization systems and may also result in interference between systems. Rain may also limit signaling bandwidth.

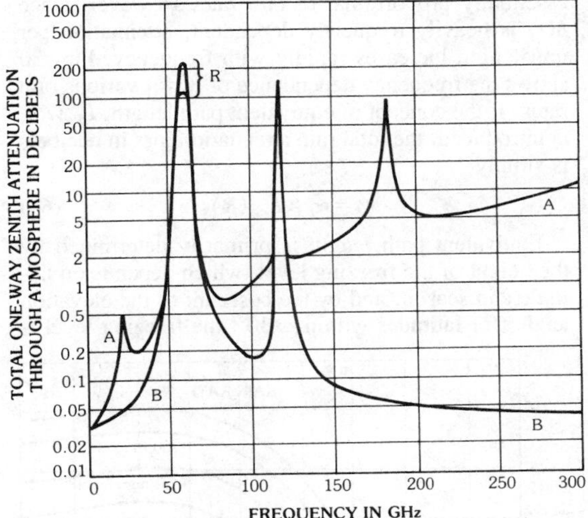

Fig. 29. Total clear sky loss versus frequency. (*From K. Miya, ed.,* Satellite Communications Technology. *Tokyo: KDD Engineering and Consulting, Inc., 1982.*)

Clear Sky Noise Temperature

In the absence of hydrometeors, the absorption by O_2 and H_2O molecules contributes to clear sky noise temperature T_{cs}

$$T_{cs} = T_{gal}/L_{cs} + [(L_{cs} - 1)/L_{cs}]T_m \qquad (61)$$

where,

$T_{gal} \cong 26/f_{GHz}$ kelvins,

$L_{cs} = 10^{L_a/10}$, clear sky loss factor,

$T_m \cong 1.12\, T_{surface} - 50$, mean raindrop temperature along the path.

The dependence of clear sky noise temperature on frequency and elevation angle is shown in Fig. 6.

Above 10 GHz, the term T_{gal}/L_{cs}, which represents the galactic noise contribution after passage through the clear sky, can be neglected. Then

$$T_{cs} \cong [(L_{cs} - 1)/L_{cs}]\, T_m \qquad (62)$$
$$= T_m(1 - 10^{-L_a/10})$$

Rain Attenuation

The amount of attenuation depends on the type of rain (stratiform, convective, cyclonic) and intensity (rain rate). As raindrop size distribution is a function of rain type and intensity, various models have been proposed to best fit specific situations. Rainfall data are available for most parts of the world; eight different types of climates have been defined and boundaries of their existence regions identified. Cumulative rain statistics provide information about probability of exceedance, i.e., the total time that a specific rain rate will be exceeded over a sufficiently long observation period. Cumulative statistics do not provide information about the frequency of occurrence and the duration of the periods of exceedance.

The attenuation per unit length (specific attenuation), α_r (dB/km), is tied to the rain rate, R (mm/hr), by the empirically derived relationship

$$\alpha_r = a(f)R^{b(f)} \qquad (63)$$

where $a(f)$ and $b(f)$ are frequency dependent coefficients. The approximate analytic expressions given in Table 4 are adequate for engineering use. Up to 50

TABLE 4. PROPAGATION COEFFICIENTS

Frequency f(GHz)	$a(f)$	$b(f)$
8.5 – 25	$4.21 \times 10^{-5}(f)^{2.42}$	$1.41(f)^{-0.0779}$
25 – 54	$4.21 \times 10^{-5}(f)^{2.42}$	$2.63(f)^{-0.272}$
54 – 100	$4.09 \times 10^{-2}(f)^{0.699}$	$2.63(f)^{-0.272}$

GHz, $b(f)$ is near unity; hence specific attenuation is essentially proportional to rain rate. However, since $a(f)$ is heavily frequency dependent, attenuation per unit length increases rapidly with frequency. Fig. 30 shows the frequency dependence of α_r for various rain rates. If the concept of equivalent path length, $L_{eq}(R)$, is introduced, the total rain attenuation loss in decibels is simply

$$A_r = \alpha_r \times L_{eq}(R) \tag{64}$$

Equivalent path length is primarily determined by the height of the freezing level, which depends on latitude and season, and by the cosecant of the elevation angle (for latitudes within $\pm 30°$, the freezing level is at 4.8 km). A correction coefficient is required to take into account the distribution of rain over a long path; however, except for very light or very heavy rain, the value of the correction is near unity. Equivalent path length can be expressed in terms of rain rate and elevation angle θ as follows:

$$L_{eq}(R,\theta) = [7.413 \times 10^{-3} R^{0.766} + (0.232 - 1.803 \times 10^{-4} R) \sin \theta]^{-1} \tag{65}$$

Curves of equivalent path lengths versus elevation angle and for different rain rates are shown in Fig. 31.

Rain rate can be modeled by the Rice-Holmberg distribution

$$P(R) = ae^{-0.03R} + be^{-0.258R} + ce^{-1.63R} \tag{66}$$

with

$$a = M\beta/2922$$

$$b = M(1-\beta)/438.3$$

$$c = 1.86\beta$$

where,

M = total mean yearly rainfall in millimeters,
β = ratio of thunderstorm rain accumulation to total accumulation.

In Eq. 66, each term represents the predominant component for specific ranges of R.

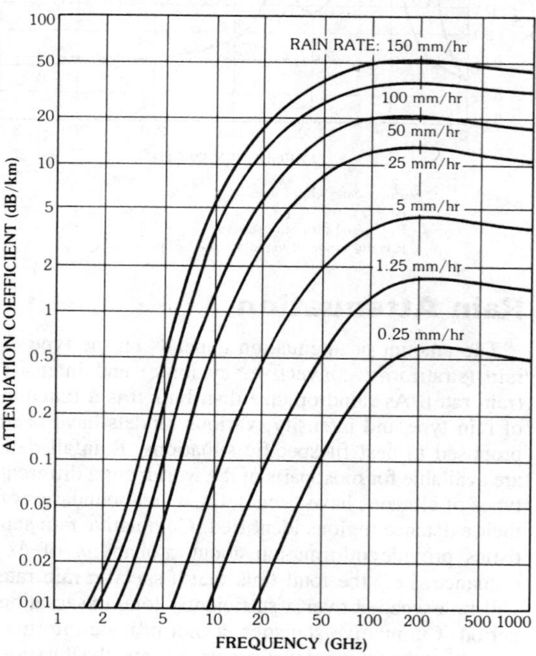

Raindrop size distribution: Laws and Parsons, 1943
Terminal velocity of raindrops: Gunn and Kinzer, 1949
Dielectric constant of water at 20°C: Ray, 1972

Fig. 30. Attenuation per unit length versus frequency and rain rate. (*From K. Miya, ed.,* Satellite Communications Technology. *Tokyo: KDD Engineering and Consulting, Inc., 1982.*)

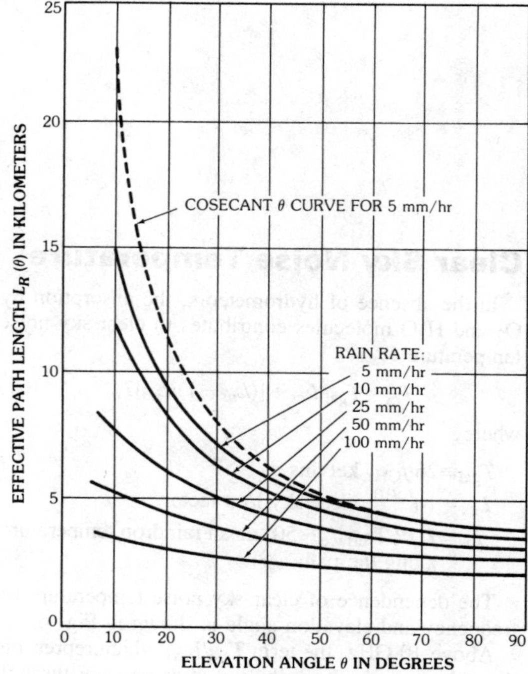

Fig. 31. Equivalent path length versus rain rate and elevation angle. (*From K. Miya, ed.,* Satellite Communications Technology. *Tokyo: KDD Engineering and Consulting, Inc., 1982.*)

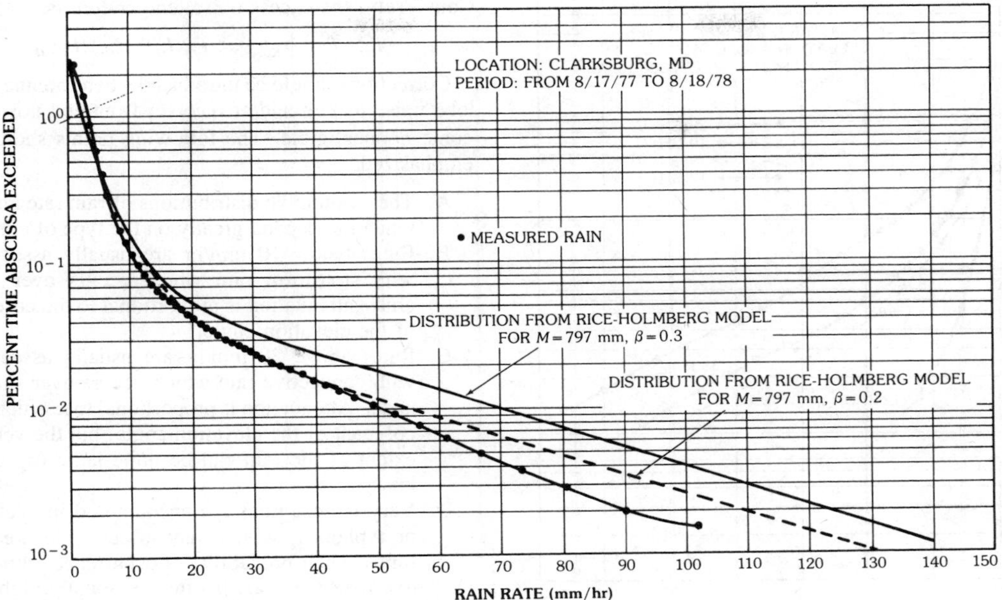

Fig. 32. Comparison of measured surface rain rate and theoretical model distributions. (*From* COMSAT Technical Review. *Vol. 12, No. 1, p. 7; Spring 1982.*)

Comparative data of measured surface rain rate and theoretical model distributions are shown in Fig. 32. Cumulative attenuation statistics are shown in Figs. 33, 34, and 35.

The comparison of rain attenuation at different frequencies leads to a scaling relationship of the form

$$A_r(f_2) = A_r(f_1)\ (f_2/f_1)^{\alpha} \tag{67}$$

with $1.7 < \alpha < 2.0$.

Water-cloud attenuation is about the same as that produced by rainfall less than 5 mm/hr. Ice-cloud attenuation is at least two orders of magnitude lower and hence negligible.

Sky Noise Temperature With Rain

Neglecting the galactic contribution, the sky noise temperature in the presence of rain, T_s, is:

$$T_s = T_{cs}/L_r + [(L_r - 1)/L_r]T_m \tag{68}$$

If the first term, which represents the contribution of the clear sky through the lossy rainy medium, is negligible compared to the second term, one can write

$$T_s \cong [(L_r - 1)/L_r]T_m = T_m(1 - 10^{-A_r/10}) \tag{69}$$

with all symbols as previously defined. Then the decibel difference (excess loss) between the attenuation due to rain and clear sky attenuation is

$$(A_r - A_a)\mid_{\mathrm{dB}} = 10 \log [(T_m - T_{cs})/(T_m - T_s)] \tag{70}$$

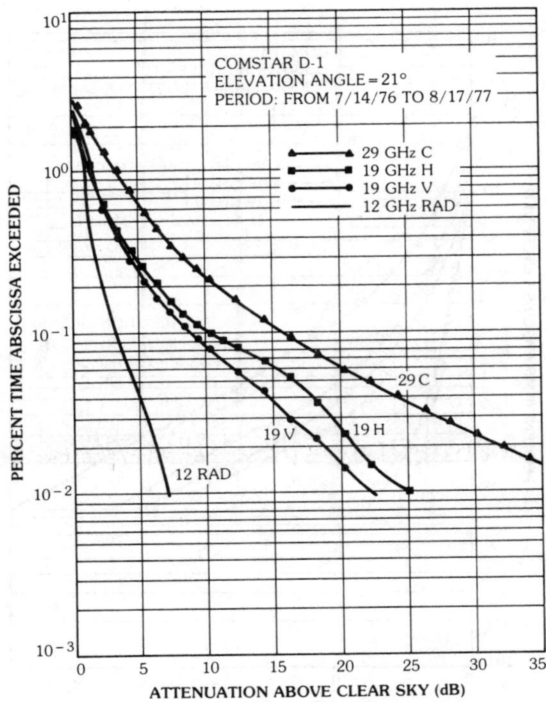

Fig. 33. Cumulative attenuation statistics (COMSTAR D-1). (*From* COMSAT Technical Review, *Vol. 12, No. 1, p. 10; Spring 1982.*)

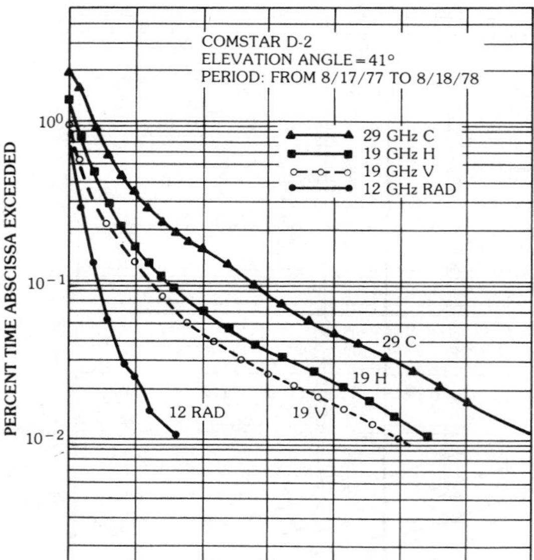

Fig. 34. Cumulative attenuation statistics (COMSTAR D-2). (*From* COMSAT Technical Review, *Vol. 12, No. 1, p. 12; Spring 1982.*)

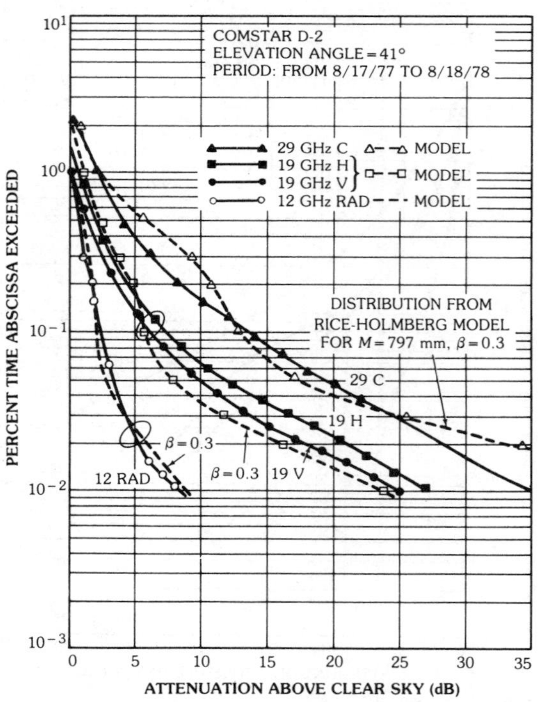

Fig. 35. Comparison of measured attenuation with theoretical distributions. (*From* COMSAT Technical Review, *Vol. 12, No. 1, p. 13; Spring 1982.*)

Conversely, the excess noise temperature is:

$$(T_s - T_{cs}) \mid_{\text{kelvin}} = T_m(L_r - L_a)/L_r L_a \qquad (71)$$

Corrections should be introduced when antenna sidelobe noise pickup and/or receiver front-end noise become nonnegligible. The following points should be emphasized:

A. The cumulative distributions of rain rate and attenuation depend greatly on the type of climate.
B. Rates below 10 mm/hr are usually associated with stratiform rain which occurs over wide areas; attenuation is proportional to the cosecant of the elevation angle.
C. Rates above 20 mm/hr are usually associated with convective rain which occurs over smaller areas. Attenuation is proportional to less than the cosecant of the elevation angle, but the vertical extent of the rain can be quite large (up to 10 km).
D. Very intense rates are encountered in cyclones or typhoons, which may affect wide areas at times. Their probability of occurrence, although low over the year, is often seasonally high.
E. Rain attenuation that is not a serious obstacle below 10 GHz becomes important at the higher frequencies. Up to 11–14 GHz, continuity of service can be maintained by allowing adequate power margins (3–10 dB) in the link design. Higher margins are either impractical or uneconomical.

Thus, especially at the higher frequencies (20/30 GHz) at which rain-induced attenuation can be quite severe, unless a lower value of service continuity is acceptable, means other than power margins may be needed.

Diversity

The most commonly proposed alternate solution is site diversity based on having two earth stations jointly operating with a given satellite. By placing the two stations at a distance for which the rain statistics along the two paths become sufficiently uncorrelated, and selecting at any time the best path, diversity gain over single-site operation can be obtained. Orientation of the baseline of the two sites with respect to the prevalent direction of weather fronts is also important. Other forms of diversity such as angle or time/bandwidth can be envisaged, but site diversity remains the most common technique.*

With reference to Fig. 36, once the attenuation statistics for a single and a double path have been acquired, diversity gain G_{div} can be defined as

$$G_{\text{div}} = A_{r_1} - A_{r_{1,2}} \mid_p \qquad (72)$$

when the single and double subscripts refer to the same

*Another solution is frequency diversity, which implies shifting traffic to frequency bands not affected by rain.

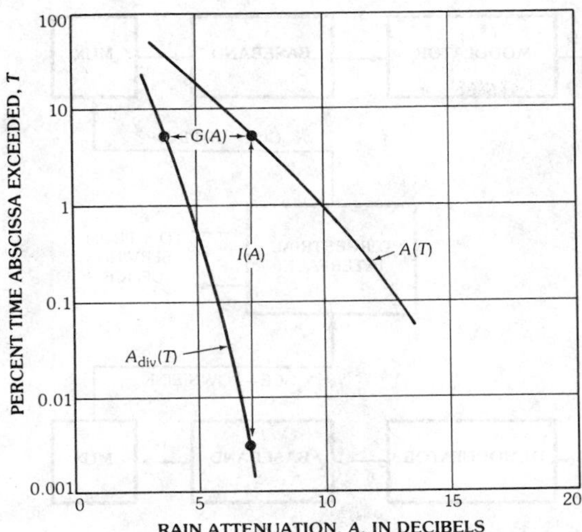

Fig. 36. Diversity and advantage gain definitions. (*From L. J. Ippolito, R. D. Kaul, and R. G. Wallace,* Propagation Effects Handbook for Satellite Systems Design. *NASA Reference Publication 1082, Dec. 1981.*)

value of single and joint event probabilities. Otherwise, for a given value of the abscissa A_r, the ratio

$$I_{div} = p_1/p_{1,2} \mid A_r \qquad (73)$$

defines diversity advantage, i.e., the factor by which exceedance time is improved.

Experiments have shown that both diversity gain and advantage increase with increasing spacing of the two sites, but at a decreasing rate until leveling off occurs at the distance for which the path statistics become independent.

Depolarization

While very small raindrops such as those encountered in light drizzle are spherical, the larger drops associated with heavier rain depart substantially from spherical shape under the combined effects of gravity, hydrostatic forces, and aerodynamic forces. Nonspherical drops and their canting angle cause differential attenuation and phase shift for different wave components and hence depolarization.

For linear- or circular-polarized waves, the power ratio of the copolarized and cross-polarized components

$$XPD = 10 \log_{10} (P_{copol}/P_{xpol}) \qquad (74)$$

gives a measure of polarization purity or cross-polarization discrimination. Once differential attenuation A and differential phase shift are determined, cross-polarization discrimination can be computed. For circular polarization, it is

$$XPD = 20 \log_{10} \mid (1 + e^{A+jB})/(1 - e^{A-jB}) \mid \qquad (75)$$

and for linear polarization

$$XPD = 20 \log_{10} \mid (1 + \tan^2\chi\, e^{A+jB})/(1 - e^{A+jB} \tan\chi) \mid \qquad (76)$$

where χ is the angle between the incident wave polarization plane and the major axis plane of the raindrops. With linear polarization, when $\chi = 0°$ or $90°$ XPD becomes infinite; for $\chi = \pm 45°$ XPD takes a minimum value. With circular polarization, XPD is always at a minimum.

Proper operation of frequency reuse systems with dual polarization requires XPD of 30 dB or more. As a residual coupling usually originating in the antennas and feeds always exists, the presence of rain worsens the situation and reduces further the value of XPD. Depolarization compensation networks can be used at the earth stations to restore orthogonality.

EARTH STATIONS

At the earth stations that form the ground segment of space and satellite communications systems, a variety of equipment is needed depending on: (A) function of the station, (B) type of service, (C) frequency bands used, (D) transmitter, (E) receiver, and (F) antenna characteristics.

Three categories can be distinguished: (1) transmit and receive, (2) receive only, and (3) transmit only stations. The first is encountered in two-way communications systems. Receive-only stations are presently used in catv systems and will be used in great numbers in a few years in direct tv broadcast satellite systems. Transmit-only earth stations are found in data collection systems. Types of service include fixed, mobile (maritime, aeronautical, and terrestrial), broadcasting, and others. Spectrum use is regulated by international allocations with actual bandwidth occupancy as needed for different services, type of traffic, and modulation.

In satellite communications systems, earth-station transmitter power ranges from a few tens of watts to ten to twelve kilowatts generated by vacuum tubes (klystrons or traveling-wave tubes). Transmitters up to 400 kW are used in the deep-space network. Receivers cover a wide range of sensitivity and bandwidth with noise temperature from a few tens of kelvins achieved with cryogenically cooled amplifiers to hundreds of kelvins in uncooled amplifiers. A highly simplified block diagram of a typical transmit and receive earth-station layout is shown in Fig. 37. The station has six major subsystems:

A. The power subsystem
B. The terrestrial interface
C. The transmit chain
D. The receive chain
E. The antenna subsystem
F. The control subsystem

Signal paths and functions are as indicated. The value of the intermediate frequency (if) in the receive and transmit chains is usually the same, 70 MHz being

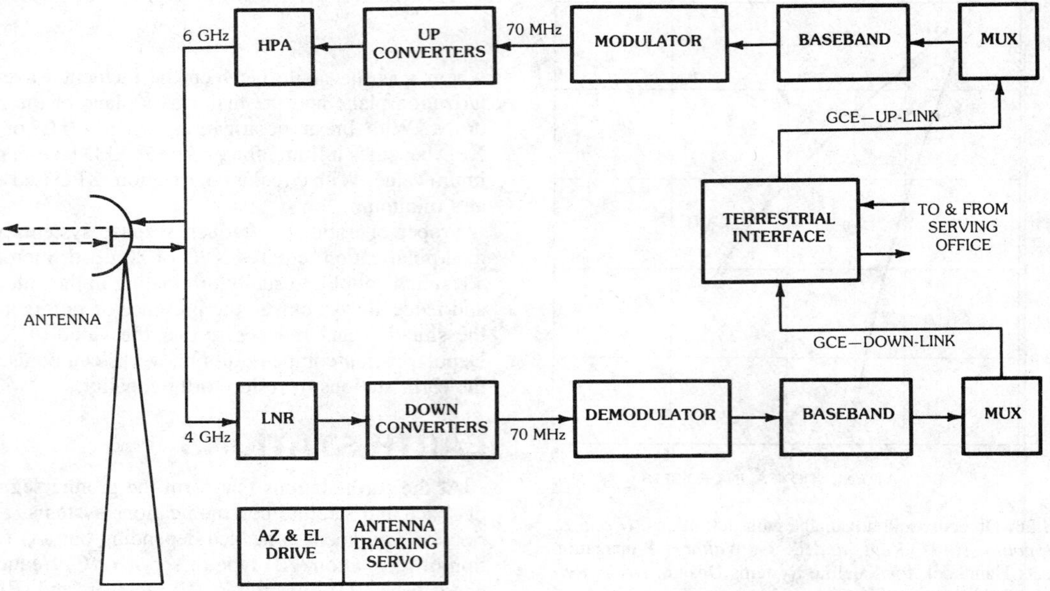

Fig. 37. Earth-station block diagram.

commonly used. The up-link signals may be transmitted by means of waveguides to high-power amplifiers located in the upper antenna room, or in if form via coaxial cables to the up-converters and high-power amplifiers in the upper antenna room. It is also possible to locate all the equipment at the base of the antenna and to extend the input and output feeds via cascaded reflectors. High-power-amplifier (HPA) outputs can be combined through bandpass filters and circulators or by means of hybrids. In the receive chain, the weak signals from the satellite are accepted by the same feed that carries the transmitter output. These two signals, which differ in power by several orders of magnitude, are kept separate in the frequency domain as they are assigned to the up-link and down-link bands, and in addition by means of orthogonal polarization. Diplexers can be used to enhance the separation in the frequency domain. Orthomode transducers and polarizers are employed, respectively, to couple orthogonally polarized signals into a single waveguide and to convert linear into circular polarization and vice versa. After preamplification, the received signals are down-converted to if and demodulated to baseband.

The antenna is a very important part of an earth station, since it enhances eirp, receive sensitivity, and interference. Antenna size varies considerably: Parabolic reflectors with a diameter of 1 meter or less are to be used in receive-only earth stations of forthcoming broadcast satellite systems that will provide tv programs direct to private homes. Much larger transmit and receive antennas (up to 32-meter diameter parabolas) are used in high-capacity systems, and even larger antennas (up to 64-meter diameter) are used in the deep-space network. In addition to parabolic reflectors, which are widely used, other forms of antennas such as horns, torus-shaped reflectors, Yagis, helices, and phased arrays have been proposed or used.

In terms of the general relationship,

$$G=(4\pi/\lambda^2)A_{\text{eff}}=(4\pi/\lambda^2)\cdot\eta\cdot A_{\text{geom}} \qquad (77)$$

where

$$\eta=A_{\text{eff}}/A_{\text{geom}} \qquad (78)$$

is the antenna efficiency. In the case of parabolic reflectors,

$$G=\pi^2(D/\lambda)^2\cdot\eta \qquad (79)$$

In practice, values of D/λ ranging from about 20 to 700 are common in communications satellite systems, and values up to 2000 are encountered in deep space systems. With efficiency in the range from 0.5 to 0.8, gains between 30 and 75 dBi have been achieved with corresponding half-power beamdwidths from a few degrees to a few hundredths of a degree. Very high gain values imply electrically as well as mechanically large antennas and consequently large and costly structure because cost is approximately proportional to aperture area.

Various forms of mounts have been used with limited or full steerability as required by the specific system characteristics. Antenna-mount types can be 2-axes (x, y, or azimuth, elevation), 3-axes (x, y, y', or azimuth, elevation, and cross-elevation), or 4-axes for maritime applications.

Reflector antennas can be classified according to the number of reflecting surfaces and/or the type of feed positioning. The overall antenna efficiency factor, η, in Eq. 78 can be broken into four parts:

$$\eta = \eta_1 \times \eta_2 \times \eta_3 \times \eta_4 \tag{80}$$

where,

η_1 = illumination factor,
η_2 = spillover factor,
η_3 = blockage factor,
η_4 = surface tolerances factor.

A decrease in main-lobe gain usually brings in an increase of side-lobe energy and consequently a higher noise temperature and a decrease of interference-rejection capability. The term η_4 is due to departure of the reflector surface from the ideal. Assuming randomly distributed profile errors over the surface, efficiency is reduced by a factor

$$\eta' = e^{-(4\pi\epsilon/\lambda)^2} \tag{81}$$

where ϵ is the rms surface tolerance. As the gain loss in decibels is

$$\eta'|_{dB} = 685.8\ (\epsilon/\lambda)^2 \tag{82}$$

if the maximum permissible gain loss is set at 1 dB, the rms surface tolerance, ϵ, must be down to $\lambda/25$ or less. The ratio D/ϵ represents a measure of the quality of a parabolic antenna. For the above-mentioned values of the D/λ ratio, the corresponding range of the D/ϵ ratio is between 500 and 25 000. In electrically larger antennas intended for radio astronomy, the higher D/ϵ values have been achieved with protective means against the environment. In satellite communications systems, aside from early designs of large horn-type antennas with radome protection against the weather, open-air antennas are generally used.* The only weather protection device is electric heaters for melting snow and ice buildups.

The other three losses—blockage, spillover, and illumination—apply to the widely used Cassegrain configuration, which is characterized by a hyperbolic subreflector and parabolic main reflector. Since aperture illumination and far-field radiation are related by Fourier transforms, uniform illumination leads to a sin x/x radiation pattern, unity illumination factor but relatively high-level side lobes. With tapering of the illumination, the illumination factor falls below unity, but better side-lobe control can be achieved.

The crowding of the geostationary orbit resulting from the expansion of satellite systems requires very strict control of antenna side lobes in order to satisfy the opposite requirements of minimum mutual interference and decreased in-orbit spacing (from 5° to 4° to 3° and even 2°).

The side-lobe envelope of most electrically large Cassegrain antennas can be approximated by an expression of the form

$$G_{dBi} = A - B \log \theta \tag{83}$$

where A and B are constants and θ is the off-boresight angle. In 1965, the CCIR adopted the rule

$$G_{dBi} = \begin{cases} 32 - 25 \log \theta & 1° \le \theta \le 48° \\ 10 & \theta > 48° \end{cases} \tag{84}$$

The antennas of the INTELSAT system have been standardized to follow the CCIR rule. Until 1977, side lobes beyond 1° from boresight were not to exceed the −29 dB level, but after 1977 a new rule was introduced requiring that no more than 10 percent of the side-lobe peaks exceed the envelope as defined above. A more recent recommendation would change the value of the constant from 32 to 29 in the above expression.

The parameter that characterizes system performance is the G/T ratio, usually expressed in dBi/K. All contributions to the system noise temperature must be properly taken into account once the measurement port is chosen. Antenna-feed design is of great importance with regard to both gain and side-lobe considerations.

The amounts of spillover and illumination taper for both the subreflector and main reflector depend on the design of the feed system. Much progress has occurred in the transition from pyramidal to conical horns of various types such as single mode (TE_{11}), multimode types ($TE_{11} + TM_{11}$), and hybrid mode (EH_{11}). In the last mentioned case, corrugated horns with $\lambda/4$ grooves have made it possible to increase bandwidth, improve symmetry of the radiation pattern, reduce the side lobes, and achieve better off-axis polarization characteristics, a feature of special importance in dual-polarization systems. In this case, the isolation between the two orthogonally polarized cofrequency channels should be as high as possible. Experience in systems using linear or circular orthogonal polarizations has confirmed that 30-dB isolation is a representative design goal.

Sophisticated orthomode junctions and an arrangement of cascaded polarizers are required to satisfy the above-mentioned requirements.

The actual isolation depends on the polarization purity of the signal source and of the antenna system. Since rain along the signal path depolarizes the signal, automatic means of depolarization correction have been successfully introduced at 6/4 GHz.

Monopulse tracking systems are employed to correct continuously the pointing of the antenna in the direction of the satellite. To maintain continuity of service, an auxiliary power source is provided as well as redundant communications equipment.

REFERENCES

1. Feher, K. *Digital Communications, Satellite/Earth Stations Engineering*. Englewood Cliffs, N.J.: Prentice-Hall Inc., 1983.
2. Miya, K., ed. *Satellite Communications Technology*. Tokyo: KDD Engineering and Consulting, Inc., 1982.
3. Van Trees, Harry L., ed. *Satellite Communications*. New York: IEEE Press, 1979.
4. Spilker, J. J. *Digital Communications by Satel-*

* Shipboard antennas of maritime satellite communications systems are an exception.

lite. Englewood Cliffs, N. J.: Prentice-Hall Inc., 1977.

5. *Proceedings of the IEEE*, Special Issue on Satellite Communications, Vol. 65, No. 3, March 1977.
6. *COMSAT Technical Review*. A periodical published twice a year since 1971 by the Communications Satellite Corp., Washington, D.C.
7. Ippolito, Louis J., Kaul, R. D., and Wallace, R. G. *Propagation Effects Handbook for Satellite Systems Design*. NASA Reference Publication 1082, Dec. 1981.

28 Digital Signal Processing and Surface-Acoustic-Wave Filters

W. Kenneth Jenkins and Bill J. Hunsinger

During the last decade, there has been rapid advancement in the theory and application of digital signal processing (DSP) in various engineering disciplines. Interest has grown in digital signal processing because, not only has the general-purpose computer become more readily available, but digital integrated circuits have become more highly integrated and cheaper, a trend that will continue into the foreseeable future. New very large scale integration (VLSI) techniques have produced high-density read-only memories (ROM) and microprocessors that provide enormous flexibilities in the design of digital hardware systems.

Digital filters offer distinct advantages over analog (continuous-time) filters in many applications, although they are not good substitutes for all analog filters. The major advantages are good numerical accuracy, programmability, stability in the presence of changing environmental conditions, suitability for multiplexing, and convenience for processing data that is directly available in binary form. Some of their disadvantages are the relatively high per-unit costs, frequency limitations imposed by the speed of the digital hardware, and the necessity for a significant amount of clocking and control circuitry to sequence the binary operations properly.

The *design* of a digital filter involves determining either a set of time-domain difference equations or a z-domain digital transfer function that satisfies given specifications. A digital filter can be obtained by first designing an analog prototype and then transforming it into a discrete-time system by a sampled-data transformation. Another approach is to use a computer optimization to place the z-domain poles and zeros so the discrete-time system will meet specifications directly. The first approach takes advantage of well known analog design techniques, while the second provides greater flexibility because it does not depend on an analog design step. Digital filter *implementation* involves choosing a network topology and hardware modules for the final network. At this stage, the designer must analyze the effects of quantization error, because error performance and network topology are closely related. Digital *hardware design* consists of designing the individual circuit elements (adders, multipliers, shift registers, etc.). If the system is to be integrated, it also includes the IC layouts. Finally, the term *architecture* refers to the arrangement and interconnection of the various functional modules. This involves the accessing of memories, multiplexing of functional units, and control and intercommunication among units executing sequential operations.

In many applications, implementation is ultimately accomplished in software on a general-purpose computer. In these cases, the emphasis is on the design and implementation stages, since hardware design and system architecture are dictated by the general computer system. However, with the rapid advances that are now being made in the automated design and manufacture of VLSI monolithic circuits, it appears that engineers of the future will enjoy more freedom to specify custom designed digital functions and have them quickly fabricated in low-cost silicon devices. This new capability will result in digital signal processing becoming less dependent on the general computer. New techniques for improving data rates, reducing circuit complexity, and improving reliability will become increasingly important as more custom-designed VLSI digital systems come into common usage.

FUNDAMENTALS FOR DISCRETE-TIME SYSTEMS

Basic Definitions

A *continuous-time* (CT) signal is a function, $s(t)$, that is defined for all time t contained in some interval on the real line. For historical reasons, CT signals are often called *analog signals*. If the domain of definition for $s(t)$ is restricted to a set of discrete points $t_n = nT$, where n is an integer and T is the sampling period, the signal $s(t_n)$ is called a *discrete-time* (DT) signal. Often, if the sampling interval is well understood within the context of the discussion, the sampling period is normalized by $T = 1$, and a DT signal is represented simply as a sequence $s(n)$. If the values of the sequence $s(n)$ are to be represented with a finite number of bits (as required in a finite state machine), then $s(n)$ can take on only a discrete set of values. In this case, $s(n)$ is called a *digital signal*. Much of the theory that is used in DSP is actually the theory of DT signals and DT systems, in that no amplitude quantization is assumed in the mathematics. However, all signals processed in binary machines are truly digital signals. One important question that arises in virtually every application is the question of how many bits are required in the representation of the digital signals to guarantee that the performance of the digital systems is acceptably close to the performance of the ideal DT system.

Linear CT systems are characterized by the familiar mathematics of differential equations, continuous convolution operators, Laplace transforms, and Fourier transforms. Similarly, linear DT systems are described by the mathematics of difference equations, discrete convolution operators, Z-transforms, and discrete Fourier transforms. It appears that for every major concept in CT systems, there is a similar concept for DT systems (e.g., differential equations and difference equations, continuous convolution and discrete convolution, etc.). However, in spite of this duality of concepts, it is impossible to apply directly the mathematics of CT systems to DT systems, or vice versa.

Many modern systems consist of both analog and digital subsystems, with appropriate analog-to-digital (A/D) and digital-to-analog (D/A) devices at the interfaces. For example, it is common to use a digital computer in the control loop of an analog plant. Analytical difficulties often occur at the boundaries between the analog and digital portions of the system because the mathematics used on the two sides of the interface must be different. It is often useful to assume that a sequence $s(n)$ is derived from an analog signal $s_a(t)$ by ideal sampling, i.e.

$$s(n) = s_a(t)\Big|_{t=nT} \tag{1}$$

An alternative model for the sampled signal is denoted by $s^*(t)$ and defined by

$$s^*(t) = \sum_{n=-\infty}^{+\infty} s_a(t)\, \delta_a\,(t-nT) \tag{2}$$

where $\delta_a(t)$ is an analog impulse function. Both $s(n)$ and $s^*(t)$ are used throughout the literature to represent an ideal sampled signal. Note that even though $s(n)$ and $s^*(t)$ represent the same essential information, $s(n)$ is a DT signal and $s^*(t)$ is a CT signal. Hence, they are not mathematically identical. In fact, $s(n)$ is a "DT-world" model of a sampled signal, whereas $s^*(t)$ is a "CT-world" model of the same phenomenon.

Finite Convolution and Difference Equations

Let $y(n) = \mathcal{S}\{x(n)\}$ define the input-output relation for a discrete-time system with input $x(n)$ and output $y(n)$. The following definitions are commonly used to define the properties of linearity, shift invariance, causality, and stability:

Linearity—$\mathcal{S}$ is linear if and only if $\mathcal{S}\{ax_1(n) + bx_2(n)\} = a\mathcal{S}\{x_1(n)\} + b\mathcal{S}\{x_2(n)\}$, where a and b are scalar constants and $x_1(n)$ and $x_2(n)$ are two arbitrary input sequences.

Shift-Invariance—$\mathcal{S}$ is shift-invariant if and only if $y(n-n_o) = \mathcal{S}\{x(n-n_o)\}$ for all integer values of n_o.

Causality—$\mathcal{S}$ is causal if and only if $y(n)$ depends on samples of $x(k)$ at times $k \leq n$.

Stability (BIBO)—$\mathcal{S}$ is stable in the bounded-input–bounded-output (BIBO) sense if and only if $y(n)$ remains bounded for all $x(n)$ that are bounded.

Whenever $\mathcal{S}$ is linear and shift-invariant, it is possible to express the zero state response, $y(n)$, due to an arbitrary input, $x(n)$, in terms of a discrete convolution of $h(n)$ and $x(n)$, where $h(n) = \mathcal{S}\{\delta(n)\}$, and $\delta(n) = 0$ for $n \neq 0$ and $\delta(n) = 1$ when $n = 1$. The function $\delta(n)$ is called a unit pulse, $h(n)$ is the unit-pulse response* of S, and the finite convolution is expressed by

$$y(n) = \sum_{k=-\infty}^{+\infty} h(k)\, x(n-k) \tag{3}$$

If $\mathcal{S}$ is causal, $h(k) = 0$ for $k<0$, and the lower limit on the summation in Eq. 3 becomes zero.

* The term "impulse response" is often used interchangeably with the term "unit pulse response."

In certain types of DT systems, the unit-pulse response is zero outside of a finite interval containing N samples; i.e., $h(n) \equiv 0$ for $n<0$ and $n \geq N$. This type of system is a finite impulse response (FIR) DT system. If $h(n)$ is supported over an infinite length interval, then $\mathcal{S}$ is called an infinite impulse response (IIR) DT system. These two classes constitute the two important types of digital filters, with each class having distinct advantages and disadvantages with regard to stability, quantization error performance, and computational efficiency.

An IIR system has an infinite memory because, in general, the output depends on the input all the way into the infinite past. Such a system can also be characterized by an Nth order linear difference equation, as given by

$$\begin{aligned} & y(n) + a_1 y(n-1) \\ & \quad + \cdots + a_{N-1}\, y(n-N+1) \\ & \quad = b_o\, x(n) + \cdots + b_{M-1}\, x(n-M+1) \end{aligned} \tag{4}$$

A linear difference equation is a recursive relation that can be realized quite easily with digital multipliers, adders, and unit delay registers. Hence, in some of the older literature, IIRs are referred to as recursive filters. Similarly, FIRs were often referred to as nonrecursive (or transversal) filters. However, the terms recursive and nonrecursive refer more specifically to the implementation of the filter, rather than the mathematical structure. This fact is illustrated by the frequency sampling structure,† which is a recursive realization of an FIR system. The terms IIR and FIR are used in the modern literature to eliminate any ambiguity that was caused by the older terminology.

In general, IIR systems are cheaper to implement than FIR systems because the iteration of a difference equation requires fewer arithmetic operations per output sample, as compared to calculating a finite convolution. However, IIR systems have "poles," whereas FIR systems do not. This implies that IIR systems must be carefully designed to ensure stability. Also, IIR systems suffer from limit cycle oscillations and quantization error accumulation because quantization errors are recycled through the inherent feedback in the recursion.

The Z-Transform

The Z-transform occupies the same position of importance in DT system theory as the Laplace transform does in CT system theory. They are very similar transforms that share many common properties. However, it is important to emphasize that they cannot be used interchangeably because they apply to different types of systems.

The 2-sided Z-transform of a DT signal, $s(n)$, is defined by

† Reference 26.

$$S(z) = \mathfrak{Z}\{s(n)\} = \sum_{n=-\infty}^{+\infty} s(n)z^{-n} \qquad (5)$$

where $S(z)$ is said to exist for all $z \epsilon R$ such that the infinite summation converges. Symbol R represents the region of convergence (R.O.C.) of $S(z)$. If the summation is taken from $n = 0$, rather than $n = -\infty$, the result is called the 1-sided Z-transform. It is clear that for all $s(n)$ such that $s(n) \equiv 0$ for $n < 0$, the 1-sided and 2-sided Z-transforms are equivalent.

Historically, the Z-transform arose from attempts to apply the Laplace transform to an ideally sampled signal, $s^*(t)$, as previously defined in Eq. 2. Since $s^*(t)$ is in fact a CT signal, it is proper to apply the Laplace transform, provided it converges. Then, using Eq. 2 results in

$$\mathfrak{L}\{s^*(t)\} = \sum_{n=-\infty}^{+\infty} s(nT)e^{-sTn} \qquad (6)$$

Comparing Eqs. 5 and 6 shows that taking the Laplace transform of $s^*(t)$ with $z = e^{sT}$ and $T = 1$ results in the Z-transform of $s(n)$, as defined directly in the DT space. Therefore the historical approach and the modern approach defining the Z-transform by Eq. 5 are entirely equivalent. Most authors now prefer to define the Z-transform in DT space according to Eq. 5 so that the Z-transform is strictly a DT space concept that does not depend on the existence of a related CT space. This approach is particularly useful in problems that are inherently discrete, such as population growth, or the compounding of monthly interest rates, where there is no underlying CT process.

The inverse Z-transform is defined by

$$s(n) = (1/2\pi j) \oint_c S(z)\, z^{n-1} dz \qquad (7)$$

where c is a closed contour that encircles the origin and lies entirely within the region of convergence for $S(z)$. Eq. 7 is valid for both positive and negative values of n. In many applications, the inverse Z-transform can be found from a table of Z-transform pairs, and therefore it is seldom necessary to find the inverse Z-transform from its definition. However, for cases where it is desired to apply the definition of Eq. 7, the contour integral can be evaluated by means of the well known residue theorem.

$$(1/2\pi j) \oint_c S(z)\, z^{n-1}\, dz = \sum (\text{residues of } S(z)\, z^{n-1} \text{ at the poles inside } c) \qquad (8)$$

The normal technique for finding the residues of $S(z)\, z^{n-1}$ is by the same method used in expanding $S(z)\, z^{n-1}$ in a partial fraction expansion. It is important to note that the residue theorem is applicable only when $S(z)\, z^{n-1}$ has singularities (poles) that occur at isolated points in the z-plane. Fortunately, virtually all one-dimensional DT systems and DT signals of practical interest produce Z-transforms that are ratios of polynomials in powers of z. The singular behavior of such systems is characterized by poles and zeros, and hence the residue theorem can be applied in virtually all cases of practical interest. Table 1 summarizes a number of Z-transform pairs that are frequently encountered in practical problems. More complete tables of Z-transforms can be found in references 8 and 18. Note that most of the sequences in Table 1 are right-sided; i.e., $x(n) = 0$ for $n < 0$. The exceptions are entry 6, which is left-sided, and entry 7, which is two-sided. For a given $X(z)$, the corresponding $x(n)$ may be right-sided, left-sided, or two-sided, depending on the nature of the region of convergence, R. Fig. 1 shows the three general types of regions of convergence for these three cases. A more complete discussion of left-sided and two-sided sequences is presented by Gabel and Roberts,* where they also describe in detail the preferred forms for partial fraction expansion when inverting these three different cases.

Table 2 summarizes some important properties of the Z-transform. Although the table is specifically given for the two-sided Z-transform, most of the properties hold for the 1-sided Z-transform also. A notable exception is property 4, the shift theorem. Since the shift theorem for 1-sided sequences is often needed for solving difference equations, it is stated here:

If $X(z)$ is the 1-sided Z-transform of $x(n)$ with a region of convergence R, then for $n_0 > 0$,

$$\text{(i)} \qquad \mathfrak{Z}\{x(n+n_0)\} = z^{n_0}\, \{X(z) - \sum_{k=0}^{n_0-1} x(k)\, z^{-k}\}$$

$$\text{(ii)} \qquad \mathfrak{Z}\{x(n-n_0)\} = z^{-n_0}\, \{X(z) + \sum_{k=-n_0}^{-1} x(k)\, z^{-k}\}$$

with region of convergence R.†

The Discrete-Time Fourier Transform

It is important to distinguish between the concepts of the discrete-time Fourier transform (DTFT) and the discrete Fourier transform (DFT). The DTFT is a transform-pair relationship between a DT signal and its continuous-frequency transform that is used extensively in the analysis and design of DT systems. In contrast, the DFT is a transform-pair relationship between a DT signal and its discrete-frequency transform that is used in practical digital processing.

The DTFT of a sequence $x(n)$ is defined by

$$X(e^{j\omega}) = \sum_{n=-\infty}^{+\infty} x(n)\, e^{-jn\omega} \qquad (9)$$

for all $x(n)$'s for which the sum converges. The spectrum $X(e^{j\omega})$ is a continuous frequency function which

* Reference 8.

† Reference 26 describes pathological cases where R for the shifted sequence may no longer include the points $z = 0$ or $z = \infty$.

TABLE 1. Z-TRANSFORMS

	$x(n)$	$X(z)$
1.	$\delta(n)$	1, for all z
2.	$u(n)$	$1/(1-z^{-1}),\quad 1<\|z\|$
3.	$n\,u(n)$	$1/(1-z^{-1})^2,\quad 1<\|z\|$
4.	$n^r a^n\,u(n)$ (r = integer)	$-z(d^r/dz^r)\,\{1/(1-az^{-1})\},\quad \|a\|<\|z\|$
5.	$a^n\,u(n)$	$1/(1-az^{-1}),\quad \|a\|<\|z\|$
6.	$n^r a^n\,u(-n)$ (r = integer)	$-z(d^r/dz^r)\,\{1/(1-az^{-1})\},\quad \|z\|<\|a\|$
7.	$a^{\|n\|}$	$(a-a^2)/(1-az)\,(1-az^{-1}),\quad \|a\|<\|z\|<\|1/a\|$
8.	$n^{-1}\,u(n)$	$-\ln\,(1-t^{-1}),\quad 1<\|z\|$
9.	$\cos\,(an)\,u(n)$	$\dfrac{(1-z^{-1}\cos a)}{(1-2z^{-1}\cos a+z^{-2})},\quad 1<\|z\|$
10.	$\sin\,(an)\,u(n)$	$\dfrac{z^{-1}\sin a}{(1-2z^{-1}\cos a+z^{-2})},\quad 1<\|z\|$
11.	$(K_1\cos an+K_2\sin an)u(n)$	$\dfrac{K_1+z^{-1}(K_2\sin a-K_1\cos a)}{(1-2z^{-1}\cos a+z^{-2})}$
12.	$(K_1\cos an+G\sin an)u(n)$ where, $G=(K_2+K_2\cos a)/\sin a$	$\dfrac{K_1+K_2z^{-1}}{(1-2z^{-1}\cos a+z^{-2})},\quad 1<\|z\|$
13.	$(n+1)a^n\,u(n)$	$1/(1-az^{-1})^2,\quad \|a\|<\|z\|$
14.	$(1/2)\,(n+1)\,(n+2)a^n\,u(n)$	$1/(1-az^{-1})^3,\quad \|a\|<\|z\|$
15.	$[1/(r-1)!]\,(n+1)\,(n+2)\ldots(n+r-1)a^n\,u(n)$	$1/(1-az^{-1})^r,\quad \|a\|<\|z\|$

is periodic, with period 2π. Eq. 9 is equivalent to evaluating the 2-sided Z-transform $X(z)=\mathcal{Z}\{x(n)\}$ on the unit circle $z=e^{j\omega}$, provided the unit circle lies within R, the region of convergence of $X(z)$. A sequence $x(n)$ can be recovered from its spectrum by

$$x(n)=(1/2\pi)\int_{-\pi}^{+\pi} X(e^{j\omega})e^{jn\omega}d\omega \qquad (10)$$

which can be interpreted on the inverse-DTFT (IDTFT). An alternate interpretation is that Eq. 9 is a Fourier series expansion of the periodic function $X(e^{j\omega})$, the $x(n)$'s are the Fourier coefficients, and Eq. 10 is an expression for finding the Fourier coefficients.

The DTFT obeys the same properties as those given for the Z-transform in Table 2, as long as the unit circle is contained within the various regions of convergence. In particular, the DTFT obeys the convolutional property (entry 11), which makes it useful for frequency analyses of DT systems. If $h(n)$ is the unit pulse response of a linear shift-invariant DT system, then $H(e^{j\omega})$ = DTFT $\{h(n)\}$ is intepreted as the frequency response of the DT system. Therefore, if a discrete sinusoidal waveform $x(n)=A\cos\,(n\omega_0)$ is applied as an input to a linear shift-invariant DT system, the sinusoidal steady-state output is given by $y_{ss}(n)=A\,|\,H(e^{j\omega_0})\,|\cos\,[n\omega_0+\measuredangle H(e^{j\omega_0})]$.

Sampling and Reconstruction

The traditional CT Fourier transform can be applied to the ideal sampled signal, $s^*(t)$, as follows:

$$\begin{aligned}\mathcal{F}\{s^*(t)\} &= \sum_{n=-\infty}^{+\infty} s(nT)\mathcal{F}\{\delta_a(t-nt)\}\\ &= \sum_{n=-\infty}^{+\infty} s_a(nT)e^{-jnT\omega} \qquad (11)\\ &= \text{DTFT}\{s_a(nT)\}\end{aligned}$$

This verifies that the traditional Fourier transform of $s^*(t)$ is identical to the DTFT of $s(n)$, where $s(n)\equiv s_a(t)|_{t=nT}$; i.e., sequence $s(n)$ is derived from $s_a(t)$ by ideal sampling. This proves that $s^*(t)$ and $s(n)$ are really different models of the same phenomenon,

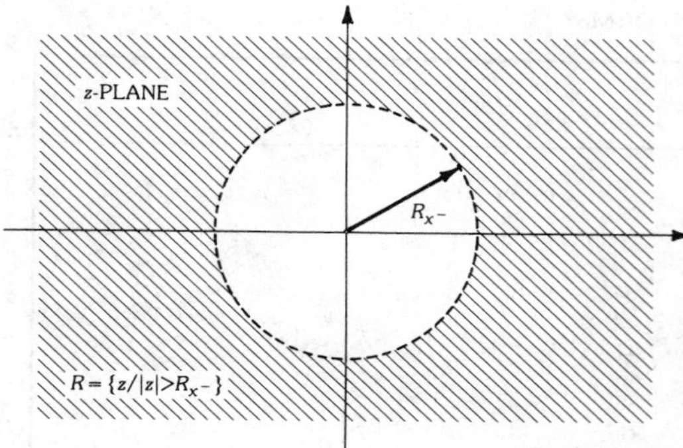

(A) Right-sided sequence.

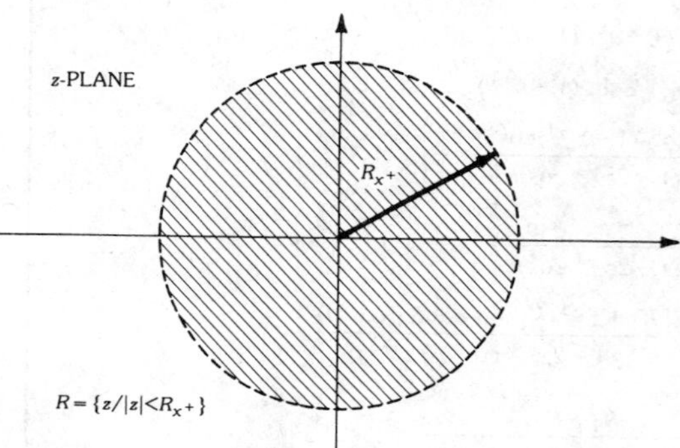

(B) Left-sided sequence.

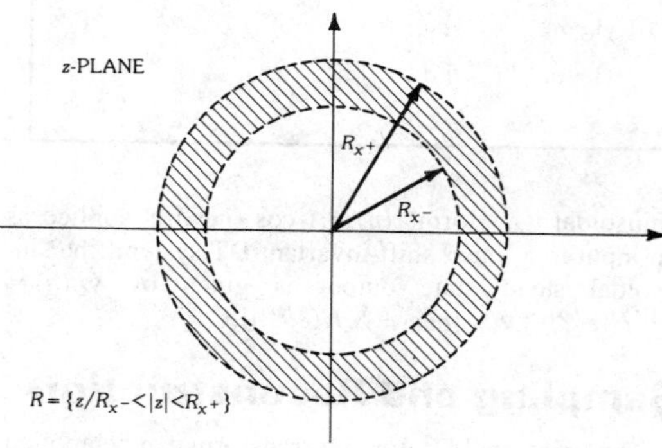

(C) Two-sided sequence.

Fig. 1. Typical regions of convergence for Z-transforms.

since their spectra (as computed with appropriate transforms) are identical. Suppose that $S_a(j\Omega)$ is the spectrum of $s_a(t)$ and $S(e^{j\omega})$ is the spectrum of $s(n)$. It can be shown that

$$S(e^{j\omega})=(1/T)\sum_{r=-\infty}^{+\infty} S_a(j[\Omega-2\pi r/T]) \qquad (12)$$

where $\omega=\Omega T$ is often referred to as the "normalized digital frequency." Eq. 12 shows that the DT spectrum is formed from a superposition of an infinite number of replicas of the analog signal, as illustrated in Fig. 2. As long as the sampling frequency, $\Omega_s=2\pi/T$, is chosen so that $\Omega_s>2\Omega_B$, where Ω_B is the highest frequency component contained in $s_a(t)$, then each period of $S(e^{j\omega})$ contains a perfect copy of $S_a(j\Omega)$, and $s_a(t)$ can be recovered exactly from $s(n)$ by ideal low-pass filtering. Sampling under these conditions is said to satisfy the Nyquist sampling criterion, since the sampling frequency exceeds the Nyquist rate, $2\Omega_B$. If the sampling rate does not satisfy the Nyquist criterion, the adjacent periods of the analog spectrum will overlap, causing a distorted spectrum (see Fig. 2). This effect, called *aliasing distortion*, is rather serious because it cannot be easily corrected once it has occurred. In general, an analog signal should be prefiltered with an analog low-pass filter prior to sampling so that aliasing distortion does not occur.

If the Nyquist criterion has been satisfied, it is always possible to reconstruct an analog signal from its samples according to

$$s_a(t)=\sum_{k=-\infty}^{+\infty} s_a(kT)\ \mathrm{sinc}(\ (\pi/T)\ (t-kT)\ ) \qquad (13)$$

This reconstruction formula results by filtering $s^*(t)$ with an ideal low-pass analog filter with a bandwidth of $\Omega_B=\pi/T$ (see Fig. 2). In general, exact reconstruction requires an infinite number of samples, although a good approximation can be obtained by using a large but finite number of terms in Eq. 13.

Most practical systems use a digital-to-analog converter for reconstruction, which results in an analog staircase approximation to the true analog signal; i.e.

$$\hat{s}_a(t)=\sum_{k=-\infty}^{+\infty} s_a(kT)[u(t-kT)-u(t-(k+1)T)] \qquad (14)$$

It can be shown that $\hat{s}_a(t)$ is obtained by filtering $s_a{}^*(t)$ with an analog filter whose frequency response is

$$H_a(j\Omega)=2Te^{-j\Omega T/2}\ \mathrm{sinc}(\Omega T/2) \qquad (15)$$

The approximation $\hat{s}_a(t)$ is said to contain "sin x/x distortion," which occurs because $H_a(j\Omega)$ is not an ideal low-pass filter. The $H_a(j\Omega)$ response distorts the signal by causing a droop near the band edge, as well as passing high-frequency distortion terms that "leak" through the side lobes of $H_a(j\Omega)$. Therefore, a practical D/A converter is normally followed by a postfilter

$$H_p(j\Omega)=\begin{Bmatrix} H_a^{-1}(j\Omega),\ 0\le|\Omega|\le\pi/T \\ 0\ ,\ \Omega \text{ otherwise} \end{Bmatrix} \qquad (16)$$

which compensates for the distortion and produces the correct $s_a(t)$ analog output. Notice, however, that $H_p(j\Omega)$ can only be approximated in practice, so that the best reconstruction is necessarily an approximation. Fig. 3 shows a digital processor complete with sampling and reconstruction devices at the input and output.

TABLE 2. SOME IMPORTANT PROPERTIES OF THE 2-SIDED Z-TRANSFORM*

Sequence	Z-Transform	
1. $x(n)$	$X(z)$	$R_{x-} < \lvert z \rvert < R_{x+}$
2. $y(n)$	$Y(z)$	$R_{y-} < \lvert z \rvert < R_{y+}$
3. $ax(n)+by(n)$	$aX(z)+bY(z)$	$\max\,[R_{x-}, R_{y-}] < \lvert z \rvert < \min\,[R_{z+}, R_{y+}]$
4. $x(n+n_0)$	$z^{n_0}X(z)$	$R_{x-} < \lvert z \rvert < R_{x+}$
5. $a^n x(n)$	$X(a^{-1}z)$	$\lvert a \rvert R_{x-} < \lvert z \rvert < \lvert a \rvert R_{x+}$
6. $nx(n)$	$-z(dX(z)/dz)$	$R_{x-} < \lvert z \rvert < R_{x-}$
7. $x^*(n)$	$X^*(z^*)$	$R_{x-} < \lvert z \rvert < R_{x+}$
8. $x(-n)$	$X(1/z)$	$1/R_{x+} < \lvert z \rvert < 1/R_{x-}$
9. $\mathrm{Re}[x(n)]$	$(1/2)\ [X(z)+X^*(z^*)]$	$R_{x-} < \lvert z \rvert < R_{x+}$
10. $\mathrm{Im}[x(n)]$	$(1/2j)[X(z)-X^*(z^*)]$	$R_{x-} < \lvert z \rvert < R_{x+}$
11. $x(n) * y(n)$	$X(z)Y(z)$	$\max\,[R_{x-}, R_{y-}] < \lvert z \rvert < \min\,[R_{x+}, R_{y+}]$
12. $x(n)y(n)$	$(1/2\pi j) \oint X(v)Y(z/v)v^{-1}dv$	$R_{x-}R_{y-} < \lvert z \rvert < R_{z+}R_{y+}$

* From A.V. Oppenheim and R.W. Schafer. *Digital Signal Processing*. Englewood Cliffs, N.J.: Prentice-Hall, Inc., 1975.

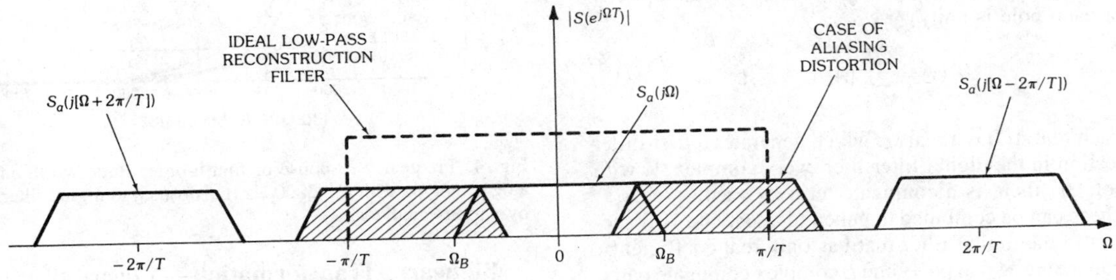

Fig. 2. Relationship between the spectrum of an analog signal and the spectrum of the ideally sampled signal.

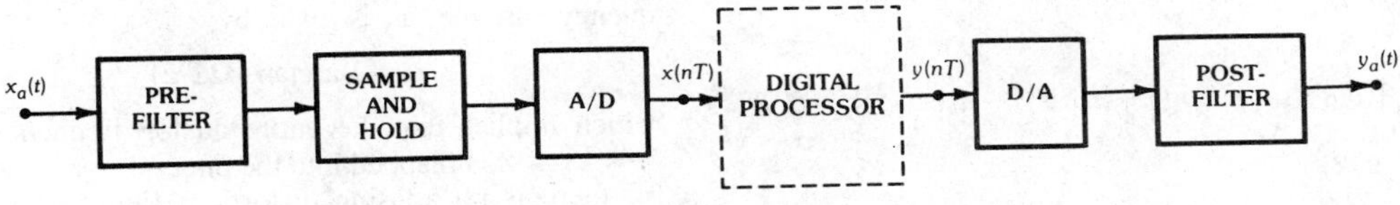

Fig. 3. Elements required for the digital processing of an analog signal.

DIGITAL-FILTER DESIGN

A common approach to designing IIR digital* filters is first to design an analog prototype, which is then transformed into a digital filter by one of several analog-to-digital mappings. This approach allows the designer to take advantage of many well known techniques for analog-filter design. Computer-based design techniques are also used for IIRs, as evidenced by the availability of several IIR design programs.† Usually, FIRs are designed either by windowing techniques or by computer optimization algorithms. This section

* It is customary to refer to these as "digital" filters, although at the design stage they really are discrete-time filters.

† Reference 30.

summarizes some of the more common design techniques.

IIR Filters

In the following discussion, $h_a(t)$ and $H_a(j\Omega)$ denote the impulse response and frequency response of an analog prototype, respectively. Similarly, $h(n)$ and $H(e^{j\omega})$ denote the unit pulse response and the frequency response of the digital filter to be designed. The objective is to obtain $H(e^{j\omega})$ (or $h(n)$) from $H_a(j\Omega)$ (or $h_a(t)$) so that the desirable features of the prototype are preserved.

Impulse Invariance—A digital filter is produced that has a unit pulse response which is exactly equal to the impulse response of the analog filter at the sampling instants; i.e., $h(n) \equiv h_a(t)|_{t=nT}$. The frequency responses are related by

$$H(e^{j\Omega T}) = \sum_{k=-\infty}^{+\infty} H_a(j(\Omega + k2\pi/T)) \qquad (17)$$

Hence, this technique preserves the frequency response when $H_a(j\Omega)$ is band-limited and T is chosen to satisfy the Nyquist criterion.

A. The design procedure is:
Design the prototype to obtain $H_a(s)$. Then expand $H_a(s)$ into partial fractions (assuming the multiplicity of each pole is unity), i.e.

$$H_a(s) = \sum_{k=1}^{N} [A_k/(s - s_k)] \qquad (18)$$

Each real A_k has a real s_k, which generates a first-order section in the digital filter. For every complex A_k with pole s_k, there is a conjugate term A_k^* with pole s_k^*. These can be combined to generate a second-order section in the digital filter that has only real coefficients. If there are M real poles and L complex conjugate pairs, the resulting digital filter becomes

$$H(z) = \sum_{l=1}^{M} [a_l/(1 + b_l z^{-1})] + \sum_{k=1}^{L} [(c_{k0} + c_{k1} z^{-1})/(1 + d_{k1} z^{-1} + d_{k2} z^{-2})] \qquad (19)$$

where,

$$\begin{aligned} a_l &= A_l \quad \text{(real)} \\ b_l &= -e^{\alpha_l T} \\ c_{k0} &= -2T \operatorname{Re}\{A_k\} \\ c_{k1} &= -2T e^{\alpha_k T} \operatorname{Re}\{A_k e^{-j\beta_k T}\} \\ d_{k1} &= -2 e^{\alpha_k T} \cos(\beta_k T) \\ d_{k2} &= e^{2\alpha_k T} \end{aligned} \qquad (20)$$

and $s_k = \alpha_k + j\beta_k$ is the kth pole of $H_a(s)$. Note that $H(z)$ is completely defined in terms of A_k, s_k, and T.

B. Features:

(1.) Good only for band-limited $H_a(j\Omega)$.
(2.) Results initially in a parallel form.
(3.) $\omega = \Omega T \rightarrow$ linear relationship between the analog and digital frequency variables.
(4.) The impulse invariant technique is *not* an algebraic mapping; i.e., $H(z)$ *cannot* be obtained from $H_a(s)$ by a substitution of variable.

C. Example:
The magnitude of the frequency response is shown in Fig. 4 for a fourth-order IIR digital filter designed by the impulse invariant technique with a sampling frequency of $f_s = 16$ kHz. The analog prototype is a fourth-order Butterworth low-pass function with a −3-dB cutoff frequency of $f_c = 2000$ Hz.

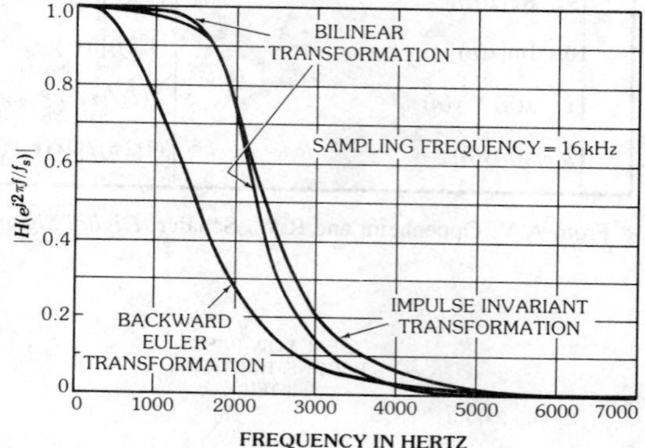

Fig. 4. Frequency response of fourth-order Butterworth analog prototype with $f_c = 2000$ Hz transformed to digital filters by various methods.

Bilinear z-Transformation—A digital filter is produced by substituting $s = (2/T)[(1 - z^{-1})/(1 + z^{-1})]$ into $H_a(s)$ to produce $H(z)$. The relation between the analog frequency variable, Ω, and the digital frequency variable, ω, is given by

$$\omega = 2 \arctan[\Omega T/2] \qquad (21)$$

which implies that the entire analog frequency axis $-\infty < \Omega < \infty$ is mapped into the interval $-\pi < \omega < \pi$ so there is no aliasing distortion. Fig. 5 shows the important features of how the s-plane is mapped into the z-plane: (A) the $j\Omega$ axis is mapped into the unit circle in the z-plane; (B) the left half s-plane is mapped inside the unit circle; and (C) the right half z-plane is mapped outside the unit circle. Since the bilinear z-transform warps the frequency axis, it is necessary to prewarp the analog prototype so that the critical frequencies end up at the right places. Suppose the final digital filter has N critical frequencies (corner frequencies, band edges, zeros, etc.), denoted ω_i, $i = 1, \ldots, N$. The prewarped analog frequencies, Ω_i, $i = 1, \ldots, N$ are generated by

$$\Omega_i = (2/T) \tan [\omega_i/2],\ i = 1, \ldots, N \quad (22)$$

and are then used to design the analog prototype.

A. Features:

(1.) Eliminates aliasing distortion, but causes nonlinear warping of the frequency axis.
(2.) Requires prewarping of the analog prototype.
(3.) Transformation is algebraic, preserves the prototype order, and can be applied to $H_a(s)$ in any form (parallel, cascade, direct, ladder, etc.).
(4.) Particularly useful for wideband filters with piecewise constant magnitude specifications.

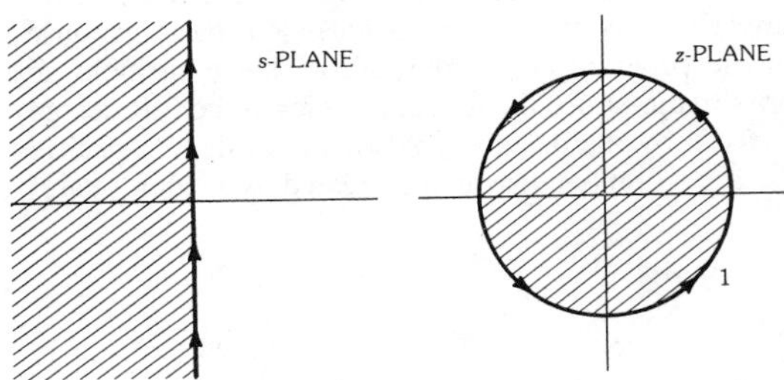

Fig. 5. Mapping of the s-plane into the z-plane by the bilinear z-transformation.

B. Example:

Fig. 4 shows the magnitude of the frequency response for a fourth-order Butterworth IIR filter designed with the bilinear z-transform for a sampling frequency $f_s = 16$ kHz. Prewarping has caused the -3-dB cutoff of the digital filter to occur at 0.707. The bilinear z-transform design does not exhibit aliasing distortion, and hence its transition regions fall off faster than for the impulse invariant design.

Euler Transformations—Although the Euler transformations are, in general, inferior to the above two design techniques for digital filters, they are mentioned here because they occur frequently for the discrete-time solution of differential equations. The backward and forward Euler transformations are algebraic transformations characterized in the frequency domain by $s = (1/T)(1 - z^{-1})$ and $s = (1/T)(z - 1)$, respectively. The backward Euler transformation results from a backward difference approximation to the first derivative, whereas the forward Euler transformation results from a forward difference approximation.* Both of the Euler transformations are known to distort the frequency response unless the sampling rate is very high with respect to the upper band edge of the analog prototype. The backward Euler transformation always maps stable analog prototypes into stable digital filters. In contrast, the forward Euler transformation has the potential to map a stable prototype into an unstable digital filter unless the sampling rate is chosen sufficiently high.

* Reference 26.

In Fig. 4, a backward Euler design for the fourth-order Butterworth IIR example is shown for the sampling rate $f_s = 16$ kHz. It is apparent that the backward Euler design results in considerable distortion in comparison with the other two designs.

FIR Filters

Two general approaches are available for the design of FIR digital filters: (1) designs by windowing techniques, and (2) optimal designs by computer optimization techniques. In general window designs can be carried out with the aid of a hand calculator and a table of well known window functions. In contrast, computer optimization requires a sophisticated computer program and a considerable amount of computer time. Filters obtained by method 2 are, in general, higher quality filters. Also, more elaborate filter specifications can be approximated because the computer does most of the work.

Window Designs—Let $h(n)$ be the unit pulse response that corresponds to some desired frequency response, $H(e^{j\omega})$. If $H(e^{j\omega})$ has sharp discontinuities, such as the low-pass example shown in Fig. 6, then $h(n)$ will represent an IIR function. The objective is to time-limit $h(n)$ in such a way as to not disturb $H(e^{j\omega})$ any more than necessary. If $h(n)$ is simply truncated, a ripple* occurs around discontinuities, resulting in a distorted filter, as illustrated in Fig. 6.

Suppose that $\omega(n)$ is a window function that time-limits $h(n)$ to create an FIR approximation, $\hat{h}(n)$; i.e., $\hat{h}(n) = \omega(n)h(n)$. Then if $W(e^{j\omega})$ is the spectrum of $\omega(n)$, $\hat{h}(n)$ has a Fourier transform given by $\hat{H}(e^{j\omega}) = W(e^{j\omega}) \circledast H(e^{j\omega})$. From this it can be seen that the ripples in $\hat{H}(e^{j\omega})$ result from the side lobes of $W(e^{j\omega})$. Ideally, $W(e^{j\omega})$ should be similar to an impulse so that $\hat{H}(e^{j\omega}) \cong H(e^{j\omega})$.

A. Special case (simple FIRs):

Let $h(n) = \cos n\omega_0$, $-\infty < n < +\infty$. Then $\hat{h}(n) = \omega(n) \cos n\omega_0$, and

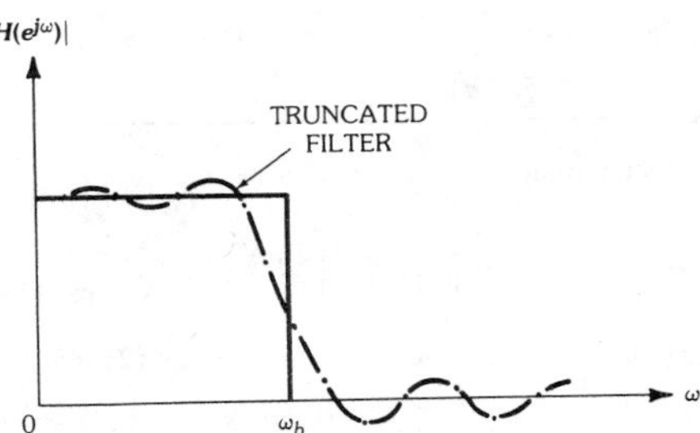

Fig. 6. Gibbs effect in a low-pass filter caused by truncating the unit pulse response.

* This ripple, known as the Gibbs phenomenon, occurs around points of discontinuity due to the truncation of the Fourier series expansion of $H(e^{j\omega})$.

$$\hat{H}(e^{j\omega}) = (1/2)W(e^{j[\omega+\omega_0]}) + (1/2)W(e^{-j[\omega-\omega_0]})$$

as depicted in Fig. 7. For this simple class, the center of the passband is controlled by ω_0, and both the shape of the passband and the side-lobe structure are strictly determined by the choice of the window. While this simple class of FIRs does not allow very flexible designs, it is a simple technique for obtaining relatively good low-pass, bandpass, and high-pass FIRs.

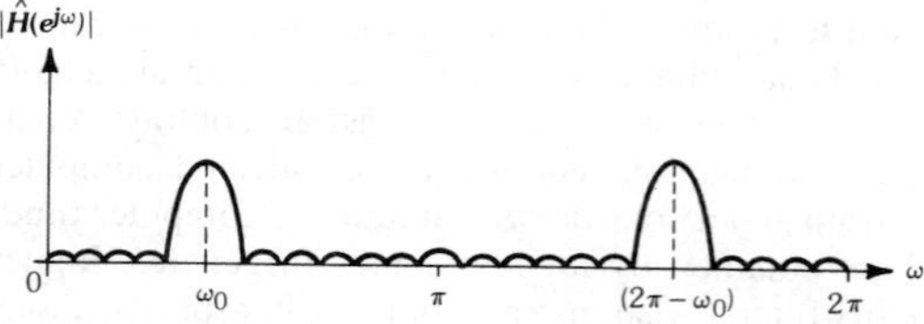

Fig. 7. Design of a simple FIR filter by windowing.

B. General case (design procedure):

Specify an ideal frequency response, $H(e^{j\omega})$, and choose samples from the curve. Use a long inverse FFT of length $\hat{N}$ to find an approximation to $h(n)$, where if N is the desired length of the FIR, then $\hat{N} >> N$. Then use a carefully selected window to truncate $h(n)$ according to $\hat{h}(n) = \omega(n)h(n)$. Finally, use an FFT of length $\hat{N}$ to find $\hat{H}(e^{j\omega})$. If $\hat{H}(e^{j\omega})$ is a satisfactory approximation to $H(e^{j\omega})$, the design is finished. If not, choose a new $H(e^{j\omega})$ or a new $\omega(n)$ and repeat. Throughout the procedure, it is important to choose $\hat{N} = kN$, with k an integer in the range $[4, \ldots, 10]$. The k should be made as large as possible within the limits of the computer resources. This design technique is a trial-and-error procedure. The quality of the result will depend to some degree on the skill and experience of the designer.

Table 3 lists a few well known window functions that can be used in this procedure. An extensive list of windows and their figures of merit has been published in reference 11.

Computer Design by Equiripple Approximation—There are quite a few publications and numerous computer programs that deal with optimized designs for FIRs using equiripple approximations in the frequency domain. One popular program that has been widely distributed is the one published by McClellan, et. al.† This program is capable of designing bandpass (including low-pass and high-pass), differentiator, and Hilbert-transform FIR filters under the constraint of linear phase. Fig. 8 illustrates an equiripple approximation for an ideal low-pass response. In the program, the order, N, of the equiripple approximation (also the resulting order of the FIR design) is fixed by the designer. Also, the designer specifies the transition band, which is treated as a "don't care"

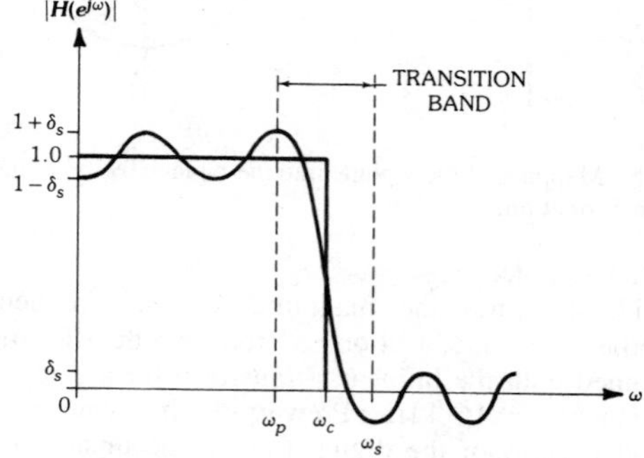

Fig. 8. Equiripple approximation to an ideal low-pass response.

† Reference 24.

TABLE 3. WINDOW FUNCTIONS

Name	Function	Peak Side-Lobe Amplitude (dB)	Main-Lobe Width	Minimun Stop-Band Attenuation (dB)
Rectangular	$\omega(n) = 1,\ 0 \le n \le N-1$	−13	$4\pi/N$	−21
Bartlett	$\omega(n) = \begin{cases} 2n/N, & 0 \le n \le (N-1)/2 \\ 2 - 2n/N, & (N-1)/2 \le n \le N-1 \end{cases}$	−25	$8\pi/N$	−25
Hanning	$\omega(n) = (1/2)[1 - \cos(2\pi n/N)],\ 0 \le n \le N-1$	−31	$8\pi/N$	−44
Hamming	$\omega(n) = 0.54 - 0.46\cos(2\pi n/N),\ 0 \le n \le N-1$	−43	$8\pi/N$	−53
Blackman	$\omega(n) = 0.42 - 0.5\cos(2\pi n/N) + 0.08\cos(4\pi n/N),\ 0 \le n \le N-1$	−57	$12\pi/N$	−74

interval during the computer optimization; i.e., the optimization is unconstrained in this interval. This particular program uses the Remez Exchange Algorithm to find the best equiripple approximation to the specified response, returning to the user the FIR coefficients and the stopband and passband ripple factors. If the result is unsatisfactory, the user can change the order, N, the transition width, $\omega_s - \omega_p$, or the relative weighting factors in the various bands to improve the result. In general, this design program is easy to use, executes efficiently, and produces excellent results. In certain circumstances, the design for multiple-band bandpass filters results in undesirable resonance peaks in the "don't-care" bands, a problem that has been addressed in the literature. Fig. 9 shows a 64th-order dual bandpass FIR filter that was designed with the program of McClellan, et. al.†

A number of other FIR design approaches have been reported in the literature,** and in some cases design programs are also available on computer tape. The IEEE Acoustics, Speech, and Signal Processing Society's collection of programs for digital signal processing‡ contains several digital-filter design programs, as well as many other useful DSP programs. This collection is available from the IEEE and makes an excellent starting point for those interested in building a library of DSP programs.

† Reference 24.

** See for example references 12, 13, 14, 23, and 31.

‡ Reference 30.

DIGITAL-FILTER IMPLEMENTATION

The fundamental building blocks for implementing digital filters are multipliers, adders, unit delay registers, and semiconductor memories. This section addresses some of the issues concerning the selection of a network topology, arithmetic number codes, and the analysis of quantization effects in digital signal processors. Much of the discussion is directed toward IIR digital filters, because the feedback in IIR structures causes the performance to be rather sensitive to network topology and the effects of quantization error.

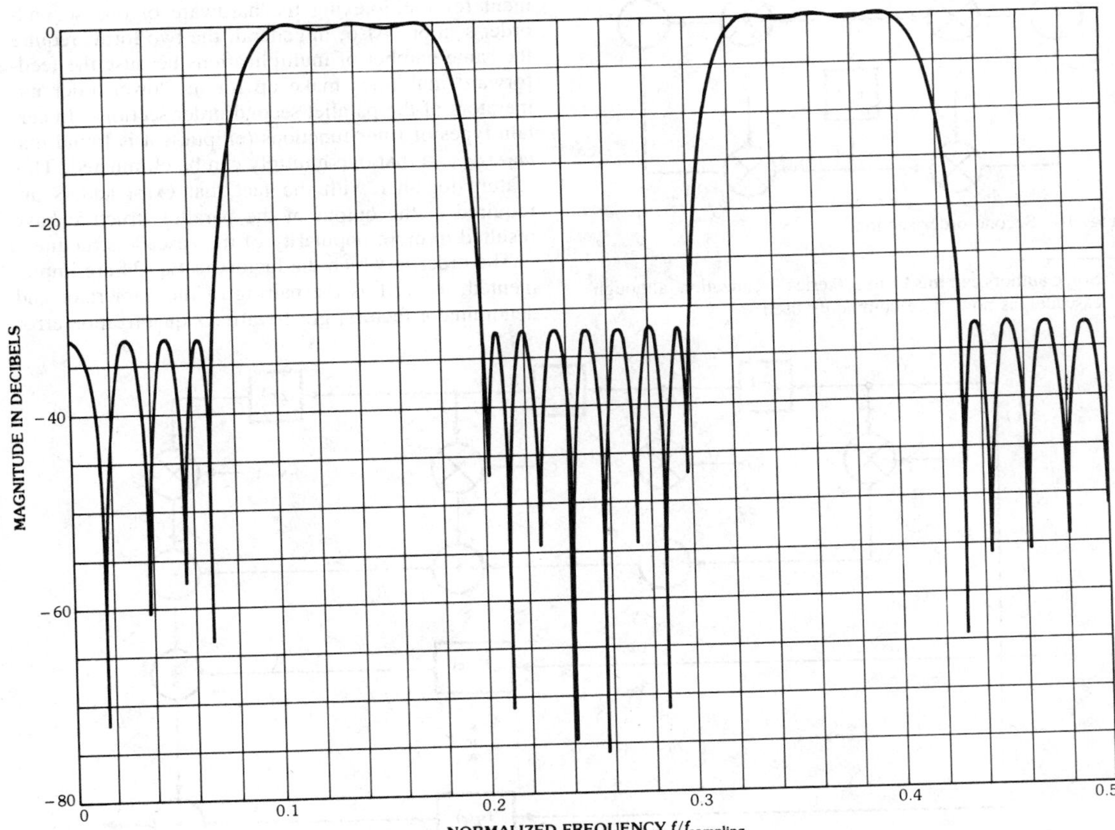

Fig. 9. Response of a 64th-order filter designed with the program of McClellan et. al.

Network Structures

The generic building block for IIR filters is the second-order section (Fig. 10). The canonical-form second-order section requires two delay registers, five multipliers, and four adders and is capable of realizing a pair of complex poles and complex zeros with real-valued multiplier coefficients. It is well known that the sensitivity of the zeros of a polynomial with respect to incremental changes in the coefficients grows large rapidly as the polynomial order increases. Therefore, with respect to transfer function sensitivity and round-off error accumulation, it is always preferable to decompose a high-order transfer function into lower-order subsystems. This results in a parallel or cascade* realization.

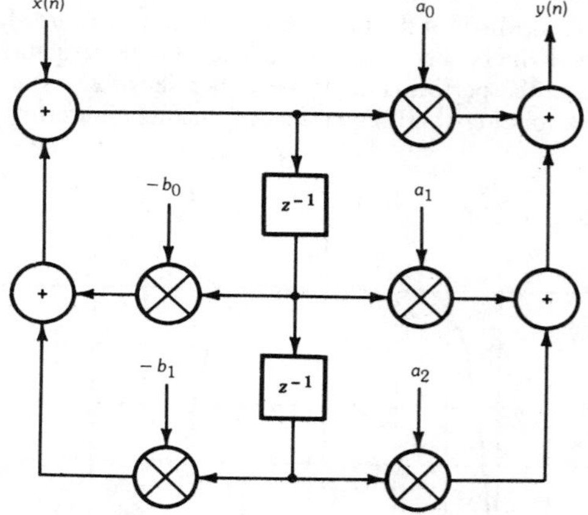

Fig. 10. Second-order section.

* Some authors call this form a "series" realization, although "cascade" is now more commonly used.

In the parallel realization, $H(z)$ is expanded as a partial fraction expansion of the form

$$H(z)=\sum_{k=0}^{M-1} c_k z^{-k}+\sum_{j=1}^{L} H_j(z) \tag{23}$$

where the first M terms are called the feed forward terms, and each $H_j(z)$ has a first-order numerator and a second-order denominator of the form

$$H_j(z)=(\gamma_{0j}+\gamma_{1j}z^{-1})/(1+b_{1j}z^{-1}+b_{2j}z^{-2})$$

The network structure takes the form shown in Fig. 11. The C_k's result by dividing the denominator of $H(z)$ into the numerator until the remainder is in proper form to expand into partial fractions.

In the cascade realization, $H(z)$ is factored into numerator and denominator factors that are at most second order, i.e.

$$H(z)=H_1(z)\cdot H_2(z)\ .\ .\ .\ H_L(z) \tag{24}$$

where each factor has the form

$$H_j(z)=(a_{0j}+a_{1j}z^{-1}+a_{2j}z^{-2})/(1+b_{1j}z^{-1}+b_{2j}z^{-2})$$

The cascade structure is shown in Fig. 12.

Both the parallel and cascade structures are convenient for multiplexing the hardware of one second-order section. Also, in general, the two forms require the same number of multiplications because the feed-forward multiplies make up for the lower-order numerators of the parallel second-order sections. In certain types of filter functions (elliptic), it is found that $a_{0j}=a_{2j}$, so that one multiply can be eliminated. This feature together with the fact that extra adders are required at the output of the parallel structure have resulted in more popularity of the cascade structure.

The order in which the factors in Eq. 24 are implemented, as well as the pairing of the numerator and denominator factors, greatly affects quantization error

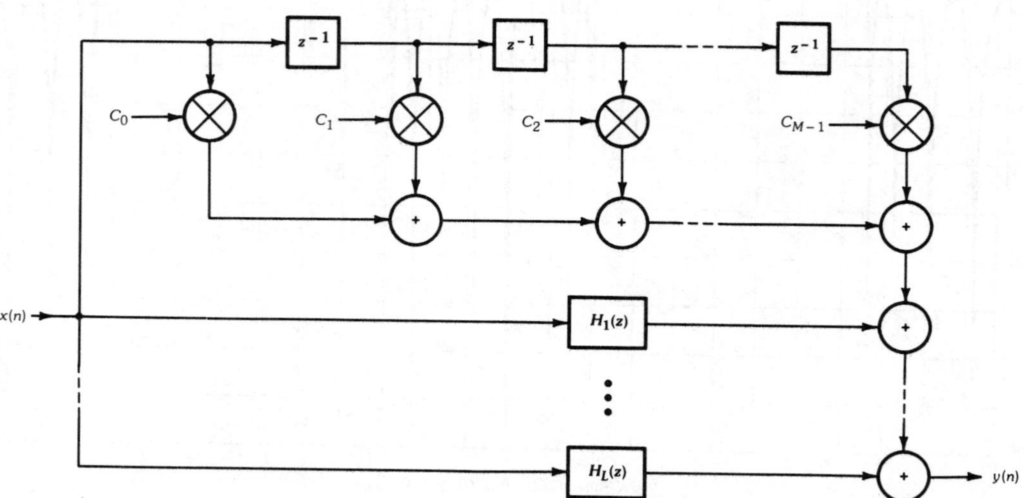

Fig. 11. Parallel realization.

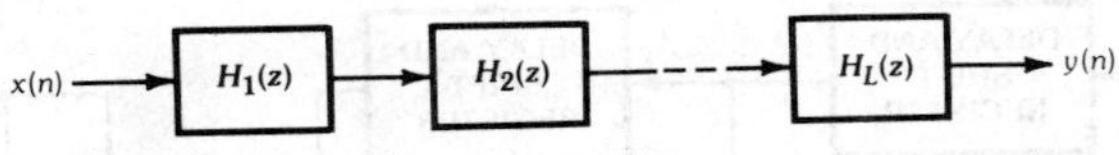

Fig. 12. Cascade realization.

accumulation and represents an additional design parameter for cascade filters. The problem has been called pole-zero pairing, and at the present time there does not appear to be a known analytical technique to determine the correct pairing for minimum-error performance. Jackson* studied the problem and suggested some "rules of thumb," but the final result is usually achieved by computer simulation.

Two other realizations for the second-order section have been proposed for special situations. The coupled-form second-order section, shown in Fig. 13, realizes a complex pair of poles, $p_k = re^{\pm j\theta}$ with the transfer function

$$H(z) = r \sin \theta \, z^{-1}/[1 - (2r \cos \theta)z^{-1} + r^2 z^{-2}] \quad (25)$$

The coupled form is a low-sensitivity second-order section that is preferable for very narrow-band low-pass filters, which are well known to be quite sensitive with respect to coefficient quantization. A second structure that has gained popularity in the last few years is a multiplierless ROM-accumulator realization that was popularized by Peled and Liu.† Assume that the second-order section is characterized by the difference equation

$$\begin{aligned} y(n) = a_0 x(n) &+ a_1 x(n-1) \\ &+ a_2 x(n-2) \\ &- b_1 y(n-1) - b_2 y(n-2) \end{aligned} \quad (26)$$

and that each data sample is encoded as a $(b+1)$-bit twos-complement binary word, e.g.

$$x(n) = x^0(n) + \sum_{j=1}^{b} x^j(n)\, 2^{-j} \quad (27)$$

If Eq. 27 is used in Eq. 26 for both the $x(n)$'s and the $y(n)$'s and the order of summations is interchanged, the output of the section can be expressed as

$$y(n) = \sum_{j=1}^{b} 2^{-j}\, \phi(A_j(n)) - \phi(A_0(n)) \quad (28)$$

where $A_j(n) = x^j(n),\ x^j(n-1),\ x^j(n-2),\ y^j(n-1),\ y^j(n-2)$ is a five-bit binary address that is used to address the stored function

$$\begin{aligned} \phi(A_j(n)) = a_0 x^j(n) &+ a_1 x^j(n-1) \\ &+ a_2 x^j(n-2) - b_1 y^j(n-1) - b_2 y^j(n-2) \end{aligned}$$

The structure for the resulting filter is shown in Fig. 14. The output, $y(n)$, is computed by a sequence of memory fetches, shifts, and adds. Therefore, multiplication has been entirely eliminated. This architecture is very appealing for high-speed real-time operation, as well as for filters that may be integrated as part of a VLSI system implementation. For a second-order section, the ROM must have 32 words of memory, and the generation of each $y(n)$ requires $b+1$ memory fetches and b add-shift cycles. The major disadvantage of this structure is that the filter coefficients are fixed in the ROM according to the function $\phi(\cdot)$. Although several ROMs can be used to select among a predetermined set of filter functions, the ROM-accumulator structure is not well suited for adaptive filters, where the a_i's and b_i's are changed in a continual time-varying fashion. This is due to the fact that a change of only a few coefficients would require a complete recalculation of the stored $\phi(\cdot)$ function, thereby nullifying the efficiencies of using a stored function in the first place.

The most common network structure for FIR filters is shown in Fig. 15. A length-N filter requires N multipliers, $N-1$ delay registers, and $N-1$ two-input adders. Normally, one hardware multiplier and one adder are time-multiplexed to minimize hardware costs, resulting in a low-cost/low-speed realization. When higher speed is required, more computational elements are used in a pipe-lined configuration, resulting in a high-speed/high-cost realization. Since there is no feedback, limit cycles cannot occur, and quantization

* Reference 15.

† Reference 28.

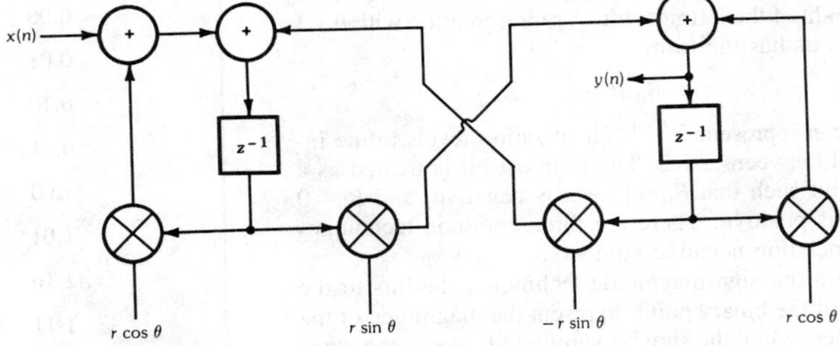

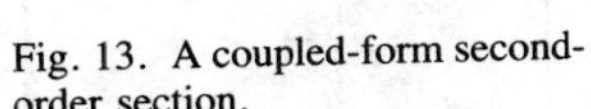
Fig. 13. A coupled-form second-order section.

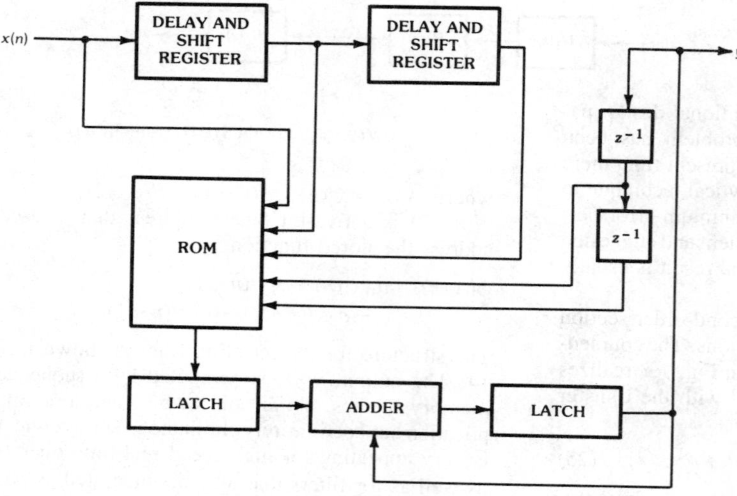

Fig. 14. ROM-accumulator realization of a second-order section.

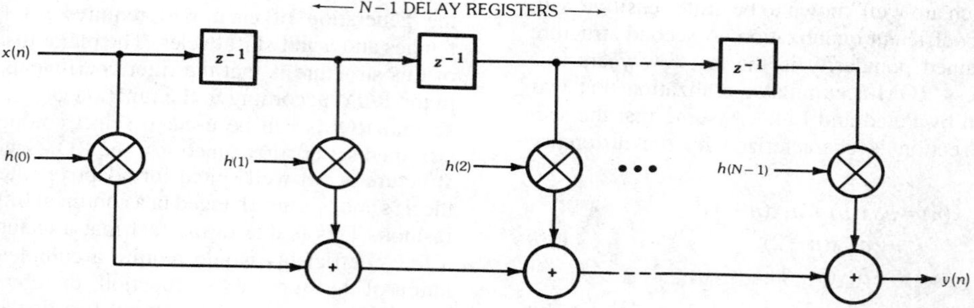

Fig. 15. Common FIR filter structure.

error in the output is simply the additive result of all the quantization error in the multipliers.

Arithmetic Number Codes

A complete discussion of binary data formats, arithmetic number codes, and popular algorithms for binary addition, multiplication, and division is presented in Chapter 42. The following summarizes important features of fixed-point binary representation that is commonly used in signal-processing hardware.

The most common fixed-point format is the binary fraction, in which the binary point is assumed to lie to the right of the leftmost bit. A representation with $b+1$ total bits has the form

$$x \cong B_0.B_{-1} \ldots B_{-b}$$

and can represent $2^{(b+1)}$ quantization levels in the interval between $\pm 1/2$. The leftmost bit is treated as a sign bit such that $B_0 = 1 \rightarrow x$ is negative, and $B_0 = 0 \rightarrow x$ is positive. There are three common techniques for encoding negative numbers.

With the sign-magnitude technique, the bits to the right of the binary point represent the magnitude of the number, while the sign bit simply indicates if the number is positive ($B_0 = 0$) or negative ($B_0 = 1$). For example, with three bits, seven states can be represented (± 0 are the same state), as listed in Table 4. Sign-magnitude representation is convenient for sign detection and magnitude comparison, although it requires

TABLE 4. STATES FOR THREE BITS (SIGN-MAGNITUDE)

Binary	Decimal
0.00	0
0.01	$^1/_4$
0.10	$^1/_2$
0.11	$^3/_4$
1.00	-0
1.01	$-^1/_4$
1.10	$-^1/_2$
1.11	$-^3/_4$

that subtraction be treated as a separate operation (i.e., subtraction cannot be implemented as the addition of an additive inverse).

With the twos-complement technique, a negative fraction in the range $-1 \leq x < 0$ is represented by $\overline{X}_2 = 2 - |x|$. A system with three bits can represent eight states (Table 5). The complement, $\overline{X}_2$, can be obtained from x by complementing each bit of x and adding 2^{-b} to this result. In this system, B_0 serves as

TABLE 5. STATES FOR THREE BITS (TWOS-COMPLEMENT)

Binary	Decimal
0.00	0
0.01	1/4
0.10	1/2
0.11	3/4
1.00	−1
1.01	−3/4
1.10	−1/2
1.11	−1/4

a sign bit, and subtraction is implemented as the addition of an additive inverse. When two numbers are added, $z = x + y$, and the result, z, falls in the range $[-1,1]$ carry overflow from the sign bit can be simply discarded. For example,

$$-\tfrac{1}{4} \cong 1.11$$
$$\underline{-\tfrac{1}{2} \cong 1.10}$$
$$-\tfrac{3}{4} \cong \not{1}1.01 \rightarrow 1.01 \text{ (correct result)}$$

discard overflow bit.

With the ones-complement technique, a negative fraction in the range $-1 < x < 0$ is represented by $\overline{x}_1 = (2 - 2^{-b}) - |x|$. A system with three bits can represent seven states (Table 6). The complement, $\overline{x}_1$, can be obtained from x by simply complementing each bit of x. The codes 0.00 and 1.11 are redundant representations of the zero state. Bit B_0 serves as a sign bit, and subtraction is implemented as the addition of an additive inverse. In this system, carry overflow from the sign bit during addition is added onto the least significant bit (end-around carry):

$$-\tfrac{1}{4} \cong 1.10$$
$$\underline{-\tfrac{1}{2} \cong 1.01}$$
$$-\tfrac{3}{4} \cong 10.11 \rightarrow 1.00$$

The most popular number code presently used in

TABLE 6. STATES FOR THREE BITS (ONES-COMPLEMENT)

Binary	Decimal
0.00	0
0.01	1/4
0.10	1/2
0.11	3/4
1.00	−3/4
1.01	−1/2
1.10	−1/4

special-purpose digital hardware, in commercially available microprocessors, and in programmable signal-processing chips is twos complement. This is due largely to the simplicities provided by ignoring carry overflow and due to the unique representation for the zero state. Twos-complement adders are constructed by interconnecting a number of full adders, and in many cases, carry look-ahead circuits are provided to improve speed. There are many different types of digital multipliers, ranging from the slow shift-add types to the high-speed array structures.†

All of the above codes have a particular feature in common: they are weighted number codes. This means that there are varying weights attached to the various bits, so that some bits are more important than others. An interesting nonweighted number code is the residue number system (RNS) code, which has attracted considerable interest among researchers during recent years. The basis of an RNS code is a finite set of mutually prime integers $\mathfrak{M} = \{m_1, m_2, \ldots, m_L\}$. The code represents $M = \prod_{i=1}^{L} m_i$ integer states in the interval $\Re = [-(M-1)/2, (M-1)/2]$ if M is odd, or $\Re = [-M/2, (M/2)-1]$ if M is even. An integer $X \in \Re$ is encoded by L residue digits $X \cong x_1 x_2 \ldots x_L$, where

$$x_i = \begin{Bmatrix} (X) \bmod m_i, & X < 0 \\ m_i - |X| \bmod m_i, & X \geq 0 \end{Bmatrix}$$

If $* \in \{+, \cdot, -\}$, then $Z = X*Y$ is defined by $z_i = (x_i * y_i) \bmod m_i$, $i = 1, \ldots, L$. Therefore RNS arithmetic does not require a carry between the residue digits during these algebraic operations. Since the m_i's are generally small integers, modular arithmetic can easily be implemented by small look-up tables stored in read-only memories (ROMs). This results in a system architecture with many parallel data paths, and

† An excellent review of logic families, commercial logic packages, multipliers, memories, and multiplexers is given in chapter 8 of reference 32.

which is well suited for high-speed applications. The inherent parallelism prevents the propagation of errors from one digit to the next, so that the number code also provides an inherent error-isolation property that is useful for fault-tolerant designs.

Certain operations such as division, sign detection, and magnitude comparison are relatively difficult to carry out within the RNS code. This has prevented RNS arithmetic from being used in commercial general-purpose computers. However, RNS techniques appear to be useful for special computation-bound digital processing functions such as FIR filtering.‡ Also, the inherent parallelism in the RNS code provides interesting properties that are currently being studied for VLSI implementation.**

Finite Wordlength Effects

Three basic sources of quantization error occur in digital filters: (1) errors due to quantizing an input signal in the A/D converter, (2) errors introduced by rounding during arithmetic operations, and (3) errors in the filter response due to a finite number of bits used in the filter coefficients. The following describes quantization in fixed-point systems. (Floating-point systems are treated in references 20 and 26.)

Input Error—Quantization that occurs in the A/D converter does not corrupt the filter itself, but rather acts as an additive noise component at the input of the ideal system. This noise cannot cause filter instabilities, although it does create inaccuracy in the time-domain response. A common model for A/D quantization noise is shown in Fig. 16, where $e(n)$ is usually assumed to be white noise that is uncorrelated with the signal, $x_a(nT)$, and which is uniformly distributed over the smallest quantization interval. If the probability

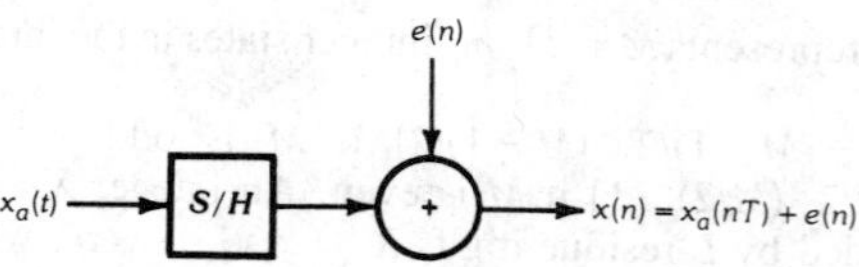

Fig. 16. Noise model for A/D quantization error.

density function has the form shown in Fig. 17, then the mean and variance of the noise at the filter input are $m_e=0$ and $\sigma_e^2=[2^{-2b}/12]$. Then the mean and variance of the noise in the filter output, $\epsilon(n)$, is given by

$$m_\epsilon=0$$

$$\sigma_\epsilon^2=[2^{-2b}/12]\sum_{n=-\infty}^{+\infty}|h(n)|^2 \qquad (29)$$

$$=[2^{-2b}/24\pi]\int_{-\pi}^{\pi}|H(e^{j\omega})|^2 d\omega$$

‡ Reference 17.

** Reference 16.

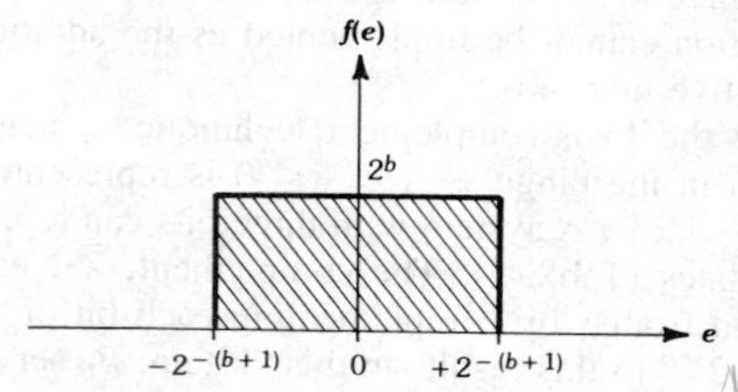

Fig. 17. Pdf for rounded A/D quantization.

where $h(n)$ is the unit pulse response and $H(e^{j\omega})$ is the frequency response of the DT system. The statistics of $e(n)$ depend on the specific hardware details of the A/D converter; i.e., different assumptions would be made depending on whether the quantization is implemented by rounding or truncation, and whether the digital code is ones-complement, twos-complement, or sign-magnitude.

Coefficient Error—Filter coefficients are represented by a finite number of bits, commonly in the range of 8 to 16 bits. This results in the implementation of a filter that differs slightly from the original design. The effect is particularly noticeable in IIR filters, mainly because the poles are moved when the coefficients are quantized. If the unquantized design has a pole very close to the unit circle (or on the unit circle, as in an iterative sine generator), slight coefficient changes may move a pole across the unit circle, causing instability. Pole movement also affects frequency response, so the final result must be carefully checked to ensure that coefficient quantization does not cause intolerable magnitude (phase) distortion.

Coefficient errors are fixed at the time of implementation. These errors have no inherently random properties, although one technique of predicting their effect is to treat them as random variables.* This approach has been quite successful in predicting the number of bits required to keep the frequency response between specified tolerance limits.

Uncorrelated Roundoff Errors—A product resulting from the multiplication of two $b+1$ bit twos-complement numbers requires $2b+1$ bits for its representation. Therefore a product must be rounded (truncated) to maintain an N-bit data stream. Roundoff is an inherently random process when viewed over a great number of computations. It has been rather successfully modeled as a random noise entering the system through a summation node immediately following a multiplier. If the assumption holds that the error from each multiplier is a white-noise source that is uncorrelated with other quantization errors in the structure, roundoff simply causes a jitter in the output, which can be well described by a statistical noise analysis.†

For example, a second-order section with two real

* References 6 and 19.

† Reference 26.

poles at $z=1/2$ and $z=1/3$ is shown in Fig. 18 as a cascade of two first-order sections. The errors $e_1(n)$, and $e_2(n)$, and $e_3(n)$ are assumed to be uncorrelated white-noise sources that represent the quantization errors due to the multipliers 1/2, −5/6, and 1/3. (Note that the multiplier 2 does not require quantization.) If it is also assumed that the number code is twos-complement with $b+1$ bits and that the quantization is rounding, then the statistics are $m_{e_i}=0$ and $\sigma_{e_i}^2=[2^{-2b}/12]$, for $i=1$, 2, and 3. If the e_i's are uncorrelated with the data stream, the output can be expressed as $\hat{y}(n)=y(n)+\epsilon(n)$ with

$$m_\epsilon=0$$

and

$$\sigma_\epsilon^2=\sigma_{e_1}^2\sum_{n=0}^{\infty}|h(n)|^2+(\sigma_{e_2}^2+\sigma_{e_3}^2)\sum_{n=0}^{\infty}|h_2(n)|^2$$

where,

$h(n)=[(1/2)^n+(1/3)^n]u(n)$is the overall system unit pulse response,

$h_2(n)=(1/3)^n u(n)$ is the unit pulse response of the second section.

Furthermore, it is found that $\sigma_\epsilon^2=[7.26/12]2^{-2b}$. If the order of the sections is interchanged, a similar analysis reveals that $\sigma_\epsilon^2=[9.89/12]2^{-2b}$. This demonstrates that the error accumulation in the output of a cascade filter depends on the ordering of the sections, and that in general there is a preferred ordering (the first in this case).

This general approach to the analysis of arithmetic quantization errors can be extended to higher-order filters realized in any configuration, provided the transfer functions (unit pulse responses) from each noise source to the filter output can be determined. Noise analysis is particularly well suited for computer-aided analysis routines that obtain the required transfer function as a step during a frequency-domain analysis. One such computer program is called DINAP*. Figs. 19 and 20 show the roundoff noise spectrum at the output of a fourth-order Butterworth filter realized with 18-bit twos-complement arithmetic in the parallel and cascade form, as generated by DINAP. The noise spectrum for the parallel structure is somewhat smaller than that of the cascade structure, although for this filter the difference is not very great.

In filters implemented with floating-point arithmetic, a roundoff error can also occur during addition. When two floating-point numbers are added, the exponents must be made equal, usually by scaling the smaller exponent until it equals the larger. This requires a right shift of the mantissa, resulting in a loss of bits on the lower end of the word and the necessity for rounding (truncating). Adder roundoff is also an inherently random process, although there is evidence that the process is not zero mean, and elementary statistical assumptions are more difficult to justify than in the case of multiplicative roundoff error.†

There arise situations in which roundoff errors become correlated. This often occurs during zero input response, when all internal states are decaying toward the zero state. Roundoff forces the filter states to take on one of a finite number of distinct levels. Forcing the state into these unnatural quanta by the nonlinear quantization scheme often results in limit cycles. These small amplitude limit cycles, which have been called *deadbands*, are a manifestation of highly correlated roundoff errors. In most filters using fixed-point hardware, and in some using floating-point hardware, these deadbands result in idle channel noise, which can be unsatisfactory. Since it is a tedious job to discover all possible limit cycles, a great amount of attention has been focused on the computation of an absolute bound

* *D*igital *N*etwork *A*nalysis *P*rogram (DINAP) was written by S. Better and S. C. Bass at Purdue University. The program performs transient analysis, frequency analysis, and statistical noise analysis for any network composed of digital adders, multipliers, and unit delays.

† Reference 21.

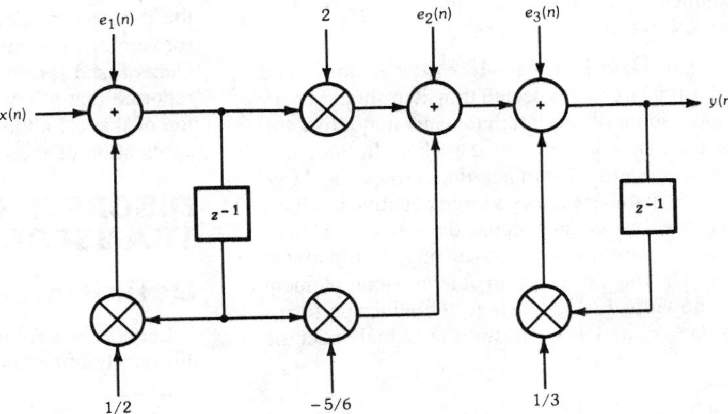

Fig. 18. Second-order section with roundoff error sources $e_1(n)$, $e_2(n)$, and $e_3(n)$.

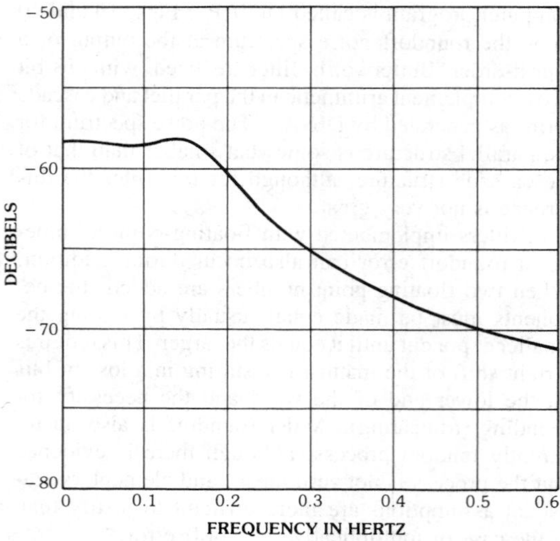

Fig. 19. Roundoff noise spectrum for fourth-order Butterworth low-pass filter in parallel form.

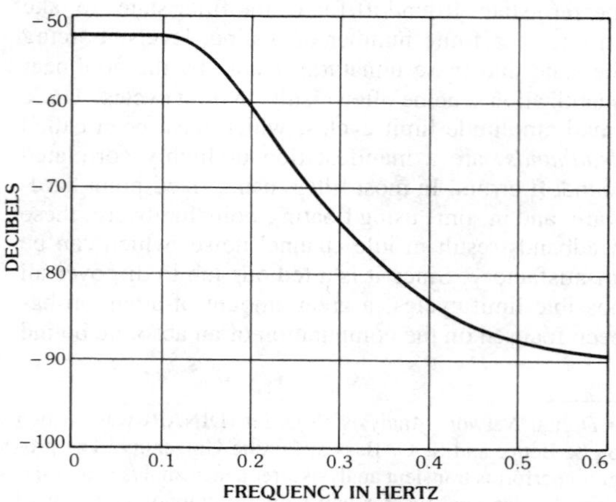

Fig. 20. Roundoff noise spectrum for fourth-order Butterworth low-pass filter in cascade form.

that is tight enough to be useful, and yet which bounds all possible limit cycles.**

Overflow Oscillations—If a filter is not scaled properly or if a large unexpected transient should occur at the input, some of the internal states may grow too large for the dynamic range of the filter. In the twos-complement system, a number that grows too large reenters the dynamic range as a large negative number. This phenomenon occurs because the range of a twos-complement system forms a closed ring, as illustrated in Fig. 21. The transfer function of a twos-complement adder is shown in Fig. 22, where the adder performs $[x+y]=f(x+y)$ and $[\cdot]$ means the actual machine

** Reference 27.

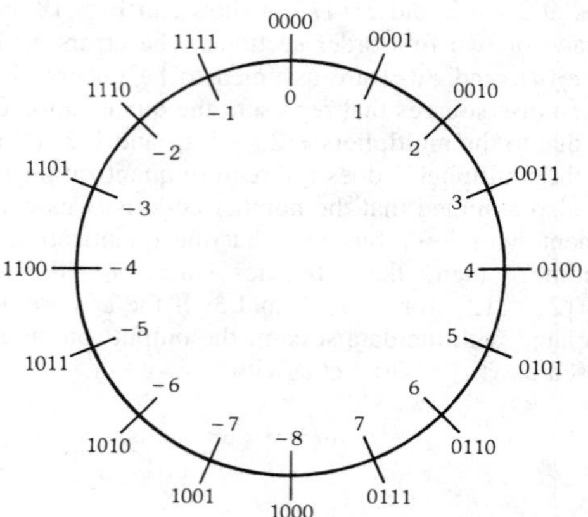

Fig. 21. Dynamic range of twos-complement number system.

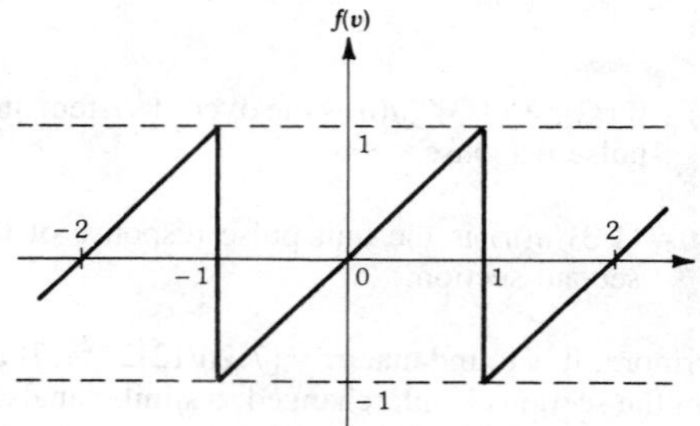

Fig. 22. The transfer function of a twos-complement adder.

result. The sudden change in sign due to range overflow can result in large self-sustained limit cycles, which are called *overflow oscillations*. These oscillations are characterized as being large and disabling, contrary to the deadband effect, which is usually small and primarily a source of annoyance.

Overflow oscillations cannot be tolerated. One way to prevent them is to scale the input so that overflow at internal summation points cannot occur. A second method is to alter the adders as shown in Fig. 23 so that they saturate (or zero) when the result is outside the dynamic range. Such a scheme, called saturating (or zeroing) arithmetic, will stop overflow oscillations. Generalized forms of saturation arithmetic have been reported that allow $f(v)$ to take on a continuum of different shapes, although there appears to be no practical application of these generalizations to date.**

DISCRETE FOURIER TRANSFORM

Definitions and Properties

Let $x(n)$ be a finite-length sequence of length N. The discrete Fourier transform (DFT) pair is defined by

** Reference 27.

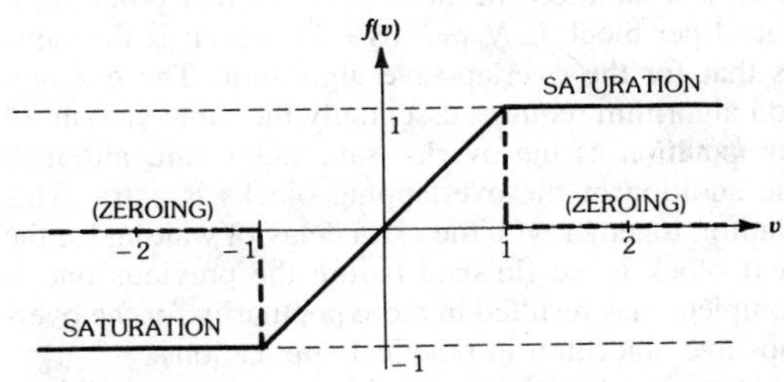

Fig. 23. Modified twos-complement adder.

$$X(k) = \sum_{n=0}^{N-1} x(n)\, e^{-j(2\pi/N)\, nk} \qquad (30a)$$

$$k = 0, \ldots, N-1$$

and

$$x(n) = (1/N) \sum_{k=0}^{N-1} X(k) e^{j(2\pi/N)\, nk} \qquad (30b)$$

$$n = 0, \ldots, N-1$$

The DFT treats $x(n)$ as though it were one period of a periodic sequence. This is an important feature that must be handled properly in signal processing to prevent the introduction of artifacts. Important properties of the DFT are summarized in Table 7. The notation $((k))_N$ denotes the least positive residue of k modulo N, and $R_N(n)$ is a rectangular window function such that $R_N(n) = 1$ for $n = 0, \ldots, N-1$, and $R_N(n) = 0$ for $n \leq 0$ and $n \geq N$.

Most of the properties given in Table 7 for the DFT are the same as those of the z-transform, although there are some important differences. Property 4, the shift property, holds for *circular* shifts of the finite-length sequence $x(n)$. Also, the multiplication of two DFTs results in the *circular convolution* of the corresponding time-domain sequences (property 6).

Suppose it is desired to implement an FIR filter $y(n) = \sum_{k=0}^{N-1} h(k)x(n-k)$ by transforming $h(n)$ and $x(n)$ into $H(k)$ and $X(k)$ using a DFT, multiplying pointwise to obtain $Y(k) = H(k)X(k)$, and then using the inverse-FFT to obtain $y(n) = \text{IFFT}\{Y(k)\}$. If $x(n)$ is a finite-length sequence of length M, then the results of the

TABLE 7. PROPERTIES OF THE DFT*

Finite-Length Sequence (length N)	DFT
1. $x(n)$	$X(k)$
2. $y(n)$	$Y(k)$
3. $ax(n) + by(n)$	$aX(k) + bY(k)$
4. $x((n+m))_N \mathcal{R}_N(n)$	$W_N^{-km} X(k)$
5. $W_N^{ln} x(n)$	$X((k+l))_N \mathcal{R}_N(k)$
6. $\left[\sum_{m=0}^{N-1} x((m))_N y((n-m))_N\right] \mathcal{R}_N(n)$	$X(k)Y(k)$
7. $x(n)y(n)$	$(1/N)\left[\sum_{l=0}^{N-1} X((l))_N Y((k-l))_N\right] \mathcal{R}_N(k)$
8. $x^*(n)$	$X^*((-k))_N \mathcal{R}_N(k)$
9. $x^*((-n))_N \mathcal{R}_N(n)$	$X^*(k)$
10. Re $[x(n)]$	$X_{ep}(k) = \frac{1}{2}[X((k))_N + X^*((-k))_N]\mathcal{R}_N(k)$
11. j Im $[x(n)]$	$X_{op}(k) = \frac{1}{2}[X((k))_N - X^*((-k))_N]\mathcal{R}_N(k)$
12. $x_{ep}(n)$	Re $[X(k)]$
13. $x_{op}(n)$	j Im $[X(k)]$
The following properties apply only when $x(n)$ is real:	
14. Any real $x(n)$	$X(k) = X^*((-k))_N \mathcal{R}_N(k)$ Re $[X(k)]$ = Re $[X((-k))_N]\mathcal{R}_N(k)$ Im $[X(k)] = -$Im $[X((-k))_N]\mathcal{R}_N(k)$ $\lvert X(k) \rvert = \lvert X((-k))_N \rvert \mathcal{R}_N(k)$ arg $[X(k)] = -$arg $[X((-k))_N]\mathcal{R}_N(k)$
15. $x_{ep}(n)$	Re $[X(k)]$
16. $x_{op}(n)$	j Im $[X(k)]$

* From A. V. Oppenheim and R. W. Schafer. *Digital Signal Processing*. Englewood Cliffs, N. J.: Prentice-Hall, Inc., 1975.

circular convolution implemented by the FFT will correspond to the desired linear convolution if and only if $N_{FFT} \geq N+M-1$, where N_{FFT} is the block length of the FFT and $h(n)$ and $x(n)$ are each padded with zeros to create blocks of length N_{FFT}. In some applications, either the value of M is too large for the memory available, or $x(n)$ may not be finite in length, but rather it may be a continual stream of data that arrives at the filter for real-time processing. Two algorithms are available that partition $x(n)$ into smaller blocks and process the individual blocks with a smaller-length FFT: (1) overlap-save partitioning and (2) overlap-add partitioning. These are summarized below.

Overlap-Save—In this algorithm, N_{FFT} is chosen to be some convenient value with $N_{FFT}>N$. The signal, $x(n)$, is partitioned into blocks which are of length N_{FFT} and which overlap by $N-1$ data points. Hence, the kth block is $x_k(n)=x(n+k(N_{FFT}-N+1))$, $n=0, \ldots, N-1$. The filter is augmented with $N_{FFT}-N$ zeros to produce

$$\hat{h}(n)=\begin{cases} h(n) & n=0, \ldots, N-1 \\ 0 & N \leq n < N_{FFT} \end{cases}$$

The FFT is then used to obtain $\hat{Y}(k)=\mathrm{FFT}\{\hat{h}(n)\} \cdot \mathrm{FFT}\{x_k(n)\}$, and $\hat{y}_k(n)=\mathrm{IFFT}\{\hat{Y}(k)\}$. From the $\hat{y}(n)$ array, the values that correctly correspond to the linear convolution are selected and saved; values that are "contaminated by wrap-around error" are discarded; i.e.,

$$y_k(n)=\begin{cases} \hat{y}_k(n), & N-1 \leq n \leq N_{FFT}-1 \\ 0 \quad , & 0 \leq n \leq N-1 \end{cases}$$

The blocks of output data $y_k(n)$ are then assembled to produce the correct output by

$$y(n)=\sum_{k=0}^{\infty} y_k(n-k(N_{FFT}-N+1))$$

For the overlap-save algorithm, each time a block is processed there are $N_{FFT}-N+1$ points saved and $N-1$ points discarded. Each block moves forward by $N_{FFT}-N+1$ data points and overlaps the previous block by $N-1$ points.

Overlap-Add—This algorithm is similar to the previous one except that the kth input block is defined to be

$$x_k(n)=\begin{cases} x(n+k\,L), \; n=0, \ldots, L-1 \\ 0 \qquad\quad , \; n=L, \ldots, N_{FFT}-1 \end{cases}$$

where $L=N_{FFT}-N+1$. The filter function $\hat{h}(n)$ is augmented with zeros, as before, to create $\hat{h}(n)$, and the DFT processing is executed exactly as before. In each block $\hat{y}_k(n)$ that is obtained at the output, the first $N-1$ points are "bad," the last $N-1$ points are "bad," and the middle $N_{FFT}-2(N-1)$ points are "good." However, if the last $N-1$ points from block k are overlapped with the first $N-1$ points of block $k+1$ and added pointwise, correct results corresponding to linear convolution are obtained in these positions. Hence, after this addition the number of correct points produced per block is $N_{FFT}-(N-1)$, which is the same as that for the overlap-save algorithm. The overlap-add algorithm requires essentially the same amount of computation as the overlap-save algorithm, although the addition of the overlapping blocks is extra. This feature, together with the extra delay of waiting for the next block to be finished before the previous one is complete, has resulted in more popularity for the overlap-save algorithm in practical applications.

These block filtering algorithms make it possible to filter continual data streams in real time because efficiency of the FFT minimizes the total computation time and can achieve reasonably high overall data rates. However, block filtering generates data in bursts; i.e., there is a delay during which no filtered data appears, and then suddenly an entire block is generated. In real-time systems, buffering must be used. The block algorithms are particularly effective for filtering very long sequences of data that are prerecorded on tape or disk.

FFT Algorithms

"Fast Fourier transform"(FFT) is a generic name for a class of algorithms that efficiently compute the DFT (see Eqs. 30). The FFT is easily understood by examining a radix-2 FFT for the case $N=2^3$. First, each of the indices k and n can be expressed in binary form, $k=k_2 4+k_1 2+k_0$ and $n=n_2 4+n_1 2+n_0$, where k_i and n_i are bits that take the values of either 0 or 1. If these expressions are substituted into Eq. 30a, all terms in the exponent that contain the factor $N=8$ can be simply deleted, because $e^{j2\pi l}=1$ for any integer l. Upon deleting such terms and regrouping, the product nk can be expressed in one of two ways:

$$nk=(4k_0)n_2+(4k_1+2k_0)n_1 + (4k_2+2k_1+k_0)\,n_0 \qquad (31a)$$

$$nk=(4n_0)k_2+(4n_1+2n_0)k_1 + (4n_2+2n_1+n_0)\,k_0 \qquad (31b)$$

Substituting Eq. 31a into Eq. 30a leads to the decimation-in-time (D-I-T) FFT, whereas substituting Eq. 31b into Eq. 30a leads to the decimation-in-frequency (D-I-F) FFT. Only the D-I-T FFT is discussed further here. The D-I-F FFT and various related forms are treated in detail in reference 4.

The D-I-T FFT decomposes into $\log_2 N$ stages of computation, plus a stage of bit reversal. (Let $W_N= e^{j2\pi/N}$.)

$$x_1(k_0,n_1,n_0)=\sum_{n_2=0}^{1} x(n_2,n_1,n_0)W_8^{4k_0 n_2} \qquad \text{(stage 1)}$$

$$x_2(k_0,k_1,n_0)=\sum_{n_1=0}^{1} x_1(k_0,n_1,n_0)W_8^{(4k_1+2k_0)n_1} \qquad \text{(stage 2)}$$

$$x_3(k_0,k_1,k_2)=\sum_{n_0=0}^{1} x_2(k_0,k_1,n_0)W_8^{(4k_2+2k_1+k_0)n_0} \quad \text{(stage 3)}$$

$$X(k_2,k_1,k_0)=x_3(k_0,k_1,k_2) \quad \text{(bit reversal)}$$

In each summation above, one of the n_i's is summed out of the expression, while at the same time a new k_i is introduced. The notation is chosen to reflect this. For example, in stage 3, n_0 is summed out, k_2 is introduced as a new variable, and n_0 is replaced by k_2 in the result. The last operation, called bit reversal, is necessary to correctly locate the frequency samples $X(k)$ in the memory. It is easy to show that if the samples are paired correctly, an in-place computation can be done by a sequence of butterfly computations. For example, in stage 3 the $k=6$ and $k=7$ samples should be paired, yielding a butterfly computation that requires one complex multiply, one complex add, and one complex subtract.

$$x_3(1,1,0)=x_2(1,1,0)+W_8^3\ x_2(1,1,1)$$

$$x_3(1,1,1)=x_2(1,1,0)-W_8^3\ x_2(1,1,1)$$

Therefore samples $x_2(6)$ and $x_2(7)$ are read from memory, the butterfly is executed, and $x_3(6)$ and $x_3(7)$ are written back to memory, thereby destroying the original values of $x_2(6)$ and $x_2(7)$. In general, there are $N/2$ butterflies per stage and $\log_2 N$ stages, so the total number of butterflies is $(N/2)\log_2 N$. Since there is at most one complex multiply per butterfly, the total number of multiplies is bounded by $(N/2)\log_2 N$ (some of the multiplies involve factors of 1, and should not be counted). A direct computation of the DFT requires on the order of N^2 complex multiplies. It is obvious that for large N, the order of complexity of the FFT $0\{(N/2)\log_2 N\}$ is much less than the order of complexity $0\{N^2\}$ of the DFT. The saving is so dramatic for large N that the FFT has made possible the solution of many DSP problems that are compute-bound and impossible to solve with the direct DFT.

Fig. 24 shows the signal-flow graph for the D-I-T FFT with $N=8$, which is referred to as an in-place FFT with normally ordered input and bit reversed output. Minor variations that include bit reversed input and normally ordered output, and non-in-place algorithms with normally ordered inputs and outputs are possible. Also, when N is not a power of 2, a mixed-radix algorithm can be used to reduce computation.* The mixed-radix FFT is most efficient when N is highly composite, i.e., $N=\prod_{i=1}^{L} P_i^{r_i}$ where the p_i's are small primes and the r_i's are positive integers. It can be shown that the order of complexity of the mixed-radix FFT is $0(N\sum_{i=1}^{L} r_i(p_i-1))$. Because of the lack of uniformity of structure among stages, this algorithm has not received much attention for hardware implementation. However, the mixed-radix FFT is often used in software applications, especially for processing data recorded in laboratory experiments where it is not convenient to restrict the block lengths to be powers of 2. The most widely used mixed-radix FFT program was published by Singleton.* Singleton's program is also contained in the collection of DSP programs published by the IEEE Acoustics, Speech, and Signal Processing Society.†

* Reference 33.

† Reference 30.

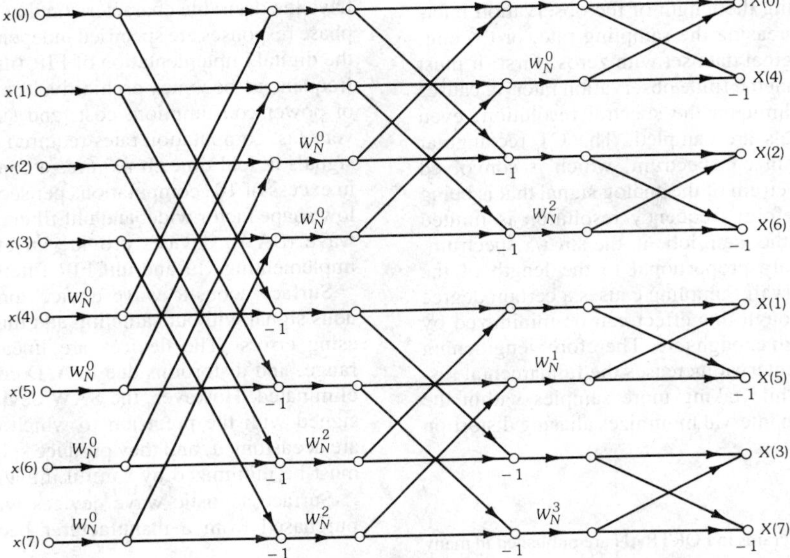

Fig. 24. Decimation-in-time FFT algorithm with normally ordered inputs and bit reversed output.

A radix-2 decimation-in-time FFT program was published by Peled and Liu.** This program, written in PL/1, is listed in Chart 1 for reference.‡

The FFT in Spectral Analysis

An FFT program is often used to perform spectral analysis on signals that are sampled and recorded as part of laboratory experiments, or in certain types of data acquisition systems. There are several issues that should be addressed when spectral analysis is performed on (sampled) analog waveforms that are observed over a finite interval of time.

Windowing—The FFT treats the block of data as though it is one period of a periodic sequence. If the underlying waveform is not periodic, then harmonic distortion may occur because the periodic waveform "created by the FFT" may have sharp discontinuities. This effect is minimized by removing the mean of the data (it can always be reinserted) and by windowing the data so the ends of the block are smoothly tapered to zero and discontinuities do not occur when the FFT treats the windowed block as one period of a periodic sequence. A good rule of thumb is to taper 10 percent of the data on each end of the block using either a cosine taper or one of the other common windows from Table 3.§ An alternate interpretation of this phenomenon is that the finite length observation has already windowed the true waveform with a rectangular window that has large spectral side lobes. Hence, applying an additional window results in a more desirable window that minimizes frequency-domain distortion.

Zero-Padding—It would appear that more accuracy is produced in the spectral domain if the block length of the FFT is increased. This can be done by (1) taking more samples within the observation interval, (2) increasing the length of the observation interval, without increasing the sampling rate, or (3) augmenting the original data set with zeros. First, it must be recognized that the finite observation interval causes a fundamental limit on the spectral resolution, even before the signals are sampled. The CT rectangular window has a $\sin x/x$ spectrum, which is convolved with the true spectrum of the analog signal that is being observed. Therefore, frequency resolution is limited by the width of the main lobe in the $\sin x/x$ spectrum, which is inversely proportional to the length of the observation interval. Sampling causes a certain degree of aliasing, although this effect can be minimized by sampling at a high enough rate. Therefore, lengthening the observation interval increases the fundamental resolution limit, while taking more samples within the same observation interval minimizes aliasing distortion and provides a better definition (more sample points) on the underlying spectrum.

Padding the data with zeros and computing a longer FFT does give more frequency domain points, but it does not improve the fundamental resolution limit, nor does it alter the effects of aliasing. The resolution limits are established by the observation interval and the sampling rate. No amount of zero padding can improve these basic limits. However, zero padding is a useful tool for providing more spectral definition; i.e., it enables one to get a better look at the (distorted) spectrum that exists after the observation and sampling effects have occurred.

Leakage and the Picket-Fence Effect—An FFT with block length N can accurately resolve only frequencies $\omega_k = (2\pi/N)k$, $k = 0, \ldots, N-1$ that are harmonics of the fundamental, $\omega_1 = 2\pi/N$. An analog waveform that is sampled and subjected to spectral analysis may have frequency components between the harmonics. A component at frequency $\omega_{k+1/2} = (2\pi/N)(k+\frac{1}{2})$ will appear to be scattered throughout the spectrum. The effect is illustrated in Fig. 25 for a sinusoid that is observed through a rectangular window and then sampled at N points. The "picket-fence effect" means that not all frequencies can be seen by the FFT. Harmonic components are seen accurately. Other components "slip through the picket fence" while their energy is "leaked" in the harmonics.* These effects produce artifacts in the spectral domain that must be carefully monitored to assure that an accurate spectrum is obtained from FFT processing.

SURFACE-ACOUSTIC-WAVE (SAW) FIR FILTERS

Finite impulse response (FIR) digital bandpass filters have the desirable characteristic that the amplitude and phase responses are specified independently. However, the digital implementation of FIR filters at vhf and uhf frequencies becomes prohibitively expensive in terms of power consumption, cost, and size because of the very fast computation rates required to process the uhf signals in real time. It requires a computer with speed in excess of 10^9 computations per second to implement low shape factor wideband uhf filters. Surface acoustic wave (SAW) devices provide an efficient means of implementing vhf and uhf FIR filters.

Surface acoustic wave devices process the continuous signal without sampling and thus produce no aliasing errors. The devices are linear over an 80-dB range, and distortions due to A/D conversion are also eliminated. However, the SAW devices cannot be designed with the precision to which digital FIR users are accustomed, and they produce spurious signals that must be minimized by careful implementation.

Surface acoustic wave devices will most likely be purchased from a manufacturer because there are a

** Reference 28.

‡ Radix-2 FFT programs in FORTRAN are published in many different places. One can be found in reference 26.

§ Reference 10.

* Reference 3.

CHART 1. AN FFT PROGRAM*

The following is a listing for a PL/1 program that computes the DFT of a complex sequence of N points, $X(*)$, using the radix 2 decimation-in-time algorithm. The calculation is done in-place, and the resulting DFT is returned in $X(*)$. The vector SINT is a sine table for determining the twiddle factors. INV is a character: if INV = "1," the inverse transform is performed; otherwise the direct transform is performed.

```
DFT:PROCEDURE(X,INV,N,SINT);
 /*THIS PROCEDURE COMPUTES THE DFT OF THE COMPLEX INPUT SEQUENCE X.
  X(0:N) - GIVEN COMPLEX INPUT SEQUENCE, RESULTING COMPLEX OUTPUT
     SEQUENCE.
  N - NUMBER OF ELEMENTS IN INPUT ARRAY - 1.
  SINT(0:(N+1)/4 - GIVEN INPUT SINE TABLE, QUARTER PERIOD.
  INV = '0' FORWARD TRANSFORM IS COMPUTED, INV = '1' - INVERSE
  TRANSFORM IS COMPUTED. */
DECLARE
   (X(*),R,R1,S)
   COMPLEX FLOAT DECIMAL (16),
   (SINT(*),S1,S2)
   FLOAT DECIMAL (16),
   (N, I, J, K, L, M, NT, NH, NQ, IR, IT)
   BINARY FIXED (15),
   (ID,ND) BINARY FIXED (31),
   INV CHAR(1);
NT=N;
NQ=NT/4+1;
NH=NQ+NQ;  /*REORDER INITIAL VECTOR X*/
J =0;      /*USING THE BIT REVERSAL TECHNIQUE*/
 DO I =0 TO NT-2;
 IF J> I THEN DO;
    R=X(J); X(J)=X(I); X(I)=R;
    END;
 K =NH;
    DO WHILE (J>=K);
    J =J-K;
    K =K/2;
    END;
 J =J+K;
 END;
ID=1;       /*CALCULATE THE DISCRETE FOURIER TRANSFORM*/
 IR=1; M=0; IT=0;

 TRAN:I=ID;
  ID=ID+ID;
     DO J=0 TO I-1;
     S2=SINT(M);
     IF INV-='1' THEN S2=-S2;
     S1=SINT(NQ-M);
     IF J>=IR THEN DO;
        M=M-IT: S1=-S1;
        END;
     ELSE M=M+IT;
        DO ND=J TO NT BY ID;
        K =ND;
        L=K+I;
        S=CPLX(S1,S2);
        R1=X(L);      /*BUTTERFLY COMPUTATION*/
        R=S*R1;
        R1=X(K);
        X(K)=R1+R;
        X(L)=R1-R;
        END;
     END;
 IR=I;
 IT=NQ/I;
  IF I<NH THEN GO TO TRAN;
 RETURN ;
END DFT;
```

* From A. Peled and B. Liu. "A New Hardware Realization of Digital Filters." *IEEE Trans. Acoust., Spch., and Sig. Proc.*, Vol. ASSP-22, No. 6, December 1974, pp. 456–462.

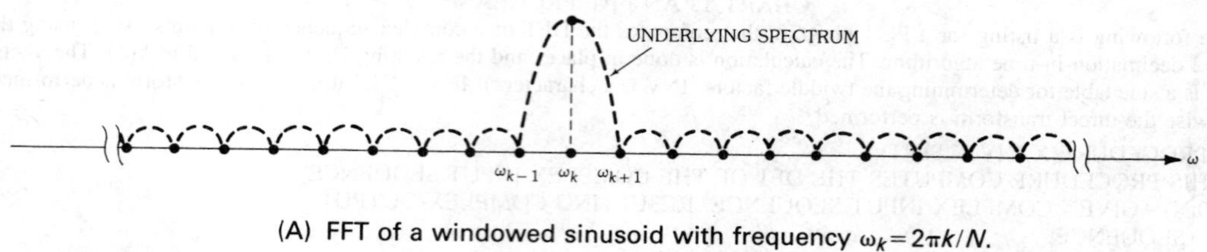

(A) FFT of a windowed sinusoid with frequency $\omega_k = 2\pi k/N$.

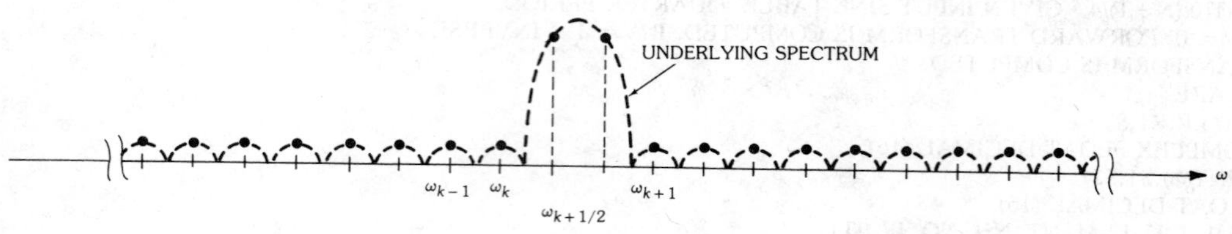

(B) Leakage for a nonharmonic sinusoidal component.

Fig. 25. Picket-fence effect.

considerable number of proprietary techniques involved in the fabrication of a high-quality SAW filter. Most filters will have to be specifically designed for the application. It is important to the system designer to know whether it is possible to realize a SAW filter that meets the specifications and what the characteristics of such a device will be before getting involved with a manufacturer. Fortunately, the characteristics of a SAW filter can be estimated on the basis of performance requirements regardless of the manufacturer involved.

The following subsections describe the SAW filter and its operation, provide an overview of the performance features, and describe how to apply the specifications to a series of graphs and estimate the admittances and insertion loss of a SAW filter.

Description

A SAW device consists of an input transducer that generates a surface-acoustic-wave replica of the signal, a polished piezoelectric substrate, which acts as the propagation path for the waves, and an output transducer that converts the delayed acoustic waves back to an electrical signal (Fig. 26). The input and output transducers each have an impulse response with a finite duration, and the construction is such that the transducer electrodes have a one-to-one correspondence with individual cycles in the impulse response. The impulse-response center frequency is set by electrode spacing, and its envelope is set by the electrode length. The frequency dependent transfer function of each transducer is the Fourier transform of its impulse response. The surface-acoustic-wave device connected between a voltage source and a low-impedance load operates as a filter with an overall transfer function that is the product of the input and output transducer frequency responses and a delay proportional to the separation between the centers of the transducers.

Consider the SAW device shown in Fig. 26, which uses a single electrode placed between a pair of grounded electrodes as the SAW generation transducer. When a pulse is applied, it creates a uniform wavefront acoustic disturbance that propagates along the surface both to the left and to the right. The wave propagating to the left is absorbed, to prevent it from bouncing back and creating an echo. Since the substrate is piezoelectric, the disturbance propagating to the right carries an electric field with it. The output transducer in this example consists of seven electrode pairs with center-to-center spacing p. The electrodes vary in length and are connected interdigitally to two bus bars, one of which is grounded and the other connected to a load. As the SAW pulse passes through the output transducer, it creates an oscillating impulse response, $h_o(t)$, with an overall duration, T_o, equal to the output transducer length, L_o, divided by the SAW velocity, v_s (Fig. 26B). The center frequency, f_o, is equal to the SAW velocity divided by the electrode-to-electrode spacing, p, and the amplitude of each cycle, h_i, is proportional to the corresponding transducer electrode length, l_i. The delay, T_D, is equal to the center-to-center distance of the input and output transducers, L_D, divided by the SAW velocity. The frequency response of the output transducer, $H_o(f)$, is equal to the Fourier transform of the impulse response (Fig. 26C). The device has a transition bandwidth, BW_T, that is approximately equal to the reciprocal of the impulse response duration as measured at the 3-dB points, T_{3dB}.

SAW FIR Overview

Response—The SAW-filter performance limits are summarized in Table 8. Operation of SAW filters is limited to a specific range of frequencies, fractional bandwidths, transition bandwidths, and delays because the SAW propagation velocity is approximately 3000

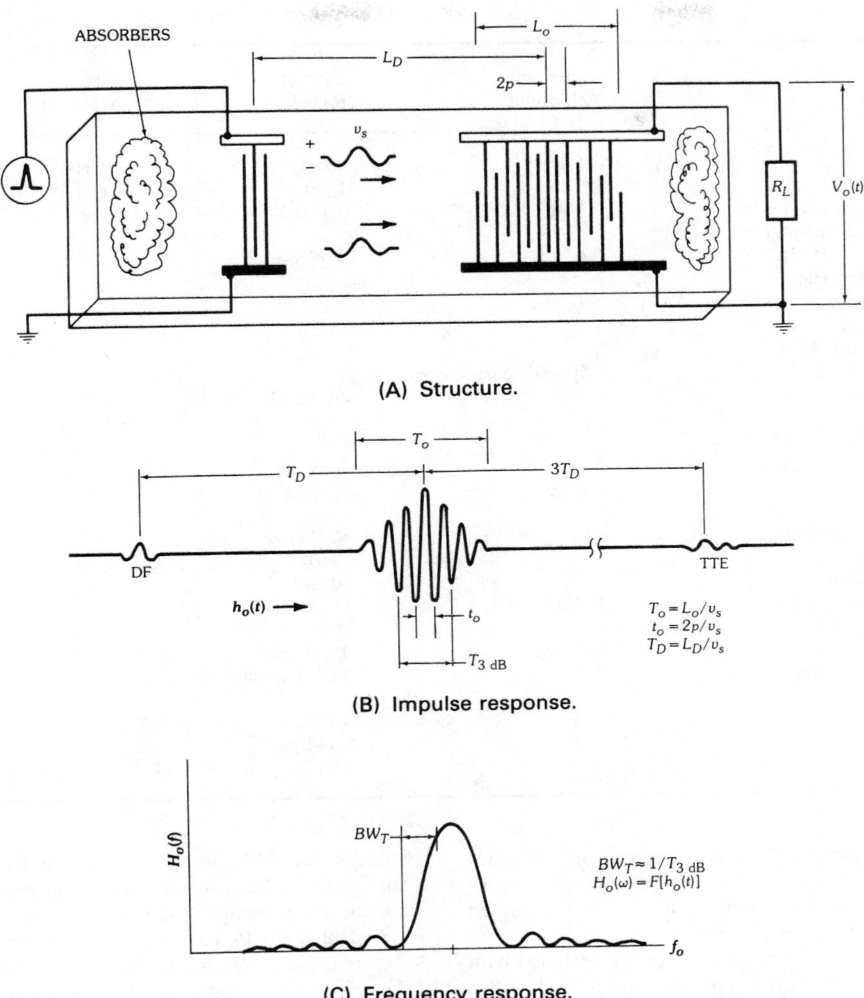

Fig. 26. Surface-acoustic-wave FIR filter.

meters per second (plus or minus 20 percent) for most useful materials.

Frequency: The highest center frequency achievable with the SAW device is dependent on the ability to print fine lines. If an electrode-to-electrode spacing of 0.75 micron can be achieved, then the devices will operate with a 2-GHz center frequency. At low frequencies, the wavelengths become very long (3 mm at 1 MHz). Since the device must be hundreds or thousands of wavelengths in length and be much thicker than a wavelength, SAW devices become impractical below 5 MHz.

Delay: The substrate size for practical devices is limited to approximately 6 cm, which means that the maximum usable delay is in the order of 20 microseconds. If the application calls for very long delays and cost is not a particularly strong consideration, then delays can be expanded to the 100-microsecond range by using long substrates. The problem of achieving very short delays is the problem of getting the transducers physically close to each other. The minimum possible delay that can be achieved in a filter is twice the reciprocal of the transition bandwidth.

Fractional Bandwidth: The fractional bandwidth of a SAW filter is inversely related to the number of electrodes contained in the main lobe of the transducers. The maximum fractional bandwidth that can be achieved is approximately 30 percent because interdigital transducers with less then three electrode pairs are inefficient. Fractional bandwidths less than 0.1 percent are relatively difficult to implement because this requires more than 1000 electrode pairs. Narrow fractional bandwidth can be achieved with SAW resonator filters, but these filters are not FIR filters, and the amplitude and phase response cannot be designed independently.*

* Reference 7.

TABLE 8. SAW FIR FILTER PARAMETERS

	Minimum Achievable	Typical Range	Max Achievable
Frequency	5 MHz	50–500 MHz	2 GHz
Delay	2/TBW	1–20 s	100 s
Fractional Bandwidth	0.01%	0.1–0.30%	50%
Transition Bandwidth (TBW) (3–30 dB)	0.05 MHz	0.5–100 MHz	
Response Precision Inband Ripple	±0.2dB	±0.5 dB	
Inband Phase	1°	5°	
Inband Group Delay Deviation	"Special specification required"		
Out of Band Rejection		40–55 dB	80 dB
Echo Levels		45 dB	60 dB
Temperature Coefficient of Frequency Delay		1–38 ppm/°C	
Insertion Loss	2.0 dB	8–30 dB	
VSWR		1.0–4	
Modulation Products		Negligible	
Maximum Power		−10–30 dB	40 dB
Dynamic Range		30–70	100 dB
Noise Figure	1–4 dB	15–35 dB	
Input Capacitive Susceptance	0.1 mmho	0.5–5 mho	50 mmho
Input Conductance		0.01–1 mmho	
Length		0.1–3 cm	10 cm
Width		0.1–1 cm	
Thickness		0.01–0.1 cm	

Transition Bandwidth: The impulse response duration is limited by substrate size to periods less than 10 microseconds so that transition bandwidths are limited to values greater than 100 kHz.

Inband Ripple: The SAW devices are designed with the same mathematical techniques used to design digital FIR filters. In the case of SAW devices, the design is realized by adjusting the length and position of the electrodes, and this can be done with an accuracy of approximately one percent. In addition, internally generated echoes add ripples to the frequency response. The measured amplitude-vs-frequency response of SAW bandpass filters generally matches the theoretical design specifications within ±0.5 dB. In special cases where the application can tolerate higher insertion losses and warrants extra development costs, the response deviations can be reduced to ±0.2 dB.

Group Delay: The group-delay variations that affect the rise time, fall time, undershoot, and overshoot in the step responses are caused by inaccurate implementation of the specified impulse response. But these variations are generally overshadowed by group-delay ripples caused by echoes. It is not uncommon for SAW devices to have group-delay variations that are in excess of those specified for LC filters and still have a perfectly acceptable response. The group-delay specification is useful for LC filters but is excessively restrictive for SAW devices; SAW devices should be specified with an echo-level specification together with a truncated response group-delay specification.†

Out-of-Band Rejection: The SAW out-of-band rejection depends on the accuracy with which the transducers are implemented and the transducer isolation built into the device. Typical state of the art SAW filters provide approximately 50–70 dB of out-of-band rejection. In certain circumstances, the out-of-band rejection can be raised to 80 dB. It is also possible to cascade SAW filters to improve rejection, if the cost, losses, and inband deviations of two filters can be tolerated.

Echo Levels: The main source of SAW-filter-generated noise is signals (called echoes) that precede and/or follow the main impulse response of the device. The two echoes in a SAW filter are the direct feed signals that are coupled directly to the output without any delay and the triple-transit echo that emerges with a delay three times that of the main signal. These echoes are highly dependent on the method by which the devices are interfaced into the supporting electronic circuitry and are typically 45 dB below the main signal.

Temperature Dependence: Both the center frequency and the delay of SAW filters are temperature dependent

† Reference 34.

because the substrate dimensions and the SAW velocity vary with temperature. These two effects are combined into a single term called the *SAW temperature coefficient of delay* (TCD). The temperature coefficient of delay is approximately 90 ppm/°C (parts per million per degree Celsius) for the $LiNbO_3$ substrates used in large-percentage-bandwidth devices, and less than 1 ppm/°C for the quartz material used in low-percentage-bandwidth filters.

Signal Power—Insertion Loss: Surface-acoustic-wave filters are typically lossy devices having an 8–30-dB midband insertion loss. Insertion loss is defined as the ratio of the actual power provided to the load through a SAW filter to the power that would have been transferred if the load were matched directly to the signal source. The insertion loss is dependent on the type of circuitry used to interface the SAW device with the rest of the electronics. Low-shape-factor, flat-response filters typically have higher losses, and the system designer must trade off losses for response accuracy. In certain cases, the insertion losses can be reduced to 2 dB at the cost of additional complexity by the use of unidirectional transducers.*

Modulation Products: The SAW devices produce negligible levels of cross-modulation products because the devices are passive and linear. In addition, the transducers are made to discriminate against signals produced by nonlinear interactions.

Maximum Power: The largest signal that the device can handle is limited by transducer burnout to a level of +20 to +36 dBm. Power-handling capability is a tradeoff among size, fractional bandwidth, and temperature stability.

Dynamic Range: A 50–70-dB dynamic range is typical, and it can be expanded to 100 dB in certain cases. The largest possible signal is limited by burnout, and the smallest input signal that a SAW device can process is limited by the transducer losses and the thermal noise generated in the load.

Noise Figure: The noise figure of a SAW filter is equal to the noise figure of the output amplifier plus the device insertion loss. The effective noise figure is high (typically 15–35 dB), and that is the reason SAW devices are presently being used primarily in the if sections of receivers.

Admittances—The SAW finite impulse response bandpass filter is modeled as a lossy capacitor.

Capacitance: The capacitance of a SAW transducer depends on the percentage bandwidth and the power-handling requirement. Typically, vhf and uhf SAW bandpass filters have a capacitive susceptance ranging from 0.5 to 5 mhos, and occasionally very special devices may have susceptances as low as 0.1 mmho.

Radiation Conductance: Part of the conductance that accounts for the power absorbed from the source and converted to the surface acoustic waves is termed "radiation conductance." It is dependent on the percentage bandwidth, temperature stability requirement, size, and power-handling specifications of the device. The typical value of conductance runs between 0.01 and 1 millimho for most devices.

Vswr: The SAW filters are inclined to produce a large vswr, particularly at frequencies that fall out of band. The major part of filter conductance accounts for the generation of surface acoustic waves, and that goes to zero out of band when no acoustic waves are generated. If one matches the device conductance to that of the generator in order to produce a low vswr at the center of the band, the vswr will increase dramatically for out-of-band frequencies where the SAW radiation conductance goes to a much lower value. Additional conductance is needed in the matching network in order to maintain a low vswr at all frequencies.

Size Considerations—The size of SAW bandpass filters depends on the transition bandwidth and the delay required for the application. Most SAW devices have long, thin geometry with a length of approximately 0.1 to 3 cm, a width ranging from 0.1 to 1 cm, and a thickness between 0.01 and 0.1 cm.

SAW FIR Parameter Estimation

Admittance—The dominant admittance of a SAW transducer is due to the frequency-independent dielectric capacitance, C_T, between the transducer electrodes, represented as a capacitor in Fig. 27. The acoustic admittance is defined as the ratio of current

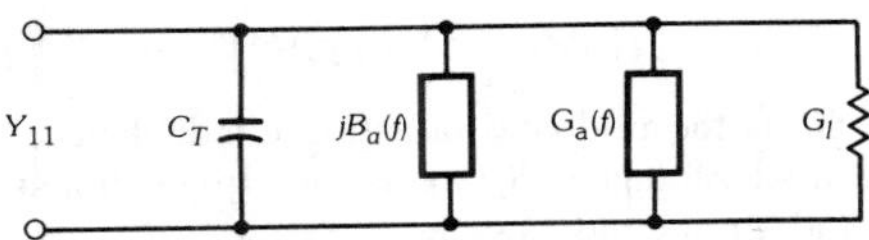

Fig. 27. Transducer equivalent circuit.

into the transducer as a result of piezoelectric interactions with the surface wave to the voltage applied at the input port. The imaginary part of the acoustic admittance, B_a, shown as deviation from the straight-line dielectric susceptance in Fig. 28A, is frequency dependent and represents the quadrature phase currents induced by the potential carried in the newly generated surface acoustic waves as they pass under the electrodes on their way out of the transducer. The radiation susceptance is related to the radiation conductance (discussed in the next paragraph), and its magnitude is zero at midband because all SAW induced currents are in phase with the driving voltage. It has a low value out of band where small-amplitude SAWs are generated (see Fig. 28B). The maximum value is significantly smaller than the dielectric susceptance, C_T.

The radiation conductance, $G_a(f)$, represents the power converted to surface acoustic waves and is pro-

* Reference 22.

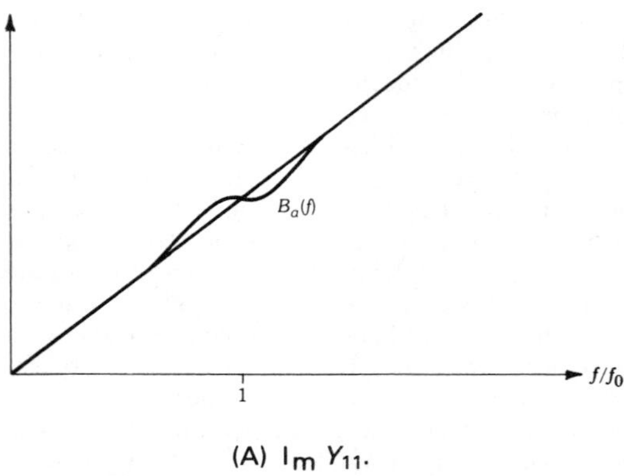

(A) $I_m\ Y_{11}$.

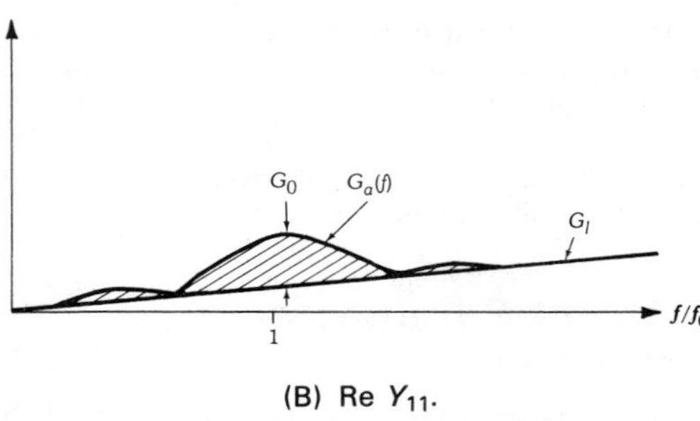

(B) Re Y_{11}.

Fig. 28. Basic SAW delay line transfer function and admittances.

portional to the transducer transfer function, $H(f)$, squared (shaded area in Fig. 28B):

$$G_a(f) = G_0\ [H(f)]^2 \tag{32}$$

where G_0 is the midband radiation conductance. The radiation susceptance, $B_a(f)$, is the Hilbert transform of the radiation conductance:

$$B_a(f) = \text{Hil}[G_a(f)] \tag{33}$$

The resistive, dielectric, and other losses are represented by the loss conductance, G_l. This conductance usually varies slowly with frequency and can be measured by subtracting the frequency-dependent component of the measured conductance of the transducer.

Admittance Levels—The dielectric capacitance of a well designed SAW-filter transducer is a function of the specified fractional bandwidth, temperature stability, shape factor, and power-handling requirements.

Quartz and lithium niobate are the two materials most commonly used for commercial devices. Quartz is good for low percentage bandwidth applications where temperature stability is important, and lithium niobate is used for wide fractional bandwidth applications where larger variations in center frequency as a function of temperature can be tolerated. The surface acoustic wave basic delay line (SAW BDL) has an equal number of uniform-length electrode pairs on both transducers, and a triangular shaped impulse response envelope (convolution of two equal-length rectangles). The frequency response is the Fourier transform of the impulse response:

$$H(f) = \{[\sin (f-f_0)N]/(f-f_0)N\}^2 \tag{34}$$

where,

N is the number of electrode pairs in each transducer,

f_0 is the center frequency $= v_s/2p$.

The input transducers for a SAW BDL have admittance B_{BDL} that is *independent* of center frequency but highly dependent on fractional bandwith. The center frequency dielectric capacitive admittance for a typical width low-power SAW BDL is plotted for ST quartz (TC<1 ppm/°C) and YZ $LiNbO_3$ (TC = 90 ppm/°C) in Fig. 29.* If the device must handle high power, it may be necessary to make the transducers wider than normal. In that case, the admittance increases along the branch curves as shown for the different power levels. The curves extend only over the useful range of fractional bandwidths for each material and show the practical options that are available.

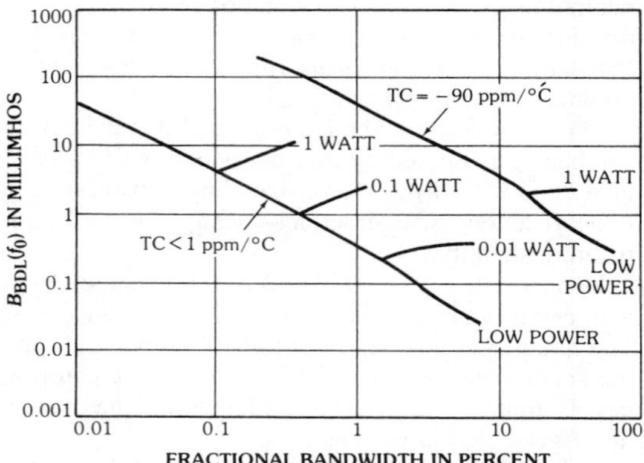

Fig. 29. Basic delay line admittance B_{BDL}.

If the out-of-band rejection must be better than 26 dB and if the response shape must be more rectangular than the $(\sin x/x)^2$ shape, then additional electrodes must be added to the transducers to shape the response. These electrodes add dielectric admittance but do not contribute to the midband SAW generation; therefore, low-shape-factor, high-rejection filters have higher admittance than a basic SAW delay line. A useful rule of thumb for estimating the admittance of the actual filters, $B(f)$, is to use the SAW BDL value, $B_{\text{BDL}}(f)$,

* The curves of Fig. 29 are calculated for a minimum beam width of 20 acoustic wavelengths. Double electrodes are used whenever the fractional bandwidth is less than 6 times the material coupling coefficient. The maximum allowable electrode-to-electrode potential that can be used to generate the specified power is set at 75 volts.

and increase it by a factor related to the transition bandwidth:

$$B(f_0) = B_{BDL}(f_0)[1 + 0.7 \log (BW_{3dB}/2TBW)] \quad (35)$$

where,

TBW is the transition bandwidth,

BW_{3dB} is the 3-dB bandwidth.

The band-center radiation conductance of a low-power SAW FIR, G_0, is dependent only on the temperature stability requirements presently measured and the fractional bandwidth. The response shape and the out-of-band rejection have a relatively minor impact on the center-frequency radiation-conductance magnitude. The estimated values of G_0 for a SAW FIR are:

$$G_0 = 1.5 \times 10^{-2}/(BWF)^2 \text{ mmho-(Hz)}^2 \ @ \ T_C = -90 \text{ ppm/°C} \quad (36a)$$

$$G_0 = 1.5 \times 10^{-8}/(BWF)^2 \text{ mmho-(Hz)}^2 \ @ \ T_C = 1 \text{ ppm/°C} \quad (36b)$$

Loss Conductance—The loss conductance, G, is primarily a function of manufacturing techniques, and good SAW devices have loss conductances much smaller than G_0.

Insertion Loss—The center-frequency insertion loss of a SAW filter driven by a generator with a conductance G_G and connected to a load with a conductance G_L is

$$IL = 10 \log \frac{4\, G_{01}\, G_G\, G_{02}\, G_L}{[(G_{01} + G_2 + G_G)^2 + B_{01}^2]\,[(G_{02} + G_2 + G_L)^2 + B_{02}^2]} \quad (37)$$

where,

G_{01} is the center-frequency conductance of the input transducer,

G_{02} is the center-frequency conductance of the output transducer,

B_{01} is the center-frequency susceptance of the input transducer,

B_{02} is the center-frequency susceptance of the output transducer,

G_1 is the loss conductance of the input transducer,

G_2 is the loss conductance of the output transducer,

G_G is the generator conductance,

G_L is the load conductance.

The triple transit echo (TTE) transit echo level measured relative to the main signal is:

$$TTE = IL + 10 \log (4\, G_L G_G / G_{01} G_{02}) \quad (38)$$

When the transducer is operated with a low-impedance generator and load ($G_G >> B_{01}$ $G_L >> B_{02}$):

$$IL = 10 \log (4\, G_{01} G_{02} / G_G G_L) \quad (39)$$

$$TTE = 2IL - 6 \text{ dB} \quad (40)$$

The FIR frequency response is approximately equal to the product of the transducer transfer functions

$$H_{12}(f) = H_1(f) \cdot H_2(f) \quad (41)$$

If the generator and load conductances are matched in magnitude to B_{01} and B_{02}, respectively (that is, $G_G = |\, B_{01} \,|$ and $G_L = |\, B_{02} \,|$),

$$IL = 10 \log (2 G_{01} G_{02} / B_{01} B_{02}) \quad (42)$$

$$TTE = 2IL \quad (43)$$

The frequency response is approximately equal to the product of the transfer functions multiplied by the response of two isolated and cascaded single-pole low-pass filters with corners at f_0:

$$H(f) = H_1(f)\, H_2(f)\, [1/(1 + j\, f/f_0)]^2 \quad (44)$$

If the generator and load contain series inductive components that tune the transducers, there is a significant impact on the device frequency response as shown in Fig. 30. The series tuned transducer response is altered because the radiation conductance, $G_a(f)$, is frequency dependent and the insertion loss, which is primarily mismatch loss, is actually reduced at the band edges when there is a better match to the generator and load. Tuned SAW filters are not precise FIR devices, and phase distortions are suffered at the band edges.

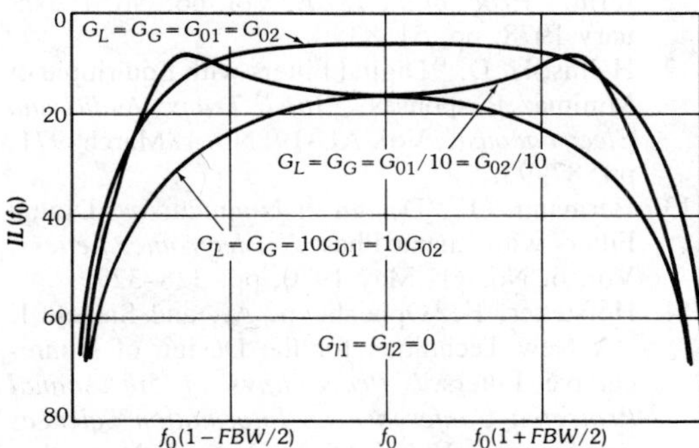

Fig. 30. Tuned SAW BDL insertion loss versus frequency.

When the transducer and the generator are conjugate-matched, the center IL and TTE are significantly dependent on loss conductance:

$$G_G = G_{01} + G_1 \quad (45)$$

$$G_L = G_{02} + G_2 \quad (46)$$

$$IL = 10 \log [G_{01} G_{02} / 4(G_{01} + G_1)(G_{02} + G_2)] \quad (47)$$

$$TTE = (G_{01}\, G_{02})^2 = 2IL \quad (48)$$

REFERENCES

1. Agarwal, A. C. and Burrus, C. S. "Number Theoretic Transforms to Implement Fast Digital Convolution." *IEEE Proc.*, Vol. 63, April 1975, pp. 555–560.
2. Antoniou, A. *Digital Filters: Analysis and Design.* New York: McGraw-Hill Book Co., 1979.
3. Bergland, G. D. "A Guided Tour of the Fast Fourier Transform." *IEEE Spectrum*, Vol. 6, July 1969, pp. 41–52.
4. Brigham, E. O. *The Fast Fourier Transform.* Englewood Cliffs, N.J.: Prentice-Hall, Inc., 1974.
5. Chen, C. T. *Digital Signal Processing.* New York: Marcel Dekker, Inc., 1980.
6. Crochiere, R. E. "A New Statistical Approach to the Coefficient Word Length Problem for Digital Filters." *IEEE Trans. Cir. and Sys.*, Vol. CAS-22, No. 3, March 1975, pp. 190–196.
7. Elliott, S., Mierzwinski, M., and Planting, P. "The Production of Surface Acoustic Wave Resonators." *1981 IEEE Ultrasonics Symposium*, Vol. 1 81CH1689-9, pp. 89–93.
8. Gabel, R. A. and Roberts, R. A. *Signals and Linear Systems.* 2nd ed. New York: John Wiley & Sons, Inc., 1980.
9. Gold, B. and Rader, C. *Digital Processing of Signals.* New York: McGraw-Hill Book Co., 1969.
10. Hamming, R. W. *Digital Filters.* Englewood Cliffs, N.J.: Prentice-Hall, Inc., 1977.
11. Harris, F. J. "On the Use of Windows for Harmonic Analysis with the Discrete Fourier Transform." *Proc. of the IEEE*, Vol. 66, No. 1, January 1978, pp. 51–83.
12. Helms, H. D. "Digital Filters with Equiripple or Minimax Responses." *IEEE Trans. Audio and Electroacoust.*, Vol. AU-19, No. 1, March 1971, pp. 87–93.
13. Hermann, O. "Design of Nonrecursive Digital Filters with Linear Phase." *Electronics Letters*, Vol. 6, No. 11, May 1970, pp. 328–329.
14. Hofstetter, E., Oppenheim, A., and Siegel, J. "A New Technique for the Design of Nonrecursive Filters." *Proceedings of 5th Annual Princeton Conference on Information Sciences and Systems*, Vol. CAS-25, No. 11, November 1978, pp. 893–902.
15. Jackson, L. B. "Roundoff Noise Analysis for Fixed-Point Digital Filters Realized in Cascade or Parallel Form." *IEEE Trans. Audio and Electroacoust.*, Vol. AU-18, No. 2, June 1970, pp. 107–122.
16. Jenkins, W. K. "The Design of Error Checkers for Self-Checking Residue Number Arithmetic." *IEEE Trans. on Computers*, Special Issue on Computer Arithmetic, Vol. C-32, No. 4, April 1983, pp. 388–396.
17. Jenkins, W. K. and Leon, B. J. "The Use of Residue Number Systems in the Design of Finite Impulse Response Digital Filters." *IEEE Trans. on Circuits and Systems*, Vol. CAS-24, No. 4, April 1977, pp. 191–201.
18. Jury, E. I. *Theory and Application of the Z-Transform Method.* Huntington, N.Y.: Krieger Publishing Co., 1964.
19. Knowles, J. B. and Olcayto, E. M. "Coefficient Accuracy and Digital Filter Response." *IEEE Trans. on Circuit Theory*, Vol. CT-15, No. 1, March 1968, pp. 31–41.
20. Liu, B. "Effects of Finite Word Length on the Accuracy of Digital Filters—A Review." *IEEE Trans. on Circuit Theory*, Vol. CT-18, No. 6, November 1971, pp. 670–677.
21. Liu, B. and Kaneko, T. "Error Analysis of Digital Filters Realized with Floating-Point Arithmetic." *Proc. IEEE*, Vol. 57, No. 10, October 1969, pp. 1735–1747.
22. Malocha, D. C. "Surface Wave Devices Using Low Loss Filter Technologies." *1981 IEEE Ultrasonics Symposium*, Vol. 1, 81CH1689-9, pp. 83–88.
23. McCallig, M. T. and Leon, B. J. "Constrained Ripple Design of FIR Digital Filters." *IEEE Trans. Circuits and Systems*, Vol. AU-20, No. 5, October 1972, pp. 280–288.
24. McClellan, J. H., Parks, T. W., and Rabiner, L. R. "A Computer Program for Designing Optimum FIR Linear Phase Digital Filters." *IEEE Trans. Audio and Electroacoust.*, Vol. AU-21, No. 6, December 1973, pp. 506–526.
25. Oppenheim, A. V, ed. *Applications of Digital Signal Processing.* Englewood Cliffs, N.J.: Prentice-Hall, Inc., 1977.
26. Oppenheim, A. V. and Schafer, R. W. *Digital Signal Processing.* Englewood Cliffs, N.J.: Prentice-Hall, Inc., 1975.
27. Parker, S. R. "Limit Cycles and Correlated Noise in Digital Filters," in *Digital Signal Processing*, J. K. Aggarwal, ed. North Hollywood, Cal.: Point Lobos Press, 1979.
28. Peled, A. and Liu, B. "A New Hardware Realization of Digital Filters." *IEEE Trans. Acoust., Spch., and Sig. Proc.*, Vol. ASSP-22, No. 6, December 1974, pp. 456–462.
29. Peled, A. and Liu, B. *Digital Signal Processing.* New York: John Wiley & Sons, Inc., 1976.
30. *Programs for Digital Signal Processing.* Ed. by the Digital Signal Processing Committee of the IEEE Acoustics, Speech and Signal Processing Society. New York: IEEE Press, 1979.
31. Rabiner, L. R. "Linear Program Design of Finite Impulse Response (FIR) Digital Filters." *IEEE Trans. Audio and Electroacoust.*, Vol. AU-21, No. 5, October 1973, pp. 456–460.
32. Rabiner, L. R. and Gold, B. *Theory and Applications of Digital Signal Processing.* Englewood Cliffs, N.J.: Prentice-Hall, Inc., 1975.
33. Singleton, R. C. "An Algorithm for Computing

the Mixed Radix Fast Fourier Transform.'' *IEEE Trans. Audio and Electroacoust.*, Vol. AU-17, June 1969, pp. 93–103.

34. Stribling, Sidney Nira. ''Surface Acoustic Wave Filter for a Television Transmitter.'' MS Thesis, University of Illinois, Urbana, 1975.
35. Tretter, S. A. *Introduction to Discrete-Time Signal Processing.* New York: John Wiley & Sons, Inc., 1976.
36. Trick, T. N. and Jenkins, W. K. ''Uncorrelated Roundoff Noise in Digital Filters,'' in *Digital Signal Processing*, J. K. Aggarwal, ed. North Hollywood, Cal.: Point Lobos Press, 1979.

29 Transmission Lines

Revised by
Tatsuo Itoh

The equations and charts of this chapter are for transmission lines operating in the TEM mode.* At the beginning of several of the sections (e.g., "Fundamental Quantities and Line Parameters," "Voltage and Current," "Impedance and Admittance," "Voltage Reflection Coefficient and Standing-Wave Ratio") there are accurate equations, according to conventional transmission-line theory. These are applicable from the lowest power and communication frequencies, including direct current, up to the frequency where a higher mode begins to appear on the line.

Following the accurate equations are others that are specially adapted for use in radio-frequency problems. In cases of small attenuation, the terms α^2x^2 and higher powers in the expansion of $\exp\alpha x$, etc., are neglected. Thus, when $\alpha x = (\alpha/\beta)\theta = 0.1$ neper (or about 1 decibel), the error in the approximate equations is of the order of 1 percent.

Much of the information is useful also in connection with special lines that function in a quasi-TEM mode (e.g., microstrip).

It should be observed that Z_0 and Y_0 are complex quantities and the imaginary part cannot be neglected in the accurate equations, unless preliminary examination of the problems indicates the contrary. Even when attenuation is small, $Z_0 = 1/Y_0$ must often be taken at its complex value, especially when the standing-wave ratio is high. In the first few pages of equations, the symbol R_0 is used frequently. However, in later charts and special applications, the conventional symbol Z_0 is used where the context indicates that the quadrature component need not be considered for the moment.

RULE OF SUBSCRIPTS AND SIGN CONVENTIONS

The equations for voltage, impedance, etc., are generally for the quantities at the input terminals of the line in terms of those at the output terminals (Fig. 1). In case it is desired to find the quantities at the output in terms of those at the input, it is simply necessary to interchange the subscripts 1 and 2 in the equations and to place a minus sign before x or θ. The minus sign may then be cleared through the hyperbolic or circular functions; thus

$$\sinh(-\gamma x) = -\sinh\gamma x$$

and so on.

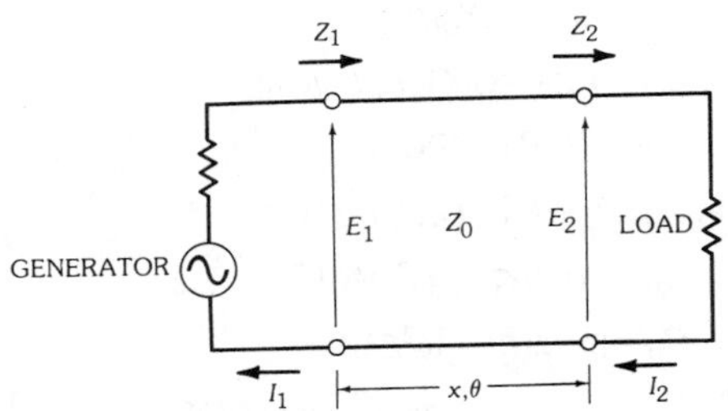

Fig. 1. Transmissionn line with generator and load.

* The information through the section "Length of Transmission Line" (page 29-19) is valid for single-mode wavequides in general, except for equations where the symbols R, L, G, or C per unit length are involved.

SYMBOLS

Voltage and current symbols usually represent the alternating-current complex sinusoid, with magnitude equal to the root-mean-square value of the quantity.

Certain quantities, namely C, c, f, L, T, v, and ω are shown with an optional set of units in parentheses. Either the standard units or the optional units may be used, provided the same set is used throughout. (For the physical significance of C, G, L, and R, refer to Fig. 2.)

$A = 10 \log_{10}(1/\eta)$ = dissipation loss in a length of line in decibels

$A_0 = 8.686\alpha x$ = normal or matched-line attenuation of a length of line in decibels

B_0 = susceptive component of Y_0 in mhos

C = capacitance of line in farads/unit length (microfarads/unit length)

c = velocity of light in vacuum in units of length/second (units of length/microsecond). See table of physical constants in Chapter 3.

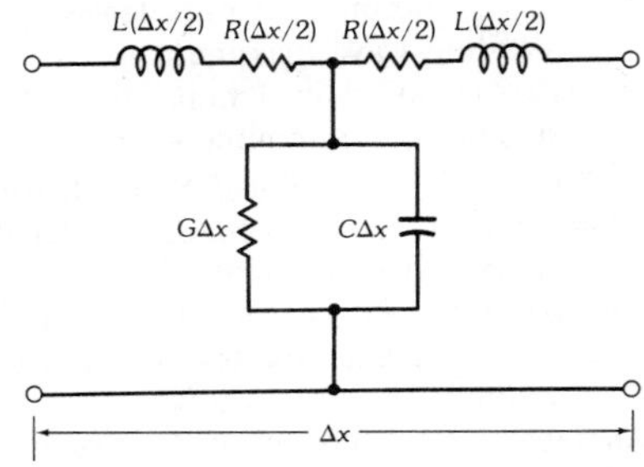

(A) T network.

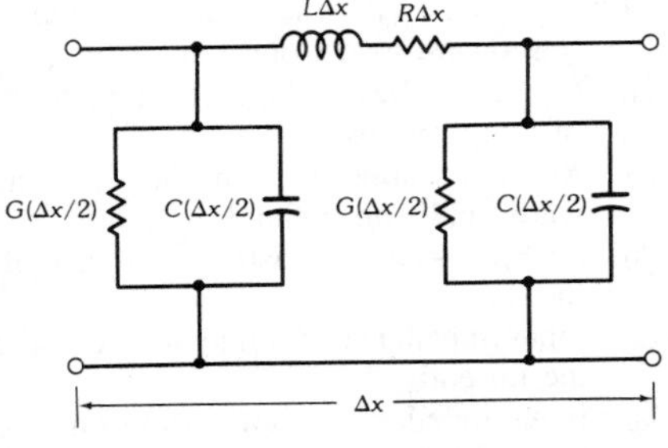

(B) π network.

Fig. 2. Equivalent networks of a short section, Δx, of a transmission line.

E = voltage (root-mean-square complex sinusoid) in volts
${}_fE$ = voltage of forward wave, traveling toward load
${}_rE$ = voltage of reflected wave
$|E_{flat}|$ = root-mean-square voltage when standing-wave ratio = 1.0
$|E_{max}|$ = root-mean-square voltage at crest of standing wave
$|E_{min}|$ = root-mean-square voltage at trough of standing wave
e = instantaneous voltage
$F_p = G/\omega C$ = power factor of dielectric
f = frequency in hertz (megahertz)
G = conductance of line in mhos/unit length
G_0 = conductive component of Y_0 in mhos
$g_a = Y_a/Y_0$ = normalized admittance at voltage standing-wave maximum
$g_b = Y_b/Y_0$ = normalized admittance at voltage standing-wave minimum
I = current (root-mean-square complex sinusoid) in amperes
${}_fI$ = current of forward wave, traveling toward load
${}_rI$ = current of reflected wave
i = instantaneous current
L = inductance of line in henries/unit length (microhenries/unit length)
P = power in watts
R = resistance of line in ohms/unit length
R_0 = resistive component of Z_0 in ohms
$r_a = Z_a/Z_0$ = normalized impedance at voltage standing-wave maximum
$r_b = Z_b/Z_0$ = normalized impedance at voltage standing-wave minimum
$S = |E_{max}/E_{min}|$ = voltage standing-wave ratio
T = delay of line in seconds/unit length (microseconds/unit length)
v = phase velocity of propagation in units of length/second (units of length/microsecond)
X_0 = reactive component of Z_0 in ohms
x = distance between points 1 and 2 in units of length (also used for normalized reactance = X/Z_0)
$Y_1 = G_1 + jB_1 = 1/Z_1$ = admittance in mhos looking toward load from point 1
$Y_0 = G_0 + jB_0 = 1/Z_0$ = characteristic admittance of line in mhos
$Z_1 = R_1 + jX_1$ = impedance in ohms looking toward load from point 1
$Z_0 = R_0 + jX_0$ = characteristic impedance of line in ohms
Z_{oc} = input impedance of a line open-circuited at the far end
Z_{sc} = input impedance of a line short-circuited at the far end
α = attenuation constant = nepers/unit length = 0.1151 × decibels/unit length
β = phase constant in radians/unit length
$\gamma = \alpha + j\beta$ = propagation constant
ϵ = base of natural logarithms = 2.718; or dielectric constant of medium (relative to air), according to context
$\eta = P_2/P_1$ = efficiency (fractional)
$\theta = \beta x$ = electrical length or angle of line in radians
$\theta° = 57.3\theta$ = electrical angle of line in degrees
λ = wavelength in units of length
λ_0 = wavelength in free space
$\rho = |\rho| \angle 2\psi$ = voltage reflection coefficient
$\rho_{dB} = -20 \log_{10}(1/\rho)$ = voltage reflection coefficient in decibels
ϕ = time phase angle of complex voltage at voltage standing-wave maximum
ψ = half the angle of the reflection coefficient = electrical angle to nearest voltage standing-wave maximum on the generator side
$\omega = 2\pi f$ = angular velocity in radians/second (radians/microsecond).

FUNDAMENTAL QUANTITIES AND LINE PARAMETERS

$$dE/dx = -(R + j\omega L)I$$

$$d^2E/dx^2 = \gamma^2 E$$

$$dI/dx = -(G + j\omega C)E$$

$$d^2I/dx^2 = \gamma^2 I$$

$$\gamma = \alpha + j\beta = [(R + j\omega L)(G + j\omega C)]^{1/2}$$

$$= j\omega(LC)^{1/2} \times [(1 - jR/\omega L)(1 - jG/\omega C)]^{1/2}$$

$$\alpha = (\tfrac{1}{2}\{[(R^2 + \omega^2 L^2)(G^2 + \omega^2 C^2)]^{1/2} + RG - \omega^2 LC\})^{1/2}$$

$$\beta = (\tfrac{1}{2}\{[(R^2 + \omega^2 L^2)(G^2 + \omega^2 C^2)]^{1/2} - RG + \omega^2 LC\})^{1/2}$$

$$Z_0 = 1/Y_0 = [(R + j\omega L)/(G + j\omega C)]^{1/2}$$

$$= (L/C)^{1/2} [(1 - jR/\omega L)/(1 - jG/\omega C]^{1/2}$$

$$= R_0(1 + jX_0/R_0)$$

$$Y_0 = 1/Z_0 = G_0(1 + j B_0/G_0)$$

$$\alpha = \tfrac{1}{2}(R/R_0 + G/G_0)$$

$$\beta B_0/G_0 = \tfrac{1}{2}(R/R_0 - G/G_0)$$

$$R_0 = [M/2(G^2 + \omega^2 C^2)]^{1/2}$$

$$G_0 = [M/2(R^2 + \omega^2 L^2)]^{1/2}$$

$$B_0/G_0 = -X_0/R_0 = (\omega RC - \omega LG)/M$$

where

$$M = [(R^2 + \omega^2 L^2)(G^2 + \omega^2 C^2)]^{1/2} + RG + \omega^2 LC$$

$$1/T = v = f\lambda = \omega/\beta$$

$$\beta = \omega/v = \omega T = 2\pi/\lambda$$

$$\gamma x = \alpha x + j\beta x = (\alpha/\beta)\theta + j\theta$$
$$\theta = \beta x = 2\pi\ x/\lambda = 2\pi f T x$$
$$\theta° = 57.3\theta = 360x/\lambda = 360fTx$$

(A) Special case—distortionless line: When $R/L = G/C$, the quantities Z_0 and α are independent of frequency.

$$X_0 = 0$$
$$\alpha = R/R_0$$
$$Z_0 = R_0 + j0 = (L/C)^{1/2}$$
$$\beta = \omega(LC)^{1/2}$$

(B) For small attenuation: $R/\omega L$ and $G/\omega C$ are small.

$$\gamma = j\omega(LC)^{1/2}\{1 - j[(R/2\omega L) + (G/2\omega C)]\}$$
$$= j\beta(1 - j\alpha/\beta)$$
$$\beta = \omega(LC)^{1/2} = \omega L/R_0 = \omega C R_0$$
$$T = 1/v = (LC)^{1/2} = R_0 C$$
$$\alpha/\beta = (R/2\omega L) + (G/2\omega C) = (R/2\omega L) + \tfrac{1}{2}F_p$$
$$= (Rv/2\omega R_0) + \tfrac{1}{2}F_p$$

= attenuation in nepers/radian

= (decibels per 100 feet)(wavelength in line in meters)/1663

$$\alpha = \tfrac{1}{2}R(C/L)^{1/2} + \tfrac{1}{2}G(L/C)^{1/2}$$
$$= (R/2R_0) + \pi(F_p/\lambda)$$
$$= (R/2R_0) + \tfrac{1}{2}(F_p\beta)$$

where R and G vary with frequency, while L and C are nearly independent of frequency.

$$Z_0 = 1/Y_0$$
$$= (L/C)^{1/2}\{1 - j[(R/2\omega L) - (G/2\omega C)]\}$$
$$= R_0(1 + jX_0/R_0)$$
$$= 1/[G_0(1 + j\,B_0/G_0)]$$
$$= (1/G_0)\,(1 - jB_0/G_0)$$
$$R_0 = 1/G_0 = (L/C)^{1/2}$$
$$B_0/G_0 = -(X_0/R_0) = (R/2\omega L) - \tfrac{1}{2}F_p = (\alpha/\beta) - F_p$$
$$X_0 = -[R/2\omega(LC)^{1/2}] + (G/2\omega C)\,(L/C)^{1/2}$$
$$= -(R\lambda/4\pi) + (\tfrac{1}{2}F_p)R_0.$$

(C) With certain exceptions, the following few equations are for ordinary lines (e.g., not spiral delay lines) with the field totally immersed in a uniform dielectric of dielectric constant ϵ (relative to air). The exceptions are all the quantities not including the symbol ϵ, these being good also for special types such as spiral delay lines, microstrip, etc.

$$L = 1.016R_0(\epsilon^{1/2}) \times 10^{-3} \text{ microhenries/foot}$$
$$= \tfrac{1}{3}R_0(\epsilon^{1/2}) \times 10^{-4} \text{ microhenries/centimeter}$$
$$C = 1.016[(\epsilon^{1/2})/R_0] \times 10^{-3} \text{ microfarads/foot}$$
$$= [(\epsilon^{1/2})/3R_0] \times 10^{-4} \text{ microfarads/centimeter}$$
$$v/c = 1016/R_0C' = \epsilon^{-1/2}$$

= velocity factor (with capacitance C' in picofarads/foot)

$$\lambda = \lambda_0 v/c = c/f(\epsilon^{1/2}) = \lambda_0/(\epsilon^{1/2})$$
$$T = 1/v = R_0C' \times 10^{-6} = 1.016 \times 10^{-3}/(v/c)$$

$= 1.016 \times 10^{-3}\epsilon^{1/2}$ microseconds/foot (with capacitance C' in picofarads/foot)

The line length is

$$x/\lambda = xf(\epsilon^{1/2})/984 \text{ wavelengths}$$
$$\theta = 2\pi x/\lambda = xf(\epsilon^{1/2})/156.5 \text{ radians}$$

where xf is the product of feet times megahertz.

VOLTAGE AND CURRENT

$$E_1 = {}_fE_1 + {}_rE_1 = {}_fE_2\epsilon^{\gamma x} + {}_rE_2\epsilon^{-\gamma x}$$
$$= E_2\{[(Z_2 + Z_0)/2Z_2]\epsilon^{\gamma x} + [(Z_2 - Z_0)/2Z_2]\epsilon^{-\gamma x}\}$$
$$= \tfrac{1}{2}(E_2 + I_2Z_0)\epsilon^{\gamma x} + \tfrac{1}{2}(E_2 - I_2Z_0)\epsilon^{-\gamma x}$$
$$= E_2[\cosh\gamma x + (Z_0/Z_2)\,\sinh\gamma x]$$
$$= E_2\,\cosh\gamma x + I_2Z_0\,\sinh\gamma x$$
$$= [E_2/(1 + \rho_2)](\epsilon^{\gamma x} + \rho_2\epsilon^{-\gamma x})$$
$$I_1 = {}_fI_1 + {}_rI_1 = {}_fI_2\epsilon^{\gamma x} + {}_rI_2\epsilon^{-\gamma x}$$
$$= Y_0({}_fE_2\epsilon^{\gamma x} - {}_rE_2\epsilon^{-\gamma x})$$
$$= I_2\{[(Z_0 - Z_2)/2Z_0]\epsilon^{\gamma x} + [(Z_0 + Z_2)/2Z_0]\epsilon^{-\gamma x}\}$$
$$= \tfrac{1}{2}(I_2 + E_2Y_0)\epsilon^{\gamma x} + \tfrac{1}{2}(I_2 - E_2Y_0)\epsilon^{-\gamma x}$$
$$= I_2[\cosh\gamma x + (Z_2/Z_0)\,\sinh\gamma x]$$
$$= I_2\,\cosh\gamma x + E_2Y_0\,\sinh\gamma x$$
$$= [I_2/(1 - \rho_2)]\,(\epsilon^{\gamma x} - \rho_2\epsilon^{-\gamma x})$$
$$E_1 = AE_2 + BI_2$$
$$I_1 = CE_2 + DI_2$$

where the general circuit parameters are $A = \cosh\gamma x$, $B = Z_0\,\sinh\gamma x$, $C = Y_0\,\sinh\gamma x$, and $D = \cosh\gamma x$.

Refer to section entitled "Matrix Algebra" in Chapter 46.

(A) When point 2 is at a voltage maximum or minimum, x' is measured from voltage maximum and x'' from voltage minimum (similarly for currents).

$$E_1 = E_{max}(\cosh\gamma x' + S^{-1}\,\sinh\gamma x')$$
$$= E_{min}(\cosh\gamma x'' + S\,\sinh\gamma x'')$$

$$I_1 = I_{max}(\cosh\gamma x' + S^{-1}\sinh\gamma x') = I_{min}(\cosh\gamma x'' + S\sinh\gamma x'')$$

When attenuation is neglected

$$E_1 = E_{max}(\cos\theta' + jS^{-1}\sin\theta') = E_{min}(\cos\theta'' + jS\sin\theta'')$$

(B) Letting Z_l = impedance of load, l = distance from load to point 2, and x_l = distance from load to point 1

$$E_1 = E_2 \frac{\cosh\gamma x_l + (Z_0/Z_l)\sinh\gamma x_l}{\cosh\gamma l + (Z_0/Z_l)\sinh\gamma l}$$

$$I_1 = I_2 \frac{\cosh\gamma x_l + (Z_l/Z_0)\sinh\gamma x_l}{\cosh\gamma l + (Z_l/Z_0)\sinh\gamma l}$$

(C) $e_1 = \sqrt{2}\,|{}_fE_2|\,\epsilon^{\alpha x}\sin[\omega t + 2\pi(x/\lambda) - \psi_2 + \phi] + \sqrt{2}\,|{}_rE_2|\,\epsilon^{-\alpha x}\sin[\omega t - 2\pi(x/\lambda) + \psi_2 + \phi]$

$i_1 = \sqrt{2}\,|{}_fI_2|\,\epsilon^{\alpha x}\times\sin[\omega t + 2\pi(x/\lambda) - \psi_2 + \phi + \tan^{-1}(B_0/G_0)] + \sqrt{2}\,|{}_rI_2|\,\epsilon^{-\alpha x} \times\sin[\omega t - 2\pi(x/\lambda) + \psi_2 + \phi + \tan^{-1}(B_0/G_0)]$

(D) For small attenuation

$$E_1 = E_2\{[1 + (Z_0/Z_2)\alpha x]\cos\theta + j[(Z_0/Z_2) + \alpha x]\sin\theta\}$$

$$I_1 = I_2\{[1 + (Z_2/Z_0)\alpha x]\cos\theta + j[(Z_2/Z_0) + \alpha x]\sin\theta\}$$

(E) When attenuation is neglected

$$E_1 = E_2\cos\theta + jI_2Z_0\sin\theta = E_2[\cos\theta + j(Y_2/Y_0)\sin\theta] = {}_fE_2\epsilon^{j\theta} + {}_rE_2\epsilon^{-j\theta}$$

$$I_1 = I_2\cos\theta + jE_2Y_0\sin\theta = I_2[\cos\theta + j(Z_2/Z_0)\sin\theta] = Y_0({}_fE_2\epsilon^{j\theta} - {}_rE_2\epsilon^{-j\theta})$$

General circuit parameters are

$$A = \cos\theta$$

$$B = jZ_0\sin\theta$$

$$C = jY_0\sin\theta$$

$$D = \cos\theta$$

IMPEDANCE AND ADMITTANCE

$$\frac{Z_1}{Z_0} = \frac{Z_2\cosh\gamma x + Z_0\sinh\gamma x}{Z_0\cosh\gamma x + Z_2\sinh\gamma x}$$

$$\frac{Y_1}{Y_0} = \frac{Y_2\cosh\gamma x + Y_0\sinh\gamma x}{Y_0\cosh\gamma x + Y_2\sinh\gamma x}$$

(A) The input impedance of a line at a position of maximum or minimum voltage has the same phase angle as the characteristic impedance.

$Z_1/Z_0 = Z_b/Z_0 = Y_0/Y_b = r_b + j0 = S^{-1}$ at a voltage minimum (current maximum).

$Y_1/Y_0 = Y_a/Y_0 = Z_0/Z_a = g_a + j0 = S^{-1}$ at a voltage maximum (current minimum).

(B) When attenuation is small

$$\frac{Z_1}{Z_0} = \frac{[(Z_2/Z_0) + \alpha x] + j[1 + (Z_2/Z_0)\alpha x]\tan\theta}{[1 + (Z_2/Z_0)\alpha x] + j[(Z_2/Z_0) + \alpha x]\tan\theta}$$

For admittances, replace Z_0, Z_1, and Z_2 by Y_0, Y_1, and Y_2, respectively.

(C) When attenuation is neglected

$$\frac{Z_1}{Z_0} = \frac{Z_2/Z_0 + j\tan\theta}{1 + j(Z_2/Z_0)\tan\theta} = \frac{1 - j(Z_2/Z_0)\cot\theta}{Z_2/Z_0 - j\cot\theta}$$

and similarly for admittances.

(D) When attenuation $\alpha x = \theta\alpha/\beta$ is small and the standing-wave ratio is large (say>10)(*Note*: The complex value of Z_0 or Y_0 must be used in computing the resistive component of Z_1 or Y_1.): For θ measured from a voltage minimum

$$Z_1/Z_0 = [r_b + (\alpha/\beta)\theta](1 + \tan^2\theta) + j\tan\theta = [r_b + (\alpha/\beta)\theta](\cos^2\theta)^{-1} + j\tan\theta$$

(See Note 1)

$$Z_0/Z_1 = Y_1/Y_0 = [r_b + (\alpha/\beta)\theta](1 + \cot^2\theta) - j\cot\theta = [r_b + (\alpha/\beta)\theta](\sin^2\theta)^{-1} - j\cot\theta$$

(See Note 2)

For θ measured from a voltage maximum

$$Z_0/Z_1 = Y_1/Y_0 = [g_a + (\alpha/\beta)\theta](1 + \tan^2\theta) + j\tan\theta$$

(See Note 1)

$$Z_1/Z_0 = [g_a + (\alpha/\beta)\theta](1 + \cot^2\theta) - j\cot\theta$$

(See Note 2)

Note 1: Not valid when $\theta \approx \pi/2, 3\pi/2$, etc., due to approximation in denominator $1 + (r_b + \theta\alpha/\beta)^2\tan^2\theta = 1$ (or with g_a in place of r_b).

Note 2: Not valid when $\theta \approx 0, \pi, 2\pi$, etc., due to approximation in denominator $1 + (r_b + \theta\alpha/\beta)^2\cot^2\theta = 1$ (or with g_a in place of r_b). For open- or short-circuited line, valid at $\theta = 0$.

(E) When x is an integral multiple of $\lambda/2$ or $\lambda/4$: For $x = n\lambda/2$, or $\theta = n\pi$

$$\frac{Z_1}{Z_0} = \frac{(Z_2/Z_0) + \tanh n\pi(\alpha/\beta)}{1 + (Z_2/Z_0)\tanh n\pi(\alpha/\beta)}$$

For $x = n\lambda/2 + \lambda/4$, or $\theta = (n + \frac{1}{2})\pi$

$$\frac{Z_1}{Z_0} = \frac{1 + (Z_2/Z_0)\tanh(n + \frac{1}{2})\pi(\alpha/\beta)}{(Z_2/Z_0) + \tanh(n + \frac{1}{2})\pi(\alpha/\beta)}$$

(F) For small attenuation, with any standing-wave ratio: For $x = n\lambda/2$, or $\theta = n\pi$, where n is an integer

$$\frac{Z_1}{Z_0} = \frac{(Z_2/Z_0) + n\pi(\alpha/\beta)}{1 + (Z_2/Z_0)n\pi(\alpha/\beta)}$$

$$g_{a1} = \frac{g_{a2} + \alpha n\lambda/2}{1 + g_{a2}\alpha n\lambda/2} = S_1^{-1}$$

For $x = (n+\frac{1}{2})\lambda/2$, or $\theta = (n+\frac{1}{2})\pi$, where n is an integer or zero

$$\frac{Z_1}{Z_0} = \frac{1 + (Z_2/Z_0)\ (n+\frac{1}{2})\alpha(\lambda/2)}{(Z_2/Z_0) + (n+\frac{1}{2})\alpha(\lambda/2)}$$

$$g_{b1} = \frac{1 + g_{a2}(n+\frac{1}{2})\ (\alpha/\beta)\pi}{g_{a2} + (n+\frac{1}{2})\ (\alpha/\beta)\pi} = S_1$$

Subscript a refers to the voltage-maximum point and b to the voltage minimum. In the above equations, subscripts a and b may be interchanged, and/or r may be substituted in place of g, except for the relationships to standing-wave ratio.

LINES OPEN- OR SHORT-CIRCUITED AT THE FAR END

Point 2 is the open- or short-circuited end of the line, from which x and θ are measured.

(A) Voltages and Currents: Use the equations of the "Voltage and Current" section, with the following conditions.

Open circuited line:

$$\rho_2 = 1.00\angle 0° = 1.00$$

$${}_rE_2 = {}_fE_2 = E_2/2$$

$${}_rI_2 = -{}_fI_2$$

$$I_2 = 0$$

$$Z_2 = \infty$$

Short-circuited line:

$$\rho_2 = 1.00\angle 180° = -1.00$$

$${}_rE_2 = -{}_fE_2$$

$$E_2 = 0$$

$${}_rI_2 = {}_fI_2 = I_2/2$$

$$Z_2 = 0$$

(B) Impedances and admittances:

$$Z_{oc} = Z_0 \coth\gamma x$$

$$Z_{sc} = Z_0 \tanh\gamma x$$

$$Y_{oc} = Y_0 \tanh\gamma x$$

$$Y_{sc} = Y_0 \coth\gamma x$$

(C) For small attenuation: Use the equations for large swr in **(D)** of the preceding section, with the following conditions.

Open-circuited line:

$$g_a = 0$$

Short-circuited line:

$$r_b = 0$$

(D) When attenuation is neglected:

$$Z_{oc} = -jR_0 \cot\theta$$

$$Z_{sc} = jR_0 \tan\theta$$

$$Y_{oc} = jG_0 \tan\theta$$

$$Y_{sc} = -jG_0 \cot\theta$$

(E) Relationships between Z_{oc} and Z_{sc}:

$$(Z_{oc}Z_{sc})^{1/2} = Z_0$$

$$\pm(Z_{sc}/Z_{oc})^{1/2} = \tanh\gamma x$$

$$\pm(Z_{oc}/Z_{sc})^{1/2} = \coth\gamma x$$

(F) When attenuation is small (except for $\theta \approx n\pi/2$, $n = 1, 2, 3, \cdots$)

$$\pm(Z_{sc}Z_{oc})^{1/2} = \pm(Y_{oc}/Y_{sc})^{1/2}$$
$$= \pm j[-(C_{oc}/C_{sc})]^{1/2}$$
$$\times[1 - j\tfrac{1}{2}(G_{oc}/\omega C_{oc} - G_{sc}/\omega C_{sc})]$$

where $Y_{oc} = G_{oc} + j\omega C_{oc}$ and $Y_{sc} = G_{sc} + j\omega C_{sc}$. The + sign is to be used before the radical when C_{oc} is positive, and the − sign when C_{oc} is negative.

(G) $R/|X|$ component of input impedance of low-attenuation nonresonant line:

Short-circuited line (except when $\theta \approx \pi/2$, $3\pi/2$, etc.)

$$R_1/|X_1| = G_1/|B_1|$$
$$= |(\alpha/\beta)\theta(\tan\theta + \cot\theta) + (B_0/G_0)|$$
$$= |(\alpha/\beta)(2\theta/\sin 2\theta) + (B_0/G_0)|$$

Open-circuited line (except when $\theta \approx \pi$, 2π, etc.)

$$R_1/|X_1| = G_1/|B_1|$$
$$= |(\alpha/\beta)\theta(\tan\theta + \cot\theta) - (B_0/G_0)|$$
$$= |(\alpha/\beta)(2\theta/\sin 2\theta) - (B_0/G_0)|$$

VOLTAGE REFLECTION COEFFICIENT AND STANDING-WAVE RATIO

$$\rho = {}_rE/{}_fE = -{}_rI/{}_fI = (Z - Z_0)/(Z + Z_0)$$
$$= (Y_0 - Y)/(Y_0 + Y) = |\rho| \angle 2\psi$$

where ψ is the electrical angle to the nearest voltage maximum on the generator side of the point where ρ is measured (Figs. 3 and 4).

$$\rho_1 = \rho_2\epsilon^{-2\alpha x} \angle -2\theta$$

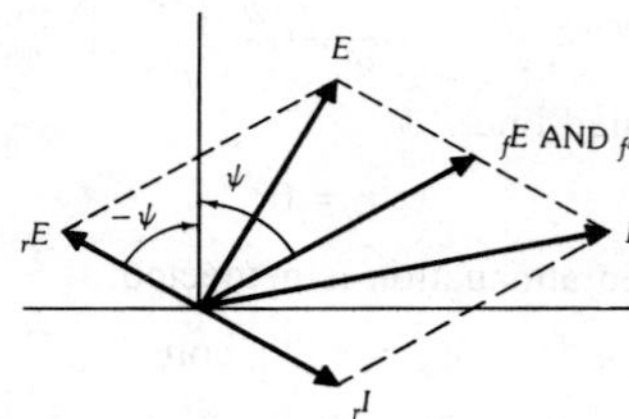

Fig. 3. Voltages and currents at time $t=0$ at point ψ electrical degrees toward the load from a voltage standing-wave maximum.

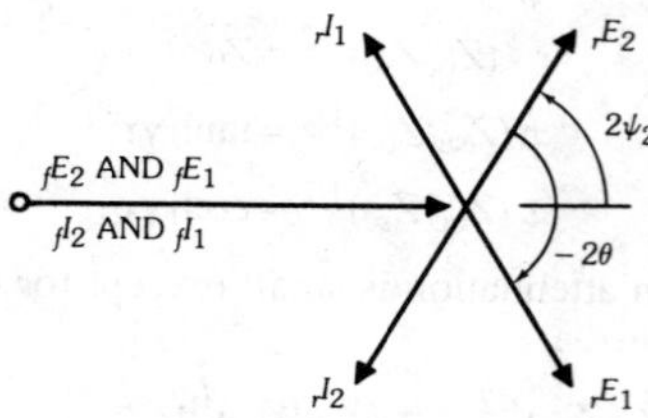

Fig. 4. Abbreviated diagram of a line with zero attenuation.

$$|\rho_1| = |\rho_2|/10^{A_0/10}$$

Voltage reflection coefficient in decibels

$$\rho_{dB} = -20 \log_{10} |1/\rho|$$

The minus sign is frequently omitted.

$$|\rho_{dB} \text{ at input}| = |\rho_{dB} \text{ at load}| + 2A_0$$

These two relationships and standing-wave ratio versus reflection coefficient in decibels are shown in Figs. 5 and 6.

$$Z = E/I = ({}_fE + {}_rE)/({}_fI + {}_rI)$$
$$= Z_0[(1+\rho)/(1-\rho)]$$
$$Z/Z_0 = (1+\rho)/(1-\rho)$$
$$= \frac{1+jS\cot\psi}{S+j\cot\psi}$$
$$S = |E_{max}/E_{min}| = |I_{max}/I_{min}|$$
$$= \frac{|{}_fE| + |{}_rE|}{|{}_fE| - |{}_rE|} = \frac{|{}_fI| + |{}_rI|}{|{}_fI| - |{}_rI|}$$
$$= \frac{1+|\rho|}{1-|\rho|} = r_a = g_a^{-1} = g_b = r_b^{-1}$$
$$|\rho| = (S-1)/(S+1)$$
$$1/S_1 = \tanh[\alpha x + \tanh^{-1}(1/S_2)]$$
$$= \tanh[0.1151A_0 + \tanh^{-1}(1/S_2)]$$

(A) For high standing-wave ratio: When the ratio S_1 is greater than 6/1, then with 1 percent accuracy or better

$$1/S_1 = 1/S_2 + \alpha x = 1/S_2 + 0.115A_0$$
$$|\rho_{dB}| = 17.4/S$$

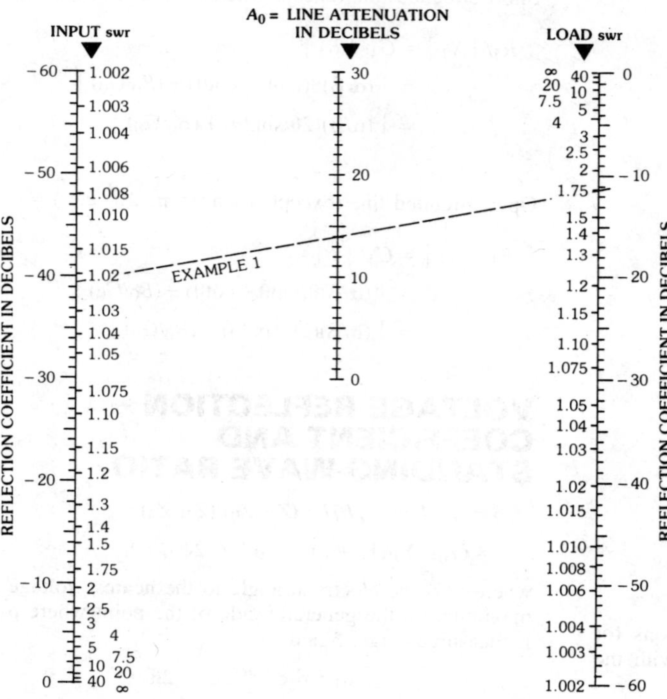

Fig. 5. Line attenuation and voltage reflection coefficient for low swr.

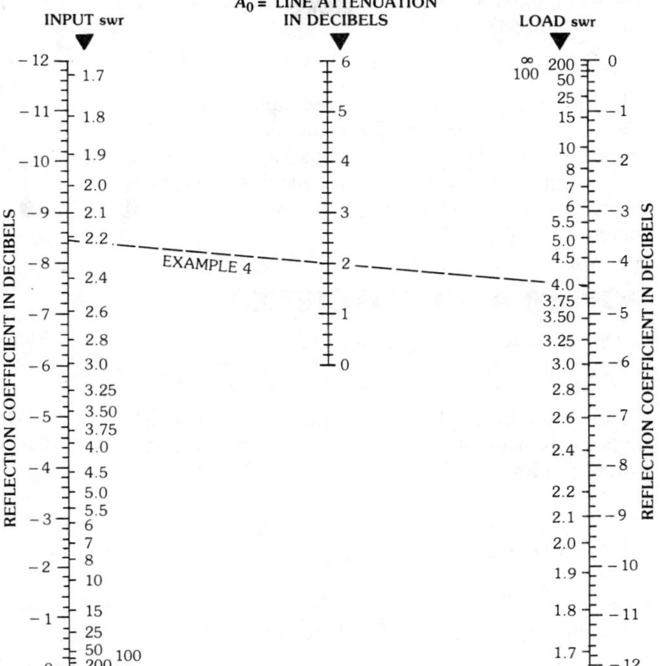

Fig. 6. Line attenuation and voltage reflection coefficient for high swr.

Subject to the conditions below, the standing-wave ratio is given by one or the other of

$$S \approx (1+x^2)/r$$

$$S \approx (1+b^2)/g$$

where

$$r+jx = Z/Z_0 = (1/R_0)[R-(B_0/G_0)X+jX]$$

$$g+jb = Y/Y_0 = (1/G_0)[G+(B_0/G_0)B+jB]$$

Conditions, for 1-percent accuracy

$$r<0.1\,|\,x+1/x\,| \quad \text{when } |\,x\,|>0.3$$

$$g<0.1\,|\,b+1/b\,| \quad \text{when } |\,b\,|>0.3$$

The boundary of the 1-percent-error region can be

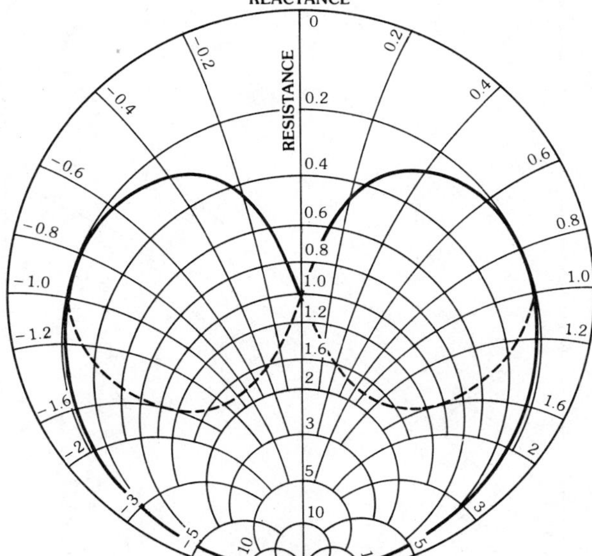

Fig. 7. Permitted region for use of equation $S \approx (1+x^2)/r$. *(From W. W. Macalpine, "Computation of Impedance and Efficiency of Transmission Line with High Standing-Wave Ratio,"* Trans. of the AIEE, *Vol. 72, Part 1, p. 336, Fig. 2; July 1953.)*

plotted on the Smith chart (Fig. 7) by use of the equation (for impedances)

$$|\cot\psi| = 0.1S^2/(S^2-1)^{1/2}$$

The same boundary line on the chart holds when reading admittances. The area outside the solid heart-shaped curve is where the swr equation is accurate to within 1 percent. The area outside the dashed curve is where the reciprocal of $r+jx$ lies in the permitted region.

POWER AND EFFICIENCY

The net power flowing toward the load is

$$P=|{}_fE|^2G_0[1-|\rho|^2+2|\rho|(B_0/G_0)\sin 2\psi]$$

(A) When the angle B_0/G_0 of the characteristic admittance is negligibly small, the net power flowing toward the load is given by

$$P=G_0(|{}_fE|^2-|{}_rE|^2)$$
$$=|{}_fE|^2G_0(1-|\rho|^2)$$
$$=|E_{max}E_{min}|/R_0$$
$$P_1=|{}_fE_2|^2G_0(\epsilon^{2(\alpha/\beta)\theta}-|\rho_2|^2\epsilon^{-2(\alpha/\beta)\theta})$$

(B) Efficiency, when B_0/G_0 is negligibly small:

$$\eta=P_2/P_1=\frac{1-|\rho_2|^2}{\epsilon^{2(\alpha/\beta)\theta}-|\rho_2|^2\epsilon^{-2(\alpha/\beta)\theta}}$$
$$=\eta_{max}\frac{1-|\rho_2|^2}{1-|\rho_2|^2\eta_{max}^2}=\frac{1-|\rho_2|^2}{1-|\rho_1|^2}\epsilon^{-2\alpha x}$$
$$=\frac{1/|\rho_2|-|\rho_2|}{1/|\rho_1|-|\rho_1|}=\frac{S_1-1/S_1}{S_2-1/S_2}$$

The maximum error in the above expressions is

$$\pm 100(S_2-1/S_2)B_0/G_0 \quad \text{percent}$$
$$\pm 4.34(S_2-1/S_2)B_0/G_0 \quad \text{decibels}$$

When the ratio S_1 is greater than 6/1:

$$\eta\approx S_1/S_2\approx(1+0.115A_0S_2)^{-1}$$

When the load matches the line, $\rho_2=0$ and the efficiency is accurately

$$\eta_{max}=\exp[-2(\alpha/\beta)\theta]=\exp(-2\alpha x)=10^{-A_0/10}$$
$$A-A_0=10\log_{10}(\eta_{max}/\eta)$$

Fig. 8 is drawn from the expressions in this paragraph.

(C) Efficiency, when swr is high:

$$\eta=\frac{P_2}{P_1}=\frac{R_2}{R_1}\left(\frac{1+x_1^2}{1+x_2^2}\right)=\frac{G_2}{G_1}\left(\frac{1+b_1^2}{1+b_2^2}\right)$$

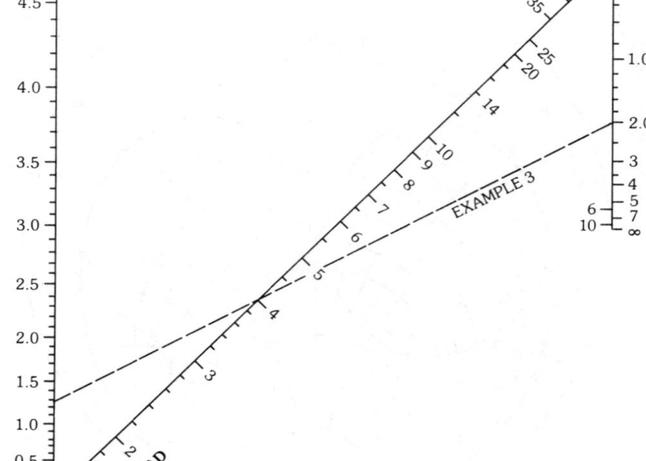

Fig. 8. Standing-wave loss factor. Due to load mismatch, an increase of loss in decibels as read from this figure must be added to normal line attenuation to give total dissipation loss in the line. This does not include mismatch loss due to any difference of line *input* impedance from the conjugate of the generator impedance [Paragraph **(B)** in section headed "Power and Efficiency"].

$$=\frac{R_2}{R_0{}^2G_1}\left(\frac{1+b_1{}^2}{1+x_2{}^2}\right)=\frac{R_0{}^2G_2}{R_1}\left(\frac{1+x_1{}^2}{1+b_2{}^2}\right)$$

where R is the ohmic resistance and x is the normalized reactance, and similarly for G and b. It is important that the Rs and Gs be computed properly, using equations in the following section, headed "Transformation of Impedance on Lines with High SWR." Note the identity of the efficiency equations with the left-hand terms of the impedance equations. The conditions for accuracy are the same as stated for the impedance equations for high standing-wave ratio.

Example: Physical significance of the equation for efficiency at high standing-wave ratio: Subject to stated conditions, approximately, $x=\cot\psi$ and $I=I_{max}\sin\psi$. I_{max} = current standing-wave maximum, practically constant along line when standing-wave ratio>6. Then

$$P=I^2R=I_{max}{}^2R/(1+x^2)$$

When line length is greater than $\frac{1}{3}$ wavelength, then

$$\eta\approx[1+0.115A_0(1+x_2{}^2)\,(R_0/R_2)]^{-1}$$

(D) Loss in nepers $=\frac{1}{2}\log_e(P_1/P_2)=0.1151\times$ (loss in decibels).

For a matched line, loss = attenuation $=(\alpha/\beta)\theta=\alpha x$ nepers.

Loss in decibels $=10\log_{10}(P_1/P_2)=8.686\times$ (loss in nepers).

When $2(\alpha/\beta)\theta$ is small

$$P_1/P_2=1+2(\alpha/\beta)\theta\,\frac{1+|\rho_2|^2}{1-|\rho_2|^2}$$

and

decibels/wavelength

$$=10\log_{10}\left(1+4\pi(\alpha/\beta)\,\frac{1+|\rho_2|^2}{1-|\rho_2|^2}\right)$$

(E) For the same power flowing in a line with standing waves as in a matched or "flat" line:

$$P=|E_{flat}|^2/R_0$$

$$|E_{max}|=|E_{flat}|\,S^{1/2}$$

$$|E_{min}|=|E_{flat}|\,/S^{1/2}$$

$$|{}_fE|=\tfrac{1}{2}\,|E_{flat}|\,(S^{1/2}+S^{-1/2})$$

$$|{}_rE|=\tfrac{1}{2}\,|E_{flat}|\,(S^{1/2}-S^{-1/2})$$

When the loss is small, so that S is nearly constant over the entire length, then per half wavelength

$$\text{(power loss)/(loss for flat line)}\approx\tfrac{1}{2}(S+1/S)$$

(F) The power dissipation per unit length, for unity standing-wave ratio, is

$$\Delta P_d/\Delta x=2\alpha P$$

$$\frac{\text{(dissipation in watts/foot)}}{\text{(line power in kilowatts)}}=2.30\text{ (decibels/100 feet)}$$

where the decibels/100 feet is the normal attenuation for a matched line.

When swr>1, the dissipation at a current maximum is S times that for swr = 1, assuming the attenuation to be due to conductor loss only. The multiplying factor for local heating reaches a minimum value equal to $(S+1/S)/2$ all along the line when conductor loss and dielectric loss are equal.

(G) Further considerations on power and efficiency are given in the section headed "Mismatch and Transducer Loss" (p. 29-12).

TRANSFORMATION OF IMPEDANCE ON LINES WITH HIGH SWR*

When standing-wave ratio is greater than 10 or 20, resistance cannot be read accurately on the Smith chart, although it is satisfactory for reactance.

Use the equation (Fig. 9)

$$R_1=R_2\,\frac{1+x_1{}^2}{1+x_2{}^2}+R_0(1+x_1{}^2)$$

$$\times\left[(\alpha/\beta)\theta+(B_0/G_0)\left(\frac{x_1}{1+x_1{}^2}-\frac{x_2}{1+x_2{}^2}\right)\right]$$

where,

R = ohmic resistance,

$x=X/R_0$ = normalized reactance.

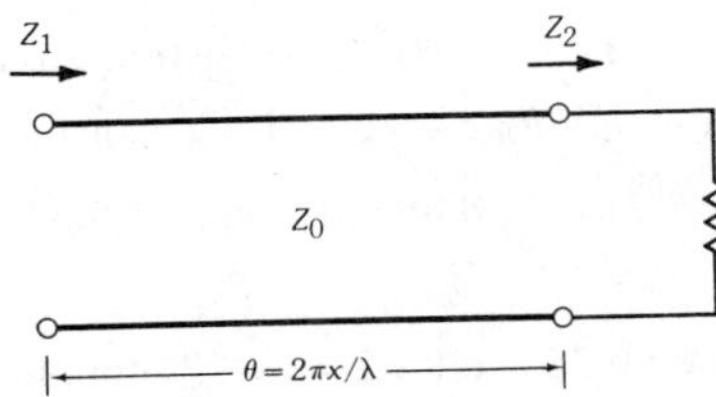

Fig. 9. Transmission-line impedances.

When admittance is given or required, similar equations can be written with the aid of the following tabulation. The top row shows the terms in the above equation.

R_1	R_2	$x_1{}^2$	$x_2{}^2$	R_0	x_1	$-x_2$
G_1	G_2	$b_1{}^2$	$b_2{}^2$	$1/R_0$	$-b_1$	b_2
R_1	$G_2R_0{}^2$	$x_1{}^2$	$b_2{}^2$	R_0	x_1	b_2
G_1	$R_2/R_0{}^2$	$b_1{}^2$	$x_2{}^2$	$1/R_0$	$-b_1$	$-x_2$

* W. W. Macalpine, "Computation of Impedance and Efficiency of Transmission Lines with High Standing-Wave Ratio," *Transactions of the AIEE*, Vol. 72, Part I, pp. 334-339; July 1953; also *Electrical Communication*, Vol. 30, pp. 238-246; September 1953.

For transforming R to G or vice versa:

$$R = R_0^2 G \mid x/b \mid$$

where x and b are read on the Smith chart in the usual manner for transforming impedances to admittances.

The conditions for roughly 1-percent accuracy of the equations are:

Standing-wave ratio is greater than 6/1 at input; $|B_0/G_0| < 0.1$; $r+jx$ or $g+jb$ (whichever is used, at each end of line) meet the requirements stipulated in paragraph **(A)** of the section headed "Voltage Reflection Coefficient and Standing-Wave Ratio" (p. 29-7); and the line parameters and given impedance are known to 1-percent accuracy.

When line length is greater than $\frac{1}{3}$ wavelength, then

$$R_1 \approx R_2[(1+x_1^2)/(1+x_2^2)] + 0.115 A_0 R_0 (1+x_1^2)$$

$$\frac{R_1/R_0}{1+x_1^2} \approx \frac{R_2/R_0}{1+x_2^2} + (\alpha/\beta)\theta$$

The equation for resistance transformation is derived from expressions for high swr in paragraph **(A)**, just referred to.

Example: A load of $0.4 - j2000$ ohms is fed through a length of RG-218/U cable at a frequency of 2.0 megahertz. What are the input impedance and the efficiency for a 24-foot length of cable and for a 124-foot length?

For RG-218/U, the attenuation at 2.0 megahertz is 0.095 decibel/100 feet (see Fig. 29). The dielectric constant $\epsilon = 2.26$, and F_p is negligibly small. Then, by equations in **(B)** and **(C)**, p. 29-5:

$$B_0/G_0 = \alpha/\beta = (\text{dB}/100 \text{ ft})(\lambda_{\text{meters}})/1663$$

$$= [0.095 \times 150/(2.26)^{1/2}]/1663 = 0.0057$$

$$x/\lambda = xf\epsilon^{1/2}/984 = 24 \times 2.0 \times 1.5/984 = 0.073$$

$$\theta = 2\pi x/\lambda = 0.46 \text{ radian for 24-foot length}$$

while

$$x/\lambda = 0.38 \text{ and } \theta = 2.4 \text{ for 124-foot length}$$

$$Z_2/Z_0 \approx (0.4 - j2000)/50 = 0.008 - j40$$

For the 24-foot length, by the Smith chart

$$x_1 = X_1/Z_0 = -1.9, \text{ or } X_1 = -95 \text{ ohms}$$

The conditions for accuracy of the resistance transformation equation are satisfied. Now

$$1 + x_1^2 = 1 + (1.9)^2 = 4.6$$

$$1 + x_2^2 = 1 + (40)^2 = 1600$$

$$R_1 = 0.4(4.6/1600) + 50 \times 4.6 \times 0.0057 \times [0.46 - (1.9/4.6) + (40/1600)]$$

$$= 0.0012 + 0.105 = 0.106 \text{ ohm}$$

The efficiency equation in paragraph **(C)** on p. 29-10 gives

$$\eta = 0.0012/0.106 = 0.0113, \text{ or } 1.1 \text{ percent}$$

where the 0.0012 figure is taken directly from the first quantity on the right-hand side of the computation of the value of R_1.

Similarly, for the 124-foot length, $x_1 = 1.1$, $X_1 = 55$ ohms, $1 + x_1^2 = 2.21$, $R_1 = 0.00055 + 1.83 = 1.83$ ohms

$$\eta = 0.00055/1.83 = 3.1 \times 10^{-4}, \text{ or } 0.03 \text{ percent}$$

Tabulating the results,

Length (feet)	Input Impedance (ohms)	Efficiency (%)	Loss (dB)
24	$0.106 - j95$	1.1	19.6
124	$1.8 + j55$	0.03	35

The considerably greater loss for 124 feet compared with 24 feet is because the transmission passes through a current maximum where the loss per unit length is much higher than at a current minimum.

MISMATCH AND TRANSDUCER LOSS

Figs. 5, 6, and 8, plus the equations in this section, permit the calculation of loss when impedance mismatch exists in a transmission-line system; also, the change in standing-wave ratio along a line due to attenuation can be determined.

One End Mismatched

When either generator or load impedance is mismatched to the Z_0 of the line and the other is matched (Fig. 10)

$$(\text{mismatch loss}) = P_m/P = 1/(1 - |\rho|^2) = (S+1)^2/4S \quad (1)$$

where,

P = power delivered to load,

P_m = power that would be delivered were system matched,

S = standing-wave ratio of mismatched impedance referred to Z_0.

Compared with an ideal transducer (ideal matching network between generator and load)

$$(\text{transducer loss}) = A_0 + 10 \log_{10}(P_m/P) \text{ decibels} \quad (2)$$

where A_0 = normal attenuation of line.

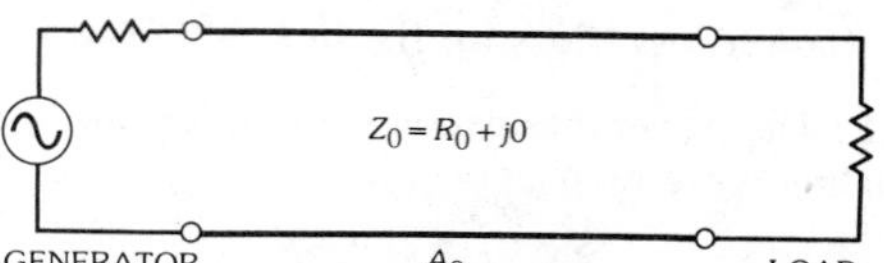

Fig. 10. Transmission line with generator and load.

Generator and Load Mismatched

$$| X_0/R_0 | << 1$$

When mismatches exist at both ends of the system (Fig. 11)

(mismatch loss at input)

$$= P_m/P$$

$$= [(R_g + R_1)^2 + (X_g + X_1)^2]/4R_gR_1 \quad (3)$$

$$\text{(transducer loss)} = (A - A_0) + A_0$$

$$+ 10 \log_{10}(P_m/P) \text{ decibels} \quad (4)$$

where $(A - A_0)$ = standing-wave loss factor obtained from Fig. 8 for S = standing-wave ratio at load.

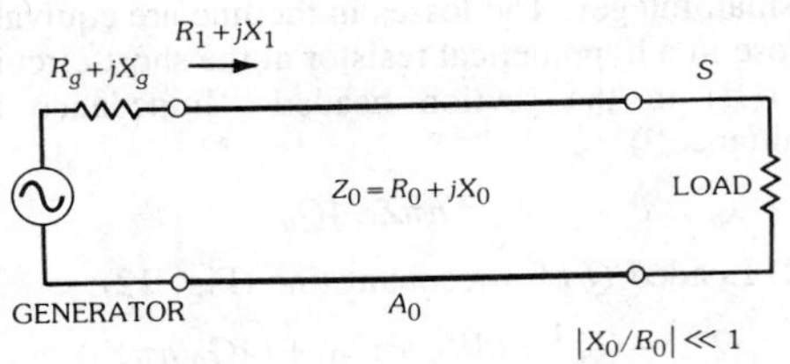

Fig. 11. Transmission line with mismatch at both ends.

Notes on Equation (3)

Equation (3) reduces to Eq. (1) when X_g and/or X_1 is zero.

In Eq. (3), the impedances can be either ohmic or normalized with respect to any convenient Z_0.

When determining input impedance $R_1 + jX_1$ on a Smith chart, adjust the radius arm for S at the input, determined from that at the output by the aid of Figs. 5 and 6.

For the junction of two admittances, use Eq. (3) with G and B substituted for R and X, respectively.

Equation (3) is valid for a junction in any linear passive network. The same is true of Eq. (1) when at least one of the impedances concerned is purely resistive. Determine S as if one impedance were that of a line.

Examples

Example 1: The swr at the load is 1.75, and the line has an attenuation of 14 decibels. What is the input swr?

Using Fig. 5, set a straightedge through the 1.75 division on the "load swr" scale and the 14-decibel point on the middle scale. Read the answer on the "input swr" scale, which the straightedge intersects at 1.022.

Example 2: Readings on a reflectometer show the reflected wave to be 4.4 decibels below the incident wave. What is the swr?

Using Fig. 6, locate the reflection coefficient 4.4 (or −4.4) decibels on either outside scale. Beside it, on the same horizontal line, read swr = 4.0+.

Example 3: A 50-ohm line is terminated with a load of $200 + j0$ ohms. The normal attenuation of the line is 2.00 decibels. What is the loss in the line?

Using Fig. 8, align a straightedge through the points $A_0 = 2.0$ and swr = 4.0. Read $A - A_0 = 1.27$ decibels on the left-hand scale. Then the transmission loss in the line is

$$A = 1.27 + 2.00 = 3.27 \text{ decibels.}$$

This is the dissipation, or heat loss, as opposed to the mismatch loss at the input (example 4).

Example 4: In the preceding example, suppose the generator impedance is $100 + j0$ ohms, and the line is 5.35 wavelengths long. What is the mismatch loss between the generator and the line?

According to example 3, the load swr = 4.0 and the line attenuation is 2.0 decibels. Then, using Fig. 6, the input swr is found to be 2.22. On the Smith chart, locate the point corresponding to 0.35 wavelength toward the generator from a voltage maximum, and swr = 2.22. Read the input normalized impedance as $0.62 + j0.53$ with respect to $Z_0 = 50$ ohms. Now the mismatch loss at the input can be determined by use of Eq. (3). However, since the generator impedance is nonreactive, Eq. (1) can be used if desired. Refer to the following paragraph and to the "Notes on Equation (3)" above.

With respect to $100 + j0$ ohms, the normalized impedance at the line input is $0.31 + j0.265$, which gives swr = 3.5 according to the Smith chart. Then by Eq. (1), $P_m/P = 1.45$, giving a mismatch loss of 1.62 decibels. The transducer loss is found by using the results of examples 3 and 4 in Eq. (4). This is

$$1.27 + 2.00 + 1.62 = 4.9 \text{ decibels}$$

ATTENUATION AND RESISTANCE OF TRANSMISSION LINES AT ULTRAHIGH FREQUENCIES

The normal or matched-line attenuation in decibels/100 feet is

$$A_{100} = 4.34R_t/Z_0 + 2.78f\epsilon^{1/2}F_p$$

where the total line resistance/100 feet (for perfect surface conditions of the conductors) is, for copper coaxial line

$$R_t = 0.1(1/d + 1/D)f^{1/2}$$

and for copper 2-wire open line

$$R_t = (0.2/d)f^{1/2}$$

where,

D = diameter of inner surface of outer coaxial conductor in inches,

d = diameter of conductors (coaxial-line center conductor) in inches,

f = frequency in megahertz,

ϵ = dielectric constant relative to air,

F_p = power factor of dielectric at frequency f.

For other conductor materials, the resistance of a conductor of diameter d (and similarly for D) is

$$0.1(1/d)\,(f\mu_r\rho/\rho_{Cu})^{1/2} \text{ ohms/100 feet}$$

where,

μ_r = relative permeability of material (1 for nonmagnetic materials),

ρ = resistivity of material at any temperature,

ρ_{Cu} = resistivity of copper at 20°C (1.724 microhms-centimeter),

f = frequency in megahertz.

RESONANT LINES

Symbols:

f_0 = resonance frequency in megahertz
G_a = conductance load in mhos at voltage standing-wave maximum, equivalent to some or all of the actual loads
k = coefficient of coupling
n = integral number of quarter wavelengths
$p = k^2Q_{1s}Q_{2s}$ = load transfer coefficient or matching factor
P_c = power converted into heat in resonator
P_m = power available from generator in watts $= E_{oc}{}^2/4R_{gen}$
P_x = power transferred when load is directly connected to generator (for single resonators); or an analogous hypothetical power (for two coupled resonators)
Q = figure of merit of a resonator as it exists, whether loaded or unloaded
Q_d = doubly loaded Q (all loads being included)
Q_s = singly loaded Q (all loads included except one). For a pair of coupled resonators, Q_{1s} is the value for the first resonator when isolated from the other. (Similarly for Q_{2s})
Q_u = unloaded Q
R_b = resistance load in ohms at voltage standing-wave minimum, equivalent to some or all of the actual loads
R_u = resistance similar to R_b except for unloaded resonator
R_1 = generator resistance, referred to short-circuited end
R_2 = load resistance
$S_x = R_1/R_2$ or R_2/R_1 = mismatch factor between generator and load
Z_{10} = characteristic impedance of the first of a pair of resonators
θ_1 = electrical angle from a voltage standing-wave minimum point

(A) Q of a resonator (electrical, mechanical, or any other) is

$$Q = 2\pi \frac{\text{(energy stored)}}{\text{(energy dissipated per cycle)}}$$

$$= 2\pi f \frac{\text{(energy stored)}}{\text{(power dissipation)}}$$

In a freely oscillating system, the amplitude decays exponentially.

$$I = I_0 \exp(-\pi f t/Q)$$

(B) Unloaded Q of a resonant line:

$$Q_u = \beta/2\alpha$$

the line length being n quarter-wavelengths, where n is a small integer. The losses in the line are equivalent to those in a hypothetical resistor at the short-circuited end ((**D**) in the section headed "Impedance and Admittance")

$$R_u = n\pi Z_0/4Q_u$$

(C) Loaded Q of a resonant line (Fig. 12):

$$Q^{-1} = Q_u^{-1} + (4R_b/n\pi Z_0) + (4G_a/n\pi Y_0)$$

$$= (4/n\pi Z_0)\,(R_u + R_b + G_a/Y_0{}^2)$$

All external loads can be referred to one end and represented by either R_b or G_a as in Fig. 13.

The total loading is the sum of all the individual loadings.

General conditions:

$$R_b/Z_0 = G_a/Y_0 << 1.0$$

or, roughly, $Q>5$.

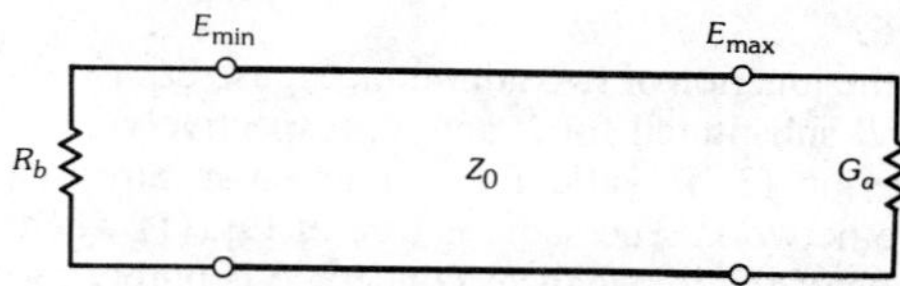

Fig. 12. Quarter-wave line with loadings at nominal short-circuit and open-circuit points.

(D) Input admittance and impedance:

The converse of the equations for Fig. 13 can be used at the resonance frequency. Then R or G is the input impedance or admittance, while

$$R_b = n\pi Z_0/4Q_s$$

where Q_s = singly loaded Q with the losses and all the loads considered except that at the terminals where input R or G is being measured.

In the vicinity of the resonance frequency, the input admittance when looking into a line at a tap point θ_1 in Fig. 14 is approximately

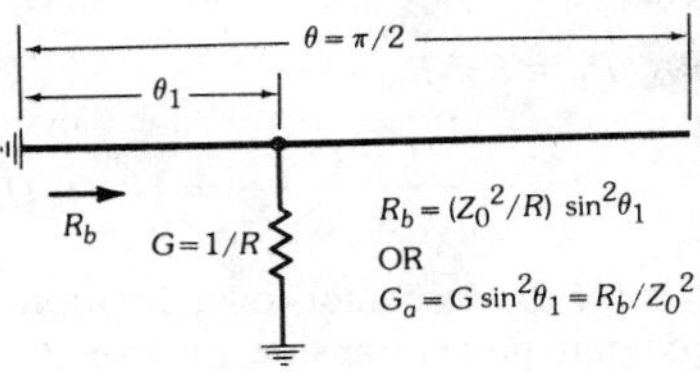

(A) Shunt or tapped load.

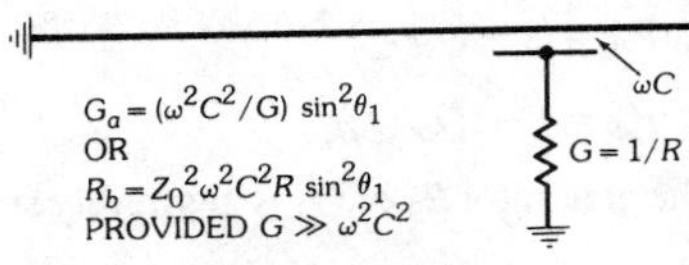

(B) Probe coupling.

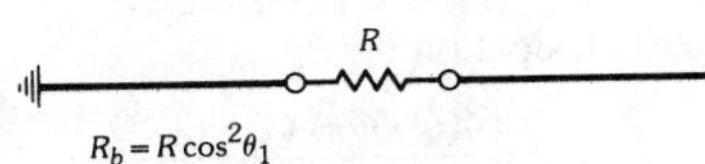

(C) Series load.

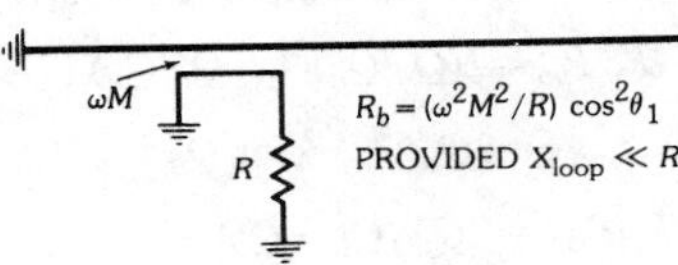

(D) Loop coupling.

Fig. 13. Typical loaded quarter-wave sections with apparent R_b equivalent to the loading at distance θ_1 from voltage-minimum point of the line. Outer conductor not shown.

$$Y = G + jB = \frac{n\pi Y_0}{4\sin^2\theta_1}\left(Q_s^{-1} + j2\frac{f-f_0}{f_0}\right)$$

provided

$$|f-f_0|/f_0 << 1.0$$

and

$$|[\theta(f-f_0)/f_0]\cot\theta_1| << 1.0$$

where $\theta = n\pi/2$ = length of line at f_0. The equation is not valid when $\theta_1 \approx 0$, π, 2π, etc., except that it is good near the short-circuited end when $f-f_0 \approx 0$.

Such a resonant line is approximately equivalent to a lumped LCG parallel circuit, where

$$\omega_0{}^2 L_1C_1 = (2\pi f_0)^2 L_1C_1 = 1$$

Admittance of the equivalent circuit is

$$Y = G + j[\omega C_1 - (1/\omega L_1)]$$

$$\approx \omega_0 C_1\{Q_s^{-1} + j2[(f-f_0)/f_0]\}$$

Then, subject to the conditions stated above

$$L_1 = (4\sin^2\theta_1)/n\pi\omega_0 Y_0$$

$$C_1 = n\pi Y_0/(4\omega_0\sin^2\theta_1) = nY_0/(8f_0\sin^2\theta_1)$$

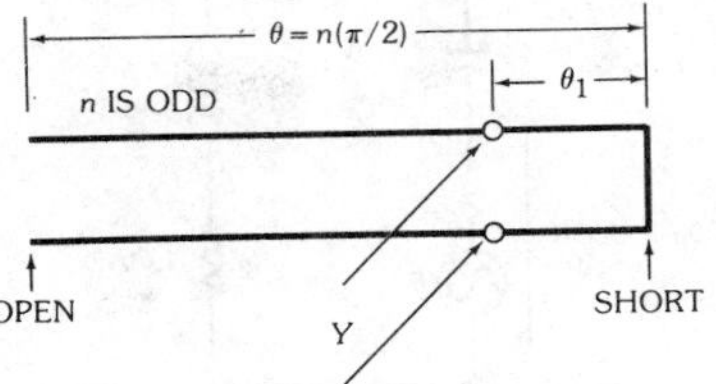

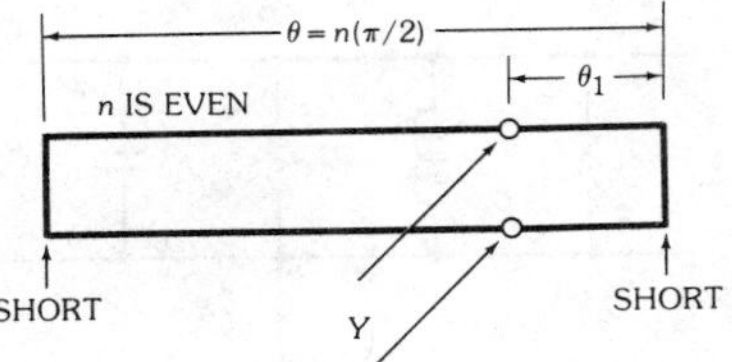

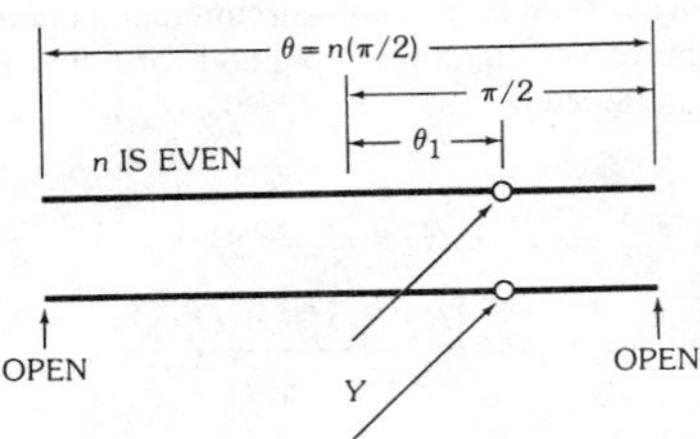

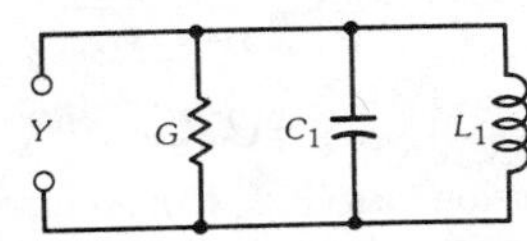

Fig. 14. Resonant transmission lines and their equivalent lumped circuit.

$$G = n\pi Y_0/(4Q_s\sin^2\theta_1)$$

$$Q_s = \omega_0 C_1/G = 1/\omega_0 L_1 G$$

Similarly, the input impedance at a point in series with the line (Figs. 13C and D) is

$$Z = R + jX = \frac{n\pi Z_0}{4\cos^2\theta_1}\left(Q_s^{-1} + j2\frac{f-f_0}{f_0}\right)$$

provided

$$|f-f_0|/f_0 << 1.0$$

and

$$|\theta[(f-f_0)/f_0]\tan\theta_1| << 1.0$$

The equation is not valid when $\theta_1 \approx \pi/2$, $3\pi/2$, etc.

The voltage standing-wave ratio at resonance, on the generator (Fig. 15), is

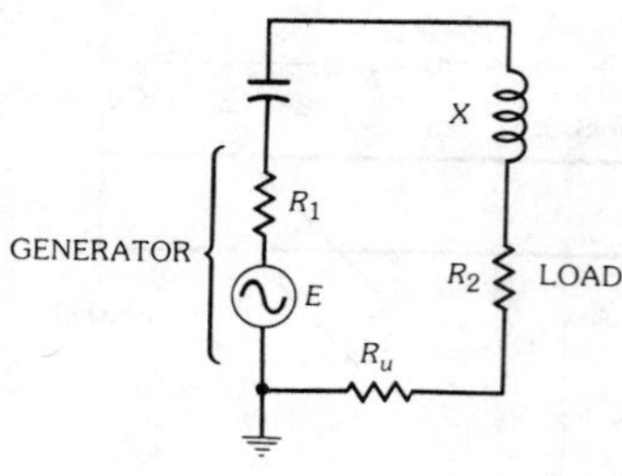

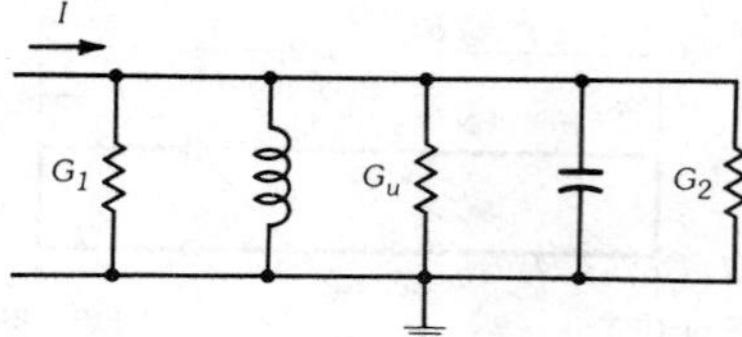

Fig. 15. Equivalent circuits of a resonant line (or a lumped tuned circuit) as seen at the short-circuited and open-circuited ends. All the power equations are good for either lumped or distributed parameters.

$$S=(R_2+R_u)/R_1$$

$$=\frac{(R_2/R_1)Q_u+Q_d}{Q_u-Q_d}$$

When $R_1=R_2$

$$S=\frac{1+Q_d/Q_u}{1-Q_d/Q_u}$$

$$\rho=Q_d/Q_u$$

(E) Insertion loss (Fig. 15): At resonance, for either a distributed or a lumped-constant device

(dissipation loss)

$$=10\ \log_{10}(P_x/P_{\text{out}})$$

$$=\ 20\ \log_{10}[1/(1-Q_d/Q_u)]$$

$$\approx 20\ \log_{10}(1+Q_d/Q_u)$$

$$\approx 8.7Q_d/Q_u \text{ decibels}$$

(mismatch loss)

$$=10\ \log_{10}(P_m/P_x)$$

$$=10\ \log_{10}[(1+S_x)^2/4S_x] \text{ decibels}$$

The dissipation loss also includes a small additional mismatch loss due to the presence of the resonator. The error in the form $20\log_{10}(1+Q_d/Q_u)$ is about twice that of the form $8.7Q_d/Q_u$. The last expression ($8.7Q_d/Q_u$) is in error compared with the first, $20\log_{10}[1/(1-Q_d/Q_u)]$, by roughly $-50(Q_d/Q_u)$ percent for $(Q_d/Q_u)<0.2$.

The selectivity is given on page 8-7, where $Q=Q_d$. That equation is accurate over a smaller range of $(f-f_0)$ for a resonant line than it is for a single tuned circuit.

At resonance*

$$P_{\text{out}}/P_{\text{in}}=R_2/(R_u+R_2)$$

$$=\frac{Q_u-Q_d}{Q_u+(R_1/R_2)Q_d}=1-Q_s/Q_u$$

where Q_s is for the resonator loaded with R_2 only.

The maximum power transfer, for fixed Q_u, Q_d, and Z_0 occurs when $R_1=R_2$. Then

$$P_{\text{out}}/P_{\text{in}}=(Q_u-Q_d)/Q_u+Q_d)=1-Q_s/Q_u$$

$$P_{\text{out}}/P_m=(1-Q_d/Q_u)^2$$

$$P_{\text{in}}/P_m=1-(Q_d/Q_u)^2$$

When the generator R_1 or G_1 is negligibly small (then $Q=Q_s=Q_d$)

$$(P_{\text{in}}/P_{\text{out}})_s=Q_u/(Q_u-Q)$$

(F) Power dissipation ($=P_c$):

$$P_c/P_m=\frac{4(Q_d/Q_u)\ (1-Q_d/Q_u)}{1+R_2/R_1}$$

For matching input and output ($R_1=R_2$)

$$P_c/P_m=2(Q_d/Q_u)(1-Q_d/Q_u)$$

$$\approx 2Q_d/Q_u \quad (\text{for } Q_d<<Q_u)$$

$$P_c/P_{\text{out}}=2Q_d/(Q_u-Q_d)$$

$$P_c/P_{\text{in}}=2Q_d/(Q_u+Q_d)$$

For generator matched by load plus cavity

$$P_c/P_m=2Q_d/Q_u$$

When the generator R_1 or G_1 is negligibly small

$$(P_c/P_{\text{out}})_s=Q/(Q_u-Q)$$

$$(P_c/P_{\text{in}})_s=Q_s/Q_u$$

(G) Voltage and current: At the current-maximum point of an n-quarter-wavelength resonant line

$$I_{sc}=4\left[\frac{P_mQ_d(1-Q_d/Q_u)}{(1+R_2/R_1)n\pi Z_0}\right]^{1/2}, \text{ rms amperes}$$

$$=4\left[\frac{P_mQ_d}{\{1+[(R_2+R_u)/R_1]\}n\pi Z_0}\right]^{1/2}$$

When the generator R_1 or G_1 is negligibly small

$$I_{sc}=2\left[\frac{P_sQ_s}{n\pi Z_0(R_2+R_u)/R_s}\right]^{1/2}$$

where,

P_s = rated power of generator,

* When the line is resonated by a reactive load ($\theta \neq n\pi/2$), it is frequently preferable to use the resistance form of the equation. Compute R_u by the method in the section "Transformation of Impedance on Lines With High SWR," or the section "Impedance and Admittance," where $Z_0=R_0(1-jB_0/G_0)$.

R_s = rated load impedance as transformed into current-maximum point of cavity.

$$I = I_{sc} \cos\theta_1$$

$$E = Z_0 I_{sc} \sin\theta_1$$

The voltage and current are in quadrature time phase. When $R_1 = R_2 + R_u$ and $n = 1$

$$I_{sc} \approx (8P_m Q_d / \pi Z_0)^{1/2}$$

In a lumped-constant tuned circuit

$$I = 2\left[\frac{P_m Q_d(1 - Q_d/Q_u)}{(1 + R_2/R_1)X}\right]^{1/2}$$

(H) Pair of coupled resonators (Fig. 16):
With inductive coupling near the short-circuited end of a pair of quarter-wave resonant lines

$$k = (4/\pi)\omega M/(Z_{10}Z_{20})^{1/2}$$

For coupling through a lossless quarter-wavelength line, inductively coupled near the short-circuited ends of the resonators (Fig. 16D):

$$k = \frac{4\omega^2 M_1 M_2}{\pi Z_0 (Z_{10}Z_{20})^{1/2}}$$

Probe coupling near top (Fig. 16C):

$$k = (4/\pi)\omega C_{12}(Z_{10}\, Z_{20})^{1/2} \sin\theta_1 \sin\theta_2$$

For lumped-constant coupled circuits, p and k are defined on pp. 8-6 and 8-2, respectively.
In either lumped or distributed resonators

(dissipation loss)

$$= 10 \log_{10}(P_x/P_{out})$$

$$= 10 \log_{10}[1/(1 - Q_{1s}/Q_{1u})\ (1 - Q_{2s}/Q_{2u})]$$

$$\approx 20 \log_{10}[1/(1 - Q_s/Q_u)]$$

$$\approx 20 \log_{10}(1 + Q_s/Q_u)$$

$$\approx 8.7 Q_s/Q_u \text{ decibels}$$

where

$$Q_s/Q_u = [(Q_{1s}/Q_{1u})\ (Q_{2s}/Q_{2u})]^{1/2}$$

provided (Q_{1s}/Q_{1u}) and (Q_{2s}/Q_{2u}) do not differ by a ratio of more than 4 to 1, and neither exceeds 0.2.

(mismatch loss at f_0)

$$= 10 \log_{10}(P_m/P_x)$$

$$= 10 \log_{10}[(1 + p)^2/4p] \text{decibels.}$$

Equations and curves for selectivity are given on pp. 8-7 and 8-8, where $Q = Q_s$.

At the peaks, when $p \geq 1$, the mismatch loss is zero, except for some that is included in the dissipation loss.

Input voltage standing-wave ratio at f_0 for equal or unequal resonators:

$$S = \frac{p + Q_{1s}/Q_{1u}}{1 - Q_{1s}/Q_{1u}}$$

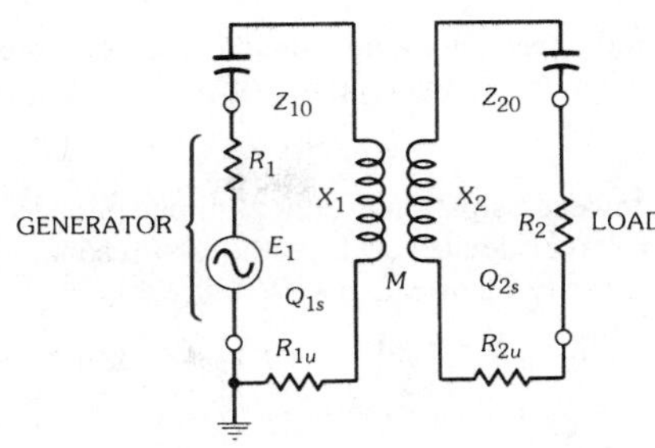

(A) Equivalent circuit with resistances as seen at the short-circuited end.

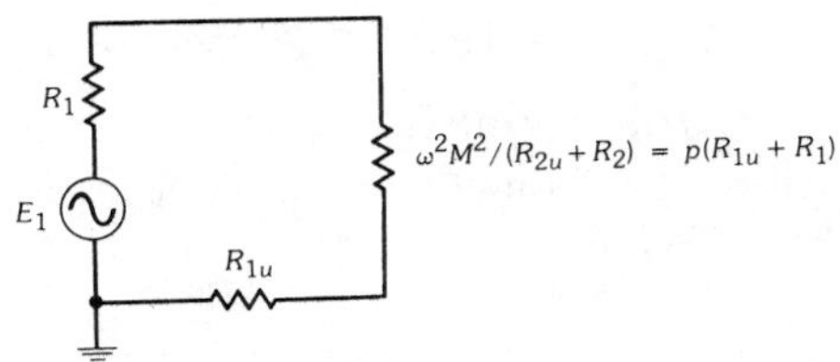

(B) Equivalent circuit of first resonator at resonance frequency.

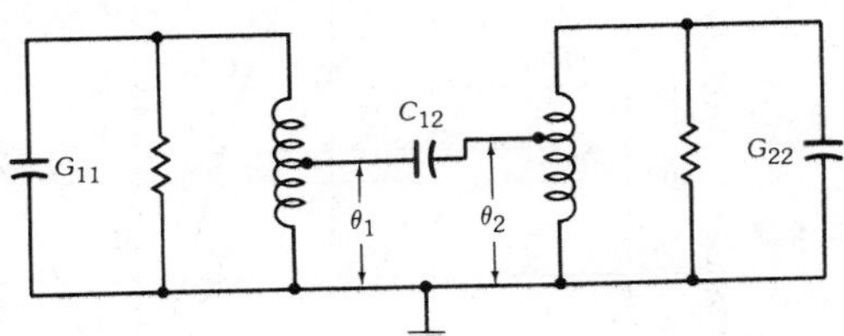

(C) Probe-coupled or aperture-coupled resonators.

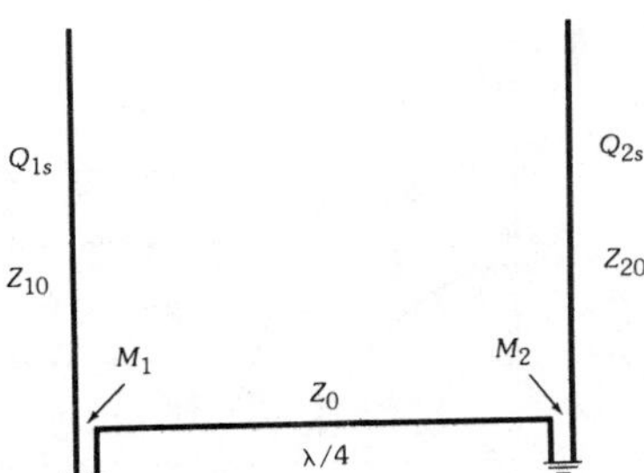

(D) Quarter-wavelength line coupling.

Fig. 16. Two coupled resonators.

At the peak frequencies ($p \geq 1$) for equal or nearly equal resonators:

$$S = \frac{1 + Q_{1s}/Q_{1u}}{1 - Q_{1s}/Q_{1u}}$$

Similar equations, using subscript 2 instead of 1, apply at the output.

When the resonators are isolated, each one presents to the generator or load an swr of

$$S=(Q_u/Q_s)-1$$

The power dissipation in either lumped or distributed (quarter-wave) devices, where the two resonators are not necessarily identical, but $Q_s << Q_u$, is

$$P_{1c}=I_{1sc}{}^2 R_{1u}[4/(1+p)^2]P_m Q_{1s}/Q_{1u}$$

$$P_{2c}=[4p/(1+p)^2]P_m Q_{2s}/Q_{2u}$$

These equations and those below for the currents assume that P_m is concentrated at f_0.

The currents in quarter-wave resonant lines, when $Q_s << Q_u$:

$$I_{1sc}=[4/(1+p)](P_m Q_{1s}/\pi Z_{10})^{1/2}$$

$$I_{2sc}/I_{1sc}=(p Z_{10} Q_{2s}/Z_{20} Q_{1s})^{1/2}$$

Similarly, for a pair of tuned circuits at resonance, when $Q_s << Q_u$:

$$I_1=[2/(1+p)]\ (P_m Q_{1s}/X_1)^{1/2}$$

$$I_2/I_1=(p X_1 Q_{2s}/X_2 Q_{1s})^{1/2}$$

QUARTER-WAVE MATCHING SECTIONS

Fig. 17 shows how voltage-reflection coefficient and standing-wave ratio (swr) vary with frequency f when quarter-wave matching lines are inserted between a line of characteristic impedance Z_0 and a load of resistance R. The symbol f_0 represents the frequency for which the matching sections are exactly one-quarter wavelength ($\lambda/4$) long.

IMPEDANCE MATCHING WITH SHORTED STUB

The use of a shorted stub for impedance matching is illustrated in Fig. 18.

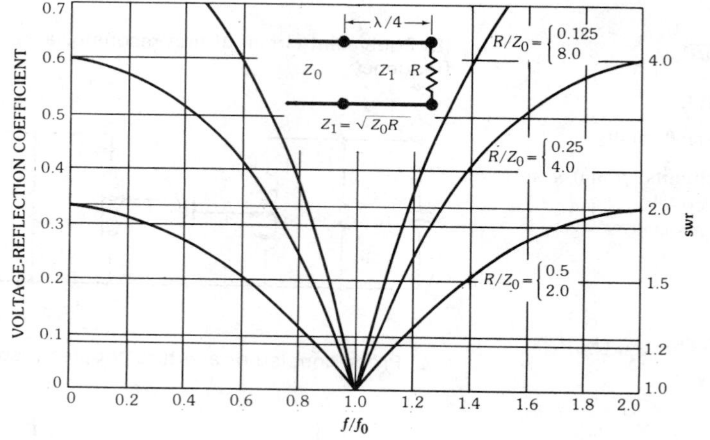

(A) One section.

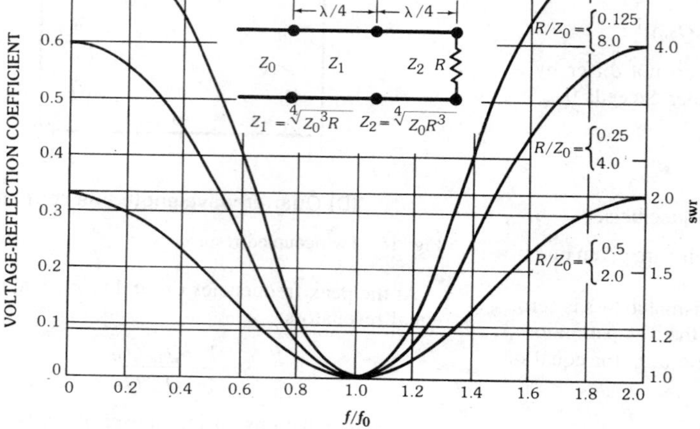

(B) Two sections.

Fig. 17. Quarter-wave matching sections.

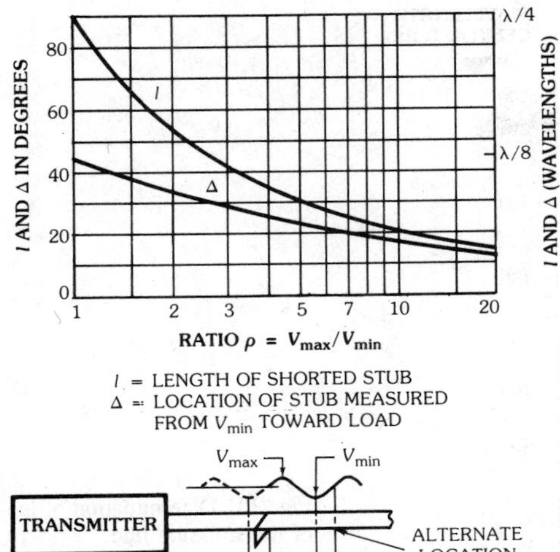

Fig. 18. Impedance matching with shorted stub.

IMPEDANCE MATCHING WITH OPEN STUB

The use of an open stub for impedance matching is illustrated in Fig. 19.

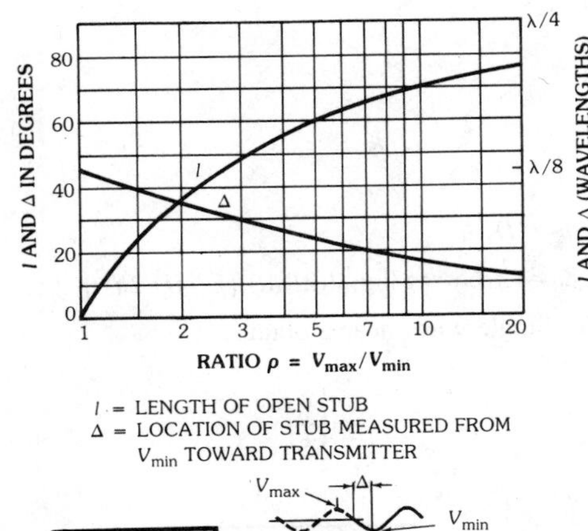

Fig. 19. Impedance matching with open stub.

LENGTH OF TRANSMISSION LINE

Fig. 20 relates the actual length of line in centimeters and inches to the length in electrical degrees and the frequency, provided that the velocity of propagation on the transmission line is equal to that in free space. The length is equal to that in free space. The length is given on the L-scale intersection by a line between λ and $l°$, where

$$l° = \frac{360L \text{ in centimeters}}{\lambda \text{ in centimeters}}$$

Example: $f = 600$ megahertz, $l° = 30$, length $L = 1.64$ inches or 4.2 centimeters.

CHARACTERISTIC IMPEDANCE OF LINES

A. Single coaxial line (See also Fig. 21.)

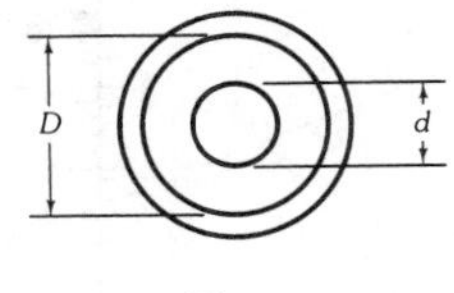

$$Z_0 = (138/\epsilon^{1/2}) \log_{10} (D/d)$$

$$= (60/\epsilon^{1/2}) \log_e (D/d)$$

ϵ = dielectric constant

= 1 in air

B. Balanced shielded line

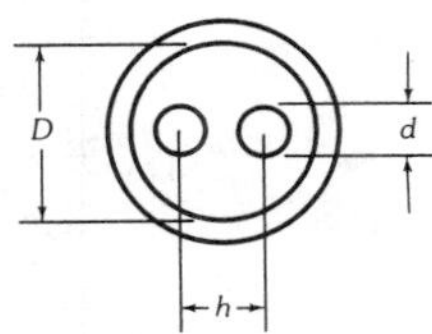

For $D >> d$, $h >> d$

$$Z_0 = (276/\epsilon^{1/2}) \log_{10}\{2v[(1-\sigma^2)/(1+\sigma^2)]\}$$

$$= (120/\epsilon^{1/2}) \log_e\{2v[(1-\sigma^2)/(1+\sigma^2)]\}$$

$v = h/d \qquad \sigma = h/D$

C. Beads—dielectric ϵ_1

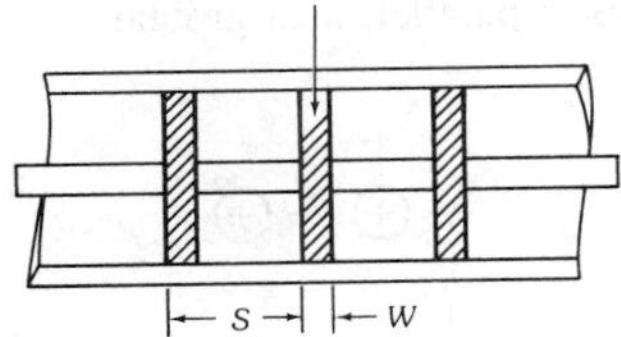

For lines *A* and *B*, if insulating beads are used at frequent intervals, call new characteristic impedance Z_0'

$$Z_0' = Z_0/\{1 + [(\epsilon_1/\epsilon) - 1](W/S)\}^{1/2}$$

$$W << S << \lambda/4$$

D. Open two-wire line in air (See also Fig. 21.)

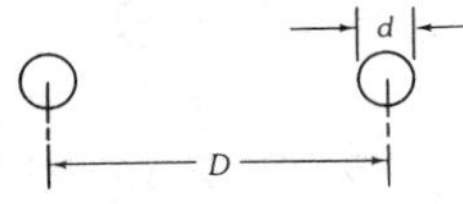

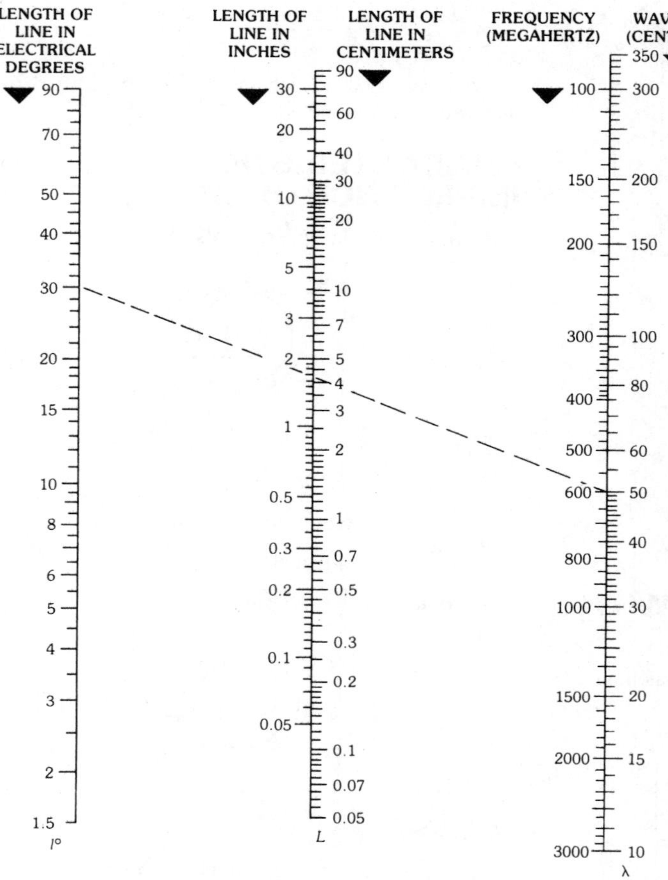

Fig. 20. Determination of length of transmission line.

$$Z_0 = 120 \cosh^{-1}(D/d)$$

$$\approx 276 \log_{10}(2D/d)$$

$$\approx 120 \log_e(2D/d)$$

E. Wires in parallel, near ground

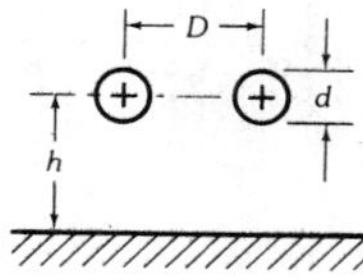

For $d << D, h$

$$Z_0 = (69/\epsilon^{1/2}) \log_{10}\{(4h/d)[1+(2h/D)^2]^{1/2}\}$$

F. Balanced, near ground

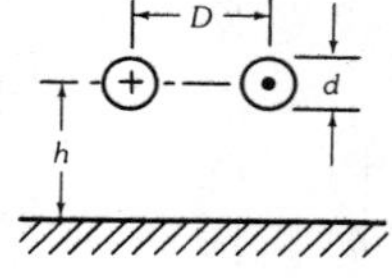

For $d << D, h$

$$Z_0 = (276/\epsilon^{1/2}) \log_{10}\{(2D/d)\,[1+(D/2h)^2]^{-1/2}\}$$

G. Single wire, near ground

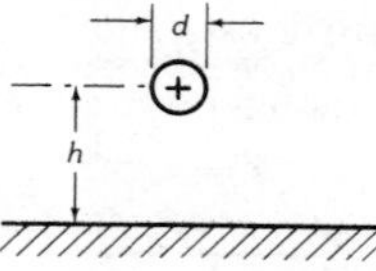

For $d << h$

$$Z_0 = (138/\epsilon^{1/2}) \log_{10}(4h/d)$$

H. Single wire, square enclosure

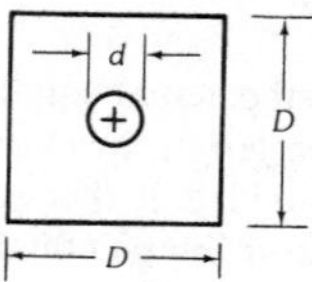

$$Z_0 \approx [138 \log_{10}\rho + 6.48 - 2.34A - 0.48B - 0.12C]\epsilon^{-1/2}$$

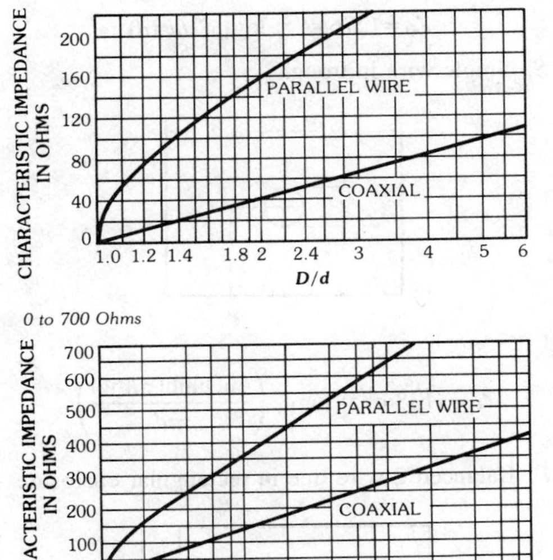

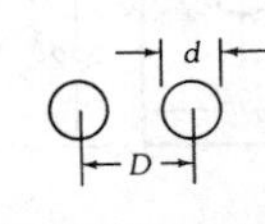
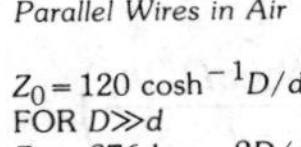
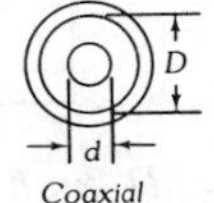

$Z_0 = 120 \cosh^{-1} D/d$
FOR $D \gg d$
$Z_0 \approx 276 \log_{10} 2D/d$

$Z_0 = (138/\sqrt{\epsilon}) \log_{10} D/d$
CURVE IS FOR
$\epsilon = 1.00$

Fig. 21. Characteristic impedance of transmission lines.

where $\rho = D/d$

$$A = (1 + 0.405\rho^{-4})/(1 - 0.405\rho^{-4})$$
$$B = (1 + 0.163\rho^{-8})/(1 - 0.163\rho^{-8})$$
$$C = (1 + 0.067\rho^{-12})/(1 - 0.067\rho^{-12})$$

I. Balanced 4-wire

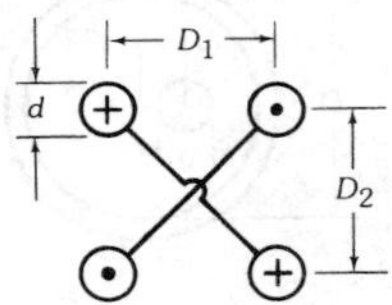

For $d \ll D_1, D_2$

$$Z_0 = (138/\epsilon^{1/2}) \log_{10}\{(2D_2/d)[1 + (D_2/D_1)^2]^{-1/2}$$

J. Parallel-strip line

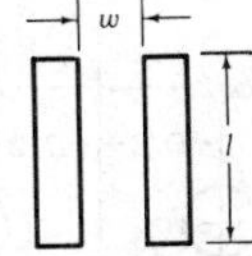

$$w/l < 0.1$$
$$Z_0 \approx 377(w/l)$$

K. Five-wire line

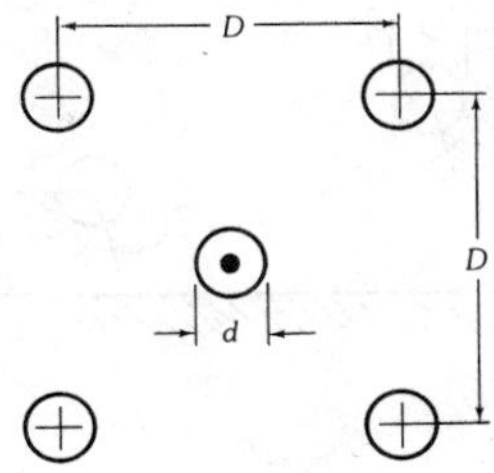

For $d \ll D$

$$Z_0 = (173/\epsilon^{1/2}) \log_{10}(D/0.933d)$$

L. Wires in parallel—sheath return

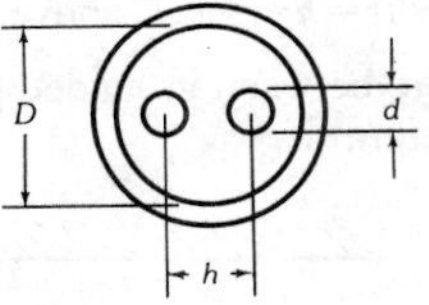

For $d \ll D, h$

$$Z_0 = (69/\epsilon^{1/2}) \log_{10}[(v/2\sigma^2)(1 - \sigma^4)]$$
$$\sigma = h/D$$
$$v = h/d$$

M. Air coaxial with dielectric supporting wedge

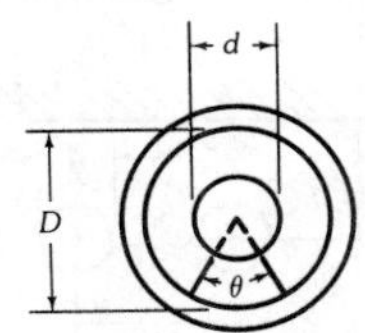

$$Z_0 \approx \frac{138 \log_{10}(D/d)}{[1 + (\epsilon - 1)(\theta/360)]^{1/2}}$$

ϵ = dielectric constant of wedge

θ = wedge angle in degrees

N. Balanced 2-wire—unequal diameters

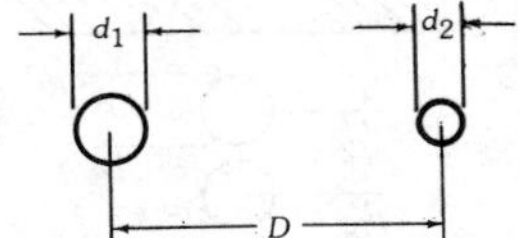

$$Z_0 = (60/\epsilon^{1/2}) \cosh^{-1} N$$

$$N = \tfrac{1}{2}[(4D^2/d_1 d_2) - (d_1/d_2) - (d_2/d_1)]$$

O. Balanced 2-wire near ground

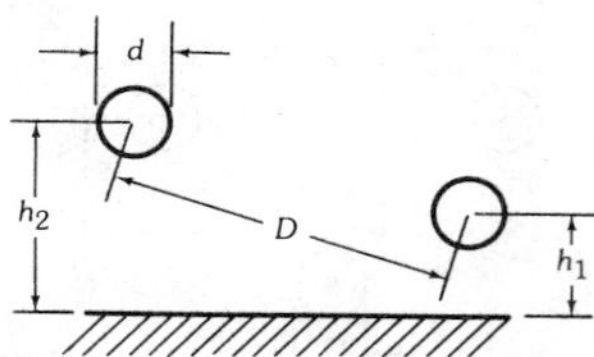

For $d << D,\ h_1,\ h_2$

$$Z_0 = (276/\epsilon^{1/2}) \log_{10}\{(2D/d)[1 + (D^2/4h_1 h_2)]^{-1/2}\}$$

Holds also in either of the following special cases:

$$D = \pm(h_2 - h_1)$$

or

$$h_1 = h_2 \text{ (see } F \text{ above)}$$

P. Single wire between grounded parallel planes—ground return

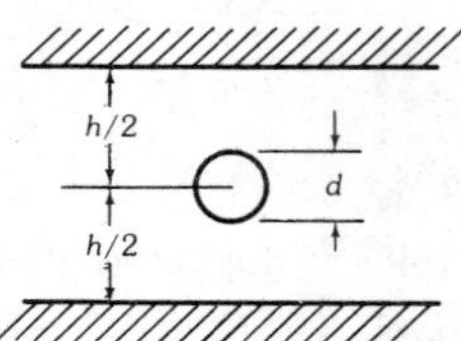

For $d/h < 0.75$

$$Z_0 = (138/\epsilon^{1/2}) \log_{10}(4h/\pi d)$$

Q. Balanced line between grounded parallel planes

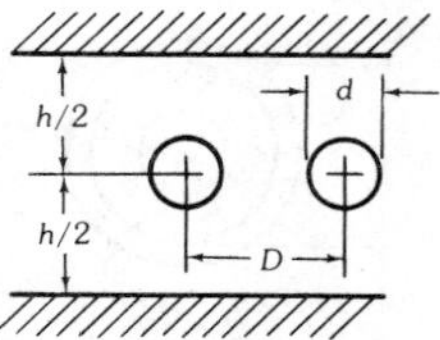

For $d << D,\ h$

$$Z_0 = (276/\epsilon^{1/2}) \log_{10}\left(\frac{4h \tanh(\pi D/2h)}{\pi d}\right)$$

R. Balanced line between grounded parallel planes

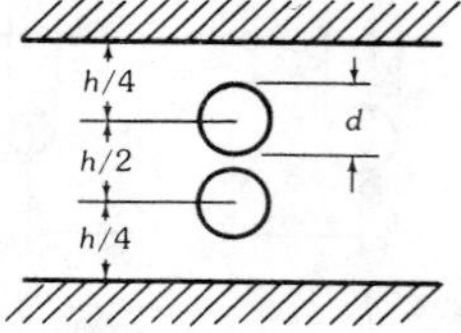

For $d << h$

$$Z_0 = (276/\epsilon^{1/2}) \log_{10}(2h/\pi d)$$

S. Single wire in trough

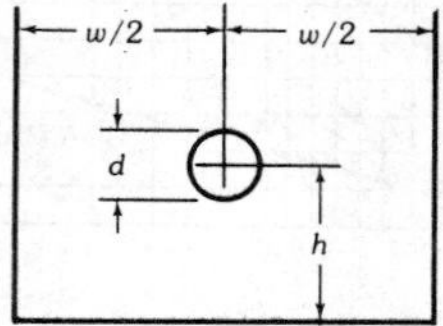

For $d << h,\ w$

$$Z_0 = (138/\epsilon^{1/2}) \log_{10}\left(\frac{4w \tanh(\pi h/w)}{\pi d}\right)$$

T. Balanced 2-wire line in rectangular enclosure

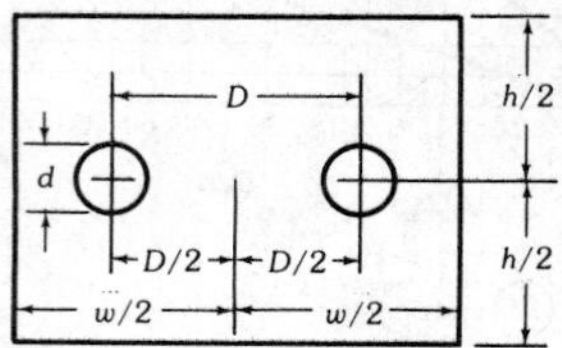

For $d << D,\ w,\ h$

$$Z_0 = (276/\epsilon^{1/2}) \left[\log_{10}\left(\frac{4h \tanh(\pi D/2h)}{\pi d}\right) - \sum_{m=1}^{\infty} \log_{10}\left(\frac{1 + u_m^2}{1 - v_m^2}\right)\right]$$

where

$$u_m = \frac{\sinh(\pi D/2h)}{\cosh(m\pi w/2h)} \qquad v_m = \frac{\sinh(\pi D/2h)}{\sinh(m\pi w/2h)}$$

U. Eccentric line

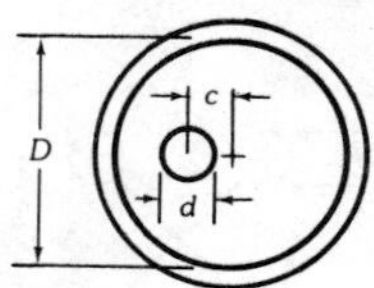

$$Z_0 = (60/\epsilon^{1/2}) \cosh^{-1} U$$

$$U = \tfrac{1}{2}[(D/d) + (d/D) - (4c^2/dD)]$$

V. Balanced 2-wire line in semi-infinite enclosure

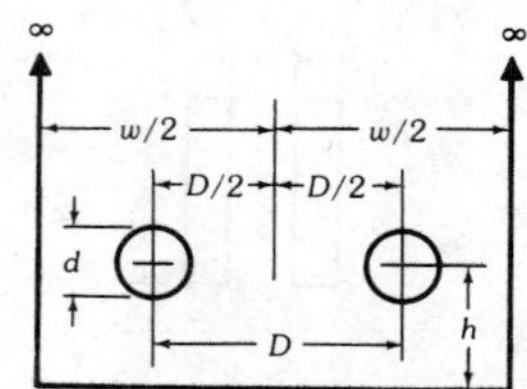

For $d << D, w, h$

$$Z_0 = (276/\epsilon^{1/2}) \log_{10}[2w/\pi d(A^{1/2})]$$

where

$$A = \mathrm{cosec}^2(\pi D/w) + \mathrm{cosech}^2(2\pi h/w)$$

W. Outer wires grounded, inner wires balanced to ground

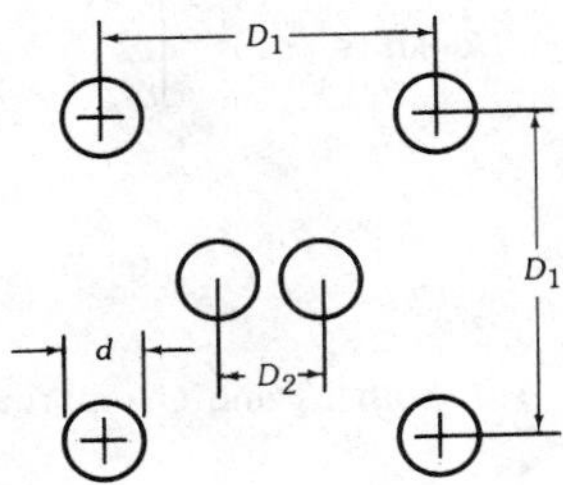

$$Z_0 \approx (276/\epsilon^{1/2}) \left\{ \log_{10}(2D_2/d) - \left[\log_{10} \frac{1+(1+D_2/D_1)^2}{1+(1-D_2/D_1)^2}\right]^2 \left[\log_{10}(2D_1\sqrt{2}/d)\right]^{-1} \right\}$$

X. Split thin-walled cylinder

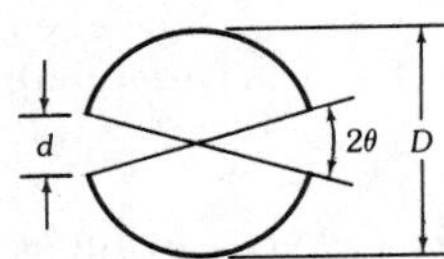

$$Z_0 \approx \frac{129}{\log_{10}[\cot\frac{1}{2}\theta + (\cot^2\frac{1}{2}\theta - 1)^{1/2}]}$$

For θ small:

$$Z_0 \approx 129/\log_{10}(4D/d)$$

Courtesy of Electronic Engineering

Y. Slotted air line

When a slot is introduced into an air coaxial line for measuring purposes, the increase in characteristic impedance in ohms, compared with a normal coaxial line, is less than

$$\Delta Z = 0.03\theta^2$$

where θ is the angular opening of the slot in radians.

MICROSTRIP LINES

Microstrip line consists of a conductor strip placed on a dielectric substrate (relative dielectric constant ϵ_r), which is in turn backed by a conducting ground plane (Fig. 22). At lower microwave frequencies, the modal field is considered almost TEM. However, as the frequency is increased, the dispersion effect becomes more obvious, and the characteristic impedance and the phase velocity defined under the quasi-TEM analysis must be modified.

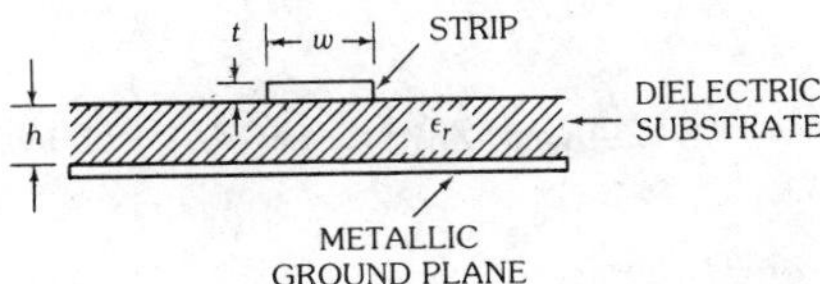

Fig. 22. Cross section of microstrip line.

Quasi-TEM Characteristics*

The characteristic impedance Z_0 and the effective dielectric constant ϵ_e are functions of structure and the dielectric constant ϵ_r.

$$Z_0(w,t,h,\epsilon_r) = Z_{01}(U_r)/[\epsilon_e(U_r, \epsilon_r)]^{1/2}$$

$$\beta = \frac{Z_{01}(U_1)}{Z_{01}(U_r)} [\epsilon_e(U_r, \epsilon_r)]^{1/2} \omega/c$$

where,

$$Z_{01}(x) = (376.73/2\pi) \ln\left\{\frac{f(x)}{x} + [1+(2/x)^2]^{1/2}\right\}$$

$$f(x) = 6 + (2\pi - 6)\exp\left[-(30.666/x)^{0.7528}\right]$$

* These formulas are from E. Hammerstadt and O. Jensen, *1980 IEEE International Microwave Symposium Digest*, pp. 407–409, June 1980. For the effects of an additional ground plane over the structure, see S. March, *Microwaves*, pp. 83-94, December 1981.

$$\epsilon_e(x, \epsilon_r) = [(\epsilon_r + 1)/2] + [(\epsilon_r - 1)/2](1 + 10/x)^y$$
$$y = -a(x)b(\epsilon_r)$$
$$a(x) = 1 + (1/49)\ \ln\left[\frac{x^4 + (x/52)^2}{x^4 + 0.432}\right] + (1/18.7)\ \ln\left[1 + (x/18.1)^3\right]$$
$$b(\epsilon_r) = 0.564\ [(\epsilon_r - 0.9)/(\epsilon_r + 3)]^{0.053}$$

x is either

$$U_r = w/h + (t/\pi)\ \ln\left[1 + \frac{4\ \exp(1)}{t\ \coth^2(6.517\ w/h)^{1/2}}\right]\left[1 + 1/\cosh\ (\epsilon_r - 1)^{1/2}\right]/2$$

or

$$U_1 = w/h + (t/\pi)\ \ln\left[1 + \frac{4\ \exp(1)}{t\ \coth^2(6.517\ w/h)^{1/2}}\right]$$

depending on which is called for in the above equations. Note that $t \to 0$, both U_r and U_1 approach w/h.

Fig. 23 shows Z_0 as a function of w/h.

Attenuation

Dielectric loss (dB/unit length)

$$\alpha_D = 4.34\ \beta\ \tan\delta/[1 + (1/\epsilon_r)(F - 1)/(F + 1)]$$
$$F = (1 + 10h/w)^{1/2}$$

The factor tanδ is the loss tangent of the substrate material.** The formula for β is given in the preceding subsection, "Quasi-TEM Characteristics."

Conductor loss is†

$$\alpha_c = (4.34R_s/\pi hZ_0)[1 - (w'/4h)^2]\{1 + h/w' + (h/\pi w')\ [\ln(4\pi w/t + 1) - (1 - t/w)/(1 + t/4\pi w)]\} \qquad w/h \le 1/2\pi$$

$$= (4.34R_s/\pi hZ_0)\ [1 - (w'/4h)^2]\ \{1 + h/w' + (h/\pi w')[\ln(2h/t + 1) - (1 + t/h)/(1 + t/2h)]\} \qquad 1/2\pi < w/h \le 2$$

$$= \frac{8.68\ R_s\left[\dfrac{w'}{h} + \dfrac{w'/\pi h}{w'/2h + 0.94}\right]}{hZ_0\{w'/h + (2/\pi)\ \ln\ [2\pi\epsilon\ (w'/2h + 0.94)]\}^2}\{1 + h/w' + (h/\pi w')[\ \ln\ (2h/t + 1) - (1 + t/h)/(1 + t/2h)]\} \qquad w/h > 2$$

where

$w' = w + \Delta w$

and for $2t/h < w/h,\ 1/2\pi$

$$\Delta w = (t/\pi)\ \ln\ (4\pi w/t + 1) \qquad w/h \le 1/2\pi$$
$$= (t/\pi)\ \ln\ (2h/t + 1) \qquad w/h \ge 1/2\pi$$

The quantity Z_0 is the characteristic impedance discussed before, and R_s is the surface resistivity. Typically, $R_s = 2.61 \times 10^{-7}\sqrt{f}$ for copper.

** M. V. Schneider, "Dielectric Loss in Hybrid Integrated Circuits," *Proc. IEEE*, Vol. 57, pp. 1206–1207, 1969.

† R. A. Pucel, D. J. Massé, and C. P. Hartwig, "Losses in Microstrip," *IEEE Trans. Microwave Theory and Techniques*, Vol. MTT-16, pp. 342–350, 1968.

Frequency-Dependent Characteristics

As the microstrip mode is not purely TEM, both Z_0 and ϵ_e are functions of frequency.‡

$$Z_0(f) = Z_0(0)[\epsilon_e(0)/\epsilon_e(f)]^{1/2}[\epsilon_e(f) - 1]/[\epsilon_e(0) - 1]$$
$$\epsilon_e(f) = \epsilon_r - [\epsilon_r - \epsilon_e(0)]/[1 + G(f/f_p)^2]$$

where,

$f_p = Z_0(0)/(2\mu_0 h)$

μ_0 = free-space permeability.

$$G = (\pi^2/12)[(\epsilon_r - 1)/\epsilon_e(0)][Z_0(0)/60]^{1/2}$$

‡ E. Hammerstadt and O. Jensen, *1980 IEEE International Microwave Symposium Digest*, pp. 407–409, June 1980.

In the above, $Z_0(0)$ and $\epsilon_e(0)$ are dc values and are obtained from the formulas in "Quasi-TEM Characteristics."

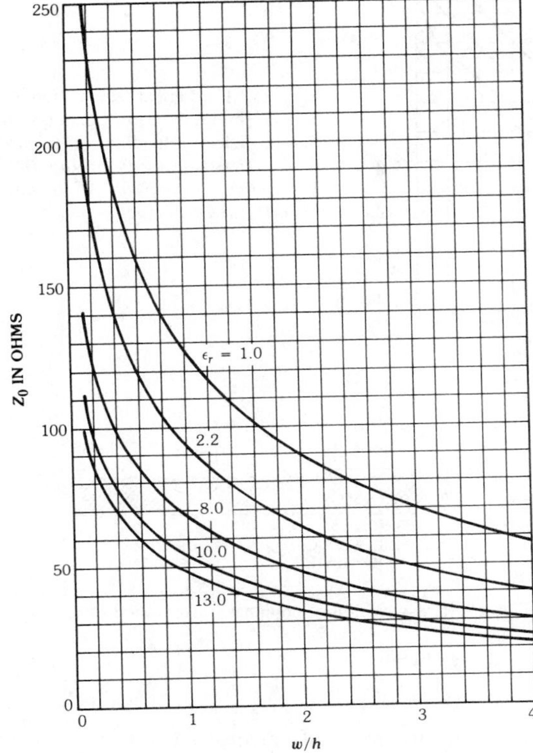

Fig. 23. Characteristic impedance of microstrip line. Calculated from quasi-tem formulas by Hammerstadt and Jensen. (Curves assume t is negligible; *i.e.*, $U_r = U_e = w/h$.)

Power-Handling Capacity

For a microstrip line composed of a strip 7/32-inch wide on a Teflon-impregnated Fibreglas base 1/16-inch thick:

(A) At 3000 megahertz with 300 watts cw, the temperature under the strip conductor has been measured at 50° Celsius rise above 20° Celsius ambient.

(B) Under pulse conditions, corona effects appear at the edge of the strip conductor for pulse power of roughly 10 kilowatts at 9000 megahertz.

STRIP TRANSMISSION LINES*

Strip transmission lines differ from microstrip in that a second ground plane is placed above the strip and the space between the two ground planes is filled completely with a homogeneous dielectric (Fig. 24). The characteristic impedance is shown in Fig. 25 and the attenuation in Fig. 26.

Dielectric loss in decibels/unit length

$$\alpha_d = 27.3F_p(\epsilon_r)^{1/2}/\lambda_0$$

where,

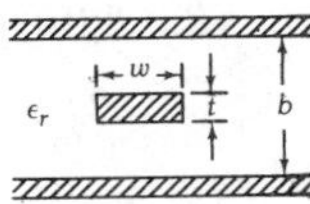

Fig. 24. Cross section of a strip transmission line.

* S. B. Cohn, "Problems in Strip Transmission Lines," *Transactions of the IRE Professional Group on Microwave Theory and Techniques*, Vol. MTT3, pp. 119–126, March 1955. Other papers on strip-type lines also appear in that issue of the journal.

Fig. 25. Plot of strip-transmission-line Z_0 versus w/b for various values of t/b. For lower-left family of curves, refer to left-hand ordinate values; for upper-right curves, use right-hand scale. (*Courtesy of Transactions of the IRE Professional Group on Microwave Theory and Techniques.*)

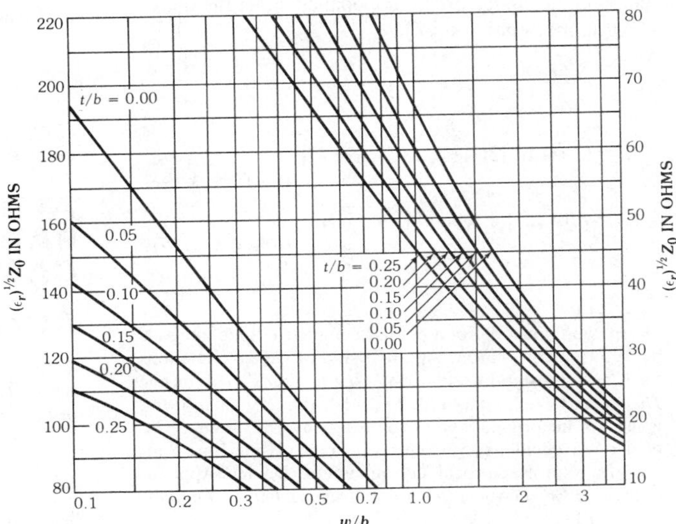

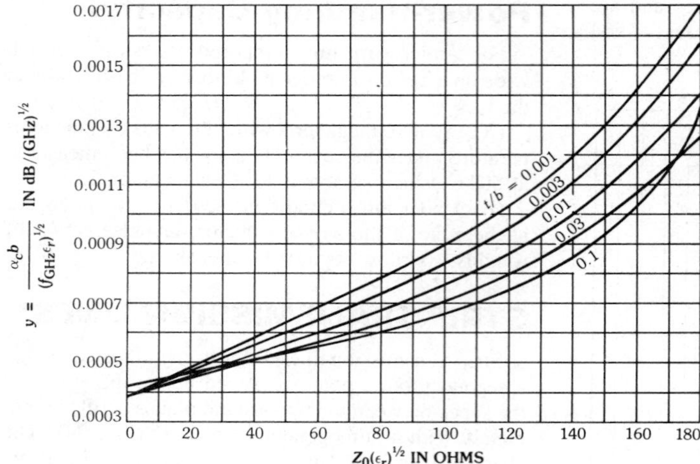

Fig. 26. Theoretical attenuation of copper-shielded strip transmission line in dielectric medium ϵ_r. (*Courtesy of Transactions of the IRE Professional Group on Microwave Theory and Techniques.*)

λ_0 = free-space wavelength,

F_p = power factor or loss angle.

Conductor loss in decibels/unit length

$$\alpha_c = (y/b)\ (f_{GHz}\epsilon_r\mu_r\rho/\rho_{Cu})^{1/2}$$

where,

y = ordinate from Fig. 26,

ρ/ρ_{Cu} = resistivity relative to copper.

The unit of length in α_d is that of λ_0, and in α_c it is that of b.

COPLANAR TRANSMISSION LINES

A quasi-TEM propagation takes place in a coplanar transmission line (Fig. 27). When $t=0$, the phase constant and the characteristic impedance under the quasi-TEM approximation are**

$$\beta = \sqrt{\epsilon_e}\ \omega/c$$

$$Z_0 = (30\pi/\sqrt{\epsilon_e})\ K_r$$

$$K_r = (1/\pi)\ \ln\ [2(1+\sqrt{k})/(1-\sqrt{k})] \qquad 0.707 \le k \le 1$$

$$K_r = \pi/\ln\ [2(1+\sqrt{k'})/(1-\sqrt{k'})] \qquad 0 \le k \le 0.707$$

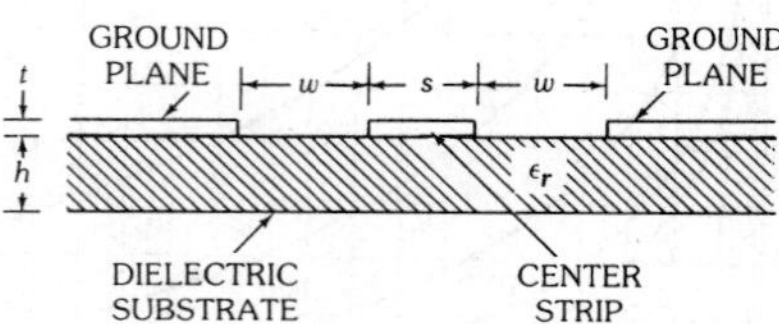

Fig. 27. Cross section of coplanar transmission line.

$$k = s/(2w+s)$$

$$k' = (1-k^2)^{1/2}$$

where for the range of practical interest $1 \le h/w < \infty$. The value of ϵ_e is given by†

$$\epsilon_e = [(\epsilon_r+1)/2]\ \{\tanh[0.775\ \ln\ (h/w)+1.75]$$
$$+\,k(w/h)[0.04-0.7k+0.01(1-0.1\ \epsilon_r)\ (0.25+k)]\}$$

If t is not negligible, k and ϵ_e in the expressions for $t=0$ must be replaced with†

$$k^e = k+(1-k^2)\ (1.25t/2w\pi)[1+\ \ln\ (4\pi s/t)]$$

$$\epsilon_e^t = \epsilon_e - 0.7(\epsilon_e - 1)\ (t/w)/(K_r + 0.7t/w)$$

Fig. 28 is a graph of characteristic impedance calculated from the formulas in this section for $t=0$.

ATTENUATION AND POWER RATING OF LINES AND CABLES

Attenuation

Fig. 29 illustrates the attenuation of general-purpose radio-frequency lines and cables up to their practical upper frequency limit. Most of these are coaxial-type

** An approximation for a ratio of complete elliptic integrals by W. Helberg, "From Approximation to Exact Relations for Characteristic Impedances," *IEEE Trans. Microwave Theory and Technique*, Vol. MTT-17, pp. 259–265, 1969, is used in the original expression for Z_0 by C.P. Wen, "Coplanar Waveguide: A Surface Strip Transmission Line Suitable for Non-Reciprocal Gyromagnetic Device Applications," *IEEE Trans. Microwave Theory and Techniques*, Vol. MTT-17, pp. 1087–1090, 1969.

† K. C. Gupta, et. al., *Microstrip Lines and Slotlines*, (Dedham, Mass.: Artech House, 1969).

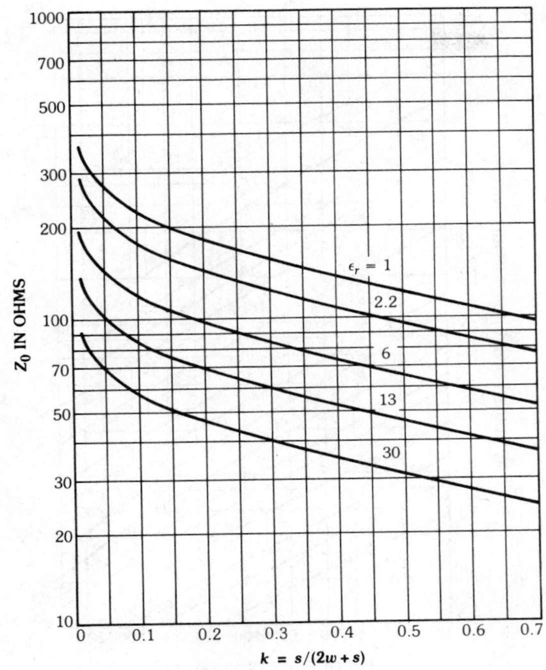

Fig. 28. Graph showing characteristic impedance of coplanar transmission line.

lines, but waveguide and microstrip are included for comparison.

The following notes are applicable to this figure.

(A) For the RG-type cables, only the number is given (for instance, the curve for RG-218/U is labeled 218. Refer to the table of radio-frequency cables.) The data on RG-type cables are taken mostly from "RF Transmission Lines and Fittings," MIL–HDBK–216, 4 January 1962, revised 18 May 1965, and from "Solid Dielectric Transmission Lines," Electronic Industries Association Standard RS–199, December 1957.

Some approximation is involved in order to simplify the figure. Thus, where a single curve is labeled with several type numbers, the actual attenuation of each individual type may be slightly different from that shown by the curve.

(B) The curves for rigid copper coaxial lines are labeled with the diameter of the line only, as $^7/_8''C$. These have been computed for the lines listed in "Rigid Coaxial Transmission Lines, 50 Ohms," Electronic Industries Association Standard RS–225, August 1959. The computations considered the copper losses only, on the basis of a resistivity $\rho = 1.724$ microhm-centimeters; a derating of 20 percent has been applied to allow for imperfect surface, presence of fittings, etc., in long installed lengths. Relative attenuations of the different sizes are

$$A_{6^1/_8''} \approx 0.13 A_{^7/_8''}$$

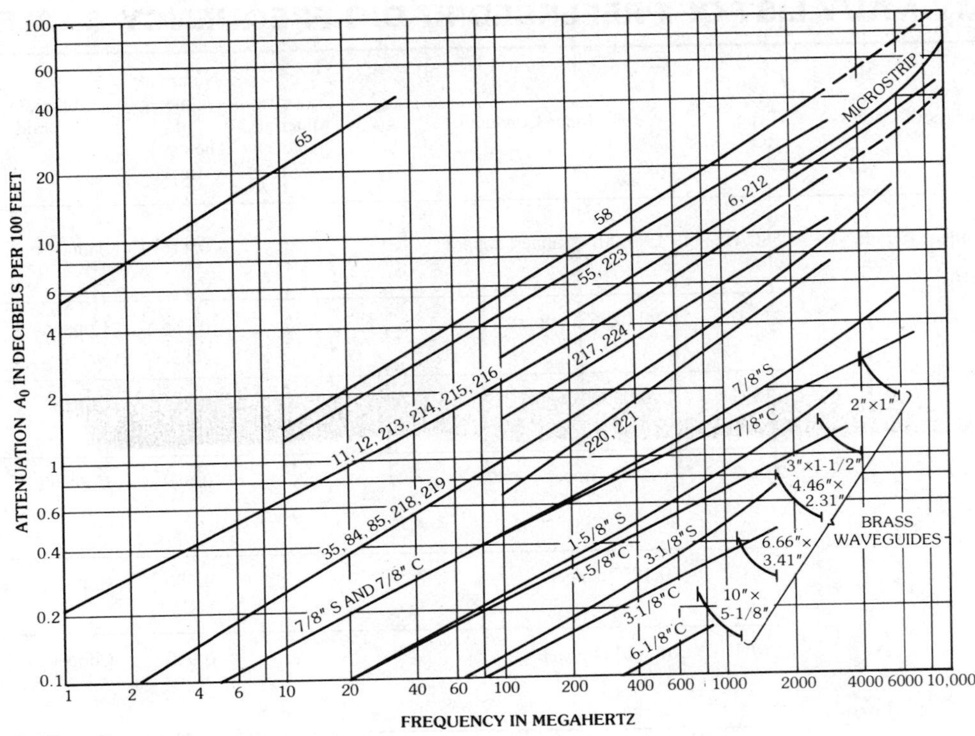

Fig. 29. Attenuation of cables.

$$A_{3^1/_8''} \approx 0.26 A_{^7/_8''}$$

$$A_{1^5/_8''} \approx 0.51 A_{^7/_8''}$$

(C) Typical curves are shown for three sizes of 50-ohm semirigid cables such as Styroflex, Spiroline, Heliax, Alumispline, etc. These are labeled by size in inches, as $^7/_8''S$.

(D) The microstrip curve is for Teflon-impregnated Fibreglas dielectric $^1/_{16}$-inch thick and conductor strip $^7/_{32}$-inch wide.

(E) Shown for comparison is the attentuation in the $TE_{1,0}$ mode of five sizes of brass waveguide. The resistivity of brass was taken as $\rho = 6.9$ microhm-centimeters, and no derating was applied. For copper or silver, attenuation is about half that for brass. For aluminum, attenuation is about two-thirds that for brass.

Power Rating

Fig. 30 shows the approximate power transmitting capabilities of various coaxial-type lines. The following notes are applicable.

(A) Identification of the curves for the RG-type cables is the same as in Fig. 29. The data for these cables are from the same sources. For polyethylene cables, an inner-conductor maximum temperature of 80 de-

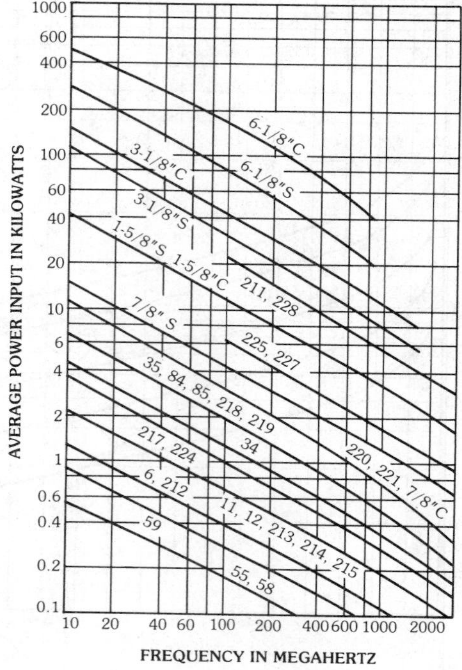

Fig. 30. Power rating of cables.

ARMY-NAVY LIST OF PREFERRED RADIO-FREQUENCY CABLES*

Class of Cables		JAN Type RG-	Inner Conductor†	Dielectric Material (Note 1)	Nominal Diameter of Dielectric (in.)	Shielding Braid
50 ohms	Single braid	58C/U	19/0.0071″ tinned copper	A	0.116	Tinned copper
		213/U	7/0.0296″ copper	A	0.285	Copper
		215/U	7/0.0296″ copper	A	0.285	Copper
		218/U	0.195″ copper	A	0.680	Copper
		219/U	0.195″ copper	A	0.680	Copper
		220/U	0.260″ copper	A	0.910	Copper
		221/U	0.260″ copper	A	0.910	Copper

grees Celsius is specified (**G**). For high-temperature cables (types 211, 228; 225, 227) the inner-conductor temperature is 250 degrees Celsius.

(**B**) The curves for 50-ohm rigid coaxial line are labeled with the diameter of the line only, as 7/8″*C*. These are rough estimates based largely on miscellaneous charts published in catalogs.

(**C**) For Styroflex, Spiroline, Heliax, Alumispline, etc., cables, refer to (**C**) above.

(**D**) The curves are for unity voltage standing-wave ratio. Safe operating power is inversely proportional to swr expressed as a numerical ratio greater than unity. Do not exceed maximum operating voltage (see table of radio-frequency cables).

(**E**) An ambient temperature of 40 degrees Celsius is assumed.

(**F**) The four curves meeting the 100-watt ordinate may be extrapolated: at 3000 megahertz for 55, 58, power is 28 watts; for 59, power is 44 watts; and for 6, 212, power is 58 watts.

(**G**) Electronic Industries Association Standard RS–199 states that operation of a polyethylene dielectric cable at a center-conductor temperature in excess of 80 degrees Celsius is likely to cause permanent damage to the cable. Where practicable, and particularly where continuous flexing is required, it is recommended that a cable be selected which, in regular operation, will produce a center-conductor temperature not greater than 65 degrees Celsius. Rating factors for various operating temperatures are given in Table 1. Multiply points on the power-rating curve by the factors in the table to determine power rating at operating conditions.

TABLE 1. DERATING FACTOR

Ambient Temperature (°C)	Maximum Allowable Center-Conductor Temperature (°C)			
	80	75	70	65
40	1.0	0.86	0.72	0.59
50	0.72	0.59	0.46	0.33
60	0.46	0.33	0.22	0.10
70	0.20	0.09	0	—
80	0	—	—	—

Protective Covering (Note 2)	Nominal Overall Diameter (in.)	Weight (lb/ft)	Nominal Capacitance (pF/ft)	Maximum Operating Voltage (rms)	Remarks
IIa	0.195	0.029	28.5	1900	Small-size flexible cable
IIa	0.405	0.120	29.5	5000	Medium-size flexible cable (formerly RG-8A/U
IIa, with armor	0.475 max.	0.160	29.5	5000	Same as RG-213/U, but with armor (formerly RG-10A/U)
IIa	0.870	0.491	29.5	11,000	Large-size low-attenuation high-power transmission line (formerly RG-17A/U)
IIa, with armor	0.945 max.	0.603	29.5	11,000	Same as RG-218/U, but with armor (formerly RG-18A/U)
IIa	1.120	0.745	29.5	14,000	Very-large low-attenuation high-power transmission cable (formerly RG-19A/U
IIa, with armor	1.195 max.	0.925	29.5	14,000	Same as RG-220/U, but with armor (formerly RG-20A/U)

Continued on next page.

Army-Navy List of Preferred Radio-Frequency Cables—cont.

Class of Cables		JAN Type RG-	Inner Conductor†	Dielectric Material (Note 1)	Nominal Diameter of Dielectric (in.)	Shielding Braid
50 ohms (*Continued*)	Double braid	55B/U	0.032″ silvered copper	A	0.116	Silvered copper
		212/U	0.0556″ silvered copper	A	0.185	Silvered copper
		214/U	7/0.0296″ silvered copper	A	0.285	Silvered copper
		217/U	0.106″ copper	A	0.370	Copper
		223/U	0.035″ silvered copper	A	0.116	Silvered copper
		224/U	0.106″ copper	A	0.370	Copper
75 ohms	Single braid	11A/U	7/26 AWG tinned copper	A	0.285	Copper
		12A/U	7/26 AWG tinned copper	A	0.285	Copper
		34B/U	7/0.0249″ copper	A	0.460	Copper
		35B/U	0.1045″ copper	A	0.680	Copper
		59B/U	0.0230″ copper-covered steel	A	0.146	Copper
		84A/U	0.1045″ copper	A	0.680	Copper
		85A/U	0.1045″ copper	A	0.680	Copper
		164/U	0.1045″ copper	A	0.680	Copper
		307A/U	17/0.0058″ silvered copper	A Foamed	0.029	Silvered copper
	Double braid	6A/U	21 AWG copper-covered steel	A	0.185	Inner: silvered copper. Outer: copper
		216/U	7/0.0159″ tinned copper	A	0.285	Copper

Protective Covering (Note 2)	Nominal Overall Diameter (in.)	Weight (lb/ft)	Nominal Capacitance (pF/ft)	Maximum Operating Voltage (rms)	Remarks
IIIa	0.206	0.032	28.5	1900	Small-size flexible cable
IIa	0.332	0.093	28.5	3000	Small-size microwave cable (formerly RG-5B/U)
IIa	0.425	0.158	30.0	5000	Special medium-size flexible cable (formerly RG-9B/U)
IIa	0.545	0.236	29.5	7000	Medium-size power transmission line (formerly RG-14A/U)
IIa	0.216	0.036	28.5	1900	Small-size flexible cable (formerly RG-55A/U)
IIa, with armor	0.615 max.	0.282	29.5	7000	Same as RG-217/U, but with armor (formerly RG-74A/U)
IIa	0.412	—	20.5	5000	Medium-size flexible video cable
IIa, with armor	0.475	—	20.5	5000	Similar to RG-11A/U, but with armor
IIa	0.630	0.231	21.5	6500	Large-size high-power low-attenuation flexible cable
IIa, with armor	0.945 max.	0.480	21.5	10,000	Large-size high-power low-attenuation video and communication cable
IIa	0.242	—	21.0	2300	General-purpose small-size video cable
IIa, with lead sheath	1.000	1.325	21.5	10,000	Same as RG-35B/U, except lead sheath instead of armor for underground installations
IIa, with lead sheath and special armor	1.565 max.	2.910	21.5	10,000	Same as RG-84A/U, with special armor for underground installations
IIa	0.870	0.490	21.5	10,000	Same as RG-35B/U, except without armor
IIIa	0.270	—	20	400	
IIa	0.332	—	20.0	2700	Small-size video and communication cable
IIa	0.425	0.121	20.5	5000	Medium-size flexible video and communication cable (formerly RG-13A/U)

Continued on next page.

Army-Navy List of Preferred Radio-Frequency Cables—cont.

Class of Cables		JAN Type RG-	Inner Conductor†	Dielectric Material (Note 1)	Nominal Diameter of Dielectric (in.)	Shielding Braid
High temperature	Single braid	144/U	7/0.0179″ silvered copper-covered steel	F1	0.285	Silvered copper
		178B/U	7/0.004″ silvered copper-covered steel	F1	0.034	Silvered copper
		179B/U	Same as above	F1	0.063	Silvered copper
		180B/U	Same as above	F1	0.102	Silvered copper
		187A/U	7/0.004″ annealed silvered copper-covered steel	F1	0.060	Silvered copper
		195A/U	Same as RG-187A/U	F1	0.102	Silvered copper
		196A/U	Same as RG-187A/U	F1	0.034	Silvered copper
		211A/U	0.190″ copper	F1	0.620	Copper
		228A/U	0.190″ copper	Fl	0.620	Copper
		302/U	0.025″ silvered copper-covered steel	Fl	0.146	Silvered copper
		303/U	0.039″ silvered copper-covered steel	Fl	0.116	Silvered copper
		304/U	0.059″ silvered copper-covered steel	Fl	0.185	Silvered copper
		316/U	7/0.0067″ annealed silvered copper-covered steel	F1	0.060	Silvered copper
	Double braid	115/U	7/0.028″ silvered copper	F2	0.250	Silvered copper
		142B/U	0.039″ silvered copper-covered steel	F1	0.116	Silvered copper
		225/U	7/0.0312″ silvered copper	F1	0.285	Silvered copper

Protective Covering (Note 2)	Nominal Overall Diameter (in.)	Weight (lb/ft)	Nominal Capacitance (pF/ft)	Maximum Operating Voltage (rms)	Remarks
Teflon-tape moisture seal with double-braid type-V jacket	0.410	0.120	20.5	5000	Similar to RG-11A/U, except cable core is Teflon. $Z=75$ ohms
IX	0.075 max.	—	29.0	1000	$Z=50$ ohms
IX	0.105	—	20.0	1200	
IX	0.145	—	15.5	1500	$Z=95$ ohms
VII	0.110	—	—	1200	Miniaturized cable. $Z=75$ ohms
VII	0.155	—	—	1500	Miniaturized cable. $Z=95$ ohms
VII	0.080	—	—	1000	Miniaturized cable. $Z=50$ ohms
Same as RG-144/U	0.730	0.450	29.0	7000	Semiflexible cable operating at −55°C to +200°C (formerly RG-117A/U). $Z=50$ ohms
Teflon-tape moisture seal with double-braid type-V jacket, with armor	0.795	0.600	29.0	7000	Same as RG-211A/U, but with armor (formerly RG-118A/U). $Z=50$ ohms)
IX	0.206	—	21.0	2300	$Z=75$ ohms
IX	0.170	—	28.5	1900	$Z=50$ ohms
IX	0.280	—	28.5	3000	$Z=50$ ohms
IX	0.102	—	—	1200	Miniaturized cable. $Z=50$ ohms
Same as RG-144/U	0.375	—	29.5	5000	Medium-size cable for use where expansion and contraction are a major problem. $Z=$ 50 ohms
IX	0.195	—	28.5	1900	Small-size flexible cable. $Z=50$ ohms
Same as RG-144/U	0.430	0.176	29.5	5000	Semiflexible cable operating at −55°C to +200°C (formerly RG-87A/U). $Z=50$ ohms

Continued on next page.

Army-Navy List of Preferred Radio-Frequency Cables—cont.

Class of Cables		JAN Type RG-	Inner Conductor†	Dielectric Material (Note 1)	Nominal Diameter of Dielectric (in.)	Shielding Braid
High temperature (*Continued*)	Double braid	226/U	19/0.0254″ silvered copper wire	F2	0.370	Copper
		227/U	7/0.0312″ silvered copper	F1	0.285	Silvered copper
Pulse	Single braid	26A/U	19/0.0117″ tinned copper	E	0.288	Tinned copper
		27A/U	19/0.0185″ tinned copper	D	0.455	Tinned copper
	Double braid	25A/U	19/0.0117″ tinned copper	E	0.288	Tinned copper
		28B/U	19/0.0185″ tinned copper	D	0.455	Inner: tinned copper. Outer: galvanized steel
		64A/U	19/0.0117″ tinned copper	E	0.288	Tinned copper
		156/U	7/21 AWG tinned copper	First layer A; second layer H	0.285	Inner: tinner copper. Outer: galvanized steel.
		157/U	19/24 AWG tinned copper	First layer H; second layer A; third layer H	0.455	
		158/U	37/21 AWG tinned copper		0.455	Tinned copper outer shield
		190/U	19/0.0117″ tinned copper	First layer H; second layer J; third layer H	0.380	Same as above
		191/U	30 AWG tinned copper; single braid over supporting elements; 0.485″ max.		1.065	Same as above
	Four braids	88/U	19/0.0117″ tinned copper	E	0.288	Tinned copper
Low capacitance	Single braid	62A/U	0.0253″ solid copper-covered steel	A	0.146	Copper

Protective Covering (Note 2)	Nominal Overall Diameter (in.)	Weight (lb/ft)	Nominal Capacitance (pF/ft)	Maximum Operating Voltage (rms)	Remarks
Same as RG-144/U	0.500	0.247	29.0	7000	Medium-size cable for use where expansion and contraction are a major problem (formerly RG-94A/U). $Z=50$ ohms
Same as RG-228A/U	0.490	0.224	29.5	5000	Same as RG-225/U, but with armor (formerly RG-116/U). $Z=50$ ohms
IV, with armor	0.505	0.189	50.0	10,000	High-voltage cable. $Z=48$ ohms
IV, with armor	0.670	0.304	50.0	15,000 peak	Large-size cable. $Z=48$ ohms
IV	0.505	0.205	50.0	10,000	High-voltage cable. $Z=48$ ohms
IV	0.750	0.370	50.0	15,000 peak	Large-size cable. $Z=48$ ohms
IV	0.475 max.	0.205	50.0	10,000	Medium-size cable. $Z=48$ ohms
IIa	0.540	0.211	30.0	10,000	Taped inner layers, first layer type K and second layer type A-1R, between the outer braid of the outer conductor and the tinned copper shield Triaxial pulse cables. $Z=50$ ohms
IIa	0.725	0.317	38.0	15,000	
IIa	0.725	0.380	78.0	15,000	Same as above, except $Z=25$ ohms
VIII over one wrap of type K	0.700	0.353	50.0	15,000	Taped inner layers, two wraps of type K and two wraps of type L between the outer braid and the tinned copper shield. Pulse cable. $Z=50$ ohms
Same as above	1.460	1.469	85.0	15,000	Same as RG-190/U, except $Z=25$ ohms
IIa	0.515	—	50.0	10,000	Medium-size multishielded high-voltage cable. $Z=48$ ohms
IIa	0.242	0.382	14.5	750	$Z=93$ ohms

Continued on next page.

Army-Navy List of Preferred Radio-Frequency Cables—cont.

Class of Cables		JAN Type RG-	Inner Conductor†	Dielectric Material (Note 1)	Nominal Diameter of Dielectric (in.)	Shielding Braid
Low capacitance (*Continued*)	Single braid	63B/U	0.0253″ copper-covered steel	A	0.285	Copper
		79B/U	0.0253″ copper-covered steel	A	0.285	Copper
	Double braid	71B/U	0.0253″ copper-covered steel	A	0.146	Tinned copper
High attenuation	Single braid	301/U	7/0.0203″ Karma wire	F1	0.185	Karma wire
High delay	Single braid	65A/U	No. 32 Formex F. Helix diameter 0.128″	A	0.285	Copper
Twin conductor	Single braid	57A/U	Each conductor 7/0.0285″ plain copper	A	0.472	Tinner copper
		130/U		A	0.472	Tinned copper
		131/U		A	0.472	Tinned copper
	Double braid	22B/U	Each conductor 7/0.0152″ copper	A	0.285	Tinned copper
		111A/U		A	0.285	Tinned copper
	Twin coaxial	181/U	Each conductor 7/26 AWG copper	A	0.210	Copper inner braids and common braid

* From "RF Transmission Lines and Fittings," MIL–HDBK–216, 4 January 1962, revised 18 May 1965. Requirements for listed cables are in Specification MIL–C–17.

† Diameter of strands given in inches, as, 7/0.0296″ = 7 strands, each 0.0296-inch diameter.

Note 1–Dielectric materials: A = Polyethylene, *D* = Layer of synthetic rubber between two layers of conducting rubber, E = Layer of conducting rubber plus two layers of synthetic rubber, F1 = Solid polytetrafluoroethylene (Teflon), F2 = Semisolid or taped polytetrafluoroethylene (Teflon), H = Conducting synthetic rubber, and J = Insulating butyl rubber.

Protective Covering (Note 2)	Nominal Overall Diameter (in.)	Weight (lb/ft)	Nominal Capacitance (pF/ft)	Maximum Operating Voltage (rms)	Remarks
IIa	0.405	0.082	10.0	1000	Medium-size low-capacitance air-spaced cable. $Z=125$ ohms
IIa, with armor	0.475 max.	0.138	10.0	1000	Same as RG-63B/U, but with armor. $Z=125$ ohms
IIIa	0.250 max.	—	14.5	750	Low-capacitance cable. $Z=93$ ohms
IX	0.245	—	29.0	3000	High-attenuation cable. $Z=50$ ohms
IIa	0.405	0.096	44.0	1000	High-impedance video cable; high-delay line. $Z=950$ ohms. (Refer to Note 3.)
IIa	0.625	0.225	17.0	3000	$Z=95$ ohms
I	0.625	0.220	17.0	8000	Same as RG-57A/U, except inner conductors are twisted to improve flexibility. $Z=95$ ohms
I, with aluminum armor	0.710	0.295	17.0	8000	Same as RG-130/U, but with armor. $Z=95$ ohms
IIa	0.420	0.116	16.0	1000	Small-size balanced twin-conductor cable. $Z=95$ ohms
IIa, with armor	0.490 max.	0.145	16.0	1000	Same as RG-22B/U, but with armor. $Z=95$ ohms
IIa	0.640	—	12	3500	Filled-to-round, unbalanced transmission cable. Twin coaxial. $Z=125$ ohms

Note 2—Jacket types: I = Polyvinyl chloride (colored black), IIa = Noncontaminating synthetic resin, IIIa = Noncontaminating synthetic resin (colored black), IV = Chloroprene, V = Fibreglas, silicone-impregnated varnish, VII = Polytetrafluoroethylene, VIII = Polychloroprene, and IX = Fluorinated ethylene propylene.

Note 3—For RG-65A/U, delay = 0.042 microsecond per foot at 5 megahertz; dc resistance = 7.0 ohms/foot.

30 Waveguides and Resonators

Revised by
Tatsuo Itoh

PROPAGATION OF ELECTROMAGNETIC WAVES IN HOLLOW WAVEGUIDES

For propagation of energy at microwave frequencies through a hollow metal tube under fixed conditions, the following different types of waves are available.

TE Waves: Transverse-electric waves, sometimes called H waves, characterized by the fact that the electric vector (E vector) is always perpendicular to the direction of propagation. This means that

$$E_z \equiv 0$$

where z is the direction of propagation.

TM Waves: Transverse-magnetic waves, also called E waves, characterized by the fact that the magnetic vector (H vector) is always perpendicular to the direction of propagation. This means that

$$H_z \equiv 0$$

where z is the direction of propagation.

Note—TEM Waves: Transverse-electromagnetic waves. These waves are characterized by the fact that both the electric vector (E vector) and the magnetic vector (H vector) are perpendicular to the direction of propagation. This means that

$$E_z \equiv H_z \equiv 0$$

where z is the direction of propagation. This is the mode commonly excited in coaxial and open-wire lines. It cannot be propagated in a waveguide.

The solutions for the field configurations in waveguides are characterized by the presence of the integers m and n, which can take on separate values from 0 or 1 to infinity. Only a limited number of these different m,n modes can be propagated, depending on the dimensions of the guide and the frequency of excitation. For each mode there is a definite lower limit or cutoff frequency below which the wave is incapable of being propagated. Thus, a waveguide is seen to exhibit definite properties of a high-pass filter.

The propagation constant, $\gamma_{m,n}$, determines the amplitude and phase of each component of the wave as it is propagated along the length of the guide. With $z =$ (direction of propagation) and $\omega = 2\pi \times$ (frequency), the factor for each component is

$$\exp[j\omega t - \gamma_{m,n}z]$$

Thus if $\gamma_{m,n}$ is real, the phase of each component is constant, but the amplitude decreases exponentially with z. When $\gamma_{m,n}$ is real, it is said that no propagation takes place. The frequency is considered below cutoff. Actually, propagation with high attenuation does take place for a small distance, and a short length of guide below cutoff is often used as a calibrated attenuator.

When $\gamma_{m,n}$ is imaginary, the amplitude of each component remains constant, but the phase varies with z. Hence, propagation takes place. The value of $\gamma_{m,n}$ is purely imaginary only in a lossless guide. In the practical case, $\gamma_{m,n}$ usually has both a real part, $\alpha_{m,n}$, which is the attenuation constant, and an imaginary part, $\beta_{m,n}$, which is the phase propagation constant. Then $\gamma_{m,n} = \alpha_{m,n} + j\beta_{m,n}$.

RECTANGULAR WAVEGUIDES

Fig. 1 shows a rectangular waveguide and a rectangular system of coordinates, disposed so that the origin falls on one of the corners of the waveguide; z is the direction of propagation along the guide, and the cross-sectional dimensions are y_0 and x_0.

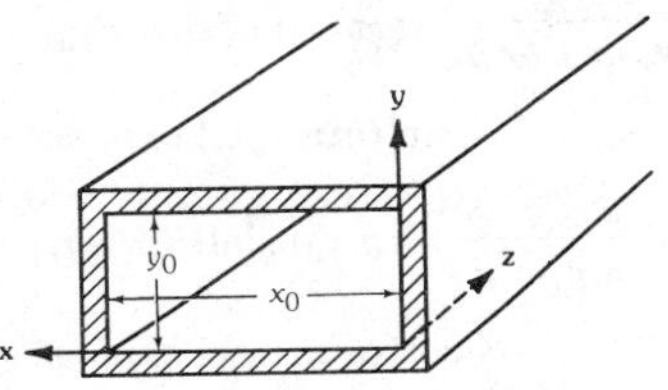

Fig. 1. Rectangular waveguide.

For the case of perfect conductivity of the guide walls with a nonconducting interior dielectric (usually air), the equations for the $\mathrm{TM}_{m,n}$ or $E_{m,n}$ waves in the dielectric are

$$E_x = -A \frac{\gamma_{m,n}}{\gamma_{m,n}z + \omega^2\mu\epsilon} (m\pi/x_0) \sin[(n\pi/y_0)y] \times \cos[(m\pi/x_0)x] \exp(j\omega t - \gamma_{m,n}z)$$

$$E_y = -A \frac{\gamma_{m,n}}{\gamma_{m,n}z + \omega^2\mu\epsilon} (n\pi/y_0) \cos[(n\pi/y_0)y] \times \sin[(m\pi/x_0)x] \exp(j\omega t - \gamma_{m,n}z)$$

$$E_z = A \sin[(n\pi/y_0)y] \sin[(m\pi/x_0)x] \times \exp(j\omega t - \gamma_{m,n}z)$$

$$H_x = -A \frac{j\omega\epsilon}{\gamma_{m,n}z + \omega^2\mu\epsilon} (n\pi/y_0) \cos[(n\pi/y_0)y] \times \sin[(m\pi/x_0)x] \exp(j\omega t - \gamma_{m,n}z)$$

$$H_y = A \frac{j\omega\epsilon}{\gamma_{m,n}z + \omega^2\mu\epsilon} (m\pi/x_0) \sin[(n\pi/y_0)y] \times \cos[(m\pi/x_0)x] \exp(j\omega t - \gamma_{m,n}z)$$

$$H_z \equiv 0$$

where ϵ is the dielectric constant and μ the permeability of the dielectric material in meter-kilogram-second (rationalized) units.

Constant A is determined solely by the exciting voltage. It has both amplitude and phase. Integers m and n may individually take values from 1 to infinity. No TM waves of the 0,0 type or 1,0 type are possible in a rectangular guide, so neither m nor n may be 0.

Equations for the $\mathrm{TE}_{m,n}$ waves or $H_{m,n}$ waves in a dielectric are

$$E_x = -B \frac{j\omega\mu}{\gamma_{m,n}z + \omega^2\mu\epsilon} (n\pi/y_0) \sin[(n\pi/y_0)y] \times \cos[(m\pi/x_0)x] \exp(j\omega t - \gamma_{m,n}z)$$

$$E_y = B \frac{j\omega\mu}{\gamma_{m,n}z + \omega^2\mu\epsilon} (m\pi/x_0) \cos[(n\pi/y_0)y] \times \sin[(m\pi/x_0)x] \exp(j\omega t - \gamma_{m,n}z)$$

$$E_z \equiv 0$$

$$H_z = B \frac{\gamma_{m,n}}{\gamma_{m,n}z + \omega^2\mu\epsilon} (m\pi/x_0) \cos[(n\pi/y_0)y] \times \sin[(m\pi/x_0)x] \exp(j\omega t - \gamma_{m,n}z)$$

$$H_y = B \frac{\gamma_{m,n}}{\gamma_{m,n}z + \omega^2\mu\epsilon} (n\pi/y_0) \sin[(n\pi/y_0)y] \times \cos[(m\pi/x_0)x] \exp(j\omega t - \gamma_{m,n}z)$$

$$H_z = B \cos[(n\pi/y_0)y] \cos[(m\pi/x_0)x] \times \exp(j\omega t - \gamma_{m,n}z)$$

Constant B depends only on the original exciting voltage and has both magnitude and phase; m and n individually may assume any integer value from 0 to infinity. The 0,0 type of wave where both m and n are 0 is not possible; all other combinations are.

As stated previously, propagation takes place only when propagation constant $\gamma_{m,n}$ is imaginary.

$$\gamma_{m,n} = [(m\pi/x_0)^2 + (n\pi/y_0)^2 - \omega^2\mu\epsilon]^{1/2}$$

This means, for any m,n mode, propagation takes place when

$$\omega^2\mu\epsilon > (m\pi/x_0)^2 + (n\pi/y_0)^2$$

or, in terms of frequency f and velocity of light c, when

$$f > \frac{c}{2\pi(\mu_1\epsilon_1)^{1/2}} [(m\pi/x_0)^2 + (n\pi/y_0)^2]^{1/2}$$

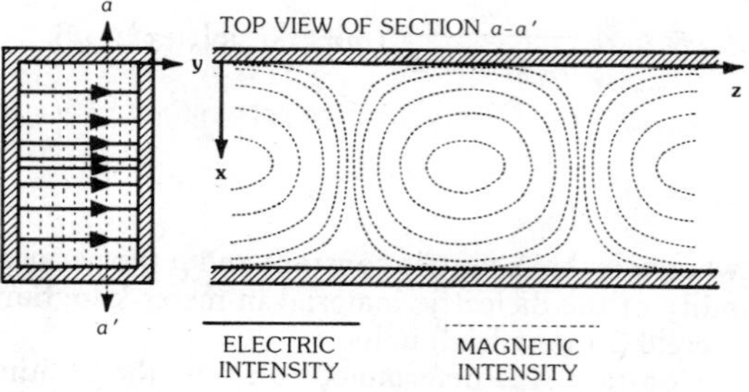

Fig. 2. Field configuration for a $\mathrm{TE}_{1,0}$ wave.

where μ_1 and ϵ_1 are the relative permeability and relative dielectric constant, respectively, of the dielectric material with respect to free space.

The wavelength in the air-filled waveguide is always greater than the wavelength in free space. The wavelength in the dielectric-filled waveguide may be less than the wavelength in free space. If λ is the wavelength in free space and the medium filling the waveguide has a relative dielectric constant ϵ,

$$\lambda_{g\,(m,n)} = \frac{\lambda}{[\epsilon - (m\lambda/2x_0)^2 - (n\lambda/2y_0)^2]^{1/2}} = \frac{\lambda}{[\epsilon - (\lambda/\lambda_c)^2]^{1/2}}$$

where $(1/\lambda_c)^2 = (m/2x_0)^2 + (n/2y_0)^2$.

The phase velocity within the guide is also always greater than in an unbounded medium. The phase velocity, v, and group velocity, u, are related by

$$u = c^2/v$$

where the phase velocity is given by $v = c\lambda_g/\lambda$, and the group velocity is the velocity of propagation of the energy.

To couple energy into waveguides, it is necessary to understand the configuration of the characteristic electric and magnetic lines. Fig. 2 shows the field configuration for a $\mathrm{TE}_{1,0}$ wave. Fig. 3 shows the instantaneous field configuration for a higher mode, a $\mathrm{TE}_{2,1}$ wave.

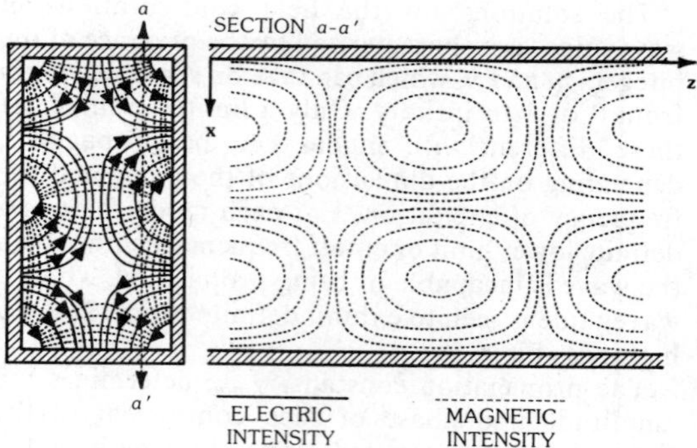

Fig. 3. Field configuration for a $\mathrm{TE}_{2,1}$ wave.

In Fig. 4 are shown only the characteristic E lines for the $\mathrm{TE}_{1,0}$, $\mathrm{TE}_{2,0}$, $\mathrm{TE}_{1,1}$, and $\mathrm{TE}_{2,1}$ waves. The arrows on the lines indicate their instantaneous relative directions. To excite a TE wave, it is necessary to insert a probe to coincide with the direction of the E lines. Thus, for a $\mathrm{TE}_{1,0}$ wave, a single probe projecting from the side of the guide parallel to the E lines would be sufficient to couple into it. Two ways of coupling from a coaxial line to a rectangular waveguide to excite the $\mathrm{TE}_{1,0}$ mode are shown in Fig. 5. With structures such as these, it is possible to

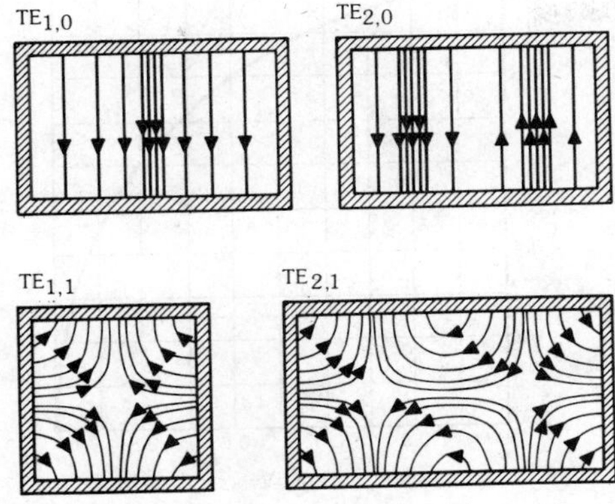

Fig. 4. Characteristic E lines for TE waves.

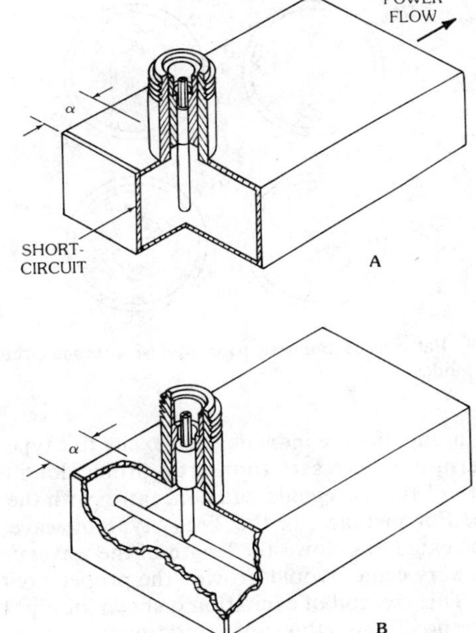

Fig. 5. Methods of coupling to $TE_{1,0}$ mode ($\alpha \approx \lambda_g/4$).

make the standing-wave ratio due to the junction less than 1.15 over a 10- to 15-percent frequency band.

Fig. 6 shows the instantaneous configuration of a $TM_{1,1}$ wave; Fig. 7 shows the instantaneous field configuration for a $TM_{2,1}$ wave. Coupling to this type of wave may be accomplished by inserting a probe that is parallel to the E lines, or by means of a loop oriented to link the lines of flux.

CIRCULAR WAVEGUIDES

The usual coordinate system is ρ, θ, z, where ρ is the radial direction, θ is the angle, and z is in the longitudinal direction.

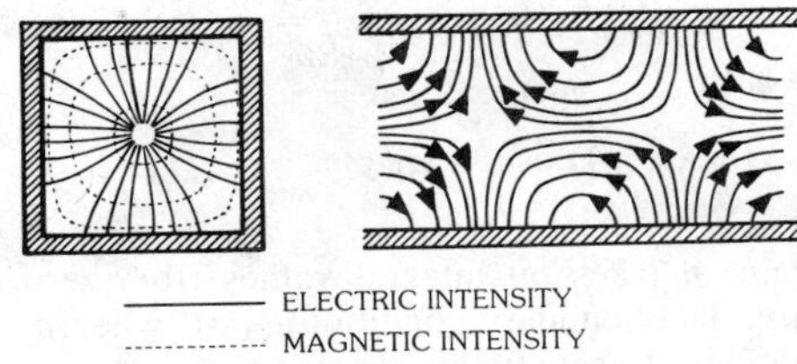

Fig. 6. Instantaneous field configuration, $TM_{1,1}$ wave.

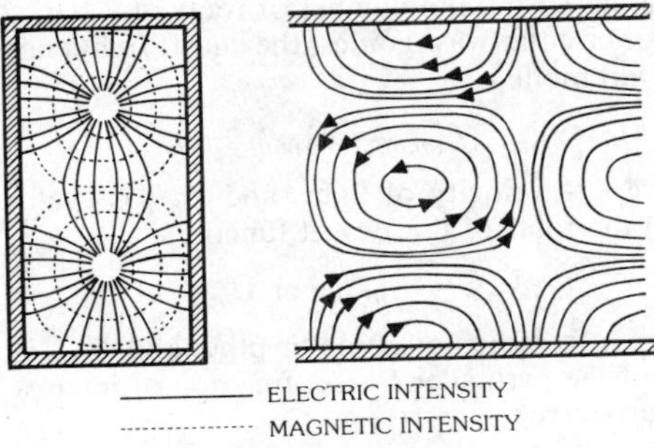

Fig. 7. Instantaneous field configuration, $TM_{2,1}$ wave.

TM Waves (E Waves): $H_z \equiv 0$

$$E_\rho = H_\theta \eta(\lambda/\lambda_{g(m,n)}) \exp(j\omega t - \gamma_{m,n}z)$$

$$E_\theta = -H_\rho \eta(\lambda/\lambda_{g(m,n)}) \exp(j\omega t - \gamma_{m,n}z)$$

$$E_z = AJ_n(k_{m,n}\rho) \cos n\theta \exp(j\omega t - \gamma_{m,n}z)$$

$$H_\rho = -jA(2\pi n/\lambda k_{m,n}{}^2\eta\rho)J_n(k_{m,n}\rho) \sin n\theta \times \exp(j\omega t - \gamma_{m,n}z)$$

$$H_\theta = -jA(2\pi/\lambda k_{m,n}\eta)J_n{}'(k_{m,n}\rho) \cos n\theta \times \exp(j\omega t - \gamma_{m,n}z)$$

where $\eta = (\mu/\epsilon)^{1/2}$, with μ and ϵ in absolute units.

By the boundary conditions, $E_z = 0$ when $\rho = a$, the radius of the guide. Thus, the only permissible values of k are those for which $J_n(k_{m,n}a) = 0$, because E_z must be zero at the boundary.

The numbers m,n take on all integral values from zero to infinity. The waves are seen to be characterized by the numbers m and n, where n gives the order of the Bessel functions, and m gives the order of the root of $J_n(k_{m,n}a)$. The Bessel function has an infinite number of roots, so there are an infinite number of k's that make $J_n(k_{m,n}a) = 0$.

TE Waves (H Waves): $E_z \equiv 0$

$$E_\rho = jB(2\pi n\eta/\lambda k_{m,n}{}^2\rho)J_n(k_{m,n}\rho) \sin n\theta \times \exp(j\omega t - \gamma_{m,n}z)$$

$$E_\theta = jB(2\pi\eta/\lambda k_{m,n})J_n{}'(k_{m,n}\rho) \cos n\theta \times \exp(j\omega t - \gamma_{m,n}z)$$

$$H_\rho = -E_\theta(\lambda_{g(m,n)}/\eta\lambda)\exp(j\omega t - \gamma_{m,n}z)$$
$$H_\theta = E_\rho(\lambda_{g(m,n)}/\eta\lambda)\exp(j\omega t - \gamma_{m,n}z)$$
$$H_z = BJ_n(k_{m,n}\rho)\cos n\theta \exp(j\omega t - \gamma_{m,n}z)$$

Again, n takes on integral values from zero to infinity. The boundary condition $E_\theta = 0$ when $\rho = a$ still applies. To satisfy this condition, k must be such as to make $J_n'(k_{m,n}a)$ equal to zero [where the superscript indicates the derivative of $J_n(k_{m,n}a)$]. It is seen that m takes on values from 1 to infinity, since there are an infinite number of roots of $J_n'(k_{m,n}a)$.

For circular waveguides, the cutoff frequency for the m,n mode is

$$f_{c(m,n)} = ck_{m,n}/2\pi$$

where c = velocity of light, and $k_{m,n}$ is evaluated from the roots of the Bessel functions.

$$k_{m,n} = U_{m,n}/a \text{ or } U_{m,n}'/a$$

where a = radius of guide or pipe, and $U_{m,n}$ is the root of the particular Bessel function of interest (or its derivative).

The wavelength in any guide filled with a homogeneous dielectric ϵ (relative) is

$$\lambda_g = \lambda_0/[\epsilon - (\lambda_0/\lambda_c)^2]^{1/2}$$

where λ_0 is the wavelength in free space, and λ_c is the free-space cutoff wavelength for any mode under consideration.

Tables 1 and 2 are useful in determining the values of k. For TE waves, the cutoff wavelengths are given in Table 1, and for TM waves the cutoff wavelengths are given in Table 2, where n is the order of the Bessel function and m is the order of the root.

TABLE 1. TE WAVES, VALUES OF λ_c/a (WHERE a = RADIUS OF GUIDE)

n \ m	0	1	2
1	1.640	3.414	2.057
2	0.896	1.178	0.937
3	0.618	0.736	0.631

TABLE 2. TM WAVES, VALUES OF λ_c/a

n \ m	0	1	2
1	2.619	1.640	1.224
2	1.139	0.896	0.747
3	0.726	0.618	0.541

Fig. 8 shows λ_0/λ_g as a function of λ_0/λ_c. From this, λ_g may be determined when λ_0 and λ_c are known.

The pattern of magnetic force of TM waves in a circular waveguide is shown in Fig. 9. Only the maximum lines are indicated. To excite this type of pattern, it is necessary to insert a probe along the length of the waveguide and concentric with the H lines. For instance, in the $TM_{0,1}$ type of wave, a probe extending down the length of the waveguide at its very center would provide the proper excitation. This method of excitation is shown in Fig. 10. Corresponding methods of excitation may be used for the other types of TM waves shown in Fig. 9.

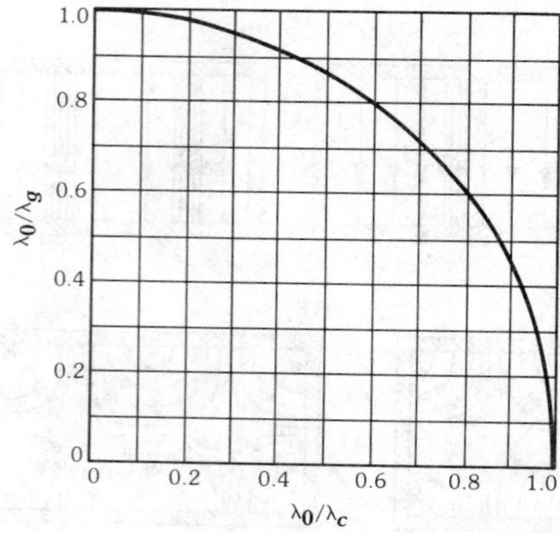

Fig. 8. Chart for determining guide wavelength.

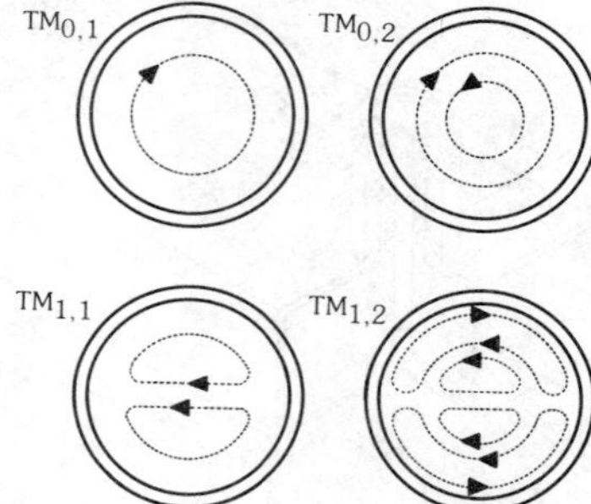

Fig. 9. Patterns of magnetic force of TM waves in circular waveguides.

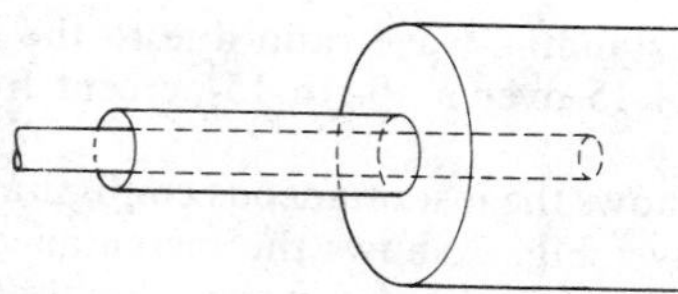

Fig. 10. Method of coupling to circular waveguide for $TM_{0,1}$ wave.

Fig. 11 shows the patterns of electric force for TE waves. Again, only the maximum lines are indicated. This type of wave may be excited by an antenna that is parallel to the electric lines of force. The $TE_{1,1}$ wave may be excited by means of an

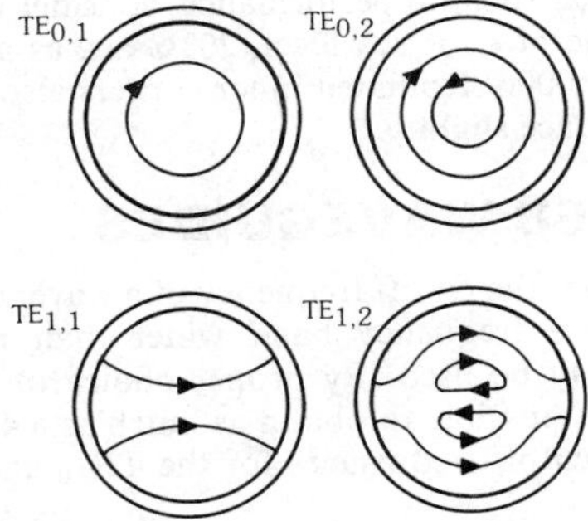

Fig. 11. Patterns of electric force of TE waves in circular waveguides.

antenna extending across the waveguide. This is illustrated in Fig. 12.

Propagating E waves have a minimum attenuation at $(3)^{1/2}f_c$. The $H_{1,1}$ wave has minimum attenuation at the frequency $2.6\ (3)^{1/2}\ f_c$.

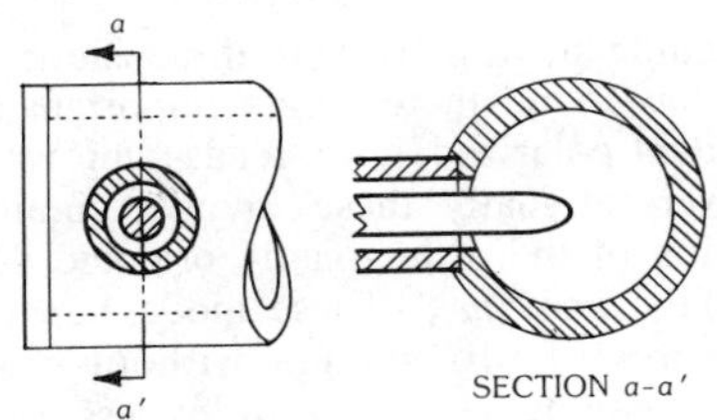

Fig. 12. Method of coupling to circular waveguide for $TE_{1,1}$ wave.

The $H_{0,1}$ wave has the interesting and useful property that attenuation decreases as the frequency increases. This has made the $H_{0,1}$ mode exceedingly useful in the transmission of microwave signals over long distances.

Table 3 presents some of the important equations for various guides.

TABLE 3. CUTOFF WAVELENGTHS AND ATTENUATION FACTORS*

Type of Guide (Copper)†	Cutoff Wavelength λ_c $f_c = (c/\lambda_c)$	Attenuation Constant α (dB/100 ft)
Coaxial line‡ TEM (2a, 2b)	0	$\dfrac{3.58 \times 10^{-6}(f)^{1/2}(1/b)[1 + (b/a)]}{\ln (b/a)}$ Note: Fig. 29 in Chapter 29 includes a derating factor of 20% applied to calculated α.
Rectangular pipe $TE_{m,0}$ or $H_{m,0}$ (a, b)	$2a/m$	$\dfrac{1.107}{a^{3/2}} \times \dfrac{{}^1\!/_2(a/b)(f/f_c)^{3/2} + (f/f_c)^{-1/2}}{[(f/f_c)^2 - 1]^{1/2}}$
Circular pipe: (2a)		
$TM_{0,1}$ or $E_{0,1}$	$2.613a$	$\dfrac{0.485}{a^{3/2}} \times \dfrac{(f/f_c)^{3/2}}{[(f/f_c)^2 - 1]^{1/2}}$
$TE_{1,1}$ or $H_{0,1}$	$3.412a$	$\dfrac{0.423}{a^{3/2}} \times \dfrac{(f/f_c)^{-1/2} + (1/2.38)(f/f_c)^{3/2}}{[(f/f_c)^2 - 1]^{1/2}}$
$TE_{0,1}$ or $H_{0,1}$	$1.640a$	$\dfrac{0.611}{a^{3/2}} \times \dfrac{(f/f_c)^{-1/2}}{[(f/f_c)^2 - 1]^{1/2}}$

* Dimensions are in inches and frequencies in hertz; vacuum dielectric.
† For other metals multiply α by the square root of ratio of resistivity relative to that of copper.
‡ Inner and outer conductors same material.

SQUARE WAVEGUIDES

Waveguide having interior dimensions $x_0 = y_0$ (Fig. 1) has found increasingly important application in dual-polarized horn feeds and waveguide multiplexers. Usually these involve simultaneous propagation of the orthogonally oriented dominant modes, $TE_{1,0}$ and $TE_{0,1}$. These modes are theoretically capable of propagation without cross coupling, at the same frequency, in lossless waveguide of square cross section. In practice, wall losses, surface irregularities, and unequal transverse interior dimensions give rise to $TE_{1,0}$ and $TE_{0,1}$ mode cross-conversion. This occurs continually along the waveguide so that unless special care is taken, long lengths of dual-polarized waveguide exhibit a deteriorated mode isolation as a function of length of guide.

Most important in establishing the initial mode isolation is proximity of the operating frequency to the cutoff frequency of the $TE_{1,0}$ mode in the square waveguide so that the total operating frequency band is above $TE_{1,0}$ cutoff and well below $TE_{1,1}$ cutoff. The lowest operating frequency should be approximately 25% above the $TE_{1,0}$ cutoff frequency. Thus, a dual-polarized feed propagating a 4400-MHz signal should use a square waveguide having internal dimensions of about 1.68 inches. If the internal dimensions are arrived at by using the 1.87 internal dimension of WR 187 waveguide, the dual-mode isolation will probably not exceed 35 dB. By operating about 25% above the $TE_{1,0}$ mode cutoff frequency and well below $TE_{1,1}$ cutoff, the isolation can exceed 50 dB.*

ATTENUATION IN A WAVEGUIDE BEYOND CUTOFF

When a waveguide is used at a wavelength greater than the cutoff wavelength, there is no real propagation and the fields are attenuated exponentially. The attenuation, L, in a length, d, is given by

$$L = 54.5(d/\lambda_c)[1 - (\lambda_c/\lambda)^2]^{1/2} \text{ decibels}$$

where λ_c = cutoff wavelength and λ = operating wavelength.

Note that for $\lambda >> \lambda_c$, attenuation is essentially independent of frequency and

$$L = 54.5d/\lambda_c \text{ decibels}$$

where λ_c is a function of geometry.

STANDARD WAVEGUIDES

Table 4 lists some properties and dimensions of standard rectangular waveguides. For other than theoretical vacuum performance, consider the relative value of ϵ for sea level, 20°C air, as approximately 1.0006. Rounded inner corners also modify performance slightly.*

RIDGED WAVEGUIDES

To lower the cutoff frequency of a waveguide for use over a frequency band wider than normal, ridges may be used. By proper choice of dimensions, it is possible to obtain as much as a 4:1 ratio between cutoff frequencies for the $TE_{2,0}$ and $TE_{1,0}$ modes.

Tables 5 and 6 and Figs. 13 and 14 give the essential characteristics of single- and double-ridged guides. Figs. 15 and 16 show the relationship between the cutoff wavelength and the critical dimensions. Figs. 17 and 18 describe the bandwidth (ratio of cutoff wavelengths of the 0,1 and 0,2 modes). The price paid for increased bandwidth is an increase in attenuation relative to the equivalent rectangular guide (Figs. 19 and 20).

Coaxial line can be coupled to either the single- or double-ridged guide by partly inserting the probe into the open space between the ridge and the opposite wall or ridge (Fig. 5A) or connecting it directly across the gap.

FLEXIBLE WAVEGUIDES

Flexible waveguide is used to join rigid sections or components that cannot be accurately dimensioned, positioned, or rendered immobile in space. Thus, rather than attempt to specify and hold precisely every bend, twist, and straight section of a long waveguide run between fixed points, it is often adequate to leave a short run or bend to be filled by a flexible guide insert. Flexible sections are also used to permit thermally induced relative movement and to insulate portions of a waveguide run from shock and vibration. Flexible waveguide should not be treated as a link between a cabinet and its frequently opened doors and drawers, unless the cabinet is specified and/or designed for that type of service. Most flexible-waveguide structures are susceptible to cracking under these conditions if flexure is repeated often.

Flexible waveguide is available in many different forms. It may be made from flat ribbons wound on a rectangular mandrel with the edges convoluted or folded in and interlocked. The convoluted guide may be soldered or unsoldered, since the bending and twisting results from a flexure of each turn and not from a relative sliding as in the case of the interlocked guide. If soldered, it is more difficult to flex and essentially loses twist capability.

---

* D. J. LeVine and W. Sichak, "Dual-Mode Horn Feed for Microwave Multiplexing." *Electronics,* September 1954.

* M. M. Brady, "Cutoff Wavelengths and Frequencies of Standard Rectangular Waveguides." *I.E.E. Electronics Letters,* Vol. 5, No. 17, 21 August 1969.

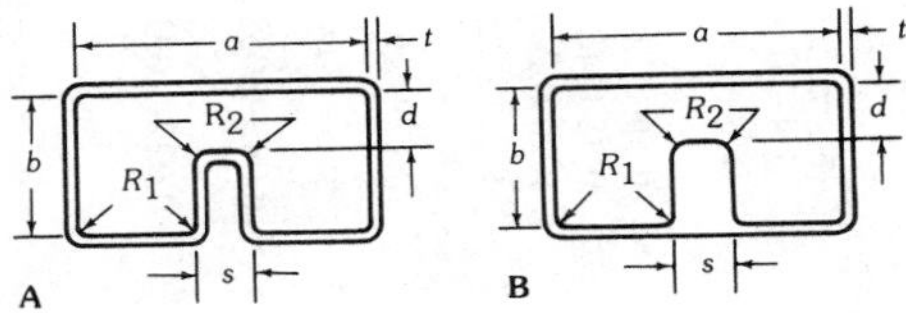

Fig. 13. Single-ridged waveguides (refer to Table 5).

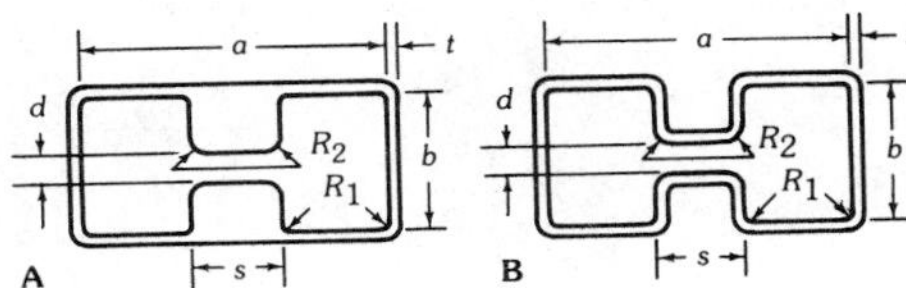

Fig. 14. Double-ridged waveguides (refer to Table 6).

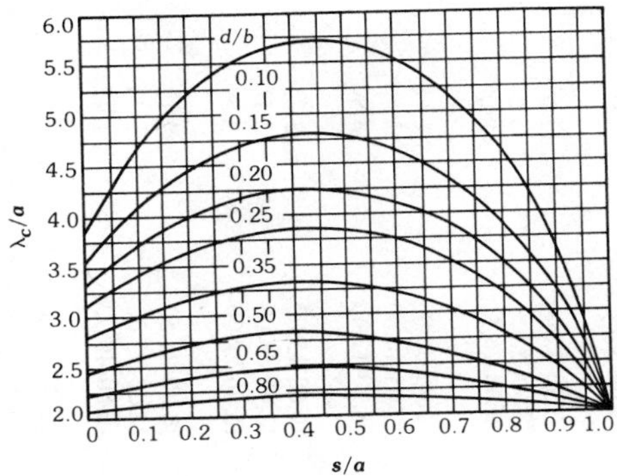

Fig. 15. Cutoff wavelength, single-ridged guide. (*From S. Hopfer, "The Design of Ridged Waveguides,"* Transactions of the I.R.E., *Vol. MTT-3, No. 5, 1955; Fig. 5.*)

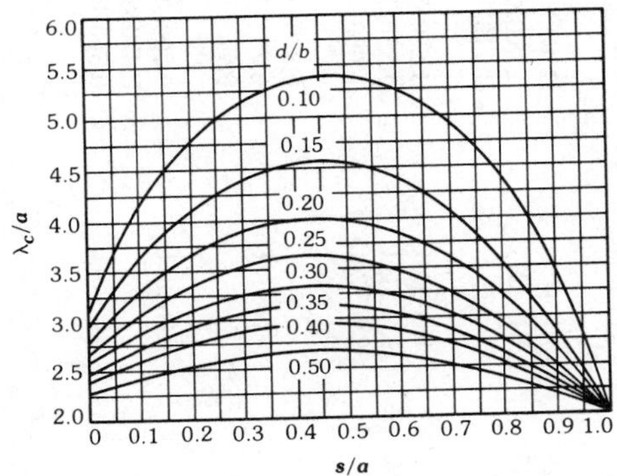

Fig. 16. Cutoff wavelength, double-ridged guide. (*From S. Hopfer, "The Design of Ridged Waveguides,"* Transactions of the I.R.E., *Vol. MTT-3, No. 5, 1955; Fig. 2.*)

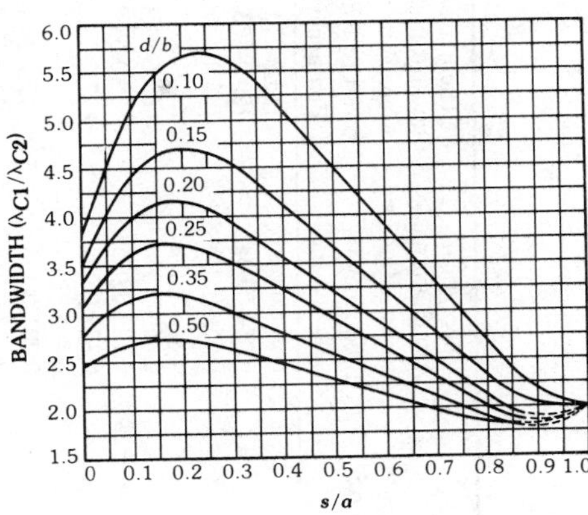

Fig. 17. Bandwidth curves, single-ridged guide. (*From S. Hopfer, "The Design of Ridged Waveguides,"* Transactions of the I.R.E., *Vol. MTT-3, No. 5, 1955; Fig. 9.*)

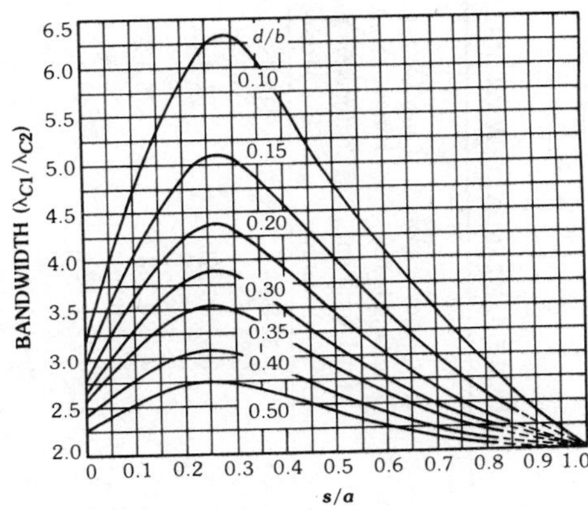

Fig. 18. Bandwidth curves, double-ridged guide. (*From S. Hopfer, "The Design of Ridged Waveguides,"* Transactions of the I.R.E., *Vol. MTT-3, No. 5, 1955; Fig. 10.*)

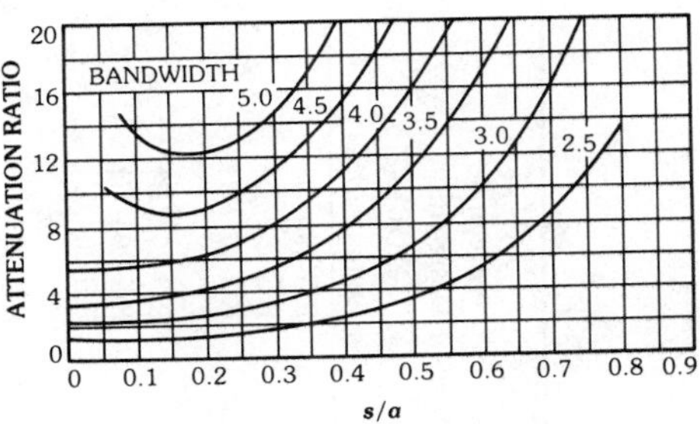

Fig. 19. Attenuation ratio, parametric in bandwidth, single-ridged guide. (*From S. Hopfer, "The Design of Ridged Waveguides,"* Transactions of the I.R.E., *Vol. MTT-3, No. 5, 1955; Fig. 11.*)

TABLE 4. STANDARD WAVEGUIDES

EIA Waveguide Designation (Standard RS-261-A)	JAN Waveguide Designation (MIL-HDBK-216, 4 January 1962)	Outer Dimensions and Wall Thickness (in inches)	Frequency Range in Gigahertz for Dominant ($TE_{1,0}$) Mode	Cutoff Wavelength, λ_c, in Centimeters for $TE_{1,0}$ Mode	Cutoff Frequency, f_c, in Gigahertz for $TE_{1,0}$ Mode	Theoretical Attenuation, Lowest to Highest Frequency in dB/100 ft	Theoretical Power Rating in Megawatts for Lowest to Highest Frequency*
WR-2300	RG-290/U†	23.250×11.750×0.125	0.32-0.49	116.8	0.256	0.051-0.031	153.0-212.0
WR-2100	RG-291/U†	21.250×10.750×0.125	0.35-0.53	106.7	0.281	0.054-0.034	120.0-173.0
WR-1800	RG-201/U†	18.250×9.250×0.125	0.425-0.620	91.4	0.328	0.056-0.038	93.4-131.9
WR-1500	RG-202/U†	15.250×7.750×0.125	0.49-0.740	76.3	0.393	0.069-0.050	67.6-93.3
WR-1150	RG-203/U†	11.750×6.000×0.125	0.64-0.96	58.4	0.514	0.128-0.075	35.0-53.8
WR-975	RG-204/U†	10.000×5.125×0.125	0.75-1.12	49.6	0.605	0.137-0.095	27.0-38.5
WR-770	RG-205/U†	7.950×4.100×0.125	0.96-1.45	39.1	0.767	0.201-0.136	17.2-24.1
WR-650	RG-69/U	6.660×3.410×0.080	1.12-1.70	33.0	0.908	0.317-0.212	11.9-17.2
WR-510	—	5.260×2.710×0.080	1.45-2.20	25.9	1.16	—	—
WR-430	RG-104/U	4.460×2.310×0.080	1.70-2.60	21.8	1.375	0.588-0.385	5.2-7.5
WR-340	RG-112/U	3.560×1.860×0.080	2.20-3.30	17.3	1.735	0.877-0.572	—
WR-284	RG-48/U	3.000×1.500×0.080	2.60-3.95	14.2	2.08	1.102-0.752	2.2-3.2
WR-229	—	2.418×1.273×0.064	3.30-4.90	11.6	2.59	—	—
WR-187	RG-49/U	2.000×1.000×0.064	3.95-5.85	9.50	3.16	2.08-1.44	1.4-2.0
WR-159	—	1.718×0.923×0.064	4.90-7.05	8.09	3.71	—	—
WR-137	RG-50/U	1.500×0.750×0.064	5.85-8.20	6.98	4.29	2.87-2.30	0.56-0.71
WR-112	RG-51/U	1.250×0.625×0.064	7.05-10.00	5.70	5.26	4.12-3.21	0.35-0.46
WR-90	RG-52/U	1.000×0.500×0.050	8.20-12.40	4.57	6.56	6.45-4.48	0.20-0.29
WR-75	—	0.850×0.475×0.050	10.00-15.00	3.81	7.88	—	—
WR-62	RG-91/U	0.702×0.391×0.040	12.40-18.00	3.16	9.49	9.51-8.31	0.12-0.16
WR-51	—	0.590×0.335×0.040	15.00-22.00	2.59	11.6	—	—
WR-42	RG-53/U	0.500×0.250×0.040	18.00-26.50	2.13	14.1	20.7-14.8	0.043-0.058
WR-34	—	0.420×0.250×0.040	22.00-33.00	1.73	17.3	—	—
WR-28	RG-96/U‡	0.360×0.220×0.040	26.50-40.00	1.42	21.1	21.9-15.0	0.022-0.031

WR-22	RG-97/U‡	0.304×0.192×0.040	33.00-50.00	1.14	26.35	31.0-20.9	0.014-0.020
WR-19	—	0.268×0.174×0.040	40.00-60.00	0.955	31.4	—	—
WR-15	RG-98/U‡	0.228×0.154×0.040	50.00-75.00	0.753	39.9	52.9-39.1	0.0063-0.0090
WR-12	RG-99/U‡	0.202×0.141×0.040	60.00-90.00	0.620	48.4	93.3-52.2	0.0042-0.0060
WR-10	—	0.180×0.130×0.040	75.00-110.00	0.509	59.0	—	—
WR-8	RG-138/U§	0.140×0.100×0.030	90.00-140.00	0.406	73.84	152-99	0.0018-0.0026
WR-7	RG-136/U§	0.125×0.0925×0.030	110.00-170.00	0.330	90.84	163-137	0.0012-0.0017
WR-5	RG-135/U§	0.111×0.0855×0.030	140.00-220.00	0.259	115.75	308-193	0.00071-0.00107
WR-4	RG-137/U§	0.103×0.0815×0.030	170.00-260.00	0.218	137.52	384-254	0.00052-0.00075
WR-3	RG-139/U§	0.094×0.0770×0.030	220.00-325.00	0.173	173.28	512-348	0.00035-0.00047

* For these computations, the breakdown strength of air was taken as 15 000 volts per centimeter. A safety factor of approximately 2 at sea level has been allowed.

† Aluminum, 2.83×10^{-6} ohm-cm resistivity. ‡ Silver, 1.62×10^{-6} ohm-cm resistivity. § JAN types are silver, with a circular outer diameter of 0.156 inch and a rectangular bore matching EIA types. All other types are of a Cu-Zn alloy, 3.9×10^{-6} ohm-cm resistivity.

Note: Equivalent designations of waveguides follow.

EIA	British	IEC	EIA	British	IEC	EIA	British	IEC
WR-2300	00	-R3	WR-340	9A	-R26	WR-51	19	-R180
WR-2100	0	-R4	WR-284	10	-R32			
WR-1800	1	-R5				WR-42	20	-R220
			WR-229	11A	-R40	WR-34	21	-R260
WR-1500	2	-R6	WR-187	12	-R48	WR-28	22	-R320
WR-1150	3	-R8	WR-159	13	-R58			
WR-975	4	-R9				WR-22	23	-R400
			WR-137	14	-R70	WR-19	24	-R500
WR-770	5	-R12	WR-112	15	-R84	WR-15	25	-R620
WR-650	6	-R14	WR-90	16	-R100			
WR-510	7	-R18				WR-12	26	-R740
			WR-75	17	-R120	WR-10	27	-R900
WR-430	8	-R22	WR-62	18	-R140	WR-8	28	-R1200

TABLE 5. CHARACTERISTICS OF SINGLE-RIDGED WAVEGUIDES*

Frequency Range (GHz)	$f_{c1,0}$ (GHz)	$\lambda_{c1,0}$ (in)	$f_{c2,0}$ (GHz)	Dimensions in Inches							At $f=(3)^{1/2}f_{c1,0}$	
				a	b	d	s	t	R_1 (max)	R_2	Atten** (dB/ft)	Power Rating† (kW)
						Bandwidth 2.4:1						
0.175-0.42	0.148	79.803	0.431	28.129	12.658	5.278	4.360	—	—	1.056	0.00024	32 870.
0.267-0.64	0.226	52.260	0.658	18.421	8.289	3.457	2.855	—	—	0.691	0.00045	14 100.
0.42-1.0	0.356	33.177	1.036	11.695	5.263	2.195	1.813	0.125	0.047	0.439	0.00087	5 682.
0.64-1.53	0.542	21.792	1.577	7.682	3.457	1.442	1.191	0.125	0.047	0.288	0.00164	2 451.
0.84-2.0	0.712	16.588	2.072	5.847	2.631	1.097	0.906	0.080	0.047	0.219	0.00248	1 421.
1.5-3.6	1.271	9.293	3.699	3.276	1.474	0.615	0.508	0.080	0.047	0.123	0.00591	445.8
2.0-4.8	1.695	6.968	4.933	2.456	1.105	0.461	0.381	0.080	0.047	0.092	0.00908	250.6
3.5-8.2	2.966	3.982	8.632	1.404	0.632	0.264	0.218	0.064	0.031	0.053	0.0212	81.87
4.75-11.0	4.025	2.934	11.714	1.034	0.465	0.194	0.160	0.050	0.031	0.039	0.0333	44.43
7.5-18.0‡	6.356	1.858	18.498	0.655	0.295	0.123	0.1015	0.050	0.015	0.025	0.0661	17.82
11.0-26.5‡	9.322	1.267	27.130	0.4466	0.2010	0.0838	0.0692	0.040	0.015	0.017	0.117	8.285
18.0-40.0‡	15.254	0.7743	44.393	0.2729	0.1228	0.0512	0.0423	0.040	0.015	0.010	0.246	3.035
						Bandwidth 3.6:1						
0.108-0.39	0.092	128.37	0.404	31.218	14.048	2.402	5.307	—	—	0.480	0.0016	14 550.
0.27-0.97	0.229	51.572	1.006	12.542	5.644	0.965	2.132	—	—	0.193	0.0065	2 348.
0.39-1.4	0.331	35.680	1.454	8.677	3.905	0.668	1.475	0.125	0.047	0.134	0.0112	1 124.
0.97-3.5	0.822	14.367	3.611	3.494	1.572	0.269	0.594	0.080	0.047	0.054	0.0438	182.2
1.4-5.0	1.186	9.958	5.210	2.422	1.090	0.186	0.412	0.080	0.047	0.037	0.0758	87.56
3.5-12.4	2.966	3.982	13.030	0.968	0.436	0.075	0.165	0.050	0.031	0.015	0.300	13.99
5.0-18.0‡	4.237	2.787	18.613	0.678	0.305	0.052	0.115	0.050	0.015	0.010	0.513	6.857
12.4-40.0‡	10.508	1.124	46.162	0.273	0.123	0.021	0.046	0.040	0.015	0.004	2.008	1.115

* From MIL-HDBK-216, *RF Transmission Lines and Fittings*, 4 January 1962.

** Copper.

† Based on breakdown of air—15000 volts per cm (safety factor of approx 2 at sea level). Corner radii considered.

‡ Fig. 13B in these frequency ranges only.

TABLE 6. CHARACTERISTICS OF DOUBLE-RIDGED WAVEGUIDES*

Frequency Range (GHz)	$f_{c1,0}$ (GHz)	$\lambda_{c1,0}$ (in)	$f_{c2,0}$ (GHz)	Dimensions in Inches							At $f=(3)^{1/2}f_{c1,0}$	
				a	b	d	s	t	R_1 (max)	R_2	Atten** (dB/ft)	Power Rating† (kW)
				Bandwidth 2.4:1								
0.175-0.42				29.667	13.795	5.863	7.417	—	—	1.173		
0.267-0.64				19.428	9.034	3.839	4.857	—	—	0.768		
0.42-1.0				12.333	5.737	2.437	3.083	0.125	0.050	0.487		
0.64-1.53				8.100	3.767	1.601	2.025	0.125	0.050	0.320		
0.84-2.0				6.167	2.868	1.219	1.542	0.125	0.050	0.244		
1.5-3.6				3.455	1.607	0.683	0.864	0.080	0.050	0.137		
2.0-4.8				2.590	1.205	0.512	0.648	0.080	0.050	0.102		
3.5-8.2				1.480	0.688	0.292	0.370	0.064	0.030	0.058		
4.75-11.0				1.090	0.506	0.215	0.272	0.050	0.030	0.043		
7.5-18.0				0.691	0.321	0.136	0.173	0.050	0.020	0.027		
11.0-26.5‡				0.471	0.219	0.093	0.118	0.040	0.015	0.019		
18.0-40.0‡				0.288	0.134	0.057	0.072	0.040	0.015	0.011		
				Bandwidth 3.6:1								
0.108-0.39	0.092	128.37	0.401	34.638	14.894	2.904	8.660	—	—	0.581	0.0014	28 830.
0.27-0.97	0.229	51.572	0.999	13.916	5.984	1.167	3.479	—	—	0.233	0.0055	4 653.
0.39-1.4	0.331	35.680	1.444	9.628	4.140	0.807	2.407	0.125	0.050	0.161	0.0097	2 227.
0.97-3.5	0.822	14.367	3.587	3.877	1.667	0.325	0.969	0.080	0.050	0.065	0.0378	361.2
1.4-5.0	1.186	9.958	5.176	2.687	1.155	0.225	0.672	0.080	0.050	0.045	0.0656	173.5
3.5-12.4	2.966	3.982	12.944	1.074	0.462	0.090	0.269	0.050	0.030	0.018	0.259	27.74
5.0-18.0	4.237	2.787	18.490	0.752	0.323	0.063	0.188	0.050	0.020	0.013	0.443	13.59
12.4-40.0‡	10.508	1.124	45.857	0.303	0.130	0.025	0.076	0.040	0.015	0.005	1.730	2.210

* From MIL-HDBK-216, *RF Transmission Lines and Fittings*, 4 January 1962.

** Copper.

† Based on breakdown of air—15000 volts per cm (safety factor of approx 2 at sea level). Corner radii considered.

‡ Fig. 14B in these frequency ranges only.

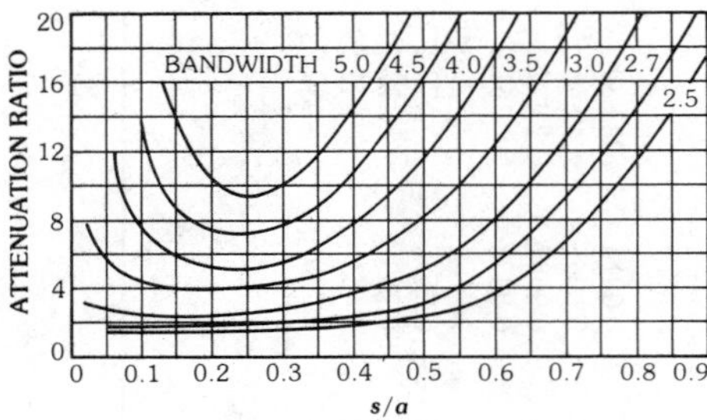

Fig. 20. Attenuation ratio, parametric in bandwidth, double-ridged guide. (*From S. Hopfer, "The Design of Ridged Waveguides,"* Transactions of the I.R.E., *Vol. MTT-3, No. 5, 1955; Fig. 12.*)

Corrugated flexible guide may be made by properly shaping thin-wall seamless rectangular tubing, or by bending and soldering corrugated sheet metal (with due consideration to current flow so that a low-loss joint results).

A bellows-type guide is produced from a group of radial chokes in tandem configuration and made of a flexible alloy.

Vertebral guide is made from a tandem chain of choke-cover sections contained within a neoprene or rubber jacket.

In general, all types except the seamless corrugated waveguide should be jacketed with neoprene or rubber. The unsoldered convoluted, interlocking, and vertebral guide must be jacketed to be pressurized.

Table 7 gives the properties of soldered convoluted flexible waveguide.

The wide variety of manufacturing techniques, the jacketing material and thickness, the length dependence, and other characteristics make it impossible to limit the (±) stretch, twist, and centerline displacement ranges of flexible waveguide. These are usually described in terms of maximum acceptable vswr or loss of the section as a function of the (±) stretch, twist, or displacement, and are best established on advice from the manufacturer selected.

WAVEGUIDE LOSSES

Hollow, enclosed single-conductor waveguides, propagating in the interior space, exhibit losses via dissipation in the waveguide walls and the dielectric material filling the space, leakage through the walls and connections to the guide, and localized power absorption (and heating) at the connections (flanges) because of poor contact or fabrication. The following discussion assumes that the dielectric is air, with zero loss tangent, and that the depth of penetration into the walls is very much less than the wall thickness, so that no appreciable wall leakage occurs.

Waveguide Material and Modes

Fig. 21 shows attenuation as a function of percent conductivity for WR-112 waveguide at 9.0 GHz.* Table 8 relates material composition and percent conductivity.

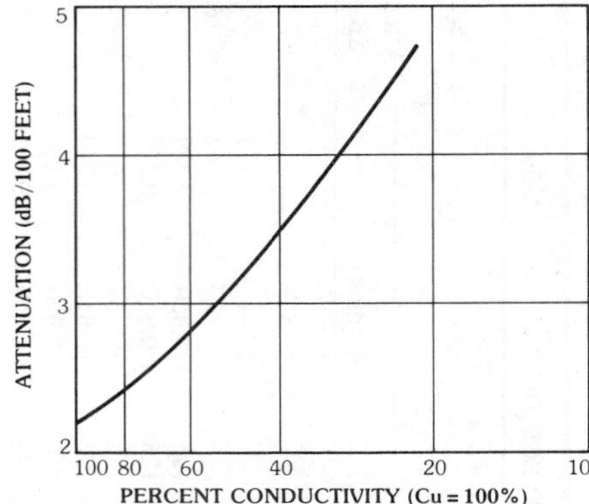

Fig. 21. Attenuation as a function of percent conductivity for WR-112 waveguide at 9.0 GHz.

To obtain the lowest attenuation where a choice of waveguide size and cross section is possible, consideration should be given to mode selection. Fig. 22 shows the relation among various waveguides for different modes and cross sections over a broad frequency band. Fig. 23 relates the attenuation rates of several circular and square guides (in the $TE_{0,1}$, $TE_{1,1}$, and $TE_{1,0}$ modes) over a limited frequency band.*

Waveguide Flange Leakage**

It is extremely difficult to apply quantitative measurements to flange leakage without direct reference to a specific set of measuring procedures, equipment, and test environment. However, in general, measurements of flange fields, made with a probe at the flanges, have indicated that the leakage fields exhibit sharp peaks distributed around the edge of the flange connection. The levels of the peaks are of the order of −130 decibels relative to the guide power, and may be higher or lower depending on the bolt tension and rfi gaskets employed.

* R. M. Cox and W. E. Rupp, "Fight Waveguide Losses 5 Ways." *Microwaves,* Vol. 5, No. 8, August 1966, pp. 32–40.

** S. Galagan. *Electrical Characteristics of Waveguide Seals with EMI Supplement.* Unpublished report prepared for the Parker Seal Co., Culver City, Cal.; January 1964.

TABLE 7. PROPERTIES OF SOLDERED CONVOLUTED FLEXIBLE WAVEGUIDES*

Dimensions (inches)		Minimum Bending Radii (inches)				Equivalent Rectangular Waveguide Type	Weight (lb/ft)	Nominal Attenuation (dB/100 ft)	Nominal Power Rating (MW)	Maximum Operating Pressure (psi)
		Standard Molded Assembly		Unjacketed or Special Molded Assembly						
Inside	Outside	*H* Plane	*E* Plane	*H* Plane	*E* Plane					
6.500×3.250	6.660×3.410	27	13	17	8 1/2	RG-69/U	2.88	0.50	10	15
4.300×2.150	4.460×2.310	18	9	11 1/2	5 3/4	RG-104/U	1.46	0.80	8.0	20
2.840×1.340	3.000×1.500	14	7	9	4 1/2	RG-48/U	0.530	1.50	2.0	30
1.872×0.872	2.000×1.000	8	4	5	2 1/2	RG-49/U	0.332	3.0	1.0	30
1.372×0.622	1.500×0.750	5	2 1/2	3 1/4	1 5/8	RG-50/U	0.266	4.7	0.50	30
1.122×0.497	1.250×0.625	3 1/2	1 3/4	2 1/4	1 1/8	RG-51/U	0.200	5.7	0.40	45
0.900×0.400	1.000×0.500	3	1 1/2	2	1	RG-52/U	0.112	9.0	0.25	60
0.622×0.311	0.702×0.391	3	1 1/2	2	1	RG-91/U	0.085	15.0	0.20	60
0.420×0.170	0.500×0.250	2 1/2	1 1/4	1 1/2	3/4	RG-53/U	0.050	29.0	0.10	60
0.280×0.140	0.360×0.220	2 1/2	1 1/4	1 1/2	3/4	RG-96/U	0.039	35.0	0.05	60

* From MIL-HDBK-216, *RF Transmission Lines and Fittings*, 4 January 1962.

TABLE 8. COMPOSITION AND CONDUCTIVITY OF WAVEGUIDE MATERIAL*

Material	Composition (%) Cu	Zn	P	Ag	Al	Mg	% Conductivity**
Copper (oxygen free)	99.95† min	—	—	—	—	—	97.6 min
Copper DLP (deoxidized, low phosphorus)	99.90† min	—	0.004–0.012	—	—	—	96.1 min
Commercial Bronze	89–91	9–11	—	—	—	—	44.2 min
Silver (fine)	0.08 max	—	—	99.90 min	—	—	100.0 min
Coin Silver	9–10.4	0.06	—	89.6–91.0	—	—	82.0 min
Aluminum 1100	0.2	0.10	—	—	99.0 min	—	59.5 min
Aluminum 6061	0.15–0.40	0.25	—	—	95	0.8–1.2	40.0 min
Magnesium	—	0.05	0.6–1.4	—	2.5–3.5	94.0	37.5‡

* From R. M. Cox and W. E. Rupp, "Fight Waveguide Losses 5 Ways," *Microwaves*, Vol. 5, No. 8, August 1966, p. 34.
** International Annealed Copper Standard.
† Any silver present is counted in the copper content.
‡ MIL-HDBK-216, Military Standardization Handbook, *RF Transmission Lines and Fittings*, 4 January 1962.

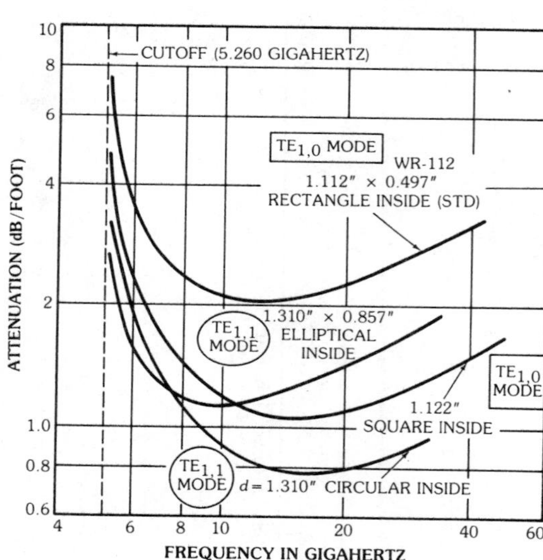

Fig. 22. Attenuation as a function of frequency for various waveguides. (*From R. M. Cox and W. E. Rupp, "Fight Waveguide Losses 5 Ways,"* Microwaves, *Vol. 5, No. 8, August 1966; p. 36.*)

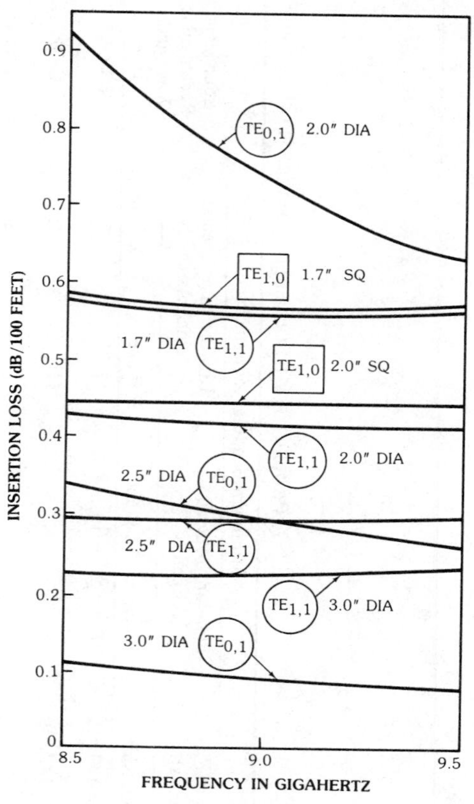

Fig. 23. Attenuation curves for various square and circular waveguides in range 8.5–9.5 GHz. (*From R. M. Cox and W. E. Rupp, "Fight Waveguide Losses 5 Ways,"* Microwaves, *Vol. 5, No. 8, August 1966; p. 37.*)

Flange Resistance and Bolt Torque†

The equivalent series resistance of a die-cast sealed rectangular aluminum *S*-band flange pair as a function of bolt torque (each bolt) is shown in Fig. 24, as an indication of the importance of proper bolt tightening at a flange connection. Table 9 gives the bolt torque for several bolt sizes to meet the recommended value of about 1000 lb/linear

† Handbook Catalog No. W5460, *Waveguide Flange and EMI Sealing*, Culver City, Calif.: Parker Seal Co., Copyright 1967.

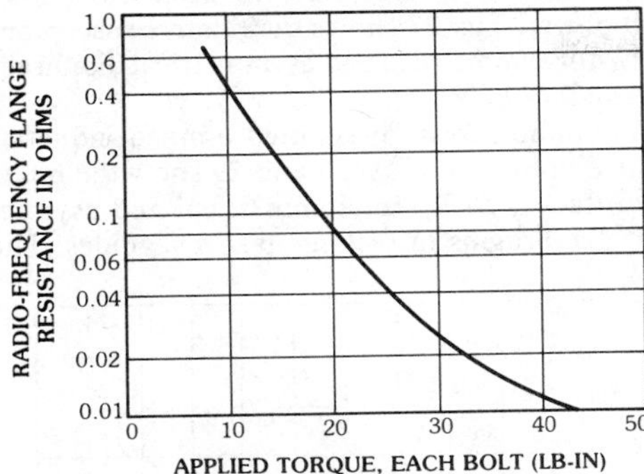

Fig. 24. Typical flange radio-frequency resistance as a function of bolt torque. (*From Handbook Catalog W5460,* Waveguide Flange and EMI Sealing, *Culver City, Calif.: Parker Seal Co., © 1967.*)

inch of flange connection, which is estimated to give a satisfactory waveguide seal for high-power applications.

Fig. 25 shows typical flange resistance as a function of frequency for a UG-53/U plus UG-54/U choke-cover combination, with and without a mesh gasket seal. A combination of two cover flanges, without seals, may have about 10 times the choke-cover resistance.

Flange Insertion Loss**

The relationship between the flange resistance and insertion loss is

$$L(\text{decibels}) = 10 \log(1 - R_F/Z_0)$$

where R_F = radio-frequency flange resistance (measured), and

$$Z_0 = 593\, b/a\,[1-(f_c/f)^2]^{-1/2} \text{ (ohms)}‡$$

The approximate value of flange insertion loss calculated from the above equation may be scaled to other flange sizes and frequencies by

$$R_{F1}/R_{F2} \sim (A_1/A_2)(f_1/f_2)^{1/2}$$

where A = flange area, f = frequency, and the subscripts refer to the two conditions.

Losses and Noise Temperature

In radio telescopes, satellite ground antennas, and other loss-sensitive waveguide systems, the noise-temperature contributions must be controlled and accounted for in the system design. The waveguide losses may be converted to noise temperature* by

$$T = T_R(1 - a)$$

where,

T = temperature in kelvins,
T_R = temperature of the lossy insert in kelvins,
a = power transmission coefficient.

For A (decibels) = $-10 \log_{10} a$

$$T = T_R(1 - 10^{-A/10})$$

TABLE 9. RECOMMENDED TORQUE TABLE*

Screw Size	Threads per Inch	Recommended Torque (lb-in)	Tension** (lb)
No. 4	40	4.5	235
	80	5.5	280
No. 6	32	8.5	360
	40	10	410
No. 8	32	18	625
	36	20	685
No. 10	24	23	705
	32	32	940
1/4"	20	80	1800
	28	100	2200
5/16"	18	140	2540
	24	150	2620
3/8"	16	250	3740
	24	275	3950
7/16"	14	400	4675
	20	425	4700
1/2"	13	550	6110
	20	575	6140

* From Handbook Catalog No. W5460, *Waveguide Flange and EMI Sealing.* © 1967, Parker Seal Co., Culver City, Cal.
** Tension (lb) = torque (lb-in)/0.2×diameter of bolt (in).

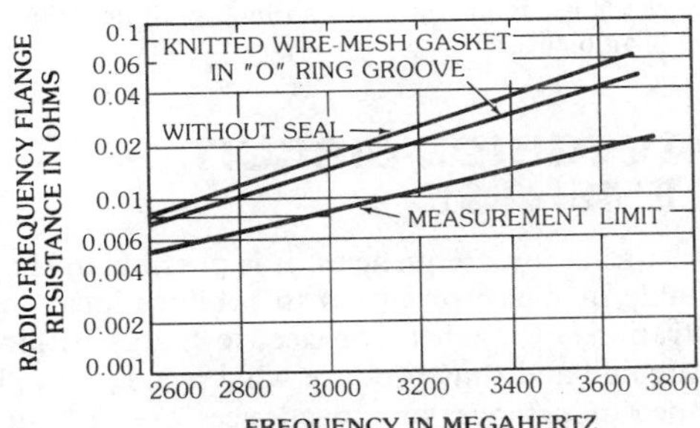

Fig. 25. Typical flange resistance as a function of frequency for choke-cover combination (UG-53/U plus UG-54/U, measured). *From Handbook Catalog W5460,* Waveguide Flange and EMI Sealing, Culver City, Calif.: *Parker Seal Co., © 1967.*)

** S. Galagan. *Electrical Characteristics of Waveguide Seals with EMI Supplement.* Unpublished report prepared for the Parker Seal Co., Culver City, Cal.; January 1964.

‡ This is the characteristic impedance, defined as the maximum transverse voltage divided by the total longitudinal current.

* A. J. Giger, S. Pardee, Jr., and P. R. Wickliffe, Jr., "The Ground Transmitter and Receiver." *Bell System Technical Journal,* Vol. 42, No. 4, Part 1, July 1963, p. 1096.

Fig. 26 relates the noise temperature added to a lossless waveguide system at 290 K by inserting a 290 K pad having A decibels of loss between the measuring point (or input to the receiver) and a 0 K load, as a function of insertion loss A (decibels).

Some typical values of insertion loss measured for a specific configuration are:

23-decibel directional coupler*	0.03 decibel
Flexible waveguide*	0.023 decibel
Flanges (UG-53/U plus UG-53/U)†	0.0017 decibel

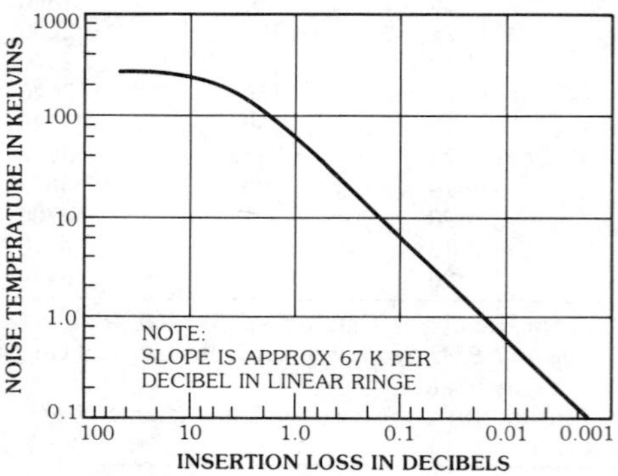

Fig. 26. Noise temperature as a function of insertion loss (290 K ambient).

WAVEGUIDE CIRCUIT ELEMENTS‡

Just as at low frequencies, it is possible to shape metallic or dielectric pieces to produce local concentrations of magnetic or electric energy within a waveguide and thus produce what are, essentially, lumped inductances or capacitances over a limited frequency bandwidth.

This behavior as a lumped element will be evident only at some distance from the obstacle in the guide, since the fields in the immediate vicinity are disturbed.

Capacitive elements are formed from electric-field concentrating devices, such as screws or thin diaphragms inserted partially along electric-field lines. These are susceptible to breakdown under high power. Fig. 27 shows the relative susceptance, B/Y_0, for symmetrical and asymmetrical diaphragms for small b/λ_g.

A common form of shunted lumped inductance is the diaphragm. Figs. 28 and 29 show the relative susceptance, B/Y_0, for symmetrical and asymmetrical diaphragms in rectangular waveguides. These

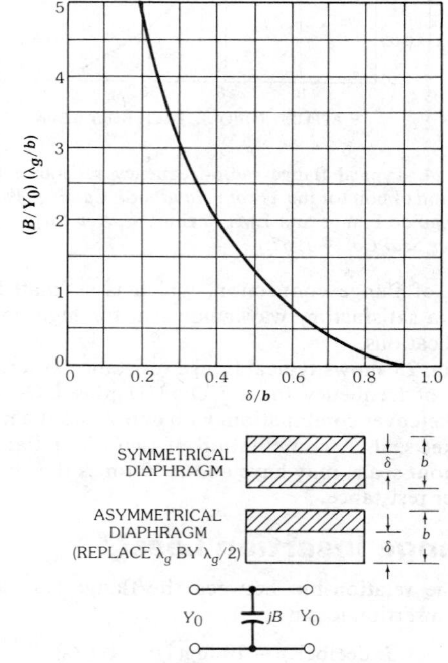

Fig. 27. Normalized susceptance of capacitive diaphragms.

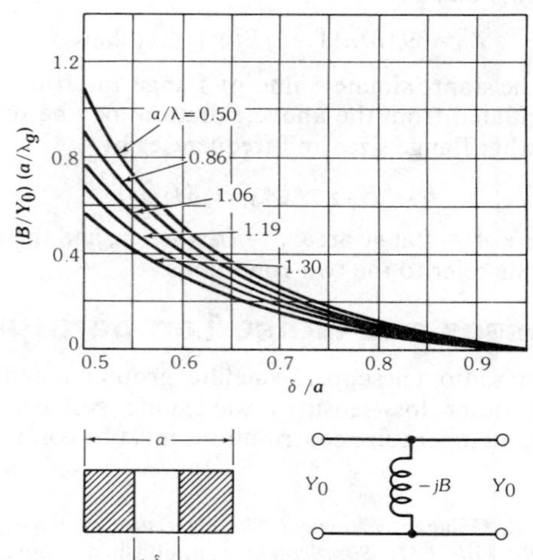

Fig. 28. Normalized susceptance of a symmetrical inductive diaphragm. (*Reprinted from* Microwave Transmission Circuits, *by George L. Ragan, 1st ed., 1948; by permission, McGraw-Hill Book Co., N.Y.*)

---

* A. J. Giger, S. Pardee, Jr., and P. R. Wickliffe, Jr., "The Ground Transmitter and Receiver." *Bell System Technical Journal,* Vol. 42, No. 4, Part 1, July 1963, p. 1096.

† Handbook Catalog No. W5460, *Waveguide Flange and EMI Sealing*, Culver City, Calif.: Parker Seal Co., Copyright 1967.

‡ C. G. Montgomery, R. H. Dicke, and E. M. Purcell. *Principles of Microwave Circuits*. New York: McGraw-Hill Book Co., 1948; Chapters 1 and 6. Also N. Marcuvitz. *Waveguide Handbook*. New York: McGraw-Hill Book Co., 1951.

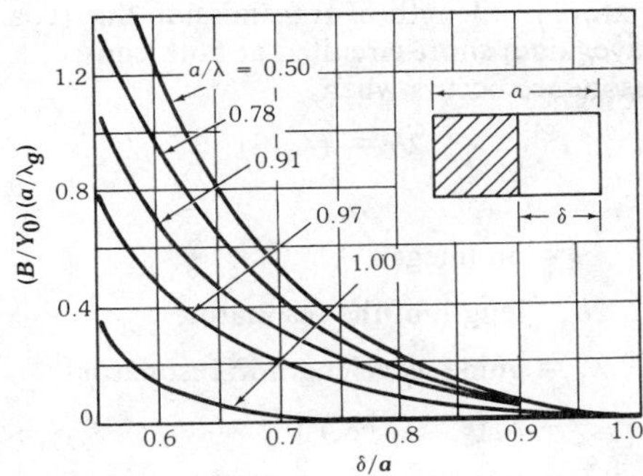

Fig. 29. Normalized susceptance of an asymmetrical inductive diaphragm. (*Reprinted from* Microwave Transmission Circuits, *by George L. Ragan, 1st ed., 1948; by permission, McGraw-Hill Book Co., N.Y.*)

are computed for infinitely thin diaphragms. Finite thicknesses result in an increase in B/Y_0.

Another form of shunt inductance that is useful because of mechanical simplicity is a round post completely across the narrow dimension of a rectangular guide (for $TE_{1,0}$ mode). Fig. 30 gives the normalized values of the elements of the equivalent four-terminal network for several post diameters.

Frequency dependence of waveguide susceptances may be given approximately as

$$\text{Inductive} = B/Y_0 \propto \lambda_g$$
$$\text{Capacitive} = B/Y_0 \propto 1/\lambda_g \text{ (distributed)}$$
$$= B/Y_0 \propto \lambda_g/\lambda^2 \text{ (lumped)}$$

Distributed capacitances are found in junctions and slits, whereas tuning screws act as lumped capacitances.

HYBRID JUNCTIONS*

The hybrid junction is illustrated in various forms in Fig. 31. An ideal junction is characterized by the fact that there is no direct coupling between arms 1 and 4 or between 2 and 3. Power flows from 1 to 4 only by virtue of reflections in arms 2 and 3. Thus, if arm 1 is excited, the voltage arriving at arm 4 is

$$E_4 = \tfrac{1}{2}E_1[\Gamma_2 \exp(j2\theta_2) - \Gamma_3 \exp(j2\theta_3)]$$

and the reflected voltage in arm 1 is

$$E_{r1} = \tfrac{1}{2}E_1[\Gamma_2 \exp(j2\theta_2) + \Gamma_3 \exp(j2\theta_3)]$$

where,

E_1 is the amplitude of the incident wave,

Γ_2 and Γ_3 are the reflection coefficients of the terminations of arms 2 and 3,

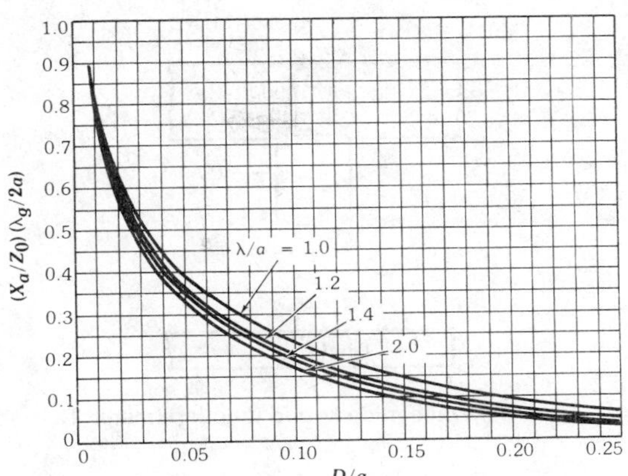

(A) Shunt reactance characteristic.

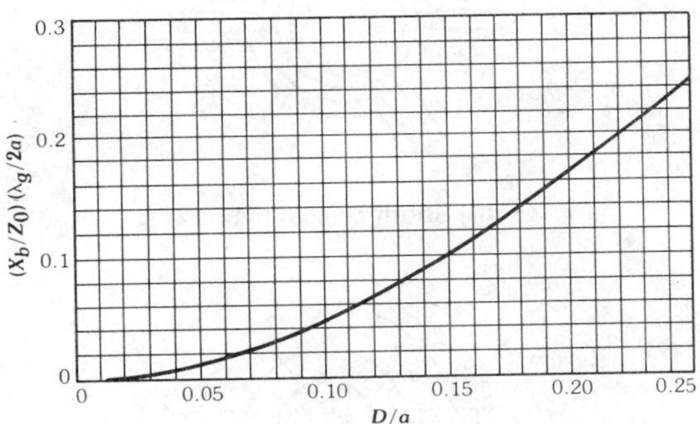

(B) Series reactance characteristic.

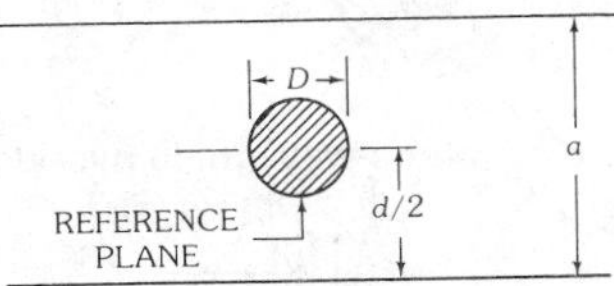

(C) Physical dimensions.

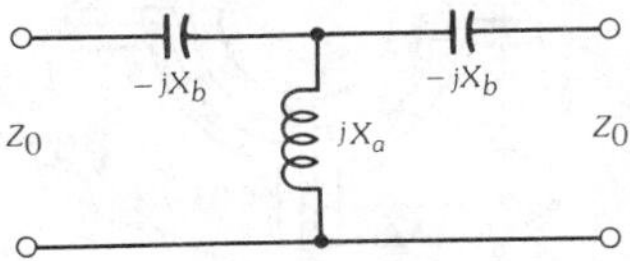

(D) Electrical equivalent.

Fig. 30. Equivalent circuit for inductive cylindrical post.

θ_2 and θ_3 are the respective distances of the terminations from the junctions.

In the case of the rings, θ is the distance between the arm-and-ring junction and the termination.

If the decoupled arms of the hybrid junction are independently matched and the other arms are ter-

* C. G. Montgomery, R. H. Dicke, and E. M. Purcell. *Principles of Microwave Circuits*. New York: McGraw-Hill Book Co., 1948; Chapter 9.

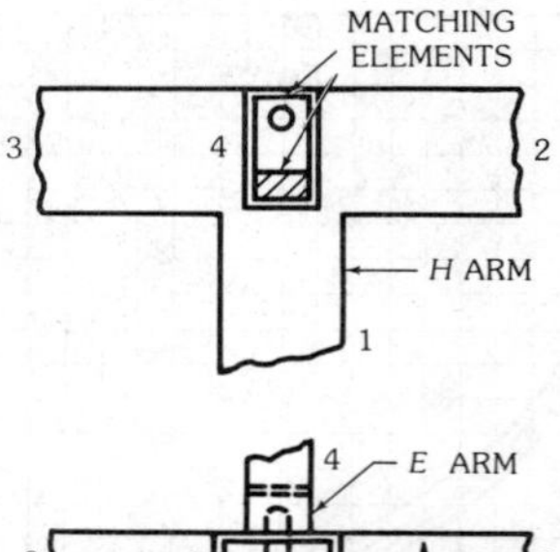

(A) Waveguide hybrid junction (magic T).

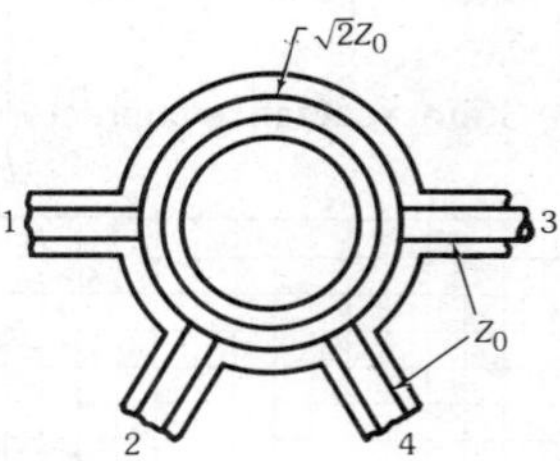

(B) Shunt coaxial hybrid ring.

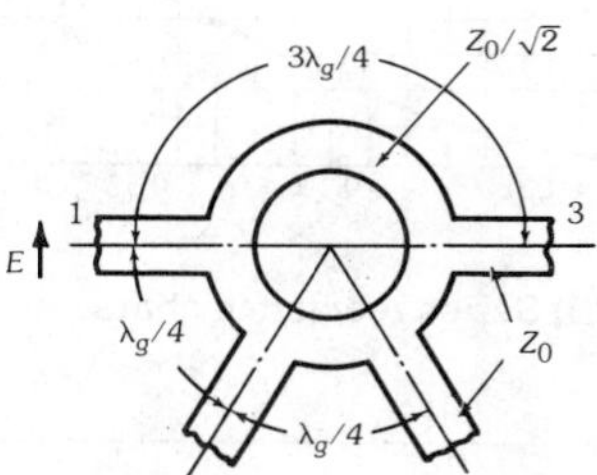

(C) *E*-plane waveguide hybrid ring.

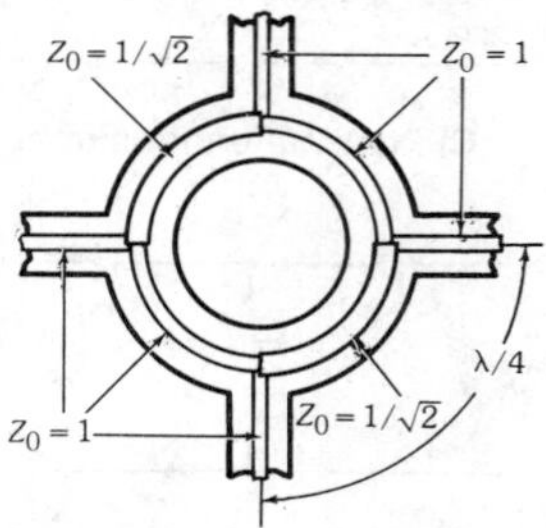

(D) Symmetrical coaxial hybrid.

Fig. 31. Hybrid junctions.

minated in their characteristic impedances, then all four arms are matched at their inputs.

RESONANT CAVITIES

A cavity enclosed by metal walls has an infinite number of natural frequencies at which resonance will occur. One of the more common types of cavity resonators is a length of transmission line (coaxial or waveguide) short-circuited at both ends.

Resonance occurs when

$$2h = l(\lambda_g/2)$$

where,

l = an integer,

$2h$ = length of the resonator,

λ_g = guide wavelength in resonator $= \lambda/[\epsilon - (\lambda/\lambda_c)^2]^{1/2}$

λ = free-space wavelength,

λ_c = guide cutoff wavelength,

ϵ = relative dielectric constant of medium in cavity.

For $TE_{m,n}$ or $TM_{m,n}$ waves in a rectangular cavity with cross section a, b

$$\lambda_c = 2/[(m/a)^2 + (n/b)^2]^{1/2}$$

where m and n are integers.

For $TE_{m,n}$ waves in a cylindrical cavity

$$\lambda_c = 2\pi a/U_{m,n}'$$

where a is the guide radius and $U_{m,n}'$ is the mth root of the equation $J_n'(U) = 0$.

For $TM_{m,n}$ waves in a cylindrical cavity

$$\lambda_c = 2\pi a/U_{m,n}$$

where a is the guide radius and $U_{m,n}$ is the mth root of the equation $J_n(U) = 0$.

For TM waves, $l = 0, 1, 2. \ . \ . \ .$

For TE waves, $l = 1, 2, \ . \ . \ . \ ,$ but not 0.

Rectangular Cavity of Dimensions *a*, *b*, 2*h*

$$\lambda = 2/[(l/2h)^2 + (m/a)^2 + (n/b)^2]^{1/2}$$

where only one of l, m, n may be zero.

Cylindrical Cavities of Radius *a* and Length 2*h*

$$\lambda = 1/[(l/4h)^2 + (1/\lambda_c)^2]^{1/2}$$

where λ_c is the guide cutoff wavelength.

Spherical Resonators of Radius *a*

$$\lambda = 2\pi a/U_{m,n} \text{ for a TE wave}$$

$$\lambda = 2\pi a/U_{m,n}' \text{ for a TM wave}$$

Values of $U_{m,n}$:

$$U_{1,1} = 4.5,\ U_{2,1} = 5.8,\ U_{1,2} = 7.64$$

Values of $U_{m,n}'$:

$$U_{1,1}' = 2.75 = \text{lowest-order root}$$

Additional Cavity Equations

Note that resonant modes are characterized by three subscripts in the mode designations of Table 10, Chart 1, and Fig. 32.

Fig. 32 is a mode chart for a right-circular-cylindrical resonator, showing the distribution of resonant modes with frequency as a function of cavity shape. With the aid of such a chart, the various possible resonances can be predicted as the length ($2h$) of the cavity is varied by a movable piston.

Effect of Temperature and Humidity on Cavity Tuning

The resonant frequency of a cavity changes with temperature and humidity (due to changes in dielectric constant of the atmosphere) and with thermal expansion of the cavity. A homogeneous cavity made of one kind of metal will have a thermal-tuning coefficient equal to the linear coefficient of expansion of the metal (Table 11), since the frequency is inversely proportional to the linear dimension of the cavity.

The relative dielectric constant of air (vacuum =1) is given by

$$k_e = 1 + 210 \times 10^{-6}(P_a/T) + 180 \times 10^{-6}[1 + (5580/T)](P_w/T)$$

where P_a and P_w are partial pressures of air and water vapor in millimeters of mercury and T is the absolute temperature. Fig. 33 is a nomograph showing change of cavity tuning relative to conditions at 25 degrees Celsius and 60 percent relative humidity (expansion is not included).

Coupling to Cavities and Loaded Q

Near resonance, a cavity may be represented as a simple shunt-resonant circuit, characterized by a loaded $Q = Q_l$, where $1/Q_l = (1/Q_0) + (1/Q_{ext})$, Q_0 is the unloaded Q characteristic of the cavity itself, and $1/Q_{ext}$ is the loading due to the external circuits. The variation of Q_{ext} with size of the coupling is approximately as given in Table 12.

Equations for Coupling Through a Cavity

Table 13 summarizes some of the useful relationships in a four-terminal cavity (transmission type) for three conditions of coupling: matched input (input resistance at resonance equals Z_0 of input line), equal coupling ($1/Q_{in} = 1/Q_{out}$), and matched output (resistance seen looking into output terminals at resonance equals output-load resistance). A matched generator is assumed.

In the table, g_c' is the apparent conductance of the cavity at resonance, with no output load; the transmission T is the ratio of the actual output-circuit power delivered to the available power from the matched generator. The loaded Q is Q_l and unloaded Q is Q_0.

Cavity Coupling Techniques*

To couple power into or out of a resonant cavity, either waveguide or coaxial loops, probes, or apertures may be used.

The essentially inductive loop (a certain amount of electric-field coupling exists) is inserted in the resonator at a desired point where it can couple to a strong magnetic field. The degree of coupling may be controlled by rotating the loop so that more or less loop area links this field. For a fixed location of the loop, the loaded Q of a loop-coupled coaxial resonator varies as the square of the effective loop area and inversely as the square of the distance of the loop center from the resonator axis of revolution.

The off-resonance input impedance of the loop is low, a feature that sometimes is helpful in series connections.

The capacitive probe is inserted in the resonator at a point where it is parallel to and can couple to strong electric fields. The degree of coupling is controlled by adjusting the length of the probe relative to the electric field.

*C. G. Montgomery, R. H. Dicke, and E. M. Purcell. *Principles of Microwave Circuits.* New York: McGraw-Hill Book Co., 1948; Chapter 7.

TABLE 10. EQUATIONS FOR A RIGHT-CIRCULAR-CYLINDRICAL CAVITY

Mode	λ_0 Resonant Wavelength	Q (all dimensions in same units)
$TM_{0,1,1}$ (E_0)	$\dfrac{4}{[(1/h)^2 + (2.35/a^2)]^{1/2}}$	$(\lambda_0/\delta)(a/\lambda_0)[1 + (a/2h)]^{-1}$
$TE_{0,1,1}$ (H_0)	$\dfrac{4}{[(1/h)^2 + (5.93/a^2)]^{1/2}}$	$(\lambda_0/\delta)(a/\lambda_0)\left[\dfrac{1 + 0.168\,(a/h)^2}{1 + 0.168\,(a/h)^3}\right]$
$TE_{1,1,1}$ (H_1)	$\dfrac{4}{[(1/h)^2 + (1.17/a^2)]^{1/2}}$	$(\lambda_0/\delta)(h/\lambda_0)\left[\dfrac{2.39h^2 + 1.73a^2}{3.39\,(h^3/a) + 0.73ah + 1.73a^2}\right]$

CHART 1. CHARACTERISTICS OF VARIOUS TYPES OF RESONATORS

Square Prism $TE_{1,0,1}$

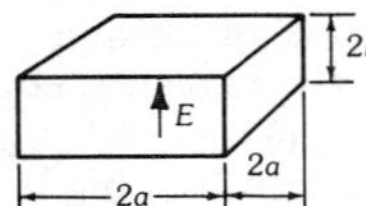

$\lambda_0 = 2.83a$
$Q = (0.353\lambda/\delta)[1+(0.177\lambda/h)]^{-1}$

Circular Cylinder $TM_{0,1,0}$

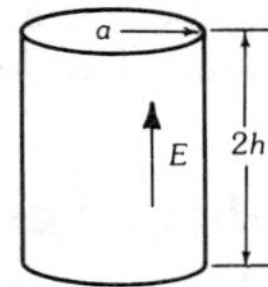

$\lambda_0 = 2.61a$
$Q = (0.383\lambda/\delta)[1+(0.192\lambda/h)]^{-1}$

Sphere

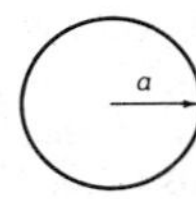

$\lambda_0 = 2.28a$
$Q = 0.318(\lambda/\delta)$

Sphere With Cones

$\lambda_0 = 4a$
Optimum Q for $\theta = 34°$
$Q = 0.1095(\lambda/\delta)$

Coaxial TEM

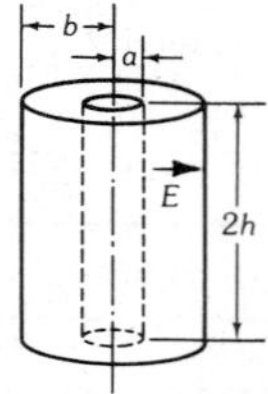

$\lambda_0 = 4h$
Optimum Q for $(b/a) = 3.6$
$(Z_0 = 77 \text{ ohms})$
$\lambda/[4\delta+7.2(h\delta/b)]$

Skin depth in meters $= \delta = (10^7/2\pi\omega\sigma)^{1/2}$, where σ = conductivity of wall in mhos/meter and $\omega = 2\pi \times$ frequency.

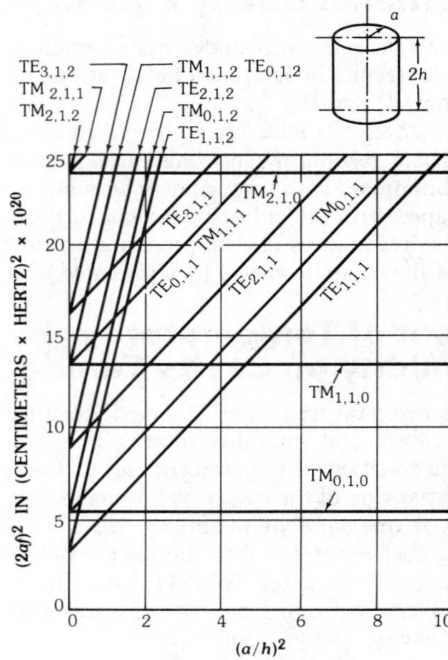

Fig. 32. Mode chart for right-circular-cylindrical cavity. (*Reprinted from* Techniques of Microwave Measurements, *by Carol G. Montgomery, 1st ed., 1947; by permission, McGraw-Hill Book Co., N.Y.*)

TABLE 11. LINEAR COEFFICIENTS OF EXPANSION

Metal	Linear Coefficient of Expansion/°C
Yellow brass	20×10^{-6}
Copper	17.6×10^{-6}
Mild steel	12×10^{-6}
Invar	1.1×10^{-6}

The off-resonance input impedance of the probe-coupled resonator is high; this property is useful in parallel connections.

Aperture coupling is suitable when coupling waveguides to resonators or in coupling resonators together. In this case, the aperture must be located and shaped to excite the proper propagating modes.

For all means of coupling, the input impedance at resonance and the loaded Q may be adjusted by proper selection of the point of coupling and the degree of coupling.

Simple Waveguide Cavity*

A cavity may be made by enclosing a section of waveguide between a pair of large shunt susceptances, as shown in Fig. 34. Its loaded Q is given by

* G. L. Ragan. *Microwave Transmission Circuits.* New York: McGraw-Hill Book Co., 1948; Chapter 10.

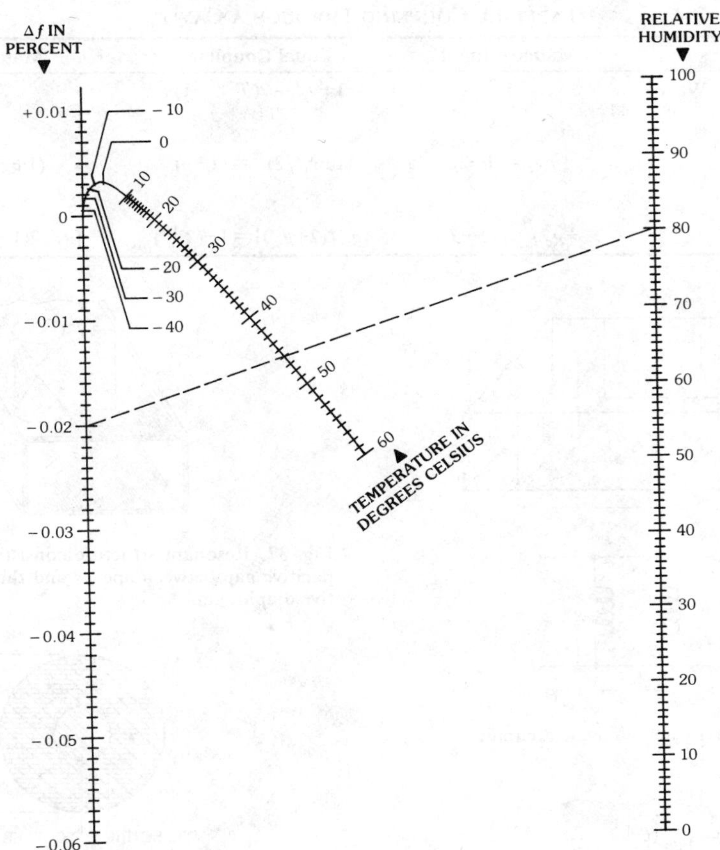

Fig. 33. Effect of temperature and humidity on cavity tuning. (*Reprinted from* Techniques of Microwave Measurements, *by Carol G. Montgomery, 1st ed., 1947; by permission, McGraw-Hill Book Co., N.Y.*)

TABLE 12. VARIATION OF Q_{ext}

Coupling	$1/Q_{ext}$ is Proportional to
Small round hole	$(\text{Diameter})^6$
Symmetrical inductive diaphragm	$(\delta)^4$ (See Fig. 28)
Small loop	$(\text{Diameter})^4$

$$Q_l = \tfrac{1}{4}(\lambda_g/\lambda)^2(b^4 + 4b^2)^{1/2}\tan^{-1}(2/b)$$

and the resonant guide wavelength, λ_{g0}, is obtained from

$$2\pi l/\lambda_{g0} = \tan^{-1}(2/b)$$

Resonant Irises

Resonant irises may be used to obtain low values of loaded Q (< 30). The simplest type is shown in Fig. 35. It consists of an inductive diaphragm and a capacitive screw located in the same plane across the waveguide. For $Q_l < 50$, the losses in the resonant circuit may be ignored and

$$1/Q_l \approx 1/Q_{ext}$$

To a good approximation, the loaded Q (matched load and matched generator) is given by

$$Q_l = (B_l/2Y_0)(\lambda_{g0}/\lambda)^2$$

where B_l is the susceptance of the inductive diaphragm. This value may be taken from charts such as Figs. 28 and 29 as a starting point, but because of the proximity of the elements, the susceptance value is modified. Exact Qs must be obtained experimentally. Other resonant structures are given in Figs. 36 and 37. These are often designed so that the capacitive gap will break down under high power levels for use as transmit-receive (tr) switches in radar systems.

SURFACE-WAVE TRANSMISSION LINE*

The surface-wave transmission line is a single-conductor line having a relatively thick dielectric

* G. Goubau. "Designing Surface-Wave Transmission Lines." *Electronics*, Vol. 27, April 1954, pp.180–184.

TABLE 13. COUPLING THROUGH A CAVITY

	Matched Input	Equal Coupling	Matched Output
Input Standing-Wave Ratio	1	$1+g_c' = 2(T^{-1/2}-1)$	$1+2g_c'$
Transmission Ratio $= T$	$1-g_c' = 1-2\rho$	$(1+g_c'/2)^{-2} = (1-\rho)^2$	$(1+g_c')^{-1} = 1-2\rho$
$Q_l/Q_0 = \rho$	$\frac{1}{2}g_c' = \frac{1}{2}(1-T)$	$[g_c'/(2+g_c')] = 1-(T^{1/2})$	$[g_c'/2(1+g_c')] = \frac{1}{2}(1-T)$

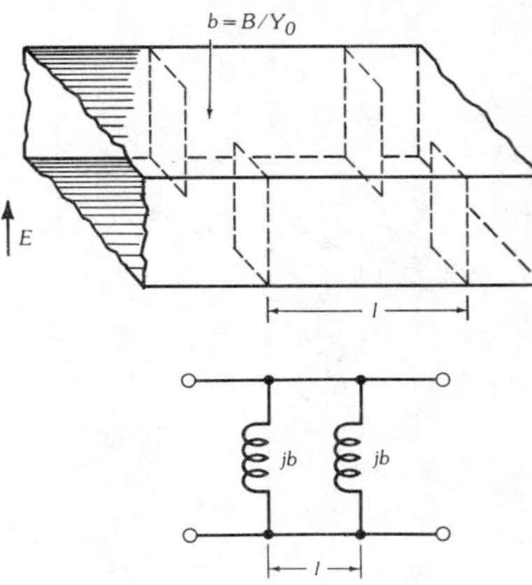

Fig. 34. Waveguide cavity and equivalent circuit.

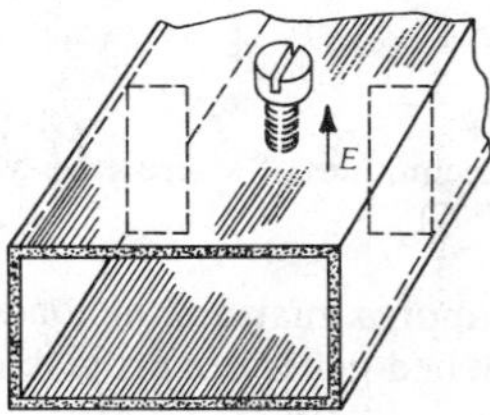

Fig. 35. Resonant iris in waveguide. Capacitive screw is tuned to resonance with inductive diaphragm.

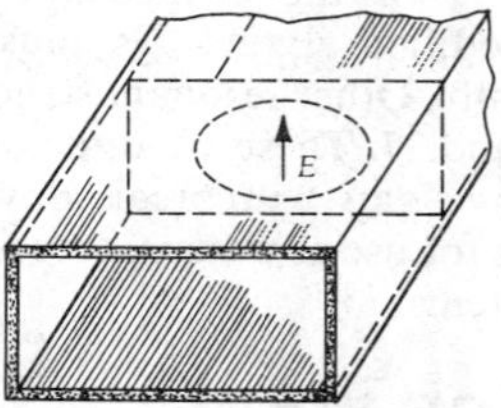

Fig. 36. Resonant element consisting of an oblong aperture in a thin transverse diaphragm.

sheath (Fig. 38). The sheath diameter is often three or more times the conductor diameter. A mode of propagation that is practically nonradiating is excited on the line by means of a conical horn at

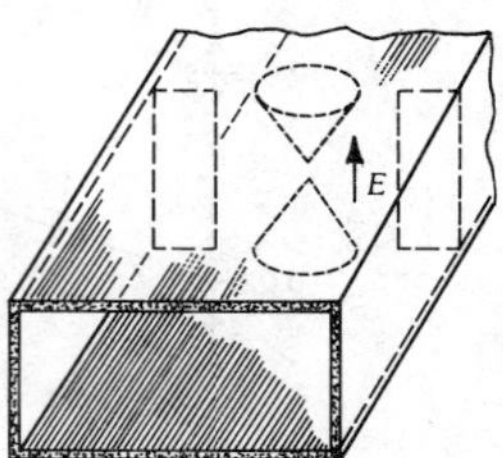

Fig. 37. Resonant structure consisting of cones with capacitive gap between apexes and thin symmetrical inductive diaphragm.

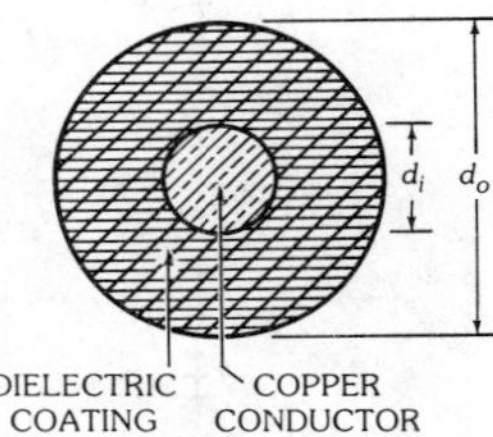

Fig. 38. Cross section of surface-wave transmission line.

each end as shown in Fig. 39. The mouth of the horn is roughly one-quarter to one wavelength in diameter. Losses are about half those of a two-wire line, but the surface-wave line has a practical lower frequency limit of about 50 megahertz. Design charts are given in Figs. 40, 41, and 42.

Fig. 39. Surface-wave transmission line with launchers at each end. (*Courtesy* Electronics.)

The losses in the two launchers combined vary from less than 0.5 decibel to a little more than 1.0 decibel, according to their design.

Conductor loss L_c by the equation below is 5 percent over the theoretical value for pure copper. Dielectric loss L_p for polyethylene at 100 megahertz is shown in Fig. 41. For other dielectrics and frequencies, find L_i by the equation.

$$L_c = 0.455 f^{1/2}/Zd_i \text{ decibels/100 feet}$$

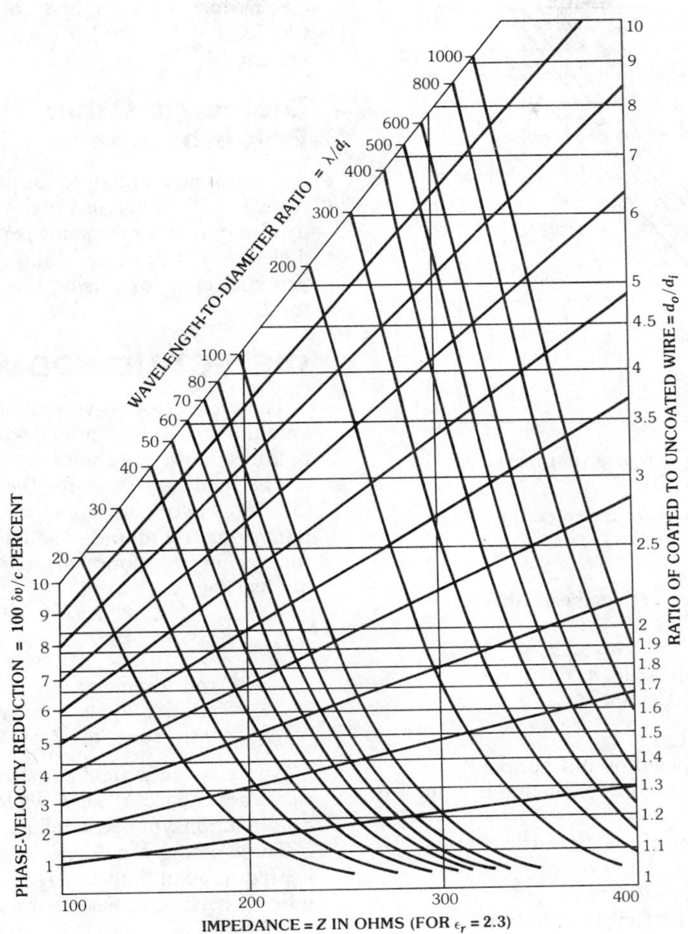

Fig. 40. Relationship among wire diameter, dielectric layer, phase-velocity reduction, and impedance (for brown polyethylene). (*Courtesy* Electronics.)

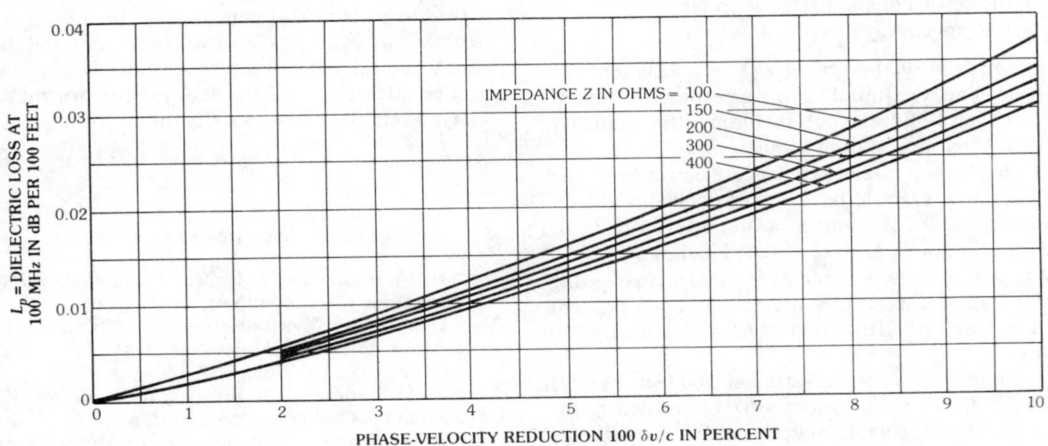

Fig. 41. Dielectric loss at 100 MHz for brown polyethylene ($\epsilon_r = 2.3$ and $F_p = 5 \times 10^{-4}$). (*Courtesy* Electronics.)

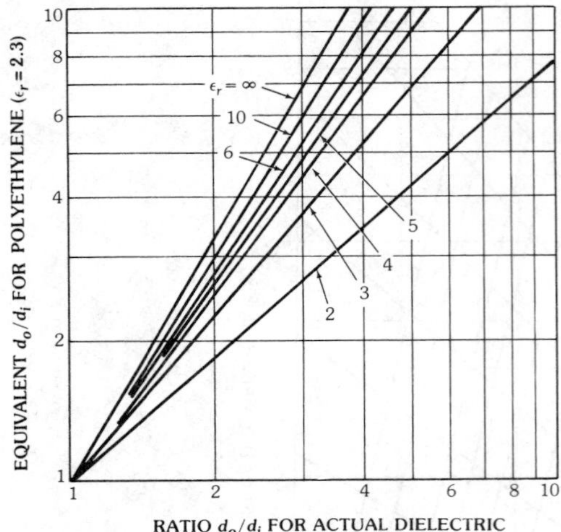

Fig. 42. Conversion chart for dielectric other than polyethylene. (*Courtesy* Electronics.)

$$L_i = 26fF_pL_p/(\epsilon_r - 1) \text{ decibels}/100 \text{ feet}$$

$$L_i = L_pf/100$$

for brown polyethylene (Fig. 41).

Symbols

c = velocity of propagation in free space
d_i = diameter of the conductor (inches in equation for L_c)
d_o = outside diameter of the dielectric coating
f = frequency in megahertz
F_p = power factor of dielectric
L_c = conductor loss in decibels/100 feet
L_i = dielectric loss in decibels/100 feet
L_p = dielectric loss shown in Fig. 41
Z = waveguide impedance in ohms
δv = reduction in phase velocity
ϵ_r = dielectric constant relative to air
λ = free-space wavelength

Example: At 900 megahertz ($\lambda = 0.333$ meter), a 200-foot line is required having a permissible loss of 1.0 decibel/100 feet (not including the launcher losses). What are its dimensions?

If 20 percent is allowed for dielectric loss, the conductor loss would be $L_c = 0.8$ decibel/100 feet. Assuming $Z = 250$ ohms as a first approximation, the formula for L_c gives $d_i = 0.068$ inch. Use No. 14 AWG wire ($d_i = 0.064$ and $\lambda/d_i = 204$). Now going to Fig. 40 and assuming that $100\delta v/c = 6$ percent is adequate, we find that $d_o/d_i = 3$ and $Z = 270$ ohms.

Recomputing, $L_c = 0.79$ decibel/100 feet. By Fig. 41, $L_p = 0.017$ at 100 megahertz for brown polyethylene. For the same material at 900 megahertz, the loss is $L_i = 0.15$ decibel/100 feet.

For 200 feet, the combined conductor and dielectric loss is 1.9 decibels, to which must be added the loss of 0.5 to 1.0 decibel total for the two launchers.

Dielectric Other Than Polyethylene

Determine Z and $\delta v/c$ for polyethylene ($\epsilon_r = 2.3$) from Fig. 40. Then use Fig. 42 to find the value of d_o/d_i required for the same performance with actual dielectric constant ϵ_r. Make a computation of the new dielectric loss, using Fig. 41 and the equation for L_i.

DIELECTRIC-ROD WAVEGUIDES

The dielectric-rod waveguide has applications in antenna structures, laser devices, fiber optics, and millimetric-wave techniques.

The field structures for the nonradiating modes fall into two classes—circularly symmetric and nonsymmetric modes. The cutoff wavelengths (λ_c) for symmetric modes are (for the $E_{0,m}$ and $H_{0,m}$ modes)**

$$\lambda_c = \pi d(\epsilon - 1)^{1/2}/j_{0,m}$$

where,

d = rod diameter,
ϵ = relative dielectric constant,
$j_{0,m}$ = mth root of $J_0(X)$.

Analysis of the field equations reveals the necessary coexistence of an E wave with an H wave to obtain a nonsymmetric field structure.†

These modes are described as HE if the H mode is predominant, and as EH if the E mode predominates. The special case of the $HE_{1,1}$ mode, referred to as the "dipole" mode because of the resemblance of the transverse-electric-field pattern to that of the electrostatic dipole, is of special interest because it has zero cutoff frequency.‡

Fig. 43 describes the relation between λ/λ_0 and d/λ_0 for rods of different ϵ (λ = operating wavelength, λ_0 = free-space wavelength, d = rod diameter), and the field structure is shown in Fig. 44.

The attenuation of the $HE_{1,1}$ mode, for material having relatively low loss,* is found from

$$\alpha = 27.3(\epsilon/\lambda_0)R \tan\delta \text{ (dB/cm)}$$

where,

ϵ = relative dielectric constant,

** H. M. Barlow and J. Brown. *Radio Surface Waves.* Oxford at the Clarendon Press, 1962; p. 71.

† D. G. Kieley. "Dielectric Aerials," *Methuen's Monographs on Physical Subjects.* New York: John Wiley & Sons, Inc., 1953; pp. 7–29.

‡ H. M. Barlow and J. Brown. *Radio Surface Waves.* Oxford at the Clarendon Press, 1962; p. 69.

* W. M. Elsasser. "Attenuation in a Dielectric Rod." *Journal of Applied Physics*, Vol. 20, December 1949, pp. 1193–1196.

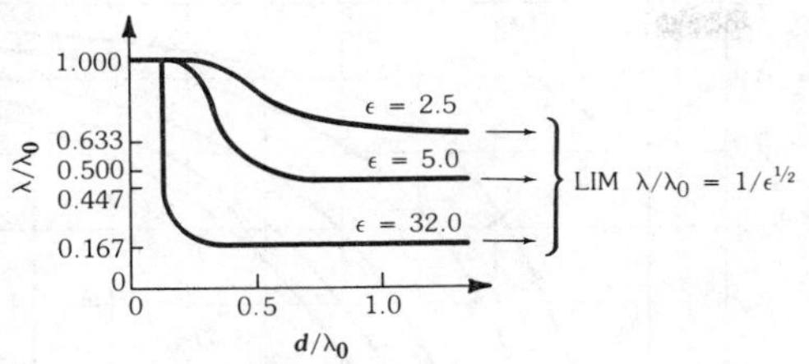

Fig. 43. Wavelength of $HE_{1,1}$ mode as a function of d/λ_0. *(From D. G. Kieley, "Dielectric Aerials,"* Methuen's Monographs on Physical Subjects. *New York: John Wiley & Sons, Inc., 1953; p. 27.)*

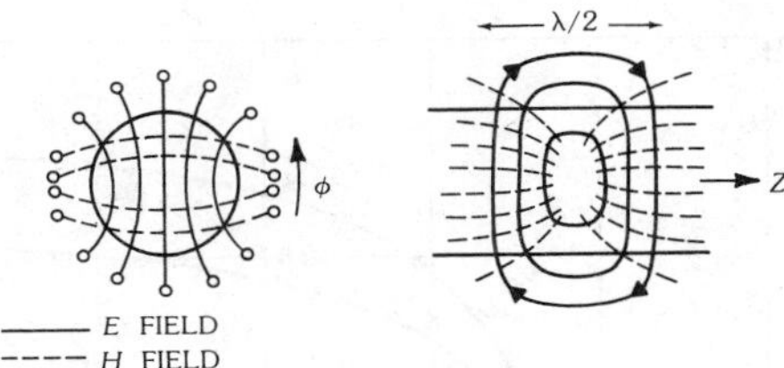

Fig. 44. $HE_{1,1}$ mode, field distribution. *(From D. G. Kieley, "Dielectric Aerials,"* Methuen's Monographs on Physical Subjects. *New York: John Wiley & Sons, Inc., 1953; p. 27.)*

$\tan\delta$ = loss tangent of dielectric,
R = attenuation factor (dimensionless).

Note that this closely resembles the expression for TEM-mode propagation in a low-loss dielectric medium,† given by

$$\alpha = [27.3(\epsilon)^{1/2}/\lambda] \tan\delta \text{ (dB/cm)}$$

For d/λ_0 larger than 0.8, $R \approx [1/(\epsilon)^{1/2}]$. For other values see Elsasser.*

The HE_{11}-mode waveguide is impractical for many frequencies in the uhf and lower bands because of the size of the rod and field spread adjacent to the rod outside of the dielectric. At higher microwave and millimeter-wave frequencies, lack of suitable material prevents use of this rod waveguide for transmission over long distances. Some attempts are being made to use this structure as a flexible waveguide by covering the rod with a cladding dielectric having a lower dielectric constant and further with a lossy jacket. The structure then resembles that of an optical fiber.

The optical fiber is now widely used in practice due to its extremely low propagation loss. More detailed information on optical fibers can be found in Chapter 22.

† G. L. Ragan. *Microwave Transmission Circuits,* 1st ed. New York: McGraw-Hill Book Co., 1948; p. 29.

* W. M. Elsasser. "Attenuation in a Dielectric Rod." *Journal of Applied Physics,* Vol. 20, December 1949, pp. 1193–1196.

RECTANGULAR DIELECTRIC GUIDES AND IMAGE GUIDES

Rectangular dielectric waveguides (Fig. 45) find use in integrated optics and in millimeter-wave integrated circuits. In the latter case, image guides (Fig. 46) are more frequently used. Modes are classified into E^y_{pq} and E^x_{pq}. The former has a principal E field in the y direction and the latter in the x direction. The subscripts p and q indicate the number of transverse field maxima in the x and y directions.

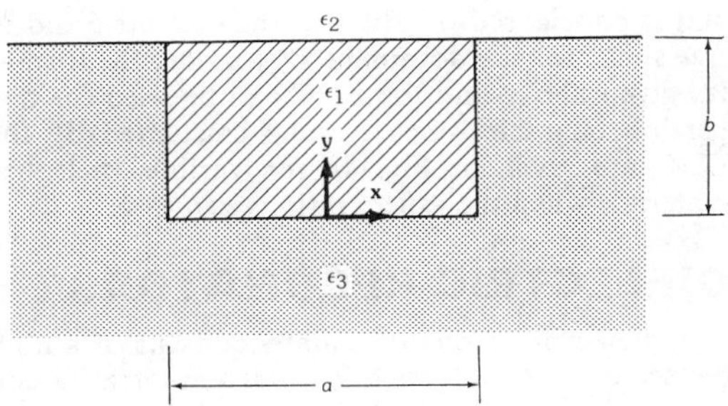

Fig. 45. Cross section of rectangular dielectric waveguide.

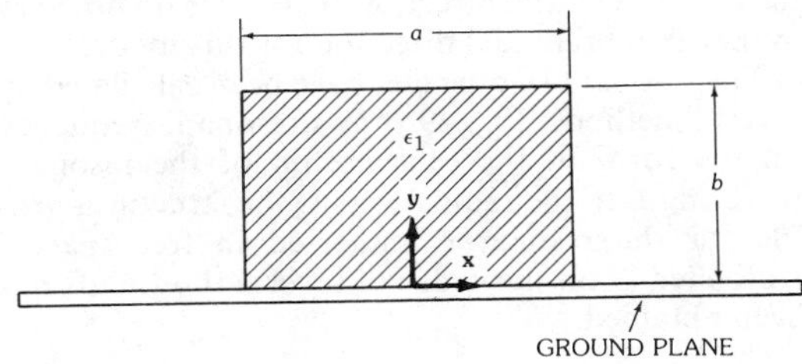

Fig. 46. Cross section of image guide.

For a well-guided mode, approximate expressions for these modes in the guide in Fig. 45 are*

$$\beta = \{\epsilon_1 k^2 - (p\pi/a)^2/(1 + 2A_3/\pi a)^2 - (q\pi/b)^2/[1 + (\epsilon_2 A_2 + \epsilon_3 A_3)/\epsilon_1 \pi b]^2\}^{1/2}$$

for the E^y_{pq} mode; $p, q = 1, 2 \ldots$

$$\beta = \{\epsilon_1 k^2 - (p\pi/a)^2/(1 + 2\epsilon_3 A_3/\epsilon_1 \pi a)^2 - (q\pi/b)^2/[1 + (A_2 + A_3)/\pi b]^2\}^{1/2}$$

for the E^x_{pq} mode; $p, q = 1, 2 \ldots$

where,

$$A_2 = \lambda_0/2\sqrt{\epsilon_1 - \epsilon_2}$$
$$A_3 = \lambda_0/2\sqrt{\epsilon_1 - \epsilon_3}$$

* E. A. J. Marcatili. "Dielectric Rectangular Waveguide and Directional Coupler for Integrated Optics." *Bell Syst. Tech. J.,* Vol. 48, 1969, pp. 2071–2102.

λ_0 is the free space wavelength,

ϵ_1, ϵ_2, and ϵ_3 are the relative dielectric constants of the materials involved.

Dispersion characteristics are provided for a rectangular guide created in a glass substrate in normalized form (Fig. 47). These curves were derived with the above equations.

The above formulas are not very accurate near the cutoff frequencies. More accurate data are obtainable by using a number of techniques, the simplest of which is the effective dielectric constant (EDC) method.**

As the image guide is usually surrounded by air and is considered one half of the dielectric rod in free space for the dominant E^y_{11} mode, the formulas given above and the EDC can be used for dispersion characteristics. Results calculated by the EDC are given in Fig. 48 for various dielectric materials when the aspect ratio is $a/b = 2$.

DIELECTRIC RESONATORS

A dielectric resonator usually consists of a high dielectric material such as barium-tetratitanate formed into a pillbox shape. The dominant resonant mode is $TE_{01\delta}$ and has magnetic fields in the radial and axial directions and an electric field in the circumferential direction. The fields do not vary in the circumferential direction for this mode.

The resonant frequencies have been calculated by several methods. Fig. 49 shows resonant frequency (in the form of the relationship of the resonant wavelength to the radius) versus the structural profile for the resonator immersed in free space.* Unloaded Q on the order of several thousands has been obtained.†

SLOT LINES

Slot line consists of a narrow gap (or slot) in a conductive coating on a dielectric substrate, as shown in Fig. 50. The slot line normally uses a dielectric substrate of sufficiently high permittivity (e.g., $\epsilon_r = 16$) that the guide wavelength is much smaller than the free-space wavelength, and the fields are closely confined to the slot with negligible radiation loss. The nature of the slot-mode configuration is such that the electric field extends across the slot while the magnetic field is in a plane perpendicular to the slot and forms closed loops at half-wave intervals.

Since the slot mode is a non-TEM wave, the def-

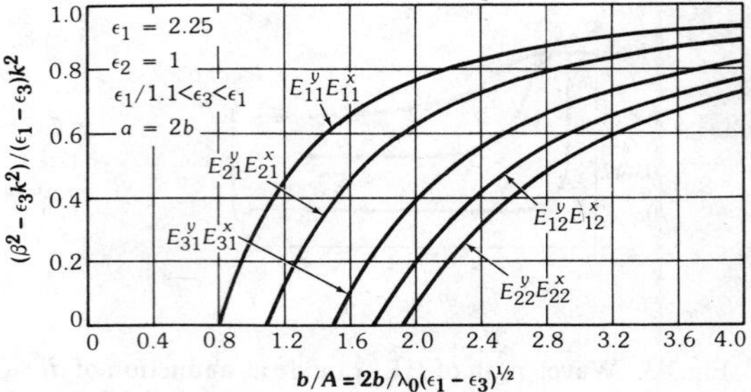

Fig. 47. Approximated dispersion characteristics of a rectangular dielectric waveguide.

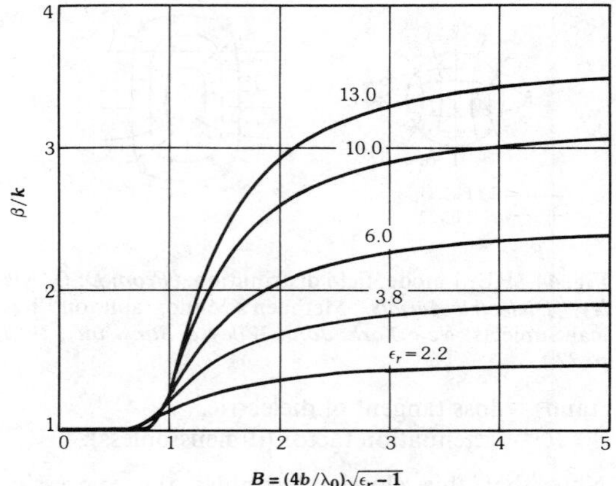

Fig. 48. Dispersion characteristics of image guide.

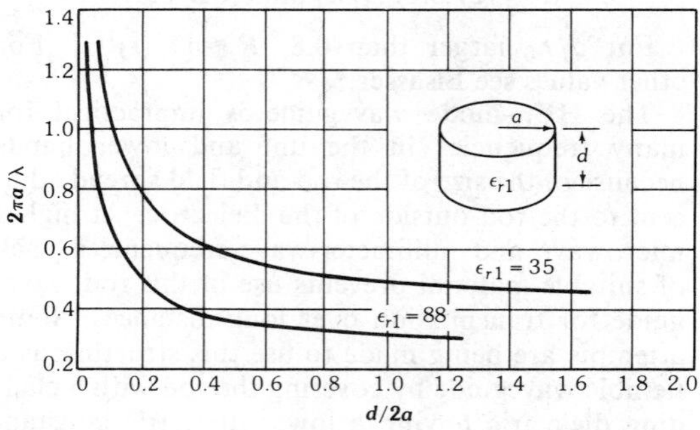

Fig. 49. Resonant frequency of the $TE_{01\delta}$ mode of a pillbox dielectric resonator. (*After T. Itoh and R. Rudokus, "New Method for Computing the Resonant Frequencies of Dielectric Resonators."* IEEE Trans. Microwave Theory and Techniques, *Vol. MTT-25, Jan. 1977; pp. 32-54.*)

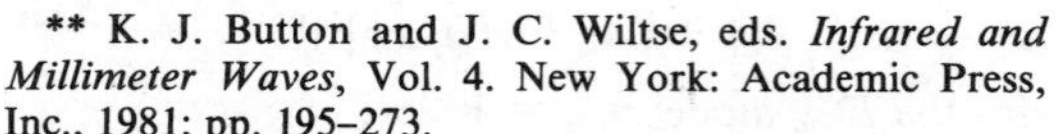

** K. J. Button and J. C. Wiltse, eds. *Infrared and Millimeter Waves*, Vol. 4. New York: Academic Press, Inc., 1981; pp. 195–273.

* Calculated by the method reported by T. Itoh and R. Rudokus in "New Method for Computing the Resonant Frequencies of Dielectric Resonators," *IEEE Trans. Microwave Theory and Techniques*, Vol. MTT-25, Jan. 1977, pp. 32–54.

† For the resonant frequencies and Q of the dielectric resonator placed in a microwave integrated circuit configuration, see M. Dydyk, "Apply High-Q Resonators to mm-Wave Microstrip," *Microwaves*, Vol. 19, Dec. 1980, pp. 62–63.

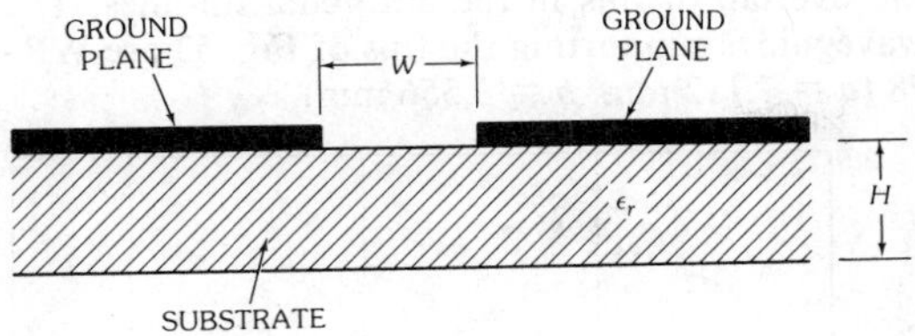

Fig. 50. Cross section of slot line.

inition of characteristic impedance is not unique. One definition could be

$$Z_0 = V^2/2P$$

where,

V is the peak voltage amplitude across the slot,
P is the average power flow of the wave.

For the intervals given by $9.6 \leq \epsilon_r \leq 20$, $0.02 \leq W/H \leq 2.0$, and $0.015 \leq H/\lambda \leq 0.08$ and an infinitesimally thin ground plane, a closed-form approximation to λ_g/λ (where λ_g is the guide wavelength and λ is the free-space wavelength) can be expressed as *

$$\lambda_g/\lambda = f_1(\epsilon_r)\{f_2(W/h) + f_3(W/h)(H/\lambda)^{f_4(W/H)} + f_5(W/H)\}$$

$$f_1(\epsilon_r) = 3.549\,\epsilon_r^{-0.56}$$

$$f_2(W/H) = 0.5632\,(W/H)^{0.104(W/H)^{0.266}}$$

$$f_3(W/H) = -0.8777\,(W/H)^{0.81} + 0.4233\,(W/H) - 0.2492$$

$$f_4(W/H) = -1.269 \times 10^{-2}\,[\ln(50\,W/H)]^{1.7} + 0.0674\ln(50\,W/H) + 0.20$$

$$f_5(W/H) = 1.906 \times 10^{-3}\,[\ln(50\,W/H)]^{2.9} - 7.203 \times 10^{3}\ln(50\,W/H) + 0.1223$$

The accuracy of the expression is $\pm$ 3.7%

A closed-form approximation for Z_0 is

$$Z_0 = (11/\epsilon_r)^{p(W/H,\,H/\lambda)}\,g(W/H,\,H/\lambda)$$

$$p(W/H,H/\lambda) = [-30.21\ln(W/H) - 46.03](H/\lambda)^2 + [0.5073\ln(W/H) + 3.358\,(W/H) + 6.492]\,(H/\lambda) - 2.013 \times 10^{-2}\ln(W/H) - 0.1374(W/H) + 0.2365$$

$$g(W/H,H/\lambda) = [-1.176 \times 10^4\,(W/H)^{0.502} - 6.311\times10^3\,(W/H) - 162.7]\,(H/\lambda)^2 + [900.5(W/H)^{0.28} + 1262(W/H) - 123.8]\,(H/\lambda) + 1.637\ln(50\,W/H) + 40.99(W/H)^{0.46} + 30.96$$

Over the same parameter ranges as above, the accuracy of Z_0 is ±14.5%. For $W/H \geq 0.2$, it is better than 4%. Data calculated by these formulas are plotted in Fig. 51.

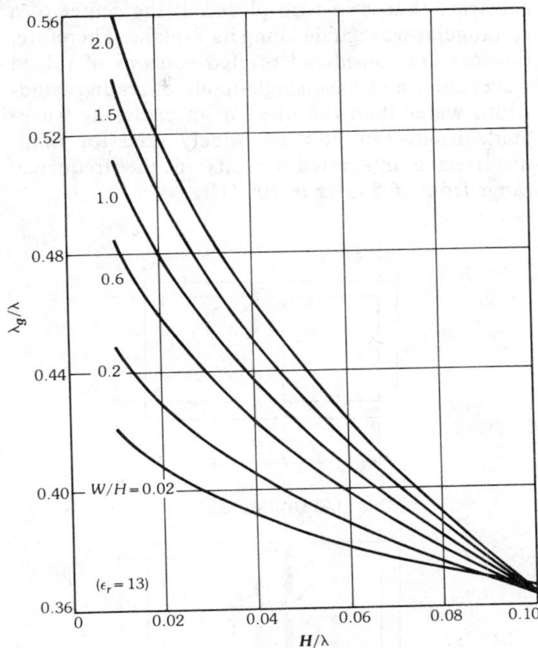

(A) Relative guide wavelength.

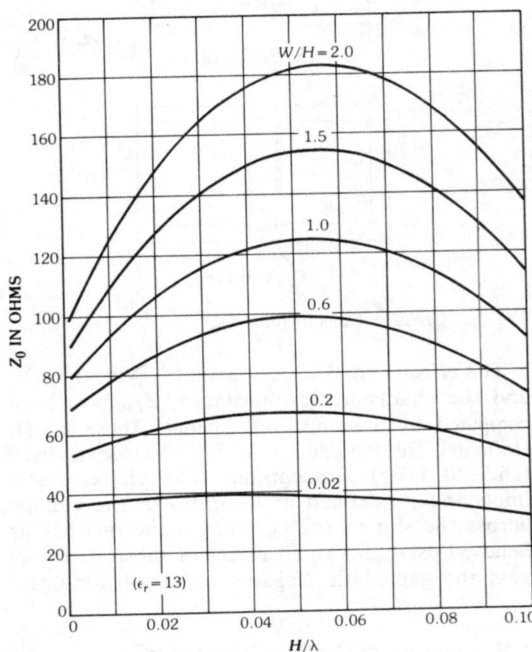

(B) Characteristic impedance.

Fig. 51. Characteristics of slot line.

---

* C. M. Krowne. "Approximations to Hybrid Mode Slot Line Behavior." *Electronics Letters*, Vol. 14, 13 April 1978, pp. 258–259.

FIN-LINES

Fin-lines (Fig. 52) consist of fins separated by a gap printed on one or both sides of a dielectric substrate that is in turn placed at the center of a rectangular waveguide along its *E*-plane. Therefore, fin-lines are considered printed versions of ridged waveguides and have single-mode operating bandwidth wider than the one for an enclosing waveguide itself. Fin-lines are widely used for millimeter-wave integrated circuits in the frequency range from 26.5 GHz to 100 GHz.

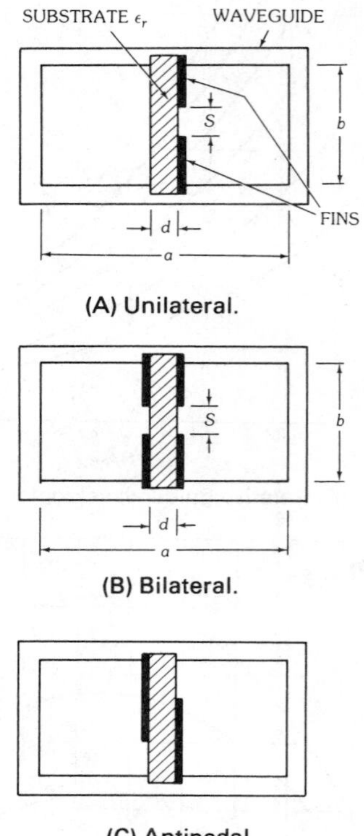

Fig. 52. Cross sections of fin-lines.

The effective dielectric constant, $\epsilon_{eff} = (\lambda/\lambda_g)^2$, and the characteristic impedance, Z_L, have been computed by a number of authors. Those by H. Hofmann are reproduced in Fig. 53 for K_a-band (26.5–40 GHz) application.* The characteristic impedance is defined as the ratio of the voltage across the slot to the current on the fins and is believed useful for small values of substrate thickness and gap width. Negative values of *S* indicate the overlap of fins in the antipodal fin-lines. The waveguides supporting the fins of Fig. 53 are WR-28 ($a = 7.112$ mm, $b = 3.556$ mm).

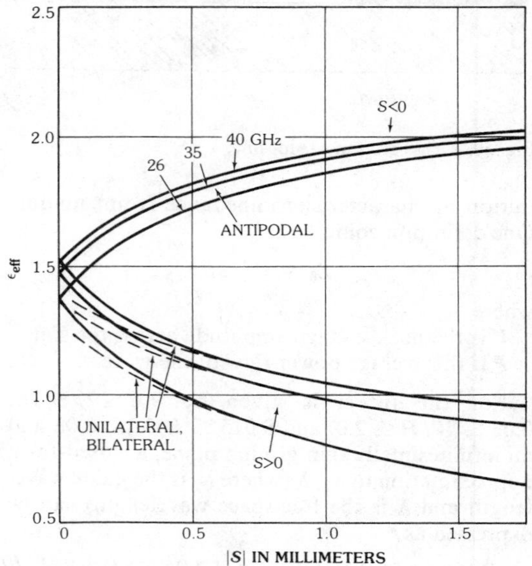

(A) Dielectric constant.

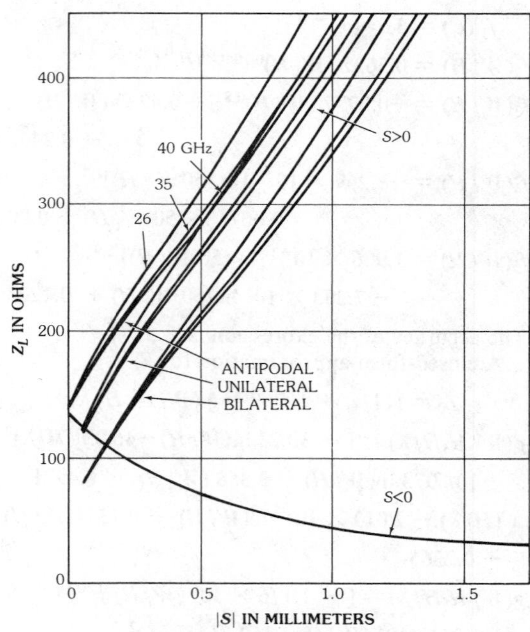

(B) Characteristic impedance.

Fig. 53. Effective dielectric constant, $\epsilon_{eff} = (\lambda/\lambda_g)^2$, and characteristic impedance of fin-lines. (*From Figs. 3 and 4 of H. Hofmann, "Calculation of Quasi-Planar Lines for mm-Wave Application,"* 1977 IEEE MTT-S International Microwave Symposium Digest, *June 1977, pp. 381-384. ©1977 IEEE.*)

---

* H. Hofmann. "Calculation of Quasi-Planar Lines for mm-Wave Application." 1977 IEEE MTT-S International Microwave Symposium Digest, San Diego, June 1977, pp. 381–384.

31 Scattering Matrices

Georges A. Deschamps and John D. Dyson

Microwave structures are characterized by dimensions that are of the order of the wavelength of the propagated signal. The notions of current, voltage, and impedance, useful at lower frequencies, have been successfully extended to these structures, but these quantities are not as directly available for measurement; there are no voltmeters or ammeters and no apparent "terminal pair" between which to connect them. The electromagnetic field itself, distributed throughout a region, becomes the relevant quantity.

Within uniform structures, which are the usual form of waveguides, the *power flow* and the *phase* of the field at a cross section are the quantities of importance. The most usual form of measurement, that of the standing-wave pattern in a slotted section, is easily interpreted in terms of *traveling* waves and gives directly the *reflection coefficient*. The scattering description of waveguide junctions was introduced* to express this point of view. It is not, however, restricted to microwaves; a low-frequency network can be considered as a "waveguide junction" between transmission lines connected to its terminal pairs, and the scattering matrix is a useful complement to the impedance and admittance descriptions.

AMPLITUDE OF A TRAVELING WAVE

In a uniform transmission line a traveling wave is characterized, for a given mode and frequency, by the electromagnetic-field distribution in a transverse cross section and by a propagation constant h. The field in any other cross section, at a distance z in the direction of propagation, has the same pattern but is multiplied by $\exp(-jhz)$. A wave propagating in the opposite direction, for the same mode and frequency, varies with z as $\exp(jhz)$. When losses are negligible, h is real.

The *amplitude* of a traveling wave, at a given cross section in the waveguide, is a complex number a defined as follows. The square $|a|^2$ of the magnitude of a is the power flow,** that is, the integral of the Poynting vector over the waveguide cross section. The phase angle of a is that of the transverse field in the cross section.†

The amplitude of a given traveling wave varies with z as $\exp(-jhz)$.

The wave amplitude has the dimensions of the square root of a power. The meter-kilogram-second unit is therefore the $(\text{watt})^{1/2}$.

REFLECTION COEFFICIENT

Definition

At a cross section in a waveguide, the reflection coefficient W (also often represented by Γ) is the ratio of the amplitudes of the waves traveling respectively in the negative and positive directions.

The positive direction must be specified and is usually taken as toward the load. To give a definite phase to the reflection coefficient, a convention is necessary that describes how the phases of waves traveling in opposite directions are to be compared. The usual convention is to compare in the two waves the phases of the transverse electric-field vectors.‡

For a short-circuit, produced, for instance, by a perfect conducting plane placed across the waveguide, the reflection coefficient is $W = -1$. For an open circuit, it is $W = +1$; and for a matched load, it is $W = 0$.

When the cross section is displaced by z in the positive direction, the reflection coefficient W becomes

$$W' = W \exp(2jhz) \qquad (1)$$

Measurement

In a slotted waveguide equipped with a sliding voltage probe,* the position of a maximum is one where the phase of the reflection coefficient is zero.

The ratio of the maximum to the minimum (the standing-wave ratio, or swr) is

$$(\text{swr}) = (1 + |W|)/(1 - |W|)$$

Therefore

$$W = [(\text{swr}) - 1]/[(\text{swr}) + 1] \qquad (2)$$

is the value of W at the position of a maximum. At the position of a minimum, which is easier to locate in practice, the reflection coefficient is $[1 - (\text{swr})]/[1 + (\text{swr})]$.

At any other position, the value of W is obtained by applying Eq. (1). If the reflection coefficient is wanted in some waveguide connected to the slotted section, a good match must obtain at the transition, or a correction must be applied as explained later in problems A and B (pages 31-6 to 31-8).

Reflectometers that give the reflection coefficient

---

* C. G. Montgomery, R. H. Dicke, and E. M. Purcell, *Principles of Microwave Circuits* (New York: McGraw-Hill Book Co., 1948).

** The amplitude is sometimes defined to make the power flow equal to $\frac{1}{2}|a|^2$ rather than to $|a|^2$. This would correspond to the use of peak values instead of root-mean-square values.

† This phase is well defined for a pure mode, since the field has the same phase everywhere in the cross section.

‡ The dual convention, based on the magnetic-field vector, would give the "current" reflection coefficient, equal to minus the "voltage" reflection coefficient. The latter is used almost exclusively, and the "voltage" is implicit.

* A probe that gives a reading proportional to the electric field.

by direct reading, or display it on a Smith chart, are in current use.

SCATTERING MATRIX OF A JUNCTION

Definition

To define accurately the waves incident on a waveguide junction and those reflected (or scattered) from it, some reference locations must be chosen in the waveguides. These locations are called the ports† of the junction. In a waveguide that can support several propagating modes, there should be as many ports as there are modes. (These ports may or may not have the same physical location in the multimode waveguide.)

At each port i of a junction, consider the amplitude a_i of the incident wave traveling toward the junction, and the amplitude b_i of the scattered wave, traveling away from it. As a consequence of Maxwell's equations, there exists a linear relation between the b_i and the a_i. Considering the a_i (where i varies from 1 to n) as the components of a vector $\mathbf{a}$, and the b_i as the components of a vector $\mathbf{b}$, this relation can be expressed by

$$\mathbf{b} = \mathbf{S}\mathbf{a}$$

where $\mathbf{S} = (s_{ij})$ as an $n \times n$ matrix called the *scattering matrix* of the junction.

The s_{ii} is the *reflection coefficient* looking into port i, and s_{ij} is the *transmission coefficient* from j to i, all other ports being terminated in matching impedances.

Properties

For a *reciprocal* junction, the transmission coefficient from i to j equals that from j to i; the matrix S is symmetrical.

$$\mathbf{S} = \tilde{\mathbf{S}}$$

where $\tilde{\mathbf{S}}$ denotes the transpose of S.

The total power incident on the junction is

$$|\mathbf{a}|^2 = \sum_{i=1}^{n} |a_i|^2$$

The total power scattered is

$$|\mathbf{b}|^2 = \sum_{i=1}^{n} |b_i|^2$$

For a lossless junction, these two powers are equal

$$|\mathbf{a}|^2 = |\mathbf{b}|^2$$

† At lower frequencies, for a network connecting transmission lines, a port is a terminal pair.

This implies that the matrix S is unitary (see "Matrix Algebra" in Chapter 46).

$$\mathbf{S}^\dagger = \mathbf{S}^{-1}$$

For a *passive* junction with losses, $|\mathbf{b}|^2 < |\mathbf{a}|^2$; hence the matrix $\mathbf{1} - \mathbf{S}\mathbf{S}^\dagger$ is positive definite.

Change of Terminal Plane

If the port in arm i is moved away from the junction by ϕ_i electrical radians, the scattering matrix becomes

$$\mathbf{S}' = \Phi \mathbf{S} \Phi \tag{3}$$

where

$$\Phi = \begin{bmatrix} e^{-j\phi_1} & 0 & 0 & 0 & \cdots \\ 0 & e^{-j\phi_2} & 0 & 0 & \cdots \\ 0 & 0 & e^{-j\phi_3} & 0 & \cdots \\ \cdot & \cdot & \cdot & \cdot & \cdots \\ \cdot & \cdot & \cdot & \cdot & \cdots \\ \cdot & \cdot & \cdot & \cdot & \cdots \end{bmatrix} \tag{4}$$

TWO-PORT JUNCTIONS

The two-port junction includes the case of an obstacle or discontinuity placed in a waveguide as well as that of two essentially different waveguides connected to each other.

If reciprocity applies, the scattering matrix

$$\mathbf{S} = \begin{bmatrix} s_{11} & s_{12} \\ s_{21} & s_{22} \end{bmatrix} \tag{5}$$

is symmetrical

$$s_{21} = s_{12}$$

For a lossless junction, the scattering coefficients can be expressed by

$$\begin{aligned} s_{11} &= +\tanh(u/2)\exp(-2j\alpha) \\ s_{22} &= -\tanh(u/2)\exp(-2j\beta) \\ s_{12} &= +\operatorname{sech}(u/2)\exp[-j(\alpha+\beta)] \end{aligned} \tag{6}$$

in terms of three parameters, u, α, and β.

This corresponds to the representation of the junction by an ideal transformer with transformer ratio $n = \exp(-u/2)$, of hyperbolic amplitude u, placed between two sections of transmission line with electrical lengths α and β, respectively.

The quantity $-20 \log_{10} |s_{12}|$ is the insertion loss.

TRANSFORMATION MATRIX

To find the effect of successive obstacles in a waveguide or to combine two-port junctions placed in cascade, it is convenient to introduce the wave transformation matrix **T**. This matrix **T** relates the traveling waves on one side of the junction to those on the other side. Using the notations of Fig. 1

$$\begin{bmatrix} A_1 \\ B_1 \end{bmatrix} = \mathrm{T} \begin{bmatrix} A_2 \\ B_2 \end{bmatrix} \qquad (7)$$

The 2×2 transformation matrix **T** may be deduced from the scattering matrix **S**

$$\mathbf{T} = s_{21}{}^{-1} \begin{bmatrix} 1 & -s_{22} \\ s_n & -\det \mathbf{S} \end{bmatrix} \qquad (8)$$

Conversely, if $\mathbf{T} = (t_{ij})$, the scattering matrix is

$$\mathbf{S} = t_{11}{}^{-1} \begin{bmatrix} t_{21} & \det \mathbf{T} \\ 1 & -t_{12} \end{bmatrix} \qquad (9)$$

When reciprocity applies to the junction

$$\det \mathbf{T} = s_{12}/s_{21} \qquad (10)$$

becomes unity.

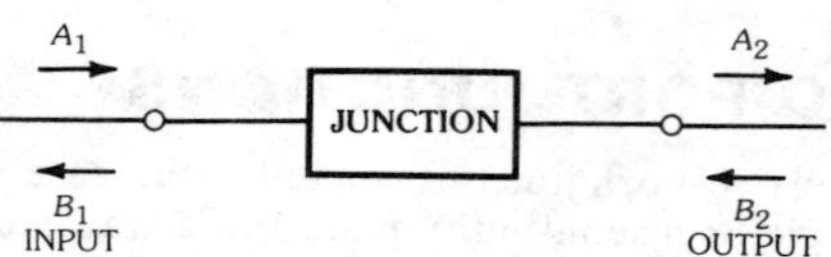

Fig. 1. Convention for wave transformation matrix **T**.

The input reflection coefficient $W' = B_1/A_1$ is related to the load reflection coefficient $W = B_2/A_2$ by

$$W' = (t_{21} + t_{22}W)/(t_{11} + t_{12}W) \qquad (11)$$
$$= s_{11} + [s_{12}{}^2 W/(1 - s_{22}W)] \qquad (12)$$

When a number of junctions, 1, 2, 3, are placed in cascade (Fig. 2), the output port of each of them being the input port of the following one, the resulting junction has the transformation matrix

$$\mathbf{T} = \mathbf{T}_1\mathbf{T}_2\mathbf{T}_3$$

If n similar junctions with transformation matrix **T** are placed in cascade, the resulting transformation matrix is $\mathbf{T}^n$.

Letting trace $\mathbf{T} = t_{11} + t_{22} = 2\cos\theta$

$$\mathbf{T}^n = (\sin n\theta/\sin\theta)\,\mathbf{T} - [\sin(n-1)\,\theta/\sin\theta] \qquad (13)$$

MEASUREMENT OF THE SCATTERING MATRIX

The measurement of the scattering parameters of the junction by conventional techniques is covered

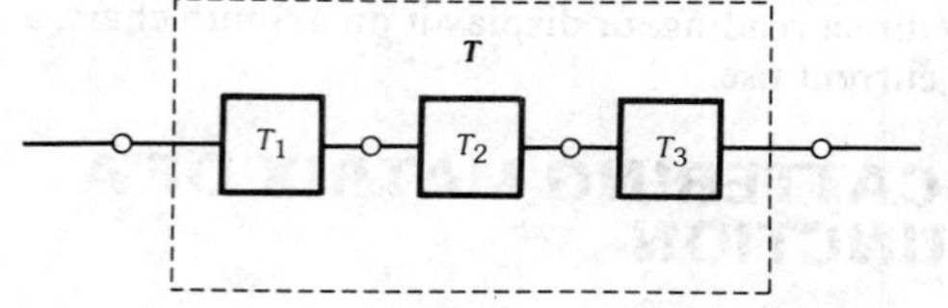

Fig. 2. Junctions in cascade.

in Chapter 12. However, there may be occasions when it is not possible, or not desirable, to connect equipment to the output ports for the measurement of the transmission coefficients s_{ij}. Under these conditions, it is still possible to determine the coefficients s_{ii} and s_{ij} from measurements of the reflection coefficient at one port.

For any load with reflection coefficient W, placed on side 2 of a junction, the input reflection coefficient W' on side 1 can be measured. Coefficient W' is called the *image* of W, and they are related by the transformation

$$W' = s_{11} + [s_{12}s_{21}W/(1 - s_{22}W)] \qquad (14)$$

The measurement of the images of three known loads, W_1, W_2, W_3, will provide sufficient information to determine, either analytically or graphically, the coefficients s_{11} and s_{22} and the product $s_{12}s_{21}$.

Considering an analytical approach first, Eq. 14 can be expressed as:

$$W_i' = s_{11} + W_iW_i'\,s_{22} - W_i\Delta s \qquad (15)$$

where,

$$\Delta s = s_{11}s_{22} - s_{12}s_{21}$$
$$i = 1,\ 2,\ 3$$

Equation 15 represents three linear equations in s_{11}, s_{22}, and Δs with solutions:

$$s_{11} = [W'_1\,W_2\,W_3\,(W'_2 - W'_3) + W'_2W_3W_1\,(W'_3 - W'_1) + W'_3W_1\,W_2(W'_1 - W'_2)]/D$$
$$s_{22} = -[W_1\,(W'_2 - W'_3) + W_2\,(W'_3 - W'_1) + W_3\,(W'_1 - W'_2)]/D$$
$$\Delta s = -[W'_1\,W_1\,(W'_2 - W'_3) + W'_2W_2(W'_3 - W'_1) + W'_3\,W_3\,(W'_1 - W'_2)]/D$$
$$D = [W_2\,W_3(W'_2 - W'_3) + W_3W_1(W'_3 - W'_1) + W_1W_2\,(W'_1 - W'_2)] \qquad (16)$$

Once s_{11}, s_{22}, and $s_{12}s_{22}$ are known, the reflection coefficient W_L of an unknown termination can be determined from the measured image W_M by rewriting Eq. (14):

$$W_L = (W_M - s_{11})/[s_{12}s_{21} + s_{22}(W_M - s_{11})] \qquad (17)$$

Solutions to Eq. 14 can also be obtained graphically.* Doing so provides an insight into the trans-

* G. A. Deschamps, "Determination of the Reflection Coefficients and Insertion Loss of a Waveguide Junction," *Journal of Applied Physics,* Vol. 24, pp. 1046–1050; August 1953: Also, *Electrical Communication,* Vol. 31, pp. 57–62; March 1954.

formation of W into its image W'. The images of various known loads can be plotted on a reflection chart and the scattering coefficients deduced by the following procedures.

(A) With a matched load, one obtains directly s_{11} plotted as O' on Fig. 3. Point O' is called the iconocenter.

(B) With a sliding short-circuit on side 2, or any variable reactive load, the input reflection coefficient describes a circle Γ', image of the unit circle Γ. This circle can be deduced from three or more measurements. Let C be its center and R its radius (Fig. 3). The magnitudes of the scattering coefficients result:

$$\begin{aligned} |s_{11}| &= OO' \\ |s_{22}| &= O'C/R \\ |s_{12}|^2 &= R(1-|s_{22}|^2) \end{aligned} \tag{18}$$

The phases of these coefficients all follow from one more measurement as described in **(C)**.

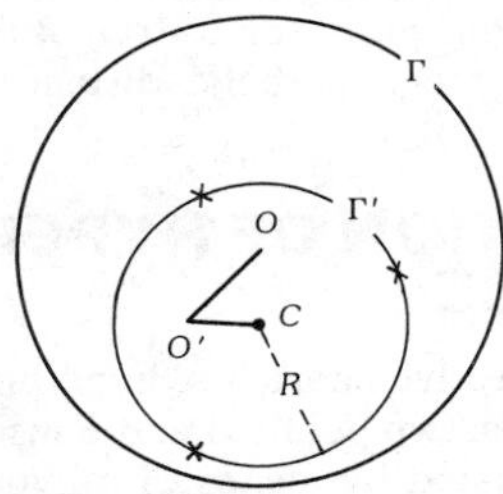

Fig. 3. Construction for the magnitudes of the scattering coefficients.

(C) The input reflection coefficient is measured with an open-circuit load placed at port 2, or for a short-circuit placed a quarter wave away from it. This may be one of the measurements taken in step **(B)**. It gives (Fig. 4) the point P', image of the point $P(W=+1)$.

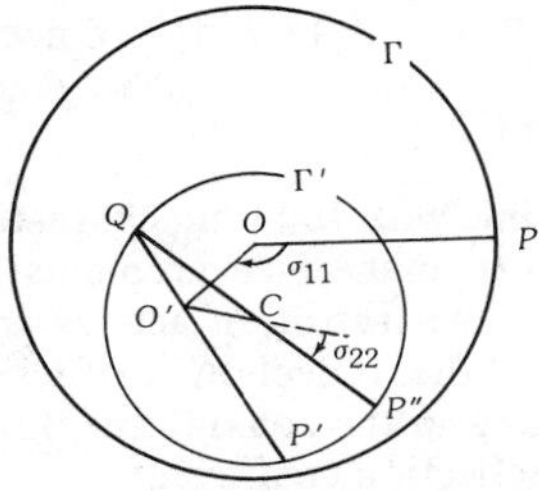

Fig. 4. Construction for the phases of the scattering coefficients.

A point P'' is constructed by projecting P' through O' onto Q on Γ', then Q through C onto P'' on Γ' (Fig. 4). Then

$$\text{Phase of } s_{11} = \text{angle }(OP,\ OO')$$
$$\text{Phase of } s_{22} = \text{angle }(O'C,\ CP'')$$
$$\text{Phase of } s_{12} = \tfrac{1}{2}\text{ angle }(OP,\ CP'') \tag{19}$$

(D) When no matched load is available, as was assumed in **(A)**, the iconocenter O' may be obtained as in Fig. 5. Let P_1, P_2, P_3, P_4 represent the input reflection coefficients when a short circuit is placed successively at port 2 and at distances $\lambda/8$, $\lambda/4$, and $3\lambda/8$ from it. These points define the circle Γ' [as in **(B)**], and the intersection I (the crossover point) of P_1P_3 and P_2P_4 may be used to find O': draw perpendiculars to CI at points C and I up to their intersections with Γ' and get C' and I'; then O' is the intersection of CI and $C'I'$.

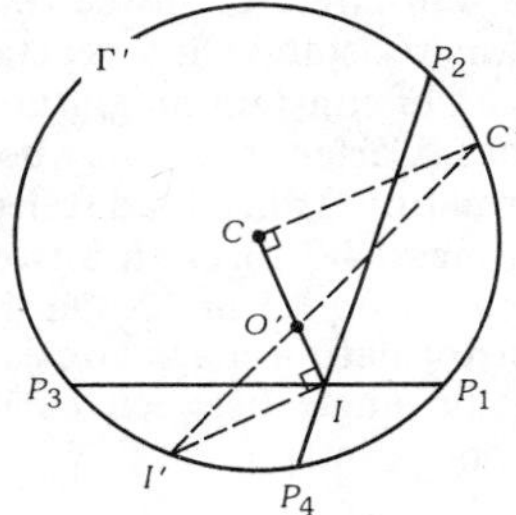

Fig. 5. Determination of O' from four measurements.

The point P_3 is identical to P' in **(C)** above; hence the four measurements give the complete scattering matrix by constructing P'' and applying Eqs. 18 and 19.

(E) The construction of O' in **(D)** above is valid with any sliding load not necessarily reactive. Taking a load with small standing-wave ratio increases the accuracy of the construction.

(F) When exact measurements of the displacements of the sliding load are difficult to make, for instance if the wavelength is very short, the point O' may be obtained as follows. Using a reactive load, construct the circle Γ' as in **(B)** above; then using a sliding load as in **(E)** above, construct a circle Γ'' (see Fig. 6). The iconocenter O' is the hyperbolic midpoint of the diameter of Γ'' (through C) with respect to Γ'. It may be constructed by means of the

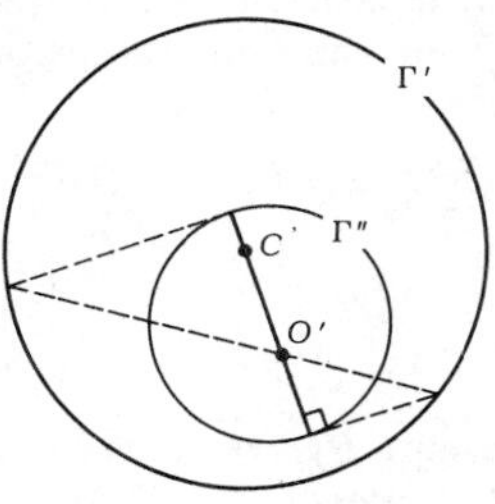

Fig. 6. Use of circles Γ'' and Γ' for determination of O'.

hyperbolic protractor* (Fig. 8) or by means of the dotted-line construction (Fig. 5).

GEOMETRY OF REFLECTION CHARTS

The following brief outline is complemented by the section on hyperbolic trigonometry contained in Chapter 46.

Conformal Charts

A reflection coefficient can be represented by a point in a plane just as any complex number is represented on the Argand diagram.

The passive loads, $|W| \leq 1$, are represented by points inside a unit circle Γ. Inside this circle, the lines of constant resistance and reactance (Smith chart) or the lines of constant magnitude and phase of the impedance (Carter chart) may be drawn.

The transformation from a load reflection coefficient W to its image W' through a two-port junction is bilinear as in Eq. 11 or 12. On the reflection chart, this transformation maps circles into circles and preserves the angle between curves and the cross ratio of four points; if

$$[W_1, W_2, W_3, W_4] = \frac{W_1 - W_3}{W_1 - W_4} : \frac{W_2 - W_3}{W_2 - W_4}$$

denotes the cross ratio of four reflection coefficients W_1, W_2, W_3, and W_4, then

$$[W_1', W_2', W_3', W_4'] = [W_1, W_2, W_3, W_4]$$

The transformation through a lossless junction preserves also the unit circle Γ and therefore leaves invariant the *hyperbolic distance* defined in Chapter 46. The hyperbolic distance to the origin of the chart is the *mismatch*, that is, the standing-wave ratio expressed in decibels. It may be evaluated by means of the proper graduation on the radial arm of the Smith chart. For two arbitrary points W_1, W_2, the hyperbolic distance between them may be interpreted as the mismatch that results from the load W_2 seen through a lossless network that matches W_1 to the input waveguide.

Projective Chart

The reflection coefficient W is represented by the point $\bar{W}$ (Fig. 7) on the same radius of the circle Γ but at a distance

$$O\bar{W} = 2OW/(1 + OW^2) \quad (20)$$

from the origin.

* G. A. Deschamps, *Hyperbolic Protractor for Microwave Impedance Measurements and Other Purposes* (New York: International Telephone and Telegraph Corp., 1953).

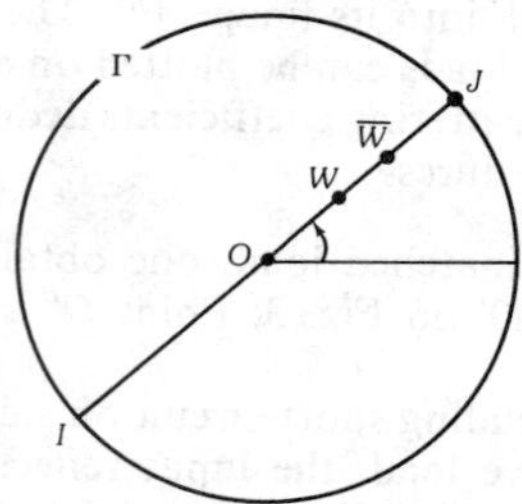

Fig. 7. Representation of a reflection coefficient by W on a Smith chart and $\bar{W}$ on the projective chart.

This is equivalent to using the standing-wave ratio squared instead of the direct ratio:

$$\bar{W}J/\bar{W}I = (WJ/WI)^2 \quad (21)$$

The transformation of Eqs. 11, 12, when the junction is lossless, is represented on this chart by a projective transformation, that is, one that maps straight lines into straight lines and preserves the cross ratio of four points on a straight line. It therefore preserves the hyperbolic distance defined in Chapter 46.

EVALUATION OF HYPERBOLIC DISTANCE

On the projective chart, the hyperbolic distance $\langle AB \rangle$ between two points A and B inside the circle Γ can be evaluated by means of a hyperbolic protractor as shown in Fig. 8. The line AB is extended to its intersections I and J with Γ. The protractor is placed so that the sides OX, OY of the right angle go through I and J. (This can be done in many ways but does not affect the result.) The numbers read on the radial lines of the protractor going through A and B, respectively, are added if A and B are on opposite sides of the radial line marked O; they are subtracted otherwise. This result divided by 2 is the distance $\langle AB \rangle$. In Fig. 8, for instance

$$\langle AB \rangle = \tfrac{1}{2}(12 + 4) = 8 \text{ decibels}$$

Problem A

A slotted line with 100-ohm characteristic impedance is used to make measurements on a 60-ohm coaxial line. The transition acts as an ideal transformer. Find the reflection coefficient W of an obstacle placed in the coaxial line, knowing that it produces a reflection coefficient

$$W' = 0.5 \exp(j\pi/2)$$

in the slotted line.

A match in the coaxial line appears in the slotted line as a normalized impedance of 0.6; hence the mismatch (standing-wave ratio in decibels) is 4.5 decibels. The corresponding point $\bar{O}'$ is plotted on the projective chart as in Fig. 9 at the distance

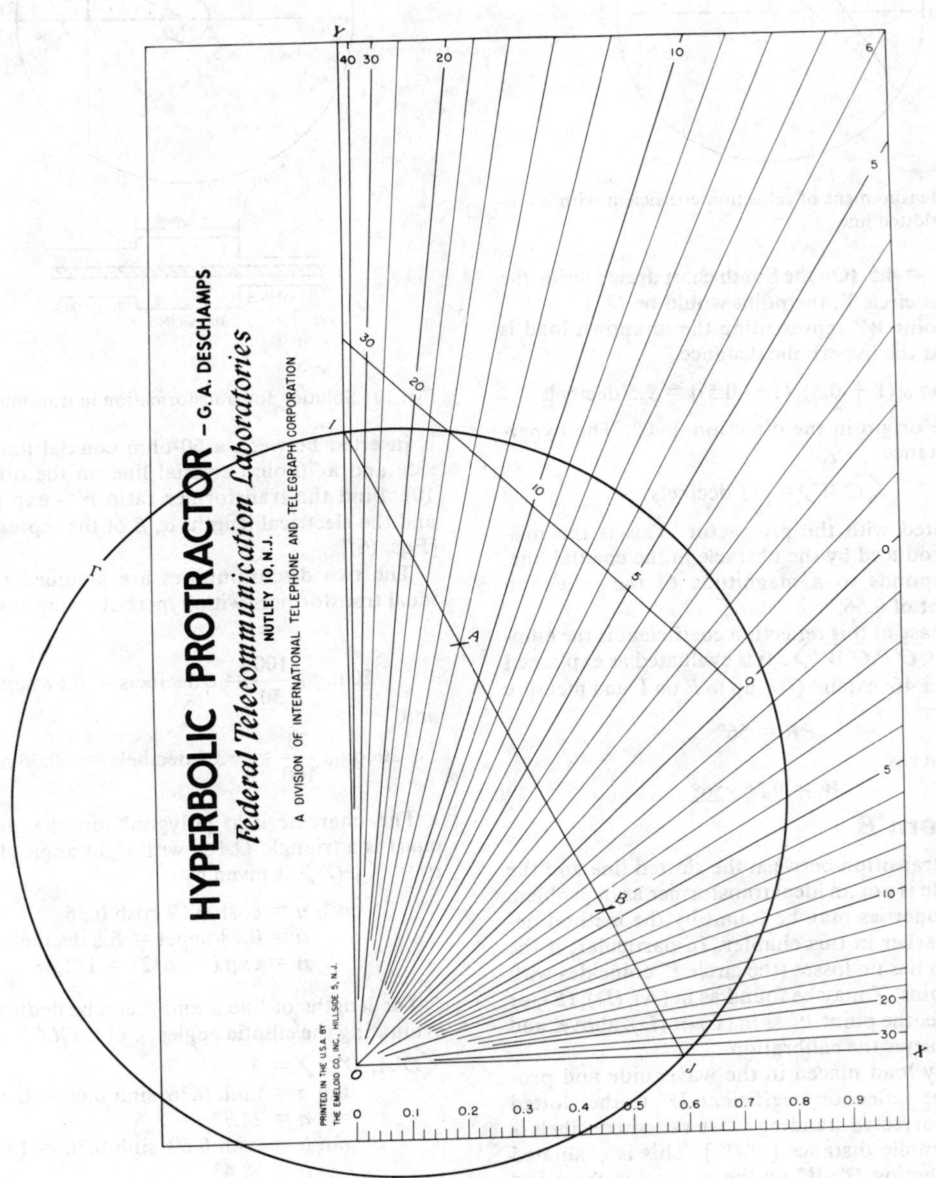

Fig. 8. Definition and evaluation of hyperbolic distance $\langle AB \rangle$ using hyperbolic protractor.

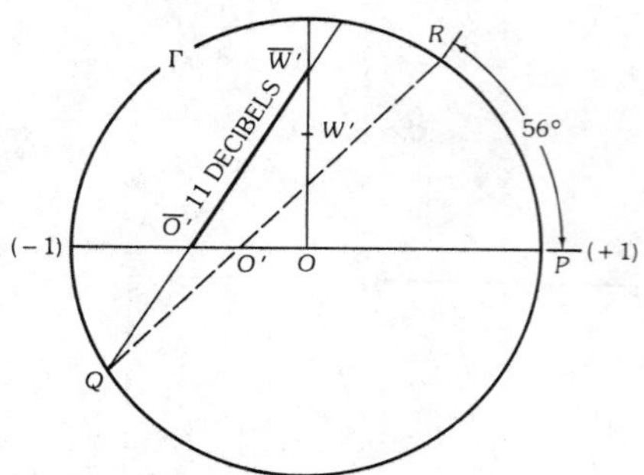

Fig. 9. Measurement of reflection coefficient with a mismatched slotted line.

$\langle O\bar{O}'\rangle = 4.5$. (On the Smith chart drawn inside the same unit circle Γ, the point would be O'.)

The point $\bar{W}'$ representing the unknown load is plotted at the hyperbolic distance

$$20 \log_{10}[(1 + 0.5)/(1 - 0.5)] = 9.5 \text{ decibels}$$

from the origin in the direction +90°. The hyperbolic distance

$$\langle \bar{O}'\bar{W}'\rangle = 11 \text{ decibels}$$

is measured with the protractor. This is the mismatch produced by the obstacle in the coaxial line. It corresponds to a magnitude of the reflection coefficient of 0.56.

The phase of this reflection coefficient is the elliptic angle $\langle \bar{O}'P, \bar{O}'\bar{W}'\rangle$. It is evaluated as explained in Chapter 46: extend QO' up to R on Γ and measure the arc

$$PR = 56°$$

The answer is

$$W = 0.56 \angle 56°$$

Problem B

If the transition between the slotted line and the waveguide is not an ideal transformer as in problem A, its properties may be found by the method described earlier in this chapter. In particular, if the transition has no losses (the circle Γ' coincides with Γ), the point O' may be found as in **(A)**, **(D)**, **(E)**, or **(F)**, above, the point P' as in **(C)** or **(D)**, above, and this completes the calibration.

For any load placed in the waveguide and producing the reflection coefficient W' in the slotted line, the corrected standing-wave ratio in decibels is the hyperbolic distance $[\bar{O}'\bar{W}']$. This is evaluated by constructing $\bar{O}'$, $\bar{W}'$ on the projective chart and measuring $\langle \bar{O}'\bar{W}'\rangle$ with the protractor. The phase angle is the elliptic angle $\langle \bar{O}'P', \bar{O}'\bar{W}'\rangle$ (see Chapter 46).

Problem C

A section of coaxial line 90 electrical degrees in length and with 100-ohm characteristic impedance is inserted between a 50-ohm coaxial line on one side and a 70-ohm coaxial line on the other (Fig. 10). Find the transformer ratio $n = \exp(-u/2)$ and the electrical lengths α, β of the representation [Eqs. (6)].

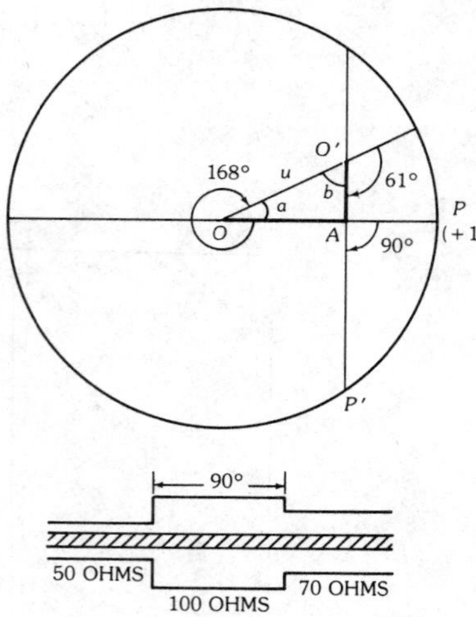

Fig. 10. Solution for transformation in transmission line.

The two discontinuities are assumed to act as ideal transformers with hyperbolic amplitudes

$$20 \log_{10} \frac{100}{50} = 6 \text{ decibels} = 0.67 \text{ neper}$$

and

$$20 \log_{10} \frac{70}{100} = -3.1 \text{ decibels} = -0.36 \text{ neper}$$

The characteristic polygon* on the projective chart is a triangle OAO' with right angle A; hence, $u = \langle OO'\rangle$ is given by

$$\cosh u = \cosh 0.69 \cosh 0.36$$
$$u = 0.78 \text{ neper} = 6.8 \text{ decibels}$$
$$n = \exp(-u/2) = 1/1.48$$

The lengths of line α and β can be deduced from evaluating the elliptic angles $\langle OA, OO'\rangle = a$ and $\langle O'A, O'O\rangle = b$

$$\tan a = \tanh 0.36/\sinh 0.69 = 0.46$$
$$a = 24.7°$$
$$\tan b = \tanh 0.69/\sinh 0.36 = 1.62$$
$$b = 58.4°$$
$$\alpha = \tfrac{1}{2}(360° - 24.7°) = 167.6°$$
$$\beta = \tfrac{1}{2}(180° - 58.4°) = 60.8°$$

* G. A. Deschamps, *Hyperbolic Protractor for Microwave Impedance Measurements and Other Purposes* (New York: International Telephone and Telegraph Corp., 1953; pp. 15–16 and p. 41).

The resulting equivalent network is shown in Fig. 11. It could also have been obtained by geometrical evaluation of the distance $\langle OO' \rangle$ with the hyperbolic protractor and the elliptic angles a and b by constructions (page 31-6 and Chapter 46).

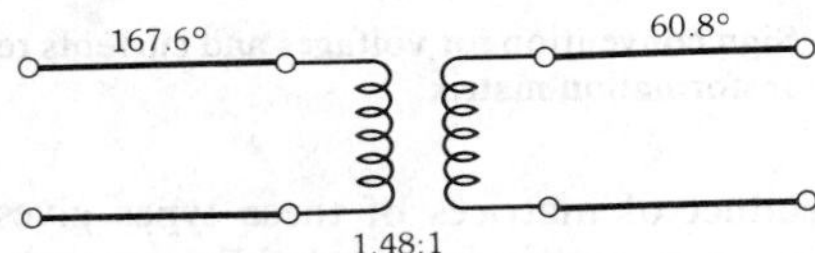

Fig. 11. Equivalent circuit for Fig. 10.

CORRESPONDENCES WITH CURRENT, VOLTAGE, AND IMPEDANCE VIEWPOINTS

Normalized Current and Voltage

In a waveguide, at a point where the amplitudes of the waves traveling in the positive and negative directions are, respectively, a and b, the normalized voltage v and the normalized current i are defined by the expressions

$$\begin{aligned} v &= a + b \\ i &= a - b \end{aligned} \tag{22}$$

The *net power flow* at that point in the positive direction is

$$|a|^2 - |b|^2 = \text{Re } vi^* \tag{23}$$

Current and Voltage Not Normalized

A more general definition for current and voltage becomes possible when a meaning has been assigned to the characteristic impedance Z_0 of the waveguide

$$\begin{aligned} V &= vZ_0{}^{1/2} \\ I &= iY_0{}^{1/2} \end{aligned} \tag{24}$$

where,

$Y_0 = 1/Z_0$ is the characteristic admittance,
v and i are the normalized values defined above.

Conversely, if by some convention the voltage (or the current) has been defined, a characteristic impedance will result from Eqs. (24). This is the case for a two-conductor waveguide supporting the tem mode; the characteristic impedance is the ratio of voltage to current in a traveling wave.

If V and I are the voltage and the current at a point in a waveguide of characteristic impedance $Z_0 = 1/Y_0$, the amplitudes of the waves traveling in both directions at that point are

$$\begin{aligned} a &= \tfrac{1}{2}\,(V\,Y_0{}^{1/2} + IZ_0{}^{1/2}) \\ b &= \tfrac{1}{2}\,(V\,Y_0{}^{1/2} + IZ_0{}^{1/2}) \end{aligned} \tag{25}$$

Normalized Impedance and Admittance

At a point in a waveguide, the normalized impedance is $Z = v/i$, and the normalized admittance is the inverse, $Y = 1/Z$.

They are related to the reflection coefficient $W = b/a$ by

$$\begin{aligned} Z &= (1 + W)/(1 - W) \\ Y &= (1 - W)/(1 + W) \end{aligned} \tag{26}$$

hence

$$W = (1 - Y)/(1 + Y) = (Z - 1)/(Z + 1) \tag{27}$$

Impedance and Admittance Matrix of a Junction

The **Z** and **Y** matrices of a junction are defined in terms of the scattering matrix **S** by

$$\begin{aligned} \mathbf{Y} &= (\mathbf{1} - \mathbf{S})(\mathbf{1} + \mathbf{S})^{-1} \\ \mathbf{Z} &= (\mathbf{1} + \mathbf{S})(\mathbf{1} - \mathbf{S})^{-1} \end{aligned} \tag{28}$$

The matrices **Y** and **Z** do not always exist, since **S** may have eigenvalues + 1 or −1, which means that det (**1** − **S**) or det (**1** + **S**) may be zero.

Conversely

$$\mathbf{S} = (\mathbf{1} - \mathbf{Y})(\mathbf{1} + \mathbf{Y})^{-1} = (\mathbf{Z} - \mathbf{1})\,(\mathbf{Z} + \mathbf{1})^{-1} \tag{29}$$

These equations may be used as definitions for the scattering matrix of lumped-constant networks with n terminal pairs. This is equivalent to considering the network as a junction between n transmission lines of unit characteristic impedance.

If the network or the junction is reciprocal, **Y** and **Z** are purely imaginary.

For a two-port junction, Eq. 28 becomes

$$\begin{aligned} \mathbf{Y} &= (\mathbf{1} - \mathbf{S})(\mathbf{1} + \mathbf{S})^{-1} \\ &= [\det(\mathbf{1} + \mathbf{S})]^{-1} \\ &\times \begin{bmatrix} 1 - \det \mathbf{S} + (s_{22} - s_{11}) & -2s_{12} \\ -2s_{21} & 1 - \det \mathbf{S} - (s_{22} - s_{11}) \end{bmatrix} \end{aligned} \tag{30}$$

and

$$\begin{aligned} \mathbf{Z} &= (\mathbf{1} + \mathbf{S})(\mathbf{1} - \mathbf{S})^{-1} \\ &= [\det(\mathbf{1} - \mathbf{S})]^{-1} \\ &\times \begin{bmatrix} 1 - \det \mathbf{S} - (s_{22} - s_{11}) & 2s_{12} \\ 2s_{21} & 1 - \det \mathbf{S} + (s_{22} - s_{11}) \end{bmatrix} \end{aligned} \tag{31}$$

$$\det(\mathbf{1}+\mathbf{S}) = 1 + \operatorname{tr}\mathbf{S} + \det\mathbf{S} = 1 + (s_{11}+s_{22}) + (s_{11}s_{22} + s_{12}s_{21})$$

$$\det(\mathbf{1}-\mathbf{S}) = 1 - \operatorname{tr}\mathbf{S} + \det\mathbf{S} = 1 - (s_{11}+s_{22}) + (s_{11}s_{22} - s_{12}s_{21})$$

The matrices **Y** and **Z** relate normalized voltages and currents at both ports (Fig. 12) as follows:

$$\begin{bmatrix} v_1 \\ v_2 \end{bmatrix} = \mathbf{Z} \begin{bmatrix} i_1 \\ i_2 \end{bmatrix}$$

$$\begin{bmatrix} i_1 \\ i_2 \end{bmatrix} = \mathbf{Y} \begin{bmatrix} v_1 \\ v_2 \end{bmatrix}$$

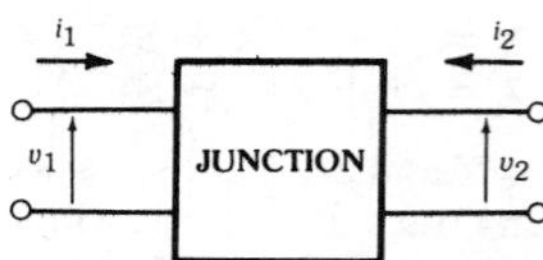

Fig. 12. Sign convention for defining the impedance and admittance of a two-port junction.

TRANSFORMATION MATRIX

A transformation matrix useful for composing two-port junctions in cascade relates the voltage and current on one side of the junction to the same quantities on the other side. With the notation in Fig. 13

$$\begin{bmatrix} v' \\ i' \end{bmatrix} = \mathbf{U} \begin{bmatrix} v \\ i \end{bmatrix} \tag{32}$$

The matrix **U**, sometimes called the *ABCD* matrix, has the same properties as **T** described earlier.

For a series element that has a normalized impedance Z

$$\mathbf{U} = \begin{bmatrix} 1 & Z \\ 0 & 1 \end{bmatrix}$$

and for a shunt element with normalized admittance Y

$$\mathbf{U} = \begin{bmatrix} 1 & 0 \\ Y & 1 \end{bmatrix}$$

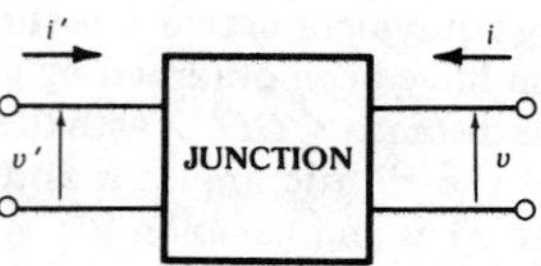

Fig. 13. Sign convention for voltages and currents related by the transformation matrix.

A product of matrices of these types gives the transformation matrix for any ladder network.

For the shunt element Y, the scattering matrix is

$$\mathbf{S} = (2+Y)^{-1} \begin{bmatrix} -Y & 2 \\ 2 & -Y \end{bmatrix} \tag{33}$$

hence

$$s_{11} = s_{22}$$
$$s_{12} = 1 + s_{11} \tag{34}$$

For the series element Z, the scattering matrix is

$$\mathbf{S} = (2+Z)^{-1} \begin{bmatrix} Z & 2 \\ 2 & Z \end{bmatrix} \tag{35}$$

hence

$$s_{11} = s_{22}$$
$$s_{12} = 1 - s_{11} \tag{36}$$

Relations 34 and 36 are characteristic, respectively, of a shunt and a series obstacle in a waveguide.

The matrix **T** can be deduced from **U** and vice versa:

$$\mathbf{T} = \frac{1}{2} \begin{bmatrix} 1 & 1 \\ 1 & -1 \end{bmatrix} \mathbf{U} \begin{bmatrix} 1 & 1 \\ 1 & -1 \end{bmatrix}$$

$$= \frac{1}{2} \begin{bmatrix} u_{11}+u_{12}+u_{21}+u_{22} & u_{11}-u_{12}+u_{21}-u_{22} \\ u_{11}+u_{12}-u_{21}-u_{22} & u_{11}-u_{12}-u_{21}+u_{22} \end{bmatrix} \tag{37}$$

A similar equation will transform **T** into **U**, since

$$\mathbf{U} = \frac{1}{2} \begin{bmatrix} 1 & 1 \\ 1 & -1 \end{bmatrix} \mathbf{T} \begin{bmatrix} 1 & 1 \\ 1 & -1 \end{bmatrix} \tag{38}$$

32 Antennas

Robert C. Hansen

GENERAL

Introduction

The field of antennas is sufficiently broad to be beyond the scope of this chapter, even for succinct design information. Thus the intent is to be eclectic rather than inclusive. Fortunately, several excellent books on antennas have recently appeared. For more detailed design data, the reader should consult the two-volume *Handbook of Antenna Design* edited by Rudge, et al.* Thorough texts on antenna theory, including extensive coverage of the powerful geometric theory of diffraction (GTD) and moment method analytical approaches, as well as exemplary computer programs, are by Balanis† and by Stutzman and Thiele.§ A third excellent text, which has extensive coverage on fixed beam array design, is by Elliott.** Finally, the most extensive treatment of phased arrays (electronic scanning) is still the three-volume *Microwave Scanning Antennas* by Hansen.‡

Six parts comprise this chapter. Basic antenna behavior, including definitions, fields, near-field power density, antenna noise temperature, and polarization coupling are in this general part. The next two parts cover, respectively, low- and medium-gain antennas, where dipoles, slots, loops, and microstrip patches are low-gain, and horns, Yagi-Udas, helices, spirals, and log-periodics are medium-gain. Arrays of all types are covered next. Aperture distributions, because of their common importance, occupy an entire part. Finally, reflector-type antennas are the subject of the last part. Some old favorites have been left out, and this reflects somewhat the changing antenna usage.

Because of the power and ready availability of calculators and computers, tables of calculated functions have largely been omitted. Instead, key performance indices have been quantified, with design formulas given so that the designer can implement them directly.

Definitions

Directivity and gain are measures of how well energy is concentrated in a given direction. Directivity is the ratio of power density (PD) in that direction to the power density that would be produced if the power were radiated isotropically. The reference can be linearly or circularly polarized, and directivity is often given in dBi, decibels above isotropic. Some early literature refers to gain above a dipole; this usage is deprecated as it is confusing and unnecessary. Directivity, then, is given by

$$D = 4\pi \, PD \Big/ \iint |E|^2 \, d\Omega$$

Gain includes antenna losses; thus gain is the field intensity produced in the given direction by a fixed input power to the antenna. Gain is related to directivity by efficiency η, and is

$$G = D\eta$$

$$G = 4\pi \, PD/P_{in}$$

Through reciprocity, directivity is independent of transmission or reception, as is gain.

Effective area is defined by:

$$A_e = \lambda^2 G/4\pi$$

where λ is the free-space wavelength. (All through this chapter, commonly used symbols are employed.) For an antenna matched to a load, the load power is $P_{load} = PD \cdot A_e$, where PD is the power density at the antenna in watts per square meter.

Path loss is part of the range equation, where received power is related to transmitted power as

$$P_r = P_t \, G_r G_t \, \lambda^2/(4\pi R)^2$$

The distance between antennas is R, and the path loss is given by

$$\text{Path loss} = (4\pi R/\lambda)^2$$

Effective length relates the ability of a receiving antenna to produce open-circuit voltage. It is

$$l_e = V/E$$

where V is the open-circuit voltage for an incident field strength E. The early usage "effective height," is deprecated, as it also has a meaning for antennas over earth. For any antenna, the preceding parameters are related through

$$30\pi l_e^2 = R_r A_e$$

Here, R_r is radiation resistance, where the radiated power of a current-driven antenna is $P_r = I^2 R_r$. Table 1 gives often used parameters for short dipoles and for half-wave dipoles. Dipole half-length and monopole length are h, and θ_3 is the half-power beamwidth, i.e., the width of the pattern between -3-dB points. For loops, the diameter is d, N is the number of turns, $k = 2\pi/\lambda$, and μ_e is the effective permeability.

Bandwidth may be defined through pattern characteristics, efficiency, impedance, etc. The latter is often used, with the bandwidth being the range between the two frequencies where the radiated power falls to half (the 3-dB, or half-power, bandwidth) or where the vswr reaches a fixed value, e.g., 2. Note that half-power bandwidth occurs when the input $R = |X|$, and for vswr $= 5.828$. Since the product of antenna Q and fractional bandwidth, BW, is:

---

* Rudge, A. W., et al, eds. *Handbook of Antenna Design.* London: Peter Peregrinus Ltd., 1983.

† Balanis, C. A. *Antenna Theory: Analysis and Design.* New York: Harper & Row Publishers, 1982.

§ Stutzman, W. L., and Thiele, G. A. *Antenna Theory and Design.* New York: John Wiley & Sons, Inc., 1981.

** Elliott, R. S. *Antenna Theory and Design.* Englewood Cliffs, N.J.: Prentice-Hall, Inc., 1981.

‡ Hansen, R. C. *Microwave Scanning Antennas.* New York: Academic Press, Inc.; Vol. 1, 1964; Vols. 2 and 3, 1966.

TABLE 1. LOW-GAIN-ANTENNA PARAMETERS

Type	D	l_e	A_e	θ_3	R_r
Isotropic	1	—	$\lambda^2/4\pi = 0.0796\lambda^2$	360°	—
Short Dipole	1.5	h	$3\lambda^2/8\pi = 0.1194\lambda^2$	90°	$20k^2h^2$
$\lambda/2$ Dipole	1.6409	λ/π	$30\lambda^2/\pi R_r = 0.1306\lambda^2$	78.078°	73.13
$\lambda/4$ Monopole	3.2818	λ/π	$30\lambda^2/\pi R_r = 0.2612\lambda^2$	78.078°	36.56
Small Loop	1.5	$\pi Nkd^2\mu_e/4$	$3\lambda^2/8\pi = 0.1194\lambda^2$	90°	$5\pi^2N^2k^4d^4\mu_e^2/4$

$$Q \cdot BW = (\text{vswr} - 1)/\sqrt{\text{vswr}}$$

the bandwidth at one vswr is related to that at another vswr by:

$$BW_1/BW_2 = [(\text{vswr}_1 - 1)/(\text{vswr}_2 - 1)](\text{vswr}_2/\text{vswr}_1)^{1/2}$$

For example, the 3-dB bandwidth is 2.828 times larger than the vswr = 2 bandwidth. These results include the effect of a matched load (generator); for half-power $Q \cdot BW = 2$.

Field regions must be defined carefully, as the early optical terms are inadequate for modern antennas. The terms Fraunhofer and Fresnel, which refer to specific field integral approximations in optics, are obsolete and deprecated. For example, with focused antennas, a Fraunhofer-type pattern may exist well within the usual D^2/λ near-far field boundary distance, while a Fresnel-type field may exist for smaller and for larger distances. Some antennas have no Fraunhofer-type field anywhere, as they have no phase center, e.g., some horns and annular slots. Finally, the Fresnel approximation itself is unambiguous only in one dimension; for the more common and useful area sources, the Fresnel results vary widely with coordinate system and formulation.*

Thus the following definitions have evolved. Space is divided into three regions as follows. That region of space immediately surrounding the antenna in which the reactive components predominate is known as the *reactive near-field region*. The size of this region varies for different antennas. For most antennas, however, the outer limit is on the order of a few wavelengths or less. For the particular case of an electrically small dipole, the reactive field predominates to a distance of approximately $\lambda/2\pi$, where the radiating and reactive fields are equal. Beyond the reactive near-field region, the radiating field predominates. The radiating region is divided into two subregions, the *radiating near-field region* and the *far-field region*. In the radiating near-field region, the relative angular distribution of the field (the usual radiation pattern) is dependent on the distance from the antenna. The reason for this behavior is twofold: the relative phase relationship of field contributions from different elements of the antenna changes with distance, and the relative amplitudes of these field contributions also change with distance. As the observation point in space moves away from the antenna, the amplitude of the field first oscillates and then decays monotonically. This variation in the limit is given by the reciprocal of the first power of distance. Furthermore, the relative phase and amplitude relationships between the field contributions from different elements of the antenna asymptotically approach a fixed relationship, and the relative angular distribution of the field becomes independent of the distance. This occurs in the far-field region; patterns are essentially independent of distance. For most antennas, the transition distance is D^2/λ, where D is the width of the equivalent uniformly excited aperture. Precision gain measurements or measurements of nulls may require a distance of $2D^2/\lambda$ or more. Low sidelobe antennas will require multiples of D^2/λ, depending on the sidelobe level and allowable error.† Fig. 1 shows the reactive near-field boundary at $R = \lambda$ and the D^2/λ far-field boundary for several planar apertures.

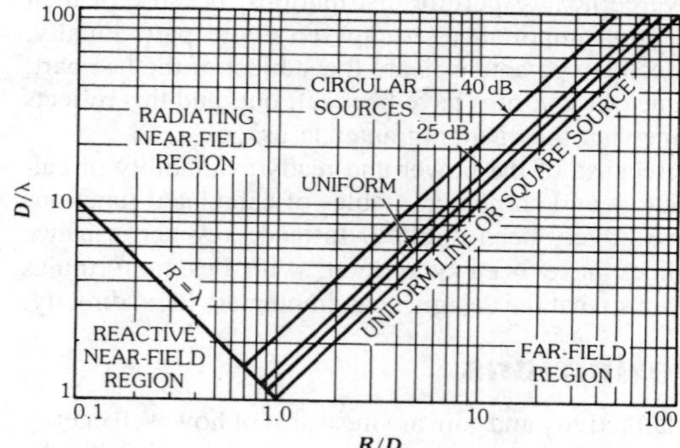

Fig. 1. Field regions.

Small Antennas

Antennas that are small in wavelength are conceptually simple, and difficult to use, as will appear below. It has been observed by Wheeler§ that all small antennas are dipoles, loops, or combinations of these two canonical types. When the dipole (or monopole over

* Hansen, R. C. *Microwave Scanning Antennas*, Vol. 1. New York: Academic Press, Inc., 1964; Chap. 1.

† See Aperture Distributions in array chapters by R. C. Hansen in Rudge, A. W., et al, eds., *Handbook of Antenna Design*, Vol. 2, London: Peter Peregrinus Ltd., 1983.

§ Wheeler, H. A., "Fundamental Limitations of Small Antennas," *Proc. IRE*, Vol. 35, Dec. 1947, pp. 1479–1484. Wheeler, H. A., "Small Antennas," *IEEE Trans. Antennas Propagat.*, Vol. AP-23, 1975, pp. 462–469.

a large ground plane) is short, the current is essentially linear from feed to end, and the fields produced, when the dipole is along the z-axis of a standard spherical coordinate system, are

$$E_\theta = j\,(60\pi h I_0 e^{-jkr} \sin\theta)/r\lambda,$$

$$H_\phi = j\,(h I_0 e^{-jkr} \sin\theta)/2r\lambda$$

The radiation resistance is $R_r = 20k^2h^2$, where $k = 2\pi/\lambda$ and h is the dipole half-length. Similarly, a small loop of diameter d carries essentially a constant current, and the fields with the loop axis along z are analogously

$$E_\phi = (30\pi^3 d^2 I_0 e^{-jkr} \sin\theta)/r\lambda^2$$

$$H_\theta = -(\pi^2 d^2 I_0 e^{-jkr} \sin\theta)/4r\lambda^2$$

The radiation resistance is given by:

$$R_r = (5/4)\,\pi^2 N^2 k^4 d^4 \mu_e^2$$

See the section on low-gain antennas for details of μ_e. When $h << \lambda$ or $d << \lambda$, the radiation resistance is very small, often smaller than the loss resistance of the conductors. Efficiency then can be small, and although the directivity is 1.5 the gain is also small. Reactance, on the other hand, is high; for short dipoles it varies as $1/kh$, whereas for small loops it varies as $1/k^2d^2$. The Q is therefore high, and the bandwidth, which is approximately $(h/\lambda)^3$ for $Q >> 1$, is small. It is useful to know what bandwidth can be achieved; the Wheeler papers referred to above give practical answers. A theoretical fundamental limitation was derived by Chu** and refined by Harrington.‡ See also Hansen§§ for a discussion of fundamental limitations in antennas. Since any radiating field can be written as a sum of spherical modes, the antenna, of whatever type it happens to be, is enclosed in a sphere of radius a. The radiated power can be calculated from propagating modes within the sphere. All modes contribute to the reactive power. When the sphere is sufficiently large to support several propagating modes, this approach is of little value because the modal coefficients are difficult to calculate. With only one propagating mode, the radiated power arises primarily from that mode. The utility of the Chu work becomes apparent when the sphere is too small to allow a propagating mode; all modes are then evanescent (below cutoff), and the Q becomes large, as the evanescent modes contribute little real power. Note that, unlike the case of a closed waveguide, there is a real part of each evanescent mode. Each mode has a Q_n based on the ratio of stored energy to radiated energy, and the Q_n rises rapidly when kr drops below the mode number. For all modes well below cutoff, the result is

$$Q = (1 + 3k^2a^2)/[k^3a^3(1 + k^2a^2)]$$

When the antenna contains loss, the Q is reduced by the efficiency, η. Fig. 2 gives fundamental limitation curves for several values of efficiency. This Q is for the lowest TM mode. When both a TM mode and a TE mode are excited, the value of Q is halved. The importance of the Chu result is that it relates the lowest achievable Q to the maximum dimension of an electrically small antenna, and this result is independent of the art that is used to construct the antenna within the hypothetical sphere, except in determining whether a pure TE or pure TM mode, or both, is excited. Since the Q grows rapidly (inverse cube) as size decreases, this indeed represents a fundamental limit that has only been approached but never equaled, much less exceeded. Bandwidth is derived from Q by assuming that the antenna equivalent is a resonant circuit with fixed values. Then the fractional bandwidth is

$$\text{Bandwidth} = \frac{f_{upper} - f_{lower}}{f_{center}} = 1/Q$$

A matched load is not included here. For $Q >> 1$ this relationship is meaningful, as the fixed resonant circuit is a good approximation to the antenna. But for $Q < 2$, the representation is no longer accurate. However, the curves are still useful for low Q even though imprecise. An octave bandwidth, for example, requires

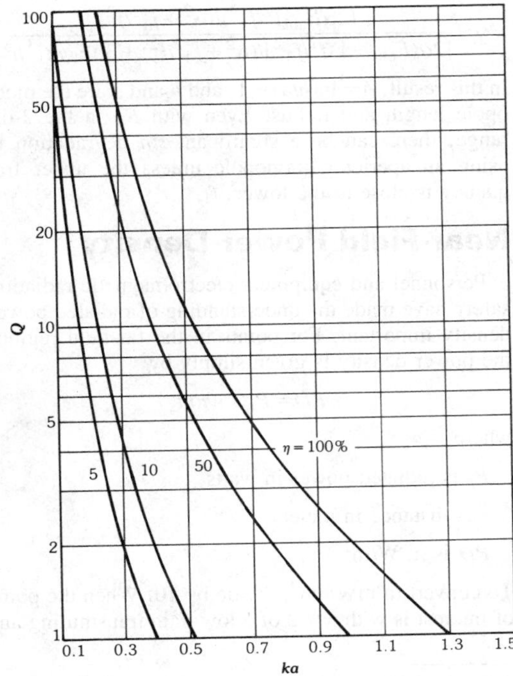

Fig. 2. Chu-Harrington fundamental limitations for single-mode antenna, various efficiencies.

** Chu, L. J. "Physical Limitations of Omnidirectional Antennas." *J. Appl. Phys.*, Vol. 19, Dec. 1948, pp. 1163–1175.

‡ Harrington, R. F. "Effect of Antenna Size on Gain, Bandwidth, and Efficiency." *J. Res. Nat. Bur. Stand.*, Vol. 64D, Jan.-Feb. 1960, pp. 1–12.

§§ Hansen, R. C. "Fundamental Limitations in Antennas." *Proc. IEEE*, Vol. 69, Feb. 1981, pp. 170–182.

$Q=\sqrt{2}$, and with no losses this requires a minimum antenna length of 0.365λ. Since most small antennas are loops or dipoles, which do not use the spherical volume efficiently, an actual octave antenna is significantly larger, often larger than $\lambda/2$.

Electrically small antennas can be broadbanded by introducing loss as indicated, or by utilization of a large mismatch. A small loop with a very low-impedance preamplifier at its terminals or a short monopole (dipole) with a high-impedance preamplifier at its terminals is called an "aperiodic loop" or "aperiodic monopole." The term "integrated antenna" has been used, but it is confusing and is now obsolete. If the amplifier impedance is much larger than the monopole reactance at the lowest frequency of interest, the output signal is nearly constant over a bandwidth of one or more octaves. The large mismatch reduces the signal and external noise. Aperiodic antennas are useful at those frequencies where the external noise is so large that even after the large mismatch loss the antenna noise is still larger than the preamp noise; thus the system is external-noise limited. Let the ratio of preamp input resistance to lowest antenna reactance be $\alpha=R/X_1$, and use F_{ant} as the ratio of external antenna noise temperature to $T_0=290$ K, and F_N as the preamp noise figure. A key system performance factor, γ, is the ratio of actual s/n to s/n if the system were external-noise limited. That is, γ determines s/n degradation compared to a narrow-band matched antenna-preamp system at frequency f. The s/n degradation factor for a monopole that is short over the band of interest is*

$$\gamma=\frac{(2\alpha F_{ant}f^2/f_1{}^2)/(\alpha^2+f_1{}^2/f^2)}{[2\alpha(F_{ant}-1)f^2/f_1{}^2]/(\alpha^2+f_1{}^2/f^2)+3AF_N/k_1{}^3h^3}$$

In this result, $A=\ln(h/a)-1$, and h and a are the monopole length and radius. Even with F_N in the 2-dB range, there can be a significant s/n degradation in using an aperiodic monopole unless the upper frequency is close to the lower, f_1.

Near-Field Power Density

Personnel and equipment electromagnetic radiation safety have made the understanding of radiated power density important. For points in the far-field region, the power density is given simply by

$$PD=P_rG/4\pi r^2$$

where,

P_r is radiated power in watts,

r is distance in meters,

PD is in W/m^2.

To convert to mW/cm^2, divide by 10. When the point of interest is within $\lambda/2$ of a low-gain transmitting antenna, such as a single vertical tower, the near-field dipole formulas can be used for approximate results. For dipoles at greater distances, $PD=30I^2/\pi R_r{}^2$.

Near fields of larger antennas can be accurately estimated if the aperture (or array) distribution is known. Both line sources and planar apertures will be discussed, because the near-field behavior of these two is quite different. For a line source, the beam is usually designed to be narrow in one plane while it is broad or omnidirectional in the other plane. In the near field, the phase errors degrade the pencil beam pattern, effectively decollimating it so that $1/r$ of the field intensity variation with distance is effectively cancelled. The other $1/r$ still remains, with the result that near-field power density for a line source oscillates about a $1/r$ variation. The oscillations are reduced for tapered amplitude distributions, and there is, of course, a transition region from the far-field $1/r^2$ behavior to the near-field $1/r$. Because most high-power antennas are planar sources instead of line sources, only limited information is given on line sources. For uniform excitation the near-field power density is*

$$PD=(4.05/\Delta)[C^2\ (1/2\sqrt{\Delta})+S^2\ (1/2\sqrt{\Delta})]$$

Here, the distance from the antenna is normalized in terms of $2L^2/\lambda$, where $\Delta=r/(2L^2/\lambda)$; C and S are conventional Fresnel integrals. For a square uniformly excited aperture, the power density is

$$PD=16.4\ [C^2(1/2\sqrt{\Delta})+S^2\ (1/2\sqrt{\Delta})]^2$$

Fig. 3 shows power density for both the line source and the square source.

Of more interest is the near-field behavior of circular apertures, either arrays or reflectors. The Hansen one-parameter circular space factor† is a good fit to the

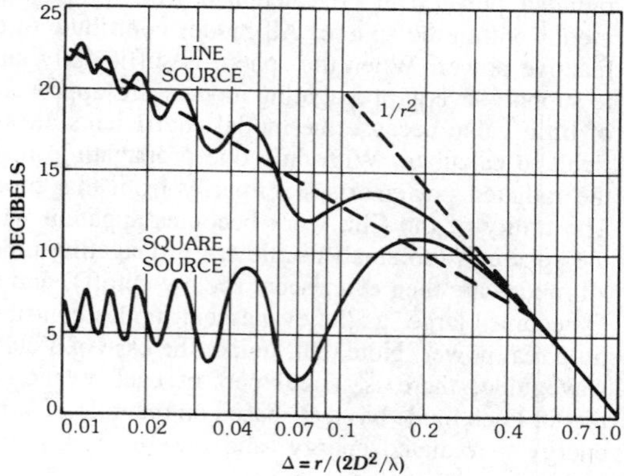

Fig. 3. On-axis power density.

* Radjy, A. H., and Hansen, R. C. "S/N Performance of Aperiodic Monopoles." *Trans. IEEE*, Vol. AP-27, March 1979, pp. 259–261.

*Ricardi, L. J., and Hansen, R. C. "Comparison of Line and Square Source Near Fields." *Trans. IEEE*, Vol AP-11, Nov. 1963, pp. 711–712.

† Hansen, R. C. "A One-Parameter Circular Aperture Distribution with Narrow Beamwidth and Low Sidelobes." *Trans. IEEE*, Vol. AP-24, July 1976, pp. 477–480.

main beam and near sidelobes of most reflector antennas if the parameter is properly chosen. Similarly, it is an excellent representation for near-field power density near the axis. These robust low-Q distributions are also used for arrays. The axial power density is obtained from numerical integration with only the results given here.§ Figs. 4 and 5 give axial power density for circular apertures with sidelobe ratios from 20 to 35 dB. Power density on-axis for a uniformly excited circular aperture is shown in Fig. 6.** Thus, it can be observed that the near-field phase errors do decollimate the beam in both planes, thereby producing a constant power density value around which the near field oscillates. All of these data are normalized to a value of unity at a distance of $2D^2/\lambda$, and the nominal power density at that distance is found simply from

$$PD = PG/[4\pi(2D^2/\lambda)^2]$$

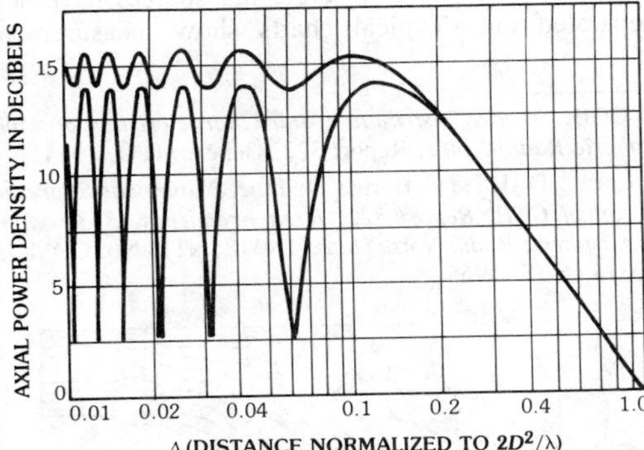

Fig. 4. Axial power density for SLR of 20 and 30 dB, one-parameter circular aperture.

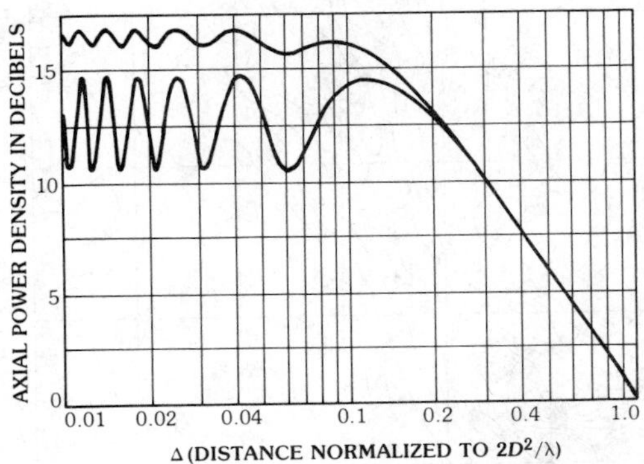

Fig. 5. Axial power density for SLR of 25 and 35 dB, one-parameter circular aperture.

§ Hansen, R. C. "Circular-Aperture Axial Power Density." *Microwave J.*, Vol. 19, Feb. 1976, pp. 50–52.

** Hansen, R. C. "Antenna Power Densities in the Fresnel Region." *Proc. IRE*, Vol. 47, Dec. 1959, pp. 2119–2120.

where,

P is the input power,

G is the antenna gain.

Calculation of near fields off-axis is extremely difficult because the small-angle approximate techniques quickly become useless. At large angles, geometric theory of diffraction techniques are useful.

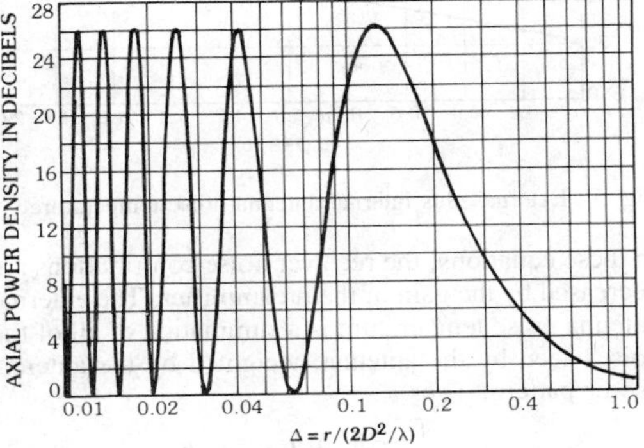

Fig. 6. Axial power density of uniform circular aperture.

Antenna Noise Temperature

Antenna noise contributes to system noise and thus will affect system performance. Antenna noise consists of external noise and internal noise, with the latter caused by loss in the antenna, feed, cable, etc. It is customary and convenient to refer noise measurements to the input terminals of the preamplifier or receiver. If there is loss within the antenna or in components between the antenna and the preamplifier input, this loss will change the antenna contribution to system noise. If L is the loss ratio (P_{in}/P_{out}) and T_a is the external noise temperature, the total antenna temperature at the preamplifier input is given by

$$T_{ant} = [290(L-1) + T_a]/L$$

Fig. 7 shows this in graphical form. It is assumed that the lossy components are at an ambient temperature of 290 K. For other ambient temperatures, the coefficient in the formula is, of course, changed. The loss applies only to actual dissipative loss and not to virtual losses such as aperture excitation efficiency. The systems temperature is

$$T_{sys} = [T_0(L-1) + T_a]/L + T_0(F_N - 1) + T_0(F_{rec} - 1)/G_{pre}$$

where the preamplifier noise figure is F_N and the receiver noise figure is F_{rec}. Note that the preamplifier noise temperature contribution involves $(F_N - 1)$; this is because the measurement of amplifier noise figure includes an input standard temperature of $T_0 = 290$ K. An alternate form of the systems temperature is:

$$T_{sys} = T_a/L - T_0(F_N - 1/L) + T_0(F_{rec} - 1)/G_{pre}$$

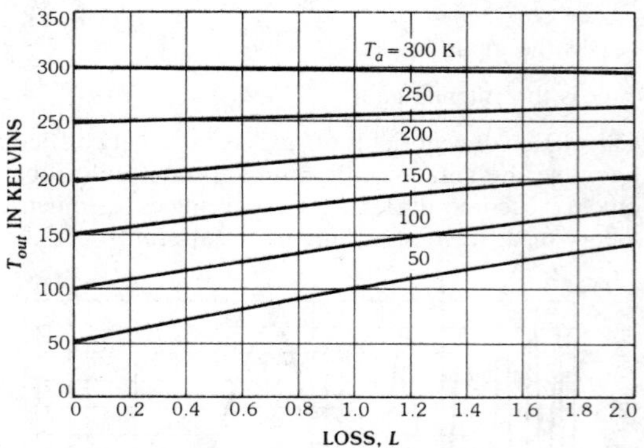

Fig. 7. External plus internal antenna noise temperature.

In these equations, the receiver noise contributions are decreased by the gain of the preamplifier. The external antenna noise temperature is a summation of all of the noises seen by the antenna, weighted by the antenna power pattern. Thus

$$T_a = \frac{\iint T(\theta, \phi)P(\theta, \phi)d\Omega}{\iint P(\theta, \phi)d\Omega}$$

In this equation, $T(\theta,\phi)$ is the spatial temperature pattern in kelvins, $P(\theta,\phi)$ is the antenna power pattern, and the integration is over all space. Next, specific external noise sources will be reviewed.

Below 10 MHz, the noise is primarily atmospheric and very large. This noise is mostly Gaussian, but there are impulsive tails at low probabilities of occurrence. Extensive data are given by the CCIR,* and a revision of this document is in press. The highest noise generally occurs in spring in the 2000–2400 time block. The CCIR data have been computerized by Lucas and Harper.† Broadly, atmospheric noise is omnidirectional and independent of distance. A minor exception occurs in the vlf absorption notch around 4 kHz in that noise may be higher in some directions due to rapid attenuation with distance. The atmospheric noise contribution becomes negligible between 35 and 40 MHz. Galactic noise generally consists of noise from a few hot stars plus a varying background that results from many, many stars. The background results, of course, depend on how many discrete star sources have been separated out. Typical charts show measurements

* CCIR. *World Distribution and Characteristics of Atmospheric Radio Noise*, Report 322. Geneva: ITU, 1964.

† Lucas, D. L., and Harper, J. D. *A Numerical Representation of CCIR Report 322: High Frequency (3–30 Mcps) Atmospheric Radio Noise Data*, TN318, NTIS No. COM-75-10374. NBS, 1965.

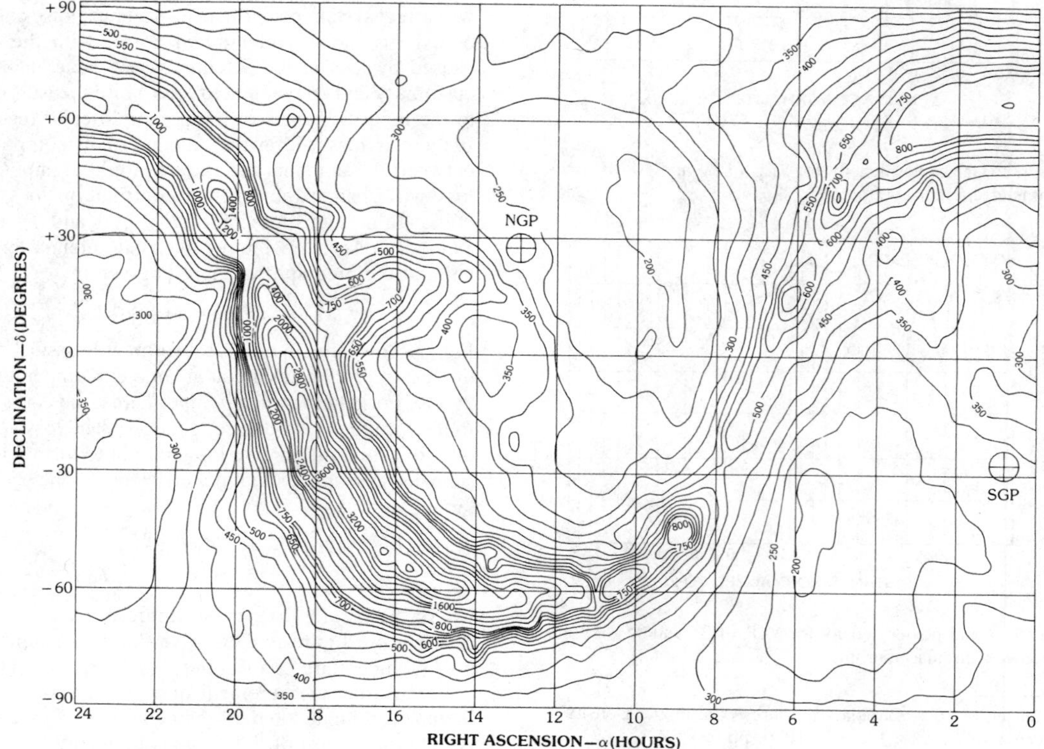

Fig. 8. Noise temperature (kelvins), 136 MHz. (*After Taylor.*)

made by Taylor§ at 136 and 400 MHz; see Figs. 8 and 9. A composite fit to these and other measurements is:

$$T \simeq (3.068 \times 10^8)/f_{MHz}2.3$$

This line is shown on the composite noise plot, Fig. 10. The sun is a special case because it subtends a finite angle and is much closer. Sun noise temperature measurements by Hogg and Mumford** show a slight oscillation about a linear frequency dependence. The latter is

$$T_s = 1.958 \times 10^8/f_{MHz}$$

For antenna beamwidths larger than 0.5°, the sun noise temperature contribution is

$$T_{sun} = T_s \theta_s^2 G/4\pi$$

where,

θ_s is the subtended angle of the sun,

G is the antenna gain in that direction.

For those portions of the antenna pattern that see the earth, the earth will contribute noise at a blackbody temperature of roughly 310 K. In the range from 20 to 600 MHz, all of these noise sources are usually dwarfed by man-made noise. This consists of power-line noise caused by arcs and corona and by user-generated pulses; noise from various transmitters including tv, fm, and push-to-talk; and noise generated by vehicles. Extensive studies and measurements of urban noise have been made, and the reader is referred to books by Skomal‡ and Herman.§§ The most appropriate division into cultural areas now seems to be "business," "residential," and "rural." These divisions and the data of Spaulding and Disney* are used here as seen in Fig. 10. However, it must be recognized that these urban noise envelopes apply away from vehicular traffic streams. That is to say, vehicle noise is included, but the vehicles are not closer than several hundred feet. A vehicle stream at 50 feet can generate noise 20 to 40 dB higher than these envelopes. A single "supernoisy" vehicle can generate as much noise as

§ Taylor, R. E. "136/400 MHz Radio Sky Maps." *Proc. IEEE*, Vol. 61, April 1973, pp. 469-472.

** Hogg, D. C., and Mumford, W. W. "The Effective Noise Temperature of the Sky." *Microwave J.*, Vol. 3, March 1960, pp. 80–84.

‡Skomal, E. N. *Man-Made Radio Noise*. New York: Van Nostrand Reinhold Co., 1978.

§§ Herman, J. R. *Electromagnetic Ambients and Man-Made Noise*. Gainesville, Va.: Don White Consultants, Inc., 1979.

* Spaulding, A. D., and Disney, R. T. *Man-Made Radio Noise. Part I: Estimates for Business, Residential, and Rural Areas*, OTR Report 74-38, June 1974.

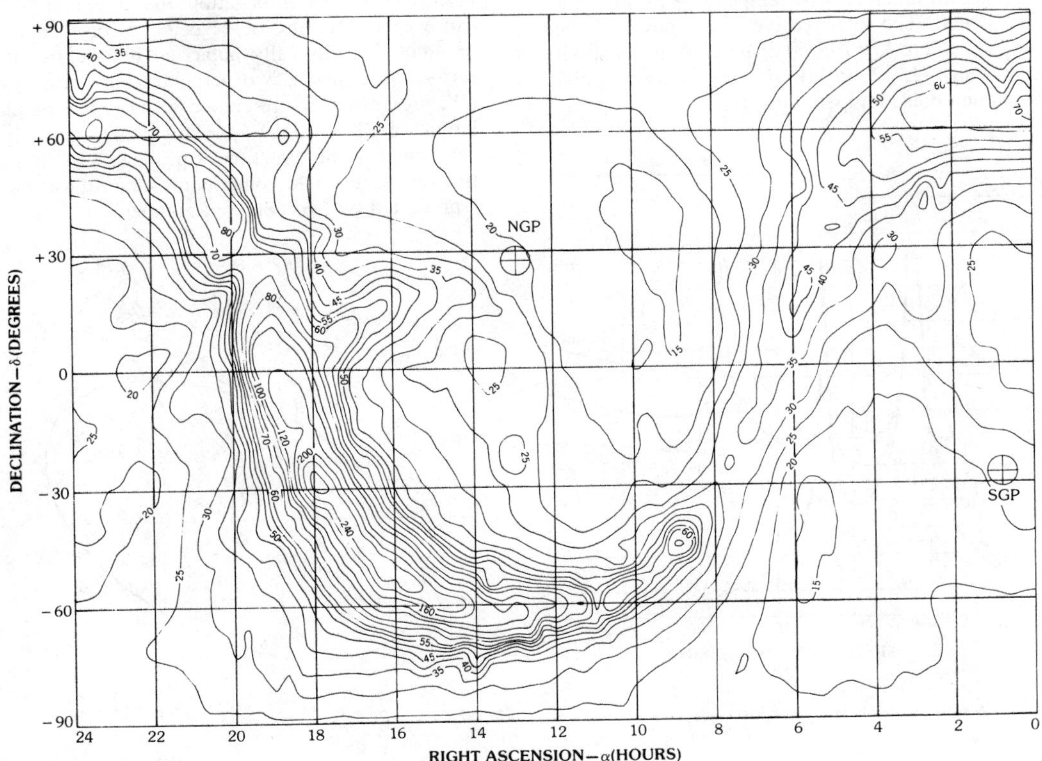

Fig. 9. Noise temperature (kelvins), 400 MHz. (*After Taylor*.)

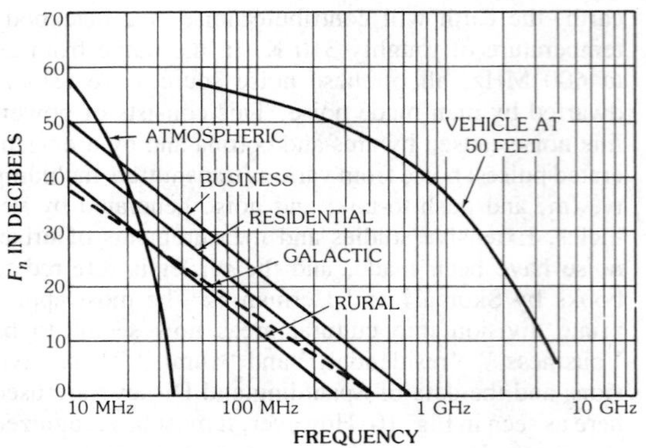

Fig. 10. Composite noise data. (*After Hansen.*)

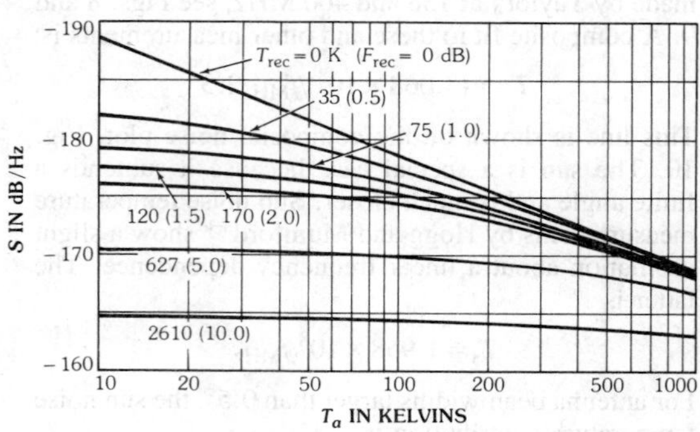

Fig. 12. Absolute sensitivity $S=K(T_a+T_{rec})$.

a stream of vehicles, and at close distances the noise levels can be extremely large. Although few data are available, vehicular noise can be significant at frequencies as high as 5 – 10 GHz, again for short distances. Atmospheric noise is produced by gaseous absorption, with the lowest molecular resonance being that for water at 22.235 GHz. Fig. 11 from Smith† gives a composite of absorption data, and the transmission windows around 33, 95, 140, and 230 GHz may be observed.

From the two-dimensional antenna pattern and these noise data, the antenna noise temperature may be readily calculated. The total receiver noise power is then $N=KT_{sys}$, where $K=198.60$ dBm. System sensitivity is shown in Fig. 12, in dBm/Hz for various receiver and antenna noise temperatures.

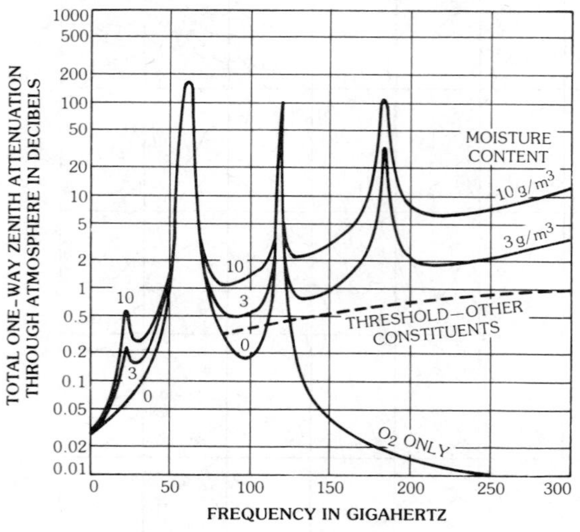

Fig. 11. Zenith atmospheric absorption.

Elliptical and Circular Polarization

An electromagnetic wave is linearly polarized when the electric field lies wholly in one plane containing the direction of propagation. A plane electromagnetic wave, at a given frequency, is elliptically polarized when the extremity of the electric vector describes an ellipse in a plane perpendicular to the direction of propagation, making one complete revolution during one period of the wave. If the rotation is clockwise looking in the direction of propagation, the sense is right-hand. More generally, any field vector, electric, magnetic, or other, is elliptically polarized if its extremity describes an ellipse. Two perpendicular axes, OX and OY, are chosen for reference in the plane of the polarization ellipse (Fig. 13). This plane is usually perpendicular to the direction of propagation. At a given frequency, the field components along these axes are represented by two complex numbers

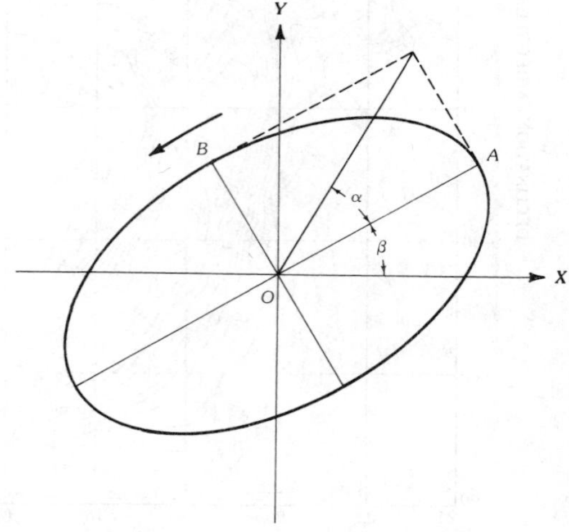

Fig. 13. Polarization ellipse.

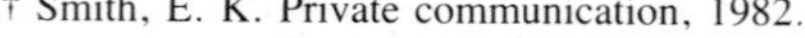

† Smith, E. K. Private communication, 1982.

$$X = |X| \exp j\theta_1$$

$$Y = |Y| \exp j\theta_2$$

The axial ratio, r, is the ratio of the minor axis, OB, to the major axis, OA, of the polarization ellipse. The relative power coupling, η, between two elliptically polarized receiving antennas is given by

$$\eta = \frac{1}{2}\left[1 + \frac{(1-r_1^2)(1-r_2^2)\cos 2\theta \pm 4r_1r_2}{(1+r_1^2)(1+r_2^2)}\right]$$

In this equation, r_1 and r_2 are the axial ratios of the antennas, and θ is the angle between the major axes of the polarization ellipses. The plus sign is used if both antennas produce the same sense of polarization, and the minus sign is used for opposite senses.* When a typical circularly polarized antenna is measured by a perfect rotating linear antenna, the circularly polarized gain is calculated from

$$G_{cir} = \frac{1}{2}[\sqrt{G_{lin\ max}} + \sqrt{G_{lin\ min}}]^2$$

Figs. 14 and 15 show transmission efficiency from a nearly linear antenna to a nearly circular antenna for both senses of polarization and for both zero and 90-degree orientation. When antenna gain is measured with a nominally linear standard-gain antenna where the orthogonal powers for two linear polarizations are summed, there is an error if the standard-gain antennas are imperfect. If the antenna under test has axial ratio r_1 and the standard-gain antenna has axial ratio r_2, the error in gain measurement is as follows:

$$\text{Error} = 1 \pm 4\, r_1r_2/(1+r_1^2)(1+r_2^2)$$

Figs. 16 and 17 show these errors for same sense and opposite sense of polarization.

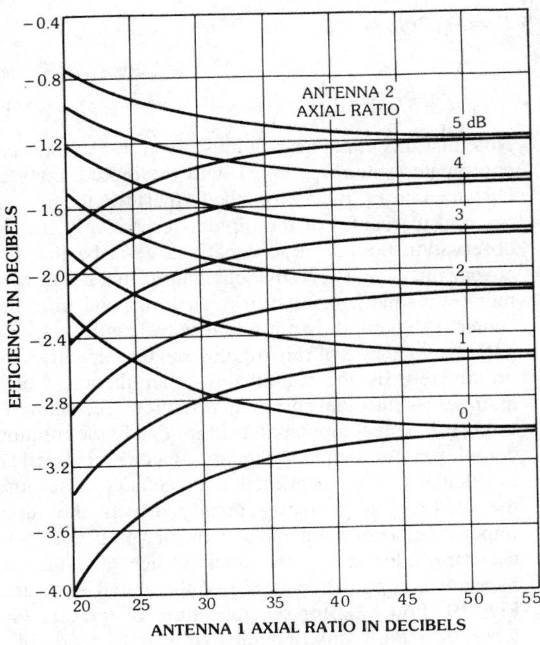

Fig. 14. Near-linear to near-circular efficiency, orientation 0°. (*Courtesy J. J. Murphy, Aerospace Corp.*)

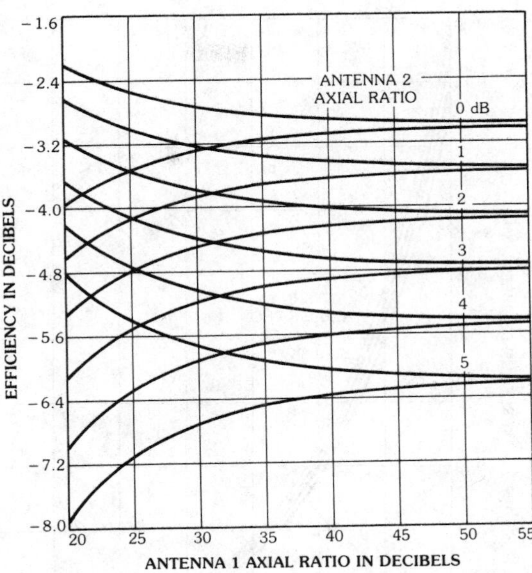

Fig. 15. Near-linear to near-circular efficiency, orientation 90°. (*Courtesy J. J. Murphy, Aerospace Corp.*)

LOW-GAIN ANTENNAS

Half-Wave Dipole

The resonant dipole, with length near a half wavelength, is a principal canonical antenna. The half-wave dipole has a rotationally symmetric pattern of $\cos[(\pi/2)\cos\theta]/\sin\theta$, where $\theta = 90°$ gives the dipole axis. Half-power beamwidth is 78.1°, directivity is $1.641 = 2.15$ dB, effective area is $A_e = 0.131\lambda^2$, and effective length is 0.318λ. When the dipole is not oriented along a spherical-coordinate-system z axis, but instead along x (Fig. 18), the pattern becomes

$$E(\theta,\phi) = \frac{\cos[(\pi/2)\sin\theta\cos\phi]}{\sqrt{1-\sin^2\theta\cos^2\phi}}$$

The far-field half-wave dipole field expressions, with I_0 the feed current, are

$$E_\theta = j\frac{60\, I_0 e^{-jkr}\cos[(\pi/2)\cos\theta]}{r\sin\theta}$$

$$H_\phi = j\frac{I_0 e^{-jkr}\cos[(\pi/2)\cos\theta]}{2\pi r\sin\theta}$$

* Hatkin, L. "Elliptically Polarized Waves." *Proc. IRE*, Vol. 38, Dec. 1950, p. 1455.

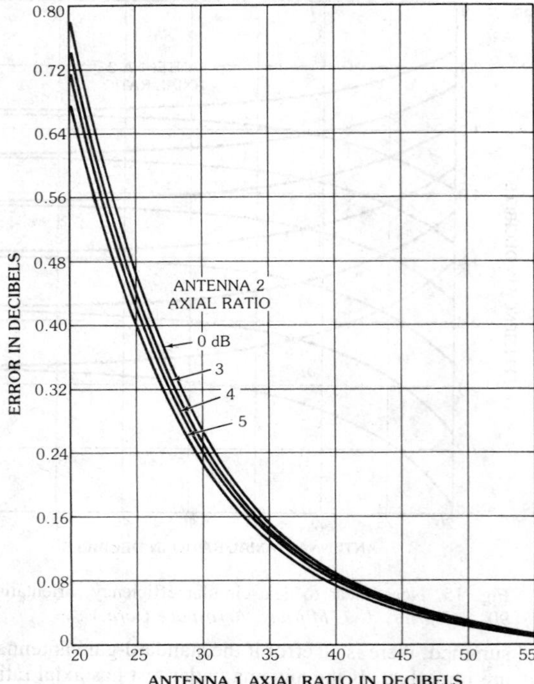

Fig. 16. Error in gain measurement, same sense. (*Courtesy J. J. Murphy, Aerospace Corp.*)

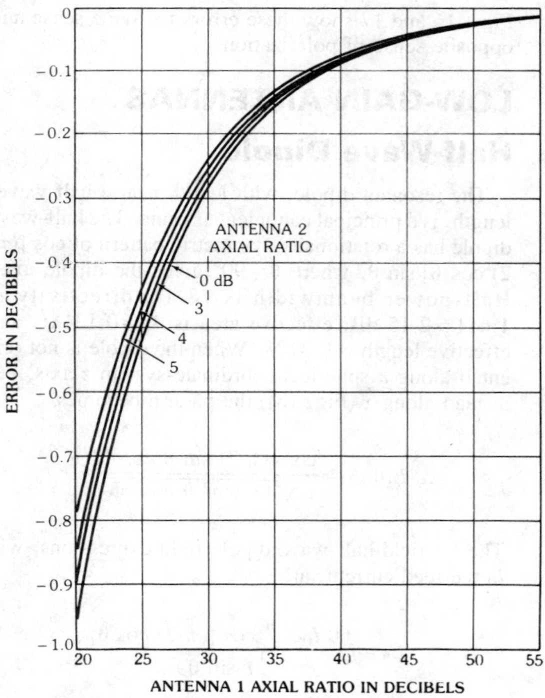

Fig. 17. Error in gain measurement, opposite sense. (*Courtesy J. J. Murphy, Aerospace Corp.*)

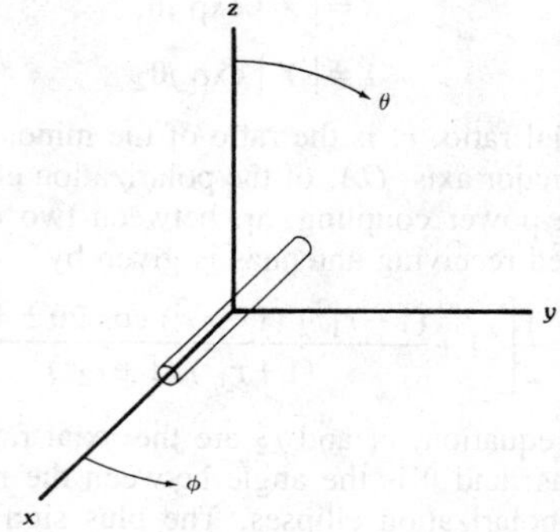

Fig. 18. Dipole along x-axis.

These formulas can be used to calculate power densities from $PD = E^2/120\pi$ as close as several wavelengths from the antenna. At closer distances, use the near-field results given below. When the dipole length is not $\lambda/2$, the pattern becomes

$$E(\theta) = \frac{\cos(kh\cos\theta) - \cos kh}{\sin\theta}$$

Exact near-fields for thin dipoles are:

$$E_\rho = j(30I_0/\rho)[(z-h)\, e^{-jkr_1}/r_1 + (z+h)e^{-jkr_2}/r_2 - (2z\cos kh)e^{-jkr_0}/r_0]$$

$$E_z = -j\,30I_0\,[e^{-jkr_1}/r_1 + e^{-jkr_2}/r_2 - (2\cos kh)e^{-jkr_0}/r_0]$$

$$H_\phi = -j(I_0/4\pi\rho)\,[e^{-jkr_1} + e^{-jkr_2} - (2\cos kh)e^{-jkr_0}]$$

Note that this exact near-field is given in a cylindrical coordinate system (ρ, ϕ, z) with the dipole along z. Distances r_1, r_2, r_0 are from the farthest tip, the nearest tip, and the center of the dipole, respectively, to the observation point. These equations describe the discovery many years ago by Schelkunoff that a thin wire radiates a spherical wave from each tip, and from the center if the length is not a half wavelength.

Self-impedance of thin dipoles can be calculated approximately by the Carter zero-order theory, which assumes a sinusoidal current distribution. See Hansen* for an efficient computer algorithm. Fat dipoles require complex integral-equation theory, which is impractical to calculate, or moment method calculations. Because the feed-region geometry greatly affects the input impedance, a common and satisfactory procedure is to use zero-order theory for quick values, and then to measure the input impedance of the actual structure. Fig. 19 shows Z_{in} for several values of h/a and l/λ, where a is the radius. Dipoles are usually made of a solid or tubular cylinder; when a flat (strip) dipole is

* Hansen, R. C. "Formulation of Echelon Dipole Mutual Impedance for Computer." *IEEE Trans.*, AP-20, 1972, pp. 780–781.

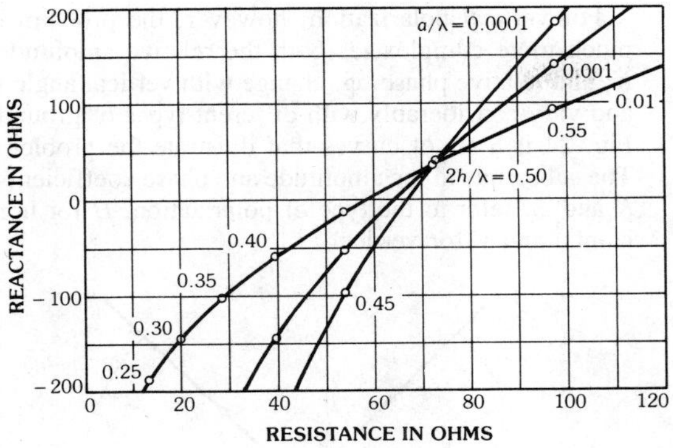

Fig. 19. Dipole self-impedance, zero order.

used, it is closely equivalent to a cylindrical dipole with $a = w/4$, where w is the strip width. Again, when zero-order theory is used, Fig. 20 gives shortening of resonant length versus h/a. Only a vanishingly thin

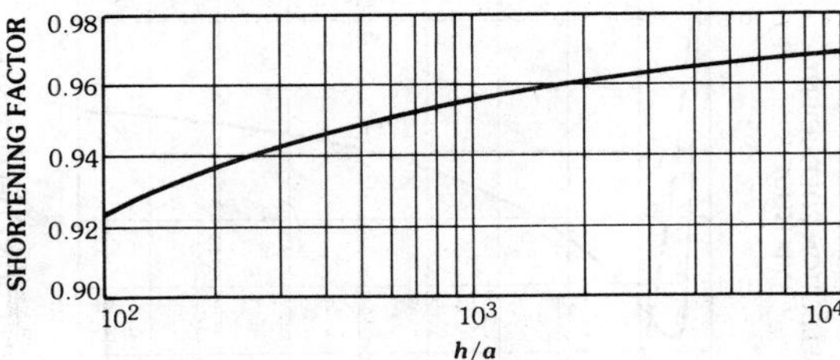

Fig. 20. Dipole shortening due to diameter.

dipole is resonant at half-wave. The resonant input resistance changes only slowly with a/λ, as seen in Table 2. In the limit of $a = 0$, the resonant resistance is 73.13 ohms.

Dipoles are frequently used parallel to a back screen, especially in arrays. If the dipole-screen spacing is s, the overall pattern, assuming the screen extends past the dipole by several wavelengths in each direction, is the dipole pattern multiplied by the screen array factor:

$$P(\theta) = P_{\text{dipole}}(\theta) \times \sin(ks \cos\theta)$$

where θ is zero normal to the screen. From this, the beamwidths in the E- and H-planes are 72.67° and 120°. Directivity is determined from the self- and mutual impedance:

$$D = 480/(R_{11} - R_{12})$$

Using zero-order mutual impedance, $R_1 = 73.13$ ohms, and a spacing $s = \lambda/4$ gives $D = 5.603 = 7.48$ dB. Actual directivities will be different because the R_{11} for actual dipoles is not 73 ohms. In principle, use of smaller spacing should increase gain; in practice, when the dipole and its image in the screen are close together, the impedance drops so that loss resistance limits the gain.† See Fig. 21.

† Kraus, J. D. *Antennas*. New York: McGraw-Hill Book Co., 1950.

TABLE 2. RESONANT RESISTANCE VS a/λ

a/λ	R_{res}(ohms)
0	73.13
0.00001	68.26
0.0001	66.79
0.001	64.11
0.01	58.16

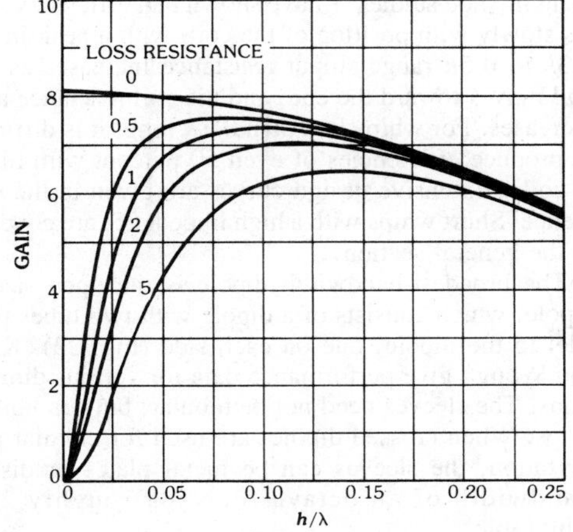

Fig. 21. Effect of loss on dipole over screen.

Sometimes when space is limited, the dipole ends are bent, forming a U dipole. This affects both resonant length and resistance; these have been calculated using moment methods§ for $a/\lambda = 0.01$. Fig. 22 shows the

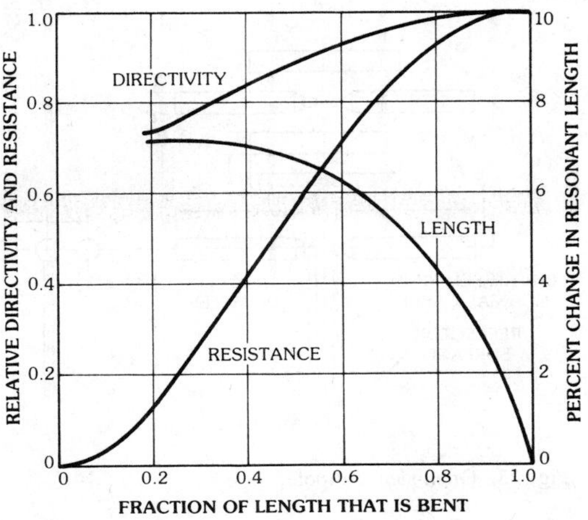

Fig. 22. Characteristics of resonant U dipoles.

§ Hansen, R. C. Array chapters in A. W. Rudge, et al, eds. *Handbook of Antenna Design*, Vol. 2. London: Peter Peregrinus Ltd., 1983.

change in resonant length and the reduction in radiation resistance as more of the length is in the arms.

Monopoles (whips) are widely used for communications, e.g., on vehicles. When a quarter wavelength cannot be accommodated, a shorter whip can be loaded with a series inductance. The loading coil keeps the current distribution roughly constant from the feed to (and slightly past) the coil, thus increasing radiation resistance by the square of the current moment over the triangular current distribution of a short whip. Moment method studies** have shown that efficiency varies slowly with position of the coil, with a peak in the $0.3h$ to $0.5h$ range. Input resistance increases as the coil moves toward the end, and the coil reactance also increases. For whips less than 0.1λ long, it is difficult to produce efficiencies of even 50 percent with high-Q coils. Extensive design curves are given in the reference. Short whips with a high impedance are covered in the general section.

The broadest-bandwidth dipole is the open sleeve dipole, which consists of a dipole with two tubes parallel to the dipole, one on each side (Fig. 23). King and Wong‡ give performance data for various dimensions. The sleeves need not be tubular, but can be flat strips. When crossed dipoles are used for circular polarization, the sleeves can be metal plates or disks. Bandwidth of an octave for vswr roughly 2 is achievable.

The behavior of the earth as a reflecting surface is considerably different for horizontal than for vertical polarization. For horizontal polarization, the earth may be considered a perfect conductor; that is, the reflected wave at all vertical angles β is substantially equal to the incident wave and 180° out of phase with it. The approximation is good for practically all types of ground.

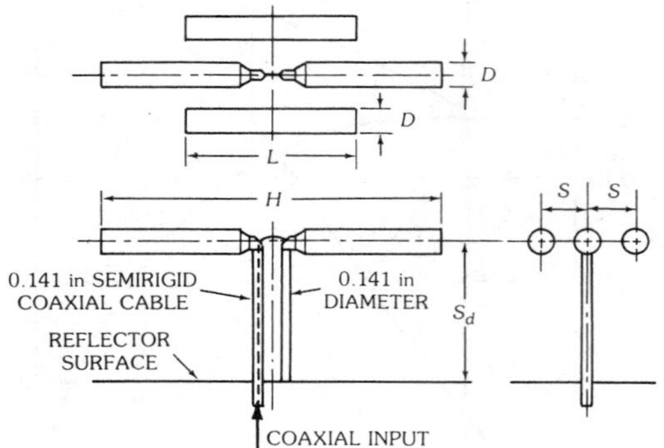

Fig. 23. Open-sleeve dipole.

For vertical polarization, however, the problem is much more complex, as both the relative amplitude, K, and relative phase, ϕ, change with vertical angle β and vary considerably with different types of ground. Fig. 24 is a set of curves that illustrate the problem. The subscripts to the amplitude and phase coefficients, K and ϕ, refer to the type of polarization, H for horizontal and V for vertical.

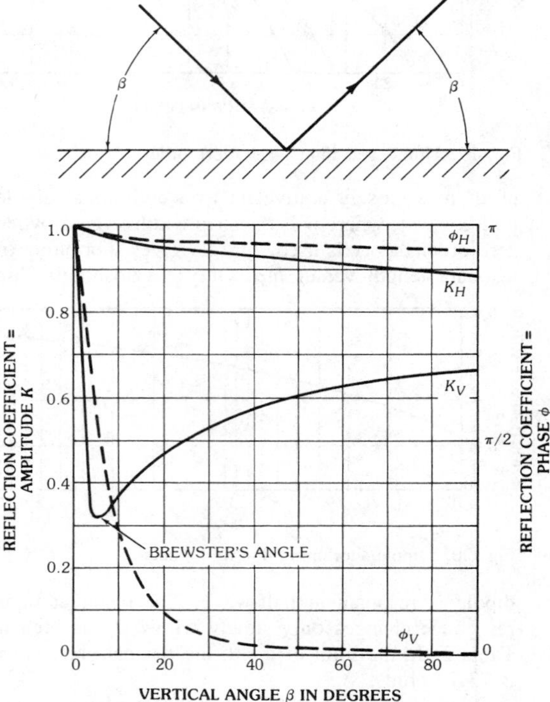

Fig. 24. Typical ground-reflection coefficients for horizontal and vertical polarizations.

Radial-Wire Ground Systems

A vertical tower antenna for mf and below usually requires a ground system to act as a low-loss "spreader" of current along the surface of the earth and to increase efficiency. Typically, the antenna base is surrounded by a metallic disk or closely spaced wire mesh, with radial wires emanating from this. It is not always possible or economical to make the radials sufficiently long and dense that the monopole performance is independent of the earth parameters. There is some evidence that surface wire radials, if sufficiently long, give higher efficiency than buried radials. Extensive calculations have been made by Maley and King,* but

** Hansen, R. C. "Efficiency and Matching Tradeoffs for Inductively Loaded Short Antennas." *IEEE Trans.* Vol. COM-23, April 1975, pp. 430-435.

‡ King, H. E., and Wong, J. L. "An Experimental Study of a Balun-Fed Open-Sleeve Dipole in Front of a Metallic Reflector." *IEEE Trans.*, Vol. AP-20, 1972, pp. 201–204.

* Maley, S. W., and King, R. J. "Impedance of a Monopole Antenna With a Radial-Wire Ground System on an Imperfectly Conducting Half Space." *Rad. Sci.*, Part 1, March-April 1962, pp. 175–180; Part 2, February 1964, pp. 159–165; Part 3, March 1964, pp. 297–301.

comparisons with measurements are yet not made. Fig. 25 shows experimental measurements at 3 MHz made by Brown et al† and Christman.§ Efficiency is given versus monopole height, and number and length of radials. For different earth, the absolute efficiencies will change, but the relative results are useful.

Printed-Circuit Antennas

Printed-circuit antennas include stripline slots, printed-circuit-board dipoles, and microstrip patches. A stripline slot consists of a narrow rectangular slot cut in the top stripline ground plane, with the slot excited by proper positioning of the center conductor below the slot. A linear resonant array of collinear slots may be fed by a single center conductor, with this conductor centered under each slot at the proper angle.** Boxed stripline is used to reduce higher modes.

Dipoles may be printed for low-cost fabrication, either as an array of dipoles on a single dielectric substrate, with each dipole fed by a balun which is normal to the dipole array face, or with each dipole and balun on a separate dielectric substrate. The array comprises a stack of these sheets.

The most widely used printed-circuit antenna is the microstrip patch, which in its simplest form is a rectangular or circular patch of metal fed by the microstrip upper conductor; see Fig. 26. Thus the element and feed line, and usually other elements, power dividers, etc., can all be prepared as a single etched pc board. A patch can also be fed from below by a coaxial probe, but this in large part removes the cost advantage. Early work on microstrip antennas was done by Deschamps and Sichak.‡

Since microstrip is thin, the patch and ground plane act as a TM resonator, with the side walls (the gaps) acting approximately as magnetic walls. Treating the patch as a lowest mode resonator gives useful results. More recent methods of analysis make use of a series of cavity modes to match a side wall admittance that includes radiation conductance and discontinuity susceptance.§§ The major disadvantage of patches is their narrow bandwidth, due to the close spacing between patch and ground plane. Fig. 27 shows bandwidth versus substrate height in free space wavelengths, for vswr < 2.* Increasing the substrate height does increase the bandwidth, but the coupling to surface wave modes and the cross-polarized radiation also increase, with undesirable pattern effects. These also allow bandwidth above the straight line, for large heights. Most techniques for extending the bandwidth of patches have done so at the expense of efficiency, i.e., through use of matching networks with high standing waves. However, the parasitic patch† avoids these problems by keeping the substrate thickness low, with a parasitic patch above the driven patch increasing the effective radiation height. Bandwidth can be doubled, with the parasitic patch dimensions and height above the driven patch adjusted to give a symmetrical impedance curve. Fig. 28 sketches the parasitic patch configuration. These may be arrayed as are ordinary patches. In practice, the parasitic patches are printed on a thin dielectric substrate, with a foam layer used to support this substrate above the microstrip.

Rectangular or circular patches, as described, are mostly linearly polarized radiators, with patch widths roughly a third of a wavelength. Accordingly, the patterns are between those of a short dipole and a half-wave dipole. Although a square patch could be fed on two adjacent sides with the proper phases to obtain circularly polarized radiation, simpler circularly polarized patches have been developed.§ The key is to modify dimensions to allow the two cross-polarized modes to be of equal amplitude and 90 degrees out of phase. Fig. 29 sketches four ways of accomplishing this. A simple analysis has been given by Lo and Richards.**

Loops

Loops small in wavelengths are equivalent to short magnetic dipoles (see the general section at the beginning of this chapter). When the circumference of a loop is small in wavelengths and the area is A, the patterns are given by

$$E_\phi = (120\pi^2 I_0 A \mu_e \sin\theta)/r\lambda^2$$

$$H_\theta = (\pi I_0 A \mu_e \sin\theta)/r\lambda^2$$

---

† Brown, G. H., Lewis, R. F., and Epstein, J. "Ground Systems as a Factor in Antenna Efficiency." *Proc. IRE,* Vol. 25, June 1937, pp. 753–787.

§ Christman, A. M. "Ground Systems for Vertical Antennas," *Ham Radio*, August 1979, pp. 31–33.

** Park, P. K., and Elliott, R. S. "Design of Collinear Longitudinal Slot Arrays Fed by Boxed Stripline." *IEEE Trans.*, Vol. AP-29, pp. 135–140.

‡ Deschamps, G., and Sichak, W. "Microstrip Microwave Antenna." Proc. 1953 Allerton Antenna Symposium, University of Illinois.

§§Lo, Y. T., et al. "Theory and Experiment on Microstrip Antennas." *Trans. IEEE*, Vol. AP-27, 1979, pp. 137–145. Carver K. R. "Practical Analytical Techniques for the Microstrip Antenna." *Printed Circuit Antenna Technology Workshop*, New Mexico State Univ., 1979. Hammerstad, E. O. "Equations for Microstrip Circuit Design." *Proc. European Microwave Conf.*, 1975, pp. 268–272.

* Derneryd, A. G., and Lind, A. G. "Extended Analysis of Rectangular Microstrip Resonant Antennas." *Trans. IEEE*, Vol. AP-27, 1979, pp. 846–849.

† Hall, P. S., et al. "Wide Bandwidth Microstrip Antennas for Circuit Integration." *Electron. Lett.*, 15, 1979, pp. 458–460.

§ Kerr, J. L. "Microstrip Antenna Developments." *Proc. Printed Circuit Antenna Technology Workshop,* October 1979, New Mexico State University.

** Lo, Y. T., and Richards, W. F. "Perturbation Approach to Design of Circularly Polarized Microstrip Antennas." *Elec. Letters*, Vol. 17, 28 May 1981, pp. 383–385.

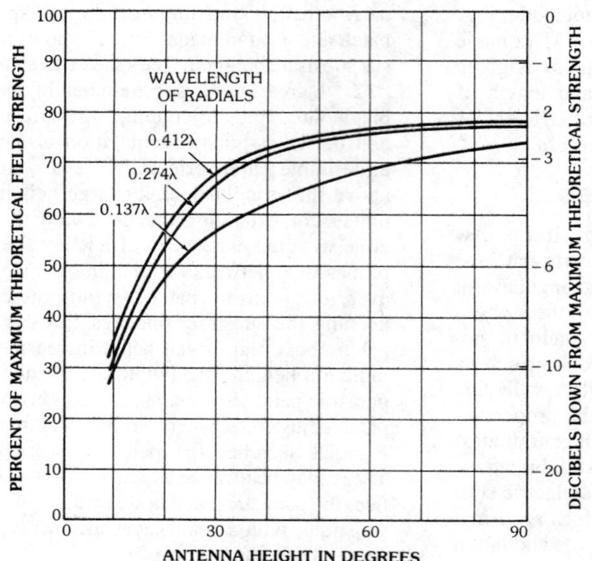

(A) 15 radials.

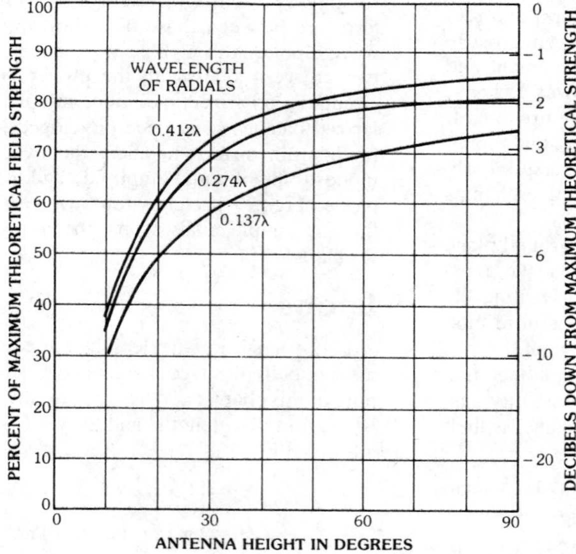

(B) 30 radials.

Fig. 25. Efficiency vs monopole height for

where,

I_0 is the loop current,

θ is measured from the loop axis.

Radiation resistance is

$$R_{\text{rad}} = 20N^2k^4A^2\mu_e^2$$

where μ_e is the effective permeability of a magnetic core. Effective length is

$$l_e = NkA\mu_e$$

The effective area and directivity are those of a short dipole: $3\lambda^2/8\pi$ and 1.5. Magnetically cored loops almost universally use ferrite cores with the highest permeability available for the frequencies of interest. Typically, the coil diameter is small in wavelengths, and the core is long compared with the diameter. Whether the coil is distributed over the core or concentrated near the center is important mostly for practical factors such as distributed capacitance. Tape cores can be used below a few hundred kilohertz with permeabilities of many tens of thousands; ferrite cores extend

(C) 60 radials.

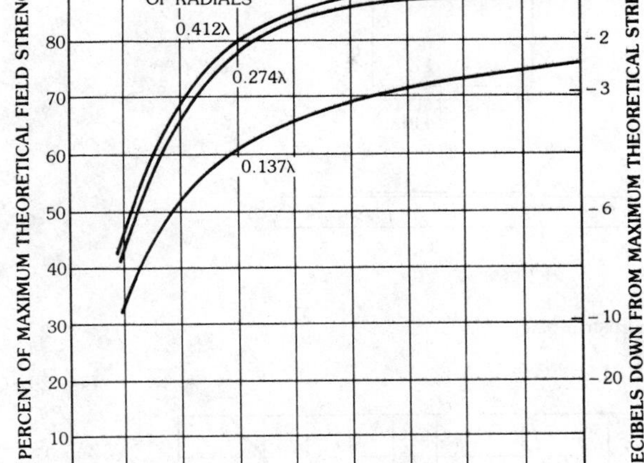

(D) 113 radials.

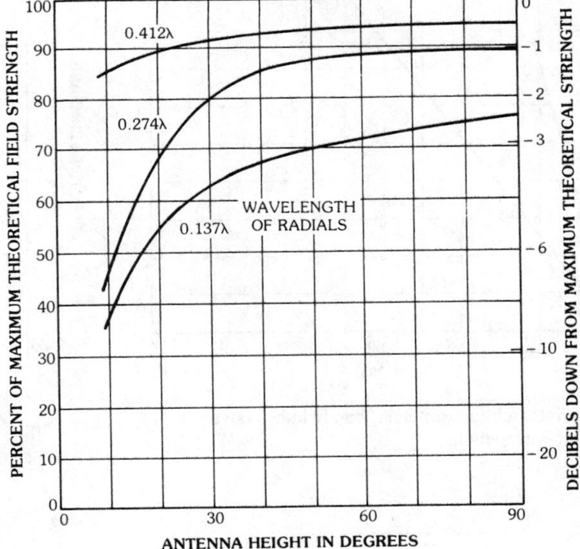

various numbers and lengths of radials.

into several hundred megahertz, with permeabilities of thousands at lower frequencies and on the order of 20 at upper frequencies. Effective permeability can be calculated by assuming the core is a prolate spheroid. With the formulas of Wait,‡ calculations of μ_e in terms of L/d, the core length/diameter ratio, have been made. Fig. 30 shows these data, which come from:

$$\mu_e/\mu = 1/[1 + (\mu - 1)\,(\zeta^2 - 1)\,Q_1]$$

where,

$$\zeta = (L/d)/[(L^2/d^2) - 1]$$

and

$$Q_1 = (\zeta/2)\ \ln\,[(\zeta + 1)/(\zeta - 1)] - 1$$

Higher permeabilities require larger L/d to realize μ_e as a significant fraction of μ. The core need not be solid, as much of the magnetic field is concentrated

‡ Wait, J. R. "Receiving Properties of a Wire Loop With a Spheroidal Core." *Can. J. Tech.*, Vol. 31, Jan. 1953, pp. 9–14.

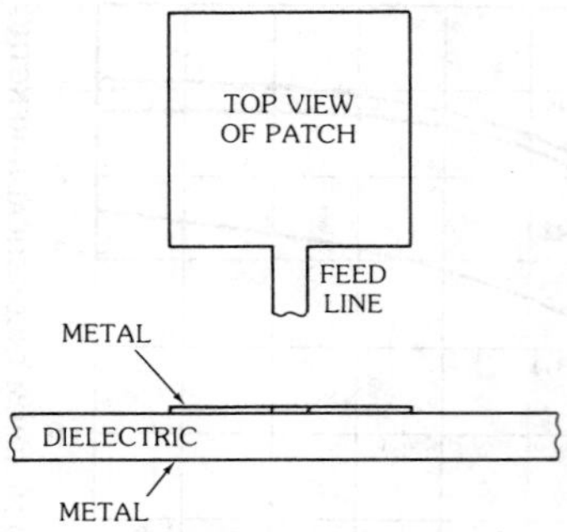

Fig. 26. Microstrip patch.

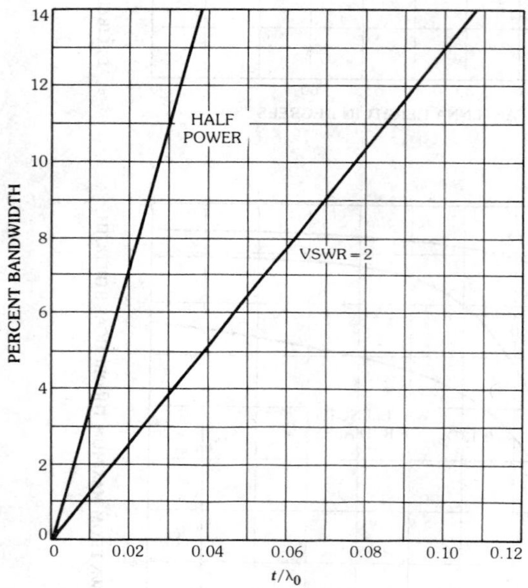

Fig. 27. Printed-circuit antenna bandwidth, vswr = 2, vs thickness/air wavelength.

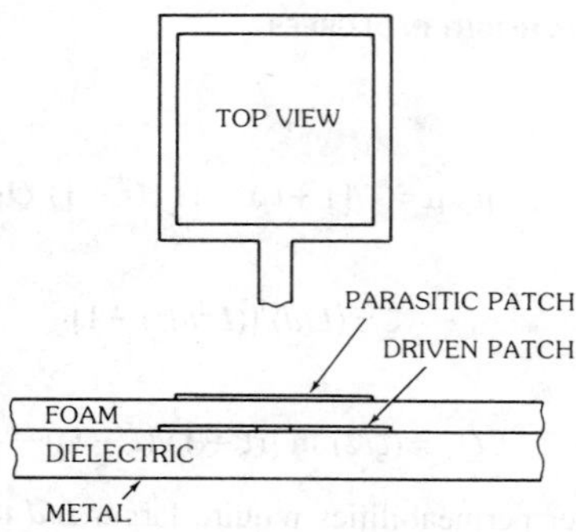

Fig. 28. Parasitic microwave patch.

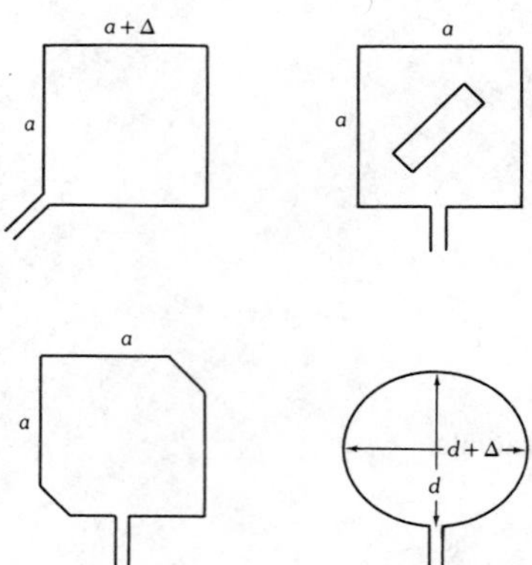

Fig. 29. Circularly polarized patches.

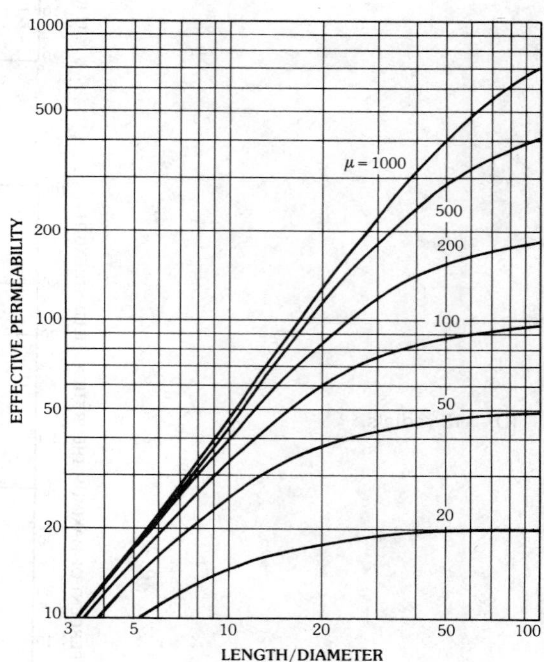

Fig. 30. Effective permeability of spheroidal core.

near the surface. Again, the hollow cylindrical core can be approximated by two prolate spheroids, and a formula for μ_e was developed by Wait.§§ Figs. 31, 32, and 33 show hollow-core effective permeability for $\mu = 20$, 100, and 500, respectively.* It is apparent that a significant part of the solid core may be elimi-

§§ Wait, J. R. "The Receiving Loop With a Hollow Prolate Spheroidal Core." *Can. J. Tech.*, Vol. 31, June 1953, pp. 132–137.

* Hansen, R. C. In *Microwave Engineers Handbook*, Vol. 2. T. S. Saad, ed. Dedham, Mass.: Horizon House, 1971; pp. 44–46 and 89–96.

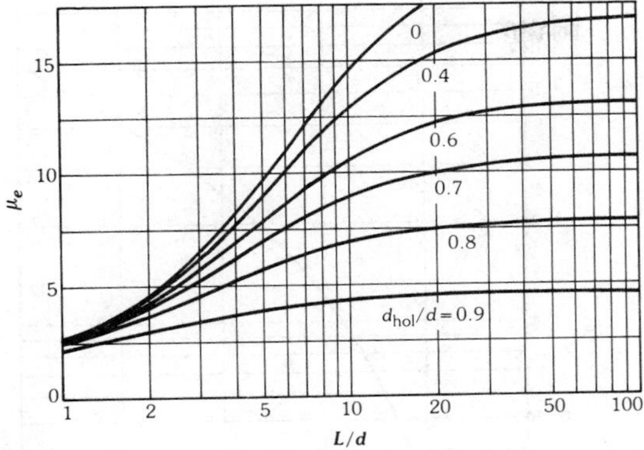

Fig. 31. Effective permeability of hollow spheroidal core, for $\mu = 20$.

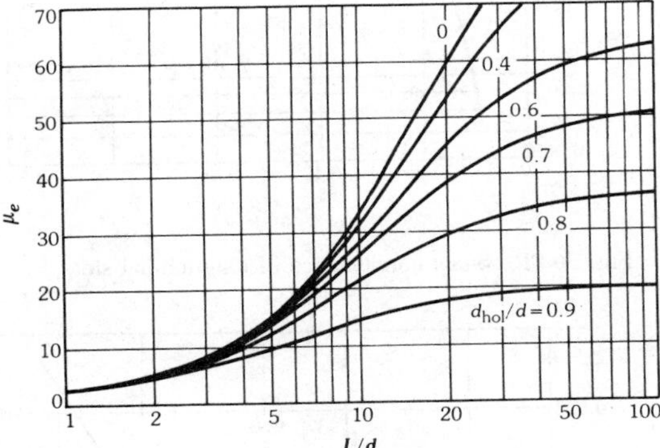

Fig. 32. Effective permeability of hollow spheroidal core, for $\mu = 100$.

nated. Such a hollow core can be assembled from parallel rods or strips around a foam block.

Slot Antennas

The properties of many slot antennas can be deduced from the properties of the complementary metallic antenna as enunciated by Booker in his extension of the optical work of Babinet. The admittance, Y_s, of the slot antenna is related to the impedance, Z_d, of the metallic antenna by

$$Z_d/Y_s = (120\pi)^2/4$$

The magnitude of the electric field, E_s, produced by the slot is proportional to the magnitude of the magnetic field, H_m, of the metallic antenna, and H_s is proportional to E_m. The electric- and magnetic-plane patterns of the slot are similar to the magnetic- and electric-plane patterns, respectively, of the metallic antenna. In Fig. 34, a rectangular slot antenna in an infinite metallic plane is shown with the complementary strip

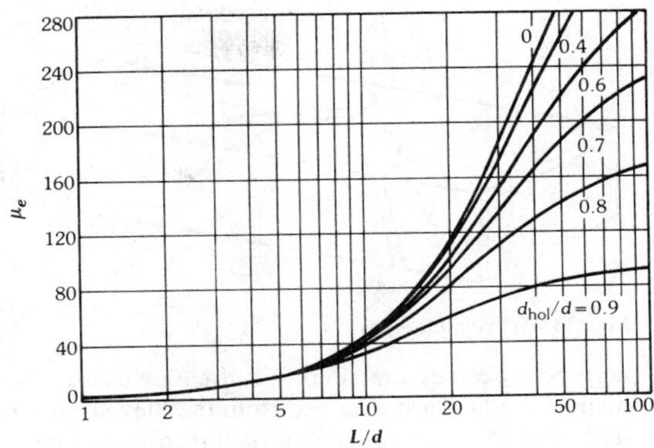

Fig. 33. Effective permeability of hollow spheroidal core, for $\mu = 500$.

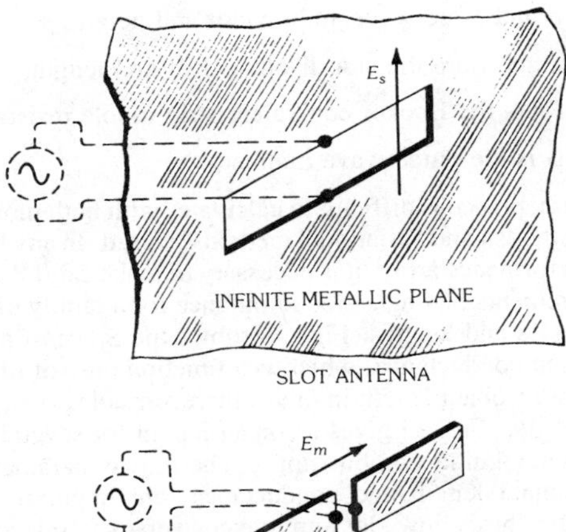

Fig. 34. Slot antenna and strip dipole.

dipole. If the slot radiates on only one side, the relationship is

$$Z_d/Y_s = (120\pi)^2/2$$

The *E*-plane pattern of the slot and *H*-plane pattern of the dipole are omnidirectional, while the slot *H*-plane pattern is the same as the dipole *E*-plane pattern.

Due to the ease of feeding, most slots are located in waveguide. Fig. 35 shows displaced longitudinal, inclined, and edge slots; the first and last are widely used for linear and planar arrays. The pattern and mutual admittance of slots in an array are close to those of the Booker complementary strip dipole array, and dipole mutual impedances are used in the design of such arrays. The slot self-admittance, however, contains a term involving reactive energy in the waveguide in the vicinity of the slot. Longitudinal slots are covered

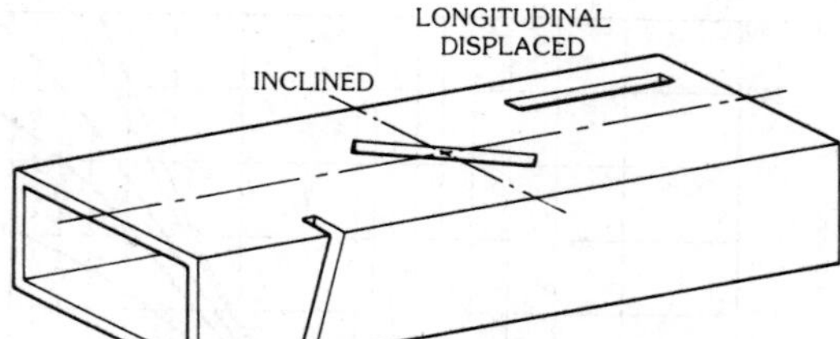

Fig. 35. Waveguide slots.

here because they are simpler and easier to use. Radiation conductance obtained from the Stevenson formula and for a sinusoidal slot field distribution is†

$$\frac{G_{rad}}{Y_0}=\frac{480(a/b)[\sin^2(\pi x/a)]\ (\cos\beta l-\cos kl)^2}{\pi\beta/k\ R_0}$$

where,

a and b are waveguide width and height,

x and l are the slot displacement and length,

R_0 is the Booker equivalent strip dipole resistance,

β is the guide wave number.

Susceptance is difficult to calculate, with both moment methods§ and variational methods** used. In any high-performance array, it is necessary to make careful measurements of single-slot admittance for a family of slot offsets and lengths. Fig. 36 shows the Stevenson resonant conductance, which is a function only of offset. Slot resonant length involves three variables: x/a, a/b, and β/k. Fig. 37 gives resonant length for several frequency-guide combinations. These two parameters, resonant length and conductance, are important because they allow slot admittances versus frequency to be plotted in the Kaminow and Stegen "universal" form.‡ Slot length is given in terms of resonant length, and both conductance and susceptance are given in terms of resonant conductance. The result was originally thought to be a single symmetric curve for G/Y_0 and a single asymmetric curve for B/Y_0. As Fig. 38 shows, it now appears that although all longitudinal slots fit these curves approximately, there are differences that vary with x/a, a/b, and β/k. The precise

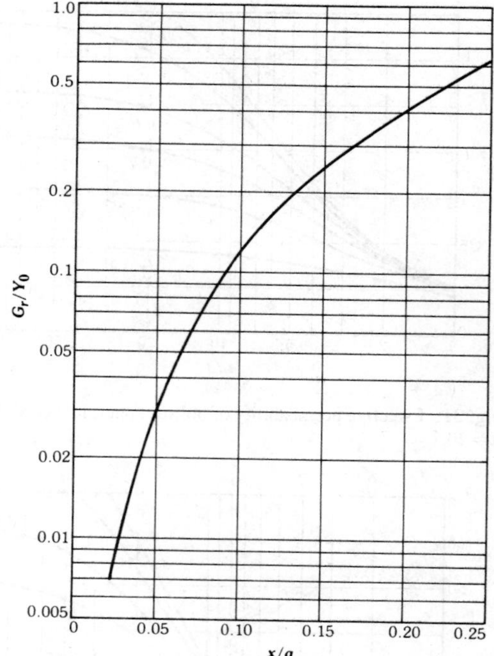

Fig. 36. Resonant conductance of longitudinal slot.

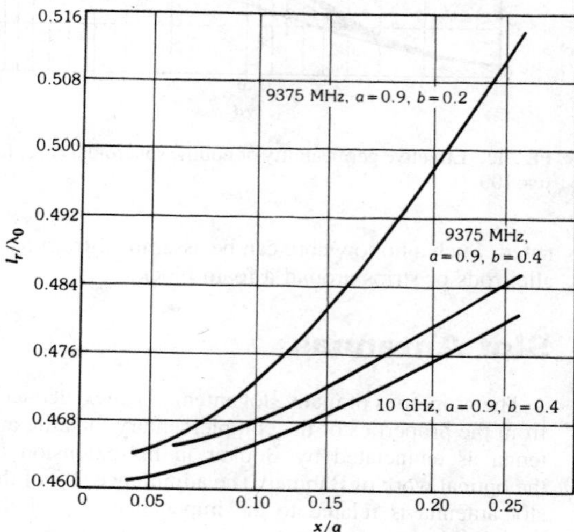

Fig. 37. Resonant length of longitudinal slot.

control of aperture distribution allowed by waveguide slots and the use of computer design have allowed high-performance low sidelobe or tailored sidelobe arrays to be constructed.§§

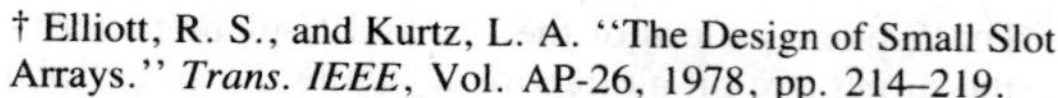

† Elliott, R. S., and Kurtz, L. A. "The Design of Small Slot Arrays." *Trans. IEEE*, Vol. AP-26, 1978, pp. 214–219.

§ Khac, T. V., and Carson, C. T. "Impedance Properties of a Longitudinal Slot Antenna in the Broad Face of a Rectangular Waveguide." *Trans. IEEE*, Vol. AP-21, 1973, pp. 708–710.

** Yee, H. Y. "Impedance of a Narrow Longitudinal Shunt Slot in a Slotted Waveguide Array." *Trans. IEEE*, Vol. AP-22, 1974, pp. 589–592. Hansen, R. C. Array chapters in *Handbook of Antenna Design*, Vol. 2. A. W. Rudge, et al, eds. London: Peter Peregrinus Ltd., 1983.

‡ Blass, J. "Slot Antennas," Chapter 8 of *Antenna Engineering Handbook*. H. Jasik, ed. New York: McGraw Hill Book Co., 1961.

§§ Elliott, R. S. *Antenna Theory and Design*. Englewood Cliffs, N.J.: Prentice-Hall, Inc., 1981. Hansen, R. C. Array chapters in *Handbook of Antenna Design*, Vol. 2. A. W. Rudge, et al, eds. London: Peter Peregrinus Ltd., 1983.

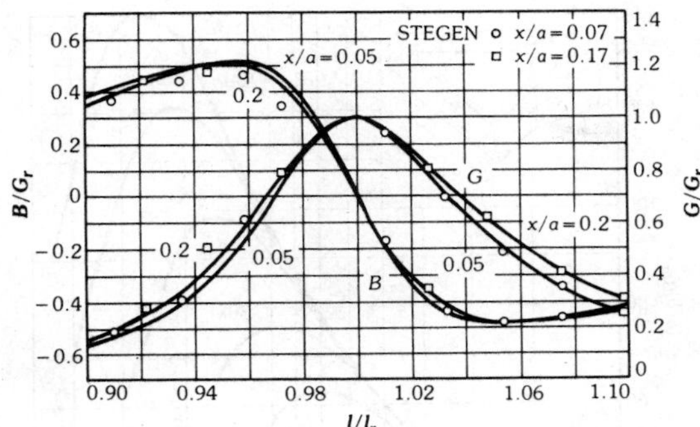

Fig. 38. Longitudinal-slot admittance, $a=0.9$, $b=0.4$, frequency = 9375 MHz.

Slots are sometimes used on cylinders, in either an axial or circumferential configuration. A circumferential half-wave slot has a smooth pattern, with the cylinder reducing the back lobes. Axial half-wave slots also exhibit shadowing, with pattern ripples around $\phi=\pi$. As ka increases, the number of ripples increases, but their depth and angular extent decrease. For both types of slots, larger cylinders give more shadowing and lower field at $\phi=\pi$. Fig. 39 shows typical axial and circumferential slot patterns in the azimuth

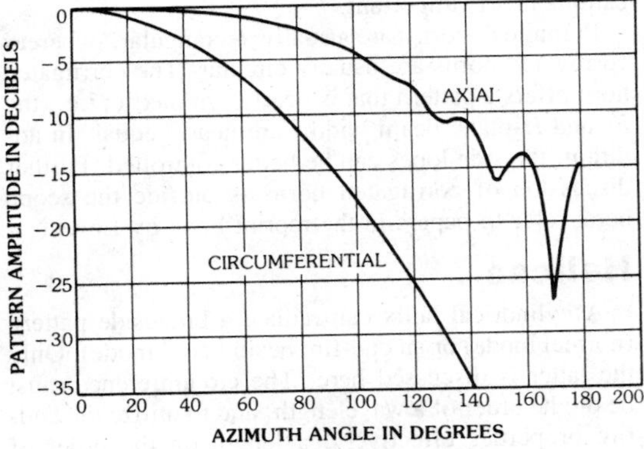

Fig. 39. Half-wave slot on cylinder, $ka=8$.

plane for $ka=8$. These were calculated with the formulas of Wait:*

Axial

$$E_\phi=\frac{1}{\pi ka}\sum_{n=0}^{\infty}\frac{\epsilon_n j^n \cos n\phi}{H_n^{(2)\prime}(ka)}$$

Circumferential

$$E_\theta=\frac{ka}{\pi}\sum_{n=0}^{\infty}\frac{\epsilon_n j^n[\cos(n\pi/2ka)]\cos n\phi}{(k^2a^2-n^2)\,H_n^{(2)}(ka)}$$

* Wait, J. R. *Electromagnetic Radiation from Cylindrical Structures*. Elmsford, N.Y.: Pergamon Press, Inc., 1959.

Here $\epsilon_n=1$ for $n=0$, $\epsilon_n=2$ for $n>0$, and $H_n^{(2)}$ is the Hankel function of the second kind. Three-dimensional patterns can be calculated by replacing ka by $ka \sin\theta$, except that finite length cylinders will have end effects that require other calculation methods such as geometric theory of diffraction.†

MEDIUM-GAIN ANTENNAS

Horns

Horns are often used as gain standards, as feeds for dish antennas, and as microwave antennas directly. The simplest type of horn is a flared waveguide. For a flare only in the E-plane or only in the H-plane, directivity can be accurately and simply calculated:§

$$G_E=(64\,a\,R_e/\pi\lambda b)\,[C^2(w)+S^2(w)]$$
$$=32\,ab\,F_E/\pi\lambda^2$$

where,

$$w=b/\sqrt{2\lambda R_e}$$

$$G_H=(4\pi bR_m/\lambda a)\,\{[(C(u)-C(v)]^2+[S(u)-S(v)]^2\}$$
$$=32\,ab\,F_H/\pi\lambda^2$$

where,

$$\sqrt{2}\,u=\sqrt{\lambda R_m}/a+a/\sqrt{\lambda R_m}$$
$$\sqrt{2}\,v=\sqrt{\lambda R_m}/a-a/\sqrt{\lambda R_m}$$

In the formulas above, R_e and R_m are the slant lengths, and C and S are conventional Fresnel integrals.** The formulas are also written in terms of directivity reduction factors F_E and F_H, which are shown in Fig. 40. Figs. 41 and 42 show normalized directivity for each of these, and it may be seen that for each slant length there is an optimum width or height. More commonly, a pyramidal horn is used with dimensions selected to be optimum in both planes. However, this may not have coincident phase centers; see below. The pyramidal horn directivity can be calculated from the constituent sectoral horn directivities by

$$G_{\text{pyramidal}}=\pi\lambda^2\,G_E\,G_H/32ab=32\,ab\,F_E\,F_H/\pi\lambda^2$$

This can also be written in terms of E- and H-plane gain reduction factors as shown in the formula above. Half-power beamwidths are given approximately by $\theta_3=51\lambda/b$ for the E-plane and $70\lambda/a$ for the H-plane.

† Hansen, R. C., ed. *Geometric Theory of Diffraction*. IEEE Press/Wiley, 1981.

§ Schelkunoff, S. A., and Friis, H. T., eds. *Antennas, Theory and Practice*. New York: John Wiley & Sons, Inc., 1952.

** Abramowitz, M., and Stegun, I. A., eds. *Handbook of Mathematical Functions*, NBS Applied Math Series, 1970.

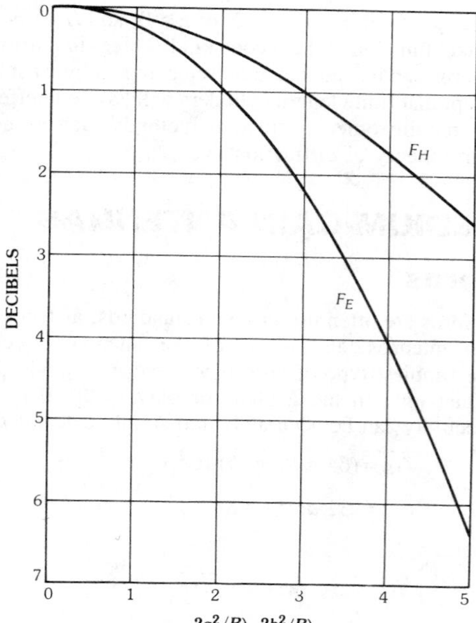

Fig. 40. Sectoral-horn directivity factors.

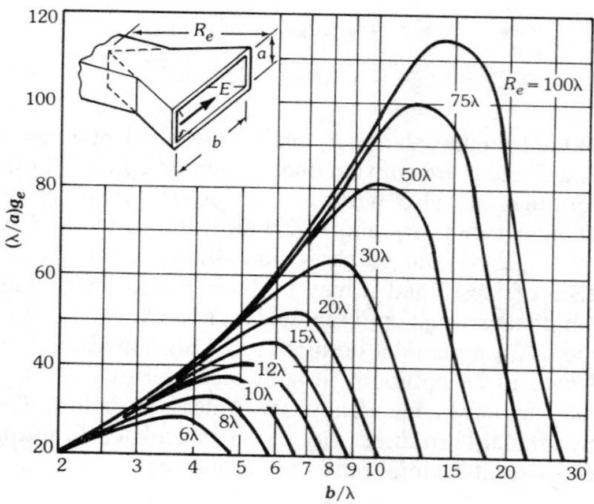

Fig. 41. Directivity of a large horn flared in the electric plane. (*After Schelkunoff.*)

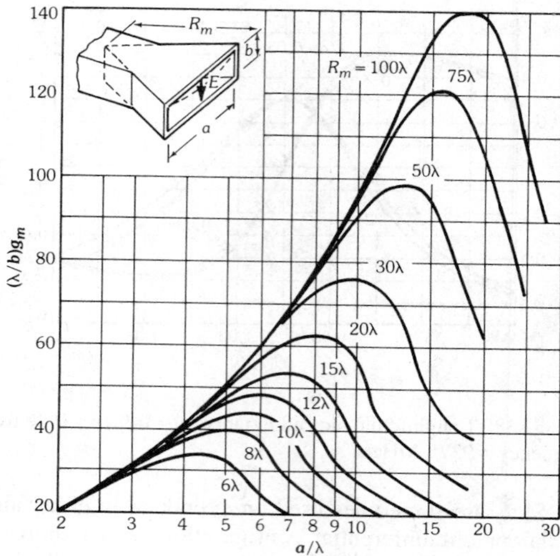

Fig. 42. Directivity of a large horn flared in the magnetic plane. (*After Schelkunoff.*)

The beamwidths in each plane are essentially independent of dimensions in the other plane. Fig. 43 shows experimental patterns in both planes for a variety of radial lengths and flare angles.‡‡ Wide-angle patterns may be computed by using geometric theory of diffraction; see reprint book by Hansen.§§ Phase centers of pyramidal horns may also be calculated by using Fresnel integrals.* The phase center is generally different for each plane, and it is located toward but not at the horn apex. The optimum rectangular pyramidal horn has slightly different dimensions from one designed with coincident phase centers. For some applications this is important.

Pyramidal horns are usually rectangular, whereas corrugated horns are usually circular. The corrugated horn offers a pattern that is nearly symmetric; i.e., the *E*- and *H*-plane beam widths are nearly equal. In addition, the side lobes can be better controlled. Further discussion of corrugated horns is outside the scope here; refer to papers in the reprint book by Love.†

Helices

A cylindrical helix can radiate a broadside pattern (normal mode) or an end-fire beam (axial mode). Only the latter is discussed here. The circumference must be on the order of a wavelength, and to utilize the end-fire properties effectively, a length on the order of several wavelengths is usually used. Fig. 44 shows beamwidth versus length for several circumferences. The axial-mode helix is broadband but is limited to roughly a 2.5:1 frequency ratio.§ A wider bandwidth can be covered by use of a long multifilar helix such as a quadhelix.** In this configuration, the helix con-

‡‡ Rhodes, D. R. "An Experimental Investigation of the Radiation Patterns of Electromagnetic Horn Antennas." *Proc. IRE*, Vol. 36, Sept. 1948; pp. 1101–1105.

§§ Hansen, R. C., ed. *Geometric Theory of Diffraction*. IEEE Press/Wiley, 1981.

* Muehldorf, E. I. "The Phase Center of Horn Antennas." *Trans. IEEE*, Vol. AP-18, Nov. 1970, pp. 753–760.

† Love, A. W., ed. *Electromagnetic Horn Antennas*. IEEE Press/Wiley, 1976.

§ Kraus, J. D. *Antennas*. New York: McGraw-Hill Book Co., 1950.

** Adams, A. T., Greenough, R. K., Wallenberg, R. F., Mendelovicz, A., and Lumjiak, C. "The Quadrifilar Helix Antenna." *Trans. IEEE*, Vol. AP-22, March 1974, pp. 173–178.

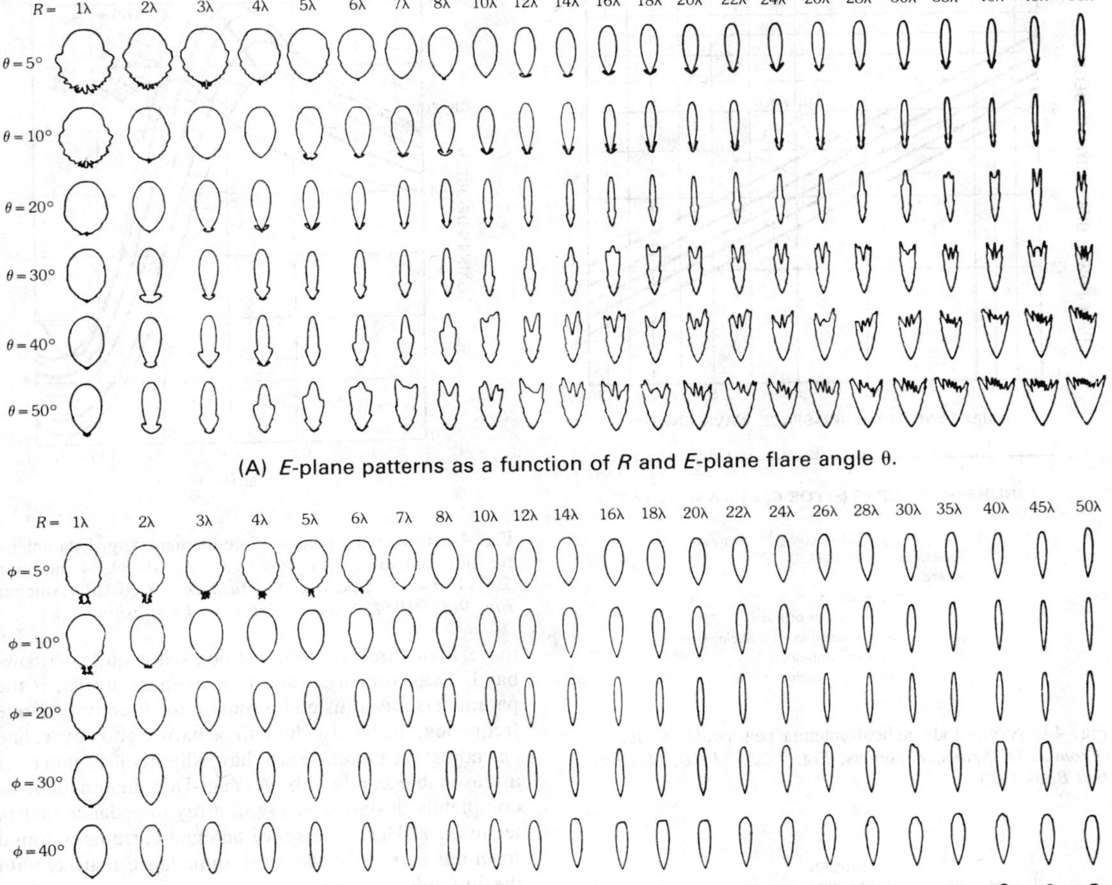

Fig. 43. Patterns of a rectangular horn antenna with radial horn length *R*. (*From Rhodes, D. R. "An Experimental Investigation of the Radiation Patterns of Electromagnetic Horn Antennas."* Proc. IRE, *Vol. 36, Sept. 1948, pp. 1101–1105.*)

sists of four windings spaced 90 degrees with the wires fed with a 90-degree phase progression at the base. Although the feed is complicated, this type of helix does not require a ground plane, although a ground plane may improve performance. A short resonant version of the quadhelix that radiates an omnidirectional circularly polarized pattern is also available.‡ This quadhelix utilizes a fractional turn but operates without ground plane and can be small in size.

Yagi-Uda Antennas

The Yagi-Uda antenna is an end-fire array constructed usually of a single driven dipole with a reflector dipole behind and one or more parasitic director dipoles in front. Fig. 45 sketches a three-element Yagi-Uda antenna. Extensive design curves for three- and four-element Yagi-Uda antennas are given by Uda and Mushiake.§§ Fig. 46 is extracted from that work. One of the difficulties of designing Yagi-Uda antennas of many elements is the large number of variables. Typically, the spacing between directors is kept constant, and the lengths are adjusted. Table 3 from Stutzman and Thiele* shows performance of equally spaced Yagi-Uda antennas with up to seven elements. When the number of elements becomes large, the directors

‡ Kilgus, C. C. "Shaped-Conical Radiation Pattern Performance of the Backfire Quadrifilar Helix." *Trans. IEEE*, Vol. AP-23, May 1975, pp. 392–397.

§§ Uda, S., and Mushiake, Y. *Yagi-Uda Antenna*. Research Inst. of Electrical Communication, 1954.

* Stutzman, W. L., and Thiele, G. A. *Antenna Theory and Design*. New York: John Wiley & Sons, Inc., 1981.

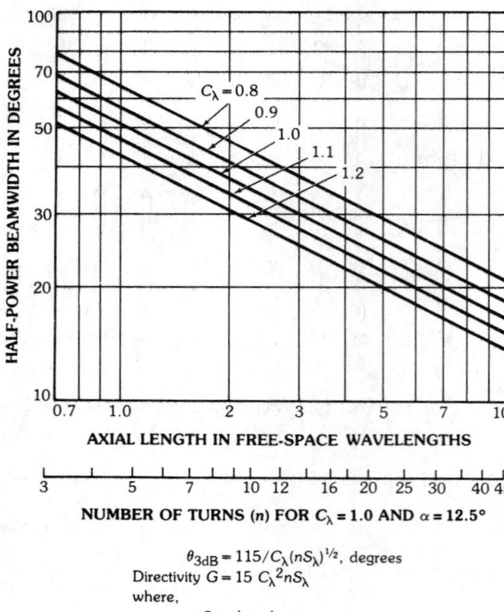

$\theta_{3dB} = 115/C_\lambda(nS_\lambda)^{1/2}$, degrees

Directivity $G = 15\ C_\lambda{}^2 nS_\lambda$

where,

nS_λ = length,
n = number of turns,
S_λ = spacing in air wavelengths,
C_λ = circumference,
α = pitch angle.

Fig. 44. Axial-mode helical-antenna beamwidth vs length. (*From J. D. Kraus,* Antennas, *Fig. 7-22, ©1950, McGraw-Hill Book Co.*)

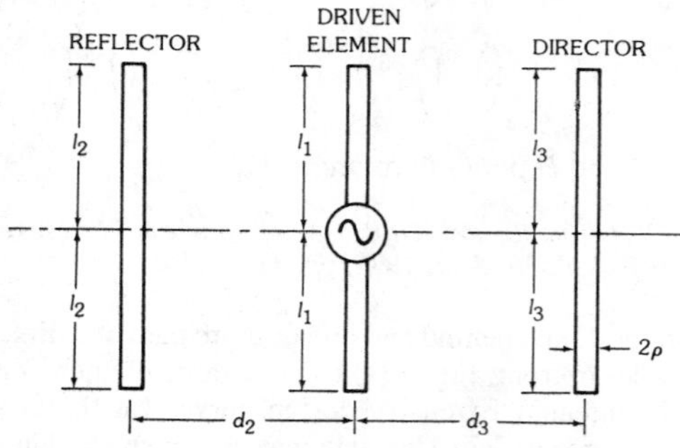

Fig. 45. Three-element Yagi-Uda antenna.

act as a slow-wave transmission line,† and the directivity can be approximately calculated from this.‡ Unfortunately, as the number of elements increases, the bandwidth decreases significantly so that arrays of

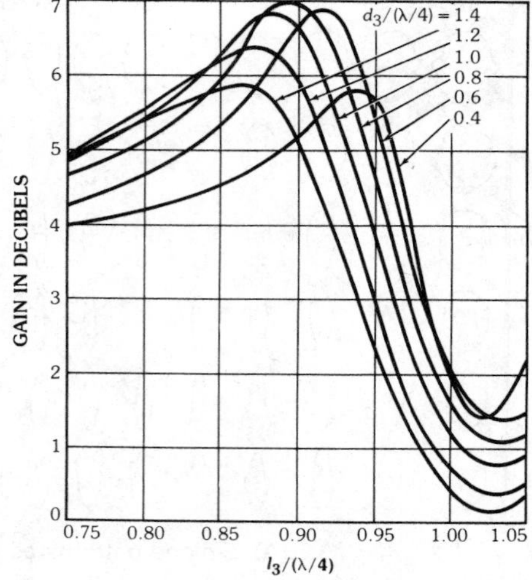

Fig. 46. Calculated gain of three-element Yagi-Uda antenna for indicated values of $d_3/(\lambda/4)$: $l_1 = l_2 = d_2 = \lambda/4$ and $\rho = \lambda/200$. (*From S. Uda and Y. Mushiake,* Yagi-Uda Antenna, *Fig. 9-3, ©1954, Sasaki Printing and Publishing Co.*)

more than three or four elements are quite narrow-band. Even for three- and four-element arrays, if the parameters are adjusted for maximum directivity at one frequency, the bandwidth will be narrow; however, one can adjust the parameters to have slightly less gain over a reasonable bandwidth. A Yagi-Uda antenna can be completely designed by use of array impedance matrix techniques. Here, the set of antenna currents is found from the inverse of the array impedance matrix, with the individual mutual impedance terms calculated by using the zero-order theory of Carter. Extensive calculations have been made by Lawson in a series of articles,§ using this technique. Fig. 47 shows gain and front-to-back ratio for various three-element Yagi-Uda antennas. Table 4 gives the parasitic lengths and relative resonances for the six curves in each graph. It should be noted that although a single Yagi-Uda can give excellent performance, arraying of Yagi-Udas where the booms are parallel is only partially satisfactory. Surface-wave antennas do not array well because their near fields interact in an undesirable way, and thus the gain realized from an array of Yagi-Uda arrays is always significantly less than the product of the gain of a single Yagi-Uda array times the number of arrays.

A related antenna is the short backfire antenna of Ehrenspeck.** As shown in Fig. 48, it consists of a

† Serracchioli, F., and Levis, C. A. "The Calculated Phase Velocity of Long End-Fire Uniform Dipole Arrays," *Trans. AP*, December 1959, pp. S424–S434.

‡ Ehrenspeck, H. W., and Poehler, H. "A New Method of Obtaining Maximum Gain From Yagi Antennas." *Trans. IEEE*, Vol. AP-7, October 1959, pp. 379–386.

§ Lawson, J. L. "Yagi Antenna Design." *Ham Radio*, Jan., Feb., May, June, July, 1980.

** Ehrenspeck, H. W. "A New Class of Medium-Size High-Efficiency Reflector." *Trans. IEEE*, Vol. AP-22, March 1974, pp. 329–332. Large, A. C. "Short Backfire Antennas With Waveguide and Linear Fields." *Microwave J.*, Vol. 19, August 1976, pp. 49–52. Kumar, A. "Backfire Antennas Aim at Direct Broadcast TV." *Microwaves Magazine*, April 1978, pp. 106–112.

TABLE 3. CHARACTERISTICS OF EQUALLY SPACED YAGI-UDA ANTENNAS*

N, No. of Elements	Spacing (wavelengths)	Element Lengths: Reflector, L_R (wavelengths)	Element Lengths: Driver, L (wavelengths)	Element Lengths: Directors, L_D (wavelengths)	Gain (dB)	Front-to-Back Ratio (dB)	Input Impedance (ohms)	H-Plane: HP_H (degrees)	H-Plane: SLL_H (dB)	E-Plane: HP_E (degrees)	E-Plane: SLL_E (dB)
3	0.25	0.479	0.453	0.451	9.4	5.6	$22.3+j15.0$	84	−11.0	66	−34.5
4	0.15	0.486	0.459	0.453	9.7	8.2	$36.7+j9.6$	84	−11.6	66	−22.8
4	0.20	0.503	0.474	0.463	9.3	7.5	$5.6+j20.7$	64	−5.2	54	−25.4
4	0.25	0.486	0.463	0.456	10.4	6.0	$10.3+j23.5$	60	−5.8	52	−15.8
4	0.30	0.475	0.453	0.446	10.7	5.2	$25.8+j23.2$	64	−7.3	56	−18.5
5	0.15	0.505	0.476	0.456	10.0	13.1	$9.6+j13.0$	76	−8.9	62	−23.2
5	0.20	0.486	0.462	0.449	11.0	9.4	$18.4+j17.6$	68	−8.4	58	−18.7
5	0.25	0.477	0.451	0.442	11.0	7.4	$53.3+j6.2$	66	−8.1	58	−19.1
5	0.30	0.482	0.459	0.451	9.3	2.9	$19.3+j39.4$	42	−3.3	40	−9.5
6	0.20	0.482	0.456	0.437	11.2	9.2	$51.3-j1.9$	68	−9.0	58	−20.0
6	0.25	0.484	0.459	0.446	11.9	9.4	$23.2+j21.0$	56	−7.1	50	−13.8
6	0.30	0.472	0.449	0.437	11.6	6.7	$61.2+j7.7$	56	−7.4	52	−14.8
7	0.20	0.489	0.463	0.444	11.8	12.6	$20.6+j16.8$	58	−7.4	52	−14.1
7	0.25	0.477	0.454	0.434	12.0	8.7	$57.2+j1.9$	58	−8.1	52	−15.4
7	0.30	0.475	0.455	0.439	12.7	8.7	$35.9+j21.7$	50	−7.3	46	−12.6

Conductor diameter $= 0.005\lambda$.

* From Stutzman, W. L., and Thiele, G. A., *Antenna Theory and Design*. New York: John Wiley & Sons, Inc., 1981.

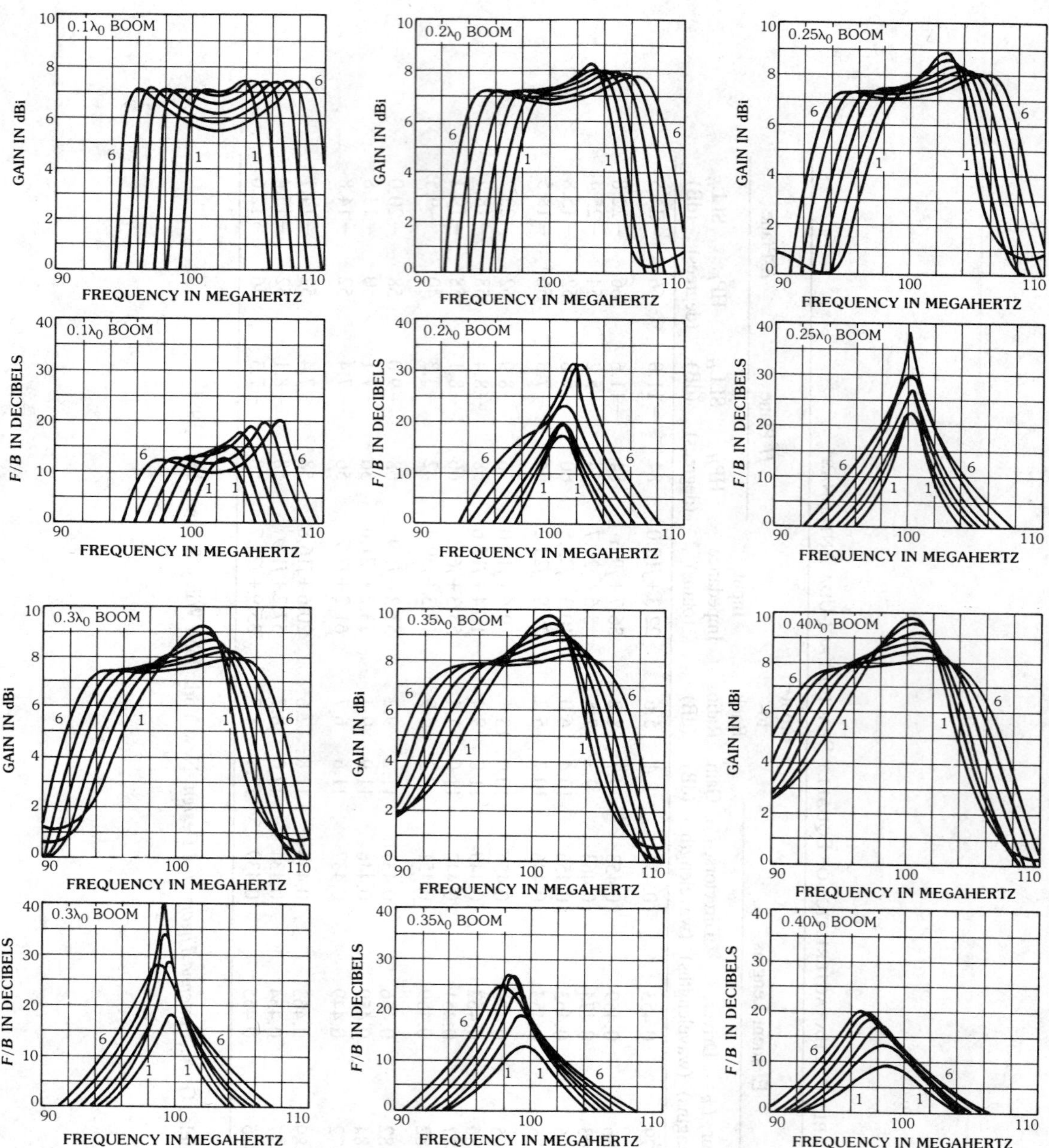

Fig. 47. Gain and front-to-back (F/B) ratio for three-element Yagi beams with varying boom lengths and for changing reflector and director lengths. (*From Lawson, J. L. "Yagi Antenna Design."* Ham Radio, *Jan., Feb., May, June, July, 1980.*)

flat plate with a dipole in front of the plate, and with a cylindrical rim around the plate. The dipole feed typically has a dipole and small plate reflector. Performance of the antenna is affected by both the plate diameter and the depth of the rim. A single feed can produce gains in the region of 10 dB. Several of these may be arrayed, all over a flat reflector with a common rim. However, when more than four feed elements are used, a parabolic reflector may be competitive. Because the antenna has dimensions that are in the resonance region, analyses have only been partially successful. Precise analysis will probably require patch moment methods. Information on bandwidth is sketchy, but the aperture efficiencies tend to be well above those for conventional parabolic-reflector antennas. Higher gain may be realized by feeding the reflecting plate by a feed dipole and dipole reflector, with a number of director dipoles between the feed

TABLE 4. PARASITIC LENGTHS AND RELATIVE RESONANT FREQUENCIES

Curve	Reflector Length/λ	Reflector Resonance	Director Length/λ	Director Resonance
1	0.49150	0.98	0.47223	1.02
2	0.49657	0.97	0.46764	1.03
3	0.50174	0.96	0.46314	1.04
4	0.50702	0.95	0.45873	1.05
5	0.51241	0.94	0.45441	1.06
6	0.51792	0.93	0.45016	1.07

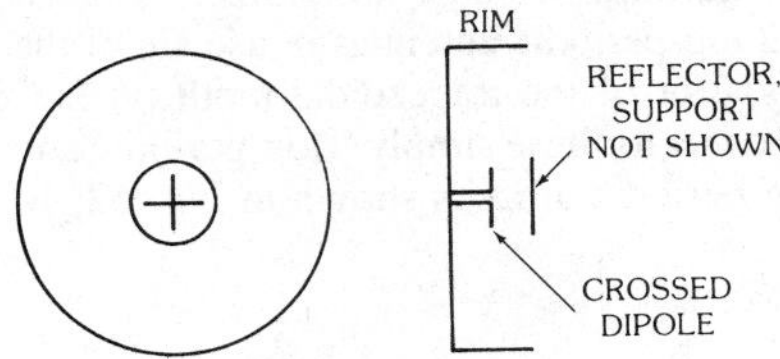

Fig. 48. Backfire antenna.

dipole and the plate. In this scheme, the Yagi-Uda performance is combined with that of the backfire.

Frequency-Independent Antennas

The principle of frequency-independent antennas was established circa 1957 through the recognition that conventional antenna bandwidth limitations occur because critical dimensions change in wavelengths. An antenna specified only in angles should then be frequency independent. Of course, all antennas are of finite size, so an absolute low-frequency cutoff must exist, but within the maximum size the angular prescription can be followed.†† A commonly used frequency-independent antenna is the spiral, which exists in two forms. The equiangular spiral is shown in Fig. 49. The conductor edges are formed in polar coordinates by

$$r_1 = r_0 e^{a\phi}, \qquad r_2 = c r_1$$

Another version of the spiral is the Archimedean shown in Fig. 50. Although this spiral is not prescribed only by angles, it gives excellent performance and is widely used. Both spirals operate through an "active region" from which the radiation takes place. At the highest frequency, the active region is contiguous with the feed, and as the frequency increases, the active region moves out to the edge. Corresponding with the active-region movement is a rotation of the pattern. The pattern is circularly polarized corresponding to the direction of the spiral arms, and the pattern exhibits a peak on the axis. Typically, the pattern is broad and the gain slightly more than that of a dipole. Since the circumference of the active region is a wavelength, the spiral must have a diameter of more than λ/π at the lowest frequency. The spiral may be used for azimuth direction finding by utilizing four, six, or eight arms. A four-arm spiral, for example, would have four interleaved arms fed with 90-degree phase rotation by a simple network of hybrids. The higher mode utilized in the four-arm spiral requires a diameter twice as large

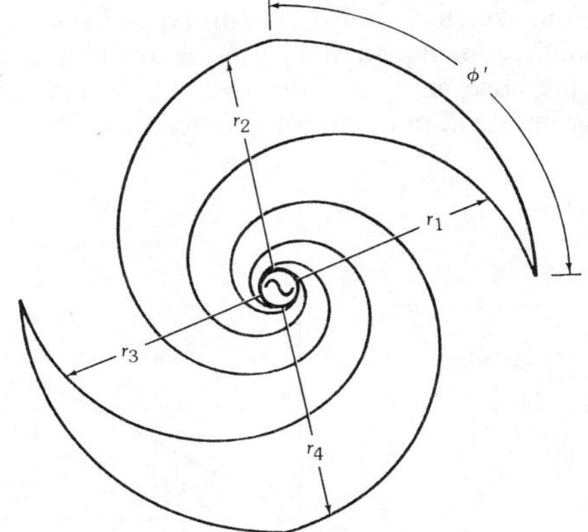

Fig. 49. Two-arm equiangular spiral. (*Courtesy Wolff.*)

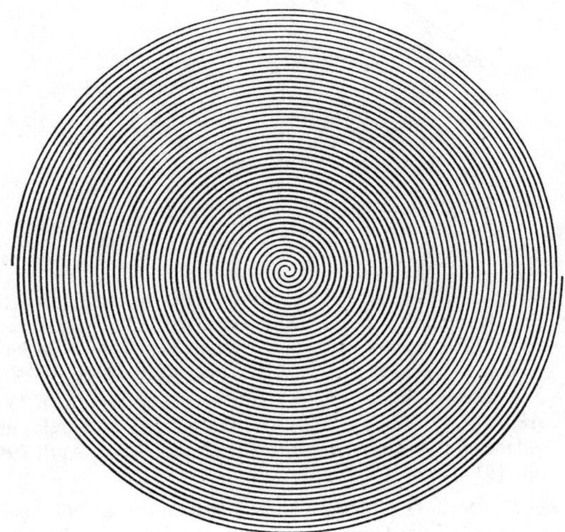

Fig. 50. Archimedean spiral. (*Courtesy Wolff.*)

†† Rumsey, V. H. *Frequency Independent Antennas*. New York: Academic Press, Inc., 1966.

as that of the two-arm spiral so that the diameter now approaches a wavelength. Spirals with more arms require even larger diameters. The spiral by itself radiates on both sides, and a cavity is often used to allow surface mounting. Unfortunately, the cavity usually significantly reduces the bandwidth, and in order to achieve several octaves, it is necessary to accept some gain degradation via the use of absorbing materials and other techniques. Further information may be found in several references.‡‡

The conical spiral consists of a two-arm spiral deployed on a cone with the feed point at the truncated apex as sketched in Fig. 51. This type of antenna gives a unidirectional pattern with the beamwidth controlled by the cone angle and the size. Further information may be found in a number of references.§§

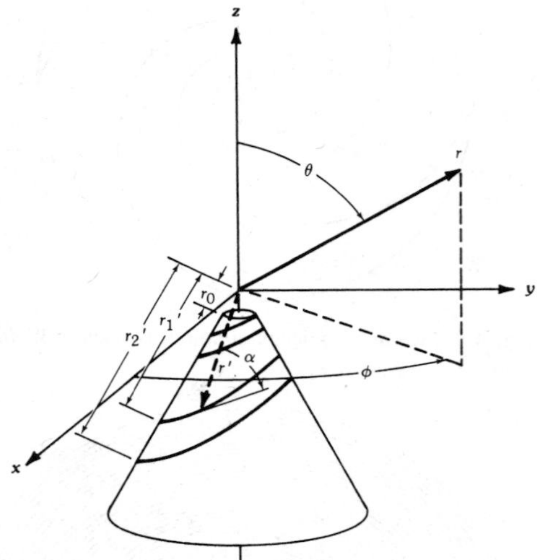

Fig. 51. Conical spiral. (*Courtesy Wolff.*)

‡‡ Sivan-Sussman, R. "Various Modes of the Equiangular Spiral Antenna." *Trans. IEEE,* Vol. AP-11, Sept. 1963, pp. 533–539. Kaiser, J. A. "The Archimedean Two-Wire Spiral Antenna." *Trans. IEEE,* Vol. AP-8, May 1960, pp. 312–323. Curtis, W. L. "Spiral Antennas." *Trans. IEEE,* Vol. AP-8, May 1960, pp. 298–306. Dyson, J. D. "The Equiangular Spiral Antenna." *Trans. IEEE,* Vol. AP-7, April 1959, pp. 181–187.

§§ Dyson, J. D. "Measuring the Capacitance per Unit Length of Two Infinite Cones of Arbitrary Cross Section." *Trans. IEEE,* Vol. AP-7, January 1959, pp. 102–103. Dyson, J. D. "Characteristics and Design of the Conical Log-Spiral Antenna." *Trans. IEEE,* Vol. AP-13, July 1965, pp. 488–499. Dyson, J. D., and Mayes, P. E. "New Circularly Polarized Frequency-Independent Antennas With Conical Beam or Omnidirectional Patterns." *Trans. IEEE,* Vol. AP-9, July 1961, pp. 334–342. Tang, C. H. "A Class of Modified Log-Spiral Antennas." *Trans. IEEE,* Vol. AP-11, July 1963, pp. 422–427. Atia, A. E., and Mei, K. K. "Analysis of Multiple-Arm Conical Log-Spiral Antennas." *Trans. IEEE,* Vol. AP-19, May 1971, pp. 320–331.

Another way of achieving a frequency-independent antenna is to design an active region that moves not continuously as in the case of the spiral, but in a discrete fashion. The antenna must now have a performance that is repetitive over a number of suitably related frequency subbands, and this, of course, is the log-periodic antenna discussed next.

Log-Periodic Antennas

Although the term "log-periodic" can be applied to any antenna designed with a structure that is periodic in the logarithm of some normalized dimensions, almost all log-periodic antennas in use are of the dipole-array type or of the trapezoidal-tooth type. Common usage now calls these simply "log-periodic antennas." The log-periodic array is shown in Fig. 52, where the

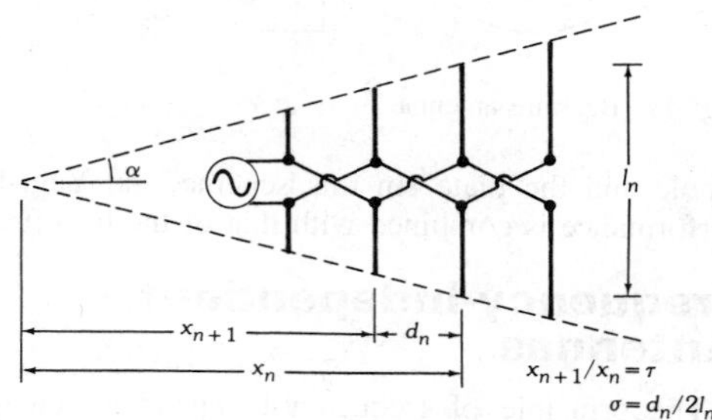

Fig. 52. Log-periodic antenna. (*Courtesy Weeks.*)

critical parameters σ and τ are described. It can be seen that the dipoles fit within an envelope of fixed angle and that the ratio of spacing is constant. The log-periodic antenna operates through an "active region" from which the radiation takes place. This active region consists of those elements that are around $\lambda/2$ in length. Obviously, the closer τ is to unity, the larger the active region and the higher the gain. The low-frequency cutoff occurs when the active region has reached the large end of the antenna. Conversely, the high-frequency cutoff occurs with the active region at the small end. In principle, the log-periodic antenna may be designed to cover many octaves, but, in practice, the upper frequency is limited by the precision required in constructing the small elements, feed lines, and support structure, and by the feeder loss, since the antenna is fed from the small end. Fig. 53 from Carrel* shows directivity versus σ and τ. Fig. 54 shows typical construction using a two-wire feed line. The dipole diameter is not critical, although it is generally desirable to use smaller diameter dipoles at the small end than at the large end when the bandwidth is large. Log-periodic antennas may be printed on a dielectric substrate, but care must be exercised in the design because

* Carrel, R. L. "The Design of Log-Periodic Dipole Antennas." *1961 IRE Conv. Rec.*, Part 1, pp. 61–75.

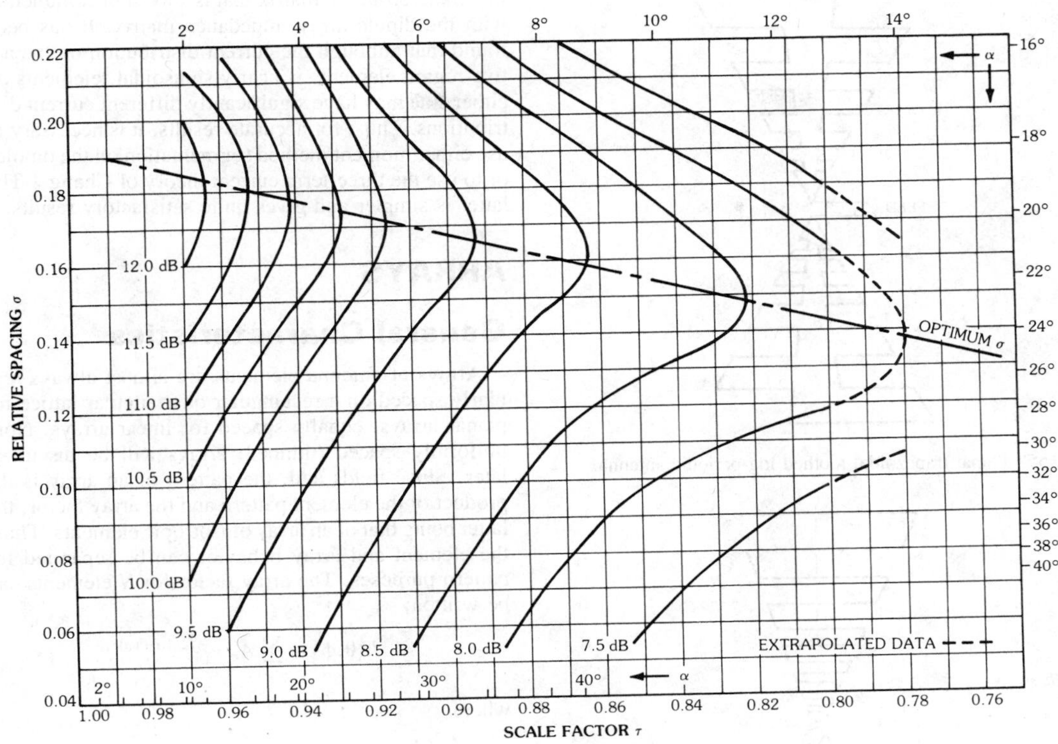

Fig. 53. Constant-directivity contours. (*After Carrel.*)

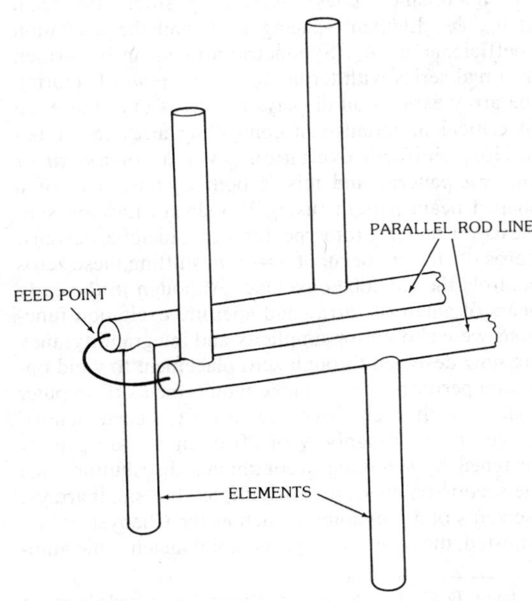

Fig. 54. Log-periodic construction. (*Courtesy Weeks.*)

substrate thickness will limit performance at the upper frequencies. In principle, one can taper the substrate thickness to maintain scaling along the log-periodic, but this, of course, is not practical. The length of the dipoles can be reduced somewhat by loading techniques such as bending with a modest reduction in performance. Reduction of the spacing between elements does not appear practical. Care must be exercised to ensure that resonances are not set up in the feed line; swept impedance and gain measurements are essential for any new log-periodic design. There are several useful log-periodic references.†

The trapezoidal-toothed log-periodic can be in planar form as shown in Fig. 55, but it is more commonly in pyramidal form where the feed is at the apex of a rectangular pyramid and the toothed arms form the sides of the pyramid. Fig. 56 shows a wire version,

† Isbell, D. E. "Log-Periodic Dipole Arrays," *Trans. IEEE*, Vol. AP-8, May 1960, pp. 260–267. Carrel, R. L. "The Design of Log-Periodic Dipole Antennas." TR 52, Antenna Lab, Univ. of Illinois, 1961. Smith, C. E., ed. *Log-Periodic Antenna Design Handbook*. Cleveland, Ohio: Smith Electronics, Inc., 1966. DeVito, G., and Stracca, G. B. "Comments on the Design of Log-Periodic Dipole Antennas." *Trans. IEEE*, Vol. AP-21, May 1973, pp. 303–308; see also Butson, P. C., and Thompson, G. T. "A Note on the Calculation of the Gain of Log-Periodic Dipole Antennas." *Trans. IEEE*, Vol. AP-24, January 1976, pp. 105–106. Balmain, K., and Nkeng, J. N. "Asymmetry Phenomenon of Log-Periodic Dipole Antennas." *Trans. IEEE*, Vol. AP-24, July 1976, pp. 402–410.

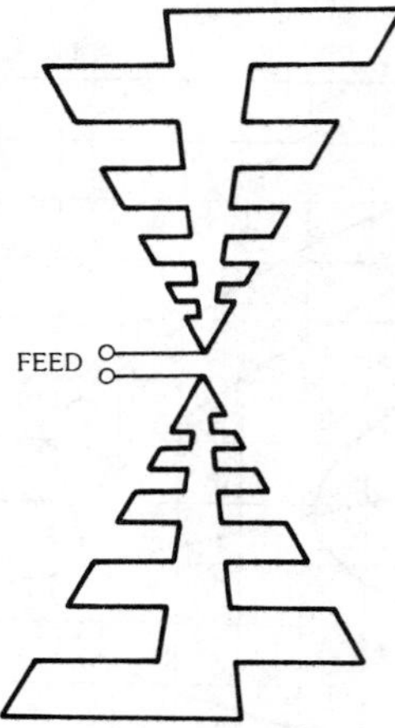

Fig. 55. Planar trapezoidal toothed log-periodic antenna.

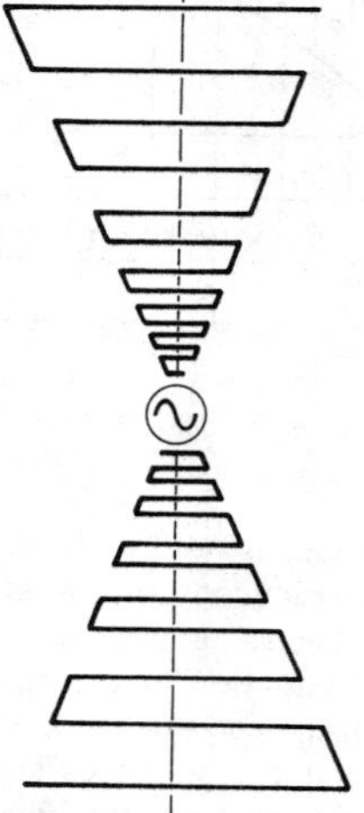

Fig. 56. Log-periodic trapezoid wire antenna (wires are connected to booms).

which, again, is usually used in pyramidal form. In the pyramidal form, the beam is unidirectional. Cross polarized levels are sometimes high.§

The log-periodic antenna can be analyzed and designed with the impedance matrix techniques described in the section on arrays, except that the transmission feed line plays an inseparable role in antenna performance. Thus the transmission line is modeled as a series of cascaded short sections with each section loaded by the impedance of the dipole at that point on the feeder. Of course, the dipole impedance includes all mutual impedances. This results in a transmission line transformation matrix that is solved in conjunction with the dipole array impedance matrix. It has been found that although the current distribution on the active-region elements is nearly sinusoidal, elements on either side may have significantly different current distributions. Thus, for accurate results, it is necessary to use either moment method segmentation on the dipoles or to use the three-term current theory of Chang.‡ The latter is simpler and gives quite satisfactory results.

ARRAYS

General Characteristics

Arrays of antenna elements are almost always regularly spaced on a rectangular or triangular lattice for planar arrays, equally spaced for linear arrays. Nonuniformly spaced (thinned) arrays will be discussed later. Since fields add, the pattern of an array is the product of the element pattern and the array factor, the latter being that of an array of isotropic elements. Thus, the element and array behavior can be separated for pattern purposes. The array factor for N elements can be written:

$$E(\theta,\phi) = \sum_{n=1}^{N} A_n \, e^{-j2n(u-u_0)}$$

where

$$u = \pi(d/\lambda) \sin\theta \cos\phi$$

$$u_0 = \pi(d/\lambda) \sin\theta_0 \cos\phi_0$$

The interelement phase shift, u_0, positions the beam at θ_0, ϕ_0. Element spacing is d, and the excitation coefficients are A_n. Symmetric arrays can be written as a real series with terms $A_n \cos n(u-u_0)$. Factoring the array expression displays the zeros in u; these are of critical importance in controlling array-factor behavior. Uniform excitation gives a $\sin \pi u/\pi u$ or sinc πu pattern, and this is both a constituent of a shaped beam pattern (using Woodward-Lawson synthesis**) and a prototype for low sidelobe designs. Zeros of sinc πu occur at $u = \pm n$; shifting these zeros controls the sidelobe envelope. Although in the early years of antennas, array and aperture excitation functions were chosen for simplicity and integrability, they are now designed through zero placement to yield optimum performance given the requirements. Computer codes are then employed to furnish needed details. Large arrays, roughly N of 20 or more, are usually designed by sampling a continuous distribution. See the section on aperture distributions. For small arrays, the zeros of a polynomial, such as the Chebyshev, are adjusted; the order of the polynomial matches the num-

§ DuHamel, R. H., and Ore, F. R. "Logarithmically Periodic Antenna Design." *IRE Convention Record*, Part I, 1958, pp. 139–151. DuHamel, R. H., and Isbell, D. E. "Broadband Logarithmically Periodic Antenna Structures." *IRE Convention Record*, Part I, 1957, pp. 119–128.

‡ Chang, D. C., Lee, S. W., and Rispin, L. "Simple Formula for Current on a Cylindrical Receiving Antenna," *Trans. IEEE*, Vol. AP-26, Sept. 1978, pp. 683–690.

** Balanis, C. A. Antenna Theory, Analysis and Design. New York: Harper & Row Publishers, Inc., 1982.

ber of elements in the array.†† For all types of arrays, design using physical principles of zero placement should be used to obtain optimum performance.§§ This also obviates comparing various well-established but meretricious array/aperture distributions.

Examples of two-element arrays for various spacings and phasings have been given by Southworth as shown in Fig. 57.

The 3-dB beamwidth is given by

$$\theta_3 = \arc\sin(\sin\theta_0 + 0.443\lambda/Nd) - \arc\sin(\sin\theta_0 - 0.443\lambda/Nd)$$

For $N \geq 6$ this is

$$\theta_3 \simeq 0.886\lambda/Nd \cos\theta_0$$

These formulas show the beamwidth broadening with scan angle θ_0. With a linear array factor, the pattern is rotationally symmetric about the array axis, with a disk-shaped pattern at broadside. As the beam is moved toward end-fire, the disk becomes a cone, and the inner −3 dB points coalesce along the axis for an angle of:

$$\theta_0 = \arc\sin(1 - 0.443\lambda/Nd)$$

†† Villeneuve, A. T. "Taylor Patterns for Discrete Arrays." *Trans. IEEE*, Vol. AP-32, Oct. 1984; pp. 1089–1093.

§§ Hansen, R. C. Array chapters in *Handbook of Antenna Design*, Vol. 2. A. W. Rudge, et al, eds. London: Peter Peregrinus Ltd., 1983.

For beam angles beyond this angle, a single end-fire beam exists. At end-fire, the beamwidth is larger than broadside beamwidth by the factor:

$$\theta_{3\ \mathrm{endfire}}/\theta_{3\ \mathrm{broadside}} \simeq 2\sqrt{0.886\lambda/Nd}$$

When analog phasers are used to produce the inter-element phase shift, u_0, required to scan the beam, any beam position may be reached if the phasers can be set precisely. Digital phasers, however, have a least count of phase shift, and the array positioning will similarly have a least position change. An m-bit phaser has a least count phase shift of

$$\phi_{\mathrm{l.c.}} = 2\pi/2^m$$

and this produces a least beam shift of $kd\sin\theta_{\mathrm{l.c.}}$, $k = 2\pi/\lambda$. These are related by

$$kd\sin\theta_{\mathrm{l.c.}} = \pi/2^{m-1}$$

For the common $\lambda/2$-spaced array, the relationship is $\sin\theta_{\mathrm{l.c.}} = 2^{1-m}$. Table 5 shows the fineness of beam steering available versus the number of bits per phaser.

Positioning the beam of an array at one frequency requires only a phaser per element. Positioning the beam over a large bandwidth requires the proper time delay, i.e., line length, at each element. The bandwidth allowed with the use of phasers depends on array size; bandwidth is taken as beam movement to the −3-dB point:

$$BW = (f_2 - f_1)/f_0 \simeq (f_2 - f_1)/f_2$$
$$= (\sin\theta_2 - \sin\theta_1)/\sin\theta_1 \simeq \lambda/2L\sin\theta_1$$

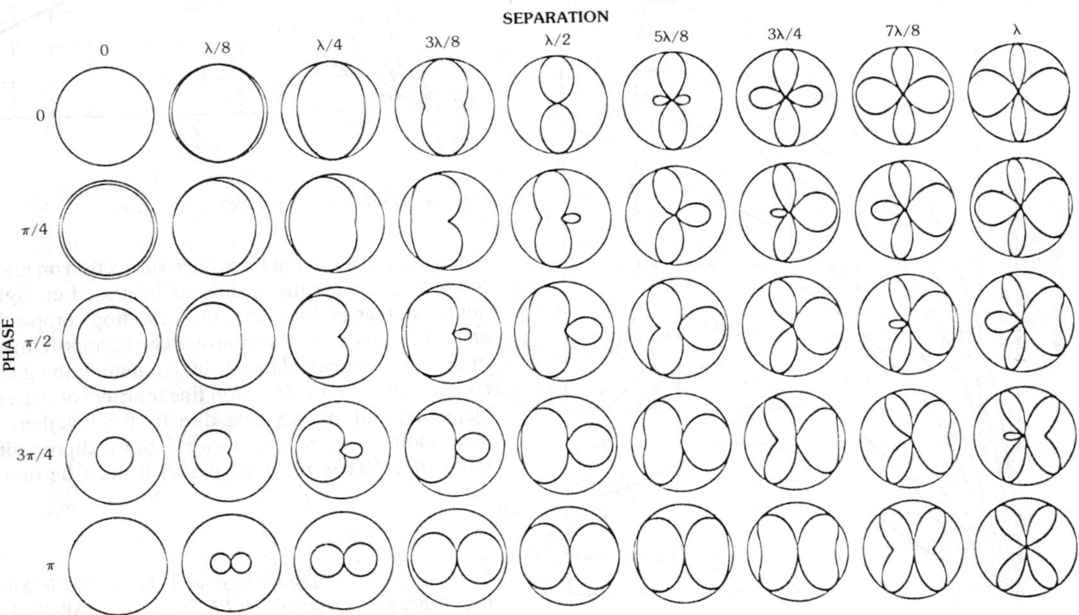

Fig. 57. Patterns of two-element arrays. (*From Southworth, G. C. "Certain Factors Affecting the Gain of Directive Antenna Arrays."* Proc. IRE, *Vol. 18, Sept. 1930, pp. 1502–1536.*)

TABLE 5. FINENESS OF BEAM STEERING

m	$\theta_{l.c.}$ (degrees)
2	30
3	14.48
4	7.18
5	3.58
6	1.79

Assumed here is a not small array length. For a beam angle of 30 degrees, for example, the fractional bandwidth is

$$BW \simeq \lambda/L$$

Thus, long arrays have less bandwidth in terms of beam shift at the band edges.

Directivity of arrays, unlike fields, cannot be found by combining element and array-factor directivity. The latter for a linear array is given by:

$$G = N \Big/ \left[1 + (2/N) \sum_{n=1}^{N-1} (N-n) \operatorname{sinc} nkd \right]$$

Fig. 58 shows this versus spacing. Linear arrays of half-wave dipoles are often used; directivity is obtained from mutual resistances (for an efficient computer algorithm, see Hansen*) with the proper values used for parallel or collinear dipoles:

$$G = 120N \Big/ \left[R_0 + (2/N) \sum_{n=1}^{N-1} (N-n) R_n \right]$$

Here R_0 is the self-resistance and R_n is the mutual resistance between the first and nth dipoles. See Figs. 59 and 60. These directivity curves are for a uniform array with a broadside beam. The directivity drops at a spacing somewhat less than a wavelength due to the emergence of a grating lobe. (See subsection on grating lobes.) Each time the spacing is increased enough to admit another grating lobe, the gain drops proportionately. Directivity of planar arrays can be approximated, in lieu of exact calculations, by computing the gain of constituent x- and y-direction linear arrays of the exact elements, multiplying these directivities together, and then adding a correction factor.† Since dipole directivity of 1.604 is included twice, it must be divided out.

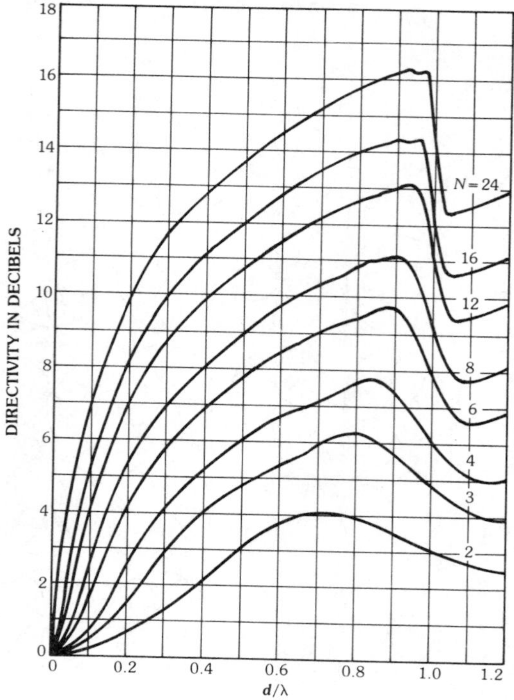

Fig. 58. Array factor directivity.

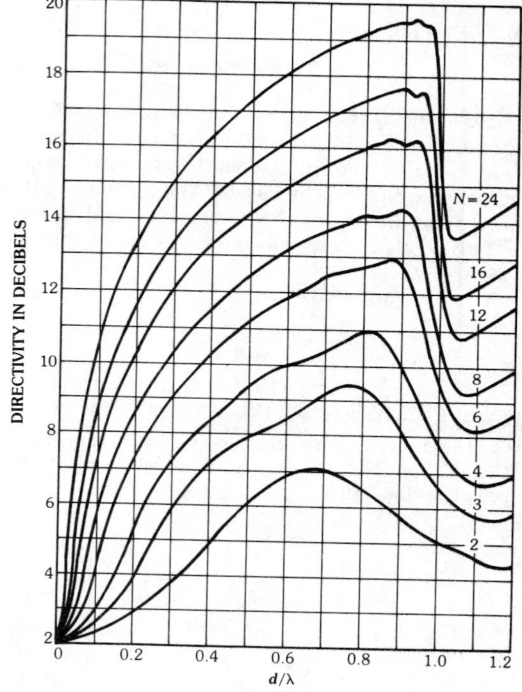

Fig. 59. Directivity of parallel dipole array.

* Hansen, R. C. "Formulation of Echelon Dipole Mutual Impedance for Computer," *IEEE Trans.*, Vol. AP-20, 1972, pp. 780–781.

† Hansen, R. C. "Comparison of Square Array Directivity Formulas." *IEEE Trans.*, Vol. AP-20, 1972, pp. 100–102.

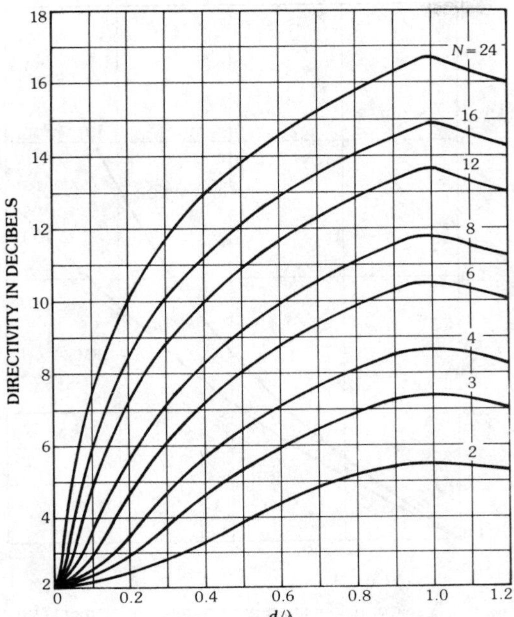

Fig. 60. Directivity of collinear dipole array.

Grating and Quantization Lobes

The uniform array equation indicates that a maximum of unity occurs whenever $u=n$. Grating lobes for a small array are shown in Fig. 61. For another array, the pattern is as shown in Fig. 62, which is for a spacing of 0.707λ. For a broadside main beam, a spacing of one wavelength produces a grating lobe at $-90°$ with a symmetric lobe at $+90°$. Similarly, a main beam scanned to 90° will produce a grating lobe at $-90°$ for half-wave spacing. Fig. 63 shows the grating-lobe angles for various spacings and main beam angles. Here $\theta_0=0$ represents broadside. These curves allow the designer to select a spacing to minimize grating-lobe effects. However, they are for appearance of the grating lobe at $\pm$ 90° so that half of the grating

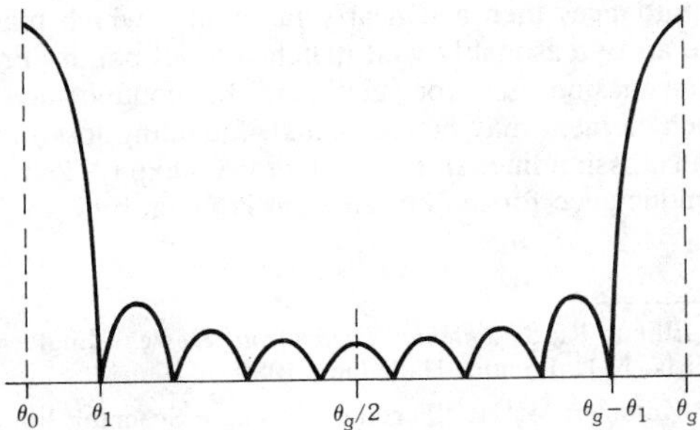

Fig. 61. Array pattern with grating lobe.

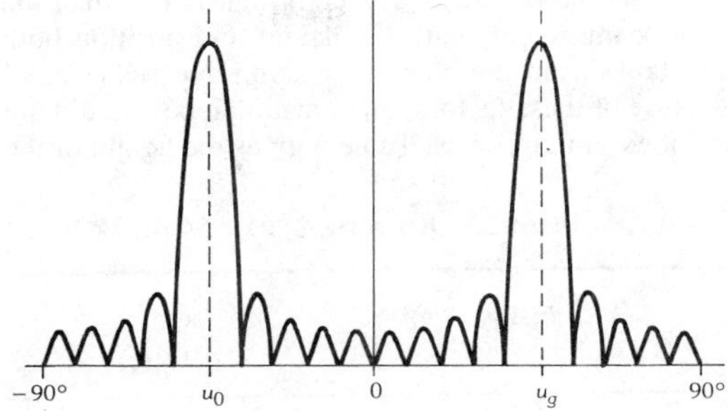

Fig. 62. Array pattern scanned to 45°.

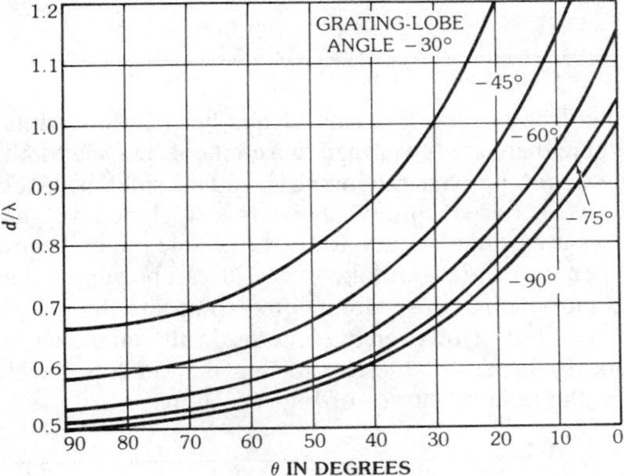

Fig. 63. Element spacing vs grating-lobe angle.

lobe is visible. In some cases, it is desirable to reduce the spacing such that the entire grating lobe is precluded, with the grating-lobe null at $\pm$ 90°. This spacing reduction has been calculated for the general Taylor one-parameter line source, which includes the uniform as a special case.§ The spacing reduction factor is

$$d/d_{gl}=(N-1+B^2)/N$$

The B is the Taylor parameter; see the section on aperture distributions. Grating lobes for arrays whose elements are on a rectangular lattice can be analyzed by using two linear-array cases. Arrays using hexagonal or triangular lattices are more complex.§

Phaser quantization, as mentioned, affects the beam position least count, and it also distorts the pattern. The desired phase versus position along the array is a straight line whose slope is related to the beam angle, but digital phasers produce a stair-step approximation to this. The effects of the sawtooth error depend on the number of elements that fit within a stair step. If there are two or more elements per step, one or more discrete lobes called quantization lobes will be pro-

---

§ Hansen, R. C. Array chapters in *Handbook of Antenna Design,* Vol. 2. A. W. Rudge, et al, eds. London: Peter Peregrinus Ltd., 1983.

duced. These have a fixed height that is less than the main beam height, with the height and position both calculable from the element spacing and scan angle. Because of these factors, quantization lobes should not be called grating lobes. Table 6 gives the height of the

TABLE 6. PHASER QUANTIZATION LOBE HEIGHT

Phaser Bits	QL (dB)
2	−10.5
3	−17.1
4	−23.6
5	−29.8
6	−36.0

quantization lobe in terms of number of phaser bits. Where there are fewer than two elements per sawtooth, the errors approach randomness, and the sidelobe level, if low, is raised. Fig. 64 shows rms sidelobe level due to phaser quantization for a broadside main beam. When a quantization lobe exists, it can be suppressed by introducing a pseudorandomization into the phaser control bits. However, the energy in the quantization lobe is dispersed, thereby raising the sidelobe level, and the random curves of Fig. 64 apply.

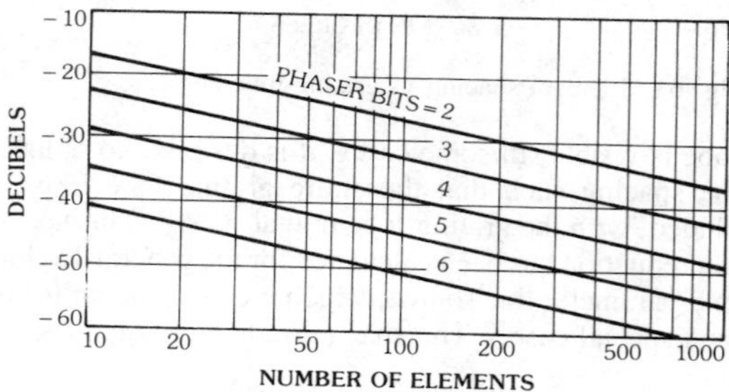

Fig. 64. Rms sidelobe level due to random phaser quantization error, uniform array at broadside.

Linear Array Feeds

Resonant arrays have elements spaced along a transmission line feed at half-guide wavelength intervals. The transmission line ends are shorted so that a standing wave is set up in the feed. At center frequency, each element is resonant, and the input conductance is the sum of the element conductances. The bandwidth of resonant arrays is narrow, because at frequencies off resonance the conductances are out of phase. The feed admittance can be calculated by using a cascade of A-B-C-D matrices with the results shown in Fig. 65. The abscissa is the number of elements times percent bandwidth. Within the range shown, the vswr is monotonic, but for larger frequency excursions, vswr

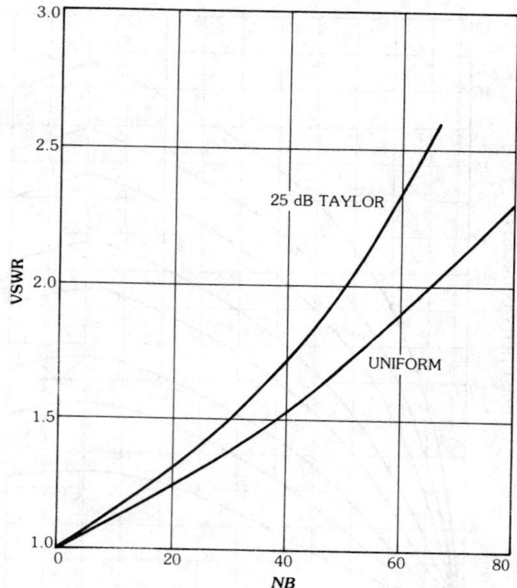

Fig. 65. Vswr of resonant array vs number of elements times percent bandwidth.

oscillates about a value of two for the uniformly illuminated array. In these results, mutual coupling has not been included; thus the bandwidth indicated is optimistic. A design procedure for resonant arrays including the effects of mutual impedance has been developed by Elliott.‡ To avoid splitting of the main beam, arrays should be designed with a vswr well below 2.0.**

A traveling-wave array consists of a transmission line with equally spaced elements coupled to it, with spacings that are not any multiple of half-guide wavelength. Power is fed into the feed at one end, and a load is placed at the other end. A portion of the feed energy is coupled to each element, with the power remaining after the last element dissipated in the load. This array is thus nonresonant, and the beam angle will be either forward or backward and will change with frequency. The broadside resonant condition must be avoided. For traveling-wave arrays of many elements, the coupling of each element is small. The element admittances then add nearly randomly, which makes the array reasonably well matched at all points. From transmission line formulations, the conductance of each element may be calculated, including loss in the transmission line. If the loss factor $s = \exp(-2\alpha)$, the conductance for the nth element is given by

‡ Elliott, R. S. *Antenna Theory and Design*. Englewood Cliffs, N.J.: Prentice-Hall, Inc., 1981.

** Kummer, W. H. "Feeding and Phase Scanning" in *Microwave Scanning Antennas*, Vol. III. R. C. Hansen, ed. New York: Academic Press, Inc., 1966.

$$G_n = F_n s^{-n} / [(1 - Ls^{1-N})^{-1} \sum_{m=1}^{N} F_m s^{-m} - \sum_{m=1}^{n} F_m s^{-m}]$$

The F_n excitation coefficients are usually those of the Taylor one-parameter or Taylor $\bar{n}$ distributions described elsewhere in this chapter. A continuous version of the conductance formula is sometimes used, but unless the number of elements is very large, the discrete form above is preferable. Fig. 66 gives slot conductance values for a 29-element array with uniform excitation for several fractions of power dissipated in the load. As expected, the element conductances increase along the feed as there is less power to which to couple. Fig. 67 gives similar curves for a Taylor one-parameter distribution, and Fig. 68 is for the Taylor $\bar{n}$ distribution.

The beam position of a traveling-wave array is easily determined from the phase equation:

$$kd \sin \theta_0 + 2n\pi = \beta d - \pi$$

where,

θ is the beam angle measured from broadside,

d is the element spacing,

$k = 2\pi/\lambda$,

β is the waveguide wave number.

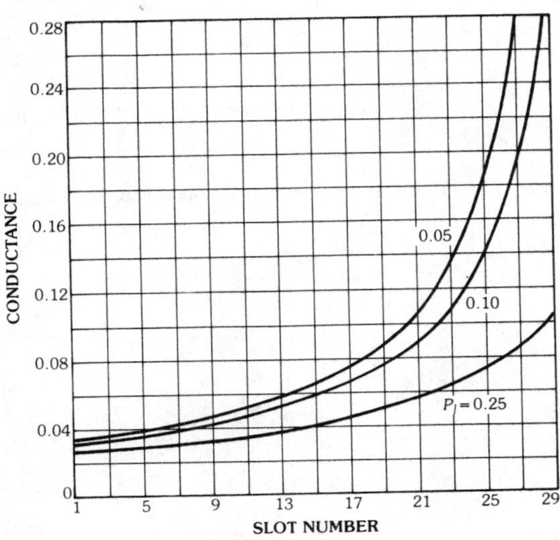

Fig. 66. Slot conductance values, uniform excitation, $N = 29$.

For a given waveguide velocity ratio, a curve such as that in Fig. 69 can be plotted; this curve is important in showing the number of beams versus beam angles. For example, for spacings smaller than 0.9375λ, only one beam exists, and for most spacings it is a backward-directed beam. Multiple beams are, of course, undesirable; this type of plot enables the single-beam range to be quickly discovered. It is necessary to inspect beam behavior over the entire frequency band of interest, which can be done with other types of plots.†† Change of beam angle with frequency is important and can readily be written in a normalized form:

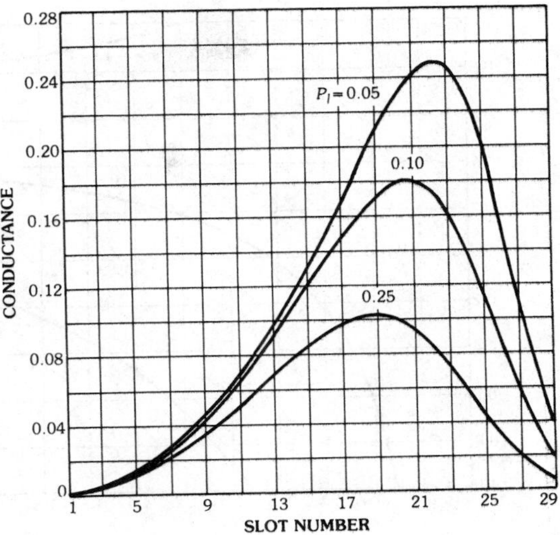

Fig. 67. Slot conductance values, Taylor one-parameter, SLR = 25, $N = 29$.

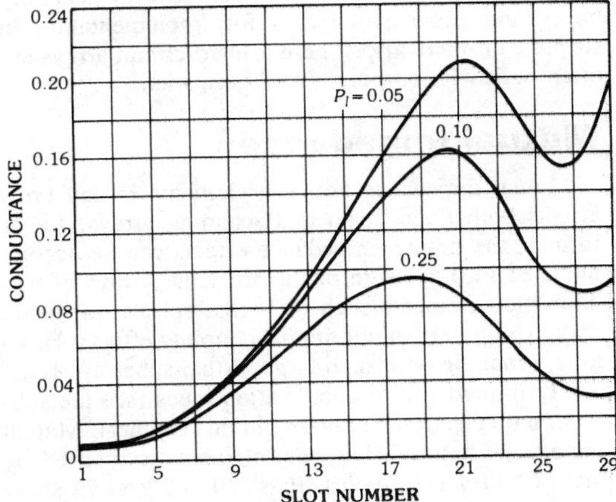

Fig. 68. Slot conductance values, Taylor $\bar{n} = 5$, SLR = 25 dB, $N = 29$.

$$f[(d/df)(\sin \theta_0)] = 1/(\beta/k) - \sin \theta_0$$

It can be seen that the normalized slope depends simply on β/k and the beam angle.

Corporate feeds can be used with a linear array; power dividers or hybrid junctions can usually provide significantly greater bandwidth than the array ele-

†† Hansen, R. C. Array Chapters in *Handbook of Antenna Design*, Vol. 2. A. W. Rudge, et al, eds. London: Peter Peregrinus Ltd., 1983.

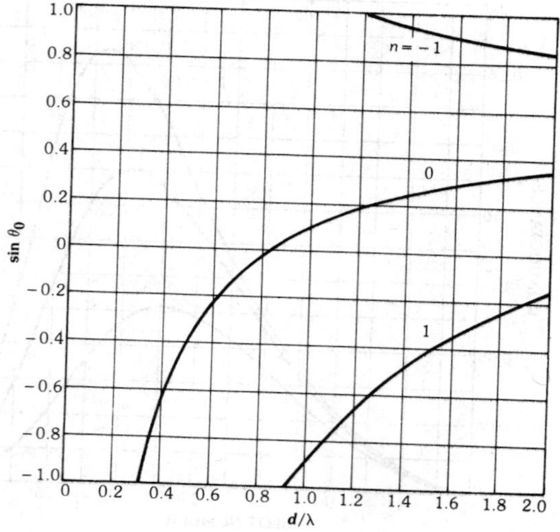

Fig. 69. Tw array beam angle vs spacing, $\beta/k = 0.6$.

ments. Thus, for wideband arrays corporate feeds are recommended. However, corporate feeds are generally bulky and expensive except for implementation in stripline or microstrip. These printed-circuit arrays are often built with corporate feed networks.

Mutual Impedance

Mutual impedance between elements in any array affects both fixed beam and scanning arrays. In the former, the mutual impedance effects can be accommodated in the design of the array.§§ Arrays of medium-gain elements such as horns, spirals, or dishes may exhibit only minor mutual coupling effects. However, scanning with such large virtual subarrays is severely limited due to quantization lobes (see the subsection on electronic scanning in this chapter). Mutual impedance between low-gain elements is typified by that of half-wave dipoles. Figs. 70, 71, and 72 show complex-plane spiral plots of mutual impedance between parallel, collinear, and 45° echelon dipoles. The circles on the graph represent different values of dipole center-to-center spacing. It is important to note that the *H*-plane coupling is strong while the *E*-plane coupling is weak. Accurate and efficient computer codes in Fortran 4 for the calculation of mutual impedance have been given by Hansen‡‡ for equal lengths and Hansen and Brunner* for unequal dipoles. These results and

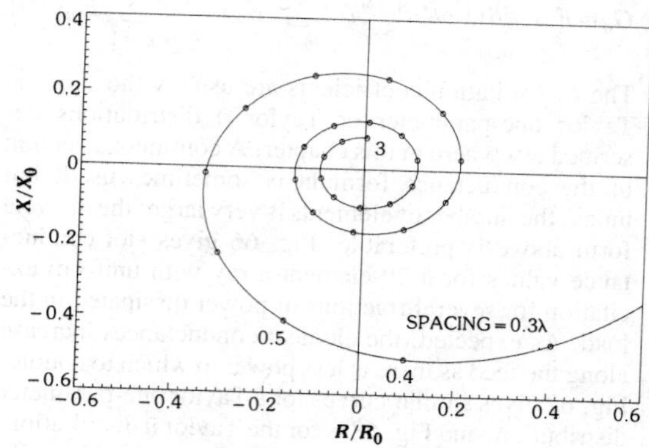

Fig. 70. Parallel dipole mutual impedance/R_0.

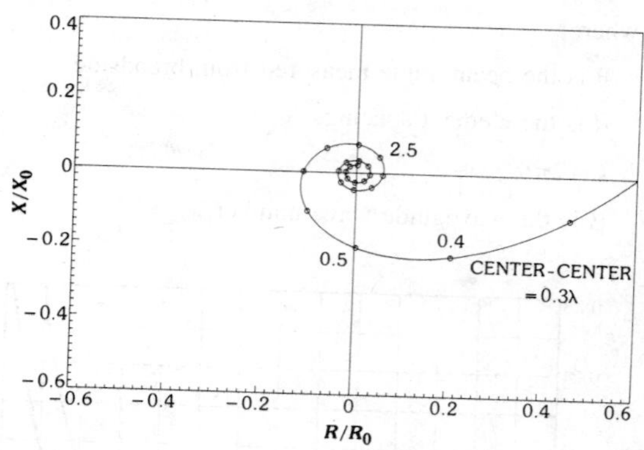

Fig. 71. 45° echelon dipole mutual impedance/R_0.

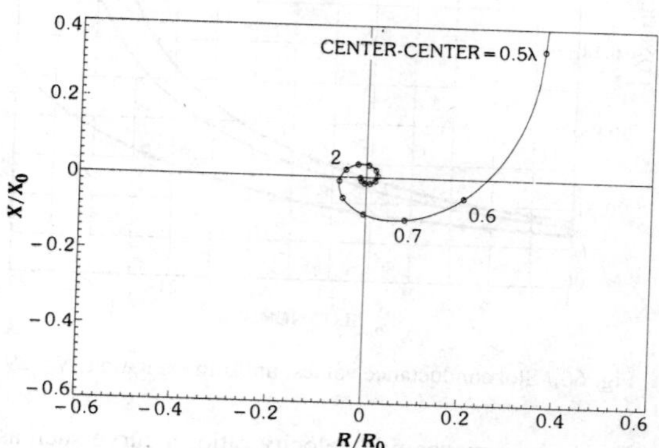

Fig. 72. Collinear dipole mutual impedance/R_0.

codes can be used for slots by replacing the dipole with a Booker equivalent Babinet dipole. The slot admittance and dipole impedance are related by $2Z_{12} = \eta^2 Y_{12}$ where $\eta = 120\pi$. If the slot radiates on both sides, the factor 2 is replaced by 4.

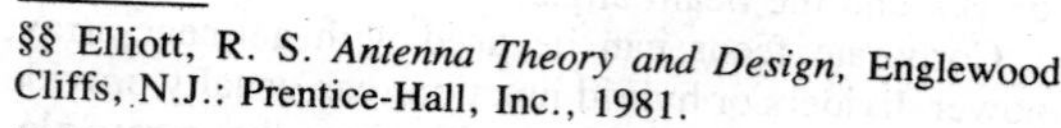

§§ Elliott, R. S. *Antenna Theory and Design*, Englewood Cliffs, N.J.: Prentice-Hall, Inc., 1981.

‡‡ Hansen, R. C. "Formulation of Echelon Dipole Mutual Impedance for Computer." *IEEE Trans.*, AP-20, 1972, pp. 780–781.

* Hansen, R. C., and Brunner, G. "Dipole Mutual Impedance for Design of Slot Arrays." *Microwave J.*, Vol. 22, Dec. 1979, pp. 54–56.

The strong mutual coupling could be reduced if the element power pattern could be transformed into the ideal symmetric cos θ pattern. One way of approximating this is to use a round or square open-end waveguide radiator with dielectric plugs, etc., to produce equal TE and TM modes.† A simpler way is to use a slot straddled by a pair of monopoles, sometimes called a Clavin slot. The monopole length and spacing can be adjusted to produce a slot *E*-plane pattern similar to the narrow *H*-plane pattern. Fig. 73 shows a Clavin slot. These slots were empirically designed by Clavin§; an analytical design was given by Papierz, et at.‡ Mutual coupling calculations have also been developed by Elliott.** Figs. 74 and 75 show complex-plane plots of mutual coupling between Clavin slots for parallel and collinear geometries. It can be noted that these couplings are both small and are more alike. Thus, this type of element should give improved scanning performance in electronically scanned arrays.

Mutual coupling effects in electronic scanning are difficult to calculate and outside the scope here. Only a few general remarks will be made. An array with a feed network that provides fixed drive voltages (using dipoles as an example) can be simply analyzed from an impedance matrix equation. That is, the dipole currents are found by multiplying the array voltage factor by the inverse of the mutual impedance matrix. Since the scan angles are contained in the drive vector, it is simple to find the active impedance of each element and the resulting array pattern for a given scan angle. Many arrays do not have a constant voltage feed network but are of the constant-available-power type. Here an element impedance mismatch reduces the applied voltage. This approach is more suited to a scattering equation. For details, see Hansen.†† Lengthy formulas have been developed for active impedance of dipole arrays,† †† and these can be translated to slot arrays. The active impedance is the element input impedance as modified by mutual coupling from neighboring elements, and this active impedance, of course, varies with scan angle. The array gain versus scan can

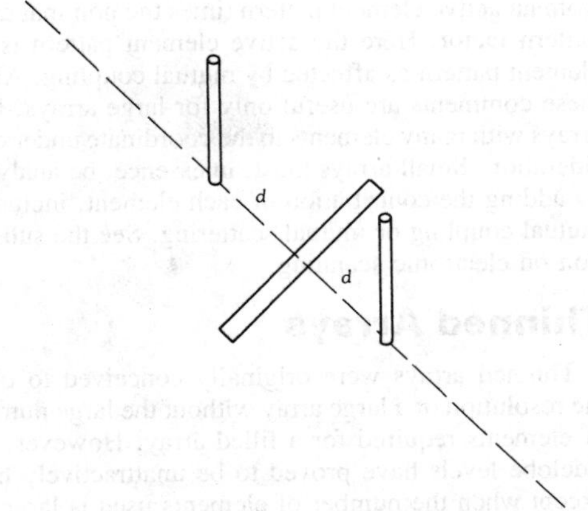

Fig. 73. Slot with parasitic monopoles (Clavin slot).

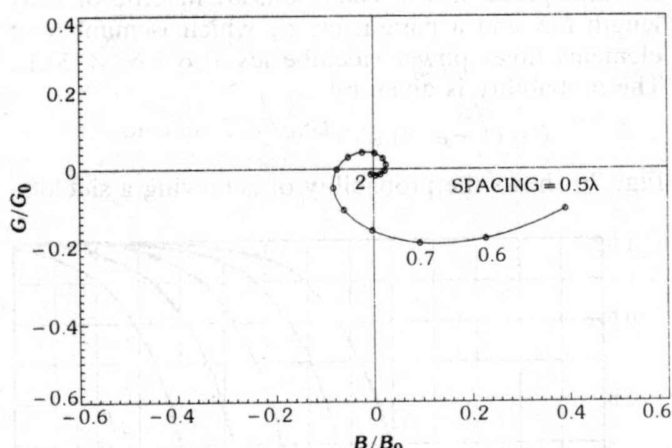

Fig. 74. Parallel Clavin-slot mutual admittance/G_0.

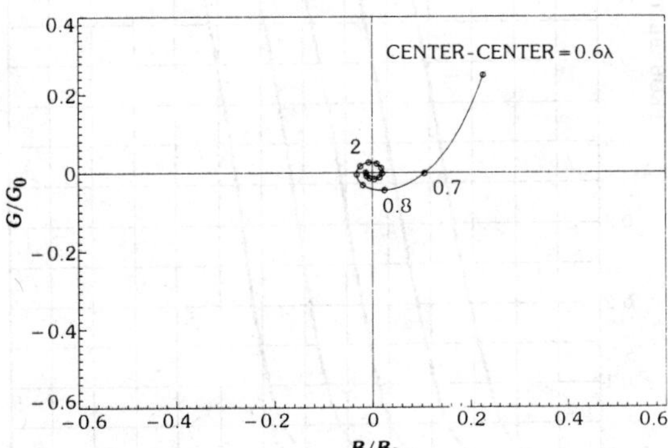

Fig. 75. Collinear Clavin-slot mutual admittance/G_0.

be calculated by using the isolated element pattern times the nominal array pattern factor times one minus the active power reflection coefficient. The active reflection coefficient is, of course, determined from the active impedance. The array gain can also be calculated

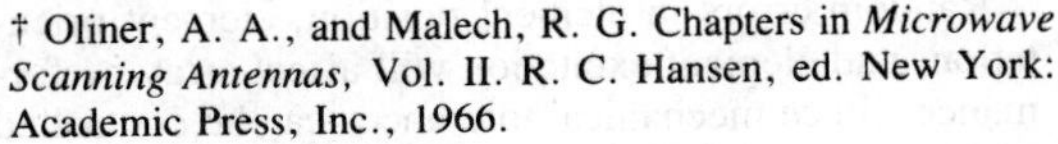

† Oliner, A. A., and Malech, R. G. Chapters in *Microwave Scanning Antennas,* Vol. II. R. C. Hansen, ed. New York: Academic Press, Inc., 1966.

§ Clavin, A., Huebner, D. A., and Kilburg, F. J. "An Improved Element for Use in Array Antennas." *Trans. IEEE,* Vol. AP-22, July 1974, pp. 521–526.

‡ Papierz, M. S., et al. "Analysis of Antenna Structure With Equal *E*- and *H*-Plane Patterns." *IEE Proc.*, Vol. 124, Jan. 1977, pp. 25–30.

** Elliott, R. S. "On the Mutual Admittance Between Clavin Elements." *Trans. IEEE,* Vol. AP-28, Nov. 1980, pp. 864–870.

†† Hansen, R. C. Array chapters in *Handbook of Antenna Design,* Vol. 2. A. W. Rudge, et al, eds. London: Peter Peregrinus Ltd., 1983.

from an active element pattern times the nominal array pattern factor. Here the active element pattern is the element pattern as affected by mutual coupling. All of these comments are useful only for large arrays, i.e., arrays with many elements in the coordinate under consideration. Small arrays must, in essence, be analyzed by adding the contribution of each element, including mutual coupling or mutual scattering. See the subsection on electronic scanning.

Thinned Arrays

Thinned arrays were originally conceived to offer the resolution of a large array without the large number of elements required for a filled array. However, the sidelobe levels have proved to be unattractively high except when the number of elements used is large. A probabilistic analysis was given by Agrawal and Lo‡‡ with results that compare closely with Monte Carlo simulations for both short and long arrays and for low and high probabilities. The results are in terms of array length L/λ and a parameter, α, which is number of elements times power sidelobe level: $\alpha = N \times \text{SLL}$. The probability is given by

$$P = (1 - e^{-\alpha})\, e^{-(2L/\lambda)\sqrt{\pi\alpha/3}\,\exp(-\alpha)}$$

Fig. 76 shows the probability of achieving a sidelobe

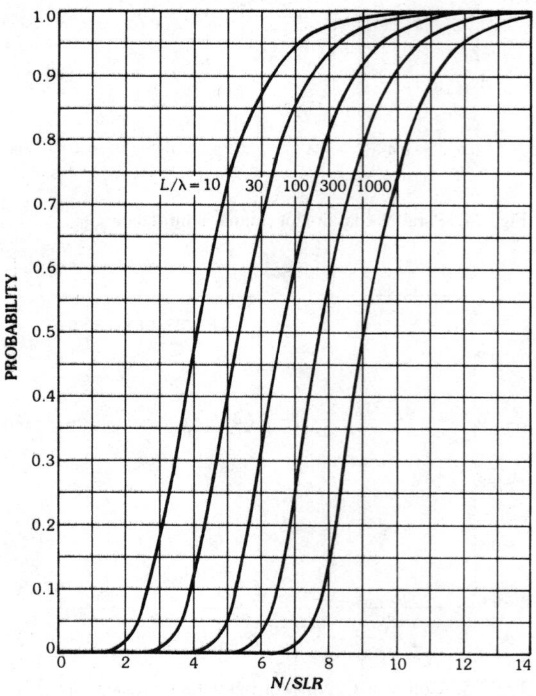

Fig. 76. Probability of random array sidelobe ratio greater than SLR.

‡‡ Agrawal, V. D., and Lo, Y. T. "Mutual Coupling in Phased Arrays of Randomly Spaced Antennas." *Trans. IEEE*, Vol. AP-20, May 1972, pp. 288–295.

level versus parameter α for different length arrays. For example, a 300-wavelength aperture with 150 elements, giving an average element spacing of 2 wavelengths, has only a 40% probability of achieving a -13 dB sidelobe level. Table 7 shows the α values for several different lengths of arrays for a 90% probability. Also shown is the limit of sidelobe ratio that can be achieved when the number of elements is sufficient to fill the array. Of course, if this number of elements were rearranged with constant half-wave spacing, the sidelobe level would be -13.26 dB. The filling ratio necessary to achieve a 90% probability of sidelobes below -10 dB is also shown. From this, it can be seen that either high filling ratios or large arrays are needed to achieve even modest sidelobe levels.

TABLE 7. 90% SIDELOBE PROBABILITY

L/λ	$\alpha = N \cdot \text{SLL}$	Maximum SLR (dB)	Filling Ratio for 10 dB SLL
10	6.201	5.1	—
30	7.373	9.1	—
100	8.652	13.6	43%
300	9.813	17.9	16%

A special type of thinned array is the space-tapered array. Here, the spacing of uniformly excited elements is tapered to match a low sidelobe distribution of the type covered in the section on aperture distributions. This scheme is attractive for arrays that have distributed transmitter/receiver modules because it allows all modules and elements to be alike. Experience has shown that the fall-off of sidelobes is not quite as optimistic as the theory predicts. With any space-tapered array, thorough calculations are needed to ensure that no high sidelobes exist, and it may be necessary to adjust the filling ratio and spacing to ensure sidelobe control over the bandwidth and range of scan angles. For information on degree of thinning versus sidelobe level, see Hansen.*

Tolerances

Random errors in element position, element orientation, and element excitation will affect array performance. Since mechanical tolerances can be controlled much more tightly than excitation, only phase and amplitude excitation errors are of concern. They are assumed to have normal (Gaussian) distribution with zero mean and variance σ^2. These errors reduce the gain a small amount, and on the average add a small constant amount to the sidelobes.† The effect on sidelobes is

* Hansen, R. C. Array chapters in *Handbook of Antenna Design*, Vol. 2. A. W. Rudge, et al, eds. London: Peter Peregrinus Ltd., 1983

† Ruze, J. "Antenna Tolerance Theory—a Review." *Proc. IEEE*, Vol. 54, 1966; pp. 633–640.

less for large arrays and more for low sidelobes. The gain, with respect to error-free gain, is given by

$$G/G_0 \simeq \exp(-\sigma_{ampl}{}^2 - \sigma_{ph}{}^2)$$
$$\simeq 1/(1 + \sigma_{ampl}{}^2 + \sigma_{ph}{}^2)$$

Thus, errors reduce gain or directivity only to second order, and gain reduction can generally be neglected. For sidelobes, the effect, however, is first order. The mean sidelobe level is related to the error-free sidelobe level by

$$SL = SL_0 \sqrt{1 + \sigma^2/G\, SL_0}$$

where,

G is the antenna gain,

σ^2 is the total error variance,

SL_0 is the power sidelobe level.

Fig. 77 shows mean sidelobe level versus the universal factor $\sigma/\sqrt{G}$. From this, it is apparent that sidelobe designs from 13 to 30 dB are quite robust, whereas low sidelobe designs have sensitive tolerances. Similarly, these curves clearly show that for a given sidelobe level and sidelobe degradation, larger arrays allow larger errors. The probability density function of the pattern with random errors is a modified Rayleigh, and the probability can be written in terms of the Marcum Q function. Curves of probability of certain sidelobes exceeding specified values are beyond the scope here, and are given in Hansen.§

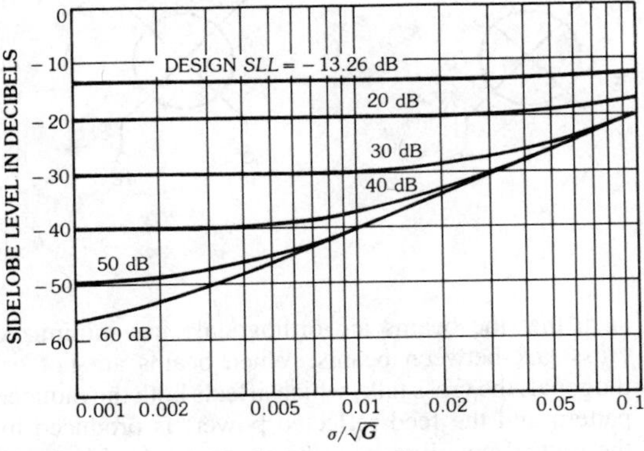

Fig. 77. Mean sidelobe level vs universal factor.

Multiple-Beam Arrays

Multiple-beam arrays usually utilize a beam-forming network (BFN) such as the Butler matrix or Rotman-Gent lens. The BFN, in essence, takes the discrete Fourier transform of the array distribution and produces the transform (space factor) at each beam port. The Butler matrix is, as has been pointed out innumerable times, the microwave equivalent of the FFT. An eight-element, eight-beam Butler BFN is shown in Fig. 78. The beams produced by this network are sketched in Fig. 79. The beam position for any spacing is

$$\sin\theta_i = \pm\,(2i-1)\lambda/2Nd, \qquad i = 1, 2, \cdots N/2$$

The crossover level is independent of spacing; when the element spacing is increased, the beamwidths become narrower and the beams move closer together. Beam coverage from the center of the leftmost beam to the center of the rightmost beam is

$$\theta_{cov} = 2 \text{ arc sin } (N-1)\lambda/2Nd$$

When $d = \lambda/2$, the space $-90°$ to $+90°$ is just filled with beams, as indicated in Fig. 79. Larger spacing moves the beams closer together so that the coverage angle of the N beams is less. However, the space will be filled with beams, and thus there is aliasing and directivity loss due to the extra beams, which are essentially grating lobes. The foregoing assumes isotropic or low-gain elements. When moderate-gain elements such as horns are used with larger spacing, the grating lobes are replaced by subarray quantization lobes; i.e., the grating-lobe amplitude is reduced by the element pattern. The sidelobe ratio and crossover level of a Butler BFN approach those of a uniform line source for large N. Table 8 shows these parameters for popular-size Butlers. The performance of a Butler BFN over a frequency band corresponds to that which is the result of change in spacing. At frequencies above the design frequency, the beams become narrow, and additional beams are added at each side. Conversely, at lower frequency, the beams spread out, and one or more beams on each side may disappear.

The Rotman-Gent BFN or lens was originally designed with circular arcs for mechanical scanning. However, they are now usually implemented in microstrip with fixed feeds, and the input-output faces are now allowed to follow the optimum curves. See Fig. 80. When the lens is symmetric, i.e., the input and output curves are the same, the lens can be specified with a single parameter.‡ This type of lens admits a three-point correction, where the center feed and two other symmetrically located feeds can produce a perfectly collimated beam. Note that the bootlace lines connecting the output face to the radiating arrays are an integral part of the lens and are specified by the lens design. The Rotman-Gent lens, being an optical device, has a fixed set of beam positions controlled by the design of the lens. Accordingly, the crossover level at center frequency is adjustable and is a tradeoff with feed mutual coupling and other design factors. At frequencies above or below the design frequency, the

§ Hansen, R. C. Array chapters in *Handbook of Antenna Design*, Vol. 2. A. W. Rudge, et al, eds. London: Peter Peregrinus Ltd., 1983.

‡ Shelton, J. P. "Focusing Characteristics of Symmetrically Configured Bootlace Lenses." *Trans. IEEE*, Vol. AP-26, July 1978, pp. 513–518.

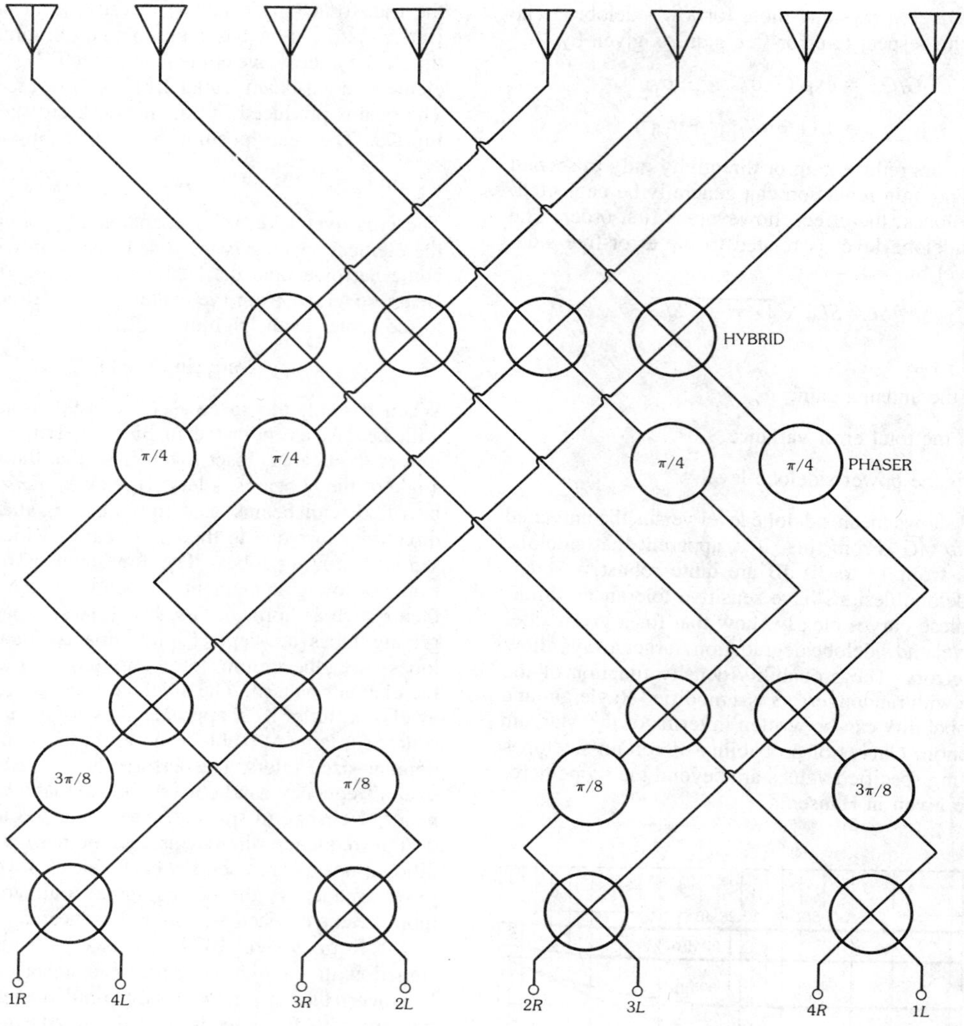

Fig. 78. Butler matrix BFN.

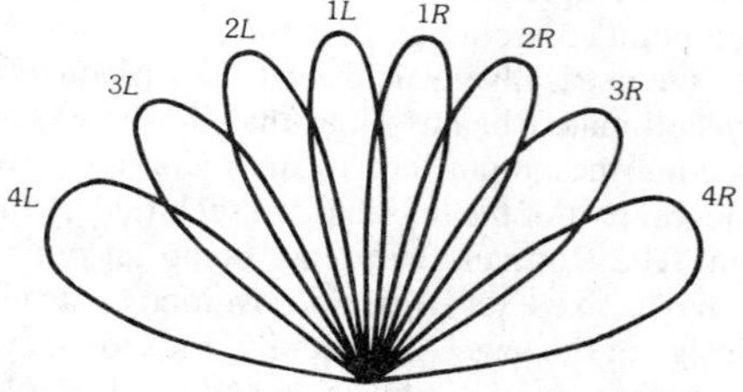

Fig. 79. Butler BFN beams.

beams become narrower or broader, and the crossover level changes accordingly. Rotman-Gent lenses tend to be lossy, and information has not been published on the minimum loss of which this design is capable.

It is known that a multiple-beam antenna designed such that the beams are orthogonal* has minimized cross talk between beams. When beams are not orthogonal, the cross talk, which affects both the radiated pattern and the feed reflected power, is produced by the mutual coupling between feed ports and between array ports. When the spacing, d, and excitation are such as to provide orthogonal beams, mutual coupling still exists, of course, but its effects cancel out in the proper directions. Unfortunately, although many configurations have approximately orthogonal beams, only a uniform line source and line sources with a cosine

---

*

$$\int_{-1/2}^{1/2} F_i(\pi v)\, F_j(\pi v)\, dv = \begin{cases} 1 \text{ for } i=j \\ 0 \text{ for } i \neq j \end{cases}$$

where $v=(d/\lambda)\sin\theta$

TABLE 8. BUTLER-ARRAY PARAMETERS

N	SLR	Crossover Level
4	11.30 dB	−3.70 dB
8	12.80 dB	−3.87 dB
16	13.15 dB	−3.91 dB
32	13.23 dB	−3.92 dB

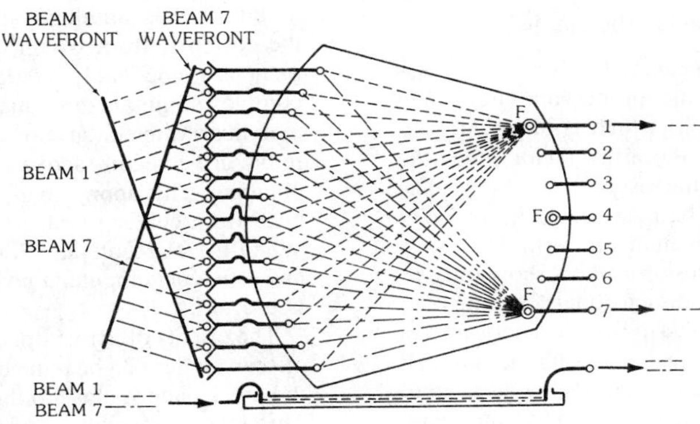

Fig. 80. Rotman BFN.

to an integral power distribution are exactly orthogonal. Contrary to older literature, a cosine-on-a-pedestal does not produce orthogonal beams. A discrete array of isotropic elements can approach orthogonality if N is large, but for small N the condition is not satisfied. The addition of practical elements such as dipoles or slots destroys the orthogonality condition. In spite of all this, arrays with as few as eight elements show excellent sidelobes for collinear dipoles and modest degradation for parallel dipoles, both at half-wave spacing. Table 9 shows the drawbacks of using purely orthogonal distributions in that only the uniform line source has acceptable efficiency and crossover level.

TABLE 9. ORTHOGONAL SPACE FACTORS

Distribution	SLR	Efficiency	Crossover
Unif	13.26 dB	1	−3.92 dB
cos	23.00 dB	0.8106	−9.54 dB
$\cos^2$	31.47 dB	0.6667	−15.40 dB

The effects of nonorthogonality can be reduced by those techniques used to reduce mutual coupling between elements in an array. For example, the use of monopole pins astride a slot reduces E-plane coupling.

Low sidelobes in multiple beams can be produced in several ways. However, the only way that is efficient and simple involves the superposition of beams. For example, the Taylor $\bar{n}$ distribution (see the section on aperture distributions) has been written as a sum over $\bar{n}$ of sinc beams. Since a Butler matrix or Rotman-Gent lens produces sinc beams if uniformly excited, several beam ports can be combined with suitable weighting to produce a single beam with low sidelobes. Using this technique, Thomas† achieved −28-dB sidelobes over a 10% bandwidth with a Rotman-Gent lens. All of the above techniques and devices have a one-dimensional radiating array. A two-dimensional array may be handled by cascading rows of Butler BFNs or of Rotman-Gent BFNs, but this requires much hardware. There is no rotationally symmetrical three-dimensional equivalent of the Rotman-Gent lens; the three-dimensional lens has four perfect points but is not symmetric.§ There is, however, a Butler matrix for two-dimensional arrays. This applies to an array disposed on a regular hexagonal lattice. The resulting BFN is a three-dimensional microwave network with multiple-arm hybrid junctions but is simpler than cascading two-dimensional Butler BFNs.‡

Electronic Scanning

Some of the design tradeoffs for electronic scanning, such as grating lobes, have been addressed above. A large electronic scanning array, or phased array, can be characterized in terms of either active impedance or active element pattern; the modifier "active" im-

† Thomas, D. T. "Multiple Beam Synthesis of Low Sidelobe Patterns in Lens Fed Arrays." *Trans. IEEE*, Vol. AP-26, November 1978, pp. 883–886.

§ Rao, J. B. L. "Multifocal Three-Dimensional Bootlace Lenses." *Trans. IEEE*, Vol. AP-30, Nov. 1982, pp. 1050–1056.

‡ Williams, W. F., and Schroeder, K. G. "Performance Analysis of Planar Hybrid Matrix Arrays." *Trans. IEEE*, Vol. AP-27, July 1969, pp. 526–528.

plies a quantity that varies with scan angles. In the active impedance approach, the pattern is written as the product of three factors. First is the power array factor (of isotropic elements), which is h_N; the second is the isolated element power pattern, g_i; the last term involves the active impedance. These combine to give a gain pattern of

$$G(\theta) = (h_N\, g_i\, R_i/R_a)\ [1 - |\, \Gamma_a\, |^2]$$

where,

R_i is the isolated element resistance,

R_a is the active element resistance,

Γ_a is the active reflection coefficient.

Active impedance can be measured only when all elements are radiating with the proper amplitude and phase. The second approach, which is identical, represents the gain pattern as the array factor times the active element pattern with the result $G(\theta) = h_N\, g_a$. The active element pattern can be measured with one element excited and all other elements terminated in a matched load. Although these formulas show only θ, they apply in general to two-dimensional scanning over θ, ϕ. For such large arrays, a powerful analysis was developed by Oliner and colleagues using the unit cell concept. Here, one element in a large array is seen to be closely equivalent to an element in an infinite array; in the latter, the periodic nature of the array allows a single cell to contain the entire characteristics of the array. This cell, called the unit cell, consists of a virtual waveguide with suitably chosen impedance walls with the waveguide symmetrically located about the element (typically a slot) in the element lattice. The wall boundary conditions are derived from the scan angles. These unit cells are normal to the array face and are contiguous. Interior modes are LSE and/or LSM. If no grating lobes exist, the conductance is given by the single propagating mode in the unit cell waveguide, and thus a closed form result occurs. All modes contribute susceptance, and thus many series terms must be computed. This method has proved to be powerful and perceptive in the understanding of scanning behavior of arrays. For more detailed information, refer to Oliner and Malech** and Hansen.††

A blind spot can be produced in an array at a particular angle; the array radiation is zero at this angle. This occurs when the dominant mode is cancelled by a higher mode, and can occur because of external or internal structure. External structure often used for matching or for protective purposes includes dielectric sheets over or near the array face and dielectric plugs protruding from waveguide-type elements. Internal structure includes dielectric loaded waveguide elements, monopole loaded slots, etc. Blind spots can be severe for large arrays but usually are negligible for small arrays. In case a blind spot is experienced, the array design can usually be changed to remove it. Typical measures are reducing lattice dimensions, changing design to improve vswr, and altering the periodicity of the external structure. Simulators have also proved useful. For more detailed information, see Oliner and Malech** and Hansen.††

Small arrays are usually analyzed with the element-by-element approach. Here, a matrix equation is set up relating the drive voltage to the element currents times the impedance matrix, using dipoles as an example. These simultaneous equations are solved for the currents, from which the impedance of each element is immediately available. Note that solution of complex simultaneous equations is several times faster than matrix inversion. Since the scan direction appears in the drive vector, active element impedance is easily obtained. This approach allows the variation of impedance from center to edge elements to be determined. When the elements are slots or dipoles, a convenient and efficient computer algorithm for mutual impedance is useful.§§

The effects of active impedance change over a range of scan angles can be reduced by designing the element to have a pattern close to the "ideal" element pattern. This is a power pattern of cos θ (for half-wavelength spacing); it is a rotationally symmetric cosine pattern.‡‡ Compensation always involves establishing both TE and TM modes (to make the pattern equal in both planes), and perhaps additional modes for pattern shape control. This can be done with open-ended waveguides with appropriately chosen dielectric plugs and slabs.* A simpler but less effective technique uses monopoles astride a slot, the "Clavin pins."†

Adaptive Arrays

Adaptive arrays are sometimes used to suppress interfering sources; the array is automatically adjusted

** Oliner, A. A., and Malech, R. G. Chapters 3 and 4 in *Microwave Scanning Antennas*, Vol. II. R. C. Hansen, ed. New York: Academic Press, Inc., 1966.

†† Hansen, R. C. Array chapters in *Handbook of Antenna Design*, Vol. 2. A. W. Rudge, et al, eds. London: Peter Peregrinus Ltd., 1983.

** Oliner, A. A., and Malech, R. G. Chapters 3 and 4 in *Microwave Scanning Antennas*, Vol. II. R. C. Hansen, ed. New York: Academic Press, Inc., 1966.

†† Hansen, R. C. Array chapters in *Handbook of Antenna Design*, Vol. 2. A. W. Rudge, et al, eds. London: Peter Peregrinus Ltd., 1983.

§§ Hansen, R. C. "Formulation of Echelon Dipole Mutual Impedance for Computer." *IEEE Trans.*, Vol. AP-20, Nov. 1972, pp. 780–781.

‡‡ Wheeler, H. A. "Simple Relations Derived From a Phased-Array Antenna Made of an Infinite Current Sheet." *IEEE Trans.*, Vol. AP-13, July 1965, pp. 506–514.

* Knittel, G. H. "Wide-Angle Impedance Matching of Phased-Array Antennas: A Survey of Theory and Practice," in Oliner, A. A., and Knittel, G. H., eds. *Phased Array Antennas*. Dedham, Mass.: Artech House, Inc., 1972.

† Elliott, R. S. "On the Mutual Admittance Between Clavin Elements." *Trans. IEEE*, Vol. AP-28, Nov. 1980, pp. 864–870.

to provide a pattern null in the direction of the interferer. The simplest type is the sidelobe canceller in which an auxiliary element is used with the main antenna to provide a single null. A feedback loop is used to adjust the phase and amplitude of the single element to produce a subtraction of the interfering signal. A more powerful configuration is the adaptive array in which each element is controlled. There are three general types of adaptive arrays, which differ in the reference signal used for driving the control loops that provide the weights (amplitude and phase) for each element. The first type uses an externally provided steering vector that indicates the desired signal direction. This type of array is called Applebaum-Howells (Fig. 81). The second type uses a signal replica that correlates with the desired signal. This replica may be from a subcarrier oscillator, a binary sequence of the proper bit rate, etc. This type of adaptive array is associated with Widrow and is frequently called an LMS array. Fig. 82 shows a simplified configuration. The control loops produce a least-mean-square error between the desired signal and the replica, hence the LMS name. The third type of adaptive array does not have an externally supplied reference signal. This array is called a power inversion array and has been associated with Compton. It will form a null on the strongest signal with additional nulls formed on weaker signals. All of these adaptive arrays have $N-1$ degrees of freedom, where N is the number of elements in one plane, and $N-1$ nulls can, in general, be placed in this plane. If there are more degrees of freedom than interferers, the power inversion array tends to null out the desired signal as well, since it cannot tell the desired signal from interfering signals. The improvement available, essentially independent of the type of adaption, is related to the interference-to-noise ratio. To illustrate, a single interferer is assumed. The ratio of signal to interference plus noise before and after adaption indicates the improvement (Fig. 83). For large I/N, the improvement in $S/(I+N)$ is essentially just I/N, but for small values the improvement rapidly decreases. Convergence depends on the ratio of the strongest signal to the desired signal; strong interferers require reduced loop gain to avoid instability, and this slows convergence. Methods of accelerating convergence can utilize orthogonalization of Eigenvalues, or a modified control law. The control loops and weighting can be performed either with analog or digital hardware. Thus, there are several configurations possible. Analog control loops are simple, but some control laws such as accelerated convergence are difficult to implement other than digitally. Further information is in several sources.§

§ Widrow, B., et al "Stationary and Nonstationary Learning Characteristics of the LMS Adaptive Filter." *Proc. IEEE*, Vol. 64, August 1976, pp. 1151–1162. Gabriel, W. F. "Adaptive Arrays—An Introduction." *Proc. IEEE*, Vol. 64, February 1976, pp. 239–272. Gabriel, W. F., ed. Special Issue on Adaptive Antennas, *Trans. IEEE*, Vol. AP-24, September 1976. Monzingo, R. A., and Miller, T. W. *Introduction to Adaptive Arrays*. New York: John Wiley & Sons, Inc., 1980. Hudson, J. E. *Adaptive Array Principles*. London: Peter Peregrinus Ltd., 1981.

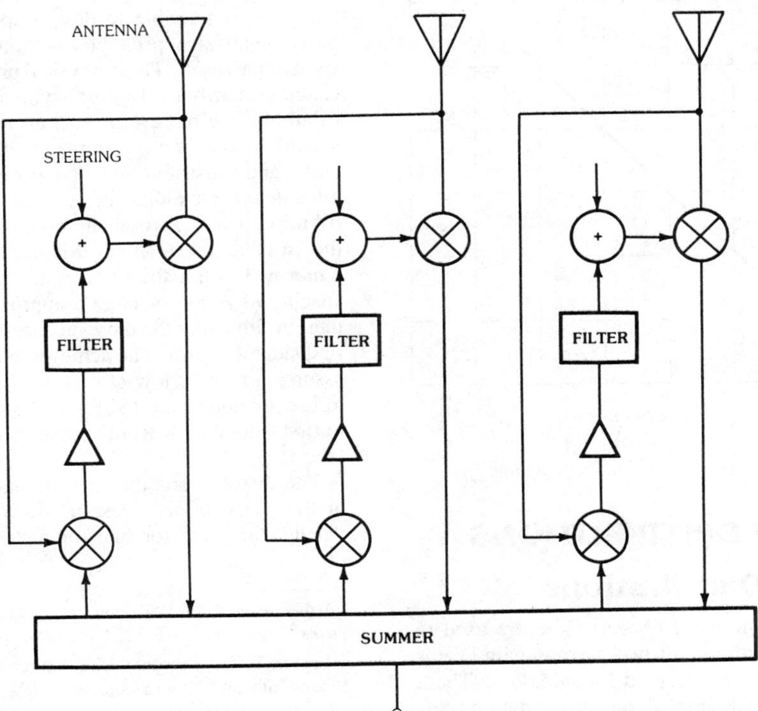

Fig. 81. Applebaum-Howells adaptive array.

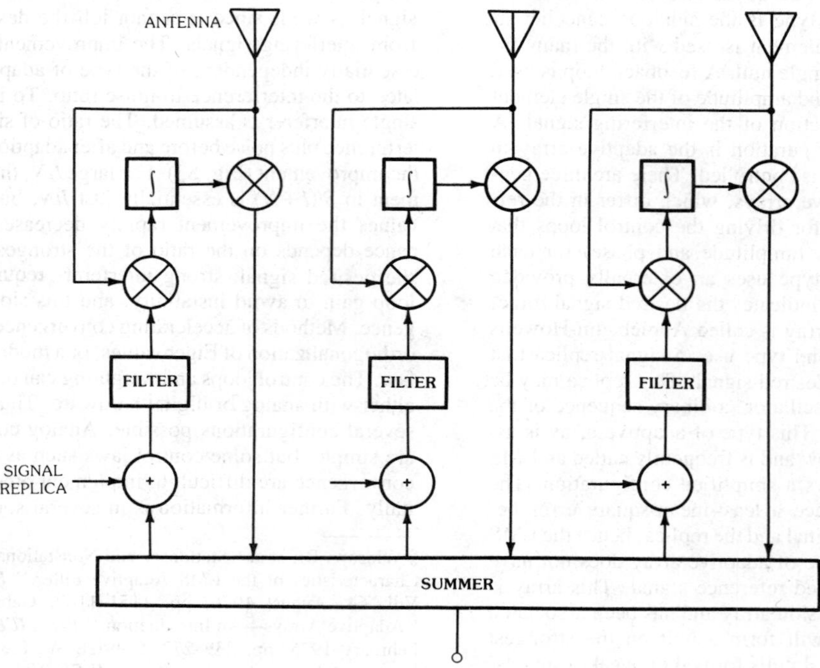

Fig. 82. Widrow LMS adaptive array.

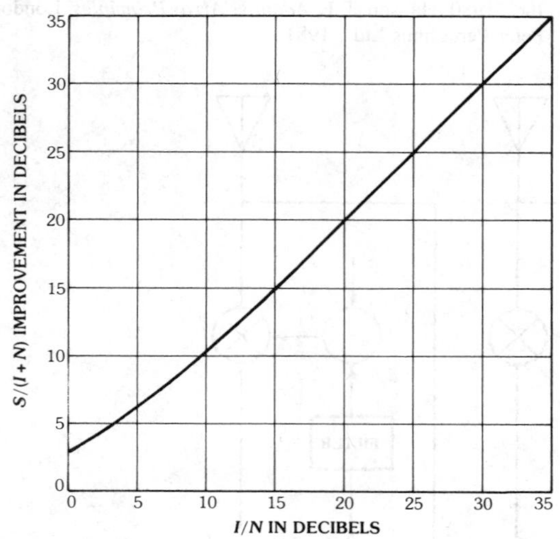

Fig. 83. Improvement in $S/(I+N)$ due to adaption.

APERTURE DISTRIBUTIONS

Design of Distributions

Aperture distributions of concern here are used to produce patterns with one or two narrow main beams with the rest of space occupied by sidelobes. These distributions are often sampled to yield excitation coefficients for arrays of many elements. For arrays of a few elements, adjustment of zeros of a polynomial is used.‡ In earlier years, aperture distributions were chosen for easy integrability. With the advent of computers, it is possible to design aperture distributions based on physical principles without concern about the form of the result. These physical principles were enunciated primarily by Taylor:* symmetric amplitude distributions are more efficient; array polynomial zeros should be real; far-out zeros should be separated by unity; and a distribution with a pedestal gives a $1/u$ far-out sidelobe envelope and is more efficient than a distribution that is zero at the ends. In all that follows, the variable $u=(L/\lambda)\sin\theta$ is used. In general, then, a narrow beam distribution is designed by making the spacing of zeros for large u approximately unity, and then by adjusting the close-in zeros to give the desired first sidelobe level. The achievement of a $1/u$ envelope assures a robust, low-Q distribution. It is convenient to use sidelobe ratio (SLR) as the ratio of beam peak to first sidelobe; SLR, of course, is the inverse of sidelobe level.

The distributions that will be discussed here differ in the shape of the close-in sidelobe envelope, and whether they are for line (or rectangular) or circular

‡ Villeneuve, A. T. "Taylor Patterns for Discrete Arrays." *Trans. IEEE*, Vol. AP-32, Oct. 1984; pp. 1089–1093.

* Taylor, T. T. "Design of Line-Source Antennas for Narrow Beamwidth and Low Sidelobes." *IRE Trans.*, AP-3, Jan. 1955, pp. 16–28.

sources. These distributions are now widely used and have superseded such distributions as cosine^n-on-a-pedestal, Gaussian, etc. The latter distributions not only have inefficient zero placement, but they involve two or more parameters. For example, a practical Gaussian must be truncated, and two parameters must be specified to determine performance indices. Incidentally, when the Gaussian is truncated, its advantages of a simple transform disappear. For these distributions, it is difficult to perform optimization: for example, determination of parameters to give maximum efficiency for a given sidelobe level is difficult. In contrast, the Taylor one-parameter distribution yields all performance indices as a function of that single parameter, while the Taylor $\bar{n}$ distribution is easily optimized because of the physical meaning of the parameters.

Taylor One-Parameter Line Source Distribution

In this distribution, the far-out zeros are placed at integer values of u, while close-in zeros are shifted to reduce close-in sidelobes to suitable values. Taylor accomplished this by using zeros

$$u=\sqrt{n^2+B^2}$$

where B is the single parameter. The pattern is written in two forms, depending on whether u is less or greater than B. These are

$$F(u)=(\sinh \pi\sqrt{B^2-u^2})/\pi\sqrt{B^2-u^2}, \qquad u\leq B$$

$$F(u)=(\sin \pi\sqrt{u^2-B^2})/\pi\sqrt{u^2-B^2}, \qquad u\geq B$$

The sidelobe ratio is given by

$$\text{SLR}=20\log(\sinh \pi B/\pi B)+13.26\text{ dB}$$

Taking the inverse transform of the pattern gives the Taylor one-parameter line source distribution:

$$g(p)=I_0(\pi B\sqrt{1-p^2})$$

where,

p is 0 at the aperture center and 1 at each end,

I_0 is the modified Bessel function.

A special case occurs for $B=0$. This is the uniformly excited line source (constant amplitude), which has a pattern of simply $\sin \pi u/\pi u$. The Taylor pattern is a modified $\sin \pi u/\pi u$ pattern, with a transition from that to the hyperbolic form at $u=B$, on the side of the main beam. The hyperbolic form provides the central part of the main beam as indicated in the formula above for SLR. Fig. 84 shows a typical one-parameter pattern or space factor. Typical aperture distributions are shown in Fig. 85. Only half of a symmetric distribution is shown. Not unexpectedly, the pattern with lower sidelobes has a smaller pedestal at the end of the aperture. The integral for aperture effeciency can be reduced to a tabulated function, I_0.†

† Abramowitz, M., and Stegun, L. *Handbook of Mathematical Functions*. NBS, 1970.

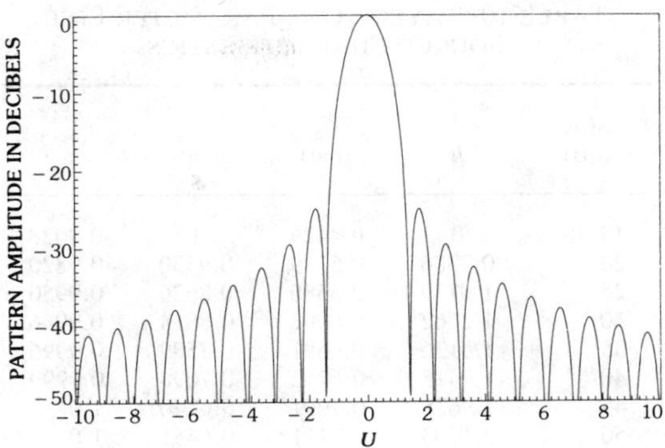

Fig. 84. Taylor one-parameter line source pattern for SLR = 25 dB.

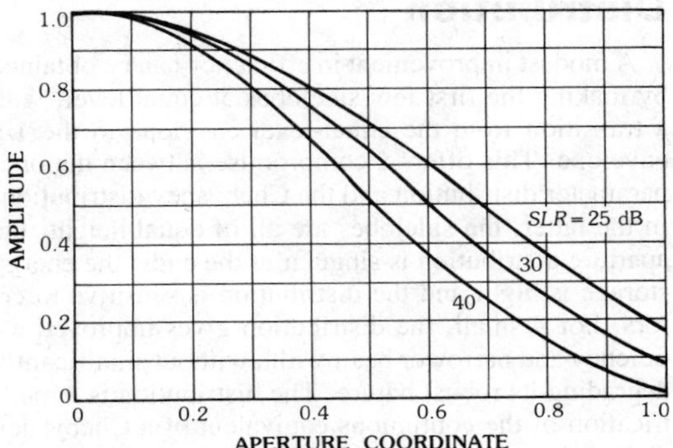

Fig. 85. Taylor one-parameter aperture distributions.

$$\eta=(2\sinh^2\pi B)/\pi B\bar{I}_0(2\pi B)$$

Although all of the performance indices can be given in terms of a table of equally spaced values of B, it is more convenient to specify sidelobe ratio. Table 10 gives beamwidth, efficiency, and beam efficiency. The beamwidth is normalized and is converted to an actual angle by

$$\theta_3/(\lambda/L)=(2L/\lambda)\arcsin[u_3/(L/\lambda)]\simeq 2u_3$$

The efficiency relates the directivity to that of a uniformly excited source. Beam efficiency is the fraction of radiated energy contained in the main beam, null to null.§ Given a desired performance index, the formulas allow B to be found, and from B all other indices can be determined. Taylor described this work in a Hughes Aircraft report; a review and tables are contained in Hansen.‡

§ Hansen, R. C. Array chapters in *Handbook of Antenna Design*, Vol. 2. A. W. Rudge, et al, eds. London: Peter Peregrinus Ltd., 1983.

‡ Hansen, R. C. *Microwave Scanning Antennas*, Vol. 1. New York: Academic Press, Inc., 1964; Chap. 1.

TABLE 10. TAYLOR ONE-PARAMETER LINE SOURCE CHARACTERISTICS

SLR (dB)	B	u_3 (rad)	η	η_b
13.26	0	0.4429	1	0.9028
20	0.7386	0.5119	0.9330	0.9820
25	1.0229	0.5580	0.8626	0.9950
30	1.2762	0.6002	0.8014	0.9986
35	1.5136	0.6391	0.7509	0.9996
40	1.7415	0.6752	0.7090	0.9999
45	1.9628	0.7091	0.6740	1.0
50	2.1793	0.7411	0.6451	1.0

Taylor $\bar{n}$ Line Source Distribution

A modest improvement in efficiency can be obtained by making the first few sidelobes at equal level, with a transition from the equal-level envelope to the $1/u$ envelope. This offers a compromise between the one-parameter distribution and the Chebyshev distribution. In the latter, the sidelobes are all of equal height, the aperture distribution is singular at the ends, the energy storage is high, and the distribution is sensitive to errors. For $\bar{n}$ small, the distribution gives improved efficiency and narrower beamwidth without significantly degrading its robust nature. The distribution is a modification of the continuous equivalent of a Chebyshev distribution, with a dilation factor used to modify the first $\bar{n}$ zeros. The pattern is given by a canonical product on zeros.**

$$F(u) = [(\sin \pi u)/\pi u] \prod_{n=1}^{\bar{n}-1} (1 - u^2/z_n^2)/(1 - u^2/n^2)$$

$$= \sum_{n=-(\bar{n}-1)}^{n=\bar{n}-1} F(n, A, \bar{n}) \text{ sinc } \pi(u+n)$$

In the second form, the pattern is a superposition of $\bar{n}$ sinc beams. The pattern coefficients are

$$F(n, A, \bar{n}) = \frac{[(\bar{n}-1)!]^2}{(\bar{n}-1+n)!(\bar{n}-1-n)!} \prod_{m=1}^{\bar{n}-1} (1 - \bar{n}^2/z_m^2)$$

with $F(0, A, \bar{n}) = 1$. Here the zeros are:

$$z_n = \pm\sigma\sqrt{A^2 + (n-1/2)^2} \qquad 1<n<\bar{n}$$

$$z_n = \pm n \qquad \bar{n} \leq n$$

where,

A is the sidelobe parameter,

σ is the dilation index.

The dilation index is

$$\sigma = \bar{n}/\sqrt{A^2 + (\bar{n}-1/2)^2}$$

The corresponding aperture distribution is

$$g(p) = 1 + 2 \sum_{n=1}^{\bar{n}-1} F(n, A, \bar{n}) \cos n\pi p$$

Tables of aperture distribution are not given because of the ease of calculating these with a computer. Values for check purposes can be obtained from Hansen.†† Table 11 gives parameter A and beamwidth factor u_3 for several values of SLR. Also shown are σ values for $\bar{n}$. The beamwidth in u is approximately $2\sigma u_3$; this may be converted to angle by using the formula in the previous subsection. Fig. 86 shows a typical Taylor $\bar{n}$ pattern; Fig. 87 shows typical aperture distributions. The value of $\bar{n}$ must be carefully chosen. If $\bar{n}$ is too small, the zero spacing in the transition region will oscillate, thereby producing oscillations in the sidelobe envelope. If $\bar{n}$ is large, the aperture distribution will not be monotonic; for suitably large $\bar{n}$ the end value may be larger than the center value, which is clearly disastrous. There is a value of $\bar{n}$ that gives maximum efficiency. Table 12 gives these values and the largest $\bar{n}$ that maintains a monotonic distribution. Note that the maximum efficiency values require an aperture distribution with peaks at the ends and so are generally not desirable. Aperture efficiency is easily calculated from

** Taylor, T. T. "Design of Line-Source Antennas for Narrow Beamwidth and Low Sidelobes." *IRE Trans.* Vol. AP-3, Jan. 1955, pp. 16–28.

†† Hansen, R. C. *Microwave Scanning Antennas*, Vol. 1. New York: Academic Press, Inc., 1964; Chap 1.

TABLE 11. TAYLOR $\bar{n}$ LINE SOURCE CHARACTERISTICS

SLR (dB)	A	u_3	σ: $\bar{n}=2$	4	6	8	10
20	0.9528	0.4465	1.1255	1.1027	1.0749	1.0582	1.0474
25	1.1366	0.4890		1.0870	1.0683	1.0546	1.0452
30	1.3200	0.5284		1.0693	1.0608	1.0505	1.0426
35	1.5032	0.5653			1.0523	1.0459	1.0397
40	1.6865	0.6000			1.0430	1.0407	1.0364
45	1.8697	0.6328				1.0350	1.0328
50	2.0530	0.6639					1.0289

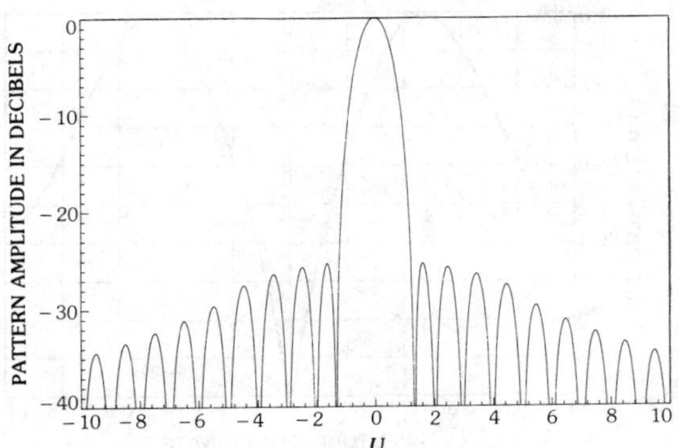

Fig. 86. Taylor line source pattern for SLR = 25 dB, $\bar{n}=5$.

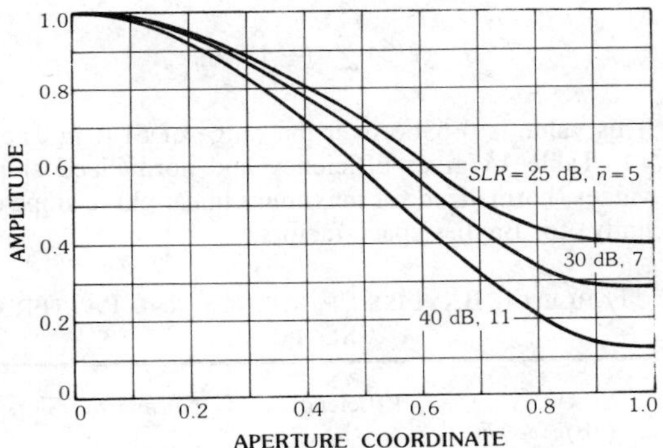

Fig. 87. Taylor $\bar{n}$ aperture distributions.

TABLE 12. TAYLOR $\bar{n}$ EFFICIENCIES

	Max η Values		Monotonic $\bar{n}$	
SLR	$\bar{n}$	η	$\bar{n}$	η
25	12	0.9252	5	0.9105
30	23	0.8787	7	0.8619
35	44	0.8326	9	0.8151
40	81	0.7899	11	0.7729

$$\eta = 1/[1+2\sum_{n=1}^{\bar{n}-1} F^2(n,A,\bar{n})]$$

The difference in efficiency between the maximum $\bar{n}$ case and the monotonic $\bar{n}$ case is roughly 1%, so the monotonic $\bar{n}$ should be used as an upper limit. The best $\bar{n}$ can be selected by comparing computer runs of patterns for various values of $\bar{n}$. For additional information, see Hansen.§§

Bayliss $\bar{n}$ One-Parameter Difference Line Source

The Bayliss space factor is a difference pattern (for tracking purposes) constructed to have the same features as the Taylor $\bar{n}$ space factor. That is, it has $\bar{n}$ roughly equal-level sidelobes adjacent to each difference beam, with a $1/u$ envelope beyond. Taylor started with an "ideal" pattern, $\cos \pi \sqrt{(u^2-A^2)}$, which has equal-level sidelobes. Then a space factor was constructed with $\bar{n}$ close-in zeros matching those of the "ideal" and with remaining zeros to give a $1/u$ envelope. Finally, a dilation factor was used to make a smooth transition around $u=\bar{n}$. A good starting point to get a difference pattern is to differentiate the "ideal" sum pattern. This gives $u \operatorname{sinc} \pi \sqrt{u^2-A^2}$. Unfortunately, not all sidelobes are of equal level; the first several are tapered. Bayliss used an iterative procedure to adjust these zeros to yield equal-level sidelobes. It was necessary to adjust only four. Since these four zeros (and A) depend on the sidelobe ratio, results for each were given in terms of fourth-order polynomials in SLR. Table 13 gives the results, for various values of SLR, for z_1, z_2, z_3, z_4, A, and u_0. The latter is the value of u at the difference peak. These zeros are used for both Bayliss line sources and circular sources.‡‡ For $n>\bar{n}$, $z_n=\sqrt{A^2+n^2}$. Next, the Taylor procedure is followed, where the envelope is made to approach $1/u$ for large u, and with a smooth transition. The difference pattern is

§§ Hansen, R. C. Array chapters in *Handbook of Antenna Design*, Vol. 2. A. W. Rudge, et al, eds. London: Peter Peregrinus Ltd., 1983.

‡‡ Bayliss, E. T. "Design of Monopulse Antenna Difference Patterns With Low Sidelobes." *BSTJ*, Vol. 47, May-June 1968, pp. 623–650.

TABLE 13. BAYLISS LINE SOURCE PARAMETERS

	SLR, Decibels					
	15	20	25	30	35	40
A	1.00790	1.22474	1.43546	1.64126	1.84308	2.04154
z_1	1.51240	1.69626	1.88266	2.07086	2.26025	2.45039
z_2	2.25610	2.36980	2.49432	2.62754	2.76748	2.91231
z_3	3.16932	3.24729	3.33506	3.43144	3.53521	3.64518
z_4	4.12639	4.18544	4.25273	4.32758	4.40934	4.49734
u_0	0.66291	0.71194	0.75693	0.79884	0.83847	0.87649

$$F(u) = u \cos \pi u \frac{\prod_{n=1}^{\bar{n}-1} (1 - u^2/\sigma^2 z_n^2)}{\prod_{n=0}^{\bar{n}-1} (1 - u^2/(n+1/2)^2)}$$

As before,

$$\sigma = (\bar{n} + 1/2)/\sqrt{A^2 + \bar{n}^2}$$

Fig. 88 shows a Bayliss line source pattern for SLR = 25 dB, $\bar{n} = 5$. The aperture distribution for $-1 \le p \le 1$ is

$$g(p) = \sum_{n=0}^{\bar{n}-1} B_n \sin \pi(n + 1/2)p$$

where

$$B_m = -(-1)^m (m + 1/2)^2 \cdot \frac{\prod_{n=1}^{\bar{n}-1} [1 - (m+1/2)^2/\sigma^2 z_n^2]}{\prod_{\substack{n=0 \\ n \neq m}}^{\bar{n}-1} [1 - (m+1/2)^2/(n+1/2)^2]}$$

These formulas allow rapid calculation of Bayliss space factors and aperture distributions. Fig. 89 gives the aperture distribution corresponding to the space factor of Fig. 88. Fig. 89 shows only amplitude; each half

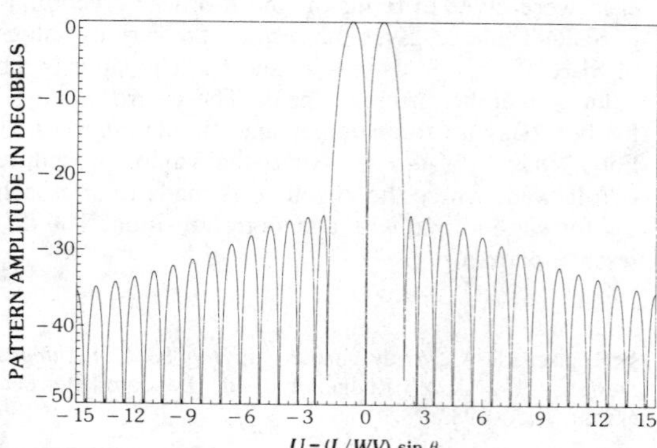

Fig. 88. Bayliss space factor for SLR = 25 dB, $\bar{n} = 5$.

of the aperture has constant phase, with the two halves out of phase. Excitation efficiency, which is the ratio of directivity (at one difference peak) to that of a uniformly excited line source, is given by

$$\eta = \frac{2u_0^2 \cos^2 \pi u_0 \left\{ \sum_{n=0}^{\bar{n}-1} (-1)^n B_n/[u_0^2 - (n+1/2)^2] \right\}^2}{\sum_{n=0}^{\bar{n}-1} B_n^2}$$

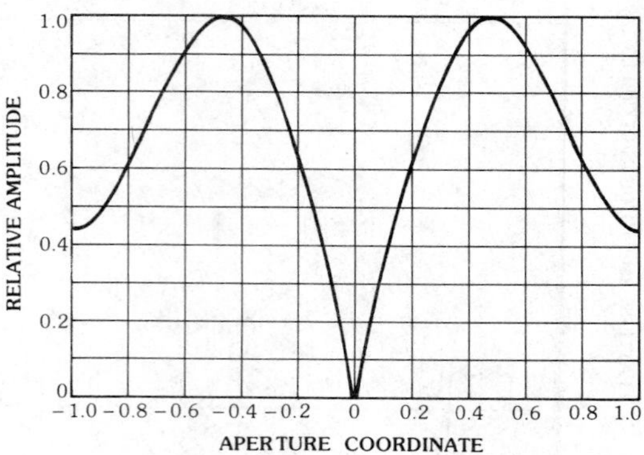

Fig. 89. Bayliss distribution for SLR = 25 dB, $\bar{n} = 5$.

The slope at $u = 0$ is given by

$$S = (2/\pi) \sum_{n=0}^{\bar{n}-1} (-1)^n B_n/z_n^2$$

This value is 0.6366 over the range of SLR in Table 14. Table 14 gives efficiency and normalized slope values (normalized for maximum linear phase slope of unity) for Bayliss space factors.

TABLE 14. BAYLISS EFFICIENCY AND PATTERN SLOPE

SLR (dB)	$\bar{n}$	Efficiency η	Normalized Slope S
15	4	0.5959	0.9567
20	4	0.5846	0.8974
25	5	0.5633	0.8427
30	6	0.5393	0.7912
35	7	0.5162	0.7448
40	8	0.4951	0.7037

Low Sidelobe Distributions

High sidelobe ratios, in the range from 30 to 60 dB, have recently attracted more interest. The one-parameter space factors have very low pedestals for this range of SLR, and thus the Taylor $\bar{n}$ space factors are more appropriate. Again, efficient design requires properly spaced space-factor zeros. In contrast, most of the popular distributions were coined before the computer allowed performance evaluations to be easily performed. The Taylor $\bar{n}$ distribution is, in essence, perfect for low sidelobes because it is low-Q, which eases the tolerance problem, and because the zeros are monotonically and smoothly spaced. This can be made evident by comparing the Taylor with the popular Hamming distribution. The latter* is

* Blackman, R. B., and Tukey, J. W. *Measurement of Power Spectra*. New York: Dover Publications, Inc., 1958.

$$g(p)=0.54+0.46 \cos \pi p=a+b \cos \pi p$$

The excitation efficiency of the Hamming is

$$\eta_t=2a^2/(2a^2+b^2)=0.7338$$

The space factor has zeros at $u=2, 3, 4, \ldots$ and, in addition, a zero at $u=\sqrt{a/(a-b)}=2.5981$. This close spacing of the first three zeros produces an uneven sidelobe envelope with the fourth sidelobe the highest at -42.7 dB. Thus, the Hamming is compared with a Taylor $\bar{n}$ distribution with SLR=42.7 dB. Table 15 shows the zeros, and it can be observed that both of the Taylor space factors have a smoothly increasing zero spacing from the first zero out to the transition point, beyond which the spacing is unity. The Hamming, on the other hand, has a first spacing of 0.598, a second spacing of 0.402, and the remaining spacings all unity. This is obviously not as good a design, as the following will show.

TABLE 15. ZEROS OF DISTRIBUTIONS FOR SLR=42.7 dB

n	Taylor One-Parameter	Taylor $\bar{n}=6$	Taylor $\bar{n}=10$	Hamming
1	2.112	1.894	1.897	2
2	2.732	2.398	2.396	2.598
3	3.550	3.173	3.166	3
4	4.412	4.069	4.056	4
5	5.335	5.020	5.002	5
6	6.282	6	5.978	6
7	7.243	7	6.970	7
8	8.214	8	7.974	8
9	9.190	9	8.984	9
10	10.172	10	10	10

Table 16 gives efficiency and normalized beamwidth for the three Taylor cases and for the Hamming. It may be seen that the Hamming beamwidth is roughly 3% broader, while the efficiency is roughly 3% lower. Thus, by use of appropriate space factors such as the Taylor $\bar{n}$, improved performance can be obtained at no additional cost in complexity, hardware, or tolerances.

TABLE 16. COMPARISON OF DISTRIBUTIONS FOR SLR=42.7 dB

	Taylor One-Parameter	Taylor $\bar{n}=6$	Taylor $\bar{n}=10$	Hamming
u_3	0.694	0.637	0.635	0.651
η	0.690	0.754	0.755	0.734

For rectangular arrays with separable aperture distributions, Taylor distributions may, of course, be used along each coordinate. For circular disk apertures, the circular Taylor $\bar{n}$ distribution is similarly excellent.

Measurement of Low-Sidelobe Patterns

In addition to the obvious problem of providing a pattern range with a low background level, the measurement of low sidelobe patterns imposes restrictions on the measurement distance. A distance of $2D^2/\lambda$ is adequate for patterns with modest sidelobe ratios (25 dB or less), but larger distances are needed to measure lower sidelobes accurately, due to the quadratic phase error produced by a finite distance. As the measurement distance is reduced, the first sidelobe and first null rise, and at the same distance the sidelobe becomes a shoulder on the main beam. (The second sidelobe is only slightly raised at this point.) At closer distances, the main beam broadens, and the second sidelobe and second null rise, and again the main beam eventually absorbs the sidelobe. Calculations have been made of the change in sidelobe level versus measurement distance for the universal Taylor $\bar{n}$ linear space factor.† Fig. 90 gives these results for a uniform line source, and for 20 (10) 60-dB Taylor $\bar{n}$ line sources. Distance is normalized to $2D^2/\lambda$, where

$$\gamma=R/(2D^2/\lambda)$$

Each curve terminates when the sidelobe becomes a shoulder on the main beam, without any dip.

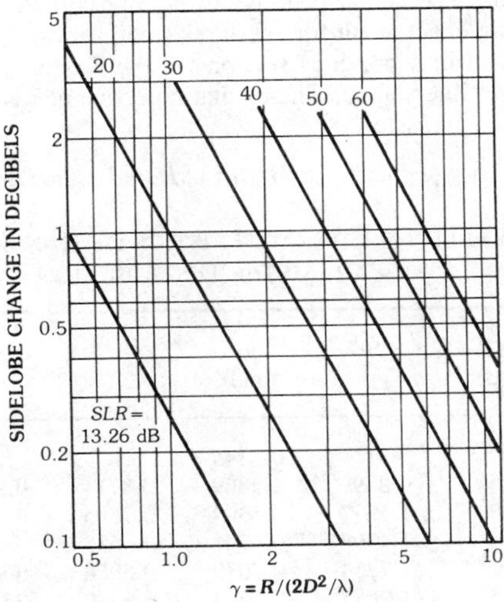

Fig. 90. Sidelobe increase due to measurement distance.

† Hansen, R. C. "Measurement Distance Effects on Low Sidelobe Patterns." *Trans. IEEE*, Vol. AP-32, June 1984; pp. 591–594.

Hansen One-Parameter Circular Source Distribution

A symmetric distribution for circular disk apertures analogous to the Taylor one-parameter line source distribution has been developed by Hansen.§ This is a modified $2J_1(\pi u)/\pi u$ pattern just as the Taylor one-parameter was a modified sinc πu. The close-in zeros are shifted to produce the desired SLR. The single parameter is H, and the space factor is again written in two forms depending on whether u is less or greater than H. These are

$$F(u)=2J_1(\pi\sqrt{u^2-H^2})/\pi\sqrt{u^2-H^2} \qquad u\geq H$$

$$F(u)=2I_1(\pi\sqrt{H^2-u^2})/\pi\sqrt{H^2-u^2} \qquad u\leq H$$

The sidelobe ratio is given by

$$\text{SLR}=20\log[2I_1(\pi H)/\pi H]+17.57\text{ dB}$$

In these formulas, J_1 and I_1 are the usual Bessel and modified Bessel functions of first kind and order one. Again, the top portion of the main beam is provided by the modified Bessel form, while the remainder of the main beam and the sidelobe structure are provided by the Bessel form. The aperture distribution is

$$g(p)=I_0(H\sqrt{\pi^2-p^2})$$

where p is zero at the center of the aperture and the aperture radius is π. The patterns and aperture distributions are much like those of the Taylor $\bar{n}$ line source and so have not been included. Table 17 gives parameter H, beamwidth u_3, the aperture excitation efficiency, and the beam efficiency. The latter is the fraction of energy that is contained in the main beam, null to null. The actual beamwidth is approximately $2u_3$. A more exact result for apertures small in wavelengths is found in a previous section for the Taylor one-parameter line source. The excitation efficiency is given by the expression

$$\eta=4I_1^2(\pi H)/\pi^2H^2[I_0^2(\pi H)-I_1^2(\pi H)]$$

TABLE 17. CHARACTERISTICS OF HANSEN ONE-PARAMETER DISTRIBUTION

SLR (dB)	H	u_3 (rad)	η	η_b
17.57	0	0.5145	1	—
25	0.8899	0.5869	0.8711	0.9745
30	1.1977	0.6304	0.7595	0.9930
35	1.4708	0.6701	0.6683	0.9981
40	1.7254	0.7070	0.5964	0.9994
45	1.9681	0.7413	0.5390	0.9998
50	2.2026	0.7737	0.4923	1.0000

§ Hansen, R. C. "A One-Parameter Circular Aperture Distribution With Narrow Beamwidth and Low Sidelobes." *Trans. IEEE*, Vol. AP-24, July 1976, pp. 477–480.

The I_0 used above is the modified Bessel function of order zero. This distribution, like the Taylor distribution, is low-Q and robust, and has been found to fit well to many reflector antenna patterns, for the main beam and close-in sidelobes.

Taylor $\bar{n}$ Circular Source Distribution

The Taylor $\bar{n}$ circular source distribution offers an improvement in efficiency and beamwidth over the Hansen one-parameter circular source distribution, just as the Taylor $\bar{n}$ line source distribution shows an improvement over the Taylor one-parameter source distribution. The circular aperture is a disk with rotationally symmetric excitation, and, again, the starting point is the uniform $2J_1(\pi u)/\pi u$ space factor. On each side of the main beam, $\bar{n}$ zeros are modified by moving them to produce the desired sidelobe ratio.‡ Again, a dilation factor, σ, is used to provide a smooth transition between the roughly equal-level sidelobes and the tapered-envelope sidelobes. The pattern is given by a canonical product on zeros:

$$F(u)=[2J_1(\pi u)/\pi u]\prod_{n=1}^{\bar{n}-1}(1-u^2/u_n^2)/(1-u^2/\mu_n^2)$$

where μ_n are the zeros of $J_1(\pi u)$. The close-in pattern zeros are given by

$$u_n=\pm\sigma\sqrt{A^2+(n-1/2)^2} \qquad 1\leq n\leq\bar{n}$$

while

$$\sigma=\mu_{\bar{n}}/\sqrt{A^2+(\bar{n}-1/2)^2}$$

Table 18 gives the sidelobe parameter, A, the beamwidth, u_3, and values of σ for various values of $\bar{n}$. The actual beamwidth for large apertures is $2\sigma u_3$. The patterns and aperture distributions are much like those of the Taylor $\bar{n}$ line source and so have not been included. The aperture distribution is

$$g(p)=(2/\pi^2)\sum_{m=0}^{\bar{n}-1}F_mJ_0(p\mu_m)/[J_0(\pi\mu_m)]^2$$

where $p=2\pi\rho/D$ and

$$F_m=-J_0(\pi\mu_m)\frac{\prod_{n=1}^{\bar{n}-1}(1-\mu_m^2/u_n^2)}{\prod_{n=1,\,n\neq m}^{\bar{n}-1}(1-\mu_m^2/\mu_n^2)}, \qquad F_0=1$$

All of these formulas are easily computed, so no tables have been included. For check purposes, reference may

‡ Taylor, T. T. "Design of Circular Apertures for Narrow Beamwidths and Low Sidelobes." *IRE Trans.*, AP-8, 1960, pp. 17–22.

TABLE 18. TAYLOR $\bar{n}$ CIRCULAR SOURCE CHARACTERISTICS

SLR (dB)	A	u_3	σ $\bar{n}=4$	5	6	7	8	9	10
20	0.9528	0.4465	1.1692	1.1398	1.1186	1.1028	1.0906	1.0810	1.0732
25	1.1366	0.4890	1.1525	1.1296	1.1118	1.0979	1.0870	1.0782	1.0708
30	1.3200	0.5284	1.1338	1.1180	1.1039	1.0923	1.0827	1.0749	1.0683
35	1.5032	0.5653	1.1134	1.1050	1.0951	1.0859	1.0779	1.0711	1.0653
40	1.6865	0.6000	1.0916	1.0910	1.0854	1.0789	1.0726	1.0670	1.0620

TABLE 19. EXCITATION EFFICIENCY VERSUS $\bar{n}$

SLR (dB)	η $\bar{n}=4$	5	6	8	10
20	0.9723	0.9356	0.8808	0.7506	0.6238
25	0.9324	0.9404	0.9379	0.9064	0.8526
30	0.8482	0.8623	0.8735	0.8838	0.8804
35	0.7708	0.7779	0.7880	0.8048	0.8153
40	0.7056	0.7063	0.7119	0.7252	0.7365

be made to tables of Hansen.** The aperture excitation efficiency is given by

$$\eta = 1/\left[1 + \sum_{n=1}^{\bar{n}-1} F_n^2/J_0^2(\pi\mu_n)\right]$$

Table 19 gives the aperture excitation efficient for several combinations of SLR and $\bar{n}$.

REFLECTORS

Parabolic Reflectors

The parabolic reflector commonly exists in both focal feed and Cassegrain form (Figs. 91 and 92). Offset reflectors are covered later. For a front-fed reflector, the reflector f/D must be matched to the feed pattern. Reflectors with pattern sidelobes roughly −25 dB and good efficiency typically have illumination edge tapers of −10 to −11 dB. For lower sidelobes, the edge taper must be lower. The tradeoff between edge taper and sidelobe level can be accurately determined from the circular one-parameter distributions in the section on aperture distributions. Fig. 93 shows the total reflector subtended angle versus f/D, where

$$\theta = 2 \text{ arc tan } [(8f/D)/(16f^2/D^2 - 1)]$$

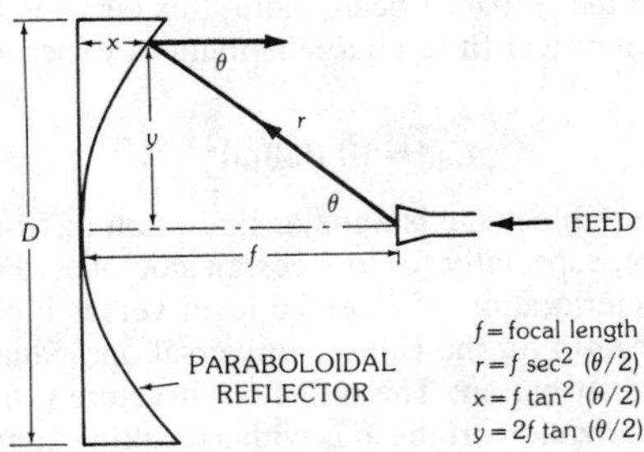

Fig. 91. Paraboloidal-reflector design.

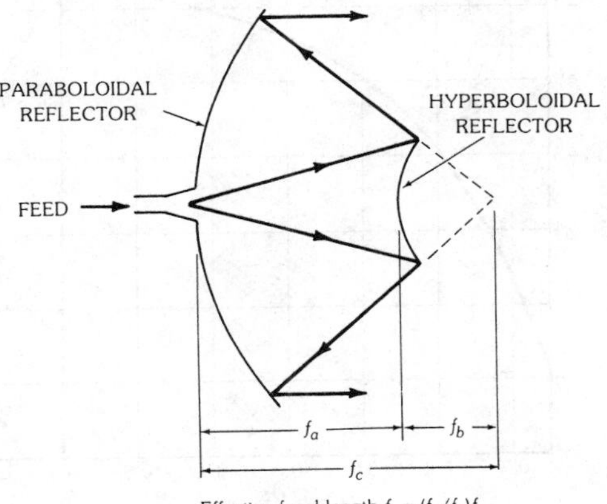

Fig. 92. Cassegrain reflector system.

** Hansen, R. C. "Tables of Taylor Distributions for Circular Aperture Antennas." *IRE Trans.*, Vol. AP-8, Jan. 1960, pp. 23–26. Hansen, R. C. *Microwave Scanning Antennas*, Vol. 1. New York: Academic Press, Inc., 1964; Chap. 1.

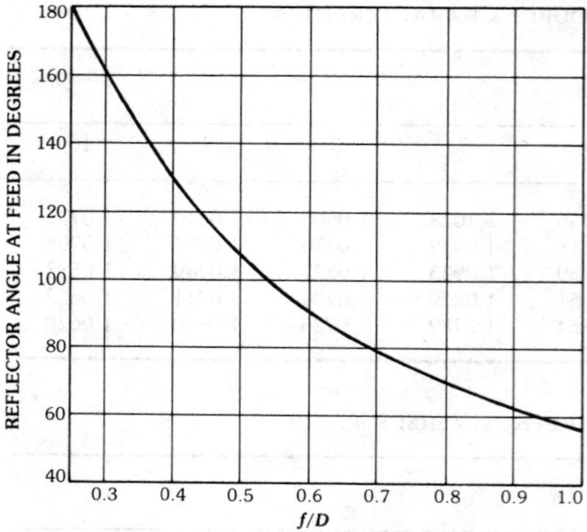

Fig. 93. Angle of reflector at feed.

Fig. 94 gives the differential path loss between the edge ray and the apex ray, also as a function of f/D:

$$R_{\text{edge}}/f = 1 + 1/(16f^2/D^2)$$

The edge illumination is, of course, the feed pattern value at the edge angle plus the differential path loss. Fig. 95 allows horn beamwidth at any level to be converted to the -10-dB beamwidth; this curve is an excellent empirical fit to a large amount of experimental data:

$$E_{\text{dB}} = 10\ (\theta/\theta_{10})^2$$

Aperture blockage is a major limitation of front-fed reflectors, especially for low design sidelobes. Fig. 96 shows deterioration of sidelobe level versus blockage diameter ratio for the Hansen universal one-parameter circular distribution. The reflector curvature produces cross polarization off the axis with a resulting gain loss.

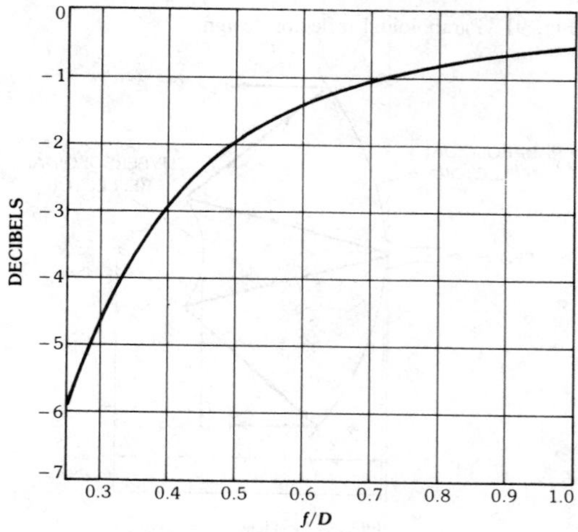

Fig. 94. Edge-center space loss.

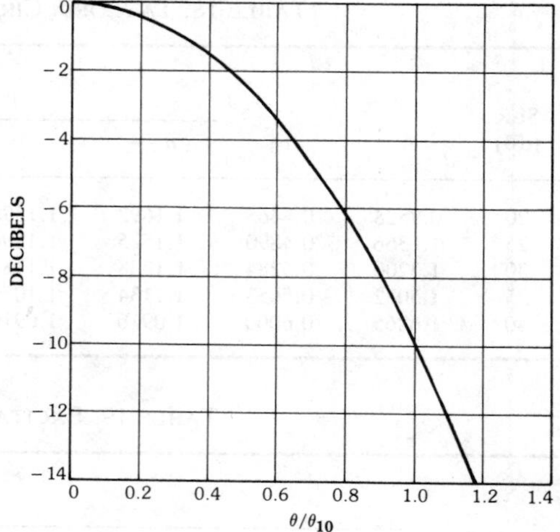

Fig. 95. Universal horn beam width conversion.

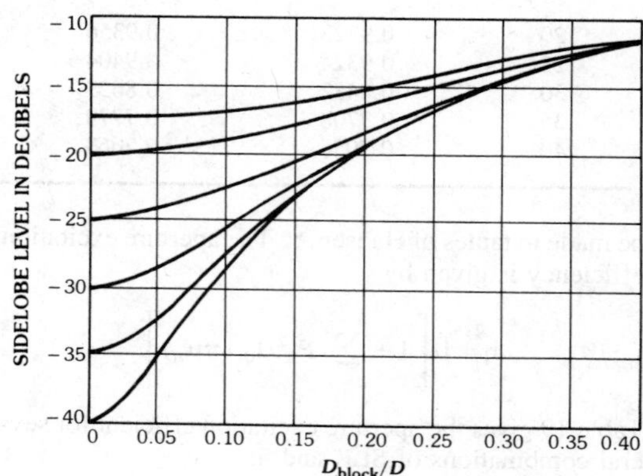

Fig. 96. Sidelobe level vs the blockage diameter ratio for Hansen one-parameter circular distribution.

Determination of aperture efficiency or gain is complex and involves calculation of aperture taper efficiency, feed spillover, feed cross polarization loss, blockage loss, and reflector cross polarization loss. For more information, refer to Rusch et al.* Random errors in the reflector surface raise the sidelobes and reduce the gain. The effect on sidelobes can be estimated by using the universal curve in Fig. 77, where sidelobe degradation is plotted against $\sigma/\sqrt{G}$. Note, however, that the σ measured for a reflector surface must be doubled before being used in that figure, due to the reflective operation of the surface. Directivity loss has been calculated by Ruze† with a modification added by

* Rusch, W. V. T., et al. "Quasi-Optical Antenna Design and Applications," Chapter 3 in the *Handbook of Antenna Design*, Vol. 1. A. W. Rudge, et al, eds. London: Peter Peregrinus Ltd., 1983.

† Ruze, J. "Antenna Tolerance Theory—A Review." *Proc. IEEE*, Vol. 54, April 1966, pp. 633–640.

Wested§ to take into account f/D. His result, which is

$$G/G_0 = (A+1)/[A + \exp(4\pi\epsilon/\lambda)^2]$$

where,

$$A = 1/[16f^2/D^2 \ln(1 + 1/16f^2/D^2)] - 1$$

is shown in Fig. 97, which gives gain decrease against one sigma tolerance in wavelengths for several values of f/D. It can be seen that for loss of more than a few tenths of a decibel, f/D needs to be included. Most narrow-band reflector antennas have aperture efficiency (gain compared to that of a uniformly excited area) in the 55–65% range. Wideband antennas using log-periodic or conical spiral feeds may have efficiencies anywhere from 20 to 50%.

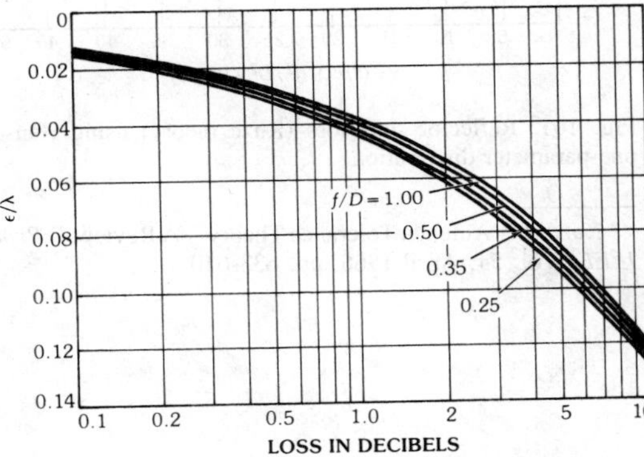

Fig. 97. Reflector directivity loss vs rms tolerance.

The Cassegrain antenna, through use of a hyperboloidal subreflector, allows the feed to protrude through a hole in the dish. Thus, low-noise and/or high-power components can be conveniently located behind the dish. The minimum blockage condition exists when the feed blockage equals the subreflector blockage. Fig. 98 shows gain decrease for equal-blockage Cassegrain reflectors versus beamwidth. From this and previous material, it is apparent that front-fed reflectors are attractive for moderate beamwidths, whereas Cassegrains are attractive for beamwidths of several degrees or less. Cross-polarization effects of a Cassegrain reflector are shown in Fig. 99. A high-efficiency Cassegrain may be designed by shaping the two reflector surfaces; only small changes on the order of a wavelength at the edge are needed.‡

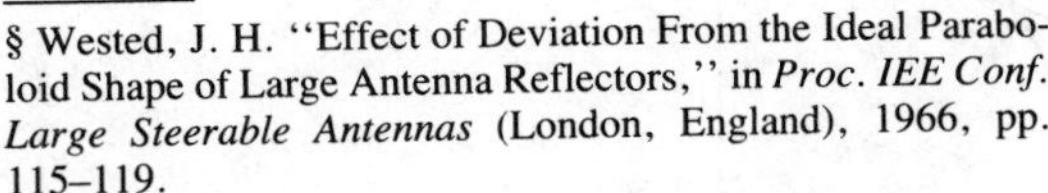

§ Wested, J. H. "Effect of Deviation From the Ideal Paraboloid Shape of Large Antenna Reflectors," in *Proc. IEE Conf. Large Steerable Antennas* (London, England), 1966, pp. 115–119.

‡ Galindo, V. "Design of Dual-Reflector Antennas With Arbitrary Phase and Amplitude Distribution." *Trans. IEEE,* Vol. AP-12, July 1964, pp. 403–408. Williams, W. F. "High Efficiency Antenna Reflector." *Microwave J.,* Vol. 8, July 1965, pp. 79–82.

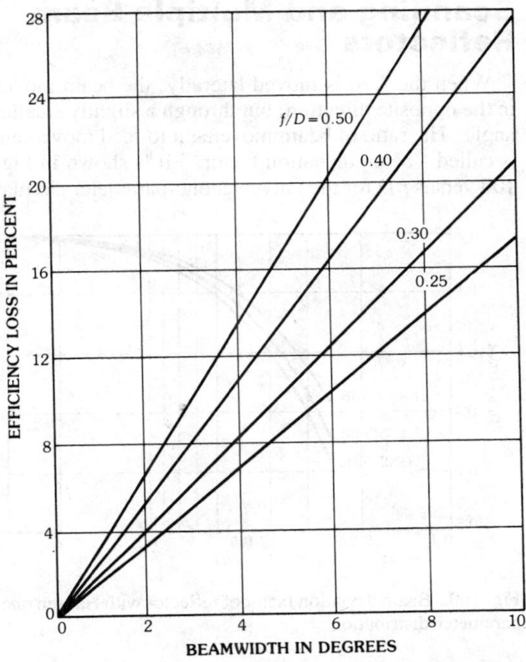

Fig. 98. Minimum blockage loss vs antenna beamwidth. (*From Rusch, W. V. T., et al. "Quasi-Optical Antenna Design and Applications," Chapter 3 in the* Handbook of Antenna Design, *Vol. 1. A. W. Rudge, et al, eds. London: Peter Peregrinus Ltd., 1983.*)

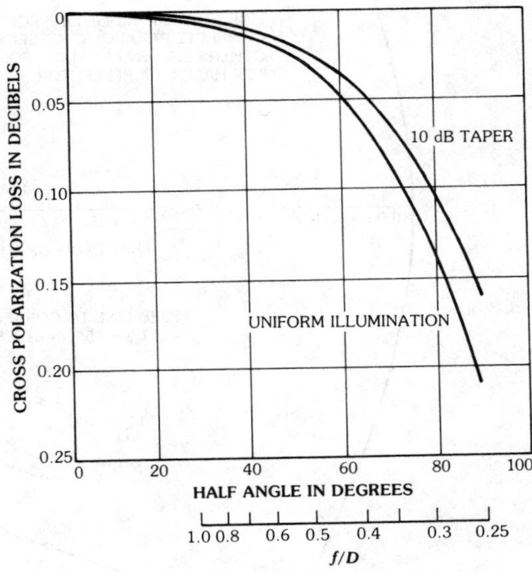

Fig. 99. Cross polarization vs subtended half angle due to reflector curvature. (*From Rusch, W. V. T., et al. "Quasi-Optical Antenna Design and Applications," Chapter 3 in the* Handbook of Antenna Design, *Vol. 1. A. W. Rudge, et al, eds. London: Peter Peregrinus Ltd., 1983.*)

Scanning and Multiple-Beam Reflectors

When the feed is moved laterally, the beam moves in the opposite direction, but through a slightly smaller angle. The ratio of beam movement to feed movement is called "beam deviation factor." It is shown in Fig. 100 versus f/D for the universal one-parameter circular distribution. There is a concomitant loss of directivity due to coma, and this was calculated by Ruze** for several distributions. A recalculation using the circular one-parameter distribution is shown in Fig. 101. Ruze

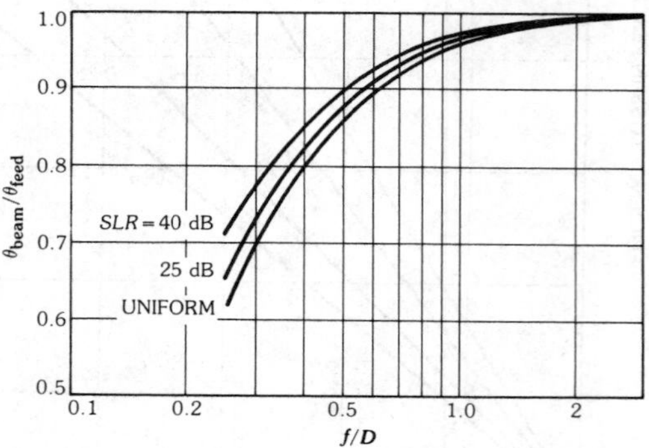

Fig. 100. Beam deviation factor of reflector with Hansen one-parameter distribution.

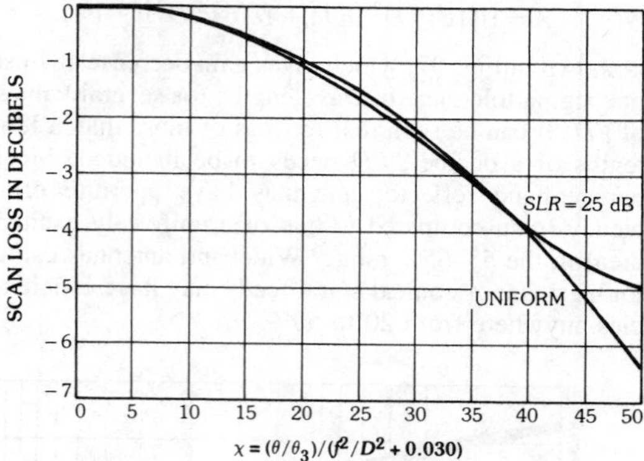

Fig. 101. Reflector scan loss (Ruze theory) using Hansen one-parameter distribution.

** Ruze, J. "Antenna Tolerance Theory—A Review." *Proc. IEEE*, Vol. 54, April 1966, pp. 633-640.

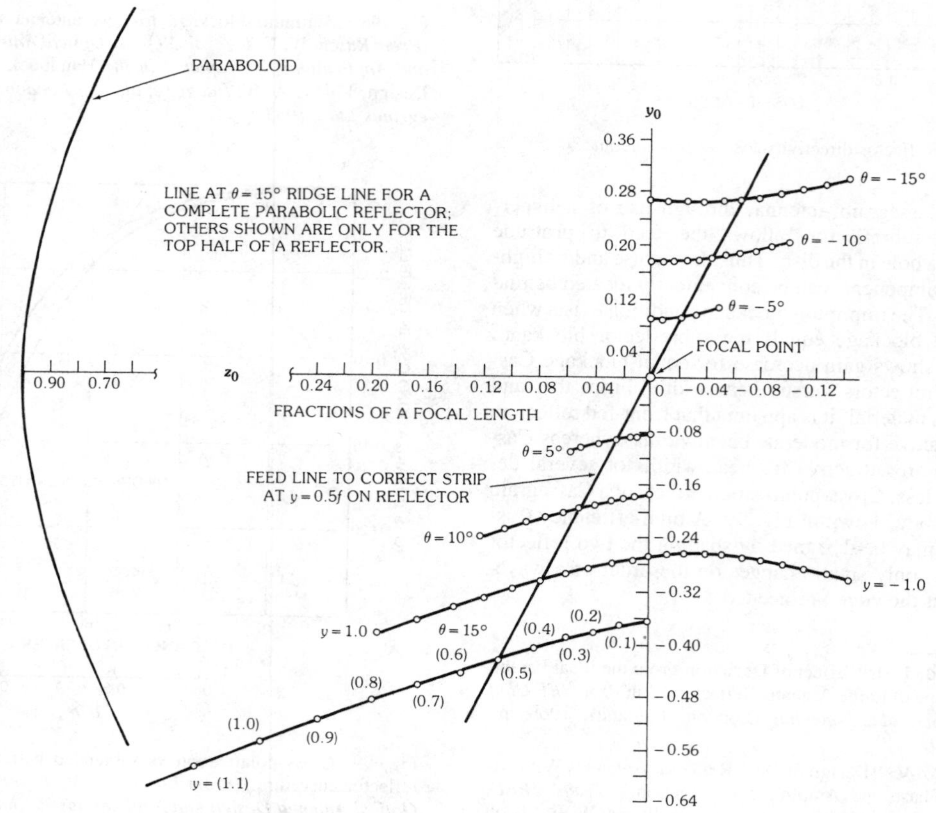

Fig. 102. Ridge lines produced by plane waves incident on paraboloidal section inclined at angles θ. (*Courtesy C. J. Sletten.*)

found that a normalized parameter, χ, could be used that closely fitted results from a wide range of f/D, and this parameter also fits these new data. Parameter χ is

$$\chi = (\theta/\theta_3)/(f_2/D^2 + 0.030)$$

From this, it can be seen that a one-dB loss occurs for values shown in Table 20.

TABLE 20. BEAM MOVEMENT FOR 1-dB LOSS

f/D	θ/θ_3
0.25	1.8
0.3	2.3
0.35	2.9
0.4	3.6
0.5	5.3
0.6	7.4
1	19.6

When the coma loss is unacceptably large, a feed array can be used in the transverse focal region. One coma correction scheme sums the array element outputs with weighting that is the conjugate of the focal field. Another requires only phase correction and utilizes a beam-forming network connected to the feed array. The phases at the BFN output are then corrected (in transform space) and summed.†† When a single horn is moved off axis, the surface giving the smallest-diameter circle of confusion is not plane and is called the Petzval surface. The maximum-gain surface also is not plane and deviates slightly from the Petzval surface. Fig. 102 shows focal ridge lines for various offsets. Generally, the precise focal surface used is not important until coma correction becomes necessary. Multiple-beam antennas typically employ a cluster of horns with each horn producing a separate beam. These configurations are usually offset to reduce the large blockage that would occur with a front-fed reflector. For further information, see Rusch, et al.§§

†† Rudge, A. W., and Davies, D. E. N. "Electronically Controllable Primary Feed for Profile-Error Compensation of Large Parabolic Reflectors." *Proc. IEE*, Vol. 117, Feb. 1970, pp. 351–357.

§§ Rusch, W. V. T., et al. "Quasi-Optical Antenna Design and Applications." Chapter 3 in the *Handbook of Antenna Design*, Vol. 1. A. W. Rudge, et al, eds. London: Peter Peregrinus Ltd., 1983.

33 Electromagnetic-Wave Propagation

Revised by
Douglass D. Crombie

Radio waves may be propagated* from the transmitting antenna to the receiving antenna through or along the surface of the earth, through the atmosphere, or by reflection or scattering from natural or artificial reflectors. The conductivity and dielectric constant of the ground vary considerably from those of the atmosphere. At very-low frequencies, ground waves may be satisfactorily propagated for distances of several thousand kilometers. At high frequencies, however, the losses are so great that signals can be propagated for only a few hundred kilometers by ground wave. Propagation in the medium- and high-frequency bands is chiefly by ground wave and by reflection from the ionosphere, and severe fading is caused in these frequency bands by the interference between ground and ionospheric waves.

The refractive index of the atmosphere is an important factor in radio propagation. At frequencies between about 100 and 8000 megahertz, scattering of radio waves by inhomogeneities in the electromagnetic characteristics of the atmosphere can be used to provide satisfactory wideband communication up to several times the line-of-sight distance. New techniques are being developed for generating coherent high-power waves in the optical spectrum. Atmospheric absorption at these frequencies is high, but the large bandwidths and small antenna beam widths may make such frequencies practical for certain applications.

VERY-LOW FREQUENCIES—UP TO 30 KILOHERTZ

The propagation of long radio waves is of considerable importance in reliable communication, long-range navigation, and the detection of nuclear explosions. Considerable progress has been made in recent years in understanding the propagation of such waves in the earth-ionosphere waveguide.†

At short distances from a transmitter, the received signal is chiefly by a ground or surface wave, and at very-low frequencies its intensity is essentially inversely proportional to distance. At greater distances, the field intensity falls at a higher rate because of losses in the ground and because of the curvature of the earth. These losses increase with frequency. At sufficiently great distances, the received level is chiefly due to sky waves reflected from the ionosphere. At intermediate distances, the field is a combination of sky waves and ground waves that result in an interference pattern. The total field at the receiver may be obtained in two distinct ways. The first method, which leads to the geometric-optics theory, directly sums the contributions at the receiver from the primary source and each of its images. The second method treats the source and its images as self-illuminating diffraction gratings, one above the earth and one below, and leads to the waveguide mode theory. The advantages of mode theory are restricted to very-low frequencies, where relatively few modes can be supported in the earth-ionosphere waveguide. For example, when the height of the ionosphere is 80 kilometers and the wavelength is 20 kilometers, only the first eight modes can be supported. When the wavelength is 2 kilometers, however, all modes up to the 80th can be supported.

Thus at very-low frequencies, and for distances greater than say 3000 kilometers, it is simpler to use the mode of lowest order to obtain the received field. At distances less than about 1000 kilometers, it is simpler to use ray theory. The above is based on an idealized condition, since it has been assumed that the earth is flat, that the ionosphere is sharply bounded, and that the effect of the magnetic field of the earth can be ignored. Even with these simplifying assumptions, the results are useful. A full treatment of the general case can be obtained by reference to several sources.*

The results of calculations made by Wait and Spies, taking into account the curvature of the earth and the conductivities of the earth and the ionosphere, are shown in Figs. 1 and 2, where σ_g = ground conductivity, ω_r = ionospheric conductivity parameter, and n = mode order. It is seen that the lowest attenuation of mode 1 in the daytime ($h = 70$ km) is at about 18 kilohertz and at night ($h = 90$ km) it is at about 15 kilohertz.

The phase velocity ratio of mode 1 in the daytime is greater than 1 for frequencies less than about 13 kilohertz and less than 1 for higher frequencies. At night the crossover frequency is about 9 kilohertz.

In daylight, the attenuation of mode 1 is always less than that of mode 2. At night, the attenuation for the two modes may be of the same order.

LOW AND MEDIUM FREQUENCIES—30 TO 3000 KILOHERTZ†

For low and medium frequencies of approximately 30 to 3000 kilohertz with a short vertical antenna over perfectly reflecting ground

$$E = 186.4(P_r)^{1/2} \text{ millivolts rms/meter at 1 mile}$$

$$E = 300(P_r)^{1/2} \text{ millivolts rms/meter at 1 kilometer}$$

where P_r = radiated power in kilowatts.

* CCIR XVth Plenary Assembly, Geneva, 1982; Vol. V, *Propagation in Non-Ionized Media*, and Vol. VI, *Propagation in Ionized Media*.

† Watt, A. D. *VLF Radio Engineering*, Vol. 14, International Series of Monographs in Electromagnetic Waves. New York: Pergamon Press, Inc., 1967.

* Wait, J. R., *Electromagnetic Waves in Stratified Media*, New York: Pergamon Press, Inc., 1962. Budden, K. G., *The Wave-Guide Mode Theory of Wave Propagation*, New York: Prentice-Hall, Inc., 1962. Johler, J. R., "Propagation of the Low-Frequency Radio Signal," *Proceedings of the IRE*, Vol. 50, No. 4, 1962; pp. 404-427. CCIR XVth Plenary Assembly, Geneva, 1982, Vol. VI, Report 895.

† *Radio Spectrum, Utilization*. New York: Joint Technical Advisory Committee (IEEE and EIA), 1964. CCIR XVth Plenary Assembly, Geneva, 1982, Vol. VI.

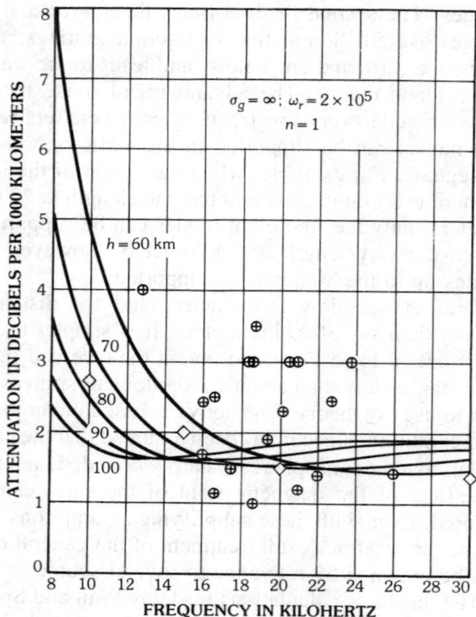

Fig. 1. Relation among attenuation, frequency, and height of the ionosphere. The diamonds represent some experimental observations by Taylor on the average daytime attenuation for west-to-east propagation over sea water. Attenuation rate in the opposite direction is greater by about 1 decibel per 1000 kilometers. (*From* Radio Spectrum Utilization, *Joint Technical Advisory Committee (IEEE and EIA), IEEE, New York, 1964; p. 104*)

Actual inverse-distance fields at 1 kilometer for a given transmitter output power depend on the height and power radiation efficiency of the antenna and associated circuit losses.

Typical values found in practice for well-designed stations are:

Small *L* or *T* antennas as on ships:

$40(P_t)^{1/2}$ millivolts/meter at 1 kilometer

Vertical radiators 0.15 to 0.25λ high:

$290(P_t)^{1/2}$ millivolts/meter at 1 kilometer

Vertical radiators 0.25 to 0.40λ high:

$322(P_t)^{1/2}$ millivolts/meter at 1 kilometer

Vertical radiators 0.40 to 0.60λ high or top-loaded vertical radiators:

$386(P_t)^{1/2}$ millivolts/meter at 1 kilometer

where P_t = transmitter output power in kilowatts. These values can be increased by directive antenna systems.

It has been found that the concept of basic transmission loss, also called path loss, is convenient for the analysis of radio communication systems. Basic transmission loss is the dimensionless ratio P_R/P_A, where P_R is the power radiated from a lossless, isotropic transmitting antenna and P_A is the power available from a lossless, isotropic receiving antenna in a matched load. The isotropic antennas are at the same physical locations and operate in the same band of frequencies as the actual antennas.

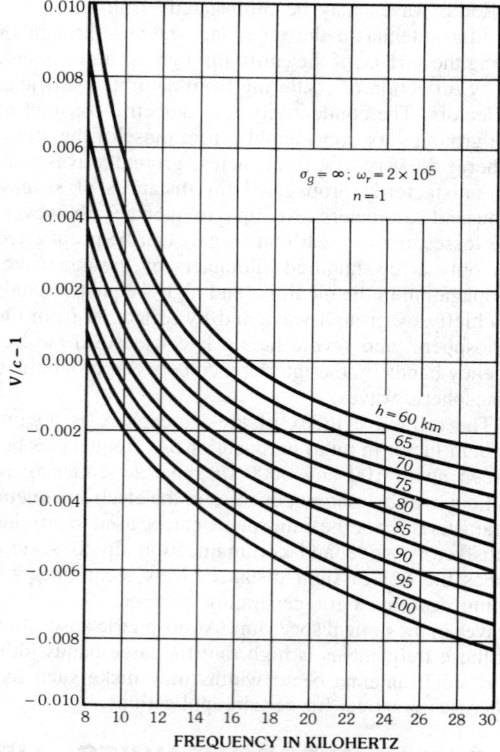

Fig. 2. Phase velocity *V* as a function of ionospheric height *h* and frequency, relative to the velocity in free space, *c*. (*From* Radio Spectrum Utilization, *Joint Technical Advisory Committee (IEEE and EIA), IEEE, New York, 1964; p. 105.*)

Surface-wave (commonly called ground-wave) basic transmission loss is plotted in Fig. 3 for vertically polarized propagation over land having a representative conductivity and dielectric constant and in Fig. 4 for vertically polarized propagation over sea water. Both antennas are 30 feet above the surface in both figures.

In the low-frequency and medium-frequency ranges, propagation losses for horizontally polarized transmission between antennas on the surface of the earth are impractically high. Ground constants typical of various terrain types are listed in Table 1.

Under the conditions used in Figs. 3 and 4, the surface of the earth behaves like a nearly perfect reflector for the isotropic antennas that are only a small fraction of a wavelength from it and that are used to calculate the basic transmission loss. As a result, each isotropic antenna together with the surface of the earth has a gain of nearly 3.01 dB in the general direction of the horizon. By contrast, a lossless quarter-wave monopole erected over a good ground screen would have a gain of 5.16 dB, and a short lossless monopole would have

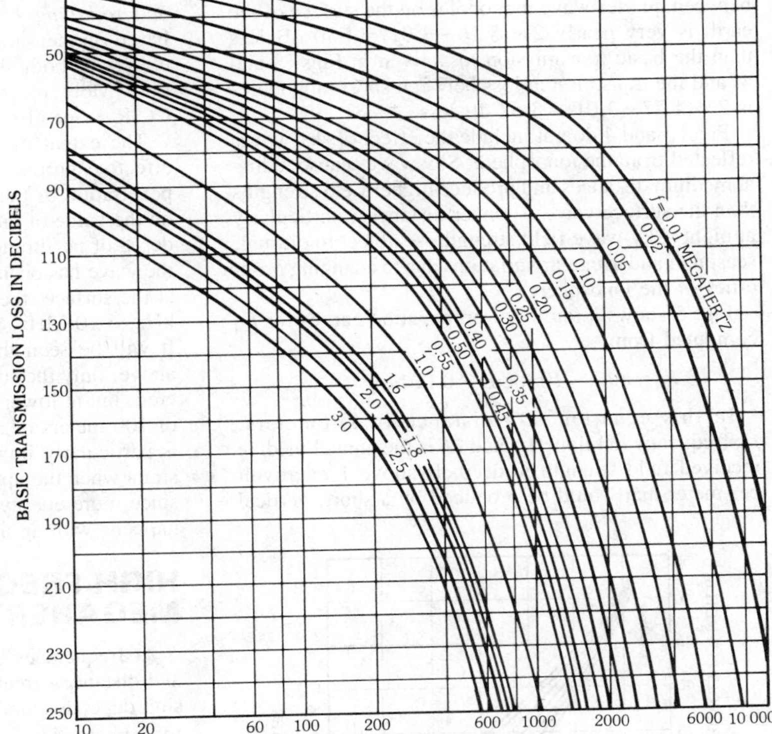

Fig. 3. Basic transmission loss expected for surface waves propagated over a smooth spherical earth. Over land: $\sigma = 0.005$ mho/meter, $\epsilon = 15$. Lossless isotropic antennas 30 feet above the surface. Vertical polarization. (*Adapted from K.A. Norton,* Transmission Loss in Radio Propagation: II, *National Bureau of Standards Technical Note 12, June 1959; Fig. 7.*)

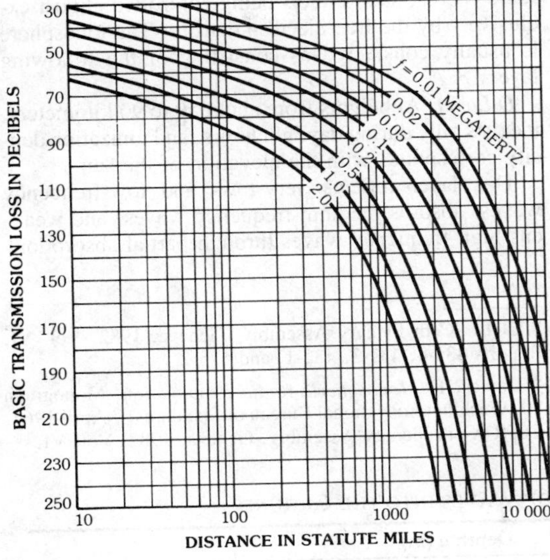

Fig. 4. Basic transmission loss expected for surface waves propagated over a smooth spherical earth. Over sea water: $\sigma = 5$ mhos/meter, $\epsilon = 80$. Lossless isotropic antenna 30 feet above the surface. Vertical polarization. (*Adapted from K. A. Norton,* Transmission Loss in Radio Propagation: II, *National Bureau of Standards Technical Note 12, June 1959; Fig. 8.*)

TABLE 1. GROUND CONDUCTIVITY AND DIELECTRIC CONSTANT FOR MEDIUM- AND LONG-WAVE PROPAGATION TO BE USED WITH NORTON, BURROWS, BREMMER, OR OTHER DEVELOPMENTS OF SOMMERFELD PROPAGATION EQUATION

Terrain	Conductivity σ (mhos/meter)	Dielectric Constant ϵ (esu)
Sea water	5	80
Fresh water	8×10^{-3}	80
Dry, sandy, flat coastal land	2×10^{-3}	10
Marshy, forested flat land	8×10^{-3}	12
Rich agricultural land, low hills	1×10^{-2}	15
Pastoral land, medium hills and forestation	5×10^{-3}	13
Rocky land, steep hills	2×10^{-3}	10
Mountainous (hills up to 3000 feet)	1×10^{-3}	5
Cities, residential areas	2×10^{-3}	5
Cities, industrial areas	1×10^{-4}	3

a gain of 4.77 dB. It follows that the transmission loss between quarter-wave monopoles on the surface of the earth is very nearly $2\times(5.16-3.01)=4.30$ dB less than the basic transmission loss given in Figs. 3 and 4, and the transmission loss between short monopoles is $2\times(4.77-3.01)=3.52$ dB less.

Figs. 3 and 4 do not include the effect of sky waves reflected from the ionosphere. Sky waves cause fading at medium distances and produce higher field strengths than the surface wave at longer distances, particularly at night. Sky-wave field strength is subject to diurnal, seasonal, and irregular variations due to changing properties of the ionosphere.

Fig. 5 shows a family of propagation curves for F_0 computed from

$$F_0=80.2-10\log D-0.00176f^{0.26}D$$

In this equation, D = distance in kilometers, f = frequency in kilohertz, and F_0 is the annual median received field strength in decibels above 1 microvolt per meter that would be produced by a short, vertical transmitting dipole at or near the surface of the earth and *radiating* 1 kilowatt. The empirical equation is based on measured data. Fig. 5 therefore includes the effects of both sky-wave and surface-wave propagation. More recent work has been described by the CCIR.* See also Fig. 7 of Chapter 35.

The extent to which the lower strata influence the effective ground constants depends on the depth of penetration of the radio energy. This in turn depends on the value of the constants and the frequency. If the depth of penetration is defined as that depth in which the wave has been attenuated to $1/e$ (37%) of its value at the surface, then over the frequency range from 10 kHz to 10 MHz, δ has the values shown in Table 2. It will be seen that, at frequencies of 10 MHz and above, only the surface of the ground need be considered, but at lower frequencies, strata down to a depth of 100 meters or more must be taken into account. It is particularly important to take account of the lower strata when the upper strata are of lower conductivity, since more energy penetrates to the lower levels than happens with an upper layer of higher conductivity.

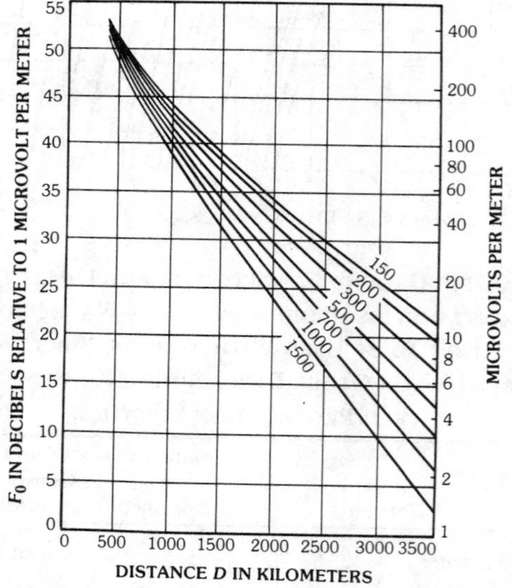

Fig. 5. Family of basic curves of F_0 to be used to determine the annual median value of the field strength for the frequencies (in kilohertz) on the curves. (*From CCIR XIIIth Plenary Assembly, Geneva, 1974, Vol. VI, Report 264-3; p. 108.*)

HIGH FREQUENCIES—3 TO 30 MEGAHERTZ†

At frequencies between about 3 and 25 megahertz and distances greater than about 100 miles, transmission depends chiefly on sky waves reflected from the ionosphere. This is a region high above the surface of the earth where the rarefied air is sufficiently ionized (primarily by ultraviolet sunlight) to reflect or absorb radio waves, such effects being controlled almost exclusively by the free-electron density. The ionosphere is usually considered as consisting of the following layers.

D Layer: At heights from about 50 to 90 kilometers, it exists only during daylight hours, and ionization density corresponds with the elevation of the sun.

This layer reflects very-low- and low-frequency waves, absorbs medium-frequency waves, and weakens high-frequency waves through partial absorption.

* CCIR XVth Plenary Assembly, Geneva, 1982, Vol. VI, Reports 435-4, 431-3, 432-1, and 575-2.

† Davies, K., *Ionospheric Radio Propagation*, Monograph 80, Washington: National Bureau of Standards, 7 April 1965. CCIR XVth Plenary Assembly, Geneva, 1982, Vol. VI.

TABLE 2. DEPTH OF PENETRATION OF WAVES INTO THE GROUND

	Depth δ (m)		
Frequency	$\sigma=4$ mho/m, $\epsilon=80$	$\sigma=10^{-2}$ mho/m, $\epsilon=10$	$\sigma=10^{-3}$ mho/m, $\epsilon=5$
10 kHz	2.5	50	150
100 kHz	0.80	15	50
3 MHz	0.14	5	17
10 MHz	0.08	2	9

E Layer: At a height of about 110 kilometers, this layer is important for high-frequency daytime propagation at distances less than 1000 miles, and for medium-frequency nighttime propagation at distances in excess of about 100 miles. Ionization density corresponds closely with the elevation of the sun. Irregular cloud-like areas of unusually high ionization, called sporadic *E*, may occur up to more than 50 percent of the time on certain days or nights. Sporadic *E* occasionally prevents frequencies that normally penetrate the *E* layer from reaching higher layers and also causes occasional long-distance transmission at very-high frequencies. Some portion (perhaps the major part) of the sporadic-*E* ionization is ascribable to visible- and subvisible-wavelength bombardment of the atmosphere.

F_1 Layer: At heights of about 175 to 250 kilometers, it exists only during daylight. This layer occasionally is the reflecting region for high-frequency transmission, but usually oblique-incidence waves that penetrate the *E* layer also penetrate the F_1 layer and are reflected by the F_2 layer. The F_1 layer introduces additional absorption of such waves.

F_2 Layer: At heights of about 250 to 400 kilometers, F_2 is the principal reflecting region for long-distance high-frequency communication. Height and ionization density vary diurnally, seasonally, and over the sunspot cycle. Ionization does not follow the elevation of the sun in any fashion, since (at such extremely low air densities and molecular-collision rates) the medium can store received solar energy for many hours, and, by energy transformation, can even detach electrons during the night. At night, the F_1 layer merges with the F_2 layer at a height of about 300 kilometers. The absence of the F_1 layer, and reduction in absorption of the *E* layer, causes nighttime field intensities and noise to be generally higher than during daylight.

As indicated to the right on Fig. 6, these layers are contained in a thick region throughout which ionization generally increases with height. The layers are said to exist where the ionization gradient is capable of refracting waves back to earth. Obliquely incident waves follow a curved path through the ionosphere due to gradual refraction or bending of the wave front. When attention need be given only to the end result, the process can be assimilated to a reflection.

Depending on the ionization density at each layer, there is a critical or highest frequency, f_c, at which the layer reflects a vertically incident wave. Frequencies higher than f_c pass through the layer at vertical incidence. At oblique incidence, and distances such that the curvature of the earth and ionosphere can be neglected, the maximum usable frequency is given by

$$\text{muf} = f_c \sec \phi$$

where,

muf is the maximum usable frequency for the particular layer and distance,

ϕ is the angle of incidence at the reflecting layer.

At greater distances, curvature is taken into account by the modification

$$\text{muf} = kf_c \sec \phi$$

where k is a correction factor that is a function of distance and vertical distribution of ionization.

Both f_c and height, and hence ϕ for a given distance, vary for each layer with local time of day, season, latitude, and throughout the 11-year sunspot cycle. The various layers change in different ways with these parameters. In addition, ionization is subject to frequent abnormal variations.

Ionospheric losses are a minimum near the maximum usable frequency and increase rapidly for lower frequencies during daylight.

High frequencies travel from the transmitter to the receiver by reflection from the ionosphere and earth in one or more hops as indicated in Figs. 6 and 7. Additional reflections may occur along the path between the bottom edge of a higher layer and the top edge of a lower layer, the wave finally returning to earth near the receiver.

Fig. 6 indicates transmission on a common fre-

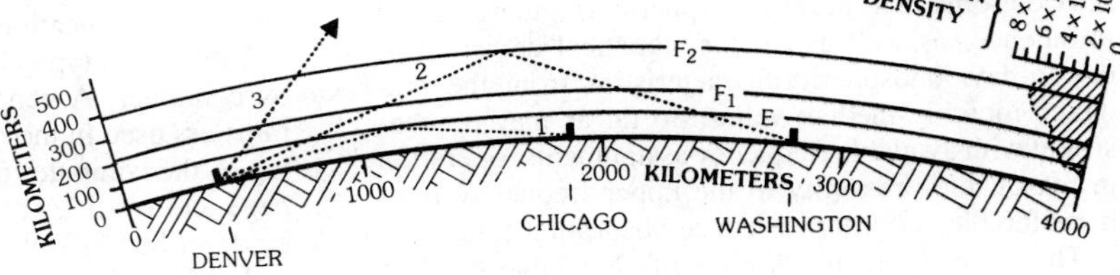

Fig. 6. Schematic explanation of skip-signal zones.

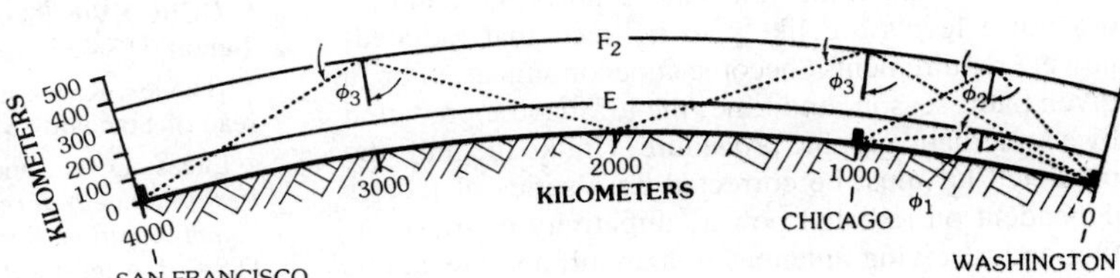

Fig. 7. Single-hop and two-hop transmission paths due to *E* and F_2 layers.

quency, (1) single-hop via E layer, Denver to Chicago, and (2) single-hop via F_2, Denver to Washington, with (3) the wave failing to reflect at higher angles, thus producing a skip region of no signal between Denver and Chicago. Fig. 7 illustrates single-hop transmission, Washington to Chicago, via the E layer (ϕ_1). At higher frequencies over the same distance, single-hop transmission would be obtained via the F_2 layer (ϕ_2). Fig. 7 also shows two-hop transmission, Washington to San Francisco, via the F_2 layer (ϕ_3).

Actual transmission over long distances is more complex than indicated by Figs. 6 and 7, because the layer heights and critical frequencies differ with time (and hence longitude) and with latitude. Further, scattered reflections occur at the various surfaces.

Typical values of critical frequency for Washington, D.C., are shown in Fig. 8.

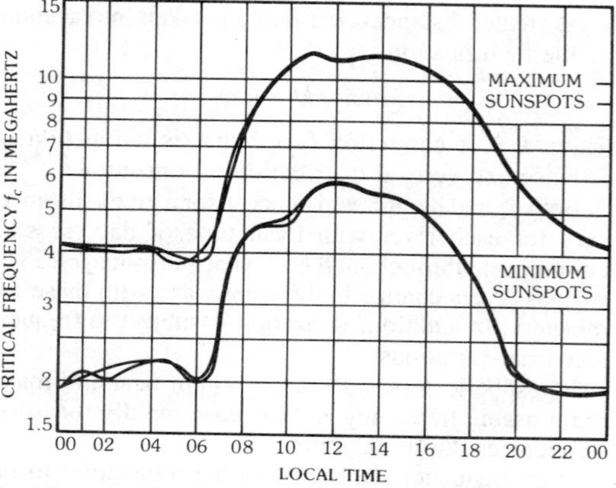

Fig. 8. Critical frequency for Washington, D.C. (*From National Bureau of Standards Circular 462.*)

Preferably, operating frequencies should be selected from a specific frequency band that is bounded above and below by limits that are systematically determinable for the transmission path under consideration. The recommended upper limit is called the *optimum working frequency* (fot) and is selected below the muf to provide some margin for ionospheric irregularities and turbulence, as well as for the statistical deviation of day-to-day ionospheric characteristics from the predicted monthly median value. So far as may be consistent with available frequency assignments, operation in reasonable proximity to the upper frequency limit is preferable, in order to reduce absorption loss.

The lower limit of the normally available band of frequencies is called the *lowest useful high frequency* (luf). Below this limit, ionospheric absorption and radio noise levels are likely to be such that radiated-power requirements become uneconomical. For a given path, season, and time, the luf may be predicted by a systematic graphic procedure. Unlike the muf, the predicted luf must be corrected by a series of factors dependent on radiated power, directivity of transmitting and receiving antennas in azimuth and elevation, class of service, and presence of local noise sources. Available data include atmospheric-noise maps, transmission-loss charts, antenna diagrams, and nomograms facilitating the computation. The procedure is formidable but worthwhile.

The upper and lower frequency limits change continuously throughout the day, whereas it is ordinarily impracticable to change operating frequencies correspondingly. Each operating frequency, therefore, should be selected to fall within the above limits for a substantial portion of the daily operating period.

Angles of Departure and Arrival

Angles of departure and arrival are of importance in the design of high-frequency antenna systems. These angles, for single-hop transmission, are obtained from the geometry of a triangular path over a curved earth with the apex of the triangle placed at the virtual height assumed for the altitude of the reflection. Fig. 9 is a family of curves showing radiation angle for different distances.

D = great-circle distance in statute miles
H = virtual height of ionosphere layer in kilometers
Δ = radiation angle in degrees
ϕ = semiangle of reflection at ionosphere

Forecasts of High-Frequency Propagation

The CCIR publishes "Basic Indices for Ionospheric Propagation" predictions several months in advance in the *Telecommunications Journal* (ITU, Geneva). A list of organizations concerned with issuing forecasts of propagation conditions is published by the CCIR.*

In designing a high-frequency communication circuit, it is necessary to determine the optimum traffic frequencies, system loss, signal-to-noise ratio, angle of arrival, and circuit reliability. Manual methods for calculating the values of these factors have been described,** as has the use of electronic computers for predicting the performance of high-frequency sky-wave communication circuits.†

Table 3 is a typical performance prediction prepared by computer. A general description of the circuit parameters used in the calculations is shown in the heading of the computer printout. Starting at the top of the page and reading from left to right, the heading may be described as follows.

The first line contains the month, the solar activity

* CCIR XVth Plenary Assembly, Geneva, 1982, Vol. VI, Report 313-4.

** Davies, K. *Ionospheric Radio Propagation*. National Bureau of Standards Monograph 80, 1 April 1965.

† Lucas, D. L., and Haydon, G. W. *Predicting Statistical Performance Indexes for High-Frequency Ionospheric Telecommunication Systems*. ESSA Technical Report IER1-ITSA, 1 August 1966.

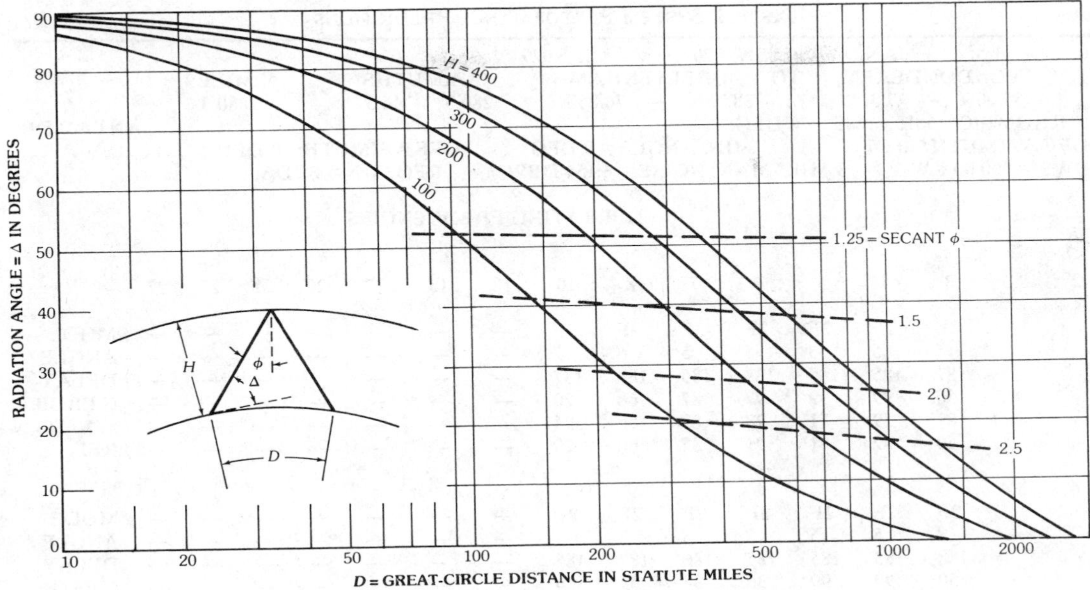

Fig. 9. Single-reflection radiation angle and great-circle distance.

level in 12-month moving average Zurich sunspot number, and a circuit identification number. The second and third lines contain the transmitter and receiver locations, the bearings, and the distance. The fourth and fifth lines contain the antenna system for each terminal and their orientation relative to the great-circle path. The minimum angle indicates the lowest vertical angle considered in the mode selection.

The sixth line is the power delivered to the transmitting antenna, the man-made noise level assumed for the area in dBW in a 1-hertz bandwidth at 3 megahertz, and the hourly median signal-to-noise ratio required to provide the service requested. The signal is in the same units as the transmitter power, and the noise is in a 1-hertz bandwidth. The seventh line contains the heading for the operating frequencies, which are given in megahertz in the eighth line. In addition to the operating frequencies, the eighth line also contains the time heading (GMT) and the classically defined maximum-usable-frequency heading (muf), i.e., frequency which has a 50% probability of having a sky-wave path.

For each time and operating frequency the body of the tabulation contains: (*A*) the mode having the greatest probability (MODE), (*B*) the median vertical angle associated with this mode (ANGLE), (*C*) the propagation time in tenths of milliseconds (DELAY), (*D*) the percentage of days that any sky-wave mode is expected to exist, circuit probability (C.PROB.), (*E*) the median of the hourly median signal-to-noise ratios for the days sky-wave modes exist (S/N..DB), and (*F*) the percentage of days within the month that the median required signal-to-noise ratio is expected to be equalled or exceeded (REL.).

Bandwidth Limitations*

In high-frequency transmission, the communication bandwidth is limited by multipath propagation. The greatest limitation occurs when two or more paths exist with a different number of hops. The bandwidth may then be as small as 100 hertz, but such multipath may be minimized by operating near the muf. Operation at a frequency within approximately 10% of the muf is necessary for paths less than about 600 kilometers to obtain bandwidths greater than, say, 1 kilohertz. The multipath reduction factor (mrf) is defined as the smallest ratio of muf to operating frequency for which the range of multipath propagation time difference is less than a specified value. The mrf thus defines the frequency above which a specified minimum protection against multipath is provided. Fig. 10 shows the mrf for various lengths of path.**

Diversity†

It has been shown that if two or more high-frequency radio channels are sufficiently separated in space, fre-

* *Multipath Propagation Over High-Frequency Radio Circuits*. CCIR, Geneva, 1982, Vol. III, Report 203-1.

** Salaman, R. K. "A New Ionospheric Multipath Reduction Factor (MRF)." *IRE Transactions on Communication Systems*, Vol. CS-10, June 1962; pp. 220–222.

† "Bandwidth and Signal-to-Noise Ratios in Complete Systems," CCIR, Geneva, 1982, Vol. III, Report 195. Grisdale, G. L., Morriss, J. H., and Palmer, D. S. "Fading of Long-Distance Radio Signals and a Comparison of Space and Polarization Diversity Reception in the 6–18 Mc Range," *Proceedings of the IEE*, Part B, No. 13, Jan. 1957, pp. 39–51.

TABLE 3. SYSTEM PERFORMANCE PREDICTIONS

1 JAN SSN = 20 CH 5.029
LONDONDERRY TO CHELTENHAM AZIMUTHS N.MILES
55.00N — 7.31W 38.75N — 76.85W 280.5 46.3 2880.3
RHOMBIC 50H 168L 70 DEG. ANT = 0 DB
OFF AZIMUTH 0 DEG. MIN. ANGLE = 0 DEG. OFF AZIMUTH 0 DEG.
PWR = 200.00 KW 3 MHZ MAN. NOISE = −154 DBW REQ. S/N = 61 DB

OPERATING FREQUENCIES

GMT	MUF	3	4	5	6	7	8	10	12	15	17	20	22	25	27	
2	8.6															
		2F	2F	2F	2F	2F	2F	2F	—	—	—	—	—	—	—	MODE
		7	5	5	5	5	6	7	—	—	—	—	—	—	—	ANGLE
		187	185	185	185	185	186	187	—	—	—	—	—	—	—	DELAY
		50	99	99	97	87	66	20	—	—	—	—	—	—	—	C.PROB.
		89	59	71	78	83	86	94	—	—	—	—	—	—	—	S/N..DB
		50	42	81	91	85	66	20	—	—	—	—	—	—	—	REL.
4	8.4															
		2F	2F	2F	2F	2F	2F	2F	—	—	—	—	—	—	—	MODE
		8	5	5	5	6	7	7	—	—	—	—	—	—	—	ANGLE
		188	185	185	185	186	187	188	—	—	—	—	—	—	—	DELAY
		50	99	99	96	84	61	16	—	—	—	—	—	—	—	C.PROB.
		89	58	69	77	82	87	95	—	—	—	—	—	—	—	S/N..DB
		50	34	77	89	82	60	16	—	—	—	—	—	—	—	REL.
6	8.3															
		2F	2F	2F	2F	2F	2F	2F	—	—	—	—	—	—	—	MODE
		7	5	5	5	6	7	7	—	—	—	—	—	—	—	ANGLE
		187	185	185	185	186	187	187	—	—	—	—	—	—	—	DELAY
		50	99	99	94	81	57	10	—	—	—	—	—	—	—	C.PROB.
		89	56	68	76	82	87	97	—	—	—	—	—	—	—	S/N..DB
		50	28	76	89	80	57	10	—	—	—	—	—	—	—	REL.
8	7.3															
		2F	2F	2F	2F	2F	2F	—	—	—	—	—	—	—	—	MODE
		7	5	5	5	6	7	—	—	—	—	—	—	—	—	ANGLE
		187	185	185	185	186	187	—	—	—	—	—	—	—	—	DELAY
		50	99	97	85	59	27	—	—	—	—	—	—	—	—	C.PROB.
		84	56	69	78	83	87	—	—	—	—	—	—	—	—	S/N..DB
		48	29	71	76	56	26	—	—	—	—	—	—	—	—	REL.
10	7.9															
		2F	2F	2F	2X	2X	2F	—	—	—	—	—	—	—	—	MODE
		6	4	4	1	1	6	—	—	—	—	—	—	—	—	ANGLE
		186	184	184	181	182	186	—	—	—	—	—	—	—	—	DELAY
		50	99	99	98	87	47	—	—	—	—	—	—	—	—	C.PROB.
		87	56	70	77	84	87	—	—	—	—	—	—	—	—	S/N..DB
		49	30	75	86	84	46	—	—	—	—	—	—	—	—	REL.
12	14.5															
		2F	3E	4F	3F	2F	2F	2F	2F	2F	2F	—	—	—	—	MODE
		4	3	15	10	4	3	3	3	4	4	—	—	—	—	ANGLE
		184	181	190	187	184	183	183	183	184	184	—	—	—	—	DELAY
		50	99	99	99	99	99	99	90	39	8	—	—	—	—	C.PROB.
		101	5	53	67	75	81	91	93	101	103	—	—	—	—	S/N..DB
		50	0	14	67	84	93	98	90	39	8	—	—	—	—	REL.

quency, angle of arrival, time, or polarization, the fading on the various channels is more or less independent. Diversity systems make use of this fact to improve the overall performance, combining or selecting separate radio channels on a single high-frequency circuit.

Satisfactory diversity improvement can be obtained if the correlation coefficient of the fading on the various channels does not exceed about 0.6, and experiments have indicated that a frequency separation of the order of 400 hertz gives satisfactory diversity performance

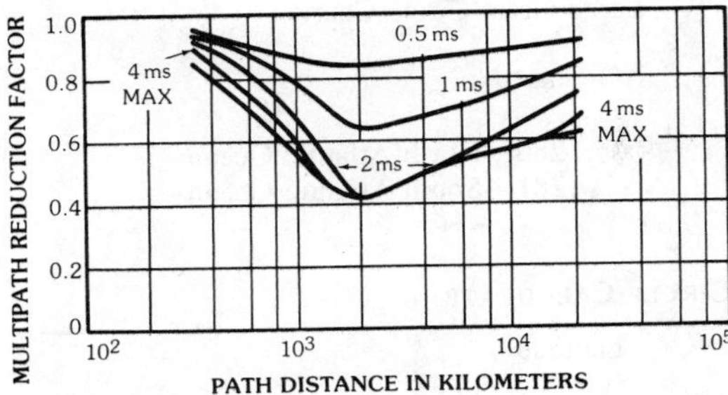

Fig. 10. Multipath reduction factor as a function of path distance. (*From Salaman, R. K. "A New Ionospheric Multipath Reduction Factor (MRF)."* © *1962, Institute of Radio Engineers.*)

on long high-frequency paths. Spacing between antennas at right angles to the direction of propagation should be about 10 wavelengths. Polarization diversity has been found to be about equivalent to space diversity in the high-frequency band. Measurements have indicated that times varying from 0.05 to 95 seconds may be necessary to obtain fading correlation coefficients as low as 0.6 in high-frequency time-diversity systems. Angle-of-arrival diversity requires the use of large antennas so as to obtain the required vertical directive characteristics. Differences in the angle of arrival of 2° have been shown to give satisfactory diversity improvement on high-frequency circuits.

GREAT-CIRCLE CALCULATIONS

With reference to Fig. 11, A and B are two places on the surface of the earth the latitudes and longitudes of which are known. In the figure, B = place of greater latitude (nearer the pole), L_A = latitude of A, L_B = latitude of B, and C = difference of longitude between A and B. Angles X and Y at A and B of the great circle passing through the two places and the distance, Z, between A and B along the great circle can be calculated as follows:

$$\tan \tfrac{1}{2}(Y-X) = \cot \tfrac{1}{2}C \frac{\sin \frac{1}{2}(L_B-L_A)}{\cos \frac{1}{2}(L_B+L_A)}$$

and

$$\tan \tfrac{1}{2}(Y+X) = \cot \tfrac{1}{2}C \frac{\cos \frac{1}{2}(L_B-L_A)}{\sin \frac{1}{2}(L_B+L_A)}$$

give the values of $\frac{1}{2}(Y-X)$ and $\frac{1}{2}(Y+X)$ from which

$$\tfrac{1}{2}(Y+X)+\tfrac{1}{2}(Y-X)=Y$$

and

$$\tfrac{1}{2}(Y+X)-\tfrac{1}{2}(Y-X)=X$$

In the above equations, north latitudes are taken as positive and south latitudes as negative. For example, if B is latitude 60° N and A is latitude 20° S,

$$\frac{L_B+L_A}{2}=\frac{60+(-20)}{2}=\frac{60-20}{2}=\frac{40}{2}=20°$$

$$\frac{L_B-L_A}{2}=\frac{60-(-20)}{2}=\frac{60+20}{2}=\frac{80}{2}=40°$$

If both places are in the southern hemisphere and L_B+L_A is negative, it is simpler to call the place of greater south latitude B and to use the above method for calculating bearings from true south and to convert the results afterward to bearings east of north.

The distance Z (in degrees) along the great circle between A and B is given by the following:

$$\tan \tfrac{1}{2}Z = \tan \tfrac{1}{2}(L_B-L_A) \times [\sin \tfrac{1}{2}(Y+X)]/[\sin \tfrac{1}{2}(Y-X)]$$

The angular distance Z (in degrees) between A and B may be converted to linear distance as follows.

Z (in degrees) × 111.12 = kilometers
Z (in degrees) × 69.05 = statute miles
Z (in degrees) × 60.00 = nautical miles

In multiplying, the minutes and seconds of arc must be expressed in decimals of a degree. For example, Z = 37°45′36″ becomes 37.755°.

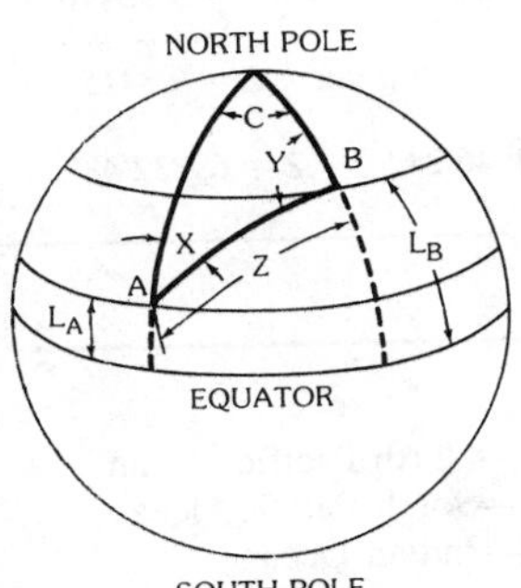

(A) Both points located in northern hemisphere.

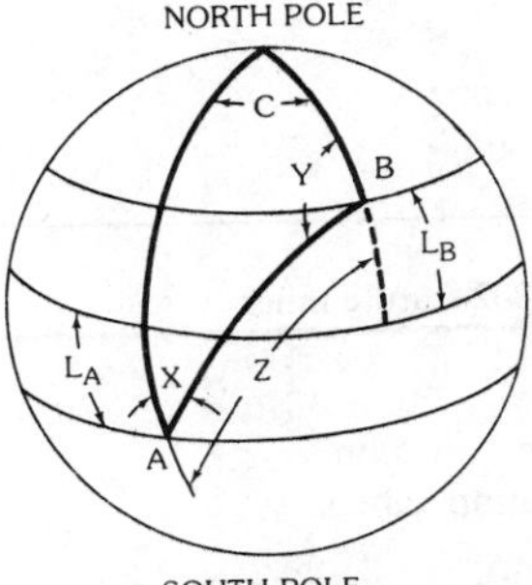

(B) In opposite hemispheres.

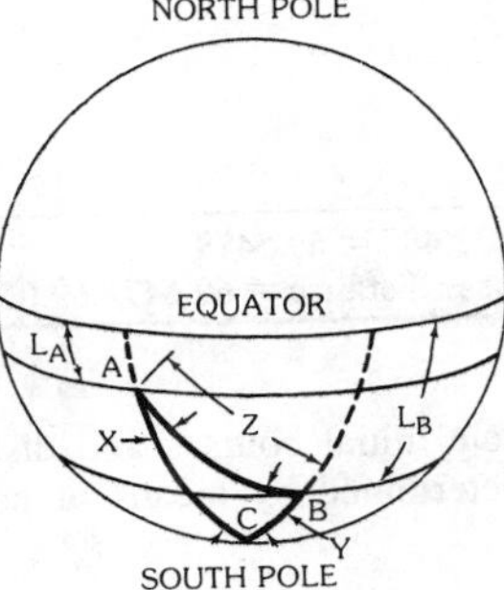

(C) Both points located in southern hemisphere.

Fig. 11. Three globes representing points A and B.

Example: Find the great-circle bearings at Brentwood, Long Island, longitude 73°15′10″W, latitude 40°48′40″N, and at Rio de Janeiro, Brazil, longitude 43°22′07″W, latitude 22°57′09″S; and the great-circle distance in statute miles between the two points. Refer to Chart 1.

CHART 1. EXAMPLE OF GREAT-CIRCLE CALCULATIONS

	Longitude	Latitude	
Brentwood	73°15′10″W	40°48′40″N	L_B
Rio de Janeiro	43°22′07″W	(−)22°57′09″S	L_A
	29°53′03″		C
		17°51′31″	L_B+L_A
		63°45′49″	L_B-L_A

$\frac{1}{2}C = 14°56'31''$ $\frac{1}{2}(L_B+L_A) = 8°55'45''$ $\frac{1}{2}(L_B-L_A) = 31°52'54''$

log cot 14°56′31″ = 10.57371	log cot 14°56′31″ = 10.57371
plus log cos 31°52′54″ = 9.92898	plus log sin 31°52′54″ = 9.72277
0.50269	0.29648
minus log sin 8°55′45″ = 9.19093	minus log cos 8°55′45″ = 9.99471
log tan $\frac{1}{2}(Y+X)$ = 1.31176	log tan $\frac{1}{2}(Y-X)$ = 0.30177
$\frac{1}{2}(Y+X) = 87°12'26''$	$\frac{1}{2}(Y-X) = 63°28'26''$

Bearing at Brentwood = $\frac{1}{2}(Y+X)+\frac{1}{2}(Y-X) = Y = 150°40'52''$ East of North

Bearing at Rio de Janeiro = $\frac{1}{2}(Y+X)-\frac{1}{2}(Y-X) = X = 23°44'00''$ West of North

$\frac{1}{2}(L_B-L_A) = 31°52'54''$	log tan 31°52′54″ = 9.79379
$\frac{1}{2}(Y+X) = 87°12'26''$	plus log sin 87°12′26″ = 9.99948
	9.79327
$\frac{1}{2}(Y-X) = 63°28'26''$	minus log sin 63°28′26″ = 9.95170
	log tan $\frac{1}{2}Z$ = 9.84157
	$\frac{1}{2}Z = 34°46'24''$ $Z = 69°32'48''$

69°32′48″ = 69.547°

Linear distance = 69.547×69.05 = 4802 statute miles

Great-circle initial courses and distances are conveniently determined by means of navigation tables such as:

(A) Navigation Tables for Navigators and Aviators—HO No. 206.

(B) Dead-Reckoning Altitude and Azimuth Table—HO No. 211.

(C) Large Great-Circle Charts:

HO Chart No.

1280—North Atlantic Ocean
1281—South Atlantic Ocean
1282—North Pacific Ocean
1283—South Pacific Ocean
1284—Indian Ocean

The above tables and charts may be obtained at a nominal charge from the United States Navy Department Hydrographic Office, Washington, D.C.

EFFECT OF NUCLEAR EXPLOSIONS ON RADIO PROPAGATION*

Nuclear explosions below an altitude of about 15 kilometers have little effect on radio transmission. However, a detonation occurring at an altitude between 15 and 60 kilometers can produce blackout in the low-frequency, medium-frequency, and high-frequency bands over a radius of several hundred kilometers. This effect lasts only for a few minutes except in an area close to the site of the explosion. In general, it can be said that the effect of nuclear explosions is greatest near the site of the detonation, but the effects of ionization and shock waves do not last longer than a few minutes at distances greater than a few hundred kilometers from the site of the explosion.

IONOSPHERIC SCATTER PROPAGATION†

This type of transmission permits communication in the frequency range from approximately 30 to 60 megahertz and over distances from about 1000 to 2000 kilometers. It is believed that this type of propagation is due to scattering from the lower *D* region of the ionosphere and that the useful bandwidth is restricted to less than 10 kilohertz. The greatest use for this type of transmission has been for printing-telegraph channels, particularly in the auroral regions where conventional high-frequency ionospheric transmission is often unreliable.

The median attenuation over paths between 800 and 1000 miles in length is about 80 decibels greater than the free-space path attenuation at 30 megahertz and about 90 decibels greater than the free-space value at 50 megahertz.

METEOR-BURST PROPAGATION‡

Frequencies in the very-high- and ultra-high-frequency bands may be propagated by reflection from columns of ionization produced by meteors entering the lower *E* region. Experimental single-channel two-way telegraph circuits have been operated in the frequency range from 30 to 40 megahertz over distances of 600 to 1300 kilometers with transmitter powers of 1 to 3 kilowatts. One-way transmission of voice and facsimile have also been made with transmitter powers of 1 kilowatt and 20 kilowatts, respectively.

The frequency range from about 50 to 80 megahertz has been found best suited for meteor-burst transmission.

* Glasstone, S. *The Effects of Nuclear Weapons*. Washington: US Government Printing Office, 1962.

† "Ionospheric Scatter Transmission," *Proceedings of the IRE*, Vol. 48, No. 1, 1960; pp. 5–29. CCIR XVth Plenary Assembly, Geneva, 1982, Vol. VI, Report 260-3, and Vol. III, Report 109-2.

‡ "Communication by Meteor-Burst Propagation," CCIR XVth Plenary Assembly, Geneva, 1982, Vol. VI, Report 251-3. Oetting, J. D., "An Analysis of Meteor Burst Communications for Military Applications," *IEEE Transactions on Communications*, Vol. Com-28, Sept. 1980.

PROPAGATION ABOVE 30 MEGAHERTZ, LINE-OF-SIGHT CONDITIONS†

Radio Refraction‡

Under normal propagation conditions, the refractive index of the atmosphere decreases with height so that radio rays travel more slowly near the ground than at higher altitudes. This variation in velocity with height results in bending of the radio rays. Uniform bending may be represented by straight-line propagation, but with the radius of the earth modified so that the relative curvature between the ray and the earth remains unchanged. The new radius of the earth is known as the effective earth radius, and the ratio of the effective earth radius to true earth radius is usually denoted by *K*. The average value of *K* in temperate climates is about 1.33; however, values from about 0.6 to 5.0 are to be expected.

The decrease in the refractive index with height may at times be so great that the ray is bent down with a radius equal to that of the earth so that the earth may then be considered to be flat. A further increase in the refractive-index gradient results in the radio ray being bent down sufficiently to be reflected from the earth. The ray then appears to be trapped in a duct between the earth and the maximum height of the radio path.

Under certain atmospheric conditions, the refractive index may increase with height, causing the radio rays to bend upward. Such inverse bending results in a decrease in path clearance on line-of-sight paths.

The distance to the radio horizon over smooth earth, when the height, h, is very small compared with the radius of the earth, is given with a good approximation by the expression

$$d=(3Kh/2)^{1/2}$$

where,

h = height in feet above the earth,
d = distance to radio horizon in miles,
K = ratio of the effective to the true radius of the earth.

† Bullington, K., "Radio Propagation at Frequencies Above 30 Megacycles," *Proceedings of the IRE*, Vol. 35, October 1947, pp. 1122–1136. Kerr, D. E., *Propagation of Short Radio Waves*, New York: McGraw-Hill Book Co., 1951. CCIR XVth Plenary Assembly, Geneva, 1982, Vol. V.

‡ Bean, B. R., and Dutton, E. J. *Radio Meteorology*. Monograph 92, Institute for Telecommunication Sciences and Aeronomy, Environmental Science Services Administration (ESSA). Washington: Superintendent of Documents, 1966.

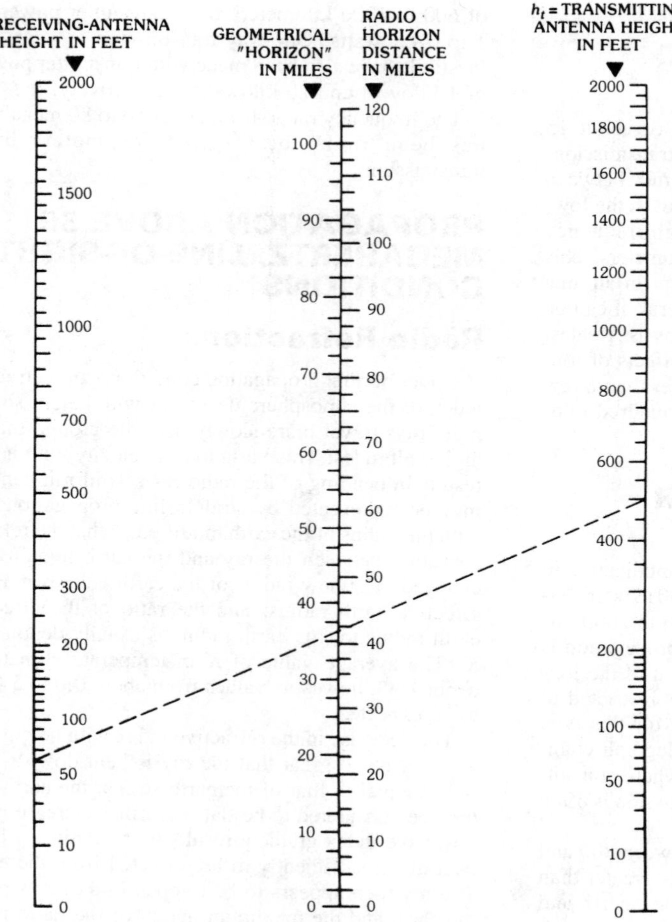

Fig. 12. Nomogram giving radio-horizon distance in miles when h_r and h_t are known. Example shown: Height of receiving antenna 60 feet; height of transmitting antenna 500 feet; maximum radio-path length = 41.5 miles. ($K = 1.33$)

Over a smooth earth, a transmitter antenna at height h_t (feet) and a receiving antenna at height h_r (feet) are in radio line-of-sight provided the spacing in miles is less than $(2h_t)^{1/2} + (2h_r)^{1/2}$ (assuming $K = 1.33$).

The nomogram in Fig. 12 gives the radio-horizon distance between a transmitter at height h_t and a receiver at height h_r. Fig. 13 extends the first nomogram to give the maximum radio-path length between two airplanes whose altitudes are known. Both figures assume a value of $K = 1.33$.

Path Plotting and Profile-Chart Construction

Path Plotting—When laying out a microwave system, it is usually convenient to plot the path on a profile chart. Such charts are scaled to indicate the departure of the curvature of the earth from a straight line. With reference to Fig. 14,

$$D^2 + R^2 = (h + R)^2 = h^2 + 2Rh + R^2$$

$$D^2 = h^2 + 2Rh$$

where,

D = distance,
R = radius of earth (3960 miles),
h = altitude.

Since $h << R$, $D = (2Rh)^{1/2}$, and inserting the true earth radius with R and D in statute miles and h in feet

$$D = \left(\frac{2 \times 3960}{5280} h\right)^{1/2}$$

$$D = [(3/2)h]^{1/2}$$

$$h = (2/3)D^2$$

for true earth, where D is in miles and h in feet. For a value of $K = 1.33$

$$D = [(3/2)h]^{1/2}(4/3)^{1/2} = (2h)^{1/2}$$

$$h = D^2/2$$

Or more generally

$$h = 2D^2/3K$$

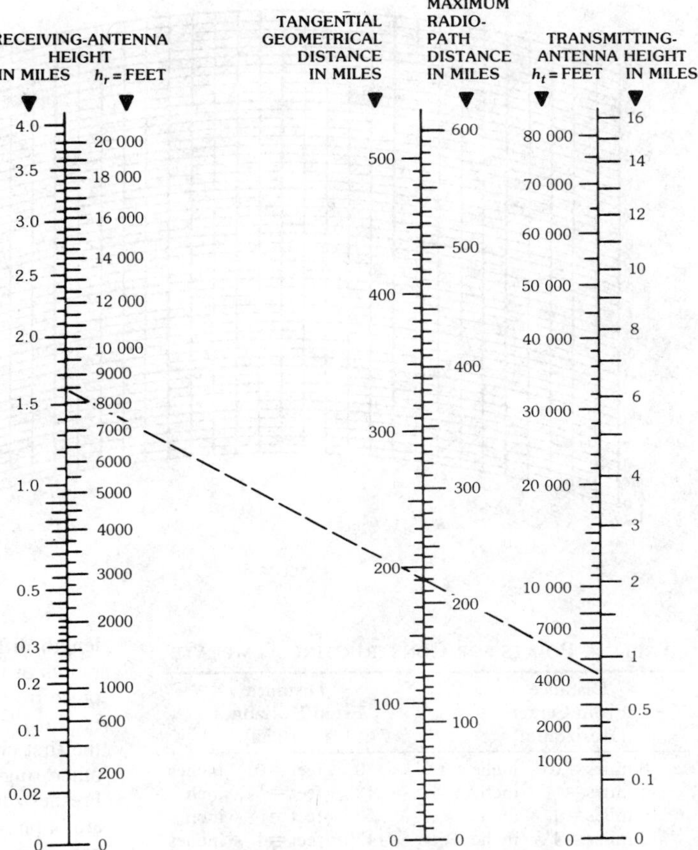

Fig. 13. Nomogram giving radio-path length and tangential distance for transmission between two airplanes at heights h_r and h_t. Example shown: Height of receiving-antenna airplane 8500 feet (1.6 miles); height of transmitting-antenna airplane 4250 feet (0.8 mile); maximum radio-path distance = 220 miles. (K = 1.33)

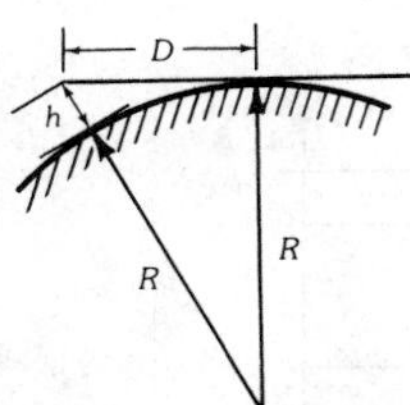

Fig. 14. Straight line tangent to surface of earth.

Profile Paper—If a 4/3 effective-radius factor is used, the departure from a horizontal tangent line is

$$h = D^2/2$$

where symbols are as above. By using this equation, a template can be made for convenient drawing of profile paper (Fig. 15). For instance, if the horizontal scale is 10 miles/inch, the vertical scale 100 feet/inch, and a width corresponding to 40 miles is desired, the points in Table 4 may be plotted.

A typical example of a template constructed according to these figures is given in Fig. 16. If a different scale is desired than is provided on available profile-chart paper (for example, if a 50-mile hop is to be plotted on 30-mile paper), then the scale of miles may be doubled to extend the range of the paper to 60 miles. The vertical scale in feet must then be quadrupled; i.e., 100-foot divisions become 400-foot divisions, as on the right in Fig. 15.

Fresnel Zones

The Fresnel-Kirchhoff theory was originally developed to account for the diffraction of light when obstructed by diaphragms, and when transmitting through apertures of various shapes and sizes. This theory may be applied to radio and sound waves and is based on the concept that any small element of space in the path of a wave may be considered as the source of a secondary wavelet, and that the radiated field can be built up by the superposition of all these wavelets (Huygens principle).

Consider a transparent screen between a distant transmitter, T, and a receiver, R, with the distance from screen to transmitter being at least 10 times the distance from screen to receiver, and with the plane of the screen perpendicular to direction T-R. Concentric circles may be drawn on this screen, with the centers at the point

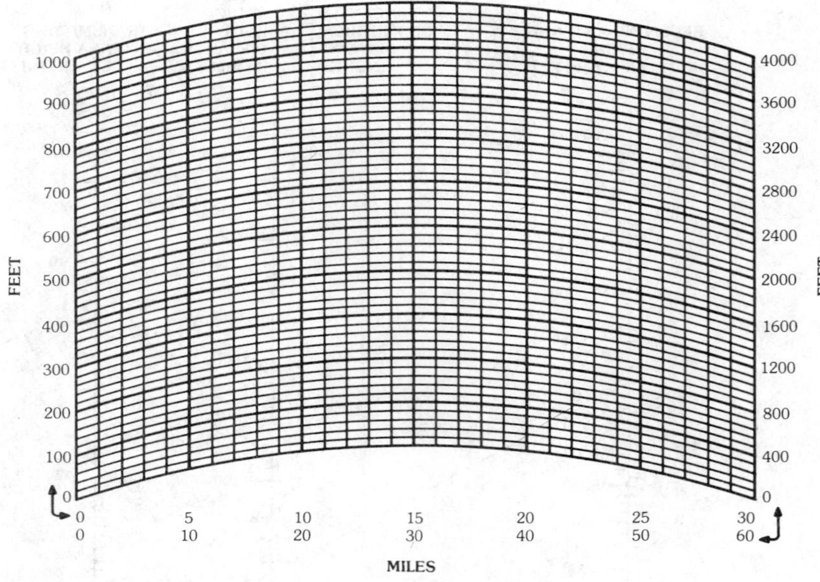

Fig. 15. Typical 4/3-earth profile paper, 1000-foot scale.

TABLE 4. POINTS FOR CONSTRUCTING TEMPLATE

Distance From Center (Horizontal)	Distance From Horizontal (Vertical)
0 miles = 0 inches	0 feet = 0 inches
5 miles = 1/2 inch	12 1/2 feet = 1/8 inch
10 miles = 1 inch	50 feet = 1/2 inch
15 miles = 1 1/2 inches	112 1/2 feet = 1 1/8 inches
20 miles = 2 inches	200 feet = 2 inches

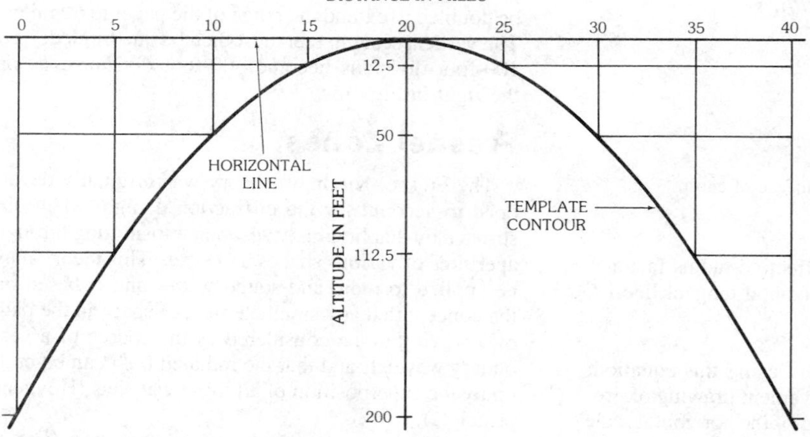

Fig. 16. Construction of a template for profile charts.

where line T-R intersects the screen at O, the radius of the first circle being such that the difference in length between path O-R and the path from the circumference of this circle to R is $\frac{1}{2}$ wavelength (λ). The radii of the other circles are such that the corresponding path-length differences are integral multiples of $\frac{1}{2}\lambda$. The radius of the first circle is $(d\lambda)^{1/2}$, where d is the distance from O to R, and the radius of the second circle is $(2d\lambda)^{1/2}$, of the third $(3d\lambda)^{1/2}$, etc. The area within the first circle is called the first Fresnel zone, and the other ring-shaped areas are the second, third, etc., Fresnel zones. The fields from the odd-number zones are in phase at R, and the fields from the even-number zones are also in phase at R but are opposite in phase to the fields from the odd-number zones. It can be shown that the effect at R of each zone is nearly equal. If an infinitely absorbing screen is provided with an aperture of the same diameter as the first Fresnel zone, it will be found that the field at R is twice as great as the unobstructed or free-space field. If the aperture is increased to include the second zone, the field at R will then be nearly zero, since the fields from zones 1 and 2 are nearly equal in amplitude and opposite in phase. With a continued increase in the diameter of the aperture, further maxima and minima appear; the amplitude of these oscillations decreases very gradually until eventually the field at R approaches the free-space value, which is half that due to the first Fresnel zone. If the distance from the screen to the transmitter is d_1, and from the screen to the receiver is d_2, then the general expression for the radius of the nth Fresnel zone is

$$\{n\lambda[(d_1 \times d_2)/(d_1 + d_2)]\}^{1/2}$$

Required Path Clearance

A criterion to determine whether the earth is sufficiently removed from the radio line-of-sight ray to allow mean free-space propagation conditions to apply is to have the first Fresnel zone clear all obstacles in the path of the rays. This first zone is bounded by points for which the transmission path from transmitter to receiver is greater by one-half wavelength than the direct path. Let d be the length of the direct path and d_1 and d_2 be the distances to the transmitter and receiver from a point, P. The radius of the first Fresnel zone at P is approximately given by

$$R_1{}^2 = \lambda(d_1 d_2/d)$$

where all quantities are expressed in the same units.

If d is in miles and frequency F in megahertz, the first Fresnel-zone radius in feet at P is given by

$$R_{1m} = 2280(d_1 d_2/Fd)^{1/2}$$

The maximum occurs when $d_1 = d_2$ and is equal to

$$R_{1m} = 1140(d/F)^{1/2}$$

While a fictitious earth of 4/3 of true earth radius is generally accepted for determining first Fresnel-zone clearance under normal refraction conditions, unusual conditions that occur in the atmosphere may make it desirable to allow first Fresnel clearance of an effective earth radius of 0.7 to 0.5 of the true radius.

Fig. 17 shows the effect of path clearance on radio transmission.*

Interference Between Direct and Reflected Rays

Where there is one reflected ray combining with the direct ray at the receiving point (Fig. 18), the resulting field strength (neglecting the difference in angles of arrival, and assuming perfect reflection at T) is related to the free-space intensity, irrespective of the polarization, by

$$E = 2E_d \sin 2\pi(\delta/2\lambda)$$

where,

E = resulting field strength, } same
E_d = direct-ray field strength, } units
δ = geometrical length difference between direct and reflected paths, which is given to a close approximation by $\delta = 2h_{at}h_{ar}/d$,

* Bullington, K. "Radio Propagation Fundamentals," *Bell System Tech. J.*, Vol. 36, No. 3, 1957, pp. 593–626.

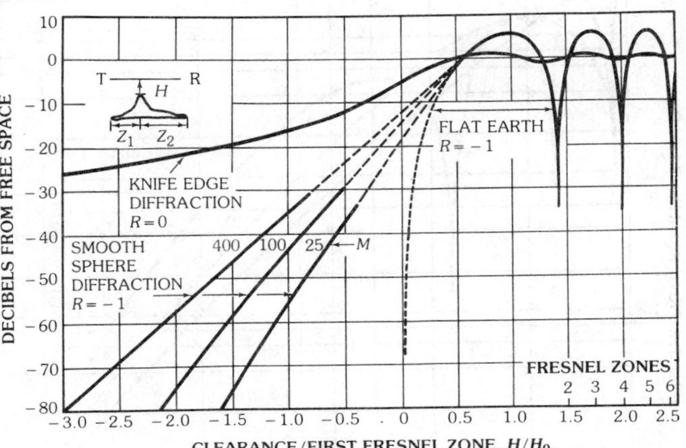

Fig. 17. Effect of path clearance on radio transmission. (*From Bullington, K. "Radio Propagation Fundamentals."* Bell System Tech. J., *Vol. 36, No. 3, Fig. 8, ©1957 American Telephone and Telegraph Co.*)

R = reflection coefficient of surface
H = clearance
H_0 = first Fresnel zone radius = $[(\lambda Z_1 \cdot Z_2)/(Z_1 + Z_2)]^{1/2}$
$M = (H_1/K^{1/3})\{[1 + (H_2/H_1)^{1/2}]/2\}^2(F/4000)^{2/3}$
H_1, H_2 = antenna height in feet above a smooth sphere
F = frequency in megahertz
K = (effective earth radius)/(true earth radius)

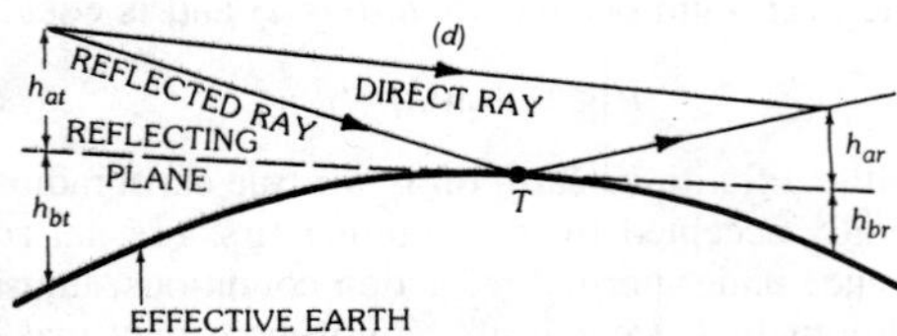

Fig. 18. Interference between direct and reflected rays.

where h_{at} and h_{ar} are the heights of the antennas above a reflecting plane tangent to the effective earth. (See Fig. 18.)

The following cases are of interest:

$E=0$ for $h_{at}h_{ar}=d\lambda/2$

$E=2E_d$ for $h_{at}h_{ar}=d\lambda/4$

$E=E_d$ for $h_{at}h_{ar}=d\lambda/12$

In case $h_{at}=h_{ar}=h$,

$E=0$ for $h=(d\lambda/2)^{1/2}$

$E=2E_d$ for $h=(d\lambda/4)^{1/2}$

$E=E_d$ for $h=(d\lambda/12)^{1/2}$

All these equations are written with the same units for all quantities.

Space-Diversity Reception

When h_{ar} is varied, the field strength at the receiver varies approximately according to the preceding equation. The use of two antennas at different heights provides a means of compensating to a certain extent for changes in electrical-path differences between direct and reflected rays (space-diversity reception).

The antenna spacing at the receiver should be approximately such as to give a $\lambda/2$ variation between geometrical-path differences in the two cases. An approximate value of the spacing is given by $\lambda d/4h_{at}$ when all quantities are in the same units.

The spacing in feet for d in miles, h_{at} in feet, λ in centimeters, and f in megahertz is given by

$$\text{spacing}=43.4\lambda d/h_{at}$$
$$=1.3\times10^6 d/fh_{at}$$

Example: $\lambda=3$ centimeters, $d=20$ miles, and $h_{at}=50$ feet; therefore spacing = 52 feet.

Variation of Field Strength With Distance

Fig. 19 shows the variation of resultant field strength with distance and frequency; this effect is due to interference between the free-space wave and the ground-reflected wave as these two components arrive in or out of phase.

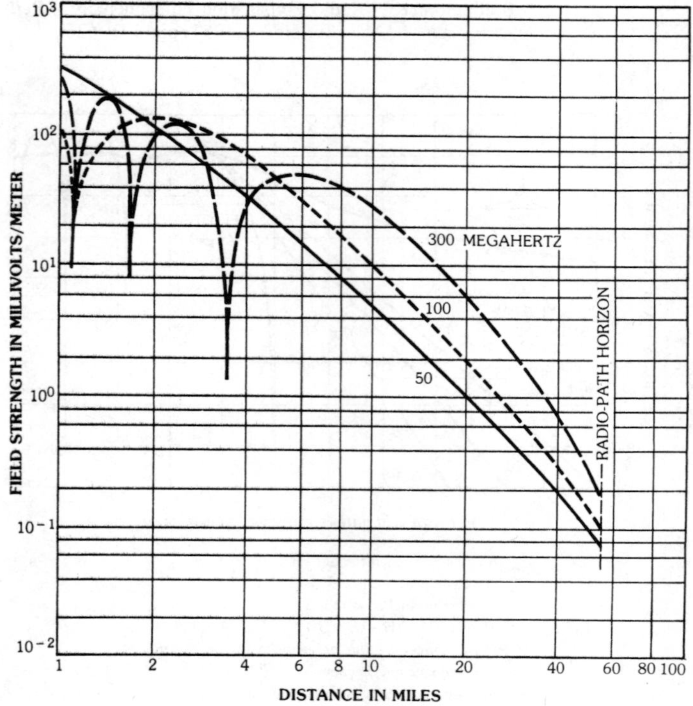

Fig. 19. Variation of resultant field strength with distance and frequency. Antenna heights: 1000 feet, 30 feet; power: 1 kilowatt; ground constants: $\sigma=5\times10^{-14}$ emu, $\epsilon=15$ esu; polarization: horizontal.

To compute the field accurately under these conditions, it is necessary to calculate the two components separately and to add them in the correct phase relationship. The phase and amplitude of the reflected ray are determined by the geometry of the path and the change in magnitude and phase at ground reflection. For horizontally polarized waves, the reflection coefficient can be taken as approximately 1, and the phase shift at reflection as 180 degrees, for nearly all types of ground and angles of incidence. For vertically polarized waves, the reflection coefficient and phase shift vary appreciably with the ground constants and angle of incidence. (See Fig. 24 of Chapter 32.)

Measured field strengths usually show large deviations from point to point because of reflections from ground irregularities, buildings, trees, etc.

For transmission paths of the order of 30 miles and for frequencies up to about 6000 megahertz, good engineering practice should allow for possible increases of signal strength of + 10 decibels with respect to free-space propagation and should allow a fading margin depending on the degree of reliability desired in accordance with the following:

10 decibels—90 percent

20 decibels—99 percent

30 decibels—99.9 percent

40 decibels—99.99 percent

Fading and Diversity*

Line-of-sight propagation at ultrahigh frequencies is affected both by signal-strength variation due to multipath transmission and by bending of the beam due to abnormal variation of refractive index with height in the lower atmosphere.

At frequencies below about 8000 megahertz, and on paths having adequate clearance, the fading on line-of-sight paths is due to multipath transmission. Multipath fading may be divided into two main types; the first is relatively rapid and is caused by interference between two or more rays arriving by slightly different paths; this is known as *atmospheric-multipath*. The second type of fading is less rapid and is due to interference between direct and reflected rays; this is referred to as *reflection-multipath*. In general, the number of fades per unit time due to atmospheric-multipath increases with path length; however, the duration of a fade of a given depth tends to decrease with increasing path length. Fig. 20 shows the typical fading characteristics of a terrestrial line-of-sight path. See Chapter 27 for earth-to-space paths.

* Bullington, K., "Radio Propagation Fundamentals," *Bell System Technical Journal*, Vol. 36, No. 3, 1957; pp. 593–626. Pearson, K. W., "Method for the Prediction of the Fading Performance of a Multisection Microwave Link," *Proceedings of the IEE*, Vol. 112, No. 7, July 1965; pp. 1291-1300. CCIR XVth Plenary Assembly, Geneva, 1982, Vol. V, Report 338-4; pp. 279–314.

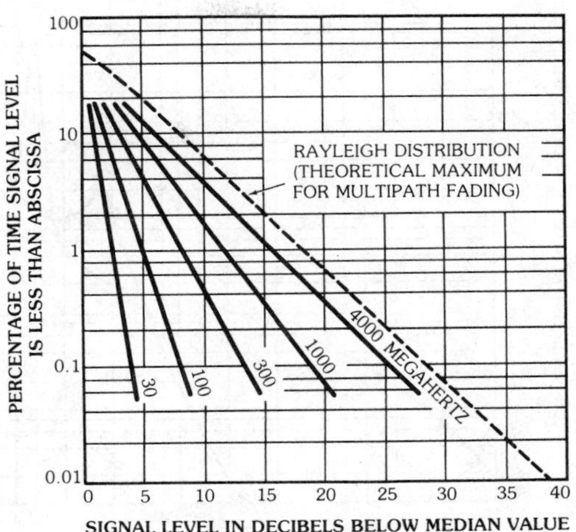

Fig. 20. Typical fading characteristics in the worst month on line-of-sight paths of 30 to 40 miles with clearance of 50 to 100 feet. (*From Bullington, K. "Radio Propagation Fundamentals,"* Bell System Technical Journal, *Vol. 36, No. 3, Fig. 4, ©1957 American Telephone and Telegraph Co.*)

Either frequency or space diversity may be used to reduce the amplitude of multipath fading. In the case of atmospheric-multipath fading on line-of-sight paths, it has been found that considerable diversity improvement can usually be obtained with a frequency difference of 100 to 200 megahertz or with a vertical antenna spacing of between 100 and 200 wavelengths.

Atmospheric Absorption*

Oxygen and water vapor may absorb energy from a radio wave by virtue of the permanent electric dipole moment of the water molecule and the permanent magnetic dipole moment of the oxygen molecule. Fig. 21 shows water-vapor absorption γ_{wo} and oxygen absorption γ_{oo} as a function of frequency.

The attenuation due to rain increases with frequency and with increasing rate of precipitation. Fig. 22 shows the frequency dependence of attenuation due to precipitation. Typical rainfall rates in a temperate climate are shown in Fig. 23. In temperate climates, rainfall rates exceeding 1 inch (25.4 millimeters) per hour are unlikely to occur over an area larger than about 4 miles in diameter.

Free-Space Transmission Equations

If the incoming wave is a plane wave having a power flow per unit area equal to P_0, the available power at the output terminals of a receiving antenna may be expressed as

* *Transmission Loss Predictions for Tropospheric Communication Circuits*. National Bureau of Standards Technical Note No. 101.

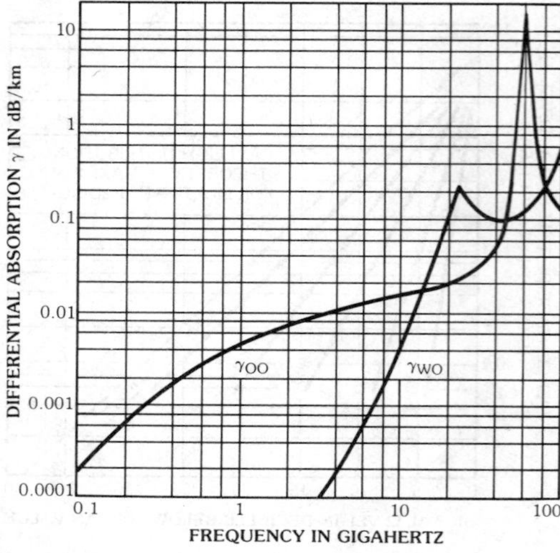

Fig. 21. Atmospheric absorption versus wavelength. (*From CCIR XIIIth Plenary Assembly, Geneva, 1974, Vol. V, Report 233-3.*)

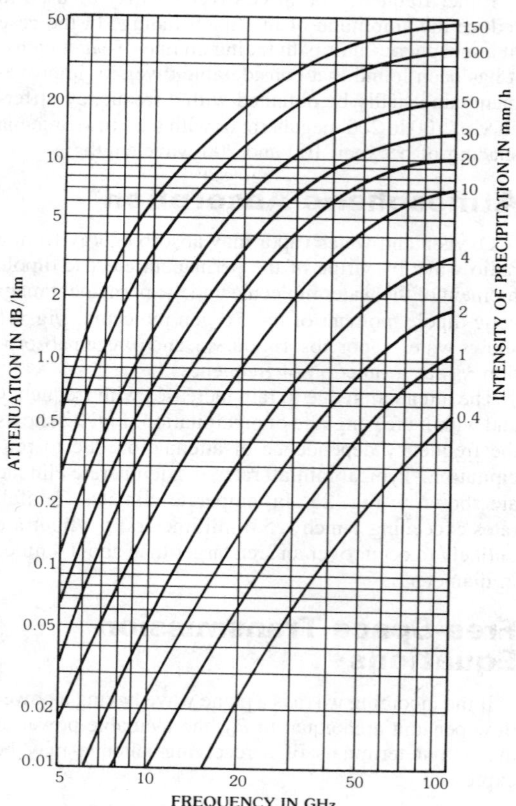

Fig. 22. Attenuation due to precipitation. (*From CCIR XIIIth Plenary Assembly, Geneva, 1974, Vol. V, Report 233-3.*)

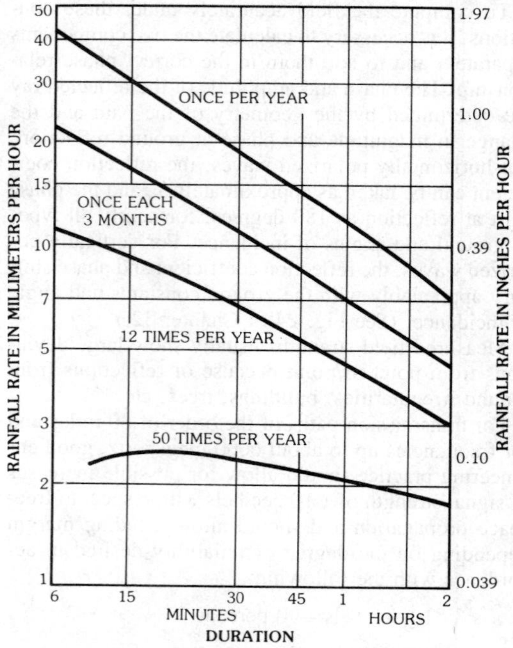

Fig. 23. Rainfall duration in England. (*From Bilham, E. G. "Climate of British Isles," Toronto: Macmillan Company, 1938.*)

$$P_r = A_r P_0$$

where A_r is the effective area of the receiving antenna.

The free-space path attenuation is given by

$$\text{attenuation} = 10 \log (P_t/P_r)$$

where P_t is the power radiated from the transmitting antenna (same units as for P_r). Then

$$P_r/P_t = A_r A_t / d^2 \lambda^2$$

where,

A_r = effective area of receiving antenna,
A_t = effective area of transmitting antenna,
λ = wavelength,
d = distance between antennas.

The length and surface units in the equation should be consistent. This is valid provided $d >> 2a^2/\lambda$, where a is the largest linear dimension of either of the antennas.

Path attenuation between isotropic antennas is

$$P_t/P_r = 4.56 \times 10^3 f^2 d^2$$

where f is in megahertz and d is in miles.

Path attenuation α (in decibels) is

$$\alpha = 36.6 + 20 \log f + 20 \log d$$

A nomogram for determining the solution of α is given in Fig. 24.

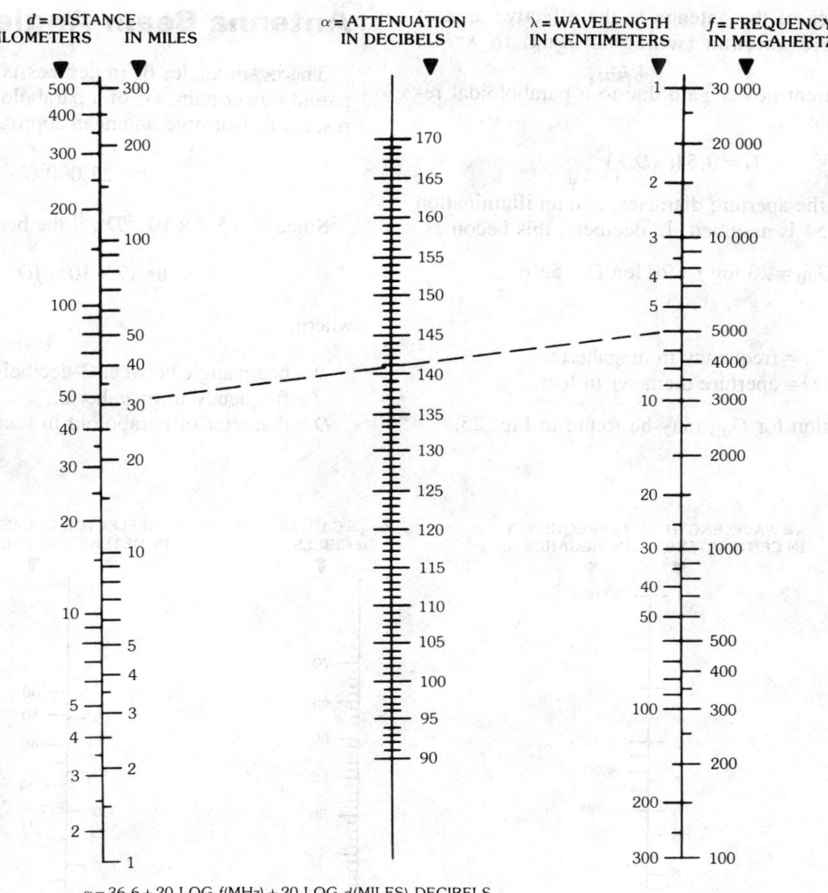

Fig. 24. Nomogram for solution of free-space path attenuation α between isotropic antennas. Example shown: distance 30 miles; frequency 5000 megahertz; attenuation = 141 decibels.

Effective Areas of Typical Antennas*

Hypothetical isotropic antenna (no heat loss)

$$A = (1/4\pi)\lambda^2 \approx 0.08\lambda^2$$

Small uniform-current dipole, short compared with wavelength (no heat loss)

$$A = (3/8\pi)\lambda^2 \approx 0.12\lambda^2$$

Half-wavelength dipole (no heat loss)

$$A \approx 0.13\lambda^2$$

Parabolic reflector of aperture area S (here, the factor 0.54 is due to nonuniform illumination of the reflector)

$$A \approx 0.54S$$

Very long horn with small aperture dimensions compared with length

$$A = 0.81S$$

Horn producing maximum field for given horn length

$$A = 0.45S$$

The aperture sides of the horn are assumed to be large compared with the wavelength.

Antenna Gain Relative to Hypothetical Isotropic Antennas

If directive antennas are used in place of isotropic antennas, the transmission equation becomes

$$P_r/P_t = G_t G_r [P_r/P_t]_{\text{isotropic}}$$

where G_t and G_r are the power gains due to the directivity of the transmitting and receiving antennas, respectively.

The apparent power gain is equal to the ratio of the

* Refer to Chapter 32.

effective area of the antenna to the effective area of the isotropic antenna (which is equal to $\lambda^2/4\pi \approx 0.08\lambda^2$).

The apparent power gain due to a paraboloidal reflector is thus

$$G = 0.54(\pi D/\lambda)^2$$

where D is the aperture diameter, and an illumination factor of 0.54 is assumed. In decibels, this becomes

$$G_{dB} = 20 \log f + 20 \log D - 52.6$$

where,

f = frequency in megahertz,
D = aperture diameter in feet.

The solution for G_{dB} may be found in Fig. 25.

Antenna Beam Angle

The beam angle, θ, in degrees is related to the apparent power gain, G, of a paraboloidal reflector with respect to isotropic antennas approximately by

$$\theta^2 \approx 27\ 000/G$$

Since $G = 5.5 \times 10^{-6} D^2 f^2$, the beam angle becomes

$$\theta \approx (7 \times 10^4)/fD$$

where,

θ = beam angle between 3-decibel points in degrees,
f = frequency in megahertz,
D = diameter of paraboloid in feet.

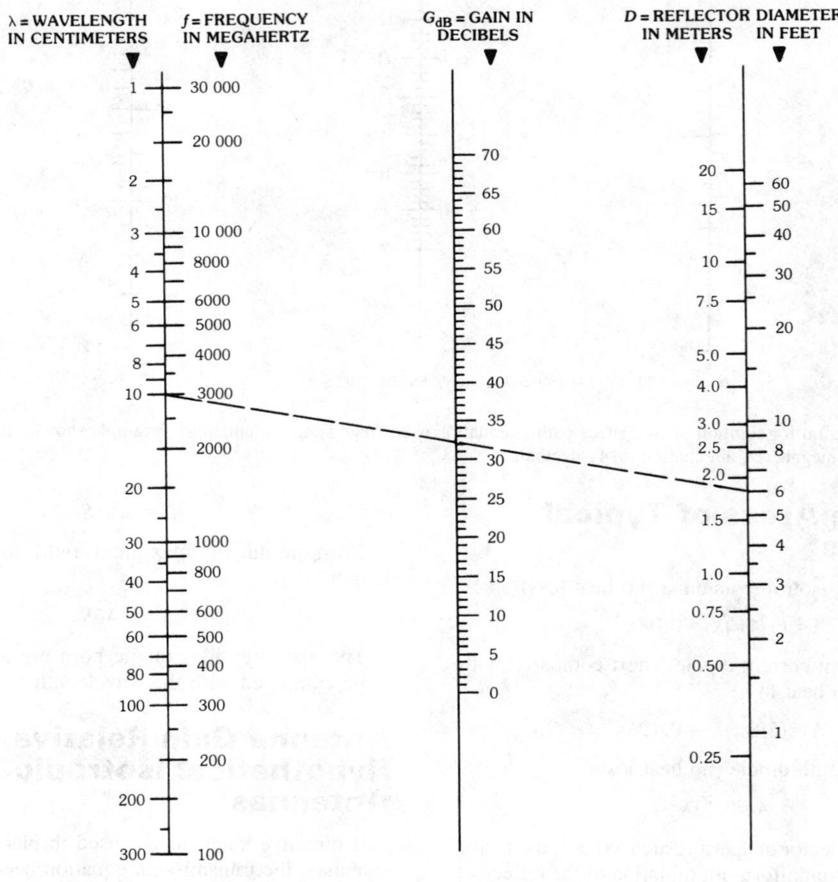

Fig. 25. Nomogram for determination of apparent power gain G_{dB} (in decibels) of a paraboloidal reflector. Example shown: frequency 3000 megahertz; diameter 6 feet; gain = 32 decibels.

Transmitter Power for a Required Output Signal/Noise Ratio

Based on the above expressions for path attenuation and reflector gain, the ratio of transmitted power to theoretical receiver noise, in decibels, is given by

$$10 \log (P_t/P_n) = A_p + (S/N) + (nf) - G_t - G_r - (\overline{ni}f)$$

where,

S/N = required signal/noise ratio at receiver in decibels,
(nf) = noise figure of receiver in decibels,
$(\overline{ni}f)$ = noise improvement factor in decibels due to modulation methods where extra bandwidth is used to gain noise reduction,
P_n = theoretical noise power in receiver,
P_t = radiated transmitter power,
G_t = gain of transmitting antenna in decibels,
G_r = gain of receiving antenna in decibels,
A_p = path attenuation in decibels.

An equivalent way to compute the transmitter power for a required output signal/noise ratio is given below directly in terms of reflector dimensions and system parameters.

(A) Normal free-space propagation

$$P_t = \frac{\beta_1 \beta_2}{40} \frac{BL^2}{f^2 r^4} \frac{F}{K} \frac{S}{N}$$

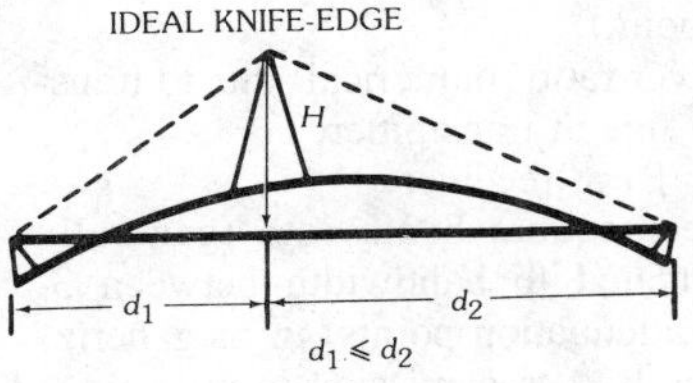

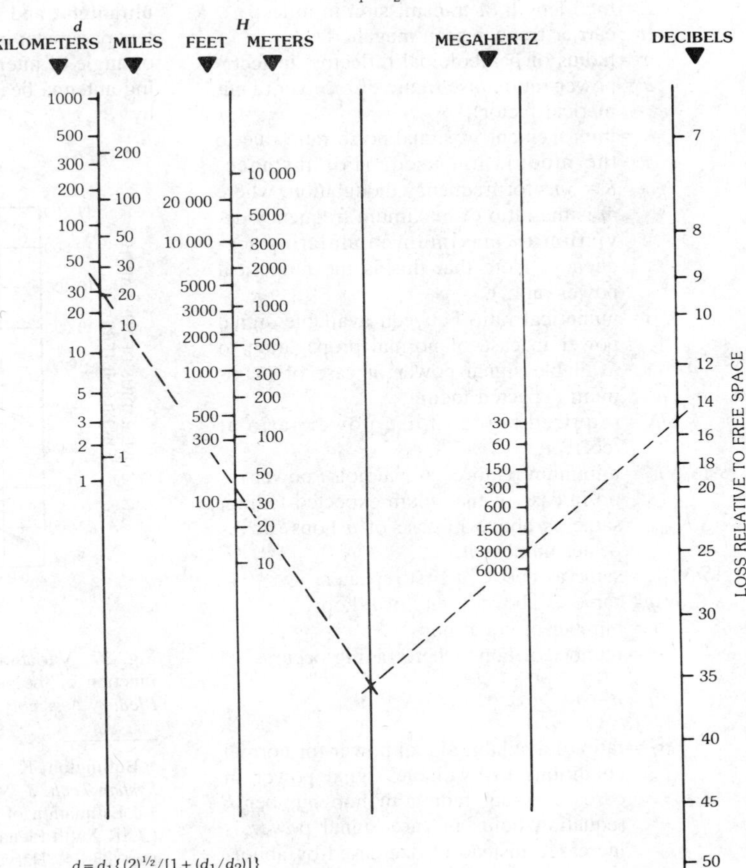

Fig. 26. Knife-edge diffraction loss relative to free space. (*From Bullington, K. "Radio Propagation Fundamentals." Bell System Technical Journal, Vol. 36, No. 3, Fig. 7, ©1957 American Telephone and Telegraph Co.*)

(B) With allowance for fading

$$P_t = \frac{\beta_1\beta_2}{40}\,\frac{BL^2}{f^2r^4}\,\frac{F}{K}\,\sigma\left(\frac{S}{N}\right)_m$$

(C) For multirelay transmission in n equal hops

$$P_t = \frac{\beta_1\beta_2}{40}\,\frac{BL^2}{f^2r^4}\,\frac{F}{K}\,\sigma\left(\frac{S}{N}\right)_{nm}$$

(D) Signal/noise ratio for nonsimultaneous fading is

$$10 \log (S/N)_n = 10 \log \sigma\, (S/N)_{1m} - 10 \log \bar{n}$$

where,

P_t = power in watts available at transmitter output terminals (kept constant at each repeater point),
β_1 = loss power ratio (numerical) due to transmission line at transmitter,
β_2 = same as β_1 at receiver,
B = root-mean-square bandwidth (generally approximated to bandwidth between 3-decibel attenuation points) in megahertz,
L = total length of transmission in miles,
f = carrier frequency in megahertz,
r = radius of paraboloidal reflectors in feet,
F = power-ratio noise figure of receiver (a numerical factor),
K = improvement in signal/noise ratio due to the modulation used. (For instance, $K = 3m^2$ for frequency modulation, where m is the ratio of maximum frequency deviation to maximum modulating frequency. Note that this is the numerical power ratio.),
σ = numerical ratio between available signal power in case of normal propagation to available signal power in case of maximum expected fading,
S/N = required signal/noise power ratio at receiver,
$(S/N)_m$ = minimum required signal/noise power ratio in case of maximum expected fading,
$(S/N)_{nm}$ = same as above in case of n hops, at repeater number n,
$(S/N)_{1m}$ = same as above at first repeater,
$(S/N)_n$ = same as above at end of n hops,
n = number of equal hops,
m = number of hops where fading occurs,

$$\bar{n} = n - m + \sum_{1}^{m} \sigma_k$$

σ_k = ratio of available signal power for normal conditions to available signal power in case of actual fading in hop number k (equation holds in case signal power is increased instead of decreased by abnormal propagation or reduced hop distance).

KNIFE-EDGE DIFFRACTION PROPAGATION*

Diffraction loss at an ideal knife-edge can be estimated from Fig. 26. However, the transmission loss over a practical knife-edge diffraction path depends critically on the shape of the diffracting edge. Since a natural obstacle, such as a mountain ridge, may depart considerably from an ideal knife-edge, the diffraction loss in practice is usually 10 to 20 decibels greater than that estimated for the ideal case.

A nonuniform transverse profile of the diffracting edge, or reflections on the transmission paths each side of the diffracting edge, may result in multipath transmission causing variations in the received level as a function of frequency, space, and time. The amplitude of such variations may be reduced by either space or frequency diversity and by the use of narrow-beam-width antennas.

TROPOSPHERIC SCATTER PROPAGATION†

Weak but reliable fields are propagated several hundred miles beyond the horizon in the very-high-, ultrahigh-, and superhigh-frequency bands. An important parameter in scatter propagation is the scatter angle or angle of intersection of the transmitting and receiving antenna beams. This angle, θ, in radians is given by

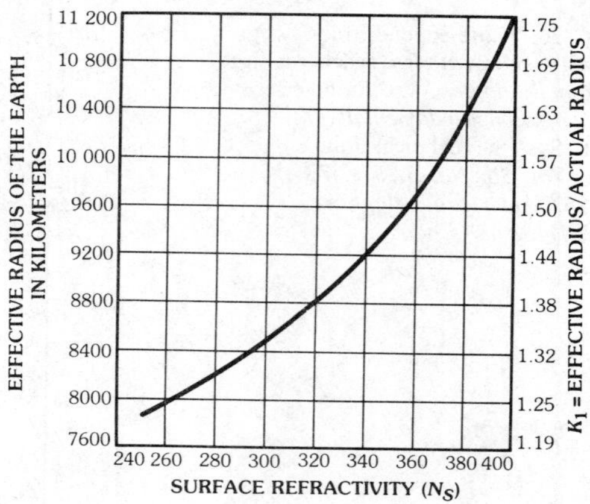

Fig. 27. Variation of the effective radius of the earth as a function of the surface refractivity, N_s. (*From CCIR XVth Plenary Assembly, Geneva, 1982, Vol. V, Report 338-4.*)

---

* Bullington, K., "Radio Propagation Fundamentals." *Bell System Tech. J.* Vol. 36, No. 3, 1957; pp. 593–626.
† "Estimation of Tropospheric-Wave Transmission Loss," CCIR XVth Plenary Assembly, Geneva, 1982, Vol. V, Report 238-4. Harvey, A. F., *Microwave Engineering*, New York: Academic Press, Inc., 1963.

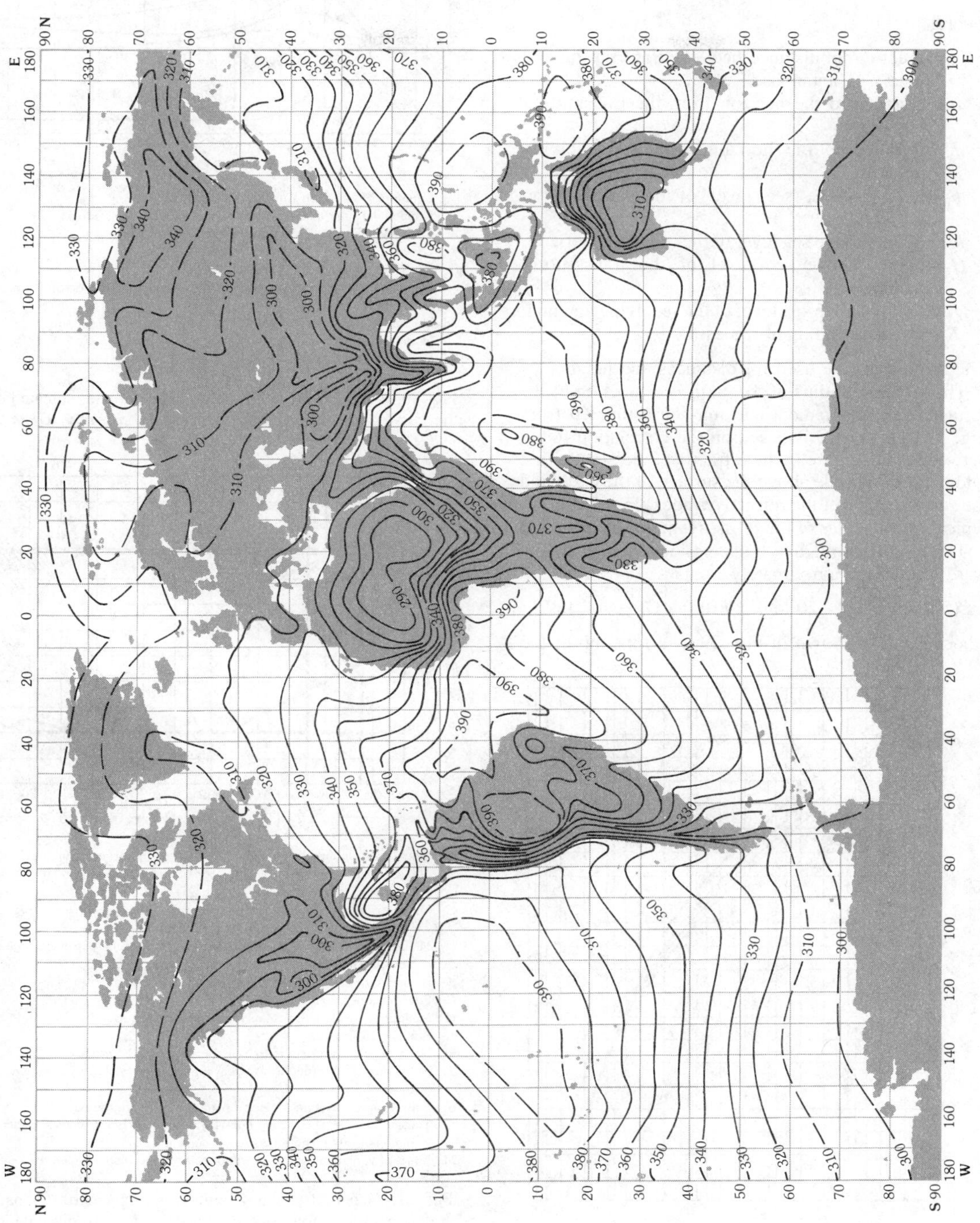

Fig. 28. Worldwide mean value of N_0 for February. (*From CCIR XVth Plenary Assembly, Geneva, 1982, Vol. V, Report 563-2, "Influence of the Atmosphere on Wave Propagation."*)

$$\theta = \frac{2d - d_t - d_v}{2R} + \frac{h_t - H_t}{d_t} + \frac{h_v - H_v}{d_v}$$

where,

d = great-circle distance between transmitting and receiving antennas,
d_t = distance to the horizon from the transmitting antenna,
d_v = distance to the horizon from the receiving antenna,
h_t = height above sea level of the transmitting horizon,
h_v = height above sea level of the receiving horizon,
H_t = height above sea level of the transmitting antenna,
H_v = height above sea level of the receiving antenna,
R = effective radius of the earth.

The same units are used for distances and heights.

The effective radius of the earth is a function of the refractive index gradient and may be estimated from Fig. 27. This curve is based on the correlation found between the decrease in the refractive index in the first kilometer of altitude above the surface of the earth and the surface value of the refractive index. Fig. 28 shows typical mean values of the refractive index at sea level.

The long-term median transmission loss due to forward scatter is approximately

$$L(50) = 30 \log f - 20 \log d + F(\theta d) - G_p - V(d_e) \text{ dB}$$

where $F(\theta d)$ is shown in Fig. 29 as a function of the

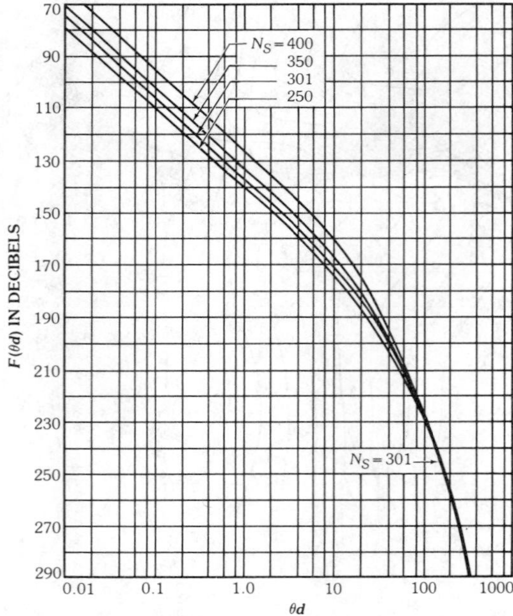

Fig. 29. Attenuation function $F(\theta d)$, where d is in kilometers and θ is in radians, for indicated values of surface refractivity N_s. (*From CCIR XVth Plenary Assembly, Geneva, 1982, Vol. V, Report 238-4.*)

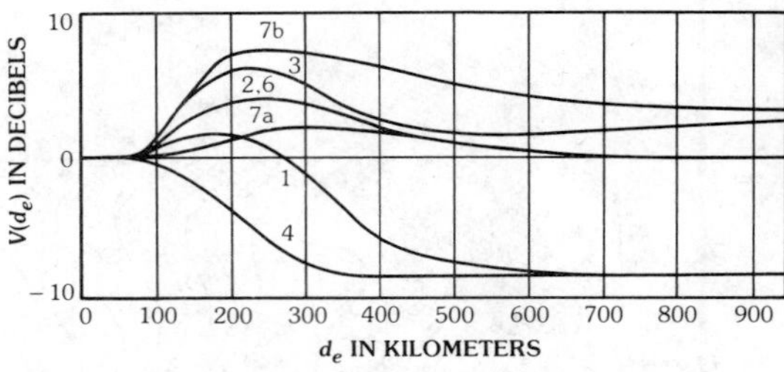

1. Equatorial (data from Congo and Ivory Coast)
2. Continental subtropical (Sudan)
3. Maritime subtropical (data from West Coast of Africa)
4. Desert (Sahara)
5. Mediterranean (no curves available)
6. Continental temperate (data from France, Federal Republic of Germany, and U.S.A.)

7a. Maritime temperate, over land (data from U.K.)
7b. Maritime temperate, over sea (data from U.K.)
8. Polar (no curves available)

Fig. 30. Function $V(d_e)$ for the types of climate indicated on the curves. (*From CCIR XVth Plenary Assembly, Geneva, 1982, Vol. V, Report 238-4.*)

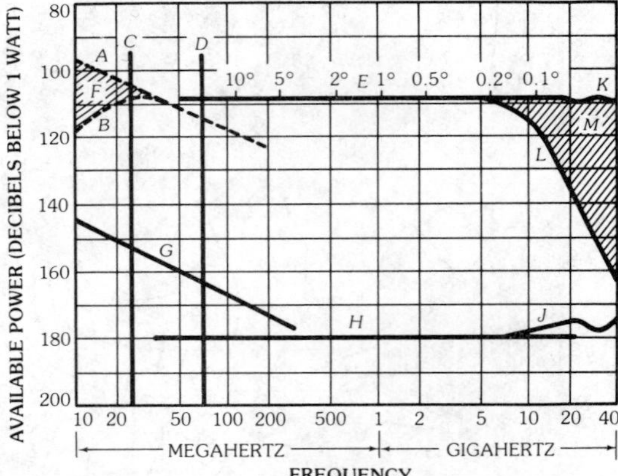

A: Signal level during ideal nighttime conditions (no absorption).
B: Typical signal level during daytime conditions, assuming an angle of elevation of 5°.
C: Minimum frequency to assure penetration of ionosphere: polar region, oblique path; tropical region, vertical path.
D: Minimum frequency to assure penetration of ionosphere: tropical region, oblique path.
E: Beam width of paraboloid between half-power points.
F: Effect of ionospheric absorption.
G: Minimum cosmic noise. Maximum value will be found to be higher by about 15 decibels.
H: Noise level corresponding to a temperature of 70 kelvins.
J: Noise due to absorption in a clear atmosphere, assuming an elevation angle of 5°.
K: Typical signal level for a vertical path in a clear atmosphere.
L: Typical signal level in heavy rain (16 millimeters/hour), vertical depth 1 kilometer, assuming an elevation angle of 5°.
M: Effect of varying atmospheric conditions and elevation angles.

Fig. 31. General frequency limits in a simple earth-to-spacecraft communication system. (Spacecraft: isotropic antenna; transmitter power, 1 watt; bandwidth, 1 kilohertz; distance, 1000 kilometers. Earth station antenna: paraboloid; diameter, 20 meters; efficiency, 55%—. 15-decibel gain above isotropic antenna - - -.) (*From CCIR Xth Plenary Assembly, Geneva, 1963, Vol. IV, Report 205, p. 194.*)

product θd. Angular distance θ is the angle between radio horizon rays in the great-circle plane containing the antennas, and d is the distance between antennas.

A semiempirical estimate of the path antenna gain, G_p, is provided by

$$G_p = G_t + G_r - 0.07 \exp[0.055(G_t + G_r)] \text{ dB}$$

for values of G_t and G_r each less than 50 dB.

Fig. 30 shows $V(d_e)$, an adjustment for the indicated types of climate.

This division is, of course, rather crude, and local geographical conditions may require serious modifications. A brief description of these climates is given in Annex 1 of CCIR Report 238-2, Geneva, 1974.

Fast and slow fading is experienced on tropospheric scatter paths. Fast fading is due to multipath transmission, is in general Rayleigh distributed, and can be considerably reduced by diversity, an antenna spacing of 60 wavelengths usually being adequate. Slow fading, with periods of hours or days, is caused by changes in the gradient of the refractive index of the atmosphere along the transmission path and is little affected by diversity.

The plane-wave gains of large antennas are not fully realized on tropospheric scatter paths. The power on such a path is received, not from a single point source, but from a volume in the atmosphere that subtends a solid angle at the receiving antenna. If the antenna beam angles are such as to limit the available scattering volume, then the received power will be correspondingly limited, and the antennas are said to suffer an antenna-to-medium coupling loss. The resulting median loss of received power is likely to be about 5 decibels for two 40-decibel-gain plane-wave antennas, and 17 decibels for two 50-decibel-gain antennas. The extent to which the path antenna gain is a function of the scatter angle, θ, or the height of the scatter volume has not yet been established.

Multipath transmission limits the communication bandwidth that can be used on a single carrier; however, useful bandwidths of several megahertz have been shown to be available on some 200-mile scatter paths. Narrow-beam antennas and diversity reduce the effects of multipath transmission.

EARTH-SPACE COMMUNICATION

Communication between earth and outer space (see Chapter 27, Satellite and Space Communications) must pass through the atmosphere of the earth, so that the optimum frequencies for this service are those that pass through the atmosphere with minimum attenuation. A range of frequency little attenuated by the atmosphere is known as a window; one such window occurs between the critical frequency of the ionosphere and the frequency absorbed by rainfall and oxygen. This frequency range extends from about 10 to 10 000 megahertz. Another window exists in the optical and infrared region of 10^6 to 10^9 megahertz. Fig. 31 shows the general frequency limits for earth-space communication.

34 Radio Noise and Interference

Revised by
Wilbur R. Vincent

Radio noise and interference limit the performance of all communications systems by restricting the operating range, generating errors in messages, and in extreme cases preventing the successful operation of receivers. At locations where man-made noise is low, natural noise sources determine receiver performance. When man-made noise encroaches upon receiving sites, the performance of receiving equipment is degraded below design levels.

When noise from sources external to a receiver is involved, the gain and orientation of the antenna must be considered. For narrow-band receivers, noise is usually flat in amplitude across the bandwidth of a receiver. For such cases, noise power affecting receiver performance is proportional to the bandwidth. For wide-band receivers, the noise may not be flat across the receiver bandwidth, and the determination of effective noise power requires further consideration.

Noise level can be expressed in terms of voltage or power at the terminals of a receiver, the strength of an electromagnetic field at an antenna location, or thermal noise power at a temperature referenced to 290 kelvins. Noise that is flat in amplitude across the bandwidth of a receiver is often expressed in terms of an effective antenna noise factor, f_a, which is defined as

$$f_a = P_n/kT_0B = T_a/T_0 \tag{1}$$

where,

P_n = noise power in watts from an equivalent lossless antenna,

k = Boltzmann's constant,

T_0 = reference temperature (290 kelvins),

B = receiver noise bandwidth in hertz,

T_a = antenna noise temperature in the presence of external noise.

NATURAL NOISE

Natural noise consists of thermal noise, atmospheric noise, and cosmic noise. These noise sources usually determine the minimum detectable signal level of a receiver operated in an environment free of man-made noise sources.

Thermal Noise

For many years, thermal noise in the first stage of a receiver was usually the main factor limiting the sensitivity of radar and microwave receivers. Recent advances in low-noise amplifier performance have reduced thermal noise of microwave amplifiers to very low levels, and thermal radiation from nearby objects, ground, and the sky are now major factors that must be considered in choosing sites for satellite receivers and radars. Fig. 1 shows noise temperatures for various devices (as of 1984) and natural limits at microwave frequencies. Receivers operated at frequencies below about 20 MHz usually encounter noise from other sources that is considerably above the thermal noise of conventional amplifiers; hence, low-noise amplifier performance is not usually a factor in the design of low-frequency receivers. An exception occurs for vlf receivers operating in the Arctic and Antarctic, where atmospheric noise is extremely low and cosmic noise is screened by the ionosphere.

Atmospheric Noise

Lightning from thunderstorms produces bursts of impulsive noise. At low frequencies, these bursts are propagated to distant receivers by normal ionospheric modes. The noise is dependent on the weather, time of day, season, location of the receiver with respect to storm areas, and ionospheric propagation conditions. Atmospheric noise generally decreases with increasing latitude and increases in high-latitude equatorial areas. Atmospheric noise sources are particularly active during the rainy season in the Caribbean, the East Indies, equatorial Africa, northern India, and the Far East. An excellent summary of worldwide atmospheric noise levels is contained in CCIR Report 322.* An example of a CCIR developed map of atmospheric noise levels in the summer during daytime hours is shown in Fig. 2. The map shows the median noise level in decibels above kTB at a frequency of 1 MHz as received on a short vertical rod antenna installed over ground (k is Boltzmann's constant, T is 290 kelvins, and B is the receiver bandwidth in hertz). This parameter is related to noise field strength by

$$E_n = F_a + 20 \log_{10} f_{\mathrm{MHz}} - 65.5 \tag{2}$$

where,

E_n = rms noise field strength in a 1-kHz bandwidth in decibels above 1 microvolt/meter,

F_a = noise level in decibels above kTB,

f_{MHz} = frequency in megahertz.

The level of atmospheric noise at a receiver site decreases with increasing frequency. Fig. 3 shows the frequency dependence of atmospheric noise for the data shown in Fig. 2. These levels represent the median amplitude of the noise bursts. Individual bursts can vary in amplitude from insignificant to very strong. An example of an individual atmospheric burst propagated at hf frequencies over a one-hop ionospheric mode is shown in Fig. 4. The data were obtained with a rapidly scanning receiver and a time memory display. The noise burst originated from a lightning event that was accompanied by numerous discharges lasting about one second. The wide-bandwidth impulsive energy produced by the lightning discharge was modified by the ionospheric propagation path so that only a portion of the discharge was seen at the distant receiver (2000 km from the storm to the receiver). The maximum prop-

* CCIR Report 322, 10th Plenary Assembly, Geneva; 1963.

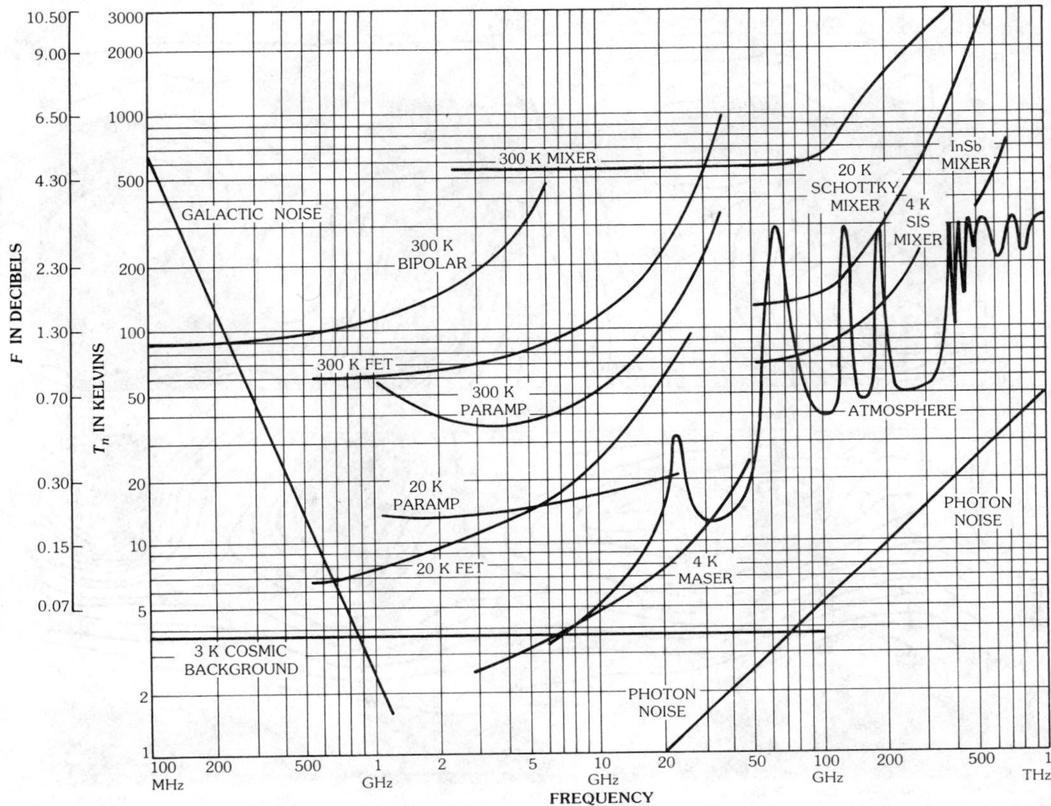

Fig. 1. Noise figure (F) and noise temperature (T_n) for various devices and natural limits—1984. (*From S. Weinreb, "Low-Noise GASFET Amplifiers."* IEEE Trans. on MTT, *Vol. MTT-28, No. 10, October 1980, pp. 1041–1054. Material updated to 1984 by Weinreb.*)

agating frequency and the minimum propagating frequency of the ionosphere are shown.

Cosmic Noise

The sources of cosmic noise are external to the atmosphere of the earth; primary sources are the sun, the Milky Way galaxy, and other discrete cosmic sources. Radio astronomy activities have identified a very large number of sources of cosmic noise. Radio noise from cosmic sources must penetrate the atmosphere of the earth to reach antennas located on the surface of the earth. Ionospheric absorption limits the reception of cosmic noise at frequencies below about 20 MHz, while molecular absorption processes limit the reception of extraterrestrial noise at frequencies above about 10 GHz. Satellite-borne receivers above about 1000 km do not encounter these limitations.

Recent advances in low-noise receiver design (see Fig. 1) and the widespread deployment of satellites and space probes have increased the importance of cosmic noise. Satellite communications systems, the broadcasting of television from satellites, and the need for data links between the space vehicles and the earth have increased the number of skyward-pointed antennas equipped with sensitive receivers that are capable of receiving cosmic noise. Cosmic noise often limits the performance of such systems.

Figs. 5 and 6 show detailed radio-sky maps** of the celestial sphere for the 136-megahertz and 400-megahertz space research satellite frequency bands. Also shown are the well-known discrete sources of cosmic radio waves, known as radio stars, including the intense source Cassiopeia A. The 136-megahertz and 400-megahertz radio-sky maps in Figs. 5 and 6 are a composite of data obtained through the use of high-gain antennas with solid-angle beams ranging in size from 2° to 5° half-power beam width (HPBW) at 136 megahertz, and 7° to 16° HPBW at 400 megahertz.

Fig. 7 shows the level of galactic noise in decibels relative to a noise temperature of 290 K when receiving on a half-wave dipole. The noise levels shown in this figure assume no atmospheric absorption and refer to the following sources of cosmic noise.

** R. F. Taylor, "136 MHz/400 MHz Radio-Sky Maps," *Proceedings of the IEEE,* Vol. 61, No. 4, "Proceedings Letters," pp. 469–472; April 1973.

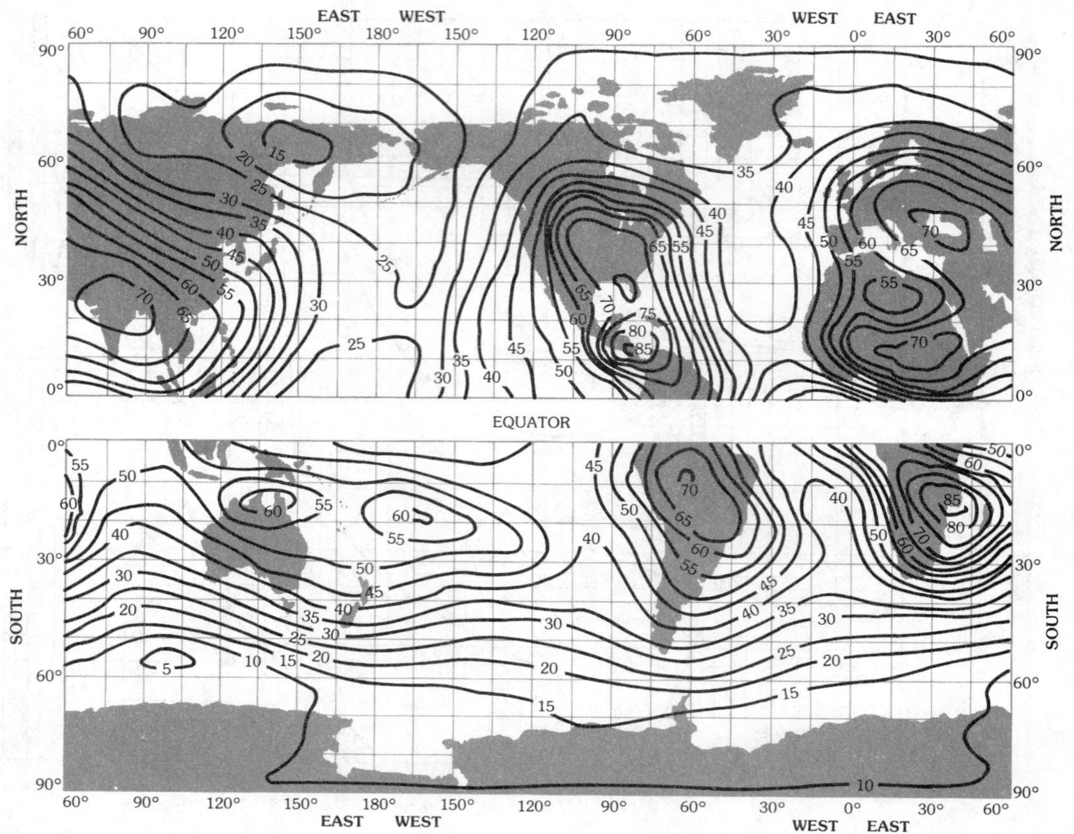

Fig. 2. Atmospheric noise levels in northern and southern hemispheres, summer, 1200–1600 hours local time. The maps show the expected values of F_a at 1 MHz, in decibels above kT_0B. (*From CCIR Report 322, 10th Plenary Assembly, Geneva; 1963.*)

Galactic Plane: Galactic noise from the galactic plane in the direction of the center of the galaxy. The noise levels from other parts of the galactic plane can be as much as 12 to 15 decibels below the levels given in Fig. 7.

Quiet Sun: Noise from the "quiet" sun; that is, solar noise at times when there is little or no sunspot activity.

Disturbed Sun: Noise from the "disturbed" sun. The term "disturbed" refers to times of sunspot and solar-flare activity.

Cassiopeia A: Noise from a high-intensity discrete source of cosmic noise known as Cassiopeia A. This is one of thousands of known discrete sources. Cassiopeia A subtends a solid angle at the surface of the earth of only about 5 arc minutes.

The levels of cosmic noise received by a highly directive antenna with main lobe pointed along the galactic plane can be obtained from equations given by Kraus† for the antenna-noise temperature (T_A) at the output terminals of an ideal, loss-free, antenna as

$$T_A = \frac{\int_0^{\theta=90^\circ-\theta_0} \int_0^{\phi=2\pi} T(\theta,\phi)G(\theta,\phi)\sin\theta d\theta d\phi}{\int_0^{\theta=90^\circ-\theta_0} \int_0^{\phi=2\pi} G(\theta,\phi)\sin\theta d\theta d\phi} K$$

where,

$\theta = 0°$ at zenith,

$\phi = 360°$ azimuth angle,

$T(\theta,\phi)$ = brightness-noise temperature distribution from radio-sky map, kelvins,

$G(\theta,\phi)$ = antenna radiation pattern gain distribution, assumed symmetrical,

θ_0 = minimum elevation angle between antenna main-lobe axis and the horizon, degrees.

However, for a practical antenna, Taylor and Stocklin‡ give a simplified approximation for T_A including contributions from the main lobe, side lobes, and back lobe as

† J. D. Kraus, *Radio Astronomy*. (New York: McGraw-Hill Book Co., 1966.)

‡ R. F. Taylor and F. J. Stocklin, "VHF/UHF Stellar Calibration Error Analysis," *Proceedings International Telemetering Conference*, Washington, D.C., Vol. VII, pp. 553-566; September 27–29, 1971.

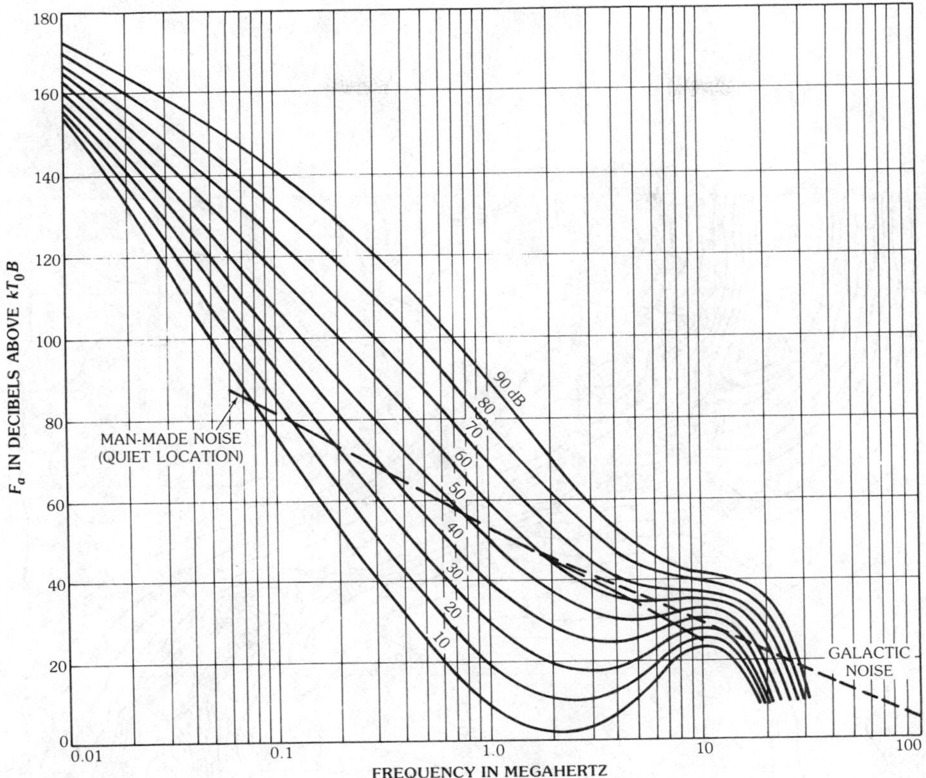

Fig. 3. Variation of radio noise with frequency, for data given in Fig. 2 legend. (*From CCIR Report 322, 10th Plenary Assembly, Geneva; 1963.*)

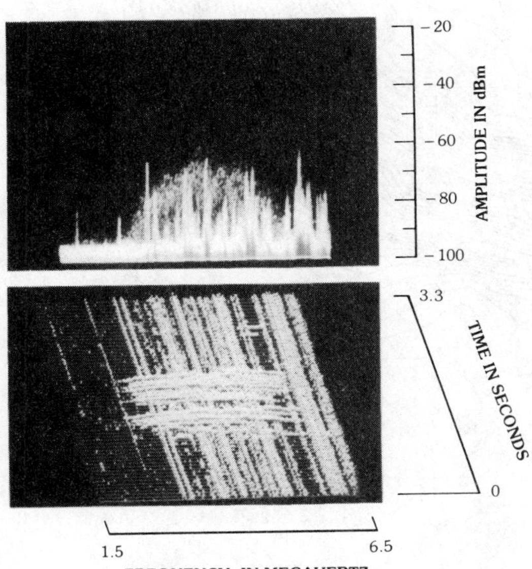

Fig. 4. Atmospheric noise burst.

$T_A \simeq 0.82\ T_{sky} + 0.13(\overline{T}_{sky} + T_E)\ K$, for a solid-angle beam, $\theta_{HPBW} = \phi_{HPBW} \leq 25°$

where,

T_{sky} = mean value of sky-brightness temperature within main-lobe HPBW, in kelvins,

$\overline{T}_{sky}$ = mean value of sky-brightness temperature within antenna side lobes, in kelvins,

$T_E \simeq T_0 = 290$ K, effective noise temperature of earth.

For example, a 136-megahertz, phased-array, directive antenna with main-lobe HPBW equal to 12°, pointed near Cassiopeia A, has a value of T_A equal to approximately 870 K, for T_{sky} equal to 950 K and $\overline{T}_{sky}$ equal to 400 K obtained from Fig. 5.

MAN-MADE RADIO NOISE

Man-made radio noise frequently limits the performance of receivers. This is particularly true for land-mobile communications, television reception, high-frequency radio, and other radio services below uhf. Man-made radio noise originates from a wide variety of sources; some examples are noise from ignition systems of gasoline engines, corona noise from high-voltage power lines, gap noise from utility distribution lines, noise from radio-frequency stabilized welders, and noise produced by many other electrical devices found in homes and businesses. Fig. 8 shows the frequency range that is affected by several common types of man-made noise. Sources that affect the perfor-

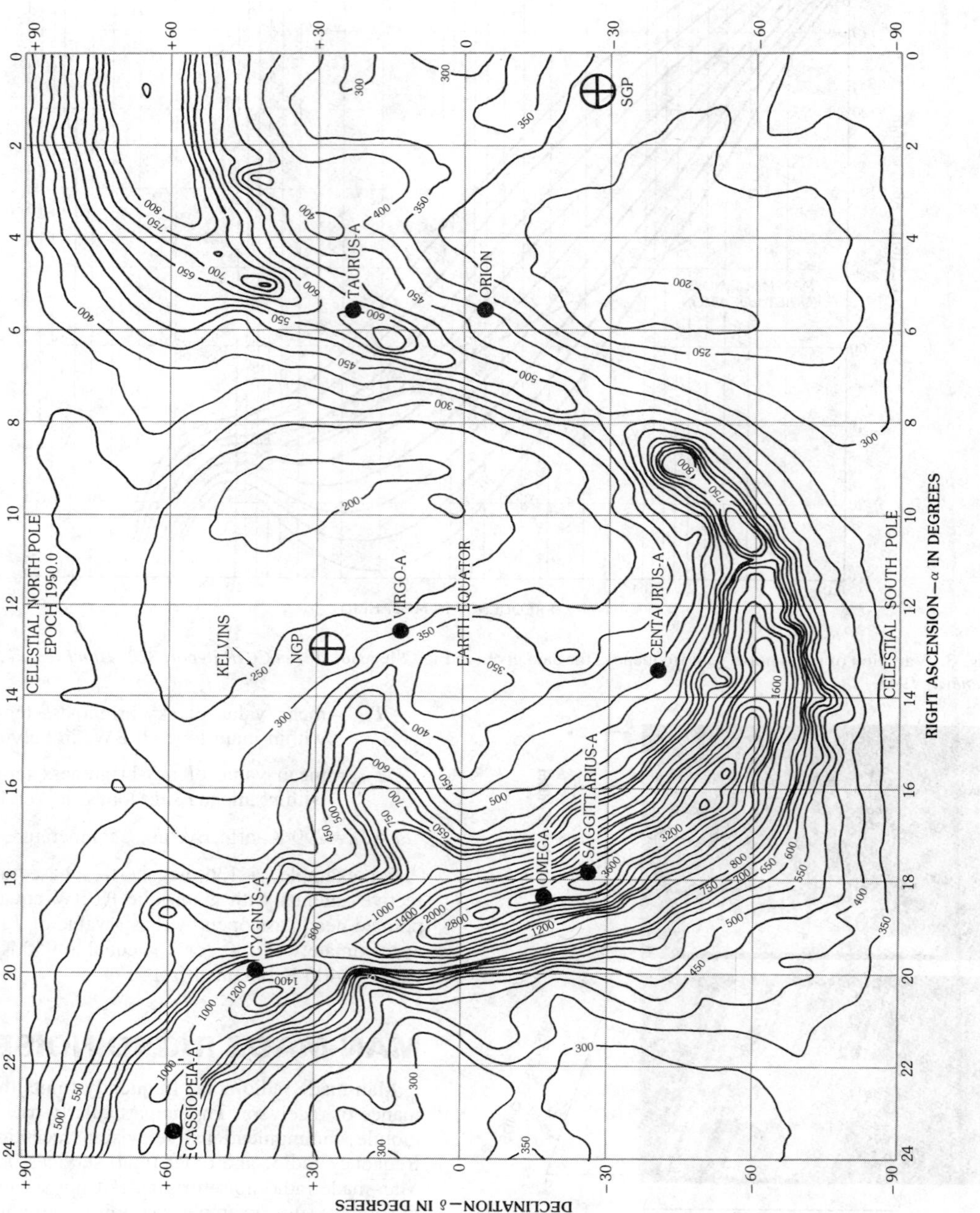

Fig. 5. Radio-sky map, 136-megahertz brightness temperature (kelvins).

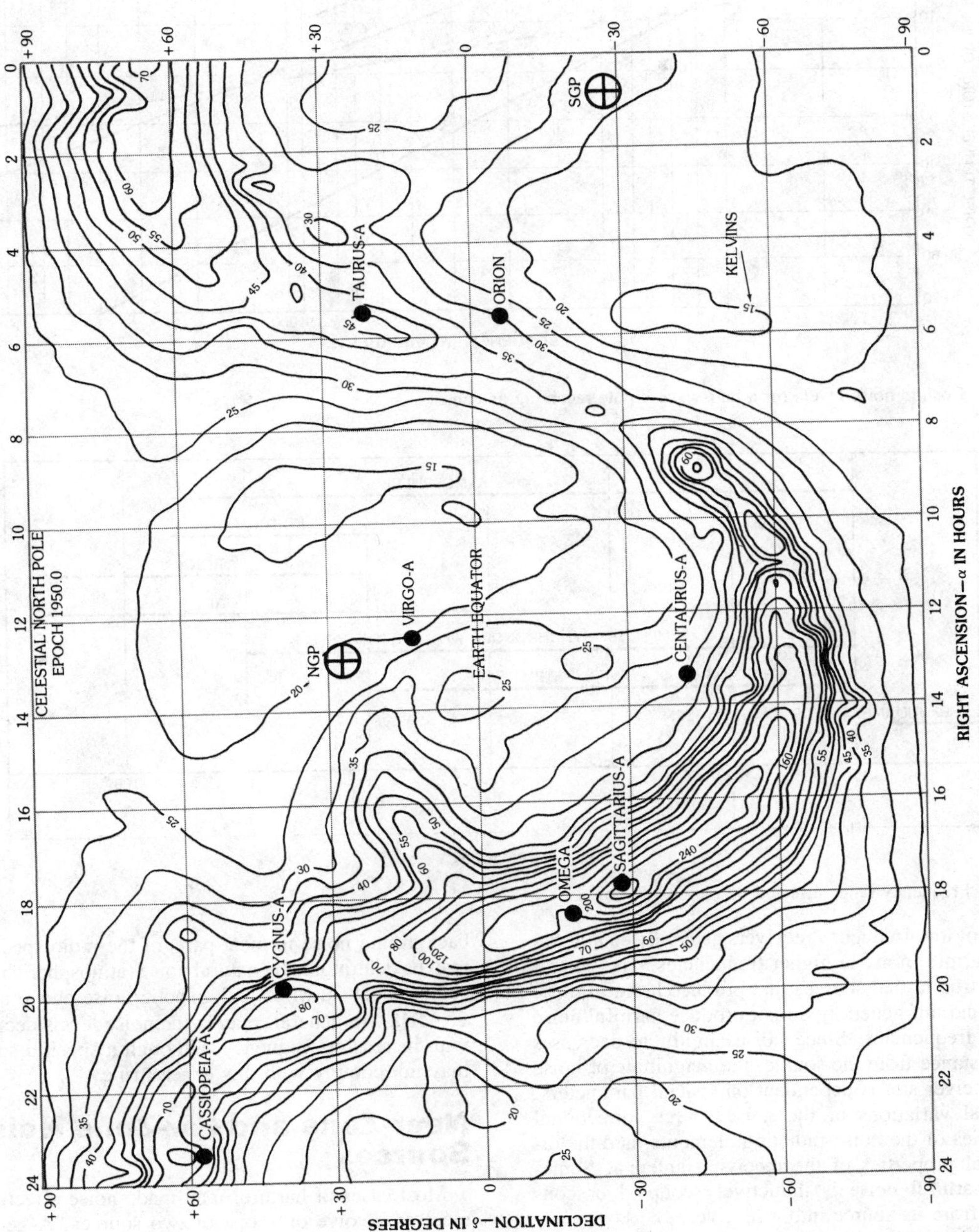

Fig. 6. Radio-sky map, 400-megahertz brightness temperature (kelvins).

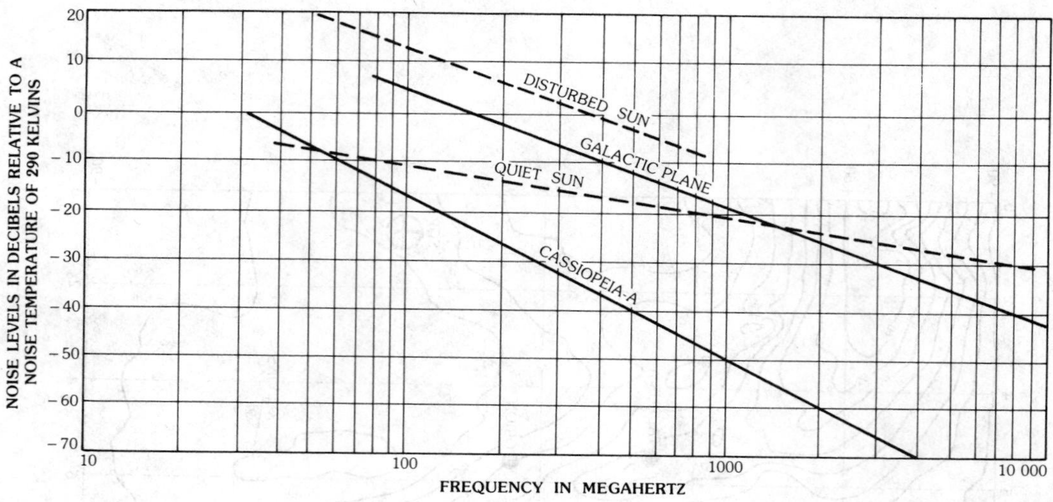

Fig. 7. Cosmic noise levels for a half-wave-dipole receiving antenna.

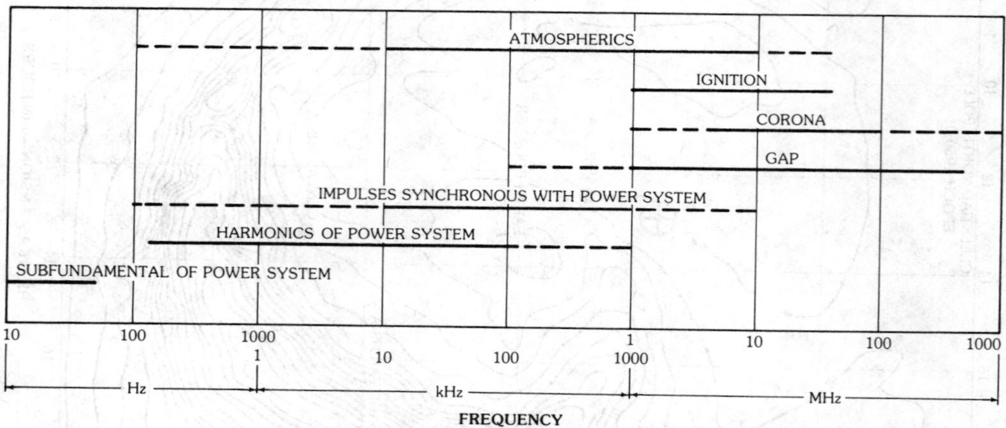

Fig. 8. Frequency range affected by man-made noise.

mance of low-frequency receivers generally do not produce harmful noise at higher frequencies. The reverse is also true in that sources that produce harmful noise at vhf and uhf generally do not produce harmful noise at low frequencies. Since noise amplitude decreases with distance from the source, the magnitude of noise at a receiver site is dependent on spacial parameters, temporal variations of the noise source, directional properties of the noise-radiating elements, and the directional properties of the receiving antenna. Sometimes harmful noise is inductively coupled or conducted from its source into a receiver.

The average levels of man-made noise are higher in urban and suburban areas than in rural areas because of the larger number of sources in areas of higher population. Fig. 9 shows median values of average noise power in urban and suburban areas. In remote quiet locations, man-made noise can be extremely low, and background noise in most parts of the radio spectrum may be determined by natural noise (atmospheric noise below about 20 MHz and galactic noise above about 20 MHz). In general, man-made noise levels decrease with increasing frequency, although a specific source may not comply with this general rule.

Near-Zone and Far-Zone Noise Sources

Most cases of harmful man-made noise affecting a receiver involve only one or two sources. A general background level of man-made noise from numerous sources is seldom encountered. Frequently, the sources of harmful noise are relatively close to a receiver. The propagation from the noise source to the receiver sometimes involves the near-zone field rather than the far-zone field of the source. For such cases, both the elec-

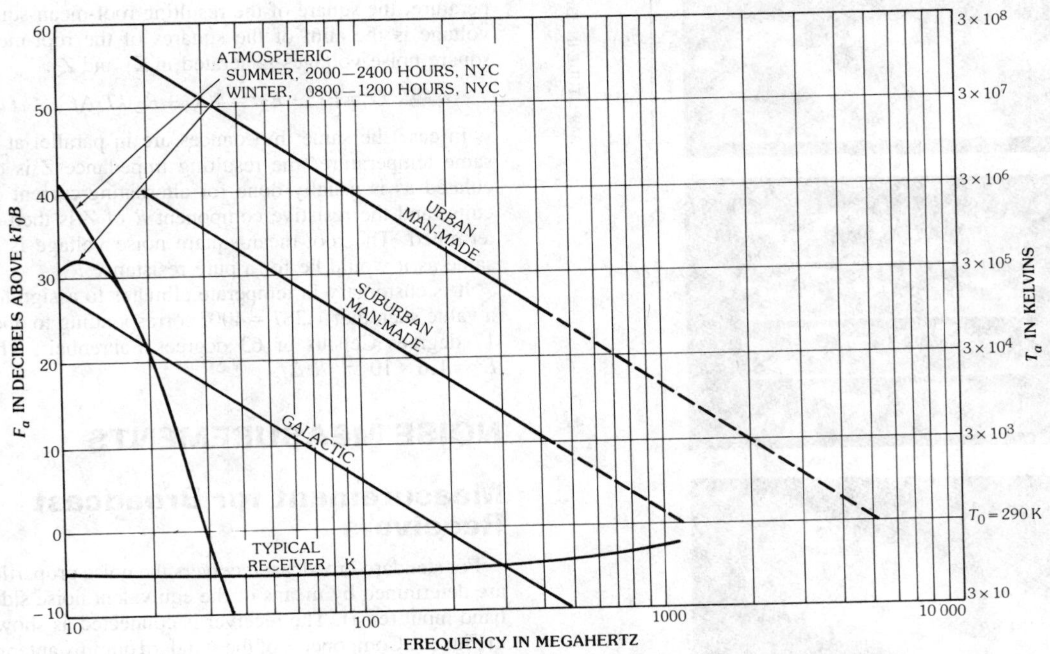

Fig. 9. Median values of average noise power expected from various sources (omnidirectional antenna near surface).

tric- and magnetic-field components of the noise source must be measured in order to define fully the ability of noise to affect receivers. Table 1 provides a general set of rules for the measurement of noise. When the path from the noise source to the receiver is greater than one wavelength, the far-field approximation applies, and the noise can be measured with either an electric-field or a magnetic-field antenna. The electric- and magnetic-field strengths can be related by the free-space impedance of $120\pi = 377$ ohms. This applies to measurements of man-made noise at frequencies above about 30 MHz. At lower frequencies, the noise source may be electrically close to the receiving antenna. At distances from the source to the affected receiver of one sixth of a wavelength or less, i.e., $<\lambda/2\pi$, the receiver is in the near field of the source. At these short distances, the ratio of E to H is no longer equal to 377 ohms, and both the electric and magnetic fields must be measured in order to define the impact of a noise source on a receiver.

TABLE 1. GENERAL RULES FOR MEASUREMENT OF NOISE

Distance Source to Sensor	Zone	Measure
Greater than λ	Distant	Either E or B
Less than λ/6	Inductive	Must measure both E and B
Between λ/6 and λ	Mixed	Recommend E and B

Power Line Noise

The temporal and spectral properties of man-made noise vary considerably from one type of source to another. Noise associated with a power line usually contains bursts of noise at intervals determined by the fundamental frequency of the power line. The temporal properties of gap noise (formed by the minute electrical breakdown between two pieces of metallic hardware exposed to a strong electric field) are shown in Fig. 10. Multiple bursts of noise occur each time the power-line voltage reaches a maximum. Individual impulses within a burst are spaced very close together (less than 1 ms apart), while the bursts are spaced 8.33 ms apart for a power line operating at 60 Hz. Individual bursts have the same amplitude. A careful inspection of the waveform in Fig. 10 shows that two gap-noise sources are present in the view.

PRECIPITATION STATIC

Precipitation static is produced by rain, hail, snow, or dust storms in the vicinity of the receiving antenna and is important chiefly at frequencies below 10 MHz. This form of interference can be reduced by eliminating sharp points from the antenna and its surroundings, and also by providing means for dissipating the charges

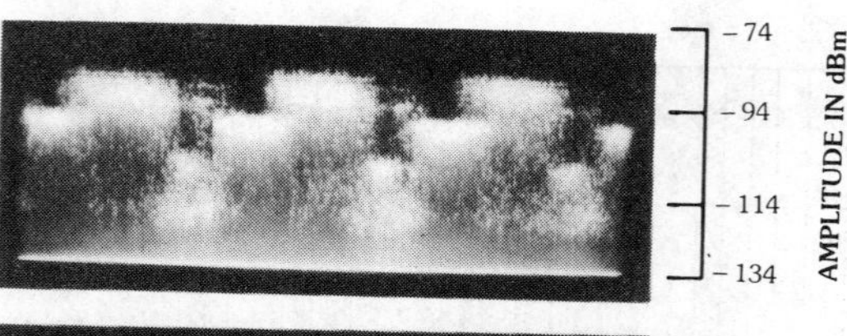

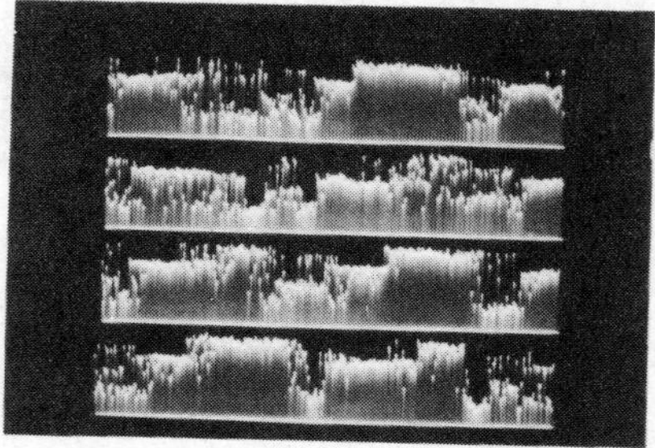

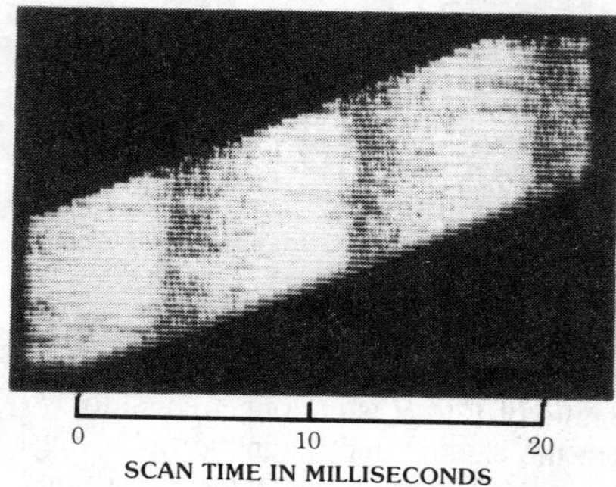

Fig. 10. Temporal structure of gap noise from 138-kV transmission line.

that build up on an antenna and on its surroundings during electrical storms.

THERMAL NOISE CALCULATIONS

Thermal noise is caused by the thermal agitation of electrons in resistances. Let R be the resistive component in ohms of an impedance Z. The mean-square value of thermal-noise voltage is given by

$$E^2 = 4RkT \cdot \Delta f$$

where,

k is Boltzmann's constant (1.38×10^{-23} joules/kelvin),

T is the absolute temperature in kelvins,

Δf is the bandwidth in hertz,

E is the root-mean-square noise voltage.

The equation given above assumes that thermal noise has a uniform distribution of power through the bandwidth Δf.

In case two impedances Z_1 and Z_2 with resistive components R_1 and R_2 are in series at the same temperature, the square of the resulting root-mean-square voltage is the sum of the squares of the root-mean-square noise voltages generated in Z_1 and Z_2:

$$E^2 = E_1{}^2 + E_2{}^2 = 4(R_1 + R_2)kT \cdot \Delta f$$

In case the same impedances are in parallel at the same temperature, the resulting impedance Z is calculated as is usually done for alternating-current circuits, and the resistive component R of Z is then determined. The root-mean-square noise voltage is the same as it would be for a pure resistance R.

It is customary in temperate climates to assign to T a value such that $1.38T = 400$, corresponding to about 17 degrees Celsius or 63 degrees Fahrenheit. Then $E^2 = 1.6 \times 10^{-20}\ R \cdot \Delta f$.

NOISE MEASUREMENTS

Measurement for Broadcast Receivers

For standard broadcast receivers the noise properties are determined by means of the equivalent noise sideband input (ensi). The receiver is connected as shown in Fig. 11. Components of the standard dummy antenna are $C_1 = 200$ picofarads, $C_2 = 400$ picofarads, $L = 20$ microhenries, and $R = 400$ ohms.

The equivalent noise sideband input is

$$(\text{ensi}) = mE_s(P'_n/P'_s)^{1/2}$$

where,

E_s = root-mean-square unmodulated carrier-input voltage,

m = degree of modulation of signal carrier at 400 hertz,

P'_s = root-mean-square signal-power output when signal is applied,

P'_n = root-mean-square noise-power output when signal input is reduced to zero.

It is assumed that no appreciable noise is transferred from the signal generator to the receiver, and that m is small enough for the receiver to operate without distortion.

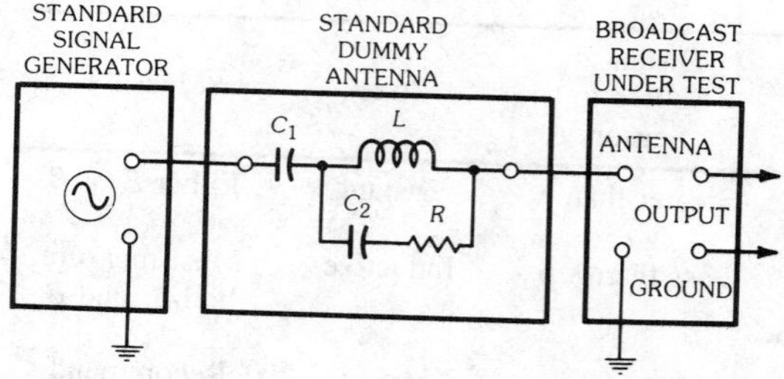

Fig. 11. Measurement of equivalent noise sideband input of a broadcast receiver.

Noise Factor of a Receiver

A more precise evaluation of the quality of a receiver as far as noise is concerned is obtained by means of its noise factor.

It should be clearly realized that the noise factor evaluates only the linear part of the receiver, i.e., up to the demodulator.

The equipment used for measuring noise factor is shown in Fig. 12. The incoming signal (applied to the receiver) is replaced by an unmodulated signal generator with R_0 = internal resistive component, E_i = root-mean-square open-circuit carrier voltage, and E_n = root-mean-square open-circuit noise voltage produced in signal generator. Then

$$E_n{}^2 = 4kT_0R_0\Delta f'$$

where,

k is Boltzmann's constant (1.38×10^{-23} J/K),

T_0 is the temperature in kelvins,

$\Delta f'$ is the effective bandwidth of the receiver (determined as below).

If the receiver does not include any other source of noise, the ratio $E_i^2/E_n{}^2$ is equal to the power carrier/noise ratio measured by the indicator:

$$E_i^2/E_n{}^2 = (E_i^2/4R_0)/kT_0\Delta f' = P_i/N_i$$

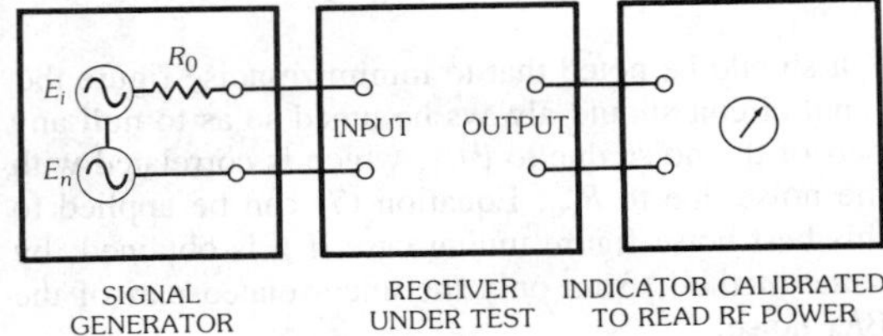

Fig. 12. Measurement of the noise factor of a receiver. The receiver is considered as a 4-terminal network. Output refers to last intermediate-frequency stage.

The quantities $E_i^2/4R_0$ and $kT_0\Delta f'$ are called the *available* carrier and noise powers, respectively.

The output carrier/noise power ratio measured in a resistance R may be considered as the ratio of an available carrier-output power P_0 to an available noise-output power N_0.

The noise factor, F, of the receiver is defined by

$$P_0/N_0 = F^{-1}(P_i/N_i)$$

$$F = (N_0/N_i)(P_0/P_i)^{-1}$$

$$= E_{i1:1}^2/4kT_0R_0\Delta f' = P_{i1:1}/kT_0\Delta f'$$

where,

P_0/P_i = available gain G of the receiver,

$P_{i1:1}$ = available power from the generator required to produce a carrier-to-noise ratio of one at the receiver output.

Noise figure is the noise factor expressed in decibels:

$$F_{dB} = 10 \log_{10} F$$

Effective bandwidth $\Delta f'$ of the receiver is

$$\Delta f' = G^{-1} \int G_f\, df$$

where G_f is the differential available gain. Generally, $\Delta f'$ is approximated to the bandwidth of the receiver between those points of the response showing a 3-dB attenuation with respect to the center frequency.

Measurement of Noise Figure With a Thermal Noise Source

For the case where the spurious responses of the receiver are negligible, receiver noise figure can be conveniently measured by using the noise output of a thermal noise source having an equivalent generator resistance equal to that specified for use with the receiver.

With the noise source off, but still possessing the correct output resistance, receiver gain is adjusted for a convenient amount of noise power output; then with the noise source on, and still possessing the correct output resistance, the noise power output is increased by a convenient power ratio (N_2/N_1). The measured noise figure is then given by

$$\text{NF} = (\text{excess})_{dB} - 10 \log[(N_2/N_1) - 1]$$

For a thermal diode operating in the temperature limited emission mode

$$(\text{excess})_{dB} = 10 \log(20R_dI_d)$$

where,

R_d is the noise source output resistance,

I_d is the diode current in amperes.

When the receiver has appreciable spurious responses, the correction factor that must be used with the above simple equation is a complex function of the spurious response ratios, and of the percentage of total internal receiver noise produced by the circuits preceding the mixer causing the spurious responses. For the simple case of no preselection and a diode mixer having negligible excess noise, 3 dB must be added to the measured noise figure to obtain the true noise figure.

A thermal noise source designed for a given generator impedance R_1 can be used to measure the noise figure of a receiver designed for a higher generator R_2 by adding a resistor ($R_2 - R_1$) between noise source and receiver input and using

$$\text{NF} = \text{NF}_{read} - 10 \log (R_2/R_1)$$

Conversion of receiver noise temperature to noise factor:

$$F = 1 + (T_R/T_0)$$

where,

T_R = receiver noise temperature in kelvins,

T_0 = 290 K,

F = noise factor of receiver (power ratio).

Conversely,

$$T_R = (F-1)T_0$$

Determination of effective noise temperature of receiving system (i.e., antenna, transmission line, and receiver):

$$T_E = T_A + (LF-1)T_0$$

where,

T_E = effective noise temperature of receiving system,

T_A = antenna noise temperature,

L = transmission line loss (power ratio),

F = noise factor of receiver (power ratio),

T_0 = 290 K.

Determination of the effective input noise power of the receiving system:

$$N_i = kBT_E$$

where,

N_i = effective input noise power of the receiving system,

k = Boltzmann's constant (1.38×10^{-23} joules/kelvin),

B = bandwidth in hertz,

T_E = effective noise temperature in Kelvins.

$$\text{dBm}_i = -198.6 + 10\log B + 10\log T_E$$

Calculation of Noise Figure

The active device can be defined for noise-figure calculations as in Fig. 13.

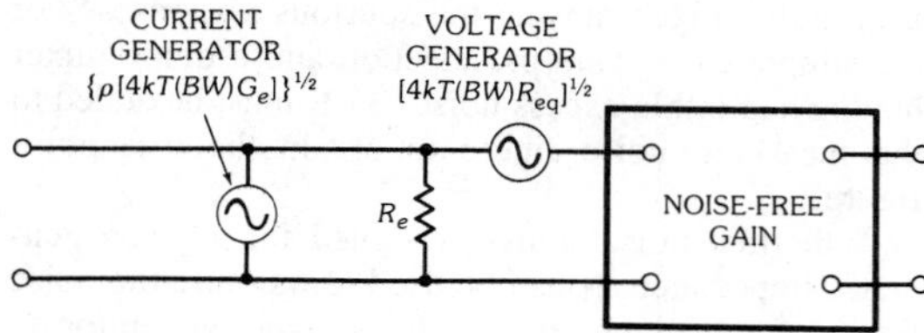

Fig. 13. Calculation of the noise figure of a receiver.

The value of R_{eq} can be obtained experimentally by measuring, with a "zero impedance" generator, the equivalent microvolts of noise V_{sc}, in a bandwidth B, in series with the input terminals, with an almost-short-circuit on the output terminals. Then R_{eq} is given by

$$R_{eq} = |V_{sc}|^2/1.64\times10^{-20}\langle BW\rangle \tag{3}$$

The value of R_e is obtained straightforwardly by input impedance measurements with a short-circuit on the output terminals.

The value of ρ can be obtained experimentally by approximately open-circuiting the input terminals at the frequency of interest with a tuned circuit of parallel resonant resistance R_0, and measuring the total equivalent microvolts of noise produced across the input terminals, with an almost-short-circuit on the output terminals. Then, assuming negligible correlation

$$\rho = [1+(R_e/R_0)]^2\{[|V_{oc}|^2/(1.64\times10^{-20}\langle BW\rangle R_e)] - (R_{eq}/R_e)\} - (R_e/R_0) \tag{4}$$

When the above characterized device is used with an input transforming circuit of parallel resonant resistance R_r, the resulting noise factor can be calculated as follows: First calculate R_1 and β from

$$R_1^{-1} = R_r^{-1} + R_e^{-1} \tag{5}$$

$$\beta = [1+\rho(R_r/R_e)]/[1+(R_r/R_e)] \tag{6}$$

In terms of the above quantities and the transformed generator resistance R_s seen by the input terminals of the active device, the resulting noise factor is given by

$$F = 1 + 2(R_{eq}/R_1) + (R_{eq}/R_s) + (R_s/R_1) \times[\beta + (R_{eq}/R_1)] \tag{7}$$

It should be noted that to minimize noise figure the input circuit should always be tuned so as to null any part of the noise due to βR_1, which is correlated with the noise due to R_{eq}. Equation (7) can be applied to this best noise figure tuning case if ρ is obtained, by some method, from only the uncorrelated part of the βR_1 noise.

This resulting noise figure is minimized when the transformed generator resistance has the value

$$R_{s\ \text{opt}} = \{(R_1R_{eq}/\beta)/[1+(R_{eq}/\beta R_1)]\}^{1/2} \tag{8}$$

and with this optimum source resistance

$$F_{\text{opt}} = 1 + 2\beta(R_{eq}/R_1)^{1/2} \times\{[1+(R_{eq}/R_1)]^{1/2} + (R_{eq}/R_1)^{1/2}\} \tag{9}$$

Noise Factor of Cascaded Networks

The overall noise factor of two networks, a and b, in cascade (Fig. 14) is

$$F_{ab} = F_a + [(F_b - 1)/G_a]$$

provided $\Delta f_b' \leq \Delta f_a'$.

The additional noise due to external sources influencing real antennas (such as galactic noise) may be accounted for by an apparent antenna temperature, bringing the available noise-power input to $kT_a\Delta f'$ instead of $N_i = kT_0\Delta f'$ (the physical antenna resistance at temperature T_0 is generally negligible in high-frequency systems). The internal noise sources contribute

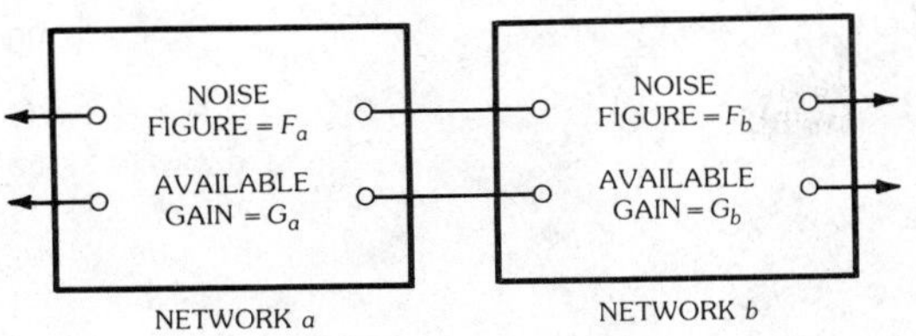

Fig. 14. Overall noise figure F_{ab} of two networks, *a* and *b*, in cascade.

$(F-1)N_i$ as before, so that the new noise factor is given by

$$F'N_i = (F-1)N_i + kT_0\Delta f'$$
$$F' = (F-1) + (T_a/T_0)$$

The average temperature of the antenna for a 6-MHz equipment is found to be 3000 kelvins, approximately. The contribution of external sources is thus of the order of 10, compared with a value of $(F-1)$ equal to 1 or 2, and becomes the limiting factor of reception. At 3000 MHz, however, values of T_a may fall below T_0.

35 Broadcasting, Cable Television, and Recording System Standards

F. M. Remley, F. Davidoff, J. F. X. Browne, and G. Jacobs

In this age of rapidly evolving communication systems, it is no longer sufficient to identify communications with the population at large exclusively with "broadcasting" in its historical form. More and more of the general public is served by, for example, cable television. During the decade of the 1980s, it is likely that direct broadcasting to the home from satellites will become a reality. Hence, this chapter not only will deal with conventional broadcasting of radio and television programs, and with cable television, but also will focus on the elements that enter into program recording for the several transmission media that serve the public. In general, the various transmission methods, and the signals that they convey to the public, must meet certain technical specifications. Much of this chapter will consist of summaries of such specifications. Sources for current versions of relevant specifications are also identified.

In the United States, broadcasting is regulated by the Federal Communications Commission (FCC), which assigns frequencies and establishes technical standards.* Three general classes of broadcast stations have traditionally been identified. These are standard broadcast stations (amplitude modulation in the band 535–1605 kHz), fm broadcast stations (frequency modulation in the band 88–108 MHz), and television broadcast stations (operating in the bands 54–72, 76–88, 174–216, and 470–806 MHz with vestigial-sideband amplitude modulation of the visual carrier and frequency modulation of the aural carrier). Technical specifications for these broadcast services are summarized in this chapter. Cable television systems are also subject to FCC technical control; the channel assignments and other specifications for this service are also summarized in this chapter.

This chapter also discusses certain technical aspects of international broadcasting, a service utilizing frequencies between 5950 and 26 100 kHz in accordance with international agreements.

Other technical information related to broadcasting is covered under auxiliary services, intercity transmission, and terminal facilities.

This chapter provides a comprehensive reference for standards applicable to audio recording and video recording. United States standards are listed, and reference is made to comparable international standards (listed separately in a later part of the chapter). Digital recording systems are also described.

The current situation in the rapidly evolving field of digital television systems is summarized in a separate section, which describes the present international agreements in this field.

A listing of revelant standards promulgated by the CCIR, ISO, and IEC concludes the chapter.

* Federal Communications Commission Rules and Regulations, Volume III, etc. These documents are available from the Superintendent of Documents, U.S. Government Printing Office, Washington, D.C. 20402.

BROADCAST AND CABLE TRANSMISSION SYSTEMS

Standard Broadcasting*

Standard-broadcast stations are licensed for operation on channels spaced by 10 kHz and occupying the band from 535 to 1605 kHz. The major classifications are clear channel, regional channel, and local channel. A clear-channel station renders service over wide areas and is protected from objectionable interference within its prescribed primary and secondary service areas. A regional-channel station renders primary service to larger cities and the surrounding rural areas; a channel may be occupied by several stations, and the primary service area may be limited by interference. A local station is designed to render service primarily to a city or town and its nearby suburban or rural areas. Its primary service area may also be limited by interference.

Field-Strength Requirements

Primary Service:

- City business, factory areas—10 to 50 millivolts/meter, ground wave
- City residential areas—2 to 10 millivolts/meter, ground wave
- Rural, all areas during winter or northern areas during summer—0.1 to 0.5 millivolt/meter, ground wave
- Rural, southern areas during summer—0.25 to 1.0 millivolt/meter, ground wave

Secondary Service: All areas having sky-wave field strength equal to or greater than 500 microvolts/meter for 50% or more of the time.

For stations employing a directional antenna, all determinations of service and interference are based on the inverse field of a "standard pattern" for that station. When applied to nighttime operation, this includes the radiation pattern in the horizontal plane and at angles above the horizontal plane (vertical radiation pattern).

Table 1 outlines generally the protected contours and permissible interference for the various classes of stations. There are additional details and some exceptions in Sections 73.21–73.29 and 73.181–73.190 of Part 73 of the FCC Rules and Regulations.

Coverage Data—Figs. 1, 2, and 3 show computed values of ground-wave field strength as a function of the distance from the transmitting antenna. These are used to determine coverage and interference. They were computed for the frequencies indicated, for a dielectric constant equal to 15 for land and 80 for sea water (referred to air as unity), and for the surface conductivities noted. The curves are for radiation from a short vertical antenna at the surface of a uniformly

* FCC Rules and Regulations, Part 73, Subpart A.

TABLE 1. CLASSIFICATION OF STANDARD-BROADCAST STATIONS

Class of Channel	Class of Station	Permissible Power (kW)	Signal-Intensity Contour of Area Protected from Objectionable Interference (microvolts/meter)		Permissible Interfering Signal on Same Channel (microvolts/meter)	
			Day[1]	Night	Day[1]	Night[3]
Clear	I-A	50	SC = 100 AC = 500	SC = 500[2] AC = 500[1]	5	25
	I-B	10–50	SC = 100 AC = 500	SC = 500[2] AC = 500[1]	5	25
	II-A	0.25–50 day 10–50 night	500	500[1]	25	25
	(II-B, II-D)	0.25–50	500	2500[1]	25	125
	(II-B, II-D)[4]	0.25–1	500	10 000	25	500
Regional	III-A	1–5	500	2500[1]	25	125
	III-B	0.5–5 day 0.5–1 night	500	4000[1]	25	200
Local	IV	0.25–1 day 0.25 night	500	not prescribed	25	not prescribed

Notes:
SC—same channel, AC—adjacent channel.
[1] Ground wave.
[2] 50% sky wave.
[3] 10% sky wave.
[4] Applies to those Class II-B and II-D stations described in Part 73 of FCC Rules and Regulations.

conductive spherical earth with an antenna power and efficiency such that the inverse-distance field is 100 millivolts/meter at one mile. (Twenty such charts, for frequencies at intervals throughout the standard broadcast band, are contained in Section 73.184 of the FCC Rules and Regulations.) Fig. 4 shows the estimated effective field for vertical omnidirectional antennas of various heights. Figs. 5 and 6 show the effective ground conductivity for various parts of the U.S. and Canada, and Fig. 7 shows the sky-wave fields for 10% and 50% of the time.

Transmission System Requirements

Modulation: 85% to 95% at authorized operating power without exceeding required audio-frequency distortion limits.

Audio-Frequency Response: Transmission characteristics between 100 and 5000 Hz within 2 dB, referenced 1000 Hz, from microphone to antenna output.

Audio-Frequency Distortion: Harmonics less than 5% (voltage measurement of arithmetic sum or root sum squared) from 0 to 84%; not over 7.5% when modulating 85% to 95% from 50 to 7500 Hz.

System Noise: At least 45 dB, unweighted root-sum squared, below 400 Hz 100% modulation for frequencies from 30 to 20 000 Hz.

Carrier Amplitude: Carrier shift less than 5% at any level of modulation percentage.

Out-of-Band Radiation: Referenced to unmodulated carrier level, attenuation of emissions removed from the carrier 15 kHz to 30 kHz greater than 25 dB, 30 kHz to 75 kHz greater than 35 dB, and more than 75 kHz 43 + 10 $\log_{10}$ (power in watts) or 80 dB whichever is less.

AM Stereo Transmission—Three basic emission methods have been proposed for adding stereo to the standard am broadcast transmission system. The three methods are:

Mixed Mode: The carrier is amplitude modulated with the (L + R) signal and phase or frequency modulated with the (L − R) signal.

Independent Sideband: The upper and lower side-

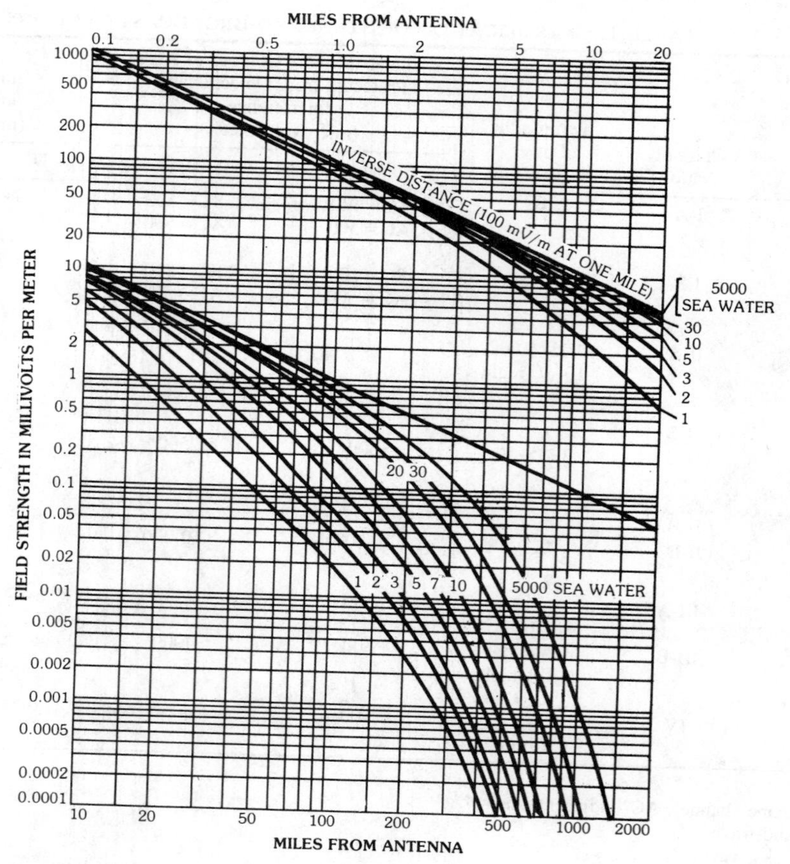

Fig. 1. Ground-wave field strength plotted against distance. Computed for 550 kilohertz. Dielectric constant = 15. Ground-conductivity values are in millimhos/meter.

bands of the carrier are modulated with the L and R signals, respectively.

Quadrature: Two phase-locked carriers are amplitude modulated and combined at a fixed phase angle.

The marketplace approach has been employed to select the method, and the manufacturer's technique will become the standard.

Frequency-Modulation Broadcasting*

Frequency-modulation (fm) broadcasting stations are authorized for operation on 101 allocated channels, each 200 kHz wide, extending consecutively from channel 200 on 87.9 MHz to channel 300 on 107.9 MHz. Commercial broadcasting is authorized on channels 221 (92.1 MHz) through 300. Noncommercial educational broadcasting is permitted on any channel, but channels 200 through 220 are reserved for such use (see Part 73 of the FCC Rules and Regulations).

Station Service Classification—Class A stations render service primarily to a relatively small community, city, or town and the rural surroundings. The coverage will not exceed the equivalent of 3 kW effective radiated power† at an antenna height above average terrain‡ of 300 feet. Minimum effective radiated power is 100 watts. Class A channels are 221, 224, 228, 232, 237, 240, 244, 249, 252, 257, 261, 265, 269, 272, 276, 280, 285, 288, 292, and 296. (A proposal to change the Rules to permit Class A station operation on any channel was pending as of December, 1982.)

* FCC Rules and Regulations, Part 73, Subparts B and C.

† Effective radiated power is the product of antenna gain and antenna input power. Antenna input power is transmitter power minus transmission-line loss.

‡ Average terrain is defined as the average of the elevations between 2 and 10 miles from the antenna along eight radials evenly spaced by 45°.

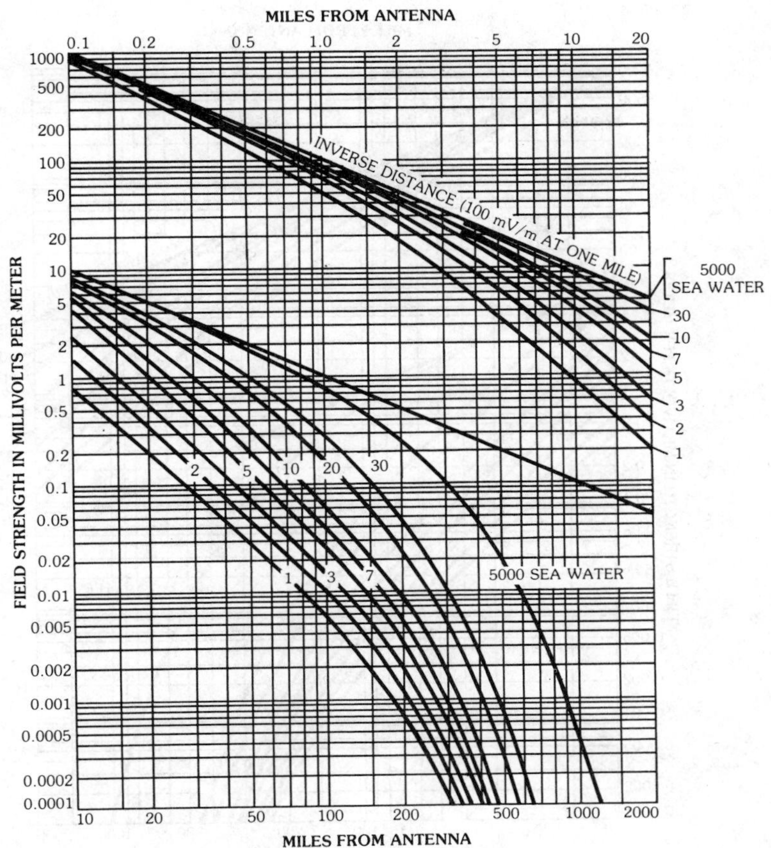

Fig. 2. Ground-wave field strength plotted against distance. Computed for 1000 kilohertz. Dielectric constant = 15. Ground-conductivity values are in millimhos/meter.

Class B stations render service to a large community. These stations operate in Zone I or Zone IA,§ and their coverage will not exceed the equivalent of an effective radiated power of 50 kW at an antenna height of 500 feet above average terrain. Minimum effective radiated power is 5 kW.

Class C stations render service to a large community. These stations operate in Zone II,** and their coverage will not exceed the equivalent of an effective radiated power of 100 kW at an antenna height of 2000 feet above average terrain. Minimum effective radiated power is 25 kW.

Class B and C stations are authorized on channels not designated for Class-A use only.

Class D stations are noncommercial stations limited to a maximum transmitter output power of 10 watts. These stations are authorized on a secondary basis to all other classes of stations and are intended to render an unprotected service to a very small area.

If the actual antenna height above average terrain for a station operating on channels 217 through 300 exceeds the maximum for the class to which the station belongs, the effective radiated power must be reduced to limit the coverage to the maximum permitted for that class (see Section 73.211 and Fig. 3 of Section 73.333).†† Stations operating on channels 201 through 216 are not necessarily limited to maximum height and power restrictions.

Channel Availability—Channels 221 through 300 are assigned to specific communities in accordance with the Table of Assignments (see Section 73.202). The number and type of channels assigned are based on population and limits imposed by cochannel and adjacent-channel interference. The Table of Assignments may be amended to add channels upon approval of a petition showing need and conformance with the

§ Generally speaking, Zone I is the northeastern part of the U.S., and Zone IA is Puerto Rico, the Virgin Islands, and California south of 40° latitude. For exact boundaries, refer to Section 73.205 of the FCC Rules and Regulations.

** Zone II includes Alaska, Hawaii, and other parts of the U.S. not in Zone I or IA.

†† Section numbers refer to the FCC Rules and Regulations.

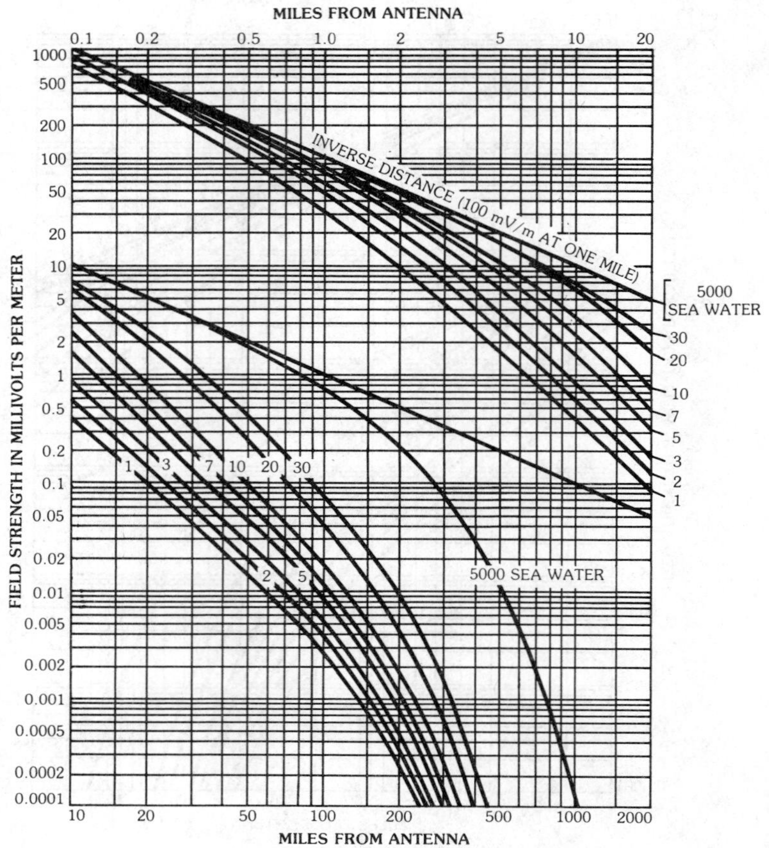

Fig. 3. Ground-wave field strength plotted against distance. Computed for 1600 kilohertz. Dielectric constant = 15. Ground-conductivity values are in millimhos/meter.

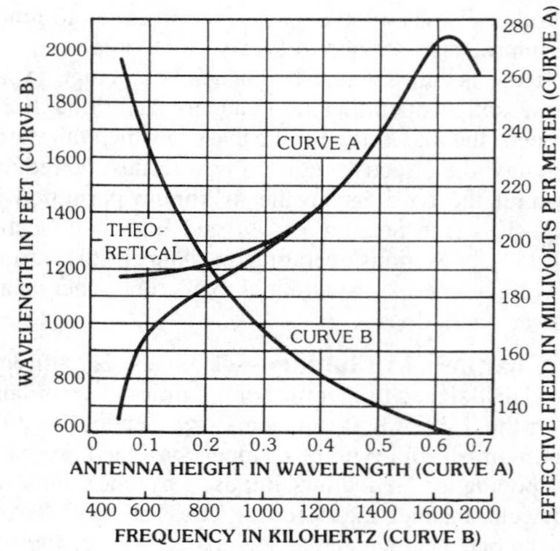

Fig. 4. Effective field at 1 mile for 1 kilowatt (curve *A*). Use for simple omnidirectional vertical antenna with ground system of at least 120 radials λ/4. *From FCC Rules and Regulations, Vol. III, Part 73, March 1980.*

separation standards shown in Table 2. Noncommercial stations operating on channels 217–220 must conform with Table 2 with respect to stations operating on channels 221–223. Stations operating on channels 201 through 216 are authorized based on mutual interference standards in lieu of a Table of Assignments. These standards are expressed in terms of the following interference ratios:

Cochannel: 10:1, or 20 dB
1st Adjacent Channel: 2:1, or 6 dB
2nd Adjacent Channel: 1:10, or −20 dB
3rd Adjacent Channel: 1:100, or −40 dB

The ratios are the magnitude of the field strength from the undesired station, determined from the *F*50,10 curves (see Section 73.333, Fig. 1a), at the limit of the coverage area of the desired station (see "Coverage" below).

Channel 200 is available only for Class D use in areas not served by a television broadcast station operating on channel 6. Stations on channels 201 through 220 are subject to maximum power restrictions if lo-

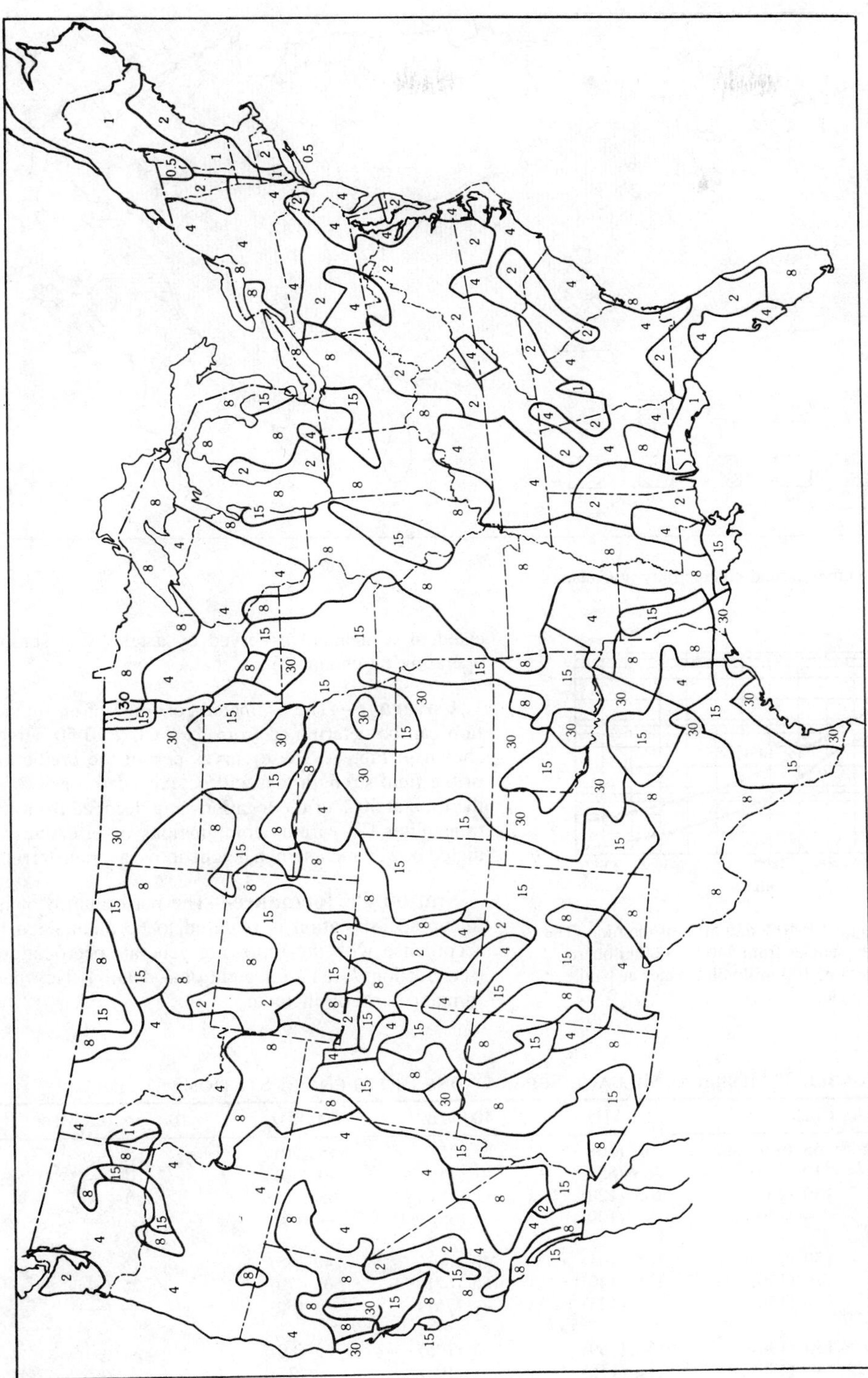

Fig. 5. Estimated effective ground conductivity in the United States. The numbers are in millimhos/meter. The conductivity of sea water (not shown) is assumed to be 5000 millimhos/meter.

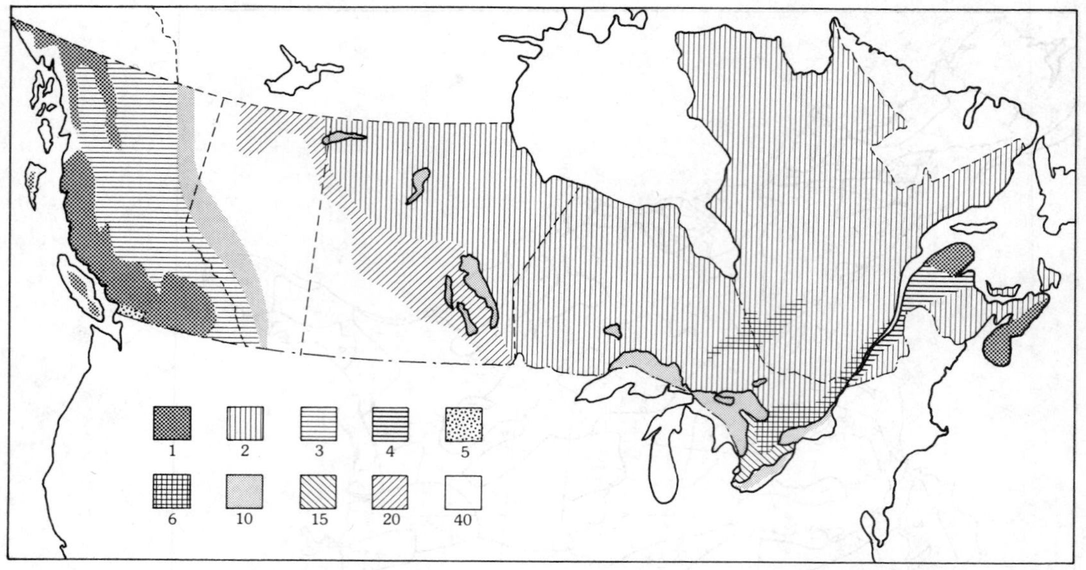

Fig. 6. Estimated effective ground conductivity in Canada.

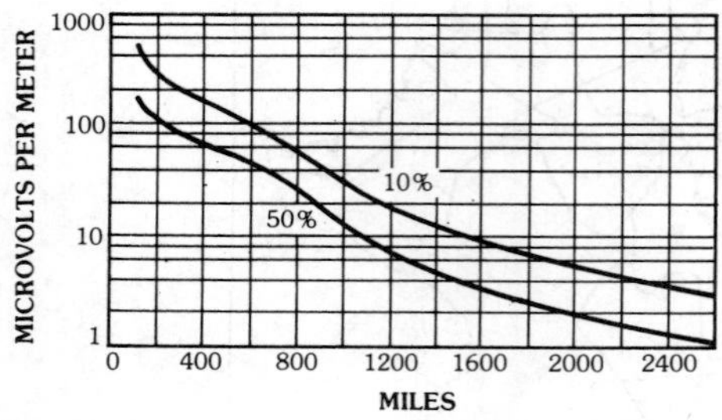

Fig. 7. Sky-wave signals for 10% and 50% of the time. The sky-wave range for frequencies from 540 to 1600 kilohertz is based on a radiated field of 100 millivolts/meter at 1 mile at the pertinent vertical angle.

cated in or near areas served by a television station operating on channel 6.

Coverage—The estimated coverage of an fm station can be determined from the FCC *F*50,50 curves shown in Fig. 8. These curves permit the prediction of the field strength that will be exceeded for 50% of the time at 50% of the locations at a distance from the transmitter for various combinations of effective radiated power and antenna height above average terrain.

Station Performance—The performance of an fm broadcast station is required to be maintained in accordance with the following general specifications (see Section 73.317 for additional details). Key performance criteria include:

TABLE 2. MINIMUM MILEAGE SEPARATIONS BETWEEN FM STATIONS*

Relation	Cochannel	200 kHz	400 kHz	600 kHz	10.6 or 10.8 MHz
A to A	65 (90)	40 (50)	15 (25)	15 (20)	5 (—)
A to B	110 (135)	65 (85)	40 (45)	40 (40)	10 (—)
A to C	130 (150)	105 (120)	65 (75)	65 (70)	20 (—)
A to C1	— (150)	— (100)	— (65)	— (60)	— (—)
B to B	150 (155)	105 (105)	40 (60)	40 (45)	15 (—)
B to C	170 (170)	135 (140)	65 (85)	65 (70)	25 (—)
B to C1	— (170)	— (125)	— (75)	— (60)	— (—)
C to C	180 (190)	150 (160)	65 (105)	65 (80)	30 (—)
C to C1	— (190)	— (155)	— (105)	— (75)	— (—)
C1 to C1	— (190)	— (140)	— (90)	— (70)	— (—)

* Canada-U.S. separations are in parentheses.

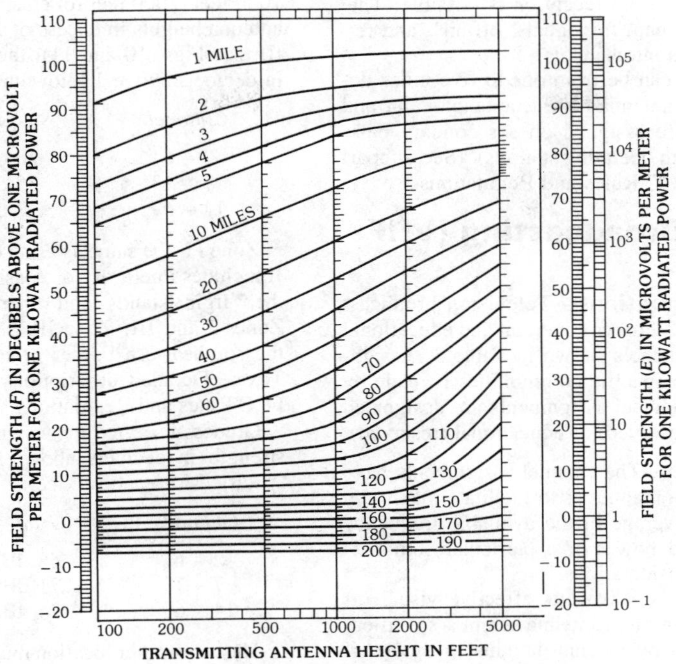

Fig. 8. $F(50,50)$ for fm channels, showing estimated field strength exceeded at 50% of the potential receiver locations for at least 50% of the time at a receiving antenna height of 30 feet. *From FCC Rules and Regulations, Vol. III, Part 73, March 1980.*

Modulation Capability: ±75 kHz deviation of carrier for 100% modulation.

Audio-Frequency Response: 50–15 000 Hz within limits shown in Section 73.333, Fig. 2 for 75 microseconds preemphasis.

Audio Distortion: Less than 2.5% to 3.5% depending on modulation frequency.

AM Noise: 50 dB below 100% amplitude modulation.

FM Noise: 60 dB below 100% frequency modulation.

Polarization: Typically right-hand circular, horizontal component required, elliptical or other circular modes acceptable.

Stereophonic Transmission—Most fm stations broadcast in a stereophonic mode that is compatible with monophonic receivers. The stereophonic signal consists of a main channel and a subchannel. The main channel is modulated by the sum of the left (L) and right (R) stereo signals. The subchannel consists of the sidebands of a 38-kHz suppressed carrier that is modulated with an L minus R (difference) signal. Since the modulating frequencies can be as high as 15 kHz, the sidebands occupy the baseband spectrum from 23 kHz to 53 kHz (Fig. 9). A pilot carrier of 19 kHz is also transmitted to allow the receiver to generate a phase-locked 38-kHz carrier in order to demodulate the L − R signal. A matrix in the stereo receiver recovers the L and R signals by algebraic operation on the L plus R and L minus R signals. A monophonic receiver demodulates only the L plus R (main channel) signal.

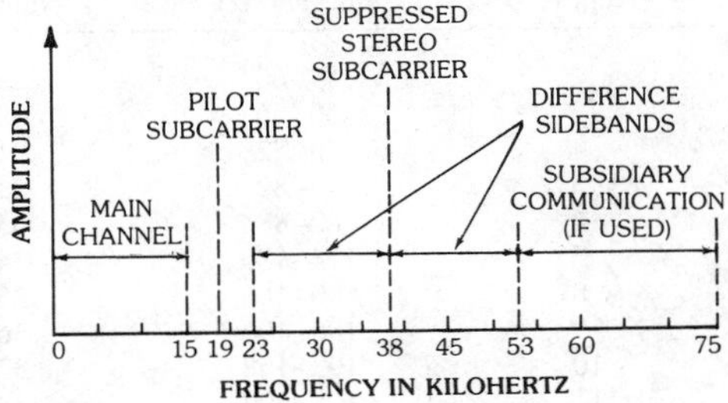

Fig. 9. Resulting stereophonic frequency spectrum.

Other FM Services—A Subsidiary Communications Authorization (SCA) permits the addition of one or more subcarriers on the main channel; these subcarriers may be used for the transmission of voice or data signals related or unrelated to the broadcast-station operation. The subcarrier is amplitude modulated and is bandwidth limited. The total modulation of the main carrier by the sum of all subcarriers may not exceed 10% (assuming stereophonic transmission). (See Section 73.319.)

FM Translators—The FCC permits the use of very low power (1 watt east of the Mississippi and 10 watts west) translators to rebroadcast the signals of fm stations in areas where no reception is possible. The stations receive the main fm signal ''off air'' and rebroadcast it on different channels.

Coverage ranges can be from one to five miles depending on power, antenna height, antenna gain, and terrain. Channels are assigned on a secondary basis (noninterference with operating stations). (See Subpart L, Part 74 of the FCC Rules and Regulations.)

Television Broadcasting (VHF and UHF)

Channel Designations—Television broadcast stations are authorized for commercial and educational operation on the channels shown in Table 3. Assignment of channels to specific communities is made by the FCC, and the channel assignments are designated as commercial (unreserved) or educational (reserved).

Coverage Data—The channel assignments have been made in such a manner as to facilitate maximum interference-free coverage in the available frequency bands. The radiated power of a particular station is fixed by several considerations.

Minimum power is 100 watts effective visual radiated power. No minimum antenna height is specified.

Except as limited by antenna heights in excess of 1000 feet (2000 feet for channels 14–69) in Zone I and antenna heights in excess of 2000 feet in Zones II and III (see Figs. 10 and 11), the maximum visual power in decibels above 1 kilowatt (dBk) is:

Channel	*Maximum Power*
2 – 6	20 dBk = 100 kilowatts
7 – 13	25 dBk = 316 kilowatts
14 – 69	37 dBk = 5000 kilowatts

Zone I is the same as Zone I for fm allocations. Zone II includes Puerto Rico, Alaska, the Hawaiian Islands, the Virgin Islands, and other parts of the U.S. not in Zones I and III. Zone III is essentially a strip along the southeastern border of the U.S. from Florida to Texas. Detailed descriptions of the zones are in the FCC Rules and Regulations, Vol. III, Section 73.609.

Grades of service are designated A and B. The signal strength in decibels above 1 microvolt/meter (dBu) specified for each service is:

Channel	*Grade A*	*Grade B*
2 – 6	68 dBu	47 dBu
7 – 13	71 dBu	56 dBu
14 – 69	74 dBu	64 dBu

The transmitter location must be so chosen that, with

TABLE 3. NUMERICAL DESIGNATION OF TELEVISION CHANNELS

Channel Number	Band (megahertz)	Channel Number	Band (megahertz)	Channel Number	Band (megahertz)
2	54–60	29	560–566	57	728–734
3	60–66	30	566–572	58	734–740
4	66–72	31	572–578	59	740–746
5	76–82	32	578–584	60	746–752
6	82–88	33	584–590	61	752–758
7	174–180	34	590–596	62	758–764
8	180–186	35	596–602	63	764–770
9	186–192	36	602–608	64	770–776
10	192–198	37	608–614	65	776–782
11	198–204	38	614–620	66	782–788
12	204–210	39	620–626	67	788–794
13	210–216	40	626–632	68	794–800
14	470–476	41	632–638	69	800–806
15	476–482	42	638–644	70*	806–812
16	482–488	43	644–650	71*	812–818
17	488–494	44	650–656	72*	818–824
18	494–500	45	656–662	73*	824–830
19	500–506	46	662–668	74*	830–836
20	506–512	47	668–674	75*	836–842
21	512–518	48	674–680	76*	842–848
22	518–524	49	680–686	77*	848–854
23	524–530	50	686–692	78*	854–860
24	530–536	51	692–698	79*	860–866
25	536–542	52	698–704	80*	866–872
26	542–548	53	704–710	81*	872–878
27	548–554	54	710–716	82*	878–884
28	554–560	55	716–722	83*	884–890
		56	722–728		

* The frequencies between 806 and 890 MHz, formerly allocated to television broadcasting, are now allocated to the land mobile services. Operation, on a secondary basis, of some television translators may continue on these frequencies.

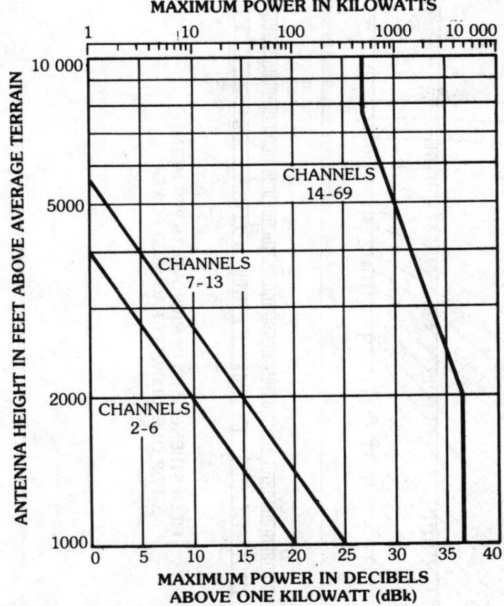

Fig. 10. Maximum television-station power versus antenna height for Zone I. *From FCC Rules and Regulations, Vol. III, Part 73, March 1980.*

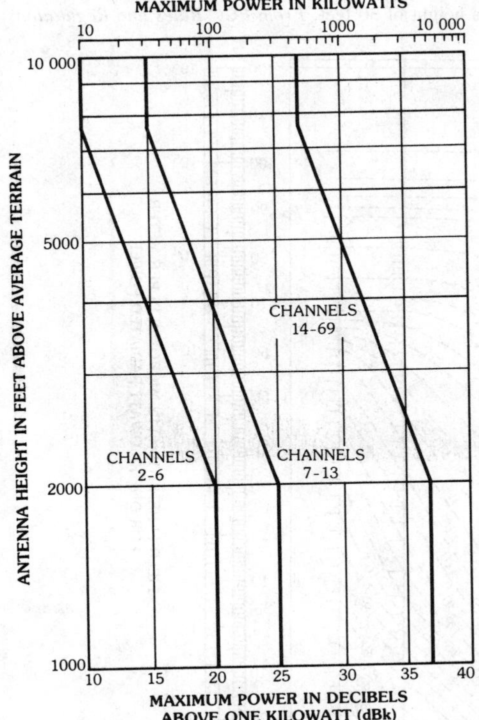

Fig. 11. Maximum television-station power versus antenna height for Zones II and III. *From FCC Rules and Regulations, Vol. III, Part 73, March 1980.*

the effective radiated power and antenna height used, the following minimum field strength in decibels above 1 microvolt/meter will be provided over the principal community to be served (sometimes referred to as "City Grade").

Channel	*Signal*
2 – 6	74 dBu
7 – 13	77 dBu
14 – 83	80 dBu

The curves of Figs. 12 and 13 give estimated field strengths for the television channels at different heights and powers. The antenna height is the height of the radiation center of the antenna above average terrain. Average terrain is determined by the elevations between 2 and 10 miles from the antenna site, taken along eight radials separated by 45° in azimuth. Effective radiated power is the product of the antenna gain and the antenna input power. Antenna input power is the peak visual output power of the transmitter less transmission-line and diplexer losses. The procedures to be followed in determining the effective radiated power to be used in the prediction of coverage are detailed in Section 73.684 of the FCC Rules and Regulations.

Directional antennas may be employed to improve service. The ratio of maximum to minimum radiation in the horizontal plane shall not exceed 10 dB for channels 2–13, and 15 dB for channels 14–69 if the transmitter power is more than 1 kW. There is no restriction for channels 14–69 if the transmitter power is 1 kW or less.

Transmission Standards—The standards for television transmission in the U.S., as defined by the FCC, are:

Channel Width: 6 MHz.

Picture Carrier Location: 1.25 MHz ± 1000 Hz above lower boundary of the channel.

Aural Center Frequency: 4.5 MHz ± 1000 Hz above visual carrier.

Polarization of Radiation: Horizontal required; right-hand circular optional.

Modulation: Amplitude-modulated composite picture and synchronizing signal on visual carrier, together with frequency-modulated audio signal on aural carrier. (See Figs. 14 and 15.)

Scanning Lines: 525 lines/frame interlaced two to one.

Scanning Sequence: Horizontally from left to right, vertically from top to bottom.

Horizontal Scanning Frequency: 2/455 times chrominance subcarrier frequency (15 734 Hz).

Vertical Scanning Frequency: 2/525 times the horizontal scanning frequency (59.94 Hz).

Chrominance Subcarrier Frequency: 3.579545 MHz ± 10 Hz.

Polarity of Transmission: Negative—a decrease in initial light intensity causes an increase in radiated power.

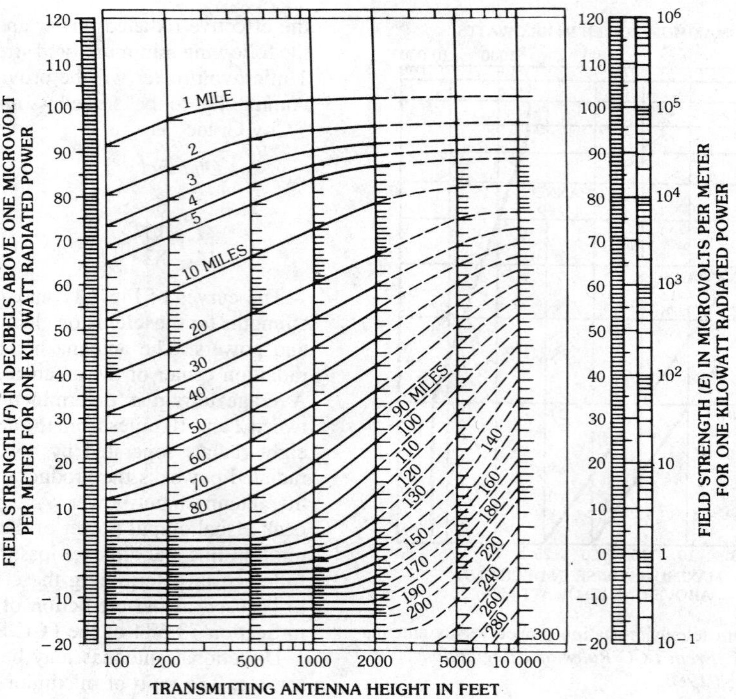

Fig. 12. $F(50,50)$ for television channels 2–6 and 14–69, showing estimated field strength exceeded at 50% of the potential receiver locations for at least 50% of the time at a receiving antenna height of 30 feet. *From FCC Rules and Regulations, Vol. III, Part 73, March 1980.*

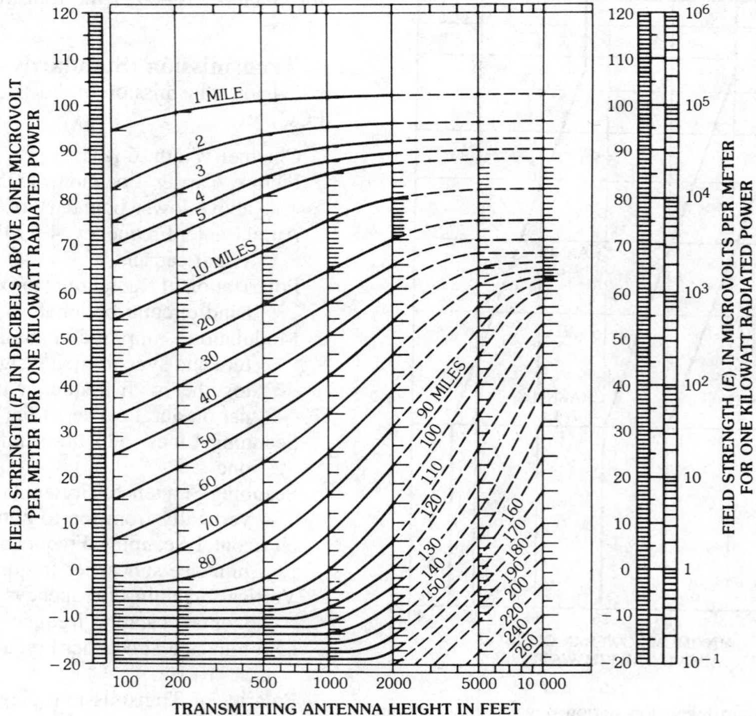

Fig. 13. $F(50,50)$ for television channels 7–13, showing estimated field strength exceeded at 50% of the potential receiver locations for at least 50% of the time at a receiving antenna height of 30 feet. *From FCC Rules and Regulations, Vol. III, Part 73, March 1980.*

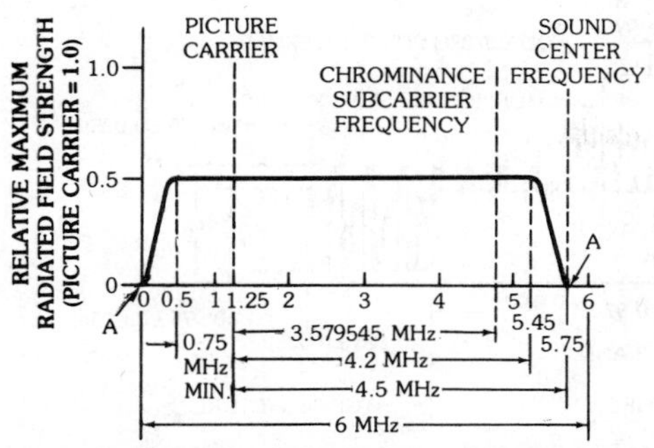

Fig. 14. Radio-frequency amplitude characteristics of television picture transmission. (Drawing not to scale.)

Transmitter Brightness Response: For luminance signal, radio-frequency output varies in an inverse logarithmic relation to the brightness of the scene.

Aural-Transmitter Power: Maximum radiated power is 20% (minimum, 10%) of peak visual transmitter power.

For color transmission, the luminance component shall be transmitted as amplitude modulation of the picture carrier and the chrominance components as amplitude-modulation sidebands of a pair of suppressed subcarriers in quadrature (Fig. 16).

The interval beginning with line 17 and continuing through line 20 of the vertical-blanking interval of each field may be used for the transmission of data, test signals, and cue and control signals. Test signals may include signals designed to check the performance of the overall transmission system or its individual components. Test signals or cue and control signals may not be transmitted during that portion of each line devoted to horizontal blanking. Line 19 in each field may be used only for transmission of the standard vertical interval reference (VIR) signal. Line 21 may be used for transmission data such as closed captioning for the deaf. Lines 15–18 may be used for data and text transmission (Teletext).

Multiplexing of the aural carrier may be employed for the purpose of transmitting telemetry and alerting signals from the transmitter site to the control point of a television broadcast station.

Visual Performance—For color transmission, between 2.1 and 4.1 MHz the amplitude shall be within ±2 dB of its value at 3.58 MHz, and the amplitude shall be no more than 4 dB below that value at 4.18 MHz. For modulating frequencies of 1.25 MHz or greater, lower-sideband radiation must be 20 dB below carrier level.

Aural Transmitter

Modulation: A frequency swing of ±25 kHz is defined as 100% modulation.

Audio-Frequency Response: 50 to 15 000 Hz within limits of 75-microsecond preemphasis curve.

Audio-Frequency Distortion: Maximum combined audio-frequency harmonics (rms) at the system output shall be less than 3.5% depending on frequency.

FM Noise: −55 dB.

Cable Television

Cable television (catv) systems were originally called community antenna television systems. They had their origin in mountainous regions of Pennsylvania where "off-air" reception of broadcast television stations was difficult if not impossible. Beginning with crude antennas and amplifiers that delivered a few channels to homes in sparsely populated valleys, catv systems have evolved into a broad-band technology utilizing coaxial cables of 3/4-inch and 7/8-inch diameter to carry signals in the spectrum from 5 MHz to 450 MHz.

Some modern systems employ multiple cables to separate subscriber and institutional networks, each with two-way capability. Cable tv systems are developing carriers of more than entertainment programming from broadcast stations. It is not unusual for a modern urban cable system to carry more television entertainment signals emanating from satellites and local sources than from "off-air" broadcast stations.

Converters are used to change nonstandard channel frequencies to a standard vhf channel (usually between channels 2 and 6 inclusive) so that ordinary television receivers can demodulate the signals. Newer receivers are being manufactured with tuners capable of tuning more than one hundred channels. The cable channel frequency assignments generally used are given in Table 4.

Bidirectional cable systems normally employ a low-band split (i.e., downstream channels above 50 MHz and upstream channels below 50 MHz) or a midband split with the division occurring at 150 MHz. As the upper limit of usable cable-system frequencies increases, the frequency where the split occurs is also raised.

Cable systems are not regulated to any great extent by the FCC,* and the technology has far exceeded the minimum requirements set by the Commission. Of particular regulatory concern, however, is the unwanted leakage of signals from poorly constructed or maintained catv systems, particularly spurious signals radiated in bands used for services crucial to the safety of life. The Commission has restricted operation on catv frequencies which coincide with aviation services (108–136 MHz) and other governmental/public-safety bands.

Some services delivered by catv systems are tiered to varying levels of subscription costs, and access to the higher-level tiers is limited by the use of scrambled signals and addressable converters and decoders.

---

* FCC Rules and Regulations, Part 76.

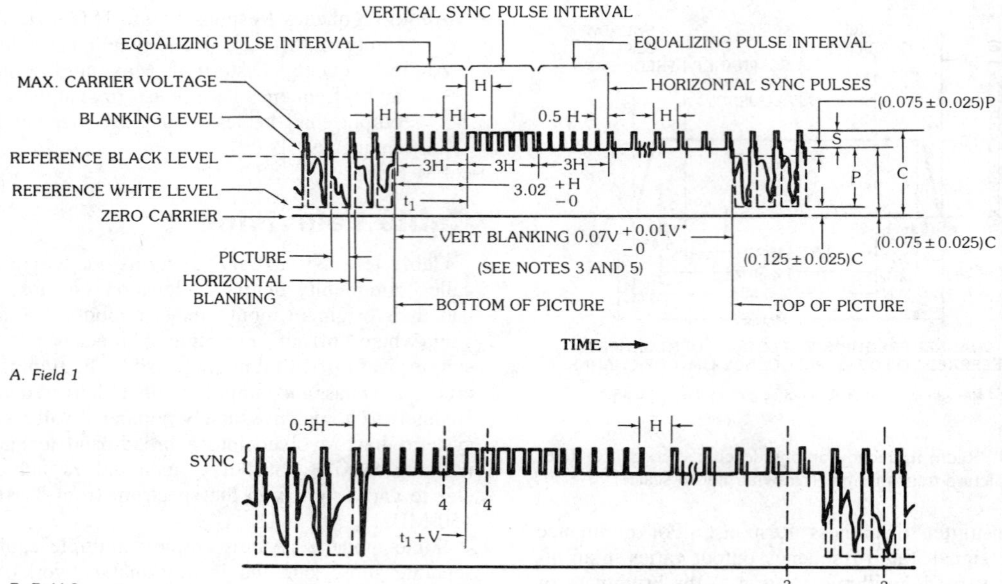

A. Field 1

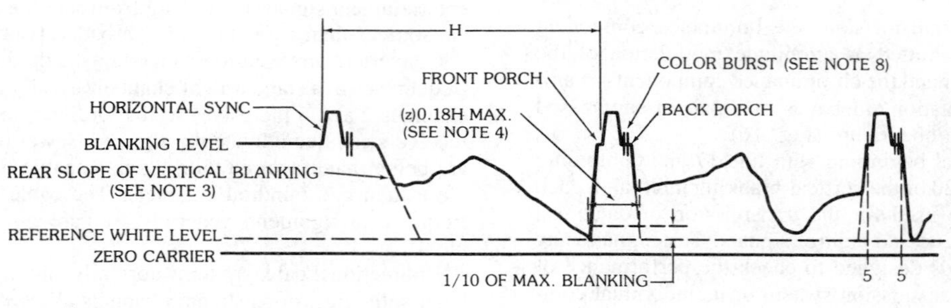

B. Field 2

C. Detail between 3-3 in B

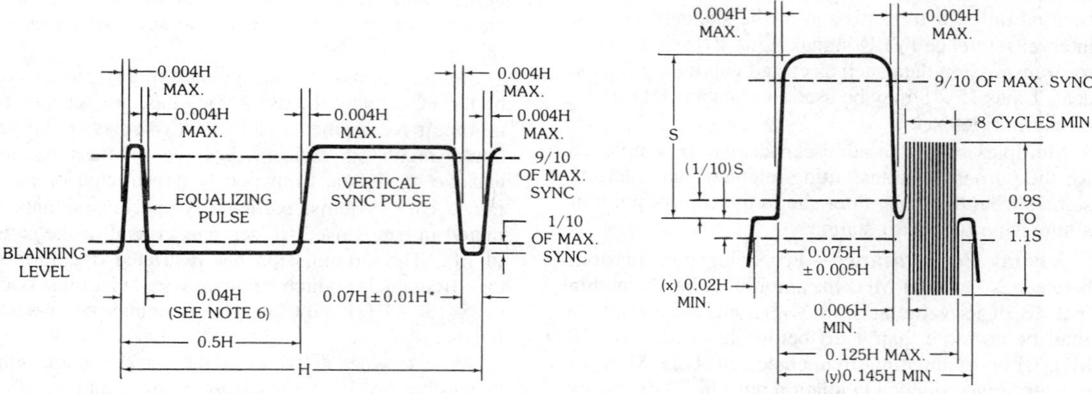

D. Detail between 4-4 in B

E. Detail between 5-5 in C

Fig. 15. Television composite-signal waveform data. (See notes on facing page.)

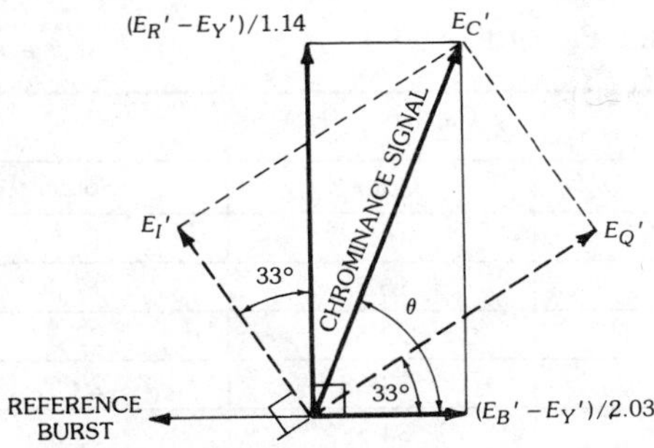

Fig. 16. Phases of color signal.

One of the limitations on the number of channels a cable system can carry is the distortion due to intermodulation and cross-modulation products of the numerous carriers. Composite triple beat is a term applied to third† and higher-order products, which can greatly degrade the preformance of a system. One method used to reduce these effects is the employment of coherent oscillators to regenerate (and convert) carrier frequencies such that they are harmonically related (HRC). This intermodulation problem increases with amplifier output levels and builds as amplifiers are cascaded. This then becomes the limiting factor on the length of a system (number of amplifiers in cascade), and as the number of channels is increased the amplifier separation must be reduced to compensate for lowered output levels. Extension of systems and system interconnection into large networks are also limited by this constraint.

Fiber optics or light-wave cable systems are being developed which promise to improve many of the transmission characteristics of coaxial cable (broad-band) amplifier systems. While there are no operating catv systems delivering signals directly to subscribers via fiber optics, this technology is being employed to deliver signals to headends or as "super trunks" to deliver signals to multiple miniheadends in large urban systems. However, coaxial-cable hardware development has advanced to the point where it is the most practical method of achieving economical broad-band systems, and fiber-optic technology is still in the developmental stage for this application.

† Products of the general type $(2f_1 \mp f_2)$ or $(f_1 + f_2 \pm f_3)$.

Notes for Fig. 15:

1. H = time from start of one line to start of next line.
2. V = time from start of one field to start of next field.
3. Leading and trailing edges of vertical blanking should be complete in less than $0.1H$.
4. Leading and trailing slopes of horizontal blanking must be steep enough to preserve minimum and maximum values of $(x+y)$ and z under all conditions of picture content.
5. Dimensions marked with an asterisk indicate that tolerances given are permitted only for long-time variations, and not for successive cycles.
6. Equalizing-pulse area shall be between 0.45 and 0.5 of the area of a horizontal-synchronizing pulse.
7. Color burst follows each horizontal pulse, but is omitted following the equalizing pulses and during the broad vertical pulses.
8. Color bursts to be omitted during monochrome transmission.
9. The burst frequency shall be 3.579545 megahertz. The tolerance on the frequency shall be ±10 hertz with a maximum rate of change of frequency not to exceed 1/10 hertz per second.
10. The horizontal scanning frequency shall be 2/455 times the burst frequency.
11. The dimensions specified for the burst determine the times of starting and stopping the burst but not its phase. The color burst consists of amplitude modulation of a continuous sine wave.
12. Dimension P represents the peak excursion of the luminance signal from blanking level but does not include the chrominance signal. Dimension S is the synchronizing amplitude above blanking level. Dimension C is the peak carrier amplitude.
13. Refer to FCC standards for further explanations and tolerances.
14. Horizontal dimensions not to scale in A, B, and C.

Other Television Services

Translators and LPTV—Television broadcast translators and low-power television (LPTV) stations operate under Subpart G of Part 74 of the FCC Rules. A television broadcast translator is a station that rebroadcasts the programming of a broadcast station in an essentially unaltered form. The signals from the originating station may be delivered by direct off-air pickup, microwave relay, satellite transmission, or rebroadcast of another translator. An LPTV station is similar to a translator station in terms of equipment and service area, but it is permitted to originate programming from virtually any source. The technical rules governing translators and LPTV are essentially identical.

For vhf stations, the transmitter output power (peak visual) may be up to 10 watts unless the station operates on an unoccupied channel assigned to its community for regular television broadcast use; in this case, 100 watts may be employed. A uhf station may be authorized on any channel with transmitter power up to 1 kW. There is no limitation on effective radiated power for either type of station.

The technical standards for operation of these stations are generally the same as those for broadcast stations (A5/F3 modulation), with the principal exceptions relating to carrier-frequency tolerances, spurious emissions, and lower-sideband attenuation characteristics.

Channel assignments are made on the assumption that these facilities are secondary to broadcast stations and must protect existing facilities from objectionable interference. While the service area of a regular broadcast station is defined as the area within the predicted Grade B contour, the service area of an LPTV station is much smaller and generally equivalent to the area

TABLE 4. CABLE TV CHANNEL FREQUENCIES

Channel	Freq. Range (MHz)	Carriers (MHz)		
		Video	Color	Sound
T- 7	5.75–11.75	7	10.58	11.5
T- 8	11.75–17.75	13	16.58	17.5
T- 9	17.75–23.75	19	22.58	23.5
T-10	23.75–29.75	25	28.58	29.5
T-11	29.75–35.75	31	34.58	35.5
T-12	35.75–41.75	37	40.58	41.5
T-13	41.75–47.55	43	46.58	47.5
2	54–60	55.25	58.83	59.75
3	60–66	61.25	64.83	65.75
4	66–72	67.25	70.83	71.75
5	76–82	77.25	80.83	81.75
6	82–88	83.25	86.83	87.75
7	174–180	175.25	178.83	179.75
8	180–186	181.25	184.83	185.75
9	186–192	187.25	190.83	191.75
10	192–198	193.25	196.83	197.75
11	198–204	199.25	202.83	203.75
12	204–210	205.25	208.83	209.75
13	210–216	211.25	214.83	215.75
FM	88–108	—	—	
14	120–126	121.25	124.83	125.75
15	126–132	127.25	130.83	131.75
16	132–138	133.25	136.83	137.75
17	138–144	139.25	142.83	143.75
18	144–150	145.25	148.83	149.75
19	150–156	151.25	154.83	155.75
20	156–162	157.25	160.83	161.75
21	162–168	163.25	166.83	167.75
22	168–174	169.25	172.83	173.75
23	216–222	217.25	220.83	221.75
24	222–228	223.25	226.83	227.75
25	228–234	229.25	232.83	233.75
26	234–240	235.25	238.83	239.75
27	240–246	241.25	244.83	245.75
28	246–252	247.25	250.83	251.75
29	252–258	253.25	256.83	257.75
30	258–264	259.25	262.83	263.75
31	264–270	265.25	268.83	269.75
32	270–276	271.25	274.83	275.75
33	276–282	277.25	280.83	281.75
34	282–288	283.25	286.83	287.75
35	288–294	289.25	292.83	293.75

Continued on next page.

TABLE 4 (CONT). CABLE TV CHANNEL FREQUENCIES

Channel	Freq. Range (MHz)	Carriers (MHz)		
		Video	Color	Sound
36	294–300	295.25	298.83	299.75
37	300–306	301.25	304.83	305.75
38	306–312	307.25	310.83	311.75
39	312–318	313.25	316.83	317.75
40	318–324	319.25	322.83	323.75
41	324–330	325.25	328.83	329.75
42	330–336	331.25	334.83	335.75
43	336–342	337.25	340.83	341.75
44	342–348	343.25	346.83	347.75
45	348–354	349.25	352.83	353.75
46	354–360	355.25	358.83	359.75
47	360–366	361.25	364.83	365.75
48	366–372	367.25	370.83	371.75
49	372–378	373.25	376.83	377.75
50	378–384	379.25	382.83	383.75
51	384–390	385.25	388.83	389.75
52	390–396	391.25	394.83	395.75
53	396–402	397.25	400.83	401.75
54	72–78	73.25	76.83	77.75
55	78–84	79.25	82.83	83.75
56	84–90	85.25	88.83	89.75
57	90–96	91.25	94.83	95.75
58	96–102	97.25	100.83	101.75
59	102–108	103.25	106.83	107.75
60	108–114	109.25	112.83	113.75
61	114–120	115.25	118.83	119.75

within its Grade A contour. Existing LPTV stations are protected from interference from newly proposed LPTV stations.

ITFS—Instructional Television Fixed Service (ITFS) stations operate under Subpart I of Part 74 of the FCC Rules in the band 2500–2686 MHz. These stations are licensed to eligible educational entities for the distribution of program material to students enrolled in instructional curricula. Public broadcast stations rendering such services are also eligible. An additional 4 MHz (2686–2690 MHz) is assigned for ITFS response stations. Response channels are intended for use as return links to the originating point for aural information such as student questions and responses. These stations employ very low power and narrowband equipment (250 mW and 125 kHz). Channels are six megahertz wide and have a standard television broadcast signal format. The channels are presently grouped in bands of four alternately spaced channels (e.g., group A is A-1, 2500–2506 MHz; A-2, 2512–2518 MHz; A-3, 2524–2530 MHz; etc.), with adjacent groups interleaved as shown in Table 5. Licensees are permitted to use multiple channels based on need and availability. Stations are usually limited to transmitter output power of 10 watts unless an adequate technical showing is made of need for a higher power (such as 20 W or 50 W). Effective radiated power is limited by coverage requirements and considerations of interference to other stations in the band (some ''grandfathered'' facilities of other services remain in the ITFS band). The band is also shared with the broadcast satellite service, and protection to those facilities is afforded.

Reception of ITFS signals usually requires the use of a small parabolic antenna, which is coupled to a down-converter. The down-converter incorporates a low-noise front end (NF<5 dB) and changes the signal(s) to vhf channels. A block of four channels (group) can be converted in a single device to yield four alternately spaced vhf channels (e.g., 7, 9, 11, 13). Four-foot antennas are generally used at distances up to ten miles for received carrier-to-noise ratios of approximately 50 dB.

TABLE 5. FREQUENCY ASSIGNMENTS OF INSTRUCTIONAL TELEVISION FIXED STATIONS

Channel	Band Limits (megahertz)
Group A	
A-1	2500–2506
A-2	2512–2518
A-3	2524–2530
A-4	2536–2542
Group B	
B-1	2506–2512
B-2	2518–2524
B-3	2530–2536
B-4	2542–2548
Group C	
C-1	2548–2554
C-2	2560–2566
C-3	2572–2578
C-4	2584–2590
Group D	
D-1	2554–2560
D-2	2566–2572
D-3	2578–2584
D-4	2590–2596
Group E	
E-1	2596–2602
E-2	2608–2614
E-3	2620–2626
E-4	2632–2638
Group F	
F-1	2602–2608
F-2	2614–2620
F-3	2626–2632
F-4	2638–2644
Group G	
G-1	2644–2650
G-2	2656–2662
G-3	2668–2674
G-4	2680–2686
Group H*	
H-1	2650–2656
H-2	2662–2668
H-3	2674–2680

* These frequencies shared with other stations.

MDS—The Multipoint Distribution Service (MDS) is a common-carrier service but is included in this chapter on broadcasting because the service is used primarily for the distribution of television program material to subscribers (''pay tv''). It is governed under Subpart K of Part 21 of the FCC Rules.

This service is assigned 10 MHz of spectrum in the band 2150–2160 MHz, except in the top 50 market areas where the band is extended to 2162 MHz. Channel 1 is designated as 2150–2156 MHz, and channel 2 as 2156–2162 MHz. Channel 1 has ''inverted'' carrier relationships (i.e., aural carrier below visual carrier) from channel 2 in order to minimize adjacent-channel interference in those locations where two channels are authorized. The band 2156–2160 MHz in smaller markets is too narrow to be used for standard television picture transmission.

In all other aspects MDS stations are very much akin to ITFS stations, and the hardware is usually common. As in ITFS, reception is accomplished with a down-converter that changes the 2-GHz signal to a standard vhf channel. The higher power generally employed in the MDS service permits the use of smaller receiving antennas and higher-noise-figure down-converters, making home reception practical.

(A pending rulemaking proceeding, if adopted, would divide the ITFS band among ITFS, MDS, and OFS* users.)

Network Distribution of Broadcast Program Signals

Terrestrial—Most network radio and television programming is distributed by terrestrial microwave systems owned and operated by common carriers (see Chapter 38). Transmission standards for aural programming vary with user requirements but generally are limited to 5-kHz circuits. Television program transmission is governed by the Network Transmission Committee Standard, NTC-7, which establishes reasonable requirements for a system that is presumed to have over 100 microwave relay stations in tandem. While most programming is carried over facilities belonging to AT&T, other smaller carriers provide extensions of the AT&T service to states or regions. The principal advantage of the terrestrial system is its infrastructure, which provides for many routing alternatives to alleviate outages due to failures and to make capacity available for occasional users (specials, sporting events, etc.). Its primary disadvantage is the poorer performance of long cascades (coast-to-coast) as compared to satellite transmission circuits.

Satellite Program Distribution—Space stations in the Domestic Satellite Service are being used extensively for the distribution of television and radio programming. Space stations are located in the geo-

* Operational Fixed Service, Private Microwave, Part 94, FCC Rules and Regulations.

stationary (or geosynchronous) orbit located approximately 23 000 miles in space above the equator. Of interest to continental US locations are satellites located between 70° and 143° west longitude, which can be "seen" from most sites except those in the most northerly latitudes. One network, the Public Broadcasting Service (PBS), distributes its television programming almost exclusively by the use of satellite channels. Its sister organization, National Public Radio (NPR), also distributes its programming almost exclusively by satellite. Cable television systems are receiving an ever increasing proportion of their programming from satellites.

When a satellite system is viewed as a microwave system with only one heterodyne repeater, it is readily apparent that vastly improved performance can be obtained over the terrestrial system if the basic carrier-to-noise limitations can be overcome. With very-low-noise amplifiers (NF<1.5 dB or 120 K), the weak signals from the satellites can be processed to produce signal-to-noise ratios in excess of 55 dB with antennas 7 to 10 meters in diameter (depending on satellite EIRP at the particular receiving location).

The present domestic satellite system uses mostly C-band transponders (3.7–4.2 GHz down-link and 5.925–6.425 GHz up-link). These bands are shared with terrestrial common carriers on a coequal basis. Thus, satellite terminals must be coordinated with existing facilities to assure interference-free operation. For this reason, it is frequently difficult to coordinate satellite earth stations in urban areas where terrestrial frequency congestion abounds. Techniques such as antenna shielding and if filtering can sometimes be employed to resolve interference problems.

Satellites with Ku-band (12/14 GHz) transponders hold promise of eliminating the interference problems while permitting the use of smaller antennas. Interference will not be a problem since these bands are not shared with terrestrial users; smaller antennas will be usable because of the higher effective gain of both satellite transmitting antennas and earth-terminal receiving antennas. The disadvantages of this band include the much higher potential for signal degradation due to precipitation and the generally higher cost of antennas and electronics. Distributed, interconnected, earth stations may solve the former and technological advances will solve the latter. There is presently no general use of Ku band for television program transmission.

DBS—Direct Broadcast Satellite transmission has been proposed and approved by the FCC. The bands proposed (subject to confirmation at the 1984 Space WARC Conference) are 12.2–12.7 GHz for down-link and 17.3–17.8 GHz for up-link. Various schemes have been proposed for channelization and modulation formats. Some proponents of high-definition television (hdtv) have proposed the use of DBS as the only practical means of broadcasting such programming to the public. Direct Broadcast Satellite transmission would further improve on the Ku-band advantage (use of small antennas) by having high-power transponder amplifiers that could permit a reduction in antenna size to less than one meter for practical home reception. It is predicted that as many as 20 DBS satellites could be in operation by 1990, offering over 200 hdtv channels for home reception.

Auxiliary Broadcast Services

In Part 74 of its Rules and Regulations, the FCC has made provision for various auxiliary broadcast services. These are generally channels allocated for delivering programming and associated communications from remote locations to the station. The bands and services include the following.

Remote Pickup Broadcast Stations—Stations in this service are used for the transmission of aural program material and associated cues and data. Assigned frequencies are in bands at 1.6 MHz, 26 MHz, 153 MHz, 161 MHz, 166 MHz, 170 MHz, 450 MHz, and 455 MHz. (See Subpart D of Part 74, FCC Rules and Regulations, for exact frequency assignments and limitations on use.)

Aural Intercity and STL Stations—Stations in this service are to be used for relay of aural program material from studio to transmitter and between fixed facilities in other locations. Assigned frequencies are in the band 947–951 MHz. (See Subpart E, Part 74, FCC Rules and Regulations.)

Television Auxiliary Stations—Stations in this service are used for transmission of television programming (aural and visual) between studio and transmitter, for intercity relay, and for remote pick-ups. Various bands are available at 2 GHz, 7 GHz, 13 GHz, and 22 GHz. (See Subpart F, Part 74 of the FCC Rules and Regulations for specific channel frequencies and limitations.)

Low Power Auxiliary Stations—Stations in this service are intended for use over very short distances for audio, cues, control, etc., associated with broadcast programs (for example, wireless microphones). Frequency bands are at 26 MHz, 161 MHz, 174–216 MHz, 450 MHz, and 950 MHz. (See Subpart H, Part 74 of the FCC Rules and Regulations for specific frequency assignments and limitations.)

Frequency Sharing—All users of Broadcast Auxiliary services must share frequencies. Users must coordinate use of specific frequencies with other local broadcasters to eliminate interference problems.

International Broadcasting Service in the United States

Transmissions from international broadcasting stations located within the United States are intended to be received directly by the general public in foreign countries. International broadcasting is conducted from both government (Voice of America) and privately owned stations. Public Law 80-402, the *United States Information and Educational Exchange Act of 1948*, encourages the participation of the private sector in international broadcasting. Nongovernment international broadcasting stations are licensed by the Federal Communications Commission in accordance with Part 73, Subpart F of the FCC Rules and Regulations, March 1980. A license for an international broadcasting station will be issued only after a satisfactory showing has been made in regard to the following, among others:

1. That there is a need for the international broadcasting service proposed to be rendered.
2. That the necessary program sources are available to the applicant to render the international service proposed.
3. That the production of the program service and the technical operation of the proposed station will be conducted by qualified persons.
4. That the applicant is legally, technically, and financially qualified and possesses adequate technical facilities to carry forward the service proposed.
5. That the public interest, convenience, and necessity will be served through the operation of the proposed station.

International broadcasting stations employ frequencies in bands between 5950 and 26 100 KHz. Frequencies authorized by the FCC fall within the following bands, which are allocated exclusively for broadcasting:

Meter Band	*Frequency in Kilohertz*
49	5950 – 6200
31	9500 – 9775
25	11 700 – 11 975
19	15 100 – 15 450
16	17 700 – 17 900
13	21 450 – 21 750
11	25 600 – 26 100

The band 7100–7300 kHz is also allocated for broadcasting, except in the western hemisphere.

The carrier frequencies assigned begin 5 kHz above the frequency specified for the beginning of each band, and are in successive steps of 5 kHz ending 5 kHz below the frequency specified as the end of each band.

Since international broadcasts cross frontiers, bridge oceans, and span continents, frequency assignments must be regulated internationally. Article 17 of the *Radio Regulations*, International Telecommunication Union, Geneva, 1979, specifies a frequency coordination procedure that member administrations must follow. The FCC has incorporated this procedure into its Rules and Regulations.

Schedules for international broadcasting stations are prepared seasonally and are implemented at 0100 UTC on the first Sunday of March, May, September, and November. The FCC requires licensees to submit their tentative schedules to the Commission six months prior to the start of each season, indicating for the season the frequency or frequencies desired for transmission to each zone or area of reception specified in the license; the specific hours of transmission to such zones or areas on each frequency; and the power, antenna gain, and antenna bearing to be used. The geographical areas to which broadcasts are directed should be designated in accordance with the ITU Geographical Zones for Broadcasting shown in Fig. 17.

Frequencies proposed should be as close as possible to the optimum working frequency, which is defined as that frequency which is returned to the surface of the earth for a specific transmission path and time of day on 90% of the days of the month, and should be chosen so that a given frequency will provide the largest period of reliable transmission to the selected zone or area of reception. The minimum transmitter power permitted for an international broadcasting station licensed by the FCC is 50 kW; the antenna power gain toward the intended reception zone must be at least 10; and the field strength incident in the reception zone, either measured or calculated, should exceed 150 microvolts/meter for 50% for the time. For purposes of calculating interference under practical operating conditions, it is assumed that the field strength in directions other than the main lobe of the antenna is not less than 222 microvolts/meter at 1 kilometer for 1 kilowatt supplied to the antenna.

Frequencies for short-wave broadcasting are very much in demand throughout the world. It has been estimated that between two and three stations often compete for an available channel during prime listening hours in many areas of the world. For this reason, international broadcasting is often subjected to a high level of cochannel and adjacent-channel interference.

International broadcasts may include commercial or sponsored programs, provided that commercial program continuities give no more than the name of the sponsor of the program and the name and general character of the commodity, utility or service, or attraction advertised and that the commodity is regularly sold or is being promoted for sale on the open market in the foreign country or countries to which the program is directed. An international broadcast station may transmit the program of a standard broadcast station or network system provided that the above commercial restrictions are met.

The 1979 World Administrative Radio Conference of the ITU allocated the following additional shortwave spectrum to broadcasting, but the new allocations are not expected to become effective before 1990:

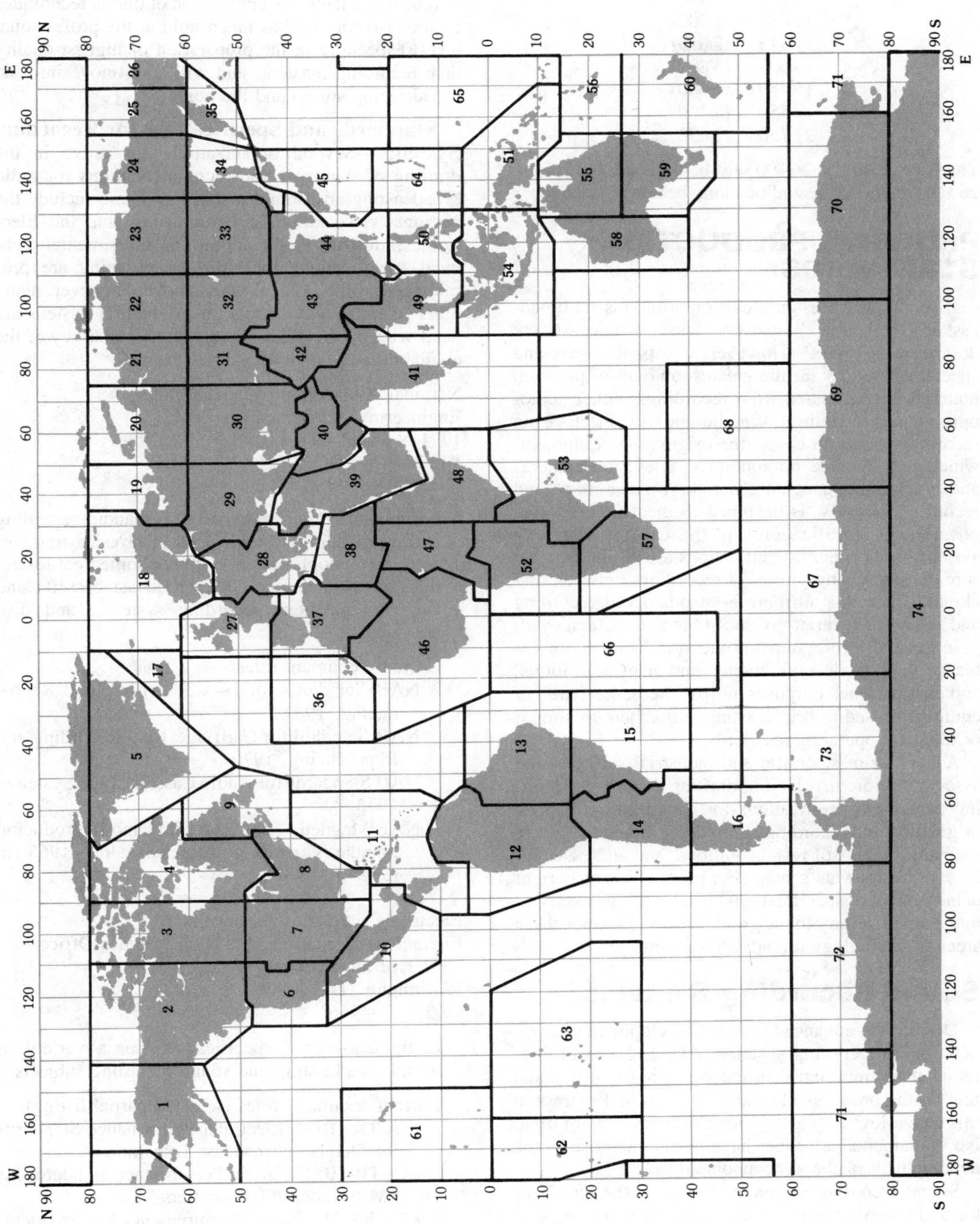

Fig. 17. Geographical zones for broadcasting.

kHz

9775 – 9900
11650 – 11700
11975 – 12050
13600 – 13800
15450 – 15600
17550 – 17700
21750 – 21850

The band 25600 – 26100 will be reduced to 25670 – 26100 when the new allocations become effective.

PROGRAM PRODUCTION STANDARDS

If sound programs or television programs are broadcast in "real time," such broadcasts or telecasts are described as "live." However, except for news and special events, by far the greatest amount of program material is transmitted from recordings, either audio-only (sound recording) or audio and video (television recording). In both cases, the origination equipment, which may include microphones, television cameras and synchronizing signal generators, transmitters, and recording systems, is designed to meet industry and government specifications. In the case of recording systems, the primary specifications are designed to assure the interchangeability of recordings between machines, since this attribute is mandatory if recording and replay in a variety of locations is to be achieved.

In general, professional production facilities for cable television, teleconferencing, and many educational and instructional purposes utilize the same types of equipment used in broadcasting and adhere to similar or identical operating standards.

A number of scientific and industrial societies and associations are involved in drafting documents defining agreed-to specifications for sound and television origination and recording equipment. Listed below are the names, areas of responsibilities, and addresses for most of the relevant groups. For information on current details of these specifications, it is often necessary to make direct inquiry to the group concerned, since these areas of technology are advancing rapidly.

Sound Recording Systems

Despite the advanced state of development of magnetic audio recording systems, disk audio recordings retain their importance in the recording industry and thus in the broadcasting industry as well. Progress in this technology is largely one of refinement with time. No fundamental advances have been evident since the development of the stereophonic disk.

Sound recording on magnetic tape is the dominant form of audio recording. Most magnetic recording systems use conventional analog techniques—gamma-ferric iron oxide coated onto a plastic tape backing material, recording heads of iron or ferritic material making use of high-frequency ac recording bias, playback heads having narrow gaps and driving the input stages of low-noise analog amplifiers. However, as will be noted later, the application of digital techniques to sound recording has taken hold at the professional level, especially in the preparation of highest-quality disk recording masters, and will soon move into the broadcasting studio and the home.

Standards and Specifications for Recording Systems—Several organizations are active in the drafting of standards and recommendations for radio broadcasting and sound recording. These include the National Association of Broadcasters and the Electronic Industries Association. The fundamental technical specifications for radio broadcasting are promulgated by the FCC, as noted above. However, many other aspects of the program production system are dealt with by the other organizations. A survey of the available specifications is listed below.

National Association of Broadcasters
Engineering Department
1771 N Street, NW
Washington, DC 20036
(202) 293-3500

Equipment-related parameters for audio recording and reproducing systems used in broadcasting, including monophonic and stereophonic reel-to-reel magnetic sound recorders, broadcast cassette and cartridge magnetic recording systems, and disk recordings.

Current document titles:

- NAB Standards for Disk Recording and Reproducing, 1964
- NAB Standard for Cartridge Tape Recording and Reproducing, 1976
- NAB Standards for Audio Cassette Tape Systems, 1976
- NAB Magnetic Tape Recording and Reproducing Standards for Reel-to-Reel Systems, 1965 (in revision, 1982)

Electronic Industries Association
Engineering Department, Standards Sales Office
2001 Eye Street, NW
Washington, DC 20006
(202) 457-4966

Equipment-related specifications for a variety of sound broadcasting and sound recording subjects.

Current document titles (selected partial listing):

- EIA TR-101-A Electrical Performance Standards for Standard Broadcast Transmitters
- EIA TR-107 Electrical Performance Standards for FM Broadcast Transmitters
- EIA RS-215 Basic Requirements for Broadcast Microphone Cables
- EIA RS-219 Audio Facilities for Radio Broadcasting Systems
- EIA RS-221-A Polarity or Phase of Microphones

for Broadcasting, Recording and Sound Reinforcement

EIA RS-288 Audio Magnetic Playback Characteristic at $7^1/2$ IPS

EIA RS-297-A Cable Connectors for Audio Facilities for Radio Broadcasting

EIA RS-298 Audio Transmitter Input Impedances for Single Input Transmitters

EIA RS-332-A Dimensional Standard—Endless Loop Magnetic Tape Cartridges (EIA Type III)

EIA RS-339-A Dimensional Standard Coplanar Magnetic Tape Cartridge, Type CP II (Compact Cassette)

EIA RS-400 EIA Reproducer Test Tape: Full-Track 1/4-inch (6.3 mm) Width, Open-Reel (for Tape Speeds of 7.5 in/s = 190.5 mm/s and 3.75 in/s = 95.3 mm/s)

EIA RS-490 Standard Test Methods of Measurement for Audio Amplifiers

NOTE: In addition to this selected list of EIA publications, the association publishes an extensive range of standards and other engineering publications concerned with electronics and electronic engineering. A catalog may be obtained from the address noted above.

International Sound-Recording Standards—The contents of most of the specifications and standards listed above are duplicated in international standards. In sound recording, most international standardization is performed by the International Electrotechnical Commission (IEC), a subdivision of the International Standards Organization (ISO). Responsibility for international sound-recording standards rests with IEC Technical Committee 60—Recording (TC 60) and its Subcommittee SC 60A. Information on IEC standards may be obtained from the American National Standards Institute, 1430 Broadway, New York, NY 10018. International exchange of broadcast sound programs recorded on tape or disk is dealt with by the International Radio Consultative Committee (CCIR). Copies of CCIR Volume 10—Sound Broadcasting and other CCIR and ITU publications may be obtained from the National Technical Information Service, US Department of Commerce, 5285 Port Royal Road, Box 1553, Springfield, VA 22161. A comprehensive listing of international standards affecting sound recording is given in the section "International Broadcasting Standards" at the end of this chapter.

Digital Audio Recording—The use of digital techniques to record audio signals has the following advantages when compared to conventional analog techniques:

A. The dynamic range of the recording can be greater. This parameter is primarily determined by the number of sampling bits per digital word used in the analog-digital conversion process. In professional audio applications, it is usual to specify 14 to 16 bits, thus providing a dynamic range of from 84 to 96 decibels. The amplitude/frequency response of digital systems is primarily determined by the digital sampling rate. For professional use, this is usually a rate of 48000 bits per second (48 kb/s)

B. Deterioration of quality when original tapes are copied is virtually absent. The situation is quite different in analog recording, where each "generation" of tape-copying results in increased distortion and noise as compared to the original recording.

C. Through digital error detecting, correcting, and concealment techniques, the effects of magnetic oxide defects ("dropouts") in the tape used for making the recording are eliminated.

In addition to more-or-less conventional tape-transport systems using fixed heads and capstan tape-pulling mechanisms, digital audio recorders are available that use video-tape-recorder tape transports. Other designs use magnetic disk mechanisms that are similar to diskette equipment used in computer applications. It is anticipated that this latter format, the magnetic disk, will have a major effect in both broadcast and home use of digital audio recording systems in the future.

The professional sound recording industry is in the process of defining final digital sound recording specifications. In the United States, this work is organized by the Audio Engineering Society (60 E. 42nd Street, New York, NY 10017). It has been agreed that the sampling rate for professional applications is to be 48 kb/s. Other parameters, such as equipment interface specifications and studio operational practices, are under active study.

For some home applications, a different rate has been chosen, largely because of the use of digital video recorders for high-quality audio recording. A rate of 44.3 kb/s functions well with the conventional home video recording system. A consortium of Japanese and European manufacturers has proposed a compact audio disk recorded digitally. This system will also utilize the 44.3 kb/s sampling rate.

Television Recording Systems

The magnetic recording of television signals, variously described as video tape recording (vtr) and/or television recording, is a technology that has been in professional use for more than 25 years. Various recording formats are now in service, from the original broadcast format using 2-inch-wide magnetic tape, through formats using 1-inch tape or $^3/4$-inch tape, to the millions of home-use systems using $^1/2$-inch tape in cassettes and still other home units utilizing $^1/4$-inch tape.

In professional studio applications, the most widely used systems include the 2-inch format, usually called either quadruplex ("quad") or transverse-scan, and the newer 1-inch broadcast formats identified either as

SMPTE (Society of Motion Picture and Television Engineers) Type C or, less frequently used in the United States, SMPTE Type B. For use in news gathering and production outside the studio (field production), both the 1-inch SMPTE Type C and another format, the 3/4-inch U-format video cassette (SMPTE Type E), are frequently placed into service. In the near future, professional-quality 1/2-inch video cassette recorders will be placed into service; these will use one or more proposed but as yet unstandardized recording systems that will record the component parts of the color signal on separate tracks, rather than recording the composite color video signal. They may displace the 3/4-inch format for portable professional applications.

For home recording purposes, the most common video recording formats are the two popular video cassette systems known commercially as VHS and Beta. These formats, identified as SMPTE Types F and G, respectively, are in the final stages of national and international standardization at the time of this writing.

All of these formats achieve US standardization through the work of the Engineering Committees of the SMPTE. A list of relevant documents, including SMPTE Recommended Practices and ANSI (American National Standards Institute) standards, is given below.

As highly advanced as the magnetic recording of television signals may be, it is true that many television programs are produced by photographic processes. Motion picture technology is also standardized in the USA by the SMPTE, and the list of SMPTE standards printed below includes many documents defining the parameters for motion picture films intended for television and other uses.

Standards for Video Magnetic Tape and Motion Picture Film Recording Systems for Television—The standards documents listed below are available from the Society of Motion Picture and Television Engineers, 862 Scarsdale Ave., Scarsdale, New York 10583.

SMPTE STANDARDS AND RECOMMENDED PRACTICES

Subject	*No.*	*Journal*	
Film Dimensions			
8 mm, Perforated Super 8, 1R	PH22.149-1981	Dec.	1981
16 mm, Perforated Regular 8, 2R-1500	PH22.17-1982	Aug.	1982
16 mm, Perforated Super 8, (1-3)	PH22.151-1981	Dec.	1981
(1-4)	PH22.168-1973 R1980	Aug.	1973
16 MM, 1R	PH22.109-1980	May	1981
16 mm, 2R	PH22.110-1980	May	1981
35 mm, Perforated Super 8 2R-1664 (1-0)	PH22.169-1980	May	1981
5R	PH22.165-1981	Oct.	1981
35 mm, Perforated 16 mm, 3R (1-3-0)	PH22.171-1980	May	1981
35 mm, Perforated 32 mm, 2R	PH22.73-1981	Oct.	1981
35 mm, BH	PH22.93-1980	Apr.	1981
35 mm, CS-1870	PH22.102-1980	Apr.	1981
35 mm, DH-1870	PH22.1-1981	Dec.	1981
35 mm, KS	PH22.139-1980	Apr.	1981
65 mm, KS	PH22.145-1981	Dec.	1981
70 mm, Perforated 65 mm, KS-1870	PH22.119-1981	Dec.	1981
Film Usage, Camera			
Regular 8	PH22.21M-1981	Mar.	1982
Super 8	PH22.156-1976	Dec.	1976
16 mm	PH22.9-1976	Feb.	1977
35 mm	PH22.2-1979	Sept.	1979
Film Usage, Projector			
Regular 8	PH22.22-1975 R1981	Apr.	1976
Super 8	PH22.155-1976	Dec.	1976
16 mm	PH22.10-1980	July	1981
Image Areas, Camera			
Regular 8	PH22.19-1976	Oct.	1976
Super 8	PH22.157-1971 R1977	June	1971
16 mm	PH22.7-1976	Oct.	1976
Super 16	PH22.201M-1981	Nov.	1981
35 mm	PH22.59-1974 R1981	June	1974
Image Areas, Printers			
Super 8 on 16 mm (1-3)	PH22.181-1973 R1979	Apr.	1973
(1-4)	PH22.153-1979	Sept.	1979
Super 8 on 35 mm	PH22.179-1980	Nov.	1980
16 mm Contact (Positive From Negative and Reversal)	PH22.48-1976	Oct. Oct.	1976 1982[1]
16 mm to 35 mm Enlargement Ratio	RP 66-1982	Dec.	1982
Super 16 to 35 Enlargement Ratio	PH22.201M-1981	Nov.	1981
35 mm to 16 mm Prints and Dupe Negatives	RP65-1982	Dec.	1982
35 mm Release Picture-Sound Continuous Contact	PH22.111-1982	Aug.	1982
Image Areas, Projectable			
8 mm Release Prints	RP56-1974 R1979	Jan.	1975
Regular 8	PH22.20-1981	Feb.	1982
Super 8	PH22.154-1976	Dec.	1976
16 mm	PH22.8-1981	Feb.	1982
16 & 35 mm TV Review Room	PH22.148-1967	Dec.	1967
35 mm	PH22.195-1977	Sept.	1977
70 mm	PH22.152-1969 R1976	Dec.	1969

Subject	No.	Journal	
Sound			
Photographic Record			
Super 8	PH22.182-1978	Nov.	1978
16 mm	PH22.41-1975	Mar.	1976
		Feb.	1982[1]
2-track	PH22.204-1981	May	1982
35 mm	PH22.40-1978	Aug.	1978
		Nov.	1982[1]
2-track	PH22.203-1981	May	1982
Magnetic Record			
Regular 8	PH22.135-1982	Oct.	1982
Super 8	PH22.164-1975	Oct.	1975
16 mm 100 mil	PH22.112-1977	Nov.	1977
200 mil	PH22.97-1982	Oct.	1982
Recorded Characteristics	PH22.209M	Nov.	1982[2]
35 mm 3 track	PH22.86-1981	Mar.	1982
4 track	PH22.108-1974 R1980	Mar.	1975
Release	PH22.137-1981	Nov.	1981
6 Track	PH22.186-1974 R1980	Feb.	1975
Recorded Characteristics	PH22.208M	Nov.	1982[2]
70 mm	PH22.185-1980		1981
Modulation, Sound Level	PH22.183	Oct.	1972[2]
Spectral Response, Super 8 Reproducers	RP 109	Sept.	1981[2]
Use and Care, Test Films	RP 45-1972 R1977	Aug.	1972
Stripe			
Regular 8	PH22.88-1976	Nov.	1976
Super 8	PH22.161-1980	Feb.	1981
Regular 8 on 16 mm	PH22.136-1976	Nov.	1976
Super 8 on 16 mm			
(1-3)	PH22.176-1976	Dec.	1976
(1-4)	PH22.162-1980	Feb.	1981
Super 8 on 35 mm (5R)	PH22.163-1980	Feb.	1981
16 mm 30 mil	PH22.101-1976	Nov.	1976
50 mil	PH22.127-1976	Nov.	1976
100 mil	PH22.87-1966 R1977	Aug.	1966
35 mm 4-Track Release	PH22.177-1970 R1976	Jan.	1971
Television			
Alignment Color Bar Signal	ECR 1-1978	Oct.	1978
Color Temperature, Monitors	RP 37-1969 R1976	Sept.	1969
Review Room Screens	RP 41-1974	Nov.	1974
		Sept.	1982[1]
Comparator	RP 52-1974	Nov.	1974
WITHDRAWN 1981		Apr.	1982[3]
Density, Monochrome, Films and Slides	RP 7-1970 R1976	Apr.	1970
Density, Color	RP 46-1972 R1977	Sept.	1972
Illuminator for Test Pattern Transparencies	RP 72-1977	June	1977
Image Area			
16 mm Film	PH22.96-1982	Dec.	1982
35 mm Film	PH22.95-1963 R1975	Oct.	1963
Review Rooms	PH22.148-1967 R1978	Dec.	1967
Slides and Opaques	PH22.94-1973 R1979	Nov.	1973
Monitors, Setting of White for	RP 71-1977	June	1977
2×2 Slide Mount	RP 9-1977	Dec.	1977
Test Patterns			
Alignment	RP 27.1-1977	May	1978
		Sept.	1972
Cameras, Studio and Field	RP 27.6-1972	Dec.	1982[3]
Telecine	RP 27.7-1972 R1977	Sept.	1972
Linearity	RP 38.1-1971 R1976	Apr.	1972
Mid-Frequency Response	RP 27.5-1977	May	1978
Picture Steadiness	RP 27.4-1972 R1977	June	1972
Registration	RP 27.2-1971 R1976	Apr.	1972
Safe Areas	RP 27.3-1972 R1977	June	1972
Test Films			
Photographic			
Regular 8 Registration	RP 19-1965	Jan.	1966
	R1975	Mar.	1982[1]
Super 8 Registration	RP 32-1969 R1975	Sept.	1969
Sound Projector	RP 18-1981	Apr.	1982
16 mm Buzz-Track	RP 67-1976	Mar.	1977
		July	1982[1]
Flutter	RP 70-1977	Aug.	1977
Projector Alignment	RP 82-1978	Mar.	1979
Registration	RP 20-1965	Jan.	1966
	R1975	Mar.	1982[1]
Scanning Beam	RP 81-1978	Feb.	1979
Sound Focusing	RP 63-1976	Aug.	1976
Sound Projector	RP 18-1981	Apr.	1982
Theater Test	RP 35-1969 R1977	June	1969
35 mm Buzz-Track	RP 68-1976	Mar.	1977
Flutter	RP 97-1981	Sept.	1981
Projector Alignment	RP 40-1971	Aug.	1971
	R1977	May	1982[1]
Anamorphic Attachments	RP 110	Apr.	1982[2]
Projector Test	RP 33-1968 R1977	Dec.	1968
Scanning Beam	RP 69-1977	Aug.	1977
Sound Focusing	RP 64-1981	July	1982
Theater Test	RP 35-1969 R1977	June	1969
70 mm Projector Alignment	RP 91-1981	Sept.	1981
Magnetic			
Super 8 Azimuth		May	1976
Alignment	RP 61-1975	May	1982[1]
Flutter	RP 62-1975	May	1976

Subject	No.	Journal	
Multifrequency	RP 92-1980	Apr.	1981
16 mm Azimuth			
Alignment	RP 78-1978	Aug.	1978
Flutter	RP 76-1977	July	1978
Multifrequency	RP 90-1979	Jan.	1980
35 mm Azimuth			
Alignment	RP 77-1978	Aug.	1978
4-Track	RP 80-1977	June	1978
D Flutter	RP 75-1977	July	1978
4-Track	RP 79-1977	June	1978
Video Magnetic Tape Recording			
Digital Control Interface			
Electrical and Mechanical Characteristics	PH22.207M	Sept.	1982[2]
Supervisory Protocol	RP 113	Sept.	1982[2]
Tape Care and Handling	RP 103-1982	Oct.	1982
Helical Scan			
Code, Time and Control, Recording Requirements	RP 93-1980	Apr.	1981
Raw Stock, Reference Tape	V98.26M	Oct.	1980[2]
Reels, 1-in	V98.24M	July	1980[2]
Tape, 1-in	V98.25M	July	1980[2]
Type B 1-in			
Basic Parameters	C98.15M-1980	Apr.	1980
Carrier Frequencies and Preemphasis	RP 84-1980	Apr.	1980
Frequency Response and Operating Level	C98.17M-1980	Apr.	1980
Record Dimensions	C98.16M-1980	Apr.	1980
Reference Tapes			
Video and Audio	RP 107	June	1981[2]
Record Dimensions	V98.30M	June	1981[2]
Recorder Parameters	V98.29M	June	1981[2]
Tracking-Control Record	RP 83-1980	Apr.	1980
Type C 1-in			
Basic Parameters	C98.18M-1979	Feb.	1980
		Jan.	1982[1]
Frequency Response and Reference Level	C98.20M-1979	Feb.	1980
Record		Feb.	1980
Dimensions	C98.19M-1979	Oct.	1982[1]
Recorder Parameters	RP 86-1979	Feb.	1980
Reference Tapes			
Interchange	RP 100	Oct.	1980[2]
Video and Audio	RP 99	Oct.	1980[2]
Record Dimensions	V98.28M	Oct.	1980[2]
Recorder Parameters	V98.27M	Oct.	1980[2]
Tracking-Control Record	RP 85-1979	Feb.	1980
Type E $^3/_4$-in			
Carrier Frequencies, Preemphasis, Audio and Control Signals	RP 87-1980	May	1980
Cassette Dimensions	C98.22M-1980	May	1980
Receiver/Monitor		Oct.	1979[2]
Test Tape	RP 96	July	1982[1]
Record Dimensions	C98.21M-1980	May	1980
Small Cassette	V98.31M	Nov.	1981[2]
Type F $^1/_2$-in			
Carrier Frequencies and Preemphasis	RP 88-1980	May	1980
Records and Parameters	C98.23M-1980	May	1980
Type H $^1/_2$-in			
Carrier Frequencies, Preemphasis, Audio and Control Signals	RP 112	July	1982[2]
Records	V98.32M	July	1982[2]
Tape and Cassette	V98.33M	July	1982[2]
Quadruplex			
Audio 2			
Level/Response	RP 102-1981	Jan.	1982
Code, Time and Control	V98.12M-1981	Aug.	1981
Recording Requirements	RP 101-1981	Jan.	1982
Dropout Detection	RP 47-1972	Dec.	1972
	R1978		
Dual-Program Audio	RP 89-1978	Mar.	1979
Headwheel and Guides	RP 36-1978	Feb.	1979
Labels	RP 26-1981	Apr.	1982
Leader, Monochrome	V98.2-1982	Dec.	1982
Color	C98.9-1967	Sept.	1967
	R1975	Jan.	1982
Modulation Practices	RP 6-1979	Jan.	1980
Patch Splices	RP 5-1976	May	1977
Records, Characteristics of Audio	V98.3-1980	July	1981
Record Dimensions, Video, Audio and Tracking Control	V98.6-1981	Oct.	1981
Record, Tracking Control	RP 16-1977	Dec.	1977
Reels, 2 in.	C98.5-1970	Feb.	1971
	R1976		
$^1/_2$ in.	C98.14-1975	Mar.	1976
	R1981		
Speed	C98.4-1970	Feb.	1971
	R1976		
Spools, Cartridge	V98.13-1981	Oct.	1981
Labels	RP 60-1975	May	1975
	R1980		
Tape Dimensions	C98.1-1978	Aug.	1978
Tape Vacuum Guide	RP 11-1968	Aug.	1968
Tape Usage, Cartridge/ Cassette Spools	EG 6	July	1981[2]
Test Tapes			
Multifrequency			
15 in/s	C98.8-1977	Jan.	1978
		Mar.	1981[2]
7.5 in/s	C98.11-1977	Jan.	1978
		Mar.	1981[2]
Video Frequency			
15 in/s, LBM	RP 29-1968	Sept.	1968
WITHDRAWN 1981	R1980	Apr.	1982[3]
15 inb/s, HB	RP 43-1975	June	1976
		Sept.	1982[1]
7.5 in/s, HB	RP 44-1975	June	1976
WITHDRAWN 1982		Aug.	1982[3]
Vertical Interval Signal	RP 57-1974	Jan.	1975
	R1980		
Time and Control Code	RP 108	Sept.	1981[2]

Subject	No.	Journal
MISCELLANEOUS		
Cartridge, Super 8 Camera		
Notches	PH22.166-1981	Nov. 1981
Silent		
50 ft.		
Model I		
Aperture, Profile, Pressure Pad, Film Position	PH22.159.2-1980	Nov. 1980
Camera Run Length, Perforation Cut-Out, End of Run Notch	PH22.200M	Nov. 1981[2]
Cartridge, Cartridge-Camera Interface, Take-Up Core Drive	PH22.159.1-1980	Feb. 1981
Model II		
Cartridge, Cartridge-Camera Fit, Core	PH22.190-1976	Sept. 1976
Film Length, Camera Run	PH22.188-1976	Sept. 1976
Position	PH22.189-1976	Sept. 1976
Speed, Color Balance, Identification	PH22.191-1976	Sept. 1976
Sound		
50 Ft.		
Model I		
Aperture, Pressure Pad, Film Position	PH22.198-1980	Aug. 1980
Camera-Run Length, Perforation Cut-Out, End-of-Run Notch	PH22.200M-1980	Aug. 1980 Nov. 1981[2]
Cartridge, Cartridge-Camera Interface, Core Drive	PH22.197-1980	Aug. 1980
Pressure Pad Flatness, Aperture Profile	PH22.199-1980	Aug. 1980
200 Ft.		
Model I		
Aperture, Profile, Film Position, Pressure Pad, Flatness	PH22.206	Aug. 1981[2]
Camera-Run Length, Perforation Cut-Out, End-of-Run Notch	PH22.200M	Nov. 1981[2]
Cartridge, Cartridge-Camera Interface, Core Drive	PH22.205	Aug. 1981[2]
Conference		
Audio Reinforcement	ECR 4	June 1979[2]
Projector	ECR 3	June 1979[2]
Cores for Raw Stock Film	PH22.37-1975 R1981	Oct. 1975
Density Measurements		
Calibration of Densitometers	RP 15-1982	Dec. 1982
Special Diffuse	PH22.117-1968 R1979	Mar. 1969
Edge Numbering, 16 mm Film	PH22.83-1972 R1978	Dec. 1972
16 mm Release Prints	RP 54-1974 R1979	July 1974
Electro-Acoustic Response		
Control Rooms and Theaters	PH22.202	Jan. 1979[2]
Emulsion Orientation		
Print Winding	RP 39-1970 R1982	Apr. 1970
Raw Stock Winding	PH22.75-1975 R1982	Mar. 1976
Super 8 Release Prints	RP 42-1970	Dec. 1970
WITHDRAWN 1982		Aug. 1982[3]
Film Length, 8 mm Camera Spool (25 ft Capacity)	PH22.143-1975 R1981	Feb. 1976
Graph Paper	RP 22-1966 R1976	Dec. 1966 Sept. 1980[2]
Image Quality, 70, 35, 16 mm	EG 5	July 1981[2]
Jump and Weave, 70, 35, 16 mm	RP 105-1981	June 1982
Leaders		
Preprint, 8 mm	RP 49-1973 R1979	Mar. 1973
Universal	PH22.55-1975	Mar. 1976 Feb. 1982[1]
Lenses		
Aperture Calibration	PH22.90-1964 R1977	June 1964
Focal Lengths, Markings, 35 & 70 mm	PH22.28-1982	Nov. 1982
Focus Scales, 16 mm and 8 mm Cameras	Ph22.74-1965 R1981	May 1965
Lens Mounts		
16 mm & 8 mm Cameras	PH22.76-1960 R1979	Feb. 1960
Lubrication 16 & 8 mm Prints	RP 48-1973 R1979	Mar. 1973
Nomenclature		
Cartridge/Cassette	RP 58-1974 R1980	Jan. 1975
Film	PH22.56-1978	Nov. 1978
Notching, Scene Change, 35 mm	RP 53-1974	July 1974 Aug. 1982[2]
Raw Stock Identification	PH22.184-1973 R1980	Nov. 1973
Container Edge	ECR 2-1979	Sept. 1979
Reels		
Regular 8	PH22.23-1975 R1981	Apr. 1976
Super 8	PH22.160-1977	July 1977
16 mm	PH22.11-1981	Feb. 1982
35 mm	PH22.4-1976	Apr. 1977
Large Capacity	PH22.193-1976	Apr. 1977
Shipping	PH22.192-1976	Apr. 1977
70/35 mm	PH22.147-1976	Apr. 1977
Reversal Color Film Speed	PH22.146M-1980	May 1981
Safety Film	PH22.31M-1980	July 1981
Screens		
Gain Determination	RP 94-1980	June 1981

Subject	No.	Journal	
Installation	RP 95-1980	June	1981
Luminance			
Drive-in Theaters	RP 12-1972	Dec.	1972
	R1980	Aug.	1982[2]
Indoor Theaters	PH22.196-1978	Aug.	1978
Measurement	RP 98-1981	Sept.	1981
Review Rooms, 8 mm	RP 51-1974	May	1974
	R1979		
Slides & Film Strips	RP 591-975	May	1975
	R1980		
Sensitometric Strips	RP 14-1982	Dec.	1982
Spindles			
Super 8 Projector	RP 50-1974	May	1974
	R1979		
16 mm Camera	RP24-1967	July	1967
	R1978		
16 mm Projector	RP 34-1968	Dec.	1968
	R1978		
35 mm Rewind	RP 21-1976	May	1977
Splices			
16 & Regular 8	PH22.24-1982	Aug.	1982
Super 8			
Cemented	PH22.172.1-1969	Mar.	1970
	R1975	June	1982[3]
Tape	PH22.172.2-1976	Dec.	1976
70, 65 & 35 mm	P 111	Apr.	1982[2]
		June	1982[3]
70 mm Reinforcement	RP 23-1979	Jan.	1980
Spools			
8 mm, 25-ft Capacity	PH22.107-1975	Feb.	1976
	R1981		
Double 8, 100-ft Capacity	PH22.173-1975	Feb.	1976
	R1981		
16 mm, daylight-loading, 50- to 400-ft Capacity	PH22.174-1981	Mar.	1982
Sprockets			
Regular 8	RP 73-1977	Jan.	1978
Super 8	RP 55-1974	Jan.	1975
	R1979		
16 mm	RP 74-1977	Jan.	1978
35 mm	PH22.35-1982	Nov.	1982
Synchronization, Sound-Picture	RP25-1968	Mar.	1968
	R1978		
Tension, 35 mm Systems	RP 106-1982	Oct.	1982
Test Methods, Sound Distortion			
Cross Modulation, Variable-Area	RP 104-1981	June	1982
Intermodulation, Variable-Density	PH22.51-1961	July	1961
	R1975	June	1982[3]
Unsteadiness, High-Speed Camera	RP 17-1964	May	1964
	R1982		

R—Reaffirmed.
[1]Proposed editorial revision.
[2]Proposal.
[3]Withdrawal notice.

International Standards for Television Recording—Most of the SMPTE documents listed above have equivalent specifications in the publications of international standardizing bodies. Video recording standards are produced by IEC Subcommittee 60B and broadcast program exchange specifications by the CCIR (published in CCIR Volume 11—Television Broadcasting). A listing of relevant IEC and CCIR standards is given later in this chapter. In addition, a set of technical publications concerning video recording of 625-line PAL and SECAM color television signals is available from the European Broadcasting Union, Technical Centre, ave. Albert Lancaster 32, B-1180 Brussels, Belgium.

Standards defining international specifications for motion picture film are drafted by ISO Technical Committee 36. A listing of these documents is contained in the last part of this chapter. In addition, the EBU publishes specifications for motion picture film intended for television use by members of the EBU, and the CCIR publishes specifications, in CCIR Volume 11—Television, for motion picture films to be used for the international exchange of television programs.

Video Disk Recording—The sale and use of video disk systems is relatively new as this chapter is being written. Currently, two systems that differ markedly from each other are generally available in North America. The first of these is an "optical" system, making use of finely etched pits arrayed in a continuous spiral on the surface of an aluminized reflective plastic disk. The pits modulate the focused beam of a low-power laser. The modulated light beam is then reflected by the aluminized core of the disk through a precise optical system to a photodetector that drives a video amplifier. The resulting video and audio outputs can be applied to a monitor or used to modulate an rf carrier, which can, in turn, be delivered to the antenna terminals of a conventional television receiver. This system is capable of very high quality audio and video reproduction and can offer the advantages of still-frame and slow-motion image display. The second commercially available system makes use of an opaque plastic disk embossed with microscopic ridges in a continuous spiral, which deflect a diamond stylus. Through changes in electrical capacitance produced as the stylus deflects, video and audio signals are reproduced. This system is also capable of high quality performance, but by its basic design it does not normally permit still-frame and slow-motion operation.

Standardization of video disk systems seems to lie in the future. Both the SMPTE in the United States and IEC SC 60B internationally have active study groups examining a variety of proposals. It is likely that some progress in defining standards for video disk systems will be apparent by the middle of the 1980s. Relatively little material specifically produced for disk systems has yet appeared.

Selected Lists of Television Standards

Listed below are two groups of industrial standards that are useful in television engineering. The EIA list is abbreviated; readers are encouraged to contact the source listed in the sub-subsection "Standards and Specifications for Recording Systems" earlier in this chapter for up-to-date comprehensive listings.

EIA Television Standards (Partial List)

EIA Television Test Charts:
- Logarithmic Reflectance Chart
- Linear Reflectance Chart
- Color Registration Chart
- Resolution Chart
- Linearity (Ball) Chart
- CRL (Color Registration and Logarithmic Reflectance Charts)
- LR (Linearity and Resolution Charts)
- Color Calibration Chart
- Multi-Burst Chart
- Window Chart

Note: The test charts listed above are to be ordered from:

Hale Color Consultants
1220 Bolton Street
Baltimore, MD 21217
(301) 669-8631

EIA RS-312-A	Engineering Specifications Outline for Monochrome CCTV Camera Equipment
EIA RS-330	Electrical Performance Standards for Closed Circuit Television Camera 525/60 Interlaced 2:1
EIA RS-343-A	Electrical Performance Standards for High Resolution Monochrome Closed Circuit Television Camera
EIA RS-439	Engineering Specification for Color CCTV Camera Equipments

SMPTE Television Standards and Practices—See "Standards for Video Magnetic Tape and Motion Picture Film Recording Systems for Television" earlier in this chapter for a complete listing of SMPTE documents.

DIGITAL TELEVISION SYSTEMS

Basic Concepts

In the context of this section, digital television is taken to refer to the digitizing of the video (picture) signal. The principal application of digital television equipment in broadcasting at the present time is in the form of digital "black boxes." As shown in Fig. 18, a typical digital black box receives an analog video signal input and delivers an analog video signal output. The digital equipment is inserted into an otherwise analog video signal environment without any special attention being given to the digital nature of the box.

The internal functions of the black box of Fig. 18 are common to all digital systems. The input signal is applied to an analog-to-digital (a/d) converter. This device is generally called a "coder" and will be discussed in more detail shortly. The central portion performs some form of digital processing depending on the design function of the equipment. The last portion is a digital-to-analog (d/a) converter generally called a "decoder." The occurrence of coders and decoders in digital equipment is so common that the pair is generally referred to as a "codec." For each digital black box used in a broadcast operation, the video signal will pass through one codec.

The a/d converter discussed above involves several fundamental digital television concepts, namely sampling, quantizing, and coding. Each of these steps is discussed below.

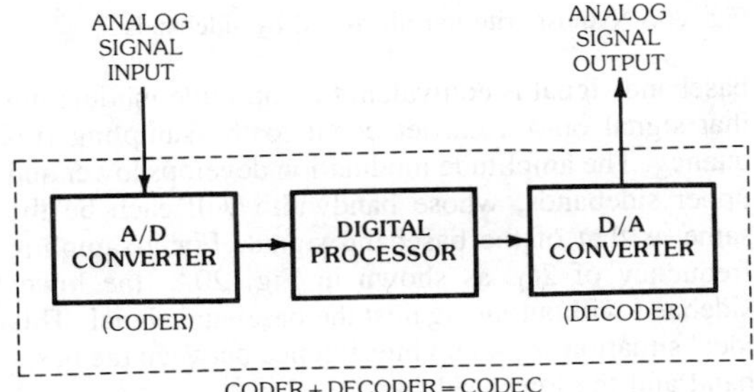

Fig. 18. Typical black-box digital device.

Sampling—Fig. 19 illustrates the sampling process. An input analog signal is shown as a graph of amplitude versus time. Periodically, the amplitude of the input signal is sampled, with the period between samples designated as t. The sampling frequency, f_s, equals $1/t$. The Nyquist criterion states that for a signal with a bandwidth of f_0 the sampling frequency should be equal to or greater than $2f_0$, i.e., twice the bandwidth of the signal being sampled. The Nyquist criterion can be simply illustrated by considering the sidebands of the sampled signal.

Fig. 20A shows a frequency spectrum in which the baseband signal has a bandwidth of f_0. Sampling the

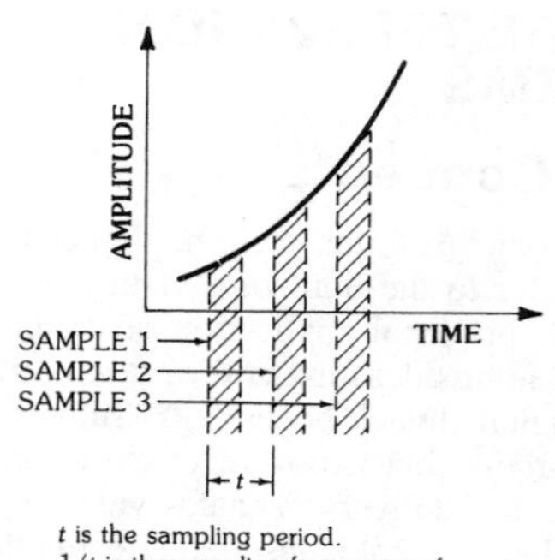

t is the sampling period.
$1/t$ is the sampling frequency $= f_s$.

Fig. 19. Sampling an analog signal.

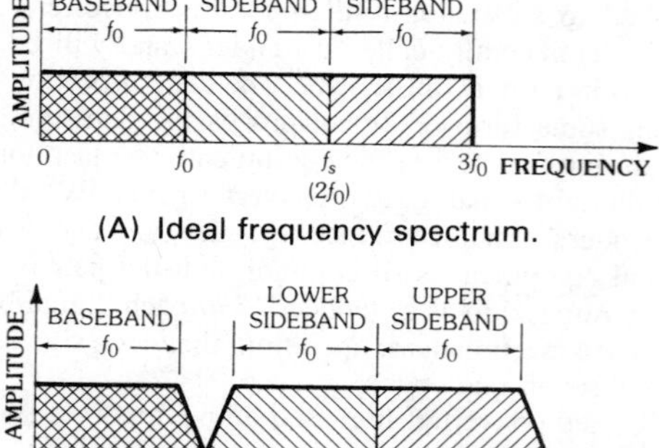

(A) Ideal frequency spectrum.

(B) Practical frequency spectrum.

Fig. 20. Nyquist criterion illustrated by sidebands.

baseband signal is equivalent to amplitude modulating that signal onto a carrier equal to the sampling frequency. The amplitude modulation develops lower and upper sidebands, whose bandwidth will each be the same as that of the baseband signal. For a sampling frequency of $2f_0$, as shown in Fig. 20A, the lower sideband will butt up against the baseband signal. This ideal situation shows no interference between the baseband and the lower sideband.

In actual equipment, a filter is used to control the bandwidth of the baseband signal. Since practical filters have finite attenuation slopes, the lower-sideband and baseband signals would overlap and cause what is called "aliasing" distortion. What is typically done, as shown in Fig. 20B, is to use a sampling frequency at least as great as $2.5f_0$. Even with a finite filter attenuation slope, the lower sideband will then not extend into the baseband spectrum. In much current North American digital equipment, the sampling frequency has been conveniently chosen to be either three or four times the television color subcarrier frequency (10.7 or 14.3 MHz, corresponding to 2.5 or 3.4 times the baseband bandwidth of 4.2 MHz).

The value of the sampling frequency is the fundamental digital television parameter. A small value lowers the digital bit rate of the system and minimizes the amount of storage and video tape required in broadcast operations. However, it causes filters to be complex and costly. A high value of sampling frequency has the opposite effects and also provides higher picture quality and ease of processing.

Quantizing—The second step in digitizing the analog signal is shown in Fig. 21. The single sample shown has found the analog signal to be at a specific level. Quantizing converts each of the various amplitude levels of a continuously varying analog signal into one of a group of selected values. All signals within a certain range of amplitude (shown by the brackets in the figure) are converted to one particular quantizing level.

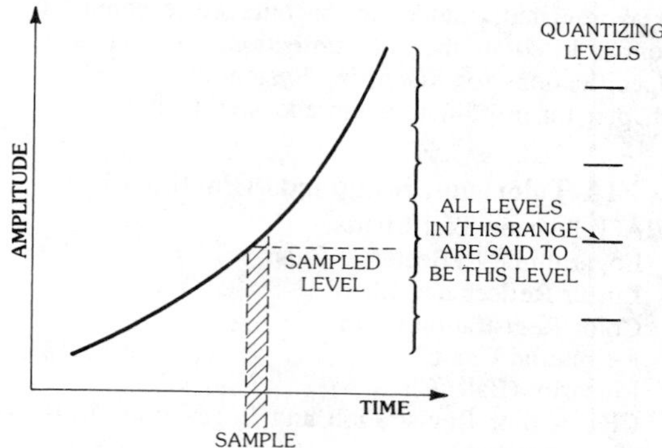

Fig. 21. The quantizing process.

Various subjective tests made by different organizations* have shown that 256 quantizing levels (equal to 2 to the eighth power) are required to digitize a television signal properly. Special situations may warrent a different number of levels.

When a smaller number of levels is used, two types of distortion may appear. "Contour" distortion may become apparent in areas of the picture having a relatively constant level. Slight variations in signal amplitude cause this area to assume one quantizing level or the next. The small but abrupt change in amplitude will be more noticeable than would a continuous analog variation.

Another type of distortion that becomes greater with smaller numbers of quantizing levels is called "quantizing noise." The fact that a particular sample of an analog signal has been converted to a slightly different value means that an error in amplitude occurs. These amplitude errors will generally be random and will appear as noise in the picture.

Usually the quantizing levels are evenly spaced.

---

* Devereux, V. G., "Pulse Code Modulation of Video Signals: Subjective Study of Coding Parameters," BBC Research Dept., Report No. 1971/40. Goldberg, A. A., "PCM Encoded NTSC Color Television Subjective Tests," SMPTE *Journal*, 82: 649–654, August 1973.

However, for special applications, nonlinear quantization in which the spacing between levels varies according to a prescribed law may be used.

Coding—The third step in digitizing an analog signal is shown in Fig. 22. For simplicity, only four quantizing levels are shown at the right side of the figure. Coding is the process of assigning a binary code to designate each of the quantizing levels. For the simplified case shown, two-bit coding may be used to obtain a unique designation for each level. If the levels shown are considered to be part of the 256 levels normally used, then the 8-bit binary codes also shown in the figure could typically be the code words for those four levels.

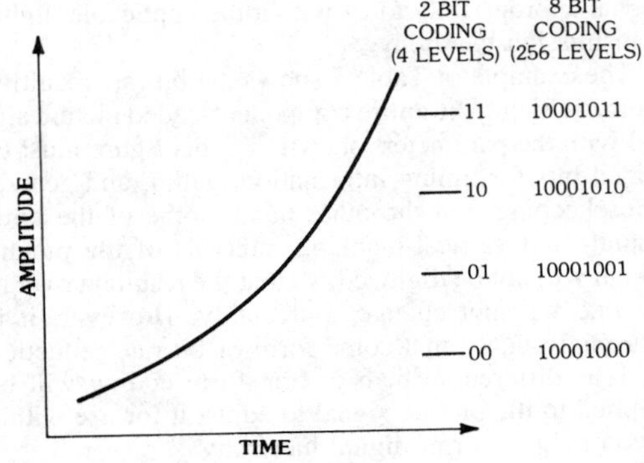

Fig. 22. Coding using binary numbers.

Composite and Component Signal Coding—The above discussion has assumed that the analog picture signal was in the conventional composite format; i.e., the signal contained the luminance component, the two chrominance signals (I and Q) modulated on a color subcarrier, and setup and synchronizing pulses. This format is typically the one that is distributed about a television studio, recorded on video tape recorders, transmitted to affiliated stations, etc. Fig. 23A shows composite coding.

For situations where a digital picture signal is distributed between equipment having digital input and output interfaces, component signal coding is generally preferred. In this picture-signal format, the red (R), green (G), and blue (B) output signals of the camera are converted by a linear matrix into a luminance (Y) signal and two color-difference (R – Y, B – Y) signals. The three component signals are then individually digitized by separate a/d converters as shown in Fig. 23B.

The three digital component signals may be combined together with timing and housekeeping information by means of a multiplexer unit into a serial bit stream for distribution over a single conductor. For some situations, a serial-parallel combination may be desirable. One arrangement currently being considered is to use eight signal conductors, each containing one bit position for the three video signal components. One or two additional conductors would distribute a clock and other information.

Composite signal coding has been chosen in a majority of the digital black boxes in use today by broadcasters because it provides a simple interface to associated analog equipment and does not require decoding and reencoding of the analog signal. It has the advantage of requiring a lower data rate than that needed for component coding; for a digital tape recorder, this means less tape usage. Its principal disadvantages are that it cannot easily be used for digital signal processing such as special effects and expanded pictures. Composite coding maintains the quality compromises inherent in the NTSC color system, such as restricted chrominance bandwidth, cross color, and edge effects. It leads to subcarrier phase problems and the four-field sequence problems of video tape program editing.

Component signal coding can provide wider luminance and chrominance bandwidths, the latter being essential for good chroma key and expanded pictures.

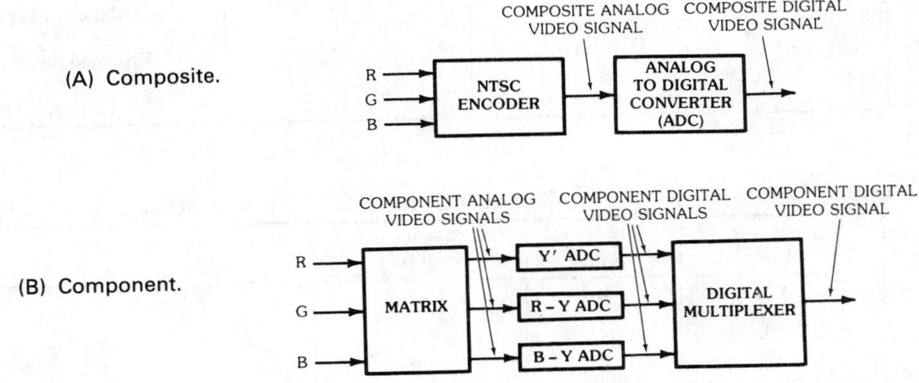

Fig. 23. Composite and component coding.

It also has the advantages of ease in complex picture signal processing and the elimination of subcarrier and four-field sequence complications in editing and other broadcast operations. Finally, adoption of component coding of the television picture signal has made possible the worldwide adoption of basic digital television parameters as listed in CCIR Recommendation 601, to be described later in this chapter.

Extensible Family of Digital Codes—Component coding schemes have generally made use of a simple ratio to relate the sampling frequencies of the luminance and color-difference signals. A simple ratio has the advantage of cositing the luminance and chrominance samples in the image plane. This characteristic makes for a stable image and ease of processing. A typical ratio is expressed as 4:2:2. The first digit, "4," is a symbol for the luminance-signal sampling frequency. The next two digits represent the sampling frequencies of the $R - Y$ and $B - Y$ color-difference signals, respectively. Thus, for the above ratio, if the luminance sampling frequency were 14 MHz, the color-difference sampling frequencies would each be 7 MHz. Although different-bandwidth digital color-difference signals have been proposed, only equal-bandwidth signals have been seriously considered.

Use of this ratio scheme allows a family of related ratios to be developed in a binary manner. Each member of the family can be associated with a particular quality level of the program. Table 6 lists several family members using CCIR sampling frequencies, together with possible program applications.

Bit-Rate Requirements—Table 7 shows the bit-rate requirements for the typical case of a 4:2:2 component-coded digital picture signal with a luminance sampling frequency of 13.5 MHz. This will probably be the most common input signal to a source coder for application to a distribution or transmission network. This network will frequently be a common-carrier, post office, or PTT system. The signal will have to be adjusted to fit into one of the standard bit rates of a system. The various carriers throughout the world have specific digital hierarchies of bit rates, and many of these hierarchies are different and nonrelated. The source coder will have to adjust the coded television signal appropriately to fit within the applicable digital transmission hierarchy.

The example of Table 7 shows the bit rate resulting from digitizing the entire component-coded picture signal with the parameters shown. To this figure must be added bits for timing information, audio, and general housekeeping. On the other hand, some of the horizontal- and vertical-blanking intervals of the picture signal will not be digitized, so that the resultant overall bit rate will not change appreciably. However, it is extremely likely that some form of bit-rate reduction such as differential pcm or transform coding will be applied to the picture signal to adjust it for use within the common-carrier digital hierarchy.

CCIR Recommendation 601

The Society of Motion Picture and Television Engineers (SMPTE)* in North America and the European Broadcast Union (EBU)† in Europe submitted papers

* SMPTE *Journal*, Volume 90, No. 10, October 1981 (entire issue).

† *EBU Technical Review*, No. 187, June 1981 (entire issue).

TABLE 6. EXTENSIBLE FAMILY OF DIGITAL CODES

Ratio	Sampling Frequency, MHz		Possible Application
	Y	R–Y, B–Y	
4:4:4	13.5	13.5	Original top-quality production
4:2:2	13.5	6.75	Studio distribution and exchange
4:1:1	13.5	3.375	ENG and documentaries
2:1:1	6.75	3.375	ENG and documentaries

TABLE 7. BIT-RATE CALCULATIONS

Component Signal	Ratio	Sampling Frequency	Quantization Bits	Bit Rate
Y	4	13.5	8	108
R − Y	2	6.75	8	54
B − Y	2	6.75	8	54
				Total 216 Mb/s

to the CCIR that resulted in Recommendation 601, approved by the CCIR Plenary Assembly in February 1982. Document 601 recommends that the following four sections "be used as a basis for digital coding standards for television studios in both the 525-line and 625-line areas of the world." Verbatim extracts from the document are enclosed in quotation marks.

Section 1: Component Coding

"The digital coding should be based on the use of one luminance and two colour-difference signals (or, if used, the red, green, and blue signals)." As stated in a later section, the red, green, and blue signals are intended only for optional use with the 4:4:4 member of the family.

Section 2: Extensible Family of Compatible Digital Coding Standards

"The digital coding should allow the establishment and evolution of an extensible family of compatible digital coding standards. The member of the family to be used for the standard digital interface between main digital studio equipment, and for international programme exchange (i.e. for the interface with video recording equipment and for the interface with the transmission system) should be that in which the luminance and colour-difference sampling frequencies are related in the ratio 4:2:2." This in effect defines the 4:2:2 coding digital scheme as the main studio system.

Section 3: Specifications Applicable to any Member of the Family

"Sampling structures should be spatially static. This is the case, for example, for the orthogonal sampling structure specified in Section 4 of the present recommendation for the 4:2:2 member of the family." Fig. 24 illustrates an orthogonal sampling structure with spatially static picture elements that do not change their horizontal position from line to line or from field to field. Nonorthogonal structures that had been considered are the "interleaved" or "quincunx" type, in which the picture elements of a line are shifted one half element spacing with respect to the elements of the previous line.

"If the samples represent luminance and two simultaneous colour-difference signals, each pair of colour-difference samples should be spatially cosited. If samples representing red, green, and blue signals are used, they should be cosited." Fig. 24 illustrates the cositing of samples. This characteristic is desirable to facilitate signal processing.

"The digital standard adopted for each member of the family should permit worldwide acceptance and application in operation; one condition to achieve this goal is that, for each member of the family, the number of samples per line specified for 625-line and 525-line systems shall be compatible (preferably the same number of samples per line)." This goal has been achieved for the 4:2:2 member of the family; Section 4 below specifies 720 samples per active line for both 525-line and 625-line television systems.

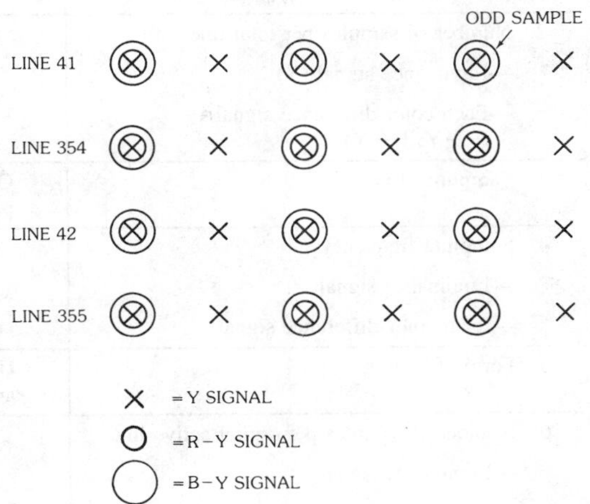

Fig. 24. Section of 625-line 4:2:2 digital sampling array showing orthogonal arrangement and cositing of luminance and color-difference signals.

Section 4: Encoding Parameter Values for the 4:2:2 Member of the Family

"The following specification (Table I) applies to the 4:2:2 member of the family, to be used for the standard digital interface between main digital studio equipment and for international programme exchange." Table I is here reproduced as Table 8. Note that the sampling frequency of 13.5 MHz fortuitously yields an integer number of samples per total horizontal scanning line for both the 525-line and 625-line television systems. All multiples of 2.25 MHz have this characteristic. Fig. 25 shows the relationships between this base frequency and other significant frequencies in the two systems.

The implementation of an all-digital television studio is not expected for some years. It may not be economically practical to completely convert an existing analog studio into a digital one. Perhaps creation of newly designed and constructed digital studios may be more feasible.

One gradual approach to the all-digital studio is the introduction of the "digital postproduction editing suite." Within this suite, the switching, signal processing, special effects, and successive tape generations of program editing would all be done with digital video signals. The well known advantages of a digital format would provide essentially distortion-free multigeneration recordings. Fig. 26 shows one embodiment of a digital editing suite with digital input and output.* This particular configuration is based on the

* Davidoff, Frank. "The All-Digital Television Studio," SMPTE *Journal*, 89, June 1980, pp. 445–449.

TABLE 8. ENCODING PARAMETER VALUES FOR THE 4:2:2 MEMBER OF THE FAMILY

Parameters	525-line, 60-field/s Systems	625-line, 50-field/s Systems
1. Coded signals	Y, R − Y, B − Y	
2. Number of samples per total line		
—Luminance signal (Y)	858	864
—Each color-difference signal (R − Y, B − Y)	429	432
3. Sampling structure	Orthogonal, line, field, and picture repetitive. R − Y and B − Y samples cosited with odd (1st, 3rd, 5th, etc.) Y samples in each line.	
4. Sampling frequency		
—Luminance signal	13.5 MHz	
—Each color-difference signal	6.75 MHz	
5. Form of coding	Uniformly quantized pcm, 8 bits per sample, for the luminance signal and each color-difference signal.	
6. Number of samples per digital active line		
—Luminance signal	720	
—Each color-difference signal	360	
7. Correspondence between video signal levels and quantization levels		
—Luminance signal	220 quantization levels with the black level corresponding to level 16 and the peak white level corresponding to level 235.	
—Each color-difference signal	224 quantization levels in the center part of the quantization scale with zero signal corresponding to level 128.	

assumption that each taking camera of a studio has its own dedicated digital video tape recorder (dvtr). If the cost of dvtr's follows the trend of analog vtr's, this appears very likely in the future. The digital tapes would be the direct input to the editing suite. The digital suite itself would not be very different functionally from existing analog suites, except that everything would be done digitally with component-coded signals. One exception would be the provision for the occasional analog or outside broadcast signal that might be an input to the suite. The line marked "Other Inputs" symbolizes the conversion means necessary for such signals to become compatible with the suite. The output of the suite could be a digital tape. This tape would be distributed to the broadcaster, who would play it back on a dvtr that would have a digital-to-analog converter at its output for distribution to other stations or for other purposes.

INTERNATIONAL BROADCASTING STANDARDS

CCIR Documents

Background—The CCIR (International Radio Consultative Committee) is a branch of the International Telecommunication Union of which most countries of the world are members through membership in the United Nations. The CCIR is concerned with preparing documents dealing with the preparation, transmission, and reception of all kinds of information by the use of "radio" signals, with radio being taken in its broadest sense. The CCIR has established several study groups, each dealing with a specific aspect of the overall terms of reference.

The study groups of interest in radio and television broadcasting are Study Group 10 on the Sound Broadcasting Service and Study Group 11 on the Television Broadcasting Service. Also of interest to broadcasters is the CMTT (Joint Committee on Television Transmission), a group set up jointly by the CCIR and the CCITT (International Telephone and Telegraph Consultative Committee). The latter organization is the branch of the International Telecommunication Union dealing with telegraph and telephone transmission. The CMTT is concerned with the transmission of signals over long distances through facilities that are common to both CCIR and CCITT disciplines.

The CCIR and CMTT produce documents of all kinds. Categories are Recommendations and Reports (the two most important), Questions, Study Programs, Decisions, Resolutions, and Opinions. The documents

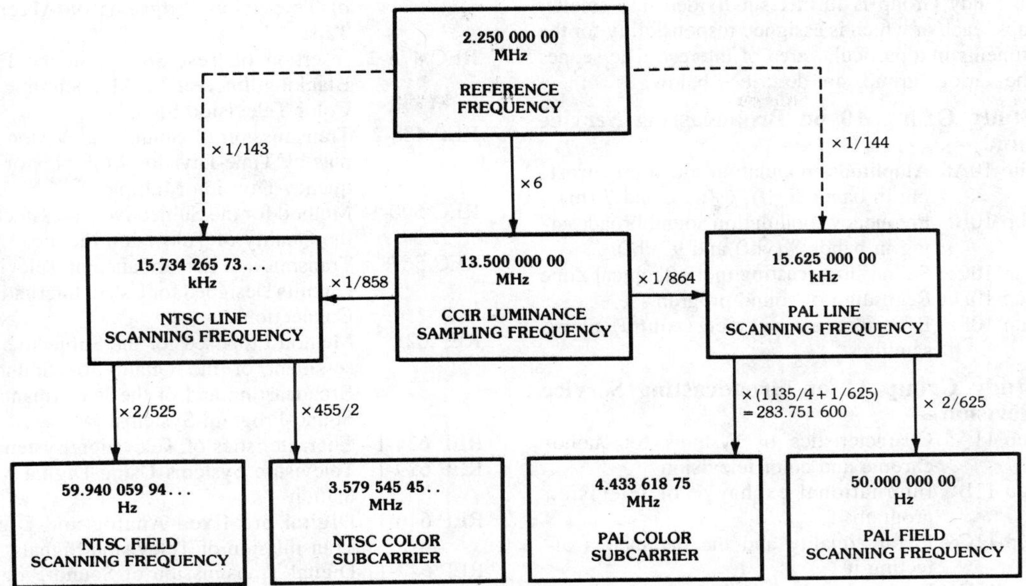

Fig. 25. Frequency relationships in NTSC and PAL color television systems.

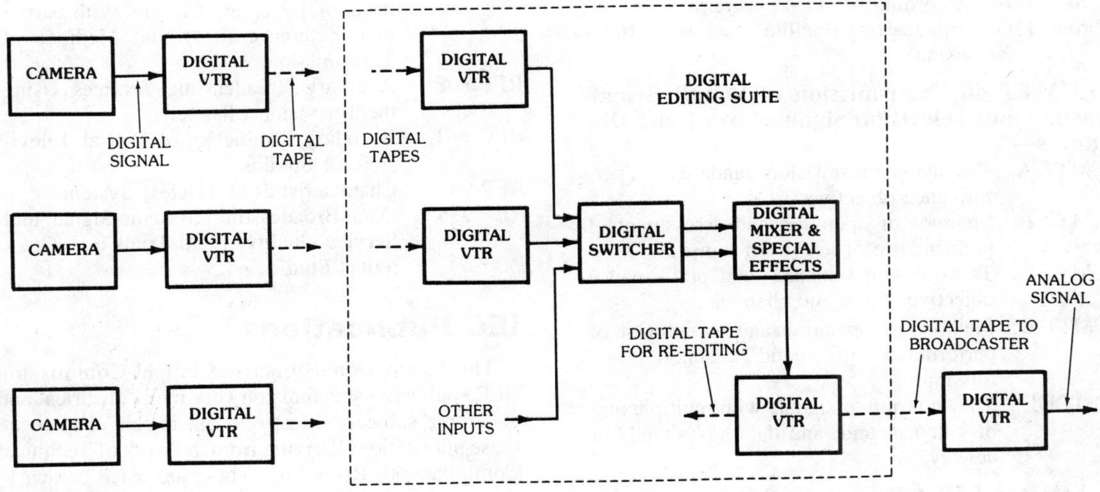

Fig. 26. Digital editing suite with digital input and output.

are reviewed every four years. During this period, there are two preliminary meetings of the Study Groups to review proposals submitted by the members. These proposals may take the form of superseding or modifying existing documents or submitting new ones. The agreements of the Study Groups after the second preliminary meeting are reviewed and generally approved at the final Plenary Assembly of the CCIR. The last Plenary Assembly took place in February 1982 and is commonly referred to as the CCIR XVth Plenary Assembly Geneva 1982.

The CCIR Plenary Assembly documents are published toward the end of the Plenary year in thirteen volumes, generally one for each Study Group. The volumes are bound in paper, with a green cover for the English text, and are colloquially known as the "Green Books." They are available from the International Telecommunication Union in Geneva, Switzerland and in the United States from the United Nations bookstore in New York City and the National Technical Information Service (see "International Sound-Recording Standards," above, for address).

Each Study Group is further subdivided into smaller groups, each of which is assigned responsibility for the documents in a particular area of interest. The scopes of the various groups are described below.

Study Group 10 on Broadcasting Service (Sound)—

Group 10A: Amplitude-modulation sound broadcasting in bands 5 (lf), 6 (mf), and 7 (hf)
Group 10B: Frequency-modulation sound broadcasting in bands 8 (vhf) and 9 (uhf)
Group 10C: Sound broadcasting in the Tropical Zone
Group 10D: Recording of sound programs
Group 10E: Broadcasting service (sound) using satellites

Study Group 11 on Broadcasting Service (Television)—

Group 11A: Characteristics of systems for monochrome and color television
Group 11B: International exchange of television programs
Group 11C: Picture quality and the parameters affecting it
Group 11D: Elements and methods for planning
Group 11E: Television systems using digital modulation
Group 11F: Recording of video programs
Group 11G: Broadcasting-satellite service (television)

CMTT on Transmission of Sound Broadcasting and Television Signals Over Long Distances—

CMTT A: Television transmission standards and performance objectives
CMTT B: Methods of operation and assessment of performance of television transmissions
CMTT C: Transmission standards and performance objectives for sound channels
CMTT D: Methods of operation and assessment of performance of sound channel transmissions
CMTT E: Transmission of signals with multiplexing of video, sound, and data and signals of new systems.

Selected CCIR Reports and Recommendations—Listed below are some of the more important Reports (''REP'') and Recommendations (''REC'') issued by the CCIR in the broadcasting and transmission areas.

REP 215-4 Systems for the Broadcasting-Satellite Service (Sound and Television)
REC 265-3 Standards for the International Exchange of Monochrome and Color-Television Programs on Film
REP 306-3 Ratio of Wanted-to-Unwanted Signal for Color Television
REP 418-3 Ratio of Wanted-to-Unwanted Signal in Monochrome Television
REC 469-2 Standards for the International Exchange of Television Programs on Magnetic Tape
REC 473-2 Insertion of Test Signals in the Field-Blanking Interval of Monochrome and Color Television Signals
REP 488-2 Transmission of Sound and Vision Signals by Time-Division Multiplex or Frequency-Division Multiplex
REC 500-1 Method for the Subjective Assessment of the Quality of Television Pictures
REC 567 Transmission Performance of Television Circuits Designed for Use in International Connections
REP 623 Method Proposed for the Subjective Assessment of the Quality of Sound in Broadcasting and of the Performance of Sound-Program Systems
REP 624-1 Characteristics of Television Systems
REP 629-1 Television Systems Using Digital Modulation
REP 646-1 Digital or Mixed Analog-and-Digital Transmission of Television Signals
REP 647-1 Digital Transmission of Sound-Program Signals
REP 649 Transmission Performance of the Hypothetical Reference Circuit for High Quality Sound Program Circuits With Particular Reference to Digital Methods of Transmission
REP 802 Ancillary Broadcasting Services Using the Television Channel
REC 601 Encoding Parameters of Digital Television for Studios
REP 957 Characteristics of Teletext Systems
REP 956 Data Broadcasting Systems-Signal and Service Quality, Field Trials and Theoretical Studies

IEC Publications

The International Electrotechnical Commission (IEC) publishes standards on very many electrical and electronic subjects. Most relevant to this chapter are those publications deriving from the work of Technical Committee 60, Recording. These are listed below:

94—Magnetic Tape Sound Recording and Reproducing Systems

94 (1968) Magnetic Tape Recording and Reproducing Systems
Amendment No. 1 (1971)
Amendment No. 2 (1973)
Amendment No. 3 (1976)
Amendment No. 4 (1978)

94A (1972) First Supplement: Cassette for Commercial Tape Records and Domestic Use. Dimensions and Characteristics.
Amendment No. 1 (1972)
Amendment No. 2 (1976)

94B (1974) Second Supplement: Eight-Track Endless Loop Magnetic Tape Cartridge. Dimensions and Characteristics.

94-1 (1981) Part 1: General Conditions and Requirements

94-2 (1975) Part 2: Calibration Tapes

94-3 (1979) Part 3: Methods of Measuring the Characteristics of Recording and Reproducing Equipment for Sound on Magnetic Tape
Amendment No. 1 (1980)

98 (1964) Processed Disk Records and Reproducing Equipment
Amendment No. 1 (1967)
Amendment No. 2 (1971)
Amendment No. 3 (1972)
Amendment No. 4 (1976)

98A (1972) First Supplement: Methods of Measuring the Characteristics of Disk Record Playing Units

347 (1972) Transverse Track Recorders
Amendment No. 1 (1980)

386 (1972) Method of Measurement of Speed Fluctuations in Sound Recording and Reproducing Equipment

461 (1974) Time and Control Code for Video Tape Recordings

503 (1975) Spools for 1 in (25.4 mm) Video Magnetic Tape

511 (1975) Helical-Scan Video-Tape Cassette System Using 0.5 in (12.70 mm) Magnetic Tape (50 Hz-625 Lines)

511A (1977) First Supplement

558 (1982) Type C Helical Video Tape Recorders

574—Audio-Visual, Video and Television Equipment and Systems

574-1 (1977) Part 1: General

574-2 (1977) Part 2: Explanation of General Terms

574-5 (1980) Part 5: Chapter 1: Synchronized Tape/Visual Operating Practice

574-8 (1979) Part 8: Symbols and Identification

574-10 (1977) Part 10: Audio Cassette Systems

602 (1980) Type B Helical Video Recorders

608 (1977) Interconnections Between Video-Tape Recorders and Television Receivers for 50 Hz, 625 Lines Systems

698 (1982) Measuring Methods for Television Tape Machines

712 (1982) Helical-Scan Video-Tape Cassette System Using 19 mm ($^3/_4$ in) Magnetic Tape, Known as U-Format

ISO Recommendations

The International Organization for Standardization (ISO) is primarily involved in the specification of mechanical systems. Electrical systems are the province of the IEC. Accordingly, standards for motion picture films are drafted by ISO Technical Committee 36, Cinematography. A list of the few ISO TC 36 documents relevant to television broadcasting appears below:

ISO 23-1976 Camera Usage of 35 mm Motion Picture Film—Specifications

ISO 26-1976 Camera Usage of 16 mm Motion Picture Film—Specifications

ISO 1188-1974 Recording Characteristic for Magnetic Sound Record on 16 mm Motion Picture Film—Specifications

ISO 1189-1975 Recorded Characteristic for Magnetic Sound Records on 35 mm Motion Picture Film—Specifications

ISO 3640-1976 Motion Picture Prints and Sound Records for International Exchange of Television Programs—Specifications

36 Radar

Merrill I. Skolnik

Radar is an electromagnetic device for the detection and location of reflecting objects such as aircraft, ships, satellites, and the natural environment. It operates by transmitting a known waveform, usually a series of narrow pulses, and observing the nature of the echo signal reflected by the target back to the radar. In addition to determining the presence of targets within its coverage, the basic measurements made by a radar are *range* (distance) and *angular location*. The Doppler frequency shift of the echo from a moving target is sometimes extracted as a measure of the *relative velocity*. The Doppler shift is also important in cw (continuous wave), MTI (moving target indication), and PD (pulse Doppler) radars for separating desired moving targets (such as aircraft) from large undesired fixed echoes (such as ground clutter). In addition to the usual measurements of range, angular location, and relative velocity, radar sometimes can obtain information about the size, shape, symmetry, surface roughness, and surface dielectric constant of a target.

Many microwave radars use a magnetron oscillator as the transmitter; but when high-average-power or controlled-modulation waveforms are required, the transmitter is often a power amplifier like the klystron, traveling-wave tube, grid-control tube, or crossed-field amplifier. The solid-state transmitter, usually based on the microwave transistor, has potential as a radar transmitter with good reliability and maintainability. Low-noise receivers, such as the transistor or the parametric amplifier, have also found their way into modern radar. Sophisticated doppler processing techniques for MTI radar have been reduced to practice because of the availability of low-cost, small-size digital circuitry. The small size and low cost of digital computers make it possible to automatically detect and accurately track many hundreds of targets simultaneously so as to present to the operator fully processed tracks rather than raw radar data. The parabolic-reflector antenna has been the antenna most commonly employed with operational radar systems. The phased-array antenna with agile beam-steering controlled by electronic phase shifters is of interest for radar applications because of the ease and rapidity with which its beam can be pointed anywhere within its coverage. However, the cost and complexity of electronically steered phased-array antennas that cover a wide angular region tend to restrict their application. On the other hand, phased arrays whose beam is scanned only over limited angular regions are more practical and have seen wide use as 3D air-surveillance radars, aircraft landing radars, and hostile-weapons location radars.

The range resolution of a radar can be of the order of a fraction of a meter, if desired; but the beamwidths that are practical with a conventional microwave antenna limit the resolution in the angle coordinate (or cross-range dimension) to much larger values. It has been possible to synthesize the effect of a large antenna and thus overcome the cross-range limitation by employing the radar on a moving vehicle (such as an aircraft) and coherently storing the received echoes in an electronic or photographic memory for a time duration equivalent to the length of a large antenna. This technique for achieving cross-range resolution comparable to the resolution that can be obtained in the range dimension is called *synthetic aperture radar* (SAR). The output of such a radar is a map or image of the target scene. The use of a stationary radar to image a moving or rotating target by using resolution in the doppler domain is called *inverse SAR*.

Radar is generally found within what is known as the microwave region of the electromagnetic spectrum. It is possible, however, to apply the radar principle at hf frequencies to obtain the advantage of long "over-the-horizon" ranges by refraction of the radar waves by the ionosphere. Aircraft can be detected by one-hop ionospheric propagation out to ranges of about 2000 nmi. Radar has also been considered for use at frequencies higher than the microwave region, at millimeter wavelengths. Laser radars are found in the IR and optical regions of the spectrum, where they offer precision range and relative-velocity measurement.

PREDICTION OF RADAR RANGE*

The simple form of the radar range equation is

$$P_r = P_t G A_e \sigma / (4\pi)^2 R^4 \qquad (1a)$$

where,

P_r = received echo signal power in watts,
P_t = transmitted signal power in watts,
G = antenna gain,
A_e = antenna effective area in square meters,
σ = radar cross section of the target in square meters,
R = range to the target in meters.

If a single antenna is used for both transmitting and receiving, as is usually the case, $G = 4\pi A_e/\lambda^2$, where λ is the radar wavelength in meters. Then

$$P_r = P_t G^2 \lambda^2 \sigma / (4\pi)^3 R^4 = P_t A_e^2 \sigma / 4\pi\lambda^2 R^4 \qquad (1b)$$

The maximum range R_{max} of a radar occurs when $P_r = S_{min}$, the minimum detectable signal. The minimum detectable signal is a statistical quantity limited by receiver noise. It can be written as

$$S_{min} = kT_o B F_n (S/N)_1 \qquad (2)$$

where,

k = Boltzmann's constant,
T_o = absolute temperature (290 K),
B = receiver bandwidth in hertz,
F_n = receiver noise figure,
$(S/N)_1$ = minimum signal-to-noise ratio required for reliable detection.

* Reference 1.

The received echo signal power can be increased by integrating (adding) a number of echo signal pulses n. This can be incorporated into the radar equation by dividing S_{min} by $nE_i(n)$, where $E_i(n)$ is the efficiency with which the n pulses can be integrated. Since the average power P_{av} is more indicative of radar capability than is the peak power, it is introduced via the relation

$$P_{av} = P_t \tau f_p \quad (3)$$

where,

τ = pulse width in seconds,
f_p = pulse repetition frequency in hertz.

With the above, the form of the radar equation suitable for calculating the range is

$$R_{max} = \left[\frac{P_{av} G^2 \lambda^2 \sigma \, nE_i(n)}{(4\pi)^3 kT_o F_n \, (B\tau) f_p (S/N)_1 L_s} \right]^{1/4} \quad (4)$$

The radar system losses L_s (number greater than one) have been included. For most radars designed with a matched filter receiver (a filter that maximizes the output signal-to-noise ratio), the product $B\tau \simeq 1$. At room temperature, $kT_o = 4 \times 10^{-21}$ W/Hz. [In Eq. 4, $(S/N)_1/nE_i(n)$ is the required signal-to-noise ratio per *pulse* $(S/N)_n$.]

Fig. 1 shows the relationship of the required signal-to-noise ratio $(S/N)_1$ to the probability of detection and the probability of false alarm. The probability of detection, also sometimes called the blip-scan ratio, is

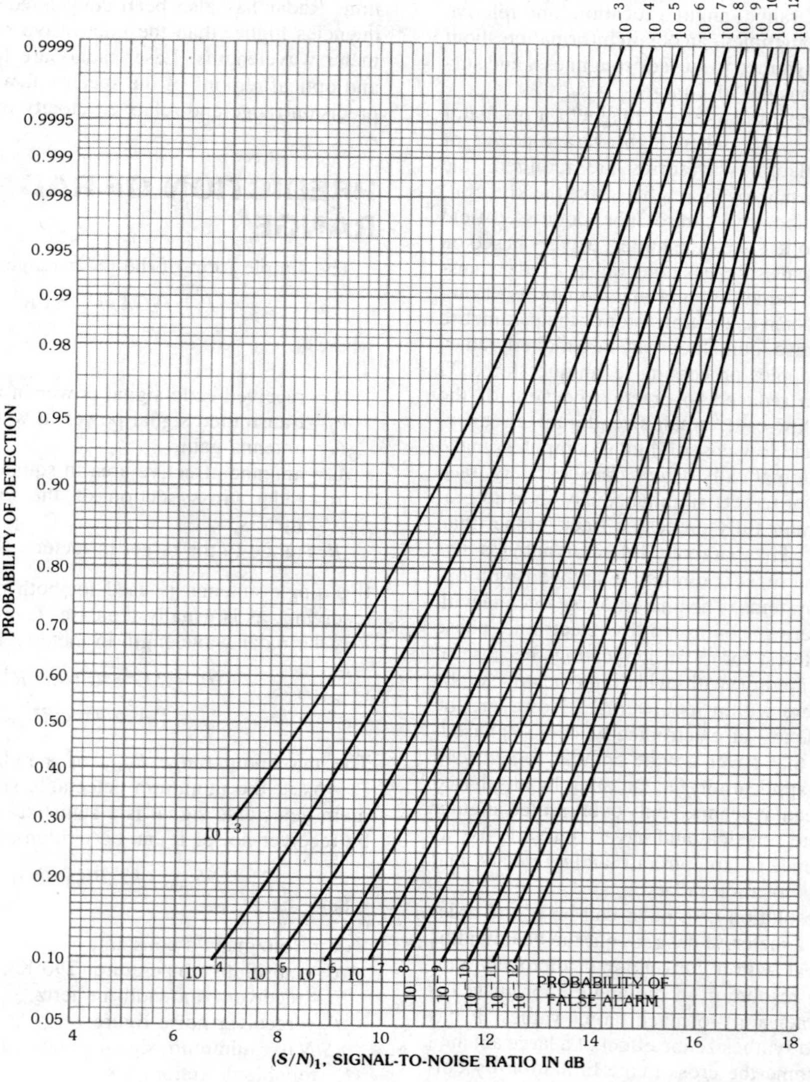

Fig. 1. Probability of detection for a sine wave in noise as a function of the signal-to-noise (power) ratio and the probability of false alarrm. (*Courtesy McGraw-Hill Book Co.*)

usually taken as 0.90, but its choice is usually the prerogative of the customer. The probability of a false alarm for a single pulse is

$$P_{fa} = 1/BT_{fa}$$

where,

B = receiver bandwidth in hertz,
T_{fa} = average time between false alarms.

The reciprocal of P_{fa} is n_f, the false alarm number. However, it is the false alarm time T_{fa} that is usually specified for radar performance rather than the probability of false alarm.

Fig. 2 is a plot of the integration improvement factor $nE_i(n)$ as a function of n. The number of pulses returned from a target when an antenna of beamwidth θ_B degrees rotates at a rate of ω_m revolutions per minute, with a pulse repetition rate of f_p Hz is

$$n = \theta_B f_p / 6\omega_m \qquad (5)$$

Failure to include the many factors that contribute to the system losses L_s can result in considerable difference between the calculated and actual range. Losses include:

Loss in the transmission line connecting the antenna to the transmitter and receiver

Loss in duplexer, rotary joint, and other microwave components

Beam-shape loss, to account for the fact that the radar equation employs the maximum gain rather than a gain that varies pulse to pulse as the antenna is scanned past the target

Collapsing loss, when additional noise samples are integrated along with the wanted signal-to-noise pulses

Loss due to degradation of transmitter power and receiver noise figure through use

FLUCTUATING TARGET MODELS

The complex nature of most radar targets causes the radar cross section to vary with changing aspect. The fluctuations in the radar cross section are difficult to specify precisely; however, four simple statistical models first described by P. Swerling (Ref. 2) are often used for computing the radar range. The statistical nature of the radar cross section σ for cases 1 and 2 is described by the probability density function

$$p(\sigma) = (1/\sigma_{av})\ e^{-(\sigma/\sigma_{av})} \qquad (6a)$$

where σ_{av} is the average cross section. Cases 3 and 4 are described by

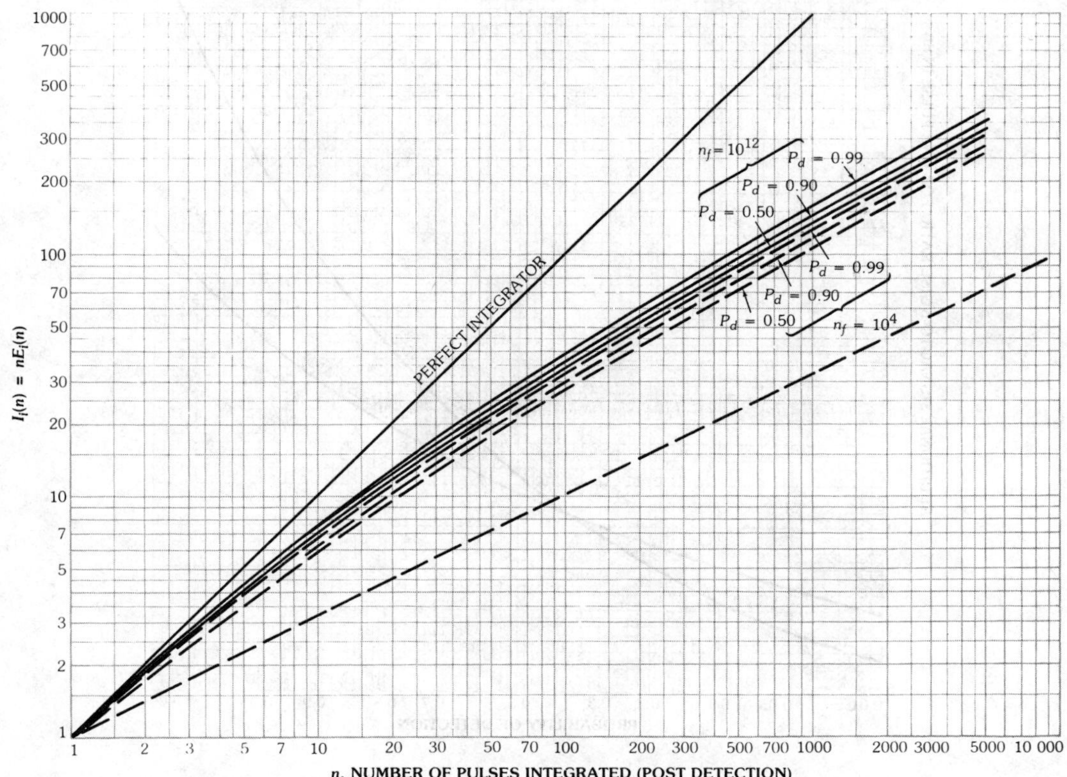

Fig. 2. Integration-improvement factor, square-law detector, P_d = probability of detection, $n_f = T_{fa}B$ = false alarm number, T_{fa} = average time between false alarms, B = bandwidth. (*Courtesy McGraw-Hill Book Co.*)

$$p(\sigma) = (4\sigma/\sigma_{av})\ e^{-(2\sigma/\sigma_{av})} \qquad (6b)$$

In cases 1 and 3, it is assumed that the echo pulses received from a target on any one scan are of constant amplitude throughout the entire scan, but are independent (uncorrelated) from scan to scan. In cases 2 and 4, the cross section is assumed to be independent from pulse to pulse. When these models are used, the signal-to-noise ratios and the integration improvement factors inserted in the radar equation are different from those used for constant cross section. The required values can be found in several references (Refs. 3–6); however, a simple approximate method (Ref. 1) valid for most purposes can be obtained using Figs. 1, 3, and 4 with the following procedures:

A. Find the signal-to-noise ratio from Fig. 1 corresponding to the desired values of detection probability P_d and false-alarm probability $P_{fa} = 1/BT_{fa}$.
B. From Fig. 3, determine for the desired Swerling case the correction factor to be applied to the signal-to-noise ratio found from Fig. 1.
C. Find the integration improvement factor $nE_i(n)$ from Fig. 4.

A more general expression for describing the statistical properties of radar target cross section is the chi-square distribution of degree $2m$, whose probability density function is

$$p(\sigma) = [m/(m-1)!\sigma_{av}](m\sigma/\sigma_{av})^{m-1}\ \exp\ (-m\sigma/\sigma_{av}) \qquad (7)$$

The Swerling cases 1 and 2 are given by the chi-square distribution with $m = 1$; the other two cases are given with $m = 2$.

RADAR CROSS SECTION OF TARGETS

Expressions for the radar cross sections of simple target shapes are given in Table 1. The radar cross

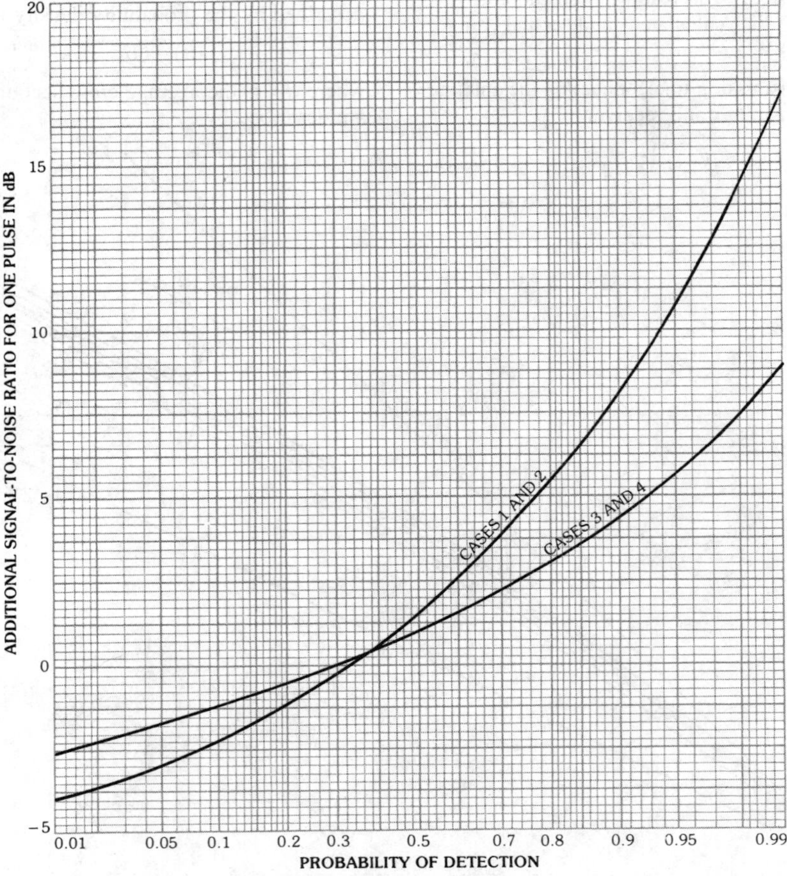

Fig. 3. Additional signal-to-noise ratio required to achieve a particular probability of detection, when the target cross section fluctuates according to the Swerling models, as compared with a nonfluctuating target; single hit, $n = 1$. To be used with Fig. 1 to find $(S/N)_1$. (*Courtesy McGraw-Hill Book Co.*)

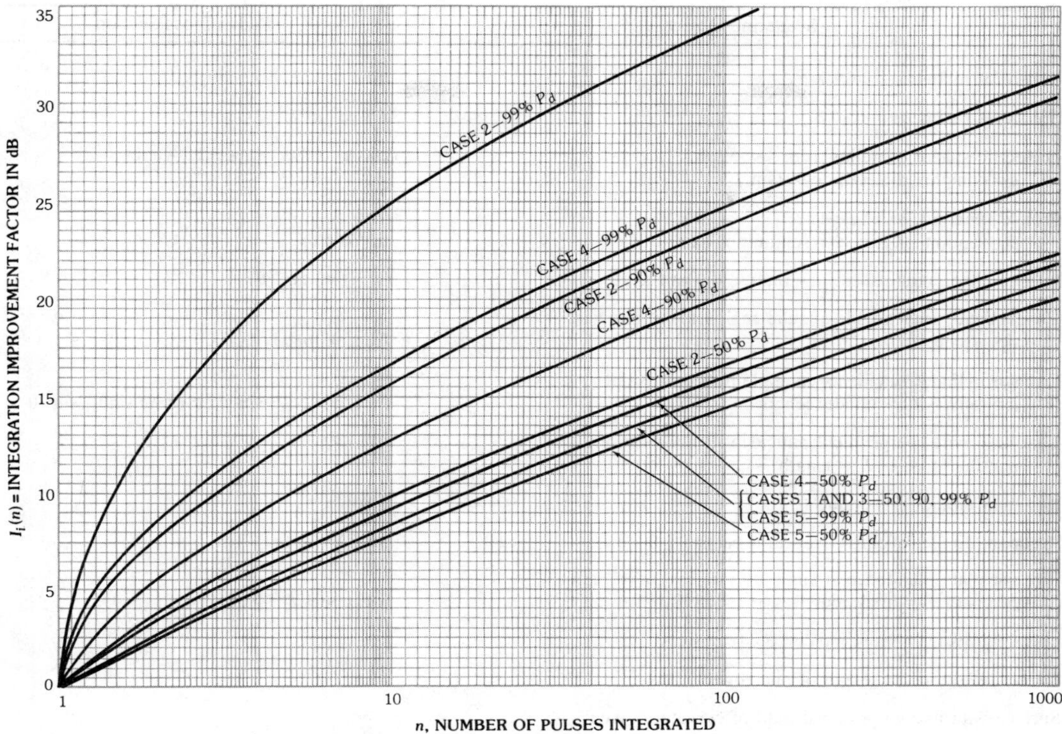

Fig. 4. Integration-improvement factor as a function of the number of pulses integrated for the four Swerling cases (cases 1 to 4) and the constant cross section (case 5). (*Courtesy McGraw-Hill Book Co.*)

section of a complex target, however, varies considerably with change in aspect or change in frequency so that a single number cannot adequately describe the radar cross section of a target. Nevertheless, Table 2 lists "example" values for various targets at microwave frequencies. These are for illustrative purposes to show the relative "sizes" of common targets as "seen" by radar.

OTHER FORMS OF THE RADAR EQUATION

The radar equation is not used only for the calculation of range; it can also provide a basis for assessing tradeoffs in radar design. The simple forms of the radar equation given above (Eqs. 1 and 2) are seldom sufficiently complete, however, and they must usually be extended. Each specific radar application has some particular requirements or constraints that can result in a slightly different form of the radar equation. Examples are presented below. All symbols used in the equations are defined at the end of this section.

Surveillance Radar Equation

$$R_{max}^4 = \frac{P_{av} A_e \, \sigma \, E_i(n)}{4\pi k T_o F (S/N)_1 \, L_s} \cdot \frac{t_s}{\Omega} \qquad (8)$$

This equation applies to a radar that must observe all targets within an angular region of solid angle Ω steradians once every t_s seconds. When the surveillance radar utilizes a conventional rotating fan beam whose elevation beamwidth is θ_e, the solid angle Ω equals $2\pi \sin \theta_e$, and t_s is the azimuth rotation period.

Tracking Radar Equation

Equation 4 is basically the tracking radar equation, where $n/f_p = t_o$ is the signal integration time. (It has also been called the searchlight equation.)

Surface Clutter Range Equation

$$R_{max} = \sigma \, n_e / (S/C)_o \, \sigma^\circ \, \theta_a \, (c\tau/2) \sec \psi \qquad (9)$$

This equation describes the detection of a target when viewed at a low grazing angle ψ in the presence of surface clutter of radar cross section per unit area σ°. It assumes that the received clutter echo power is much greater than receiver noise. The effective number of pulses integrated, n_e, will depend on the decorrelation time of the clutter. (With stationary clutter, there is no integration improvement, and n_e = 1.)

TABLE 1. RADAR CROSS SECTION OF SIMPLE SHAPES

Scattering Object	Radar Cross Section
Sphere of radius a, $a >> \lambda$	πa^2
Sphere of radius a, $a << \lambda$	$144\ \pi^5\ a^6/\lambda^4$
Dielectric sphere of radius a, $a << \lambda$, ϵ = complex dielectric constant	$256\ \pi^5\ (a^6/\lambda^4)\ \lvert (\epsilon-1)/(\epsilon+1) \rvert^2$
Prolate spheroid, axial aspect, a = semimajor axis, b = semiminor axis	$\pi b^4/a^4$
Infinite cone, axial aspect, a = radius, θ = half angle	$(\lambda^2/16\pi)\ \tan^4\theta$
Cone-sphere, over a wide angular region looking toward tip of core	Approximately $0.1\ \lambda^2$
Circular cylinder, incidence at angle θ to broadside, a = radius, L = cylinder length	$(a\lambda/2\pi)\ \{\cos\theta\ \sin^2\ [2\pi\ (L/\lambda)\ \sin\theta]\}/\sin^2\theta$
Long thin wire, perpendicular incidence, polarization parallel to wire, L = length	L^2/π
Resonant dipole, max value	$0.866\ \lambda^2$
Resonant dipole, average over all aspects	$0.15\ \lambda^2$
Circular plate, incidence at angle θ to normal, a = radius of plate, J_1 = first-order Bessel function	$\pi a^2\ \cot^2\theta\ J_1{}^2[4\pi(a/\lambda)\sin\theta]$
Large flat plate, normal incidence, A = area	$4\pi\ A^2/\lambda^2$
Corner reflector, a = length of one edge	$4\pi\ a^4/3\lambda^2$
Surface with its two principal radii of curvature given by R_1, R_2	$\pi\ R_1\ R_2$

Volume Clutter Radar Equation

$$R^2_{\max} = \sigma\ G\ n_e/(S/C)_o\ \eta\ (\pi^3/4)\ (c\tau/2) \qquad (10)$$

The reflectivity η is the radar cross section of the clutter per unit volume.

Noise Jamming Radar Equation (Surveillance)

$$R^2_{\max} = \frac{P_{\text{av}}\ G\ E_i(n)}{G_{SL}L_s} \cdot \frac{\sigma}{(S/N)_1} \cdot \frac{t_s}{\Omega} \cdot \frac{Bj}{P_j\ G_j} \qquad (11)$$

This equation assumes that the jamming noise enters the antenna side lobes whose gain is G_{SL}. When the jamming enters the main beam, $G_{SL}=G$. The jammer power P_j is spread over a bandwidth B_j and is radiated by an antenna gain G_j.

Noise Jamming Radar Equation (Tracking)

$$R^2{}_{\max} = \frac{P_{\text{av}}\ G^2\ E_i(n)\ t_o}{4\pi\ G_{SL}\ L_s} \cdot \frac{\sigma}{(S/N)_1} \cdot \frac{Bj}{P_j\ G_j} \qquad (12)$$

When the jamming noise enters the radar via the main beam, $G_{SL}=G$.

Self-Screening Range Equation

This is the range at which the radar echo signal S received from a target exceeds the received jamming noise power J by the amount S/J. It is also called the *cross-over range*. The self-screening range is found from either Eq. 11 or 12 (depending on the application) by setting $G_{SL}=G$, setting $(S/N)_1=S/J$, and calling $R_{\max}$ the self-screening range R_{ss}. The value of required S/J is often taken to be the same as $(S/N)_1$ found for receiver noise.

Weather Radar Equation

$$\overline{P}_r = 2.4\ P_t\ G\ \tau\ r^{1.6}/R^2\lambda^2 L_s \qquad (13)$$

This equation is employed by radar meteorologists to relate the average echo signal power $\overline{P}_r$ to the rainfall rate r (mm/hr). It assumes that rain uniformly fills the radar resolution cell.

Synthetic Aperture Radar Equation

$$\frac{S}{N} = \frac{2\ P_{\text{av}}\ \rho_a^2\ \sigma^\circ\ \delta_{cr}\ \delta_r}{\pi f\ kT_oF_n\ R\ S_w\ L_s\ \sin^2\psi} \qquad (14)$$

TABLE 2. "EXAMPLE" VALUES OF RADAR CROSS SECTION

Target	σ (square meters)
Conventional unmanned winged missile	0.5
Small single-engine aircraft	1
Small fighter, or 4-passenger jet	2
Large fighter	6
Medium bomber or medium jet airliner	20
Large bomber or large jet airliner	40
Jumbo jet	100
Small open boat	0.02
Small pleasure boat	2
Cabin cruiser	10
Ship, grazing angle greater than zero	Displacement tonnage expressed in m^2
Pickup truck	200
Automobile	100
Bicycle	2
Man	1
Bird	0.01
Insect	10^{-5}

This equation relates the signal-to-noise ratio of a resolution cell (sometimes called a pixel) with range resolution δ_r and cross-range resolution δ_{cr} located within a swath S centered at a range R. The above takes account of the combined restriction on cross-range resolution and swath necessary to avoid ambiguities in either range or cross range. Another form of the equation which does not account for this restriction is

$$S/N = P_{av}\, A^2_e\, \sigma^\circ \sec \psi / 8\pi\, \lambda k T_o F_n\, v\, R^3 \qquad (15)$$

The velocity of the vehicle carrying the radar is v.

HF Over-the-Horizon Radar Equation

$$R^2_{max} = \frac{P_{av} G_t\, G_r\, \lambda^2\, \sigma F_p^2\, T_c}{(4\pi)^3\, N_o\, (S/N)_1\, L_s} \qquad (16)$$

The transmitting antenna gain G_t and the receiving antenna gain G_r are shown separate since two different antennas are often used for transmit and receive. The propagation loss is accounted for by F_p (number less than unity), and T_c is the coherent processing time. The noise power per unit bandwidth N_o (W/Hz) at the receiver is determined by external noise.

Laser Radar Equation

$$R^4_{max} = P_t\, A_e\, \eta_q\, \sigma / 32\, \theta^2_B\, n_p\, hfB\, L_s \qquad (17)$$

The quantum efficiency of the detector is η_q, the required number of signal photoelectrons is n_p, and h = Planck's constant. The above assumes the laser beam is larger than the target. When the target is larger than the laser beam

$$R^2_{max} = \pi P_t\, A_e\, \eta_q\, \rho \sin \psi / 32\, n_p hfB\, L_s \qquad (18)$$

where ρ = surface reflection coefficient.

Bistatic Radar Equation

$$P_r = P_t G_t G_r \lambda^2 \sigma_b / (4\pi)^3 D_t^2 D_r^2 L_s \qquad (19)$$

The transmitter-to-target distance is D_t, and the target-to-receiver distance is D_r. The bistatic cross section is designated σ_b.

Symbol Definitions

The symbols used in the above radar equations are defined as follows.

A_e = antenna effective aperture in square meters
B = receiver bandwidth in hertz
B_j = jammer bandwidth in hertz
c = velocity of propagation in meters/second
D_r = range from target to receiver in meters
D_t = range from transmitter to target in meters
δ_{cr} = cross-range resolution in meters
δ_r = range resolution in meters
$E_i(n)$ = efficiency in integrating n pulses
η = volume clutter of reflectivity, or radar cross section of clutter per unit volume, in meters^{-1}
η_q = quantum efficiency
f = frequency in hertz
f_p = pulse repetition frequency in hertz
F_n = receiver noise figure
F_p = propagation factor
G = antenna gain
G_j = jammer antenna gain
G_r = receiving antenna gain
G_{SL} = antenna side lobe gain
G_t = transmitting antenna gain
h = Planck's constant, 6.62×10^{-34} joule-seconds
k = Boltzmann's constant = 1.38×10^{-23} joules/degree
L_s = system losses
λ = wavelength in meters
n = number of echo pulses received per target
n_e = effective number of pulses integrated
n_p = number of signal photoelectrons
N_o = noise power per unit bandwidth
Ω = solid angular region (steradians) of radar coverage
P_{av} = average power in watts

P_j = jammer power in watts
$\bar{P}_r$ = received signal power in watts
P_t = peak power in watts
ψ = grazing angle
r = rainfall rate in millimeters/hour
R = range in meters
R_{max} = maximum radar range in meters
ρ = surface reflection coefficient
ρ_a = antenna efficiency
$(S/C)_o$ = minimum signal-to-clutter ratio necessary to detect a target with a specified probability of detection and probability of false alarm, for a single pulse
S/N = signal-to-noise ratio in a SAR resolution cell
$(S/N)_1$ = minimum signal-to-noise ratio necessary to detect a target with a specified probability of detection and probability of false alarm, for a single pulse
S_w = swath width in meters
σ = radar cross section of target in square meters
σ_b = bistatic radar cross section in square meters
σ° = radar cross section of surface clutter per unit area
t_s = scan time, or revisit time, in seconds
$t_o = n/f_p$ = signal integration time in seconds
T_c = coherent processing time in seconds
T_o = standard temperature = 290 K
τ = pulse width in seconds
θ_a = azimuth beamwidth in radians
θ_B = antenna beamwidth in degrees
θ_e = elevation beamwidth in radians
v = velocity in meters/second

RADAR LETTER BANDS

The frequency bands in which radar operates have traditionally been designated by letters of the alphabet. This nomenclature is given in Table 3 along with the International Telecommunication Union frequency allocations for "radiolocation." There are other letter-band nomenclatures that have been employed in the past for various purposes, but those of Table 3 are the only ones that should be employed for *radar*.

RADAR ANTENNAS

The half-power beamwidth of a radar antenna of dimension D is

$$\theta_B = \beta\lambda/D \qquad (20)$$

where beamwidth θ_B is in degrees and wavelength λ and dimension D are in the same units. The constant β depends on the shape of the illumination (current distribution) across the aperture. It is sometimes called the normalized beamwidth. The choice of aperture illumination determines the normalized beamwidth and the peak side-lobe level. Equation 20 is plotted in Fig. 5 for $\beta = 65$, a typical value for horn-fed parabolic-reflector antennas. The Taylor illumination is often

TABLE 3. RADAR LETTER-BAND NOMENCLATURE

Band Designation	Nominal Frequency Range	Specific Radiolocation (Radar) Bands Based on ITU Assignments for Region 2
hf	3 – 30 MHz	
vhf	30 – 300 MHz	138 – 144 MHz
		216 – 225
uhf	300 – 1000 MHz	420 – 450 MHz
		890 – 942
L	1 – 2 GHz	1215 – 1400 MHz
S	2 – 4 GHz	2300 – 2500 MHz
		2700 – 3700
C	4 – 8 GHz	5250 – 5925 MHz
X	8 – 12 GHz	8500 – 10680 MHz
K_u	12 – 18 GHz	13.4 – 14.0 GHz
		15.7 – 17.7
K	18 – 27 GHz	24.05 – 24.25 GHz
K_a	27 – 40 GHz	33.4 – 36.0 GHz
V	40 – 75 GHz	59 – 64 GHz
W	75 – 110 GHz	76 – 81 GHz
		92 – 100
mm	110 – 300 GHz	126 – 142 GHz
		144 – 149
		231 – 235
		238 – 248

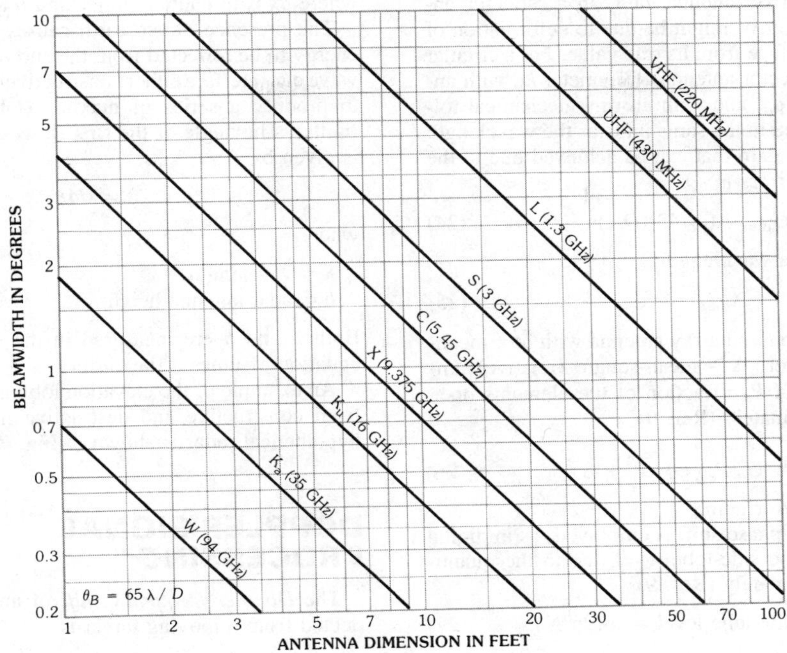

Fig. 5. Antenna beamwidth as a function of the antenna size, for the various radar letter bands, as given by Eq. 20 with $\beta = 65$.

used in radar-antenna design. It is characterized as having its first $\bar{n}-1$ side lobes equal, beyond which the side-lobe amplitude decreases with increasing angle. The variation of the beamwidth as a function of the side-lobe level of a planar antenna designed with a Taylor circular aperture illumination is shown in Fig. 6 for various $\bar{n}$ (Refs. 7,8). For a Taylor line source, the beamwidths of Fig. 6 are reduced by the (approximate) factor $\bar{n}/(\bar{n}+0.25)$.

Antenna Gain

The gain of "typical" reflector antennas can sometimes be approximated by

$$G_D \simeq 20{,}000/\theta_B \phi_B \qquad (21)$$

where θ_B and ϕ_B (in degrees) are the half-power beamwidths measured in the two principal (orthogonal) planes.

Cosecant-Squared Antenna Loss

The cosecant-squared antenna has a conventional beam shape from $\phi = 0$ to $\phi = \phi_o$ (ϕ = elevation angle), but its gain is proportional to $\csc^2\phi/\csc^2\phi_o$ from $\phi = \phi_o$ to $\phi = \phi_m$. For small ϕ_o and large ϕ_m, the loss in gain of a cosecant-squared antenna relative to the gain of a conventional antenna of beamwidth ϕ_o (in radians) is approximately

$$L_c \simeq 2 - \phi_o \cot \phi_m \qquad (22)$$

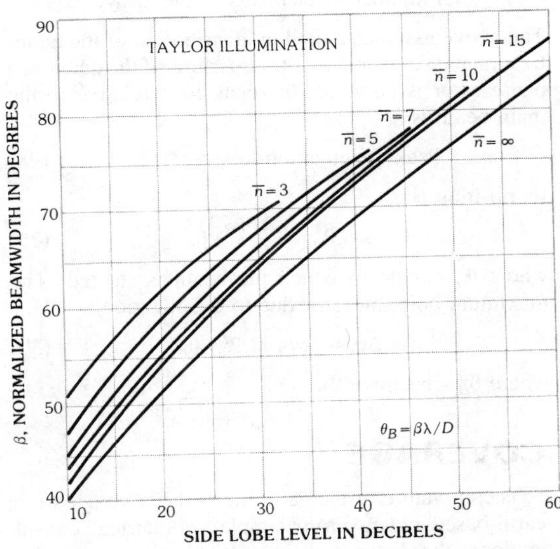

Fig. 6. Normalized beamwidth, β, as a function of the peak side-lobe level for the Taylor circular aperture illumination in which the first $\bar{n}-1$ side lobes are all equal. (After Ref. 8.)

Antenna Errors

The gain of an antenna when the rms phase error of the aperture illumination is $(\overline{\delta^2})^{1/2}$ is approximately

$$G = G_o \exp(-\overline{\delta^2}) \qquad (23)$$

where G_o = no-error antenna gain. In a reflector antenna, the phase error might be due to deformation of the reflecting surface from its true value. For a circular paraboloidal reflector antenna of diameter D, with antenna efficiency ρ_a, and with an rms mechanical tolerance ϵ measured in the same units as the wavelength λ, the maximum gain that can be achieved due to the resulting phase errors is

$$G_{max} = (\rho_a/43)(D/\epsilon)^2 \quad (24)$$

which occurs at a wavelength

$$\lambda_{max} = 4\pi\epsilon \quad (25)$$

The gain of a phased array antenna with $\overline{\delta}^2$ = mean-square phase error, $\overline{\Delta}^2$ = mean-square (relative) amplitude error, and P_e = fraction of the elements operative, is approximately (Ref. 9)

$$G \simeq G_o P_e/(1 + \overline{\Delta}^2 + \overline{\delta}^2) \quad (26)$$

where G_o = no-error gain.

When discrete phase shift is employed, as in digital phase shifters, the side-lobe level due to the quantization is approximately (Ref. 9)

$$\text{rms side-lobe level} \simeq 5/(2^{2B}N) \quad (27)$$

where,

B = number of bits in the phase shifter,
N = total number of elements in the array.

The above assumes a random distribution of the quantization phase error across the aperture. If the quantized phase error is periodic, the peak lobe relative to the main beam is

$$\text{Peak quantization lobe} = 1/2^{2B} \quad (28)$$

Its position is

$$\sin\theta_q \simeq (1-2^B)\,\theta_o \quad (29)$$

where θ_o = angle to which the beam is steered. The maximum pointing error due to quantization is

$$\Delta\theta_o = (\pi/4)(1/2^B)\,\theta_B \quad (30)$$

where θ_B = beamwidth.

COVERAGE

The curvature of the earth limits the coverage of an earth-based radar. From simple geometrical considerations, the distance d (in nautical miles) to the horizon from a radar at height h (in feet) is

$$d = 1.23\sqrt{h} \quad (31)$$

This assumes that the refraction of the radar energy by the atmosphere can be represented by an effective earth radius 4/3 times the actual radius. If the target is at a height h_t, the distance between the radar and target when the line of sight just grazes the surface of the earth is

$$d_o = 1.23\ (\sqrt{h} + \sqrt{h_t}) \quad (32)$$

where d_o is in nautical miles and h and h_t are in feet.

The presence of the earth causes part of the radar energy to be reflected from the surface. This reflected wave can interfere either constructively or destructively to produce a series of maxima (lobes) and minima (nulls). The angle of the first (lowest) lobe in radians is given by

$$\theta_l = \lambda/4h \quad (33)$$

where,

λ = wavelength
h = radar antenna height.

Both λ and h are measured in the same units. This equation assumes a flat earth.

An example of the elevation lobing pattern resulting from constructive and destructive interference for a hypothetical radar is shown in Fig. 7.

DOPPLER SIGNAL PROCESSING

The *Doppler frequency shift* of an echo signal reflected from a moving target is

$$f_d = (2v\cos\theta)/\lambda \quad (34)$$

where,

f_d is the frequency shift in hertz,
v is the velocity of the target in meters/second,
λ is the wavelength in meters,
θ is the angle defined by the direction of target travel and the radar line of sight to the target.

The relative velocity is $v_r = v\cos\theta$. The Doppler frequency shift per knot of relative velocity (f_d/v_r) is plotted in Fig. 8. When v_r is in knots and λ in meters, an approximate expression for the Doppler frequency shift is the following:

$$f_d\ (\text{Hz}) \approx v_r\ (\text{kn})/\lambda\ (\text{m}) \quad (35)$$

In a pulse radar, the measurement of the Doppler frequency shift is ambiguous if twice the pulse repetition frequency (prf) is less than the Doppler frequency. When the prf is equal to Doppler frequency f_d or some multiple, the target velocity cannot be distinguished from stationary clutter; *i.e.*, it will appear to have no Doppler shift. The relative velocity that produces a Doppler frequency equal to the prf or some multiple thereof is called a *blind speed,* since in an MTI radar the signal from a target at a blind speed is rejected along with the clutter. The blind speeds v_b are given by

$$v_b = n\lambda f_p/2 \quad (36)$$

where,

v_b is the blind speed in meters/second,
λ is the wavelength in meters,
f_p is the pulse repetition frequency in hertz,
n is an integer.

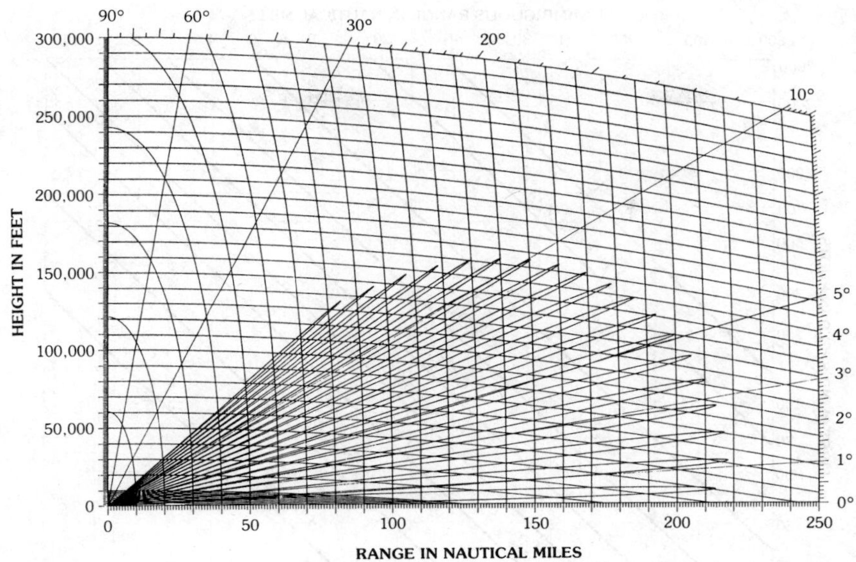

Fig. 7. Theoretical vertical lobing of the radar coverage diagram for an example uhf radar with its antenna located 25 meters over a perfectly reflecting plane surface.

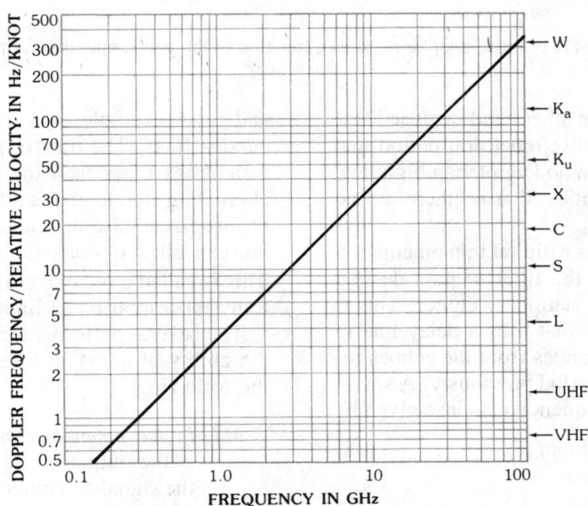

Fig. 8. Doppler frequency shift per unit relative velocity (hertz per knot) as a function of radar frequency.

An approximate expression for the first blind speed in knots (Fig. 9) is

$$v_1 \text{ (kn)} \approx \lambda \text{ (m)} f_p \text{(Hz)} \qquad (37)$$

MTI and Pulse Doppler Radars

There are two types of pulse radars that extract the Doppler frequency shift, or relative velocity, in addition to the range information. One is called MTI (Moving Target Indication) radar, and the other is the pulse Doppler (PD) radar. The difference between the two depends on the manner in which range and Doppler ambiguities are handled. An MTI radar is one that generally operates with a pulse repetition frequency that results in an unambiguous range measurement, but which produces ambiguities in the Doppler measurement. Thus, the first blind speed is less than the highest

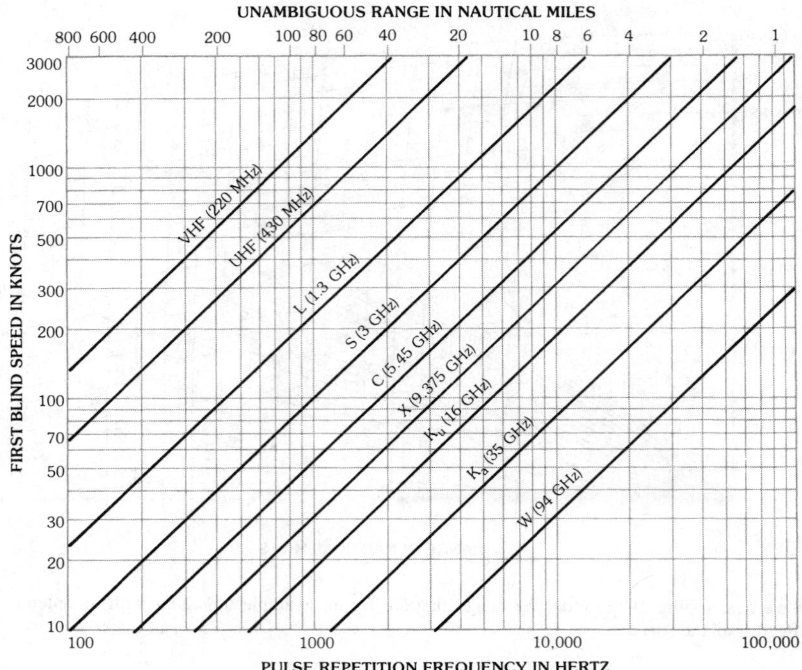

Fig. 9. First blind speed in an MTI or pulse doppler radar as a function of the pulse repetition frequency and the radar frequency.

target speed expected. On the other hand, a pulse Doppler radar operates with a pulse repetition period that is sufficiently high to result in no Doppler ambiguities, but it accepts range ambiguities; that is, there are no blind speeds.

The MTI radar usually uses a digital implementation of a delay-line canceller as the filter to pass desired moving targets, but to reject stationary clutter. This is a form of time-domain filter that uses a delay line to subtract the received radar echoes from the echoes received one pulse repetition period previously. A single delay-line canceller has a frequency response given by

$$H_1(f) = 2 \sin \pi f_d T_p \tag{38}$$

where,

f_d = Doppler frequency,
T_p = pulse repetition period.

When n single delay-line cancellers are used in cascade the frequency response is

$$H_n(f) = 2^n \sin^n \pi f_d T_p \tag{39}$$

An arrangement of n delay lines in cascade with $n+1=N$ taps (one at the input of each delay line and one at the output of the last line), whose outputs are weighted by the binomial coefficients with alternating sign before summing, produces a response equivalent to that of Eq. 39 for the n cascaded delay-line cancellers. This is sometimes called an *N-pulse canceller* and is an example of a *transversal* filter, or *nonrecursive* filter. The transversal filter may be used with weightings other than the binomial coefficients with alternating sign to give a different frequency response. Feedback can also be employed in a cascade of delay-line cancellers to shape the frequency response further. This is called a recursive filter. Its "infinite" duration transient response can limit its usefulness, however.

The several performance measures used to describe the ability of an MTI to see targets in clutter include the following:

MTI Improvement Factor: The signal-to-clutter ratio at the output of the MTI processor divided by the signal-to-clutter ratio at the input, averaged uniformly over all target relative velocities of interest. It is the basic measure now used to describe MTI performance.

Clutter Visibility Factor: The signal-to-noise clutter ratio, after cancellation or Doppler filtering, that provides stated probabilities of detection and false alarm. When the MTI is limited by noiselike system instabilities, the clutter visibility factor should be chosen as is the signal-to-noise ratio previously described in the section headed "Prediction of Radar Range."

Subclutter Visibility: The ratio by which the target echo power may be weaker than the clutter echo power and still be detected with specified detection and false alarm probabilities. All target relative velocities are assumed equally likely.

Clutter Attenuation: The ratio of the clutter power at the canceller input to the clutter residue at the output, normalized to the attenuation of a single pulse passing through the unprocessed channel of the canceller.

Note that:

Improvement Factor = Subclutter Visibility × Clutter Visibility Factor

or in decibels:

$$I = SCV + V_{oc} \quad (40)$$

The performance of an MTI radar is limited by clutter fluctuations, antenna scanning, and equipment instabilities.

Clutter Fluctuations—The improvement of an N-pulse canceller with $n = N - 1$ delay lines, or n cascaded delay-line cancellers, is

$$I_N = (2^n/n!)\ (f_p/2\pi\sigma_c)^{2n} \quad (41)$$

where,

f_p = pulse repetition frequency,
σ_c = standard deviation of the clutter spectrum which is assumed to be of gaussian shape.

This is plotted in Fig. 10. The standard deviation of the clutter spectrum in hertz is related to the standard deviation of the velocity spread in meters/second by the expression $\sigma_c = 2\sigma_v/\lambda$. "Typical" values of σ_v are (Ref. 10):

Heavily wooded hills, 20 mph wind: 0.2 m/s
Sparsely wooded hills, calm day: 0.02 m/s
Sea echo, windy day: 0.9 m/s
Rain clouds: 2 m/s
Chaff: 1 m/s

Antenna Scanning—The limitation to the improvement factor for a single delay-line canceller is

$$I_s = n_B^2/1.388 \quad (42)$$

and for a double delay-line canceller it is

$$I_s = n_B^4/3.853 \quad (43)$$

where n_B = number of pulses received within the half-power beamwidth as the antenna scans by the target.

Equipment Instabilities—The limits to the improvement factor due to pulse-to-pulse instability are (Ref. 11):

Transmitter frequency	$(\pi\Delta f\tau)^{-2}$
Stalo or coho frequency	$(2\pi\Delta fT)^{-2}$
Transmitter phase shift	$(\Delta\phi)^{-2}$
Pulse width	$\tau^2/(\Delta\tau)^2$
Pulse amplitude	$(A/\Delta A)^2$

where,

Δf = interpulse frequency change,
τ = pulse width,
T = transmission time to and from target,
$\Delta\phi$ = interpulse phase change,
$\Delta\tau$ = pulse-width jitter,
A = pulse amplitude,
ΔA = interpulse amplitude change.

Digital Quantization—The limit on the improvement factor in a digital MTI due to the finite size

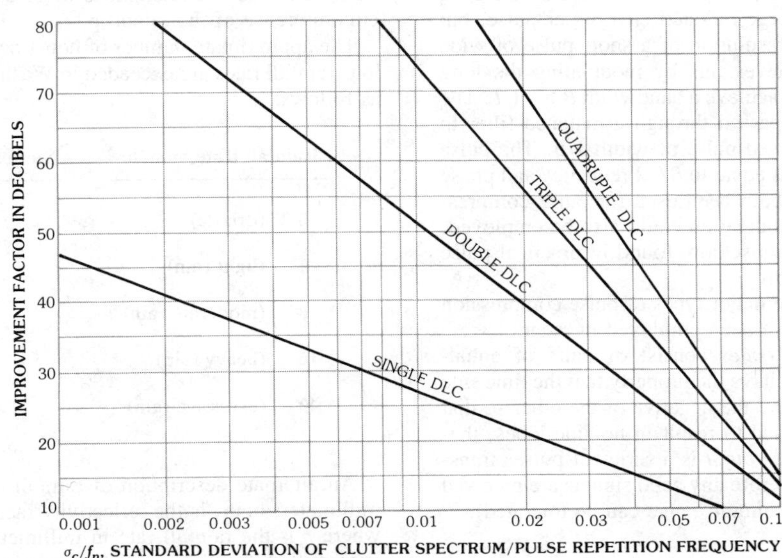

Fig. 10. The improvement factor as a function of σ_c/f_p (ratio of the standard deviation of the clutter and the pulse repetition frequency) for single and cascaded delay-line cancellers (DLC) as given by Eq. 41.

of the quantization interval of the A/D converter is approximately

$$I_q \simeq 6 \times \text{number of bits} \quad (44)$$

where I_q is in decibels. For example, an eight-bit A/D converter limits the improvement factor to 48 dB.

Limiting

Limiters employed in MTI radar can result in significantly less improvement factor than would be predicted from linear analysis (Ref. 11).

Staggered PRF MTI Radar

A staggered or multiple PRF MTI radar is one that uses more than one pulse repetition period to increase the first blind speed beyond that given by Eq. 36. If the individual periods of the multiple PRF waveform are designated by $T_1, T_2, \ldots, T_N$, and if they are related by the ratios $n_1/T_1 = n_2/T_2 = \ldots = n_N/T_N$, where $n_1, n_2, \ldots, n_N$ are integers, then the first blind speed when N different periods are used is

$$v_1 = [(n_1 + n_2 + \cdots + n_N)/N]v_B \quad (45)$$

where v_B is the first blind speed as given by Eq. 36 for a nonstaggered waveform with a constant period equal to the average of the N periods, or $(T_1 + T_2 + \cdots + T_N)/N$. For example, if $n_1 = 25$, $n_2 = 30$, $n_3 = 27$, and $n_4 = 31$, the first blind speed of the staggered waveform is 28.25 times that of a waveform with constant pulse repetition period.

PULSE COMPRESSION

Pulse compression allows a radar to utilize a long pulse to achieve large radiated energy per pulse, but to obtain a range resolution of a short pulse of wide bandwidth. It achieves this by modulating the long pulse of width T to achieve a bandwidth $B >> 1/T$. The received signal is passed through a matched filter to produce a compressed pulse of width $1/B$. The pulse compression ratio is equal to BT. Frequency and phase modulations have both been used for pulse compression. Amplitude modulation could also be employed, in principle, but it is seldom found in practical pulse compression systems.

Table 4 gives the major types of pulse compression waveforms that have been employed in radar.

Complementary codes consist of pairs of equal-length codes which have the property that the time side lobes of one code are the negative of the other so that they can be cancelled to result in no side lobes, theoretically. The *pulse burst* is a series of pulses transmitted as a group before any echo signals are received (as when a long minimum range can be tolerated).

CLUTTER

Clutter consists of those radar echo signals that are received from undesired scatterers such as land, sea, rain, snow, chaff, clouds, birds, insects, aurora, and meteors. Good radar practice requires that these clutter echoes be eliminated or minimized so as to prevent degraded detection of desired target echoes. It is especially important to limit the amount of clutter echoes presented to automatic detection and tracking (ADT) systems.

Surface clutter is described by the dimensionless parameter $\sigma°$, defined as the radar clutter cross section per unit area illuminated. Fig. 11 is a plot of the mean value of $\sigma°$ for sea clutter as a function of the grazing angle. It applies for a medium sea (10–20 knot winds). At the lower grazing angles, sea clutter increases slowly with increasing winds above 15 to 20 knots (a few tenths of a decibel per knot). Below about 5 knots wind velocity, sea clutter decreases rapidly with decreasing wind.

Land clutter is difficult to describe because of the many factors that affect its value. Fig. 12 is a plot of the mean value of clutter measured over North America in the summer for X and L bands (Ref. 12).

The radar echo from volume clutter, such as rain, is described by the volume reflectivity, η, defined as the radar cross section per unit volume. The volume reflectivity of rain in the microwave region is

$$\eta = 7f^4 r^{1.6} \times 10^{-12} \quad (46)$$

where,

η is the volume reflectivity in meters^2/meter^3,
f is the radar frequency in gigahertz,
r is the rainfall rate in millimeters/hour.

This expression assumes that the backscatter from rain is described by Rayleigh scattering, which is applicable when the radar wavelength is large compared to the circumference of the raindrop.

The approximate number of hours per year that various rainfall rates are exceeded in Washington, DC are as follows:

Rainfall Rate, mm/hr	Number of hrs/yr
0.25 (drizzle)	500
1 (light rain)	230
4 (moderate rain)	65
16 (heavy rain)	10
40 (excessive rain)	2

An alternate description of rainfall rate, instead of millimeters/hour, is the reflectivity factor $Z = 200r^{1.6}$, where r is the rainfall rate in millimeters/hour and Z is in millimeters^6/meter^3. This measure is often used by radar meteorologists. It is usually expressed in decibels and abbreviated dBz. For example, a rainfall rate of 4 mm/hr is equivalent to 32.6 dBz.

TABLE 4. PULSE COMPRESSION WAVEFORMS

Type	Description	Pulse Compression Ratio	Peak (Time) Side Lobe	Comments
Linear FM (chirp)	Linear modulation of the frequency over the range Δf in a time T.	$\Delta f \cdot T$	-13.2 dB; with weighting in the frequency domain on receive, side lobes can be -30 dB with loss in snr of about 1 dB.	Widely used. Doppler tolerant. Especially applicable to very high resolution.
Binary Phase (also known as linear recursive sequences, pseudorandom sequences, P-N sequences, and maximal length sequence)	A pulse of width T is divided into N subpulses of width τ, $N = T/\tau$. The phase of each subpulse is either 0 or π radians, chosen at random or pseudorandom.	N, the number of subpulses.	Approximately $0.5/N$	Side lobes are not low, especially with doppler shift. Not well suited when employed with multiple filters for doppler processing, because of high side lobes.
Barker	A form of binary phase code that has all its time side lobes equal to $1/N^2$ that of the compressed pulse peak power.	N	$1/N^2$	Limited to $N \leq 13$. Longer pulse compression ratios can be had with *compound Barker codes* in which each segment of a Barker code is modulated by a Barker code, but no decrease in side lobes is obtained.
Frank Polyphase (variants due to Lewis and Kretschmer are known as P codes)	A pulse of width T is divided into N^2 equal subpulses. The phase of each subpulse is chosen so that a linear fm waveform is approximated with a phase quantization equal to $2\pi/N$ radians.	N^2	$1/\pi^2 N^2$	Doppler tolerant for aircraft velocities

Continued on next page.

TABLE 4 (CONT). PULSE COMPRESSION WAVEFORMS

Type	Description	Pulse Compression Ratio	Peak (Time) Side Lobe	Comments
Discrete Frequency Shift (also called time-frequency coded)	A pulse of width T is divided into N subpulses, and the carrier frequency is changed subpulse to subpulse, with frequency steps separated by the reciprocal of the subpulse width. Frequencies are selected in random order from N frequency steps over a band $\Delta f = N/\tau$, where $\tau = T/N$.	$\Delta f \cdot T = N^2$	$1/N^2$	Produces a thumbtack ambiguity diagram. (When the frequencies are stepped uniformly from one end of the band to the other, the result is similar to linear fm.)
Nonlinear FM	Frequency is varied nonlinearly over a band Δf in a time T.	$\Delta f \cdot T$	Depends on the nonlinear waveform used.	Low side lobes can be obtained with matched filter processing without a loss in snr. Can obtain a thumbtack ambiguity diagram.
Stretch	Transmits a linear fm waveform of bandwidth Δf over a time T, but processes the received signal with circuitry of bandwidth $\Delta f/a$ $(a > 1)$ over a restricted range interval $1/a$ that of conventional fm pulse compression.	$\Delta f \cdot T$	As with linear fm pulse compression.	The time waveform is "stretched" by a factor of a so as to achieve a reduction of processing bandwidth by $1/a$.

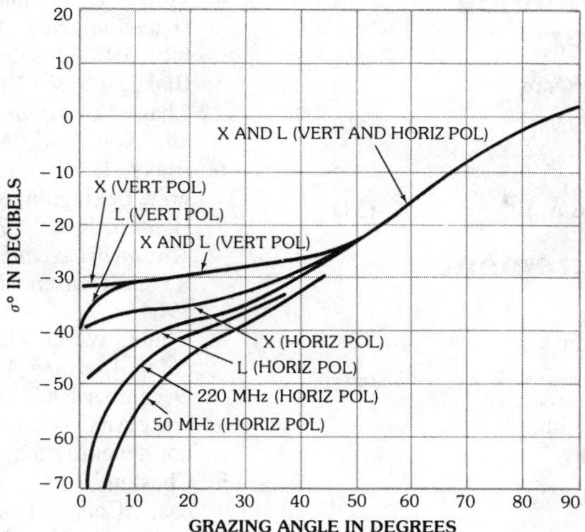

Fig. 11. Composite of σ° data for a "medium" sea, as a function of the grazing angle.

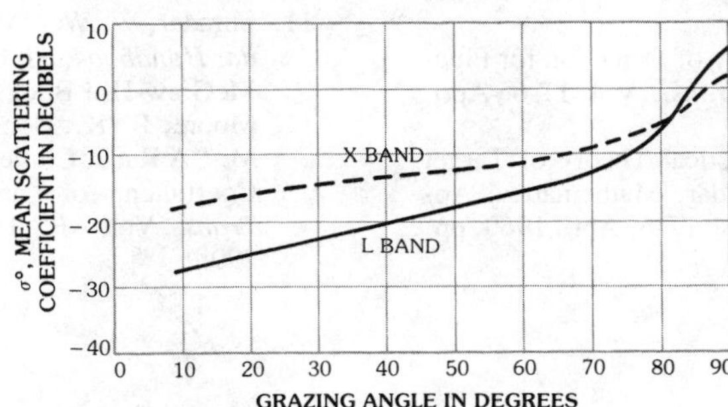

Fig. 12. Mean value of clutter σ° as a function of grazing angle, measured over North America in the summer. (Adapted from Reference 12.)

RADAR MEASUREMENT ACCURACY

The rms error, δM, in making a measurement of a radar parameter M in the presence of Gaussian noise and in the absence of any bias can be written conceptually in the form

$$\delta M = kM/(2E/N_o)^{1/2} \qquad (47)$$

where,

E is the received signal energy,
N_o is the noise power per unit bandwidth,
k is a constant of the order of unity.

(Note that $2E/N_o$ is also the peak-signal–to–mean-noise ratio at the output of a matched filter.) For a range measurement (time delay), k depends on the shape of the signal spectrum, and M can be taken as the rise time of the pulse. For a Doppler frequency measurement, k depends on the shape of the time waveform, and M is the spectral resolution (or the reciprocal of the observation time). For an angle measurement, k depends on the shape of the aperture illumination, and M is the beamwidth.

RMS Range Measurement Error

Band-limited rectangular pulse:

$$\delta R = (c/2)[\tau/(4BE/N_o)]^{1/2} \qquad (48)$$

where,

τ = pulse width,
B = bandwidth.

Gaussian-shaped pulse:

$$\delta R = (c/2)\tau/1.18(2E/N_o)^{1/2} \qquad (49)$$

where τ = half-power pulse width.

RMS Doppler Frequency Measurement Error

Rectangular pulse:

$$\delta f = \sqrt{3}/\pi\tau(2E/N_o)^{1/2} \quad (50)$$

Gaussian pulse:

$$\delta f = 1.18/\pi\tau(2E/N_o)^{1/2} \quad (51)$$

RMS Angle Measurement Accuracy

Uniform aperture illumination:

$$\delta\theta = 0.628\theta_B/(2E/N_o)^{1/2} \quad (52)$$

where θ_B = half-power beamwidth.

Cosine aperture illumination:

$$\delta\theta = 0.73\theta_B/(2E/N_o)^{1/2} \quad (53)$$

REFERENCES

1. Skolnik, M. I. *Introduction to Radar Systems.* 2d ed. New York: McGraw-Hill Book Co., 1980, Chap. 2.
2. Swerling, P. "Probability of Detection for Fluctuating Targets." *IRE Trans.*, Vol. IT-6, April 1960, pp. 269–308.
3. Marcum, J. I. "A Statistical Theory of Target Detection by Pulsed Radar, Mathematical Appendix," *IRE Trans.*, Vol. IT-6, April 1960, pp. 145–267.
4. Meyer, D. P., and Mayer, H. A. *Radar Target Detection.* New York, N.Y.: Academic Press, Inc., 1973.
5. Blake, L. V. "Prediction of Radar Range." Chap. 2 of *Radar Handbook,* ed. by M. I. Skolnik. New York: McGraw-Hill Book Co., 1970.
6. Blake, L. V. *Radar Range-Performance Analyses.* Lexington, Mass.: Lexington Books, 1980.
7. Hansen, R. C. "Aperture Theory." Chap. 1 of *Microwave Scanning Antennas,* Vol. 1, ed. by R. C. Hansen. New York: Academic Press, 1964.
8. White, W. D. *Desirable Illuminations for Circular Aperture Arrays.* Research Paper P-351, December 1967. Arlington, Va.: Institute for Defense Analyses. (Approved for public release, but not generally available.)
9. Cheston, T. C., and Frank, J. "Array Antennas." Chap. 11 of *Radar Handbook*, ed. by M. I. Skolnik. New York: McGraw-Hill Book Co., 1970.
10. Barlow, E. J. "Doppler Radar." *Proc. IRE*, Vol. 37, April 1949, pp. 340–355.
11. Shrader, W. W. "MTI Radar." Chap. 17 of *Radar Handbook*, ed. by M. I. Skolnik. New York: McGraw-Hill Book Co., 1970.
12. Moore, R. K., Soofi, K. A., and Purduski, S. M. "A Radar Clutter Model: Average Scattering Coefficients of Land, Snow, and Ice." *IEEE Trans.*, Vol. AES-16, November 1980, pp. 783–799.

37 Radio Navigation Aids

Revised by
E. A. Robinson

The speed of propagation of radio waves (300 000 kilometers per second) permits measurement of distance as a function of time and of direction as a function of differential distance to two or more known points. In free space, radio navigation has the fundamental capability of considerable accuracy. Along the surface of the earth, however, accuracy is reduced by the effects of multiple propagation paths between transmitter and receiver. Most navigation systems are, therefore, a compromise between service areas, accuracy, and convenience of use. Thus, the radio navigation aids that have found general use have been the result of these tradeoffs by the various user groups. In general, the complexity of the navigation equipment has been minimized at the expense of greater complexity and cost at the source of the navigation signal generation. A large degree of international standardization has resulted for these navigation systems due to the large number of multinational users. These standards, once established, change slowly and then primarily due to the development of a requirement of the navigation community. Many countries have established additional military systems, of which some are compatible with the international civil systems.

MAJOR NAVIGATION AGENCIES

Airlines Electronic Engineering Committee (AEEC), Annapolis Science Center, Annapolis, Maryland: A division of Aeronautical Radio, Inc. (ARINC) and owned by the scheduled US airlines. Publishes technical standards for avionics purchased by the scheduled airlines.

Department of Transportation, United States Coast Guard (USCG), Washington, D.C.: Operates the Loran-C and Omega navigation systems for marine and aeronautical navigation.

Federal Aviation Administration (FAA), Washington, D.C.: Operates navigation aids and air traffic control systems for both civil and military aircraft in the US and its possessions.

Federal Communications Commission (FCC), Washington, D.C.: The agency that licenses transmitters and operators in the United States and aboard US registered ships and aircraft.

International Air Transport Association (IATA), Montreal, Canada: The international association representing scheduled airlines.

International Civil Aviation Organization (ICASO), Montreal, Canada: A United Nations agency that formulates standards and recommended practices, including navigation aids, for all civil aviation.

International Telecommunication Union (ITU), Geneva, Switzerland: An agency of the United Nations that allocates frequencies for best use of the radio spectrum.

Radio Technical Commission for Aeronautics (RTCA), Washington, D.C.: Supported by contributions from industry and government agencies. Participation by manufacturers, users, and others in the recommended standards for aviation electronics. Many of these standards are adopted, at least in part, by the ICAO and the FAA.

Radio Technical Commission for Marine (RTCM), Washington, D.C.: Functions similar to those of RTCA; however, addresses primarily marine issues.

REDUCTION OF PROPAGATION ERRORS

In low-frequency navigation systems, complex propagation models may be used to correct for nonhomogeneous propagation paths. In addition, in most radio navigation systems, errors may result from the contamination of the signals with those that have traveled by a nondirect path that is often variable. To reduce such multipath effects, the following techniques are commonly used in the navigation system.

Pulse Transmission

Through the use of an appropriate pulse length and repetition rate combined with a means in the receiver to recognize the leading edges of the pulses, the desired direct signal may be separated from a signal that has traveled a longer path (skywave for low-frequency systems and multipath for higher-frequency systems). Pulse transmission techniques are effectively used in Loran-C, DME, transponder, and radar systems.

Space Diversity

The larger the aperture of the antenna system, the greater is the statistical probability that the desired signals will add linearly while the multipath signals add randomly. This approach is effectively used in doppler VOR systems. Antenna directivity is frequently used to reduce interference from undesired multipath signals. Horizontal directivity is used in the ILS system, and vertical directivity is used in VOR and DME systems.

Frequency Diversity

While the line-of-sight path remains the same at all radio frequencies, indirect paths may vary with frequency. In such cases, spectrum-spreading techniques may achieve the same result as space diversity.

RADIO NAVIGATION AIDS

The following subsections provide information on radio navigation aids ranging from the low-frequency Omega and Loran-C systems through the current Transit satellite navigation system and the Navstar Global navigation system being developed for operation in the 1986–87 time frame. For the specific extent to which these navigation aids are currently implemented throughout the world, see the navigational facility maps issued at frequent intervals by the US Coast and Geodetic Survey, Washington Science Center, Rockville, Maryland.

Omega

The Omega system is a worldwide vlf navigation system used for marine and enroute air navigation. The system comprises eight cw transmitting stations. Each station sequentially transmits long, but precisely timed, pulses at four frequencies: 10.2 kHz, 11.3 kHz, 13.6 kHz, and 11.05 kHz. Position information is obtained by measuring the relative phase difference of the received signals. The inherent accuracy of the Omega system is limited by propagation corrections that must be applied to the receiver. These corrections vary depending on location and the time of day. In many cases, accuracies of 2 nautical miles (rms) day and 4 nautical miles (rms) night are being achieved in most of the coverage areas.

Loran-C

Loran-C is a long-range hyperbolic radio navigation system that possesses an inherent high degree of accuracy at ranges of 800 to 1000 nautical miles. The Loran-C system transmits synchronized, phase-coded pulses from a master station and two or more secondary stations at 100 kHz. The transmitting stations form a chain characterized by the group repetition interval (GRI) in which the pulses are repeated. A GRI starts with the master station transmitting eight pulses, each spaced one millisecond apart, followed by a ninth pulse two milliseconds later. The master-station transmission is followed at a prescribed coding delay by transmissions from each of the secondary stations in the chain, each transmitting eight pulses at one-millisecond intervals. Phase coding is used to differentiate the master pulses from those of the secondaries. The pulse spacing and phase code allow the ground wave (direct propagation path) to be differentiated from the varying sky wave. Fig. 1 shows the Loran-C signal format and chain parameters.

Currently, there are 17 chains consisting of 50 transmitting stations. Table 1 lists the Loran-C chains and their group repetition intervals (GRI). A typical Loran-C receiver makes use of a microprocessor for signal processing, navigation computation, and control. Through the use of microprocessors and other signal-processing and timing integrated circuits, the cost of high-quality navigation has continued to decrease. This has caused the system to be accepted by a large number of users, and this coupled with the accuracy of the system (one quarter of a mile absolute and less than 100 feet relative) assures its continued use, improvement, and expansion, through the year 2000. A typical receiver may be divided into the sections shown in Fig. 2.

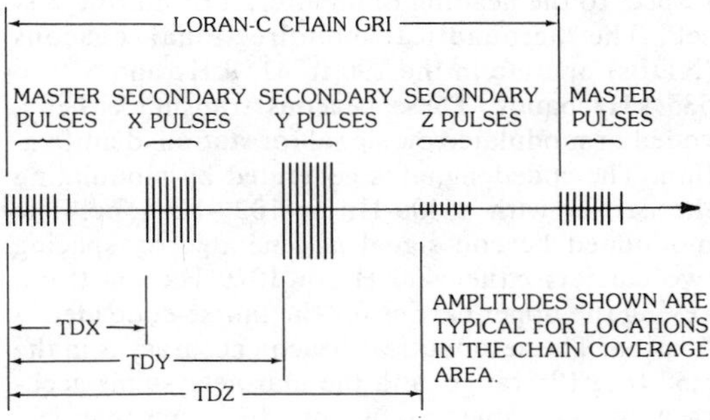

(A) Pulse groups.

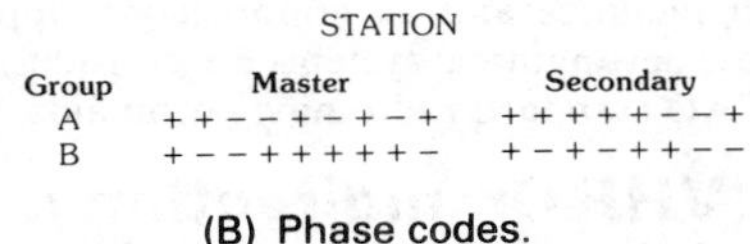

(B) Phase codes.

Fig. 1. Loran-C signal format.

TABLE 1. LORAN-C CHAINS

Area/Chain	GRI
Continental US	
Pacific Northwest	5990
Southeastern US	7980
Great Lakes	8970
Western US	9940
Northeastern US	9960
Alaska	
Western Canada	5990
Gulf of Alaska	7960
Northern Pacific	9990
Hawaii	
Central Pacific	4990
Canada	
Eastern Canada	5930
Western Canada	5990
Northern Atlantic	7930
Europe	
Norwegian Sea	7970
Mediterranean Sea	7990
Far East	
Far East	5970
Northwestern Pacific	9970

Radio Beacons

Radio beacons are nondirectional transmitters that operate in the low-frequency and medium-frequency bands. A radio direction finder is used to measure the relative bearing to the transmitter with respect to the heading of an aircraft or marine vessel. The aeronautical nondirectional beacons (NDBs) operate in the 190 to 415 kHz and 510 to 535 kHz bands. These beacons transmit either a coded or modulated cw signal for station identification. The coded signal is generated by modulating the carrier with a 400-Hz or 1020-Hz tone. The modulated beacon signal is generated by spacing two carriers either 400 Hz or 1020 Hz apart and keying the upper carrier for the morse-code identification. The aeronautical-beacon accuracy is in the ±3° to ±10° range, and the marine-systems accuracy is maintained to within ±3°. Aeronautical NDBs are used to supplement the VOR-DME system for transition from enroute to precision approach facilities and as nonprecision approach aids. These aeronautical systems are considered one of the ICAO standard radio navigation aids.

VOR (VHF Omnidirectional Range)

The VOR transmits continuous-wave signals on one of the 20 assigned channels in the 108 to 118 MHz band with 100-kHz channel separation. A nondirectional 30-Hz reference signal with a ±480-Hz frequency modulation on a 9960-Hz subcarrier is transmitted along with a carrier radiating from a rotating antenna with a horizontal cardioid pattern. The cardioid antenna pattern rotates at a 30-Hz rate, allowing the airborne receiver to determine its bearing from the station as a function of phase between the reference and the rotating signal. The VOR system has line-of-sight limitations in that at altitudes above 5000 feet the range is approximately 100 nautical miles and above 20 000 feet the range is approximately 200 nautical miles. The enroute VOR stations are rated at 50 watts. The accuracy of the VOR ground station is better than ±1.4°. However, the station magnetic declination is usually allowed to increase to 2° before being reset. The total system error (rss of the ground and airborne system plus flight technical error) is less than ±4.5°.

Distance measuring equipments (DMEs) are often colocated with the VOR stations to provide ranging information. In the US and other countries, TACAN (Tactical Air Navigation) installations are colocated with the VORs to provide a navigation system utilized primarily by the military.

ILS (Instrument Landing System)

At present, the instrument landing system operating in the 108 to 112 MHz band is the primary worldwide, ICAO approved, precision landing system. The system has limitations in siting, frequency allocation, and performance. An alternate system, the microwave landing system (MLS) is scheduled to be implemented in the 1980s and eventually replace the ILS by the year 2000. An ILS normally consists of two or three marker beacons, a localizer, and a glide slope to provide both vertical and horizontal guidance information. The localizer, operating in the 108 to 112 MHz band, is normally located 1000 feet beyond the stop end of the runway. The glide slope is normally positioned 1000 feet after the approach end of the runway and operates in the 328.6 to 335.4 MHz band. Marker beacons operating along the extension of the runway centerline at 75 MHz are used to indicate decision height points for the approach or distance to the threshold of the runway.

Azimuth guidance provided by the localizer is accomplished by use of a 90-Hz–modulated left-hand antenna pattern and a 150-Hz–modulated right-hand pattern as viewed from the aircraft on an approach. A 90-Hz signal detected by the aircraft receiver will cause the course deviation indicator (CDI) to deviate to the right. A 150-Hz signal will drive the CDI vertical needle to the left when the aircraft is right of the centerline course. When the aircraft is on the center line, the CDI vertical needle will be centered. A total of 40 channels is provided by the ILS localizer system, each being paired with a possible glide-slope channel.

Vertical guidance is provided by the glide-slope facility that is normally located to the side of the approach end of the runway. A total of 40 channels is provided in the 328.6 to 335.4 MHz band; each is paired with one of 40 ILS localizer channels. The carrier radiated by the antenna pattern below the glide slope is amplitude modulated with a 150-Hz signal. The pattern above the glide slope produces a signal with 90-Hz amplitude modulation. When the approaching aircraft is on the glide slope, the CDI horizontal (glide slope) needle will be centered.

The marker-beacon facilities along the course provide vertical fan markers to mark the key positions along the approach. The inner marker is normally at the runway threshold; the middle marker is about 3500 feet from the threshold; and the outer marker is usually 5 miles from the runway. A DME on one of the 20 paired channels with the localizer channels may also be used for indicating position during the approach.

The quality of the ILS installations varies depending of the equipment, terrain, and calibration. The ICAO has established the categories in Table 2 for minimum approach ceiling and forward visibility.

DME (Distance Measuring Equipment)

The airborne equipment (interrogator) generates a pulsed signal that is recognized by the ground

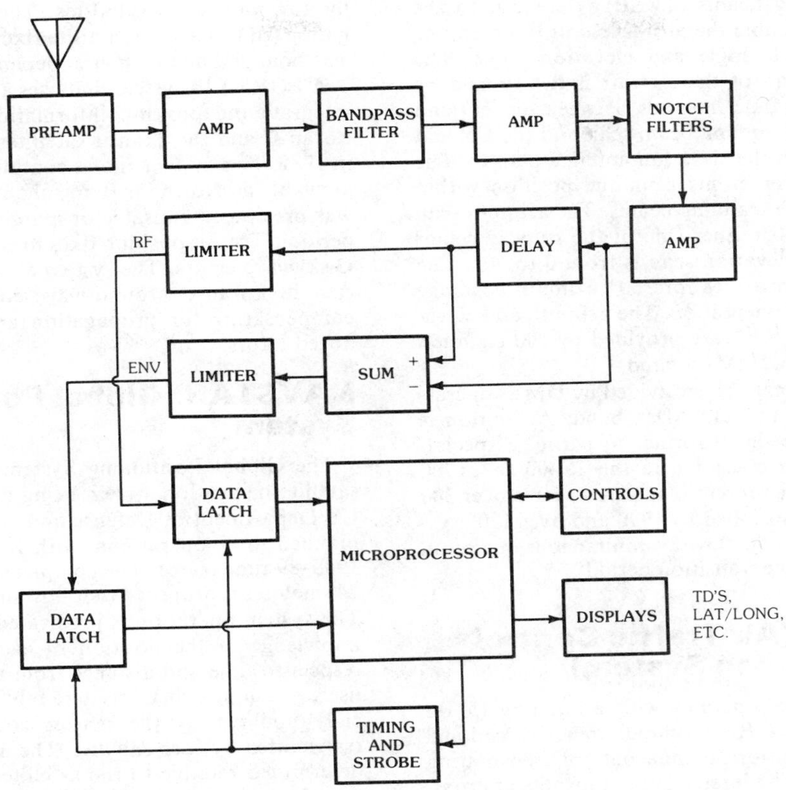

Fig. 2. Typical Loran-C receiver.

TABLE 2. CATEGORIES FOR MINIMUM APPROACH CEILING AND FORWARD VISIBILITY

Category	Min Ceiling (feet)	Fwd Visibility (feet)
I	200	2600
II	100	1200
III-A	50	700
III-B	35	150
III-C	0	0

equipment (transponder); the transponder then transmits a reply that is identified by the tracking circuit in the interrogator. The distance is computed by measuring the total round-trip time of interrogation, reply, and fixed delay introduced by the ground transponder. The airborne interrogator transmits about 30 pulse pairs per second on one of the 126 allocated channels between 1025 and 1150 MHz. The ground transponder replies on one of the paired channels in the 962 to 1024 MHz band or 1151 to 1213 MHz band. A DME and a colocated VOR constitute the ICAO standard rho-theta system.

TACAN (Tactical Air Navigation)

The TACAN system provides both omnibearing and distance-measuring capability. The rotating directional horizontal-plane radiation pattern produces the azimuth signal, which contains a coarse (15 Hz) and a fine (135 Hz) azimuth element. The rotation of the pattern at 15 Hz results in a modulation of the carrier with a composite 15-Hz sine wave. Reference signals are transmitted by coded pulse trains to provide the phase reference. Bearing is obtained by the airborne receiver by comparing the 15-Hz and 135-Hz sine waves with the reference pulse groups. The TACAN system operates in the 960 to 1215 MHz band with 1-MHz channel separations.

MLS (Microwave Landing System)

The microwave landing system is the ICAO-approved replacement for the current ILS system. The MLS is being developed to meet the full range of user operational requirements to the year 2000 and beyond. The system is based on time-referenced scanning beams (TRSB) referenced to the runway that enable the airborne unit to determine precise azimuth angle and elevation angle. The angular position of the aircraft is determined by measuring the time intervals between the TO and FRO azimuth antenna beam scan and the UP and DOWN scan of the elevation antenna pattern. The time interval represents a unique position within the range of the scanning beams. The azimuth scan is typically ±60° either side of the runway center line, and the elevation scan is from 0 to 30°. The signal format provides for 360° azimuth coverage for future implementation. The azimuth and elevation angle functions are provided by 200 channels in the 5000 to 5250 MHz band.

Range information is provided by DMEs operating in the 960 to 1215 MHz band. An option is included in the signal format to permit a special-purpose system operating in the 15400 to 15700 MHz band. The present plans are to have over 300 MLS facilities installed by 1990 and over 1200 systems by the year 2000 with simultaneous operation of ILS during the transition period.

ATCRBS (Air Traffic Control Radar Beacon System)

The ATCRBS operates with a rotating (5 Hz enroute and 2.5 Hz terminal areas) directional ground interrogator antenna pattern transmitting at 1030 MHz. The interrogator transmits approximately 400 pulse pairs per second and receives replies from aircraft transponders that are within the beam of the antenna pattern. The airborne transponder replies at 1090 MHz with one of the 4096 pulse codes available. The decoded replies are displayed on the surveillance radar ppi indicator along with primary radar returns. An omnidirectional pulsed pattern is also radiated from the ground to suppress unwanted side-lobe replies. This system is often referred to as secondary surveillance radar.

Transit

Transit is a satellite navigation system consisting of four or more satellites in approximately 600–nautical-mile poloar orbits. The system is operated by the U.S. Navy. The satellites broadcast ephemeris information continuously at 150 MHz and 400 MHz. A receiver measures successive Doppler shifts of the signal as the satellite approaches or passes the user. The geographic position of the receiver is then calculated from the satellite position information (transmitted from the satellite every two minutes) and the measurement of the Doppler shift. Normally only one frequency (400 MHz) is used for navigation, and the other is used when ionosphere correction (for improved accuracy) is performed. Coverage is worldwide but not continuous due to the relatively low altitude and the low number of satellites. The update of navigation (time between satellite fixes) can be as short as 1 hour and up to 8 hours, depending on latitude.

A LORSAT* system that uses a Kalman filter to integrate the position information derived from Loran-C and the Transit satellites was introduced in 1982. This system allows continuous coverage in areas of the Northern Hemisphere where Loran-C was previously unusable or marginal, by using the periodic Transit position fixes to correct the Loran-C skywave errors. The system also improves accuracy in Loran-C ground-wave coverage areas by compensating for propagation and geometry induced errors.

NAVSTAR/Global Positioning System

The Global Positioning System is a worldwide satellite navigation system being developed by the US Department of Defense and is currently being planned to be operational with 18 satellites in the 1988–89 time frame. The complete constellation of 24 satellites is proposed to follow in the early 1990s. The system operation is predicated upon accurate knowledge of the position of each satellite with respect to time and distance from a satellite to the user. A unique ephemeris data table that is periodically updated by the master control station is transmitted by each satellite. The user's position is determined relative to the satellites by processing signals received from at least four satellites to solve for the time-of-arrival difference to obtain the distance to the satellites. A time correction then relates the satellite system to earth coordinates.

The satellite signals are transmitted at two L-band frequencies (L1 of 1575.42 MHz and L2 of 1227.6 MHz) to permit corrections for ionospheric delays in propagation. The signals are modulated with both a P and a C/A pseudorandom noise code that are in phase quadrature. The L1 and L2 signals are also continuously modulated with the navigation data-bit stream at 50 b/s. These codes allow identification of the satellites and measurement of the transit time through measurement of the phase shift required to match the codes.

The P code is a long precision code operating at 10.23 Mb/s, and the C/A code is a short code, readily acquired but operating at 1.023 Mb/s, which provides a less accurate measurement. The 50 b/s navigation message contains the data that the user's receiver requires to perform the operations for successful position determination and navigation.

The navigation message is formatted in five subframes, each of six seconds duration, which make up a data frame that is 30 seconds long and contains 1500 bits. The data are nonreturn to zero (NRZ)

*LORSAT is a trademark of Texas Instruments Inc.

at 50 b/s and are common to the P and C/A signals on both the L1 and L2 frequencies. Each data subframe starts with a telemetry word and the C/A to P code handover word. The initial eight bits of the telemetry word contain a preamble that facilitates the acquisition of the data message.

The accuracy of a position fix varies with the capability of the user's equipment and with the user-to-satellite geometry. The most sophisticated equipment using both the C/A and P codes will provide positioning accuracy of approximately 20 meters horizontally and 30 meters vertically. The presently projected accuracy that will be made available to the civil user is 500 meters. The Global Position System is illustrated in Fig. 3. Fig. 4 is a block diagram of a multiplexed GPS receiver.

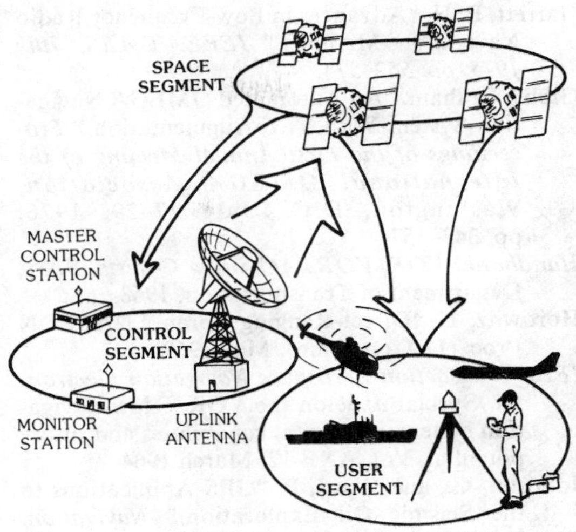

Fig. 3. Global Positioning System.

REFERENCES

Beck, G. E. *Navigation Systems—A Survey of Modern Electronic Aids.* London: Van Nostrand Reinhold Co., 1971.

Beser, J., and Parkinson, B. W. "The Application of NAVSTAR Differential GPS in the Civilian Community." *Navigation,* Vol. 29, No. 2, Summer 1982.

Blackbane, W. T., ed. "Advanced Navigation Techniques." *AGARD Conference Proceeding 28.* Slough, England: Technivision Services, 1970.

Borty, J. E., Gupta, R. R., Scull, D. C., and Morris, P. B. "OMEGA Signal Coverage Prediction." *Navigation,* Vol. 23, No. 1, Spring 1976.

Bowditch, N. *American Practical Navigator.* US Navy Hydrographic Office, Publ. 9, 1958.

Copps, E. M., Geier, G. J., Fidler, W. C., and Grundy, P. A. "Optimal Processing of GPS Signals." *Navigation,* Vol. 27, No. 3, Fall 1980.

Frank, Robert L. "History of Loran C." *Navigation,* Vol. 29, No. 1, Spring 1982.

Fritch, V. J., and Sanders, L. J. "Instrument Landing Systems." *IEEE Communications,* May 1973, pp. 435–454.

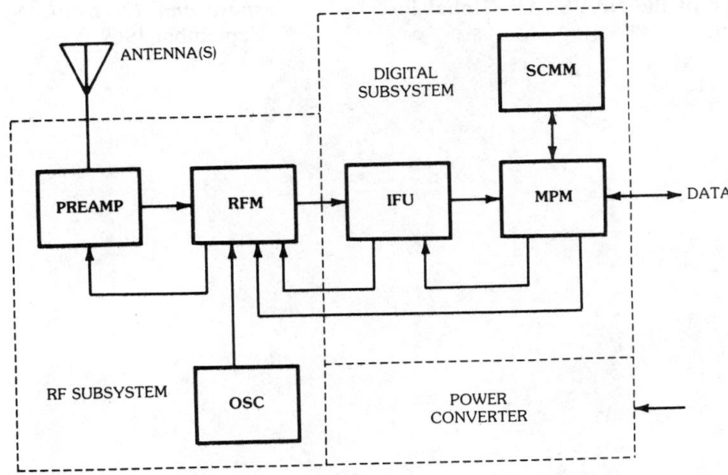

NOTE:
RFM—RF Module (Dual-conversion from L-band to 10.28 MHz 1st if)
IFU—IF Module (If signal processing, code correlation, and generation of I/Q signals along with carrier/frequency track)
MPM—Microprocessor Module (Two microprocessors for signal processing, control, and display functions)
SCMM—Semiconductor Memory Module

Fig. 4. Global Positioning System receiver.

Garrett, P. H. "Advances in Low-Frequency Radio Navigation Methods." *IEEE, T-AES, July 1975, pp. 562*–574.

Gibbs, Graham. "An Automated OMEGA Navigation System With VLF Augmentation," *Proceedings of the First Annual Meeting of the International OMEGA Association,* Washington, D.C., July 27–29, 1976, pp. 343–357.

Handbook: VOR/VORTAC Siting Criteria. FAA, Department of Transportation, 1968.

Horowitz, L. "Direct-Ranging Loran." NAECON Proc. (Dayton, Ohio), May 1969.

IEEE Transaction Aerospace Navigation Electronics. Special issue on the VOR/DME navigation system–its present capabilities and future potential, Vol. ANE-12, March 1964.

Johnson, C., and Ward, P. "GPS Applications to the Seismic Oil Exploration." *Navigation,* Vol. 26, No. 2, Summer 1979.

Kayton, M., and Fried, W. R. *Avionics Navigation Systems.* New York: John Wiley & Sons, Inc., 1969.

Kelly, R. J. "Time Reference Microwave Landing System Multipath Control Techniques." *Navigation,* Vol. 23, No. 1, Spring 1976.

Loran C User Handbook COMDTINST M16562.3, US Coast Guard (G-NRN-3), Washington, D.C.

Microwave Landing System Signal Format and System Level Functional Requirements, FAA-ER-700-08C. April 1979.

Navigation: Journal of the Institute of Navigation. Vol. 25, No. 2, Summer 1978; Special issue with coverage of the NAVSTAR Global Position System.

Pierce, J. A. "OMEGA." *IEEE Transaction Aerospace Electronic Systems,* Vol. AES-1, December 1965, pp. 206–215.

Pierce, J. A., and Woodward, R. H. "The Development of Long Range Hyperbolic Navigation in the United States." *Navigation,* Vol. 18, No. 1, Spring 1971.

Pisacane, V. L., Holland, B. B., and Black, H. D. "Recent Improvements in the Navy Navigational Satellite System." *Navigation,* Vol. 20, 1973.

Poppe, Martin. "The Loran C Receiver, A Functional Description." *Navigation,* Vol. 29, No. 1, Spring 1982.

Robinson, E. A. "Certification of Loran C in the National Air Space by Texas Instruments." Wild Goose Association Technical Symp., October 1982.

Stansell, T. A., Jr. "The Navy Navigation Satellite System; Description and Status." *Navigation,* Vol. 15, No. 3, Fall 1968.

Swanson, E. R. "OMEGA Possibilities: Limitations, Options, and Opportunities." *Navigation,* Vol. 26, No. 3, Fall 1979.

Vass, E. R. "OMEGA Navigation System: Present Status and Plans 1977–1980." *Navigation,* Vol. 25, No. 1, Spring 1978.

Watt, A. D. *VLF Radio Engineering.* Elmsford, N.Y.: Pergamon Press, Inc., 1967.

Winnick, A. B., and Brandewie, D. M. "Recent VOR/DME System Improvements." *Proc. IEEE,* Vol. 58, March 1970.

Zimmerman, W. "Optimum Integration of Aircraft Navigation Systems." *IEEE Transaction Aerospace and Electronic Systems,* Vol. AES-5, September 1969.

38 Common Carrier Transmission

Revised by
Richard A. McDonald,
M. R. Aaron, and L. C. J. Roscoe

DEFINITIONS OF COMMONLY USED TERMS

System Reference Level Point: A point in a communication circuit arbitrarily chosen as a reference point for signal level measurements. Common equivalent terms are "0 dB transmission level point," "zero level," "zero level point," "0 dB TL," "0 dB TLP," and "0 TLP."

Relative Level: The relative level at any point in a circuit is the power gain or loss in decibels between the 0 TLP and the point under consideration. Relative level is expressed with terms such as "3 dB TLP." Signal and/or interference powers may be referred to the 0 TLP with a phrase such as "a signal power of −16 dBm0," which indicates the power of the signal had it been measured at the 0 TLP. Note that the 0 TLP may not be accessible for measurement and, in fact, need not even exist in a given system. Reference of signal and interference powers to the 0 TLP is convenient in system design. Present practice is to define the outgoing side of a toll switch to be −2 or − 3 dB TLP, and the outgoing side of an end office switch to be 0 dB TLP.

Volume: Volume is a method of expressing the amplitude of complex nonperiodic signals such as speech. Volume is expressed in volume units (vu) and is defined as the reading obtained on a specified meter when read in a prescribed manner. The volume indicator is not frequency weighted in its response.*

Noise: Noise, in its broadest definition, consists of any undesired signal in a communication channel. Noise may be classified as thermal, or white, noise; impulse noise; cross talk; tone interference; and miscellaneous.

When noise is measured on voice communication channels, a weighting network is often inserted in front of the detector to account for the different subjective annoyance from noise of different frequencies when modern telephone sets are used.

Thermal Noise: A form of noise arising from random electron motion. It occurs on all transmission media and in all communications apparatus. It is characterized by uniform energy distribution over the frequency spectrum and a normal or Gaussian distribution of voltage or current.

Impulse Noise: Noncontinuous noise consisting of irregular pulses of short duration and relatively high amplitude. Some sources of impulse noise in voice communication channels are: induced interference by transients due to relay and switch operation, transients due to switching or lightning in adjacent power circuits, and cross talk from high-level telegraph circuits.

Cross Talk: Interference from other communication channels is called cross talk. It is classified as near-end and far-end cross talk and as intelligible and unintelligible cross talk. Near-end cross talk is measured on a channel at a receiving point near the sending point of the interfering channel; far-end cross talk is measured on a channel at a receiving point near the receiving point of the interfering channel. (In far-end cross talk, the two channels transmit in the same direction.)

Intelligible cross talk can be understood by the listener, and, because it diverts his attention, it has more interfering effect than unintelligible cross talk. Cross talk into a voice-frequency circuit from adjacent voice-frequency circuits or between groups and supergroups in frequency-multiplexed systems is generally intelligible. Cross talk due to incomplete suppression of sidebands, to intermodulation of two or more frequency-multiplexed channels, or to otherwise intelligible cross talk between frequency-multiplexed channels having offset frequency spectra is generally unintelligible. Such cross talk is often classed as miscellaneous noise. Intermodulation cross talk in wideband frequency-multiplexed systems approaches thermal noise in its spectral distribution.

Tone Interference: Interference due to single tones or complex periodic waveforms.

Miscellaneous Noise: Interferences that cannot readily be placed in any of the preceding categories.

Reference Noise: 1 picowatt (10^{-12} watt) of power. Also commonly stated as −90 dBm.

dBrn: Decibels above reference noise. The unit of measurement of noise power used in the Western Electric 3A noise measuring set is the dBrn. For measuring noise on voice communication channels, the 3A noise measuring set may be equipped with a C-message weighting filter. When this filter is used, the unique network response causes the reading to deviate from what would be obtained with some other instrument and filter. Readings in dBrn taken with the C-message weighting filter are designated "dBrnc." The calibration tone is 0 dBm (1 milliwatt) at 1000 Hz, which reads 90 dBrn with or without the C-message weighting filter. Refer to Table 1.

pWp: Picowatts of noise psophometrically weighted. Units of noise power derived from measurements with the CCITT recommended psophometer. The psophometer is frequency weighted by a curve having a shape similar to the F1A weighting curve. The reference tone is −90 dBm (1 picowatt) at 800 hertz. Refer to Table 1.

Net Loss: The net loss of a transmission channel is the ratio of the signal power at the input and the output of the channel. By custom, the net loss of a channel is understood to be measured at 1000 hertz in the American and Canadian plant and at 800 hertz in European practice (CCITT). The power ratio is typically expressed in decibels (dB).

Insertion Loss: The ratio of the power delivered from a source to a load to the power delivered from

---

* Chinn, H. A., Gannett, D. K., and Morris, R. M. "A New Standard Volume Indicator and Reference Level." *BSTJ*, Vol. 19, Jan. 1940, pp. 94-137.

TABLE 1. COMPARISON OF TWO NOISE MEASURING SETS

Noise Measuring Set	Reading Due to 0 dBM of	
	1000 Hertz	White Noise Limited to 0–3 Kilohertz Band
Western Electric 3A(C-message weighting)	90 dBrn	88 dBrn
CCITT Psophometer	+1 dBm*	−2 dBm*

*The psophometer is defined as measuring the internal (open-circuit) voltage of an equivalent noise generator having an impedance of 600 ohms and delivering noise power to a 600-ohm load. For convenience in comparison, the psophometric electromotive force has been converted to dBm.

the same source through a transducer (network, channel) to the same load. The definition may be applied to the loss or gain effect caused by the insertion of a gain (repeater) or equalization element into a two- or four-wire transmission channel. The concept is equally applicable to the channel as a whole.

Transducer Loss: The ratio of the maximum power available from a source to the power delivered from that source through a transducer (network, channel) to a load. The transducer loss of a channel is different from the insertion loss only because the maximum power from the source is substituted for the power from that source to that load.

Singing Margin: The singing margin of a circuit is defined as the amount by which the loss of the two directions of transmission may be reduced before oscillation (singing) occurs. Inadequate singing margin results in distortion that is often described as sounding hollow or like a "rain barrel."

Return Loss: A measure of the match between the two impedances on either side of a junction point. Return loss is defined by

$$RL\,(\text{dB}) = 20 \log_{10} |(Z_1 + Z_2)/(Z_1 - Z_2)|$$

where Z_1 and Z_2 are the complex impedances of the two halves of the circuit. When the impedances are not matched, the junction becomes a reflection point. The return loss expresses the ratio of incident to reflected signal power.

Echo Return Loss: The weighted power averaged return loss at a reflection point. The echo return loss expresses in decibels the ratio of the power of a broad-band incident signal to that of the correspondingly broad-band reflected signal. Both powers are measured through a weighting network, typically covering approximately the band 500 Hz to 2500 Hz.

Singing Return Loss: The same as echo return loss but over a considerably narrower band near an edge of the voice band, e.g., 200 Hz to 500 Hz or 2500 Hz to 3200 Hz.

Talker Echo: A signal returned to the talker after making one or more round trips between the talker and a distant reflection point (Fig. 1). The first talker echo is generally the most important. Echos that are sufficiently loud and sufficiently delayed can be annoying and can interfere with the normal speech process.

Listener Echo: A signal first returned toward the talker at a distant reflection point and then reflected again toward the listener (Fig. 1). The listener echo mixes with the original signal, adding to its strength at some frequencies and diminishing it at others. It can result in amplitude distortion and reduced singing margin.

Via Net Loss: A loss factor proportional to echo path delay, used to prescribe loss objectives for interoffice trunks. Via net loss (*VNL*) can be calculated as:

$$VNL = VNLF \times L + 0.4 \text{ dB}$$

where,

L = one-way circuit length in miles,

VNLF is the via net loss factor, proportional to echo path delay per mile.

Commonly accepted values of *VNLF* are given in Table 2.

THE SWITCHED TELECOMMUNICATIONS SYSTEM

The telecommunications system consists of many elements that work together to provide a variety of services. Switched voice-band channels may be set up as needed by giving appropriate instructions at the subscriber's telephone set. Every subscriber is connected to a local end-office switching system via a local loop. Therefore, at each end of a switched connection there is a local loop. To connect between two loops served by different end offices, it is

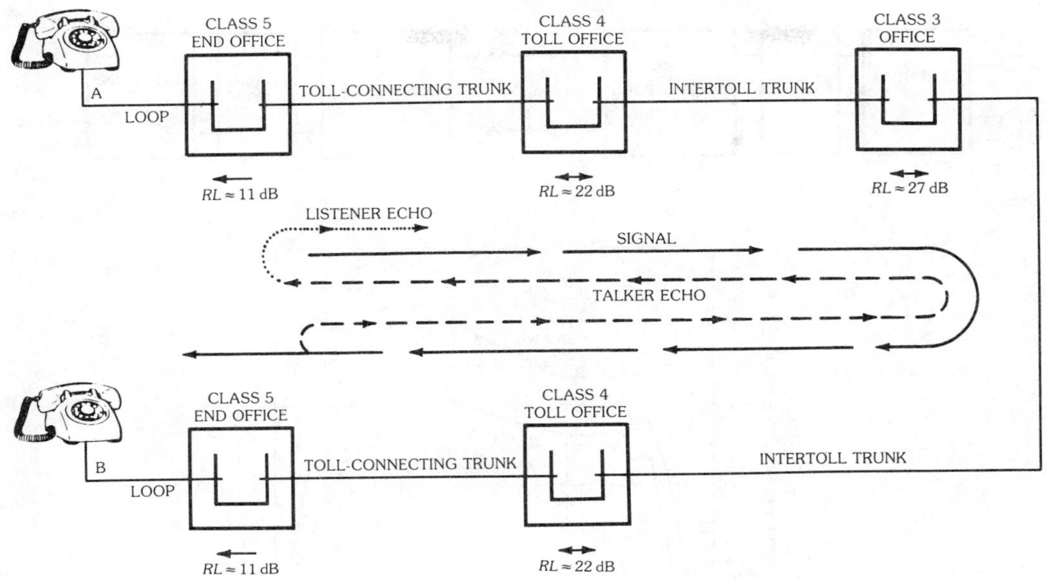

Fig. 1. Return losses and echo paths in a toll connection. (*From* Transmission Systems for Communication, *fifth edition, p. 185. © 1982 Bell Telephone Laboratories, Inc.*)

necessary to use a trunk route. The trunk route may be a single trunk connected directly between end-office switching systems. Alternatively, the trunk route may consist of several trunks routed through one or more toll or tandem switching systems. Direct trunks and trunks to tandem switching systems are generally limited to the exchange area, covering tens of miles or occasionally a few hundred miles. When a connection must be made over greater distances, the trunk route passes through toll connecting and intertoll trunks and is switched by a hierarchy of toll switching systems. The following paragraphs give information about the various transmission elements (e.g., loops and trunks) and about the signaling systems used to set up, hold, and disconnect the channels.

THE EXCHANGE PLANT

Subscriber Sets

Figs. 2 and 3 illustrate transmission performance of a 500-type subscriber set on some hypothetical connections. Typical station-set dc internal resistance is 200 to 300 ohms up to a maximum of 400 ohms.

Station Loops

Resistance Design— Present practice is to design most subscriber loops to use wire facilities with maximum loop resistance limited to a value dependent on requirements for supervisory signaling or transmission, whichever governs. Bridged tap limits and loading rules are also included to control transmission. Resistance objectives in current use are as follows.

For step-by-step offices: Maximum loop resistance of 1300 ohms, limited by supervisory signaling and dial-pulsing requirements. The maximum resistance may be less in some older installations.

For No. 5 crossbar and ESS offices: Maximum loop resistance of 1300 ohms, based primarily on transmission considerations. The 1300-ohm limit applies provided that loops of 18 000 feet and greater are loaded.

For longer loops: Higher resistance is permitted provided supervisory signaling range extension and transmission range extension are provided as required.

Subscriber carrier: The declining relative cost of electronics and new technology has made subscrib-

TABLE 2. VIA-NET-LOSS FACTORS*

Facility	*VNLF* (decibels per mile)
4-W VF Cable	0.017
2-W VF Cable	0.04
Carrier Systems (all types)	0.0015

*From *Transmission Systems for Communication*, fifth edition, p. 154. © 1982 Bell Telephone Laboratories, Inc.

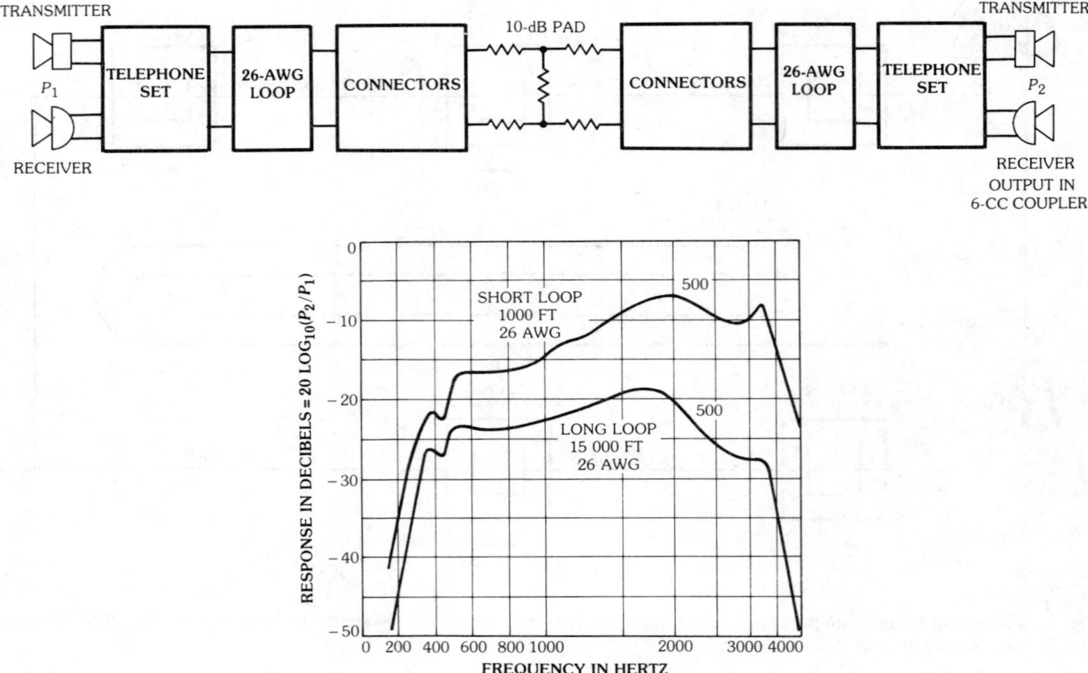

Fig. 2. Comparison of overall response. (*From W. F. Tuffnell, "500-Type Telephone Set,"* Bell Laboratories Record, *Vol. 29, September 1951, pp. 414–418. © 1951 Bell Telephone Laboratories, Inc.*)

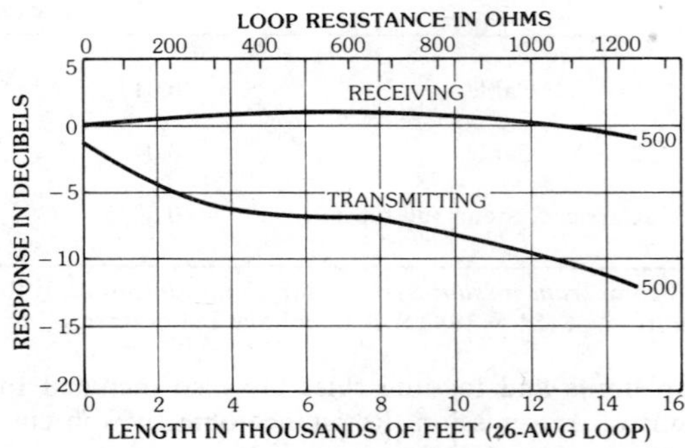

Fig. 3. Relative volume levels. (*From W. F. Tuffnell, "500-Type Telephone Set,"* Bell Laboratories Record, *Vol. 29, September 1951, pp. 414–418. © 1951 Bell Telephone Laboratories, Inc.*)

er carrier more attractive economically. Where subscriber carrier is applied, the loop transmission and signaling are generally equivalent to or better than loops on wire facilities.

Loss— The loss of a station loop should be less than 9.0 dB. The objective of the resistance design method is to achieve a distribution of loss values clustered broadly in the vicinity of 3 to 4 dB.

Noise— The noise objective for station loops is typically 20 dBrnc at the customer interface.

Current— For satisfactory transmission and signaling performance, station loop resistance limits are set in conjunction with line supervision source designs to assure a typical minimum loop current of 20 milliamperes.

Local Interoffice Trunks

Local interoffice trunks interconnect end-office switching systems and tandem switching systems. Most local trunks are so short that talker echo is not a problem. Fixed loss design objectives are usually used, rather than VNL objectives. Telephone administrations use objectives that depend on the type of trunk. A direct trunk between end-office switching systems is always the only trunk in a connection and may have loss objectives between 3 dB and 6 dB. Trunks between end-office switching systems and tandem switching systems typically have a 3-dB objective. Trunks between tandem switching systems usually have a 0-dB objective.

THE TOLL TRANSMISSION PLANT

Toll Connecting Trunks

Toll connecting trunks interconnect an end-office switching system and a toll switching system. Present transmission design objectives require a minimum loss of 2 dB in each toll connecting trunk to help mask the generally poor impedance match to the subscriber loops at the end office. The match is poor because the variety of impedances of different loops cannot be matched well by a single compromise impedance. When necessary, impedance-correcting networks are added at the toll end. Maximum loss of toll connecting trunks is set at 4 dB. Loss objectives for toll connecting trunks are typically $VNL + 2.5$ dB.

Intertoll Trunks

Overall Connection Loss— Intertoll trunks interconnect toll switching systems. Intertoll trunks are typically designed for *VNL* dB. Objectives for toll connecting trunks together with those for intertoll trunks yield an overall interoffice toll connection loss of:

$$OCL = VNLF \times L + 0.4N + 4 \text{ dB}$$

where,

N is the number of trunks in the connection,
L is the route length between the end offices.

Delay—Delay, by itself, is seldom annoying in speech communication until the delay has reached a value of approximately 600 milliseconds. Delays encountered in the modern toll plant seldom reach this value, but delay of this magnitude can be expected in circuits operating via synchronous-orbit satellites.

Optimum Loss—Talker echo (Fig. 1) without delay appears as sidetone and is not detrimental unless quite excessive. Delayed echoes, on the other hand, are particularly annoying. By increasing the overall connection loss of longer connections, the subjective annoyance of the talker echo may be reduced to the point of optimum tradeoff against impairment due to loss. Fig. 4 shows the optimum loss in a connection as a function of its length, compared to the value of *OCL* provided by the *VNL* plan.

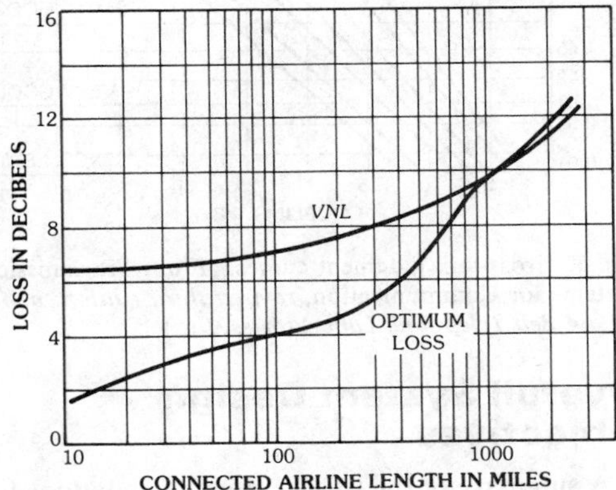

Fig. 4. Optimum loss compared with *VNL*. (*From* Transmission Systems for Communication, *fifth edition, p. 153. © 1982 Bell Telephone Laboratories, Inc.*)

Echo Objectives—To assure satisfactory echo performance of toll facilities, requirements for echo return loss (ERL) must be met. For two- to four-wire interfaces at toll switching points between intertoll trunks, an objective of 27 dB (minimum 21 dB) is typical. For two-wire interfaces at toll offices to toll connecting trunks, the objective is typically 22 dB (minimum 16 dB). The ERL objective for loops at the end office is typically 11 dB.

Echo Suppressors—If a trunk is longer than 1800 miles, American practice is to insert echo suppressors and then set the net loss of the trunk to 0 dB. Echo suppressors are voice-operated devices that, when one party is talking, insert a high loss in the opposite direction of transmission. Though modern designs minimize problems, echo suppressors must be used carefully because they can cause clipping of the start and finish phonemes of words. If two or more circuits containing echo suppressors are switched in tandem, a phenomenon known as lock-out may occur when both parties attempt to talk simultaneously. In this case, neither party will be heard by the other.

Echo performance may also be improved by the application of an echo canceller. The technique involves formulating an estimate of the echo that will result from an incoming signal on the four-wire side of a hybrid. That estimate is subtracted from the signal on the return path. The echo-canceller device produces the echo estimate by means of a signal processor (filter) for which parameters are set by correlation techniques.

CCITT RECOMMENDATIONS

The CCITT currently recommends* that international toll circuits in the new transmission plan be given an insertion loss of 0.5 dB for each 500 kilometers of length or fraction thereof, assuming that the international circuit is derived from carrier systems in coaxial cable or radio relay. Table 3 compares the losses of long-haul switching trunks designed in accordance with CCITT Recommendation G.131 and with *VNL*. For the *VNL* design, a *VNLF* of 0.0015 dB per statute mile is used. The 0.4 dB term is included.

Echo Control

The CCITT recommends* that the overall loss of a connection may be adjusted so that echo signals are sufficiently attenuated or, alternatively, an echo suppressor may be fitted if the loss adjustment results in an excessive insertion loss.

* CCITT Recommendaion G.131, Yellow Book, Vol. III-1, Seventh Plenary Assembly, Geneva, 1980.

TABLE 3. CIRCUIT NET LOSS IN DECIBELS

Circuit Length		CCITT Rec. G.131 (dB)	*VNL* Design (dB)
Kilometers	Statute Miles		
0	0	0.5	0.40
250	155.3	0.5	0.63
500	310.7	0.5	0.87
750	466.0	1.0	1.10
1000	621.4	1.0	1.33
1250	776.7	1.5	1.56
1500	932.1	1.5	1.80
1750	1087.4	2.0	2.03
2000	1242.8	2.0	2.26
2250	1398.1	2.5	2.50
2500	1553.5	2.5	2.73

Noise Objectives

Noise objectives in present use are:
Long-haul circuits: 44 dBrnC at 0 TLP.
Short- and medium-haul circuits: 38 dBrnC at 0 TLP.

Received Volume Objectives

Speech volume in the commercial American plant has been the subject of continuing study over the years. A study by Ahern, Duffy, and Maher† is a recent definitive study of speech volume and related parameters expressed as average power and equivalent peak level.

Cross-Talk Objectives

If coupling paths between transmission systems give rise to intelligible or nearly intelligible cross-talk, normal practice is to design the system so that the probability that a customer will hear a "foreign" conversation does not exceed 1 percent. In measuring cross talk, a commonly used unit is the dBx, which is defined as the difference between 90 dB of loss and the transmission of the coupling path. Fig. 5 illustrates the relationship between the cross-talk coupling in dBx and the cross-talk index, which is defined as the chance of encountering intelligible cross-talk.

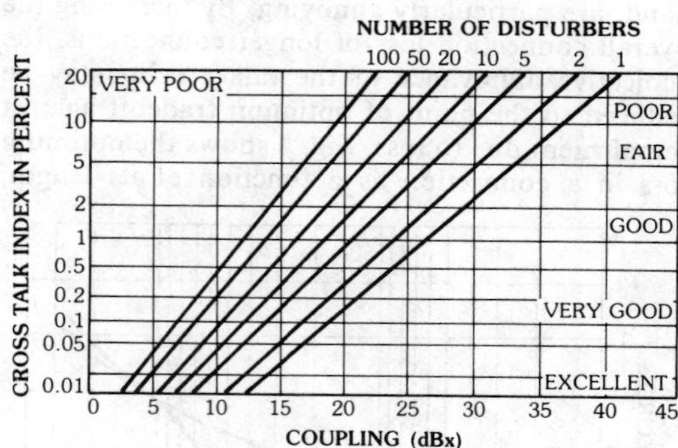

Fig. 5. Cross-talk judgment curves. (*From* Transmission Systems for Communication, *revised third edition, p. 48.* *© 1964 Bell Telephone Laboratories, Inc.*)

Overall System Design Objectives

A summary of the CCITT recommendations for system objectives and design criteria for circuits used in the international telephone service is given in Chapter 2.

†W. C. Ahern, F. P. Duffy, and J. A. Maher, "Speech Signal Power in the Switched Message Network," *Bell System Technical Journal*, Vol. 57, No.7, Sept. 1978.

WIRE FACILITIES

Open-wire facilities are generally obsolete and limited to rural areas. Multiconductor cable with a variety of insulation types is widely used in pair counts up to 2700 or higher. Paper pulp and polyethylene insulated cable are most common for voice-band transmission and are also used for frequencies up to several megahertz. At higher frequencies and for carrier transmission, single or multiple coaxial cable facilities may also be used. The transmission characteristics and parameters of some typical exchange area cables are given in Table 4. The same cable types may be used either for loops or for local interoffice trunks. Trunk cables and longer loops are normally loaded to reduce loss and minimize the need for repeaters.

CABLE TRANSMISSION REPEATERS

If cable loss exceeds the allowable maximum loss, it is necessary to add gain to the circuit. Amplifiers designed for this purpose are termed "voice-frequency repeaters." The advent of carrier transmission systems has made obsolete the use of cable and voice-frequency repeaters for intertoll trunks. However, in the exchange plant, local interoffice and toll-connecting trunks frequently use voice-frequency-repeatered cable. Voice-frequency repeaters are also used for transmission range extension on some local loops. Three types of repeaters are typically used: negative-impedance, hybrid, and four-wire repeaters.

Negative-Impedance Repeaters—Two-Wire

A negative-impedance telephone repeater is a voice-frequency repeater that provides gain by inserting a series or shunt negative impedance into the line. Between lines having reasonably similar impedances, the bridged-T configuration combination repeater may be used (Fig. 6). Negative-impedance repeaters require special impedance-matching techniques to avoid instability at or near maximum gain. Negative-impedance repeaters have fallen into disuse in favor of hybrid repeaters.

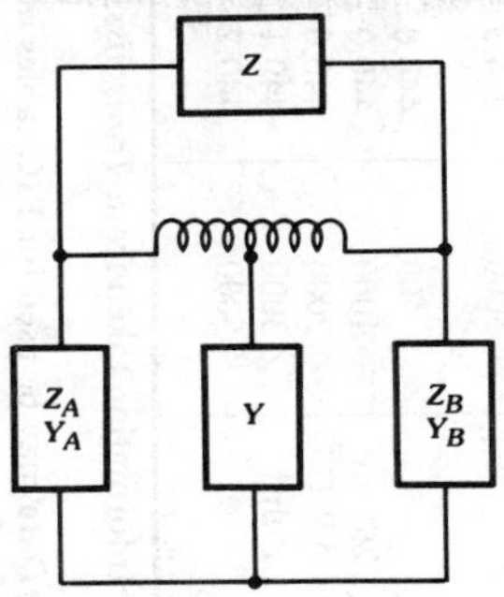

Fig. 6. Series-shunt repeater.

Hybrid Repeaters

Hybrid repeaters provide independent gain and equalization for each direction of transmission by splitting the path with four-wire hybrids. A typical hybrid repeater configuration is shown in Fig. 7. To assure stability of the internal gain loop, hybrid balance must be carefully controlled by proper design and adjustment of the balance networks. For optimal perfomance, the networks (often active) should exactly match the impedance seen at the cable-facility interface. To the extent that impedance is mismatched, gain must be limited to prevent instability.

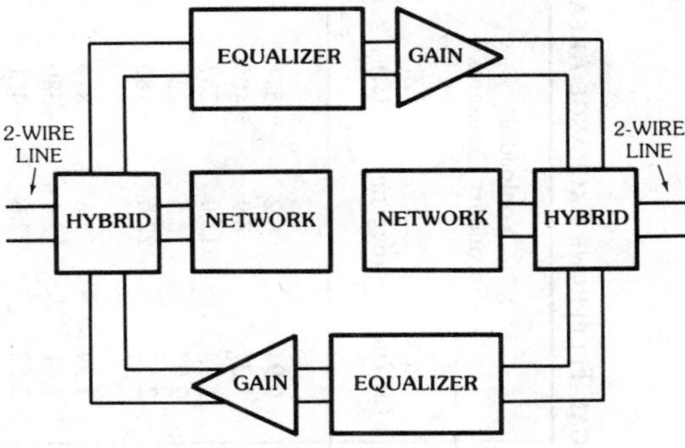

Fig. 7. A typical hybrid repeater configuration.

Four-Wire Voice-Frequency Repeaters

If voice-frequency trunks must include more gain than can be obtained from two-wire repeaters, four-wire operation is employed (Fig. 8). Repeaters for four-wire circuits are conventional one-way amplifiers. The facility is converted from two-wire to four-wire at the terminations by the use of four-wire terminating sets consisting of resistance or transformer hybrids with appropriate balancing networks to match the connecting channels. Low-pass filters may be included in the terminating sets and/or in the repeaters to limit the possibility of the circuit having sufficient gain to sing at frequencies outside the voice band.

CARRIER SYSTEMS

When many telecommunications channels are needed between two points, it is usually economical to use a carrier system to multiplex many channels over one medium. Media used for carrier telephony include wire pairs, coaxial cables, microwave radio, waveguides, synchronous-orbit satellites, and glass fibers.

TABLE 4. CHARACTERISTICS OF TYPICAL TELEPHONE EXCHANGE AREA CABLES*

	Frequency	Primary Constants				Nonloaded Secondary Constants				H 88 Loaded** Secondary Constants			
	f (Hz)	R (Ω/mi)	L (mH/mi)	G (μmho/mi)	C (μF/mi)	α (dB/mi)	β (deg/mi)	$\|Z\|$ (Ω)	$\angle Z$ (deg)	α (dB/mi)	β (deg/mi)	$\|Z\|$ (Ω)	$\angle Z$ (deg)
	500	86.1	0.887	0.506	0.084	0.91	6.2	571	−44.0	0.42	14.9	1009	−10.3
19	1000	86.2	0.886	1.219	0.084	1.27	8.9	405	−43.1	0.42	29.8	1017	−5.3
AWG	2000	86.4	0.884	3.133	0.084	1.74	13.0	287	−41.3	0.43	62.2	1183	−2.7
Pulp†	3000	86.6	0.883	5.520	0.084	2.07	16.5	236	−39.5	0.52	105.5	1914	−3.4
	5000	87.1	0.881	11.35	0.084	2.52	22.6	186	−36.1	13.78	158.7	926	−88.5
	500	173.1	0.871	0.494	0.082	1.29	8.6	819	−44.5	0.76	15.3	1098	−18.0
22	1000	173.2	0.870	1.190	0.082	1.81	12.3	580	−44.0	0.79	29.7	1051	−9.9
AWG	2000	173.3	0.869	3.058	0.082	2.52	17.6	411	−43.1	0.79	61.5	1196	−5.1
Pulp†	3000	173.5	0.867	5.388	0.082	3.04	21.9	336	−42.2	0.88	103.5	1871	−5.6
	5000	173.8	0.866	11.08	0.082	3.81	29.2	262	−40.4	13.54	159.5	960	−87.4
	500	274.0	0.951	0.506	0.084	1.64	10.9	1019	−44.6	1.13	16.4	1201	−24.3
24	1000	274.0	0.950	1.219	0.084	2.31	15.6	721	−44.3	1.21	30.7	1079	−14.6
AWG	2000	274.1	0.949	3.133	0.084	3.24	22.2	510	−43.7	1.23	62.7	1200	−7.7
Pulp†	3000	274.2	0.948	5.520	0.084	3.92	27.5	417	−43.0	1.35	105.6	1906	−8.7
	5000	274.3	0.947	11.35	0.084	4.95	36.3	324	−41.8	13.81	160.8	927	−86.2
	500	440.0	0.995	0.476	0.079	2.03	13.4	1330	−44.7	1.58	17.6	1444	−30.5
26	1000	440.0	0.995	1.146	0.079	2.85	19.0	942	−44.5	1.79	31.0	1199	−20.8
AWG	2000	440.0	0.995	2.946	0.079	4.01	27.1	666	−44.1	1.86	61.3	1247	−11.8
Pulp†	3000	440.1	0.994	5.191	0.079	4.87	33.4	544	−43.7	1.97	101.0	1820	−11.8
	5000	440.2	0.993	10.67	0.079	6.20	43.7	422	−42.9	13.21	161.7	1018	−84.4

*Information taken from *Transmission Data-Exchange Area Cables*, © 1962 AT&T.

†Data may be used for PIC cables in the voice band.

**H 88 loading implies use of 88-mH loading coils at intervals of 6000 ft. Data apply to cables with 3000-ft end sections on both ends.

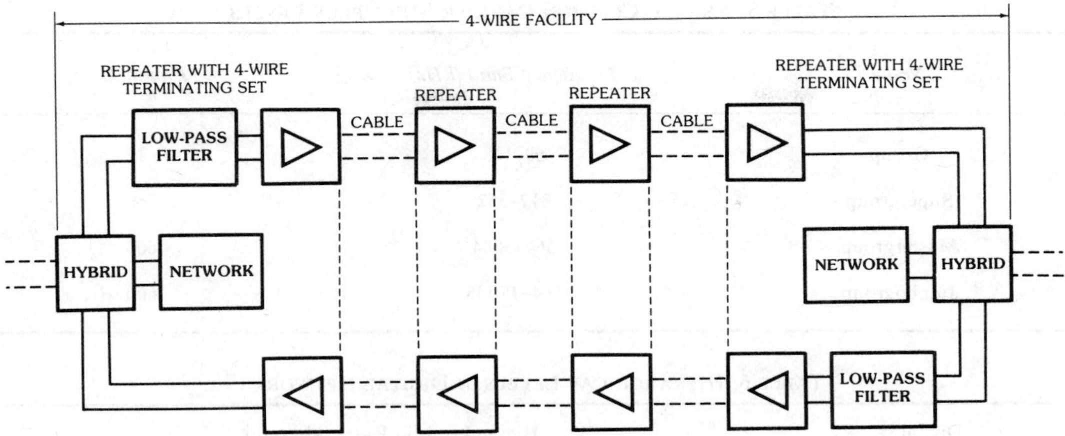

Fig. 8. Four-wire method of operation. (*From* Transmission Systems for Communication, *revised third edition. p. 74 ©1964 Bell Telephone Laboratories, Inc.*)

Multiplexing Techniques

Two basic techniques are used for the transmission of a plurality of telephone channels over a single transmission medium:

A. Frequency-division (analog) systems, in which a unique band of frequencies within the wideband frequency spectrum of the transmission medium is allotted to each communication channel on a continuous time basis.

B. Time-division (digital) systems, in which each communication channel is allotted a discrete time slot within a sampling frame, with a theoretical occupancy of the entire wideband frequency spectrum for the allotted time.

Several types of modulation techniques may be employed with each of the multiplexing techniques.

Modulation Techniques

Several frequency-division modulation techniques are in common use, including amplitude modulation and frequency modulation, double and single sideband. Pulse-code modulation is the most common time-division modulation technique, although delta modulation and other techniques are also found. See Chapters 23 and 24 for a more detailed discussion of modulation techniques.

Subscriber Carrier

Although first introduced to serve customers at great distances from an end-office switching system, subscriber carrier systems have begun to be economical on shorter routes. One carrier terminal is placed at the end office and the other in the general vicinity of a group of customers. Ordinary wire facilities are used to carry derived voice-frequency channels between the remote terminal and the customers. In some cases, the remote terminal may be placed on the premises of a customer requiring many lines. Both frequency-division and time-division techniques are used. The overwhelming majority of new systems now use time division.

Trunk Carrier

Trunk carrier systems operate between carrier terminals at switching offices. Trunk systems may be further categorized as short- or medium-haul systems for use between switching systems within an exchange area, and long-haul systems having performance specifications suitable for transcontinental or intercontinental connections.

Common-Carrier Hierarchies

There are several agreed-upon bundle sizes or multiplex levels for carrying telephone channels in common-carrier telephony. Most of the standards are covered in a variety of available CCITT documents. Table 5 gives the common-carrier multiplex levels used in North America for analog transmission of single-sideband channels (nominally 4 kHz wide) stacked in frequency-division multiplex.

In digital transmission, the basic entity is a 64 kb/s channel in which the nominal 4 kHz signal has been filtered (to minimize aliasing) and sampled at 8 kHz, and each sample is represented by eight bits. Of course, digital data at up to 64 kb/s can be substituted for the digitized voice-band signal. Table 6 gives the digital hierarchy levels corresponding to bit rates and number of usable 64-kb/s time slots recommended by CCITT. As noted below, other levels are being introduced.

At level 1, wire pairs are used to carry the bit stream. Indeed, in the United States more than half

TABLE 5. ANALOG COMMON-CARRIER MULTIPLEX LEVELS

Name	*Frequency Band (kHz)*	*Channels*
Group	60–108	12
Supergroup	312–552	60
Mastergroup	564–3084	600
Jumbogroup	564–17 548	3600

TABLE 6. HIERARCHICAL LEVELS IN DIGITAL NETWORKS

Digital Hierarchy Levels	Hierarchical Bit Rates (kb/s) for Networks With the Digital Hierarchy Based on a First Level Bit Rate of: 1544 kb/s*	2048 kb/s*
0	64	64
1	1544	2048
2	6312	8448
3	32 064 44 736	34 368
4		139 264

*Typically, 1544 kb/s is used at the first level in the US and Canada, and 2048 kb/s is used in Europe.

of the local interoffice trunks are digital, and by 1990 virtually all toll connecting trunks are planned to be digital due to the synergy between digital transmission and switching. Radio and optical-fiber media have been used for higher bit rates such as 44 736 kb/s and 90 254 kb/s. Rapid technology advances such as the realization of low-loss fiber, both multimode and single mode, is expected to lead to the introduction of other bit rates in the digital hierarchy. Table 7 lists some of the levels in the US digital hierarchy used or to be introduced soon. Figs. 9 and 10 diagram the levels, the associated multiplexes, digital line systems, and references to CCITT documents that give more detailed specifications for the two generic hierarchies.

NETWORK CONTROL SIGNALING

Subscriber-Loop Signaling

Subscriber-loop signaling generally uses control of direct current in the subscriber loop to provide both supervisory and address signals. Supervisory on-hook (open-loop) and off-hook (closed-loop) signals are used to detect when the calling subscriber is demanding service and when a called subscriber answers. Numerics are transmitted as dial pulses obtained by opening and closing the loop at a rate on the order of 8 to 12 pulses per second. The number of pulses in a train represents the dialed digit, except for the digit 0, which is represented by 10 pulses.

A ringing signal for summoning the called subscriber to the telephone is commonly transmitted as a high-voltage, low-frequency signal (typically at least 40 V rms, 20 Hz, at the subscriber terminal) to actuate directly a bell in the subscriber terminal. Another approach has been to transmit a tone at normal speech level, this tone being used to actuate an audible device.

Dual-Tone Multifrequency Signaling (DTMF)

To increase the speed of service and reduce holding time on registers in the end-office switching system, a tone signaling technique for subscriber lines has been introduced. At the subscriber set, the conventional dial is replaced with a set of push-button keys that, when pressed by the subscriber, cause transmission to the central office of combinations of two audio-frequency tones, one combination

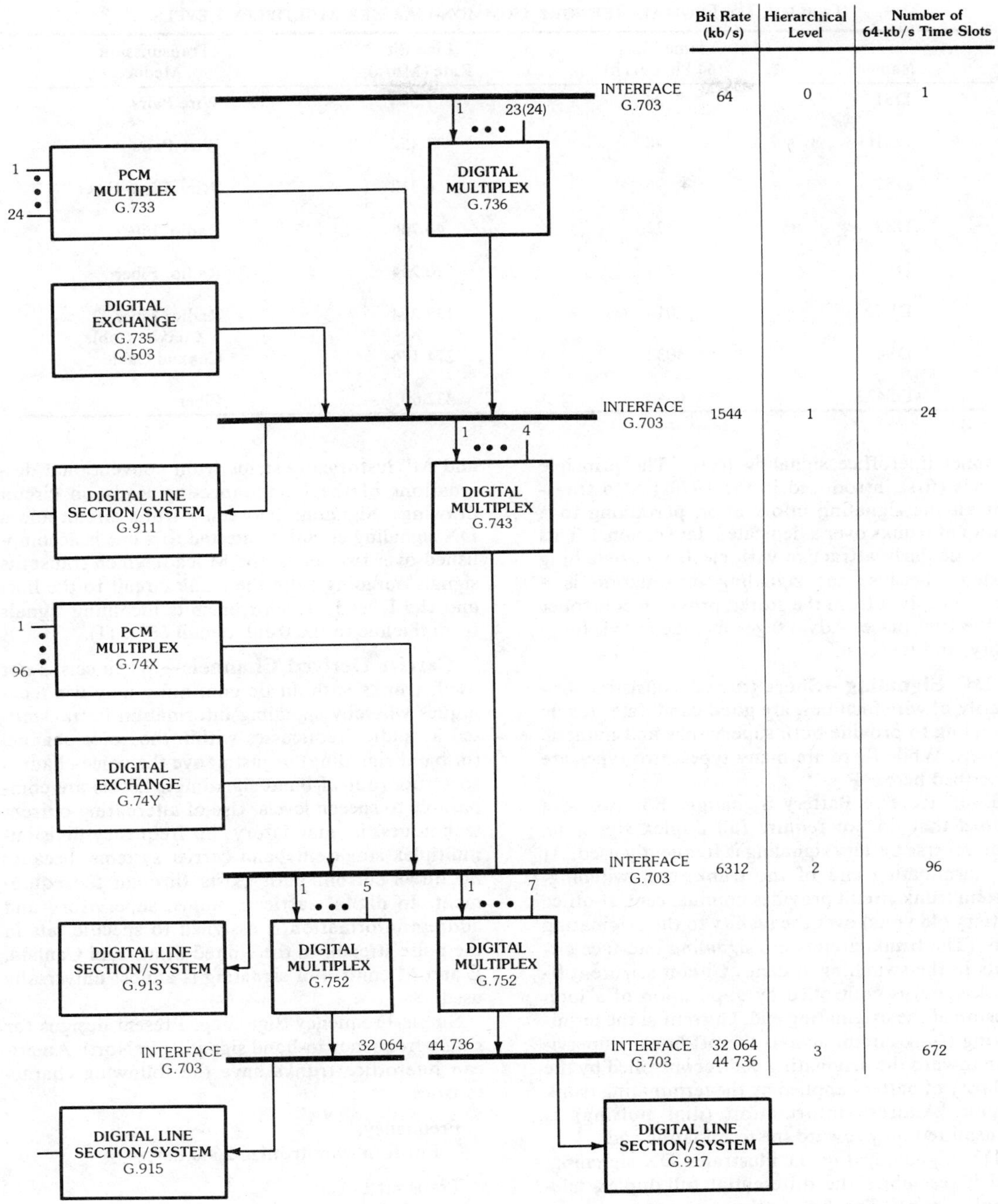

Fig. 9. Hierarchical bit rates for networks with the digital hierarchy based on the first-level bit rate of 1544 kb/s (including references to related CCITT Recommendations).

representing each numerical digit. A total of eight tone frequencies is provided, the excess combinations over the 10 required for numerics being reserved for special signals. The dtmf signaling code is given in Table 8.

Interoffice Signaling

Common-Channel Interoffice Signaling— Signaling between switching systems (whether exchange or toll) may be either per-trunk or common-

TABLE 7. US DIGITAL-NETWORK COMMON-CARRIER MULTIPLEX LEVELS

Name	Time Slots (64 kb/s each)	Line Bit Rate (Mb/s)	Transmission Media
DS1	24	1.544	Wire Pairs
DS1C	48	3.152	Wire Pairs
DS2	96	6.312	Wire Pairs, Fiber
DS3	672	44.736	Radio, Fiber
DS3C	1344	90.254	Radio, Fiber
DS4E	2016	139.264	Radio, Fiber, Coaxial Cable
DS4	4032	274.176	Coaxial Cable
DS432	6048	432.00	Fiber

channel interoffice signaling (ccis). The principle of ccis (first introduced in the 1970s) is to transmit all the signaling information pertaining to a group of trunks over a dedicated data channel. This is particularly attractive with electronic switching systems because the signaling information is a stream of data from the stored program controller in the first place. Advantages include speed, flexibility, and low cost.

DC Signaling—Short trunks, consisting primarily of wire facilities, are good candidates for dc signaling to provide both supervisory and numeric signals. While there are many types, two types are described here.

Loop Reverse Battery Signaling: For two-wire trunks that do not require full duplex signaling, loop reverse battery signaling is frequently used. At the terminating end of the trunk, the switching-system trunk circuit provides nominal central-office battery (48 volts) over the facility to the originating end. (The trunk circuits are signaling interface circuits in the switching system.) Circuit seizure (off-hook state) is controlled by application of a loop closure at the originating end. Current at the terminating trunk circuit signals the off-hook. Supervision toward the originating end is controlled by the polarity of battery applied by the terminating trunk circuit. Address information (dial pulsing) is transmitted only toward the terminating end.

DX Signaling: Fig. 11 illustrates DX signaling, which resembles the differential full-duplex telegraph circuit. The full-duplex operation permits independent two-way signal transmission. Signaling bypasses may be required at repeaters. The diagram shows that E and M control is used between the DX circuit and the trunk circuits.

E and M Lead Control—Most signaling systems other than loop signaling are separated from the trunk circuit and generally are introduced between the trunk circuit and the line. The name "E and M" historically stems from conventional designations of the interconnecting leads on circuit drawings. Signaling between a trunk circuit and a DX signaling circuit connected to a line is accomplished over two leads, the M lead, which transmits signals outgoing from the trunk circuit to the line, and the E lead, which transmits incoming signals from the line to the trunk circuit (Fig. 11).

Carrier Derived Channels—When ccis is not used, trunks without dc continuity may use techniques whereby signaling information is transmitted at audio frequencies within the voice channel (in-band signaling) or just above the voice-channel spectrum (out-of-band signaling). Levels are comparable to speech levels. Use of alternating-current techniques is mandatory in frequency-division-multiplex single-sideband carrier systems, because no direct-current path exists through the equipment. In digital carrier systems, supervisory and address information is assigned to specific bits in the pulse stream. In the United States and Canada, E and M control of signaling is almost universally used.

Single-Frequency Signaling: Present designs for single-frequency in-band signaling in North American interoffice trunks have the following characteristics:

Frequency:
For four-wire trunks, 2600 hertz

Transmit Level:
− 8 dBm0 during pulsing
− 20 dBm0 for continuous tone (on-hook state)

Provision is made to minimize mutual interference between signal and speech circuits by the use of a guard circuit that inhibits operation of the sf receiver when frequencies other than signal tone are present. Additional arrangements insert a tone-blocking filter in the receiving speech path when

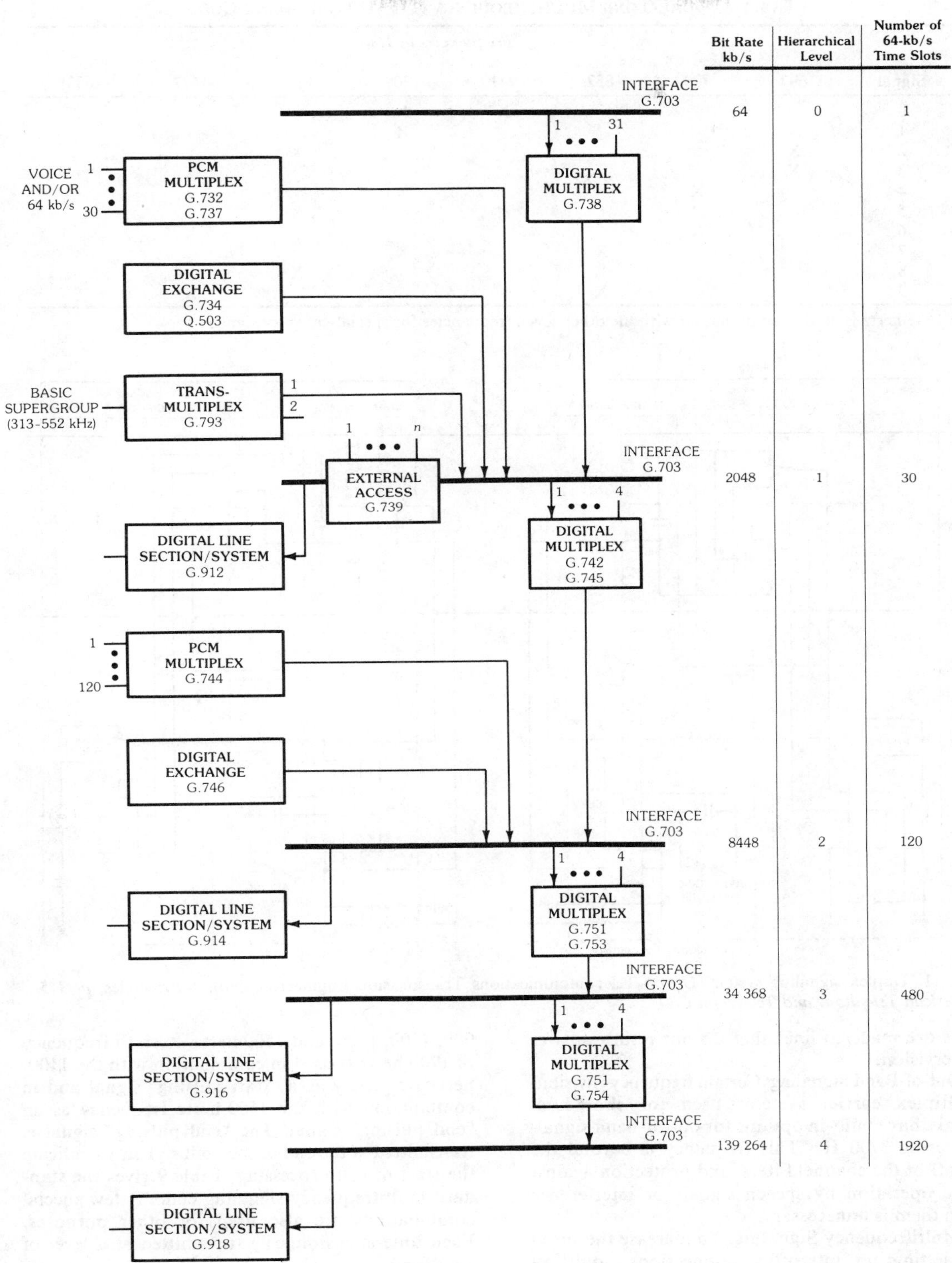

Fig. 10. Hierarchical bit rates for networks with the digital hierarchy based on the first-level bit rate of 2048 kb/s (including references to related CCITT Recommendations).

TABLE 8. DUAL-TONE MULTIFREQUENCY (DTMF) SIGNALING CODE

	Frequencies in Hertz							
Signal	697	770	852	941	1209	1336	1477	1633*
0				X	X			
1	X				X			
2	X					X		
3	X						X	
4		X			X			
5		X				X		
6		X					X	
7			X		X			
8			X			X		
9			X				X	

* 1633 hertz is used in combination with the other seven frequencies for special-category signals.

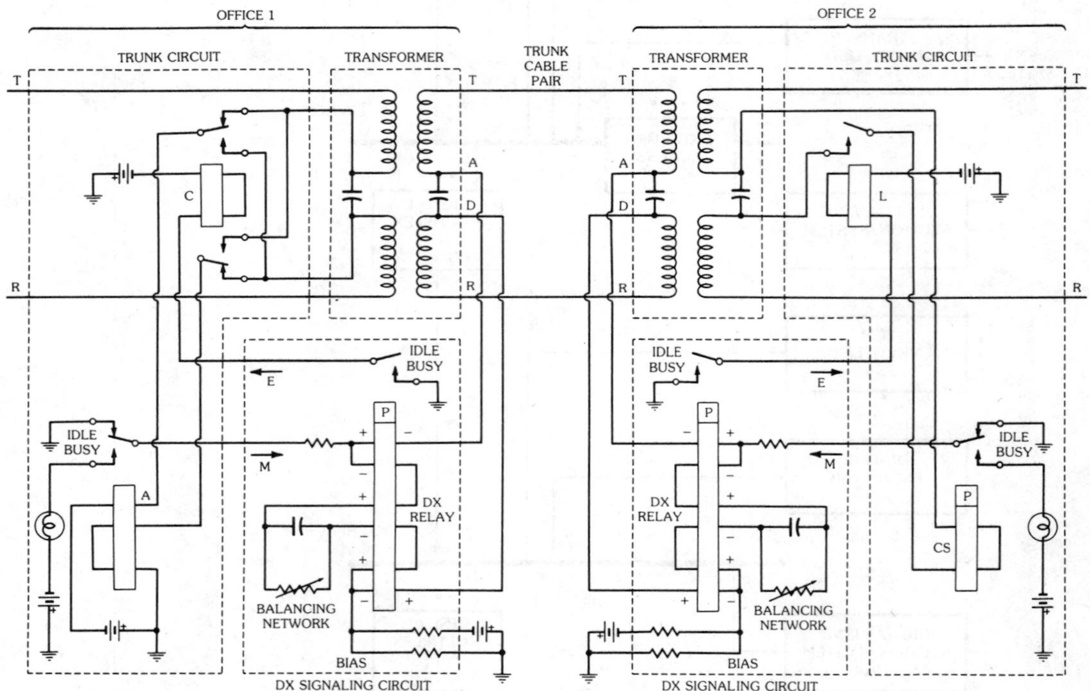

Fig. 11. Duplex signaling system. (*From* Telecommunications Transmission Engineering, Vol. 1 Principles, *p. 345.* © *American Telephone and Telegraph Co.*)

calls are made to lines that do not return answer supervision.

Out-of-Band Signaling: Certain frequency-division-multiplex carrier systems used for short-haul trunks have built-in options for out-of-band signaling using 3700 Hz. This frequency is beyond the cutoff of the channel filters, and protection against false operation by speech signals or interference with them is unnecessary.

Multifrequency Signaling: To increase the speed of setting up interoffice connections, multifrequency signaling is often applied to trunk circuits. Digital information is transmitted by combinations of two of the following five audio frequencies: 700, 900, 1100, 1300, and 1500 hertz. A sixth frequency of 1700 hertz is used in combination with the 1100-hertz frequency as a "start pulsing" signal and in combination with the 1500-hertz frequency as an "end pulsing" signal. The "end pulsing" signal is transmitted at the end of the digit signals to indicate the start of call processing. Table 9 gives the standard multifrequency signaling code. A few special combinations are also used for other purposes. Each tone is customarily transmitted at a level of −6 dBm0.

Signaling Systems—CCITT—For CCITT Recommendations, refer to Chapter 2.

TABLE 9. MULTIFREQUENCY TRUNK SIGNALING CODE

	Frequencies in Hertz					
Signal	700	900	1100	1300	1500	1700
Start Pulsing			X			X
End Pulsing					X	X
0				X	X	
1	X	X				
2	X		X			
3		X	X			
4	X			X		
5		X		X		
6			X	X		
7	X				X	
8		X			X	
9			X		X	

REFERENCES

Andrews, F. T. "Loop Plant Electronics—Overview." *Bell System Tech. J.*, Vol. 57, Apr. 1978, pp. 1025–1034.

AT&T Network Planning Division, Fundamental Network Planning Section. *Notes on the Network*. American Telephone and Telegraph Co. Inc., 1980.

AT&T Co., Bell Telephone Companies, and Bell Laboratories; technical personnel. *Telecommunications Transmission Engineering*. American Telephone and Telegraph Co., Inc., 1974.

Bennett, A. F. "An Improved Circuit for the Telephone Set." *Bell System Tech. J.*, Vol. 32, May 1953, pp. 611–626.

Bonner, A. L., Garrison, J. L., and Kopp, W. J. "The E6 Negative Impedance Repeater." *Bell System Tech. J.*, Vol. 39, Nov. 1960, pp. 1445–1504.

Chinn, H. A., Gannett, D. K., and Morris, R. M. "A New Standard Volume Indicator and Reference Level." *Bell System Tech. J.*, Vol. 19, Jan. 1940, pp. 94–137.

Drechsler, R. C. "Echo Suppressor Terminal for No. 4 ESS." Conference Record, ICC '76, Vol. III, pp. 36–9, 36–12.

Drechsler, R. C., Koehler, D. C., and Royer, R. D. "Suppressing Echoes Digital Style." *Bell Laboratories Record*, Vol. 56, June 1978, pp. 142–146.

Duffy, F. P., Ahern, W. C., and Maher, J. A. "Speech Signal Power in the Switched Message Network." *Bell System Tech. J.*, Vol. 57, Sept. 1978, pp. 2695–2726.

Duttweiler, D. L. "A Single-Chip VLSI Echo Canceller." *Bell System Tech. J.*, Vol. 59, Feb. 1980, pp. 149–160.

Gresh, P. A. "Physical and Transmission Characteristics of Customer Loop Plant." *Bell System Tech. J.*, Vol. 48, Dec. 1969, pp. 3337–3385.

Holman, E. W., and Suhocki, V. P. "A New Echo Suppressor." *Bell Laboratories Record*, April 1966.

Jacobs, I. "Fiber Optics Communications Technology and Application Trends in the Bell System." 4th World Telecommunications Forum, Speaker's Papers, Pt. II, Vol. III, pp. 3.16.5.1–3.16.5.6.

Manhire, L. M. "Physical and Transmission Characteristics of Customer Loop Plant." *Bell System Tech. J.*, Vol. 57, Jan. 1978, pp. 35–39.

Park, K. I. "Intelligible Crosstalk Performance of Voice-Frequency Customer Loops." *Bell System Tech. J.*, Vol. 57, Oct. 1978, pp. 3001–3029.

Ritchie, A. E., and Menard, J. Z. "Common Channel Interoffice Signaling: An Overview." *Bell System Tech. J.*, Vol. 57, Feb. 1978, pp. 221–224.

Sondhi, M. M. "An Adaptive Echo Canceller." *Bell System Tech. J.*, Vol. 46, Mar. 1967, pp. 497–511.

Spang, T. C. "Loss Noise Echo Study of the Direct Distance Dialing Network." *Bell System Tech. J.*, Vol. 55, Jan. 1976.

Technical staff, members of. *Transmission Systems for Communication*, 5th ed. Bell Telephone Laboratories, Inc., 1982. (See also 3rd ed., 1964, and 4th ed., 1970, 1971.)

Transmission Data—Exchange Area Cables. American Telephone and Telegraph Co., Inc., 1962.

Tuffnell, W. F. "500-Type Telephone Set." *Bell Laboratories Record*, Vol. 29, Sept. 1951, pp. 414–418.

39 Switching Networks and Traffic Concepts

Revised by
Walter S. Hayward

PART 1; COORDINATE SWITCHING NETWORKS

DEFINITIONS OF TERMS

Concentration: The function associated with a switching network having fewer outlet than inlet terminals.

Coordinate switch: A rectangular array of crosspoints in which one side of the crosspoint is multipled in rows and the other side in columns.

Crosspoint: A two-state switching device containing one or more elements that have a low transmission impedance in one state and a very high one in the other.

Expansion: The function associated with a switch or switching network having more outlet than inlet terminals.

Folded network: A network in which each terminal can serve as either inlet or outlet and which is capable of completing a path between any pair of inlet-outlets.

Full availability: Property of a switch or switching network capable of providing a path from every inlet terminal to every outlet terminal.

Internal blocking: The inability to interconnect an idle inlet to an idle outlet because all possible paths between them are already in use.

Nonblocking network: A network in which there is at all times at least one available path between any idle inlet and any idle outlet, regardless of the number of paths already occupied.

Nonfolded network: A network in which inlets and outlets are separate and which is capable of completing a path between any inlet and outlet.

Single-linkage array: The mesh or spread of interconnections between the stages of a switching network whereby every switch of one stage has one connection to every switch of the adjacent stage.

Space-division switching network: A switching network in which the transmission paths are physically distinct.

Switching network: That part of a switching system that establishes transmission paths between pairs of terminals.

Switching stage: Those switches in a switching network that have identical parallel functions.

Time-division switching network: A switching network in which the transmission paths are separated in time.

SPACE- AND TIME-DIVISION SWITCHING

Most of the switching systems of the world have utilized space-division analog switching. Recently, combinations of space and time division have become almost universal in new switching systems, employing primarily pulse-code modulation on multiplexed lines. Because there is duality between time-division and space-division switching and the principles are more easily understood in terms of space division, space division will be used in describing switching-network principles. This will be followed by a description of how to map time-division into space-division networks.

PROPERTIES OF COORDINATE SWITCHING NETWORKS

The simplest coordinate switch has a number of "crosspoints" that can give an inlet to the switch a connection to an outlet. (In most applications, only one connection will be desired at a time, but this is not always a requirement.) It is convenient to give a number of inlets identical access to a number of outlets so that the network can be arranged in a single piece of apparatus (switch) as shown in Fig. 1A. Such a switch is known as a rectangular coordinate switch.

In electromechanical coordinate switches, the connecting devices may be individual contact-making relays, in which case the number of complete relays (coil and a set of contacts) required is the product of the numbers of inlets and outlets. Alternatively, the whole crosspoint array may be provided by a crossbar switch in which a single relay coil is associated with each row and column of the switch, and the concurrent energizing of a row coil and column coil closes an individual set of contacts. In electronic switches, the crosspoints may be solid-state elements, usually transistors.

The rectangular switch discussed so far is nonblocking—any idle inlet can reach an idle outlet. For small numbers of inlets and outlets, this is an efficient arrangement, but when the inlets and outlets are more than twenty to thirty, the number of crosspoints can be reduced significantly by replacing the single switch by a number of interconnected smaller coordinate switches, arranged in a multistage coordinate switching network in one of many possible ways, as will be described. If blocking is acceptable, the number of crosspoints can be reduced even more.

SINGLE-STAGE COORDINATE SWITCHES

Fig. 1A shows a rectangular-coordinate switch interconnecting inlets from N lines and outlets to M

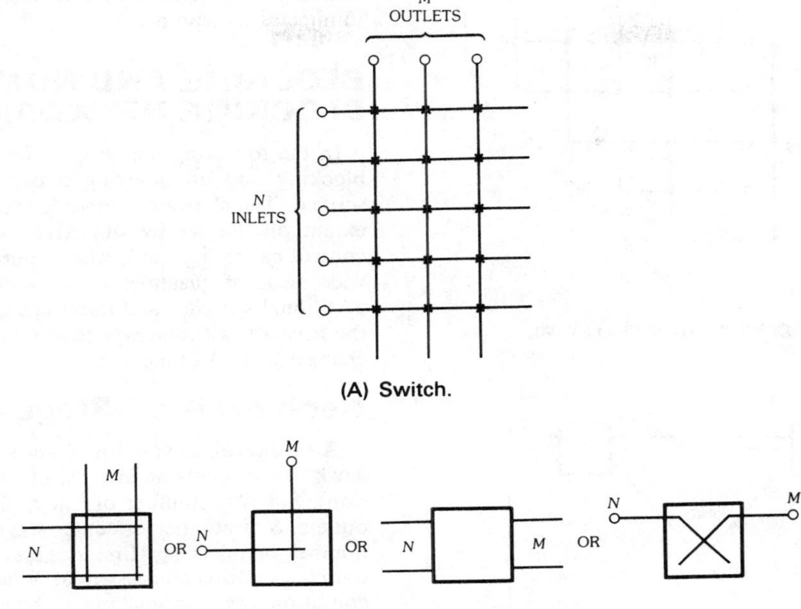

(A) Switch.

(B) Conventional equivalent symbols.

Fig. 1. Single-stage rectangular coordinate switch (full availability).

other lines. When interconnection is possible at every crosspoint, the switch provides full availability and is said to be nonblocking. The particular switch shown in Fig. 1A acts as a concentrating switch so that, although the switch is nonblocking, an idle inlet may be blocked because there is no idle outlet. Such concentration switches are often used to bring higher traffic loads per circuit to the subsequent stages of a multistage system in order to use switches more efficiently. A rectangular switch with full availability requires NM crosspoints.

If full availability is not a requirement, an economy of crosspoints can be achieved by use of a limited-availability (restricted-access) coordinate switch. Fig. 2 shows a square switch with five inlets and five outlets in which every inlet has access to only three of the outlets. The grading is said to be homogeneous when, as shown, each set of three outlets is unique.

When the outlets from a switch are connected to the same lines or trunks as are the inlets to the switch, a triangular "folded" arrangement may be used. Fig. 3 shows a triangular switch for N inlets-outlets. The switch provides full availability and is nonblocking.

$$\text{Number of crosspoints required} = N(N-1)/2$$

Fig. 4 shows a symmetrical two-stage network with square switches. This formed the basic frame arrangement of many electromechanical switching systems in which the limited size of the crossbar switch was compensated for by introducing two

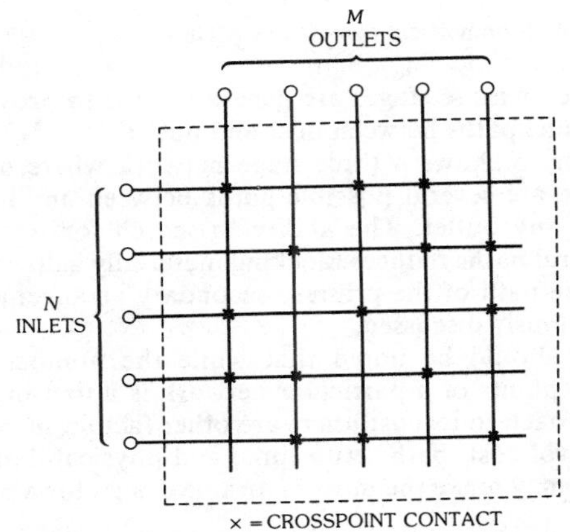

Fig. 2. Single-stage square coordinate switch (limited availability).

stages of switching—in the decimal planned system, expanding the access of a ten-by-ten crossbar switch to 100. This increase is achieved at the cost of introducing blocking; that is, an idle inlet can no longer reach every idle outlet regardless of the number of calls in progress. In fact, it is easy to see that every call from an inlet on primary switch 1 to an outlet on secondary switch 1 blocks all possible calls from other inlets on primary switch 1 to other outlets on secondary switch 1. Such an arrangement would usually give intolerable service, so

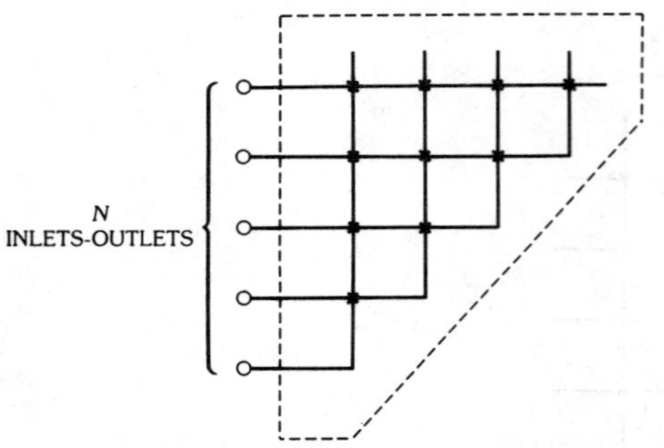

Fig. 3. Single-stage folded (triangular) switch.

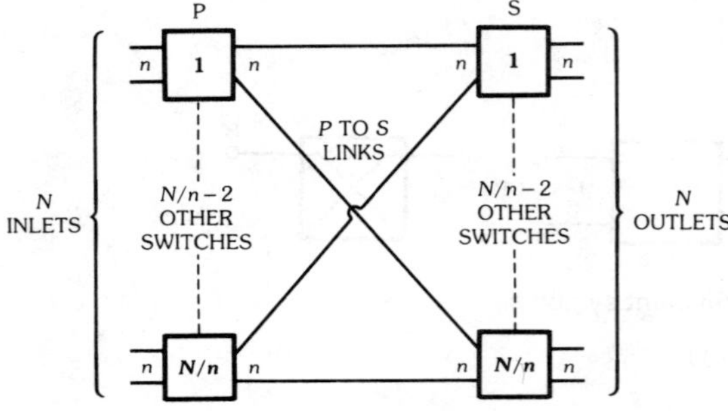

Fig. 4. Symmetrical two-stage switch.

three or more stages are generally used to provide parallel paths between inlet and outlet.

Fig. 5 shows a three-stage network where now there are several possible paths between any inlet and any outlet. The ability to search for one of several paths reduces blocking materially below the single path of the primary-secondary arrangement previously discussed.

It should be noted that while the number of crosspoints of a particular network is a first-order approach to its cost, there are other factors such as control cost, path setup time, and physical design that may make the most favorable design for a particular application quite different from the one that minimizes crosspoints.

BLOCKING AND NON-BLOCKING NETWORKS

In the following paragraphs, configurations for blocking and nonblocking networks will be described. The choice of network depends to a large extent on the service objectives, cost objectives, control capability, and, where a network covers a wide range of quantities of inlets and outlets, how additional switches and terminals can be added to the network without expensive rearrangement and changes in control methods.

Nonblocking 3-Stage Networks

A. General Case—Fig. 6 shows part of a network that presents no internal blocking (see definition). Let N = number of inlets, M = number of outlets, S = number of center-stage switches, n = number of inlets per first stage switch, and m = outlets per final-stage switch. When $N \neq M$, the condition for nonblocking is given by $S = n + m - 1$.

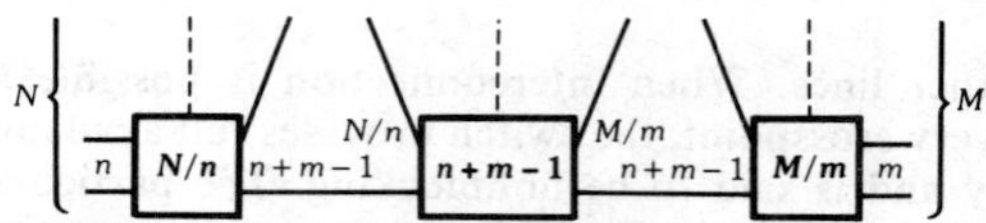

Fig. 6. Nonblocking three-stage network.

The number of crosspoints required is given by

$$X = (n + m - 1)[N + M + (NM/nm)]$$

A minimum number of crosspoints is obtained when $m = n$ and when n satisfies the equation

$$NM/(N + M) = n^3/(n - 1) \qquad (1)$$

Columns 1 and 2 of Table 1 show corresponding values of n and $NM/(N + M)$ that satisfy equation 1, from which the optimum value of n for given

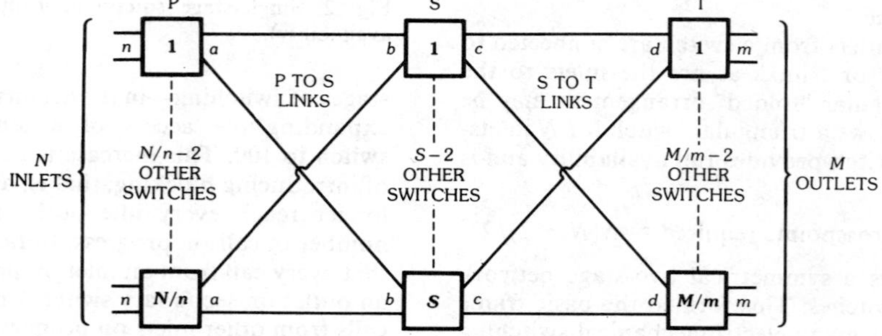

Fig. 5. Three-stage network—general case.

TABLE 1. RELATIONSHIP OF n, N, AND M IN NONBLOCKING 3-STAGE NETWORKS

n	$NM/(N+M)$ (Eq. 1)	N (Eq. 2)	N (Eq. 3)
2	8.0	16*	8
3	13.5	27*	18
4	21.3	42.7	32
5	32.2	62.5	50
6	43.2	86.4	72
7	57.2	114.3	98
8	73.1	146.3	128
9	91.1	182.2	162
10	111.1	222.2	200
11	133.1	266.2	242
12	157.1	314.2	288

*The only two integral solutions of equation 2.

values of N and M may be selected. Fig. 7 shows part of the network.

The number of crosspoints required is then given by

$$X_{\min} = (2n - 1)[N + M + (NM/n^2)]$$

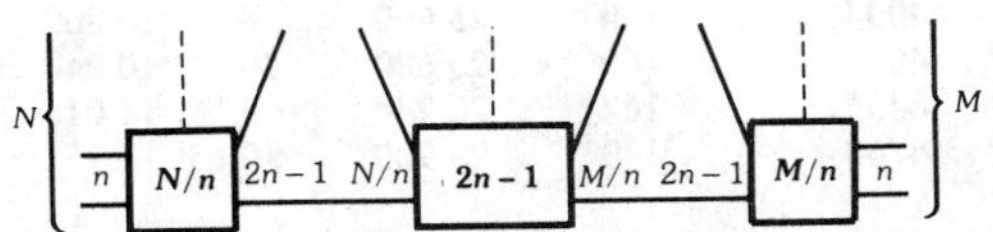

Fig. 7. Nonblocking three-stage network with minimum crosspoints.

B. Symmetrical Case—Fig. 8 shows part of a symmetrical nonblocking network, where $M = N$ and $m = n$. The condition for nonblocking is given by

$$S = 2n - 1$$

The number of crosspoints required is given by

$$X = N(2n - 1)[2 + (N/n^2)]$$

A minimum number of crosspoints is obtained when n satisfies the equation

$$N = 2n^3/(n - 1) \tag{2}$$

Columns 1 and 3 of Table 1 show corresponding values of n and N satisfying equation 2, from which the optimum value of n may be selected.

For large values of n, beyond the range of practical three-stage networks (see **C** below)

$$2n^2 \rightarrow N$$

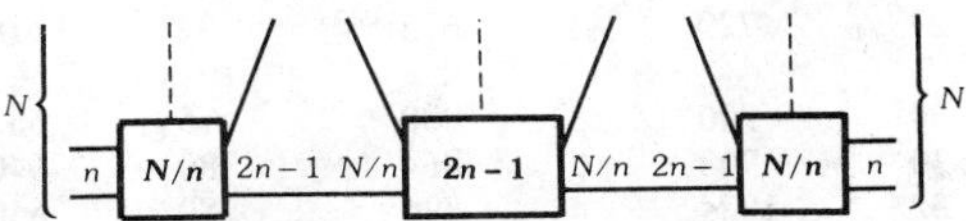

Fig. 8. Symmetrical nonblocking network with minimum crosspoints.

Columns 1 and 4 of Table 1 show corresponding values of n and N satisfying the equation

$$N = 2n^2 \tag{3}$$

The optimum value of n selected by using column 3 is either equal to or one less than the value indicated by column 4.

C. Comparison of 3-Stage Network and Single-Stage Switch— A single-stage, full-access switch is inherently nonblocking. The most favorable nonblocking three-stage network requires fewer than NM crosspoints if

$$NM/(N + M) > n^2(2n - 1)/(n - 1)^2$$

Table 2 shows, for some practical values of n, the limiting value of $NM/(N + M)$ below which a single-stage switch requires fewer crosspoints.

Table 3 compares single-stage switches and three-stage symmetrical nonblocking networks for typical values of N and n, where N/n is the integral (see **D** below), to illustrate the trends of design choices.

D. Practical Nonblocking 3-Stage Network with Minimum Crosspoints— When a nonblocking network with a minimum number of crosspoints is sought, the indicated optimum value of

TABLE 2. LOWER LIMITS OF $NM/(N+M)$ FOR NONBLOCKING 3-STAGE NETWORKS

n	2	3	4	5	6
$NM/(N+M) >$	12	11.25	12.45	14.06	15.84

TABLE 3. COMPARISON OF TYPICAL 1-STAGE AND 3-STAGE NONBLOCKING COORDINATE NETWORKS ($M = N$, $m = n$, N/n INTEGRAL)

	1-Stage	3-Stage			1-Stage	3-Stage			1-Stage	3-Stage	
N	*X*	*n*	*X*	*N*	*X*	*n*	*X*	*N*	*X*	*n*	*X*
8*	64	2	96	50*	2500	5	1800	100	10 000	5	5 400
15	225	3	275	54	2916	6	2079	100	10 000	10	5 700
16	256	2	288	55	3025	5	2079	105	11 025	7	5 655
16	256	4	336	56	3136	4	2156	108	11 664	6	5 940
18*	324	2	351	56	3136	7	2236	108	11 664	9	6 120
18	324	3	360	60	3600	5	2376	110	12 100	5	6 336
20	400	2	420	60	3600	6	2420	110	12 100	10	6 479
20	400	4	455	64	4096	4	2688	120	14 400	6	7 040
24	576	3	560	64	4096	8	2880	120	14 400	8	6 975
24	576	4	588	70	4900	5	3024	128*	16 384	8	7 680
25	625	5	675	70	4900	7	3120	130	16 900	10	8 151
27	729	3	675	72*	5184	6	3168	140	19 600	7	8 840
30	900	3	800	75	5625	5	3375	140	19 600	10	9 044
30	900	5	864	80	6400	5	3744	144	20 736	8	9 180
32*	1024	4	896	80	6400	8	3900	144	20 736	9	9 248
35	1225	5	1071	81	6561	9	4131	150	22 500	10	9 975
36	1296	4	1071	84	7056	6	4004	160	25 600	8	10 800
36	1296	6	1188	84	7056	7	4056	160	25 600	10	10 944
40	1600	4	1260	90	8100	6	4455	162*	26 244	9	11 016
40	1600	5	1296	90	8100	9	4760	170	28 900	10	11 951
45	2025	5	1539	91	8281	7	4563	180	32 400	9	12 920
48	2304	4	1680	96	9216	6	4928	180	32 400	10	12 996
48	2304	6	1760	96	9216	8	5040	190	36 100	10	14 089
49	2401	7	1911	98*	9604	7	5096	200*	40 000	10	15 200

Legend:

N = Inputs, outputs per network
n = Inputs per *P*-switch, outputs per *T*-switch
X = Crosspoints required
*** = Optimum configuration (where $N = 2n^2$, except when $n = 3$).**

inlets, n, per primary-stage switch may be such that N/n is not integral. The desired result may be achieved by providing some of the primary-stage and tertiary-stage switches with $(n - 1)$ inlets and outlets, respectively, by adjusting the sizes of the secondary-stage switches, and by superimposing two sets of interstage linkages.

The method is illustrated in Fig. 9, a nonblocking network for 100 inlets and 100 outlets requiring a minimum number of crosspoints. The value $n = 6$ is selected from column 3 of Table 1. The nearest multiple of 6 that exceeds 100 is 102. Thus, 17 primary-stage switches are required, 15 with 6 inlets each and 2 with 5 inlets. The larger switches require $2n - 1 = 11$ outlets each, cross-linked to 11 secondary-stage switches (links shown by continuous lines). The smaller switches require $2n - 2 = 10$ outlets each, cross-linked to 10 only of the secondary-stage switches (links shown by dashed lines). Thus, the secondary-stage switches also are of two sizes. The total number of crosspoints is 5291 (as compared with 5423 for a nonblocking network for 102 inlets and outlets, with $n = 6$). The crosspoint saving may not be worth the added control complexity.

E. Extension to 5-Stage and 7-Stage Networks—If the number of inlets and outlets on each secondary-stage switch of a three-stage network is large, it is advantageous to use a five-stage network. One possible arrangement is shown in Fig. 10, in which each secondary switch of a symmetrical three-stage network conforming to Fig. 8 is expanded into a nonblocking three-stage subnetwork.

If $N > 160$, a five-stage nonblocking network

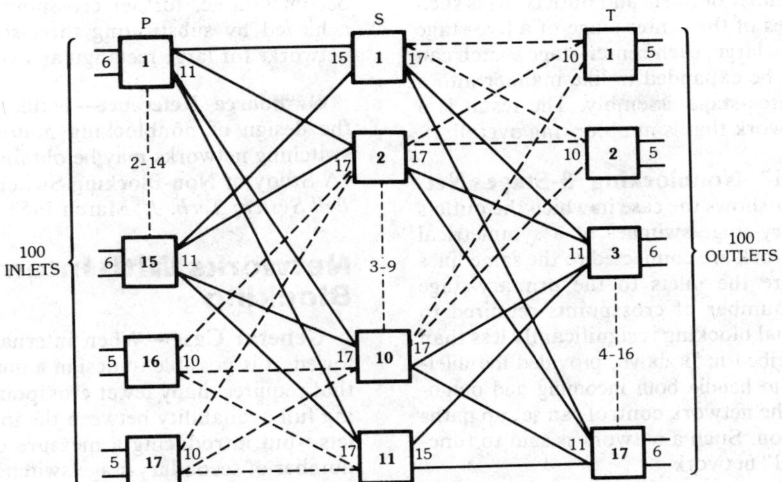

Fig. 9. Nonblocking three-stage network for 100 inlets, 100 outlets.

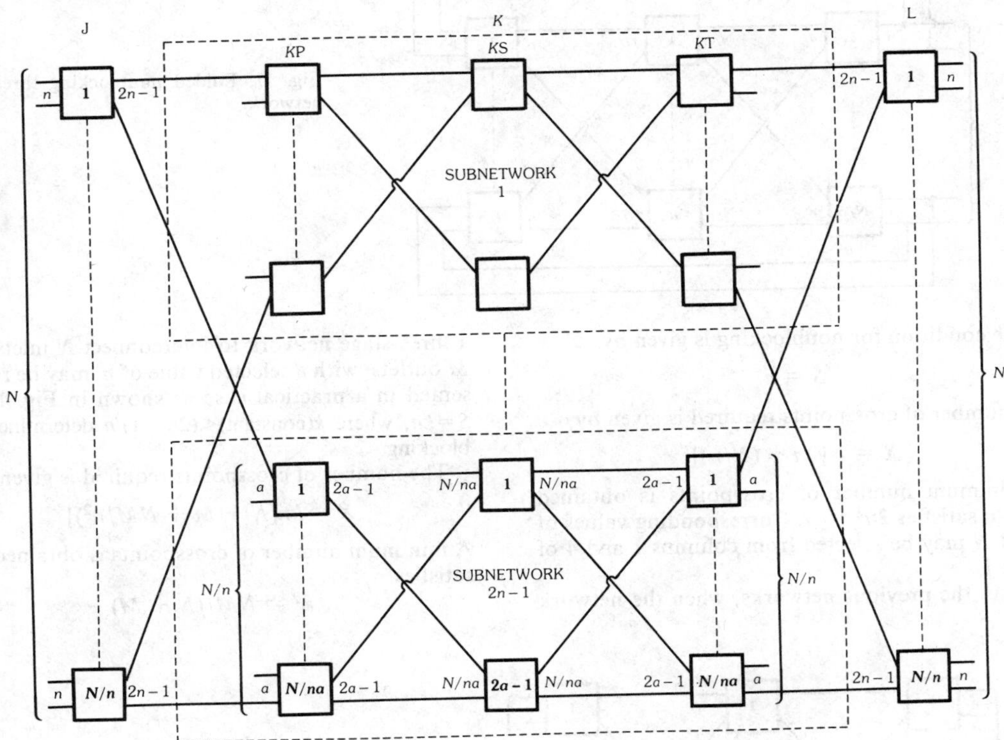

Fig. 10. Typical nonblocking five-stage network.

can be designed—by judicious selection of parameters n and a —which requires fewer crosspoints than the most favorable nonblocking three-stage network. The advantage increases slowly with N; at $N = 240$, the advantage is less than 5%.

The number of crosspoints is given by

$$X = N(2n - 1)\{2 + (2a - 1)[(2/n) + (N/n^2a^2)]\}$$

A minimum number of crosspoints is obtained when n and a satisfy the equations

$$N = 2na^3/(a - 1)$$
$$= [na^2(2n^2 + 2a - 1)]/[(2a - 1)(n - 1)]$$

When the number of inlets and outlets, N, is such that the switches of the center stage of a five-stage network become large, each center-stage switch can with advantage be expanded in like manner into a nonblocking three-stage assembly. The result is a seven-stage network that is nonblocking overall.

F. "Folded" Nonblocking 3-Stage Networks—Fig. 11 shows the case in which the outlets from the tertiary-stage switches of a symmetrical three-stage network are connected to the same lines or trunks as are the inlets to the primary-stage switches. The number of crosspoints required to ensure no internal blocking is significantly less than in the case described in **B** above, provided the inlet-outlets are able to handle both incoming and outgoing traffic and the network control can set up paths in either direction. Such a network is said to function as a "folded" network.

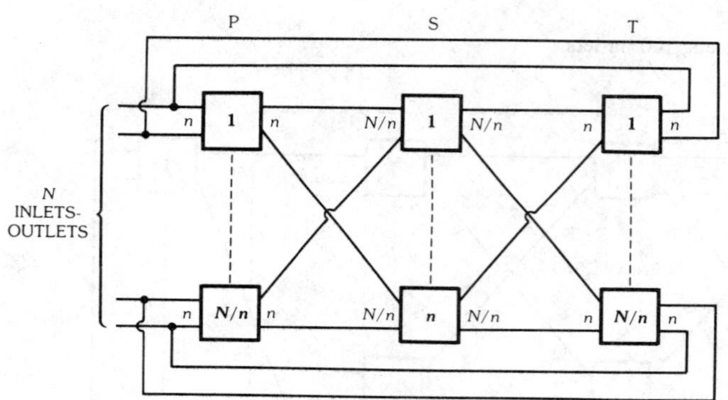

Fig. 11. Folded nonblocking three-stage network.

The condition for nonblocking is given by

$$S = n$$

The number of crosspoints required is given by

$$X = N[2n + (N/n)]$$

A minimum number of crosspoints is obtained when n satisfies $2n^2 = N$. Corresponding values of n and N may be selected from columns 1 and 4 of Table 1.

As in the previous networks, when the network becomes large, further crosspoint savings can be achieved by substituting three-stage nonblocking networks for large rectangular switches.

G. Source Reference—Further information for the design of nonblocking multistage coordinate switching networks may be obtained from C. Clos, "A Study of Non-Blocking Switching Networks," *Bell System Tech. J.*, March 1953.

Networks With Internal Blocking

General Case—When internal blocking is allowed, it is possible to design a multistage network that requires many fewer crosspoints while providing full availability between the inlets and the outlets—but introducing a measure of blocking. The number of secondary-stage switches, S, required in a three-stage network to interconnect N inlets and M outlets, with a selected value of n, may be represented in a practical case as shown in Fig. 12 by $S = kn$, where k(constant)$<(2n-1)/n$ determines the blocking.

The number of crosspoints required is given by

$$X = kn[N + M + NM/n^2)]$$

A minimum number of crosspoints is obtained if n satisfies

$$n^2 = NM/(N + M)$$

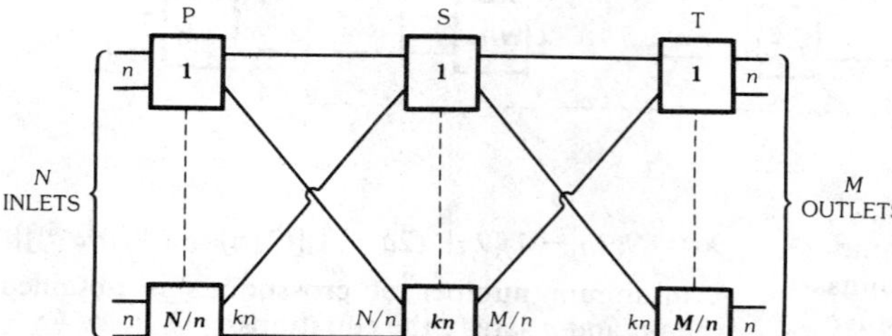

Fig. 12. Three-stage network with internal blocking.

The number of crosspoints is then given by

$$X_{min} = 2k[NM(N+M)]^{1/2}$$

When the inlets and outlets of a symmetrical network are connected to the same lines or trunks, it is advantageous to arrange the network to function as a "folded" network (see **F** above).

Traffic-Carrying Capability of Blocking Networks—Exact calculation of the probability of blocking in a multistage network is difficult. Order-of-magnitude estimates can be made by using the method described by C. Y. Lee, in which the number of parallel paths, r, possible for a connection is used. It a_1, a_2, and a_{s-1} designate the occupancy of links leaving stages 1, 2, and $s - 1$ of an s-stage network, the probability, P, of blocking is approximated by:

$$P = [1 - (1 - a_1)(1 - a_2) \ldots (1 - a_{s-1})]^r$$

More detailed methods of calculating blocking in switching networks may be found in the publications of the Institute of Switching and Data Technics, University of Stuttgart, Stuttgart, Germany.

TIME-DIVISION SWITCHING

Time division with pulse-code modulation has for many years been used as a method of multiplexing a number of telephone conversations on a digital carrier. With the same methods, a great reduction in active elements in a switching network can be achieved; in fact, it is advantageous to switch channels from digital carriers without demodulating them. The prevalence of digital transmission systems and the rapid advance of digital technology have made the present trend completely toward digital switching.

Conceptually, each time slot in a digital carrier can be thought of as a separate traffic path separated by the dimension of time rather than that of space as in the networks discussed so far. There are three usual time-division network configurations. In the first, the time slots in one digital stream are inserted into similarly positioned time slots in an output stream. This forms the basis of the "space division" part of many time-division networks, as will be discussed.

In the second configuration, the information in one time-slot position of the input digital stream is inserted in a different position in the output digital stream (which may have a different bit rate).

Hybrid between the first two configurations is the taking of a sample of an analog inlet signal, encoding it and placing it in a selected time slot in an output digital stream, and the converse. This configuration is usual in the first stage of a time-division line-switching system.

The last two configurations require storage of signals—one for each time slot in the case of the second and one for each analog inlet in the case of the third.

The first configuration is analogous to a set of crosspoints arranged to connect n inlets to n outlets but with each inlet having access to only one outlet. It is achieved, of course, with only one physical crosspoint. This crosspoint can be combined with others in a rectangular array to provide the equivalent of n rectangular switches in which all the same-positioned time slots of a number of inlet digital streams have access to like-positioned slots in a number of outlet streams.

The second and third configurations correspond to a rectangular switch in which any inlet, path, or time slot has access to any outlet.

Because digital transmission is generally unidirectional, two paths must be established, one for each direction of transmission. This may be done by pairing the transmitter and receiver for each inlet and outlet. A connection is set up through parallel networks from the transmitter of the inlet to the receiver of the outlet and from the receiver of the inlet to the transmitter of the outlet. Crosspoints in the parallel networks are controlled in common.

The above arrangement cannot be used on folded networks where the ability is needed to connect any inlet-outlet to any other, because connections would be made from transmitter to transmitter. It can, however, be implemented by placing all the transmitters on one side of a nonfolded network and all the receivers on the other. Independent paths are then set up for each direction of transmission.

References

Bellamy, J. *Digital Telephony*. New York: John Wiley & Sons, Inc., 1982.

Lee, C. Y. "Analysis of Switching Networks." *Bell Syst. T. J.*, Vol. 3, No. 6, Nov. 1955.

Talley, D. *Basic Telephone Switching Systems*. Rochelle Park, N.J.: Hayden Book Co., Inc., 1979.

PART 2: TRAFFIC CONCEPTS

DEFINITIONS OF TERMS

Busy hour: The continuous one-hour period that, on consecutive days in the busy part of the year, contains the maximum average traffic intensity.

Call: A discrete engagement or occupation of a traffic path.

Calling rate: The average number of calls placed during the busy hour.

Call-second, call-minute, call-hour: Units of traffic quantity representing the occupation of a circuit for a second, minute, or hour.

Equated busy-hour call (EBHC): A European unit of traffic intensity equal to 1/30 of an erlang.

Erlang: The international dimensionless unit of traffic intensity. One erlang is the traffic intensity represented by an average of one circuit busy out of a group of circuits over some period of time.

Full-availability group: A group of traffic-carrying trunks or circuits in which every circuit is accessible to all the traffic sources.

Grade of service: A measure of the probability that, during a specified period of peak traffic, a call offered to a group of trunks or circuits will fail to find an idle circuit at the first attempt. Usually applied to the busy hour of traffic.

Holding time: The duration of occupancy of a traffic path by a call. Sometimes used to mean the average duration of occupancy of one or more paths by calls.

Hundred call-second per hour (CCS): A unit of traffic intensity equal to 1/36 of an erlang.

Infinite sources: The assumption that the number of sources offering traffic to a group of trunks or circuits is large in comparison with the traffic load. A ratio of ten or higher is often considered "infinite."

Limited-access group: A group of traffic-carrying trunks or circuits in which only a fraction of the circuits is accessible to any one group of the traffic sources.

Limited sources: The assumption that the number of sources offering traffic to a group of trunks or circuits must be included in loss-probability calculations. Used in the binomial and Engset loss-probability equations.

Lost calls cleared: The assumption that calls not immediately satisfied at the first attempt are cleared from the system and do not reappear during the period under consideration. Used in the Erlang B loss-probability equation.

Lost calls delayed: The assumption that calls not immediately satisfied at the first attempt are held in the system until satisfied. Used in the Erlang C delay-probability equation.

Lost calls held: The assumption that calls not immediately satisfied at the first attempt are held in the system until served or abandoned. The sum of waiting and service time is assumed equal to what the service time would have been without delay. Used in the Poisson loss-probability equation. (When the service time is a negative exponential, this assumption has the same effect if holding and service times are independent and the distribution of the caller's holding time to abandon is a negative exponential with the mean equal to that of the service time.)

Occupancy: The traffic intensity per traffic path. One hundred percent occupancy implies all paths busy.

Traffic concentration: The average ratio of the traffic quantity during the busy hour to the traffic quantity during the day.

Traffic density: *See* Traffic intensity.

Traffic flow: *See* Traffic intensity.

Traffic intensity: The average number of calls present on a group of traffic paths over a period of time.

Traffic load: *See* Traffic intensity.

Traffic path: A channel, time slot, frequency band, line, trunk, switch, or circuit over which individual communications pass in sequence.

Traffic quantity: The aggregate engagement time or occupancy time of one or more traffic paths.

TRAFFIC MEASURES

In the terminology of traffic, there are three concepts that often cause confusion and need to be differentiated.

Traffic Quantity

Traffic quantity represents the total number of channel uses in units of time. It is of particular use for billing or other volume calculations but is not useful by itself. Where only this gross daily quantity is available, an estimate can be made of the busy-hour traffic by making an assumption about the proportion of daily traffic appearing in the busy hour. Direct busy-hour measurements are preferable, however, for calculating the traffic performance of a traffic-carrying facility. The latter is a function of traffic intensity, which is the quantity of traffic carried during a period of time.

Traffic Intensity

Traffic intensity, often called traffic load, is the average number of calls present during a period of time. It can be measured by averaging periodic counts of the number of calls present during the period or, alternatively, by summing the duration of calls within the period and dividing by the period length.

Call Intensity

For many traffic-carrying elements, the number of calls making up the total traffic load is immaterial; the load represented by two calls of ten minutes duration has the same impact as one call of twenty minutes duration. This is the case for switching-system processors and controllers, where the duration of calls on the traffic-carrying paths is irrelevant. Call intensity, or call rate, in calls per time unit, is frequently used, therefore, in estimating traffic loads on control equipment. Often the time unit is an hour and is implied, as in "a line calling rate of 3."

TRAFFIC EQUATIONS AND TABLES

Grade of Service

The overall "grade of service" of switching system or trunk system refers to the ratio of calls that are not completed at the first attempt to the total number of attempts to establish a connection through the system during a specific period of time, usually the busy hour. The traffic capacity of a switch, switching network, or trunk group is the traffic load that will, on the average, provide the grade of service that is chosen as the service objective.

For measurement purposes, it is sometimes easier to observe the proportion of time that all circuits are busy than to count the total of offered and carried calls (or carried and lost calls). For the usual assumptions of random calling, infinite sources, and a level value of offered traffic, the proportion of calls blocked and the proportion of time all circuits are busy are equal. This relation does not persist for assumptions of peaked traffic, smooth traffic, or limited sources.

Choice of Formula

The most important factors determining the choice of formula for estimating grade of service are as follows:

A. The statistical nature of the call originations
   - Random from infinite sources
   - Random from finite sources
   - Peaked from infinite sources
   - Smooth from infinite sources
   - Periodic

B. The probability distribution of the call holding time

C. The availability of the circuits serving the traffic
   - Full access
   - Limited access in a graded multiple
   - Limited access through a switching network

D. The behavior of calls when blocked
   - Lost calls cleared
   - Lost calls delayed
   - Lost calls held
   - Lost calls retried at a later time

Traffic Equations

The two most commonly used equations in trunk service estimation are the Erlang B and the Poisson. In general, the Erlang B is a good estimator of single-hour service, while the Poisson is a good estimator of the service given on the average over a busy season during which the offered load varies over a wider range than would be expected from the assumptions of random offered traffic. In general, the use of the Poisson equation is not recommended for grades of service that exceed 0.10 (10% blocking).

In small networks, it may be necessary to take limited sources into account. This results in the Engset and binomial equations as indicated below.

The Erlang B equation is given by

$$B(c,a) = (a^c/c!)/\sum_{x=0}^{x=c} a^x/x!$$

The corresponding Engset equation for limited sources is solved only be iterative techniques. Tables are available for this purpose.*

The Poisson equation is given by

$$P(c,a) = e^{-a}\sum_{x=c}^{\infty}(a^x/x!)$$

The corresponding binomial equation is given by

$$P(c,n+1,r) = \sum_{x=c}^{n}\binom{n}{x} r^x(1-r)^{n-x}$$

Finally, the Erlang C equation is given by

$$C(c,a) = (a^c/c!)[c/(c-a)]/\left\{\sum_{x=0}^{c-1} a^x/x! + (a^c/c!)[c/(c-a)]\right\}$$

where,

a = offered load in erlangs,
c = number of circuits,
n = number of limited sources less one,
r = load offered by an idle source.

Although the above equations are derived on the assumption of a negative exponential distribution of holding times, they are reasonable approximations for most distributions encountered in practice.

The Erlang B equation is particularly useful in estimating the load that will be carried on and overflowing from a direct group of trunks to which random traffic is offered. Calls that find all direct circuits busy are directed to an alternate route. The traffic carried on the direct route will be $a[1-B(c,a)]$, and the traffic overflowing will be $aB(c,a)$, where a=the random offered traffic. The overflow

*_Telephone Traffic Theory, Tables and Charts._ Berlin-Munich: Siemens Aktiengesellschaft, 1970.

TABLE 4. OFFERED LOAD TO A TRAFFIC-CARRYING FACILITY FOR OBJECTIVE GRADE OF SERVICE

Trunks	Grade of Service 0.001 *B*	0.001 *P*	0.005 *B*	0.005 *P*	0.010 *B*	0.010 *P*	0.02 *B*	0.02 *P*	0.05 *B*	0.05 *P*	0.10 *B*	0.10 *P*
1	0.001	0.001	0.005	0.005	0.01	0.01	0.02	0.02	0.05	0.05	0.11	0.11
2	0.05	0.05	0.11	0.10	0.16	0.15	0.22	0.21	0.38	0.36	0.60	0.53
3	0.19	0.19	0.35	0.34	0.46	0.44	0.60	0.57	0.90	0.82	1.3	1.1
4	0.44	0.43	0.70	0.67	0.87	0.82	1.1	1.0	1.5	1.4	2.0	1.7
5	0.76	0.74	1.1	1.1	1.4	1.3	1.7	1.5	2.2	2.0	2.9	2.4
6	1.1	1.1	1.6	1.5	1.9	1.8	2.3	2.1	3.0	2.6	3.8	3.2
7	1.6	1.5	2.2	2.0	2.5	2.3	2.9	2.7	3.7	3.3	4.7	3.9
8	2.1	2.0	2.7	2.6	3.1	2.9	3.6	3.3	4.5	4.0	5.6	4.7
9	2.6	2.4	3.3	3.1	3.8	3.5	4.3	4.0	5.4	4.7	6.6	5.4
10	3.1	3.0	4.0	3.7	4.5	4.1	5.1	4.6	6.2	5.4	7.5	6.2
11	3.7	3.5	4.6	4.3	5.2	4.8	5.8	5.3	7.1	6.2	8.5	7.0
12	4.2	4.0	5.3	4.9	5.9	5.4	6.6	6.0	8.0	6.9	9.5	7.8
13	4.8	4.6	6.0	5.6	6.6	6.1	7.4	6.7	8.8	7.7	10.5	8.6
14	5.4	5.2	6.7	6.2	7.4	6.8	8.2	7.4	9.7	8.5	11.5	9.5
15	6.1	5.8	7.4	6.9	8.1	7.5	9.0	8.2	10.6	9.2	12.5	10.3
16	6.7	6.4	8.1	7.6	8.9	8.2	9.8	8.9	11.5	10.0	13.5	11.1
17	7.4	7.0	8.8	8.3	9.6	8.9	10.7	9.6	12.5	10.8	14.5	12.0
18	8.0	7.7	9.6	8.9	10.4	9.6	11.5	10.4	13.4	11.6	15.5	12.8
19	8.7	8.3	10.3	9.6	11.2	10.3	12.3	11.2	14.3	12.4	16.5	13.7
20	9.4	9.0	11.1	10.4	12.0	11.1	13.3	11.9	15.2	13.2	17.6	14.5
30	16.7	15.9	19.0	17.8	20.3	18.7	21.9	19.8	24.8	21.6	28.1	23.2
40	24.4	23.3	27.4	25.6	29.0	26.8	31.0	28.1	34.6	30.2	38.8	32.1
50	32.5	30.0	36.0	33.7	37.9	35.0	40.3	36.6	44.5	39.0	49.6	41.2
60	40.8	38.9	44.8	41.9	46.9	43.5	49.6	45.2	54.6	47.9	60.4	50.3
70	49.2	47.0	53.7	50.3	56.1	52.0	59.1	53.9	64.7	56.8	71.3	59.5
80	57.8	55.2	62.7	58.8	65.4	60.7	68.7	62.7	74.8	65.9	82.2	68.8
90	66.5	63.5	71.8	67.4	74.7	69.4	78.3	71.6	85.0	75.0	93.1	78.1
100	75.2	71.9	80.9	76.1	84.1	78.2	88.0	80.6	95.2	84.1	104.1	87.4

Legend:

All loads in erlangs

B = Lost calls cleared, Erlang B

P = Lost calls held, Poisson

traffic is no longer random, and other approximations are necessary to estimate the grade of service given by an alternate trunk group to overflow traffic. For this purpose, it is necessary to estimate the variance of the overflow traffic, or more conveniently, the ratio of the variance to the mean. This latter quantity is known as the peakedness of the traffic. (Random traffic from an infinite source has a peakedness of one.) Peakedness, *z*, of traffic overflowing from *c* trunks offered *a* erlangs under the LCC assumptions is given by:

$$z = 1 - aB(c,a) + a/[c + 1 + aB(c,a) - a]$$

When a number of overflow loads from a number of direct groups are combined, the peakedness of the total is given closely by

$$z = \sum_{x=1}^{n} z_x a'_x / \sum_{x=1}^{n} a'_x$$

where a'_x is the overflow load from the *x*th source.

An approximation to the grade of service given by a group of trunks to peaked traffic may be made with the equation

$$P = B(c/z, y/z)$$

(See D. W. Hill and S. R. Neal, "Traffic Capacity of a Probability Engineered Trunk Group," *Bell Syst. T. J.*, Vol. 55, No. 7, September, 1976, for closer approximations.)

Delays

Many control systems operate on a delay basis, and it is convenient to have an equation for estimating delays. Under the Erlang C assumptions, the delay has a mean of

$$d(c,a,h) = [c/(c-a)]C(c,a)h$$

If calls are served in order of arrival, the length of delay for delayed calls will be negative exponential.

Approximations to other service time distributions are given by

$$d(c,a,h) = [c/(c-a)]C(c,a)h(1 + V/h^2)/2$$

where,

V is the variance,
h is the average of the holding time distribution.

For calls served in order of arrival, the negative exponential is an approximation to the delay distribution of delayed calls.

Table 4 gives values of loads for objective losses over a range of loads and trunks for the Poisson and Erlang B formulas. For computer use, advantage can be taken of the following recurrence relation for computing the infinite source loss formulas:

Let $E(c+1,a) = (c/a)[E(c,a) + 1]$

and $E(0,a) = 1$

Then for lost calls cleared

$$B(c,a) = 1/E(c,a)$$

and for lost calls delayed

$$C(c,a) = 1/\{[(c-a)/c]\,[E(c,a) - 1] + 1\}$$

For lost calls held:

Let $D(c,a) = [(a/c) + 1]\,D(c-1,a) - (a/c)\,D(c-2,a)$

and $D(-1, a) = 0$

$D(0,a) = 1$

Then

$$P(c,a) = 1 - e^{-a}D(c-1,a)$$

References

Telephone Traffic Theory, Tables and Charts. Berlin-Munich: Siemens Aktiengesellschaft, 1970. (In English and German.)

Bear, D. *Principles in Telecommunications—Traffic Engineering*. Peter Peregrinus, Ltd., 1976 (Repr. 1980).

Mina, R. *Introduction to Teletraffic Engineering*. Telephony Publishing Co., 1974.

40 Electroacoustics

Revised and Expanded by
Paul A. Schomer

THEORY OF SOUND WAVES*

Sound (or a sound wave) is an alteration in pressure, stress, particle displacement, or particle velocity that is propagated in an elastic material, or the superposition of such propagated alterations. Sound (or sound sensation) is also the sensation produced through the ear by the above alterations.

Wave Equation

The behavior of small-amplitude sound waves is given by the wave equation

$$\nabla^2 p = (1/c^2)(\partial^2 p/\partial t^2) \qquad (1)$$

where,

p is the instantaneous pressure increment above and below a steady pressure (newtons/meter^2),
t is the time in seconds,
c is the velocity of propagation in meters/second,
∇^2 is the Laplacian.

The quantity p is a function of time and of the three coordinates of space. For the particular case of rectangular coordinates x, y, and z (in meters), the Laplacian is given by

$$\nabla^2 \equiv (\partial^2/\partial x^2) + (\partial^2/\partial y^2) + (\partial^2/\partial z^2) \qquad (2)$$

Plane Waves—For a plane wave of sound, where variations with respect to y and z are zero, $\nabla^2 p = \partial^2 p/\partial x^2 = d^2p/dx^2$; the latter is approximately equal to the curvature of the plot of p versus x at some instant. Equation 1 states simply that, for variations in x only, the acceleration in pressure p (which is the second time derivative of p) is proportional to the curvature in p (which is the second space derivative of p).

Sinusoidal variations in time are usually of interest. For this case, the standard procedure is to put p=(real part of $\bar{p}e^{j\omega t}$), where the phasor $\bar{p}$ now satisfies the equation

$$\nabla^2\bar{p} + (\omega/c)^2\bar{p} = 0 \qquad (3)$$

The velocity phasor $\bar{v}$ of the sound wave in the medium is related to the complex pressure phasor $\bar{p}$ by

$$\bar{v} = -(1/j\omega\rho_0)\,\text{grad}\,\bar{p} \qquad (4)$$

The specific acoustic impedance $\bar{Z}$ at any point in the medium is the ratio of the pressure phasor to the velocity phasor, or

$$\bar{Z} = \bar{p}/\bar{v} \qquad (5)$$

Spherical Waves—The solutions of Eqs.1 and 3 take particularly simple and instructive forms for the case of one-dimensional plane and spherical waves in one direction. Table 1 summarizes the pertinent information.

For example, the acoustic impedance for spherical waves has an equivalent electrical circuit comprising a resistance shunted by an inductance. In this form, it is obvious that a small spherical source (r is small) cannot radiate efficiently since the radiation resistance $\rho_0 c$ is shunted by a small inductance $\rho_0 r$. Efficient radiation begins approximately at the frequency where the resistance $\rho_0 c$ equals the inductive (mass) reactance $\rho_0 r$. This is the frequency at which the period ($1/f$) equals the time required for the sound wave to travel the distance $2\pi r$.

Sound in Gases

The acoustic behavior of a medium is determined by its physical characteristics and, in the case of gases, by the density, pressure, temperature, specific heat, coefficients of viscosity, and the amount of heat exchange at the boundary surfaces.

The velocity of propagation in a gas is a function of the equation of state ($PV = RT$ plus higher-order terms), the molecular weight, and the specific heat.*

For small displacements relative to the wavelength of sound, the velocity is given by

$$c = (\gamma p_0/\rho_0)^{1/2} \qquad (6)$$

where,

γ = ratio of the specific heat at constant pressure to that at constant volume,
p_0 = the steady pressure of the gas in newtons/meter^2,
ρ_0 = the steady or average density of the gas in kilograms/meter^3.

The values of the velocity in a few gases are given in Table 2 for 0 degrees Celsius and 760 millimeters of mercury barometric pressure.

The velocity of sound, c, in dry air is given by the experimentally verified equation

$$c = 331.45 \pm 0.05 \text{ meters/second}$$
$$= 1087.42 \pm 0.16 \text{ feet/second}$$

for the audible-frequency range, at 0 degrees Celsius and 760 millimeters of mercury with 0.03-mole-percent content of CO_2.

The velocity in air for a range of about 20 degrees Celsius change in temperature is given by

$$c = 331.45 + 0.607T_c \text{ meters/second}$$
$$= 1052.03 + 1.106T_f \text{ feet/second}$$

where T_c is the temperature in degrees Celsius and

---

* Rayleigh, Lord. *Theory of Sound*. Vols. 1 and 2. New York: Dover Publications, 1945. Morse, P. M. *Vibration & Sound*. 2d ed. New York: McGraw-Hill Book Co., 1948.

* Hardy, H. C., Telfair, D., and Pielemeier, W. H. "The Velocity of Sound in Air." *Journal of the Acoustical Society of America*, Vol. 13, January 1942, pp. 226-233. See also Beranek, L. *Acoustic Measurements*. New York: John Wiley & Sons, Inc., 1949, p. 46.

TABLE 1. SOLUTIONS FOR VARIOUS PARAMETERS

Factor	Type of Sound Wave: Plane Wave	Type of Sound Wave: Spherical Wave
Equation for p	$\partial^2 p/\partial x^2 = (1/c^2)(\partial^2 p/\partial t^2)$	$(\partial^2 p/\partial x^2)+(2/r)(\partial p/\partial r) = (1/c^2)(\partial^2 p/\partial t^2)$
Equation for $\overline{p}$	$(d^2\overline{p}/dx^2)+(\omega/c)^2\overline{p}=0$	$(d^2\overline{p}/dx^2)+(2/r)(d\overline{p}/dt)+(\omega/c)^2 p=0$
Solution for p	$p=F[t-(x/c)]$	$p=(1/r)\,F[t-(x/c)]$
Solution for $\overline{p}$	$\overline{p}=\overline{P}\exp(-j\omega x/c+j\theta)$	$\overline{p}=(1/r)\,\overline{P}\exp(-j\omega r/c+j\theta)$
Solution for $\overline{v}$	$\overline{v}=(p/\rho_0 c)\exp(-j\omega x/c+j\theta)$	$\overline{v}=(p/\rho_0 cr)\,[1+(c/j\omega r)]\exp(-j\omega\, r/c+j\theta)$
$\overline{Z}$	$\overline{Z}=\rho_0 c$	$\overline{Z}=\rho_0 c/[1+(c/j\omega r)]$
Equivalent Electrical Circuit for $\overline{Z}$	$\overline{Z}\rightarrow$ $\rho_0 c$ (resistor)	$\overline{Z}\rightarrow$ $\rho_0 c$ (resistor) in parallel with $\rho_0 r$ (inductor)

p = excess pressure in newtons/meter^2

$\overline{p}$ = complex excess pressure in newtons/meter^2
t = time in seconds
x = space coordinate for plane wave in meters
r = space coordinate for spherical wave in meters

$\overline{v}$ = complex velocity in meters/second

$\overline{Z}$ = specific acoustic impedance in newton-seconds/meter^3
c = velocity of propagation in meters/second
$\omega = 2\pi f$; f = frequency in hertz
F = an arbitrary function
θ = phase constant

ρ_0 = density of medium in kilograms/meter^3

$\overline{P}$ = peak amplitude of the phasor (units dependent on wave type)

T_f is the temperature in degrees Fahrenheit. For values of T_c greater than 20 degrees, the following equation may be used:

$$c = 331.45 \times (T_k/273)^{1/2} \text{ meters/second}$$

where T_k is the temperature in kelvins.

For other corrections, if extreme accuracy is desired, reference should be made to the literature.*

From Eq. 5 and Table 1, characteristic impedance is equal to the ratio of the sound pressure to the particle velocity.

$$\overline{Z} = \overline{p}/\overline{v} = \rho_0 c \cos\phi$$

For plane waves, $\phi = 0$ and $\cos\phi = 1$. For spherical waves, $\tan\phi = \lambda/2\pi r$, where λ = wavelength of the acoustic wave, and r = distance from the sound source. For r greater than a few wavelengths, $\cos\phi \approx 1$.

Characteristic impedance $\rho_0 c$ in newton-seconds/meter^3 for several gases at 0 degrees Celsius and 760 millimeters of mercury is given in Table 3.

Sound in Liquids

In liquids, the velocity of sound is given by

$$c = (1/K\rho_0)^{1/2}$$

* Hardy, H. C., Telfair, D., and Pielemeier, W. H. "The Velocity of Sound in Air." *Journal of the Acoustical Society of America,* Vol. 13, January 1942, pp. 226–233.

TABLE 2. GASES AND VAPORS

Substance	Formula	Density (gm/L)	Velocity (m/s)	$\Delta v/\Delta t$ (m/s°C)
Gases (0°C)				
Air, dry		1.293	331.45	0.59
Ammonia	NH_3	0.771	415	
Argon	A	1.783	319	0.56
Carbon dioxide	CO_2	1.977	259	0.4
Carbon monoxide	CO	1.25	338	0.6
Chlorine	Cl_2	3.214	206	
Deuterium	D_2		890	1.6
Ethane (10°C)	C_2H_6	1.356	308	
Ethylene	C_2H_4	1.260	317	
Helium	He	0.178	965	0.8
Hydrogen	H_2	0.0899	1284	2.2
Hydrogen bromide	HBr	3.50	200	
Hydrogen chloride	HCl	1.639	296	
Hydrogen iodide	HI	5.66	157	
Hydrogen sulfide	H_2S	1.539	289	
Illuminating (coal) gas			453	
Methane	CH_4	0.7168	430	
Neon	Ne	0.900	435	0.8
Nitric oxide (10°C)	NO	1.34	324	
Nitrogen	N_2	1.251	334	0.6
Nitrous oxide	N_2O	1.977	263	0.5
Oxygen	O_2	1.429	316	0.56
Sulfur dioxide	SO_2	2.927	213	0.47
Vapors (97.1°C)				
Acetone	C_3H_6O		239	0.32
Benzene	C_6H_6		202	0.3
Carbon Tetrachloride	CCl_4		145	
Chloroform	$CHCl_3$		171	0.24
Ethanol	C_2H_6O		269	0.4
Ethyl ether	$C_4H_{10}O$		206	0.3
Methanol	CH_4O		335	0.46
Water Vapor (134°C)	H_2O		494	

TABLE 3. CHARACTERISTIC IMPEDANCE $\rho_0 c$ FOR GASES

Gas	Symbol	$\rho_0 c$ (N·s/m³)
Air		428.6
Argon	A	569
Carbon dioxide	CO_2	511
Carbon monoxide	CO	421
Helium	He	173.2
Hydrogen	H_2	114
Neon	Ne	383
Nitric oxide	NO	435
Nitrogen	N_2	418
Nitrous oxide	N_2O	518
Oxygen	O_2	453

where,

c is the velocity in meters/second,
K is the compressibility in meter-seconds²/kilogram and may be regarded as constant.

For most liquids,

$$K = (47 \times 10^{-8})/981$$

Figures for the velocity of sound in meters/second through some liquids are given in Table 4.

TABLE 4. VELOCITY OF SOUND IN LIQUIDS

Liquid	Temperature in °C	Velocity in (m/s) × 10^3
Alcohol, ethyl	12.5	1.24
	20	1.17
Benzene	20	1.32
Carbon disulfide	20	1.16
Chloroform	20	1.00
Ether, ethyl	20	1.01
Glycerin	20	1.92
Mercury	20	1.45
Pentane	18	1.05
	20	1.02
Petroleum	15	1.33
Turpentine	3.5	1.37
	27	1.28
Water, fresh	17	1.43
Water, sea (36 parts/thousand salinty)	15	1.505

Sound in Solids

See Chapter 4, Tables 17, 18, and 19.

Sound Intensity

The sound intensity is the average rate of sound energy transmitted in a specified direction through a unit area normal to this direction at the point considered. In the case of a plane or spherical wave (several wavelengths from the source) in a fluid, the intensity in the direction of propagation is given by

$$I = p^2/\rho c \tag{7}$$

The units of I are newtons/second/meter^2.

Sound-Pressure Level (SPL)

It is customary (and convenient) to use the decibel scale to express the ratio between any two sound pressures. Since sound pressure is usually proportional to the square root of the corresponding sound intensity, sound-pressure level (spl) is defined as

$$\text{spl} = 20 \log_{10} (p/p_0) \tag{8}$$

where p_0 is 20 micronewtons/meter^2 in air.*

At times, sound pressure is measured in units other than newtons/meter^2. Table 5 lists the spl for a quantity of one in these other units (e.g., 1 atmosphere).

ELECTRICAL ANALOGIES FOR ACOUSTICAL AND/OR MECHANICAL SYSTEMS**

Analysis of electromechanical-acoustical transducers, mechanical systems, or acoustical systems can be facilitated by the use of electrical analogies. Two forms of electrical analogies are possible: (1) force or pressure-voltage analogue and (2) force or pressure-current analogue. Tables 6 and 7 tabulate the pertinent quantities in these two analogies.

In a specific problem in which it is desired to represent an acoustical or mechanical system by an

* The reference of 1 micronewton/meter^2 is usually used for liquids and all other media.

** Swenson, George W., *Principles of Modern Acoustics,* New York: Van Nostrand, 1953.

TABLE 5. SOUND-PRESSURE LEVELS FOR SEVERAL UNITS USED TO MEASURE SOUND PRESSURE IN AIR (RE 20 MICRONEWTONS/METER^2)

SPL	Unit	
94	1 newton/meter^2	(N/m^2)
127.6	1 pound/foot^2	(psf)
170.8	1 pound/in^2	(psi)
194.1	1 atmosphere	(atm)
74	1 microbar	(μbar)

TABLE 6. FORCE/PRESSURE-VOLTAGE ANALOGY

Mechanical Quantity	Acoustical Quantity	Electrical Quantity
Force (newtons)	Sound Pressure (newtons meter^{-2})	Voltage (volts)
Velocity (meters/second)	Volume Velocity (meters^3 second^{-1})	Current (amperes)
Displacement (meters)	Volume Displacement (meters^3)	Charge (coulombs)
Mass (kilograms)	Acoustic Mass (kilograms meter^{-4})	Inductance (henries)
Compliance (stiffness^{-1}) (meters/newton)	Acoustic Stiffness (kilograms meter^{-4} second^{-2})	Capacitance^{-1} (farads^{-1})
Viscous Friction (newtons/meter/second)	Acoustic Resistance (kilograms meter^{-4} second^{-1})	Resistance (ohms)
Mechanical Impedance (mechanical ohms) or (force/velocity)	Acoustic Impedance (kilograms meter^{-4} second^{-1})	Impedance (ohms)

TABLE 7. FORCE/PRESSURE-CURRENT ANALOGY

Mechanical Quantity	Acoustical Quantity	Electrical Quantity
Force (newtons)	Sound Pressure (newtons meter^{-2})	Current (amperes)
Velocity (meters/second)	Volume Velocity (meters^3 second^{-1})	Voltage (volts)
Displacement (meters)	Volume Displacement (meters^3)	Impulse $\int v\,dt$ (volt-seconds)
Mass (kilograms)	Acoustic Mass (kilograms meter^{-4})	Capacitance (farads)
Compliance (stiffness^{-1}) (meters/newton)	Acoustic Stiffness (kilograms meter^{-4} second^{-2})	Inductance^{-1} (henries^{-1})
Viscous Friction (newtons/meter/second)	Acoustic Resistance (kilograms meter^{-4} second^{-1})	Conductance (mhos)
Mechanical Impedance (mechanical ohms) or (force/velocity)	Acoustic Impedance (kilograms meter^{-4} second^{-1})	Admittance (mhos)

electrical circuit, a choice must be made between the force-voltage and force-current analogies. If the system includes electrical as well as mechanical or acoustical elements, as in a motor or a loudspeaker, for example, the choice is dictated by the type of coupling between the electrical and mechanical parts. However, if the system is entirely mechanical or acoustical, the choice is arbitrary and depends upon personal preference. In any event, given the equivalent circuit derived on the basis of one analogy, the equivalent circuit corresponding to the other analogy can be derived by taking the dual of the given circuit.

In a system that combines acoustical and

mechanical quantities, it is necessary to establish a common frame of reference. Usually, it is easiest to convert acoustical quantities to mechanical quantities, as for example, integrating the pressure over a piston (loudspeaker or microphone) to calculate the force, etc.

HEARING*

The auditory system consists of the periphery sensors, acoustic neurological transducers, the ears, the eighth cranial nerve leading to a programming and priority switching center at various levels of the brain stem, and finally to the auditory area of the cortex located near the Sylvian fissure of the frontal-lobe convolution.

The auditory system does much more than detect minute sounds. Among other functions, it preferentially places more weight on certain preprogrammed characteristic sounds, localizes the direction of most sounds by a variety of ingenious techniques, and initiates involuntary actions for visual acquisition of the source.

The hearing mechanism was probably evolved to help man survive in a hostile environment, and not for linguistic communication or for musical entertainment.

The apparent loudness attributed to a sound varies not only with the sound pressure, but also with

* Flanagan, J. L., *Speech Analysis, Synthesis and Perception.* New York: Academic Press, 1965. Also Richardson, E. G., ed. *Technical Aspects of Sound.* New York: Elsevier Press, 1953.

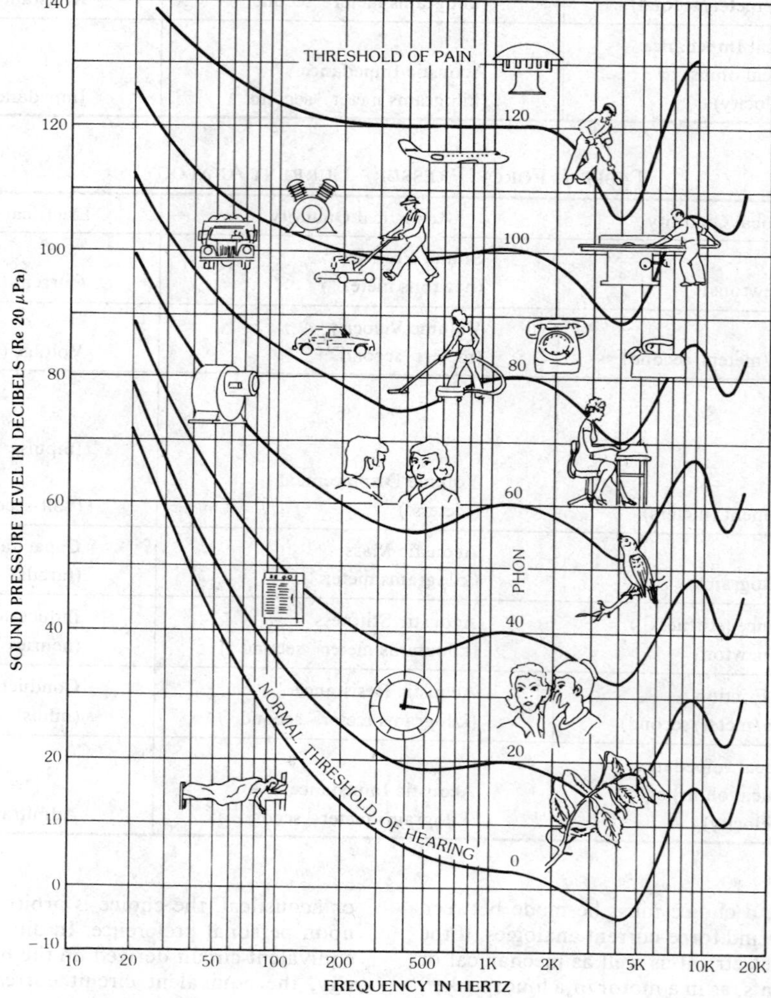

Fig. 1. Equal loudness contours. (*Ginn. K. B.* Architechtural Acoustics. *Brüel & Kjaer.*)

the frequency (or pitch) of the sound. In addition, the way it varies with frequency depends on the pressure. Fig. 1 illustrates experimentally determined equal loudness contours (spl) as a function of frequency.

MEASUREMENT OF SOUND*

Sound Level Meters

A Sound Level Meter (SLM) is an instrument designed to measure directly sound-pressure levels while incorporating frequency-weighting networks that approximate the inverse of equal loudness contours at various levels. The current American National Standard Specification for Sound Level Meters (ANSI S1.4-1983) designates three alternate frequency-weighting networks (Fig. 2). In addition, some SLMs incorporate a *D*-weighting (Fig. 2), which corresponds to the inverse of an experimentally determined moderate level "equal noisiness contour," and a "flat" or unweighted measure. (In the latter case, the measurement bandwidth is determined by the microphone and electrical network characteristics.)

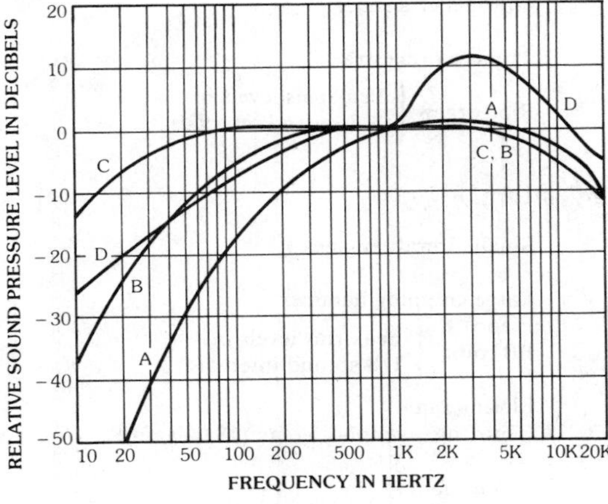

Fig. 2. IEC Standard A, B, and C weighting curves for Sound Level Meters. (*EPA*. Public Health and Welfare Criteria for Noise. *550/9-73-002. p. 2-2.*)

An SLM measures the "quasi-root-mean-square" (rms) value of a time-varying sound pressure. Table 8 lists the rms detector time constant and corresponding label as given in ANSI S1.4-1983. In addition, some SLMs incorporate a peak-hold (*p*), which captures the instantaneous largest sound pressure level.

In text, complete description must be given for any measurement made (*e.g.,* 55 dB *B*-weighted slow). In equations, spl's are abbreviated as L_{AF} for *A*-weighted fast, L_{BI} for *B*-weighted impulse, L_{PC} for *C*-weighted peak, etc. Maximum should not be confused with peak. Peak refers to an instantaneous measure, whereas maximum refers to a detected level [*e.g.,* max *A*-weighted slow ($L_{AS\max}$)].

TABLE 8. SOUND LEVEL METER TIME CONSTANTS

Label	RMS Detector Time Constant
Slow (S)	1 s
Fast (F)	0.1 s
Impulse (I)	35 ms

Measurement and/or Laboratory Microphones and Their Calibration

Measurement and/or laboratory microphones are generally capacitor, electret, or ceramic (piezoelectic). These are chosen because of their flat frequency response, low noise, and stability. One-inch and 1/2-inch microphones are the most common with smaller sizes also available. Capacitor microphones require a polarization voltage but are not potentially as susceptible to aging or temperature effects as the others. All three are very high-impedance devices requiring an FET follower close-coupled for impedance transformation. For a given size, the ceramic microphone is on the order of 20 dB less sensitive than the capacitor or electret microphone.

Calibration is normally performed with a close-coupled single-frequency acoustical source. The acoustical signal may be generated mechanically (pistons driven by a cam and motor) or electromechanically. Typical frequencies are 1000 Hz and 250 Hz. Other frequencies, high pressure, and reciprocity calibrators are also available as needs dictate, but generally are not required.

Sound Level

In recent years, it has become customary to measure most environmental or industrial work-place (occupational) noise with the *A*-weighting. In the United States, *A*-weighted spl is commonly designated "sound level." (Special high-amplitude impulse sounds, such as sonic booms or explosions, that can noticeably vibrate structures are measured with *C*-weighting.) Fig. 1 also illustrates sound levels for some common noise sources.

Power Level

One is usually concerned with the acoustic power radiated by some source of noise or radiated purposefully by a transducer such as a loudspeaker.

* See, for example, reference 4.2.

Typically, these sources have a directivity pattern, Q, that is a function of angle. Radiated power is the normal component of the intensity integrated over an imaginary surface encompassing the source. (Sufficiently far from the source, the intensity is proportional to the pressure squared.) Power level (pwl), symbolized by L_w, is defined as

$$L_w = 10 \log W/W_0 \tag{9}$$

where,

W is the radiated acoustic power in watts,
W_0 is the reference power of 10^{-12} watts.

Table 9 lists the pwl for some common noise sources. Note that there is no *a priori* relation between the spl's in Fig. 1 and pwl's in Table 9.

ENVIRONMENTAL NOISE

Environmental-noise regulations typically take one of two forms. Either they are specified as simple spl limits at some measurement location, or they are specified by time-integrated quantities. The former are similar in concept to "speed limits" or "discharge limits" and are relatively easy to enforce. They may take the form of property-line limits such as so many decibels A-weighted slow or may even be expressed as octave-band or $^1/_3$-octave–band

TABLE 9. ACOUSTIC POWER AND SOUND POWER LEVELS OF TYPICAL NOISE SOURCES* (A-WEIGHTED)

Power (Watts)	Power Level (dB re 10^{-12} Watts)	Source
100,000	170	Ram jet Turbojet engine with afterburner
10,000	160	Turbojet engine, 7000 lb thrust
1000	150	4-propeller airliner
100	140	75-piece orchestra Pipe organ { Peak rms levels in 1/8-second intervals
10	130	
3	125	Small aircraft engine
1.0	120	Large chipping hammer Piano, BB^b tuba { Peak rms levels in 1/8-second intervals
0.1	110	Blaring radio Centrifugal ventilating fan (13,000 cfm)
0.01	100	4-foot loom Auto on highway
0.001	90	Vanaxial ventilating fan (1500 cfm) Voice—shouting (average long-time rms)
0.0001	80	
0.00001	70	Voice—conversational level (average long-time rms)
0.000001	60	
0.0000001	50	
0.00000001	40	
0.000000001	30	Voice—very soft whisper

*Space average sound pressure level at 10 meters = Power level − 28 dB

limits. (Some SLMs include an octave-band filter set; others have octave and 1/3-octave filter sets available as attachments. The filters are specified by ANSI S1.11-1966.) For vehicular sources, the limits are usually specified in terms of a maximum *A*-weighted fast (or slow) level at a reference distance (usually 15 m). Airports, highways, railways, and other large, distributed time-varying sources are usually regulated and assessed with the time-integrated measures. These better correlate with human and community response than do the single event limits, but they are harder to enforce. The time-integrated measures are typically portrayed by contours or zone maps and are easily utilized in land-use planning and zoning.

Day-night average sound level (dnl) is the preferred integrated noise measure in the United States.* Symbolized by L_{dn}, dnl is defined as

$$L_{dn} = 10 \log \frac{1}{p_0{}^2 T} \times \left[\int_{7\,\text{am}}^{10\,\text{pm}} p^2 dt + 10 \int_{10\,\text{pm}}^{7\,\text{am}} p^2 dt \right] \qquad (10)$$

where,

$p(t)$ is the time-varying *A*-weighted sound pressure,

p_0 is the reference pressure (20 μN/m^2) defined earlier,

T is 86 400, the number of seconds in a day.

Note the tenfold, or 10-dB, penalty applied to nighttime to better reflect the community response that is thought to result from altered nighttime activity and lower ambient levels.

Normally, dnl is based on the yearly average day and can be thought of as yearly average "noise fall" (like rainfall) but divided by the seconds in a year, expressed in decibels and with a 10× nighttime penalty. Fig. 3 illustrates the dnl level found in various areas, and several references list land uses commonly considered compatible with several dnl zones.** Special large-amplitude impulse sounds such as sonic booms are assessed with a *C*-weighted dnl and a similar, but not identical, land-use table. Reference 3.5 offers greater detail on these special sources.

ASSESSMENT OF WORKPLACE-INDUSTRIAL NOISE

A Sound Level Meter (SLM) set on the *A*-scale can be used to assess industrial noise for hearing

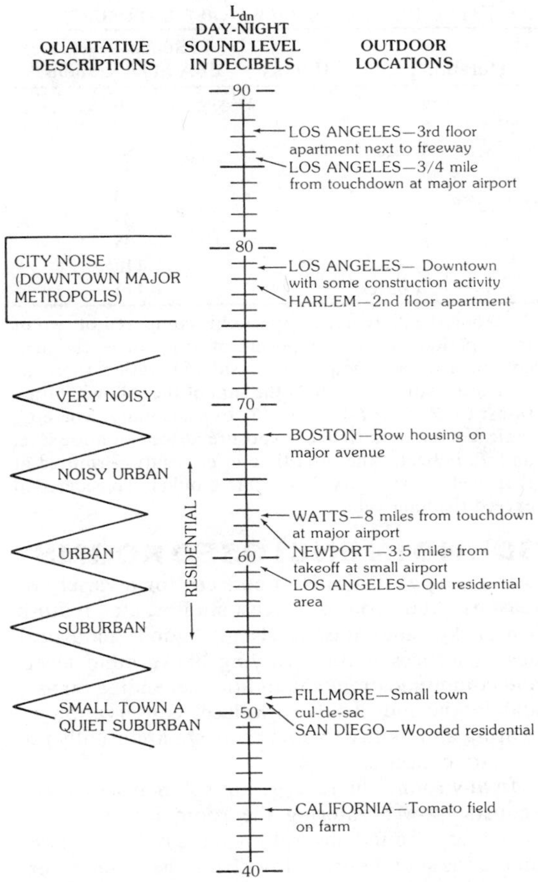

Fig. 3. Outdoor day-night sound level in decibels relative to 20 micronewtons/meter^2 at various locations. (*EPA.* Information on Levels of Environmental Noise Requisite to Protect Health and Welfare With an Adequate Margin of Safety. *550/9-74-004.*)

hazard. For enforcement purposes, the US Department of Labor (DOL) has set 90 dB (*A*) as the limit for eight hours of exposure per day. Most industrial noise is not uniform and continuous, and so the regulations of DOL include procedures for determining the equivalent exposure. These procedures incorporate a 5-dB trading rule in that halving the exposure time allows the noise limit to rise by 5 dB.* The present limits are given in Table 10.

As an alternative to using an SLM to measure industrial noise exposure, several personnel dosimeters are available that directly measure a worker's dose (usually as a percent of the eight-hour limit). These devices directly incorporate the 5-dB trading rule in their electronics.

Exposure to impulsive or impact noise should not exceed 140 dB peak sound pressure level.

* EPA. *Information on Levels of Environmental Noise Requisite to Protect Health and Welfare With an Adequate Margin of Safety.* 550/9-74-004, March 1974.

** For greater detail, see references 3.2, 3.3, 3.4, and 3.6.

* Most European countries use a 3-dB or equal energy rule, and the USAF uses a 4-dB rule.

TABLE 10. PERMISSIBLE NOISE EXPOSURES[1]

Duration per Day, Hours	Sound Level, dBA Slow Response
8	90
6	92
4	95
3	97
2	100
$1\frac{1}{2}$	102
1	105
$\frac{1}{2}$	110
$\frac{1}{4}$ or less	115

[1]When the daily noise exposure is composed of two or more periods of noise exposure of different levels, their combined effect should be considered, rather than the individual effect of each. If the sum of the following fractions: $C_1/T_1 + C_2/T_2 + \ldots C_n/T_n$ exceeds unity, where C_n indicates the total time of exposure at the nth noise level and T_n indicates the total time of exposure permitted at that level, then the mixed exposure should be considered to exceed the limit value.

SOUND IN ENCLOSED ROOMS*

Indoors, acoustics is of concern for a variety of reasons. Auditoria, broadcast studios, etc., require low background noise levels and "good" acoustical design; offices require varying background levels and communications capabilities depending on use; and in the industrial workplace, conservation of hearing and ability to communicate are usually the primary concerns.

In any room, the reverberant spl increases as the acoustic power entering the room increases and decreases as sound absorption in the room increases and/or as sound power flows from the room. Sources of sound may be internal to rooms, such as machinery or loudspeakers, or may be external. In the latter case, the sound flows through walls, windows, ventilation ducts, etc., from other building spaces or components, or from outdoors. In either case, the sound field in a room is composed of two parts, the direct field and the reverberant field. The direct field of a source is

$$p = \rho ckDQ^{1/2}/2\pi^{1/2}r \qquad (11)$$

where,

$k = (2\pi f/c)$ (rad/m),
$p =$ pressure (n/m^2),
$c =$ speed of sound (m/s),
$\rho =$ density of air (kg/m^3)
$D =$ source strength (m^3/s·rad),
$Q =$ directivity (rad)**,
$r =$ distance from the source (m).

* See, for example, ref. 2.5, 2.6, or 2.8.

** The directivity factor of a source is the ratio of the sound pressure squared, at some fixed distance and specified direction (such as the axis of a loudspeaker), to the mean-square sound pressure at the same distance averaged over all directions from the source. The distance must be great enough so that the sound appears to diverge spherically from the effective acoustic center of the source.

Near a source, the sound field may be very complicated; the pressure and velocity may not be in phase, and hence the power radiated is not proportional to the pressure squared. This is particularly a problem when determining the noise power radiated by machinery in the industrial environment.*

By substitution in Eq. 11, the direct field intensity (I_d) of a source can be written as

$$I_d = WQ/4\pi r^2 \qquad (12)$$

where,

$W =$ radiated source power (watts),
$Q =$ directivity (rad),
$r =$ distance from the source (meters).

The total intensity (I) in an enclosed room (direct plus reverberant field) is

$$I = WQ/4\pi r^2 + 4W/R \qquad (13)$$

The quantity R is the room constant defined by

$$R = S\bar{\alpha}/(1 - \bar{\alpha}) \qquad (14)$$

where,

$S =$ total surface area of the room,
$\bar{\alpha} =$ average room absorption coefficient.

Fig. 4 illustrates the sound fields in a room. In terms of the sound pressure level,

$$\text{spl} = \text{pwl} + 10 \log (Q/4\pi r^2 + 4/R) + K \qquad (15)$$

where,

K = 0.2 dB if metric units are used,
K = 10.5 dB if English units are used.

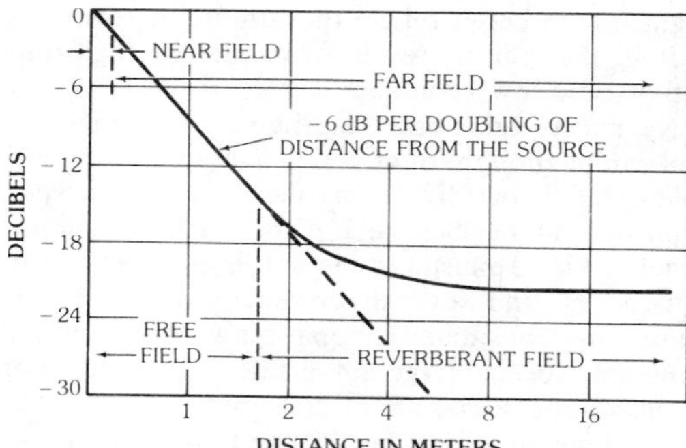

Fig. 4. Description of the sound field around a sound source in a reverberant room. (*From Ginn, K. B.* Architectural Acoustics. *Brüel and Kjaer, 1978.*)

Standing Sound Waves

Resonant conditions in sound studios cause standing waves by reflections from opposing paral-

* Newly developed sound-intensity meters are a valuable aid in these situations.

lel surfaces, such as ceiling-floor and parallel walls, resulting in serious peaks in the reverberation-time/frequency curve. Standing sound waves in a room can be considered comparable to standing electrical waves in an improperly terminated transmission line where the transmitted power is not fully absorbed by the load.

In properly proportioned rooms, resonances can be effectively reduced and standing waves practically eliminated by introducing numerous surfaces disposed obliquely. Thus, large-order reflections can be avoided by breaking them up into numerous smaller reflections. The object is to prevent sound reflection back to the point of origin until after several reflections.

The most desirable ratios of dimensions for broadcast studios are given in Fig. 5.

Reverberation Time

Reverberation time varies with frequency and is measured by the time required for a sound, when suddenly interrupted, to die away or decay to a level 60 decibels below the original sound. Measurement of reverberation time is the easiest way to find the average absorption coefficient ($\bar{\alpha}$) and its associated value of reflectivity ($1 - \bar{\alpha}$) in an acoustic environment. Conversely, absorption coefficients and room sizes are used to predict the reverberation time for new construction or remodeling.

Measurement of Reverberation Time

The reverberation time of an enclosed space that already exists is an important quantity that is relatively easy to measure. The degree of accuracy required is determined by the use to which the data will be put. The majority of day-to-day reverberation measurements are taken with quite satisfactory results by using the interrupted-noise method (Fig. 6). This is the most widely used method and yields excellent field results; only in the most critical concert halls is this method found wanting. It has been stated by Atal that the subjective assessment of reverberation time is governed by the early decay time (EDT, which is the time it takes the early decay to change by 15 dB). Good agreement between this EDT and subjective estimates of the presence or absence of excessive reverberation has been obtained in a number of concert halls. This first 15 dB of decay is difficult to obtain with the interrupted-noise method because of the statistical variations that occur in the test signal over short time periods.

The Schroeder-Kuttruff method (Fig. 7) employs

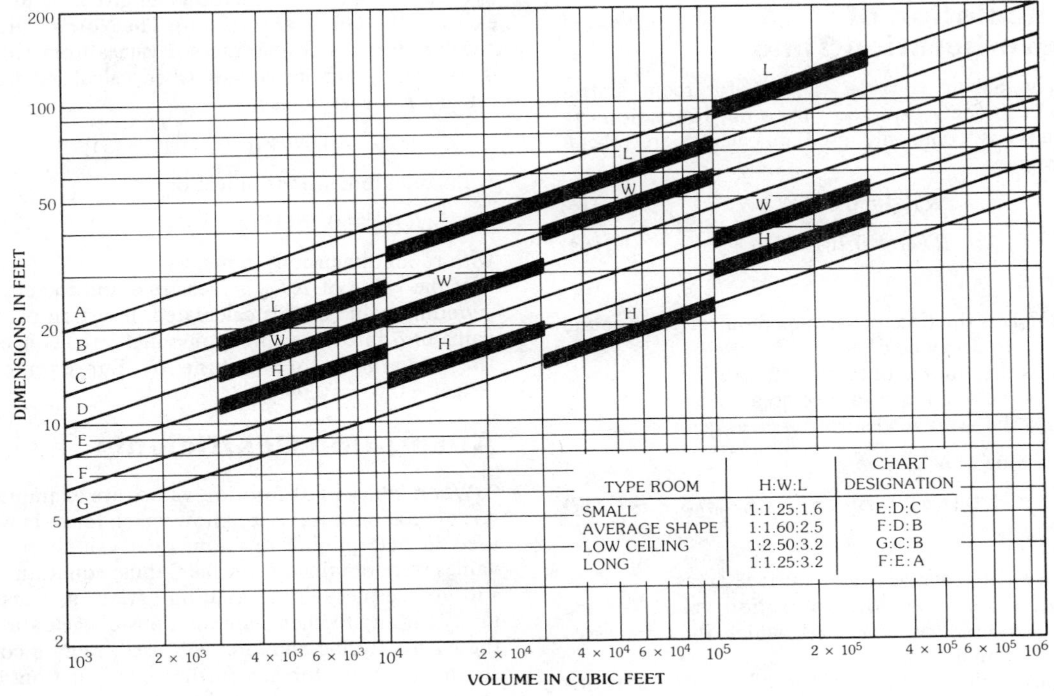

Fig. 5. Preferred room dimensions based on $2^{1/3}$ ratio. Permissible deviation is 5 percent. (*Courtesy Acoustical Society of America and RCA.*)

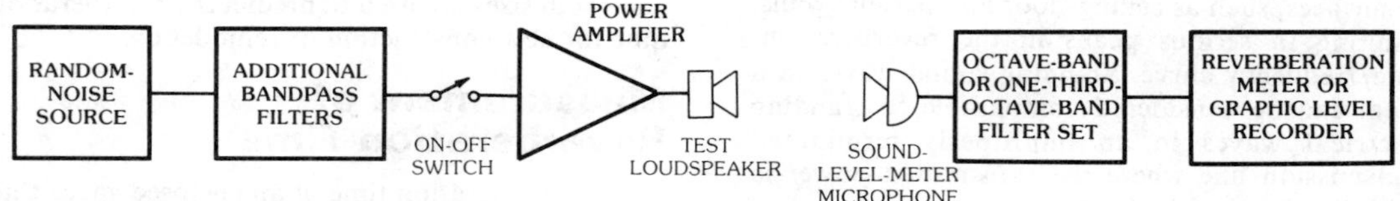

Fig. 6. The interrupted-noise method.

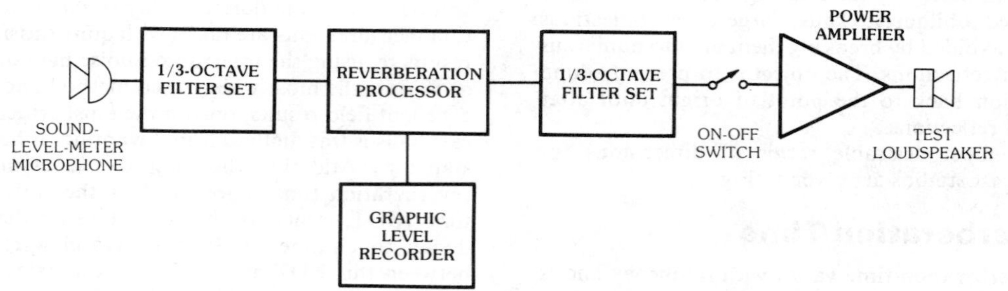

Fig. 7. The Schroeder-Kuttruff method.

a 2.7-ms rectangular pulse that is used to excite a standard 1/3-octave bandpass filter. The resulting "ringing" of the filter provides a statistically reliable signal that allows highly repeatable decay recordings and excellent resolution of the EDT.

Calculation of Reverberation Time

The Sabine equation (Eq. 16 for room dimensions in feet, Eq. 16a for room dimensions in meters) is commonly used to calculate reverberation time.

$$RT_{60} = 0.049V/S\bar{\alpha} \quad (16)$$

$$RT_{60} = 0.161V/S\bar{\alpha} \quad (16a)$$

where,

RT_{60} is the time in seconds required for a sound to decay 60 dB,
V is the volume of the room,
S is the boundary surface area,
$\bar{\alpha}$ is the average absorption coefficient.

The value of $\bar{\alpha}$ is:

$$\bar{\alpha} = (s_1\alpha_1 + s_2\alpha_2 + \ldots s_n\alpha_n)/S \quad (17)$$

where,

s_1, s_2, etc., are boundary surface areas,
α_1, α_2, etc., are the absorption values for the boundary areas with which they are associated,
$s_n\alpha_n$ is the total absorption of the people, furniture, etc., present in the room.

Note that $S\bar{\alpha}$ can be replaced by A, the total absorption in the room. This concept is useful when considering the effects of surfaces plus objects or people in the room.

In the limiting case, the Sabine equation predicts a finite reverberation time in a room with 100% absorption present, and for true absorption values in excess of 0.63, this equation can give α values in excess of 1.0 (100% absorption). The Norris-Eyring equation (Eq. 18 or 18a) gives $\bar{\alpha}$ values from 1.0 to 0 for true absorption values when calculated from actual RT_{60} measurements.

$$RT_{60} = 0.049V/[-S\ln(1-\bar{\alpha})] \quad (18)$$

with room dimensions in feet or

$$RT_{60} = 0.161V/[-S\ln(1-\bar{\alpha})] \quad (18a)$$

with room dimensions in meters.

If the value of RT_{60} is measured and the corresponding value of $\bar{\alpha}$ is calculated, insertion of this value of $\bar{\alpha}$ into the expression $-\ln(1-\alpha)$ converts the $\bar{\alpha}$ value into a Sabine $\bar{\alpha}$. For example, $-\ln(1-0.63)=0.99$.

Absorption Coefficients

When tables of absorption values are examined, it is of vital importance to know which formula was used in determining the numerical values. If the values were obtained from the Sabine equation, be sure to use the Sabine equation variations consistently for any further manipulations of the data. If the Norris-Eyring equation was used, remain consistent in its use for any further manipulations of the data. The Sabine formula is used in measuring and specifying most building materials. Hence, the Sabine equation must be used unless the absorption values are converted (see above). Table 11 gives

TABLE 11. COEFFICIENTS OF GENERAL BUILDING MATERIALS AND FURNISHINGS

Complete tables of coefficients of the various materials that normally constitute the interior finish of rooms may be found in the various books on architectural acoustics. This short list will be useful in making simple calculations.

Materials	Coefficients					
	125 Hz	250 Hz	500 Hz	1000 Hz	2000 Hz	4000 Hz
Brick, unglazed	0.03	0.03	0.03	0.01	0.05	0.07
Brick, unglazed, painted	0.01	0.01	0.02	0.02	0.02	0.03
Carpet, heavy, on concrete	0.02	0.06	0.14	0.37	0.60	0.65
Same, on 40 oz hairfelt or foam rubber	0.08	0.24	0.57	0.69	0.71	0.73
Same, with impermeable latex backing on 40 oz hairfelt or foam rubber	0.08	0.27	0.39	0.34	0.48	0.63
Concrete block, coarse	0.36	0.44	0.31	0.29	0.39	0.25
Concrete block, painted	0.10	0.05	0.06	0.07	0.09	0.08
Fabrics (draperies)						
Light velour, 10 oz per sq yd, hung straight, in contact with wall	0.03	0.04	0.11	0.17	0.24	0.35
Medium velour, 14 oz per sq yd, draped to half area	0.07	0.31	0.49	0.75	0.70	0.60
Heavy velour, 18 oz per sq yd, draped to half area	0.14	0.35	0.55	0.72	0.70	0.65
Floors						
Concrete or terrazzo	0.01	0.01	0.015	0.02	0.02	0.02
Linoleum, asphalt, rubber, or cork tile on concrete	0.02	0.03	0.03	0.03	0.03	0.02
Wood	0.15	0.11	0.10	0.07	0.06	0.07
Wood parquet in asphalt on concrete	0.04	0.04	0.07	0.06	0.06	0.07
Glass						
Large panes of heavy plate glass	0.18	0.06	0.04	0.03	0.02	0.02
Ordinary window glass	0.35	0.25	0.18	0.12	0.07	0.04
Gypsum board, ½″ nailed to 2×1's, 16″ o.c.	0.29	0.10	0.05	0.04	0.07	0.09
Marble or glazed tile	0.01	0.01	0.01	0.01	0.02	0.02
Openings						
Stage, depending on furnishings			0.25 – 0.75			
Deep balcony, upholstered seats			0.50 – 1.00			
Grills, ventilating			0.15 – 0.50			
Plaster, gypsum or lime, smooth finish on tile or brick	0.013	0.015	0.02	0.03	0.04	0.05
Plaster, gypsum or lime, rough finish on lath	0.14	0.10	0.06	0.05	0.04	0.03
Same, with smooth finish	0.14	0.10	0.06	0.04	0.04	0.03
Plywood paneling, ⅜″ thick	0.28	0.22	0.17	0.09	0.10	0.11
Water surface, as in a swimming pool	0.008	0.008	0.013	0.015	0.020	0.025
Air, Sabins per 1000 cubic feet @ 50% RH				0.9	2.3	7.2

Absorption of Seats and Audience
Values given are in Sabins per square foot of seating area per unit.

	125 Hz	250 Hz	500 Hz	1000 Hz	2000 Hz	4000 Hz
Audience, seated in upholstered seats, per sq ft of floor area	0.60	0.74	0.88	0.96	0.93	0.85
Unoccupied cloth-covered upholstered seats, per sq ft of floor area	0.49	0.66	0.80	0.88	0.82	0.70
Unoccupied leather-covered upholstered seats, per sq ft of floor area	0.44	0.54	0.60	0.62	0.58	0.50
Wooden pews, occupied, per sq ft of floor area	0.57	0.61	0.75	0.86	0.91	0.86
Chairs, metal or wood seats, each, unoccupied	0.15	0.19	0.22	0.39	0.38	0.30

absorption coefficients of some typical building materials.

Optimum Reverberation Time

Optimum, or most desirable, reverberation time varies with room size and use, such as music, speech, etc. (see Fig. 8).

Fig. 9 shows the desirable ratio of the reverberation time for various frequencies to the reverberation time for 500 hertz. The desirable reverberation time for any frequency between 50 Hz and 10 kHz may be found by multiplying the reverberation time at 500 hertz (from Fig. 8) by the desirable ratio in Fig. 9 that corresponds to the frequency chosen.

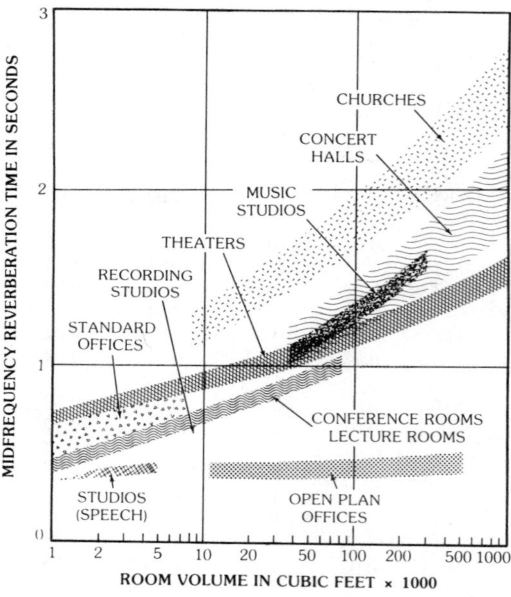

Fig. 8. Optimum midfrequency reverberation times.

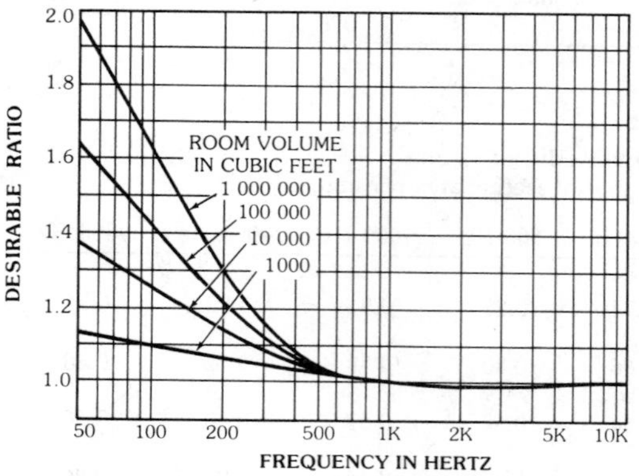

Fig. 9. Desirable relative reverberation time versus frequency for various structures.

INDOOR DESIGN NOISE LEVELS AND SPEECH COMMUNICATION

Articulation Index

The concept and use of articulation index is obtained from Fig. 10. The abscissa is divided into 20 bandwidths of unequal frequency intervals. Each of these bands will contribute 5 percent to the articulation index when the speech spectrum is not masked by noise and is sufficiently loud to be above the threshold of audibility. The ordinates give the root-mean-square peaks and minimums (in $^1/_8$-second intervals) and the average sound pressure levels created at 1 meter from a speaker's mouth in an anechoic (echo-free) chamber. The units are in decibels of pressure level per hertz. (For example, for a bandwidth of 100 hertz, rather than 1 hertz, the pressure level would be that indicated plus 20 decibels; the latter figure is obtained by taking 10 times the logarithm to the base 10 of the ratio of the 100-hertz band to the indicated band of 1 hertz.)

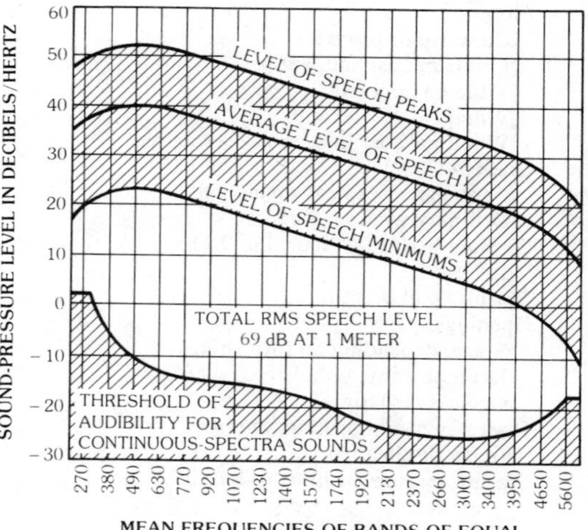

Fig. 10. Bands of equal articulation index; 0 decibels = 20 micronewtons/meter^2. (*Courtesy* Proceedings of the IRE.)

An articulation index of 5 percent results in any of the 20 bands when a full 30-decibel range of speech-pressure level peaks to speech-pressure level minimums is obtained in that band. If the speech minimums are masked by noise of a higher pressure level, the contribution to articulation is accordingly reduced to a value given by $^1/_6$ [(decibels level of speech peaks) − (decibels level of average noise)]. Thus, if the average noise is 30 decibels under the speech peaks, this expression gives 5 percent. If the noise is only 10 decibels below the speech peaks, the contribution to articulation index reduces to $^1/_6 \times 10 = 1.67$ percent. If the noise is more than 30 decibels below the speech peaks, a value of 5 percent is used for the articulation index. Such a computation is made for each of the 20 bands of Fig. 10, and the results are added to give the expected articulation index.

A number of important results follow from Fig. 10. For example, in the presence of a large white (thermal-agitation) noise having a flat spectrum, an improvement in articulation results if pre-emphasis is used. A pre-emphasis rate of about 8 decibels/octave is sufficient.

Preferred Speech Interference Levels/ Speech Interference Levels

As discussed above, noise interference with speech is usually a masking process. For many noises, the measurement and calculation of articulation index can be simplified further by the use of a three-band analysis.* The bands chosen are the octave bands centered on 500, 1000, and 2000 Hz. The arithmetic average of these three spl's in these three bands gives the quantity called the three-band preferred octave speech-interference level (psil). One can use this level for determining when telephone use or speech communication is easy or difficult and what changes in levels are required to improve the situation (Figs. 11 and 12).

* Beranek, L. L. "The Design of Speech Communications Systems." *Proc. IEEE,* Vol. 35, No. 9, Sept. 1947, pp. 880–890.

For satisfactory intelligibility of difficult speech material, maximum permissible values of speech-interference levels for men with average voice strengths are given in Fig. 12, which is an extension by Webster of Beranek's work.** It is assumed in this chart that there are no reflecting surfaces nearby, that the speaker is facing the listener, and that the spoken material is not already familiar to the listener. For example, the speech-interference level in a factory might be 80 dB, which is high. The chart indicates that the two people must ordinarily be no more than two feet apart in order to be understood satisfactorily. If the words spoken are carefully selected and limited in number, intelligible speech will be possible at greater distances.

If a number of conversations are to be held in the same reverberant room, the procedure is more complicated. This chart cannot be used on the basis of the background-noise level before the conversations are in progress, because a given conversation will be subject to interference from the noise produced by all the other conversations. The general procedure for calculating a speech-interference level under those conditions has not been completely worked out.

Not only background, equipment, and other occupant noise affects speech intelligibility. In a live room (too much reverberation), speech syllables are

** Webster, J. C. "Effects of Noise on Speech Intelligibility." *Proc. Conference Noise as a Public Health Hazard,* Washington, D. C., June 13–14, 1968, ASHA Rpt. 4. Washington, DC: The American Speech and Hearing Association, 1969, pp. 49–73.

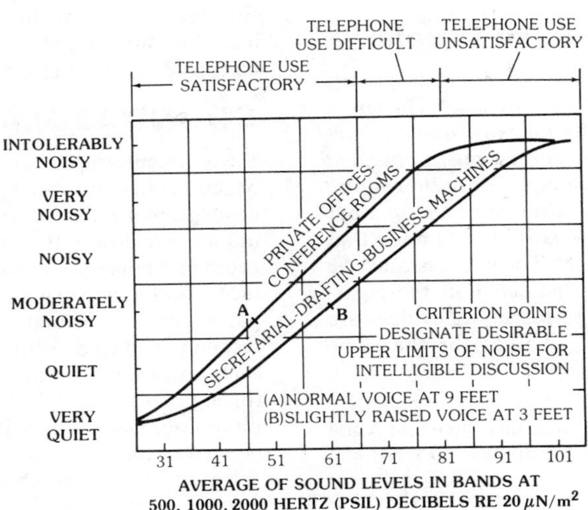

Fig. 11. Rating chart for office noises. Data were determined by an octave-band analysis and correlated with subjective tests. (*From Peterson, A. and Gross, E.* Handbook of Noise Measurement. *General Radio, 1972. Modified from Beranek and Newman for preferred bands.*)

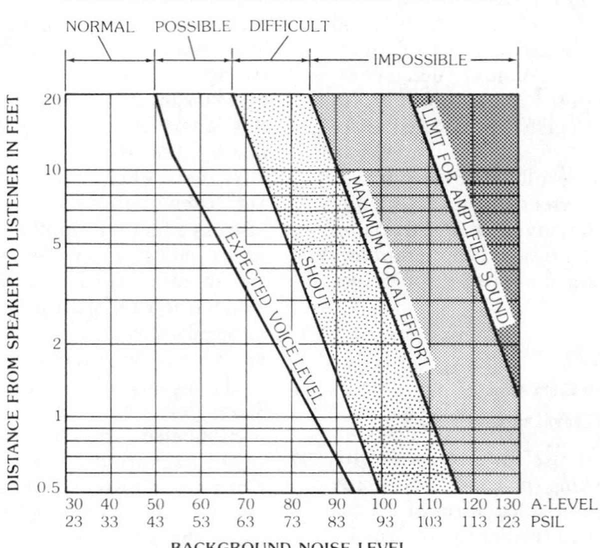

Fig. 12. Degree of face-to-face speech communication between speaker and listener as a function of distance and background noise level (A-levels or psil). (*After Contractor Report to Army for TW-5-838-2.*)

smeared by reflected noise at the other end, even though the noise is not very great. That is, if the distance from speaker (or loudspeaker) to listener is too great or the room too reverberant, then the speaker's own sound energy may make the speech unintelligible. As a rule of thumb, the direct field from the speaker (or loudspeaker) must exceed his (its) own reverberant field by 6 dB in order to maintain good speech intelligibility and prevent feedback.

Interior Design Noise Levels

Table 12 contains a list of planning levels for activities conducted in interior spaces.* The (interior) planning levels for *exterior noise and for interior equipment that is not continuously operated* are given in terms of L_{eq} values. *Continuous noise sources,* for example ventilating systems or other mechanical equipment, emit steady-state noise that is measured in terms of L_s.** The sources must be considered separately. As indicated in the right-hand column of Table 12, permissible L_s values are 5 and 10 dB less than L_{eq} values for the same activity.

Interior steady-state noise levels more than 5–10 dB below the levels specified in Table 12 are not desirable. Annoyance will actually increase with the lowered background noise levels because individuals will hear intruding sounds that normally would be masked by the steady-state noise. Occasionally, where adequate noise insulation cannot be provided, increasing the continuous background noise levels over the values shown in Table 12 will provide better masking of intruding intermittent sounds. For such occasions, the characteristics of both the intruding noises and the background noise should be considered during the design of the facility.

SOUND ISOLATION*

Sound entering a room from external areas can be controlled by quieting the source, by adding absorption to the source room (if the source is indoors) to lower the ambient level in the source room and hence the energy striking the party** surface, and by increasing the transmission loss of the party surface. Sound travels from sources into receiving rooms directly through building elements such as walls, windows, or doors or indirectly through "sound leaks" such as cracks, HVAC, plumbing, electrical facilities, plenums above suspended ceilings, corridors, etc.

Sound isolation should be contrasted with absorption. Efficient sound absorbers such as glass

* In offices or office-type environments, Preferred Noise Criteria Curves (PNC) are frequently used to select and specify the ambient.

** L_s is the A-weighted noise level produced by the ventilation or mechanical systems (or other interior noise sources) which operate more or less continuously. The L_s value for design should be the noise level produced in the space during the time of occupancy while the equipment is at the typical mode of operation.

* See, for example, references 2.1 through 2.11.

** The party surface is the surface (e.g., wall or ceiling) common to two spaces.

TABLE 12. INTERIOR NOISE ENVIRONMENT PLANNING LEVELS

Activity	All Noise Source L_{eq}(dB)	Continuous Interior Sources* L_s(dB)**
Sleeping	45	40
Other residential activities (conversations, radio, tv, listening, etc.)	50	40
Classrooms, libraries, churches, hospitals	45	40
Offices — private, conference	45	40
Offices/work spaces, telephone use satisfactory	55	45
Work spaces — occasional speech communication or telephone use	60	55
Work spaces — infrequent speech communication, telephone use infrequent	70	60
Concert halls, large auditoriums, theaters, or churches	30	25
Broadcast, tv, and recording studios	35	30
Small auditoriums, theaters, or churches	45	40

*Typically, ventilation systems and mechanical equipment in near-continuous operations.
**The L_s value is given in terms of A-weighted noise level. The approximate noise criteria (NC) curve values are 8 dB less than the A-level values.

fiber or open-cell polyurethane foam do not contain or isolate sound. Efficient sound isolators or containers such as heavy walls do not absorb sound.

In the United States, the sound isolating value of walls and other elements is frequently specified in terms of STC (see ASTM E413-73). More recently, it is also specified in terms of ΔL_A, the A-weighted sound reduction between typically furnished small rooms (see ASTM E597-77T) . Where usable, ΔL_A has the advantage of easier field measurement than STC.

Walls

Figs. 13–15 illustrate many building details and their corresponding STC and ΔL_A values. The total noise reduction between a source room and a receiving room is a function of the area of the party surface and the absorption in the receiving room. Small amounts of energy entering a very reverberant room will build up to high levels. Large wall areas with adequate STC being used to isolate a recording studio from a machine room will likely fail. Thus, party boundaries between noise-critical and noisy spaces should be minimized or eliminated.

Doors

Solid-core wood doors and hollow metal doors filled with insulation are generally better sound isolators than hollow-core doors; however, the amount of air space around the edges of the door is usually the controlling sound path. (See also the section on sound leaks.) For maximum sound isolation there must be a soft gasketing or weather stripping around the door to provide an airtight seal. The gasket must not be so stiff that it is difficult to close the door.

Resilient Channels

Frequent mention is made of the use of *resilient channels* to improve the noise-isolation performance of typical constructions. Typically, these channels are made of light-gauge metal in a Z-section, with one flange of the channel firmly attached to a stud or joist, and the other flange of the channel providing support for one or two layers of gypsum wall board. Because of the spring-like action of the metal between the two flanges, the gypsum board is able to vibrate independently of the wall supports.

In normal rigid constructions, sound energy from the source room passes directly from one surface

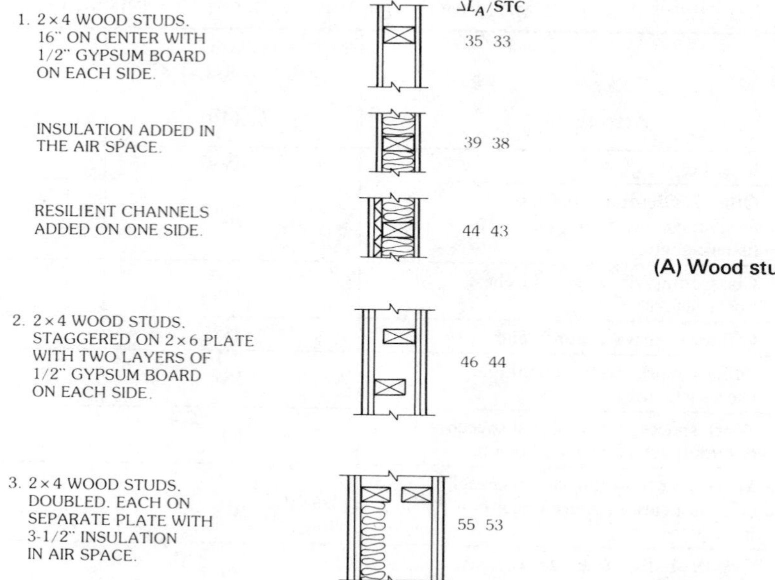
ΔL_A/STC
1. 2×4 WOOD STUDS, 16" ON CENTER WITH 1/2" GYPSUM BOARD ON EACH SIDE.
35 33
INSULATION ADDED IN THE AIR SPACE.
39 38
RESILIENT CHANNELS ADDED ON ONE SIDE.
44 43
2. 2×4 WOOD STUDS, STAGGERED ON 2×6 PLATE WITH TWO LAYERS OF 1/2" GYPSUM BOARD ON EACH SIDE.
46 44
3. 2×4 WOOD STUDS, DOUBLED, EACH ON SEPARATE PLATE WITH 3-1/2" INSULATION IN AIR SPACE.
55 53

(A) Wood stud framing.

THE EFFECTS ON THE ΔL_A RATING FOR METAL-STUD PARTITIONS DUE TO VARIOUS ADDITIONS OR CHANGES IN THE COMPONENTS OF THE BASIC PARTITION ARE SUMMARIZED BELOW.

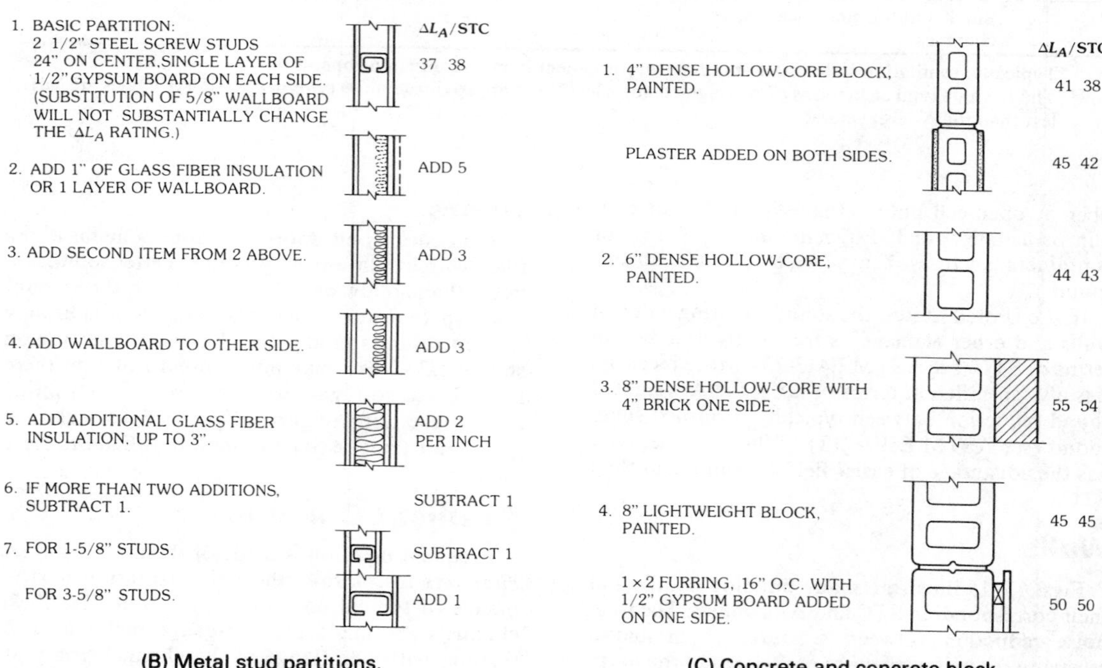
ΔL_A/STC
1. BASIC PARTITION: 2 1/2" STEEL SCREW STUDS 24" ON CENTER,SINGLE LAYER OF 1/2" GYPSUM BOARD ON EACH SIDE. (SUBSTITUTION OF 5/8" WALLBOARD WILL NOT SUBSTANTIALLY CHANGE THE ΔL_A RATING.)
37 38
2. ADD 1" OF GLASS FIBER INSULATION OR 1 LAYER OF WALLBOARD.
ADD 5
3. ADD SECOND ITEM FROM 2 ABOVE.
ADD 3
4. ADD WALLBOARD TO OTHER SIDE.
ADD 3
5. ADD ADDITIONAL GLASS FIBER INSULATION, UP TO 3".
ADD 2 PER INCH
6. IF MORE THAN TWO ADDITIONS, SUBTRACT 1.
SUBTRACT 1
7. FOR 1-5/8" STUDS.
SUBTRACT 1
FOR 3-5/8" STUDS.
ADD 1
ΔL_A/STC
1. 4" DENSE HOLLOW-CORE BLOCK, PAINTED.
41 38
PLASTER ADDED ON BOTH SIDES.
45 42
2. 6" DENSE HOLLOW-CORE, PAINTED.
44 43
3. 8" DENSE HOLLOW-CORE WITH 4" BRICK ONE SIDE.
55 54
4. 8" LIGHTWEIGHT BLOCK, PAINTED.
45 45
1×2 FURRING, 16" O.C. WITH 1/2" GYPSUM BOARD ADDED ON ONE SIDE.
50 50

(B) Metal stud partitions.

(C) Concrete and concrete block.

Fig. 13. Interior-wall sound isolation.

(wall or floor), through the structural support (studs or joists), to the surface on the opposite side, causing this second surface to vibrate and thus reradiate sound energy from the source room. However, resilient channels reduce this structural continuity. Now, most of the sound energy from one side of the wall or floor will pass through the airspace between the studs or joists. This structural discontinuity greatly improves the sound isolation of the construction. Because the sound path in this

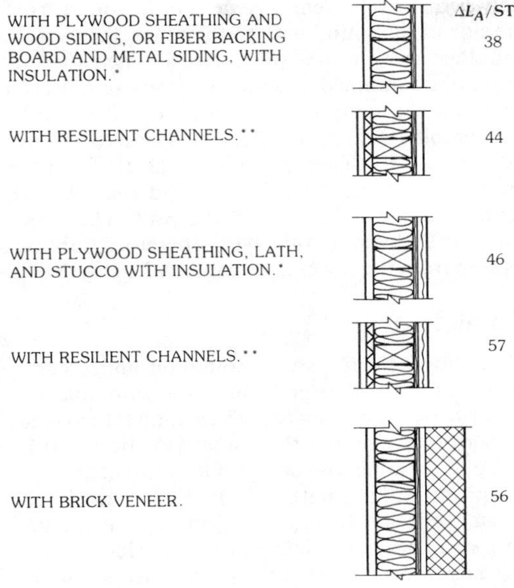
ΔL_A/STC
WITH PLYWOOD SHEATHING AND WOOD SIDING, OR FIBER BACKING BOARD AND METAL SIDING, WITH INSULATION.*
38
WITH RESILIENT CHANNELS.**
44
WITH PLYWOOD SHEATHING, LATH, AND STUCCO WITH INSULATION.*
46
WITH RESILIENT CHANNELS.**
57
WITH BRICK VENEER.
56

* It is assumed that all new wood frame construction would have glass fiber or mineral wool insulation in the stud cavity.
** Always use sound-absorptive insulation with resilient channels.

(A) Wood stud framing.

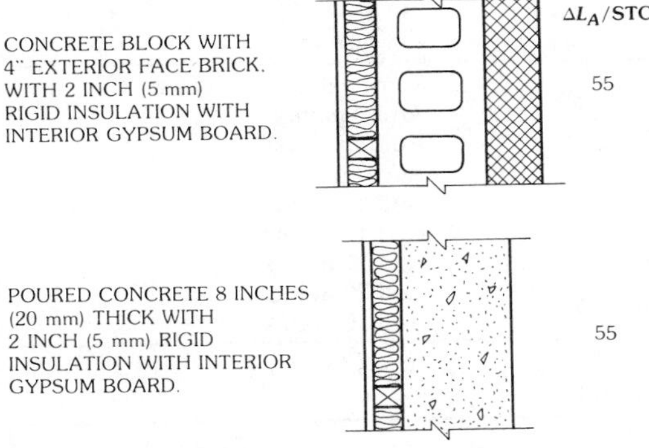
ΔL_A/STC
CONCRETE BLOCK WITH 4" EXTERIOR FACE BRICK, WITH 2 INCH (5 mm) RIGID INSULATION WITH INTERIOR GYPSUM BOARD.
55
POURED CONCRETE 8 INCHES (20 mm) THICK WITH 2 INCH (5 mm) RIGID INSULATION WITH INTERIOR GYPSUM BOARD.
55

(B) Concrete and masonry.

Fig. 14. Exterior-wall sound isolation.

case is through the wall or floor cavity, it is especially useful to have a sound-absorbing blanket of glass fiber or mineral wool insulation in the cavity air space. When resilient channels are not used, this insulation offers very little benefit, but when the resilient channels are used, this insulation should always be added.

It is very easy to short-circuit the isolation provided by the resilient channel. For example, wall-hung cabinets require rigid contact with the wall; they must be fastened directly to the studs, thus eliminating the isolation of the resilient channel. In

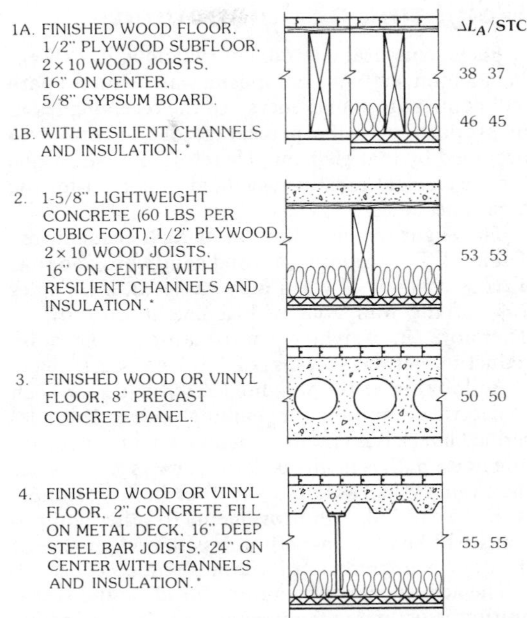
ΔL_A/STC
1A. FINISHED WOOD FLOOR, 1/2" PLYWOOD SUBFLOOR, 2 x 10 WOOD JOISTS, 16" ON CENTER, 5/8" GYPSUM BOARD.
38 37
1B. WITH RESILIENT CHANNELS AND INSULATION.*
46 45
2. 1-5/8" LIGHTWEIGHT CONCRETE (60 LBS PER CUBIC FOOT), 1/2" PLYWOOD, 2 x 10 WOOD JOISTS, 16" ON CENTER WITH RESILIENT CHANNELS AND INSULATION.*
53 53
3. FINISHED WOOD OR VINYL FLOOR, 8" PRECAST CONCRETE PANEL.
50 50
4. FINISHED WOOD OR VINYL FLOOR, 2" CONCRETE FILL ON METAL DECK, 16" DEEP STEEL BAR JOISTS, 24" ON CENTER WITH CHANNELS AND INSULATION.*
55 55

* Insulation and resilient channels should always be used together.

Fig. 15. Floor/ceiling-system sound isolation.

general, use of the channels for wall constructions is not recommended; however, in special instances and for ceilings, which are out of direct contact and abuse, resilient channels can be very helpful.

For ceilings and walls, resilient channels might typically be attached 24 inches on center at right angles to the floor joists or studs. Gypsum board is then attached to the channels with screws (Fig. 16).

For wall installations, manufacturers often recommend that a 3-inch-wide continuous filler strip be applied at the base. This will provide good mechanical support for the baseboard and gypsum board against damage without sacrificing much resilience.

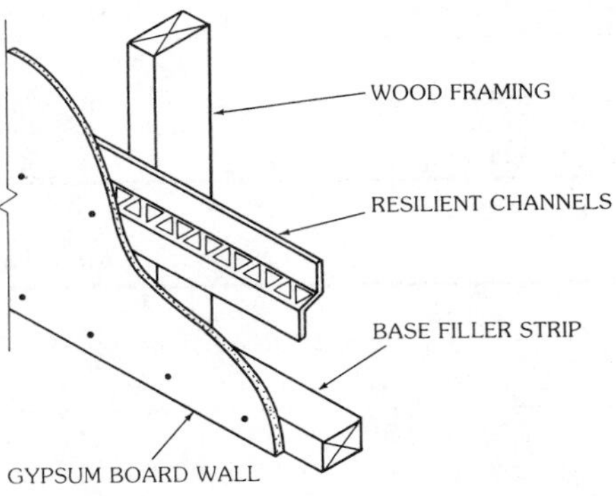
WOOD FRAMING
RESILIENT CHANNELS
BASE FILLER STRIP
GYPSUM BOARD WALL

Fig. 16. Installation of resilient channels.

Weighting of Components

Each structural component (window, door, ceiling plenum, etc.) in a composite wall or sound path will contribute sound energy to the receiving room in proportion to the percentage of the total area occupied by that element. Therefore the ΔL_A value for the element must be weighted to correct for the area it occupies.

To weight the contribution of each element, use Table 13. For example, a wood solid-core door has a ΔL_A rating of 27 when fully gasketed; it occupies 20% of the wall area with a neighboring office. Therefore, its weighted contribution to the noise reduction between rooms is $27 + 7 = 34\ \Delta L_A$.

By looking at the weighted ΔL_A value for each structural component or flanking path or leak (described later), it is readily apparent what the controlling noise path is in any isolation analysis. Consider the example in Table 14. The door in this example is the "weak link." Improvements to other components will be of no benefit unless the ΔL_A value of the door is upgraded or its area is reduced.

The acoustical contributions for all sound-transmitting structural components (wall, door, window, etc.) must be combined to determine the performance of the composite wall. Since ΔL_A values for each element represent a *reduction* in sound from one side of the sound path to the other, the overall, combined ΔL_A value will decrease when the elements are combined (Table 15). For example: To determine the composite ΔL_A for the wall example, first combine ΔL_A 36 and ΔL_A 34. These two values differ by 2 dB; therefore, subtract 2 dB from the lower ΔL_A value: ΔL_A 34 less 2 dB $= \Delta L_A$ 32. Next, combine ΔL_A 32 and ΔL_A 42. These two values differ by more than 9 dB; therefore, subtract nothing from the lower ΔL_A value: ΔL_A stays at 32.

Flanking

Flanking is the transfer of sound through secondary sound paths that go around the common wall barrier between two spaces. These indirect passages can seriously degrade the sound isolation performance of an otherwise acceptable construction.

Typical flanking paths, A, B, C, D, and E, are shown in Fig. 17. In addition, flanking paths F and G are structural connections, which are less suitable for analysis with specific ΔL_A sound-isolation performance ratings. In these cases, a sound from a source room will be transmitted to a receiver room by way of a common wall or floor surface.

It is most important to break the structural paths

TABLE 13. WEIGHTING FACTORS FOR COMBINING PARTY SURFACE COMPONENTS

Percent of Total Area Occupied by Component	Weighting Factor for ΔL_A/STC
100	Same value
50	Add 3 points
33	Add 5 points
20	Add 7 points
10	Add 10 points
5	Add 13 points
2	Add 17 points
1	Add 20 points
0.1	Add 30 points

TABLE 14. WALL EXAMPLE

Component	Intrinsic STC/ΔL_A Value	Percent of Area	Weighted STC/ΔL_A Value
Window	26	10	36
Door	27	20	34
Wall	40	70	42
Flanking (floor)	50	—	50

TABLE 15. COMBINED ΔL_A VALUE

Where two ΔL_A values differ by	Subtract the following amount from the lower value ΔL_A
0–1	3
2–3	2
4–8	1
9 or more	0

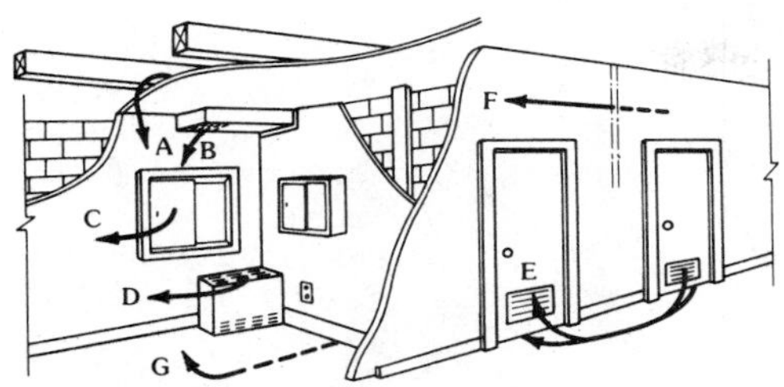

A. OPEN PLENUM ABOVE WALLS
B. UNBAFFLED DUCTS
C. OUTDOOR PATH, WINDOW TO WINDOW
D. CONTINUOUS UNBAFFLED INDUCTOR UNITS
E. CORRIDOR PATH
F. COMMON WALL
G. COMMON FLOOR

Fig. 17. Flanking paths.

that connect two spaces. Floors, rafters, and joists should not span between separate dwelling units. Interior wallboard skins should not span between noise-sensitive spaces.

Leaks

Leaks are sound paths through open air passages caused by poor detailing, inadequate field supervision, and sloppy workmanship. They represent a major cause for poor performance of partitions and can be easily resolved by proper caulking and sealing of all openings and penetrations between spaces. Typical locations of leaks are shown in Fig. 18.

The noise isolation value of an untreated hole in a wall is, of course, $\Delta L_A = 0$. Depending on the size of the leak, the noise isolation performance of an entire wall or other structural component can be severely degraded. Fig. 19 plots this effect.

For example, consider a perfectly good composite construction with $\Delta L_A = 50$ and a typical area of 100 square feet (9 square meters). A crack only 0.01 inch (2.5 mm) high along the base of a wall 12 feet (4 meters) long would be a total opening of about 1.4 square inches. The ΔL_A performance would be degraded from 50 to 39. Also consider that a typical properly installed door with a 1-inch undercut for rugs or ventilation has an opening of about 40 square inches.

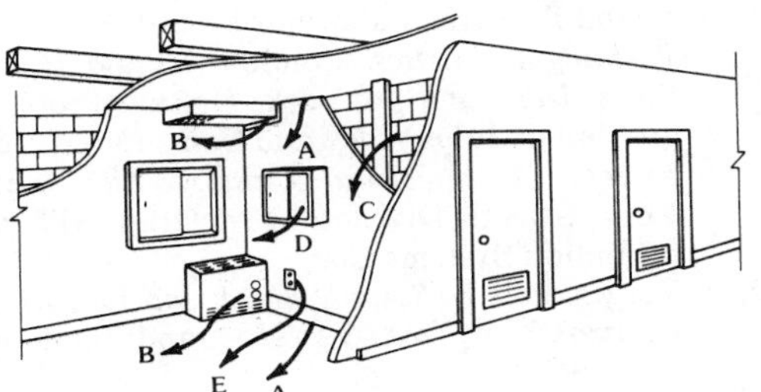

A. POOR SEAL AT EDGES OF CEILING, FLOOR, AND WALLS
B. POOR SEAL AROUND DUCT PENETRATIONS; GAPS AND HOLES AT PIPE PENETRATIONS
C. POOR MORTAR JOINTS AND POROUS MASONRY
D. BACK-TO-BACK CABINETS (AND BATH TUBS) WITHOUT WALL BEHIND
E. BACK-TO-BACK ELECTRICAL OUTLETS

Fig. 18. Sound leaks.

SOUND-REINFORCEMENT SYSTEMS*

Listeners should be almost unaware of the operation of a sound-reinforcement system. There should be adequate level, high intelligibility, and natural sound.

The purpose of a sound system is to boost the message or program sufficiently above the ambient level. Ideally, the amplification at the receiver should be unity. That is, the sound level at a receiver's ear should be about 65–70 dB, or the level generated by a speaker at a listening distance of 1 meter. In most auditoria or lecture rooms, the background level is not a problem, but in sports arenas, spectator noise can raise the ambient such that the direct sound amplification at the receiver must be much greater than unity. It has been found that about 9 dB of amplification at the receiver is the most that can go undetected—the sports-arena situation frequently requires more.

In general, a sound-reinforcement system may contain a single loudspeaker cluster or a distribution of loudspeakers. If the sound originates from a stage or performance arena, then a single cluster should be used if the sound is to be natural. An arrangement of two loudspeakers flanking the stage is inappropriate; the two speakers create interference patterns and an unnatural sound. If one speaker cluster cannot maintain sufficient level all the way to the rear of an audience, then additional speakers can be added closer to the rear with appropriate time-delay hardware so that the sound remains realistic and in correct time phase with the source. Distributed loudspeaker systems are used when there is no identifiable source. Situations requiring distributed systems include paging, such as in offices, and background music, such as in food stores.

A rule of thumb for placing loudspeakers in a distributed system is:

$$D = 2(H - 4)$$

where,

D = distance between speaker centers in feet,
H = ceiling height in feet.

or

$$D = 2(H - 1)$$

where D and H are in meters.

* Yerges, Lyle F. *Sound, Noise, and Vibration Control.* New York: Van Nostrand-Reinhold Co., 1969.

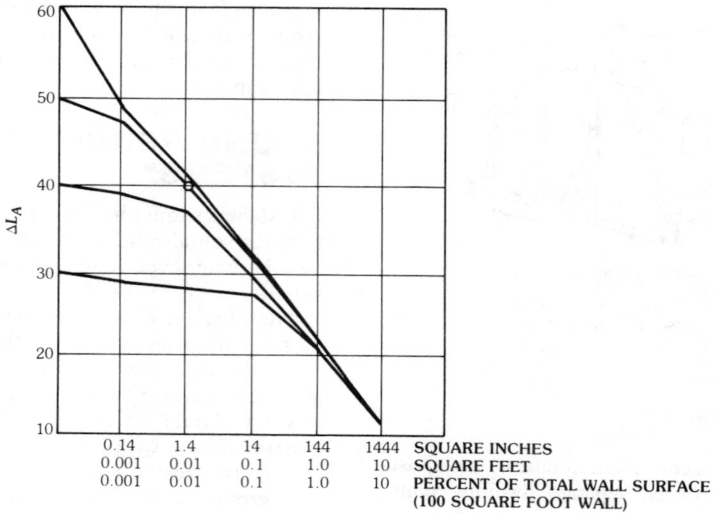

Fig. 19. The effect of leaks on ΔL_A value of partition system.

The center of a single speaker should be aimed at the most distant centrally located receiver. The directivity pattern (vertically and horizontally) should be chosen to provide a uniform sound field to the audience given the elevation and plan of the room.*

SELECTED BIBLIOGRAPHY

1. General Acoustics

1.1 Kinsler and Frey. *Fundamentals of Acoustics.* New York: John Wiley and Sons, Inc., 1965.

1.2 Morse and Ingard. *Theoretical Acoustics.* New York: McGraw-Hill Book Co., 1968.

1.3 Rayleigh, J. *The Theory of Sound.* New York: Dover Publications, 1945 (from the 1894 edition).

1.4 Swenson, George W. *Principles of Modern Acoustics.* New York: Van Nostrand, 1953.

2. Building Acoustics

2.1 Department of Housing and Urban Development (HUD), *A Guide to Airborne, Impact and Structure Borne Noise Control in Multi-Family Dwellings,* FT/TS-24, 1967.

2.2 Dept. of Commerce, *Acoustical and Thermal Performance of Exterior Residential Walls, Doors and Windows,* NBS Building Science Series No. 77, Nov., 1975.

2.3 *Acoustical Manual: Apartment and Home Construction.* Rockville, Md.: National Association of House Builders (NAHB), Research Foundation, No. 315.04, June, 1971.

2.4 *Guide and Data Book.* New York: American Society of Heating, Refrigeration and Air-Conditioning Engineers (ASHRAE).

2.5 Ginn, K. *Architectural Acoustics.* Denmark: Bruel and Kjaer, 1978.

2.6 Knudsen and Harris. *Acoustical Design in Architecture.* New York: John Wiley and Sons, Inc., 1950.

2.7 Newman, R., et. al. *Acoustics, Time Saver Standard.* New York: McGraw-Hill Book Co., 1973.

2.8 Dept. of Commerce, *Quieting: A Practical Guide to Noise Control,* NBS Handbook No. 119, July, 1971.

2.9 Sound Research Laboratories, Ltd. *Practical Building Acoustics.* London: E. and F. N. Spon, Ltd.; or New York: Halsted Press, a division of John Wiley and Sons, Inc.; 1976.

2.10 *Sweet's Architectural Catalog File.* New York: Sweet's Division of McGraw-Hill Information Systems Co.

2.11 Yerges, Lyle F. *Sound, Noise, and Vibration Control.* New York: Van Nostrand Reinhold Co., 1969.

3. Environmental Noise Assessment and Land Use Planning With Respect to Noise

3.1 Beranek, Leo L., ed. *Noise and Vibration Control.* New York: McGraw-Hill Book Co., 1971.

* Sound systems in large and/or critical areas (*e.g.,* playhouses) are dealt with in reference 4.1.

3.2 Dept. of Housing and Urban Development (HUD), Office of Policy Development and Research, *Criteria for Acceptable Outdoor and Indoor Acoustical Environments for Schools, Hospitals, Nursing and Convalescent Homes, Churches, Business Offices and Other Community Facilities,* June, 1976.

3.3 Dept of Commerce (NBS), *Design Guide for Reducing Transportation Noise in and Around Buildings,* Building Science Series 84, April, 1978.

3.4 Dept. of the Army, *Environmental Protection: Planning in the Noise Environment,* Technical Manual (TM) 5-803-2, 15 June 1978.

3.5 *Environmental Assessment of High Energy Impulsive Sounds,* Report of Working Group 84, Committee on Hearing, Bioacoustics and Biomechanics (CHBAB), Assembly of Behavioral and Social Sciences, National Research Council, National Academy of Science, 1981.

3.6 Dept. of Transportation, Federal Interagency Committee on Urban Noise, *Guidelines for Considering Noise in Land Use Planning and Control,* June, 1980.

3.7 *Guidelines for Preparing Environmental Impact Statements on Noise,* Committee on Hearing, Bioacoustics and Biomechanics, (CHBAB), Assembly of Behavioral and Social Sciences, National Research Council, National Academy of Science, 1977.

3.8 EPA, *Information on Levels of Environmental Noise Requisite to Protect Public Health and Welfare with an Adequate Margin of Safety,* 550/9-74-004, March, 1974.

3.9 Kryter, Karl. *The Effects of Noise on Man.* New York: Academic Press, 1970.

3.10 von Gierke, H. "Noise—How Much Is Too Much." *Noise Control Engineering,* Vol. 5, No. 1, 1975.

4. Noise Measurement and Instrumentation

4.1 Davis, Don and Carolyn. *Sound System Engineering,* Indianapolis: Howard W. Sams & Co., Inc., 1975.

4.2 Hassall, J. R. and Zaveri, K. *Acoustic Noise Measurement,* Denmark: Bruel and Kjaer, 1979.

4.3 Peterson, A. G. and Gross, E. E. Jr. *Handbook of Noise Measurement.* Concord, Mass.: General Radio Co., 1972.

41 Lasers

Joseph T. Verdeyen and Thomas A. DeTemple

The term "laser" is an acronym for *L*ight *A*mplification by *S*timulated *E*mission of *R*adiation. The first demonstration of an electronic amplifier using stimulated emission was in the microwave frequency domain where the acronym "maser" is used. Stimulated emission has been observed in amorphous and single crystal solids, semiconductors, and dyes and in atomic, molecular, and ionized gases. Fig. 1 shows the spectral region of operation of the various types of lasers, a smoothed version of atmospheric transmission, common detector classes, and transmissive materials for various optical functions. Since the operation of lasers is dependent on quantum interactions, they are known as quantum electronic devices.

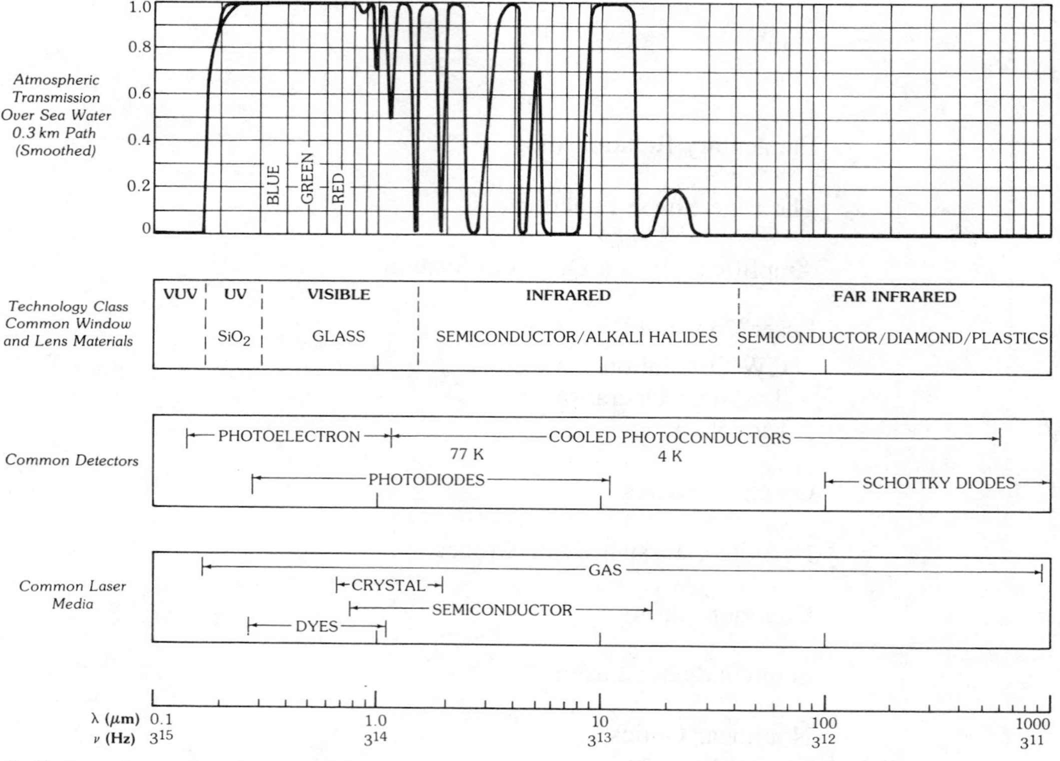

Fig. 1. Regions of operation of various types of laser and examples of the technology required in the various domains.

ELEMENTARY RELATIONSHIPS

In order for a quantum electronic device to amplify (or attenuate) at a frequency ν_{21}, there must be a set of allowed energy levels (or bands) E_2 or E_1 in the material separated by the Planck spacing.

$$E_2 - E_1 = h\nu_{21} \quad \text{(Planck's law)} \qquad (1)$$

An atom can make discontinuous jumps between these energy levels, accompanied by either the emission or absorption of a photon by one of three processes (Fig. 2).

A. *Stimulated emission* occurs when an external applied wave causes an atom to yield its potential energy given by Eq. 1 to the inducing wave. This added energy (photon) propagates in the same direction, at the same frequency and phase, and in the same sense of polarization as the stimulating field. This results in amplification of the wave.

B. *Absorption* occurs when the atom increases its potential energy by Eq. 1, thereby decreasing the wave energy by the same amount, and results in attenuation.

C. *Spontaneous emission* is the name for a 2→1 radiative transition for which there is no apparent external influence and which yields a photon at the frequency given by Eq. 1 in any of 4π directions. It is the cause of noise in quantum elec-

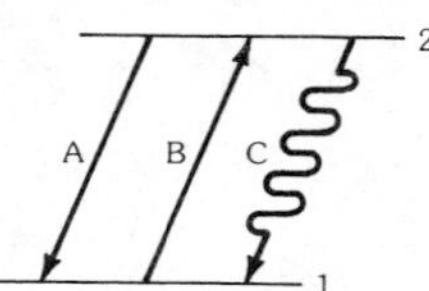

Fig. 2. Radiative processes between two levels in an atom. (The letters A, B, and C refer to stimulated emission, absorption, and spontaneous emission of radiation.)

tronic amplifiers; however, discharge lamps produce most of their radiation by this process.

The rate of production of electromagnetic energy by the above processes is given by

$$\text{Spontaneous: } h\nu_{21}A_{21}N_2 \quad (2)$$
$$\text{Stimulated: } h\nu_{21}B_{21}N_2 g(\nu)I_\nu/c' \quad (3)$$
$$\text{Absorption: } h\nu_{21}B_{12}N_1 g(\nu)I_\nu/c' \quad (4)$$

where,

$N_{2,(1)}$ is the density of atoms in energy state 2 (1),
A_{21}, B_{21}, and B_{12} are the Einstein coefficients,
$g(\nu)$ is the line shape function,
I_ν is the intensity (watts/area) of the externally applied wave,
$c' = c/n$ is the velocity of propagation of the wave,
n is the index of refraction of the material.

The Einstein coefficients are interrelated.

$$A_{21}/B_{21} = 8\pi n^3 \nu^3/c^3 \qquad B_{21}/B_{12} = g_1/g_2 \quad (5)$$

where $g_{2(1)}$ is the degeneracy of the state, i.e., number of quantum levels with the same energy.

At thermal equilibrium, the ratio of the population of the two energy states, E_2 and E_1, which are separated by Eq. 1 is given by the Boltzmann distribution:

$$N_2/N_1 = (g_2/g_1) \exp[-h\nu_{21}/kT] \quad (6)$$

THE LINE SHAPE FUNCTION

The function $g(\nu)d\nu$ has the following physical interpretations:

A. It is the fraction of the spontaneous emission from state 2 to state 1 that appears in the frequency interval ν to $\nu + d\nu$.
B. It is the relative strength of interaction between the medium and the wave for stimulated emission or absorption.
C. It is the line profile of an optically thin transition.

Since it is a probability function, it obeys a normalization condition

$$\int_0^\infty g(\nu)\, d\nu = 1 \quad (7)$$

Transitions in the near ir through the uv in low-pressure gases (a few Torr) are broadened by the thermal motion of the atoms, which leads to a Doppler broadened (Gaussian) line shape:

$$g_D(\nu) = [(4 \ln 2)/\pi]^{1/2}(1/\Delta\nu_D) \exp\{(-4 \ln 2)[(\nu - \nu_{21})/\Delta\nu_D]^2\} \quad (8)$$

with

$$\Delta\nu_D = (8kT/Mc^2)^{1/2}\nu_{21} \quad (9)$$

where,

M is the mass of the atom (or molecule),
T is the temperature,
$\Delta\nu_D$ is the *f*ull-*w*idth-at-*h*alf *m*aximum (FWHM) of the transition centered at ν_{21}.

At higher pressures and/or longer wavelengths (lower frequency), elastic gas collisions broaden the transition, and the homogeneous (Lorentzian) line shape function becomes:

$$g(\nu) = \Delta\nu_h/2\pi[(\nu - \nu_{21})^2 + (\Delta\nu_h/2)^2] \quad (10)$$

with

$$\Delta\nu_h = (1/2\pi)[A_2 + k_2 + A_1 + k_1 + 2\nu_{\text{col.}}] \quad (11)$$

In Eq. 11, the As are the total radiative decay rates of levels 2 and 1, the k's are total quenching rates of the two states, and $\nu_{\text{col.}}$ is the elastic collision rate of the active atoms or molecules with the other gas constituents. Usually, the elastic-collision term dominates, and one refers to the resulting line profile as "pressure broadened."

Seldom does one attempt to describe the line shape of a transition in a solid or band-to-band transition in semiconductors by elementary functions such as those used in Eqs. 8 and 10. Rather, one depends on the experimental determination of the line profile (e.g., C above) and often approximates $g(\nu)$ by

$$g(\nu) \simeq 1/\Delta\nu \quad (12)$$

where $\Delta\nu$ is the FWHM of the spontaneous emission of an optically thin line.

AMPLIFICATION IN A QUANTUM SYSTEM

The net rate of increase of an externally applied electromagnetic wave $I\nu$ (watts/area) per unit of length of active material is the difference between the stimulated rate of Eq. 3 and the absorptive rate given by Eq. 4:

$$dI_\nu/dz = h\nu_{21} B_{21}[N_2 - (g_2/g_1)N_1]g(\nu)I_\nu/c' \quad (13)$$

If $N_2 < (g_2/g_1)N_1$, one has a "normal" population with a positive "temperature" specified by Eq. 6; such a medium will attenuate an electromagnetic wave. If $N_2 > (g_2/g_1)N_1$, one has an "inverted" population with a "negative" temperature, and this medium will amplify an electromagnetic wave. By combining Eqs. 5 and 13, one obtains the *small-signal gain coefficient* given by

$$\gamma_o(\nu) \triangleq (1/I_\nu)(dI_\nu/dz)$$
$$= A_{21}(\lambda^2/8\pi)\, g(\nu)[N_2 - (g_2/g_1)N_1] \quad (14)$$

where $\lambda = \lambda_o/n$. The collection of terms multiplying N_2 is often referred to as the *stimulated emission cross section* (i.e., "area")

$$\sigma_{se}(\nu) = A_{21}(\lambda^2/8\pi)\, g(\nu_{21}) \quad (15)$$

The absorption cross-section refers to the factors multiplying N_1 in Eq. 14.

$$\sigma_{abs}(\nu) = (g_2/g_1)\,\sigma_{se}(\nu) \quad (16)$$

The small-signal gain of the medium of length l is given by the expression

$$G_o = \exp \gamma_o(\nu) l \quad (17)$$

The subscript "o" in Eqs. 14 and 17 implies that the intensity of the stimulating wave is sufficiently small so as to create negligible perturbation on the density of atoms in the two states.

If the intensity is large enough, the populations, $N_{2,1}$, in Eq. 14 are functions of the intensity, I_ν, of the stimulating wave. For homogeneously broadened cw systems, the *saturated gain coefficient* is given by

$$\gamma(\nu) = \gamma_o(\nu)/[1 + (I_\nu/I_s)\ \bar{g}(\nu)] \triangleq (1/I_\nu)\ dI_\nu/dz \quad (18)$$

where,

$\gamma_o(\nu)$ is the small-signal value,

$\bar{g}(\nu)$ is the line shape function normalized to unity at line center,

I_s is the saturation intensity given by

$$I_s = h\nu_{21}/\tau_2\ \sigma_{se}(\nu_{21}) \quad (19)$$

In Eq. 19, τ_2^{-1} is the decay rate of state 2 due to all causes—radiation plus quenching, and $\sigma_{se}(\nu d_{21})$ is the stimulated-emission cross section given by Eq. 15 evaluated at line center.

The overall gain, G, of the amplifier is thus a function of the input intensity and is found by integrating Eq. 18 from 0 to l.

$$\ln G + (I_{in}/I_s)\ \bar{g}(\nu)\ (G-1) = \gamma_o(\nu) l \quad (20)$$

where $G = I_{out}/I_{in}$.

The maximum intensity that can be extracted from an amplifier by stimulated emission is given by

$$(I_{out} - I_{in}) \Big|_{max} = \gamma_o(\nu_{21})\ I_s l \quad (21)$$

For inhomogeneously broadened transitions the saturated law is modified:

$$\gamma(\nu) = \gamma_o(\nu)/(1 + I_\nu/I_s)^{1/2} \quad (22)$$

where all factors are as defined previously, and Eqs. 20 and 21 must be suitably modified.

LASER THRESHOLD

Oscillation with an inverted population is possible provided sufficient feedback is provided, usually in the form of an open cavity. In general, the necessary condition for oscillation is that net round-trip gain must exceed 1 so that an initially small wave can grow in amplitude as it passes around the feedback path. For a simple system such as that shown in Fig. 3, the threshold condition is

$$R_1 R_2 T_a^2 T_b^2 \exp[2\ \gamma_o(\nu) l] \geq 1$$

or

$$\gamma_o \geq (1/l) \ln (1/\sqrt{R_1 R_2}) + (1/l)\ ln\ (1/T_a\ T_b) \triangleq \alpha \quad (23)$$

In Fig. 3, the power transmission coefficients for the windows at the ends of the laser cell are T_a and T_b, and those of the mirrors are $T_1 = 1 - R_1$ and $T_2 = 1 - R_2$. Implicit in Eq. 23 is the assumption that the population inversion exists long enough to allow many transits back and forth between mirrors.

If the transition is homogeneously broadened, then oscillation will occur on a cavity mode that has the highest gain-to-loss coefficient ratio γ_o/α. For the simple laser shown in Figs. 3 and 4, this would correspond to the cavity mode nearest line center. If the transition is inhomogeneously broadened, multimode oscillation is possible on those cavity modes at frequencies ν_q such that $\gamma_o(\nu_q)/\alpha > 1$.

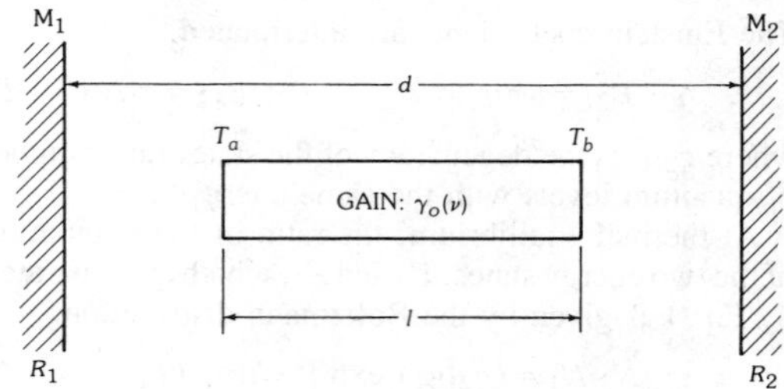

Fig. 3. A simple laser.

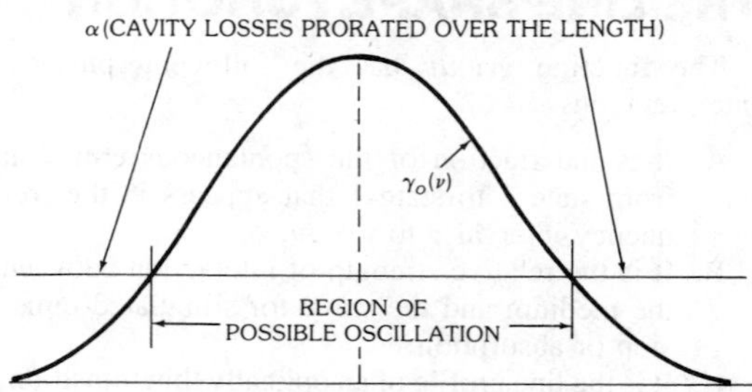

Fig. 4. Spectral bandwidth available for oscillation.

CW Oscillation

For the simple laser shown in Fig. 3, there are three types of losses: internal discrete losses represented by the imperfect transmission by the windows, a residual loss due to imperfections in the medium, and external or coupling losses by virtue of the transmission through the mirrors. The photon flux inside the cavity builds up to a large intensity so that the gain coefficient (Eq. 18 or 21) saturates at the threshold value given by Eq. 23. If the mirrors are highly reflecting, little power escapes the cavity; if the mirrors have too large a transmission coefficient, the laser fails to oscillate. The transmission coefficient of M_2 that provides maximum output power from the laser illustrated in Fig. 3 is

$$T_2 = -L + (g_o L)^{1/2}; \qquad (T_1 = 0) \quad (24)$$

where,

$g_o = \gamma_o \cdot 2l$ is the gain coefficient integrated over the double pass through the active medium,

$L = \alpha_{int} \cdot 2l$ is the integrated internal losses given by

$$\alpha_{int} = (1/2l)\ [\ln(1/R_1) + \ln(1/T_a^2 T_b^2)] + \alpha_{residual} \quad (25)$$

The output intensity (through M_2) is given by

$$I_{out} = (I_s/2) \cdot [g_o/(L+T_2) - 1]\ T_2 \quad (26)$$

For T_2 given by Eq. 24, the output intensity is given by the expression

$$I_{out} = (I_s/2) \cdot [\sqrt{g_o} - \sqrt{L}]^2 \quad (27)$$

If $\sqrt{g_o} >> \sqrt{L}$, then $I_{out} \simeq (I_s\ g_o)/2 \simeq \gamma_o l I_s$.

Transient Operation

The previous discussion presumes steady-state or cw laser oscillation. However, some lasers are operated in a pulsed mode with rather spectacular results, as discussed below.

If the upper laser level has a lifetime that is long compared to the pumping duration, then it is a candidate for Q-switching. The scheme is similar to that used for radar modulators in which energy is slowly stored in a capacitor, transmission line, or PFN and then is rapidly discharged into the magnetron load. In the laser scheme, the energy is stored in the upper quantum state and is then extracted by stimulated emission.

The scheme is illustrated in Fig. 5. The Q of the laser cavity is spoiled (intentionally) during the time the upper state is being populated by the external pump, thus preventing lasing. When the upper state has reached an equilibrium with the pump, the cavity Q is restored to its high value. The population inversion is larger than threshold for cw oscillation, and thus stimulated emission rapidly amplifies the photons initially present in the cavity. This buildup occurs so rapidly that one can neglect any pumping on the time scale of the output pulse; in fact, it is so rapid and intense that the population inversion is driven below threshold for cw oscillation.

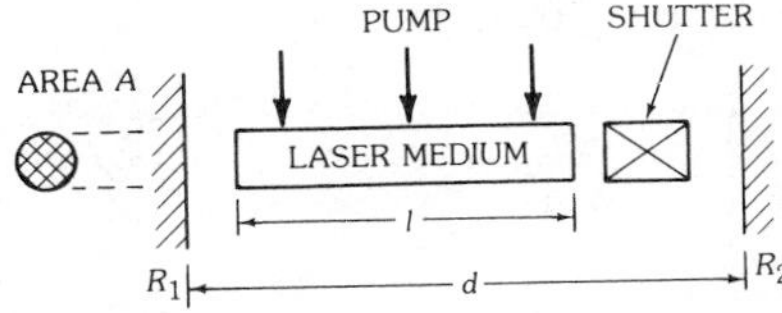

(A) The optical cavity.

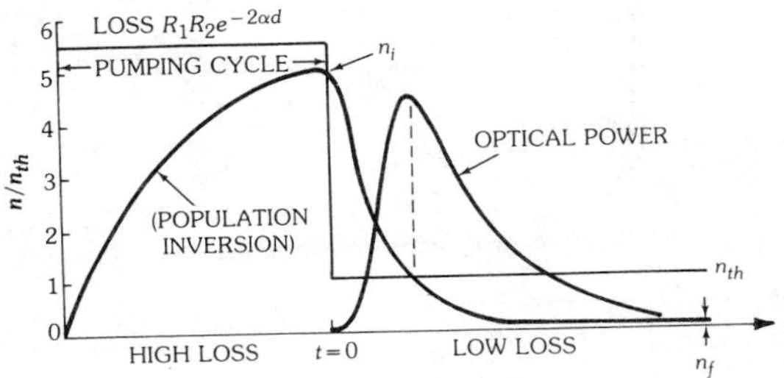

(B) Time evolution of the population diference and the optical pulse.

Fig. 5. Q-switching of a simple laser. (*From J. T. Verdeyen,* Laser Electronics. *Englewood Cliffs, N. J.: Prentice-Hall, Inc., 1981; Fig. 9-5.*)

The parameters that characterize the Q-switched pulse are:

n_i = initial number of inverted atoms or molecules in the cavity

n_{th} = the inversion required for cw oscillation in the high-Q cavity

n_f = the final inversion in population (see Fig. 5)

$T = t/\tau_c$, time divided by the passive cavity lifetime

The peak power produced by such a laser is given by:

$$P_p = k(h\nu/\tau_c)\ [(n_i - n_{th}/2) - (n_{th}/2)\ \ln(n_i/n_{th})] \quad (28)$$

where k is the coupling factor on the order of 1/2.

The output energy of the Q-switch pulse is given by

$$W_{out} = k \cdot [(n_i - n_f)/2]h\nu \quad (29)$$

and the final inversion is found from a solution to

$$n_f/n_i = \exp\ [-(n_i - n_f)/n_{th}] \quad (30)$$

Inasmuch as the maximum energy that could be extracted is $n_i \cdot h\nu/2$, Eq. 29 can be written as:

$$W_{out} = k \cdot [(n_i - n_f)/n_i] \cdot (n_i h\nu/2) \quad (31)$$

The quantity $(n_i - n_f)/n_i$ is called the energy extraction efficiency and is plotted in Fig. 6A. The time evolution of this Q-switch pulse requires a numerical solution of the coupled differential equation. A typical solution is shown in Fig. 6B for various values of the ratio n_i/n_{th}. A rough estimate for the FWHM of the pulse can be found by combining Eqs. 28 and 31:

$$\Delta t_{1/2} \simeq W_{out}/P_p \quad (32)$$

Modelocking

If the system is inhomogeneously broadened, the laser may oscillate on many cavity modes within the spectral bandwidth available for oscillation (Fig. 4). If these modes can be locked together in a definite phase sequence, then the time-domain representation of the output consists of a series of pulses at a repetition rate equal to the round-trip transit time of a photon in the cavity.

Each individual pulse is very short; typically, 0.1–2 ns is common, with 30 fs (30×10^{-15} s) being the shortest presently recorded. The time evolution of these pulses is related to the spectral distribution of modes by the Fourier transform pair (see Chapter 7). Rough rules of thumb for the pulse characteristics are:

$$\text{Peak power} = N \times \text{average power} \quad (33a)$$

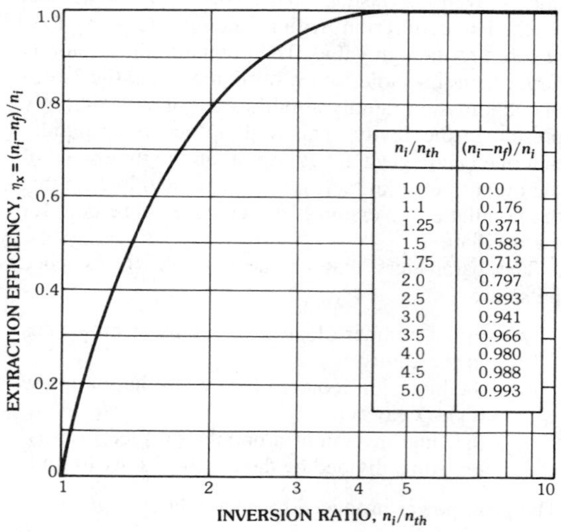

n_i/n_{th}	$(n_i-n_f)/n_i$
1.0	0.0
1.1	0.176
1.25	0.371
1.5	0.583
1.75	0.713
2.0	0.797
2.5	0.893
3.0	0.941
3.5	0.966
4.0	0.980
4.5	0.988
5.0	0.993

(A) The extraction efficiency for a *Q*-switched laser.

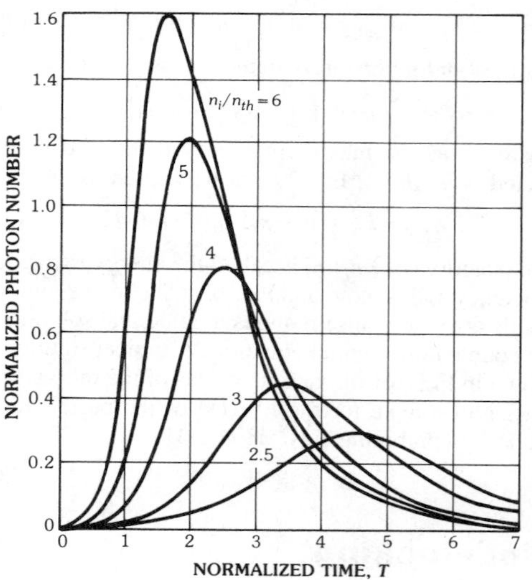

(B) The time evolution of the *Q*-switched pulse, time normalized to the passive cavity lifetime.

Fig. 6. Characteristic parameters of *Q*-switching. (*From J. T. Verdeyen,* Laser Electronics. *Englewood Cliffs, N. J.: Prentice-Hall, Inc., 1981; Figs. 9-7 and 9-8.*)

Pulse repetition rate $= c'/2d$ (33b)

Pulse width $= (2d/c')/N \approx 1/\Delta\nu$ (33c)

where,

N is the number of modes locked,
$2d/c'$ is the round-trip transit time,
$\Delta\nu$ is the FWHM of the transition.

OPTICAL CAVITIES

Optical cavities perform the essential function of providing positive feedback for the quantum electronic amplifier so as to obtain oscillation. The spontaneous emission from the upper state of the quantum system is reflected by the mirrors and amplified by the inverted population. The cavity also selects the oscillation frequency within the spectral bandwidth defined in Fig. 4. The laser depends upon stimulated emission for amplification, and that rate is a maximum at the resonant frequencies of the cavity since the electromagnetic energy is a maximum under these conditions.

The resonant buildup of electromagnetic energy in any cavity is due to the near-perfect phase addition of the various partial waves bouncing back and forth between the mirrors. In general then, the resonant frequency of any cavity is determined by

Round-trip phase shift
= Integral multiple of 2π radians (34)

If one assumes a uniform plane wave for the normal mode of the simple Fabry-Perot cavity shown in Fig. 7A, then Eq. 34 reduces to

$$\nu_q = q\, c'/2d \quad (35)$$

where,

c' is the velocity of light in the medium between the mirrors,
q is an integer.

Typically $d \approx 0.1$ to 1.0 m for solid-state and gas lasers, and thus $c'/2d \approx 150$ MHz to 1.5 GHz, considerably smaller than the line width of many transitions (approximately 5 to 100 GHz).

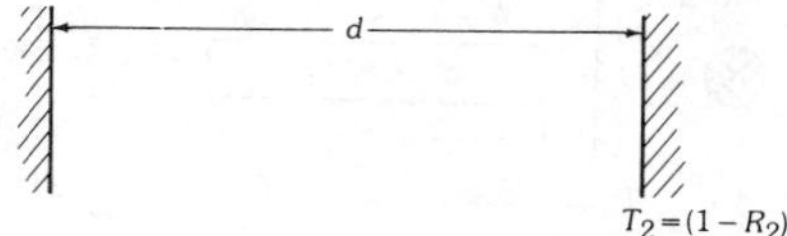

(A) Fabry-Perot cavity.

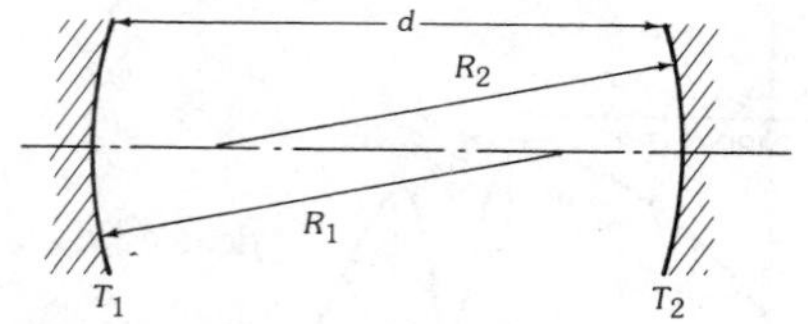

(B) A stable optical cavity.

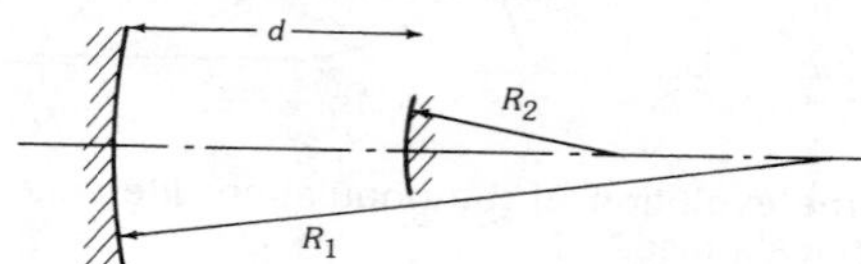

(C) An unstable optical cavity.

Fig. 7. Typical cavities used with lasers.

Most optical cavities use one or more focusing elements in the manner indicated in Figs. 7B and C. The cavity shown in Fig. 7B is *stable* in the sense that the maximum excursion of a paraxial ray from the axis of the cavity is bounded. However, a ray will "walk off" the surfaces of the feedback mirrors of Fig. 7C; hence, this cavity is classified as *unstable*. Simple cavities such as those shown in Fig. 7 are stable if

$$0 \le g_1 g_2 \le 1 \tag{36}$$

where,

$$g_{1,2} = 1 - d/R_{1,2} \tag{37}$$

If the equality is not satisfied, the cavity is unstable. The stability criterion given by Eq. 36 is quite often embodied in a graphical format such as shown in Fig. 8. The dashed curves illustrate the equiloss contours for the unstable cavity. The quantity Γ is related to the mean fractional power lost per pass by $\Gamma = 1 - L$. For stable cavities, the diffraction loss (i.e., loss around edges of the mirrors) is very small; most of the loss of electromagnetic energy is transmission through the mirrors.

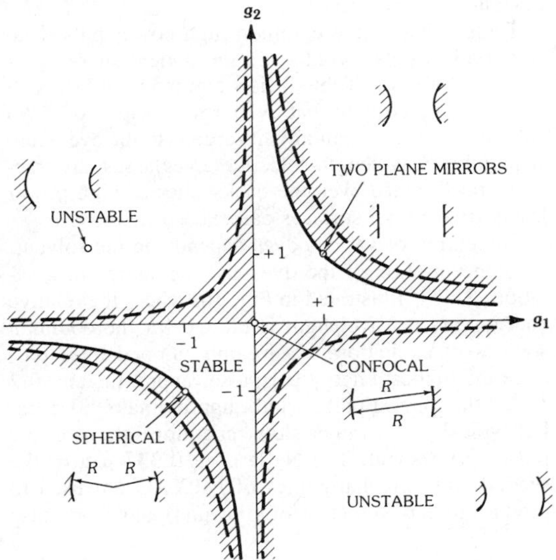

Fig. 8. Graphical display of the stability of optical cavities.

Unstable cavities have a very large mode volume and thus can utilize the stimulated emission from a large number of atoms. The losses from around the edges of the mirrors are usually large compared to the loss from direct coupling—usually by design. One can relate Γ, the mean fraction of the power that survives a single pass through the cavity, to the cavity g parameters by

$$\Gamma = \pm \frac{1 - [1 - (g_1 g_2)^{-1}]^{1/2}}{1 + [1 - (g_1 g_2)^{-1}]^{1/2}} \tag{38}$$

where the choice of the sign is determined by the requirement that $0 < \Gamma < 1$. This leads to the equiloss contours shown in Fig. 8.

If all external sources of electromagnetic energy for the resonant mode of a cavity are suddenly removed, the energy will decay with time constant τ_p, called the photon lifetime:

$$dW/dt = -W/\tau_p \tag{39a}$$

$$\tau_p = \frac{\text{Time for a round trip}}{\text{Fraction of energy lost per round trip}} \tag{39b}$$

where W is the energy stored in the cavity.

For the simple cavity shown in Fig. 7B, this leads to the expression

$$\tau_p = (2d/c)/(1 - R_1 R_2) \tag{40}$$

where all losses except those due to imperfect mirror reflectivities have been neglected.

The photon lifetime is directly related to the cavity Q by the equation

$$Q = 2\pi \times \frac{\text{Energy stored}}{\text{Energy lost in a cycle}} \tag{41a}$$

or

$$Q = \omega_o W/(-dW/dt) \tag{41b}$$

where ω_o is the (angular) resonant frequency of the cavity. The differential equation of Eq. 41b is the same as Eq. 39a and relates Q and τ_p by

$$\tau_p = Q/\omega_o \tag{42a}$$

The full-width at half maximum (FWHM) of the cavity resonance is given by the familiar formula

$$\Delta\omega = \omega_o/Q \quad \text{or} \quad \Delta\nu = \nu_o/Q \tag{42b}$$

Hence

$$\Delta\omega\tau_p = 1 \tag{42c}$$

HERMITE-GAUSSIAN BEAM MODES

Stable cavities such as that shown in Fig. 7B produce a beam whose electric field is described by

$$\begin{aligned} E_{m,n}(x,y,z) = E_o\, H_m(\sqrt{2}x/w)\, H_n(\sqrt{2}y/w) \\ \cdot \exp[-(x^2+y^2)/w^2] \\ \cdot \exp[-jk(x^2+y^2)/2R(z)] \\ \cdot \exp\{-j[kz - (1+m+n)\tan^{-1} z/z_o]\} \end{aligned} \tag{43}$$

where H_m is a Hermite polynomial and $k = \omega/c$. The mode indices, m and n, refer to the variation of the tem field in the x and y directions similar to the convention used for waveguides. The interrelationships between the parameters in Eq. 43 are listed below.

$$w^2(z) = w_o^2\,[1 + (z/z_o)^2] \tag{44a}$$

$$R(z)=z\ [1+(z_o/z)^2] \quad (44b)$$

$$z_o=\pi w_o^2/\lambda \quad (44c)$$

$$H_n(u)=(-1)^n\ (\exp u^2)\ (d^n/du^n)\exp(-u^2) \quad (44d)$$

The plane $z=0$ is defined to be the location where the spot size, w, is a minimum and the wave-front curvature becomes planar ($R=\infty$). The spot size of the beam will expand by $\sqrt{2}$ when it has propagated a distance z_o from $z=0$. The interrelationships are plotted in Fig. 9 for the $\text{tem}_{o,o}$ mode.

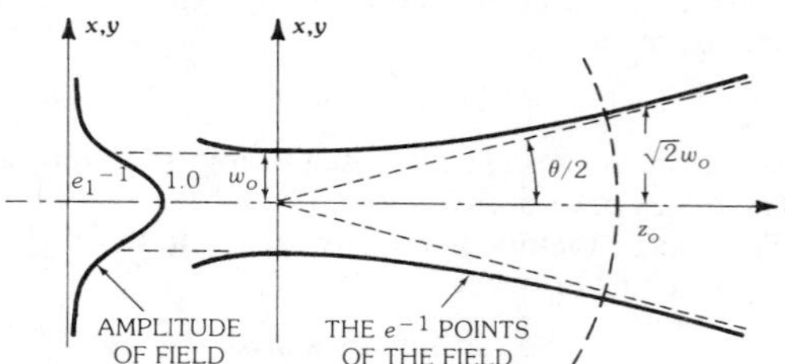

Fig. 9. The expansion of a $\text{tem}_{o,o}$ mode.

Notice that in the far field the beam expands as if it were a spherical wave of limited angular extent and originating at $z=0$. Angle θ is the angular divergence of the beam given by

$$\theta=2\lambda/\pi w_o \quad (45)$$

and is an important laser characteristic.

The characteristic parameter, z_o, and the location of the plane $z=0$ are related to the radius of curvature of the mirrors and the spacing for the cavity shown in Fig. 7B by:

$$z_o^2=(\pi w_o^2/\lambda)^2 = \frac{d(R_1-d)\ (R_2-d)\ (R_1+R_2-d)}{(R_1+R_2-2d)^2} \quad (46a)$$

$$z_1=d(R_2-d)/(R_1+R_2-2d) \quad (46b)$$

$$z_2=d(R_1-d)/(R_1+R_2-2d) \quad (46c)$$

The resonant frequencies of the $\text{tem}_{m,n}$ modes differ slightly from those predicted by Eq. 35:

$$\nu_{m,n,q}=(c'/2d)\ \{q+[(1+m+n)/\pi]\ \cos^{-1}(g_1 g_2)^{1/2}\} \quad (47)$$

COMMON LASERS

Gases, impurity doping in crystalline and amorphous solids, direct bandgap semiconductors, and dyes dissolved in various solvents are the common materials used for lasers. All have a common theoretical framework, but the details of the excitation route differ from one type to another as illustrated in Table 1.

The excitation can be either pulsed or cw, resulting in the corresponding temporal characteristic of the laser radiation. With the exception of the semiconductors, the wall-plug efficiency of most lasers is typically less than 1%; the cw CO_2 gas laser system is an exception with efficiencies of 10–20% being common. Therefore, many very high power lasers are operated in a pulsed mode so as to dissipate the waste heat associated with the excitation. Many lasers, such as N_2 at 0.337 μm, require pulsed operation since only a transient inversion can be created.

Table 2 lists some common high-power pulsed lasers; typical pulse widths and the optical energy produced are shown. Table 3 lists common cw lasers. It should be noted here that even a milliwatt (10^{-3} W) coherent laser is potentially hazardous to the eye. Consequently, wavelength-selective eyeglasses are recommended, since even the backscatter of high-power lasers from rough surfaces can exceed this level.

The output of various dyes depends on the solvent, the concentration of the dye, and the source of excitation. This is illustrated in Fig. 10, where the relative energy versus wavelength is plotted for the common sources of excitation. Flash-lamp-pumped dyes produce the highest energy per pulse, approximately 0.2 J, and the longest pulse width, approximately 0.2 μs, but typically at a rather slow repetition rate of a few pulses per second. The N_2 laser at 0.337 μm or the second and third harmonics of Nd:YAG are used to produce high power (tens of kilowatts) and short pulse

TABLE 1. COMMON EXCITATION MECHANISMS FOR VARIOUS LASERS

Material	Excitation Route
Gases	Electric discharge. Optical pumping, chemical reactions.
Impurity doping in solids	Optical pumping usually with incoherent flash lamps, halogen lamps, or semiconductor LEDs.
Semiconductor	Carrier injection.
Dyes	Coaxial flash lamps, N_2 laser, excimer, or harmonics of YAG (0.532 μm, 0.355 μm) for pulsed operation; argon or krypton ion lasers for cw operation.

TABLE 2. PULSED LASERS — HIGH POWER

λ*	Atom or Molecule	Materials	Pulse Width (ns)	Energy (mJ)	Rate (pps)	Comments
157	F_2	g	6	10	50	RGH excimers: mixtures of rare gas and halide donors; 10^5–10^6 laser pulses per gas fill. Can also be used with N_2, HF, CO_2 in a typical laser system.
193	ArF	g	14	200	50	
222	KrCl	g	9	30	50	
249	KrF	g	16	250	50	
282	XeBr	g	8	17	50	
308	XeCl	g	6	150	50	
351	XeF	g	14	400	50	
266	Nd:YAG, Glass	s	4	50	0.02–20	Fourth harmonic of 1.06 μm Nd
337	N_2	g	6	16	100	
347	Cr:Al_2O_3	s	25	100	0.1	Second harmonic of 694.3 nm (ruby)
355	Nd:YAG	s	5	100	0.1–20	Third harmonic of 1.06 μm Nd
502	HgBr	g	50	100	5–100	New, high temperature (>150°C) req.
532	Nd:YAG, Glass	s	7	200	20	Second harmonic of 1.06 μm Nd
510.6, 578.2	Cu	g	30	2.5	6 kHz	High temperature, high average power
694.3	Cr:Al_2O_3	s	20	1–10 J	0.02	*Q* Switched
850.0	GaAs	Semi	100	0.01	1 kHz	Semiconductor diode array
1.06 μm	Nd:YAG	s	15	0.1–1 J	10	Other transitions are possible.
1.06	Nd:Glass	s	20	1–20 J	0.03	Other transitions are possible.
2.64–3.01 μm	HF	g	500	300	2	Chemical lasers, discrete line spectra associated with VR bands
5–6 μm	CO	g	1000	8 J	5	Discrete ~50 line spectra associated with VR structure
9.4, 10.6 μm	CO_2	g	100	10^2 J	0.1–10^2	Discrete ~50 line spectra associated with VR structure
12–13 μm	NH_3	g	100	0.1 J	1–10	Optically pumped with CO_2
385 μm	D_2O	g	100	0.1 J	1–10	Optically pumped with CO_2
496.1 μm	CH_3F	g	100	10	1–10	Optically pumped with CO_2

*Nanometers unless noted otherwise.

TABLE 3. COMMON CW LASERS

λ (nm)	Atom or Molecule	Materials	Power	Comments
325.0	Cd^+	g	8 mW	Small compact
441.6			40mW	
290–305	Dye	l	50 mW	SHG of R6G dye
351.1, 363.8	Ar^{++}	g	2.5 W	1. Many other lines possible 2. Very intense discharge 3. Dominant lines denoted by * 4. Water cooling usually required
454.5			1.1 W	
457.9			1.35 W	
476.5			2.7 W	
488.0*	Ar^+	g	6.5W	
496.5			2.5 W	
501.7			1.5 W	
514.5*			7.5 W	
528.7			1.0 W	
405–805 nm	Dye	l	0.05–2.0 W	Optically pumped by Ar^+ or K^+ lasers (See Fig. 10 for more detail.)
413.1	Kr^{++}	g	0.5 W	1. Many other lines possible 2. Very intense discharge 3. Dominant lines denoted by * 4. Water cooling required
468.0			0.3 W	
476.2			0.40 W	
482.5			0.40 W	
520.8	Kr^+	g	0.7 W	
530.9*			1.5 W	
568.2*			1.1 W	
647.1*			3.5 W	
676.4*			0.9 W	
752.5			1.2 W	
799.0*			0.3 W	
632.8	Ne	g	1–50 mW	Common "red" laser
1152.3	Ne	g	10 mW	First gas laser
3391.3	Ne	g	10 mW	Frequency Standard
750–900	$Al_xGa_{1-x}As$	Semi	10 mW	Typical drive current = 150 mA (See Table 4.)
864–904	GaAs	Semi	1–1000 mW	λ is temperature dependent.

Continued on next page.

TABLE 3. (CONT). COMMON CW LASERS

λ (nm)	Atom or Molecule	Materials	Power	Comments
1.06 μm	Nd^{3+}(YAG)	s	1–10 W	Optically pumped (flash lamp)
1.1–1.3 μm	$In_{1-x}Ga_xP_{1-z}$	Semi	0.1–10 mW	
2.3–3.3 μm	F-center	s	1–10 mW	Color center
2.6–3 μm	HF	g	1–50 W	Chemical laser, gas flow required
3.6–4.1 μm	DF	g		
5–6.5 μm	CO	g	10–20 W	Line selectable
9.2–11.2 μm	CO_2	g	4 W	Line selectable, sealed
10.6 μm	CO_2	g	50 W–8.5 kW	Flowing gas
28, 78, 118 μm	H_2O	g	10 mW	Electrically excited
118.1 μm	CH_3OH	g	100 mW	Optically pumped (See Fig. 11.)
311, 377 μm	HCN	g	100 mW	Electrically excited
496.1 μm	CH_3F	g	25 mW	Optically pumped with CO_2

widths (5 ns) at moderate repetition rates of 10–100 pps. Continuous-wave dye lasers are usually excited with the lines from an argon or krypton ion laser. While the conversion efficiency of the dye is quite reasonable (10–25%), the electrical efficiency of the pump laser is usually less than 0.1%.

Many of the far-infrared lasers, using various gases such as CH_3OH, are pumped by the CO_2 laser. Fig. 11 illustrates the range of powers and frequencies available when such a pumping scheme is used.

SEMICONDUCTOR LASERS

Population inversion, and thus optical gain in the sense of Eq. 14, may be obtained in a direct bandgap semiconductor exhibiting radiative recombination of electrons and holes provided there is, simultaneously in space and time, a large density of electrons in the conduction band and holes in the valence band such as that provided by the junction region of a diode under heavy forward bias where large numbers of electrons and holes exist.

Such a diode is shown in Fig. 12 for a homojunction laser consisting of, for example, heavily doped regions of GaAs to form a pn junction. Under forward bias, electrons are injected into the p region, leading to an "inverted" population of width 1 μm or so perpendicular to the junction. The light (L) output is directly proportional to the recombining carriers and thus is directly related to the injected current (I). At low currents, the device acts as a LED. Beyond a threshold current, stimulated emission of recombination takes place, and the slope of light output versus current shows a dramatic increase. Due to the short recombination lifetime (typically 10^{-9} second), one is able to amplitude modulate the light output by simple current modulation.

Semiconductor lasers that involve only one material (e.g., GaAs) are classified as *homo*junctions; those that utilize a different material on one side of the active region are called single heterostructures (SH); if both sides use different materials, the device is called a double heterostructure (DH). Usually, the different materials used to obtain a heterostructure have a different bandgap and index of refraction than that of the active region. A typical configuration for a double heterostructure is shown in Fig. 13. The higher bandgap confines the carrier, and change in index of refraction guides the electromagnetic mode. Due to the small size of the active region, the divergence angle of the mode may be on the order of 10–20° and is different in the planes parallel or perpendicular to the junction.

Table 4 lists various semiconductor lasers using combinations of column III-V and IV-VI elements. Also indicated in this table are typical threshold current densities (J_{th}). Note that the current densities required for heterostructure lasers are much less than those required for homostructures.

NONLINEAR OPTICS

As in the case of other electronic devices, under suitable conditions the ordinary linear behavior of transparent materials becomes nonlinear. These nonlinearities, which are caused by strong optical or low-frequency electric fields, are used in the frequency and amplitude control of light.

The various effects are described by a simple Taylor series expansion of the microscopic polarization

$$P = \epsilon_0 \chi^{(1)} E + \epsilon_0 \chi^{(2)} EE + \epsilon_0 \chi^{(3)} EEE + \cdots \quad (48)$$

so that the normal index of refraction is given by $n^2 = 1 + \chi^{(1)}$.

If the symmetry property of the material is such that a reversal of the sign of all coordinates causes no change in the spatial arrangement of the atoms, the material is said to be centrosymmetric. Applying the reversal to P and E, one sees that the $\chi^{(2)}$ coefficient must be zero in centrosymmetric systems but that $\chi^{(1)}$ and $\chi^{(3)}$ are not. Accordingly, the lowest order or quadratic nonlinearity is seen in noncentrosymmetric systems which comprise 21 of the 32 known crystal classes.

In general, $\chi^{(1)}$ has nine components. If one chooses a natural coordinate system that reflects the symmetry axis of the material, $\chi^{(1)}$ becomes diagonal with three components only. This is referred to as the principal axis frame. In the principal axis frame $\chi^{(2)}$ has at most

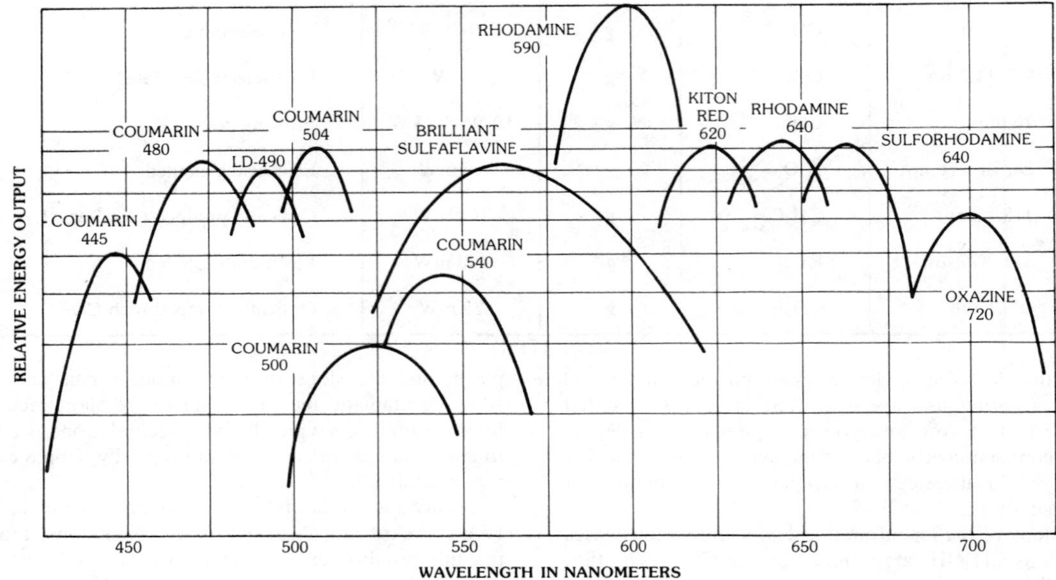

(A) Coaxial flash lamp pumped dyes (Candela Corp.).

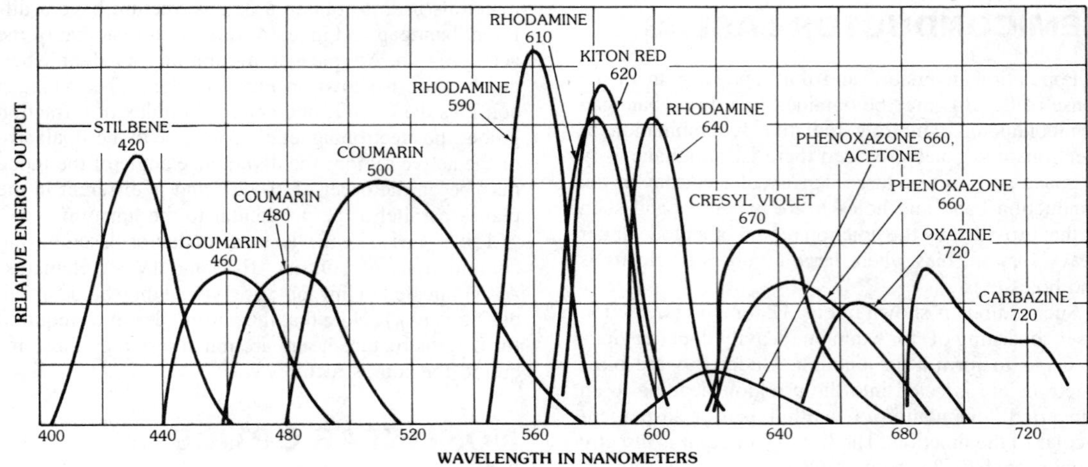

(B) Nd:YAG pumped dyes (Quanta-Ray).

Fig. 10. Typical tuning curves for various dyes when

27 components while $\chi^{(3)}$ has 3^4 components. Not all components are separately distinguishable, so it is possible to reduce the independent number down to 18 for $\chi^{(2)}$. This is done by noting that the ith component of the quadratic part of the polarization is $P_i = \sum_k \sum_l \epsilon_0 \chi^{(2)}_{ikl} E_k E_l$, so a new parameter is accordingly defined as $d_{iq} = \epsilon_0 \chi^{(2)}_{i(jk)}$ where q is a contracted index for the jk combination in parentheses (the Voight notation). Principal axes along x, y, and z are labeled 1, 2, and 3 so that q ranges from 1 to 6 according to the jk combinations 11, 22, 33, 23, or 32; 13 or 31; and 12 or 21, respectively. So, for example, $\chi^{(2)}_{x(yz)} = \chi^{(2)}_{1(23)}$ contributes to d_{14}. With this, d is a 3×6 matrix with at most 18 elements.

The symmetry of the individual system is then used to deduce which elements are nonzero. Since $\chi^{(2)}$ is the same order tensor as that used to describe the piezoelectric effect, the nonzero elements are well known and are compiled in Chapter 4. Similarly, the individual values for d_{iq} for various materials and optical frequencies are tabulated elsewhere. The same kind of symmetry-based reductions can be applied to $\chi^{(3)}$ for crystals, but none are used for amorphous materials.

A specific example is for the crystal class $\bar{4}2m$, which is appropriate to the common nonlinear material potassium dihydrogen phosphate (KDP). For this class, $\chi_1^{(1)} = \chi_2^{(1)} \neq \chi_3^{(1)}$, and

$$d = \begin{bmatrix} 0 & 0 & 0 & d_{14} & 0 & 0 \\ 0 & 0 & 0 & 0 & d_{25} & 0 \\ 0 & 0 & 0 & 0 & 0 & d_{36} \end{bmatrix} \qquad (49)$$

with $d_{14} = d_{25}$.

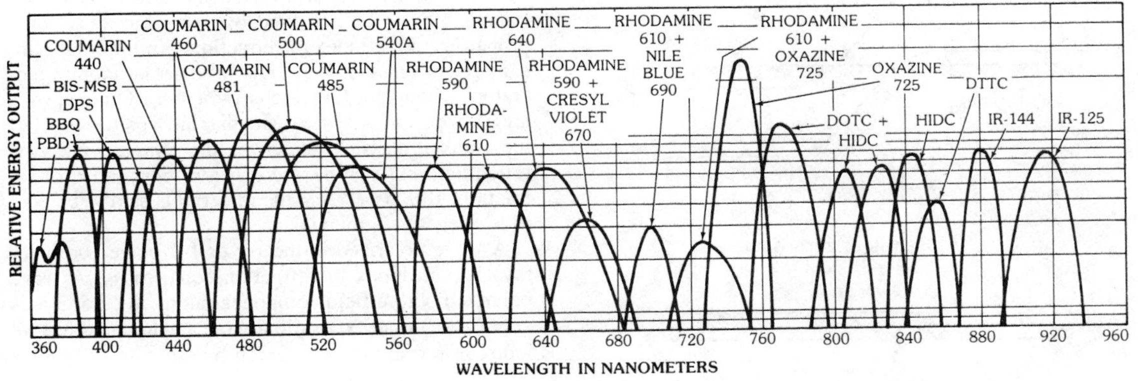

(C) N_2 pumped dyes (Molectron).

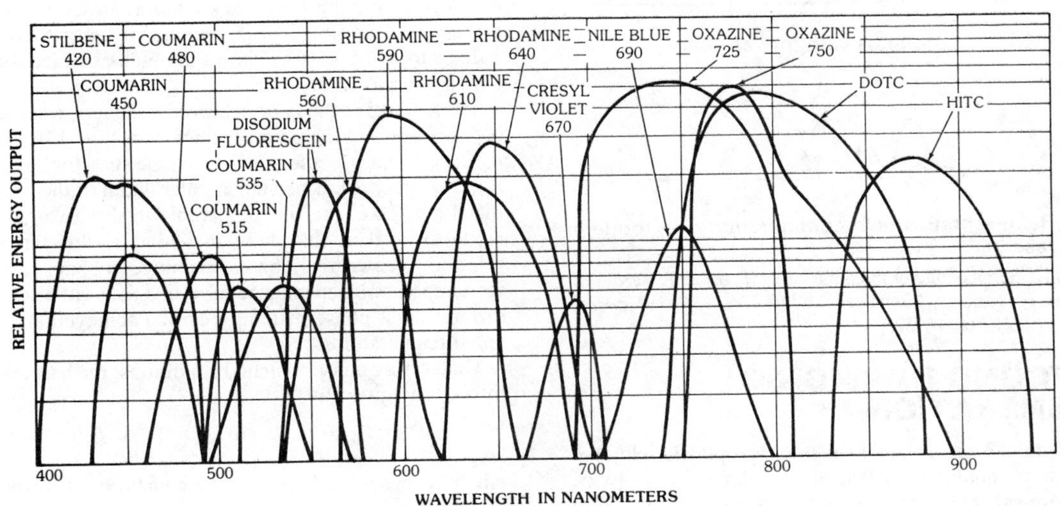

(D) Ar^+ and Kr^+ cw pumped dyes (Spectra-Physics).

excited by the indicated sources. (*Courtesy Exciton Corp.*)

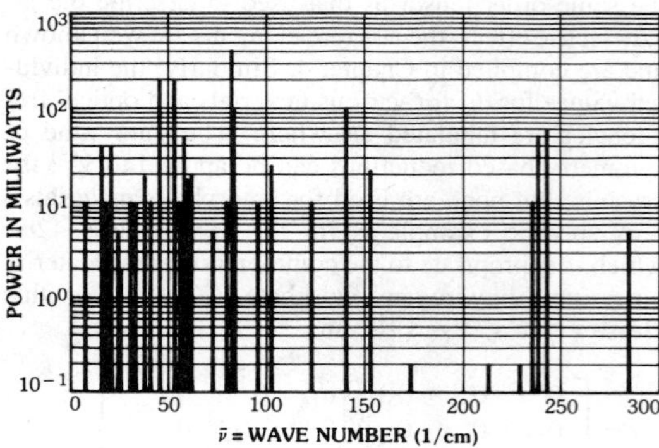

Fig. 11. Optically pumped far-infrared laser lines.

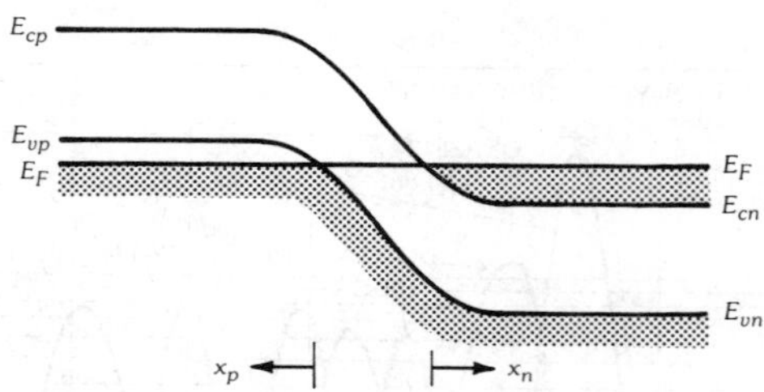

(A) Bands at zero bias.

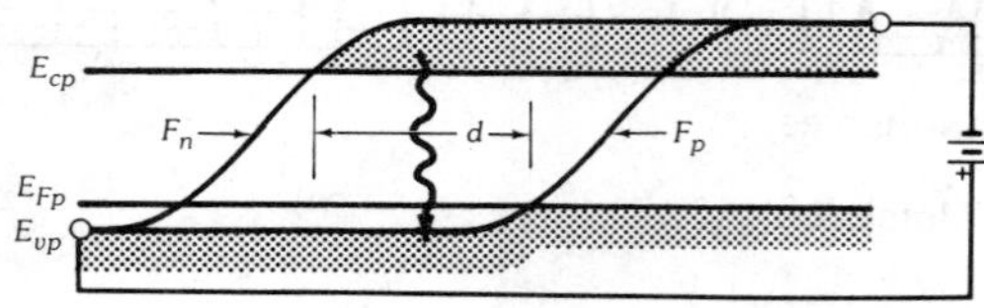

(B) Bands with forward bias.

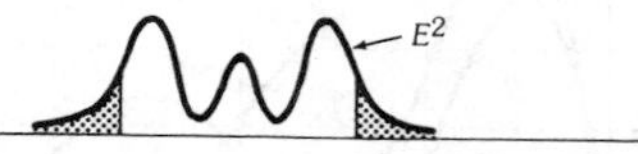

(C) Representative field configuration of a mode in the laser.

Fig. 12. Semiconductor injection laser. (*From J. T. Verdeyen,* Laser Electronics, *Englewood Cliffs, N.J.: Prentice-Hall, Inc., 1981; Fig. 10-19.*)

SECOND HARMONIC GENERATION

With the presence of an intense optical field, the quadratic nonlinearity leads to a polarization at twice the optical frequency, which may act as a driving term for a new field component at 2ω. Since the growth tends to occur on spatial scales long compared with the wavelengths, the full set of Maxwell's equations can be reduced to a first-order equation for the evolution of the field amplitude. The growth of the harmonic is governed by

$$d(E_{2\omega})/dz = A\, E_{\omega}^{2} \exp\,(i\Delta kz) \qquad (50)$$

where,

$\Delta k = k(2\omega) - 2k(\omega) = 2k_o[n(2\omega) - n(\omega)]$,

$k_o(\omega) = \omega/c$,

A = a constant proportional to the d_{iq} elements, the exact form of which depends on the field orientations and the d matrix.

If E_ω is approximately constant, $E_{2\omega}$ is seen to be harmonic in z if $\Delta k \neq 0$. If $\Delta k = 0$, then $E_{2\omega} \simeq z$, and the growth increases with distance. This latter case is clearly favorable and requires $\Delta k = 0$; it is referred to as a phase-matched condition.

Since most materials are normally dispersive, $n(2\omega) > n(\omega)$, the phase-matched condition is not automatic. It can be achieved by using the anisotropic properties of the index of refraction. An important linear optical class of materials used as harmonic generators is uniaxial, for which $n_1 = n_2 \neq n_3$. Other classes are isotropic, $n_1 = n_2 = n_3$, and biaxial, $n_1 < n_2 < n_3$. For the uniaxial class, $n_1 = n_2 = n_o$, the *ordinary* index of refraction, and $n_3 = n_e$, the *extraordinary* index of refraction. If $n_e > (<) n_o$, the material is said to be positive (negative) uniaxial.

A wave polarized in the x-y or 1-2 plane (ordinary ray) has an index of refraction equal to n_o. A wave polarized with a field component along the z or 3 axis (optic axis) is an extraordinary ray and has an effective index of refraction, n_{eff}, given by

$$1/n_{\text{eff}}^{2} = \cos^2\theta/n_o^2 + \sin^2\theta/n_2^2 \qquad (51)$$

where θ is the angle between the optic axis and the direction of propagation. A common representation of these is a polar plot of $n(\theta)$, which is referred to as an index surface. Fig. 14A shows the index surface for a negative uniaxial crystal.

Phase matching may be achieved if the index-surface curves at ω and 2ω cross as shown in Fig. 14B. For $n_e(2\omega) < n_o(\omega)$, there is an angle θ_m for which $n_{\text{eff}}(2\omega) = n_o(\omega)$. Since the angular width of the phase matching ($\Delta kL = \pi$) is comparable to the divergence angle of a laser, the phase matching of this type is referred to as critical phase matching. In some cases, the temperature dependence of n_o and n_e is sufficiently different that phase matching can be achieved at elevated temperatures.

Under fully phase-matched conditions, the harmonic intensity is given by

$$I_{2\omega}(z) = I_\omega(0)\tanh^2\,[AE_\omega(0)z] \underset{z\to\infty}{\longrightarrow} I_\omega(0) \qquad (52)$$

which shows the full conversion capability of the process, which has an initial growth of

$$I_{2\omega}(z) = A^{\,2}E_\omega^2(0)I_\omega(0)z^2 \cong \mathrm{I}^2(0)z^2 \qquad (53)$$

The term $AE_\omega(0)$ represents an inverse conversion

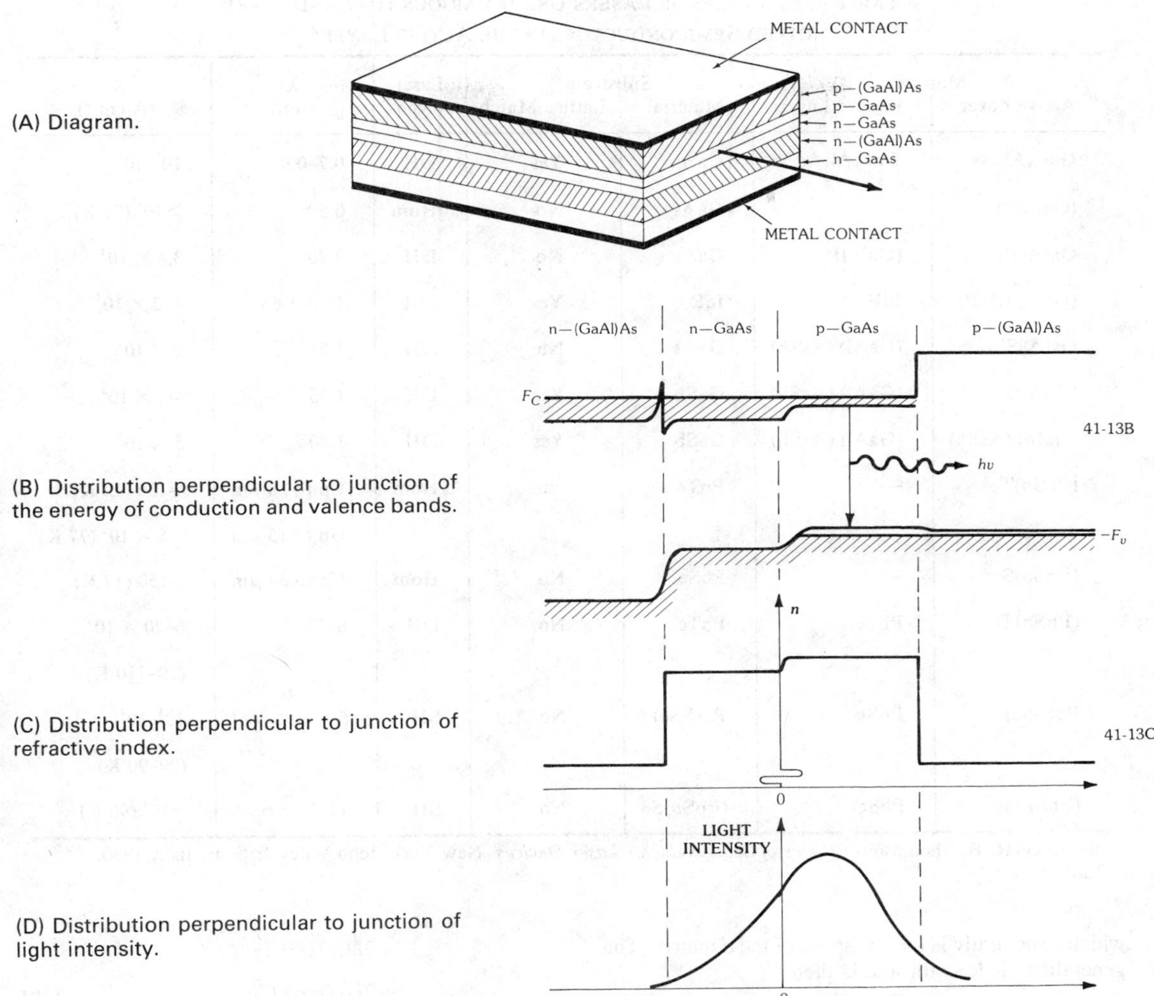

Fig. 13. Four layer separate confinement heterostructure. (*From G. H. B. Thompson,* Physics of Semiconductor La *New York: John Wiley & Sons, Inc., 1980; Figs. 3.16 and 3.17.*) 41-13D

distance, $l_{conv'}$, such that at $z = l_{conv'}$ 58% of the fundamental has been converted to a harmonic. For practically all materials with large d values, $l_{conv'}$ on the order of centimeters requires initial pumping intensities of 10–100 MW/cm^2. Although the d_{iq} values tend to be larger at lower frequencies, the equivalent A terms tend to be numerically about the same. Hence, harmonic generation by this method tends to be efficient, 50–80%, but requires large powers. Representative nonlinear materials and coefficients are listed in Table 5.

Pockels Effect

A second effect associated with $\chi^{(2)}$ is the Pockels, or linear electro-optic, effect used to amplitude modulate light. Here, the nonlinear polarization is at the optical frequency and is proportional to $E_\omega E^0$, where E^0 is a nonoptical field. The effect, which has the same basic d matrix as for the case of second-harmonic generation, leads only to a shift in the index of refraction caused by E^0. The effect is described by a three-dimensional version of the index surface called the index ellipsoid, which relates n_{eff} to the displacement vector rather than to the direction of propagation. With no applied field, the index ellipsoid is of the general form

$$1 = D_1^2/n_1^2 + D_2^2/n_2^2 + D_3^2/n_3^2 \tag{54}$$

where D_i are the normalized displacement vectors along the principal axis. The effective index of refraction is simply the length of this vector.

The effect of E^0 is to distort the index ellipsoid. The distortion is expressed in terms of a new matrix, r,

TABLE 4. EXAMPLES OF LASERS USING VARIOUS III-V AND IV-VI MIXED SEMICONDUCTORS IN THE ACTIVE LAYER*

Material		Substrate		Laser	λ	J_{th}
Active Layer	Passive Layer	Material	Lattice Match	Type	(μm)	(A cm^{-2})
$Ga_{1-x}Al_xAs$	$Ga_{1-y}Al_yAs$	GaAs	Yes	DH	0.7–0.9	10^3–10^4
(GaIn)P	—	GaAs	No	Hom	0.59	$>10^4$ (77 K)
Ga(AsP)	(GaIn)P	GaAs	No	DH	0.70	3.4×10^3
(GaIn) (AsP)	InP	InP	Yes	DH	1.15–1.65	$1\text{–}2 \times 10^3$
Ga(AsSb)	(GaAl) (AsSb)	GaAs	No	DH	1.0	2×10^3
(GaAl)Sb	(GaAl) (AsSb)	GaSb	Yes	DH	1.35	$\simeq 2 \times 10^3$
(GaIn) (AsSb)	(GaAl) (AsSb)	GaSb	Yes	DH	1.80	5×10^3
(PbSn)Te	—	PbTe	No	Hom	Up to 28 μm	≃150 (12 K)
					Up to 15 μm	$\simeq 5 \times 10^3$ (77 K)
(PbSn)Se	—	PbSe	No	Hom	Up to 34 μm	≃150 (12 K)
(PbSn)Te	PbTe	PbTe	No	DH	8–15	$6\text{–}20 \times 10^3$
						(70–110 K)
Pb(SSe)	PbSe	Pb(SSe)	No	SH	5	10^3–10^4
						(20–90 K)
(PbSn)Se	PbSe	(PbSn)Se	No	SH	11	$\simeq 10^4$ (90 K)

*From G. H. B. Thompson, *Physics of Semiconductor Laser Devices*. New York: John Wiley & Sons, Inc., 1980; Table 1.1.

which structurally is the transpose of the d matrix. The generalized index ellipsoid is then

$$1=\sum_{i=1}^{3}\left[D_i^2/n_i^2+D_i^2\sum_{j=1}^{3}\Delta B_{(ii)j}\right]$$
$$+\sum_{i=1}^{3}\sum_{j\neq i}^{3}D_iD_j\sum_{k}\Delta B_{(ij)k} \quad (55)$$

where $\Delta B_{(ij)k}=r_{(ij)k}E_k^0$, and the indices in parentheses are contracted according to the Voight notation. Continuing with the example of $\bar{4}2m$ crystals,

$$r=\begin{bmatrix}0&0&0\\0&0&0\\0&0&0\\r_{41}&0&0\\0&r_{52}&0\\0&0&r_{63}\end{bmatrix}$$

and $r_{41}=r_{52}$. The resulting index elipsoid is

$$1=D_1^2/n_o^2+D_2^2/n_o^2+D_3^2/n_e^2$$
$$+2D_2D_3r_{41}E_1^0$$
$$+2D_1D_3r_{52}E_2^0$$
$$+2D_1D_2r_{63}E_3^0 \quad (56)$$

The problem then reduces to finding a new principal axis frame for which the index elipsoid is diagonal.

Two special cases exist. For longitudinal modulators, the optical field is perpendicular to E^0, whereas for transverse modulators, it is parallel to E^0. Fig. 15A illustrates the former, for which the index ellipsoid is diagonal in the indicated primed axis for $\bar{4}2m$. The index ellipsoid is then

$$1=D_1'^2[1/n_o^2-r_{63}E_3^0]+$$
$$D_2'^2[1/n_o^2+r_{63}E_3^0]+D_3^2/n_e^2 \quad (57)$$
$$=D_1'^2/n_1'^2+D_2'^2/n_2'^2+D_3^2/n_e^2$$

For all materials, the product $r_{qk}E_k{}^0$ is small compared with the indices of refraction, so in the primed axis the new indices of refraction are approximately $n_1'\cong n_o+(1/2)\,n_o^3r_{63}E_3^0$ and $n_2'\cong n_o-(1/2)\,n_o^3r_{63}E_3^0$. A wave originally polarized along the 1 axis will have equal field components along the 1′ and 2′ axes, each

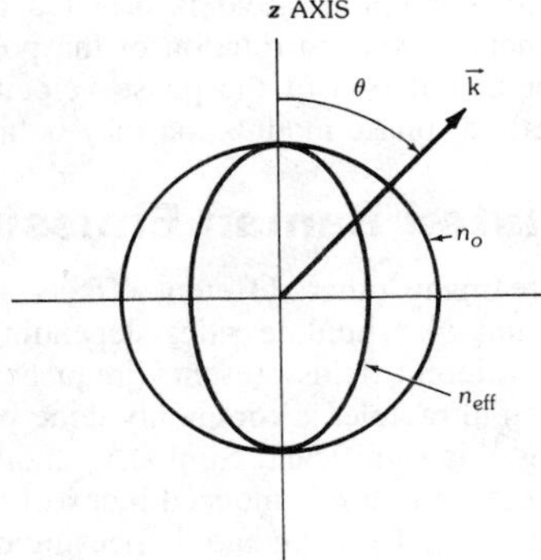

(A) Negative uniaxial crystal.

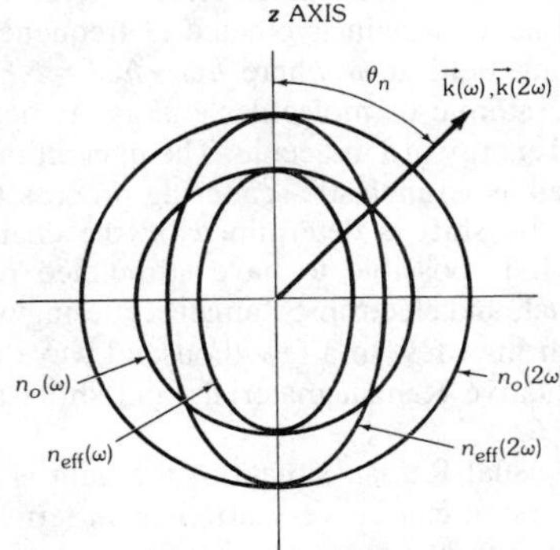

(B) Phase matching.

Fig. 14. Wave vectors in a uniaxial crystal.

of which propagates with a different phase velocity. After exiting from the crystal, the net optical polarization will be rotated from the incident polarization by an amount proportional to $n_1' - n_2' = n_o^3 r_{63} E_3^0$. Amplitude modulation is accomplished by placing a polarizer after the crystal. For this orientation, the transmission, T, through the crystal-polarizer combination is given by

$$T = \sin^2 (\Gamma/2) \quad (58)$$

where,

Γ is the retardation given by $\omega(n'_1 - n'_2)L/c$

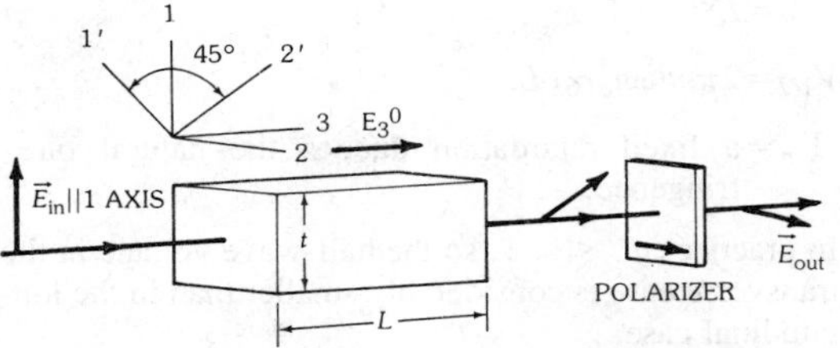

(A) Longitudinal modulator.

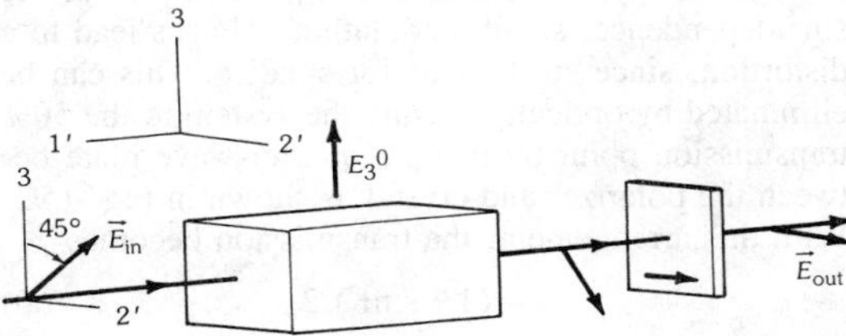

(B) Transverse modulator.

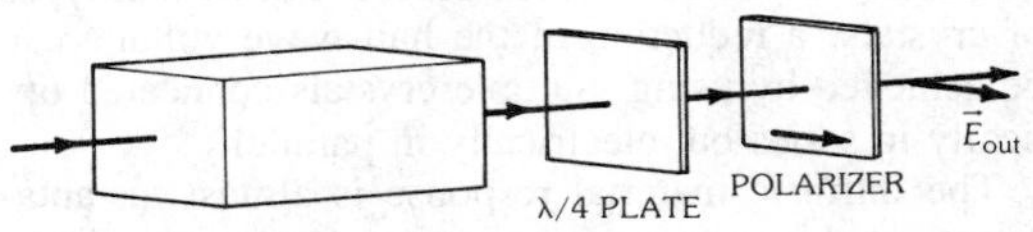

(C) Bias with λ/4 plate.

Fig. 15. Configurations for amplitude modulation of light.

$$= \omega n_o^3 r_{63} E_3^0 L/c = \pi V/V_{1/2},$$

$$V = E_3^0 L,$$

$V_{1/2} = \pi c/\omega n_o^3 r_{63}$, the half-wave voltage.

The half-wave voltage is the most important parameter in specifying electro-optic modulators.

For the transverse case shown in Fig. 15B, the retardation becomes

$$\Gamma = (\omega/c)\ (n_o - n_e - \tfrac{1}{2}\ n_o^3 r_{63} E_3^0)L \quad (59a)$$

$$= \Gamma_m - \pi V/V_{1/2} \quad (59b)$$

where,

TABLE 5. NONLINEAR OPTICAL COEFFICIENTS

Material	Class	λ (μm)	r_{qk} (10^{-12}m/V)	d_{iq}/ϵ_o (10^{-12}m/V)	Class
KDP	$\bar{4}2m$	1	$r_{63} = -9$	$d_{14} = 0.49$	Uniaxial
			$r_{41} = 8.6$	$d_{36} = 0.47$	
GaAs	$\bar{4}3m$	10	$r_{41} = 1.6$	$d_{14} = 134$	Isotropic
CdTe	$\bar{4}3m$	10	$r_{41} = 6.8$	$d_{14} = 16.7$	Isotropic

$V = E_3^0 t$,

$V_{1/2} = 2\pi ct/\omega n_o^3 r_{63} L$,

Γ_m = a fixed retardation due to the natural birefringence.

In practice, $t/L << 1$, so the half-wave voltage in the transverse case is considerably smaller than in the longitudinal case.

Half-wave voltages of 1–10 kV are not uncommon so that for high-frequency modulation, operational voltages are typically less than $V_{1/2}$. Because of the $\sin^2$ dependence, small modulating voltages lead to a distortion, since $\sin^2\theta \simeq \theta^2$ for small θ. This can be eliminated by optically biasing the system to the 50% transmission point by using a quarter-wave plate between the polarizer and crystal as shown in Fig. 15C. With this arrangement, the transmission becomes

$$T = (1 + \sin\Gamma)/2 \tag{60}$$

and results in a change in transmission linear in $V(\Gamma)$ for small modulation. Hence, the standard amplitude modulator is a three-element device. For certain types of crystals, a reduction of the half-wave voltage can be achieved by using multiple crystals connected optically in series but electrically in parallel.

The intrinsic material response is almost instantaneous, so the system response time tends to be determined simply by circuit effects. If these are made small, a second frequency limitation occurs because of the finite optical transit time through the crystal. Operation above 1 GHz generally requires a traveling-wave modulation wherein a certain type of phase matching is used.

If the incident optical field is oriented along the primed principal axis, no rotation of the polarization occurs, but modulation of the phase velocity exists. By this method, phase modulation may be achieved.

Stimulated Raman Emission

There are many other different effects associated with $\chi^{(2)}$ and $\chi^{(3)}$ nonlinearities depending on the number of different fields present. Frequency tripling into the deep ultraviolet is commonly done with gases for which $\chi^{(3)}$ is significant. Similarly, another third-order effect leads to a self-induced index of refraction change and causes focusing and defocusing of a beam by virtue of its radial intensity variation.

One important $\chi^{(3)}$ effect used to frequency shift light is the stimulated Raman effect. Here, a polarization and hence a gain is created at frequency ω' due to a strong field at ω where $\hbar\omega - \hbar\omega' = \Delta E$, a characteristic atomic or molecular energy, typically a vibrational energy in a molecule. The interaction can also be viewed as an inelastic scattering process (Fig. 16) wherein the shift is determined by the characteristic energy. It is possible to have stimulated rotational, vibrational, and electronic Raman scattering with shifts ranging from a few to a few thousand wave numbers. Representative Raman materials and shifts are listed in Table 6.

In the usual Raman situation, the gain is large, so the conversion can be very efficient in terms of photons. In terms of power, the conversion efficiency is limited by the Manley-Rowe limit of $\hbar\omega'/\hbar\omega$. Typically, the requirements for pump intensities lie in the range 10–100 kW/cm^3.

TABLE 6. RAMAN MATERIALS AND SHIFTS

Substance	States	Stokes Shift (cm^{-1})
H_2	Vibrational	4155
N_2	Vibrational	2331
D_2	Vibrational	2991
Carbon Tetrachloride	Vibrational	460
Carbon Disulfide	Vibrational	656
Benzene	Vibrational	992
Nitrobenzene	Vibrational	1344
Acetone	Vibrational	2921
Water	Vibrational	3651
H_2	Rotational	365
Lead	Electronic	10600

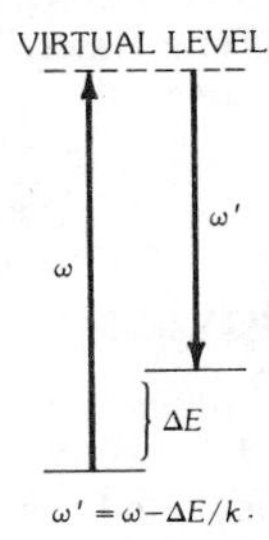

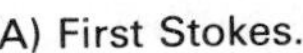

(A) First Stokes.

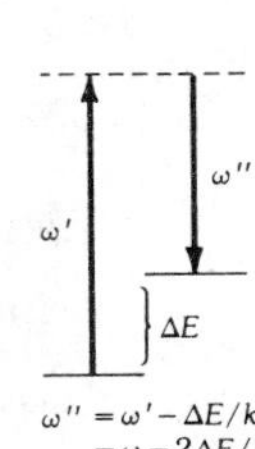

(B) Second Stokes.

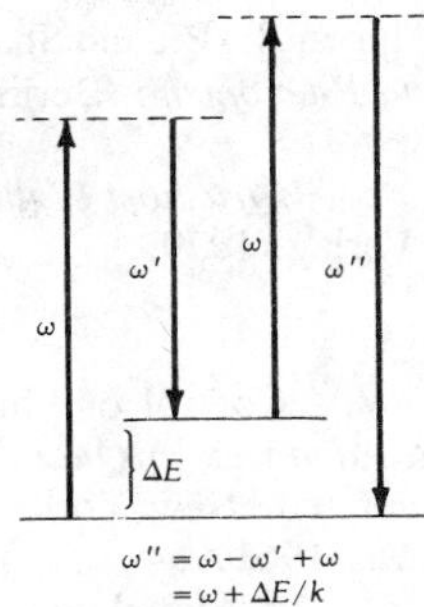

(C) First anti-Stokes.

Fig. 16. Raman effect.

The normal Raman effect occurs when $\omega' < \omega$ and is called a Stokes wave. The growing first Stokes can reach a sufficient intensity to act as a pump for a second Stokes, and so on. With the presence of a strong pump and first Stokes, an anti-Stokes wave may be generated at $\omega'' = \omega + \Delta E/\hbar$. Hence, under normal conditions a number of Stokes and anti-Stokes waves may be created as shown in Fig. 16. A Raman cell is usually constructed locally because of its simplicity.

REFERENCES

General Laser Theory

Elementary Laser Theory

Siegman, A. *An Introduction to Lasers and Masers*. New York: McGraw-Hill Book Co., 1976.

Verdeyen, J. T. *Laser Electronics*. Englewood Cliffs, N.J.: Prentice-Hall, Inc.,1981.

Yariv, A. *Optical Electronics*. 2nd ed. New York: Holt, Rinehart & Winston, Inc., 1976.

Advanced Laser Theory

Pantell, R., and Puthoff, H. *Fundamentals of Quantum Electronics*. New York: John Wiley & Sons, Inc., 1969.

Sargent, M. III, Scully, M. O., and Lamb, W. E., Jr. *Laser Physics*. Reading, Mass.: Addison-Wesley Publishing Co., Inc., 1974.

Yariv, A. *Quantum Electronics*. 2nd ed. New York: John Wiley & Sons, Inc., 1975.

Specific Lasers

Solid State and Gas Lasers

Brown, D. C., ed. *High-Peak-Power ND: Glass Laser Systems*. Springer Series in Optical Sciences. Berlin: Springer-Verlag, 1979.

Pressley, R. J., ed. *Handbook of Lasers With Selected Data on Optical Technology*. Cleveland, O.: Chemical Rubber Co., 1971.

Weber, M. J., ed. *Handbook of Laser Science and Technology*. Vol. I, *Lasers and Masers*; Vol. II, *Gas Lasers*. Cleveland, O.: Chemical Rubber Co., 1982.

Willett, C. S. *Introduction to Gas Lasers*. New York: Pergamon Press, Inc., 1974.

Semiconductor Lasers

Kressel, H., ed. *Semiconductor Devices for Optical Communications*. Vol. 39 in Topics of Applied Physics. Berlin: Springer-Verlag, 1980.

Kressel, H., and Butler, J. K. *Semiconductor Lasers and Heterojunction LED's*. New York: Academic Press, Inc., 1977.

Thompson, G. H. B. *Physics of Semiconductor Laser Devices*. New York: John Wiley & Sons, Inc., 1980.

Dye Lasers

Hansch, T. W. "Repetitively Pulsed Tunable Dye Laser for High Resolution Spectroscopy." *Applied Optics*, 4, 1972, pp. 895–898.

Wallenstein, R. "Pulse Dye Lasers," in *Laser Handbook*, Vol. 3. M. L. Stitch, ed. New York: North-Holland Publishing Co., 1979.

Pulse Behavior

Modelocking

Haus, H. A. "Theory of Mode Locking With a Slow Saturable Absorber." *IEEE J. of Quant. Electr.*, QE-11, 1975, pp. 736–746.

IEEE J. of Quant. Electr., QE-19, 1983. Special issue on Picosecond Phenomena and references cited therein.

Kuzenga, D. J., and Siegman, A. E. "FM and AM Mode Locking of the Homogeneous Laser." Part I, Theory. *IEEE J. of Quant. Electr.*, QE-6, 1970, p. 694.

Laubereau, A., and Eisenthal, K. B., eds. *Picosecond Phenomena III*. Berlin: Springer-Verlag, 1982.

Shank, C. V., Ippen, E. P., and Shapiro, S. L., eds. *Picosecond Phenomena*. Berlin: Springer-Verlag, 1978.

Shapiro, S. L., ed. *Ultrashort Light Pulses*. Berlin: Springer-Verlag, 1979.

Q-Switching

Hellwarth, R. W. "Control of Fluorescent Pulsations," in *Advances in Quantum Electronics*, J. R. Singer, ed. New York: Columbia University Press, 1961.

Hellwarth, R. W. "A Modulation of Lasers," in *Lasers* I, A. K. Levine, ed. New York: Marcel Dekker Inc., 1966.

Wagner, W. G., and Lengyel, B. A. "Evolution of a Giant Pulse in a Laser." *J. of Appl. Phys.*, 34, 1963, p. 2044.

Nonlinear Optics

Bloembergen, N. *Nonlinear Optics*. New York: W. A. Benjamin Inc., 1965.

Hellwege, K. H. and Hellwege, A. M., eds. *Landolt-Bornstein New Series Group III Vol. 2*. Berlin: Springer-Verlag, 1979. (See also A. Yariv, *Quantum Electronics*, Chapters 14–19.)

42 Computer Organization and Programming

John Wakerly

INTRODUCTION

Computers come in all shapes and sizes, from large corporate data-processing machines that fill a room to tiny microcontrollers buried deep in automobiles, stereo systems, and toasters. Even though computers span a wide range of sizes, capabilities, and cost, they share a great many characteristics and operating principles. The goal of this chapter is to describe general principles of computer organization and programming that apply to computers of any category.

Taken as a whole, a computer is an incredibly complex system, with many more levels of detail than, say, the noncomputer part of a radio. Fortunately, a computer system *is* structured into many levels, so that it is easy to understand if taken one level at a time; see Fig. 1.

USERS			
APPLICATION PROGRAMS		COMMAND INTERPRETERS	SOFTWARE
HIGH-LEVEL LANGUAGES	OPERATING SYSTEM	FILE SYSTEM	
ASSEMBLY LANGUAGE			
MACHINE LANGUAGE			
ARCHITECTURE			HARDWARE
IMPLEMENTATION			
HARDWARE REALIZATION			

Fig. 1. Levels of detail of a computer.

Hardware

Central to the organization, or indeed existence, of a computer is its *architecture*. The architecture of a computer is really an abstract concept—it is just a definition of the functionality of the computer as seen by a machine-language programmer. A user of a particular computer architecture does not care how its functions are achieved, only that they match their definitions. An example of a particular computer architecture, the Motorola 68000, is given later in the chapter.

At the next level down is *implementation*, the logic structures used to accomplish the functions defined by the architecture. An implementation consists of a block diagram of the system hardware modules, their interconnections, and further refinements of the internal logic structure of each module. Note that implementation is still a somewhat abstract concept in that an implementation can exist on paper without the machine actually having been built. Chapter 43 discusses logic design—the tools and techniques for designing hardware modules.

The lowest level of detail in a computer is its *hardware realization*—the electronic circuits and other components from which the computer is built. Realization encompasses integrated-circuit technologies, such as those discussed in Chapter 20, as well as packaging, interconnections, and all other physical details of the machine.

The bottom three levels of a computer just described are often grouped together and simply called "the hardware." The higher levels in Fig. 1 are usually grouped together as "software."

Software

Computer *software* consists of the instructions and data that the computer hardware manipulates to perform useful work. A sequence of instructions for a computer is called a *program*. The data manipulated by a program is called a *data base*, a *file*, *input*, or simply *data*, depending on its nature and extent.

The most primitive instructions that can be given to a computer are those interpreted directly by hardware, in the *machine language* of the computer. Machine-language instructions are encoded as strings of bits in the computer memory, often one instruction per memory location. The processor fetches machine instructions from memory and executes them one by one.

Since it is difficult for humans to read and recognize strings of bits, machine-language programs are written in *assembly language* and translated into bit strings by an *assembler*. Assembly language represents machine instructions by mnemonic names and allows memory addresses and other constants to be represented by symbols rather than bit strings.

Most programs are written in *high-level languages* that allow common operations such as expression evaluation, repetition, assignment, and conditional action to be invoked in a single high-level *statement*. Popular high-level languages include BASIC, FORTRAN, and Pascal.

Few computers execute a high-level language directly. Therefore, a *compiler* is needed to translate a high-level-language program into a sequence of machine instructions that performs the desired task.

Assemblers and compilers are not the only *software tools* that a programmer may encounter. Other useful tools related to program development are interpreters, simulators, and on-line debuggers. Like a compiler, an *interpreter* processes a high-level-language program. Unlike a compiler, an interpreter actually exe-

---

The material in this chapter is adapted, with permission, from portions of Chapters 1, 4, 5, 10, and 14 of *Microcomputer Architecture and Programming* by John F. Wakerly, ©1981 by John Wiley & Sons, Inc., New York; and from the 68000 edition of the same book, to be published in 1986.

cutes the high-level-language program one statement at a time, rather than translating each statement into a sequence of machine instructions to be run later. Most BASIC environments use an interpreter.

A *simulator* is a program that simulates individual machine instructions, usually on a machine other than the one being simulated. A typical use of a simulator is to test programs to be run on a processor before the processor hardware is available. An *on-line debugger* executes a program on a machine one or a few instructions at a time, allowing the programmer to see the effects of small pieces of the program and thereby isolate programming errors (*bugs*).

Text editors are used to enter and edit text in a general-purpose computer, whether the text is a letter, a report, or a computer program. *Text formatters* read text with imbedded formatting commands and produce formatted documents such as this book. Text editors and formatters belong to the area of computing known as *word processing*.

In a medium to large computer system, cooperating programs run under the control of an *operating system*. An operating system schedules programs for execution, controls the use of i/o devices, and provides utility functions for all of the programs that run on the computer. Programs and text stored on disks and other mass-storage devices are managed by a *file system*, a collection of programs for reading, writing, and structuring such information in "files." The operating systems in most computers include file systems. Even a very small computer with no mass-storage or file system has a simple operating system, at least to monitor inputs and accept commands from the outside world.

Computer Data Types

The basic unit of information storage in a digital computer is the *bit*, which has a value of either 0 or 1. Obviously a single-bit data type is not very useful for numeric computation, since it only allows us to count from 0 to 1.

By assembling two or more bits into a string, we can represent more than two values or conditions. The bits in a string of n bits can take on 2^n different combinations of values.

Strings of eight bits are usually referred to as *bytes*. The name "byte" was invented at IBM in the early days of electronic computers. The name for a string of four bits is fancifully derived from the byte—the *nibble*.

A bit string manipulated by a computer in one operation is usually called a *word*. Some computers have a *word length* as short as four bits, others as long as 64 bits or more. For many minicomputers and microcomputers, the nomenclature has been standardized to use "word" to describe 16-bit strings and "double word" or "long word" to describe 32-bit strings.

Bits, nibbles, bytes, and words are easy data types to classify because they require differing amounts of storage in the computer memory. There are other data types that are classified not by how much storage they take, but by how they are interpreted and used by the computer hardware and software. For example, the following three data types, all using an eight-bit byte, might be defined for a microcomputer:

Unsigned integer: The byte represents an unsigned integer between 0 and 255.

Signed integer: The byte represents a signed integer between −128 and +127.

Character: The byte represents an ASCII character.

Even though all three data types are stored exactly the same way in the computer memory, they may be manipulated differently by the hardware and software.

Classification of Computers

Computers are classified as supercomputers, maxicomputers, midicomputers, minicomputers, and microcomputers according to system size:

Supercomputer: A very high-performance computer for large scientific "number-crunching" applications, costing millions of dollars. Examples: CRAY-II, ILLIAC IV.

Maxicomputer: A large, high-performance, general-purpose computer system costing over a million dollars. Examples: CYBER 76, IBM 3033.

Midicomputer: A general-purpose computer lying between minis and maxis in performance and price. Example: VAX-11/780.

Minicomputer: A general-purpose computer, often tailored for a specific, dedicated application, costing between $20 000 and $200 000. Examples: HP 1000, PDP-11/70.

Microcomputer: A computer whose CPU is a microprocessor, usually configured for a specific, dedicated application and costing well under $20 000. Examples: Macintosh, TRS-80, PDP-11/23.

There are four other relevant terms that use the prefix "micro-":

Microprocessor: A complete processor (CPU) contained in one or a few LSI circuits, used to build a microcomputer in the context above. Examples: MC68000, i80286, MC6809.

Microcomputer: In another context, a processor, memory, and i/o system contained in one LSI circuit, often referred to as a *single-chip microcomputer* for clarity. Examples: Z8, MCS-51, MC6801.

Microcontroller: A processor, memory, and i/o system contained in one or a few LSI circuits, tailored to an application that hides the general-purpose data-processing capabilities of the computer from the users. Example: TMS1000.

Microprogrammed processor: A particular type of processor hardware that executes each machine instruction as a sequence of primitive *microinstructions*. "Microprogramming" refers to writing sequences of microinstructions, *not* the programming of microprocessors. Most programmers never see a microprogram. Examples of microprogrammed processors: MC68000, LSI-11/2, iAPX432.

The definitions of computer classes are vague for a number of reasons. Advances in technology tend to blur classifications based on hardware design or performance. For example, many contemporary minicomputers outperform the maxicomputers of 10 years ago.

NUMBER SYSTEMS AND ARITHMETIC

Decimal and Binary Positional Number Systems

Positional number systems are used in all computers and almost all day-to-day business of people. In a *positional number system*, a number is represented by a string of digits in which each digit position has an associated weight. For example, the value D of a 4-digit decimal number, $d_3d_2d_1d_0$, is

$$D = d_3 \cdot 10^3 + d_2 \cdot 10^2 + d_1 \cdot 10^1 + d_0 \cdot 10^0$$

Each digit d_i has a weight of 10^i. Thus, the value of 6851 is computed as follows:

$$6851 = 6 \cdot 1000 + 8 \cdot 100 + 5 \cdot 10 + 1 \cdot 1$$

A decimal point is used to allow negative as well as positive powers of 10 in a decimal number representation. Thus, $d_1d_0 \,.\, d_{-1}d_{-2}$ has the value

$$D = d_1 \cdot 10^1 + d_0 \cdot 10^0 + d_{-1} \cdot 10^{-1} + d_{-2} \cdot 10^{-2}$$

In a general positional number system, each digit position has an associated weight of b^i, where b is called the *base*, or *radix*, of the number system. The general form of a number in such a system is

$$d_{p-1}d_{p-2} \cdots d_1d_0 \,.\, d_{-1}d_{-2} \cdots d_{-n}$$

where there are p digits to the left of the point and n digits to the right of the point, called the *radix point*. The value of the number is

$$D = \sum_{p-1 \geq i \geq -n} d_i \cdot b^i$$

the summation of each digit times the corresponding power of the radix. If the radix point is missing, it is assumed to be to the right of the rightmost digit. Except for possible leading and trailing zeros, the representation of a number in a positional number system is unique. (Obviously, 34.85 equals 034.85000, and so on.)

The *binary radix* is used in almost all computers. The allowable digits, 0 and 1, are called *bits*, and each bit d_i has weight 2^i. In the examples below, subscripts distinguish between radix-2 and radix-10 numbers.

$$10001_2 = 1 \cdot 16 + 0 \cdot 8 + 0 \cdot 4 + 0 \cdot 2 + 1 \cdot 1 = 17_{10}$$
$$110.011_2 = 1 \cdot 4 + 1 \cdot 2 + 0 \cdot 1 + 0 \cdot 0.5 + 1 \cdot 0.25 + 1 \cdot 0.125 = 6.375_{10}$$

The leftmost bit of a binary number is called the *high-order* or *most significant bit (msb)*; the rightmost is the *low-order* or *least significant bit (lsb)*.

Octal and Hexadecimal Numbers

Most computer software uses either the *octal number system* (radix 8) or the *hexadecimal number system* (radix 16) to represent binary numbers. Table 1 shows the binary numbers from 0 through 1111 and their octal, decimal, and hexadecimal equivalents. The octal system uses eight digits, 0–7. The hexadecimal system requires 16 digits, so the letters A–F are used in addition to the digits 0–9 of the decimal system.

The octal and hexadecimal number systems are useful for representing binary numbers because their radices are powers of two. Since a string of three bits can take on eight different combinations, it follows that each 3-bit string is uniquely represented by one octal

TABLE 1. BINARY, DECIMAL, OCTAL, AND HEXADECIMAL NUMBERS

Binary	Decimal	Octal	3-Bit String	Hexadecimal	4-Bit String
0	0	0	000	0	0000
1	1	1	001	1	0001
10	2	2	010	2	0010
11	3	3	011	3	0011
100	4	4	100	4	0100
101	5	5	101	5	0101
110	6	6	110	6	0110
111	7	7	111	7	0111
1000	8	10	—	8	1000
1001	9	11	—	9	1001
1010	10	12	—	A	1010
1011	11	13	—	B	1011
1100	12	14	—	C	1100
1101	13	15	—	D	1101
1110	14	16	—	E	1110
1111	15	17	—	F	1111

digit, according to the third and fourth columns of Table 1. Likewise, a 4-bit string is represented by one hexadecimal digit according to the fifth and sixth columns of the table.

Thus, it is very easy to convert a binary integer to octal (or hexadecimal). Starting at the binary point and working left, we simply separate the bits into groups of three (or four) and replace each group with the corresponding octal (or hexadecimal) digit:

$$101011000110_2 = 101\ 011\ 000\ 110_2 = 5306_8$$
$$= 1010\ 1100\ 0110_2 = AC6_{16}$$
$$11011001110101001_2 = 011\ 011\ 001\ 110\ 101\ 001_2$$
$$= 331651_8$$
$$= 0001\ 1011\ 0011\ 1010\ 1001_2$$
$$= 1B3A9_{16}$$

Conversion from octal or hexadecimal to binary is easy, too. We simply replace each octal or hexadecimal digit with the corresponding 3- or 4-bit string. And to convert from octal to hexadecimal or vice versa, we first convert to binary:

$$1573_8 = 001\ 101\ 111\ 011_2$$
$$= 0011\ 0111\ 1011_2$$
$$= 37B_{16}$$

General Positional Number System Conversions

In the more general case, conversion between two bases cannot be done by simple substitutions; base-10 arithmetic operations are generally required.

The base-10 value of a number in any base can be found from the expansion formula given earlier:

$$D = \sum_{p-1 \geq i \geq -n} d_i \cdot b^i$$

We convert each digit d_i of the number to its base-10 equivalent and expand the formula using base-10 arithmetic. For example:

$$1BE8_{16} = 1 \cdot 16^3 + 11 \cdot 16^2 + 14 \cdot 16^1 + 8 \cdot 16^0$$
$$= 7144_{10}$$
$$3176_8 = 3 \cdot 8^3 + 1 \cdot 8^2 + 7 \cdot 8^1 + 6 \cdot 8^0$$
$$= 1662_{10}$$

In the opposite direction, to convert a decimal number D to base b, we perform a series of divisions by b. In the first step, we divide D by b; the remainder is used as the least significant digit (lsd) of the desired result, and the quotient is used as the input to the next step. At each succeeding step, we divide the input number by b, use the remainder as the next-higher-order digit of the result, and use the quotient as the input to the next step. Since a remainder is always less than b, it is always representable as a single base-b digit. The process eventually terminates when a zero quotient is produced. Examples are given below.

$89 \div 2 = 44$ remainder 1 (lsb)

$\div 2 = 22$ remainder 0

$\div 2 = 11$ remainder 0

$\div 2 = 5$ remainder 1

$\div 2 = 2$ remainder 1

$\div 2 = 1$ remainder 0

$\div 2 = 0$ remainder 1 (msb)

$89_{10} = 1011001_2$

$3417 \div 16 = 213$ remainder 9 (lsd)

$\div 16 = 13$ remainder 5

$\div 16 = 0$ remainder 13 (msd)

$3417_{10} = D59_{16}$

Addition and Subtraction of Nondecimal Numbers

Addition and subtraction of nondecimal numbers by hand uses the same technique that we learned in grammar school for decimal numbers; the only catch is that the addition and subtraction tables are different. Table 2 gives the addition and subtraction tables for binary

TABLE 2. BINARY ADDITION AND SUBTRACTION TABLE

c_{in}	0	0	0	0	1	1	1	1
x	0	0	1	1	0	0	1	1
y	0	1	0	1	0	1	0	1
$x+y+c_{in}$	0	1	1	0	1	0	0	1
c_{out}	0	0	0	1	0	1	1	1
b_{in}	0	0	0	0	1	1	1	1
x	0	0	1	1	0	0	1	1
y	0	1	0	1	0	1	0	1
$x-y-b_{in}$	0	1	1	0	1	0	0	1
b_{out}	0	1	0	0	1	1	0	1

numbers. To add two binary numbers X and Y, we add together the least significant bits with an initial carry (c_{in}) of 0, producing sum ($x+y+c_{in}$) and carry (c_{out}) bits according to the table. We continue processing bits from right to left, including the carry out of each column in the sum for the next column. Two examples of decimal additions and the corresponding binary additions are shown below, with the carries shown as a bit string, C.

C		101111000
X	190	10111110
Y	+141	+10001101
$X+Y$	331	101001011

C		001011000
X	173	10101101
Y	+ 44	+00101100
$X+Y$	217	11011001

Subtraction is performed similarly, except with borrows (b_{in} and b_{out}) instead of carries:

B		001111100
X	229	11100101
Y	− 46	−00101110
$X-Y$	183	10110111

Addition and subtraction tables can be developed for octal and hexadecimal numbers, or any other desired base. However, rather than memorize tables, many computer engineers use hand-held "hex calculators" to perform arithmetic on binary, octal, and hexadecimal numbers.

Representation of Negative Numbers

So far, we have dealt only with positive numbers. There are many ways to represent negative numbers.

The representation of decimal numbers used in everyday business is called the *signed-magnitude system*. In this system, a number consists of a magnitude and a symbol indicating whether the magnitude is positive or negative. The signed-magnitude system can be applied to binary numbers quite easily by using an extra bit position to represent the sign. Traditionally, the most significant bit (msb) is used to represent the sign (0 = plus, 1 = minus), and the lower-order bits contain the magnitude. In this system, there are two representations of zero, "plus zero" and "minus zero."

To add two signed-magnitude numbers, we follow the rules that we learned in grammar school. If the signs are the same, we add the magnitudes and give the result the same sign. If the signs are different, we subtract the smaller magnitude from the larger and give the result the sign of the larger. To subtract signed-magnitude numbers, we change the sign of the subtrahend and proceed as in addition.

Despite the conceptual simplicity of the signed-magnitude system, for performance and efficiency reasons negative numbers in a computer are usually represented in *complement number systems*. In such a system, a number is negated not by inverting a sign bit, but rather by "taking the complement."

Complement number systems can be defined for any base or radix, but we shall concentrate on the base-2 radix complement, commonly called the *twos-complement* system. In this system, we assume that a binary number, B, has n bits,

$$B = b_{n-1}b_{n-2} \cdots b_1b_0.$$

so that the radix point is on the right and the number is an integer. If any operation produces a result that requires more than n bits, we throw away the extra high-order bit(s).

By definition, the twos complement of an n-bit number, B, is obtained by subtracting B from 2^n. If B is between 1 and 2^n-1, this subtraction will result in another number between 1 and 2^n-1. If B is 0, the result of the subtraction is b^n; this has the form $100 \cdots 00$, where there are a total of $n+1$ bits. We throw away the extra high-order bit and get the result $00 \cdots 00 = 0$. Thus there is only one representation of zero in the twos-complement system.

From the foregoing, it might seem that a subtraction operation is needed to form the twos complement of a number B. However, this subtraction is avoided by rewriting 2^n as $(2^n-1)+1$ and $b^n - B$ as $((2^n-1)-B)+1$. Since 2^n-1 has the form $11 \cdots 11$, any n-bit number B may be subtracted from it by simply complementing each bit of B. Thus, the twos complement of B is obtained by complementing the individual bits of B and adding 1. Some examples of 8-bit numbers and their twos complements are shown below.

$17_{10} =$	00010001_2	
	11101110	(complement bits)
	+1	
	$11101111_2 = -17_{10}$	
$119_{10} =$	01110111_2	
	10001000	(complement bits)
	+1	
	$10001001_2 = -119_{10}$	
$0_{10} =$	00000000_2	
	11111111	(complement bits)
	+1	
	$1\vert\ 00000000_2 = 0_{10}$	
$-99_{10} =$	10011101_2	
	01100010	(complement bits)
	+1	
	$01100011_2 = 99_{10}$	

The decimal value of a twos-complement number can be obtained from the same expansion formula that we showed for unsigned numbers, except that the weight of the msb is -2^{n-1} instead of $+2^{n-1}$ (for example, $-119 = -128+8+1$ above).

Some computers use the *ones-complement* system, in which the complement of an n-bit number, B, is defined to be $2^n - 1 - B$. Thus, the ones complement of a number is obtained by simply inverting its bits. Although taking the complement is faster, addition may be slower than in the twos-complement system, requiring an "end-around carry" operation. Also, there are two representations of zero, 00 · · · 00 and 11 · · · 11.

Twos-Complement Addition and Subtraction

Twos-complement numbers, regardless of sign, are added by ordinary binary addition, ignoring any carries beyond the msb. The result is always the correct sum as long as the range of the number system is not exceeded. Some examples of decimal addition and the corresponding 4-bit twos-complement additions illustrate this:

Decimal	Binary	Decimal	Binary
+3	0011	−2	1110
+ +4	+0100	+ −6	+1010
+7	0111	−8	1|1000
+6	0110	+4	0100
+ −3	+1101	+ −7	+1001
+3	1|0011	−3	1101

If an addition operation produces a result that exceeds the range of the number system, *overflow* is said to occur. Addition of two numbers with different signs can never produce overflow, but addition of two numbers of like sign can, as shown by the following examples:

Decimal	Binary	Decimal	Binary
−3	1101	+5	0101
+ −6	+1010	+ +6	0110
−9	1|0111 = +7	+11	1011 = −5
−8	1000	+7	0111
+ −8	+1000	+ +7	+0111
−16	1|0000 = +0	+14	1110 = −2

Fortunately, there is a simple rule for detecting overflow in addition: an addition overflows if the signs of the addends are the same and the sign of the sum is different. Most computers have built-in hardware for detecting overflow.

Two numbers may be subtracted by negating the subtrahend and adding the result to the minuend. A twos-complement number is negated by taking its twos complement. Thus, negating the subtrahend and adding the minuend can be accomplished with only one addition operation as follows: Perform a bit-by-bit complement of the subtrahend, and add the complemented subtrahend to the minuend with an initial carry of 1 instead of 0. Examples are the following:

Decimal	Binary	Complemented
		1—initial carry
+4	0100	0100
− +3	−0011	+1100
+1		1|0001
		1—initial carry
+3	0011	0011
− +4	− 0100	+1011
−1		1111
		1—initial carry
+3	0011	0011
− −4	−1100	+0011
+7		0111

Overflow in subtraction can be detected by examining the signs of the minuend and the *complemented* subtrahend, using the same rule as in addition.

Since twos-complement numbers are added and subtracted by the same basic binary addition and subtraction algorithms as unsigned numbers, most computers have only one type of addition or subtraction instruction that handles both signed and unsigned numbers. However, a program must interpret the results of such an instruction differently, depending on whether it thinks it is dealing with signed numbers (e.g., −8 through +7) or unsigned numbers (e.g., 0 through 15).

Thus, most computers have an *overflow* flag to indicate that the range of signed numbers has been exceeded, and a *carry* or *borrow* flag to indicate that the range of unsigned numbers has been exceeded. Addition and subtraction instructions affect both flags, but a program looks only at the flag for the number system it is using. Also, a program interprets the msb of a number as a sign bit only when dealing with signed numbers.

Binary Multiplication

In grammar school, we learned to multiply by adding a list of shifted multiplicands computed according to the digits of the multiplier. The same technique can be used to obtain the product of two unsigned binary numbers as shown below:

```
   11
  ×13
  ---
   33
  11
  ---
  143

       1011   multiplicand
      ×1101   multiplier
      -----
       1011
      0000    shifted multiplicands
     1011
    1011
   --------
   10001111   product
```

Forming the shifted multiplicands is trivial in binary multiplication, since the only possible values of the multiplier digits are 0 and 1.

Instead of listing all the shifted multiplicands and then adding, in a computer it is more convenient to add each shifted multiplicand as it is created to a *partial product*. The previous example is repeated in Chart 1, calculated with this technique.

CHART 1. MULTIPLICATION WITH PARTIAL PRODUCTS

1011	multiplicand
× 1101	multiplier
00000000	partial product
1011	shifted multiplicand
00001011	partial product
0000	shifted multiplicand
00001011	partial product
1011	shifted multiplicand
00110111	partial product
1011	shifted multiplicand
10001111	product

When we multiply an n-bit word and an m-bit word in a computer, the resulting product will take at most $n+m$ bits to express. Therefore, a typical multiplication instruction multiplies two n-bit words and produces a $2n$-bit double-word product.

Multiplication of signed numbers can be accomplished by using unsigned multiplication and the usual grammar-school rules: Perform an unsigned multiplication of the magnitudes, and make the product positive if the operands had the same sign, negative if they had different signs.

Binary Division

The simplest binary division algorithm is also based on the technique we learned in grammar school, as shown in Chart 2. In both the decimal and binary cases, we mentally compare the reduced dividend with multiples of the divisor to determine which multiple of the shifted divisor to subtract. In the decimal case, we first pick 11 as the smallest multiple of 11 less than 21, and then pick 99 as the smallest multiple less than 107. In the binary case, the choice is somewhat simpler, since the only two choices are zero and the divisor itself. Still, a comparison operation *is* needed to pick the proper shifted divisor.

CHART 2. EXAMPLE OF LONG DIVISION

Decimal	Binary	
19	10011	quotient
11 ⟌ 217	1011 ⟌ 11011001	dividend
11	1011	shifted divisor
107	0101	reduced dividend
99	0000	shifted divisor
8	1010	reduced dividend
	0000	shifted divisor
	10100	reduced dividend
	1011	shifted divisor
	10011	reduced dividend
	1011	shifted divisor
	1000	remainder

Unsigned division in a computer processor is complementary to multiplication. A typical division instruction accepts a double-word dividend and a single-word divisor, and produces single-word quotient and remainder. Such a division *overflows* if the divisor is zero or the quotient would take more than one word to express. The second situation occurs only if the divisor is less than or equal to the high-order word of the dividend.

Fixed-Point and Floating-Point Representations

All of the number systems discussed so far fix the binary point to the right of the rightmost bit. Thus, a 16-bit unsigned number lies in the range 0 through +65 535. This type of number system is most appropriate for programs that count objects or otherwise deal with integer quantities. However, many programs must deal with fractional quantities. For example, in a scientific program that computes a table of positive sines, it might be convenient to fix the binary point to the right of the *leftmost* bit of a 16-bit number, so that numbers range between 0 and 1.999 969 482 421 875 in increments of 0.000 030 517 578 125 (2^{-15}).

In either case, the binary point is fixed in a particular location for all numbers in the system. Thus, the system is called a *fixed-point representation*. Addition and subtraction may be performed directly on fixed-point numbers without regard to the location of the binary point; however, scaling may be required after multiplication and division operations.

To avoid error-prone bookkeeping and to represent a large range of numbers with relatively few bits, a *floating-point representation* can be used to encode explicitly a scale factor in each number. Similar to numbers in scientific notation, a floating-point number has two variable components, the mantissa (M) and the exponent (E). The value of the number is $M \cdot R^E$, where R is a fixed *radix*. In decimal scientific notation, the radix is 10, while in the floating-point representations of most computers, the radix is 2.

Within a computer, a floating-point number is represented by a bit string containing explicit fields for the mantissa and the exponent; the radix is implicit. The number of mantissa bits determines the precision of the number system, while the number of exponent bits determines the dynamic range of the system (the ratio between the largest and smallest nonzero numbers). Both the mantissa and the exponent are signed numbers, allowing representation of a large range of positive and negative numbers with absolute values both greater than and less than 1. Floating-point formats in typical computers use a total of 32 bits or more for the mantissa, exponent, and sign fields.

Many computers use a "single precision" floating-point format developed by the IEEE Standards Committee. This format uses a 24-bit mantissa and an 8-bit exponent to represent signed numbers with absolute values in the range of about 2^{-127} to 2^{+127} (10^{-38} to 10^{+38}). The IEEE standard also specifies a 64-bit ("double precision") format and gives guidelines for extended formats (e.g., 128-bit "quad precision" format).

Computer systems may use either of two methods to provide floating-point operations. Some computers have built-in hardware and instructions to process floating-point numbers in a prescribed format. Others have software programs that operate on floating-point numbers using sequences of primitive arithmetic instructions. The hardware approach typically performs operations 100 times faster than the software approach.

A computer that "supports" floating-point numbers is one that provides a minimum set of floating-point instructions including addition, subtraction, multiplication, division, and comparison of floating-point numbers, and conversion to and from integer fixed-point format. Special algorithms are needed to perform these instructions, including scaling or "normalization" of operands and results, as well as arithmetic manipulations of mantissas and exponents.

BCD Representation

The *binary-coded decimal (bcd)* number system encodes the digits 0 through 9 by their 4-bit unsigned binary representations, 0000 through 1001. The codes 1010 through 1111 are not used. Conversions between bcd and decimal representations are trivial, a direct substitution of four bits for each decimal digit. Two bcd digits may be packed into one byte; thus one byte may represent the values from 0 to 99 as opposed to 0 to 255 for a normal unsigned 8-bit binary number. Binary-coded decimal numbers with any desired number of digits may be obtained by using a string of bytes, one byte for each two digits.

Computers "support" bcd representation by providing instructions that perform arithmetic on packed-bcd numbers. Some computers have a "decimal adjust" instruction that corrects the result of an ordinary 8-bit add instruction, assuming that the operands were packed-bcd bytes. Others have decimal add, subtract, multiply, and divide instructions that operate directly on packed-bcd bytes, words, or arbitrary-length strings of bytes.

Character Codes

A string of bits in a computer need not represent a number; in fact most input and output of contemporary computers is nonnumeric. The most common type of nonnumeric data is *text*, strings of characters from some character set. Each character is represented in the computer by a bit string according to an established convention.

The most commonly used character code is ASCII, the American Standard Code for Information Interchange. This code represents each of 128 different characters by a 7-bit string, as shown in Table 3. Thus, the text string "Yeccch!" is represented by a rather innocent looking list of seven 7-bit numbers:

1011001 1100101 1100011 1100011 1100011
1101000 0100001

Some of the 7-bit strings in ASCII denote device control functions instead of "printing" characters. For example, CR (0001101) returns the print head or cursor on a printer or display to the first column, and LF (0001010) advances to the next line. Most of the other control characters are intended for use by data communication links, but different computer systems may use these characters for different functions.

Most computers manipulate an 8-bit quantity as a single unit, a byte, and store one character in each byte. The disposition of the extra bit when 7-bit ASCII is used depends on the system or program. Sometimes this bit is set to a particular value, sometimes it is ignored, and sometimes it is used to encode an additional 128 non-ASCII characters.

An important feature of ASCII is that the bit strings for letters and digits form a reasonable numerical sequence, so that text strings can be sorted by computer instructions that compare numerical values.

BASIC COMPUTER ORGANIZATION

A computer system consists of three major subsystems: processor, memory, and input/output (i/o), as shown in Fig. 2.

Processors

The *processor* (or *central processing unit, CPU*) is the heart of the computer. As shown in Fig. 3, a simple processor contains control circuits for fetching and executing instructions, an arithmetic logic unit (ALU) for manipulating data, and registers for storing the processor status and a small amount of data. It also has interface circuits for controlling and communicating with the memory and i/o subsystems.

Different processors have different organizations, both in their internal implementations and in their architectures as seen by programmers. Typical organizations can be roughly classified as accumulator-based processors, general-register processors, and stack machines.

The simplest processor organization is *accumulator based*. Such a processor has only one or two registers, called *accumulators*, in which arithmetic and logical operations and data transfers take place. The Intel 8051

TABLE 3. AMERICAN STANDARD CODE FOR INFORMATION INTERCHANGE (ASCII), STANDARD NO. X3.4-1968 OF THE AMERICAN NATIONAL STANDARDS INSTITUTE

		$b_6b_5b_4$ (Column)							
$b_3b_2b_1b_0$	Row (hex)	000 0	001 1	010 2	011 3	100 4	101 5	110 6	111 7
0000	0	NUL	DLE	SP	0	@	P	`	p
0001	1	SOH	DC1	!	1	A	Q	a	q
0010	2	STX	DC2	"	2	B	R	b	r
0011	3	ETX	DC3	#	3	C	S	c	s
0100	4	EOT	DC4	$	4	D	T	d	t
0101	5	ENQ	NAK	%	5	E	U	e	u
0110	6	ACK	SYN	&	6	F	V	f	v
0111	7	BEL	ETB	'	7	G	W	g	w
1000	8	BS	CAN	(	8	H	X	h	x
1001	9	HT	EM	)	9	I	Y	i	y
1010	A	LF	SUB	*	:	J	Z	j	z
1011	B	VT	ESC	+	;	K	[	k	{
1100	C	FF	FS	,	<	L	\	l	|
1101	D	CR	GS	–	=	M	]	m	}
1110	E	SO	RS	.	>	N	^	n	~
1111	F	SI	US	/	?	O	_	o	DEL

Control Codes

NUL	Null	DLE	Data link escape
SOH	Start of heading	DC1	Device control 1
STX	Start of text	DC2	Device control 2
ETX	End of text	DC3	Device control 3
EOT	End of transmission	DC4	Device control 4
ENQ	Enquiry	NAK	Negative acknowledge
ACK	Acknowledge	SYN	Synchronize
BEL	Bell	ETB	End transmitted block
BS	Backspace	CAN	Cancel
HT	Horizontal tab	EM	End of medium
LF	Line feed	SUB	Substitute
VT	Vertical tab	ESC	Escape
FF	Form feed	FS	File separator
CR	Carriage return	GS	Group separator
SO	Shift out	RS	Record separator
SI	Shift in	US	Unit separator
SP	Space	DEL	Delete or rubout

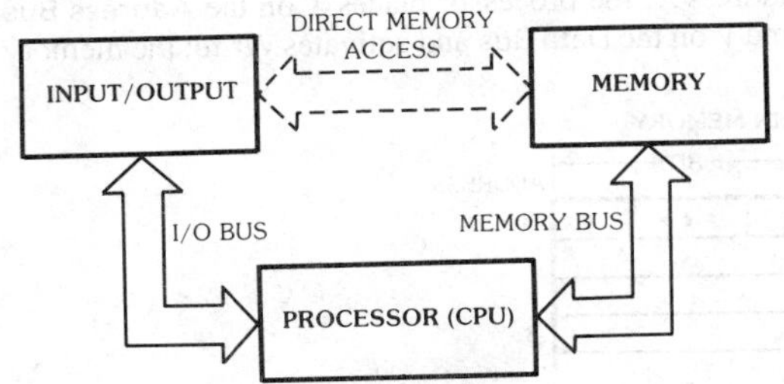

Fig. 2. Block diagram of a typical computer.

and 8085, the Zilog Z80, and the MOS Technology 6502 are single-accumulator processors; the Motorola 6809 is a two-accumulator processor. Accumulator-based processors often have other special-purpose registers, in addition to the accumulator(s), for address manipulation and other operations. The Intel 8086 is best classified as a single-accumulator machine with many special-purpose registers.

Processors with more than two registers for arithmetic and logical operations are classified as *general-register processors*. The IBM 370, the DEC PDP-11, the Zilog Z8000, the National 32000, and the Motorola 68000 are all general-register processors. Later in this chapter, a hypothetical general-register processor with a subset of the Motorola 68000 architecture will be used to illustrate computer organization.

A *stack machine* has neither general registers nor accumulators, only a stack pointer, SP, that points to a pushdown stack in memory; all operations are performed on the stack. Many of Hewlett-Packard's desktop and hand-held calculators, for example, are stack machines. Also, most accumulator-based and general-register processors have at least a few stack-oriented instructions. Among contemporary microprocessors, the Intel iAPX432 comes closest to being a pure stack machine. It has no accumulators or general registers; all operations are performed either on the stack or directly on memory locations.

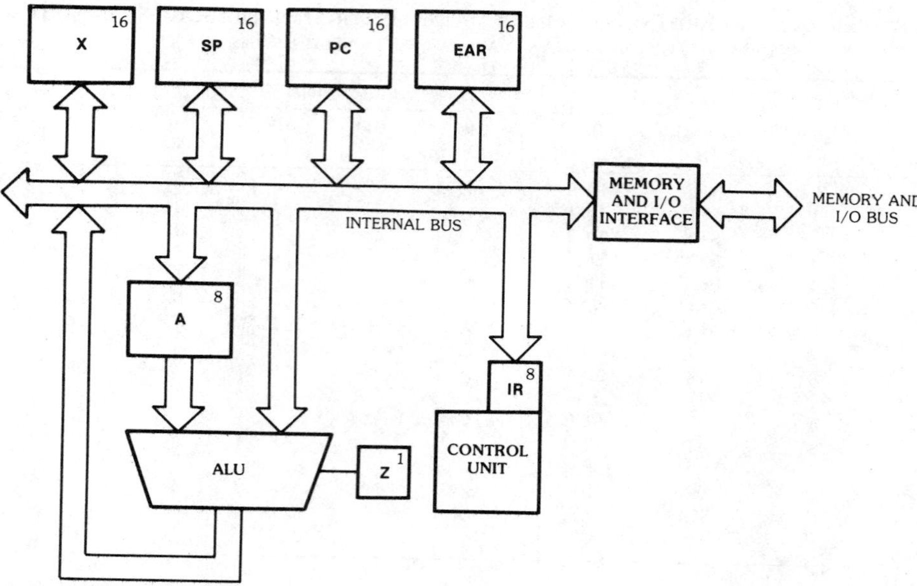

Fig. 3. Block diagram of a simple processor.

Memory

The *memory* (or *main memory*) of a computer contains storage for instructions and data, and is tied to the processor via the Memory Bus in Fig. 2. A *bus* is simply a bundle of wires or any other physical medium for transferring information. A computer memory has some number of *locations*, each of which stores a b-bit quantity. In most contemporary computers, each location stores an 8-bit "byte." Associated with each location in the memory is a unique binary number called the *address*. If there are n locations, then the addresses range from 0 to $n-1$.

The key feature that distinguishes main memory from other forms of mass storage in a computer is *random access*—the processor has equally fast access to every location in memory. Random-access memory is analogous to a wall of post-office boxes; a postal clerk can deposit mail in any box with equal ease. Compare this with the *serial access* method of a letter carrier who visits locations sequentially, in the order of the route. Magnetic tapes provide serial-access memory in computer systems.

Fig. 4 shows how the processor accesses main memory in a typical small computer system. The memory is an array of n locations of b bits each. To read the data stored at address X, the processor places the number X on the Address Bus and activates a Read control signal; the memory responds by placing the contents of address X on the Data Bus. To write a value V at address X, the processor places X on the Address Bus and V on the Data Bus and activates Write; the memory

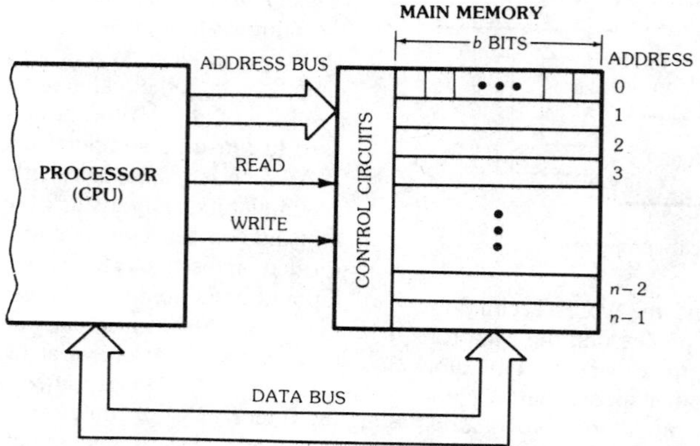

Fig. 4. Main memory in a typical computer.

immediately writes the value *V* in the specified location. Subsequent reading of address *X* will now return the value *V*.

Fig. 5A gives a conceptual view of a memory with 2^{16} bytes; a 16-bit address uniquely specifies any byte in this memory. Even if a computer must deal with 16-bit or larger quantities, it can still have a byte-addressable memory as shown in Fig. 5B. Each 16-bit "word" consists of two consecutive bytes starting on an even address (a *word boundary*). For example, bytes 2 and 3 form a word, but bytes 3 and 4 do not. Instructions that manipulate words must specify even addresses; instructions that manipulate bytes may specify any address.

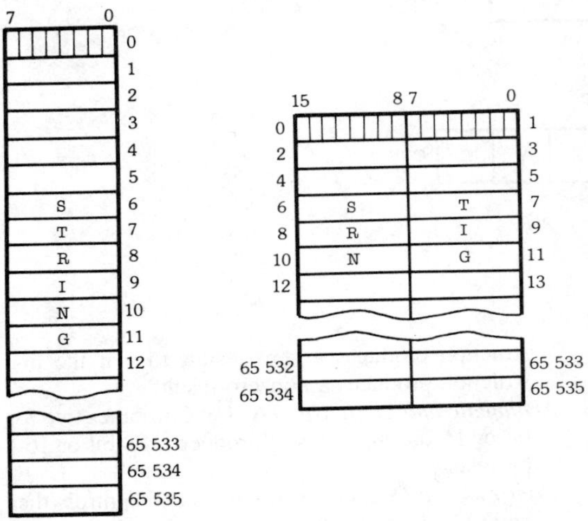

(A) Byte-addressable memory for an 8-bit processor.

(B) Byte-addressable memory for a 16-bit processor.

Fig. 5. Memory organizations.

Several types of memories are commonly used in computer main memory systems. With a *read/write memory (RWM)*, we can store data at any address and read it back at any time. With a *read-only memory (ROM)*, we can read the contents of any address at any time, but data can be stored only once, when the ROM is manufactured.

A *programmable read-only memory (PROM)* is similar to a ROM, except that the customer may store the data (i.e., "program the PROM") using a *PROM programmer*. A PROM chip is manufactured with all its bits at a particular value, say 1. The PROM programmer can be used to set desired bits to 0, typically by blowing tiny fuses inside the PROM corresponding to each bit.

An *erasable programmable read-only memory (EPROM)* is similar to a PROM, except that the EPROM can be "erased" to the all-ones state by physically exposing it to ultraviolet light. An *electrically erasable programmable read-only memory (EEPROM)* is erased electronically, so it can be erased without manual intervention.

Every computer has RWM for storing variable data. Depending on the computer and the application, the programs may also be stored in RWM, or they may be stored in ROM, PROM, or EPROM.

All of the memories described above are *random-access memories*, because all locations have equally fast access. However, computer jargon has developed so that the acronym "RAM" most commonly refers to read/write memory only.

Input/Output

The *input/output (i/o)* subsystem contains *peripheral devices* for communicating with, observing, and controlling the world outside the computer. Peripheral devices include terminals, printers, communication devices, and mechanical sensors and actuators. Also included in the i/o subsystem are *mass storage* devices such as magnetic tapes and disks. These devices are used to store information not needed in the main memory at all times, such as applications programs and text files. Not all computers have mass storage devices, but all useful computers have at least one peripheral device, since by definition a peripheral is the only means by which the computer can communicate with the outside world.

The processor writes and reads information to and from peripherals by means of i/o instructions that place commands and data on the I/O Bus. In many computers, both memory and peripherals share the same physical bus. Going one step further, some computers communicate with their peripherals by using registers that behave as if they were memory locations. In such systems, the hardware makes no distinction between accessing main memory and accessing peripherals.

In simple computer systems, there is no direct path from peripherals to main memory; the only way to transfer data between a peripheral and memory is for the processor to read it from the peripheral and store it in memory, or vice versa. However, systems requiring a higher data transfer rate incorporate *direct memory access (DMA)*, a link between a special peripheral controller and memory that allows a peripheral to read and write memory without processor intervention.

PROCESSORS AND PROGRAMMING

Organization of a General-Register Processor

Fig. 6 shows the internal organization of a general-register processor, the S68000. The hypothetical processor has a subset of the registers and instructions of the Motorola 68000.

The S68000 accesses a memory of up to 2^{16} (65 536) bytes, arranged as shown in Fig. 5B; addresses are two

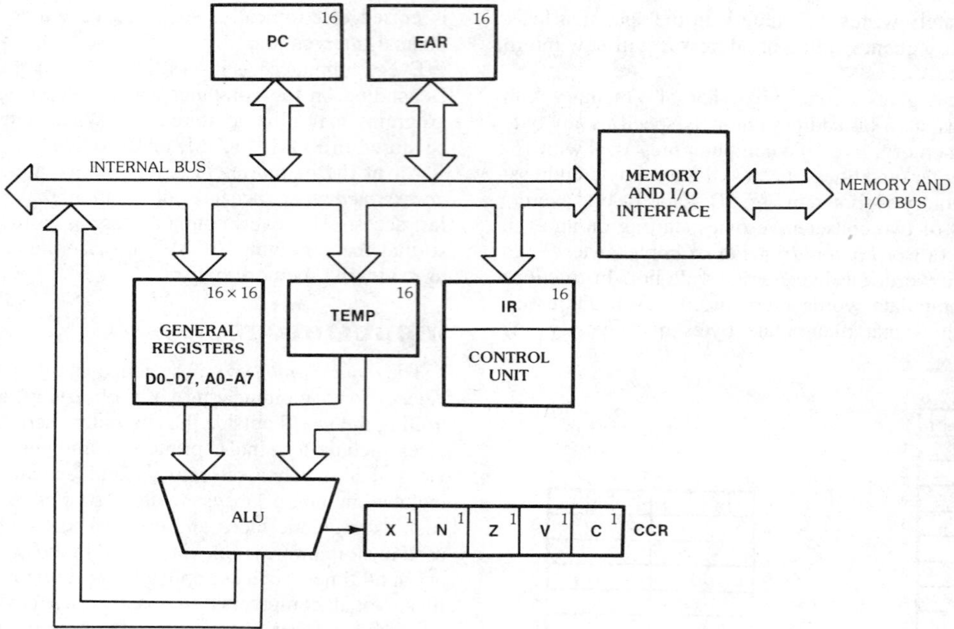

Fig. 6. General-register processor organization, the S68000.

bytes long. Most instructions can manipulate either 1-byte or 2-byte quantities. An S68000 instruction occupies 1, 2, or 3 words (2, 4, or 6 bytes) in memory. [The real 68000 accesses a memory of up to 2^{32} bytes, manipulates 1-, 2-, or 4-byte quantities, and has instructions up to 5 words long.]

The operation of the S68000 (or almost any other computer processor) consists of endless repetition of two steps: read the next instruction from memory (the *fetch cycle*), and perform the actions it requires (the *execution cycle*). Several registers and functional units are involved in instruction fetching and execution:

Program Counter (PC): A 16-bit register that holds the memory address of the next instruction to be executed.

Instruction Register (IR): A 16-bit register that holds the first word of the currently executing instruction.

Effective Address Register (EAR): A 16-bit register that, when used, holds an address at which the processor reads or writes memory during the execution of an instruction.

General Registers (D0–D7,A0–A7): A set of sixteen 16-bit registers containing data and addresses to be processed.

Temporary Register (TEMP): A 16-bit register that holds operands or intermediate results during the execution of an instruction.

Condition Code Register (CCR): A set of 1-bit flags that the processor sets or clears during the execution of each data manipulation instruction. In particular, the *Zero Bit* (Z) is set to 1 if the instruction produces a zero result, to 0 if the instruction produces a nonzero result.

Arithmetic and Logic Unit (ALU): Combines two 8-bit or 16-bit quantities to produce an 8-bit or 16-bit result.

Control Unit: Decodes instructions and controls the other blocks to fetch and execute instructions.

Memory and I/O Interface: Reads and writes memory and communicates with i/o (input/output) devices according to commands from the control unit.

Although all of the blocks above are essential to the internal operation of the S68000, only the registers PC, D0–D7, A0–A7, and CCR are explicitly manipulated by instructions and have values that are meaningful after the execution of each instruction. Such registers comprise the *processor state*, and may be shown in a *programming model* for the processor, as in Fig. 7. Only these registers are of concern to a programmer.

Instruction Formats

The instructions executed by the hypothetical S68000 are 1 to 3 words long. The first word of each instruction is called the *opcode word*. It contains an *opcode*, which uniquely specifies the operation to be performed, and *effective-address specifiers* for zero to two *operands*, which are registers or memory locations to be manipulated by the instruction. If an operand is a memory location, then a second or third instruction word contains its address.

The S68000 uses several different formats for the

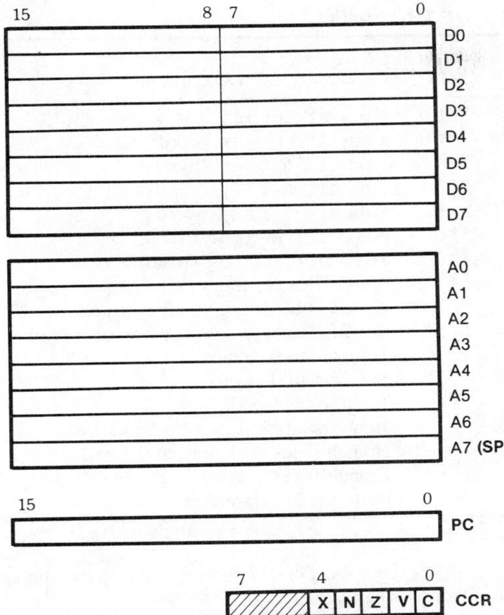

Fig. 7. Programming model for S68000.

opcode word of different instructions, as required by different operations and operand types. For example, five different formats are shown in Fig. 8, and there are many other formats in the real 68000.

To understand why a particular set of formats is used in a given architecture, one must understand the goals, optimizations, and compromises of its architects, which is beyond the scope of this chapter. If you look at S68000 instruction formats in particular, you will often ask, "Why did they do that?" This question is easier to answer in the real 68000, on which the S68000 is based, but even the instructions of the real 68000 are encoded somewhat irregularly.

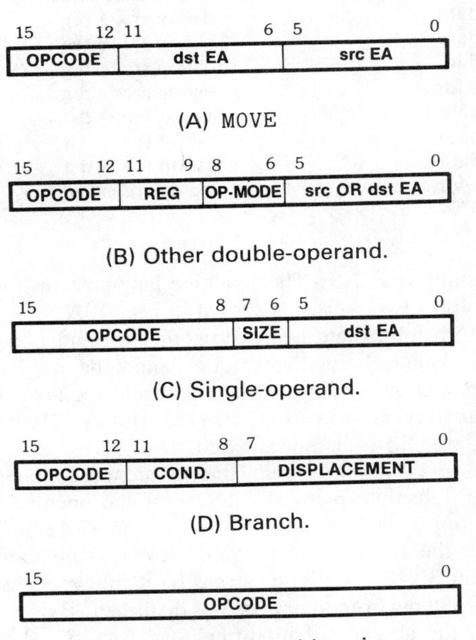

Fig. 8. Format of first word of typical S68000 instructions.

Most S68000 data manipulation instructions can operate on bytes or words. The operand size is encoded as part of the opcode in MOVE instructions (Fig. 8A), and in the size or op-mode field in other instructions (Figs. 8B and C).

Instructions that reference memory contain a 6-bit "effective-address" (EA) field that specifies the location of the operand; MOVE contains two such fields. In all these instructions, the EA field can specify one of several addressing modes, as discussed shortly. Some addressing modes require an additional word of information to be appended to the instruction, thus giving rise to 2- and 3-word instructions.

Instruction Set

Table 4 lists the machine instructions of the S68000; the real 68000 has many more instructions. Even for the few instructions in the table, the real 68000 has many more variations. While the S68000 uses only three of eight possible combinations of the op-mode field, the real 68000 uses them all, as shown in Table 5. In both the real 68000 and the S68000, there are still more variations in each instruction according to the addressing modes specified for the source (src) and destination (dst) operands, as discussed next.

Addressing Modes

In both the 68000 and the S68000, an operand is specified by a 6-bit source or destination "effective address" (EA) field in the instruction, consisting of a 3-bit addressing mode designator and a 3-bit register number, as shown in Fig. 9. In conjunction with the reg field, the eight mode combinations are used as follows:

- 0 reg specifies a data register (D0–D7) that contains the operand.
- 1 reg specifies an address register (A0–A7) that contains the operand.
- 2–6 reg specifies an address register that is used to compute the effective address of an operand in memory.
- 7 reg specifies one of five addressing modes that do not use a general register; three combinations are left over for future expansion.

A full treatment of 68000 addressing modes is beyond the scope of this chapter, but for the S68000 we shall discuss the seven modes summarized in Table 6.

In *register-direct addressing*, the operand is contained in one of the data registers, D0–D7, or in one of the address registers, A0–A7. As suggested in Fig. 7, word operations use the entire 16-bit register, while byte operations use only the eight low-order bits and never affect the high-order bits of the register. Byte

TABLE 4. INSTRUCTION SET OF THE S68000

Mnemonic	Operands	Format (Fig. 8)	Opcode Word (Binary)	XNZVC	Description
MOVE.B	src,dst	A	0001ddddddssssss	-**00	Copy src to dst (byte)
MOVE.W	src,dst	A	0011ddddddssssss	-**00	Copy src to dst (word)
ADD.B	src,dreg	B	1101rrr000ssssss	*****	Add src to dreg (byte)
ADD.W	src,dreg	B	1101rrr001ssssss	*****	Add src to dreg (word)
ADDA.W	src,areg	B	1101rrr011ssssss	-----	Add src to areg (word)
AND.B	src,dreg	B	1100rrr000ssssss	-**00	AND src to dreg (byte)
AND.W	src,dreg	B	1100rrr001ssssss	-**00	AND src to dreg (word)
CMP.B	src,dreg	B	1011rrr000ssssss	-****	Set CCR according to dreg–src (byte)
CMP.W	src,dreg	B	1011rrr001ssssss	-****	Set CCR according to dreg–src (word)
CMPA.W	src,areg	B	1011rrr011ssssss	-****	Set CCR according to areg–src (word)
CLR.B	dst	C	0100001000dddddd	-0100	Set dst to 0 (byte)
CLR.W	dst	C	0100001001dddddd	-0100	Set dst to 0 (word)
JMP	dst	C	0100111011dddddd	-----	Jump to dst address
JSR	dst	C	0100111010dddddd	-----	Jump to subroutine at dst address
BNE	offset	D	01100110oooooooo	-----	Branch if result is nonzero (Z = 0)
BEQ	offset	D	01100111oooooooo	-----	Branch if result is zero (Z = 1)
RTS		E	0100111001110101	-----	Return from subroutine

Notes:
1. src = source operand; dst = destination operand; dreg = one of the data registers, D0–D7; areg = one of the address registers, A0–A7; offset = 8-bit signed integer added to PC if branch is taken.
2. s = bit in src field; d = bit in dst field; r = bit in reg field; o = bit in offset field.
3. Effects of instructions on condition bits: –: not affected; 1: always set to 1; 0: always cleared to 0; *: set to 1 if operation caused corresponding condition, else cleared to 0.
4. Conditions: N—most significant bit of result was 1; Z—result was zero; V—operation caused twos-complement overflow; C—addition or subtraction caused a carry or borrow from most significant bit, or shift caused a 1 bit to "fall off" the end; X—same as C (but not always affected).

TABLE 5. TYPICAL USE OF THE OP-MODE FIELD IN THE MOTOROLA 68000

Value (Binary)	Operand Size	Register Type	Operation
000	byte	data	reg op src → reg
001	word	data	reg op src → reg
010	long	data	reg op src → reg
011	word	addr	reg op src → reg
100	byte	data	dst op reg → dst
101	word	data	dst op reg → dst
110	long	data	dst op reg → dst
111	long	addr	reg op src → reg

5 3	2 0
MODE	REG

Fig. 9. S68000 operand effective-adress specification (mode and reg are reversed in the dst EA field of the MOVE instruction).

operations are not allowed on address registers, since an "address" should normally be a full 16-bit quantity.

In *immediate addressing*, a 1- or 2-byte operand is contained in the second word of the instruction (with 1-byte operands, the high-order byte of the second word is unused). Thus, to place the value $1234 in register D5, we could use the instruction, "MOVE.W #$1234,D5". By convention, the notation "#" denotes an immediate operand, and "$" denotes a hexadecimal number. The machine-language instruction occupies two words as shown in Fig. 10A.

With immediate addressing, the operand is a *constant* value. Immediate mode cannot be used for a "destination" operand, since it would require the instruction to modify itself; thus "MOVE.W D5,#$1234" is not allowed.

In *absolute addressing*, the instruction contains the 16-bit absolute memory address of the operand. For example, the instruction "MOVE.W $5432,D5" reads the 16-bit value currently stored at memory location $5432 and copies (*loads*) it into register D5; the contents of memory are not disturbed. By convention, the absence of modifiers such as "#" denotes an absolute operand.

In the other direction, "MOVE.W D4,$5432" (Fig. 10B) *stores* the value of D4 into memory location

TABLE 6. ADDRESSING MODES OF THE S68000

Name	mode	reg	Notation	Operand
Data-register direct	0	0–7	Dn	Dn
Address-register direct	1	0–7	An	An
Immediate	7	4	#data	data
Absolute	7	0	addr	MEM[addr]
Address-register indirect	2	0–7	(An)	MEM[An]
Auto-increment (by 1 or 2)	3	0–7	(An)+	MEM[An], then An ← An + operand size
Auto-decrement (by 1 or 2)	4	0–7	−(An)	An ← An − operand size, then MEM[An]

Notes:
1. Dn denotes a data register: D0–D7.
2. An denotes an address register: A0–A7 or SP (same as A7).
3. data is an 8- or 16-bit value as appropriate for the size of the operation.
4. addr is the 16-bit absolute memory address of the operand.
5. MEM[x] is the 8- or 16-bit value beginning at memory address x, as appropriate for the size of the operation.

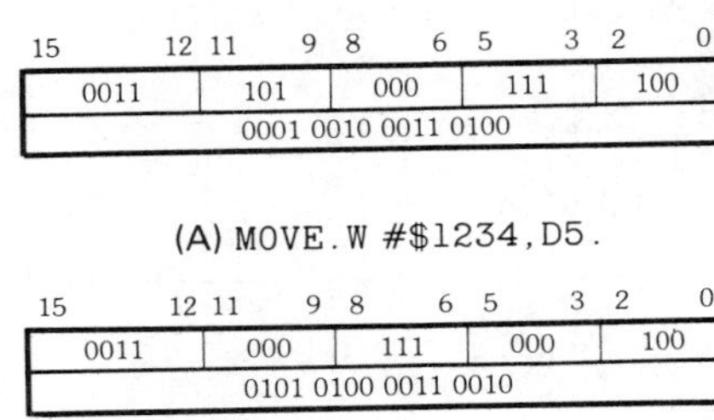

Fig. 10. Encodings of machine instructions.

$5432, without disturbing D4. Thus, an operand with absolute addressing is typically a *variable*, since the program may store different values into the specified memory location from time to time.

In *address-register indirect addressing*, the specified address register (An) contains the 16-bit address of the operand. In this mode, not only is the operand a variable, but so is its address. Each time the instruction is executed, the address register may "point to" a different memory location. This mode is useful for dealing with arrays, lists, and other data structures.

Auto-increment addressing is the same as address-register indirect, except that after being used, the address register is automatically incremented by 1 or 2 according to the size of the operand, byte or word. *Auto-decrement addressing* is similar, except that the address register is *decremented* before being used as the address. These two modes are used together to step through arrays, lists, and other data structures and to manipulate pushdown stacks.

A Machine-Language Program

Table 7 shows the values stored in memory for a sequence of instructions and data that forms a program for multiplying 123 by 456. A list of machine instructions stored in memory, as defined by the two left-hand columns of the table, is called a *machine-language program*. Even though they do not mean much to a human reader, these two columns completely specify the operations to be performed by the computer. The remaining columns of Table 7 are an *assembly-language program* that gives the same information in symbolic form much more understandable by humans. In the next two subsections, we will discuss assembly language and then discuss the actual behavior of the machine-language program.

Assembly Language

The program development process using assembly language is illustrated in Fig. 11. A typical programmer uses a *text editor* to create a text file containing an assembly-language program. The names *source file* and *source program* are often used for assembly-language and high-level-language text files. The assembler accepts a source program as input, checks for format errors, and produces an *object module* containing the machine-language program. A *loader* then reads the object module and creates a corresponding set of bit patterns (the machine-language program) in the memory of the target machine. There, the machine-language program is run, possibly with the aid of a *debugger*.

The loader, debugger, and machine-language program described above *must* run on the target machine; the text editor and assembler may run there or on different machines. An assembler that runs on one machine and produces object modules for another is called a *cross assembler*. For example, it is possible to use a text editor on a personal computer to generate an assembly-language program, carry the source file on floppy disk to a large computer that runs a cross assembler, and "download" the object module to a loader in a local microcomputer by way of a serial data link.

As a minimum, any assembly language must define:

A mnemonic for each machine-language instruction

TABLE 7. MEMORY CONTENTS FOR A SEQUENCE OF INSTRUCTIONS AND DATA

Machine Language		Assembly Language			
Addr (hex)	Contents (hex)	Label (Sym)	Opcode (Mnem)	Operand (Sym)	Comments
0000	????	*			Program to multiply MCND by MPY.
0000	????		ORG	$1000	
1000	????	MON	EQU	$E000	Address to go to when done.
1000	????	*			
1000	4240	MULT	CLR.W	D0	D0 will accumulate product.
1002	3E01		MOVE.W	MPY,D1	D1 holds loop count (multiplier).
1004	2002				
1006	670A		BEQ	DONE	Done if count is down to zero.
1008	D078	LOOP	ADD.W	MCND,D0	Else add multiplicand to product
100A	2004				
100C	D27C		ADD.W	#−1,D1	and do loop MPY (D1) times.
100E	FFFF				
1010	66F6		BNE	LOOP	
1012	3038	DONE	MOVE.W	D0,PROD	Save product.
1014	2000				
1016	4EF8		JMP	MON	Return to monitor at $E000.
1018	E000				
1018			ORG	$2000	
2000	????	PROD	DS.W	1	Storage for PROD.
2002	007B	MPY	DC.W	123	Multiplier value.
2004	01C8	MCND	DC.W	456	Multiplicand value.
2006			END	MULT	

SYMBOL TABLE:

DONE	1012	MCND	2004	MPY	2002	PROD	2000
LOOP	1008	MON	E000	MULT	1000		

Note: Addr = Address; hex = hexadecimal; sym = symbolic; mnem = mnemonic.

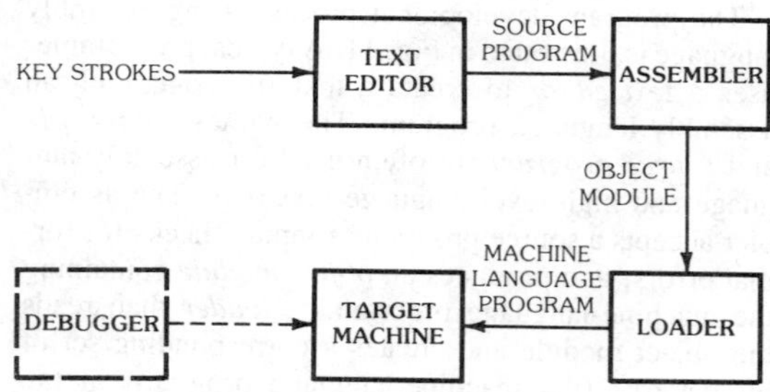

Fig. 11. Assembly-language program development.

A standard format for the lines of an assembly-language program

Formats for specifying different addressing modes and other instruction variations

Formats for specifying character constants and integer constants in different bases

Mechanisms for associating symbolic names with addresses and other numeric values

Mechanisms for defining data to be stored in memory along with the instructions when the program is loaded

Directives that specify how the program is to be assembled

In a typical assembly language, each line has four fields arranged as shown:

```
LABEL OPCODE OPERAND COMMENTS
```

The LABEL field is optional; if present it begins in column 1. Successive fields are separated by one or more spaces.

A *label* is simply a symbol, that is, a sequence of letters and digits beginning with a letter. Every symbol in an assembly-language program is assigned a value at the time that it is defined; the assembler program keeps track of labels and their values by an internal *symbol table*. For example, in the CLR.W statement in Table 7, the value of the symbol "MULT" equals the memory address at which the CLR.W instruction is stored, $1000 to be exact. In general, the value of a symbol is the memory address at which the corresponding instruction or data is stored.

Symbols only have values at the time a program is being assembled, that is, at *assembly time*. Symbols exist in the source text of the program, and in the environment of the assembler, but do not appear in the machine-language program. For instance, when our example program is loaded into memory (at *load time*) or actually run (at *run time*), there is no way to discover, simply by looking at the object module or at memory, that the programmer had associated the symbol MULT with address $1000, or even that there had ever been a symbol MULT at all.

The OPCODE field contains the mnemonic of either

a machine instruction or an assembler directive. Depending on the contents of the OPCODE field, the OPERAND field specifies zero or more operands separated by commas. An operand is an expression consisting of symbols, constants, and operators such as + and −. The simplest expression consists of a single symbol or constant.

The COMMENTS field is ignored by the assembler, but it is used by the programmer to convey a high-level explanation of the program to human readers. It is also possible to use an entire line as a comment by placing a star "*" in column 1 of the line.

Five different *assembler directives* appear in Table 7, as described below:

ORG (Origin). The operand specifies the address at which the next instruction or datum is to be deposited when the program is loaded into memory. Subsequent instructions and data are deposited in successive memory addresses.

DS.W (Define Storage—Word). The operand specifies a number of memory words to be skipped without storing any instructions or data, thereby reserving space to be used by variables in a program.

DC.W (Define Constant—Word). The specified word value is stored into memory when the program is first loaded into memory, thereby establishing a constant value that may be accessed when the program is run.

EQU (Equate). The identifier in the Label field is assigned the value in the Operand field, instead of the value in the Address field. This makes the identifier a synonym for a constant value for the duration of the assembly process.

END (End Assembly). This directive denotes the end of the text to be assembled. Its operand, if present, is the address of the first executable instruction of the program.

Of course, there are many other directives in a full assembly language, and different assembly languages may use different names and conventions for the same thing.

The input of an assembler is a text file containing the source code, while the output consists of an object module (typically an "object file" or "binary file") and a listing similar to Table 7. To understand a little more about how an assembler works, it is useful to examine the Address column in the listing. The Address column indicates the address at which the assembler "thinks" that it should assemble the instruction on the current line. By default, this address is initialized to zero.

The ORG directive tells the assembler to assemble successive instructions starting from a new address, and hence changes the Address column on the *next* line. Some directives, such as EQU, do not assemble any instructions or data and hence do not affect the Address column at all. Others, such as DS.W, advance the Address value but do not cause anything in particular to be loaded into memory at load time.

For its own and the programmer's use, the assembler creates a *symbol table* that lists the numeric value associated with each symbol. As is evident in Table 7, in most cases a symbol receives the value of the Address column in the line in which it is defined, except in the case of EQU statements, which assign the symbol the value of the Operand column. In all cases, the symbols and their values exist only at assembly time.

Operation of a Simple Program

We are now ready to explain the program in Table 7. It multiplies MCND by MPY by initializing the product to 0 and then adding MCND to it MPY times.

To begin with, notice that the values of the multiplier and multiplicand are "passed" to the program in fixed memory locations, $2002 and $2004, which are initialized to 123 and 456 when the program is loaded. If, after loading the program, we start the program at location $1000, it will indeed compute the product of 123 and 456. Alternatively, if after loading the program we place different numbers in locations $2002 and $2004, then the program will compute the product of the new numbers. In any case, the program "returns" its result by placing the result in memory location $2000, which may be examined by whoever ran the program in the first place.

The program uses registers to accumulate the product (in D0) and to keep track of the multiplier count (in D1), because registers can be accessed faster than main memory in most computers, and hence the program will run faster. In fact, we could optimize the program to run even faster by placing MCND into a register (say D2) before entering the multiplication loop.

So, the program begins by setting D0 to zero and copying MPY into D1. The MOVE instructions set the condition bits in CCR according to the value stored. In particular, the Z bit is set to 1 if a zero value (00..00) was stored; otherwise, Z is cleared to 0. Therefore, if MPY is zero, the Z bit is 1 when the BEQ DONE instruction is first reached, and the branch will be taken, leaving zero in D0. Notice that the branch instruction gives a *relative offset* of $0A, so that either the next instruction (at address $1008) or the instruction at the branch offset (address $1008+$0A) is executed.

If the BEQ branch is not taken, then the two ADD instructions add MCND to the product in D0 and subtract 1 from the loop count in D1. Like a MOVE instruction, an ADD instruction sets Z to 1 if its result is zero. Thus, if D1 has *not* been reduced to zero, the BNE instruction will branch back to address $1006. (The 8-bit signed offset $F6, interpreted in twos-complement, equals −$0A; hence the branch is to address $1012−$0A=$1008.)

Eventually, after executing the loop MPY times, the program reaches the MOVE instruction at address

$1012, which saves the accumulated product in memory. Finally, the program unconditionally jumps to address $E000, where we assume there exists a program that prints out or gives the user some way to examine the result of the computation stored at location $2000.

Indirect Addressing

The program in Table 7 manipulated only simple variables and constants. More complicated data structures such as arrays, stacks, queues, records, and lists are used in almost all programs. Consider the problem of initializing the components of an array of five bytes to zero. An assembly-language solution is shown in Table 8. Note that the operand expressions "Q+1," "Q+2," and so on are evaluated at assembly time.

The choice of a 5-component array above was very judicious—the corresponding program for a 100-component array would have 101 instructions. *Indirect addressing* avoids this problem by taking the address from an address register at run time. Thus, we can write a loop to initialize an array; an address register points to a different array element on each iteration of the loop.

The program in Table 9 solves the array initialization problem by using indirect addressing. It initializes A0 to point to the first component of Q and then executes a loop that clears successive components of Q, incrementing A0 once per iteration of the loop. The program could be further optimized by using auto-increment addressing, changing the CLR.B instruction to CLR.B (A0)+ and eliminating the ADD.W instruction.

Not only does the program in Table 9 occupy fewer bytes than the one in Table 8, but it also stays the same length for an array of any size. The program is easily modified to work on an array of different length by changing the occurrences of the length "5" to the desired length.

Subroutines

A *subroutine* is a sequence of instructions, defined and stored only once, that may be invoked (or *called*) from many places. In order to use subroutines, we need instructions to save the current value of the PC each time the subroutine is called, and restore it when the subroutine is finished. In the S68000, these instructions save and restore PC using a stack.

A *push-down stack* (or simply a *stack*) is a one-dimensional data structure in which values are entered and removed one item at a time at one end, called the *top of stack*. A register called the *stack pointer* (SP) points to the top of stack. A datum is entered by a *push* operation that points the stack pointer to the next available memory location and then stores the datum at the top of stack. Data is removed by a *pop* (or *pull*) operation that removes the datum at the top of stack and then backs up the stack pointer.

In the S68000, the JSR and RTS instructions perform subroutine calls and returns using address register A7 as a stack pointer. Any program that uses subroutines is required to reserve a small area of memory for a push-down stack for return addresses. At the beginning of such a program, A7 must be initialized to point

TABLE 8. INITIALIZING AN ARRAY THE HARD WAY

Addr	Contents	Label	Opcode	Operands	Comments
0000	????		ORG	$3000	Set components of Q to zero.
3000	4238 3100	INIT	CLR.B	Q	First component.
3004	4238 3101		CLR.B	Q+1	Second component.
3008	4238 3102		CLR.B	Q+2	Third component.
300C	4238 3103		CLR.B	Q+3	Fourth component.
3010	4238 3104		CLR.B	Q+4	Fifth component.
3014	4EF8 E000		JMP	$E000	Return to monitor.
3018			ORG	$3100	
3100	???? ????	Q	DS.B	5	Reserve 5 bytes for array.
3105	????		END	INIT	

TABLE 9. INITIALIZING AN ARRAY BY USING INDIRECT ADDRESSING

Addr	Contents	Label	Opcode	Operands	Comments
0000	????		ORG	$3000	Set components of Q to zero.
3000	307C 3100	INIT	MOVE.W	#Q,A0	A0 points to first component
3004	4210	ILOOP	CLR.B	(A0)	Clear byte that A0 points to.
3006	D0FC 0001		ADD.W	#1,A0	Point to next byte.
300A	B0FC 3105		CMPA.W	A0,#QEND	Past the end?
300E	66F4		BNE	ILOOP	No, keep clearing.
3010	4EF8 E000		JMP	$E000	Yes, return to monitor.
3014			ORG	$3100	
3100	???? ????	Q	DS.B	5	Reserve 5 bytes for array.
3105	????	QEND	EQU	Q+5	Address of byte just past Q.
3105	????		END	INIT	

at this area by means of a `MOVE.W #addr,A7` instruction. As shown in Fig. 12, A7 points to the top item in the stack, or just past the stack area if the stack is empty. Register A7 is decremented by two before each word is stored on the stack, and incremented by two after each word is popped. These operations correspond exactly to auto-decrement and auto-increment addressing as described in Table 6.

The `JSR addr` instruction saves the address of the next instruction by pushing it onto the stack and then jumps to the instruction at location `addr`, the first instruction of the subroutine. At the end of the subroutine, RTS pops an address from the stack into PC, effecting a return to the original program sequence.

A stack is the most appropriate data structure for saving return addresses, because it can store more than one return address when subroutines are *nested*, that is, when one subroutine calls another. The number of levels of nesting is limited only by the size of the memory area reserved by the programmer for the stack.

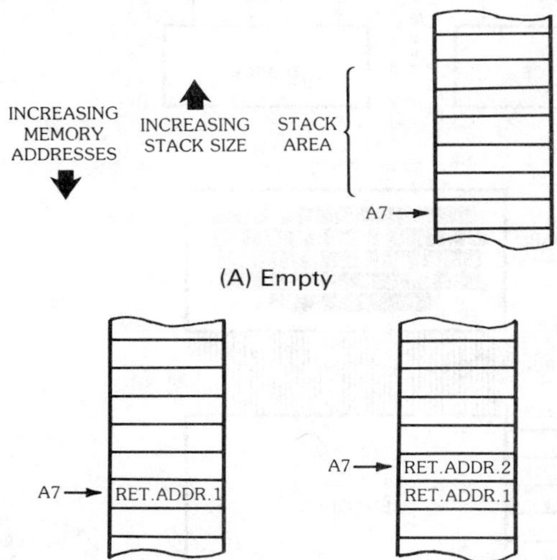

Fig. 12. S68000 return-address stack.

INPUT/OUTPUT

Of the three major subsystems of a computer, input/output has experienced the biggest evolution, because of the explosion of computer applications and the hundreds of different devices that are now part of computer systems. In the 1950s, few computer architects would have predicted that some day more computers would be used in automobiles than in any other application, and that one of the most common output devices would be a fuel-injected carburetor.

Despite the proliferation of devices, fairly standard techniques are still used to connect typical devices to a computer system, as described in this section.

Buses

Fig. 2 showed the basic organization of a computer as consisting of processor, memory, and input/output (i/o). The processor communicated with the i/o subsystem by means of an i/o bus. In Fig. 13, we expand our view. Like a memory bus, the i/o bus in Fig. 13 contains data, address, and control lines. The address lines allow a program to select among different i/o devices connected to the system, while the data lines carry the actual data being transferred.

Devices and Interfaces

The i/o subsystem in Fig. 13 contains both devices and interfaces. A *peripheral device* (or *i/o device*) performs some function for the computer. An *i/o interface* (or *device interface*) controls the operation of a peripheral device according to commands from the computer processor; it also converts computer data into whatever format is required by the device and vice versa. Also as shown in the figure, a peripheral device is often housed separately from the processor, while the interface is almost always packaged together with the processor and memory in one "CPU box."

There are many different peripheral devices that convert computer data into forms that are useful in the world outside the computer; such devices include displays, printers, plotters, digital-to-analog converters, mechanical relays, and fuel-injected carburetors. Many other devices convert data from the outside world into forms usable by the computer; examples include keyboards, text scanners, joysticks, analog-to-digital converters, mechanical switches, and crash detectors. The sole purpose of some devices is simply to store data for later retrieval; these are called *mass storage devices* and include magnetic disks and tapes.

Sometimes the dividing line between an interface and the device it controls is fuzzy. For example, Fig. 14 shows the circuitry associated with a simple mechanical keyboard. The encoder circuit converts a mechanical switch depression into a 7-bit number in the ASCII code. The bus interface can place this number on the i/o bus on demand by the processor. So it seems that the device interface consists of the Encoder and Bus Interface blocks. However, in a typical system the encoder is packaged with the keyboard; then only a small number of wires are needed between the encoder (in the keyboard package) and the bus interface (in the CPU box). Most computer designers would say that the Encoder block is part of the keyboard and the device interface consists of the Bus Interface block alone. Fortunately, the dividing line is unimportant to i/o programs that deal with the keyboard. More important is the "i/o programming model" that an i/o program sees, as discussed next.

Ports

An *i/o port* (or *i/o register*) is a part of a device interface, a group of bits accessed by the processor

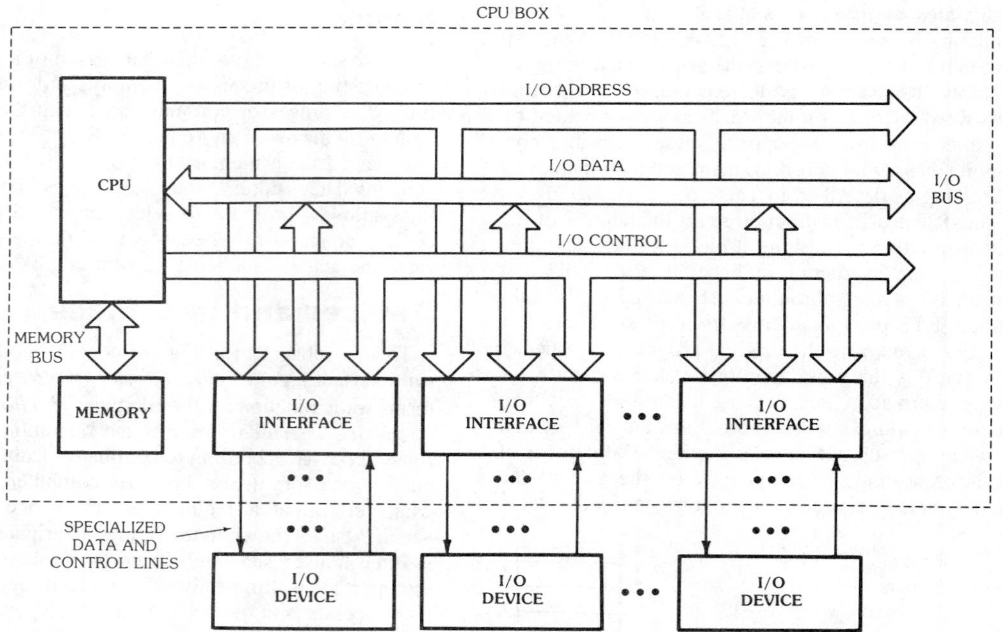

Fig. 13. Input/output (i/o) subsystem.

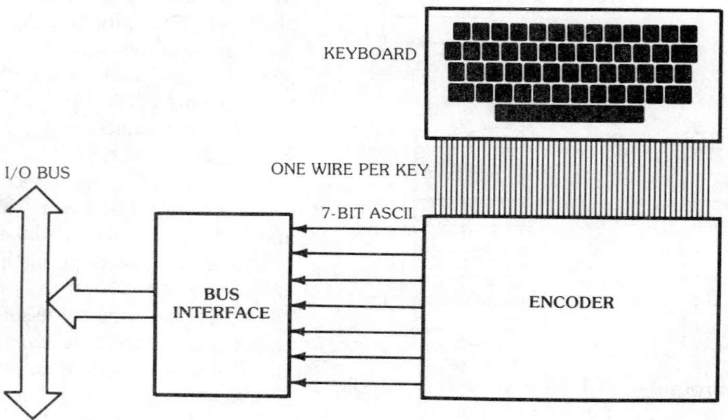

Fig. 14. Keyboard and interface.

during i/o operations. The "i/o programming model" of the keyboard in Fig. 14 contains one 8-bit i/o port named KBDATA, as shown in Fig. 15. The high-order bit of KBDATA is always 0. The low-order bits of KBDATA contain the output of the Encoder block in Fig. 14, that is, the 7-bit ASCII code for the key that is currently being depressed, or 0000000 if no key is depressed.

In order to read data from the keyboard in Fig. 14, a program must execute an instruction that transfers the contents of KBDATA into one of the processor registers. Once the data is in the processor, it can be manipulated like any other data. Although the interface "writes" keyboard data into KBDATA, the port is read-only from the point of view of the processor; any attempt by the processor to write data into KBDATA has no effect. Therefore, we can call KBDATA an "input port."

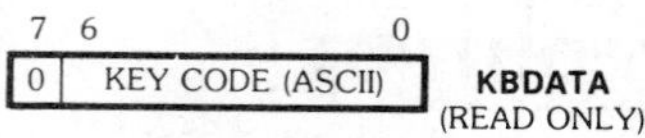

Fig. 15. Programming model for a keyboard.

Fig. 16A shows a very simple output device that interprets an 8-bit byte as two 4-bit bcd digits and displays the digits on two seven-segment displays. The i/o programming model consists of one 8-bit port, DIG-

OUT, shown in Fig. 16B. In order to display two digits, the processor must transfer an 8-bit value into DIGOUT. In this case, DIGOUT is an "output port" and is write-only from the point of view of the processor; an attempt to read it produces an undefined value.

I/O Programming

So far, nothing has been said about how i/o-port data is transferred to and from processor registers. This subsection discusses two techniques that are used in different processors to perform i/o transfers.

In *isolated i/o*, the ports are accessed by special i/o instructions. For example, the Z8000 has two instructions for transferring the contents of an 8-bit port to and from registers:

```
INB    rn,pn     REGISTER[rn] ←
                      INPUTPORT[pn];
OUTB   pn,rn     OUTPUTPORT[pn] ←
                       REGISTER[rn];
```

The INB and OUTB instructions perform simple data transfers, as do load and store instructions, except that they access an array of i/o ports instead of an array of memory bytes. Since the main memory and the i/o ports are on different buses, the "address spaces" accessed by memory reference and i/o instructions are different, even though they both may happen to use 16-bit addresses.

It has been observed that i/o buses are very similar to memory buses, and that i/o instructions are similar to load and store instructions on memory. *Memory-mapped i/o* takes advantage of the similarity by eliminating the i/o bus and i/o instructions.

Fig. 17 shows the hardware organization of a computer with memory-mapped i/o. Both the main memory

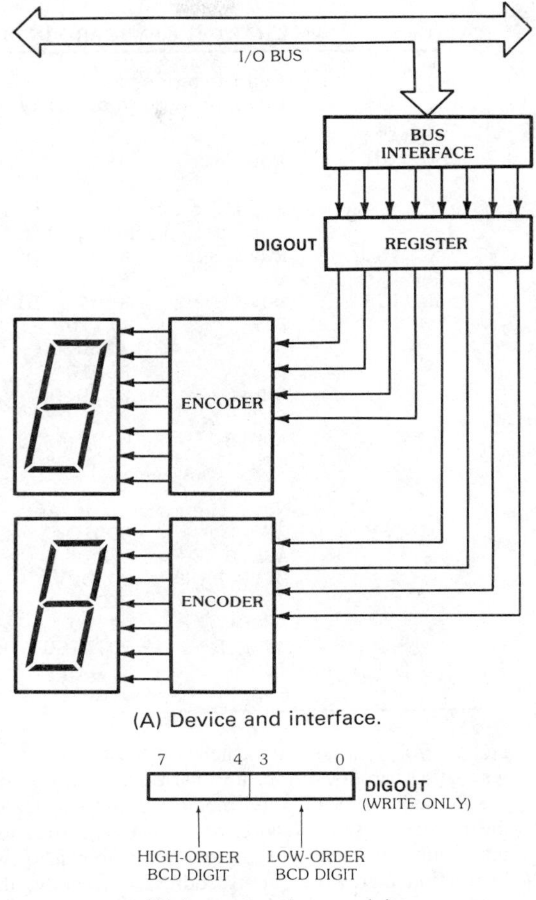

Fig. 16. Seven-segment display.

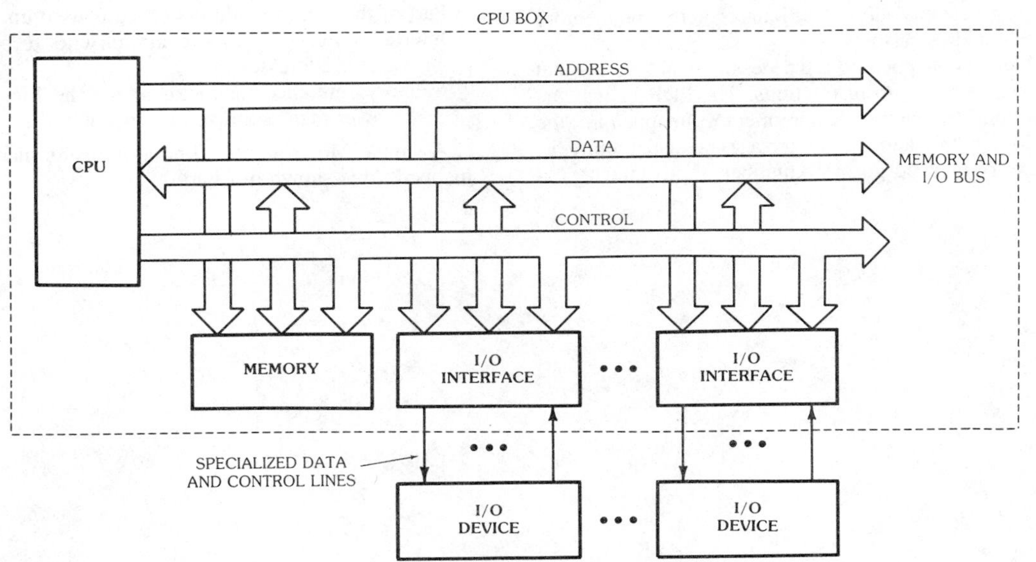

Fig. 17. Memory-mapped i/o structure.

CHART 3. KEYBOARD INPUT AND DISPLAY OUTPUT FOR THE 68000

```
*
*          Read two decimal digits from the keyboard and display them on the
*          seven-segment display. Ignore illegal characters.
KBDATA     EQU        $F004           Keyboard input port address.
DIGOUT     EQU        $F008           Seven-segment display port address.
*
*          First, a subroutine to read one decimal digit, convert it to 4-bit bcd,
*          and return the result in register D0.
RDDIG      MOVE.B     KBDATA,D0       Read current character.
           BEQ        RDDIG           Wait for a key to be pressed.
WAITUP     MOVE.B     KBDATA,D1       Now the character is in D0,
           BNE        WAITUP            wait for the key to be released.
           CMP.B      #$30,D0         Was it a valid decimal digit?
           BLT        RDDIG           Not if it is less than ASCII '0' ...
           CMP.B      #$39,D0
           BGT        RDDIG           ... or greater than ASCII '9'.
           ADD.B      D0,#-$30        Convert ASCII '0'-'9' to 0-9.
           RTS                        Done, return.
*
*          Now, the main program.
DIGDSP     JSR        RDDIG           Read high-order decimal digit into D0.
           ASL.B      #4,D0           Shift left 4 bits
           MOVE.B     D0,D2             and save in D2.
           JSR        RDDIG           Read low-order decimal digit.
           OR.B       D2,D0           Merge (logical OR) with high-order digit.
           MOVE.B     D0,DIGOUT       Send to seven-segment display.
           JMP        DIGDSP          Do another pair of digits.
           END        DIGDSP
```

and all i/o ports communicate with the processor by way of a shared Memory and I/O Bus. Each i/o port has an address in the main memory address space of the processor. An input port responds to any instruction that reads at its address; an output port responds to any instruction that writes at its address. Typically, the system designer reserves a portion of the total address space for i/o ports, for example, the top 4 kilobytes. However, theoretically a hardware designer can locate an i/o port at any address, as long as there is no memory at that address also.

Memory-mapped i/o is a necessity in processors that have no special i/o instructions. The PDP-11 was the first minicomputer to require memory-mapped i/o; the 68000, 6809, and 32000 processors also require it. Memory-mapped i/o has a number of advantages:

No opcodes or processor circuits are used up for i/o instructions.

All memory reference instructions, not just loads and stores, may be used to manipulate i/o ports.

The number of available i/o port addresses is virtually unlimited.

The hardware bus structure is simplified.

There are also disadvantages:

Part of the memory address space is used up.

Interfaces may need more circuitry to recognize longer addresses.

Memory reference instructions may be longer or slower than optimized i/o instructions.

A simple Motorola 68000 program using memory-mapped i/o is shown in Chart 3.

43 Logic Design

Edward J. McCluskey

INTRODUCTION

The distinguishing feature of the circuits to be discussed in this chapter is the use of two-valued, or binary, signals. There will be some deviation of the signals from their nominal values, but within certain limits this variation will not affect the performance of the circuit. If the variations exceed these limits, the circuit will not behave properly, and steps must be taken to confine the signals to the proper ranges. When the statement is made that the signals are two-valued, what is really meant is that the value of each signal is within one of two (nonoverlapping) continuous ranges. Since the operation of the circuit does not depend on exactly which value within a given range the signal takes on, a particular value is chosen to represent the range, and the signal is said to be equal to this value.

The exact numerical value of the signal is not important. It is possible to have two circuits perform the same function and have completely different values for their signals. In order to avoid any possible confusion that might arise because of this situation and to simplify the design procedures, it is customary to carry out the logic design without specifying the actual values of the signals. Once the logic design has been completed, actual values must be assigned to the signals in the course of designing the detailed electrical circuit. For the purposes of the logic design, arbitrary symbols are chosen to represent the two values to which the signals are to be restricted. An algebra* using these symbols is then developed as the basis for formal design techniques. The development of such an algebra will be described next.

POSTULATES

The two symbols most commonly chosen to represent the two logic values taken on by binary signals are 0 and 1. It should be emphasized that there is no numerical significance attached to these symbols. For an electronic circuit that has its signals equal to either 0 or 5 volts, it is possible to assign the symbol 1 to 0 volts and the symbol 0 to 5 volts. This choice for the correspondence between logic value and physical value in which 1 corresponds to the more negative physical value is called *negative logic*. More common is *positive logic* in which the logic 1 is assigned to the more positive physical signal value. Negative logic is often used for technologies such as PMOS in which the active signal has the lower value. Unless stated otherwise, positive logic will be used here.

* This algebra will be called *switching algebra*. It is identical with a Boolean algebra and was originally applied to switching circuits (reference 1) by reinterpreting Boolean algebra in terms of switching circuits rather than by developing a switching algebra directly, as will be done here.

Some other set of symbols such as T and F, H and L, or + and − could be used instead of 0 and 1, but there is a strong tradition behind the use of 0 and 1.

Switching variables are used in logic networks to represent the signals present at the network inputs and outputs. A switching variable is also useful in representing the state of a switch. This is shown in Fig. 1. The switch operation is defined as the two switch terminals (a and b) being connected together if and only if the control variable, X, equals 1. When $X = 0$, there is an open circuit between the two switch terminals. There is a variable, T, associated with the switch that equals 1 when the terminals are connected together and that equals 0 when there is an open circuit between the terminals. The variable, T, is called the *transmission* of the switch. (It is also possible to associate with the switch a variable that equals 1 only when the switch is open. Such a variable is called the *switch hindrance*. This was used in the very early papers on switching theory in connection with contact networks. The transmission concept is the standard usage at present.)

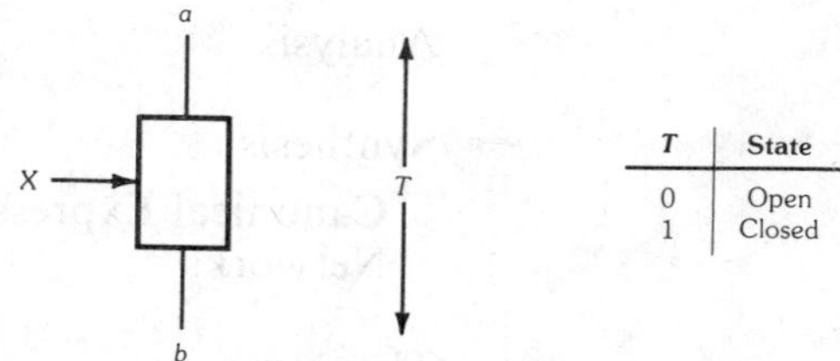

T	State
0	Open
1	Closed

Fig. 1. The transmission, T, of a switch.

The first postulate of the switching algebra can now be presented. This is merely a formal statement of the fact that switching variables are always equal to either 0 or 1. (In the statements of postulates and theorems that follow, the symbols X, Y, Z, X_1, X_2, . . . , X_n will be used to represent switching variables.)

$$\text{(P1)}\ X = 0 \text{ if } X \neq 1 \qquad \text{(P1')}\ X = 1 \text{ if } X \neq 0$$

To implement general switching networks, it is necessary to be able to obtain, for any signal representing a switching variable, a signal that has the opposite value. A circuit for realizing this function is called an *inverter*. The symbol used to represent an inverter is shown in Fig. 2.

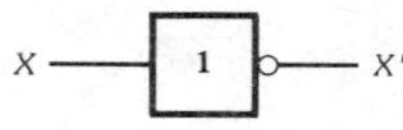

Fig. 2. Symbol for inverter.

If X represents a switching variable, the symbol X' is used to represent the signal having the opposite value. This notation is specified formally in the second switching-algebra postulate:

(P2) If X = 0, then $X' = 1$ (P2′) if $X = 1$, then $X' = 0$

The two symbols, X and X', are not two different variables, since they involve only X. In order to distinguish them, the term *literal* is used, where a literal is defined as a variable with or without an associated prime, and X and X' are different literals. The literal X' is called the *complement*† of X. Similarly, 0 is called the complement of 1, and 1 is called the complement of 0. The logical operation of the inverter can now be described in terms of switching algebra. If X_1 represents the input signal and X_0 the output signal, then $X_0 = X_1'$, since X_0 is high when X_1 is low, and vice versa.

In order to represent the action of other switching devices, it is necessary to define additional algebraic operations. A two-input AND gate will be considered first. If the two inputs are represented by X_1 and X_2 and the output by X_0, the logical performance of the circuit is represented by the table of Fig. 3A, in which 1 corresponds to a high voltage and 0 corresponds to a low voltage. This table is also correct for the transmission, X_0, of two switches, X_1 and X_2, in series. The operation represented is identical with ordinary multiplication and will be defined as multiplication in the switching algebra developed here. Thus, the equation for an AND gate, or two switches in series, is $X_0 = X_1X_2$.

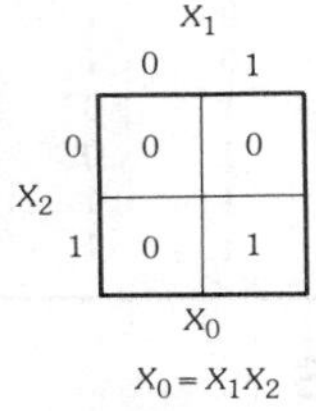

(A) Logical performance.

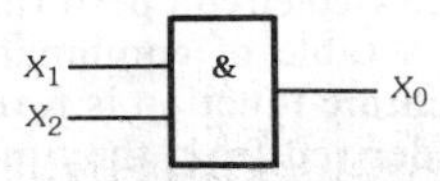

(B) AND gate.

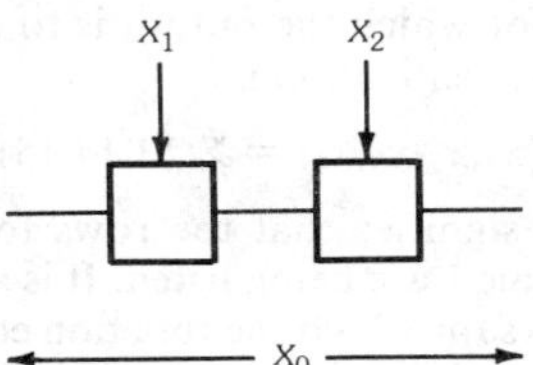

(C) Transmission gates in series.

Fig. 3. The AND function.

† Some authors use the symbols $\bar{X}$ or $\sim X$ to indicate the complement of X.

The table for an OR gate, or two switches in parallel, is shown in Fig. 4A. This table will be taken as the definition of switching-algebra addition.§ It is identical with ordinary addition except for the case $1 + 1 = 1$. The remaining postulates are merely restatements of the definitions of multiplication and addition:

(P3) $0 \cdot 0 = 0$ (P3′) $1 + 1 = 1$
(P4) $1 \cdot 1 = 1$ (P4′) $0 + 0 = 0$
(P5) $1 \cdot 0 = 0 \cdot 1 = 0$ (P5′) $0 + 1 = 1 + 0 = 1$

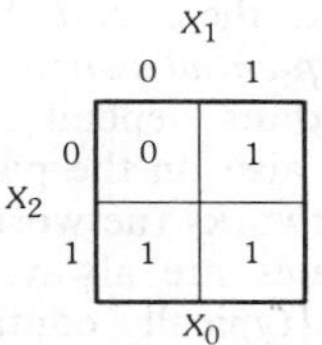

(A) Logical performance.

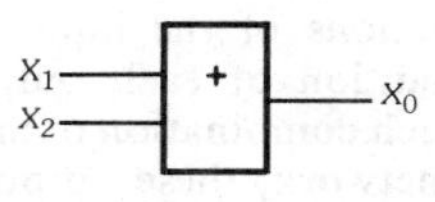

(B) OR gate.

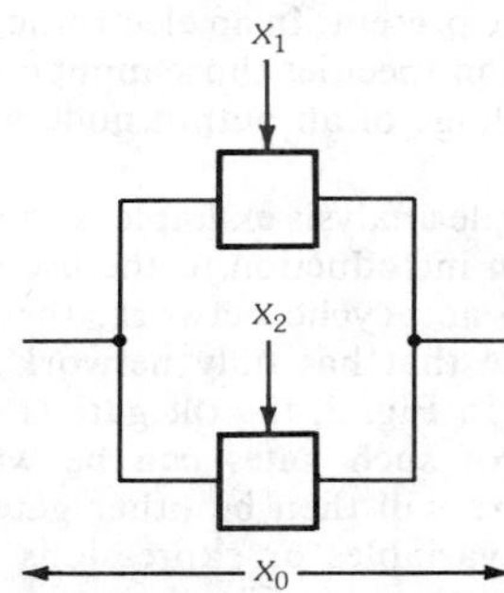

(C) Transmission gates in parallel.

Fig. 4. The OR function.

All the postulates have been stated in pairs. These pairs have the property that when (0 and 1) and (+ and ·) are interchanged one member of the pair is changed into the other. This is an example of the general *principle of duality*, which is true for all switching-algebra theorems since it is true for all the postulates. This algebraic duality arises from the fact that either 1 or 0 can be assigned to an open circuit. If 0 is chosen to represent a high voltage, the expression for an AND gate becomes $x_0 = x_1 + x_2$.

§ This operation is also called logical addition, and some writers use the notation $x_1 \vee x_2$ rather than $x_1 + x_2$.

Similarly, the transmission of two switches in series is equal to $x_1 + x_2$ if 0 is chosen to represent a closed circuit.

ANALYSIS

Analysis of a circuit consists in examining the circuit and somehow determining what the behavior of the circuit will be for all possible inputs. The switching algebra developed in the previous section is sufficient for the analysis of those switching circuits that are known as *combinational switching circuits*. A combinational switching circuit is defined as one for which the outputs depend only on the present inputs to the circuit. The other type of circuit, called a *sequential switching circuit*, is one for which the outputs depend not only on the present inputs but also on the past history of the inputs. Acyclic networks (networks with no feedback loops) of gates are always combinational. Sequential circuits typically contain latches, flip-flops, or feedback loops.

The analysis of a combinational circuit consists in writing an algebraic function for each output. These are functions of the input variables from which the condition of each output can be determined for each combination of input conditions. For a switch network, these output functions are the transmissions between each pair of external terminals. Only two-terminal networks will be considered for the present. In an electronic network, an output function specifies those input conditions for which the voltage of an output node will be at the high level.

A very simple analysis example is shown in Fig. 5 to serve as an introduction to the use of switching functions. In an acyclic network, there will be at least one gate that has only network inputs connected to it. In Fig. 5, the OR gate is such a gate. The output of such gates can be written down directly. There will then be other gates that have either input variables or expressions on all their inputs; the AND gate in Fig. 5 is such a gate. It is then possible to write expressions for the output of all these gates. By repeating this procedure, an expression will eventually be obtained for the network output.

A procedure for using these output functions to determine the circuit performance is to substitute a set of values for the input variables and then use the postulates to simplify the resulting expressions. A

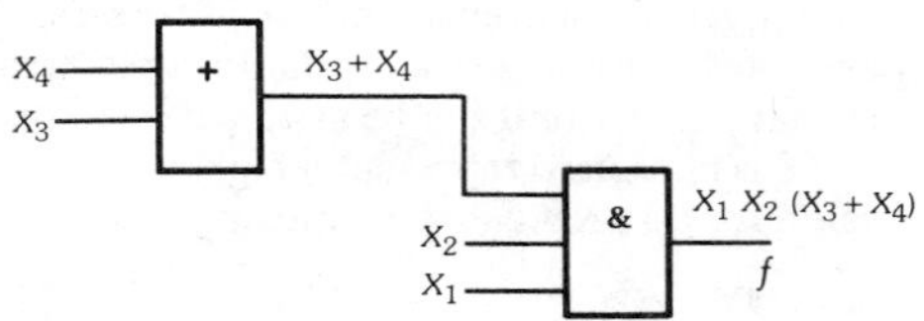

Fig. 5. Simple example of analysis of a gate network.

value for the output function will be obtained that specifies the output for the particular input combination chosen. For example, if the input combination $X_1 = 1$, $X_2 = 1$, $X_3 = 0$, $X_4 = 1$, is chosen, the output function of Fig. 5 becomes

$$f = X_1 X_2 (X_3 + X_4)$$
$$= 1 \cdot 1 (0 + 1) = 1(1) = 1$$

By carrying out this procedure for all possible input combinations, it is possible to form a table that lists the output for each input combination. Such a table describes completely the circuit performance and is called a *table of combinations.** Table 1 is the table of combinations for the circuit of Fig. 5.

TABLE 1. TABLE OF COMBINATIONS FOR $f = X_1X_2(X_3 + X_4)$

	X_1	X_2	X_3	X_4	f
0	0	0	0	0	0
1	0	0	0	1	0
2	0	0	1	0	0
3	0	0	1	1	0
4	0	1	0	0	0
5	0	1	0	1	0
6	0	1	1	0	0
7	0	1	1	1	0
8	1	0	0	0	0
9	1	0	0	1	0
10	1	0	1	0	0
11	1	0	1	1	0
12	1	1	0	0	0
13	1	1	0	1	1
14	1	1	1	0	1
15	1	1	1	1	1

SYNTHESIS

In designing a combinational circuit, it is necessary to carry out in reverse the procedure just described. The desired circuit performance is specified by means of a table of combinations. From this table, an algebraic function is formed, and finally the circuit is derived from the function. A concise means of specifying the table of combinations, called a *decimal specification*, is to list the numbers of the rows for which the output is to equal 1. For Table 1, this specification is:

$$f(x_1,x_2,x_3,x_4) = \Sigma(13,14,15)$$

where the Σ signifies that the rows for which the function equals 1 are being listed. It is also possible to list the rows for which the function equals 0, such a list being preceded by the symbol Π to indicate that it is the zero rows which are listed. This specification for the preceding table is

* In logic, the table is called a *truth table*, and some writers use this term when discussing switching circuits.

$$f(x_1,x_2,x_3,x_4) = \Pi(0,1,2,3,4,5,6,7,8,9,10,11,12)$$

In order to avoid any ambiguity in these specifications, it is necessary to adopt some rule for numbering the rows of the table of combinations. The usual procedure is to regard each row of the table as a binary number and then use the decimal equivalent of this binary number as the row number. The output-column entries are not included in forming the binary row numbers. There is nothing special about decimal numbers other than the fact that they are the most familiar; any other number base, such as octal, could be used. The reason for using a number system other than binary is simply that binary numbers take too much space to write down.

It has been pointed out that the table of combinations is a complete specification for a combinational circuit. The first step in designing a circuit is to formulate such a table. There are no general formal techniques for doing this. When a sequential circuit is being designed, it is customary to reduce the sequential-design problem to (several) combinational problems, and formal techniques exist for doing this. However, when a combinational circuit is being designed, no formal techniques are available, and it is necessary to rely on common sense. This is not too surprising, since any formal technique must start with a formal statement of the problem, and this is precisely what the table of combinations is. As an example of how this is done, the table of combinations for a circuit to check binary-coded decimal digits for the 8, 4, 2, 1 code is shown in Table 2. This circuit is to deliver an output whenever a digit having an invalid combination of bits is received.

In forming a table of combinations, there very often are rows for which it is unimportant whether the function equals 0 or 1. The usual reason for this situation is that the combinations of inputs corresponding to these rows can never occur (when the circuit is functioning properly). As an example of this, consider a circuit to translate from the 8, 4, 2, 1 bcd code to a Gray (cyclic binary) code. When the circuit is working correctly, the input combinations represented by rows 10 through 15 of the table of combinations cannot occur. Therefore, the output need not be specified for these rows. The symbol d will be used to indicate the output condition for such a situation.† The output conditions so denoted are called *don't-care conditions* (see Table 3). It is possible to include the d rows in the decimal specification of a function by listing them after the symbol d. Thus the decimal specification for g_1 of Table 3 would be

$$g_1(b_8,b_4,b_2,b_1) = \Sigma(1,2,5,6,9) + d(10,11,12,13,14,15)$$
$$= \Pi(0,3,4,7,8) + d(10,11,12,13,14,15)$$

TABLE 2. TABLE OF COMBINATIONS FOR CIRCUIT TO CHECK BINARY-CODED-DECIMAL DIGITS

	b_8	b_4	b_2	b_1	f	
0	0	0	0	0	0	
1	0	0	0	1	0	
2	0	0	1	0	0	
3	0	0	1	1	0	
4	0	1	0	0	0	
5	0	1	0	1	0	
6	0	1	1	0	0	
7	0	1	1	1	0	
8	1	0	0	0	0	
9	1	0	0	1	0	
10	1	0	1	0	1	Invalid code words
11	1	0	1	1	1	
12	1	1	0	0	1	
13	1	1	0	1	1	
14	1	1	1	0	1	
15	1	1	1	1	1	

TABLE 3. TABLE OF COMBINATIONS FOR CIRCUIT TO TRANSLATE FROM BCD 8,4,2,1 CODE TO GRAY CODE

	BCD-Code Inputs				Gray-Code Outputs			
	b_8	b_4	b_2	b_1	g_4	g_3	g_2	g_1
0	0	0	0	0	0	0	0	0
1	0	0	0	1	0	0	0	1
2	0	0	1	0	0	0	1	1
3	0	0	1	1	0	0	1	0
4	0	1	0	0	0	1	1	0
5	0	1	0	1	0	1	1	1
6	0	1	1	0	0	1	0	1
7	0	1	1	1	0	1	0	0
8	1	0	0	0	1	1	0	0
9	1	0	0	1	1	1	0	1
10	1	0	1	0	d	d	d	d
11	1	0	1	1	d	d	d	d
12	1	1	0	0	d	d	d	d
13	1	1	0	1	d	d	d	d
14	1	1	1	0	d	d	d	d
15	1	1	1	1	d	d	d	d

Canonical Expressions

After the table of combinations has been formed, the next step in designing a circuit is to write an algebraic expression for the output function. The simplest output functions to write are those that equal 1 for only one row of the table of combinations or those that equal 0 for only one row. It is possible to associate with each row two functions, one that equals 1 only for the row and one that equals 0 only for the row (Table 4). These functions are called *fundamental products or minterms* and *fundamental sums or maxterms*, respectively. Each fundamental product or sum contains all the input variables. The rule for forming the fundamental product for a given row is to prime any variables

† In the literature, the symbol ϕ is also used.

TABLE 4. TABLE OF COMBINATIONS SHOWING FUNDAMENTAL PRODUCTS AND FUNDAMENTAL SUMS

	x_1	x_2	x_3	Fundamental Products	Fundamental Sums
0	0	0	0	$x'_1x'_2x'_3$	$x_1 + x_2 + x_3$
1	0	0	1	$x'_1x'_2x_3$	$x_1 + x_2 + x'_3$
2	0	1	0	$x'_1x_2x'_3$	$x_1 + x'_2 + x_3$
3	0	1	1	$x'_1x_2x_3$	$x_1 + x'_2 + x'_3$
4	1	0	0	$x_1x'_2x'_3$	$x'_1 + x_2 + x_3$
5	1	0	1	$x_1x'_2x_3$	$x'_1 + x_2 + x'_3$
6	1	1	0	$x_1x_2x'_3$	$x'_1 + x'_2 + x_3$
7	1	1	1	$x_1x_2x_3$	$x'_1 + x'_2 + x'_3$

that equal 0 for the row and leave unprimed variables that equal 1 for the row. The fundamental product equals the product of the literals so formed. The fundamental sum is formed by a completely reverse, or dual, procedure. Each variable that equals 0 for the row is left unprimed, and each variable that equals 1 for the row is primed. The fundamental sum is the sum of the literals obtained by this process. The algebraic expression for any table for which the output is equal to 1 (or 0) for only one row can be written down directly by choosing the proper fundamental product (or sum). For example, the output function specified by $f(x_1,x_2,x_3) = \Sigma(6)$ is written algebraically as $f = x_1x_2x'_3$, and the output function $f(x_1, x_2,x_3) = \Pi(6)$ is written as $f = x'_1 + x'_2 + x_3$. The fundamental product corresponding to row i of the table of combinations will be denoted by p_i, and the fundamental sum corresponding to row i will be denoted by s_i.

The algebraic expression that equals 1 (or 0) for more than one row of the table of combinations can be written directly as a sum of fundamental products or as a product of fundamental sums. A function f that equals 1 for two rows, i and j, of the table of combinations can be expressed as a sum of the two fundamental products p_i and p_j: $f = p_i + p_j$. When the inputs correspond to row i, $p_i = 1$ and $p_j = 0$ so that $f = 1 + 0 = 1$. When the inputs correspond to row j, $p_i = 0$, $p_j = 1$, and $f = 0 + 1 = 1$. When the inputs correspond to any other row, $p_i = 0$, $p_j = 0$, $f = 0 + 0 = 0$. This shows that the function $f = p_i + p_j$ does equal 1 only for rows i and j. This argument can be extended to functions that equal 1 for any number of input combinations —they can be represented algebraically as a sum of the corresponding fundamental products (see Table 5). An algebraic expression that is a sum of fundamental products is called a *canonical sum*. An arbitrary function can also be expressed as a product of fundamental sums. This form is called a *canonical product*. The canonical product for a function that is equal to 0 only for rows i and j of the table of combinations is given by $f = s_i\, s_j$. For row i, $s_i = 0$ so that $f = 0$. For row j, $s_j = 0$ so that $f = 0$, and for any other row $s_i = s_j = 1$ so that $f = 1$. In general, the canonical product is equal to the product of all fundamental sums that correspond to input conditions for which the function is to equal 0.

TABLE 5. $f(x_1,x_2,x_3) = \Sigma(1,2,3,4)$

(A) Table of Combinations

	x_1	x_2	x_3	f
0	0	0	0	0
1	0	0	1	1
2	0	1	0	1
3	0	1	1	1
4	1	0	0	1
5	1	0	1	0
6	1	1	0	0
7	1	1	1	0

(B) Canonical Sum

$$f = x'_1x'_2x_3 + x'_1x_2x'_3 + x'_1x_2x_3 + x_1x'_2x'_3$$

(C) Canonical Product

$$f = (x_1 + x_2 + x_3)\,(x'_1 + x_2 + x'_3)\,(x'_1 + x'_2 + x_3)\,(x'_1 + x'_2 + x'_3)$$

It is possible to write a general expression for the canonical sum by making use of the following theorems:

$$\text{(T2}'\text{)}\ x\cdot 0 = 0$$
$$\text{(T1)}\ x + 0 = x \qquad \text{(T1}'\text{)}\ x\cdot 1 = x$$

If the value of the function $f(x_1,x_2, \ldots ,x_n)$ for the ith row of the table of combinations is f_i ($f_i = 0$ or 1), then the canonical sum is given by

$$f(x_1,x_2, \ldots ,x_n) = f_0p_0 + f_1p_1 + \ldots + f_{(2^n-1)}p_{(2^n-1)}$$
$$= \sum_{i=0}^{2^n-1} f_ip_i$$

For the function $f(x_1,x_2) = \Sigma(0,2)$, the values of the f_i are $f_0 = f_2 = 1, f_1 = f_3 = 0$ so that

$$\begin{aligned} f(x_1,x_2) &= 1\cdot p_0 + 0\cdot p_1 + 1\cdot p_2 + 0\cdot p_3 \\ &= p_0 + 0 + p_2 + 0 \\ &= p_0 + p_2 \\ &= x'_1x'_2 + x_1x'_2 \end{aligned}$$

In a similar fashion, a general expression for the canonical product can be obtained by using the theorems

$$\text{(T2)}\ x + 1 = 1$$
$$\text{(T1)}\ x + 0 = x \qquad \text{(T1}'\text{)}\ x\cdot 1 = x$$

The resulting expression is

$$f(x_1,x_2, \ldots ,x_n) = (f_0 + s_0)\ (f_1 + s_1)\ \ldots\ [f_{(2^n-1)} + s_{(2^n-1)}]$$

$$= \sum_{i=0}^{2^n-1} (f_i + s_i)$$

For $f(x_1,x_2) = \Sigma(0,2) = \Pi(1,3)$ this becomes

$$\begin{aligned} f(x_1,x_2) &= (1 + s_0)(0 + s_1)(1 + s_2)(0 + s_3) \\ &= 1 \cdot s_1 \cdot 1 \cdot s_3 \\ &= s_1 \cdot s_3 \\ &= (x_1 + x'_2)(x'_1 + x'_2) \end{aligned}$$

Networks

A technique for obtaining an algebraic expression from a table of combinations has just been described. A circuit can be drawn directly from this expression by reversing the procedures used to analyze series-parallel switch networks or gate networks. The circuit for a single fundamental product is just a series connection of switches or an AND gate with appropriate inputs. For a canonical sum involving more than one fundamental product, the circuit consists of a number of parallel subnetworks, each subnetwork corresponding to one fundamental product, or a number of AND gates with their outputs connected as inputs to an OR gate. This is shown in Figs. 6A and 7A. Similarly, the switch network corresponding to a canonical product consists of a number of subnetworks in series, each subnetwork corresponding to one fundamental sum and consisting of switches in parallel (Fig. 6B). The gate network corresponding to a canonical product consists of a number of OR gates with their outputs connected as the inputs of an AND gate (Fig. 7B). These conclusions can be summarized as follows:

For any arbitrary table of combinations, a network whose performance corresponds to the table of combinations can be constructed of:

1. Switches
2. AND gates, OR gates, and inverters if there are no restrictions on the number of gates available

In a certain sense, the design procedure is now complete. A method has been presented for going from a table of combinations to a circuit diagram. However, the canonical circuits so designed are usually very uneconomical and therefore unsatis-

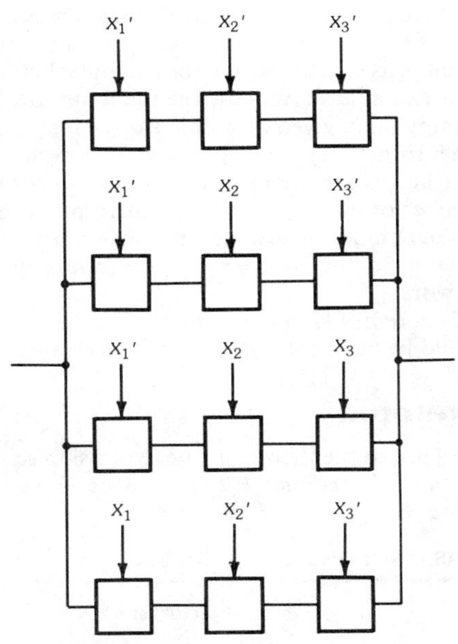

(A) Network derived from canonical sums.

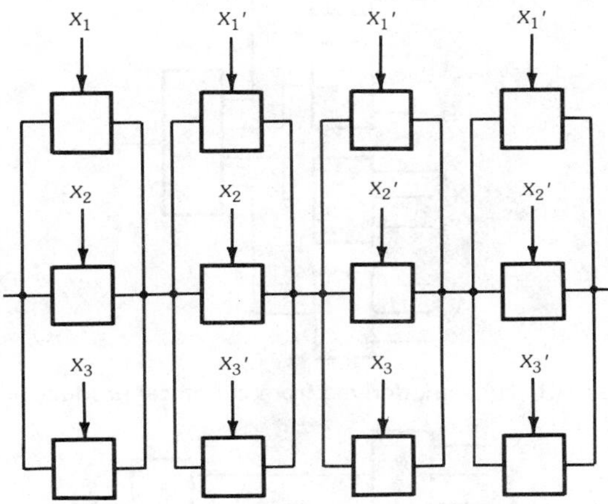

(B) Network derived from canonical product.

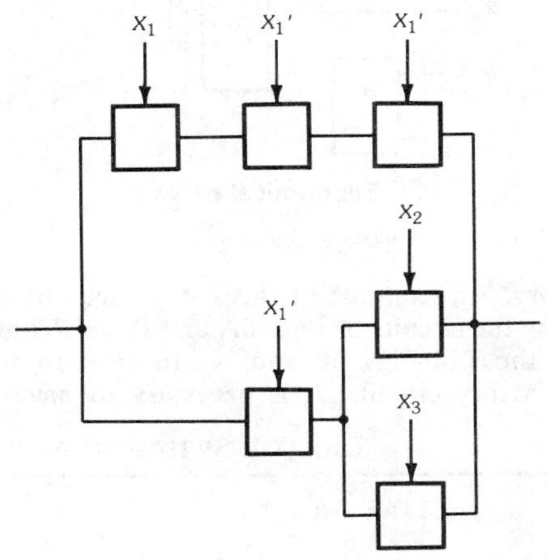

(C) Economical network.

Fig. 6. Switch networks for $\Sigma(1,2,3,4)$.

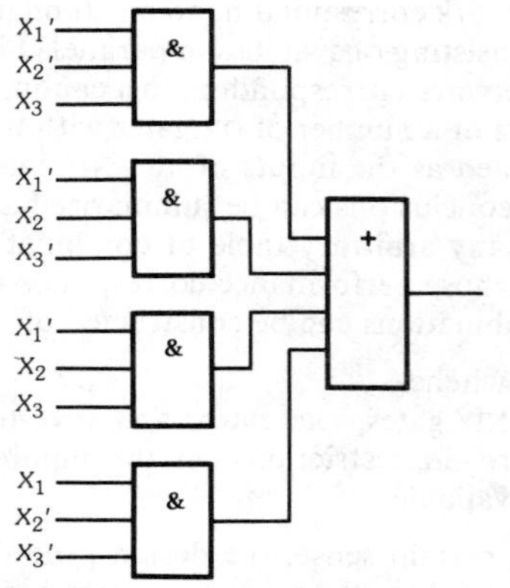

(A) Network derived from canonical sums.

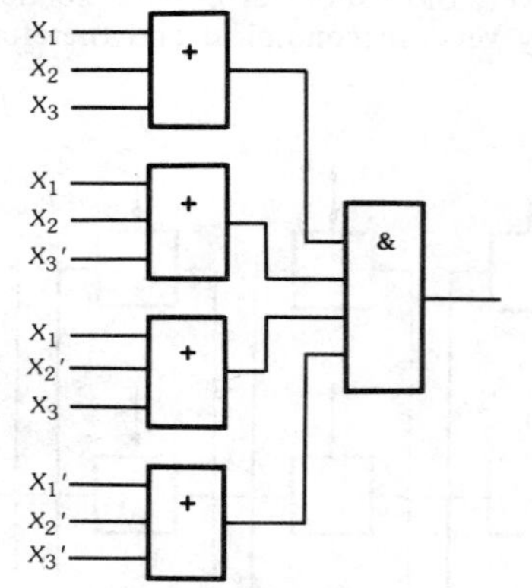

(B) Network derived from canonical product.

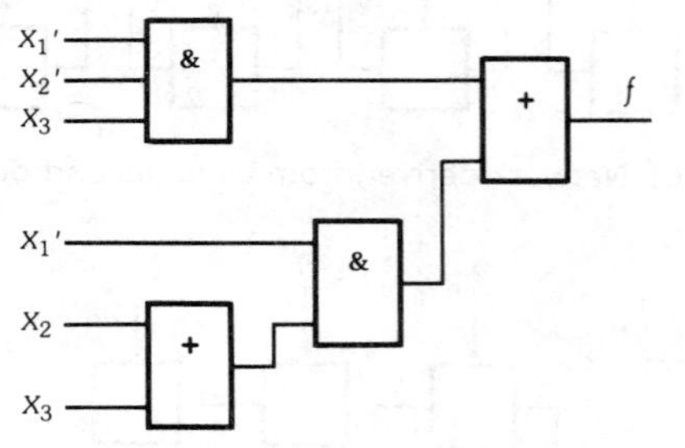

(C) Economical network.

Fig. 7. Gate networks for $\Sigma(1,2,3,4)$.

factory. An example of this can be seen by comparing the circuits of Figs. 6A and B and 7A and B with those of Figs. 6C and 7C. In order to design satisfactory circuits, it is necessary to have procedures for simplifying them so that they correspond to simpler circuits.

THEOREMS

The method of using switching algebra in designing switching circuits is to formulate the desired circuit performance as an algebraic expression and then to manipulate the expression into a form from which a desirable circuit can be arrived at directly. The manipulations are carried out by means of the theorems that will be presented in this section.

Single-Variable Theorems

The switching-algebra theorems that involve only a single variable are shown in Chart 1. Three of these theorems, T2, T3 and T3′, are especially noteworthy since they are false for ordinary algebra. Theorems T3 and T3′ can be extended to $X + X + X + \ldots + X = nX = X$ and $X \cdot X \ldots X = X^n = X$.

Two- and Three-Variable Theorems

Chart 2 lists the switching-algebra theorems that involve two or three variables. Theorems T7 and T7′ are useful in eliminating terms from algebraic expressions and thereby eliminating elements from the corresponding networks. Theorem T10′ is noteworthy in that it is not true for ordinary algebra even though its dual, T10, is.

In reducing algebraic expressions, Theorems T11 and T11′ are very important and are used frequently, as is illustrated in the example that follows. This example shows how the theorems are used to manipulate a given algebraic expression into some other form. Very frequently, the form desired is one that has as few literals occurring as possible. The number of literal occurrences corresponds directly to the number of switches in a switch network and roughly to the number of gate inputs in a gate network.

Theorem T12 has no dual (T12′), for the dual would be identical with the original theorem.

Example

By use of the theorems, the expression $(c' + abd + b'd + a'b)(c + ab + bd)$ is to be shown equal to $b(a + c)(a' + c') + d(b + c)$.

CHART 1. SWITCHING-ALGEBRA THEOREMS INVOLVING ONE VARIABLE

(T1) $X + 0 = X$	(T′) $X \cdot 1 = X$	(Identities)
(T2) $X + 1 = 1$	(T2′) $X \cdot 0 = 0$	(Null elements)
(T3) $X + X = X$	(T3′) $X \cdot X = X$	(Idempotency)
(T4) $(X')' = X$		(Involution)
(T5) $X + X' = 1$	(T5′) $X \cdot X' = 0$	(Complements)

CHART 2. SWITCHING-ALEGBRA THEOREMS INVOLVING TWO OR THREE VARIABLES

(T6)	$X+Y=Y+X$	(T6′) $XY=YX$	(Commutative)
(T7)	$X+XY=X$	(T7′) $X(X+Y)=X$	(Absorption)
(T8)	$(X+Y')Y=XY$	(T8′) $XY'+Y=X+Y$	
(T9) (T9′)	$(X+Y)+Z=X+(Y+Z)=X+Y+Z$ $(XY)Z=X(YZ)=XYZ$		(Associative)
(T10) (T10′)	$XY+XZ=X(Y+Z)$ $(X+Y)(X+Z)=X+YZ$		(Distributive)
(T11) (T11′)	$(X+Y)(X'+Z)(Y+Z)=(X+Y)(X'+Z)$ $XY+X'Z+YZ=XY+X'Z$		(Consensus)
(T12)	$(X+Y)(X'+Z)=XZ+X'Y$		

	$(c'+abd+b'd+a'b)(c+ab+bd)$
(T12)	$c'(ab+bd)+c(abd+b'd+a'b)$
(T10)	$abc'+bc'd+abcd+b'cd+a'bc$
(T6)	$abc'+abcd+bc'd+a'bc+b'cd$
(T10)	$ab(c'+cd)+bc'd+a'bc+b'cd$
(T8′)	$ab(c'+d)+bc'd+a'bc+b'cd$
(T11′)	$ab(c'+d)+bc'd+a'bc+a'bd+b'cd$
(T10)	$abc'+abd+bc'd+a'bc+a'bd+b'cd$
(T6)	$(abd+a'bd)+abc'+bc'd+a'bc+b'cd$
(T10)	$bd(a+a')+abc'+bc'd+a'bc+b'cd$
(T5)	$bd(1)+abc'+bc'd+a'bc+b'cd$
(T1′)	$bd+abc'+bc'd+a'bc+b'cd$
(T6)	$bd+bc'd+abc'+a'bc+b'cd$
(T7)	$bd+abc'+a'bc+b'cd$
(T6),(T10)	$d(b+b'c)+abc'+a'bc$
(T8′)	$d(b+c)+abc'+a'bc$
(T10)	$d(b+c)+b(ac'+a'c)$
(T12)	$d(b+c)+b(a+c)(a'+c')$

n-Variable Theorems

The switching-variable theorems that involve an arbitrary number of variables are shown in Chart 3. Three of these theorems (T13, T13′, and T14) cannot be proved by perfect induction. For these theorems, the proofs require the use of finite induction.* Theorems T13 and T13′ are proved by first letting $n = 2$ and using perfect induction to prove their validity for this special case. It is then assumed that the theorems are true for $n = k$, and this is shown to imply that they must then be true for $n = k + 1$. This completes the proof, the details of which are given in the reference. Theorem T14 is proved by using Theorems T13 and T13′ along with the fact that every function can be split into the sum of several functions or the product of several functions. By successively splitting the function into subfunctions and using T13 and T13′, it is possible to prove T14.

Theorem T14, which is a generalization of T13, forms the basis of a method for constructing complementary networks. Two networks having outputs T_1 and T_2 are said to be *complementary* if $T_1 = T'_2$. The complementary network for any given network can be designed by writing the output, T_1,

* Reference 2.

CHART 3. SWITCHING-VARIABLE THEOREMS INVOLVING *n* VARIABLES

(DeMorgan's Theorems)

(T13) $(X_1+X_2+\cdots+X_n)'=X'_1X'_2\cdots X'_n$

(T13′) $(X_1X_2\cdots X_n)'=X'_1+X'_2+\cdots+X'_n$

(Generalized DeMorgan's Theorem)

(T14) $f(X_1,X_2,\ldots,X_n,+,\cdot)'=f(X'_1,X'_2,\ldots,X'_n,\cdot,+)$

(Expansion Theorem)

(T15) $f(X_1,X_2,\ldots,X_n)=X_1f(1,X_2,\ldots,X_n)+X'_1f(0,X_2,\ldots,X_n)$

(T15′) $f(X_1,X_2,\ldots,X_n)=[X_1+f(0,X_2,\ldots,X_n)][X'_1+f(1,X_2,\ldots,X_n)]$

for the first network, then forming T'_1 by means of T14, and then designing a network having output T'_1. For example, if $T_1 = (x + y)[w(y' + z) + xy]$, then $T'_1 = x'y' + (w' + yz')(x' + y')$.

It was pointed out in connection with Figs. 6 and 7 that the canonical networks are generally uneconomical. By manipulating the canonical sum or product with the aid of the theorems just presented, it is usually possible to obtain algebraic expressions that correspond to more economical networks than the canonical networks. The following example shows how this is done for the networks of Figs. 6 and 7. The final expressions correspond to the networks of Figs. 6C and 7C.

$$
\begin{aligned}
f &= X'_1X'_2X_3 + X'_1X_2X'_3 + X'_1X_2X_3 + X_1X'_2X'_3 \\
f &= X'_1X'_2X_3 + X'_1X_2X_3 + X'_1X_2X'_3 + X_1X'_2X'_3 \\
f &= X'_1X'_2X_3 + X'_1X_2X_3 + X'_1X_2X_3 + X'_1X_2X'_3 + X_1X'_2X'_3 \\
f &= X'_1X_3(X'_2 + X_2) + X'_1X_2(X_3 + X'_3) + X_1X'_2X'_3 \\
f &= X'_1X_3(1) + X'_1X_2(1) + X_1X'_2X'_3 \\
f &= X'_1X_3 + X'_1X_2 + X_1X'_2X'_3 \\
f &= X'_1(X_3 + X_2) + X_1X'_2X'_3
\end{aligned}
$$

Many of the theorems of ordinary algebra are also valid for switching algebra. One that is not is the cancellation law. In ordinary algebra, it follows that $Y=Z$ if $X+Y=Y+Z$. In switching algebra, this is not true. For example, it is generally true that $X+XY=X+0$, but it is not necessarily true that $XY=0$. This can be easily verified by writing out the tables of combinations for $f_1(X,Y)=X+XY$, $f_2=(X,Y)=X+0$, and $f_3(X,Y)=XY$. Similar remarks apply to the situation in which $XY=XZ$ does not imply that $Y=Z$.

GENERAL GATE NETWORKS

The previous discussion of gate networks in this chapter has been concerned solely with networks constructed of AND gates and OR gates. This can be considered only an introduction to the topic of gate networks, for other types of gate are equally important. In this section, other types of gate networks will be considered.

It has been shown that any arbitrary switching function can be realized by a network of AND gates, OR gates, and inverters. A natural question to ask in this connection is whether all three types of elements are necessary. Inverters are required if the inputs to the network consist of signals representing the input variables but not of signals representing the complements of the input variables. The situation when signals representing the complements are available is called *double-rail logic*, and when the complements are not available, the term *single-rail* is used. Both techniques are employed, but for the purposes of the present discussion it will be assumed that complements are not directly available (single-rail logic). The function $f(x) = x'$ cannot be realized by a network of AND gates and OR gates only.

It is clear that inverters are required to realize arbitrary functions, but the possibility of using only AND gates and inverters still exists. That the OR gates are not necessary is easily demonstrated, for it is possible to construct a network having the function of an OR gate and using only AND gates and inverters. This is done by making use of DeMorgan's theorem—$X + Y = (X'Y')'$—as is illustrated in Fig. 8. Thus any network consisting of AND gates, OR gates, and inverters can be changed into a network containing only AND gates and inverters by using the replacement shown in Fig. 8 to remove the OR gates. By duality, a similar technique can be used to remove the AND gates instead.

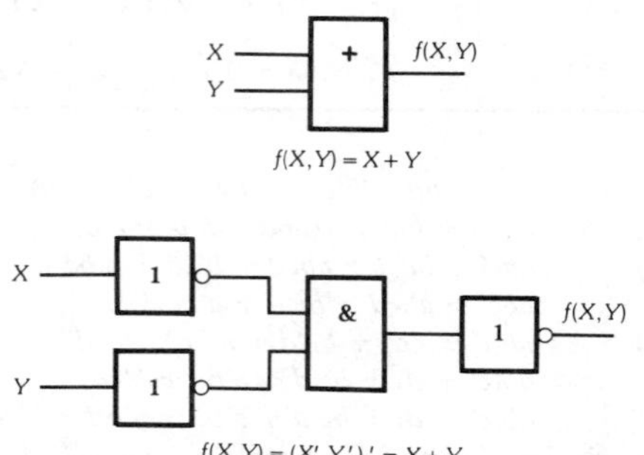

Fig. 8. Realization of an OR gate by means of an AND gate and inverters.

Since it is not possible to use only inverters to realize arbitrary functions, a minimum set of elements has now been determined. Because it is possible to construct a network containing only AND gates and inverters for any arbitrary function, the AND gate and inverter are said to form a *complete gate set*. Similarly, the OR gate and inverter form a complete gate set.

The two operations of the AND function and the complement can be combined in a single gate, the NAND gate, shown in Fig. 9. This is a very common integrated-circuit gate. It comprises a complete gate set in one gate since: (1) an inverter is obtained if all inputs are connected to the same source, as in Fig. 9C; (2) an AND gate is formed by combining two NAND gates, as in Fig. 9D.

Two symbols for the NAND-gate are shown in Fig. 9A. This is because the basic NAND-gate function (XY') can also be written as $X' + Y'$ by using DeMorgan's Theorem (T13). Use of the two symbols facilitates analysis and synthesis using these gates. The small circles ("bubbles") in Fig. 9 indicate inversion, and from a logical standpoint each bubble can be replaced by an inverter.

Another very common integrated-circuit gate is the NOR gate shown in Fig. 10. It is the dual of the NAND gate and is also a single gate that is sufficient to implement a network for any arbitrary switching function.

The other important IC gate type is the *sum*

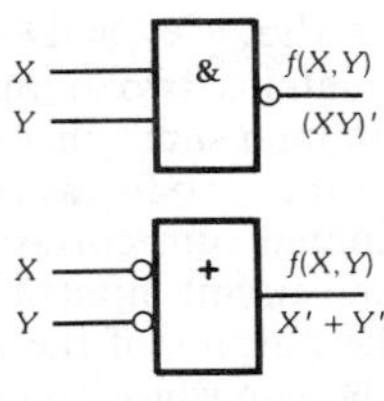

(A) Symbols.

X	Y	NAND $f(X,Y)$
0	0	1
0	1	1
1	0	1
1	1	0

(B) Table of combinations.

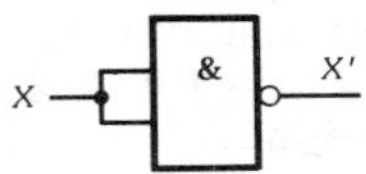

(C) Inverter connection.

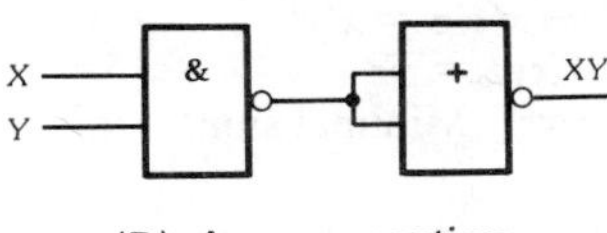

(D) AND connection.

Fig. 9. NAND gate.

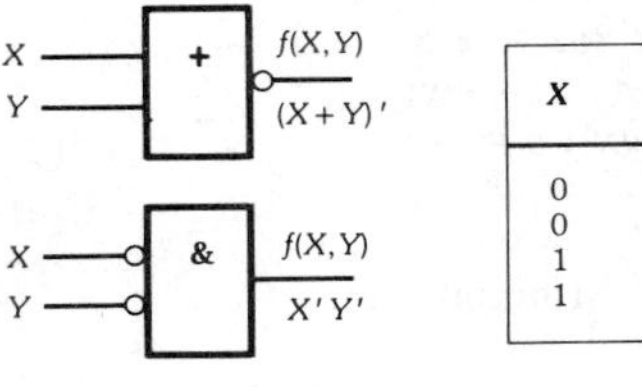

(A) Symbols.

X	Y	NOR $f(X,Y)$
0	0	1
0	1	0
1	0	0
1	1	0

(B) Table of combinations.

Fig. 10. NOR gate.

modulo two, or EXCLUSIVE OR (XOR) gate,† which has a high output only when an odd number of its inputs are high. The table of combinations for a two-input XOR gate is given in Table 6. This table shows that the output of an XOR gate with inputs x_1 and x_2 is given by $x_1 \oplus x_2 = x'_1x_2 + x_1x'_2$. It is easily demonstrated that the XOR operation is commutative and associative; that is,

$$x \oplus y = y \oplus x \text{ and } (x \oplus y) \oplus z = x \oplus (y \oplus z)$$

†The term XOR is used since a two-input gate has a 1 output if one *but not both* of the inputs is equal to 1. The OR gate (sometimes called INCLUSIVE OR) has a 1 output if one or both inputs are equal to 1.

TABLE 6. TABLE OF COMBINATIONS FOR A TWO-INPUT XOR GATE

X	Y	XOR $f(X,Y)$
0	0	0
0	1	1
1	0	1
1	1	0

If both inputs of an XOR gate are connected together to x, the output is given by $x \oplus x = xx' + x'x = 0$. The fact that this operation is associative means that any network composed only of XOR gates is equivalent to a single XOR gate with all inputs connected directly to it. This is illustrated in Fig. 11. Because of the facts that $x \oplus x = 0$ and $x \oplus 0 = x$, it is possible to remove any duplicated inputs so that the resulting network contains only one gate for which no input appears more than once. Since this gate, called the *reduced gate*, cannot function as an inverter, it follows that no network containing only XOR gates can be equivalent to an inverter. Thus the XOR gate is not a complete set by itself.

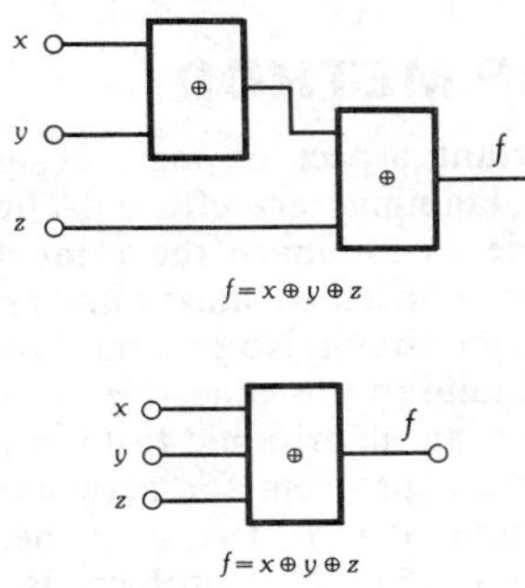

Fig. 11. Equivalent networks of XOR gates.

The XOR gate can perform as an inverter if a signal representing a constant 1 is available, since $x \oplus 1 = x1' + x'1 = x'$ (see Fig. 12). It is still not possible to construct an OR gate by using XOR gates and a 1 signal, for the output of the reduced gate will still be of the form $X \oplus Y \oplus 1$, which does not equal 1 when either or both of X and Y are equal to 1. A complete set can be formed by using both AND gates and XOR gates. As demonstrated above, any arbitrary function can be realized by a network containing only AND gates and inverters. By using XOR gates as in Fig. 12 to replace the inverters, it is possible to obtain a network containing only AND gates and XOR gates.

Canonical expressions involving AND, OR, and inverter operations have been derived. It is possible

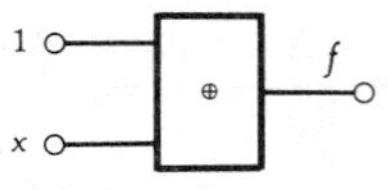

Fig. 12. Use of a constant 1 signal to form an inverter from an XOR gate.

to obtain similar canonical expressions for any complete set. The canonical expression using AND and XOR for two-variable functions is

$$f(x,y) = g_0 \oplus g_1 \cdot x \oplus g_2 \cdot y \oplus g_3 \cdot x \cdot y$$

where,

$$g_0 = f_0$$
$$g_1 = f_0 \oplus f_2$$
$$g_2 = f_0 \oplus f_1$$
$$g_3 = f_0 \oplus f_1 \oplus f_2 \oplus f_3$$

Thus for $f(x,y) = \Sigma(0,3) = x'y' + xy$

$$g_0 = 1$$
$$g_1 = 1$$
$$g_2 = 1$$
$$g_3 = 0$$

so that

$$f(x,y) = 1 \oplus x \oplus y$$

THE MAP METHOD

One important aspect of logic design is concerned with obtaining very efficient circuits. It is often desirable to minimize the total number of gates used for a specified maximum propagation delay through the circuit. No general algorithms are known for obtaining this objective. The best that can be done in an algorithmic fashion is to solve this minimization problem for networks that are limited to having at most two gates between any input and any output. This problem is important because: (1) the solution corresponds directly to a minimum PLA (programmable logic array) realization, (2) it is a minimum-delay solution, and (3) the resulting expressions are good starting places from which to derive efficient multistage networks. This section presents an introduction to the topic of two-stage minimization. The simplest sum-of-product-terms form of a function will be called a *minimal sum.** The sum-of-products form that has the fewest terms will be taken as the minimal sum. If there is more than one sum-of-products form having the minimum number of terms, and if these forms do not all contain the same total number of literals, then only the form(s) with the fewest literals will be called the minimal sum(s). For example, the function $f = x'yz + xyz + xyz'$ can be written as $f = yz + xyz'$, $f = x'yz + xy$, and $f = yz + xy$. Each of these forms contains two terms, but only the third form is a minimal sum, since it contains four literals, while the other two forms contain five literals each. The minimal sum corresponds to a gate circuit in which the circuit inputs are connected to AND gates and the outputs of the AND gates form the inputs to an OR gate whose output is the circuit output. Such a circuit is called a *two-stage circuit*, since there are two gates connected in series between the circuit inputs and output. It is also possible to have two-stage circuits in which the circuit inputs are connected to OR gates and the circuit output is obtained from an AND gate. The minimal sum just defined corresponds to the two-stage circuit in which the output is derived from an OR gate and which contains the minimum number of gates. The basic method for obtaining the minimal sum is to apply the theorem $XY + X'Y = Y$ to as many terms as possible and then to use the theorem $XY + X'Z + YZ = XY + X'Z$ to eliminate as many terms as possible.

* This is called "minimal" rather than "minimum," since there may be more than one such form.

Example:

$$f = x'y'z' + x'y'z + xy'z + xyz$$
$$x'y'z' + x'y'z = x'y'$$
$$xy'z + xyz = xz$$
$$f = x'y' + xz \quad \text{Minimal sum.}$$

Example:

$$f = w'x'y'z + w'x'yz + w'xy'z + w'xyz + wxy'z' + wxy'z + wx'y'z' + wx'y'z$$
$$w'x'y'z + w'x'yz = w'x'z$$
$$w'xy'z + w'xyz = w'xz$$
$$wxy'z' + wxy'z = wxy'$$
$$wx'y'z' + wx'y'z = wx'y'$$
$$w'x'z + w'xz = w'z$$
$$wxy' + wx'y' = wy'$$
$$f = w'z + wy' \quad \text{Minimal sum.}$$

Example:

$$f = xyz + x'yz + xy'z$$
$$xyz = xyz + xyz$$
$$f = (xyz + x'yz) + (xyz + xy'z)$$
$$f = yz + xz \quad \text{Minimal sum.}$$

The last example illustrates the fact that it may be necessary to apply the theorem $XY + X'Y = Y$ several times, the number of literals in the terms being reduced each time. A single term may be paired with more than one other term, as shown in this example. The process of comparing pairs of terms to determine whether or not the theorem $XY + X'Y = Y$ applies can become very tedious for large functions. This comparison process can be simplified by using an "*n*-cube map," which is called a *Karnaugh map* when used for minimization.

Maps for Two, Three, and Four Variables

A map for a function of two variables, as shown in Fig. 13, is a square of four cells, or a 2-cube map. The value 0 or 1 that the function is to equal when $x = 1, y = 0$ (the entry in the 10 location or 2 row of the table of combinations) is placed in the cell having coordinates $x = 1, y = 0$. In general, the scheme for filling in the map is to place a 1 in all cells whose coordinates form a binary number that corresponds to one of the fundamental products included in the function and to place a 0 in all cells whose binary numbers correspond to fundamental products not included in the function. This is done very simply by writing a 1 in each cell whose decimal designation (decimal equivalent of the binary number formed by the coordinates) occurs in the decimal specification of the function and writing 0s in the remaining cells.

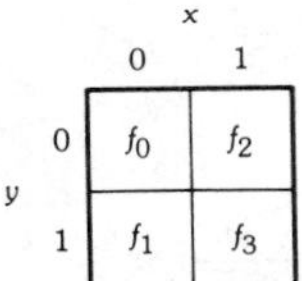

(A) General form.

y \ x	0	1
0	1	1
1	0	0

(B) Map for $f = x'y' + xy' = \Sigma\,(0,2) = \Pi\,(1,3)$.

Fig. 13. Two-variable map.

The maps for functions of three and four variables are direct extensions of the two-variable map and are shown in Figs. 14 and 15. Maps for more than four variables are possible, but they are much more difficult to use and will not be discussed here.

z \ xy	00	01	11	10
0	f_0	f_2	f_6	f_4
1	f_1	f_3	f_7	f_5

(A) General form.

z \ xy	00	01	11	10
0	1	1	0	0
1	0	0	1	0

(B) Map for $f = x'y'z' + x'yz' + xyz$; $f = \Sigma(0,2,7) = \Pi(1,3,4,5,6)$.

Fig. 14. Three-variable map.

Prime Implicants

Two fundamental products can be "combined" by means of the theorem $XY + XY' = X$ if their corresponding binary numbers differ in only 1 bit. For the fundamental products *wxyz* and *wxyz'*,

$$wxyz + wxyz' = wxy$$

The corresponding binary numbers are 1111 and 1110, which differ only in the lowest-order bit position. The fundamental products *wxyz* and *w'xyz'* cannot combine, and their corresponding numbers, 1111 and 0110, differ in the first and last bit positions.

Fundamental products that can be combined correspond to adjacent cells on the *n*-cube map. Thus cells that represent fundamental products that can be combined can be determined very quickly by inspection. In carrying out this inspection process, it must be remembered that cells such as f_4 and f_6 or f_1 and f_9 in Fig. 15 must be considered to be adjacent.

yz \ wx	00	01	11	10
00	f_0	f_4	f_{12}	f_8
01	f_1	f_5	f_{13}	f_9
11	f_3	f_7	f_{15}	f_{11}
10	f_2	f_6	f_{14}	f_{10}

(A) General form.

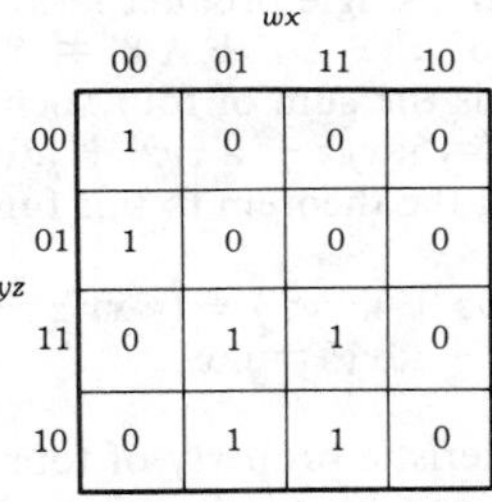

(B) Map for $f = w'x'y'z' + w'x'y'z + w'xyz' + w'xyz + wxyz' + wxyz$; $f = \Sigma\,(1,6,7,14,15) = \Pi\,(2,3,4,5,8,9,10,11,12,13)$.

Fig. 15. Four-variable map.

In a four-variable map, each cell is adjacent to four other cells, corresponding to the four bit positions in which two binary numbers can differ. In inspecting a map to determine which fundamental products can be combined, only cells with 1 entries (1 cells) need be considered, since these correspond to the fundamental products included in the function. Fig. 16 shows a four-variable map with adjacent 1 cells encircled. Notice that the 0111 cell is adjacent to two 1 cells. The rule for writing down the algebraic expression corresponding to a map is that there will be one product term for each pair of adjacent 1 cells and a fundamental product for each 1 cell that is not adjacent to any other 1 cell. The fundamental products are written down according to the rule: Any variable corresponding to a 0 in the binary number formed by the coordinates of the corresponding 1 cell is primed; the variables corre-

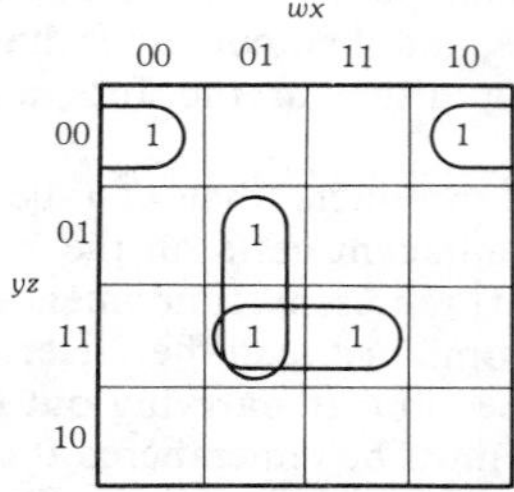

Fig. 16. Four-variable map with adjacent 1 cells encircled. $f=\Sigma(0,5,6,8,15)=x'y'z'+w'xz+xy$.

sponding to 1s are left unprimed. The product terms corresponding to pairs of adjacent 1 cells are obtained by the same rule, with the exception that one variable is not included in the product. The variable excluded is that corresponding to the bit position in which the coordinates of the two 1 cells differ (see Fig. 16).

The situation in which it is possible to combine two of the terms obtained from pairs of the fundamental products must be considered next. In such a situation, four of the fundamental products can be combined into a single product term by successive applications of the $XY + XY' = X$ theorem. A function that is the sum of four such fundamental products is $f = wxyz + wxyz' + wxy'z + wxy'z'$. Application of the theorem to this function yields

$$\begin{aligned} f &= (wxyz + wxyz') + (wxy'z + wxy'z') \\ &= wxy + wxy' = wx \end{aligned}$$

The characteristic property of four fundamental products that can be combined in this fashion is that all but two of the variables are the same (either primed or unprimed) in all four terms. The corresponding four binary numbers are identical in all but two bit positions. The corresponding cells on a map form "squares" (Fig. 17A) or "lines" (Fig. 17B) of four adjacent cells. For such a group of four cells on the map of a function, the corresponding product term is written just as for two adjacent cells, except that two variables corresponding to the two bit positions for which the cell coordinates change must be omitted.

It is also possible that eight of the fundamental products can be combined. In this case, all but three of the variables are identical (either primed or unprimed) in all eight terms.

The general rule is that, if in 2^i fundamental products all but i of the variables are identical (primed or unprimed), then the 2^i products can be combined and the i variables that change can be dropped.

In searching for a minimal sum for a function by means of a map, the first step is to encircle all sets of cells corresponding to fundamental products that can be combined (see Fig. 18). If one such set is contained in a larger set, only the larger set is

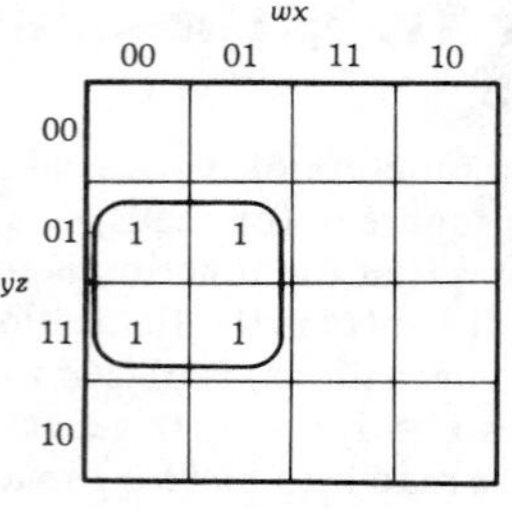

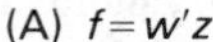

(A) $f=w'z$.

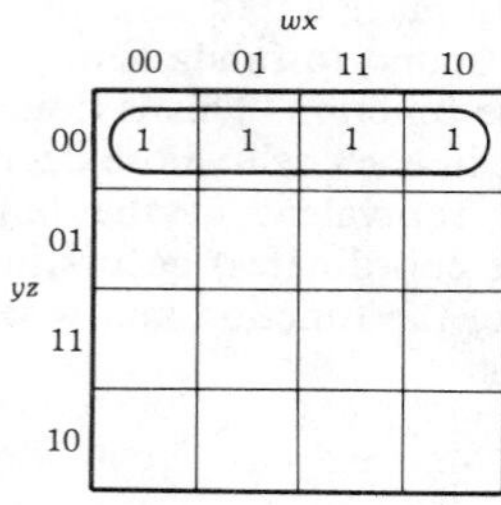

(B) $f=y'z'$.

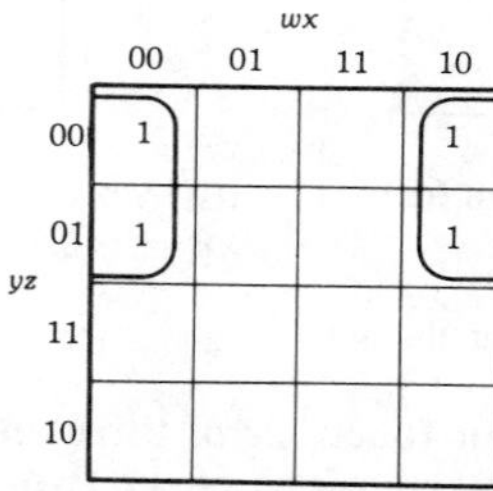

(C) $f=x'y'$.

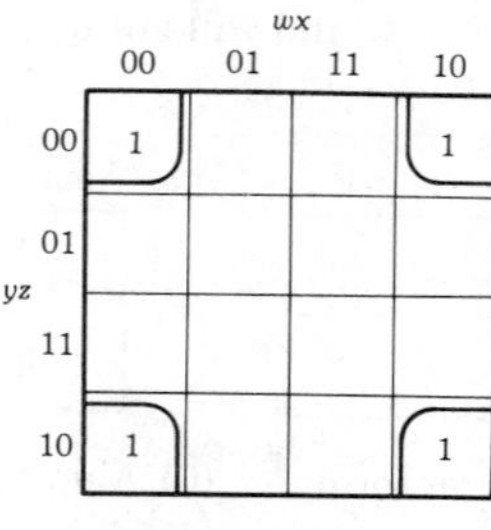

(D) $f=x'z'$.

Fig. 17. Four-variable maps showing sets of cells corresponding to four fundamental products that can be combined.

encircled.* In Fig. 18 the set (0101,0111) is not encircled. The encircled sets and the corresponding product terms will be called *prime implicants*.† These are exactly the terms that would result from

*This corresponds to using the theorem $X + XY = X$.
†This term was introduced by W. V. Quine. It is derived from the terminology of mathematical logic, but it has received widespread use in connection with logic design.

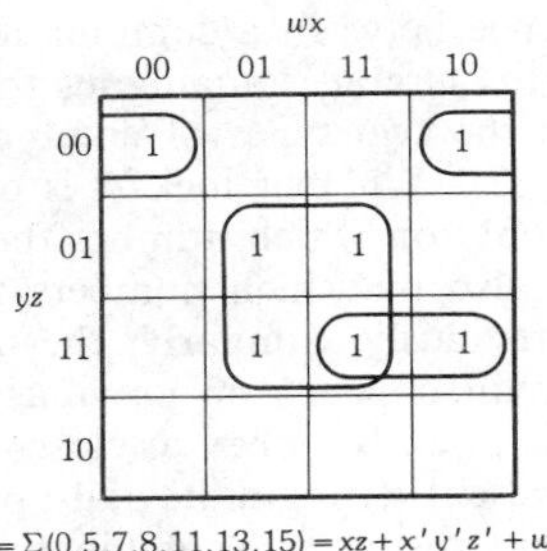

$f = \Sigma(0,5,7,8,11,13,15) = xz + x'y'z' + wyz$

Fig. 18. Map showing prime implicants.

repeated applications of the theorem $XY + XY' = X$. The terms appearing in the minimal sum will be some or all of the prime implicants.

Formation of Minimal Sums

Once the prime implicants have been formed, the minimal sum can be determined directly from the map. The rule that must be followed in choosing the prime implicants that are to correspond to terms of the minimal sum is that each 1 cell must be included in at least one of the chosen prime implicants. The problem of obtaining a minimal sum is equivalent to that of selecting the fewest prime implicants. This rule is based on the fact that, for each combination of values of the input variables for which the function is to equal 1, the minimal sum must equal 1, and therefore at least one of its terms must equal 1. More simply, the map corresponding to the minimal sum must have the same 1 cells as the map of the original function.

A procedure for determining the minimal sum is first to determine whether any 1 cells are included in only one prime implicant. In Fig. 19, an asterisk has been placed in each 1 cell that is included in only one prime implicant. A 1 cell that is included in only one prime implicant is called a *distinguished 1 cell.*

A prime implicant that includes a 1 cell that is not included in any other prime implicant is called an *essential prime implicant* and must be included in the corresponding minimal sum.

In Fig. 19A, both the prime implicants are essential and must be included in the minimal sum. A minimal sum does not always consist only of essential prime implicants. In Fig. 19B, only the essential prime implicants are shown. Cell 7 is not included in any of these, so another prime implicant that includes cell 7 must be present in the minimal sum. Fig. 19C shows the function of Fig. 19B after removal of the essential prime implicants. One of the two prime implicants shown must be included in the minimal sum, and the larger is chosen because the corresponding term contains fewer literals. The final minimal sum is $f = y'z + wz + w'yz' + xz$. There are some functions, such as that shown in

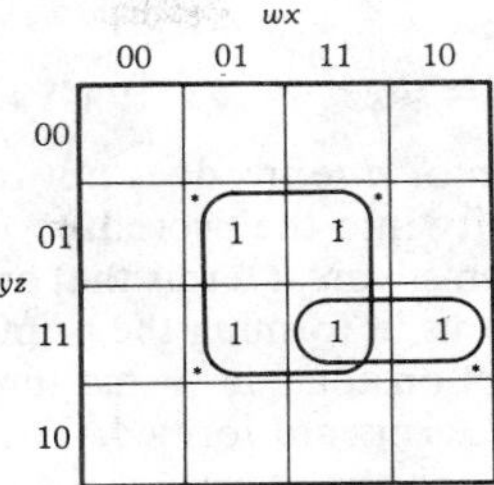

(A) $f = \Sigma(5,7,11,12,15)$, minimal sum: $f = xz + wyz$.

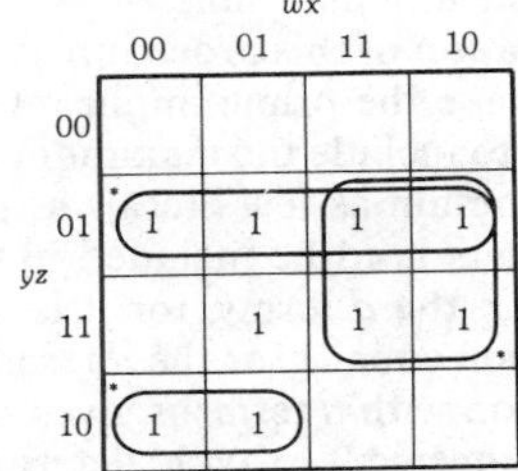

(B) $f = \Sigma(1,2,5,6,7,9,11,13,15)$, essential prime implicants.

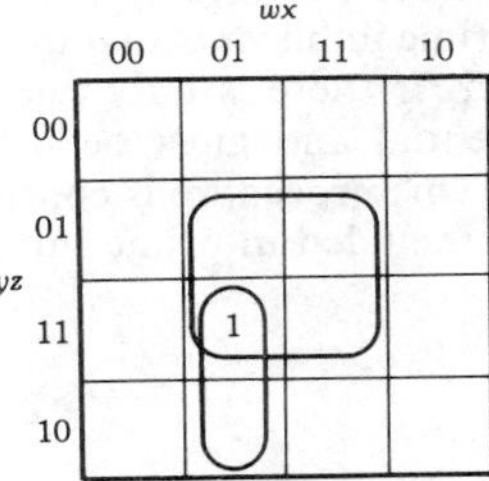

(C) After removal of the essential prime implicants.

yz \ wx	00	01	11	10
00			1	
01		1	1	
11		1		
10		1	1	

(D) $f = \Sigma(5,6,7,12,13,14)$, no essential prime implicants present.

Fig. 19. Determination of minimal sums.

Fig. 19D, that do not contain any essential prime implicants. For such functions, the minimum number of prime implicants required in the minimal sum can be determined by trial and error. The function of Fig. 19D has two minimal sums,

$$f = wxy' + w'xy + xyz'$$

and

$$f = wxz' + xy'z + w'xy$$

The addition of d terms does not introduce any extra complexity into the procedure for determining minimal sums. Any d terms that are present are treated as 1 terms in forming the prime implicants, with the exception that no prime implicants containing only d terms are formed. The d terms are disregarded in choosing terms of the minimal sum. No prime implicants are included in order to ensure that each d term is contained in at least one prime implicant of the minimal sum.

The explanation of this procedure is that d terms are used to make the prime implicants as large as possible so as to include the maximum number of 1 cells and to contain as few literals as possible. No prime implicants need be included in the minimal sum because of the d terms, for it is not required that the function equal 1 for the d terms. An example of a function with d terms is given in Fig. 20.

It is often convenient to avoid determining all the prime implicants. This can sometimes be done by searching for 1 cells that are contained in only one prime implicant and thus determining the essential prime implicants. A 1 cell is selected, and the prime implicant or prime implicants that include the 1 cell are determined. If there is only one prime implicant, it is essential and must be included in the minimal sum. This procedure is continued until all the 1 cells are included in prime implicants of the minimal sum.

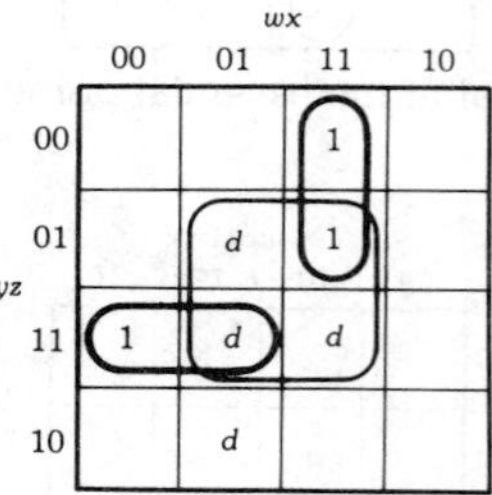

$f = \Sigma(3,12,13) + d(5,6,7,15)$ Minimal sum: $f = wxy' + w'yz$

Fig. 20. Determination of minimal sum for a function with d terms. Prime implicants used in minimal sums are shown darkened.

SEQUENTIAL CIRCUITS

The methods developed above are applicable to combinational circuits—circuits whose outputs are determined completely by their present inputs. Many digital circuits satisfy this restriction; however, there are also many circuits that do not. Circuits whose outputs depend not only on the present inputs but also on previous inputs are called *sequential circuits.*

The difference between a combinational circuit and a sequential circuit is analogous to the difference between the two types of combination lock shown in Fig. 21. Whether lock A is open or not depends not only on which number the pointer is selecting but also on which numbers the pointer stopped at previously. Similarly, the output of a sequential circuit depends on previous as well as present inputs. Lock B is open or closed depending only on the present setting of its dials; past settings are unimportant, just as, in a combinational circuit, past inputs are unimportant in determining the present circuit outputs.

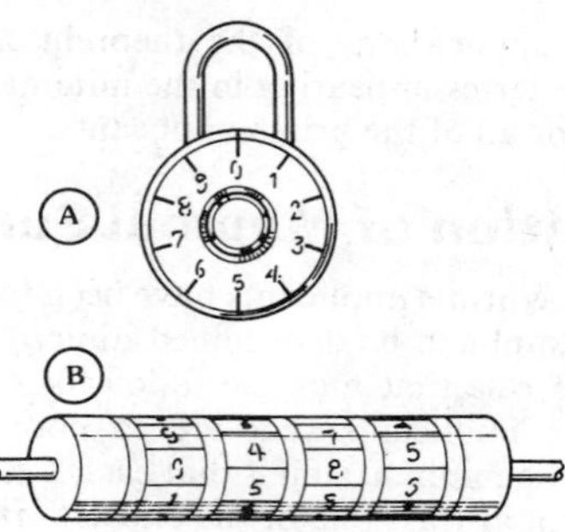

Fig. 21. Two types of combination lock.

In order for the output of a sequential circuit to depend on past inputs, the circuit must have some mechanism to retain information about previous inputs. This mechanism is some type of memory element. The first memory element used in electronic logic circuits is now called a *latch.* A latch constructed of NOR gates is shown in Fig. 22. This circuit is called a *set-reset latch* or *SR latch* (S represents the set input and R represents the reset input).

The operation of the circuit is illustrated in the figure. When S = 1 and R = 0 as in Fig. 22A, the circuit is "set," and the output values are Q = 1, $\bar{Q}$ = 0. The asterisks in the top NOR gate of this figure indicate that the output of that gate is "forced" by an external signal (S in this case). If S is then changed to 0, the circuit remains in the set condition, as shown in Fig. 22B. Changing S back to 1 returns the circuit to the situation of Fig. 22A. If, instead, R becomes 1, the circuit becomes "reset," and the conditions of Fig. 22C are present. The circuit remains in the reset state if R is returned to 0 as in Fig. 22D. Thus, when S and R are both 0, the output indicates which of S or R was last equal to 1—the circuit "remembers" the last nonzero input condition. Waveforms illustrating the latch operation are shown in Fig. 23. Fig. 22 includes tables showing the circuit conditions, and the logic symbol for a set-reset latch with the appropriate signal values. A table typical of those used by manufacturers to summarize the circuit operation is shown in Table 7.

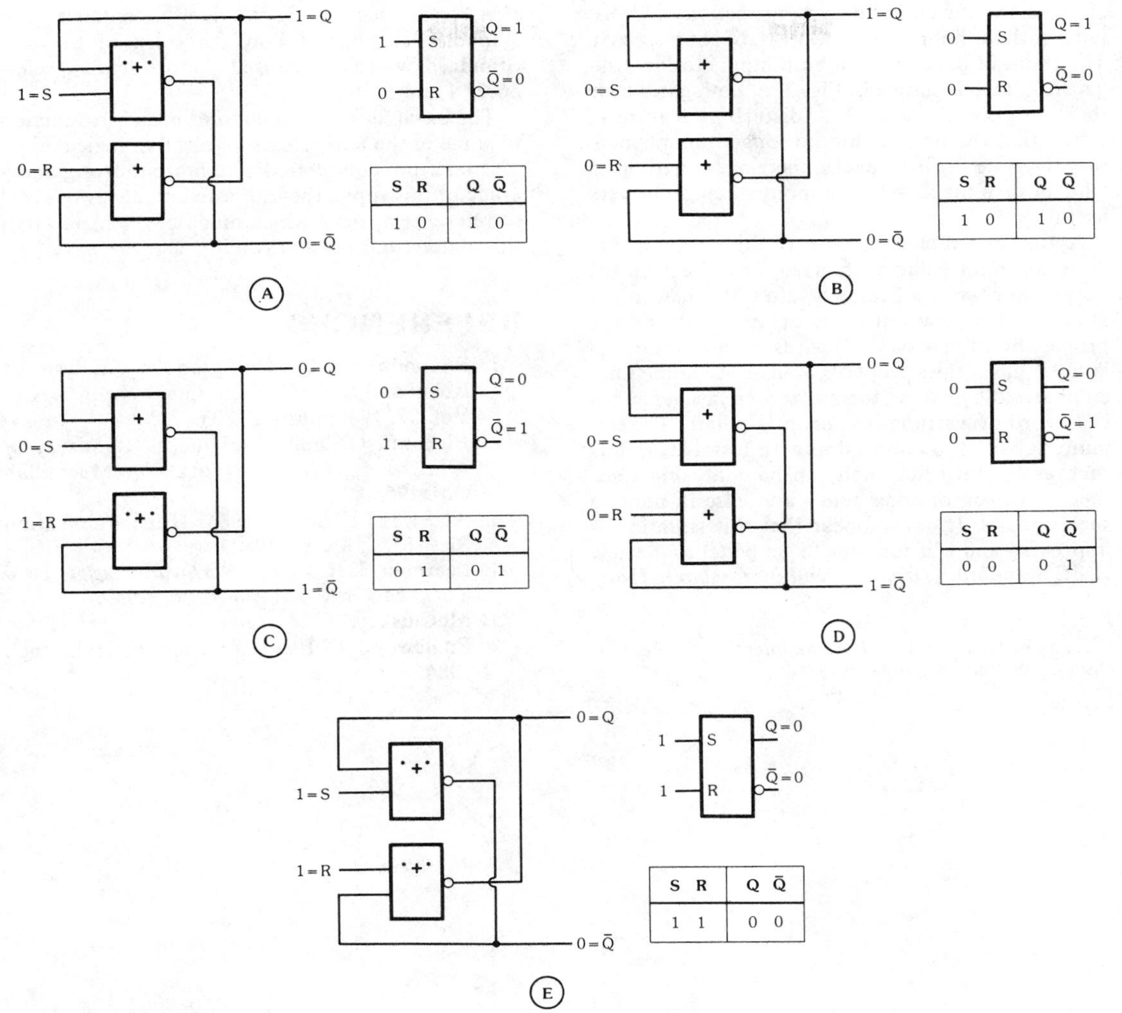

A

S R	Q $\bar{Q}$
1 0	1 0

B

S R	Q $\bar{Q}$
1 0	1 0

C

S R	Q $\bar{Q}$
0 1	0 1

D

S R	Q $\bar{Q}$
0 0	0 1

E

S R	Q $\bar{Q}$
1 1	0 0

Fig. 22. Set-reset latch constructed of NOR gates.

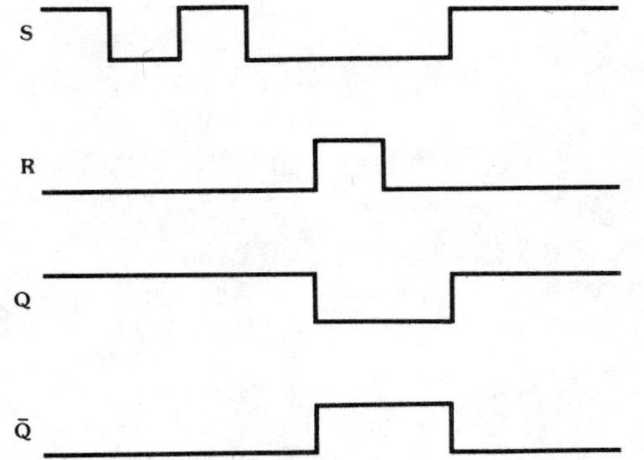

Fig. 23. Waveforms of S-R latch.

TABLE 7. SR LATCH OPERATION

S	R	Q_n
0	0	Q_{n-1}
1	0	1
0	1	0
1	1	F

Q_n is present output state.

Q_{n-1} is last previous output state.

F is a forbidden input condition.

Definitions of the SR latch do not specify the circuit operation with both inputs at 1. This is because the circuit action with S = R = 1 is very dependent on the specific implementation. Usually, networks containing SR latches are designed so that the S = R = 1 input condition never occurs.

For a specific circuit design, the action of an SR latch with both inputs equal to 1 is, of course, fixed. The result of placing 1s on both inputs to the NOR-gate SR latch is shown in Fig. 22E: both gates have their outputs forced to 0. A disturbing feature of this is that the outputs are no longer complementary. Despite the drawbacks, there are situations in which use of the S = R = 1 input state can be very useful.

An important characteristic of this circuit is that when an input value is changed, any effect on the output appears immediately after the new input appears. The new output is delayed only by the propagation times of the devices in the circuit. All latches have this property, but it is sometimes emphasized by calling them *transparent latches.**

A word of warning is appropriate here. The terminology used in connection with bistable circuits such as the set-reset latch can be confusing since there is a lack of consistency and care in naming such circuits. It does appear that this situation is improving and will continue to get better as the new IEEE Standard—*Graphic Symbols for Logic Functions* (reference 3)—gains wider acceptance. The symbols used here follow the spirit of the new Standard, which is discussed in more detail in references 4 and 5.

The discussion in this chapter is an introduction to some of the basic issues involved in logic design. It is far from complete. For a more thorough coverage of this topic, the interested reader is referred to reference 5, from which much of the material of this chapter has been drawn.

* Texas Instruments Inc. *1981 Supplement to The TTL Data Book.* 2nd ed., 1981; p. 338.

REFERENCES

1. Shannon, C. E. "A Symbolic Analysis of Relay and Switching Circuits." *Trans. AIEE,* Vol. 57, 1938, pp. 713–723.
2. Birkhoff, G., and MacLane, S. *A Survey of Modern Algebra.* New York: The Macmillan Co., 1955.
3. ANSI/IEEE Std 91-1984, IEEE Publication SH08615, July 13, 1984.
4. Peatman, J. B. *Digital Hardware Design.* New York: McGraw-Hill Book Co., 1980.
5. McCluskey, E. J. *Logic Design Principles.* Englewood Cliffs, N.J.: Prentice-Hall, Inc., 1986.

44 Probability and Statistics

Georges A. Deschamps

GENERAL

A *random experiment* is one that can be repeated a large number of times, under similar circumstances, but which yields unpredictable results at each trial. For example, rolling of a die is a random experiment where the result is one of the numbers 1, 2, 3, 4, 5, or 6. Observing the noise voltage across a resistor is another random experiment that gives a number V dependent on the instant of observation. A random experiment may consist of observing or measuring elements taken from a set that is then known as a *population*. A real number associated with the result of a random experiment is called a *random variable*, or *variate*. A variate may be *discrete*, as in the case of the die, or *continuous*, as in the case of a noise voltage.

Fluctuations of the result of a random experiment are due to causes that cannot be entirely controlled. However, if the conditions of the experiment are sufficiently uniform (for instance, if the same die is used in successive throws or if the resistor is at a constant temperature), some statistical regularity will be observed when results of a large number of experiments are considered. This regularity is expressed by the law that gives the probability of obtaining a given result or a result falling within a given range of values. The law of probability is assumed to be the same for each performance of the experiment, independent of the results of other trials.

A discrete variate which may take values $x_1, x_2, \cdots, x_n$ from a finite or denumerable set is described by p_k, its probability function; p_k is the probability of obtaining x_k as the result of one trial.

$$0 \leq p_k \leq 1$$

$$\sum_{\text{all } k} p_k = 1$$

The *cumulative probability function*

$$P(x) = \sum_{x_k \leq x} p_k$$

also describes the variate. The p_k are the jumps of this function.

For a discrete variate of more than one dimension, several possibilities arise. For example, in two dimensions let (x_j, y_k) be the coordinates of a point in the (x,y) plane. The quantity $p(x_j, y_k)$ is the *joint* probability that $x = x_j$ and $y = y_k$.

$$p_1(x_j) = \sum_{\text{all } k} p(x_j, y_k)$$

and

$$p_2(y_k) = \sum_{\text{all } j} (x_j, y_k)$$

are, respectively, the probabilities that $x = x_j$ independent of y, and $y = y_k$ independent of x.

$$p(x_j | y_k) = p(x_j, y_k)/p_2(y_k), \quad p_2(y_k) > 0$$

and

$$p(y_k | x_j) = p(x_j, y_k)/p_1(x_j), \quad p_1(x_j) > 0$$

are, respectively, the probabilities that $x = x_j$ given that $y = y_k$ has already occurred, and that $y = y_k$ given that $x = x_j$ has already occurred.

Quantities x_j and y_k are said to be *independent* if $p(x_j, y_k) = p_1(x_j) p_2(y_k)$ for all x_j and y_k.

A continuous variate is one that takes values from a nondenumerable set. The probability that one trial of the experiment gives a result between x and $x + dx$ is $p(x)dx$, where $p(x)$ is known as the *probability density function*. The *cumulative distribution function* is

$$P(x) = \int_{-\infty}^{x} p(s) ds$$

where $P(x)$ is the probability that the variate is less than or equal to x.

$$P(x_1) \geq P(x_2) \text{ if } x_1 \geq x_2$$

$$P(-\infty) = 0$$

$$P(+\infty) = \int_{-\infty}^{+\infty} p(s) ds = 1$$

$$p(x) \geq 0$$

$$p(x) = dP/dx$$

For a continuous random variable of more than one dimension, or *multivariate*, the probability density function, p, and the cumulative distribution function, P, can also be defined. For instance, if (x,y) are the coordinates of a point in the plane, then $p(x,y)dxdy$ is the probability that the multivariate has its abscissa between x and $x + dx$ and its ordinate between y and $y + dy$.

The *marginal* probability density functions are

$$p_1(x) = \int_{-\infty}^{\infty} p(x,y) dy$$

$$p_2(y) = \int_{-\infty}^{\infty} p(x,y) dx$$

For example, $p_1(x)\, dx$ is the probability that the variate x lies between x and $x + dx$ independent of y.

The *conditional* probability functions are

$$p(x|y) = p(x,y)/p_2(y), \text{ for } p_2(y) > 0$$

and

$$p(y|x) = p(x,y)/p_1(x), \text{ for } p_1(x) > 0$$

For example, $p(x|y_0)dx$ is the probability that the variate x lies between x and $x + dx$, knowing that $y = y_0$.

Two variates x and y are said to be independent if

$$p(x,y) = p_1(x)\, p_2(y)$$

for all x and y.

The cumulative distribution function is

$$P(x,y) = \int_{-\infty}^{x} ds \int_{-\infty}^{y} p(s,t) dt$$

where $P(x,y)$ is the probability that the variates are less than or equal to x and y, respectively.

DEFINITIONS

Quantities used to describe the main features of a distribution are listed below. The mean and median locate the center. Geometrically, the mean is the abscissa of the center of gravity of the probability (density) function, and the median divides that function into two equal parts. The rms, variance, standard deviation, and mean absolute deviation are measures of the spread about the mean. The moments of order three and four about the mean describe asymmetry and peakedness, respectively, of the probability (density) function.

The equations containing Σ and $\int$ refer to the discrete and continuous cases, respectively. Some authors combine these cases by means of a Stieltjes integral, for example

$$\mu = \int_{-\infty}^{\infty} x dP(x)$$

Expected Value or Mathematical Expectation: For any variable y equal to a given function $g(x)$ of the random variable x, the expected value is

$$E(y) = \sum_{\text{all } k} g(x_k) p_k$$

$$= \int_{-\infty}^{+\infty} g(x) p(x) dx$$

Average or Mean:

$$\mu = E(x) = \sum_{\text{all } k} p_k x_k$$

$$= \int_{-\infty}^{+\infty} x p(x) dx$$

Root-Mean-Square:

$$r = [E(x^2)]^{1/2} = \left[\sum_{\text{all } k} p_k x_k^2\right]^{1/2}$$

$$= \left[\int_{-\infty}^{\infty} x^2 p(x) dx\right]^{1/2}$$

Moment of Order r, About the Origin:

$$\nu_r = E(x^r) = \sum_{\text{all } k} p_k x_k^r$$

$$= \int_{-\infty}^{+\infty} x^r p(x) dx$$

$$\nu_1 = \mu$$

Moment of Order r, About the Mean:

$$\mu_r = E[(x-\mu)^r] = \sum_{\text{all } k} p_k (x_k - \mu)^r$$

$$= \int_{-\infty}^{+\infty} (x-\mu)^r p(x) dx$$

Variance:

$$\sigma^2 = \mu_2 = E[(x-\mu)^2] = \sum_{\text{all } k} p_k (x_k - \mu)^2$$

$$= \int_{-\infty}^{+\infty} (x-\mu)^2 p(x) dx$$

Standard Deviation or RMS Deviation From the Mean:

$$\sigma = \{E[(x-\mu)^2]\}^{1/2} = \left[\sum_{\text{all } k} p_k (x_k - \mu)^2\right]^{1/2}$$

$$= \left[\int_{-\infty}^{+\infty} (x-\mu)^2 p(x) dx\right]^{1/2}$$

Mean Absolute Deviation or Mean Absolute Error:

$$\text{mae} = E(|x-\mu|) = \sum_{\text{all } k} p_k |x_k - \mu|$$

$$= \int_{-\infty}^{+\infty} |x-\mu| p(x) dx$$

Median: A value m such that the variate x_k (or x) has equal probabilities of being larger or smaller than m. For the continuous case

$$\int_{-\infty}^{m} p(x) dx = \int_{m}^{+\infty} p(x) dx = {}^1/_2$$

Mode: A value of x (or x_k) where the probability density $p(x)$ (or p_k) is largest. There may be more than one mode.

p-Percent Value or Percentile: A value of x exceeded only p percent of the time; that is, with probability $p/100$. For continuous distributions, the p-percent value denoted by x_p satisfies

$$1 - P(x_p) = \int_{x_p}^{+\infty} p(x) dx = p/100$$

The median is the 50-percent value.

Quartile: The 25- and 75-percent values are called the lower and upper quartiles, respectively.

Chebishev Inequality:

$$\text{Prob}(|x-\mu| \geq \epsilon) \leq \sigma^2/\epsilon^2$$

CHARACTERISTIC FUNCTION

The characteristic function, $C(u)$, for a distribution defined by its probability (density) function, p, or by its cumulative distribution function, P, is

$$C(u) = E[\exp(jux)] = \int_{-\infty}^{+\infty} \exp(jux) dP(x)$$

$$= \int_{-\infty}^{+\infty} \exp(jux) p(x) dx$$

$$= \sum_{\text{all } k} p_k \exp(jux_k)$$

Properties:

$$C(0)=1$$

$$|C(u)| \leq 1, \text{ for } u \text{ real}$$

the moment

$$\nu_r = \left. \frac{j^{-r} d^r C(u)}{du^r} \right|_{u=0}$$

$$C(-u)=C^*(u), \quad \text{for } u \text{ real}$$

where the asterisk denotes the complex conjugate. Function $C(u)$ can be expanded in terms of the moments ν_r:

$$C(u)=1+\sum_{r=1}^{\infty} \nu_r (ju)^r/r!$$

Function C is the Fourier transform of p in the continuous case; hence

$$p(x)=(1/2\pi)\int_{-\infty}^{\infty} \exp(-jux)C(u)\,du$$

The moment generating function is

$$C(-jt)=E(e^{tx})$$

For a multivariate $\mathbf{x}=(x_1, x_2, \cdots, x_n)$, the characteristic function is

$$C(u_1, u_2, \cdots, u_n)$$
$$=E\{\exp[j(u_1x_1+u_2x_2+\cdots+u_nx_n)]\}$$

or

$$C(\mathbf{u})=E[\exp(j\mathbf{u}\tilde{\mathbf{x}})]$$

ADDITION OF STATISTICALLY INDEPENDENT VARIABLES

If two continuous independent variates x_1, x_2 have probability densities $p_1(x_1)$ and $p_2(x_2)$, the probability density function for their sum $x=x_1+x_2$ is the convolution integral

$$p(x)=\int_{-\infty}^{+\infty} p_1(x-\xi)p_2(\xi)\,d\xi$$

or in shortened form

$$p=p_1*p_2$$

Similarly, the cumulative distribution function for the sum is

$$P(x)=P_1*p_2=\int_{-\infty}^{+\infty} P_1(x-\xi)\,dP_2(\xi)$$

$$=\int_{-\infty}^{+\infty} P_1(x-\xi)p_2(\xi)\,d\xi$$

Instead of computing these convolutions, it is sometimes simpler to use the corresponding property of the characteristic functions

$$C(u)=C_1(u)C_2(u)$$

and to deduce $p(x)$ as the Fourier transform of $C(u)$. This property extends to the sum of n independent variates.

DISTRIBUTIONS

Normal Distribution

The *normal*, or *Gaussian*, distribution is often found in practice because it occurs whenever (1) a large number of independent random causes produces additive effects and (2) an appreciable fraction of the causes produces effects of nearly maximum variance. This is know as the *Central limit theorem*: If the variates x_i are independent and have a mean μ and a standard deviation σ, then the sum $x_1+x_2+\cdots+x_n$ has a distribution that is approximately normal with mean $n\mu$ and standard deviation $\sigma n^{1/2}$. (The x_i may be discrete or continuous.)

The normal probability density function, for a mean of zero and a standard deviation σ, is

$$\varphi_\sigma(x)=[1/\sigma(2\pi)^{1/2}]\exp[-\tfrac{1}{2}(x/\sigma)^2]$$

(See Fig. 1 and Table 10 in Chapter 47.) When the mean value is μ instead of 0, the probability density becomes $\varphi_\sigma(x-\mu)$.

The cumulative distribution function

$$\Phi(x)=\int_{-\infty}^{x} \varphi_\sigma(x)\,dx$$

is given by scale C on Fig. 1 and more accurately by Table 10 in Chapter 47.) Related to Φ are the error function erf t and the complementary error function erfc t.

$$\text{erf } t=(2/\pi^{1/2})\int_0^t \exp(-t^2)\,dt=2\Phi[t2^{1/2}]-1$$

$$\text{erfc } t=1-\text{erf } t$$

The distribution of the absolute deviation from the mean $|x-\mu|$, sometimes called the error, is given by scale E on Fig. 1. The median value of the error, equal to 0.6745σ, is called the *probable error*. It is exceeded 50 percent of the time. The average of $|x-\mu|$, equal to 0.7979σ, is the *mean absolute error*. The 3σ error is exceeded with probability of about 0.3 percent.

Additive Property: The linear combination, with constant coefficients of n normal independent random variables, is also a normal random variable. If

$$y=c_1x_1+c_2x_2+\cdots+c_nx_n$$

where x_i has mean μ_i and variance σ_i^2, then y has a mean

$$\mu=\sum_{i=1}^{n} c_i\mu_i$$

and a variance

$$\sigma^2=\sum_{i=1}^{n} c_i^2\sigma_i^2$$

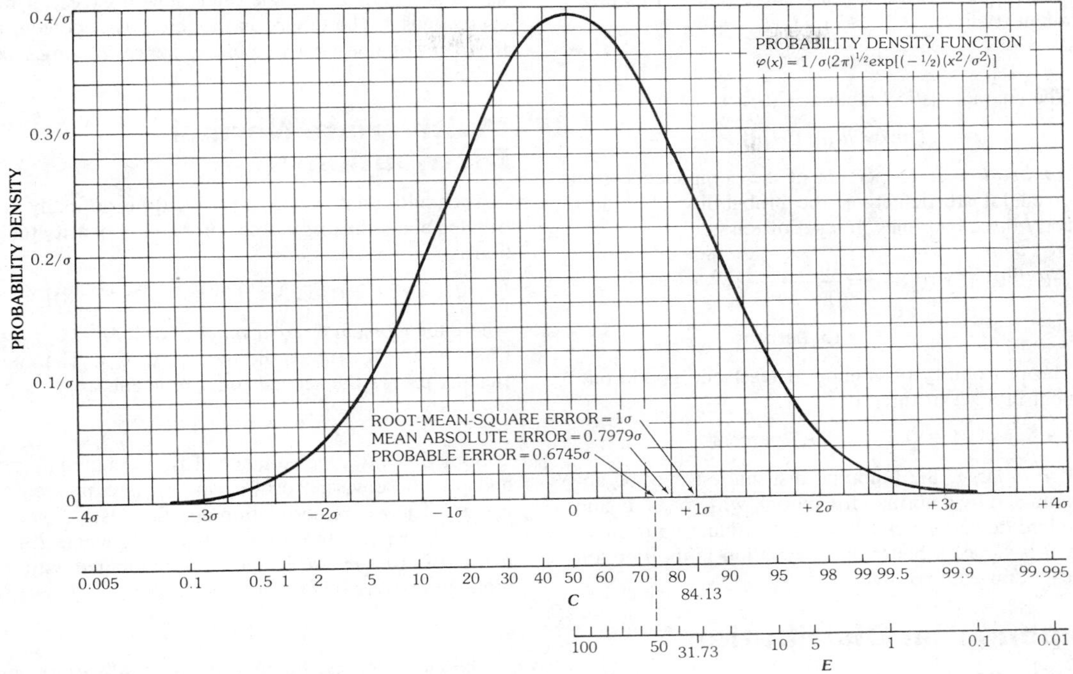

Fig. 1. The normal distribution; σ is the standard deviation. Scale *C* is the cumulative distribution function in percent = $100\Phi(x)$. For example, the probability of finding *x* between $-\sigma$ and $+2\sigma$ is $97-16=81$ percent. Scale *E* is the probability that the error (absolute deviation) exceeds the value read on the axis. For example, if the deviation is larger than 2σ in either direction, probability is 4.5 percent.

Poisson Distribution

A random experiment that leads to the Poisson distribution might consist of counting during a given time, *T*, the electrons emitted by a cathode, the telephone calls received at a central office, or the noise pulses exceeding a threshold value. In all these cases, the events are, in general, independent of each other, and there is a constant probability, νdt, that one of them will occur during a short interval, dt.

The probability that exactly *k* events will occur during time interval *T* is given by the Poisson frequency function

$$p_k = (m^k/k!)\exp(-m), \quad k=0,\ 1,\ 2,\ \cdots$$

where the parameter $m=\nu T$ is the average number of events during interval *T*.

The variance of *k* is

$$E[(k-\nu T)^2]=m$$

The standard deviation is

$$m^{1/2}$$

The characteristic function, $C(u)$, is

$$\exp\{m[\exp(ju)-1]\}$$

Binomial Distribution

If the result of a random experiment is one of two alternatives, the statistics are completely defined by the probability, *p*, of one of the alternatives. The trial may be the flipping of a coin or the testing of a transistor taken at random. The preferred alternative or "success" could be a head in the first case, an acceptable transistor in the second case. The probability of failure in one trial is

$$q=1-p$$

In *n* independent trials, the probability of exactly *k* "successes" is given by

$$p_k = C_k^n p^k (1-p)^{n-k}$$
$$= [n!/k!(n-k)!]p^k(1-p)^{n-k}, \quad 0 \le k \le n$$

This is called the binomial distribution because p_k is the *k*th term in the development of the binomial $(p+q)^n$.

The average of *k* is np and the variance is

$$E[(k-np)^2]=npq$$

The standard deviation is

$$(npq)^{1/2}$$

The probability of at least one success in n independent trials is

$$1-(1-p)^n$$

The characteristic function, $C(u)$, is

$$[p\exp(ju)+1-p]^n$$

Application: If 15 percent of the components from a given lot are defective, the probability of finding exactly three bad ones in a set of ten is

$$C_3{}^{10}(0.15)^3(0.85)^7=\frac{10\times 9\times 8}{1\times 2\times 3}15^3 85^7 10^{-20}$$

$$=13 \text{ percent}$$

The probability of finding at least one good component in a set of three is

$$1-(0.15)^3=99.7 \text{ percent}$$

If n is large, the binomial distribution may be approximated by a normal distribution with mean np and standard deviation $(npq)^{1/2}$. If the product np is small and n is large, a better approximation is the Poisson distribution with parameter $m=np$.

Exponential Distribution

An important case where exponential distribution arises is the following. In a Poisson process, the probability that the interval between two consecutive events lies between t and $t+dt$ is

$$\nu\exp(-\nu t)\,dt=d[1-\exp(-\nu t)]$$

with $t\geq 0$. The average interval is

$$E[t]=1/\nu$$

The probability density function is

$$p(t)=\nu\exp(-\nu t),\quad t\geq 0$$

The root-mean-square is

$$[E(t^2)]^{1/2}=\sqrt{2}/\nu$$

The standard deviation is

$$\{E[(t-1/\nu)^2]\}^{1/2}=1/\nu$$

The median is

$$(\log_e 2)/\nu=0.6931/\nu$$

The cumulative distribution function is

$$1-\exp(-\nu t)$$

The probability that an interval is larger than t is

$$\exp(-\nu t)$$

Problem: Pulses of noise above a certain level occur with an average density of 2 per millisecond. A device is triggered every time two pulses occur within the same 5-microsecond interval. How often does this happen? Since $\nu t=0.01$, then

$$\exp(-0.01)=0.990$$

is the probability that one interval will exceed 5 microseconds.* The device is triggered by 1 percent of the pairs of consecutive pulses, hence 20 times per second.

Multivariate Normal Distribution

The multivariate $\mathbf{x}=(x_1,x_2,\cdots,x_n)$ is normally distributed about the origin if the probability density function is

$$\varphi_M(\mathbf{x})=[(2\pi)^n\det M]^{-1/2}\exp[-\tfrac{1}{2}(\mathbf{x}M^{-1}\tilde{\mathbf{x}})]$$

where the *moment* or *covariance matrix* $M=(\mu_{ij})$ is of order n, and $\tilde{\mathbf{x}}$ denotes the transpose of $\mathbf{x}$. The coefficients μ_{ij} are the second-order moments

$$\mu_{ij}=E[x_ix_j]$$

Sometimes μ_{ii}, the variance of x_i, is denoted by $\sigma_i{}^2$, and μ_{ij}, the covariance of x_i and x_j, is expressed by $\sigma_i\sigma_j r_{ij}$. The r_{ij} are correlation coefficients.

Any linear function of $\tilde{\mathbf{x}}$, say $\tilde{\mathbf{y}}=L\tilde{\mathbf{x}}$, where L is a matrix of order $m\times n$, is normally distributed with the moment matrix

$$N=LM\tilde{L}$$

The characteristic function of the multivariate normal distribution is

$$C(\mathbf{u})=E[\exp(j\mathbf{u}\tilde{\mathbf{x}})]=\exp[-\tfrac{1}{2}(\mathbf{u}M\tilde{\mathbf{u}})]$$

The sum of two independent, normally distributed multivariates, $\mathbf{x}$ and $\mathbf{y}$, with covariance matrices M and N, respectively, is normally distributed with covariance matrix $M+N$

$$\varphi_M*\varphi_N=\varphi_{M+N}$$

Normal Distribution in Two Dimensions: The probability density function at the point (x,y) is

$$\varphi(x,y)=[2\pi\sigma_1\sigma_2(1-\rho^2)^{1/2}]^{-1}$$
$$\times\exp\left[-[2(1-\rho^2)]^{-1}\left(\frac{x^2}{\sigma_1{}^2}-\frac{2\rho xy}{\sigma_1\sigma_2}+\frac{y^2}{\sigma_2{}^2}\right)\right]$$

where $\sigma_1{}^2$ and $\sigma_2{}^2$ are the variances of x and y, and ρ is their correlation coefficient.

Circular Case—Rayleigh Distribution: When the two normally distributed variates have the same variance ($\sigma_1=\sigma_2=\sigma$) and are not correlated ($\rho=0$)

$$\varphi(x,y)=(2\pi\sigma^2)^{-1}\exp[-\tfrac{1}{2}(x^2+y^2)/\sigma^2]$$

The distance R to the origin, $R=(x^2+y^2)^{1/2}$, is distributed according to the probability density function

$$p(R)=(R/\sigma^2)\exp(-R^2/2\sigma^2),\ \text{for } R\geq 0$$

This is sometimes called the Rayleigh distribution. When a large number of small independent random

* The value $\exp(-0.01)=0.990$ was obtained from Table 9 in Chapter 47.

phasors with equiprobable phases are added, the extremity of the vector sum is distributed according to the circular normal bivariate distribution. The magnitude, R, of the sum has therefore the probability density $p(R)$. This applies to the electromagnetic field scattered by a large number of small scatterers. It also describes the distribution of the envelope of a narrow band of Gaussian noise.

Fig. 2 shows function $p(R)$, and scale C gives the probability that some given level will be exceeded. The rms of R is $\sigma(2)^{1/2}$. The average $\sigma(\pi/2)^{1/2}=1.2533\sigma$ is the *mean radial error*. The *median*, or 50-percent value, 1.1774σ, is also called cep (circular error probable), because it is the radius of the 50-percent-probability circle in the x, y plane.

Using $X=\frac{1}{2}R^2$(power) as the variable

$$p(R)\,dR=[\exp(-X/X_0)]\,d(X/X_0)$$

with $X_0=\sigma^2$ (an exponential distribution).

When the circular normal distribution has its center at a distance S from the origin, the distance R to the origin is distributed according to

$$p(R)\,dR=(R/\sigma^2)\times\exp[-(R^2+S^2)/2\sigma^2]I_0(RS/\sigma^2)\,dR$$

where I_0 = Bessel function of zero order with imaginary argument. This is the distribution of the envelope of a sine wave plus some Gaussian noise. It also represents the distribution of the amplitude of a field that results from the addition of a fixed vector and a random component obtained, for instance, by scattering from a large number of small independent scatterers. See Fig. 3.

Chi-Square Distribution

The distribution of the sum of the squares of n independent normal variates, each having mean zero and variance unity, is called the chi-square distribution of n degrees of freedom.

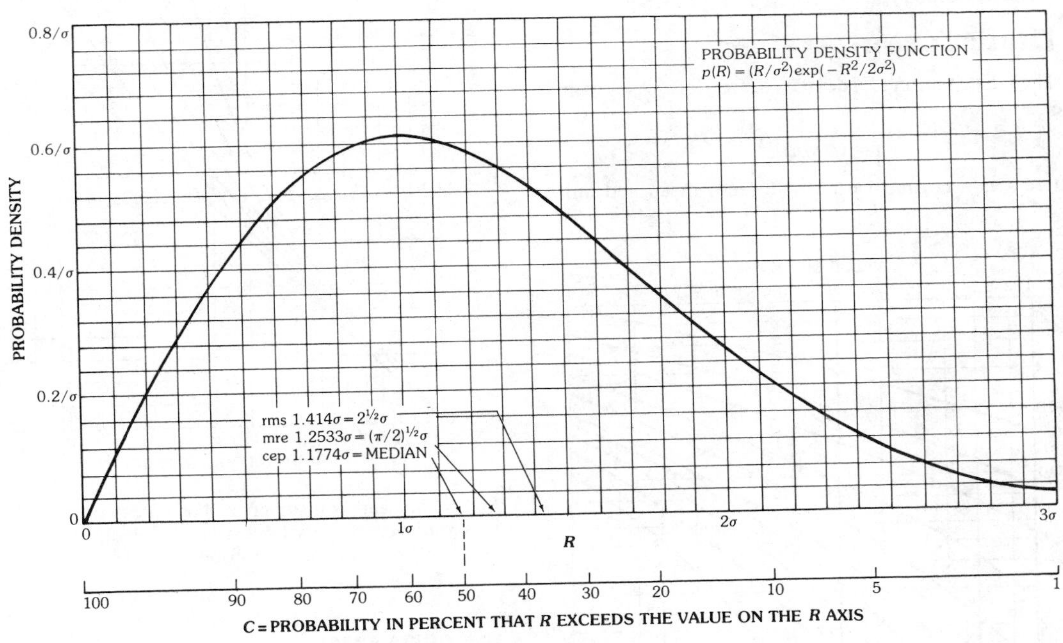

Fig. 2. Rayleigh distribution. R is the distance to the origin in a bivariate normal distribution. σ is the standard deviation for either component of the normal distribution.

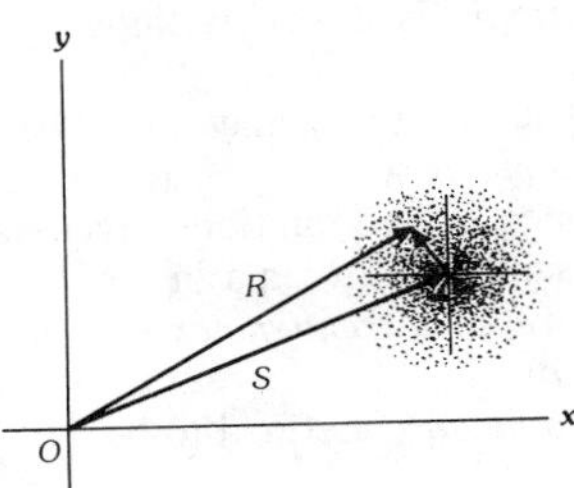

Fig. 3.

The probability density function for this sum x is

$$k_n(x)=\frac{x^{(n/2)-1}}{2^{n/2}\Gamma(n/2)}\exp(-x/2)$$

(x, being the sum of n squares, is nonnegative). Parameter n is called the *number of degrees of freedom*. The mean of x is n, its variance is $2n$, and its characteristic function is $(1-j2t)^{-n/2}$.

The p-percent value of x (exceeded p percent of the time) is denoted, for n degrees of freedom, by $\chi_p^{\,2}(n)$:

$$\int_{\chi_p^2}^{\infty} k_n(x)dx = p/100$$

Curves of $\chi_p^{\,2}(n)$ versus p are shown in Fig. 4. For values of n greater than 30, $(2x)^{1/2}$ is approximately normal with mean $(2n-1)^{1/2}$ and unit variance.

The first functions k_n are:

$$k_1(x) = (2\pi x)^{-1/2}\exp(-x/2)$$

In this case, x is the square of a normal variate.

$$k_2(x) = \tfrac{1}{2}\exp(-x/2)$$

In this case, x corresponds to R^2/σ^2 in the Rayleigh distribution.

$$k_3(x) = (x/2\pi)^{1/2}\exp(-x/2)$$

In this case, x is the square of the distance to the origin of a point in space having a normal distribution with spherical symmetry.

t Distribution

The t distribution of n degrees of freedom arises from the ratio

$$t = x/(y/n)^{1/2}$$

where x is a normal variate with zero mean and unit variance; y has a chi-square distribution with n degrees of freedom.

The probability density function is

$$p(t) = (n\pi)^{-1/2}\frac{\Gamma[(n+1)/2]}{\Gamma(n/2)}(1+t^2/n)^{-(n+1)/2},$$

$$\text{for } -\infty < t < \infty$$

where Γ refers to the gamma function. The mean of t is zero, and its variance is $n/(n-2)$ for $n \geq 3$.

Values of the cumulative distribution of $|t|$ may be read from Fig. 5.

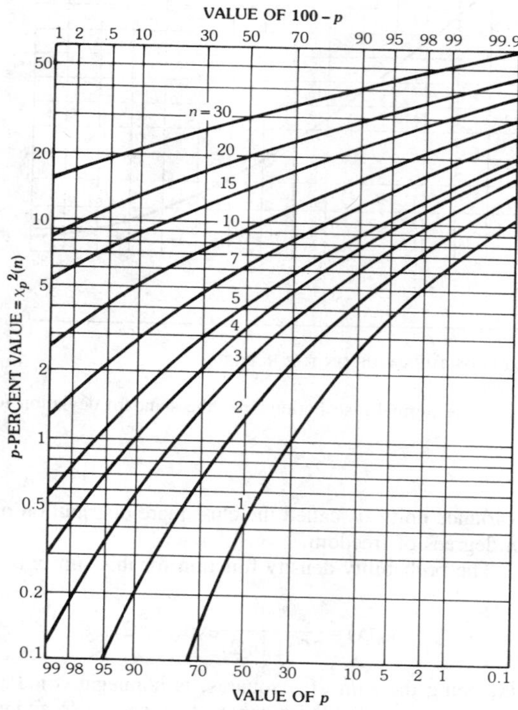

Fig. 4. Chi-square distribution. The function $\chi_p^{\,2}(n)$ vs p.

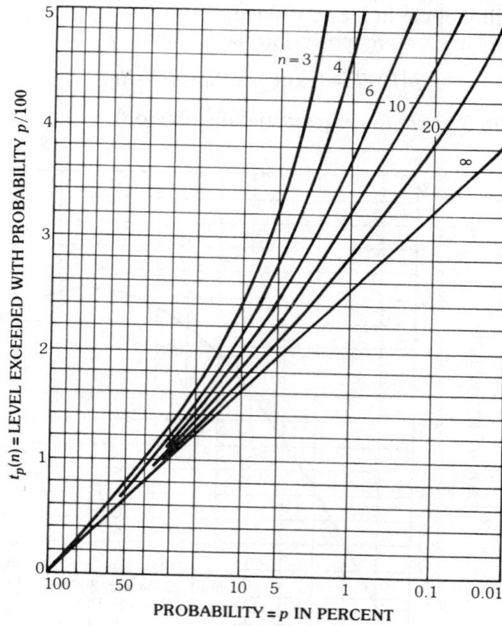

Fig. 5. Student's t distribution. For n degrees of freedom, the ordinate on the curve labeled n is the value $t_p(n)$ exceeded, in either direction, with a probability $p/100$.

SAMPLING

If a random experiment is repeated independently n times, the set of variates $x_1, x_2, \cdots, x_n$ is called a random sample of size n. The distribution of x from which the sample is drawn is known as the *parent distribution*.

The numbers $x_1, \cdots, x_n$ may not all be different and may form a smaller set $x_1, \cdots, x_k, \cdots, x_m$, where x_k occurs n_k times. The definitions given earlier can be applied to a sample (or to an arbitrary set of numbers) by using the *relative frequencies* n_k/n in place of the probabilities p_k.

The *sample mean* is defined to be

$$\bar{x} = (1/n)(x_1 + x_2 + \cdots + x_n)$$

This estimate is unbiased: $E(\bar{x}) = \mu$. It usually has the least variance among unbiased estimators.

The *sample variance* is defined to be

$$s^2 = (n-1)^{-1} \sum_{i=1}^{n} (x_i - \bar{x})^2$$

The factor $(n-1)$ makes s^2 unbiased: $E(s^2) = \sigma^2$.

For a normal variate, the standard deviation σ can also be deduced from the sample range, that is, from the difference between the largest number and the smallest number in the sample. For a sample of size n, σ is obtained by dividing the range by the number c_n in Table 1. Usually this estimator will have a larger variance than the sample variance.

Let the x_k be in such order that

$$x_1 \leq x_2 \leq x_3 \leq \cdots \leq x_n$$

The sample median is defined to be

$$\xi = x_{(n+1)/2}$$

if n is odd and

$$\xi = \tfrac{1}{2}[x_{n/2} + x_{(n/2)+1}]$$

if n is even. A sample may always be so ordered.

Interval Estimation of the Mean and Variance of a Normal Variate

A p-percent *confidence interval* is such that the quantity estimated falls within that interval p percent of the time. Intervals of this type can be deduced from a given sample for the mean μ and for the variance σ of the parent population.

Mean, σ known:

$$\bar{x} - a\sigma/n^{1/2} < \mu < \bar{x} + a\sigma/n^{1/2}$$

where a is chosen so that $\phi(a) = 0.5 + p/200$.

Mean, σ unknown:

$$\bar{x} - (s/n^{1/2})t_{100-p}(n-1) < \mu < \bar{x} + (s/n^{1/2})t_{100-p}(n-1)$$

Example: For a sample of size 5, a 99% confidence interval for μ, σ unknown, is

$$\bar{x} - 2.06s < \mu < \bar{x} + 2.06s$$

Refer to Fig. 5 for the value of $t_{100-p}(4)$.

Variance, μ known:

$$\sum_{i=1}^{n} (x_i - \mu)^2/\chi_{(100-p)/2}^2(n) < \sigma^2 < \sum_{i=1}^{n} (x_i - \mu)^2/\chi_{(100+p)/2}^2(n)$$

Variance, μ unknown:

$$(n-1)s^2/\chi_{(100-p)/2}^2(n-1) < \sigma^2 < (n-1)s^2/\chi_{(100+p)/2}^2(n-1)$$

Example: For a random sample of size 5, a 90% confidence interval for σ^2, μ unknown, is

$$0.42s^2 < \sigma^2 < 5.7s^2$$

Refer to Fig. 4 for the values of $\chi_5^2(4)$ and $\chi_{95}^2(4)$.

CHI-SQUARE TEST

The problem is to find how well a sample taken from a population agrees with some distribution function assumed for that population.

Let the random sample be divided into m disjoint sets, and let the number of sample points falling within the ith set be f_i. From the assumed distribution and the size of the sample, the expected number, g_i, of points in the ith set is computed. The deviation between this and the actual result is expressed by

$$D = \sum_{i=1}^{m} (f_i - g_i)^2/g_i$$

If the f_i are sufficiently large, this deviation is distributed according to the chi-square distribution with $m-1$ degrees of freedom, approximately. The curves of Fig. 4 can be used to evaluate in percent the significance of a given deviation.

TABLE 1. VALUES OF c_n*

n	c_n
5	2.33
10	3.08
20	3.73
30	4.09
100	5.02

* From E. S. Pearson, "Percentage Limits for the Distribution of Range in Samples for a Normal Population," *Biometrika*, Vol. 24, November 1932, pp. 404–417; see p. 416. See also, E. S. Pearson and H. O. Hartley, *Biometrika Tables for Statisticians*, Vol. 1, London: Cambridge University Press, 1954; Table 22.

If the assumed parent distribution is not completely known and r parameters defining it have been determined to fit the sample, the number of degrees of freedom is reduced to $m-1-r$. For example, in the following application, ν is unknown; hence $r=1$.

Application: During three successive one-hour periods, the number of telephone calls received at a station was 11, 15, and 23, while during two nonoverlapping two-hour periods, it was 40 and 37. How does this agree with a Poisson process?

Since the density ν (the number of calls per hour) has not been specified, it is deduced from the sample:

$$\nu = (11 + 15 + 23 + 40 + 37)/7 = 18$$

The deviation from the expected number is

$$7^2/18+3^2/18+5^2/18+4^2/36+1^2/36=5.1$$

For $5-2=3$ degrees of freedom, this deviation is exceeded about 15 percent of the time. The assumption of a Poisson process is therefore very good. It would have been significantly doubtful only if the deviation obtained was exceeded as rarely as 5 percent or less of the time, which corresponds to D larger than 7.8 in this application.

MONTE CARLO METHOD

The Monte Carlo method consists of solving statistical problems, or problems that can be interpreted as such, by substituting for the actual random experiment a simpler one where the desired probability laws are obtained by drawing random numbers.

Reading in order the digits in Table 3 of Chapter 47 is equivalent to successive trials where the result is one out of 10 equiprobable eventualities. Taking pairs of digits simulates 100 equiprobable eventualities. An event with probability of 63 percent may be simulated by the reading of successive pairs, considering as a "success" any pair from 00 to 62. The successive pairs divided by 100 approximate a random variable uniformly distributed over the interval (0, 1). For a smoother approximation, three or four consecutive digits could be used.

Any continuous variate x with cumulative distribution function $P(x)$ can be transformed into a new variate r that is uniformly distributed between zero and one by means of $r=P(x)$. Conversely, the variate x can be simulated by solving $P(x)=r_i$ for x, where the r_i are successive random numbers. For example, from Table 3 in Chapter 47, we have 0.49, 0.31, 0.97, 0.45, 0.80, etc. Table 10 in Chapter 47 gives the corresponding values of x that will be normally distributed with mean zero and variance one: 0.0, −0.5, 1.9, −0.1, 0.8. This simulates the result of successive shots aimed at the point $x=0$.

To obtain accurate numerical results by the Monte Carlo method, a large number of trials, N, may be necessary since the accuracy increases roughly as the square root of N. There are cases, however, where only a crude evaluation is needed, and it may be obtained even with a short table such as Table 3 in Chapter 47.

When a computer is used, it becomes essential to avoid the looking up in a table by generating directly in the computer a sequence of pseudorandom digits, i.e., a sequence that although deterministic approximates closely the statistical properties of a truly random sequence of uncorrelated digits. A number of algorithms are available to generate such sequences. See for instance M. Abramowitz and I. A. Stegen, eds., *Mathematical Functions*, New York: Dover Publications, Inc., 1965, p. 949.

Problem: Airplanes arrive over an airfield at random, independently of each other, at the average rate of one per minute. The landing operation takes 3/4 minute, and only one airplane can be handled at a time. Will many airplanes have to wait before landing? The cumulative distribution function for the interval t minutes between arrival of successive airplanes is $1-\exp(-t)$.* The successive intervals, during an imaginary experiment, may therefore be taken as $t_i=-\log_e(1-r_i)$, where r_i are the random numbers uniformly distributed between 0 and 1. This is equivalent to $t_i=-\log_e r_i$. Starting at the top left of Table 3 in Chapter 47 gives 0.71, 1.17, 0.03, 0.80, 0.22, 0.13, 0.25, 0.40, 0.37, 0.46, 0.17, 0.15, 0.37, 0.65, 3.91, 2.21, 0.17, · · · for the successive intervals in minutes. It is apparent that after a few minutes airplanes will be waiting. A few other trials using other parts of the table show that this situation is not exceptional. The traffic density is too high. The problem could be made more realistic by assuming a normal distribution of the landing times, simulated for instance as explained above.

RANDOM OR STOCHASTIC PROCESSES

A random or stochastic process is a family of random variables $\{y_t\}$ where the index t is usually considered to be time. If t assumes discrete values (for example, all the positive integers), the process is called *discrete*. If t takes on all the values from an interval, the process is said to be *continuous*.

The noise voltage, y_t, across a resistor is an example of a stochastic process. Suppose the general shape of a typical recording of this noise voltage does not appear to change: The oscillations show no tendency to become larger or smaller, do not change in rapidity, etc. This type of process is called *stationary*. The probability density functions of stationary stochastic processes are independent of time:

$$p(y_{t1})=p(y_{t2}), \text{ for any } t_1 \text{ and } t_2$$

The process is called *Gaussian* or *normal* if the joint distribution

$$p(y_{t1}, y_{t2}, y_{t3}, \cdots), \text{ for any } t_1, t_2, t_3, \cdots$$

is a multidimensional normal distribution.

Suppose the random variables in the family exhibit similar statistical properties and one member could be taken as representative of the entire family. This type of process is called *ergodic*. For ergodic processes, time averages are equivalent to ensemble averages, e.g.

$$\lim_{T\to\infty} T^{-1}\int_0^T y_t dt=\int_{-\infty}^{\infty} y_t p(y_t) dy_t$$

Power Density Spectrum

In arriving at the power density spectrum of a stationary random function, let

---

* See Table 9 in Chapter 47.

$$F_T(\nu) = \int_0^T f(t)\exp(-2\pi j\nu t)\,dt$$

be the Fourier transform of the given random function $f(t)$ limited to the interval 0 to T.

The power density spectrum is defined by

$$W(\nu) = \lim_{T\to\infty} T^{-1}|F_T(\nu)|^2$$

Function W is an even function of frequency

$$W(-\nu) = W(\nu)$$

since for a real function f

$$F_T(-\nu) = F_T{}^*(\nu)$$

Sometimes the spectrum is limited to positive frequencies by considering

$$W'(\nu) = 2W(\nu) \quad \text{for } \nu > 0$$
$$= 0 \qquad \text{for } \nu < 0$$

The power in a band extending from ν_1 to ν_2 is

$$\int_{\nu_1}^{\nu_2} W'(\nu)\,d\nu$$

Correlation Functions

The autocorrelation function is defined by

$$\varphi(\tau) = \lim_{T\to\infty} T^{-1}\int_0^T f(t)f(t+\tau)\,dt$$

It is particularly useful because its Fourier transform is the power density spectrum

$$W(\nu) = \int_{-\infty}^{+\infty} \varphi(t)\exp(-j2\pi\nu t)\,dt$$

$$\varphi(t) = \int_{-\infty}^{+\infty} W(\nu)\exp(j2\pi\nu t)\,d\nu$$

or also

$$W'(\nu) = 4\int_0^{\infty} \varphi(t)\cos(2\pi\nu t)\,dt$$

$$\varphi(t) = \int_0^{\infty} W'(\nu)\cos(2\pi\nu t)\,d\nu$$

The cross-correlation function is defined by

$$\varphi_{f0} = \lim_{T\to\infty} T^{-1}\int_0^T f(t)g(t+\tau)\,dt$$

The mean square of $f(t)$ is

$$\varphi(0) = \int_{-\infty}^{+\infty} W(\nu)\,d\nu = \int_0^{\infty} W'(\nu)\,d\nu$$

If the process is Gaussian, it is entirely specified by its second-order properties: power spectrum or autocorrelation function. For instance, $p(y_t, y_{t+\tau})$ is a bivariate normal probability density function with $\mu_{11} = \mu_{22} = \varphi(0)$, and $\mu_{12} = \varphi(\tau)$.

Effect of a Linear Filter

A linear filter is defined by its impulse response, $h(t)$, or by its transfer function, $H(\nu)$, which is the Fourier transform of $h(t)$.

If the input to the filter is the random function $f_1(t)$, the output is the random function

$$f_2(t) = h * f_1$$

$$= \int_{-\infty}^{+\infty} h(t-\tau)f_1(\tau)\,d\tau$$

Introducing the power gain

$$G(\nu) = |H(\nu)|^2$$

the power density spectrum of f_2 is

$$W_2 = GW_1$$

The autocorrelation function of f_2 is

$$\varphi_2 = g * \varphi_1$$

where g is the Fourier transform of G or

$$g(t) = h(t) * h(-t) = \int_{-\infty}^{+\infty} h(\tau)h(\tau+t)\,d\tau$$

45 Reliability and Life Testing

Revised by
Douglas L. Marriott

DEFINITIONS AND TERMINOLOGY

Availability: Probability of a system subject to repair operating satisfactorily on demand.

Average Life: The mean value for a normal distribution of lives. Generally applied to mechanical failures resulting from "wear-out."

Burn-In (also *Initial Failure*): Initially high failure rate encountered on first placing a component on test. Burn-in failures are usually associated with manufacturing defects and the debugging phase of early service.

Component: (Normally used interchangeably with the term "unit.") A component is defined as an article which is normally a combination of parts, subassemblies, or assemblies, and is a self-contained element of a complete operating equipment and performs a function necessary to the operation of that equipment. *Examples*: indicator unit, power unit, receiver, transmitter, rotating antenna, modulator unit, amplifier unit.

Confidence Level (Coefficient): The degree of desired trust or assurance in a given result. A confidence level is always associated with some assertion and measures the probability that a given assertion is true. For example, it could be the probability that a particular characteristic will fall within specified limits, i.e., the chance that the true value of P lies between $P=a$ and $P=b$.

Configuration Management: Management of and knowledge of where all specifications, procedures, and associated test results are located and assigned, so that it is possible to produce these controlled items and all reliability evaluations and predictions pertaining to the system.

Cumulative Distribution Function: If x is a random variable, then the cumulative distribution function of x is defined to be the function F such that for every real number t, $F(t)$ is the probability that a given outcome of x will not exceed t; in symbols:

$$F(t)=\Pr\ (x \leq t)$$

Defect: Any deviation of a unit of product from specified requirements. A unit of product may contain more than one defect.

Degradation Failure: A failure that results from a gradual change in performance characteristics of an equipment or part with time.

Design Reviews:

(A) Preliminary design review: As soon as possible after a contract has been signed, a breadboard model should be built and its reliability estimated. Reliability engineering shall re-evaluate all parts and components and determine and recommend improvements in the design.

(B) Intermediate design reviews: While developing the system, conduct design reviews on a formal basis at all suppliers as well as at the prime contractor. This program should be coordinated with the reliability growth program. Account must be taken of the contract requirements for reliability goals at scheduled points in the production schedule.

(C) Critical design review: When engineering believes the design is ready to be "frozen" and also when a satisfactory prototype has met the qualification and other reliability tests, a final design review shall be scheduled. This formal review takes into account all contract demands as modified by the most recent changes in the contract. If the product is adjudged to be satisfactory, the final design may be approved for production. The block system should be used and authorization should be given to production to make x units per the specifications and blueprints without any change.

Downtime: Time during which equipments are not capable of doing useful work because of malfunction. This does not include preventive-maintenance time. In other words, downtime is measured from the occurrence of a malfunction to the correction of that malfunction.

Equipment: Material or articles (such as sets) comprising an outfit. The term "equipment" sometimes is used instead of "set," i.e., one or more assemblies or a combination of items capable of independently performing a complete function. *Examples*: Radio receiver, digital computer, automobile. An equipment may contain several sets as components. An example would be two radio receivers assembled for dual-diversity reception. The combination would constitute the equipment.

Failure: A failure is a detected cessation of ability to perform a specified function or functions within previously established limits on the area of interest. It is beyond adjustment by the operator by means of controls normally accessible to him during the routine operation of the device. This requires that measurable limits be established to define satisfactory performance of the function.

Failure Mode Analysis: Research, development, and production engineers as well as the reliability engineers analyze the basic design and determine by simulation and logistics what possible failures might occur. Corrective measures for eliminating and preventing failures are built into the basic design. Standard forms for the failure mode analysis are made available, and the results must be given to the prime contractors and production engineers for evaluation.

Failure Modes and Effects Analysis (FMEA): Extension of failure mode analysis which also considers the consequences of the failure mode, such as "revealed fault," "fail-safe," etc.

Failure Rate and Hazard Rate: Failure rate is generally the rate at which failure occurs during an interval of time (given that it has not occurred before the start of that interval) as a function of the total interval length. Hazard rate is an instantaneous failure rate and is defined as the limit of the failure rate as the time interval approaches 0. An example might be: A family takes an automobile trip of 120 miles and completes the trip in three hours. The average rate was 40 mph, although they drove at various rates of speed. The rate at any given instant could be determined by reading the speedometer at that instant. Therefore, the 40 mph is equivalent to the failure rate and the speed at any instant of time equals the hazard rate.

Inherent Reliability: The basic or generic failure rates of components have often been compiled by several companies as well as by some government agencies. A library covering failure rates should be part of a good reliability program. Such a listing gives concretely the reliability that can be guaranteed.

Lot Size: A specific quantity of similar material or collection of similar units from a common source; in inspection work, the quantity offered for inspection and acceptance at any one time. It may be a collection of raw material, parts, subassemblies inspected during production, or a consignment of finished products to be sent out for service.

Maintainability: The maintainability of an equipment in a specified maintenance environment is the probability that a failure will be repaired within a specified time after the failure occurs.

Mean Time Before Failure (*MTBF*): The total measured operating time of a population of equipments divided by the total number of failures of a repairable equipment is defined as the ratio of the total operating time to the total number of failures. The measured operating time of the equipments of the population that did not fail must be included. This measurement is normally made during that period of time between the early life and wear-out failures. In the case of a constant hazard rate, this ratio is the reciprocal of the failure rate.

The MTBF can be determined by dividing the product of the number of equipments tested, N, and the test time, t, by the number of failures, f, which occur during that time; i.e., MTBF, or often just m, is equal to Nt/f. The quantity m is the reciprocal of λ, i.e., $m = 1/\lambda$, and is related to the probability of survival by the exponential law $P_s = e(-t/m)$. The figure of merit m (sometimes expressed as $\bar{t}$) is convenient for use in determining if the reliability of an equipment is likely to be adequate for missions of specific lengths.

Mean Time to Failure (*MTTF*): The measured operating time of a single piece of equipment divided by the total number of failures of the equipment during the measured period of time. This measurement is normally made during that period of time between the early life and wear-out failures.

Mean Time to Repair (*MTTR*): The measured repair time divided by the total number of failures of the equipment.

Mission Success Rate: That percentage of the total missions uninterrupted by failure of the equipment. This figure of merit is more closely dependent on the reliability of the parts included in the system and on the design of the system than are either maintenance ratio or in-commission rate. However, this measure of reliability is valuable primarily to a using agency that has a regular schedule of missions. A mission success rate obtained by one agency is not typical of the equipment in general and will not necessarily apply for other agencies with different operating schedules.

Mode of Failure: The physical description of the manner in which a failure occurs and the operating condition of the equipment or part at the time of the failure.

Part Failure Rate: The rate at which a part fails to perform its intended function.

Probability: The limiting relative frequency in an infinite random series. If an event can occur in n ways and its failure in m ways, and if these $m+n$ ways are equally likely, then the mathematical probability that the event will occur in any one trial is the ratio $n/(n+m)$.

In other words, the probability of an event is the theoretical relative frequency with which it will occur, such relative frequency being the ratio of the number of times the event is observed under experimental conditions to the total of a great number of observations made under those conditions.

Common notation is:

$$\Pr(X) = \text{Probability of event } X$$
$$\Pr(X/Y) = \text{Probability of event } X \text{ given } Y \text{ has occurred}$$

Probability Density Function (*PDF*): The relative frequency of a continuous random variable, obtained from the cumulative distribution function by differentiation.

$$f(t) = dF(t)/dt$$

Probability of Survival: A numerical expression of reliability with the accepted nomenclature of P_s and a range from 0 to 1.0 indicating the extremes of "impossibility" and "certainty."

In other words, the probability of a given equipment performing its intended function or the given use cycle is

$$R(t) = 1 - F(t)$$

Product Effectiveness: The entire reliability program must be tied in with the quality engineering programs for securing the most effective operation possible. Product effectiveness includes all the elements for securing at minimum cost a product with maximum effectiveness. Programs for quality assurance contain schedules and procedures that encompass within them reliability, preventive maintenance, value engineering, human engineering, quality control, inspection, and tests that result in systems and products that will prove most effective when in operation.

Quality Assurance (QA): All those activities, including surveillance, inspection, control, and documentation, aimed at ensuring that the product will meet its performance specifications. Quality assurance generally has an independent role from production and reports directly to senior management.

Reliability Allocations: With an overall system reliability goal and where the confidence level is known, reliability values may be allocated to every component in the system by the use of available failure rates and weighting factors.

Reliability Demonstration: Critical tests must be programmed to provide sufficient valid data to determine the reliability of the system and all critical component parts. Provision should be made for processing all such data as expeditiously as possible to speed up all phases of the program with truly reliable materials and parts.

Reliability Goals: Requests for bids, specifications, and contracts requiring quality assurance and reliability generally describe completely the reliability goals. The reliability of the system with specified confidence levels is the principal goal. Values to be secured at specified points in time are listed on the growth curves. A reliability demonstration program is detailed to show that the reliability goals have been attained. How are such reliability goals expressed? The simplest statement covers only an expected value. The quality assurance paragraph in the specification will state simply: "The desired reliability is 99.7%." This is too simple, as the length of the mission or number of cycles of operation has not been detailed. The statement should be, "For 100 hours of operation the specified reliability is 99.7%."

Many contracts also introduce confidence levels. For example, after the initial design review, the reliability of the system must be 99% with a confidence level of 0.60. After qualifying, the system must have a reliability of 99.7% for a mission time of 10 hours with a confidence level of 0.99. Thus expressed, this goal has within it a final goal plus some information concerning the desired growth curve for reliability. In many programs, provision for the reliability demonstration program has not been made, or in making final arrangements for the finalized contract it is cancelled because of lack of funds. The reliability engineer should establish a very modest program to check the achievement of the reliability goals by means of an economic reliability demonstration program. Thus, it must be emphasized that the simplest possible reliability demonstration test should be made a part of the final program covered by contract and funds. This provides vital evidence that the reliability goals have been achieved and that the customer reliability requirements have been met.

Reliability Growth Curves: Periodic reports, such as monthly or quarterly, should be prepared containing up-to-date predictions of the system's reliability. These should be presented graphically on the reliability growth curves for this system and its various components and parts.

Reliability Predictions: Many agencies and companies have compiled failure rates for parts, components, subassemblies, assemblies, and systems. These generic failure rates are used as basic data to predict a value for reliability.

Sample: One or more sample units selected at random from a quantity of product to represent that quantity of product for inspection purposes.

Sequential Sampling: Sampling inspection in which, after each unit is inspected, the decision is made to accept, to reject, or to inspect another unit. *Note*: Sequential sampling as defined here is sometimes called "unit sequential sampling" or "multiple sampling." Multiple sampling is preferred, to differentiate from sequential testing.

Significance Level: The level of confidence (i.e., the probability of being correct) at which a hypothesis, such as "goodness of fit," is to be accepted or rejected.

Specification Limits: The specification limit(s) is the requirement that a quality characteristic should meet. This requirement may be expressed as an upper or a lower specification limit (called a single specification limit) or both upper and lower specification limits (called a double specification limit).

System (General): A combination of parts, assemblies, and sets joined together to perform a specific operational function or functions. *Examples*: Piping system, refrigeration system, air conditioning system.

Test to Failure: Testing conducted on one or more items until a predetermined number of failures have been observed. Failures are induced by increasing electrical, mechanical, and/or environmental stress levels, usually in contrast to life tests in which failures occur after extended exposure to predetermined stress levels. A life test can be considered a test to failure using age as the stress.

Value Engineering: One feature of a good reliability program is a concurrent value engineering program. Many programs include incentive provisions. If an operation is unusually expensive, provisions should be made for a series of improvements that fall in with the value engineering and incentive programs. Provisions for successive cost reduction programs as well as basic value engineering improvements should be added to each contract. It may be made a part of the incentive programs that are now usually included in military and government contracts for procurement.

Weapon System: A missile and all the necessary support equipment (either ground or airborne) necessary to launch and properly operate a missile.

Wear-Out Period: The wear-out period of an equipment is that period of equipment life, following the normal operating period, during which the equipment failure rate increases above the normal rate.

RELIABILITY DEFINITIONS

Reliability is the probability that a given product, system, or action will achieve its designated goal successfully under the specified environmental conditions and for a prescribed period of time or for the number of cycles of operation required for the mission or task.

Reliability involves three distinct concepts:

1. Attaining a specified level of performance
2. The probability of achieving that level
3. Maintaining that level for a specified time

ORGANIZATION FOR RELIABILITY

Reliability is one of the disciplines that form the assurance sciences. These disciplines govern the quality, safety, and dependability of a product throughout its life cycle. Whereas many of the aspects of assurance science, such as quality control, are aimed at ensuring that a product has been manufactured within acceptable limits, reliability is also related to aspects of design and development, in that its objective is to predict and assure an acceptable level of performance in service. For this reason, provision must be made in the structure of any manufacturing organization to integrate reliability activities with the product evolution process.

There is no single answer to the question of where the reliability function should exist in any organization. To some extent, the solution depends on the form of product and the organization itself. Comprehensive discussions of organization for product assurance and reliability are contained in References 1 and 2. Guidance is also available from professional bodies. For instance, the Institute of Electrical and Electronics Engineers and the American Society for Quality Control have defined a structure as shown in Fig. 1.

Today, reliability seems to fit most naturally into the product-assurance or product-effectiveness section of major space equipment manufacturers.

The most important organizational aspect is the fact that reliability and well-versed quality personnel use one tool which, when shared jointly, adds to the "experience-retention" capability; that tool is statistics. If the intent of any organization is to get the most out of the "fall of data," the organization of Fig. 1 is effective. Because of the problem of decision making, reliability is best placed at the staff level.

In a small organization, reliability is usually shared by engineering and by quality assurance. Some firms, to better satisfy the requirements of the Department of Defense, put reliability in the quality organization spectrum of operations, viz.,

Quality and reliability assurance tasks
Data reduction
Planning
Inspection, etc.
Test plan
Vendor survey
Quality-control engineering

Reliability planning, to be effective, must fit into the total program. One of the best ways to gain ac-

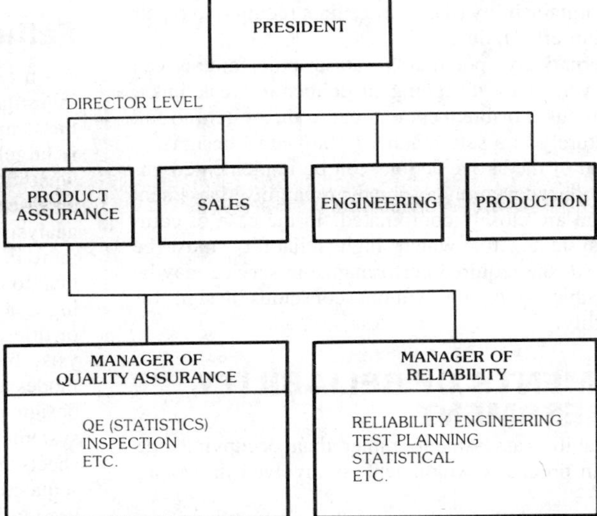

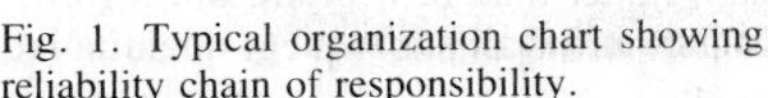
Fig. 1. Typical organization chart showing reliability chain of responsibility.

ceptance of the reliability or maintainability effort is to establish a proper sequence for its insertion into a master operating plan. For companies without a particular project, this plan usually is discussed with the engineering head and the production head. For corporations working on a program that requires reliability by specified intent, the customer invariably requires a reliability plan. The major items (or task delineators) in a reliability plan are noted in Chart 1.

CHART 1. TASK DELINEATION

1	Program plan update	11	Change and configuration control
2	Education and manuals	12	Reports and project review
3	Design to specified reliability and maintainability	13	Corrective action control
4	Apportionment	14	Supplier control
5	Models and prediction	15	Manufacturing reliability and maintainability control
6	Cost-effectiveness analysis	16	Failure diagnosis
7	Failure modes and effects analysis	17	Data acquisition and reduction
8	Human factors	18	Verification
9	Stress/strength analysis	19	Summary
10	Design review	20	References

In the plan itself, the time relationships of the tasks to reliability and overall project tasks are noted as milestones. Some of these milestones are then placed on more sophisticated PERT* or CPM† charts for management review of major effort.

The recognition of reliability as an integral part of product performance is more than simply a matter of early prediction of in-service behavior. It also identifies a number of major economic and safety implications. The realization that failures are inevitable adds an extra dimension to conventional design, consideration being given to the consequences of failures and actions which must be taken to mitigate them. One possible action, for instance, is to provide redundant or standby systems. Others are to give early attention to availability (the percentage of time a system is in a working state) and maintainability (the average time required to repair a system after failure).

Alternatively, potentially catastrophic failures can be prevented by designing in deliberate weak links, such as fuses or breakers that cause the system to fail prematurely in a safe manner (''fail-safe'' design).

None of the above actions can be implemented unless product development and the reliability assessment program are closely coordinated. In the case of complex systems, for which high reliability must be achieved, the required performance in service may be impossible to obtain without conscious design for reliability.

ELEMENTS OF RELIABILITY ASSESSMENT

Reliability assessment, rather than occupying one place in product development, is involved in a more or less continuous basis from the early conceptual stages through to the monitoring of service performance. As a result, reliability analysts may be called on to take part in the following activities:

Design reviews
Failure mode analysis
Quantitative reliability assessment
Failure data collection and assessment

* Program Evaluation and Review Technique.
† Critical Path Method.

Design Reviews

Several reviews of a system design are performed, usually by an interdisciplinary team of designers, production engineers, safety and reliability analysts, and others. The objectives are generally to ensure that the product is capable of meeting performance requirements, conforms with necessary standards, optimizes production facilities, and contains no unanticipated flaws. Guidelines for performing a design review have been set out in Reference 2. These emphasize careful documentation and the use of check lists prepared from past experience to ensure coverage of all relevant factors. The list of reliability terms at the beginning of this chapter gives a brief description of three common design reviews.

Failure Mode Analysis

An essential step in controlling reliability is a full investigation of all potential failure mechanisms and their causes, effects, and possible consequences. This is largely a qualitative exercise aimed at finding out what may happen, rather than how often, but it forms an essential foundation for all subsequent quantitative analysis. In fact, there are many instances where it is more important to have all the failure modes identified than to make accurate analyses of some while neglecting others altogether. This task is known by a variety of titles, one alternative being cause and effects analysis, but the most frequently used name is failure modes and effects analysis (FMEA). As in the case of design reviews, FMEA is helped considerably by a systematic approach using check lists and analysis sheets (See Reference 2 for examples). In addition, the sequential structure of many forms of failure can be used to identify the failure itself, using logic diagrams or decision trees. Two of the many variants available are shown in Figs. 2 and 3. Fig. 2 illustrates an event tree. This type of diagram is commonly used to map the end results of a sequence of safety device operations. Each branch point represents two possibilities, e.g., "device operates" or "device fails to operate." The event tree starts from an initiating event and progresses forward through subsequent events to a final state which may or may not be a significant failure state for the system. It is capable of pointing to undesirable event sequences which might not be identified by a less systematic approach. Unfortunately, event trees tend to generate many spurious branches and rapidly reach unmanageable proportions. Their main use is in mapping sequences where there is a clear chain of events, for instance, as in certain human operations (see Reference 3).

used in logic-circuit theory. They are not confined to only two branches at each branch point as is normally the case in event-tree construction. In theory, there is a problem with fault-tree analysis, in that the top event must first be identified by some other means. In practice, it is usually possible to identify failure events at a sufficiently high level to ensure exhaustive coverage of all possibilities. For instance, the top events for reliability analysis of a radio receiver would include:

Fail to operate at all
Fail to receive any signal
Fading or hunting reception
Drift
Excessive interference and noise levels

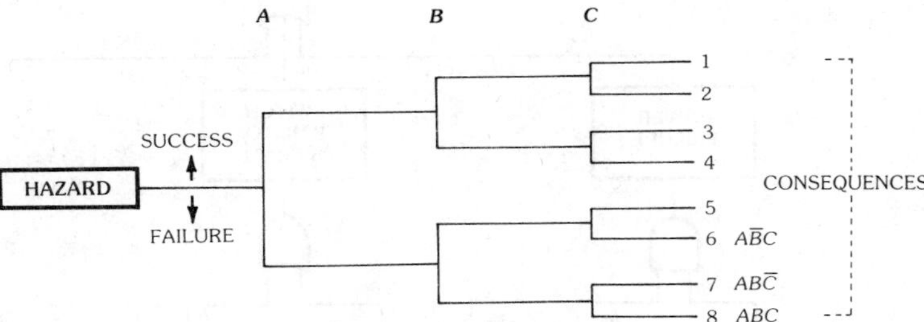

Fig. 2. A simple event tree.

The most common form of diagram is the fault tree (Fig. 3). This diagram employs a top-down logic, expanding out from a known end result to identify events, or combinations of events, that lead to failure. Fault trees use the same conventions for logic gates that are

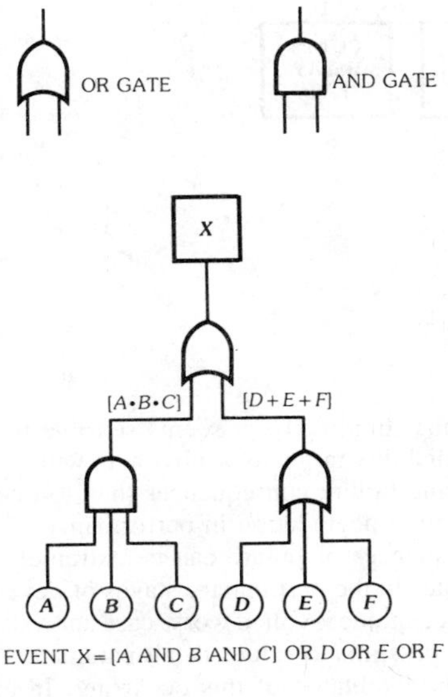

OR GATE
AND GATE
X
[$A \cdot B \cdot C$]
[$D+E+F$]
A B C D E F

EVENT X = [A AND B AND C] OR D OR E OR F

Fig. 3. A simple fault tree.

More detailed failure mechanisms can then be identified within each category, on the next lower level of detail. Fig. 4 is a partially developed fault tree for the top event, "fail to operate." To illustrate the technique more fully, it has been assumed that the receiver is powered from a mains (power-line) supply with an automatic switching backup battery. This diagram is by no means complete.

Fault trees can grow rapidly beyond easy manipulation by hand, and, for formal analysis, it is advisable to resort to computer codes, of which several are currently available for fault tree analysis. (See Reference 4 for an extensive summary of currently available software.) On a less formal level, the fault tree is an extremely useful aid in that it imposes a disciplined approach to the problem of failure-mode identification and it provides a structure whereby details of potential failures can be recorded in a sequential manner, without getting lost in the process.

In addition to identifying individual events that lead to failure, a fault tree also contains the logic governing the necessary sequences of events that lead to failure. These sequences are known as "cut sets." In Fig. 5, any one of the combinations *A* and *D*, *B* and *C*, *D* and *G*, or *J* and *B* is a cut set. In more complicated problems, the cut sets may be very complex themselves, and not necessarily independent of one another. It can be shown that cut sets can be reduced to an equivalent set of minimal independent requirements for failure called "minimum cut sets." Two simple examples are shown in Fig. 5. In Fig. 5A, four events, A, B, C, and D contribute to some failure event X. Combining these events by the rules of Boolean logic algebra, X occurs if A and C or A and D or B and C or B and D occur. These sequences are cut sets and, in this case, minimum cut sets. In Fig. 5B, a similar fault tree contains a common element B in both branches. In this case, the top event, Y, is only dependent on the occurrence of A and C or B. Here, B represents an important class of failures called "common mode" or "common cause" failures. It can be easily seen that if A, B, C, and D all have similar probabilities of occurrence, Y has a much higher probability of occurrence that X, because it only requires a single event instead of two simultaneous events. Failure to identify common modes or dependencies between different failure mechanisms is a frequent cause of excessively optimistic reliability predictions. It can be seen from the above simple example how fault tree analysis can be used to avoid this problem.

Quantitative Reliability Assessment

Quantitative reliability assessment is one of the most important tasks of the reliability analyst. Reliability quantification covers two dimensions—probability of occurrence and magnitude of the consequence of failure. Assessment of failure consequence is probably more the task of design. Nevertheless, it cannot be neglected in reliability assessment, particularly when assigning reliability targets, or allocating reliability to subsystems. Intuitively, it seems sensible to assign a higher reliability target to a subsystem with potentially catastrophic failure consequences than to one that involves only a degradation in performance.

Consequences of failure can be extremely difficult to estimate. In the preliminary stages of development, a useful technique is called worst-case analysis. It takes the most pessimistic values of all parameters, regardless of the likelihood of this occurring. In electronic circuits, for instance, all components are assumed to

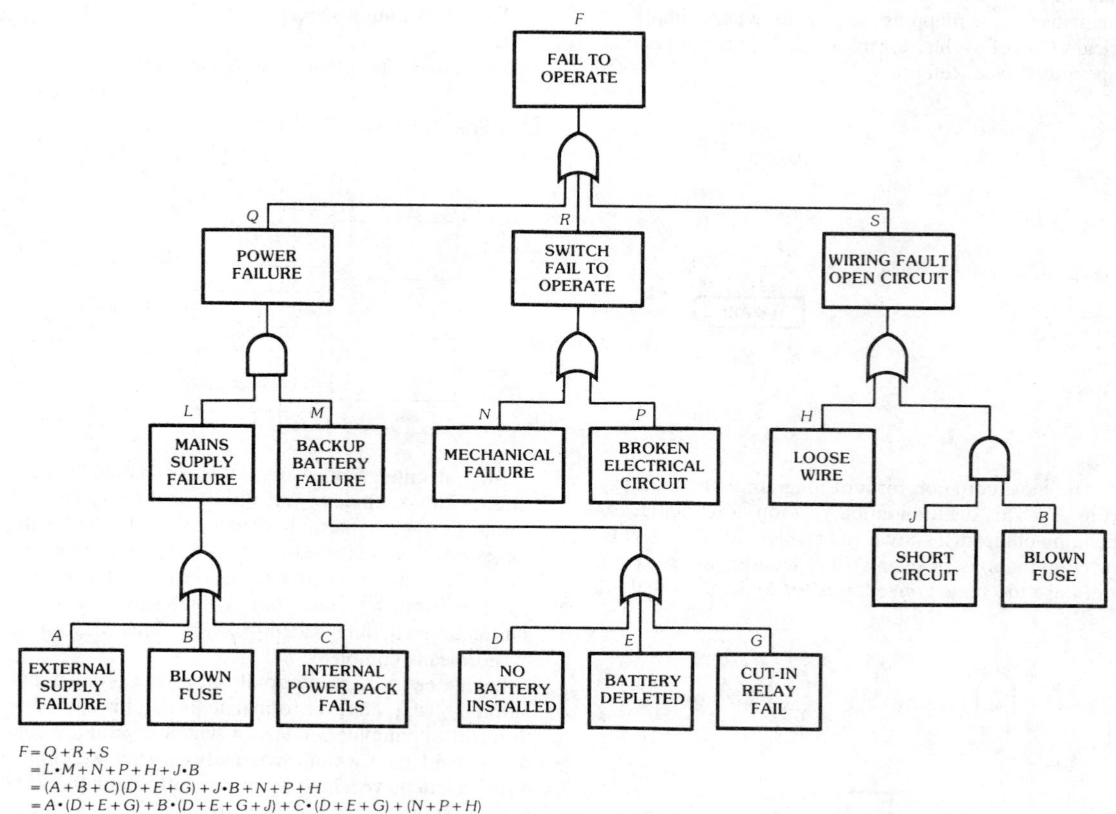

Fig. 4. Fault tree for radio-receiver failure mode "fail to operate on demand."

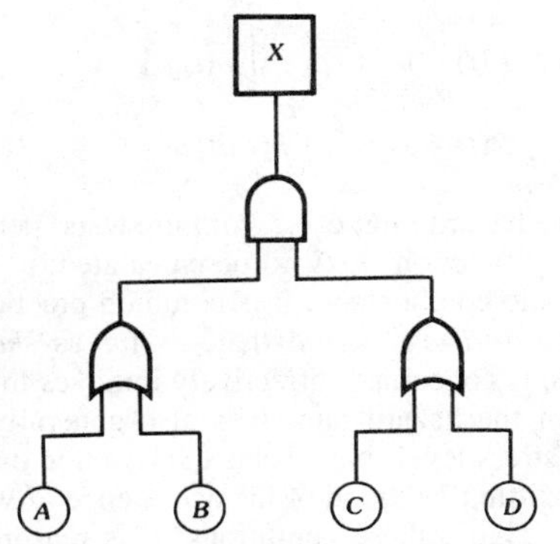

EVENT $X = (A \text{ AND } C) \text{ OR } (A \text{ AND } D) \text{ OR } (B \text{ AND } C) \text{ OR } (B \text{ AND } D)$
$\Pr(A) = A \cdot C + A \cdot D + B \cdot C + B \cdot D$
IF $\Pr(A) = \Pr(B) = \Pr(C) = \Pr(D) = 0.01$
$\Pr(X) \doteq \Pr(A) \cdot \Pr(C) + \Pr(A) \cdot \Pr(D) + \Pr(B) \cdot \Pr(C) + \Pr(B) \cdot \Pr(D)$
$= 4 \times 10^{-4}$

(A) Independent events.

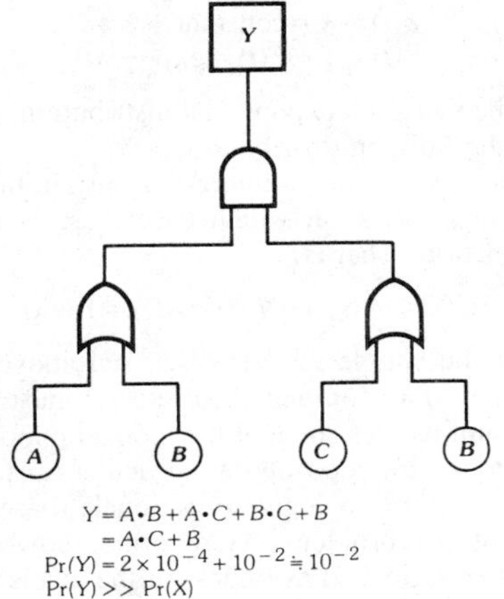

$Y = A \cdot B + A \cdot C + B \cdot C + B$
$= A \cdot C + B$
$\Pr(Y) = 2 \times 10^{-4} + 10^{-2} \doteq 10^{-2}$
$\Pr(Y) >> \Pr(X)$

(B) Dependent events.

Fig. 5. Calculation of event probability.

be at the extreme limits of their tolerance bands, including any drift due to environmental factors. By this procedure, it is possible to identify undesirable modes of operation that might not be considered possible otherwise. The detection of potentially catastrophic modes of failure is probably the most useful outcome of this type of analysis. Regardless of their likelihood, failures of this type may require designing out of the system completely. It is only in those cases where the cost of failure can be reasonably balanced against the cost of improved design that the question of probability of occurrence becomes important.

The area of quantification most closely identified with the reliability analyst is the assessment of failure probability. This is a natural result of defining reliability in probabilistic terms. A wide range of mathematical techniques have been developed in the general fields of probability and statistics, supplemented with methods specifically directed toward reliability modeling. These methods will be reviewed in a special section later in this chapter.

Failure Data Collection and Assessment

The collection of failure data is part of a continuous feedback loop that is essential to reliability growth. The process begins in the early stages of product development, where failures are likely to be common due to design faults and as yet uncontrolled production processes, and extends through to collection of data under normal service conditions. The reliability analyst should be closely involved in this process. In addition to receiving basic failure-rate data from this source, the reliability analyst also has an active role in evaluating failure reports, assessing whether the failures are random or represent a significant deviation from expected conditions, providing backup to quality control, and offering recommendations for remedial action.

For failure analysis to be successful, the data collection system must be carefully designed and controlled. Considerable thought has been given to this problem in recent years, and a number of reporting systems have been developed. A comprehensive review of data banks and their methods of operation is contained in Reference 4. Some of these are appropriate models for in-house reporting systems. This subject will be discussed further in a later section of this chapter on sources of data. One reporting system of particular relevance to integrated circuit reliability was developed by Lockheed Corporation on behalf of NASA (Reference 5). This report is both an excellent source of information and a model for setting up an in-house reporting and analysis system.

COMPONENT RELIABILITY

A population of components on test or in service under nominally similar conditions will decrease in time as a result of failures as shown in Fig. 6. The numbers failed and surviving after time t are $N_f(t)$ and $N_s(t)$, respectively. For a sufficiently large population number N, the relative frequencies represent the cumulative probability of failure and the probability of survival (i.e., the reliability), respectively:

$$\hat{F}(t) = N_f(t)/N$$
$$\hat{R}(t) = N_s(t)/N$$

As N tends to infinity, estimates $\hat{F}(t)$ and $\hat{R}(t)$ tend to the true failure and reliability functions, $F(t)$ and $R(t)$. Here, and subsequently in this chapter, a "hat" (^) will be used to represent an estimate of a random variable. These functions are shown also in Fig. 6. By differentiating $F(t)$, the probability density function of failures, $f(t)$, is obtained (Fig. 7). This represents the relative frequency of failures with respect to the orig-

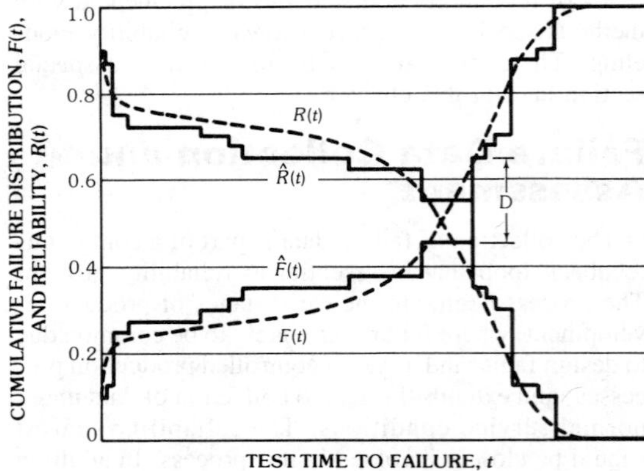

Fig. 6. Cumulative failure distribution and reliability plots for a test with 40 samples—experimental and theoretical curves shown.

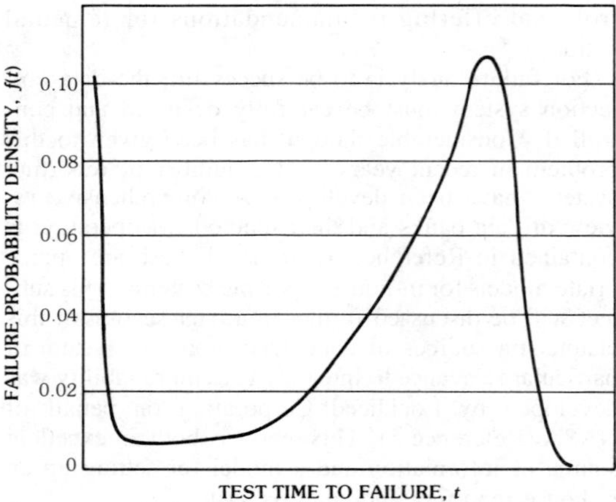

Fig. 7. Failure probability density derived from Fig. 6.

inal population. Three zones can be identified in this figure. They are an initial period of high failure density, corresponding to initial manufacturing defects, an intermediate period of relatively low failure rate, where the only cause is external and random, and a final peak of high density corresponding to wearout or material deterioration. These zones can be more readily distinguished by calculating the instantaneous failure rate, or hazard rate, $h(t)$, which is the failure rate expressed as a proportion of the population surviving at time t.

$$h(t)=f(t)/[1-F(t)]$$

The hazard rate has a characteristic shape, illustrated in Fig. 8, known as the "bathtub curve." Here, initial, or burn-in, failures; random failures; and wearout failures are clearly identified by having decreasing, constant, and increasing hazard rates, respectively. This characteristic of the hazard rate is an important diagnostic tool, as will be demonstrated later in this chapter.

It can be shown (see Reference 4 for details) that

$$F(t)=1-\exp[-\int_0^t h(t)dt]$$

$$R(t)=\exp[-\int_0^t h(t)dt]$$

Knowing the hazard rate, e.g., from life tests, permits the reliability function, $R(t)$, to be calculated.

In electronic components, it is common practice to require a burn-in test period that, as far as service performance is concerned, effectively removes the initial stage of the hazard rate. It is also generally assumed that stress levels have been so chosen in design that there is little likelihood of the occurrence of wearout failures. Under these conditions, it is reasonable to use a restricted version of reliability theory that assumes a constant hazard rate and is relatively easy to manipulate mathematically. In fact, the majority of formal reliability theory is based on the constant-hazard-rate model (sometimes referred to loosely and erroneously in much of the literature as a "constant-failure-rate" theory). In this model,

$$h(t)=\lambda=\text{constant}$$
$$R(t)=1-F(t)=\exp(-\lambda t)$$

This is the so-called exponential distribution, a special case of the Poisson distribution.

In many cases, one is interested only in high component reliabilities, in which case there is a simple approximation. That is,

$$\text{if } \lambda t << 1,\ 1-F(t)=R(t)=1-\lambda t$$

Where the simple models discussed above are not valid, other more complex distributions must be used. Table 1 summarizes some of the more commonly used distributions. In choosing a particular distribution, there is sometimes a degree of rationale, based on the physics of the problem. More often, however, the choice is empirical so as to best fit the available data. One distribution of predominant interest in reliability studies is the Weibull distribution (see Table 1). This distribution has some justification in that it is one of a class of extreme distributions, i.e., distributions of extreme values of groups of components. Its main attribute is its ability to represent any one of the three zones in the bathtub curve (Fig. 8) by choosing the shape parameter, β, to be less than, equal to, or greater than unity (see Fig. 9). This property of the Weibull distribution will be shown later to be a useful method of evaluating service data.

SYSTEM RELIABILITY

System reliability can be dealt with in two different ways, depending on the advancement of the project.

In the early stages of making a bid, the eventual configuration will not be known precisely, but a reliability estimate is required for the purpose of making the proposal. This estimate is inevitably crude and must draw on previous experience and generic data for its quantitative basis. The simplest approach, once the

TABLE 1. LIST OF COMMON FREQUENCY DISTRIBUTIONS

Name	Mass/Density Function		Mean	Variance
Binomial	$P_X(x) = \binom{n}{x} p^x (1-p)^{n-x}$	$x = 0, 1, 2, \ldots$	np	$np(1-p)$
Poisson	$p_X(x) = (\theta^x / x!) e^{-\theta}$	$x = 0, 1, 2, \ldots$	θ	θ
Normal	$f_X(x) = (1/\sqrt{2\pi}\sigma) \exp[(-1/2\sigma^2)(x-\mu)^2]$	$-\infty < x < \infty$	μ	σ^2
Lognormal	$f_X(x) = (1/\sqrt{2\pi}\sigma x) \exp[(-1/2\sigma^2)(\ln x - \mu)^2]$	$x > 0$	$\exp(\mu + \sigma^2/2)$	$\exp(2\mu + \sigma^2) [\exp(\sigma^2) - 1]$
Student-t	$f_X(t) = [\Gamma(m/2 + 1/2)/\sqrt{m\pi}\Gamma(m/2)] (1 + t^2/m)^{-(m+1)/2}$	$-\infty < t < \infty$	0	$m/(m-2)$
Gamma	$f_X(x) = [\theta^n / \Gamma(n)] x^{n-1} e^{-\theta x}$	$x > 0$	n/θ	n/θ^2
Exponential	$f_X(x) = \theta e^{-\theta x}$	$x > 0$	$1/\theta$	$1/\theta^2$
Weibull	$f_X(x) = \beta(x^{\beta-1}/\eta^\beta) \exp[-(x/\eta)^\beta]$	$x > 0$	$\eta\Gamma(\beta+1)/\beta$	$\eta^2\{\Gamma(\beta+2)/\beta - [\Gamma(\beta+1)/\beta]^2\}$
Chi-Square	$f_X(x) = [1/2^{n/2}\Gamma(n/2)] x^{n/2-1} e^{-x/2}$	$x > 0$	n	$2n$

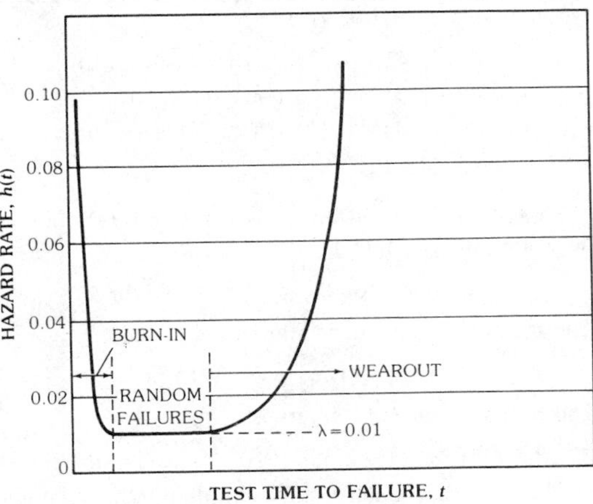

Fig. 8. Hazard rate for data given in Fig. 6.

number and types of components are known approximately, is to assume the individual failure rates to be additive:

$$\lambda_s = \sum \lambda_i$$

The system reliability, $R_s(t)$, is obtained as

$$R_s(t) = \exp(-\lambda_s t)$$

This simple approach can also be used in reverse to assign component reliabilites so as to achieve a target system reliability that may have been set as a design requirement. Illustrating this simple method, if an assembly is made up of two components of equal weight and is to have an end reliability of 0.99, then each component should have a reliability of at least 0.995. The product $0.995 \times 0.995 = 0.990025$. If the assembly is made up of three units of equal weight, then each should have a reliability of 0.997, since $(0.997)^3 = 0.991026973$. The same procedure is used even if the parts have widely different importance and hence extremely different reliability values.

Formalized variants of this approach are given in the AGREE report (Reference 6) and in a more recent version in Military Handbook MIL-HDBK-217B (Reference 7).

The AGREE report uses a more complex method, detailed on pages 192–195 of Reference 18. The AGREE report cautions against the use of its method for items of low importance. Its basis is the complexity of the units rather than their failure rates. Consideration is given to the number of hours the jth unit will be required to operate in T system hours and the total number of modules in the system. The importance factor of each unit must also be considered. All units do not operate the same length of time; hence, these times of operation must be considered, and many of these t_j values will be less than T hours. The AGREE report gives the allocated rate of the jth unit as

$$\tilde{\lambda}_j = n_j[-\log R^*(T)]/E_j t_j N$$

This is used in many instances to allocate reliabilities among the components.

MIL-HDBK-217B describes two methods of failure-rate prediction, "part stress analysis" and "parts count." The latter is the simpler of the two, and more appropriate to early design and bidding. For full explanations, reference is made to the handbook itself and to an expanded explanation in Reference 1. A parts count failure rate prediction lumps components into generic part types and assigns part quality levels to each type. The inputs are therefore

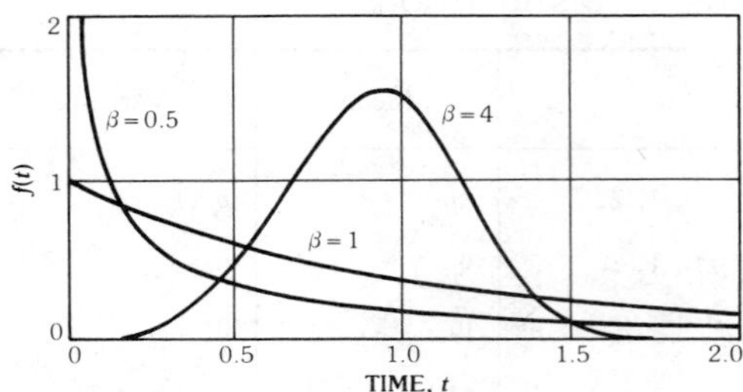

(A) Probability density function.

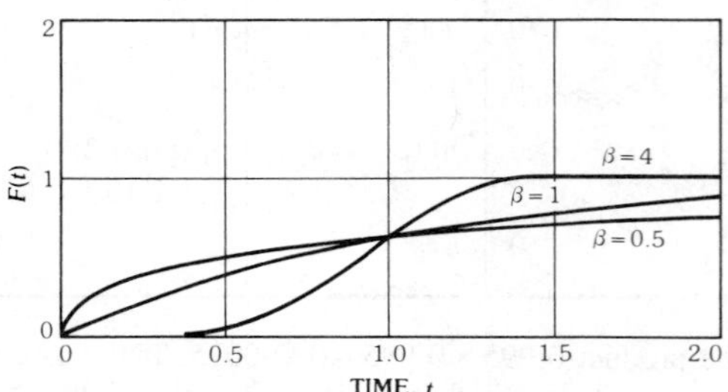

(B) Probability distribution.

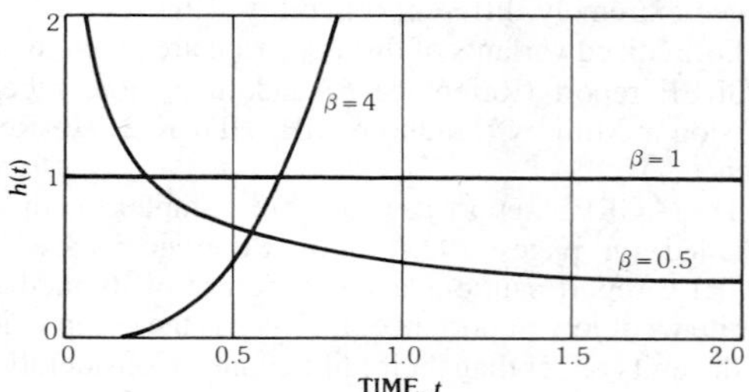

(C) Hazard function rate or instantaneous failure rate.

Fig. 9. The Weibull distribution, $F(t)=1-\exp[-(t/\eta)^{\beta}]$.

Generic part types
Quantities of parts in each type
Part quality levels
Equipment environment

The total failure rate, λ_s, is then calculated from

$$\lambda_s=\sum_i(\lambda_{pc})_i$$

where $(\lambda_{pc})_i$ is the ith part category failure rate:

$$\lambda_{pc}=n\cdot\lambda_g\cdot\pi_Q$$

where,

n = number of parts in category i,

λ_g = generic failure rate, provided in MIL-HDBK-217B, for a range of components, taking into account environmental factors such as "ground, benign," "space, flight," "naval, unsheltered," up to the most severe, "missile, launch,"

π_Q = quality factor, specifying the standard of selection, e.g., "above spec.," "MIL-spec.," "below MIL-spec."

EXAMPLE 1. A preliminary design of an RC network for a ground mobile (G_M) environment gives the information shown in Table 2.

TABLE 2. DATA FOR EXAMPLE 1

Part Count	Generic Part Type*	Number of Parts	Generic Failure Rate, $\lambda_G/10^6$ hr*	Quality Factor, π_Q*
1	Film Resistor RNR 55182	32	0.042	0.1 (R)
2	Wire Resistor RWR 35007	5	0.110	0.1 (R)
3	Capacitor CKR 39014	15	0.044	0.1 (R)

* Data from MIL-HDBK-217B

$$(\lambda_{pc})_1=32\times0.042\times10^{-6}\times0.1/\text{hr}=0.1344\times10^{-6}/\text{hr}$$
$$(\lambda_{pc})_2=5\times0.11\times10^{-6}\times0.1/\text{hr}=0.055\times10^{-6}/\text{hr}$$
$$(\lambda_{pc})_3=15\times0.044\times10^{-6}\times0.1/\text{hr}=0.066\times10^{-6}/\text{hr}$$
$$\lambda_s=0.2554\times10^{-6}/\text{hr}$$

EXAMPLE 2. A system consisting of 1000 units as described in Example 1.

$$\lambda_{\text{system}}=1000\lambda_s=2.554\times10^{-4}/\text{hr}$$

The mean time between failures is

$$\text{MTBF}=1/\lambda_{\text{system}}=3915\text{ hr}$$

The reliability at 1000 hours is

$$R(1000)=\exp(-2.554\times10^{-4}\times10^3)=0.775$$

The above approach assumes implicitly that all components in a system are connected in series and that a single failure causes system failure (weakest-link theory). This is conservative and unrealistic in large systems where backups are necessary if the system is to work at all. A more accurate assessment can be made once the system configuration has been finalized. In general, this may contain parallel, or redundant, elements in addition to series elements. A number of examples of redundant systems are given in Fig. 10. Note particularly that a series system is less reliable than its components, whereas a parallel system is more reliable. Series connections are generally forced on the designer by functional requirements, and redundancy is one method of retaining an acceptable level of reliability. Other techniques, such as standby and voting logic systems, are also possible (see Reference 4 for further details).

In practice, reliability of a system is not simply a function of the inherent failure rates of its components. The system will be subject to repair and scheduled maintenance. While periodic repair ensures that the reliability does not decay continuously to zero, the system will not be available on demand while it is being repaired. For complex systems with repair, therefore, it is more convenient to think in terms of "availability," A.

Product availability is defined as the probability that the system will operate satisfactorily at any point in time, where time includes not only operating life, but also active repair time and administrative and logistic time. An equation for availability is

$$A = MTBF/(MTBF + MTTR)$$

where,

A = availability,
MTBF = mean time before failure,
MTTR = mean time to repair.

The calculation of MTTR is related to repair hours, while the calculation of MTBF is related to component operating hours. Fig. 11 is a graph of the above equation.

It is evident from the figure that the effect of maintainability on availability increases as the ratio of MTBF to MTTR decreases. If an item has an inherently low MTBF, the MTTR must be very low to sustain a good level of availability.

In the design of any complex system, an optimum relationship should be established between reliability and maintainability, so that reliability is not increased beyond the point where very little availability gain is obtained because of lack of consideration of the effect of maximum maintainability.

To look at this another way, manufacturers of microelectronic devices claim that MTBF is very high and repair time is nil or very low. Look again at Fig. 11 and the ratio of MTBF/MTTR in the region to the right of 100. Little is to be gained by designing a module that can be repaired extremely quickly, if to do the job special tools and costs are involved. In other words, the throwaway concept in this case is clearly justifiable. NOTE: There is some ambiguity in the use of the terms MTBF and MTTF. Some references use the same definitions as accepted in this chapter. Others, such as Reference 4, use MTTF as the inverse of failure rate, with MTBF to be given by

$$MTBF = MTTF + MTTR$$

Hence, by this convention, $A = MTTF/MTBF$

SOURCES OF RELIABILITY DATA

There are two sources of failure-rate data:

Generic data banks
Experimental determination by life testing

Generic Data Sources

Generic data are usually used for initial design and evaluation, whereas life testing is more appropriate to

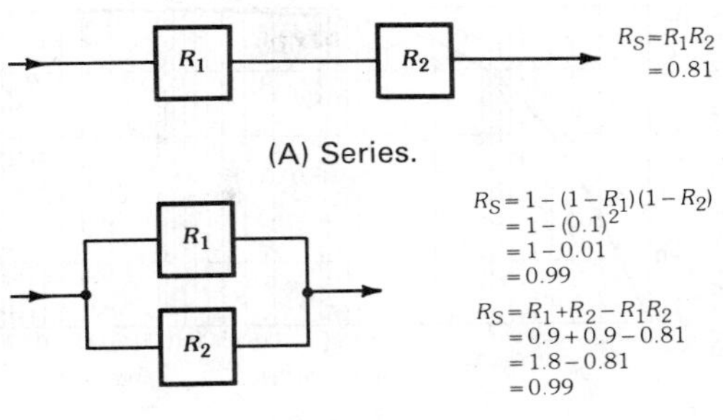

(A) Series.

(B) Parallel.

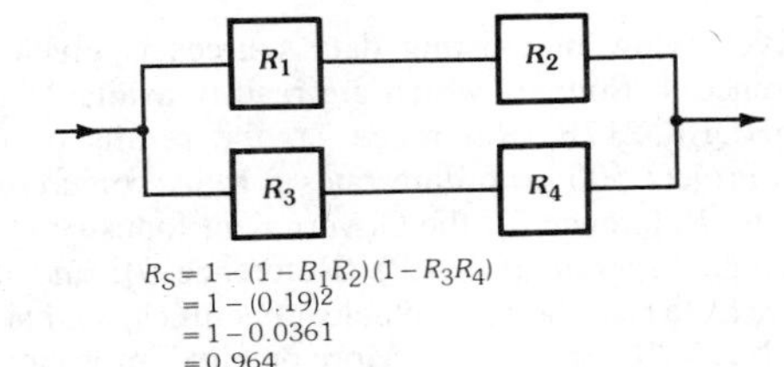

(C) Series-parallel.

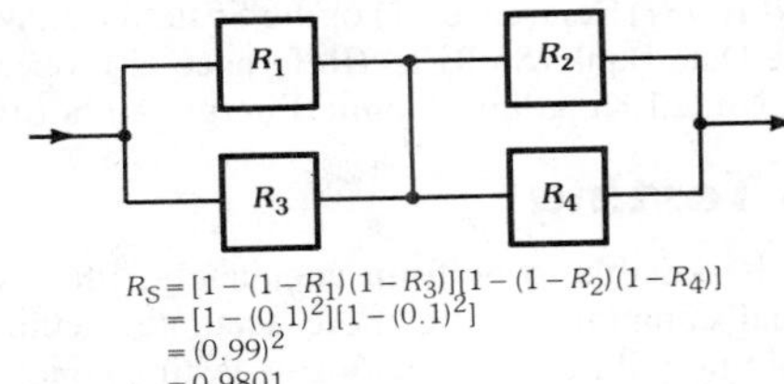

R1
R2
R3
R4
RS = [1 − (1 − R1)(1 − R3)][1 − (1 − R2)(1 − R4)]
= [1 − (0.1)²][1 − (0.1)²]
= (0.99)²
= 0.9801

(D) Improved series-parallel.

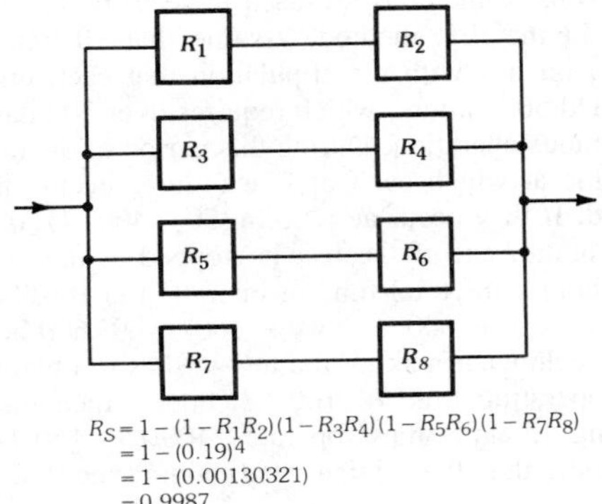

(E) Further improved series-parallel.

Fig. 10. Diagrams of redundant systems with relationships for computing system reliabilities. All parts are assumed to have equal reliability ($R_C = 0.90$).

the latter stages of system development, where questions such as reliability growth and system substantiation are important.

Data banks and other sources of failure-rate data have grown considerably in recent years. A compre-

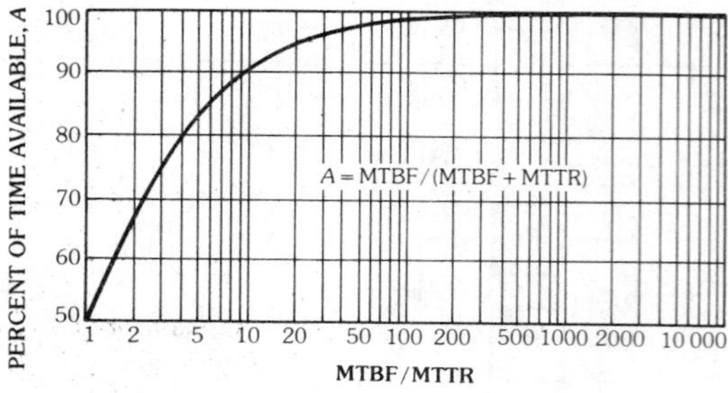

Fig. 11. Product availability.

hensive review of existing data sources is given in Reference 4. Sources which are readily available are MIL-HDBK-217B (Reference 7), the results of the IEEE Project 500 on failure rates for electronic components (Reference 8), the Government/Industry Data Exchange Program (GIDEP) (Reference 9), and the Failure Data Handbook for Nuclear Facilities, available from NTIS (Reference 10). More detailed information can be obtained by subscribing to one of several data banks, such as the Nuclear Plant Reliability Data System (NPRDS) (Reference 11) or the System Reliability Service Data Bank (SYREL) (Reference 12), operated by the United Kingdom Atomic Energy Authority.

Life Testing

If it is possible to perform adequately, life testing of actual components is a more accurate method of determining failure rates, as long as testing conditions are chosen to simulate actual service conditions as realistically as possible.

Two basic methods are used in securing reliability data. In the first method, assume that 20 units are placed on test with the stipulation that each unit be operated 5000 hours, which requires over 200 days of continuous operation. During this period of operation, the time at which each of five failures occurs is recorded. If they occur at 4200, 4350, 4400, 4750, and 4900 hours, and if the test is stopped at the end of 5000 hours, the total time of operation is 100 000 − (800 + 650 + 600 + 250 + 100) = 97 600 hours. If the technician making the test wishes to obtain the total operating time of 100 000 hours, then the remaining 15 units must operate 2400/15 = 160 hours each more than the original 5000 hours specified with no additional failures. Otherwise, the estimate of the failure rate should be $r = 5/97\,600 = 5.12/100\,000 = 5.12\%/1000$ hours. The value of $m = \text{MTBF}$ is $m = 97\,600/5 = 19\,520$ hours.

The second basic method for determining the failure rate is to operate all 20 units until they fail. This makes it difficult to plan a testing program, as some equipment may be tied up for two or even several times the period designated for the original test. However, failure rates found by the second method should be more representative of actual field failure rates. The data secured under these conditions for the 20 units might be as given in Chart 2.

This gives a failure rate much larger than the first estimate, i.e., $r = 20/153\,200 = 13.05\%/1000$ hours. If the 5% value is near the actual central value, then the total operating time should be about 400 000 hours, or 20 000 hours (833 days) on the average for each unit.

It is desirable to establish confidence levels for reliability values. A distribution of probability values is used to give either a one-sided confidence value or a two-sided confidence value with both upper and lower limits. When determining failure rates, a one-sided level applies. A confidence level of 0.95 might be stipulated for a reliability of 0.95 or better (the goal stipulated in the contract). In the first set of data, the rate of 5%/1000 hours may be expressed as a reliability of 95% where the mission time is 1000 hours. If the mission time is 10 hours, then the corresponding reliability is 99.95%. These values are associated with expected values rather than having a range of values associated with a designated confidence-level value. Where a confidence limit is used, a limit is established that is associated with this limit. It will be expressed as a failure rate or an MTBF value.

Fixed time testing is covered in a useful publication of the Department of Defense (DOD), Handbook H108 (Reference 13). Both fixed and sequential testing are covered by a military standard, MIL-STD-781B (Reference 14). These references and their applications are well illustrated in Reference 1.

PROBABILITY AND STATISTICAL INFERENCE

Confidence Limits

If 100 components are tested and two fail, the failure probability can be estimated approximately as

CHART 2. TIME IN HOURS TO FAILURE FOR RELIABILITY TESTS (n = 20 UNITS)

	6500		8000		10200		5500	
	4200		7000		4750		8900	
	9400		4350		4900		10500	
	4400		9100		8750		9150	
	7800		8100		9200		12500	
Sum =	32300	+	36550	+	37800	+	46550	= 153 200

$$\hat{P}_F = 0.02$$

The corresponding reliability is therefore

$$\hat{R} = 1 - P_F = 0.98$$

However, these are only estimates based on a single test. If the test is repeated, a different estimate will be obtained, as shown in Table 3, which summarizes 20 such tests, each of 100 components. Empirically, it can be determined that the mean observed reliability is $\overline{R} = 0.975$, with a standard deviation of $\sigma_R = 0.01775$. By fitting a theoretical distribution such as a normal or Poisson distribution to this data, it is possible to determine the probability of the number of failures in some subsequent batch exceeding some value, say 6. The relative frequencies of different observed reliability estimates have been calculated in Table 4 using a normal distribution with sample mean and standard deviation, and using a Poisson distribution with the same observed overall failure rate, 0.025. Predictions are shown in Fig. 12. It appears that the normal distribution gives the better fit to the data.

This problem shows that the true reliability may be different from any observed estimate. However, it is possible to use the data summarized in Table 3 to calculate limits that will contain the true value a specified percentage of the time. These limits are known as *confidence limits*, and the percentage is called the *confidence level*.

In practice, the observed values of $\hat{R}$ will be dispersed around the true reliability, R, but R is unknown. If it can be assumed that R is normally distributed with a true standard deviation equal to the estimate from the 20 tests, then it can be shown that the mean of n tests, $\overline{R}$, will be normally distributed with a standard deviation $\sigma_R = \sigma_{\hat{R}}/\sqrt{n}$, where $n = 20$ in this case; further-

TABLE 3. RESULTS OF 20 LIFE TEST GROUPS OF 100 COMPONENTS EACH

Test	Units on Test n_i	Failures d_i	Successes s_i	Reliability $R_i = s_i/n_i$
1	100	0	100	1.000
2	100	1	99	0.990
3	100	3	97	0.970
4	100	0	100	1.000
5	100	2	98	0.980
6	100	4	96	0.960
7	100	3	97	0.970
8	100	5	95	0.950
9	100	1	99	0.990
10	100	2	98	0.980
11	100	0	100	1.000
12	100	0	100	1.000
13	100	3	97	0.970
14	100	6	94	0.940
15	100	4	96	0.960
16	100	5	95	0.950
17	100	2	98	0.980
18	100	2	98	0.980
19	100	3	97	0.970
20	100	4	96	0.960
Total	2000	50	1950	(0.975)

$$m = 20,\ n = n_i = 100,\ \sum d_i = 50,\ \sum n_i = 2000.$$

$$R = (\sum n_i - \sum d_i)/\sum n_i = \sum s_i/\sum n_i$$

Average: $\overline{R} = 19.500/20 = 0.9750$

$\sigma_R{}^2 = (\sum f_j R_j{}^2/\sum f_j) - \overline{R}^2 = (19.0188/20) - (0.975)^2 = 0.950940 - 0.950625 = 0.000315.$

Also $\sigma_R{}^2 = \sum f_j \langle \Delta R_j \rangle^2 / \sum f_j$ (where $\Delta R_j = R_j - \overline{R}_j$) and $\sigma_R = 0.017748$; hence

$\sigma_R{}^2 = \frac{1}{20}(0.001225 + 0.001250 + 0.000675 + 0.000100 + 0.000100 + 0.000450 + 0.002500)$

$= \frac{1}{20}(0.006300) = 0.000315$, or using relative weights, $w_j f_j = 1.00$, $\sigma_R{}^2 = w_j f_j \langle \Delta R_j \rangle^2 = 0.000315$, given in last column above. Hence $\sigma_R = (0.000315)^{1/2} = 0.017748$.

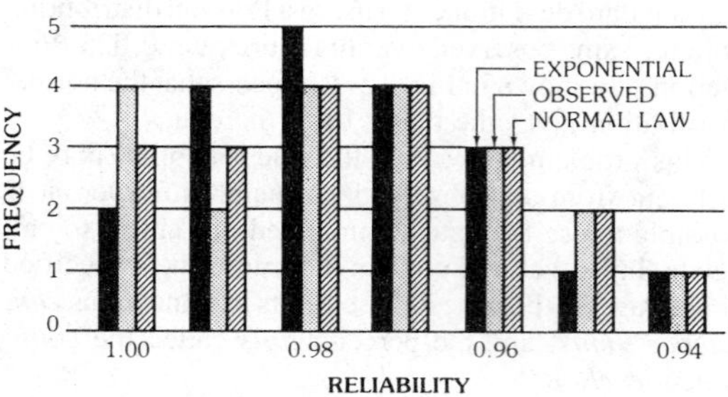

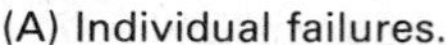
(A) Individual failures.

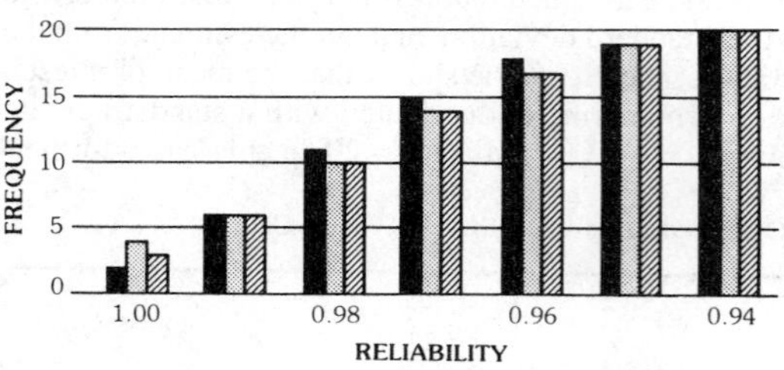

(B) Cumulated failures.

Fig. 12. Observed reliability distributions of individual and cumulated failures for 20 sets with $n = 100$ on test for 1000 hours each. The corresponding theoretical frequencies for exponential and normal law are also shown. Failure rate $r = 0.025$ and $rn = 2.5$ for exponential; $\overline{R} = 0.975$ and $\sigma_R = 0.01775$ for normal law.

more $\overline{R}$ tends to the true reliability as n increases. If we imagine many sets of n tests to be performed, the results will be distributed around the true (unknown) reliability R, as shown in Fig. 13. The value $\overline{R} = 0.975$ obtained from Table 3 is also shown.

The probability that $\overline{R}$ is greater than some amount L larger than R is

$$\Pr(\overline{R} > R + L) = \int_{R+L}^{\infty} n(R;\sigma_{\hat{R}}/\sqrt{n})dR$$

where $n(R;\sigma_{\hat{R}}/\sqrt{n})$ is the normal distribution with mean R and standard deviation $\sigma_{\hat{R}}/\sqrt{n}$.

This can be expressed more conveniently in terms of the standard normal variate, Z_X.

$$Z_X = (X - R)/(\sigma_{\hat{R}}/\sqrt{n}),\ Z_L = L/(\sigma_{\hat{R}}/\sqrt{n})$$

$$\Pr[(\overline{R} - R)/(\sigma_{\hat{R}}/\sqrt{n}) > Z_L] = \int_{Z_L}^{\infty} n(0;1)dZ$$

If as is usual $\overline{R}$ is known and R is unknown, we need to determine the probability that R is less than $\overline{R}$ by some amount L. From the previous equation

$$\Pr(R < \bar{R} - Z_L/\sigma_{\hat{R}}/\sqrt{n}) = \int_{Z_L}^{\infty} n(0;1)dZ = \alpha$$

The quantity α can be expressed in terms of the error function that is tabulated in any standard text for the standardized normal distribution, $n(0;1)$.

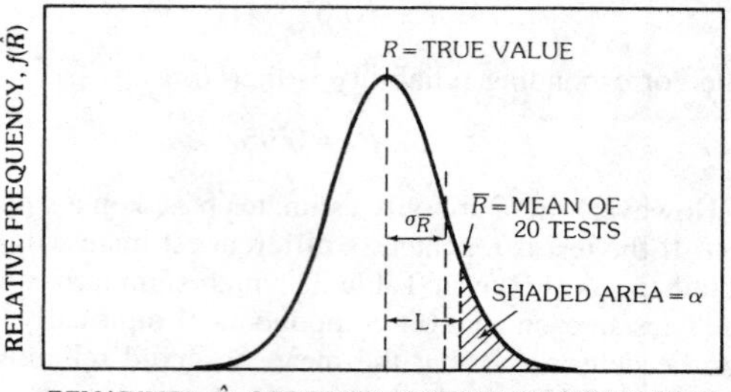

Fig. 13. Calculation of lower confidence limit when standard deviation of sample is known.

The quantity $(\overline{R} - L)$ is the lower confidence limit (LCL) for a confidence level of $(1 - \alpha)$. Table 5 gives a short list of α against standardized L, derived from tables of the error function.

EXAMPLE 2 (*continued*). Find the 97.5% LCL for the reliability, $R_{L,97.5}$.

$$\alpha = 0.025$$

Hence,

$$L/(\sigma_{\hat{R}}/\sqrt{n}) = 1.96 \text{ (from Table 5)}$$
$$\sigma_{\hat{R}}/\sqrt{n} = 0.01775/\sqrt{20}$$
$$L = 1.96 \times 0.01775/\sqrt{20} = 7.77 \times 10^{-3}$$
$$R_{L,97.5} = 0.975 - 7.77 \times 10^{-3} = 0.967$$

Similarly, an upper confidence level can be obtained by postulating $\overline{R}$ to be to the left of R (see Fig. 14). Hence,

$$R_{U,97.5} = 0.975 + 7.77 \times 10^{-3} = 0.983$$

The above two quantities represent the one-sided LCL and UCL at the $(1 - \alpha)$ confidence level, respectively. Taken together, they constitute a two-sided confidence band at the $(1 - 2\alpha)$ confidence level. That is,

$$0.967 < R < 0.983 \text{ at } 95\% \text{ confidence level}$$

When the standard deviation of the population is unknown, it is no longer valid to use the above approach based on a normal distribution. Instead, the student-t distribution must be used. The t distribution is flatter than the normal and gives wider confidence limits for the same data.

Confidence limits are calculated from the following relation:

$$\text{UCL,LCL} = R \pm t_\alpha S/\sqrt{n}$$

where,

t_α = confidence coefficient for level $(1 - \alpha)$ (See Table 6),
S = sample standard deviation,
n = number of components in sample,
r = number of degrees of freedom $(= n - 1)$.

TABLE 4. CALCULATION TO FIT THEORETICAL DISTRIBUTIONS TO DATA OF TABLE 3

Derivation of Normal Law Theoretical Probabilities

	Deviations from $\overline{R}$				Theoretical Frequency		
					Individual		
Boundary Values	Numerical	z Values σ_R Units	Normal Law Probabilities Corresponding to z	Reliability Values	Prob.	No.	Cumulated No.
				1.000	0.1299	3	3
0.995	+0.020	+1.127	0.3701	0.990	0.1567	3	6
0.985	+0.010	+0.5634	0.2134	0.980	0.2134	4	10
0.975	0	0	0	0.970	0.2134	4	14
0.965	−0.010	−0.5634	0.2134	0.960	0.1567	3	17
0.955	−0.020	−1.127	0.3701	0.950	0.0844	2	19
0.945	−0.030	−1.690	0.4545	0.940	0.0455	1	20

Observed and Theoretical Individual and Cumulated Values for Failure Rate $r=0.025$
Exponential and Normal Laws

	Observed Values		$m=20$ Sets					
			Experimental, $rn=2.5$				Normal Law $\overline{R}=0.975$, $\sigma_R=0.01775$	
			Probabilities		Number			
Reliability Values	Ind.	Cum.	Ind.	Cum.	Ind.	Cum.	Individual	Cumulated
1.00	4	4	0.082	0.082	2	2	3	3
0.99	2	6	0.205	0.287	4	6	3	6
0.98	4	10	0.257	0.544	5	11	4	10
0.97	4	14	0.214	0.758	4	15	4	14
0.96	3	17	0.134	0.892	3	18	3	17
0.95	2	19	0.067	0.959	1	19	2	19
0.94	1	20	0.041	1.000	1	20	1	20
Sum =	20				20		20	

TABLE 5. SHORT LIST OF CONFIDENCE LEVELS VERSUS STANDARD NORMAL VARIATE FOR THE NORMAL DISTRIBUTION

Confidence Level, $(1-\alpha)$	Standard Normal Variate, Z
0.900	1.282
0.925	1.440
0.950	1.645
0.960	1.751
0.965	1.812
0.970	1.881
0.975	1.960
0.980	2.054
0.985	2.170
0.990	2.326
0.995	2.580

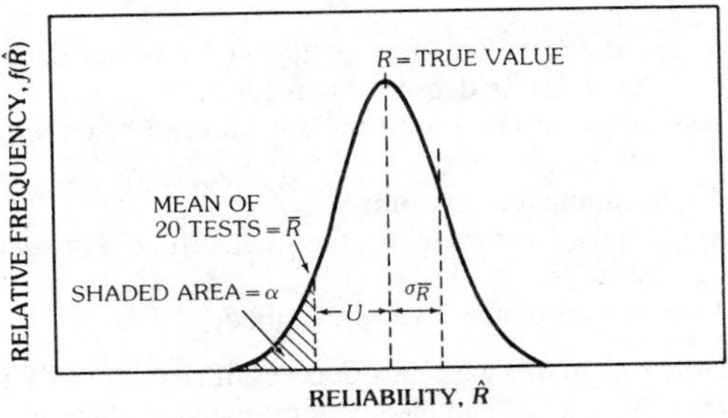

Fig. 14. Calculation of upper confidence limit when standard deviation of sample is known.

TABLE 6. SHORT LIST OF STUDENT-t DISTRIBUTION

r	$t_{0.1}$	$t_{0.05}$	$t_{0.025}$	$t_{0.01}$
1	3.078	6.314	12.706	31.82
2	1.886	2.920	4.303	6.965
4	1.533	2.132	2.776	3.747
9	1.383	1.833	2.262	2.821
19	1.328	1.729	2.093	2.539
29	1.311	1.699	2.045	2.462
∞	1.282	1.645	1.96	2.326

EXAMPLE 3. Same problem as example 1, but with no assumption on standard deviation. The number of components in the sample is 20; therefore the number of degrees of freedom is 19. From Table 6, $t_{0.025}$ = 2.09.

$$\text{UCL} = 0.975 + 2.09 \times 0.01775/\sqrt{20} = 0.9833$$

$$\text{LCL} = 0.975 - 2.09 \times 0.01775/\sqrt{20} = 0.9667$$

There is not much difference between this result and the previous one. This is because the sample size is relatively large. In general, the t distribution need only be used if the sample is smaller than 30.

In calculating confidence limits for other quantities, such as the MTBF, neither the normal nor t distribution is valid because the distribution of the random variable differs too much from a normal distribution (exponential in the case of MTBF). In the case of the exponential distribution, the appropriate relation for determining confidence limits is the chi-squared distribution.

When it is desired to specify that the true mean time between failures must exceed a given minimum value with a confidence level of $(1-\alpha)$, the procedure for a one-sided confidence limit is applied. This provides a tail area α and means there is a probability α that the m value actually observed by test will be smaller than the specified minimum and a probability of $1-\alpha$ that it will be larger. Reference 15 denotes the one-sided confidence limit by the notation C_L to distinguish it from the two-sided lower limit L. Its value is given by

$$C_L = (2r/\chi^2_{\alpha;2r})\hat{m} = 2T/\chi^2_{\alpha;2r}$$

where,

$\chi^2_{\alpha;2r}$ is the value of χ^2 at the $(1-\alpha)$ confidence level for $2r$ degrees of freedom,
tests are continued until the rth failure occurs with $r = 1, 2, \cdots, d$,
T = accumulated test time = Σt_i,
$m = T/r$ = an estimate of the mean time between failures,
$1-\alpha$ = confidence level prescribed.

Note that in this case $2r$ = degrees of freedom (d.f.).

However, a test can also be terminated at some preselected test time without a failure occurring exactly at that time. For such a case, Reference 16 has shown that for the accumulated hours of operating time $T = \Sigma t_i$, then

$$m \geq 2T/\chi^2_{\alpha;2r+2}$$

where d.f. $= 2r+2$ and the case where $r=0$ is covered. For $r=0$, then

$$C_L = 2T/\chi^2_{\alpha;2}$$

In the percent survival method, the accumulated operating time T is not measured, and only the straight test duration time t_d is known, at which time r failures of n units on test are counted. In this method, confusion may exist between chance failures and failures due to actual wearout. The time to wearout must be known, and it is necessary to design and select parts from manufacturers that can be made so that their respective wearout time is many hours past the time of the mission. Again referring to both Reference 15 and to Epstein, for a one-sided confidence level of $1-\alpha$, the lower-limit estimated reliability for t_d hours is

$$\hat{R}(t_d) = \frac{1}{1 + [(r+1)/(n-r)]F_{\alpha;2r+2;2n-2r}}$$

where F is the upper α percentage point of the Fisher distribution (termed the F distribution) with the two corresponding degrees of freedom, $2r+2$ and $2n-2r$. For this estimate of reliability there is a probability of $1-\alpha$ that the true reliability for t_d hours is equal to or larger than $R(t_d)$. It must be noted that this reliability estimate is nonparametric and is valid for the exponential as well as the nonexponential case.

A general mathematical approach is used in many cases to determine the confidence levels for either one-sided or two-sided distributions for various density functions. Confidence levels and reliability values are related by the two following general relations, where P_b = degree of belief, equivalent to the confidence level. One relation covers continuous distributions and makes use of the area under the density function secured by integration, while the second relation covers summations for integral values. These relations are

$$P_b = \int_0^{x^*} f(x)dx \Big/ \int_0^{x=\infty} f(x)dx$$

$$P_b = \sum_0^{x^*} F(x) / \sum_0^{x=\infty} F(x)$$

For the exponential density function, use of these relations gives

$$P_b = \int_0^{t^*} \lambda \exp(-\lambda t)dt \Big/ \int_0^{\infty} \lambda \exp(-\lambda t)dt$$

$$= -\exp(-\lambda t)_0^{t^*} / -\exp(-\lambda t)_0^{\infty}$$

$$= [-\exp(-\lambda t^*) + 1] / (0+1)$$

$$= 1 - \exp(-\lambda t^*)$$

The reliability $R(t)$ for time t for P_b (the one-sided confidence level) is derived from the term λt^*, where no failures have been observed in time $T = \Sigma t_i$. From an exponential table determine $\lambda t^* = a$, corresponding to $1-P_b$. Then $\lambda = a/t^* = a/T$, since t^* corresponds to the total time required for the test. The final reliability value is determined from

$$R(t)=\exp(-\lambda t)=\exp(-at/T)$$

If a test is terminated when the rth failure has occurred, the ratio $2r(\hat{m}/m)$ has a chi-square distribution with $2r$ degrees of freedom. The two-sided confidence interval at a confidence level of $(1-\alpha)$ is

$$\hat{m}(2r/\chi^2_{\alpha/2;2r}) \leq m \leq \hat{m}(2r/\chi^2_{1-\alpha/2;2r})$$

Here $\hat{m}$ represents the estimate of m derived from the samples tested and is the MTBF. The lower limit L is given by

$$L=(2r/\chi^2_{\alpha/2;2r})\hat{m}=2T/\chi^2_{\alpha/2;2r}$$

while the upper confidence limit is given by

$$U=(2r/\chi^2_{1-\alpha/2;2r})\hat{m}=2T/\chi^2_{1-\alpha/2;2r}$$

Herein $\hat{m}=T/r$ and can be derived from either a replacement or a nonreplacement test, while $T=\sum t_i$, the sum of the operating times accumulated by all the components during the test. When the test is terminated at time t_d without a failure occurring exactly at that time, then the degrees of freedom for the lower limit are changed from $2r$ to $2r+2$. The upper and lower limits are given by

$$2T/\chi^2_{\alpha/2;2r+2} \leq m \leq 2T/\chi^2_{1-\alpha/2;2r}$$

From these limits giving lower and upper limits, L and U, in terms of mean time before failure, for any mission time t, then lower and upper limiting values for the reliability $R(t)$ may be readily computed from

$$L{::}R_L{:}R_L(t)=\exp(-t/L)$$

$$U{::}R_U{:}R_U(t)=\exp(-t/U)$$

When the Gaussian (normal law) distribution applies or is used as a means of determining upper and lower limits for either m or $R(t)$, where $\overline{m}$, σ_m and $\overline{R}$, σ_R are known, either symmetric or nonsymmetric confidence limits may be determined from

$$L{:}\overline{m}-z_\alpha\sigma_m;\ \overline{R}-z_\alpha\sigma_R$$

$$U{:}\overline{m}+z_\beta\sigma_m;\ \overline{R}+z_\beta\sigma_R$$

where $\alpha+\beta=\gamma=$ probability for the specified confidence band.

As an aid to the calculation of confidence bands under certain stated conditions, several of the military specifications listed in Table 7 allow an easy calculation of these limits. One specification to note is MIL-R-22973.

EXAMPLE 4. Estimation of reliability for times different from test time.

In most cases, reliability values are associated with life usage or with the time of storage. A mission may require t hours to be accomplished. For example, it may require 10 hours to drive an automobile from Los Angeles to San Francisco, a distance of approximately 420 miles. What is $R(10)$, the reliability of accomplishing this mission in 10 hours at any time? In a prior example, $p=2.5\%$ where it was assumed that each unit was tested 1000 hours. The failure rate then may be expressed as 2.5%/1000 hours. If the mission time is 10 hours, the reliability $R(t)$, assumed to be based on the exponential, is determined as follows. For $t=10$ hours, $v=2.5\%/1000$ hours $=0.000025$/hr, and for this case

$$R(10)=\exp[-0.000025(10)]$$
$$=\exp(-0.00025)=0.99975$$

This value of reliability is based on the expected value. For the exponential, the variance is equal to the expected value. Hence, since for this 10-hour mission $t=0.00025$, $\sigma_t=(0.00025)^{1/2}=0.01581$. For a 90% confidence level using the proper multiplying factor based on the normal law

$$t_{0.90}=0.00025+1.282(0.01581)=0.00025$$
$$+0.02026842=0.02051842$$

The corresponding reliability for $t=10$ hours is $R_{0.90}(t=10)=\exp(-0.02052)$, $R(10)_{0.90}=0.97968$. For a 95% confidence level based on the normal law:

$$(\lambda t)_{0.95}=0.00025+1.645(0.01581)$$
$$0.00025+0.0260075=0.0262575$$

For this expected value of λt with $t=10$ hours, $R(10)=\exp(-0.02626)=0.97408$. For a 99% confidence level based on the normal law:

$$(\lambda t)_{0.99}=0.00025+2.326(0.01581)$$
$$=0.00025+0.03677406=0.03702406$$

Hence $R(t=10)_{0.99}=\exp(-0.03702)=0.96366$. For the six confidence levels often used, the reliability values for a one-tailed confidence level may be obtained from Table 8.

According to the data in Table 3, there existed only one set of 100 units $=n$ out of 10 000 that had a reliability observed of 0.94. Associated with this value is a confidence level of $(10\,000-100)/10\,000=9900/10\,000=0.99$ (Table 9). This reliability value is determined for an assumed operating period of 1000 hours. Hence this gives $R(1000)=0.94$ when the confidence level $P_C=0.99$. Hence the value of λ is thus computed

$$R(1000)=0.94=\exp[-\lambda(1000)]=\exp(-0.0619)$$

Then $1000\lambda=0.0619$, and $\lambda=6.19\%/1000$ hours. For $t=10$ hours, then

$$R(10)=\exp(-\lambda t)=\exp[-0.0619(10)/1000]$$
$$=\exp(-0.000619)=0.99938$$

Thus the actual data provide more optimistic estimates of the reliability based on field test results. Since

TABLE 7. RELIABILITY SPECIFICATIONS

Specification	Title
MIL-A-8866	Airplane Strength and Rigidity Reliability Requirements, Repeated Loads and Fatigue
MIL-R-19610	General Specifications for Reliability of Production Electronic Equipment
MIL-R-22732	Reliability Requirements for Shipboard and Ground Electronic Equipment
MIL-R-22973	General Specification for Reliability Index Determination for Avionic Equipment Models
MIL-R-23094	General Specification for Reliability Assurance for Production Acceptance of Avionic Equipment
MIL-R-26484	Reliability Requirements for Development of Electronic Subsystems for Equipment
MIL-R-26667	General Specification for Reliability and Longevity Requirements, Electronic Equipment
MIL-R-27173	Reliability Requirements for Electronic Ground Checkout Equipment
M-REL-M-131-62	Reliability Engineering Program Provisions for Space System Contractors
NASA NPC 250-1	Reliability Program Provisions for Space System Contractors
NASA Circular No. 293	Integration of Reliability Requirements into NASA Procurements
LeRC-REL-1	Reliability Program Provisions for Research and Development Contracts
WR-41 (BUWEPS)	Naval Weapons Requirements, Reliability Evaluation
NAVSHIPS 900193	Reliability Stress Analysis for Electronic Equipment
NAVSHIPS 93820	Handbook for Prediction of Shipboard and Shore Electronic Equipment Reliability
NAVSHIPS 94501	Bureau of Ships Reliability Design Handbook
NAVWEPS 16-1-519	Handbook Preferred Circuits—Naval Aeronautical Electronic Equipment
PB 181080	Reliability Analysis Data for Systems and Components Design Engineers
PB 131678	Reliability Stress Analysis for Electronic Equipment, TR-1100
TR-80	Techniques for Reliability Measurement and Prediction Based on Field Failure Data
TR-98	A Summary of Reliability Prediction and Measurement Guidelines for Shipboard Electronic Equipment
AD-DCEA	Reliability Requirements for Production Ground Electronic Equipment
AD 114274	(ASTIA) Reliability Factors for Ground Electronic Equipment
AD 131152	(ASTIA) Air Force Ground Electronic Equipment-Reliability Improvement Program
AD 148556	(ASTIA) Philosophy and Guidelines—Prediction on Ground Electronic Equipment
Ad 148801	(ASTIA) Methods of Field Data Acquisition, Reduction and Analysis
AD 148977	(ASTIA) Prediction and Measurement of Air Force Ground Electronic Reliability
MIL-HDBK-217B	Reliability Prediction of Electronic Equipment
RADC 2623	Reliability Requirements for Ground Electronic Equipment
USAF BLTN 2629	Reliability Requirements for Ground Electronic Equipment
AR-705-25	Reliability Program for Material and Equipment
OP 400	General Instructions: Design, Manufacture and Inspection of Naval Ordnance Equipment
MIL-STD-105D	Sampling Procedures and Tables for Inspection by Attributes
MIL-STD-414	Sampling Procedures and Tables for Inspection by Variables of Percent Defective
MIL-STD-721	Definitions for Reliability Engineering
MIL-STD-756	Procedures for Prediction and Reporting Prediction of Reliability of Weapon Systems
MIL-STD-781B	Reliability Tests Exponential Distribution
MIL-STD-785	Requirements for Reliability Program (for Systems and Equipments)
DOD H-108	Sampling Procedure and Table for Life and Reliability Testing

TABLE 8. RELIABILITY VALUES FOR A MISSION OF $t = 10$ HOURS FOR $\lambda = 2.5\%/1000$ HOURS FOR 6 ONE-TAILED CONFIDENCE LEVELS FOR EXPONENTIAL: $R(\lambda t) = R[\lambda t + z(\lambda t)^{1/2}]$, z GIVEN IN NORMAL LAW (GAUSSIAN) TABLES FOR P_z TABULATED FOR CONFIDENCE LEVEL

Confidence Level	Normal Law z Values	Upper Limit for $(\lambda t)_z = 0.00025 + z(0.00025)^{1/2}$	$R(\lambda t)_z = \exp[-(\lambda t)_z]$
0.90	1.282	0.02052	0.97968
0.95	1.645	0.02626	0.97408
0.96	1.751	0.02794	0.97245
0.97	1.881	0.02999	0.97045
0.98	2.054	0.03272	0.96780
0.99	2.326	0.03702	0.96366

TABLE 9. RELIABILITIES ASSOCIATED WITH ONE-TAILED CONFIDENCE LEVELS

Confidence Level	Multiplying Factor for Normal Law, z	$n = 10\ 000$ units Observed Reliability	Normal Law Theoretical Reliability
0.90	1.282	95.07%	95.69%
0.95	1.645	94.50%	95.18%
0.96	1.751	94.38%	95.03%
0.97	1.881	94.25%	94.85%
0.98	2.054	94.12%	94.60%
0.99	2.326	94.00%	94.22%

the distribution as graphed appears to be almost rectangular in Fig. 12, the assumption of normality is pessimistic.

In these life tests, each failure must be carefully analyzed to determine whether it is a *chance* failure or a *wearout* failure. These results must be fed back to the design engineers to make certain that corrective measures for improving the life characteristics are taken and established as standard procedures.

Fitting a Distribution Using Chi-Squared Test

In a sample of n observed values of a random variable, $n_1, n_2, \ldots, n_k$ frequencies are observed in each of k intervals (Fig. 12). An assumed distribution predicts $e_1, e_2, \ldots, e_k$ for the same intervals. The goodness-of-fit is tested by the criterion

$$\sum (n_i - e_i)^2/e_i < \chi^2_{1-\alpha;r}$$

In this expression, $\chi^2_{1-\alpha;r}$ is the value of the χ^2 distribution at the cumulative probability level $(1-\alpha)$ and with degrees of freedom $r = k-1$ if the mean is known independently and $r = k-2$ if the sample mean is used.

EXAMPLE 5. Using data from example 1, test the goodness of fit for normal and Poisson distributions.

From Table 4 for $k = 7$, obtain the data in Table 10. For the normal distribution,

$$\chi^2 = (1/3 + 1/3 + 0 + 0 + 0 + 0 + 0) = 0.667$$

For $\alpha = 0.05$ significance level and $r = k - 2 = 5$, from standard tables of χ^2, $\chi^2_{0.95;5} = 11.1$, which is greater than χ^2 for the normal distribution. Hence the normal distribution is a valid model at the 5% level.

For the Poisson distribution,

$$\chi^2 = (4/2 + 4/4 + 1/5 + 0 + 0 + 1/1 + 0) = 4.2$$

Hence, the Poisson distribution is also valid at the 5% significance level, but it is a poorer fit to the data than the normal distribution.

TABLE 10. DATA FOR EXAMPLE 5

i	n	e_i(normal)	e_i(Poisson)
1	4	3	2
2	2	3	4
3	4	4	5
4	4	4	4
5	3	3	3
6	2	2	1
7	1	1	1

Probability Paper

An alternative method of fitting a distribution is to use probability plotting paper. This procedure attempts to produce the best fit to the cumulative distribution rather than the probability density function. A number of results obtained in a life test are tabulated in Table 11. The cumulative failure distribution, $F(t)$, can be approximated by

$$F(t_i) \doteq i/n$$

where i is the number of failures at time t_i.

The principle in using probability paper is to transform the axes of the cumulative plot to give a straight line if the data conform with the assumed theoretical distribution. For instance, an exponential distribution gives a straight line if log $[F(t)]$ is plotted against t, with a slope equal to the hazard rate, λ. In other cases, such as the normal distribution, special plotting papers must be prepared. These can be obtained for the more common distributions from suppliers of drawing materials.

Weibull Analysis

By far the most commonly used plotting paper in reliability studies is Weibull paper (see Fig. 15). A Weibull distribution has the attribute that it can describe distributions with decreasing, constant, or increasing hazard rate, simply by appropriate choice of the shape parameter, β. Furthermore, the slope of the Weibull plot gives β directly.

EXAMPLE 6. Use of Weibull paper to analyze the data in Table 11. It is first necessary to place the results

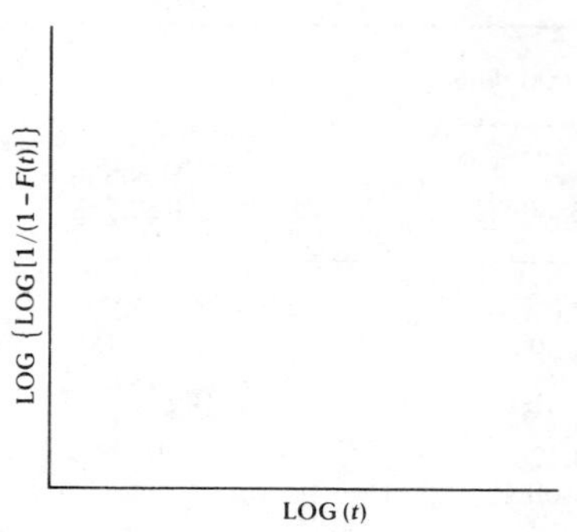

Fig. 15. Weibull plotting transformation.

in increasing rank order m. Next the cumulative failure distribution $F(t)$ is approximated by the median rank

$$\hat{F}_n(t) = m/(n+1) \doteqdot F(t)$$

This is slightly different from the approximation to $F(t)$ quoted earlier, i.e., $F(t_i) = m/n$. The difference is that the median rank is a minimally biased estimate of $F(t)$ and avoids some plotting problems near $m = 1$ and $m = n$ as well. Similarly, the reliability $R(t)$ can be estimated by

$$\hat{R}_n(t) = (n-m)/(n+1)$$

Table 12 shows the ordered list with the median ranks calculated. The results are plotted in Fig. 16. Note the following:

1. Fitting a distribution using probability paper requires relatively few points compared with the number needed to produce a histogram to approximate a probability density function. As few as five or six can be used to fit a cumulative distribution, whereas more than 20 are needed for a histogram.
2. The data in the Weibull plot clearly indicate three distinct zones, suggesting that three modes of failure are occurring.
3. From the estimates of the shape parameter, it can be deduced that the first two or three failures are burn-in ($\beta = 0.6$), the last five are clearly wearout ($\beta = 3.6$, which is incidentally very close to a normal distribution in shape), while the remainder are random failures with external cause ($\beta = 1.08$, approximately 1).

The Weibull plot is therefore not just a means of

TABLE 11. RESULTS OF COMPONENT LIFE TEST ON 15 UNITS

Test Number	Time to Failure (Hours)
1	8.35×10^4
2	2.24×10^4
3	4.84×10^4
4	0.33×10^4
5	9.90×10^4
6	8.85×10^4
7	0.83×10^4
8	10.60×10^4
9	6.55×10^4
10	1.80×10^4
11	4.30×10^4
12	8.45×10^4
13	9.27×10^4
14	3.15×10^4
15	6.02×10^4

TABLE 12. CALCULATION OF MEDIAN RANKS FOR WEIBULL ANALYSIS OF DATA IN TABLE 11

Rank Order, m	Median Rank, $m/(n+1)$	Time to Failure, (10^4 hr)
1	0.0625	0.33
2	0.125	0.83
3	0.1875	1.80
4	0.25	2.24
5	0.3125	3.15
6	0.375	4.30
7	0.4375	4.84
8	0.50	6.02
9	0.5625	6.55
10	0.625	8.35
11	0.6875	8.45
12	0.75	8.85
13	0.8125	9.27
14	0.875	9.90
15	0.9375	10.60

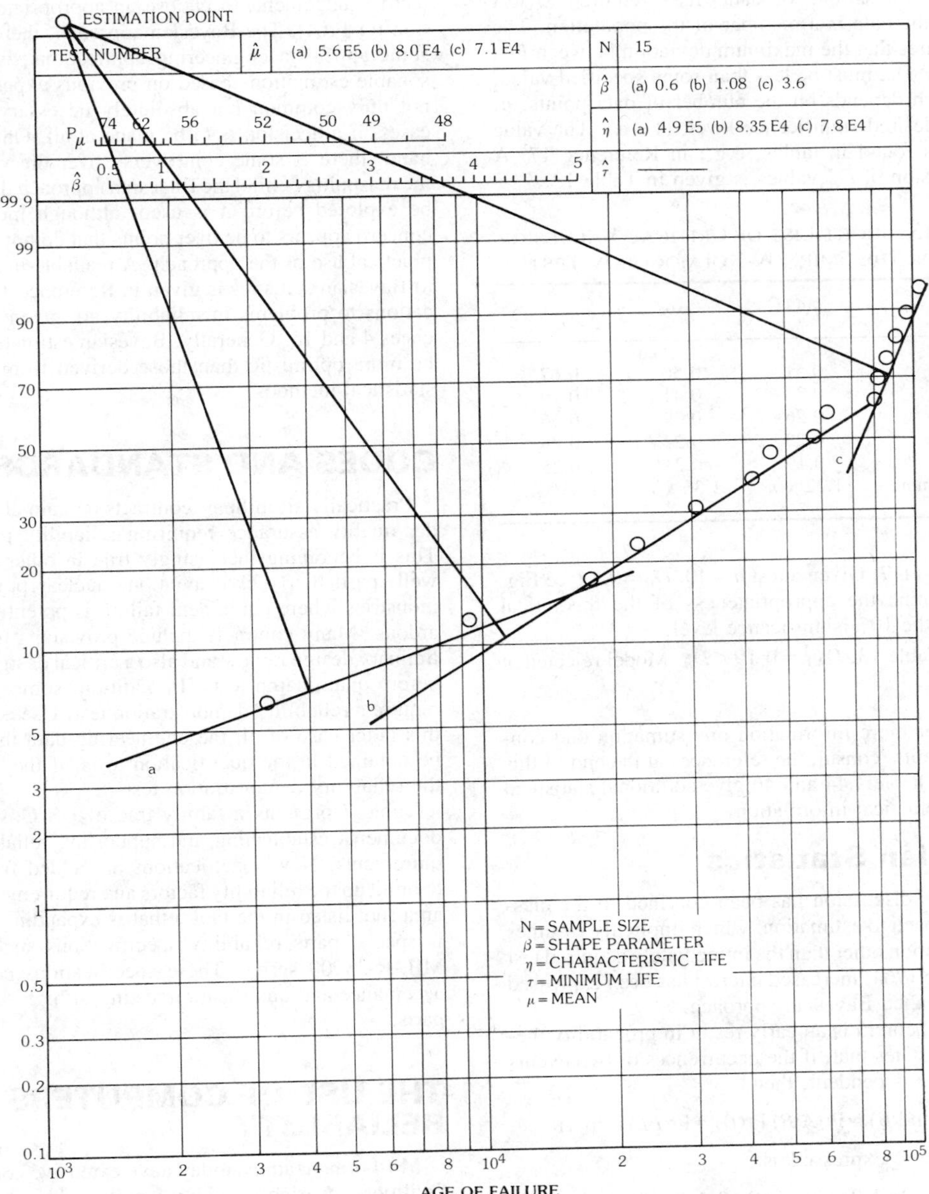

Fig. 16. Standardized Weibull paper with example.

fitting a distribution to data. It is also a useful diagnostic tool, indicating the number and nature of failure modes present.

Distribution-Free Tests of Goodness of Fit

As with all data-fitting procedures, probability-paper plots provide only an estimate of the true distribution. Confidence limits should be placed on any parameter estimates made. This is more difficult to do for several parameters simultaneously than for a single variable such as failure rate. Further details of this type of analysis are to be found in any standard text on statistics, e.g., Reference 17. One technique that will be illustrated here is a check to find out whether a set of data belongs to an already known and characterized population. This test is known as the Smirnov-Kolmogorov test. The procedure is illustrated in Fig. 6, which shows the actual $F(t)$ of the population and a median-rank plot for a typical sample of data. It is required to test whether the data is a member of the population. The test requires that the maximum deviation between $F(t)$ and $F_n(t)$, D_n, must be less than some specified value, D_n^α, which depends on the number of data points, n, and the desired significance level, $(1-\alpha)$. The value for D_n^α is found in tables, e.g., in Reference 17. A short version of D_n^α values is given in Table 13.

TABLE 13. SHORT LIST OF CRITICAL VALUES OF D_n FOR THE SMIRNOV-KOLMOGOROV TEST

n	$D_n^{0.1}$	$D_n^{0.05}$	$D_n^{0.01}$
5	0.51	0.56	0.67
10	0.37	0.41	0.49
20	0.26	0.29	0.36
30	0.22	0.24	0.29
40	0.19	0.21	0.25
50 and more	$1.22/\sqrt{n}$	$1.36/\sqrt{n}$	$1.63/\sqrt{n}$

EXAMPLE 7. Given a test $n=40$, $D_n=0.2$ (see Fig. 6), determine the appropriateness of the theoretical model at the 10% significance level.

From Table 13, $D_{40}^{0.1}=0.19<D_n$. Model rejected at 10% level.

NOTE. For more information on estimation and confidence limits, consult the references at the end of this chapter. Chapters 44 and 46 give additional statistical and mathematical information.

Bayesian Statistics

Previous discussion has been confined to the classical approach to estimation, which implicitly assumes no information other than the immediately acquired test data. Since 1964, increased interest has been expressed in the so-called Bayesian approach.

Bayes' theorem is an early result in probability theory which states that, if the occurrences of two events A and B are dependent, then

$$\Pr(A \text{ and } B)=\Pr(A/B)\cdot\Pr(B)=\Pr(B/A)\cdot\Pr(A)$$

This can be re-expressed as

$$\Pr(A/B)=\Pr(A)\cdot\Pr(B/A)/\Pr(B)$$

This equation can be used to infer a result using both test data and previous experience. For instance,

$A=\lambda$, the actual (unknown) failure rate
$B=\hat{\lambda}$, the observed failure rate in a test

The question is posed, what is the actual failure rate given the observed rate in the test?

The focal point of the method is the so-called "prior" distribution, $\Pr(A)$. In classical statistics, A would be considered a constant, but unknown. In Bayesian statistics, A is considered to be a random variable, and use is made of previous experience, or even subjective judgement, to choose an appropriate distribution for $\Pr(A)$. The Bayesian approach, therefore, has some appeal in engineering applications, where reasonable estimations based on previous experience are not only common but absolutely necessary in some cases, if a decision is to be made at all. On the other hand, there is some controversy over the validity of the reasoning behind the Bayesian approach that should be explored before it is used, although most of the concern appears to be over points that do not affect the practical use of the approach. A readable introduction to Bayesian statistics is given in Reference 17. Applications to problems in reliability are given in References 4 and 15. Generally, Bayesian estimates tend to be more optimistic than those derived from classical statistical methods.

CODES AND STANDARDS

Practically all military contracts contain clauses under quality assurance requiring reliability programs. This is becoming increasingly true in other fields as well, particularly civil aviation, nuclear power, and industries where equipment failure is potentially hazardous. Many contracts include provisions for a preliminary design review and also a critical design review before qualificaton tests. In addition, some contracts require a reliability demonstration test. Use is made in this latter case of all the engineering data that might be obtained in the qualification tests, if they precede the reliability demonstration test.

Table 7 presents a family tree of US Government documents establishing and supporting reliability requirements. New specifications are added frequently to build up the reliability factors and requirements. One area (not listed in the table) that is expanding steadily is special parts reliability specifications such as the MIL-R-38000 series. These specifications cover the acceptance and qualification testing of high-reliability parts.

THE USE OF COMPUTERS IN RELIABILITY

Most organizations today have extensive computing facilities. Invariably, such a facility will include software packages for performing statistical calculations. Even the most modest of desk-top computers has a package of this type, or one can be obtained as an option. The majority of standard calculations, such as curve fitting, evaluation of confidence limits, etc., can be done as a routine exercise, and cumbersome, error-prone tasks like referring to sets of tables for values of chi-square and the t distribution are now obsolete, except for educational demonstration.

Of particular interest to reliability analysts is an ongoing series in *Quality Progress*, the journal of the American Society for Quality Control, which offers computer programs for use on small desk-top models and hand-held programmable calculators, covering a range of reliability-related topics.

Several large computer codes are now generally available for various aspects of reliability analysis of complex systems. These are reviewed in Reference 4. Many of the codes described in Reference 4 are available from the authors of the reference. Some examples of the range of capabilities are

KITT — Fault-tree analysis of complex systems (several versions).

BACFIRE — Search routine for common-mode failures.

FAMULS — Calculates cut sets for systems with multiple control loops given signal flow graph representation of system.

SCHE — Converts reliability block diagrams (see Fig. 10) into fault trees.

HEUR — Reliability optimization under constraints of cost, weight, etc.

REFERENCES

1. Halpern, S. *The Assurance Sciences—An Introduction to Quality Control and Reliability*. Englewood Cliffs, N.J.: Prentice-Hall, Inc., 1978.
2. Marguglio, B. W. *Quality Systems in the Nuclear Industry and Other High Technology Industries*. ASTM/STP 616, 1977.
3. Swain, A. D., and Guttmann, H. E. *Handbook of Human Reliability Analysis with Emphasis on Nuclear Power Plant Applications*. USNRC Report NUREG/CR-1278, October 1980.
4. Hensley, E. J., and Kumamoto, H. *Reliability Engineering and Risk Assessment*. Englewood Cliffs, N.J.: Prentice-Hall, Inc., 1981.
5. NASA. *Parts Materials, and Process Experience Summary*. Report CR114391, Feb. 1972.
6. Advisory Group on Reliability of Electronic Equipment (AGREE), Office of the Assistant Secretary of Defense (Research and Engineering). *Reliability of Military Electronic Equipment*. Washington, D.C.: Supt. of Documents, US Government Printing Office, 4 June 1957.
7. MIL-HDBK-217B, *Military Standardization Handbook, Reliability Prediction of Electronic Equipment*. September 20, 1974.
8. IEEE Std. 500-1977, *Guide to the Collection and Presentation of Electrical, Electronic and Sensing Component Reliability Data for Nuclear Power Generating Stations*. New York: IEEE.
9. Government/Industry Data Exchange Program (GIDEP), Fleet Missile Systems Analysis and Evaluation Group, Corona, California.
10. *Failure Data Handbook for Nuclear Facilities*, LNEC-Memo-69-7. Available from NTIS.
11. Nuclear Plant Reliability Data Systems (NPRDS), operated by Southwest Research Institute, San Antonio, Texas.
12. SYREL—System Reliability Service Data Bank, UKAEA, Culcheth, Warrington, United Kingdom.
13. H108, Quality Control and Reliability Handbook, *Sampling Procedures and Tables for Life and Reliability Testing*. Washington, DC: Office of the Assistant Secretary of Defense, April 19, 1960.
14. MIL-STD-781B, *Military Standard, Reliability Tests Exponential Distribution*. Washington, DC: Department of Defense, November 15, 1967.
15. Bazovsky, Igor. *Reliability Theory and Practice*. Englewood Cliffs, N.J.: Prentice-Hall, Inc., 1961.
16. Epstein, B. ''Estimation From Life Test Data.'' *IRE Trans on Reliability and Quality Control*, Vol. RQC-9, April 1960.
17. Ang, A. H-S., and Tang, W. H. *Probability Concepts in Engineering Planning and Design*. New York: John Wiley and Sons, Inc., 1975.
18. Von Alven, William H., ed. *Reliability Engineering*. Englewood Cliffs, N.J.: Prentice-Hall, Inc., 1964 (23 contributors).

Other Useful References

Abramowitz, Milton, and Stegun, Irene A. *Applied Mathematics Series AMS 55, Handbook of Mathematical Functions With Formulas, Graphs and Mathematical Tables* (third printing with corrections). Washington, DC: Supt. of Documents, US Government Printing Office, 1965.

Ad Hoc Study Group on Parts Specification Management for Reliability, Office of the Director of Defense Research and Engineering and Office of the Assistant Secretary of Defense Supply and Logistics. *Parts Specification Management for Reliability*, Vols. 1 and 2, PSMR-1. Washington, DC: Supt. of Documents, US Government Printing Office, May 1960.

Calabro, S. R. *Reliability Principles and Practices*. New York: McGraw-Hill Book Co., 1962.

Staff of the Computation Laboratory. *Tables of the Error Function and of Its First Twenty Derivatives*. Cambridge, Mass.: Harvard University Press, 1952.

Goldman, A. S., and Slattery, T. B. *Maintainability*. New York: John Wiley & Sons, Inc., 1964. (Contributions by S. Firstman, Rand Corp.; and J. Rigney, University of Southern California): Chapter on General Electric Co. ''TEMPO.''

Gryna, Frank M. Jr., McAfee, Naomi J., Ryerson, Clifford M., and Zwerling, Stanley, eds. *Reliability Training Text*, 2nd ed. Sponsored by ASQC and IEEE, March 1960.

Hald, A. *Statistical Tables and Formulas*. New York: John Wiley & Sons, Inc., 1952.

Ireson, W. Grant, ed. *Reliability Handbook*. New

York: McGraw-Hill Book Co., 1966. (19 contributors)

Johnson, Norman L., and Leone, Fred C. *Statistics and Experimental Design in Engineering and the Physical Sciences*, Vols. 1 and 2. New York: John Wiley & Sons, Inc., 1964.

Lambe, C. G. *Elements of Statistics*. London and New York: Longmans, Green and Co., 1952.

Landers, Richard R. *Reliability and Product Assurance, A Manual for Engineering and Management*. Englewood Cliffs, N.J.: Prentice-Hall, Inc., 1963.

Lloyd, David K., and Lipow, Myron. *Reliability: Management Methods and Mathematics*. Englewood Cliffs, N.J.: Prentice-Hall, Inc., 1962.

Lowan, Arnold N., and Staff. *Applied Mathematics Series AMS 23, Tables of Normal Probability Function*. Washington, DC: Supt. of Documents, US Government Printing Office, 1953.

Lowan, Arnold N., and Staff. *Applied Mathematics Series AMS 14, Tables of the Exponential Function,* e^x, 4th ed. Washington, DC: Supt. of Documents, US Government Printing Office, 1961.

Molina, E. C. *Poisson's Exponential Binomial Limit, Table 1—Individual Terms; Table 2—Cumulated Terms*. New York: D. Van Nostrand Co., Inc., 1949.

Zelen, Marvin, ed. ''Statistical Theory of Reliability.'' Proceedings of an Advanced Seminar Conducted by the Mathematics Research, United States Army, at the University of Wisconsin, Madison, 8–10 May 1962. Madison, Wis.: The University of Wisconsin Press, 1963.

46 Mathematical Equations

MENSURATION EQUATIONS

Areas and Lengths Associated With Plane Figures

Parallelogram:

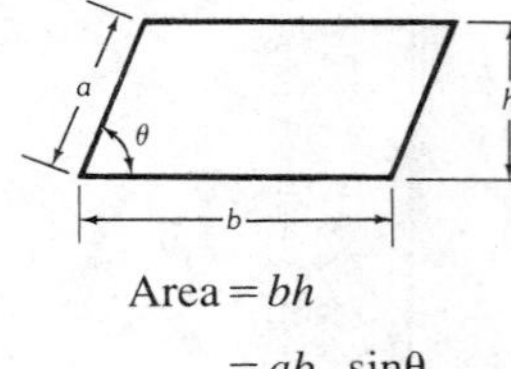

$$\text{Area} = bh$$
$$= ab \sin\theta$$

Trapezoid:

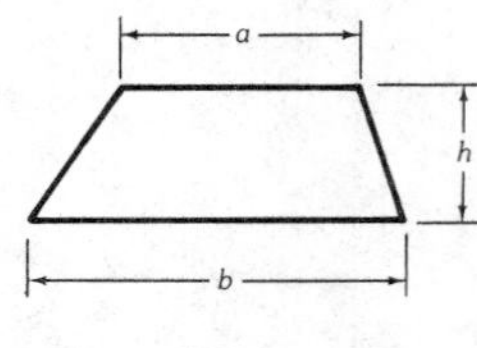

$$\text{Area} = \tfrac{1}{2}h(a+b)$$

Triangle:

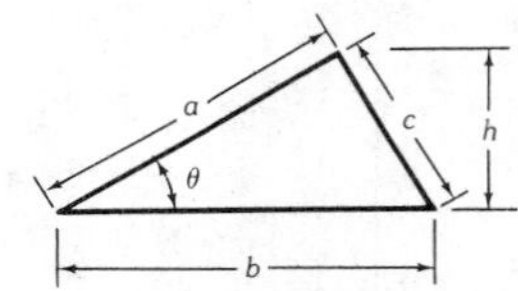

$$\text{Area} = \tfrac{1}{2}bh$$
$$= \tfrac{1}{2}ab \sin\theta$$
$$= [s(s-a)(s-b)(s-c)]^{1/2}$$

where $s = \tfrac{1}{2}(a+b+c)$

Regular Polygon:

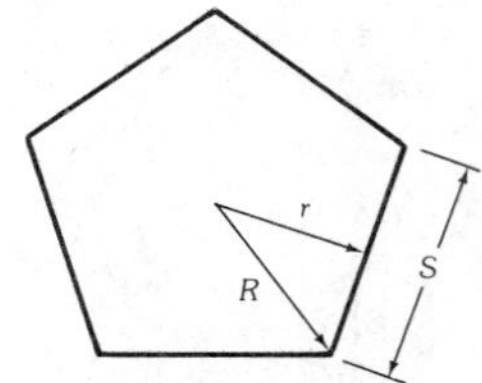

$$\text{Area} = \tfrac{1}{2}nrS$$
$$= nr^2 \tan(180°/n)$$
$$= \tfrac{1}{4}nS^2 \cot(180°/n)$$
$$= \tfrac{1}{2}nR^2 \sin(360°/n)$$

where,

n = number of sides,

S = length of one side,

R = long radius,

r = short radius $= R\cos(180°/n) = \tfrac{1}{2}S\cot(180°/n)$.

Circle:

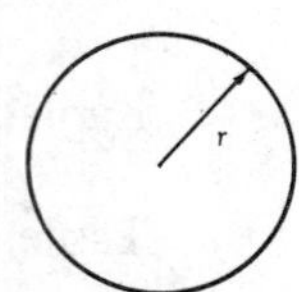

$$\text{Area} = \pi r^2$$
$$\text{Circumference} = 2\pi r$$

where,

r = radius,

$\pi = 3.1416$.

Segment of Circle:

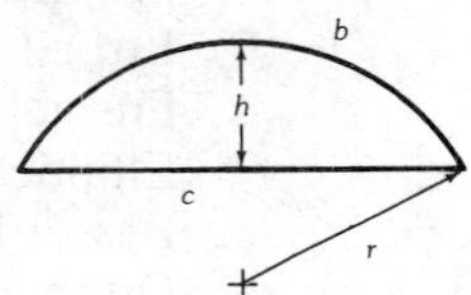

$$\text{Area} = \tfrac{1}{2}[br - c(r-h)]$$

where,

b = length of arc,

c = length of chord $= [4(2hr - h^2)]^{1/2}$.

Sector of Circle:

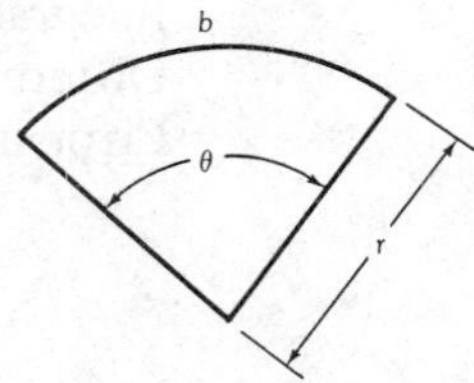

$$\text{Area} = br/2 = \pi r^2(\theta/360°)$$

where,

$b = (\pi r\theta/180°)$,

θ is in degrees.

Parabola:

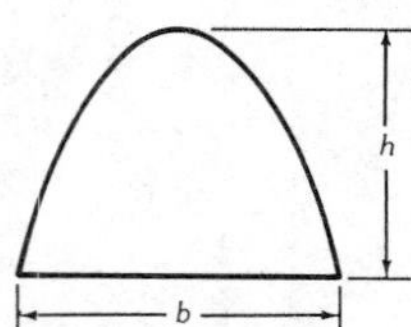

$$\text{Area} = \tfrac{2}{3}bh$$

$$\text{Arc length} = [4h^2 + (b^2/4)]^{1/2} + (b^2/8h)\ \ln\ \{4h + 2[4h^2 + (b^2/4)]^{1/2}/b\}$$

Ellipse:

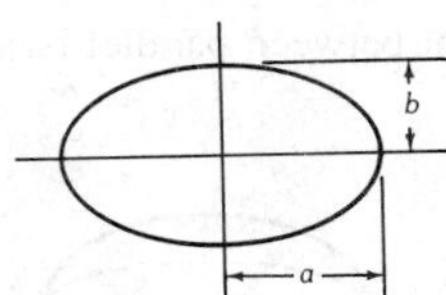

$$\text{Area} = \pi ab$$

$$\text{Circumference} = 4aE(k)$$

where $E(k)$ is a complete elliptic integral with $k = (a^2 - b^2)^{1/2}/a$ of the second kind, $a > b$.

Trapezium:

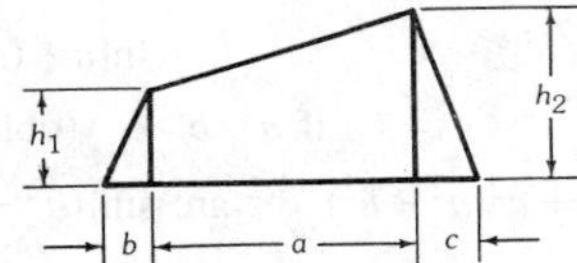

$$\text{Area} = \tfrac{1}{2}[a(h_1 + h_2) + bh_1 + ch_2]$$

Surface Areas and Volumes of Solid Figures

Sphere:

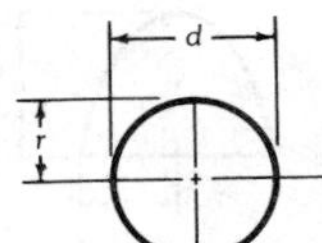

$$\text{Surface} = 4\pi r^2 = 12.5664r^2 = \pi d^2$$

$$\text{Volume} = (4\pi r^3/3) = 4.1888r^3$$

Sector of Sphere:

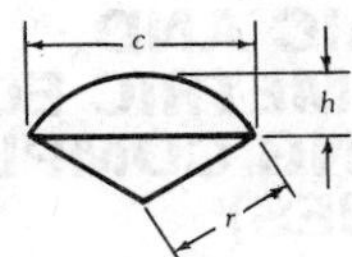

$$\text{Total surface} = (\pi r/2)(4h + c)$$

$$\text{Volume} = (2\pi r^2h/3) = 2.0944r^2h$$

$$= (2\pi r^2/3)[r - (r^2 - \tfrac{1}{4}c^2)^{1/2}]$$

$$c = [4(2hr - h^2)]^{1/2}$$

Segment of Sphere:

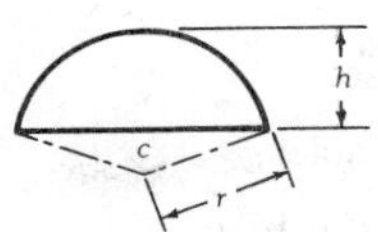

$$\text{Spherical surface} = 2\pi rh = \tfrac{1}{4}\pi(c^2 + 4h^2)$$

$$\text{Volume} = \pi h^2(r - \tfrac{1}{3}h)$$

$$= \pi h^2[(c^2 + 4h^2)/8h - \tfrac{1}{3}h]$$

$$c = [4(2hr - h^2)]^{1/2}$$

Cylinder:

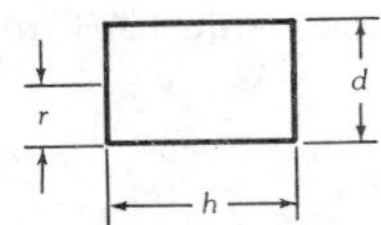

$$\text{Cylindrical surface} = \pi dh = 3.1416dh$$

$$\text{Total surface} = 2\pi r(r + h)$$

$$\text{Volume} = \pi r^2h = 0.7854d^2h$$

$$= c^2h/4\pi = 0.0796c^2h$$

$$c = \text{circumference}$$

Torus or Ring of Circular Cross Section:

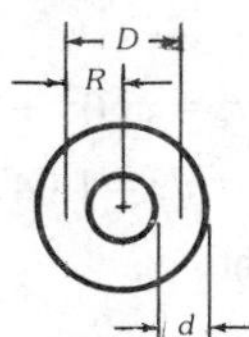

$$\text{Surface} = 4\pi^2Rr = 39.4784Rr = 9.8696Dd$$

$$\text{Volume} = 2\pi^2Rr^2 = 19.74Rr^2 = 2.467Dd^2$$

where,

$D = 2R$ = diameter to centers of cross-section of torus,

$r = d/2$.

Pyramid:

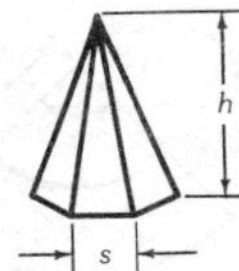

$$\text{Volume} = Ah/3$$

When base is a regular polygon

$$\text{Volume} = \tfrac{1}{3}h\{nr^2[\tan(360°/2n)]\}$$

$$= \tfrac{1}{3}h\{\tfrac{1}{4}(ns^2)[\cot(360°/2n)]\}$$

where,

A = area of base,

n = number of sides,

r = short radius of base.

See "Regular Polygon" in subsection on plane figures.

Pyramidal frustum:

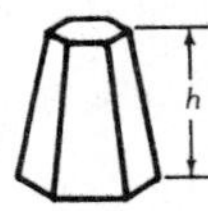

$$\text{Volume} = \tfrac{1}{3}h[a + A + (aA)^{1/2}]$$

where,

A = area of base,

a = area of top.

Cone With Circular Base:

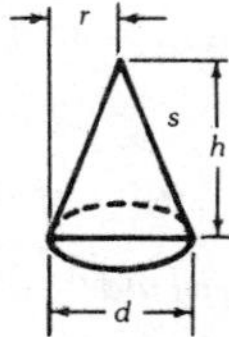

$$\text{Conical area} = \pi rs = \pi r(r^2 + h^2)^{1/2}$$

$$\text{Volume} = \pi r^2 h/3 = 1.047 r^2 h = 0.2618 d^2 h$$

where s = slant height.

Conic Frustum:

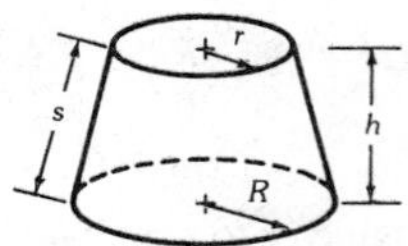

$$\text{Volume} = (\pi h/3)(R^2 + Rr + r^2)$$

$$\text{Area of conic surface} = \pi s(R + r)$$

Development of Conic Surface:

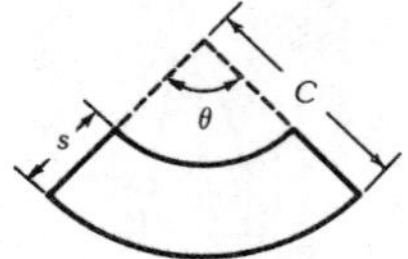

$$C = sR/(R - r)$$
$$\theta = 360R/C$$

where θ is in degrees.

Wedge Frustum:

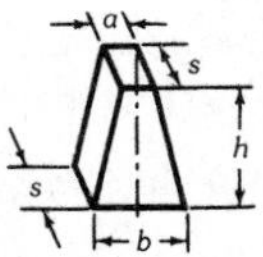

$$\text{Volume} = \tfrac{1}{2}hs(a + b)$$

where h = height between parallel bases.

Ellipsoid:

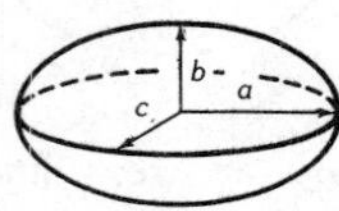

$$\text{Volume} = (4\pi abc/3)$$

$$\text{Surface} = 2\pi\{c^2 + [b/(a^2 - c^2)^{1/2}][c^2F(\phi,k) + (a^2 - c^2)E(\phi,k)]\} \quad \text{if } a>b>c$$

$$= 2\pi a\{a + [c^2/(a^2 - c^2)^{1/2}] \times \ln[a + (a^2 - c^2)^{1/2}]/c\} \quad \text{if } a=b>c \text{ (oblate ellipsoid)}$$

$$= 2\pi c\{c + a^2(a^2 - c^2)^{-1/2} \operatorname{arc\,sin}[(a^2 - c^2)^{1/2}/a]\} \quad \text{if } a>b=c \text{ (prolate ellipsoid)}$$

$F(\phi, k)$ and $E(\phi, k)$ are incomplete elliptic integrals of the first and second kinds, respectively.

$$\phi = \operatorname{arc\,sin}\,[(a^2 - c^2)^{1/2}/a]$$
$$k = (a/b)[(b^2 - c^2)/(a^2 - c^2)]^{1/2}$$

Paraboloid:

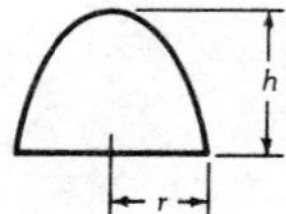

$$\text{Volume} = (\pi r^2 h/2)$$

$$\text{Curved surface} = (\pi r/6h^2)[(r^2 + 4h^2)^{3/2} - r^3]$$

ALGEBRAIC AND TRIGONOMETRIC EQUATIONS (INCLUDING COMPLEX QUANTITIES)

Quadratic Equation

If $ax^2 + bx + c = 0$, then

$$x = [-b \pm (b^2 - 4ac)^{1/2}]/2a$$
$$= 2c/[-b \mp (b^2 - 4ac)^{1/2}]$$

Solution of Cubic Equations*

Given $z^3+a_2z^2+a_1z+a_0=0$, let

$$q=\tfrac{1}{3}a_1-\tfrac{1}{9}a_2^2;\quad r=\tfrac{1}{6}(a_1a_2-3a_0)-\tfrac{1}{27}a_2^3$$

If $q^3+r^2>0$, one real root and a pair of complex conjugate roots

$q^3+r^2=0$, all roots real and at least two are equal

$q^3+r^2<0$, all roots real (irreducible case)

Let

$$s_1=[r+(q^3+r^2)^{1/2}]^{1/3}$$

$$s_2=[r-(q^3+r^2)^{1/2}]^{1/3}$$

then

$$z_1=(s_1+s_2)-(a_2/3)$$

$$z_2=-\tfrac{1}{2}(s_1+s_2)-(a_2/3)+(j\sqrt{3}/2)(s_1-s_2)$$

$$z_3=-\tfrac{1}{2}(s_1+s_2)-(a_2/3)-(j\sqrt{3}/2)(s_1-s_2)$$

If z_1, z_2, z_3 are the roots of the cubic equation

$$z_1+z_2+z_3=-a_2$$

$$z_1z_2+z_1z_3+z_2z_3=a_1$$

$$z_1z_2z_3=-a_0$$

Solution of Quartic Equations*

Given $z^4+a_3z^3+a_2z^2+a_1z+a_0=0$, find the real root u_1 of the cubic equation

$$u^3-a_2u^2+(a_1a_3-4a_0)u-(a_1^2+a_0a_3^2-4a_0a_2)=0$$

and determine the four roots of the quartic as solutions of the two quadratic equations

$$v^2+\{(a_3/2)\mp[(a_3^2/4)+u_1-a_2]^{1/2}\}v + (u_1/2)\mp[(u_1/2)^2-a_0]^{1/2}=0$$

If all roots of the cubic equation are real, use the value of u_1 that gives real coefficients in the quadratic equation.

Complex Quantities

In the following equations, all quantities are real except $j=(-1)^{1/2}$.

$$(A+jB)+(C+jD)=(A+C)+j(B+D)$$

$$(A+jB)(C+jD)=(AC-BD)+j(BC+AD)$$

$$\frac{A+jB}{C+jD}=\frac{AC+BD}{C^2+D^2}+j\frac{BC-AD}{C^2+D^2}$$

$$\frac{1}{A+jB}=\frac{A}{A^2+B^2}-j\frac{B}{A^2+B^2}$$

Polar Form:

$$A+jB=\rho(\cos\theta+j\sin\theta)=\rho e^{j\theta}$$

De Moivre's Equation:

$$(A+jB)^\nu=\rho^\nu(\cos\nu\theta+j\sin\nu\theta)$$

where,

$\rho=(A^2+B^2)^{1/2}>0$,

$\cos\theta=A/\rho$,

$\sin\theta=B/\rho$.

For nonintegral ν, this quantity is many-valued.

Complex Conjugate:

$$\langle A+jB\rangle=(A+jB)^*=A-jB$$

Analytic Function: Let $f(z)$ be a function of the complex variable $z=x+jy$. Function $f(z)$ is analytic at a point $z=z_0$, if

$$\lim_{\Delta z\to 0}[f(z_0+\Delta z)-f(z_0)]/\Delta z$$

exists independent of the manner in which Δz approaches zero. Analyticity is equivalent to the existence of a Taylor series about the point in question. In addition, $f(z)$ is analytic if and only if the Cauchy-Riemann equations hold:

$$\partial u/\partial x=\partial v/\partial y$$

$$\partial v/\partial x=-\partial u/\partial y$$

where,

$u=\text{Re}f(z)$,

$v=\text{Im}f(z)$.

If $f(z)$ is analytic for all z, it is called an entire function.

Properties of *e*

$$e=\lim_{n\to\infty}(1+n^{-1})^n=\sum_{k=0}^{\infty}(k!)^{-1}=2.71828$$

$$e^{\pm jx}=\cos x\pm j\sin x=\exp(\pm jx)$$

Properties of Logarithms

If $\log_a x=N$, then $a^N=x$

$\log_a x=\log_a b\ \log_b x$

$\log_a xy=\log_a x+\log_a y$

$\log_a x/y=\log_a x-\log_a y$

* Abramovitz, M. and Stegun, I. A. *Handbook of Mathematical Functions*. Washington, D. C.: National Bureau of Standards; p. 17.

$\log_a x^y = y \log_a x$

$\log_a b = 1/\log_b a$

$a^{\log_a x} = x$

$\log_a 1 = 0$

$\log_a a = 1$

$\log_e x = \ln x = \log_e 10 \log_{10} x = 2.30259 \log_{10} x$

$\log_{10} x = \log_{10} e \log_e x = 0.43429 \log_e x$

Sums

In this subsection, the following symbols will be used:

$\Gamma(\alpha)$ = Gamma function of α

$\binom{n}{k} = \frac{n!}{k!(n-k)!}$ in which n and k are positive integers, $n \geqslant k$, and

$\binom{n}{0} = 1$

Arithmetic Progression:

$$\sum_{k=0}^{n-1} (a+kd) = a + (a+d) + (a+2d) + \cdots + [a+(n-1)d]$$

$$= \tfrac{1}{2}n[2a+(n-1)d]$$

Geometric Progression:

$$\sum_{k=0}^{n-1} ar^k = a + ar + ar^2 + \cdots + ar^{n-1}$$

$$= [a(r^n - 1)/(r-1)], \quad \text{for } r \neq 1$$

$$= na, \quad \text{for } r = 1$$

Sums of Powers of Integers:

$$\sum_{k=1}^{n} k^2 = 1^2 + 2^2 + 3^2 + \cdots + n^2$$

$$= [n(n+1)(2n+1)/6]$$

$$\sum_{k=1}^{n} k^3 = 1^3 + 2^3 + 3^3 + \cdots + n^3$$

$$= [n^2(n+1)^2/4]$$

$$\sum_{k=1}^{n} k^4 = 1^4 + 2^4 + 3^4; \cdots + n^4$$

$$= \tfrac{1}{30} n(n+1)(2n+1)(3n^2+3n-1)$$

$$\sum_{k=1}^{n} k^r = 1^r + 2^r + 3^r + \cdots + n^r$$

$$= \frac{n^{r+1}}{r+1} + \tfrac{1}{2}n^r + \sum_{k=1}^{[r/2]} (2k)^{-1} \binom{r}{2k-1} B_{2k} n^{r-2k+1}$$

where,

r is a positive integer,

$[r/2]$ is the largest integer less than or equal to $r/2$,

B_{2k} is the $2k$th Bernoulli number.

These sums are tabulated for $n = 1, 2, 3, \cdots, 100$ and for $r = 1, 2, 3, \cdots, 10$.*

Sums of integral powers of odd integers may be obtained from the above. For example,

$$\sum_{k=0}^{n} (2k+1)^2 = 1^2 + 3^2 + 5^2 + \cdots + (2n+1)^2$$

$$= 1^2 + 2^2 + 3^2 + \cdots + (2n+1)^2 - 2^2 - 4^2 - 6^2 - \cdots - (2n)^2$$

$$= \sum_{k=1}^{2n+1} k^2 - 2^2 \sum_{k=1}^{n} k^2$$

$$= \tfrac{1}{3}(n+1)(2n+1)(2n+3)$$

Sums of Powers of Reciprocals of Integers:

$$\zeta(z) = \sum_{k=1}^{\infty} (1/k^z) = (1/1^z) + (1/2^z) + (1/3^z) + \cdots$$

for $\mathrm{Re}\, z > 1$, $\zeta(z)$ is the Riemann zeta function of z

$$\sum_{k=1}^{\infty} (1/k^2) = (1/1^2) + (1/2^2) + (1/3^2) + \cdots$$

$$= \pi^2/6 = 1.64493$$

$$\sum_{k=1}^{\infty} (1/k^3) = (1/1^3) + (1/2^3) + (1/3^3) + \cdots$$

$$= 1.20206$$

$$\sum_{k=1}^{\infty} (1/k^4) = (1/1^4) + (1/2^4) + (1/3^4) + \cdots$$

$$= \pi^4/90 = 1.08232$$

$$\sum_{k=1}^{\infty} (1/k^{2r}) = (1/1^{2r}) + (1/2^{2r}) + (1/3^{2r}) + \cdots$$

$$= [2^{2r-1}\pi^{2r}|B_{2r}|/(2r)!], \quad r \text{ integral}$$

where B_{2r} is the $2r$th Bernoulli number. Values of $\zeta(z)$ are tabulated for $z = 2, 3, 4, \cdots, 42$, to 20 decimal places in the *Handbook of Mathematical Functions*. The truncated sums

$$\sum_{k=1}^{n} (1/k^r)$$

are related to the polygamma functions.

---

* Abramovitz, M. and Stegun, I. A. *Handbook of Mathematical Functions*. Washington, D. C.: National Bureau of Standards.

Finite Sums of Binomial Coefficients:

$$\sum_{k=0}^{n}\binom{m+k}{m}=\binom{m}{m}+\binom{m+1}{m}+\cdots+\binom{m+n}{m}$$

$$=\binom{m+n+1}{m+1}$$

$$\sum_{k=0}^{n}\binom{r}{k}\binom{s}{n-k}=\binom{r}{0}\binom{s}{n}$$

$$+\binom{r}{1}\binom{s}{n-1}+\cdots+\binom{r}{n}\binom{s}{0}$$

$$=\binom{r+s}{n}\quad (r\geq n \text{ and } s\geq n)$$

Binomial Theorem:

Nonnegative integral exponent:

$$(a+b)^n=a^n+na^{n-1}b+\tfrac{1}{2}[n(n-1)]a^{n-2}b^2$$

$$+\cdots+b^n=\sum_{k=0}^{n}\binom{n}{k}a^{n-k}b^k$$

For other n, integral negative and nonintegral, the series will be infinite. For convergence it is assumed $|a|>|b|$. (If $|a|<|b|$, interchange a and b.)

Negative integral exponent:

$$(a+b)^{-n}=a^{-n}-na^{-n-1}b$$

$$+\tfrac{1}{2}[n(n+1)]a^{-n-2}b^2-\cdots$$

$$=\sum_{k=0}^{\infty}(-1)^k\binom{n+k-1}{k}a^{-n-k}b^k$$

Nonintegral exponent:

$$(a+b)^{\alpha}=a^{\alpha}+\alpha a^{\alpha-1}b+\tfrac{1}{2}\alpha(\alpha-1)a^{\alpha-2}b^2+\cdots$$

$$=\sum_{k=0}^{\infty}\frac{\Gamma(\alpha+1)}{k!\,\Gamma(\alpha-k+1)}a^{\alpha-k}b^k$$

Multinomial Series:

$$(x_1+x_2+\cdots+x_r)^n$$

$$=\sum_{n_1=0}^{n}\ \sum_{n_2=0}^{n-n_1}\ \sum_{n_3=0}^{n-n_1-n_2}\cdots\sum_{n_{r-1}=0}^{n-n_1-n_2-\cdots-n_{r-2}} n!\prod_{k=1}^{r}\frac{x_k^{\,n_k}}{n_k!}$$

where the interpretation $n_r=n-n_1-n_2-\cdots-n_{r-1}$ is to be used in the final product.

Combinations and Permutations

A combination is a selection from a number of things in which the order of the selected objects is disregarded, whereas a permutation is a selection in which the order is taken into consideration. For example, if from the letters a, b, and c a group of two is selected, then ab, bc, ac are the combinations and ab, ba, bc, cb, ac, ca are the permutations.

The number of different combinations of n (dissimilar) things taken r at a time is

$$\binom{n}{r}=C_r^n=\frac{n!}{r!(n-r)!}$$

The number of different permutations of n (dissimilar) things taken r at a time is

$$P_r^n=n!/(n-r)!=n\times(n-1)\times\cdots\times(n-r+1)$$

Bernoulli Numbers

Definition:

$$B_n=(d^n/dx^n)[x/(e^x-1)]|_{x=0}$$

Values for Small n:

$$B_0=1,\ B_1=-\tfrac{1}{2},\ B_2=\tfrac{1}{6},\ B_4=-\tfrac{1}{30},$$

$$B_6=\tfrac{1}{42},\ B_8=-\tfrac{1}{30},\ B_{10}=\tfrac{5}{66},$$

$$B_{2n+1}=0,\ \text{for all integral } n>0$$

Identities:

$$B_{2n}=(-1)^{n-1}\frac{2\cdot(2n)!}{(2\pi)^{2n}}\sum_{k=1}^{\infty}\frac{1}{k^{2n}},\ n\geq 1$$

$$B_n=\sum_{k=0}^{n}\binom{n}{k}B_k,\ n>1$$

The latter identity may be used for recursive calculation of the B_n's.

Trigonometric Identities

$$1=\sin^2A+\cos^2A=\sin A\csc A=\tan A\cot A$$

$$=\cos A\sec A$$

$$\sin A=\cos A/\cot A=1/\csc A=\cos A\tan A$$

$$=\pm(1-\cos^2A)^{1/2}$$

$$\cos A=\sin A/\tan A=1/\sec A=\sin A\cot A$$

$$=\pm(1-\sin^2A)^{1/2}$$

$$\tan A=\sin A/\cos A=1/\cot A=\sin A\sec A$$

$$\sin A=(e^{jA}-e^{-jA})/2j$$

$$\cos A=(e^{jA}+e^{-jA})/2$$

$$\sin(A\pm B)=\sin A\cos B\pm\cos A\sin B$$

$$\cos(A\pm B)=\cos A\cos B\mp\sin A\sin B$$

$$\tan(A\pm B)=(\tan A\pm\tan B)/(1\mp\tan A\tan B)$$

$$=(\tan A\cot B\pm 1)/(\cot B\mp\tan A)$$

$$\cot(A \pm B) = (\cot A \cot B \mp 1)/(\cot B \pm \cot A)$$
$$= (\cot A \mp \tan B)/(1 \pm \cot A \tan B)$$

$$\sin 2A = 2 \sin A \cos A$$

$$\cos 2A = \cos^2 A - \sin^2 A$$

$$\tan 2A = (2 \tan A)/(1 - \tan^2 A)$$

$$\sin 3A = 3 \sin A - 4 \sin^3 A$$

$$\cos 3A = -3 \cos A + 4 \cos^3 A$$

$$\tan 3A = (3 \tan A - \tan^3 A)/(1 - 3 \tan^2 A)$$

$$\cos nA = \mathrm{Re}(\cos A + j \sin A)^n$$

$$\sin nA = \mathrm{Im}(\cos A + j \sin A)^n$$

$$\sin\tfrac{1}{2}A = \pm[(1 - \cos A)/2]^{1/2}$$

$$\cos\tfrac{1}{2}A = \pm[(1 + \cos A)/2]^{1/2}$$

$$\tan\tfrac{1}{2}A = \sin A/(1 + \cos A) = (1 - \cos A)/\sin A$$

$$\sin A \pm \sin B = 2 \sin\tfrac{1}{2}(A \pm B)\cos\tfrac{1}{2}(A \mp B)$$

$$\cos A + \cos B = 2 \cos\tfrac{1}{2}(A + B)\cos\tfrac{1}{2}(A - B)$$

$$\cos B - \cos A = 2 \sin\tfrac{1}{2}(A + B)\sin\tfrac{1}{2}(A - B)$$

$$\tan A \pm \tan B = [\sin(A \pm B)/\cos A \cos B]$$

$$\cot A \pm \cot B = [\sin(B \pm A)/\sin A \sin B]$$

$$\sin^2 A - \sin^2 B = \sin(A + B)\sin(A - B)$$

$$\cos^2 A - \sin^2 B = \cos(A + B)\cos(A - B)$$

$$\tan\tfrac{1}{2}(A \pm B) = (\sin A \pm \sin B)/(\cos A + \cos B)$$

$$\cot\tfrac{1}{2}(A \mp B) = (\sin A \pm \sin B)/(\cos B - \cos A)$$

$$\cos^2 A = \tfrac{1}{2}(\cos 2A + 1)$$

$$\cos^3 A = \tfrac{1}{4}(\cos 3A + 3 \cos A)$$

$$\cos^4 A = \tfrac{1}{8}(\cos 4A + 4 \cos 2A + 3)$$

$$\sin^2 A = \tfrac{1}{2}(-\cos 2A + 1)$$

$$\sin^3 A = \tfrac{1}{4}(-\sin 3A + 3 \sin A)$$

$$\sin^4 A = \tfrac{1}{8}(\cos 4A - 4 \cos 2A + 3)$$

$$\sin A \cos B = \tfrac{1}{2}[\sin(A + B) + \sin(A - B)]$$

$$\cos A \cos B = \tfrac{1}{2}[\cos(A + B) + \cos(A - B)]$$

$$\sin A \sin B = \tfrac{1}{2}[\cos(A - B) - \cos(A + B)]$$

$$\sin A + m \sin B = \rho \sin C$$

with

$$\rho^2 = 1 + m^2 + 2m\cos(B - A)$$

and

$$\tan(C - A) = [m\sin(B - A)]/[1 + m\cos(B - A)]$$

$$\sum_i A_i \exp(j\theta_i) = \rho e^{j\psi}$$

with

$$\tan\psi = \left(\sum_i A_i \sin\theta_i / \sum_i A_i \cos\theta_i\right)$$

and

$$\rho = \left[\sum_i A_i^2 + \sum_{i<j}\sum A_i A_j \cos(\theta_i - \theta_j)\right]^{1/2}$$

In the previous notation

$$\sum_i A_i \cos\theta_i = \rho \cos\psi$$

$$\sum_i A_i \sin\theta_i = \rho \sin\psi$$

apply.

$$\sin x + \sin 2x + \sin 3x + \cdots + \sin mx$$
$$= [\sin\tfrac{1}{2}mx \sin\tfrac{1}{2}(m+1)x / \sin\tfrac{1}{2}x]$$

$$\cos x + \cos 2x + \cos 3x + \cdots + \cos mx$$
$$= [\sin\tfrac{1}{2}mx \cos\tfrac{1}{2}(m+1)x / \sin\tfrac{1}{2}x]$$

$$\sin x + \sin 3x + \sin 5x + \cdots + \sin(2m-1)x$$
$$= (\sin^2 mx / \sin x)$$

$$\cos x + \cos 3x + \cos 5x + \cdots + \cos(2m-1)x$$
$$= (\sin 2mx / 2 \sin x)$$

$$\tfrac{1}{2} + \cos x + \cos 2x + \cdots + \cos mx$$
$$= [\sin(m + \tfrac{1}{2})x / 2 \sin\tfrac{1}{2}x]$$

Angle (degrees)	Sine	Cosine	Tangent
0	0	1	0
30	$\frac{1}{2}$	$\frac{1}{2}\sqrt{3}$	$\frac{1}{3}\sqrt{3}$
45	$\frac{1}{2}\sqrt{2}$	$\frac{1}{2}\sqrt{2}$	1
60	$\frac{1}{2}\sqrt{3}$	$\frac{1}{2}$	$\sqrt{3}$
90	1	0	$\pm\infty$
180	0	−1	0
270	−1	0	$\pm\infty$
360	0	1	0
0–90	+	+	+
90–180	+	−	−
180–270	−	−	+
270–360	−	+	−

versine: $\mathrm{vers}\ \theta = 1 - \cos\theta$
haversine: $\mathrm{hav}\ \theta = \tfrac{1}{2}(1 - \cos\theta) = \sin^2\tfrac{1}{2}\theta$

Approximations for Small Angles

$$\left.\begin{array}{l} \sin\theta = (\theta - \theta^3/6 \cdots) \\ \tan\theta = (\theta + \theta^3/3 \cdots) \\ \cos\theta = (1 - \theta^2/2 \cdots) \end{array}\right\} \theta \text{ in radians}$$

$$\sin\theta = \theta$$

with less than 1-percent error up to $\theta = 0.24$ radian $= 14.0°$

with less than 10-percent error up to
$\theta = 0.78$ radian $= 44.5°$

$\tan\theta = \theta$

with less than 1-percent error up to
$\theta = 0.17$ radian $= 10.0°$

with less than 10-percent error up to
$\theta = 0.54$ radian $= 31.0°$

Inequalities

$$\sin x \le x \le \tan x, \quad \text{for} \quad 0 \le x < \pi/2$$
$$\sin x \ge (2/\pi)x, \quad \text{for} \quad 0 \le x \le \pi/2$$
$$\cos x < \sin x/x \le 1, \quad \text{for} \quad 0 < x \le \pi$$

where x is in radians.

PLANE TRIGONOMETRY

Right Triangles

Refer to Fig. 1.

$$C = 90°$$
$$B = 90° - A$$
$$\sin A = \cos B = a/c$$
$$\tan A = a/b$$
$$c^2 = a^2 + b^2$$
$$\text{area} = \tfrac{1}{2}ab$$

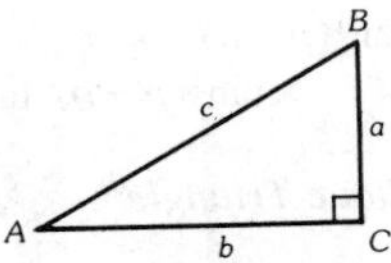

Fig. 1. Right triangle.

Oblique Triangles

Refer to Fig. 2.
Sum of Angles:

$$A + B + C = 180° \qquad (1)$$

Law of Cosines:

$$a^2 = b^2 + c^2 - 2bc \cos A \qquad (2)$$

Law of Sines:

$$a/\sin A = b/\sin B = c/\sin C \qquad (3)$$

Law of Tangents:

$$\frac{a-b}{a+b} = \frac{\tan\frac{1}{2}(A-B)}{\tan\frac{1}{2}(A+B)} \qquad (4)$$

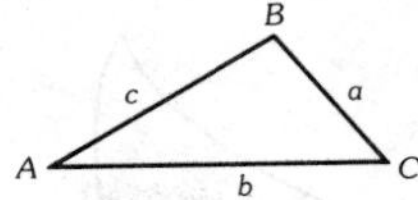

Fig. 2. Oblique triangle.

Half-Angle Equation (Fig. 3):

$$\tan\tfrac{1}{2}A = r/(p-a) \qquad (5)$$

where,

$$2p = a + b + c$$
$$r = [(p-a)(p-b)(p-c)/p]^{1/2}$$

Solving an Oblique Triangle:

Given	Use Eq.	To Obtain
aBC	(1) (3)	A bc
Abc	(1) (4)	$B+C$ hence $B-C$ } B,C
abc	(5) or (2)	ABC
abA ambiguous case	(3) and (1)	BCc

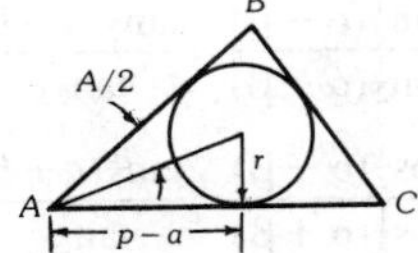

Fig. 3. Half angle.

SPHERICAL TRIGONOMETRY

Spherical triangles are bounded by the arcs of great circles. These are circles formed by the intersection of a sphere with planes passing through the center of the sphere. In the following equations, α, β, γ are the angles, and a, b, c are the corresponding opposite sides, respectively. The sides are measured by the angles subtended by the arcs; for example, a side extending from the Equator to the North Pole is a 90° side.

Right Spherical Triangles ($\gamma = 90°$)

Refer to Fig. 4.

$$\cos c = \cos a \ \cos b = \cot\alpha \ \cot\beta$$
$$\cos\alpha = \sin\beta \ \cos a = \tan b \ \cot c$$
$$\cos\beta = \sin\alpha \ \cos b = \tan a \ \cot c$$
$$\sin a = \sin c \ \sin\alpha = \tan b \ \cot\beta$$
$$\sin b = \sin c \ \sin\beta = \tan a \ \cot\alpha \qquad (6)$$

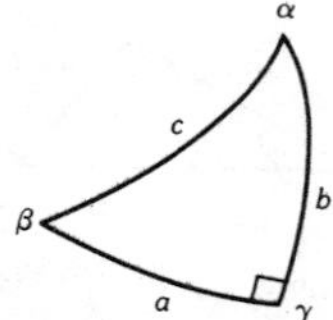

Fig. 4. Right spherical triangle.

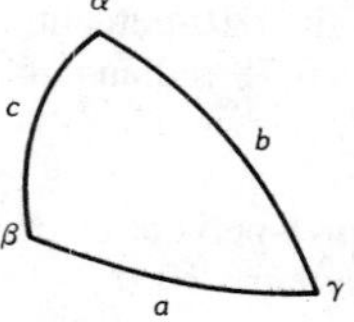

Fig. 5. Oblique spherical triangle.

Oblique Triangles

Refer to Fig. 5.

Law of Cosines for Sides:

$$\cos a = \cos b \ \cos c + \sin b \ \sin c \ \cos\alpha$$
$$\cos b = \cos c \ \cos a + \sin c \ \sin a \ \cos\beta$$
$$\cos c = \cos a \ \cos b + \sin a \ \sin b \ \cos\gamma \quad (7a)$$

Law of Cosines for Angles:

$$\cos\alpha = -\cos\beta \ \cos\gamma + \sin\beta \ \sin\gamma \ \cos a$$
$$\cos\beta = -\cos\gamma \ \cos\alpha + \sin\gamma \ \sin\alpha \ \cos b$$
$$\cos\gamma = -\cos\alpha \ \cos\beta + \sin\alpha \ \sin\beta \ \cos c \quad (7b)$$

Law of Sines:

$$\sin a/\sin\alpha = \sin b/\sin\beta = \sin c/\sin\gamma \quad (8)$$

Napier's Analogies:

$$\frac{\sin\frac{1}{2}(\alpha-\beta)}{\sin\frac{1}{2}(\alpha+\beta)} = \frac{\tan\frac{1}{2}(a-b)}{\tan\frac{1}{2}c} \quad (9a)$$
$$\frac{\cos\frac{1}{2}(\alpha-\beta)}{\cos\frac{1}{2}(\alpha+\beta)} = \frac{\tan\frac{1}{2}(a+b)}{\tan\frac{1}{2}c} \quad (9b)$$
$$\frac{\sin\frac{1}{2}(a-b)}{\sin\frac{1}{2}(a+b)} = \frac{\tan\frac{1}{2}(\alpha-\beta)}{\cot\frac{1}{2}\gamma} \quad (9c)$$
$$\frac{\cos\frac{1}{2}(a-b)}{\cos\frac{1}{2}(a+b)} = \frac{\tan\frac{1}{2}(\alpha+\beta)}{\cot\frac{1}{2}\gamma} \quad (9d)$$

Half-Angle Equations:

$$\tan\tfrac{1}{2}\alpha = \tan r/\sin(p-a)$$
$$\tan\tfrac{1}{2}\beta = \tan r/\sin(p-b)$$
$$\tan\tfrac{1}{2}\gamma = \tan r/\sin(p-c) \quad (10a)$$

where $2p = a+b+c$ and

$$\tan^2 r = \frac{\sin(p-a) \ \sin(p-b) \ \sin(p-c)}{\sin p}$$
$$\sin^2 \tfrac{1}{2}\alpha = [\sin(p-b) \ \sin(p-c)/\sin b \ \sin c]$$
$$\cos^2 \tfrac{1}{2}\alpha = [\sin p \ \sin(p-a)/\sin b \ \sin c]$$
$$\tan^2 \tfrac{1}{2}\alpha = [\sin(p-b) \ \sin(p-c)/\sin p \ \sin(p-a)] \quad (10b)$$

and equations obtained by cyclical permutation for β and γ.

Half-Side Equations:

$$\tan\tfrac{1}{2}a = \tan R \ \sin(\alpha - E)$$
$$\tan\tfrac{1}{2}b = \tan R \ \sin(\beta - E)$$
$$\tan\tfrac{1}{2}c = \tan R \ \sin(\gamma - E) \quad (11a)$$

where $2E = \alpha+\beta+\gamma-\pi$ is the spherical excess and

$$\tan^2 R = \frac{\sin E}{\sin(\alpha-E) \ \sin(\beta-E) \ \sin(\gamma-E)}$$
$$\sin^2 \tfrac{1}{2}a = -[\sin E \ \sin(E-\alpha)/\sin\beta \ \sin\gamma]$$
$$\cos^2 \tfrac{1}{2}a = [\sin(E-\beta) \ \sin(E-\gamma)/\sin\beta \ \sin\gamma]$$
$$\tan^2 \tfrac{1}{2}a = -[\sin E \ \sin(E-\alpha)/\sin(E-\beta) \ \sin(E-\gamma)] \quad (11b)$$

and equations obtained by cyclical permutation for b and c.

Area: On a sphere of radius one, the area of a triangle is equal to the spherical excess

$$2E = \alpha+\beta+\gamma-\pi$$

L'Huilier's Theorem:

$$\tan^2 \tfrac{1}{2}E = \tan\tfrac{1}{2}p \ \tan\tfrac{1}{2}(p-a) \times \tan\tfrac{1}{2}(p-b) \ \tan\tfrac{1}{2}(p-c) \quad (12)$$

*Solving an Oblique Triangle**:

Given	Use Eq.	To Obtain
abc	(10)	$\alpha\beta\gamma$
$\alpha\beta\gamma$	(11)	abc
$ab\gamma$	(9)	$\alpha\pm\beta$, hence α,β, then c
$\alpha\beta c$	(9)	$a\pm b$, hence a,b, then γ
$ab\alpha$ ambiguous	(8)	β
case	(9)	$c\gamma$
$\alpha\beta a$ ambiguous	(8)	b
case	(9)	$c\gamma$

HYPERBOLIC FUNCTIONS†

$$\sinh x = (e^x - e^{-x})/2$$
$$\cosh x = (e^x + e^{-x})/2$$

* See also great-circle calculations in Chapter 33.

† Tables of hyperbolic functions appear in Chapter 47.

$\tanh x = \sinh x/\cosh x$

$= [1 - \exp(-2x)]/[1 + \exp(-2x)]$

$= 1/\coth x$

$\text{sech} x = 1/\cosh x$

$\text{csch} x = 1/\sinh x$

$\sinh(-x) = -\sinh x$

$\cosh(-x) = \cosh x$

$\tanh(-x) = -\tanh x$

$\coth(-x) = -\coth x$

$\sinh jx = j \sin x$

$\cosh jx = \cos x$

$\tanh jx = j \tan x$

$\coth jx = -j \cot x$

$\cosh^2 x - \sinh^2 x = 1$

$1 - \tanh^2 x = 1/\cosh^2 x$

$\coth^2 x - 1 = 1/\sinh^2 x$

$\sinh 2x = 2 \sinh x \cosh x$

$\cosh 2x = \cosh^2 x + \sinh^2 x$

$\sinh(x \pm jy) = \sinh x \cos y \pm j \cosh x \sin y$

$\cosh(x \pm jy) = \cosh x \cos y \pm j \sinh x \sin y$

$\tanh(x \pm y) = (\tanh x \pm \tanh y)/(1 \pm \tanh x \tanh y)$

If $y = \text{gd} x$ (gudermannian of x) is defined by

$$x = \log_e \tan(\tfrac{1}{4}\pi + \tfrac{1}{2}y)$$

then

$$\sinh x = \tan y$$

$$\cosh x = \sec y$$

$$\tanh x = \sin y$$

$$\tanh(x/2) = \tan(y/2)$$

HYPERBOLIC TRIGONOMETRY

Hyperbolic (or pseudospherical) trigonometry applies to triangles drawn in the hyperbolic type of non-Euclidean space. Reflection charts, used in transmission-line theory and waveguide analysis, are models of this hyperbolic space.*

Conformal Model

The space is limited to the inside of a unit circle, Γ. Geodesics (or "straight lines" for the model) are arcs of circle orthogonal to Γ as shown in Fig. 6. The

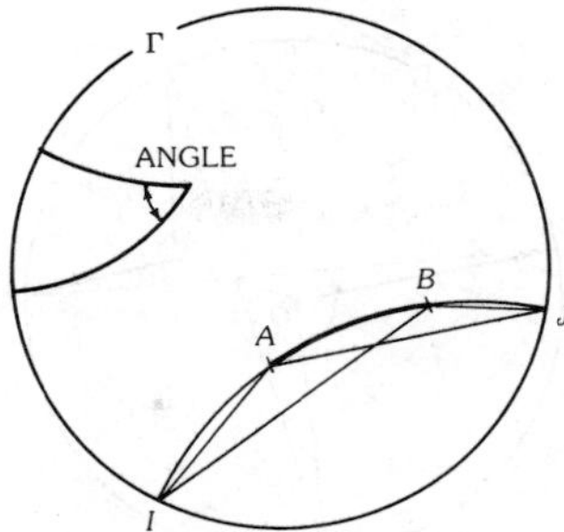

Fig. 6. Conformal model.

hyperbolic distance between two points A and B is defined by

$$[AB] = \log_e[(BI/BJ):(AI/AJ)]$$

where I and J are the intersections with Γ of geodesic AB. The distance $[AB]$ is expressed in nepers. For engineering purposes, a unit corresponding to the decibel and equal to 1/8.686 neper is sometimes used.

As this model is conformal, the angle between two lines is the ordinary angle between the tangents at their common point.

Projective Model

The space is again composed of the points inside a circle, Γ. Geodesics are straight-line segments limited to the inside of Γ (JI in Fig. 7).

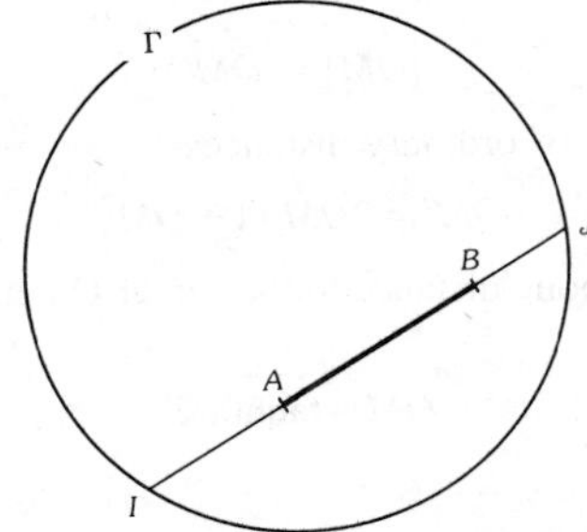

Fig. 7. Projective model.

The hyperbolic distance $\langle AB \rangle$ is defined by

$$\langle AB \rangle = \tfrac{1}{2}\log_e[(BI/BJ):(AI/AJ)]$$

and can be measured directly by means of a hyperbolic protractor. The angles for this model do not appear in true size except when at the center of Γ. An angle such as BAC, when it is considered in reference to the projective model, will be called an *elliptic* angle. It can be evaluated, as shown in Fig. 8, by projecting B and C through the hyperbolic midpoint of OA onto B' and C' on circle Γ, then measuring $B'OC'$ as in Euclidean geometry.

The two models in Fig. 9 drawn inside the same circle Γ can be set into a distance-preserving corre-

---

* G.A. Deschamps, *Hyperbolic Protractor for Microwave Impedance Measurements and Other Purposes*. New York: International Telephone and Telegraph Corp., 1953.

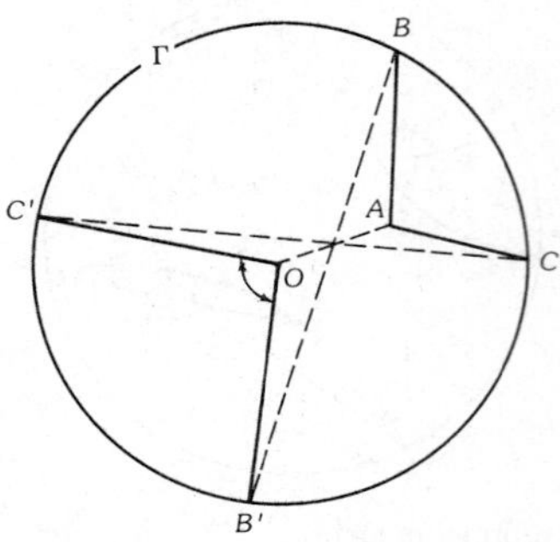

Fig. 8. Construction of angle on projective model.

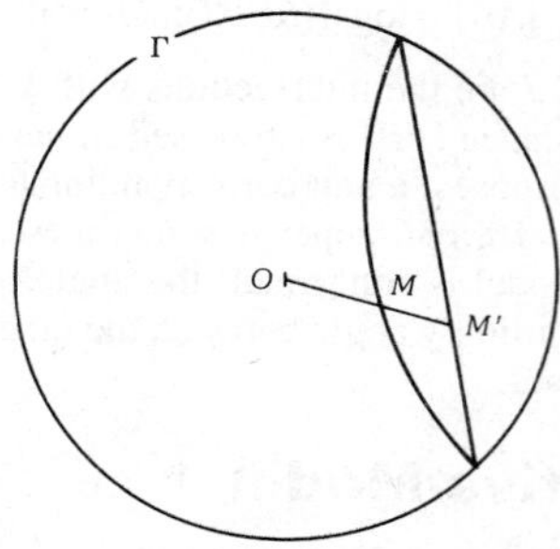

Fig. 9. Correspondence between the two models.

spondence by the transformation: $\mathfrak{B}(M)=M'$ defined by

$$[OM]=\langle OM'\rangle$$

or in terms of ordinary distances

$$OM'=2OM/(1+OM^2)$$

The hyperbolic distance to the center O being denoted by u

$$OM=\tanh(u/2)$$

and

$$OM'=\tanh u$$

The points on Γ are at an infinite distance from any point inside Γ.

In the following equations, the sides are expressed in nepers, the angles in radians. The three points A,B,C are assumed to be inside circle Γ.

Right Hyperbolic Triangles ($\gamma=90°$)

Refer to Figs. 10 and 11.

$$\cosh c=\cosh a\ \cosh b$$
$$=\cot\alpha\ \cot\beta$$
$$\cos\alpha=\sin\beta\ \cosh a$$
$$=\tanh b\ \coth c$$

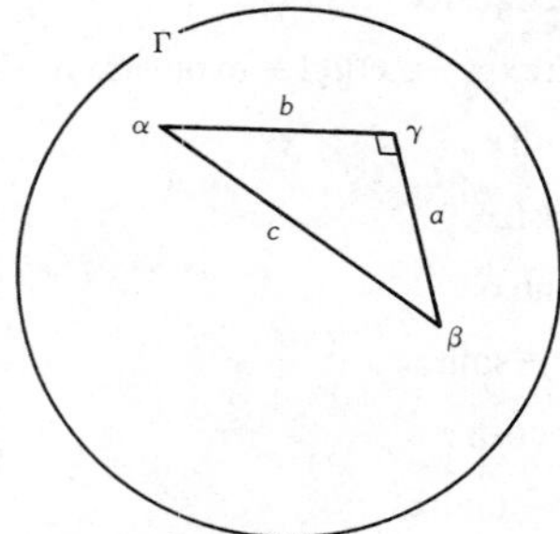

Fig. 10. Projective representation of right hyperbolic triangle.

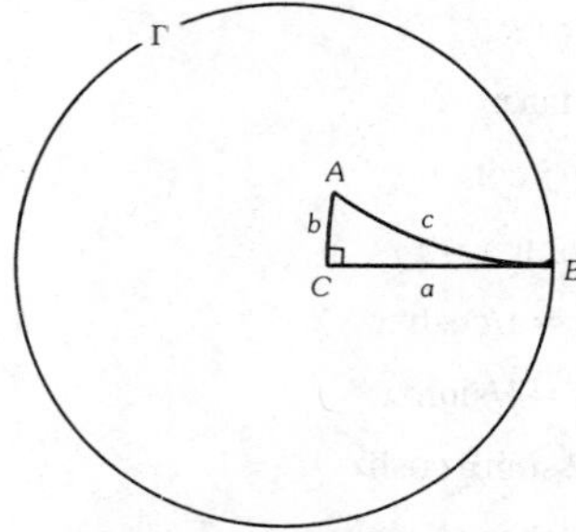

Fig. 11. Conformal representation of right hyperbolic triangle with B at infinity.

$$\cos\beta=\sin\alpha\ \cosh b$$
$$=\tanh a\ \coth c$$

When B is at infinity, i.e., on Γ

$$\cos A=\tanh b$$
$$\cot A=\sinh b$$
$$\csc A=\cosh b$$
$$\tan\tfrac{1}{2}A=\exp b$$

or

$$(\pi/2)-A=\mathrm{gd}\,b$$

(See definition of gd in section "Hyperbolic Functions.")

CB and AB are "parallel." A is also called angle of parallelism and is noted by

$$A=\Pi(b)$$
$$=\pi/2-\mathrm{gd}\,b$$

Oblique Hyperbolic Triangles

Law of Cosines:

$$\cosh a=\cosh b\ \cosh c-\sinh b\ \sinh c\ \cos\alpha$$

and permutations (13a)

$$\cos\alpha=-\cos\beta\ \cos\gamma+\sin\beta\ \sin\gamma\ \cosh a$$

and permutations (13b)

Law of Sines (Fig. 12):

$$\sinh a/\sin\alpha = \sinh b/\sin\beta = \sinh c/\sin\gamma \qquad (14)$$

Napier's Analogies:

$$\frac{\sin\frac{1}{2}(\alpha-\beta)}{\sin\frac{1}{2}(\alpha+\beta)} = \frac{\tanh\frac{1}{2}(a-b)}{\tanh\frac{1}{2}c} \qquad (15a)$$

$$\frac{\cos\frac{1}{2}(\alpha-\beta)}{\cos\frac{1}{2}(\alpha+\beta)} = \frac{\tan\frac{1}{2}(a+b)}{\tanh\frac{1}{2}c} \qquad (15b)$$

$$\frac{\sinh\frac{1}{2}(a-b)}{\sinh\frac{1}{2}(a+b)} = \frac{\tan\frac{1}{2}(\alpha-\beta)}{\cot\frac{1}{2}\gamma} \qquad (15c)$$

$$\frac{\cosh\frac{1}{2}(a-b)}{\cosh\frac{1}{2}(a+b)} = \frac{\tan\frac{1}{2}(\alpha+\beta)}{\cot\frac{1}{2}\gamma} \qquad (15d)$$

Half-Angle Equations:

$$\tan\tfrac{1}{2}\alpha = \tanh r/\sinh(p-\alpha)$$

and permutations where $2p = a+b+c$ and

$$\tanh^2 r = \frac{\sinh(p-a)\,\sinh(p-b)\,\sinh(p-c)}{\sinh p} \qquad (16a)$$

$$\sin^2\tfrac{1}{2}\alpha = \frac{\sinh(p-b)\,\sinh(p-c)}{\sinh b\,\sinh c}$$

$$\cos^2\tfrac{1}{2}\alpha = \frac{\sinh p\,\sinh(p-a)}{\sinh b\,\sinh c}$$

$$\tan^2\tfrac{1}{2}\alpha = \frac{\sinh(p-b)\,\sinh(p-c)}{\sinh p\,\sin(p-a)} \qquad (16b)$$

Half-Side Equations:

$$\coth\tfrac{1}{2}a = \coth R/\sin(\Delta+\alpha)$$

and permutations where $2\Delta = \pi-\alpha-\beta-\gamma$ is the hyperbolic defect and

$$\tanh^2 R = \frac{\sin\Delta}{\sin(\Delta+\alpha)\,\sin(\Delta+\beta)\,\sin(\Delta+\gamma)} \qquad (17a)$$

$$\sinh^2\tfrac{1}{2}a = \frac{\sin\Delta\,\sin(\Delta+\alpha)}{\sin\beta\,\sin\gamma}$$

$$\cosh^2\tfrac{1}{2}a = \frac{\sin(\Delta+\beta)\,\sin(\Delta+\gamma)}{\sin\beta\,\sin\gamma}$$

$$\tanh^2\tfrac{1}{2}a = \frac{\sin\Delta\,\sin(\Delta+\alpha)}{\sin(\Delta+\beta)\,\sin(\Delta+\gamma)} \qquad (17b)$$

Area: The hyperbolic area of a triangle is equal to the hyperbolic defect.

$$2\Delta = \pi - (\alpha+\beta+\gamma) \qquad (18)$$

Solving an Oblique Hyperbolic Triangle: Solution of an oblique hyperbolic triangle is analogous to that for an oblique spherical triangle, as follows.

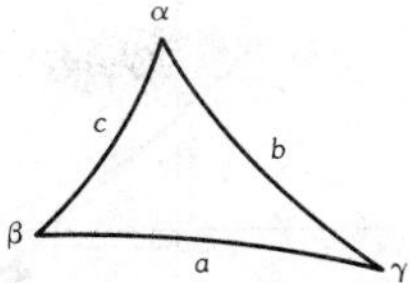

Fig. 12. Oblique hyperbolic triangle.

Given	Use Eq.	To Obtain
abc	(16)	$\alpha\beta\gamma$
$\alpha\beta\gamma$	(17)	abc
$ab\gamma$	(15)	$\alpha\pm\beta$, hence α, β, then c
$\alpha\beta c$	(15)	$a\pm b$, hence a, b, then γ
$ab\alpha$ ambiguous case	(14) (15)	β $c\gamma$
$\alpha\beta a$ ambiguous case	(14) (15)	b $c\gamma$

PLANE ANALYTIC GEOMETRY

In the following, x and y are coordinates of a variable point in a rectangular-coordinate system.

Straight Line

General Equation:

$$Ax+By+C=0$$

A, B, and C are constants.

Slope-Intercept Form (Fig. 13):

$$y = sx+b$$

$$b = y\text{-intercept}$$

$$s = \tan\theta$$

$$= \text{slope}$$

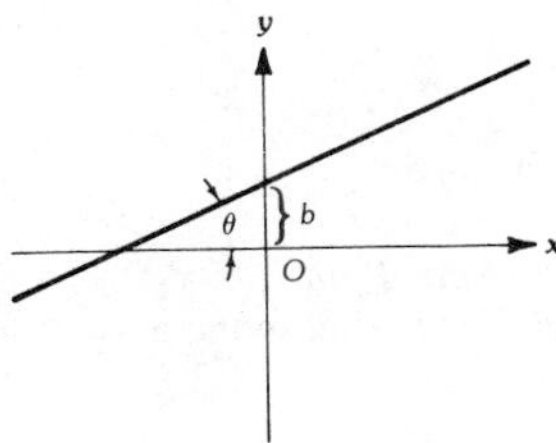

Fig. 13. Slope-intercept.

Intercept-Intercept Form (Fig. 14):

$$(x/a)+(y/b)=1$$

$$a = x\text{-intercept}$$

$$b = y\text{-intercept}$$

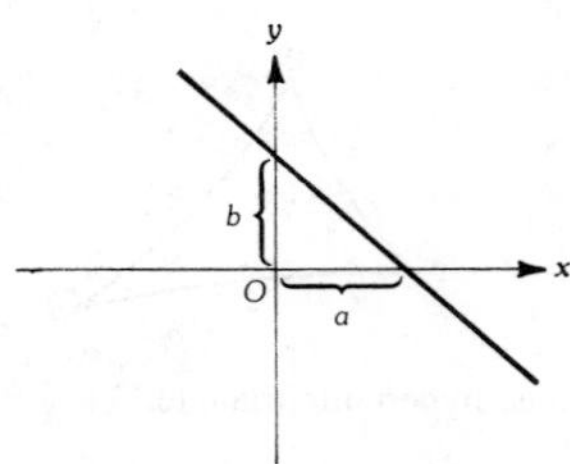

Fig. 14. Intercept-intercept.

Point-Slope Form (Fig. 15):

$$y-y_1=s(x-x_1)$$

$$s=\tan\theta$$

(x_1,y_1) = coordinates of known point on line.

Point-Point Form:

$$(y-y_1)/(y_1-y_2)=(x-x_1)/(x_1-x_2)$$

where (x_1, y_1) and (x_2, y_2) are coordinates of two different points on the line.

Normal Form:

$$\frac{A}{\pm(A^2+B^2)^{1/2}}x+\frac{B}{\pm(A^2+B^2)^{1/2}}y+\frac{C}{\pm(A^2+B^2)^{1/2}}=0$$

The sign of the radical is chosen so that

$$\frac{C}{\pm(A^2+B^2)^{1/2}}<0$$

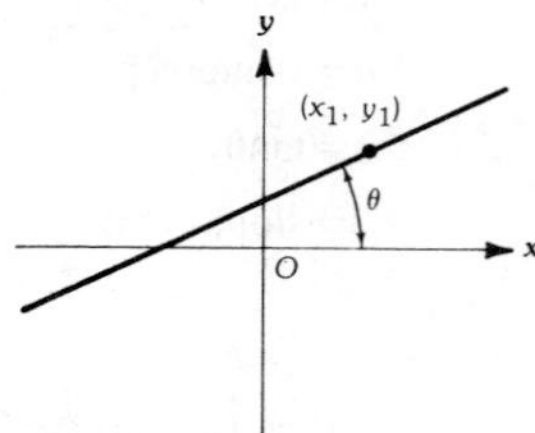

Fig. 15. Point-slope.

Distance From Point (x_1,y_1) to a Line: Substitute coordinates of the point in the normal form of the line. Thus

$$\text{distance}=\frac{A}{\pm(A^2+B^2)^{1/2}}x_1+\frac{B}{\pm(A^2+B^2)^{1/2}}y_1+\frac{C}{\pm(A^2+B^2)^{1/2}}$$

Angle Between Two Lines:

$$\tan\phi=(s_1-s_2)/(1+s_1s_2)$$

where,

ϕ = angle between the lines,

s_1 = slope of one line,

s_2 = slope of other line.

When the lines are mutually perpendicular, $\tan\phi=\pm\infty$, whence

$$s_1=-1/s_2$$

Transformation of Rectangular Coordinates

Translation:

$$x_1=h+x_2$$
$$y_1=k+y_2$$
$$x_2=x_1-h$$
$$y_2=y_1-k$$

(h,k) = coordinates of new origin referred to old origin

Rotation (Fig. 16):

$$x_1=x_2\cos\theta-y_2\sin\theta$$
$$y_1=x_2\sin\theta+y_2\cos\theta$$
$$x_2=x_1\cos\theta+y_1\sin\theta$$
$$y_2=-x_1\sin\theta+y_1\cos\theta$$

(x_1, y_1) = "old" coordinates

(x_2, y_2) = "new" coordinates

θ = counterclockwise angle of rotation of axes

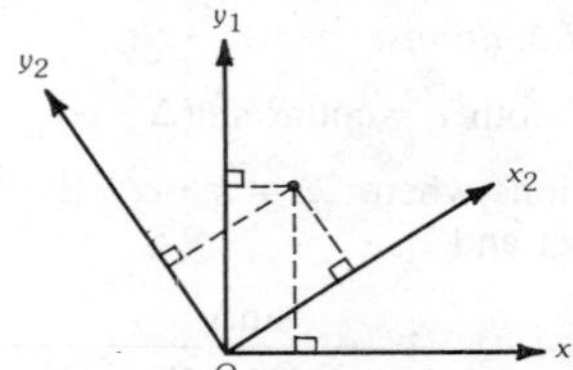

Fig. 16. Rotation of axes.

Circle

The equation of a circle of radius r with center at (m, n) is

$$(x-m)^2+(y-n)^2=r^2$$

Tangent Line to a Circle: At (x_1, y_1) is

$$y-y_1=-[(x_1-m)/(y_1-n)](x-x_1)$$

Normal Line to a Circle: At (x_1, y_1) is

$$y-y_1=[(y_1-n)/(x_1-m)](x-x_1)$$

Parabola

Fig. 17 shows an x-parabola centered at the origin open to the right.

Focus: F
Directrix: D
Vertex: O
Latus rectum: AA'
e = eccentricity = 1
$MP/FP = 1$ for any point P on the parabola

x-Parabola:

$$(y-k)^2 = \pm 2p(x-h)$$

where (h, k) are the coordinates of the vertex, and the sign used is plus or minus when the parabola is open to the right or to the left, respectively. The semilatus rectum is p.

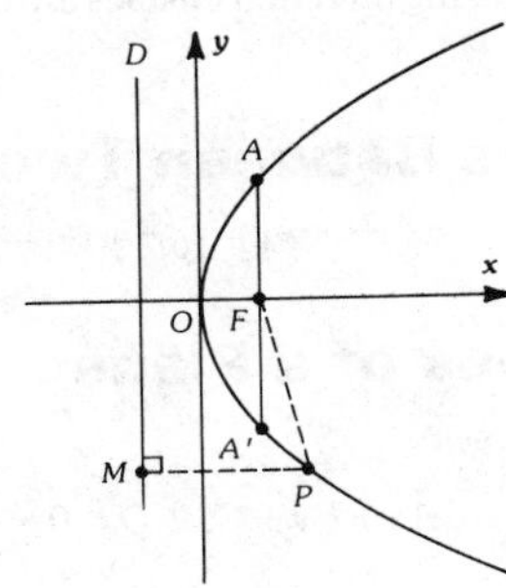

Fig. 17. Parabola.

y-Parabola:

$$(x-h)^2 = \pm 2p(y-k)$$

where (h, k) are the coordinates of the vertex. Use plus sign if parabola is open above, and minus sign if open below.

Tangent Lines to a Parabola:

(x_1, y_1) = point of tangency

For x-parabola

$$y - y_1 = \pm [p/(y_1 - k)](x - x_1)$$

Use plus sign if parabola is open to the right, minus sign if open to the left.

For y-parabola

$$y - y_1 = \pm [(x_1 - h)/p](x - x_1)$$

Use plus sign if parabola is open above, minus sign if open below.

Normal Lines to a Parabola:

(x_1, y_1) = point of contact

For x-parabola

$$y - y_1 = \mp [(y_1 - k)/p](x - x_1)$$

Use minus sign if parabola is open to the right, plus sign if open to the left.

For y-parabola

$$y - y_1 = \mp [p/(x_1 - h)](x - x_1)$$

Use minus sign if parabola is open above, plus sign if open below.

Ellipse

Fig. 18 shows an ellipse centered at the origin. If the ellipse is centered at (h, k) instead, the equations that follow must be modified by replacing x, x_1, y, y_1 by $x-h$, x_1-h, $y-k$, y_1-k, respectively.

Foci: F, F'
Directrices: D, D'

$$e = \text{eccentricity} < 1$$

$$2a = A'A = \text{major axis}$$

$$2b = BB' = \text{minor axis}$$

$$2c = FF' = \text{focal distance}$$

Then

$$OC = a/e$$

$$BF = a$$

$$FC = ae$$

$$1 - e^2 = b^2/a^2$$

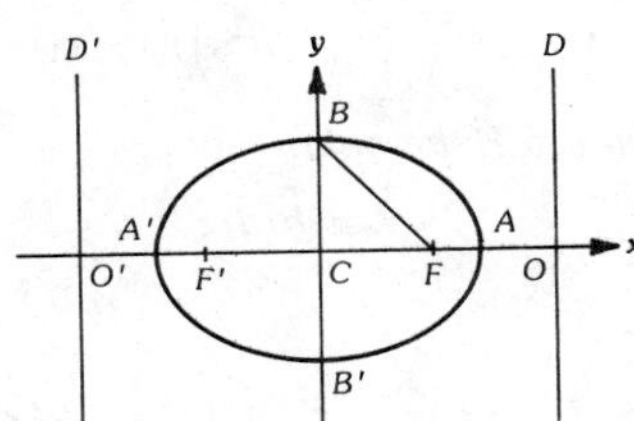

Fig. 18. Ellipse.

Equation of Ellipse:

$$(x^2/a^2) + (y^2/b^2) = 1$$

Sum of the Focal Radii:

To any point on ellipse = $2a$

Equation of Tangent Line to Ellipse:

(x_1, y_1) = point of tangency

$$(xx_1/a^2) + (yy_1/b^2) = 1$$

Equation of Normal Line to an Ellipse:

$$y - y_1 = (a^2 y_1/b^2 x_1)(x - x_1)$$

Hyperbola

Fig. 19 shows an x-hyperbola centered at the origin. If the hyperbola is centered at (h, k) instead, the equations that follow must be modified by replacing x, x_1, y, y_1 by $x-h$, x_1-h, $y-k$, y_1-k, respectively.

Foci: F, F'
Directrices: D, D'

$$e = \text{eccentricity} > 1$$

$$2a = \text{transverse axis} = A'A$$

$$CO = a/e$$

$$CF = ae$$

Equation of x-Hyperbola:

$$(x^2/a^2) - (y^2/b^2) = 1$$

where,

$$b^2 = a^2(e^2 - 1)$$

Equation of Conjugate (y-) Hyperbola:

$$(y^2/b^2) - (x^2/a^2) = 1$$

Tangent Line to x-Hyperbola:

$$(x_1, y_1) = \text{point of tangency}$$

$$a^2y_1y - b^2x_1x = -a^2b^2$$

Normal Line to x-Hyperbola:

$$y - y_1 = -(a^2y_1/b^2x_1)(x - x_1)$$

Asymptotes to Hyperbola:

$$y = \pm(b/a)x$$

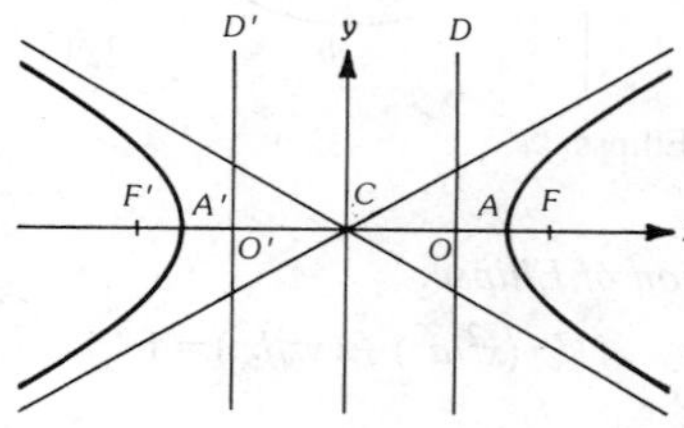

Fig. 19. Hyperbola.

SOLID ANALYTIC GEOMETRY

In the following, x, y, and z are the coordinates of a variable point in space in a right-handed rectangular-coordinate system. See Fig. 20.

Coordinates

The coordinates of a point are given by a triplet, e.g., (x_0, y_0, z_0).

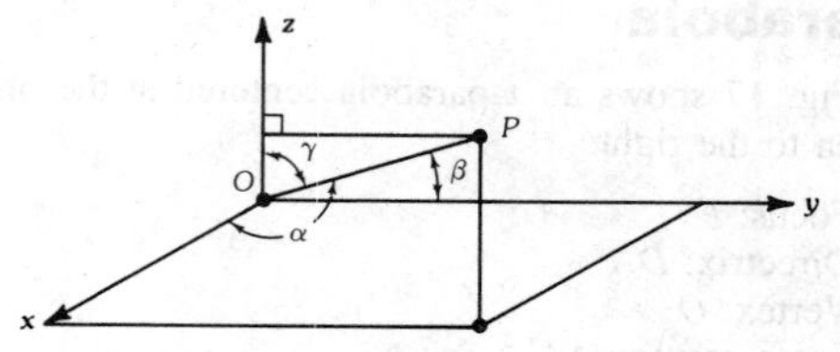

Fig. 20. Coordinates of point P.

Direction Cosines and Numbers

In Fig. 20, α, β, and γ are the angles line OP makes with the x, y, and z axes, respectively. $\cos\alpha$, $\cos\beta$, and $\cos\gamma$ are the direction cosines of line OP. Numbers proportional to the direction cosines are called direction numbers.

Distance Between Two Points

$$d = [(x_1 - x_0)^2 + (y_1 - y_0)^2 + (z_1 - z_0)^2]^{1/2}$$

Equations of a Plane

General Form:

$$Ax + By + Cz + D = 0$$

Three-Point Form:

$$\begin{vmatrix} x - x_0 & y - y_0 & z - z_0 \\ x_1 - x_0 & y_1 - y_0 & z_1 - z_0 \\ x_2 - x_0 & y_2 - y_0 & z_2 - z_0 \end{vmatrix} = 0$$

where (x_0, y_0, z_0), (x_1, y_1, z_1), and (x_2, y_2, z_2) are three points on the plane.

Intercept Form:

$$(x/a) + (y/b) + (z/c) = 1$$

where a, b, and c are the intercepts on the x, y, and z axes, respectively.

Point-Direction Form:

$$A(x - x_1) + B(y - y_1) + C(z - z_1) = 0$$

where,

A, B, and C are the direction numbers of a normal to the plane,
(x_1, y_1, z_1) is a point on the plane.

Normal Form:

$$x \cos\alpha + y \cos\beta + z \cos\gamma - p = 0$$

where,

$\cos\alpha$, $\cos\beta$, and $\cos\gamma$ are the direction cosines of a normal to the plane,

p is the distance of the plane to the origin.

Equations of a Straight Line

Point-Direction Form:

$$(x-x_1)/A=(y-y_1)/B=(z-z_1)/C$$

or

$$x=At+x_1$$

$$y=Bt+y_1$$

$$z=Ct+z_1 \text{ (parametric)}$$

where,

A, B, and C are direction numbers of the line,

(x_1, y_1, z_1) is a point on the line.

Two-Point Form:

$$(x-x_1)/(x_1-x_0)=(y-y_1)/(y_1-y_0)$$
$$=(z-z_1)/(z_1-z_0)$$

or

$$x=x_0+(x_1-x_0)t$$

$$y=y_0+(y_1-y_0)t$$

$$z=z_0+(z_1-z_0)t \text{ (parametric)}$$

where (x_0, y_0, z_0) and (x_1, y_1, z_1) are two points on the line.

Ellipsoid

$$(x^2/a^2)+(y^2/b^2)+(z^2/c^2)=1$$

where a, b, c are the semiaxes of the ellipsoid or the intercepts on the x, y, and z axes, respectively.

Prolate Spheroid

$$a^2(y^2+z^2)+b^2x^2=a^2b^2$$

where $a>b$, and x-axis = axis of revolution.

Oblate Spheroid

$$b^2(x^2+z^2)+a^2y^2=a^2b^2$$

where $a>b$, and y-axis = axis of revolution.

Paraboloid of Revolution

$$y^2+z^2=2px$$

x-axis = axis of revolution

Hyperboloid of Revolution

Revolving an x-hyperbola about the x-axis results in the hyperboloid of two sheets

$$a^2(y^2+z^2)-b^2x^2=-a^2b^2$$

Revolving an x-hyperbola about the y-axis results in the hyperboloid of one sheet

$$b^2(x^2+z^2)-a^2y^2=a^2b^2$$

DIFFERENTIAL CALCULUS

List of Derivatives

In the following, u, v, w are differentiable functions of x, and c is a constant.

General Equations:

$$dc/dx=0$$

$$dx/dx=1$$

$$(d/dx)(u+v-w)=(du/dx)+(dv/dx)-(dw/dx)$$

$$(d/dx)(cv)=c(dv/dx)$$

$$(d/dx)(uv)=u(dv/dx)+v(du/dx)$$

$$(d/dx)(v^c)=cv^{c-1}(dv/dx)$$

$$\frac{d}{dx}\left(\frac{u}{v}\right)=\frac{v(du/dx)-u(dv/dx)}{v^2}$$

$$dy/dx=(dy/dv)(dv/dx), \quad \text{if } y=y(v)$$

$$dy/dx=(dx/dy)^{-1}, \quad \text{if } dx/dy\neq 0$$

Transcendental Functions:

$$(d/dx)\ \ln v=v^{-1}(dv/dx)$$

$$(d/dx)\ (c^v)=c^v \ln c(dv/dx)$$

$$(d/dx)\ (e^v)=e^v(dv/dx)$$

$$(d/dx)\ (u^v)=vu^{v-1}(du/dx)+(\ln u)u^v(dv/dx)$$

$$(d/dx)\ (\sin v)=\cos v(dv/dx)$$

$$(d/dx)\ (\cos v)=-\sin v(dv/dx)$$

$$(d/dx)\ (\tan v)=\sec^2 v(dv/dx)$$

$$(d/dx)\ (\cot v)=-\csc^2 v(dv/dx)$$

$$(d/dx)\ (\sec v)=\sec v\ \tan v(dv/dx)$$

$$(d/dx)\ (\csc v)=\csc v\ \cot v(dv/dx)$$

$$(d/dx)\ (\arcsin v)=(1-v^2)^{-1/2}(dv/dx)$$

$$(d/dx)\ (\arccos v)=-(1-v^2)^{-1/2}(dv/dx)$$

$$(d/dx)\ (\arctan v)=(1+v^2)^{-1}(dv/dx)$$

$$(d/dx)\ (\text{arccot}\, v)=-(1+v^2)^{-1}(dv/dx)$$

$$(d/dx)\ (\text{arcsec}\, v)=[v(v^2-1)^{1/2}]^{-1}(dv/dx)$$

$$(d/dx)\ (\text{arccsc}\, v)=-[v(v^2-1)^{1/2}]^{-1}(dv/dx)$$

$$(d/dx)\ (\sinh v)=\cosh v(dv/dx)$$

$$(d/dx)\ (\cosh v)=\sinh v(dv/dx)$$

$$(d/dx)\ (\tanh v)=\text{sech}^2 v(dv/dx)$$

TABLE OF INTEGRALS

Indefinite Integrals

General Equations:

$$\int af(x)\,dx = a\int f(x)\,dx$$

$$\int [f(x)+g(x)]\,dx = \int f(x)\,dx + \int g(x)\,dx$$

$$(d/dx)\int f(x)\,dx = f(x)$$

$$(d/dx)\int_u^v f(y,x)\,dy$$
$$= f(v,x)\cdot(dv/dx) - f(u,x)\cdot(du/dx) + \int_u^v \frac{\partial f(y,x)}{\partial x}\,dy$$

$$\int f'(x)\,g(x)\,dx = f(x)g(x) - \int f(x)g'(x)\,dx$$

$$\int f(x)\,dx = \int f[h(y)]\,h'(y)\,dy$$

$$\int_{x_1}^{x_2} f(x)\,dx = \int_{-x_2}^{-x_1} f(-x)\,dx$$

$$\int f(\sin x, \cos x)\,dx = \int f\left(\frac{2z}{1+z^2}, \frac{1-z^2}{1+z^2}\right)\frac{dz}{1+z^2}$$

Elementary Forms:

$$\int x^m\,dx = x^{m+1}/(m+1),\ m \neq -1$$

$$\int (dx/x) = \ln|x|$$

$$\int e^x\,dx = e^x$$

$$\int \ln x\,dx = x\ln x - x$$

$$\int \sin x\,dx = -\cos x$$

$$\int \cos x\,dx = \sin x$$

$$\int \tan x\,dx = -\ln|\cos x|$$

$$\int \cot x\,dx = \ln|\sin x|$$

$$\int \csc x\,dx = \ln|\tan\tfrac{1}{2}x|$$

$$\int \sec x\,dx = \ln|\sec x + \tan x|$$

$$\int \sinh x\,dx = \cosh x$$

$$\int \cosh x\,dx = \sinh x$$

$$\int \tanh x\,dx = \ln|\cosh x|$$

$$\int \coth x\,dx = \ln|\sinh x|$$

$$\int \operatorname{sech} x\,dx = 2\tan^{-1}e^x$$

$$\int \operatorname{csch} x\,dx = \ln|\tanh\tfrac{1}{2}x|$$

$$\int dx/(1+x^2) = \tan^{-1}x$$

$$\int dx/(1-x^2) = \tfrac{1}{2}\ln[(1+x)/(1-x)]$$

$$\int dx/(x^2-1) = \tfrac{1}{2}\ln[(x-1)/(x+1)]$$

$$\int dx/(1-x^2)^{1/2} = \sin^{-1}x$$

$$\int dx/(x^2\pm 1)^{1/2} = \ln|x+(x^2\pm 1)^{1/2}|$$

$$\int (x^2\pm 1)^{1/2}\,dx = \tfrac{1}{2}[x(x^2\pm 1)^{1/2} \pm \ln|x+(x^2\pm 1)^{1/2}|]$$

$$\int (1-x^2)^{1/2}\,dx = \tfrac{1}{2}[x(1-x^2)^{1/2} + \sin^{-1}x]$$

Forms Containing $(ax+b)$, $a \neq 0$, $b \neq 0$:

$$\int (ax+b)^n\,dx = [(ax+b)^{n+1}/a(n+1)],\ n \neq -1$$

$$\int (ax+b)^{-1}\,dx = (1/a)\ln|ax+b|$$

$$\int x\,dx/(ax+b) = a^{-2}[ax+b-b\ln(ax+b)]$$

$$\int x dx/(ax+b)^2 = a^{-2}[b/(ax+b) + \ln (ax+b)]$$

$$\int dx/[x(ax+b)] = b^{-1} \ln [x/(ax+b)]$$

$$\int dx/[x(ax+b)^2] = [b(ax+b)]^{-1} + b^{-2} \ln [x/(ax+b)]$$

$$\int dx/[x^2(ax+b)] = -(bx)^{-1} + (a/b^2) \ln [(ax+b)/x]$$

$$\int dx/[x^2(ax+b)^2] = -(2ax+b)/[b^2x(ax+b)] + (2a/b^3) \ln [(ax+b)/x]$$

$$\int x^m (ax+b)^n dx = [x^{m+1}(ax+b)^n/(m+n+1)] + [bn/(m+n+1)]\int x^m(ax+b)^{n-1} dx$$

$$\int \frac{x^m dx}{(ax+b)^n} = [b(1-n)]^{-1}\left[\frac{-x^{m+1}}{(ax+b)^{n-1}} + (m-n+2)\int \frac{x^m}{(ax+b)^{n-1}}\, dx\right], \quad n \neq 1$$

$$\int \frac{x^m dx}{ax+b} = \frac{x^m}{am} + (b/a)\int \frac{x^{m-1}}{ax+b}\, dx, \quad m \neq 0$$

Forms Containing $(ax+b)^{1/2}$, $a \neq 0$, $b \neq 0$:

$$\int x(ax+b)^{1/2} dx = \frac{2(3ax-2b)[(ax+b)^3]^{1/2}}{15a^2}$$

$$\int x^m(ax+b)^{1/2} dx = [2/a(2m+3)]\left\{x^m[(ax+b)^3]^{1/2} - mb\int x^{m-1}(ax+b)^{1/2} dx\right\}$$

$$\int \frac{(ax+b)^{1/2}}{x} dx = 2(ax+b)^{1/2} + (b)^{1/2} \ln \left|\frac{(ax+b)^{1/2} - (b)^{1/2}}{(ax+b)^{1/2} + (b)^{1/2}}\right|, \quad b > 0$$

$$= 2(ax+b)^{1/2} - 2(-b)^{1/2} \tan^{-1}[-(ax+b)/b]^{1/2}, \quad b < 0$$

$$\int \frac{(ax+b)^{1/2}}{x^m} dx = -[(m-1)b]^{-1}\frac{[(ax+b)^3]^{1/2}}{x^{m-1}} + \tfrac{1}{2}[(2m-5)a]\int \frac{(ax+b)^{1/2} dx}{x^{m-1}}, \quad m \neq 1$$

$$\int \frac{x dx}{(ax+b)^{1/2}} = [2(ax-2b)/3a^2](ax+b)^{1/2}$$

$$\int \frac{x^m dx}{(ax+b)^{1/2}} = \frac{2x^m(ax+b)^{1/2}}{(2m+1)a} - \frac{2bm}{(2m+1)a}\int \frac{x^{m-1} dx}{(ax+b)^{1/2}}$$

$$\int \frac{dx}{x(ax+b)^{1/2}} = (b)^{-1/2} \ln \left|\frac{(ax+b)^{1/2} - (b)^{1/2}}{(ax+b)^{1/2} + (b)^{1/2}}\right|, \quad b > 0$$

$$= [2/(-b)^{1/2}]\tan^{-1}[(ax+b)/-b]^{1/2}, \quad b < 0$$

$$\int \frac{dx}{x^n(ax+b)^{1/2}} = -\frac{(ax+b)^{1/2}}{(n-1)bx^{n-1}} - \frac{(2n-3)a}{(2n-2)b}\int \frac{dx}{x^{n-1}(ax+b)^{1/2}}, \quad n \neq 1$$

Forms Containing $R = ax^2 + bx + c$, $a \neq 0$, $x > 0$: *Let* $q = 4ac - b^2$.

$$\int \frac{dx}{R} = [2/(q)^{1/2}]\tan^{-1}[(2ax+b)/(q)^{1/2}], \quad q > 0$$

$$= (-q)^{-1/2} \ln \frac{2ax+b-(-q)^{1/2}}{2ax+b+(-q)^{1/2}}, \quad q < 0$$

(If $q=0$, R is a perfect square.)

$$\int \frac{dx}{R^n} = \frac{2ax+b}{(n-1)qR^{n-1}} + \frac{2(2n-3)a}{q(n-1)} \int \frac{dx}{R^{n-1}}$$

$$\int \frac{x^m}{R^n}\,dx = \frac{x^{m-1}}{(2n-m-1)aR^{n-1}} - \frac{n-m}{2n-m-1}(b/a)\int \frac{x^{m-1}dx}{R^n} + \frac{m-1}{2n-m-1}(c/a)\int \frac{x^{m-2}dx}{R^n}$$

$$\int \frac{dx}{x^m R^n} = -[(m-1)cx^{m-1}R^{n-1}]^{-1} - \frac{m+n-2}{m-1}(b/c)\int \frac{dx}{x^{m-1}R^n} - \frac{m+2n-3}{m-1}(a/c)\int \frac{dx}{x^{m-2}R^n}$$

Forms Containing $(R)^{1/2} = (ax^2+bx+c)^{1/2}$, $a \neq 0$: *Let* $q = 4ac - b^2$.

$$\int \frac{dx}{(R)^{1/2}} = (a)^{-1/2} \ln\left[(R)^{1/2} + x(a)^{1/2} + \frac{b}{2(a)^{1/2}}\right], \quad a>0$$

$$= \frac{-1}{(-a)^{1/2}} \sin^{-1}\left[\frac{2ax+b}{(-q)^{1/2}}\right], \quad a<0$$

$$\int (R)^{1/2}dx = \frac{(2ax+b)(R)^{1/2}}{4a} + \frac{q}{8a}\int \frac{dx}{(R)^{1/2}}$$

$$\int \frac{dx}{R^n(R)^{1/2}} = \frac{2(2ax+b)(R)^{1/2}}{(2n-1)qR^n} + \frac{8a(n-1)}{(2n-1)q}\int \frac{dx}{R^{n-1}(R)^{1/2}}$$

$$\int R^n(R)^{1/2}dx = \frac{(2ax+b)R^n(R)^{1/2}}{4(n+1)a} + \frac{(2n+1)q}{8(n+1)a}\int \frac{R^n dx}{(R)^{1/2}}$$

$$\int \frac{x\,dx}{(R)^{1/2}} = \frac{(R)^{1/2}}{a} - \frac{b}{2a}\int \frac{dx}{(R)^{1/2}}$$

$$\int x(R)^{1/2}dx = \frac{R(R)^{1/2}}{3a} - \frac{b}{2a}\int (R)^{1/2}dx$$

$$\int \frac{x^m dx}{R^n(R)^{1/2}}\,dx = a^{-1}\int \frac{x^{m-2}dx}{R^{n-1}(R)^{1/2}} - (b/a)\int \frac{x^{m-1}dx}{R^n(R)^{1/2}} - (c/a)\int \frac{x^{m-2}dx}{R^n(R)^{1/2}}$$

$$\int \frac{x^m R^n}{(R)^{1/2}}\,dx = \frac{x^{m-1}R^n(R)^{1/2}}{(2n+m)a} - \frac{(2n+2m+1)b}{2a(2n+m)}\int \frac{x^{m-1}R^n dx}{(R)^{1/2}} - \frac{(m-1)c}{(2n+m)a}\int \frac{x^{m-2}R^n dx}{(R)^{1/2}}$$

$$\int \frac{dx}{x^m R^n(R)^{1/2}} = -\frac{(R)^{1/2}}{(m-1)cx^{m-1}R^n} - \frac{(2n+2m-3)b}{2c(m-1)}\int \frac{dx}{x^{m-1}R^n(R)^{1/2}} - \frac{(2n+m-2)a}{(m-1)c}\int \frac{dx}{x^{m-2}R^n(R)^{1/2}}$$

$$\int \frac{R^n dx}{x^m(R)^{1/2}} = -\frac{R^{n-1}(R)^{1/2}}{(m-1)x^{m-1}} + \frac{(2n-1)b}{2(m-1)}\int \frac{R^{n-1}dx}{x^{m-1}(R)^{1/2}} + \frac{(2n-1)a}{m-1}\int \frac{R^{n-1}dx}{x^{m-2}(R)^{1/2}}$$

Logarithmic Integrands

$$\int \ln ax\,dx = x(\ln ax - 1)$$

$$\int \log_b x\,dx = \log_b e(\ln x - 1)x = [(\ln x - 1)x/\ln b]$$

$$\int (\ln x)^n dx = x(\ln x)^n - n\int (\ln x)^{n-1}dx$$

$$\int x^m \ln x dx = x^{m+1}[(m+1)^{-1} \ln x - (m+1)^{-2}],\ m \neq -1$$

$$\int \frac{\ln x}{x} dx = \tfrac{1}{2}(\ln x)^2$$

$$\int x^m (\ln x)^n dx = [x^{m+1}(\ln x)^n/(m+1)] - [n/(m+1)] \int x^m (\ln x)^{n-1} dx,\ m \neq -1$$

$$\int [(\ln x)^n/x] dx = (\ln x)^{n+1}/(n+1),\ n \neq -1$$

$$\int dx/(x \ln x) = \ln|\ln x|$$

$$\int \frac{x^m dx}{(\ln x)^n} = -\frac{x^{m+1}}{(n-1)(\ln x)^{n-1}} + \frac{m+1}{n-1} \int \frac{x^m dx}{(\ln x)^{n-1}},\ n \neq 1$$

$$\int (x^m/\ln x) dx = \mathrm{Ei}[(n+1)\ln x],\quad m \neq -1$$

Exponential Integrands

$$\int a^{bx} dx = a^{bx}/(b \ln a)$$

$$\int x^m e^x dx = x^m e^x - m \int x^{m-1} e^x dx$$

$$\int dx/(a + be^{mx}) = (x/a) - (am)^{-1} \ln|a + be^{mx}|,\quad a \text{ and } m \neq 0$$

$$\int \frac{dx}{(a+be^{mx})^{1/2}} = [m(a)^{1/2}]^{-1} \ln \frac{(a+be^{mx})^{1/2} - (a)^{1/2}}{(a+be^{mx})^{1/2} + (a)^{1/2}},\quad a > 0$$

$$= \{2/[m(-a)^{1/2}]\} \arctan[(a+be^{mx})^{1/2}/(-a)^{1/2}],\quad a < 0$$

Trigonometric Integrands

$$\int \sin^2 x dx = \tfrac{1}{2}(x - \sin x \cos x)$$

$$\int \sin^n x dx = -[(\sin^{n-1} x \cos x)/n] + (n-1)/n \int \sin^{n-2} x dx$$

$$\int \cos^2 x dx = \tfrac{1}{2}(x + \sin x \cos x)$$

$$\int \cos^n x dx = [(\cos^{n-1} x \sin x)/n] + (n-1)/n \int \cos^{n-2} x dx$$

$$\int \sin x \cos^m x dx = -\cos^{m+1} x/(m+1)$$

$$\int \sin^m x \cos x dx = \sin^{m+1} x/(m+1)$$

$$\int \sin^n x \cos^m x = \frac{\cos^{m-1} x \sin^{n+1} x}{m+n} + \frac{m-1}{m+n} \int \cos^{m-2} x \sin^n x dx$$

$$= -\frac{\sin^{n-1} x \cos^{m+1} x}{m+n} + \frac{n-1}{m+n} \int \cos^m x \sin^{n-2} x dx$$

$$\int \frac{\sin^n x dx}{\cos^m x} = (m-1)^{-1}\left[\frac{\sin^{n-1}x}{\cos^{m-1}x} - (n-1)\int \frac{\sin^{n-2}x dx}{\cos^{m-2}x}\right]$$

$$\int \frac{\cos^m x}{\sin^n x} dx = -(n-1)^{-1}\left[\frac{\cos^{m-1}x}{\sin^{n-1}x} + (m-1)\int \frac{\cos^{m-2}x dx}{\sin^{n-2}x}\right]$$

$$\int \frac{dx}{\sin^m x \cos^n x} = [(n-1)\sin^{m-1}x \cos^{n-1}x]^{-1} + \frac{m+n-2}{n-1}\int \frac{dx}{\sin^m x \cos^{n-2}x}$$

$$= -[(m-1)\sin^{m-1}x \cos^{n-1}x]^{-1} + \frac{m+n-2}{m-1}\int \frac{dx}{\sin^{m-2}x \cos^n x}$$

$$\int \tan^n x dx = [\tan^{n-1}x/(n-1)] - \int \tan^{n-2}x dx$$

$$\int \cot^n x dx = -[\cot^{n-1}x/(n-1)] - \int \cot^{n-2}x dx$$

$$\int \sec^2 x dx = \tan x$$

$$\int \sec^n x dx = [\sin x/(n-1)\cos^{n-1}x] + [(n-2)/(n-1)]\int \sec^{n-2}x dx, \ n \neq 1$$

$$\int \frac{dx}{a+b\cos x + c\sin x} = \frac{2}{(a^2-b^2-c^2)^{1/2}}\tan^{-1}\frac{(a-b)\tan(x/2)+c}{(a^2-b^2-c^2)^{1/2}}, \ a^2 > b^2+c^2$$

$$= (b^2+c^2-a^2)^{-1/2}\ln\left|\frac{(a-b)\tan(x/2)+c-(b^2+c^2-a^2)^{1/2}}{(a-b)\tan(x/2)+c+(b^2+c^2-a^2)^{1/2}}\right|, \ a^2 < b^2+c^2, \ a \neq b$$

$$= c^{-1}\ln|a+c\tan(x/2)|, \ a=b$$

$$= \frac{-2}{c+(a-b)\tan(x/2)}, \quad a^2 = b^2+c^2$$

$$\int x^n \sin ax dx = -\sum_{k=0}^{n} k!\binom{n}{k}\frac{x^{n-k}}{a^{k+1}}\cos[ax+(k\pi/2)], \ n \text{ nonnegative integer}$$

$$\int x^n \cos ax dx = \sum_{k=0}^{n} k!\binom{n}{k}\frac{x^{n-k}}{a^{k+1}}\sin[ax+(k\pi/2)], \ n \text{ nonnegative integer}$$

Inverse Trigonometric Integrals, x and a > 0:

$$\int \sin^{-1}(x/a)dx = x\sin^{-1}(x/a) + (a^2-x^2)^{1/2}$$

$$\int \cos^{-1}(x/a)dx = x\cos^{-1}(x/a) - (a^2-x^2)^{1/2}$$

$$\int \tan^{-1}(x/a)dx = x\tan^{-1}(x/a) - \tfrac{1}{2}a\ln(a^2+x^2)$$

$$\int \cot^{-1}(x/a)dx = x\cot^{-1}(x/a) + \tfrac{1}{2}a\ln(a^2+x^2)$$

$$\int \sec^{-1}(x/a)dx = x\sec^{-1}(x/a) - a\ln[x+(x^2-a^2)^{1/2}]$$

$$\int \csc^{-1}(x/a)dx = x\csc^{-1}(x/a) + a\ln[x+(x^2-a^2)^{1/2}]$$

$$\int x \sin^{-1} x dx = \tfrac{1}{4}[(2x^2-1)\sin^{-1} x + x(1-x^2)^{1/2}]$$

$$\int x \cos^{-1} x dx = \tfrac{1}{4}[(2x^2-1)\cos^{-1} x - x(1-x^2)^{1/2}]$$

$$\int x^n \sin^{-1} x dx = [x^{n+1}\sin^{-1} x/(n+1)] - (n+1)^{-1}\int [x^{n+1}/(1-x^2)^{1/2}]dx$$

$$\int x^n \cos^{-1} x dx = [x^{n+1}\cos^{-1} x/(n+1)] + (n+1)^{-1}\int [x^{n+1}/(1-x^2)^{1/2}]dx$$

Miscellaneous Integrals:

$$\int \frac{dx}{ax^3+b} = \frac{p}{3b}\left[\tfrac{1}{2}\ln\frac{(x+p)^2}{x^2-px+p^2} + \sqrt{3}\arctan\frac{x\sqrt{3}}{2p-x}\right],\ p=(b/a)^{1/3}$$

$$\int \frac{dx}{x^4+a^4} = (4a^3\sqrt{2})^{-1}\left[\ln\left(\frac{x^2+ax\sqrt{2}+a^2}{x^2-ax\sqrt{2}+a^2}\right) + 2\tan^{-1}\left(\frac{ax\sqrt{2}}{a^2-x^2}\right)\right]$$

$$\int \frac{dx}{x^4-a^4} = (1/4a^3)\left[\ln\left(\frac{x-a}{x+a}\right) - 2\tan^{-1}(x/a)\right]$$

Definite Integrals

$$\int_0^1 x^u(1-x)^v dx = [\Gamma(u+1)\Gamma(v+1)/\Gamma(u+v+2)],\ u > -1,\ v > -1$$

$$= [u!v!/(u+v+1)!],\ \text{if } u,v \text{ nonnegative integers}$$

$$\int_0^\infty x^u dx/(1+x^v) = (\pi/v)\csc[(u+1)\pi/v],\ 0<u+1<v$$

$$\int_0^1 \frac{x^{2n+1}dx}{(1-x^2)^{1/2}} = \frac{(2n)\ (2n-2)\cdots 6\cdot4\cdot2}{(2n+1)\ (2n-1)\cdots 5\cdot3\cdot1},\ n \text{ positive integer}$$

$$\int_0^1 \frac{x^{2n}dx}{(1-x^2)^{1/2}} = \frac{(2n-1)\ (2n-3)\cdots 5\cdot3\cdot1}{(2n)\ (2n-2)\cdots 6\cdot4\cdot2}\cdot\tfrac{1}{2}\pi,\ n \text{ positive integer}$$

$$\int_0^\infty x^b e^{-ax} dx = \Gamma(b+1)/a^{b+1},\ b>-1,\ a>0$$

$$= b!/a^{b+1},\ b \text{ nonnegative integer}$$

$$\int_0^\infty e^{-ax^2} dx = \tfrac{1}{2}(\pi/a)^{1/2},\ a>0$$

$$\int_0^\infty x^b e^{-ax^2} dx = \frac{\Gamma[(b+1)/2]}{2a^{(b+1)/2}},\ b>-1$$

$$\int_0^\infty \exp[-(ax^2+bx+c)]dx = \tfrac{1}{2}(\pi/a)^{1/2}\exp[(b^2-4ac)/4a][1-\operatorname{erf}(\tfrac{1}{2}ba^{1/2})]$$

$$\int_0^\infty \exp\{-[x^2+(a^2/x^2)]\}\ dx = \tfrac{1}{2}\exp(-2|a|)\ (\pi)^{1/2}$$

$$\int_0^{\pi/2} \sin^n x dx = \int_0^{\pi/2} \cos^n x dx = \tfrac{1}{2}(\pi)^{1/2}\frac{\Gamma[\frac{1}{2}(n+1)]}{\Gamma(\frac{1}{2}n+1)},\ n>-1$$

$$\int_0^\infty (\sin mx/x)\,dx = \tfrac{1}{2}\pi,\ m>0$$

$$= 0,\ m=0$$

$$= -\tfrac{1}{2}\pi,\ m<0$$

$$\int_0^\infty (\sin x \cdot \cos mx/x)\,dx = 0,\ |m|>1$$

$$= \tfrac{1}{4}\pi,\ m=\pm 1$$

$$= \tfrac{1}{2}\pi,\ -1<m<1$$

$$\int_0^\infty (\sin x/x)^2\,dx = \tfrac{1}{2}\pi$$

$$\int_0^\infty \cos(x^2)\,dx = \int_0^\infty \sin(x^2)\,dx = \tfrac{1}{2}(\tfrac{1}{2}\pi)^{1/2}$$

$$\int_0^\infty [\cos mx/(1+x^2)]\,dx = \tfrac{1}{2}\pi e^{-|m|}$$

$$\int_0^\infty (\cos x/x^{1/2})\,dx = \int_0^\infty (\sin x/x^{1/2})\,dx = (\tfrac{1}{2}\pi)^{1/2}$$

$$\int_0^\infty \exp(-a^2x^2)\cos bx\,dx = [(\pi)^{1/2}/2\,|\,a\,|]\exp(-b^2/4a^2)$$

$$\int_0^1 [\ln x/(1-x)]\,dx = -\tfrac{1}{6}\pi^2$$

$$\int_0^1 [\ln x/(1+x)]\,dx = -\tfrac{1}{12}\pi^2$$

$$\int_0^1 [\ln x/(1-x^2)]\,dx = -\tfrac{1}{8}\pi^2$$

$$\int_0^1 [\ln x/(1-x^2)^{1/2}]\,dx = -\tfrac{1}{2}\pi \ln 2$$

$$\int_0^1 (\ln x)^n\,dx = (-1)^n n!$$

$$\int_0^1 x^m(\ln x^{-1})^n\,dx = [\Gamma(n+1)/(m+1)^{n+1}],\ m>-1,\ n>-1$$

$$\int_0^{\pi/2} \ln \sin x\,dx = -\tfrac{1}{2}\pi \ln 2$$

$$\int_0^{\pi/2} \ln \cos x\,dx = -\tfrac{1}{2}\pi \ln 2$$

$$\int_0^{\pi} x \ln \sin x\,dx = -\tfrac{1}{2}\pi^2 \ln 2$$

$$\int_0^{\pi} \ln(a+b\cos x)\,dx = \pi \ln\tfrac{1}{2}[a+(a^2-b^2)^{1/2}],\ a \geq b$$

SERIES

Taylor's Series for a Single Variable

$$f(z)=f(a)+f'(a)\,(z-a)+\tfrac{1}{2}[f''(a)]\,(z-a)^2+\cdots+[f^{(n)}(a)/n!]\,(z-a)^n+R_n$$

$$=\sum_{k=0}^{n}(k!)^{-1}f^{(k)}(a)\,(z-a)^k+R_n$$

where the remainder is bounded by $|R_n|\le[M/(n+1)!]\,|z-a|^{n+1}$ in which

$$M=\max_{0\le\theta\le1}|f^{(n+1)}[a+\theta(z-a)]|$$

If a, z, and f are real, then there exists a real θ, $0<\theta<1$, such that the remainder is

$$R_n=\{f^{(n+1)}[a+\theta(z-a)]/(n+1)!\}\,(z-a)^{n+1}$$

When $a=0$, this series is often called MacLaurin's series.

Taylor's Series for Two Variables

$$f(x,y)=f(a,b)+\left[\frac{\partial f(x,y)}{\partial x}\bigg|_{x=a,y=b}(x-a)+\frac{\partial f(x,y)}{\partial y}\bigg|_{x=a,y=b}(y-b)\right]$$

$$+\tfrac{1}{2}\left[\frac{\partial^2 f(x,y)}{\partial x^2}\bigg|_{x=a,y=b}(x-a)^2+2\frac{\partial^2 f(x,y)}{\partial x\partial y}\bigg|_{x=a,y=b}(x-a)(y-b)+\frac{\partial^2 f(x,y)}{\partial y^2}\bigg|_{x=a,y=b}(y-b)^2\right]+\cdots$$

$$=\sum_{k=0}^{n}(k!)^{-1}\{[(x-a)(\partial/\partial\xi)+(y-b)(\partial/\partial\eta)]^k f(\xi,\eta)\}_{\xi=a,\eta=b}+R_n$$

where the remainder is bounded by $|R_n|\le[M/(n+1)!](|x-a|+|y-b|)^{n+1}$ in which

$$M=\max_{\substack{k=0,1,2,\cdots,n+1\\0\le\theta\le1}}|\partial^{n+1}f(\xi,\eta)/\partial^k\xi\,\partial^{n+1-k}\eta|,\ \xi=a+\theta(x-a),\eta=b+\theta(y-b)$$

If $f(x,y)$ is a real function of two variables, there exists a number θ, $0<\theta<1$ such that the remainder is

$$R_n=[(n+1)!]^{-1}\{[(x-a)(\partial/\partial\xi)+(y-b)(\partial/\partial\eta)]^{n+1}f(\xi,\eta)\},\ \xi=a+\theta(x-a),\ \eta=b+\theta(y-b)$$

Miscellaneous Series

$$\ln(1+x)=x-\tfrac{1}{2}x^2+\tfrac{1}{3}x^3-\tfrac{1}{4}x^4+\cdots=-\sum_{k=1}^{\infty}(-1)^k\frac{x^k}{k},\ |x|<1$$

$$e^x=1+x+\frac{x^2}{2!}+\frac{x^3}{3!}+\cdots=\sum_{k=0}^{\infty}\frac{x^k}{k!},\ |x|<\infty$$

$$\sin x=x-\frac{x^3}{3!}+\frac{x^5}{5!}-\frac{x^7}{7!}+\cdots=\sum_{k=0}^{\infty}(-1)^k\frac{x^{2k+1}}{(2k+1)!},\ |x|<\infty$$

$$\cos x=1-\frac{x^2}{2!}+\frac{x^4}{4!}-\frac{x^6}{6!}+\cdots=\sum_{k=0}^{\infty}(-1)^k\frac{x^{2k}}{(2k)!},\ |x|<\infty$$

$$\sinh x=x+\frac{x^3}{3!}+\frac{x^5}{5!}+\frac{x^7}{7!}+\cdots=\sum_{k=0}^{\infty}\frac{x^{2k+1}}{(2k+1)!},\ |x|<\infty$$

$$\cosh x=1+\frac{x^2}{2!}+\frac{x^4}{4!}+\frac{x^6}{6!}+\cdots=\sum_{k=0}^{\infty}\frac{x^{2k}}{(2k)!},\ |x|<\infty$$

$$\tan x=x+\frac{x^3}{3}+\frac{2x^5}{15}+\frac{17x^7}{315}+\frac{62x^9}{2835}+\cdots,\ |x|<\tfrac{1}{2}\pi$$

$$\cot x = \frac{1}{x} - \frac{x}{3} - \frac{x^3}{45} - \frac{2x^5}{945} - \frac{x^7}{4725} - \cdots, \quad |x| < \pi$$

$$\arcsin x = x + \frac{1}{2}\frac{x^3}{3} + \frac{1 \cdot 3}{2 \cdot 4}\frac{x^5}{5} + \frac{1 \cdot 3 \cdot 5}{2 \cdot 4 \cdot 6}\frac{x^7}{7} + \cdots, \quad |x| < 1$$

$$\arctan x = x - \tfrac{1}{3}x^3 + \tfrac{1}{5}x^5 - \tfrac{1}{7}x^7 + \cdots, \quad |x| < 1$$

$$\operatorname{arc\,sinh} x = x - \frac{1}{2}\frac{x^3}{3} + \frac{1 \cdot 3}{2 \cdot 4}\frac{x^5}{5} - \frac{1 \cdot 3 \cdot 5}{2 \cdot 4 \cdot 6}\frac{x^7}{7} + \cdots, \quad |x| < 1$$

$$\operatorname{arc\,tanh} x = x + \tfrac{1}{3}x^3 + \tfrac{1}{5}x^5 + \tfrac{1}{7}x^7 + \cdots, \quad |x| < 1$$

MATRIX ALGEBRA

Notation

A matrix of order $m \times n$ is a rectangular array of numbers, real or complex, consisting of m rows and n columns:

$$\mathbf{A} = \begin{bmatrix} a_{11} & a_{12} & a_{13} & \cdots & a_{1n} \\ a_{21} & a_{22} & a_{23} & \cdots & a_{2n} \\ \cdots & \cdots & \cdots & \cdots & \cdots \\ a_{m1} & a_{m2} & a_{m3} & \cdots & a_{mn} \end{bmatrix} = [a_{ij}]$$

A row (column) vector is a $1 \times n$ $(n \times 1)$ matrix. An $n \times n$ matrix is called a square matrix. A matrix with all entries equal to zero is called a zero matrix; it is denoted by $\mathbf{0}$. A square zero matrix with the elements on the main diagonal, that is, the diagonal extending from the upper left corner to the lower right corner, replaced by ones is called an identity matrix. It is denoted by $\mathbf{I}_n$ or $\mathbf{I}$. A square matrix with all entries zero above or below the main diagonal is called triangular. For example,

$$(1,2,5), \quad \begin{bmatrix} 2 \\ 3 \end{bmatrix}, \quad \begin{bmatrix} 0 & 0 \\ 0 & 0 \\ 0 & 0 \end{bmatrix},$$

$$\begin{bmatrix} 1 & 4 & 3 & 4 \\ 0 & 3 & -1 & 5 \\ 0 & 0 & 4 & -2 \\ 0 & 0 & 0 & 2 \end{bmatrix}, \quad \begin{bmatrix} 1 & 0 & 0 \\ 0 & 1 & 0 \\ 0 & 0 & 1 \end{bmatrix}$$

are row, column, zero, triangular, and identity matrices, respectively.

Operations

Addition and Subtraction: If $\mathbf{A}$ and $\mathbf{B}$ are matrices of the same order with elements a_{ij} and b_{ij}, respectively, the matrix

$$\mathbf{C} = \mathbf{A} \pm \mathbf{B}$$

has elements

$$c_{ij} = a_{ij} \pm b_{ij}, \quad i = 1, 2, \cdots, m; j = 1, 2, \cdots, n$$

Multiplication by a Number: If k is a number, real or complex, the matrix

$$\mathbf{C} = k\mathbf{A}$$

has elements

$$c_{ij} = ka_{ij}, \quad i = 1, 2, \cdots, m; j = 1, 2, \cdots, n$$

Multiplication of Two Matrices: Let $\mathbf{A}$ and $\mathbf{B}$ be two matrices of orders $m \times n$ and $n \times p$, respectively. The matrix

$$\mathbf{C} = \mathbf{AB}$$

will have elements

$$c_{ij} = \sum_{k=1}^{n} a_{ik} b_{kj}, \; i = 1, 2, \cdots, m; j = 1, 2, \cdots, p$$

Matrix $\mathbf{C}$ has order $m \times p$. Note that the product $\mathbf{BA}$ is defined only when $m = p$. In general $\mathbf{AB} \neq \mathbf{BA}$ even when $m = n = p$, so it is necessary to distinguish between premultiplication and postmultiplication.

For a square matrix $\mathbf{A}$ of order n, powers are defined

$$\mathbf{A}^0 = \mathbf{I}, \; \mathbf{A}^1 = \mathbf{A}, \; \mathbf{A}^2 = \mathbf{A} \cdot \mathbf{A}, \; \mathbf{A}^3 = \mathbf{A} \cdot \mathbf{A}^2, \text{ etc.}$$

A polynomial function of $\mathbf{A}$ is a square matrix of order n given by

$$P(\mathbf{A}) = a_n\mathbf{A}^n + a_{n-1}\mathbf{A}^{n-1} + \cdots + a_1\mathbf{A}^1 + a_0\mathbf{A}^0$$

where the a_i are real or complex numbers.

Division of Two Matrices: Not defined.

Determinant

Definition: The determinant of a square matrix $\mathbf{A}$ of order n is usually defined

$$|\mathbf{A}| = \sum \pm \; a_{1i}a_{2j} \cdots a_{nr}$$

where the second subscripts $i, j, \cdots, r$ form a permutation (rearrangement) of the integers $1, 2, \cdots, n$. The sum is taken over all permutations with a plus (minus) sign if the permutation is even (odd). A permutation is called even (odd) if an even (odd) number of inversions is necessary to attain the natural or ascending order. For example, 4132→1432→1342→1324→1234; therefore, 4132 is an even permutation.

A square matrix is said to be singular if its determinant is zero and nonsingular otherwise.

Laplace's Development: By an expansion known as Laplace's development, the determinant of a matrix **A** of order n can be expressed in terms of determinants of matrices of order $n-1$.

$$|\mathbf{A}| = \sum_{j=1}^{n} (-1)^{i+j} a_{ij} M_{ij} \qquad \text{(expansion by the } i\text{th row)}$$

$$= \sum_{i=1}^{n} (-1)^{i+j} a_{ij} M_{ij} \qquad \text{(expansion by the } j\text{th column)}$$

The notation M_{ij} represents the determinant of the matrix formed by deleting the ith row and jth column of **A**. Expanding by any row or column will lead to the same value of $|\mathbf{A}|$. The M_{ij} are in turn evaluated in terms of determinants of order $n-2$. This process is continued until, say, second-order determinants are obtained.

$$\begin{vmatrix} b_{11} & b_{12} \\ b_{21} & b_{22} \end{vmatrix} = b_{11}b_{22} - b_{12}b_{21}$$

A first-order determinant has a value equal to its only entry. M_{ij} is called the minor and $(-1)^{i+j} \times M_{ij} = A_{ij}$ is called the cofactor of the element a_{ij}.

Laplace's development is valuable for a literal expansion. For numerical evaluation of determinants of large order, say greater than four, the Gauss algorithm described in the next subsection requires less effort.

Linear Transformations

The linear transformation or set of equations

$$\begin{aligned} a_{11}x_1 + a_{12}x_2 + \cdots + a_{1n}x_n &= y_1 \\ a_{21}x_1 + a_{22}x_2 + \cdots + a_{2n}x_n &= y_2 \\ &\vdots \\ a_{m1}x_1 + a_{m2}x_2 + \cdots + a_{mn}x_n &= y_m \end{aligned}$$

may be compactly written in matrix form

$$\mathbf{AX} = \mathbf{Y}$$

where $\mathbf{A} = [a_{ij}]$ is a matrix of order $m \times n$, and **X** and **Y** are column vectors.

Inverse Matrix: If $m = n$ and $|\mathbf{A}| \neq 0$, then there exists an inverse matrix denoted by $\mathbf{A}^{-1}$ such that $\mathbf{AA}^{-1} = \mathbf{A}^{-1}\mathbf{A} = \mathbf{I}$. The inverse transformation expressing the x's in terms of the y's may then be compactly written

$$\mathbf{X} = \mathbf{A}^{-1}\mathbf{Y}$$

This inverse transformation may be effected using *Cramer's rule*

$$x_i = (1/|\mathbf{A}|) \sum_{k=1}^{n} A_{ki} y_k, \quad i = 1, 2, \cdots, n$$

where A_{ki} denotes the cofactor associated with the element a_{ki} in the original matrix **A**. Cramer's rule provides a useful literal expansion of the solution. However, numerical evaluation of the determinants involved, say Laplace's development, requires of the order of $n!$ operations.

The following *Gauss algorithm* requires only of the order of n^3 operations and is therefore preferred for numerical evaluation when n is large: Renumber the x_k's if necessary to make $a_{11} \neq 0$. Normalize the first equation by dividing it by a_{11}. If $a_{21} = 0$, leave the second equation intact. If $a_{21} \neq 0$, eliminate x_1 by subtracting the normalized first equation multiplied by a_{21} from the second equation. Similarly, eliminate x_1 from the remaining $n-2$ equations. The result is

$$\begin{aligned} x_1 + (a_{12}/a_{11})x_2 + (a_{13}/a_{11})x_3 + \cdots + (a_{1n}/a_{11})x_n &= (y_1/a_{11}) \\ [a_{22} - a_{21}(a_{12}/a_{11})]x_2 + [a_{23} - a_{21}(a_{13}/a_{11})]x_3 + \cdots + [a_{2n} - a_{21}(a_{1n}/a_{11})]x_n &= y_2 - a_{21}(y_1/a_{11}) \\ &\vdots \\ [a_{n2} - a_{n1}(a_{12}/a_{11})]x_2 + [a_{n3} - a_{n1}(a_{13}/a_{11})]x_3 + \cdots + [a_{nn} - a_{n1}(a_{1n}/a_{11})]x_n &= y_n - a_{n1}(y_1/a_{11}) \end{aligned}$$

The entire process is now repeated with the first equation omitted. There then results a set of the form

$$\begin{aligned} x_1 + b_{12}x_2 + b_{13}x_3 + \cdots + b_{1n}x_n &= c_{11}y_1 \\ x_2 + b_{23}x_3 + \cdots + b_{2n}x_n &= c_{21}y_1 + c_{22}y_2 \\ b_{33}x_3 + \cdots + b_{3n}x_n &= c_{31}y_1 + c_{32}y_2 + c_{33}y_3 \\ &\vdots \\ b_{n3}x_3 + \cdots + b_{nn}x_n &= c_{n1}y_1 + c_{n2}y_2 + c_{nn}y_n \end{aligned}$$

Again the process is repeated with the first two equations omitted. Continuing in this manner yields a triangular form

$$x_1 + b_{12}x_2 + b_{13}x_3 + b_{14}x_4 + \cdots + b_{1n}x_n = c_{11}y_1$$

$$
\begin{aligned}
x_2 + b_{23}x_3 + b_{24}x_4 + \cdots + b_{2n}x_n &= c_{21}y_1 + c_{22}y_2 \\
x_3 + d_{34}x_4 + \cdots + d_{3n}x_n &= e_{31}y_1 + e_{32}y_2 + e_{33}y_3 \\
x_4 + \cdots + d_{4n}x_n &= e_{41}y_1 + e_{42}y_2 + e_{43}y_3 + e_{44}y_4 \\
&\vdots \\
x_n &= e_{n1}y_1 + e_{n2}y_2 + e_{n3}y_3 + \cdots + e_{nn}y_n
\end{aligned}
$$

Note that the last equation gives the value of x_n. This may be substituted in the next to the last equation to obtain x_{n-1} and so on. If the y's are literal as shown above, the process will yield the inverse transformation

$$\mathbf{X} = \mathbf{A}^{-1}\mathbf{Y}$$

where $\mathbf{A}^{-1}$ is the inverse of the original matrix. If the y's are numerical, labor is saved by combining the values on the right side of each equation at each step of the algorithm.

Since the determinant of a triangular matrix is equal to the product of the elements on the main diagonal

$$|\mathbf{A}| = a_{11}a_{22}'a_{33}' \cdots a_{nn}'$$

where a_{kk}' is the quantity the kth equation is divided by in the above Gauss algorithm. This is useful for evaluating determinants of large order since it requires only of the order of n^3 operations.

A matrix $\mathbf{A}$ may be viewed as consisting of column or row vectors. The largest number of linearly independent column vectors (which is the same as the largest number of linearly independent row vectors) is called the *rank* of the matrix, $\rho(\mathbf{A})$. A set of vectors $\mathbf{V}_i$ is linearly independent if

$$\sum_i a_i \mathbf{V}_i = \mathbf{0}$$

implies that $a_i = 0$ for $i = 1, 2 \cdots$.

The rank is equal to the order of the largest nonvanishing determinant of the submatrix by deleting rows and columns of the original matrix. Consider the matrices

$$\mathbf{A} = \begin{bmatrix} a_{11} & a_{12} & \cdots & a_{1n} \\ a_{21} & a_{22} & \cdots & a_{2n} \\ & & \vdots & \\ a_{m1} & a_{m2} & \cdots & a_{mn} \end{bmatrix}$$

and

$$\mathbf{B} = \begin{bmatrix} a_{11} & a_{12} & \cdots & a_{1n} & y_1 \\ a_{21} & a_{22} & \cdots & a_{2n} & y_2 \\ & & \vdots & & \\ a_{m1} & a_{m2} & \cdots & a_{mn} & y_m \end{bmatrix}$$

The equations

$$\mathbf{AX} = \mathbf{Y}$$

have a solution if and only if

$$\rho(\mathbf{A}) = \rho(\mathbf{B})$$

in which case the equations are said to be consistent.

If $\rho(\mathbf{A}) < n = m$, that is, $|\mathbf{A}| = 0$, and if the equations are consistent, then the Gauss algorithm will terminate before n steps. That is, the coefficients of all the x_k's will be zero in the remaining $n - \rho(\mathbf{A})$ equations. Therefore, among the x_k's there will be certain ones, $n - \rho(\mathbf{A})$ in number, which may be assigned arbitrary values. Similarly, if $m \neq n$, the Gauss algorithm will yield an equivalent set of $\rho(\mathbf{A})$ equations that has the same solution as the original set. Again $n - \rho(\mathbf{A})$ (possibly zero) of the x_k's may be assigned arbitrary values.

It is not necessary to know beforehand whether the equations are inconsistent. If they are inconsistent, the algorithm will yield an "equation" in which the coefficients of the x_k's on the left side are zero but there is a nonzero combination of the y_k's on the right side. Since the right side of the "equation" may contain accumulated round-off errors, an analysis of the error propagation in the Gauss algorithm may be necessary to determine whether a small right side is caused by inconsistent equations or by round-off errors.

Eigenvectors and Eigenvalues

An eigenvector of the square matrix $\mathbf{A}$ of order n is a nonzero vector $\mathbf{X}$ such that

$$\mathbf{AX} = \lambda\mathbf{X}$$

The scalar λ is called an eigenvalue of $\mathbf{A}$, and $\mathbf{X}$ is called an eigenvector corresponding to or associated with λ. The eigenvalues may be determined from the characteristic equation

$$|\mathbf{A} - \lambda\mathbf{I}| = 0$$

The corresponding eigenvectors $\mathbf{X}$ may then be found by solving

$$(\mathbf{A} - \lambda\mathbf{I})\mathbf{X} = \mathbf{0}$$

The solution may be obtained by the Gauss algorithm. If the eigenvalues λ are distinct, an explicit solution may be obtained by taking a nontrivial row of cofactors from $\mathbf{A} - \lambda_i\mathbf{I}$. This is possible since the rank of $\mathbf{A} - \lambda_i\mathbf{I}$ is $n-1$ and, therefore, there exists a nonvanishing subdeterminant of order $n-1$.

Note that eigenvectors are determined only to within a multiplicative constant.

Further Definitions and Properties

The matrix whose elements are a_{ji}^* is called the conjugate transpose of $\mathbf{A} = [a_{ij}]$; it is denoted by $\mathbf{A}\dagger$. The conjugate transpose of a product is

$$(AB)\dagger = B\dagger A\dagger$$

If $\mathbf{A}=\mathbf{A}^\dagger$ ($\mathbf{A}=-\mathbf{A}^\dagger$), $\mathbf{A}$ is said to be Hermitian (skew-Hermitian). If $\mathbf{A}^\dagger\mathbf{A}=\mathbf{I}$, $\mathbf{A}$ is called unitary. If $\mathbf{A}^\dagger\mathbf{A}=\mathbf{A}\mathbf{A}^\dagger$, $\mathbf{A}$ is called normal. Diagonal, unitary, Hermitian, and skew-Hermitian matrices are special cases of normal matrices. If the matrix $\mathbf{A}$ is real (that is, all entries are real), the terms Hermitian, skew-Hermitian, and unitary are usually replaced by symmetric, skew-symmetric, and unitary, respectively. The eigenvalues of Hermitian, skew-Hermitian, and unitary matrices are real, pure imaginary, and of unit absolute value, respectively.

The inner product of vectors x and y is

$$\mathbf{X}^\dagger\mathbf{Y}=\sum_{i=1}^{n} x_i^* y_i$$

where x_i^* is the complex conjugate of x_i. (A square matrix of order one is considered here as a scalar.)

The length of a vector is given by

$$\|\mathbf{X}\|=(\mathbf{X}^\dagger\mathbf{X})^{1/2}=\left(\sum_{i=1}^{n}|x_i|^2\right)^{1/2}$$

Two vectors are said to be orthogonal if

$$\mathbf{X}^\dagger\mathbf{Y}=0$$

Orthogonality corresponds to perpendicularity in two and three dimensions. Inner products obey the following inequalities

$$|\mathbf{X}^\dagger\mathbf{Y}| \le \|\mathbf{X}\|\|\mathbf{Y}\| \qquad \text{(Schwarz)}$$

$$\|\mathbf{X}+\mathbf{Y}\| \le \|\mathbf{X}\|+\|\mathbf{Y}\| \qquad \text{(Triangle)}$$

Two matrices $\mathbf{A}$ and $\mathbf{B}$ are called similar if there exists a nonsingular matrix $\mathbf{C}$ such that

$$\mathbf{B}=\mathbf{CAC}^{-1}$$

Of particular interest is the case in which matrix $\mathbf{B}$ is diagonal

$$\mathbf{B}=\begin{bmatrix} \lambda_1 & 0 & 0 & \cdots & 0 \\ 0 & \lambda_2 & 0 & \cdots & 0 \\ 0 & 0 & \lambda_3 & \cdots & 0 \\ & & & \vdots & \\ 0 & 0 & 0 & \cdots & \lambda_n \end{bmatrix}$$

The diagonal elements λ_i are the eigenvalues of the matrix $\mathbf{A}$. Not all matrices are similar to diagonal ones. However, if a matrix of order n has n linearly independent eigenvectors, as is the case with normal matrices or when all its eigenvalues are distinct, it can be diagonalized. This may be done as follows: Let

$$\mathbf{A}\begin{bmatrix} x_{1i} \\ x_{2i} \\ \vdots \\ x_{ni} \end{bmatrix}=\lambda_i\begin{bmatrix} x_{1i} \\ x_{2i} \\ \vdots \\ x_{ni} \end{bmatrix}$$

Then

$$\mathbf{A}\begin{bmatrix} x_{11} & x_{12} & \cdots & x_{1n} \\ x_{21} & x_{22} & \cdots & x_{2n} \\ & & \vdots & \\ x_{n1} & x_{n2} & \cdots & x_{nn} \end{bmatrix}$$

$$=\begin{bmatrix} x_{11} & x_{12} & \cdots & x_{1n} \\ x_{21} & x_{22} & \cdots & x_{2n} \\ & \vdots & & \\ x_{n1} & x_{n2} & \cdots & x_{nn} \end{bmatrix}\begin{bmatrix} \lambda_1 & 0 & 0 & \cdots & 0 \\ 0 & \lambda_2 & 0 & \cdots & 0 \\ & & & \vdots & \\ 0 & 0 & 0 & \cdots & \lambda_n \end{bmatrix}$$

The matrix formed from the eigenvectors is nonsingular since the eigenvectors are linearly independent. Hence

$$\mathbf{C}^{-1}=\begin{bmatrix} x_{11} & x_{12} & \cdots & x_{1n} \\ x_{21} & x_{22} & \cdots & x_{2n} \\ & & \vdots & \\ x_{n1} & x_{n2} & \cdots & x_{nn} \end{bmatrix}$$

If a matrix $\mathbf{A}$ is reducible to diagonal form, polynomial functions of $\mathbf{A}$ are readily calculated

$$f(\mathbf{A})=f(\mathbf{C}^{-1}\mathbf{BC})$$

$$=\mathbf{C}^{-1}\begin{bmatrix} f(\lambda_1) & 0 & 0 & \cdots & 0 \\ 0 & f(\lambda_2) & 0 & \cdots & 0 \\ & & & \vdots & \\ 0 & 0 & 0 & \cdots & f(\lambda_n) \end{bmatrix}\mathbf{C}$$

Matrices can always be reduced to the Jordan canonical form, in which the eigenvalues are on the main diagonal and there are ones in certain places just above the main diagonal, and zeros elsewhere. In this form, operations such as polynomial functions are simplified.

Hermitian Forms

A Hermitian form is a polynomial

$$\mathbf{X}^\dagger\mathbf{AX}=\sum_{j=1}^{n}\sum_{i=1}^{n} a_{ij}x_i^* x_j$$

where $\mathbf{A}$ is a Hermitian matrix. Hermitian forms are real-valued and satisfy the inequality

$$\lambda_n\|\mathbf{X}\|^2 \le \mathbf{X}^\dagger\mathbf{AX} \le \lambda_1\|\mathbf{X}\|^2$$

in which λ_1 and λ_n are the largest and smallest eigenvalues of $\mathbf{A}$. By a suitable change of variable

$$\mathbf{Y}=\mathbf{TX}$$

any Hermitian form $\mathbf{X}^\dagger\mathbf{AX}$ may be reduced.

$$\mathbf{X}^\dagger\mathbf{AX}=\mathbf{Y}^\dagger\mathbf{TAT}^\dagger\mathbf{Y}=\sum_{i=1}^{n}\lambda_i|y_i|^2$$

by choosing **T** to be that unitary matrix which diagonalizes **A**.

The maximum value of **X†AX** for all unit vectors **X**, that is, $\|\mathbf{X}\|=1$, is the largest eigenvalue of **A**, say λ_1. A vector yielding this largest value will be a corresponding eigenvector, say $\mathbf{X}_1$. The maximum of **X†AX** overall unit vectors orthogonal to $\mathbf{X}_1$ will be another eigenvalue, say λ_2. A vector yielding λ_2 will be a corresponding eigenvector, say $\mathbf{X}_2$. The process is repeated considering unit vectors orthogonal to both $\mathbf{X}_1$ and $\mathbf{X}_2$. In this way, the eigenvalues of any Hermitian matrix may be found.

VECTOR-ANALYSIS EQUATIONS

Rectangular Coordinates

(In the following, vectors are indicated in bold-faced type.)

Notation:

$$\mathbf{a}=a\hat{\mathbf{a}}$$

a = magnitude of **a**

$\hat{\mathbf{a}}$ = unit vector in direction of **a**

Associative Law: For addition

$$\mathbf{a}+(\mathbf{b}+\mathbf{c})=(\mathbf{a}+\mathbf{b})+\mathbf{c}=\mathbf{a}+\mathbf{b}+\mathbf{c}$$

Commutative Law: For addition

$$\mathbf{a}+\mathbf{b}=\mathbf{b}+\mathbf{a}$$

Scalar or "Dot" Product (Fig. 21):

$$\mathbf{a}\cdot\mathbf{b}=\mathbf{b}\cdot\mathbf{a}$$
$$=ab\cos\theta$$

where θ = angle included by **a** and **b**.

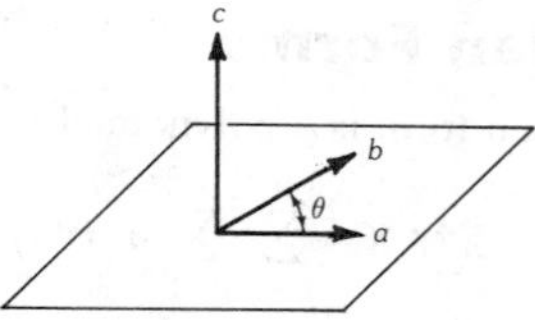

Fig. 21. "Dot" product.

Vector or "Cross" Product:

$$\mathbf{a}\times\mathbf{b}=-\mathbf{b}\times\mathbf{a}$$
$$=ab\sin\theta\hat{\mathbf{c}}$$

where,

θ = smallest angle swept in rotating **a** into **b**,
$\hat{\mathbf{c}}$ = unit vector perpendicular to plane of **a** and **b**, and directed in the sense of travel of a right-hand screw rotating from **a** to **b** through the angle θ.

Distributive Law for Scalar Multiplication:

$$\mathbf{a}\cdot(\mathbf{b}+\mathbf{c})=\mathbf{a}\cdot\mathbf{b}+\mathbf{a}\cdot\mathbf{c}$$

Distributive Law for Vector Mulitplication:

$$\mathbf{a}\times(\mathbf{b}+\mathbf{c})=\mathbf{a}\times\mathbf{b}+\mathbf{a}\times\mathbf{c}$$

Scalar Triple Product:

$$\mathbf{a}\cdot(\mathbf{b}\times\mathbf{c})=(\mathbf{a}\times\mathbf{b})\cdot\mathbf{c}=\mathbf{c}\cdot(\mathbf{a}\times\mathbf{b})=\mathbf{b}\cdot(\mathbf{c}\times\mathbf{a})$$

Vector Triple Product:

$$\mathbf{a}\times(\mathbf{b}\times\mathbf{c})=(\mathbf{a}\cdot\mathbf{c})\mathbf{b}-(\mathbf{a}\cdot\mathbf{b})\mathbf{c}$$
$$(\mathbf{a}\times\mathbf{b})\cdot(\mathbf{c}\times\mathbf{d})=(\mathbf{a}\cdot\mathbf{c})(\mathbf{b}\cdot\mathbf{d})-(\mathbf{a}\cdot\mathbf{d})(\mathbf{b}\cdot\mathbf{c})$$
$$(\mathbf{a}\times\mathbf{b})\times(\mathbf{c}\times\mathbf{d})=(\mathbf{a}\times\mathbf{b}\cdot\mathbf{d})\mathbf{c}-(\mathbf{a}\times\mathbf{b}\cdot\mathbf{c})\mathbf{d}$$

Del Operator:

$$\nabla\equiv\mathbf{i}(\partial/\partial x)+\mathbf{j}(\partial/\partial y)+\mathbf{k}(\partial/\partial z)$$

where **i**, **j**, **k** are unit vectors in the directions of the x, y, z coordinate axes, respectively.

Gradient:

$$\text{grad}\phi=\nabla\phi$$
$$=\mathbf{i}(\partial\phi/\partial x)+\mathbf{j}(\partial\phi/\partial y)+\mathbf{k}(\partial\phi/\partial z),$$

in Cartesian coordinates

$$\text{grad}(\phi+\psi)=\text{grad}\phi+\text{grad}\psi$$
$$\text{grad}(\phi\psi)=\phi\,\text{grad}\psi+\psi\,\text{grad}\phi$$

Divergence:

$$\text{div}\mathbf{a}=\nabla\cdot\mathbf{a}=(\partial a_x/\partial x)+(\partial a_y/\partial y)+(\partial a_z/\partial z),$$

in Cartesian coordinates

where a_x, a_y, a_z are components of **a** in the directions of the x, y, z coordinate axes, respectively.

$$\text{div}(\mathbf{a}+\mathbf{b})=\text{div}\mathbf{a}+\text{div}\mathbf{b}$$
$$\text{div}(\phi\mathbf{a})=\phi\,\text{div}\mathbf{a}+\mathbf{a}\cdot\text{grad}\phi$$

Curl:

$$\text{curl}\mathbf{a}=\nabla\times\mathbf{a}$$
$$=\mathbf{i}\left(\frac{\partial a_z}{\partial y}-\frac{\partial a_y}{\partial z}\right)+\mathbf{j}\left(\frac{\partial a_z}{\partial z}-\frac{\partial a_z}{\partial x}\right)$$
$$+\mathbf{k}\left(\frac{\partial a_y}{\partial x}-\frac{\partial a_x}{\partial y}\right)$$

$$=\begin{vmatrix}\mathbf{i} & \mathbf{j} & \mathbf{k}\\ \partial/\partial x & \partial/\partial y & \partial/\partial z\\ a_x & a_y & a_z\end{vmatrix},\quad \text{in Cartesian coordinates}$$

$\text{curl}(\mathbf{a}+\mathbf{b}) = \text{curl}\mathbf{a} + \text{curl}\mathbf{b}$

$\text{curl}(\phi\mathbf{a}) = \text{grad}\phi \times \mathbf{a} + \phi\text{curl}\mathbf{a}$

$\text{curl grad}\phi = 0$

$\text{div curl}\mathbf{a} = 0$

$\text{div}(\mathbf{a} \times \mathbf{b}) = \mathbf{b} \cdot \text{curl}\mathbf{a} - \mathbf{a} \cdot \text{curl}\mathbf{b}$

$\text{Laplacian} \equiv \nabla^2 = \nabla \cdot \nabla$

$$\nabla^2\phi = (\partial^2\phi/\partial x^2) + (\partial^2\phi/\partial y^2) + (\partial^2\phi/\partial z^2),$$

in Cartesian coordinates

$$\text{curl curl}\mathbf{a} = \text{grad div}\mathbf{a} - (\mathbf{i}\nabla^2 a_x + \mathbf{j}\nabla^2 a_y + \mathbf{k}\nabla^2 a_z)$$
$$= \nabla(\nabla \cdot \mathbf{a}) - \nabla^2\mathbf{a}$$

Directional Derivative: Derivative of ϕ in the direction of **s**

$$d\phi/ds = \hat{\mathbf{s}} \cdot \nabla\phi$$

Integral Relations: In the following equations, τ is a volume bounded by a closed surface, S. The unit vector **n** is normal to the surface and is directed outward. The symbol dS represents an element of surface area. If the surface is represented by $z = f(x,y)$ then

$$dS = [1 + (\partial f/\partial x)^2 + (\partial f/\partial y)^2]^{1/2} dx\, dy$$

$$\int_\tau \nabla\phi\, d\tau = \int_S \phi\mathbf{n}\, dS$$

$$\int_\tau \nabla \cdot \mathbf{a}\, d\tau = \int_S \mathbf{a} \cdot \mathbf{n}\, dS \text{ (Gauss' theorem)}$$

$$\int_\tau \nabla \times \mathbf{a}\, d\tau = \int_S \mathbf{n} \times \mathbf{a}\, dS$$

$$\int_\tau (\psi\nabla^2\phi - \phi\nabla^2\psi)d\tau; \int_S [\psi(\partial\phi/\partial n) - \phi(\partial\psi/\partial n)]\, dS$$

where $\partial/\partial n$ is the derivative in the direction of **n** (Green's theorem).

In the two following equations, S is an open surface bounded by a contour C, with distance along C represented by s.

$$\int_S \mathbf{n} \times \nabla\phi\, dS = \int_C \phi\, ds$$

$$\int_S (\nabla \times \mathbf{a}) \cdot \mathbf{n}\, dS = \int_C \mathbf{a} \cdot d\mathbf{s} \text{ (Stokes' theorem)}$$

where,

$\mathbf{s} = s\hat{\mathbf{s}}$,

$\hat{\mathbf{s}}$ is the unit tangent vector along C.

Gradient, Divergence, Curl, and Laplacian in Coordinate Systems Other Than Rectangular

Cylindrical Coordinates: (ρ, ϕ, z), unit vectors $\hat{\boldsymbol{\rho}}$, $\hat{\boldsymbol{\phi}}$, $\mathbf{k}$, respectively

$$\text{grad}\psi = \nabla\psi = (\partial\psi/\partial\rho)\hat{\boldsymbol{\rho}} + \rho^{-1}(\partial\psi/\partial\phi)\hat{\boldsymbol{\phi}} + (\partial\psi/\partial z)\mathbf{k}$$

Let $\mathbf{a} = a_\rho\hat{\boldsymbol{\rho}} + a_\phi\hat{\boldsymbol{\phi}} + a_z\mathbf{k}$. Then

$$\text{div}\mathbf{a} = \nabla \cdot \mathbf{a} = \rho^{-1}(\partial/\partial\rho)(\rho a_\rho) + \rho^{-1}(\partial a_\phi/\partial\phi) + (\partial a_z/\partial z)$$

$$\text{curl}\mathbf{a} = \nabla \times \mathbf{a} = [\rho^{-1}(\partial a_z/\partial\phi) - (\partial a_\phi/\partial z)]\hat{\boldsymbol{\rho}} + [(\partial a_\rho/\partial z) - (\partial a_z/\partial\rho)]\hat{\boldsymbol{\phi}} + [\rho^{-1}(\partial/\partial\rho)(\rho a_\phi) - \rho^{-1}(\partial a_\rho/\partial\phi)]\mathbf{k}$$

$$\nabla^2\psi = \rho^{-1}(\partial/\partial\rho)[\rho(\partial\psi/\partial\rho)] + \rho^{-2}(\partial^2\psi/\partial\phi^2) + (\partial^2\psi/\partial z^2)$$

Spherical Coordinates: (r, θ, ϕ), unit vectors $\hat{\mathbf{r}}$, $\hat{\boldsymbol{\theta}}$, $\hat{\boldsymbol{\phi}}$

r = distance to origin

θ = polar angle

ϕ = azimuthal angle

$$\text{grad}\psi = \nabla\psi = (\partial\psi/\partial r)\hat{\mathbf{r}} + r^{-1}(\partial\psi/\partial\theta)\hat{\boldsymbol{\theta}} + (r\sin\theta)^{-1}(\partial\psi/\partial\phi)\hat{\boldsymbol{\phi}}$$

Let $\mathbf{a} = a_r\hat{\mathbf{r}} + a_\theta\hat{\boldsymbol{\theta}} + a_\phi\hat{\boldsymbol{\phi}}$. Then

$$\text{div}\mathbf{a} = \nabla \cdot \mathbf{a} = r^{-2}(\partial/\partial r)(r^2 a_r) + (r\sin\theta)^{-1}(\partial/\partial\theta)(a_\theta\sin\theta) + (r\sin\theta)^{-1}(\partial a_\phi/\partial\phi)$$

$$\text{curl}\mathbf{a} = \nabla \times \mathbf{a} = (r\sin\theta)^{-1}[(\partial/\partial\theta)(a_\phi\sin\theta) - (\partial a_\theta/\partial\phi)]\hat{\mathbf{r}}$$
$$+ r^{-1}[(\sin\theta)^{-1}(\partial a_r/\partial\phi) - (\partial/\partial r)(r a_\phi)]\hat{\boldsymbol{\theta}} + r^{-1}[(\partial/\partial r)(r a_\theta) - (\partial a_r/\partial\theta)]\hat{\boldsymbol{\phi}}$$

$$\nabla^2\psi = r^{-2}(\partial/\partial r)[r^2(\partial\psi/\partial r)] + (r^2\sin\theta)^{-1}(\partial/\partial\theta)[\sin\theta(\partial\psi/\partial\theta)] + (r^2\sin^2\theta)^{-1}(\partial^2\psi/\partial\phi^2)$$

Orthogonal Curvilinear Coordinates:

Coordinates: u_1, u_2, u_3

Metric coefficients: h_1, h_2, h_3 $(ds^2 = h_1^2 du_1^2 + h_2^2 du_2^2 + h_3^2 du_3^2)$

Unit vectors: $\mathbf{i}_1, \mathbf{i}_2, \mathbf{i}_3$ $(ds=\mathbf{i}_1h_1du_1+\mathbf{i}_2h_2du_2+\mathbf{i}_3h_3du_3)$

$$\text{grad}\psi=\nabla\psi=h_1^{-1}(\partial\psi/\partial u_1)\mathbf{i}_1+h_2^{-1}(\partial\psi/\partial u_2)\mathbf{i}_2+h_3^{-1}(\partial\psi/\partial u_3)\mathbf{i}_3$$

$$\text{div}\mathbf{a}=\nabla\cdot\mathbf{a}=(h_1h_2h_3)^{-1}[(\partial/\partial u_1)(h_2h_3a_1)+(\partial/\partial u_2)(h_3h_1a_2)+(\partial/\partial u_3)(h_1h_2a_3)]$$

$$\text{curl}\mathbf{a}=\nabla\times\mathbf{a}=(h_2h_3)^{-1}[(\partial/\partial u_2)(h_3a_3)-(\partial/\partial u_3)(h_2a_2)]\mathbf{i}_1+(h_3h_1)^{-1}[(\partial/\partial u_3)(h_1a_1)-(\partial/\partial u_1)(h_3a_3)]\mathbf{i}_2+(h_1h_2)^{-1}[(\partial/\partial u_1)(h_2a_2)-(\partial/\partial u_2)(h_1a_1)]\mathbf{i}_3$$

$$=(h_1h_2h_3)^{-1}\begin{vmatrix} h_1\mathbf{i}_1 & h_2\mathbf{i}_2 & h_3\mathbf{i}_3 \\ \partial/\partial u_1 & \partial/\partial u_2 & \partial/\partial u_3 \\ h_1a_1 & h_2a_2 & h_3a_3 \end{vmatrix}$$

$$\nabla^2\psi=(h_1h_2h_3)^{-1}\left[\frac{\partial}{\partial u_1}\left(\frac{h_2h_3}{h_1}\frac{\partial\psi}{\partial u_1}\right)+\frac{\partial}{\partial u_2}\left(\frac{h_3h_1}{h_2}\frac{\partial\psi}{\partial u_2}\right)+\frac{\partial}{\partial u_3}\left(\frac{h_1h_2}{h_3}\frac{\partial\psi}{\partial u_3}\right)\right]$$

Space Curves

A curve may be represented vectorially as $\mathbf{r}=\mathbf{r}(s)$. See Fig. 22.

A unit tangent $\mathbf{t}$ is then given by

$$\mathbf{t}=d\mathbf{r}/ds$$

The principal normal $\mathbf{n}$ is given by

$$\mathbf{n}=(1/k)(d\mathbf{t}/ds)$$

where k is the curvature. The radius of curvature $R=1/k$. For a plane curve $y=f(x)$, the curvature may be computed from

$$k=|y''|/[1+(y')^2]^{3/2}$$

The binormal is defined by

$$\mathbf{b}=\mathbf{t}\times\mathbf{n}$$

These vectors satisfy Frenet's equations

$$d\mathbf{n}/ds=-k\mathbf{t}+\tau\mathbf{b}$$
$$d\mathbf{b}/ds=-\tau\mathbf{n}$$

where τ is the torsion. The torsion is zero everywhere if and only if the curve lies in a plane.

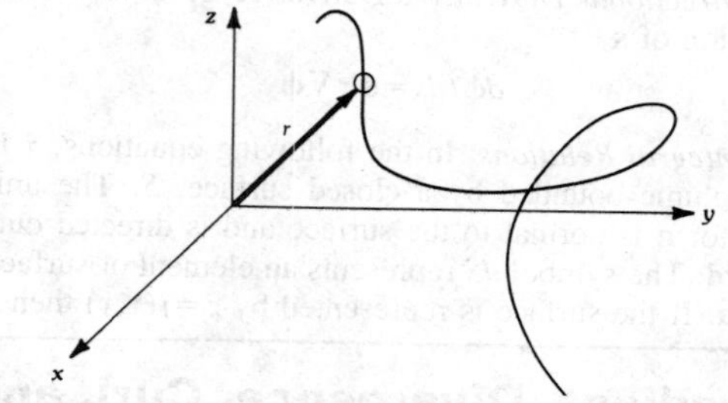

Fig. 22. Space curve.

LAPLACE TRANSFORM

The Laplace transform of a function $f(t)$ is defined by the expression

$$F(p)=\int_0^\infty f(t)e^{-pt}dt$$

If this integral converges for some $p=p_0$, real or complex, then it will converge for all p such that $\text{Re}p>\text{Re}p_0$.

The inverse transform may be found by

$$f(t)=(j2\pi)^{-1}\int_{c-j\infty}^{c+j\infty}F(z)e^{tz}dz,\ t>0$$

where there are no singularities to the right of the path of integration.

TABLE OF LAPLACE TRANSFORMS

General Equations

	Function	Transform*
Shifting theorem	$f(t-a),\quad f(t)=0,\quad t<0$	$e^{-ap}F(p),\quad a>0$
Convolution	$\int_0^t f_1(\lambda)f_2(t-\lambda)d\lambda$	$F_1(p)F_2(p)$
Linearity	$a_1f_1(t)+a_2f_2(t),\quad (a_1,\ a_2\ \text{const})$	$a_1F_1(p)+a_2F_2(p)$
Derivative	$df(t)/dt$	$-f(0)+pF(p)$

Continued on next page.

General Equations (cont)

	Function	Transform*
Integral	$\int f(t)dt$	$p^{-1}\left[\left[\int f(t)dt\right]_{t=0} + [F(p)/p]\right], \quad \mathrm{Re}p>0$
Periodic function	$f(t)=f(t+r)$	$\int_0^r f(\lambda)e^{-p\lambda}d\lambda/(1-e^{-pr}), \quad r>0$
	$f(t)=-f(t+r)$	$\int_0^r f(\lambda)e^{-p\lambda}d\lambda/(1+e^{-pr}), \quad r>0$
	$f(at), \quad a>0$	$F(p/a)/a$
	$e^{at}f(t)$	$F(p-a), \quad \mathrm{Re}p>\mathrm{Re}a$
	$t^n f(t)$	$(-1)^n[d^nF(p)/dp^n]$
Final-value theorem	$f(\infty)$	$\lim_{p\to 0} pF(p)$
Initial-value theorem	$f(0+)$	$\lim_{p\to\infty} pF(p)$

* $F(p)$ denotes the Laplace transform of $f(t)$.

Miscellaneous Functions*

	Function	Transform
Step	$u(t-a)=0, \quad 0\le t<a$ $=1, \quad t\ge a$	e^{-ap}/p
Impulse	$\delta(t)$	1
	$t^a, \quad \mathrm{Re}a>-1$	$\Gamma(a+1)/p^{a+1}$
	e^{at}	$1/(p-a), \quad \mathrm{Re}p>\mathrm{Re}a$
	$t^a e^{bt}, \quad \mathrm{Re}a>-1$	$\Gamma(a+1)/(p-b)^{a+1}, \quad \mathrm{Re}p>\mathrm{Re}b$
	$\cos at$	$p/(p^2+a^2)$ } $\mathrm{Re}p>\lvert\mathrm{Im}a\rvert$
	$\sin at$	$a/(p^2+a^2)$ } $\mathrm{Re}p>\lvert\mathrm{Im}a\rvert$
	$\cosh at$	$p/(p^2-a^2)$ } $\mathrm{Re}p>\lvert\mathrm{Re}a\rvert$
	$\sinh at$	$1/(p^2-a^2)$ } $\mathrm{Re}p>\lvert\mathrm{Re}a\rvert$
	$\ln t$	$-(\gamma+\ln p)/p, \quad \gamma$ is Euler's constant $=0.57722$
	$1/(t+a), \quad a>0$	$e^{ap}E_1(ap)$
	e^{-at^2}	$\frac{1}{2}(\pi/a)^{1/2}e^{p^2/4a}\,\mathrm{erfc}[p/2(a)^{1/2}]$
Bessel function $J_\nu(at), \quad \mathrm{Re}\nu>-1$		$r^{-1}[(r-p)/a]^\nu, \quad r=(p^2+a^2)^{1/2}, \quad \mathrm{Re}p>\lvert\mathrm{Re}a\rvert$
Bessel function $I_\nu(at), \quad \mathrm{Re}\nu>-1$		$R^{-1}[(R-p)/a]^\nu, \quad R=(p^2-a^2)^{1/2}, \quad \mathrm{Re}p>\lvert\mathrm{Re}a\rvert$

* For an extensive listing, refer to A. Erdéyli, ed., *Tables of Integral Transforms*, Vol. 1, Bateman Manuscript Project, New York: McGraw-Hill Book Co., 1954.

Inverse Transforms*

Transform	Function
1	$\delta(t)$
$1/(p+a)$	e^{-at}
$1/(p+a)^{\nu}$, $\mathrm{Re}\nu>0$	$t^{\nu-1}e^{-at}/\Gamma(\nu)$
$1/[(p+a)\,(p+b)]$	$(e^{-at}-e^{-bt})/(b-a)$
$p/[(p+a)\,(p+b)]$	$(ae^{-at}-be^{-bt})/(a-b)$
$1/(p^2+a^2)$	$a^{-1}\sin at$
$1/(p^2-a^2)$	$a^{-1}\sinh at$
$p/(p^2+a^2)$	$\cos at$
$p/(p^2-a^2)$	$\cosh at$
$1/(p^2+a^2)^{1/2}$	$J_o(at)$
e^{-ap}/p	$u(t-a)$
e^{-ap}/p^{ν}, $\mathrm{Re}\nu>0$	$(t-a)^{\nu-1}u(t-a)/\Gamma(\nu)$
$(1/p)e^{-a/p}$	$J_0[2(at)^{1/2}]$
$(1/p^{\nu})e^{-a/p}$ } $\mathrm{Re}\nu>0$	$(t/a)^{(\nu-1)/2}J_{\nu-1}[2(at)^{1/2}]$
$(1/p^{\nu})e^{a/p}$ } $\mathrm{Re}\nu>0$	$(t/a)^{(\nu-1)/2}I_{\nu-1}[2(at)^{1/2}]$
$(1/p)\ln p$	$-\gamma-\ln t$, $\gamma=0.57722$

* Refer to A. Erdéyli, ed., *Tables of Integral Transforms*, Vol. 1, Bateman Manuscript Project, New York: McGraw-Hill Book Co., 1954.

SELECTED FUNCTIONS

Exponential Integrals

Definitions:

$$E_1(z)=\int_z^{\infty}(e^{-t}/t)dt,\ |\arg z|<\pi$$

in which the path of integration does not cross the negative t-axis and also excludes the origin. For $\arg z=\pi$ the following function is used:

$$E_i(x)=-\mathrm{pv}\int_{-x}^{\infty}(e^{-t}/t)dt,\ x>0$$

where pv stands for the Cauchy principal value.

$$E_1(-x\pm j0)=Ei(x)\mp j\pi$$

Series Expansions:

$$E_1(z)=-\gamma-\ln z-\sum_{n=1}^{\infty}[(-1)^n z^n/(n\cdot n!)],\quad |\arg z|<\pi$$

$$Ei(x)=\gamma+\ln x+\sum_{n=1}^{\infty}x^n/(n\cdot n!)$$

in which γ is Euler's constant, $\gamma=0.57722$.

Asymptotic Expansions:

$$E_1(z)\sim(e^{-z}/z)[1-z^{-1}+(2!/z^2)-(3!/z^3)+\cdots],\quad |\arg z|<\pi$$

$$Ei(x)\sim(e^x/x)[1+x^{-1}+(2!/x^2)+(3!/x^3)+\cdots]$$

Cosine and Sine Integrals

Definitions:

$$Si(z)=\int_0^z(\sin t/t)dt$$

$$Ci(z)=\gamma+\ln z+\int_0^z[(\cos t-1)/t]dt,\ |\arg z|<\pi$$

where γ is Euler's constant, $\gamma=0.57722$.

Series Expansions:

$$Si(z)=\sum_{n=0}^{\infty}\frac{(-1)^n z^{2n+1}}{(2n+1)(2n+1)!}$$

$$Ci(z)=\gamma+\ln z+\sum_{n=1}^{\infty}\frac{(-1)^n z^{2n}}{2n(2n)!}$$

Asymptotic Expansions:

$$Si(z)\sim\tfrac{1}{2}\pi-\left(1-\frac{2!}{z^2}+\frac{4!}{z^4}-\frac{6!}{z^6}+\cdots\right)\frac{\cos z}{z}-\left(1-\frac{3!}{z^2}+\frac{5!}{z^4}-\frac{7!}{z^6}+\cdots\right)\frac{\sin z}{z^2}$$

$$Ci(z)\sim\left(1-\frac{2!}{z^2}+\frac{4!}{z^4}-\frac{6!}{z^6}+\cdots\right)\frac{\sin z}{z}-\left(1-\frac{3!}{z^2}+\frac{5!}{z^4}-\frac{7!}{z^6}+\cdots\right)\frac{\cos z}{z^2}$$

Gamma Function

Definition:

$$\Gamma(z)=\int_0^{\infty}t^{z-1}e^{-t}dt$$

$$\Gamma(-z)=-\pi/[\Gamma(z+1)\sin\pi z],\ \mathrm{Re}z>0$$

The function $\Gamma(z)$ is an analytic function everywhere except at the negative integers.

Identities:

$$z\Gamma(z)=\Gamma(z+1)$$

$$\Gamma(\tfrac{1}{2}+z)\Gamma(\tfrac{1}{2}-z)=\pi\sec\pi z$$

Special Values:

$$\Gamma(z)=(z-1)!\quad z=1,2,3,\cdots$$

$$\Gamma(z+\tfrac{1}{2})=[(2z)!/2^{2z}z!](\pi)^{1/2}$$

n	1	2	3	4	5	6	7	8	9	10
$\Gamma(n)=(n-1)!$	1	1	2	6	24	120	720	5040	40 320	362 880

Asymptotic Expansion (Stirling's Equation):

$$\Gamma(z)\sim e^{-z}z^{z-1/2}(2\pi)^{1/2}[1+(1/12z)+(1/288z^2)+\cdots]$$

$$\text{Note } z!=z\Gamma(z)\sim e^{-z}z^{z+1/2}(2\pi)^{1/2}[1+(1/12z)+(1/288z^2)+\cdots]$$

Psi and Polygamma Functions

Definitions:

$$\psi^{(n)}(z)=\frac{d^{n+1}}{dz^{n+1}}[\ln\Gamma(z)]\quad n=0,1,2,\cdots$$

$\psi^{(0)}(z)=\psi(z)$ is known as the psi function, and $\psi^{(n)}(z)$ as the polygamma function of order n, $n=1,2,\cdots$

Special Values:

$$\psi^{(n)}(1)=(-1)^{n+1}n!\zeta(n+1),\quad n=1,2,\cdots$$

$$\psi(1)=-\gamma$$

$$\psi^{(n)}(\tfrac{1}{2})=(-1)^{n+1}n!(2^{n+1}-1)\zeta(n+1),\quad n=1,2,\cdots$$

$$\psi(\tfrac{1}{2})=-\gamma-2\ln 2$$

in which $\zeta(n+1)$ is the Riemann zeta function of $n+1$.

Series Expansions:

$$\psi^{(n)}(1+z)=\sum_{k=0}^{\infty}\frac{(-1)^{n+k+1}(n+k)!\zeta(n+k+1)z^k}{k!},\quad |z|<1,\ n=1,2,\cdots$$

$$\psi(1+z)=-\gamma-\sum_{k=1}^{\infty}(-1)^k\zeta(k+1)z^k,\ |z|<1$$

Asymptotic Expansions:

$$\psi^{(n)}(z)\sim[(-1)^n(n-1)!/z^n][1+(n/2z)+\cdots],\quad \text{as } z\to\infty,\ |\arg z|<\pi,\ n=1,2,\cdots$$

$$\psi(z)\sim\ln z-(1/2z)+\cdots$$

Error Function

Definitions:

$$\text{erf}z=(2/\pi^{1/2})\int_0^z\exp(-t^2)\,dt$$

$$\text{erfc}z=(2/\pi^{1/2})\int_z^{\infty}\exp(-t^2)\,dt=1-\text{erf}z$$

The path of integration for large t must remain within $|\arg t|<\pi/4$ in the latter integral.

Series Expansion:

$$\text{erf}z=(2/\pi^{1/2})\sum_{n=0}^{\infty}[(-1)^nz^{2n+1}/n!(2n+1)],\ |z|<\infty$$

Asymptotic Expansion:

$$\text{erf}z\sim1-[\exp(-z^2)/z\pi^{1/2}]\times[1-(1/2z^2)+(3/4z^4)-\cdots],\quad \text{as } z\to\infty,\ |\arg z|<3\pi/4$$

Inequality:

$$\frac{(2/\pi^{1/2})\exp(-x^2)}{x+(x^2+2)^{1/2}}<\text{erfc}x\le\frac{(2/\pi^{1/2})\exp(-x^2)}{x+[x^2+(4/\pi)]^{1/2}},\ x\ge0$$

Derivatives:

$$(d^{(n+1)}/dz^{(n+1)})\text{erf}z=(-1)^n(2/\pi^{1/2})H_n(z)e^{-z^2},\ n=0,1,2,\cdots$$

where $H_n(z)$ is the Hermite polynomial of order n.

Relation to Gaussian Distribution:

$$\text{Prob}(X\le x)=[\sigma(2\pi)^{1/2}]^{-1}\int_{-\infty}^{x}\exp\left(-\frac{(t-\mu)^2}{2\sigma^2}\right)dt=\tfrac{1}{2}\left[1+\text{erf}\left(\frac{x-\mu}{\sigma\sqrt{2}}\right)\right]$$

Fresnel Integrals

Definitions:

$$C(z)=\int_0^z\cos[(\pi/2)t^2]\,dt$$

$$S(z)=\int_0^z\sin[(\pi/2)t^2]\,dt$$

Series Expansions:

$$C(z)=\sum_{n=0}^{\infty}\frac{(-1)^n(\pi/2)^{2n}}{(2n)!(4n+1)}z^{4n+1}$$

$$S(z)=\sum_{n=0}^{\infty}\frac{(-1)^n(\pi/2)^{2n+1}}{(2n+1)!(4n+3)}z^{4n+3},\ |z|<\infty$$

Asymptotic Expansions:

$$C(z) \sim \tfrac{1}{2} + \left(1 - \frac{1\cdot 3}{(\pi z^2)^2} + \frac{1\cdot 3\cdot 5\cdot 7}{(\pi z^2)^4} - \cdots\right)\frac{\sin\frac{1}{2}\pi z^2}{\pi z} - \left[(1/\pi z^2) - \frac{1\cdot 3\cdot 5}{(\pi z^2)^3} + \frac{1\cdot 3\cdot 5\cdot 7\cdot 9}{(\pi z^2)^5} - \cdots\right]\frac{\cos\frac{1}{2}\pi z^2}{\pi z}$$

$$S(z) \sim \tfrac{1}{2} - \left[1 - \frac{1\cdot 3}{(\pi z^2)^2} + \frac{1\cdot 3\cdot 5\cdot 7}{(\pi z^2)^4} - \cdots\right]\frac{\cos\frac{1}{2}\pi z^2}{\pi z} - \left[(1/\pi z^2) - \frac{1\cdot 3\cdot 5}{(\pi z^2)^3} + \frac{1\cdot 3\cdot 5\cdot 7\cdot 9}{(\pi z^2)^5} - \cdots\right]\frac{\sin\frac{1}{2}\pi z^2}{\pi z}$$

Elliptic Integrals

First Kind:

$$F(\phi, k) = \int_0^{\phi} \frac{d\theta}{(1 - k^2\sin^2\theta)^{1/2}}$$

Second Kind:

$$E(\phi, k) = \int_0^{\phi} (1 - k^2\sin^2\theta)^{1/2} d\theta$$

If $\phi = \pi/2$, the elliptic integrals are said to be complete, and they are denoted by K or $K(k)$ and E or $E(k)$.

Bessel Functions*

Refer to Fig. 23.

Definitions: Bessel functions are solutions to Bessel's differential equation

$$z^2(d^2w/dz^2) + z(dw/dz) + (z^2 - \nu^2)w = 0$$

They are divided into

First kind

$$J_\nu(z) = (z/2)^\nu \sum_{k=0}^{\infty} \frac{(-1)^k (z/2)^{2k}}{k!\,\Gamma(k+\nu+1)}$$

Second kind

$$Y_\nu(z) = \frac{J_\nu(z)\cos\nu\pi - J_{-\nu}(z)}{\sin\pi\nu}, \quad \nu \text{ not an integer}$$

$$Y_n(z) = \lim_{\nu\to n} Y_\nu(z), \quad n \text{ integral}$$

Third kind

$$H_\nu^{(1)}(z) = J_\nu(z) + jY_\nu(z)$$

$$H_\nu^{(2)}(z) = J_\nu(z) - jY_\nu(z)$$

The second and third kinds are sometimes called Neumann and Hankel functions, respectively.

The modified Bessel functions are solutions to Bessel's differential equation with z replaced by jz.

* For an extensive treatment of Bessel functions, see G. N. Watson, *A Treatise on the Theory of Bessel Functions*, New York: Cambridge University Press, 1943.

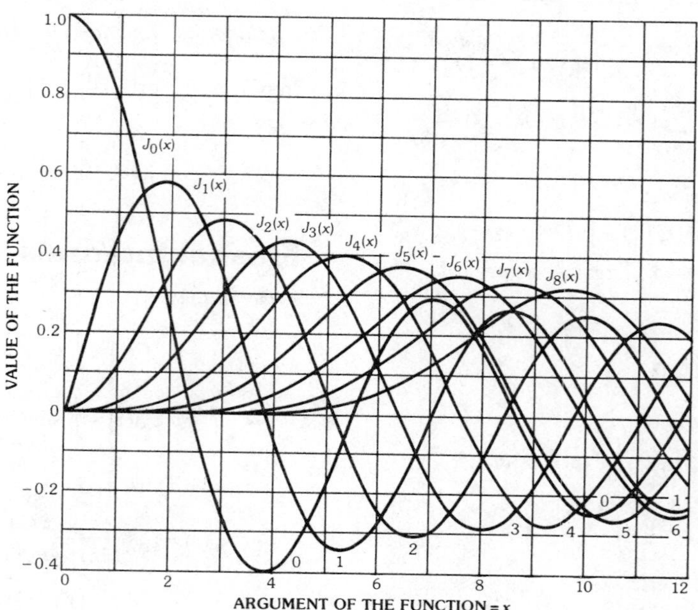

Fig. 23. Bessel functions for the first eight orders.

$$I_\nu(z) = \exp(-j\pi\nu/2) J_\nu[z\exp(j\pi/2)], \quad -\pi < \arg z \le \pi/2$$
$$= \exp(j3\pi\nu/2) J_\nu[z\exp(-j3\pi/2)], \quad \pi/2 < \arg z \le \pi$$
$$K_\nu(z) = (j\pi/2)\exp(j\pi\nu/2) H_\nu^{(1)}[z\exp(j\pi/2)], \quad -\pi < \arg z \le \pi/2$$
$$= -(j\pi/2)\exp(-j\pi\nu/2) \times H_\nu^{(2)}[z\exp(-j\pi/2)], \quad -\pi/2 < \arg z \le \pi$$

Integral Representations:

$$J_\nu(z) = \frac{2(z/2)^\nu}{\Gamma(\nu+\frac{1}{2})\pi^{1/2}} \int_0^{\pi/2} \cos(z\sin\phi)(\cos\phi)^{2\nu} d\phi, \quad \mathrm{Re}\,\nu > -\tfrac{1}{2}$$
$$H_\nu^{(1)}(z) = -(j/\pi)e^{-j\pi\nu/2} \int_{-\infty}^{\infty} \exp(jz\cosh t - \nu t)dt, \quad 0 < \arg z < \pi$$
$$H_\nu^{(2)}(z) = (j/\pi)e^{j\pi\nu/2} \int_{-\infty}^{\infty} \exp(-jz\cosh t - \nu t)dt, \quad -\pi < \arg z < 0$$

Asymptotic Expressions: For $|z| \to \infty$

$$J_\nu(z) \sim [(2/\pi z)]^{1/2} \cos(z - \nu\pi/2 - \pi/4), \quad |\arg z| < \pi$$
$$Y_\nu(z) \sim [(2/\pi z)]^{1/2} \sin(z - \nu\pi/2 - \pi/4), \quad |\arg z| < \pi$$
$$H_\nu^{(1)}(z) \sim [(2/\pi z)]^{1/2} \exp[j(z - \nu\pi/2 - \pi/4)], \quad -\pi < \arg z < 2\pi$$
$$H_\nu^{(2)}(z) \sim [(2/\pi z)]^{1/2} \exp[-j(z - \nu\pi/2 - \pi/4)], \quad -2\pi < \arg z < \pi$$

Recurrence Relations:

$$C_{\nu-1}(z) + C_{\nu+1}(z) = (2\nu/z)C_\nu(z)$$
$$C_{\nu-1}(z) - C_{\nu+1}(z) = 2C_\nu'(z)$$

where C denotes J, Y, $H^{(1)}$ or $H^{(2)}$

Series Containing Bessel Functions:

$$\exp(-ju\sin x) = \sum_{n=-\infty}^{\infty} J_n(u)\exp(-jnx)$$
$$\cos(u\sin x) = J_0(u) + 2\sum_{n=1}^{\infty} J_{2n}(u)\cos 2nx$$
$$\sin(u\sin x) = 2\sum_{n=1}^{\infty} J_{2n-1}(u)\sin(2n-1)x$$
$$\cos(u\cos x) = J_0(u) + 2\sum_{n=1}^{\infty} (-1)^n J_{2n}(u)\cos 2nx$$
$$\sin(u\cos x) = 2\sum_{n=1}^{\infty} (-1)^{n+1} J_{2n-1}(u) \times \cos(2n-1)x$$
$$1 = \sum_{n=-\infty}^{\infty} J_n^2(x)$$

Orthogonal Polynomials

Any set of polynomials $\{f_n(x)\}$ with the property

$$\int_a^b w(x)f_n(x)f_m(x)dx = 0, \quad \text{for } m \ne n$$
$$= h_n, \quad \text{for } m = n$$

is called a set of orthogonal polynomials on the interval (a, b) with respect to the weight function $w(x)$. These functions occur in the Gauss quadrature equations among other places. Chebishev polynomials are involved in the theory of the Chebishev filter; Hermite polynomials arise in the refinements of the central limit theorem, the so-called Edgeworth series, etc. The important properties are summarized in Table 1.

NUMERICAL ANALYSIS

Algorithms for Solving *F*(*x*) = 0

Bisection Method and Regula Falsi (*Rule of False Position*): First determine x_1 and x_2 such that $F(x_1)F(x_2) < 0$, i.e., x_1 and x_2 are points at which the function has opposite signs.

Bisection Method: Calculate

$$x_3 = (x_1 + x_2)/2$$

Regula Falsi: Calculate

$$x_3 = [x_1F(x_2) - x_2F(x_1)]/[F(x_2) - F(x_1)]$$

To obtain the next approximation, take x_3 and x_i, $i = 1$ or 2, such that $F(x_3)F(x_i) < 0$, and repeat the procedure.

Newton-Raphson: Take some initial value x_1 and calculate successively

$$x_{n+1} = x_n - [F(x_n)/F'(x_n)], \quad n = 1, 2, 3 \cdots$$

This method may not converge. When it converges, the rate of convergence is generally faster than the bisection method or the regula falsi.

Algorithm for Solving *F*(*x*, *y*) = *G*(*x*, *y*) = 0

The following is an extension of the Newton-Raphson method described above. Take some initial values x_1 and y_1 and calculate successively

TABLE 1. PROPERTIES OF ORTHOGONAL POLYNOMIALS

$f_n(x)$	Name	a	b	$w(x)$	h_n	Explicit Expression*
$T_n(x)$	Chebishev	-1	$+1$	$(1-x^2)^{-1/2}$	$\frac{1}{2}\pi,\ n\neq 0;\ \pi,\ n=0$	$\frac{1}{2}n\sum_{m=0}^{[n/2]}(-1)^m\frac{(n-m-1)!}{m!(n-2m)!}(2x)^{n-2m}$
$H_n(x)$	Hermite	$-\infty$	∞	e^{-x^2}	$2^n n!\pi^{1/2}$	$n!\sum_{m=0}^{[n/2]}(-1)^m\frac{(2x)^{n-2m}}{m!(n-2m)!}$
$L_n(x)$	Laguerre	0	∞	e^{-x}	1	$\sum_{m=0}^{n}(-1)^m\binom{n}{m}\frac{x^m}{m!}$
$P_n(x)$	Legendre	-1	1	1	$2/(2n+1)$	$2^{-n}\sum_{m=0}^{[n/2]}(-1)^m\binom{n}{m}\binom{2n-2m}{n}x^{n-2m}$

* $[n/2]$ denotes the largest integer less than or equal to $n/2$.

$$x_{n+1}=x_n+\left[\left(\frac{\partial F}{\partial y}G-F\frac{\partial G}{\partial y}\right)\Big/\left(\frac{\partial F}{\partial x}\frac{\partial G}{\partial y}-\frac{\partial F}{\partial y}\frac{\partial G}{\partial x}\right)\right]\Bigg|_{(x,y)=(x_n,y_n)}$$

$$y_{n+1}=y_n+\left[\left(F\frac{\partial G}{\partial x}-\frac{\partial F}{\partial x}G\right)\Big/\left(\frac{\partial F}{\partial x}\frac{\partial G}{\partial y}-\frac{\partial F}{\partial y}\frac{\partial G}{\partial x}\right)\right]\Bigg|_{(x,y)=(x_n,y_n)}$$

for $n=1,2,3,\cdots$.

By using Taylor's series to first derivatives and Cramer's rule, this algorithm may be further extended to the case of m simultaneous equations in m variables.

Interpolation Polynomial

The polynomial of lowest degree that passes through n points (x_i, y_i), $i=1,2,\cdots,n$, is given by

$$P(x)=\sum_{i=1}^{n}\left(y_i\prod_{k=1\ k\neq i}^{n}\frac{x-x_k}{x_i-x_k}\right)$$

where $x_i\neq x_k$ for $i\neq k$.

Interpolation at Equidistant Points

Let

$$f_i=f(x_0+ih)$$

$$g_{ij}=g(x+ih,\ y+jk)$$

Then $f(x_0+ph)$ may be approximated by

$$(1-p)f_0+pf_1,\ \text{given two points}$$

$$[p(p-1)/2]f_{-1}+(1-p^2)f_0+[p(p+1)/2]f_1,\ \text{given three points}$$

$$[-p(p-1)(p-2)/6]f_{-1}+[(p^2-1)(p-2)/2]f_0-[p(p+1)(p+2)/2]f_1+[p(p^2-1)/6]f_2,\ \text{given four points}$$

$g(x_0+ph,\ y_0+qk)$ may be approximated by

$$(1-p-q)g_{00}+pg_{10}+qg_{01},\ \text{given three points}$$

$$(1-p)(1-q)g_{00}+p(1-q)g_{10}+q(1-p)g_{01}+pqg_{11},\ \text{given four points}$$

Integration

Refer to Fig. 24. Let

$$f_i=f(x_0+ih)$$

$$\int_{x_0}^{x_0+mh}f(x)dx\approx h(\tfrac{1}{2}f_0+f_1+f_2+\cdots+f_{n-1}+\tfrac{1}{2}f_n)\quad\text{(Trapezoidal Rule)}$$

$$\int_{x_0}^{x_0+2nh}f(x)dx\approx\tfrac{1}{3}h(f_0+4f_1+2f_2+4f_3+2f_4+\cdots+2f_{2n-2}+4f_{2n-1}+f_{2n})\quad\text{(Simpson's Rule)}$$

Differentiation

If a function $f(x)$ is known at two points a and b, then

$$f'(a)\approx[f(b)-f(a)]/(b-a)$$

If $f(x)$ is given at a discrete set of points, one may differentiate the interpolation polynomial stated above. The equation simplifies when the derivative is calcu-

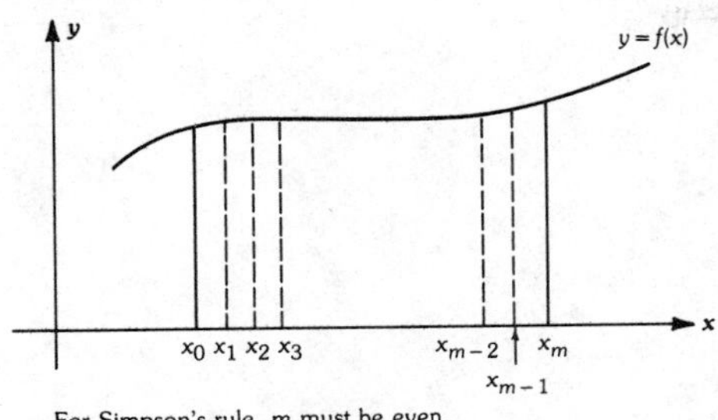

Fig. 24. Integration.

lated at interpolation points that are equidistant, for example

$$f_{-1}' \approx (1/2h)(-3f_{-1}+4f_0-f_1)$$

$$f_0' \approx (1/2h)(-f_{-1}+f_1)$$

$$f_1' \approx (1/2h)(f_{-1}-4f_0+3f_1)$$

where,

$$f_i = f(x_0+ih), \quad i=-1,0,1$$

Error in Arithmetic Operations

Let the quantities A and B be known to within errors of a and b, respectively, where a and b are small relative to A and B. Then

Operation	Maximum Error (to first order in a and b)
$A \pm B$	$a+b$
$A \cdot B$	$\lvert AB \rvert(\lvert a/A \rvert + \lvert b/B \rvert)$
A/B	$\lvert A/B \rvert(\lvert a/A \rvert + \lvert b/B \rvert)$

The magnitude, q, of small random errors often has a normal distribution

$$(h/\pi^{1/2})\exp(-h^2q^2)$$

where h is called the index of precision. The rms of q, denoted by σ, is, in terms of h

$$\sigma = 1/h\sqrt{2}$$

Let $f(x_1, x_2, \cdots, x_n)$ be a function of n variables x_i, which have normally distributed independent errors q_i with indices of precision h_i. Then $f(x_1, x_2, \cdots, x_n)$ will have an error Q, which is also normally distributed with index of precision H. For small errors

$$H \approx \{[h_1^{-1}(\partial f/\partial q_1)]^2 + [h_2^{-1}(\partial f/\partial q_2)]^2 + \cdots + [h_n^{-1}(\partial f/\partial q_n)]^2\}^{-1/2}$$

47 Mathematical Tables

Hyperbolic Sines

Hyperbolic Cosines

Hyperbolic Tangents

Multiples of 0.4343

Multiples of 2.3026

Logarithms to Base 2 and Powers of 2

Random Digits

Exponentials

Normal or Gaussian Distribution

Bessel Functions

HYPERBOLIC SINES

TABLE 1. HYPERBOLIC SINES [$\sinh x = \frac{1}{2}(e^x - e^{-x})$]

x	0	1	2	3	4	5	6	7	8	9	avg diff
0.0	0.0000	0.0100	0.0200	0.0300	0.0400	0.0500	0.0600	0.0701	0.0801	0.0901	100
.1	0.1002	0.1102	0.1203	0.1304	0.1405	0.1506	0.1607	0.1708	0.1810	0.1911	101
.2	0.2013	0.2115	0.2218	0.2320	0.2423	0.2526	0.2629	0.2733	0.2837	0.2941	103
.3	0.3045	0.3150	0.3255	0.3360	0.3466	0.3572	0.3678	0.3785	0.3892	0.4000	106
.4	0.4108	0.4216	0.4325	0.4434	0.4543	0.4653	0.4764	0.4875	0.4986	0.5098	110
0.5	0.5211	0.5324	0.5438	0.5552	0.5666	0.5782	0.5897	0.6014	0.6131	0.6248	116
.6	0.6367	0.6485	0.6605	0.6725	0.6846	0.6967	0.7090	0.7213	0.7336	0.7461	122
.7	0.7586	0.7712	0.7838	0.7966	0.8094	0.8223	0.8353	0.8484	0.8615	0.8748	130
.8	0.8881	0.9015	0.9150	0.9286	0.9423	0.9561	0.9700	0.9840	0.9981	1.012	138
.9	1.027	1.041	1.055	1.070	1.085	1.099	1.114	1.129	1.145	1.160	15
1.0	1.175	1.191	1.206	1.222	1.238	1.254	1.270	1.286	1.303	1.319	16
.1	1.336	1.352	1.369	1.386	1.403	1.421	1.438	1.456	1.474	1.491	17
.2	1.509	1.528	1.546	1.564	1.583	1.602	1.621	1.640	1.659	1.679	19
.3	1.698	1.718	1.738	1.758	1.779	1.799	1.820	1.841	1.862	1.883	21
.4	1.904	1.926	1.948	1.970	1.992	2.014	2.037	2.060	2.083	2.106	22
1.5	2.129	2.153	2.177	2.201	2.225	2.250	2.274	2.299	2.324	2.350	25
.6	2.376	2.401	2.428	2.454	2.481	2.507	2.535	2.562	2.590	2.617	27
.7	2.646	2.674	2.703	2.732	2.761	2.790	2.820	2.850	2.881	2.911	30
.8	2.942	2.973	3.005	3.037	3.069	3.101	3.134	3.167	3.200	3.234	33
.9	3.268	3.303	3.337	3.372	3.408	3.443	3.479	3.516	3.552	3.589	36
2.0	3.627	3.665	3.703	3.741	3.780	3.820	3.859	3.899	3.940	3.981	39
.1	4.022	4.064	4.106	4.148	4.191	4.234	4.278	4.322	4.367	4.412	44
.2	4.457	4.503	4.549	4.596	4.643	4.691	4.739	4.788	4.837	4.887	48
.3	4.937	4.988	5.039	5.090	5.142	5.195	5.248	5.302	5.356	5.411	53
.4	5.466	5.522	5.578	5.635	5.693	5.751	5.810	5.869	5.929	5.989	58
2.5	6.050	6.112	6.174	6.237	6.300	6.365	6.429	6.495	6.561	6.627	64
.6	0.695	6.763	6.831	6.901	6.971	7.042	7.113	7.185	7.258	7.332	71
.7	7.406	7.481	7.557	7.634	7.711	7.789	7.868	7.948	8.028	8.110	79
.8	8.192	8.275	8.359	8.443	8.529	8.615	8.702	8.790	8.879	8.969	87
.9	9.060	9.151	9.244	9.337	9.431	9.527	9.623	9.720	9.819	9.918	96
3.0	10.02	10.12	10.22	10.32	10.43	10.53	10.64	10.75	10.86	10.97	11
.1	11.08	11.19	11.30	11.42	11.53	11.65	11.76	11.88	12.00	12.12	12
.2	12.25	12.37	12.49	12.62	12.75	12.88	13.01	13.14	13.27	13.40	13
.3	13.54	13.67	13.81	13.95	14.09	14.23	14.38	14.52	14.67	14.82	14
.4	14.97	15.12	15.27	15.42	15.58	15.73	15.89	16.05	16.21	16.38	16
3.5	16.54	16.71	16.88	17.05	17.22	17.39	17.57	17.74	17.92	18.10	17
.6	18.29	18.47	18.66	18.84	19.03	19.22	19.42	19.61	19.81	20.01	19
.7	20.21	20.41	20.62	20.83	21.04	21.25	21.46	21.68	21.90	22.12	21
.8	22.34	22.56	22.79	23.02	23.25	23.49	23.72	23.96	24.20	24.45	24
.9	24.69	24.94	25.19	25.44	25.70	25.96	26.22	26.48	26.75	27.02	26
4.0	27.29	27.56	27.84	28.12	28.40	28.69	28.98	29.27	29.56	29.86	29
.1	30.16	30.47	30.77	31.08	31.39	31.71	32.03	32.35	32.68	33.00	32
.2	33.34	33.67	34.01	34.35	34.70	35.05	35.40	35.75	36.11	36.48	35
.3	36.84	37.21	37.59	37.97	38.35	38.73	39.12	39.52	39.91	40.31	39
.4	40.72	41.13	41.54	41.96	42.38	42.81	43.24	43.67	44.11	44.56	43
4.5	45.00	45.46	45.91	46.37	46.84	47.31	47.79	48.27	48.75	49.24	47
.6	49.74	50.24	50.74	51.25	51.77	52.29	52.81	53.34	53.88	54.42	52
.7	54.97	55.52	56.08	56.64	57.21	57.79	58.37	58.96	59.55	60.15	58
.8	60.75	61.36	61.98	62.60	63.23	63.87	64.51	65.16	65.81	66.47	64
.9	67.14	67.82	68.50	69.19	69.88	70.58	71.29	72.01	72.73	73.46	71
5.0	74.20										

If $x > 5$, $\sinh x = \frac{1}{2}(e^x)$ and $\log_{10} \sinh x = (0.4343)x + 0.6990 - 1$, correct to four significant figures.

HYPERBOLIC COSINES

TABLE 2. HYPERBOLIC COSINES [$\cosh x = \frac{1}{2}(e^x + e^{-x})$]

x	0	1	2	3	4	5	6	7	8	9	avg diff
0.0	1.000	1.000	1.000	1.000	1.001	1.001	1.002	1.002	1.003	1.004	1
.1	1.005	1.006	1.007	1.008	1.010	1.011	1.013	1.014	1.016	1.018	2
.2	1.020	1.022	1.024	1.027	1.029	1.031	1.034	1.037	1.039	1.042	3
.3	1.045	1.048	1.052	1.055	1.058	1.062	1.066	1.069	1.073	1.077	4
.4	1.081	1.085	1.090	1.094	1.098	1.103	1.108	1.112	1.117	1.122	5
0.5	1.128	1.133	1.138	1.144	1.149	1.155	1.161	1.167	1.173	1.179	6
.6	1.185	1.192	1.198	1.205	1.212	1.219	1.226	1.233	1.240	1.248	7
.7	1.255	1.263	1.271	1.278	1.287	1.295	1.303	1.311	1.320	1.329	8
.8	1.337	1.346	1.355	1.365	1.374	1.384	1.393	1.403	1.413	1.423	10
.9	1.433	1.443	1.454	1.465	1.475	1.486	1.497	1.509	1.520	1.531	11
1.0	1.543	1.555	1.567	1.579	1.591	1.604	1.616	1.629	1.642	1.655	13
.1	1.669	1.682	1.696	1.709	1.723	1.737	1.752	1.766	1.781	1.796	14
.2	1.811	1.826	1.841	1.857	1.872	1.888	1.905	1.921	1.937	1.954	16
.3	1.971	1.988	2.005	2.023	2.040	2.058	2.076	2.095	2.113	2.132	18
.4	1.151	2.170	2.189	2.209	2.229	2.249	2.269	2.290	2.310	2.331	20
1.5	2.352	2.374	2.395	2.417	2.439	2.462	2.484	2.507	2.530	2.554	23
.6	2.577	2.601	2.625	2.650	2.675	2.700	2.725	2.750	2.776	2.802	25
.7	2.828	2.855	2.882	2.909	2.936	2.964	2.992	3.021	3.049	3.078	28
.8	3.107	3.137	3.167	3.197	3.228	3.259	3.290	3.321	3.353	3.385	31
.9	3.418	3.451	3.484	3.517	3.551	3.585	3.620	3.655	3.690	3.726	34
2.0	3.762	3.799	3.835	3.873	3.910	3.948	3.987	4.026	4.065	4.104	38
.1	4.144	4.185	4.226	4.267	4.309	4.351	4.393	4.436	4.480	4.524	42
.2	4.568	4.613	4.658	4.704	4.750	4.797	4.844	4.891	4.939	4.988	47
.3	5.037	5.087	5.137	5.188	5.239	5.290	5.343	5.395	5.449	5.503	52
.4	5.557	5.612	5.667	5.723	5.780	5.837	5.895	5.954	6.013	6.072	58
2.5	6.132	6.193	6.255	6.317	6.379	6.443	6.507	6.571	6.636	6.702	64
.6	6.769	6.836	6.904	6.973	7.042	7.112	7.183	7.255	7.327	7.400	70
.7	7.473	7.548	7.623	7.699	7.776	7.853	7.932	8.011	8.091	8.171	78
.8	8.253	8.335	8.418	8.502	8.587	8.673	8.759	8.847	8.935	9.024	86
.9	9.115	9.206	9.298	9.391	9.484	9.579	9.675	9.772	9.869	9.968	95
3.0	10.07	10.17	10.27	10.37	10.48	10.58	10.69	10.79	10.90	11.01	11
.1	11.12	11.23	11.35	11.46	11.57	11.69	11.81	11.92	12.04	12.16	12
.2	12.29	12.41	12.53	12.66	12.79	12.91	13.04	13.17	13.31	13.44	13
.3	13.57	13.71	13.85	13.99	14.13	14.27	14.41	14.56	14.70	14.85	14
.4	15.00	15.15	15.30	15.45	15.61	15.77	15.92	16.08	16.25	16.41	16
3.5	16.57	16.74	16.91	17.08	17.25	17.42	17.60	17.77	17.95	18.13	17
.6	18.31	18.50	18.68	18.87	19.06	19.25	19.44	19.64	19.84	20.03	19
.7	20.24	20.44	20.64	20.85	21.06	21.27	21.49	21.70	21.92	22.14	21
.8	22.36	22.59	22.81	23.04	23.27	23.51	23.74	23.98	24.22	24.47	23
.9	24.71	24.96	25.21	25.46	25.72	25.98	26.24	26.50	26.77	27.04	26
4.0	27.31	27.58	27.86	28.14	28.42	28.71	29.00	29.29	29.58	29.88	29
.1	30.18	30.48	30.79	31.10	31.41	31.72	32.04	32.37	32.69	33.02	32
.2	33.35	33.69	34.02	34.37	34.71	35.06	35.41	35.77	36.13	36.49	35
.3	36.86	37.23	37.60	37.98	38.36	38.75	39.13	39.53	39.93	40.33	39
.4	40.73	41.14	41.55	41.97	42.39	42.82	43.25	43.68	44.12	44.57	43
4.5	45.01	45.47	45.92	46.38	46.85	47.32	47.80	48.28	48.76	49.25	47
.6	49.75	50.25	50.75	51.26	51.78	52.30	52.82	53.35	53.89	54.43	52
.7	54.98	55.53	56.09	56.65	57.22	57.80	58.38	58.96	59.56	60.15	58
.8	60.76	61.37	61.99	62.61	63.24	63.87	64.52	65.16	65.82	66.48	64
.9	67.15	67.82	68.50	69.19	69.89	70.59	71.30	72.02	72.74	73.47	71
5.0	74.21										

If $x > 5$, $\cosh x = \frac{1}{2}(e^x)$, and $\log_{10} \cosh x = (0.4343)x + 0.6990 - 1$, correct to four significant figures.

HYPERBOLIC TANGENTS

TABLE 3. HYPERBOLIC TANGENTS [$\tanh x = (e^x - e^{-x})/(e^x + e^{-x}) = \sinh x/\cosh x$]

x	0	1	2	3	4	5	6	7	8	9	avg diff
0.0	.0000	.0100	.0200	.0300	.0400	.0500	.0599	.0699	.0798	.0898	100
.1	.0997	.1096	.1194	.1293	.1391	.1489	.1587	.1684	.1781	.1878	98
.2	.1974	.2070	.2165	.2260	.2355	.2449	.2543	.2636	.2729	.2821	94
.3	.2913	.3004	.3095	.3185	.3275	.3364	.3452	.3540	.3627	.3714	89
.4	.3800	.3885	.3969	.4053	.4136	.4219	.4301	.4382	.4462	.4542	82
0.5	.4621	.4700	.4777	.4854	.4930	.5005	.5080	.5154	.5227	.5299	75
.6	.5370	.5441	.5511	.5581	.5649	.5717	.5784	.5850	.5915	.5980	67
.7	.6044	.6107	.6169	.6231	.6291	.6352	.6411	.6469	.6527	.6584	60
.8	.6640	.6696	.6751	.6805	.6858	.6911	.6963	.7014	.7064	.7114	52
.9	.7163	.7211	.7259	.7306	.7352	.7398	.7443	.7487	.7531	.7574	45
1.0	.7616	.7658	.7699	.7739	.7779	.7818	.7857	.7895	.7932	.7969	39
.1	.8005	.8041	.8076	.8110	.8144	.8178	.8210	.8243	.8275	.8306	33
.2	.8337	.8367	.8397	.8426	.8455	.8483	.8511	.8538	.8565	.8591	28
.3	.8617	.8643	.8668	.8693	.8717	.8741	.8764	.8787	.8810	.8832	24
.4	.8854	.8875	.8896	.8917	.8937	.8057	.8977	.8996	.9015	.9033	20
1.5	.9052	.9069	.9087	.9104	.9121	.9138	.9154	.9170	.9186	.9202	17
.6	.9217	.9232	.9246	.9261	.9275	.9289	.9302	.9316	.9329	.9342	14
.7	.9354	.9367	.9379	.9391	.9402	.9414	.9425	.9436	.9447	.9458	11
.8	.9468	.9478	.9488	.9498	.9508	.9518	.9527	.9536	.9545	.9554	9
.9	.9562	.9571	.9579	.9587	.9595	.9603	.9611	.9619	.9626	.9633	8
2.0	.9640	.9647	.9654	.9661	.9668	.9674	.9680	.9687	.9693	.9699	6
.1	.9705	.9710	.9716	.9722	.9727	.9732	.9738	.9743	.9748	.9753	5
.2	.9757	.9762	.9767	.9771	.9776	.9780	.9785	.9789	.9793	.9797	4
.3	.9801	.9805	.9809	.9812	.9816	.9820	.9823	.9827	.9830	.9834	4
.4	.9837	.9840	.9843	.9846	.9849	.9852	.9855	.9858	.9861	.9863	3
2.5	.9866	.9869	.9871	.9874	.9876	.9879	.9881	.9884	.9886	.9888	2
.6	.9890	.9892	.9895	.9897	.9899	.9901	.9903	.9905	.9906	.9908	2
.7	.9910	.9912	.9914	.9915	.9917	.9919	.9920	.9922	.9923	.9925	2
.8	.9926	.9928	.9929	.9931	.9932	.9933	.9935	.9936	.9937	.9938	1
.9	.9940	.9941	.9942	.9943	.9944	.9945	.9946	.9947	.9949	.9950	1
3.0	.9951	.9959	.9967	.9973	.9978	.9982	.9985	.9988	.9990	.9992	4
4.0	.9993	.9995	.9996	.9996	.9997	.9998	.9998	.9998	.9999	.9999	1
5.0	.9999										

If $x > 5$, $\tanh x = 1.0000$ to four decimal places.

MULTIPLES OF 0.4343

TABLE 4. MULTIPLES OF 0.4343 ($0.43429448 = \log_{10} e$)

x	0	1	2	3	4	5	6	7	8	9
0.0	0.0000	0.0434	0.0869	0.1303	0.1737	0.2171	0.2606	0.3040	0.3474	0.3909
1.0	0.4343	0.4777	0.5212	0.5646	0.6080	0.6514	0.6949	0.7383	0.7817	0.8252
2.0	0.8686	0.9120	0.9554	0.9989	1.0423	1.0857	1.1292	1.1726	1.2160	1.2595
3.0	1.3029	1.3463	1.3897	1.4332	1.4766	1.5200	1.5635	1.6069	1.6503	1.6937
4.0	1.7372	1.7806	1.8240	1.8675	1.9109	1.9543	1.9978	2.0412	2.0846	2.1280
5.0	2.1715	2.2149	2.2583	2.3018	2.3452	2.3886	2.4320	2.4755	2.5189	2.5623
6.0	2.6058	2.6492	2.6926	2.7361	2.7795	2.8229	2.8663	2.9098	2.9532	2.9966
7.0	3.0401	3.0835	3.1269	3.1703	3.2138	3.2572	3.3006	3.3441	3.3875	3.4309
8.0	3.4744	3.5178	3.5612	3.6046	3.6481	3.6915	3.7349	3.7784	3.8218	3.8652
9.0	3.9087	3.9521	3.9955	4.0389	4.0824	4.1258	4.1692	4.2127	4.2561	4.2995

MULTIPLES OF 2.3026

TABLE 5. MULTIPLES OF 2.3026 (2.3025851 = 1/0.4343 = $\log_e 10$)

x	0	1	2	3	4	5	6	7	8	9
0.0	0.0000	0.2303	0.4605	0.6908	0.9210	1.1513	1.3816	1.6118	1.8421	2.0723
1.0	2.3026	2.5328	2.7631	2.9934	3.2236	3.4539	3.6841	3.9144	4.1447	4.3749
2.0	4.6052	4.8354	5.0657	5.2959	5.5262	5.7565	5.9867	6.2170	6.4472	6.6775
3.0	6.9078	7.1380	7.3683	7.5985	7.8288	8.0590	8.2893	8.5196	8.7498	8.9801
4.0	9.2103	9.4406	9.6709	9.9011	10.131	10.362	10.592	10.822	11.052	11.283
5.0	11.513	11.743	11.973	12.204	12.434	12.664	12.894	13.125	13.355	13.585
6.0	13.816	14.046	14.276	14.506	14.737	14.967	15.197	15.427	15.658	15.888
7.0	16.118	16.348	16.579	16.809	17.039	17.269	17.500	17.730	17.960	18.190
8.0	18.421	18.651	18.881	19.111	19.342	19.572	19.802	20.032	20.263	20.493
9.0	20.723	20.954	21.184	21.414	21.644	21.875	22.105	22.335	22.565	22.796

LOGARITHMS TO BASE 2 AND POWERS OF 2

TABLE 6.

x	$\log_2 x$
0.1	−3.32193
0.2	−2.32193
0.3	−1.73697
0.4	−1.32193
0.5	−1.00000
0.6	−0.73697
0.7	−0.51457
0.8	−0.32193
0.9	−0.15200
1.0	0.00000
1.1	0.13750
1.2	0.26303
1.3	0.37851
1.4	0.48543
1.5	0.58496
1.6	0.67807
1.7	0.76553
1.8	0.84800
1.9	0.92600
2.0	1.00000
10	3.32193
100	6.64386
1000	9.96578
2^y	y

$$\log_2 x = \log_2 10 \log_{10} x = \log_2 e \log_e x$$

$$2^y = e^{y \log_e 2} = 10^{y \log_{10} 2}$$

$$\log_2 10 = 3.32193 = 1/\log_{10} 2$$

$$\log_{10} 2 = 0.30103 = 1/\log_2 10$$

$$\log_2 e = 1.44269 = 1/\log_e 2$$

$$\log_e 2 = 0.69315 = 1/\log_2 e$$

TABLE 7.

y	2^y
0.1	1.072
0.2	1.149
0.3	1.231
0.4	1.320
0.5	1.414
0.6	1.515
0.7	1.625
0.8	1.741
0.9	1.866
1	2
2	4
3	8
4	16
5	32
6	64
7	128
8	256
9	512
10	1 024
11	2 048
12	4 096
13	8 192
14	16 384
15	32 768
16	65 536
17	131 072
18	262 144
19	524 288
20	1 048 576
21	2 097 152
22	4 194 304
23	8 388 608
24	16 777 216
25	33 554 432
26	67 108 864
27	134 217 728
28	268 435 456
29	536 870 912
30	1 073 741 824
31	2 147 483 648
32	4 294 967 296
$\log_2 x$	x

RANDOM DIGITS

TABLE 8. RANDOM DIGITS

49 31 97 45 80	57 47 01 46 00	57 16 83 04 58	23 89 20 78 25	18 53 20 38 74	66 22 07 90 50	29 22 37 05 41	67 11 58 45 84
88 78 67 69 63	12 12 72 50 14	71 88 66 53 34	38 01 30 93 79	22 93 62 20 58	49 17 11 10 27	22 68 18 01 10	31 59 50 92 46
84 86 69 52 02	43 98 37 26 55	40 41 85 95 04	52 38 30 72 32	66 39 77 65 10	81 15 00 07 04	74 58 09 03 54	43 74 42 21 78
11 84 92 64 82	20 46 19 94 50	28 83 37 66 61	47 27 79 29 35	89 73 02 32 72	65 42 03 50 91	69 09 37 13 64	08 10 79 69 52
54 96 61 75 94	57 39 37 32 67	37 88 36 21 24	62 19 94 95 42	81 82 17 53 23	96 06 89 17 24	40 45 69 12 34	58 09 06 53 42
10 95 93 33 49	80 71 99 67 51	44 88 23 35 92	66 23 41 38 21	94 37 78 25 54	53 58 61 14 32	72 92 76 73 49	83 96 25 89 12
22 78 40 77 83	35 90 30 00 91	19 08 21 38 73	07 18 42 15 66	68 48 54 99 91	53 16 51 98 65	61 86 93 30 93	81 12 90 64 81
86 03 76 17 91	33 81 56 39 68	45 31 62 92 83	89 31 85 58 06	07 33 00 71 84	86 78 86 45 77	40 04 81 65 20	07 63 81 07 97
80 03 76 50 89	85 91 97 43 91	22 78 85 54 33	31 18 87 48 82	10 99 31 49 30	35 07 23 64 29	68 77 39 76 69	28 65 68 99 38
72 75 18 43 59	15 76 91 36 15	08 29 38 61 93	05 02 62 12 55	20 80 11 51 78	64 45 38 33 57	09 77 43 07 51	49 74 01 13 85
79 24 13 53 47	66 85 17 92 47	46 13 93 66 89	82 58 71 35 86	93 36 91 30 44	69 68 67 81 62	66 37 80 29 19	34 01 25 00 80
43 59 33 95 55	97 34 55 84 94	26 56 69 53 23	32 99 38 99 88	19 36 05 50 49	94 95 17 63 41	84 01 93 06 90	25 65 67 29 96
29 52 26 27 13	33 70 11 71 86	06 76 55 71 41	48 61 71 82 82	47 79 88 98 90	06 89 36 54 83	17 70 12 12 92	14 88 01 53 86
88 83 64 72 90	67 27 47 83 62	35 38 49 03 80	12 31 78 97 02	69 22 33 20 07	03 51 36 11 49	32 54 69 20 72	62 52 22 15 04
65 90 56 62 53	91 48 23 06 89	49 33 37 84 82	36 19 91 13 55	34 51 15 07 21	84 85 03 41 59	97 13 86 19 19	97 78 92 85 75
44 79 86 93 71	07 86 59 17 56	45 59 51 40 44	56 80 69 91 26	54 03 15 93 29	58 96 35 22 20	35 29 22 79 24	55 46 74 30 36
35 51 09 91 39	32 03 12 79 25	79 81 91 50 54	76 17 41 22 06	66 72 28 55 15	04 72 39 24 11	02 73 70 81 68	30 04 36 34 50
50 12 59 32 23	64 20 94 97 14	11 97 16 22 34	74 85 74 64 01	71 05 90 74 96	38 40 41 81 26	28 26 13 78 44	12 54 31 43 98
25 17 39 00 38	63 87 14 04 18	11 45 28 93 18	53 08 42 19 93	45 47 88 60 66	31 13 53 32 43	80 57 33 06 06	48 64 45 30 08
68 45 99 00 94	44 99 59 37 18	38 74 68 12 71	96 26 09 81 37	97 24 69 11 21	89 43 72 03 93	77 15 38 85 52	26 84 31 28 44
22 98 22 59 36	96 41 73 48 45	85 14 95 75 04	15 05 93 68 49	84 98 36 83 12	25 51 95 61 58	86 30 00 76 89	14 00 67 77 53
48 24 36 29 93	47 13 28 52 48	35 22 97 28 37	36 75 27 16 55	35 55 40 29 35	72 88 96 87 72	19 85 03 96 50	65 22 21 55 63
93 51 41 49 15	67 96 08 22 03	40 11 72 43 46	32 18 98 70 74	04 36 81 76 32	50 96 27 19 08	94 46 46 64 32	62 24 31 36 74
69 70 79 83 03	93 06 91 62 16	60 87 59 75 45	68 65 29 21 60	81 31 16 04 79	69 98 53 09 52	23 92 14 97 30	21 71 89 23 14
87 46 79 17 94	70 81 41 27 43	03 76 93 25 51	74 80 14 16 92	03 82 38 98 87	55 82 87 44 52	72 77 52 37 16	42 85 37 47 93
81 00 68 14 98	59 37 53 05 02	94 07 79 22 09	31 50 66 96 06	80 42 26 54 37	38 79 75 62 61	27 81 64 67 04	82 73 50 33 39
15 45 88 14 81	50 18 74 33 75	94 37 60 06 66	94 14 52 23 99	61 30 74 94 68	43 34 44 37 00	20 20 77 70 88	17 16 72 45 31
33 46 91 25 10	23 09 54 80 16	42 35 41 13 47	90 92 00 38 64	83 87 38 25 57	10 00 28 00 93	59 28 30 44 94	60 72 52 14 31
67 19 80 71 76	65 99 61 83 17	81 14 94 32 91	10 81 74 43 48	38 11 01 68 55	28 92 29 37 58	88 73 13 63 16	51 38 35 76 19
58 03 79 22 61	85 50 45 56 90	10 63 17 82 38	00 15 74 62 59	43 89 29 11 89	87 22 65 69 35	84 76 26 79 36	75 00 00 17 95
93 68 30 96 64	53 92 74 98 85	20 75 49 23 55	57 95 51 09 40	14 95 42 22 99	40 15 65 26 85	29 22 33 83 83	30 31 57 09 99
32 74 80 21 21	11 97 29 69 14	28 06 56 95 64	06 83 55 68 45	01 71 19 84 39	09 44 63 39 37	49 09 54 02 38	81 69 71 24 74
49 21 19 29 63	38 62 56 53 12	62 17 57 33 53	84 97 21 77 26	62 32 85 53 28	45 73 89 39 40	27 46 62 69 27	53 34 51 13 79
63 36 56 42 24	69 47 55 75 12	11 04 45 04 83	68 82 19 74 26	73 00 46 21 09	81 90 77 10 77	57 46 37 00 45	65 12 34 90 70
63 57 62 63 73	44 61 04 37 48	00 33 16 34 22	99 62 27 67 57	34 21 88 94 45	05 60 95 23 36	50 55 89 22 42	52 73 28 15 02
41 07 84 70 36	65 52 46 84 66	67 15 72 64 19	37 97 81 65 11	99 15 90 19 68	45 88 68 68 75	28 41 39 59 18	44 15 64 69 59
70 84 68 95 58	64 17 31 53 81	87 71 35 08 41	46 27 02 65 08	92 85 82 99 49	15 81 79 33 72	56 65 74 31 93	58 13 05 42 73
68 80 06 44 92	20 16 23 27 07	10 28 18 25 25	74 15 58 67 49	27 39 69 74 77	65 55 47 16 01	13 12 16 88 67	95 76 35 96 67
44 97 78 95 25	51 26 96 37 47	91 36 77 40 33	67 02 06 90 92	37 10 34 53 09	30 12 94 33 80	96 99 68 93 56	22 78 46 01 84
79 35 46 38 47	24 39 55 36 79	40 56 03 69 14	69 17 63 19 18	57 34 79 70 12	48 42 82 06 06	60 74 22 22 26	89 99 32 45 97

EXPONENTIALS

TABLE 9. EXPONENTIALS (e^n AND e^{-n})

n	e^n	diff	n	e^n	diff	n	e^n (*)	n	e^{-n}	diff	n	e^{-n}	n	e^{-n}(*)
0.00	1.000	10	0.50	1.649	16	1.0	2.718	0.00	1.000	−10	0.50	.607	1.0	.368
.01	1.010	10	.51	1.665	17	.1	3.004	.01	0.990	−10	.51	.600	.1	.333
.02	1.020	10	.52	1.682	17	.2	3.320	.02	.980	−10	.52	.595	.2	.301
.03	1.030	11	.53	1.699	17	.3	3.669	.03	.970	− 9	.53	.589	.3	.273
.04	1.041	10	.54	1.716	17	.4	4.055	.04	.961	−10	.54	.583	.4	.247
0.05	1.051	11	0.55	1.733	18	1.5	4.482	0.05	.951	− 9	0.55	.577	1.5	.223
.06	1.062	11	.56	1.751	17	.6	4.953	.06	.942	−10	.56	.571	.6	.202
.07	1.073	10	.57	1.768	18	.7	5.474	.07	.932	− 9	.57	.566	.7	.183
.08	1.083	11	.58	1.786	18	.8	6.050	.08	.923	− 9	.58	.560	.8	.165
.09	1.094	11	.59	1.804	18	.9	6.686	.09	.914	− 9	.59	.554	.9	.150
0.10	1.105	11	0.60	1.822	18	2.0	7.389	0.10	.905	− 9	0.60	.549	2.0	.135
.11	1.116	11	.61	1.840	19	.1	8.166	.11	.896	− 9	.61	.543	.1	.122
.12	1.127	12	.62	1.859	19	.2	9.025	.12	.887	− 9	.62	.538	.2	.111
.13	1.139	11	.63	1.878	18	.3	9.974	.13	.878	− 9	.63	.533	.3	.100
.14	1.150	12	.64	1.896	20	.4	11.02	.14	.869	− 8	.64	.527	.4	.0907
0.15	1.162	12	0.65	1.916	19	2.5	12.18	0.15	.861	− 9	0.65	.522	2.5	.0821
.16	1.174	11	.66	1.935	19	.6	13.46	.16	.852	− 8	.66	.517	.6	.0743
.17	1.185	12	.67	1.954	20	.7	14.88	.17	.844	− 9	.67	.512	.7	.0672
.18	1.197	12	.68	1.974	20	.8	16.44	.18	.835	− 8	.68	.507	.8	.0608
.19	1.209	12	.69	1.994	20	.9	18.17	.19	.827	− 8	.69	.502	.9	.0550
0.20	1.221	13	0.70	2.014	20	3.0	20.09	0.20	.819	− 8	0.70	.497	3.0	.0498
.21	1.234	12	.71	2.034	20	.1	22.20	.21	.811	− 8	.71	.492	.1	.0450
.22	1.246	13	.72	2.054	21	.2	24.53	.22	.803	− 8	.72	.487	.2	.0408
.23	1.259	12	.73	2.075	21	.3	27.11	.23	.795	− 8	.73	.482	.3	.0369
.24	1.271	13	.74	2.096	21	.4	29.96	.24	.787	− 8	.74	.477	.4	.0334
0.25	1.284	13	0.75	2.117	21	3.5	33.12	0.25	.779	− 8	0.75	.472	3.5	.0302
.26	1.297	13	.76	2.138	22	.6	36.60	.26	.771	− 8	.76	.468	.6	.0273
.27	1.310	13	.77	2.160	21	.7	40.45	.27	.763	− 7	.77	.463	.7	.0247
.28	1.323	13	.78	2.181	22	.8	44.70	.28	.756	− 8	.78	.458	.8	.0224
.29	1.336	14	.79	2.203	23	.9	49.40	.29	.748	− 7	.79	.454	.9	.0202
0.30	1.350	13	0.80	2.226	22	4.0	54.60	0.30	.741	− 8	0.80	.449	4.0	.0183
.31	1.363	14	.81	2.248	22	.1	60.34	.31	.733	− 7	.81	.445	.1	.0166
.32	1.377	14	.82	2.270	23	.2	66.69	.32	.726	− 7	.82	.440	.2	.0150
.33	1.391	14	.83	2.293	23	.3	73.70	.33	.719	− 7	.83	.436	.3	.0136
.34	1.405	14	.84	2.316	24	.4	81.45	.34	.712	− 7	.84	.432	.4	.0123
0.35	1.419	14	0.85	2.340	23	4.5	90.02	0.35	.705	− 7	0.85	.427	4.5	.0111
.36	1.433	15	.86	2.363	24			.36	.698	− 7	.86	.423		
.37	1.448	14	.87	2.387	24	5.0	148.4	.37	.691	− 7	.87	.419	5.0	.00674
.38	1.462	15	.88	2.411	24	6.0	403.4	.38	.684	− 7	.88	.415	6.0	.00248
.39	1.477	15	.89	2.435	25	7.0	1097.	.39	.677	− 7	.89	.411	7.0	.000912
0.40	1.492	15	0.90	2.460	24	8.0	2981.	0.40	.670	− 6	0.90	.407	8.0	.000335
.41	1.507	15	.91	2.484	25	9.0	8103.	.41	.664	− 7	.91	.403	9.0	.000123
.42	1.522	15	.92	2.509	26	10.0	22026.	.42	.657	− 6	.92	.399	10.0	.000045
.43	1.537	16	.93	2.535	25			.43	.651	− 7	.93	.395		
.44	1.553	15	.94	2.560	26	π/2	4.810	.44	.644	− 6	.94	.391	π/2	.208
						2π/2	23.14						2π/2	.0432
0.45	1.568	16	0.95	2.586	26	3π/2	111.3	0.45	.638	− 7	0.95	.387	3π/2	.00898
.46	1.584	16	.96	2.612	26	4π/2	535.5	.46	.631	− 6	.96	.383	4π/2	.00187
.47	1.600	16	.97	2.638	26	5π/2	2576.	.47	.625	− 6	.97	.379	5π/2	.000388
.48	1.616	16	.98	2.664	27	6π/2	12392.	.48	.619	− 6	.98	.375	6π/2	.000081
.49	1.632	17	.99	2.691	27	7π/2	59610.	.49	.613	− 6	.99	.372	7π/2	.000017
						8π/2	286751.						8π/2	.000003
0.50	1.649		1.00	2.718				0.50	0.607		1.00	.368		

*Note: Do not interpolate in this column.

NORMAL OR GAUSSIAN DISTRIBUTION

TABLE 10. NORMAL OR GAUSSIAN DISTRIBUTION

x	$p(x)=(2\pi)^{-1/2}\exp(-x^2/2)$	$P(x)=\int_{-\infty}^{x} p(t)\,dt$	x	$p(x)=(2\pi)^{-1/2}\exp(-x^2/2)$	$P(x)=\int_{-\infty}^{x} p(t)\,dt$
0.00	0.39894	0.50000	2.55	0.01545	$0.9^{2}4614$
0.05	0.39844	0.51994	2.60	0.01358	$0.9^{2}5339$
0.10	0.39695	0.53983	2.65	0.01191	$0.9^{2}5975$
0.15	0.39448	0.55962	2.70	0.01042	$0.9^{2}6533$
0.20	0.39104	0.57926	2.75	$0.0^{2}9094$	$0.9^{2}7020$
0.25	0.38667	0.59871	2.80	$0.0^{2}7915$	$0.9^{2}7445$
0.30	0.38139	0.61791	2.85	$0.0^{2}6873$	$0.9^{2}7814$
0.35	0.37524	0.63683	2.90	$0.0^{2}5953$	$0.9^{2}8134$
0.40	0.36827	0.65542	2.95	$0.0^{2}5143$	$0.9^{2}8411$
0.45	0.36053	0.67364	3.00	$0.0^{2}4432$	$0.9^{2}8650$
0.50	0.35207	0.69146	3.05	$0.0^{2}3810$	$0.9^{2}8856$
0.55	0.34294	0.70884	3.10	$0.0^{2}3267$	$0.9^{3}0324$
0.60	0.33322	0.72575	3.15	$0.0^{2}2794$	$0.9^{3}1836$
0.65	0.32297	0.74215	3.20	$0.0^{2}2384$	$0.9^{3}3129$
0.70	0.31225	0.75804	3.25	$0.0^{2}2029$	$0.9^{3}4230$
0.75	0.30114	0.77337	3.30	$0.0^{2}1723$	$0.9^{3}5166$
0.80	0.28969	0.78814	3.35	$0.0^{2}1459$	$0.9^{3}5959$
0.85	0.27798	0.80234	3.40	$0.0^{2}1232$	$0.9^{3}6631$
0.90	0.26609	0.81594	3.45	$0.0^{2}1038$	$0.9^{3}7197$
0.95	0.25406	0.82894	3.50	$0.0^{3}8727$	$0.9^{3}7674$
1.00	0.24197	0.84134	3.55	$0.0^{3}7317$	$0.9^{3}8074$
1.05	0.22988	0.85314	3.60	$0.0^{3}6119$	$0.9^{3}8409$
1.10	0.21785	0.86433	3.65	$0.0^{3}5105$	$0.9^{3}8689$
1.15	0.20594	0.87493	3.70	$0.0^{3}4248$	$0.9^{3}8922$
1.20	0.19419	0.88493	3.75	$0.0^{3}3526$	$0.9^{4}1158$
1.25	0.18265	0.89435	3.80	$0.0^{3}2919$	$0.9^{4}2765$
1.30	0.17137	0.90320	3.85	$0.0^{3}2411$	$0.9^{4}4094$
1.35	0.16038	0.91149	3.90	$0.0^{3}1987$	$0.9^{4}5190$
1.40	0.14973	0.91924	3.95	$0.0^{3}1633$	$0.9^{4}6092$
1.45	0.13943	0.92647	4.00	$0.0^{3}1338$	$0.9^{4}6833$
1.50	0.12952	0.93319	4.05	$0.0^{3}1094$	$0.9^{4}7439$
1.55	0.12001	0.93943	4.10	$0.0^{4}8926$	$0.9^{4}7934$
1.60	0.11092	0.94520	4.15	$0.0^{4}7263$	$0.9^{4}8338$
1.65	0.10226	0.95053	4.20	$0.0^{4}5894$	$0.9^{4}8665$
1.70	0.09405	0.95543	4.25	$0.0^{4}4772$	$0.9^{4}8931$
1.75	0.08628	0.95994	4.30	$0.0^{4}3854$	$0.9^{5}1460$
1.80	0.07895	0.96407	4.35	$0.0^{4}3104$	$0.9^{5}3193$
1.85	0.07206	0.96784	4.40	$0.0^{4}2494$	$0.9^{5}4587$
1.90	0.06562	0.97128	4.45	$0.0^{4}1999$	$0.9^{5}5706$
1.95	0.05959	0.97441	4.50	$0.0^{4}1598$	$0.9^{5}6602$
2.00	0.05399	0.97725	4.55	$0.0^{4}1275$	$0.9^{5}7318$
2.05	0.04879	0.97982	4.60	$0.0^{4}1014$	$0.9^{5}7888$
2.10	0.04398	0.98214	4.65	$0.0^{5}8047$	$0.9^{5}8340$
2.15	0.03955	0.98422	4.70	$0.0^{5}6370$	$0.9^{5}8699$
2.20	0.03547	0.98610	4.75	$0.0^{5}5030$	$0.9^{5}8983$
2.25	0.03174	0.98778	4.80	$0.0^{5}3961$	$0.9^{6}2067$
2.30	0.02833	0.98928	4.85	$0.0^{5}3112$	$0.9^{6}3827$
2.35	0.02522	$0.9^{2}0613$	4.90	$0.0^{5}2439$	$0.9^{6}5208$
2.40	0.02239	$0.9^{2}1802$	4.95	$0.0^{5}1907$	$0.9^{6}6289$
2.45	0.01984	$0.9^{2}2857$	5.00	$0.0^{5}1487$	$0.9^{6}7133$
2.50	0.01753	$0.9^{2}3790$			

Note: $0.0^{2}9094 = 0.009094$ $\quad 0.9^{3}0324 = 0.9990324$

$P(-x) = 1 - P(x)$ $\quad \int_{-x}^{x} p(t)\,dt = 2P(x) - 1$

BESSEL FUNCTIONS

TABLE 11. $J_0(z)$

z	0	0.1	0.2	0.3	0.4	0.5	0.6	0.7	0.8	0.9
0	1.0000	0.9975	0.9900	0.9776	0.9604	0.9385	0.9120	0.8812	0.8463	0.8075
1	0.7652	0.7196	0.6711	0.6201	0.5669	0.5118	0.4554	0.3980	0.3400	0.2818
2	0.2239	0.1666	0.1104	0.0555	0.0025	−0.0484	−0.0968	−0.1424	−0.1850	−0.2243
3	−0.2601	−0.2921	−0.3202	−0.3443	−0.3643	−0.3801	−0.3918	−0.3992	−0.4026	−0.4018
4	−0.3971	−0.3887	−0.3766	−0.3610	−0.3423	−0.3205	−0.2961	−0.2693	−0.2404	−0.2097
5	−0.1776	−0.1443	−0.1103	−0.0758	−0.0412	−0.0068	+0.0270	0.0599	0.0917	0.1220
6	0.1506	0.1773	0.2017	0.2238	0.2433	0.2601	0.2740	0.2851	0.2931	0.2981
7	0.3001	0.2991	0.2951	0.2882	0.2786	0.2663	0.2516	0.2346	0.2154	0.1944
8	0.1717	0.1475	0.1222	0.0960	0.0692	0.0419	0.0146	−0.0125	−0.0392	−0.0653
9	−0.0903	−0.1142	−0.1367	−0.1577	−0.1768	−0.1939	−0.2090	−0.2218	−0.2323	−0.2403
10	−0.2459	−0.2490	−0.2496	−0.2477	−0.2434	−0.2366	−0.2276	−0.2164	−0.2032	−0.1881
11	−0.1712	−0.1528	−0.1330	−0.1121	−0.0902	−0.0677	−0.0446	−0.0213	+0.0020	0.0250
12	0.0477	0.0697	0.0908	0.1108	0.1296	0.1469	0.1626	0.1766	0.1887	0.1988
13	0.2069	0.2129	0.2167	0.2183	0.2177	0.2150	0.2101	0.2032	0.1943	0.1836
14	0.1711	0.1570	0.1414	0.1245	0.1065	0.0875	0.0679	0.0476	0.0271	0.0064
15	−0.0142	−0.0346	−0.0544	−0.0736	−0.0919	−0.1092	−0.1253	−0.1401	−0.1533	−0.1650

TABLE 12. $J_1(z)$

z	0	0.1	0.2	0.3	0.4	0.5	0.6	0.7	0.8	0.9
0	0.0000	0.0499	0.0995	0.1483	0.1960	0.2423	0.2867	0.3290	0.3688	0.4059
1	0.4401	0.4709	0.4983	0.5220	0.5419	0.5579	0.5699	0.5778	0.5815	0.5812
2	0.5767	0.5683	0.5560	0.5399	0.5202	0.4971	0.4708	0.4416	0.4097	0.3754
3	0.3391	0.3009	0.2613	0.2207	0.1792	0.1374	0.0955	0.0538	0.0128	−0.0272
4	−0.0660	−0.1033	−0.1386	−0.1719	−0.2028	−0.2311	−0.2566	−0.2791	−0.2985	−0.3147
5	−0.3276	−0.3371	−0.3432	−0.3460	−0.3453	−0.3414	−0.3343	−0.3241	−0.3110	−0.2951
6	−0.2767	−0.2559	−0.2329	−0.2081	−0.1816	−0.1538	−0.1250	−0.0953	−0.0652	−0.0349
7	−0.0047	+0.0252	0.0543	0.0826	0.1096	0.1352	0.1592	0.1813	0.2014	0.2192
8	0.2346	0.2476	0.2580	0.2657	0.2708	0.2731	0.2728	0.2697	0.2641	0.2559
9	0.2453	0.2324	0.2174	0.2004	0.1816	0.1613	0.1395	0.1166	0.0928	0.0684
10	0.0435	0.0184	−0.0066	−0.0313	−0.0555	−0.0789	−0.1012	−0.1224	−0.1422	−0.1603
11	−0.1768	−0.1913	−0.2039	−0.2143	−0.2225	−0.2284	−0.2320	−0.2333	−0.2323	−0.2290
12	−0.2234	−0.2157	−0.2060	−0.1943	−0.1807	−0.1655	−0.1487	−0.1307	−0.1114	−0.0912
13	−0.0703	−0.0489	−0.0271	−0.0052	+0.0166	0.0380	0.0590	0.0791	0.0984	0.1165
14	0.1334	0.1488	0.1626	0.1747	0.1850	0.1934	0.1999	0.2043	0.2066	0.2069
15	0.2051	0.2013	0.1955	0.1879	0.1784	0.1672	0.1544	0.1402	0.1247	0.1080

TABLE 13. $J_2(z)$

z	0	0.1	0.2	0.3	0.4	0.5	0.6	0.7	0.8	0.9
0	0.0000	0.0012	0.0050	0.0112	0.0197	0.0306	0.0437	0.0588	0.0758	0.0946
1	0.1149	0.1366	0.1593	0.1830	0.2074	0.2321	0.2570	0.2817	0.3061	0.3299
2	0.3528	0.3746	0.3951	0.4139	0.4310	0.4461	0.4590	0.4696	0.4777	0.4832
3	0.4861	0.4862	0.4835	0.4780	0.4697	0.4586	0.4448	0.4283	0.4093	0.3879
4	0.3641	0.3383	0.3105	0.2811	0.2501	0.2178	0.1846	0.1506	0.1161	0.0813

TABLE 14. $J_3(z)$

z	0	0.1	0.2	0.3	0.4	0.5	0.6	0.7	0.8	0.9
0	0.0000	0.0000	0.0002	0.0006	0.0013	0.0026	0.0044	0.0069	0.0102	0.0144
1	0.0196	0.0257	0.0329	0.0411	0.0505	0.0610	0.0725	0.0851	0.0988	0.1134
2	0.1289	0.1453	0.1623	0.1800	0.1981	0.2166	0.2353	0.2540	0.2727	0.2911
3	0.3091	0.3264	0.3431	0.3588	0.3734	0.3868	0.3988	0.4092	0.4180	0.4250
4	0.4302	0.4333	0.4344	0.4333	0.4301	0.4247	0.4171	0.4072	0.3952	0.3811

TABLE 15. $J_4(z)$

z	0	0.1	0.2	0.3	0.4	0.5	0.6	0.7	0.8	0.9
0	0.0000	0.0000	0.0000	0.0000	0.0001	0.0002	0.0003	0.0006	0.0010	0.0016
1	0.0025	0.0036	0.0050	0.0068	0.0091	0.0118	0.0150	0.0188	0.0232	0.0283
2	0.0340	0.0405	0.0476	0.0556	0.0643	0.0738	0.0840	0.0950	0.1067	0.1190
3	0.1320	0.1456	0.1597	0.1743	0.1891	0.2044	0.2198	0.2353	0.2507	0.2661
4	0.2811	0.2958	0.3100	0.3236	0.3365	0.3484	0.3594	0.3693	0.3780	0.3853

48 Miscellaneous Data

Summary of Military Nomenclature System

EFFECT OF ALTITUDE ON BREAKDOWN VOLTAGES

Pressure-Altitude Graph

Design of electrical equipment for aircraft is somewhat complicated by the requirement of additional insulation for high voltages as a result of the decrease in atmospheric pressure. The extent of this effect may be determined from Figs. 1 and 2 and Table 1. (1 inch mercury = 25.4 millimeters mercury = 0.4912 pound/inch^2 = 3.38×10^3 pascals.)

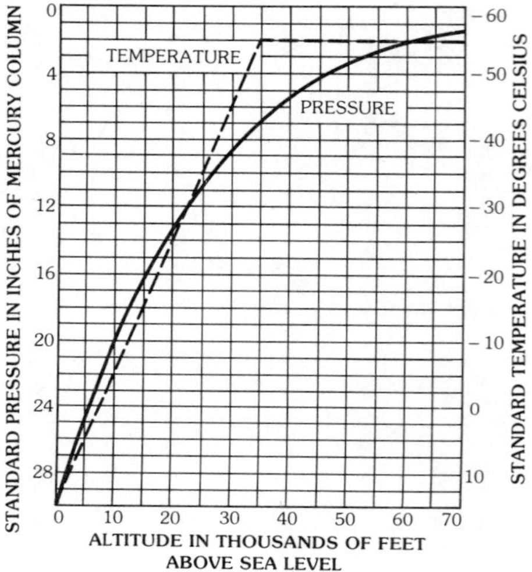

Fig. 1. Pressure as a function of altitude.

TABLE 1. MULTIPLYING FACTORS

Pressure in Hg	Pressure mm Hg	Temperature in Degrees Celsius −40	−20	0	20	40	60
5	127	0.26	0.24	0.23	0.21	0.20	0.19
10	254	0.47	0.44	0.42	0.39	0.37	0.34
15	381	0.68	0.64	0.60	0.56	0.53	0.50
20	508	0.87	0.82	0.77	0.72	0.68	0.64
25	635	1.07	0.99	0.93	0.87	0.82	0.77
30	762	1.25	1.17	1.10	1.03	0.97	0.91
35	889	1.43	1.34	1.26	1.19	1.12	1.05
40	1016	1.61	1.51	1.42	1.33	1.25	1.17
45	1143	1.79	1.68	1.58	1.49	1.40	1.31
50	1270	1.96	1.84	1.73	1.63	1.53	1.44
55	1397	2.13	2.01	1.89	1.78	1.67	1.57
60	1524	2.30	2.17	2.04	1.92	1.80	1.69

Spark-Gap Breakdown Voltages

Fig. 2 is for a voltage that is continuous or at a frequency low enough to permit complete deionization between cycles, between needle points or clean smooth spherical surfaces (electrodes ungrounded) in dust-free dry air. Temperature is 25 degrees Celsius and pressure is 760 millimeters (29.9 inches) of mercury. Peak kilovolts shown in the figure should be multiplied by the factors given in Table 1 for atmospheric conditions other than the above.

An approximate rule for uniform fields at all frequencies up to at least 300 megahertz is that the breakdown gradient of air is 30 peak kilovolts/centimeter or 75 peak kilovolts/inch at sea level (760 millimeters of mercury) and normal temperature (25 degrees Celsius). The breakdown voltage is approximately proportional to pressure and inversely proportional to absolute (Kelvin) temperature.

Certain synthetic gases have higher dielectric strengths than air. Two such gases that appear to be useful for electrical insulation are sulfur hexafluoride (SF_6) and Freon 12 (CCl_2F_2), which both have about 2.5 times the dielectric strength of air. Mixtures of sulfur hexafluoride with helium and of perfluoromethylcyclohexane (C_7F_{14}) with nitrogen have good dielectric strength as well as other desirable properties.

WEATHER DATA*

Temperature Extremes

United States (contiguous):

Lowest temperature:
Rodgers Pass, Montana (January 20, 1954)—−57°C (−70°F).

Highest temperature:
Greenland Ranch, Death Valley, California (July 10, 1913)—57°C (134°F).

Alaska:

Lowest temperature:
Prospect Creek Camp (January 23, 1971)—−62°C (−79.8°F).

Highest temperature:
Fort Yukon (June 27, 1915)—38°C (100°F).

World:

Lowest temperature:
Vostok, Antarctica (July 21, 1983)—−89°C (−129°F).

---

* Compiled in part from "Climate and Man," *Yearbook of Agriculture*, US Dept. of Agriculture. Obtainable from Superintendent of Documents, Government Printing Office, Washington, DC 20402. See also "Weather Extremes Around the World," a world map compiled (1984) by the Geographic Sciences Laboratory, US Army Engineer Topographic Laboratories, Belvoir, Virginia 22060.

For a comprehensive summary of available climatological information, refer to *Selective Guide to Climatic Data Sources* (No. 4.11), by Warren L. Hatch, July 1983, available from the National Climatic Data Center, Asheville, North Carolina 28801-2696; phone: (704) 259-0682.

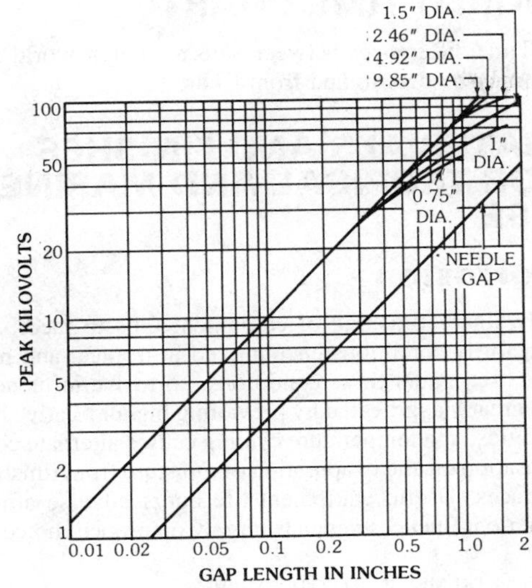

Fig. 2. Spark-gap breakdown voltages.

Highest temperature:
Azizia, Libya, North Africa (September 13, 1922)—58°C (136°F).

Lowest mean temperature (annual):
Framheim, Antarctica— −26°C (−14°F).

Highest mean temperature (annual):
Dallol Ethiopia—34°C (94°F).

World Temperatures

Territory	Maximum °C	Maximum °F	Minimum °C	Minimum °F
NORTH AMERICA				
Alaska	38	100	−62	−79.8
Canada	45	113	−63	−81
Canal Zone	36	97	−17	63
Greenland	30	86	−43	−46
Mexico	48	118	−12	11
USA	57	134	−57	−70
West Indies	39	102	7	45
SOUTH AMERICA				
Argentina	46	115	−33	−27
Bolivia	28	82	−4	25
Brazil	42	108	−6	21
Chile	37	99	−7	19
Venezuela	39	102	7	45
EUROPE				
British Isles	38	100	−16	4
France	42	107	−26	−14
Germany	38	100	−27	−16
Iceland	22	71	−21	−6
Italy	46	114	−16	4
Norway	35	95	−32	−26
Spain	51	124	−12	10
Sweden	33	92	−45	−49
Turkey	38	100	−8	17
USSR (Russia)	43	110	−52	−61
ASIA				
Saudi Arabia	51	123	2	35
China	44	111	−23	−10
East Indies	38	101	16	60
India	49	120	−28	−19
Iraq	52	125	−7	19
Japan	38	101	−22	−7
Malaysia	36	97	19	66
Philippine Islands	38	101	14	58
Thailand	41	106	11	52
Tibet	29	85	−29	−20
Turkey	44	111	−30	−22
USSR (Russia)	43	109	−68	−90
Vietnam	45	113	1	33
AFRICA				
Algeria	56	133	−17	1
Angola	33	91	1	33
Egypt	51	124	−1	31
Ethiopia	44	111	0	32
Libya	58	136	2	35
Morocco	48	119	−15	5
Rhodesia	44	112	−8	18
Somalia	34	93	16	61
Sudan	52	126	−2	28
Tunisia	50	122	−2	28
Union of South Africa	44	111	−6	21
Zaire	36	97	1	34
AUSTRALASIA				
Australia	53	127	−7	19
Hawaii	33	91	11	51
New Zealand	34	94	−5	23
Samoan Islands	36	96	16	61
Solomon Islands	36	97	21	70

Precipitation Extremes

United States:

Wettest state:
Louisiana—average annual rainfall 57.34 inches.

Driest state:
Nevada—average annual rainfall 8.60 inches.

Maximum recorded:
Camp Leroy, California (January 22–23, 1943)—26.12 inches in 24 hours.

Minimums recorded:
Bagdad, California (1909–1913)—3.93 inches in 5 years.
Greenland Ranch, California—1.76 inches annual average.

World:

Maximums recorded:

Cherrapunji, India (July, 1861)—366 inches in 1 month. (Average annual rainfall of Cherrapunji is 450 inches).

Baguio, Luzon, Philippines, July 14–15, 1911—46 inches in 24 hours.

Minimums recorded:

Wadi Halfa (Sudan) and Aswan (Egypt) are in the "rainless" area; average annual rainfall is too small to be measured.

Wind-Velocity and Temperature Extremes in North America

Data regarding extremes of temperature and wind velocity for several locations in North America may be found in Table 2.

WORLD TIME CHART

Time differences between selected major world cities may be determined from Chart 1.

MATERIALS AND FINISHES FOR TROPICAL AND MARINE USE

Corrosion

Ordinary finishing of equipment fails to meet satisfactorily conditions encountered in tropical and marine use. Under these conditions, corrosive influences are greatly aggravated by prevailing higher relative humidities, and temperature cycling causes alternate condensation on and evaporation of moisture from finished surfaces. Useful equipment life under adverse atmospheric influences depends largely on proper choice of base materials and finishes applied. Especially important in tropical and marine applications is avoidance of electrical contact between dissimilar metals.

Dissimilar metals widely separated in the galvanic series should not be bolted, riveted, etc., without separation by insulating material at the facing surfaces. The only exception occurs when both surfaces have

TABLE 2. WIND-VELOCITY AND TEMPERATURE EXTREMES IN NORTH AMERICA

Station	Wind* (miles/hour)	Temperature, Degrees Fahrenheit: Maximum	Temperature, Degrees Fahrenheit: Minimum
UNITED STATES, 1871–1955			
Albany, New York	71	104	−26
Amarillo, Texas	84	108	−16
Buffalo, New York	91	99	−21
Charleston, South Carolina	76	104	7
Chicago, Illinois	87	105	−23
Bismarck, North Dakota	72	114	−45
Hatteras, North Carolina	110	97	8
Miami, Florida	132	95	27
Minneapolis, Minnesota	92	108	−34
Mobile, Alabama	87	104	−11
Mt. Washington, New Hampshire	188**	71	−46
Nantucket, Massachusetts	91	95	−6
New York, New York	99	102	−14
North Platte, Nebraska	72	112	−35
Pensacola, Florida	114	103	7
Washington, D.C.	62	106	−15
San Juan, Puerto Rico	149†	94	62
CANADA, 1955			
Banff, Alberta	52‡	97	−60
Kamloops, British Columbia	34‡	107	−37
Sable Island, Nova Scotia	64‡	86	−12
Toronto, Ontario	48‡	105	−46

*Maximum corrected wind velocity (fastest single mile).
**Gusts were recorded at 231 miles/hour (corrected).
†Estimated.
‡For a period of 5 minutes.

Celsius Table of Relative Humidity or Percent of Saturation

To find the relative humidity when the wet-bulb and dry-bulb temperatures are known, consult Table 3.

TABLE 3. CELSIUS TABLE OF RELATIVE HUMIDITY OR PERCENT OF SATURATION

Dry Bulb °C	Difference Between Readings of Wet and Dry Bulbs in Degrees Celsius																																
	0.5	1.0	1.5	2.0	2.5	3.0	3.5	4.0	4.5	5	6	7	8	9	10	11	12	13	14	15	16	18	20	22	24	26	28	30	32	34	36	38	40
4	93	85	77	70	63	56	48	41	34	28	15																						
8	94	87	81	74	68	62	56	50	45	39	28	17																					
12	94	89	84	78	73	68	63	58	53	48	38	30	21	12	4																		
16	95	90	85	81	76	71	67	62	58	54	45	37	29	21	14	7																	
20	96	91	87	82	78	74	70	66	62	58	51	44	36	30	23	17	11																
22	96	92	87	83	79	75	72	68	64	60	53	46	40	34	27	21	16	11															
24	96	92	88	85	81	77	74	70	66	63	56	49	43	37	31	26	21	14	10														
26	96	92	89	85	81	77	74	71	67	64	57	51	45	39	34	28	23	18	13														
28	96	92	89	85	82	78	75	72	68	65	59	53	47	42	37	31	26	21	17	13													
30	96	93	89	86	82	79	76	73	70	67	61	55	50	44	39	35	30	24	20	16	12												
32	96	93	90	86	83	80	77	74	71	68	62	56	51	46	41	36	32	27	23	19	15												
34	97	93	90	87	84	81	77	74	71	69	63	58	53	48	43	38	34	30	26	22	18	10											
36	97	93	90	87	84	81	78	75	72	70	64	59	54	50	45	41	36	32	28	24	21	13											
38	97	94	90	87	84	81	79	76	73	70	65	60	56	51	46	42	38	34	30	26	23	16	10										
40	97	94	91	88	85	82	79	76	74	71	66	61	57	52	48	44	40	36	32	29	25	19	13										
44	97	94	91	88	86	83	80	77	75	73	68	63	59	54	50	47	43	39	36	32	29	23	17	12									
48	97	94	92	89	86	84	81	78	76	74	69	65	61	56	53	49	45	42	39	35	33	27	21	16	12								
52	97	94	92	89	87	84	82	79	77	75	70	66	62	58	55	51	48	44	41	38	35	30	25	20	16	11							
56	97	95	92	90	87	85	83	80	78	76	72	68	64	60	57	53	50	46	43	40	38	32	27	23	19	15	11						
60	98	95	93	90	88	86	83	81	79	77	73	69	65	62	58	55	52	48	45	43	40	35	30	26	21	18	14	11					
70	98	96	93	91	89	87	85	83	81	79	75	71	68	65	61	58	55	52	50	47	44	40	35	31	27	23	20	17	14	11			
80	98	96	94	92	90	88	86	84	83	81	77	74	71	67	64	61	58	56	53	50	48	43	39	35	31	28	24	22	19	16	14	11	
90	98	97	95	93	91	89	87	85	84	82	79	76	73	69	67	64	61	58	56	53	51	47	42	39	35	32	28	26	23	20	18	16	14
100	99	97	95	93	92	90	88	86	85	83	80	77	74	71	68	66	63	60	58	56	54	49	45	42	38	35	32	29	26	24	22	19	17

Example: Assume dry-bulb reading (thermometer exposed directly to atmosphere) is 20°C and wet-bulb reading is 17°C, or a difference of 3°C. The relative humidity at 20°C is then 74%.

CHART 1. WORLD TIME CHART

Aleutian Islands, Tutuila, Samoa	Anchorage, Fairbanks, Hawaiian Islands, Tahiti	Los Angeles, San Francisco, Seattle, Juneau	Chicago, Central America (except Panama), Mexico, Winnipeg	New York, Montreal, Miami, Havana, Panama, Bogota, Lima, Quito	Bermuda, Puerto Rico, Caracas, La Paz, Asuncion	Buenos Aires,* Rio de Janeiro, Santos, Sao Paulo, Montevideo	Iceland	Lisbon, Dublin, Algiers, Dakar, Ascension Island	Greenwich Civil Time (GCT) or Universal Time (UT)	London,* Paris,* Madrid,* Brussels, Rome, Berlin, Vienna, Oslo, Stockholm, Copenhagen, Amsterdam, Tunis, Warsaw	Athens, Israel, Ankara, Cairo, Capetown	Moscow,* Ethiopia, Iraq, Malagasy Republic	Bombay, Sri Lanka, New Delhi	Bangkok, Chungking, Chengtu, Kunming	Hong Kong, Manila, Shanghai, Saigon, Taipeh, Celebes	Japan, Adelaide, Korea, Manchuria	Sydney, Melbourne, Brisbane, Guam, New Guinea, Khabarovsk	Solomon Islands, New Caledonia	Wellington,* Auckland*
1:00	**2:00**	**4:00**	**6:00**	**7:00**	**8:00**	**9:00**	**11:00**	Midnite	0000	1:00	2:00	3:00	5:30	7:00	8:00	9:00	10:00	11:00	11:30
2:00	**3:00**	**5:00**	**7:00**	**8:00**	**9:00**	**10:00**	Midnite	1:00	0100	2:00	3:00	4:00	6:30	8:00	9:00	10:00	11:00	Noon	**12:30**
3:00	**4:00**	**6:00**	**8:00**	**9:00**	**10:00**	**11:00**	1:00	2:00	0200	3:00	4:00	5:00	7:30	9:00	10:00	11:00	Noon	**1:00**	**1:30**
4:00	**5:00**	**7:00**	**9:00**	**10:00**	**11:00**	Midnite	2:00	3:00	0300	4:00	5:00	6:00	8:30	10:00	11:00	Noon	**1:00**	**2:00**	**2:30**
5:00	**6:00**	**8:00**	**10:00**	**11:00**	Midnite	1:00	3:00	4:00	0400	5:00	6:00	7:00	9:30	11:00	Noon	**1:00**	**2:00**	**3:00**	**3:30**
6:00	**7:00**	**9:00**	**11:00**	Midnite	1:00	2:00	4:00	5:00	0500	6:00	7:00	8:00	10:30	Noon	**1:00**	**2:00**	**3:00**	**4:00**	**4:30**
7:00	**8:00**	**10:00**	Midnite	1:00	2:00	3:00	5:00	6:00	0600	7:00	8:00	9:00	11:30	**1:00**	**2:00**	**3:00**	**4:00**	**5:00**	**5:30**
8:00	**9:00**	**11:00**	1:00	2:00	3:00	4:00	6:00	7:00	0700	8:00	9:00	10:00	**12:30**	**2:00**	**3:00**	**4:00**	**5:00**	**6:00**	**6:30**
9:00	**10:00**	Midnite	2:00	3:00	4:00	5:00	7:00	8:00	0800	9:00	10:00	11:00	**1:30**	**3:00**	**4:00**	**5:00**	**6:00**	**7:00**	**7:30**
10:00	**11:00**	1:00	3:00	4:00	5:00	6:00	8:00	9:00	0900	10:00	11:00	Noon	**2:30**	**4:00**	**5:00**	**6:00**	**7:00**	**8.00**	**8.30**
11:00	Midnite	2:00	4:00	5:00	6:00	7:00	9:00	10:00	1000	11:00	Noon	**1:00**	**3:30**	**5:00**	**6:00**	**7:00**	**8:00**	**9:00**	**9:30**
Midnite	1:00	3:00	5:00	6:00	7:00	8:00	10:00	11:00	1100	Noon	**1:00**	**2:00**	**4:30**	**6:00**	**7:00**	**8:00**	**9:00**	**10:00**	**10:30**
1:00	2:00	4:00	6:00	7:00	8:00	9:00	11:00	Noon	1200	**1:00**	**2:00**	**3:00**	**5:30**	**7:00**	**8:00**	**9:00**	**10:00**	**11:00**	**11:30**
2:00	3:00	5:00	7:00	8:00	9:00	10:00	Noon	**1:00**	1300	**2:00**	**3:00**	**4:00**	**6:30**	**8:00**	**9:00**	**10:00**	**11:00**	Midnite	12:30
3:00	4:00	6:00	8:00	9:00	10:00	11:00	**1:00**	**2:00**	1400	**3:00**	**4:00**	**5:00**	**7:30**	**9:00**	**10:00**	**11:00**	Midnite	1:00	1:30
4:00	5:00	7:00	9:00	10:00	11:00	Noon	**2:00**	**3:00**	1500	**4:00**	**5:00**	**6:00**	**8:30**	**10:00**	**11:00**	Midnite	1:00	2:00	2:30
5:00	6:00	8:00	10:00	11:00	Noon	**1:00**	**3:00**	**4:00**	1600	**5:00**	**6:00**	**7:00**	**9:30**	**11:00**	Midnite	1:00	2:00	3:00	3:30
6:00	7:00	9:00	11:00	Noon	**1:00**	**2:00**	**4:00**	**5:00**	1700	**6:00**	**7:00**	**8:00**	**10:30**	Midnite	1:00	2:00	3:00	4:00	4:30
7:00	8:00	10:00	Noon	**1:00**	**2:00**	**3:00**	**5:00**	**6:00**	1800	**7:00**	**8:00**	**9:00**	**11:30**	1:00	2:00	3:00	4:00	5:00	5:30
8:00	9:00	11:00	**1:00**	**2:00**	**3.00**	**4:00**	**6:00**	**7:00**	1900	**8:00**	**9:00**	**10:00**	12:30	2:00	3:00	4:00	5:00	6:00	6:30
9:00	10:00	Noon	**2:00**	**3:00**	**4:00**	**5:00**	**7:00**	**8:00**	2000	**9:00**	**10:00**	**11:00**	1:30	3:00	4:00	5:00	6:00	7:00	7:30
10:00	11:00	**1:00**	**3:00**	**4:00**	**5:00**	**6:00**	**8:00**	**9:00**	2100	**10:00**	**11:00**	Midnite	2:30	4:00	5:00	6:00	7:00	8:00	8:30
11:00	Noon	**2:00**	**4:00**	**5:00**	**6:00**	**7:00**	**9:00**	**10:00**	2200	**11:00**	Midnite	1:00	3:30	5:00	6:00	7:00	8:00	9:00	9:30
Noon	**1:00**	**3:00**	**5:00**	**6:00**	**7:00**	**8:00**	**10:00**	**11:00**	2300	Midnite	1:00	2:00	4:30	6:00	7:00	8:00	9:00	10:00	10:30
1:00	**2:00**	**4:00**	**6:00**	**7:00**	**8:00**	**9:00**	**11:00**	Midnite	2400	1:00	2:00	3:00	5:30	7:00	8:00	9:00	10:00	11:00	11:30

Notes:

(1) Lightface figures designate am, bold figures pm.

(2) Time is that used at places indicated. In general, this is standard time, but for places marked with asterisks it is permanent daylight saving time. Temporary daylight saving time is commonplace but not indicated above.

(3) When passing the heavy line going down or to the right, add 1 day. When passing the heavy line going up or to the left, subtract 1 day.

been coated with the same protective metal, e.g., electroplating, hot dipping, galvanizing, etc.

Aluminum, steel, zinc, and cadmium should never be used bare. Electrical contact surfaces should be given copper-nickel-chromium or copper-nickel finish, and, in addition, they should be silver plated. Adjustable-capacitor plates should be silver plated.

An additional 0.000015-inch to 0.000020-inch electro-plating of hard, bright gold over the silver greatly improves resistance to tarnish and oxidation and to attack by most chemicals, lowers electrical resistance, and provides long-term solderability.

Fungus and Decay

The value of fungicidal coatings or treatments is controversial. If equipment is to operate under tropical conditions, greater success can be achieved by the use of materials that do not provide a nutrient medium for fungus and insects. The following types or kinds of materials are examples of nonnutrient mediums that are generally considered acceptable.

- Metals
- Glass
- Ceramics (steatite, glass-bonded mica)
- Mica
- Polyamide
- Cellulose acetate
- Rubber (natural or synthetic)
- Plastic materials using glass, mica, or asbestos as a filler
- Polyvinylchloride
- Polytetrafluoroethylene
- Monochlortrifluoroethylene

The following types or kinds of materials should not be used, except where such materials are fabricated into completed parts and their use is acceptable to the customer.

- Linen
- Cellulose nitrate
- Regenerated cellulose
- Wood
- Jute
- Leather
- Cork
- Paper and cardboard
- Organic fiberboard
- Hair or wool felts
- Plastic materials using cotton, linen, or wood flour as a filler

Wood should not be used as an electrical insulator, and its use for other purposes should be restricted to those parts for which a superior substitute is not known. When used, it should be pressure-treated and impregnated to resist moisture, insects, and decay with a waterborne preservative (as specified in Federal Specification TT-W-571), and it should also be treated with a suitable fire-retardant chemical.

Finish Application Table

For information regarding the application of finishes, see Table 4.

PRINCIPAL LOW-VOLTAGE POWER SUPPLIES IN THE WORLD

Territory (Frequency) Voltage

North America:

Alaska (60) 120/240
Belize (60) 110/220
Bermuda (60) 115/230; some 120/208
Canada (60) 120/240; some 115/230
Costa Rica (60) 110/220
El Salvador (60) 110/220
Guatemala (60) 110/240; some 220, 120/208
Honduras (60) 110/220
Mexico (50, 60) 127/220 and other voltages
  Mexico City (50) 125/216
Nicaragua (60) 120
Panama (60) 110/220; some 120/240, 115/230
United States (60) 120/240 and 120/208

West Indies:

Antigua (60) 230/400
Bahamas (60) 115/200; some 115/220
Barbados (50) 120/208; some 110/200
Cuba (60) 115/230; some 120/208
Dominican Republic (60) 115/230
Guadeloupe (50) 127/220
Jamaica (50, some 60) 110/220
Martinique (50) 127/220
Puerto Rico (60) 120/240
Trinidad (60) 115/230
Virgin Islands (60) 120/240

South America:

Argentina (50) 220/380; also 220/440 dc
Bolivia (50, also 60) 220 and other voltages
Brazil (50, 60) 110, 220; also other voltages and dc
  Rio de Janeiro (50) 125/216
Chile (50) 220/380; some 220 dc
Colombia (60) 110/220; also 120/240 and others
Ecuador (60) 120/208; also 110/220 and others
French Guiana (50) 127/220
Guyana (50, 60) 110/220
Paraguay (50) 220/440; some 220/440 dc
Peru (60) 220; some 110
Surinam (50, 60) 127/220; some 115/230
Uruguay (50) 220
Venezuela (60, some 50) 120/208, 120/240

Europe:

Austria (50) 220/380; Vienna also has 220/440 dc
Azores (50) 220/380
Belgium (50) 220/380 and many others; some dc

TABLE 4. FINISH APPLICATION TABLE*

Material	Finish	Remarks
Aluminum alloy	Anodizing	An electrochemical-oxidation surface treatment, for improving corrosion resistance; not an electroplating process. For riveted or welded assemblies, specify chromic acid anodizing. Do not anodize parts with nonaluminum inserts. Colors vary: yellow-green, gray, or black.
	"Alrok"	Chemical-dip oxide treatment. Cheap. Inferior in abrasion and corrosion resistance to the anodizing process, but applicable to assemblies of aluminum and nonaluminum materials.
Copper and zinc alloys	Bright acid dip	Immersion of parts in acid solution. Clear lacquer applied to prevent tarnish.
Brass, bronze, zinc die-casting alloys	Brass, chrome, nickel, tin	As discussed under steel.
Magnesium alloy	Dichromate treatment	Corrosion-preventive dichromate dip. Yellow color.
Stainless steel	Passivating treatment	Nitric-acid immunizing dip.
Steel	Cadmium	Electroplate, dull white color, good corrosion resistance, easily scratched, good thread antiseize. Poor wear and galling resistance.
	Chromium	Electroplate, excellent corrosion resistance and lustrous appearance. Relatively expensive. Specify hard chrome plate for exceptionally hard abrasion-resistive surface. Has low coefficient of friction. Used to some extent on nonferrous metals particularly when die-cast. Chrome-plated objects usually receive a base electroplate of copper, then nickel, followed by chromium. Used for buildup of parts that are undersized. Do not use on parts with deep recesses.
	Blueing	Immersion of cleaned and polished steel into heated saltpeter or carbonaceous material. Part then rubbed with linseed oil. Cheap. Poor corrosion resistance.
	Silver plate	Electroplate, frosted appearance; buff to brighten. Tarnishes readily. Good bearing lining. For electrical contacts, reflectors.
	Zinc plate	Dip in molten zinc (galvanizing) or electroplate of low-carbon or low-alloy steels. Low cost. Generally inferior to cadmium plate. Poor appearance. Poor wear resistance: Electroplate has better adherence to base metal than hot-dip coating. For improving corrosion resistance, zinc-plated parts are given special inhibiting treatments.
	Nickel plate	Electroplate, dull white. Does not protect steel from galvanic corrosion. If plating is broken, corrosion of base metal will be hastened. Finishes in dull white, polished, or black. Do not use on parts with deep recesses.
	Black-oxide dip	Nonmetallic chemical black oxidizing treatment for steel, cast iron, and wrought iron. Inferior to electroplate. No buildup. Suitable for parts with close dimensional requirements as gears, worms, and guides. Poor abrasion resistance.
	Phosphate treatment	Nonmetallic chemical treatment for steel and iron products. Suitable for protection of internal surfaces of hollow parts. Small amount of surface buildup. Inferior to metallic electroplate. Poor abrasion resistance. Good paint base.

Continued on next page.

TABLE 4 (CONT.) FINISH APPLICATION TABLE*

Material	Finish	Remarks
	Tin plate	Hot dip or electroplate. Excellent corrosion resistance, but if broken will not protect steel from galvanic corrosion. Also used for copper, brass, and bronze parts that must be soldered after plating. Tin-plated parts can be severely worked and deformed without rupture of plating.
	Brass plate	Electroplate of copper and zinc. Applied to brass and steel parts where uniform appearance is desired. Applied to steel parts when bonding to rubber is desired.
	Copper plate	Electroplate applied before nickel or chrome plates. Also for parts to be brazed or protected against carburization. Tarnishes readily.

* By Z. Fox. Reprinted by permission from *Product Engineering*, Vol. 19, January 1948, p. 161.

Canary Islands (50) 127/220
Denmark (50) 220/380; also 220/440 dc
Finland (50) 220/380
France (50) 120/240, 220/380, and many others
Germany (Federal Republic) (50) 220/380; also others, some dc
Gibraltar (50) 240/415
Greece (50) 220/380; also others, some dc
Iceland (50) 220; some 220/380
Ireland (50) 220/380; some 220/440 dc
Italy (50) 127/220, 220/380 and others
Luxembourg (50) 110/190, 220/380
Madeira (50) 220/380; also 220/440 dc
Malta (50) 240/415
Monaco (50) 127/220, 220/380
Netherlands (50) 220/380; also 127/220
Norway (50) 230
Portugal (50) 220/380; some 110/190
Spain (50) 127/220; also 220/380, some dc
Sweden (50) 127/220, 220/380; some dc
Switzerland (50) 220/380
Turkey (50) 220/380; some 110/190
United Kingdom (50) 240/415 and others, some dc
Yugoslavia (50) 220/380

Asia:

Afghanistan (50) 220/380
Burma (50) 230
Cambodia (50) 120/208; some 220/380
Cyprus (50) 240
Hong Kong (50) 200/346
India (50) 230/400 and others, some dc
Indonesia (50) 127/220
Iran (50) 220/380
Iraq (50) 220/380
Israel (50) 230/400
Japan (50, 60) 100/200
Jordan (50) 220/380
Korea (60) 100/200
Kuwait (50) 240/415
Laos (50) 127/220; some 220/380
Lebanon (50) 110/190; some 220/380
Malaysia (50) 230/400; some 240/415
Nepal (50) 110/220
Okinawa (60) 120/240
Pakistan (50) 230/400 and others, some dc
Philippines (60) 110, 220, and others
Saudi Arabia (50, 60) 120/208; also 220/380, 230/400
Singapore (50) 230/400
Sri Lanka (50) 230/400
Syria (50) 115/200; some 220/380
Taiwan (60) 100/200
Thailand (50) 220/380; also 110/190
Vietnam (50) 220/380 future standard
Yemen Arab Republic (50) 220
Yemen, Peoples Democratic Republic (50) 230/400

Africa:

Algeria (50) 127/220, 220/380
Angola (50) 220/380
Dahomey (50) 220/380
Egypt (50) 110, 220 and others; some dc
Ethiopia (50) 220/380; some 127/220
Guinea (50) 220/380; some 127/220
Kenya (50) 240/415
Liberia (60) 120/240
Libya (50) 125/220; some 230/400
Malagasy Republic (50) 220/380; some 127/220
Mauritius (50) 230/400
Morocco (50) 115/200; also 230/400 and others
Mozambique (50) 220/380
Niger (50) 220/380
Nigeria (50) 230/400
Rhodesia (50) 220/380; also 230/400
Senegal (50) 127/220
Sierra Leone (50) 230/400
Somalia (50) 220/440; also 110, 230
South Africa (50) 220/380; also others, some dc
Sudan (50) 240/415
Tanganyika (50) 230/400
Tunisia (50) 220/380; also others
Uganda (50) 240/415
Upper Volta (50) 220/380
Zaire (50) 220/380

Oceania:

Australia (50) 240/415; also others and dc
Fiji Islands (50) 240/415
Hawaii (60) 120/240
New Caledonia (50) 220/440
New Zealand (50) 230/400

Notes:

1. Abstracted from *Electric Power Abroad*, issued 1963 by the Bureau of International Commerce of the US Department of Commerce. This pamphlet is obtainable from the Superintendent of Documents, US Government Printing Office, Washington, D.C. 20402.
2. The listings show electric (residential) power supplied in each country; as indicated, in very many cases other types of supply also exist to a greater or lesser extent. Therefore, for specific characteristics of the power supply of particular cities, reference should be made to *Electric Power Abroad*. This pamphlet also gives additional details such as number of phases, number of wires to the residence, frequency stability, grounding regulations, and some data on types of commercial service.
3. In the United States in urban areas, the usual supply is 60-hertz 3-phase 120/208 volts; in less densely populated areas, it is usually 120/240 volts, single phase, to each customer. Any other supplies, including dc, are rare and are becoming more so. Additional information for the US is given in the current edition of *Directory of Electric Utilities*, published by McGraw-Hill Book Co., New York, N.Y.
4. All voltages in the table are ac except where specifically stated as dc. The latter are infrequent and in most cases are being replaced by ac. The lower voltages shown for ac, wye or delta ac, or for dc distribution lines, are used mostly for lighting and small appliances; the higher voltages are used for larger appliances.

POWER SUPPLY WIRING

Electric power supply (mains) wiring is usually controlled for public safety by local or state government boards, based in the USA primarily on the National Electrical Code (NEC)* and the National Electric Safety Codes.† Brief extracts from some NEC requirements are given here for convenient reference.

Many products such as wire and cable, fuses, outlet boxes, appliances, etc., are governed within the USA by Underwriters Laboratories (UL) Standards, which specify the terminology used for the various classes of an item as well as the safety requirements that must be met by UL approved items. Note that the overall performance of assemblies such as appliances, motors, radio equipment, or television equipment is not covered by NEC or UL standards, which are primarily for personnel safety.

The following tables are provided.

Tables 5 and 6: NEC standard types of insulated wires and cables.

Tables 7 and 8: Allowable currents for conductors.

Table 8: Derating factors to be applied for ambient temperatures above 30°C (86°F) and for more than three conductors in a cable or conduit.

Table 9: Motor starting currents, which determine the overcurrent protection requirements during the starting period.

Table 10: Motor full-load operating currents for usual conditions and speeds.

Guide to Use of Tables

Determine the total equipment load by adding the loads of the various individual items, estimating motor currents according to Table 10 if specific operating-current information is not otherwise available. Any load substantially bigger than this should be interrupted by an overload protective device; normally the next-larger standard fuse or circuit breaker is considered satisfactory.

Determine the total starting load by using the locked-rotor currents computed from Table 9 and the steady-state currents for resistive devices. Make an additional allowance for any large quantities of tungsten lamps, starting transients, and high-inertia loads that will increase the duration of the starting period. The circuit overload protection must be designed to carry this load for the entire starting period. Time-lag fuses or time-delay circuit breakers are usually desirable.

Using the starting currents, determine the voltage drops in the supply circuit; thus be sure that the terminal voltage of the motor or other device will be adequate at start. Increase the size of the supply conductor or reduce the source impedance if necessary. From the starting and running currents, determine the required size of supply conductors.

WIRING OF ELECTRONIC EQUIPMENT AND CHASSIS

There are few official standards for the internal wiring of electronic equipment and chassis. Nevertheless, the following points should be considered.

(A) Probable maximum continuous ambient temperature where the wiring is located.

(B) Allowable temperature rise of conductor surface under full-load conditions (determines minimum wire size).

(C) Maximum voltage to ground or to surrounding metal parts (determines required insulation thickness).

(D) Possibility of corona on high-voltage leads;

* American National Standards Institute, Inc., ANSI Standard C1, prepared by the National Fire Protection Association.

† ANSI Standard C2.

TABLE 5. NEC CONDUCTOR APPLICATIONS AND INSULATIONS

Trade Name	Type Letter	Maximum Operating Temperature	Application	Insulation	Outer Covering
Heat-resistant rubber	RH	75°C 167°F	Dry locations	Heat-resistant rubber	Moisture-resistant flame-retardant non-metallic covering
Heat-resistant rubber	RHH	90°C 194°F	Dry locations	Heat-resistant rubber	Moisture-resistant flame-retardant non-metallic covering
Moisture- and heat-resistant rubber	RHW	75°C 167°F	Dry and wet locations	Moisture- and heat-resistant rubber	Moisture-resistant flame-retardant non-metallic covering
Heat-resistant latex rubber	RUH	75°C 167°F	Dry locations	90% unmilled grainless rubber	Moisture-resistant flame-retardant non-metallic covering
Moisture-resistant latex rubber	RUW	60°C 140°F	Dry and wet locations	90% unmilled grainless rubber	Moisture-resistant flame-retardant non-metallic covering
Thermoplastic	T	60°C 140°F	Dry locations	Flame-retardant thermoplastic compound	None
Moisture-resistant thermoplastic	TW	60°C 140°F	Dry and wet locations	Flame-retardant moisture-resistant thermoplastic	None
Moisture- and heat-resistant thermoplastic	THW	75°C 167°F	Dry and wet locations	Flame-retardant moisture- and heat-resistant thermoplastic	None
Heat-resistant thermoplastic	THHN	90°C 194°F	Dry locations	Flame-retardant heat-resistant thermoplastic	Nylon jacket
Moisture- and heat-resistant thermoplastic	THWN	75°C 107°F	Dry and wet locations	Flame-retardant, moisture- and heat-resistant thermoplastic	Nylon jacket
Moisture- and heat-resistant cross-linked synthetic polymer	XHHW	90°C 194°F	Dry locations	Flame-retardant cross-linked synthetic polymer	None
Extruded polytetra-fluoroethylene	TFE	250°C 482°F	Dry locations	Extruded polytetra-fluoroethylene	None
Silicone asbestos	SA	90°C 194°F	Dry locations	Silicone rubber	Asbestos
Fluorinated Ethylene Polypylene	FEP	90°C 194°F	Dry locations	Fluorinated Ethylene Propylene	None
Varnished Cambric	V	85°C 185°F	Dry locations	Varnished Cambric	Nonmetallic Covering or lead sheath

some insulating materials deteriorate rapidly under corona conditions.

(E) Need for shield braid on some conductors to reduce noise pickup. Shields must be insulated if positive single-point grounding is to be attained.

(F) Skin effect on conductors carrying high radio-frequency currents.

(G) Vibration, shock, or relative motion of conductors during normal use of the equipment. Stranded or flexible conductors and adequate clamping or other tie-down of conductors and cables may be essential.

(H) Wiring should be shielded from the direct heat radiation of high-temperature parts such as electron tubes and power resistors.

(I) Wire identification may be required for convenience in manufacture, installation, or servicing.

As a matter of expediency, most electronic equipment employs the smallest conveniently handled wire size (usually 20 to 24 AWG) for most wiring, with the larger conductors being installed only for circuits carrying currents greater than that permitted for the "general-use" wire size. With the trend toward compact solid-state integrated-circuit equipment, smaller wire sizes are being used. However, the reduction in wiring bundle size may be small unless the conductor insulation thickness (determined by voltage considerations) can be reduced.

Table 11 gives recommended current ratings for copper and some aluminum based on a 45°C (40°C for wires smaller than 22 AWG) conductor temperature

TABLE 6. NEC FLEXIBLE-CORD DATA

Trade Name*	Type Letter†	Size Range (AWG)	No. of Conductors	Insulation	Outer Covering
All-rubber parallel cord	SP-3	18 – 12	2 or 3	Rubber	Rubber
All-plastic parallel cord	SPT-3	18 – 10	2 or 3	Thermoplastic	Thermoplastic
Lamp cord	C	18 – 10	2 or more	Rubber	None
Twisted portable cord	PD	18 – 10	2 or more	Rubber	Cotton or rayon
Vacuum-cleaner cord	SV	18	2 or 3	Rubber	Rubber
Vacuum-cleaner cord	SVT	18 – 17	2 or 3	Thermoplastic	Thermoplastic
Junior hard-service cord	SJ	18 – 14	2 – 4	Rubber	Rubber
Junior hard-service cord	SJO	18 – 14	2 – 4	Rubber	Oil-resistant compound
Junior hard-service cord	SJT	18 – 14	2 – 4	Rubber or thermoplastic	Thermoplastic
Junior hard-service cord	SJTO	18 – 14	2 – 4	Rubber or thermoplastic	Oil-resistant thermo-plastic
Hard-service cord	S	18 – 2	2 or more	Rubber	Rubber
Hard-service cord	SO	18 – 2	2 or more	Rubber	Oil-resistant compound
Hard-service cord	ST	18 – 2	2 or more	Rubber or thermoplastic	Thermoplastic
Hard-service cord	STO	18 – 2	2 or more	Rubber or thermoplastic	Oil-resistant thermo-plastic
Rubber-jacketed heat-resistant cord	AFSJ	18 – 16	2 or 3	Impregnated asbestos	Rubber
Rubber-jacketed heat-resistant cord	AFS	18 – 14	2 or 3	Impregnated asbestos	Rubber
Heater cord	HPD	18 – 12	2 – 4	Rubber or thermoplastic with asbestos or all neoprene	Cotton or rayon
Rubber-jacketed heater cord	HSJ	18 – 16	2 – 4	Rubber or thermoplastic and asbestos or all neoprene	Cotton and rubber
Jacketed heater cord	HSJO	18 – 16	2 – 4	Rubber with asbestos or all neoprene	Cotton and oil-resistant compound
Jacketed heater cord	HS	14 – 12	2 – 4	Rubber with asbestos or all neoprene	Cotton and rubber or neoprene
Jacketed heater cord	HSO	14 – 12	2 – 4	Rubber with asbestos or all neoprene	Cotton and oil-resistant compound
Parallel heater cord	HPN	18 – 12	2 or 3	Thermosetting	Thermosetting

*All types shown are recommended for use in damp locations.
†"S" series cords may also be used in pendant applications.

rise due to load current. Table 11 may be used for the temperature conditions given in Table 12.

A 60°C ambient temperature around the wiring (20°C internal temperature rise from 40°C [104°F] ambient around the equipment) is typical of some electronic equipment. If higher ambient temperatures, high power, or compact designs with electron tubes or magnetic-core components (except for very low power) are a factor, the temperature in the wiring space should be specifically determined.

"Wiring confined" ratings are based on 15 or more wires in a bundle, with the sum of all the actual load currents of the bundled wires not exceeding 20% of the permitted "wiring confined" sum total carrying capacity of the bundled wires. These ratings approximate 60% of the free-air ratings (with some variations due to rounding). They should be used for wire in harnesses, cable, conduit, and general chassis conditions. Bundles of fewer than 15 wires may have the allowable sum of the load currents increased as the bundle approaches the single-wire condition.

RESISTANCE CHANGE WITH TEMPERATURE

The resistance of most conductor materials changes with temperature. Table 11 shows the copper-wire resistance at 100°C. Correction factors must be applied to determine the resistance at other temperatures, or for other materials. Thus, from Table 11 determine the copper-wire resistance at 100°C (multiply by conversion factor m of Table 13 for other materials). Use the equation

$$R_t = R_r m[1 + K(t - 100)]$$

where,

R_t = resistance at desired temperature t,

R_r = resistance at 100°C for copper (Table 11),

m = material factor for 100°C resistance value (Table 13),

K = correction factor (Table 13),

t = desired temperature (°C).

TABLE 7. NEC CURRENT-CARRYING CAPACITY, IN AMPERES, OF FLEXIBLE CORDS

Size AWG	Rubber TP,TS Thermoplastic TPT, TSP	Rubber C, PD, E, EO, EN, S, SO, SRD, SS, SSO, SV, SVO, SP Thermoplastic ET, ETT, ETLB, ETP, ST, STO, SRDT, SVT, SVTO, SPT		AFS, AFSJ, HPD, HSJ, HSJO, HS HSO, HPN	Cotton* CFPD. Asbestos* AFC, AFPD
27†	0.5	—	—	—	—
18		7‡	10§	10	6
16		10‡	13§	15	8
14		15‡	18§	20	17
12		20‡	25§	30	23
10		25‡	30§	35	28
8		35‡	40§	—	—
6		45‡	55§	—	—
4		60‡	70§	—	—
2		80‡	95§	—	—

* Generally used in fixtures exposed to high temperatures, derated accordingly.
† Tinsel.
‡ Three-conductor and other multiconductor cords connected so only three conductors are current-carrying.
§ Two-conductor and other multiconductor cords connected so only two conductors are current-carrying.

Notes:
1. For not more than three current-carrying conductors in a cord. If four to six conductors are used, allowable capacity of each conductor shall be reduced to 80% of values for not more than three current-carrying conductors.
2. A conductor used for equipment grounding and a neutral conductor which carries only the unbalanced current from other conductors shall not be considered as current-carrying conductors.
3. Based on room temperature of 30°C (86°F).

WIRE IDENTIFICATION

In a complex wiring assembly, or if both ends of a wire cannot be seen from one station, a means of identifying each lead simplifies manufacture, installation, and servicing. Common identification methods are:

(A) Tag each end of a lead with an assigned designation (an alternative method is to print the designation at frequent intervals along the wire insulation).

(B) Color code the wires. The wire insulation may be a solid color, color stripes may be spiraled around the wire, or the name of the color (or its numerical code equivalent) may be stamped at frequent intervals along the wire.

Color Coding

The commonly used colors and their numerical codes are:

0	Black	5	Green
1	Brown	6	Blue
2	Red	7	Violet (purple)
3	Orange	8	Gray (slate)
4	Yellow	9	White

While spiral stripes can be applied on top of any basic insulation color, under less favorable viewing conditions it is difficult to distinguish some colors from the basic insulation color. Identification may be slow and subject to error. The preferred combination consists of one or two (sometimes three) colored stripes on a white basic insulation. To minimize identification errors, the first stripe is made wider than the second (or third), and some rules require that the second stripe be of higher numerical code than the first stripe. If the required variety of wire color codes is not great, the preceding guides should be followed.

Table 14 gives a standard color code used to distinguish by function the various leads in electronic circuits.

In manufacturing practice, it is preferred that, at any harness breakout point, all wires of the same color code be connected to the same terminal at that location. When this rule and the wire color coding of Table 14 are both applicable, additional tracers may be used to supplement the primary coding of Table 14.

DIAMETER OF CIRCLE ENCLOSING A GIVEN NUMBER OF SMALLER CIRCLES*

Four of many possible compact arrangements of circles within a circle are shown in Fig. 3. To determine the diameter of the smallest enclosing circle for a particular number of enclosed circles all of the same size, three factors that influence the size of the enclosing circle should be considered, as follows.

* J. Dutka. "How Many Wires Can Be Packed Into a Circular Conduit," *Machinery's Handbook*. New York: Industrial Press, Inc., 1956.

TABLE 8. NEC ALLOWABLE CURRENT-CARRYING CAPACITIES, IN AMPERES, OF CONDUCTORS

	Copper-Conductor Insulation				Aluminum or Copper-Clad Aluminum Conductor Insulation		
Size AWG	RUW (14 – 2) T, TW, UF	RH, RHW, RUH, (14 – 2) THW, THWN, XHHW	TA, TBS, SA, FEP, FEPB, RHH, THHN, XHHW*	TFE†	RUW (12 – 2) T, TW, UF	RH, RHW, RUH (12 – 2), THW, THWN, XHHW	TA, TBS, SA, RHH, THHN, XHHW*
14	15	15	25‡	40	—	—	—
12	20	20	30‡	55	15	15	25§
10	30	30	40‡	75	25	25	30§
8	40	45	50	95	30	40	40
6	55	65	70	120	40	50	55
4	70	85	90	145	55	65	70
3	80	100	105	170	65	75	80
2	95	115	120	195	75	90	95
1	110	130	140	220	85	100	110
0	125	150	155	250	100	120	125
00	145	175	185	280	115	135	145
000	165	200	210	315	130	155	165
0000	195	230	235	370	155	180	185
Correction Factors for Higher Room Temperatures °C °F							
40 104	0.82	0.88	0.91	—	0.82	0.88	0.91
45 113	0.71	0.82	0.87	—	0.71	0.82	0.87
50 122	0.58	0.75	0.82	—	0.58	0.75	0.82
55 131	0.41	0.67	0.76	—	0.41	0.67	0.76
60 140	—	0.58	0.71	.95	—	0.58	0.71

*Dry locations only.
† Nickel or nickel-coated copper only.
‡ For types FEP, FEPB, RHH, THHN, and XHHW, sizes 14, 12, 10 shall be the same as designated for RH, RHW, etc.
§ For types RHH, THHN, and XHHW, sizes 12 and 10 shall be the same as designated for RH, RHW, etc.

Notes:

1. Not more than three conductors in raceway or cable.
2. Based on room temperature of 30°C (86°F). See correction factors for higher temperatures.
3. Derating factors—more than three conductors in raceway or cable:

Number of conductors	4 – 6	7 – 24	25 – 42	>42
% of current capacity	80	70	60	50

Arrangement of Center or Core Circles

The four most common arrangements of center or core circles are shown in cross section in Fig. 3. It may seem that Fig. 3A would require the smallest enclosing circle for a given number of enclosed circles, but this is not always the case since the most compact arrangement will depend in part on the number of circles to be enclosed.

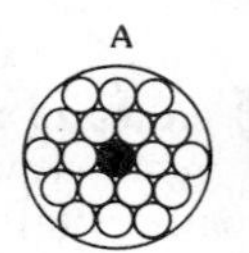

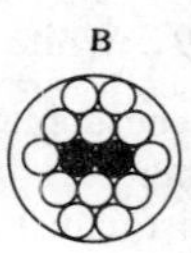

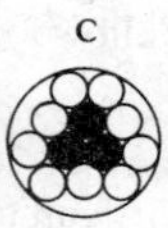

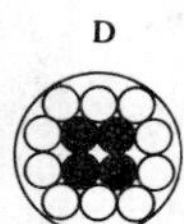

Fig. 3. Arrangements of circles within a circle. (*Reprinted with permission from* Machinery's Handbook, *17th Edition, Industrial Press, Inc., New York.*)

TABLE 9. NEC MOTOR STARTING-CURRENT DATA*

Code Letter	Kilovolt-Amperes per Horsepower with Locked Rotor
A	0 – 3.14
B	3.15 – 3.54
C	3.55 – 3.99
D	4.00 – 4.49
E	4.50 – 4.99
F	5.00 – 5.59
G	5.60 – 6.29
H	6.30 – 7.09
J	7.10 – 7.99
K	8.00 – 8.99
L	9.00 – 9.99
M	10.00 – 11.19
N	11.20 – 12.49
P	12.50 – 13.99
R	14.00 – 15.99
S	16.00 – 17.99
T	18.00 – 19.99
U	20.00 – 22.39
V	22.40 up

* Locked-rotor currents of motors are useful in determining branch-circuit overcurrent protection requirements and voltage drop at start. These values are indicated by code letters on motor nameplate. Note NEMA standard.

plete "layers" around a central core consisting of one circle; this agrees with the data shown in the left half of Table 15 for $n=2$.

To determine the diameter of the enclosing circle, the data in the right half of Table 15 are used. Thus, for $n=2$ and an "A" pattern, diameter D is 5 times the diameter, d, of the enclosed circles.

Diameter of Enclosing Circle When Outer Layer of Circles Is Not Complete

In most cases, it is possible to reduce the size of the enclosing circle from that required if the outer layer were complete. Thus, for example, Fig. 3B shows that the central core consisting of two circles is surrounded by one complete layer of eight circles and one partial outer layer of four circles so that the total number of circles enclosed is 14. If the outer layer were complete, then (from Table 15) the total number of enclosed circles would be 24 and the diameter of the enclosing circle would be $6d$; however, since the outer layer is composed of only four circles out of a possible 14 for a complete second layer, a smaller diameter of enclosing circle may be used. Table 16 shows that for a total of 14 enclosed circles arranged in a "B" pattern with the outer layer of circles incomplete, the diameter for the enclosing circle is $4.606d$.

TABLE 10. NEC MOTOR FULL-LOAD RUNNING CURRENTS IN AMPERES (USUAL CONDITIONS AND SPEEDS)

Horsepower	Single-Phase AC		3-Phase AC†			DC	
	115 V	230 V*	115 V	230 V*	460 V	120 V	240 V
1/6	4.4	2.2	—	—	—	—	—
1/4	5.8	2.9	—	—	—	3.1	1.6
1/3	7.2	3.6	—	—	—	4.1	2.0
1/2	9.8	4.9	4	2	1	5.4	2.7
3/4	13.8	6.9	5.6	2.8	1.4	7.6	3.8
1	16	8	7.2	3.6	1.8	9.5	4.7
1½	20	10	10.4	5.2	2.6	13.2	6.6
2	24	12	13.6	6.8	3.4	17	8.5
3	34	17	—	9.6	4.8	25	12.2
5	56	28	—	15.2	7.6	40	20
7½	80	40	—	22	11	58	29
10	100	50	—	28	14	76	38

* For 208 V, multiply by 1.1; for 200 V, multiply by 1.15.
† Induction type, squirrel cage and wound rotor.

Diameter of Enclosing Circle When Outer Layer of Circles Is Complete

Successive, complete "layers" of circles may be placed around each of the central cores of 1, 2, 3, or 4 circles. The number of circles contained in arrangements of complete "layers" around a central core of circles, as well as the diameter of the enclosing circle, may be obtained from Table 15. Thus, for example, Fig. 3A has a total of 18 circles arranged in two com-

Table 16 can be used to determine the smallest enclosing circle for a given number of circles to be enclosed by direct comparison of the "A," "B," and "C" columns. For data outside the range of Table 16, use the equations in Dr. Dutka's article.*

*J. Dutka. "How Many Wires Can Be Packed Into a Circular Conduit," *Machinery's Handbook*. New York: Industrial Press, Inc., 1956.

TABLE 11. RECOMMENDED CURRENT RATINGS (CONTINUOUS DUTY) FOR ELECTRONIC EQUIPMENT AND CHASSIS WIRING*

Wire Size		Copper Conductor (100°C) Nominal Resistance (Ohms/1000 ft)	Maximum Current in Amperes			
			Copper Wire		Aluminum Wire	
AWG	Circular Mils		Wiring in Free Air	Wiring Confined	Wiring in Free Air	Wiring Confined
32	63.2	188.0	0.53	0.32		
30	100.5	116.0	0.86	0.52		
28	159.8	72.0	1.4	0.83		
26	254.1	45.2	2.2	1.3		
24	404.0	28.4	3.5	2.1		
22	642.4	22.0	7.0	5.0		
20	1022	13.7	11.0	7.5		
18	1624	6.50	16	10		
16	2583	5.15	22	13		
14	4107	3.20	32	17		
12	6530	2.02	41	23		
10	10 380	1.31	55	33		
8	16 510	0.734	73	46	60	36
6	26 250	0.459	101	60	83	50
4	41 740	0.290	135	80	108	66
2	66 370	0.185	181	100	152	82
1	83 690	0.151	211	125	174	105
0	105 500	0.117	245	150	202	123
00	133 100	0.092	283	175	235	145
000	167 800	0.074	328	200	266	162
0000	211 600	0.059	380	225	303	190

* See Table 12.

TABLE 12. TEMPERATURE CONDITIONS FOR TABLE 11

Maximum Allowable Conductor Temperature °C	Maximum Ambient Temperature Around Wire °C	Typical Conductor and Insulation
105	60	Bare or tinned copper or aluminum; polyvinyl-chloride insulation
200	155	Silver-coated copper; FEP or PTFE with FEP jacket insulation*
260	215	Nickel-coated copper; PTFE insulation*

* FEP = Fluorinated ethylene propylene.
PTFE = Polytetrafluoroethylene.

TABLE 13. TEMPERATURE AND MATERIALS CORRECTION FACTORS

Conductor Material	Material Factor m	Correction Factor K
Soft copper	1.00	0.0039
Hard copper	1.03	0.0038
Copper-clad steel:		
30% conductivity	3.47	0.0044
40% conductivity	2.56	0.0041
Aluminum	1.64	0.0039
Nickel	5.28	0.0047
Nickel-clad copper:		
10%	1.07	0.0038
30%	1.35	0.0036
Silver	0.94	0.0038

Approximate Equation When Number of Enclosed Circles is Large

When a large number of circles is to be enclosed, the arrangement of the center circles has little effect on the diameter of the enclosing circle. For numbers of circles greater than 10 000, the diameter of the enclosing circle may be calculated within 2 percent from the equation

$$D = d[1 + (N/0.907)^{1/2}]$$

where,

D = diameter of enclosing circle,

d = diameter of enclosed circles,

N = number of enclosed circles.

TORQUE AND HORSEPOWER

Torque varies directly with power and inversely with rotating speed of the shaft, or

$$T = KP/N$$

where,

T = torque in pound-inches,

TABLE 14. COLORS FOR WIRE IDENTIFICATION BY FUNCTION

Function	Color	Identification No.
Grounds, grounded elements	Black	0
Heaters or filaments	Brown	1
Power supply B +	Red	2
Screen grids	Orange	3
Cathodes and transistor emitters*†	Yellow	4
Control grids and transistor bases†	Green	5
Anodes (plates) and transistor collectors*†	Blue	6
Power supply, negative (−)	Violet (purple)	7
Ac power lines	Gray (slate)	8

* Applies to diodes, semiconductor elements, photoelectric cells, mercury-arc rectifiers, and other elements with operation similar to vacuum tubes and transistors.
† Applies to all types of gas tubes with operation similar to vacuum tubes.

take the form of a parabola instead of their actual form of a catenary. The error is negligible, and the computations are much simplified. In calculating sags, the changes in cables due to variations in load and temperature must be considered.

Supports at Same Level

Refer to Fig. 4. The equations used in calculating sags are

$$H = WL^2/8S$$

$$S = WL^2/8H = [(L_c - L)3L/8]^{1/2}$$

$$L_c = L + 8S^2/3L$$

where,

L = length of span in feet,

L_c = length of cable in feet,

S = sag of cable at center of span in feet,

H = tension in cable at center of span in pounds = horizontal component of the tension at any point,

TABLE 15. NUMBER OF CIRCLES CONTAINED IN COMPLETE LAYERS OF CIRCLES AND DIAMETER OF ENCLOSING CIRCLE

No.	Number of Circles in Center Pattern: 1	2	3	4	1	2	3	4
	Arrangement of Circles in Center Pattern (see Fig. 3)							
Complete Layers Over Core, n	"A"	"B"	"C"	"D"	"A"	"B"	"C"	"D"
	Number of Circles, N, Enclosed				Diameter, D, of Enclosing Circle*			
0	1	2	3	4	d	$2d$	$2.155d$	$2.414d$
1	7	10	12	14	$3d$	$4d$	$4.055d$	$4.386d$
2	19	24	27	30	$5d$	$6d$	$6.033d$	$6.379d$
3	37	44	48	52	$7d$	$8d$	$8.024d$	$8.375d$
4	61	70	75	80	$9d$	$10d$	$10.018d$	$10.373d$
5	91	102	108	114	$11d$	$12d$	$12.015d$	$12.372d$
n	†	†	†	†	†	†	†	†

* Diameter D is given in terms of d, the diameter of the enclosed circles.
† For n complete layers over core, the number of enclosed circles N for "A" center pattern is $3n^2+3n+1$; for "B," $3n^2+5n+2$; for "C," $3n^2+6n+3$; for "D," $3n^2+7n+4$; while the diameter D of the enclosing circle for "A" center pattern is $(2n+1)d$; for "B," $(2n+2)d$; for "C," $[1+2(n^2+n+\frac{1}{3})^{1/2}]d$; and for "D," $[1+(4n^2+5.644n+2)^{1/2}]d$.
Reprinted with permission from *Machinery's Handbook*, 17th edition. New York: Industrial Press, Inc.

P = horsepower,

N = revolutions/minute,

K = 63 000 (constant).

TRANSMISSION-LINE SAG CALCULATIONS†

For transmission-line work, with towers on the same or slightly different levels, the cables are assumed to

† Reprinted by permission from *Transmission Towers*. Pittsburgh, Pa.: American Bridge Co., 1923; p. 70.

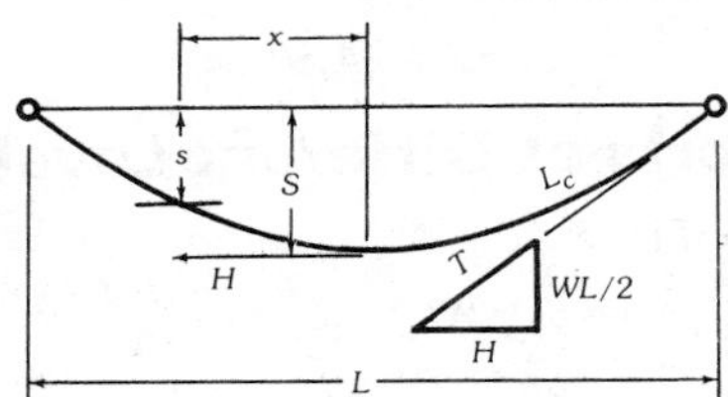

Fig. 4. Supports at same elevation.

TABLE 16. FACTORS FOR DETERMINING DIAMETER, D, OF SMALLEST ENCLOSING CIRCLE FOR VARIOUS NUMBERS, N, OF ENCLOSED CIRCLES*

No. N	Center Circle Pattern "A"	Center Circle Pattern "B"	Center Circle Pattern "C"	No. N	Center Circle Pattern "A"	Center Circle Pattern "B"	Center Circle Pattern "C"	No. N	Center Circle Pattern "A"	Center Circle Pattern "B"	Center Circle Pattern "C"
	Diameter Factor K				Diameter Factor K				Diameter Factor K		
2	3	2	—	34	7	7.083	7.110	66	9.718	9.544	9.326
3	3	2.732	2.155	35	7	7.245	7.110	67	9.718	9.544	9.326
4	3	2.732	3.309	36	7	7.245	7.110	68	9.718	9.544	9.326
5	3	3.646	3.309	37	7	7.245	7.429	69	9.718	9.660	9.326
6	3	3.646	3.309	38	7.928	7.245	7.429	70	9.718	9.660	10.018
7	3	3.646	4.055	39	7.928	7.557	7.429	71	9.718	9.888	10.018
8	4.464	3.646	4.055	40	7.928	7.557	7.429	72	9.718	9.888	10.018
9	4.464	4	4.055	41	7.928	7.557	7.429	73	9.718	9.888	10.018
10	4.464	4	4.055	42	7.928	7.557	7.429	74	10.165	9.888	10.018
11	4.464	4.606	4.055	43	7.928	8	8.024	75	10.165	10	10.018
12	4.464	4.606	4.055	44	8.211	8	8.024	76	10.165	10	10.238
13	4.464	4.606	5.163	45	8.211	8	8.024	77	10.165	10.539	10.238
14	5	4.606	5.163	46	8.211	8	8.024	78	10.165	10.539	10.238
15	5	5.359	5.163	47	8.211	8	8.024	79	10.165	10.539	10.452
16	5	5.359	5.163	48	8.211	8	8.024	80	10.165	10.539	10.452
17	5	5.359	5.163	49	8.211	8.550	8.572	81	10.165	10.539	10.452
18	5	5.359	5.163	50	8.211	8.550	8.572	82	10.165	10.539	10.452
19	5	5.583	5.619	51	8.211	8.550	8.572	83	10.165	10.539	10.452
20	6.292	5.583	5.619	52	8.211	8.550	8.572	84	10.165	10.539	10.452
21	6.292	5.583	5.619	53	8.211	8.810	8.572	85	10.165	10.644	10.866
22	6.292	5.583	6.033	54	8.211	8.810	8.572	86	11	10.644	10.866
23	6.292	6	6.033	55	8.211	8.810	9.083	87	11	10.644	10.866
24	6.292	6	6.033	56	9	8.810	9.083	88	11	10.644	10.866
25	6.292	6.196	6.033	57	9	8.937	9.083	89	11	10.849	10.866
26	6.292	6.196	6.033	58	9	8.937	9.083	90	11	10.849	10.866
27	6.292	6.568	6.033	59	9	8.937	9.083	91	11	10.849	11.214
28	6.292	6.568	6.773	60	9	8.937	9.083	92	11.392	10.849	11.214
29	6.292	6.568	6.773	61	9	9.185	9.083	93	11.392	11.149	11.214
30	6.292	6.568	6.773	62	9.718	9.185	9.083	94	11.392	11.149	11.214
31	6.292	7.083	7.110	63	9.718	9.185	9.083	95	11.392	11.149	11.214
32	7	7.083	7.110	64	9.718	9.185	9.326	96	11.392	11.149	11.214
33	7	7.083	7.110	65	9.718	9.544	9.326	97	11.392	11.440	11.214

* The diameter, D, of the enclosing circle is equal to the diameter factor, K, multiplied by d, the diameter of the enclosed circles, or $D = K \times d$. For example, if the number of circles to be enclosed, N, is 12, and the center circle arrangement is "C," then for $d = 1\frac{1}{2}$ inches, $D = 4.005 \times 1\frac{1}{2} = 6.083$ inches. Reprinted with permission from *Machinery's Handbook*, 17th Edition. New York: Industrial Press, Inc.

W = weight of cable in pounds per lineal foot.

If cables are subject to wind and ice loads, W = the algebraic sum of the loads. That is, for ice on cables, W = weight of cables plus weight of ice; for wind on bare or ice-covered cables, W = the square root of the sum of the squares of the vertical and horizontal loads.

For any intermediate point at a distance x from the center of the span, the sag is

$$S_x = S(1 - 4x^2/L^2)$$

Supports at Different Levels

Refer to Fig. 5.

$$S = S_0 = WL_0^2 \cos a/8T$$

$$= WL^2/8T \cos a$$

$$S_1 = WL_1^2/8H$$

$$S_2 = WL_2^2/8H$$

$$L_1/2 = L/2 - (hH \cos a/WL)$$

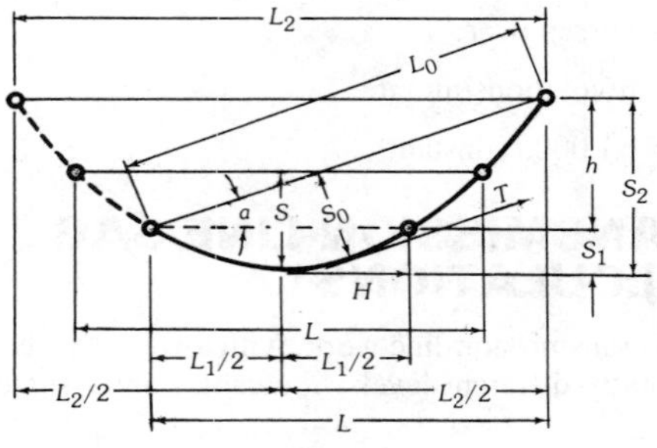

Fig. 5. Supports at different elevations.

$$L_2/2 = L/2 + (hH \cos a/WL)$$

$$L_c = L + \tfrac{4}{3}[(S_1^2/L_1) + (S_2^2/L_2)]$$

where,

W = weight of cable in pounds per lineal foot between supports or in direction of L_0,

T = tension in cable direction parallel with line between supports.

The change, l, in length of cable, L_c, for varying temperature is found by multiplying the number of degrees, n, by the length of the cable in feet times the coefficient of linear expansion per foot per degree Fahrenheit, c. This is*

$$l = L_c \times n \times c$$

A short approximate method for determining sags under varying temperatures and loadings that is close enough for all ordinary line work is as follows:

(A) Determine sag of cable with maximum stress under maximum load of lowest temperature occurring at the time of maximum load, and find length of cable with this sag.

(B) Find length of cable at the temperature for which the sag is required.

(C) Assume a certain reduced tension in the cable at the temperature and under the loading combination for which the sag is required; then find the decrease in length of the cable due to the decrease of the stress from its maximum.

(D) Combine the algebraic sum of **(B)** and **(C)** with **(A)** to get the length of the cable under the desired conditions; from this length, the sag and tension can be determined.

(E) If this tension agrees with that assumed in **(C)**, the sag in **(D)** is correct. If it does not agree, another assumption of tension in **(C)** must be made and the process repeated until **(C)** and **(D)** agree.

STRUCTURAL STANDARDS FOR STEEL RADIO TOWERS†

Material

(A) Structural steel shall conform to American Society for Testing Materials *Standard Specifications for Steel for Bridges and Buildings*, Serial Designation A-7, as amended to date.

(B) Steel pipe shall conform to American Society for Testing Materials standard specifications either for electric-resistance welded steel pipe, Grade A or Grade B, Serial Designation A-135, or for welded and seamless steel pipe, Grade A or Grade B, Serial Designation A-53, each as amended to date.

Loading

(A) 20-pound design: Structures up to 600 feet in height (unless they are to be located within city limits) shall be designed for a horizontal wind pressure of 20 pounds/foot^2 on flat surfaces and 13.3 pounds/foot^2 on cylindrical surfaces.

(B) 30-pound design: Structures more than 600 feet in height and those of any height to be located within city limits shall be designed for a horizontal wind pressure of 30 pounds/foot^2 on flat surfaces and 20 pounds/foot^2 on cylindrical surfaces.

(C) Other designs: Certain structures may be designed to resist loads greater than those described in **(A)** and **(B)**. Figure 1 of American Standard A58.1-1955 shows sections of the United States where greater wind pressures may occur. In all such cases, the pressure on cylindrical surfaces shall be computed as being $\tfrac{2}{3}$ of that specified for flat surfaces.

(D) For open-face (latticed) structures of square cross section, the wind pressure normal to one face shall be applied to 2.20 times the normal projected area of all members in one face, or 2.40 times the normal projected area of one face for wind applied to one corner. For open-face (latticed) structures of triangular cross section, the wind pressure normal to one face shall be applied to 2.00 times the normal projected area of all members in one face, or 1.50 times the normal projected area for wind parallel to one face. For closed-face (solid) structures, the wind pressure shall be applied to 1.00 times the normal projected area for square or rectangular shape, 0.80 for hexagonal or octagonal shape, and 0.60 for round or elliptical shape.

(E) Provisions shall be made for all supplementary loadings caused by the attachment of guys, antennas, transmission and power lines, ladders, etc. The pressure shall be as described for the respective designs and shall be applied to the projected area of the construction.

(F) The total load specified above shall be applied to the structure in the directions that will cause the maximum stress in the various members.

(G) The dead weight of the structure, and all materials attached thereto, shall be included.

Unit Stresses

(A) All parts of the structure shall be so designed that the unit stresses resulting from the specified loads shall not exceed the following values in pounds/inch^2:

Axial tension on net section = 20 000 pounds/inch^2.

Axial compression on gross section:

---

* Temperature coefficient of linear expansion is given in Chapter 4, Table 28.

† Abstracted from *American Standard Minimum Design Loads in Buildings and Other Structures*, A58.1-1955. New York: American National Standards Institute, Inc. Also from Electronic Industries Association Standard TR-116, October 1949. Sections on manufacture and workmanship, finish, and plans and marking of the standard are not reproduced here. The section "Wind Velocities and Pressures" is not part of the standard.

For members with values of L/R not greater than 120,

$= 17\,000 - 0.485L^2/R^2$ pounds/inch^2

For members with values of L/R greater than 120,

$$= \frac{18\,000}{1 + L^2/18\,000R^2} \text{ pounds/inch}^2$$

where L = unbraced length of the member, and R = corresponding radius of gyration, both in inches.

Maximum L/R for main leg members = 140
Maximum L/R for other compression members with calculated stress = 200
Maximum L/R for members with no calculated stress = 250
Bending on extreme fibres = 20 000 pounds/inch^2
Single shear on bolts = 13 500 pounds/inch^2
Double shear on bolts = 27 000 pounds/inch^2
Bearing on bolts (single shear) = 30 000 pounds/inch^2
Bearing on bolts (double shear) = 30 000 pounds/inch^2
Tension on bolts and other threaded parts, on nominal area at root of thread = 16 000 pounds/inch^2

Members subject to both axial and bending stresses shall be so designed that the calculated unit axial stress divided by the allowable unit axial stress, plus the calculated unit bending stress, divided by the allowable unit bending stress, shall not exceed unity.

(B) Minimum thickness of material for structural members:

Painted structural angles and plates = $\frac{3}{16}$ inch

Hot-dip galvanized structural angles and plates = $\frac{1}{8}$ inch

Other structural members to mill minimum for standard shapes.

(C) Where materials of higher quality than specified under "Material" above are used, the above unit stresses may be modified. The modified unit stresses must provide the same factor of safety based on the yield point of the materials.

Foundations

(A) Standard foundations shall be designed for a soil pressure not to exceed 4000 pounds/foot^2 under the

TABLE 17. WIND VELOCITIES AND PRESSURES

Actual Velocity V_a* (miles/hour)	Indicated Velocity V_i (miles/hour): 3-cup Anemometer	Indicated Velocity V_i (miles/hour): 4-cup Anemometer	Pressure P (pounds/foot^2) Projected Areas†: Cylindrical Surfaces ($P=0.0025V_a^2$)	Pressure P (pounds/foot^2) Projected Areas†: Flat Surfaces ($P=0.0042V_a^2$)
10	9	10	0.25	0.42
20	20	23	1.0	1.7
30	31	36	2.3	3.8
40	42	50	4.0	6.7
50	54	64	6.3	10.5
60	65	77	9.0	15.1
70	76	91	12.3	20.6
80	88	105	16.0	26.8
90	99	119	20.3	34.0
100	110	132	25.0	42.0
110	121	146	30.3	50.8
120	133	160	36.0	60.5
130	144	173	42.3	71.0
140	155	187	49.0	82.3
150	167	201	56.3	94.5

* Although wind velocities are measured with cup anemometers, all data published by the US Weather Bureau since January 1932 includes instrumental corrections and are actual velocities. Prior to 1932, indicated velocities were published.

In calculating pressures on structures, the "fastest-single-mile velocities" published by the Weather Bureau should be multiplied by a gust factor of 1.3 to obtain the maximum instantaneous actual velocities. See "Wind-Velocity and Temperature Extremes in North America" in this chapter for fastest-single-mile records at various places in the United States and Canada.

† The American Bridge Company equations given here are based on a ratio of 25/42 for pressures on cylindrical and flat surfaces, respectively, while the Electronic Industries Association specifies a ratio of 2/3. The actual ratio varies in a complex manner with Reynolds number, shape, and size of the exposed object.

specified loading. In uplift, the foundations shall be designed to resist 100 percent more than the specified loading, assuming that the base of the pier will engage the frustum of an inverted pyramid of earth whose sides form an angle of 30 degrees with the vertical. Earth shall be considered to weigh 100 pounds/foot^3 and concrete 140 pounds/foot^3.

(B) Foundation plans shall ordinarily show standard foundations as defined in **(A)**. Where the actual soil conditions are not normal, requiring some modification in the standard design, and complete soil information is provided to the manufacturer by the purchaser, the foundation plan shall show the required design.

(C) Under conditions requiring special engineering such as pile construction, roof installations, etc., the manufacturer shall provide the necessary information so that proper foundations can be designed by the purchaser's engineer or architect.

(D) In the design of guy anchors subject to submersion, the upward pressure of the water should be taken into account.

Wind Velocities and Pressures

Data regarding wind velocities and pressures are contained in Table 17.

VIBRATION AND SHOCK ISOLATION

Symbols

b = damping factor
d = static deflection in inches
E = relative transmissibility
= (force transmitted by isolators)/(force transmitted by rigid mountings)
F = force in pounds
F_0 = peak force in pounds
f = frequency in hertz
f_0 = resonant frequency of system in hertz
G = acceleration of gravity
$\approx$ 386 inches per second^2
g = peak acceleration in dimensionless gravitational units
= $\ddot{X}_0/G$
$j = (-1)^{1/2}$, vector operator
k = stiffness constant; force required to compress or extend isolators unit distance in pounds per inch
r = coefficient of viscous damping in pounds per inch per second
t = time in seconds
W = weight in pounds
x = displacement from equilibrium position in inches
X_0 = peak displacement in inches
$\dot{x}$ = velocity in inches per second
= dx/dt
$\dot{X}_0$ = peak velocity in inches per second
$\ddot{x}$ = acceleration in inches per second^2
= d^2x/dt^2
$\ddot{X}_0$ = peak acceleration in inches per second^2
ϕ = phase angle in radians
ω = angular velocity in radians per second
= $2\pi f$

Equations

The following relations apply to simple harmonic motion in systems with one degree of freedom. Although actual vibration is usually more complex, the equations provide useful approximations for practical purposes.

$$F = W(\ddot{x}/G) \tag{1}$$

$$F_0 = Wg \tag{2}$$

$$x = X_0 \sin(\omega t + \phi) \tag{3}$$

$$X_0 = 9.77g/f^2 \tag{4}$$

$$\dot{X}_0 = \omega X_0 = 6.28fX_0 = 61.4g/f \tag{5}$$

$$\ddot{X}_0 = \omega^2 X_0 = 39.5f^2 X_0 = 386g \tag{6}$$

$$E = \left| \frac{r - j(k/\omega)}{r + j[(\omega W/G) - k/\omega]} \right| \tag{7}$$

$$f_0 = 3.13\ (k/W)^{1/2} \tag{8}$$

$$b = 9.77r/(kW)^{1/2} \tag{9}$$

For critical damping, $b = 1$.

Neglecting dissipation ($b = 0$), or at $f/f_0 = (2)^{1/2}$ for any degree of damping

$$E = \left| \frac{1}{(f/f_0)^2 - 1} \right| \tag{10}$$

When damping is neglected

$$k = W/d \tag{11}$$

$$f_0 = 3.13/d^{1/2} \tag{12}$$

$$E = 9.77/(df^2 - 9.77) \tag{13}$$

Acceleration

The intensity of vibratory forces is often defined in terms of g values. From Eq. 2, it is apparent, for example, that a peak acceleration of $10g$ on a body will result in a reactionary force by the body equal to 10 times its weight.

If an object is mounted on vibration isolators, the accelerations of the vehicle are transmitted to the object (or vice versa) in an amplitude and phase that depend on the elastic flexing of the isolators in the direction in which the accelerations (dynamic forces) are applied.

Magnitudes

The relations among X_0, $\dot{X}_0$, $\ddot{X}_0$, and f are shown in Fig. 6. Any two of these parameters applied to the

TABLE 18. VIBRATION IN TYPICAL VEHICLES

Vehicle	Range of Frequencies (hertz)	Approximate Peak Amplitude (inches)	Nature of Excitation	Usual Choice of Isolator Resonant Frequency
Ships	0 to 15	0.02	Engine vibration in diesel or reciprocating steam drive	6 hertz for vibration isolation in commercial vessels. 27 to 30 hertz for shock isolation on naval vessels. These latter mounts amplify most vibrations to some extent.
	0 to 33	0.01	Propeller-blade frequency $=$(propeller rpm)$\times$(number of blades)/60	
Piston-engine aircraft	0 to 60	0.01	Engine vibrations	Above 20 hertz. Amplitude of vibrations varies with location in aircraft. Landing shock can be neglected.
	0 to 100	0.01	Propeller vibrations. Aerodynamic vibrations due to buffeting	
Turboprop aircraft	0 to 60	0.01	Engine vibrations$=$(engine rpm)/60. Also aerodynamic vibrations due to buffeting and turbulence	9 hertz
	0 to 100	0.01	Propeller vibrations	
Jet aircraft	Up to 500	0.001	Audible noise frequencies due to jet wake and combustion turbulence; very little engine vibration	9 hertz
Passenger automobiles	1	6	Suspension resonance	25 hertz will usually avoid resonance with wheel hop and suspension resonant frequencies.
	8 to 12	0.02	Unsprung weight resonance (wheel hop)	
	20+	0.002	Irregular transient vibrations due to resonances of structural members with road roughnesses	
Automobile trucks	4	5	Suspension resonance	Above 20 hertz and should not correspond with any structural resonance. It is not advisable to attempt to isolate suspension and unsprung weight resonances.
	20	0.05	Unsprung weight resonance	
	80+	0.005	Structural resonances	
Military tanks	1 to 3	2	Suspension resonance	Similar to automobile truck
	Depends on speed	—	Track-laying frequency $\approx$17.6(speed in mph)/(tread spacing in inches)	
	100+	0.001	Structural resonances	
Railroad trains	Broad and erratic		Similar to automobiles with additional excitations from rail joints and from side slop in rail trucks and draft gear	20 hertz has been successful in railroad applications. Shock with velocity changes up to 100 inches/second in direction of train occurs when coupling cars or starting freight trains.

graph locate the other two. For example, suppose $f = 10$ hertz and peak displacement $X_0 = 1$ inch. From Fig. 6, peak velocity $\dot{X}_0 = 63$ inches per second and peak acceleration $\ddot{X}_0 = 10g$.

Natural Frequency

If damping is neglected, the natural frequency, f_0, of vibration of an isolated system in the vertical direction can be calculated from Eq. 12 from the static deflection of the mounts. For example, suppose an object at rest causes a 0.25-inch deflection of its supporting springs. Then

$$f_0 = 3.13/(0.25)^{1/2} = 6.3 \text{ hertz}$$

Resonance

In Fig. 7, E is plotted against f/f_0 for various damping factors. Note that resonance occurs when $f_0 \approx f$ and that the vibratory forces are then increased by the isolators. To reduce vibration, f_0 must be less than $0.7f$, and it should be as small as $0.3f$ for good isolation.

It is not possible to secure good isolation at all vibrational frequencies in vehicles and similar environments where several different and varying exciting frequencies are present and where the isolators may have to withstand shock as well as vibration. In such cases, f_0 is often selected as about 1.5 to 2 times the predominant f. Information regarding vibration in typical vehicles is shown in Table 18.

Although all supporting structures have compliance and may reduce the effects of vibration and shock, the apparent stiffness of many "rigid" mountings is merely a matter of degree, and in conjunction with the supported mass, they can also give rise to resonance effects, thus magnifying the amplitude of certain vibrations.

Damping

Damping is desirable to reduce vibration amplitude when the exciting frequency is in the vicinity of f_0. This occurs occasionally in most installations. Any isolator that absorbs energy provides damping.

It is seldom practical to introduce damping as an

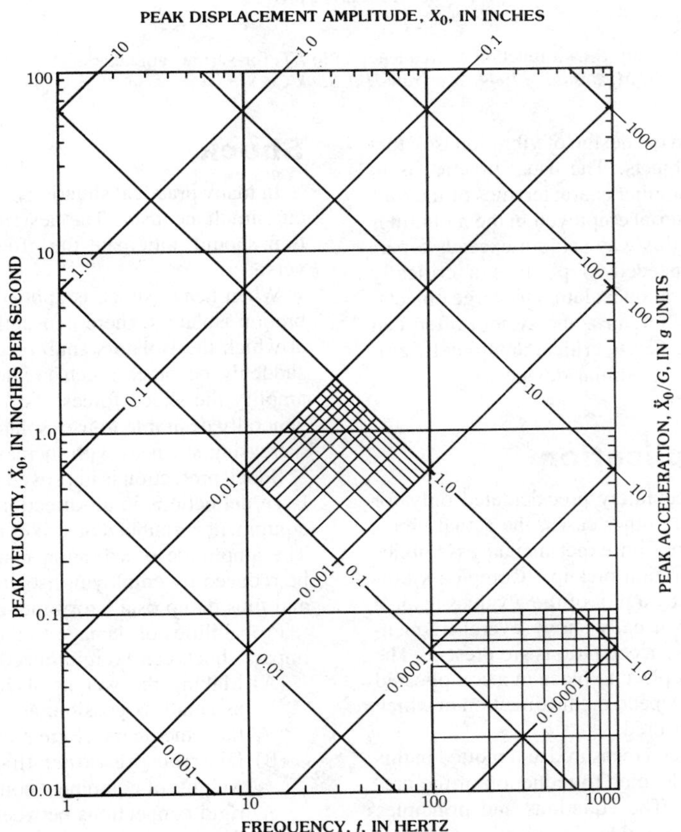

Fig. 6. Relation of frequency and peak values of velocity, displacement, and acceleration.

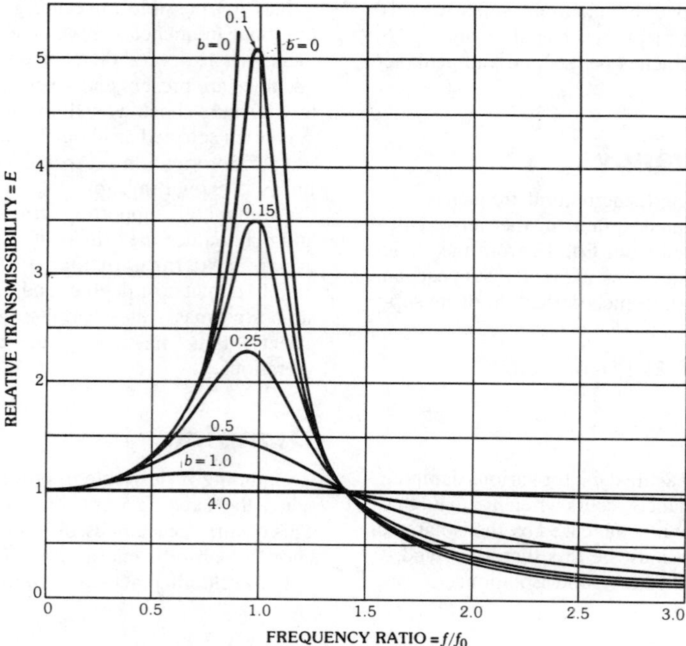

Fig. 7. Relative transmissibility E as a function of frequency ratio f/f_0 for various amounts of damping b. (*By permission from* Vibration Analysis, *by N. O. Myklestad. ©1944, McGraw-Hill Book Co.*)

independent variable in the design of vibration isolators for relatively small objects. The usual practice is to rely on the inherent damping characteristics of the rubber or other elastic material employed in the mounting. Damping achieved in this way seldom exceeds 5 percent of the amount needed to produce a critically damped system. In vibration isolators for large objects, such as variable-speed engines, the system often can be designed to produce nearly critical damping by employing fluid dashpots or similar devices.

Practical Application

Vibration can be accurately precalculated only for the simplest systems. In other cases, the actual vibration should be measured on experimental assemblies by using electrical vibration pickups. Complex vibration is often described by a plot of the g values against frequency. These plots usually show several frequencies at which the largest accelerations are present. The patterns will vary from place to place in a complicated structure and will also depend on the direction in which the acceleration is measured.

After vibration has been measured and plotted in this way, attention can be devoted to reduction of the predominant components. The equations and principles given above are used as guides in selecting the size, stiffness, damping characteristics, and location of isolators.

Shock

In many practical situations, vibration and shock occur simultaneously. The design of isolators for vibration should anticipate the effects of shock and vice versa.

When heavy shock is applied to a system using vibration isolators, there is usually a definite deflection at which the isolators snub or at which their stiffness suddenly becomes much greater. These actions may amplify the shock forces. To reduce this effect, it is generally desirable to use isolators that have smoothly increasing stiffness with increasing deflection.

Shock protection is improved by isolators that permit large deflections in all directions before the protected equipment is snubbed or strikes neighboring apparatus. The amplitude of vibration resulting from shock can be reduced by employing isolators that absorb energy and thus damp oscillatory movement.

Probabilities of damage to the apparatus itself from impact shock can be minimized by:

(A) Making the weight of equipment components as small as possible and the strength of structural members as great as possible.

(B) Distributing rather than concentrating the weights of equipment components and avoiding rigid connections between components.

(C) Employing structural members that have high ratios of stiffness to weight, such as tubes, *I* beams, etc.

(D) Avoiding, so far as is practical, stress concentrations at joints, supports, discontinuities, etc.
(E) Using materials such as steel that yield rather than rupture under high stress.

US GRAPHIC SYMBOLS

The USA standard graphic symbols for electrical and electronics diagrams cover both the communication and power fields. Symbols of primary interest to communications workers are in Fig. 8. They have been abstracted from IEEE Standard 315-1975.* Symbols that also agree with Recommendation No. 117 of the International Electrotechnical Commission are indicated by "*IEC*."

Diagram Types

Block diagrams consist of simple rectangles and circles with names or other designations within or adjacent to them to show the general arrangement of apparatus to perform desired functions. The direction of power or signal flow is often indicated by arrows near the connecting lines or arrowheads on the lines.

Schematic diagrams show all major components and their interconnections. A single-line diagram, as indicated by that name, uses single lines to interconnect components even though two or more conductors are actually required. It is a shorthand form of schematic diagram. It is always used for waveguide diagrams.

Wiring diagrams are complete in that all conductors are shown and all terminal identifications are included. The contact numbers on electron-tube sockets, colors of transformer leads, rotors of adjustable capacitors, and other terminal markings are shown so that a workman having no knowledge of the operation of the equipment can wire it properly.

Orientation

Graphic symbols are not considered to be coarse pictures of specific pieces of equipment but are true symbols representing the functions of parts in the circuit. Consequently, they may be rotated to any orientation with respect to each other without changing their meanings. Ground, chassis, and antenna symbols, for instance, may "point" in any direction that is convenient for drafting purposes.

Graphic symbols may be correlated with parts lists, descriptions, or instructions by means of reference designations (MIL-STD-16).

Detached Elements

Switches and relays often have many sets of contacts, and these sets may be separated and placed in the parts of the drawing to which they apply. Each separated element should be clearly labeled as part of the basic switch or relay.

Terminals

The terminal symbol need not be used unless it is needed. Thus, it may be omitted from relay and switch symbols. In particular, the terminal symbol often shown at the end of the movable element of a relay or switch should not be considered as the fulcrum or bearing but only as a terminal.

Associated or Future Equipment

Associated equipment, such as for measurement purposes, or additions that may be made later are identified as such by using broken lines for both symbols and connections.

BRITISH GRAPHIC SYMBOLS

Commonly used British block-diagram graphic symbols are in Fig. 9. They have been abstracted with permission from British Standard 530: 1948 and Supplement No. 5 (1962), and superseding BS 3939 (1966–1969). The issuing organization is the British Standards Institution, British Standards House, 2 Park Street, London, W. 1, England.

STANDARDIZED GRAPHIC SYMBOLS FOR LOGIC DESIGN

Symbols for logic design are so varied and in such a state of flux that many authors use their own symbols. Fig. 10 shows a short set of standardized graphic symbols for logic design selected from a revised draft of ANSI/IEEE Standard 91-1982, soon to be published.

SAFETY LEVELS FOR EXPOSURE TO EM FIELDS*

The American National Standards Institute (ANSI) has published recommendations concerning safety levels with respect to human exposure to electromagnetic fields in the frequency range from 300 kHz to 100 GHz.† The ANSI radio-frequency protection guides (RFPG) are intended to apply to both occupational and nonoccupational exposure, but not to exposure of patients for medical purposes. These guides specify maximum allowable levels as a function of frequency in terms of the mean squared electric and magnetic fields, E and H, respectively, and the equivalent free-space plane-wave power density. To determine adherence to this RFPG, the power density or squares of the field strengths should be averaged over any 0.1-h period with all measurements to be made at a distance 5 cm

* Published by the Institute of Electrical and Electronics Engineers, 345 East 47th Street, New York, N.Y.

* This section contributed by Charles A. Cain.

† Reference 1.

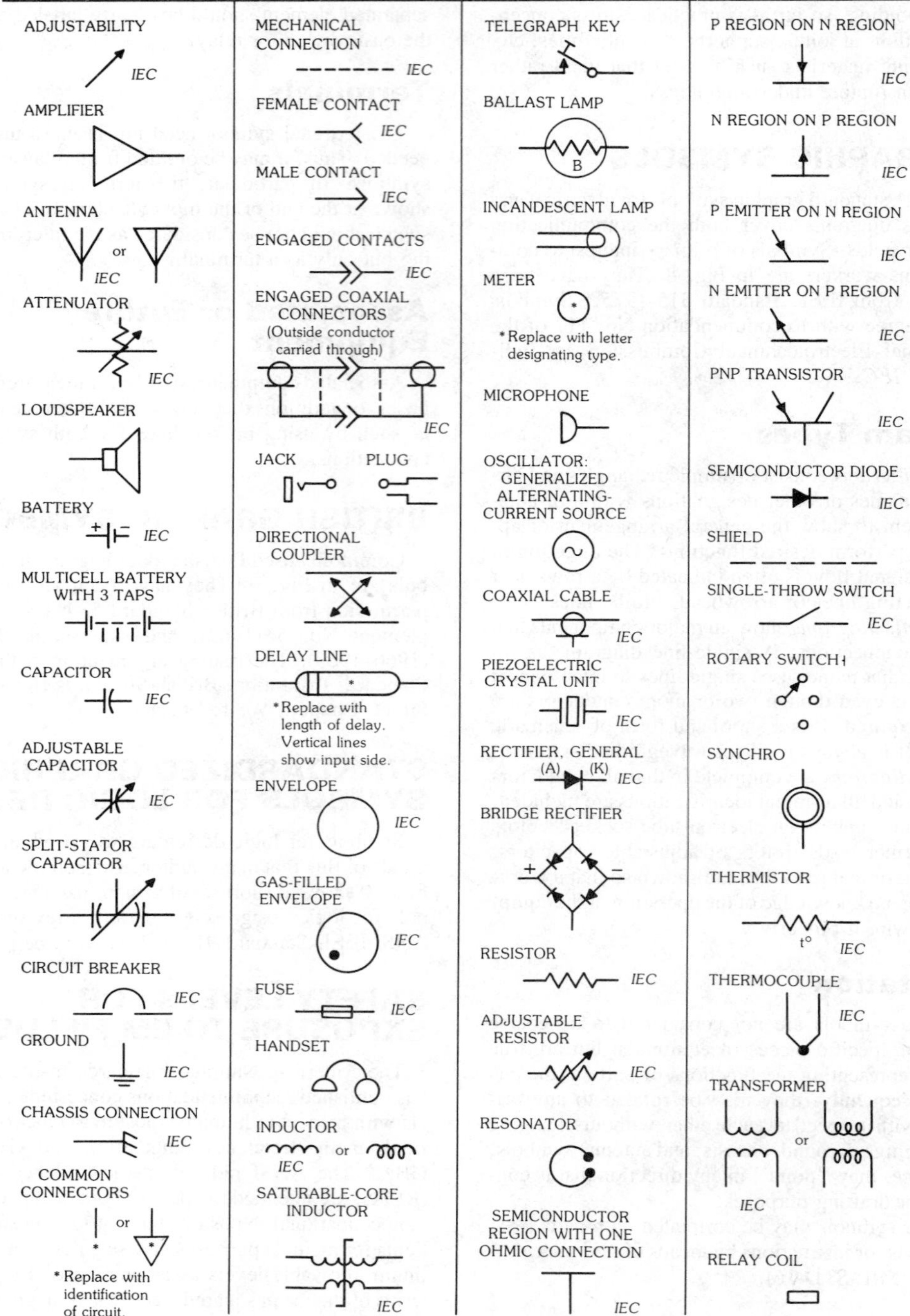

Fig. 8. Selected graphic symbols from IEEE Standard 315-1975.

or greater from any object. Table 19 is a summary of the ANSI C95.1-1982 RFPG.

A detailed rationale for the ANSI C95.1-1982 RFPG is given in reference 1. The frequency dependency of the maximum allowable equivalent plane-wave power density and of the mean squared field strengths is based on research results which show that the whole-body-average specific absorption rates§ approach maximum

§ Specific absorption rate: the time rate per unit mass of radio-frequency electromagnetic energy deposition in biological tissues.

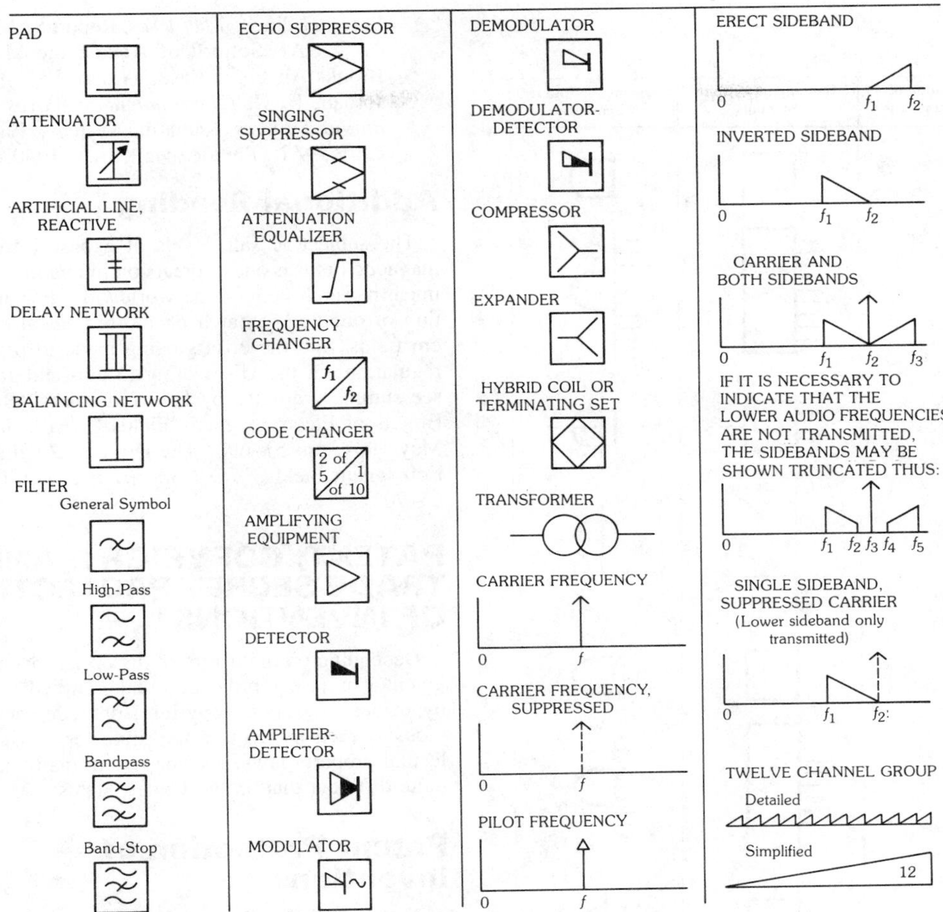

Fig. 9. Selected British block-diagram graphic symbols.

values when the long axis of the body is parallel to the E-field vector and is approximately 0.4λ in length where λ is the free-space wavelength of the incident field.** This resonance occurs at a frequency of about 70 MHz for a man of average size. To account for the range of body dimensions from small infants to large adults, the RFPG specifies a maximum allowable power density of 1 mW/cm^2 over the frequency range from 30 MHz to 300 MHz. As indicated in Table 19, the allowable incident intensities increase below 30 MHz and above 300 MHz.

The ANSI C95.1-1982 standard specifies an exclusion from the recommended protection guides for low-power devices. The levels in Table 19 may be exceeded if the input power to a radiating device at radio frequencies between 300 kHz and 1 GHz is seven watts or less. The RFPG may also be exceeded between 300 kHz and 100 GHz if the exposure conditions result in specific absorption rates below 0.4 W/kg as averaged over the whole body and spatial peak specific absorption rates below 8 W/kg as averaged over any one gram of tissue for any 0.1-h time interval.

The type of situation that this exclusion is meant to address is illustrated by Table 20, which shows the field strengths and power densities in the near field of a 50-cm monopole antenna radiating at 150 MHz and 2 W of input power. It is clear from these data that the RFPG maximum recommended limits of field strength or of plane-wave equivalent power density are exceeded at locations near the antenna. However, this device radiates less than 7 watts and may be excluded.

References

1. *American National Standard Safety Levels with Respect to Human Exposure to Radio Frequency Electromagnetic Fields, 300 kHz to 100 GHz*, ANSI C95.1-1982. Institute of Electrical and Electronics Engineers, 1982.
2. Durney, C. H., Johnson, C. C., Barber, P. W., Massoudi, H., Iskander, M. F., Lords, J. L., Ryser, D. K., Allen, S. J., and Mitchell, J. C. *Radiofrequency Radiation Dosimetry Handbook*,

** Reference 2.

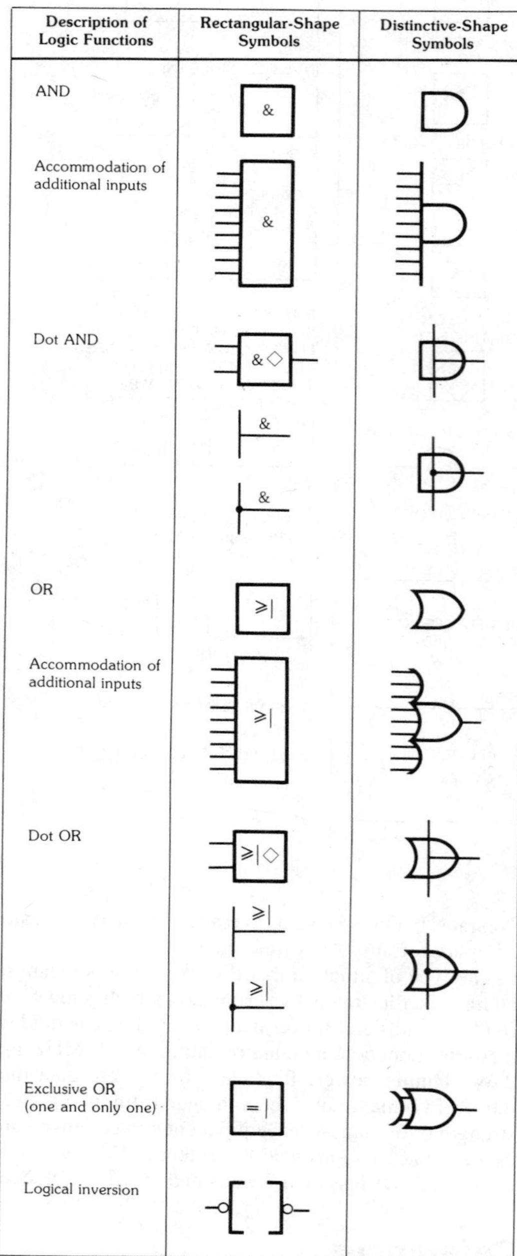

Description of Logic Functions
Rectangular-Shape Symbols
Distinctive-Shape Symbols
AND
&
Accommodation of additional inputs
&
Dot AND
&◇
&
&
OR
≥1
Accommodation of additional inputs
≥1
Dot OR
≥1◇
≥1
≥1
Exclusive OR (one and only one)
=1
Logical inversion

Fig. 10. Selected standardized graphic symbols for logic diagrams. (*Excerpted with permission from the revised draft of* IEEE Standard Graphic Symbols for Logic Diagrams (Two State Devices). *For a complete listing of all standardized symbols for logic diagrams, including the relationship between the new and superseded standards, see ANSI/IEEE Std. 91-1982, Appendix C, a revised draft of which is soon to be published.*)

Second Edition, May 1978, Report SAM-TR-78-22. USAF School of Aerospace Medicine, Brooks Air Force Base, Texas.

3. Jordan, E. C. *Electromagnetic Waves and Radiating Systems.* Radiating Systems. Englewood Cliffs, N.J.: Prentice-Hall, Inc., 1950; p. 323.

Additional Reading

The subject of safe levels of exposure to electromagnetic fields is one of great continuing interest, with much research being done worldwide. For an indication of ongoing research on the biological effects of em fields, and the efforts being made to promulgate regulations in the US relating to em-field standards, see summary articles by Eric J. Lerner: "Biological Effects of Electromagnetic Fields," *IEEE Spectrum*, May 1984, pp. 57–69; "The Drive to Regulate Electromagnetic Fields," *IEEE Spectrum*, March 1984, pp. 63–70.

PATENT, COPYRIGHT, AND TRADE SECRET PROTECTION OF INVENTIONS*

Depending on the nature of his or her invention, an inventor may seek protection under United States law by means of patent, copyright, or trade secret. Obviously, each nation is responsible for its own intellectual property laws, and the law in many nations is quite different than in the United States.

Patent Protection of Inventions

Nature of the Patent Grant—A United States patent grants to its owner the right to exclude others from making, using, or selling the claimed invention in the United States for 17 years. It does not, however, confer on the owner the right to practice that invention, for one or more other patents may also cover the invention. So, for example, if a patent for a carburetor coexisted with a patent for a dual-barrelled carburetor, the owner of the latter could not lawfully manufacture his invention without a license from the owner of the former. However, 17 years after a patent issues, the owner loses all right to exclude.

What Can Be Patented?—A patent may be obtained on any new and useful process, machine, manufacture, or composition of matter, or any new and useful improvement thereof. The patent standard of novelty is a strict one. Among the novelty requirements is the requirement that the inventor have made the invention before it was used or known by others in the United States, and before it was patented or described by others in a printed publication anywhere on earth.

* This section contributed by David Bender.

TABLE 19. RADIO-FREQUENCY PROTECTION GUIDES*

Frequency Range (MHz)	E^2 (V^2/m^2)	H^2 (A^2/m^2)	Power Density (mW/cm^2)
0.3— 3	400 000	2.5	100
3 — 30	4000 $(900/f^2)$	0.025 $(900/f^2)$	$900/f^2$
30 — 300	4000	0.025	1.0
300 — 1500	4000 $(f/300)$	0.025 $(f/300)$	$f/300$
1500 —100 000	20 000	0.125	5.0

* From ANSI C95.1-1982.
Note: f = frequency in megahertz.

TABLE 20. FIELD STRENGTHS (E AND H) AND POWER DENSITIES IN PROXIMITY TO A CURRENT-FED, QUARTER-WAVE, RADIATING MONOPOLE ANTENNA*

Distance in Centimeters				
0	10	20	30	40
E = 64.3 V/m H = 0.00 A/m PD = 1.10 mW/cm² *	48.7 0.0217 0.628 *	32.7 0.0317 0.287 *	24.2 0.0341 0.156 *	19.3 0.0348 0.0985 *
	95.1 0.0524 2.40 *	45.7 0.0524 0.553 *	29.8 0.0489 0.236 *	22.4 0.0456 0.133 *
	121 0.136 3.88 *	53.4 0.0856 0.757 *	33.3 0.0682 0.295 *	24.5 0.0533 0.159 *
50-cm Monopole Antenna	110 0.229 3.72 *	51.7 0.124 0.708 *	33.3 0.089 0.300 *	25.0 0.0716 0.193 *
	83.6 0.306 3.54 *	43.0 0.157 0.931 *	30.2 0.108 0.437 *	24.1 0.0531 0.261 *
	50.1 0.358 4.82 *	32.0 0.180 1.22 *	26.2 0.120 0.548 *	22.7 0.0910 0.312 *
	27.7 0.375 5.31 *	26.3 0.188 1.33 *	24.3 0.125 0.589 *	22.1 0.0938 0.332 *

* From ANSI C95.1-1982

A 50-cm antenna operating at 150 MHz and 2 W of input power is assumed, with an input impedance of 36 Ω. Each asterisk associated with a set of field measurements denotes the spatial point of measurement in relation to the antenna. The vertical and horizontal distance between adjacent points of measurement is 10 cm. (Calculations based on reference 3.)

Even beyond novelty itself, the claimed invention must be nonobvious to a person of ordinary skill in the area of technology to which it pertains.

Computer Programs—In recent years, many inventors of program-related inventions have sought patents on their inventions. Much doubt existed as to whether such inventions were in fact processes or machines—and therefore patentable subject matter—or were rather akin to laws of nature or mathematical equations—and therefore unpatentable. Since 1981, it has been clear that some program-related inventions are patentable, and the law is in an early stage of development on the issue of which such inventions are patentable.

Recognizing Inventions—If an original development appears to be of commercial significance (or if there is some other reason why a patent is desired), the inventor may wish to transmit the pertinent details to a patent attorney or patent agent in order to have a patent search done to assess whether the invention meets the patent standards of novelty and nonobviousness. The attorney or agent will generally compare the invention with issued US patents in reaching this determination.

Who Is the Inventor?—Patent law recognizes that invention comprises two stages: (1) conception of the idea, and (2) its reduction to practice (which amounts to either constructing the invention or making a detailed written description or model). The person(s) actually responsible for the conception and reduction to practice is (are) the inventor(s). If a person's contribution to the invention does not rise above the level of ordinary mechanical skill, that person is not an inventor. Supervisors should not simply as a matter of course be named on patent applications as inventors. They should be named only where they actually participated in the conception or reduction to practice.

Once an application for a patent is filed, it is freely assignable, as an item of personal property. Similarly, an issued patent is also freely assignable.

The Inventive Process—The question of whether an invention is patentable does not depend on the manner in which it is made. In particular, the fact that an invention is made by accident, or serendipity, will not negate patentability. However, often the inventive process comprises the following steps: someone identifies either a problem to be solved or a result to be achieved; the inventor conceives a way of solving the problem or effecting the result; and the inventor reduces the conception to reality, either by actually building or practicing the invention, or by making a detailed written description of it.

Obtaining a Patent—One seeks a patent by filing and prosecuting a patent application with the US Patent and Trademark Office ("PTO"). There are many requirements that must be met in the application. The application must include: a written description of the invention, and the manner and process of making and using it, in such full, clear, concise, and exact terms as to enable any person skilled in the appropriate technology to make and use it; a statement of the best mode contemplated by the inventor; and one or more claims pointing out the subject matter regarded as the invention. The preparation and prosecution of a patent application is a very technical and detailed task. An application can be filed and prosecuted only by the inventor, or by a registered patent attorney or patent agent. Each year, roughly 100 000 patent applications are filed and some 70 000 patents issued. The average time between filing of the application and issuance of the patent is currently about $2\frac{1}{2}$ years.

Interferences—On occasion, different inventors may independently file patent applications claiming the same invention. This possibility is enhanced by the fact that the PTO maintains all patent applications in secrecy, so that when an applicant files, he or she is unaware of other applications. When the PTO notes applications by different inventors claiming substantially the same inventions, the PTO declares an "interference." This is a proceeding in the PTO to determine the first inventor. If two interfering patents actually issue, a resolution may be had by suit in federal court.

Publication—For a valid United States patent to issue, an application must be filed within one year of publication or public use of the invention. However, it is generally unwise publicly to disclose an invention before a patent application is filed. The law in some foreign countries is such that publication before filing for a patent in those countries will extinguish the right to obtain a patent there.

Importance of Records—In certain instances, the patent owner may be called upon to prove various facts associated with his patent rights. In an interference, he will be expected to prove when the inventor conceived the invention, when it was reduced to practice, and perhaps that due diligence was exercised in the interim. In licensing negotiations, the owner may be challenged to show similar facts. And in litigation, the owner will be expected to show these and other facts relating to the invention. The interferences, licensing negotiations, or litigation may occur years—even decades—after the invention was made. Memories dim and witnesses die. With clear and detailed contemporaneous documents and models, it is less difficult to reconstruct precisely what happened, and when.

Such documents may consist of the lab or engineering notebooks of the inventors and others, memoranda, and sketches. The notebooks in particular are often of prime importance, and certain characteristics may render these notebooks more credible as evidence: (1) They should be bound, rather than looseleaf. (2) Entries should be in chronological order, written clearly and completely in ink, without leaving blank pages or spaces. (3) Changes should be made by drawing one line through material to be deleted (but not obliterating it). (4) Erasures and removal of pages should be avoided. (5) Where other materials must be included in the notebook (e.g., small lab samples, graphs), they should be fastened securely to the appropriate page, referenced on that page or as nearly as practical, dated, signed, and witnessed. (6) Each entry should be made at or shortly after the events recorded, and should be dated and signed when made. (7) When an event of perceived significance is recorded, the entry should be shown to someone of discretion who will understand it. This person should then write the date and his or her signature below the entry, subscribing above the signature "read and understood."

Copyright Protection of Inventions

Copyright law protects expression, as opposed to concept or idea. For this reason, copyright is not generally an appropriate vehicle for protecting technology against misappropriation. The single major exception to this generality is the computer program. Amendments made to the copyright law in 1980 contemplate that copyright provide protection against the unlawful copying of programs. Many owners of programs have sought protection under the copyright laws, especially where their goal is widely to market a relatively inexpensive program package, and most particularly where the market consists of individuals rather than companies. This method of seeking to protect programs was not used on a widespread basis until relatively recently. Accordingly, the law is still unsettled as to such issues as the form in which programs may be protected by copyright and the scope of copyright protection of programs.

Trade Secret Protection of Inventions

What Is a Trade Secret?—A trade secret is "any formula, pattern, device, or compilation of information which is used in one's business and which gives him an opportunity to obtain an advantage over competitors who do not know or use it." Basically, when a business has secret information giving it a commercial edge over competitors who don't have it, the business may seek to protect the information as a trade secret. Although trade secret protection often extends to such commercial items as customer lists and cost data, much of the information protected by trade-secret law is technological.

For example, one may maintain as trade secrets chemical processes used in the trade secret owner's plant, or the structure and functioning of machines used by licensees who have agreed to keep these items confidential. And, at least until the recent surge of interest in copyright protection, trade secret law was the most common way of seeking to protect computer programs. However, a device marketed to consumers and embodying a "trade secret" whose characteristics are obvious from a casual inspection, would not be a good candidate for trade secret protection after marketing commenced, because there would then be no secret.

The Law of Trade Secrets—The basic rule of trade secret law is as easy to state as it is difficult to apply: a person may not make unauthorized use or disclosure of another's trade secret. Such conduct constitutes misappropriation of the trade secret. It is not necessary that any tangible object be taken for misappropriation to occur; rather, misappropriation can occur on the basis of use or disclosure of information existing in a person's memory.

Unlike patent and copyright law, trade secrets are governed largely by state (rather than federal) law. One result of this is that trade secret law differs from state to state. Under certain circumstances, in some states trade secret misappropriation may constitute a crime. Aside from the general larceny statutes in each state (which may or may not apply), about half the states have specific trade secret theft statutes.

However, unlike patent rights, trade secret rights do not extend to preclude another from use or disclosure of information independently developed or otherwise lawfully acquired free of restrictions. Accordingly, a trade secret owner is continually subject to loss of the trade secret through the lawful conduct of others.

General Requirements for a Trade Secret—Generally, six factors are important in determining whether one has a trade secret:

(1) The extent to which the information is known outside the owner's business. Obviously, if everyone in the industry knows it, there is no secret. But if one or two others know of it, and maintain it as a secret, it may still be a trade secret.

(2) The extent to which it is known by employees and others involved in the owner's business. A business which restricts internal proliferation of its secrets on a "need to know" basis is in a better posture to claim that it has trade secrets.

(3) The extent of measures taken by the owner to guard the secrecy of the information. The owner will rarely be required to duplicate the security features of Fort Knox. But if the owner permits people to walk in off the street and roam around in areas where this information is exposed, then it will be difficult to convince a court that it constitutes a trade secret.

(4) The value of the information to the owner and his competitors. Value embodies the commercial importance and desirability of the information. Without such value, it is difficult to see how the information gives its owner an advantage over competitors.

(5) The amount of money or effort expended by the owner in developing the information. This may be looked to as one indicator of value. Moreover, as an equitable matter, a court or jury may be reluctant to permit a defendant to reap where he has not sown.

(6) The ease or difficulty with which the information could properly be acquired or duplicated by others. For example, as indicated above, if the "secret" is obvious from simply looking at the marketed product, it is generally unworthy of protection as a trade secret, for it is not secret.

Relationships Giving Rise to Trade Secret Relationship—The owner of a trade secret has rights, not against the world, but only against one unlawfully acquiring the trade secret or making an unauthorized use or disclosure of the trade secret lawfully acquired through some relationship with the owner. Typically, the owner's right to restrain use or disclosure arises from a "confidential relationship" that exists between the owner and the other party, whereby the other agrees to maintain the information as confidential. This confidential relationship may arise from a contract expressly creating it, or it may be implied from the relationship of the parties. For example, a confidential relationship arises generally from the owner's employment of an employee to work on developing a process or product.

Also, it is not uncommon for employers to require certain employees to sign "covenants not to compete." In such a covenant, the employee agrees that for a specified period following his employment by the employer, he will not, in a particular geographic region, undertake activities of the type in which he engaged for his present employer.

Who Owns the Rights in a Trade Secret?—It is not always easy to determine who owns rights in a trade secret. For example, in the employer-employee situation, some factors that a court would look to in adjudicating that issue might be: the terms of any contract between the two; whether the employee was hired to do research and/or development work; whether the secrets in question are the result only of the employee's efforts; whether the work resulting in the trade secret was assigned to the employee by the employer; and whether the work done by the employee was done on the employer's premises, during normal working hours, and with the use of the employer's facilities and materials.

SUMMARY OF MILITARY NOMENCLATURE SYSTEM*

In the Joint Electronics Type Designation System (JETDS), formerly called the "AN" system, nomenclature for electronic equipment consists of a name, followed by a type number.

A type designation assignment for equipment such as a definitive system, subsystem, center, central, set, etc., shall consist of at least an AN, a slant bar, a three-letter equipment designation (Table 21), a dash, and a number. *Example:* AN/VRC-12 would be a radio communication set installed in a vehicle designed for functions other than carrying electronic equipment.

All groups, including commercial off-the-shelf equipment, are identified by a two-letter indicator from Table 22. Applicable equipment indicator letters (Table 21) follow the slant bar to indicate the potential of the group for multiple or peculiar application. *Example:* OE-162/ARC indicates an antenna for aircraft radio-communication equipment. Equipment indicators with a specific model number (e.g., OK-450/TRC-26) are used following the slant bar when the group is peculiar to specific equipment (e.g., AN/TRC-26) with no known potential for other use.

The type designation for units having one end use consists of an indicator (Table 23), a dash, a number, a slant bar, and the equipment the unit is a part of or used with. *Example:* the receiver portion of the AN/VRC-12 is identified as R-40/VRC-12. If the unit has multiple usage, only those indicators that are common or appropriate are included after the slant bar. *Examples:* A power supply, part of or used with the AN/VRC-12 and AN/VRC-19 would be identified as PP-50/VRC. A power supply, "part of" the AN/VRC-12 and "used with" the AN/VRR-40 would be identified as PP-60/VR.

The system indicator (AN) does not mean that the Army, Navy, and Air Force use the equipment, but simply that the type number was assigned in the JETDS system.

Nomenclature Policy

JETDS nomenclature will be assigned to:

(A) Complete systems, subsystems, centers, centrals, sets, groups, kits, and units of military design, either definitive or variable in configuration.
(B) Groups of articles of either commercial or military design that are grouped for a military purpose.
(C) Electronic articles of military design that are part of or used with an item not identified in the JETDS.
(D) Commercial articles requiring military identification for use by US Government.
(E) Electronic materials of military design which are not part of or used with a set.

JETDS nomenclature will not be assigned to:

(A) Articles cataloged commercially except in accordance with paragraph (D) above.
(B) Minor components of military design for which other adequate means of identification are available.
(C) Small parts such as capacitors and resistors.
(D) Articles having other adequate identification in joint military specifications.

Nomenclature assignments will remain unchanged regardless of later changes in installation and/or application.

Modification Letters

Component modification suffix letters will be assigned for each modification of a component when detail parts and subassemblies used therein are no longer interchangeable, but the component itself is interchangeable physically, electrically, and mechanically.

Set modification letters will be assigned for each modification not affecting interchangeability of the sets or equipment as a whole, except that in some special cases they will be assigned to indicate functional interchangeability and not necessarily complete electrical and mechanical interchangeability. Modification letters will only be assigned if the frequency coverage of the unmodified equipment is maintained.

The suffix letters X, Y, and Z will be used only to

* Adapted from MIL-STD-196C, *Joint Electronics Type Designation System*, 22 April 1971; Notice 1, 20 April 1972; Notice 2, 8 June 1972; Notice 3, 14 November 1972; and Notice 4, 17 July 1977. Available from the Superintendent of Documents, Washington, D.C. 20402.

TABLE 21. SET OR EQUIPMENT INDICATOR LETTERS

1st Letter (Type of Installation)		2nd Letter (Type of Equipment)		3rd Letter (Purpose)	
A	Piloted aircraft	A	Invisible light, heat radiation	A	Auxiliary assemblies (not complete operating sets used with or part of two or more sets or sets series) (inactivated, do not use)
B	Underwater mobile, submarine	B	Pigeon (do not use)	B	Bombing
C	Air transportable (inactivated, do not use)	C	Carrier	C	Communications (receiving and transmitting)
D	Pilotless carrier	D	Radiac	D	Direction finder, reconnaissance, and/or surveillance
		E	Nupac (inactivated, do not use)	E	Ejection and/or release
F	Fixed ground	F	Photographic*		
G	General ground use (includes two or more ground-type installations)	G	Telegraph or Teletype	G	Fire control or searchlight directing
				H	Recording and/or reproducing (graphic meteorological and sound)
		I	Interphone and public address		
		J	Electromechanical or inertial wire covered		
K	Amphibious	K	Telemetering	K	Computing
		L	Countermeasures	L	Searchlight control (inactivated, use "G")
M	Ground, mobile (installed as operating unit in a vehicle which has no function other than transporting the equipment)	M	Meteorological	M	Maintenance and/or test assemblies (including tools)
		N	Sound in air	N	Navigational aids (including altimeters, beacons, compasses, racons, depth sounding, approach, and landing)
P	Pack or portable (animal or man)	P	Radar	P	Reproducing (inactivated, use "H")
		Q	Sonar and underwater sound	Q	Special, or combination of purposes
		R	Radio	R	Receiving, passive detecting
S	Water surface craft	S	Special types, magnetic, etc., or combinations of types	S	Detecting and/or range and bearing, search
T	Ground, transportable	T	Telephone (wire)	T	Transmitting
U	General utility (includes two or more general installation classes, airborne, shipboard, and ground)				
V	Ground, vehicular (installed in vehicle designed for functions other than carrying electronic equipment, etc., such as tanks)	V	Visual and visible light		
W	Water surface and underwater combination	W	Armament (peculiar to armament, not otherwise covered)	W	Automatic flight or remote control
		X	Facsimile or television	X	Identification and recognition
		Y	Data processing	Y	Surveillance (search, detect, and multiple target tracking) and control (both fire control and air control)
Z	Piloted and pilotless airborne vehicle combination				

* Not for US use except for assigning suffix letters to previously nomenclatured items.

designate a set or equipment modified by changing the power input voltage, phase, or frequency. They will be used as follows: X will indicate the first change, Y the second, Z the third, XX the fourth, etc. These letters will be in addition to other modification letters applicable.

Developmental Indicators

Experimental Sets—To identify a set or equipment of an experimental nature with the development organization concerned, the following indicators are used within the parentheses:

TABLE 22. GROUP INDICATORS

Indicator	Family Name	Indicator	Family Name
OA	Miscellaneous groups	ON	Interconnecting groups
OB	Multiplexer and/or demultiplexer groups	OP	Power supply groups
OD	Indicator groups	OQ	Test set groups
OE	Antenna groups	OR	Receiver groups
OF	Adapter groups	OT	Transmitter groups
OG	Amplifier groups	OU	Converter groups
OH	Simulator groups	OV	Generator groups
OJ	Consoles and console groups	OW	Terminal groups
OK	Control groups	OX	Coder, decoder, interrogator, transponder groups
OL	Data analysis and data processing groups	OY	Radar set groups
OM	Modulator and/or demodulator groups	OZ	Radio set groups

XA	Aeronautical Systems Division, Wright-Patterson Air Force Base, Ohio.
XB	Naval Research Laboratory, Washington, DC.
XC	US Army Signal Engineering Laboratories, The Hexagon, Fort Monmouth, NJ (inactivated, use XE).
XD	Electronic Systems Division, Laurence G. Hanscom Field, Bedford, Massachusetts.
XE	US Army Electronics Laboratories, Fort Monmouth, NJ.
XF	Frankford Arsenal, Philadelphia, Pa.
XG	USN Electronics Laboratory, San Diego, California.
*XH	Aerial Reconnaissance Laboratory, Wright-Patterson Air Force Base, Ohio.
XI	Air Force Armament Laboratory, Eglin Air Force Base, Florida.
XJ	Naval Air Development Center, Johnsville, Pa.
*XK	Flight Control Laboratory, Wright-Patterson Air Force Base, Ohio.
XL	US Army Signal Electronics Research Unit, Mountain View, Cal.
XM	US Army Signal Engineering Laboratories, The Hexagon, Fort Monmouth, NJ (inactivated, use XE).
XN	Department of the Navy, Washington, DC.
XO	US Army Missile Command, Redstone Arsenal, Alabama.
XP	Canadian Department of National Defence, Ottawa, Ontario, Canada.
*XQ	Aeronautical Accessories Laboratory, Wright-Patterson Air Force Base, Ohio.
XR	National Security Agency, Fort George G. Meade, Maryland.
*XS	Electronic Components Laboratory, Wright-Patterson Air Force Base, Ohio.
XT	US Army Security Agency, Arlington Hall Station, Arlington, Va.

* Not for Air Force use except for assigning additional developmental designations to previously type-designated items. Use XA for all new equipments.

XU	USN Underwater Sound Laboratory, Fort Trumbull, New London, Conn.
XV	Air Force Weapons Laboratory, Kirtland Air Force Base, New Mexico.
XW	Rome Air Development Center, Rome, New York.
*XY	Weapons Guidance Laboratory, Wright-Patterson Air Force Base, Ohio.
XZ	USN Bureau of Naval Weapons Activities.
XAA	Air Force Ballistic Systems Division, Norton Air Force Base, Cal.
XAE	US Army Electronics Research and Development Activity, Fort Huachuca, Arizona.
XAN	Naval Avionics Facility, Indianapolis, Ind.
XBB	US Army Electronics Command, Proc and Prod Div., Fort Monmouth, NJ.
XCA	US Naval Ammunition Depot, Crane, Ind.
XCC	Air Force Missile Test Center, Patrick Air Force Base, Florida.
XCL	Naval Weapons Center, China Lake, Calif.
XCR	Naval Weapons Center, Corona Laboratory, Corona, Calif.
XDD	US Army Signal Air Defense Engineering Agency, Fort George G. Meade, Md.
XDV	US Naval Weapons Laboratory, Dahlgren, Va.
XGS	Grand Support Equipment Division, Naval Air Engineering Center, Philadelphia, Pennsylvania.
XIH	US Naval Ordnance Station, Indianhead, Maryland.
XLW	US Army Limited War Laboratory, Aberdeen Proving Ground, Md.
XMG	Naval Missile Center, Point Mugu, Calif.
XPM	US Army, Project Michigan, Ypsilanti, Michigan.

* Not for Air Force use except for assigning additional developmental designations to previously type-designated items. Use XA for all new equipments.

TABLE 23. UNIT INDICATORS

Indicator	Family Name	Indicator	Family Name
AB	Supports, antenna	OC*	Oceanographic devices
AM	Amplifiers	OS	Oscilloscopes, test
AS	Antennas, complex and simple	PD*	Prime drivers
AT*	Antennas, simple	PF*	Fittings, pole
BA	Batteries, primary type	PG*	Pigeon articles
BB	Batteries, secondary type	PH*	Photographic articles
BZ	Alarm units	PL	Plug-in units
C	Controls	PP	Power supplies
CA*	Commutator assemblies, sonar	PT	Mapping and plotting units
CB*	Capacitor banks	PU	Power equipments
CG	Cable assemblies, rf	R	Receivers
CK*	Crystal kits	RC*	Reels
CM	Comparators	RD	Recorder-reproducers
CN	Compensators	RE	Relay assemblies
CP	Computers	RF*	Radio-frequency components
CR*	Crystals	RG*	Cables, rf bulk
CU	Couplers	RL	Reeling machines
CV	Converters (electronic)	RO	Recorders
CW	Radomes	RP	Reproducers
CX	Cable assemblies, nonrf	RR	Reflectors
CY	Cases and cabinets	RT	Receiver and transmitter
D	Dispensers	S	Shelters
DA	Loads, dummy	SA	Switching units
DT	Detecting heads	SB	Switchboards
DY*	Dynamotors	SG	Generators, signal
E*	Hoists	SM	Simulators
F	Filter units	SN	Synchronizers
FN*	Furniture	ST*	Straps
FR	Frequency-measuring devices	SU	Optical devices
G	Generators, power	T	Transmitters
GO*	Goniometers	TA	Telephone apparatus
GP*	Ground rods	TB	Towed bodies
H	Head, hand, and chest sets	TC*	Towed cables
HC*	Crystal holders	TD	Timing devices
HD	Environmental apparatus (heating, cooling, etc.)	TF	Transformers
		TG	Positioning devices
ID	Indicators, noncathode-ray tube	TH	Telegraph apparatus
IL*	Insulators	TK*	Tool kits
IM	Intensity-measuring devices	TL*	Tools
IP	Indicators, cathode-ray tube	TN	Tuning units
J	Interface units	TR	Transducers
KY	Keying devices	TS	Test units
LC*	Tools, line-construction	TT	Teletypewriter and facsimile apparatus
LS	Loudspeakers		
M	Microphones	TV*	Testers, tube
MA*	Magazines	TW	Tape units
MD	Modulators, demodulators, discriminators	U*	Connectors, audio and power
		UG*	Connectors, rf
ME	Meters	V	Vehicles
MF*	Magnets or magnetic-field generators	VS*	Signaling equipment, visual
MK	Miscellaneous kits	WD*	Cables, two-conductor
ML	Meteorological devices	WF*	Cables, four-conductor
MT	Mountings	WM*	Cables, multiple-conductor
MU	Memory units	WS*	Cables, single-conductor
MX	Miscellaneous	WT*	Cables, three-conductor
O	Oscillators	ZM	Impedance-measuring devices

* Not for US use except for assigning suffix letters to previously nomenclatured items.

XSC US Army Satellite Communications Agency, Fort Monmouth, NJ.

XWH US Naval Ammunition Depot Earle, Naval Weapons Handling Laboratory, Colts Neck, NJ.

XWO Naval Ordnance Laboratory, White Oak, Silver Spring, Md.

Example—Radio Set AN/ARC–3 ( ) might be assigned for a new airborne radio communication set under development. The cognizant development organization might then assign AN/ARC–3(XA–1), AN/ARC–3(XA–2), etc., type numbers to the various sets developed for test. When the set was considered satisfactory for use, the experimental indicator would be dropped, and procurement nomenclature, AN/ARC–3 would be officially assigned thereto.

Training Sets—A set or equipment designed for training purposes will be assigned type numbers as follows:

(A) A set to train for a specific basic set will be assigned the basic-set type number followed by a dash, the letter T, and a number. *Example:* Radio Training Set AN/ARC–6A–T1 would be the first training set for Radio Set AN/ARC–6A.

(B) A set to train for general types of sets will be assigned the usual set indicator letters followed by a dash, the letter T, and a number. *Example:* Radio Training Set AN/ARC–T1 would be the first training set for general airborne radio communication sets.

Parenthesis Indicators—A series of a basic item, i.e., all production and/or nonproduction versions, may be identified by a type designation with an empty parenthesis. *Examples:* AN/APS–25 ( ) or R–275( )/APS–25. Such an assignment is all inclusive and does not refer to any specific version within the series.

Systems, subsystems, centers, centrals, sets, groups, or units with variable parts lists are assigned type designations in the same manner, except a parenthetical V (V) is added to the type designation.

Units designed to accept plug-ins that change the function, frequency, or technical characteristics of the type-designated unit will designate with a (P) preceding the slant bar. The plug-in is not considered a part of the unit.

Examples of JETDS Type Numbers

AN/SRC–3( )	General reference set nomenclature for water surface craft radio communication set number 3.
AN/SRC–3	Original procurement set nomenclature applied against AN/SRC–3( ).
AN/SRC–3A	Modification set nomenclature applied against AN/SRC–3.
AN/APQ–13–T1( )	General reference training set nomenclature for the AN/APQ–13 set.
AN/APQ–13–T1	Original procurement training set nomenclature applied against AN/APQ–13–T1( ).
AN/APQ–13–T1A	Modification training set nomenclature applied against AN/APQ–13–T1.
AN/UPT–T3( )	General reference training set nomenclature for general utility radar transmitting training set number 3.
AN/UPT–T3	Original procurement training set nomenclature applied against AN/UPT–T3( ).
AN/UPT–T3A	Modification training set nomenclature applied against AN/UPT–T3.
T–51( )/ARQ–8	General reference component nomenclature for transmitter number 51, part of or used with airborne radio special set 8.
T–51/ARQ–8	Original procurement component nomenclature applied against T–51( )/ARQ–8.
T–51A/ARQ–8	Modification component nomenclature applied against T–51/ARQ–8.
RD–31( )/U	General reference component nomenclature for recorder-reproducer number 31 for general utility use, not part of a specific set.
RD–31/U	Original procurement component nomenclature applied against RD–31( )/U.
RD–31A/U	Modification component nomenclature applied against RD–31/U.

Index

A

C

INDEX

INDEX

F

G

J

K

L

N

O

P

Q

R

S

T

U

V

W

X

Y

Z

Other Reference Books Available from Sams

Computer Dictionary and Handbook (4th Edition) Charles J. Sippl
994 pages, 5¼×8½, hardbound, ©1985, No. 22206

Data Communications, Networks, and Systems Thomas Bartee, Editor
416 pages, 6½×9¼, hardbound, ©1985, No. 22235

Electronics: Circuits and Systems Swaminathan Madhu
800 pages, 6×9, hardbound, ©1985, No. 21984

Gallium Arsenide Technology David K. Ferry, Editor
256 pages, 7½×9¾, hardbound, ©1985, No. 11274

Handbook for Sound Engineers Glen Ballou, Editor
1200 pages, 7½×9¾, hardbound, ©1985, No. 21983

Handbook of Antenna Theory and Design Lee and Lo, Editors
700 pages, 7½×9¾ hardbound, ©1985, No. 22054

Handbook of Electronic Tables and Formulas (6th Edition)
Robert G. Middleton, 288 pages, 7½×9¾, hardbound, ©1985, No. 22469

Image Tubes Illes P. Csorba
448 pages, 5½×8½, hardbound, ©1985, No. 22023

Landmobile and Marine Radio Technology Handbook Edward M. Noll
464 pages, 7½×9¾, hardbound, ©1985, No. 22427

Mobile Communications Fundamentals William C. Lee
352 pages, 7½×9¾, hardbound, ©1985, No. 22305

Optical Signal Processing Pankaj K. Das
480 pages, 7½×9¾, hardbound, ©1985, No. 21961

Optical Waveguide Communications E. E. "Bert" Basch, Editor
554 pages, 7½×9¾, hardbound, ©1985, No 22301

Practical Microwaves Thomas S. Laverghetta
496 pages, 6½×9½, hardbound, ©1983, No. 21945

Principles of Solid-State Power Conversion Ralph E. Tarter
608 pages, 7½×9¾, hardbound, ©1985, No. 22018

Semiconductor Device Technology Malcolm E. Goodge
481 pages, 6×9, hardbound, ©1983, No. 22074

Sound System Engineering (2nd Edition) Don and Carolyn Davis
448 pages, 7½×9¾, hardbound, ©1985, No. 21857

To order these and other Sams books, call toll-free 1-800-428-SAMS.